National Electrical Code®
Handbook

NINTH EDITION

International Electrical Code® Series

Mark W. Earley, P.E.
Editor-in-Chief

Joseph V. Sheehan, P.E.
Senior Editor

Jeffrey S. Sargent
Editor

John M. Caloggero
Editor

Timothy M. Croushore, P.E.
Editor

With the complete text of the 2002 edition of the *National Electrical Code®*

National Fire Protection Association, Quincy, Massachusetts

Product Manager: Charles Durang
Developmental Editor: Jean Peck
Project Editor: Joyce Grandy
Copy Editors: Joyce Grandy, Dana Richards, Nancy Wirtes
Editorial Assistant: Carol A. Henderson
Text Processing: Louise Grant
Composition: Modern Graphics
Art Coordinator: Cheryl Langway
Illustration: Rollin Graphics, George Nichols, J. Philip Simmons
Interior Design: The Davis Group, Inc.
Cover Design: Corey, McPherson, Nash
Manufacturing Manager: Ellen Glisker
Printer: R. R. Donnelley/Willard

Notice Concerning Liability: Publication of this handbook is for the purpose of circulating information and opinion among those concerned for fire and electrical safety and related subjects. While every effort has been made to achieve a work of high quality, neither the NFPA nor the contributors to this handbook guarantee the accuracy or completeness of or assume any liability in connection with the information and opinions contained in this handbook. The NFPA and the contributors shall in no event be liable for any personal injury, property, or other damages of any nature whatsoever, whether special, indirect, consequential, or compensatory, directly or indirectly resulting from the publication, use of, or reliance upon this handbook.

This handbook is published with the understanding that the NFPA and the contributors to this handbook are supplying information and opinion but are not attempting to render engineering or other professional services. If such services are required, the assistance of an appropriate professional should be sought.

Notice Concerning Code Interpretations: This ninth edition of the *National Electrical Code® Handbook* is based on the 2002 edition of NFPA 70, *National Electrical Code*. All NFPA codes, standards, recommended practices, and guides are developed in accordance with the published procedures of the NFPA by technical committees comprised of volunteers drawn from a broad array of relevant interests. The handbook contains the complete text of NFPA 70 and any applicable Formal Interpretations issued by the Association. These documents are accompanied by explanatory commentary and other supplementary materials.

The commentary and supplementary materials in this handbook are not a part of the *Code* and do not constitute Formal Interpretations of the NFPA (which can be obtained only through requests processed by the responsible technical committees in accordance with the published procedures of the NFPA). The commentary and supplementary materials, therefore, solely reflect the personal opinions of the editor or other contributors and do not necessarily represent the official position of the NFPA or its technical committees.

®Registered Trademark National Fire Protection Association, Inc.

NFPA No.: 70HB02
ISBN: 0-87765-462-x
Library of Congress Card Catalog No.: 2001093271

Printed in the United States of America
03   04   05   06   5   4   3

Contents

Chapter 4   Equipment for General Use   419

Chapter 5   Special Occupancies   603

Chapter 6   Special Equipment   837

Chapter 7   Special Conditions   981

Chapter 8   Communications Systems   1045

Chapter 9   Tables   1087

Annexes

Index   1161

About the Editors   1191

On March 18–19, 1896, a group of 23 persons, representing a wide variety of organizations, met at the headquarters of the American Society of Mechanical Engineers in New York City. Their purpose was to develop a national code of rules for electrical construction and operation. (It is interesting to note that this meeting took place a mere 17 years after the invention of the incandescent light bulb.) This was not the first attempt to establish consistent rules for electrical installations, but it was the first national effort. The number of electrical fires was increasing, and the need for standardization was becoming urgent. One insurer reported electrical fires in 65 textile mills in New England by 1881.

The major problem was the lack of an authoritative, nationwide electrical installation standard. As one of the early participants noted, "We were without standards and inspectors, while manufacturers were without experience and knowledge of real installation needs. The workmen frequently created the standards as they worked, and rarely did two men think and work alike."

By 1895, five electrical installation codes had come into use in the United States, causing considerable controversy and confusion. It was difficult to manufacture products that met the requirements of all five codes. Something had to be done to develop a single, national code. The committee that met in 1896 recognized that the five existing codes should be used collectively as the basis for the new code. In the first known instance of international harmonization, the group also referred to the German code, the code of the British Board of Trade, and the Phoenix Rules of England. The importance of industry consensus was immediately recognized; before the committee met again in 1897, the new code was reviewed by 1200 individuals in the United States and Europe. Shortly thereafter, the first standardized U.S. electrical code, the *National Electrical Code*®, was published.

The *National Electrical Code* has become the most widely adopted code in the United States—it is the standard used in all 50 states and all U.S. territories. Moreover, it has grown well beyond the borders of the United States and is now used in numerous other countries. Because the code is a living document, constantly changing to reflect changes in technology, its use continues to grow.

Some things have not changed. The *National Electrical Code* continues to offer an open-consensus process. Anyone can submit a proposal for change or a public comment, and all proposals and comments are subject to a rigorous public review process. The *NEC*® still provides the best technical information, ensuring the practical safeguarding of persons and property from the hazards arising from the use of electricity.

Throughout its history, the National Electrical Code Committee has been guided by giants in the electrical industry. The names are too numerous to mention. Certainly the first chairman, William J. Hammer, should be applauded for providing the leadership necessary to get the *Code* started. More recently, the *Code* has been chaired by outstanding leaders such as Richard L. Loyd, Richard W. Osborne, Richard G. Biermann, and D. Harold Ware. Each of these men devoted many years to the National Electrical Code Committee.

The publication of the 2002 *National Electrical Code* marks the end of the tenure of D. Harold Ware as chair of the Technical Correlating Committee. Harold has devoted countless hours of quality personal time in his 35-plus years as a committee member. We could not let this occasion pass without noting that Harold has been one of the truly outstanding leaders in the 104-year history of the *Code*. We thank him for his dedication and his leadership.

The editors dedicate this book to the memory of W. Terry Lindsay, who represented the National Electrical Contractors Association on the Technical Correlating Committee and on CMP 8. Terry participated in several task groups, all of which focused on improving the *Code* or the functioning of the Committee structure. He is sorely missed.

The editors also wish to note the passing of several other long-term committee members who made numerous contributions to the *National Electrical Code*: Richard W. Osborne, former chair of the Correlating Committee and

CMP 11; William Wusinich, former chair of CMP 8 and member of CMP 1; Arthur Buxbaum, CMPs 7 and 18; Norman H. Davis III, CMP 5; Melvin J. Schiff, CMP 5; Walter Short, CMP 14; John Mangan, CMP 16; George Schuck, former chair of CMP 17; and William C. Boteler, CMP 19.

The editors have conferred closely with members of the National Electrical Code Committee in developing the revisions incorporated into the 2002 edition of the *Code*. The assistance and cooperation of code-making panel chairs and various committee members are herein gratefully acknowledged.

This edition of the *NEC Handbook* would not have been possible without the invaluable technical assistance of NFPA staff: Merton W. Bunker, former Chief Electrical Engineer; Ken Mastrullo, Senior Electrical Specialist; and Lee F. Richardson, Senior Electrical Engineer. Their contributions are greatly appreciated.

The editors acknowledge with thanks the manufacturers and their representatives who generously supplied photographs, drawings, and data upon request. Special thanks also to the editors of and contributors to past editions. Their work provided an excellent foundation on which to build.

The editors express special thanks to Joyce Grandy for her long hours and extraordinary effort in attending to all of the editorial details that we technical types often overlook. Special thanks are also due to Sylvia Dovner, an outstanding manager who kept this project on track. Without the efforts of Joyce and Sylvia, this new and improved edition of the *NEC Handbook* would not have been possible.

We also wish to thank the electrical support staff, Carol Henderson, Mary Warren-Pilson, Kathleen Stevens, and Elizabeth Schaffer, along with their leader, Jean O'Connor, for their support on this project.

The editors express their sincere appreciation to Richard Berman, Philip H. Cox, Allan Manche, Dean K. Wilson, Brian Phelan, David Kendall, Lori Tennant, Ray C. Mullin, James T. Pauley, Vincent Saporita, Peter J. Schram, and John C. Wiles for special help on specific articles. Finally, we also thank the following for contributing photos and graphics for this edition:

3M Co., Electrical Products Division
AFC Cable Systems, Inc.
Agfa Corporation

Alcoa Inc.
Allied Tube & Conduit, a Tyco International Co.
Appleton Electric Co.
Appleton Electric Co., EGS Electrical Group
Bose Corp.
Bussmann Division, Cooper Industries
Cable Tray Institute
Carlon®, Lamson & Sessions
Caterpillar
Cincinnati Milacron
Colortran, Inc.
Cooper Crouse-Hinds
Daniel Woodhead Co.
Dranetz-BMI
Dual-Lite, Inc.
Electronic Theatre Controls, Inc.
Fire Control Instruments
Ford Motor Co.
General Electric Co.
Hubbell, Inc.
Hubbell Inc., Kellems Division
Kliegl Bros.
L. E. Mason Co.
Lithonia Lighting, Reloc Wiring Systems
MPHusky Corp.
NAPCO Security Systems, Inc.
O.Z./Gedney
Pass & Seymour/Legrand®
Production Arts Lighting, Inc.
Pyrotenax Cables, Ltd.
Raco, Inc.
Reading Municipal Light Department
Rockbestos-Suprenant Cable Corp.
S & C Electric Co.
Smart House
Solar Design Associates, Inc.
Square D Co.
The Wiremold Co.
Thomas & Betts Corp.
Tocco Division, Park Ohio Industries
Tyco Electronics Corp.
Underwriters Laboratories Inc.
Union Connector Co., Inc.
Walker Systems, a Wiremold Co.

Administration and Enforcement

ARTICLE 80
Administration and Enforcement

Contents

This article is informative unless specifically adopted by the local jurisdiction adopting the National Electrical Code®. (See 80.5.)

Article 80 is new for the 2002 edition. The purpose of this administrative article is to assist jurisdictions that do not have formalized electrical inspection procedures but want to amend their inspection laws. It is intended to serve as a guide to the adoption of the *National Electrical Code® (NEC®)*. Because Article 80 is offered as a guide to the adoption process, its application is not mandatory unless specifically adopted by local law, as stated in the note after the title. See 80.5 for adoption information.

Article 80 has an interesting history. Originally, it was a model law that provided for inspection of electrical installations. It was prepared by the Electrical Field Service Advisory Committee of NFPA as a guide for those jurisdictions that either did not have formalized electrical inspection procedures or desired to amend their electrical inspection laws and to serve as a guide for adoption of the *NEC*. The model law was intended for use by states as well as municipalities. The first edition was adopted by NFPA on May 15, 1973. A second edition was approved on March 27, 1987.

Article 80 continues to cover such issues as creation of an electrical board, plan review, and inspection. The professional qualifications of the electrical inspector and the investigation of fires attributed to electrical installations have been added.

Adoption of the *National Electrical Code* can occur in two ways. It can be incorporated in a law, or a law can be enacted authorizing a governmental agency or board to adopt it. To facilitate the drafting of such laws, as well as establishment of the accompanying inspection and enforcement procedures, NFPA offers Article 80, Administration and Enforcement. Article 80 may require modification in order to comply with the structure-writing rules of the adopting political jurisdiction.

Circumstances in a particular jurisdiction determine which alternative is most appropriate. Provisions that are less comprehensive may be adequate for smaller political subdivisions.

80.1 Scope.

The following functions are covered:

(1) The inspection of electrical installations as covered by 90.2
(2) The investigation of fires caused by electrical installations
(3) The review of construction plans, drawings, and specifications for electrical systems
(4) The design, alteration, modification, construction, maintenance, and testing of electrical systems and equipment
(5) The regulation and control of electrical installations at special events including but not limited to exhibits, trade shows, amusement parks, and other similar special occupancies

80.2 Definitions.

Authority Having Jurisdiction. The organization, office, or individual responsible for approving equipment, materials, an installation, or a procedure.

Chief Electrical Inspector. An electrical inspector who either is the authority having jurisdiction or is designated by the authority having jurisdiction and is responsible for administering the requirements of this *Code*.

Electrical Inspector. An individual meeting the requirements of 80.27 and authorized to perform electrical inspections.

80.3 Purpose.

The purpose of this article shall be to provide requirements for administration and enforcement of the *National Electrical Code*.

80.5 Adoption.

Article 80 shall not apply unless specifically adopted by the local jurisdiction adopting the *National Electrical Code*.

80.7 Title.

The title of this *Code* shall be NFPA 70, *National Electrical Code®*, of the National Fire Protection Association. The short title of this *Code* shall be the *NEC®*.

80.9 Application.

(A) New Installations. This *Code* applies to new installations. Buildings with construction permits dated after adoption of this *Code* shall comply with its requirements.

(B) Existing Installations. Existing electrical installations that do not comply with the provisions of this *Code* shall be permitted to be continued in use unless the authority having jurisdiction determines that the lack of conformity

with this *Code* presents an imminent danger to occupants. Where changes are required for correction of hazards, a reasonable amount of time shall be given for compliance, depending on the degree of the hazard.

(C) Additions, Alterations, or Repairs. Additions, alterations, or repairs to any building, structure, or premises shall conform to that required of a new building without requiring the existing building to comply with all the requirements of this *Code*. Additions, alterations, installations, or repairs shall not cause an existing building to become unsafe or to adversely affect the performance of the building as determined by the authority having jurisdiction. Electrical wiring added to an existing service, feeder, or branch circuit shall not result in an installation that violates the provisions of the *Code* in force at the time the additions are made.

80.11 Occupancy of Building or Structure.

(A) New Construction. No newly constructed building shall be occupied in whole or in part in violation of the provisions of this *Code*.

(B) Existing Buildings. Existing buildings that are occupied at the time of adoption of this *Code* shall be permitted to remain in use provided the following conditions apply:

(1) The occupancy classification remains unchanged
(2) There exists no condition deemed hazardous to life or property that would constitute an imminent danger

80.13 Authority.

Where used in this article, the term *authority having jurisdiction* shall include the chief electrical inspector or other individuals designated by the governing body. This *Code* shall be administered and enforced by the authority having jurisdiction designated by the governing authority as follows.

(1) The authority having jurisdiction shall be permitted to render interpretations of this *Code* in order to provide clarification to its requirements, as permitted by 90.4.
(2) When the use of any electrical equipment or its installations is found to be dangerous to human life or property, the authority having jurisdiction shall be empowered to have the premises disconnected from its source of electric supply, as established by the Board. When such equipment or installation has been so condemned or disconnected, a notice shall be placed thereon listing the causes for the condemnation, the disconnection, or both and the penalty under 80.23 for the unlawful use thereof. Written notice of such condemnation or disconnection and the causes therefor shall be given within 24 hours to the owners, the occupant, or both, of such building, structure, or premises. It shall be unlawful for

any person to remove said notice, to reconnect the electric equipment to its source of electric supply, or to use or permit to be used electric power in any such electric equipment until such causes for the condemnation or disconnection have been remedied to the satisfaction of the inspection authorities.

(3) The authority having jurisdiction shall be permitted to delegate to other qualified individuals such powers as necessary for the proper administration and enforcement of this *Code*.

(4) Police, fire, and other enforcement agencies shall have authority to render necessary assistance in the enforcement of this *Code* when requested to do so by the authority having jurisdiction.

(5) The authority having jurisdiction shall be authorized to inspect, at all reasonable times, any building or premises for dangerous or hazardous conditions or equipment as set forth in this *Code*. The authority having jurisdiction shall be permitted to order any person(s) to remove or remedy such dangerous or hazardous condition or equipment. Any person(s) failing to comply with such order shall be in violation of this *Code*.

(6) Where the authority having jurisdiction deems that conditions hazardous to life and property exist, he or she shall be permitted to require that such hazardous conditions in violation of this *Code* be corrected.

(7) To the full extent permitted by law, any authority having jurisdiction engaged in inspection work shall be authorized at all reasonable times to enter and examine any building, structure, or premises for the purpose of making electrical inspections. Before entering a premises, the authority having jurisdiction shall obtain the consent of the occupant thereof or obtain a court warrant authorizing entry for the purpose of inspection except in those instances where an emergency exists. As used in this section, *emergency* means circumstances that the authority having jurisdiction knows, or has reason to believe, exist and that reasonably can constitute immediate danger to persons or property.

(8) Persons authorized to enter and inspect buildings, structures, and premises as herein set forth shall be identified by proper credentials issued by this governing authority.

(9) Persons shall not interfere with an authority having jurisdiction carrying out any duties or functions prescribed by this *Code*.

(10) Persons shall not use a badge, uniform, or other credentials to impersonate the authority having jurisdiction.

(11) The authority having jurisdiction shall be permitted to investigate the cause, origin, and circumstances of any fire, explosion, or other hazardous condition.

(12) The authority having jurisdiction shall be permitted to require plans and specifications to ensure compliance with this *Code*.

(13) Whenever any installation subject to inspection prior to use is covered or concealed without having first been inspected, the authority having jurisdiction shall be permitted to require that such work be exposed for inspection. The authority having jurisdiction shall be notified when the installation is ready for inspection and shall conduct the inspection within _____ days.

(14) The authority having jurisdiction shall be permitted to order the immediate evacuation of any occupied building deemed unsafe when such building has hazardous conditions that present imminent danger to building occupants.

(15) The authority having jurisdiction shall be permitted to waive specific requirements in this *Code* or permit alternative methods where it is assured that equivalent objectives can be achieved by establishing and maintaining effective safety. Technical documentation shall be submitted to the authority having jurisdiction to demonstrate equivalency and that the system, method, or device is approved for the intended purpose.

(16) Each application for a waiver of a specific electrical requirement shall be filed with the authority having jurisdiction and shall be accompanied by such evidence, letters, statements, results of tests, or other supporting information as required to justify the request. The authority having jurisdiction shall keep a record of actions on such applications, and a signed copy of the authority having jurisdiction's decision shall be provided for the applicant.

80.15 Electrical Board.

(A) Creation of the Electrical Board. There is hereby created the Electrical Board of the _____ of _____, hereinafter designated as the Board.

(B) Appointments. Board members shall be appointed by the Governor with the advice and consent of the Senate (or by the Mayor with the advice and consent of the Council, or the equivalent).

(1) Members of the Board shall be chosen in a manner to reflect a balanced representation of individuals or organizations. The Chair of the Board shall be elected by the Board membership.

(2) The Chief Electrical Inspector in the jurisdiction adopting this Article authorized in 80.15(B)(3)(a) shall be the nonvoting secretary of the Board. Where the Chief Electrical Inspector of a local municipality serves a Board at a state level, he or she shall be permitted to serve as a voting member of the Board.

(3) The board shall consist of not fewer than five voting members. Board members shall be selected from the following:

a. Chief Electrical Inspector from a local government (for State Board only)
b. An electrical contractor operating in the jurisdiction
c. A licensed professional engineer engaged primarily in the design or maintenance of electrical installations
d. A journeyman electrician

(4) Additional membership shall be selected from the following:

a. A master (supervising) electrician
b. The Fire Marshal (or Fire Chief)
c. A representative of the property/casualty insurance industry
d. A representative of an electric power utility operating in the jurisdiction
e. A representative of electrical manufacturers primarily and actively engaged in producing materials, fittings, devices, appliances, luminaires (fixtures), or apparatus used as part of or in connection with electrical installations
f. A member of the labor organization that represents the primary electrical workforce
g. A member from the public who is not affiliated with any other designated group
h. A representative of a telecommunications utility operating in the jurisdiction

(C) Terms. Of the members first appointed, _____ shall be appointed for a term of 1 year, _____ for a term of 2 years, _____ for a term of 3 years, and _____ for a term of 4 years, and thereafter each appointment shall be for a term of 4 years or until a successor is appointed. The Chair of the Board shall be appointed for a term not to exceed _____ years.

(D) Compensation. Each appointed member shall receive the sum of _____dollars ($_____) for each day during which the member attends a meeting of the Board and, in addition thereto, shall be reimbursed for direct lodging, travel, and meal expenses as covered by policies and procedures established by the jurisdiction.

(E) Quorum. A quorum as established by the Board operating procedures shall be required to conduct Board business. The Board shall hold such meetings as necessary to carry out the purposes of Article 80. The Chair or a majority of the members of the Board shall have the authority to call meetings of the Board.

(F) Duties. It shall be the duty of the Board to:

(1) Adopt the necessary rules and regulations to administer and enforce Article 80.
(2) Establish qualifications of electrical inspectors.

(3) Revoke or suspend the recognition of any inspector's certificate for the jurisdiction.
(4) After advance notice of the public hearings and the execution of such hearings, as established by law, the Board is authorized to establish and update the provisions for the safety of electrical installations to conform with the current edition of the *National Electrical Code* (NFPA 70) and other nationally recognized safety standards for electrical installations.
(5) Establish procedures for recognition of electrical safety standards and acceptance of equipment conforming to these standards.

(G) Appeals.

(1) Review of Decisions. Any person, firm, or corporation may register an appeal with the Board for a review of any decision of the Chief Electrical Inspector or of any Electrical Inspector, provided that such appeal is made in writing within fifteen (15) days after such person, firm, or corporation shall have been notified. Upon receipt of such appeal, said Board shall, if requested by the person making the appeal, hold a public hearing and proceed to determine whether the action of the Board, or of the Chief Electrical Inspector, or of the Electrical Inspector complies with this law and, within fifteen (15) days after receipt of the appeal or after holding the hearing, shall make a decision in accordance with its findings.
(2) Conditions. Any person shall be permitted to appeal a decision of the authority having jurisdiction to the Board when it is claimed that any one or more of the following conditions exist:

a. The true intent of the codes or ordinances described in this *Code* has been incorrectly interpreted.
b. The provisions of the codes or ordinances do not fully apply.
c. A decision is unreasonable or arbitrary as it applies to alternatives or new materials.

(3) Submission of Appeals. A written appeal, outlining the *Code* provision from which relief is sought and the remedy proposed, shall be submitted to the authority having jurisdiction within 15 calendar days of notification of violation.

(H) Meetings and Records. Meetings and records of the Board shall conform to the following:

(1) Meetings of the Board shall be open to the public as required by law.
(2) Records of meetings of the Board shall be available for review during normal business hours, as required by law.

80.17 Records and Reports.

The authority having jurisdiction shall retain records in accordance with 80.17(A) and (B).

(A) Retention. The authority having jurisdiction shall keep a record of all electrical inspections, including the date of such inspections and a summary of any violations found to exist, the date of the services of notices, and a record of the final disposition of all violations. All required records shall be maintained until their usefulness has been served or as otherwise required by law.

(B) Availability. A record of examinations, approvals, and variances granted shall be maintained by the authority having jurisdiction and shall be available for public review as prescribed by law during normal business hours.

80.19 Permits and Approvals.

Permits and approvals shall conform to 80.19(A) through (H).

(A) Application.

(1) Activity authorized by a permit issued under this *Code* shall be conducted by the permittee or the permittee's agents or employees in compliance with all requirements of this *Code* applicable thereto and in accordance with the approved plans and specifications. No permit issued under this *Code* shall be interpreted to justify a violation of any provision of this *Code* or any other applicable law or regulation. Any addition or alteration of approved plans or specifications shall be approved in advance by the authority having jurisdiction, as evidenced by the issuance of a new or amended permit.
(2) A copy of the permit shall be posted or otherwise readily accessible at each work site or carried by the permit holder as specified by the authority having jurisdiction.

(B) Content. Permits shall be issued by the authority having jurisdiction and shall bear the name and signature of the authority having jurisdiction or that of the authority having jurisdiction's designated representative. In addition, the permit shall indicate the following:

(1) Operation or activities for which the permit is issued
(2) Address or location where the operation or activity is to be conducted
(3) Name and address of the permittee
(4) Permit number and date of issuance
(5) Period of validity of the permit
(6) Inspection requirements

(C) Issuance of Permits. The authority having jurisdiction shall be authorized to establish and issue permits, certificates, notices, and approvals, or orders pertaining to electrical safety hazards pursuant to 80.23, except that no permit

shall be required to execute any of the classes of electrical work specified in the following:

(1) Installation or replacement of equipment such as lamps and of electric utilization equipment approved for connection to suitable permanently installed receptacles. Replacement of flush or snap switches, fuses, lamp sockets, and receptacles, and other minor maintenance and repair work, such as replacing worn cords and tightening connections on a wiring device
(2) The process of manufacturing, testing, servicing, or repairing electric equipment or apparatus

(D) Annual Permits. In lieu of an individual permit for each installation or alteration, an annual permit shall, upon application, be issued to any person, firm, or corporation regularly employing one or more employees for the installation, alteration, and maintenance of electric equipment in or on buildings or premises owned or occupied by the applicant for the permit. Upon application, an electrical contractor as agent for the owner or tenant shall be issued an annual permit. The applicant shall keep records of all work done, and such records shall be transmitted periodically to the Electrical Inspector.

(E) Fees. Any political subdivision that has been provided for electrical inspection in accordance with the provisions of Article 80 may establish fees that shall be paid by the applicant for a permit before the permit is issued.

(F) Inspection and Approvals.

(1) Upon the completion of any installation of electrical equipment that has been made under a permit other than an annual permit, it shall be the duty of the person, firm, or corporation making the installation to notify the Electrical Inspector having jurisdiction, who shall inspect the work within a reasonable time.
(2) Where the Inspector finds the installation to be in conformity with the statutes of all applicable local ordinances and all rules and regulations, the Inspector shall issue to the person, firm, or corporation making the installation a certificate of approval, with duplicate copy for delivery to the owner, authorizing the connection to the supply of electricity and shall send written notice of such authorization to the supplier of electric service. When a certificate of temporary approval is issued authorizing the connection of an installation, such certificates shall be issued to expire at a time to be stated therein and shall be revocable by the Electrical Inspector for cause.
(3) When any portion of the electrical installation within the jurisdiction of an Electrical Inspector is to be hidden from view by the permanent placement of parts of the building, the person, firm, or corporation installing the equipment shall notify the Electrical Inspector, and such

equipment shall not be concealed until it has been approved by the Electrical Inspector or until _____ days have elapsed from the time of such notification, provided that on large installations, where the concealment of equipment proceeds continuously, the person, firm, or corporation installing the equipment shall give the Electrical Inspector due notice in advance, and inspections shall be made periodically during the progress of the work.

(4) At regular intervals, the Electrical Inspector having jurisdiction shall visit all buildings and premises where work may be done under annual permits and shall inspect all electric equipment installed under such permits since the date of the previous inspection. The Electrical Inspector shall issue a certificate of approval for such work as is found to be in conformity with the provisions of Article 80 and all applicable ordinances, orders, rules, and regulations, after payments of all required fees.

(5) If, upon inspection, any installation is found not to be fully in conformity with the provisions of Article 80, and all applicable ordinances, rules, and regulations, the Inspector making the inspection shall at once forward to the person, firm, or corporation making the installation a written notice stating the defects that have been found to exist.

(G) Revocation of Permits. Revocation of permits shall conform to the following:

(1) The authority having jurisdiction shall be permitted to revoke a permit or approval issued if any violation of this *Code* is found upon inspection or in case there have been any false statements or misrepresentations submitted in the application or plans on which the permit or approval was based.

(2) Any attempt to defraud or otherwise deliberately or knowingly design, install, service, maintain, operate, sell, represent for sale, falsify records, reports, or applications, or other related activity in violation of the requirements prescribed by this *Code* shall be a violation of this *Code*. Such violations shall be cause for immediate suspension or revocation of any related licenses, certificates, or permits issued by this jurisdiction. In addition, any such violation shall be subject to any other criminal or civil penalties as available by the laws of this jurisdiction.

(3) Revocation shall be constituted when the permittee is duly notified by the authority having jurisdiction.

(4) Any person who engages in any business, operation, or occupation, or uses any premises, after the permit issued therefor has been suspended or revoked pursuant to the provisions of this *Code*, and before such suspended permit has been reinstated or a new permit issued, shall be in violation of this *Code*.

(5) A permit shall be predicated upon compliance with the requirements of this *Code* and shall constitute written authority issued by the authority having jurisdiction to install electrical equipment. Any permit issued under this *Code* shall not take the place of any other license or permit required by other regulations or laws of this jurisdiction.

(6) The authority having jurisdiction shall be permitted to require an inspection prior to the issuance of a permit.

(7) A permit issued under this *Code* shall continue until revoked or for the period of time designated on the permit. The permit shall be issued to one person or business only and for the location or purpose described in the permit. Any change that affects any of the conditions of the permit shall require a new or amended permit.

(H) Applications and Extensions. Applications and extensions of permits shall conform to the following:

(1) The authority having jurisdiction shall be permitted to grant an extension of the permit time period upon presentation by the permittee of a satisfactory reason for failure to start or complete the work or activity authorized by the permit.

(2) Applications for permits shall be made to the authority having jurisdiction on forms provided by the jurisdiction and shall include the applicant's answers in full to inquiries set forth on such forms. Applications for permits shall be accompanied by such data as required by the authority having jurisdiction, such as plans and specifications, location, and so forth. Fees shall be determined as required by local laws.

(3) The authority having jurisdiction shall review all applications submitted and issue permits as required. If an application for a permit is rejected by the authority having jurisdiction, the applicant shall be advised of the reasons for such rejection. Permits for activities requiring evidence of financial responsibility by the jurisdiction shall not be issued unless proof of required financial responsibility is furnished.

80.21 Plans Review.

Review of plans and specifications shall conform to 80.21(A) through (C).

(A) Authority. For new construction, modification, or rehabilitation, the authority having jurisdiction shall be permitted to review construction documents and drawings.

(B) Responsibility of the Applicant. It shall be the responsibility of the applicant to ensure the following:

(1) The construction documents include all of the electrical requirements.

(2) The construction documents and drawings are correct and in compliance with the applicable codes and standards.

(C) Responsibility of the Authority Having Jurisdiction. It shall be the responsibility of the authority having jurisdiction to promulgate rules that cover the following:

(1) Review of construction documents and drawings within established time frames for the purpose of acceptance or to provide reasons for nonacceptance

(2) Review and approval by the authority having jurisdiction shall not relieve the applicant of the responsibility of compliance with this *Code*.

(3) Where field conditions necessitate any substantial change from the approved plan, the authority having jurisdiction shall be permitted to require that the corrected plans be submitted for approval.

80.23 Notice of Violations, Penalties.

Notice of violations and penalties shall conform to 80.23(A) and (B).

(A) Violations.

(1) Whenever the authority having jurisdiction determines that there are violations of this *Code*, a written notice shall be issued to confirm such findings.

(2) Any order or notice issued pursuant to this *Code* shall be served upon the owner, operator, occupant, or other person responsible for the condition or violation, either by personal service or mail or by delivering the same to, and leaving it with, some person of responsibility upon the premises. For unattended or abandoned locations, a copy of such order or notice shall be posted on the premises in a conspicuous place at or near the entrance to such premises and the order or notice shall be mailed by registered or certified mail, with return receipt requested, to the last known address of the owner, occupant, or both.

(B) Penalties.

(1) Any person who fails to comply with the provisions of this *Code* or who fails to carry out an order made pursuant to this *Code* or violates any condition attached to a permit, approval, or certificate shall be subject to the penalties established by this jurisdiction.

(2) Failure to comply with the time limits of an abatement notice or other corrective notice issued by the authority having jurisdiction shall result in each day that such violation continues being regarded as a new and separate offense.

(3) Any person, firm, or corporation who shall willfully violate any of the applicable provisions of this article shall be guilty of a misdemeanor and, upon conviction

thereof, shall be punished by a fine of not less than _____dollars (\$_____) or more than _____dollars (\$_____) for each offense, together with the costs of prosecution, imprisonment, or both, for not less than _____ (_____) days or more than _____ (_____) days.

80.25 Connection to Electricity Supply.

Connections to the electric supply shall conform to 80.25(A) through (E).

(A) Authorization. Except where work is done under an annual permit and except as otherwise provided in 80.25, it shall be unlawful for any person, firm, or corporation to make connection to a supply of electricity or to supply electricity to any electric equipment installation for which a permit is required or that has been disconnected or ordered to be disconnected.

(B) Special Consideration. By special permission of the authority having jurisdiction, temporary power shall be permitted to be supplied to the premises for specific needs of the construction project. The Board shall determine what needs are permitted under this provision.

(C) Notification. If, within _____ business days after the Electrical Inspector is notified of the completion of an installation of electric equipment, other than a temporary approval installation, the Electrical Inspector has neither authorized connection nor disapproved the installation, the supplier of electricity is authorized to make connections and supply electricity to such installation.

(D) Other Territories. If an installation or electric equipment is located in any territory where an Electrical Inspector has not been authorized or is not required to make inspections, the supplier of electricity is authorized to make connections and supply electricity to such installations.

(E) Disconnection. Where a connection is made to an installation that has not been inspected, as outlined in the preceding paragraphs of this section, the supplier of electricity shall immediately report such connection to the Chief Electrical Inspector. If, upon subsequent inspection, it is found that the installation is not in conformity with the provisions of Article 80, the Chief Electrical Inspector shall notify the person, firm, or corporation making the installation to rectify the defects and, if such work is not completed within fifteen (15) business days or a longer period as may be specified by the Board, the Board shall have the authority to cause the disconnection of that portion of the installation that is not in conformity.

80.27 Inspector's Qualifications.

(A) Certificate. All electrical inspectors shall be certified by a nationally recognized inspector certification program accepted by the Board. The certification program shall specifically qualify the inspector in electrical inspections. No person shall be employed as an Electrical Inspector unless that person is the holder of an Electrical Inspector's certificate of qualification issued by the Board, except that any person who on the date on which this law went into effect was serving as a legally appointed Electrical Inspector of _____ shall, upon application and payment of the prescribed fee and without examination, be issued a special certificate permitting him or her to continue to serve as an Electrical Inspector in the same territory.

(B) Experience. Electrical inspector applicants shall demonstrate the following:

(1) Have a demonstrated knowledge of the standard materials and methods used in the installation of electric equipment

(2) Be well versed in the approved methods of construction for safety to persons and property

(3) Be well versed in the statutes of _____ relating to electrical work and the *National Electrical Code*, as approved by the American National Standards Institute

(4) Have had at least _____ years' experience as an Electrical Inspector or _____ years in the installation of electrical equipment. In lieu of such experience, the applicant shall be a graduate in electrical engineering or of a similar curriculum of a college or university considered by the Board as having suitable requirements for graduation and shall have had two years' practical electrical experience.

(C) Recertification. Electrical inspectors shall be recertified as established by provisions of the applicable certification program.

(D) Revocation and Suspension of Authority. The Board shall have the authority to revoke an inspector's authority to conduct inspections within a jurisdiction.

80.29 Liability for Damages.

Article 80 shall not be construed to affect the responsibility or liability of any party owning, designing, operating, controlling, or installing any electric equipment for damages to persons or property caused by a defect therein, nor shall the _____ or any of its employees be held as assuming any such liability by reason of the inspection, reinspection, or other examination authorized.

80.31 Validity.

If any section, subsection, sentence, clause, or phrase of Article 80 is for any reason held to be unconstitutional, such decision shall not affect the validity of the remaining portions of Article 80.

80.33 Repeal of Conflicting Acts.

All acts or parts of acts in conflict with the provisions of Article 80 are hereby repealed.

80.35 Effective Date.

Article 80 shall take effect _____ (_____) days after its passage and publication.

Introduction

ARTICLE 90
Introduction

Contents

90.1 Purpose.

(A) Practical Safeguarding. The purpose of this *Code* is the practical safeguarding of persons and property from hazards arising from the use of electricity.

The *NEC* is prepared by the National Electrical Code Committee, which consists of a Technical Correlating Committee and 20 code-making panels. The code-making panels have specific subject responsibility within the *Code*. The scope of the National Electrical Code Committee follows:

This committee shall have primary responsibility for documents on minimizing the risk of electricity as a source of electric shock and as a potential ignition source of fires and explosions. It shall also be responsible for text to minimize the propagation of fire and explosions due to electrical installations.

In addition to its overall responsibility for the *National Electrical Code*, the Technical Correlating Committee is responsible for the *Electrical Code for One- and Two-Family Dwellings* (NFPA 70A) and for the correlation of *Recommended Practice for Electrical Equipment Maintenance* (NFPA 70B), *Standard for Electrical Safety Requirements for Employee Workplaces* (NFPA 70E), *Electrical Inspection Code for Existing Dwellings* (NFPA 73), and *Electrical Standard for Industrial Machinery* (NFPA 79).

(B) Adequacy. This *Code* contains provisions that are considered necessary for safety. Compliance therewith and proper maintenance will result in an installation that is essentially free from hazard but not necessarily efficient, convenient, or adequate for good service or future expansion of electrical use.

> FPN: Hazards often occur because of overloading of wiring systems by methods or usage not in conformity with this *Code*. This occurs because initial wiring did not provide for increases in the use of electricity. An initial adequate installation and reasonable provisions for system changes will provide for future increases in the use of electricity.

Consideration should always be given to future expansion of the electrical system. Future expansion might be unlikely in some occupancies, but for others it is wise to plan an initial installation comprising service-entrance conductors and equipment, feeder conductors, and panelboards that allows for future additions, alterations, designs, and so on.

(C) Intention. This *Code* is not intended as a design specification or an instruction manual for untrained persons.

The *NEC* is intended for use by capable engineers and electrical contractors in the design and/or installation of electrical equipment; by inspection authorities exercising legal jurisdiction over electrical installations; by property insurance inspectors; by qualified industrial, commercial, and residential electricians; and by instructors of electrical apprentices or students.

(D) Relation to International Standards. The requirements in this *Code* address the fundamental principles of protection for safety contained in Section 131 of International Electrotechnical Commission Standard 60364–1, *Electrical Installations of Buildings*.

> FPN: IEC 60364-1, Section 131, contains fundamental principles of protection for safety that encompass protection against electric shock, protection against thermal effects, protection against overcurrent, protection against fault currents, and protection against overvoltage. All of these potential hazards are addressed by the requirements in this *Code*.

The *NEC* has been widely adopted in many countries. This section makes it clear that the *NEC* is compatible with international safety principles. Added as a Tentative Interim Amendment (TIA) to the 1999 *Code*, this section calls attention to the fact that installations meeting the requirements of the *NEC* are also in compliance with the fundamental principles outlined in IEC 60364-1, *Electrical Installations of Buildings*, Section 131. This TIA allows countries that do not have any formalized rules for electrical installations to adopt the *NEC* and by so doing to be fully compatible with the safety principles of IEC 60364-1, Section 131. The addition of this section will promote acceptance and adoption of the *NEC* internationally.

The *NEC* is an essential part of the safety system of the Americas. Its future will be enhanced by increased international acceptance of the *Code*.

90.2 Scope.

(A) Covered. This *Code* covers the installation of electric conductors, electric equipment, signaling and communications conductors and equipment, and fiber optic cables and raceways for the following:

(1) Public and private premises, including buildings, structures, mobile homes, recreational vehicles, and floating buildings
(2) Yards, lots, parking lots, carnivals, and industrial substations

> FPN: For additional information concerning such installations in an industrial or multibuilding complex, see the ANSI C2-1997, *National Electrical Safety Code*.

Requirements for locations such as these are found throughout the *Code*. Specific items such as outside feeders and branch circuits can be found in Article 225, grounding in Article 250, surge arresters in Article 280, switches in Article 404, outside lighting in Article 410, transformers in Article 450, and carnivals in Article 525.

(3) Installations of conductors and equipment that connect to the supply of electricity

Often, but not always, the source of supply of electricity is the serving electric utility. The point of connection from a premises wiring system to a serving electric utility system is, by definition, referred to as the *service point*. The conductors on the premises side of the service point are, by definition, referred to as *service conductors*. These definitions are found in Article 100. The requirements for service conductors as well as for service-related equipment are found in Article 230. Article 230 applies only where the source of supply of electricity is from a utility.

Where the source of supply of electricity is not the serving electric utility, the source may be a generator, a battery system, a solar photovoltaic system, a fuel cell, or a combination of those sources. Requirements for such sources of supply are found in Article 445, and in Articles 700 through 702 for generators, Article 480 for storage batteries, Article 690 for solar photovoltaic systems, and Article 692 for fuel cells. The associated delivery wiring requirements are found in Chapters 2 and 3 (except Article 230) and in Articles 700 through 702 for emergency, legally required, and optional standby power system circuits.

(4) Installations used by the electric utility, such as office buildings, warehouses, garages, machine shops, and recreational buildings, that are not an integral part of a generating plant, substation, or control center

For the 2002 *Code*, 90.2(A) was rewritten. It provides order and clarity concerning the portions of electric utility facilities covered by the *NEC*. See 90.2(B) and the related commentary for information on facilities and specific lighting not covered by the *NEC*. Exhibit 90.1 illustrates the distinction between electric utility facilities to which the *NEC* does and does not apply.

Industrial and multibuilding complexes and campus-style wiring often include substations and other installations that employ construction and wiring similar to that of electric utility installations. Although such nonutility installations are within the scope of the *NEC*, the *NEC* requirements may not always be all-inclusive, for example, in clearances of conductors or in clearances from buildings or structures for nominal voltages over 600 volts. In such cases, the user can find additional information in the *National Electrical Safety Code (NESC)*, published by the Institute of Electrical and Electronics Engineers, Inc., P.O. Box 1331, 445 Hoes Lane, Piscataway, NJ 08855-1331.

(B) Not Covered. This *Code* does not cover the following:

(1) Installations in ships, watercraft other than floating buildings, railway rolling stock, aircraft, or automotive vehicles other than mobile homes and recreational vehicles

Installation requirements for floating buildings are found in Article 553.

> FPN: Although the scope of this *Code* indicates that the *Code* does not cover installations in ships, portions of

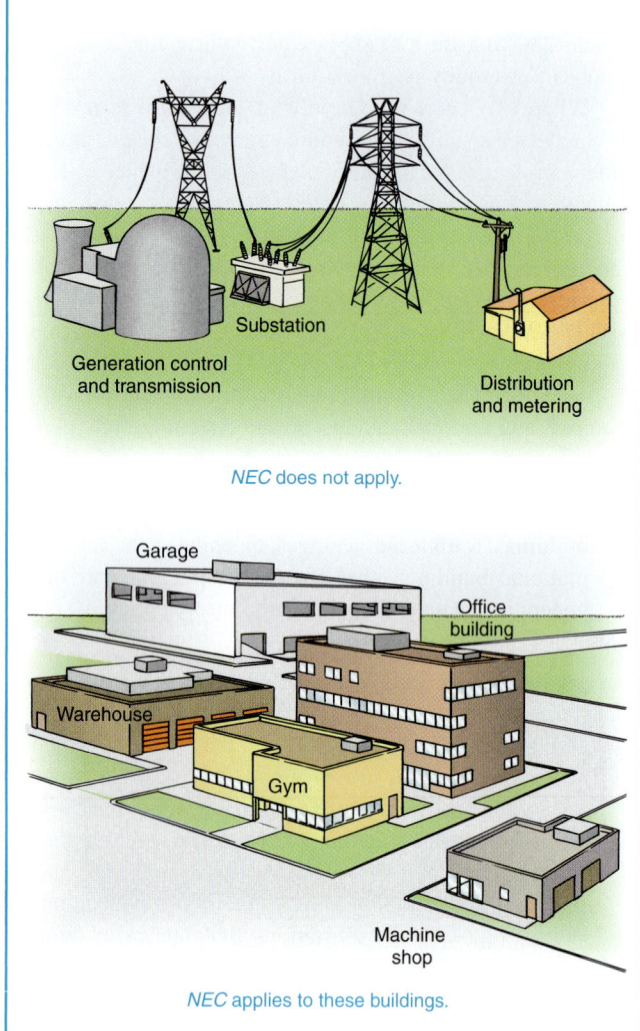

Exhibit 90.1 Typical electric utility complexes showing examples of facilities covered and not covered by the provisions of the *NEC*.

this *Code* are incorporated by reference into Title 46, *Code of Federal Regulations*, Parts 110–113.

The *NEC* does not specifically cover shipboard wiring. Title 46 of the *Code of Federal Regulations*, Parts 110–113, however, does contain many specific *NEC*-referenced requirements. These requirements, which originated in the *NEC*, are enforced by the U.S. Coast Guard.

(2) Installations under ground in mines and self-propelled mobile surface mining machinery and its attendant electrical trailing cable

(3) Installations of railways for generation, transformation, transmission, or distribution of power used exclusively for operation of rolling stock or installations used exclusively for signaling and communications purposes

(4) Installations of communications equipment under the exclusive control of communications utilities located outdoors or in building spaces used exclusively for such installations

(5) Installations under the exclusive control of an electric utility where such installations

   a. Consist of service drops or service laterals, and associated metering, or

   b. Are located in legally established easements, rights-of-way, or by other agreements either designated by or recognized by public service commissions, utility commissions, or other regulatory agencies having jurisdiction for such installations, or

   c. Are on property owned or leased by the electric utility for the purpose of communications, metering, generation, control, transformation, transmission, or distribution of electric energy.

In the 2002 *Code*, 90.2(B)(5) was changed. The term *associated metering* was added to declare that the *Code* does not cover metering equipment associated with service drops and laterals. Further, the purpose of new wording added to identify access by "easements, right-of-ways, or by other agreements" associated with the authority of "public service commissions, utility commissions, or other regulatory agencies having jurisdiction" is to clarify that those agencies generally have authority over those types of installations and establish the rules that govern such installations.

It is not the intent of this section to exclude the *NEC* as an installation regulatory document. After all, the *NEC* is fully capable of being utilized for electrical installations in most cases, and 90.2(B)(5) does not pertain to areas where portions of the *NEC* could not be used. Rather 90.2(B)(5) lists specific areas where the nature of the installation requires specialized rules or where the use of other installation rules, standards, and guidelines has been developed for specific uses and industries. For example, the electrical utility industry uses the *NESC* as its primary requirement in the generation, transmission, distribution, and metering of electrical energy.

(C) Special Permission. The authority having jurisdiction for enforcing this *Code* may grant exception for the installation of conductors and equipment that are not under the exclusive control of the electric utilities and are used to connect the electric utility supply system to the service-entrance conductors of the premises served, provided such installations are outside a building or terminate immediately inside a building wall.

90.3 Code Arrangement.

This *Code* is divided into the introduction and nine chapters, as shown in Figure 90.3. Chapters 1, 2, 3, and 4 apply generally; Chapters 5, 6, and 7 apply to special occupancies, special equipment, or other special conditions. These latter chapters supplement or modify the general rules. Chapters 1 through 4 apply except as amended by Chapters 5, 6, and 7 for the particular conditions.

Chapter 8 covers communications systems and is not subject to the requirements of Chapters 1 through 7 except where the requirements are specifically referenced in Chapter 8.

Chapter 9 consists of tables.

Annexes are not part of the requirements of this *Code* but are included for informational purposes only.

The reference to the Introduction is intended to include Article 90 in the application of this *Code*. Chapters 1 through 4 apply generally, except as amended or specifically referenced in Chapters 5, 6, and 7 (Articles 500 through 780). For example, 300.22 (Chapter 3) is modified by 725.3(C) and 760.3(B) and is specifically referenced in 800.52(C) and 830.3(B). A graphic explanation of the *NEC* arrangement, Figure 90.3, has been added for the 2002 *Code*.

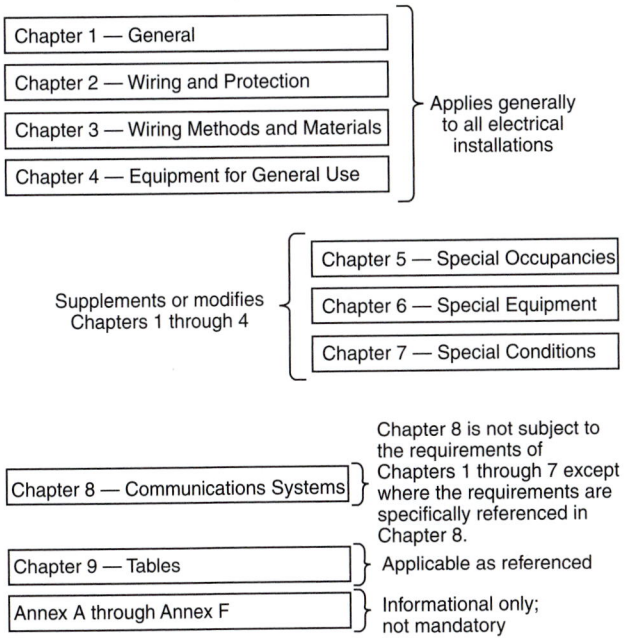

Figure 90.3 Code arrangement.

90.4 Enforcement.

This *Code* is intended to be suitable for mandatory application by governmental bodies that exercise legal jurisdiction over electrical installations, including signaling and communications systems, and for use by insurance inspectors. The authority having jurisdiction for enforcement of the *Code* has the responsibility for making interpretations of the rules, for deciding on the approval of equipment and materials, and for granting the special permission contemplated in a number of the rules.

Some localities do not adopt the *NEC*, but even in those localities, installations that comply with the current *Code* are prima facie evidence that the electrical installation is safe.

Section 90.4 advises that all materials and equipment used under the requirements of this *Code* are subject to the approval of the *authority having jurisdiction*. The text of 90.7, 110.2, and 110.3, along with the definitions of the terms *approved*, *identified*, *labeled*, and *listed*, is intended to provide a basis for the authority having jurisdiction to make the judgments that fall within that particular area of responsibility.

The phrase *including signaling and communication systems* was added for the 2002 *Code* to emphasize that, indeed, these systems are also subject to enforcement.

By special permission, the authority having jurisdiction may waive specific requirements in this *Code* or permit alternative methods where it is assured that equivalent objectives can be achieved by establishing and maintaining effective safety.

It is the responsibility of the *authority having jurisdiction* to interpret the specific rules of the *Code*. This paragraph empowers the authority having jurisdiction, using special permission (written consent), to permit alternative methods where specific rules are not established in the *Code*. For example, the authority having jurisdiction may waive specific requirements in industrial occupancies, research and testing laboratories, and other occupancies where the specific type of installation is not covered in the *Code*.

This *Code* may require new products, constructions, or materials that may not yet be available at the time the *Code* is adopted. In such event, the authority having jurisdiction may permit the use of the products, constructions, or materials that comply with the most recent previous edition of this *Code* adopted by the jurisdiction.

This paragraph of 90.4 permits the authority having jurisdiction to waive a new *Code* requirement during the interim period between acceptance of a new edition of the *NEC* and the availability of a new product, construction, or material redesigned to comply with the increased safety required by the latest edition. It is difficult to establish a viable future effective date within each section of the *NEC* because the time needed to change existing products and standards, as well as to develop new materials and test methods, usually is not known at the time the latest edition of the *Code* is adopted.

90.5 Mandatory Rules, Permissive Rules, and Explanatory Material.

(A) Mandatory Rules. Mandatory rules of this *Code* are those that identify actions that are specifically required or prohibited and are characterized by the use of the terms *shall* or *shall not.*

Section 90.5 (revised and reorganized in the 1999 *Code*) clarifies that two distinctive types of rules are stated in the *Code*. Mandatory rules, characterized by the terms *shall* and *shall not,* are covered in 90.5(A).

(B) Permissive Rules. Permissive rules of this *Code* are those that identify actions that are allowed but not required, are normally used to describe options or alternative methods, and are characterized by the use of the terms *shall be permitted* or *shall not be required.*

Permissive rules (added in the 1999 *Code*) are simply options or alternative methods of achieving equivalent safety — they are not requirements. A close reading of permissive terms is important. Permissive rules are often misinterpreted. For example, the frequently used permissive term *shall be permitted* can be mistaken for a requirement. Substituting "the inspector must allow [item A or method A]" for "[item A or method A] shall be permitted" generally clarifies the interpretation.

(C) Explanatory Material. Explanatory material, such as references to other standards, references to related sections of this *Code,* or information related to a *Code* rule, is included in this *Code* in the form of fine print notes (FPNs). Fine print notes are informational only and are not enforceable as requirements of this *Code.*

Fine print notes (FPNs) do not contain "statements" of intent or recommendations. They present additional supplementary material that aids in the application of the requirement. In addition to printing explanatory material in fine print (small type), the material is further identified in the *Code* by the abbreviation *FPN* preceding the paragraph. Fine print notes are not requirements of the *NEC* and are not enforceable.

Footnotes to tables, although also in fine print, are not explanatory material unless they are identified by the abbre-

viation *FPN.* However, the footnotes are part of the tables and are necessary for proper use of the tables. For example, the footnotes at the end of Table 310.13 are necessary for the use of the table and therefore are mandatory and enforceable *Code* text.

Additional explanatory material is also found in the annexes at the back of this handbook. Annex A is a reference list of product safety standards used for product listing where that listing is required by the *Code.* Annex B provides guidance on the use of the general formula for ampacity found in 310.15(C). Annex C consists of wire fill tables for conduit and tubing. Annex D contains example calculations. Annex E, a table showing fire resistance ratings for Types I–V construction, has been added to this edition to correlate with the expanded use of Type NM cable, as permitted in 334.10. In the 1999 *NEC,* Appendix E was a cross-reference of section numbers to the reorganized Article 250. That cross-reference has been removed for the 2002 edition, and in its place, in Annex F, is a cross-reference for the renumbering of the previous Chapter 3 articles.

FPN: The format and language used in this *Code* follows guidelines established by NFPA and published in the *NEC Style Manual.* Copies of this manual can be obtained from NFPA.

This fine print note informs the user that a style manual is available for the *NEC.* A style manual is basically a "how-to" pamphlet for editors. The *NEC Style Manual* contains a list of rules and regulations used by the panels and editors who prepare the *NEC.* The *NEC Style Manual,* which was revised for the 2002 edition of the *Code,* is available from NFPA.

90.6 Formal Interpretations.

To promote uniformity of interpretation and application of the provisions of this *Code,* formal interpretation procedures have been established and are found in the NFPA Regulations Governing Committee Projects.

The procedures for implementing Formal Interpretations of the provisions of the *NEC* are outlined in "NFPA Regulations Governing Committee Projects." These regulations are included in the *NFPA Directory,* which is published annually and can be obtained from the Secretary of the NFPA Standards Council. The Formal Interpretations procedure can be found in Section 6 of the Regulations.

The National Electrical Code Committee cannot be responsible for subsequent actions of authorities enforcing the *NEC* that accept or reject its findings. The authority having jurisdiction is responsible for interpreting *Code* rules and should attempt to resolve all disagreements at the local level.

Two general forms of Formal Interpretations are recognized: (1) those that are interpretations of the literal text and (2) those that are interpretations of the intent of the Committee at the time the particular text was issued.

Interpretations of the *NEC* not subject to processing are those that involve (a) a determination of compliance of a design, installation, product, or equivalency of protection; (b) a review of plans or specifications or judgment or knowledge that can be acquired only as a result of on-site inspection; (c) text that clearly and decisively provides the requested information; or (d) subjects not previously considered by the Technical Committee or not addressed in the document.

Formal Interpretations of *Code* rules are published in the NFPA Electrical Section News Bulletin, "Current Flashes," and in the *National Fire Codes* subscription service and are sent to interested trade publications.

Most interpretations of the *NEC* are rendered as the personal opinions of NFPA Electrical Engineering staff or of an involved member of the National Electrical Code Committee because the request for interpretation does not qualify for processing as a Formal Interpretation in accordance with "NFPA Regulations Governing Committee Projects." Such opinions are rendered in writing only in response to written requests. The correspondence contains a disclaimer indicating that it is not a Formal Interpretation issued pursuant to NFPA Regulations and that any opinion expressed is the personal opinion of the author and does not necessarily represent the official position of NFPA or the National Electrical Code Committee.

90.7 Examination of Equipment for Safety.

For specific items of equipment and materials referred to in this *Code*, examinations for safety made under standard conditions provide a basis for approval where the record is made generally available through promulgation by organizations properly equipped and qualified for experimental testing, inspections of the run of goods at factories, and service-value determination through field inspections. This avoids the necessity for repetition of examinations by different examiners, frequently with inadequate facilities for such work, and the confusion that would result from conflicting reports on the suitability of devices and materials examined for a given purpose.

It is the intent of this *Code* that factory-installed internal wiring or the construction of equipment need not be inspected at the time of installation of the equipment, except to detect alterations or damage, if the equipment has been listed by a qualified electrical testing laboratory that is recognized as having the facilities described in the preceding paragraph and that requires suitability for installation in accordance with this *Code*.

FPN No. 1: See requirements in 110.3.
FPN No. 2: *Listed* is defined in Article 100.
FPN No. 3: Annex A contains an informative list of product safety standards for electrical equipment.

Testing laboratories, inspection agencies, and other organizations concerned with product evaluation publish lists of equipment and materials that have been tested and meet nationally recognized standards or that have been found suitable for use in a specified manner. The *Code* does not contain detailed information on equipment or materials but refers to the products as "listed," "labeled," or "identified." See Article 100 for definitions of these terms.

NFPA does not approve, inspect, or certify any installations, procedures, equipment, or materials; nor does it approve or evaluate testing laboratories. In determining the acceptability of installations or procedures, equipment, or materials, the authority having jurisdiction may base acceptance on compliance with NFPA or other appropriate standards. In the absence of such standards, the authority may require evidence of proper installation, procedures, or use. The authority having jurisdiction may also refer to the listing or labeling practices of an organization concerned with product evaluations that is able to determine compliance with appropriate standards for the current production of listed items.

New FPN No. 3 for the 2002 edition points the user to Annex A, which is a nonmandatory list of product safety standards used for product listing. The list includes only product safety standards for which a listing is required by the *Code*. For example, 344.6 requires that rigid metal conduit, Type RMC, be listed. By using Annex A, the user finds that the listing standard for rigid metal conduit is UL 6, *Rigid Metal Conduit*. Because associated conduit fittings are also required to be listed, UL 514B, *Fittings for Cable and Conduit*, is found in Annex A also.

90.8 Wiring Planning.

(A) Future Expansion and Convenience. Plans and specifications that provide ample space in raceways, spare raceways, and additional spaces allow for future increases in the use of electricity. Distribution centers located in readily accessible locations provide convenience and safety of operation.

The requirement for providing the exclusively dedicated equipment space mandated by 110.26(F) supports the intent of 90.8(A) regarding future increases in the use of electricity.

Distribution centers should contain additional space and capacity for future additions and should be conveniently located for easy accessibility.

Where distribution equipment is installed so that easy access cannot be achieved, a spare raceway(s) or pull line(s) should be run at the initial installation, as illustrated in Exhibit 90.2.

(B) Number of Circuits in Enclosures. It is elsewhere provided in this *Code* that the number of wires and circuits confined in a single enclosure be varyingly restricted. Limiting the number of circuits in a single enclosure minimizes the effects from a short circuit or ground fault in one circuit.

These limitations will minimize the heating effects inherently present wherever current-carrying conductors are grouped together. See 408.15 for restrictions on the number of overcurrent devices on one panelboard.

90.9 Units of Measurement.

(A) Measurement System of Preference. For the purpose of this *Code*, metric units of measurement are in accordance with the modernized metric system known as the International System of Units (SI).

According to a recent report titled "A Metric for Success" by the National Institute of Standards and Technology (NIST), most U.S. industries that do business abroad are predominantly metric already because of global sourcing of parts, service, components, and production. However, quite a few domestic industries still use U.S. Customary units. The NIST report warns that domestic industries that ignore global realities and continue to design and manufacture with nonmetric measures will find that they risk increasing their costs. Nonmetric modular products (the building construction industry uses great quantities of modular parts) and those that interface with outside industry products are especially vulnerable to the added costs of adapting to a metric environment.

Metric standards are beginning to appear in the domestic building construction industry because our national standards are being harmonized with international standards.

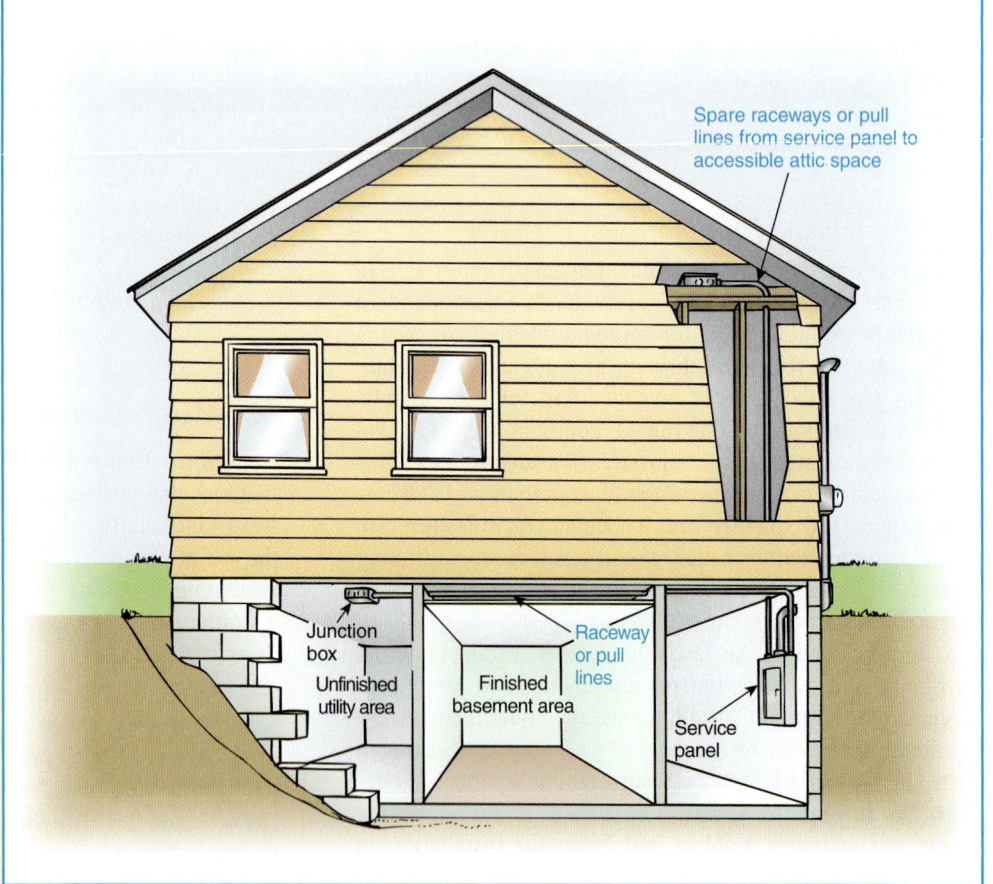

Exhibit 90.2 A residential distribution system showing spare raceways or pull lines that allow for future circuits and loads.

The *National Electrical Code* is an important building construction standard and, with this edition, has moved one more step in the metric direction.

(B) Dual System of Units. The SI units shall appear first, and the inch-pound units shall immediately follow in parentheses. The conversion from the inch-pound units to SI units shall be based on hard conversion except as provided in 90.9(C).

Hard conversion is explained in FPN No. 1 following 90.9(D). Calculations to convert measurements from inch-pound units to metric units must be made using "hard" conversion. The hard-conversion method is mandatory except for trade sizes [e.g., raceway sizes in Table 300.1(C)], extracted material (e.g., class and zone measurements from other NFPA documents), and safety calculations (e.g., minimum distances taken from Table 110.31).

Example

Using the hard-conversion method, determine the equivalent metric conversion for 24 in., generally the minimum cover requirements for direct burial cables and conductors in nonspecific locations taken from row 1 of Table 300.5.

Solution

STEP 1. $24 \text{ in.} \times \dfrac{25.4 \text{ mm}}{1 \text{ in.}} = 609.6 \text{ mm}$

STEP 2. Because the calculation is being performed as a hard conversion, the 609.6 mm dimension may be changed. For the 2002 Code, the selected equivalent cover requirement is 600 mm.

For the 2002 Code, the measurements of 600 mm and 24 in. appear in Table 300.5 for the minimum cover requirements for direct burial cables and conductors in nonspecific locations. For the 1999 NEC, the selected SI unit of measure was required to be 609.6 mm. However, the 2002 Code permits much more latitude for the final selected dimension, and so the equivalent minimum cover requirement of 600 mm is a more practical solution. Basically, a hard conversion permits a change in a dimension or allows rounding up or down to better fit the physical constraints of the installation.

(C) Permitted Uses of Soft Conversion. The cases given in 90.9(C)(1) through (4) shall not be required to use hard conversion and shall be permitted to use soft conversion.

(1) Trade Sizes. Where the actual measured size of a product is not the same as the nominal size, trade size designators shall be used rather than dimensions. Trade practices shall be followed in all cases.

Metric trade sizes (metric designators) of conduits were added in the 1996 *Code* as fine print notes in each raceway article. In the 2002 *Code*, these metric designators appear in the *Code* text, preceding the trade size equivalents, in the raceway articles.

For example, in 350.20(A) of the 2002 *NEC*, the size requirement is now stated as follows: "LFMC smaller than metric designator 16 (trade size ½) shall not be used." In 351-5(a) of the 1999 *NEC*, the size requirement was stated as follows: "Liquidtight flexible metallic conduit smaller than ½-in. electrical trade size shall not be used."

This change does not reflect a technical change but rather provides acceptable language to both domestic and international users of the *NEC*. For ease of use, in Table 4 of Chapter 9, metric designators are now separate columns.

(2) Extracted Material. Where material is extracted from another standard, the context of the original material shall not be compromised or violated. Any editing of the extracted text shall be confined to making the style consistent with that of the *NEC*.

(3) Industry Practice. Where industry practice is to express units in inch-pound units, the inclusion of SI units shall not be required.

(4) Safety. Where a negative impact on safety would result, hard conversion shall not be required.

The following examples illustrate conversions from U.S. Customary units to SI units. Example 1 shows the process of converting a dimension from feet to meters, where safety is a concern. Table 110.31 contains minimum permitted distances from a fence to a live part for voltages 601 and greater. Example 1 calculates the equivalent metric conversion for 10 ft using the minimum distance of 10 ft in Table 110.31 where the measurement is from a fence to a live part from 601 volts to 13,799 volts.

Example 1

Determine the equivalent metric conversion for 10 ft where the calculation could have a negative impact on safety, such as the minimum distance of 10 ft from Table 110.31, and where the measurement is from a fence to a live part from 601 volts to 13,799 volts.

Solution

STEP 1. $10 \text{ ft} \times \dfrac{0.3048 \text{ m}}{1 \text{ ft}} = 3.048 \text{ m}$

STEP 2. Round up the calculation to 3.05 m, because a distance less than 3.048 could have a negative impact on safety. The answer, 3.05 m, matches the minimum distance in Table 110.31 from a fence to a live part from 601 volts to 13,799 volts.

Because safety is a concern for this conversion calculation, the original *Code* distance (the U.S. Customary units, for this example) remains the shortest permitted distance. The final metric equivalent ends up slightly larger. The exact difference is of no practical concern, however, since 0.2 mm is less than $\frac{1}{32}$ in. From a practical point of view, a variance of $\frac{1}{32}$ in. in a length of 10 ft is insignificant.

Example 2

Using the "soft-conversion" method, determine the equivalent metric conversion for 30 in. where the calculation could have a negative impact on safety, such as a 30-in. minimum horizontal working space requirement in the rear of equipment that requires access to nonelectrical parts in 110.26(A)(1)(a).

Solution

STEP 1. $30 \text{ in.} \times \frac{25.4 \text{ mm}}{1 \text{ in.}} = 762 \text{ mm}$

STEP 2. Do not round off the calculation, because even a slight reduction in the original distance could have a negative impact on safety. The answer is 762 mm, and this metric dimension matches the minimum distance of 110.26(A)(1)(a) for a minimum horizontal working space.

(D) **Compliance.** The conversion from inch-pound units to SI units shall be permitted to be an approximate conversion. Compliance with the numbers shown in either the SI system or the inch-pound system shall constitute compliance with this *Code*.

> FPN No. 1: Hard conversion is considered a change in dimensions or properties of an item into new sizes that might or might not be interchangeable with the sizes used in the original measurement. Soft conversion is considered a direct mathematical conversion and involves a change in the description of an existing measurement but not in the actual dimension.
>
> FPN No. 2: SI conversions are based on IEEE/ASTM SI 10-1997, *Standard for the Use of the International System of Units (SI): The Modern Metric System.*

Table 90.1 offers some examples of the hard-conversion process used for the 2002 *NEC* revision cycle. U.S. Customary units were used in the 1993, 1996, and 1999 *Code* and are still valid for the 2002 *Code*. Soft-conversion SI units were used in the 1996 and 1999 *Code*. The hard-conversion SI units, which are new to the 2002 *Code*, are listed with their equivalent U.S. Customary units. The equivalent U.S. units are given only to show the small variance between customary units and the hard-conversion units of the 2002 *Code*.

Warning signs that state specific clearances, such as required in 513.10(B), permit distance measurements in either inch-pound units or metric units.

Table 90.1 Conversions Using the Hard-Conversion Method

U.S. Customary Unit	Soft Conversion, SI Unit	Hard Conversion, SI Unit	Equivalent U.S. Customary Unit
½ in.	12.7 mm	13 mm	0.51 in.
¾ in.	19 mm	19 mm	0.75 in.
1 in.	25.4 mm	25 mm	0.98 in.
4 in.	102 mm	100 mm	3.94 in.
12 in.	305 mm	300 mm	11.81 ft
2 ft	610 mm	600 mm	1.97 ft
3 ft	914 mm	900 mm	2.95 ft
6 ft	1.83 m	1.8 m	5.91 ft
15 ft	4.57 m	4.5 m	14.76 ft

ARTICLE 100
Definitions

Contents

Scope. This article contains only those definitions essential to the proper application of this *Code*. It is not intended to include commonly defined general terms or commonly defined technical terms from related codes and standards. In general, only those terms that are used in two or more articles are defined in Article 100. Other definitions are included in the article in which they are used but may be referenced in Article 100.

Part I of this article contains definitions intended to apply wherever the terms are used throughout this *Code*. Part II contains definitions applicable only to the parts of articles specifically covering installations and equipment operating at over 600 volts, nominal.

Commonly defined general terms include those terms defined in general English language dictionaries and terms that are not used in a unique or restricted manner in the *NEC*. Commonly defined technical terms such as *volt* (abbreviated V) and *ampere* (abbreviated A) are found in the *IEEE Standard Dictionary of Electrical and Electronic Terms*.

Definitions that are not listed in Article 100 are included in their appropriate article. For articles that follow the common format according to the *NEC Style Manual*, the section number is generally XXX.2 Definition(s). For example, the definition of *nonmetallic-sheathed cable* is found in 334.2, Definition. This edition of the *Code* contains some isolated exceptions to this general rule because the *NEC* has not been entirely converted to a common numbering system.

I. General

Accessible (as applied to equipment). Admitting close approach; not guarded by locked doors, elevation, or other effective means.

Exhibit 100.1 illustrates a few examples of equipment considered accessible (as applied to equipment). The main rule for switches and circuit breakers used as switches is shown in (a) and is according to 404.8(A). In (b), the busway installation is according to 368.12. The exceptions to the

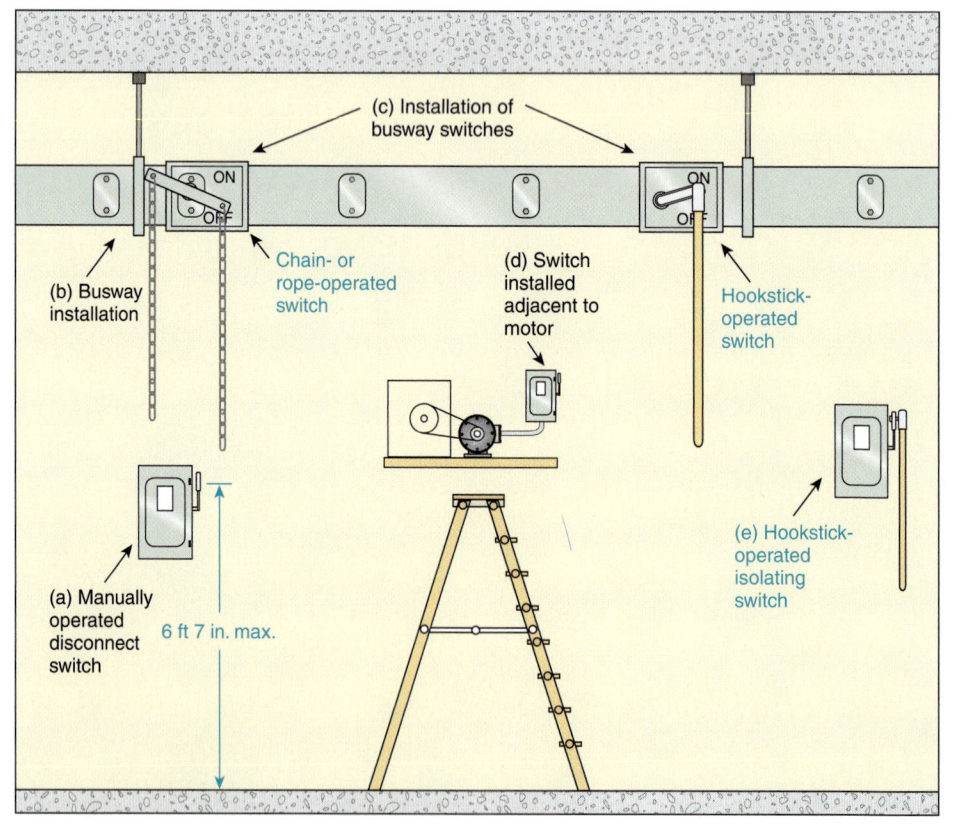

Exhibit 100.1 Example of busway and of switches considered accessible even if located above 6 ft 7 in.

(c) Installation of busway switches

(b) Busway installation

Chain- or rope-operated switch

(d) Switch installed adjacent to motor

Hookstick-operated switch

(a) Manually operated disconnect switch

6 ft 7 in. max.

(e) Hookstick-operated isolating switch

main rule are illustrated in (c) the installation of busway switches installed according to 404.8(A), Exception No. 1; (d) a switch installed adjacent to a motor according to 404.8(A), Exception No. 2; and (e) a hookstick-operated isolating switch installed according to 404.8(A), Exception No. 3.

Accessible (as applied to wiring methods). Capable of being removed or exposed without damaging the building structure or finish or not permanently closed in by the structure or finish of the building.

Wiring methods located behind removable panels designed to allow access are not considered permanently enclosed and are considered exposed as applied to wiring methods. See 300.4(C) regarding cables located in spaces behind accessible panels.

Exhibit 100.2 illustrates examples of wiring methods and equipment that are considered accessible.

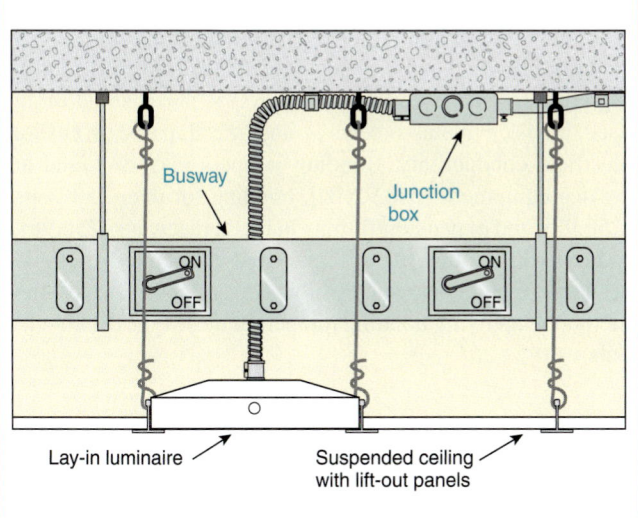

Exhibit 100.2 Examples of busways and junction boxes considered accessible even if located behind hung ceilings having lift-out panels.

Accessible, Readily (Readily Accessible). Capable of being reached quickly for operation, renewal, or inspections without requiring those to whom ready access is requisite to climb over or remove obstacles or to resort to portable ladders, and so forth.

The definition of *readily accessible* does not preclude the use of a locked door for service equipment or rooms containing service equipment, provided those for whom ready access is necessary have a key (or lock combination) available.

For example, 230.70(A)(1) and 230.205(A) require service-disconnecting means to be readily accessible. Section 225.32 requires that feeder disconnecting means for separate buildings be readily accessible. A commonly used, permitted practice is to locate the disconnecting means in the electrical equipment room of an office building or large apartment building and to keep the door to that room locked to prevent access by unauthorized persons. Section 240.24(A) requires that overcurrent devices be so located as to be readily accessible.

Ampacity. The current, in amperes, that a conductor can carry continuously under the conditions of use without exceeding its temperature rating.

The definition of the term *ampacity* states that the maximum current a conductor carries continuously varies with the conditions of use as well as with the temperature rating of the conductor insulation. For example, ambient temperature is a condition of use. A conductor with insulation rated at 60°C, installed near a furnace where the ambient temperature is continuously maintained at 60°C, has no current-carrying capacity. Any current flowing through the conductor will raise its temperature above the 60°C insulation rating. Therefore, the ampacity of this conductor, regardless of its size, is zero. See the ampacity correction factors for temperature at the bottom of Tables 310.16 through 310.20, or see Annex B. The temperature limitations on conductors is further explained, and examples are given, in 310.10 and in the commentary following that section.

Another condition of use is the number of conductors in a raceway or cable. [See 310.15(B)(2).]

Appliance. Utilization equipment, generally other than industrial, that is normally built in standardized sizes or types and is installed or connected as a unit to perform one or more functions such as clothes washing, air conditioning, food mixing, deep frying, and so forth.

Approved. Acceptable to the authority having jurisdiction.

See the definition of *authority having jurisdiction* and 110.2 for a better understanding of the approval process. Understanding *NEC* terms such as *listed*, *labeled*, and *identified* (as applied to equipment) will also assist the user in understanding the approval process.

Askarel. A generic term for a group of nonflammable synthetic chlorinated hydrocarbons used as electrical insulating media. Askarels of various compositional types are used. Under arcing conditions, the gases produced, while consisting predominantly of noncombustible hydrogen chloride, can include varying amounts of combustible gases, depending on the askarel type.

Attachment Plug (Plug Cap) (Plug). A device that, by insertion in a receptacle, establishes a connection between the conductors of the attached flexible cord and the conductors connected permanently to the receptacle.

Standard attachment caps are also available with built-in options, such as switching, fuses, or even ground-fault circuit-interrupter protection.

Attachment plug contact blades have specific shapes, sizes, and configurations so that a receptacle or cord connector will not accept an attachment plug of a different voltage or current rating than that for which the device is intended. Configuration charts from NEMA Standard WD 6 for general-purpose nonlocking and specific-purpose locking plugs and receptacles are shown in Exhibits 406.2 and 406.3, respectively.

Authority Having Jurisdiction. The organization, office, or individual responsible for approving equipment, materials, an installation, or a procedure.

> FPN: The phrase "authority having jurisdiction" is used in NFPA documents in a broad manner, since jurisdictions and approval agencies vary, as do their responsibilities. Where public safety is primary, the authority having jurisdiction may be a federal, state, local, or other regional department or individual such as a fire chief; fire marshal; chief of a fire prevention bureau, labor department, or health department; building official; electrical inspector; or others having statutory authority. For insurance purposes, an insurance inspection department, rating bureau, or other insurance company representative may be the authority having jurisdiction. In many circumstances, the property owner or his or her designated agent assumes the role of the authority having jurisdiction; at government installations, the commanding officer or departmental official may be the authority having jurisdiction.

The important role of the authority having jurisdiction (AHJ) cannot be overstated in the current North American safety system. The basic role of the AHJ is to verify that an installation complies with the *Code*. The definition of *authority having jurisdiction* and the accompanying explanation (the FPN) bring a sense of uniformity to the *Code*, since this exact definition has appeared in many other NFPA documents for quite some time and now is incorporated into the 2002 *NEC*. This definition is very helpful in understanding *Code* enforcement, the inspection process, the definition of *approved*, and 110.2 and 90.7.

Automatic. Self-acting, operating by its own mechanism when actuated by some impersonal influence, as, for exam-

ple, a change in current, pressure, temperature, or mechanical configuration.

Bathroom. An area including a basin with one or more of the following: a toilet, a tub, or a shower.

Bonding (Bonded). The permanent joining of metallic parts to form an electrically conductive path that ensures electrical continuity and the capacity to conduct safely any current likely to be imposed.

The purpose of bonding is to establish an effective path for fault current that, in turn, facilitates the operation of the overcurrent protective device. This is explained in 250.4(A)(3) and (4) and 250.4(B)(3) and (4). Specific bonding requirements are found in Part V of Article 250 and in other sections of the *Code* as referenced in 250.3.

Bonding Jumper. A reliable conductor to ensure the required electrical conductivity between metal parts required to be electrically connected.

Both concentric- and eccentric-type knockouts can impair the electrical conductivity between the metal parts and may actually introduce unnecessary impedance into the grounding path. Installing bonding jumper(s) is one method often used between metal raceways and metal parts to ensure electrical conductivity. Bonding jumpers may be found at service equipment [250.92(B)], bonding for over 250 volts (250.97), and expansion fittings in metal raceways (250.98). Exhibit 100.3 shows the difference between concentric- and eccentric-type knockouts. Exhibit 100.3 also illustrates one method of applying bonding jumpers at these types of knockouts.

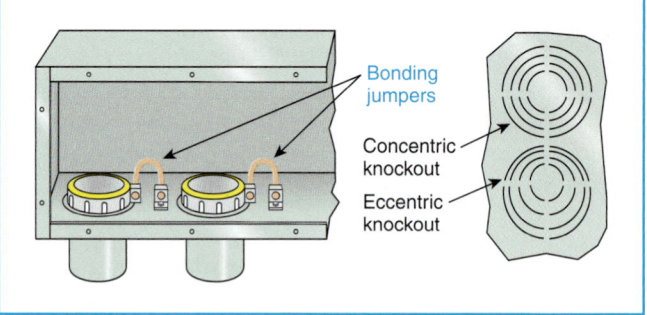

Bonding jumpers

Concentric knockout

Eccentric knockout

Exhibit 100.3 Bonding jumpers installed around concentric or eccentric knockouts.

Bonding Jumper, Equipment. The connection between two or more portions of the equipment grounding conductor.

Bonding Jumper, Main. The connection between the grounded circuit conductor and the equipment grounding conductor at the service.

Exhibit 100.4 shows a main bonding jumper used to provide the connection between the grounded service conductor and the equipment grounding conductor at the service. Bonding jumpers may be located throughout the electrical system, but a main bonding jumper is located only at the service. Main bonding jumper requirements are found in 250.28.

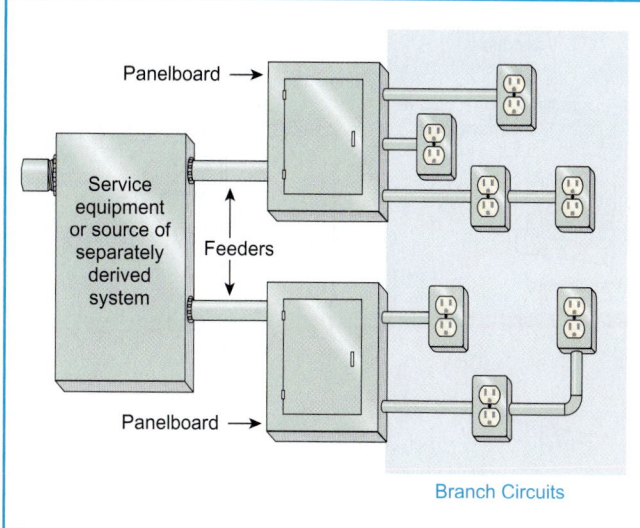

Exhibit 100.5 Feeder (circuits) and branch circuits.

210.52(B)(1) requires that these circuits supply receptacle outlets located in such rooms as the kitchen, pantry, and so on. These small-appliance branch circuits are not permitted to supply other outlets or permanently connected lighting fixtures. (See 210.52 for exact details.)

Branch Circuit, General-Purpose. A branch circuit that supplies two or more receptacles or outlets for lighting and appliances.

Branch Circuit, Individual. A branch circuit that supplies only one utilization equipment.

An individual branch circuit is a circuit that supplies only one piece of utilization equipment (e.g., one range, one space heater, one motor). See 210.23 regarding permissible loads for branch circuits.

An individual branch circuit supplies only one single receptacle for the connection of a single attachment plug. This single receptacle is required to have an ampere rating not less than that of the branch circuit, as stated in 210.21(B)(1).

Exhibit 100.6 illustrates an individual branch circuit with a single receptacle intended for the connection of one piece of utilization equipment. A branch circuit that supplies one duplex receptacle that can accommodate two cord-and-plug-connected appliances or similar equipment is not an individual branch circuit.

Branch Circuit, Multiwire. A branch circuit that consists of two or more ungrounded conductors that have a voltage between them, and a grounded conductor that has equal voltage between it and each ungrounded conductor of the

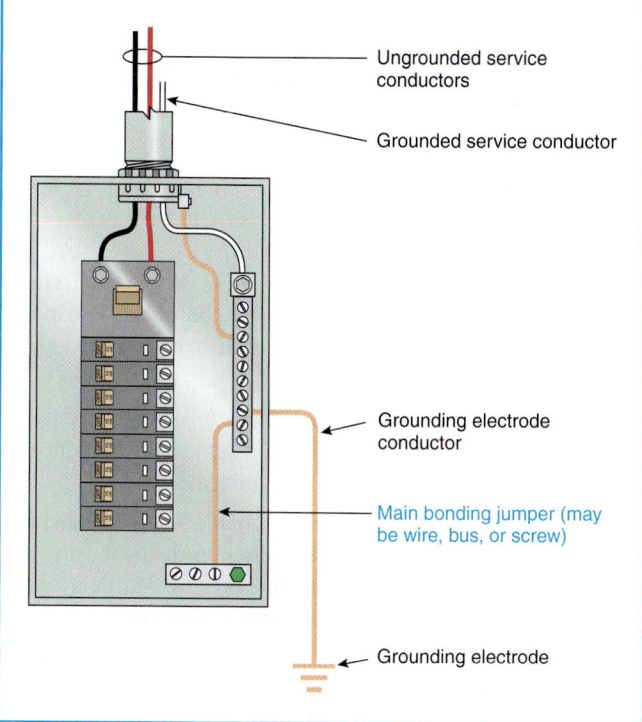

Exhibit 100.4 A main bonding jumper installed at the service between the grounded service conductor and the equipment grounding conductor.

Branch Circuit. The circuit conductors between the final overcurrent device protecting the circuit and the outlet(s).

Exhibit 100.5 shows the difference between branch circuits and feeders. Conductors between the overcurrent devices in the panelboards and the duplex receptacles are branch-circuit conductors. Conductors between the service equipment or source of separately derived systems and the panelboards are feeders.

Branch Circuit, Appliance. A branch circuit that supplies energy to one or more outlets to which appliances are to be connected and that has no permanently connected luminaires (lighting fixtures) that are not a part of an appliance.

Two or more 20-ampere small-appliance branch circuits are required by 210.11(C)(1) for dwelling units. Section

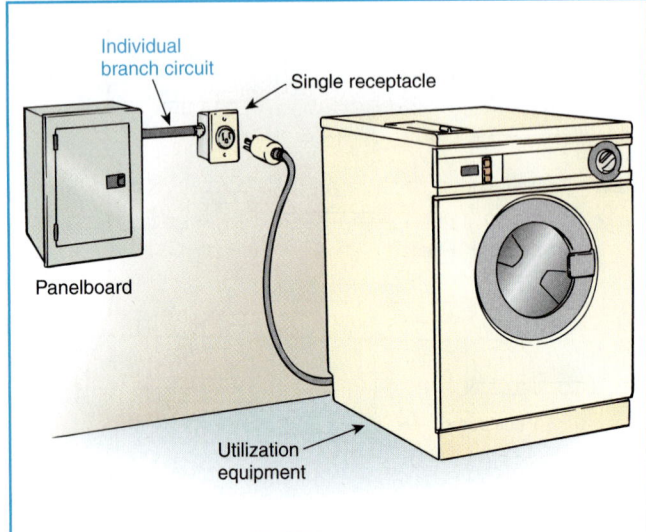

Individual
branch circuit

Single receptacle

Panelboard

Utilization
equipment

Exhibit 100.6 An individual branch circuit, which supplies only
one utilization equipment via a single receptacle.

circuit and that is connected to the neutral or grounded
conductor of the system.

For the 2002 edition, this definition was editorially modified
by substituting the word *voltage* for the term *potential difference*. See 210.4, 240.20(B)(1), and 300.13(B) for specific
information about multiwire branch circuits.

Building. A structure that stands alone or that is cut off
from adjoining structures by fire walls with all openings
therein protected by approved fire doors.

A building is generally considered to be a roofed or walled
structure that may be used or intended for supporting or
sheltering any use or occupancy. However, it may also be
a separate structure such as a pole, billboard sign, or water
tower.
    Definitions of the terms *fire walls* and *fire doors* are
the responsibility of building codes. Generically, a fire wall
may be defined as a wall that separates buildings or sub-
divides a building to prevent the spread of fire and that has
a fire resistance rating and structural stability. Fire doors
(and fire windows) are used to protect openings in walls,
floors, and ceilings against the spread of fire and smoke
within, into, or out of buildings.

Cabinet. An enclosure that is designed for either surface
mounting or flush mounting and is provided with a frame,
mat, or trim in which a swinging door or doors are or can
be hung.

Both cabinets and cutout boxes are covered in Article 312.
Cabinets are designed for surface or flush mounting with a
trim to which a swinging door(s) is hung. Cutout boxes
are designed for surface mounting with a swinging door(s)
secured directly to the box. Panelboards are electrical assemblies designed to be placed in a cabinet or cutout box. (See
definitions of *cutout box* and *panelboard*.)

Circuit Breaker. A device designed to open and close a
circuit by nonautomatic means and to open the circuit automatically on a predetermined overcurrent without damage
to itself when properly applied within its rating.

> FPN: The automatic opening means can be integral, di-
> rect acting with the circuit breaker, or remote from the
> circuit breaker.

Adjustable (as applied to circuit breakers). A qualifying
term indicating that the circuit breaker can be set to trip at
various values of current, time, or both, within a predetermined range.

Instantaneous Trip (as applied to circuit breakers). A qualifying term indicating that no delay is purposely introduced
in the tripping action of the circuit breaker.

Inverse Time (as applied to circuit breakers). A qualifying
term indicating that there is purposely introduced a delay in
the tripping action of the circuit breaker, which delay decreases as the magnitude of the current increases.

Nonadjustable (as applied to circuit breakers). A qualifying
term indicating that the circuit breaker does not have any
adjustment to alter the value of current at which it will trip
or the time required for its operation.

Setting (of circuit breakers). The value of current, time, or
both, at which an adjustable circuit breaker is set to trip.

Concealed. Rendered inaccessible by the structure or finish
of the building. Wires in concealed raceways are considered
concealed, even though they may become accessible by
withdrawing them.

Raceways and cables supported or located within hollow
frames or permanently closed in by the finish of buildings
are considered concealed. Open-type work — such as raceways and cables in exposed areas; in unfinished basements;
in accessible underfloor areas or attics; attached to the surface of finished areas; or behind, above, or below panels
designed to allow access and that may be removed without
damage to the building structure or finish — is not considered concealed. [See definition of *exposed (as applied to
wiring methods)*.]

Conductor, Bare. A conductor having no covering or electrical insulation whatsoever.

Conductor, Covered. A conductor encased within material of composition or thickness that is not recognized by this *Code* as electrical insulation.

Typical covered conductors are the green-covered equipment grounding conductors contained within a nonmetallic-sheathed cable or the uninsulated grounded system conductors within the overall exterior jacket of a Type SE cable. Covered conductors should always be treated as bare conductors for working clearances, since they are really uninsulated conductors.

Conductor, Insulated. A conductor encased within material of composition and thickness that is recognized by this *Code* as electrical insulation.

For the covering on a conductor to be considered insulation, the conductor with the covering material generally is required to pass minimum testing required by a product standard. One such product standard is UL 83, *Thermoplastic-Insulated Wires and Cables*. To meet the requirements of UL 83, specimens of finished single-conductor wires must pass specified tests that measure (1) resistance to flame propagation, (2) dielectric strength, even while immersed, and (3) resistance to abrasion, cracking, crushing, and impact. Only wires and cables that meet the minimum fire, electrical, and physical properties required by the applicable standards are permitted to be marked with the letter designations found in Tables 310.13 and 310.61. See 310.13 for the exact requirements of insulated conductor construction and applications.

Conduit Body. A separate portion of a conduit or tubing system that provides access through a removable cover(s) to the interior of the system at a junction of two or more sections of the system or at a terminal point of the system.

Boxes such as FS and FD or larger cast or sheet metal boxes are not classified as conduit bodies.

Conduit bodies are a portion of a raceway system with removable covers to allow access to the interior of the system. They include the short-radius type as well as capped elbows and service-entrance elbows.

Some conduit bodies are referred to in the trade as "condulets" and include the LB, LL, LR, C, T, and X designs. (See 300.15 and Article 314 for rules on the usage of conduit bodies.)

Types FS and FD boxes are not classified as conduit bodies; they are listed with boxes in Table 314.16(A).

Connector, Pressure (Solderless). A device that establishes a connection between two or more conductors or between one or more conductors and a terminal by means of mechanical pressure and without the use of solder.

Continuous Load. A load where the maximum current is expected to continue for 3 hours or more.

Controller. A device or group of devices that serves to govern, in some predetermined manner, the electric power delivered to the apparatus to which it is connected.

A controller may be a remote-controlled magnetic contactor, switch, circuit breaker, or device that is normally used to start and stop motors and other apparatus and, in the case of motors, is required to be capable of interrupting the stalled-rotor current of the motor. Stop-and-start stations and similar control circuit components that do not open the power conductors to the motor are not considered controllers.

Cooking Unit, Counter-Mounted. A cooking appliance designed for mounting in or on a counter and consisting of one or more heating elements, internal wiring, and built-in or mountable controls.

Copper-Clad Aluminum Conductors. Conductors drawn from a copper-clad aluminum rod with the copper metallurgically bonded to an aluminum core. The copper forms a minimum of 10 percent of the cross-sectional area of a solid conductor or each strand of a stranded conductor.

Cutout Box. An enclosure designed for surface mounting that has swinging doors or covers secured directly to and telescoping with the walls of the box proper.

Dead Front. Without live parts exposed to a person on the operating side of the equipment.

Demand Factor. The ratio of the maximum demand of a system, or part of a system, to the total connected load of a system or the part of the system under consideration.

Device. A unit of an electrical system that is intended to carry but not utilize electric energy.

Components (such as switches, circuit breakers, fuseholders, receptacles, attachment plugs, and lampholders) that distribute or control but do not consume electricity are considered devices.

Disconnecting Means. A device, or group of devices, or other means by which the conductors of a circuit can be disconnected from their source of supply.

For disconnecting means for service equipment, see Part VI of Article 230; for fuses, see Part IV of Article 240; for circuit breakers, see Part VII of Article 240; for appliances, see Part III of Article 422; for space-heating equipment, see Part III of Article 424; for motors and controllers, see Part IX of Article 430; and for air-conditioning and refrigerating equipment, see Part II of Article 440. (See also references for *disconnecting means* in the index.)

Dusttight. Constructed so that dust will not enter the enclosing case under specified test conditions.

Table 430.91, Motor Controller Enclosure Selection, provides a basis for selecting enclosure types that are dusttight. (See also the commentary following the definition of *enclosure*.)

Note that the term *dustproof* was removed from the *Code* for the 2002 edition because it is no longer applicable or used in the *Code*.

Duty, Continuous. Operation at a substantially constant load for an indefinitely long time.

Duty, Intermittent. Operation for alternate intervals of (1) load and no load; or (2) load and rest; or (3) load, no load, and rest.

Duty, Periodic. Intermittent operation in which the load conditions are regularly recurrent.

Duty, Short-Time. Operation at a substantially constant load for a short and definite, specified time.

Duty, Varying. Operation at loads, and for intervals of time, both of which may be subject to wide variation.

Information on the protection of intermittent, periodic, short-time, and varying-duty motors against overload can be found in 430.33.

Dwelling Unit. One or more rooms for the use of one or more persons as a housekeeping unit with space for eating, living, and sleeping, and permanent provisions for cooking and sanitation.

A mobile home may be considered to be a dwelling unit. Where dwelling units are referenced throughout the *Code*, it is important to note that rooms in motels, hotels, and similar occupancies could be classified as dwelling units if they satisfy the requirements of the definition. For example, the motel or hotel room illustrated in Exhibit 100.7 clearly meets the definition because it has eating, living, and sleeping space and permanent areas for cooking and sanitation.

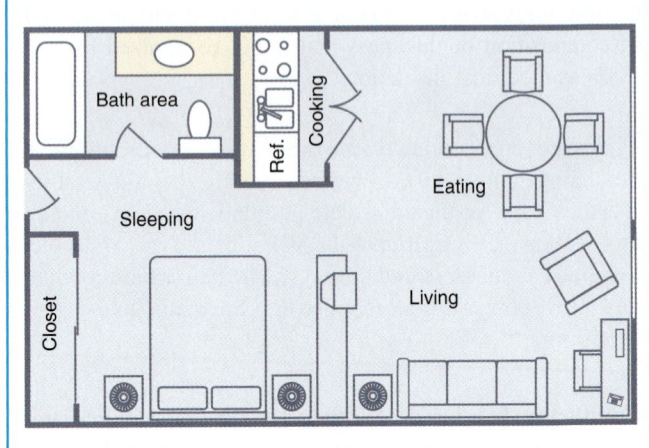

Exhibit 100.7 Example of motel or hotel room considered to be a dwelling unit.

Dwelling, One-Family. A building that consists solely of one dwelling unit.

Dwelling, Two-Family. A building that consists solely of two dwelling units.

Dwelling, Multifamily. A building that contains three or more dwelling units.

Electric Sign. A fixed, stationary, or portable self-contained, electrically illuminated utilization equipment with words or symbols designed to convey information or attract attention.

Enclosed. Surrounded by a case, housing, fence, or wall(s) that prevents persons from accidentally contacting energized parts.

Enclosure. The case or housing of apparatus, or the fence or walls surrounding an installation to prevent personnel from accidentally contacting energized parts or to protect the equipment from physical damage.

FPN: See Table 430.91 for examples of enclosure types.

The information in Table 1.1 is taken from the 2002 UL *General Information Directory* (White Book), category AALZ, "Electrical Equipment for Use in Ordinary Locations." It summarizes the intended uses of the various types of enclosures for nonhazardous locations.

Enclosures that comply with the requirements for more than one type of enclosure may be marked with multiple designations. Enclosures marked with a type may also be marked as follows:

A Type 1 may be marked "Indoor Use Only."

A Type 3, 3S, 4, 4X, 6, or 6P may be marked "Raintight."

Table 1.1 Environmental Protections for Nonhazardous Locations, by Type of Enclosure

Enclosure Type Number	Provides a Degree of Protection Against the Following Environmental Conditions*
1	Indoor use
2	Indoor use, limited amounts of falling water
3R	Outdoor use, undamaged by the formation of ice on the enclosure**
3	Same as 3R plus windblown dust
3S	Same as 3R plus windblown dust; external mechanisms remain operable while ice laden
4	Outdoor use, splashing water, windblown dust, hose-directed water, undamaged by the formation of ice on the enclosure**
4X	Same as 4 plus resists corrosion
5	Indoor use to provide a degree of protection against settling airborne dust, falling dirt, and dripping noncorrosive liquids
6	Same as 3R plus entry of water during temporary submersion at a limited depth
6P	Same as 3R plus entry of water during prolonged submersion at a limited depth
12, 12K	Indoor use, dust, dripping noncorrosive liquids
13	Indoor use, dust, spraying water, oil, and noncorrosive coolants

*All enclosure types provide a degree of protection against ordinary corrosion and against accidental contact with the enclosed equipment when doors or covers are closed and in place. All types of enclosures provide protection against a limited amount of falling dirt.

**All outdoor-type enclosures provide a degree of protection against rain, snow, and sleet. Outdoor enclosures are also suitable for use indoors if they meet the environmental conditions present.

Source: Underwriters Laboratories, *General Information Directory*, 2001 edition.

A Type 3R may be marked "Rainproof."

A Type 4, 4X, 6, or 6P may be marked "Watertight."

A Type 4X or 6P may be marked "Corrosion Resistant."

A Type 2, 5, 12, 12K, or 13 may be marked "Driptight."

A Type 3, 3S, 5, 12K, or 13 may be marked "Dusttight."

For equipment designated "Raintight," testing designed to simulate exposure to a beating rain will not result in entrance of water. For equipment designated "Rainproof," testing designed to simulate exposure to a beating rain will not interfere with the operation of the apparatus or result in wetting of live parts and wiring within the enclosure. "Watertight" equipment is so constructed that water does not enter the enclosure when subjected to a stream of water. "Corrosion resistant" equipment is constructed so that it provides a degree of protection against exposure to corrosive agents such as salt spray. "Driptight" equipment is constructed so that falling moisture or dirt does not enter the enclosure. "Dusttight" equipment is constructed so that circulating or airborne dust does not enter the enclosure.

Energized. Electrically connected to a source of voltage.

This definition was editorially revised for the 2002 *Code* by substituting the term *voltage* for *potential difference*. For a more thorough understanding of *energized*, also see the definitions of *exposed (as applied to live parts)* and *live parts*.

Equipment. A general term including material, fittings, devices, appliances, luminaires (fixtures), apparatus, and the like used as a part of, or in connection with, an electrical installation.

Explosionproof Apparatus. Apparatus enclosed in a case that is capable of withstanding an explosion of a specified gas or vapor that may occur within it and of preventing the ignition of a specified gas or vapor surrounding the enclosure by sparks, flashes, or explosion of the gas or vapor within, and that operates at such an external temperature that a surrounding flammable atmosphere will not be ignited thereby.

> FPN: For further information, see ANSI/UL 1203-1999, *Explosion-Proof and Dust-Ignition-Proof Electrical Equipment for Use in Hazardous (Classified) Locations.*

Exposed (as applied to live parts). Capable of being inadvertently touched or approached nearer than a safe distance by a person. It is applied to parts that are not suitably guarded, isolated, or insulated.

For a more thorough understanding of *exposed (as applied to live parts)*, also see the definitions of *energized* and *live parts*. Requirements for guarding of live parts may be found in 110.27.

Exposed (as applied to wiring methods). On or attached to the surface or behind panels designed to allow access.

See Exhibit 100.2, where wiring methods located behind a suspended ceiling with lift-out panels are considered exposed (as applied to wiring methods).

Externally Operable. Capable of being operated without exposing the operator to contact with live parts.

Feeder. All circuit conductors between the service equipment, the source of a separately derived system, or other power supply source and the final branch-circuit overcurrent device.

See the commentary following the definition of *branch circuit,* including Exhibit 100.5, which illustrates the difference between branch circuits and feeders.

Festoon Lighting. A string of outdoor lights that is suspended between two points.

The general requirements for festoon lighting are located in 225.6(B). Use the index to find specific requirements.

Fitting. An accessory such as a locknut, bushing, or other part of a wiring system that is intended primarily to perform a mechanical rather than an electrical function.

Items such as condulets, conduit couplings, EMT connectors and couplings, and threadless connectors are considered fittings.

Garage. A building or portion of a building in which one or more self-propelled vehicles can be kept for use, sale, storage, rental, repair, exhibition, or demonstration purposes.

> FPN: For commercial garages, repair and storage, see Article 511.

Revised for the 2002 *Code*, this definition was simplified and includes the garages for electric vehicles covered in Article 625.

Ground. A conducting connection, whether intentional or accidental, between an electrical circuit or equipment and the earth or to some conducting body that serves in place of the earth.

Grounded. Connected to earth or to some conducting body that serves in place of the earth.

Grounded, Effectively. Intentionally connected to earth through a ground connection or connections of sufficiently low impedance and having sufficient current-carrying capacity to prevent the buildup of voltages that may result in undue hazards to connected equipment or to persons.

Grounded Conductor. A system or circuit conductor that is intentionally grounded.

Grounding Conductor. A conductor used to connect equipment or the grounded circuit of a wiring system to a grounding electrode or electrodes.

Grounding Conductor, Equipment. The conductor used to connect the non–current-carrying metal parts of equipment, raceways, and other enclosures to the system grounded conductor, the grounding electrode conductor, or both, at the service equipment or at the source of a separately derived system.

See 250.118 for types of equipment grounding conductors. Proper sizing of the equipment grounding conductor is found in 250.122 and its associated Table 250.122.

Grounding Electrode Conductor. The conductor used to connect the grounding electrode(s) to the equipment grounding conductor, to the grounded conductor, or to both, at the service, at each building or structure where supplied from a common service, or at the source of a separately derived system.

The grounding electrode conductor is covered extensively in Article 250, Part III. The grounding electrode conductor is required to be copper, aluminum, or copper-clad aluminum. It is used to connect the equipment grounding conductor or the grounded conductor (at the service or at the separately derived system) to the grounding electrode or electrodes for either grounded or ungrounded systems. Refer to Exhibits 100.4 and 250.1, which show the grounding electrode conductor in a typical grounding system for a single-phase, 3-wire service. The grounding electrode conductor is sized according to the requirements of 250.66 and the accompanying Table 250.66.

Ground-Fault Circuit Interrupter. A device intended for the protection of personnel that functions to de-energize a circuit or portion thereof within an established period of time when a current to ground exceeds the values established for a Class A device.

> FPN: Class A ground-fault circuit interrupters trip when the current to ground has a value in the range of 4 mA to 6 mA. For further information, see UL 943, *Standard for Ground-Fault Circuit Interrupters.*

The commentary following 210.8 contains a list of applicable cross-references for ground-fault circuit interrupters (GFCIs). Exhibits 210.7 through 210.15 contain specific information regarding the requirements for GFCIs.

This definition was revised for the 2002 *Code* by adding an FPN that describes how personal protection is achieved.

Ground-Fault Protection of Equipment. A system intended to provide protection of equipment from damaging line-to-ground fault currents by operating to cause a disconnecting means to open all ungrounded conductors of the faulted circuit. This protection is provided at current levels less than those required to protect conductors from damage through the operation of a supply circuit overcurrent device.

See the commentary on 230.95, 426.28, and 427.22(2).

Guarded. Covered, shielded, fenced, enclosed, or otherwise protected by means of suitable covers, casings, barriers, rails, screens, mats, or platforms to remove the likelihood of approach or contact by persons or objects to a point of danger.

Hoistway. Any shaftway, hatchway, well hole, or other vertical opening or space in which an elevator or dumbwaiter is designed to operate.

See Article 620 for the installation of electrical equipment and wiring methods in hoistways.

Identified (as applied to equipment). Recognizable as suitable for the specific purpose, function, use, environment, application, and so forth, where described in a particular *Code* requirement.

> FPN: Some examples of ways to determine suitability of equipment for a specific purpose, environment, or application include investigations by a qualified testing laboratory (listing and labeling), an inspection agency, or other organizations concerned with product evaluation.

In Sight From (Within Sight From, Within Sight). Where this *Code* specifies that one equipment shall be "in sight from," "within sight from," or "within sight," and so forth, of another equipment, the specified equipment is to be visible and not more than 15 m (50 ft) distant from the other.

Exhibit 430.20 depicts requirements for the placement of a disconnecting means that is not in sight.

Interrupting Rating. The highest current at rated voltage that a device is intended to interrupt under standard test conditions.

> FPN: Equipment intended to interrupt current at other than fault levels may have its interrupting rating implied in other ratings, such as horsepower or locked rotor current.

Interrupting ratings are essential in the coordination of electrical systems so that available fault currents can be properly controlled. Other sections specifically dealing with interrupting ratings are 110.9, 240.60(C), 240.83(C), and 240.86.

Isolated (as applied to location). Not readily accessible to persons unless special means for access are used.

See the definition of *accessible, readily.*

Labeled. Equipment or materials to which has been attached a label, symbol, or other identifying mark of an organization that is acceptable to the authority having jurisdiction and concerned with product evaluation, that maintains periodic inspection of production of labeled equipment or materials, and by whose labeling the manufacturer indicates compliance with appropriate standards or performance in a specified manner.

Equipment and conductors required or permitted by this *Code* are acceptable only if they have been approved for a specific environment or application by the authority having jurisdiction, as stated in 110.2. See 90.7 regarding the examination of equipment for safety. Listing or labeling by a qualified testing laboratory provides a basis for approval.

Lighting Outlet. An outlet intended for the direct connection of a lampholder, a luminaire (lighting fixture), or a pendant cord terminating in a lampholder.

Listed. Equipment, materials, or services included in a list published by an organization that is acceptable to the authority having jurisdiction and concerned with evaluation of products or services, that maintains periodic inspection of production of listed equipment or materials or periodic evaluation of services, and whose listing states that the equipment, material, or services either meets appropriate designated standards or has been tested and found suitable for a specified purpose.

> FPN: The means for identifying listed equipment may vary for each organization concerned with product evaluation, some of which do not recognize equipment as listed unless it is also labeled. Use of the system employed by the listing organization allows the authority having jurisdiction to identify a listed product.

The definition of *listed* was revised for the 2002 *Code*. This slightly reworded definition is now in line with the definition of *listed* found in the NFPA Regulations Governing Committee Projects. Reviewing other *NEC*-defined terms such as *approved, authority having jurisdiction, labeled,* and *identified (as applied to equipment)* will help the user understand the approval process.

Live Parts. Energized conductive components.

The definition of *live parts* was simplified for the 2002 *Code*. The condition of a shock hazard has been removed from this new definition. *Live parts* can now be associated with all voltage levels, not simply voltage levels that present a shock hazard.

Location, Damp. Locations protected from weather and not subject to saturation with water or other liquids but subject to moderate degrees of moisture. Examples of such locations include partially protected locations under canopies, marquees, roofed open porches, and like locations, and interior locations subject to moderate degrees of moisture, such as some basements, some barns, and some cold-storage warehouses.

Location, Dry. A location not normally subject to dampness or wetness. A location classified as dry may be temporarily subject to dampness or wetness, as in the case of a building under construction.

Location, Wet. Installations under ground or in concrete slabs or masonry in direct contact with the earth; in locations subject to saturation with water or other liquids, such as vehicle washing areas; and in unprotected locations exposed to weather.

It is intended that the inside of a raceway in a wet location or a raceway installed under ground be considered a wet location. Therefore, any conductors contained therein would be required to be suitable for wet locations.

See 300.6(C) for some examples of wet locations and 410.4(A) for information on luminaires installed in wet locations.

See *patient care area* in 517.2 for a definition of wet locations in a patient care area.

Luminaire. A complete lighting unit consisting of a lamp or lamps together with the parts designed to distribute the light, to position and protect the lamps and ballast (where applicable), and to connect the lamps to the power supply.

The term *luminaire* first appeared in the 1996 *NEC* in a fine print note following 410-1. In the 2002 *Code*, the term *luminaire* is used throughout in place of the term *lighting fixture*.

Although new lighting techniques such as light pipe and glass fiber optics are sometimes referred to as "lighting systems," the definition of *luminaire* does not necessarily preclude such systems, because light pipe and fiber optics are actually "parts designed to distribute the light."

Metal-Enclosed Power Switchgear. A switchgear assembly completely enclosed on all sides and top with sheet metal (except for ventilating openings and inspection windows) containing primary power circuit switching, interrupting de-

vices, or both, with buses and connections. The assembly may include control and auxiliary devices. Access to the interior of the enclosure is provided by doors, removable covers, or both.

Motor Control Center. An assembly of one or more enclosed sections having a common power bus and principally containing motor control units.

Multioutlet Assembly. A type of surface, flush, or freestanding raceway designed to hold conductors and receptacles, assembled in the field or at the factory.

The definition of *multioutlet assembly* now includes a reference to a freestanding assembly with multiple outlets, commonly called a power pole. In dry locations, metallic and nonmetallic multioutlet assemblies are permitted; however, they are not permitted to be installed if concealed. See Article 380 for details on recessing these multioutlet assemblies. Exhibit 100.8 shows a multioutlet assembly used for countertop appliances.

Exhibit 100.8 Multioutlet assembly installed to serve countertop appliances. (Courtesy of The Wiremold Co.)

Nonautomatic. Action requiring personal intervention for its control. As applied to an electric controller, nonautomatic control does not necessarily imply a manual controller, but only that personal intervention is necessary.

Nonincendive Circuit. A circuit, other than field wiring, in which any arc or thermal effect produced under intended operating conditions of the equipment is not capable, under specified test conditions, of igniting the flammable gas–air, vapor–air, or dust–air mixture.

A nonincendive circuit employs a protection technique that prevents electrical circuits from causing a fire or explosion in a hazardous (classified) location. A careful reading of 500.7(F) points out that this protection technique is permitted only for Division 2 areas of Class I or Class II locations. Nonincendive circuits are not permitted in Division 1 areas of Class I or Class II locations. Nonincendive circuits and equipment are tested or evaluated essentially in the same way that intrinsically safe circuits and equipment are tested and evaluated, except that abnormal conditions are not considered. Because of its definition, a nonincendive circuit is a low-energy circuit.

> FPN: For test conditions, see ANSI/ISA-S12.12-1994, *Nonincendive Electrical Equipment for Use in Class I and II, Division 2 and Class III, Divisions 1 and 2 Hazardous (Classified) Locations.*

Nonincendive Field Wiring. Wiring that enters or leaves an equipment enclosure and, under normal operating conditions of the equipment, is not capable, due to arcing or thermal effects, of igniting the flammable gas–air, vapor–air, or dust–air mixture. Normal operation includes opening, shorting, or grounding the field wiring.

The definition of *nonincendive field wiring* was added to the 1999 *Code* to alert users that, although the circuits and equipment may have been evaluated and approved as nonincendive, field wiring is not generally approved by a testing laboratory. Field wiring meeting this definition would require limitations of energy on the wiring under conditions such as opening, shorting, or grounding. For example, stored energy in the form of mutual inductance or capacitance could be released during an opening, shorting, or grounding of nonincendive field wiring, thus defeating the purpose of this protection technique. Further information regarding the nonincendive protection techniques can be found in the definition of *nonincendive circuit* in Article 100 and also in 500.7(F).

Nonlinear Load. A load where the wave shape of the steady-state current does not follow the wave shape of the applied voltage.

> FPN: Electronic equipment, electronic/electric-discharge lighting, adjustable-speed drive systems, and similar equipment may be nonlinear loads.

Nonlinear loads are a major cause of harmonic currents in modern circuits. Additional conductor heating is just one of the undesirable operational effects often associated with harmonic currents. The FPN following 310.10 points out that harmonic current, as well as fundamental current, should be used in determining the heat generated internally in a conductor. Also, 310.4, Exception No. 4, permits limited use of parallel neutrals.

Actual circuit measurements of current for nonlinear loads should be made using only true rms-measuring ammeter instruments. Averaging ammeters produce inaccurate values if used to measure nonlinear loads. [See 310.15(B)(4)(c) and associated commentary.]

Outlet. A point on the wiring system at which current is taken to supply utilization equipment.

An example is a lighting outlet or a receptacle outlet.

Outline Lighting. An arrangement of incandescent lamps or electric-discharge lighting to outline or call attention to certain features such as the shape of a building or the decoration of a window.

See Article 600 for details on outline lighting.

Overcurrent. Any current in excess of the rated current of equipment or the ampacity of a conductor. It may result from overload, short circuit, or ground fault.

> FPN: A current in excess of rating may be accommodated by certain equipment and conductors for a given set of conditions. Therefore the rules for overcurrent protection are specific for particular situations.

Overload. Operation of equipment in excess of normal, full-load rating, or of a conductor in excess of rated ampacity that, when it persists for a sufficient length of time, would cause damage or dangerous overheating. A fault, such as a short circuit or ground fault, is not an overload.

Panelboard. A single panel or group of panel units designed for assembly in the form of a single panel, including buses and automatic overcurrent devices, and equipped with or without switches for the control of light, heat, or power circuits; designed to be placed in a cabinet or cutout box placed in or against a wall, partition, or other support; and accessible only from the front.

See Article 408 for details on panelboards.

Plenum. A compartment or chamber to which one or more air ducts are connected and that forms part of the air distribution system.

Power Outlet. An enclosed assembly that may include receptacles, circuit breakers, fuseholders, fused switches, buses, and watt-hour meter mounting means; intended to supply and control power to mobile homes, recreational vehicles, park trailers, or boats or to serve as a means for distributing power required to operate mobile or temporarily installed equipment.

Premises Wiring (System). That interior and exterior wiring, including power, lighting, control, and signal circuit wiring together with all their associated hardware, fittings, and wiring devices, both permanently and temporarily installed, that extends from the service point or source of power, such as a battery, a solar photovoltaic system, or a generator, transformer, or converter windings, to the outlet(s). Such wiring does not include wiring internal to appliances, luminaires (fixtures), motors, controllers, motor control centers, and similar equipment.

Qualified Person. One who has skills and knowledge related to the construction and operation of the electrical equipment and installations and has received safety training on the hazards involved.

Raceway. An enclosed channel of metal or nonmetallic materials designed expressly for holding wires, cables, or busbars, with additional functions as permitted in this *Code.* Raceways include, but are not limited to, rigid metal conduit, rigid nonmetallic conduit, intermediate metal conduit, liquidtight flexible conduit, flexible metallic tubing, flexible metal conduit, electrical nonmetallic tubing, electrical metallic tubing, underfloor raceways, cellular concrete floor raceways, cellular metal floor raceways, surface raceways, wireways, and busways.

Rainproof. Constructed, protected, or treated so as to prevent rain from interfering with the successful operation of the apparatus under specified test conditions.

Raintight. Constructed or protected so that exposure to a beating rain will not result in the entrance of water under specified test conditions.

Receptacle. A receptacle is a contact device installed at the outlet for the connection of an attachment plug. A single receptacle is a single contact device with no other contact device on the same yoke. A multiple receptacle is two or more contact devices on the same yoke.

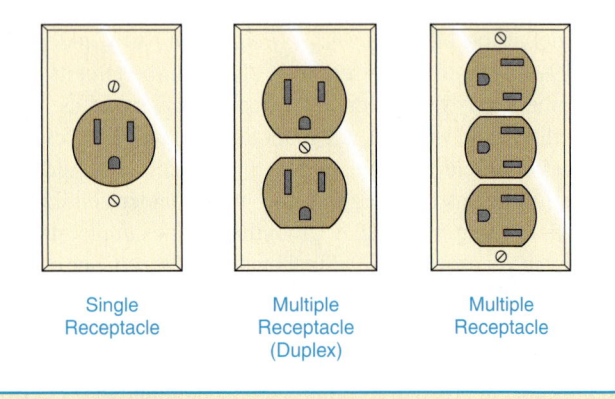

| Single Receptacle | Multiple Receptacle (Duplex) | Multiple Receptacle |

Exhibit 100.9 Receptacles.

Receptacle Outlet. An outlet where one or more receptacles are installed.

Remote-Control Circuit. Any electric circuit that controls any other circuit through a relay or an equivalent device.

Exhibit 100.10 illustrates a remote-control circuit that starts and stops an electric motor.

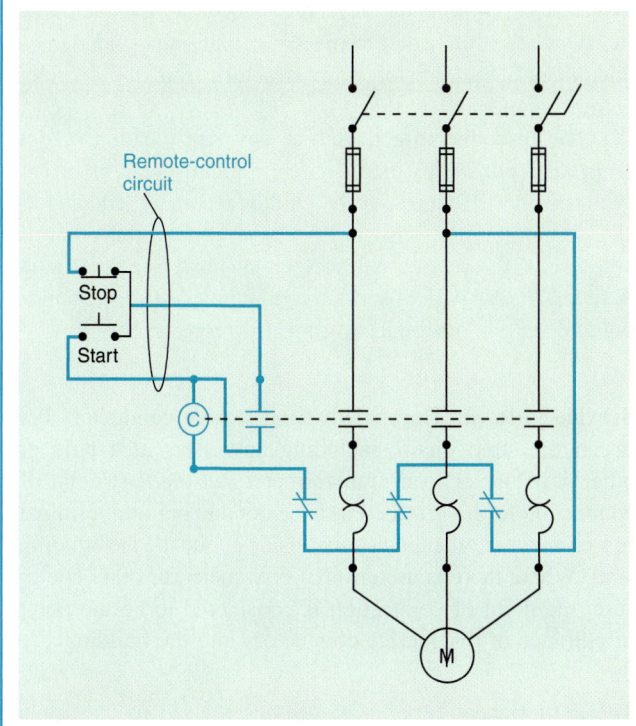

Exhibit 100.10 Remote-control circuit for starting and stopping an electric motor.

Sealable Equipment. Equipment enclosed in a case or cabinet that is provided with a means of sealing or locking so that live parts cannot be made accessible without opening the enclosure. The equipment may or may not be operable without opening the enclosure.

Separately Derived System. A premises wiring system whose power is derived from a battery, from a solar photovoltaic system, or from a generator, transformer, or converter windings, and that has no direct electrical connection, including a solidly connected grounded circuit conductor, to supply conductors originating in another system.

Service. The conductors and equipment for delivering electric energy from the serving utility to the wiring system of the premises served.

The definition of *service* was modified for the 1999 *Code* to state that electric energy to a service can be supplied only by the serving utility. If electric energy is supplied by other than the serving utility, the supplied conductors and equipment are considered feeders, not a service.

Service Cable. Service conductors made up in the form of a cable.

Service Conductors. The conductors from the service point to the service disconnecting means.

The definition of *service conductors* was revised in the 1999 *Code* to be more precise. The phrase "or other source of power" was deleted. The service conductors originate at the service point (where the serving utility ends) and end at the service disconnect. Due to the associated 1999 *Code* change in the definition of *service,* these service conductors may originate only from the serving utility.

Service conductors is a broad term and may include service drops, service laterals, and service-entrance conductors. But this term specifically excludes any wiring on the supply side (serving utility side) of the service point.

If the utility has specified that the service point is at the utility pole, then the service conductors from an overhead distribution system originate at the utility pole and terminate at the service disconnecting means.

If the utility has specified that the service point is at the utility manhole, then the service conductors from an underground distribution system originate at the utility manhole and terminate at the service disconnecting means. Where utility-owned primary conductors are extended to outdoor pad-mounted transformers on private property, the service conductors originate at the secondary connections of the transformers only if the utility has specified that the service point is at the secondary connections.

See Article 230, Part VIII, and the commentary following 230.200 for service conductors exceeding 600 volts, nominal.

Service Drop. The overhead service conductors from the last pole or other aerial support to and including the splices, if any, connecting to the service-entrance conductors at the building or other structure.

In Exhibit 100.11, the overhead service-drop conductors run from the utility pole and connect to the service-entrance conductors at the service point. Conductors on the utility side of the service point are not covered by the *NEC*. The utility specifies the location of the service point. Exact locations of the service point may vary from utility to utility, as well as from occupancy to occupancy.

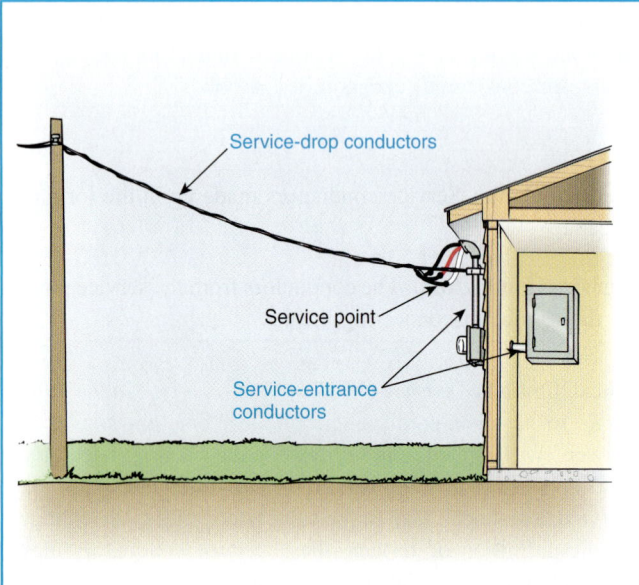

Exhibit 100.11 Overhead system showing a service drop from a utility pole to attachment on a house and service-entrance conductors from point of attachment (spliced to service-drop conductors), down the side of the house, through the meter socket, and terminating in the service equipment.

Service-Entrance Conductors, Overhead System. The service conductors between the terminals of the service equipment and a point usually outside the building, clear of building walls, where joined by tap or splice to the service drop.

See Exhibit 100.11 for an illustration of service-entrance conductors in an overhead system.

Service-Entrance Conductors, Underground System. The service conductors between the terminals of the service equipment and the point of connection to the service lateral.

See Exhibit 100.12 for an illustration of service-entrance conductors in an underground system.

FPN: Where service equipment is located outside the building walls, there may be no service-entrance conductors, or they may be entirely outside the building.

Service Equipment. The necessary equipment, usually consisting of a circuit breaker(s) or switch(es) and fuse(s) and their accessories, connected to the load end of service conductors to a building or other structure, or an otherwise designated area, and intended to constitute the main control and cutoff of the supply.

The definition of *service equipment* was revised for the 1999 *Code*. In addition to editorial corrections, the phrase "connected to the load end of service conductors" was substituted for the phrase "located near the point of entrance." This change is in concert with the other associated changes made to the definitions of *service* and *service conductors*.

Service equipment may consist of circuit breakers or fused switches provided to disconnect all ungrounded conductors in a building or other structure from the service-entrance conductors.

The disconnecting means at any *one* location is not allowed to consist of more than six circuit breakers or six switches and is required to be readily accessible either outside the building or structure or inside nearest the point of entrance of the service-entrance conductors. See 230.6 and Article 230, Part VI, for service conductors outside the building and service disconnecting means, respectively.

Service Lateral. The underground service conductors between the street main, including any risers at a pole or other structure or from transformers, and the first point of connection to the service-entrance conductors in a terminal box or meter or other enclosure, inside or outside the building wall. Where there is no terminal box, meter, or other enclosure, the point of connection is considered to be the point of entrance of the service conductors into the building.

The definition of *service lateral* was revised for the 1999 *Code,* eliminating the references to adequate space.

As Exhibit 100.12 shows, the underground service laterals may be run from poles or from transformers and with or without terminal boxes, provided they begin at the service point. Conductors on the utility side of the service point are not covered by the *NEC*. The utility specifies the location of the service point. Exact locations of the service point may vary from utility to utility, as well as from occupancy to occupancy.

Service Point. The point of connection between the facilities of the serving utility and the premises wiring.

The service point is the point of demarcation between the serving utility and the premises wiring. The service point is the point on the wiring system where the serving utility ends and the premises wiring begins. The serving utility generally specifies the location of the service point.

Because the location of the service point is generally determined by the utility, the service-drop conductors and the service-lateral conductors may or may not be part of the service as covered by the *NEC*. For these types of conductors

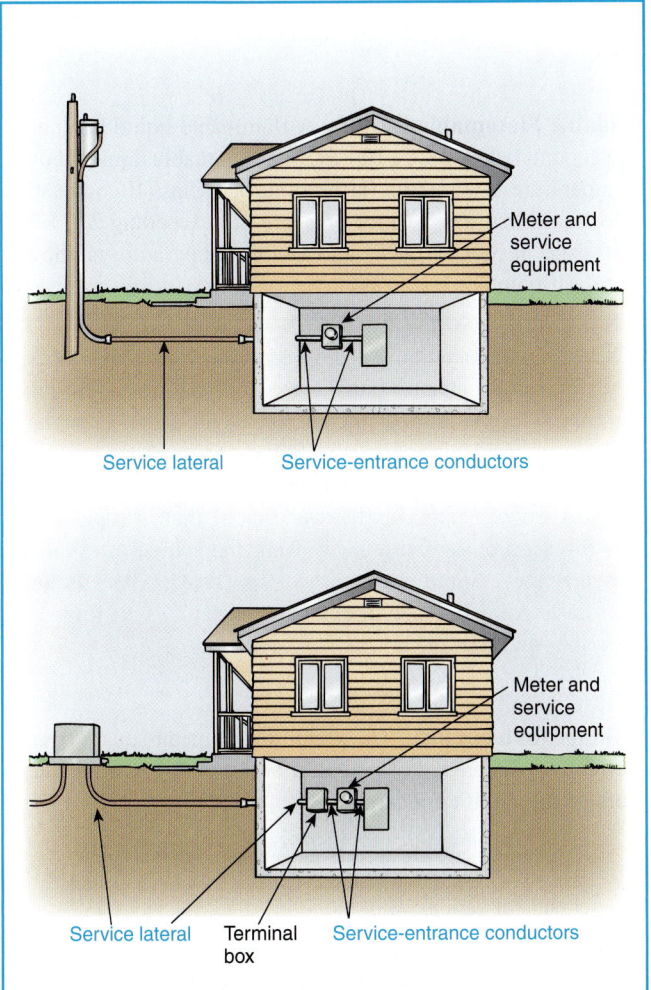

Exhibit 100.12 Underground systems showing service laterals run from a pole and from a transformer.

to be covered, they must be physically located on the premises wiring side of the service point. If these conductors are located on the utility side of the service point, they are not covered in the definition of *service conductors* and are therefore not covered by the *NEC*.

Generally, based on the definitions of the terms *service point* and *service conductors,* any conductor on the serving utility side of the service point is not covered by the *NEC.* For example, a typical suburban residence has an overhead service drop from the utility pole to the house. If the utility specifies that the service point is at the point of attachment of the service drop to the house, then the service-drop conductors are not considered service conductors because the service drop is not on the premises wiring side of the service point. Alternatively, if the service point is specified as "at the pole" by the utility, then the service-drop conductors are considered service conductors, and the *NEC* would apply to the service drop.

Exact locations for a service point may vary from utility to utility, as well as from occupancy to occupancy.

Show Window. Any window used or designed to be used for the display of goods or advertising material, whether it is fully or partly enclosed or entirely open at the rear and whether or not it has a platform raised higher than the street floor level.

See 220.3(B)(7), 220.12(A), and Exhibit 220.1 for show-window lighting load requirements.

Signaling Circuit. Any electric circuit that energizes signaling equipment.

Solar Photovoltaic System. The total components and subsystems that, in combination, convert solar energy into electrical energy suitable for connection to a utilization load.

See Article 690 for solar photovoltaic system requirements.

Special Permission. The written consent of the authority having jurisdiction.

The authority having jurisdiction for enforcement of the *Code* is responsible for making interpretations and granting special permission contemplated in a number of the rules, as stated in 90.4. For specific examples of special permission, see 110.26(A)(1)(b), 230.2(B), and 426.14.

Structure. That which is built or constructed.

Added for the 2002 *Code*, this definition allows the architect, electrical engineer, general contractor, electrical contractor, and all building officials to use the same definition for *structure*.

Switch, Bypass Isolation. A manually operated device used in conjunction with a transfer switch to provide a means of directly connecting load conductors to a power source and of disconnecting the transfer switch.

See 700.6(B) and 701.7(B) for further information on bypass isolation transfer switches.

Switch, General-Use. A switch intended for use in general distribution and branch circuits. It is rated in amperes, and it is capable of interrupting its rated current at its rated voltage.

Switch, General-Use Snap. A form of general-use switch constructed so that it can be installed in device boxes or on box covers, or otherwise used in conjunction with wiring systems recognized by this *Code.*

Switch, Isolating. A switch intended for isolating an electric circuit from the source of power. It has no interrupting rating, and it is intended to be operated only after the circuit has been opened by some other means.

Switch, Motor-Circuit. A switch rated in horsepower that is capable of interrupting the maximum operating overload current of a motor of the same horsepower rating as the switch at the rated voltage.

Switch, Transfer. An automatic or nonautomatic device for transferring one or more load conductor connections from one power source to another.

Switchboard. A large single panel, frame, or assembly of panels on which are mounted on the face, back, or both, switches, overcurrent and other protective devices, buses, and usually instruments. Switchboards are generally accessible from the rear as well as from the front and are not intended to be installed in cabinets.

Busbars are required to be arranged to avoid inductive overheating. Service busbars are required to be isolated by barriers from the remainder of the switchboard. Most modern switchboards are totally enclosed to minimize the probability of spreading fire to adjacent combustible materials and to guard live parts. See Article 408 for more information regarding switchboards.

Thermal Protector (as applied to motors). A protective device for assembly as an integral part of a motor or motor-compressor that, when properly applied, protects the motor against dangerous overheating due to overload and failure to start.

> FPN: The thermal protector may consist of one or more sensing elements integral with the motor or motor-compressor and an external control device.

Thermally Protected (as applied to motors). The words *Thermally Protected* appearing on the nameplate of a motor or motor-compressor indicate that the motor is provided with a thermal protector.

Utilization Equipment. Equipment that utilizes electric energy for electronic, electromechanical, chemical, heating, lighting, or similar purposes.

Ventilated. Provided with a means to permit circulation of air sufficient to remove an excess of heat, fumes, or vapors.

See the commentary following 110.13(B).

Volatile Flammable Liquid. A flammable liquid having a flash point below 38°C (100°F), or a flammable liquid whose temperature is above its flash point, or a Class II combustible liquid that has a vapor pressure not exceeding 276 kPa (40 psia) at 38°C (100°F) and whose temperature is above its flash point.

The *flash point* of a liquid is defined as the minimum temperature at which it gives off sufficient vapor to form an ignitible mixture with the air near the surface of the liquid or within the vessel used to contain the liquid. An *ignitible mixture* is defined as a mixture within the explosive or flammable range (between upper and lower limits) that is capable of the propagation of flame away from the source of ignition when ignited. Some emission of vapors takes place below the flash point but not in sufficient quantities to form an ignitible mixture.

Voltage (of a circuit). The greatest root-mean-square (rms) (effective) difference of potential between any two conductors of the circuit concerned.

> FPN: Some systems, such as 3-phase 4-wire, single-phase 3-wire, and 3-wire direct current, may have various circuits of various voltages.

Common 3-phase, 4-wire wye systems are 480/277 volts and 208/120 volts. The voltage of the circuit is the highest voltage between any two conductors (i.e., 480 volts or 208 volts). The voltage of the circuit of a 2-wire feeder or branch circuit (single phase and the grounded conductor) derived from these systems would be the voltage between the two conductors at the lower voltage (i.e., 277 volts or 120 volts). The same applies to dc or single-phase, 3-wire systems where there are two voltages.

Voltage, Nominal. A nominal value assigned to a circuit or system for the purpose of conveniently designating its voltage class (e.g., 120/240 volts, 480Y/277 volts, 600 volts).

The actual voltage at which a circuit operates can vary from the nominal within a range that permits satisfactory operation of equipment.

See 220.2(A) for a list of nominal voltages used in computing branch-circuit and feeder loads.

> FPN: See ANSI C84.1-1995, *Voltage Ratings for Electric Power Systems and Equipment (60 Hz).*

Voltage to Ground. For grounded circuits, the voltage between the given conductor and that point or conductor of the circuit that is grounded; for ungrounded circuits, the greatest voltage between the given conductor and any other conductor of the circuit.

The voltage to ground of a 277/480-volt wye system would be 277 volts; of a 120/208-volt wye system, 120 volts; and of a 3-phase, 3-wire ungrounded 480-volt system, 480 volts.

For a 3-phase, 4-wire delta system with the center of one leg grounded, there are two voltages to ground. For example, on a 240-volt system, two legs would each have 120 volts to ground and the third, or "high" leg, would have 208 volts to ground. See 215.8, 230.56, and 408.3(E) for special marking and arrangements on such circuit conductors.

Watertight. Constructed so that moisture will not enter the enclosure under specified test conditions.

Unless the enclosure is hermetically sealed, it is possible for moisture to enter the enclosure. See the commentary following the definition of *enclosure* and following Table 430.91.

Weatherproof. Constructed or protected so that exposure to the weather will not interfere with successful operation.

See the commentary following the definition of *enclosure*. Industry standards for enclosures are found in the commentary following 430.91.

FPN: Rainproof, raintight, or watertight equipment can fulfill the requirements for weatherproof where varying weather conditions other than wetness, such as snow, ice, dust, or temperature extremes, are not a factor.

II. Over 600 Volts, Nominal

Whereas the preceding definitions are intended to apply wherever the terms are used throughout this *Code*, the following definitions are applicable only to parts of the article specifically covering installations and equipment operating at over 600 volts, nominal.

Electronically Actuated Fuse. An overcurrent protective device that generally consists of a control module that provides current sensing, electronically derived time–current characteristics, energy to initiate tripping, and an interrupting module that interrupts current when an overcurrent occurs. Electronically actuated fuses may or may not operate in a current-limiting fashion, depending on the type of control selected.

Although they are called fuses because they interrupt current by melting a fusible element, electronically actuated fuses respond to a signal from an electronic control rather than from the heat generated by actual current passing through a fusible element. Electronically actuated fuses have controls similar to those of electronic circuit breakers.

Fuse. An overcurrent protective device with a circuit-opening fusible part that is heated and severed by the passage of overcurrent through it.

FPN: A fuse comprises all the parts that form a unit capable of performing the prescribed functions. It may or may not be the complete device necessary to connect it into an electrical circuit.

Controlled Vented Power Fuse. A fuse with provision for controlling discharge circuit interruption such that no solid material may be exhausted into the surrounding atmosphere.

FPN: The fuse is designed so that discharged gases will not ignite or damage insulation in the path of the discharge or propagate a flashover to or between grounded members or conduction members in the path of the discharge where the distance between the vent and such insulation or conduction members conforms to manufacturer's recommendations.

Expulsion Fuse Unit (Expulsion Fuse). A vented fuse unit in which the expulsion effect of gases produced by the arc and lining of the fuseholder, either alone or aided by a spring, extinguishes the arc.

Nonvented Power Fuse. A fuse without intentional provision for the escape of arc gases, liquids, or solid particles to the atmosphere during circuit interruption.

Power Fuse Unit. A vented, nonvented, or controlled vented fuse unit in which the arc is extinguished by being drawn through solid material, granular material, or liquid, either alone or aided by a spring.

Vented Power Fuse. A fuse with provision for the escape of arc gases, liquids, or solid particles to the surrounding atmosphere during circuit interruption.

Multiple Fuse. An assembly of two or more single-pole fuses.

Switching Device. A device designed to close, open, or both, one or more electric circuits.

Circuit Breaker. A switching device capable of making, carrying, and interrupting currents under normal circuit conditions, and also of making, carrying for a specified time, and interrupting currents under specified abnormal circuit conditions, such as those of short circuit.

Cutout. An assembly of a fuse support with either a fuse-holder, fuse carrier, or disconnecting blade. The fuseholder or fuse carrier may include a conducting element (fuse link) or may act as the disconnecting blade by the inclusion of a nonfusible member.

Disconnecting (or Isolating) Switch (Disconnector, Isolator). A mechanical switching device used for isolating a circuit or equipment from a source of power.

Disconnecting Means. A device, group of devices, or other means whereby the conductors of a circuit can be disconnected from their source of supply.

Interrupter Switch. A switch capable of making, carrying, and interrupting specified currents.

Oil Cutout (Oil-Filled Cutout). A cutout in which all or part of the fuse support and its fuse link or disconnecting blade is mounted in oil with complete immersion of the contacts and the fusible portion of the conducting element (fuse link) so that arc interruption by severing of the fuse link or by opening of the contacts will occur under oil.

Oil Switch. A switch having contacts that operate under oil (or askarel or other suitable liquid).

Regulator Bypass Switch. A specific device or combination of devices designed to bypass a regulator.

ARTICLE 110
Requirements for Electrical Installations

Contents

I. General

110.1 Scope.

This article covers general requirements for the examination and approval, installation and use, access to and spaces about electrical conductors and equipment, and tunnel installations.

110.2 Approval.

The conductors and equipment required or permitted by this *Code* shall be acceptable only if approved.

> FPN: See 90.7, Examination of Equipment for Safety, and 110.3, Examination, Identification, Installation, and Use of Equipment. See definitions of *Approved, Identified, Labeled,* and *Listed*.

All electrical equipment is required to be approved as defined in Article 100 and, as such, to be acceptable to the authority having jurisdiction (also defined in Article 100). Section 110.3 provides guidance for the evaluation of equipment and recognizes listing or labeling as a means of establishing suitability.

Approval of equipment is the responsibility of the electrical inspection authority, and many such approvals are based on tests and listings of testing laboratories.

110.3 Examination, Identification, Installation, and Use of Equipment.

(A) Examination. In judging equipment, considerations such as the following shall be evaluated:

(1) Suitability for installation and use in conformity with the provisions of this *Code*

> FPN: Suitability of equipment use may be identified by a description marked on or provided with a product to identify the suitability of the product for a specific purpose, environment, or application. Suitability of equipment may be evidenced by listing or labeling.

(2) Mechanical strength and durability, including, for parts designed to enclose and protect other equipment, the adequacy of the protection thus provided

(3) Wire-bending and connection space

(4) Electrical insulation

(5) Heating effects under normal conditions of use and also under abnormal conditions likely to arise in service

(6) Arcing effects

(7) Classification by type, size, voltage, current capacity, and specific use

(8) Other factors that contribute to the practical safeguarding of persons using or likely to come in contact with the equipment

For wire-bending and connection space in cabinets and cut-out boxes, see 312.6, Tables 312.6(A) and (B), and 312.7, 312.9, and 312.11. For wire-bending and connection space in other equipment, see the appropriate *NEC* article and section. For example, see 314.16 and 314.28 for outlet, device, pull, and junction boxes, as well as for conduit bodies; 404.3 and 404.18 for switches; 408.3(F) for switchboards and panelboards; and 430.10 for motors and motor controllers.

(B) Installation and Use. Listed or labeled equipment shall be installed and used in accordance with any instructions included in the listing or labeling.

Manufacturers usually supply installation instructions with equipment for use by general contractors, erectors, electrical contractors, electrical inspectors, and others concerned with an installation. It is important to follow the listing or labeling installation instructions. For example, 210.52, second paragraph, permits permanently installed electric baseboard heaters to be equipped with receptacle outlets that meet the requirements for the wall space utilized by such heaters. The installation instructions for such permanent baseboard heaters indicate that these heaters should not be mounted beneath a receptacle. In dwelling units, it is common to use low-density heating units that measure in excess of 12 ft in length. Therefore, to meet the provisions of 210.52(A) and also the installation instructions, a receptacle must either be part of the heating unit or be installed in the floor close to the wall but not above the heating unit. (See 210.52, FPN, and Exhibit 210.23 for more specific details.)

In itself, 110.3 does not require listing or labeling of equipment. It does, however, require considerable evaluation of equipment. Section 110.2 requires that equipment be acceptable only if approved. The term *approved* is defined in Article 100 as acceptable to the authority having jurisdiction. Before issuing approval, the authority having jurisdiction (also defined in Article 100) may require evidence of compliance with 110.3(A). The most common form of this evidence considered acceptable by authorities having jurisdiction is a listing or labeling by a third party.

Some sections in the *Code* require listed or labeled equipment, such as 250.8, which includes the phrase "listed pressure connectors, listed clamps, or other listed means."

110.4 Voltages.

Throughout this *Code*, the voltage considered shall be that at which the circuit operates. The voltage rating of electrical equipment shall not be less than the nominal voltage of a circuit to which it is connected.

Voltages used for computing branch-circuit and feeder loads are nominal voltages as listed in 220.2. See the definitions of *voltage (of a circuit); voltage, nominal;* and *voltage to ground* in Article 100. See also 300.2 and 300.3, which specify the voltage limitations of conductors of circuits rated 600 volts, nominal, or less, and over 600 volts, nominal.

110.5 Conductors.

Conductors normally used to carry current shall be of copper unless otherwise provided in this *Code*. Where the conductor material is not specified, the material and the sizes given in this *Code* shall apply to copper conductors. Where other materials are used, the size shall be changed accordingly.

FPN: For aluminum and copper-clad aluminum conductors, see 310.15.

See 310.14 for aluminum conductor material.

110.6 Conductor Sizes.

Conductor sizes are expressed in American Wire Gage (AWG) or in circular mils.

For copper, aluminum, or copper-clad aluminum conductors up to size 4/0 AWG, this *Code* uses the American Wire Gage (AWG) for size identification, which is the same as the Brown and Sharpe (BS) Gage. Changed for the 2002 *Code*, wire sizes up to size 4/0 AWG are now expressed as XX AWG, XX being the size wire. For example, a wire size expressed as *No. 12* in prior editions of the *Code* is now expressed as *12 AWG*. The resulting expression would therefore appear as *six 12 AWG conductors* instead of *6 No. 12 conductors*.

Conductors larger than 4/0 AWG are sized in circular mils, beginning with 250,000 circular mils. Prior to the 1990

edition, a 250,000-circular-mil conductor was labeled *250 MCM*. The term *MCM* was defined as 1000 circular mils (the first *M* being the Roman numeral designation for 1000). Beginning in the 1990 edition, the notation was changed to *250 kcmil* to recognize the accepted convention that *k* indicates 1000. UL standards and IEEE standards also use the notation *kcmil* rather than *MCM*.

The circular mil area of a conductor is equal to its diameter in mils squared (1 in. = 1000 mils). For example, the circular mil area of an 8 AWG solid conductor that has a 0.1285-in. diameter is calculated as follows:

$$0.1285 \text{ in.} \times 1000 = 128.5 \text{ mils}$$

$$128.5 \times 128.5 = 16,512.25 \text{ circular mils}$$

or 16,510 circular mils (rounded off)

According to Table 8 in Chapter 9, this rounded value represents the circular mil area for one conductor. Where stranded conductors are used, the circular mil area of each strand must be multiplied by the number of strands to determine the circular mil area of the conductor.

110.7 Insulation Integrity.

Completed wiring installations shall be free from short circuits and from grounds other than as required or permitted in Article 250.

Insulation is the material that prevents the flow of electricity between points of different potential in an electrical system. Failure of the insulation system is one of the most common causes of problems in electrical installations, in both high-voltage and low-voltage systems.

Insulation tests are performed on new or existing installations to determine the quality or condition of the insulation of conductors and equipment. The principal causes of insulation failures are heat, moisture, dirt, and physical damage (abrasion or nicks) occurring during and after installation. Insulation can also fail due to chemical attack, sunlight, and excessive voltage stresses.

Insulation integrity must be maintained during overcurrent conditions. Overcurrent protective devices must be selected and coordinated using tables of insulation thermal-withstand ability to ensure that the damage point of an insulated conductor is never reached. These tables, entitled "Allowable Short-Circuit Currents for Insulated Copper (or Aluminum) Conductors," are contained in the Insulated Cable Engineers Association's publication ICEA P-32-382. See 110.10 for other circuit components.

In an insulation resistance test, an applied voltage ranging from 100 to 5000 (usually 500 to 1000 volts for systems of 600 volts or less), supplied from a source of constant potential, is applied across the insulation. A megohmmeter is usually the potential source, and it indicates the insulation resistance directly on a scale calibrated in megohms (MΩ). The quality of the insulation is evaluated based on the level of the insulation resistance.

The insulation resistance of many types of insulation varies with temperature, so the field data obtained should be corrected to the standard temperature for the class of equipment being tested. The megohm value of insulation resistance obtained is inversely proportional to the volume of insulation tested. For example, a cable 1000 ft long would be expected to have one-tenth the insulation resistance of a cable 100 ft long, if all other conditions are identical.

The insulation resistance test is relatively easy to perform and is useful on all types and classes of electrical equipment. Its main value lies in the charting of data from periodic tests, corrected for temperature, over a long period so that deteriorative trends can be detected.

Manuals on this subject are available from instrument manufacturers. Thorough knowledge in the use of insulation testers is essential if the test results are to be meaningful. Exhibit 110.1 shows a typical megohmmeter insulation tester.

Exhibit 110.1 A manual multivoltage, multirange insulation tester.

110.8 Wiring Methods.

Only wiring methods recognized as suitable are included in this *Code*. The recognized methods of wiring shall be permitted to be installed in any type of building or occupancy, except as otherwise provided in this *Code*.

The scope of Article 300 applies generally to all wiring methods, except as amended, modified, or supplemented by other *NEC* chapters. The application statement is found in 90.3, Code Arrangement.

110.9 Interrupting Rating.

Equipment intended to interrupt current at fault levels shall have an interrupting rating sufficient for the nominal circuit voltage and the current that is available at the line terminals of the equipment.

Equipment intended to interrupt current at other than fault levels shall have an interrupting rating at nominal circuit voltage sufficient for the current that must be interrupted.

In the 1999 *Code*, 110.9 was changed by substituting the word *interrupt* for the word *break* in two places.

The interrupting rating of overcurrent protective devices is determined under standard test conditions. It is important that the test conditions match the actual installation needs. Section 110.9 states that all fuses and circuit breakers intended to interrupt the circuit at fault levels must have an adequate interrupting rating wherever they are used in the electrical system. Fuses or circuit breakers that do not have adequate interrupting ratings could rupture while attempting to clear a short circuit.

Interrupting ratings should not be confused with short-circuit current ratings. Short-circuit current ratings are further explained in the commentary following 110.10.

110.10 Circuit Impedance and Other Characteristics.

The overcurrent protective devices, the total impedance, the component short-circuit current ratings, and other characteristics of the circuit to be protected shall be selected and coordinated to permit the circuit-protective devices used to clear a fault to do so without extensive damage to the electrical components of the circuit. This fault shall be assumed to be either between two or more of the circuit conductors or between any circuit conductor and the grounding conductor or enclosing metal raceway. Listed products applied in accordance with their listing shall be considered to meet the requirements of this section.

In the 1999 *Code*, 110.10 was changed by substituting the word *current* for *withstand*. That change correlated the *Code* language with the standard marking language used on equipment. Withstand ratings are not marked on equipment; short-circuit current ratings are marked on the equipment. This marking appears on many pieces of equipment, such as panelboards, switchboards, busways, contactors, and starters. Additionally, a new last sentence was added in the 1999 *Code* to address concerns of what exactly constitutes "extensive damage." Since, under product safety requirements, electrical equipment is evaluated for indications of extensive damage, listed products used within their ratings are considered to have met the requirements of this section.

The basic purpose of overcurrent protection is to open the circuit before conductors or conductor insulation is damaged when an overcurrent condition occurs. An overcurrent condition can be the result of an overload, a ground fault, or a short circuit and must be eliminated before the conductor insulation damage point is reached.

Overcurrent protective devices (such as fuses and circuit breakers) should be selected to ensure that the short-circuit current rating of the system components is not exceeded should a short circuit or high-level ground fault occur.

System components include wire, bus structures, switching, protection and disconnect devices, and distribution equipment, all of which have limited short-circuit ratings and would be damaged or destroyed if those short-circuit ratings were exceeded. Merely providing overcurrent protective devices with sufficient interrupting rating would not ensure adequate short-circuit protection for the system components. When the available short-circuit current exceeds the short-circuit current rating of an electrical component, the overcurrent protective device must limit the let-through energy to within the rating of that electrical component.

Utility companies usually determine and provide information on available short-circuit current levels at the service equipment. Literature on how to calculate short-circuit currents at each point in any distribution generally can be obtained by contacting the manufacturers of overcurrent protective devices or by referring to IEEE 141-1993, *IEEE Recommended Practice for Electric Power Distribution for Industrial Plants* (Red Book).

For a typical one-family dwelling with a 100-ampere service using 2 AWG aluminum supplied by a 37½ kVA transformer with 1.72 percent impedance located at a distance of 25 ft, the available short-circuit current would be approximately 6000 amperes.

Available short-circuit current to multifamily structures, where pad-mounted transformers are located close to the multimetering location, can be relatively high. For example, the line-to-line fault current values close to a low-impedance transformer could exceed 22,000 amperes. At the secondary of a single-phase, center-tapped transformer, the line-to-neutral fault current is approximately one and one-half times that of the line-to-line fault current. The short-circuit current rating of utilization equipment located and connected near the service equipment should be known. For example, HVAC

equipment is tested at 3500 amperes through a 40-ampere load rating and at 5000 amperes for loads rated more than 40 amperes.

Adequate short-circuit protection can be provided by fuses, molded-case circuit breakers, and low-voltage power circuit breakers, depending on specific circuit and installation requirements.

110.11 Deteriorating Agents.

Unless identified for use in the operating environment, no conductors or equipment shall be located in damp or wet locations; where exposed to gases, fumes, vapors, liquids, or other agents that have a deteriorating effect on the conductors or equipment; or where exposed to excessive temperatures.

> FPN No. 1: See 300.6 for protection against corrosion.
> FPN No. 2: Some cleaning and lubricating compounds can cause severe deterioration of many plastic materials used for insulating and structural applications in equipment.

Equipment identified only as "dry locations," "Type 1," or "indoor use only" shall be protected against permanent damage from the weather during building construction.

110.12 Mechanical Execution of Work.

Electrical equipment shall be installed in a neat and workmanlike manner.

The regulation in 110.12 calling for "neat and workmanlike" installations has appeared in the *NEC* as currently worded for more than a half-century. It stands as a basis for pride in one's work and has been emphasized by persons involved in the training of apprentice electricians for many years.

Many *Code* conflicts or violations have been cited by the authority having jurisdiction based on the authority's interpretation of "neat and workmanlike manner." Many electrical inspection authorities use their own experience or precedents in their local areas as the basis for their judgments.

Examples of installations that do not qualify as "neat and workmanlike" include exposed runs of cables or raceways that are improperly supported (e.g., sagging between supports or using improper support methods); field-bent and kinked, flattened, or poorly measured raceways; or cabinets, cutout boxes, and enclosures that are not plumb or not properly secured.

(A) Unused Openings. Unused cable or raceway openings in boxes, raceways, auxiliary gutters, cabinets, cutout boxes, meter socket enclosures, equipment cases, or housings shall be effectively closed to afford protection substantially equivalent to the wall of the equipment. Where metallic plugs or plates are used with nonmetallic enclosures, they shall be recessed at least 6 mm (¼ in.) from the outer surface of the enclosure.

In the 2002 *Code*, 110.12(A) has been revised by changing the phrase *unused openings* to *unused cable or raceway openings* to clarify that openings such as weep holes are not required to be closed up. In addition, text from 370.18 of the 1999 *NEC* has been moved to the end of 110.12(A). (The reorganization of Chapter 3 in the 2002 *NEC* moved Article 370 to Article 314.)

(B) Subsurface Enclosures. Conductors shall be racked to provide ready and safe access in underground and subsurface enclosures into which persons enter for installation and maintenance.

(C) Integrity of Electrical Equipment and Connections. Internal parts of electrical equipment, including busbars, wiring terminals, insulators, and other surfaces, shall not be damaged or contaminated by foreign materials such as paint, plaster, cleaners, abrasives, or corrosive residues. There shall be no damaged parts that may adversely affect safe operation or mechanical strength of the equipment such as parts that are broken; bent; cut; or deteriorated by corrosion, chemical action, or overheating.

110.13 Mounting and Cooling of Equipment.

(A) Mounting. Electrical equipment shall be firmly secured to the surface on which it is mounted. Wooden plugs driven into holes in masonry, concrete, plaster, or similar materials shall not be used.

(B) Cooling. Electrical equipment that depends on the natural circulation of air and convection principles for cooling of exposed surfaces shall be installed so that room airflow over such surfaces is not prevented by walls or by adjacent installed equipment. For equipment designed for floor mounting, clearance between top surfaces and adjacent surfaces shall be provided to dissipate rising warm air.

Ventilation for motor locations is covered in 430.14(A) and 430.16. Ventilation for transformer locations is covered in 450.9 and 450.45.

Electrical equipment provided with ventilating openings shall be installed so that walls or other obstructions do not prevent the free circulation of air through the equipment.

For example, a ventilated busway must be located where there are no walls or other objects that might interfere with the natural circulation of air and convection principles for cooling. See the definition of *ventilated* in Article 100.

Panelboards, transformers, and other types of equipment are adversely affected if enclosure surfaces normally exposed to room air are covered or tightly enclosed. Ventilating openings in equipment are provided to allow the circulation of room air around internal components of the equipment; the blocking of such openings can cause dangerous overheating.

110.14 Electrical Connections.

Because of different characteristics of dissimilar metals, devices such as pressure terminal or pressure splicing connectors and soldering lugs shall be identified for the material of the conductor and shall be properly installed and used. Conductors of dissimilar metals shall not be intermixed in a terminal or splicing connector where physical contact occurs between dissimilar conductors (such as copper and aluminum, copper and copper-clad aluminum, or aluminum and copper-clad aluminum), unless the device is identified for the purpose and conditions of use. Materials such as solder, fluxes, inhibitors, and compounds, where employed, shall be suitable for the use and shall be of a type that will not adversely affect the conductors, installation, or equipment.

FPN: Many terminations and equipment are marked with a tightening torque.

Section 110.3(B) applies where terminations and equipment are marked with tightening torques.

For the testing of wire connectors for which the manufacturer has not assigned another value appropriate for the design, Tables 1.2 through 1.5 provide data on the tightening torques that Underwriters Laboratories uses. These tables should be used for guidance only if no tightening information on the specific wire connector is available. They should *not* be used to replace the manufacturer's instructions, which should always be followed.

The information in the tables was taken from the edition of UL Standard 486B, *Wire Connections for Use with Aluminum Conductors,* in effect at the time of the printing of the 2002 edition of this handbook. Similar information can be found in UL 486A, *Wire Connections and Solder Lugs for Use with Copper Conductors.*

(A) Terminals. Connection of conductors to terminal parts shall ensure a thoroughly good connection without damaging the conductors and shall be made by means of pressure connectors (including set-screw type), solder lugs, or splices to flexible leads. Connection by means of wire-binding

Table 1.2 Tightening Torques for Screws* in Pound-Inches

Wire Size (AWG or kcmil)	Slotted Head No. 10 and Larger		Hexagonal Head–External Drive Socket Wrench	
	Slot Width to ³⁄₆₄ in. or Slot Length to ¼ in.**	Slot Width Over ³⁄₆₄ in. or Slot Length Over ¼ in.**	Split-Bolt Connectors	Other Connectors
30–10	20	35	80	75
8	25	40	80	75
6	35	45	165	110
4	35	45	165	110
3	35	50	275	150
2	40	50	275	150
1	—	50	275	150
1/0	—	50	385	180
2/0	—	50	385	180
3/0	—	50	500	250
4/0	—	50	500	250
250	—	50	650	325
300	—	50	650	325
350	—	50	650	325
400	—	50	825	325
500	—	50	825	375
600	—	50	1000	375
700	—	50	1000	375
750	—	50	1000	375
800	—	50	1100	500
900	—	50	1100	500
1000	—	50	1100	500
1250	—	—	1100	600
1500	—	—	1100	600
1750	—	—	1100	600
2000	—	—	1100	600

*Clamping screws with multiple tightening means. For example, for a slotted hexagonal head screw, use the torque value associated with the tool used in the installation. UL uses both values when testing.

**For values of slot width or length other than those specified, select the largest torque value associated with conductor size.

screws or studs and nuts that have upturned lugs or the equivalent shall be permitted for 10 AWG or smaller conductors.

Terminals for more than one conductor and terminals used to connect aluminum shall be so identified.

(B) Splices. Conductors shall be spliced or joined with splicing devices identified for the use or by brazing, welding, or soldering with a fusible metal or alloy. Soldered splices shall first be spliced or joined so as to be mechanically and electrically secure without solder and then be soldered. All splices and joints and the free ends of conductors shall be covered with an insulation equivalent to that of the conductors or with an insulating device identified for the purpose.

Wire connectors or splicing means installed on conductors for direct burial shall be listed for such use.

Field observations and trade magazine articles indicate that electrical connection failures have been determined to be the cause of many equipment burnouts and fires. Many of these failures are attributable to improper terminations, poor workmanship, the differing characteristics of dissimilar metals, and improper binding screws or splicing devices.

UL's requirements for listing solid aluminum conductors in 12 AWG and 10 AWG and for listing snap switches and receptacles for use on 15- and 20-ampere branch circuits incorporate stringent tests that take into account the factors listed in the preceding paragraph. For further information regarding receptacles and switches using CO/ALR-rated terminals, refer to 404.14(C) and 406.2(C).

Screwless pressure terminal connectors of the conductor

Table 1.3 Torques in Pound-Inches for Slotted Head Screws* Smaller Than No. 10, for Use with 8 AWG and Smaller Conductors

Screw-Slot Length (in.)**	Screw-Slot Width Less Than ³⁄₆₄ in.	Screw-Slot Width ³⁄₆₄ in. and Larger
To ⁵⁄₃₂	7	9
⁵⁄₃₂	7	12
³⁄₁₆	7	12
⁷⁄₃₂	7	12
¹⁄₄	9	12
⁹⁄₃₂	—	15
Above ⁹⁄₃₂	—	20

*Clamping screws with multiple tightening means. For example, for a slotted hexagonal head screw, use the torque value associated with the tool used in the installation. UL uses both values when testing.
**For slot lengths of intermediate values, select torques pertaining to next-shorter slot length.

Table 1.4 Torques for Recessed Allen Head Screws

Socket Size Across Flats (in.)	Torque (lb-in.)
¹⁄₈	45
⁵⁄₃₂	100
³⁄₁₆	120
⁷⁄₃₂	150
¹⁄₄	200
⁵⁄₁₆	275
³⁄₈	375
¹⁄₂	500
⁹⁄₁₆	600

Table 1.5 Lug-Bolting Torques for Connection of Wire Connectors to Busbars

Bolt Diameter	Tightening Torque (lb-ft)
No. 8 or smaller	1.5
No. 10	2.0
¹⁄₄ in. or less	6
⁵⁄₁₆ in.	11
³⁄₈ in.	19
⁷⁄₁₆ in.	30
¹⁄₂ in.	40
⁹⁄₁₆ in. or larger	55

push-in type are for use with solid copper and copper-clad aluminum conductors only.

Instructions that describe proper installation techniques and emphasize the need to follow those techniques and prac-

tice good workmanship are required to be included with each coil of 12 AWG and 10 AWG insulated aluminum wire or cable. See also the commentary on tightening torque that follows 110.14, FPN.

New product and material designs that provide for increased levels of safety of aluminum wire terminations have been developed by the electrical industry. To assist all concerned parties in the proper and safe use of solid aluminum wire in making connections to wiring devices used on 15- and 20-ampere branch circuits, the following information is presented. Understanding and using this information is essential for proper application of materials and devices now available.

For New Installations

The following commentary is based on a report prepared by the Ad Hoc Committee on Aluminum Terminations prior to the 1975 *Code*. This information is still pertinent today and is necessary to comply with 110.14(A) when using aluminum wire in new installations.

New Materials and Devices. For direct connection, only 15- and 20-ampere receptacles and switches marked "CO/ALR," and connected as follows in Installation Method, should be used.

The "CO/ALR" marking is on the device mounting yoke or strap. The "CO/ALR" marking means the devices have been tested to stringent heat-cycling requirements to determine their suitability for use with UL-labeled aluminum, copper, or copper-clad aluminum wire.

Listed solid aluminum wire, 12 AWG or 10 AWG, marked with the aluminum insulated wire label should be used. The installation instructions that are packaged with the wire should be used.

Installation Method. Exhibit 110.2 illustrates the following correct method of connection to be used:

1. The freshly stripped end of the wire should be wrapped two-thirds to three-quarters of the distance around the wire-binding screw post, as shown in Step A of Exhibit 110.2. The loop is made so that rotation of the screw during tightening will tend to wrap the wire around the post rather than unwrap it.
2. The screw should be tightened until the wire is snugly in contact with the underside of the screw head and with the contact plate on the wiring device, as shown in Step B of Exhibit 110.2.
3. The screw should be tightened an additional half-turn, thereby providing a firm connection, as shown in Step C of Exhibit 110.2. Where torque screwdrivers are used, the screw should be tightened to 12 lb-in., as shown in Step C of Exhibit 110.2.

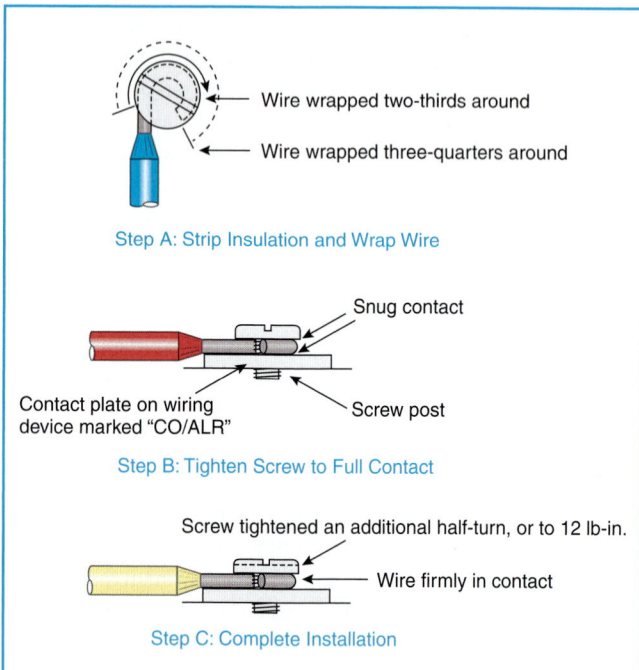

Step A: Strip Insulation and Wrap Wire
- Wire wrapped two-thirds around
- Wire wrapped three-quarters around

Step B: Tighten Screw to Full Contact
- Snug contact
- Contact plate on wiring device marked "CO/ALR"
- Screw post

Step C: Complete Installation
- Screw tightened an additional half-turn, or to 12 lb-in.
- Wire firmly in contact

Exhibit 110.2 Correct method of terminating aluminum wire at wire-binding screw terminals of receptacles and snap switches. (Redrawn from Underwriters Laboratories Inc.)

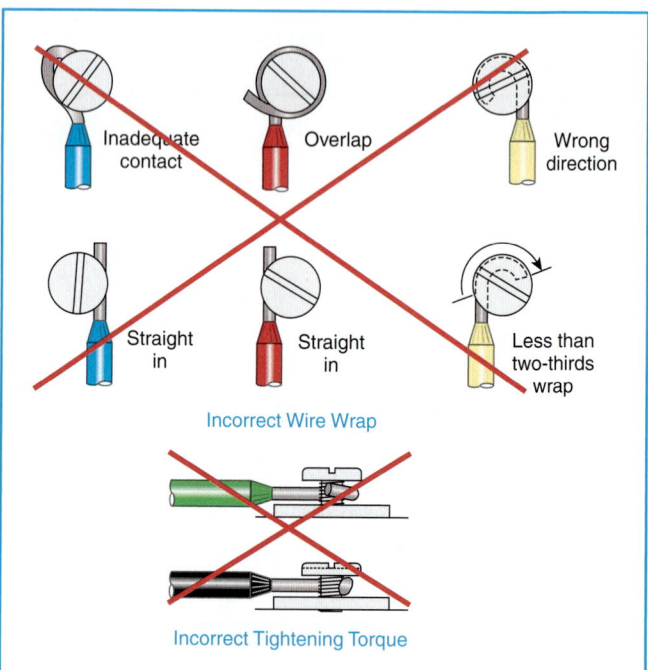

Incorrect Wire Wrap
- Inadequate contact
- Overlap
- Wrong direction
- Straight in
- Straight in
- Less than two-thirds wrap

Incorrect Tightening Torque

Exhibit 110.3 Incorrect methods of terminating aluminum wire at wire-binding screw terminals of receptacles and snap switches. (Redrawn from Underwriters Laboratories Inc.)

4. The wires should be positioned behind the wiring device to decrease the likelihood of the terminal screws loosening when the device is positioned into the outlet box.

Exhibit 110.3 illustrates incorrect methods for connection. These methods should *not* be used.

Existing Inventory. Labeled 12 AWG or 10 AWG solid aluminum wire that does not bear the new aluminum wire label should be used with wiring devices marked "CO/ALR" and connected as described in the preceding Installation Method. This is the preferred and recommended method for using such wire.

In the following types of devices, the terminals should not be directly connected to aluminum conductors but may be used with labeled copper or copper-clad conductors:

1. Receptacles and snap switches marked "AL-CU"
2. Receptacles and snap switches having no conductor marking
3. Receptacles and snap switches that have back-wired terminals or screwless terminals of the push-in type

For Existing Installations

If examination discloses overheating or loose connections, the recommendations described under Existing Inventory should be followed.

Twist-On Wire Connectors

Because 110.14(B) requires conductors to be spliced with "splicing devices identified for the use," wire connectors are required to be marked for conductor suitability. Twist-on wire connectors are not suitable for splicing aluminum conductors or copper-clad aluminum to copper conductors unless it is so stated and marked as such on the shipping carton. The marking is typically "AL-CU (dry locations)." Presently, one style of wire nut and one style of crimp-type connector have been listed as having met these requirements.

On February 2, 1995, Underwriters Laboratories announced the listing of a twist-on wire connector suitable for use with aluminum-to-copper conductors, in accordance with UL 486C, *Splicing Wire Connectors*. That was the first listing of a twist-on type connector for aluminum-to-copper conductors since 1987. The UL listing does *not* cover aluminum-to-aluminum combinations. However, more than one aluminum or copper conductor is allowed when used in combination.

These listed wire-connecting devices are available for pigtailing short lengths of copper conductors to the original aluminum branch-circuit conductors, as shown in Exhibit 110.4. Primarily, these pigtailed conductors supply 15- and 20-ampere wiring devices. Pigtailing is permitted, provided there is suitable space within the enclosure.

(C) Temperature Limitations. The temperature rating associated with the ampacity of a conductor shall be selected

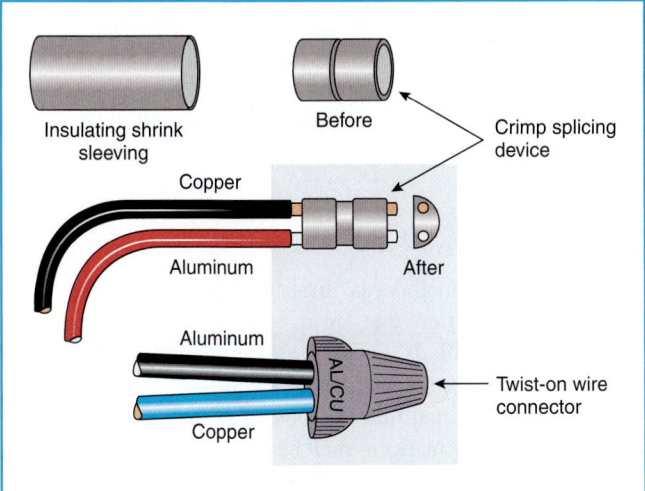

Exhibit 110.4 Pigtailing copper to aluminum conductors using two listed devices.

and coordinated so as not to exceed the lowest temperature rating of any connected termination, conductor, or device. Conductors with temperature ratings higher than specified for terminations shall be permitted to be used for ampacity adjustment, correction, or both.

(1) Equipment Provisions. The determination of termination provisions of equipment shall be based on 110.14(C)1(a) or (b). Unless the equipment is listed and marked otherwise, conductor ampacities used in determining equipment termination provisions shall be based on Table 310.16 as appropriately modified by 310.15(B)(1) through (6).

   (a) Termination provisions of equipment for circuits rated 100 amperes or less, or marked for 14 AWG through 1 AWG conductors, shall be used only for one of the following:

(1) Conductors rated 60°C (140°F)
(2) Conductors with higher temperature ratings, provided the ampacity of such conductors is determined based on the 60°C (140°F) ampacity of the conductor size used
(3) Conductors with higher temperature ratings if the equipment is listed and identified for use with such conductors
(4) For motors marked with design letters B, C, D, or E, conductors having an insulation rating of 75°C (167°F) or higher shall be permitted to be used provided the ampacity of such conductors does not exceed the 75°C (167°F) ampacity.

   (b) Termination provisions of equipment for circuits rated over 100 amperes, or marked for conductors larger than 1 AWG, shall be used only for one of the following:

(1) Conductors rated 75°C (167°F)

(2) Conductors with higher temperature ratings, provided the ampacity of such conductors does not exceed the 75°C (167°F) ampacity of the conductor size used, or up to their ampacity if the equipment is listed and identified for use with such conductors

(2) Separate Connector Provisions. Separately installed pressure connectors shall be used with conductors at the ampacities not exceeding the ampacity at the listed and identified temperature rating of the connector.

> FPN: With respect to 110.14(C)(1) and (2), equipment markings or listing information may additionally restrict the sizing and temperature ratings of connected conductors.

Revised for the 2002 *Code*, 110.14(C)(1) now states that where conductors are terminated in equipment, the selected conductor ampacities must be based on Table 310.16, unless the equipment is specifically listed and marked otherwise. The intent of this requirement is to clarify which ampacities are used to determine the proper conductor size at equipment terminations.

When 600-V-and-less equipment is evaluated relative to the appropriate temperature characteristics of the terminations, conductors sized according to Table 310.16 are required to be used. The UL *General Information Directory* (pages 1 and 2; UL White Book) clearly indicates that the 60°C and 75°C provisions for equipment have been determined using conductors from Table 310.16. However, installers or designers unaware of the UL guide card information might attempt to select conductors based on a table other than 310.16. That is especially true if a wiring method that allows the use of ampacities such as those in Table 310.17 is used. That use can result in overheated terminations at the equipment. Clearly, the ampacities shown in other tables (such as Table 310.17) could be used for various conditions to which the wiring method is subject (ambient, ampacity correction, etc.), but the conductor size at the termination must be based on ampacities from Table 310.16. This change introduces no new impact on the equipment or the wiring methods; it simply adds a rule from the listing information into the *Code* because it is an installation and equipment selection issue.

Section 110.14(C)(1)(a) requires that conductor terminations, as well as conductors, be rated for the operating temperature of the circuit. For example, the load on an 8 AWG THHN, 90°C copper wire is limited to 40 amperes where connected to a disconnect switch with terminals rated at 60°C. This same 8 AWG THHN, 90°C wire is limited to 50 amperes where connected to a fusible switch with terminals rated at 75°C. The conductor ampacities were selected from Table 310.16. Not only does this requirement apply to conductor terminations of breakers and fusible switches, but the equipment enclosure must also permit terminations above 60°C. Exhibit 110.5 shows an example of termination temperature markings.

Exhibit 110.5 An example of termination temperature markings on a main circuit breaker. (Courtesy of Square D Co.)

110.15 High-Leg Marking.

On a 4-wire, delta-connected system where the midpoint of one phase winding is grounded to supply lighting and similar loads, the conductor or busbar having the higher phase voltage to ground shall be durably and permanently marked by an outer finish that is orange in color or by other effective means. Such identification shall be placed at each point on the system where a connection is made if the grounded conductor is also present.

Added for the 2002 *Code*, this section now contains a requirement that appeared in 384-3(e) of the 1999 *NEC*. This requirement was moved to Article 110, where the application becomes a more general requirement.

The high leg is common on a 240/120-volt 3-phase, 4-wire delta system. It is typically designated as "B phase." The high-leg marking is required to be the color orange or other similar effective means and is intended to prevent

problems due to the lack of complete standardization where metered and nonmetered equipment are installed in the same installation. Electricians should always test each phase relative to ground with suitable equipment to determine exactly where the high leg is located in the system.

110.16 Flash Protection.

Switchboards, panelboards, industrial control panels, and motor control centers that are in other than dwelling occupancies and are likely to require examination, adjustment, servicing, or maintenance while energized shall be field marked to warn qualified persons of potential electric arc flash hazards. The marking shall be located so as to be clearly visible to qualified persons before examination, adjustment, servicing, or maintenance of the equipment.

> FPN No. 1: NFPA 70E-2000, *Electrical Safety Requirements for Employee Workplaces*, provides assistance in determining severity of potential exposure, planning safe work practices, and selecting personal protective equipment.
> FPN No. 2: ANSI Z535.4-1998, *Product Safety Signs and Labels*, provides guidelines for the design of safety signs and labels for application to products.

This requirement is new in the 2002 *Code*. Field marking that warns electrical workers of potential electrical arc flash hazards is now required because significant numbers of electricians are being seriously burned or killed from accidental electrical arc flash while working equipment "hot" (energized). Most of these accidents could be prevented or their severity significantly reduced if electricians wore the proper type of protective clothing. Requiring switchboards, panelboards, and motor control centers to be individually field marked with proper warning labels will raise the level of awareness of electrical arc flash hazards and thereby decrease the number of accidents.

In Exhibit 110.6, an electrical employee is working inside the flash protection boundary and in front of a large capacity service type switchboard that has not been de-energized and is not under the lockout/tagout procedure. The worker is wearing personal protective equipment (PPE) considered appropriate flash protection clothing for the flash hazard involved. Suitable PPE appropriate to a particular hazard can be determined by using NFPA 70E, *Standard for Electrical Safety Requirements for Employee Workplaces*.

Accident reports continue to confirm the fact that workers responsible for the installation or maintenance of electrical equipment often do not turn off the power source before working on the equipment. Working electrical equipment energized is a major safety concern in the electrical industry. The real purpose of this additional code requirement is to alert electrical contractors, electricians, facility owners and

Exhibit 110.6 Electrical worker clothed in personal protective equipment (PPE) appropriate for the hazards involved.

managers, and other interested parties to some of the hazards of working on or near energized equipment and to emphasize the importance of turning off the power before working on electrical circuits.

The information in these fine print notes is not mandatory. Employers can be assured that they are providing a safe workplace for their employees if safety-related work practices required by NFPA 70E have been implemented and are being followed.

In addition to the standards referenced in the fine print notes and their individual bibliographies, additional information on this subject can be found in the 1997 report "Hazards of Working Electrical Equipment Hot," published by the National Electrical Manufacturers Association.

110.18 Arcing Parts.

Parts of electric equipment that in ordinary operation produce arcs, sparks, flames, or molten metal shall be enclosed or separated and isolated from all combustible material.

> FPN: For hazardous (classified) locations, see Articles 500 through 517. For motors, see 430.14.

Examples of electrical equipment that may produce sparks during ordinary operation include open motors having a centrifugal starting switch, open motors with commutators, and collector rings. Adequate separation from combustible material is essential if open motors with these features are used.

110.19 Light and Power from Railway Conductors.

Circuits for lighting and power shall not be connected to any system that contains trolley wires with a ground return.

Exception: Such circuit connections shall be permitted in car houses, power houses, or passenger and freight stations operated in connection with electric railways.

110.21 Marking.

The manufacturer's name, trademark, or other descriptive marking by which the organization responsible for the product can be identified shall be placed on all electric equipment. Other markings that indicate voltage, current, wattage, or other ratings shall be provided as specified elsewhere in this *Code*. The marking shall be of sufficient durability to withstand the environment involved.

The *Code* requires that equipment ratings be marked on the equipment and that such markings be located so as to be visible or easily accessible during or after installation.

110.22 Identification of Disconnecting Means.

Each disconnecting means shall be legibly marked to indicate its purpose unless located and arranged so the purpose is evident. The marking shall be of sufficient durability to withstand the environment involved.

Where circuit breakers or fuses are applied in compliance with the series combination ratings marked on the equipment by the manufacturer, the equipment enclosure(s) shall be legibly marked in the field to indicate the equipment has been applied with a series combination rating. The marking shall be readily visible and state the following:

> CAUTION — SERIES COMBINATION SYSTEM
> RATED ____ AMPERES. IDENTIFIED
> REPLACEMENT COMPONENTS REQUIRED.

> FPN: See Section 240.86(A) for interrupting rating marking for end-use equipment.

Proper identification needs to be specific. For example, the marking should indicate not merely "motor" but rather "motor, water pump"; not merely "lights" but rather "lights, front lobby." Consideration also should be given to the form of identification. Marking often fades or is covered by paint after installation.

The second paragraph of 110.22 requires series-rated overcurrent devices to be legibly marked. The equipment manufacturer can mark the equipment to be used with series combination ratings. If the equipment is installed in the field at its marked series combination rating, the equipment must have an additional label, as specified in 110.22, to indicate that the series combination rating has been used.

110.23 Current Transformers.

Unused current transformers associated with potentially energized circuits shall be short-circuited.

Because Article 450 specifically exempts current transformers, the practical solution to prevent damage to current transformers not connected to a load or for unused current transformers has been placed here as a new requirement for the 2002 *Code*.

II. 600 Volts, Nominal, or Less

110.26 Spaces About Electrical Equipment.

Sufficient access and working space shall be provided and maintained about all electric equipment to permit ready and safe operation and maintenance of such equipment. Enclosures housing electrical apparatus that are controlled by lock and key shall be considered accessible to qualified persons.

Key to understanding 110.26 is the division of requirements for spaces about electrical equipment in two separate and distinct categories: working space and dedicated equipment space. Working space generally applies to the protection of the worker, and dedicated equipment space applies to the space reserved for future access to electrical equipment and to protection of the equipment from intrusion by nonelectrical equipment. The performance requirements for all spaces about electrical equipment are set forth in the first sentence. Storage of materials that blocks access or prevents safe work practices must be avoided at all times.

(A) Working Space. Working space for equipment operating at 600 volts, nominal, or less to ground and likely to require examination, adjustment, servicing, or maintenance while energized shall comply with the dimensions of 110.26(A)(1), (2), and (3) or as required or permitted elsewhere in this *Code*.

The intent of 110.26(A) is to provide enough space for personnel to perform any of the operations listed without jeopardizing worker safety. These operations include examination, adjustment, servicing, and maintenance of equipment. Examples of such equipment include panelboards, switches, circuit breakers, controllers, and controls on heating and air-conditioning equipment. It is important to understand that the word *examination,* as used in 110.26(A), includes such tasks as checking for the presence of voltage using a portable voltmeter.

Minimum working clearances are not required if the equipment is such that it is not likely to require examination, adjustment, servicing, or maintenance while energized. However, "sufficient" access and working space are still required by the opening paragraph of 110.26.

(1) Depth of Working Space. The depth of the working space in the direction of live parts shall not be less than that

specified in Table 110.26(A)(1) unless the requirements of 110.26(A)(1)(a), (b), or (c) are met. Distances shall be measured from the exposed live parts or from the enclosure or opening if the live parts are enclosed.

Table 110.26(A)(1) Working Spaces

Nominal Voltage to Ground	Minimum Clear Distance		
	Condition 1	Condition 2	Condition 3
0–150	900 mm (3 ft)	900 mm (3 ft)	900 mm (3 ft)
151–600	900 mm (3 ft)	1 m (3½ ft)	1.2 m (4 ft)

Note: Where the conditions are as follows:
Condition 1 — Exposed live parts on one side and no live or grounded parts on the other side of the working space, or exposed live parts on both sides effectively guarded by suitable wood or other insulating materials. Insulated wire or insulated busbars operating at not over 300 volts to ground shall not be considered live parts.
Condition 2 — Exposed live parts on one side and grounded parts on the other side. Concrete, brick, or tile walls shall be considered as grounded.
Condition 3 — Exposed live parts on both sides of the work space (not guarded as provided in Condition 1) with the operator between.

Included in these clearance requirements is the step-back distance from the face of the equipment. Table 110.26(A)(1) provides requirements for clearances away from the equipment, based on the circuit voltage to ground and whether there are grounded or ungrounded objects in the step-back space or exposed live parts across from each other. The voltages to ground consist of two groups: 0 to 150 and 151 to 600, inclusive. Remember, where an ungrounded system is utilized, the voltage to ground is the greatest voltage between the given conductor and any other conductor of the circuit. For example, the voltage to ground for a 480-volt ungrounded delta system is 480 volts. See Exhibit 110.7 for general working clearance requirements for each of the three conditions expressed in Table 110.26(A)(1).

(a) Dead-Front Assemblies. Working space shall not be required in the back or sides of assemblies, such as dead-front switchboards or motor control centers, where all connections and all renewable or adjustable parts, such as fuses or switches, are accessible from locations other than the back or sides. Where rear access is required to work on nonelectrical parts on the back of enclosed equipment, a minimum horizontal working space of 762 mm (30 in.) shall be provided.

The intent of this section is to point out that work space is required only from the side(s) of the enclosure that requires access. The general rule still applies: Equipment that requires front, rear, or side access for electrical activities described

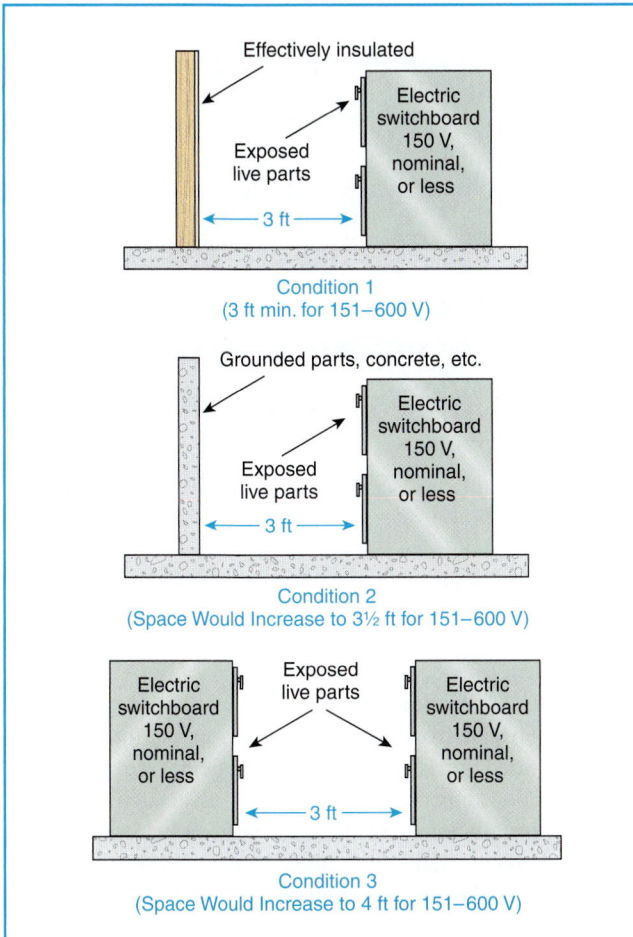

Condition 1
(3 ft min. for 151–600 V)

Condition 2
(Space Would Increase to 3½ ft for 151–600 V)

Condition 3
(Space Would Increase to 4 ft for 151–600 V)

Exhibit 110.7 Distances measured from the live parts if the live parts are exposed or from the enclosure front if the live parts are enclosed. If any assemblies, such as switchboards or motor-control centers, are accessible from the back and expose live parts, the working clearance dimensions would be required at the rear of the equipment, as illustrated. Note that for Condition 3, where there is an enclosure on opposite sides of the working space, the clearance for only one working space is required.

in 110.26(A) must meet the requirements of Table 110.26(A)(1). In many cases, equipment of "dead-front" assemblies requires only front access. For equipment that requires rear access for nonelectrical activity, however, a reduced working space of at least 762 mm (30 in.) must be provided. Exhibit 110.8 shows a reduced working space of 30 in. at the rear of equipment to allow work on nonelectrical parts.

(b) Low Voltage. By special permission, smaller working spaces shall be permitted where all uninsulated parts operate at not greater than 30 volts rms, 42 volts peak, or 60 volts dc.

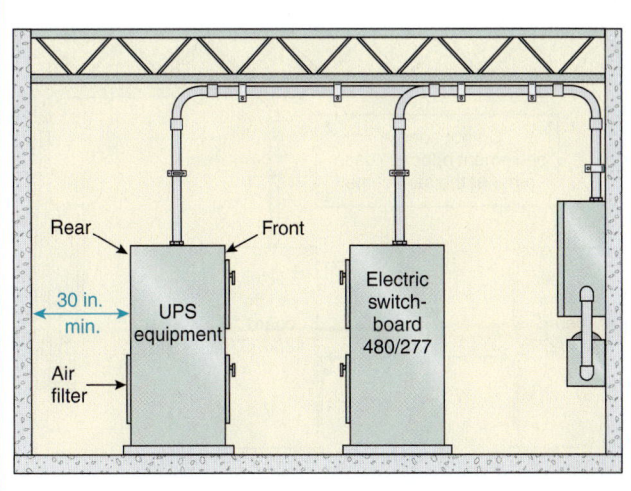

Exhibit 110.8 Example of the 30-in. minimum working space at the rear of equipment to allow work on nonelectrical parts, such as the replacement of an air filter.

(c) Existing Buildings. In existing buildings where electrical equipment is being replaced, Condition 2 working clearance shall be permitted between dead-front switchboards, panelboards, or motor control centers located across the aisle from each other where conditions of maintenance and supervision ensure that written procedures have been adopted to prohibit equipment on both sides of the aisle from being open at the same time and qualified persons who are authorized will service the installation.

This section permits some relief for installations being upgraded. When dead-front switchboards, panelboards, or motor-control centers are replaced in an existing building, the working clearance allowed is that required by Table 110.26(A)(1), Condition 2. The reduction from a Condition 3 to a Condition 2 clearance is allowed only where a written procedure prohibits facing doors of equipment from being open at the same time and where only authorized and qualified persons service the installation. Exhibit 110.9 illustrates this relief for existing buildings.

(2) Width of Working Space. The width of the working space in front of the electric equipment shall be the width of the equipment or 750 mm (30 in.), whichever is greater. In all cases, the work space shall permit at least a 90 degree opening of equipment doors or hinged panels.

Regardless of the width of the electrical equipment, the working space cannot be less than 30 in. wide. This allows an individual to have at least shoulder-width space in front of the equipment. This 30-in. measurement can be made

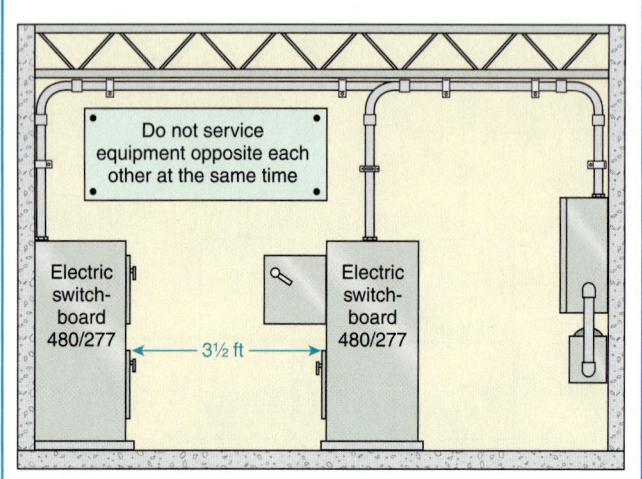

Exhibit 110.9 Permitted reduction from a Condition 3 to a Condition 2 clearance according to 110.26(A)(1)(c).

from either the left or the right edge of the equipment and can overlap other electrical equipment, provided the other equipment does not extend beyond the clearance required by Table 110.26(A)(1). If the equipment is wider than 30 in., the left-to-right space must be equal to the width of the equipment. See Exhibit 110.10 for an explanation of the 30-in. width requirement.

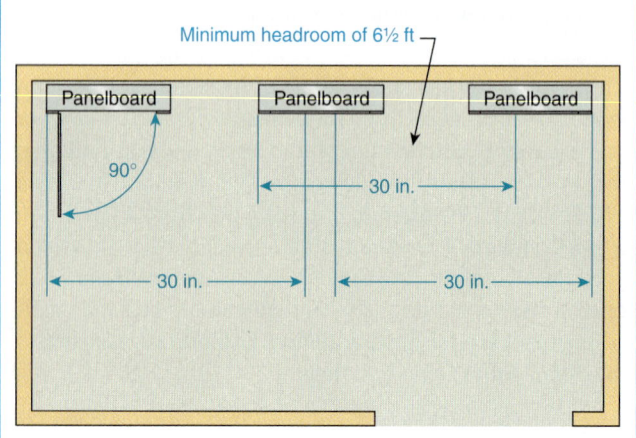

Exhibit 110.10 The 30 in. wide front working space not required to be directly centered on the electrical equipment if space is sufficient for safe operation and maintenance of such equipment.

Sufficient depth in the working space also must be provided to allow a panel or door to open at least 90 degrees. If doors or hinged panels are wider than 3 ft, more than a 3 ft deep working space must be provided to allow a full 90-degree opening. (See Exhibit 110.11.)

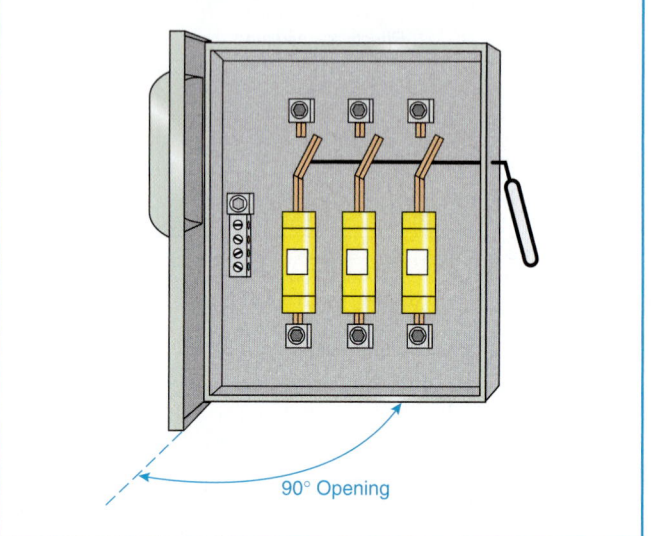

Exhibit 110.11 Equipment doors required to open a full 90 degrees to ensure a safe working space.

(3) Height of Working Space. The work space shall be clear and extend from the grade, floor, or platform to the height required by 110.26(E). Within the height requirements of this section, other equipment that is associated with the electrical installation and is located above or below the electrical equipment shall be permitted to extend not more than 150 mm (6 in.) beyond the front of the electrical equipment.

In addition to requiring a working space to be clear from the floor up to a height of 6½ ft or to the height of the equipment, whichever is greater, this section permits electrical equipment located above or below other electrical equipment to extend into the "working space" not more than 6 in. This requirement allows the placement of a 12 in. × 12 in. wireway on the wall directly above or below a 6 in. deep panelboard without impinging into the working space or compromising practical working clearances. The requirement continues to prohibit large differences in depth of equipment below or above other equipment that specifically requires working space. Electrical equipment that produces heat or that otherwise requires ventilation also must comply with 110.3(B) and 110.13.

(B) Clear Spaces. Working space required by this section shall not be used for storage. When normally enclosed live parts are exposed for inspection or servicing, the working space, if in a passageway or general open space, shall be suitably guarded.

Section 110.26(B), as well as the rest of 110.26, does not prohibit the placement of panelboards in corridors or passageways. Thus, while the covers of corridor-mounted panelboards are removed for servicing or other work, access to the area around the panelboard should be guarded or limited, to protect unqualified persons using the corridor.

(C) Entrance to Working Space.

(1) Minimum Required. At least one entrance of sufficient area shall be provided to give access to working space about electrical equipment.

(2) Large Equipment. For equipment rated 1200 amperes or more and over 1.8 m (6 ft) wide that contains overcurrent devices, switching devices, or control devices, there shall be one entrance to the required working space not less than 610 mm (24 in.) wide and 2.0 m (6½ ft) high at each end of the working space. Where the entrance has a personnel door(s), the door(s) shall open in the direction of egress and be equipped with panic bars, pressure plates, or other devices that are normally latched but open under simple pressure.

For the 2002 *Code,* a new requirement provides for the installation of panic or push-type hardware on personnel door(s) used for egress from electrical rooms containing large electrical equipment (such as switchboards, panelboards, and the like that are over 6 ft wide and rated 1200 amperes or more).

This new requirement affords safety for workers exposed to energized conductors by allowing an injured worker to safely and quickly exit an electrical room without having to turn knobs or pull doors open.

For a graphical explanation of access and entrance requirements to a working space, see Exhibits 110.12 and 110.13. Notice the unacceptable and hazardous situation shown in Exhibit 110.14.

A single entrance to the required working space shall be permitted where either of the conditions in 110.26(C)(2)(a) or (b) is met.

(a) Unobstructed Exit. Where the location permits a continuous and unobstructed way of exit travel, a single entrance to the working space shall be permitted.

(b) Extra Working Space. Where the depth of the working space is twice that required by 110.26(A)(1), a single entrance shall be permitted. It shall be located so that the distance from the equipment to the nearest edge of the entrance is not less than the minimum clear distance specified in Table 110.26(A)(1) for equipment operating at that voltage and in that condition.

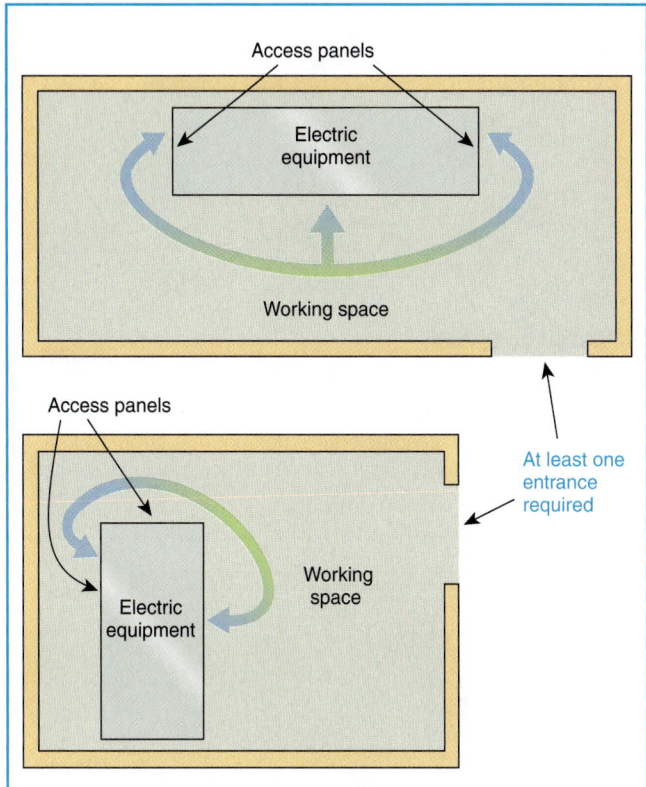

Exhibit 110.12 Basic Rule, first paragraph. At least one entrance is required to provide access to the working space around electrical equipment [110.26(C)(1)]. The lower installation would not be acceptable for a switchboard over 6 ft wide and rated 1200 amperes or more.

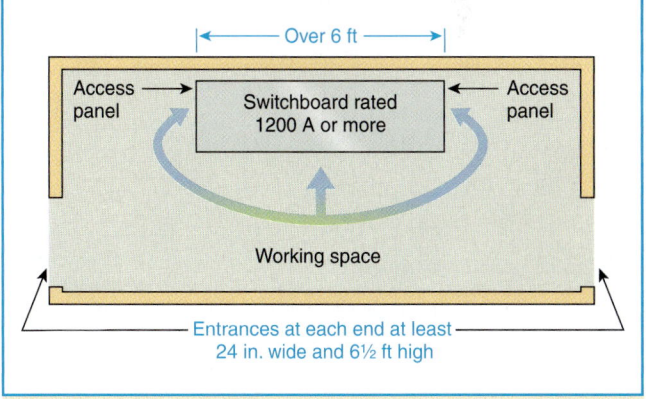

Exhibit 110.13 Basic Rule, second paragraph. For equipment rated 1200 amperes or more and over 6 ft wide, one entrance not less than 24 in. wide and 6½ ft high is required at each end [110.26(C)(2)].

For an explanation of 110.26(C)(2)(a) and (b), see Exhibits 110.15 and 110.16.

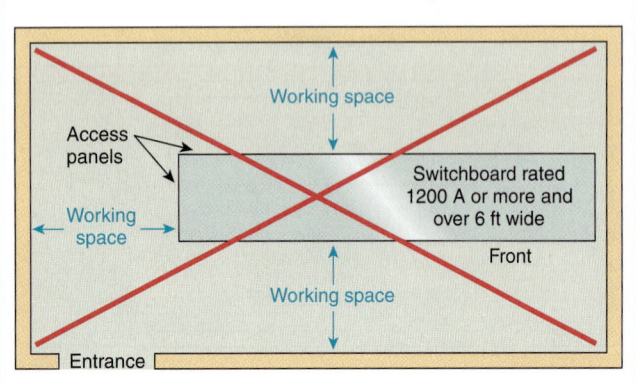

Exhibit 110.14 Unacceptable arrangement of a large switchboard. A person could be trapped behind arcing electrical equipment.

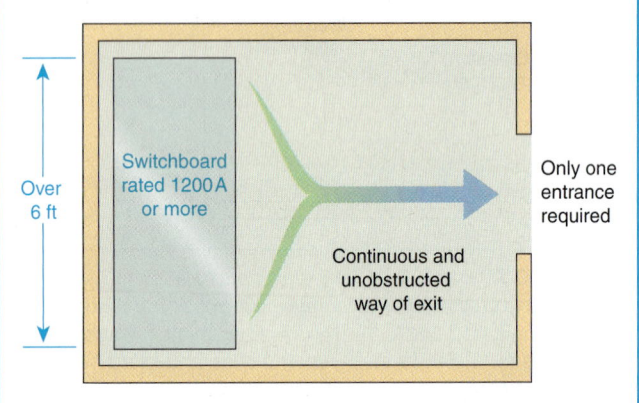

Exhibit 110.15 Equipment location allowing a continuous and unobstructed way of exit travel.

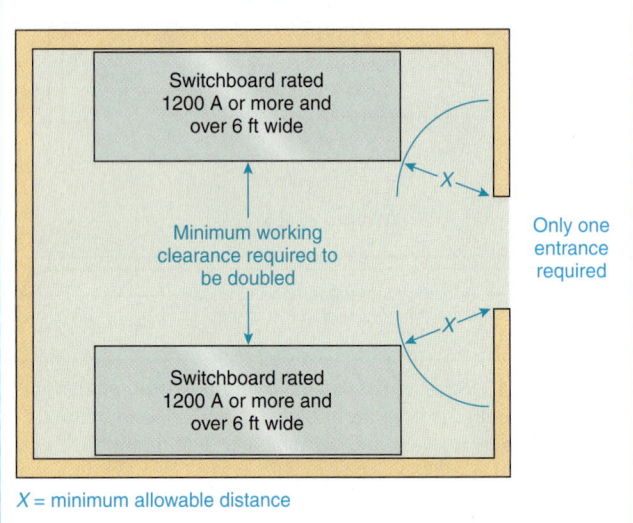

X = minimum allowable distance

Exhibit 110.16 Working space with one entrance. Only one entrance is required if the working space required by 110.26(A) is doubled. See Table 110.26(A)(1) for permitted dimensions for *X*.

(D) Illumination. Illumination shall be provided for all working spaces about service equipment, switchboards, panelboards, or motor control centers installed indoors. Additional lighting outlets shall not be required where the work space is illuminated by an adjacent light source or as permitted by 210.70(A)(1), Exception No. 1, for switched receptacles. In electrical equipment rooms, the illumination shall not be controlled by automatic means only.

(E) Headroom. The minimum headroom of working spaces about service equipment, switchboards, panelboards, or motor control centers shall be 2.0 m (6½ ft). Where the electrical equipment exceeds 2.0 m (6½ ft) in height, the minimum headroom shall not be less than the height of the equipment.

Exception: In existing dwelling units, service equipment or panelboards that do not exceed 200 amperes shall be permitted in spaces where the headroom is less than 2.0 m (6½ ft).

(F) Dedicated Equipment Space. All switchboards, panelboards, distribution boards, and motor control centers shall be located in dedicated spaces and protected from damage.

Exception: Control equipment that by its very nature or because of other rules of the Code must be adjacent to or within sight of its operating machinery shall be permitted in those locations.

(1) Indoor. Indoor installations shall comply with 110.26(F)(1)(a) through (d).

(a) Dedicated Electrical Space. The space equal to the width and depth of the equipment and extending from the floor to a height of 1.8 m (6 ft) above the equipment or to the structural ceiling, whichever is lower, shall be dedicated to the electrical installation. No piping, ducts, leak protection apparatus, or other equipment foreign to the electrical installation shall be located in this zone.

Exception: Suspended ceilings with removable panels shall be permitted within the 1.8-m (6-ft) zone.

(b) Foreign Systems. The area above the dedicated space required by 110.26(F)(1)(a) shall be permitted to contain foreign systems, provided protection is installed to avoid damage to the electrical equipment from condensation, leaks, or breaks in such foreign systems.

(c) Sprinkler Protection. Sprinkler protection shall be permitted for the dedicated space where the piping complies with this section.

(d) Suspended Ceilings. A dropped, suspended, or similar ceiling that does not add strength to the building structure shall not be considered a structural ceiling.

The dedicated electrical space includes the space defined by extending the footprint of the switchboard or panelboard from the floor to a height of 6 ft above the height of the equipment or to the structural ceiling, whichever is lower. This reserved space permits busways, conduits, raceways, and cables to enter the equipment. The dedicated electrical space must be clear of any piping, ducts, leak protection apparatus, or equipment foreign to the electrical installation. Plumbing, heating, ventilation, and air-conditioning piping, ducts, and equipment must be installed outside the width and depth zone.

Foreign systems installed directly above the dedicated space reserved for electrical equipment must include protective equipment that ensures that occurrences such as leaks, condensation, and even breaks do not damage the electrical equipment located below.

Sprinkler protection is permitted for the dedicated spaces as long as it complies with this section. A dropped, suspended, or similar ceiling is permitted to be located directly in the dedicated space, as are building structural members.

The electrical equipment also must be protected from physical damage. Damage can be caused by activities occurring near this equipment, such as material handling by personnel or the operation of a forklift or other mobile equipment. See 110.27(B) for other provisions relating to the protection of electrical equipment.

Exhibits 110.17, 110.18, and 110.19 illustrate the two distinct indoor installation spaces required in 110.26(A) and (F), that is, the working space and the dedicated electrical space.

In Exhibit 110.17, the dedicated electrical space required by 110.26(F) is the space outlined by the width and depth of the equipment (the footprint) and extending from the floor to 6 ft above the equipment or to the structural ceiling (whichever is lower). The dedicated electrical space is reserved for the installation of electrical equipment and for the installation of any conduits, cable trays, and so on, entering or exiting that equipment. The outlined area in front of the electrical equipment in Exhibit 110.17 is the working space required by 110.26(A). Note that sprinkler protection is afforded the entire dedicated electrical space and working space without actually entering either space. Also, note that the exhaust duct is not located in or directly above the dedicated electrical space. Although not specifically required to be located here, this duct location may be a cost-effective solution that avoids the substantial physical protection requirements of 110.26(F)(1)(b).

Exhibit 110.18 illustrates the working space required in front of the panelboard by 110.26(A). No equipment, electrical or otherwise, is allowed in the working space.

Exhibit 110.19 illustrates the dedicated electrical space required over and under the panelboard by 110.26(F)(1).

This space is for the cables, raceways, and so on, that run to and from the panelboard.

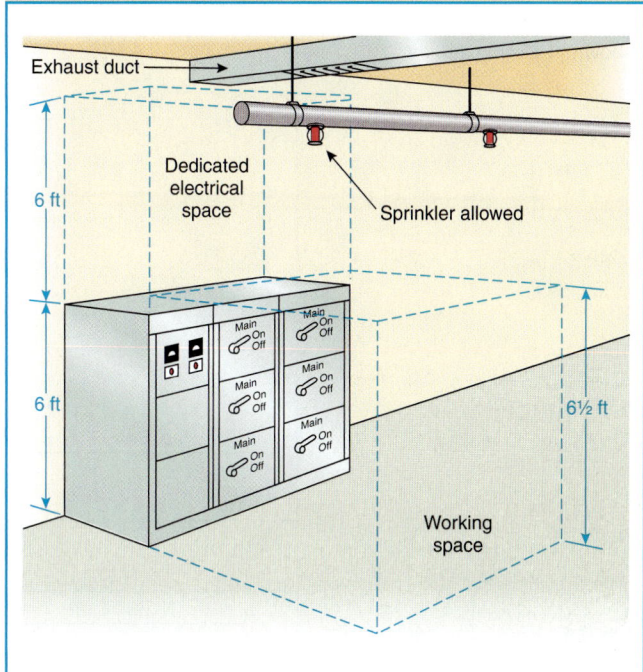

Exhibit 110.17 The two distinct indoor installation spaces required by 110.26(A) and 110.26(F): the working space and the dedicated electrical space.

(2) Outdoor. Outdoor electrical equipment shall be installed in suitable enclosures and shall be protected from accidental contact by unauthorized personnel, or by vehicular traffic, or by accidental spillage or leakage from piping systems. The working clearance space shall include the zone described in 110.26(A). No architectural appurtenance or other equipment shall be located in this zone.

Extreme care should be taken where protection from unauthorized personnel or vehicular traffic is added to existing installations in order to comply with this requirement. Any excavation or driving of steel into the ground for the placement of fencing, vehicle stops, or bollards should be done only after a thorough investigation of the belowgrade wiring.

110.27 Guarding of Live Parts.

(A) Live Parts Guarded Against Accidental Contact. Except as elsewhere required or permitted by this *Code,* live parts of electrical equipment operating at 50 volts or more shall be guarded against accidental contact by approved enclosures or by any of the following means:

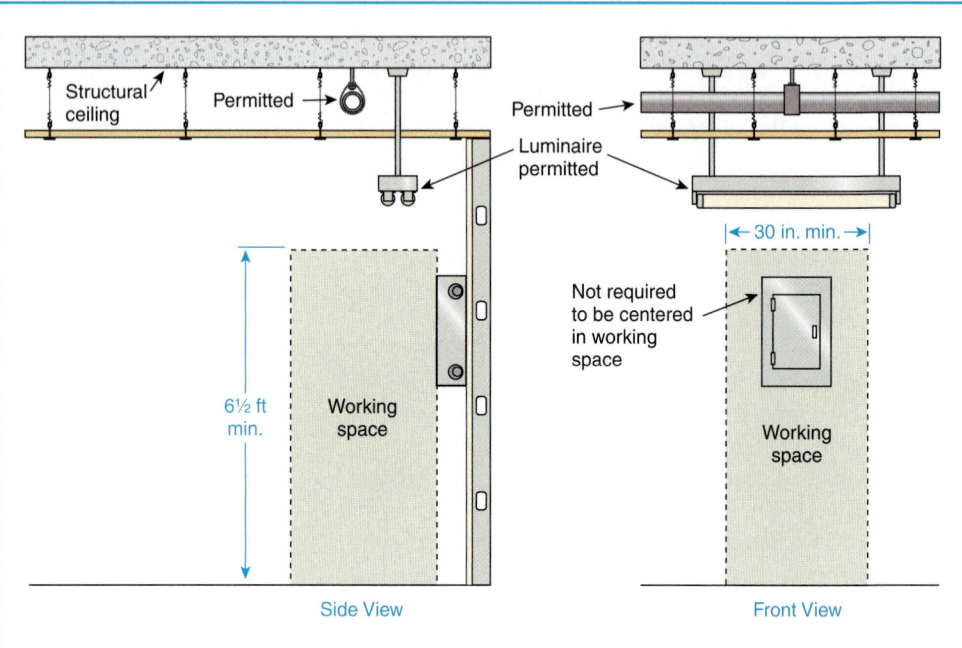

Exhibit 110.18 The working space in front of a panelboard as required by 110.26(A). This illustration supplements the dedicated electrical space shown in Exhibit 110.17.

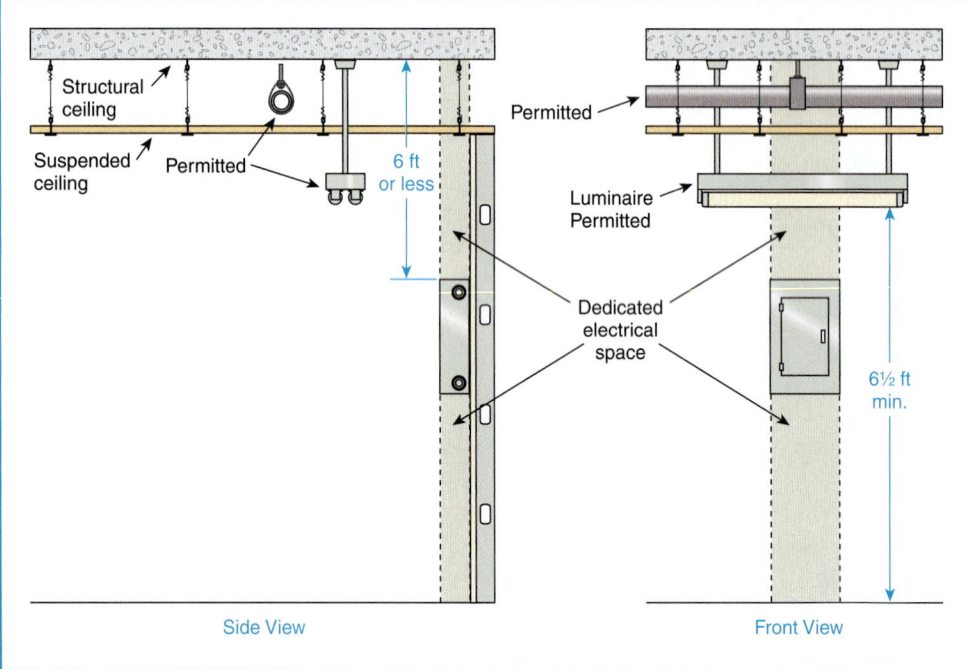

Exhibit 110.19 The dedicated electrical space above and below a panelboard as required by 110.26(F)(1).

(1) By location in a room, vault, or similar enclosure that is accessible only to qualified persons.

(2) By suitable permanent, substantial partitions or screens arranged so that only qualified persons have access to the space within reach of the live parts. Any openings in such partitions or screens shall be sized and located so that persons are not likely to come into accidental contact with the live parts or to bring conducting objects into contact with them.

(3) By location on a suitable balcony, gallery, or platform elevated and arranged so as to exclude unqualified persons.

(4) By elevation of 2.5 m (8 ft) or more above the floor or other working surface.

Contact conductors used for traveling cranes are permitted to be bare by 610.13(B) and 610.21(A). Although contact conductors obviously have to be bare for contact shoes on the moving member to make contact with the conductor, it is possible to place guards near the conductor to prevent its accidental contact with persons and still have slots or spaces through which the moving contacts can operate. Note that the *Code* also recognizes the guarding of live parts by elevation.

(B) Prevent Physical Damage. In locations where electric equipment is likely to be exposed to physical damage, enclosures or guards shall be so arranged and of such strength as to prevent such damage.

(C) Warning Signs. Entrances to rooms and other guarded locations that contain exposed live parts shall be marked with conspicuous warning signs forbidding unqualified persons to enter.

FPN: For motors, see 430.132 and 430.133. For over 600 volts, see 110.34.

Live parts of electrical equipment should be covered, shielded, enclosed, or otherwise protected by covers, barriers, mats, or platforms to prevent the likelihood of contact by persons or objects. See the definitions of *dead front* and *isolated (as applied to location)* in Article 100.

III. Over 600 Volts, Nominal

110.30 General.

Conductors and equipment used on circuits over 600 volts, nominal, shall comply with Part I of this article and with the following sections, which supplement or modify Part I. In no case shall the provisions of this part apply to equipment on the supply side of the service point.

See "Over 600 volts" in the index to this *Handbook* for various articles, sections, and parts that include requirements for installations over 600 volts.

Equipment on the supply side of the service point is outside the scope of the *NEC*. Such equipment is covered by ANSI C2, *National Electrical Safety Code*, published by the Institute of Electrical and Electronics Engineers (IEEE).

110.31 Enclosure for Electrical Installations.

Electrical installations in a vault, room, or closet or in an area surrounded by a wall, screen, or fence, access to which is controlled by lock and key or other approved means, shall be considered to be accessible to qualified persons only. The type of enclosure used in a given case shall be designed

and constructed according to the nature and degree of the hazard(s) associated with the installation.

For installations other than equipment as described in 110.31(D), a wall, screen, or fence shall be used to enclose an outdoor electrical installation to deter access by persons who are not qualified. A fence shall not be less than 2.1 m (7 ft) in height or a combination of 1.8 m (6 ft) or more of fence fabric and a 300-mm (1-ft) or more extension utilizing three or more strands of barbed wire or equivalent. The distance from the fence to live parts shall be not less than given in Table 110.31.

Table 110.31 Minimum Distance from Fence to Live Parts

Nominal Voltage	Minimum Distance to Live Parts	
	m	**ft**
601 – 13,799	3.05	10
13,800 – 230,000	4.57	15
Over 230,000	5.49	18

Note: For clearances of conductors for specific system voltages and typical BIL ratings, see ANSI C2-1997, *National Electrical Safety Code.*

FPN: See Article 450 for construction requirements for transformer vaults.

(A) Fire Resistivity of Electrical Vaults. The walls, roof, floors, and doorways of vaults containing conductors and equipment over 600 volts, nominal, shall be constructed of materials that have adequate structural strength for the conditions, with a minimum fire rating of 3 hours. The floors of vaults in contact with the earth shall be of concrete that is not less than 4 in. (102 mm) thick, but where the vault is constructed with a vacant space or other stories below it, the floor shall have adequate structural strength for the load imposed on it and a minimum fire resistance of 3 hours. For the purpose of this section, studs and wallboards shall not be considered acceptable.

(B) Indoor Installations.

(1) In Places Accessible to Unqualified Persons. Indoor electrical installations that are accessible to unqualified persons shall be made with metal-enclosed equipment. Metal-enclosed switchgear, unit substations, transformers, pull boxes, connection boxes, and other similar associated equipment shall be marked with appropriate caution signs. Openings in ventilated dry-type transformers or similar openings in other equipment shall be designed so that foreign objects inserted through these openings are deflected from energized parts.

(2) In Places Accessible to Qualified Persons Only. Indoor electrical installations considered accessible only to qualified persons in accordance with this section shall comply with 110.34, 110.36, and 490.24.

(C) Outdoor Installations.

(1) In Places Accessible to Unqualified Persons. Outdoor electrical installations that are open to unqualified persons shall comply with Article 225.

> FPN: For clearances of conductors for system voltages over 600 volts, nominal, see ANSI C2-1997, *National Electrical Safety Code.*

(2) In Places Accessible to Qualified Persons Only. Outdoor electrical installations that have exposed live parts shall be accessible to qualified persons only in accordance with the first paragraph of this section and shall comply with 110.34, 110.36, and 490.24.

(D) Enclosed Equipment Accessible to Unqualified Persons. Ventilating or similar openings in equipment shall be designed so that foreign objects inserted through these openings are deflected from energized parts. Where exposed to physical damage from vehicular traffic, suitable guards shall be provided. Nonmetallic or metal-enclosed equipment located outdoors and accessible to the general public shall be designed so that exposed nuts or bolts cannot be readily removed, permitting access to live parts. Where nonmetallic or metal-enclosed equipment is accessible to the general public and the bottom of the enclosure is less than 2.5 m (8 ft) above the floor or grade level, the enclosure door or hinged cover shall be kept locked. Doors and covers of enclosures used solely as pull boxes, splice boxes, or junction boxes shall be locked, bolted, or screwed on. Underground box covers that weigh over 45.4 kg (100 lb) shall be considered as meeting this requirement.

110.32 Work Space About Equipment.

Sufficient space shall be provided and maintained about electric equipment to permit ready and safe operation and maintenance of such equipment. Where energized parts are exposed, the minimum clear work space shall not be less than 2.0 m (6½ ft) high (measured vertically from the floor or platform) or less than 900 mm (3 ft) wide (measured parallel to the equipment). The depth shall be as required in 110.34(A). In all cases, the work space shall permit at least a 90 degree opening of doors or hinged panels.

110.33 Entrance and Access to Work Space.

(A) Entrance. At least one entrance not less than 610 mm (24 in.) wide and 2.0 m (6½ ft) high shall be provided to give access to the working space about electric equipment. Where the entrance has a personnel door(s), the door(s) shall

open in the direction of egress and be equipped with panic bars, pressure plates, or other devices that are normally latched but open under simple pressure.

(1) Large Equipment. On switchboard and control panels exceeding 1.8 m (6 ft) in width, there shall be one entrance at each end of the equipment. A single entrance to the required working space shall be permitted where either of the conditions in 110.33(A)(1)(a) or (b) is met.

(a) Unobstructed Exit. Where the location permits a continuous and unobstructed way of exit travel, a single entrance to the working space shall be permitted.

(b) Extra Working Space. Where the depth of the working space is twice that required by 110.34(A), a single entrance shall be permitted. It shall be located so that the distance from the equipment to the nearest edge of the entrance is not less than the minimum clear distance specified in Table 110.34(A) for equipment operating at that voltage and in that condition.

(2) Guarding. Where bare energized parts at any voltage or insulated energized parts above 600 volts, nominal, to ground are located adjacent to such entrance, they shall be suitably guarded.

> Section 110.33(A) contains requirements similar to those of 110.26(C). Because of the high voltages involved, 110.33(A) differs slightly in that a minimum working space entrance dimension is required even for switchboards of 6 ft or less in width. For further information, see the commentary following 110.26(C)(2), since most of it remains valid for over-600-volt installations as well.

(B) Access. Permanent ladders or stairways shall be provided to give safe access to the working space around electric equipment installed on platforms, balconies, or mezzanine floors or in attic or roof rooms or spaces.

110.34 Work Space and Guarding.

(A) Working Space. Except as elsewhere required or permitted in this *Code,* the minimum clear working space in the direction of access to live parts of electrical equipment shall not be less than specified in Table 110.34(A). Distances shall be measured from the live parts, if such are exposed, or from the enclosure front or opening if such are enclosed.

Exception: Working space shall not be required in back of equipment such as dead-front switchboards or control assemblies where there are no renewable or adjustable parts (such as fuses or switches) on the back and where all connections are accessible from locations other than the back. Where rear access is required to work on de-energized parts

Table 110.34(A) Minimum Depth of Clear Working Space at Electrical Equipment

Nominal Voltage to Ground	Minimum Clear Distance		
	Condition 1	Condition 2	Condition 3
601–2500 V	900 mm (3 ft)	1.2 m (4 ft)	1.5 m (5 ft)
2501–9000 V	1.2 m (4 ft)	1.5 m (5 ft)	1.8 m (6 ft)
9001–25,000 V	1.5 m (5 ft)	1.8 m (6 ft)	2.8 m (9 ft)
25,001V–75 kV	1.8 m (6 ft)	2.5 m (8 ft)	3.0 m (10 ft)
Above 75 kV	2.5 m (8 ft)	3.0 m (10 ft)	3.7 m (12 ft)

Note: Where the conditions are as follows:
Condition 1— Exposed live parts on one side and no live or grounded parts on the other side of the working space, or exposed live parts on both sides effectively guarded by suitable wood or other insulating materials. Insulated wire or insulated busbars operating at not over 300 volts shall not be considered live parts.
Condition 2— Exposed live parts on one side and grounded parts on the other side. Concrete, brick, or tile walls shall be considered as grounded surfaces.
Condition 3— Exposed live parts on both sides of the work space (not guarded as provided in Condition 1) with the operator between.

on the back of enclosed equipment, a minimum working space of 750 mm (30 in.) horizontally shall be provided.

(B) Separation from Low-Voltage Equipment. Where switches, cutouts, or other equipment operating at 600 volts, nominal, or less are installed in a room or enclosure where there are exposed live parts or exposed wiring operating at over 600 volts, nominal, the high-voltage equipment shall be effectively separated from the space occupied by the low-voltage equipment by a suitable partition, fence, or screen.

Exception: Switches or other equipment operating at 600 volts, nominal, or less and serving only equipment within the high-voltage vault, room, or enclosure shall be permitted to be installed in the high-voltage enclosure, room, or vault if accessible to qualified persons only.

(C) Locked Rooms or Enclosures. The entrances to all buildings, rooms, or enclosures containing exposed live parts or exposed conductors operating at over 600 volts, nominal, shall be kept locked unless such entrances are under the observation of a qualified person at all times.

Where the voltage exceeds 600 volts, nominal, permanent and conspicuous warning signs shall be provided, reading as follows:

DANGER — HIGH VOLTAGE — KEEP OUT

Equipment used on circuits over 600 volts, nominal, and containing exposed live parts or exposed conductors is required to be located in a locked room or an enclosure. The provisions for locking are not required if the location is

under observation at all times, as in the case of some engine rooms.

(D) Illumination. Illumination shall be provided for all working spaces about electrical equipment. The lighting outlets shall be arranged so that persons changing lamps or making repairs on the lighting system are not endangered by live parts or other equipment.

The points of control shall be located so that persons are not likely to come in contact with any live part or moving part of the equipment while turning on the lights.

(E) Elevation of Unguarded Live Parts. Unguarded live parts above working space shall be maintained at elevations not less than required by Table 110.34(E).

Table 110.34(E) Elevation of Unguarded Live Parts Above Working Space

Nominal Voltage Between Phases	Elevation	
	m	ft
601–7500 V	2.8	9
7501–35,000 V	2.9	9
Over 35 kV	2.9 m + 9.5 mm/kV above 35	9 ft + 0.37 in./kV above 35

(F) Protection of Service Equipment, Metal-Enclosed Power Switchgear, and Industrial Control Assemblies. Pipes or ducts foreign to the electrical installation and requiring periodic maintenance or whose malfunction would endanger the operation of the electrical system shall not be located in the vicinity of the service equipment, metal-enclosed power switchgear, or industrial control assemblies. Protection shall be provided where necessary to avoid damage from condensation leaks and breaks in such foreign systems. Piping and other facilities shall not be considered foreign if provided for fire protection of the electrical installation.

110.36 Circuit Conductors.

Circuit conductors shall be permitted to be installed in raceways; in cable trays; as metal-clad cable, as bare wire, cable, and busbars; or as Type MV cables or conductors as provided in 300.37, 300.39, 300.40, and 300.50. Bare live conductors shall conform with 490.24.

Insulators, together with their mounting and conductor attachments, where used as supports for wires, single-conductor cables, or busbars, shall be capable of safely withstanding the maximum magnetic forces that would prevail when two or more conductors of a circuit were subjected to short-circuit current.

Open runs of insulated wires and cables that have a bare lead sheath or a braided outer covering shall be supported in a manner designed to prevent physical damage to the braid or sheath. Supports for lead-covered cables shall be designed to prevent electrolysis of the sheath.

110.40 Temperature Limitations at Terminations.

Conductors shall be permitted to be terminated based on the 90°C (194°F) temperature rating and ampacity as given in Table 310.67 through Table 310.86, unless otherwise identified.

IV. Tunnel Installations Over 600 Volts, Nominal

Prior to the 1999 *NEC*, the requirements for systems operating over 600 volts, nominal, were placed throughout the *Code,* so housekeeping of these requirements was determined to be in order. For that reason, tunnel installation requirements for systems operating over 600 volts, nominal, have been moved from 1999 *NEC*, Article 710, Part F, to Article 110, Part IV.

110.51 General.

(A) Covered. The provisions of this part shall apply to the installation and use of high-voltage power distribution and utilization equipment that is portable, mobile, or both, such as substations, trailers, cars, mobile shovels, draglines, hoists, drills, dredges, compressors, pumps, conveyors, and underground excavators, and the like.

(B) Other Articles. The requirements of this part shall be additional to, or amendatory of, those prescribed in Articles 100 through 490 of this *Code*. Special attention shall be paid to Article 250.

(C) Protection Against Physical Damage. Conductors and cables in tunnels shall be located above the tunnel floor and so placed or guarded to protect them from physical damage.

110.52 Overcurrent Protection.

Motor-operated equipment shall be protected from overcurrent in accordance with Article 430. Transformers shall be protected from overcurrent in accordance with Article 450.

110.53 Conductors.

High-voltage conductors in tunnels shall be installed in metal conduit or other metal raceway, Type MC cable, or other approved multiconductor cable. Multiconductor portable cable shall be permitted to supply mobile equipment.

110.54 Bonding and Equipment Grounding Conductors.

(A) Grounded and Bonded. All non–current-carrying metal parts of electric equipment and all metal raceways and cable sheaths shall be effectively grounded and bonded to all metal pipes and rails at the portal and at intervals not exceeding 300 m (1000 ft) throughout the tunnel.

(B) Equipment Grounding Conductors. An equipment grounding conductor shall be run with circuit conductors inside the metal raceway or inside the multiconductor cable jacket. The equipment grounding conductor shall be permitted to be insulated or bare.

110.55 Transformers, Switches, and Electrical Equipment.

All transformers, switches, motor controllers, motors, rectifiers, and other equipment installed below ground shall be protected from physical damage by location or guarding.

110.56 Energized Parts.

Bare terminals of transformers, switches, motor controllers, and other equipment shall be enclosed to prevent accidental contact with energized parts.

110.57 Ventilation System Controls.

Electrical controls for the ventilation system shall be arranged so that the airflow can be reversed.

110.58 Disconnecting Means.

A switch or circuit breaker that simultaneously opens all ungrounded conductors of the circuit shall be installed within sight of each transformer or motor location for disconnecting the transformer or motor. The switch or circuit breaker for a transformer shall have an ampere rating not less than the ampacity of the transformer supply conductors. The switch or circuit breaker for a motor shall comply with the applicable requirements of Article 430.

110.59 Enclosures.

Enclosures for use in tunnels shall be dripproof, weatherproof, or submersible as required by the environmental conditions. Switch or contactor enclosures shall not be used as junction boxes or as raceways for conductors feeding through or tapping off to other switches, unless special designs are used to provide adequate space for this purpose.

Wiring and Protection

ARTICLE 200
Use and Identification of Grounded Conductors

Contents

200.1 Scope.

This article provides requirements for the following:

(1) Identification of terminals
(2) Grounded conductors in premises wiring systems
(3) Identification of grounded conductors

> FPN: See Article 100 for definitions of *Grounded Conductor* and *Grounding Conductor*.

Article 200 contains requirements for grounded conductor use and identification. The grounded circuit conductor is referred to throughout the *Code* as the *grounded conductor*. The grounded conductor is often, but not always, the neutral conductor. For example, in corner-grounded delta systems, the grounded conductor is not the neutral conductor.

With the 1999 edition, the *Code* began providing an alternative for identifying the grounded conductor. According to 200.6, three continuous white stripes along the entire length of conductor insulation that is colored other than green may be used to identify a conductor as the grounded conductor. Continuous white or gray coloring is also used to identify the grounded conductor. The three white stripes method of identification is permitted for all conductor sizes.

200.2 General.

All premises wiring systems, other than circuits and systems exempted or prohibited by 210.10, 215.7, 250.21, 250.22, 250.162, 503.13, 517.63, 668.11, 668.21, and 690.41, Exception, shall have a grounded conductor that is identified in accordance with 200.6.

The grounded conductor, where insulated, shall have insulation that is (1) suitable, other than color, for any ungrounded conductor of the same circuit on circuits of less than 1000 volts or impedance grounded neutral systems of 1 kV and over, or (2) rated not less than 600 volts for solidly grounded neutral systems of 1 kV and over as described in 250.184(A).

200.3 Connection to Grounded System.

Premises wiring shall not be electrically connected to a supply system unless the latter contains, for any grounded conductor of the interior system, a corresponding conductor that is grounded. For the purpose of this section, *electrically connected* shall mean connected so as to be capable of carrying current, as distinguished from connection through electromagnetic induction.

Grounded conductors of premises wiring (other than separately derived systems) must be connected to the supply system grounded conductor to ensure a common, continuous, grounded system.

200.6 Means of Identifying Grounded Conductors.

(A) **Sizes 6 AWG or Smaller.** An insulated grounded conductor of 6 AWG or smaller shall be identified by a continuous white or gray outer finish or by three continuous white stripes on other than green insulation along its entire length. Wires that have their outer covering finished to show a white or gray color but have colored tracer threads in the braid identifying the source of manufacture shall be considered as meeting the provisions of this section. Insulated grounded conductors shall also be permitted to be identified as follows:

(1) The grounded conductor of a mineral-insulated, metal-sheathed cable shall be identified at the time of installation by distinctive marking at its terminations.
(2) A single-conductor, sunlight-resistant, outdoor-rated cable used as a grounded conductor in photovoltaic power systems as permitted by 690.31 shall be identified

at the time of installation by distinctive white marking at all terminations.

(3) Fixture wire shall comply with the requirements for grounded conductor identification as specified in 402.8

(4) For aerial cable, the identification shall be as above, or by means of a ridge located on the exterior of the cable so as to identify it.

The general rule of 200.6(A) requires insulated conductors to be white or gray for their entire length. In the 1999 *Code,* however, the concept of marking the grounded conductor with three continuous white stripes along the entire length of the insulated conductor was added.

Other methods of identification are also permitted within 200.6(A). For example, the grounded conductor of MI cable, due to its unique construction, is permitted to be identified at the time of installation. Aerial cable may have its grounded conductor identified by a ridge along its insulated surface, and fixture wires are permitted to have the grounded conductor identified by various methods, including colored insulation, stripes on the insulation, colored braid, colored separator, and tinned conductors. These identification methods are found in 402.8 and explained in detail in 400.22(A) through (E).

For 6 AWG or smaller, identification of the grounded conductor solely by distinctive white marking at the time of installation is not permitted except as described for multiconductor cables and cords in 200.6(C) and (E) and in outdoor photovoltaic power installations.

(B) Sizes Larger Than 6 AWG. An insulated grounded conductor larger than 6 AWG shall be identified either by a continuous white or gray outer finish or by three continuous white stripes on other than green insulation along its entire length or at the time of installation by a distinctive white marking at its terminations. This marking shall encircle the conductor or insulation.

The general rule of 200.6(B) requires the insulated conductors to be white or gray for their entire length or to be identified by three continuous white stripes along the entire length of the insulated conductor. Another permitted method for these larger conductors is field-applied distinctive white markings applied at the time of installation at all the conductor termination points. If field applied, the white marking must completely encircle the conductor in order to be clearly visible. This method of identification is shown in Exhibit 200.1.

(C) Flexible Cords. An insulated conductor that is intended for use as a grounded conductor, where contained within a flexible cord, shall be identified by a white or gray

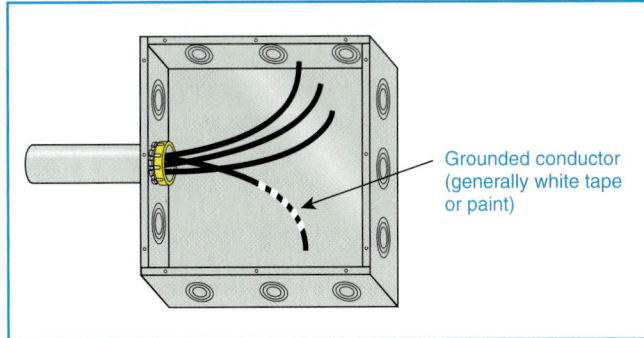

Exhibit 200.1 Field-applied identification permitted by 200.6(B) to a 4 AWG conductor to identify it as the grounded conductor.

outer finish or by three continuous white stripes on other than green insulation or by methods permitted by 400.22.

(D) Grounded Conductors of Different Systems. Where conductors of different systems are installed in the same raceway, cable, box, auxiliary gutter, or other type of enclosure, one system grounded conductor, if required, shall have an outer covering conforming to 200.6(A) or 200.6(B). Each other system grounded conductor shall have an outer covering of white with a readily distinguishable, different colored stripe other than green running along the insulation, or shall have other and different means of identification as allowed by 200.6(A) or (B) that will distinguish each system grounded conductor.

The requirements found in 200.6(D) are essentially the same since the 1987 edition of the *NEC.* However, these requirements are often misapplied. As Exhibit 200.2 shows, if grounded conductors of different systems are present in the same enclosure, these grounded conductors must be distinguishable, such as by different colors. Careful study of 200.6(D) reveals what Exhibit 200.2 shows, that is, where

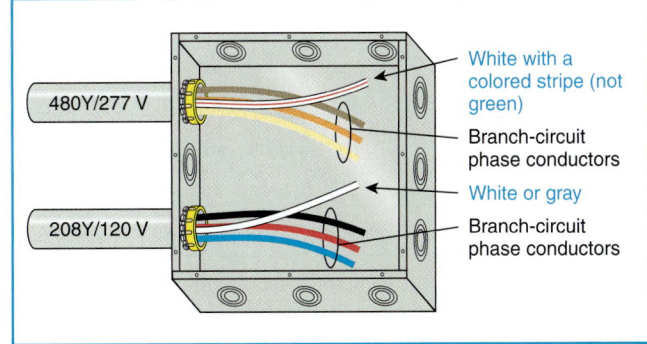

Exhibit 200.2 Grounded conductors of different systems in the same enclosure. The insulation of the grounded conductors of the different systems is of a color as prescribed by 200.6(D).

one system uses a white or gray insulation for the grounded conductor, the other system must use an identification different from the color selected for the first system colored insulation. For example, where one system uses white for the grounded conductor, the second system must use a different color or marking such as gray or white with a strip.

(E) Grounded Conductors of Multiconductor Cables. The insulated grounded conductors in a multiconductor cable shall be identified by a continuous white or gray outer finish or by three continuous white stripes on other than green insulation along its entire length. Multiconductor flat 4 AWG or larger shall be permitted to employ an external ridge on the grounded conductor.

Exception No. 1: Where the conditions of maintenance and supervision ensure that only qualified persons service the installation, grounded conductors in multiconductor cables shall be permitted to be permanently identified at their terminations at the time of installation by a distinctive white marking or other equally effective means.

Exception No. 1 to 200.6(E) introduces a concept for identifying grounded conductors of multiconductor cables. This exception allows identification of a grounded conductor of a multiconductor cable, as illustrated in Exhibit 200.3, at the time of installation by use of a distinctive white marking or other equally effective means, such as numbering, lettering, or tagging. Exception No. 1 to 200.6(E) is intended to apply in locations where a regulated system of maintenance and supervision ensures that only qualified persons will service the installation.

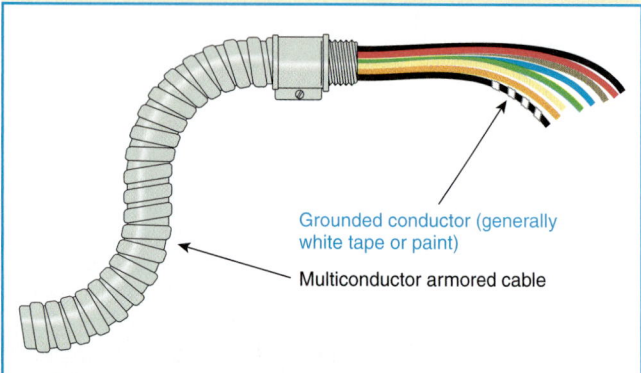

Grounded conductor (generally white tape or paint)

Multiconductor armored cable

Exhibit 200.3 Field-applied identification to the conductor of a multiconductor armored cable that will be used as the grounded conductor as permitted by 200.6(E), Exception No. 1.

Exception No. 2: The grounded conductor of a multiconductor varnished-cloth-insulated cable shall be permitted to be

identified at its terminations at the time of installation by a distinctive white marking or other equally effective means.

> FPN: The color gray may have been used in the past as an ungrounded conductor. Care should be taken when working on existing systems.

In the 2002 *Code*, the term *natural gray* has been changed to *gray* because natural gray outer finish is no longer manufactured and because varying shades of a color are sometimes difficult to distinguish. The new FPN following 200.6 warns the user to exercise caution when working on existing systems because gray may have been used on those existing systems.

200.7 Use of Insulation of a White or Gray Color or with Three Continuous White Stripes.

(A) General. The following shall be used only for the grounded circuit conductor, unless otherwise permitted in 200.7(B) and (C):

(1) A conductor with continuous white or gray covering
(2) A conductor with three continuous white stripes on other than green insulation
(3) A marking of white or gray color at the termination

(B) Circuits of Less Than 50 Volts. A conductor with white or gray color insulation or three continuous white stripes or having a marking of white or gray at the termination for circuits of less than 50 volts shall be required to be grounded only as required by 250.20(A).

(C) Circuits of 50 Volts or More. The use of insulation that is white or gray or that has three continuous white stripes for other than a grounded conductor for circuits of 50 volts or more shall be permitted only as in (1) through (3).

(1) If part of a cable assembly and where the insulation is permanently reidentified to indicate its use as an ungrounded conductor, by painting or other effective means at its termination, and at each location where the conductor is visible and accessible.
(2) Where a cable assembly contains an insulated conductor for single-pole, 3-way or 4-way switch loops and the conductor with white or gray insulation or a marking of three continuous white stripes is used for the supply to the switch but not as a return conductor from the switch to the switched outlet. In these applications, the conductor with white or gray insulation or with three continuous white stripes shall be permanently reidentified to indicate its use by painting or other effective means at its terminations and at each location where the conductor is visible and accessible.

Previous editions of the *Code* permitted switch loops that contained a white insulated conductor that was used to supply the switch and that was not used to supply the luminaires to remain white. Prior to the 1999 *NEC*, re-identification of this particular ungrounded conductor was not required. However, many electronic automation devices requiring a grounded conductor are now available for installation into switch outlets. Therefore, re-identification of all ungrounded conductors that are white or any permitted white coloring is now required at each and every termination point. The required re-identification must be effective, permanent, and suitable for the environment, in order to clearly identify the insulated conductor as an ungrounded conductor. Proper re-identification should eliminate the possibility of miswiring new electronic automation devices during installation.

(3) Where a flexible cord, having one conductor identified by a white or gray outer finish or three continuous white stripes or by any other means permitted by 400.22, is used for connecting an appliance or equipment permitted by 400.7. This shall apply to flexible cords connected to outlets whether or not the outlet is supplied by a circuit that has a grounded conductor.

FPN: The color gray may have been used in the past as an ungrounded conductor. Care should be taken when working on existing systems.

In the 2002 *Code*, the term *natural gray* has been changed to *gray* throughout Article 200 because natural gray outer finish is no longer manufactured and because varying shades of a color are sometimes difficult to distinguish. The new FPN following 200.7 warns the user to exercise caution when working on existing systems because gray may have been used on those existing systems.

200.9 Means of Identification of Terminals.

The identification of terminals to which a grounded conductor is to be connected shall be substantially white in color. The identification of other terminals shall be of a readily distinguishable different color.

Exception: Where the conditions of maintenance and supervision ensure that only qualified persons service the installations, terminals for grounded conductors shall be permitted to be permanently identified at the time of installation by a distinctive white marking or other equally effective means.

200.10 Identification of Terminals.

(A) Device Terminals. All devices, excluding panelboards, provided with terminals for the attachment of conductors and intended for connection to more than one side of the circuit shall have terminals properly marked for identification, unless the electrical connection of the terminal intended to be connected to the grounded conductor is clearly evident.

Exception: Terminal identification shall not be required for devices that have a normal current rating of over 30 amperes, other than polarized attachment plugs and polarized receptacles for attachment plugs as required in 200.10(B).

(B) Receptacles, Plugs, and Connectors. Receptacles, polarized attachment plugs, and cord connectors for plugs and polarized plugs shall have the terminal intended for connection to the grounded conductor identified as follows:

(1) Identification shall be by a metal or metal coating that is substantially white in color or by the word *white* or the letter *W* located adjacent to the identified terminal.

(2) If the terminal is not visible, the conductor entrance hole for the connection shall be colored white or marked with the word *white* or the letter *W*.

FPN: See 250.126 for identification of wiring device equipment grounding conductor terminals.

Section 200.10(B) requires that terminals of receptacles, plugs, and connectors that are intended for the connection of the grounded conductor be marked by one of several methods, including the word *white,* the letter *W,* or a distinctive white color. The variety of these methods allows the plating of all screws and terminals to meet other requirements of specific applications, such as corrosion-resistant devices.

(C) Screw Shells. For devices with screw shells, the terminal for the grounded conductor shall be the one connected to the screw shell.

(D) Screw Shell Devices with Leads. For screw shell devices with attached leads, the conductor attached to the screw shell shall have a white or gray finish. The outer finish of the other conductor shall be of a solid color that will not be confused with the white or gray finish used to identify the grounded conductor.

FPN: The color gray may have been used in the past as an ungrounded conductor. Care should be taken when working on existing systems.

In the 2002 *Code*, the term *natural gray* has been changed to *gray* because natural gray outer finish is no longer manufactured and because varying shades of a color are sometimes difficult to distinguish. The new FPN following 200.10 warns the user to exercise caution when working on existing systems because gray may have been used on those existing systems.

(E) Appliances. Appliances that have a single-pole switch or a single-pole overcurrent device in the line or any line-connected screw shell lampholders, and that are to be connected by (1) a permanent wiring method or (2) field-installed attachment plugs and cords with three or more wires (including the equipment grounding conductor), shall have means to identify the terminal for the grounded circuit conductor (if any).

200.11 Polarity of Connections.

No grounded conductor shall be attached to any terminal or lead so as to reverse the designated polarity.

ARTICLE 210
Branch Circuits

Contents

I. General Provisions
   210.1 Scope
   210.2 Other Articles for Specific-Purpose Branch Circuits
   210.3 Rating
   210.4 Multiwire Branch Circuits
     (A) General
     (B) Dwelling Units
     (C) Line-to-Neutral Loads
     (D) Identification of Ungrounded Conductors
   210.5 Identification for Branch Circuits
     (A) Grounded Conductor
     (B) Equipment Grounding Conductor
   210.6 Branch-Circuit Voltage Limitations
     (A) Occupancy Limitation
     (B) 120 Volts Between Conductors
     (C) 277 Volts to Ground
     (D) 600 Volts Between Conductors
     (E) Over 600 Volts Between Conductors
   210.7 Branch Circuit Receptacle Requirements
     (A) Receptacle Outlet Location
     (B) Receptacle Requirements
     (C) Multiple Branch Circuits
   210.8 Ground-Fault Circuit-Interrupter Protection for Personnel
     (A) Dwelling Units
     (B) Other Than Dwelling Units
   210.9 Circuits Derived from Autotransformers
   210.10 Ungrounded Conductors Tapped from Grounded Systems
   210.11 Branch Circuits Required
     (A) Number of Branch Circuits
     (B) Load Evenly Proportioned Among Branch Circuits
     (C) Dwelling Units

   210.12 Arc-Fault Circuit-Interrupter Protection
     (A) Definition
     (B) Dwelling Unit Bedrooms
II. Branch-Circuit Ratings
   210.19 Conductors—Minimum Ampacity and Size
     (A) Branch Circuits Not More Than 600 Volts
     (B) Branch Circuits Over 600 Volts
   210.20 Overcurrent Protection
     (A) Continuous and Noncontinuous Loads
     (B) Conductor Protection
     (C) Equipment
     (D) Outlet Devices
   210.21 Outlet Devices
     (A) Lampholders
     (B) Receptacles
   210.23 Permissible Loads
     (A) 15- and 20-Ampere Branch Circuits
     (B) 30-Ampere Branch Circuits
     (C) 40- and 50-Ampere Branch Circuits
     (D) Branch Circuits Larger Than 50 Amperes
   210.24 Branch-Circuit Requirements—Summary
   210.25 Common Area Branch Circuits
III. Required Outlets
   210.50 General
     (A) Cord Pendants
     (B) Cord Connections
     (C) Appliance Outlets
   210.52 Dwelling Unit Receptacle Outlets
     (A) General Provisions
     (B) Small Appliances
     (C) Countertops
     (D) Bathrooms
     (E) Outdoor Outlets
     (F) Laundry Areas
     (G) Basements and Garages
     (H) Hallways
   210.60 Guest Rooms
     (A) General
     (B) Receptacle Placement
   210.62 Show Windows
   210.63 Heating, Air-Conditioning, and Refrigeration Equipment Outlet
   210.70 Lighting Outlets Required
     (A) Dwelling Units
     (B) Guest Rooms
     (C) Other Than Dwelling Units

I. General Provisions

210.1 Scope.

This article covers branch circuits except for branch circuits that supply only motor loads, which are covered in Article 430. Provisions of this article and Article 430 apply to branch circuits with combination loads.

According to 668.3(C)(1), electrolytic cell line conductors, cells, cell line attachments, and the wiring of auxiliary equipment and devices within the cell line working zone are not required to comply with the provisions of Article 210.

210.2 Other Articles for Specific-Purpose Branch Circuits.

Branch circuits shall comply with this article and also with the applicable provisions of other articles of this *Code*. The provisions for branch circuits supplying equipment in Table 210.2 amend or supplement the provisions in this article and shall apply to branch circuits referred to therein.

210.3 Rating.

Branch circuits recognized by this article shall be rated in accordance with the maximum permitted ampere rating or setting of the overcurrent device. The rating for other than individual branch circuits shall be 15, 20, 30, 40, and 50 amperes. Where conductors of higher ampacity are used for any reason, the ampere rating or setting of the specified overcurrent device shall determine the circuit rating.

To compensate for voltage drop in a long circuit, larger conductors with a higher ampacity are commonly used. For example, a branch circuit of 10 AWG, Type TW copper conductors has a 30-ampere ampacity. However, if a 20-ampere overcurrent device protects this branch circuit, it is *rated* as a 20-ampere branch circuit.

Exception: Multioutlet branch circuits greater than 50 amperes shall be permitted to supply nonlighting outlet loads on industrial premises where conditions of maintenance and supervision ensure that only qualified persons service the equipment.

It is common in industrial establishments to provide several single receptacles of 50-ampere or higher rating on a single branch circuit to allow quick relocation of equipment for production and/or maintenance use, such as in the case of electric welders. Generally, only one piece of equipment is operated at a time. The type of receptacle used in this situation is generally a configuration known as a pin-and-sleeve receptacle, although the *Code* does not so limit the design. Pin-and-sleeve receptacles may or may not be horsepower rated.

210.4 Multiwire Branch Circuits.

(A) General. Branch circuits recognized by this article shall be permitted as multiwire circuits. A multiwire branch circuit shall be permitted to be considered as multiple

Table 210.2 Specific-Purpose Branch Circuits

Equipment	Article	Section
Air-conditioning and refrigerating equipment		440.6, 440.31, 440.32
Busways		368
Circuits and equipment operating at less than 50 volts	720	
Central heating equipment other than fixed electric space-heating equipment		422.12
Class 1, Class 2, and Class 3 remote-control, signaling, and power-limited circuits	725	
Closed-loop and programmed power distribution	780	
Cranes and hoists		610.42
Electric signs and outline lighting		600.6
Electric welders	630	
Elevators, dumbwaiters, escalators, moving walks, wheelchair lifts, and stairway chair lifts		620.61
Fire alarm systems	760	
Fixed electric heating equipment for pipelines and vessels		427.4
Fixed electric space-heating equipment		424.3
Fixed outdoor electric deicing and snow-melting equipment		426.4
Information technology equipment		645.5
Infrared lamp industrial heating equipment		422.48, 424.3
Induction and dielectric heating equipment	665	
Marinas and boatyards		555.19
Mobile homes, manufactured homes, and mobile home parks	550	
Motion picture and television studios and similar locations	530	
Motors, motor circuits, and controllers	430	
Pipe organs		650.7
Recreational vehicles and recreational vehicle parks	551	
Sound-recording and similar equipment		640.8
Switchboards and panelboards		408.32
Theaters, audience areas of motion picture and television studios, and similar locations		520.41, 520.52, 520.62
X-ray equipment		660.2, 517.73

circuits. All conductors shall originate from the same panelboard.

> FPN: A 3-phase, 4-wire, wye-connected power system used to supply power to nonlinear loads may necessitate that the power system design allow for the possibility of high harmonic neutral currents.

The power supplies for equipment such as computers, printers, and adjustable-speed motor drives may introduce harmonic currents in the system neutral conductor. The resulting total harmonic distortion current could exceed the load current of the device itself. See the commentary following 310.15(B)(4)(c) for a discussion of neutral conductor ampacity.

(B) Dwelling Units. In dwelling units, a multiwire branch circuit supplying more than one device or equipment on the same yoke shall be provided with a means to disconnect simultaneously all ungrounded conductors at the panelboard where the branch circuit originated.

Multiwire branch circuits can be dangerous. Section 210.4(B) specifically requires simultaneous disconnection of all ungrounded conductors and requires that it take place at the panelboard of origin. The reason for this requirement is to reduce the risk of shock should a worker fail to disconnect all of the ungrounded circuits to the equipment mounted on a single yoke or strap of a device. Most commonly, receptacles are the focus of this requirement. However, equipment mounted on a yoke can include devices such as receptacles, switches, and lampholders, as well as other items such as dimmers, pilot lights, and home automation controls.

Many 125-volt, 15- and 20-ampere duplex receptacles have a break-off tab that permits each of the two receptacles to be supplied from different circuits or a 3-wire (multiwire) branch circuit. This arrangement is commonly called a *split-wired receptacle* (i.e., one circuit supplies half the duplex receptacle and another circuit supplies the other half). The simultaneous opening of both "hot" conductors at the panelboard effectively protects personnel from inadvertent contact during servicing with an energized conductor or device terminal. The simultaneous disconnection can be achieved by a 2-pole circuit breaker or by two single-pole circuit breakers with an approved handle tie, as shown in Exhibit 210.1 (bottom). If fuses are used, a 2-pole disconnect switch is required.

(C) Line-to-Neutral Loads. Multiwire branch circuits shall supply only line-to-neutral loads.

Exception No. 1: A multiwire branch circuit that supplies only one utilization equipment.

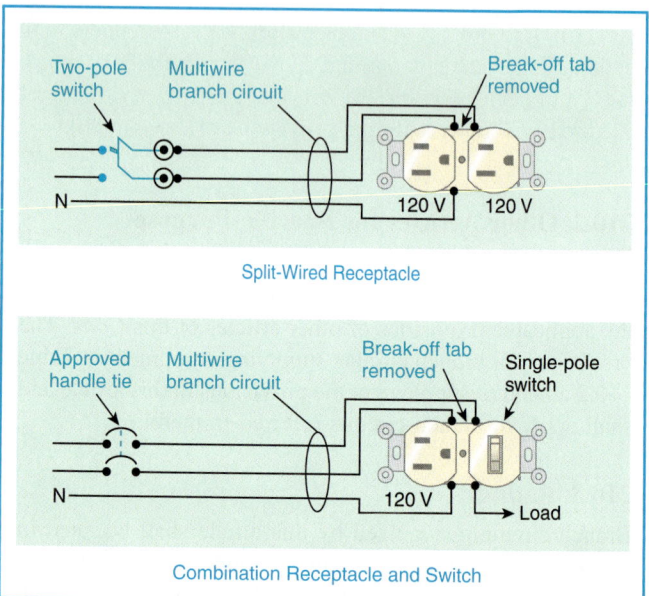

Split-Wired Receptacle

Combination Receptacle and Switch

Exhibit 210.1 An example where 210.4(B) requires the simultaneous disconnection of all ungrounded conductors to multiwire branch circuits supplying more than one device or equipment on the same yoke.

Exception No. 2: Where all ungrounded conductors of the multiwire branch circuit are opened simultaneously by the branch-circuit overcurrent device.

> FPN: See 300.13(B) for continuity of grounded conductor on multiwire circuits.

The term *multiwire branch circuit* is defined in Article 100 as a branch circuit that consists of two or more ungrounded conductors that have a voltage between them and a grounded conductor that has equal voltage between it and each ungrounded conductor of the circuit and that is connected to the neutral or grounded conductor of the system.

The circuit most commonly used as a multiwire branch circuit is illustrated in Exhibit 210.2 and consists of two ungrounded conductors and one grounded conductor supplied from a 120/240-volt, single-phase, 3-wire system. Multiwire branch circuits have many advantages, such as three wires doing the work of four (in place of two 2-wire circuits), less raceway fill, easier balancing and phasing of a system, and less voltage drop. See the commentary following 215.2(A)(4), FPN No. 3, for further information on voltage drop for branch circuits.

Multiwire branch circuits may be derived from a 120/240-volt, single-phase; a 208Y/120-volt and 480Y/277-volt, 3-phase, 4-wire; or a 240/120-volt, 3-phase, 4-wire delta system. Section 210.11(B) requires multiwire branch circuits to be properly balanced. If two ungrounded conductors and a common neutral are used as a multiwire branch circuit

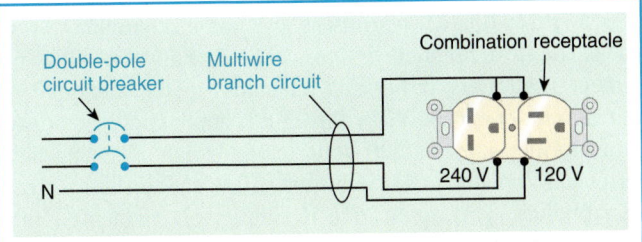

Exhibit 210.2 An example where 210.4(C), Exception No. 2, permits a multiwire branch circuit to supply line-to-neutral loads provided the ungrounded conductors are opened simultaneously by the branch-circuit overcurrent device.

supplied from a 208Y/120-volt, 3-phase, 4-wire system, the neutral carries the same current as the phase conductor with the highest current and, therefore, should be the same size. The neutral for a 2-phase, 3-wire or a 2-phase, 5-wire circuit must be sized to carry 140 percent of the ampere rating of the circuit as required by 220.22. See the commentary following 210.4(A), FPN, for further information on 3-phase, 4-wire system neutral conductors.

If loads are connected line to line (i.e., utilization equipment connected between 2 or 3 phases), 2-pole or 3-pole circuit breakers are required to disconnect all ungrounded conductors simultaneously. In testing 240-volt equipment, it is quite possible not to realize that the circuit is still energized with 120 volts if one pole of the overcurrent device is open. See 210.10 and 240.20(B) for further information on circuit breaker overcurrent protection of ungrounded conductors. Other precautions concerning device removal on multiwire branch circuits are found in the commentary following 300.13(B).

(D) Identification of Ungrounded Conductors. Where more than one nominal voltage system exists in a building, each ungrounded conductor of a multiwire branch circuit, where accessible, shall be identified by phase and system. This means of identification shall be permitted to be by separate color coding, marking tape, tagging, or other approved means and shall be permanently posted at each branch-circuit panelboard.

Exhibit 210.3 shows an example of two different nominal voltage systems in a building. Each ungrounded system conductor is identified by color-coded marking tape. A notice indicating the means of the identification is permanently located at each panelboard. It should be noted that this requirement applies only to multiwire branch circuits.

210.5 Identification for Branch Circuits.

(A) Grounded Conductor. The grounded conductor of a branch circuit shall be identified in accordance with 200.6.

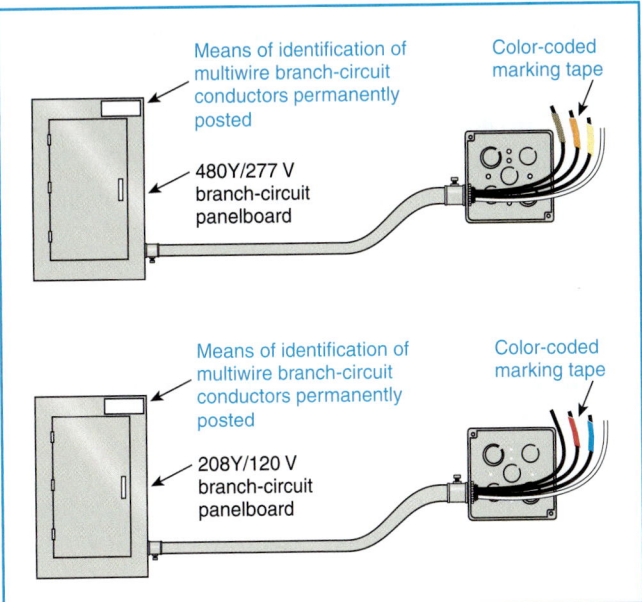

Exhibit 210.3 An example of accessible (ungrounded) phase conductors identified by marking tape.

(B) Equipment Grounding Conductor. The equipment grounding conductor shall be identified in accordance with 250.119.

210.6 Branch-Circuit Voltage Limitations.

The nominal voltage of branch circuits shall not exceed the values permitted by 210.6(A) through (E).

(A) Occupancy Limitation. In dwelling units and guest rooms of hotels, motels, and similar occupancies, the voltage shall not exceed 120 volts, nominal, between conductors that supply the terminals of the following:

(1) Luminaires (lighting fixtures)
(2) Cord-and-plug-connected loads 1440 volt-amperes, nominal, or less or less than ¼ hp

The term *similar occupancies* in 210.6(A) refers to sleeping rooms in dormitories, fraternities, sororities, nursing homes, and other such facilities.

Small loads, such as those of 1440 volt-amperes or less and motors of less than ¼ horsepower, are limited to 120-volt circuits. High-wattage cord-and-plug-connected loads, such as electric ranges, clothes dryers, and some window air conditioners, may be connected to a 208-volt or 240-volt circuit.

(B) 120 Volts Between Conductors. Circuits not exceeding 120 volts, nominal, between conductors shall be permitted to supply the following:

(1) The terminals of lampholders applied within their voltage ratings

(2) Auxiliary equipment of electric-discharge lamps

(3) Cord-and-plug-connected or permanently connected utilization equipment

(C) 277 Volts to Ground. Circuits exceeding 120 volts, nominal, between conductors and not exceeding 277 volts, nominal, to ground shall be permitted to supply the following:

(1) Listed electric-discharge luminaires (lighting fixtures)

(2) Listed incandescent luminaires (lighting fixtures), where supplied at 120 volts or less from the output of a stepdown autotransformer that is an integral component of the luminaire (fixture) and the outer shell terminal is electrically connected to a grounded conductor of the branch circuit

(3) Luminaires (lighting fixtures) equipped with mogul-base screw shell lampholders
(4) Lampholders, other than the screw shell type, applied within their voltage ratings
(5) Auxiliary equipment of electric-discharge lamps
(6) Cord-and-plug-connected or permanently connected utilization equipment

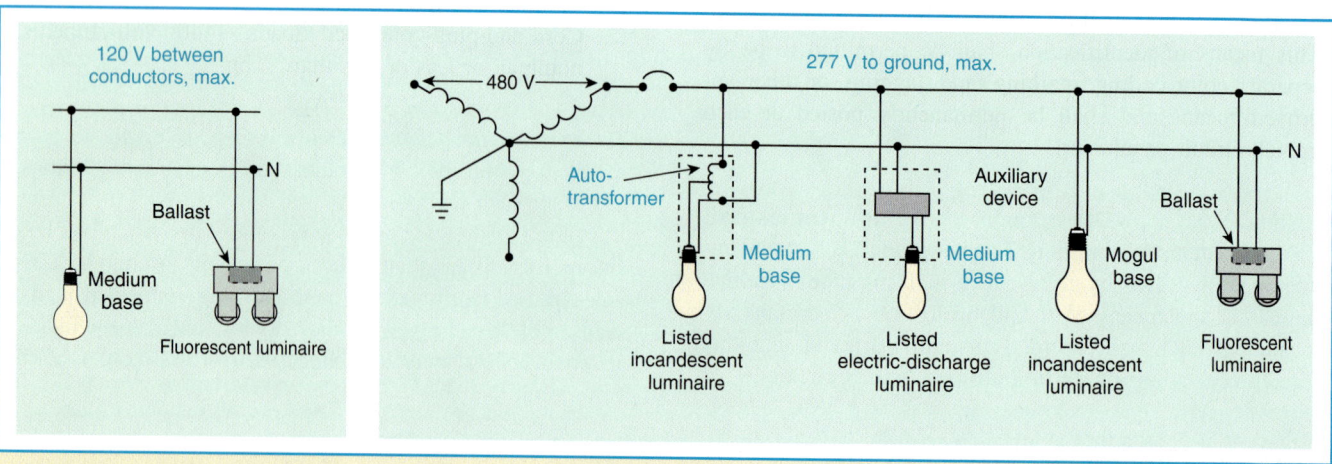

Exhibit 210.4 Examples of luminaires permitted by 210.6 (B) and (C) to be connected to branch circuits.

210.6(A), (B), (D), and (E) describe voltage as "volts between conductors." Luminaires listed for and connected to a 480-volt source may be used in applications permitted by 210.6(C) provided that the 480-volt system is in fact a grounded wye system that contains a grounded conductor (thus limiting the system "voltage to ground" to the 277-volt level).

(D) 600 Volts Between Conductors. Circuits exceeding 277 volts, nominal, to ground and not exceeding 600 volts, nominal, between conductors shall be permitted to supply the following:

(1) The auxiliary equipment of electric-discharge lamps mounted in permanently installed luminaires (fixtures) where the luminaires (fixtures) are mounted in accordance with one of the following:

   a. Not less than a height of 6.7 m (22 ft) on poles or similar structures for the illumination of outdoor areas such as highways, roads, bridges, athletic fields, or parking lots

   b. Not less than a height of 5.5 m (18 ft) on other structures such as tunnels

The minimum mounting heights required by 210.6(D)(1) are for circuits that exceed 277 volts to ground and do not exceed 600 volts phase to phase. These circuits supply the auxiliary equipment of electric-discharge lamps. Exhibit

210.5 (left) shows the minimum mounting height of 18 ft for luminaires installed in tunnels and similar structures. Exhibit 210.5 (right) illustrates the minimum mounting height of 22 ft for luminaires in outdoor areas such as parking lots.

(2) Cord-and-plug-connected or permanently connected utilization equipment

   FPN: See 410.78 for auxiliary equipment limitations.

Exception No. 1 to (B), (C), and (D): For lampholders of infrared industrial heating appliances as provided in 422.14.

Exception No. 2 to (B), (C), and (D): For railway properties as described in 110.19.

(E) Over 600 Volts Between Conductors. Circuits exceeding 600 volts, nominal, between conductors shall be permitted to supply utilization equipment in installations where conditions of maintenance and supervision ensure that only qualified persons service the installation.

210.7 Branch Circuit Receptacle Requirements.

(A) Receptacle Outlet Location. Receptacle outlets shall be located in branch circuits in accordance with Part III of Article 210.

(B) Receptacle Requirements. Specific requirements for receptacles are covered in Article 406.

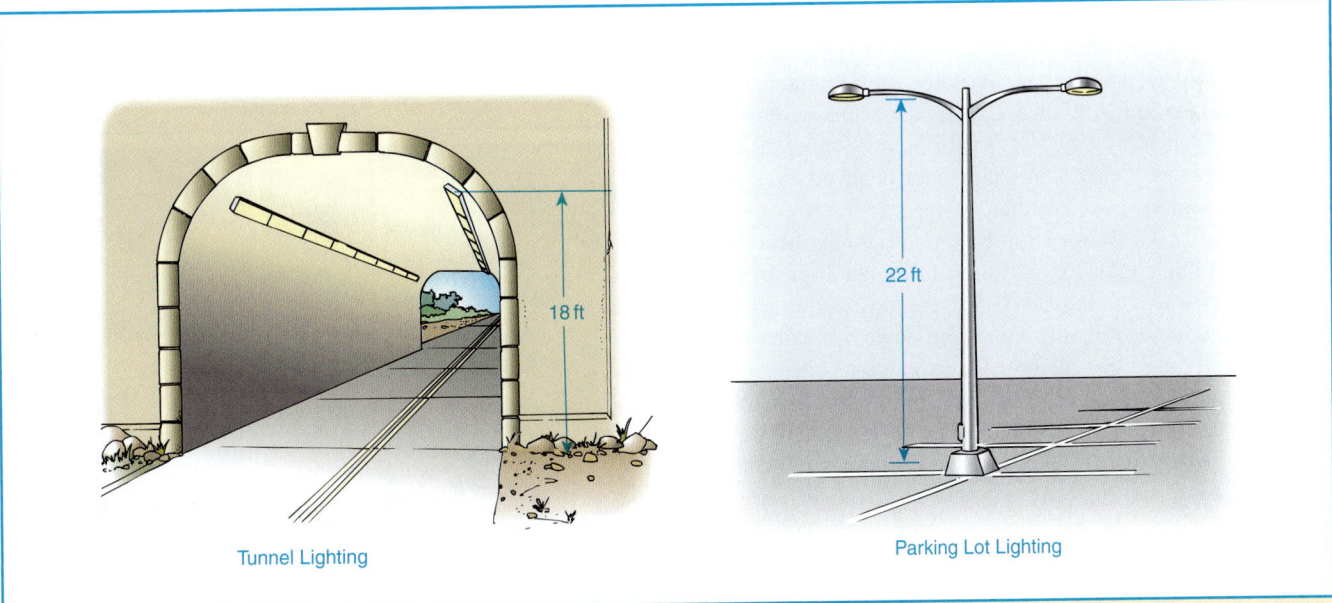

Tunnel Lighting

18 ft

Parking Lot Lighting

22 ft

Exhibit 210.5 Minimum mounting heights for tunnel and parking lot lighting as required by 210.6(D)(1) for circuits exceeding 277 volts to ground and not exceeding 600 volts between conductors supplying auxiliary equipment of electric-discharge lampholders.

Section 210.7 in the 1999 *NEC* is now new Article 406, Receptacles, Cord Connectors, and Attachment Plugs (Caps), for the 2002 *Code*. Requirements for replacement receptacles are now located in 406.3(D).

(C) Multiple Branch Circuits. Where more than one branch circuit supplies more than one receptacle on the same yoke, a means to simultaneously disconnect the ungrounded conductors supplying those receptacles shall be provided at the panelboard where the branch circuits originated.

In 210.7(C), specifying a means to simultaneously disconnect the ungrounded conductors is a safety issue that applies to devices (actually, the single yoke) where more than one branch circuit is involved.

210.8 Ground-Fault Circuit-Interrupter Protection for Personnel.

> FPN: See 215.9 for ground-fault circuit-interrupter protection for personnel on feeders.

Section 210.8 is the main rule for the application of ground-fault circuit interrupters (GFCIs). Since the introduction of the GFCI in the 1971 *Code,* these devices have proved to their users and to the electrical community that they are worth the added cost during construction or remodeling. Published data from the Consumer Product Safety Commission show a decreasing trend in the number of electrocutions in the United States since the introduction of GFCI devices. Unfortunately, no statistics are available for the actual number of lives saved by GFCI devices or the actual number of injuries prevented by GFCI devices. However, most experts in the field would agree that the number of saved lives and prevented injuries is substantial.

Exhibit 210.6 shows a typical circuit arrangement of a GFCI. The line conductors are passed through a sensor and are connected to a shunt-trip device. As long as the current in each conductor remains equal, the device remains in a closed position. If one of the conductors comes in contact with a grounded object, either directly or through a person's body, some of the current returns by an alternative path, resulting in an unbalanced current. The toroidal coil senses the unbalanced current, and a circuit is established to the shunt-trip mechanism that reacts and opens the circuit. Note that the circuit design does not require the presence of an equipment grounding conductor, which is the reason 406.3(D)(3)(b) permits the use of GFCIs as replacements for receptacles where a grounding means does not exist.

GFCIs operate on currents of 5 mA. Listing standards

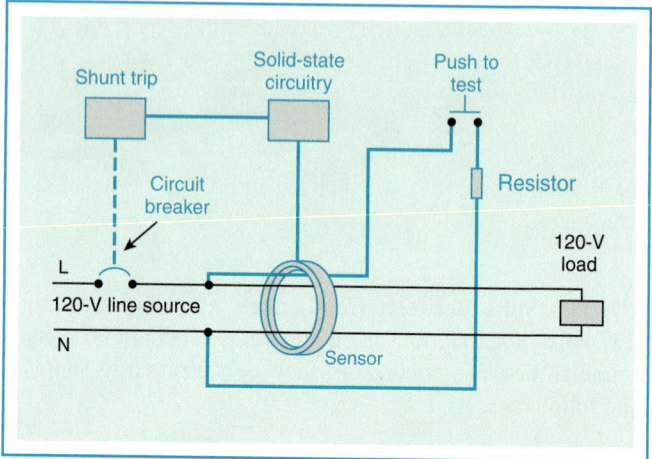

Exhibit 210.6 The circuitry and components of a typical GFCI.

permit a differential of 4 to 6 mA. At trip levels of 5 mA (the instantaneous current could be much higher), a shock can be felt during the time of the fault. The shock can lead to involuntary reactions that may cause secondary accidents such as falls. GFCIs will not protect persons from shock hazards where contact is between phase and neutral or between phase-to-phase conductors.

A variety of GFCIs are available, including portable and plug-in types and circuit-breaker types, types built into attachment plug caps, and receptacle types. Each type has a test switch so that units can be checked periodically to ensure proper operation. See Exhibits 210.7 and 210.8.

Although 210.8 is the main rule for GFCIs, other specific applications require the use of GFCIs. These additional specific applications are listed in Table 210.1.

Exhibit 210.7 A portable plug-in type of GFCI. (Courtesy of Pass & Seymour/Legrand®)

Exhibit 210.8 A 15-ampere duplex receptacle with integral GFCI that also protects downstream loads. (Courtesy of Pass & Seymour/Legrand®)

Table 210.1 Additional Requirements for the Application of Ground-Fault Circuit-Interrupter Protection

Location	Applicable Section(s)
Audio system equipment	640.10(A)
Boathouses	555.19(B)(1)
Carnivals, circuses, fairs, and similar events	525.23
Commercial garages	511.12
Electric vehicle charging systems	625.22
Electronic equipment, sensitive	647.7(A)
Elevators, escalators, and moving walkways	620.85
Feeders	215.9
Fountains	680.51(A)
Health care facilities	517.20(A), 517.21
High-pressure spray washers	422.49
Hydromassage bathtubs	680.71
Marinas	555.19(B)(1)
Mobile and manufactured homes	550.13(B) and (E), 550.32(E)
Park trailers	552.41(C)
Pools, permanently installed	680.22(A)(1), (A)(5), and (B)(4); 680.23(A)(3)
Pools, storable	680.32
Sensitive electronic equipment	647.7(A)
Signs with fountains	680.57(B)
Signs, mobile or portable	600.10(C)(2)
Recreational vehicles	551.40(C), 551.41(C)
Recreational vehicle parks	551.71
Replacement receptacles	406.3(D)(2)
Temporary installations	527.6

(A) Dwelling Units. All 125-volt, single-phase, 15- and 20-ampere receptacles installed in the locations specified in (1) through (8) shall have ground-fault circuit-interrupter protection for personnel.

(1) Bathrooms

GFCI receptacles in bathrooms prevent accidents. Therefore, 210.8(A)(1) requires that all 125-volt, single-phase, 15- and 20-ampere receptacles in bathrooms are to have GFCI protection, including receptacles that are integral with luminaires and, of course, the wall-mounted receptacles adjacent to the basin.

A *bathroom* is defined in Article 100 as an area that includes a basin with one or more of the following: a toilet, a tub, or a shower. The term applies to the entire area, whether a separating door, as illustrated in Exhibit 210.9, is present or not. Note that 210.52(D) requires that a receptacle be located on the wall or partition adjacent to each basin location. However, if the basins are adjacent and in close proximity, then one receptacle outlet may satisfy the requirement, as shown in Exhibit 210.9 (top).

(2) Garages, and also accessory buildings that have a floor located at or below grade level not intended as habitable

rooms and limited to storage areas, work areas, and areas of similar use

Exception No. 1: Receptacles that are not readily accessible.

Exception No. 2: A single receptacle or a duplex receptacle for two appliances located within dedicated space for each appliance that, in normal use, is not easily moved from one place to another and that is cord-and-plug connected in accordance with 400.7(A)(6), (A)(7), or (A)(8).

Receptacles installed under the exceptions to 210.8(A)(2) shall not be considered as meeting the requirements of 210.52(G).

The requirement for GFCI receptacles in garages and sheds, as illustrated in Exhibit 210.10, improves safety for persons using portable hand-held tools, gardening appliances, lawn

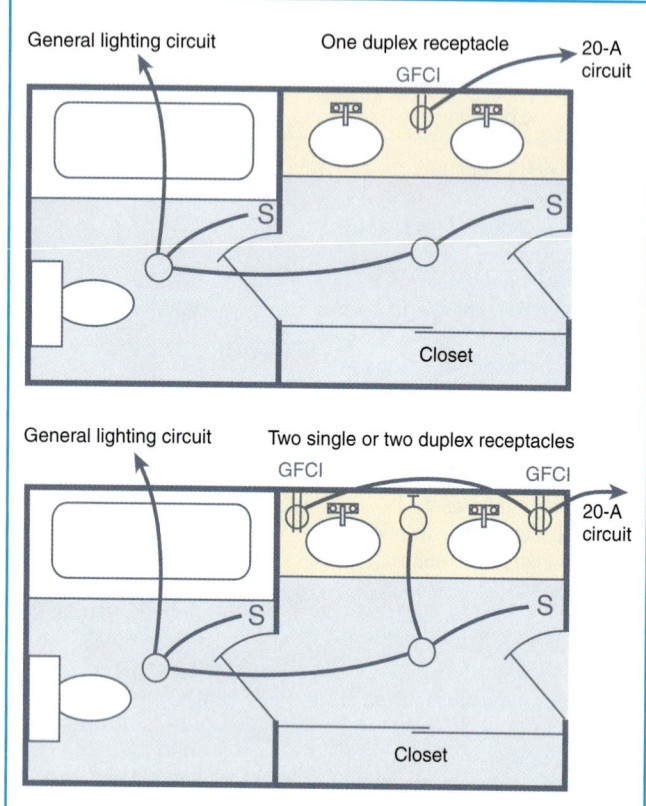

Exhibit 210.9 GFCI-protected receptacles in accordance with 210.8(A)(1) in bathrooms.

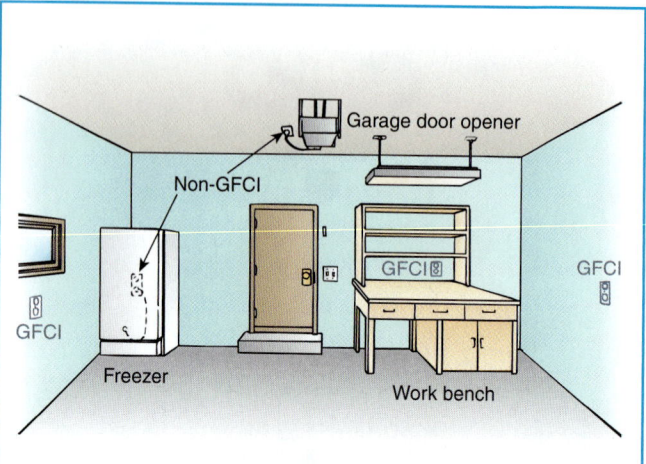

Exhibit 210.10 Examples of receptacles in a garage that are required by 210.8(A)(2) to have GFCI protection. Some receptacles are exempt because they are not readily accessible or are located for an appliance that occupies dedicated space.

mowers, string trimmers, snow blowers, and so on, that might be connected to these receptacles, which are often the closest ones available. GFCI protection is also required in garage areas where auto repair work and general workshop electrical tools are used.

Exception No. 1 to 210.8(A)(2) permits a ceiling-mounted receptacle that is installed for connection of a garage door opener to be exempt from the GFCI requirement. Exception No. 2 to 210.8(A)(2) allows a duplex receptacle located where two cord-and-plug-connected appliances occupy a dedicated space to be exempt from the GFCI requirement. If only a single cord-and-plug-connected appliance, such as a food freezer, occupies the dedicated space, then a single receptacle must be used.

(3) Outdoors

Exception: Receptacles that are not readily accessible and are supplied by a dedicated branch circuit for electric snow-melting or deicing equipment shall be permitted to be installed in accordance with the applicable provisions of Article 426.

The dwelling unit shown in Exhibit 210.11 has four outdoor receptacles. Three of these receptacles are considered to be at direct grade level access and must have GFCI protection for personnel. The fourth receptacle located adjacent to the gutter for the roof-mounted snow-melting cable is not readily accessible and, therefore, is exempt from the GFCI requirements of 210.8(A)(3). However, this receptacle is covered by the equipment protection requirements of 426.28. See the commentary following 210.52(E) and 406.8(B) regarding the installation of outdoor receptacles subject to moisture.

(4) Crawl spaces—at or below grade level
(5) Unfinished basements—for purposes of this section, unfinished basements are defined as portions or areas of the basement not intended as habitable rooms and limited to storage areas, work areas, and the like

Exception No. 1: Receptacles that are not readily accessible.

Exception No. 2: A single receptacle or a duplex receptacle for two appliances located within dedicated space for each appliance that, in normal use, is not easily moved from one place to another and that is cord-and-plug connected in accordance with 400.7(A)(6), (A)(7), or (A)(8).

Exception No. 3: A receptacle supplying only a permanently installed fire alarm or burglar alarm system shall not be required to have ground-fault circuit-interrupter protection.

Receptacles installed under the exceptions to 210.8(A)(5) shall not be considered as meeting the requirements of 210.52(G).

Exhibit 210.11 A dwelling unit with three receptacles that are required by 210.8(A)(3) to have GFCI protection and one that is exempt because it is not readily accessible.

An unfinished portion of a basement is limited to storage areas, work areas, and the like. The receptacles in the work area of the basement shown in Exhibit 210.12 must have

GFCI protection. Section 210.8(A)(5) does not apply to finished areas in basements, such as sleeping rooms or family rooms, and GFCI protection of receptacles in those areas is not required. In addition, freezer and laundry receptacles do not require GFCI protection, in accordance with 210.8(A)(5), Exception No. 2.

Exception No. 3 was added for the 2002 *Code* to permit the omission of GFCI protection for outlets that serve burglar and fire alarm systems, thus adding a degree of reliability to these important systems.

(6) Kitchens—where the receptacles are installed to serve the countertop surfaces

Many countertop kitchen appliances are ungrounded, and the presence of water and grounded surfaces contributes to a hazardous environment, leading to the requirement in 210.8(A)(6) for GFCI protection around a kitchen sink. See Exhibit 210.13 and Exhibit 210.26. The requirement is intended for receptacles serving the countertop. Receptacles installed for disposals, dishwashers, and trash compactors are not required to be protected by a GFCI. According to 406.4(E), receptacles installed to serve countertops cannot be installed in the countertop in the face-up position because liquid, dirt, and other foreign material can enter the receptacle.

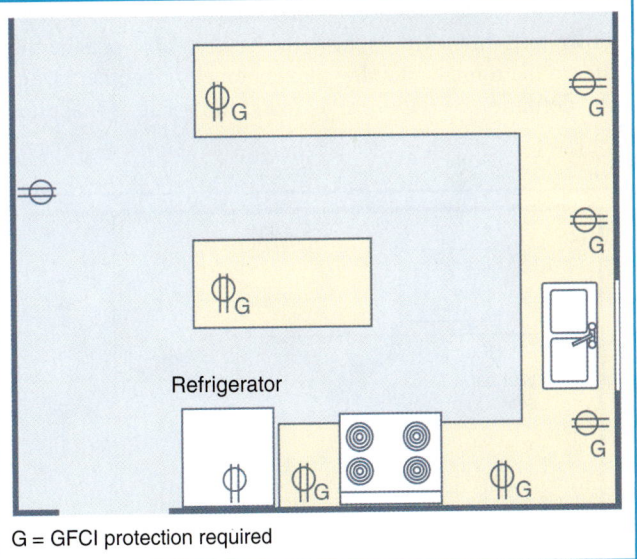

G = GFCI protection required

Exhibit 210.13 GFCI-protected receptacles shown in accordance with 210.8(A)(6) to serve countertop surfaces in dwelling unit kitchens.

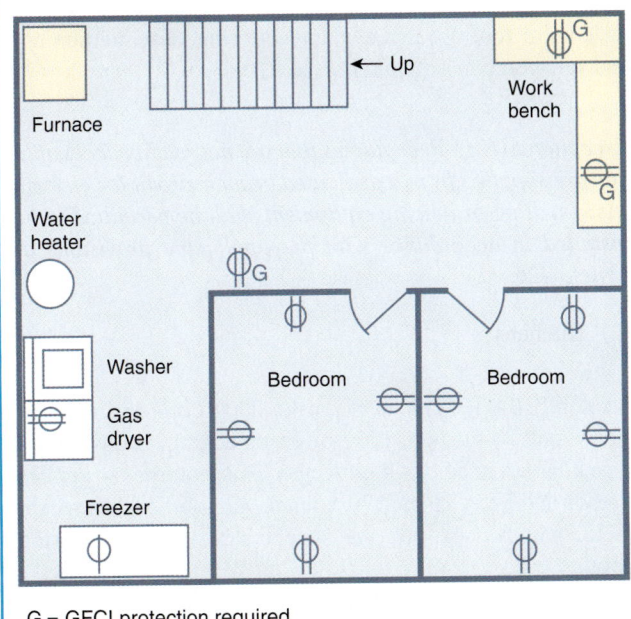

G = GFCI protection required

Exhibit 210.12 A basement floor plan with GFCI-protected receptacles, in accordance with 210.8(A)(5), in the work area and non-GFCI receptacles elsewhere.

(7) Wet bar sinks—where the receptacles are installed to serve the countertop surfaces and are located within 1.8 m (6 ft) of the outside edge of the wet bar sink.

Many countertop wet bar appliances are ungrounded, and the presence of water and grounded surfaces contributes to a hazardous environment, leading to this requirement for GFCI protection around a wet bar sink. As illustrated in Exhibit 210.14, receptacles within 6 ft of a wet bar sink are required to be GFCI protected. According to 406.4(E), these receptacles cannot be installed in the countertop in the face-up position because liquid, dirt, and other foreign material can enter the receptacle.

(8) Boathouses

(B) Other Than Dwelling Units. All 125-volt, single-phase, 15- and 20-ampere receptacles installed in the locations specified in (1), (2), and (3) shall have ground-fault circuit-interrupter protection for personnel:

(1) Bathrooms

If receptacles are provided in bathroom areas of hotels and motels, GFCI-protected receptacles are required. Lavatories in airports, commercial buildings, industrial facilities, and other nondwelling occupancies are required to have *all* their receptacles GFCI protected. The only exception to this rule is found in 517.21, which permits receptacles in hospital critical care areas to be non-GFCI if the toilet and basin are installed in the patient room rather than in a separate bathroom. Some motel and hotel bathrooms, like the one shown in Exhibit 210.15, have the basin located outside the door to the room containing the tub, toilet, or another basin. The definition of *bathroom* as found in Article 100 applies

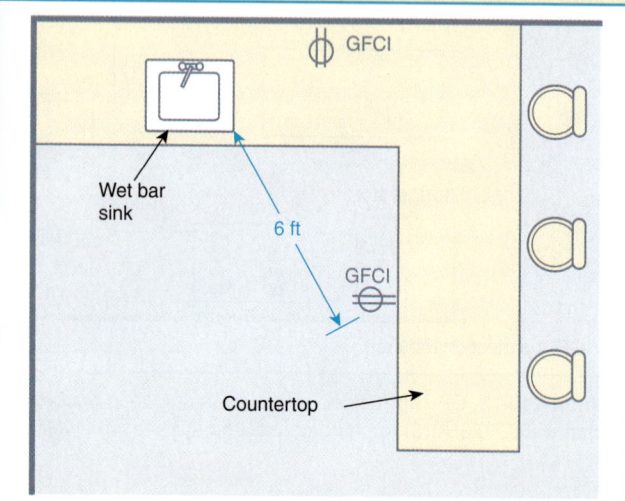

Exhibit 210.14 GFCI-protected receptacles in accordance with 210.8(A)(7) that are located near a wet bar sink and serve the countertop.

to motel and hotel bathrooms, as does the GFCI requirement of 210.8(B)(1).

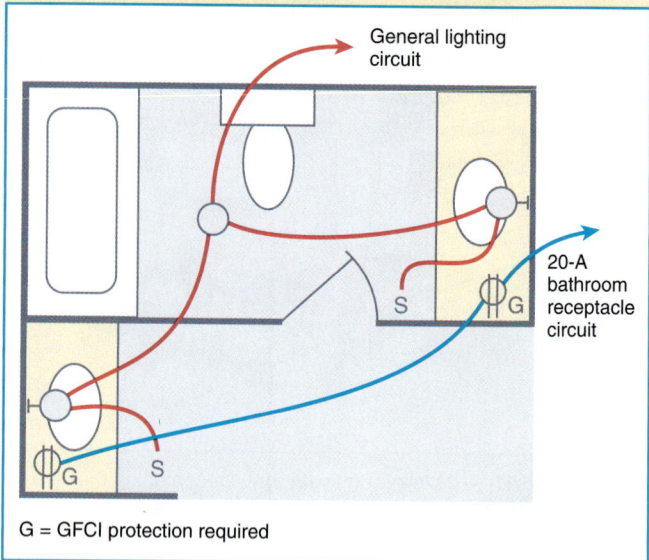

G = GFCI protection required

Exhibit 210.15 GFCI-protected receptacles shown in accordance with 210.8(B)(1) in a motel/hotel bathroom where one basin is located outside the door to the rest of the bathroom area.

(2) Rooftops

Section 210.8(B)(2) requires all rooftop 15- and 20-ampere receptacles in nondwelling occupancies to be GFCI protected. For rooftops that also have heating, air-conditioning, and refrigeration equipment, see 210.63.

Exception to (2): Receptacles that are not readily accessible and are supplied from a dedicated branch circuit for electric snow-melting or deicing equipment shall be permitted to be installed in accordance with the applicable provisions of Article 426.

(3) Kitchens

Section 210.8(B)(3) is new for the 2002 *Code* and requires all 15- and 20-ampere, 125-volt receptacles in nondwelling-type kitchens to be GFCI protected. This requirement applies to each and every 15- and 20-ampere, 125-volt kitchen receptacle, whether or not the receptacle serves countertop appliances.

Accident data related to electrical incidents in nondwelling kitchens reveal the presence of many hazards, including poorly maintained electrical apparatus, damaged electrical cords, wet floors, and employees without proper electrical safety training. Mandating some limited form of GFCI protection for high-hazard areas such as nondwelling kitchens should help prevent electrical accidents.

210.9 Circuits Derived from Autotransformers.

Branch circuits shall not be derived from autotransformers unless the circuit supplied has a grounded conductor that is electrically connected to a grounded conductor of the system supplying the autotransformer.

Exhibits 210.16 through 210.19 illustrate typical applications of autotransformers. In Exhibit 210.16, a 120-volt supply is derived from a 240-volt system. The grounded conductor of the primary system is electrically connected to the grounded conductor of the secondary system.

A buck-boost transformer is classified as an autotransformer. A buck-boost transformer provides a means of raising or lowering (boosting or bucking) a supply line voltage by a small amount (usually no more than 20 percent). A buck-boost is a transformer with two primary windings (H1-H2 and H3-H4) and two secondary windings (X1-X2 and X3-X4). Its primary and secondary windings are connected together so that the electrical characteristics are changed from a transformer that has its primary and secondary windings insulated from each other to one that has primary and secondary windings connected together to buck or boost the voltage as an autotransformer, correcting voltage by up to 20 percent.

A single unit is used to boost or buck single-phase voltage, but two or three units are used to boost or buck 3-phase voltage. An autotransformer requires little physical space, is economical, and, above all, is efficient.

One common application of a boost transformer is to derive a single-phase, 240-volt supply system for ranges, air conditioners, heating elements, and motors from a 3-phase, 208Y/120-volt source system. The boosted leg should not be used to supply line-to-neutral loads because the boosted line-to-neutral voltage will be higher than 120 volts.

Another common boost transformer application is to increase a single-phase, 240-volt source to a single-phase, 277-volt supply for lighting systems. One common 3-phase application is to boost 440 volts to 550 volts for power equipment.

Other common applications of a buck transformer include transforming 240 volts to 208 volts for use with 208-volt appliances and converting a 480Y/277-volt source to a 416Y/240-volt supply system.

Literature containing diagrams for connection and application of autotransformers is available from manufacturers.

Exception No. 1: An autotransformer shall be permitted without the connection to a grounded conductor where trans-

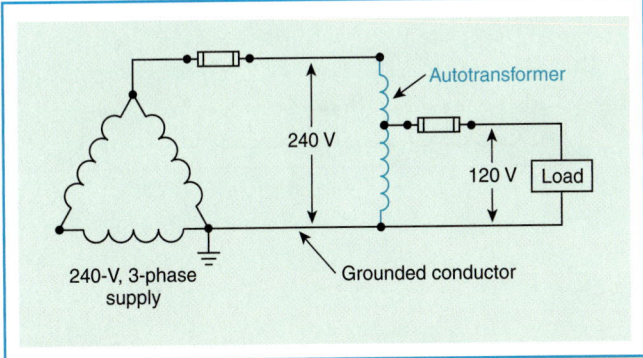

Exhibit 210.16 The circuitry for an autotransformer used to derive a 2-wire, 120-volt system for lighting or convenience receptacles from a 240-volt corner-grounded delta system.

forming from a nominal 208 volts to a nominal 240-volt supply or similarly from 240 volts to 208 volts.

Exception No. 1 to 210.9 allows an autotransformer (without an electrical connection to a grounded conductor) to extend or add an individual branch circuit in an existing installation where transforming (boosting) 208 volts to 240 volts, as shown in Exhibit 210.17. Exhibits 210.18 and 210.19 illustrate typical single-phase and 3-phase buck and boost transformers connected as autotransformers to change 240 volts to 208 volts and vice versa.

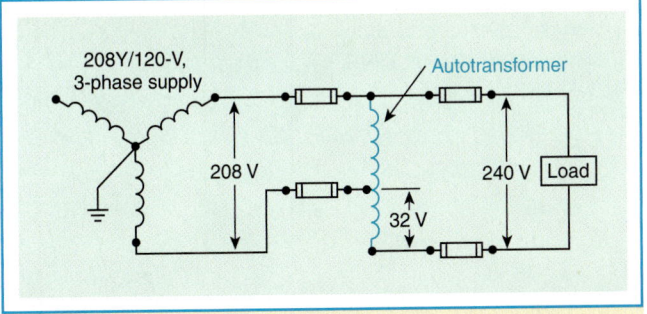

Exhibit 210.17 The circuitry for an autotransformer used to derive a 240-volt system for appliances from a 208Y/120-volt source, in accordance with 210.9, Exception No. 1.

Exception No. 2: In industrial occupancies, where conditions of maintenance and supervision ensure that only qualified persons service the installation, autotransformers shall be permitted to supply nominal 600-volt loads from nominal 480-volt systems, and 480-volt loads from nominal 600-volt systems, without the connection to a similar grounded conductor.

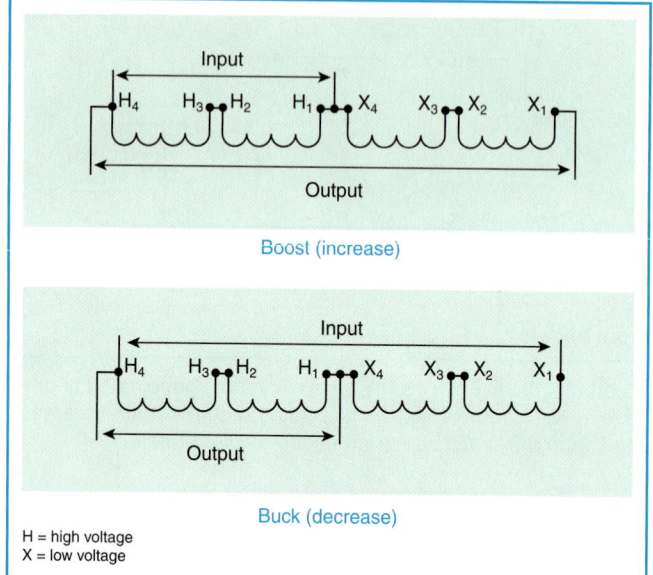

Boost (increase)

Buck (decrease)

H = high voltage
X = low voltage

Exhibit 210.18 Typical single-phase connection diagrams for buck or boost transformers connected as autotransformers to change 240 volts single-phase to 208 volts and vice versa.

In industrial locations, Exception No. 2 to 210.9 allows the use of an autotransformer to supply a 600-volt load from 480-volt systems, provided there are qualified personnel to service the installation. It also allows 480-volt loads to be supplied through an autotransformer supplied by a 600-volt system.

210.10 Ungrounded Conductors Tapped from Grounded Systems.

Two-wire dc circuits and ac circuits of two or more ungrounded conductors shall be permitted to be tapped from the ungrounded conductors of circuits that have a grounded neutral conductor. Switching devices in each tapped circuit shall have a pole in each ungrounded conductor. All poles of multipole switching devices shall manually switch together where such switching devices also serve as a disconnecting means as required by the following:

(1) 410.48 for double-pole switched lampholders
(2) 410.54(B) for electric-discharge lamp auxiliary equipment switching devices
(3) 422.31(B) for an appliance
(4) 424.20 for a fixed electric space-heating unit
(5) 426.51 for electric deicing and snow-melting equipment
(6) 430.85 for a motor controller
(7) 430.103 for a motor

Exhibit 210.19 Typical connection diagrams for buck or boost transformers connected in 3-phase open delta as autotransformers to change 240 volts to 208 volts and vice versa.

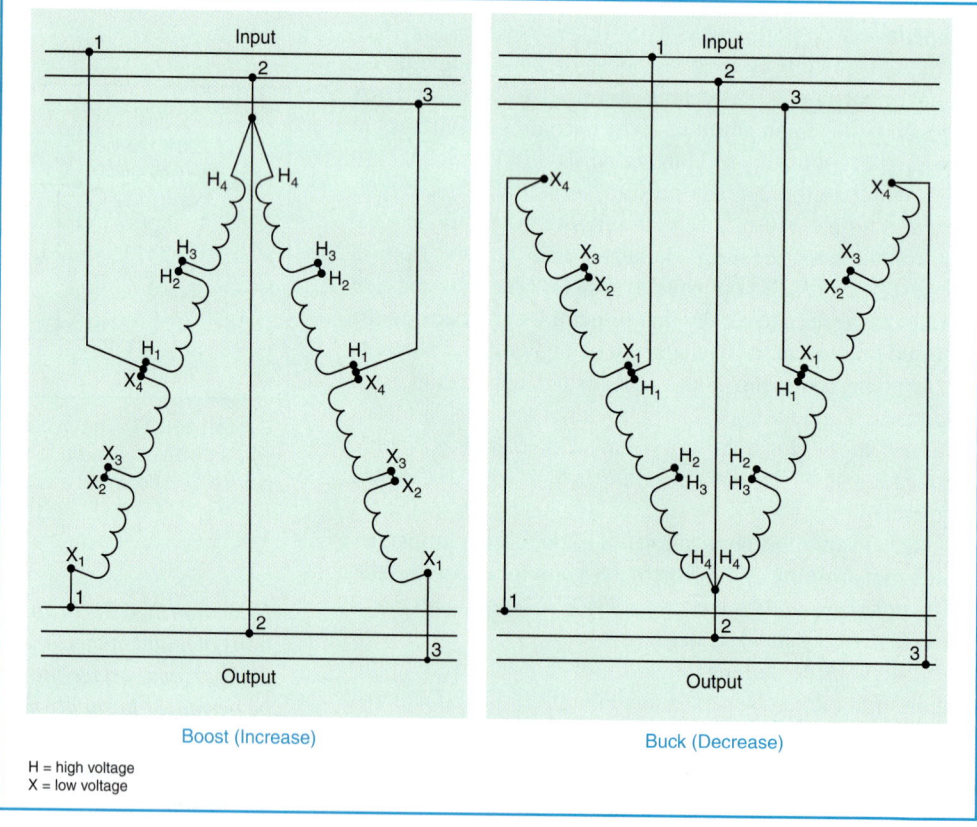

Boost (Increase)

Buck (Decrease)

H = high voltage
X = low voltage

Two-wire ungrounded branch circuits may be tapped from ac or dc circuits of two or more ungrounded conductors that have a grounded neutral conductor. Exhibit 210.20 (top) illustrates an ungrounded 2-wire branch circuit tapped from the ungrounded conductors of a dc or single-phase system to supply a small motor. Exhibit 210.20 (bottom) illustrates a 3-phase, 4-wire wye system.

Circuit breakers or switches that are used as the disconnecting means for a branch circuit must open all poles simultaneously using only the manual operation of the disconnecting means. Therefore, if switches and fuses are used and one fuse blows, or if circuit breakers (two single-pole circuit breakers with a handle tie) are used and one breaker trips, one pole could possibly remain closed. The intention is not to require a common trip of fuses or circuit breakers but rather to disconnect (manually) the ungrounded conductors of the branch circuit with one manual operation. See 240.20(B) for information on handle ties.

210.11 Branch Circuits Required.

Branch circuits for lighting and for appliances, including motor-operated appliances, shall be provided to supply the loads computed in accordance with 220.3. In addition, branch circuits shall be provided for specific loads not covered by 220.3 where required elsewhere in this *Code* and for dwelling unit loads as specified in 210.11(C).

(A) Number of Branch Circuits. The minimum number of branch circuits shall be determined from the total computed load and the size or rating of the circuits used. In all installations, the number of circuits shall be sufficient to supply the load served. In no case shall the load on any circuit exceed the maximum specified by 220.4.

(B) Load Evenly Proportioned Among Branch Circuits. Where the load is computed on a volt-amperes/square meter or square foot basis, the wiring system up to and including the branch-circuit panelboard(s) shall be provided to serve not less than the calculated load. This load shall be evenly proportioned among multioutlet branch circuits within the panelboard(s). Branch-circuit overcurrent devices and circuits shall only be required to be installed to serve the connected load.

(C) Dwelling Units.

(1) Small-Appliance Branch Circuits. In addition to the number of branch circuits required by other parts of this section, two or more 20-ampere small-appliance branch circuits shall be provided for all receptacle outlets specified by 210.52(B).

(2) Laundry Branch Circuits. In addition to the number of branch circuits required by other parts of this section, at least one additional 20-ampere branch circuit shall be provided to supply the laundry receptacle outlet(s) required by 210.52(F). This circuit shall have no other outlets.

(3) Bathroom Branch Circuits. In addition to the number of branch circuits required by other parts of this section, at least one 20-ampere branch circuit shall be provided to supply the bathroom receptacle outlet(s). Such circuits shall have no other outlets.

Exception: Where the 20-ampere circuit supplies a single bathroom, outlets for other equipment within the same bathroom shall be permitted to be supplied in accordance with 210.23(A).

FPN: See Examples D1(A), D1(B), D2(B), and D4(A) in Annex D.

210.12 Arc-Fault Circuit-Interrupter Protection.

(A) Definition. An *arc-fault circuit interrupter* is a device intended to provide protection from the effects of arc faults by recognizing characteristics unique to arcing and by functioning to de-energize the circuit when an arc fault is detected.

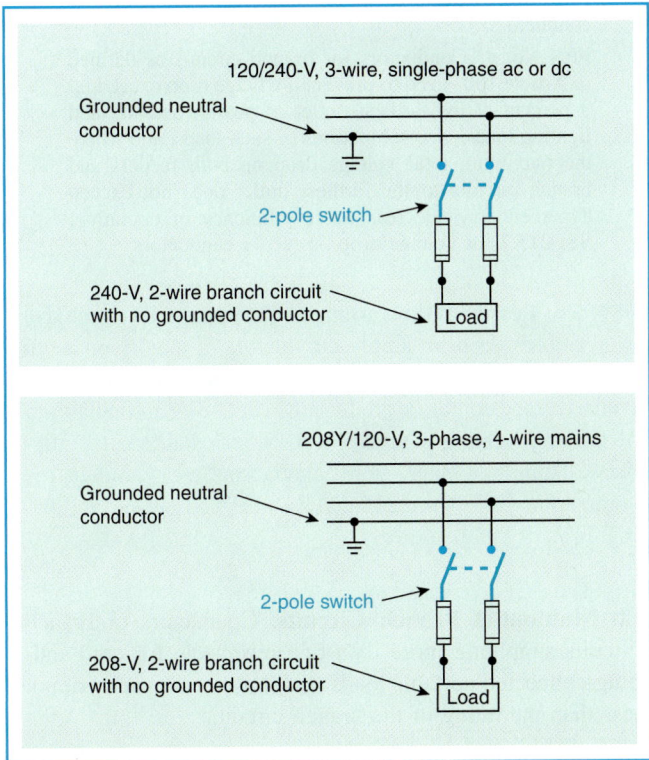

Exhibit 210.20 Branch circuits tapped from ungrounded conductors of multiwire systems.

(B) Dwelling Unit Bedrooms. All branch circuits that supply 125-volt, single-phase, 15- and 20-ampere outlets installed in dwelling unit bedrooms shall be protected by an arc-fault circuit interrupter listed to provide protection of the entire branch circuit.

The definition of *arc-fault circuit interrupter* given in 210.12(A) explains its function. The basic objective is to de-energize the branch circuit when an arc-fault is detected.

Arc-fault circuit interrupters are evaluated to UL 1699, *Safety Standard for Arc-Fault Circuit Interrupter*, using testing methods that create or simulate arcing conditions to determine the product's ability to detect and interrupt arcing faults. These devices are also tested to verify that arc detection is not unduly inhibited by the presence of loads and circuit characteristics that may mask the hazardous arcing condition. In addition, these devices are evaluated to determine resistance to unwanted tripping due to the presence of arcing that occurs in control and utilization equipment under normal operating conditions or to a loading condition that closely mimics an arcing fault, such as a solid-state electronic ballast or a dimmed load.

UL 1699 is the standard covering arc-fault devices that have a maximum rating of 20 amperes intended for use in 120-volt ac, 60 Hz circuits. These devices may also include the capability to perform other functions such as overcurrent protection, ground-fault circuit-interruption, and surge suppression. UL 1699 currently recognizes five types of arc-fault circuit interrupters: the branch/feeder AFCI, combination AFCI, cord AFCI, outlet AFCI, and portable AFCI. Placement of the device in the circuit and a review of the UL guide information must be considered when complying with 210.12. The *NEC* is clear that the objective is to provide protection of the entire branch circuit. (See Article 100 for the definition of *branch circuit*.) For instance, a cord AFCI could not be used to comply with the requirement of 210.12 to protect the entire branch circuit.

Section 210.12 requires that AFCI protection be provided on branch circuits that supply outlets (receptacle, lighting, etc.) in dwelling unit bedrooms. The requirement is limited to 15- and 20-ampere 125-volt circuits. There is no prohibition against providing AFCI protection on other circuits or in locations other than bedrooms. Because circuits are often shared between a bedroom and other areas such as closets and hallways, providing AFCI protection on the complete circuit would comply with 210.12.

II. Branch-Circuit Ratings

210.19 Conductors—Minimum Ampacity and Size.

(A) Branch Circuits Not More Than 600 Volts.

(1) General. Branch-circuit conductors shall have an ampacity not less than the maximum load to be served. Where a branch circuit supplies continuous loads or any combination of continuous and noncontinuous loads, the minimum branch-circuit conductor size, before the application of any adjustment or correction factors, shall have an allowable ampacity not less than the noncontinuous load plus 125 percent of the continuous load.

Exception: Where the assembly, including the overcurrent devices protecting the branch circuit(s), is listed for operation at 100 percent of its rating, the allowable ampacity of the branch circuit conductors shall be permitted to be not less than the sum of the continuous load plus the noncontinuous load.

Conductors of branch circuits not more than 600 volts must be able to supply power to loads without overheating. The requirements in 210.19(A)(1) establish minimum size and ampacity requirements to allow this to happen. The requirements for the minimum size of overcurrent protection devices are found in 210.20. An example showing these minimum-size calculations is found in the commentary following 210.20(A), Exception.

> FPN No. 1: See 310.15 for ampacity ratings of conductors.
> FPN No. 2: See Part II of Article 430 for minimum rating of motor branch-circuit conductors.
> FPN No. 3: See 310.10 for temperature limitation of conductors.
> FPN No. 4: Conductors for branch circuits as defined in Article 100, sized to prevent a voltage drop exceeding 3 percent at the farthest outlet of power, heating, and lighting loads, or combinations of such loads, and where the maximum total voltage drop on both feeders and branch circuits to the farthest outlet does not exceed 5 percent, provide reasonable efficiency of operation. See 215.2 for voltage drop on feeder conductors.

FPN No. 4 expresses a warning about improper voltage due to a voltage drop in supply conductors, a major source of trouble and inefficient operation in electrical equipment. Undervoltage conditions reduce the capability and reliability of motors, lighting sources, heaters, and solid-state equipment. Sample voltage-drop calculations are found in the commentary following 215.2(A)(4), FPN No. 3, and following Table 9 in Chapter 9.

(2) Multioutlet Branch Circuits. Conductors of branch circuits supplying more than one receptacle for cord-and-plug-connected portable loads shall have an ampacity of not less than the rating of the branch circuit.

Because the loading of branch-circuit conductors that supply receptacles for cord-and-plug-connected portable loads is unpredictable, it is safest simply to require such circuits to have an ampacity that is not less than the rating of the branch circuit. According to 210.3, the rating of the branch circuit is actually the rating of the overcurrent device.

(3) Household Ranges and Cooking Appliances. Branch-circuit conductors supplying household ranges, wall-mounted ovens, counter-mounted cooking units, and other household cooking appliances shall have an ampacity not less than the rating of the branch circuit and not less than the maximum load to be served. For ranges of 8 kW or more rating, the minimum branch-circuit rating shall be 40 amperes.

A minimum 40-ampere branch-circuit rating would be, for example, 8 AWG, Type TW copper or 6 AWG, Type TW aluminum. See Table 310.16 for other applications.

Exception No. 1: Tap conductors supplying electric ranges, wall-mounted electric ovens, and counter-mounted electric cooking units from a 50-ampere branch circuit shall have an ampacity of not less than 20 and shall be sufficient for the load to be served. The taps shall not be longer than necessary for servicing the appliance.

As illustrated in Exhibit 210.21, Exception No. 1 to 210.19(A)(3) permits a 20-ampere tap conductor from a range, oven, or cooking unit to be connected to a 50-ampere branch circuit, if the following four conditions are met:

1. The taps are not longer than necessary to service or permit access to the junction box.
2. The taps to each unit are properly spliced.
3. The junction box is adjacent to each unit.
4. The taps are of sufficient size for the load to be served.

Exception No. 2: The neutral conductor of a 3-wire branch circuit supplying a household electric range, a wall-mounted oven, or a counter-mounted cooking unit shall be permitted to be smaller than the ungrounded conductors where the maximum demand of a range of 8¾ kW or more rating has been computed according to Column C of Table 220.19, but shall have an ampacity of not less than 70 percent of the branch-circuit rating and shall not be smaller than 10 AWG.

Column C of Table 220.19 indicates that the maximum demand for one range (not over 12 kW rating) is 8 kW (8 kW = 8000 volt-amperes; 8000 volt-amperes ÷ 240 volts = 33.3 amperes). The allowable ampacity of an 8 AWG, Type TW copper conductor is 40 amperes (see Table 310.16) and may be used for the range branch circuit. According to this computation, the neutral of this 3-wire circuit can be

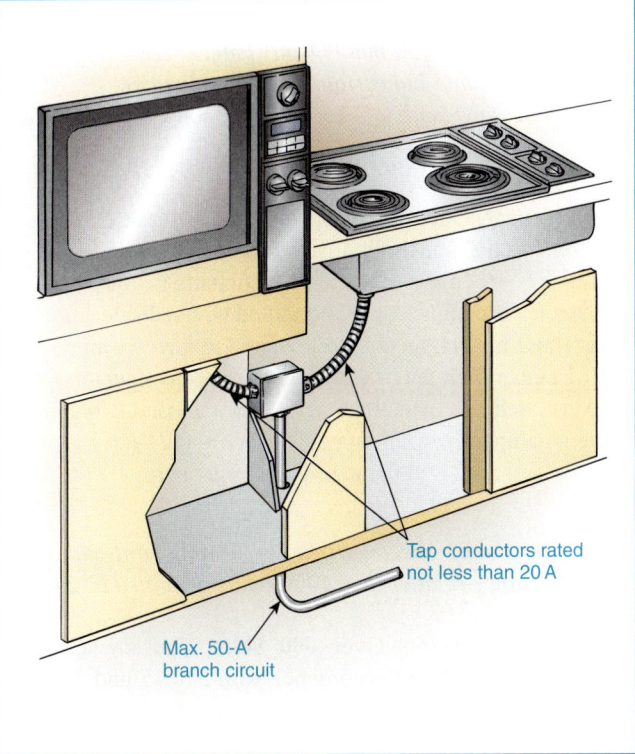

Exhibit 210.21 Tap conductors permitted by 210.19(A)(3), Exception No. 1, to be sized smaller than the branch-circuit conductors and not to be longer than necessary for servicing.

smaller than 8 AWG but not smaller than 10 AWG, which has an allowable ampacity of 30 amperes (30 amperes is more than 70 percent of 40 amperes, per Exception No. 2). The maximum demand for a neutral of an 8-kW range circuit seldom exceeds 25 amperes, since current is drawn from the neutral only for lights, clocks, timers, and some heating elements when in the low-heating position.

(4) Other Loads. Branch-circuit conductors that supply loads other than those specified in 210.2 and other than cooking appliances as covered in 210.19(A)(3) shall have an ampacity sufficient for the loads served and shall not be smaller than 14 AWG.

Exception No. 1: Tap conductors shall have an ampacity sufficient for the load served. In addition, they shall have an ampacity of not less than 15 for circuits rated less than 40 amperes and not less than 20 for circuits rated at 40 or 50 amperes and only where these tap conductors supply any of the following loads:

(a) Individual lampholders or luminaires (fixtures) with taps extending not longer than 450 mm (18 in.) beyond any portion of the lampholder or luminaire (fixture).

(b) A fixture having tap conductors as provided in 410.67.

(c) Individual outlets, other than receptacle outlets, with taps not over 450 mm (18 in.) long.

(d) Infrared lamp industrial heating appliances.

(e) Nonheating leads of deicing and snow-melting cables and mats.

Tap conductors are generally required to have the same ampacity as the branch-circuit overcurrent device. Exception No. 1 lists specific applications in subparts (a) through (e) where the tap conductors are permitted with reduced ampacities. These tap conductors are required to have an ampacity of 15 amperes or more (14 AWG copper conductors) for circuits rated less than 40 amperes. The tap conductors must have an ampacity of 20 amperes or more (12 AWG copper conductors) for circuits rated 40 or 50 amperes.

Exception No. 2: Fixture wires and flexible cords shall be permitted to be smaller than 14 AWG as permitted by 240.5.

(B) Branch Circuits Over 600 Volts. The ampacity of conductors shall be in accordance with 310.15 and 310.60 as applicable. Branch-circuit conductors over 600 volts shall be sized in accordance with 210.19(B)(1) or (B)(2).

(1) General. The ampacity of branch-circuit conductors shall not be less than 125 percent of the designed potential load of utilization equipment that will be operated simultaneously.

(2) Supervised Installations. For supervised installations, branch-circuit conductor sizing shall be permitted to be determined by qualified persons under engineering supervision. Supervised installations are defined as those portions of a facility where all of the following conditions are met:

(1) Conditions of design and installation are provided under engineering supervision.

(2) Qualified persons with documented training and experience in over 600-volt systems provide maintenance, monitoring, and servicing of the system.

Part II of Article 210 has been revised for the 2002 *Code* by including requirements for the branch circuits over 600 volts in 210.19(B). Basically, branch circuits over 600 volts must be sized at 125 percent of the combined simultaneous load, unless the branch circuits over 600 volts are located at facilities that qualify as supervised installations.

210.20 Overcurrent Protection.

Branch-circuit conductors and equipment shall be protected by overcurrent protective devices that have a rating or setting that complies with 210.20(A) through (D).

(A) Continuous and Noncontinuous Loads. Where a branch circuit supplies continuous loads or any combination of continuous and noncontinuous loads, the rating of the overcurrent device shall not be less than the noncontinuous load plus 125 percent of the continuous load.

An example calculation for a continuous load only is illustrated in Exhibit 210.22.

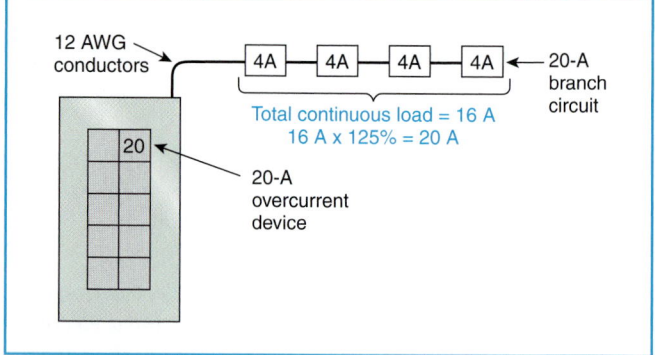

Exhibit 210.22 A continuous load (store lighting) calculated at 125 percent to determine the ampacity of the conductor and the branch-circuit size.

Exception: Where the assembly, including the overcurrent devices protecting the branch circuit(s), is listed for operation at 100 percent of its rating, the ampere rating of the overcurrent device shall be permitted to be not less than the sum of the continuous load plus the noncontinuous load.

According to 210.20, an overcurrent device that supplies continuous and noncontinuous loads must have a rating that is not less than the sum of 100 percent of the noncontinuous load plus 125 percent of the continuous load, calculated in accordance with Article 210.

In addition, 210.19(A)(1) requires that the circuit conductors, chosen from the ampacity tables, must have an initial ampacity of not less than the sum of 100 percent of the noncontinuous load plus 125 percent of the continuous load, the same as calculated for the overcurrent device.

The rating of the overcurrent device cannot exceed the final ampacity of the circuit conductors after all the derating or correction factors have been applied, such as for temperature or number of conductors.

Example

Determine the minimum-size overcurrent protective device and the minimum conductor size for the following circuit:

- 25 amperes of continuous load
- 60°C overcurrent device terminal rating

- Type THWN conductors
- Four current-carrying copper conductors in a raceway

Solution

STEP 1. Determine the size of the overcurrent protective device (OCPD). Referring to 210.20(A), 125 percent of 25 amperes is 31.25 amperes. Thus, the minimum standard-size overcurrent device, according to 240.6(A), is 35 amperes.

STEP 2. Determine the minimum conductor size. The ampacity of the conductor must not be less than 125 percent of the 25-ampere continuous load, which results in 31.25 amperes. The conductor must have an allowable ampacity of not less than 31.25 amperes before any adjustment or correction factors are applied. Because there are four current-carrying conductors in the raceway, Table 310.15(B)(2)(a) applies. First calculate the ampacity of the conductor using the ampacity value calculated above:

$$\text{Conductor ampacity} = \frac{\text{computed load}}{\text{percent adjustment factor from Table 310.15(B)(2)(a)}}$$

$$\frac{31.25 \text{ amperes}}{0.80} = 39.06 \text{ amperes}$$

Because of the 60°C rating of the overcurrent device terminal, it is necessary to choose a conductor based on the ampacities in the 60° column of Table 310.16. The calculated load must not exceed the conductor ampacity. Therefore, an 8 AWG copper, Type TW conductor with an ampacity of 40 amperes is the minimum allowed computed load. However, conductors with a higher temperature rating such as THWN or THHN may be used, but only at their 60°C ampacity.

(B) Conductor Protection. Conductors shall be protected in accordance with 240.4. Flexible cords and fixture wires shall be protected in accordance with 240.5.

(C) Equipment. The rating or setting of the overcurrent protective device shall not exceed that specified in the applicable articles referenced in 240.3 for equipment.

(D) Outlet Devices. The rating or setting shall not exceed that specified in 210.21 for outlet devices.

210.21 Outlet Devices.

Outlet devices shall have an ampere rating that is not less than the load to be served and shall comply with 210.21(A) and (B).

(A) Lampholders. Where connected to a branch circuit having a rating in excess of 20 amperes, lampholders shall be of the heavy-duty type. A heavy-duty lampholder shall have a rating of not less than 660 watts if of the admedium type and not less than 750 watts if of any other type.

The intent of 210.21(A) is to restrict a fluorescent lighting branch-circuit rating to not more than 20 amperes because most lampholders manufactured for use with fluorescent lights are not of the heavy-duty type and are rated at 660 watts or 250 watts.

Branch-circuit conductors for fluorescent electric-discharge lighting are usually connected to a ballast rather than to lampholders, and, by specifying a wattage rating for these lampholders, a limit of 20 amperes is applied to ballast circuits.

Only the admedium-base lampholder is recognized as heavy duty at the rating of 660 watts. Other lampholders are required to have a rating of not less than 750 watts to be recognized as heavy duty. The requirement of 210.21(A) prohibits the use of medium-base screw shell lampholders on branch circuits that are in excess of 20 amperes.

(B) Receptacles.

(1) Single Receptacle on an Individual Branch Circuit. A single receptacle installed on an individual branch circuit shall have an ampere rating not less than that of the branch circuit.

Exception No. 1: A receptacle installed in accordance with 430.81(C).

Exception No. 2: A receptacle installed exclusively for the use of a cord-and-plug-connected arc welder shall be permitted to have an ampere rating not less than the minimum branch-circuit conductor ampacity determined by 630.11(A) for arc welders.

FPN: See definition of *receptacle* in Article 100.

(2) Total Cord-and-Plug-Connected Load. Where connected to a branch circuit supplying two or more receptacles or outlets, a receptacle shall not supply a total cord-and-plug-connected load in excess of the maximum specified in Table 210.21(B)(2).

Table 210.21(B)(2) Maximum Cord-and-Plug-Connected Load to Receptacle

Circuit Rating (Amperes)	Receptacle Rating (Amperes)	Maximum Load (Amperes)
15 or 20	15	12
20	20	16
30	30	24

(3) Receptacle Ratings. Where connected to a branch circuit supplying two or more receptacles or outlets, receptacle ratings shall conform to the values listed in Table

210.21(B)(3), or where larger than 50 amperes, the receptacle rating shall not be less than the branch-circuit rating.

Table 210.21(B)(3) Receptacle Ratings for Various Size Circuits

Circuit Rating (Amperes)	Receptacle Rating (Amperes)
15	Not over 15
20	15 or 20
30	30
40	40 or 50
50	50

Exception No. 1: Receptacles for one or more cord-and-plug-connected arc welders shall be permitted to have ampere ratings not less than the minimum branch-circuit conductor ampacity permitted by 630.11(A) or (B) as applicable for arc welders.

Exception No. 2: The ampere rating of a receptacle installed for electric discharge lighting shall be permitted to be based on 410.30(C).

A single receptacle installed on an individual branch circuit must have an ampere rating not less than that of the branch circuit. For example, a single receptacle on a 20-ampere individual branch circuit must be rated at 20 amperes; however, two or more 15-ampere receptacles or duplex receptacles are permitted on a 20-ampere general-purpose branch circuit. This requirement does not apply to specific types of cord-and-plug-connected arc welders.

(4) Range Receptacle Rating. The ampere rating of a range receptacle shall be permitted to be based on a single range demand load as specified in Table 220.19.

210.23 Permissible Loads.

In no case shall the load exceed the branch-circuit ampere rating. An individual branch circuit shall be permitted to supply any load for which it is rated. A branch circuit supplying two or more outlets or receptacles shall supply only the loads specified according to its size as specified in 210.23(A) through (D) and as summarized in 210.24 and Table 210.24.

The requirements of 210.23 are often misunderstood. An individual (single-outlet) branch circuit can supply any load within its rating. On the other side, the load, of course, cannot be greater than the branch-circuit rating.

(A) 15- and 20-Ampere Branch Circuits. A 15- or 20-ampere branch circuit shall be permitted to supply lighting units or other utilization equipment, or a combination of both, and shall comply with 210.23(A)(1) and (A)(2).

Exception: The small appliance branch circuits, laundry branch circuits, and bathroom branch circuits required in a dwelling unit(s) by 210.11(C)(1), (2), and (3) shall supply only the receptacle outlets specified in that section.

Section 210.23(A) permits a 15- or 20-ampere branch circuit for lighting to also supply utilization equipment fastened in place, such as an air conditioner. The equipment load must not exceed 50 percent of the branch-circuit ampere rating (7.5 amperes on a 15-ampere circuit and 10 amperes on a 20-ampere circuit). However, according to 210.52(B), such fastened-in-place equipment is not permitted on the small-appliance branch circuits required in the kitchen, dining room, and so on.

(1) Cord-and-Plug-Connected Equipment. The rating of any one cord-and-plug-connected utilization equipment shall not exceed 80 percent of the branch-circuit ampere rating.

(2) Utilization Equipment Fastened in Place. The total rating of utilization equipment fastened in place, other than luminaires (lighting fixtures), shall not exceed 50 percent of the branch-circuit ampere rating where lighting units, cord-and-plug-connected utilization equipment not fastened in place, or both, are also supplied.

(B) 30-Ampere Branch Circuits. A 30-ampere branch circuit shall be permitted to supply fixed lighting units with heavy-duty lampholders in other than a dwelling unit(s) or utilization equipment in any occupancy. A rating of any one cord-and-plug-connected utilization equipment shall not exceed 80 percent of the branch-circuit ampere rating.

(C) 40- and 50-Ampere Branch Circuits. A 40- or 50-ampere branch circuit shall be permitted to supply cooking appliances that are fastened in place in any occupancy. In other than dwelling units, such circuits shall be permitted to supply fixed lighting units with heavy-duty lampholders, infrared heating units, or other utilization equipment.

A branch circuit that supplies two or more outlets is permitted to supply only the loads specified according to its size, in accordance with 210.23(A) through (C) and as summarized in 210.24 and Table 210.24. Other circuits are not permitted to have more than one outlet and are considered individual branch circuits. However, 517.71 and 660.4(B) do not require individual branch circuits for portable, mobile, and transportable medical X-ray equipment requiring a capacity of not over 60 amperes.

(D) Branch Circuits Larger Than 50 Amperes. Branch circuits larger than 50 amperes shall supply only nonlighting outlet loads.

See the commentary following 210.3, Exception, regarding multioutlet branch circuits greater than 50 amperes that are permitted to supply nonlighting outlet loads in industrial establishments.

210.24 Branch-Circuit Requirements—Summary.

The requirements for circuits that have two or more outlets or receptacles, other than the receptacle circuits of 210.11(C)(1) and (2), are summarized in Table 210.24. This table provides only a summary of minimum requirements. See 210.19, 210.20, and 210.21 for the specific requirements applying to branch circuits.

Table 210.24 summarizes branch-circuit requirements of conductors, overcurrent protection, outlet devices, maximum load, and permissible load where two or more outlets are supplied.

If the branch circuit serves a fixture load and supplies two or more fixture outlets, 210.23 requires the branch circuit to have a specific ampere rating that is based on the rating of the overcurrent device, as stated in 210.3. Thus, if the

circuit breaker that protects the branch circuit is rated 20 amperes, the conductors supplying this circuit must have an ampacity not less than 20 amperes.

Note that in accordance with the Article 100 definition of *ampacity*, the ampacity is determined after applying all derating (adjustment and correction) factors, such as those in 310.15(B)(2)(a). If seven to nine such conductors are in one conduit, a 12 AWG, Type THHN copper conductor (30 amperes, per Table 310.16) adjusted to 70 percent, per Table 310.15(B)(2)(a), would have an ampacity of 21 amperes and would be suitable for a load of 20 amperes. Thus, this conductor would be acceptable for use on the 20-ampere multioutlet branch circuit.

210.25 Common Area Branch Circuits.

Branch circuits in dwelling units shall supply only loads within that dwelling unit or loads associated only with that dwelling unit. Branch circuits required for the purpose of lighting, central alarm, signal, communications, or other needs for public or common areas of a two-family or multi-family dwelling shall not be supplied from equipment that supplies an individual dwelling unit.

Not only does 210.25 prohibit branch circuits from feeding more than one dwelling unit, it also prohibits the sharing of systems, equipment, or common lighting if that equipment is fed from any of the dwelling units. The systems, equipment, or lighting for public or common areas is required to be supplied from a separate "house load" panelboard. This requirement permits access to the branch-circuit discon-

Table 210.24 Summary of Branch-Circuit Requirements

Circuit Rating	15 A	20 A	30 A	40 A	50 A
Conductors (min. size):					
Circuit wires[1]	14	12	10	8	6
Taps	14	14	14	12	12
Fixture wires and cords—See 240.5					
Overcurrent Protection	**15 A**	**20 A**	**30 A**	**40 A**	**50 A**
Outlet devices:					
Lampholders permitted	Any type	Any type	Heavy duty	Heavy duty	Heavy duty
Receptacle rating[2]	15 max. A	15 or 20 A	30 A	40 or 50 A	50 A
Maximum Load	**15 A**	**20 A**	**30 A**	**40 A**	**50 A**
Permissible load	See 210.23(A)	See 210.23(A)	See 210.23(B)	See 210.23(C)	See 210.23(C)

[1]These gauges are for copper conductors.
[2]For receptacle rating of cord-connected electric-discharge luminaires (lighting fixtures), see 410.30(C).

necting means without the need to enter the space of any tenants. The requirement also prevents a tenant from turning off important circuits that may affect other tenants.

III. Required Outlets

210.50 General.

Receptacle outlets shall be installed as specified in 210.52 through 210.63.

(A) Cord Pendants. A cord connector that is supplied by a permanently connected cord pendant shall be considered a receptacle outlet.

(B) Cord Connections. A receptacle outlet shall be installed wherever flexible cords with attachment plugs are used. Where flexible cords are permitted to be permanently connected, receptacles shall be permitted to be omitted for such cords.

Flexible cords are permitted to be permanently connected to boxes or fittings where specifically permitted by the *Code*. However, plugging a cord into a lampholder by inserting a screw-plug adapter is not permitted, because 410.47 requires lampholders of the screw shell type to be installed for use as lampholders only.

(C) Appliance Outlets. Appliance receptacle outlets installed in a dwelling unit for specific appliances, such as laundry equipment, shall be installed within 1.8 m (6 ft) of the intended location of the appliance.

See 210.52(F) and 210.11(C)(2) for requirements regarding laundry receptacle outlets and branch circuits.

210.52 Dwelling Unit Receptacle Outlets.

This section provides requirements for 125-volt, 15- and 20-ampere receptacle outlets. Receptacle outlets required by this section shall be in addition to any receptacle that is part of a luminaire (lighting fixture) or appliance, located within cabinets or cupboards, or located more than 1.7 m (5½ ft) above the floor.

Permanently installed electric baseboard heaters equipped with factory-installed receptacle outlets or outlets provided as a separate assembly by the manufacturer shall be permitted as the required outlet or outlets for the wall space utilized by such permanently installed heaters. Such receptacle outlets shall not be connected to the heater circuits.

FPN: Listed baseboard heaters include instructions that may not permit their installation below receptacle outlets.

The requirements of 210.52 apply to dwelling unit receptacles that are rated 125 volts and 15 or 20 amperes and that are not part of a luminaire or an appliance. These receptacles are normally used to supply lighting and general-purpose electrical equipment and are in addition to the ones that are 5½ ft above the floor and within cupboards and cabinets.

According to listing requirements [see 110.3(B)], permanent electric baseboard heaters may not be located beneath wall receptacles. If the receptacle is part of the heater, appliance or lamp cords are less apt to be exposed to the heating elements, as might occur should the cords fall into convector slots. Many electric baseboard heaters are of the low-density type and are longer than 12 ft. To meet the spacing requirements of 210.52(A)(1), the required receptacle may be located as a part of the heater unit, as shown as Exhibit 210.23.

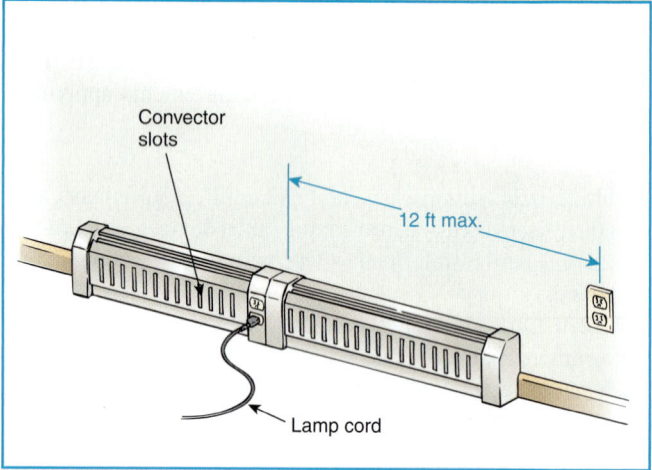

Exhibit 210.23 Permanent electric baseboard heater equipped with a receptacle outlet to meet the spacing requirements of 210.52(A).

(A) General Provisions. In every kitchen, family room, dining room, living room, parlor, library, den, sunroom, bedroom, recreation room, or similar room or area of dwelling units, receptacle outlets shall be installed in accordance with the general provisions specified in 210.52(A)(1) through (A)(3).

(1) Spacing. Receptacles shall be installed so that no point measured horizontally along the floor line in any wall space is more than 1.8 m (6 ft) from a receptacle outlet.

Receptacles are required to be located so that no point in any wall space is more than 6 ft from a receptacle. This rule intends that an appliance or lamp with a flexible cord attached may be placed anywhere in the room near a wall and be within 6 ft of a receptacle, thus eliminating the need for extension cords. Although not an enforceable requirement, receptacles may be placed equal distances apart where there is no specific room layout for the general use of electrical equipment. Section 210.52(A)(1) does not prohibit a receptacle layout designed for intended utilization equipment or practical room use. For example, receptacles in a living room, family room, or den that are intended to serve home entertainment equipment or home office equipment may be placed in corners, may be grouped, or may be placed in a convenient location. Receptacles that are intended for window-type holiday lighting may be placed under windows. In any event, even if more receptacles than the minimum are installed in a room, no point in any wall space is permitted to be more than 6 ft from a receptacle.

(2) Wall Space. As used in this section, a wall space shall include the following:

(1) Any space 600 mm (2 ft) or more in width (including space measured around corners) and unbroken along the floor line by doorways, fireplaces, and similar openings
(2) The space occupied by fixed panels in exterior walls, excluding sliding panels
(3) The space afforded by fixed room dividers such as freestanding bar-type counters or railings

A *wall space* is a wall unbroken along the floor line by doorways, fireplaces, archways, and similar openings and may include two or more walls of a room (around corners), as illustrated in Exhibit 210.24.

Fixed room dividers, such as bar-type counters and railings, are to be included in the 6 ft measurement. Fixed panels in exterior walls are counted as regular wall space, and a floor-type receptacle close to the wall can be used to meet the required spacing. Isolated, individual wall spaces 2 ft or more in width are considered usable for the location of a lamp or appliance, and a receptacle outlet is required to be provided.

(3) Floor Receptacles. Receptacle outlets in floors shall not be counted as part of the required number of receptacle outlets unless located within 450 mm (18 in.) of the wall.

(B) Small Appliances.

Section 210.52(B) requires two or more 20-ampere circuits for all receptacle outlets for the small-appliance loads, including refrigeration equipment, in the kitchen, dining room, pantry, and breakfast room of a dwelling unit. The countertop receptacle outlets in kitchens must be supplied by no fewer than two small-appliance branch circuits. These circuits may also supply receptacle outlets in the pantry, dining room,

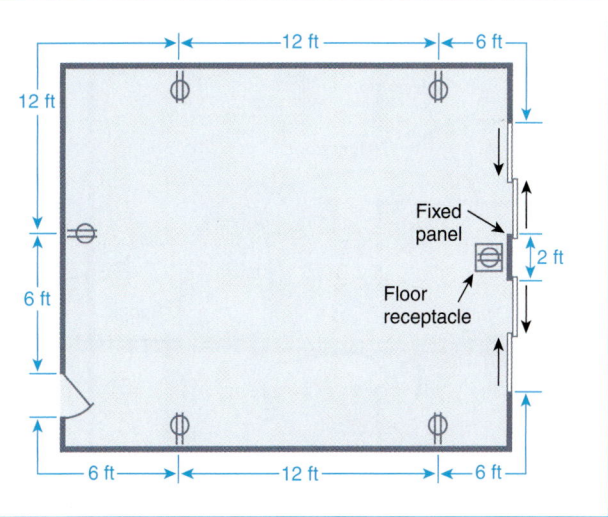

Exhibit 210.24 Typical room plan view of the location of dwelling unit receptacles meeting the requirements of 210.52(A).

and breakfast room, as well as an electric clock receptacle and electric loads associated with gas-fired appliances, but these circuits are to have no other outlets. See 210.8(A)(6) and (7) for GFCI requirements for receptacles serving countertop surfaces.

No restriction is placed on the number of outlets connected to a general-lighting or small-appliance branch circuit. The minimum number of receptacle outlets in a room is determined by 210.52(A) based on the room perimeter. It may be desirable to provide more than the minimum number of receptacle outlets required, thereby further reducing the need for extension cords.

Exhibit 210.25 illustrates the application of the requirements of 210.52(B)(1), (2), and (3). The small-appliance branch circuits illustrated are not permitted to serve any other outlets, such as might be connected to exhaust hoods or fans, disposals, or dishwashers. The countertop receptacles are also required to be supplied by these two circuits. Receptacles installed to serve countertop surfaces are required to be GFCI protected in accordance with 210.8(A)(6). The dining room switched receptacle on a 15-ampere general-purpose branch circuit is permitted according to 210.52(B)(1), Exception No. 1. The refrigerator receptacle supplied by a 15-ampere individual branch circuit (Exhibit 210.25, bottom) is permitted by 210.52(B)(1), Exception No. 2.

(1) Receptacle Outlets Served. In the kitchen, pantry, breakfast room, dining room, or similar area of a dwelling unit, the two or more 20-ampere small-appliance branch circuits required by 210.11(C)(1) shall serve all receptacle outlets covered by 210.52(A) and (C) and receptacle outlets for refrigeration equipment.

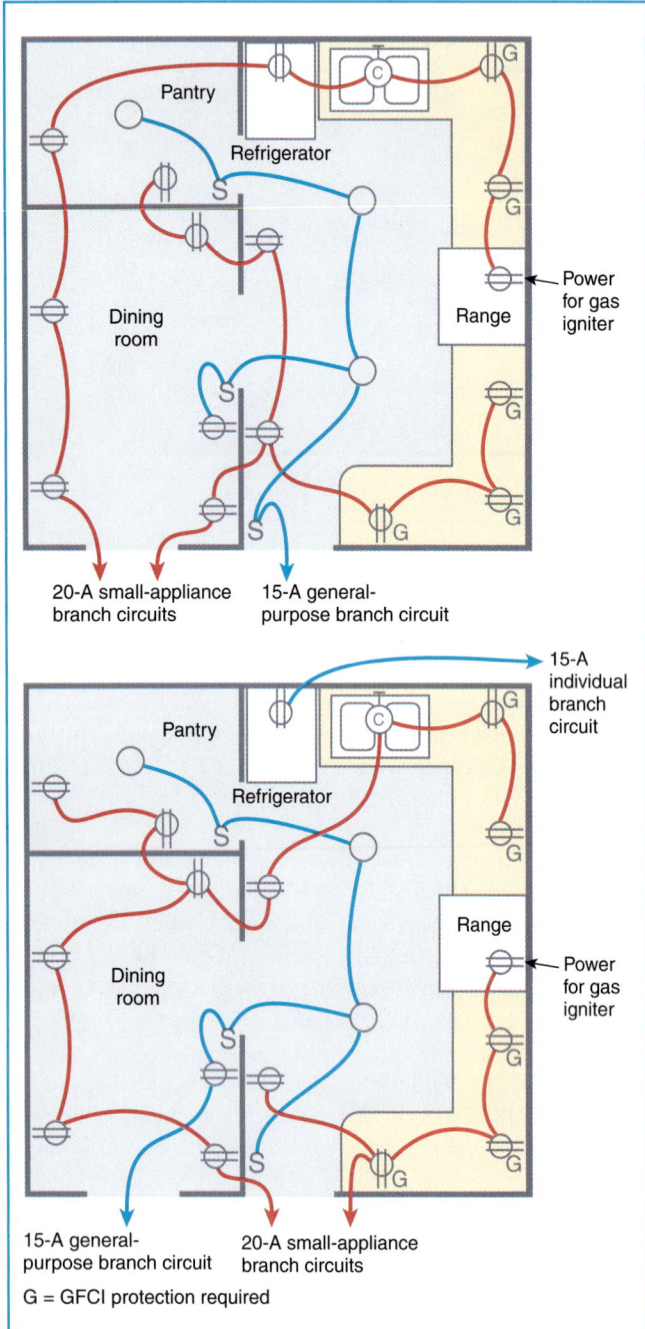

20-A small-appliance branch circuits

15-A general-purpose branch circuit

Pantry

Refrigerator

Dining room

Range

Power for gas igniter

15-A individual branch circuit

Pantry

Refrigerator

Dining room

Range

Power for gas igniter

15-A general-purpose branch circuit

20-A small-appliance branch circuits

G = GFCI protection required

Exception No. 1: In addition to the required receptacles specified by 210.52, switched receptacles supplied from a general-purpose branch circuit as defined in 210.70(A)(1), Exception No. 1, shall be permitted.

Exception No. 1 to 210.52(B)(1) permits switched receptacles supplied from general-purpose 15-ampere branch cir-

cuits to be located in kitchens, pantries, breakfast rooms, and similar areas. See 210.70(A) and Exhibit 210.25 for details.

Exception No. 2: The receptacle outlet for refrigeration equipment shall be permitted to be supplied from an individual branch circuit rated 15 amperes or greater.

Exception No. 2 to 210.52(B)(1) allows a choice for refrigeration equipment receptacle outlets located in a kitchen or similar area. An individual 15-ampere or larger branch circuit may serve this equipment, or it may be included in the 20-ampere small-appliance branch circuit. Refrigeration equipment is also exempt from the GFCI requirements of 210.8. See Exhibit 210.25 for an illustration.

(2) No Other Outlets. The two or more small-appliance branch circuits specified in 210.52(B)(1) shall have no other outlets.

Exception No. 1: A receptacle installed solely for the electrical supply to and support of an electric clock in any of the rooms specified in 210.52(B)(1).

Exception No. 2: Receptacles installed to provide power for supplemental equipment and lighting on gas-fired ranges, ovens, or counter-mounted cooking units.

Exception No. 2 to 210.52(B)(2) allows the small electrical loads associated with gas-fired appliances to be connected to small-appliance branch circuits. See Exhibit 210.25 for an illustration.

(3) Kitchen Receptacle Requirements. Receptacles installed in a kitchen to serve countertop surfaces shall be supplied by not fewer than two small-appliance branch circuits, either or both of which shall also be permitted to supply receptacle outlets in the same kitchen and in other rooms specified in 210.52(B)(1). Additional small-appliance branch circuits shall be permitted to supply receptacle outlets in the kitchen and other rooms specified in 210.52(B)(1). No small-appliance branch circuit shall serve more than one kitchen.

(C) Countertops. In kitchens and dining rooms of dwelling units, receptacle outlets for counter spaces shall be installed in accordance with 210.52(C)(1) through (5).

(1) Wall Counter Spaces. A receptacle outlet shall be installed at each wall counter space that is 300 mm (12 in.) or wider. Receptacle outlets shall be installed so that no point along the wall line is more than 600 mm (24 in.) measured horizontally from a receptacle outlet in that space.

(2) Island Counter Spaces. At least one receptacle outlet shall be installed at each island counter space with a long dimension of 600 mm (24 in.) or greater and a short dimension of 300 mm (12 in.) or greater.

(3) Peninsular Counter Spaces. At least one receptacle outlet shall be installed at each peninsular counter space with a long dimension of 600 mm (24 in.) or greater and a short dimension of 300 mm (12 in.) or greater. A peninsular countertop is measured from the connecting edge.

(4) Separate Spaces. Countertop spaces separated by range tops, refrigerators, or sinks shall be considered as separate countertop spaces in applying the requirements of 210.52(C)(1), (2), and (3).

(5) Receptacle Outlet Location. Receptacle outlets shall be located above, but not more than 500 mm (20 in.) above, the countertop. Receptacle outlets rendered not readily accessible by appliances fastened in place, appliance garages, or appliances occupying dedicated space shall not be considered as these required outlets.

Exception: To comply with the conditions specified in (a) or (b), receptacle outlets shall be permitted to be mounted not more than 300 mm (12 in.) below the countertop. Receptacles mounted below a countertop in accordance with this exception shall not be located where the countertop extends more than 150 mm (6 in.) beyond its support base.

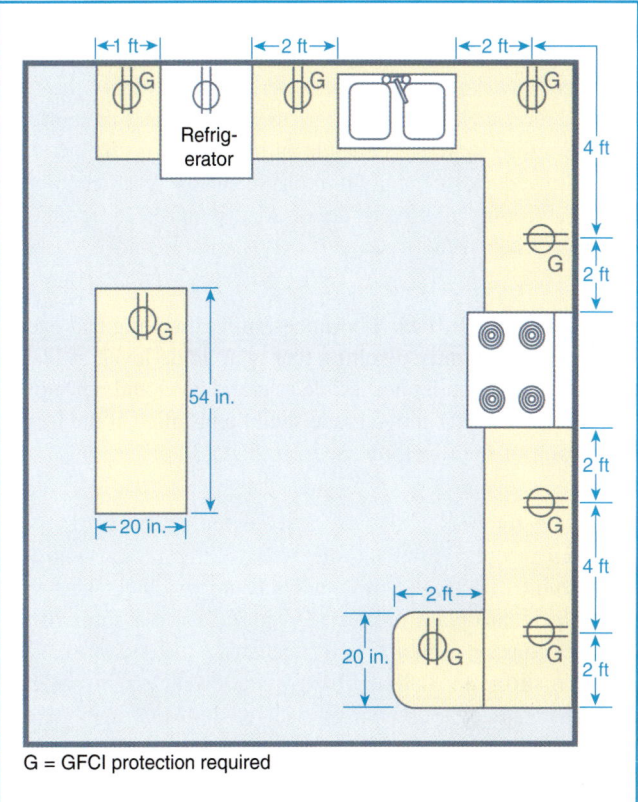

Exhibit 210.26 Dwelling unit receptacles serving countertop spaces in a kitchen and installed in accordance with 210.52(C).

Dwelling unit receptacles that serve countertop spaces in kitchens, dining areas, and similar rooms, as illustrated in Exhibit 210.26, are required to be installed as follows:

1. In each wall space wider than 12 in. and spaced so that no point along the wall line is more than 24 in. from a receptacle
2. Not more than 20 in. above the countertop [According to 406.4(E), receptacles cannot be installed in a face-up position. Receptacles installed in a face-up position in a countertop could collect crumbs, liquids, and other debris, resulting in a potential fire or shock hazard.]
3. At each countertop island and peninsular countertop with a short dimension of at least 12 in. and a long dimension of at least 24 in. (The measurement of a peninsular-type countertop is from the edge connecting to the nonpeninsular counter.)
4. Accessible for use and not blocked by appliances occupying dedicated space or fastened in place
5. Fed from two or more of the required 20-ampere small-appliance branch circuits and GFCI protected according to 210.8(A)(6)

For the 2002 *Code,* the maximum permitted height of a receptacle serving a countertop was revised in 210.52(C)(5) upward from 18 in. to 20 in. as a practical consideration.

(a) *Construction for the physically impaired.*
(b) *On island and peninsular countertops where the countertop is flat across its entire surface (no backsplashes, dividers, etc.) and there are no means to mount a receptacle within 500 mm (20 in.) above the countertop, such as an overhead cabinet.*

(D) Bathrooms. In dwelling units, at least one wall receptacle outlet shall be installed in bathrooms within 900 mm (3 ft) of the outside edge of each basin. The receptacle outlet shall be located on a wall or partition that is adjacent to the basin or basin countertop.

Section 210.52(D) requires one wall receptacle in each bathroom of a dwelling unit to be installed adjacent (within 36 in.) to the washbasin. This receptacle is required in addition to any receptacle that may be part of any luminaire or medicine cabinet. If there is more than one washbasin, a receptacle outlet is required adjacent to each basin location. If the basins are in close proximity, one receptacle outlet installed between the two basins might satisfy this requirement. See 410.57(D), which prohibits installation of a receptacle inside bathtub and shower spaces. See Exhibit 210.9 for a sample electrical layout of a bathroom.

Section 210.11(C)(3) requires the receptacle outlets to be supplied from a 20-ampere branch circuit with no other outlets. However, this circuit is permitted to supply the required receptacles in more than one bathroom. If the circuit supplies the required receptacle outlet in only one bathroom, it is allowed to also supply lighting and an exhaust fan within that bathroom. This receptacle is also required to be GFCI protected according to 210.8(A)(1).

(E) Outdoor Outlets. For a one-family dwelling and each unit of a two-family dwelling that is at grade level, at least one receptacle outlet accessible at grade level and not more than 2.0 m (6½ ft) above grade shall be installed at the front and back of the dwelling. See 210.8(A)(3).

The rule for one- and two-family dwellings requires two outdoor receptacle outlets for each dwelling unit, as shown in Exhibit 210.27. When installing outdoor receptacles the receptacle faceplate must rest securely on the supporting surface to prevent moisture from entering the enclosure. On uneven surfaces, such as brick, stone, or stucco, it may be necessary to close openings with caulking compound or mastic. See 406.8 for further information on receptacles installed in damp or wet locations.

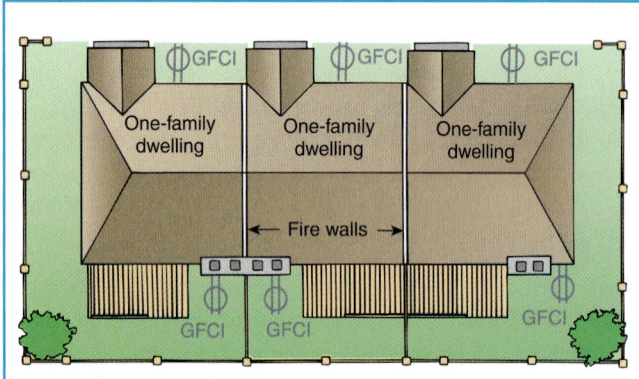

Exhibit 210.27 Row housing with GFCI-protected receptacles located, as required by 210.52(E), at the front and back of each one-family dwelling.

(F) Laundry Areas. In dwelling units, at least one receptacle outlet shall be installed for the laundry.

Exception No. 1: In a dwelling unit that is an apartment or living area in a multifamily building where laundry facilities are provided on the premises and are available to all building occupants, a laundry receptacle shall not be required.

Exception No. 2: In other than one-family dwellings where laundry facilities are not to be installed or permitted, a laundry receptacle shall not be required.

A laundry receptacle outlet(s) is supplied by a 20-ampere branch circuit. This circuit can have no other outlets. See 210.11(C)(2) for further information.

(G) Basements and Garages. For a one-family dwelling, at least one receptacle outlet, in addition to any provided for laundry equipment, shall be installed in each basement and in each attached garage, and in each detached garage with electric power. See 210.8(A)(2) and (A)(5). Where a portion of the basement is finished into one or more habitable rooms, each separate unfinished portion shall have a receptacle outlet installed in accordance with this section.

In a one-family dwelling, a receptacle must be installed in the basement (in addition to the laundry receptacle), in each attached garage, and in each detached garage with electric power.

Section 210.8(A)(5) requires receptacles in unfinished basements to be protected by GFCIs. Section 210.8(A)(2) requires receptacles that are installed in garages to be protected by GFCIs. If no electrical power is provided to detached garages, receptacles do not have to be installed.

(H) Hallways. In dwelling units, hallways of 3.0 m (10 ft) or more in length shall have at least one receptacle outlet.

As used in this subsection, the hall length shall be considered the length along the centerline of the hall without passing through a doorway.

The requirement in 210.52(H) is intended to minimize strain or damage to cords and receptacles for dwelling unit receptacles. The requirement does not apply to common hallways of hotels, motels, apartment buildings, condominiums, and so on.

210.60 Guest Rooms.

(A) General. Guest rooms in hotels, motels, and similar occupancies shall have receptacle outlets installed in accordance with 210.52(A) and 210.52(D). Guest rooms meeting the definition of a dwelling unit shall have receptacle outlets installed in accordance with all of the applicable rules in 210.52.

(B) Receptacle Placement. In applying the provisions of 210.52(A), the total number of receptacle outlets shall not be less than the minimum number that would comply with the provisions of that section. These receptacle outlets shall be permitted to be located conveniently for permanent furniture layout. At least two receptacle outlets shall be readily accessible. Where receptacles are installed behind the bed, the receptacle shall be located to prevent the bed from contacting any attachment plug that may be installed, or the receptacle shall be provided with a suitable guard.

Section 210.60(B) permits the receptacles in guest rooms of hotels and motels to be placed in accessible locations that are compatible with permanent furniture. However, the minimum number of receptacles required by 210.52 is not permitted to be reduced. The minimum number of receptacle outlets should be determined by assuming there is no furniture in the room. The practical locations of that minimum number of receptacles are then determined based on the permanent furniture layout.

Hotel and motel rooms are commonly used as remote offices for businesspeople who use laptop computers and other plug-in devices. The *Code* requires *two* receptacle outlets to be available without moving furniture to access those receptacles. To reduce the risk of fire to bedding material, receptacles located behind beds must include guards if attachment plugs might contact the bed.

Bathroom areas for guest rooms in hotels and motels are required to be provided with a receptacle outlet adjacent to the basin location, in accordance with 210.8(B)(1).

Exhibit 210.28 shows receptacle outlets in a hotel guest room located conveniently with respect to the permanent furniture layout. Some spaces that are 2 ft or more in width have no receptacle outlets because 210.60(B) permits the required number of outlets to be placed in convenient locations that are compatible with the permanent furniture layout. In Exhibit 210.28, the receptacle outlet adjacent to the permanent dresser is needed because 210.60(B) applies only to the location of receptacle outlets, not to the minimum number of receptacle outlets.

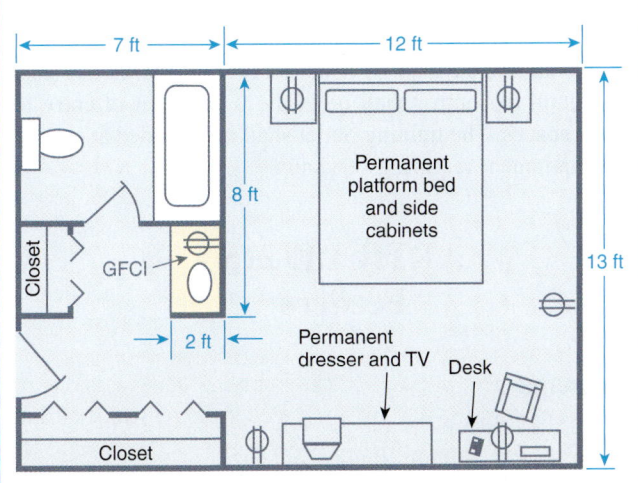

Exhibit 210.28 Floor plan of a hotel guest room with receptacles located as permitted by 210.60(B) with respect to permanent furniture.

210.62 Show Windows.

At least one receptacle outlet shall be installed directly above a show window for each 3.7 linear m (12 ft) or major fraction thereof of show window area measured horizontally at its maximum width.

Show windows usually extend from floor to ceiling for maximum display. To discourage floor receptacles and unsightly extension cords likely to cause physical injury, receptacles must be installed directly above a show window, and one receptacle is required for every 12 linear ft or "major fraction thereof" (6 ft or more). See 220.3(B)(7) and 220.12(A) for information regarding load computations for show windows.

210.63 Heating, Air-Conditioning, and Refrigeration Equipment Outlet.

A 125-volt, single-phase, 15- or 20-ampere-rated receptacle outlet shall be installed at an accessible location for the servicing of heating, air-conditioning, and refrigeration equipment. The receptacle shall be located on the same level and within 7.5 m (25 ft) of the heating, air-conditioning, and refrigeration equipment. The receptacle outlet shall not be connected to the load side of the equipment disconnecting means.

> FPN: See 210.8 for ground-fault circuit-interrupter requirements.

Section 210.63 is intended to prevent makeshift methods of obtaining 125-volt power for servicing and troubleshooting heating, air-conditioning, and refrigeration equipment. The reference to 210.8 in the fine print note to 210.63 reminds the *Code* user of the GFCI requirements for these receptacle outlets. The requirements in 210.52(E) for outdoor dwelling unit receptacles located within 25 ft of this equipment will meet the requirements of 210.63.

The requirements of 210.63 were expanded for the 2002 *Code* to improve worker safety. As a result, a receptacle outlet is now required for troubleshooting heating, air-conditioning, and refrigeration equipment at grade-accessible outdoor equipment and at rooftop units associated with one- and two-family dwelling units.

210.70 Lighting Outlets Required.

Lighting outlets shall be installed where specified in 210.70(A), (B), and (C).

(A) Dwelling Units. In dwelling units, lighting outlets shall be installed in accordance with 210.70(A)(1), (2), and (3).

(1) Habitable Rooms. At least one wall switch-controlled lighting outlet shall be installed in every habitable room and bathroom.

Exception No. 1: In other than kitchens and bathrooms, one or more receptacles controlled by a wall switch shall be permitted in lieu of lighting outlets.

A receptacle is not permitted to be switched as a lighting outlet on a small-appliance branch circuit. A receptacle can be switched as a lighting outlet (in the dining room, for example) supplied by a branch circuit other than a small-appliance branch circuit. See Exhibit 210.26, which shows a dining room switched receptacle on a 15-ampere general-purpose branch circuit.

Exception No. 2: Lighting outlets shall be permitted to be controlled by occupancy sensors that are (1) in addition to wall switches or (2) located at a customary wall switch location and equipped with a manual override that will allow the sensor to function as a wall switch.

(2) Additional Locations. Additional lighting outlets shall be installed in accordance with (a), (b), and (c).

(a) At least one wall switch-controlled lighting outlet shall be installed in hallways, stairways, attached garages, and detached garages with electric power.

(b) For dwelling units, attached garages, and detached garages with electric power, at least one wall switch–controlled lighting outlet shall be installed to provide illumination on the exterior side of outdoor entrances or exits with grade level access. A vehicle door in a garage shall not be considered as an outdoor entrance or exit.

(c) Where one or more lighting outlet(s) are installed for interior stairways, there shall be a wall switch at each floor level, and landing level that includes an entry way, to control the lighting outlet(s) where the stairway between floor levels has six risers or more.

Exception to (a), (b), and (c): In hallways, stairways, and at outdoor entrances, remote, central, or automatic control of lighting shall be permitted.

Section 210.70 points out that adequate lighting and proper control and location of switching are as essential to the safety of occupants of dwelling units, hotels, motels, and so on, as are proper wiring requirements. Proper illumination ensures safe movement for persons of all ages, thus preventing many accidents.

Although the requirement in 210.7(A)(2)(b) calls for a switched lighting outlet at outdoor entrances and exits, it does not prohibit a single lighting outlet, if suitably located, from serving more than one door.

A wall switch-controlled lighting outlet is required in the kitchen and bathroom. A receptacle outlet controlled by a wall switch is not permitted to serve as a lighting outlet in these rooms. Occupancy sensors are permitted to be used for switching these lighting outlets, provided they are equipped with a manual override or are used in addition to regular switches.

(3) Storage or Equipment Spaces. For attics, underfloor spaces, utility rooms, and basements, at least one lighting outlet containing a switch or controlled by a wall switch shall be installed where these spaces are used for storage or contain equipment requiring servicing. At least one point of control shall be at the usual point of entry to these spaces. The lighting outlet shall be provided at or near the equipment requiring servicing.

Installation of lighting outlets in attics, underfloor spaces or crawl areas, utility rooms, and basements is required when these spaces are used for storage (e.g., holiday decorations or luggage).

If such spaces contain equipment that requires servicing (e.g., air-handling units, cooling and heating equipment, water pumps, and sump pumps), 210.70(C) requires that a lighting outlet be installed in these spaces.

(B) Guest Rooms. At least one wall switch–controlled lighting outlet or wall switch–controlled receptacle shall be installed in guest rooms in hotels, motels, or similar occupancies.

(C) Other Than Dwelling Units. For attics and underfloor spaces containing equipment requiring servicing, such as heating, air-conditioning, and refrigeration equipment, at least one lighting outlet containing a switch or controlled by a wall switch shall be installed in such spaces. At least one point of control shall be at the usual point of entry to these spaces. The lighting outlet shall be provided at or near the equipment requiring servicing.

ARTICLE 215
Feeders

Contents

215.1 Scope.

This article covers the installation requirements, overcurrent protection requirements, minimum size, and ampacity of conductors for feeders supplying branch-circuit loads as computed in accordance with Article 220.

Exception: Feeders for electrolytic cells as covered in 668.3(C)(1) and (4).

The scope statement for Article 215 includes a reference to overcurrent protection requirements for feeders, which encompasses both the fact that Article 215 references Article 240 for overcurrent protection and the fact that 215.3 includes the 125 percent sizing rules for overcurrent devices.

The total connected load to be supplied by the feeder must be calculated to accurately determine feeder conductor ampacity. The sum of the computed and connected loads supplied by a feeder is multiplied by the demand factor to determine the load that the feeder conductors must be sized to serve. See Article 100 for the definition of *demand factor*.

When the total connected load is operated simultaneously, the demand factor is 100 percent; that is, the maximum demand is equal to the total connected load. Due to diversity, the maximum operating load carried at any time may be only three-quarters of the total connected load; thus, the demand factor is 75 percent.

On a new installation, a minimum value for the demand factor can be determined by applying the requirements and tables of Article 220, Branch-Circuit, Feeder, and Service Calculations.

Feeder conductor sizes are determined by calculating the total volt-amperes (VA) of the feeder load at the nominal voltage of the feeder circuit. See 220.2 for the nominal system voltages used in computing branch-circuit and feeder loads. See 310.15 for allowable ampacities and sizes of insulated conductors.

Feeder circuits must have sufficient ampacity for safety. Wiring systems that do not provide for increases in the use of electricity often create hazards. It is a good practice to allow for future expansion and convenience increases, as stated in 90.8(A).

215.2 Minimum Rating and Size.

(A) Feeders Not More Than 600 Volts.

(1) General. Feeder conductors shall have an ampacity not less than required to supply the load as computed in Parts II, III, and IV of Article 220. The minimum feeder-circuit conductor size, before the application of any adjustment or correction factors, shall have an allowable ampacity not less than the noncontinuous load plus 125 percent of the continuous load.

Exception: Where the assembly, including the overcurrent devices protecting the feeder(s), is listed for operation at 100 percent of its rating, the allowable ampacity of the feeder conductors shall be permitted to be not less than the sum of the continuous load plus the noncontinuous load.

Additional minimum sizes shall be as specified in (2), (3), and (4) under the conditions stipulated.

(2) For Specified Circuits. The ampacity of feeder conductors shall not be less than 30 amperes where the load supplied consists of any of the following number and types of circuits:

(1) Two or more 2-wire branch circuits supplied by a 2-wire feeder
(2) More than two 2-wire branch circuits supplied by a 3-wire feeder
(3) Two or more 3-wire branch circuits supplied by a 3-wire feeder
(4) Two or more 4-wire branch circuits supplied by a 3-phase, 4-wire feeder

(3) Ampacity Relative to Service-Entrance Conductors. The feeder conductor ampacity shall not be less than that of the service-entrance conductors where the feeder conductors carry the total load supplied by service-entrance conductors with an ampacity of 55 amperes or less.

(4) Individual Dwelling Unit or Mobile Home Conductors. Feeder conductors for individual dwelling units or mobile homes need not be larger than service-entrance conductors. Paragraph 310.15(B)(6) shall be permitted to be used for conductor size.

For example, according to Table 310.16, a 3/0 AWG, Type THW copper wire has an ampacity of 200 amperes. However, for a 3-wire, single-phase dwelling service, as shown in Exhibit 215.1, Table 310.15(B)(6) permits 2/0 AWG, Type THW copper conductors or 4/0 AWG, Type THW aluminum conductors for services or feeders rated at 200 amperes. Feeder conductors carrying the total load supplied by the service are not required to be sized larger than the service-entrance conductors.

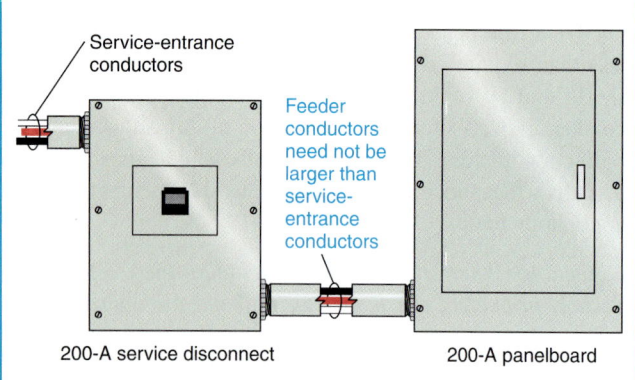

Service-entrance conductors

Feeder conductors need not be larger than service-entrance conductors

200-A service disconnect

200-A panelboard

Exhibit 215.1 A 3-wire, single-phase dwelling service with an ampacity of 200 amperes for 2/0 AWG copper or 4/0 AWG aluminum conductors used as service-entrance conductors and feeder conductors, according to 215.2(A)(4).

FPN No. 1: See Examples D1 through D10 in Annex D.

FPN No. 2: Conductors for feeders as defined in Article 100, sized to prevent a voltage drop exceeding 3 percent at the farthest outlet of power, heating, and lighting loads, or combinations of such loads, and where the maximum total voltage drop on both feeders and branch circuits to the farthest outlet does not exceed 5 percent, will provide reasonable efficiency of operation.

FPN No. 3: See 210.19(A), FPN No. 4, for voltage drop for branch circuits.

Reasonable operating efficiency is achieved if the voltage drop of a feeder or the voltage drop of a branch circuit is limited to 3 percent. However, the total voltage drop of a branch circuit plus a feeder can reach 5 percent and still achieve reasonable operating efficiency. See Article 100 for definitions of *feeder* and *branch circuit*.

The 5 percent voltage-drop value is explanatory material and, as such, appears as a fine print note. Fine print notes are not mandatory (see 90.5). However, where circuit conductors are increased due to voltage drop, 250.122(B) requires an increase in circular mil area for the associated equipment grounding conductors.

The resistance or impedance of conductors may cause a substantial difference between voltage at service equipment and voltage at the point-of-utilization equipment. Excessive voltage drop impairs the starting and the operation of electrical equipment. Undervoltage can result in inefficient operation of heating, lighting, and motor loads. An applied voltage of 10 percent below rating can result in a decrease in effi-

ciency of substantially more than 10 percent. For example, fluorescent light output would be reduced by 15 percent, and incandescent light output would be reduced by 30 percent. Induction motors would run hotter and produce less torque. With an applied voltage of 10 percent below rating, the running current would increase 11 percent, and the operating temperature would increase by 12 percent. At the same time, torque would be reduced by 19 percent.

In addition to resistance or impedance, the type of raceway or cable enclosure, the type of circuit (ac, dc, single-phase, 3-phase), and the power factor should be considered to determine voltage drop.

The following basic formula can be used to determine the voltage drop in a 2-wire dc circuit, a 2-wire ac circuit, or a 3-wire ac single-phase circuit, all with a balanced load at 100 percent power factor and where reactance can be neglected.

$$VD = \frac{2 \times L \times R \times I}{1000}$$

where:

VD = voltage drop (based on conductor temperature of 75°F)

L = one-way length of circuit (ft)

R = conductor resistance in ohms (Ω) per thousand feet (from Chapter 9, Table 8)

I = load current (amperes)

For 3-phase circuits (at 100 percent power factor), the voltage drop between any two phase conductors is 0.866 times the voltage drop calculated by the preceding formula.

Example

Determine the voltage drop in a 240-volt, 2-wire heating circuit with a load of 50 amperes. The circuit size is 6 AWG, Type THHN copper, and the one-way circuit length is 100 ft.

Solution

STEP 1. Find the conductor resistance from Table 8 in Chapter 9.

STEP 2. Substitute values into the following voltage-drop formula:

$$VD = \frac{2 \times L \times R \times I}{1000}$$

$$= \frac{2 \times 100 \times 0.491 \times 50}{1000} = 4.91 \text{ volts}$$

STEP 3. Determine the percentage of the voltage drop:

$$\% \, VD = \frac{4.91 \text{ V}}{240 \text{ V}} = 0.02 \quad \text{or} \quad 2\%$$

A 12-volt drop on a 240-volt circuit is a 5 percent drop. A 4.91-volt drop falls within this percentage. If the total voltage drop exceeds 5 percent, or 12 volts, larger-size conductors should be used, the circuit length should be shortened, or the circuit load should be reduced.

See the commentary following Chapter 9, Table 9, for an example of voltage-drop calculation using ac reactance and resistance. Voltage-drop tables and calculations are also available from various manufacturers.

(B) Feeders Over 600 Volts. The ampacity of conductors shall be in accordance with 310.15 and 310.60 as applicable. Feeder conductors over 600 volts shall be sized in accordance with 215.2(B)(1), (2), or (3).

(1) Feeders Supplying Transformers. The ampacity of feeder conductors shall not be less than the sum of the nameplate ratings of the transformers supplied when only transformers are supplied.

(2) Feeders Supplying Transformers and Utilization Equipment. The ampacity of feeders supplying a combination of transformers and utilization equipment shall not be less than the sum of the nameplate ratings of the transformers and 125 percent of the designed potential load of the utilization equipment that will be operated simultaneously.

(3) Supervised Installations. For supervised installations, feeder conductor sizing shall be permitted to be determined by qualified persons under engineering supervision. Supervised installations are defined as those portions of a facility where all of the following conditions are met:

(1) Conditions of design and installation are provided under engineering supervision.
(2) Qualified persons with documented training and experience in over 600-volt systems provide maintenance, monitoring, and servicing of the system.

Section 215.2(B) is new for the 2002 *NEC*. It sets the minimum requirements for feeders over 600 volts. Unless the circuit is part of a supervised installation [defined in 215.2(B)(3)], the minimum ampacity for feeder circuit conductors over 600 volts can be no less than 100 percent of the transformer nameplate load plus 125 percent of any additional utilization equipment. The overcurrent protection requirements for feeders over 600 volts must be in accordance with Article 240, Part IX.

215.3 Overcurrent Protection.

Feeders shall be protected against overcurrent in accordance with the provisions of Part I of Article 240. Where a feeder supplies continuous loads or any combination of continuous and noncontinuous loads, the rating of the overcurrent device shall not be less than the noncontinuous load plus 125 percent of the continuous load.

Exception No. 1: Where the assembly, including the overcurrent devices protecting the feeder(s), is listed for operation at 100 percent of its rating, the ampere rating of the overcurrent device shall be permitted to be not less than the sum of the continuous load plus the noncontinuous load.

The feeder overcurrent protection requirements in 215.3 are somewhat similar to the branch-circuit overcurrent protection requirements in 210.20(A).

Exception No. 2: Overcurrent protection for feeders over 600 volts, nominal, shall comply with Part IX of Article 240.

215.4 Feeders with Common Neutral.

(A) Feeders with Common Neutral. Two or three sets of 3-wire feeders or two sets of 4-wire or 5-wire feeders shall be permitted to utilize a common neutral.

(B) In Metal Raceway or Enclosure. Where installed in a metal raceway or other metal enclosure, all conductors of all feeders using a common neutral shall be enclosed within the same raceway or other enclosure as required in 300.20.

If feeder conductors carrying ac current, including the neutral, are installed in metal raceways, the conductors are required to be grouped together to avoid induction heating of the surrounding metal. If it is necessary to run parallel conductors through multiple metal raceways, conductors from each phase plus the neutral must be run in each raceway. See 250.102(E), 250.134(B), 300.3, 300.5(I), and 300.20 for requirements associated with conductor grouping of feeder circuits.

A 3-phase, 4-wire (208Y/120-volt, 480Y/277-volt) system is often used to supply both lighting and motor loads. The 3-phase motor loads are typically not connected to the neutral and, therefore, will not cause current to flow in the neutral conductor. The maximum current on the neutral, therefore, is due to lighting loads or circuits where the neutral is used. On this type of system (3-phase, 4-wire), a demand factor of 70 percent is permitted by 220.22 for that portion of the neutral load in excess of 200 amperes.

For example, if the maximum possible unbalanced load is 500 amperes, the neutral would have to be large enough to carry 410 amperes (200 amperes plus 70 percent of 300 amperes, or 410 amperes). No reduction of the neutral capacity for that portion of the load consisting of electric-discharge lighting is permitted.

Section 310.15(B)(4)(c) points out that a neutral conductor must be counted as a current-carrying conductor if the load it serves consists of harmonic currents. See 220.22 for other systems in which the 70 percent demand factor may be applied. The maximum unbalanced load for feeders supplying clothes dryers, household ranges, wall-mounted ovens, and counter-mounted cooking units is required to be considered 70 percent of the load on the ungrounded conductors. See Examples D1(a) through D5(b) of Annex D.

215.5 Diagrams of Feeders.

If required by the authority having jurisdiction, a diagram showing feeder details shall be provided prior to the installation of the feeders. Such a diagram shall show the area in square feet of the building or other structure supplied by each feeder, the total computed load before applying demand factors, the demand factors used, the computed load after applying demand factors, and the size and type of conductors to be used.

215.6 Feeder Conductor Grounding Means.

Where a feeder supplies branch circuits in which equipment grounding conductors are required, the feeder shall include or provide a grounding means, in accordance with the provisions of 250.134, to which the equipment grounding conductors of the branch circuits shall be connected.

215.7 Ungrounded Conductors Tapped from Grounded Systems.

Two-wire dc circuits and ac circuits of two or more ungrounded conductors shall be permitted to be tapped from the ungrounded conductors of circuits having a grounded neutral conductor. Switching devices in each tapped circuit shall have a pole in each ungrounded conductor.

Section 215.7 does not require a common trip or simultaneous opening of circuit breakers or fuses, but rather requires a switching device to manually disconnect the ungrounded conductors of the feeder. See 210.10 for similar requirements related to the ungrounded conductors of the branch circuit.

215.8 Means of Identifying Conductor with the Higher Voltage to Ground.

On a 4-wire, delta-connected secondary where the midpoint of one phase winding is grounded to supply lighting and similar loads, the phase conductor having the higher voltage to ground shall be identified by an outer finish that is orange in color or by tagging or other effective means. Such identification shall be placed at each point where a connection is made if the grounded conductor is also present.

Where the midpoint of one phase winding is grounded in order to supply 120-volt lighting and similar loads from a

delta-connected, 3-phase secondary, one phase conductor will have a higher voltage to ground. An orange finish, orange tape, or other effective means identifies this phase conductor at any point, such as junction or pull boxes or panelboards, where connections may be made and the grounded conductor is also present. The orange high leg of a 3-phase, 4-wire 240/120-volt delta system is 208 volts to ground (120 volts multiplied by 1.73 equals 208 volts) and should obviously not be used for 120-volt circuits. See 110.15, 230.56, and 408.3(E) for details on high-leg marking and phase arrangement. The conductor with the higher voltage to ground is identified in Exhibit 215.2 as having an orange finish. The identification must be visible at every point where a connection is made if the grounded conductor (neutral) is present.

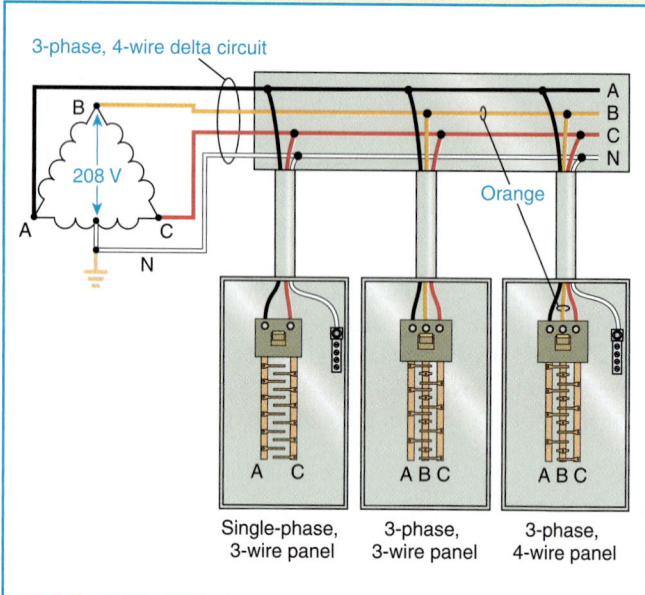

Exhibit 215.2 Identification of the high-leg (orange) conductor of a 3-phase, 4-wire, 240/120-volt delta circuit.

215.9 Ground-Fault Circuit-Interrupter Protection for Personnel.

Feeders supplying 15- and 20-ampere receptacle branch circuits shall be permitted to be protected by a ground-fault circuit interrupter in lieu of the provisions for such interrupters as specified in 210.8 and Article 527.

Several manufacturers offer double-pole 120/240-volt circuit-breaker-type ground-fault circuit interrupters (GFCIs) for application to a feeder, thereby protecting all branch circuits supplied by that feeder. This type of GFCI installation is in lieu of provisions of 210.8 for outdoor, bathroom, garage, kitchen, basement, and boathouse receptacles. GFCI protection in the feeder can also be used to protect construction-site receptacles, as covered in 527.6(A), provided that the feeder supplies no lighting branch circuits.

It may be more economical or convenient to install GFCIs for feeders. However, consideration should be given to the fact that a ground-fault circuit interrupter may be monitoring several branch circuits and will de-energize all branch circuits in response to a line-to-ground fault from one branch circuit. As stated in 90.1(B), the installation may be "free from hazard but not necessarily efficient, convenient, or adequate for good service" where GFCIs are substituted for GFP-type protection.

215.10 Ground-Fault Protection of Equipment.

Each feeder disconnect rated 1000 amperes or more and installed on solidly grounded wye electrical systems of more than 150 volts to ground, but not exceeding 600 volts phase-to-phase, shall be provided with ground-fault protection of equipment in accordance with the provisions of 230.95.

Exception No. 1: The provisions of this section shall not apply to a disconnecting means for a continuous industrial process where a nonorderly shutdown will introduce additional or increased hazards.

Exception No. 2: The provisions of this section shall not apply to fire pumps.

Exception No. 3: The provisions of this section shall not apply if ground-fault protection of equipment is provided on the supply side of the feeder.

The intent of 215.10 is to require ground-fault protection of equipment for feeder disconnects that are rated 1000 amperes or more at 480Y/277 volts. A similar requirement for services is found in 230.95. The reason for the requirement is the unusually high number of burndowns reported on feeders and services in this voltage range. Prior to being put into service, each ground-fault protection system must be performance tested and documented according to the requirements of 230.95(C).

It should be noted that ground-fault protection of feeder equipment is not required if protection is provided on an upstream feeder or at the service. However, additional levels of ground-fault protection on feeders may be desired so that a single ground fault does not de-energize the whole electrical system. See 230.95 for further commentary on ground-fault protection of services. Also, see 517.17, which requires an additional level of ground-fault protection for health care facilities.

For emergency feeders, according to Article 700, the ground-fault protection requirements are different. See 700.26 for further details.

215.11 Circuits Derived from Autotransformers.

Feeders shall not be derived from autotransformers unless the system supplied has a grounded conductor that is electrically connected to a grounded conductor of the system supplying the autotransformer.

Exception No. 1: An autotransformer shall be permitted without the connection to a grounded conductor where transforming from a nominal 208 volts to a nominal 240-volt supply or similarly from 240 volts to 208 volts.

Exception No. 2: In industrial occupancies, where conditions of maintenance and supervision ensure that only qualified persons service the installation, autotransformers shall be permitted to supply nominal 600-volt loads from nominal 480-volt systems, and 480-volt loads from nominal 600-volt systems, without the connection to a similar grounded conductor.

Section 215.11 addresses autotransformers for feeders and is similar to the requirements in 210.9 for branch circuits. See the commentary following 210.9 for further information on autotransformers that supply branch circuits.

ARTICLE 220
Branch-Circuit, Feeder, and Service Calculations

Contents

I. General

220.1 Scope.

This article provides requirements for computing branch-circuit, feeder, and service loads.

Exception: Branch-circuit and feeder calculations for electrolytic cells as covered in 668.3(C)(1) and (4).

The scope of Article 220 was previously revised to clearly indicate that the article deals with load computation for branch-circuit, feeder, and service loads and not with requirements for determining the number of branch circuits needed. The determination for the number of branch circuits is contained in Article 210.

220.2 Computations.

(A) Voltages. Unless other voltages are specified, for purposes of computing branch-circuit and feeder loads, nominal system voltages of 120, 120/240, 208Y/120, 240, 347, 480Y/277, 480, 600Y/347, and 600 volts shall be used.

(B) Fractions of an Ampere. Where computations result in a fraction of an ampere that is less than 0.5, such fractions shall be permitted to be dropped.

For uniform calculation of load, nominal voltages, as listed in 220.2(A), are required to be used in computing the ampere load on the conductors. To select conductor sizes, refer to 310.15(A) and (B).

Loads are computed on the basis of volt-amperes (VA) or kilovolt-amperes (kVA), rather than watts or kilowatts (kW), to calculate the true ampere values. However, the rating of equipment is given in watts or kilowatts for noninductive loads. Such ratings are considered to be the equivalent of the same rating in volt-amperes or kilovolt-amperes. See, for example, 220.19. This concept recognizes that load calculations determine conductor and circuit sizes, that the power factor of the load is often unknown, and that the conductor "sees" the circuit volt-amperes only, not the circuit power (watts).

See Examples D1(a) through D5(b) of Annex D. The results of these examples are generally expressed in amperes. Unless the computations result in a major fraction of an ampere (0.5 or larger), such fractions (less than 0.5) may be dropped.

220.3 Computation of Branch Circuit Loads.

Branch-circuit loads shall be computed as shown in 220.3(A) through (C).

(A) Lighting Load for Specified Occupancies. A unit load of not less than that specified in Table 220.3(A) for occupancies specified therein shall constitute the minimum lighting load. The floor area for each floor shall be computed from the outside dimensions of the building, dwelling unit, or other area involved. For dwelling units, the computed floor area shall not include open porches, garages, or unused or unfinished spaces not adaptable for future use.

Table 220.3(A) General Lighting Loads by Occupancy

Type of Occupancy	Unit Load	
	Volt-Amperes per Square Meter	Volt-Amperes per Square Foot
Armories and auditoriums	11	1
Banks	39[b]	3½[b]
Barber shops and beauty parlors	33	3
Churches	11	1
Clubs	22	2
Court rooms	22	2
Dwelling units[a]	33	3
Garages — commercial (storage)	6	½
Hospitals	22	2
Hotels and motels, including apartment houses without provision for cooking by tenants[a]	22	2
Industrial commercial (loft) buildings	22	2
Lodge rooms	17	1½
Office buildings	39	3½[b]
Restaurants	22	2
Schools	33	3
Stores	33	3
Warehouses (storage)	3	¼
In any of the preceding occupancies except one-family dwellings and individual dwelling units of two-family and multifamily dwellings:		
Assembly halls and auditoriums	11	1
Halls, corridors, closets, stairways	6	½
Storage spaces	3	¼

[a]See 220.3(B)(10).
[b]In addition, a unit load of 11 volt-amperes/m² or 1 volt-ampere/ft² shall be included for general-purpose receptacle outlets where the actual number of general-purpose receptacle outlets is unknown.

FPN: The unit values herein are based on minimum load conditions and 100 percent power factor and may not provide sufficient capacity for the installation contemplated.

General lighting loads determined by 220.3(A) are in fact minimum lighting loads, and there are no exceptions to these requirements. Therefore, energy saving-type calculations are not permitted to be used to determine the minimum calculated lighting load, if they produce loads less than the load calculated according to 220.3(A). On the other hand, energy saving-type calculations can be a useful tool to reduce the connected lighting load.

Examples of unused or unfinished spaces for dwelling units are some attics, cellars, and crawl spaces.

(B) Other Loads—All Occupancies. In all occupancies, the minimum load for each outlet for general-use receptacles and outlets not used for general illumination shall not be less than that computed in 220.3(B)(1) through (11), the loads shown being based on nominal branch-circuit voltages.

Exception: The loads of outlets serving switchboards and switching frames in telephone exchanges shall be waived from the computations.

(1) Specific Appliances or Loads. An outlet for a specific appliance or other load not covered in (2) through (11) shall be computed based on the ampere rating of the appliance or load served.

(2) Electric Dryers and Household Electric Cooking Appliances. Load computations shall be permitted as specified in 220.18 for electric dryers and in 220.19 for electric ranges and other cooking appliances.

(3) Motor Loads. Outlets for motor loads shall be computed in accordance with the requirements in 430.22, 430.24, and 440.6.

(4) Recessed Luminaires (Lighting Fixtures). An outlet supplying recessed luminaire(s) [lighting fixture(s)] shall be computed based on the maximum volt-ampere rating of the equipment and lamps for which the luminaire(s) [fixture(s)] is rated.

Recessed luminaires are included in the 3 volt-amperes per square foot calculation used for dwellings if they are used for general lighting.

(5) Heavy-Duty Lampholders. Outlets for heavy-duty lampholders shall be computed at a minimum of 600 volt-amperes.

(6) Sign and Outline Lighting. Sign and outline lighting outlets shall be computed at a minimum of 1200 volt-amperes for each required branch circuit specified in 600.5(A).

Section 220.3(B)(6) assigns 1200 volt-amperes as a minimum circuit load for the signs and outline lighting outlets required by 600.5(A). If the specific load is known to be larger, then, according to 220.3(B)(1), the actual load is used.

(7) Show Windows. Show windows shall be computed in accordance with either of the following:

(1) The unit load per outlet as required in other provisions of this section
(2) At 200 volt-amperes per 300 mm (1 ft) of show window

The following two options are permitted for the load calculations for branch circuits serving show windows:

1. 180 volt-amperes per receptacle according to 210.62, which requires one receptacle per 12 linear feet
2. 200 volt-amperes per linear foot of show-window space

As shown in Exhibit 220.1, the linear-foot calculation method is permitted in lieu of the specified unit load per outlet for branch circuits serving show windows.

200 VA per linear ft x 10 ft = 2000 VA

Exhibit 220.1 An example of the linear-foot load calculation for branch circuits serving a show window.

(8) Fixed Multioutlet Assemblies. Fixed multioutlet assemblies used in other than dwelling units or the guest rooms of hotels or motels shall be computed in accordance with (1) or (2). For the purposes of this section, the computation shall be permitted to be based on the portion that contains receptacle outlets.

(1) Where appliances are unlikely to be used simultaneously, each 1.5 m (5 ft) or fraction thereof of each separate and continuous length shall be considered as one outlet of not less than 180 volt-amperes.
(2) Where appliances are likely to be used simultaneously, each 300 mm (1 ft) or fraction thereof shall be considered as an outlet of not less than 180 volt-amperes.

Fixed multioutlet assemblies are commonly used in commercial and industrial locations. The use of multioutlet assemblies is divided into two broad areas. The first area of use is light use, which means that not all the cord-connected equipment is expected to be used at the same time, as noted in 220.3(B)(8)(1). An example of light use is a workbench area where one worker uses one electrical tool at a time. The second area of use is heavy use, which is characterized by all the cord-connected equipment generally operating at the same time, as noted in 220.3(B)(8)(2). An example of heavy use is a retail outlet displaying television sets, where most, if not all, sets are operating simultaneously.

As shown in Exhibit 220.2, the requirements of 220.3(B)(8) state that each 5 ft of a fixed multioutlet assembly must be considered as one outlet rated 180 volt-amperes and that, if appliances are likely to be used simultaneously, each 1 ft must be considered as one outlet rated 180 volt-amperes.

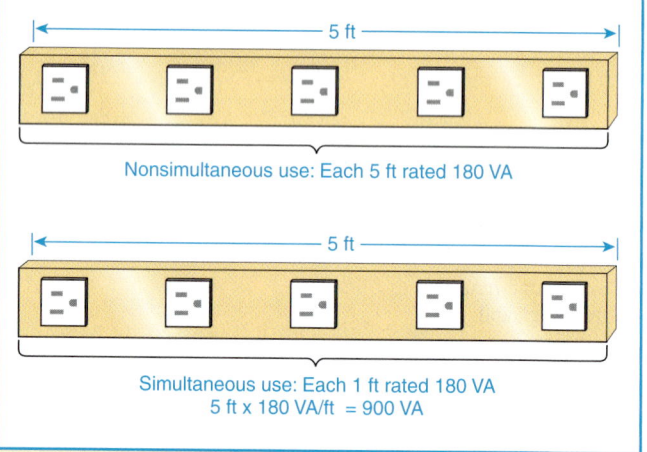

Exhibit 220.2 The requirements of 220.3(B)(8) as applied to fixed multioutlet assemblies.

(9) Receptacle Outlets. Except as covered in 220.3(B)(10), receptacle outlets shall be computed at not less than 180 volt-amperes for each single or for each multiple receptacle on one yoke. A single piece of equipment consisting of a multiple receptacle comprised of four or more receptacles shall be computed at not less than 90 volt-amperes per receptacle.

This provision shall not be applicable to the receptacle outlets specified in 210.11(C)(1) and (2).

As illustrated in Exhibit 220.3, the 180-volt-ampere load is applied to single and multiple receptacles mounted on a single yoke or strap, and a 360-volt-ampere load is applied to each receptacle that consists of four receptacles. These are considered receptacle outlets, in accordance with 220.3(B)(9). The receptacle outlets are not the lighting outlets installed for general illumination or the small-appliance branch circuits, as indicated in 220.3(B)(10). The receptacle load for outlets for general illumination in one- and two-family and multifamily dwellings and in guest rooms of hotels and motels is included in Table 220.3(A). The load requirement for the small-appliance branch circuits is 1500 volt-amperes per circuit, as described in 220.16(A).

Note in Exhibit 220.3 that the last outlet of the top circuit consists of two duplex receptacles on separate straps. That outlet is calculated at 360 volt-amperes because each duplex receptacle is on one yolk. The multiple receptacle supplied from the bottom circuit that comprises four or more receptacles is calculated at 90 volt-amperes per receptacle (4 × 90 VA = 360 VA). For example, single-strap and multiple-receptacle devices are calculated as follows:

Device	Computed Load
Duplex receptacle	180 VA
Triplex receptacle	180 VA
Double duplex receptacle	360 VA (180 × 2)
Quad or four-plex-type receptacle	360 VA (90 × 4)

A load of 180 volt-amperes is not required to be considered for outlets supplying recessed lighting fixtures, lighting outlets for general illumination, and small-appliance branch circuits. To apply the 180-volt-ampere requirement in those cases would be unrealistic, because it would unnecessarily restrict the number of lighting or receptacle outlets on branch

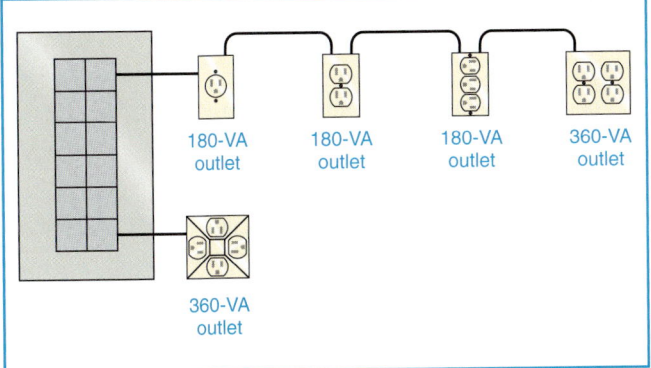

Exhibit 220.3 The 180-volt-ampere load requirement of 220.3(B)(9) as applied to single- and multiple-receptacle outlets on single straps and the 360-volt-ampere load applied to each receptacle that consists of four receptacles.

circuits in dwelling units. See the note below Table 220.3(A) that references 220.3(B)(10). This note indicates that the 180-volt-ampere requirement does not apply to most receptacle outlets in dwellings.

In Exhibit 220.4, the maximum number of outlets permitted on 15- and 20-ampere branch circuits is 10 and 13 outlets, respectively. This restriction does not apply to outlets connected to general lighting or small-appliance branch circuits in dwelling units.

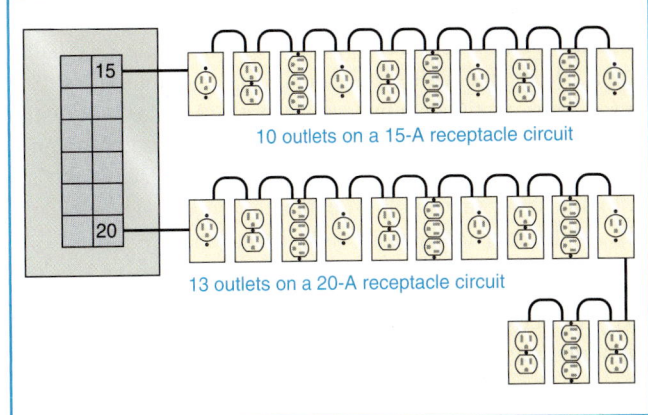

Exhibit 220.4 Maximum number of outlets permitted on 15- and 20-ampere branch circuits.

(10) Dwelling Occupancies. In one-family, two-family, and multifamily dwellings and in guest rooms of hotels and motels, the outlets specified in (1), (2), and (3) are included in the general lighting load calculations of 220.3(A). No additional load calculations shall be required for such outlets.

(1) All general-use receptacle outlets of 20-ampere rating or less, including receptacles connected to the circuits in 210.11(C)(3)

(2) The receptacle outlets specified in 210.52(E) and (G)

(3) The lighting outlets specified in 210.70(A) and (B)

(11) Other Outlets. Other outlets not covered in 220.3(B)(1) through (10) shall be computed based on 180 volt-amperes per outlet.

(C) Loads for Additions to Existing Installations.

(1) Dwelling Units. Loads added to an existing dwelling unit(s) shall comply with the following as applicable:

(1) Loads for structural additions to an existing dwelling unit or for a previously unwired portion of an existing dwelling unit, either of which exceeds 46.5 m² (500 ft²), shall be computed in accordance with 220.3(A) and (B).

(2) Loads for new circuits or extended circuits in previously wired dwelling units shall be computed in accordance with either 220.3(A) or (B), as applicable.

(2) Other Than Dwelling Units. Loads for new circuits or extended circuits in other than dwelling units shall be computed in accordance with either 220.3(A) or (B), as applicable.

220.4 Maximum Loads.

The total load shall not exceed the rating of the branch circuit, and it shall not exceed the maximum loads specified in 220.4(A) through (C) under the conditions specified therein.

(A) Motor-Operated and Combination Loads. Where a circuit supplies only motor-operated loads, Article 430 shall apply. Where a circuit supplies only air-conditioning equipment, refrigerating equipment, or both, Article 440 shall apply. For circuits supplying loads consisting of motor-operated utilization equipment that is fastened in place and has a motor larger than 1/8 hp in combination with other loads, the total computed load shall be based on 125 percent of the largest motor load plus the sum of the other loads.

(B) Inductive Lighting Loads. For circuits supplying lighting units that have ballasts, transformers, or autotransformers, the computed load shall be based on the total ampere ratings of such units and not on the total watts of the lamps.

(C) Range Loads. It shall be permissible to apply demand factors for range loads in accordance with Table 220.19, including Note 4.

II. Feeders and Services

220.10 General.

The computed load of a feeder or service shall not be less than the sum of the loads on the branch circuits supplied, as determined by Part I of this article, after any applicable demand factors permitted by Parts II, III, or IV have been applied.

> FPN: See Examples D1(A) through D10 in Annex D. See 220.4(B) for the maximum load in amperes permitted for lighting units operating at less than 100 percent power factor.

In the example shown in Exhibit 220.5, each panel serves a computed load of 80 amperes. The main feeder is sized to carry the total computed load of 240 amperes (3 multiplied by 80 amperes). The feeder tap conductors from the meter enclosure to the panelboards are sized to supply a computed load of 80 amperes. The main feeder is not intended to be sized to carry 300 amperes based on the sum of the panelboards.

See Exhibit 230.13 for a similar example for service conductors. The ungrounded service conductors are no longer required to be sized for the sum of the main overcur-

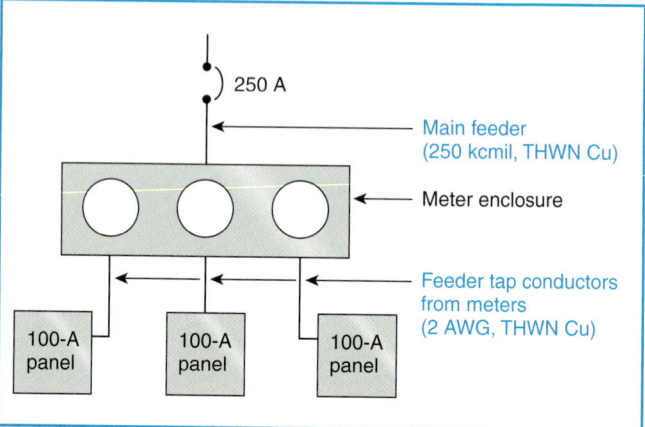

Exhibit 220.5 Feeder conductors sized in accordance with 220.10.

rent device rating of 300 amperes. Service conductors are required to have sufficient ampacity to carry the loads computed in accordance with Article 220, with appropriate demand factors applied. See 230.23, 230.31, and 230.42 for specifics on size and rating of conductors.

Part II of Article 220 contains the requirements for calculating feeder and service loads. Part III provides optional methods for calculating feeder and service loads in dwelling units and multifamily dwellings.

Except as permitted in 240.4 and 240.6, the rating of the overcurrent device cannot exceed the final ampacity of the circuit conductors after all the correction and adjustment factors have been applied, such as for temperature or number of conductors.

Example

Determine the minimum-size overcurrent protective device and the minimum conductor size for a feeder circuit with the following characteristics:

- 3-phase, 4-wire feeder (full-size neutral)
- 125-ampere noncontinuous load
- 200-ampere continuous load
- 75°C overcurrent device terminal rating
- Type THWN insulated conductors
- Four current-carrying conductors in a raceway
- A major portion of the load is nonlinear

Solution

STEP 1. Select the feeder overcurrent protective device (OCPD) rating by first totaling the continuous and noncontinuous loads according to 215.3:

$$OCPD\ rating = 125\%\ of\ continuous\ load$$
$$+\ noncontinuous\ load$$
$$= (200\ A \times 1.25) + 125\ A$$
$$= 250\ A + 125\ A$$
$$= 375\ A$$

Using 240.4(B) and 240.6(A), adjust the minimum standard-size OCPD to 400 amperes.

STEP 2. Select the feeder conductor size before derating by first summing the continuous and noncontinuous loads according to 215.2(A)(1).

$$\text{Feeder size (before derating)} = 125\% \text{ of continuous load} + \text{noncontinuous load}$$
$$= (200 \text{ A} \times 1.25) + 125 \text{ A}$$
$$= 250 \text{ A} + 125 \text{ A}$$
$$= 375 \text{ A}$$

Using Table 310.16 and using the 75°C column (because of the overcurrent device terminal), the minimum-size Type THWN copper conductor that can supply a computed load of 375 amperes is 500-kcmil copper, which has an ampacity of 380 amperes.

STEP 3. Apply the derating factors to the feeder conductor size. Section 310.15(B)(4)(c) requires that the neutral conductor be counted as a current-carrying conductor because a major portion of the load consists of fluorescent and HID luminaires. Therefore, this feeder circuit consists of four current-carrying conductors in the same raceway. Section 310.15(B)(2) requires an 80 percent adjustment factor for four current-carrying conductors in the same raceway. According to Table 310.16, 500-kcmil, Type THWN conductors have an ampacity of 380 amperes. The adjustment factors are applied to this ampacity as follows:

$$\text{Adjusted ampacity} = \text{table ampacity} \times \text{adjustment factor}$$
$$= 380 \text{ A} \times 0.80$$
$$= 304 \text{ A}$$

According to 240.4(B) and 240.6(A), a conductor with a computed ampacity of 304 amperes is not allowed to be protected by a 400-ampere overcurrent protective device. Therefore, the 500-kcmil, Type THWN copper conductor *cannot* be used.

STEP 4. Revise the feeder conductor selection and perform a check. The next standard size conductor listed in Table 310.16 is a 600-kcmil copper conductor in the 90°C column. If higher-temperature insulations are used, adjustment factors can be applied to the higher ampacity. Because Type THWN is a 75°C insulation, a 90°C Type THHN is selected. If a 600-kcmil Type THHN copper conductor is used, perform the check as follows. According to Table 310.16, a 600-kcmil conductor has an ampacity of 475 amperes in the Type THHN 90°C column:

$$\text{Adjusted ampacity} = \text{table ampacity} \times \text{adjustment factor}$$
$$= 475 \text{ A} \times 0.80$$
$$= 380 \text{ A}$$

A conductor with a computed ampacity of 380 amperes is allowed to be protected by a 400-ampere overcurrent protective device, in accordance with 240.4(B). However, because the overcurrent protective device terminations are rated 75°C, the load current cannot exceed the ampacity of a 600-kcmil conductor in the 75°C column of Table 310.16, which has a value of 420 amperes.

STEP 5. Evaluate the circuit. The calculation in Step 4 results in four 600-kcmil Type THHN copper conductors in one raceway, each with an ampacity of 380 amperes, supplying a 375-ampere continuous load, and protected by a 400-ampere overcurrent protective device. It is important to note here that a 90°C, 600-kcmil copper conductor with a final computed ampacity of 380 amperes is permitted to terminate on a 75°C-rated terminal, according to 110.14(C)(1)(b)(2).

220.11 General Lighting.

The demand factors specified in Table 220.11 shall apply to that portion of the total branch-circuit load computed for general illumination. They shall not be applied in determining the number of branch circuits for general illumination.

220.12 Show-Window and Track Lighting.

(A) Show Windows. For show-window lighting, a load of not less than 660 volt-amperes/linear meter or 200 volt-amperes/linear foot shall be included for a show window, measured horizontally along its base.

> FPN: See 220.3(B)(7) for branch circuits supplying show windows.

The 200-volt-ampere calculation for each linear foot of a show window is required to determine the *feeder* load. See the commentary following 220.3(B)(7) for load calculations for branch circuits in show windows.

(B) Track Lighting. For track lighting in other than dwelling units or guest rooms of hotels or motels, an additional load of 150 volt-amperes shall be included for every 600 mm (2 ft) of lighting track or fraction thereof. Where multicircuit track is installed, the load shall be considered to be divided equally between the track circuits.

Example

A lighting plan shows 62.5 linear ft of single-circuit track lighting for a small department store featuring clothing. Because the actual track lighting fixtures are owner supplied, neither the quantity of track lighting fixtures or the lamp size is specified. What is the minimum calculated load associated with the track lighting that must be added to the service or feeder supplying this store?

Table 220.11 Lighting Load Demand Factors

Type of Occupancy	Portion of Lighting Load to Which Demand Factor Applies (Volt-Amperes)	Demand Factor (Percent)
Dwelling units	First 3000 or less at From 3001 to 120,000 at Remainder over 120,000 at	100 35 25
Hospitals*	First 50,000 or less at Remainder over 50,000 at	40 20
Hotels and motels, including apartment houses without provision for cooking by tenants*	First 20,000 or less at From 20,001 to 100,000 at Remainder over 100,000 at	50 40 30
Warehouses (storage)	First 12,500 or less at Remainder over 12,500 at	100 50
All others	Total volt-amperes	100

*The demand factors of this table shall not apply to the computed load of feeders or services supplying areas in hospitals, hotels, and motels where the entire lighting is likely to be used at one time, as in operating rooms, ballrooms, or dining rooms.

Solution

According to 220.12(B), the minimum calculated load to be added to the service or feeder supplying this track light installation is calculated as follows:

$$\frac{62.5 \text{ ft}}{2 \text{ ft}} = 31.25, \quad \text{rounded up to } 32$$

$$32 \times 150 \text{ VA} = 4800 \text{ VA}$$

Thus, the minimum load that must be added to the service or feeder calculation is 4800 volt-amperes.

It is important to note that the branch circuits supplying this installation are covered in 410.101(B). For the track lighting branch-circuit load, the maximum load on the track cannot exceed the rating of the branch circuit supplying the track. Also, the track must be supplied by a branch circuit that has a rating not exceeding the rating of the track. The track length does not enter into the branch-circuit calculation.

Section 220.12(B) is not intended to limit the number of feet of track on a single branch circuit, nor is it intended to limit the number of fixtures on an individual track. Rather, 220.12(B) is meant to be used solely for load calculations of feeders and services.

220.13 Receptacle Loads—Nondwelling Units.

In other than dwelling units, receptacle loads computed at not more than 180 volt-amperes per outlet in accordance with 220.3(B)(9) and fixed multioutlet assemblies computed

in accordance with 220.3(B)(8) shall be permitted to be added to the lighting loads and made subject to the demand factors given in Table 220.11, or they shall be permitted to be made subject to the demand factors given in Table 220.13.

Table 220.13 Demand Factors for Nondwelling Receptacle Loads

Portion of Receptacle Load to Which Demand Factor Applies (Volt-Amperes)	Demand Factor (Percent)
First 10 kVA or less at Remainder over 10 kVA at	100 50

Section 220.13 permits receptacle loads, calculated at not more than 180 volt-amperes per strap, to be computed by either of the following two methods:

1. The receptacle loads are added to the lighting load. The demand factors in Table 220.11 are then applied to the combined load.
2. The receptacle loads are calculated (without the lighting load) with demand factors from Table 220.13 applied.

220.14 Motors.

Motor loads shall be computed in accordance with 430.24, 430.25, and 430.26 and with 440.6 for hermetic refrigerant motor compressors.

220.15 Fixed Electric Space Heating.

Fixed electric space heating loads shall be computed at 100 percent of the total connected load; however, in no case shall a feeder or service load current rating be less than the rating of the largest branch circuit supplied.

Exception: Where reduced loading of the conductors results from units operating on duty-cycle, intermittently, or from all units not operating at the same time, the authority having jurisdiction may grant permission for feeder and service conductors to have an ampacity less than 100 percent, provided the conductors have an ampacity for the load so determined.

220.16 Small Appliance and Laundry Loads—Dwelling Unit.

(A) Small Appliance Circuit Load. In each dwelling unit, the load shall be computed at 1500 volt-amperes for each 2-wire small-appliance branch circuit required by 210.11(C)(1). Where the load is subdivided through two or more feeders, the computed load for each shall include not less than 1500 volt-amperes for each 2-wire small-appliance branch circuit. These loads shall be permitted to be included with the general lighting load and subjected to the demand factors provided in Table 220.11.

Exception: The individual branch circuit permitted by 210.52(B)(1), Exception No. 2, shall be permitted to be excluded from the calculation required by 220.16.

See the commentary following 210.52(B) regarding required receptacle outlets for small-appliance branch circuits.

(B) Laundry Circuit Load. A load of not less than 1500 volt-amperes shall be included for each 2-wire laundry branch circuit installed as required by 210.11(C)(2). This load shall be permitted to be included with the general lighting load and subjected to the demand factors provided in Table 220.11.

In each dwelling unit, the feeder load is required to be calculated at 1500 volt-amperes for each of the two or more (2-wire) small-appliance branch circuits and at 1500 volt-amperes for each (2-wire) laundry branch circuit. These loads are permitted to be totaled and then added to the general lighting load. The total load (i.e., small appliance, laundry, and general lighting) is subjected to the demand factors provided in Table 220.11.

220.17 Appliance Load—Dwelling Unit(s).

It shall be permissible to apply a demand factor of 75 percent to the nameplate rating load of four or more appliances fastened in place, other than electric ranges, clothes dryers, space-heating equipment, or air-conditioning equipment, that are served by the same feeder or service in a one-family, two-family, or multifamily dwelling.

For appliances fastened in place (other than ranges, clothes dryers, and space-heating and air-conditioning equipment), feeder capacity must be provided for the sum of these loads, and for a total load of four or more such appliances, a demand factor of 75 percent may be applied. See Table 430.148 for the full-load current, in amperes, for single-phase ac motors, in accordance with 220.14.

Example

Determine the feeder capacity needed for a 120/240-volt fastened-in-place appliance load in a dwelling unit for the following:

Appliance	Rating	Load
Water heater	4000 W, 240 V	4000 VA
Kitchen disposal	½ hp, 120 V	1176 VA
Dishwasher	1200 W, 120 V	1200 VA
Furnace motor	¼ hp, 120 V	696 VA
Attic fan	¼ hp, 120 V	696 VA
Water pump	½ hp, 240 V	1176 VA

Solution

STEP 1. Calculate the total of the six fastened-in-place appliances:

$$\text{Total load} = 4000 \text{ VA} + 1176 \text{ VA} + 1200 \text{ VA}$$
$$+ 696 \text{ VA} + 696 \text{ VA} + 1176 \text{ VA}$$
$$= 8944 \text{ VA}$$

STEP 2. Because the load is for more than four appliances, apply a demand factor of 75 percent:

$$8944 \text{ VA} \times 0.75 = 6708 \text{ VA}$$

Thus, 6708 volt-amperes is the load to be added to the other determined loads when calculating the size of service and feeder conductors.

220.18 Electric Clothes Dryers— Dwelling Unit(s).

The load for household electric clothes dryers in a dwelling unit(s) shall be 5000 watts (volt-amperes) or the nameplate rating, whichever is larger, for each dryer served. The use of the demand factors in Table 220.18 shall be permitted. Where two or more single-phase dryers are supplied by a 3-phase, 4-wire feeder or service, the total load shall be computed on the basis of twice the maximum number connected between any two phases.

Table 220.18 Demand Factors for Household Electric Clothes Dryers

Number of Dryers	Demand Factor (Percent)
1–4	100%
5	85%
6	75%
7	65%
8	60%
9	55%
10	50%
11	47%
12–22	% = 47 − (number of dryers − 11)
23	35%
24–42	% = 35 − [0.5 × (number of dryers − 23)]
43 and over	25%

The exact method of calculation presented in Table 220.18 was revised for the 2002 *Code* to produce a more accurate load for all quantities of dryers. To compute the load of household electric dryers, 220.18 specifies a minimum demand of 5 kVA for the calculation of feeder conductors. If the nameplate rating is known and exceeds 5 kW, the larger rating is applied.

220.19 Electric Ranges and Other Cooking Appliances—Dwelling Unit(s).

The demand load for household electric ranges, wall-mounted ovens, counter-mounted cooking units, and other household cooking appliances individually rated in excess of 1⅓ kW shall be permitted to be computed in accordance with Table 220.19. Kilovolt-amperes (kVA) shall be considered equivalent to kilowatts (kW) for loads computed under this section.

Where two or more single-phase ranges are supplied by a 3-phase, 4-wire feeder or service, the total load shall be computed on the basis of twice the maximum number connected between any two phases.

> FPN No. 1: See Example D5(A) in Annex D.
> FPN No. 2: See Table 220.20 for commercial cooking equipment.
> FPN No. 3: See the examples in Annex D.

Table 220.19 Demand Loads for Household Electric Ranges, Wall-Mounted Ovens, Counter-Mounted Cooking Units, and Other Household Cooking Appliances over 1¾ kW Rating (Column C to be used in all cases except as otherwise permitted in Note 3.)

| Number of Appliances | Demand Factor (Percent) (See Notes) | | Column C Maximum Demand (kW) (See Notes) (Not over 12 kW Rating) |
	Column A (Less than 3½ kW Rating)	Column B (3½ kW to 8¾ kW Rating)	
1	80	80	8
2	75	65	11
3	70	55	14
4	66	50	17
5	62	45	20
6	59	43	21
7	56	40	23
8	53	36	23
9	51	35	24
10	49	34	25
11	47	32	26
12	45	32	27
13	43	32	28
14	41	32	29
15	40	32	30
16	39	28	31
17	38	28	32
18	37	28	33

Table 220.19 Continued

| Number of Appliances | Demand Factor (Percent) (See Notes) | | Column C Maximum Demand (kW) (See Notes) (Not over 12 kW Rating) |
	Column A (Less than 3½ kW Rating)	Column B (3½ kW to 8¾ kW Rating)	
19	36	28	34
20	35	28	35
21	34	26	36
22	33	26	37
23	32	26	38
24	31	26	39
25	30	26	40
26–30	30	24	15 kW + 1 kW for
31–40	30	22	each range
41–50	30	20	25 kW + ¾ kW
51–60	30	18	for each range
61 and over	30	16	

1. Over 12 kW through 27 kW ranges all of same rating. For ranges individually rated more than 12 kW but not more than 27 kW, the maximum demand in Column C shall be increased 5 percent for each additional kilowatt of rating or major fraction thereof by which the rating of individual ranges exceeds 12 kW.

> For household electric ranges and other cooking appliances, the size of the conductors must be determined by the rating of the range. According to Table 220.19, for one range rated 12 kW or less, the maximum demand load is 8 kW (8 kVA per 220.19), and 8 AWG copper conductors with 60°C insulation would suffice. Note that 210.19(A)(3) does not permit the branch-circuit rating of a circuit supplying household ranges with a nameplate rating of 8¾-kW to be less than 40 amperes.

2. Over 8¾ kW through 27 kW ranges of unequal ratings. For ranges individually rated more than 8¾ kW and of different ratings, but none exceeding 27 kW, an average value of rating shall be computed by adding together the ratings of all ranges to obtain the total connected load (using 12 kW for any range rated less than 12 kW) and dividing by the total number of ranges. Then the maximum demand in Column C shall be increased 5 percent for each kilowatt or major fraction thereof by which this average value exceeds 12 kW.

> Note 2 to Table 220.19 provides for ranges larger than 8¾ kW. Note 4 covers installations where the circuit supplies multiple cooking components, which are combined and treated as a single range.

3. Over 1¾ kW through 8¾ kW. In lieu of the method provided in Column C, it shall be permissible to add the nameplate ratings of all household cooking appliances rated more than 1¾ kW but not more than 8¾ kW and multiply the sum by the demand factors specified in Column A or B for the given number of appliances. Where the rating of cooking appliances falls under both Column A and Column B,

Table 220.19 Continued

the demand factors for each column shall be applied to the appliances for that column, and the results added together.

The branch-circuit load for one range is permitted to be computed by using either the nameplate rating of the appliance or Table 220.19. If a single branch circuit supplies a counter-mounted cooking unit and not more than two wall-mounted ovens, all of which are located in the same room, the nameplate ratings of these appliances can be added together and the total treated as the equivalent of one range, according to Note 4 of Table 220.19.

Example

Calculate the load for a single branch circuit that supplies the following cooking units:

One counter-mounted cooking unit, with rating of 8 kW

One wall-mounted oven, with rating of 7 kW

A second wall-mounted oven, with rating of 6 kW

Solution

STEP 1. The combined cooking appliances can be treated as one range according to Note 4 of Table 220.19. In Table 220.19, find the maximum demand for one range not over 12 kW, which is 8 kW (from Column C).

STEP 2. According to Note 1 in Table 220.19, for ranges that are over 12 kW but not more than 27 kW, the maximum demand in Column C (8 kW) is increased 5 percent for each kW that exceeds 12 kW. Determine the additional kilowatts:

Combined unit rating = 8 kW + 7 kW + 6 kW = 21 kW

Additional kW = 21 kW − 12 kW = 9 kW

STEP 3. Calculate by how much the maximum load in Column C in Table 220.19 must be increased for the combined appliances:

Increase = 5% per kW × 9 kW = 45%

= 0.45 × 8 kW = 3.6 kW

STEP 4. Calculate the total load in amperes, as follows:

$$\text{Total load} = 8 \text{ kW} + 3.6 \text{ kW} = 11.6 \text{ kW}$$
$$= 11.6 \text{ kW} = 11,600 \text{ W} = 11,600 \text{ VA}$$
$$= \frac{11,600 \text{ VA}}{240 \text{ V}}$$
$$= 48.3 \text{ A}$$

4. Branch-Circuit Load. It shall be permissible to compute the branch-circuit load for one range in accordance with Table 220.19. The branch-circuit load for one wall-mounted oven or one counter-mounted cooking unit shall be the nameplate rating of the appliance. The branch-circuit load for a counter-mounted cooking unit and not more than two wall-mounted ovens, all supplied from a single branch circuit and located in the same room, shall be computed by adding the nameplate rating of the individual appliances and treating this total as equivalent to one range.

5. This table also applies to household cooking appliances rated over 1¾ kW and used in instructional programs.

The nameplate ratings of all household cooking appliances rated more than 1¾ kW but not more than 8¾ kW may be added together and the sum multiplied by the demand factors specified in Column A or B of Table 220.19 for the given number of appliances. For feeder demand factors for other than dwelling units—that is, commercial electric cooking equipment, dishwasher booster heaters, water heaters, and so on—see Table 220.20.

The demand factors of this *Code* are based on the diversified use of appliances, because it is unlikely that all appliances will be used simultaneously or that all cooking units and the oven of a range will be at maximum heat for any length of time.

220.20 Kitchen Equipment—Other Than Dwelling Unit(s).

It shall be permissible to compute the load for commercial electric cooking equipment, dishwasher booster heaters, water heaters, and other kitchen equipment in accordance with Table 220.20. These demand factors shall be applied to all equipment that has either thermostatic control or intermittent use as kitchen equipment. They shall not apply to space-heating, ventilating, or air-conditioning equipment.

However, in no case shall the feeder or service demand be less than the sum of the largest two kitchen equipment loads.

Table 220.20 Demand Factors for Kitchen Equipment — Other Than Dwelling Unit(s)

Number of Units of Equipment	Demand Factor (Percent)
1	100
2	100
3	90
4	80
5	70
6 and over	65

220.21 Noncoincident Loads.

Where it is unlikely that two or more noncoincident loads will be in use simultaneously, it shall be permissible to use only the largest load(s) that will be used at one time, in computing the total load of a feeder or service.

220.22 Feeder or Service Neutral Load.

The feeder or service neutral load shall be the maximum unbalance of the load determined by this article. The maximum unbalanced load shall be the maximum net computed load between the neutral and any one ungrounded conductor, except that the load thus obtained shall be multiplied by 140 percent for 3-wire, 2-phase or 5-wire, 2-phase systems. For a feeder or service supplying household electric ranges, wall-mounted ovens, counter-mounted cooking units, and electric dryers, the maximum unbalanced load shall be considered as 70 percent of the load on the ungrounded conductors, as determined in accordance with Table 220.19 for ranges and Table 220.18 for dryers. For 3-wire dc or single-phase ac; 4-wire, 3-phase; 3-wire, 2-phase; or 5-wire, 2-phase systems, a further demand factor of 70 percent shall be permitted for that portion of the unbalanced load in excess of 200 amperes. There shall be no reduction of the neutral capacity for that portion of the load that consists of nonlinear loads supplied from a 4-wire, wye-connected, 3-phase system. There shall be no reduction in the capacity of the grounded conductor of a 3-wire circuit consisting of two phase wires and the neutral of a 4-wire, 3-phase, wye-connected system.

> FPN No. 1: See Examples D1(A), D1(B), D2(B), D4(A), and D5(A) in Annex D.
>
> FPN No. 2: A 3-phase, 4-wire, wye-connected power system used to supply power to nonlinear loads may necessitate that the power system design allow for the possibility of high harmonic neutral currents.

Section 220.22 describes the basis for calculating the neutral load of feeders or services as the maximum unbalanced load that can occur between the neutral and any other ungrounded conductor.

For a household electric range or clothes dryer, the maximum unbalanced load may be assumed to be 70 percent, so the neutral may be sized on that basis. Although 220.22 permits the reduction of the feeder neutral conductor size under specific conditions of use, the last two sentences, revised for the 2002 *Code*, cite two specific cases that would prohibit reducing a neutral or grounded conductor of a feeder.

If the system also supplies nonlinear loads such as electric-discharge lighting, including fluorescent and HID, or data-processing or similar equipment, the neutral is considered a current-carrying conductor if the load of the electric-discharge lighting, data-processing, or similar equipment on the feeder neutral consists of more than half the total load. Electric-discharge lighting and data-processing equipment may have harmonic currents in the neutral that may exceed the load current in the ungrounded conductors. It would be appropriate to require a full-size or larger feeder neutral conductor, depending on the total harmonic distortion contributed by the equipment to be supplied (see 220.22, FPN No. 2).

In some instances, the neutral current may exceed the current in the phase conductors. See the commentary following 310.15(B)(4)(c) regarding neutral conductor ampacity.

III. Optional Calculations for Computing Feeder and Service Loads

220.30 Optional Calculation—Dwelling Unit.

(A) Feeder and Service Load. For a dwelling unit having the total connected load served by a single 3-wire, 120/240-volt or 208Y/120-volt set of service or feeder conductors with an ampacity of 100 or greater, it shall be permissible to compute the feeder and service loads in accordance with this section instead of the method specified in Part II of this article. The calculated load shall be the result of adding the loads from 220.30(B) and (C). Feeder and service-entrance conductors whose demand load is determined by this optional calculation shall be permitted to have the neutral load determined by 220.22.

The optional method given in 220.30 applies to a single dwelling unit, whether it is a separate building or located in a multifamily dwelling. The optional calculation permitted by 220.30 may be used only if the service-entrance or feeder conductors have an ampacity of at least 100 amperes. See Article 100 for the definition of *dwelling unit*.

Examples of the optional calculation for a dwelling unit are given in Examples D2(a), D2(b), D2(c), and D4(b) of Annex D.

(B) General Loads. The general calculated load shall be not less than 100 percent of the first 10 kVA plus 40 percent of the remainder of the following loads:

(1) 1500 volt-amperes for each 2-wire, 20-ampere small-appliance branch circuit and each laundry branch circuit specified in 220.16.

(2) 33 volt-amperes/m^2 or 3 volt-amperes/ft^2 for general lighting and general-use receptacles. The floor area for each floor shall be computed from the outside dimensions of the dwelling unit. The computed floor area shall not include open porches, garages, or unused or unfinished spaces not adaptable for future use.

(3) The nameplate rating of all appliances that are fastened in place, permanently connected, or located to be on a specific circuit, ranges, wall-mounted ovens, counter-mounted cooking units, clothes dryers, and water heaters.

Section 220.30(B)(3) includes appliances that may not be fastened in place but may be permanently connected or on a specific circuit, such as clothes dryers, dishwashers, and freezers.

(4) The nameplate ampere or kVA rating of all motors and of all low-power-factor loads.

(C) Heating and Air-Conditioning Load. The largest of the following six selections (load in kVA) shall be included:

(1) 100 percent of the nameplate rating(s) of the air conditioning and cooling.

(2) 100 percent of the nameplate ratings of the heat pump compressors and supplemental heating unless the controller prevents the compressor and supplemental heating from operating at the same time.

(3) 100 percent of the nameplate ratings of electric thermal storage and other heating systems where the usual load is expected to be continuous at the full nameplate value. Systems qualifying under this selection shall not be calculated under any other selection in 220.30(C).

(4) 65 percent of the nameplate rating(s) of the central electric space heating, including integral supplemental heating in heat pumps where the controller prevents the compressor and supplemental heating from operating at the same time.

(5) 65 percent of the nameplate rating(s) of electric space heating if less than four separately controlled units.

(6) 40 percent of the nameplate rating(s) of electric space heating if four or more separately controlled units.

> Section 220.21 states that for loads that do not operate simultaneously, the largest load being considered is used. In concert with 220.21, 220.30(C) requires that only the largest of the six choices needs to be included in the feeder or service calculation. Examples of calculations using air conditioning and heating are found in Annex D, Examples D2(b) and D3(c).

220.31 Optional Calculations for Additional Loads in an Existing Dwelling Unit.

This section shall be permitted to be used to determine if the existing service or feeder is of sufficient capacity to serve additional loads. Where the dwelling unit is served by a 120/240-volt or 208Y/120-volt, 3-wire service, it shall be permissible to compute the total load in accordance with 220.31(A) or (B).

(A) Where Additional Air-Conditioning Equipment or Electric Space-Heating Equipment Is Not to Be Installed. The following formula shall be used for existing and additional new loads.

Load (kVa)	Percent of Load
First 8 kVA of load at	100
Remainder of load at	40

Load calculations shall include the following:

(1) General lighting and general-use receptacles at 33 volt-amperes/m^2 or 3 volt-amperes/ft^2 as determined by 220.3(A)

(2) 1500 volt-amperes for each 2-wire, 20-ampere small-appliance branch circuit and each laundry branch circuit specified in 220.16

(3) Household range(s), wall-mounted oven(s), and counter-mounted cooking unit(s)

(4) All other appliances that are permanently connected, fastened in place, or connected to a dedicated circuit, at nameplate rating

(B) Where Additional Air-Conditioning Equipment or Electric Space-Heating Equipment Is to Be Installed. The following formula shall be used for existing and additional new loads. The larger connected load of air-conditioning or space-heating, but not both, shall be used.

Air-conditioning equipment	100
Central electric space heating	100
Less than four separately controlled space-heating units	100
First 8 kVA of all other loads	100
Remainder of all other loads	40

Other loads shall include the following:

(1) General lighting and general-use receptacles at 33 volt-amperes/m^2 or 3 volt-amperes/ft^2 as determined by 220.3(A)

(2) 1500 volt-amperes for each 2-wire, 20-ampere small-appliance branch circuit and each laundry branch circuit specified in 220.16

(3) Household range(s), wall-mounted oven(s), and counter-mounted cooking unit(s)

(4) All other appliances that are permanently connected, fastened in place, or connected to a dedicated circuit, including four or more separately controlled space-heating units, at nameplate rating

> The optional method described in Section 220.31(A) or (B) allows an additional load to be supplied by an existing service.
>
> ### Example
>
> An existing dwelling unit is served by a 100-ampere service. An additional load of a single 5 kVA, 240-volt air conditioning unit is to be installed. Because the exiting load does not contain heating or air-conditioning equipment, the existing load is calculated according to 220.31(A). The load of the existing dwelling unit consists of the following:
>
> | | |
> |---|---|
> | General lighting, 24 ft × 40 ft = 960 ft^2 × 3 VA per ft^2 | 2,880 VA |
> | Small-appliance circuits (3 × 1500 VA) | 4,500 VA |
> | Laundry circuit at 1500 VA | 1,500 VA |
> | Electric range rated 10.5 kW | 10,500 VA |
> | Electric water heater rated 3.0 kW | 3,000 VA |
> | Total existing load | 22,380 VA |

STEP 1. Following the requirements of 220.31(A), calculate the existing dwelling unit load before adding any equipment:

First 8 kVA of load at 100%	8,000 VA
Remainder of load at 40%	
$(22,380 - 8,000) = 14,380 \times 40\%$	5,752 VA
Total load (without air-conditioning equipment)	13,752 VA
13,752 VA ÷ 240 V	57.3 amps

STEP 2. Prepare a list of the existing and new loads of the dwelling unit.

General lighting, 24 ft × 40 ft = 960 ft² × 3 VA per ft²	2,880 VA
Small-appliance circuits (3 × 1500 VA)	4,500 VA
Laundry circuit at 1500 VA	1,500 VA
Electric range rated 10.5 kW	10,500 VA
Electric water heater rated 3.0 kW	3,000 VA
Added air-conditioning equipment	5,000 VA
Total new load	27,380 VA

STEP 3. Following the requirements of 220.31(B), calculate the dwelling unit total load after adding any new equipment.

First 8 kVA of other load at 100%	8,000 VA
Remainder of other load at 40%	
$(22,380 - 8,000) = 14,380 \times 40\%$	5,752 VA
100% of air-conditioning equipment	5,000 VA
Total load (with added air-conditioning equipment)	18,752 VA
18,752 VA ÷ 240 V	78.13 amps

The additional load contributed by the added 5-kVA air conditioning does not exceed the allowable load permitted on a 100-ampere service.

220.32 Optional Calculation— Multifamily Dwelling.

(A) Feeder or Service Load. It shall be permissible to compute the load of a feeder or service that supplies more than two dwelling units of a multifamily dwelling in accordance with Table 220.32 instead of Part II of this article where all the following conditions are met:

(1) No dwelling unit is supplied by more than one feeder.
(2) Each dwelling unit is equipped with electric cooking equipment.

Table 220.32 Optional Calculations — Demand Factors for Three or More Multifamily Dwelling Units

Number of Dwelling Units	Demand Factor (Percent)
3–5	45
6–7	44
8–10	43
11	42
12–13	41
14–15	40
16–17	39
18–20	38
21	37
22–23	36
24–25	35
26–27	34
28–30	33
31	32
32–33	31
34–36	30
37–38	29
39–42	28
43–45	27
46–50	26
51–55	25
56–61	24
62 and over	23

The method of load calculation under 220.32 is optional and applies only where one service or feeder supplies the entire load of a dwelling unit. If all the stated conditions prevail, the optional calculations in 220.32 may be used instead of those in Part II of Article 220.

Exception: When the computed load for multifamily dwellings without electric cooking in Part II of this article exceeds that computed under Part III for the identical load plus electric cooking (based on 8 kW per unit), the lesser of the two loads shall be permitted to be used.

Section 220.32(A)(2) requires that each dwelling unit be equipped with electric cooking equipment in order to use the load calculation method found in 220.32(A). The exception to 220.32(A)(2) permits load calculation for dwelling units that do not have electric cooking equipment by calculating a simulated electric cooking equipment load of 8 kW per unit and selecting the lesser of the two loads.

(3) Each dwelling unit is equipped with either electric space heating, air conditioning, or both. Feeders and service conductors whose demand load is determined by this optional calculation shall be permitted to have the neutral load determined by 220.22.

(B) House Loads. House loads shall be computed in accordance with Part II of this article and shall be in addition to the dwelling unit loads computed in accordance with Table 220.32.

(C) Connected Loads. The computed load to which the demand factors of Table 220.32 apply shall include the following:

(1) 1500 volt-amperes for each 2-wire, 20-ampere small-appliance branch circuit and each laundry branch circuit specified in 220.16.
(2) 33 volt-amperes/m^2 or 3 volt-amperes/ft^2 for general lighting and general-use receptacles.
(3) The nameplate rating of all appliances that are fastened in place, permanently connected or located to be on a specific circuit, ranges, wall-mounted ovens, counter-mounted cooking units, clothes dryers, water heaters, and space heaters. If water heater elements are interlocked so that all elements cannot be used at the same time, the maximum possible load shall be considered the nameplate load.
(4) The nameplate ampere or kilovolt-ampere rating of all motors and of all low-power-factor loads.
(5) The larger of the air-conditioning load or the space-heating load.

220.33 Optional Calculation— Two Dwelling Units.

Where two dwelling units are supplied by a single feeder and the computed load under Part II of this article exceeds that for three identical units computed under 220.32, the lesser of the two loads shall be permitted to be used.

220.34 Optional Method—Schools.

The calculation of a feeder or service load for schools shall be permitted in accordance with Table 220.34 in lieu of Part II of this article where equipped with electric space heating, air conditioning, or both. The connected load to which the demand factors of Table 220.34 apply shall include all of the interior and exterior lighting, power, water heating, cooking, other loads, and the larger of the air-conditioning load or space-heating load within the building or structure.

Feeders and service-entrance conductors whose demand load is determined by this optional calculation shall be permitted to have the neutral load determined by 220.22. Where the building or structure load is calculated by this optional method, feeders within the building or structure shall have ampacity as permitted in Part II of this article; however, the ampacity of an individual feeder shall not be required to be larger than the ampacity for the entire building.

Table 220.34 Optional Method—Demand Factors for Feeders and Service-Entrance Conductors for Schools

Connected Load	Demand Factor (Percent)
First 33 VA/m^2 (3 VA/ft^2) at	100
Plus	
Over 33 to 220 VA/m^2 (3 to 20 VA/ft^2) at	75
Plus	
Remainder over 220 VA/m^2 (20 VA/ft^2) at	25

This section shall not apply to portable classroom buildings.

Many schools add small, portable classroom buildings. The air-conditioning load in these portable classrooms must comply with Article 440, and the lighting load must be considered continuous. The demand factors in Table 220.34 do not apply to portable classrooms, because those demand factors would decrease the feeder or service size to below that required for the connected continuous load.

220.35 Optional Calculations for Determining Existing Loads.

The calculation of a feeder or service load for existing installations shall be permitted to use actual maximum demand to determine the existing load under the following conditions:

(1) The maximum demand data is available for a 1-year period.

Exception: If the maximum demand data for a 1-year period is not available, the calculated load shall be permitted to be based on the maximum demand (measure of average power demand over a 15-minute period) continuously recorded over a minimum 30-day period using a recording ammeter or power meter connected to the highest loaded phase of the feeder or service, based on the initial loading at the start of the recording. The recording shall reflect the maximum demand of the feeder or service by being taken when the building or space is occupied and shall include by measurement or calculation the larger of the heating or cooling equipment load, and other loads that may be periodic in nature due to seasonal or similar conditions.

(2) The maximum demand at 125 percent plus the new load does not exceed the ampacity of the feeder or rating of the service.
(3) The feeder has overcurrent protection in accordance with 240.4, and the service has overload protection in accordance with 230.90.

Additional loads may be connected to existing services and feeders under the following conditions:

1. The maximum demand kVA data for a minimum one-year period (or the 30-day alternate method from the exception) is available
2. The installation complies with 220.35(2) and (3).

220.36 Optional Calculation— New Restaurants.

Calculation of a service or feeder load, where the feeder serves the total load, for a new restaurant shall be permitted in accordance with Table 220.36 in lieu of Part II of this article.

The overload protection of the service conductors shall be in accordance with 230.90 and 240.4.

Feeder conductors shall not be required to be of greater ampacity than the service conductors.

Service or feeder conductors whose demand load is determined by this optional calculation shall be permitted to have the neutral load determined by 220.22.

Section 220.36 recognizes the effects of load diversity that are typical of restaurant occupancies. It also recognizes the amount of continuous loads as a percentage of the total connected load. The exact method of calculation presented in Table 220.36 was revised for the 2002 *Code* to more accurately reflect the original load study data.

The National Restaurant Association, the Edison Electric Institute, and the Electric Power Research Institute based the data for the change in 220.36 on load studies of 262 restaurants. These studies show that the demand factors were lower for restaurants with larger connected loads. Based on this information, it was determined that demand factors for restaurant loads are appropriate.

When using the optional method found in 220.36, it is important to notice that first, all loads are added together, even heating and air conditioning, and then the appropriate demand load is calculated from Table 220.36. The service or feeder size is calculated after application of the demand load factor.

Example 1

A new, all-electric restaurant has a total connected load of 348 kVA at 208Y/120 volts. Using Table 220.36, calculate the demand load and determine the size of the service-entrance conductors and the maximum-size overcurrent device for the service.

Solution

STEP 1. Use the value in Table 220.36 for a connected load of 348 kVA (Row 3, Column 2) to calculate the demand load for an all electric restaurant.

$$
\begin{aligned}
\text{Demand load} &= 50\% \text{ of amount over } 325 \text{ kVA} + 172.5 \text{ kVA} \\
&= 0.50 \times (348 \text{ kVA} - 325 \text{ kVA}) + 172.5 \text{ kVA} \\
&= (0.50 \times 23) + 172.5 \\
&= 11.5 + 172.5 \\
&= 184 \text{ kVA}
\end{aligned}
$$

STEP 2. Calculate the service size using the calculated demand load in Step 1.

$$
\begin{aligned}
\text{Service size} &= \frac{\text{kVA}_{\text{Demand load}} \times 1000}{\text{voltage} \times \sqrt{3}} \\
&= \frac{184 \text{ kVA} \times 1000}{208 \text{ V} \times \sqrt{3}} \\
&= 510.7 \quad \text{or} \quad 511 \text{ A}
\end{aligned}
$$

STEP 3. Determine the size of the overcurrent device. The next larger standard-size overcurrent device is 600 amperes. The minimum size of the conductors must be adequate to handle the load, but 240.4(B) permits the next larger standard-rated overcurrent device to be used.

Example 2

A new restaurant has gas cooking appliances plus a total connected electrical load of 348 kVA at 208Y/120 volts. Calculate the demand load using Table 220.36 and the service size. Then determine the maximum-size overcurrent device for the service.

Solution

STEP 1. Calculate the demand load for a new restaurant using the value in Table 220.36 for a connected load of 348 kVA (Column 3, Row 3) as follows:

Table 220.36 Optional Method—Permitted Load Calculations for Service and Feeder Conductors for New Restaurants

Total Connected Load (kVA)	All Electric Restaurant Calculated Loads (kVA)	Not All Electric Restaurant Calculated Loads (kVA)
0–200	80%	100%
201–325	10% (amount over 200) + 160.0	50% (amount over 200) + 200.0
326–800	50% (amount over 325) + 172.5	45% (amount over 325) + 262.5
Over 800	50% (amount over 800) + 410.0	20% (amount over 800) + 476.3

Note: Add all electrical loads, including both heating and cooling loads, to compute the total connected load. Select the one demand factor that applies from the table, and multiply the total connected load by this single demand factor.

$$\text{Demand load} = 45\% \text{ of amount over } 325 \text{ kVA} + 262.5 \text{ kVA}$$
$$= 0.45 \times (348 - 325) + 262.5$$
$$= (0.45 \times 23) + 262.5$$
$$= 10.35 + 262.5$$
$$= 272.85 \text{ kVA}$$

STEP 2. Calculate the service size using the calculated demand load in Step 1.

$$\text{Service size} = \frac{\text{kVA}_{\text{Demand load}} \times 1000}{\text{voltage} \times \sqrt{3}}$$
$$= \frac{272.85 \text{ kVA} \times 1000}{208 \text{ V} \times \sqrt{3}}$$
$$= 757.36 \quad \text{or} \quad 757 \text{ A}$$

STEP 3. Determine the size of the overcurrent device. The next higher standard rating for an overcurrent device, according to 240.6, is 800 amperes. Section 230.79 requires that the service disconnecting means have a rating that is not less than the computed load (757 amperes). The minimum size of the conductors must be adequate to handle the load, but 240.4(B) permits the next larger standard-rated overcurrent device to be used.

IV. Method for Computing Farm Loads

220.40 Farm Loads—Buildings and Other Loads.

(A) Dwelling Unit. The feeder or service load of a farm dwelling unit shall be computed in accordance with the provisions for dwellings in Part II or III of this article. Where the dwelling has electric heat and the farm has electric grain-drying systems, Part III of this article shall not be used to compute the dwelling load where the dwelling and farm load are supplied by a common service.

(B) Other Than Dwelling Unit. Where a feeder or service supplies a farm building or other load having two or more separate branch circuits, the load for feeders, service conductors, and service equipment shall be computed in accordance with demand factors not less than indicated in Table 220.40.

Table 220.40 Method for Computing Farm Loads for Other Than Dwelling Unit

Ampere Load at 240 Volts Maximum	Demand Factor (Percent)
Loads expected to operate without diversity, but not less than 125 percent full-load current of the largest motor and not less than the first 60 amperes of load	100
Next 60 amperes of all other loads	50
Remainder of other load	25

220.41 Farm Loads—Total.

Where supplied by a common service, the total load of the farm for service conductors and service equipment shall be computed in accordance with the farm dwelling unit load and demand factors specified in Table 220.41. Where there is equipment in two or more farm equipment buildings or for loads having the same function, such loads shall be computed in accordance with Table 220.40 and shall be permitted to be combined as a single load in Table 220.41 for computing the total load.

Table 220.41 Method for Computing Total Farm Load

Individual Loads Computed in Accordance with Table 220.40	Demand Factor (Percent)
Largest load	100
Second largest load	75
Third largest load	65
Remaining loads	50

Note: To this total load, add the load of the farm dwelling unit computed in accordance with Part II or III of this article. Where the dwelling has electric heat and the farm has electric grain-drying systems, Part III of this article shall not be used to compute the dwelling load.

ARTICLE 225
Outside Branch Circuits and Feeders

Contents

225.1 Scope.

This article covers requirements for outside branch circuits and feeders run on or between buildings, structures, or poles on the premises; and electric equipment and wiring for the supply of utilization equipment that is located on or attached to the outside of buildings, structures, or poles.

> FPN: For additional information on wiring over 600 volts, see ANSI C2-1997, *National Electrical Safety Code.*

225.2 Other Articles.

Application of other articles, including additional requirements to specific cases of equipment and conductors, is shown in Table 225.2.

I. General

225.3 Calculation of Loads 600 Volts, Nominal, or Less.

(A) Branch Circuits. The load on outdoor branch circuits shall be as determined by 220.3.

(B) Feeders. The load on outdoor feeders shall be as determined by Part II of Article 220.

225.4 Conductor Covering.

Where within 3.0 m (10 ft) of any building or structure other than supporting poles or towers, open individual (aerial) overhead conductors shall be insulated or covered. Conductors in cables or raceways, except Type MI cable, shall be of the rubber-covered type or thermoplastic type and, in wet locations, shall comply with 310.8. Conductors for festoon lighting shall be of the rubber-covered or thermoplastic type.

Table 225.2 Other Articles

Equipment/Conductors	Article
Branch circuits	210
Class 1, Class 2, and Class 3 remote-control, signaling, and power-limited circuits	725
Communications circuits	800
Community antenna television and radio distribution systems	820
Conductors for general wiring	310
Electrically driven or controlled irrigation machines	675
Electric signs and outline lighting	600
Feeders	215
Fire alarm systems	760
Fixed outdoor electric deicing and snow-melting equipment	426
Floating buildings	553
Grounding	250
Hazardous (classified) locations	500
Hazardous (classified) locations — specific	510
Marinas and boatyards	555
Messenger supported wiring	396
Open wiring on insulators	398
Over 600 volts, general	490
Overcurrent protection	240
Radio and television equipment	810
Services	230
Solar photovoltaic systems	690
Swimming pools, fountains, and similar installations	680
Use and identification of grounded conductors	200

Exception: Equipment grounding conductors and grounded circuit conductors shall be permitted to be bare or covered as specifically permitted elsewhere in this Code.

The exception to 225.4 and Exception No. 2 to 250.184(A) clarify that bare neutral messenger wire cable assemblies may be regrounded at additional buildings.

225.5 Size of Conductors 600 Volts, Nominal, or Less.

The ampacity of outdoor branch-circuit and feeder conductors shall be in accordance with 310.15 based on loads as determined under 220.3 and Part II of Article 220.

225.6 Conductor Size and Support.

(A) Overhead Spans. Open individual conductors shall not be smaller than the following:

(1) For 600 volts, nominal, or less, 10 AWG copper or 8 AWG aluminum for spans up to 15 m (50 ft) in length and 8 AWG copper or 6 AWG aluminum for a longer span, unless supported by a messenger wire
(2) For over 600 volts, nominal, 6 AWG copper or 4 AWG aluminum where open individual conductors and 8 AWG copper or 6 AWG aluminum where in cable

The size limitation of copper and aluminum conductors for overhead spans is based on the need for adequate mechanical strength to support the weight of the conductors and withstand wind, ice, and other similar conditions. Exhibit 225.1 illustrates overhead spans that are not messenger supported, that are run between buildings, structures, or poles, and that are 600 volts or less. If the conductors are supported on a messenger, the messenger cable provides the necessary mechanical strength. See 396.10 for wiring methods permitted to be messenger supported.

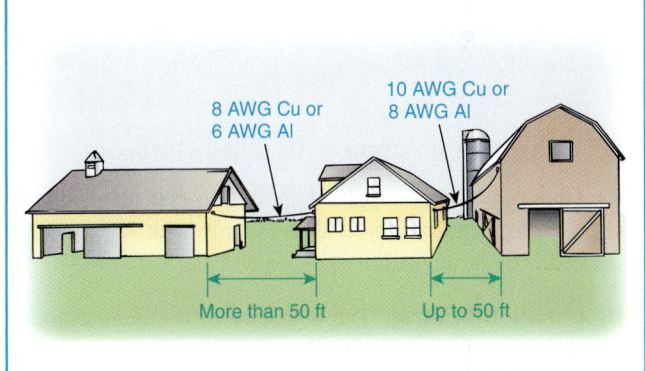

Exhibit 225.1 Minimum sizes of conductors in overhead spans as specified by 225.6(A)(1) for 600 volts, nominal, or less.

(B) Festoon Lighting. Overhead conductors for festoon lighting shall not be smaller than 12 AWG unless the conductors are supported by messenger wires. In all spans exceeding 12 m (40 ft), the conductors shall be supported by messenger wire. The messenger wire shall be supported by strain insulators. Conductors or messenger wires shall not be attached to any fire escape, downspout, or plumbing equipment.

Article 100 defines *festoon lighting* as a string of outdoor lights that is suspended between two points. The conductors for festoon lighting must be larger than 12 AWG unless a messenger wire supports them. On all spans of festoon lighting exceeding 40 ft, messenger wire is required and must be supported by strain insulators. See Exhibit 225.2. If no messenger wire is required, the 12 AWG or larger conductors are required to be supported by strain insulators.

Attachment of festoon lighting to fire escapes, plumbing

Exhibit 225.2 Messenger wire required by 225.6(B) for festoon lighting conductors in a span exceeding 40 ft.

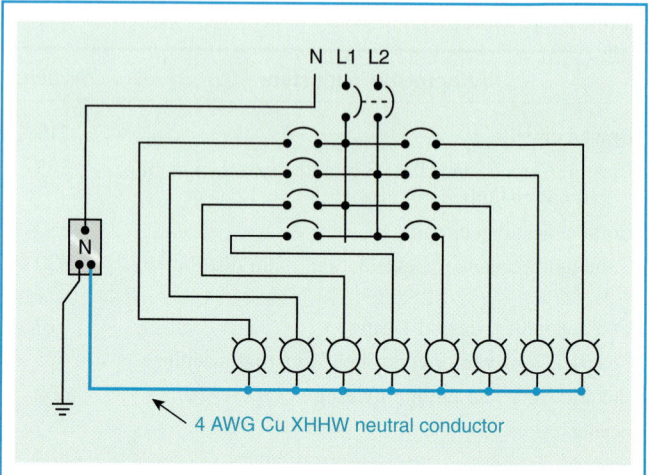

Exhibit 225.3 A 120/240-volt, single-phase, 3-wire system (branch circuits rated at 20 amperes; maximum unbalanced current of 80 amperes).

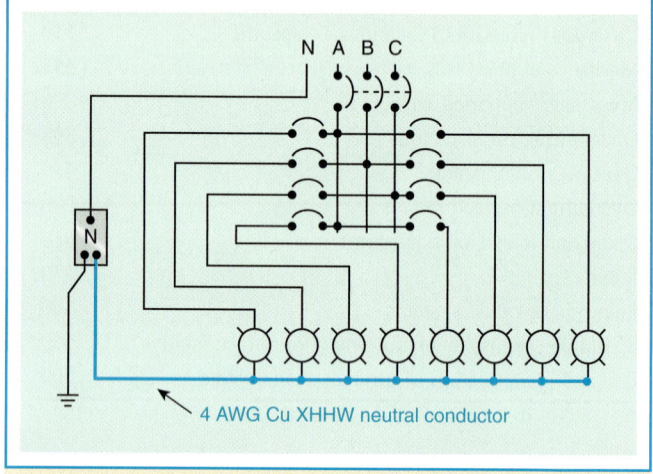

Exhibit 225.4 A 208Y/120-volt, 3-phase, 4-wire system (branch circuits rated at 20 amperes; maximum unbalanced current of 80 amperes).

equipment, or metal drain spouts is prohibited because the attachment could provide a path to ground. Moreover, such methods of attachment could not be relied on for a permanent or secure means of support.

225.7 Lighting Equipment Installed Outdoors.

(A) General. For the supply of lighting equipment installed outdoors, the branch circuits shall comply with Article 210 and 225.7(B) through (D).

(B) Common Neutral. The ampacity of the neutral conductor shall not be less than the maximum net computed load current between the neutral and all ungrounded conductors connected to any one phase of the circuit.

Multiwire branch circuits consisting of a neutral and two or more ungrounded conductors are permitted, provided the neutral capacity is not less than the total load of all ungrounded conductors connected to any one phase of the circuit.

Exhibit 225.3 illustrates a 120/240-volt, single-phase, 3-wire system, and Exhibit 225.4 illustrates a 208Y/120-volt, 3-phase, 4-wire system. In Exhibit 225.3, all branch circuits are rated at 20 amperes. The maximum unbalanced current that can occur is four times 20 amperes, or 80 amperes. In Exhibit 225.4, all branch circuits also are rated at 20 amperes. The maximum unbalanced current that can occur on a 3-phase system with the load connected as shown is 80 amperes, due to the load on phase A.

(C) 277 Volts to Ground. Circuits exceeding 120 volts, nominal, between conductors and not exceeding 277 volts, nominal, to ground shall be permitted to supply luminaires (lighting fixtures) for illumination of outdoor areas of indus-

trial establishments, office buildings, schools, stores, and other commercial or public buildings where the luminaires (fixtures) are not less than 900 mm (3 ft) from windows, platforms, fire escapes, and the like.

Branch circuits for the outdoor illumination of industrial establishments, office buildings, schools, stores, and other commercial or public buildings are permitted to operate with a voltage to ground of 277 volts. See 210.6(D) for voltages greater than 277 volts to ground. The restrictions outlined in 225.7(C) are in addition to those in 210.6(C).

(D) 600 Volts Between Conductors. Circuits exceeding 277 volts, nominal, to ground and not exceeding 600 volts, nominal, between conductors shall be permitted to supply the auxiliary equipment of electric-discharge lamps in accordance with 210.6(D)(1).

Section 210.6(D)(1) contains the minimum height requirements for circuits exceeding 277 volts, nominal, to ground but not exceeding 600 volts, nominal, between conductors for circuits that supply the auxiliary equipment of electric-discharge lamps.

225.9 Overcurrent Protection.

Overcurrent protection shall be in accordance with 210.20 for branch circuits and Article 240 for feeders.

Section 225.9 ensures that feeder overcurrent devices meet the same general accessibility requirements as service overcurrent devices. Accessibility to overcurrent protection devices is also covered in 225.40.

225.10 Wiring on Buildings.

The installation of outside wiring on surfaces of buildings shall be permitted for circuits of not over 600 volts, nominal, as open wiring on insulators, as multiconductor cable, as Type MC cable, as Type MI cable, as messenger supported wiring, in rigid metal conduit, in intermediate metal conduit, in rigid nonmetallic conduit, in cable trays, as cablebus, in wireways, in auxiliary gutters, in electrical metallic tubing, in flexible metal conduit, in liquidtight flexible metal conduit, in liquidtight flexible nonmetallic conduit, and in busways. Circuits of over 600 volts, nominal, shall be installed as provided in 300.37. Circuits for signs and outline lighting shall be installed in accordance with Article 600.

225.11 Circuit Exits and Entrances.

Where outside branch and feeder circuits leave or enter a building, the requirements of 230.52 and 230.54 shall apply.

Section 225.11 references 230.52, Individual Conductors Entering Buildings or Other Structures, and 230.54, Overhead Service Locations. Section 225.18 covers the requirements for clearances from ground (not over 600 volts), and 225.19 covers the requirements for clearances from buildings for conductors not over 600 volts. See 225.19(D) for final span clearances from windows, doors, fire escapes, and so on. Exhibit 225.5 shows an example where these requirements apply.

225.12 Open-Conductor Supports.

Open conductors shall be supported on glass or porcelain knobs, racks, brackets, or strain insulators.

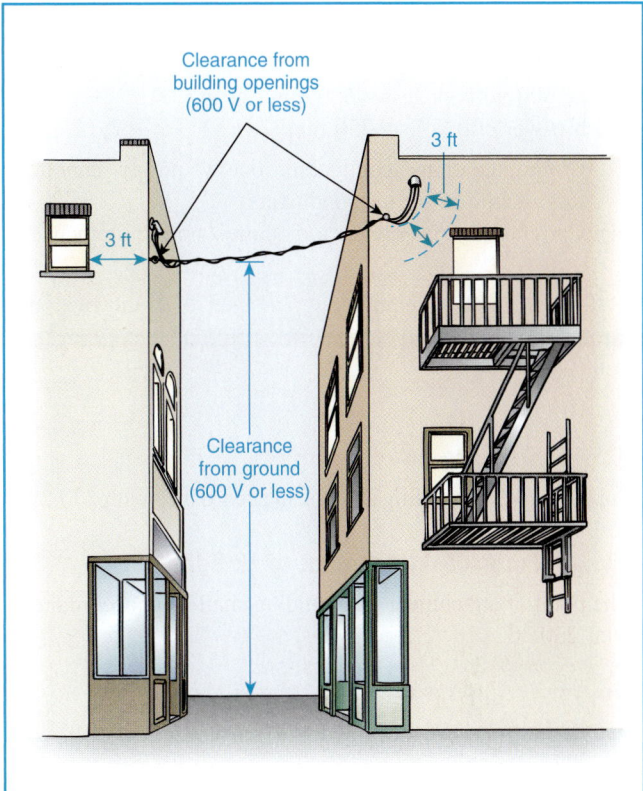

Exhibit 225.5 Section 225.11 references 230.52, Individual Conductors Entering Buildings or Other Structures, and 230.54, Overhead Service Locations. Section 225.18 covers the requirements for clearances from ground for conductors not over 600 volts and 225.19 covers the requirements for clearances from buildings for conductors not over 600 volts. See 225.19(D) for clearances from windows, doors, fire escapes, and so on.

225.14 Open-Conductor Spacings.

(A) 600 Volts, Nominal, or Less. Conductors of 600 volts, nominal, or less, shall comply with the spacings provided in Table 230.51(C).

(B) Over 600 Volts, Nominal. Conductors of over 600 volts, nominal, shall comply with the spacings provided in 110.36 and 490.24.

(C) Separation from Other Circuits. Open conductors shall be separated from open conductors of other circuits or systems by not less than 100 mm (4 in.).

(D) Conductors on Poles. Conductors on poles shall have a separation of not less than 300 mm (1 ft) where not placed on racks or brackets. Conductors supported on poles shall provide a horizontal climbing space not less than the following:

(1) Power conductors below communications conductors—750 mm (30 in.)

(2) Power conductors alone or above communications conductors:

    a. 300 volts or less—600 mm (24 in.)

    b. Over 300 volts—750 mm (30 in.)

(3) Communications conductors below power conductors—same as power conductors

(4) Communications conductors alone—no requirement

Sufficient space is required for linemen to climb over or through conductors to safely work with conductors on the pole.

225.15 Supports over Buildings.

Supports over a building shall be in accordance with 230.29.

225.16 Point of Attachment to Buildings.

The point of attachment to a building shall be in accordance with 230.26.

225.17 Means of Attachment to Buildings.

The means of attachment to a building shall be in accordance with 230.27.

225.18 Clearance from Ground.

Overhead spans of open conductors and open multiconductor cables of not over 600 volts, nominal, shall conform to the following:

(1) 3.0 m (10 ft)—above finished grade, sidewalks, or from any platform or projection from which they might be reached where the voltage does not exceed 150 volts to ground and accessible to pedestrians only

(2) 3.7 m (12 ft)—over residential property and driveways, and those commercial areas not subject to truck traffic where the voltage does not exceed 300 volts to ground

(3) 4.5 m (15 ft)—for those areas listed in the 3.7-m (12-ft) classification where the voltage exceeds 300 volts to ground

(4) 5.5 m (18 ft)—over public streets, alleys, roads, parking areas subject to truck traffic, driveways on other than residential property, and other land traversed by vehicles, such as cultivated, grazing, forest, and orchard

225.19 Clearances from Buildings for Conductors of Not Over 600 Volts, Nominal.

(A) Above Roofs. Overhead spans of open conductors and open multiconductor cables shall have a vertical clearance of not less than 2.5 m (8 ft) above the roof surface. The vertical clearance above the roof level shall be maintained for a distance not less than 900 mm (3 ft) in all directions from the edge of the roof.

Exception No. 1: The area above a roof surface subject to pedestrian or vehicular traffic shall have a vertical clearance from the roof surface in accordance with the clearance requirements of 225.18.

Exception No. 2: Where the voltage between conductors does not exceed 300, and the roof has a slope of 100 mm (4 in.) in 300 mm (12 in.) or greater, a reduction in clearance to 900 mm (3 ft) shall be permitted.

Exception No. 3: Where the voltage between conductors does not exceed 300, a reduction in clearance above only the overhanging portion of the roof to not less than 450 mm (18 in.) shall be permitted if (1) not more than 1.8 m (6 ft) of the conductors, 1.2 m (4 ft) horizontally, pass above the roof overhang and (2) they are terminated at a through-the-roof raceway or approved support.

Exception No. 4: The requirement for maintaining the vertical clearance 900 mm (3 ft) from the edge of the roof shall not apply to the final conductor span where the conductors are attached to the side of a building.

(B) From Nonbuilding or Nonbridge Structures. From signs, chimneys, radio and television antennas, tanks, and other nonbuilding or nonbridge structures, clearances—vertical, diagonal, and horizontal—shall not be less than 900 mm (3 ft).

(C) Horizontal Clearances. Clearances shall not be less than 900 mm (3 ft).

(D) Final Spans. Final spans of feeders or branch circuits shall comply with 225.19(D)(1), (2), and (3).

(1) Clearance from Windows. Final spans to the building they supply, or from which they are fed, shall be permitted to be attached to the building, but they shall be kept not less than 900 mm (3 ft) from windows that are designed to be opened, and from doors, porches, balconies, ladders, stairs, fire escapes, or similar locations.

Exception: Conductors run above the top level of a window shall be permitted to be less than the 900-mm (3-ft) requirement.

(2) Vertical Clearance. The vertical clearance of final spans above, or within 900 mm (3 ft) measured horizontally of, platforms, projections, or surfaces from which they might be reached shall be maintained in accordance with 225.18.

(3) Building Openings. The overhead branch-circuit and feeder conductors shall not be installed beneath openings through which materials may be moved, such as openings in farm and commercial buildings, and shall not be installed where they obstruct entrance to these buildings' openings.

(E) Zone for Fire Ladders. Where buildings exceed three stories or 15 m (50 ft) in height, overhead lines shall be arranged, where practicable, so that a clear space (or zone) at least 1.8 m (6 ft) wide will be left either adjacent to the buildings or beginning not over 2.5 m (8 ft) from them to facilitate the raising of ladders when necessary for fire fighting.

225.20 Mechanical Protection of Conductors.

Mechanical protection of conductors on buildings, structures, or poles shall be as provided for services in 230.50.

225.21 Multiconductor Cables on Exterior Surfaces of Buildings.

Supports for multiconductor cables on exterior surfaces of buildings shall be as provided in 230.51.

225.22 Raceways on Exterior Surfaces of Buildings or Other Structures.

Raceways on exterior surfaces of buildings or other structures shall be raintight and arranged to drain.

Exception: Flexible metal conduit, where permitted in 348.12(1), shall not be required to be raintight.

Raintight is defined in Article 100 as constructed or protected so that exposure to a beating rain will not result in the entrance of water under specified test conditions. To ensure this, all conduit bodies, fittings, and boxes used in wet locations are required to be provided with threaded hubs or other approved means. Threadless couplings and connectors used with metal conduit or electrical metallic tubing installed on the exterior of a building must be of the raintight type [see 342.42(A), 344.42(A), and 348.42].

If raceways are exposed to weather or rain through weatherhead openings, condensation is likely to occur, causing moisture to accumulate within raceways at low points of the installation and in junction boxes. Therefore, raceways should be installed to permit drainage through drain holes at appropriate locations.

225.24 Outdoor Lampholders.

Where outdoor lampholders are attached as pendants, the connections to the circuit wires shall be staggered. Where such lampholders have terminals of a type that puncture the insulation and make contact with the conductors, they shall be attached only to conductors of the stranded type.

Splices to branch-circuit conductors for outdoor lampholders of the Edison-base type or "pigtail" sockets are required to be staggered so that splices will not be in close proximity to each other. Pin-type terminal sockets must be attached to stranded conductors only and are intended for installations for temporary lighting or decorations, signs, or specifically approved applications.

225.25 Location of Outdoor Lamps.

Locations of lamps for outdoor lighting shall be below all energized conductors, transformers, or other electric utilization equipment, unless

(1) Clearances or other safeguards are provided for relamping operations, or
(2) Equipment is controlled by a disconnecting means that can be locked in the open position.

Because 225.18 requires a minimum clearance for open conductors of 10 ft above grade or platforms, it may be difficult to keep all electrical equipment above the lamps. Section 225.25(1) allows other clearances or safeguards to permit safe relamping, and 225.25(2) permits the use of a disconnecting means to de-energize the circuit.

225.26 Vegetation as Support.

Vegetation such as trees shall not be used for support of overhead conductor spans.

Where overhead conductor spans are attached to a tree, normal tree growth around the attachment device causes the mounting insulators to break and the conductor insulation to be degraded. The requirement in 225.26 reduces the likelihood of chafing of the conductor insulation and the danger of shock to tree trimmers and tree climbers. The 2002 *Code* has deleted the exception to 225.26 that appeared in previous editions, and trees are no longer permitted as support of overhead conductor spans, even on a temporary basis. However, outdoor luminaires and associated equipment are permitted by 410.16(H) to be supported by trees. To prevent the chafing damage, conductors are run up the tree from an underground wiring method. See 300.5(D) for protection of conductors.

II. More Than One Building or Other Structure

Part II covers outside branch circuits and feeders on single managed properties with more than one building or structure. Part II covers the feeders that supply buildings on college campuses, multibuilding industrial facilities, multibuilding commercial facilities, and so on, provided that these buildings are under one management. Many of these requirements for feeder disconnecting means are similar to the requirements for services and are included in Part II.

225.30 Number of Supplies.

Where more than one building or other structure is on the same property and under single management, each additional building or other structure served that is on the load side of the service disconnecting means shall be supplied by one feeder or branch circuit unless permitted in 225.30(A) through (E). For the purpose of this section, a multiwire branch circuit shall be considered a single circuit.

(A) Special Conditions. Additional feeders or branch circuits shall be permitted to supply the following:

(1) Fire pumps
(2) Emergency systems
(3) Legally required standby systems
(4) Optional standby systems
(5) Parallel power production systems

(B) Special Occupancies. By special permission, additional feeders or branch circuits shall be permitted for the following:

(1) Multiple-occupancy buildings where there is no space available for supply equipment accessible to all occupants, or
(2) A single building or other structure sufficiently large to make two or more supplies necessary.

(C) Capacity Requirements. Additional feeders or branch circuits shall be permitted where the capacity requirements are in excess of 2000 amperes at a supply voltage of 600 volts or less.

(D) Different Characteristics. Additional feeders or branch circuits shall be permitted for different voltages, frequencies, or phases or for different uses, such as control of outside lighting from multiple locations.

(E) Documented Switching Procedures. Additional feeders or branch circuits shall be permitted to supply installations under single management where documented safe switching procedures are established and maintained for disconnection.

Buildings on college campuses, multibuilding industrial facilities, and multibuilding commercial facilities are permitted to be supplied by secondary loop supply (secondary selective) networks, provided that documented switching

procedures are established. These switching procedures must establish a method to safely operate switches for the facility during maintenance and during alternate supply and emergency supply conditions. Keyed interlock systems are often used to reduce the likelihood of inappropriate switching procedures that could result in hazardous conditions.

225.31 Disconnecting Means.

Means shall be provided for disconnecting all ungrounded conductors that supply or pass through the building or structure.

225.32 Location.

The disconnecting means shall be installed either inside or outside of the building or structure served or where the conductors pass through the building or structure. The disconnecting means shall be at a readily accessible location nearest the point of entrance of the conductors. For the purposes of this section, the requirements in 230.6 shall be permitted to be utilized.

Exception No. 1: For installations under single management, where documented safe switching procedures are established and maintained for disconnection, and where the installation is monitored by qualified individuals, the disconnecting means shall be permitted to be located elsewhere on the premises.

Exception No. 2: For buildings or other structures qualifying under the provisions of Article 685, the disconnecting means shall be permitted to be located elsewhere on the premises.

Exception No. 3: For towers or poles used as lighting standards, the disconnecting means shall be permitted to be located elsewhere on the premises.

Exception No. 4: For poles or similar structures used only for support of signs installed in accordance with Article 600, the disconnecting means shall be permitted to be located elsewhere on the premises.

225.33 Maximum Number of Disconnects.

(A) General. The disconnecting means for each supply permitted by 225.30 shall consist of not more than six switches or six circuit breakers mounted in a single enclosure, in a group of separate enclosures, or in or on a switchboard. There shall be no more than six disconnects per supply grouped in any one location.

Exception: For the purposes of this section, disconnecting means used solely for the control circuit of the ground-fault protection system, or the control circuit of the power-operated supply disconnecting means, installed as part of the listed equipment, shall not be considered a supply disconnecting means.

(B) Single-Pole Units. Two or three single-pole switches or breakers capable of individual operation shall be permitted on multiwire circuits, one pole for each ungrounded conductor, as one multipole disconnect, provided they are equipped with handle ties or a master handle to disconnect all ungrounded conductors with no more than six operations of the hand.

225.34 Grouping of Disconnects.

(A) General. The two to six disconnects as permitted in 225.33 shall be grouped. Each disconnect shall be marked to indicate the load served.

Exception: One of the two to six disconnecting means permitted in 225.33, where used only for a water pump also intended to provide fire protection, shall be permitted to be located remote from the other disconnecting means.

(B) Additional Disconnecting Means. The one or more additional disconnecting means for fire pumps or for emergency, legally required standby or optional standby system permitted by 225.30 shall be installed sufficiently remote from the one to six disconnecting means for normal supply to minimize the possibility of simultaneous interruption of supply.

225.35 Access to Occupants.

In a multiple-occupancy building, each occupant shall have access to the occupant's supply disconnecting means.

Exception: In a multiple-occupancy building where electric supply and electrical maintenance are provided by the building management and where these are under continuous building management supervision, the supply disconnecting means supplying more than one occupancy shall be permitted to be accessible to authorized management personnel only.

225.36 Suitable for Service Equipment.

The disconnecting means specified in 225.31 shall be suitable for use as service equipment.

Exception: For garages and outbuildings on residential property, a snap switch or a set of 3-way or 4-way snap switches shall be permitted as the disconnecting means.

225.37 Identification.

Where a building or structure has any combination of feeders, branch circuits, or services passing through it or supplying it, a permanent plaque or directory shall be installed at each feeder and branch-circuit disconnect location denoting all other services, feeders, or branch circuits supplying that building or structure or passing through that building or structure and the area served by each.

Exception No. 1: A plaque or directory shall not be required for large-capacity multibuilding industrial installations under single management, where it is ensured that disconnection can be accomplished by establishing and maintaining safe switching procedures.

Exception No. 2: This identification shall not be required for branch circuits installed from a dwelling unit to a second building or structure.

225.38 Disconnect Construction.

Disconnecting means shall meet the requirements of 225.38(A) through (D).

Exception: For garages and outbuildings on residential property, snap switches or sets of 3-way or 4-way snap switches shall be permitted as the disconnecting means.

(A) Manually or Power Operable. The disconnecting means shall consist of either (1) a manually operable switch or a circuit breaker equipped with a handle or other suitable operating means or (2) a power-operable switch or circuit breaker, provided the switch or circuit breaker can be opened by hand in the event of a power failure.

(B) Simultaneous Opening of Poles. Each building or structure disconnecting means shall simultaneously disconnect all ungrounded supply conductors that it controls from the building or structure wiring system.

(C) Disconnection of Grounded Conductor. Where the building or structure disconnecting means does not disconnect the grounded conductor from the grounded conductors in the building or structure wiring, other means shall be provided for this purpose at the location of disconnecting means. A terminal or bus to which all grounded conductors can be attached by means of pressure connectors shall be permitted for this purpose.

In a multisection switchboard, disconnects for the grounded conductor shall be permitted to be in any of the switchboard, provided any such switchboard is marked.

(D) Indicating. The building or structure disconnecting means shall plainly indicate whether it is in the open or closed position.

225.39 Rating of Disconnect.

The feeder or branch-circuit disconnecting means shall have a rating of not less than the load to be carried, determined in accordance with Article 220. In no case shall the rating be lower than specified in 225.39(A), (B), (C), or (D).

(A) One-Circuit Installation. For installations to supply only limited loads of a single branch circuit, the branch circuit disconnecting means shall have a rating of not less than 15 amperes.

(B) Two-Circuit Installations. For installations consisting of not more than two 2-wire branch circuits, the feeder or branch-circuit disconnecting means shall have a rating of not less than 30 amperes.

(C) One-Family Dwelling. For a one-family dwelling, the feeder disconnecting means shall have a rating of not less than 100 amperes, 3-wire.

(D) All Others. For all other installations, the feeder or branch-circuit disconnecting means shall have a rating of not less than 60 amperes.

225.40 Access to Overcurrent Protective Devices.

Where a feeder overcurrent device is not readily accessible, branch-circuit overcurrent devices shall be installed on the load side, shall be mounted in a readily accessible location, and shall be of a lower ampere rating than the feeder overcurrent device.

III. Over 600 Volts

225.50 Sizing of Conductors.

The sizing of conductors over 600 volts shall be in accordance with 210.19(B) for branch circuits and 215.2(B) for feeders.

225.51 Isolating Switches.

Where oil switches or air, oil, vacuum, or sulfur hexafluoride circuit breakers constitute a building disconnecting means, an isolating switch with visible break contacts and meeting the requirements of 230.204(B), (C), and (D) shall be installed on the supply side of the disconnecting means and all associated equipment.

Exception: The isolating switch shall not be required where the disconnecting means is mounted on removable truck panels or metal-enclosed switchgear units that cannot be opened unless the circuit is disconnected and that, when removed from the normal operating position, automatically disconnect the circuit breaker or switch from all energized parts.

225.52 Location.

A building or structure disconnecting means shall be located in accordance with 225.31, or it shall be electrically operated by a similarly located remote-control device.

225.53 Type.

Each building or structure disconnect shall simultaneously disconnect all ungrounded supply conductors it controls and shall have a fault-closing rating not less than the maximum available short-circuit current available at its supply terminals.

The requirement for a disconnecting means for over 600-volt feeders to buildings or structures is similar to the requirements found in 230.205(B). This disconnect can be an air, oil, SF6, or vacuum breaker. It also can be an air, oil, SF6, or vacuum switch. Where a switch is used, fuses are permitted to help with the fault-closing capability of the switch. The common practices of using fused load-break cutouts to switch sections of overhead lines and using load-break elbows to switch sections of underground lines are permitted. However, the building disconnecting means must be gang-operated to simultaneously open and close all ungrounded supply conductors. Load-break elbows and fused cutouts cannot be used as the building disconnection means.

Where fused switches or separately mounted fuses are installed, the fuse characteristics shall be permitted to contribute to the fault closing rating of the disconnecting means.

225.60 Clearances over Roadways, Walkways, Rail, Water, and Open Land.

(A) 22 kV Nominal to Ground or Less. The clearances over roadways, walkways, rail, water, and open land for conductors and live parts up to 22 kV nominal to ground or less shall be not less than the values shown in Table 225.60.

Table 225.60 Clearances over Roadways, Walkways, Rail, Water, and Open Land

| | Clearance | |
Location	m	ft
Open land subject to vehicles, cultivation, or grazing	5.6	18.5
Roadways, driveways, parking lots, and alleys	5.6	18.5
Walkways	4.1	13.5
Rails	8.1	26.5
Spaces and ways for pedestrians and restricted traffic	4.4	14.5
Water areas not suitable for boating	5.2	17

(B) Over 22 kV Nominal to Ground. Clearances for the categories shown in Table 225.60 shall be increased by 10 mm (0.4 in.) per kV above 22,000 volts.

(C) Special Cases. For special cases, such as where crossings will be made over lakes, rivers, or areas using large vehicles such as mining operations, specific designs shall be engineered considering the special circumstances and shall be approved by the authority having jurisdiction.

FPN: For additional information, see ANSI C2-1997, *National Electrical Safety Code.*

New in the 2002 *Code*, 225.60 adds clearance requirements and specific distances that are in harmony with the *National Electrical Safety Code® (NESC®).*

225.61 Clearances over Buildings and Other Structures.

(A) 22 kV Nominal to Ground or Less. The clearances over buildings and other structures for conductors and live parts up to 22 kV, nominal, to ground or less shall be not less than the values shown in Table 225.61.

Table 225.61 Clearances over Buildings and Other Structures

Clearance from Conductors or Live Parts from:	Horizontal		Vertical	
	m	ft	m	ft
Building walls, projections, and windows	2.3	7.5	—	—
Balconies, catwalks, and similar areas accessible to people	2.3	7.5	4.1	13.5
Over or under roofs or projections not readily accessible to people	—	—	3.8	12.5
Over roofs accessible to vehicles but not trucks	—	—	4.1	13.5
Over roofs accessible to trucks	—	—	5.6	18.5
Other structures	2.3	7.5	—	—

(B) Over 22 kV Nominal to Ground. Clearances for the categories shown in Table 225.61 shall be increased by 10 mm (0.4 in.) per kV above 22,000 volts.

FPN: For additional information see ANSI C2-1997, *National Electrical Safety Code.*

New in the 2002 *Code*, 225.61 adds clearance requirements and specific distances over buildings and structures that are in harmony with the *National Electrical Safety Code (NESC).*

ARTICLE 230
Services

Contents

230.1 Scope.

This article covers service conductors and equipment for control and protection of services and their installation requirements.

FPN: See Figure 230.1.

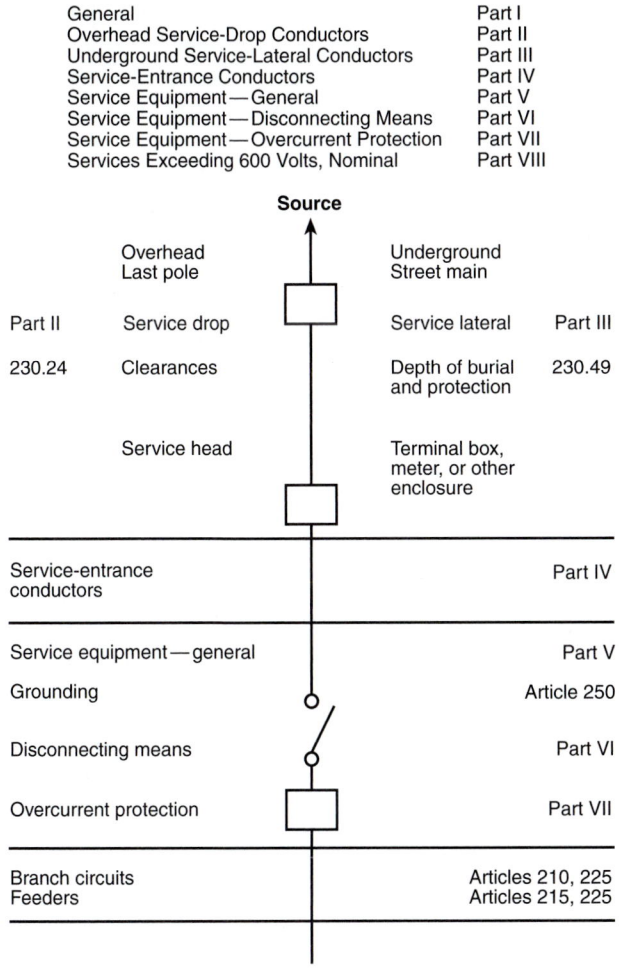

			Part I
General			Part I
Overhead Service-Drop Conductors			Part II
Underground Service-Lateral Conductors			Part III
Service-Entrance Conductors			Part IV
Service Equipment—General			Part V
Service Equipment—Disconnecting Means			Part VI
Service Equipment—Overcurrent Protection			Part VII
Services Exceeding 600 Volts, Nominal			Part VIII

Source

	Overhead Last pole		Underground Street main	
Part II	Service drop		Service lateral	Part III
230.24	Clearances		Depth of burial and protection	230.49
	Service head		Terminal box, meter, or other enclosure	

Service-entrance conductors	Part IV
Service equipment—general	Part V
Grounding	Article 250
Disconnecting means	Part VI
Overcurrent protection	Part VII
Branch circuits	Articles 210, 225
Feeders	Articles 215, 225

Figure 230.1 Services.

The requirements for the subjects covered in Parts I through VIII of Article 230 are arranged as follows:

Subject	Part	Section
General	I	230.2–230.10
Overhead Service-Drop Conductors	II	230.22–230.29
Underground Service-Lateral Conductors	III	230.30–230.33
Service-Entrance Conductors	IV	230.40–230.56
Service Equipment—General	V	230.62–230.66
Service Equipment—Disconnecting Means	VI	230.70–230.82
Service Equipment—Overcurrent Protection	VII	230.90–230.95
Services Exceeding 600 Volts, Nominal	VIII	230.200–230.212

I. General

230.2 Number of Services.

A building or other structure served shall be supplied by only one service unless permitted in 230.2(A) through (D). For the purpose of 230.40, Exception No. 2 only, underground sets of conductors, 1/0 AWG and larger, running to the same location and connected together at their supply end but not connected together at their load end shall be considered to be supplying one service.

The basic requirement of 230.2 is that a building or other structure can be supplied by only one service. However, under certain conditions, when a single service may not be adequate to supply a building or structure, 230.2 permits the use of additional services. Sections 230.2(A) through (D) describe those conditions under which more than one service is permitted. If more than one service is installed, 230.2(E) requires that a permanent plaque or directory be installed.

(A) Special Conditions. Additional services shall be permitted to supply the following:

(1) Fire pumps
(2) Emergency systems
(3) Legally required standby systems
(4) Optional standby systems
(5) Parallel power production systems

The intent of 230.2(A) is that a disruption of the main building service should not disconnect fire pump equipment or emergency systems, legally required standby systems, optional standby systems, or parallel power production systems.

(B) Special Occupancies. By special permission, additional services shall be permitted for the following:

(1) Multiple-occupancy buildings where there is no available space for service equipment accessible to all occupants, or
(2) A single building or other structure sufficiently large to make two or more services necessary

Section 230.2(B) permits additional services for certain occupancies by special permission (the written consent of the authority having jurisdiction; see the definition of *authority having jurisdiction* in Article 100). A separate service in multiple-occupancy buildings may be necessary if no space is available for service equipment accessible to all occupants. Also, if a building is large, more than one service is permitted. The expansion of buildings, shopping centers, and industrial plants often necessitates the addition of one or more services. It may, for example, be impractical or impossible to install one service for an industrial plant with sufficient capacity for any and all future loads. It is also impractical to run extremely long feeders. The serving utility and the authority having jurisdiction should be consulted before the use of this special permission is contemplated.

(C) Capacity Requirements. Additional services shall be permitted under any of the following:

(1) Where the capacity requirements are in excess of 2000 amperes at a supply voltage of 600 volts or less
(2) Where the load requirements of a single-phase installation are greater than the serving agency normally supplies through one service
(3) By special permission

Section 230.2(C) permits a building or structure to be served by two or more services if capacity requirements are in excess of 2000 amperes at a supply voltage of 600 or less. Additional services for lesser loads are also allowed by special permission.

Many electric power companies have specifications for, and have adopted special regulations covering, certain types of electrical loads and service equipment that may be energized from their lines. Consultation with the serving utility is advised to determine line and transformer capacities before electrical services for large buildings and facilities are designed.

(D) Different Characteristics. Additional services shall be permitted for different voltages, frequencies, or phases, or for different uses, such as for different rate schedules.

Section 230.2(D) permits the installation of more than one service for different characteristics, such as different voltages, frequencies, single-phase or 3-phase services, or different utility rate schedules. For example, different service characteristics exist between a 3-wire, 120/240-volt, single-phase service and a 3-phase, 4-wire, 480Y/277-volt service. For different applications, such as different rate schedules, this requirement allows a second service for supplying a second meter on a different rate. Curtailable loads, interruptible loads, electric heating, and electric water heating are examples of loads that may be on a different rate.

(E) Identification. Where a building or structure is supplied by more than one service, or any combination of branch circuits, feeders, and services, a permanent plaque or directory shall be installed at each service disconnect location denoting all other services, feeders, and branch circuits supplying that building or structure and the area served by each. See 225.37.

Section 230.2(E) states that where any combination of branch circuits, feeders, and services supplies power to a building or structure, a permanent plaque or directory must be installed at each service disconnect location to indicate where the other disconnects that feed the building are located, as illustrated in Exhibit 230.1. All the other services on or in the building or structure and the area served by each must be noted on the plaque or directory. This plaque or directory should be of sufficient durability to withstand the ambient environment.

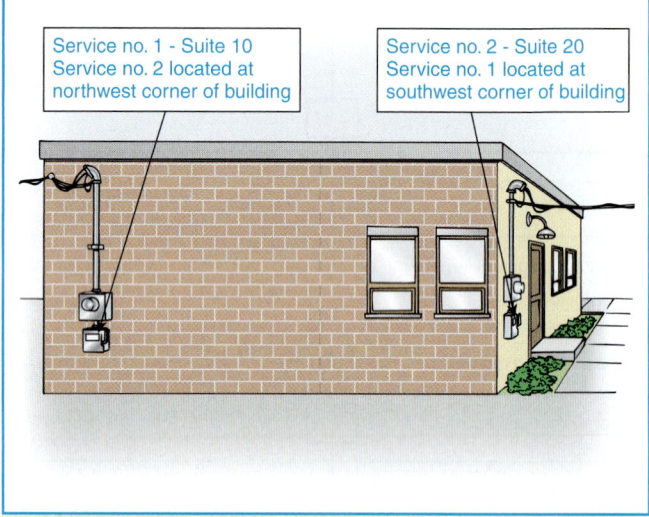

Service no. 1 - Suite 10
Service no. 2 located at northwest corner of building

Service no. 2 - Suite 20
Service no. 1 located at southwest corner of building

Exhibit 230.1 An example of more than one service installed for one building and a permanent plaque or directory denoting all other services and the area served by each.

Exhibits 230.2 through 230.13 illustrate examples of permitted service configurations. The figures are intended to clarify some of the *Code* rules that affect services but that are often misunderstood. No attempt has been made to include every type of service arrangement. It should be understood that the term *one location,* as applied to services, is determined by the authority having jurisdiction.

230.3 One Building or Other Structure Not to Be Supplied Through Another.

Service conductors supplying a building or other structure shall not pass through the interior of another building or other structure.

Service conductors are permitted to be installed along the exterior of one building to supply another building. However, service conductors supplying a building are not permitted to pass through the interior of a building. Each building served in this manner is required to be provided with a

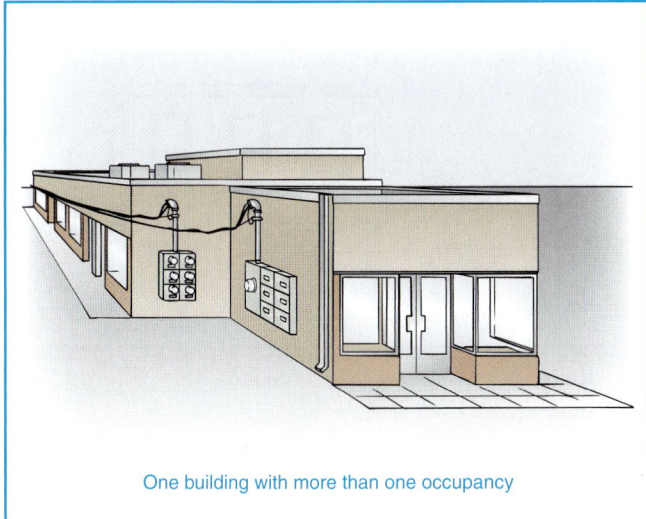

One building with more than one occupancy

Exhibit 230.2 More than one service and no available space accessible to all occupants.

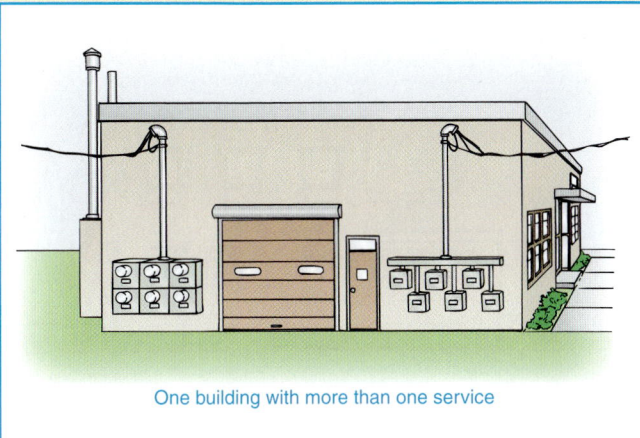

One building with more than one service

Exhibit 230.3 Two services supplying a building that has capacity requirements exceeding 2000 amperes.

One building with two groups of occupancies

Exhibit 230.4 One service with two sets of service-entrance conductors that supply two groups of disconnects in different locations.

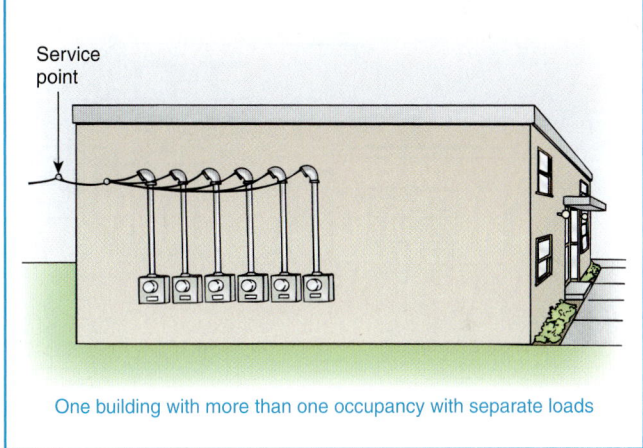

Service point

One building with more than one occupancy with separate loads

Exhibit 230.5 One service with six sets of service-entrance conductors supplying six disconnects in separate enclosures grouped in one location.

disconnecting means for all ungrounded conductors, in accordance with Part VI, Service Equipment—Disconnecting Means.

For example, in Exhibit 230.14, Building No. 2 service is *not* to be supplied through Building No. 1 service. The service disconnecting means shown for Building No. 1 and Building No. 2 are located on the exterior walls. A disconnecting means suitable for use as service equipment is required to be provided for each building.

230.6 Conductors Considered Outside the Building.

Conductors shall be considered outside of a building or other structure under any of the following conditions:

(1) Where installed under not less than 50 mm (2 in.) of concrete beneath a building or other structure
(2) Where installed within a building or other structure in a raceway that is encased in concrete or brick not less than 50 mm (2 in.) thick

Service-entrance conductors are considered to be "outside" a building if they are installed beneath the building under not less than 2 in. of concrete or concealed in a raceway within the building and encased by not less than 2 in. of concrete or brick, as illustrated in Exhibit 230.15. Service-entrance conductors are also considered outside the building

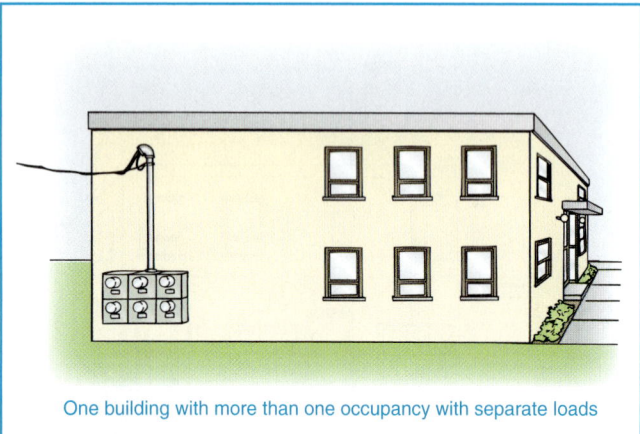

One building with more than one occupancy with separate loads

Exhibit 230.6 One service supplying six sets of service-entrance conductors supplying six disconnects all in one enclosure (optional arrangement of Exhibit 230.5).

One building with more than one occupancy with separate loads

Exhibit 230.7 One service lateral consisting of not more than six sets of conductors, 1/0 or larger.

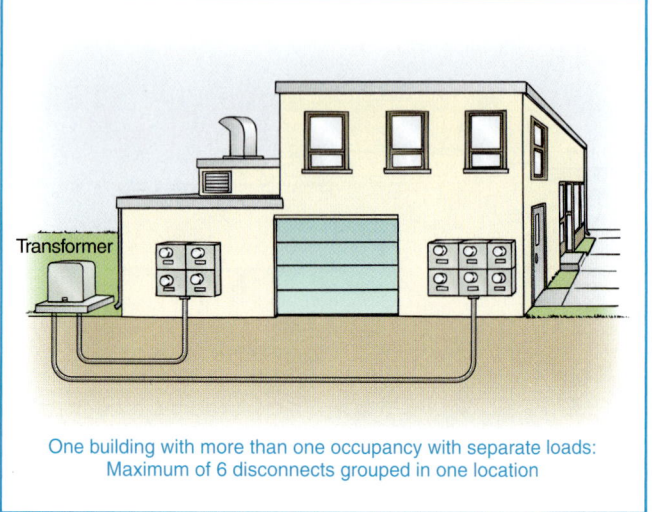

One building with more than one occupancy with separate loads: Maximum of 6 disconnects grouped in one location

Exhibit 230.8 Two service laterals consisting of not more than six sets of conductors each, 1/0 or larger, supplying two groups of loads in different locations.

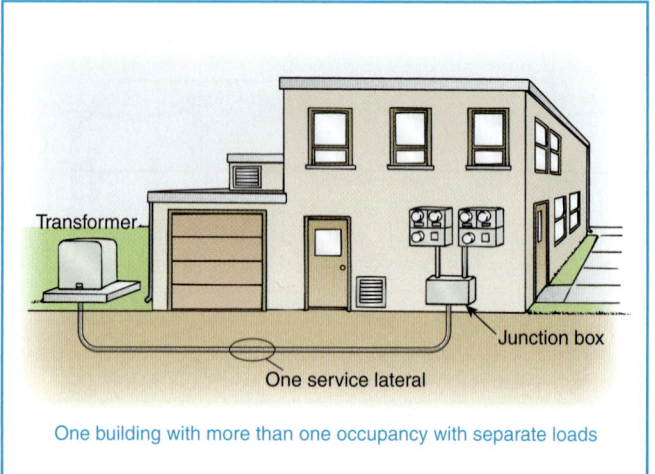

One building with more than one occupancy with separate loads

Exhibit 230.9 One service lateral supplying two sets of service-entrance conductors with not more than six disconnects grouped in one location.

if they are installed in a vault. See 450.41 through 450.48 for the construction requirements of transformer vaults. Service conductors installed under 18 in. of earth beneath the building are also considered outside the building according to 230.6(4).

(3) Where installed in any vault that meets the construction requirements of Article 450, Part III
(4) Where installed in conduit and under not less than 450 mm (18 in.) of earth beneath a building or other structure

230.7 Other Conductors in Raceway or Cable.

Conductors other than service conductors shall not be installed in the same service raceway or service cable.

All feeder and branch-circuit conductors must be separated from service conductors. Service conductors are not provided with overcurrent protection where they receive their supply; they are protected against overload conditions at their load end by the service disconnect fuses or circuit breakers. The amount of current that could be imposed on

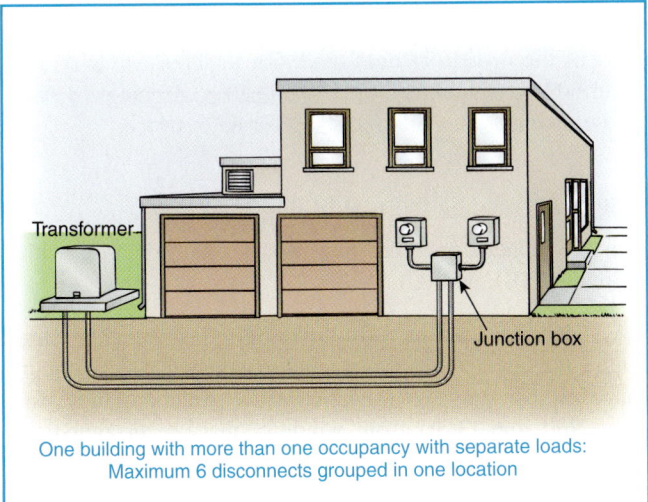

One building with more than one occupancy with separate loads:
Maximum 6 disconnects grouped in one location

Exhibit 230.10 Two service laterals, each consisting of not more than six sets of conductors, 1/0 or larger, and each supplying one set of service-entrance conductors.

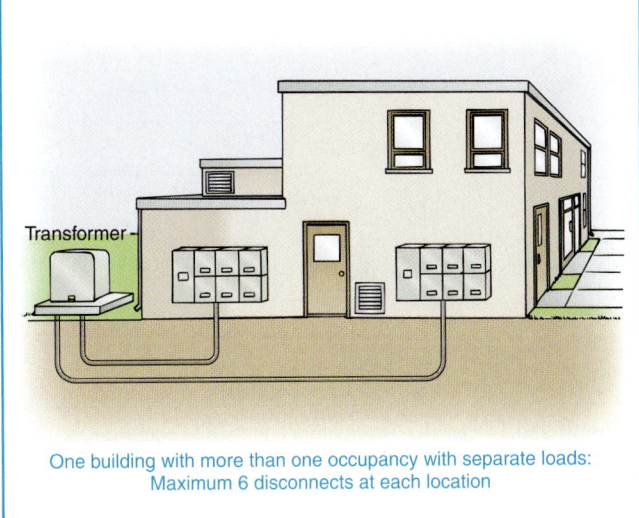

One building with more than one occupancy with separate loads:
Maximum 6 disconnects at each location

Exhibit 230.12 Two service laterals in two different locations supplying multiple loads.

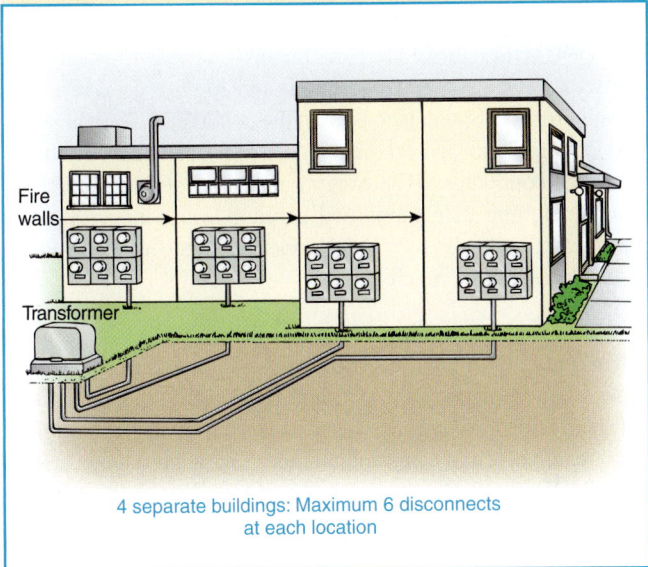

4 separate buildings: Maximum 6 disconnects
at each location

Exhibit 230.11 Four separate services consisting of four sets of service laterals, each supplying disconnects in four different locations.

One building with more than one occupancy with separate loads:
Maximum 6 disconnects allowed at each location

Exhibit 230.13 One service lateral supplying four sets of service-entrance conductors.

feeder or branch-circuit conductors, should they be in the same raceway and a fault occur, would be much higher than the ampacity of the feeder or branch-circuit conductors.

Exception No. 1: Grounding conductors and bonding jumpers.

Exception No. 2: Load management control conductors having overcurrent protection.

Because load management control conductors, control circuits, or switch leg conductors for use with special rate meters are usually short, they are allowed in the service raceway or cable. See Exhibit 230.24 for an example.

230.8 Raceway Seal.

Where a service raceway enters a building or structure from an underground distribution system, it shall be sealed in

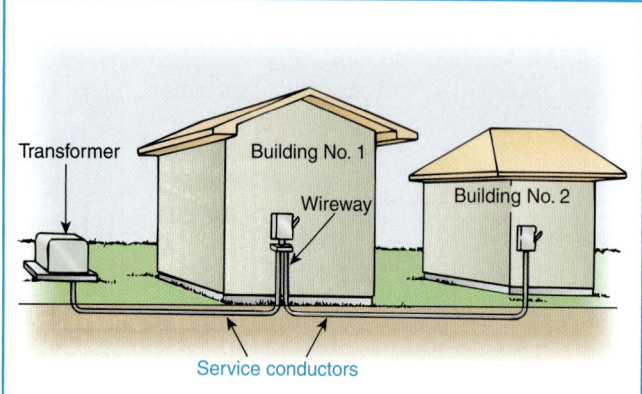

Exhibit 230.14 Service conductors installed in accordance with 230.3 so as not to pass through the interior of Building No. 1 to supply Building No. 2.

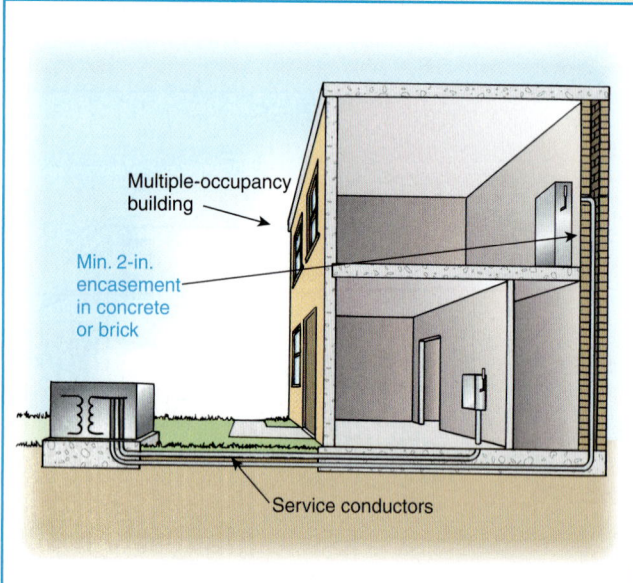

Exhibit 230.15 Service conductors considered outside a building where installed under not less than 2 in. of concrete beneath the building or in a raceway encased by 2 in. of concrete or brick within the building.

accordance with 300.5(G). Spare or unused raceways shall also be sealed. Sealants shall be identified for use with the cable insulation, shield, or other components.

Sealant, such as duct seal or a bushing incorporating the physical characteristics of a seal, must be used to seal the ends of service raceways. The intent of this requirement is to prevent water, usually the result of condensation due to temperature differences, from entering the service equipment via the raceway. The sealant material should be compatible

with the conductor insulation and should not cause deterioration of the insulation over time. For underground services over 600 volts, nominal, refer to 300.50(E) for raceway seal requirements. (See Exhibit 300.9 for an example.)

230.9 Clearance from Building Openings.

Service conductors and final spans shall comply with 230.9(A), (B), and (C).

(A) Clearance from Windows. Service conductors installed as open conductors or multiconductor cable without an overall outer jacket shall have a clearance of not less than 900 mm (3 ft) from windows that are designed to be opened, doors, porches, balconies, ladders, stairs, fire escapes, or similar locations.

Exception: Conductors run above the top level of a window shall be permitted to be less than the 900-mm (3-ft) requirement.

As illustrated in Exhibit 230.16, the clearance of 3 ft applies to open conductors, not to a raceway or cable assembly with an overall outer jacket approved for use as a service conductor. The intent is to protect the conductors from physical damage and protect personnel from accidental contact with the conductors. The exception permits service conductors, including drip loops and service-drop conductors, to be located just above window openings, because they are considered out of reach.

(B) Vertical Clearance. The vertical clearance of final spans above, or within 900 mm (3 ft) measured horizontally of, platforms, projections, or surfaces from which they might be reached shall be maintained in accordance with 230.24(B).

If service conductors are within 3 ft measured horizontally from a balcony, stair landing, or other platform, clearance to the platform of at least 10 ft must be maintained, as shown in Exhibit 230.17. See 230.24(B) for vertical clearances from ground.

(C) Building Openings. Overhead service conductors shall not be installed beneath openings through which materials may be moved, such as openings in farm and commercial buildings, and shall not be installed where they obstruct entrance to these building openings.

Elevated openings in buildings such as hay barns and silos through which materials may be moved are often high enough for the service conductors to be installed below the

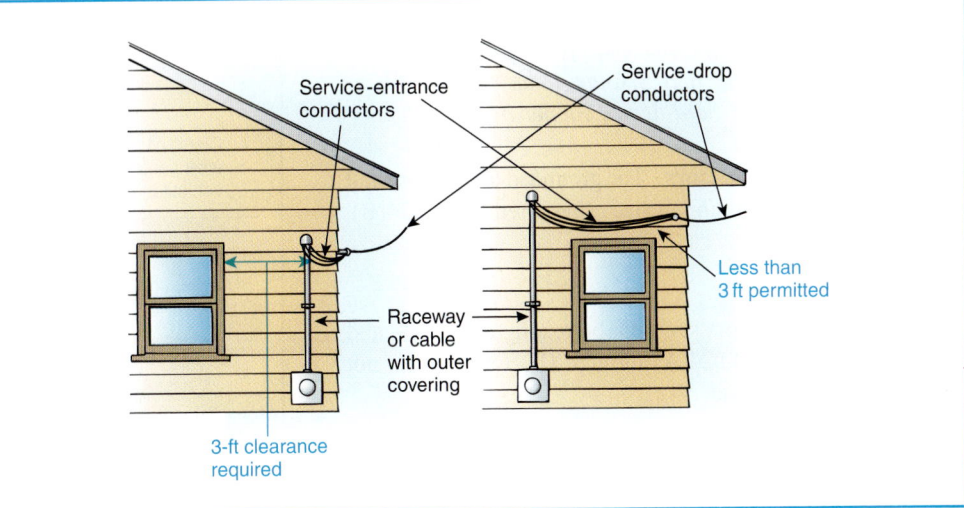

Service-entrance conductors

Service-drop conductors

Less than 3 ft permitted

Raceway or cable with outer covering

3-ft clearance required

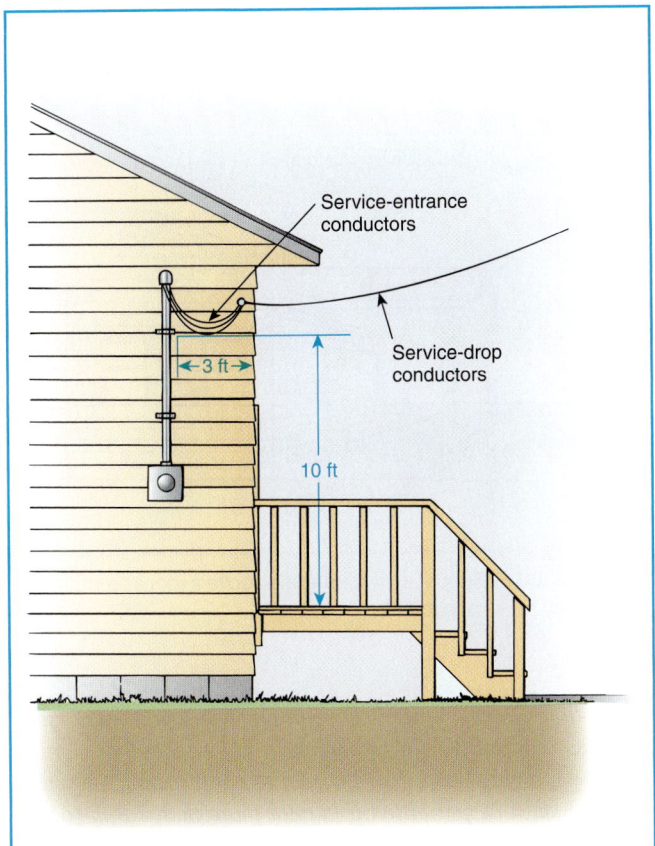

Service-entrance conductors

3 ft

Service-drop conductors

10 ft

Exhibit 230.17 Required dimensions for service conductors located above a stair landing, according to 230.9(B) and 230.24(B).

opening. However, 230.9(C) prohibits such placement in order to reduce the likelihood of damage to the service conductors and the potential for electric shock to persons using the openings.

230.10 Vegetation as Support.

Vegetation such as trees shall not be used for support of overhead service conductors.

II. Overhead Service-Drop Conductors

230.22 Insulation or Covering.

Individual conductors shall be insulated or covered.

Exception: The grounded conductor of a multiconductor cable shall be permitted to be bare.

The intent of 230.22 is to prevent problems created by weather and abrasion and other deleterious effects that reduce the insulating quality of the covering or insulation. The grounded conductor (neutral) of triplex or quadraplex service-drop conductors is often bare and used to mechanically support the other ungrounded conductors.

230.23 Size and Rating.

(A) General. Conductors shall have sufficient ampacity to carry the current for the load as computed in accordance with Article 220 and shall have adequate mechanical strength.

When a load is added to any service, the installer must be aware of all existing loads. The potential for overloading the service conductors must be governed by installer responsibility and inspector awareness; they should not rely solely on the usual method of overcurrent protection. The serving electric utility should be alerted whenever load is added to ensure that adequate power is available.

(B) Minimum Size. The conductors shall not be smaller than 8 AWG copper or 6 AWG aluminum or copper-clad aluminum.

Exception: Conductors supplying only limited loads of a single branch circuit—such as small polyphase power, controlled water heaters, and similar loads—shall not be smaller than 12 AWG hard-drawn copper or equivalent.

(C) Grounded Conductors. The grounded conductor shall not be less than the minimum size as required by 250.24(B).

230.24 Clearances.

Service-drop conductors shall not be readily accessible and shall comply with 230.24(A) through (D) for services not over 600 volts, nominal.

(A) Above Roofs. Conductors shall have a vertical clearance of not less than 2.5 m (8 ft) above the roof surface. The vertical clearance above the roof level shall be maintained for a distance of not less than 900 mm (3 ft) in all directions from the edge of the roof.

Service-drop conductors are not permitted to be readily accessible. This main rule applies to services of up to 600 volts. An 8-ft vertical clearance is required over the roof surface, extending 3 ft in all directions from the edge. Note that Exception No. 4 allows penetration of this space for the final span of the service drop.

Exception No. 1: The area above a roof surface subject to pedestrian or vehicular traffic shall have a vertical clearance from the roof surface in accordance with the clearance requirements of 230.24(B).

Exception No. 1 to 230.24(A) requires service-drop conductor clearance above a roof surface subject to vehicular or pedestrian traffic, such as the rooftop parking area shown in Exhibit 230.18, to meet the clearance requirements of 230.24(B).

Exception No. 2: Where the voltage between conductors does not exceed 300 and the roof has a slope of 100 mm (4 in.) in 300 mm (12 in.), or greater, a reduction in clearance to 900 mm (3 ft) shall be permitted.

Exception No. 2 to 230.24(A) permits a reduction in service-drop conductor clearance above the roof from 8 ft to 3 ft, as illustrated in Exhibit 230.19, if the voltage between conductors does not exceed 300 volts and the roof is sloped not less than 4 in. vertically in 12 in. horizontally. Steeply sloped roofs are less likely to be walked on. There are no restrictions on the length of the conductors over the roof.

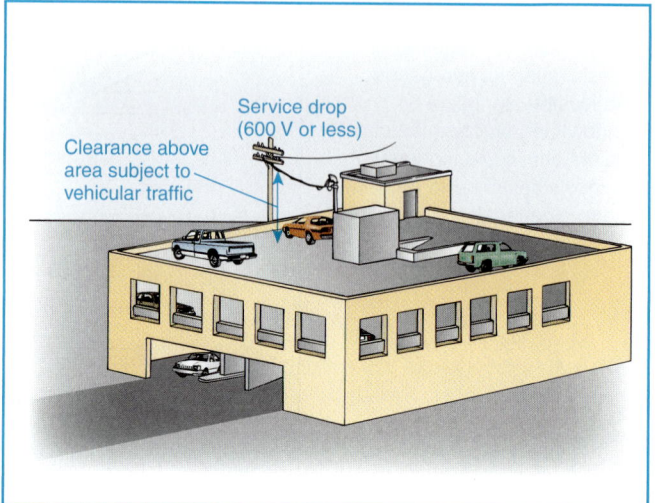

Exhibit 230.18 Service-drop conductor clearance required by 230.24(A), Exception No. 1.

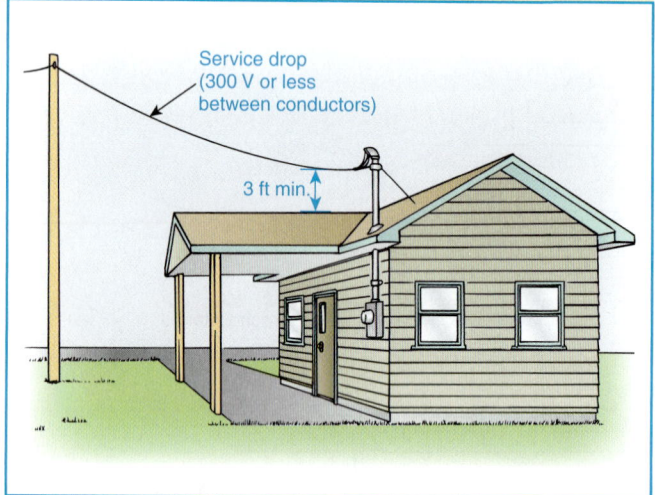

Exhibit 230.19 Reduction in clearance above a roof as permitted by 230.24(A), Exception No. 2.

Exception No. 3: Where the voltage between conductors does not exceed 300, a reduction in clearance above only the overhanging portion of the roof to not less than 450 mm (18 in.) shall be permitted if (1) not more than 1.8 m (6 ft) of service-drop conductors, 1.2 m (4 ft) horizontally, pass above the roof overhang, and (2) they are terminated at a through-the-roof raceway or approved support.

FPN: See 230.28 for mast supports.

Exception No. 3 to 230.24(A) permits a reduction of service-drop conductor clearances to 18 in. above the roof, as illustrated in Exhibit 230.20. This reduction is for service-mast (through-the-roof) installations where the voltage between conductors does not exceed 300 volts and the mast is located within 4 ft of the edge of the roof, measured horizontally. Exception No. 3 applies to either sloped or flat roofs that are easily walked on. Not more than 6 ft of conductors is permitted to pass over the roof.

Exception No. 4: The requirement for maintaining the vertical clearance 900 mm (3 ft) from the edge of the roof shall not apply to the final conductor span where the service drop is attached to the side of a building.

Section 230.24(A) applies to the vertical clearance above roofs for service-drop conductors up to 600 volts. This main rule requires a vertical clearance of 8 ft above the roof, including those areas 3 ft in all directions beyond the edge of the roof.

Exception No. 4 to 230.24(A) exempts the final span of a service drop attached to the side of a building from the 8-ft and 3-ft requirements, to allow the service conductors to be attached to the building, as illustrated in Exhibit 230.21. Exception No. 2 and Exception No. 3 permit lesser clearances for service drops of 300 volts or less, as illustrated in Exhibits 230.19 and 230.20.

If the roof is subject to pedestrian or vehicular traffic,

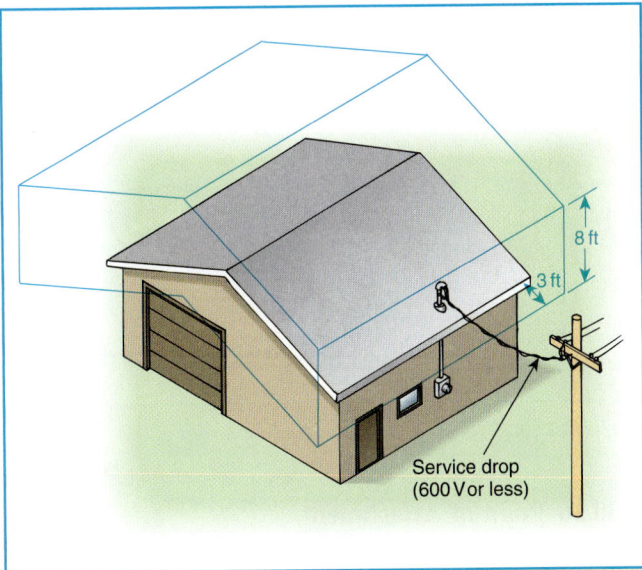

Exhibit 230.21 Clearance of the final span of a service drop, as permitted by 230.24(A), Exception No. 4.

the vertical clearance of the service drop must be the same as the vertical clearance from the ground, in accordance with 230.24(B).

(B) Vertical Clearance from Ground. Service-drop conductors, where not in excess of 600 volts, nominal, shall have the following minimum clearance from final grade:

(1) 3.0 m (10 ft)—at the electric service entrance to buildings, also at the lowest point of the drip loop of the building electric entrance, and above areas or sidewalks accessible only to pedestrians, measured from final grade or other accessible surface only for service-drop cables supported on and cabled together with a grounded bare messenger where the voltage does not exceed 150 volts to ground

(2) 3.7 m (12 ft)—over residential property and driveways, and those commercial areas not subject to truck traffic where the voltage does not exceed 300 volts to ground

(3) 4.5 m (15 ft)—for those areas listed in the 3.7 m (12 ft) classification where the voltage exceeds 300 volts to ground

(4) 5.5 m (18 ft)—over public streets, alleys, roads, parking areas subject to truck traffic, driveways on other than residential property, and other land such as cultivated, grazing, forest, and orchard

Exhibit 230.22 illustrates the 10-ft, 12-ft, 15-ft, and 18-ft vertical clearances from ground for service-drop conductors up to 600 volts, as specified by 230.24(B).

(C) Clearance from Building Openings. See 230.9.

(D) Clearance from Swimming Pools. See 680.8.

Exhibit 230.20 Reduction in clearance above a roof as permitted by 230.24(A), Exception No. 3.

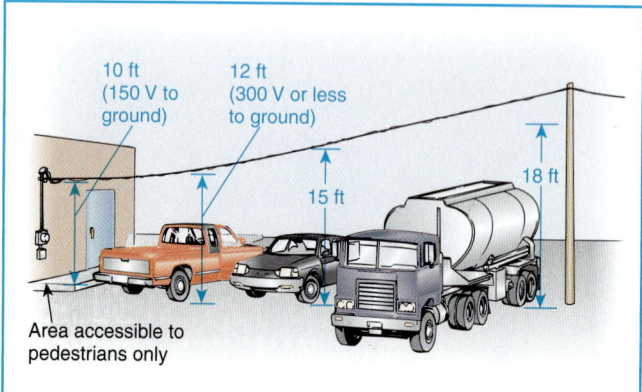

Exhibit 230.22 Clearances in accordance with 230.24(B).

230.26 Point of Attachment.

The point of attachment of the service-drop conductors to a building or other structure shall provide the minimum clearances as specified in 230.24. In no case shall this point of attachment be less than 3.0 m (10 ft) above finished grade.

230.27 Means of Attachment.

Multiconductor cables used for service drops shall be attached to buildings or other structures by fittings identified for use with service conductors. Open conductors shall be attached to fittings identified for use with service conductors or to noncombustible, nonabsorbent insulators securely attached to the building or other structure.

See 230.51 for mounting and supporting of service cables and individual open service conductors. See 230.54 for connections at service heads.

230.28 Service Masts as Supports.

Where a service mast is used for the support of service-drop conductors, it shall be of adequate strength or be supported by braces or guys to withstand safely the strain imposed by the service drop. Where raceway-type service masts are used, all raceway fittings shall be identified for use with service masts. Only power service-drop conductors shall be permitted to be attached to a service mast.

Where the service drop is secured to the mast, a guy wire may be installed to support the mast and provide adequate mechanical strength to support the service drop. Communications conductors such as those for cable TV or telephone service are not permitted to be attached to the service mast.

230.29 Supports over Buildings.

Service-drop conductors passing over a roof shall be securely supported by substantial structures. Where practicable, such supports shall be independent of the building.

III. Underground Service-Lateral Conductors

230.30 Insulation.

Service-lateral conductors shall be insulated for the applied voltage.

Exception: A grounded conductor shall be permitted to be uninsulated as follows:

(a) Bare copper used in a raceway.
(b) Bare copper for direct burial where bare copper is judged to be suitable for the soil conditions.
(c) Bare copper for direct burial without regard to soil conditions where part of a cable assembly identified for underground use.
(d) Aluminum or copper-clad aluminum without individual insulation or covering where part of a cable assembly identified for underground use in a raceway or for direct burial.

Exhibit 230.23 shows various applications of bare grounded service-lateral and service-entrance conductors for underground locations. Aluminum or copper-clad aluminum conductors must be insulated if run in a raceway or direct buried, unless they are part of a cable assembly identified for the use.

Bare copper, in raceway

Bare copper, direct buried, suitable for soil conditions

Bare copper (or aluminum), part of an identified cable assembly

Exhibit 230.23 Bare grounded service-lateral and service-entrance conductors for underground locations.

230.31 Size and Rating.

(A) General. Service-lateral conductors shall have sufficient ampacity to carry the current for the load as computed in accordance with Article 220 and shall have adequate mechanical strength.

(B) Minimum Size. The conductors shall not be smaller than 8 AWG copper or 6 AWG aluminum or copper-clad aluminum.

Exception: Conductors supplying only limited loads of a single branch circuit—such as small polyphase power, controlled water heaters, and similar loads—shall not be

smaller than 12 AWG copper or 10 AWG aluminum or copper-clad aluminum.

(C) Grounded Conductors. The grounded conductor shall not be less than the minimum size required by 250.24(B).

See 310.15(B)(3) and (B)(6) and the commentary following 230.42(C) for further information on the sizing of grounded conductors.

230.32 Protection Against Damage.

Underground service-lateral conductors shall be protected against damage in accordance with 300.5. Service-lateral conductors entering a building shall be installed in accordance with 230.6 or protected by a raceway wiring method identified in 230.43.

230.33 Spliced Conductors.

Service-lateral conductors shall be permitted to be spliced or tapped in accordance with 110.14, 300.5(E), 300.13, and 300.15.

IV. Service-Entrance Conductors

230.40 Number of Service-Entrance Conductor Sets.

Each service drop or lateral shall supply only one set of service-entrance conductors.

Exception No. 1: A building with one or more than one occupancy shall be permitted to have one set of service-entrance conductors for each service of different characteristics, as defined in 230.2(D), run to each occupancy or group of occupancies.

Exception No. 2: Where two to six service disconnecting means in separate enclosures are grouped at one location and supply separate loads from one service drop or lateral, one set of service-entrance conductors shall be permitted to supply each or several such service equipment enclosures.

See Exhibits 230.2 through 230.13 for examples of permitted service configurations. Note that in such cases the lateral conductors are considered *one* service lateral.

Exception No. 3: A single-family dwelling unit and a separate structure shall be permitted to have one set of service-entrance conductors run to each from a single service drop or lateral.

Each set of service-drop or service-lateral conductors is allowed to supply only one set of service-entrance conductors.

However, if a service drop or a service lateral supplies a building with more than one occupancy, such as multifamily dwellings, strip malls, and office buildings, each service drop or service lateral is allowed to supply more than one set of service-entrance conductors, provided they are run to each occupancy or group of occupancies.

Exception No. 3 to 230.40 allows a second set of service-entrance conductors supplied by a single service drop or lateral at a single-family dwelling unit to also supply a second building on the premises, such as a garage or storage shed.

Exception No. 4: A two-family dwelling or a multifamily dwelling shall be permitted to have one set of service-entrance conductors installed to supply the circuits covered in 210.25.

Exception No. 5: One set of service-entrance conductors connected to the supply side of the normal service disconnecting means shall be permitted to supply each or several systems covered by 230.82(4) or (5).

230.41 Insulation of Service-Entrance Conductors.

Service-entrance conductors entering or on the exterior of buildings or other structures shall be insulated.

Exception: A grounded conductor shall be permitted to be uninsulated as follows:

(a) *Bare copper used in a raceway or part of a service cable assembly.*
(b) *Bare copper for direct burial where bare copper is judged to be suitable for the soil conditions.*
(c) *Bare copper for direct burial without regard to soil conditions where part of a cable assembly identified for underground use.*
(d) *Aluminum or copper-clad aluminum without individual insulation or covering where part of a cable assembly or identified for underground use in a raceway, or for direct burial.*
(e) *Bare conductors used in an auxiliary gutter.*

Service-entrance conductors must be insulated; however, bare grounded conductors are permitted under the same conditions as underground service laterals. New in the 2002 *Code*, bare grounded conductors are permitted in an auxiliary gutter of a metering enclosure, panelboard, or switchboard. See Exhibit 230.23, which shows examples of bare grounded conductors for underground locations.

230.42 Minimum Size and Rating.

(A) General. The ampacity of the service-entrance conductors before the application of any adjustment or correction factors shall not be less than either (1) or (2). Loads shall be determined in accordance with Article 220. Ampacity shall be determined from 310.15. The maximum allowable current of busways shall be that value for which the busway has been listed or labeled.

(1) The sum of the noncontinuous loads plus 125 percent of continuous loads

(2) The sum of the noncontinuous load plus the continuous load if the service-entrance conductors terminate in an overcurrent device where both the overcurrent device and its assembly are listed for operation at 100 percent of their rating

(B) Specific Installations. In addition to the requirements of 230.42(A), the minimum ampacity for ungrounded conductors for specific installations shall not be less than the rating of the service disconnecting means specified in 230.79(A) through (D).

> The title and text of 230.42(B) were changed in the 2002 *Code* for clarity. The basic rule in 230.42(B) requires *ungrounded* service conductors to be sized large enough to carry the load and to be not smaller than the minimum size of the disconnect as specifically required in 230.79(A) through (D).

(C) Grounded Conductors. The grounded conductor shall not be less than the minimum size as required by 250.24(B).

> The maximum unbalanced load determines the size of the *grounded* service conductor. Section 220.22 allows this value to be reduced, based on the maximum unbalanced load that can occur between the ungrounded conductors and the grounded (neutral) conductor. However, the minimum size of the grounded service conductor cannot be smaller than that required by 250.24(B)(1) and (B)(2).
>
> The additional heating effect of harmonic currents, due to nonlinear loads should be considered when sizing the neutral conductor of a 3-phase, 4-wire wye system. If the service to a building is a single-phase, 3-wire, 120/240-volt system with no 240-volt loads, the maximum current in the neutral would be the same as the maximum current in the ungrounded conductor. If all loads connected to one leg are "on" and all the other loads on the other leg are "off," maximum current will flow in the neutral. In such cases, the service neutral size cannot be reduced but would be sized the same as the ungrounded conductors. See 310.15(B)(4) for more information on sizing the ungrounded conductor.

230.43 Wiring Methods for 600 Volts, Nominal, or Less.

Service-entrance conductors shall be installed in accordance with the applicable requirements of this *Code* covering the type of wiring method used and shall be limited to the following methods:

(1) Open wiring on insulators
(2) Type IGS cable
(3) Rigid metal conduit
(4) Intermediate metal conduit
(5) Electrical metallic tubing
(6) Electrical nonmetallic tubing (ENT)
(7) Service-entrance cables
(8) Wireways
(9) Busways
(10) Auxiliary gutters
(11) Rigid nonmetallic conduit
(12) Cablebus
(13) Type MC cable
(14) Mineral-insulated, metal-sheathed cable
(15) Flexible metal conduit not over 1.8 m (6 ft) long or liquidtight flexible metal conduit not over 1.8 m (6 ft) long between raceways, or between raceway and service equipment, with equipment bonding jumper routed with the flexible metal conduit or the liquidtight flexible metal conduit according to the provisions of 250.102(A), (B), (C), and (E)
(16) Liquidtight flexible nonmetallic conduit

> Where flexible metal conduit or liquidtight flexible metal conduit is installed for services, a bonding jumper must be installed between both ends within the raceway. The bonding jumper is allowed to be installed outside the raceway, but it must follow the path of the raceway and cannot exceed 6 ft in length. The bonding jumper must not be wrapped or spiraled around the flexible conduit.

230.44 Cable Trays.

Cable tray systems shall be permitted to support cable used as service-entrance conductors.

230.46 Spliced Conductors.

Service-entrance conductors shall be permitted to be spliced or tapped in accordance with 110.14, 300.5(E), 300.13, and 300.15.

> Splices are permitted in service-entrance conductors if the splice meets the requirements of 230.46. Splices must be in an enclosure or be direct buried using a listed underground splice kit. It is common to have an underground service lateral terminate at a terminal box either inside or outside the building. At this point, service conductors may be spliced or run directly to the service equipment.
>
> Splices are permitted where, for example, the cable enters a terminal box and a different wiring method, such

as conduit, continues to the service equipment. Splices are most common where metering equipment is located on the line side of service equipment, service busways, and taps for supplying up to six disconnecting means. See Exhibit 230.24 for splices permitted in metering equipment.

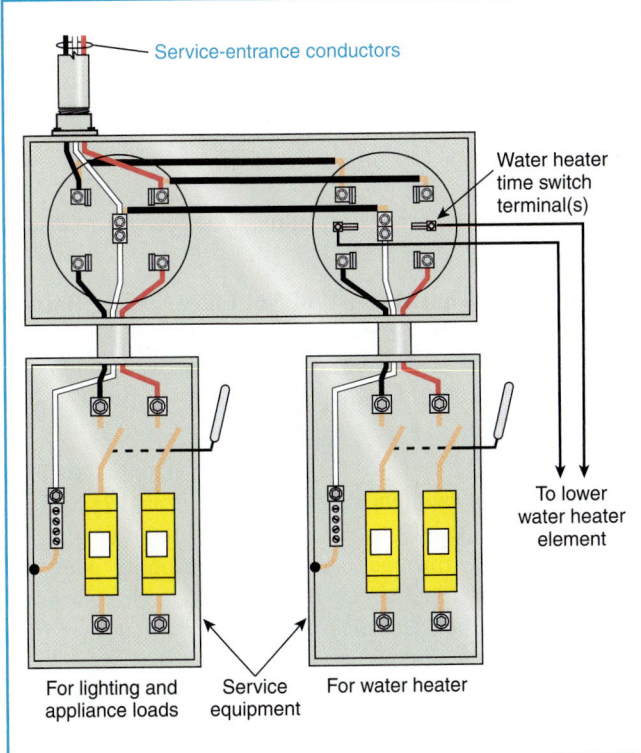

Exhibit 230.24 Time clock and control switch integral to a meter for use, generally, with water heaters.

230.49 Protection Against Physical Damage—Underground.

Underground service-entrance conductors shall be protected against physical damage in accordance with 300.5.

230.50 Protection of Open Conductors and Cables Against Damage—Above Ground.

Service-entrance conductors installed above ground shall be protected against physical damage as specified in 230.50(A) or (B).

(A) Service Cables. Service cables, where subject to physical damage, shall be protected by any of the following:

(1) Rigid metal conduit
(2) Intermediate metal conduit
(3) Schedule 80 rigid nonmetallic conduit
(4) Electrical metallic tubing
(5) Other approved means

(B) Other Than Service Cable. Individual open conductors and cables other than service cables shall not be installed within 3.0 m (10 ft) of grade level or where exposed to physical damage.

Exception: Type MI and Type MC cable shall be permitted within 3.0 m (10 ft) of grade level where not exposed to physical damage or where protected in accordance with 300.5(D).

230.51 Mounting Supports.

Cables or individual open service conductors shall be supported as specified in 230.51(A), (B), or (C).

(A) Service Cables. Service cables shall be supported by straps or other approved means within 300 mm (12 in.) of every service head, gooseneck, or connection to a raceway or enclosure and at intervals not exceeding 750 mm (30 in.).

(B) Other Cables. Cables that are not approved for mounting in contact with a building or other structure shall be mounted on insulating supports installed at intervals not exceeding 4.5 m (15 ft) and in a manner that will maintain a clearance of not less than 50 mm (2 in.) from the surface over which they pass.

(C) Individual Open Conductors. Individual open conductors shall be installed in accordance with Table 230.51(C). Where exposed to the weather, the conductors shall be mounted on insulators or on insulating supports attached to racks, brackets, or other approved means. Where not exposed to the weather, the conductors shall be mounted on glass or porcelain knobs.

230.52 Individual Conductors Entering Buildings or Other Structures.

Where individual open conductors enter a building or other structure, they shall enter through roof bushings or through the wall in an upward slant through individual, noncombustible, nonabsorbent insulating tubes. Drip loops shall be formed on the conductors before they enter the tubes.

230.53 Raceways to Drain.

Where exposed to the weather, raceways enclosing service-entrance conductors shall be raintight and arranged to drain. Where embedded in masonry, raceways shall be arranged to drain.

Exception: As permitted in 348.12(1).

The goal of 230.53 is to prevent water from entering internal electrical equipment through the raceway system. Service raceways exposed to the weather must have raintight fittings

Table 230.51(C) Supports

Maximum Volts	Maximum Distance Between Supports		Minimum Clearance			
			Between Conductors		From Surface	
	m	ft	mm	in.	mm	in.
600	2.7	9	150	6	50	2
600	4.5	15	300	12	50	2
300	1.4	4½	75	3	50	2
600*	1.4*	4½*	65*	2½*	25*	1*

*Where not exposed to weather.

and drain holes. During the installation of raceways in masonry, provisions to drain and divert water should be made to prevent the entrance of surface water, rain, or water from poured concrete.

230.54 Overhead Service Locations.

(A) Raintight Service Head. Service raceways shall be equipped with a raintight service head at the point of connection to service-drop conductors.

(B) Service Cable Equipped with Raintight Service Head or Gooseneck. Service cables shall be equipped with a raintight service head.

Exception: Type SE cable shall be permitted to be formed in a gooseneck and taped with a self-sealing weather-resistant thermoplastic.

(C) Service Heads Above Service-Drop Attachment. Service heads and goosenecks in service-entrance cables shall be located above the point of attachment of the service-drop conductors to the building or other structure.

Exception: Where it is impracticable to locate the service head above the point of attachment, the service head location shall be permitted not farther than 600 mm (24 in.) from the point of attachment.

(D) Secured. Service cables shall be held securely in place.

(E) Separately Bushed Openings. Service heads shall have conductors of different potential brought out through separately bushed openings.

Exception: For jacketed multiconductor service cable without splice.

(F) Drip Loops. Drip loops shall be formed on individual conductors. To prevent the entrance of moisture, service-entrance conductors shall be connected to the service-drop conductors either (1) below the level of the service head or

(2) below the level of the termination of the service-entrance cable sheath.

(G) Arranged That Water Will Not Enter Service Raceway or Equipment. Service-drop conductors and service-entrance conductors shall be arranged so that water will not enter service raceway or equipment.

Service raceways and service cables are required to be equipped with a raintight service (weatherhead). Type SE service-entrance cables may be installed without a service head if they are run continuously from a utility pole to metering or service equipment or if they are shaped in a downward direction (forming a "gooseneck") and sealed by taping and painting, as shown in Exhibit 230.25.

Service heads and goosenecks are required to be located above the service-drop point of attachment to the building or structure. Individual conductors should extend in a downward direction, as shown in Exhibit 230.25, or drip loops should be formed so that, where splices are made, they are at the lowest point of the drip loop.

230.56 Service Conductor with the Higher Voltage to Ground.

On a 4-wire, delta-connected service where the midpoint of one phase winding is grounded, the service conductor having the higher phase voltage to ground shall be durably and permanently marked by an outer finish that is orange in color, or by other effective means, at each termination or junction point.

Proper service connections require the service conductors having the higher voltage to ground to be durably marked by an outer finish of orange, such as by painting, colored adhesive tagging, or taping. Marking should be at both the point of connection to the service-drop conductors and the point of connection to the service disconnecting means. See 110.15, 215.8, and 408.3(E) for high-leg marking and phase arrangement requirements.

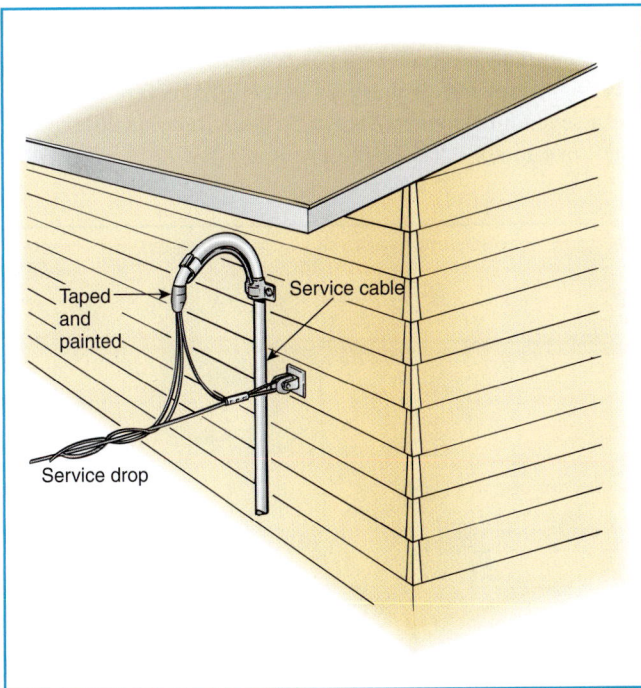

Exhibit 230.25 A service-entrance cable that terminates in a gooseneck without a raintight service weatherhead.

Labels in figure: Taped and painted; Service cable; Service drop

V. Service Equipment—General

230.62 Service Equipment—Enclosed or Guarded.

Energized parts of service equipment shall be enclosed as specified in 230.62(A) or guarded as specified in 230.62(B).

(A) Enclosed. Energized parts shall be enclosed so that they will not be exposed to accidental contact or shall be guarded as in 230.62(B).

(B) Guarded. Energized parts that are not enclosed shall be installed on a switchboard, panelboard, or control board and guarded in accordance with 110.18 and 110.27. Where energized parts are guarded as provided in 110.27(A)(1) and (2), a means for locking or sealing doors providing access to energized parts shall be provided.

230.66 Marking.

Service equipment rated at 600 volts or less shall be marked to identify it as being suitable for use as service equipment. Individual meter socket enclosures shall not be considered service equipment.

According to the listing information, panelboards with the neutral factory bonded to the enclosure will be marked "Suitable Only for Use as Service Equipment." Other types of

equipment intended for optional use, either as service equipment or as sub-distribution panelboards for feeders on the load side of the service disconnect, are required by 230.66 to be marked suitable for use as service equipment. Section 225.36 requires the feeder disconnecting means to be suitable for use as service equipment.

VI. Service Equipment— Disconnecting Means

230.70 General.

Means shall be provided to disconnect all conductors in a building or other structure from the service-entrance conductors.

(A) Location. The service disconnecting means shall be installed in accordance with 230.70(A)(1), (2), and (3).

No maximum distance is specified from the point of entrance of service conductors to a readily accessible location for the installation of a service disconnecting means. The authority enforcing this *Code* has the responsibility for, and is charged with, making the decision as to how far inside the building the service-entrance conductors are allowed to travel to the main disconnecting means. The length of service-entrance conductors should be kept to a minimum inside buildings, because power utilities provide limited overcurrent protection and, in the event of a fault, the service conductors could ignite nearby combustible materials.

Some local jurisdictions have ordinances that allow service-entrance conductors to run within the building up to a specified length to terminate at the disconnecting means. The authority having jurisdiction may permit service conductors to bypass fuel storage tanks or gas meters and the like, permitting the service disconnecting means to be located in a readily accessible location. However, if the authority judges the distance as being excessive, the disconnecting means may be required to be located on the outside of the building or near the building at a readily accessible location that is not necessarily nearest the point of entrance of the conductors. See also 230.6 and Exhibit 230.15 for conductors considered to be outside a building.

See 404.8(A) for mounting-height restrictions for switches and for circuit breakers used as switches.

(1) Readily Accessible Location. The service disconnecting means shall be installed at a readily accessible location either outside of a building or structure or inside nearest the point of entrance of the service conductors.

(2) Bathrooms. Service disconnecting means shall not be installed in bathrooms.

(3) Remote Control. Where a remote control device(s) is used to actuate the service disconnecting means, the service disconnecting means shall be located in accordance with 230.70(A)(1).

(B) Marking. Each service disconnect shall be permanently marked to identify it as a service disconnect.

(C) Suitable for Use. Each service disconnecting means shall be suitable for the prevailing conditions. Service equipment installed in hazardous (classified) locations shall comply with the requirements of Articles 500 through 517.

230.71 Maximum Number of Disconnects.

(A) General. The service disconnecting means for each service permitted by 230.2, or for each set of service-entrance conductors permitted by 230.40, Exception Nos. 1, 3, 4, or 5, shall consist of not more than six switches or sets of circuit breakers, or a combination of not more than six switches and sets of circuit breakers, mounted in a single enclosure, in a group of separate enclosures, or in or on a switchboard. There shall be no more than six sets of disconnects per service grouped in any one location. For the purpose of this section, disconnecting means used solely for power monitoring equipment, or the control circuit of the ground-fault protection system or power-operable service disconnecting means, installed as part of the listed equipment, shall not be considered a service disconnecting means.

Section 230.71(A) covers the maximum number of disconnects permitted as the disconnecting means for the service conductors that supply the building or structure. One set of service-entrance conductors, either overhead or underground, is permitted to supply two to six service disconnecting means in lieu of a single main disconnect. A single-occupancy building can have up to six disconnects for each set of service-entrance conductors. Multiple-occupancy buildings (residential or other than residential) can be provided with one main service disconnect or up to six main disconnects for each set of service-entrance conductors.

Multiple-occupancy buildings may have service-entrance conductors run to each occupancy, and each such set of service-entrance conductors may have from one to six disconnects (see 230.40, Exception No. 1).

Where service-entrance conductors are routed outside the building (see 230.6 and Exhibit 230.15), each set of service-entrance conductors is permitted to supply not more than six disconnecting means at each occupancy of a multiple-occupancy building. See Exhibits 230.2 through 230.13 for examples of permitted service configurations.

Exhibit 230.26 shows a single enclosure for grouping service equipment that consists of six circuit breakers or six fused switches. This arrangement does not require a main switch. Six separate enclosures also would be permitted as the service equipment. The last sentence of 230.71(A) makes

clear that although the disconnect for the control circuit of a ground-fault protector or a power-operable service disconnecting means installed as part of the listed equipment may be a seventh disconnect switch, it is *not* considered one of the "one to six" service disconnects.

Exhibit 230.26 An enclosure for grouping service equipment consisting of six circuit breakers or six fused switches.

(B) Single-Pole Units. Two or three single-pole switches or breakers, capable of individual operation, shall be permitted on multiwire circuits, one pole for each ungrounded conductor, as one multipole disconnect, provided they are equipped with handle ties or a master handle to disconnect all conductors of the service with no more than six operations of the hand.

> FPN: See 408.16(A) for service equipment in panelboards, and see 430.95 for service equipment in motor control centers.

230.72 Grouping of Disconnects.

(A) General. The two to six disconnects as permitted in 230.71 shall be grouped. Each disconnect shall be marked to indicate the load served.

Exception: One of the two to six service disconnecting means permitted in 230.71, where used only for a water pump also intended to provide fire protection, shall be permitted to be located remote from the other disconnecting means.

The water pump in 230.72(A), Exception, is not the fire pump covered by Article 695.

(B) Additional Service Disconnecting Means. The one or more additional service disconnecting means for fire pumps, for legally required standby, or for optional standby services permitted by 230.2 shall be installed remote from the one to six service disconnecting means for normal service to minimize the possibility of simultaneous interruption of supply.

The intent of 230.2(A) is to permit separate services, where necessary, for fire pumps (with one to six disconnects) or for emergency, legally required standby, or optional standby systems (with one to six disconnects), in addition to the one to six disconnects for the normal building service. Article 230 recognizes that a disruption of the normal building service should not disconnect the fire pump, emergency system, or other exempted systems. Because these services are in addition to the normal services, the one to six disconnects allowed for them are not included as one of the six disconnects for the normal supply.

(C) Access to Occupants. In a multiple-occupancy building, each occupant shall have access to the occupant's service disconnecting means.

Exception: In a multiple-occupancy building where electric service and electrical maintenance are provided by the building management and where these are under continuous building management supervision, the service disconnecting means supplying more than one occupancy shall be permitted to be accessible to authorized management personnel only.

A multiple-occupancy building may have any number of dwelling units, offices, and the like that are independent of each other. Unless electric service and maintenance are provided by and under continuous supervision of the building management, the occupants of a multiple-occupancy building must have ready access to their disconnecting means as required by 240.24(B).

230.74 Simultaneous Opening of Poles.

Each service disconnect shall simultaneously disconnect all ungrounded service conductors that it controls from the premises wiring system.

230.75 Disconnection of Grounded Conductor.

Where the service disconnecting means does not disconnect the grounded conductor from the premises wiring, other means shall be provided for this purpose in the service equipment. A terminal or bus to which all grounded conductors can be attached by means of pressure connectors shall be permitted for this purpose. In a multisection switchboard, disconnects for the grounded conductor shall be permitted

to be in any section of the switchboard, provided any such switchboard section is marked.

Provisions are required at the service equipment for disconnecting the grounded conductor from the premises wiring. This disconnection does not have to be by operation of the service disconnecting means. Disconnection can be, and most commonly is, accomplished by manually removing the grounded conductor from the bus or terminal bar to which it is lugged or bolted. This location is often referred to as the *neutral disconnect link.*

Manufacturers design neutral terminal bars for service equipment so that grounded conductors must be cut to be attached; that is, the grounded conductor cannot be run straight through the service equipment without means of disconnection from the premises wiring.

230.76 Manually or Power Operable.

The service disconnecting means for ungrounded service conductors shall consist of either (1) a manually operable switch or circuit breaker equipped with a handle or other suitable operating means or (2) a power-operated switch or circuit breaker, provided the switch or circuit breaker can be opened by hand in the event of a power supply failure.

230.77 Indicating.

The service disconnecting means shall plainly indicate whether it is in the open or closed position.

230.79 Rating of Service Disconnecting Means.

The service disconnecting means shall have a rating not less than the load to be carried, determined in accordance with Article 220. In no case shall the rating be lower than specified in 230.79(A), (B), (C), or (D).

Three-wire services that supply one-family dwellings are required to be installed using wire with the capacity to supply a 100-ampere service for all single-family dwellings.

A conductor ampacity of 60 amperes is permitted for other loads. Smaller sizes are permitted down to 14 AWG copper (12 AWG aluminum) for installations with one circuit. Two-circuit installations must have a rating of at least 30 amperes. Exhibit 230.27 illustrates the conductor sizing requirements of 230.79 for ungrounded service-entrance conductors. A single service disconnecting means is required to have a rating of not less than the load to be carried.

(A) One-Circuit Installation. For installations to supply only limited loads of a single branch circuit, the service disconnecting means shall have a rating of not less than 15 amperes.

(B) Two-Circuit Installations. For installations consisting of not more than two 2-wire branch circuits, the service disconnecting means shall have a rating of not less than 30 amperes.

(C) One-Family Dwelling. For a one-family dwelling, the service disconnecting means shall have a rating of not less than 100 amperes, 3-wire.

(D) All Others. For all other installations, the service disconnecting means shall have a rating of not less than 60 amperes.

230.80 Combined Rating of Disconnects.

Where the service disconnecting means consists of more than one switch or circuit breaker, as permitted by 230.71, the combined ratings of all the switches or circuit breakers used shall not be less than the rating required by 230.79.

Section 230.71(A) permits up to six individual switches or circuit breakers, mounted in a single enclosure, in a group of separate enclosures, or in or on a switchboard or several switchboards, to serve as the required service disconnecting means at any one location. Section 230.80 refers to situations in which more than one switch or circuit breaker is used as the disconnecting means and indicates that the combined rating of all the switches or circuit breakers used cannot be less than the rating required for a single switch or circuit breaker.

Section 230.90 requires an overcurrent device to provide overload protection in each ungrounded service conductor. A single overcurrent device must have a rating or setting that is not higher than the allowable ampacity of the service conductors. However, Exception No. 3 to 230.90(A) allows not more than six circuit breakers or six sets of fuses to be considered the overcurrent device. None of these individual overcurrent devices can have a rating or setting higher than the ampacity of the service conductors.

In complying with these rules, it is possible for the total of the six overcurrent devices to be greater than the rating of the service-entrance conductors. However, the size of the service-entrance conductors is required to be adequate for the computed load only, and each individual service disconnecting means is required to be large enough for the individual load it supplies. See the commentary following 230.90(A), Exception No. 3.

230.81 Connection to Terminals.

The service conductors shall be connected to the service disconnecting means by pressure connectors, clamps, or other approved means. Connections that depend on solder shall not be used.

230.82 Equipment Connected to the Supply Side of Service Disconnect.

Only the following equipment shall be permitted to be connected to the supply side of the service disconnecting means:

(1) Cable limiters or other current-limiting devices.

Cable limiters or other current-limiting devices are applied ahead of the service disconnecting means for the following reasons:

1. Faulted cable(s) are individually isolated.
2. Continuity of service is maximized even though one or more cables are faulted.
3. The possibility of severe equipment damage or burn-down as a result of a fault on the service conductors is reduced.
4. The current-limiting feature of cable limiters can be used to provide protection against high short-circuit currents for services and to provide compliance with 110.10.

(2) Meters, meter sockets, or meter disconnect switches nominally rated not in excess of 600 volts, provided all metal housings and service enclosures are grounded.

New in the 2002 *Code*, meter sockets and meter disconnect switches are permitted to be connected on the supply side of the service disconnecting means. Although the *NEC* recognizes that meter sockets have commonly been used on the supply side of the service disconnect, meter sockets have been omitted from 230.82. The meter disconnect is a load-break disconnect switch designed to interrupt the service load on 480Y/277-volt services with self-contained meter sockets. The meter disconnect is not the service disconnecting means. The purpose of the meter disconnect switch is to facilitate meter change, maintenance, or disconnect service. Self-contained meters do not have external potential transformers or current transformers. The load current of the service travels through the meter itself. Neither the self-contained meter nor the meter bypass switch in the meter socket is designed to break the load current on a 480Y/277-volt system.

Self-contained meters or meter bypass switches should not be used to break the load current of a service with over 150 volts to ground, because a hazardous arc could be generated. Arcs generated at voltages greater than 150 volts are considered self-sustaining and can transfer from the energized portions of the equipment to the grounded portions of the equipment. An arc created while breaking load current on a 480Y/277-volt system (277 volts to ground) could transfer to the grounded equipment enclosure, creating a high-energy arcing ground fault and arc flash that could develop into a 3-phase short circuit. This hazardous arcing could burn down the meter socket and injure the person performing the work.

(3) Instrument transformers (current and voltage), high-impedance shunts, load management devices, and surge arresters.

(4) Taps used only to supply load management devices, circuits for standby power systems, fire pump equipment, and fire and sprinkler alarms, if provided with service equipment and installed in accordance with requirements for service-entrance conductors.

Systems such as emergency lighting, fire alarms, fire pumps, standby power, and sprinkler alarms are permitted to be connected ahead of the normal service disconnecting means only if such systems are provided with a separate disconnecting means and overcurrent protection.

(5) Solar photovoltaic systems, fuel cell systems, or interconnected electric power production sources.

(6) Control circuits for power-operable service disconnecting means, if suitable overcurrent protection and disconnecting means are provided.

(7) Ground-fault protection systems where installed as part of listed equipment, if suitable overcurrent protection and disconnecting means are provided.

VII. Service Equipment— Overcurrent Protection

230.90 Where Required.

Each ungrounded service conductor shall have overload protection.

Service-entrance conductors, overhead or underground, are the supply conductors between the point of connection to the service-drop or service-lateral conductors and the service equipment. Service equipment is intended to constitute the main control and means of cutoff of the electrical supply to the premises wiring system. At this point, an overcurrent device, usually a circuit breaker or a fuse, must be installed in series with each ungrounded service conductor to provide overload protection only.

The service overcurrent device will not protect the service conductors under short-circuit or ground-fault conditions on the line side of the disconnect. Protection against ground faults and short circuits is provided by the special requirements for service conductor protection and the location of the conductors.

On multiwire circuits, two or three single-pole switches or circuit breakers that are capable of individual operation are permitted as one protective device. This allowance is acceptable, provided the switches or circuit breakers are equipped with handle ties or a master handle, so that all ungrounded conductors of a service can be disconnected with not more than six operations of the hand, per 230.71(B).

(A) Ungrounded Conductor. Such protection shall be provided by an overcurrent device in series with each ungrounded service conductor that has a rating or setting not higher than the allowable ampacity of the conductor. A set of fuses shall be considered all the fuses required to protect all the ungrounded conductors of a circuit. Single-pole circuit breakers, grouped in accordance with 230.71(B), shall be considered as one protective device.

Exception No. 1: For motor-starting currents, ratings that conform with 430.52, 430.62, and 430.63 shall be permitted.

If a service supplies a motor load as well as lighting or a lighting and appliance load, then the overcurrent protective device is required to have a rating that is sufficient for the lighting and/or appliance load, in accordance with Articles 210 and 220. For an individual motor, the rating is specified by 430.52; for two or more motors, the rating is specified by 430.62.

Example

Determine the minimum-size service conductors to supply a 100-ampere lighting and appliance load plus three squirrel-cage induction motors rated 460 volts, 3 phase, code letter F, service factor 1.15, 40°C, full-voltage starting one 100-hp and two 25-hp motors on a 480-volt, 3-phase system.

Solution

STEP 1. Calculate the conductor loads. The full-load current of the 100-hp motor is 124 amperes (from Table 430.150). The full-load current of each 25-hp motor is 34 amperes (from Table 430.150). The service-entrance conductors are calculated at 125 percent of 124 amperes (155 amperes) plus two motors at 34 amperes each (68 amperes), for a total of 223 amperes (see 430.24). The motor load of 223 amperes plus the lighting and appliance load of 100 amperes equals a total load of 323 amperes. Based on this calculation, the service-entrance conductors cannot be smaller than 400-kcmil copper or 600-kcmil aluminum (see Table 310.16).

STEP 2. Determine the overcurrent protection. The maximum rating of the service overcurrent protective device is based on the lighting and appliance loads calculated in accordance with Article 220 plus the largest motor branch-circuit overcurrent device plus the sum of the other full-load motor currents. Using an inverse-time circuit breaker (see Table 430.52), 250 percent of 124 amperes (100-hp motor) is 310 amperes. The next standard size allowed is 350 amperes, plus 2 times 34 amperes, for a total of 418 amperes, plus the lighting and appliance load (100 amperes), for a total of 518 amperes. Because going up to the next standard-size overcurrent device is not permitted, the next lower standard size is 450 amperes. See 240.6 and 430.63 for feeder overcurrent device rating.

Exception No. 2: Fuses and circuit breakers with a rating or setting that conform with 240.4(B) or (C) and 240.6 shall be permitted.

If the conductor rating does not correspond to the standard ampere rating of a circuit breaker or fuse, the next larger-size circuit breaker or fuse may be used, provided its rating does not exceed 800 amperes, as permitted in 240.4(B)(3). See 240.6 for standard ampere ratings of fuses and circuit breakers.

Exception No. 3: Two to six circuit breakers or sets of fuses shall be permitted as the overcurrent device to provide the overload protection. The sum of the ratings of the circuit breakers or fuses shall be permitted to exceed the ampacity of the service conductors, provided the calculated load does not exceed the ampacity of the service conductors.

Circuit breaker or fuse ampere ratings are permitted to be greater than the ampacity of the service conductors. If multiple disconnects are used as the disconnecting means, the ampacity of the service conductors must be equal to or greater than the load calculated in accordance with Article 220; however, they are not required to be sized equal to or greater than the sum of the multiple disconnects.

For example, the computed load for a service is 350 amperes. The ampacity of a 500-kcmil, Type XHHW copper conductor is 380 amperes (from Table 310.16), and the conductor is allowed to be protected by a 400-ampere fuse or circuit breaker in accordance with 240.4(B). The rating of the fuse or circuit breaker is based on the ampacity of the service conductor, not on the rating of the service disconnect switch.

In this example, a 400-ampere fuse or circuit breaker may be considered properly sized for the protection of 500-kcmil, Type XHHW copper service conductors. If the service disconnecting means [see Exception No. 3 to 230.90(A)] consists of six circuit breakers or six sets of fuses, the combined ratings must not be less than the rating required for a single switch or circuit breaker, in accordance with 230.80. See the commentary following 230.80 and Exhibit 230.27.

As Exhibit 230.27 shows, the combined ratings of the

overcurrent devices equal the size of the overcurrent device required by 240.4 and 230.90(A), Exception No. 3. However, the combined ratings of the overcurrent devices are not required to be equal to or less than this value. For example, all disconnects could be rated 100 amperes. See the commentary following 230.23(A) and the associated commentary.

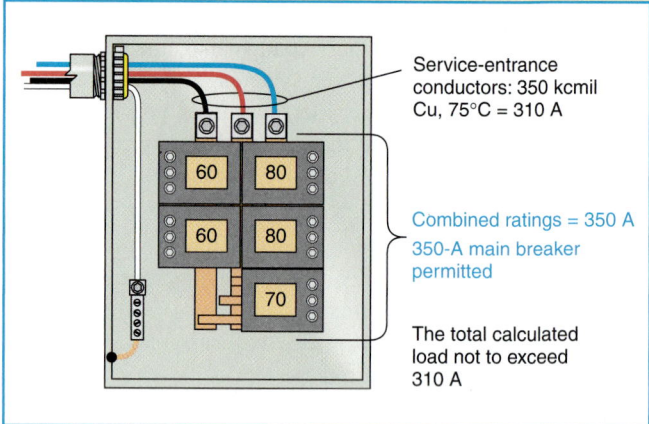

Service-entrance conductors: 350 kcmil Cu, 75°C = 310 A

Combined ratings = 350 A
350-A main breaker permitted

The total calculated load not to exceed 310 A

Exhibit 230.27 An example in which the combined ratings of the overcurrent devices are permitted to exceed the ampacity of the service conductors.

Exception No. 4: Overload protection for fire pump supply conductors shall conform with 695.4(B)(1).

Exception No. 5: Overload protection for 120/240-volt, 3-wire, single-phase dwelling services shall be permitted in accordance with the requirements of 310.15(B)(6).

(B) Not in Grounded Conductor. No overcurrent device shall be inserted in a grounded service conductor except a circuit breaker that simultaneously opens all conductors of the circuit.

230.91 Location.

The service overcurrent device shall be an integral part of the service disconnecting means or shall be located immediately adjacent thereto.

230.92 Locked Service Overcurrent Devices.

Where the service overcurrent devices are locked or sealed or are not readily accessible to the occupant, branch-circuit overcurrent devices shall be installed on the load side, shall be mounted in a readily accessible location, and shall be of lower ampere rating than the service overcurrent device.

230.93 Protection of Specific Circuits.

Where necessary to prevent tampering, an automatic overcurrent device that protects service conductors supplying only a specific load, such as a water heater, shall be permitted to be locked or sealed where located so as to be accessible.

230.94 Relative Location of Overcurrent Device and Other Service Equipment.

The overcurrent device shall protect all circuits and devices.

Exception No. 1: The service switch shall be permitted on the supply side.

Exception No. 2: High-impedance shunt circuits, surge arresters, surge-protective capacitors, and instrument transformers (current and voltage) shall be permitted to be connected and installed on the supply side of the service disconnecting means as permitted in 230.82.

Exception No. 3: Circuits for load management devices shall be permitted to be connected on the supply side of the service overcurrent device where separately provided with overcurrent protection.

Exception No. 4: Circuits used only for the operation of fire alarm, other protective signaling systems, or the supply to fire pump equipment shall be permitted to be connected on the supply side of the service overcurrent device where separately provided with overcurrent protection.

Exception No. 5: Meters nominally rated not in excess of 600 volts, provided all metal housings and service enclosures are grounded in accordance with Article 250.

Exception No. 6: Where service equipment is power operable, the control circuit shall be permitted to be connected ahead of the service equipment if suitable overcurrent protection and disconnecting means are provided.

230.95 Ground-Fault Protection of Equipment.

Ground-fault protection of equipment shall be provided for solidly grounded wye electrical services of more than 150 volts to ground but not exceeding 600 volts phase-to-phase for each service disconnect rated 1000 amperes or more.

The rating of the service disconnect shall be considered to be the rating of the largest fuse that can be installed or the highest continuous current trip setting for which the actual overcurrent device installed in a circuit breaker is rated or can be adjusted.

See the definition of *ground-fault protection of equipment* in Article 100. Ground-fault protection of equipment on services rated 1000 amperes or more operating at 480Y/277 volts was first required in the 1971 *Code* because of the unusually high number of burndowns reported on those types of service. Ground-fault protection of services does not protect the conductors on the supply side of the service disconnecting means, but it is designed to provide protection from line-to-ground faults that occur on the load side of the service

disconnecting means. An alternative to installing ground-fault protection may be to provide multiple disconnects rated less than 1000 amperes. For instance, up to six 800-ampere disconnecting means may be used, and, in that case, ground-fault protection would not be required. Fine Print Note No. 2 to 230.95(C) recognizes that ground-fault protection may be desirable at lesser amperages on solidly grounded systems for voltages exceeding 150 volts to ground but not exceeding 600 volts phase to phase.

In addition to providing ground-fault protection, engineering studies are recommended to determine the circuit impedance and short-circuit currents that would be available at the supply terminals, so that equipment and overcurrent protection of the proper interrupting rating are used. See 110.9 and 110.10 for details on interrupting rating and circuit impedance.

The two basic types of ground-fault equipment protectors are illustrated in Exhibits 230.28 and 230.29. In Exhibit 230.28, the ground-fault sensor is installed around all the circuit conductors, and a stray current on a line-to-ground fault sets up an unbalance of the currents flowing in individual conductors installed through the ground-fault sensor. When this current exceeds the setting of the ground-fault sensor, the shunt trip operates and opens the circuit breakers.

The ground-fault sensor illustrated in Exhibit 230.29 is installed around the bonding jumper only. When an unbalanced current from a line-to-ground fault occurs, the current flows through the bonding jumper and the shunt trip causes the circuit breaker to operate, removing the load from the line. See also 250.24(A)(4), which permits a grounding electrode conductor connection to the equipment grounding terminal bar or bus.

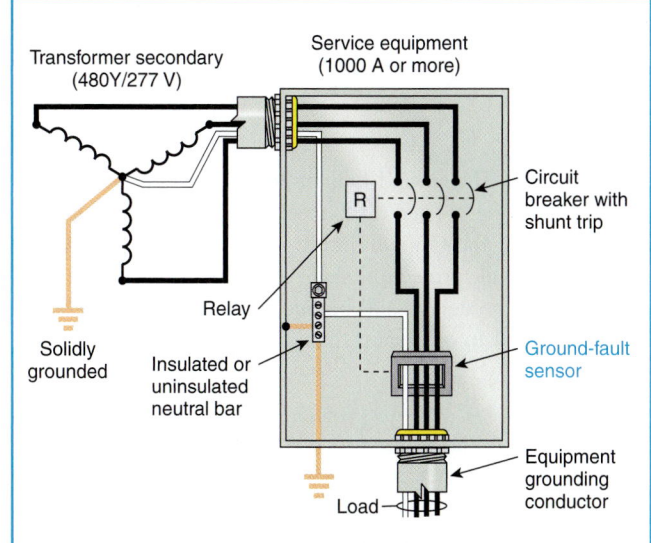

Exhibit 230.28 A ground-fault sensor encircling all circuit conductors, including the neutral.

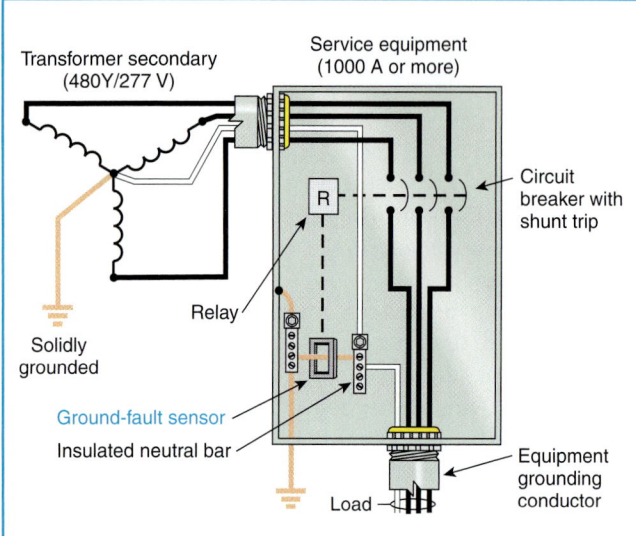

Transformer secondary
(480Y/277 V)

Service equipment
(1000 A or more)

R

Relay

Circuit
breaker with
shunt trip

Solidly
grounded

Ground-fault sensor

Insulated neutral bar

Load

Equipment
grounding
conductor

Exhibit 230.29 A ground-fault sensor encircling only the bonding jumper conductor.

Solidly Grounded—Definition. Connection of the grounded conductor to ground without inserting any resistor or impedance device.

Exception No. 1: The ground-fault protection provisions of this section shall not apply to a service disconnect for a continuous industrial process where a nonorderly shutdown will introduce additional or increased hazards.

Exception No. 2: The ground-fault protection provisions of this section shall not apply to fire pumps.

Most fire pumps rated 100 hp and over would require a disconnecting means rated at 1000 amperes or more. However, due to the emergency nature of their use, fire pumps are exempt from the provisions of 230.95.

(A) Setting. The ground-fault protection system shall operate to cause the service disconnect to open all ungrounded conductors of the faulted circuit. The maximum setting of the ground-fault protection shall be 1200 amperes, and the maximum time delay shall be one second for ground-fault currents equal to or greater than 3000 amperes.

The maximum setting for ground-fault sensors is 1200 amperes. There is no minimum, but it should be noted that settings at low levels increase the likelihood of unwanted shutdowns. The requirements of 230.95 place a restriction on fault currents greater than 3000 amperes and limit the duration of the fault to not more than 1 second. This restriction minimizes the amount of damage done by an arcing

fault, which is directly proportional to the time the arcing fault is allowed to burn.

Care should be taken to ensure that interconnecting multiple supply systems does not negate proper sensing by the ground-fault protection equipment. A careful engineering study must be made to ensure that fault currents do not take parallel paths to the supply system, thereby bypassing the ground-fault detection device. See 215.10, 240.13, 517.17, and 705.32 for further information on ground-fault protection of equipment.

(B) Fuses. If a switch and fuse combination is used, the fuses employed shall be capable of interrupting any current higher than the interrupting capacity of the switch during a time that the ground-fault protective system will not cause the switch to open.

(C) Performance Testing. The ground-fault protection system shall be performance tested when first installed on site. The test shall be conducted in accordance with instructions that shall be provided with the equipment. A written record of this test shall be made and shall be available to the authority having jurisdiction.

The requirement for ground-fault protection system performance testing is a result of numerous reports of ground-fault protection systems that were improperly wired and could not or did not perform the function for which they were intended. This *Code* and qualified testing laboratories require a set of performance testing instructions to be supplied with the equipment. Evaluation and listing of the instructions fall under the jurisdiction of those best qualified to make such judgments, the qualified electrical testing laboratory (see 90.7). If listed equipment is not installed in accordance with the instructions provided, the installation does not comply with 110.3(B).

FPN No. 1: Ground-fault protection that functions to open the service disconnect affords no protection from faults on the line side of the protective element. It serves only to limit damage to conductors and equipment on the load side in the event of an arcing ground fault on the load side of the protective element.

FPN No. 2: This added protective equipment at the service equipment may make it necessary to review the overall wiring system for proper selective overcurrent protection coordination. Additional installations of ground-fault protective equipment may be needed on feeders and branch circuits where maximum continuity of electrical service is necessary.

FPN No. 3: Where ground-fault protection is provided for the service disconnect and interconnection is made with another supply system by a transfer device, means or devices may be needed to ensure proper ground-fault sensing by the ground-fault protection equipment.

VIII. Services Exceeding 600 Volts, Nominal

230.200 General.

Service conductors and equipment used on circuits exceeding 600 volts, nominal, shall comply with all the applicable provisions of the preceding sections of this article and with the following sections, which supplement or modify the preceding sections. In no case shall the provisions of Part VIII apply to equipment on the supply side of the service point.

> FPN: For clearances of conductors of over 600 volts, nominal, see ANSI C2-1997, *National Electrical Safety Code*.

Where services rated over 600 volts supply utility-owned and utility-maintained transformers, the conductors on the line side and load side of the service point are service-lateral conductors; however, only those conductors on the load side of the service point come under the requirements of the *NEC*. The service point is a specific location where the supply conductors of the electric utility and the customer-owned (premises wiring) conductors connect.

Exhibit 230.30 depicts an installation where the transformer and service lateral conductors to the service point are owned by the electric utility. The transformer secondary conductors between the service point and the service discon-

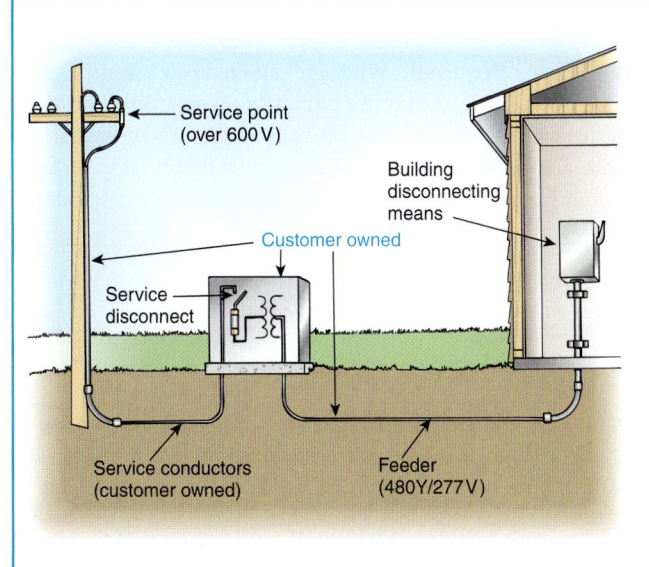

Exhibit 230.31 Service rated over 600 volts supplying a customer-owned transformer.

necting means at the building are service-entrance conductors. In the installation depicted in Exhibit 230.31, the main service disconnecting means is located at the customer-owned transformer primary. The conductors between the transformer secondary and the line side of the building disconnecting means are feeders. The connection point may be in a belowground or aboveground junction box, where a change in the wiring method might occur. Conductors on the load side of the building service disconnecting means are feeders. Each building or structure is required to have a disconnecting means, in accordance with 225.31.

Where services rated over 600 volts supply customer-owned and customer-maintained transformers, the conductors from the service point to the transformer service disconnect are service lateral conductors, as illustrated in Exhibit 230.31. The conductors between the transformer secondary and the building disconnecting means are feeders, as defined in Article 100.

230.202 Service-Entrance Conductors.

Service-entrance conductors to buildings or enclosures shall be installed to conform to 230.202(A) and (B).

(A) Conductor Size. Service-entrance conductors shall not be smaller than 6 AWG unless in multiconductor cable. Multiconductor cable shall not be smaller than 8 AWG.

(B) Wiring Methods. Service-entrance conductors shall be installed by one of the wiring methods covered in 300.37 and 300.50.

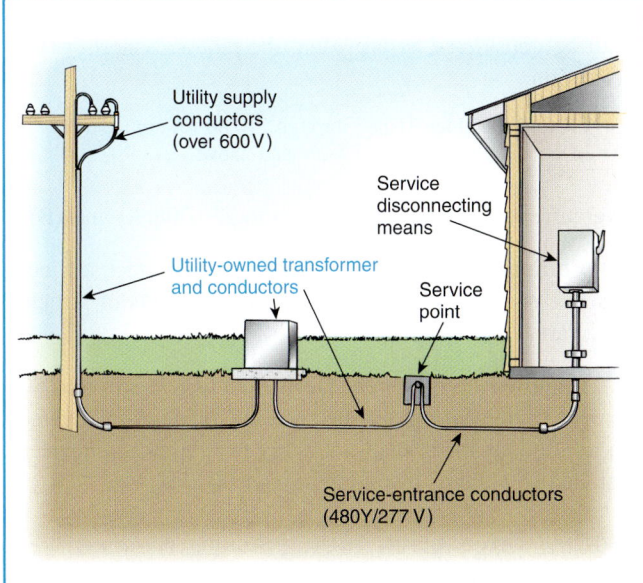

Exhibit 230.30 Service rated over 600 volts supplying a utility-owned transformer.

230.204 Isolating Switches.

(A) Where Required. Where oil switches or air, oil, vacuum, or sulfur hexafluoride circuit breakers constitute the service disconnecting means, an isolating switch with visible break contacts shall be installed on the supply side of the disconnecting means and all associated service equipment.

Exception: An isolating switch shall not be required where the circuit breaker or switch is mounted on removable truck panels or metal-enclosed switchgear units, that

(a) Cannot be opened unless the circuit is disconnected, and

(b) Where all energized parts are automatically disconnected when the circuit breaker or switch is removed from the normal operating position

(B) Fuses as Isolating Switch. Where fuses are of the type that can be operated as a disconnecting switch, a set of such fuses shall be permitted as the isolating switch.

(C) Accessible to Qualified Persons Only. The isolating switch shall be accessible to qualified persons only.

(D) Grounding Connection. Isolating switches shall be provided with a means for readily connecting the load side conductors to ground when disconnected from the source of supply.

A means for grounding the load side conductors shall not be required for any duplicate isolating switch installed and maintained by the electric supply company.

Exhibit 230.32 illustrates a two-position isolating switch for grounding a load-side conductor when it is disconnected from high-voltage line buses.

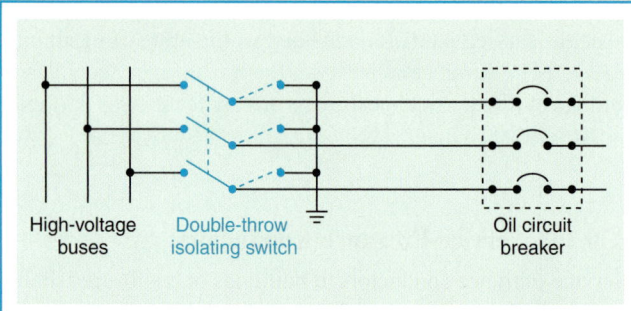

Exhibit 230.32 Two-position isolating switch for grounding a load-side conductor disconnected from high-voltage line buses.

230.205 Disconnecting Means.

(A) Location. The service disconnecting means shall be located in accordance with 230.70.

(B) Type. Each service disconnect shall simultaneously disconnect all ungrounded service conductors that it controls and shall have a fault-closing rating that is not less than the maximum short-circuit current available at its supply terminals.

Where fused switches or separately mounted fuses are installed, the fuse characteristics shall be permitted to contribute to the fault-closing rating of the disconnecting means.

(C) Remote Control. For multibuilding, industrial installations under single management, the service disconnecting means shall be permitted to be located at a separate building or structure. In such cases, the service disconnecting means shall be permitted to be electrically operated by a readily accessible, remote-control device.

230.206 Overcurrent Devices as Disconnecting Means.

Where the circuit breaker or alternative for it, as specified in 230.208 for service overcurrent devices, meets the requirements specified in 230.205, they shall constitute the service disconnecting means.

230.208 Protection Requirements.

A short-circuit protective device shall be provided on the load side of, or as an integral part of, the service disconnect, and shall protect all ungrounded conductors that it supplies. The protective device shall be capable of detecting and interrupting all values of current, in excess of its trip setting or melting point, that can occur at its location. A fuse rated in continuous amperes not to exceed three times the ampacity of the conductor, or a circuit breaker with a trip setting of not more than six times the ampacity of the conductors, shall be considered as providing the required short-circuit protection.

> FPN: See Tables 310.67 through 310.86 for ampacities of conductors rated 2001 volts and above.

Overcurrent devices shall conform to 230.208(A) and (B).

(A) Equipment Type. Equipment used to protect service-entrance conductors shall meet the requirements of Article 490, Part II.

(B) Enclosed Overcurrent Devices. The restriction to 80 percent of the rating for an enclosed overcurrent device for continuous loads shall not apply to overcurrent devices installed in systems operating at over 600 volts.

230.209 Surge Arresters (Lightning Arresters).

Surge arresters installed in accordance with the requirements of Article 280 shall be permitted on each ungrounded overhead service conductor.

230.210 Service Equipment—General Provisions.

Service equipment, including instrument transformers, shall conform to Article 490, Part I.

230.211 Metal-Enclosed Switchgear.

Metal-enclosed switchgear shall consist of a substantial metal structure and a sheet metal enclosure. Where installed over a combustible floor, suitable protection thereto shall be provided.

Exhibit 230.33 shows an assembly of metal-enclosed switchgear.

Exhibit 230.33 Metal-enclosed switchgear. (Courtesy of Square D Co.)

230.212 Over 35,000 Volts.

Where the voltage exceeds 35,000 volts between conductors that enter a building, they shall terminate in a metal-enclosed switchgear compartment or a vault conforming to the requirements of 450.41 through 450.48.

ARTICLE 240
Overcurrent Protection

Contents

I. General

240.1 Scope.

Parts I through VII of this article provide the general requirements for overcurrent protection and overcurrent protective devices not more than 600 volts, nominal. Part VIII covers overcurrent protection for those portions of supervised industrial installations operating at voltages of not more than 600 volts, nominal. Part IX covers overcurrent protection over 600 volts, nominal.

> FPN: Overcurrent protection for conductors and equipment is provided to open the circuit if the current reaches a value that will cause an excessive or dangerous temperature in conductors or conductor insulation. See also 110.9 for requirements for interrupting ratings and 110.10 for requirements for protection against fault currents.

240.2 Definitions.

Coordination. The proper localization of a fault condition to restrict outages to the equipment affected, accomplished by the choice of selective fault-protective devices.

The main goal of overcurrent protection coordination is to isolate the faulted portion of the electrical circuit quickly while at the same time maintaining normal service for the remainder of the electrical system. The electrical system overcurrent protection must guard against short circuits and ground faults to ensure that the resulting damage is minimized while other parts of the system not directly involved with the fault are kept on until other protective devices clear the fault.

Overcurrent protective devices, such as fuses and circuit breakers, have time/current characteristics that determine the time it takes to clear the fault for a given value of fault current. Selectivity occurs when the device closest to the fault opens before the next device upstream operates. For example, any fault on a branch circuit should open the branch circuit breaker rather than the feeder overcurrent protection. All faults on a feeder should open the feeder overcurrent protection rather than the service overcurrent protection. When selectivity occurs, the electrical system is considered to be coordinated.

With coordinated overcurrent protection, the faulted or overloaded circuit is isolated by the selective operation of only the overcurrent protective device closest to the overcurrent condition. This isolation prevents power loss to unaffected loads.

Current-Limiting Overcurrent Protective Device. A device that, when interrupting currents in its current-limiting range, reduces the current flowing in the faulted circuit to a magnitude substantially less than that obtainable in the same circuit if the device were replaced with a solid conductor having comparable impedance.

Most electrical distribution systems can deliver high ground-fault or short-circuit currents to components such as conductors, service equipment, and the like. These components may not be able to handle short-circuit currents; they may be damaged or destroyed, and serious burndowns and fires could result. Properly selected current-limiting overcurrent protective devices, such as the ones shown in Exhibit 240.1,

limit the let-through energy to within the rating of the components, in spite of high available short-circuit currents. A current-limiting protective device is one that cuts off a fault current in less than one-half cycle. It thus prevents short-circuit currents from building up to their full available values.

Proper selection of current-limiting devices may depend on the type of device selected. For example, a Class RK5 fuse is not as current limiting as a Class RK1 fuse. Furthermore, a Class RK1 fuse is not as current limiting as a high-speed semiconductor fuse. See 110.9 and 110.10 for details on interrupting ratings, circuit impedance, and other characteristics.

Supervised Industrial Installation. For the purposes of Part VIII, the industrial portions of a facility where all of the following conditions are met:

(1) Conditions of maintenance and engineering supervision ensure that only qualified persons monitor and service the system.
(2) The premises wiring system has 2500 kVA or greater of load used in industrial process(es), manufacturing activities, or both, as calculated in accordance with Article 220.
(3) The premises has at least one service that is more than 150 volts to ground and more than 300 volts phase-to-phase.

This definition excludes installations in buildings used by the industrial facility for offices, warehouses, garages, machine shops, and recreational facilities that are not an integral part of the industrial plant, substation, or control center.

Exhibit 240.1 Class R current-limiting fuses with rejection feature to prohibit the installation of non-current-limiting fuses. (Courtesy of Bussmann Division, Cooper Industries)

For a facility to be recognized as an industrial establishment, it must have the following: a combined process and manufacturing load greater than 2500 kVA; at least one service, 480Y/277V, nominal, or greater; and maintenance and engineering supervision that ensures that only qualified persons will monitor and service the electrical system. All process or manufacturing loads from each low-, medium-, and high-voltage system can be added together to satisfy the load requirements of Part VIII. Loads are calculated in accordance with Article 220. However, loads not associated with manufacturing or processing cannot be used to meet the 2500-kVA minimum requirement.

The provisions of Part VIII only apply to low-voltage electrical systems (600 volts, nominal, or less) used for process or manufacturing. Part VIII does not apply to electrical systems operating at over 600 volts, nominal, or to electrical systems that serve nonmanufacturing or nonprocess facilities, such as offices, warehouses, garages, machine shops, or recreational facilities. However, if a part of the process or manufacturing electrical system is used to serve an office, warehouse, garage, machine shop, or recreational facility that is an integral part of the industrial plant, control center, or substation, Part VIII can still apply to the total process or manufacturing electrical system.

Tap Conductors. As used in this article, a tap conductor is defined as a conductor, other than a service conductor, that has overcurrent protection ahead of its point of supply that exceeds the value permitted for similar conductors that are protected as described elsewhere in 240.4.

240.3 Other Articles.

Equipment shall be protected against overcurrent in accordance with the article in this *Code* that covers the type of equipment specified in Table 240.3.

240.4 Protection of Conductors.

Conductors, other than flexible cords, flexible cables, and fixture wires, shall be protected against overcurrent in accordance with their ampacities specified in 310.15, unless otherwise permitted or required in 240.4(A) through (G).

(A) Power Loss Hazard. Conductor overload protection shall not be required where the interruption of the circuit would create a hazard, such as in a material-handling magnet circuit or fire pump circuit. Short-circuit protection shall be provided.

> FPN: See NFPA 20-1999, *Standard for the Installation of Stationary Pumps for Fire Protection.*

(B) Devices Rated 800 Amperes or Less. The next higher standard overcurrent device rating (above the ampacity of the conductors being protected) shall be permitted to be used, provided all of the following conditions are met:

(1) The conductors being protected are not part of a multioutlet branch circuit supplying receptacles for cord-and-plug-connected portable loads.
(2) The ampacity of the conductors does not correspond with the standard ampere rating of a fuse or a circuit breaker without overload trip adjustments above its rating (but that shall be permitted to have other trip or rating adjustments).
(3) The next higher standard rating selected does not exceed 800 amperes.

Table 210.24 summarizes the requirements for the size of conductors and the size of the overcurrent protection for branch circuits where two or more outlets are required. The first footnote also indicates that the wire sizes are for copper conductors. Section 210.3 indicates that branch-circuit con-

Table 240.3 Other Articles

Equipment	Article
Air-conditioning and refrigerating equipment	440
Appliances	422
Audio signal processing, amplification, and reproduction equipment	640
Branch circuits	210
Busways	368
Capacitors	460
Class 1, Class 2, and Class 3 remote-control, signaling, and power-limited circuits	725
Closed-loop and programmed power distribution	780
Cranes and hoists	610
Electric signs and outline lighting	600
Electric welders	630
Electrolytic cells	668
Elevators, dumbwaiters, escalators, moving walks, wheelchair lifts, and stairway chair lifts	620
Emergency systems	700
Fire alarm systems	760
Fire pumps	695
Fixed electric heating equipment for pipelines and vessels	427
Fixed electric space-heating equipment	424
Fixed outdoor electric deicing and snow-melting equipment	426
Generators	445
Health care facilities	517
Induction and dielectric heating equipment	665
Industrial machinery	670
Luminaires (lighting fixtures), lampholders, and lamps	410
Motion picture and television studios and similar locations	530
Motors, motor circuits, and controllers	430
Phase converters	455
Pipe organs	650
Places of assembly	518
Receptacles	406
Services	230
Solar photovoltaic systems	690
Switchboards and panelboards	408
Theaters, audience areas of motion picture and television studios, and similar locations	520
Transformers and transformer vaults	450
X-ray equipment	660

ductors rated 15, 20, 30, 40, and 50 amperes must be protected at their ratings. Section 210.19(A) requires that branch-circuit conductors have an ampacity not less than the rating of the branch circuit and not less than the maximum load to be served. These specific requirements take precedence over 240.4(B), which applies generally.

Tables 310.16 through 310.86 list the ampacities of conductors. Section 240.6 lists the standard ratings of overcurrent devices. If the ampacity of the conductor in these tables does not match the rating of the standard overcurrent device, 240.4 permits the use of the next larger standard overcurrent device. All three individual conditions of 240.4(B)(3) must be met in order for this permission to apply. However, if the ampacity of a conductor matches the standard rating of 240.6, that conductor must be protected at the standard size device. For example, in Table 310.16, 3 AWG, 75°C Copper, Type THWN, the ampacity is listed as 100 amperes. That conductor would be protected by a 100-ampere overcurrent device.

Section 310.15(B)(6) allows upsizing of the ampacity of the conductor and the overcurrent devices for single-phase, 120/240-volt, residential services and certain feeders.

(C) Devices Rated Over 800 Amperes. Where the overcurrent device is rated over 800 amperes, the ampacity of the conductors it protects shall be equal to or greater than the rating of the overcurrent device defined in 240.6.

(D) Small Conductors. Unless specifically permitted in 240.4(E) through (G), the overcurrent protection shall not exceed 15 amperes for 14 AWG, 20 amperes for 12 AWG, and 30 amperes for 10 AWG copper; or 15 amperes for 12 AWG and 25 amperes for 10 AWG aluminum and copper-clad aluminum after any correction factors for ambient temperature and number of conductors have been applied.

(E) Tap Conductors. Tap conductors shall be permitted to be protected against overcurrent in accordance with 210.19(A)(3) and (4), 240.5(B)(2), 240.21, 368.11, 368.12, and 430.53(D).

(F) Transformer Secondary Conductors. Single-phase (other than 2-wire) and multiphase (other than delta-delta, 3-wire) transformer secondary conductors shall not be considered to be protected by the primary overcurrent protective device. Conductors supplied by the secondary side of a single-phase transformer having a 2-wire (single-voltage) secondary, or a three-phase, delta-delta connected transformer having a 3-wire (single-voltage) secondary, shall be permitted to be protected by overcurrent protection provided on the primary (supply) side of the transformer, provided this protection is in accordance with 450.3 and does not exceed the value determined by multiplying the secondary conductor ampacity by the secondary to primary transformer voltage ratio.

The first paragraph of 240.4 requires that conductors be protected against overcurrent in accordance with their ampacity. Section 240.4(F) permits the secondary circuit conductors from a transformer to be protected by overcurrent devices in the primary circuit conductors of the transformer only in the following two special cases:

1. A transformer with a 2-wire primary and a 2-wire secondary, provided the transformer primary is protected in accordance with 450.3
2. A 3-phase, delta-delta-connected transformer having a 3-wire, single-voltage secondary, provided its primary is protected in accordance with 450.3

Except for these two special cases, transformer secondary conductors must be protected by the use of overcurrent devices, because the primary overcurrent devices do not provide this protection. As an example, consider a single-phase transformer with a 2-wire secondary that is provided with primary overcurrent protection rated at 50 amperes. The transformer is rated 480/240 volts. Conductors supplied by the secondary have an ampacity of 100 amperes. Is the 50-ampere overcurrent protection allowed to protect the conductors that are connected to the secondary?

The secondary-to-primary voltage ratio in this example is 240 ÷ 480, a ratio of 0.5. Multiplying the secondary conductor ampacity of 100 amperes by 0.5 yields 50 amperes. Thus, the maximum rating of the overcurrent device allowed on the primary of the transformer that will also provide overcurrent protection for the secondary conductors is 50 amperes. These secondary conductors are not tap conductors, are not limited in length, and do not require overcurrent protection where they receive their supply, which is at the transformer secondary terminals.

However, if the secondary consisted of a 3-wire, 240/120-volt system, a 120-volt line-to-neutral load could draw up to 200 amperes before the overcurrent device in the primary actuated. That would be the result of the 1:4 secondary-to-primary voltage ratio of the 120-volt winding of the transformer secondary, which can cause dangerous overloading of the secondary conductors.

(G) Overcurrent Protection for Specific Conductor Applications. Overcurrent protection for the specific conductors shall be permitted to be provided as referenced in Table 240.4(G).

240.5 Protection of Flexible Cords, Flexible Cables, and Fixture Wires.

Flexible cord and flexible cable, including tinsel cord and extension cords, and fixture wires shall be protected against overcurrent by either 240.5(A) or (B).

Table 240.4(G) Specific Conductor Applications

Conductor	Article	Section
Air-conditioning and refrigeration equipment circuit conductors	440, Parts III, VI	
Capacitor circuit conductors	460	460.8(B) and 460.25(A)–(D)
Control and instrumentation circuit conductors (Type ITC)	727	727.9
Electric welder circuit conductors	630	630.12 and 630.32
Fire alarm system circuit conductors	760	760.23, 760.24, 760.41, and Chapter 9, Tables 12(A) and 12(B)
Motor-operated appliance circuit conductors	422, Part II	
Motor and motor-control circuit conductors	430, Parts III, IV, V, VI, VII	
Phase converter supply conductors	455	455.7
Remote-control, signaling, and power-limited circuit conductors	725	725.23, 725.24, 725.41, and Chapter 9, Tables 11(A) and 11(B)
Secondary tie conductors	450	450.6

(A) Ampacities. Flexible cord and flexible cable shall be protected by an overcurrent device in accordance with their ampacity as specified in Table 400.5(A) and Table 400.5(B). Fixture wire shall be protected against overcurrent in accordance with its ampacity as specified in Table 402.5. Supplementary overcurrent protection, as in 240.10, shall be permitted to be an acceptable means for providing this protection.

(B) Branch Circuit Overcurrent Device. Flexible cord shall be protected where supplied by a branch circuit in accordance with one of the methods described in 240.5(B)(1), (2), or (3).

(1) Supply Cord of Listed Appliance or Portable Lamps. Where flexible cord or tinsel cord is approved for and used with a specific listed appliance or portable lamp, it shall be permitted to be supplied by a branch circuit of Article 210 in accordance with the following:

(1) 20-ampere circuits—tinsel cord or 18 AWG cord and larger

(2) 30-ampere circuits—16 AWG cord and larger

(3) 40-ampere circuits—cord of 20-ampere capacity and over

(4) 50-ampere circuits—cord of 20-ampere capacity and over

Section 240.5(A) references Tables 400.5(A) and 400.5(B) for flexible cords and flexible cables and Table 402.5 for fixture wire ampacity. Supplementary protection, as described in 240.10, is also acceptable as an alternate for protection of either flexible cord or fixture wire.

Section 240.5(B) permits smaller conductors to be connected to branch circuits of a greater rating if the smaller conductors are approved for and used with a specific listed appliance, portable lamp, or extension cord, or if fixture wire is protected as listed in the exceptions.

(2) Fixture Wire. Fixture wire shall be permitted to be tapped to the branch circuit conductor of a branch circuit of Article 210 in accordance with the following:

(1) 20-ampere circuits—18 AWG, up to 15 m (50 ft) of run length

(2) 20-ampere circuits—16 AWG, up to 30 m (100 ft) of run length

(3) 20-ampere circuits—14 AWG and larger

(4) 30-ampere circuits—14 AWG and larger

(5) 40-ampere circuits—12 AWG and larger

(6) 50-ampere circuits—12 AWG and larger

(3) Extension Cord Sets. Flexible cord used in listed extension cord sets, or in extension cords made with separately listed and installed components, shall be permitted to be supplied by a branch circuit of Article 210 in accordance with the following:

20-ampere circuits—16 AWG and larger

Section 240.5(B)(3) also permits listed extension cord sets 16 AWG or larger to be protected by 20-ampere-or-less branch-circuit overcurrent devices.

240.6 Standard Ampere Ratings.

(A) Fuses and Fixed-Trip Circuit Breakers. The standard ampere ratings for fuses and inverse time circuit breakers shall be considered 15, 20, 25, 30, 35, 40, 45, 50, 60, 70, 80, 90, 100, 110, 125, 150, 175, 200, 225, 250, 300, 350, 400, 450, 500, 600, 700, 800, 1000, 1200, 1600, 2000, 2500, 3000, 4000, 5000, and 6000 amperes. Additional standard ampere ratings for fuses shall be 1, 3, 6, 10, and 601. The use of fuses and inverse time circuit breakers with nonstandard ampere ratings shall be permitted.

(B) Adjustable-Trip Circuit Breakers. The rating of adjustable-trip circuit breakers having external means for adjusting the current setting (long-time pickup setting), not meeting the requirements of 240.6(C), shall be the maximum setting possible.

(C) Restricted Access Adjustable-Trip Circuit Breakers.
A circuit breaker(s) that has restricted access to the adjusting means shall be permitted to have an ampere rating(s) that is equal to the adjusted current setting (long-time pickup setting). Restricted access shall be defined as located behind one of the following:

(1) Removable and sealable covers over the adjusting means
(2) Bolted equipment enclosure doors
(3) Locked doors accessible only to qualified personnel

The set long-time pickup rating (as opposed to the instantaneous trip rating) of an adjustable-trip circuit breaker can be considered the circuit breaker rating if access to the adjustment means is limited. Such limitation is due to location of the adjustment means behind sealable covers, as shown in Exhibit 240.2, or behind locked doors accessible only to qualified personnel.

Exhibit 240.2 An adjustable-trip circuit breaker with a transparent, removable and sealable cover. (Courtesy of Square D Co.)

Manufacturers of both fuses and inverse time circuit breakers have, or can make available, products with ampere ratings other than those listed in 240.6(A). Selection of such nonstandard ratings is not required by the *Code* but may permit better protection for conductors.

240.8 Fuses or Circuit Breakers in Parallel.
Fuses and circuit breakers shall be permitted to be connected in parallel where they are factory assembled in parallel and listed as a unit. Individual fuses, circuit breakers, or combinations thereof shall not otherwise be connected in parallel.

Section 240.8 prohibits the use of fuses or circuit breakers in parallel unless they are factory assembled in parallel and listed as a unit. Section 404.17 prohibits the use of fuses in parallel in fused switches.

It is not the intent of 240.8 to restore the use of standard fuses in parallel in disconnect switches. However, 240.8 recognizes parallel low-voltage circuit breakers or fuses and parallel high-voltage circuit breakers or fuses if they are tested and factory assembled in parallel and listed as a unit.

High-voltage fuses have long been recognized in parallel when they are assembled in an identified common mounting.

240.9 Thermal Devices.
Thermal relays and other devices not designed to open short circuits or ground faults shall not be used for the protection of conductors against overcurrent due to short circuits or ground faults, but the use of such devices shall be permitted to protect motor branch-circuit conductors from overload if protected in accordance with 430.40.

240.10 Supplementary Overcurrent Protection.
Where supplementary overcurrent protection is used for luminaires (lighting fixtures), appliances, and other equipment or for internal circuits and components of equipment, it shall not be used as a substitute for branch-circuit overcurrent devices or in place of the branch-circuit protection specified in Article 210. Supplementary overcurrent devices shall not be required to be readily accessible.

240.12 Electrical System Coordination.
Where an orderly shutdown is required to minimize the hazard(s) to personnel and equipment, a system of coordination based on the following two conditions shall be permitted:

(1) Coordinated short-circuit protection

With coordinated overcurrent protection, the faulted or overloaded circuit is isolated by the selective operation of only the overcurrent protective device closest to the overcurrent condition. This prevents power loss to unaffected loads. An example of noncoordinated protection and coordinated protection are illustrated in Exhibit 240.3.

(2) Overload indication based on monitoring systems or devices

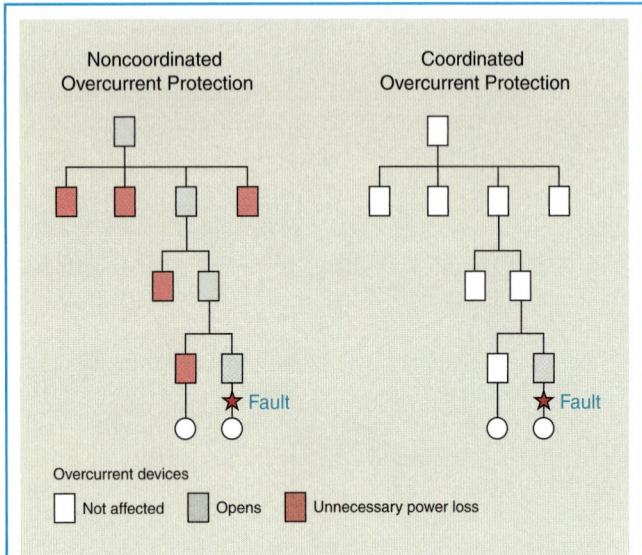

Exhibit 240.3 Noncoordinated and coordinated overcurrent protection.

FPN: The monitoring system may cause the condition to go to alarm, allowing corrective action or an orderly shutdown, thereby minimizing personnel hazard and equipment damage.

240.13 Ground-Fault Protection of Equipment.

Ground-fault protection of equipment shall be provided in accordance with the provisions of 230.95 for solidly grounded wye electrical systems of more than 150 volts to ground but not exceeding 600 volts phase-to-phase for each individual device used as a building or structure main disconnecting means rated 1000 amperes or more.

The provisions of this section shall not apply to the disconnecting means for the following:

(1) Continuous industrial processes where a nonorderly shutdown will introduce additional or increased hazards
(2) Installations where ground-fault protection is provided by other requirements for services or feeders
(3) Fire pumps installed in accordance with Article 695

Section 240.13 extends the requirement of 230.95 to building disconnects, regardless of how the disconnects are classified (service disconnects or building disconnects for feeders or even branch circuits). See 215.10 and Article 225, Part II, for the requirements for building disconnects not on the utility service.

Section 240.13 requires each building or structure disconnect that is rated 1000 amperes or more, on a solidly grounded system of more than 150 volts to ground (e.g., a 480Y/277-volt system), to be provided with ground-fault

protection for equipment. Provisions are included for fire pumps and for continuous industrial processes in which nonorderly shutdowns would introduce additional hazards.

Where ground-fault protection for equipment is installed at the service equipment and other buildings or structures are supplied by feeders or branch circuits, 250.24 requires regrounding of the grounded conductor if an equipment grounding conductor is not included in the run. However, 240.13(2) exempts the grounded conductor from being regrounded downstream from the ground-fault protected service. Regrounding of the neutral at the second building may nullify the ground-fault protection of the second building that would otherwise be provided by ground-fault protection at the main service.

II. Location

240.20 Ungrounded Conductors.

(A) Overcurrent Device Required. A fuse or an overcurrent trip unit of a circuit breaker shall be connected in series with each ungrounded conductor. A combination of a current transformer and overcurrent relay shall be considered equivalent to an overcurrent trip unit.

FPN: For motor circuits, see Parts III, IV, V, and X of Article 430.

(B) Circuit Breaker as Overcurrent Device. Circuit breakers shall open all ungrounded conductors of the circuit unless otherwise permitted in 240.20(B)(1), (B)(2), and (B)(3).

(1) Multiwire Branch Circuit. Except where limited by 210.4(B), individual single-pole circuit breakers, with or without approved handle ties, shall be permitted as the protection for each ungrounded conductor of multiwire branch circuits that serve only single-phase line-to-neutral loads.

(2) Grounded Single-Phase and 3-wire dc Circuits. In grounded systems, individual single-pole circuit breakers with approved handle ties shall be permitted as the protection for each ungrounded conductor for line-to-line connected loads for single-phase circuits or 3-wire, direct-current circuits.

(3) 3-Phase and 2-Phase Systems. For line-to-line loads in 4-wire, 3-phase systems or 5-wire, 2-phase systems having a grounded neutral and no conductor operating at a voltage greater than permitted in 210.6, individual single-pole circuit breakers with approved handle ties shall be permitted as the protection for each ungrounded conductor.

Before discussing handle ties, it is important to understand the Article 100 definition of the term *branch circuit, multiwire,* as well as 210.4(C) and its two exceptions. Multiwire branch circuits are permitted to supply line-to-line loads that consist of one piece of utilization equipment or where all ungrounded conductors are opened simultaneously by the branch-circuit overcurrent device. See the commentary following 210.4(C) for additional information.

Section 240.20(B) requires that if a circuit breaker is used, it must open all ungrounded conductors of the circuit when it trips or is manually operated. For 2-wire circuits with one conductor grounded, this rule is simple and needs no further explanation. For multiwire branch circuits of 600 volts or less, however, there are three methods of implementing this rule.

The first, and certainly the most common, method is to use a multipole circuit breaker with an internal common trip mechanism. This breaker is operated by an external single lever internally attached to two or three poles of a circuit breaker, or the external lever may be attached to multiple handles operated as one, provided the breaker is a factory-assembled unit. Underwriters Laboratories refers to these devices as multipole common trip circuit breakers. These circuit breakers are required to be used for multiwire branch circuits fed from both 3-phase and single-phase ungrounded systems. Of course, they are permitted to be used on any multiwire branch circuit within their rating.

The second method is to use two or three single-pole circuit breakers and add an approved handle tie to function as a common operating handle. This multipole circuit breaker is field assembled by externally attaching an approved common lever (handle tie) onto the two or three individual circuit breakers. It is important to understand that handle ties do not cause the circuit breaker to serve as a common trip mechanism, but rather allow for common switching only. Handle tie mechanism circuit breakers are permitted as a substitute for internal common trip mechanism circuit breakers only for limited applications. Unless specifically prohibited elsewhere, circuit breakers with approved handle ties are permitted for multiwire branch circuits only if the circuit is supplied from grounded 3-phase or grounded single-phase systems. The single-pole circuit breakers used together in this fashion must be rated for the dual voltage encountered, such as 120/240 volts.

The third method is to use individual single-pole circuit breakers without common trip mechanisms or without handle ties for multiwire branch circuits. Unless limited by other sections of the *Code,* this method is permitted for multiwire circuits, provided the multiwire branch circuit supplies only single-phase line-to-neutral loads.

Exhibits 240.4, 240.5, and 240.6 illustrate some examples of how the requirements in 240.20(B) are applied. In Exhibit 240.4, where multipole circuit breakers are required, handle ties are not permitted because the circuits are supplied from ungrounded systems. In Exhibit 240.5, in which single-

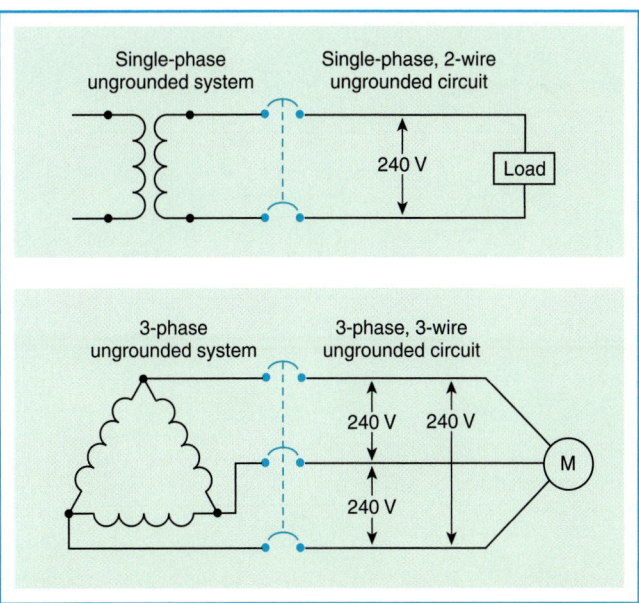

Exhibit 240.4 Examples of circuits that require multipole common trip-type circuit breakers, in accordance with 240.20(B).

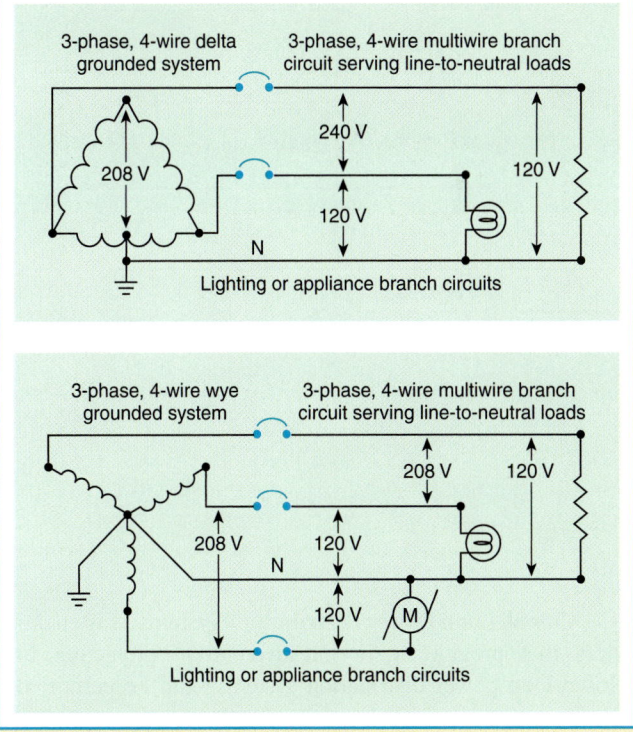

Exhibit 240.5 Examples of circuits in which single-pole circuit breakers are permitted, in accordance with 240.20(B)(1), because they open the ungrounded conductor of the circuit.

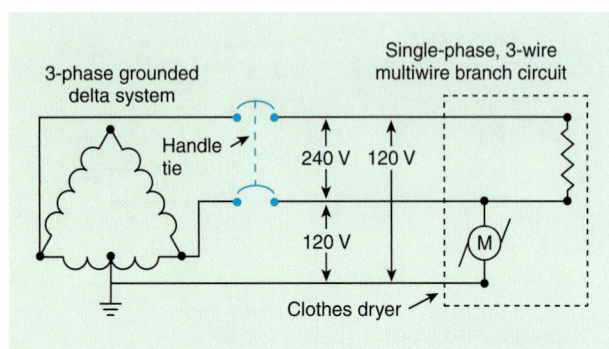

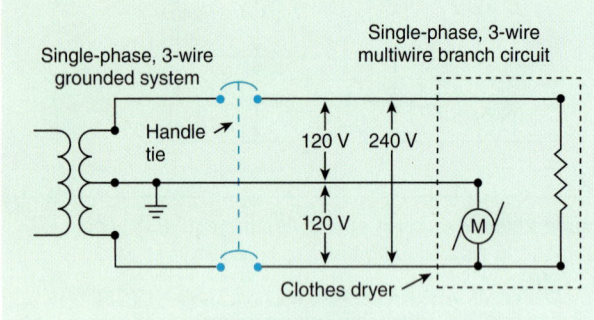

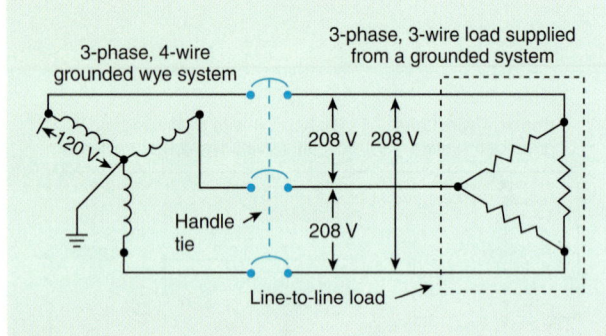

Exhibit 240.6 Examples of circuits in which approved handle ties are permitted according to 240.20(B)(2) or 240.20(B)(3).

pole circuit breakers are permitted, handle ties are not required because the circuits supply line-to-neutral loads. In Exhibit 240.6, in which line-to-line loads are supplied from single-phase or 4-wire, 3-phase systems, approved handle ties are permitted.

(C) Closed-Loop Power Distribution Systems. Listed devices that provide equivalent overcurrent protection in closed-loop power distribution systems shall be permitted as a substitute for fuses or circuit breakers.

240.21 Location in Circuit.

Overcurrent protection shall be provided in each ungrounded circuit conductor and shall be located at the point where

the conductors receive their supply except as specified in 240.21(A) through (G). No conductor supplied under the provisions of 240.21(A) through (G) shall supply another conductor under those provisions, except through an overcurrent protective device meeting the requirements of 240.4.

(A) Branch-Circuit Conductors. Branch-circuit tap conductors meeting the requirements specified in 210.19 shall be permitted to have overcurrent protection located as specified in that section.

(B) Feeder Taps. Conductors shall be permitted to be tapped, without overcurrent protection at the tap, to a feeder as specified in 240.21(B)(1) through (5).

Exhibit 240.7 illustrates how a smaller 1/0 AWG, Type THW copper conductor (150 amperes) is tapped from a larger 3/0 AWG, Type THW copper feeder conductor that is sized at 200 amperes to compensate for voltage drop and that is, in turn, protected by a 150-ampere circuit breaker equal to the ampacity of the 1/0 tap conductor. The circuit breaker protecting the feeder conductors also protects the tap conductors. *Additional overcurrent protection is not required.*

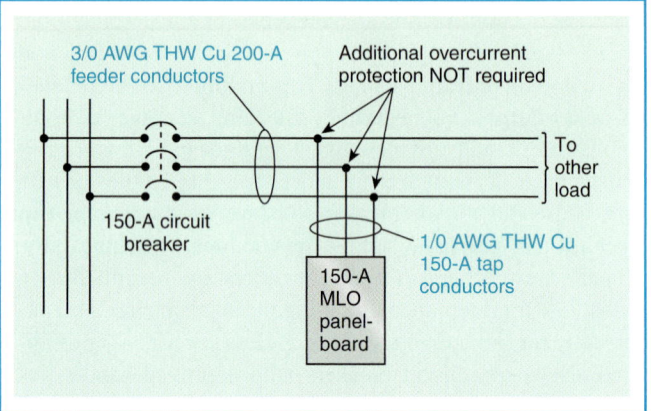

Exhibit 240.7 An example in which the circuit breaker protecting the feeder conductors is permitted by 240.21(A) to protect the tap conductors to the cabinet.

(1) Taps Not Over 3 m (10 ft) Long. Where the length of the tap conductors does not exceed 3 m (10 ft) and the tap conductors comply with all of the following:

(1) The ampacity of the tap conductors is

   a. Not less than the combined computed loads on the circuits supplied by the tap conductors, and
   b. Not less than the rating of the device supplied by the tap conductors or not less than the rating of the overcurrent-protective device at the termination of the tap conductors.

(2) The tap conductors do not extend beyond the switchboard, panelboard, disconnecting means, or control devices they supply.

(3) Except at the point of connection to the feeder, the tap conductors are enclosed in a raceway, which shall extend from the tap to the enclosure of an enclosed switchboard, panelboard, or control devices, or to the back of an open switchboard.

(4) For field installations where the tap conductors leave the enclosure or vault in which the tap is made, the rating of the overcurrent device on the line side of the tap conductors shall not exceed 10 times the ampacity of the tap conductor.

FPN: For overcurrent protection requirements for lighting and appliance branch-circuit panelboards and certain power panelboards, see 408.16(A), (B), and (E).

(2) Taps Not Over 7.5 m (25 ft) Long. Where the length of the tap conductors does not exceed 7.5 m (25 ft) and the tap conductors comply with all the following:

(1) The ampacity of the tap conductors is not less than one-third of the rating of the overcurrent device protecting the feeder conductors.

(2) The tap conductors terminate in a single circuit breaker or a single set of fuses that will limit the load to the ampacity of the tap conductors. This device shall be permitted to supply any number of additional overcurrent devices on its load side.

(3) The tap conductors are suitably protected from physical damage or are enclosed in a raceway.

Exhibit 240.8 illustrates the conditions of 240.21(B)(2), in which three 3/0 AWG, Type THW copper tap conductors are protected from physical damage in a raceway. The lengths of the tap conductors are not more than 25 ft between termina-

tions, and the conductors are tapped from 500-kcmil, Type THW copper feeders and terminate in a circuit breaker.

Note that a 3/0 AWG, Type THW copper conductor (200 amperes) is more than one-third the rating of the overcurrent device (400 amperes) protecting the feeder circuit. See Table 310.16 for the ampacity of copper conductors in conduit.

(3) Taps Supplying a Transformer [Primary Plus Secondary Not Over 7.5 m (25 ft) Long]. Where the tap conductors supply a transformer and comply with all the following conditions:

(1) The conductors supplying the primary of a transformer have an ampacity at least one-third the rating of the overcurrent device protecting the feeder conductors.

(2) The conductors supplied by the secondary of the transformer shall have an ampacity that, when multiplied by the ratio of the secondary-to-primary voltage, is at least one-third of the rating of the overcurrent device protecting the feeder conductors.

(3) The total length of one primary plus one secondary conductor, excluding any portion of the primary conductor that is protected at its ampacity, is not over 7.5 m (25 ft).

(4) The primary and secondary conductors are suitably protected from physical damage.

(5) The secondary conductors terminate in a single circuit breaker or set of fuses that limit the load current to not more than the conductor ampacity that is permitted by 310.15.

Exhibit 240.9 illustrates the conditions of 240.21(B)(3). The overcurrent protection requirements of 408.16 for panelboards and 450.3(B) for transformers also apply.

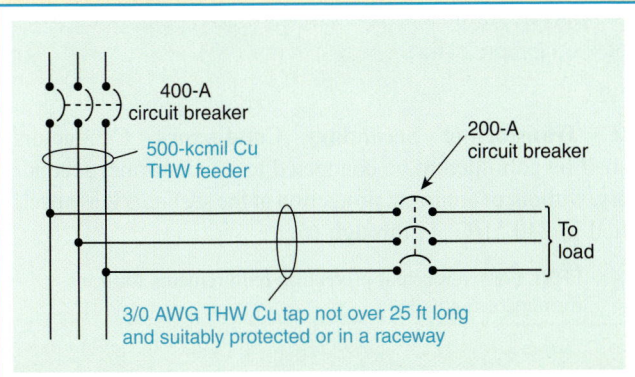

Exhibit 240.8 An example in which the feeder taps terminate in a single circuit breaker, per 240.21(B)(2).

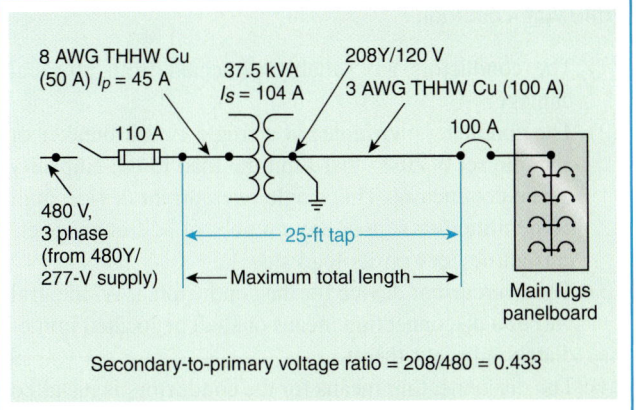

Exhibit 240.9 An example in which the transformer feeder taps (primary plus secondary) are not over 25 ft long, per 240.21(B)(3).

(4) Taps Over 7.5 m (25 ft) Long. Where the feeder is in a high bay manufacturing building over 11 m (35 ft) high at walls and the installation complies with all the following conditions:

(1) Conditions of maintenance and supervision ensure that only qualified persons service the systems.
(2) The tap conductors are not over 7.5 m (25 ft) long horizontally and not over 30 m (100 ft) total length.
(3) The ampacity of the tap conductors is not less than one-third the rating of the overcurrent device protecting the feeder conductors.
(4) The tap conductors terminate at a single circuit breaker or a single set of fuses that limit the load to the ampacity of the tap conductors. This single overcurrent device shall be permitted to supply any number of additional overcurrent devices on its load side.
(5) The tap conductors are suitably protected from physical damage or are enclosed in a raceway.
(6) The tap conductors are continuous from end-to-end and contain no splices.
(7) The tap conductors are sized 6 AWG copper or 4 AWG aluminum or larger.
(8) The tap conductors do not penetrate walls, floors, or ceilings.
(9) The tap is made no less than 9 m (30 ft) from the floor.

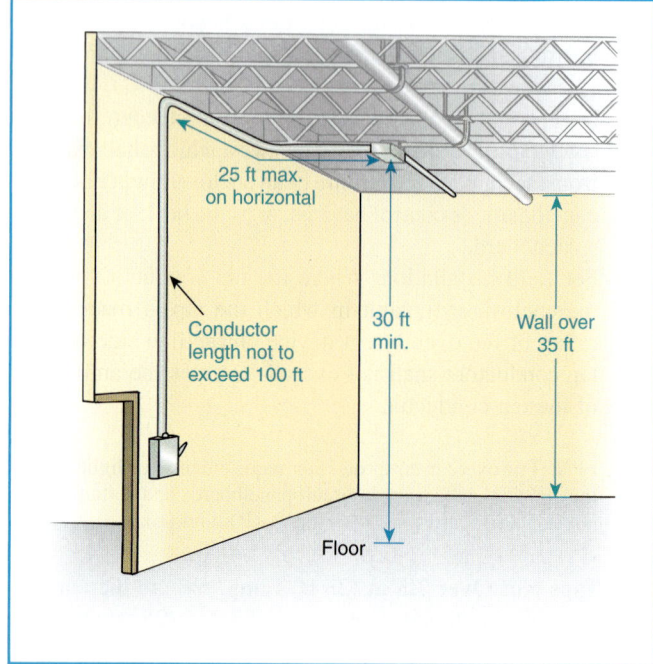

Exhibit 240.10 An example in which the feeder taps are over 25 ft long, the tap connection being not less than 30 ft from the floor, per 240.21(B)(4).

Exhibit 240.10 illustrates the requirements of 240.21(B)(4). It permits a tap of 100 ft for manufacturing buildings with walls that are over 35 ft high if the tap connection is not less than 30 ft from the floor and conditions of maintenance and supervision ensure that only qualified persons will service these systems.

(5) Outside Taps of Unlimited Length. Where the conductors are located outdoors of a building or structure, except at the point of load termination, and comply with all of the following conditions:

(1) The conductors are suitably protected from physical damage.
(2) The conductors terminate at a single circuit breaker or a single set of fuses that limit the load to the ampacity of the conductors. This single overcurrent device shall be permitted to supply any number of additional overcurrent devices on its load side.
(3) The overcurrent device for the conductors is an integral part of a disconnecting means or shall be located immediately adjacent thereto.
(4) The disconnecting means for the conductors is installed at a readily accessible location complying with one of the following:

    a. Outside of a building or structure

    b. Inside, nearest the point of entrance of the conductors
    c. Where installed in accordance with 230.6, nearest the point of entrance of the conductors

Section 240.21(B)(5) permits outside conductors to be tapped from a feeder without any limitations on the length of the tap conductors. The tap conductors must be protected and must terminate in a single, fused disconnect or a single circuit breaker rated at not more than the ampacity of the tap conductors. Also, this fused disconnect or circuit breaker must be installed at a readily accessible location either inside or outside a building or structure. Furthermore, if the fused disconnect or circuit breaker is installed inside a building or structure, it must be located nearest the point of entrance of the tap conductors.

(C) Transformer Secondary Conductors. Conductors shall be permitted to be connected to a transformer secondary, without overcurrent protection at the secondary, as specified in 240.21(C)(1) through (6).

> FPN: For overcurrent protection requirements for transformers, see 450.3.

Transformer secondary conductors are permitted without an overcurrent protective device in the secondary under any of the following four conditions:

1. The primary overcurrent protective device, as described in 240.21(C)(1), can protect single-phase (2-wire) and 3-phase (delta-delta) transformer secondary conductors.
2. The transformer secondary conductors do not exceed 10 ft.
3. The transformer secondary conductors do not exceed 25 ft.
4. The transformer secondary conductors are located outdoors.

(1) Protection by Primary Overcurrent Device. Conductors supplied by the secondary side of a single-phase transformer having a 2-wire (single-voltage) secondary, or a three-phase, delta-delta connected transformer having a 3-wire (single-voltage) secondary, shall be permitted to be protected by overcurrent protection provided on the primary (supply) side of the transformer, provided this protection is in accordance with 450.3 and does not exceed the value determined by multiplying the secondary conductor ampacity by the secondary to primary transformer voltage ratio.

Single-phase (other than 2-wire) and multiphase (other than delta-delta, 3-wire) transformer secondary conductors are not considered to be protected by the primary overcurrent protective device.

(2) Transformer Secondary Conductors Not Over 3 m (10 ft) Long. Where the length of secondary conductor does not exceed 3 m (10 ft) and complies with all of the following:

(1) The ampacity of the secondary conductors is
   a. Not less than the combined computed loads on the circuits supplied by the secondary conductors, and
   b. Not less than the rating of the device supplied by the secondary conductors or not less than the rating of the overcurrent-protective device at the termination of the secondary conductors
(2) The secondary conductors do not extend beyond the switchboard, panelboard, disconnecting means, or control devices they supply.
(3) The secondary conductors are enclosed in a raceway, which shall extend from the transformer to the enclosure of an enclosed switchboard, panelboard, or control devices or to the back of an open switchboard.

FPN: For overcurrent protection requirements for lighting and appliance branch-circuit panelboards and certain power panelboards, see 408.16(A), (B), and (E).

(3) Industrial Installation Secondary Conductors Not Over 7.5 m (25 ft) Long. For industrial installations only, where the length of the secondary conductors does not exceed 7.5 m (25 ft) and complies with all of the following:

(1) The ampacity of the secondary conductors is not less than the secondary current rating of the transformer, and the sum of the ratings of the overcurrent devices does not exceed the ampacity of the secondary conductors.
(2) All overcurrent devices are grouped.
(3) The secondary conductors are suitably protected from physical damage.

(4) Outside Secondary of Building or Structure Conductors. Where the conductors are located outdoors of a building or structure, except at the point of load termination, and comply with all of the following conditions:

(1) The conductors are suitably protected from physical damage.
(2) The conductors terminate at a single circuit breaker or a single set of fuses that limit the load to the ampacity of the conductors. This single overcurrent device shall be permitted to supply any number of additional overcurrent devices on its load side.
(3) The overcurrent device for the conductors is an integral part of a disconnecting means or shall be located immediately adjacent thereto.
(4) The disconnecting means for the conductors is installed at a readily accessible location complying with one of the following:
   a. Outside of a building or structure
   b. Inside, nearest the point of entrance of the conductors
   c. Where installed in accordance with 230.6, nearest the point of entrance of the conductors

(5) Secondary Conductors from a Feeder Tapped Transformer. Transformer secondary conductors installed in accordance with 240.21(B)(3) shall be permitted to have overcurrent protection as specified in that section.

(6) Secondary Conductors Not Over 7.5 m (25 ft) Long. Where the length of secondary conductor does not exceed 7.5 m (25 ft) and complies with all of the following:

(1) The secondary conductors shall have an ampacity that, when multiplied by the ratio of the secondary-to-primary voltage, is at least one-third of the rating of the overcurrent device protecting the primary of the transformer.
(2) The secondary conductors terminate in a single circuit breaker or set of fuses that limit the load current to not more than the conductor ampacity that is permitted by 310.15.
(3) The secondary conductors are suitably protected from physical damage.

(D) Service Conductors. Service-entrance conductors shall be permitted to be protected by overcurrent devices in accordance with 230.91.

(E) Busway Taps. Busways and busway taps shall be permitted to be protected against overcurrent in accordance with 368.10 through 368.13.

(F) Motor Circuit Taps. Motor-feeder and branch-circuit conductors shall be permitted to be protected against overcurrent in accordance with 430.28 and 430.53, respectively.

(G) Conductors from Generator Terminals. Conductors from generator terminals that meet the size requirement in 445.13 shall be permitted to be protected against overload by the generator overload protective device(s) required by 445.12.

240.22 Grounded Conductor.

No overcurrent device shall be connected in series with any conductor that is intentionally grounded, unless one of the following two conditions is met:

(1) The overcurrent device opens all conductors of the circuit, including the grounded conductor, and is designed so that no pole can operate independently.
(2) Where required by 430.36 or 430.37 for motor overload protection.

240.23 Change in Size of Grounded Conductor.

Where a change occurs in the size of the ungrounded conductor, a similar change shall be permitted to be made in the size of the grounded conductor.

Section 240.23 acknowledges that the size of a grounded conductor may be increased (because of voltage-drop problems, for example) or reduced to correspond to a change made in the size of an ungrounded conductor, as in the case of tap conductors, where all are of the same circuit.

240.24 Location in or on Premises.

(A) Accessibility. Overcurrent devices shall be readily accessible unless one of the following applies:

(1) For busways, as provided in 368.12.
(2) For supplementary overcurrent protection, as described in 240.10.
(3) For overcurrent devices, as described in 225.40 and 230.92.
(4) For overcurrent devices adjacent to utilization equipment that they supply, access shall be permitted to be by portable means.

Section 240.24(A)(4) recognizes the need for overcurrent protection in locations that are not readily accessible, such

as above suspended ceilings. It permits overcurrent devices to be located so that they are not readily accessible, as long as they are located next to the appliance, motor, or other equipment they supply and can be reached by using a ladder. See also 240.10 and 404.8(A) regarding the accessibility of supplementary overcurrent devices and other switches.

(B) Occupancy. Each occupant shall have ready access to all overcurrent devices protecting the conductors supplying that occupancy.

Exception No. 1: Where electric service and electrical maintenance are provided by the building management and where these are under continuous building management supervision, the service overcurrent devices and feeder overcurrent devices supplying more than one occupancy shall be permitted to be accessible to only authorized management personnel in the following:

(a) Multiple-occupancy buildings
(b) Guest rooms of hotels and motels that are intended for transient occupancy

Exception No. 2: Where electric service and electrical maintenance are provided by the building management and where these are under continuous building management supervision, the branch circuit overcurrent devices supplying any guest rooms shall be permitted to be accessible to only authorized management personnel for guest rooms of hotels and motels that are intended for transient occupancy.

(C) Not Exposed to Physical Damage. Overcurrent devices shall be located where they will not be exposed to physical damage.

FPN: See 110.11, Deteriorating Agents.

(D) Not in Vicinity of Easily Ignitible Material. Overcurrent devices shall not be located in the vicinity of easily ignitible material, such as in clothes closets.

Examples of locations where combustible materials may be stored are linen closets, paper storage closets, and clothes closets.

(E) Not Located in Bathrooms. In dwelling units and guest rooms of hotels and motels, overcurrent devices, other than supplementary overcurrent protection, shall not be located in bathrooms as defined in Article 100.

III. Enclosures

240.30 General.

(A) Protection from Physical Damage. Overcurrent devices shall be protected from physical damage by one of the following:

(1) Installation in enclosures, cabinets, cutout boxes, or equipment assemblies
(2) Mounting on open-type switchboards, panelboards, or control boards that are in rooms or enclosures free from dampness and easily ignitible material and are accessible only to qualified personnel

Properly selected overcurrent protective devices are designed to open a circuit before an overcurrent condition can seriously damage conductor insulation. Requirements that overcurrent devices be enclosed in cabinets or cutout boxes ensure that hot metal particles will not be ejected in the vicinity of combustible materials. Also, use of an enclosure prevents contact with live parts by personnel.

Overcurrent devices mounted on open-type switchboards, panelboards, or control boards and having exposed energized parts are to be located where accessible only to qualified persons.

(B) Operating Handle. The operating handle of a circuit breaker shall be permitted to be accessible without opening a door or cover.

240.32 Damp or Wet Locations.

Enclosures for overcurrent devices in damp or wet locations shall comply with 312.2(A).

240.33 Vertical Position.

Enclosures for overcurrent devices shall be mounted in a vertical position unless that is shown to be impracticable. Circuit breaker enclosures shall be permitted to be installed horizontally where the circuit breaker is installed in accordance with 240.81. Listed busway plug-in units shall be permitted to be mounted in orientations corresponding to the busway mounting position.

A wall-mounted vertical position for enclosures for overcurrent devices is desirable to afford easier access, natural hand operation, normal swing or closing of doors or covers, and legibility of the manufacturer's markings.

IV. Disconnecting and Guarding

240.40 Disconnecting Means for Fuses.

A disconnecting means shall be provided on the supply side of all fuses in circuits over 150 volts to ground and cartridge fuses in circuits of any voltage where accessible to other than qualified persons so that each individual circuit containing fuses can be independently disconnected from the source of power. A current-limiting device without a disconnecting means shall be permitted on the supply side of the service disconnecting means as permitted by 230.82. A single disconnecting means shall be permitted on the supply side of more than one set of fuses as permitted by 430.112, Exception, for group operation of motors and 424.22(C) for fixed electric space-heating equipment.

A single disconnect switch is allowed to serve more than one set of fuses, such as in multimotor installations or for electric space-heating equipment where the heating element load is required to be subdivided, each element with its own set of fuses.

240.41 Arcing or Suddenly Moving Parts.

Arcing or suddenly moving parts shall comply with 240.41(A) and (B).

(A) Location. Fuses and circuit breakers shall be located or shielded so that persons will not be burned or otherwise injured by their operation.

(B) Suddenly Moving Parts. Handles or levers of circuit breakers, and similar parts that may move suddenly in such a way that persons in the vicinity are likely to be injured by being struck by them, shall be guarded or isolated.

Arcing or sudden-moving parts are usually associated with switchboards or control boards that may be of the open type. Switchboards and control boards should be under competent supervision and accessible only to qualified persons. Fuses or circuit breakers must be located or shielded so that, under an abnormal condition, the subsequent arc across the opening device will not injure persons in the vicinity.

Guardrails may be provided in the vicinity of disconnecting means because sudden-moving handles may be capable of causing injury. Modern switchboards, for example, are equipped with removable handles. See Article 100 for the definition of *guarded*. See also 110.27 for the guarding of live parts (600 volts, nominal, or less).

V. Plug Fuses, Fuseholders, and Adapters

240.50 General.

(A) Maximum Voltage. Plug fuses shall be permitted to be used in the following circuits:

(1) Circuits not exceeding 125 volts between conductors
(2) Circuits supplied by a system having a grounded neutral where the line-to-neutral voltage does not exceed 150 volts

Plug fuses can be installed in circuits supplied by 120/240-volt, single-phase, 3-wire systems and by 208Y/120-volt, 3-phase, 4-wire systems.

(B) Marking. Each fuse, fuseholder, and adapter shall be marked with its ampere rating.

(C) Hexagonal Configuration. Plug fuses of 15-ampere and lower rating shall be identified by a hexagonal configuration of the window, cap, or other prominent part to distinguish them from fuses of higher ampere ratings.

Exhibit 240.11 shows some examples of plug fuses and Type S fuses. Note the hexagonal feature on the 15-ampere fuse.

Exhibit 240.11 Plug fuses and Type S fuses. (Courtesy of Bussmann Division, Cooper Industries)

(D) No Energized Parts. Plug fuses, fuseholders, and adapters shall have no exposed energized parts after fuses or fuses and adapters have been installed.

(E) Screw Shell. The screw shell of a plug-type fuseholder shall be connected to the load side of the circuit.

240.51 Edison-Base Fuses.

(A) Classification. Plug fuses of the Edison-base type shall be classified at not over 125 volts and 30 amperes and below.

(B) Replacement Only. Plug fuses of the Edison-base type shall be used only for replacements in existing installations where there is no evidence of overfusing or tampering.

240.52 Edison-Base Fuseholders.

Fuseholders of the Edison-base type shall be installed only where they are made to accept Type S fuses by the use of adapters.

240.53 Type S Fuses.

Type S fuses shall be of the plug type and shall comply with 240.53(A) and (B).

(A) Classification. Type S fuses shall be classified at not over 125 volts and 0 to 15 amperes, 16 to 20 amperes, and 21 to 30 amperes.

(B) Noninterchangeable. Type S fuses of an ampere classification as specified in 240.53(A) shall not be interchangeable with a lower ampere classification. They shall be designed so that they cannot be used in any fuseholder other than a Type S fuseholder or a fuseholder with a Type S adapter inserted.

240.54 Type S Fuses, Adapters, and Fuseholders.

(A) To Fit Edison-Base Fuseholders. Type S adapters shall fit Edison-base fuseholders.

(B) To Fit Type S Fuses Only. Type S fuseholders and adapters shall be designed so that either the fuseholder itself or the fuseholder with a Type S adapter inserted cannot be used for any fuse other than a Type S fuse.

(C) Nonremovable. Type S adapters shall be designed so that once inserted in a fuseholder, they cannot be removed.

(D) Nontamperable. Type S fuses, fuseholders, and adapters shall be designed so that tampering or shunting (bridging) would be difficult.

(E) Interchangeability. Dimensions of Type S fuses, fuseholders, and adapters shall be standardized to permit interchangeability regardless of the manufacturer.

Exhibit 240.12 shows a Type S nonrenewable plug fuse and adapter.

VI. Cartridge Fuses and Fuseholders

240.60 General.

(A) Maximum Voltage—300-Volt Type. Cartridge fuses and fuseholders of the 300-volt type shall be permitted to be used in the following circuits:

Exhibit 240.12 Type S nonrenewable plug fuse and adapter. (Redrawn from Bussmann Division, Cooper Industries)

(1) Circuits not exceeding 300 volts between conductors
(2) Single-phase line-to-neutral circuits supplied from a 3-phase, 4-wire, solidly grounded neutral source where the line-to-neutral voltage does not exceed 300 volts

(B) Noninterchangeable—0–6000-Ampere Cartridge Fuseholders. Fuseholders shall be designed so that it will be difficult to put a fuse of any given class into a fuseholder that is designed for a current lower, or voltage higher, than that of the class to which the fuse belongs. Fuseholders for current-limiting fuses shall not permit insertion of fuses that are not current-limiting.

(C) Marking. Fuses shall be plainly marked, either by printing on the fuse barrel or by a label attached to the barrel showing the following:

(1) Ampere rating
(2) Voltage rating
(3) Interrupting rating where other than 10,000 amperes
(4) Current limiting where applicable
(5) The name or trademark of the manufacturer

The interrupting rating shall not be required to be marked on fuses used for supplementary protection.

Exhibit 240.13 shows two examples of Class G fuses rated 300 volts. Note the plainly marked barrels. Class H-type cartridge fuses have an interrupting capacity (IC) rating of 10,000 amperes, which need not be marked on the fuse. However, Class CC, G, J, K, L, R, and T cartridge fuses exceed the 10,000-ampere IC rating and must be marked with the IC rating. Section 240.83(C) requires that the IC rating for circuit breakers for other than 5000 amperes be indicated on the circuit breaker. Fuses or circuit breakers used for supplementary protection of fluorescent fixtures, semiconductor rectifiers, motor-operated appliances, and so on, need not be marked for IC. See also the commentary on circuit breakers following 110.10.

Exhibit 240.13 Two Class G fuses rated 300 volts. (Courtesy of Bussmann Division, Cooper Industries)

240.61 Classification.

Cartridge fuses and fuseholders shall be classified according to voltage and amperage ranges. Fuses rated 600 volts, nominal, or less shall be permitted to be used for voltages at or below their ratings.

See 490.21(B) for application of high-voltage fuses.

VII. Circuit Breakers

240.80 Method of Operation.

Circuit breakers shall be trip free and capable of being closed and opened by manual operation. Their normal method of operation by other than manual means, such as electrical or pneumatic, shall be permitted if means for manual operation are also provided.

240.81 Indicating.

Circuit breakers shall clearly indicate whether they are in the open "off" or closed "on" position.

Where circuit breaker handles are operated vertically rather than rotationally or horizontally, the "up" position of the handle shall be the "on" position.

See 240.83(D), 404.11, and 410.81(A) for requirements for circuit breakers used as switches. To ensure that operating the handle in a downward motion turns the device off, 240.81 prohibits a circuit breaker from being inverted.

240.82 Nontamperable.

A circuit breaker shall be of such design that any alteration of its trip point (calibration) or the time required for its operation requires dismantling of the device or breaking of a seal for other than intended adjustments.

240.83 Marking.

(A) Durable and Visible. Circuit breakers shall be marked with their ampere rating in a manner that will be durable and visible after installation. Such marking shall be permitted to be made visible by removal of a trim or cover.

(B) Location. Circuit breakers rated at 100 amperes or less and 600 volts or less shall have the ampere rating molded, stamped, etched, or similarly marked into their handles or escutcheon areas.

(C) Interrupting Rating. Every circuit breaker having an interrupting rating other than 5000 amperes shall have its interrupting rating shown on the circuit breaker. The interrupting rating shall not be required to be marked on circuit breakers used for supplementary protection.

Section 240.83(C) recognizes series-rated circuit breakers and requires that the end-use equipment be marked with the series combination rating. For example, a circuit breaker with an interrupting rating of 10,000 amperes may perform safely on a circuit with an available fault current that is greater than 10,000 amperes under the following conditions:

1. It is protected on its line side by a circuit breaker with a suitable interrupting rating, and
2. The series combination has been tested and demonstrated to safely open a short-circuit current higher than the 10,000 amperes on the load side of the downstream breaker.

The UL *General Information Directory* (White Book), under the category "Switchboards, Dead Front Type (WEVZ)," provides the following information:

Short Circuit Ratings: Dead-front switchboard sections or interiors are marked with their DC or RMS symmetrical short-circuit current rating in amperes. The marking states that short-circuit ratings are limited to the lowest short-circuit rating of (1) any switchboard section connected in series or (2) the lowest short-circuit rating of any device installed or intended to be installed therein. However, for combination series-connected devices, the short-circuit current rating marked on the switchboard may be higher than the short-circuit current rating of a specific circuit breaker installed or to be installed in the switchboard. This higher rating is valid only if the specific overcurrent devices identified in the marking are used within or ahead of the switchboard in accordance with the marked instructions. In many cases, the short-circuit ratings are associated with instructions for securing supply wiring within the switchboard.

(D) Used as Switches. Circuit breakers used as switches in 120-volt and 277-volt fluorescent lighting circuits shall be listed and shall be marked SWD or HID. Circuit breakers used as switches in high-intensity discharge lighting circuits shall be listed and shall be marked as HID.

Circuit breakers marked SWD are 15- or 20-ampere breakers that have been subjected to additional endurance and temperature testing. New in the 2002 *Code*, breakers marked with HID are acceptable for switching applications. If high-intensity discharge lighting such as mercury vapor, high-pressure or low-pressure sodium, or metal halide lighting is used, the breaker used for switching must be marked HID.

(E) Voltage Marking. Circuit breakers shall be marked with a voltage rating not less than the nominal system voltage that is indicative of their capability to interrupt fault currents between phases or phase to ground.

240.85 Applications.

A circuit breaker with a straight voltage rating, such as 240V or 480V, shall be permitted to be applied in a circuit in which the nominal voltage between any two conductors does not exceed the circuit breaker's voltage rating. A two-pole circuit breaker shall not be used for protecting a 3-phase, corner-grounded delta circuit unless the circuit breaker is marked 1ϕ–3ϕ to indicate such suitability.

A circuit breaker with a slash rating, such as 120/240V or 480Y/277V, shall be permitted to be applied in a solidly grounded circuit where the nominal voltage of any conductor to ground does not exceed the lower of the two values of the circuit breaker's voltage rating and the nominal voltage between any two conductors does not exceed the higher value of the circuit breaker's voltage rating.

FPN: Proper application of molded case circuit breakers on 3-phase systems, other than solidly grounded wye, particularly on corner grounded delta systems, considers the circuit breakers' individual pole-interrupting capability.

A circuit breaker marked 480Y/277V is not intended for use on a 480-volt system with up to 480 volts to ground, such as a 480-volt circuit derived from a corner-grounded, delta-connected system. A circuit breaker marked either 480V or 600V should be used on such a system. In like manner, a circuit breaker marked 120/240V is not intended for use on a delta-connected 240-volt circuit. A 240-volt, 480-volt, or 600-volt circuit breaker should be used on such a circuit. The slash (/) between the lower and higher voltage ratings in the marking indicates that the circuit breaker has been tested for use on a circuit with the higher voltage between phases and with the lower voltage to ground.

240.86 Series Ratings.

Where a circuit breaker is used on a circuit having an available fault current higher than its marked interrupting rating by being connected on the load side of an acceptable overcurrent protective device having the higher rating, 240.86(A) and (B) shall apply.

(A) Marking. The additional series combination interrupting rating shall be marked on the end use equipment, such as switchboards and panelboards.

(B) Motor Contribution. Series ratings shall not be used where

(1) Motors are connected on the load side of the higher-rated overcurrent device and on the line side of the lower-rated overcurrent device, and

(2) The sum of the motor full-load currents exceeds 1 percent of the interrupting rating of the lower-rated circuit breaker.

A series rated system is a combination of circuit breakers, or fuses and circuit breakers, that can be applied at available short-circuit levels above the interrupting rating of the load side circuit breakers, but not above that of the main or line-side device. Series rated systems can consist of fuses that protect circuit breakers or of circuit breakers that protect circuit breakers.

Section 240.86(A) requires that, when a series rating is used, the switchboards, panelboards, and load centers be marked for use with the series rated combinations that may be used. Therefore, the enclosures must have a label affixed by the equipment manufacturer that provides the series rating of the combination(s). Because there is often not enough room in the equipment to show all the legitimate series rated combinations, UL 67 (Panelboards) allows for a bulletin to be referenced and supplied with the panelboard. These bulletins typically provide all the acceptable combinations. Note that the installer also has an additional labeling requirement (see 110.22).

One critical requirement limits the use of series rated systems in which motors are connected between the line-side (protecting) device and the load-side (protected) circuit breaker. Section 240.86(B) requires that series ratings not be used where the sum of motor full-load currents exceeds 1 percent of the interrupting rating of the load-side (protected) circuit breaker, as illustrated in Exhibit 240.14.

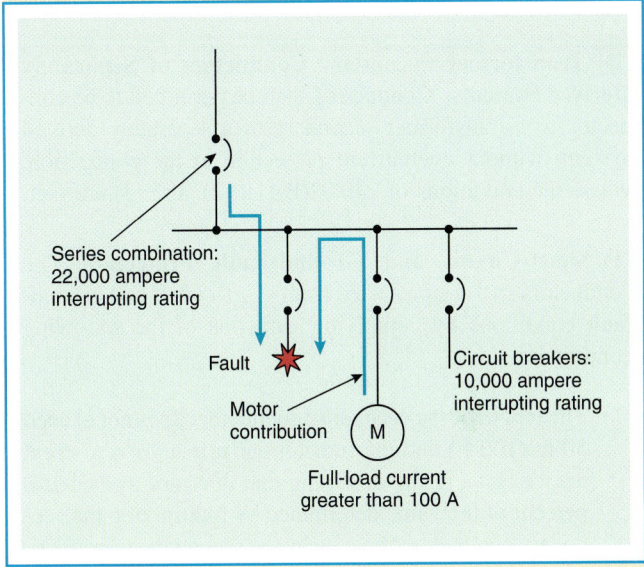

Exhibit 240.14 Example of violation of 240.86(B) due to motor contributions.

VIII. Supervised Industrial Installations

240.90 General.

Overcurrent protection in areas of supervised industrial installations shall comply with all of the other applicable provisions of this article, except as provided in Part VIII. The provisions of Part VIII shall only be permitted to apply to those portions of the electrical system in the supervised industrial installation used exclusively for manufacturing or process control activities.

Part VIII relaxes some of the location requirements for low-voltage (600 volts, nominal, and under) overcurrent protection on transformer secondary conductors and outside feeder taps for large manufacturing or process industries. Normally, 240.21 provides the requirements for the location of circuit overcurrent protection. However, Part VIII can apply to the electrical systems serving the process or manufacturing areas that meet the criteria of 240.2.

240.92 Location in Circuit.

An overcurrent device shall be connected in each ungrounded circuit conductor as required in 240.92(A) through (D).

(A) Feeder and Branch-Circuit Conductors. Feeder and branch-circuit conductors shall be protected at the point the conductors receive their supply as permitted in 240.21 or as otherwise permitted in 240.92(B), (C), or (D).

(B) Transformer Secondary Conductors of Separately Derived Systems. Conductors shall be permitted to be connected to a transformer secondary of a separately derived system, without overcurrent protection at the connection, where the conditions of 240.92(B)(1), (2), and (3) are met.

(1) Short-Circuit and Ground-Fault Protection. The conductors shall be protected from short-circuit and ground-fault conditions by complying with one of the following conditions:

(1) The length of the secondary conductors does not exceed 30 m (100 ft) and the transformer primary overcurrent device has a rating or setting that does not exceed 150 percent of the value determined by multiplying the secondary conductor ampacity by the secondary-to-primary transformer voltage ratio.
(2) The conductors are protected by a differential relay with a trip setting equal to or less than the conductor ampacity.
(3) The conductors shall be considered to be protected if calculations, made under engineering supervision, determine that the system overcurrent devices will protect the conductors within recognized time vs. current limits for all short-circuit and ground-fault conditions.

(2) Overload Protection. The conductors shall be protected against overload conditions by complying with one of the following:

(1) The conductors terminate in a single overcurrent device that will limit the load to the conductor ampacity.
(2) The sum of the overcurrent devices at the conductor termination limits the load to the conductor ampacity. The overcurrent devices shall consist of not more than six circuit breakers or sets of fuses, mounted in a single enclosure, in a group of separate enclosures, or in or on a switchboard. There shall be no more than six overcurrent devices grouped in any one location.
(3) Overcurrent relaying is connected [with a current transformer(s), if needed] to sense all of the secondary conductor current and limit the load to the conductor ampacity by opening upstream or downstream devices.

(4) Conductors shall be considered to be protected if calculations, made under engineering supervision, determine that the system overcurrent devices will protect the conductors from overload conditions.

(3) Physical Protection. The secondary conductors shall be suitably protected from physical damage.

(C) Outside Feeder Taps. Outside conductors shall be permitted to be tapped to a feeder or to be connected at a transformer secondary, without overcurrent protection at the tap or connection, where all the following conditions are met:

(1) The conductors are suitably protected from physical damage.
(2) The sum of the overcurrent devices at the conductor termination limits the load to the conductor ampacity. The overcurrent devices shall consist of not more than six circuit breakers or sets of fuses mounted in a single enclosure, in a group of separate enclosures, or in or on a switchboard. There shall be no more than six overcurrent devices grouped in any one location.
(3) The tap conductors are installed outdoors of a building or structure except at the point of load termination.
(4) The overcurrent device for the conductors is an integral part of a disconnecting means or shall be located immediately adjacent thereto.
(5) The disconnecting means for the conductors are installed at a readily accessible location complying with one of the following:
   a. Outside of a building or structure
   b. Inside, nearest the point of entrance of the conductors
   c. Where installed in accordance with 230.6, nearest the point of entrance of the conductors

(D) Protection by Primary Overcurrent Device. Conductors supplied by the secondary side of a transformer shall be permitted to be protected by overcurrent protection provided on the primary (supply) side of the transformer, provided the primary device time–current protection characteristic, multiplied by the maximum effective primary-to-secondary transformer voltage ratio, effectively protects the secondary conductors.

IX. Overcurrent Protection Over 600 Volts, Nominal

240.100 Feeders and Branch Circuits.

(A) Location and Type of Protection. Feeder and branch-circuit conductors shall have overcurrent protection in each ungrounded conductor located at the point where the conductor receives its supply or at an alternative location in the circuit when designed under engineering supervision that includes but is not limited to considering the appropriate fault studies and time–current coordination analysis of the protective devices and the conductor damage curves. The overcurrent protection shall be permitted to be provided by either 240.100(A)(1) or (A)(2).

(1) Overcurrent Relays and Current Transformers. Circuit breakers used for overcurrent protection of 3-phase circuits shall have a minimum of three overcurrent relay elements operated from three current transformers. The separate overcurrent relay elements (or protective functions) shall be permitted to be part of a single electronic protective relay unit.

On 3-phase, 3-wire circuits, an overcurrent relay element in the residual circuit of the current transformers shall be permitted to replace one of the phase relay elements.

An overcurrent relay element, operated from a current transformer that links all phases of a 3-phase, 3-wire circuit, shall be permitted to replace the residual relay element and one of the phase-conductor current transformers. Where the neutral is not regrounded on the load side of the circuit as permitted in 250.184(B), the current transformer shall be permitted to link all 3-phase conductors and the grounded circuit conductor (neutral).

(2) Fuses. A fuse shall be connected in series with each ungrounded conductor.

(B) Protective Devices. The protective device(s) shall be capable of detecting and interrupting all values of current that can occur at their location in excess of their trip-setting or melting point.

(C) Conductor Protection. The operating time of the protective device, the available short-circuit current, and the conductor used shall be coordinated to prevent damaging or dangerous temperatures in conductors or conductor insulation under short-circuit conditions.

240.101 Additional Requirements for Feeders.

(A) Rating or Setting of Overcurrent Protective Devices. The continuous ampere rating of a fuse shall not exceed three times the ampacity of the conductors. The long-time trip element setting of a breaker or the minimum trip setting of an electronically actuated fuse shall not exceed six times the ampacity of the conductor. For fire pumps, conductors shall be permitted to be protected for overcurrent in accordance with 695.4(B).

(B) Feeder Taps. Conductors tapped to a feeder shall be permitted to be protected by the feeder overcurrent device where that overcurrent device also protects the tap conductor.

ARTICLE 250
Grounding

Contents

I. General

250.1 Scope.

This article covers general requirements for grounding and bonding of electrical installations, and specific requirements in (1) through (6).

(1) Systems, circuits, and equipment required, permitted, or not permitted to be grounded
(2) Circuit conductor to be grounded on grounded systems
(3) Location of grounding connections
(4) Types and sizes of grounding and bonding conductors and electrodes
(5) Methods of grounding and bonding
(6) Conditions under which guards, isolation, or insulation may be substituted for grounding

The complete revision of Article 250 is one of the most significant changes to occur in the recent history of the *Code.* For the 1999 *Code,* this revision was a collective effort of the NEC Usability Task Group, Code-Making Panel 5, and *NEC* users who submitted proposals and comments. Similar requirements that previously appeared in different parts of Article 250 have been grouped together in the same part, and exceptions have been converted into positive code language. An overall new approach to the layout has provided a more user-friendly Article 250. See the commentary for Annex F, which provides two cross-reference lists. Exhibit F.1 references the 1996, 1999, and 2002 sections to the 1996 Article 250 topics, and Exhibit F.2 references the 2002, 1999, and 1996 sections to the 2002 Article 250 topics.

250.2 Definitions.

Effective Ground-Fault Current Path. An intentionally constructed, permanent, low-impedance electrically conductive path designed and intended to carry current under ground-fault conditions from the point of a ground fault on a wiring system to the electrical supply source.

Ground Fault. An unintentional, electrically conducting connection between an ungrounded conductor of an electrical circuit and the normally non–current-carrying conductors, metallic enclosures, metallic raceways, metallic equipment, or earth.

Ground-Fault Current Path. An electrically conductive path from the point of a ground fault on a wiring system through normally non–current-carrying conductors, equipment, or the earth to the electrical supply source.

> FPN: Examples of ground-fault current paths could consist of any combination of equipment grounding conductors, metallic raceways, metallic cable sheaths, electrical equipment, and any other electrically conductive material such as metal water and gas piping, steel framing members, stucco mesh, metal ducting, reinforcing steel, shields of communications cables, and the earth itself.

Section 250.2 is new for the 2002 *Code.* Following a common numbering sequence throughout the *NEC,* definitions that are specific to their article and not generally used elsewhere now appear in ____.2 for their respective articles. For specific examples of article-related definitions, see 240.2, 280.2, 285.2, and 517.2.

It is imperative that *Code* users be familiar with the definitions in Article 100, especially those terms associated with Article 250. Three of the most basic terms to be aware of are *grounded conductor, equipment grounding conductor,* and *grounding electrode conductor.*

250.3 Application of Other Articles.

In other articles applying to particular cases of installation of conductors and equipment, there are requirements identified in Table 250.3 that are in addition to, or modifications of, those of this article.

Table 250.3 Additional Grounding Requirements

Conductor/Equipment	Article	Section
Agricultural buildings		547.9 and 547.10
Audio signal processing, amplification, and reproduction equipment		640.7
Branch circuits		210.5, 210.6, 406.3
Cablebus		370.9
Capacitors		460.10, 460.27
Circuits and equipment operating at less than 50 volts	720	
Class 1, Class 2, and Class 3 remote-control, signaling, and power-limited circuits		725.9
Closed-loop and programmed power distribution		780.3
Communications circuits	800	
Community antenna television and radio distribution systems		820.33, 820.40, 820.41
Conductors for general wiring	310	
Cranes and hoists	610	
Electrically driven or controlled irrigation machines		675.11(C), 675.12, 675.13, 675.14, 675.15
Electric signs and outline lighting	600	
Electrolytic cells	668	
Elevators, dumbwaiters, escalators, moving walks, wheelchair lifts, and stairway chair lifts	620	
Fire alarm systems		760.9
Fixed electric heating equipment for pipelines and vessels		427.29, 427.48
Fixed outdoor electric deicing and snow-melting equipment		426.27
Flexible cords and cables		400.22, 400.23
Floating buildings		553.8, 553.10, 553.11
Grounding-type receptacles, adapters, cord connectors, and attachment plugs		406.9
Hazardous (classified) locations	500–517	
Health care facilities	517	

Table 250.3 *Continued*

Conductor/Equipment	Article	Section
Induction and dielectric heating equipment	665	
Industrial machinery	670	
Information technology equipment		645.15
Intrinsically safe systems		504.50
Luminaires (lighting fixtures) and lighting equipment		410.17, 410.18, 410.20, 410.21, 410.105(B)
Luminaires (fixtures), lampholders, lamps, and receptacles	410	
Marinas and boatyards		555.15
Mobile homes and mobile home park	550	
Motion picture and television studios and similar locations		530.20, 530.66
Motors, motor circuits, and controllers	430	
Outlet, device, pull and junction boxes, conduit bodies and fittings		314.4, 314.25
Over 600 volts, nominal, underground wiring methods		300.50(B)
Panelboards		408.20
Pipe organs	650	
Radio and television equipment	810	
Receptacles and cord connectors		406.3
Recreational vehicles and recreational vehicle parks	551	
Services	230	
Solar photovoltaic systems		690.41, 690.42, 690.43, 690.45, 690.47
Swimming pools, fountains, and similar installations	680	
Switchboards and panelboards		408.3(D)
Switches		404.12
Theaters, audience areas of motion picture and television studios, and similar locations		520.81
Transformers and transformer vaults		450.10
Use and identification of grounded conductors	200	
X-ray equipment	660	517.78

250.4 General Requirements for Grounding and Bonding.

The following general requirements identify what grounding and bonding of electrical systems are required to accomplish. The prescriptive methods contained in Article 250 shall be followed to comply with the performance requirements of this section.

> Section 250.4 provides the performance requirements for grounding and bonding of electrical systems and equipment. Performance-based requirements provide an overall objective without stating the specifics for accomplishing that objective. The first paragraph of 250.4 indicates that the performance objectives stated in 250.4(A) for grounded systems and in 250.4(B) for ungrounded systems are accomplished by complying with the prescriptive requirements found in the rest of Article 250.
>
> Section 250.2 of the 1999 *Code* was considered to be a performance requirement for the grounding path. The requirements of that section did not provide a specific rule for the sizing or connection of grounding conductors; rather it stated overall performance considerations for grounding conductors and applied to both grounded and ungrounded systems. In the 2002 *Code,* both 250.4(A)(5) and (B)(4) contain fault current path objectives that were previously found in 250.51 of the 1996 and earlier editions of the *Code*.

(A) Grounded Systems.

(1) Electrical System Grounding. Electrical systems that are grounded shall be connected to earth in a manner that will limit the voltage imposed by lightning, line surges, or unintentional contact with higher-voltage lines and that will stabilize the voltage to earth during normal operation.

(2) Grounding of Electrical Equipment. Non–current-carrying conductive materials enclosing electrical conductors or equipment, or forming part of such equipment, shall be connected to earth so as to limit the voltage to ground on these materials.

(3) Bonding of Electrical Equipment. Non–current-carrying conductive materials enclosing electrical conductors or equipment, or forming part of such equipment, shall be connected together and to the electrical supply source in a manner that establishes an effective ground-fault current path.

(4) Bonding of Electrically Conductive Materials and Other Equipment. Electrically conductive materials that are likely to become energized shall be connected together and to the electrical supply source in a manner that establishes an effective ground-fault current path.

(5) Effective Ground-Fault Current Path. Electrical equipment and wiring and other electrically conductive material likely to become energized shall be installed in a manner that creates a permanent, low-impedance circuit capable of safely carrying the maximum ground-fault current likely to be imposed on it from any point on the wiring system where a ground fault may occur to the electrical supply source. The earth shall not be used as the sole equipment grounding conductor or effective ground-fault current path.

(B) Ungrounded Systems.

(1) Grounding Electrical Equipment. Non–current-carrying conductive materials enclosing electrical conductors or equipment, or forming part of such equipment, shall be connected to earth in a manner that will limit the voltage imposed by lightning or unintentional contact with higher-voltage lines and limit the voltage to ground on these materials.

(2) Bonding of Electrical Equipment. Non–current-carrying conductive materials enclosing electrical conductors or equipment, or forming part of such equipment, shall be connected together and to the supply system grounded equipment in a manner that creates a permanent, low-impedance path for ground-fault current that is capable of carrying the maximum fault current likely to be imposed on it.

(3) Bonding of Electrically Conductive Materials and Other Equipment. Electrically conductive materials that are likely to become energized shall be connected together and to the supply system grounded equipment in a manner that creates a permanent, low-impedance path for ground-fault current that is capable of carrying the maximum fault current likely to be imposed on it.

(4) Path for Fault Current. Electrical equipment, wiring, and other electrically conductive material likely to become energized shall be installed in a manner that creates a permanent, low-impedance circuit from any point on the wiring system to the electrical supply source to facilitate the operation of overcurrent devices should a second fault occur on the wiring system. The earth shall not be used as the sole equipment grounding conductor or effective fault-current path.

> FPN No. 1: A second fault that occurs through the equipment enclosures and bonding is considered a ground fault.
>
> FPN No. 2: See Figure 250.4 for information on the organization of Article 250.

> Grounding can be divided into two areas: system grounding and equipment grounding. These two areas are kept separate

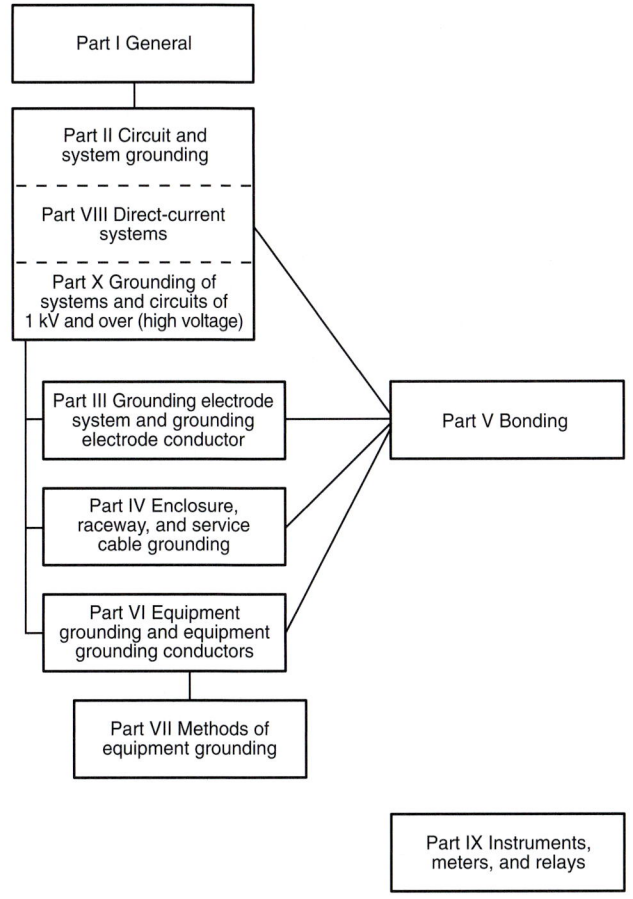

Figure 250.4 Grounding.

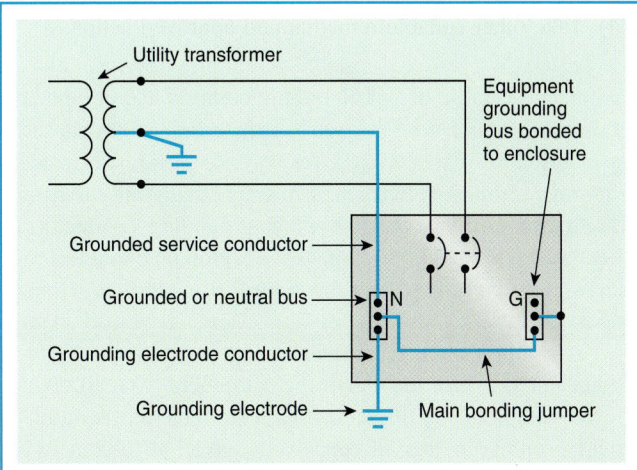

Exhibit 250.1 A typical grounding system for a single-phase, 3-wire service.

Exhibit 250.1 shows a typical grounding system for a single-phase, 3-wire service supplied from a utility transformer. Inside the service disconnecting means, the grounded conductor of the system is intentionally connected to a grounding electrode via the grounding electrode conductor. Bonding the equipment grounding bus to the grounded or neutral bus via the main bonding jumper within the service disconnecting means provides a ground reference for exposed, non–current-carrying parts of the electrical system and a circuit through the grounded service conductor back to the utility transformer (source of supply) for ground-fault current. At the utility transformer, often times, an additional connection from the grounded conductor to a separate grounding electrode is made.

from each other except at the point where they receive their source of power, such as at the service equipment or at a separately derived system.

Grounding is the intentional connection of a current-carrying conductor to ground or something that serves in place of ground. In most instances, this connection is made at the supply source, such as a transformer, and at the main service disconnecting means of the premises using the energy.

There are three basic reasons for grounding:

1. To limit the voltages caused by lightning or by accidental contact of the supply conductors with conductors of higher voltage
2. To stabilize the voltage under normal operating conditions (which maintains the voltage at one level relative to ground, so that any equipment connected to the system will be subject only to that potential difference)
3. To facilitate the operation of overcurrent devices, such as fuses, circuit breakers, or relays, under ground-fault conditions

250.6 Objectionable Current over Grounding Conductors.

(A) Arrangement to Prevent Objectionable Current. The grounding of electrical systems, circuit conductors, surge arresters, and conductive non–current-carrying materials and equipment shall be installed and arranged in a manner that will prevent objectionable current over the grounding conductors or grounding paths.

(B) Alterations to Stop Objectionable Current. If the use of multiple grounding connections results in objectionable current, one or more of the following alterations shall be permitted to be made, provided that the requirements of 250.4(A)(5) or 250.4(B)(4) are met:

(1) Discontinue one or more but not all of such grounding connections.
(2) Change the locations of the grounding connections.
(3) Interrupt the continuity of the conductor or conductive path interconnecting the grounding connections.

(4) Take other suitable remedial and approved action.

An increase in the use of electronic controls and computer equipment, which are sensitive to stray currents, has caused installation designers to look for ways to isolate electronic equipment from the effects of such stray circulating currents. Circulating currents on equipment grounding conductors, metal raceways, and building steel develop potential differences between ground and the neutral of electronic equipment.

A solution often recommended by inexperienced individuals is to isolate the electronic equipment from all other power equipment by disconnecting it from the power equipment ground. In this ill-conceived corrective action, the equipment grounding means is removed or nonmetallic spacers are installed in the metallic raceway system. The electronic equipment is then grounded to an earth ground isolated from the common power system ground. Isolating equipment in this manner creates a potential difference that is a shock hazard. The error is compounded because such isolation does not establish a low-impedance ground-fault return path to the power source, which is necessary to actuate the overcurrent protection device. Section 250.6(B) is not intended to allow disconnection of all power grounding connections to the electronic equipment. See also the commentary following 250.6(D).

(C) Temporary Currents Not Classified as Objectionable Currents. Temporary currents resulting from accidental conditions, such as ground-fault currents, that occur only while the grounding conductors are performing their intended protective functions shall not be classified as objectionable current for the purposes specified in 250.6(A) and (B).

(D) Limitations to Permissible Alterations. The provisions of this section shall not be considered as permitting electronic equipment from being operated on ac systems or branch circuits that are not grounded as required by this article. Currents that introduce noise or data errors in electronic equipment shall not be considered the objectionable currents addressed in this section.

Section 250.6(D) indicates that currents that result in noise or data errors in electronic equipment are not considered to be the objectionable currents referred to in 250.6, which limits the alterations permitted by 250.6(C). See 250.96(B), which provides methods to minimize noise and data errors.

(E) Isolation of Objectionable Direct-Current Ground Currents. Where isolation of objectionable dc ground currents from cathodic protection systems is required, a listed ac coupling/dc isolating device shall be permitted in the

equipment grounding path to provide an effective return path for ac ground-fault current while blocking dc current.

The dc ground current on grounding conductors as a result of a cathodic protection system may be considered objectionable. Because of the required grounding and bonding connections associated with metal piping systems, it is inevitable that where cathodic protection for the piping system is provided, dc current will be present on grounding and bonding conductors.

Section 250.6(E) allows the use of a listed ac coupling/dc isolating device. This device prevents the dc current on grounding and bonding conductors and allows the ground-fault return path to function properly. As part of the product testing, these devices are evaluated for proper performance under ground-fault conditions.

250.8 Connection of Grounding and Bonding Equipment.

Grounding conductors and bonding jumpers shall be connected by exothermic welding, listed pressure connectors, listed clamps, or other listed means. Connection devices or fittings that depend solely on solder shall not be used. Sheet metal screws shall not be used to connect grounding conductors to enclosures.

Section 250.8 prohibits the use of sheet metal screws as a means for attaching equipment grounding conductors to equipment. Connection means that are listed, are part of listed equipment, or are exothermically welded are required to ensure a permanent and low-resistance connection. Exhibits 250.2 and 250.3 illustrate two methods of attaching an equipment grounding conductor to a metal box.

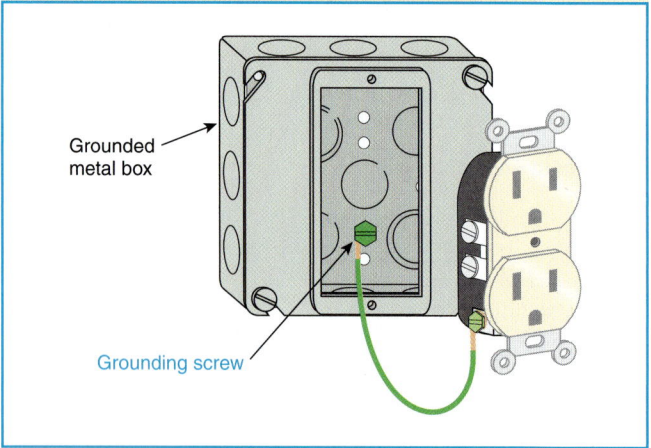

Grounded metal box

Grounding screw

Exhibit 250.2 Use of a grounding screw to attach a grounding conductor to a metal box.

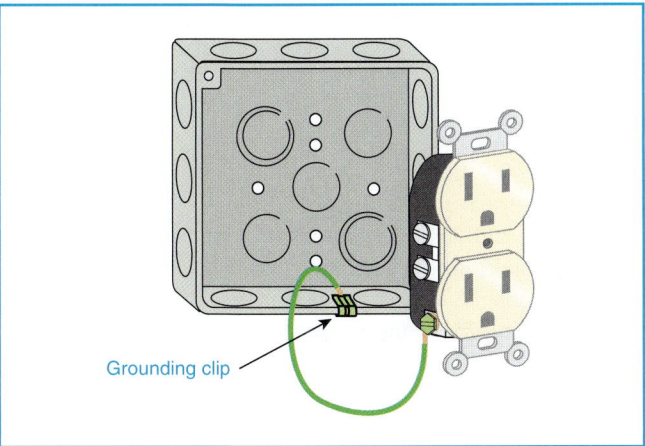

Exhibit 250.3 Use of a grounding clip to attach a grounding conductor to a metal box.

250.10 Protection of Ground Clamps and Fittings.

Ground clamps or other fittings shall be approved for general use without protection or shall be protected from physical damage as indicated in (1) or (2).

(1) In installations where they are not likely to be damaged
(2) Where enclosed in metal, wood, or equivalent protective covering

250.12 Clean Surfaces.

Nonconductive coatings (such as paint, lacquer, and enamel) on equipment to be grounded shall be removed from threads and other contact surfaces to ensure good electrical continuity or be connected by means of fittings designed so as to make such removal unnecessary.

II. Circuit and System Grounding

250.20 Alternating-Current Circuits and Systems to Be Grounded.

Alternating-current circuits and systems shall be grounded as provided for in 250.20(A), (B), (C), or (D). Other circuits and systems shall be permitted to be grounded. If such systems are grounded, they shall comply with the applicable provisions of this article.

> FPN: An example of a system permitted to be grounded is a corner-grounded delta transformer connection. See 250.26(4) for conductor to be grounded.

(A) Alternating-Current Circuits of Less Than 50 Volts.
Alternating-current circuits of less than 50 volts shall be grounded under any of the following conditions:

(1) Where supplied by transformers, if the transformer supply system exceeds 150 volts to ground

(2) Where supplied by transformers, if the transformer supply system is ungrounded
(3) Where installed as overhead conductors outside of buildings

(B) Alternating-Current Systems of 50 Volts to 1000 Volts. Alternating-current systems of 50 volts to 1000 volts that supply premises wiring and premises wiring systems shall be grounded under any of the following conditions:

(1) Where the system can be grounded so that the maximum voltage to ground on the ungrounded conductors does not exceed 150 volts

Exhibit 250.4 illustrates the grounding requirements of 250.20(B)(1) as applied to a 120-volt, single-phase, 2-wire system and to a 120/240-volt, single-phase, 3-wire system. The selection of which conductor is to be grounded is covered by 250.26.

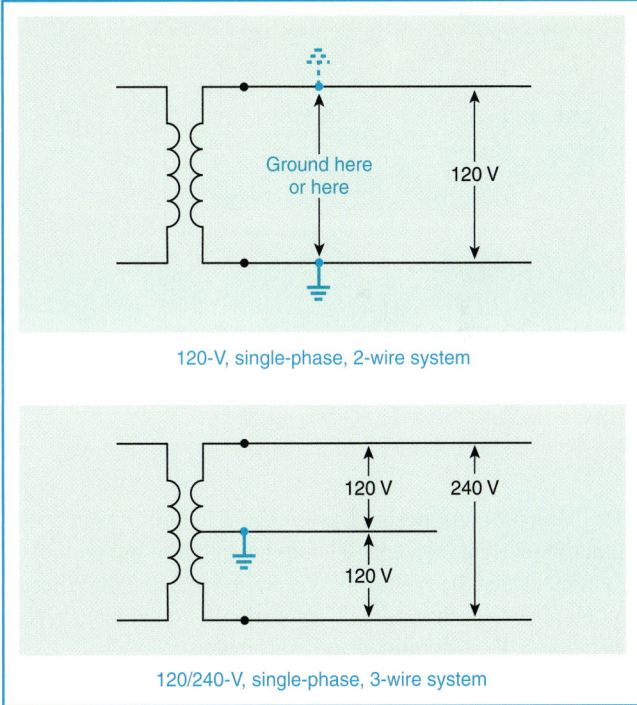

120-V, single-phase, 2-wire system

120/240-V, single-phase, 3-wire system

Exhibit 250.4 Typical systems required to be grounded by 250.20(B)(1). The conductor to be grounded is in accordance with 250.26.

(2) Where the system is 3-phase, 4-wire, wye connected in which the neutral is used as a circuit conductor
(3) Where the system is 3-phase, 4-wire, delta connected in which the midpoint of one phase winding is used as a circuit conductor

Exhibit 250.5 illustrates which conductor is required to be grounded for all wye systems if the neutral is used as a circuit conductor. Where the midpoint of one phase of a 3-phase, 4-wire delta system is used as a circuit conductor, it must be grounded and the high-leg conductor must be identified. See 250.20(B)(2) and (B)(3) as well as 250.26.

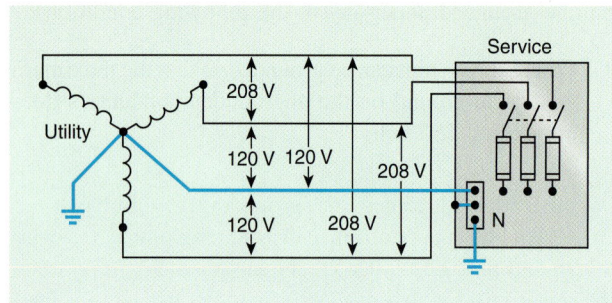

208Y/120-V, 3-phase, 4-wire wye system

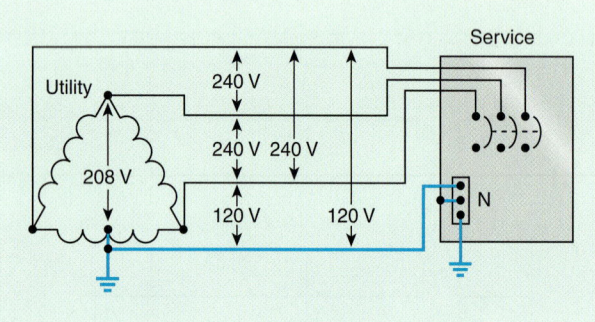

120/240-V, 3-phase, 4-wire delta system

Exhibit 250.5 Typical systems required to be grounded by 250.20(B)(2) and (B)(3). The conductor to be grounded is in accordance with 250.26.

(C) Alternating-Current Systems of 1 kV and Over. Alternating-current systems supplying mobile or portable equipment shall be grounded as specified in 250.188. Where supplying other than mobile or portable equipment, such systems shall be permitted to be grounded.

(D) Separately Derived Systems. Separately derived systems, as covered in 250.20(A) or (B), shall be grounded as specified in 250.30.

Two of the most common sources of separately derived systems in premises wiring are transformers and generators. An autotransformer or step-down transformer that is part of electrical equipment and that does not supply premises wiring is not the source of a separately derived system. See the definition of *premises wiring* in Article 100.

FPN No. 1: An alternate ac power source such as an on-site generator is not a separately derived system if the neutral is solidly interconnected to a service-supplied system neutral.

Exhibits 250.6 and 250.7 depict a 208Y/120-volt, 3-phase, 4-wire electrical service supplying a service disconnecting means to a building. The system is fed through a transfer switch connected to a generator intended to provide power for an emergency or standby system.

In Exhibit 250.6, the neutral conductor from the generator to the load is not disconnected by the transfer switch. There is a direct electrical connection between the normal grounded system conductor (neutral) and the generator neutral through the neutral bus in the transfer switch, thereby grounding the generator neutral. Because the generator is grounded by connection to the normal system ground, it is not a separately derived system, and there are no requirements for grounding the neutral at the generator. Under these conditions, it is necessary to run an equipment grounding conductor from the service equipment to the 3-pole transfer switch and from the 3-pole transfer switch to the generator. This can be in the form of any of the items listed in 250.118.

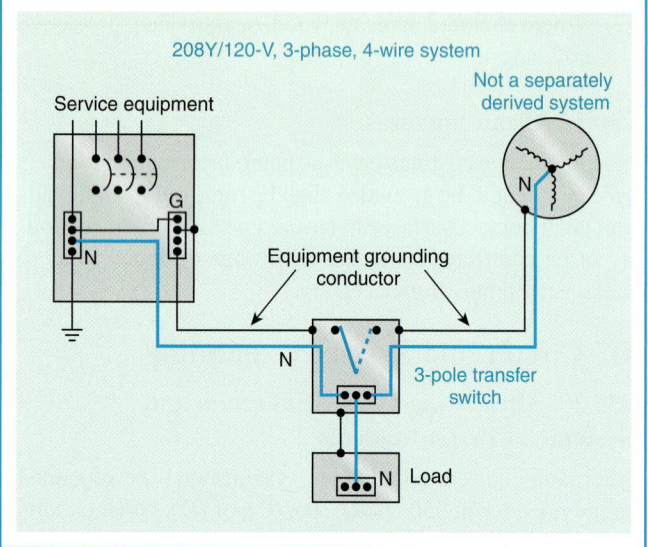

Exhibit 250.6 A 208Y/120-volt, 3-phase, 4-wire system that has a direct electrical connection of the grounded circuit conductor (neutral) to the generator and is therefore not considered a separately derived system.

In Exhibit 250.7, the grounded conductor (neutral) is connected to the switching contacts of a 4-pole transfer switch. Therefore, the generator system does not have a direct electrical connection to the other supply system grounded conductor (neutral), and the system supplied by the generator is considered separately derived. This separately derived system (3-phase, 4-wire, wye-connected system that supplies line-to-neutral loads) is required to be grounded in accordance with 250.20(B) and (D). The methods for grounding the system are specified in 250.30(A).

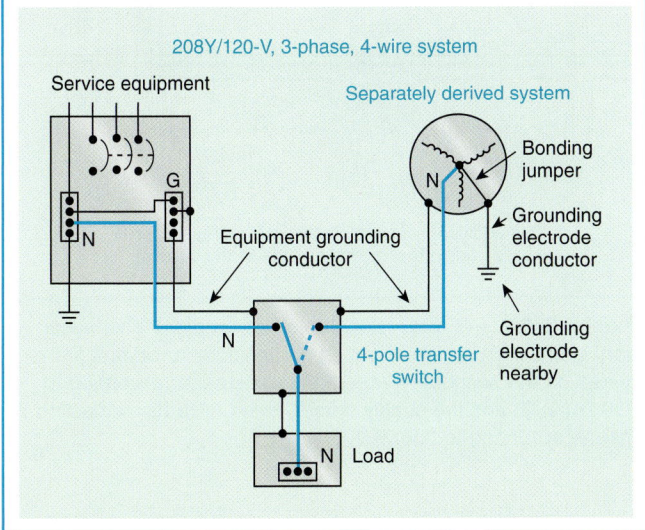

208Y/120-V, 3-phase, 4-wire system

Service equipment

Separately derived system

G

N

Bonding jumper

N

Grounding electrode conductor

Equipment grounding conductor

N

4-pole transfer switch

Grounding electrode nearby

N   Load

Exhibit 250.7 A 208Y/120-volt, 3-phase, 4-wire system that does not have a direct electrical connection of the grounded circuit conductor (neutral) to the generator and is therefore considered a separately derived system.

Section 250.30(A)(1) requires separately derived systems to have a bonding jumper connected between the generator frame and the grounded circuit conductor (neutral). The grounding electrode conductor from the generator is required to be connected to a grounding electrode. This conductor should be located as close to the generator as practicable, according to 250.30(A)(4). If the generator is in a building, the preferred grounding electrode is required to be one of the following, depending on which grounding electrode conductor is nearest the generator: (1) effectively grounded structural metal member or (2) the first 5 ft of water pipe into a building where this piping is effectively grounded. (The exception to 250.52(A)(1) permits the grounding connection to the water piping beyond the first 5 ft.) For buildings or structures in which the preferred electrodes are not available, the choice can be made from any of the grounding electrodes specified in 250.52.

FPN No. 2: For systems that are not separately derived and are not required to be grounded as specified in 250.30, see 445.13 for minimum size of conductors that must carry fault current.

250.21 Alternating-Current Systems of 50 Volts to 1000 Volts Not Required to Be Grounded.

The following ac systems of 50 volts to 1000 volts shall be permitted to be grounded but shall not be required to be grounded:

(1) Electric systems used exclusively to supply industrial electric furnaces for melting, refining, tempering, and the like
(2) Separately derived systems used exclusively for rectifiers that supply only adjustable-speed industrial drives
(3) Separately derived systems supplied by transformers that have a primary voltage rating less than 1000 volts, provided that all the following conditions are met:

   a. The system is used exclusively for control circuits.
   b. The conditions of maintenance and supervision ensure that only qualified persons service the installation.
   c. Continuity of control power is required.
   d. Ground detectors are installed on the control system.

(4) High-impedance grounded neutral systems as specified in 250.36
(5) Other systems that are not required to be grounded in accordance with the requirements of 250.20(B)

250.22 Circuits Not to Be Grounded.

The following circuits shall not be grounded:

(1) Cranes (circuits for electric cranes operating over combustible fibers in Class III locations, as provided in 503.13)
(2) Health care facilities (circuits as provided in Article 517)
(3) Electrolytic cells (circuits as provided in Article 668)
(4) Lighting systems [secondary circuits as provided in 411.5(A)]

250.24 Grounding Service-Supplied Alternating-Current Systems.

(A) System Grounding Connections. A premises wiring system supplied by a grounded ac service shall have a grounding electrode conductor connected to the grounded service conductor, at each service, in accordance with 250.24(A)(1) through (A)(5).

(1) General. The connection shall be made at any accessible point from the load end of the service drop or service lateral to and including the terminal or bus to which the grounded service conductor is connected at the service disconnecting means.

The grounded conductor of an AC service is connected to a grounding electrode system to limit the voltage to ground imposed on the system by lightning, line surges, and (unintentional) high-voltage crossovers. Another reason for requiring this connection is to stabilize the voltage to ground during normal operation, including short circuits. These performance requirements are stated in 250.4.

The actual connection of the grounded service conductor to the grounded electrode conductor is permitted to be made at various locations according to 250.24(A)(1). Allowing various locations for the connection to be made continues to meet the overall objectives for grounding while allowing the installer a variety of practical solutions. Exhibit 250.8 illustrates three possible connection point solutions to where the grounded conductor of the service could be connected to the grounding electrode conductor.

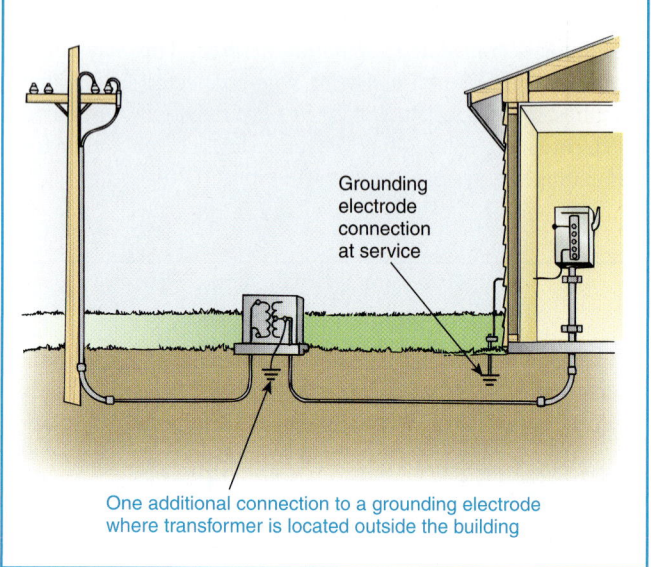

Grounding electrode connection at service

One additional connection to a grounding electrode where transformer is located outside the building

Exhibit 250.9 A 3-wire, 120/240-volt ac, single-phase, secondary distribution system in which grounding connections are required on the secondary side of the transformer according to 250.24(A)(2) and the supply side of the service disconnecting means according to 250.24(A)(1).

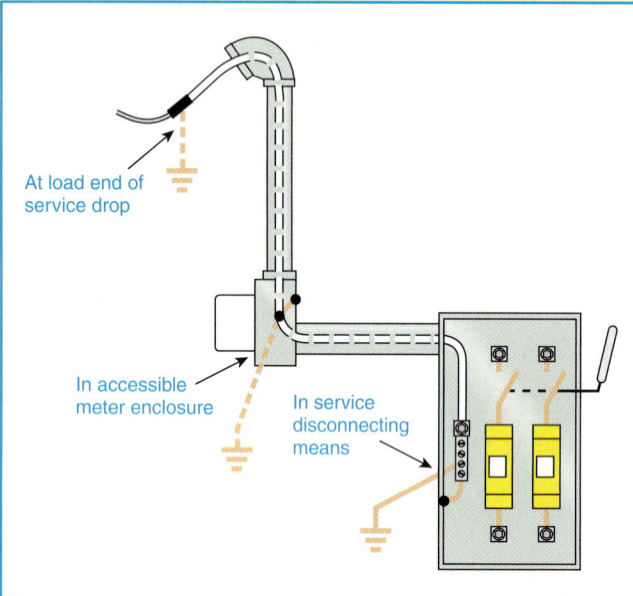

At load end of service drop

In accessible meter enclosure

In service disconnecting means

Exhibit 250.8 An AC service supplied from an overhead distribution system illustrating three accessible connection points where the grounded service conductor is connected to the grounding electrode conductor according to 250.24(A)(1).

FPN: See definitions of *Service Drop* and *Service Lateral* in Article 100.

(2) Outdoor Transformer. Where the transformer supplying the service is located outside the building, at least one additional grounding connection shall be made from the grounded service conductor to a grounding electrode, either at the transformer or elsewhere outside the building.

See Exhibit 250.9 for an illustration of an outdoor distribution system transformer connected to an additional grounding electrode.

Exception: The additional grounding connection shall not be made on high-impedance grounded neutral systems. The system shall meet the requirements of 250.36.

(3) Dual Fed Services. For services that are dual fed (double ended) in a common enclosure or grouped together in separate enclosures and employing a secondary tie, a single grounding electrode connection to the tie point of the grounded circuit conductors from each power source shall be permitted.

(4) Main Bonding Jumper as Wire or Busbar. Where the main bonding jumper specified in 250.28 is a wire or busbar and is installed from the neutral bar or bus to the equipment grounding terminal bar or bus in the service equipment, the grounding electrode conductor shall be permitted to be connected to the equipment grounding terminal bar or bus to which the main bonding jumper is connected.

(5) Load-Side Grounding Connections. A grounding connection shall not be made to any grounded circuit conductor on the load side of the service disconnecting means except as otherwise permitted in this article.

FPN: See 250.30(A) for separately derived systems, 250.32 for connections at separate buildings or structures, and 250.142 for use of the grounded circuit conductor for grounding equipment.

The power for ac premises wiring systems is either separately derived, in accordance with 250.20(D), or supplied by the service. See the definition of *service* in Article 100. Section 250.30 covers grounding requirements for separately derived ac systems. Section 250.24(A) covers system grounding requirements for service-supplied ac systems.

According to 250.24, a premises wiring system supplied by an ac service that is required to be grounded must have a grounding electrode conductor at each service connected to the grounding electrodes that meets the requirements in Part III of Article 250. Note that the grounding electrode requirements for a separately derived system are specified in 250.30(A)(4).

The grounding electrode conductor connection to the grounded conductor is very specific. The *Code* requires that the connection be made to the grounded service conductor and describes where this connection is permitted. If a transformer is installed outdoors on the load side of the service point and it supplies power to a building or structure, a grounding connection must be made at the transformer secondary under the conditions listed in 250.24(A)(2). In addition, the conductor that is grounded at the transformer is required to be grounded again at the building or structure according to 250.24(A)(1).

Section 250.24(A)(5) prohibits regrounding of the grounded conductor on the load side of the service disconnecting means. This requirement is also in concert with 250.142(B).

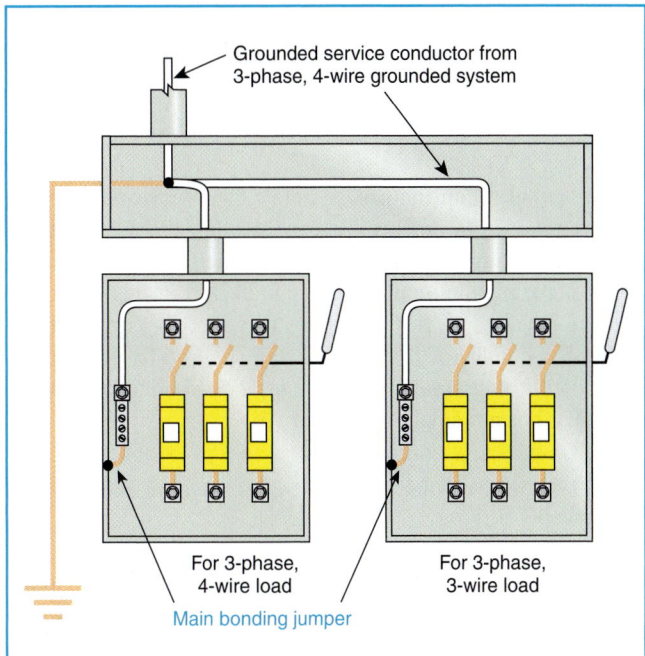

Exhibit 250.10 A grounded system in which the grounded service conductor is brought into a 3-phase, 4-wire service equipment enclosure and to the 3-phase, 3-wire service equipment enclosure, where it is bonded to each service disconnecting means.

The exception to 250.24(B) permits one bonding connection to a listed service assembly containing more than one service disconnecting means as shown in Exhibit 250.11.

(B) Grounded Conductor Brought to Service Equipment. Where an ac system operating at less than 1000 volts is grounded at any point, the grounded conductor(s) shall be run to each service disconnecting means and shall be bonded to each disconnecting means enclosure. The grounded conductor(s) shall be installed in accordance with 250.24(B)(1) through (B)(3).

Exception: Where more than one service disconnecting means are located in an assembly listed for use as service equipment, it shall be permitted to run the grounded conductor(s) to the assembly, and the conductor(s) shall be bonded to the assembly enclosure.

If the utility service that supplies premises wiring is grounded, the grounded conductor, whether or not it is used to supply a load, must be run to the service equipment, be bonded to the equipment, and be connected to a grounding electrode system. Exhibit 250.10 shows an example of the main rule in 250.24(B), which requires the grounded service conductor to be brought in and bonded to each service disconnecting means enclosure.

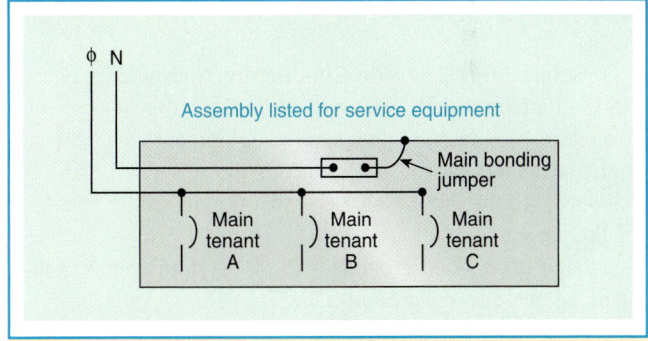

Exhibit 250.11 One bonding connection to a listed service assembly containing multiple service disconnecting means, in accordance with 250.24(B), Exception.

(1) Routing and Sizing. This conductor shall be routed with the phase conductors and shall not be smaller than the required grounding electrode conductor specified in Table 250.66 but shall not be required to be larger than the largest ungrounded service-entrance phase conductor. In addition, for service-entrance phase conductors larger than 1100 kcmil copper or 1750 kcmil aluminum, the grounded conductor shall not be smaller than 12½ percent of the area of the largest service-entrance phase conductor. The grounded service entrance conductor of a 3-phase, 3-wire delta service shall have an ampacity not less than the ungrounded conductors.

(2) Parallel Conductors. Where the service-entrance phase conductors are installed in parallel, the size of the grounded conductor shall be based on the total circular mil area of the parallel conductors as indicated in this section. Where installed in two or more raceways, the size of the grounded conductor in each raceway shall be based on the size of the ungrounded service-entrance conductor in the raceway but not smaller than 1/0 AWG.

> FPN: See 310.4 for grounded conductors connected in parallel.

For a multiple raceway or cable service installation, the minimum size for the grounded conductor in each raceway or cable where conductors are in parallel cannot be less than 1/0 AWG. Although the cumulative size of the parallel grounded conductors may be larger than is required by 250.24(B)(1), the minimum 1/0 AWG per raceway or cable correlates with the requirements for parallel conductors contained in 310.4.

(3) High Impedance. The grounded conductor on a high-impedance grounded neutral system shall be grounded in accordance with 250.36.

(C) Grounding Electrode Conductor. A grounding electrode conductor shall be used to connect the equipment grounding conductors, the service-equipment enclosures, and, where the system is grounded, the grounded service conductor to the grounding electrode(s) required by Part III of this article.

High-impedance grounded neutral system connections shall be made as covered in 250.36.

> FPN: See 250.24(A) for ac system grounding connections.

(D) Ungrounded System Grounding Connections. A premises wiring system that is supplied by an ac service that is ungrounded shall have, at each service, a grounding electrode conductor connected to the grounding electrode(s) required by Part III of this article. The grounding electrode conductor shall be connected to a metal enclosure of the service conductors at any accessible point from the load end of the service drop or service lateral to the service disconnecting means.

250.26 Conductor to Be Grounded—Alternating-Current Systems.

For ac premises wiring systems, the conductor to be grounded shall be as specified in the following:

(1) Single-phase, 2-wire—one conductor
(2) Single-phase, 3-wire—the neutral conductor
(3) Multiphase systems having one wire common to all phases—the common conductor
(4) Multiphase systems where one phase is grounded—one phase conductor
(5) Multiphase systems in which one phase is used as in (2)—the neutral conductor

Section 250.26 works in conjunction with 250.20(B), which identifies ac systems that are required to be grounded. Section 250.26 identifies which conductors in the systems in 250.20(B) must be grounded.

250.28 Main Bonding Jumper.

For a grounded system, an unspliced main bonding jumper shall be used to connect the equipment grounding conductor(s) and the service-disconnect enclosure to the grounded conductor of the system within the enclosure for each service disconnect.

Where the service equipment of a grounded system consists of multiple disconnects, a main bonding jumper for each service disconnect is required to connect the grounded service conductor, the equipment grounding conductor, and the service equipment enclosure. See Exhibits 250.10 and 250.11.

Exception No. 1: Where more than one service disconnecting means is located in an assembly listed for use as service equipment, an unspliced main bonding jumper shall bond the grounded conductor(s) to the assembly enclosure.

If multiple service disconnects are part of an assembly listed as service equipment, all grounded service conductors are required to be run to and bonded to the assembly. However, only one section of the assembly is required to have the main bonding jumper connection. See Exhibit 250.12, which accompanies the commentary following 250.28(D).

Exception No. 2: Impedance grounded neutral systems shall be permitted to be connected as provided in 250.36 and 250.186.

(A) Material. Main bonding jumpers shall be of copper or other corrosion-resistant material. A main bonding jumper shall be a wire, bus, screw, or similar suitable conductor.

(B) Construction. Where a main bonding jumper is a screw only, the screw shall be identified with a green finish that shall be visible with the screw installed.

The requirement of 250.28(B) for a green screw makes it possible to readily distinguish the main bonding jumper screw for inspection.

(C) Attachment. Main bonding jumpers shall be attached in the manner specified by the applicable provisions of 250.8.

(D) Size. The main bonding jumper shall not be smaller than the sizes shown in Table 250.66 for grounding electrode conductors. Where the service-entrance phase conductors are larger than 1100 kcmil copper or 1750 kcmil aluminum, the bonding jumper shall have an area that is not less than 12½ percent of the area of the largest phase conductor except that, where the phase conductors and the bonding jumper are of different materials (copper or aluminum), the minimum size of the bonding jumper shall be based on the assumed use of phase conductors of the same material as the bonding jumper and with an ampacity equivalent to that of the installed phase conductors.

The size of the equipment bonding jumper and the main bonding jumper on the *supply* side of a service is based on the size of the supply phase conductors. Section 250.28(D) uses Table 250.66 for sizing the main and supply side bonding jumpers. The title of Table 250.66 states that it is used for sizing the grounding electrode conductor, but it is also used to size the main and supply-side bonding jumpers. Unlike the grounding electrode conductor, bonding jumpers may be required to be larger than a 3/0 AWG copper or 250-kcmil aluminum conductor. If the size of the phase conductors exceeds 1100 kcmil copper or 1750 kcmil aluminum, the bonding jumpers cannot be less than 12½ percent of the cross-sectional area of the phase conductors.

To apply the bonding jumper requirements, each switch is treated as separate service equipment, as depicted in Exhibit 250.12. For example, to apply the bonding requirements to Exhibit 250.12, the grounded conductor would be sized for the maximum unbalance that can occur, but it cannot be smaller than that allowed by 250.24(B)(1), which has requirements similar to those in 250.24(B)(1).

Assuming a computed load of 450 amperes for the main service, the service-entrance conductors are sized at 750 kcmil copper. The bonding jumpers for the metal service conduit and trough are based on the size of the main service-entrance conductors and cannot be less than 2/0 AWG copper, based on Table 250.66.

The service-entrance conductors to the enclosures are sized 3/0 AWG and 500 kcmil copper, based on their loads.

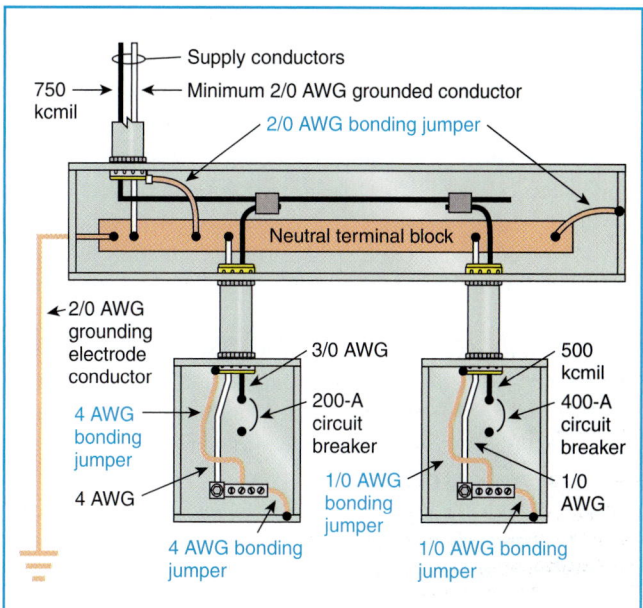

Exhibit 250.12 An example of the bonding requirements for service equipment.

The bonding jumpers for the disconnects and short nipples are sized based on the size of the phase conductors supplying each disconnect. In this case, 3/0 AWG and 500 kcmil require 4 AWG and 1/0 AWG copper bonding jumpers, respectively.

In some instances, the bonding jumper may be required to be larger than the grounding electrode conductor. Section 250.28(D) indicates that, if the service-entrance conductors are larger than 1100 kcmil copper or 1750 kcmil aluminum, the bonding jumper is required to have a cross-sectional area of not less than 12½ percent of the cross-sectional area of the largest phase conductor. For example, if a service is supplied by four 500-kcmil conductors in parallel for each phase, the minimum cross-sectional area of the bonding jumper is calculated as follows: 4 × 500 kcmil = 2000 kcmil. Therefore, the bonding jumper cannot be less than 12½ percent of 2000 kcmil, which results in a 250-kcmil copper conductor.

250.30 Grounding Separately Derived Alternating-Current Systems.

(A) Grounded Systems. A separately derived ac system that is grounded shall comply with 250.30(A)(1) through (6).

Exception: High-impedance grounded neutral system grounding connection requirements shall not be required to comply with 250.30(A)(1) and (2) and shall be made as specified in 250.36 and 250.186.

Section 250.30(A) provides the requirements for bonding and grounding the separately derived systems described in

250.20(D). A *separately derived system* is defined in Article 100 as a premises wiring system in which power is derived from a battery, a solar photovoltaic system, a generator, a transformer, or converter windings. It has no direct electrical connection, including a solidly connected grounded circuit conductor, to supply conductors originating in another system.

The requirements of 250.30 are most commonly applied to 480-volt transformers that transform a 480-volt supply to a 208Y/120-volt system for lighting and appliance loads.

These requirements provide for a low-impedance path to ground so that line-to-ground faults on circuits supplied by the transformer result in sufficient current flow to operate the overcurrent devices. These requirements also apply to generators or systems that are derived from converter windings, although these systems do not have the same wide use as separately derived systems that are derived from transformers.

(1) Bonding Jumper. A bonding jumper in compliance with 250.28(A) through (D) that is sized for the derived phase conductors shall be used to connect the equipment grounding conductors of the separately derived system to the grounded conductor. Except as permitted by 250.24(A)(3), this connection shall be made at any point on the separately derived system from the source to the first system disconnecting means or overcurrent device, or it shall be made at the source of a separately derived system that has no disconnecting means or overcurrent devices. The point of connection shall be the same as the grounding electrode conductor as required in 250.30(A)(2).

Where a separately derived system provides a grounded conductor, a bonding jumper must be installed to connect the equipment grounding conductors to the grounded conductor. Equipment grounding conductors are connected to the grounding electrode system by the grounding electrode conductor. The bonding jumper is sized according to 250.28(D) and may be located at any point between the source terminals (transformer, generator, etc.) and the first disconnecting means or overcurrent device.

Exception No. 1: A bonding jumper at both the source and the first disconnecting means shall be permitted where doing so does not establish a parallel path for the grounded circuit conductor. Where a grounded conductor is used in this manner, it shall not be smaller than the size specified for the bonding jumper but shall not be required to be larger than the ungrounded conductor(s). For the purposes of this exception, connection through the earth shall not be considered as providing a parallel path.

Exception No. 2: The size of the bonding jumper for a system that supplies a Class 1, Class 2, or Class 3 circuit, and is

derived from a transformer rated not more than 1000 volt-amperes, shall not be smaller than the derived phase conductors and shall not be smaller than 14 AWG copper or 12 AWG aluminum.

Section 250.30(A)(1) requires the bonding jumper to be not smaller than the sizes given in Table 250.66, that is, not smaller than 8 AWG copper. Exception No. 2 to 250.30(A)(1) permits a bonding jumper for a Class 1, Class 2, or Class 3 circuit to be not smaller than 14 AWG copper or 12 AWG aluminum.

(2) Grounding Electrode Conductor. The grounding electrode conductor shall be installed in accordance with (a) or (b). Where taps are connected to a common grounding electrode conductor, the installation shall comply with 250.30(A)(3).

(a) Single Separately Derived System. A grounding electrode conductor for a single separately derived system shall be sized in accordance with 250.66 for the derived phase conductors and shall be used to connect the grounded conductor of the derived system to the grounding electrode as specified in 250.30(A)(4). Except as permitted by 250.24(A)(3) or (A)(4), this connection shall be made at the same point on the separately derived system where the bonding jumper is installed.

Exception: A grounding electrode conductor shall not be required for a system that supplies a Class 1, Class 2, or Class 3 circuit and is derived from a transformer rated not more than 1000 volt-amperes, provided the system grounded conductor is bonded to the transformer frame or enclosure by a jumper sized in accordance with 250.30(A)(1), Exception No. 2, and the transformer frame or enclosure is grounded by one of the means specified in 250.134.

If a separately derived system is required to be grounded, the conductor to be grounded is allowed to be connected to the grounding electrode system at any location between the source terminals (transformer, generator, etc.) and the first disconnecting means or overcurrent device. The location of the grounding electrode conductor connection to the grounded conductor must be at the same point as where the bonding jumper is connected to the grounded conductor. By establishing a common point of connection, normal neutral current will be carried only on the system grounded conductor. Metal raceways, piping systems, and structural steel must not provide a parallel circuit. Exhibits 250.13 and 250.14 illustrate examples of grounding electrode connections for separately derived systems.

(b) Multiple Separately Derived Systems. Where more than one separately derived system is connected to a common grounding electrode conductor as provided in 250.30(A)(3), the common grounding electrode conductor shall be sized in accordance with 250.66, based on the total area of the largest derived phase conductor from each separately derived system.

(3) Grounding Electrode Conductor Taps. It shall be permissible to connect taps from a separately derived system to a common grounding electrode conductor. Each tap conductor shall connect the grounded conductor of the separately derived system to the common grounding electrode conductor.

(a) Tap Conductor Size. Each tap conductor shall be sized in accordance with 250.66 for the derived phase conductors of the separately derived system it serves.

(b) Connections. All connections shall be made at an accessible location by an irreversible compression connector listed for the purpose, listed connections to copper busbars not less than 6 mm × 50 mm (¼ in. × 2 in.), or by the exothermic welding process. The tap conductors shall be connected to the common grounding electrode conductor as specified in 250.30(A)(2)(b) in such a manner that the common grounding electrode conductor remains without a splice or joint.

(c) Installation. The common grounding electrode conductor and the taps to each separately derived system shall comply with 250.64(A), (B), (C), and (E).

(d) Bonding. Where exposed structural steel that is interconnected to form the building frame or interior metal piping exists in the area served by the separately derived system, it shall be bonded to the grounding electrode conductor in accordance with 250.104.

New for the 2002 *NEC*, a common grounding electrode conductor is now permitted as an alternative installation method, in lieu of installing separate individual grounding electrode conductors from each separately derived system to the grounding electrode system. The sizing of the common grounding electrode conductor is covered in 250.30(A)(2)(b), and the specific installations requirements are located in 250.30(A)(3)(c). The following example, together with Exhibit 250.15, illustrates this new permitted installation method.

Example

An old, large post-and-beam loft-type building is being renovated for use as an office building. The building is being furnished with four 45-kVA, 480 to 120/208-volt, 3-phase, 4-wire, wye-connected transformers. Each transformer secondary supplies an adjacent 150-ampere main circuit breaker

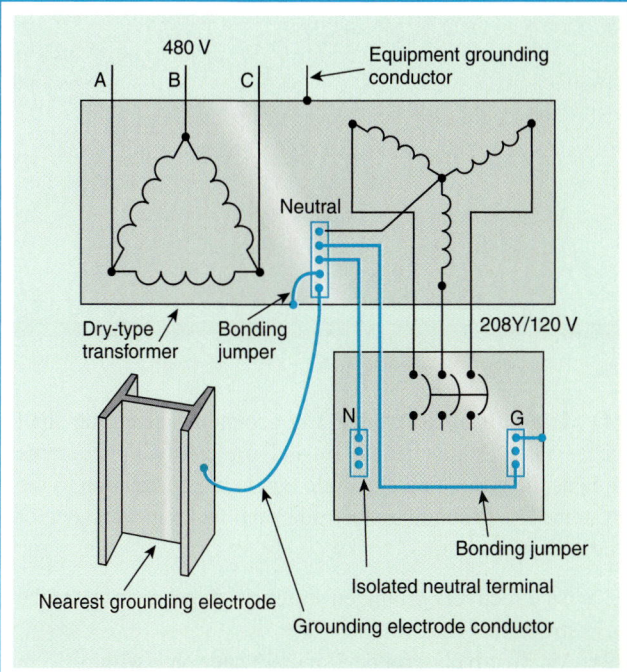

Exhibit 250.13 A grounding arrangement for a separately derived system in which the grounding electrode conductor connection is made at the transformer.

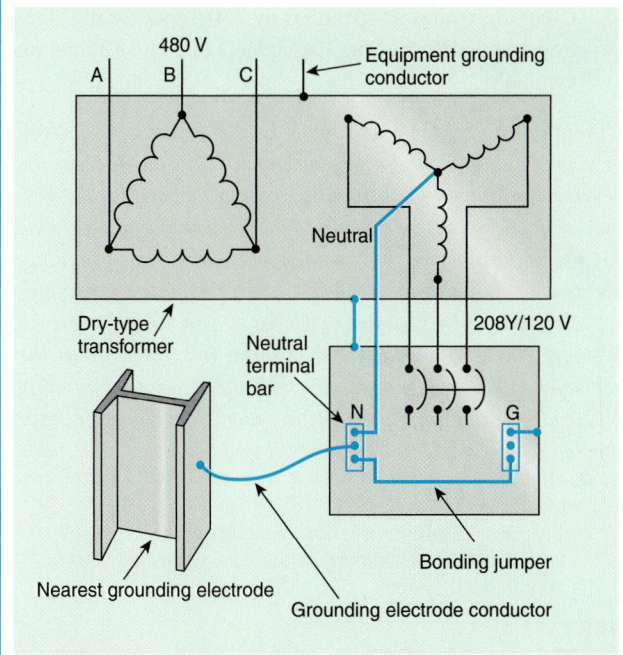

Exhibit 250.14 A grounding arrangement for a separately derived system in which the grounding electrode conductor connection is made at the first disconnecting means.

panelboard using 1/0 AWG, Type THHN copper conductors. The transformers are strategically placed throughout the building. Because the building contains no effectively grounded structural steel, each transformer secondary must be grounded to the water service electrode within the first 5 ft of entry into the building A common grounding electrode conductor has been selected as the method to ground all the transformers.

What is the minimum size common grounding electrode conductor that must be used to connect the four transformers to the grounding electrode system? What is the minimum size grounding electrode conductor to connect each of the four transformers to the common grounding electrode conductor?

Solution

STEP 1. Determine the size of the largest derived phase conductor (secondary conductor in this example) for each transformer or separately derived system that is intended to be connected to the common grounding electrode conductor. Based on the load and size of the main overcurrent device in the panelboard, the designer has already decided that each 45-kVA transformer uses a 1/0 AWG copper conductor as the largest derived phase conductor.

STEP 2. Calculate the total circular mil area for all the separately derived phase conductors. The grounding electrode conductors for each of the four transformers will tap to the common grounding electrode conductor. Each transformer uses the same size and type conductors, 1/0 AWG, Type THHN copper to supply the panelboards. According to Table 8 in Chapter 9, a 1/0 AWG has a circular mil area of 105,600 circular mils. There are four separately derived systems to be connected to the common grounding electrode conductor, so the total circular mil area is 4 times 105,600 circular mils, which equals 422,400 circular mils, or 422.4 kcmil.

STEP 3. Determine the size of the common grounding electrode conductor. The total circular mil area is calculated using the secondary phase conductors from each of the four separately derived systems (transformers) that will be connected to the common grounding electrode conductor. Because 422.4 kcmil is not a standard-size wire, the next standard-size or larger conductor must be used to determine the common grounding electrode conductor size. Using 500 kcmil as the total for the derived phase conductors and applying Table 250.66, row 5, "Over 350 through 600 kcmil," a 1/0 AWG conductor is selected as the size of the common grounding electrode conductor. This conductor is labeled "Conductor A" in Exhibit 250.15.

STEP 4. Determine the size of each individual grounding electrode tap conductor for each of the separately derived systems. According to Table 250.66, a 1/0 AWG copper derived phase conductor requires a conductor not smaller than 6 AWG copper for each transformer grounding electrode tap conductor. This individual grounding electrode conductor will be used as the permitted tap conductor and will run from the conductor to be grounded of each separately derived system to a connection point located on the common grounding electrode conductor. This conductor is labeled "Conductor B" in Exhibit 250.15. If a situation arises in which an additional load requires the installation of another 45-kVA transformer, a recalculation of this example will show that the original 1/0 AWG copper common grounding electrode conductor need not be increased in size.

(4) Grounding Electrode. The grounding electrode shall be as near as practicable to and preferably in the same area as the grounding electrode conductor connection to the system. The grounding electrode shall be the nearest one of the following:

(1) An effectively grounded structural metal member of the structure
(2) An effectively grounded metal water pipe within 1.5 m (5 ft) from the point of entrance into the building

Exception: In industrial and commercial buildings where conditions of maintenance and supervision ensure that only qualified persons service the installation and the entire length of the interior metal water pipe that is being used for the grounding electrode is exposed, the connection shall be permitted at any point on the water pipe system.

(3) Other electrodes as specified by 250.52 where the electrodes specified by 250.30(A)(4)(1) or (A)(4)(2) are not available

Exception to (1), (2), and (3): Where a separately derived system originates in listed equipment suitable for use as service equipment, the grounding electrode used for the service or feeder shall be permitted as the grounding electrode for the separately derived system, provided the grounding electrode conductor from the service or feeder to the grounding electrode is of sufficient size for the separately derived system. Where the equipment ground bus internal to the service equipment is not smaller than the required grounding electrode conductor, the grounding electrode connection for the separately derived system shall be permitted to be made to the bus.

FPN: See 250.104(A)(4) for bonding requirements of interior metal water piping in the area served by separately derived systems.

Section 250.30(A)(4) requires that the grounding electrode be as near as is practicable to the grounding conductor connection to the system to minimize the impedance to ground. If an effectively grounded structural metal member of the

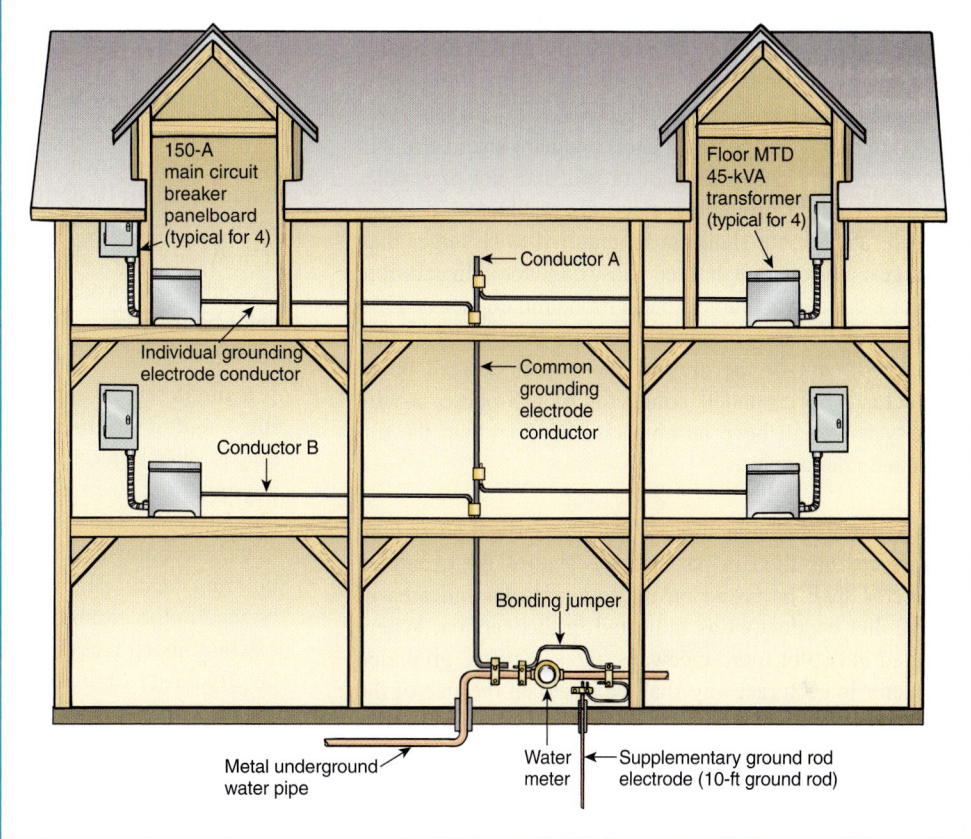

Exhibit 250.15 The grounding arrangement for multiple separately derived systems using taps from a common grounding electrode conductor according to 250.30(A)(2)(b) and 250.30(A)(3).

150-A main circuit breaker panelboard (typical for 4)

Floor MTD 45-kVA transformer (typical for 4)

Conductor A

Individual grounding electrode conductor

Common grounding electrode conductor

Conductor B

Bonding jumper

Metal underground water pipe

Water meter

Supplementary ground rod electrode (10-ft ground rod)

building structure or an effectively grounded metal water pipe is available nearby, 250.30(A)(4) requires that it be used as the grounding electrode. For example, where a transformer is installed on the fiftieth floor, the grounding electrode conductor is not required to be run to the service grounding electrode system. However, where an effectively grounded metal water pipe is used as an electrode for a separately derived system, the *Code* requires that the grounding electrode conductor connection be made within 5 ft of where the piping system enters the building. Concern over the use of nonmetallic piping or fittings is the basis for the "within 5 ft" requirement. Where the piping system is located in an industrial or commercial building that is serviced only by qualified persons and the entire length that will be used as an electrode is exposed, the connection may be made at any point on the piping system.

The practice of grounding the secondary of an isolating transformer to a ground rod or running the grounding electrode conductor back to the service ground (usually to reduce electrical noise on data processing systems) is prohibited if an item in (1) or (2) of 250.30(A)(4) is available. However, an isolation transformer that is part of a listed power supply for a data processing room is not required to be grounded in accordance with 250.30(A)(4), but it must be grounded in accordance with the manufacturer's instructions.

Exhibits 250.13 and 250.14 are typical wiring diagrams for dry-type transformers supplied from a 480-volt, 3-phase feeder to derive a 208Y/120-volt or 480Y/277-volt secondary. As indicated in 250.30(A)(1), the bonding jumper connection is required to be sized according to 250.28(D). In Exhibit 250.13, this connection is made at the source of the separately derived system, in the transformer enclosure. In Exhibit 250.14, the bonding jumper connection is made at the first disconnecting means. With the grounding electrode conductor, the bonding jumper, and the bonding of the grounded circuit conductor (neutral) connected as shown, line-to-ground fault currents are able to return to the supply source through a short, low-impedance path. A path of lower impedance is provided that facilitates the operation of overcurrent devices, in accordance with 250.4(A)(5). The grounding electrode conductor from the secondary grounded circuit conductor is sized according to Table 250.66.

(5) Equipment Bonding Jumper Size. Where a bonding jumper is run with the derived phase conductors from the source of a separately derived system to the first disconnecting means, it shall be sized in accordance with 250.28(A) through (D), based on the size of the derived phase conductors.

(6) Grounded Conductor. Where a grounded conductor is installed and the bonding jumper is not located at the source of the separately derived system, the following shall apply:

(a) *Routing and Sizing.* This conductor shall be routed with the derived phase conductors and shall not be smaller than the required grounding electrode conductor specified in Table 250.66, but shall not be required to be larger than the largest ungrounded derived phase conductor. In addition, for phase conductors larger than 1100 kcmil copper or 1750 kcmil aluminum, the grounded conductor shall not be smaller than 12 percent of the area of the largest derived phase conductor. The grounded conductor of a 3-phase, 3-wire delta system shall have an ampacity not less than the ungrounded conductors.

(b) *Parallel Conductors.* Where the derived phase conductors are installed in parallel, the size of the grounded conductor shall be based on the total circular mil area of the parallel conductors as indicated in this section. Where installed in two or more raceways, the size of the grounded conductor in each raceway shall be based on the size of the ungrounded conductors in the raceway but not smaller than 1/0 AWG.

> FPN: See 310.4 for grounded conductors connected in parallel.

(c) *High Impedance.* The grounded conductor on a high-impedance grounded neutral system shall be grounded in accordance with 250.36.

(B) Ungrounded Systems. The equipment of an ungrounded separately derived system shall be grounded as specified in 250.30(B)(1) and (2).

(1) Grounding Electrode Conductor. A grounding electrode conductor, sized in accordance with 250.66 for the derived phase conductors, shall be used to connect the metal enclosures of the derived system to the grounding electrode as specified in 250.30(B)(2). This connection shall be made at any point on the separately derived system from the source to the first system disconnecting means.

> In a separately derived system that is ungrounded, a bonding jumper must be installed to connect the disconnect enclosure and equipment grounding conductors to the grounding electrode system.

(2) Grounding Electrode. Except as permitted by 250.34 for portable and vehicle-mounted generators, the grounding electrode shall comply with 250.30(A)(4).

250.32 Two or More Buildings or Structures Supplied from a Common Service.

(A) Grounding Electrode. Where two or more buildings or structures are supplied from a common ac service by a feeder(s) or branch circuit(s), the grounding electrode(s) required in Part III of this article at each building or structure shall be connected in the manner specified in 250.32(B) or (C). Where there are no existing grounding electrodes, the grounding electrode(s) required in Part III of this article shall be installed.

> If a single service supplies more than one building, such as illustrated in Exhibit 250.16, and the feeder is installed with an equipment grounding conductor, 250.32(A) requires that a grounding electrode system be established, unless one already exists. The equipment grounding bus must be bonded to the grounding electrode system. The disconnecting means, building steel, and interior metal water piping must be bonded to the grounding electrode system. All non–current-carrying metal parts of electrical equipment are required to be grounded by connection to the equipment grounding bus.

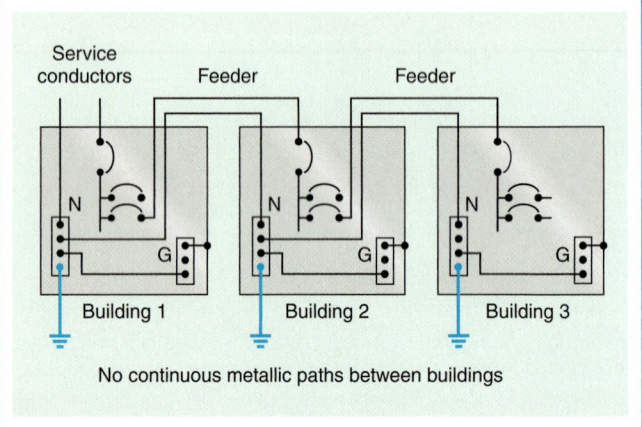

No continuous metallic paths between buildings

Exhibit 250.16 A single service (grounded system) supplying three buildings, where each building is required to have a grounding electrode installed in accordance with 250.32(A).

Exception: A grounding electrode at separate buildings or structures shall not be required where only one branch circuit supplies the building or structure and the branch circuit includes an equipment grounding conductor for grounding the conductive non–current-carrying parts of all equipment.

> If a building is supplied by only one branch circuit with an equipment grounding conductor, there is no requirement to establish a grounding electrode system or connect to one if one exists. See Exhibit 250.17 for an example of such a system.

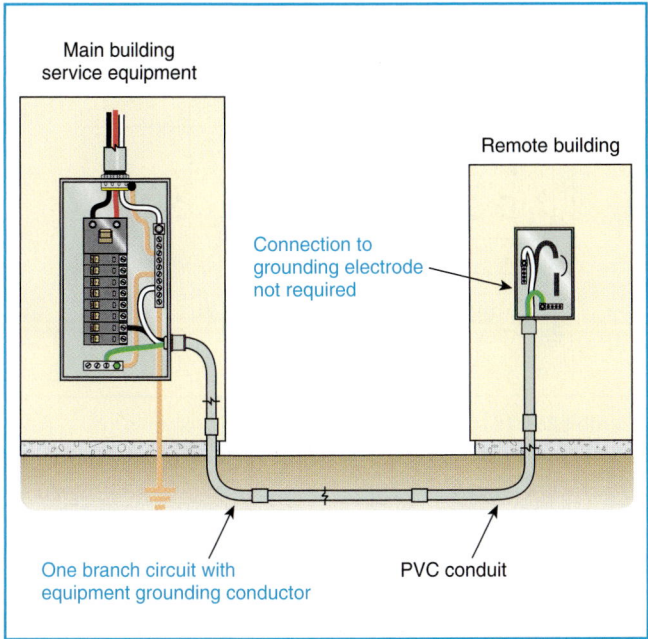

Exhibit 250.17 An installation in which connection from the enclosure of the building disconnecting means to the electrode is not required because an equipment grounding conductor is run with the circuit conductors.

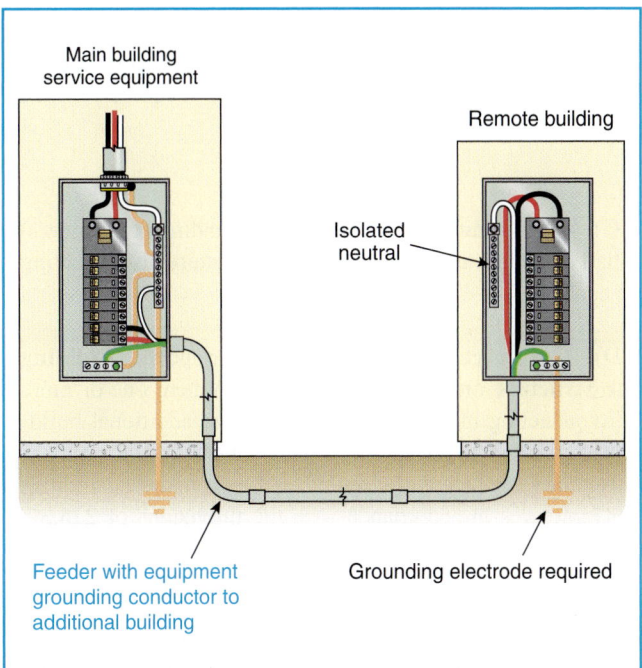

Exhibit 250.18 An installation in which connection between the grounded conductor (neutral) and equipment grounding terminal bar is not allowed. A connection to the grounding electrode is required.

(B) Grounded Systems. For a grounded system at the separate building or structure, the connection to the grounding electrode and grounding or bonding of equipment, structures, or frames required to be grounded or bonded shall comply with either 250.32(B)(1) or (2).

(1) Equipment Grounding Conductor. An equipment grounding conductor as described in 250.118 shall be run with the supply conductors and connected to the building or structure disconnecting means and to the grounding electrode(s). The equipment grounding conductor shall be used for grounding or bonding of equipment, structures, or frames required to be grounded or bonded. The equipment grounding conductor shall be sized in accordance with 250.122. Any installed grounded conductor shall not be connected to the equipment grounding conductor or to the grounding electrode(s).

If a feeder supplies another building from the same service and an equipment grounding conductor is run with the feeder, the grounded conductor (neutral) is not permitted to be connected to the equipment grounding conductor or to the grounding electrode system, as illustrated in Exhibit 250.18.

(2) Grounded Conductor. Where (1) an equipment grounding conductor is not run with the supply to the building or structure, (2) there are no continuous metallic paths

bonded to the grounding system in both buildings or structures involved, and (3) ground-fault protection of equipment has not been installed on the common ac service, the grounded circuit conductor run with the supply to the building or structure shall be connected to the building or structure disconnecting means and to the grounding electrode(s) and shall be used for grounding or bonding of equipment, structures, or frames required to be grounded or bonded. The size of the grounded conductor shall not be smaller than the larger of

(1) That required by 220.22
(2) That required by 250.122

Similar to the provisions of 250.30(A)(2), 250.32(B)(2) also eliminates the creation of parallel paths for normal neutral current on grounding conductors, metal raceways, metal piping, and other metal structures. In previous editions of the *Code,* the grounding electrode conductor and equipment grounding conductors were permitted to be connected to the grounded conductor at a separate building or structure. This multiple-location grounding arrangement could provide parallel paths for neutral current along the electrical system and along other continuous metallic piping and mechanical systems as well. Connection of the grounded conductor to a grounding electrode system at a separate building or structure is permitted only if these parallel paths are not created and if there is no common ground-fault protection of equipment provided at the service where the feeder or branch circuit originates.

(C) Ungrounded Systems. The grounding electrode(s) shall be connected to the building or structure disconnecting means.

(D) Disconnecting Means Located in Separate Building or Structure on the Same Premises. Where one or more disconnecting means supply one or more additional buildings or structures under single management, and where these disconnecting means are located remote from those buildings or structures in accordance with the provisions of 225.32, Exception Nos. 1 and 2, all of the following conditions shall be met:

(1) The connection of the grounded circuit conductor to the grounding electrode at a separate building or structure shall not be made.
(2) An equipment grounding conductor for grounding any non–current-carrying equipment, interior metal piping systems, and building or structural metal frames is run with the circuit conductors to a separate building or structure and bonded to existing grounding electrode(s) required in Part III of this article, or, where there are no existing electrodes, the grounding electrode(s) required in Part III of this article shall be installed where a separate building or structure is supplied by more than one branch circuit.
(3) Bonding the equipment grounding conductor to the grounding electrode at a separate building or structure shall be made in a junction box, panelboard, or similar enclosure located immediately inside or outside the separate building or structure.

Exhibit 250.19 illustrates an installation in which the disconnect for Building 2 is located in Building 1. Section 250.32(D) applies to separate buildings or structures that do not have a disconnect, as permitted by Exception Nos. 1 and 2 to 225.32. The feeder conductors must terminate in a panelboard, junction box, or similar enclosure that is located immediately inside or outside the building.

An equipment grounding conductor must be run with the feeder conductors, the grounded conductor must not be bonded to the enclosure or equipment grounding bus, and the equipment grounding bus must be connected to a new or existing grounding electrode system at the second building. All non–current-carrying metal parts of equipment, building steel, and interior metal piping systems must be connected to the grounding electrode system.

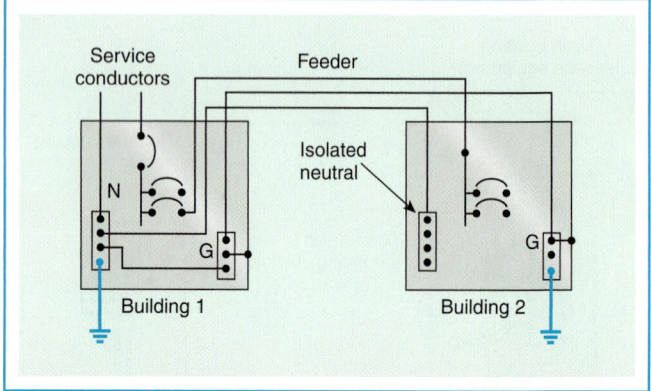

Exhibit 250.19 Grounding and bonding requirements for a separate building under single management with the disconnect remotely located from the building.

(E) Grounding Electrode Conductor. The size of the grounding electrode conductor to the grounding electrode(s) shall not be smaller than given in 250.66, based on the largest ungrounded supply conductor. The installation shall comply with Part III of this article.

A grounding electrode system is connected to the grounded conductor and/or the equipment enclosures by the *grounding electrode conductor* by definition and according to 250.24 for services, to 250.30 for separately derived systems, and to this section (250.32) for two or more buildings supplied from a common service. Each of these references directs the user to the same general requirements; that is, the grounding electrode conductor must comply with Part III of this article. For the 2002 *Code*, this revision was made to clearly state the requirement for two or more structures supplied from a common service.

250.34 Portable and Vehicle-Mounted Generators.

(A) Portable Generators. The frame of a portable generator shall not be required to be grounded and shall be permitted to serve as the grounding electrode for a system supplied by the generator under the following conditions:

(1) The generator supplies only equipment mounted on the generator, cord-and-plug-connected equipment through receptacles mounted on the generator, or both, and
(2) The non–current-carrying metal parts of equipment and the equipment grounding conductor terminals of the receptacles are bonded to the generator frame.

Portable describes equipment that is easily carried by personnel from one location to another. *Mobile* describes equipment, such as vehicle-mounted generators, that is capable of being moved, on wheels or rollers.

The frame of a portable generator is not required to be connected to earth (ground rod, water pipe, etc.) if the generator has receptacles mounted on the generator panel and the receptacles have equipment grounding terminals bonded to the generator frame.

(B) Vehicle-Mounted Generators. The frame of a vehicle shall be permitted to serve as the grounding electrode for a system supplied by a generator located on the vehicle under the following conditions:

(1) The frame of the generator is bonded to the vehicle frame, and
(2) The generator supplies only equipment located on the vehicle or cord-and-plug-connected equipment through receptacles mounted on the vehicle, or both equipment located on the vehicle and cord-and-plug-connected equipment through receptacles mounted on the vehicle or on the generator, and
(3) The non–current-carrying metal parts of equipment and the equipment grounding conductor terminals of the receptacles are bonded to the generator frame, and
(4) The system complies with all other provisions of this article.

Vehicle-mounted generators that provide a neutral conductor and are installed as separately derived systems supplying equipment and receptacles on the vehicle are required to have the neutral conductor bonded to the generator frame and to the vehicle frame. The non–current-carrying parts of the equipment must be bonded to the generator frame.

(C) Grounded Conductor Bonding. A system conductor that is required to be grounded by 250.26 shall be bonded to the generator frame where the generator is a component of a separately derived system.

> FPN: For grounding portable generators supplying fixed wiring systems, see 250.20(D).

Portable and vehicle-mounted generators that are installed as separately derived systems and that provide a neutral conductor (such as 3 phase, 4 wire wye connected; single phase 240/120 volt; or 3 phase, 4 wire delta connected) are required to have the neutral conductor bonded to the generator frame.

250.36 High-Impedance Grounded Neutral Systems.

High-impedance grounded neutral systems in which a grounding impedance, usually a resistor, limits the ground-

fault current to a low value shall be permitted for 3-phase ac systems of 480 volts to 1000 volts where all the following conditions are met:

Section 250.36 covers high-impedance grounded neutral systems of 480 to 1000 volts. Systems rated over 1000 volts are covered in 250.186. For information on the differences between solidly grounded systems and high-impedance grounded neutral systems, see "Grounding for Emergency and Standby Power Systems," by Robert B. West, *IEEE Transactions on Industry Applications,* Vol. IA-15, No. 2, March/April 1979.

As the schematic diagram in Exhibit 250.20 shows, a high-impedance grounded neutral system is designed to minimize the amount of fault current that can flow during a ground fault. The grounding impedance is usually selected to limit fault current to a value that is slightly greater than or equal to the capacitive charging current. This system is used where continuity of power is required. Therefore, a ground fault results in an alarm condition rather than in the tripping of a circuit breaker, which allows for a safe and orderly shutdown.

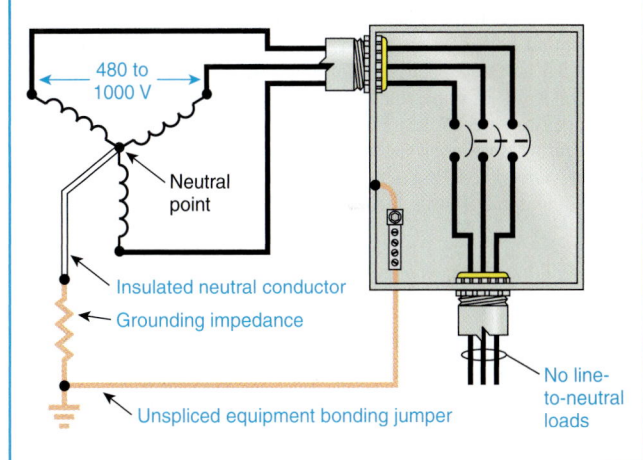

Exhibit 250.20 Schematic diagram of a high-impedance grounded neutral system.

(1) The conditions of maintenance and supervision ensure that only qualified persons service the installation.
(2) Continuity of power is required.
(3) Ground detectors are installed on the system.
(4) Line-to-neutral loads are not served.

High-impedance grounded neutral systems shall comply with the provisions of 250.36(A) through (G).

(A) Grounding Impedance Location. The grounding impedance shall be installed between the grounding electrode conductor and the system neutral. Where a neutral is not available, the grounding impedance shall be installed between the grounding electrode conductor and the neutral derived from a grounding transformer.

(B) Neutral Conductor. The neutral conductor from the neutral point of the transformer or generator to its connection point to the grounding impedance shall be fully insulated.

The neutral conductor shall have an ampacity of not less than the maximum current rating of the grounding impedance. In no case shall the neutral conductor be smaller than 8 AWG copper or 6 AWG aluminum or copper-clad aluminum.

> The current through the neutral conductor is limited by the grounding impedance. Therefore, the neutral conductor is not required to be sized to carry high-fault current. The neutral conductor cannot be smaller than 8 AWG copper or 6 AWG aluminum.

(C) System Neutral Connection. The system neutral conductor shall not be connected to ground except through the grounding impedance.

> FPN: The impedance is normally selected to limit the ground-fault current to a value slightly greater than or equal to the capacitive charging current of the system. This value of impedance will also limit transient overvoltages to safe values. For guidance, refer to criteria for limiting transient overvoltages in ANSI/IEEE 142-1991, *Recommended Practice for Grounding of Industrial and Commercial Power Systems.*

> Additional information can be found in "Charging Current Data for Guesswork-Free Design of High-Resistance Grounded Systems," by D. S. Baker, *IEEE Transactions on Industry Applications,* Vol. IA-15, No. 2, March/April 1979; and "High-Resistance Grounding," by Baldwin Bridger, Jr., *IEEE Transactions on Industry Applications,* Vol. IA-19, No. 1, January/February 1983.

(D) Neutral Conductor Routing. The conductor connecting the neutral point of the transformer or generator to the grounding impedance shall be permitted to be installed in a separate raceway. It shall not be required to run this conductor with the phase conductors to the first system disconnecting means or overcurrent device.

(E) Equipment Bonding Jumper. The equipment bonding jumper (the connection between the equipment grounding conductors and the grounding impedance) shall be an unspliced conductor run from the first system disconnecting means or overcurrent device to the grounded side of the grounding impedance.

(F) Grounding Electrode Conductor Location. The grounding electrode conductor shall be attached at any point from the grounded side of the grounding impedance to the equipment grounding connection at the service equipment or first system disconnecting means.

(G) Equipment Bonding Jumper Size. The equipment bonding jumper shall be sized in accordance with (1) or (2).

(1) Where the grounding electrode conductor connection is made at the grounding impedance, the equipment bonding jumper shall be sized in accordance with 250.66, based on the size of the service entrance conductors for a service or the derived phase conductors for a separately derived system.
(2) Where the grounding electrode conductor is connected at the first system disconnecting means or overcurrent device, the equipment bonding jumper shall be sized the same as the neutral conductor in 250.36(B).

III. Grounding Electrode System and Grounding Electrode Conductor

250.50 Grounding Electrode System.

If available on the premises at each building or structure served, each item in 250.52(A)(1) through (A)(6) shall be bonded together to form the grounding electrode system. Where none of these electrodes are available, one or more of the electrodes specified in 250.52(A)(4) through (A)(7) shall be installed and used.

Formal Interpretation 78-4

Reference: Article 250.50

Question: Is it the intent of 250.50 that reinforcing steel, if used in a building footing, must be made available for grounding?

Answer: No.

Issue Edition: 1978

Reference: 250-81

Issue Date: March 1980

> Section 250.50 introduces the important concept of a "grounding electrode system," in which all electrodes are bonded together, as illustrated in Exhibit 250.21. Rather than relying totally on a single electrode to perform its function over the life of the electrical installation, the *NEC* encourages the formation of a system of electrodes "if available on the

premises." There is no doubt that building a system of electrodes adds a level of reliability and helps ensure system performance over a long period of time.

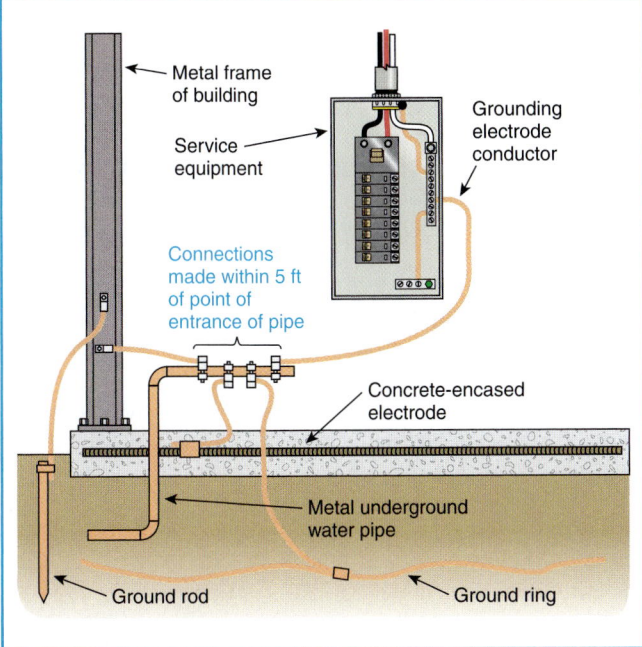

Exhibit 250.21 A grounding electrode system that uses the metal frame of a building, a ground ring, a concrete-encased electrode, a metal underground water pipe, and a ground rod.

250.52 Grounding Electrodes.

(A) Electrodes Permitted for Grounding.

(1) Metal Underground Water Pipe. A metal underground water pipe in direct contact with the earth for 3.0 m (10 ft) or more (including any metal well casing effectively bonded to the pipe) and electrically continuous (or made electrically continuous by bonding around insulating joints or insulating pipe) to the points of connection of the grounding electrode conductor and the bonding conductors. Interior metal water piping located more than 1.52 m (5 ft) from the point of entrance to the building shall not be used as a part of the grounding electrode system or as a conductor to interconnect electrodes that are part of the grounding electrode system.

Exception: In industrial and commercial buildings or structures where conditions of maintenance and supervision ensure that only qualified persons service the installation, interior metal water piping located more than 1.52 m (5 ft) from the point of entrance to the building shall be permitted as a part of the grounding electrode system or as a conductor to interconnect electrodes that are part of the grounding electrode system, provided that the entire length, other than

short sections passing perpendicular through walls, floors, or ceilings, of the interior metal water pipe that is being used for the conductor is exposed.

There has always been uncertainty as to whether metal water piping systems should be used as grounding electrodes, so many years ago the electrical industry and the waterworks industry formed a committee of all interested parties to evaluate the use of metal underground water piping systems as grounding electrodes. Based on its findings, the committee issued an authoritative report on this subject. The International Association of Electrical Inspectors published the report, *Interim Report of the American Research Committee on Grounding,* in January 1944 (reprinted March 1949).

The National Institute of Standards and Technology (NIST) has monitored the electrolysis of metal systems, because a flow of current at a grounding electrode on dc systems can cause displacement of metal. The results of this monitoring have shown that problems are minimal.

The last sentence of 250.52(A)(1) prohibits the use of that portion of the interior metal water piping system that extends more than 5 ft beyond the point of entrance into the building to interconnect grounding electrodes and the grounding electrode conductor. The exception to 250.52(A)(1), however, permits this practice, provided there is qualified maintenance and the entire length of the pipe is exposed. This 5-ft limit also applies to the replacement of nongrounding receptacles with grounding-type or branch-circuit extensions in accordance with 250.130(C). See the commentary following 250.130(C) and the illustration that accompanies that commentary, Exhibit 250.50.

(2) Metal Frame of the Building or Structure. The metal frame of the building or structure, where effectively grounded.

(3) Concrete-Encased Electrode. An electrode encased by at least 50 mm (2 in.) of concrete, located within and near the bottom of a concrete foundation or footing that is in direct contact with the earth, consisting of at least 6.0 m (20 ft) of one or more bare or zinc galvanized or other electrically conductive coated steel reinforcing bars or rods of not less than 13 mm (½ in.) in diameter, or consisting of at least 6.0 m (20 ft) of bare copper conductor not smaller than 4 AWG. Reinforcing bars shall be permitted to be bonded together by the usual steel tie wires or other effective means.

Exhibit 250.22 shows an example of a concrete-encased electrode.

(4) Ground Ring. A ground ring encircling the building or structure, in direct contact with the earth, consisting of

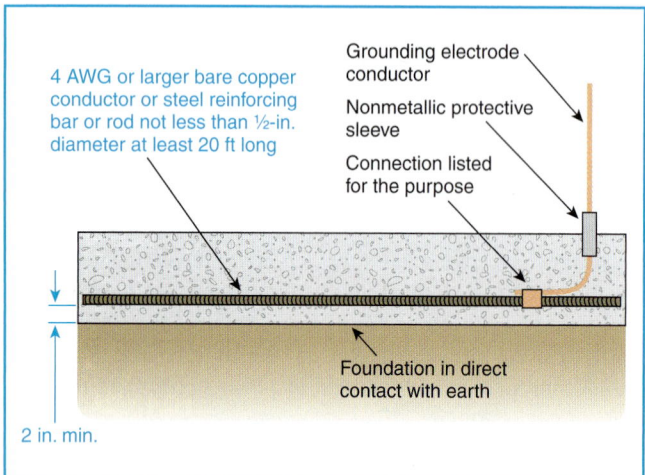

4 AWG or larger bare copper conductor or steel reinforcing bar or rod not less than ½-in. diameter at least 20 ft long

Grounding electrode conductor

Nonmetallic protective sleeve

Connection listed for the purpose

Foundation in direct contact with earth

2 in. min.

Exhibit 250.22 A concrete-encased electrode.

at least 6.0 m (20 ft) of bare copper conductor not smaller than 2 AWG.

(5) Rod and Pipe Electrodes. Rod and pipe electrodes shall not be less than 2.5 m (8 ft) in length and shall consist of the following materials.

(a) Electrodes of pipe or conduit shall not be smaller than metric designator 21 (trade size ¾) and, where of iron or steel, shall have the outer surface galvanized or otherwise metal-coated for corrosion protection.

(b) Electrodes of rods of iron or steel shall be at least 15.87 mm (⅝ in.) in diameter. Stainless steel rods less than 16 mm (⅝ in.) in diameter, nonferrous rods, or their equivalent shall be listed and shall not be less than 13 mm (½ in.) in diameter.

(6) Plate Electrodes. Each plate electrode shall expose not less than 0.186 m² (2 ft²) of surface to exterior soil. Electrodes of iron or steel plates shall be at least 6.4 mm (¼ in.) in thickness. Electrodes of nonferrous metal shall be at least 1.5 mm (0.06 in.) in thickness.

(7) Other Local Metal Underground Systems or Structures. Other local metal underground systems or structures such as piping systems and underground tanks.

(B) Electrodes Not Permitted for Grounding. The following shall not be used as grounding electrodes:

(1) Metal underground gas piping system
(2) Aluminum electrodes

250.53 Grounding Electrode System Installation.

FPN: See 547.9 and 547.10 for special grounding and bonding requirements for agricultural buildings.

(A) Rod, Pipe, and Plate Electrodes. Where practicable, rod, pipe, and plate electrodes shall be embedded below permanent moisture level. Rod, pipe, and plate electrodes shall be free from nonconductive coatings such as paint or enamel.

(B) Electrode Spacing. Where more than one of the electrodes of the type specified in 250.52(A)(5) or (A)(6) are used, each electrode of one grounding system (including that used for air terminals) shall not be less than 1.83 m (6 ft) from any other electrode of another grounding system. Two or more grounding electrodes that are effectively bonded together shall be considered a single grounding electrode system.

(C) Bonding Jumper. The bonding jumper(s) used to connect the grounding electrodes together to form the grounding electrode system shall be installed in accordance with 250.64(A), (B), and (E), shall be sized in accordance with 250.66, and shall be connected in the manner specified in 250.70.

(D) Metal Underground Water Pipe. Where used as a grounding electrode, metal underground water pipe shall meet the requirements of 250.53(D)(1) and (D)(2).

(1) Continuity. Continuity of the grounding path or the bonding connection to interior piping shall not rely on water meters or filtering devices and similar equipment.

(2) Supplemental Electrode Required. A metal underground water pipe shall be supplemented by an additional electrode of a type specified in 250.52(A)(2) through (A)(7). Where the supplemental electrode is a rod, pipe, or plate type, it shall comply with 250.56. The supplemental electrode shall be permitted to be bonded to the grounding electrode conductor, the grounded service-entrance conductor, the nonflexible grounded service raceway, or any grounded service enclosure.

Exception: The supplemental electrode shall be permitted to be bonded to the interior metal water piping at any convenient point as covered in 250.52(A)(1), Exception.

Section 250.53(D)(2) specifically requires that rod, pipe, or plate electrodes used to supplement metal water piping be installed in accordance with 250.56. This requirement clarifies that the supplemental electrode system must be installed as if it were the sole grounding electrode for the system. If 25 ohms or less of earth resistance cannot be achieved with one rod, pipe, or plate, another electrode (other than the metal piping that is being supplemented) must be provided. One of the permitted methods of bonding a supplemental grounding electrode conductor to the primary electrode system is to connect it to the service enclosure.

The requirement to supplement the metal water pipe is based on the practice of using a plastic pipe for replacement when the original metal water pipe fails. This type of replacement leaves the system without a grounding electrode unless a supplementary electrode is provided.

(E) Supplemental Electrode Bonding Connection Size. Where the supplemental electrode is a rod, pipe, or plate electrode, that portion of the bonding jumper that is the sole connection to the supplemental grounding electrode shall not be required to be larger than 6 AWG copper wire or 4 AWG aluminum wire.

Section 250.53(E) correlates with 250.52(A)(5) or (6) and 250.66(A). For example, if a metal underground water pipe or the metal frame of the building or structure is used as the grounding electrode or as part of the grounding electrode system, Table 250.66 must be used for sizing the grounding electrode conductor. The size of the grounding electrode conductor or bonding jumper for ground rod or pipe or for plate electrodes between the service equipment and the electrodes is not required to be larger than 6 AWG copper or 4 AWG aluminum.

(F) Ground Ring. The ground ring shall be buried at a depth below the earth's surface of not less than 750 mm (30 in.).

(G) Rod and Pipe Electrodes. The electrode shall be installed such that at least 2.44 m (8 ft) of length is in contact with the soil. It shall be driven to a depth of not less than 2.44 m (8 ft) except that, where rock bottom is encountered, the electrode shall be driven at an oblique angle not to exceed 45 degrees from the vertical or, where rock bottom is encountered at an angle up to 45 degrees, the electrode shall be permitted to be buried in a trench that is at least 750 mm (30 in.) deep. The upper end of the electrode shall be flush with or below ground level unless the aboveground end and the grounding electrode conductor attachment are protected against physical damage as specified in 250.10.

All rod and pipe electrodes must have at least 8 ft of length in contact with the soil, regardless of rock bottom. Where rock bottom is encountered, the electrodes must either be driven at not more than a 45 degree angle or buried in a 2½-ft deep trench. Ground clamps used on buried electrodes must be listed for direct earth burial. Ground clamps installed above ground must be protected where subject to physical damage. Exhibit 250.23 illustrates these requirements.

(H) Plate Electrode. Plate electrodes shall be installed not less than 750 mm (30 in.) below the surface of the earth.

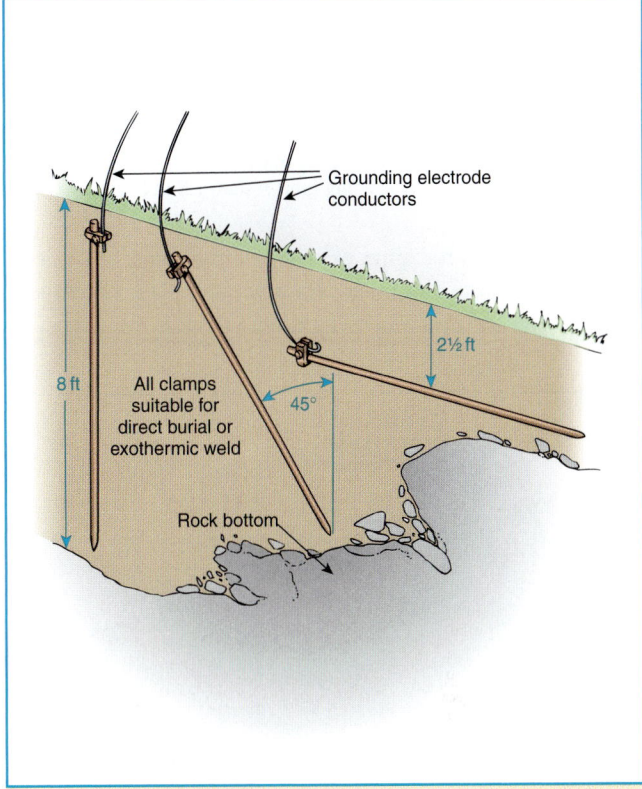

Exhibit 250.23 The installation requirements for rod and pipe electrodes as specified by 250.53(G).

250.54 Supplementary Grounding Electrodes.

Supplementary grounding electrodes shall be permitted to be connected to the equipment grounding conductors specified in 250.118 and shall not be required to comply with the electrode bonding requirements of 250.50 or 250.53(C) or the resistance requirements of 250.56, but the earth shall not be used as the sole equipment grounding conductor.

Grounding electrodes, such as ground rods, that are connected to equipment are not permitted to be used in lieu of the equipment grounding conductor, but they may be used for supplementary protection. For example, grounding electrodes may be used for lightning protection or to equalize potentials in the area of the equipment. Sections 250.4(A)(5) and 250.4(B)(4) also specify that the earth not be used as the sole equipment grounding conductor or effective (ground) fault current path.

250.56 Resistance of Rod, Pipe, and Plate Electrodes.

A single electrode consisting of a rod, pipe, or plate that does not have a resistance to ground of 25 ohms or less shall be augmented by one additional electrode of any of the types specified by 250.52(A)(2) through (A)(7). Where multiple rod, pipe, or plate electrodes are installed to meet the requirements of this section, they shall not be less than 1.8 m (6 ft) apart.

> FPN: The paralleling efficiency of rods longer than 2.5 m (8 ft) is improved by spacing greater than 1.8 m (6 ft).

A supplementary rod, pipe, or plate electrode must be spaced at least 6 ft from any other rod, pipe, and plate electrode. See Exhibit 250.24.

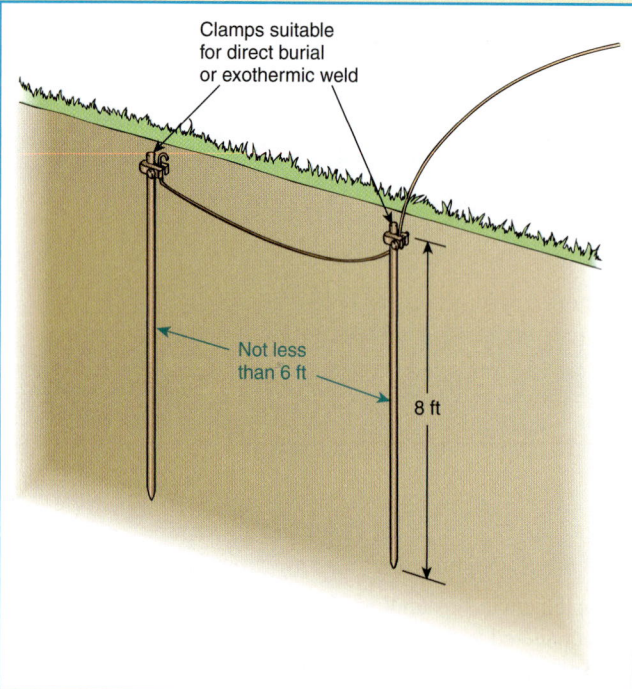

Exhibit 250.24 The 6-ft spacing between electrodes, as required by 250.53(B) and 250.56.

The resistance to ground of a driven grounding electrode can be measured by a ground tester used in the manner shown in Exhibit 250.25.

250.58 Common Grounding Electrode.

Where an ac system is connected to a grounding electrode in or at a building as specified in 250.24 and 250.32, the same electrode shall be used to ground conductor enclosures and equipment in or on that building. Where separate services supply a building and are required to be connected to a grounding electrode, the same grounding electrode shall be used.

Two or more grounding electrodes that are effectively bonded together shall be considered as a single grounding electrode system in this sense.

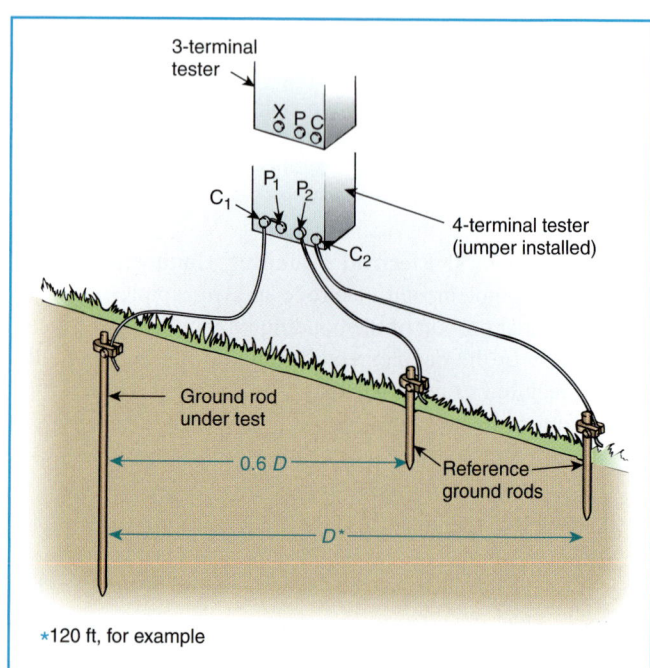

Exhibit 250.25 The resistance to ground of a ground rod being measured by a ground tester.

250.60 Use of Air Terminals.

Air terminal conductors and driven pipes, rods, or plate electrodes used for grounding air terminals shall not be used in lieu of the grounding electrodes required by 250.50 for grounding wiring systems and equipment. This provision shall not prohibit the required bonding together of grounding electrodes of different systems.

> FPN No. 1: See 250.106 for spacing from air terminals. See 800.40(D), 810.21(J), and 820.40(D) for bonding of electrodes.
>
> FPN No. 2: Bonding together of all separate grounding electrodes will limit potential differences between them and between their associated wiring systems.

250.62 Grounding Electrode Conductor Material.

The grounding electrode conductor shall be of copper, aluminum, or copper-clad aluminum. The material selected shall be resistant to any corrosive condition existing at the installation or shall be suitably protected against corrosion. The conductor shall be solid or stranded, insulated, covered, or bare.

250.64 Grounding Electrode Conductor Installation.

Grounding electrode conductors shall be installed as specified in 250.64(A) through (F).

(A) Aluminum or Copper-Clad Aluminum Conductors. Bare aluminum or copper-clad aluminum grounding conductors shall not be used where in direct contact with masonry or the earth or where subject to corrosive conditions. Where used outside, aluminum or copper-clad aluminum grounding conductors shall not be terminated within 450 mm (18 in.) of the earth.

(B) Securing and Protection from Physical Damage. A grounding electrode conductor or its enclosure shall be securely fastened to the surface on which it is carried. A 4 AWG copper or aluminum or larger conductor shall be protected if exposed to severe physical damage. A 6 AWG grounding conductor that is free from exposure to physical damage shall be permitted to be run along the surface of the building construction without metal covering or protection where it is securely fastened to the construction; otherwise, it shall be in rigid metal conduit, intermediate metal conduit, rigid nonmetallic conduit, electrical metallic tubing, or cable armor. Grounding conductors smaller than 6 AWG shall be in rigid metal conduit, intermediate metal conduit, rigid nonmetallic conduit, electrical metallic tubing, or cable armor.

See 250.64(E) for additional information on situations in which raceways enclose the grounding electrode conductor. Also see the commentary following 250.92(A)(3) and the illustration that accompanies that commentary, Exhibit 250.32, for installation requirements for metal raceways used to install and physically protect the grounding electrode conductor(s).

(C) Continuous. The grounding electrode conductor shall be installed in one continuous length without a splice or joint, unless spliced only by irreversible compression-type connectors listed for the purpose or by the exothermic welding process.

Exception: Sections of busbars shall be permitted to be connected together to form a grounding electrode conductor.

Although infrequent, it may be necessary to splice the grounding electrode conductor because of remodeling of the building or to add equipment. Section 250.64(C) permits splicing the grounding electrode conductor by irreversible compression-type fittings or exothermic welding. These methods create a permanent connection that will not become loose after installation.

For the 2002 *Code*, an exception was added to 250.64(C) to recognize bolted connections joining busbar sections as suitable connections.

(D) Grounding Electrode Conductor Taps. Where a service consists of more than a single enclosure as permitted

in 230.40, Exception No. 2, it shall be permitted to connect taps to the grounding electrode conductor. Each such tap conductor shall extend to the inside of each such enclosure. The grounding electrode conductor shall be sized in accordance with 250.66, but the tap conductors shall be permitted to be sized in accordance with the grounding electrode conductors specified in 250.66 for the largest conductor serving the respective enclosures. The tap conductors shall be connected to the grounding electrode conductor in such a manner that the grounding electrode conductor remains without a splice.

Grounding electrode (tap) conductors must be sized using Table 250.66 and are based on the size of the largest phase conductors serving each enclosure. The grounding electrode conductor is determined by the size of the largest service-entrance conductor or equivalent cross-sectional area for parallel conductors, per Table 250.66. As illustrated in Exhibit 250.26, the tap method eliminates the difficulties found in looping grounding electrode conductors from one enclosure to another. The 2 AWG grounding electrode conductor shown in Exhibit 250.26 is required to be installed without a splice or joint, except as permitted in 250.64(C).

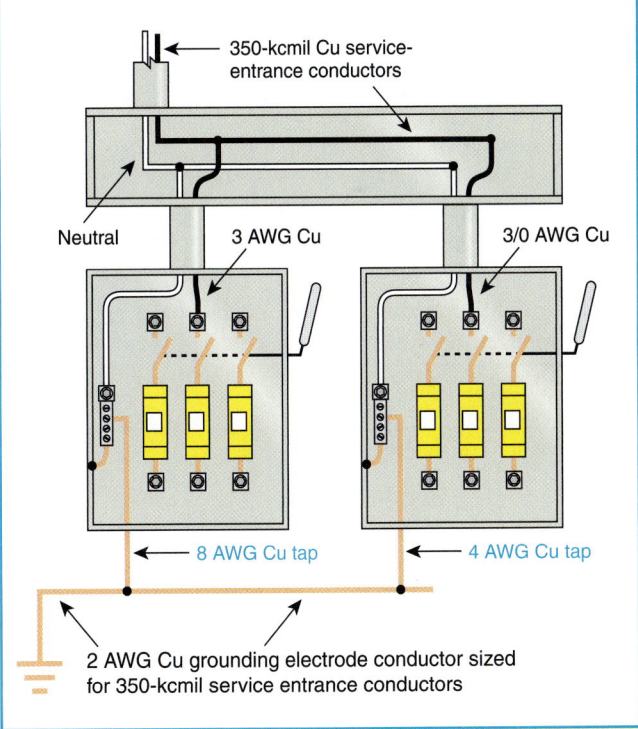

Exhibit 250.26 The tap method of connecting grounding electrode conductors from one enclosure to another.

(E) Enclosures for Grounding Electrode Conductors. Metal enclosures for grounding electrode conductors shall be electrically continuous from the point of attachment to cabinets or equipment to the grounding electrode and shall be securely fastened to the ground clamp or fitting. Metal enclosures that are not physically continuous from cabinet or equipment to the grounding electrode shall be made electrically continuous by bonding each end to the grounding electrode conductor. Where a raceway is used as protection for a grounding electrode conductor, the installation shall comply with the requirements of the appropriate raceway article.

Bonding jumpers installed to ensure the electrical continuity of metal enclosures must be sized in accordance with 250.102(C). Exhibit 250.32, which appears in the commentary following 250.92(A)(3), shows the bonding of a metal raceway to a grounding electrode conductor at both ends to ensure that the raceway and conductor are in parallel.

(F) To Electrode(s). A grounding electrode conductor shall be permitted to be run to any convenient grounding electrode available in the grounding electrode system or to one or more grounding electrode(s) individually. The grounding electrode conductor shall be sized for the largest grounding electrode conductor required among all the electrodes connected to it.

Exhibit 250.27 shows an example of a grounding electrode system. The location of the grounding electrode conductor is "to any convenient grounding electrode available."

250.66 Size of Alternating-Current Grounding Electrode Conductor.

The size of the grounding electrode conductor of a grounded or ungrounded ac system shall not be less than given in Table 250.66, except as permitted in 250.66(A) through (C).

FPN: See 250.24(B) for size of ac system conductor brought to service equipment.

Example

Apply the sizing requirements in Table 250.66 to Exhibit 250.28 to determine the size of grounding electrode conductor.

Solution

STEP 1. Using Table 8 in Chapter 9, calculate the total circular mil area of both grounded service conductors.

$$3 \text{ AWG} = 52,620 \text{ circular mils}$$
$$3/0 \text{ AWG} = \underline{167,800} \text{ circular mils}$$
$$\text{Total area} = 220,420 \text{ circular mils}$$

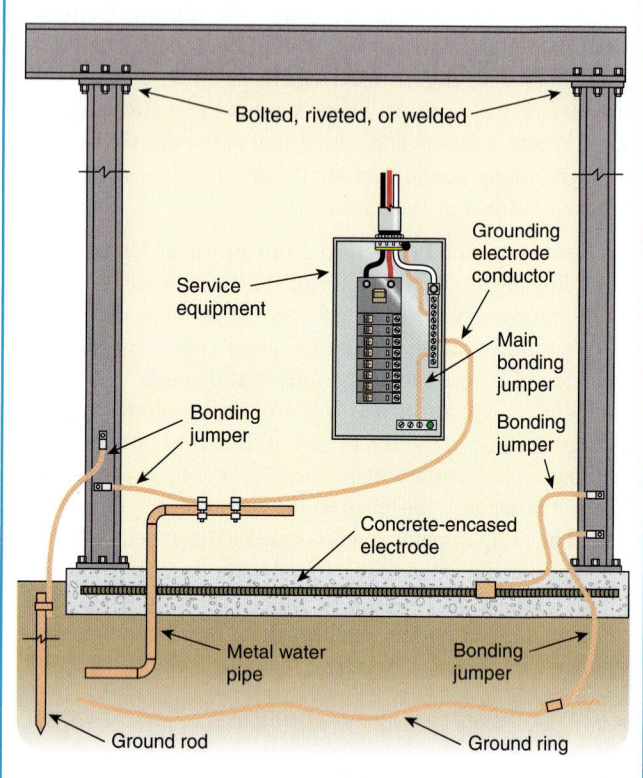

Exhibit 250.27 An example of running the grounding electrode conductor to any convenient electrode available as well as bonding electrodes together to form the grounding electrode system required by 250.50.

From Table 8, the next larger standard size is 250 kcmil.

STEP 2. Use Table 250.66 to size the grounding electrode conductor. According to the fourth row, "Over 3/0 through 350," the size should be 2 AWG.

Note that the size of the tap to the grounding electrode conductor from the disconnect on the left in Exhibit 250.28 is based on the size of the service-entrance conductors supplying the enclosure.

(A) Connections to Rod, Pipe, or Plate Electrodes. Where the grounding electrode conductor is connected to rod, pipe, or plate electrodes as permitted in 250.52(A)(5) or 250.52(A)(6), that portion of the conductor that is the sole connection to the grounding electrode shall not be required to be larger than 6 AWG copper wire or 4 AWG aluminum wire.

(B) Connections to Concrete-Encased Electrodes. Where the grounding electrode conductor is connected to a concrete-encased electrode as permitted in 250.52(A)(3), that portion of the conductor that is the sole connection to

Table 250.66 Grounding Electrode Conductor for Alternating-Current Systems

Size of Largest Ungrounded Service-Entrance Conductor or Equivalent Area for Parallel Conductors[a] (AWG/kcmil)		Size of Grounding Electrode Conductor (AWG/kcmil)	
Copper	Aluminum or Copper-Clad Aluminum	Copper	Aluminum or Copper-Clad Aluminum[b]
2 or smaller	1/0 or smaller	8	6
1 or 1/0	2/0 or 3/0	6	4
2/0 or 3/0	4/0 or 250	4	2
Over 3/0 through 350	Over 250 through 500	2	1/0
Over 350 through 600	Over 500 through 900	1/0	3/0
Over 600 through 1100	Over 900 through 1750	2/0	4/0
Over 1100	Over 1750	3/0	250

Notes:
1. Where multiple sets of service-entrance conductors are used as permitted in 230.40, Exception No. 2, the equivalent size of the largest service-entrance conductor shall be determined by the largest sum of the areas of the corresponding conductors of each set.
2. Where there are no service-entrance conductors, the grounding electrode conductor size shall be determined by the equivalent size of the largest service-entrance conductor required for the load to be served.
[a]This table also applies to the derived conductors of separately derived ac systems.
[b]See installation restrictions in 250.64(A).

the grounding electrode shall not be required to be larger than 4 AWG copper wire.

(C) Connections to Ground Rings. Where the grounding electrode conductor is connected to a ground ring as permitted in 250.52(A)(4), that portion of the conductor that is the sole connection to the grounding electrode shall not be required to be larger than the conductor used for the ground ring.

As illustrated in Exhibit 250.29, if a grounding electrode conductor is run from the service equipment or separately derived system to a water pipe or structural metal building member and from that point to one of the electrodes mentioned in 250.66(A), that portion of the grounding electrode between the service equipment or separately derived system and the water pipe or structural metal building member must be a full-size conductor, per Table 250.66. If the grounding electrode conductor from the service equipment was run, for example, to the ground rod first and then to the water pipe, the conductor to the ground rod would also have to be full size, per Table 250.66. Note that Exhibit 250.29 is

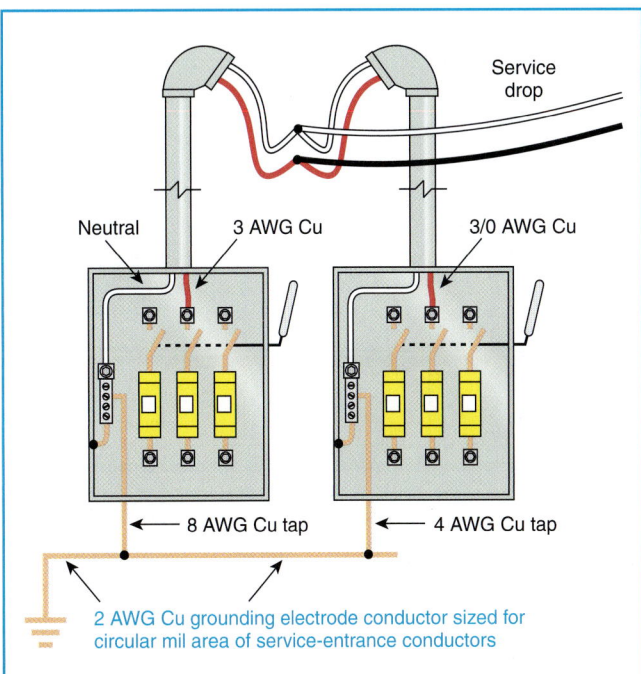

Exhibit 250.28 A grounding electrode conductor with multiple sets of service conductors, sized according to Table 250.66, Note 1.

not intended to show the physical routing and connection of the bonding jumpers.

250.68 Grounding Electrode Conductor and Bonding Jumper Connection to Grounding Electrodes.

(A) Accessibility. The connection of a grounding electrode conductor or bonding jumper to a grounding electrode shall be accessible.

Exception: An encased or buried connection to a concrete-encased, driven, or buried grounding electrode shall not be required to be accessible.

If the exposed portion of an encased, driven, or buried electrode is used for the termination of a grounding electrode conductor, the terminations must be accessible. However, if the connection is buried or encased, terminations are not required to be accessible. Ground clamps and other connectors suitable for use where buried in earth or embedded in concrete must be listed for such use, either by a marking on the connector or by a tag attached to the connector. See Exhibits 250.22 and 250.24 for illustrations of encased and buried electrodes.

(B) Effective Grounding Path. The connection of a grounding electrode conductor or bonding jumper to a

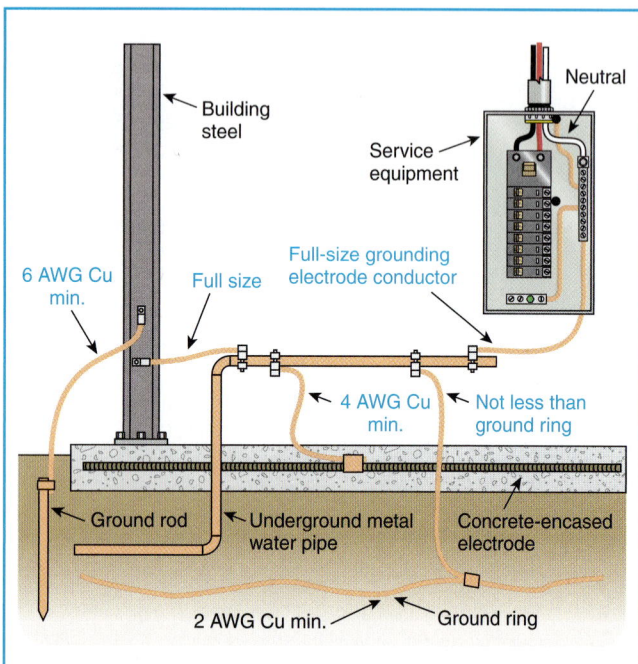

grounding electrode shall be made in a manner that will ensure a permanent and effective grounding path. Where necessary to ensure the grounding path for a metal piping system used as a grounding electrode, effective bonding shall be provided around insulated joints and around any equipment likely to be disconnected for repairs or replacement. Bonding conductors shall be of sufficient length to permit removal of such equipment while retaining the integrity of the bond.

Examples of equipment likely to be disconnected for repairs or replacement are water meters and water filter systems.

250.70 Methods of Grounding and Bonding Conductor Connection to Electrodes.

The grounding or bonding conductor shall be connected to the grounding electrode by exothermic welding, listed lugs, listed pressure connectors, listed clamps, or other listed means. Connections depending on solder shall not be used. Ground clamps shall be listed for the materials of the grounding electrode and the grounding electrode conductor and, where used on pipe, rod, or other buried electrodes, shall also be listed for direct soil burial or concrete encasement. Not more than one conductor shall be connected to the grounding electrode by a single clamp or fitting unless the clamp or fitting is listed for multiple conductors. One of the following methods shall be used:

(1) A pipe fitting, pipe plug, or other approved device screwed into a pipe or pipe fitting
(2) A listed bolted clamp of cast bronze or brass, or plain or malleable iron
(3) For indoor telecommunications purposes only, a listed sheet metal strap-type ground clamp having a rigid metal base that seats on the electrode and having a strap of such material and dimensions that it is not likely to stretch during or after installation
(4) An equally substantial approved means

Where a ground clamp is used and it terminates on a galvanized water pipe, for example, the clamp must be of a material that is compatible with steel so as to prevent galvanic corrosion. The same type of compatibility requirement applies to ground clamps on copper water pipe.

Exhibit 250.30 shows a listed ground clamp generally used with 8 AWG through 4 AWG grounding electrode conductors. Exothermic weld kits acceptable for this purpose are commercially available.

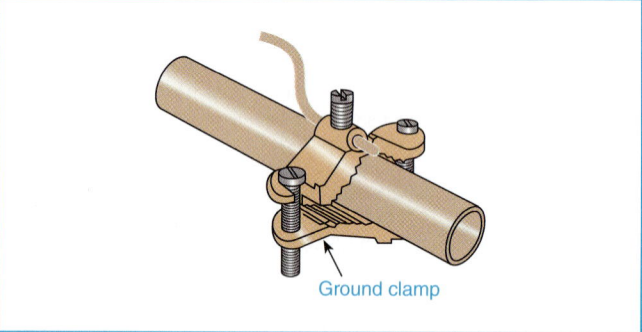

Exhibit 250.30 An application of a listed ground clamp.

Exhibit 250.31 shows a listed U-bolt ground clamp. These clamps are available for all pipe sizes and all grounding electrode conductor sizes. Where grounding electrode conductors are run in conduit, conduit hubs may be bolted to the threaded portion of the U-bolt.

IV. Enclosure, Raceway, and Service Cable Grounding

250.80 Service Raceways and Enclosures.

Metal enclosures and raceways for service conductors and equipment shall be grounded.

Exception: A metal elbow that is installed in an underground installation of rigid nonmetallic conduit and is isolated from possible contact by a minimum cover of 450 mm (18 in.) to any part of the elbow shall not be required to be grounded.

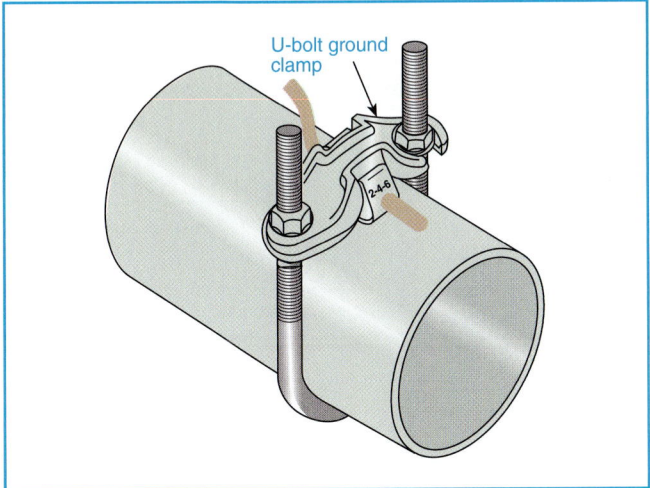

U-bolt ground clamp

Exhibit 250.31 An application of a listed U-bolt ground clamp.

The exception to 250.80 recognizes that metal sweep elbows are often installed in underground installations of rigid non-metallic conduit. The metal elbows are installed because nonmetallic elbows can be damaged by friction from the pulling ropes used during conductor installation. The elbows are isolated from physical contact by burial so that no part of the elbow is less than 18 in. below grade.

250.84 Underground Service Cable or Conduit.

(A) Underground Service Cable. The sheath or armor of a continuous underground metal-sheathed service cable system that is metallically connected to the underground system shall not be required to be grounded at the building. The sheath or armor shall be permitted to be insulated from the interior conduit or piping.

(B) Underground Service Conduit Containing Cable. An underground service conduit that contains a metal-sheathed cable bonded to the underground system shall not be required to be grounded at the building. The sheath or armor shall be permitted to be insulated from the interior conduit or piping.

250.86 Other Conductor Enclosures and Raceways.

Except as permitted by 250.112(I), metal enclosures and raceways for other than service conductors shall be grounded.

Section 250.86 requires grounding, bonding, and ensured electrical continuity of all enclosures and metal raceways. Connectors, couplings, or other similar fittings that perform

mechanical and electrical functions must ensure bonding and grounding continuity between the fitting, the metal raceway, and the enclosure. Metal enclosures must be grounded so that when a fault occurs between an ungrounded (hot) conductor and ground, the potential difference between the non–current-carrying parts of the electrical installation is minimized, thereby reducing the risk of shock.

Exception No. 1: Metal enclosures and raceways for conductors added to existing installations of open wire, knob and tube wiring, and nonmetallic-sheathed cable shall not be required to be grounded where these enclosures or wiring methods

(a) *Do not provide an equipment ground;*
(b) *Are in runs of less than 7.5 m (25 ft);*
(c) *Are free from probable contact with ground, grounded metal, metal lath, or other conductive material; and*
(d) *Are guarded against contact by persons.*

Exception No. 2: Short sections of metal enclosures or raceways used to provide support or protection of cable assemblies from physical damage shall not be required to be grounded.

Exception No. 3: A metal elbow shall not be required to be grounded where it is installed in a nonmetallic raceway and is isolated from possible contact by a minimum cover of 450 mm (18 in.) to any part of the elbow or is encased in not less than 50 mm (2 in.) of concrete.

V. Bonding

250.90 General.

Bonding shall be provided where necessary to ensure electrical continuity and the capacity to conduct safely any fault current likely to be imposed.

250.92 Services.

(A) Bonding of Services. The non–current-carrying metal parts of equipment indicated in 250.92(A)(1), (2), and (3) shall be effectively bonded together.

(1) The service raceways, cable trays, cablebus framework, auxiliary gutters, or service cable armor or sheath except as permitted in 250.84.
(2) All service enclosures containing service conductors, including meter fittings, boxes, or the like, interposed in the service raceway or armor.
(3) Any metallic raceway or armor enclosing a grounding electrode conductor as specified in 250.64(B). Bonding shall apply at each end and to all intervening raceways, boxes, and enclosures between the service equipment and the grounding electrode.

Section 250.92(A)(3) is intended to clarify that where metal raceways, boxes, or enclosures contain a grounding electrode conductor, both ends of the raceway, box, or enclosure must be bonded to the grounding electrode conductor, as illustrated in Exhibit 250.32. Bonding the raceway to the conductor reduces the impedance and minimizes the potential difference between the electrical equipment and ground. See also 250.64(E) and 250.102(A).

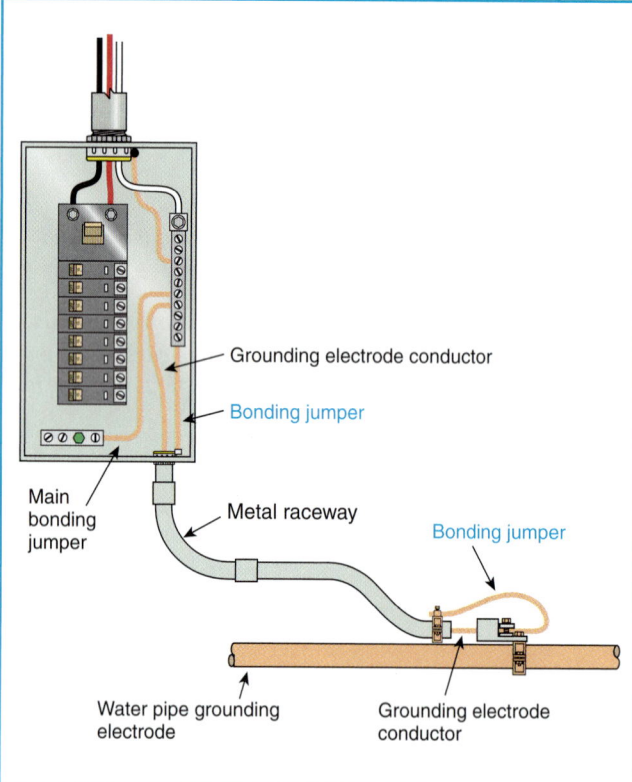

Exhibit 250.32 Bonding of a metal raceway that contains a grounding electrode conductor to the conductor at both ends, as required by 250.64(E).

(B) Method of Bonding at the Service. Electrical continuity at service equipment, service raceways, and service conductor enclosures shall be ensured by one of the following methods:

(1) Bonding equipment to the grounded service conductor in a manner provided in 250.8

Exhibit 250.33 illustrates grounding and bonding at an individual service. Exhibit 250.34 illustrates a grounding and bonding arrangement for up to six switches (three are shown) that serve as the service disconnecting means for an individual service. Section 250.24(B) clarifies that the grounded

service conductor must be run to each service disconnect and be bonded to the enclosure. Section 250.92(B)(1) permits the bonding of service equipment to be accomplished by bonding to the grounded service conductor.

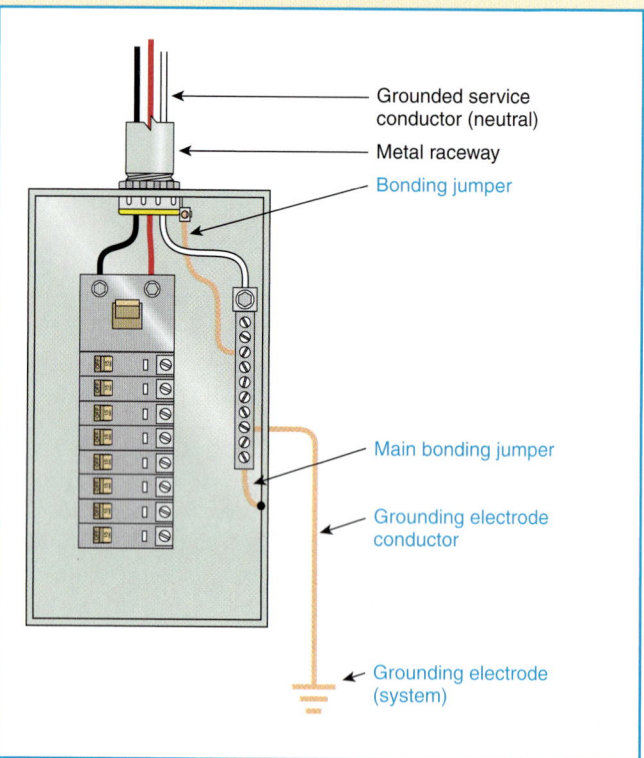

Exhibit 250.33 Grounding and bonding for a service with one disconnecting means.

(2) Connections utilizing threaded couplings or threaded bosses on enclosures where made up wrenchtight
(3) Threadless couplings and connectors where made up tight for metal raceways and metal-clad cables
(4) Other approved devices, such as bonding-type locknuts and bushings

Note that method (4) requires other similar devices, such as bonding-type locknuts or bushings. Standard locknuts or sealing locknuts are not acceptable as the "sole means" for bonding on the line side of service equipment.

Grounding and bonding bushings for use with rigid or intermediate metal conduit are provided with means (usually one or more set screws that make positive contact with the conduit) for reliably bonding the bushing and the conduit on which it is threaded to the metal equipment enclosure or box.

Grounding-and-bonding-type bushings used with rigid or intermediate metal conduit, such as those shown in Exhib-

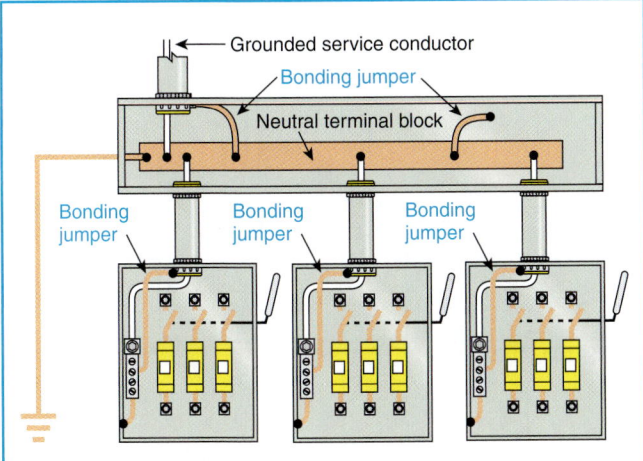

Exhibit 250.34 A grounding and bonding arrangement for up to six switches that serve as the service disconnecting means for an individual service.

its 250.35 and 250.36, have provisions for connecting a bonding jumper or have means provided by the manufacturer for use in mounting a wire connector. This type of bushing may also have means (usually one or more set screws) to reliably bond the bushing to the conduit. Exhibit 250.37 shows a bonding-type wedge lug used to connect a conduit to a box.

Bonding jumpers meeting the other requirements of this article shall be used around concentric or eccentric knock-outs that are punched or otherwise formed so as to impair the electrical connection to ground. Standard locknuts or bushings shall not be the sole means for the bonding required by this section.

For an example of concentric and eccentric knockouts, see the commentary following the definition of *bonding jumper* in Article 100 and Exhibit 100.3.

250.94 Bonding for Other Systems.

An accessible means external to enclosures for connecting intersystem bonding and grounding conductors shall be provided at the service equipment and at the disconnecting means for any additional buildings or structures by at least one of the following means:

(1) Exposed nonflexible metallic raceways
(2) Exposed grounding electrode conductor
(3) Approved means for the external connection of a copper or other corrosion-resistant bonding or grounding conductor to the grounded raceway or equipment

FPN No. 1: A 6 AWG copper conductor with one end bonded to the grounded nonflexible metallic raceway or

Exhibit 250.35 Grounding-and-bonding bushings used to connect a copper bonding or grounding wire to conduits. (Courtesy of Thomas & Betts Corp.)

Exhibit 250.36 A threaded grounding bushing with set screws used to ensure electrical and mechanical connection. (Courtesy of Thomas & Betts Corp.)

equipment and with 150 mm (6 in.) or more of the other end made accessible on the outside wall is an example of the approved means covered in 250.94(3).

Other accessible external means for intersystem bonding that comply with 250.94, FPN No. 1, are illustrated in Exhibit 250.38. On the left is an illustration of accessible means for the connection. The illustration on the right shows a method of providing the required bonding means when the panelboard is a flush type.

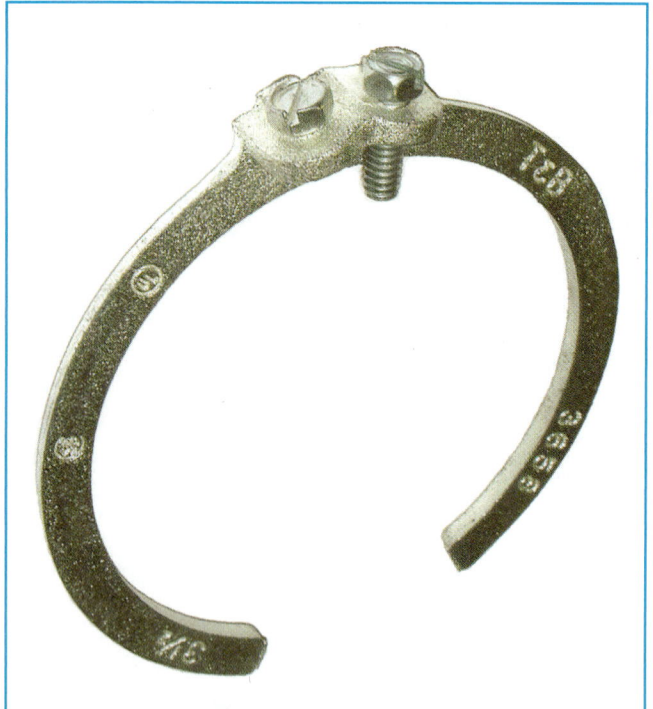

Exhibit 250.37 A grounding wedge lug used to provide an electrical connection between a conduit and a box. (Courtesy of Thomas & Betts Corp.)

FPN No. 2: See 800.40, 810.21, and 820.40 for bonding and grounding requirements for communications circuits, radio and television equipment, and CATV circuits.

An external accessible bonding means is equally important for separate buildings and mobile homes. In these occupancies, the disconnecting means enclosure on the load side of the service can be considered the equivalent of the service equipment for the purpose of intersystem bonding.

The *Code* requires that separate systems be bonded together to reduce the differences of potential between them due to lightning or accidental contact with power lines. Lightning protection systems, communications, radio and TV, and CATV systems must be bonded together to minimize the potential differences between the systems. Lack of interconnection can result in a severe shock and fire hazard.

The reason for this potential hazard is illustrated in Exhibit 250.39, which shows a CATV cable with its jacket grounded to a separate ground rod and not bonded to the power ground. The cable is connected to the cable decoder and the tuner of a television set. Also connected to the decoder and television is the 120-volt supply, with one conductor grounded at the service (the power ground). In each case, resistance to ground will be present at the grounding electrode. This resistance to ground varies widely, depending on soil conditions and the type of grounding electrode. The

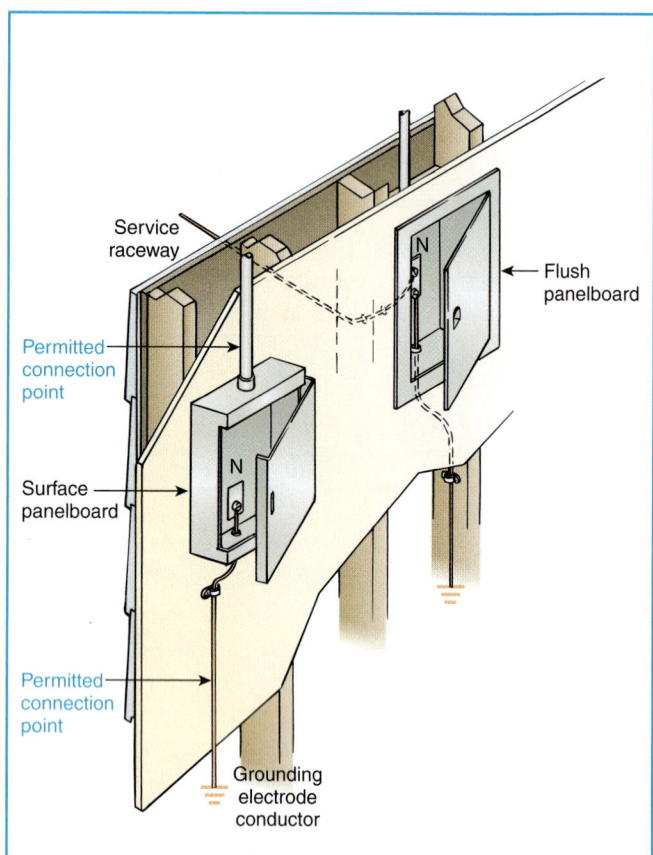

Exhibit 250.38 Examples of accessible external means for intersystem bonding, as required by 250.94.

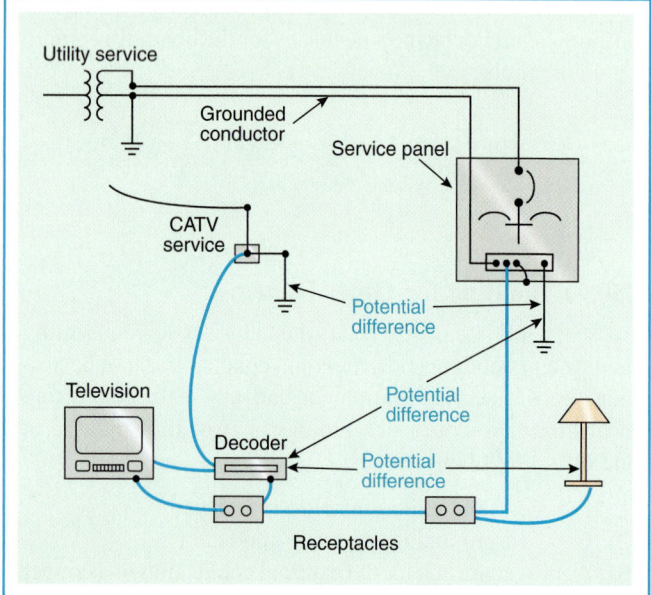

Exhibit 250.39 A CATV installation that does not comply with the *Code* that illustrates why bonding between different systems is necessary.

resistance at the CATV ground is likely to be higher than the power ground resistance, because the power ground is often an underground metal water piping system or concrete-encased electrode, whereas the CATV ground is commonly a ground rod.

For example, for the cable TV installation shown in Exhibit 250.39, assume that a current is induced in the power line by a switching surge or a nearby lightning strike, so that a momentary current of 1000 amperes flows over the power line to the power-line ground. This amount of current is not unusual under such circumstances. The amount could be, and often is, considerably higher. Also assume that the power ground has a resistance of 10 ohms, a *very* low value in most circumstances (a single ground rod in average soil has a resistance to ground in the neighborhood of 40 ohms).

According to Ohm's law, the current flowing through the equipment connected to electrical system will be raised momentarily to a potential of 10,000 volts (1000 amperes × 10 ohms). This potential of 10,000 volts would exist between the CATV system and the electrical system and between the grounded conductor within the CATV cable and the grounded surfaces in the walls of the home, such as water pipes (which are connected to the power ground), over which the cable runs. This potential could also appear across a person with one hand on the CATV cable and the other hand on a metal surface connected to the power ground (e.g., a radiator or refrigerator).

Actual voltage is likely to be many times the 10,000 volts calculated, because extremely low (below normal) values were assumed for both resistance to ground and current. Most insulation systems, however, are not designed to withstand even 10,000 volts. Even if the insulation system does withstand a 10,000-volt surge, it is likely to be damaged and breakdown of the insulation system will result in sparking.

The same situation would exist if the current surge were on the CATV cable or on a telephone line. The only difference would be the voltage involved, which would depend on the individual resistance to ground of the grounding electrodes.

The solution is to bond the two grounding electrode systems together, as shown in Exhibit 250.40 or to connect the CATV cable jacket to the power ground, which is exactly what the *Code* requires. When one system is raised above ground potential, the second system rises to the same potential, and no voltage exists between the two grounding systems.

These bonding rules are provided to address the difficulties communications and CATV installers encounter in complying with *Code* grounding and bonding requirements. These difficulties arise from the increasing use of plastic for water pipe, fittings, water meters, and service conduit. In the past, bonding between communications, CATV, and power systems was usually achieved by connecting the communications protector grounds or cable shield to an interior metallic

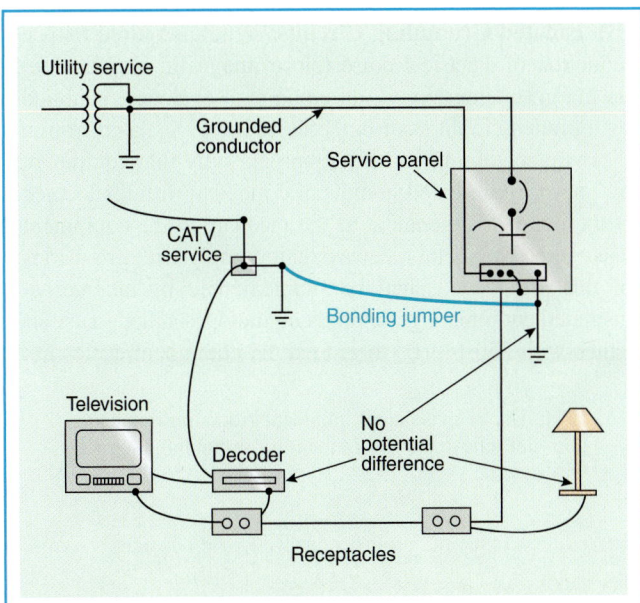

Exhibit 250.40 A cable TV installation that complies with 250.94.

water pipe, because the pipe was often used as the power grounding electrode. Thus, the requirement that the power, communications, CATV cable shield, and metallic water piping systems be bonded together was easily satisfied. If the power was grounded to one of the other electrodes permitted by the *Code,* usually by a made electrode such as a ground rod, the bond was connected to the power grounding electrode conductor or to a metallic service raceway, since at least one of these was usually accessible. With the proliferation of plastic water pipe and the increasing tendency for service equipment (often flush-mounted) to be installed in finished areas, where the grounding electrode conductor is often concealed, as well as the increased use of plastic service-entrance conduit, communications and CATV installers no longer have access to a point for connecting bonding jumpers or grounding conductors. See Exhibit 250.39 and also the commentary following 820.40(D), FPN No. 2.

250.96 Bonding Other Enclosures.

(A) General. Metal raceways, cable trays, cable armor, cable sheath, enclosures, frames, fittings, and other metal non–current-carrying parts that are to serve as grounding conductors, with or without the use of supplementary equipment grounding conductors, shall be effectively bonded where necessary to ensure electrical continuity and the capacity to conduct safely any fault current likely to be imposed on them. Any nonconductive paint, enamel, or similar coating shall be removed at threads, contact points, and contact surfaces or be connected by means of fittings designed so as to make such removal unnecessary.

(B) Isolated Grounding Circuits. Where required for the reduction of electrical noise (electromagnetic interference) on the grounding circuit, an equipment enclosure supplied by a branch circuit shall be permitted to be isolated from a raceway containing circuits supplying only that equipment by one or more listed nonmetallic raceway fittings located at the point of attachment of the raceway to the equipment enclosure. The metal raceway shall comply with provisions of this article and shall be supplemented by an internal insulated equipment grounding conductor installed in accordance with 250.146(D) to ground the equipment enclosure.

FPN: Use of an isolated equipment grounding conductor does not relieve the requirement for grounding the raceway system.

To reduce electromagnetic interference, 250.96(B) permits electronic equipment to be isolated from the raceway in a manner similar to that for cord-and-plug-connected equipment. Section 250.96(B) specifies that a metal equipment enclosure supplied by a branch circuit is the subject of the requirement and that subsequent wiring, raceways, or other equipment beyond the insulating fitting is not permitted.

Exhibits 250.41 and 250.42 show examples of installations. In Exhibit 250.41, note that the metal raceway is grounded in the usual manner, by attachment to the grounded service enclosure, satisfying the concern mentioned in the fine print note. In Exhibit 250.42, note that 408.20, Exception, permits the isolated equipment grounding conductor (which is required to be insulated) to pass through the subpanel.

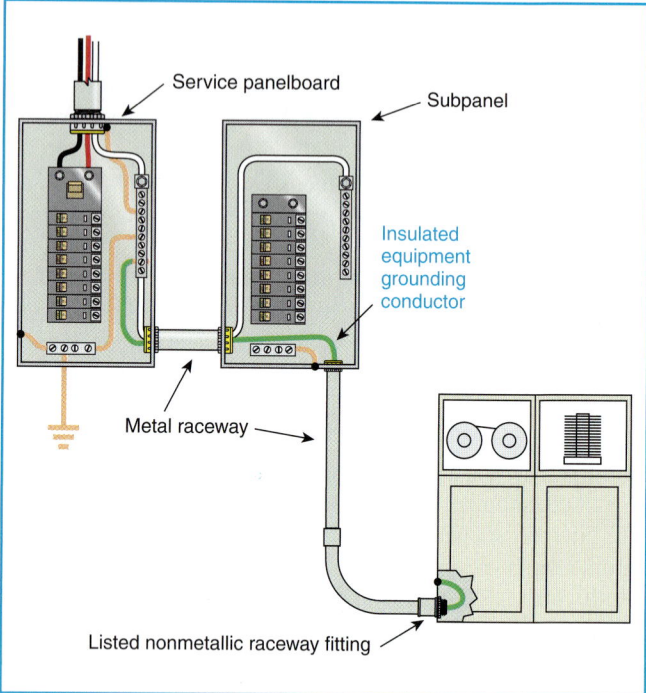

Exhibit 250.42 An installation in which the insulated equipment grounding conductor is allowed to pass through the subpanel without connecting to the grounding bus to terminate at the service grounding bus.

250.97 Bonding for Over 250 Volts.

For circuits of over 250 volts to ground, the electrical continuity of metal raceways and cables with metal sheaths that contain any conductor other than service conductors shall be ensured by one or more of the methods specified for services in 250.92(B), except for (1).

Exception: Where oversized, concentric, or eccentric knockouts are not encountered, or where a box or enclosure with concentric or eccentric knockouts is listed for the purpose, the following methods shall be permitted:

(a) *Threadless couplings and connectors for cables with metal sheaths*

(b) *Two locknuts, on rigid metal conduit or intermediate metal conduit, one inside and one outside of boxes and cabinets*

(c) *Fittings with shoulders that seat firmly against the box or cabinet, such as electrical metallic tubing connectors, flexible metal conduit connectors, and cable connectors, with one locknut on the inside of boxes and cabinets*

(d) *Listed fittings that are identified for the purpose*

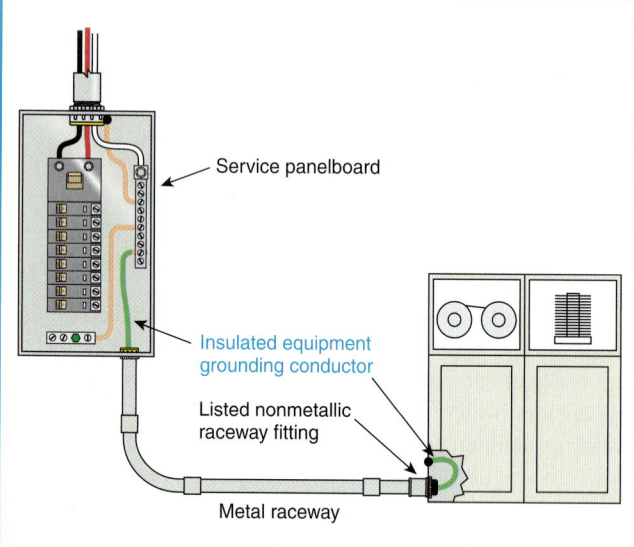

Exhibit 250.41 An installation in which the electronic equipment is grounded through the isolated grounding conductor.

Bonding around prepunched concentric or eccentric knockouts is not required if the enclosure containing the knockouts has been tested and is listed as suitable for bonding.

The methods in (a), (b), and (c) of the exception to 250.97 are permitted for circuits over 250 volts to ground only where there are no oversize, concentric, or eccentric knockouts. Note that method (c) permits fittings, such as EMT connectors, cable connectors, and similar fittings with shoulders that seat firmly against the metal of a box or cabinet, to be installed with only one locknut on the inside of the box.

250.98 Bonding Loosely Jointed Metal Raceways.

Expansion fittings and telescoping sections of metal raceways shall be made electrically continuous by equipment bonding jumpers or other means.

250.100 Bonding in Hazardous (Classified) Locations.

Regardless of the voltage of the electrical system, the electrical continuity of non–current-carrying metal parts of equipment, raceways, and other enclosures in any hazardous (classified) location as defined in Article 500 shall be ensured by any of the methods specified for services in 250.92(B) that are approved for the wiring method used.

250.102 Equipment Bonding Jumpers.

(A) Material. Equipment bonding jumpers shall be of copper or other corrosion-resistant material. A bonding jumper shall be a wire, bus, screw, or similar suitable conductor.

(B) Attachment. Equipment bonding jumpers shall be attached in the manner specified by the applicable provisions of 250.8 for circuits and equipment and by 250.70 for grounding electrodes.

(C) Size—Equipment Bonding Jumper on Supply Side of Service. The bonding jumper shall not be smaller than the sizes shown in Table 250.66 for grounding electrode conductors. Where the service-entrance phase conductors are larger than 1100 kcmil copper or 1750 kcmil aluminum, the bonding jumper shall have an area not less than 12½ percent of the area of the largest phase conductor except that, where the phase conductors and the bonding jumper are of different materials (copper or aluminum), the minimum size of the bonding jumper shall be based on the assumed use of phase conductors of the same material as the bonding jumper and with an ampacity equivalent to that of the installed phase conductors. Where the service-entrance conductors are paralleled in two or more raceways or cables, the equipment bonding jumper, where routed with the raceways or cables, shall be run in parallel. The size of the bonding jumper for each raceway or cable shall be based on the size of the service-entrance conductors in each raceway or cable.

The bonding jumper for a grounding electrode conductor raceway or cable armor as covered in 250.64(E) shall be the same size or larger than the required enclosed grounding electrode conductor.

(D) Size—Equipment Bonding Jumper on Load Side of Service. The equipment bonding jumper on the load side of the service overcurrent devices shall be sized, as a minimum, in accordance with the sizes listed in Table 250.122, but shall not be required to be larger than the largest ungrounded circuit conductors supplying the equipment and shall not be smaller than 14 AWG.

A single common continuous equipment bonding jumper shall be permitted to bond two or more raceways or cables where the bonding jumper is sized in accordance with Table 250.122 for the largest overcurrent device supplying circuits therein.

(E) Installation. The equipment bonding jumper shall be permitted to be installed inside or outside of a raceway or enclosure. Where installed on the outside, the length of the equipment bonding jumper shall not exceed 1.8 m (6 ft) and shall be routed with the raceway or enclosure. Where installed inside of a raceway, the equipment bonding jumper shall comply with the requirements of 250.119 and 250.148.

Exception: An equipment bonding jumper longer than 1.8 m (6 ft) shall be permitted at outside pole locations for the purpose of bonding or grounding isolated sections of metal raceways or elbows installed in exposed risers of metal conduit or other metal raceway.

In many applications, equipment bonding jumpers must be installed on the outside of metal raceways and enclosures. For example, it would be impractical to install the bonding jumper for a conduit expansion joint on the inside of the conduit. For some metal raceway and rigid conduit systems and conduit systems in hazardous (classified) locations, installing the bonding jumper where it is visible and accessible for inspection and maintenance is desirable. An external bonding jumper has a higher impedance than an internal bonding jumper, but by limiting the length of the bonding jumper to 6 ft and routing it with the raceway, the increase in the impedance of the equipment grounding circuit is insignificant. Exhibit 250.43 illustrates a bonding jumper run outside a length of flexible metal conduit. Because the function of a bonding jumper is readily apparent, color identification is not necessary.

250.104 Bonding of Piping Systems and Exposed Structural Steel.

(A) Metal Water Piping. The metal water piping system shall be bonded as required in (1), (2), (3), or (4) of this

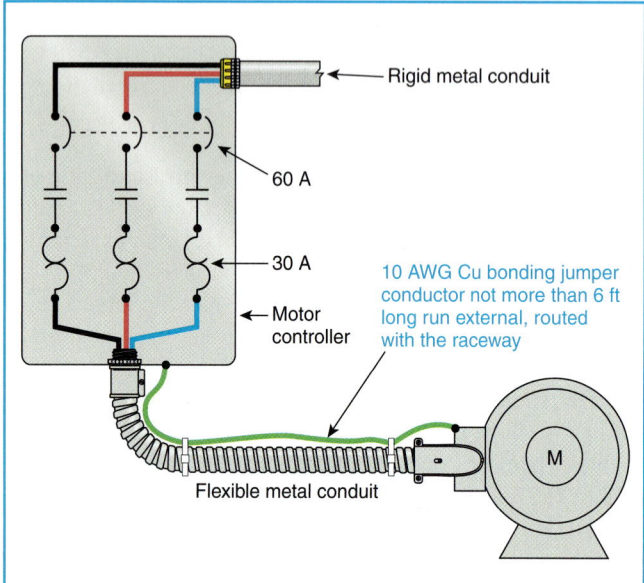

Rigid metal conduit

60 A

30 A

Motor
controller

10 AWG Cu bonding jumper
conductor not more than 6 ft
long run external, routed
with the raceway

Flexible metal conduit

M

Exhibit 250.43 A bonding jumper around the outside a flexible metal conduit.

section. The bonding jumper(s) shall be installed in accordance with 250.64(A), (B), and (E). The points of attachment of the bonding jumper(s) shall be accessible.

(1) General. Metal water piping system(s) installed in or attached to a building or structure shall be bonded to the service equipment enclosure, the grounded conductor at the service, the grounding electrode conductor where of sufficient size, or to the one or more grounding electrodes used. The bonding jumper(s) shall be sized in accordance with Table 250.66 except as permitted in 250.104(A)(2) and (A)(3).

Bonding the interior metal water piping system is not the same as using the metal water piping system as a grounding electrode. Bonding to the grounding electrode system places the bonded components at the same voltage level. For example, a current of 2000 amperes across 25 ft of 6 AWG copper conductor produces a voltage differential of approximately 26 volts. Section 250.104(A)(1) requires the interior metal water piping system and any other metal piping systems likely to become energized to be bonded to the service equipment or grounding electrode conductor.

If it cannot reasonably be concluded that the hot and cold water pipes are reliably interconnected, an electrical bonding jumper is required to ensure that this connection is made. Some judgment must be exercised for each installation. The special installation requirements provided in 250.64(A), (B), and (E) also apply to the water piping bonding jumper.

(2) Buildings of Multiple Occupancy. In buildings of multiple occupancy where the metal water piping system(s) installed in or attached to a building or structure for the individual occupancies is metallically isolated from all other occupancies by use of nonmetallic water piping, the metal water piping system(s) for each occupancy shall be permitted to be bonded to the equipment grounding terminal of the panelboard or switchboard enclosure (other than service equipment) supplying that occupancy. The bonding jumper shall be sized in accordance with Table 250.122.

The intent of 250.104(A)(2) is to recognize that the increased use of nonmetallic water piping mains causes the interior metal piping system of a multiple-occupancy building to be isolated from ground. Therefore, the water pipe is permitted to be bonded to the panelboard that serves only that particular occupancy. The bonding jumper, in this case, is permitted to be sized according to Table 250.122, based on the size of the main overcurrent device supplying the occupancy.

(3) Multiple Buildings or Structures Supplied from a Common Service. The metal water piping system(s) installed in or attached to a building or structure shall be bonded to the building or structure disconnecting means enclosure where located at the building or structure, to the equipment grounding conductor run with the supply conductors, or to the one or more grounding electrodes used. The bonding jumper(s) shall be sized in accordance with 250.66, based on the size of the feeder or branch circuit conductors that supply the building. The bonding jumper shall not be required to be larger than the largest ungrounded feeder or branch circuit conductor supplying the building.

(4) Separately Derived Systems. The grounded conductor of each separately derived system shall be bonded to the nearest available point of the interior metal water piping system(s) in the area served by each separately derived system. This connection shall be made at the same point on the separately derived system where the grounding electrode conductor is connected. Each bonding jumper shall be sized in accordance with Table 250.66.

Exception: A separate water piping bonding jumper shall not be required where the effectively grounded metal frame of a building or structure is used as the grounding electrode for a separately derived system and is bonded to the metallic water piping in the area served by the separately derived system.

Section 250.104(A)(4) requires that where a separately derived system supplies the power, the interior metal piping system in the area must be bonded to the grounded conductor at the point nearest the derived system, and this connection must be accessible.

(B) Other Metal Piping. Where installed in or attached to a building or structure, metal piping system(s), including gas piping, that may become energized shall be bonded to the service equipment enclosure, the grounded conductor at the service, the grounding electrode conductor where of sufficient size, or to the one or more grounding electrodes used. The bonding jumper(s) shall be sized in accordance with 250.122 using the rating of the circuit that may energize the piping system(s). The equipment grounding conductor for the circuit that may energize the piping shall be permitted to serve as the bonding means. The points of attachment of the bonding jumper(s) shall be accessible.

> FPN: Bonding all piping and metal air ducts within the premises will provide additional safety.

Section 250.104(B) was revised for the 2002 *Code* to state that gas piping is treated exactly the same as all "other metal piping" systems within a building.

(C) Structural Steel. Exposed structural steel that is interconnected to form a steel building frame and is not intentionally grounded and may become energized shall be bonded to the service equipment enclosure, the grounded conductor at the service, the grounding electrode conductor where of sufficient size, or the one or more grounding electrodes used. The bonding jumper(s) shall be sized in accordance with Table 250.66 and installed in accordance with 250.64(A), (B), and (E). The points of attachment of the bonding jumper(s) shall be accessible.

Section 250.104(C) requires exposed metal building framework that is not intentionally or inherently grounded to be bonded to the service equipment or grounding electrode system.

250.106 Lightning Protection Systems.

The lightning protection system ground terminals shall be bonded to the building or structure grounding electrode system.

> FPN No. 1: See 250.60 for use of air terminals. For further information, see NFPA 780-1997, *Standard for the Installation of Lightning Protection Systems*, which contains detailed information on grounding, bonding, and spacing from lightning protection systems.

> FPN No. 2: Metal raceways, enclosures, frames, and other non–current-carrying metal parts of electric equipment installed on a building equipped with a lightning protection system may require bonding or spacing from the lightning protection conductors in accordance with NFPA 780-1997, *Standard for the Installation of Lightning Protection Systems*. Separation from lightning protection conductors is typically 1.8 m (6 ft) through air or 900 mm (3 ft) through dense materials such as concrete, brick, or wood.

Section 250.106 specifies that the grounding electrode system of the lightning protection system be bonded to the electrical service grounding electrode system, as shown in Exhibit 250.44. A similar requirement is found in 3-14 of NFPA 780, *Standard for the Installation of Lightning Protection Systems*. Additional bonding between the lightning protection system and the electrical system may be necessary based on proximity and whether separation between the systems is through air or building materials.

Fine Print Note No. 2 references NFPA 780 for guidance on determining the need for additional bonding connections. Section 3-21.2 of NFPA 780 includes a method for calculating flashover distances.

Exposed, non–current-carrying metal parts of fixed equipment that are not likely to become energized are not required to be grounded. These parts include some metal nameplates on nonmetallic enclosures and small parts, such as bolts and screws, if they are located so that they are not likely to become energized.

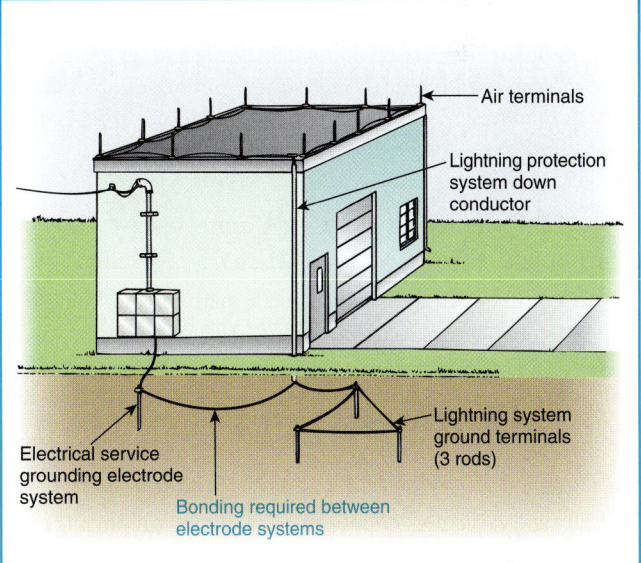

Exhibit 250.44 Bonding between the lightning system ground terminals and the electrical service grounding electrode system, in accordance with 250.106.

VI. Equipment Grounding and Equipment Grounding Conductors

250.110 Equipment Fastened in Place or Connected by Permanent Wiring Methods (Fixed).

Exposed non–current-carrying metal parts of fixed equipment likely to become energized shall be grounded under any of the following conditions:

(1) Where within 2.5 m (8 ft) vertically or 1.5 m (5 ft) horizontally of ground or grounded metal objects and subject to contact by persons
(2) Where located in a wet or damp location and not isolated
(3) Where in electrical contact with metal
(4) Where in a hazardous (classified) location as covered by Articles 500 through 517
(5) Where supplied by a metal-clad, metal-sheathed, metal-raceway, or other wiring method that provides an equipment ground, except as permitted by 250.86, Exception No. 2, for short sections of metal enclosures
(6) Where equipment operates with any terminal at over 150 volts to ground

Exception No. 1: Metal frames of electrically heated appliances, exempted by special permission, in which case the frames shall be permanently and effectively insulated from ground.

Exception No. 2: Distribution apparatus, such as transformer and capacitor cases, mounted on wooden poles, at a height exceeding 2.5 m (8 ft) above ground or grade level.

Exception No. 3: Listed equipment protected by a system of double insulation, or its equivalent, shall not be required to be grounded. Where such a system is employed, the equipment shall be distinctively marked.

250.112 Fastened in Place or Connected by Permanent Wiring Methods (Fixed)—Specific.

Exposed, non–current-carrying metal parts of the kinds of equipment described in 250.112(A) through (K), and non–current-carrying metal parts of equipment and enclosures described in 250.112(L) and (M), shall be grounded regardless of voltage.

(A) Switchboard Frames and Structures. Switchboard frames and structures supporting switching equipment, except frames of 2-wire dc switchboards where effectively insulated from ground.

Section 250.112(A) clarifies that dc switchboards insulated from ground are not required to be grounded.

(B) Pipe Organs. Generator and motor frames in an electrically operated pipe organ, unless effectively insulated from ground and the motor driving it.

(C) Motor Frames. Motor frames, as provided by 430.142.

(D) Enclosures for Motor Controllers. Enclosures for motor controllers unless attached to ungrounded portable equipment.

(E) Elevators and Cranes. Electric equipment for elevators and cranes.

(F) Garages, Theaters, and Motion Picture Studios. Electric equipment in commercial garages, theaters, and motion picture studios, except pendant lampholders supplied by circuits not over 150 volts to ground.

(G) Electric Signs. Electric signs, outline lighting, and associated equipment as provided in Article 600.

(H) Motion Picture Projection Equipment. Motion picture projection equipment.

(I) Power-Limited Remote-Control, Signaling, and Fire Alarm Circuits. Equipment supplied by Class 1 power-limited circuits and Class 1, Class 2, and Class 3 remote-control and signaling circuits, and by fire alarm circuits, shall be grounded where system grounding is required by Part II or Part VIII of this article.

(J) Luminaires (Lighting Fixtures). Luminaires (lighting fixtures) as provided in Part V of Article 410.

(K) Skid Mounted Equipment. Permanently mounted electrical equipment and skids shall be grounded with an equipment bonding jumper sized as required by 250.122.

(L) Motor-Operated Water Pumps. Motor-operated water pumps, including the submersible type.

The requirement of 250.112(L) is intended to reduce stray voltages and minimize shock hazard during maintenance, when the pump is hauled out of the well casing and might be tested in a barrel of water.

(M) Metal Well Casings. Where a submersible pump is used in a metal well casing, the well casing shall be bonded to the pump circuit equipment grounding conductor.

Section 250.112(M) is intended to prevent a shock hazard that could exist due to a potential difference between the pump, which is grounded to the system ground, and the metal well casing.

250.114 Equipment Connected by Cord and Plug.

Under any of the conditions described in (1) through (4), exposed non–current-carrying metal parts of cord-and-plug-connected equipment likely to become energized shall be grounded.

Exception: Listed tools, listed appliances, and listed equipment covered in (2) through (4) shall not be required to be grounded where protected by a system of double insulation or its equivalent. Double insulated equipment shall be distinctively marked.

The exception to 250.114 recognizes listed double-insulated appliances, motor-operated hand-held tools, stationary and fixed motor-operated tools, and light industrial motor-operated tools as not requiring equipment grounding connections.

(1) In hazardous (classified) locations (see Articles 500 through 517)

(2) Where operated at over 150 volts to ground

Exception No. 1: Motors, where guarded, shall not be required to be grounded.

Exception No. 2: Metal frames of electrically heated appliances, exempted by special permission, shall not be required to be grounded, in which case the frames shall be permanently and effectively insulated from ground.

(3) In residential occupancies:

a. Refrigerators, freezers, and air conditioners

b. Clothes-washing, clothes-drying, dish-washing machines; kitchen waste disposers; information technology equipment; sump pumps and electrical aquarium equipment

c. Hand-held motor-operated tools, stationary and fixed motor-operated tools, light industrial motor-operated tools

d. Motor-operated appliances of the following types: hedge clippers, lawn mowers, snow blowers, and wet scrubbers

e. Portable handlamps

(4) In other than residential occupancies:

a. Refrigerators, freezers, and air conditioners

b. Clothes-washing, clothes-drying, dish-washing machines; information technology equipment; sump pumps and electrical aquarium equipment

c. Hand-held motor-operated tools, stationary and fixed motor-operated tools, light industrial motor-operated tools

d. Motor-operated appliances of the following types: hedge clippers, lawn mowers, snow blowers, and wet scrubbers

e. Portable handlamps

f. Cord-and-plug-connected appliances used in damp or wet locations or by persons standing on the ground or on metal floors or working inside of metal tanks or boilers

g. Tools likely to be used in wet or conductive locations

Exception: Tools and portable handlamps likely to be used in wet or conductive locations shall not be required to be grounded where supplied through an isolating transformer with an ungrounded secondary of not over 50 volts.

Tools must be grounded by an equipment grounding conductor within the cord or cable supplying the tool, except where the tool is supplied by an isolating transformer, as permitted by the exception following 250.114(4). Portable tools and appliances protected by an approved system of double insulation must be listed by a qualified electrical testing laboratory as being suitable for the purpose, and the equipment must be distinctively marked as double insulated.

Cord-connected portable tools or appliances are not intended to be used in damp, wet, or conductive locations unless they are grounded, supplied by an isolation transformer with a secondary of not more than 50 volts, or protected by an approved system of double insulation.

Exhibit 250.45 shows an example of lighting equipment supplied through an isolating transformer operating at 6 or 12 volts that provides safe illumination for work inside boilers, tanks, and similar locations that may be metal or wet.

Exhibit 250.45 Lighting equipment supplied through an isolating transformer operating at 6 or 12 volts and therefore not required to be grounded. (Courtesy of Daniel Woodhead Co.)

250.116 Nonelectric Equipment.

The metal parts of nonelectric equipment described in this section shall be grounded.

(1) Frames and tracks of electrically operated cranes and hoists

(2) Frames of nonelectrically driven elevator cars to which electric conductors are attached

(3) Hand-operated metal shifting ropes or cables of electric elevators

> FPN: Where extensive metal in or on buildings may become energized and is subject to personal contact, adequate bonding and grounding will provide additional safety.

Because metal siding on buildings is not electrical equipment, it is outside the scope of the *Code* [see 90.2(A)]. Therefore, the *Code* cannot require that it be grounded. Quite often, however, luminaires, signs, or receptacles are installed on buildings with metal siding that could become energized. Grounding of metal siding reduces the risk of shock to persons who may come in contact with the siding.

250.118 Types of Equipment Grounding Conductors.

The equipment grounding conductor run with or enclosing the circuit conductors shall be one or more or a combination of the following:

(1) A copper, aluminum, or copper-clad aluminum conductor. This conductor shall be solid or stranded; insulated, covered, or bare; and in the form of a wire or a busbar of any shape.

(2) Rigid metal conduit.

(3) Intermediate metal conduit.

(4) Electrical metallic tubing.

(5) Flexible metal conduit where both the conduit and fittings are listed for grounding.

(6) Listed flexible metal conduit that is not listed for grounding, meeting all the following conditions:

   a. The conduit is terminated in fittings listed for grounding.

   b. The circuit conductors contained in the conduit are protected by overcurrent devices rated at 20 amperes or less.

   c. The combined length of flexible metal conduit and flexible metallic tubing and liquidtight flexible metal conduit in the same ground return path does not exceed 1.8 m (6 ft).

   d. The conduit is not installed for flexibility.

(7) Listed liquidtight flexible metal conduit meeting all the following conditions:

   a. The conduit is terminated in fittings listed for grounding.

   b. For metric designators 12 through 16 (trade sizes ⅜ through ½), the circuit conductors contained in the conduit are protected by overcurrent devices rated at 20 amperes or less.

   c. For metric designators 21 through 35 (trade sizes ¾ through 1¼), the circuit conductors contained in the conduit are protected by overcurrent devices rated not more than 60 amperes and there is no flexible metal conduit, flexible metallic tubing, or liquidtight flexible metal conduit in trade sizes metric designators 12 through 16 (trade sizes ⅜ through ½) in the grounding path.

   d. The combined length of flexible metal conduit and flexible metallic tubing and liquidtight flexible metal conduit in the same ground return path does not exceed 1.8 m (6 ft).

   e. The conduit is not installed for flexibility.

(8) Flexible metallic tubing where the tubing is terminated in fittings listed for grounding and meeting the following conditions:

   a. The circuit conductors contained in the tubing are protected by overcurrent devices rated at 20 amperes or less.

   b. The combined length of flexible metal conduit and flexible metallic tubing and liquidtight flexible metal conduit in the same ground return path does not exceed 1.8 m (6 ft).

(9) Armor of Type AC cable as provided in 320.108.

(10) The copper sheath of mineral-insulated, metal-sheathed cable.

(11) Type MC cable where listed and identified for grounding in accordance with the following:

   a. The combined metallic sheath and grounding conductor of interlocked metal tape–type MC cable

   b. The metallic sheath or the combined metallic sheath and grounding conductors of the smooth or corrugated tube type MC cable

(12) Cable trays as permitted in 392.3(C) and 392.7.

(13) Cablebus framework as permitted in 370.3.

(14) Other electrically continuous metal raceways and auxiliary gutters listed for grounding.

Exhibit 250.46 illustrates the various sizes of metal conduits that enclose the circuit conductors where the metal conduits are used as equipment grounding conductors.

250.119 Identification of Equipment Grounding Conductors.

Unless required elsewhere in this *Code,* equipment grounding conductors shall be permitted to be bare, covered, or

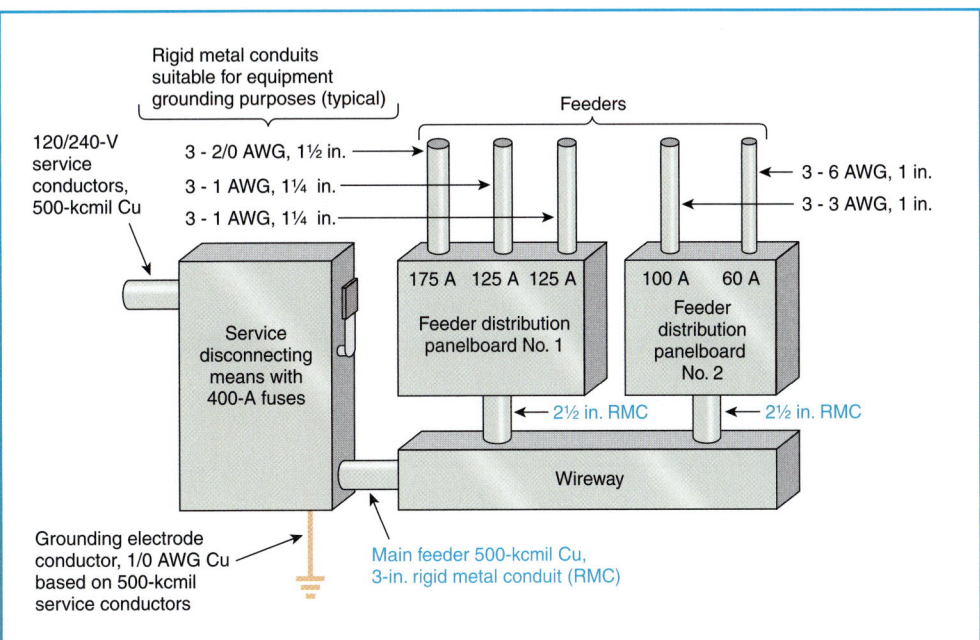

Exhibit 250.46 Various sizes of enclosing metal conduits used as equipment grounding conductors, as they apply to a service and feeder system.

insulated. Individually covered or insulated equipment grounding conductors shall have a continuous outer finish that is either green or green with one or more yellow stripes except as permitted in this section.

(A) Conductors Larger Than 6 AWG. An insulated or covered conductor larger than 6 AWG copper or aluminum shall be permitted, at the time of installation, to be permanently identified as an equipment grounding conductor at each end and at every point where the conductor is accessible. Identification shall encircle the conductor and shall be accomplished by one of the following:

(1) Stripping the insulation or covering from the entire exposed length
(2) Coloring the exposed insulation or covering green
(3) Marking the exposed insulation or covering with green tape or green adhesive labels

(B) Multiconductor Cable. Where the conditions of maintenance and supervision ensure that only qualified persons service the installation, one or more insulated conductors in a multiconductor cable, at the time of installation, shall be permitted to be permanently identified as equipment grounding conductors at each end and at every point where the conductors are accessible by one of the following means:

(1) Stripping the insulation from the entire exposed length
(2) Coloring the exposed insulation green
(3) Marking the exposed insulation with green tape or green adhesive labels

(C) Flexible Cord. An uninsulated equipment grounding conductor shall be permitted, but, if individually covered, the covering shall have a continuous outer finish that is either green or green with one or more yellow stripes.

250.120 Equipment Grounding Conductor Installation.

An equipment grounding conductor shall be installed in accordance with 250.120(A), (B), and (C).

(A) Raceway, Cable Trays, Cable Armor, Cablebus, or Cable Sheaths. Where it consists of a raceway, cable tray, cable armor, cablebus framework, or cable sheath or where it is a wire within a raceway or cable, it shall be installed in accordance with the applicable provisions in this *Code* using fittings for joints and terminations approved for use with the type raceway or cable used. All connections, joints, and fittings shall be made tight using suitable tools.

(B) Aluminum and Copper-Clad Aluminum Conductors. Equipment grounding conductors of bare or insulated aluminum or copper-clad aluminum shall be permitted. Bare conductors shall not come in direct contact with masonry or the earth or where subject to corrosive conditions. Aluminum or copper-clad aluminum conductors shall not be terminated within 450 mm (18 in.) of the earth.

(C) Equipment Grounding Conductors Smaller Than 6 AWG. Equipment grounding conductors smaller than 6 AWG shall be protected from physical damage by a raceway or cable armor except where run in hollow spaces of walls or partitions, where not subject to physical damage, or where protected from physical damage.

250.122 Size of Equipment Grounding Conductors.

(A) General. Copper, aluminum, or copper-clad aluminum equipment grounding conductors of the wire type shall not be smaller than shown in Table 250.122 but shall not be required to be larger than the circuit conductors supplying the equipment. Where a raceway or a cable armor or sheath is used as the equipment grounding conductor, as provided in 250.118 and 250.134(A), it shall comply with 250.4(A)(5) or 250.4(B)(4).

Table 250.122 Minimum Size Equipment Grounding Conductors for Grounding Raceway and Equipment

Rating or Setting of Automatic Overcurrent Device in Circuit Ahead of Equipment, Conduit, etc., Not Exceeding (Amperes)	Size (AWG or kcmil)	
	Copper	Aluminum or Copper-Clad Aluminum*
15	14	12
20	12	10
30	10	8
40	10	8
60	10	8
100	8	6
200	6	4
300	4	2
400	3	1
500	2	1/0
600	1	2/0
800	1/0	3/0
1000	2/0	4/0
1200	3/0	250
1600	4/0	350
2000	250	400
2500	350	600
3000	400	600
4000	500	800
5000	700	1200
6000	800	1200

Note: Where necessary to comply with 250.4(A)(5) or 250.4(B)(4), the equipment grounding conductor shall be sized larger than given in this table.
*See installation restrictions in 250.120.

The last sentence of 250.122(A) alerts users that, if a long distance exists between a power source and utilization equipment, some of the wiring methods allowed for grounding purposes must be evaluated further.

(B) Increased in Size. Where ungrounded conductors are increased in size, equipment grounding conductors, where installed, shall be increased in size proportionately according to circular mil area of the ungrounded conductors.

Equipment grounding conductors on the load side of the service disconnecting means and overcurrent devices are sized based on the size of the feeder or branch circuit overcurrent devices ahead of them. If the ungrounded conductors are increased in size to compensate for voltage drop, the equipment grounding conductors must also be increased proportionally.

Example

A 240-volt, single-phase, 250-ampere load is supplied from a 300-ampere breaker located in a panelboard 500 ft away. The conductors are 250 kcmil copper, installed in rigid nonmetallic conduit, with a 4 AWG copper equipment grounding conductor. If the conductors are increased to 350 kcmil, to what size must the equipment grounding conductor be increased?

Solution

STEP 1. Calculate the size ratio of the new conductors to the existing conductors:

$$\text{Size ratio} = \frac{350{,}000 \text{ circular mils}}{250{,}000 \text{ circular mils}} = 1.4$$

STEP 2. Calculate the cross-sectional area of the new equipment grounding conductor. According to Chapter 9, Table 8, 4 AWG, the size of the existing grounding conductor has a cross-sectional area of 41,740 circular mils.

STEP 3. Determine the size of the new equipment grounding conductor. Again, referring to Chapter 9, Table 8, we find that 58,436 circular mils is larger than 3 AWG. The next larger size is 66,360 circular mils, which converts to a 2 AWG copper equipment grounding conductor.

(C) Multiple Circuits. Where a single equipment grounding conductor is run with multiple circuits in the same raceway or cable, it shall be sized for the largest overcurrent device protecting conductors in the raceway or cable.

A single equipment grounding conductor must be sized for the largest overcurrent device. It is not required to be sized for the composite of all the circuits in the raceway because it is not anticipated that all circuits will develop faults at the same time. For example, three 3-phase circuits in the same raceway, protected by overcurrent devices rated 30, 60, and 100 amperes, would require only one equipment grounding conductor, sized according to the largest overcurrent device (in this case, 100 amperes). Therefore, an 8 AWG copper or 6 AWG aluminum conductor or copper-clad aluminum conductor is required, according to Table 250.122.

(D) Motor Circuits. Where the overcurrent device consists of an instantaneous trip circuit breaker or a motor short-circuit protector, as allowed in 430.52, the equipment grounding conductor size shall be permitted to be based on the rating of the motor overload protective device but not less than the size shown in Table 250.122.

(E) Flexible Cord and Fixture Wire. Equipment grounding conductors that are part of flexible cords or used with fixture wires in accordance with 240.5 shall be not smaller than 18 AWG copper and not smaller than the circuit conductors.

(F) Conductors in Parallel. Where conductors are run in parallel in multiple raceways or cables as permitted in 310.4, the equipment grounding conductors, where used, shall be run in parallel in each raceway or cable. One of the methods in 250.122(F)(1) or (2) shall be used to ensure the equipment grounding conductors are protected.

(1) Each parallel equipment grounding conductor shall be sized on the basis of the ampere rating of the overcurrent device protecting the circuit conductors in the raceway or cable in accordance with Table 250.122.

A full-sized equipment grounding conductor is needed in each raceway of an array of raceways enclosing paralleled phase conductors to prevent overloading and possible burnout of the grounding conductor should a ground fault occur along one of the parallel branches. The installation conditions for paralleled conductors prescribed in 310.4 result in proportional distribution of the current-time duty among the several paralleled grounding conductors only for overcurrent conditions downstream of the paralleled array.

Exhibit 250.47 shows a parallel array of two nonmetallic conduits installed underground. For clarity, a one-line diagram with equipment grounding conductors is shown. A ground fault at the enclosure will cause the equipment grounding conductor in the top conduit to carry more than

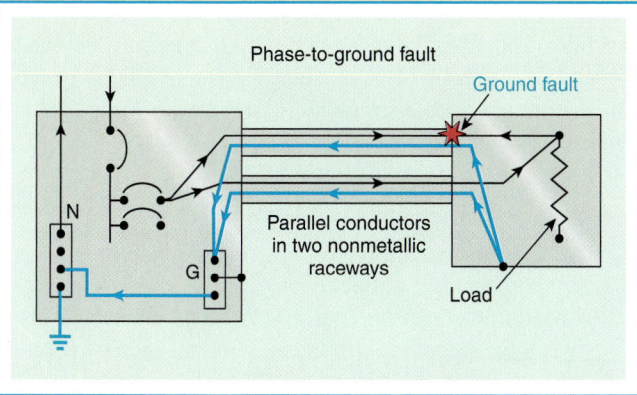

Phase-to-ground fault

Ground fault

Parallel conductors in two nonmetallic raceways

Load

Exhibit 250.47 Grounding paths for ground fault at the load supplied by parallel conductors in two nonmetallic raceways, illustrating the reason for the requirement of 250.122(F)(1).

its proportionate share of fault current. Note that the fault is fed by two different conductors of the same phase, one from the left and two from the right. The shortest and lowest-impedance path to ground from the fault to the supply panelboard is through the equipment grounding conductor in the top conduit. The grounding path from the fault through the bottom conduit is longer and of higher impedance. Therefore, the equipment grounding conductor in each raceway must be capable of carrying a major portion of the fault current without burning open.

(2) Where ground-fault protection of equipment is installed, each parallel equipment grounding conductor in a multiconductor cable shall be permitted to be sized in accordance with Table 250.122 on the basis of the trip rating of the ground-fault protection where the following conditions are met:

(1) Conditions of maintenance and supervision ensure that only qualified persons will service the installation.
(2) The ground-fault protection equipment is set to trip at not more than the ampacity of a single ungrounded conductor of one of the cables in parallel.
(3) The ground-fault protection is listed for the purpose.

Section 250.122(F)(2) applies to cables that are installed in parallel. Because cable assemblies are manufactured in standard conductor size configurations, the equipment grounding conductor in a cable is properly sized for some circuit arrangements. However, if the cable is used in large-capacity parallel circuits, the equipment grounding conductor in each cable may not be sized in accordance with Table 250.122. To address this problem, 250.122(F)(2) permits the sizing of the equipment grounding conductor within a multiconductor cable to be based on the trip setting of an equipment ground-fault device. This method of protection is only permitted where the installation will be serviced by qualified personnel and the ground-fault device is specifically listed for this purpose. The trip setting of the ground-fault protection of equipment (GFPE) cannot exceed the ampacity of a single ungrounded conductor installed in one of the parallel cables. This rule precludes the need to manufacture multiconductor cables with equipment grounding conductors that are sized for a specific parallel circuit configuration.

250.124 Equipment Grounding Conductor Continuity.

(A) Separable Connections. Separable connections such as those provided in drawout equipment or attachment plugs and mating connectors and receptacles shall provide for first-make, last-break of the equipment grounding conductor.

First-make, last-break shall not be required where inter-locked equipment, plugs, receptacles, and connectors preclude energization without grounding continuity.

(B) Switches. No automatic cutout or switch shall be placed in the equipment grounding conductor of a premises wiring system unless the opening of the cutout or switch disconnects all sources of energy.

250.126 Identification of Wiring Device Terminals.

The terminal for the connection of the equipment grounding conductor shall be identified by one of the following:

(1) A green, not readily removable terminal screw with a hexagonal head.
(2) A green, hexagonal, not readily removable terminal nut.
(3) A green pressure wire connector. If the terminal for the grounding conductor is not visible, the conductor entrance hole shall be marked with the word *green* or *ground*, the letters *G* or *GR* or the grounding symbol shown in Figure 250.126, or otherwise identified by a distinctive green color. If the terminal for the equipment grounding conductor is readily removable, the area adjacent to the terminal shall be similarly marked.

Figure 250.126 Grounding symbol.

VII. Methods of Equipment Grounding

250.130 Equipment Grounding Conductor Connections.

Equipment grounding conductor connections at the source of separately derived systems shall be made in accordance with 250.30(A)(1). Equipment grounding conductor connections at service equipment shall be made as indicated in 250.130(A) or (B). For replacement of non–grounding-type receptacles with grounding-type receptacles and for branch-circuit extensions only in existing installations that do not have an equipment grounding conductor in the branch circuit, connections shall be permitted as indicated in 250.130(C).

(A) For Grounded Systems. The connection shall be made by bonding the equipment grounding conductor to the grounded service conductor and the grounding electrode conductor.

The grounding arrangement for a grounded system is illustrated in Exhibit 250.48.

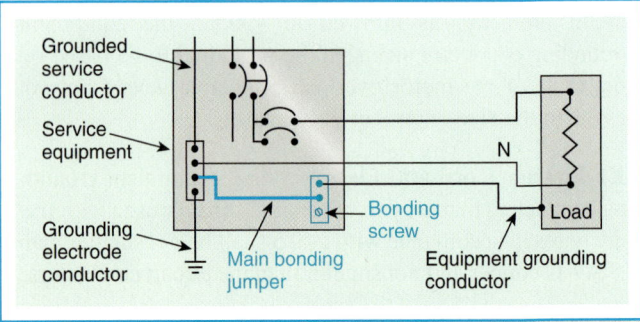

Exhibit 250.48 Grounding arrangement for grounded systems, per 250.130(A), illustrating connection of the equipment grounding conductor (bus) to the enclosures and the grounded service conductor.

(B) For Ungrounded Systems. The connection shall be made by bonding the equipment grounding conductor to the grounding electrode conductor.

(C) Nongrounding Receptacle Replacement or Branch Circuit Extensions. The equipment grounding conductor of a grounding-type receptacle or a branch-circuit extension shall be permitted to be connected to any of the following:

(1) Any accessible point on the grounding electrode system as described in 250.50
(2) Any accessible point on the grounding electrode conductor
(3) The equipment grounding terminal bar within the enclosure where the branch circuit for the receptacle or branch circuit originates
(4) For grounded systems, the grounded service conductor within the service equipment enclosure
(5) For ungrounded systems, the grounding terminal bar within the service equipment enclosure

FPN: See 406.3(D) for the use of a ground-fault circuit-interrupting type of receptacle.

Section 250.130(C) applies to both ungrounded and grounded systems. It permits a nongrounding-type receptacle to be replaced with a grounding-type receptacle under the following conditions.

1. The branch circuit does not contain an equipment ground.
2. An existing branch circuit is being extended for additional receptacle outlets.
3. An equipment grounding conductor is connected between the receptacle grounding terminal and any accessible point on the grounding electrode system.

Because of the requirements of 250.52(A)(1), an interior metal water pipe more than 5 ft from the point of entrance of the water pipe into the building is no longer allowed to serve as a connection to the grounding electrode conductor.

Exhibit 250.49 shows a branch-circuit extension made from an existing installation. This method is also permitted to ground a replacement 3-wire receptacle in the existing ungrounded box on the left, where no grounding conductor is available.

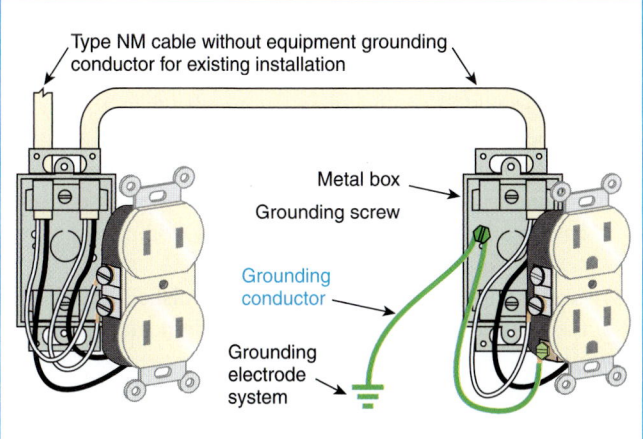

Exhibit 250.49 Branch-circuit extension to an existing installation, per 250.130(C), illustrating a separate grounding conductor connected to the main grounding electrode.

250.132 Short Sections of Raceway.

Isolated sections of metal raceway or cable armor, where required to be grounded, shall be grounded in accordance with 250.134.

250.134 Equipment Fastened in Place or Connected by Permanent Wiring Methods (Fixed)—Grounding.

Unless grounded by connection to the grounded circuit conductor as permitted by 250.32, 250.140, and 250.142, non–current-carrying metal parts of equipment, raceways, and other enclosures, if grounded, shall be grounded by one of the following methods.

Section 250.134 eliminates any conflict between 250.134(A), which requires an equipment grounding conductor to be used for equipment grounding, and 250.32, 250.140, and 250.142, which permit the grounded circuit conductor to be used for equipment grounding if certain specified conditions are met.

(A) Equipment Grounding Conductor Types. By any of the equipment grounding conductors permitted by 250.118.

(B) With Circuit Conductors. By an equipment grounding conductor contained within the same raceway, cable, or otherwise run with the circuit conductors.

One of the functions of an equipment grounding conductor is to provide a low-impedance ground-fault path between a ground fault and the electrical source. This path allows the overcurrent protective device to actuate, interrupting the current. To keep the impedance at a minimum, it is necessary to run the equipment grounding conductor within the same raceway or cable as the circuit conductor(s). This practice allows the magnetic field developed by the circuit conductor and the equipment grounding conductor to cancel, reducing their impedance.

Magnetic flux strength is inversely proportional to the square of the distance between the two conductors. By placing an equipment grounding conductor away from the conductor delivering the fault current, the magnetic flux cancellation decreases. This increases the impedance of the fault path and delays operation of the protective device.

Exception No. 1: As provided in 250.130(C), the equipment grounding conductor shall be permitted to be run separately from the circuit conductors.

Exception No. 1 to 250.134(B) permits an equipment grounding conductor to be run to the grounding electrode separately from the other conductors of the circuit. This practice applies only where a grounding-type receptacle is used on a circuit that does not include an equipment grounding conductor. See the commentary following 250.130(C) for further explanation.

Exception No. 2: For dc circuits, the equipment grounding conductor shall be permitted to be run separately from the circuit conductors.

> FPN No. 1: See 250.102 and 250.168 for equipment bonding jumper requirements.
>
> FPN No. 2: See 400.7 for use of cords for fixed equipment.

250.136 Equipment Considered Effectively Grounded.

Under the conditions specified in 250.136(A) and (B), the non–current-carrying metal parts of the equipment shall be considered effectively grounded.

(A) Equipment Secured to Grounded Metal Supports. Electrical equipment secured to and in electrical contact with a metal rack or structure provided for its support and grounded by one of the means indicated in 250.134. The structural metal frame of a building shall not be used as the required equipment grounding conductor for ac equipment.

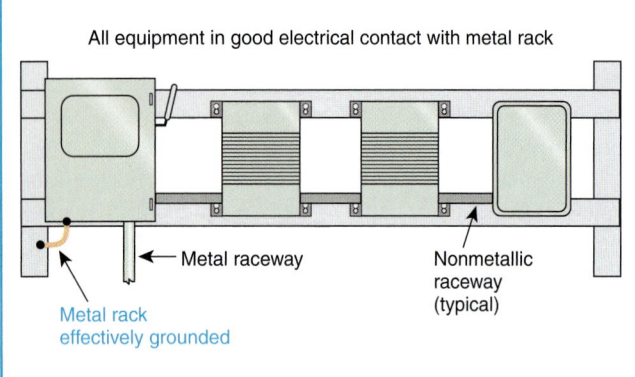

All equipment in good electrical contact with metal rack

Metal raceway

Nonmetallic raceway (typical)

Metal rack effectively grounded

Exhibit 250.50 An example of equipment grounding on a metal rack.

(B) Metal Car Frames. Metal car frames supported by metal hoisting cables attached to or running over metal sheaves or drums of elevator machines that are grounded by one of the methods indicated in 250.134.

250.138 Cord-and-Plug-Connected Equipment.

Non–current-carrying metal parts of cord-and-plug-connected equipment, if grounded, shall be grounded by one of the methods in 250.138(A) or (B).

(A) By Means of an Equipment Grounding Conductor. By means of an equipment grounding conductor run with the power supply conductors in a cable assembly or flexible cord properly terminated in a grounding-type attachment plug with one fixed grounding contact.

Exception: The grounding contacting pole of grounding-type plug-in ground-fault circuit interrupters shall be permitted to be of the movable, self-restoring type on circuits operating at not over 150 volts between any two conductors or over 150 volts between any conductor and ground.

(B) By Means of a Separate Flexible Wire or Strap. By means of a separate flexible wire or strap, insulated or bare, protected as well as practicable against physical damage, where part of equipment.

250.140 Frames of Ranges and Clothes Dryers.

This section shall apply to existing branch-circuit installations only. New branch-circuit installations shall comply

with 250.134 and 250.138. Frames of electric ranges, wall-mounted ovens, counter-mounted cooking units, clothes dryers, and outlet or junction boxes that are part of the circuit for these appliances shall be grounded in the manner specified by 250.134 or 250.138; or, except for mobile homes and recreational vehicles, shall be permitted to be grounded to the grounded circuit conductor if all the following conditions are met.

(1) The supply circuit is 120/240-volt, single-phase, 3-wire; or 208Y/120-volt derived from a 3-phase, 4-wire, wye-connected system.
(2) The grounded conductor is not smaller than 10 AWG copper or 8 AWG aluminum.
(3) The grounded conductor is insulated, or the grounded conductor is uninsulated and part of a Type SE service-entrance cable and the branch circuit originates at the service equipment.
(4) Grounding contacts of receptacles furnished as part of the equipment are bonded to the equipment.

equipment to avoid neutral current from downstream panelboards flowing on metal objects, such as pipes or ducts. Exhibit 250.51 shows an existing installation in which Type SE service-entrance cable was used for ranges, dryers, wall-mounted ovens, and counter-mounted cooking units. Junction boxes in the supply circuit were also permitted to be grounded from the grounded neutral conductor.

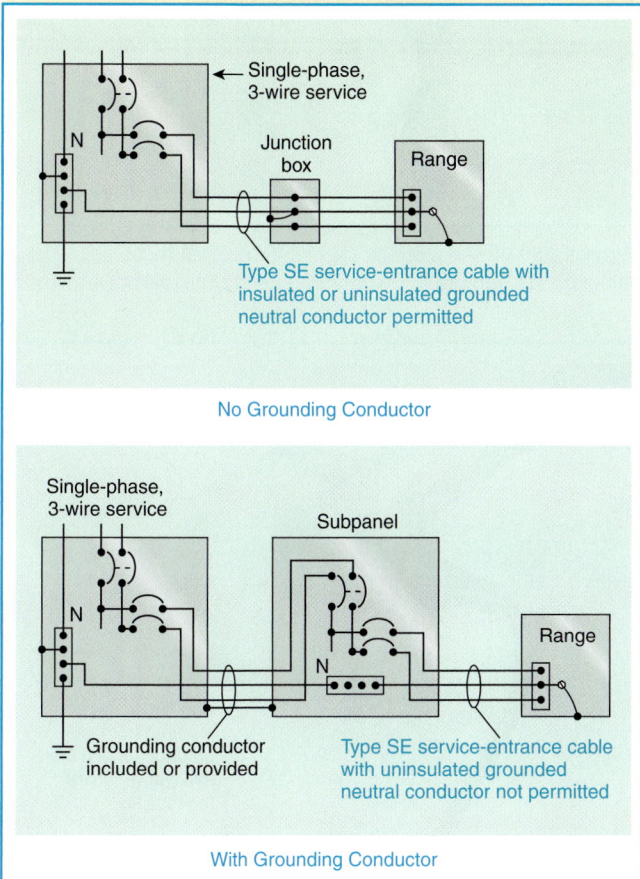

Exhibit 250.51 An existing installation in which the grounded conductor in Type SE service-entrance cable was used for grounding the appliances specified in 250.140.

250.142 Use of Grounded Circuit Conductor for Grounding Equipment.

(A) Supply-Side Equipment. A grounded circuit conductor shall be permitted to ground non–current-carrying metal parts of equipment, raceways, and other enclosures at any of the following locations:

(1) On the supply side or within the enclosure of the ac service-disconnecting means
(2) On the supply side or within the enclosure of the main disconnecting means for separate buildings as provided in 250.32(B)

(3) On the supply side or within the enclosure of the main disconnecting means or overcurrent devices of a separately derived system where permitted by 250.30(A)(1)

In separately derived systems, the grounded circuit conductor is permitted to ground non–current-carrying metal parts of equipment, raceways, and other enclosures only on the supply side of the main disconnecting means.

(B) Load-Side Equipment. Except as permitted in 250.30(A)(1) and 250.32(B), a grounded circuit conductor shall not be used for grounding non–current-carrying metal parts of equipment on the load side of the service disconnecting means or on the load side of a separately derived system disconnecting means or the overcurrent devices for a separately derived system not having a main disconnecting means.

Exception No. 1: The frames of ranges, wall-mounted ovens, counter-mounted cooking units, and clothes dryers under the conditions permitted for existing installations by 250.140 shall be permitted to be grounded by a grounded circuit conductor.

Exception No. 2: It shall be permissible to ground meter enclosures by connection to the grounded circuit conductor on the load side of the service disconnect if

(a) No service ground-fault protection is installed, and
(b) All meter enclosures are located near the service disconnecting means, and
(c) The size of the grounded circuit conductor is not smaller than the size specified in Table 250.122 for equipment grounding conductors.

Exception No. 3: Direct-current systems shall be permitted to be grounded on the load side of the disconnecting means or overcurrent device in accordance with 250.164.

Exception No. 4: Electrode-type boilers operating at over 600 volts shall be grounded as required in 490.72(E)(1) and 490.74.

One major reason the grounded circuit conductor is not permitted to be grounded on the load side of the service (except as allowed in 250.140) is that, should the grounded service conductor become disconnected at any point on the line side of the ground, the equipment grounding conductor and all conductive parts connected to it will carry the neutral current, raising the potential to ground of exposed metal parts not normally intended to carry current. This could result in arcing in concealed spaces and could pose a severe shock hazard, particularly if the path is inadvertently opened by a person servicing or repairing piping or ductwork. Even without an open grounded conductor (usually referred to as an open neutral), the equipment grounding conductor path will become a parallel path with the grounded conductor, and there will be some potential drop on exposed and concealed dead metal parts. The magnitude of this potential difference will be determined by the relative impedances of the equipment grounding path and the grounded conductor circuits. Not only would the equipment grounding conductor path be affected, but all parallel paths not intended as equipment grounding conductors would be affected as well. This could involve current flowing through metal building structures, piping, and ducts.

250.144 Multiple Circuit Connections.

Where equipment is required to be grounded and is supplied by separate connection to more than one circuit or grounded premises wiring system, a means for grounding shall be provided for each such connection as specified in 250.134 and 250.138.

250.146 Connecting Receptacle Grounding Terminal to Box.

An equipment bonding jumper shall be used to connect the grounding terminal of a grounding-type receptacle to a grounded box unless grounded as in 250.146(A) through (D).

(A) Surface Mounted Box. Where the box is mounted on the surface, direct metal-to-metal contact between the device yoke and the box shall be permitted to ground the receptacle to the box. This provision shall not apply to cover-mounted receptacles unless the box and cover combination are listed as providing satisfactory ground continuity between the box and the receptacle.

The main rule of 250.146 requires an equipment bonding jumper to be installed between the device box and the receptacle grounding terminal. However, 250.146(A) permits the equipment bonding jumper to be omitted if the metal yoke of the device is in direct contact with the metal device box ears, as illustrated in Exhibit 250.52.

Cover-mounted wiring devices, such as on 4-in. square covers, are not considered grounded. Section 250.146(A) does not apply to cover-mounted receptacles, such as the one illustrated in Exhibit 250.53. Box-cover and device combinations listed as providing grounding continuity are permitted.

(B) Contact Devices or Yokes. Contact devices or yokes designed and listed for the purpose shall be permitted in conjunction with the supporting screws to establish the

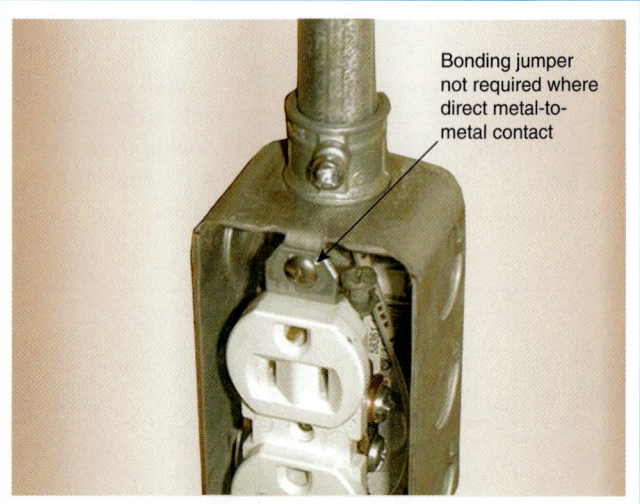

Exhibit 250.52 An example of a box-mounted receptacle attached to a surface box where a bonding jumper is not required.

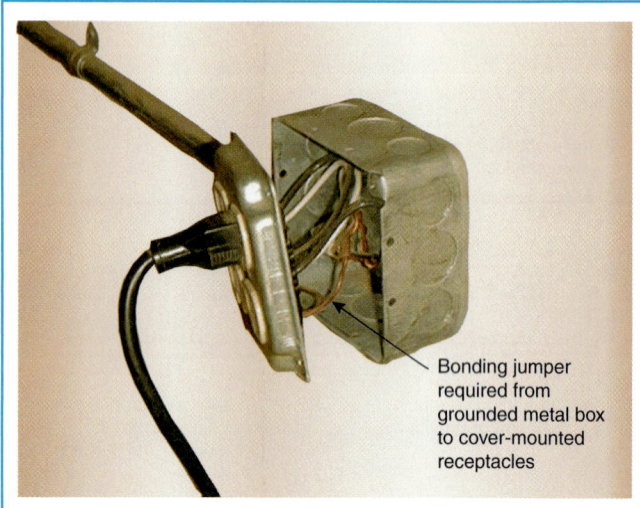

Exhibit 250.53 An example of a cover-mounted receptacle attached to a surface box where a bonding jumper is required.

grounding circuit between the device yoke and flush-type boxes.

Section 250.146(B) is illustrated by Exhibit 250.54, which shows a receptacle designed with a spring-type grounding strap for holding the mounting screw and establishing the grounding circuit so that no bonding jumper is required.

(C) Floor Boxes. Floor boxes designed for and listed as providing satisfactory ground continuity between the box and the device shall be permitted.

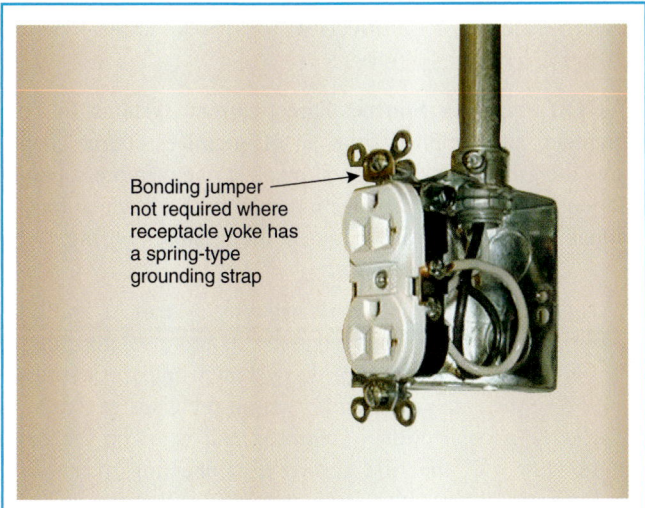

Exhibit 250.54 A receptacle designed with a listed spring-type grounding strap. The strap that holds the mounting screw captive establishes a grounding circuit and eliminates a bonding jumper to the box, in accordance with 250.146(B).

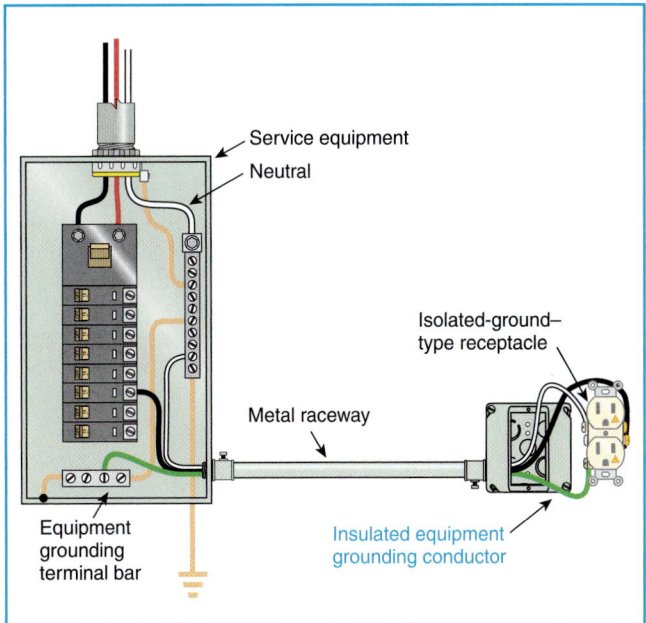

Exhibit 250.55 An isolated-ground–type receptacle with an insulated equipment grounding conductor and with the device box grounded through the metal raceway.

(D) Isolated Receptacles. Where required for the reduction of electrical noise (electromagnetic interference) on the grounding circuit, a receptacle in which the grounding terminal is purposely insulated from the receptacle mounting means shall be permitted. The receptacle grounding terminal shall be grounded by an insulated equipment grounding conductor run with the circuit conductors. This grounding conductor shall be permitted to pass through one or more panelboards without connection to the panelboard grounding terminal as permitted in 408.20, Exception, so as to terminate within the same building or structure directly at an equipment grounding conductor terminal of the applicable derived system or service.

> FPN: Use of an isolated equipment grounding conductor does not relieve the requirement for grounding the raceway system and outlet box.

Section 250.146(D) allows an isolated-ground–type receptacle to be installed without a bonding jumper between the metal device box and the receptacle grounding terminal. An insulated equipment grounding conductor, as shown in Exhibit 250.55, is installed with the branch-circuit conductors. This conductor may originate in the service panel, pass through any number of subpanels without being connected to the equipment grounding bus, and terminate at the isolated-ground–type receptacle ground terminal. However, this does not exempt the metal device box from being grounded. The metal device box must be grounded either by an equipment grounding conductor run with the circuit conductors or by a wiring method that serves as an equipment grounding

conductor. See 250.118 for types of equipment grounding conductors.

According to 250.146(D), where isolated-ground–type receptacles are used, the isolated equipment grounding conductor can terminate at an equipment grounding terminal of the applicable service or derived system within the same building as the receptacle. If the isolated equipment grounding conductor terminates at a separate building, a large voltage difference may exist between buildings during lightning transients. Such transients could cause damage to equipment connected to an isolated-ground–type receptacle and present a shock hazard between the isolated equipment frame and other grounded surfaces.

The fine print note to 250.146(D) is a reminder that metallic raceways and boxes are still required to be grounded by one of the usual required methods. This could require a separate grounding conductor, for example, to ground a metal box in a nonmetallic raceway system or to ground a metal box supplied by flexible metal conduit. When an ordinary grounding-type receptacle with an isolated-ground–type receptacle is being replaced, use of an existing equipment grounding conductor as the isolated receptacle grounding conductor could effectively defeat or seriously compromise the required box or raceway equipment ground.

250.148 Continuity and Attachment of Equipment Grounding Conductors to Boxes.

Where circuit conductors are spliced within a box, or terminated on equipment within or supported by a box, any separate equipment grounding conductors associated with those circuit conductors shall be spliced or joined within the box or to the box with devices suitable for the use. Connections depending solely on solder shall not be used. Splices shall be made in accordance with 110.14(B) except that insulation shall not be required. The arrangement of grounding connections shall be such that the disconnection or the removal of a receptacle, luminaire (fixture), or other device fed from the box will not interfere with or interrupt the grounding continuity.

Exception: The equipment grounding conductor permitted in 250.146(D) shall not be required to be connected to the other equipment grounding conductors or to the box.

(A) Metal Boxes. A connection shall be made between the one or more equipment grounding conductors and a metal box by means of a grounding screw that shall be used for no other purpose or a listed grounding device.

(B) Nonmetallic Boxes. One or more equipment grounding conductors brought into a nonmetallic outlet box shall be arranged so that a connection can be made to any fitting or device in that box requiring grounding.

VIII. Direct-Current Systems

250.160 General.

Direct-current systems shall comply with Part VIII and other sections of Article 250 not specifically intended for ac systems.

250.162 Direct-Current Circuits and Systems to Be Grounded.

Direct-current circuits and systems shall be grounded as provided for in 250.162(A) and (B).

(A) Two-Wire, Direct-Current Systems. A 2-wire, dc system supplying premises wiring and operating at greater than 50 volts but not greater than 300 volts shall be grounded.

Exception No. 1: A system equipped with a ground detector and supplying only industrial equipment in limited areas shall not be required to be grounded.

Exception No. 2: A rectifier-derived dc system supplied from an ac system complying with 250.20 shall not be required to be grounded.

Exception No. 3: Direct-current fire alarm circuits having a maximum current of 0.030 amperes as specified in Article 760, Part III, shall not be required to be grounded.

(B) Three-Wire, Direct-Current Systems. The neutral conductor of all 3-wire, dc systems supplying premises wiring shall be grounded.

250.164 Point of Connection for Direct-Current Systems.

(A) Off-Premises Source. Direct-current systems to be grounded and supplied from an off-premises source shall have the grounding connection made at one or more supply stations. A grounding connection shall not be made at individual services or at any point on the premises wiring.

As shown in the 3-wire dc distribution system in Exhibit 250.56, the neutral is grounded at the off-premises generator site. Grounding of a 2-wire dc system would be accomplished in the same manner. For an on-premises generator, a grounding connection is required and is to be located at the source of the first system disconnecting means or overcurrent device. Other equivalent means that use equipment listed and identified for such use are permitted.

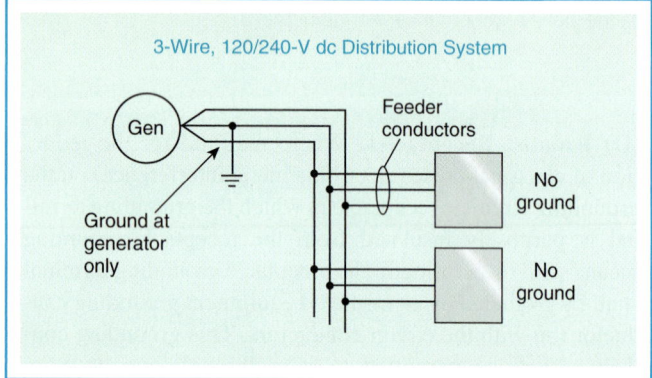

Exhibit 250.56 A 3-wire, 120/240-volt dc distribution system with the neutral grounded at the off-premises generator site.

(B) On-Premises Source. Where the dc system source is located on the premises, a grounding connection shall be made at one of the following:

(1) The source
(2) The first system disconnection means or overcurrent device
(3) By other means that accomplish equivalent system protection and that utilize equipment listed and identified for the use

250.166 Size of Direct-Current Grounding Electrode Conductor.

The size of the grounding electrode conductor for a dc system shall be as specified in 250.166(A) through (E).

(A) Not Smaller Than the Neutral Conductor. Where the dc system consists of a 3-wire balancer set or a balancer winding with overcurrent protection as provided in 445.12(D), the grounding electrode conductor shall not be smaller than the neutral conductor and not smaller than 8 AWG copper or 6 AWG aluminum.

(B) Not Smaller Than the Largest Conductor. Where the dc system is other than as in 250.166(A), the grounding electrode conductor shall not be smaller than the largest conductor supplied by the system, and not smaller than 8 AWG copper or 6 AWG aluminum.

(C) Connected to Rod, Pipe, or Plate Electrodes. Where connected to rod, pipe, or plate electrodes as in 250.52(A)(5) or 250.52(A)(6), that portion of the grounding electrode conductor that is the sole connection to the grounding electrode shall not be required to be larger than 6 AWG copper wire or 4 AWG aluminum wire.

(D) Connected to a Concrete-Encased Electrode. Where connected to a concrete-encased electrode as in 250.52(A)(3), that portion of the grounding electrode conductor that is the sole connection to the grounding electrode shall not be required to be larger than 4 AWG copper wire.

(E) Connected to a Ground Ring. Where connected to a ground ring as in 250.52(A)(4), that portion of the grounding electrode conductor that is the sole connection to the grounding electrode shall not be required to be larger than the conductor used for the ground ring.

250.168 Direct-Current Bonding Jumper.

For dc systems, the size of the bonding jumper shall not be smaller than the system grounding conductor specified in 250.166.

250.169 Ungrounded Direct-Current Separately Derived Systems.

Except as otherwise permitted in 250.34 for portable and vehicle-mounted generators, an ungrounded dc separately derived system supplied from a stand-alone power source (such as an engine–generator set) shall have a grounding electrode conductor connected to an electrode that complies with Part III to provide for grounding of metal enclosures, raceways, cables, and exposed non–current-carrying metal parts of equipment. The grounding electrode conductor connection shall be to the metal enclosure at any point on the separately derived system from the source to the first system disconnecting means or overcurrent device, or it shall be made at the source of a separately derived system that has no disconnecting means or overcurrent devices.

The size of the grounding electrode conductor shall be in accordance with 250.166.

IX. Instruments, Meters, and Relays

250.170 Instrument Transformer Circuits.

Secondary circuits of current and potential instrument transformers shall be grounded where the primary windings are connected to circuits of 300 volts or more to ground and, where on switchboards, shall be grounded irrespective of voltage.

Exception: Circuits where the primary windings are connected to circuits of less than 1000 volts with no live parts or wiring exposed or accessible to other than qualified persons.

250.172 Instrument Transformer Cases.

Cases or frames of instrument transformers shall be grounded where accessible to other than qualified persons.

Exception: Cases or frames of current transformers, the primaries of which are not over 150 volts to ground and that are used exclusively to supply current to meters.

250.174 Cases of Instruments, Meters, and Relays Operating at Less Than 1000 Volts.

Instruments, meters, and relays operating with windings or working parts at less than 1000 volts shall be grounded as specified in 250.174(A), (B), or (C).

(A) Not on Switchboards. Instruments, meters, and relays not located on switchboards, operating with windings or working parts at 300 volts or more to ground, and accessible to other than qualified persons, shall have the cases and other exposed metal parts grounded.

(B) On Dead-Front Switchboards. Instruments, meters, and relays (whether operated from current and potential transformers or connected directly in the circuit) on switchboards having no live parts on the front of the panels shall have the cases grounded.

(C) On Live-Front Switchboards. Instruments, meters, and relays (whether operated from current and potential transformers or connected directly in the circuit) on switchboards having exposed live parts on the front of panels shall not have their cases grounded. Mats of insulating rubber or other suitable floor insulation shall be provided for the operator where the voltage to ground exceeds 150.

250.176 Cases of Instruments, Meters, and Relays—Operating Voltage 1 kV and Over.

Where instruments, meters, and relays have current-carrying parts of 1 kV and over to ground, they shall be isolated by elevation or protected by suitable barriers, grounded metal, or insulating covers or guards. Their cases shall not be grounded.

Exception: Cases of electrostatic ground detectors where the internal ground segments of the instrument are connected to the instrument case and grounded and the ground detector is isolated by elevation.

250.178 Instrument Grounding Conductor.

The grounding conductor for secondary circuits of instrument transformers and for instrument cases shall not be smaller than 12 AWG copper or 10 AWG aluminum. Cases of instrument transformers, instruments, meters, and relays that are mounted directly on grounded metal surfaces of enclosures or grounded metal switchboard panels shall be considered to be grounded, and no additional grounding conductor shall be required.

X. Grounding of Systems and Circuits of 1 kV and Over (High Voltage)

250.180 General.

Where high-voltage systems are grounded, they shall comply with all applicable provisions of the preceding sections of this article and with 250.182 through 250.190, which supplement and modify the preceding sections.

250.182 Derived Neutral Systems.

A system neutral derived from a grounding transformer shall be permitted to be used for grounding high-voltage systems.

250.184 Solidly Grounded Neutral Systems.

(A) Neutral Conductor. The minimum insulation level for neutral conductors of solidly grounded systems shall be 600 volts.

Exception No. 1: Bare copper conductors shall be permitted to be used for the neutral of service entrances and the neutral of direct-buried portions of feeders.

Exception No. 2: Bare conductors shall be permitted for the neutral of overhead portions installed outdoors.

> FPN: See 225.4 for conductor covering where within 3.0 m (10 ft) of any building or other structure.

(B) Multiple Grounding. The neutral of a solidly grounded neutral system shall be permitted to be grounded at more than one point. Grounding shall be permitted at one or more of the following locations:

(1) Transformers supplying conductors to a building or other structure
(2) Underground circuits where the neutral is exposed
(3) Overhead circuits installed outdoors

(C) Neutral Grounding Conductor. The neutral grounding conductor shall be permitted to be a bare conductor if isolated from phase conductors and protected from physical damage.

(D) Multigrounded Neutral Conductor. Where a multigrounded neutral system is used, the following shall apply:

(1) The multigrounded neutral conductor shall be of sufficient ampacity for the load imposed on the conductor but not less than 33 percent of the ampacity of the phase conductors.

Exception: In industrial and commercial premises under engineering supervision, it shall be permissible to size the ampacity of the neutral conductor to not less than 20 percent of the ampacity of the phase conductor.

(2) The multigrounded neutral conductor shall be grounded at each transformer and at other additional locations by connection to a made or existing electrode.
(3) At least one grounding electrode shall be installed and connected to the multigrounded neutral circuit conductor every 400 m (1300 ft).
(4) The maximum distance between any two adjacent electrodes shall not be more than 400 m (1300 ft).
(5) In a multigrounded shielded cable system, the shielding shall be grounded at each cable joint that is exposed to personnel contact.

250.186 Impedance Grounded Neutral Systems.

Impedance grounded neutral systems in which a grounding impedance, usually a resistor, limits the ground-fault current, shall be permitted where all of the following conditions are met.

(1) The conditions of maintenance and supervision ensure that only qualified persons will service the installation.
(2) Ground detectors are installed on the system.
(3) Line-to-neutral loads are not served.

Impedance grounded neutral systems shall comply with the provisions of 250.186(A) through (D).

(A) Location. The grounding impedance shall be inserted in the grounding conductor between the grounding electrode of the supply system and the neutral point of the supply transformer or generator.

(B) Identified and Insulated. The neutral conductor of an impedance grounded neutral system shall be identified, as well as fully insulated with the same insulation as the phase conductors.

(C) System Neutral Connection. The system neutral shall not be connected to ground, except through the neutral grounding impedance.

(D) Equipment Grounding Conductors. Equipment grounding conductors shall be permitted to be bare and shall be electrically connected to the ground bus and grounding electrode conductor.

250.188 Grounding of Systems Supplying Portable or Mobile Equipment.

Systems supplying portable or mobile high-voltage equipment, other than substations installed on a temporary basis, shall comply with 250.188(A) through (F).

Portable describes equipment that is easily carried from one location to another. *Mobile* describes equipment that is easily moved on wheels, treads, and so on.

(A) Portable or Mobile Equipment. Portable or mobile high-voltage equipment shall be supplied from a system having its neutral grounded through an impedance. Where a delta-connected high-voltage system is used to supply portable or mobile equipment, a system neutral shall be derived.

(B) Exposed Non–Current-Carrying Metal Parts. Exposed non–current-carrying metal parts of portable or mobile equipment shall be connected by an equipment grounding conductor to the point at which the system neutral impedance is grounded.

(C) Ground-Fault Current. The voltage developed between the portable or mobile equipment frame and ground by the flow of maximum ground-fault current shall not exceed 100 volts.

(D) Ground-Fault Detection and Relaying. Ground-fault detection and relaying shall be provided to automatically de-energize any high-voltage system component that has developed a ground fault. The continuity of the equipment grounding conductor shall be continuously monitored so as to de-energize automatically the high-voltage circuit to the portable or mobile equipment upon loss of continuity of the equipment grounding conductor.

(E) Isolation. The grounding electrode to which the portable or mobile equipment system neutral impedance is connected shall be isolated from and separated in the ground by at least 6.0 m (20 ft) from any other system or equipment grounding electrode, and there shall be no direct connection between the grounding electrodes, such as buried pipe and fence, and so forth.

(F) Trailing Cable and Couplers. High-voltage trailing cable and couplers for interconnection of portable or mobile equipment shall meet the requirements of Part III of Article 400 for cables and 490.55 for couplers.

250.190 Grounding of Equipment.

All non–current-carrying metal parts of fixed, portable, and mobile equipment and associated fences, housings, enclosures, and supporting structures shall be grounded.

Exception: Where isolated from ground and located so as to prevent any person who can make contact with ground from contacting such metal parts when the equipment is energized.

Grounding conductors not an integral part of a cable assembly shall not be smaller than 6 AWG copper or 4 AWG aluminum.

FPN: See 250.110, Exception No. 2, for pole-mounted distribution apparatus.

ARTICLE 280
Surge Arresters

Contents

I. General

280.1 Scope.

This article covers general requirements, installation requirements, and connection requirements for surge arresters installed on premises wiring systems.

Voltage surges with peaks of several thousand volts, even on 120-volt circuits, are not uncommon. These surges occur because of induced voltages in power and transmission lines resulting from lightning strikes in the vicinity of the line.

Surges also occur as a result of switching inductive circuits on the premises. Surge arresters for installation as part of an electric service, such as the one shown in Exhibit 280.1, and for use with cord-and-plug-connected solid-state electronic equipment are commercially available. The basic standards used to investigate surge arresters are ANSI/IEEE C62.1, *Standard for Gapped Silicon-Carbide Surge Arresters for AC Power Circuits,* and ANSI/IEEE C62.11, *Standard for Metal-Oxide Surge Arresters for AC Power Circuits.*

Exhibit 280.1 A lightning surge protector suitable for service-entrance installation and for mounting in a panel knockout. (Courtesy of General Electric Co.)

280.2 Definition.

Surge Arrester. A protective device for limiting surge voltages by discharging or bypassing surge current, and it also prevents continued flow of follow current while remaining capable of repeating these functions.

280.3 Number Required.

Where used at a point on a circuit, a surge arrester shall be connected to each ungrounded conductor. A single installation of such surge arresters shall be permitted to protect a number of interconnected circuits, provided that no circuit is exposed to surges while disconnected from the surge arresters.

Means must be provided for protection of circuits that may be disconnected from the generating station bus. A switch with double-throw action used to disconnect the outside circuits from the station generator and alternatively connect these circuits to ground would satisfy the condition of a single set of arresters protecting more than one circuit.

Surge arresters are required to be installed on circuits in buildings that house explosives. For details, see Chapter 6 of NFPA 495, *Explosive Materials Code.*

280.4 Surge Arrester Selection.

(A) Circuits of Less Than 1000 Volts. The rating of the surge arrester shall be equal to or greater than the maximum

continuous phase-to-ground power frequency voltage available at the point of application.

Surge arresters installed on circuits of less than 1000 volts shall be listed for the purpose.

(B) Circuits of 1 kV and Over—Silicon Carbide Types. The rating of a silicon carbide-type surge arrester shall be not less than 125 percent of the maximum continuous phase-to-ground voltage available at the point of application.

> FPN No. 1: For further information on surge arresters, see ANSI/IEEE C62.1-1989, *Standard for Gapped Silicon-Carbide Surge Arresters for AC Power Circuits*; ANSI/IEEE C62.2-1987, *Guide for the Application of Gapped Silicon-Carbide Surge Arresters for Alternating-Current Systems*; ANSI/IEEE C62.11-1993, *Standard for Metal-Oxide Surge Arresters for Alternating-Current Power Circuits*; and ANSI/IEEE C62.22-1991, *Guide for the Application of Metal-Oxide Surge Arresters for Alternating-Current Systems.*
>
> FPN No. 2: The selection of a properly rated metal oxide arrester is based on considerations of maximum continuous operating voltage and the magnitude and duration of overvoltages at the arrester location as affected by phase-to-ground faults, system grounding techniques, switching surges, and other causes. See the manufacturer's application rules for selection of the specific arrester to be used at a particular location.

II. Installation

280.11 Location.

Surge arresters shall be permitted to be located indoors or outdoors. Surge arresters shall be made inaccessible to unqualified persons, unless listed for installation in accessible locations.

Maximum protection is achieved where the surge protective device is located as close as practicable to the equipment to be protected. When a surge passes through an arrester, a wave is reflected in both directions on the conductors connected to the surge arrester. The magnitude of the reflected wave increases as the distance from the arrester increases. If the length of the conductor between the protected equipment and the surge arrester is short, the magnitude of the wave reflected through the equipment is minimized.

280.12 Routing of Surge Arrester Connections.

The conductor used to connect the surge arrester to line or bus and to ground shall not be any longer than necessary and shall avoid unnecessary bends.

Arrester conductors should be as short and be run as straight as practicable, avoiding any sharp bends and turns, which increases the impedance.

III. Connecting Surge Arresters

280.21 Installed at Services of Less Than 1000 Volts.

Line and ground connecting conductors shall not be smaller than 14 AWG copper or 12 AWG aluminum. The arrester grounding conductor shall be connected to one of the following:

(1) Grounded service conductor
(2) Grounding electrode conductor
(3) Grounding electrode for the service
(4) Equipment grounding terminal in the service equipment

High-frequency currents, such as those common to lightning discharges, tend to reduce the effectiveness of a grounding conductor. Single-phase or 3-phase grounded or ungrounded services are permitted to have the surge arrester grounded to the equipment grounding terminal in the service equipment. Exhibit 280.2 shows three methods of grounding the ground terminals of surge arresters at service entrances. In the upper left diagram, an arrester is connected to a neutral service conductor. In the upper right diagram, an arrester is connected to a grounding electrode conductor. In the lower diagram, an arrester is connected to a grounding electrode conductor of an ungrounded system.

280.22 Installed on the Load Side Services of Less Than 1000 Volts.

Line and ground connecting conductors shall not be smaller than 14 AWG copper or 12 AWG aluminum. A surge arrester shall be permitted to be connected between any two conductors—ungrounded conductor(s), grounded conductor, grounding conductor. The grounded conductor and the grounding conductor shall be interconnected only by the normal operation of the surge arrester during a surge.

280.23 Circuits of 1 kV and Over—Surge-Arrester Conductors.

The conductor between the surge arrester and the line and the surge arrester and the grounding connection shall not be smaller than 6 AWG copper or aluminum.

280.24 Circuits of 1 kV and Over—Interconnections.

The grounding conductor of a surge arrester protecting a transformer that supplies a secondary distribution system shall be interconnected as specified in 280.24(A), (B), or (C).

(A) Metallic Interconnections. A metallic interconnection shall be made to the secondary grounded circuit conductor

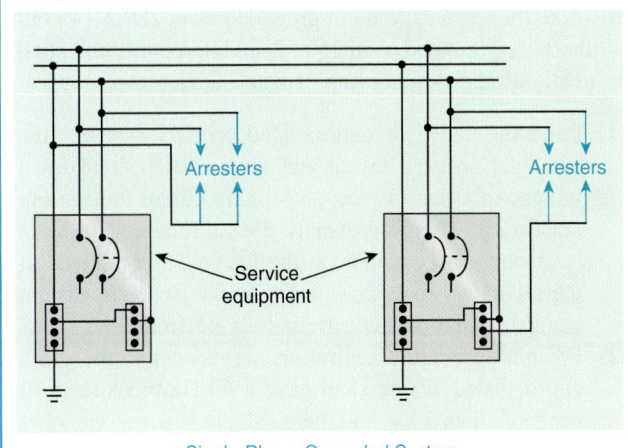

Single-Phase Grounded System

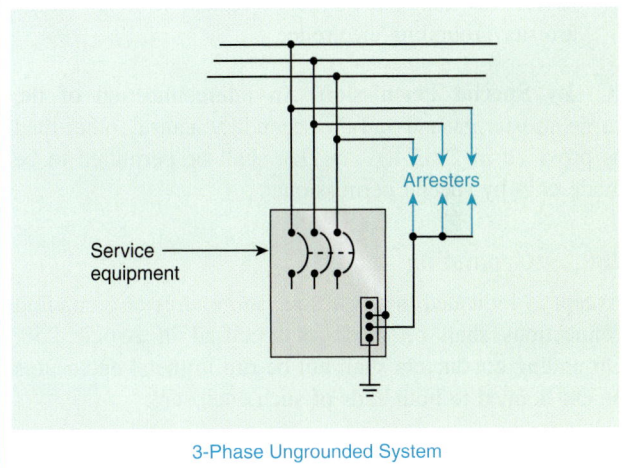

3-Phase Ungrounded System

Exhibit 280.2 Three methods of grounding surge arresters at service entrances.

or the secondary circuit grounding conductor provided that, in addition to the direct grounding connection at the surge arrester, the following occurs:

(1) The grounded conductor of the secondary has elsewhere a grounding connection to a continuous metal underground water piping system. However, in urban water-pipe areas where there are at least four water-pipe connections on the neutral and not fewer than four such connections in each mile of neutral, the metallic interconnection shall be permitted to be made to the secondary neutral with omission of the direct grounding connection at the surge arrester.

(2) The grounded conductor of the secondary system is a part of a multiground neutral system of which the primary neutral has at least four ground connections in each mile of line in addition to a ground at each service.

(B) Through Spark Gap or Device. Where the surge arrester grounding conductor is not connected as in 280.24(A) or where the secondary is not grounded as in 280.24(A) but is otherwise grounded as in 250.52, an interconnection shall be made through a spark gap or listed device as follows:

(1) For ungrounded or unigrounded primary systems, the spark gap or listed device shall have a 60-Hz breakdown voltage of at least twice the primary circuit voltage but not necessarily more than 10 kV, and there shall be at least one other ground on the grounded conductor of the secondary that is not less than 6.0 m (20 ft) distant from the surge arrester grounding electrode.

(2) For multigrounded neutral primary systems, the spark gap or listed device shall have a 60-Hz breakdown of not more than 3 kV, and there shall be at least one other ground on the grounded conductor of the secondary that is not less than 6.0 m (20 ft) distant from the surge arrester grounding electrode.

(C) By Special Permission. An interconnection of the surge arrester ground and the secondary neutral, other than as provided in 280.24(A) or (B), shall be permitted to be made only by special permission.

280.25 Grounding.

Except as indicated in this article, surge arrester grounding connections shall be made as specified in Article 250. Grounding conductors shall not be run in metal enclosures unless bonded to both ends of such enclosure.

ARTICLE 285
Transient Voltage Surge Suppressors: TVSSs

Contents

I. General

285.1 Scope.

This article covers general requirements, installation requirements, and connection requirements for transient voltage surge suppressors (TVSS) permanently installed on premises wiring systems.

> Article 285 is new to the 2002 *Code*. It was created to provide installation requirements for new technology for the protection of persons and electronic equipment.
>
> A transient voltage surge suppressor (TVSS) is a common component of an electrical system that provides a protection function similar to that of a surge arrester (see Article 280). A TVSS is generally installed to protect sensitive electronic equipment such as computers, telecommunications equipment, security systems, and electronic appliances. A TVSS should begin to divert or limit the surge current from a transient (or surge) event much closer to the operating voltage, as compared to a surge arrester.

285.2 Definition.

Transient Voltage Surge Suppressor (TVSS). A protective device for limiting transient voltages by diverting or limiting surge current; it also prevents continued flow of follow current while remaining capable of repeating these functions.

285.3 Uses Not Permitted.

A TVSS shall not be used in the following:

(1) Circuits exceeding 600 volts
(2) Ungrounded electrical systems as permitted in 250.21
(3) Where the rating of the TVSS is less than the maximum continuous phase-to-ground power frequency voltage available at the point of application

> FPN: For further information on TVSSs, see NEMA LS 1-1992, *Standard for Low Voltage Surge Suppression Devices*. The selection of a properly rated TVSS is based on criteria such as maximum continuous operating voltage, the magnitude and duration of overvoltages at the suppressor location as affected by phase-to-ground faults, system grounding techniques, and switching surges.

> UL 1449, *Safety Standard for Transient Voltage Surge Suppressors*, limits TVSS applications to 600 volts and less. Due to the voltage instability of ungrounded electrical systems, TVSS devices are not permitted to be installed on ungrounded electrical systems.

285.4 Number Required.

Where used at a point on a circuit, the TVSS shall be connected to each ungrounded conductor.

285.5 Listing.

A TVSS shall be a listed device.

285.6 Short Circuit Current Rating.

The TVSS shall be marked with a short circuit current rating and shall not be installed at a point on the system where the available fault current is in excess of that rating. This marking requirement shall not apply to receptacles.

The first TVSS device is commonly installed in the electrical system as an integral component of, or near to, the service entrance equipment in residential and commercial structures. It is imperative that the available fault current at the point of installation not exceed the short-circuit current rating of the TVSS. The installed TVSS device must match or exceed the system's available fault current at its point of installation on a system. Of course, as an alternative to the TVSS at the service, an industrial or large commercial facility may elect to install arresters (Article 280) at the service equipment and at intermediate points in the distribution system, and then install TVSS devices downstream at panelboards that serve loads susceptible to transients. See Exhibit 285.1.

Another type of TVSS is the point-of-use TVSS. These devices (for example, receptacles and power strips) may be installed at the equipment (computers, equipment with electronic controls, and so on). The function of a point-of-use TVSS is to remove any small transients that pass through the more robust surge devices located at the service. Point-of-use TVSS devices are also useful in removing small transients that have been generated within the building.

II. Installation

285.11 Location.

TVSSs shall be permitted to be located indoors or outdoors and shall be made inaccessible to unqualified persons, unless listed for installation in accessible locations.

285.12 Routing of Connections.

The conductors used to connect the the TVSS to the line or bus and to ground shall not be any longer than necessary and shall avoid unnecessary bends.

The conductor length used to connect the TVSS plays an important role in protection performance. As the length of the conductor increases, so does the impedance in the conduction path. This drives the clamping voltage higher and reduces the protection provided by the TVSS unit. Maximum protection is achieved where the TVSS is located as close as practicable to the equipment being protected, as shown in Exhibit 285.2.

III. Connecting Transient Voltage Surge Suppressors

285.21 Connection.

Where a TVSS is installed, it shall be connected as follows.

(A) Location.

(1) Service Supplied Building or Structure. The transient voltage surge suppressor shall be connected on the load side of a service disconnect overcurrent device required in 230.91.

(2) Feeder Supplied Building or Structure. The transient voltage surge suppressor shall be connected on the load side of the first overcurrent device at the building or structure.

Exception to (1) and (2): Where the TVSS is also listed as a surge arrester, the connection shall be as permitted by Article 280.

(3) Separately Derived System. The TVSS shall be connected on the load side of the first overcurrent device in a separately derived system.

Exhibit 285.1 A TVSS as an integral component of a receptacle, providing local point-of-use protection of equipment when transient events occur within the facility. (Courtesy of Pass & Seymour/ Legrand®)

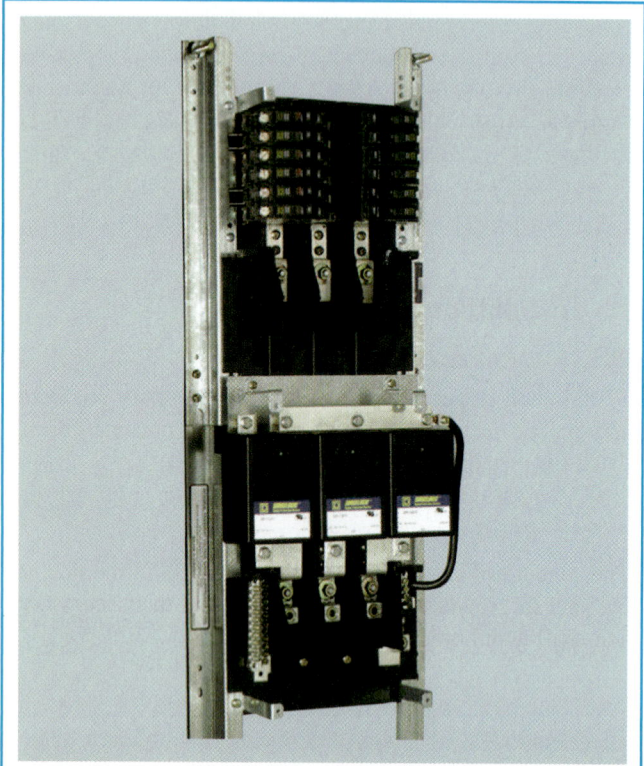

Exhibit 285.2 A TVSS mounted as an integral component of a panelboard, which minimizes conductor length between the electrical system and the TVSS. (Courtesy of Square D Co.)

UL 1449, *Safety Standard for Transient Voltage Surge Suppressors*, is used to investigate the safety of a TVSS. In accordance with the scope of UL 1449, a TVSS must be installed on the load side of the service disconnect overcurrent protection. If a service disconnect is dedicated to a TVSS unit, it becomes one of the six disconnects permitted in 230.71. Only surge arresters in accordance with Article 280 are permitted to be electrically connected ahead of the service disconnect in accordance with 230.82(3). The requirement for connection on the load side of the first overcurrent protection device in a feeder-supplied structure is necessary due to the exposure of external feeder conductors to a more hostile surge environment such as lightning.

(B) Conductor Size. Line and ground connecting conductors shall not be smaller than 14 AWG copper or 12 AWG aluminum.

(C) Connection Between Conductors. A TVSS shall be permitted to be connected between any two conductors—ungrounded conductor(s), grounded conductor, grounding conductor. The grounded conductor and the grounding conductor shall be interconnected only by the normal operation of the TVSS during a surge.

285.25 Grounding.

Grounding conductors shall not be run in metal enclosures unless bonded to both ends of such enclosure.

See 250.64(E) and 250.92(A)(3) and the associated commentary, including Exhibit 250.33, for requirements to bond both ends of a metallic raceway that encloses grounding electrode conductors.

Wiring Methods and Materials

There are two significant editorial changes within Chapter 3 for the 2002 *Code*. Both changes resulted from the collective efforts of the NEC Usability Task Group, the code making panels (CMPs) having specific subject responsibility within Chapter 3, and the users who submitted proposals and comments.

First and foremost, many common requirements for wiring methods within Chapter 3 are now aligned using a common numbering system. All of the circular raceway articles and many of the cable articles now have a common numbering format, which will assist users in locating common requirements within an article. Additionally, students of the *Code* will find it much easier to understand, learn, and compare the many common requirements of various wiring methods.

The second significant editorial change is the renumbering of many articles within Chapter 3. Achieving a true common numbering format within each article necessitated splitting a few of them into two separate articles. These additional articles forced the NEC Usability Task Group to consider and then adopt a new article renumbering scheme for Chapter 3. This new format allows for a grouping of like articles and leaves space for new wiring method articles should they be added in the future.

Annex F contains three cross-reference tables to guide the user through the Chapter 3 reorganization.

ARTICLE 300
Wiring Methods

Contents

I. General Requirements

300.1 Scope.

(A) All Wiring Installations. This article covers wiring methods for all wiring installations unless modified by other articles.

(B) Integral Parts of Equipment. The provisions of this article are not intended to apply to the conductors that form an integral part of equipment, such as motors, controllers, motor control centers, or factory assembled control equipment or listed utilization equipment.

Generally, wiring within equipment is covered by product standards. As an example, integral wiring of motors is covered by NEMA MG-1, *Motors and Generators*; integral wiring of industrial control equipment by UL 508, *Standard for Safety of Industrial Control Equipment*; and wiring that forms an integral part of industrial machinery by NFPA 79, *Electrical Standard for Industrial Machinery*.

(C) Metric Designators and Trade Sizes. Metric designators and trade sizes for conduit, tubing, and associated fittings and accessories shall be as designated in Table 300.1(C).

Table 300.1(C) Metric Designator and Trade Sizes

Metric Designator	Trade Size
12	3/8
16	1/2
21	3/4
27	1
35	1 1/4
41	1 1/2
53	2
63	2 1/2
78	3
91	3 1/2
103	4
129	5
155	6

Note: The metric designators and trade sizes are for identification purposes only and are not actual dimensions.

Using metric designators to describe circular raceways is one more step in the metrication of the *NEC*, as stated in both 90.9 and its associated commentary. Metric designators for conduits first appeared in 1989 in IEC 981, *Extra-Heavy Duty Rigid Steel Conduits for Electrical Installations*. Since then, both NEMA and the *NEC* have recognized metric designators. The NEC did so in the 1996 edition, allowing metric designators to appear as fine print notes following the mention of trade sizes of circular raceways. Assigning metric designators to traditional trade size threaded conduit does not change the physical dimensions or the traditional "NPT type" threads of the conduit. Using metric designators is simply another method of identifying the size of a circular raceway.

Table 300.1(C) identifies a distinct metric designator for each circular raceway trade size. Dimensions or descriptions of circular raceways have traditionally included an inch size or unit of measure. The unit of measure associated with a circular raceway has not been included in this table or throughout the *NEC* because it does not reflect a true measure, but rather a "modular" or "relative" measure. Many examples of modular or relative measures can be found in the building trades. For example, in North America, items such as a 2 ft × 4 ft drop-in luminaire, an 8 ft fluorescent lamp, and a 2 in. × 4 in. piece of lumber do not reflect a true dimension but rather are loosely associated dimensions common in modular construction. As stated in the footnote to Table 300.1(C), the metric designators and trade sizes are not actual dimensions.

According to Table 4 of Chapter 9, each metric designator sized circular raceway is identical in internal and external dimensions (including manufacturing tolerances) to its trade size counterpart. Therefore, Annex C wire fill tables are applicable to both metric designator and trade size circular raceways, and so for installation practices, introducing an associated metric designator for traditional circular raceways trade sizes should not be a concern.

What is a concern for installation practices, however, is the use of a circular raceway with threaded joints, where the threaded joints are not according to the product standard. For example, rigid metal conduit (RMC) is required to be listed according to 344.6, and the appropriate product standard for this listing is UL 6. Intermediate metal conduit (IMC) is required to be listed according to 342.6, and the appropriate product standard for this listing is UL 1242. Both listed conduits must be threaded in accordance with ANSI/ASME B.1.20.1-1993. Therefore, only conduits threaded to the traditional dimension of ¾-inch taper per foot are acceptable. Simply stated, an installation using metric threaded conduit is not permitted by the *NEC*. However, an installation using threads according to ANSI/ASME B1.20.1-1993 is in compliance with the *NEC*.

300.2 Limitations.

(A) Voltage. Wiring methods specified in Chapter 3 shall be used for 600 volts, nominal, or less where not specifically limited in some section of Chapter 3. They shall be permitted for over 600 volts, nominal, where specifically permitted elsewhere in this *Code*.

(B) Temperature. Temperature limitation of conductors shall be in accordance with 310.10.

See 110.14(C) and its commentary for information on temperature limitations of conductor terminations.

300.3 Conductors.

(A) Single Conductors. Single conductors specified in Table 310.13 shall only be installed where part of a recognized wiring method of Chapter 3.

Section 300.3(A) clearly states that building wire, such as individual insulated conductors identified as THHN, is prohibited from use outside of a recognized wiring method.

(B) Conductors of the Same Circuit. All conductors of the same circuit and, where used, the grounded conductor and all equipment grounding conductors and bonding conductors shall be contained within the same raceway, auxiliary gutter, cable tray, cablebus assembly, trench, cable, or cord, unless otherwise permitted in accordance with 300.3(B)(1) through (4).

This general rule remains consistent with electrical theory; that is, to reduce inductive heating and to avoid increases in overall circuit impedance, all circuit conductors of an individual circuit must be grouped. Similar requirements are found in 300.5(I).

(1) Paralleled Installations. Conductors shall be permitted to be run in parallel in accordance with the provisions of 310.4. The requirement to run all circuit conductors within the same raceway, auxiliary gutter, cable tray, trench, cable, or cord shall apply separately to each portion of the paralleled installation, and the equipment grounding conductors shall comply with the provisions of 250.122. Parallel runs in cable tray shall comply with the provisions of 392.8(D).

Exception: Conductors installed in nonmetallic raceways run underground shall be permitted to be arranged as isolated phase installations. The raceways shall be installed in close proximity, and the conductors shall comply with the provisions of 300.20(B).

(2) Grounding and Bonding Conductors. Equipment grounding conductors shall be permitted to be installed outside a raceway or cable assembly where in accordance with the provisions of 250.130(C) for certain existing installations or in accordance with 250.134(B), Exception No. 2, for dc circuits. Equipment bonding conductors shall be permitted to be installed on the outside of raceways in accordance with 250.102(E).

Section 300.3(B)(2) recognizes that some types of grounding and bonding conductors can be run as single conductors on the exterior of the raceway or outside of a cable assembly.

(3) Nonferrous Wiring Methods. Conductors in wiring methods with a nonmetallic or other nonmagnetic sheath, where run in different raceways, auxiliary gutters, cable trays, trenches, cables, or cords, shall comply with the provisions of 300.20(B). Conductors in single-conductor Type MI cable with a nonmagnetic sheath shall comply with the provisions of 332.31. Conductors of single-conductor–type MC cable with a nonmagnetic sheath shall comply with the provisions of 330.31, 330.116, and 300.20(B).

Section 300.3(B)(3) has been revised for the 2002 *Code*. The last sentence now addresses the installation of single-conductor–Type MC cable using a nonferrous (nonmagnetic) sheath.

(4) Enclosures. Where an auxiliary gutter runs between a column-width panelboard and a pull box, and the pull box includes neutral terminations, the neutral conductors of circuits supplied from the panelboard shall be permitted to originate in the pull box.

Section 300.3(B)(4) recognizes the practice of supplying narrow, column-type panelboard through an auxiliary gutter from an overhead pull box and running only the ungrounded conductors down from the pull box to the panelboard. As shown in Exhibit 300.1, the feeder and branch-circuit neutral conductors are terminated in the overhead pull box and are not carried with the ungrounded conductors. Inductive heating does not occur, because the load-carrying conductors extend down and back up within the same enclosure.

(C) Conductors of Different Systems.

(1) 600 Volts, Nominal, or Less. Conductors of circuits rated 600 volts, nominal, or less, ac circuits, and dc circuits shall be permitted to occupy the same equipment wiring enclosure, cable, or raceway. All conductors shall have an insulation rating equal to at least the maximum circuit voltage applied to any conductor within the enclosure, cable, or raceway.

Exception: For solar photovoltaic systems in accordance with 690.4(B).

FPN: See 725.55(A) for Class 2 and Class 3 circuit conductors.

Section 300.3(C)(1) makes it clear that it is the *maximum circuit voltage* in the raceway, not the maximum insulation voltage rating of the conductors in the raceway, that determines the minimum voltage rating required for the insulation of conductors for systems of 600 volts or less.

The conductors of a 3-phase, 4-wire, 208Y/120-volt ac circuit; a 3-phase, 4-wire, 480Y/277-volt ac circuit; and a 3-wire, 120/240-volt dc circuit may occupy the same equipment wiring enclosure, cable, or raceway if all of the conductors are insulated for the maximum circuit voltage of any conductor. In this case, the maximum circuit voltage would

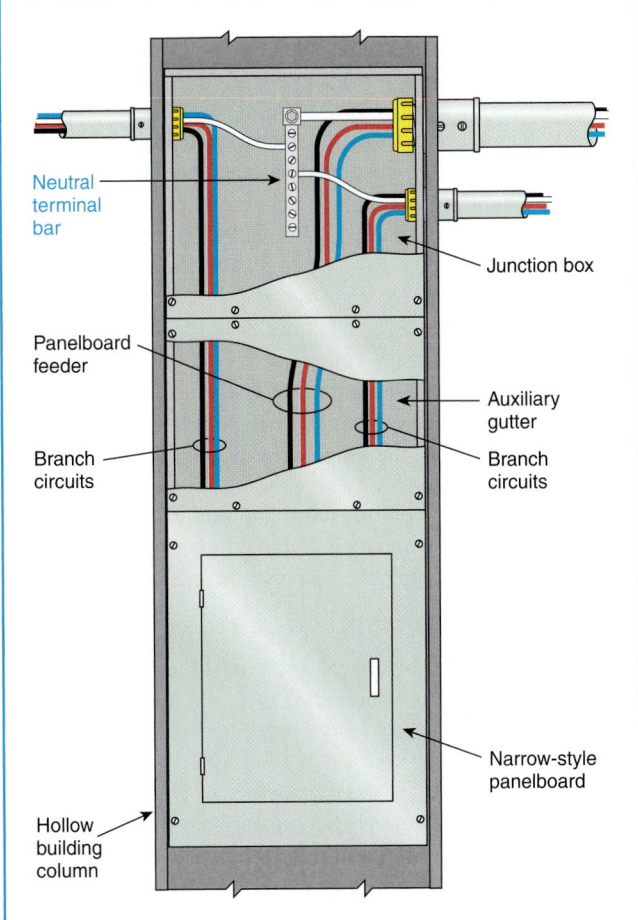

Exhibit 300.1 An installation where an auxiliary gutter extends from the panelboard up to a pull box that is used as a termination point for the feeder and branch-circuit grounded conductors (neutrals).

be 480 volts, and 600-volt insulation would be suitable for all of the conductors.

If a 2-wire, 120-volt circuit is included in the same raceway with a 3-wire, 120/240-volt circuit having 600-volt conductors, then the 2-wire, 120-volt circuit conductors could use 300-volt insulation because the maximum circuit voltage is only 240 volts.

Section 690.4(B) prohibits the location of solar photovoltaic circuits within the same enclosure as conductors of other systems unless the conductors are separated by a partition or are connected together.

Section 725.55(A) prohibits Class 2 and Class 3 circuit conductors from occupying the same enclosure, cable, or raceway as Class 1, electric light, and power conductors, unless specifically permitted in 725.55(B) through (J).

(2) Over 600 Volts, Nominal. Conductors of circuits rated over 600 volts, nominal, shall not occupy the same equipment wiring enclosure, cable, or raceway with conductors of circuits rated 600 volts, nominal, or less unless otherwise permitted in (a) through (e).

(a) Secondary wiring to electric-discharge lamps of 1000 volts or less, if insulated for the secondary voltage involved, shall be permitted to occupy the same luminaire (fixture), sign, or outline lighting enclosure as the branch-circuit conductors.

(b) Primary leads of electric-discharge lamp ballasts, insulated for the primary voltage of the ballast, where contained within the individual wiring enclosure, shall be permitted to occupy the same luminaire (fixture), sign, or outline lighting enclosure as the branch-circuit conductors.

(c) Excitation, control, relay, and ammeter conductors used in connection with any individual motor or starter shall be permitted to occupy the same enclosure as the motor-circuit conductors.

(d) In motors, switchgear and control assemblies, and similar equipment, conductors of different voltage ratings shall be permitted.

(e) In manholes, if the conductors of each system are permanently and effectively separated from the conductors of the other systems and securely fastened to racks, insulators, or other approved supports, conductors of different voltage ratings shall be permitted.

Conductors having nonshielded insulation and operating at different voltage levels shall not occupy the same enclosure, cable, or raceway.

300.4 Protection Against Physical Damage.

Where subject to physical damage, conductors shall be adequately protected.

(A) Cables and Raceways Through Wood Members.

(1) Bored Holes. In both exposed and concealed locations, where a cable- or raceway-type wiring method is installed through bored holes in joists, rafters, or wood members, holes shall be bored so that the edge of the hole is not less than 32 mm (1¼ in.) from the nearest edge of the wood member. Where this distance cannot be maintained, the cable or raceway shall be protected from penetration by screws or nails by a steel plate or bushing, at least 1.6 mm (¹⁄₁₆ in.) thick, and of appropriate length and width installed to cover the area of the wiring.

Exception: Steel plates shall not be required to protect rigid metal conduit, intermediate metal conduit, rigid nonmetallic conduit, or electrical metallic tubing.

(2) Notches in Wood. Where there is no objection because of weakening the building structure, in both exposed and concealed locations, cables or raceways shall be permitted to be laid in notches in wood studs, joists, rafters, or other wood members where the cable or raceway at those points is protected against nails or screws by a steel plate at least 1.6 mm (¹⁄₁₆ in.) thick installed before the building finish is applied.

Exception: Steel plates shall not be required to protect rigid metal conduit, intermediate metal conduit, rigid nonmetallic conduit, or electrical metallic tubing.

The intent of 300.4(A)(1) is to prevent nails and screws from being driven into cables and raceways. Keeping the edge of a drilled hole 1¼ in. from the nearest edge of a stud, as shown in Exhibit 300.2, should prevent nails from penetrating the stud far enough to injure a cable. Building codes limit the maximum size of bored or notched holes in studs, and 300.4(A)(2) indicates that consideration should be given to the size of notches in studs, so as not to affect the strength of the structure.

Exhibit 300.2 A steel plate used to protect a nonmetallic-sheathed cable within 1¼ in. of the edge of a wood stud. (Courtesy of RACO, Inc.)

The exceptions to 300.4(A)(1) and (A)(2) permit intermediate metal conduit, rigid metal conduit, rigid nonmetallic conduit, and electrical metallic tubing to be installed through bored holes or laid in notches less than 1¼ in. from the nearest edge of the stud, without a steel plate or bushing.

(B) Nonmetallic-Sheathed Cables and Electrical Nonmetallic Tubing Through Metal Framing Members.

(1) Nonmetallic-Sheathed Cable. In both exposed and concealed locations where nonmetallic-sheathed cables pass through either factory or field punched, cut, or drilled slots or holes in metal members, the cable shall be protected by listed bushings or listed grommets covering all metal edges that are securely fastened in the opening prior to installation of the cable.

The phrase "listed bushing or listed grommet covering all metal edges" is new to the 2002 *Code*. This change requires the use of listed grommets or listed bushings that completely encircle Type NM cables as they pass through holes in metal studs. These listed devices must also securely seat in the stud opening and meet pull-out requirements of the product standard. This requirement affords physical protection for nonmetallic-sheathed cables as the cables are "pulled" through the openings in metal studs. Notice, too, that this requirement mandates all metal studs to be positioned in place before cable is pulled through protected openings. Fastening the listed grommet or listed bushing in place prior to installing cable is also mandatory. Should additional metal studs become necessary after installation of a cable, the cable must be removed before the stud is added. Field notching of metal studs and then installing the stud around a nonmetallic sheathed cable already installed or in place leads to cable damage and can result in insulation failure.

(2) Nonmetallic-Sheathed Cable and Electrical Nonmetallic Tubing. Where nails or screws are likely to penetrate nonmetallic-sheathed cable or electrical nonmetallic tubing, a steel sleeve, steel plate, or steel clip not less than 1.6 mm (1/16 in.) in thickness shall be used to protect the cable or tubing.

(C) Cables Through Spaces Behind Panels Designed to Allow Access. Cables or raceway-type wiring methods, installed behind panels designed to allow access, shall be supported according to their applicable articles.

Cable- or raceway-type wiring methods installed above suspended ceilings with lift-up panels must not be laid on the suspended ceiling. They are required to be supported according to 300.11(A), 300.23, and the requirements of the article applicable to the wiring method involved. Similarly, low-voltage, optical fiber, broadband, and communications cables are not permitted to block access to equipment above the suspended ceiling. Examples of this requirement are also found in 725.5, 760.5, 770.7, 800.5, 820.5, and 830.6. For supporting of low-voltage cables, optical fiber, broadband,

and communications cables, see 720.11, 725.6, 760.6, 770.8, 800.6, 820.6, and 830.7.

(D) Cables and Raceways Parallel to Framing Members. In both exposed and concealed locations, where a cable- or raceway-type wiring method is installed parallel to framing members, such as joists, rafters, or studs, the cable or raceway shall be installed and supported so that the nearest outside surface of the cable or raceway is not less than 32 mm (1¼ in.) from the nearest edge of the framing member where nails or screws are likely to penetrate. Where this distance cannot be maintained, the cable or raceway shall be protected from penetration by nails or screws by a steel plate, sleeve, or equivalent at least 1.6 mm (1/16 in.) thick.

Exception No. 1: Steel plates, sleeves, or the equivalent shall not be required to protect rigid metal conduit, intermediate metal conduit, rigid nonmetallic conduit, or electrical metallic tubing.

The intent of 300.4(D) is to prevent mechanical damage to cables and raceways from nails and screws. The *Code* offers two means of protection. The first method is to fasten the cable or raceway so that it is 1¼ in. from the edge of the framing member, as illustrated in Exhibit 300.3. This requirement generally applies to exposed and concealed work. The second method permits the cable or raceway to be installed closer than 1¼ in. from the edge of the framing member if physical protection, such as a steel plate or a sleeve, is provided. (A steel plate is illustrated in Exhibit 300.2.) As stated in Exception No. 1, this requirement does not apply to rigid metal conduit, rigid nonmetallic conduit, intermediate metal conduit, or electrical metallic tubing wiring methods because these methods provide physical protection for the conductors.

Exception No. 2: For concealed work in finished buildings, or finished panels for prefabricated buildings where such supporting is impracticable, it shall be permissible to fish the cables between access points.

(E) Cables and Raceways Installed in Shallow Grooves. Cable- or raceway-type wiring methods installed in a groove, to be covered by wallboard, siding, paneling, carpeting, or similar finish, shall be protected by 1.6 mm (1/16 in.) thick steel plate, sleeve, or equivalent or by not less than 32 mm (1¼ in.) free space for the full length of the groove in which the cable or raceway is installed.

Exception: Steel plates, sleeves, or the equivalent shall not be required to protect rigid metal conduit, intermediate metal

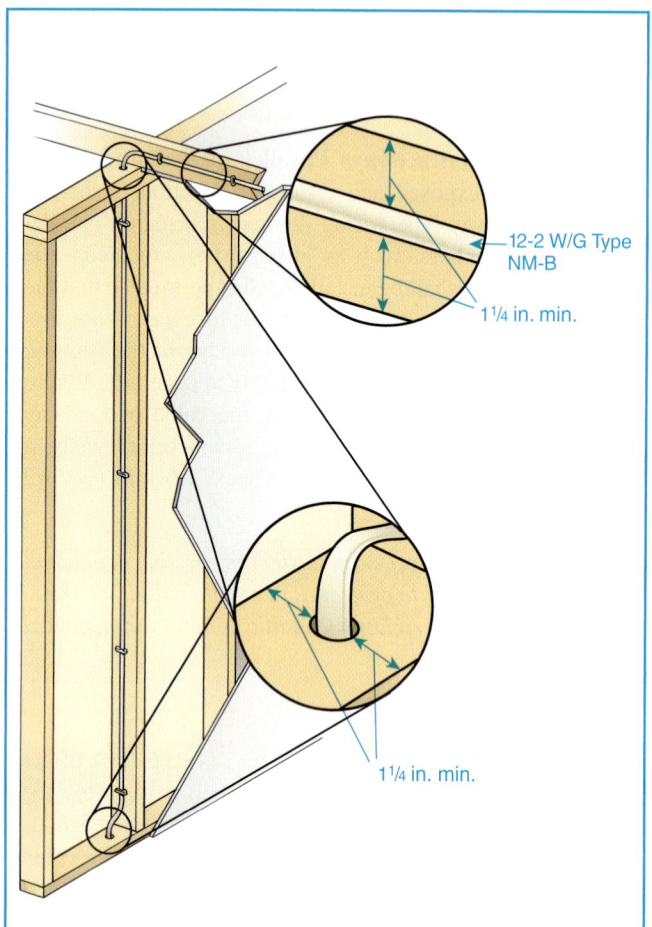

12-2 W/G Type NM-B

1¼ in. min.

1¼ in. min.

Exhibit 300.3 Cables and raceways installed parallel to framing members in accordance with 300.4(D).

conduit, rigid nonmetallic conduit, or electrical metallic tubing.

(F) Insulated Fittings. Where raceways containing ungrounded conductors 4 AWG or larger enter a cabinet, box enclosure, or raceway, the conductors shall be protected by a substantial fitting providing a smoothly rounded insulating surface, unless the conductors are separated from the fitting or raceway by substantial insulating material that is securely fastened in place.

Exception: Where threaded hubs or bosses that are an integral part of a cabinet, box enclosure, or raceway provide a smoothly rounded or flared entry for conductors.

Conduit bushings constructed wholly of insulating material shall not be used to secure a fitting or raceway. The insulating fitting or insulating material shall have a temperature rating not less than the insulation temperature rating of the installed conductors.

Heavy conductors and cables tend to stress the conductor insulation at terminating points. Providing insulated bushing or smooth rounded entries at raceway and cable terminations reduces the risk of insulation failure at conductor insulation "stress" points. The temperature ratings of insulating bushing must coordinate with the insulation of the conductor to ensure the protection remains intact over the life cycle of the insulated conductor.

Where 4 AWG or larger ungrounded conductors enter a cabinet, box enclosure, meter socket, or raceway, the conductors must be protected by a substantial fitting that provides a smoothly rounded insulating surface to protect the conductors from abrasion during and after installation. Because this requirement is located in 300.4(F), it applies generally to all wiring methods and all enclosures. See also 342.46, 344.46, and 352.46 for information relating to bushings.

Metal conduit bushings or fittings provided with insulated sleeves or linings are commonly used. A separate insulating sleeve or lining can also be used to separate the conductors from the raceway fitting.

Listed insulating bushings provided separately or as part of a fitting are colored black or brown if they are suitable for a temperature of 150°C and any other color for 90°C, unless specifically marked for a higher temperature. Exhibit 300.4 shows an insulated thermoplastic or fiber bushing that is used to protect the conductors from chafing against a metal conduit fitting. Note the use of a double locknut.

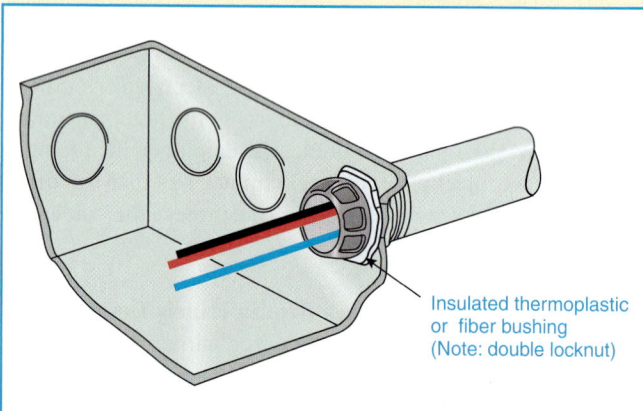

Insulated thermoplastic or fiber bushing (Note: double locknut)

Exhibit 300.4 An insulating bushing used to protect conductors from chafing against a metal conduit fitting.

300.5 Underground Installations.

(A) Minimum Cover Requirements. Direct-buried cable or conduit or other raceways shall be installed to meet the minimum cover requirements of Table 300.5.

Conductors under residential driveways must be at least 18 in. below grade. However, if the conductors are protected

Table 300.5 Minimum Cover Requirements, 0 to 600 Volts, Nominal, Burial in Millimeters (Inches)

Location of Wiring Method or Circuit	Type of Wiring Method or Circuit									
	Column 1 Direct Burial Cables or Conductors		Column 2 Rigid Metal Conduit or Intermediate Metal Conduit		Column 3 Nonmetallic Raceways Listed for Direct Burial Without Concrete Encasement or Other Approved Raceways		Column 4 Residential Branch Circuits Rated 120 Volts or Less with GFCI Protection and Maximum Overcurrent Protection of 20 Amperes		Column 5 Circuits for Control of Irrigation and Landscape Lighting Limited to Not More Than 30 Volts and Installed with Type UF or in Other Identified Cable or Raceway	
	mm	in.	mm	in.	mm	in.	mm	in.	mm	in.
All locations not specified below	600	24	150	6	450	18	300	12	150	6
In trench below 50-mm (2-in.) thick concrete or equivalent	450	18	150	6	300	12	150	6	150	6
Under a building	0 (in raceway only)	0	0	0	0	0	0 (in raceway only)	0	0 (in raceway only)	0
Under minimum of 102-mm (4-in.) thick concrete exterior slab with no vehicular traffic and the slab extending not less than 152 mm (6 in.) beyond the underground installation	450	18	100	4	100	4	150 (direct burial) 100 (in raceway)	6 4	150	6
Under streets, highways, roads, alleys, driveways, and parking lots	600	24	600	24	600	24	600	24	600	24
One- and two-family dwelling driveways and outdoor parking areas, and used only for dwelling-related purposes	450	18	450	18	450	18	300	12	450	18
In or under airport runways, including adjacent areas where trespassing prohibited	450	18	450	18	450	18	450	18	450	18

Notes:
1. Cover is defined as the shortest distance in millimeters (inches) measured between a point on the top surface of any direct-buried conductor, cable, conduit, or other raceway and the top surface of finished grade, concrete, or similar cover.
2. Raceways approved for burial only where concrete encased shall require concrete envelope not less than 50 mm (2 in.) thick.
3. Lesser depths shall be permitted where cables and conductors rise for terminations or splices or where access is otherwise required.

4. Where one of the wiring method types listed in Columns 1–3 is used for one of the circuit types in Columns 4 and 5, the shallower depth of burial shall be permitted.
5. Where solid rock prevents compliance with the cover depths specified in this table, the wiring shall be installed in metal or nonmetallic raceway permitted for direct burial. The raceways shall be covered by a minimum of 50 mm (2 in.) of concrete extending down to rock.

by an overcurrent device rated at not more than 20 amperes and provided with ground-fault circuit-interrupter (GFCI) protection for personnel, the burial depth may be reduced to 12 in. Exhibits 300.5 and 300.6 provide examples showing underground installations of 18 in. and 12 in., respectively.

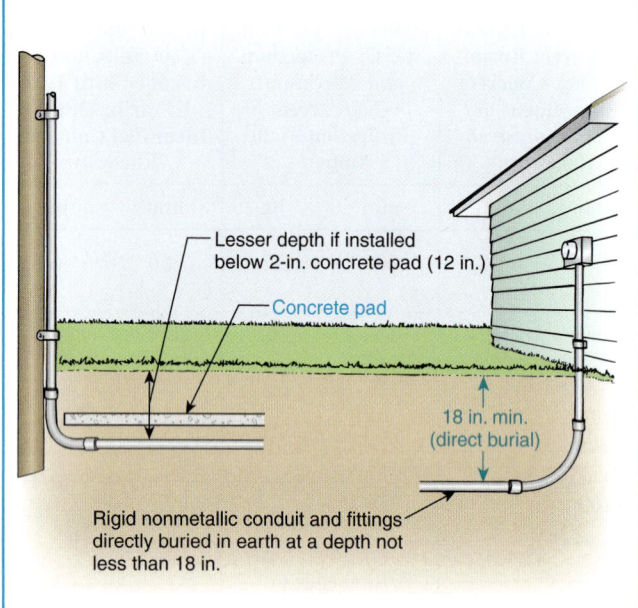

Exhibit 300.5 PVC rigid nonmetallic conduit buried in compliance with Table 300.5 and installed in accordance with 300.5(A).

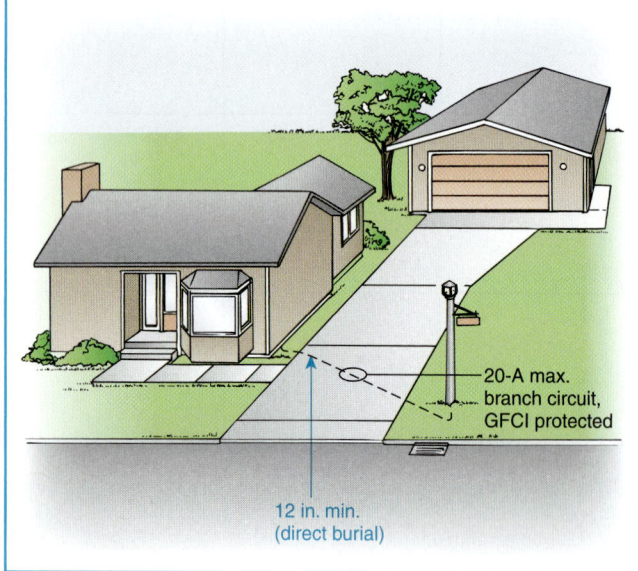

Exhibit 300.6 A 20-ampere, GFCI-protected residential branch circuit installed with a minimum burial depth of 12 in. beneath a residential driveway.

(B) Grounding. All underground installations shall be grounded and bonded in accordance with Article 250.

Rigid nonmetallic conduit elbows installed as part of a long run of conduit are often damaged in the process of pulling the conductors, due to friction at the bend. For service raceways, 250.80, Exception, permits a metal elbow to be installed without being grounded, provided it is isolated from possible contact by at least 18 in. of cover to any part of the elbow, as shown in Exhibit 300.7. For other than service raceways, Exception No. 3 to 250.86 applies.

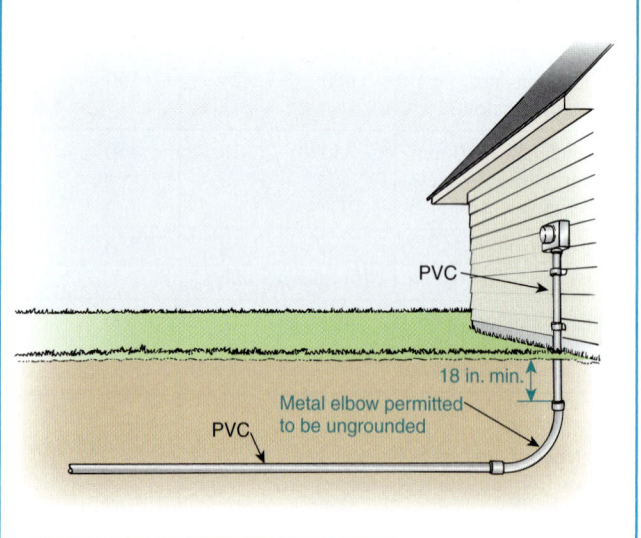

Exhibit 300.7 An application of 250.80, Exception, which permits the metal elbow to be ungrounded, provided it is isolated from contact by a minimum cover of 18 in. to any part of the elbow.

(C) Underground Cables Under Buildings. Underground cable installed under a building shall be in a raceway that is extended beyond the outside walls of the building.

(D) Protection from Damage. Direct-buried conductors and cables shall be protected from damage in accordance with (1) through (5).

(1) Emerging from Grade. Direct-buried conductors and enclosures emerging from grade shall be protected by enclosures or raceways extending from the minimum cover distance required by 300.5(A) below grade to a point at least 2.5 m (8 ft) above finished grade. In no case shall the protection be required to exceed 450 mm (18 in.) below finished grade.

(2) Conductors Entering Buildings. Conductors entering a building shall be protected to the point of entrance.

(3) Service Conductors. Underground service conductors that are not encased in concrete and that are buried 450 mm (18 in.) or more below grade shall have their location identified by a warning ribbon that is placed in the trench at least 300 mm (12 in.) above the underground installation.

Providing a warning ribbon reduces the risk of an accident or electrocution during excavation near underground service conductors that are not encased in concrete. This provision requiring a warning ribbon does not extend to feeders and branch circuits because these circuits contain short-circuit and overload protection.

(4) Enclosure or Raceway Damage. Where the enclosure or raceway is subject to physical damage, the conductors shall be installed in rigid metal conduit, intermediate metal conduit, Schedule 80 rigid nonmetallic conduit, or equivalent.

(5) Listing. Cables and insulated conductors installed in enclosures or raceways in underground installations shall be listed for use in wet locations.

Section 310.8(C) and Table 310.13 are used to determine which general wiring conductor types are permitted to be installed in wet locations.

(E) Splices and Taps. Direct-buried conductors or cables shall be permitted to be spliced or tapped without the use of splice boxes. The splices or taps shall be made in accordance with 110.14(B).

There is a difference between multiconductor cables labeled for direct burial and single conductors labeled for direct burial. Because direct-burial multiconductor cables may or may not contain individual conductors labeled for direct burial, the overall cable jacket may be the only underground protection technique for the contained conductors. Although the direct-burial splicing techniques used on multiconductor cables can differ widely from the techniques used on direct-burial single-conductor cables, the *Code* requirements are generally the same. The splicing technique should be listed for the cable type and listed for direct burial, due to the identified requirements and the listing requirements of 110.14(B) and 250.8. An example of an underground splicing method used with single-conductor direct-burial cables is shown in Exhibit 300.8.

(F) Backfill. Backfill that contains large rocks, paving materials, cinders, large or sharply angular substances, or corrosive material shall not be placed in an excavation where materials may damage raceways, cables, or other substruc-

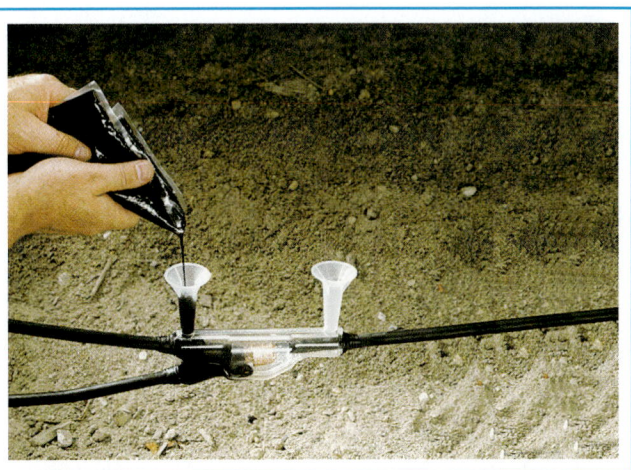

Exhibit 300.8 An underground splicing method. (Courtesy of 3M Co., Electrical Products Division)

tures or prevent adequate compaction of fill or contribute to corrosion of raceways, cables, or other substructures.

Where necessary to prevent physical damage to the raceway or cable, protection shall be provided in the form of granular or selected material, suitable running boards, suitable sleeves, or other approved means.

(G) Raceway Seals. Conduits or raceways through which moisture may contact energized live parts shall be sealed or plugged at either or both ends.

> FPN: Presence of hazardous gases or vapors may also necessitate sealing of underground conduits or raceways entering buildings.

Exhibit 300.9 shows a conduit sealing bushing used to prevent the entrance of gas or moisture. See 230.8 for sealing service raceways.

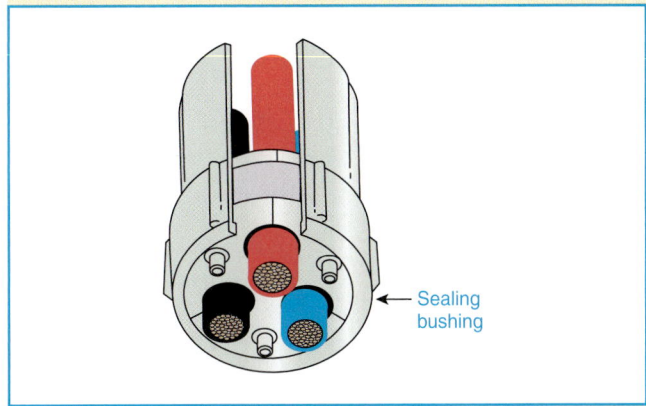

Sealing bushing

Exhibit 300.9 A conduit sealing bushing used to prevent the entrance of gas or moisture. (Redrawn from O.Z./Gedney Co.)

(H) Bushing. A bushing, or terminal fitting, with an integral bushed opening shall be used at the end of a conduit or other raceway that terminates underground where the conductors or cables emerge as a direct burial wiring method. A seal incorporating the physical protection characteristics of a bushing shall be permitted to be used in lieu of a bushing.

Exhibit 300.10 shows a Type UF cable buried in compliance with Table 300.5. Note the protective bushing where the cable is used with metal conduit. The commentary following 300.4(F) applies to this requirement also.

(I) Conductors of the Same Circuit. All conductors of the same circuit and, where used, the grounded conductor and all equipment grounding conductors shall be installed in the same raceway or cable or shall be installed in close proximity in the same trench.

Exception No. 1: Conductors in parallel in raceways or cables shall be permitted, but each raceway or cable shall contain all conductors of the same circuit including grounding conductors.

Conductors of the same circuit are also addressed in 300.3(B). Section 300.5(I), Exception No. 1, permits the installation of paralleled conductors in different raceways. See 310.4 for conductors in parallel.

Exception No. 2: Isolated phase, polarity, grounded conductor, and equipment grounding and bonding conductor instal-lations shall be permitted in nonmetallic raceways or cables with a nonmetallic covering or nonmagnetic sheath in close proximity where conductors are paralleled as permitted in 310.4, and where the conditions of 300.20(B) are met.

Isolated phase installations contain only one phase per raceway or cable. The spacing between isolated phase raceways and cables should be as small as possible and the length of the run limited to avoid increased circuit impedance and the resulting increase in voltage drop inherent in an installation involving ac circuits. Isolated phase installations may be used in ac circuits to limit available fault current at down-stream equipment.

Isolated phase installations present an inherent hazard of overheating, a risk that must be understood and carefully controlled. This hazard results from induced currents in metal surrounding a raceway that contains only one phase conductor. [See 300.20(A) and 300.20(B) for more information on induced currents in raceways.] The surrounding metal acts as a shorted transformer turn. In underground installations, a single conductor is unlikely to be installed in a metal raceway or, if it were, it is unlikely to present a fire hazard. This is not true, however, for aboveground raceways, and it is the reason isolated phase installations have limited application for aboveground installations.

See 300.3(B)(3) together with 330.31 and 332.31, which recognize single-conductor Type MI cable and single-conductor Type MC cable.

(J) Ground Movement. Where direct-buried conductors, raceways, or cables are subject to movement by settlement

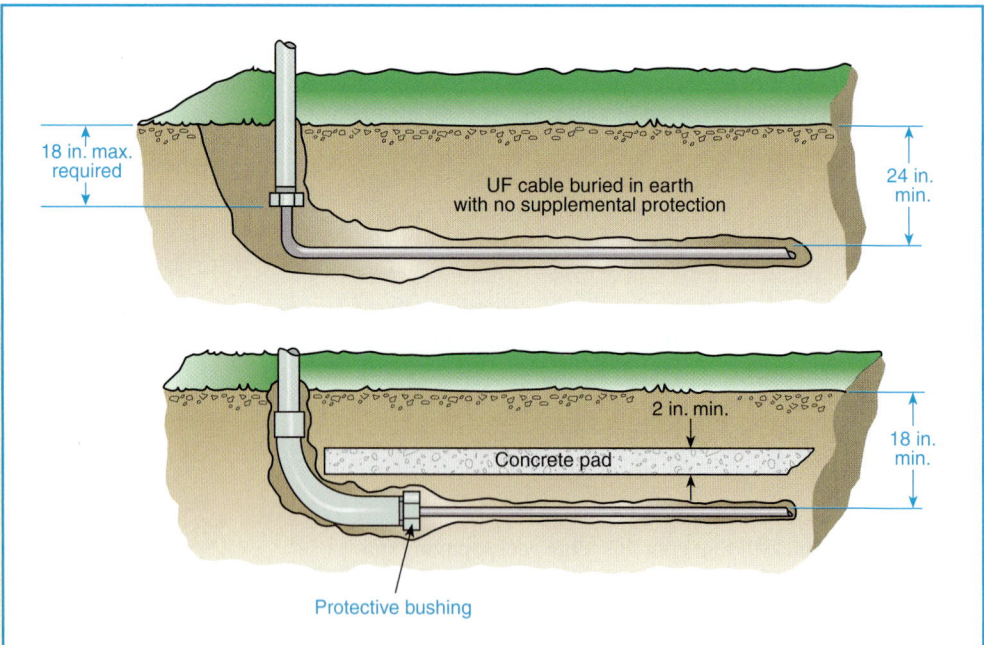

Exhibit 300.10 A Type UF cable buried in compliance with Table 300.5.

18 in. max. required

UF cable buried in earth with no supplemental protection

24 in. min.

2 in. min.

Concrete pad

18 in. min.

Protective bushing

or frost, direct-buried conductors, raceways, or cables shall be arranged to prevent damage to the enclosed conductors or to equipment connected to the raceways.

> FPN: This section recognizes "S" loops in underground direct burial to raceway transitions, expansion fittings in raceway risers to fixed equipment, and, generally, the provision of flexible connections to equipment subject to settlement or frost heaves.

Section 300.5(J) points out the practical need for installers to allow for movement of direct-buried equipment, cables, and raceways. Slack must be allowed in cables or expansion joints or other measures must be taken if ground movement due to frost or settlement is anticipated.

(K) Directional Boring. Cables or raceways installed using directional boring equipment shall be approved for the purpose.

Section 300.5(K) is new for the 2002 *Code*. Manufacturers of both metal and nonmetallic raceways suitable for underground use offer products that can endure the rigors of boring-type installation methods.

300.6 Protection Against Corrosion.

Metal raceways, cable trays, cablebus, auxiliary gutters, cable armor, boxes, cable sheathing, cabinets, elbows, couplings, fittings, supports, and support hardware shall be of materials suitable for the environment in which they are to be installed.

Section 300.6 applies generally. For specific applications, see the particular article covering the appropriate cables, raceways, or enclosures.

(A) General. Ferrous raceways, cable trays, cablebus, auxiliary gutters, cable armor, boxes, cable sheathing, cabinets, metal elbows, couplings, fittings, supports, and support hardware shall be suitably protected against corrosion inside and outside (except threads at joints) by a coating of approved corrosion-resistant material such as zinc, cadmium, or enamel. Where protected from corrosion solely by enamel, they shall not be used outdoors or in wet locations as described in 300.6(C). Where boxes or cabinets have an approved system of organic coatings and are marked "Raintight," "Rainproof," or "Outdoor Type," they shall be permitted outdoors. Where corrosion protection is necessary and the conduit is threaded in the field, the threads shall be coated with an approved electrically conductive, corrosion-resistant compound.

Zinc chromate paste is one type of electrically conductive compound that could be approved.

(B) In Concrete or in Direct Contact with the Earth. Ferrous or nonferrous metal raceways, cable armor, boxes, cable sheathing, cabinets, elbows, couplings, fittings, supports, and support hardware shall be permitted to be installed in concrete or in direct contact with the earth, or in areas subject to severe corrosive influences where made of material judged suitable for the condition, or where provided with corrosion protection approved for the condition.

Where ferrous or nonferrous metal conduit has corrosion protection and is judged suitable for the condition, it may be installed in concrete, in contact with the earth, or in areas exposed to severe corrosive influence. Special precautions are normally necessary for installing aluminum conduits in concrete, and specific approval by the authority having jurisdiction may be necessary.

Metal raceways installed in the earth can be coated with an asphalt compound, plastic sheath, or other equivalent protection to help prevent deterioration. Also, metallic raceways are available with a bonded PVC coating.

Galvanized rigid steel conduit and steel intermediate metal conduit do not generally require supplementary corrosion protection.

(C) Indoor Wet Locations. In portions of dairy processing facilities, laundries, canneries, and other indoor wet locations, and in locations where walls are frequently washed or where there are surfaces of absorbent materials, such as damp paper or wood, the entire wiring system, where installed exposed, including all boxes, fittings, conduits, and cable used therewith, shall be mounted so that there is at least a 6-mm (¼-in.) airspace between it and the wall or supporting surface.

Exception: Nonmetallic raceways, boxes, and fittings shall be permitted to be installed without the airspace on a concrete, masonry, tile, or similar surface.

> FPN: In general, areas where acids and alkali chemicals are handled and stored may present such corrosive conditions, particularly when wet or damp. Severe corrosive conditions may also be present in portions of meatpacking plants, tanneries, glue houses, and some stables; in installations immediately adjacent to a seashore and swimming pool areas; in areas where chemical deicers are used; and in storage cellars or rooms for hides, casings, fertilizer, salt, and bulk chemicals.

The exception to 300.6(C) is in harmony with 547.5(B), permitting nonmetallic boxes, fittings, conduit, and cables to be installed without the airspace in corrosive locations of agricultural buildings.

300.7 Raceways Exposed to Different Temperatures.

(A) Sealing. Where portions of a cable raceway or sleeve are known to be subjected to different temperatures and where condensation is known to be a problem, as in cold storage areas of buildings or where passing from the interior to the exterior of a building, the raceway or sleeve shall be filled with an approved material to prevent the circulation of warm air to a colder section of the raceway or sleeve. An explosionproof seal shall not be required for this purpose.

Where a raceway is used to enclose the lighting and refrigeration branch-circuit conductors within a walk-in chest, the circulation of air through the raceway from a warmer to a colder section could cause condensation within the raceway. A circulation of air can be prevented by sealing the raceway with a suitable pliable compound at a conduit body or junction box, usually installed in the raceway before it enters the colder section. Special sealing fittings, such as those used in hazardous (classified) locations, are not necessary.

(B) Expansion Fittings. Raceways shall be provided with expansion fittings where necessary to compensate for thermal expansion and contraction.

> FPN: Table 352.44(A) provides the expansion information for polyvinyl chloride (PVC). A nominal number for steel conduit can be determined by multiplying the expansion length in this table by 0.20. The coefficient of expansion for steel electrical metallic tubing, intermediate metal conduit, and rigid conduit is 11.70×10^{-6} (0.0000117 mm per mm of conduit for each °C in temperature change) [6.50×10^{-6} (0.0000065 in. per inch of conduit for each °F in temperature change)].

This fine print note provides the relationship of linear expansion of PVC rigid nonmetallic conduit to steel conduit. For example, if a calculation indicated a linear expansion of 1¼ in. for PVC conduit, the steel conduit equivalent of expansion would be only ¼ in.

300.8 Installation of Conductors with Other Systems.

Raceways or cable trays containing electric conductors shall not contain any pipe, tube, or equal for steam, water, air, gas, drainage, or any service other than electrical.

This section specifically prohibits installation of an electrical conductor in a raceway or cable tray that includes a drain, water, oil, air, or similar pipe.

300.10 Electrical Continuity of Metal Raceways and Enclosures.

Metal raceways, cable armor, and other metal enclosures for conductors shall be metallically joined together into a continuous electric conductor and shall be connected to all boxes, fittings, and cabinets so as to provide effective electrical continuity. Unless specifically permitted elsewhere in this *Code*, raceways and cable assemblies shall be mechanically secured to boxes, fittings, cabinets, and other enclosures.

Sections 250.4(A) and (B) set forth in detail what must be accomplished by grounding and bonding metal parts of the electrical system. These metal parts must form an effective low-impedance path to ground in order to safely conduct any fault current and facilitate the operation of overcurrent devices protecting the enclosed circuit conductors.

Exception No. 1: Short sections of raceways used to provide support or protection of cable assemblies from physical damage shall not be required to be made electrically continuous.

Exception No. 2: Equipment enclosures to be isolated, as permitted by 250.96(B), shall not be required to be metallically joined to the metal raceway.

300.11 Securing and Supporting.

(A) Secured in Place. Raceways, cable assemblies, boxes, cabinets, and fittings shall be securely fastened in place. Support wires that do not provide secure support shall not be permitted as the sole support. Support wires and associated fittings that provide secure support and that are installed in addition to the ceiling grid support wires shall be permitted as the sole support. Where independent support wires are used, they shall be secured at both ends. Cables and raceways shall not be supported by ceiling grids.

(1) Fire-Rated Assemblies. Wiring located within the cavity of a fire-rated floor–ceiling or roof–ceiling assembly shall not be secured to, or supported by, the ceiling assembly, including the ceiling support wires. An independent means of secure support shall be provided. Where independent support wires are used, they shall be distinguishable by color, tagging, or other effective means from those that are part of the fire-rated design.

Exception: The ceiling support system shall be permitted to support wiring and equipment that have been tested as part of the fire-rated assembly.

Wiring methods of any type and all luminaires are not allowed to be supported or secured to the support wires or T bars of a fire-rated ceiling assembly unless the assembly has been tested and listed for that use. If support wires are selected as the supporting means for the electrical system within the fire-rated ceiling cavity, they must be distinguishable from the ceiling support wires and they must be secured at both ends.

Generally, the rule for supporting electrical equipment is "securely fastened in place." This phrase means not only that vertical support for the weight of the equipment must be provided but also that the equipment must be secured to prevent horizontal movement or sway. The intention is to prevent the loss of grounding continuity provided by the raceway that could result from horizontal movement.

Sections 300.11(A)(1) and (A)(2) are quite similar. Unless the exceptions apply, these sections clearly prohibit all types of wiring from being attached in any way to the support wires of a ceiling assembly. Unless ceiling grids are part of the building structure, they, too, are prohibited from furnishing support for cables and raceways. However, if wiring and equipment are located within the ceiling cavity and rigidly supported independent of the ceiling, without the use of ceiling-type hanger wire, then the requirements of this section are met.

Refer to the appropriate wiring method article in Chapter 3 of the *Code* for cable and raceway supporting requirements. See 410.15(A) and 410.16 for the proper support of luminaires; 314.23 for the support of outlet boxes; and 725.6, 760.6, and 770.8 for various low-voltage fire alarm and optical fiber cable supports. See Chapter 8 for communications cable supports.

FPN: One method of determining fire rating is testing in accordance with NFPA 251-1999, *Standard Methods of Tests of Fire Endurance of Building Construction and Materials.*

(2) Non–Fire-Rated Assemblies. Wiring located within the cavity of a non–fire-rated floor–ceiling or roof–ceiling assembly shall not be secured to, or supported by, the ceiling assembly, including the ceiling support wires. An independent means of secure support shall be provided.

Exception: The ceiling support system shall be permitted to support branch-circuit wiring and associated equipment where installed in accordance with the ceiling system manufacturer's instructions.

(B) Raceways Used as Means of Support. Raceways shall only be used as a means of support for other raceways, cables, or nonelectric equipment under the following conditions:

(1) Where the raceway or means of support is identified for the purpose; or
(2) Where the raceway contains power supply conductors for electrically controlled equipment and is used to support Class 2 circuit conductors or cables that are solely

for the purpose of connection to the equipment control circuits; or
(3) Where the raceway is used to support boxes or conduit bodies in accordance with 314.23 or to support luminaires (fixtures) in accordance with 410.16(F)

The purpose of 300.11(B)(3) is to prevent cables from being attached to the exterior of a raceway. Electrical, telephone, and computer cables wrapped around a raceway can prevent dissipation of heat from the raceway and affect the temperature of the conductors therein. This section also prohibits the use of a raceway as a means of support for nonelectric equipment, such as suspended ceilings, water pipes, nonelectric signs, and the like, which could cause a mechanical failure of the raceway.

However, 300.11(B)(2) does allow the installation of Class 2 thermostat conductors for a boiler or air conditioner unit to be supported by the conduit supplying power to the unit, as shown in Exhibit 300.11. These Class 2 circuits are functionally associated with the branch-circuit wiring method.

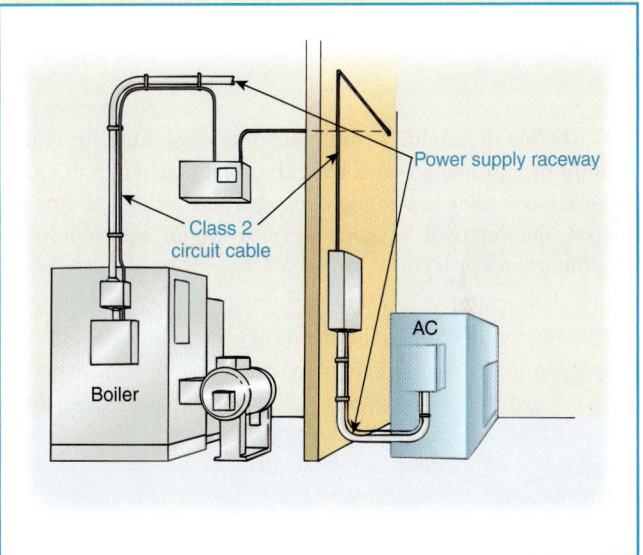

Exhibit 300.11 Raceways used to support Class 2 thermostat cables.

(C) Cables Not Used as Means of Support. Cable wiring methods shall not be used as a means of support for other cables, raceways, or nonelectrical equipment.

Added for the 2002 *Code*, this section prohibits cables from being used as a means of support for other cables, raceways, or nonelectric equipment. Taking the requirements of both 300.11(B) and (C) together, the indiscriminate practice of using one supported cable or raceway to support many other raceways and cables is now properly limited.

300.12 Mechanical Continuity—Raceways and Cables.

Metal or nonmetallic raceways, cable armors, and cable sheaths shall be continuous between cabinets, boxes, fittings, or other enclosures or outlets.

Exception: Short sections of raceways used to provide support or protection of cable assemblies from physical damage shall not be required to be mechanically continuous.

300.13 Mechanical and Electrical Continuity—Conductors.

(A) General. Conductors in raceways shall be continuous between outlets, boxes, devices, and so forth. There shall be no splice or tap within a raceway unless permitted by 300.15; 368.8(A); 376.56; 378.56; 384.56; 386.56; 388.56; or 390.6.

Splices or taps are prohibited within raceways unless the raceways are equipped with hinged or removable covers. Busway conductors are exempt from this requirement. Splices and taps must be accessible according to 300.15.

(B) Device Removal. In multiwire branch circuits, the continuity of a grounded conductor shall not depend on device connections such as lampholders, receptacles, and so forth, where the removal of such devices would interrupt the continuity.

Grounded conductors (neutrals) of multiwire branch circuits supplying receptacles, lampholders, or other such devices are not permitted to depend on terminal connections for continuity between devices. For such installations (3- or 4-wire circuits), a splice is made and a jumper is connected to the terminal, unless the neutral is looped; that is, a receptacle or lampholder could be replaced without interrupting the continuity of energized downstream line-to-neutral loads (see commentary to 300.14). Opening the neutral could cause unbalanced voltages, and a considerably higher voltage would be impressed on one part of a multiwire branch circuit, especially if the downstream line-to-neutral loads were appreciably unbalanced. This requirement does not apply to individual 2-wire circuits or other circuits that do not contain a grounded (neutral) conductor.

300.14 Length of Free Conductors at Outlets, Junctions, and Switch Points.

At least 150 mm (6 in.) of free conductor, measured from the point in the box where it emerges from its raceway or cable sheath, shall be left at each outlet, junction, and switch point for splices or the connection of luminaires (fixtures) or devices. Where the opening to an outlet, junction, or switch point is less than 200 mm (8 in.) in any dimension, each conductor shall be long enough to extend at least 75 mm (3 in.) outside the opening.

Exception: Conductors that are not spliced or terminated at the outlet, junction, or switch point shall not be required to comply with 300.14.

A conductor looping through an outlet box and intended for connection to receptacles, switches, lampholders, or other such devices requires slack so that terminal connections can be made easily. Conductors running through a box should have sufficient slack to prevent physical damage from the insertion of devices or from the use of fixture studs, hickeys, or other fixture supports within the box.

Revised for the 1999 *Code,* this section is more specific about the measurements of free conductor length required at each splice point or device outlet. For these free conductor length measurements, see Exhibit 300.12.

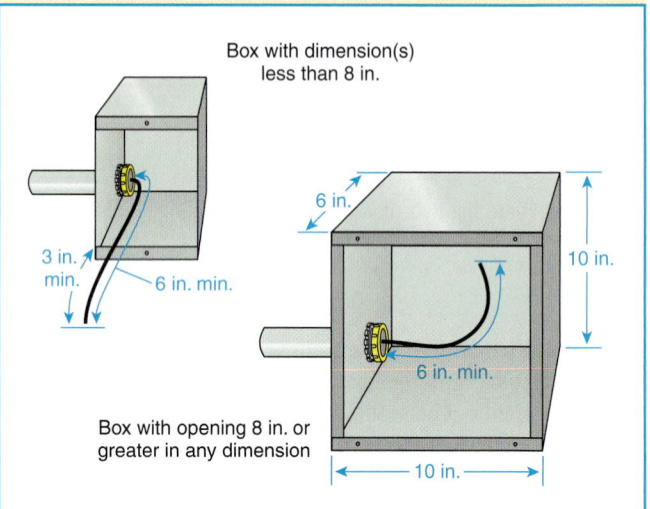

Exhibit 300.12 Two different boxes with free conductor lengths illustrated.

300.15 Boxes, Conduit Bodies, or Fittings—Where Required.

A box shall be installed at each outlet and switch point for concealed knob-and-tube wiring.

Fittings and connectors shall be used only with the specific wiring methods for which they are designed and listed.

Where the wiring method is conduit, tubing, Type AC cable, Type MC cable, Type MI cable, nonmetallic-sheathed cable, or other cables, a box or conduit body complying with Article 314 shall be installed at each conductor splice point, outlet point, switch point, junction point, termination point, or pull point, unless otherwise permitted in 300.15(A) through (M).

(A) Wiring Methods with Interior Access. A box or conduit body shall not be required for each splice, junction, switch, pull, termination, or outlet points in wiring methods with removable covers, such as wireways, multioutlet assemblies, auxiliary gutters, and surface raceways. The covers shall be accessible after installation.

(B) Equipment. An integral junction box or wiring compartment as part of approved equipment shall be permitted in lieu of a box.

(C) Protection. A box or conduit body shall not be required where cables enter or exit from conduit or tubing that is used to provide cable support or protection against physical damage. A fitting shall be provided on the end(s) of the conduit or tubing to protect the cable from abrasion.

Section 300.15(C) permits conduit or tubing to be used as support and protection against physical damage without terminating in a box. It also permits conduit or tubing to be used as physical protection for underground cables that exit from buildings or that are located outdoors on poles, without a box being required on the end of the conduit. A fitting to protect the wires or cables against physical damage is required on the ends of the conduit or tubing.

(D) Type MI Cable. A box or conduit body shall not be required where accessible fittings are used for straight-through splices in mineral-insulated metal-sheathed cable.

(E) Integral Enclosure. A wiring device with integral enclosure identified for the use, having brackets that securely fasten the device to walls or ceilings of conventional on-site frame construction, for use with nonmetallic-sheathed cable, shall be permitted in lieu of a box or conduit body.

> FPN: See 334.30(C); 545.10; 550.15(I), 551.47(E), Exception No. 1; and 552.48(E), Exception No. 1.

Section 300.15(E) applies to a device with an integral enclosure (boxless device) such as the one shown in Exhibit 300.13.

(F) Fitting. A fitting identified for the use shall be permitted in lieu of a box or conduit body where conductors are not spliced or terminated within the fitting. The fitting shall be accessible after installation.

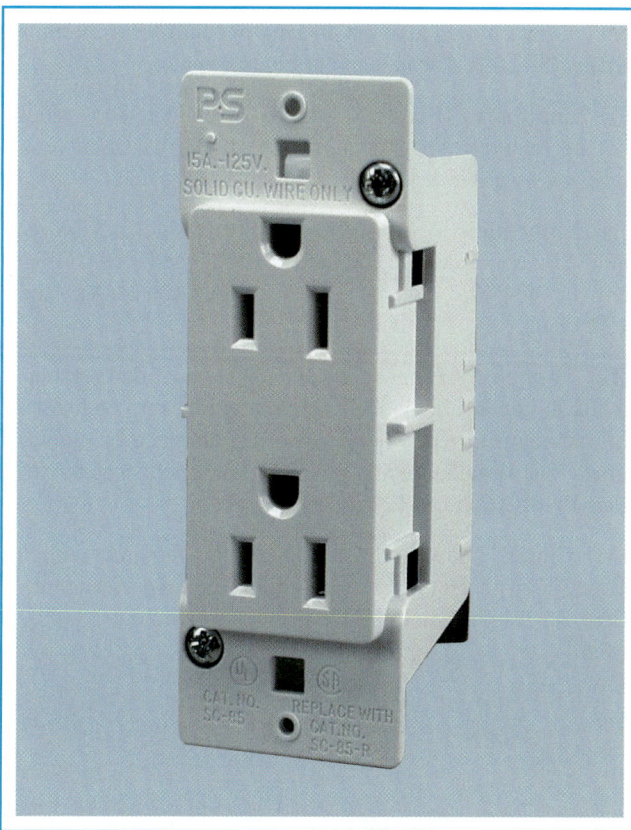

Exhibit 300.13 A self-contained device (SCD) receptacle. (Courtesy of Pass & Seymour/Legrand®)

Where a cable system makes a transition to a raceway to provide mechanical protection against damage, 300.15(F) permits the use of a fitting instead of a box. For example, where nonmetallic-sheathed cable that runs overhead on floor joists and drops down on a masonry wall to supply a receptacle needs to be protected from physical damage, a short length of raceway is installed to the outlet device box. The cable sheath is removed for the length of the raceway. The cable is then inserted in the raceway and secured by a combination fitting that is fastened to the end of the raceway.

(G) Direct-Buried Conductors. As permitted in 300.5(E), a box or conduit body shall not be required for splices and taps in direct-buried conductors and cables.

(H) Insulated Devices. As permitted in 334.40(B), a box or conduit body shall not be required for insulated devices supplied by nonmetallic-sheathed cable.

(I) Enclosures. A box or conduit body shall not be required where a splice, switch, terminal, or pull point is in a cabinet or cutout box, in an enclosure for a switch or overcurrent device as permitted in 312.8, in a motor controller as permitted in 430.10(A), or in a motor control center.

(J) Luminaires (Fixtures). A box or conduit body shall not be required where a luminaire (fixture) is used as a raceway as permitted in 410.31 and 410.32.

(K) Embedded. A box or conduit body shall not be required for splices where conductors are embedded as permitted in 424.40, 424.41(D), 426.22(B), 426.24(A), and 427.19(A).

(L) Manholes. Where accessible only to qualified persons, a box or conduit body shall not be required for conductors in manholes, except where connecting to electrical equipment. The installation shall comply with the provisions of Part IV of Article 314.

(M) Closed Loop. A box shall not be required with a closed-loop power distribution system where a device identified and listed as suitable for installation without a box is used.

See Article 780, Closed-Loop and Programmed Power Distribution.

300.16 Raceway or Cable to Open or Concealed Wiring.

(A) Box or Fitting. A box or terminal fitting having a separately bushed hole for each conductor shall be used wherever a change is made from conduit, electrical metallic tubing, electrical nonmetallic tubing, nonmetallic-sheathed cable, Type AC cable, Type MC cable, or mineral-insulated, metal-sheathed cable and surface raceway wiring to open wiring or to concealed knob-and-tube wiring. A fitting used for this purpose shall contain no taps or splices and shall not be used at luminaire (fixture) outlets.

(B) Bushing. A bushing shall be permitted in lieu of a box or terminal where the conductors emerge from a raceway and enter or terminate at equipment, such as open switchboards, unenclosed control equipment, or similar equipment. The bushing shall be of the insulating type for other than lead-sheathed conductors.

300.17 Number and Size of Conductors in Raceway.

The number and size of conductors in any raceway shall not be more than will permit dissipation of the heat and ready installation or withdrawal of the conductors without damage to the conductors or to their insulation.

> FPN: See the following sections of this *Code*: intermediate metal conduit, 342.22; rigid metal conduit, 344.22;

flexible metal conduit, 348.22; liquidtight flexible metal conduit, 350.22; rigid nonmetallic conduit, 352.22; liquidtight nonmetallic flexible conduit, 356.22; electrical metallic tubing, 358.22; flexible metallic tubing, 360.22; electrical nonmetallic tubing, 362.22; cellular concrete floor raceways, 372.11; cellular metal floor raceways, 374.5; metal wireways, 376.22; nonmetallic wireways, 378.22; surface metal raceways, 386.22; surface nonmetallic raceways 388.22; underfloor raceways, 390.5; fixture wire, 402.7; theaters, 520.6; signs, 600.31(C); elevators, 620.33; audio signal processing, amplification, and reproduction equipment, 640.23(A) and 640.24; Class 1, Class 2, and Class 3 circuits, Article 725; fire alarm circuits, Article 760; and optical fiber cables and raceways, Article 770.

Listed wire-pulling compounds are available to assist in the process of pulling wires into raceways. As pointed out in 310.9, wire-pulling compounds should not be used if they have a harmful effect on either the conductor or the conductor insulation.

300.18 Raceway Installations.

(A) Complete Runs. Raceways, other than busways or exposed raceways having hinged or removable covers, shall be installed complete between outlet, junction, or splicing points prior to the installation of conductors. Where required to facilitate the installation of utilization equipment, the raceway shall be permitted to be initially installed without a terminating connection at the equipment. Prewired raceway assemblies shall be permitted only where specifically permitted in this *Code* for the applicable wiring method.

One of the primary functions of a raceway is to provide physical protection for conductors. If raceways are incomplete at the time of conductor installation, a greater possibility exists for damage to the conductors.

Section 300.18(A) does, however, permit the installation of conductors in a raceway prior to the complete installation of the raceway, up to the point of utilization. The motor installation shown in Exhibit 300.14 is a typical application, where the motor is supplied through liquidtight flexible metal conduit that terminates in the motor terminal box through a 90 degree angle connector. Wiring a fixture whip prior to connecting a luminaire is also permitted by this section.

(B) Welding. Metal raceways shall not be supported, terminated, or connected by welding to the raceway unless specifically designed to be or otherwise specifically permitted to be in this *Code*.

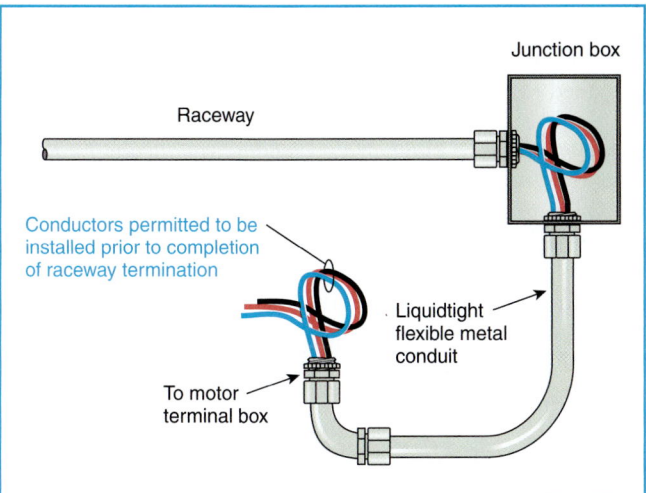

Raceway

Junction box

Conductors permitted to be installed prior to completion of raceway termination

Liquidtight flexible metal conduit

To motor terminal box

Exhibit 300.14 An application of 300.18(A), which permits the conductors supplying a motor through liquidtight flexible metal conduit to be installed prior to the connection of the raceway to the motor terminal box.

300.19 Supporting Conductors in Vertical Raceways.

(A) Spacing Intervals—Maximum. Conductors in vertical raceways shall be supported if the vertical rise exceeds the values in Table 300.19(A). One cable support shall be provided at the top of the vertical raceway or as close to the top as practical. Intermediate supports shall be provided as necessary to limit supported conductor lengths to not greater than those values specified in Table 300.19(A).

Exception: Steel wire armor cable shall be supported at the top of the riser with a cable support that clamps the steel wire armor. A safety device shall be permitted at the lower end of the riser to hold the cable in the event there is slippage of the cable in the wire-armored cable support. Additional

wedge-type supports shall be permitted to relieve the strain on the equipment terminals caused by expansion of the cable under load.

(B) Support Methods. One of the following methods of support shall be used.

(1) By clamping devices constructed of or employing insulating wedges inserted in the ends of the raceways. Where clamping of insulation does not adequately support the cable, the conductor also shall be clamped.

(2) By inserting boxes at the required intervals in which insulating supports are installed and secured in a satisfactory manner to withstand the weight of the conductors attached thereto, the boxes being provided with covers.

(3) In junction boxes, by deflecting the cables not less than 90 degrees and carrying them horizontally to a distance not less than twice the diameter of the cable, the cables being carried on two or more insulating supports and additionally secured thereto by tie wires if desired. Where this method is used, cables shall be supported at intervals not greater than 20 percent of those mentioned in the preceding tabulation.

(4) By a method of equal effectiveness.

Conductors in long vertical runs must be supported if the vertical rise exceeds the values in Table 300.19(A). This requirement prevents the weight of the conductors from damaging the insulation where they leave the conduit and prevents the conductors from being pulled out of the terminals. Supports such as those shown in Exhibits 300.15 and 300.16 may be used, in addition to many other types of grips manufactured for this purpose.

Example

A vertical raceway contains 1/0 AWG copper conductors. One cable support near the top of the run would be required

Table 300.19(A) Spacings for Conductor Supports

		Conductors			
		Aluminum or Copper-Clad Aluminum		Copper	
Size of Wire	Support of Conductors in Vertical Raceways	m	ft	m	ft
18 AWG through 8 AWG	Not greater than	30	100	30	100
6 AWG through 1/0 AWG	Not greater than	60	200	30	100
2/0 AWG through 4/0 AWG	Not greater than	55	180	25	80
Over 4/0 AWG through 350 kcmil	Not greater than	41	135	18	60
Over 350 kcmil through 500 kcmil	Not greater than	36	120	15	50
Over 500 kcmil through 750 kcmil	Not greater than	28	95	12	40
Over 750 kcmil	Not greater than	26	85	11	35

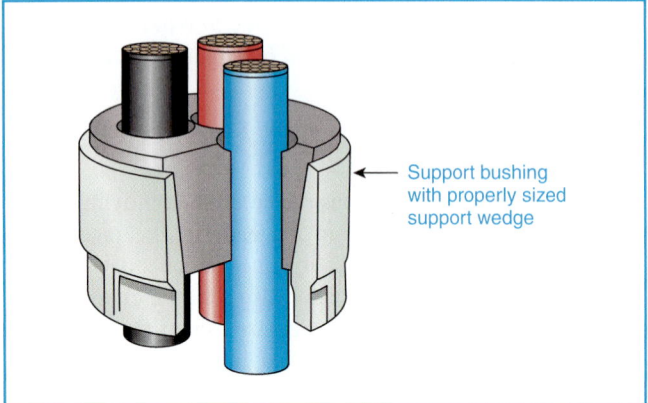

Exhibit 300.15 A support bushing, located at the top of a vertical conduit at a cabinet or pull box, used to prevent the weight of the conductors from damaging insulation or placing strain on termination points. (Redrawn from O.Z./Gedney Co.)

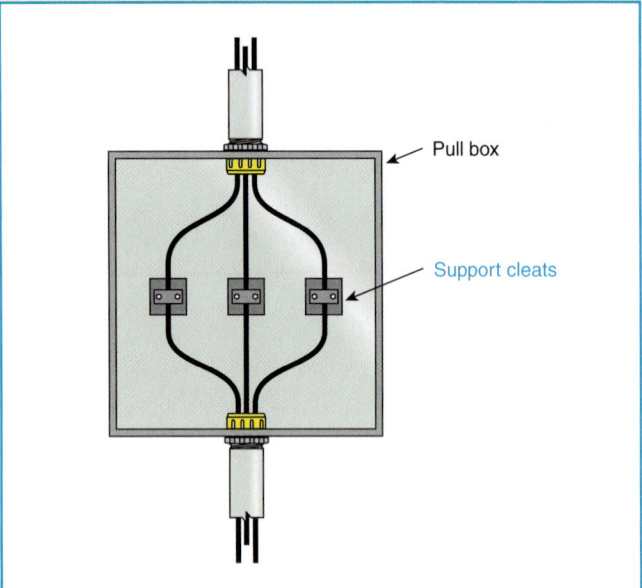

Exhibit 300.16 Support cleats used to prevent the weight of vertical conductors from damaging insulation or placing strain on termination points.

if the vertical run exceeds 100 ft. Intermediate supports may be required to limit the supported length to the table values. If the vertical run is less than 100 ft, cable supports would not be required.

300.20 Induced Currents in Metal Enclosures or Metal Raceways.

(A) Conductors Grouped Together. Where conductors carrying alternating current are installed in metal enclosures or metal raceways, they shall be arranged so as to avoid heating the surrounding metal by induction. To accomplish this, all phase conductors and, where used, the grounded conductor and all equipment grounding conductors shall be grouped together.

Exception No. 1: Equipment grounding conductors for certain existing installations shall be permitted to be installed separate from their associated circuit conductors where run in accordance with the provisions of 250.130(C).

Exception No. 2: A single conductor shall be permitted to be installed in a ferromagnetic enclosure and used for skin-effect heating in accordance with the provisions of 426.42 and 427.47.

(B) Individual Conductors. Where a single conductor carrying alternating current passes through metal with magnetic properties, the inductive effect shall be minimized by (1) cutting slots in the metal between the individual holes through which the individual conductors pass or (2) passing all the conductors in the circuit through an insulating wall sufficiently large for all of the conductors of the circuit.

Exception: In the case of circuits supplying vacuum or electric-discharge lighting systems or signs or X-ray apparatus, the currents carried by the conductors are so small that the inductive heating effect can be ignored where these conductors are placed in metal enclosures or pass through metal.

> FPN: Because aluminum is not a magnetic metal, there will be no heating due to hysteresis; however, induced currents will be present. They will not be of sufficient magnitude to require grouping of conductors or special treatment in passing conductors through aluminum wall sections.

Section 300.3(B)(3) permits single-conductor Type MI cable as well as single-conductor Type MC cable. In addition to requirements contained in their respective Articles (332 for Type MI and 330 for Type MC), the installation must conform to 300.20 regarding inductive effects.

300.21 Spread of Fire or Products of Combustion.

Electrical installations in hollow spaces, vertical shafts, and ventilation or air-handling ducts shall be made so that the possible spread of fire or products of combustion will not be substantially increased. Openings around electrical penetrations through fire-resistant-rated walls, partitions, floors, or ceilings shall be firestopped using approved methods to maintain the fire resistance rating.

> FPN: Directories of electrical construction materials published by qualified testing laboratories contain many listing installation restrictions necessary to maintain the

fire-resistive rating of assemblies where penetrations or openings are made. Building codes also contain restrictions on membrane penetrations on opposite sides of a fire-resistance–rated wall assembly. An example is the 600-mm (24-in.) minimum horizontal separation that usually applies between boxes installed on opposite sides of the wall. Assistance in complying with 300.21 can be found in building codes, fire resistance directories, and product listings.

The intent of 300.21 is that cables, cable trays, and raceways be installed through rated wall, floor, and ceiling assemblies in such a manner that they do not contribute to the spread of fire or the products of combustion. NFPA 221, *Standard for Fire Walls and Fire Barrier Walls,* defines *fire resistance rating* as "the time, in minutes or hours, that materials or assemblies have withstood a fire exposure as established in accordance with the test procedures of NFPA 251, *Standard Methods of Tests of Fire Endurance of Building Construction and Materials.*" (ASTM E 119, *Standard Test Methods for Fire Tests of Building Construction and Materials,* and ANSI/UL 263, *Fire Tests of Building Construction and Materials,* are similar to NFPA 251.)

Further, NFPA 221, Section 4.2, Penetration Seals, states the following:

All through-penetration protection systems shall be tested and rated in accordance with ASTM E 814, *Standard Test Method for Fire Tests of Through-Penetration Fire Stops,* or ANSI/UL 1479, *Fire Test of Through-Penetration Fire Stops.* The positive pressure difference between the exposed and unexposed surfaces of the test assembly shall not be less than 0.01 in. (2.5 Pa) water gauge. A through-penetration protection system shall have an F rating not less than the required fire resistance rating of the fire wall or fire barrier wall.

Exception: Concrete, mortar, or grout shall be permitted with maximum 6-in. (153-mm) nominal diameter steel or copper pipe or steel conduit. Concrete, mortar, or grout shall be the thickness required to maintain the required fire resistance rating of the wall being penetrated. The maximum opening size shall be 144 in.^{2} (0.094 m^2).

According to the UL *Fire Resistance Directory — Volume 2, Category XHEX, Through-Penetration Firestop Systems,* a firestop system is a specific construction consisting of a wall or floor assembly, a penetrating item passing through an opening in the wall or floor assembly, and the materials designed to prevent the spread of fire through the openings. The specifications for materials in a firestop system and the assembly of the materials are details that directly relate to the established ratings and are described in the individual systems. The hourly ratings apply only to the complete systems. Individual components, which are designated for use in a specific system to achieve specified ratings, are not assigned ratings and are not intended to be interchanged between systems. Additionally, the substitution or elimination of components required in a system should not be made unless specifically permitted in the individual system.

The basic standard used to investigate products in this category — ANSI/UL 1479, *Fire Tests of Through-Penetration Firestops* — defines the criteria for hourly F, T, and L ratings for firestop systems. The F rating criteria prohibit flame passage through the system and require acceptable hose stream test performance. The T rating criteria prohibit flame passage through the system and require the maximum temperature rise on the unexposed surface of the wall or floor assembly, on the penetrating item, and on the fill material not to exceed 325°F (181°C) above ambient and require acceptable hose stream test performance. The L rating criteria determine the amount of air leakage, in cubic feet per minute per square foot of opening (CFM/sq ft), through the firestop system at ambient and/or 400°F air temperatures at an air pressure differential of 0.30 in. W.C. The L ratings are intended to assist authorities having jurisdiction and others in determining the suitability of firestop systems for the protection of penetrations and miscellaneous openings in floors and smoke barriers for the purpose of restricting the movement of smoke in accordance with the NFPA *101, Life Safety Code.*

Materials used in firestop systems are to be installed in accordance with the manufacturer's instructions provided with the materials. The structural integrity of the floor or wall assembly needs to be evaluated when providing openings for the penetrating items.

A firestop system, the seals for which are shown in Exhibit 300.17, may be used to meet the requirements of 300.21.

Exhibit 300.17 Fire seals used in a through-penetration firestop system to maintain the fire resistance rating of the wall, as required by 300.21. (Courtesy of O.Z./Gedney Co.)

300.22 Wiring in Ducts, Plenums, and Other Air-Handling Spaces.

The provisions of this section apply to the installation and uses of electric wiring and equipment in ducts, plenums, and other air-handling spaces.

> FPN: See Article 424, Part VI, for duct heaters.

(A) Ducts for Dust, Loose Stock, or Vapor Removal. No wiring systems of any type shall be installed in ducts used to transport dust, loose stock, or flammable vapors. No wiring system of any type shall be installed in any duct, or shaft containing only such ducts, used for vapor removal or for ventilation of commercial-type cooking equipment.

(B) Ducts or Plenums Used for Environmental Air. Only wiring methods consisting of Type MI cable, Type MC cable employing a smooth or corrugated impervious metal sheath without an overall nonmetallic covering, electrical metallic tubing, flexible metallic tubing, intermediate metal conduit, or rigid metal conduit without an overall nonmetallic covering shall be installed in ducts or plenums specifically fabricated to transport environmental air. Flexible metal conduit and liquidtight flexible metal conduit shall be permitted, in lengths not to exceed 1.2 m (4 ft), to connect physically adjustable equipment and devices permitted to be in these ducts and plenum chambers. The connectors used with flexible metal conduit shall effectively close any openings in the connection. Equipment and devices shall be permitted within such ducts or plenum chambers only if necessary for their direct action upon, or sensing of, the contained air. Where equipment or devices are installed and illumination is necessary to facilitate maintenance and repair, enclosed gasketed-type luminaires (fixtures) shall be permitted.

Section 300.22 limits the use of materials that would contribute smoke and products of combustion during a fire in an area that handles environmental air, and 300.22(B) provides for an effective barrier against the spread of products of combustion into the ducts or plenums.

This section applies to sheet metal ducts and other ducts and plenums specifically constructed to transport environmental air. Because equipment and devices such as luminaires and motors are not normally permitted in ducts or plenums, the wiring methods in 300.22(B) differ from those permitted in paragraph 300.22(C).

(C) Other Space Used for Environmental Air. This section applies to space used for environmental air-handling purposes other than ducts and plenums as specified in 300.22(A) and (B). It does not include habitable rooms or areas of buildings, the prime purpose of which is not air handling.

> FPN: The space over a hung ceiling used for environmental air-handling purposes is an example of the type of other space to which this section applies.

Section 300.22(C) applies to other spaces that are used to transport environmental air and are not specifically manufactured as ducts or plenums, such as the space or cavity between a structural floor or roof and a suspended (hung) ceiling. Many spaces above suspended ceilings are intended to transport return air. Some spaces are also used for supply air, but they are far less common than those used for return air. This section does not apply to habitable rooms and other areas whose prime purpose is other than air handling. Such an area is shown in Exhibit 300.18. If the prime purpose of the room or space is air handling as depicted in Exhibit 300.18, then the restrictions in 300.22(C) apply, whether or not electrical equipment is located in the room.

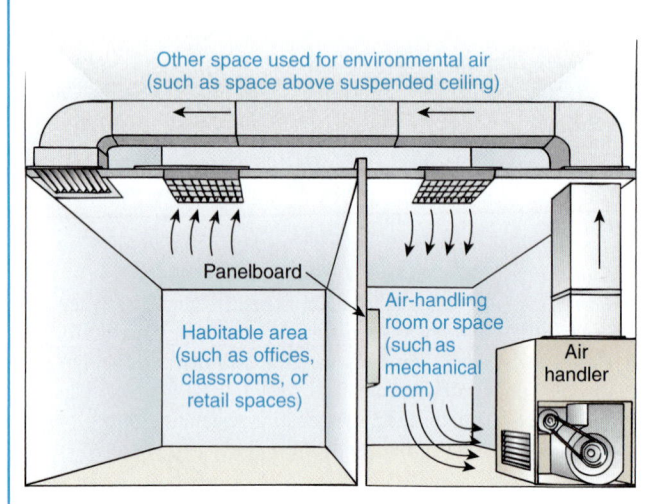

Exhibit 300.18 An application of 300.22(C). Ordinary wiring methods are permitted in habitable areas that are not used primarily for air handling. Only wiring methods described in 300.22(C)(1) are permitted in rooms or spaces that are primarily used for air handling.

Exception: This section shall not apply to the joist or stud spaces of dwelling units where the wiring passes through such spaces perpendicular to the long dimension of such spaces.

The exception to 300.22(C) permits cable to pass through joist or stud spaces of a dwelling unit, as illustrated in Exhibit 300.19. The joist space is covered with sheet metal and used as a cold-air return for a forced warm-air central heating system. Equipment such as junction boxes or device enclosures is not permitted in this location.

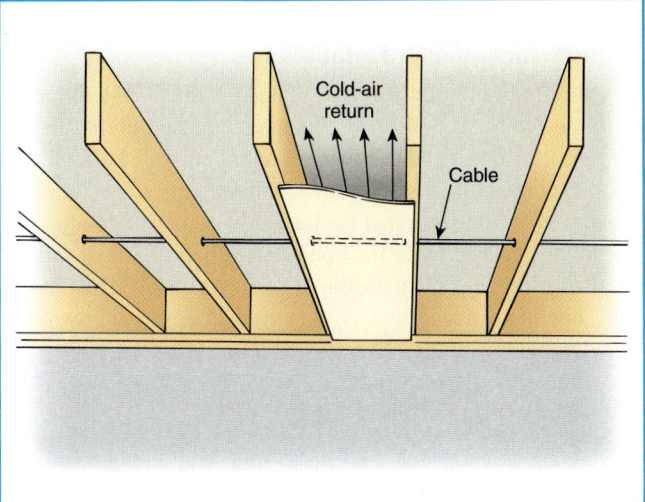

Exhibit 300.19 A cable passing through joist spaces of a dwelling unit, as permitted by 300.22(C), Exception.

(1) Wiring Methods. The wiring methods for such other space shall be limited to totally enclosed, nonventilated, insulated busway having no provisions for plug-in connections, Type MI cable, Type MC cable without an overall nonmetallic covering, Type AC cable, or other factory-assembled multiconductor control or power cable that is specifically listed for the use, or listed prefabricated cable assemblies of metallic manufactured wiring systems without nonmetallic sheath. Other types of cables and conductors shall be installed in electrical metallic tubing, flexible metallic tubing, intermediate metal conduit, rigid metal conduit without an overall nonmetallic covering, flexible metal conduit, or, where accessible, surface metal raceway or metal wireway with metal covers or solid bottom metal cable tray with solid metal covers.

Revised for the 2002 *Code*, this section no longer permits liquidtight flexible metal conduit as covered in Article 350 to be installed within "other spaces used for environmental air." The previous exception permitting this application for single lengths up to 6 ft has been removed.

(2) Equipment. Electrical equipment with a metal enclosure, or with a nonmetallic enclosure listed for the use and having adequate fire-resistant and low-smoke-producing characteristics, and associated wiring material suitable for the ambient temperature shall be permitted to be installed in such other space unless prohibited elsewhere in this *Code*.

Electrical equipment with metal enclosures is allowed within spaces used for environmental air. However, nonmetallic enclosures must be listed for this use.

Exception: Integral fan systems shall be permitted where specifically identified for such use.

It is not intended that the requirements of 300.22(B) or 300.22(C) apply to air-handling areas beneath raised floors in information technology rooms. See Article 645, Information Technology Equipment.

(D) Information Technology Equipment. Electric wiring in air-handling areas beneath raised floors for information technology equipment shall be permitted in accordance with Article 645.

300.23 Panels Designed to Allow Access.

Cables, raceways, and equipment installed behind panels designed to allow access, including suspended ceiling panels, shall be arranged and secured so as to allow the removal of panels and access to the equipment.

Section 300.23 is intended to prevent the excess accumulation of wires and cables that could limit access to electrical equipment by preventing the removal of access panels.

II. Requirements for Over 600 Volts, Nominal

Part II of this article was changed for the 1999 *Code*. Some of the requirements previously located within Article 710, Over 600 Volts, Nominal, General, in the 1996 *Code* are now in Article 300, Part II. The flow chart shown in Exhibit 300.20 shows the relocated and reorganized sections.

300.31 Covers Required.

Suitable covers shall be installed on all boxes, fittings, and similar enclosures to prevent accidental contact with energized parts or physical damage to parts or insulation.

300.32 Conductors of Different Systems.

See 300.3(C)(2).

300.34 Conductor Bending Radius.

The conductor shall not be bent to a radius less than 8 times the overall diameter for nonshielded conductors or 12 times the diameter for shielded or lead-covered conductors during or after installation. For multiconductor or multiplexed single conductor cables having individually shielded conductors, the minimum bending radius is 12 times the diameter of the individually shielded conductors or 7 times the overall diameter, whichever is greater.

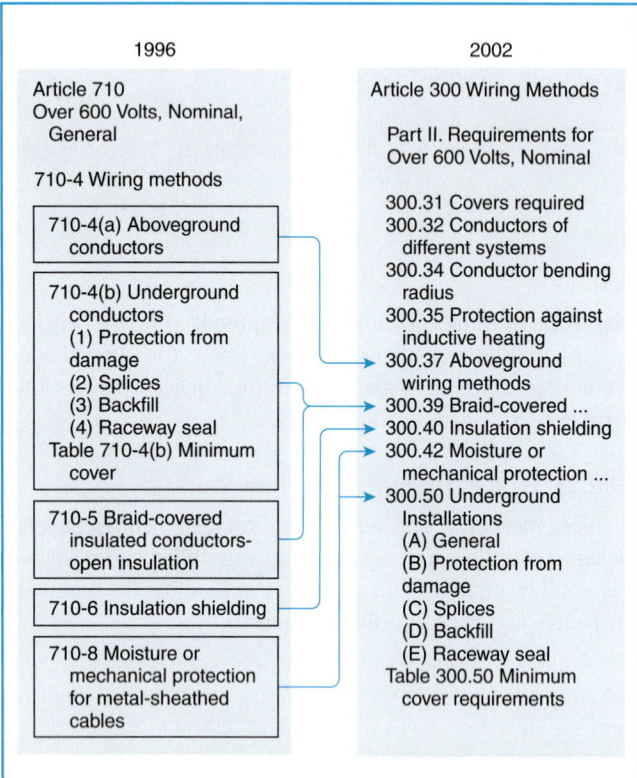

Exhibit 300.20 The relocation of Article 710, Over 600 Volts, Nominal, General (1996 *Code*) into Article 300, Part II (2002 *Code*).

300.35 Protection Against Induction Heating.

Metallic raceways and associated conductors shall be arranged so as to avoid heating of the raceway in accordance with the applicable provisions of 300.20.

300.37 Aboveground Wiring Methods.

Aboveground conductors shall be installed in rigid metal conduit, in intermediate metal conduit, in electrical metallic tubing, in rigid nonmetallic conduit, in cable trays, as busways, as cablebus, in other identified raceways, or as open runs of metal-clad cable suitable for the use and purpose. In locations accessible to qualified persons only, open runs of Type MV cables, bare conductors, and bare busbars shall also be permitted. Busbars shall be permitted to be either copper or aluminum.

In transformer vaults, switch rooms, and similar areas restricted to qualified personnel, any suitable wiring method may be used. Open wiring using bare or insulated conductors on insulators is commonly employed, as is rigid metal conduit and rigid nonmetallic conduit.

300.39 Braid-Covered Insulated Conductors—Open Installation.

Open runs of braid-covered insulated conductors shall have a flame-retardant braid. If the conductors used do not have this protection, a flame-retardant saturant shall be applied to the braid covering after installation. This treated braid covering shall be stripped back a safe distance at conductor terminals, according to the operating voltage. This distance shall not be less than 25 mm (1 in.) for each kilovolt of the conductor-to-ground voltage of the circuit, where practicable.

300.40 Insulation Shielding.

Metallic and semiconducting insulation shielding components of shielded cables shall be removed for a distance dependent on the circuit voltage and insulation. Stress reduction means shall be provided at all terminations of factory-applied shielding.

Metallic shielding components such as tapes, wires, or braids, or combinations thereof, and their associated conducting or semiconducting components shall be grounded.

300.42 Moisture or Mechanical Protection for Metal-Sheathed Cables.

Where cable conductors emerge from a metal sheath and where protection against moisture or physical damage is necessary, the insulation of the conductors shall be protected by a cable sheath terminating device.

300.50 Underground Installations.

(A) General. Underground conductors shall be identified for the voltage and conditions under which they are installed. Direct burial cables shall comply with the provisions of 310.7. Underground cables shall be installed in accordance with 300.50(A)(1) or (2), and the installation shall meet the depth requirements of Table 300.50.

Exception No. 1: Areas subject to vehicular traffic, such as thoroughfares or commercial parking areas, shall have a minimum cover of 600 mm (24 in.).

Exception No. 2: The minimum cover requirements for other than rigid metal conduit and intermediate metal conduit shall be permitted to be reduced 150 mm (6 in.) for each 50 mm (2 in.) of concrete or equivalent protection placed in the trench over the underground installation.

Exception No. 3: The minimum cover requirements shall not apply to conduits or other raceways that are located under a building or exterior concrete slab not less than 100 mm (4 in.) in thickness and extending not less than 150 mm (6 in.) beyond the underground installation. A warning ribbon or other effective means suitable for the conditions shall be placed above the underground installation.

Table 300.50 Minimum Cover Requirements

Circuit Voltage	Direct-Buried Cables		Rigid Nonmetallic Conduit Approved for Direct Burial*		Rigid Metal Conduit and Intermediate Metal Conduit	
	mm	in.	mm	in.	mm	in.
Over 600 V through 22 kV	750	30	450	18	150	6
Over 22 kV through 40 kV	900	36	600	24	150	6
Over 40 kV	1000	42	750	30	150	6

Note: *Cover* is defined as the shortest distance in millimeters measured between a point on the top surface of any direct-buried conductor, cable, conduit, or other raceway and the top surface of finished grade, concrete, or similar cover.
*Listed by a qualified testing agency as suitable for direct burial without encasement. All other nonmetallic systems shall require 50 mm (2 in.) of concrete or equivalent above conduit in addition to above depth.

Exception No. 4: Lesser depths shall be permitted where cables and conductors rise for terminations or splices or where access is otherwise required.

Exception No. 5: In airport runways, including adjacent defined areas where trespass is prohibited, cable shall be permitted to be buried not less than 450 mm (18 in.) deep and without raceways, concrete enclosement, or equivalent.

Exception No. 6: Raceways installed in solid rock shall be permitted to be buried at lesser depth where covered by 50 mm (2 in.) of concrete, which shall be permitted to extend to the rock surface.

(1) Shielded Cables and Nonshielded Cables in Metal-Sheathed Cable Assemblies. Underground cables, including nonshielded, Type MC and moisture-impervious metal sheath cables, shall have those sheaths grounded through an effective grounding path meeting the requirements of 250.4(A)(5) or 250.4(B)(4). They shall be direct buried or installed in raceways identified for the use.

(2) Other Nonshielded Cables. Other nonshielded cables not covered in 300.50(A)(1) shall be installed in rigid metal conduit, intermediate metal conduit, or rigid nonmetallic conduit encased in not less than 75 mm (3 in.) of concrete.

(B) Protection from Damage. Conductors emerging from the ground shall be enclosed in listed raceways. Raceways installed on poles shall be of rigid metal conduit, intermediate metal conduit, PVC Schedule 80, or equivalent, extending from the minimum cover depth specified in Table 300.50 to a point 2.5 m (8 ft) above finished grade. Conductors entering a building shall be protected by an approved enclosure or raceway from the minimum cover depth to the point of entrance. Where direct-buried conductors, raceways, or cables are subject to movement by settlement or frost, they shall be installed to prevent damage to the enclosed conductors or to the equipment connected to the raceways. Metallic enclosures shall be grounded.

(C) Splices. Direct burial cables shall be permitted to be spliced or tapped without the use of splice boxes, provided they are installed using materials suitable for the application. The taps and splices shall be watertight and protected from mechanical damage. Where cables are shielded, the shielding shall be continuous across the splice or tap.

Exception: At splices of an engineered cabling system, metallic shields of direct-buried single-conductor cables with maintained spacing between phases shall be permitted to be interrupted and overlapped. Where shields are interrupted and overlapped, each shield section shall be grounded at one point.

(D) Backfill. Backfill containing large rocks, paving materials, cinders, large or sharply angular substances, or corrosive materials shall not be placed in an excavation where materials can damage raceways, cables, or other substructures, or prevent adequate compaction of fill, or contribute to corrosion of raceways, cables, or other substructures.

Protection in the form of granular or selected material or suitable sleeves shall be provided to prevent physical damage to the raceway or cable.

(E) Raceway Seal. Where a raceway enters from an underground system, the end within the building shall be sealed with an identified compound so as to prevent the entrance of moisture or gases, or it shall be so arranged to prevent moisture from contacting live parts.

ARTICLE 310
Conductors for General Wiring

Contents

310.1 Scope.

This article covers general requirements for conductors and their type designations, insulations, markings, mechanical strengths, ampacity ratings, and uses. These requirements do not apply to conductors that form an integral part of equipment, such as motors, motor controllers, and similar equipment, or to conductors specifically provided for elsewhere in this *Code*.

FPN: For flexible cords and cables, see Article 400. For fixture wires, see Article 402.

310.2 Conductors.

As the electrical industry moves more toward global suppliers of insulated conductors, it is important for the user to understand that all "insulated" conductors are not equal. Only wires and cables that meet the minimum fire, electrical, and physical properties required by the applicable standards are permitted to be marked with the letter designations found in Tables 310.13, 310.61, and 310.62. See 310.13 for the requirements of insulated conductor construction and applications.

(A) **Insulated.** Conductors shall be insulated.

Exception: Where covered or bare conductors are specifically permitted elsewhere in this Code.

FPN: See 250.184 for insulation of neutral conductors of a solidly grounded high-voltage system.

(B) **Conductor Material.** Conductors in this article shall be of aluminum, copper-clad aluminum, or copper unless otherwise specified.

310.3 Stranded Conductors.

Where installed in raceways, conductors of size 8 AWG and larger shall be stranded.

Larger-size conductors are required to be stranded to provide greater flexibility. This requirement does not apply to busbars and the conductors of Type MI mineral-insulated metal-sheathed cable. In addition, the bonding conductors of a common bonding grid of a permanently installed swimming pool are required to be solid, nonstranded conductors of 8 AWG or larger, according to 680.26(C).

Exception: As permitted or required elsewhere in this Code.

310.4 Conductors in Parallel.

Aluminum, copper-clad aluminum, or copper conductors of size 1/0 AWG and larger, comprising each phase, neutral, or grounded circuit conductor, shall be permitted to be connected in parallel (electrically joined at both ends to form a single conductor).

Conductors connected in parallel, in accordance with 310.4, are considered a single conductor with a total cross-sectional area of all conductors in parallel. Therefore, if individual conductors are tapped from conductors in parallel, the tap connection must include all the conductors in parallel for that particular phase. Tapping into only one of the parallel conductors will result in unbalanced distribution of tap load current between parallel conductors. This will result in one of the conductors carrying more than its share of the load, which can cause overheating and conductor insulation failure. For example, if a 250-kcmil conductor is tapped from a set of two 500-kcmil conductors in parallel, the splicing device must include both 500-kcmil conductors and the single 250-kcmil tap conductor.

Exception No. 1: As permitted in 620.12(A)(1) .

Exception No. 2: Conductors in sizes smaller than 1/0 AWG shall be permitted to be run in parallel to supply control power to indicating instruments, contactors, relays, solenoids, and similar control devices provided

(a) *They are contained within the same raceway or cable,*

(b) *The ampacity of each individual conductor is sufficient to carry the entire load current shared by the parallel conductors, and*

(c) *The overcurrent protection is such that the ampacity of each individual conductor will not be exceeded if one or more of the parallel conductors become inadvertently disconnected.*

Exception No. 3: Conductors in sizes smaller than 1/0 AWG shall be permitted to be run in parallel for frequencies of 360 Hz and higher where conditions (a), (b), and (c) of Exception No. 2 are met.

For example, in control wiring and circuits that operate at frequencies greater than 360 Hz, it may be necessary to reduce cable capacitance or voltage drop in long lengths of wire. A 14 AWG conductor might have more than sufficient capacity to carry the load, but by installing two conductors in parallel, the voltage drop can be reduced to acceptable limits. This method is permissible, provided the safeguards listed in Exception No. 2 are followed.

Exception No. 4: Under engineering supervision, grounded neutral conductors in sizes 2 AWG and larger shall be permitted to be run in parallel for existing installations.

FPN: Exception No. 4 can be used to alleviate overheating of neutral conductors in existing installations due to high content of triplen harmonic currents.

The word *triplen* refers to a third-order harmonic current, such as the third, sixth, ninth, and so on.

The paralleled conductors in each phase, neutral, or grounded circuit conductor shall

(1) Be the same length
(2) Have the same conductor material
(3) Be the same size in circular mil area
(4) Have the same insulation type
(5) Be terminated in the same manner

Where run in separate raceways or cables, the raceways or cables shall have the same physical characteristics. Conductors of one phase, neutral, or grounded circuit conductor shall not be required to have the same physical characteristics as those of another phase, neutral, or grounded circuit conductor to achieve balance.

For example, the conductors in phases A and B may be copper, and those in phase C may be aluminum. However, all raceways or cables must be of the same size, material,

and length. Cables, in this case, means wiring method type cables such as Type MC.

FPN: Differences in inductive reactance and unequal division of current can be minimized by choice of materials, methods of construction, and orientation of conductors.

Where equipment grounding conductors are used with conductors in parallel, they shall comply with the requirements of this section except that they shall be sized in accordance with 250.122.

Conductors installed in parallel shall comply with the provisions of 310.15(B)(2)(a).

Section 310.4 permits a practical means for installing large-capacity conductors for feeders or services. The paralleling of two or more conductors in place of using one large conductor to ensure equal division of current depends on a number of factors. Therefore, several conditions must be satisfied so as not to overload any of the individual paralleled conductors. Other than as permitted in 250.122 and the exceptions to 310.4, there does not appear to be any practical need to parallel conductors smaller than 1/0 AWG.

To avoid excessive voltage drop and also to ensure equal division of current, it is essential that different phase conductors be located close together and that each phase conductor, grounded conductor, and the grounding conductor (if used) be grouped together in each raceway or cable. However, isolated phase installations are permitted underground where the phase conductors are run in nonmetallic raceways that are in close proximity.

The impedance of a circuit in an aluminum raceway or aluminum-sheathed cable will differ from the impedance of the same circuit in a steel raceway or steel-sheathed cable; therefore, separate raceways and cables must have the same physical characteristics. See 300.20 regarding induced currents in metal enclosures or raceways.

Note that all conductors of the same phase or neutral are required to be of the same conductor material. For example, if 12 conductors are paralleled for a 3-phase, 4-wire, 480Y/277-volt ac circuit, four conductors could be installed in each of three raceways. The *Code* does not intend that all 12 conductors be copper or aluminum but does intend that the individual conductors in parallel for each phase, grounded conductor, and neutral be the same material, insulation type, length, and so on. Also, it is intended that the three raceways have the same physical characteristics (e.g., three rigid aluminum conduits, three steel IMC conduits, three EMTs, or three nonmetallic conduits), not a mixture (e.g., two rigid aluminum conduits and one rigid steel conduit).

It is neither economical nor practical to use conductors

larger than 1000 kcmil in raceways unless the conductor size is governed by voltage drop. The ampacity of larger sizes would increase very little in proportion to the increase in the size of the conductor. Where the cross-sectional area of a conductor increases 50 percent (e.g., from 1000 to 1500 kcmil), a Type THW conductor ampacity increases only 80 amperes (less than 15 percent). A 100 percent increase (from 1000 to 2000 kcmil) causes an increase of only 120 amperes (approximately 2 percent). Generally, where cost is a factor, installation of two (or more) paralleled conductors per phase may be beneficial.

310.5 Minimum Size of Conductors.

The minimum size of conductors shall be as shown in Table 310.5.

Table 310.5 Minimum Size of Conductors

Conductor Voltage Rating (Volts)	Minimum Conductor Size (AWG)	
	Copper	Aluminum or Copper-Clad Aluminum
0–2000	14	12
2001–8000	8	8
8001–15,000	2	2
15,001–28,000	1	1
28,001–35,000	1/0	1/0

Exception No. 1: For flexible cords as permitted by 400.12.

Exception No. 2: For fixture wire as permitted by 402.6.

Exception No. 3: For motors rated 1 hp or less as permitted by 430.22(F).

Exception No. 4: For cranes and hoists as permitted by 610.14.

Exception No. 5: For elevator control and signaling circuits as permitted by 620.12.

Exception No. 6: For Class 1, Class 2, and Class 3 circuits as permitted by 725.27(A) and 725.51, Exception.

Exception No. 7: For fire alarm circuits as permitted by 760.27(A), 760.51, Exception, and 760.71(B).

Exception No. 8: For motor-control circuits as permitted by 430.72.

Exception No. 9: For control and instrumentation circuits as permitted by 727.6.

Exception No. 10: For electric signs and outline lighting as permitted in 600.31(B) and 600.32(B).

310.6 Shielding.

Solid dielectric insulated conductors operated above 2000 volts in permanent installations shall have ozone-resistant insulation and shall be shielded. All metallic insulation shields shall be grounded through an effective grounding path meeting the requirements of 250.4(A)(5) or 250.4(B)(4). Shielding shall be for the purpose of confining the voltage stresses to the insulation.

Exception: Nonshielded insulated conductors listed by a qualified testing laboratory shall be permitted for use up to 8000 volts under the following conditions:

(a) Conductors shall have insulation resistant to electric discharge and surface tracking, or the insulated conductor(s) shall be covered with a material resistant to ozone, electric discharge, and surface tracking.

(b) Where used in wet locations, the insulated conductor(s) shall have an overall nonmetallic jacket or a continuous metallic sheath.

(c) Where operated at 5001 to 8000 volts, the insulated conductor(s) shall have a nonmetallic jacket over the insulation. The insulation shall have a specific inductive capacity not greater than 3.6, and the jacket shall have a specific inductive capacity not greater than 10 and not less than 6.

(d) Insulation and jacket thicknesses shall be in accordance with Table 310.63.

Solid dielectric insulated conductors that are permanently installed and that operate at greater than 2000 volts must have ozone-resistant insulation and must be shielded with a grounded metallic shield (note exception). Shielding is accomplished by applying a metal tape or nonmetallic semiconducting tape around the conductor surface, to prevent corona from forming and to reduce high-voltage stresses. Corona is a faint glow adjacent to the surface of the electrical conductor at high voltage. If high-voltage stresses and a charging current are flowing between the conductor and ground (usually due to moisture), the surrounding atmosphere is ionized, and ozone—generated by an electric discharge in ordinary oxygen or air—is formed, which will attack the conductor jacket and insulation and may eventually break them down. The shield is at ground potential; therefore, no voltage above ground is present on the jacket outside the shield, thus preventing a discharge from the jacket and the subsequent formation of ozone.

Specialized training and close adherence to manufacturers' instructions are absolutely essential for high-voltage cable installations.

Example Exhibits. Exhibits 310.1, 310.2, and 310.3 show some examples of shielded cable installations: a three-conductor cable of the shielded type, a stress-relief cone for an indoor cable terminator, and a stress cone on a single-conductor shielded cable terminating inside a pothead. In Exhibit 310.3 a clamping ring provides a grounding connection between the copper shielding tape and the shield to the metallic base of the pothead.

Exhibit 310.1 A three-conductor cable of the shielded type.

Exhibit 310.2 A one-piece, premolded stress-relief cone for indoor cable terminations of up to 35 kV phase-to-phase.

310.7 Direct Burial Conductors.

Conductors used for direct burial applications shall be of a type identified for such use.

Cables rated above 2000 volts shall be shielded.

Exception: Nonshielded multiconductor cables rated 2001–5000 volts shall be permitted if the cable has an overall metallic sheath or armor.

The metallic shield, sheath, or armor shall be grounded through an effective grounding path meeting the requirements of 250.4(A)(5) or 250.4(B)(4).

FPN No. 1: See 300.5 for installation requirements for conductors rated 600 volts or less.

FPN No. 2: See 300.50 for installation requirements for conductors rated over 600 volts.

310.8 Locations.

(A) Dry Locations. Insulated conductors and cables used in dry locations shall be any of the types identified in this *Code.*

(B) Dry and Damp Locations. Insulated conductors and cables used in dry and damp locations shall be Types FEP, FEPB, MTW, PFA, RHH, RHW, RHW-2, SA, THHN, THW, THW-2, THHW, THHW-2, THWN, THWN-2, TW, XHH, XHHW, XHHW-2, Z, or ZW.

(C) Wet Locations. Insulated conductors and cables used in wet locations shall be

(1) Moisture-impervious metal-sheathed;
(2) Types MTW, RHW, RHW-2, TW, THW, THW-2, THHW, THHW-2, THWN, THWN-2, XHHW, XHHW-2, ZW; or
(3) Of a type listed for use in wet locations.

(D) Locations Exposed to Direct Sunlight. Insulated conductors and cables used where exposed to direct rays of the sun shall be of a type listed for sunlight resistance or listed and marked "sunlight resistant."

310.9 Corrosive Conditions.

Conductors exposed to oils, greases, vapors, gases, fumes, liquids, or other substances having a deleterious effect on

Exhibit 310.3 A stress cone on a single-conductor shielded cable terminating inside a pothead.

the conductor or insulation shall be of a type suitable for the application.

See the commentary following 501.13 regarding gasoline-resistant conductors. Before being used, wire pulling compounds should first be investigated to determine compliance with this section.

310.10 Temperature Limitation of Conductors.

No conductor shall be used in such a manner that its operating temperature exceeds that designated for the type of insulated conductor involved. In no case shall conductors be associated together in such a way with respect to type of circuit, the wiring method employed, or the number of conductors that the limiting temperature of any conductor is exceeded.

Most terminations are normally designed for 60°C or 75°C maximum temperatures, although some are now being designed for 90°C. Therefore, the higher-rated ampacities for conductors of 90°C, 110°C, and so on, cannot be used unless the terminals at which the conductors terminate have comparable ratings.

Tables 310.16 through 310.20 have ampacity correction factors for ambient temperatures greater or less than the ambient temperature identified in the table heading. To assign the proper ampacity to a conductor in an ambient above 30°C (86°F), the appropriate temperature correction factor must be used. This correction factor is applied in addition to any adjustment factor, such as in 310.15(B)(2)(a).

Example

Determine the adequacy of 2 AWG THHN copper conductors to be installed in a raceway in an ambient temperature of 50°C (122°F).

Solution

Table 310.16 shows that the allowable ampacity of the conductor at 30°C is 130 amperes, which is multiplied by 0.82 (taken from the correction factors at the bottom of the table). Thus, the allowable ampacity of the 2 AWG conductor at 50°C is reduced to 106.6 amperes (130 amperes × 0.82 = 106.6 amperes).

If six of these conductors were run in a raceway, 310.15(B)(2)(a) would require the allowable ampacity to be further reduced to 80 percent, which, in this case, would be 106.6 amperes × 0.8 = 85.28 amperes. Under these conditions, the 2 AWG conductors would be suitable for an 80-ampere circuit.

The basis for determining the ampacities of conductors for Tables 310.16 and 310.17 was the NEMA "Report of Determination of Maximum Permissible Current-Carrying Capacity of Code Insulated Wires and Cables for Building Purposes," dated June 27, 1938. The basis for determining the ampacities of conductors for Tables 310.18 and 310.19 and the ampacity tables in Annex B was the Neher-McGrath method. See the commentary following 310.15(C) for further explanation.

Conductors should be chosen that have a rating above the anticipated maximum ambient temperature. The operating temperature of conductors should be controlled at or below the conductor rating by coordinating conductor size, number of associated conductors, and ampacity for the particular conductor rating and ambient temperature. All tabulations should be corrected for the anticipated ambient temperature, using the correction factors at the bottom of the ampacity tables. If more than three conductors are associated together, the additional adjustment shown in 310.15(B)(2)(a) must be applied.

FPN: The temperature rating of a conductor (see Table 310.13 and Table 310.61) is the maximum temperature, at any location along its length, that the conductor can withstand over a prolonged time period without serious degradation. The allowable ampacity tables, the ampacity tables of Article 310 and the ampacity tables of Annex B, the correction factors at the bottom of these tables, and the notes to the tables provide guidance for coordinating conductor sizes, types, allowable ampacities, ampacities, ambient temperatures, and number of associated conductors.

The principal determinants of operating temperature are as follows:

(1) Ambient temperature—ambient temperature may vary along the conductor length as well as from time to time.
(2) Heat generated internally in the conductor as the result of load current flow, including fundamental and harmonic currents.
(3) The rate at which generated heat dissipates into the ambient medium. Thermal insulation that covers or surrounds conductors affects the rate of heat dissipation.
(4) Adjacent load-carrying conductors—adjacent conductors have the dual effect of raising the ambient temperature and impeding heat dissipation.

The second principal determinant explains that heating due to harmonic current should also be considered. In certain cases, this may require using larger size conductors. For existing installations, see 310.4, Exception No. 4.

The fine print note focuses attention on the necessity for derating conductors where high ambient temperatures are encountered and provides users with helpful information in coordinating ampacities, ambient temperatures, conductor size and number, and so on, to ensure operation at or below rating.

310.11 Marking.

(A) Required Information. All conductors and cables shall be marked to indicate the following information, using the applicable method described in 310.11(B):

(1) The maximum rated voltage.
(2) The proper type letter or letters for the type of wire or cable as specified elsewhere in this *Code*.
(3) The manufacturer's name, trademark, or other distinctive marking by which the organization responsible for the product can be readily identified.
(4) The AWG size or circular mil area.

FPN: See Conductor Properties, Table 8 of Chapter 9 for conductor area expressed in SI units for conductor sizes specified in AWG or circular mil area.

(5) Cable assemblies where the neutral conductor is smaller than the ungrounded conductors shall be so marked.

(B) Method of Marking.

(1) Surface Marking. The following conductors and cables shall be durably marked on the surface. The AWG size or circular mil area shall be repeated at intervals not exceeding 610 mm (24 in.). All other markings shall be repeated at intervals not exceeding 1.0 m (40 in.).

(1) Single- and multiconductor rubber- and thermoplastic-insulated wire and cable
(2) Nonmetallic-sheathed cable
(3) Service-entrance cable
(4) Underground feeder and branch-circuit cable
(5) Tray cable
(6) Irrigation cable
(7) Power-limited tray cable
(8) Instrumentation tray cable

(2) Marker Tape. Metal-covered multiconductor cables shall employ a marker tape located within the cable and running for its complete length.

Exception No. 1: Mineral-insulated, metal-sheathed cable.

Exception No. 2: Type AC cable.

Exception No. 3: The information required in 310.11(A) shall be permitted to be durably marked on the outer nonmetallic covering of Type MC, Type ITC, or Type PLTC cables at intervals not exceeding 1.0 m (40 in.).

Exception No. 4: The information required in 310.11(A) shall be permitted to be durably marked on a nonmetallic covering under the metallic sheath of Type ITC or Type PLTC cable at intervals not exceeding 1.0 m (40 in.).

Type PLTC cable is permitted to have a metallic sheath or armor over a nonmetallic jacketed cable. A second nonmetallic jacket covering the metallic sheath is optional. Exception Nos. 3 and 4 define the marking requirements for either case.

FPN: Included in the group of metal-covered cables are Type AC cable (Article 320), Type MC cable (Article 330), and lead-sheathed cable.

(3) Tag Marking. The following conductors and cables shall be marked by means of a printed tag attached to the coil, reel, or carton:

(1) Mineral-insulated, metal-sheathed cable
(2) Switchboard wires
(3) Metal-covered, single-conductor cables
(4) Type AC cable

(4) Optional Marking of Wire Size. The information required in 310.11(A)(4) shall be permitted to be marked on the surface of the individual insulated conductors for the following multiconductor cables:

(1) Type MC cable
(2) Tray cable
(3) Irrigation cable
(4) Power-limited tray cable
(5) Power-limited fire alarm cable

(6) Instrumentation tray cable

(C) Suffixes to Designate Number of Conductors. A type letter or letters used alone shall indicate a single insulated conductor. The letter suffixes shall be indicated as follows:

(1) D—For two insulated conductors laid parallel within an outer nonmetallic covering
(2) M—For an assembly of two or more insulated conductors twisted spirally within an outer nonmetallic covering

(D) Optional Markings. All conductors and cables contained in Chapter 3 shall be permitted to be surface marked to indicate special characteristics of the cable materials. These markings include, but are not limited to, markings for limited smoke, sunlight resistant, and so forth.

New cable insulations that have special characteristics are frequently developed. An example is the family of limited-smoke cables, which are permitted to be marked "LS." Other materials that have other characteristics such as sunlight resistance and low corrosivity have been developed or are in development. Section 310.11(D) allows these developments without identifying each characteristic in the *Code*.

310.12 Conductor Identification.

(A) Grounded Conductors. Insulated or covered grounded conductors shall be identified in accordance with 200.6.

Section 200.6 permits a new method of identification described as "three continuous white stripes on other than green insulation along the conductor's entire length."

(B) Equipment Grounding Conductors. Equipment grounding conductors shall be in accordance with 250.119.

(C) Ungrounded Conductors. Conductors that are intended for use as ungrounded conductors, whether used as a single conductor or in multiconductor cables, shall be finished to be clearly distinguishable from grounded and grounding conductors. Distinguishing markings shall not conflict in any manner with the surface markings required by 310.11(B)(1).

Exception: Conductor identification shall be permitted in accordance with 200.7.

Ungrounded conductors with white or gray insulation are permitted if the conductors are permanently reidentified at termination points and if the conductor is visible and accessible. The normal methods of reidentification include colored tape, tagging, or paint. Other applications where white conductors are permitted include flexible cords and circuits less than 50 volts. A white conductor used in single-pole, 3-way and 4-way switch loops also requires reidentification (a color other than white, gray, or green) if it is used as an ungrounded conductor. See 200.7(C)(2) for exact details.

310.13 Conductor Constructions and Applications.

Insulated conductors shall comply with the applicable provisions of one or more of the following: Tables 310.13, 310.61, 310.62, 310.63, and 310.64.

These conductors shall be permitted for use in any of the wiring methods recognized in Chapter 3 and as specified in their respective tables.

FPN: Thermoplastic insulation may stiffen at temperatures lower than minus 10°C (plus 14°F). Thermoplastic insulation may also be deformed at normal temperatures where subjected to pressure, such as at points of support. Thermoplastic insulation, where used on dc circuits in wet locations, may result in electroendosmosis between conductor and insulation.

Table 310.13 lists the various types of insulated conductors covered by the requirements of this *Code*. More detailed wire classification information from sizes 14 AWG through 2000 kcmil are available in standards or directories such as those published by Underwriters Laboratories Inc.

Table 310.13 also includes conductor applications and maximum operating temperatures. Some conductors have dual ratings. For example, Type XHHW is rated 90°C for dry and damp locations and 75°C for wet locations; Type THW is rated 75°C for dry and wet locations and 90°C for special applications within electric-discharge lighting equipment.

Types RHW-2, XHHW-2, and other types identified by the suffix "-2" are rated 90°C for wet locations as well as dry and damp locations. Conductors permitted to be identified by the suffix "-2," other than Types RHW-2 and XHHW-2, are identified in the table by footnote 4.

The maximum continuous ampacities for copper, copper-clad aluminum, and aluminum conductors, rated 0 through 2000 volts, are listed in Tables 310.16 through 310.20 and accompanying adjustment factors of 310.15(B)(1) through (6), or they can be calculated in accordance with 310.15(C).

Copper-clad aluminum conductors are drawn from a copper-clad aluminum rod with the copper metallurgically bonded to an aluminum core. The copper forms a minimum of 10 percent of the cross-sectional area of a solid conductor or of each strand of a stranded conductor. See the additional commentary following 110.14 for making electrical connec-

Table 310.13 Conductor Application and Insulations

Trade Name	Type Letter	Maximum Operating Temperature	Application Provisions	Insulation	Thickness of Insulation			Outer Covering[1]
					AWG or kcmil	mm	Mils	
Fluorinated ethylene propylene	FEP or FEPB	90°C 194°F	Dry and damp locations	Fluorinated ethylene propylene	14–10 8–2	0.51 0.76	20 30	None
					14–8	0.36	14	Glass braid
		200°C 392°F	Dry locations— special applications[2]	Fluorinated ethylene propylene	6–2	0.36	14	Glass braid or other suitable material
Mineral insulation (metal sheathed)	MI	90°C 194°F	Dry and wet locations	Magnesium oxide	18–16[3] 16–10 9–4 3–500	0.58 0.91 1.27 1.40	23 36 50 55	Copper or alloy steel
		250°C 482°F	For special applications[2]					
Moisture-, heat-, and oil-resistant thermo-plastic	MTW	60°C 140°F	Machine tool wiring in wet locations as permitted in NFPA 79 (See Article 670.) Machine tool wiring in dry locations as permitted in NFPA 79 (See Article 670.)	Flame-retardant moist-ure-, heat-, and oil-resistant thermo-plastic	22–12 10 8 6 4–2 1–4/0 213–500 501–1000	(A) (B) 0.76 0.38 0.76 0.51 1.14 0.76 1.52 0.76 1.52 1.02 2.03 1.27 2.41 1.52 2.79 1.78	(A) (B) 30 15 30 20 45 30 60 30 60 40 80 50 95 60 110 70	(A) None (B) Nylon jacket or equivalent
		90°C 194°F						
Paper		85°C 185°F	For underground service conductors, or by special permission	Paper				Lead sheath
Perfluoro-alkoxy	PFA	90°C 194°F	Dry and damp locations	Perfluoro-alkoxy	14–10 8–2 1–4/0	0.51 0.76 1.14	20 30 45	None
		200°C 392°F	Dry locations — special applications[2]					
Perfluoro-alkoxy	PFAH	250°C 482°F	Dry locations only. Only for leads within apparatus or within raceways connected to apparatus (nickel or nickel-coated copper only)	Perfluoro-alkoxy	14–10 8–2 1–4/0	0.51 0.76 1.14	20 30 45	None
Thermoset	RHH	90°C 194°F	Dry and damp locations		14-10 8–2 1–4/0 213–500 501–1000 1001–2000 For 601–2000 V, *see* Table 310.62.	1.14 1.52 2.03 2.41 2.79 3.18	45 60 80 95 110 125	Moisture-resistant, flame-retardant, non-metallic covering[1]

Table 310.13 *Continued*

Trade Name	Type Letter	Maximum Operating Temperature	Application Provisions	Insulation	Thickness of Insulation			Outer Covering[1]
					AWG or kcmil	mm	Mils	
Moisture-resistant thermoset	RHW[4]	75°C 167°F	Dry and wet locations	Flame-retardant, moisture-resistant thermoset	14–10 8–2 1–4/0 213–500 501–1000 1001–2000 For 601–2000 V, *see* Table 310.62.	1.14 1.52 2.03 2.41 2.79 3.18	45 60 80 95 110 125	Moisture-resistant, flame-retardant, non-metallic covering[5]
Moisture-resistant thermoset	RHW-2	90°C 194°F	Dry and wet locations	Flame-retardant moisture-resistant thermoset	14–10 8–2 1-4/0 213–500 501–1000 1001–2000 For 601–2000 V, *see* Table 310.62.	1.14 1.52 2.03 2.41 2.79 3.18	45 60 80 95 110 125	Moisture-resistant, flame-retardant, non-metallic covering[5]
Silicone	SA	90°C 194°F 200°C 392°F	Dry and damp locations For special application[2]	Silicone rubber	14–10 8–2 1–4/0 213–500 501–1000 1001–2000	1.14 1.52 2.03 2.41 2.79 3.18	45 60 80 95 110 125	Glass or other suitable braid material
Thermoset	SIS	90°C 194°F	Switchboard wiring only	Flame-retardant thermoset	14–10 8–2 1–4/0	0.76 1.14 2.41	30 45 95	None
Thermo-plastic and fibrous outer braid	TBS	90°C 194°F	Switchboard wiring only	Thermo-plastic	14–10 8 6–2 1–4/0	0.76 1.14 1.52 2.03	30 45 60 80	Flame-retardant, non-metallic covering
Extended polytetra-fluoro-ethylene	TFE	250°C 482°F	Dry locations only. Only for leads within apparatus or within raceways connected to apparatus, or as open wiring (nickel or nickel-coated copper only)	Extruded polytetra-fluoro-ethylene	14–10 8–2 1–4/0	0.51 0.76 1.14	20 30 45	None
Heat-resistant thermo-plastic	THHN	90°C 194°F	Dry and damp locations	Flame-retardant, heat-resistant thermo-plastic	14–12 10 8–6 4–2 1–4/0 250–500 501–1000	0.38 0.51 0.76 1.02 1.27 1.52 1.78	15 20 30 40 50 60 70	Nylon jacket or equivalent

Table 310.13 *Continued*

Trade Name	Type Letter	Maximum Operating Temperature	Application Provisions	Insulation	Thickness of Insulation			Outer Covering[1]
					AWG or kcmil	mm	Mils	
Moisture- and heat-resistant thermo-plastic	THHW	75°C 167°F 90°C 194°F	Wet location Dry location	Flame-retardant, moisture- and heat-resistant thermo-plastic	14–10 8 6–2 1–4/0 213–500 501–1000	0.76 1.14 1.52 2.03 2.41 2.79	30 45 60 80 95 110	None
Moisture- and heat-resistant thermo-plastic	THW[4]	75°C 167°F 90°C 194°F	Dry and wet locations Special applications within electric discharge lighting equipment. Limited to 1000 open-circuit volts or less. (size 14-8 only as permitted in 410.33)	Flame-retardant, moisture- and heat-resistant thermo-plastic	14–10 8 6–2 1–4/0 213–500 501–1000 1001–2000	0.76 1.14 1.52 2.03 2.41 2.79 3.18	30 45 60 80 95 110 125	None
Moisture- and heat-resistant thermo-plastic	THWN[4]	75°C 167°F	Dry and wet locations	Flame-retardant, moisture- and heat-resistant thermo-plastic	14–12 10 8–6 4–2 1–4/0 250–500 501–1000	0.38 0.51 0.76 1.02 1.27 1.52 1.78	15 20 30 40 50 60 70	Nylon jacket or equivalent
Moisture-resistant thermo-plastic	TW	60°C 140°F	Dry and wet locations	Flame-retardant, moisture-resistant thermo-plastic	14–10 8 6–2 1–4/0 213–500 501–1000 1001–2000	0.76 1.14 1.52 2.03 2.41 2.79 3.18	30 45 60 80 95 110 125	None
Underground feeder and branch-circuit cable — single conductor (For Type UF cable employing more than one conductor, *see* Articles 339, 340.)	UF	60°C 140°F 75°C 167°F[7]	See Article 340.	Moisture-resistant Moisture- and heat-resistant	14–10 8–2 1–4/0	1.52 2.03 2.41	60[6] 80[6] 95[6]	Integral with insulation

Table 310.13 *Continued*

Trade Name	Type Letter	Maximum Operating Temperature	Application Provisions	Insulation	Thickness of Insulation			Outer Covering[1]
					AWG or kcmil	mm	Mils	
Underground service-entrance cable — single conductor (For Type USE cable employing more than one conductor, *see* Article 338.)	USE[4]	75°C 167°F	See Article 338.	Heat- and moisture-resistant	14–10 8–2 1–4/0 213–500 501–1000 1001–2000	1.14 1.52 2.03 2.41 2.79 3.18	45 60 80 95[8] 110 125	Moisture-resistant non-metallic covering (See 338.2.)
Thermoset	XHH	90°C 194°F	Dry and damp locations	Flame-retardant thermoset	14–10 8–2 1–4/0 213–500 501–1000 1001–2000	0.76 1.14 1.40 1.65 2.03 2.41	30 45 55 65 80 95	None
Moisture-resistant thermoset	XHHW[4]	90°C 194°F 75°C 167°F	Dry and damp locations Wet locations	Flame-retardant, moisture-resistant thermoset	14–10 8–2 1–4/0 213–500 501–1000 1001–2000	0.76 1.14 1.40 1.65 2.03 2.41	30 45 55 65 80 95	None
Moisture-resistant thermoset	XHHW-2	90°C 194°F	Dry and wet locations	Flame-retardant, moisture-resistant thermoset	14–10 8–2 1–4/0 213–500 501–1000 1001–2000	0.76 1.14 1.40 1.65 2.03 2.41	30 45 55 65 80 95	None
Modified ethylene tetra-fluoro-ethylene	Z	90°C 194°F 150°C 302°F	Dry and damp locations Dry locations — special applications[2]	Modified ethylene tetra-fluoro-ethylene	14–12 10 8–4 3–1 1/0–4/0	0.38 0.51 0.64 0.89 1.14	15 20 25 35 45	None
Modified ethylene tetrafluoro-ethylene	ZW[4]	75°C 167°F 90°C 194°F 150°C 302°F	Wet locations Dry and damp locations Dry locations — special applications[2]	Modified ethylene tetra-fluoro-ethylene	14–10 8–2	0.76 1.14	30 45	None

[1] Some insulations do not require an outer covering.

[2] Where design conditions require maximum conductor operating temperatures above 90°C (194°F).

[3] For signaling circuits permitting 300-volt insulation.

[4] Listed wire types designated with the suffix "2," such as RHW-2, shall be permitted to be used at a continuous 90°C (194°F) operating temperature, wet or dry.

[5] Some rubber insulations do not require an outer covering.

[6] Includes integral jacket.

[7] For ampacity limitation, see 340.80.

[8] Insulation thickness shall be permitted to be 2.03 mm (80 mils) for listed Type USE conductors that have been subjected to special investigations. The nonmetallic covering over individual rubber-covered conductors of aluminum-sheathed cable and of lead-sheathed or multiconductor cable shall not be required to be flame retardant.

For Type MC cable, see 330.104. For nonmetallic-sheathed cable, see Article 334, Part III. For Type UF cable, see Article 340, Part III.

tions with different types of conductor material. A comparison of the characteristics of copper, copper-clad aluminum, and aluminum conductors can be made from Table 3.1.

Table 3.1 Conductor Characteristics

Characteristic	Copper	Copper-Clad Aluminum	Aluminum
Density (lb/in.³)	0.323	0.121	0.098
Density (g/cm³)	8.91	3.34	2.71
Resistivity ohms/CMF	10.37	16.08	16.78
Resistivity Microhm — CM	1.724	2.673	2.790
Conductivity (IACS %)	100	61–63	61.0
Weight % Copper	100	26.8	—
Tensile K psi — Hard	65.0	30.0	27.0
Tensile kg/mm² — Hard	45.7	21.1	19.0
Tensile K psi — Annealed	35.0	17.0	17.0*
Tensile kg/mm² — Annealed	24.6	12.0	12.0
Specific Gravity	8.91	3.34	2.71

*Semiannealed

310.14 Aluminum Conductor Material.

Solid aluminum conductors 8, 10, and 12 AWG shall be made of an AA-8000 series electrical grade aluminum alloy conductor material. Stranded aluminum conductors 8 AWG through 1000 kcmil marked as Type RHH, RHW, XHHW, THW, THHW, THWN, THHN, service-entrance Type SE Style U and SE Style R shall be made of an AA-8000 series electrical grade aluminum alloy conductor material.

Section 310.14 provides proper recognition of approved AA-8000 series electrical-grade aluminum alloy conductor materials for wire and cable products. It also provides coordination with the UL listing requirements for testing terminations, such as CO/ALR devices and other connectors, suitable for use with aluminum conductors. The electrical industry has developed AA-8000 series aluminum alloy materials and the connectors suitable for use with aluminum conductors to provide for safe and stable connections. Connections suitable for use with aluminum conductors are also generally listed as suitable for use with copper conductors and are marked accordingly, such as AL7CU or AL9CU. Numbers 7 and 9 identify the temperature ratings of 75°C and 90°C, respectively, for these connectors.

310.15 Ampacities for Conductors Rated 0–2000 Volts.

(A) General.

Section 310.15 was reorganized for the 1999 *Code*. The notes affecting the ampacity of conductors, previously enti-

tled Article 310, Notes to Ampacity Tables 0 to 2000 Volts, now appear as adjustment factors in 310.15(B)(1) through (6). Although this reorganization was a significant change, no substantive technical changes were made to the requirements.

(1) Tables or Engineering Supervision. Ampacities for conductors shall be permitted to be determined by tables or under engineering supervision, as provided in 310.15(B) and (C).

> FPN No. 1: Ampacities provided by this section do not take voltage drop into consideration. See 210.19(A), FPN No. 4, for branch circuits and 215.2(D), FPN No. 2, for feeders.
> FPN No. 2: For the allowable ampacities of Type MTW wire, see Table 11 in NFPA 79-1997, *Electrical Standard for Industrial Machinery.*

Section 310.15(A)(1) permits either of two methods of determining conductor ampacity for conductors rated 0 through 2000 volts: to select the ampacity from a table, using correction factors in the table or notes where necessary, or to calculate the ampacity. The latter method can be complex and time consuming and requires engineering supervision. It can, however, result in lower installation costs, in some cases, and if calculated properly, it provides a mathematically exact ampacity. See the commentary following 310.15(C) and accompanying Annex B for further explanation.

(2) Selection of Ampacity. Where more than one calculated or tabulated ampacity could apply for a given circuit length, the lowest value shall be used.

Exception: Where two different ampacities apply to adjacent portions of a circuit, the higher ampacity shall be permitted to be used beyond the point of transition, a distance equal to 3.0 m (10 ft) or 10 percent of the circuit length figured at the higher ampacity, whichever is less.

Example

Three 500-kcmil THW conductors in a rigid conduit are run from a motor control center for 12 ft past a heat-treating furnace to a pump motor located 150 ft from the motor control center. Where run in a 78°F to 86°F ambient, the conductors have an ampacity of 380 amperes, per Table 310.16. The ambient temperature near the furnace, where the conduit is run, is found to be 113°F, and the length of this particular part of the run is greater than 10 ft and more than 10 percent of the total length of the run at the 78°F to 86°F ambient. Determine the ampacity of total run in accordance with 310.15(A)(2).

Solution

In accordance with the correction factors for temperature at the bottom of Table 310.16, the ampacity is 0.82 × 380 amperes, or 311.6 amperes. This, therefore, is the ampacity of the total run, in accordance with 310.15(A)(2).

Had the run near the furnace at the 113°F ambient been 10 ft or less in length, the ampacity of the entire run would have been 380 amperes, in accordance with the exception to 310.15(A)(2). The heat-sinking effect of the run at the lower ambient temperature would have been sufficient to reduce the temperature of the conductor near the furnace.

FPN: See 110.14(C) for conductor temperature limitations due to termination provisions.

(B) Tables. Ampacities for conductors rated 0 to 2000 volts shall be as specified in the Allowable Ampacity Table 310.16 through Table 310.19 and Ampacity Tables 310.20 and 310.21 as modified by (1) through (6).

FPN: Tables 310.16 through 310.19 are application tables for use in determining conductor sizes on loads calculated in accordance with Article 220. Allowable ampacities result from consideration of one or more of the following:

(1) Temperature compatibility with connected equipment, especially the connection points.
(2) Coordination with circuit and system overcurrent protection.
(3) Compliance with the requirements of product listings or certifications. See 110.3(B).
(4) Preservation of the safety benefits of established industry practices and standardized procedures.

Ampacity tables, particularly Table 310.16, do not take into account all of the many factors affecting ampacity. See the commentary following 310.15(C) and accompanying Annex B for further explanation. However, experience over many years has proved the table values to be adequate for loads calculated in accordance with Article 220, because not all diversity factors and load factors found in most actual installations are specifically provided for in Article 220. If loads are not calculated in accordance with the requirements of Article 220, the table ampacities, even when corrected in accordance with ambient correction factors and the notes to the tables, might be too high. This result can be particularly true where many cables or raceways are routed close to one another underground. However, load diversity and thermal conductance fill around buried cable could result in increased ampacity. See Annex B for further information.

(1) General. For explanation of type letters used in tables and for recognized sizes of conductors for the various con-

ductor insulations, see 310.13. For installation requirements, see 310.1 through 310.10 and the various articles of this *Code*. For flexible cords, see Tables 400.4, 400.5(A), and 400.5(B).

(2) Adjustment Factors.

(a) More Than Three Current-Carrying Conductors in a Raceway or Cable. Where the number of current-carrying conductors in a raceway or cable exceeds three, or where single conductors or multiconductor cables are stacked or bundled longer than 600 mm (24 in.) without maintaining spacing and are not installed in raceways, the allowable ampacity of each conductor shall be reduced as shown in Table 310.15(B)(2)(a).

Table 310.15(B)(2)(a) Adjustment Factors for More Than Three Current-Carrying Conductors in a Raceway or Cable

Number of Current-Carrying Conductors	Percent of Values in Tables 310.16 through 310.19 as Adjusted for Ambient Temperature if Necessary
4–6	80
7–9	70
10–20	50
21–30	45
31–40	40
41 and above	35

FPN: See Annex B, Table B.310.11, for adjustment factors for more than three current-carrying conductors in a raceway or cable with load diversity.

The factors in the second column of Table 310.15(B)(2)(a) are based on no diversity, meaning that all conductors in the raceway or cable are loaded to their maximum rated load. For load diversity, the user is directed to Annex B.

Specific cross references for raceway fill and adjustment factors of 310.15(B)(2) can be found in the fine print note following 300.17. For Class 1 conductors, see 725.28(A); for fire alarm systems, 760.28 and 760.52; for optical fiber cables and raceways, 770.52; and for communications wires and cables within buildings, 800.48.

Exception No. 1: Where conductors of different systems, as provided in 300.3, are installed in a common raceway or cable, the derating factors shown in Table 310.15(B)(2)(a) shall apply to the number of power and lighting conductors only (Articles 210, 215, 220, and 230).

Exception No. 1 assumes that the watt loss (heating) from any control and signal conductors in the same raceway or cable will not be enough to significantly increase the temperature of the power and lighting conductors. See 725.26 and 725.54 for limitations on the installation of control and signal conductors in the same raceway or cable as power and lighting conductors.

Exception No. 2: For conductors installed in cable trays, the provisions of 392.11 shall apply.

Exception No. 3: Derating factors shall not apply to conductors in nipples having a length not exceeding 600 mm (24 in.).

Exception No. 4: Derating factors shall not apply to underground conductors entering or leaving an outdoor trench if those conductors have physical protection in the form of rigid metal conduit, intermediate metal conduit, or rigid nonmetallic conduit having a length not exceeding 3.05 m (10 ft) and if the number of conductors does not exceed four.

Exhibit 310.4 illustrates Exception No. 4 to 310.15(B)(2)(a), in that derating factors do not apply to the conductors because they have physical protection (conduit) that does not exceed 10 ft in length and the number of conductors does not exceed four.

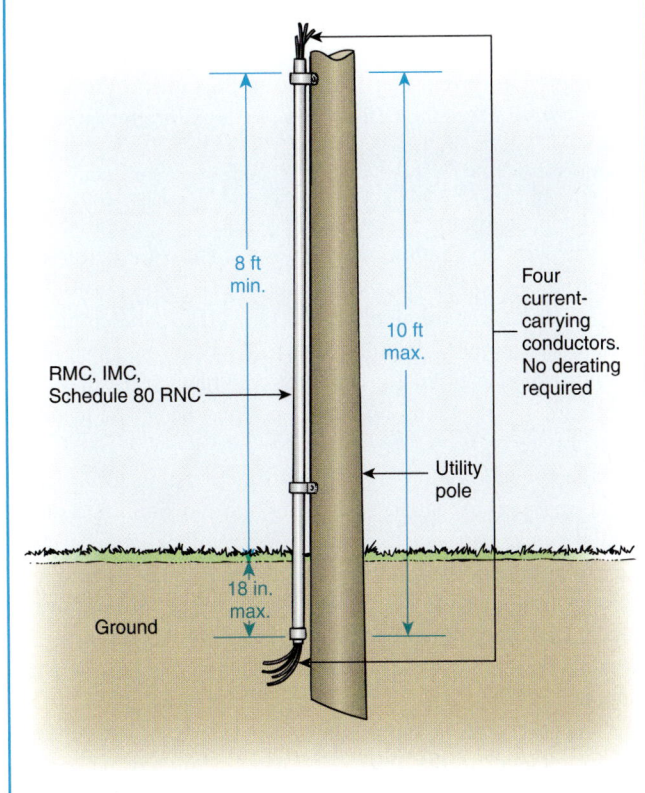

Exhibit 310.4 An application of 310.15(B)(2)(a), Exception No. 4.

Exception No. 5: Adjustment factors shall not apply to Type AC cable or to Type MC cable without an overall outer jacket under the following conditions:

(a) Each cable has not more than three current-carrying conductors.

(b) The conductors are 12 AWG copper.

(c) Not more than 20 current-carrying conductors are bundled, stacked, or supported on "bridle rings."

A 60 percent adjustment factor shall be applied where the current-carrying conductors in these cables that are stacked or bundled longer than 600 mm (24 in.) without maintaining spacing exceeds 20.

Example

A commercial office space will require fourteen 277-volt fluorescent lighting circuits to serve a single open office area. The office area lighting is assumed to be a continuous load, and the office ambient temperature will not exceed 30°C (86°F). Each circuit will be arranged so that it has a computed load not exceeding 16 amperes. The selected wiring method is Type MC cable, 3-conductor (with an additional equipment grounding conductor), 12 AWG THHN copper. Each individual MC cable will contain a 3-wire multiwire branch circuit. To serve the entire area, this arrangement will require a total of seven Type MC cables bundled for a distance of about 25 ft, without maintaining spacing between them where they leave the electrical room and enter the office area.

Determine the ampacity of each circuit conductor in accordance with 310.15, applying Exception No. 5 to 310.15(B)(2)(a) to account for the bundled cables. Then determine the maximum branch-circuit overcurrent protection permitted for these bundled MC cables.

Solution

STEP 1. To apply Exception No. 5, first determine the quantity of current-carrying conductors. According to 310.15(B)(5), equipment grounding conductors are not counted as current-carrying conductors. According to 310.15(B)(4)(c), fluorescent lighting is considered a nonlinear load, so the grounded conductor of each Type MC cable must be counted as a current-carrying conductor.

$$7 \text{ cables} \times 3 \text{ conductors each} = \frac{21 \text{ current-carrying}}{\text{conductors}}$$

Because the quantity of current-carrying conductors exceeds 20, a 60 percent adjustment factor is required by 310.15(B)(2)(a), Exception No. 5.

STEP 2. Determine the ampacity of each current-carrying conductor due to these MC cables with more than 20 current-carrying conductors being bundled.

From Table 310.16, 12 AWG THHN = 30 amps

$$30 \text{ amps} \times 0.60 = 18 \text{ amps}$$

Because the actual computed load is 16 amperes of continuous load, 210.19(A)(1) is applicable. The conductors must have an ampacity equal to or greater than the load before the adjustment factor is applied. Because the ampacity of the conductors after the adjustment factor is applied is 18 amperes, no further adjustment is necessary and the conductors are suitable for this installation.

STEP 3. Finally, determine the maximum size of overcurrent protection device permitted for these bundled MC cable branch circuits. Section 240.4(B) permits the use of the next higher standard rating of overcurrent protection device. Therefore, although the conductors have a computed ampacity of 18 amperes, a 20-ampere overcurrent protective device is permitted. In addition, and of significance, the 20-ampere overcurrent protective device is in compliance with 210.20(A), given that the actual 16-ampere continuous load would require a 20-ampere overcurrent protective device, based on the listing of the overcurrent device.

(b) More Than One Conduit, Tube, or Raceway. Spacing between conduits, tubing, or raceways shall be maintained.

Spacing is normally maintained between individual conduits in groups of conduit runs from junction box to junction box because of the need to separate the conduits where they enter the junction box, to allow room for locknuts and bushings. Field experience has indicated this degree of spacing between runs has not caused any problems.

(3) Bare or Covered Conductors. Where bare or covered conductors are used with insulated conductors, their allowable ampacities shall be limited to those permitted for the adjacent insulated conductors.

(4) Neutral Conductor.

(a) A neutral conductor that carries only the unbalanced current from other conductors of the same circuit shall not be required to be counted when applying the provisions of 310.15(B)(2)(a).

(b) In a 3-wire circuit consisting of two phase wires and the neutral of a 4-wire, 3-phase, wye-connected system, a common conductor carries approximately the same current as the line-to-neutral load currents of the other conductors and shall be counted when applying the provisions of 310.15(B)(2)(a).

(c) On a 4-wire, 3-phase wye circuit where the major portion of the load consists of nonlinear loads, harmonic currents are present in the neutral conductor; the neutral shall therefore be considered a current-carrying conductor.

During the 1996 *NEC* cycle, a task group composed of interested parties was created to recommend to the National Electrical Code Committee the direction its standards should take to improve the safeguarding of persons and property from conditions that can be introduced by nonlinear loads. This group was designated the NEC Correlating Committee Ad Hoc Subcommittee on Nonlinear Loads. The scope of this subcommittee was as follows:

1. To study the effects of electrical loads producing substantial current distortion upon electrical system distribution components including, but not limited to
   a. Distribution transformers, current transformers, and others
   b. Switchboards and panelboards
   c. Phase and neutral feeder conductors
   d. Phase and neutral branch-circuit conductors
   e. Proximate data and communications conductors
2. To study harmful effects, if any, to the system components from overheating resulting from these load characteristics
3. To make recommendations for methods to minimize the harmful effects of nonlinear loads considering all means, including compensating methods at load sources
4. To prepare proposals, if necessary, to amend the 1996 *National Electrical Code,* where amelioration to fire safety may be achieved

The subcommittee reviewed technical literature and electrical theory on the fundamental nature of harmonic distortion, as well as the requirements in and proposals for the 1993 *NEC* regarding nonlinear loads. The subcommittee concluded that, while nonlinear loads can cause undesirable operational effects, including additional heating, no significant threat to persons and property has been adequately substantiated.

The subcommittee agreed with the existing *Code* text regarding nonlinear loads. However, the subcommittee submitted many proposals for the 1996 *NEC,* including a definition of nonlinear load, revised text reflecting that definition, fine print notes calling attention to the effects of nonlinear loads, and proposals permitting the paralleling of neutral conductors in existing installations under engineering supervision.

As part of the subcommittee's final report, nine proposals for changes to the 1993 *NEC* were submitted. All were accepted without modification as changes to the 1996 *NEC.* Also included in this report and now pertinent to the 2002 *NEC* 310.15(B)(4)(c) is the following discussion.

SHOULD NEUTRAL CONDUCTORS BE OVERSIZED?

There is concern that, because the theoretical maximum neutral current is 1.73 times the balanced phase conductor current, a potential exists for neutral conductor overheating in 3-phase, 4-wire, wye-connected power systems. The subcommittee acknowledged this theoretical basis, although a review of documented information could not identify fires attributed to the use of nonlinear loads.

The subcommittee reviewed all available data regarding measurements of circuits that contain nonlinear loads. The data were obtained from consultants, equipment manufacturers, and testing laboratories, and included hundreds of feeder and branch circuits involving 3-phase, 4-wire, wye-connected systems with nonlinear loads. The data revealed that many circuits had neutral conductor current greater than the phase conductor current, and approximately 5 percent of all circuits reported had neutral conductor current exceeding 125 percent of the highest phase conductor current. One documented survey with data collected in 1988 from 146 three-phase computer power system sites determined that 3.4 percent of the sites had neutral current in excess of the rated system full-load current.

According to 384-16(C) of the 1993 *NEC* (for the 2002 *NEC*, refer to 408.16), the total continuous load on any overcurrent device located in a panelboard should not exceed 80 percent of its rating (the exception being assemblies listed for continuous operation at 100 percent of its rating). Because the neutral conductor is usually not connected to an overcurrent device, derating for continuous operation is not necessary. Therefore, neutral conductor ampacity is usually 125 percent of the maximum continuous current allowed by the overcurrent device.

Also important for gathering electrically measured data from existing installations is the following.

Measurement of Nonsinusoidal Voltages and Currents

The measurement of nonsinusoidal voltages and currents may require instruments different from the conventional meters used to measure sinusoidal waveforms. Many voltage and current meters respond only to the peak value of a waveform, and indicate a value that is equivalent to the rms value of a sinusoidal waveform. For a sinusoidal waveform the rms value will be 70.7 percent of the peak value. Meters of this type are known as "average responding meters" and will only give a true indication if the waveform being measured is sinusoidal. Both analog and digital meters may be average responding instruments. Voltages and currents that are nonsinusoidal, such as those with harmonic frequencies, cannot be accurately measured using an average responding meter. Only a meter that measures "true rms" can be used to correctly measure the rms value of a nonsinusoidal waveform.

Exhibit 310.5 shows an example of a clamp-on ammeter that uses true rms measurements. Exhibit 310.6 shows an example of a portable diagnostic analyzer used for more sophisticated power measurements, including measuring harmonic distortion.

Exhibit 310.5 A clamp-on ammeter that uses true rms measurements. (Courtesy of Fluke Corp.)

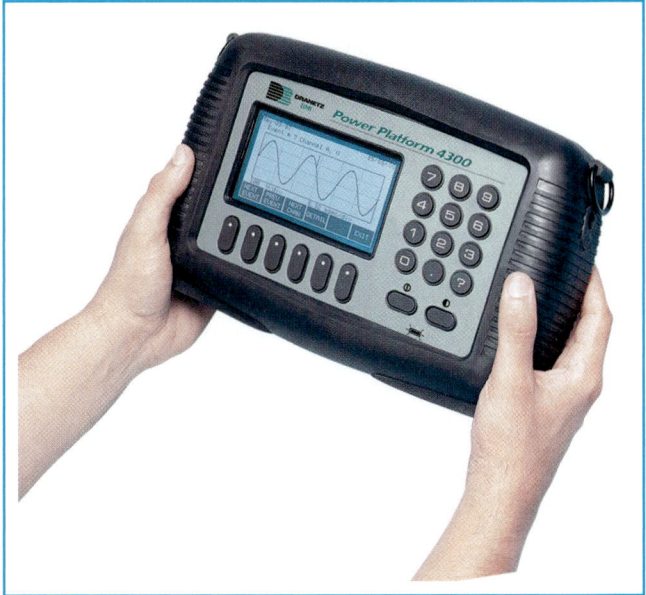

Exhibit 310.6 A portable tool for such tasks as diagnostic power analysis including harmonic distortion. (Courtesy of Dranetz-BMI)

(5) Grounding or Bonding Conductor. A grounding or bonding conductor shall not be counted when applying the provisions of 310.15(B)(2)(a).

(6) 120/240-Volt, 3-Wire, Single-Phase Dwelling Services and Feeders. For dwelling units, conductors, as listed in Table 310.15(B)(6), shall be permitted as 120/240-volt, 3-wire, single-phase service-entrance conductors, service lateral conductors, and feeder conductors that serve as the main power feeder to a dwelling unit and are installed in raceway or cable with or without an equipment grounding conductor. For application of this section, the main power feeder shall be the feeder(s) between the main disconnect and the lighting and appliance branch-circuit panelboard(s). The feeder conductors to a dwelling unit shall not be required to be larger than their service-entrance conductors. The grounded conductor shall be permitted to be smaller than the ungrounded conductors, provided the requirements of 215.2, 220.22, and 230.42 are met.

Table 310.15(B)(6) Conductor Types and Sizes for 120/240-Volt, 3-Wire, Single-Phase Dwelling Services and Feeders. Conductor Types RHH, RHW, RHW-2, THHN, THHW, THW, THW-2, THWN, THWN-2, XHHW, XHHW-2, SE, USE, USE-2.

Conductor (AWG or kcmil)		
Copper	Aluminum or Copper-Clad Aluminum	Service or Feeder Rating (Amperes)
4	2	100
3	1	110
2	1/0	125
1	2/0	150
1/0	3/0	175
2/0	4/0	200
3/0	250	225
4/0	300	250
250	350	300
350	500	350
400	600	400

If a single set of 3-wire, single-phase, service-entrance conductors in raceway or cable supplies a one-family, two-family, or multifamily dwelling, the reduced conductor size permitted by 310.15(B)(6) is applicable to the service-entrance conductors, service-lateral conductors, or any feeder conductors that supply the main power feeder to a dwelling unit.

This section permits the main feeder to a dwelling unit to be sized according to the conductor sizes in Table 310.15(B)(6) even if other loads, such as ac units and pool loads, are fed from the same service. The feeder conductors

to a dwelling unit are not required to be larger than its service-entrance conductors.

Exhibits 310.7 and 310.8 illustrate the application of 310.15(B)(6). In Exhibit 310.7, the reduced conductor size permitted is applicable to the service-entrance conductors run to each apartment from the meters. In Exhibit 310.8,

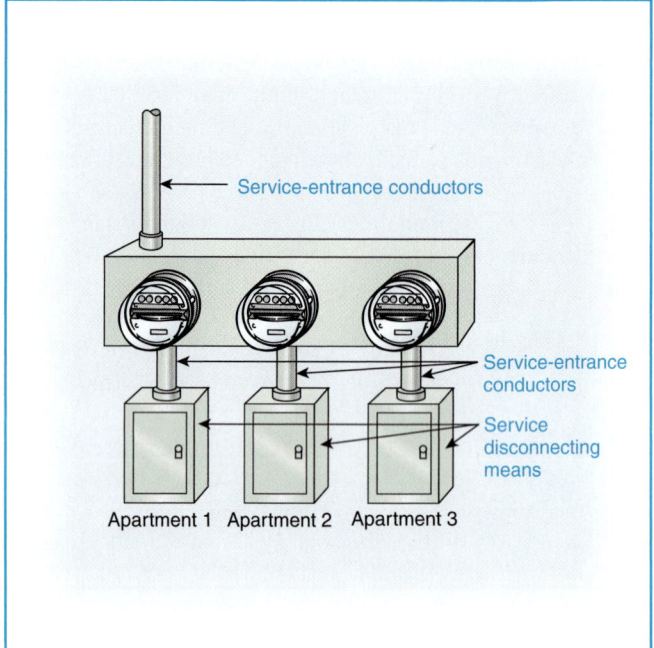

Exhibit 310.7 An application of 310.15(B)(6).

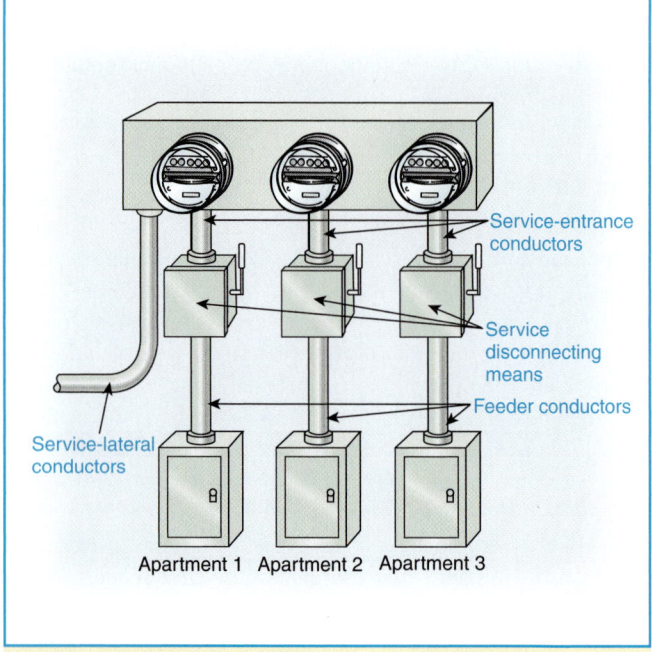

Exhibit 310.8 Another application of 310.15(B)(6).

the reduced conductor size permitted is also applicable to the feeder conductors run to each apartment from the service disconnecting means, because these feeders carry the entire load to each apartment.

Previously limited to a two-wire size reduction but changed in the 1996 *Code,* the grounded conductor is now permitted to be reduced more than two sizes. The stipulation is that the requirements from the other applicable *Code* sections are observed, including 250.24(B).

Other sections of the *Code* must also be applied for determining the size of service or feeder conductors. Section 230.42 requires service conductors to be of sufficient size to carry the load calculated in accordance with Article 220. It should not be taken for granted that the grounded (neutral) conductor can be automatically reduced. In addition to the load, the grounded (neutral) conductor must also provide a low-impedance fault path capable of conducting fault current back to the transformer. Section 250.24(B) provides the requirements for determining the minimum size grounded (neutral) conductor.

Section 250.24(B) is not applicable to the grounded (neutral) conductor of feeders; therefore, the minimum size is governed by 220.22. This section requires the grounded (neutral) conductor to be large enough to carry the maximum unbalance of the net computed load connected to the neutral and any one ungrounded conductor. In the event that there are no 240-volt loads, the neutral, under severe unbalanced conditions, carries the same current as the ungrounded conductor supplying the load.

(C) Engineering Supervision. Under engineering supervision, conductor ampacities shall be permitted to be calculated by means of the following general formula:

$$I = \sqrt{\frac{TC - (TA + \Delta TD)}{RDC(1 + YC)RCA}}$$

where:

TC = conductor temperature in degrees Celsius (°C)

TA = ambient temperature in degrees Celsius (°C)

ΔTD = dielectric loss temperature rise

RDC = dc resistance of conductor at temperature TC

YC = component ac resistance resulting from skin effect and proximity effect

RCA = effective thermal resistance between conductor and surrounding ambient

FPN: See Annex B for examples of formula applications.

The formula in 310.15(C) was developed by J. H. Neher and M. H. McGrath to determine conductor ampacity. It is actually a composite of a number of separate formulas. A description of this method of calculation was given in AIEE

paper No. 5-660, "The Calculation of the Temperature Rise and Load Capability of Cable Systems," by J. H. Neher and M. H. McGrath. This paper was presented to the AIEE general meeting in Montreal, Quebec, on June 24–28, 1956, and was published in *AIEE Transactions, Part III (Power Apparatus and Systems),* Vol. 76, October 1957, pp. 752–772. AIEE (American Institute of Electrical Engineers) is now IEEE (Institute of Electrical and Electronic Engineers).

The Neher-McGrath formula in 310.15(C) is a heat-transfer formula, composed of a series of heat-transfer calculations, that takes into account all heat sources and the thermal resistances between the heat sources and free air. The most common use for the Neher-McGrath formula is to calculate the ampacity of conductors in underground electrical ducts (raceways), although the formula is applicable to all conductor installations.

It is not the intent of the following discussion to provide instruction on the use of the Neher-McGrath method of calculation. The intent is to identify the many factors affecting the calculations. It is because of these many variables and the complexities of the many formulas involved that the *Code* requires the calculation to be made under engineering supervision.

Current passing through a conductor produces I^2R losses in the form of heat, which results from conductor losses and appears as a temperature rise in the conductor. This heat must pass through the cable insulation, the air in the raceway, and the raceway itself to the surrounding medium, usually earth or concrete, where it is dissipated into the air by radiation and convection. Unless the heat is dissipated, the temperature in the conductor will exceed the rating of the conductor insulation.

The conductor's ampacity is based on the rate of heat dissipation through the thermal resistances surrounding the conductor. Current traveling through a material with a specific resistance at a specified temperature generates this heat. Additional heat is caused by skin and proximity effects, because usually the current is ac and there are other conductors in the same duct.

For conductors in underground electrical ducts, there are several heat sources, as follows, and as illustrated in Exhibit 310.9.

1. *Conductor losses due to the load current I^2R.* These losses vary with the load current, conductor material, and conductor cross-sectional area (conductor size).
2. *Skin-effect heating if the current is alternating current.* The heat developed by the skin effect is due to the shape of the conductor and is based on the configuration of the conductors (i.e., solid, stranded, or compact).
3. *Hysteresis losses if the duct is steel or other magnetic material.* These losses are dependent upon the magnetic

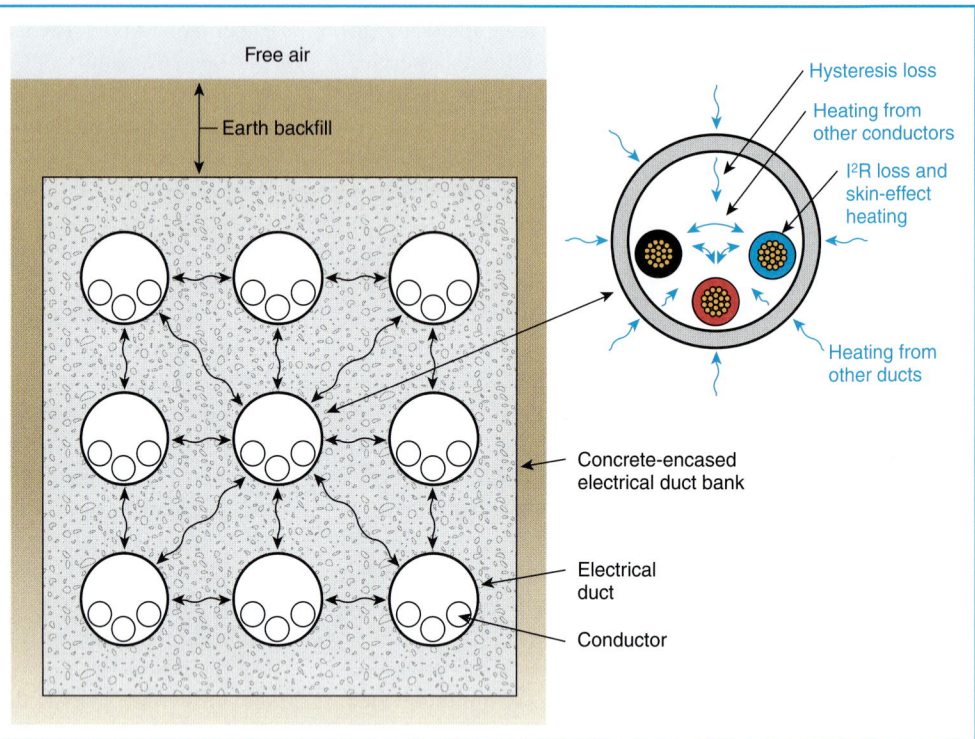

Exhibit 310.9 Heat sources.

properties of the electrical duct and the shape of the duct.

4. *Heating from other conductors in the duct.* This heating is based on the number, location, and proximity of other conductors as well as the losses in the other conductors. The more conductors in the raceway, the greater the heating effect from these conductors is likely to be. This factor replaces the adjustment factors in 310.15(B)(2)(a) to the ampacity tables.

5. *Mutual heating from other ducts, cables, etc., in the vicinity.* The closer the other heat sources and the more they surround the duct for which calculations are being made, the greater the heating effect. For example, in the case of a symmetrical 9-duct bank, 3 ducts high and 3 ducts wide, the center duct will receive the most heat as a result of mutual heating.

Heat generated by the following various types of losses is conducted through the different thermal barriers or resistances, as illustrated in Exhibit 310.10.

Conductor insulation. The conductor insulation, which is designed to perform as a good electrical insulator, also serves as a good thermal insulator. It presents a thermal resistance to heat generated by the conductor due to the I^2R losses, including any dielectric losses. This thermal resistance value depends on the thickness of the insulation and the type of insulating material used. Materials such as polyvinyl chloride, used in Type THW and other conductors; cross-linked polyethylene, used in Type XHHW and other conductors; and rubber, used in Type RHW and other conductors, have different thermal resistivities. In addition, the thickness of the conductor insulation varies from one type of insulation to another, even for the same size conductor.

Airspace. The next thermal barrier encountered by the heat flow generated in the conductor is the airspace between the conductor insulation and the surrounding wall or raceway. The thermal resistance of this airspace is based on the number of conductors in the duct, the assumed mean value of the temperature of the air in the duct, and the constants provided in the Neher-McGrath paper, which were determined from experimental data.

Duct wall. After it passes through the airspace around the conductors, the heat encounters the thermal resistance of the duct wall. This thermal resistance is based on the thermal resistivity of the type of material used and the thickness of the duct wall. Metallic materials have less thermal resistance than nonmetallic materials. The thicker the wall, the greater the thermal resistance.

Earth backfill. The thermal resistance that must be considered next is that offered by the earth or other backfill material above the duct. This incorporates not only the thermal resistivity and ambient temperature of the earth but also the number of current-carrying conductors within the duct, the outside diameter of the duct, the burial depth, a loss

Table 310.16 Allowable Ampacities of Insulated Conductors Rated 0 Through 2000 Volts, 60°C Through 90°C (140°F Through 194°F), Not More Than Three Current-Carrying Conductors in Raceway, Cable, or Earth (Directly Buried), Based on Ambient Temperature of 30°C (86°F)

Size AWG or kcmil	Temperature Rating of Conductor (See Table 310.13.)						Size AWG or kcmil
	60°C (140°F)	75°C (167°F)	90°C (194°F)	60°C (140°F)	75°C (167°F)	90°C (194°F)	
	Types TW, UF	Types RHW, THHW, THW, THWN, XHHW, USE, ZW	Types TBS, SA, SIS, FEP, FEPB, MI, RHH, RHW-2, THHN, THHW, THW-2, THWN-2, USE-2, XHH, XHHW, XHHW-2, ZW-2	Types TW, UF	Types RHW, THHW, THW, THWN, XHHW, USE	Types TBS, SA, SIS, THHN, THHW, THW-2, THWN-2, RHH, RHW-2, USE-2, XHH, XHHW, XHHW-2, ZW-2	
	COPPER			ALUMINUM OR COPPER-CLAD ALUMINUM			
18	—	—	14	—	—	—	—
16	—	—	18	—	—	—	—
14*	20	20	25	—	—	—	—
12*	25	25	30	20	20	25	12*
10*	30	35	40	25	30	35	10*
8	40	50	55	30	40	45	8
6	55	65	75	40	50	60	6
4	70	85	95	55	65	75	4
3	85	100	110	65	75	85	3
2	95	115	130	75	90	100	2
1	110	130	150	85	100	115	1
1/0	125	150	170	100	120	135	1/0
2/0	145	175	195	115	135	150	2/0
3/0	165	200	225	130	155	175	3/0
4/0	195	230	260	150	180	205	4/0
250	215	255	290	170	205	230	250
300	240	285	320	190	230	255	300
350	260	310	350	210	250	280	350
400	280	335	380	225	270	305	400
500	320	380	430	260	310	350	500
600	355	420	475	285	340	385	600
700	385	460	520	310	375	420	700
750	400	475	535	320	385	435	750
800	410	490	555	330	395	450	800
900	435	520	585	355	425	480	900
1000	455	545	615	375	445	500	1000
1250	495	590	665	405	485	545	1250
1500	520	625	705	435	520	585	1500
1750	545	650	735	455	545	615	1750
2000	560	665	750	470	560	630	2000

CORRECTION FACTORS

Ambient Temp. (°C)	For ambient temperatures other than 30°C (86°F), multiply the allowable ampacities shown above by the appropriate factor shown below.						Ambient Temp. (°F)
21–25	1.08	1.05	1.04	1.08	1.05	1.04	70–77
26–30	1.00	1.00	1.00	1.00	1.00	1.00	78–86
31–35	0.91	0.94	0.96	0.91	0.94	0.96	87–95
36–40	0.82	0.88	0.91	0.82	0.88	0.91	96–104
41–45	0.71	0.82	0.87	0.71	0.82	0.87	105–113
46–50	0.58	0.75	0.82	0.58	0.75	0.82	114–122
51–55	0.41	0.67	0.76	0.41	0.67	0.76	123–131
56–60	—	0.58	0.71	—	0.58	0.71	132–140
61–70	—	0.33	0.58	—	0.33	0.58	141–158
71–80	—	—	0.41	—	—	0.41	159–176

* See 240.4(D).

Table 310.17 Allowable Ampacities of Single-Insulated Conductors Rated 0 Through 2000 Volts in Free Air, Based on Ambient Air Temperature of 30°C (86°F)

Size AWG or kcmil	Temperature Rating of Conductor (See Table 310.13.)						Size AWG or kcmil
	60°C (140°F)	75°C (167°F)	90°C (194°F)	60°C (140°F)	75°C (167°F)	90°C (194°F)	
	Types TW, UF	Types RHW, THHW, THW, THWN, XHHW, ZW	Types TBS, SA, SIS, FEP, FEPB, MI, RHH, RHW-2, THHN, THHW, THW-2, THWN-2, USE-2, XHH, XHHW, XHHW-2, ZW-2	Types TW, UF	Types RHW, THHW, THW, THWN, XHHW	Types TBS, SA, SIS, THHN, THHW, THW-2, THWN-2, RHH, RHW-2, USE-2, XHH, XHHW, XHHW-2, ZW-2	
	COPPER			ALUMINUM OR COPPER-CLAD ALUMINUM			
18	—	—	18	—	—	—	—
16	—	—	24	—	—	—	—
14*	25	30	35	—	—	—	—
12*	30	35	40	25	30	35	12*
10*	40	50	55	35	40	40	10*
8	60	70	80	45	55	60	8
6	80	95	105	60	75	80	6
4	105	125	140	80	100	110	4
3	120	145	165	95	115	130	3
2	140	170	190	110	135	150	2
1	165	195	220	130	155	175	1
1/0	195	230	260	150	180	205	1/0
2/0	225	265	300	175	210	235	2/0
3/0	260	310	350	200	240	275	3/0
4/0	300	360	405	235	280	315	4/0
250	340	405	455	265	315	355	250
300	375	445	505	290	4350	395	300
350	420	505	570	330	395	445	350
400	455	545	615	355	425	480	400
500	515	620	700	405	485	545	500
600	575	690	780	455	540	615	600
700	630	755	855	500	595	675	700
750	655	785	885	515	620	700	750
800	680	815	920	535	645	725	800
900	730	870	985	580	700	785	900
1000	780	935	1055	625	750	845	1000
1250	890	1065	1200	710	855	960	1250
1500	980	1175	1325	795	950	1075	1500
1750	1070	1280	1445	875	1050	1185	1750
2000	1155	1385	1560	960	1150	1335	2000

Table 310.17 *Continued*

	Temperature Rating of Conductor (See Table 310.13.)						
	60°C (140°F)	75°C (167°F)	90°C (194°F)	60°C (140°F)	75°C (167°F)	90°C (194°F)	
Size AWG or kcmil	Types TW, UF	Types RHW, THHW, THW, THWN, XHHW, ZW	Types TBS, SA, SIS, FEP, FEPB, MI, RHH, RHW-2, THHN, THHW, THW-2, THWN-2, USE-2, XHH, XHHW, XHHW-2, ZW-2	Types TW, UF	Types RHW, THHW, THW, THWN, XHHW	Types TBS, SA, SIS, THHN, THHW, THW-2, THWN-2, RHH, RHW-2, USE-2, XHH, XHHW, XHHW-2, ZW-2	**Size AWG or kcmil**
	COPPER			**ALUMINUM OR COPPER-CLAD ALUMINUM**			
	CORRECTION FACTORS						
Ambient Temp. (°C)	For ambient temperatures other than 30°C (86°F), multiply the allowable ampacities shown above by the appropriate factor shown below.						**Ambient Temp. (°F)**
21–25	1.08	1.05	1.04	1.08	1.05	1.04	70–77
26–30	1.00	1.00	1.00	1.00	1.00	1.00	78–86
31–35	0.91	0.94	0.96	0.91	0.94	0.96	87–95
36–40	0.82	0.88	0.91	0.82	0.88	0.91	96–104
41–45	0.71	0.82	0.87	0.71	0.82	0.87	105–113
46–50	0.58	0.75	0.82	0.58	0.75	0.82	114–122
51–55	0.41	0.67	0.76	0.41	0.67	0.76	123–131
56–60	—	0.58	0.71	—	0.58	0.71	132–140
61–70	—	0.33	0.58	—	0.33	0.58	141–158
71–80	—	—	0.41	—	—	0.41	159–176

* See 240.4(D).

Table 310.18 Allowable Ampacities of Insulated Conductors Rated 0 Through 2000 Volts, 150°C Through 250°C (302°F Through 482°F). Not More Than Three Current-Carrying Conductors in Raceway or Cable, Based on Ambient Air Temperature of 40°C (104°F)

Size AWG or kcmil	Temperature Rating of Conductor (See Table 310.13.)				Size AWG or kcmil
	150°C (302°F)	200°C (392°F)	250 °C (482°F)	150°C (302°F)	
	Type Z	Types FEP, FEPB, PFA	Types PFAH, TFE	Type Z	
	COPPER		NICKEL OR NICKEL-COATED COPPER	ALUMINUM OR COPPER-CLAD ALUMINUM	
14	34	36	39	—	14
12	43	45	54	30	12
10	55	60	73	44	10
8	76	83	93	57	8
6	96	110	117	75	6
4	120	125	148	94	4
3	143	152	166	109	3
2	160	171	191	124	2
1	186	197	215	145	1
1/0	215	229	244	169	1/0
2/0	251	260	273	198	2/0
3/0	288	297	308	227	3/0
4/0	332	346	361	260	4/0

CORRECTION FACTORS

Ambient Temp. (°C)	For ambient temperatures other than 40°C (104°F), multiply the allowable ampacities shown above by the appropriate factor shown below.				Ambient Temp. (°F)
41–50	0.95	0.97	0.98	0.95	105–122
51–60	0.90	0.94	0.95	0.90	123–140
61–70	0.85	0.90	0.93	0.85	141–158
71–80	0.80	0.87	0.90	0.80	159–176
81–90	0.74	0.83	0.87	0.74	177–194
91–100	0.67	0.79	0.85	0.67	195–212
101–120	0.52	0.71	0.79	0.52	213–248
121–140	0.30	0.61	0.72	0.30	249–284
141–160	—	0.50	0.65	—	285–320
161–180	—	0.35	0.58	—	321–356
181–200	—	—	0.49	—	357–392
201–225	—	—	0.35	—	393–437

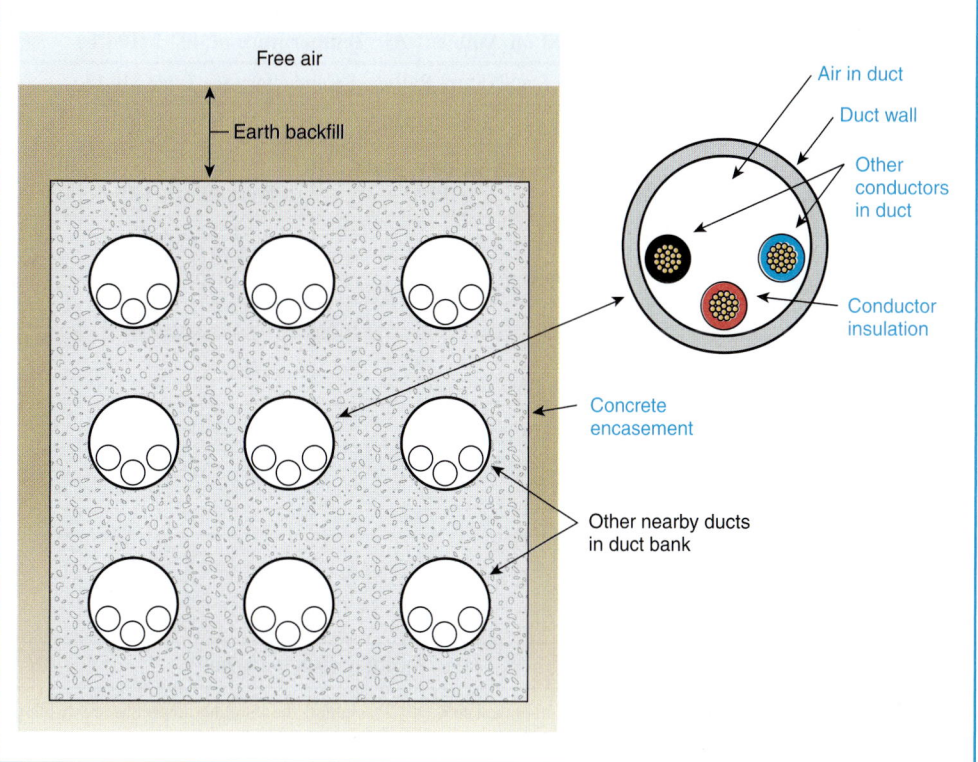

Exhibit 310.10 Thermal barriers (resistances).

factor, and the mutual heating factor caused by other nearby ducts. The deeper the duct is buried, the greater the thermal resistance.

To prevent the temperature of the conductors from exceeding the rated temperature of the insulation, heat dissipation through these thermal resistances must be equal to or greater than the heat developed. Thus, the thermal resistances of all the components of a conductor must be determined, and the allowable temperature differential above the ambient temperature and between the conductors and the surface of the earth must be known.

In addition to the Neher-McGrath paper itself, as described in the first paragraph of this commentary, the following references provide more detailed information on the use of this method of calculation:

"Power Cable Ampacity Tables," Insulated Power Cable Engineers Association, IPCEA P-46-426 (IEEE S-135-1).

"IEEE Standard Power Cable Ampacity Tables," Institute of Electrical and Electronic Engineers, IEEE Standard 835-1994.

"Neher-McGrath Calculations for Insulated Power Cables," Peter Pollak, IEEE Paper No. CH2040-4/84/0000-0172, presented at the 1984 IEEE Industrial and Commercial Power Systems Technical Conference, Atlanta, GA, May 9, 1984.

"Cable Ampacities, the *NEC* and Computerized Applications," M. T. Brown et al., IEEE Paper No. CH2207-9/85/00000-0323.

"How to Use the Neher-McGrath Method to Calculate Ampacity of Underground Conductors," John M. Caloggero, NFPA *Fire Journal*, pp. 17-18, May/June 1988.

310.60 Conductors Rated 2001 to 35,000 Volts.

(A) Definitions.

Electrical Ducts. As used in Article 310, electrical ducts shall include any of the electrical conduits recognized in Chapter 3 as suitable for use underground; other raceways round in cross section, listed for underground use, and embedded in earth or concrete.

The term *electrical duct* is used to differentiate from ducts used for air handling, for example. It is intended to include nonmetallic electrical ducts commonly used for underground wiring, as well as other raceways (e.g., rigid metal conduit, intermediate metal conduit, rigid nonmetallic conduit) listed for use underground in earth or concrete.

Thermal Resistivity. As used in this *Code*, the heat transfer capability through a substance by conduction. It is the reciprocal of thermal conductivity and is designated Rho and expressed in the units °C-cm/watt.

Table 310.19 Allowable Ampacities of Single-Insulated Conductors, Rated 0 Through 2000 Volts, 150°C Through 250°C (302°F Through 482°F), in Free Air, Based on Ambient Air Temperature of 40°C (104°F)

Size AWG or kcmil	Temperature Rating of Conductor (See Table 310.13.)				Size AWG or kcmil
	150°C (302°F)	200°C (392°F)	250 °C (482°F)	150°C (302°F)	
	Type Z	Types FEP, FEPB, PFA	Types PFAH, TFE	Type Z	
	COPPER		NICKEL OR NICKEL-COATED COPPER	ALUMINUM OR COPPER-CLAD ALUMINUM	
14	46	54	59	—	14
12	60	68	78	47	12
10	80	90	107	63	10
8	106	124	142	83	8
6	155	165	205	112	6
4	190	220	278	148	4
3	214	252	327	170	3
2	255	293	381	198	2
1	293	344	440	228	1
1/0	339	399	532	263	1/0
2/0	390	467	591	305	2/0
3/0	451	546	708	351	3/0
4/0	529	629	830	411	4/0

CORRECTION FACTORS

Ambient Temp. (°C)	For ambient temperatures other than 40°C (104°F), multiply the allowable ampacities shown above by the appropriate factor shown below.				Ambient Temp. (°F)
41–50	0.95	0.97	0.98	0.95	105–122
51–60	0.90	0.94	0.95	0.90	123–140
61–70	0.85	0.90	0.93	0.85	141–158
71–80	0.80	0.87	0.90	0.80	159–176
81–90	0.74	0.83	0.87	0.74	177–194
91–100	0.67	0.79	0.85	0.67	195–212
101–120	0.52	0.71	0.79	0.52	213–248
121–140	0.30	0.61	0.72	0.30	249–284
141–160	—	0.50	0.65	—	285–320
161–180	—	0.35	0.58	—	321–356
181–200	—	—	0.49	—	357–392
201–225	—	—	0.35	—	393–437

Table 310.20 Ampacities of Not More Than Three Single Insulated Conductors, Rated 0 Through 2000 Volts, Supported on a Messenger, Based on Ambient Air Temperature of 40°C (104°F)

	Temperature Rating of Conductor (See Table 310.13.)				
	75°C (167°F)	90°C (194°F)	75°C (167°F)	90°C (194°F)	
Size AWG or kcmil	Types RHW, THHW, THW, THWN, XHHW, ZW	Types MI, THHN, THHW, THW-2, THWN-2, RHH, RHW-2, USE-2, XHHW, XHHW-2, ZW-2	Types RHW, THW, THWN, THHW, XHHW	Types THHN, THHW, RHH, XHHW, RHW-2, XHHW-2, THW-2, THWN-2, USE-2, ZW-2	Size AWG or kcmil
	COPPER		ALUMINUM OR COPPER-CLAD ALUMINUM		
8	57	66	44	51	8
6	76	89	59	69	6
4	101	117	78	91	4
3	118	138	92	107	3
2	135	158	106	123	2
1	158	185	123	144	1
1/0	183	214	143	167	1/0
2/0	212	247	165	193	2/0
3/0	245	287	192	224	3/0
4/0	287	335	224	262	4/0
250	320	374	251	292	250
300	359	419	282	328	300
350	397	464	312	364	350
400	430	503	339	395	400
500	496	580	392	458	500
600	553	647	440	514	600
700	610	714	488	570	700
750	638	747	512	598	750
800	660	773	532	622	800
900	704	826	572	669	900
1000	748	879	612	716	1000

CORRECTION FACTORS

Ambient Temp. (°C)	For ambient temperatures other than 40°C (104°F), multiply the allowable ampacities shown above by the appropriate factor shown below.				Ambient Temp. (°F)
21–25	1.20	1.14	1.20	1.14	70–77
26–30	1.13	1.10	1.13	1.10	79–86
31–35	1.07	1.05	1.07	1.05	88–95
36–40	1.00	1.00	1.00	1.00	97–104
41–45	0.93	0.95	0.93	0.95	106–113
46–50	0.85	0.89	0.85	0.89	115–122
51–55	0.76	0.84	0.76	0.84	124–131
56–60	0.65	0.77	0.65	0.77	133–140
61–70	0.38	0.63	0.38	0.63	142–158
71–80	—	0.45	—	0.45	160–176

Table 310.21 Ampacities of Bare or Covered Conductors in Free Air, Based on 40°C (104°F) Ambient, 80°C (176°F) Total Conductor Temperature, 610 mm/sec (2 ft/sec) Wind Velocity

Copper Conductors				AAC Aluminum Conductors			
Bare		Covered		Bare		Covered	
AWG or kcmil	Amperes	AWG or kcmil	Amperes	AWG or kcmil	Amperes	AWG or kcmil	Amperes
8	98	8	103	8	76	8	80
6	124	6	130	6	96	6	101
4	155	4	163	4	121	4	127
2	209	2	219	2	163	2	171
1/0	282	1/0	297	1/0	220	1/0	231
2/0	329	2/0	344	2/0	255	2/0	268
3/0	382	3/0	401	3/0	297	3/0	312
4/0	444	4/0	466	4/0	346	4/0	364
250	494	250	519	266.8	403	266.8	423
300	556	300	584	336.4	468	336.4	492
500	773	500	812	397.5	522	397.5	548
750	1000	750	1050	477.0	588	477.0	617
1000	1193	1000	1253	556.5	650	556.5	682
—	—	—	—	636.0	709	636.0	744
—	—	—	—	795.0	819	795.0	860
—	—	—	—	954.0	920	—	—
—	—	—	—	1033.5	968	1033.5	1017
—	—	—	—	1272	1103	1272	1201
—	—	—	—	1590	1267	1590	1381
—	—	—	—	2000	1454	2000	1527

Table 310.61 Conductor Application and Insulation

Trade Name	Type Letter	Maximum Operating Temperature	Application Provision	Insulation	Outer Covering
Medium voltage solid dielectric	MV-90 MV−105*	90°C 105°C	Dry or wet locations rated 2001 volts and higher	Thermoplastic or thermosetting	Jacket, sheath, or armor

*Where design conditions require maximum conductor temperatures above 90°C.

Table 310.62 Thickness of Insulation for 601- to 2000-Volt Nonshielded Types RHH and RHW

Conductor Size (AWG or kcmil)	Column A[1]		Column B[2]	
	mm	mils	mm	mils
14–10	2.03	80	1.52	60
8	2.03	80	1.78	70
6–2	2.41	95	1.78	70
1–2/0	2.79	110	2.29	90
3/0–4/0	2.79	110	2.29	90
213–500	3.18	125	2.67	105
501–1000	3.56	140	3.05	120

[1]Column A insulations are limited to natural, SBR, and butyl rubbers.
[2]Column B insulations are materials such as cross-linked polyethylene, ethylene propylene rubber, and composites thereof.

(B) Ampacities of Conductors Rated 2001 to 35,000 Volts. Ampacities for solid dielectric-insulated conductors shall be permitted to be determined by tables or under engineering supervision, as provided in 310.60(C) and (D).

(1) Selection of Ampacity. Where more than one calculated or tabulated ampacity could apply for a given circuit length, the lowest value shall be used.

Exception: Where two different ampacities apply to adjacent portions of a circuit, the higher ampacity shall be permitted

to be used beyond the point of transition, a distance equal to 3.0 m (10 ft) or 10 percent of the circuit length figured at the higher ampacity, whichever is less.

FPN: See 110.40 for conductor temperature limitations due to termination provisions.

(C) Tables. Ampacities for conductors rated 2001 to 35,000 volts shall be as specified in the Ampacity Table 310.67 through Table 310.86. Ampacities at ambient temperatures other than those shown in the tables shall be determined by the formula in 310.60(C)(4).

FPN No. 1: For ampacities calculated in accordance with 310.60(B), reference IEEE 835-1994 (IPCEA Pub. No. P-46-426), *Standard Power Cable Ampacity Tables,* and the references therein for availability of all factors and constants.

FPN No. 2: Ampacities provided by this section do not take voltage drop into consideration. See 210.19(A), FPN No. 4, for branch circuits and 215.2(D), FPN No. 2, for feeders.

(1) Grounded Shields. Ampacities shown in Table 310.69, Table 310.70, Table 310.81, and Table 310.82 are for cable

with shields grounded at one point only. Where shields are grounded at more than one point, ampacities shall be adjusted to take into consideration the heating due to shield currents.

(2) Burial Depth of Underground Circuits. Where the burial depth of direct burial or electrical duct bank circuits is modified from the values shown in a figure or table, ampacities shall be permitted to be modified as indicated in (a) and (b).

(a) Where burial depths are increased in part(s) of an electrical duct run, no decrease in ampacity of the conductors is needed, provided the total length of parts of the duct run

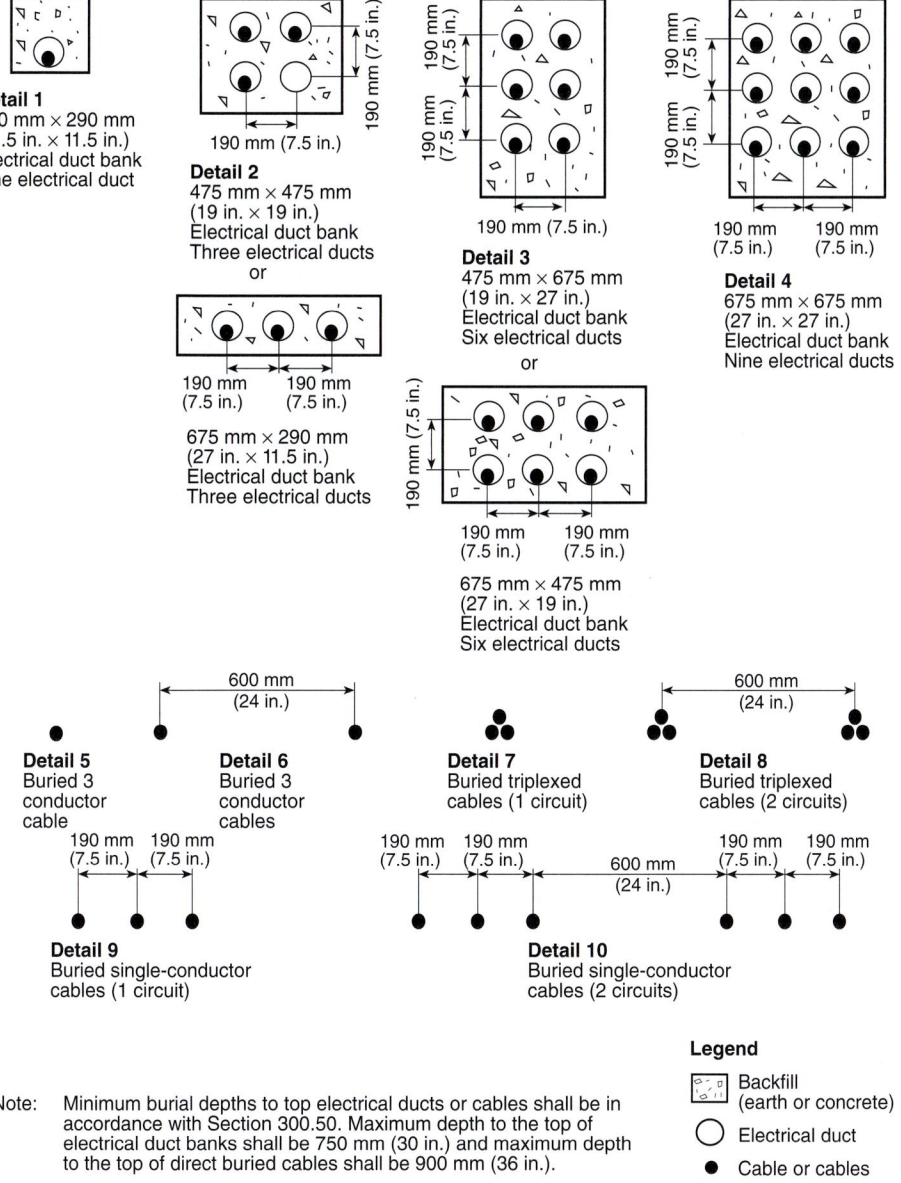

Note: Minimum burial depths to top electrical ducts or cables shall be in accordance with Section 300.50. Maximum depth to the top of electrical duct banks shall be 750 mm (30 in.) and maximum depth to the top of direct buried cables shall be 900 mm (36 in.).

Figure 310.60 Cable installation dimensions for use with Table 310.77 through Table 310.86.

Table 310.63 Thickness of Insulation and Jacket for Nonshielded Solid Dielectric Insulated Conductors Rated 2001 to 8000 Volts

Con- ductor Size (AWG or kcmil)	2001–5000 Volts												5001–8000 Volts 100 Percent Insulation Level Wet or Dry Locations					
	Dry Locations, Single Conductor						Wet or Dry Locations											
	Without Jacket Insulation		With Jacket				Single Conductor				Multi- conductor Insulation*		Single Conductor				Multi- conductor Insulation*	
			Insulation		Jacket		Insulation		Jacket				Insulation		Jacket			
	mm	mils	mm	mils	mm	mils	mm	mils	mm	mils	mm	mils	mm	mils	mm	mils	mm	mils
8	2.79	110	2.29	90	0.76	30	3.18	125	2.03	80	2.29	90	4.57	180	2.03	80	4.57	180
6	2.79	110	2.29	90	0.76	30	3.18	125	2.03	80	2.29	90	4.57	180	2.03	80	4.57	180
4–2	2.79	110	2.29	90	1.14	45	3.18	125	2.03	80	2.29	90	4.57	180	2.41	95	4.57	180
1–2/0	2.79	110	2.29	90	1.14	45	3.18	125	2.03	80	2.29	90	4.57	180	2.41	95	4.57	180
3/0–4/0	2.79	110	2.29	90	1.65	65	3.18	125	2.41	95	2.29	90	4.57	180	2.79	110	4.57	180
213–500	3.05	120	2.29	90	1.65	65	3.56	140	2.79	110	2.29	90	5.33	210	2.79	110	5.33	210
501–750	3.30	130	2.29	90	1.65	65	3.94	155	3.18	125	2.29	90	5.97	235	3.18	125	5.97	235
751–1000	3.30	130	2.29	90	1.65	65	3.94	155	3.18	125	2.29	90	6.35	250	3.56	140	6.35	250

*Under a common overall covering such as a jacket, sheath, or armor.

increased in depth is less than 25 percent of the total run length.

(b) Where burial depths are deeper than shown in a specific underground ampacity table or figure, an ampacity derating factor of 6 percent per 300-mm (1-ft) increase in depth for all values of rho shall be permitted.

No rating change is needed where the burial depth is decreased.

(3) Electrical Ducts in Figure 310.60. At locations where electrical ducts enter equipment enclosures from underground, spacing between such ducts, as shown in Figure 310.60, shall be permitted to be reduced without requiring the ampacity of conductors therein to be reduced.

(4) Ambients Not in Tables. Ampacities at ambient temperatures other than those shown in the tables shall be determined by means of the following formula:

$$I_2 = I_1 \sqrt{\frac{TC - TA_2 - \Delta TD}{TC - TA_1 - \Delta TD}}$$

where:

I_1 = ampacity from tables at ambient TA_1
I_2 = ampacity at desired ambient TA_2

TC = conductor temperature in degrees Celsius (°C)
TA_1 = surrounding ambient from tables in degrees Celsius (°C)
TA_2 = desired ambient in degrees Celsius (°C)
ΔTD = dielectric loss temperature rise

(D) Engineering Supervision. Under engineering supervision, conductor ampacities shall be permitted to be calculated by means of the following general formula:

$$I = \sqrt{\frac{TC - (TA + \Delta TD)}{RDC(1 + YC)RCA}}$$

where:

TC = conductor temperature in °C
TA = ambient temperature in °C
ΔTD = dielectric loss temperature rise
RDC = dc resistance of conductor at temperature TC
YC = component ac resistance resulting from skin effect and proximity effect
RCA = effective thermal resistance between conductor and surrounding ambient

FPN: See Annex B for examples of formula applications.

Table 310.64 Thickness of Insulation for Shielded Solid Dielectric Insulated Conductors Rated 2001 to 35,000 Volts

Conductor Size (AWG or kcmil)	2001–5000 Volts		5001–8000 Volts				8001–15,000 Volts				15,001–25,000 Volts				25,001–28,000 Volts				28,001–35,000 Volts			
			100 Percent Insulation Level 1		133 Percent Insulation Level 2		100 Percent Insulation Level 1		133 Percent Insulation Level 2		100 Percent Insulation Level 1		133 Percent Insulation Level 2		100 Percent Insulation Level 1		133 Percent Insulation Level 2		100 Percent Insulation Level 1		133 Percent Insulation Level 2	
	mm	mils	mm	mils	mm	mils	mm	mils	mm	mils	mm	mils	mm	mils	mm	mils	mm	mils	mm	mils	mm	mils
8	2.29	90	—	—	—	—	—	—	—	—	—	—	—	—	—	—	—	—	—	—	—	—
6–4	2.29	90	2.92	115	3.56	140	—	—	—	—	—	—	—	—	—	—	—	—	—	—	—	—
2	2.29	90	2.92	115	3.56	140	4.45	175	5.46	215	—	—	—	—	—	—	—	—	—	—	—	—
1	2.29	90	2.92	115	3.56	140	4.45	175	5.46	215	6.60	260	8.76	345	7.11	280	8.76	345	—	—	—	—
1/0–2000	2.29	90	2.92	115	3.56	140	4.45	175	5.46	215	6.60	260	8.76	345	7.11	280	8.76	345	8.76	345	10.67	420

[1]**100 Percent Insulation Level.** Cables in this category shall be permitted to be applied where the system is provided with relay protection such that ground faults will be cleared as rapidly as possible but, in any case, within 1 minute. While these cables are applicable to the great majority of cable installations that are on grounded systems, they shall be permitted to be used also on other systems for which the application of cables is acceptable, provided the above clearing requirements are met in completely de-energizing the faulted section.

[2]**133 Percent Insulation Level.** This insulation level corresponds to that formerly designated for ungrounded systems. Cables in this category shall be permitted to be applied in situations where the clearing time requirements of the 100 percent level category cannot be met and yet there is adequate assurance that the faulted section will be de-energized in a time not exceeding 1 hour. Also, they shall be permitted to be used where additional insulation strength over the 100 percent level category is desirable.

Table 310.67 Ampacities of Insulated Single Copper Conductor Cables Triplexed in Air Based on Conductor Temperatures of 90°C (194°F) and 105°C (221°F) and Ambient Air Temperature of 40°C (104°F)

Conductor Size (AWG or kcmil)	Temperature Rating of Conductor (See Table 310.61.)			
	2001–5000 Volts Ampacity		5001–35,000 Volts Ampacity	
	90°C (194°F) Type MV-90	105°C (221°F) Type MV-105	90°C (194°F) Type MV-90	105°C (221°F) Type MV-105
8	65	74	—	—
6	90	99	100	110
4	120	130	130	140
2	160	175	170	195
1	185	205	195	225
1/0	215	240	225	255
2/0	250	275	260	295
3/0	290	320	300	340
4/0	335	375	345	390
250	375	415	380	430
350	465	515	470	525
500	580	645	580	650
750	750	835	730	820
1000	880	980	850	950

Table 310.68 Ampacities of Insulated Single Aluminum Conductor Cables Triplexed in Air Based on Conductor Temperatures of 90°C (194°F) and 105°C (221°F) and Ambient Air Temperature of 40°C (104°F)

Conductor Size (AWG or kcmil)	Temperature Rating of Conductor (See Table 310.61.)			
	2001–5000 Volts Ampacity		5001–35,000 Volts Ampacity	
	90°C (194°F) Type MV-90	105°C (221°F) Type MV-105	90°C (194°F) Type MV-90	105°C (221°F) Type MV-105
8	50	57	—	—
6	70	77	75	84
4	90	100	100	110
2	125	135	130	150
1	145	160	150	175
1/0	170	185	175	200
2/0	195	215	200	230
3/0	225	250	230	265
4/0	265	290	270	305
250	295	325	300	335
350	365	405	370	415
500	460	510	460	515
750	600	665	590	660
1000	715	800	700	780

Table 310.69 Ampacities of Insulated Single Copper Conductor Isolated in Air Based on Conductor Temperatures of 90°C (194°F) and 105°C (221°F) and Ambient Air Temperature of 40°C (104°F)

Conductor Size (AWG or kcmil)	2001–5000 Volts Ampacity		5001–15,000 Volts Ampacity		15,001–35,000 Volts Ampacity	
	90°C (194°F) Type MV-90	105°C (221°F) Type MV-105	90°C (194°F) Type MV-90	105°C (221°F) Type MV-105	90°C (194°F) Type MV-90	105°C (221°F) Type MV-105
8	83	93	—	—	—	—
6	110	120	110	125	—	—
4	145	160	150	165	—	—
2	190	215	195	215	—	—
1	225	250	225	250	225	250
1/0	260	290	260	290	260	290
2/0	300	330	300	335	300	330
3/0	345	385	345	385	345	380
4/0	400	445	400	445	395	445
250	445	495	445	495	440	490
350	550	615	550	610	545	605
500	695	775	685	765	680	755
750	900	1000	885	990	870	970
1000	1075	1200	1060	1185	1040	1160
1250	1230	1370	1210	1350	1185	1320
1500	1365	1525	1345	1500	1315	1465
1750	1495	1665	1470	1640	1430	1595
2000	1605	1790	1575	1755	1535	1710

Table 310.70 Ampacities of Insulated Single Aluminum Conductor Isolated in Air Based on Conductor Temperatures of 90°C (194°F) and 105°C (221°F) and Ambient Air Temperature of 40°C (104°F)

Conductor Size (AWG or kcmil)	2001–5000 Volts Ampacity		5001–15,000 Volts Ampacity		15,001–35,000 Volts Ampacity	
	90°C (194°F) Type MV-90	105°C (221°F) Type MV-105	90°C (194°F) Type MV-90	105°C (221°F) Type MV-105	90°C (194°F) Type MV-90	105°C (221°F) Type MV-105
8	64	71	—	—	—	—
6	85	95	87	97	—	—
4	115	125	115	130	—	—
2	150	165	150	170	—	—
1	175	195	175	195	175	195
1/0	200	225	200	225	200	225
2/0	230	260	235	260	230	260
3/0	270	300	270	300	270	300
4/0	310	350	310	350	310	345
250	345	385	345	385	345	380
350	430	480	430	480	430	475
500	545	605	535	600	530	590
750	710	790	700	780	685	765
1000	855	950	840	940	825	920
1250	980	1095	970	1080	950	1055
1500	1105	1230	1085	1215	1060	1180
1750	1215	1355	1195	1335	1165	1300
2000	1320	1475	1295	1445	1265	1410

Table 310.71 Ampacities of an Insulated Three-Conductor Copper Cable Isolated in Air Based on Conductor Temperatures of 90°C (194°F) and 105°C (221°F) and Ambient Air Temperature of 40°C (104°F)

Conductor Size (AWG or kcmil)	2001–5000 Volts Ampacity		5001–35,000 Volts Ampacity	
	90°C (194°F) Type MV-90	105°C (221°F) Type MV-105	90°C (194°F) Type MV-90	105°C (221°F) Type MV-105
8	59	66	—	—
6	79	88	93	105
4	105	115	120	135
2	140	154	165	185
1	160	180	185	210
1/0	185	205	215	240
2/0	215	240	245	275
3/0	250	280	285	315
4/0	285	320	325	360
250	320	355	360	400
350	395	440	435	490
500	485	545	535	600
750	615	685	670	745
1000	705	790	770	860

Table 310.72 Ampacities of an Insulated Three-Conductor Aluminum Cable Isolated in Air Based on Conductor Temperatures of 90°C (194°F) and 105°C (221°F) and Ambient Air Temperature of 40°C (104°F)

Conductor Size (AWG or kcmil)	2001–5000 Volts Ampacity		5001–35,000 Volts Ampacity	
	90°C (194°F) Type MV-90	105°C (221°F) Type MV-105	90°C (194°F) Type MV-90	105°C (221°F) Type MV-105
8	46	51	—	—
6	61	68	72	80
4	81	90	95	105
2	110	120	125	145
1	125	140	145	165
1/0	145	160	170	185
2/0	170	185	190	215
3/0	195	215	220	245
4/0	225	250	255	285
250	250	280	280	315
350	310	345	345	385
500	385	430	425	475
750	495	550	540	600
1000	585	650	635	705

Table 310.73 Ampacities of an Insulated Triplexed or Three Single-Conductor Copper Cables in Isolated Conduit in Air Based on Conductor Temperatures of 90°C (194°F) and 105°C (221°F) and Ambient Air Temperature of 40°C (104°F)

Conductor Size (AWG or kcmil)	Temperature Rating of Conductor (See Table 310.61.)			
	2001–5000 Volts Ampacity		5001–35,000 Volts Ampacity	
	90°C (194°F) Type MV-90	105°C (221°F) Type MV-105	90°C (194°F) Type MV-90	105°C (221°F) Type MV-105
8	55	61	—	—
6	75	84	83	93
4	97	110	110	120
2	130	145	150	165
1	155	175	170	190
1/0	180	200	195	215
2/0	205	225	225	255
3/0	240	270	260	290
4/0	280	305	295	330
250	315	355	330	365
350	385	430	395	440
500	475	530	480	535
750	600	665	585	655
1000	690	770	675	755

Table 310.75 Ampacities of an Insulated Three-Conductor Copper Cable in Isolated Conduit in Air Based on Conductor Temperatures of 90°C (194°F) and 105°C (221°F) and Ambient Air Temperature of 40°C (104°F)

Conductor Size (AWG or kcmil)	Temperature Rating of Conductor (See Table 310.61.)			
	2001–5000 Volts Ampacity		5001–35,000 Volts Ampacity	
	90°C (194°F) Type MV-90	105°C (221°F) Type MV-105	90°C (194°F) Type MV-90	105°C (221°F) Type MV-105
8	52	58	—	—
6	69	77	83	92
4	91	100	105	120
2	125	135	145	165
1	140	155	165	185
1/0	165	185	195	215
2/0	190	210	220	245
3/0	220	245	250	280
4/0	255	285	290	320
250	280	315	315	350
350	350	390	385	430
500	425	475	470	525
750	525	585	570	635
1000	590	660	650	725

Table 310.74 Ampacities of an Insulated Triplexed or Three Single-Conductor Aluminum Cables in Isolated Conduit in Air Based on Conductor Temperatures of 90°C (194°F) and 105°C (221°F) and Ambient Air Temperature of 40°C (104°F)

Conductor Size (AWG or kcmil)	Temperature Rating of Conductor (See Table 310.61.)			
	2001–5000 Volts Ampacity		5001–35,000 Volts Ampacity	
	90°C (194°F) Type MV-90	105°C (221°F) Type MV-105	90°C (194°F) Type MV-90	105°C (221°F) Type MV-105
8	43	48	—	—
6	58	65	65	72
4	76	85	84	94
2	100	115	115	130
1	120	135	130	150
1/0	140	155	150	170
2/0	160	175	175	200
3/0	190	210	200	225
4/0	215	240	230	260
250	250	280	255	290
350	305	340	310	350
500	380	425	385	430
750	490	545	485	540
1000	580	645	565	640

Table 310.76 Ampacities of an Insulated Three-Conductor Aluminum Cable in Isolated Conduit in Air Based on Conductor Temperatures of 90°C (194°F) and 105°C (221°F) and Ambient Air Temperature of 40°C (104°F)

Conductor Size (AWG or kcmil)	Temperature Rating of Conductor (See Table 310.61.)			
	2001–5000 Volts Ampacity		5001–35,000 Volts Ampacity	
	90°C (194°F) Type MV-90	105°C (221°F) Type MV-105	90°C (194°F) Type MV-90	105°C (221°F) Type MV-105
8	41	46	—	—
6	53	59	64	71
4	71	79	84	94
2	96	105	115	125
1	110	125	130	145
1/0	130	145	150	170
2/0	150	165	170	190
3/0	170	190	195	220
4/0	200	225	225	255
250	220	245	250	280
350	275	305	305	340
500	340	380	380	425
750	430	480	470	520
1000	505	560	550	615

Table 310.77 Ampacities of Three Single-Insulated Copper Conductors in Underground Electrical Ducts (Three Conductors per Electrical Duct) Based on Ambient Earth Temperature of 20°C (68°F), Electrical Duct Arrangement per Figure 310.60, 100 Percent Load Factor, Thermal Resistance (RHO) of 90, Conductor Temperatures of 90°C (194°F) and 105°C (221°F)

Conductor Size (AWG or kcmil)	2001–5000 Volts Ampacity		5001–35,000 Volts Ampacity	
	90°C (194°F) Type MV-90	105°C (221°F) Type MV-105	90°C (194°F) Type MV-90	105°C (221°F) Type MV-105
One Circuit (See Figure 310.60, Detail 1.)				
8	64	69	—	—
6	85	92	90	97
4	110	120	115	125
2	145	155	155	165
1	170	180	175	185
1/0	195	210	200	215
2/0	220	235	230	245
3/0	250	270	260	275
4/0	290	310	295	315
250	320	345	325	345
350	385	415	390	415
500	470	505	465	500
750	585	630	565	610
1000	670	720	640	690
Three Circuits (See Figure 310.60, Detail 2.)				
8	56	60	—	—
6	73	79	77	83
4	95	100	99	105
2	125	130	130	135
1	140	150	145	155
1/0	160	175	165	175
2/0	185	195	185	200
3/0	210	225	210	225
4/0	235	255	240	255
250	260	280	260	280
350	315	335	310	330
500	375	405	370	395
750	460	495	440	475
1000	525	565	495	535
Six Circuits (See Figure 310.60, Detail 3.)				
8	48	52	—	—
6	62	67	64	68
4	80	86	82	88
2	105	110	105	115
1	115	125	120	125
1/0	135	145	135	145
2/0	150	160	150	165
3/0	170	185	170	185
4/0	195	210	190	205
250	210	225	210	225
350	250	270	245	265
500	300	325	290	310
750	365	395	350	375
1000	410	445	390	415

Table 310.78 Ampacities of Three Single-Insulated Aluminum Conductors in Underground Electrical Ducts (Three Conductors per Electrical Duct) Based on Ambient Earth Temperature of 20°C (68°F), Electrical Duct Arrangement per Figure 310.60, 100 Percent Load Factor, Thermal Resistance (RHO) of 90, Conductor Temperatures of 90°C (194°F) and 105°C (221°F)

Conductor Size (AWG or kcmil)	2001–5000 Volts Ampacity		5001–35,000 Volts Ampacity	
	90°C (194°F) Type MV-90	105°C (221°F) Type MV-105	90°C (194°F) Type MV-90	105°C (221°F) Type MV-105
One Circuit (See Figure 310.60, Detail 1.)				
8	50	54	—	—
6	66	71	70	75
4	86	93	91	98
2	115	125	120	130
1	130	140	135	145
1/0	150	160	155	165
2/0	170	185	175	190
3/0	195	210	200	215
4/0	225	245	230	245
250	250	270	250	270
350	305	325	305	330
500	370	400	370	400
750	470	505	455	490
1000	545	590	525	565
Three Circuits (See Figure 310.60, Detail 2.)				
8	44	47	—	—
6	57	61	60	65
4	74	80	77	83
2	96	105	100	105
1	110	120	110	120
1/0	125	135	125	140
2/0	145	155	145	155
3/0	160	175	165	175
4/0	185	200	185	200
250	205	220	200	220
350	245	265	245	260
500	295	320	290	315
750	370	395	355	385
1000	425	460	405	440
Six Circuits (See Figure 310.60, Detail 3.)				
8	38	41	—	—
6	48	52	50	54
4	62	67	64	69
2	80	86	80	88
1	91	98	90	99
1/0	105	110	105	110
2/0	115	125	115	125
3/0	135	145	130	145
4/0	150	165	150	160
250	165	180	165	175
350	195	210	195	210
500	240	255	230	250
750	290	315	280	305
1000	335	360	320	345

Table 310.79 Ampacities of Three Insulated Copper Conductors Cabled Within an Overall Covering (Three-Conductor Cable) in Underground Electrical Ducts (One Cable per Electrical Duct) Based on Ambient Earth Temperature of 20°C (68°F), Electrical Duct Arrangement per Figure 310.60, 100 Percent Load Factor, Thermal Resistance (RHO) of 90, Conductor Temperatures of 90°C (194°F) and 105°C (221°C)

| | Temperature Rating of Conductor (See Table 310.61.) | | | |
| | 2001–5000 Volts Ampacity | | 5001–35,000 Volts Ampacity | |
Conductor Size (AWG or kcmil)	90°C (194°F) Type MV-90	105°C (221°F) Type MV-105	90°C (194°F) Type MV-90	105°C (221°F) Type MV-105
One Circuit (See Figure 310.60, Detail 1.)				
8	59	64	—	—
6	78	84	88	95
4	100	110	115	125
2	135	145	150	160
1	155	165	170	185
1/0	175	190	195	210
2/0	200	220	220	235
3/0	230	250	250	270
4/0	265	285	285	305
250	290	315	310	335
350	355	380	375	400
500	430	460	450	485
750	530	570	545	585
1000	600	645	615	660
Three Circuits (See Figure 310.60, Detail 2.)				
8	53	57	—	—
6	69	74	75	81
4	89	96	97	105
2	115	125	125	135
1	135	145	140	155
1/0	150	165	160	175
2/0	170	185	185	195
3/0	195	210	205	220
4/0	225	240	230	250
250	245	265	255	270
350	295	315	305	325
500	355	380	360	385
750	430	465	430	465
1000	485	520	485	515
Six Circuits (See Figure 310.60, Detail 3.)				
8	46	50	—	—
6	60	65	63	68
4	77	83	81	87
2	98	105	105	110
1	110	120	115	125
1/0	125	135	130	145
2/0	145	155	150	160
3/0	165	175	170	180
4/0	185	200	190	200
250	200	220	205	220
350	240	270	245	275
500	290	310	290	305
750	350	375	340	365
1000	390	420	380	405

Table 310.80 Ampacities of Three Insulated Aluminum Conductors Cabled Within an Overall Covering (Three-Conductor Cable) in Underground Electrical Ducts (One Cable per Electrical Duct) Based on Ambient Earth Temperature of 20°C (68°F), Electrical Duct Arrangement per Figure 310.60, 100 Percent Load Factor, Thermal Resistance (RHO) of 90, Conductor Temperatures of 90°C (194°F) and 105°C (221°C)

| | Temperature Rating of Conductor (See Table 310.61.) | | | |
| | 2001–5000 Volts Ampacity | | 5001–35,000 Volts Ampacity | |
Conductor Size (AWG or kcmil)	90°C (194°F) Type MV-90	105°C (221°F) Type MV-105	90°C (194°F) Type MV-90	105°C (221°F) Type MV-105
One Circuit (See Figure 310.60, Detail 1.)				
8	46	50	—	—
6	61	66	69	74
4	80	86	89	96
2	105	110	115	125
1	120	130	135	145
1/0	140	150	150	165
2/0	160	170	170	185
3/0	180	195	195	210
4/0	205	220	220	240
250	230	245	245	265
350	280	310	295	315
500	340	365	355	385
750	425	460	440	475
1000	495	535	510	545
Three Circuits (See Figure 310.60, Detail 2.)				
8	41	44	—	—
6	54	58	59	64
4	70	75	75	81
2	90	97	100	105
1	105	110	110	120
1/0	120	125	125	135
2/0	135	145	140	155
3/0	155	165	160	175
4/0	175	185	180	195
250	190	205	200	215
350	230	250	240	255
500	280	300	285	305
750	345	375	350	375
1000	400	430	400	430
Six Circuits (See Figure 310.60, Detail 3.)				
8	36	39	—	—
6	46	50	49	53
4	60	65	63	68
2	77	83	80	86
1	87	94	90	98
1/0	99	105	105	110
2/0	110	120	115	125
3/0	130	140	130	140
4/0	145	155	150	160
250	160	170	160	170
350	190	205	190	205
500	230	245	230	245
750	280	305	275	295
1000	320	345	315	335

Table 310.81 Ampacities of Single Insulated Copper Conductors Directly Buried in Earth Based on Ambient Earth Temperature of 20°C (68°F), Arrangement per Figure 310.60, 100 Percent Load Factor, Thermal Resistance (RHO) of 90, Conductor Temperatures of 90°C (194°F) and 105°C (221°C)

| Conductor Size (AWG or kcmil) | Temperature Rating of Conductor (See Table 310.61.) | | | |
| | 2001–5000 Volts Ampacity | | 5001–35,000 Volts Ampacity | |
	90°C (194°F) Type MV-90	105°C (221°F) Type MV-105	90°C (194°F) Type MV-90	105°C (221°F) Type MV-105
One Circuit, Three Conductors (See Figure 310.60, Detail 9.)				
8	110	115	—	—
6	140	150	130	140
4	180	195	170	180
2	230	250	210	225
1	260	280	240	260
1/0	295	320	275	295
2/0	335	365	310	335
3/0	385	415	355	380
4/0	435	465	405	435
250	470	510	440	475
350	570	615	535	575
500	690	745	650	700
750	845	910	805	865
1000	980	1055	930	1005
Two Circuits, Six Conductors (See Figure 310.60, Detail 10.)				
8	100	110	—	—
6	130	140	120	130
4	165	180	160	170
2	215	230	195	210
1	240	260	225	240
1/0	275	295	255	275
2/0	310	335	290	315
3/0	355	380	330	355
4/0	400	430	375	405
250	435	470	410	440
350	520	560	495	530
500	630	680	600	645
750	775	835	740	795
1000	890	960	855	920

Table 310.82 Ampacities of Single Insulated Aluminum Conductors Directly Buried in Earth Based on Ambient Earth Temperature of 20°C (68°F), Arrangement per Figure 310.60, 100 Percent Load Factor, Thermal Resistance (RHO) of 90, Conductor Temperatures of 90°C (194°F) and 105°C (221°F)

| Conductor Size (AWG or kcmil) | Temperature Rating of Conductor (See Table 310.61.) | | | |
| | 2001–5000 Volts Ampacity | | 5001–35,000 Volts Ampacity | |
	90°C (194°F) Type MV-90	105°C (221°F) Type MV-105	90°C (194°F) Type MV-90	105°C (221°F) Type MV-105
One Circuit, Three Conductors (See Figure 310.60, Detail 9.)				
8	85	90	—	—
6	110	115	100	110
4	140	150	130	140
2	180	195	165	175
1	205	220	185	200
1/0	230	250	215	230
2/0	265	285	245	260
3/0	300	320	275	295
4/0	340	365	315	340
250	370	395	345	370
350	445	480	415	450
500	540	580	510	545
750	665	720	635	680
1000	780	840	740	795
Two Circuits, Six Conductors (See Figure 310.60, Detail 10.)				
8	80	85	—	—
6	100	110	95	100
4	130	140	125	130
2	165	180	155	165
1	190	200	175	190
1/0	215	230	200	215
2/0	245	260	225	245
3/0	275	295	255	275
4/0	310	335	290	315
250	340	365	320	345
350	410	440	385	415
500	495	530	470	505
750	610	655	580	625
1000	710	765	680	730

Table 310.83 Ampacities of Three Insulated Copper Conductors Cabled Within an Overall Covering (Three-Conductor Cable), Directly Buried in Earth Based on Ambient Earth Temperature of 20°C (68°F), Arrangement per Figure 310.60, 100 Percent Load Factor, Thermal Resistance (RHO) of 90, Conductor Temperatures of 90°C (194°F) and 105°C (221°F)

| Conductor Size (AWG or kcmil) | Temperature Rating of Conductor (See Table 310.61.) | | | |
| | 2001–5000 Volts Ampacity | | 5001–35,000 Volts Ampacity | |
	90°C (194°F) Type MV-90	105°C (221°F) Type MV-105	90°C (194°F) Type MV-90	105°C (221°F) Type MV-105
One Circuit (See Figure 310.60, Detail 5.)				
8	85	89	—	—
6	105	115	115	120
4	135	150	145	155
2	180	190	185	200
1	200	215	210	225
1/0	230	245	240	255
2/0	260	280	270	290
3/0	295	320	305	330
4/0	335	360	350	375
250	365	395	380	410
350	440	475	460	495
500	530	570	550	590
750	650	700	665	720
1000	730	785	750	810
Two Circuits (See Figure 310.60, Detail 6.)				
8	80	84	—	—
6	100	105	105	115
4	130	140	135	145
2	165	180	170	185
1	185	200	195	210
1/0	215	230	220	235
2/0	240	260	250	270
3/0	275	295	280	305
4/0	310	335	320	345
250	340	365	350	375
350	410	440	420	450
500	490	525	500	535
750	595	640	605	650
1000	665	715	675	730

Table 310.84 Ampacities of Three Insulated Aluminum Conductors Cabled Within an Overall Covering (Three-Conductor Cable), Directly Buried in Earth Based on Ambient Earth Temperature of 20°C (68°F), Arrangement per Figure 310.60, 100 Percent Load Factor, Thermal Resistance (RHO) of 90, Conductor Temperatures of 90°C (194°F) and 105°C (221°F)

| Conductor Size (AWG or kcmil) | Temperature Rating of Conductor (See Table 310.61.) | | | |
| | 2001–5000 Volts Ampacity | | 5001–35,000 Volts Ampacity | |
	90°C (194°F) Type MV-90	105°C (221°F) Type MV-105	90°C (194°F) Type MV-90	105°C (221°F) Type MV-105
One Circuit (See Figure 310.60, Detail 5.)				
8	65	70	—	—
6	80	88	90	95
4	105	115	115	125
2	140	150	145	155
1	155	170	165	175
1/0	180	190	185	200
2/0	205	220	210	225
3/0	230	250	240	260
4/0	260	280	270	295
250	285	310	300	320
350	345	375	360	390
500	420	450	435	470
750	520	560	540	580
1000	600	650	620	665
Two Circuits (See Figure 310.60, Detail 6.)				
8	60	66	—	—
6	75	83	80	95
4	100	110	105	115
2	130	140	135	145
1	145	155	150	165
1/0	165	180	170	185
2/0	190	205	195	210
3/0	215	230	220	240
4/0	245	260	250	270
250	265	285	275	295
350	320	345	330	355
500	385	415	395	425
750	480	515	485	525
1000	550	590	560	600

Table 310.85 Ampacities of Three Triplexed Single Insulated Copper Conductors Directly Buried in Earth Based on Ambient Earth Temperature of 20°C (68°F), Arrangement per Figure 310.60, 100 Percent Load Factor, Thermal Resistance (RHO) of 90, Conductor Temperatures 90°C (194°F) and 105°C (221°F)

| | Temperature Rating of Conductor (See Table 310.61.) | | | |
| | 2001–5000 Volts Ampacity | | 5001–35,000 Volts Ampacity | |
Conductor Size (AWG or kcmil)	90°C (194°F) Type MV-90	105°C (221°F) Type MV-105	90°C (194°F) Type MV-90	105°C (221°F) Type MV-105
One Circuit, Three Conductors (See Figure 310.60, Detail 7.)				
8	90	95	—	—
6	120	130	115	120
4	150	165	150	160
2	195	205	190	205
1	225	240	215	230
1/0	255	270	245	260
2/0	290	310	275	295
3/0	330	360	315	340
4/0	375	405	360	385
250	410	445	390	410
350	490	580	470	505
500	590	635	565	605
750	725	780	685	740
1000	825	885	770	830
Two Circuits, Six Conductors (See Figure 310.60, Detail 8.)				
8	85	90	—	—
6	110	115	105	115
4	140	150	140	150
2	180	195	175	190
1	205	220	200	215
1/0	235	250	225	240
2/0	265	285	255	275
3/0	300	320	290	315
4/0	340	365	325	350
250	370	395	355	380
350	445	480	425	455
500	535	575	510	545
750	650	700	615	660
1000	740	795	690	745

Table 310.86 Ampacities of Three Triplexed Single Insulated Aluminum Conductors Directly Buried in Earth Based on Ambient Earth Temperature of 20°C (68°F), Arrangement per Figure 310.60, 100 Percent Load Factor, Thermal Resistance (RHO) of 90, Conductor Temperatures 90°C (194°F) and 105°C (221°F)

| | Temperature Rating of Conductor (See Table 310.61.) | | | |
| | 2001–5000 Volts Ampacity | | 5001–35,000 Volts Ampacity | |
Conductor Size (AWG or kcmil)	90°C (194°F) Type MV-90	105°C (221°F) Type MV-105	90°C (194°F) Type MV-90	105°C (221°F) Type MV-105
One Circuit, Three Conductors (See Figure 310.60, Detail 7.)				
8	70	75	—	—
6	90	100	90	95
4	120	130	115	125
2	155	165	145	155
1	175	190	165	175
1/0	200	210	190	205
2/0	225	240	215	230
3/0	255	275	245	265
4/0	290	310	280	305
250	320	350	305	325
350	385	420	370	400
500	465	500	445	480
750	580	625	550	590
1000	670	725	635	680
Two Circuits, Six Conductors (See Figure 310.60, Detail 8.)				
8	65	70	—	—
6	85	95	85	90
4	110	120	105	115
2	140	150	135	145
1	160	170	155	170
1/0	180	195	175	190
2/0	205	220	200	215
3/0	235	250	225	245
4/0	265	285	255	275
250	290	310	280	300
350	350	375	335	360
500	420	455	405	435
750	520	560	485	525
1000	600	645	565	605

ARTICLE 312
Cabinets, Cutout Boxes, and Meter Socket Enclosures

Contents

312.1 Scope.

This article covers the installation and construction specifications of cabinets, cutout boxes, and meter socket enclosures.

See the definitions of *cabinet* and *cutout box* in Article 100. Cabinets and cutout boxes are designed with a swinging door(s) to enclose potential transformers, current transformers, switches, overcurrent devices, meters, or control equipment. Some cabinets for circuit breaker panelboards or load centers may not have doors, as permitted by 240.30(A)(2). Cabinets and cutout boxes are required to be of sufficient size to accommodate all devices and conductors without overcrowding or jamming. This condition can be prevented by the use of auxiliary gutters (Article 366).

The serving electric utility often has equipment specifications or service requirements beyond the *Code* for meter sockets, metering cabinets, and metering compartments within switchgear, switchboards, and panelboards. Consulting with the local electric utility on these requirements will help identify suitable equipment for an installation. One organization, the Electric Utility Service Equipment Requirements Committee (EUSERC), promotes uniform electric utility metering service requirements for these enclosures that meet the requirements of the *Code* and the serving utility. Many electric equipment manufacturers identify their equipment as meeting the EUSERC metering space requirements.

I. Installation

312.2 Damp, Wet, or Hazardous (Classified) Locations.

(A) Damp and Wet Locations. In damp or wet locations, surface-type enclosures within the scope of this article shall be placed or equipped so as to prevent moisture or water from entering and accumulating within the cabinet or cutout box, and shall be mounted so there is at least 6 mm (¼ in.) airspace between the enclosure and the wall or other supporting surface. Enclosures installed in wet locations shall be weatherproof.

Exception: Nonmetallic enclosures shall be permitted to be installed without the airspace on a concrete, masonry, tile, or similar surface.

   FPN: For protection against corrosion, see 300.6.

(B) Hazardous (Classified) Locations. Installations in hazardous (classified) locations shall conform to Articles 500 through 517.

312.3 Position in Wall.

In walls of concrete, tile, or other noncombustible material, cabinets shall be installed so that the front edge of the cabinet is not set back of the finished surface more than 6 mm (¼ in.). In walls constructed of wood or other combustible material, cabinets shall be flush with the finished surface or project therefrom.

312.5 Cabinets, Cutout Boxes, and Meter Socket Enclosures.

Conductors entering enclosures within the scope of this article shall be protected from abrasion and shall comply with 312.5(A) through (C).

(A) Openings to Be Closed. Openings through which conductors enter shall be adequately closed.

(B) Metal Cabinets, Cutout Boxes, and Meter Socket Enclosures. Where metal enclosures within the scope of this article are installed with open wiring or concealed knob-and-tube wiring, conductors shall enter through insulating bushings or, in dry locations, through flexible tubing extending from the last insulating support and firmly secured to the enclosure.

(C) Cables. Where cable is used, each cable shall be secured to the cabinet, cutout box, or meter socket enclosure.

The main rule of 312.5(C) prohibits the installation of several cables bunched together and run through a knockout or chase nipple. Individual cable clamps or connectors are required to be used with only one cable per clamp or connector, unless the clamp or connector is identified for more than a single cable.

Exception: Cables with entirely nonmetallic sheaths shall be permitted to enter the top of a surface-mounted enclosure through one or more nonflexible raceways not less than 450 mm (18 in.) or more than 3.0 m (10 ft) in length, provided all the following conditions are met:

(a) Each cable is fastened within 300 mm (12 in.), measured along the sheath, of the outer end of the raceway.

(b) The raceway extends directly above the enclosure and does not penetrate a structural ceiling.

(c) A fitting is provided on each end of the raceway to protect the cable(s) from abrasion and the fittings remain accessible after installation.

(d) The raceway is sealed or plugged at the outer end using approved means so as to prevent access to the enclosure through the raceway.

(e) The cable sheath is continuous through the raceway and extends into the enclosure beyond the fitting not less than 6 mm (¼ in.).

(f) The raceway is fastened at its outer end and at other points in accordance with the applicable article.

(g) Where installed as conduit or tubing, the allowable cable fill does not exceed that permitted for complete conduit or tubing systems by Table 1 of Chapter 9 of this Code and all applicable notes thereto.

FPN: See Table 1 in Chapter 9, including Note 9, for allowable cable fill in circular raceways. See 310.15(B)(2)(a) for required ampacity reductions for multiple cables installed in a common raceway.

The exception, which was added for the 1999 *NEC,* spells out the requirements that allow multiple nonmetallic cables such as Type NM, NMC, NMS, UF, SE, and USE to enter the top of a surface-mounted enclosure through a nonflexible raceway sleeve or nipple. These sleeves or nipples are permitted to be 18 in. to 10 ft in length. However, if the nipple length exceeds 24 in., ampacity adjustment factors as specified in 310.15(B)(2) apply.

312.6 Deflection of Conductors.

Conductors at terminals or conductors entering or leaving cabinets or cutout boxes and the like shall comply with 312.6(A) through (C).

Exception: Wire-bending space in enclosures for motor controllers with provisions for one or two wires per terminal shall comply with 430.10(B).

(A) Width of Wiring Gutters. Conductors shall not be deflected within a cabinet or cutout box unless a gutter having a width in accordance with Table 312.6(A) is provided. Conductors in parallel in accordance with 310.4 shall be judged on the basis of the number of conductors in parallel.

(B) Wire-Bending Space at Terminals. Wire-bending space at each terminal shall be provided in accordance with 312.6(B)(1) or (2).

(1) Conductors Not Entering or Leaving Opposite Wall. Table 312.6(A) shall apply where the conductor does not enter or leave the enclosure through the wall opposite its terminal.

(2) Conductors Entering or Leaving Opposite Wall. Table 312.6(B) shall apply where the conductor does enter or leave the enclosure through the wall opposite its terminal.

Exception No. 1: Where the distance between the wall and its terminal is in accordance with Table 312.6(A), a conductor shall be permitted to enter or leave an enclosure through the wall opposite its terminal, provided the conductor enters or leaves the enclosure where the gutter joins an adjacent gutter that has a width that conforms to Table 312.6(B) for the conductor.

Exception No. 2: A conductor not larger than 350 kcmil shall be permitted to enter or leave an enclosure containing only a meter socket(s) through the wall opposite its terminal, provided the distance between the terminal and the opposite wall is not less than that specified in Table 312.6(A) and the terminal is a lay-in type where the terminal is either of the following:

(a) Directed toward the opening in the enclosure and within a 45 degree angle of directly facing the enclosure wall

(b) Directly facing the enclosure wall and offset not greater than 50 percent of the bending space specified in Table 312.6(A)

FPN: *Offset* is the distance measured along the enclosure wall from the axis of the centerline of the terminal to a

Table 312.6(A) Minimum Wire-Bending Space at Terminals and Minimum Width of Wiring Gutters

| Wire Size (AWG or kcmil) | Wires per Terminal | | | | | | | | | |
| | 1 | | 2 | | 3 | | 4 | | 5 | |
	mm	in.	mm	in.	mm	in.	mm	in.	mm	in.
14–10	Not specified		—	—	—	—	—	—	—	—
8–6	38.1	1½	—	—	—	—	—	—	—	—
4–3	50.8	2	—	—	—	—	—	—	—	—
2	63.5	2½	—	—	—	—	—	—	—	—
1	76.2	3	—	—	—	—	—	—	—	—
1/0–2/0	88.9	3½	127	5	178	7	—	—	—	—
3/0–4/0	102	4	152	6	203	8	—	—	—	—
250	114	4½	152	6	203	8	254	10	—	—
300–350	127	5	203	8	254	10	305	12	—	—
400–500	152	6	203	8	254	10	305	12	356	14
600–700	203	8	254	10	305	12	356	14	406	16
750–900	203	8	305	12	356	14	406	16	457	18
1000–1250	254	10	—	—	—	—	—	—	—	—
1500–2000	305	12	—	—	—	—	—	—	—	—

Note: Bending space at terminals shall be measured in a straight line from the end of the lug or wire connector (in the direction that the wire leaves the terminal) to the wall, barrier, or obstruction.

line passing through the center of the opening in the enclosure.

Section 312.6(B)(2) and Table 312.6(B) provide the requirements for wire-bending space where straight-in wiring or offset (double bends) is employed at terminals. Section 312.6(B)(1) applies only to 90 degree bends.

The notes to Table 312.6(B) permit a reduction in required bending space for removable and lay-in wire terminals. The removable terminal wire connectors can be either compression type or setscrew type. However, connectors are required to be of the type intended for a single conductor (single barrel). Removable connectors designed for multiple wires are not permitted to have a reduction in bending space.

To facilitate wiring, a terminal may be placed on the stripped end of the conductor, which has been cut to the proper length. The terminal is crimped or lightly torqued on the wire as intended. The wire is bent and routed to facilitate mounting onto the stud or landing pad for proper connection. All mechanical screws, bolts, and nuts involved should then be torqued to the proper value. See the commentary following the fine print note to 110.14 regarding tightening torques.

Note that in accordance with the notes to Tables 312.6(A) and 312.6(B), when using Table 312.6(A), bending space is measured in the direction in which the wire leaves the terminal, and when using Table 312.6(B), it is measured in a direction perpendicular to the enclosure wall.

A lay-in-type terminal is a pressure wire connector in which part of the connector is removable or swings away so that the stripped end of the conductor can be laid into the fixed portion of the connector. The removable or swing-away portion is then put back in place and the connector tightened down on the conductor.

In conjunction with the following listing, Exhibit 312.1 shows the application of the rules of 312.6(B)(1) and 312.6(B)(2) and Tables 312.6(A) and 312.6(B) to the wiring for a lay-in-type terminal.

T_1, 312.6(B)(2): Table 312.6(B) applies for conductors M.

T_2, 312.6(B)(2): Table 312.6(B) applies for conductors BR_2 unless Exception No. 2 of 312.6(B)(1) applies. This exception allows Table 312.6(A) to apply to T_2 as long as BR_2 enters or leaves the enclosure where gutter G_2 joins gutter G_1, and gutter G_1 has a width conforming to Table 312.6(B) for BR_2.

T_3, 312.6(B)(2): Table 312.6(B) applies for conductors BR_3.

T_4, 312.6(B)(1): Table 312.6(A) applies for conductor N.

G_1, 312.6(A): Table 312.6(A) applies for conductors M. Table 312.6(B) applies for conductors BR_2 where T_2 does not comply with Table 312.6(B).

G_2, 312.6(A): Table 312.6(A) applies for conductors BR_2.

G_3, 312.6(A): Table 312.6(A) applies for conductors BR_3.

G_4, 312.6(A): Table 312.6(A) applies for conductor N.

Table 312.6(B) Minimum Wire-Bending Space at Terminals

Wire Size (AWG or kcmil)		Wires per Terminal							
All Other Conductors	Compact Stranded AA-8000 Aluminum Alloy Conductors (See Note 3.)	1		2		3		4 or More	
		mm	in.	mm	in.	mm	in.	mm	in.
14–10	12–8	Not specified		—	—	—		—	—
8	6	38.1	1½	—	—	—		—	—
6	4	50.8	2	—	—	—		—	—
4	2	76.2	3	—	—	—		—	—
3	1	76.2	3	—	—	—		—	—
2	1/0	88.9	3½	—	—	—		—	—
1	2/0	114	4½	—	—	—		—	—
1/0	3/0	140	5½	140	5½	178	7	—	—
2/0	4/0	152	6	152	6	190	7½	—	—
3/0	250	165[a]	6½[a]	165[a]	6½[a]	203	8	—	—
4/0	300	178[b]	7[b]	190[c]	7½[c]	216[a]	8½[a]	—	—
250	350	216[d]	8½[d]	229[d]	8½[d]	254	9[b]	254	10
300	400	254[e]	10[e]	254[d]	10[d]	279[b]	11[b]	305	12
350	500	305[e]	12[e]	305[e]	12[e]	330[e]	13[e]	356[d]	14[d]
400	600	330[e]	13[e]	330[e]	13[e]	356[e]	14[e]	381[e]	15[e]
500	700–750	356[e]	14[e]	356[e]	14[e]	381[e]	15[e]	406[e]	16[e]
600	800–900	381[e]	15[e]	406[e]	16[e]	457[e]	18[e]	483[e]	19[e]
700	1000	406[e]	16[e]	457[e]	18[e]	508[e]	20[e]	559[e]	22[e]
750	—	432[e]	17[e]	483[e]	19[e]	559[e]	22[e]	610[e]	24[e]
800	—	457	18	508	20	559	22	610	24
900	—	483	19	559	22	610	24	610	24
1000	—	508	20	—	—	—		—	
1250	—	559	22	—	—	—		—	
1500	—	610	24	—	—	—		—	
1750	—	610	24	—	—	—		—	
2000	—	610	24	—	—	—		—	

1. Bending space at terminals shall be measured in a straight line from the end of the lug or wire connector in a direction perpendicular to the enclosure wall.

2. For removable and lay-in wire terminals intended for only one wire, bending space shall be permitted to be reduced by the following number of millimeters (inches):

[a] 12.7 mm (½ in.)   [d] 50.8 mm (2 in.)
[b] 25.4 mm (1 in.)   [e] 76.2 mm (3 in.)
[c] 38.1 mm (1½ in.)

3. This column shall be permitted to determine the required wire-bending space for compact stranded aluminum conductors in sizes up to 1000 kcmil and manufactured using AA-8000 series electrical grade aluminum alloy conductor material in accordance with 310.14.

Exhibit 312.2 is an illustration of the conditions under which 312.6(B)(2), Exception No. 2, is applicable. The terminal on the left has an offset not greater than 50 percent of bending space, per condition (b) of Exception No. 2. The terminal on the right is within a 45 degree angle of the enclosure, per condition (a) of Exception No. 2. See also Exhibit 430.1, which shows an example of wire-bending space in enclosures for motor controllers.

(C) Conductors 4 AWG or Larger. Installation shall comply with 300.4(F).

312.7 Space in Enclosures.

Cabinets and cutout boxes shall have sufficient space to accommodate all conductors installed in them without crowding.

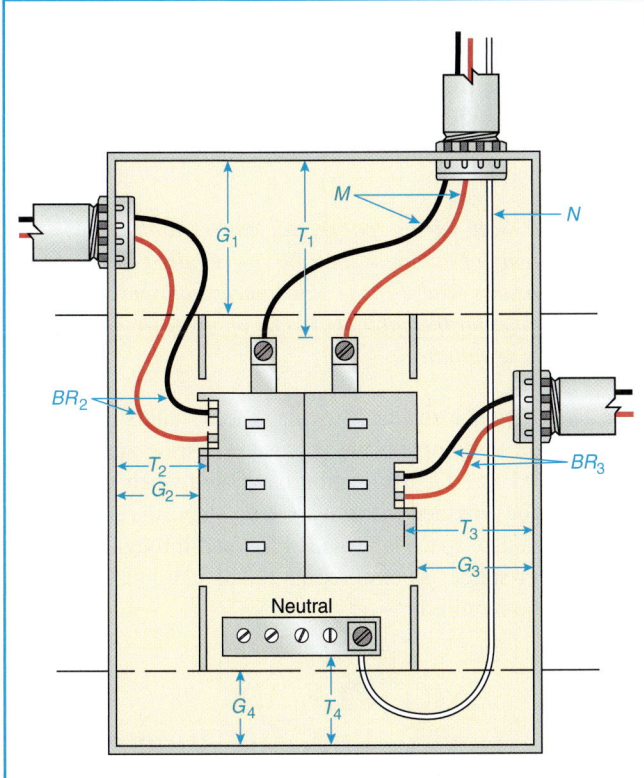

Exhibit 312.1 Wiring for a lay-in-type terminal to which the list of rules on page 297 applies.

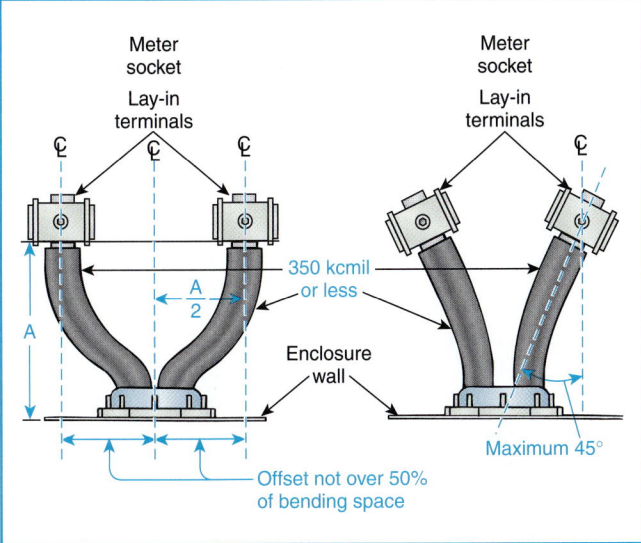

Exhibit 312.2 Conditions under which 312.6(B)(2), Exception No. 2, is applicable.

312.8 Enclosures for Switches or Overcurrent Devices.

Enclosures for switches or overcurrent devices shall not be used as junction boxes, auxiliary gutters, or raceways for conductors feeding through or tapping off to other switches or overcurrent devices, unless adequate space for this purpose is provided. The conductors shall not fill the wiring space at any cross section to more than 40 percent of the cross-sectional area of the space, and the conductors, splices, and taps shall not fill the wiring space at any cross section to more than 75 percent of the cross-sectional area of that space.

Most enclosures are intended to accommodate only those conductors that will be connected to terminals for switches or overcurrent devices within the enclosures themselves. Where adequate space is provided for additional conductors, such as control circuits, the total conductor fill in the enclosure may not exceed 40 percent of the cross section of the wiring space in the enclosure, and no more than 75 percent if splices or taps are necessary.

Example

If an enclosure has a wiring space of 4 in. by 3 in., the cross-sectional area is 12 in.^{2} Thus, the total conductor fill (see Chapter 9, Table 5 for dimensions of conductors) at any cross section cannot exceed 4.8 in.^{2} (40 percent of 12 in.^{2}), and the maximum space for conductors and splices or taps at any cross section cannot exceed 9 in.^{2} (75 percent of 12 in.^{2}).

In general, the best way to avoid overcrowding enclosures is to use properly sized auxiliary gutters (366.5 and 366.8) or junction boxes (314.16 and 314.28). See 430.10 and commentary for wiring space in enclosures for motor controllers and disconnecting means. See also 110.59 for tunnel installations over 600 volts.

312.9 Side or Back Wiring Spaces or Gutters.

Cabinets and cutout boxes shall be provided with back-wiring spaces, gutters, or wiring compartments as required by 312.11(C) and (D).

II. Construction Specifications

312.10 Material.

Cabinets, cutout boxes, and meter socket enclosures shall comply with 312.10(A) through (C).

(A) Metal Cabinets and Cutout Boxes. Metal enclosures within the scope of this article shall be protected both inside and outside against corrosion.

FPN: For information on protection against corrosion, see 300.6.

(B) Strength. The design and construction of enclosures within the scope of this article shall be such as to secure ample strength and rigidity. If constructed of sheet steel, the metal thickness shall not be less than 1.35 mm (0.053 in.) uncoated.

(C) Nonmetallic Cabinets. Nonmetallic cabinets shall be listed or they shall be submitted for approval prior to installation.

312.11 Spacing.

The spacing within cabinets and cutout boxes shall comply with 312.11(A) through (D).

(A) General. Spacing within cabinets and cutout boxes shall be sufficient to provide ample room for the distribution of wires and cables placed in them and for a separation between metal parts of devices and apparatus mounted within them as follows.

(1) Base. Other than at points of support, there shall be an airspace of at least 1.59 mm (0.0625 in.) between the base of the device and the wall of any metal cabinet or cutout box in which the device is mounted.

(2) Doors. There shall be an airspace of at least 25.4 mm (1.00 in.) between any live metal part, including live metal parts of enclosed fuses, and the door.

Exception: Where the door is lined with an approved insulating material or is of a thickness of metal not less than 2.36 mm (0.093 in.) uncoated, the airspace shall not be less than 12.7 mm (0.500 in.).

(3) Live Parts. There shall be an airspace of at least 12.7 mm (0.500 in.) between the walls, back, gutter partition, if of metal, or door of any cabinet or cutout box and the nearest exposed current-carrying part of devices mounted within the cabinet where the voltage does not exceed 250. This spacing shall be increased to at least 25.4 mm (1.00 in.) for voltages of 251 to 600, nominal.

Exception: Where the conditions in 312.11(A)(2), Exception, are met, the airspace for nominal voltages from 251 to 600 shall be permitted to be not less than 12.7 mm (0.500 in.).

(B) Switch Clearance. Cabinets and cutout boxes shall be deep enough to allow the closing of the doors when 30-ampere branch-circuit panelboard switches are in any position, when combination cutout switches are in any position, or when other single-throw switches are opened as far as their construction permits.

(C) Wiring Space. Cabinets and cutout boxes that contain devices or apparatus connected within the cabinet or box to more than eight conductors, including those of branch circuits, meter loops, feeder circuits, power circuits, and similar circuits, but not including the supply circuit or a continuation thereof, shall have back-wiring spaces or one or more side-wiring spaces, side gutters, or wiring compartments.

(D) Wiring Space—Enclosure. Side-wiring spaces, side gutters, or side-wiring compartments of cabinets and cutout boxes shall be made tight enclosures by means of covers, barriers, or partitions extending from the bases of the devices contained in the cabinet, to the door, frame, or sides of the cabinet.

Exception: Side-wiring spaces, side gutters, and side-wiring compartments of cabinets shall not be required to be made tight enclosures where those side spaces contain only conductors that enter the cabinet directly opposite to the devices where they terminate.

Partially enclosed back-wiring spaces shall be provided with covers to complete the enclosure. Wiring spaces that are required by 312.11(C) and are exposed when doors are open shall be provided with covers to complete the enclosure. Where adequate space is provided for feed-through conductors and for splices as required in 312.8, additional barriers shall not be required.

ARTICLE 314
Outlet, Device, Pull, and Junction Boxes; Conduit Bodies; Fittings; and Manholes

Contents

I. Scope and General

314.1 Scope.

This article covers the installation and use of all boxes and conduit bodies used as outlet, device, junction, or pull boxes, depending on their use, and manholes and other electric enclosures intended for personnel entry. Cast, sheet metal, nonmetallic, and other boxes such as FS, FD, and larger boxes are not classified as conduit bodies. This article also includes installation requirements for fittings used to join raceways and to connect raceways and cables to boxes and conduit bodies.

314.2 Round Boxes.

Round boxes shall not be used where conduits or connectors requiring the use of locknuts or bushings are to be connected to the side of the box.

Section 314.2 requires the use of rectangular or octagonal boxes having a flat bearing surface at each knockout for locknuts and bushings to ensure effective grounding continuity. Round boxes, however, can be used if the conduit or cable is secured by clamps within the box or if the cable does not need attachment to the box, as permitted by 314.17(C), Exception.

314.3 Nonmetallic Boxes.

Nonmetallic boxes shall be permitted only with open wiring on insulators, concealed knob-and-tube wiring, cabled wiring methods with entirely nonmetallic sheaths, flexible cords, and nonmetallic raceways.

Exception No. 1: Where internal bonding means are provided between all entries, nonmetallic boxes shall be permitted to be used with metal raceways or metal-armored cables.

Exception No. 2: Where integral bonding means with a provision for attaching an equipment bonding jumper inside the box are provided between all threaded entries in nonmetallic boxes listed for the purpose, nonmetallic boxes shall be permitted to be used with metal raceways or metal-armored cables.

The main rule of 314.3 was revised for the 2002 *Code* to permit the use of nonmetallic boxes with flexible cords. Exception No. 1 applies to nonmetallic boxes without threaded entries and permits the use of metal raceways and metal-armored cables with nonmetallic boxes. Internal bonding means must be installed to ensure ground continuity between the metal raceways or metal-armored cables. For the purposes of this exception the term *metal-armored cable* includes cables with a metal covering such as mineral-insulated, metal-sheathed cable (Type MI), metal-clad cable (Type MC), and armored cable (Type AC).

Exception No. 2 permits the use of metal raceways and metal-armored cables with listed nonmetallic boxes equipped with threaded entries. An integral bonding means is required to ensure ground continuity between the threaded entries. The requirement for means to attach a bonding jumper accommodates devices or equipment attached to the box. For the purposes of this exception, the term *metal-armored cable* includes cables with a metal covering such as mineral-insulated, metal-sheathed cable (Type MI), metal-clad cable (Type MC), and armored cable (Type AC).

314.4 Metal Boxes.

All metal boxes shall be grounded in accordance with the provisions of Article 250.

314.5 Short-Radius Conduit Bodies.

Conduit bodies such as capped elbows and service-entrance elbows that enclose conductors 6 AWG or smaller, and are only intended to enable the installation of the raceway and the contained conductors, shall not contain splices, taps, or devices and shall be of sufficient size to provide free space for all conductors enclosed in the conduit body.

Short radius conduit bodies are not permitted to contain splices.

II. Installation

314.15 Damp, Wet, or Hazardous (Classified) Locations.

(A) Damp or Wet Locations. In damp or wet locations, boxes, conduit bodies, and fittings shall be placed or equipped so as to prevent moisture from entering or accumulating within the box, conduit body, or fitting. Boxes, conduit bodies, and fittings installed in wet locations shall be listed for use in wet locations.

> FPN No. 1: For boxes in floors, see 314.27(C).
> FPN No. 2: For protection against corrosion, see 300.6.

Article 100 defines the term *weatherproof* as "constructed or protected so that exposure to the weather will not interfere with successful operation." Rainproof, raintight, or watertight equipment can fulfill the requirements for this definition where varying weather conditions other than wetness, such as snow, ice, dust, or temperature extremes, are not a factor. A weatherhead fitting is considered to be weatherproof because the openings for the conductors are placed in a downward position so that rain or snow cannot enter the fitting.

See the definitions of *damp location* and *wet location* under *location* in Article 100, as well as the commentary following the definition of *enclosure,* for further explanation.

(B) Hazardous (Classified) Locations. Installations in hazardous (classified) locations shall conform to Articles 500 through 517.

314.16 Number of Conductors in Outlet, Device, and Junction Boxes, and Conduit Bodies.

Boxes and conduit bodies shall be of sufficient size to provide free space for all enclosed conductors. In no case shall the volume of the box, as calculated in 314.16(A), be less than the fill calculation as calculated in 314.16(B). The minimum volume for conduit bodies shall be as calculated in 314.16(C).

The provisions of this section shall not apply to terminal housings supplied with motors.

> FPN: For volume requirements of motor terminal housings, see 430.12.

Boxes and conduit bodies enclosing conductors 4 AWG or larger shall also comply with the provisions of 314.28.

(A) Box Volume Calculations. The volume of a wiring enclosure (box) shall be the total volume of the assembled sections, and, where used, the space provided by plaster rings, domed covers, extension rings, and so forth, that are marked with their volume or are made from boxes the dimensions of which are listed in Table 314.16(A).

Table 314.16(A) Metal Boxes

Box Trade Size			Minimum Volume		Maximum Number of Conductors*						
mm	in.		cm³	in.³	18	16	14	12	10	8	6
100 × 32	(4 × 1¼)	round/octagonal	205	12.5	8	7	6	5	5	5	2
100 × 38	(4 × 1½)	round/octagonal	254	15.5	10	8	7	6	6	5	3
100 × 54	(4 × 2⅛)	round/octagonal	353	21.5	14	12	10	9	8	7	4
100 × 32	(4 × 1¼)	square	295	18.0	12	10	9	8	7	6	3
100 × 38	(4 × 1½)	square	344	21.0	14	12	10	9	8	7	4
100 × 54	(4 × 2⅛)	square	497	30.3	20	17	15	13	12	10	6
120 × 32	(4¹¹⁄₁₆ × 1¼)	square	418	25.5	17	14	12	11	10	8	5
120 × 38	(4¹¹⁄₁₆ × 1½)	square	484	29.5	19	16	14	13	11	9	5
120 × 54	(4¹¹⁄₁₆ × 2⅛)	square	689	42.0	28	24	21	18	16	14	8
75 × 50 × 38	(3 × 2 × 1½)	device	123	7.5	5	4	3	3	3	2	1
75 × 50 × 50	(3 × 2 × 2)	device	164	10.0	6	5	5	4	4	3	2
75 × 50 × 57	(3 × 2 × 2¼)	device	172	10.5	7	6	5	4	4	3	2
75 × 50 × 65	(3 × 2 × 2½)	device	205	12.5	8	7	6	5	5	4	2
75 × 50 × 70	(3 × 2 × 2¾)	device	230	14.0	9	8	7	6	5	4	2
75 × 50 × 90	(3 × 2 × 3½)	device	295	18.0	12	10	9	8	7	6	3
100 × 54 × 38	(4 × 2⅛ × 1½)	device	169	10.3	6	5	5	4	4	3	2
100 × 54 × 48	(4 × 2⅛ × 1⅞)	device	213	13.0	8	7	6	5	5	4	2
100 × 54 × 54	(4 × 2⅛ × 2⅛)	device	238	14.5	9	8	7	6	5	4	2
95 × 50 × 65	(3¾ × 2 × 2½)	masonry box/gang	230	14.0	9	8	7	6	5	4	2
95 × 50 × 90	(3¾ × 2 × 3½)	masonry box/gang	344	21.0	14	12	10	9	8	7	4
min. 44.5 depth	FS—single cover/gang (1¾)		221	13.5	9	7	6	6	5	4	2
min. 60.3 depth	FD — single cover/gang (2⅜)		295	18.0	12	10	9	8	7	6	3
min. 44.5 depth	FS—multiple cover/gang (1¾)		295	18.0	12	10	9	8	7	6	3
min. 60.3 depth	FD—multiple cover/gang (2⅜)		395	24.0	16	13	12	10	9	8	4

*Where no volume allowances are required by 314.16(B)(2) through 314.16(B)(5).

(1) Standard Boxes. The volumes of standard boxes that are not marked with their volume shall be as given in Table 314.16(A).

(2) Other Boxes. Boxes 1650 cm³ (100 in.³) or less, other than those described in Table 314.16(A), and nonmetallic boxes shall be durably and legibly marked by the manufacturer with their volume. Boxes described in Table 314.16(A) that have a volume larger than is designated in the table shall be permitted to have their volume marked as required by this section.

(B) Box Fill Calculations. The volumes in paragraphs 314.16(B)(1) through (5), as applicable, shall be added together. No allowance shall be required for small fittings such as locknuts and bushings.

(1) Conductor Fill. Each conductor that originates outside the box and terminates or is spliced within the box shall be counted once, and each conductor that passes through the box without splice or termination shall be counted once. The conductor fill shall be computed using Table 314.16(B). A conductor, no part of which leaves the box, shall not be counted.

Table 314.16(B) Volume Allowance Required per Conductor

Size of Conductor (AWG)	Free Space Within Box for Each Conductor	
	cm³	in.³
18	24.6	1.50
16	28.7	1.75
14	32.8	2.00
12	36.9	2.25
10	41.0	2.50
8	49.2	3.00
6	81.9	5.00

Section 314.16 provides the requirements and identifies the allowances for the number of conductors permitted to be enclosed within a box. This section requires that the *total box volume* be equal to or greater than the *total box fill*.

GENERAL REQUIREMENTS

The *total box volume* is determined by adding the individual volumes of the box components. The components include the box itself plus any attachments to it, such as a plaster ring, an extension ring, or a dome cover. The volume of each box component is determined either from the volume marking on the component itself or from the standard volumes listed in Table 314.16(A). If a box is marked with a larger volume than listed in Table 314.16(A), the larger volume can be used instead of the table value.

Adding all of the volume allowances for all items contributing to box fill determines the *total box fill*. The volume allowance for each fill item is based on the volume listed in Table 314.16(B) for the conductor size indicated. Table 314.1 summarizes the components contributing to box fill.

SPECIFIC EXAMPLES

The following three examples illustrate the applicable requirements of 314.16 and the accompanying tables.

Example 1

Using the simple method [according to the footnote of Table 314.16(A)], select a standard size box for use where all the conductors are the same size and the box does not contain any cable clamps, support fittings, devices, or equipment grounding conductors. Refer to Exhibit 314.1 as an example.

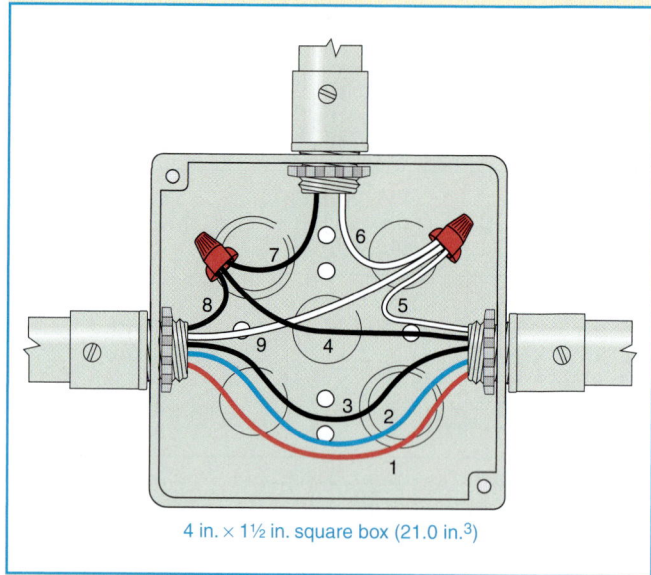

4 in. × 1½ in. square box (21.0 in.³)

Exhibit 314.1 Example 1: A standard-size 4 in. × 1½ in. square box (21.0 in.³) containing no fittings or devices, such as fixture studs, cable clamps, switches, receptacles, or equipment grounding conductors.

Table 314.1 Summary of Items Contributing to Box Fill

Items Contained Within Box	Volume Allowance	Based on [see Table 314.16(B)]
Conductors that originate outside box	One for each conductor	Actual conductor size
Conductors that pass through box without splice or connection	One for each conductor	Actual conductor size
Conductors that originate within box and do not leave box	None (These conductors are not counted.)	NA
Fixture conductors [per 314.16(B)(1), Exception]	None (These conductors are not counted.)	NA
Internal cable clamps (one or more)	One only	Largest size conductor present
Support fittings (such as fixture studs, hickeys)	One for each type of support fitting	Largest size conductor present
Devices (such as receptacles, switches)	Two for each yoke or mounting strap	Largest size conductor connected to device or equipment
Equipment grounding conductor (one or more)	One only	Largest equipment grounding conductor present
Isolated equipment grounding conductor (one or more) (see 250.74, Exception No. 4)	One only	Largest isolated and insulated equipment grounding conductor present

Solution

To determine the number of conductors permitted in this standard 4 in. × 1½ in. square box (21.0 in.³), count the conductors in the box and compare the total to the maximum number of conductors permitted by Table 314.16(A). Each unspliced conductor running through the box is counted as one conductor, and each other conductor is counted as one conductor. Therefore, the total conductor count for this box is nine conductors. Table 314.16(A) indicates that the maximum fill for this box is nine 12 AWG conductors, so the box is adequately sized.

Example 2

The standard method for determining adequate box size calculates the total box volume and then subtracts the total box fill to ensure compliance. Using this method, refer to Exhibit 314.2 and determine whether the box is adequately sized.

Solution

For a standard 3 × 2 × 3½ device box (18 in.³), Table 314.16(A) allows up to a maximum of nine 14 AWG conductors. The box fill for this situation as given in Table 314.2 is 16 in.³. Therefore, because the total box fill of 16 in.³ is less than the 18 in.³ total box volume permitted, the box is adequately sized.

Example 3

Using the standard method, determine the adequacy of the device illustrated in Exhibit 314.3. The exhibit illustrates two 3 × 2 × 3½-in. device boxes assembled to configure a single box.

Solution

The total box volume, referring to Table 314.16(A), is 36 in.³ (2 × 18 in.³). The total box fill, referring to Table 314.16(B), is determined as given in Table 314.3. With only 26 in.³ of the 36 in.³ filled, the box is adequately sized.

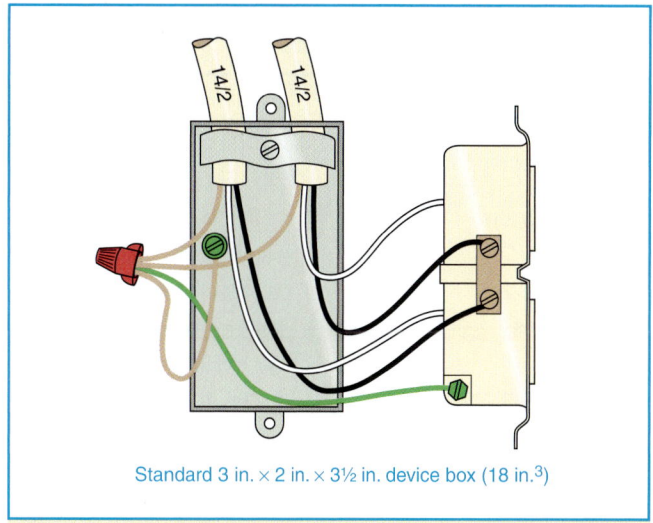

Standard 3 in. × 2 in. × 3½ in. device box (18 in.³)

Exhibit 314.2 Example 2: A device box that contains components and conductors requiring deductions in accordance with 314.16.

Exception: An equipment grounding conductor or conductors or not over four fixture wires smaller than 14 AWG, or both, shall be permitted to be omitted from the calculations where they enter a box from a domed luminaire (fixture) or similar canopy and terminate within that box.

(2) Clamp Fill. Where one or more internal cable clamps, whether factory or field supplied, are present in the box, a single volume allowance in accordance with Table 314.16(B) shall be made based on the largest conductor present in the box. No allowance shall be required for a cable connector with its clamping mechanism outside the box.

(3) Support Fittings Fill. Where one or more luminaire (fixture) studs or hickeys are present in the box, a single volume allowance in accordance with Table 314.16(B) shall be made for each type of fitting based on the largest conductor present in the box.

Table 314.2 Total Box Fill for Example No. 2

Items Contained Within Box	Volume Allowance	Unit Volume Based on Table 314.16(B) (in.³)	Total Box Fill (in.³)
4 conductors	4 volume allowances for 14 AWG conductors	2.00	8.00
1 clamp	1 volume allowance (based on 14 AWG conductors)	2.00	2.00
1 device	2 volume allowances (based on 14 AWG conductors)	2.00	4.00
Equipment grounding conductors (all)	1 volume allowance (based on 14 AWG conductors)	2.00	2.00
Total			16.00 in.³

Table 314.3 Total Box Fill for Example No. 3

Items Contained Within Box	Volume Allowance	Unit Volume Based on Table 314.16(B) (in.³)	Total Box Fill (in.³)
6 conductors	2 volume allowances for 14 AWG conductors	2.00	4.00
	4 volume allowances for 12 AWG conductors	2.25	9.00
2 clamps	1 volume allowance (based on 12 AWG conductors)	2.25	2.25
2 devices	2 volume allowances (based on 14 AWG conductors)	2.00	4.00
	2 volume allowances (based on 12 AWG conductors)	2.25	4.50
Equipment grounding conductors (all)	1 volume allowance (based on 12 AWG conductors)	2.25	2.25
Total			26.00 in.³

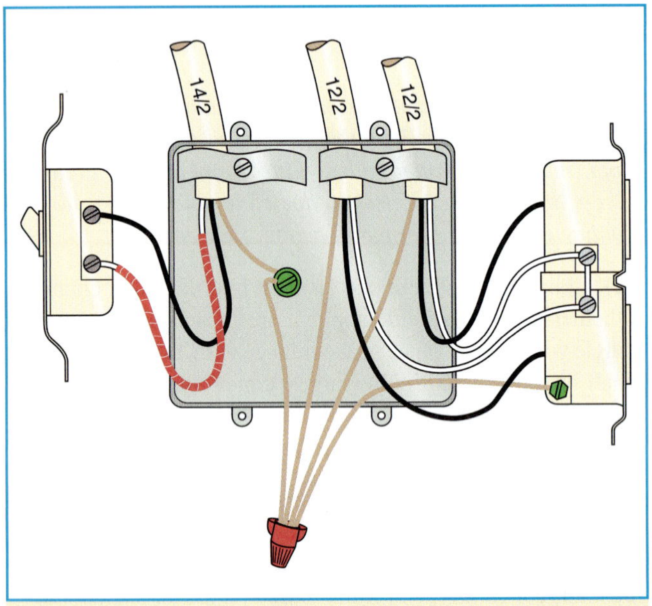

Exhibit 314.3 Example 3: Two standard gangable device boxes containing conductors of different sizes.

(4) Device or Equipment Fill. For each yoke or strap containing one or more devices or equipment, a double volume allowance in accordance with Table 314.16(B) shall be made for each yoke or strap based on the largest conductor connected to a device(s) or equipment supported by that yoke or strap.

(5) Equipment Grounding Conductor Fill. Where one or more equipment grounding conductors or equipment bonding jumpers enter a box, a single volume allowance in accordance with Table 314.16(B) shall be made based on the largest equipment grounding conductor or equipment bonding jumper present in the box. Where an additional set of equipment grounding conductors, as permitted by 250.146(D), is present in the box, an additional volume allowance shall be made based on the largest equipment grounding conductor in the additional set.

(C) Conduit Bodies.

(1) General. Conduit bodies enclosing 6 AWG conductors or smaller, other than short-radius conduit bodies as described in 314.5, shall have a cross-sectional area not less than twice the cross-sectional area of the largest conduit or tubing to which it is attached. The maximum number of conductors permitted shall be the maximum number permitted by Table 1 of Chapter 9 for the conduit or tubing to which it is attached.

(2) With Splices, Taps, or Devices. Only those conduit bodies that are durably and legibly marked by the manufacturer with their volume shall be permitted to contain splices, taps, or devices. The maximum number of conductors shall be computed in accordance with 314.16(B). Conduit bodies shall be supported in a rigid and secure manner.

As illustrated in Exhibit 314.4, conduit bodies other than the short-radius type are permitted to contain splices or taps, provided the conduit bodies are marked with their cubic inch capacity. Such conduit bodies are required to have a cross-sectional area not less than twice that of the conduit to which they are attached and are not permitted to contain more conductors than the attached raceway. The volume requirements for splicing or tapping are provided in 314.16(C).

Conduit bodies must be rigidly supported. See 314.23(D)(2) and the exception to 314.23(E), and Exception No. 1 of 314.23(F), which permit the raceway to support the conduit body, provided the conduit body is not larger than the attached raceway.

See 314.28 for requirements that apply to conduit bodies used as pull and junction boxes.

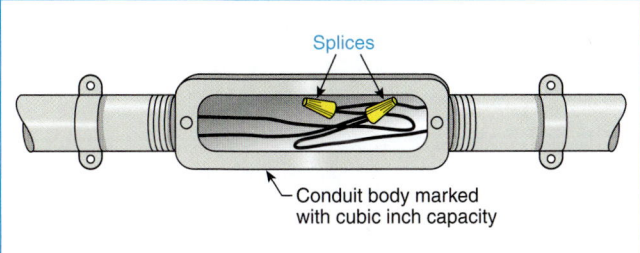

Exhibit 314.4 An example of splices in a raceway-supported conduit body.

314.17 Conductors Entering Boxes, Conduit Bodies, or Fittings.

Conductors entering boxes, conduit bodies, or fittings shall be protected from abrasion and shall comply with 314.17(A) through (D).

(A) Openings to Be Closed. Openings through which conductors enter shall be adequately closed.

(B) Metal Boxes and Conduit Bodies. Where metal boxes or conduit bodies are installed with open wiring or concealed knob-and-tube wiring, conductors shall enter through insulating bushings or, in dry locations, through flexible tubing extending from the last insulating support to not less than 6 mm (¼ in.) inside the box and beyond any cable clamps. Except as provided in 300.15(C), the wiring shall be firmly secured to the box or conduit body. Where raceway or cable is installed with metal boxes or conduit bodies, the raceway or cable shall be secured to such boxes and conduit bodies.

(C) Nonmetallic Boxes and Conduit Bodies. Nonmetallic boxes and conduit bodies shall be suitable for the lowest temperature-rated conductor entering the box. Where nonmetallic boxes and conduit bodies are used with open wiring or concealed knob-and-tube wiring, the conductors shall enter the box through individual holes. Where flexible tubing is used to enclose the conductors, the tubing shall extend from the last insulating support to not less than 6 mm (¼ in.) inside the box and beyond any cable clamp. Where nonmetallic-sheathed cable or multiconductor Type UF cable is used, the sheath shall extend not less than 6 mm (¼ in.) inside the box and beyond any cable clamp. In all instances, all permitted wiring methods shall be secured to the boxes.

Standard nonmetallic boxes are permitted for use with 90°C insulated conductors. A nonmetallic box used for splicing

a conductor of a higher temperature rating to a conductor of a lower temperature rating is required to be identified as suitable for the temperature rating of the lower-rated conductor. The intent is to avoid the necessity of giving a high temperature rating to boxes in a normal temperature location simply because high-temperature conductors enter the box from, or exit to, a high-temperature location. However, where insulated conductors rated at higher temperatures are necessary in a high-temperature environment, the box is required to be suitably identified by a marking on the box or in the listing of the box to comply with 110.3(B).

Exception: Where nonmetallic-sheathed cable or multiconductor Type UF cable is used with single gang boxes not larger than a nominal size 57 mm × 100 mm (2¼ in. × 4 in.) mounted in walls or ceilings, and where the cable is fastened within 200 mm (8 in.) of the box measured along the sheath and where the sheath extends through a cable knockout not less than 6 mm (¼ in.), securing the cable to the box shall not be required. Multiple cable entries shall be permitted in a single cable knockout opening.

Some cable-securing means is required in all single gang boxes larger than 2¼ in. by 4 in. The requirement is based on the width of the box and the likelihood that the cable will be pushed back out of the box when the conductors and device, if any, are folded back into the box during installation of receptacles, switches, dimmers, and so on.

(D) Conductors 4 AWG or Larger. Installation shall comply with 300.4(F).

> FPN: See 110.12(A) for requirements on closing unused cable and raceway knockout openings.

314.19 Boxes Enclosing Flush Devices.

Boxes used to enclose flush devices shall be of such design that the devices will be completely enclosed on back and sides and substantial support for the devices will be provided. Screws for supporting the box shall not be used in attachment of the device contained therein.

314.20 In Wall or Ceiling.

In walls or ceilings with a surface of concrete, tile, gypsum, plaster, or other noncombustible material, boxes shall be installed so that the front edge of the box will not be set back of the finished surface more than 6 mm (¼ in.).

For the 2002 *Code*, plaster, gypsum, and other construction products were added to the list of noncombustible construction materials. The intent of the section was also clarified.

The addition of the terms *surface* and *with a surface of* makes it clear that the requirements of this section apply only to the construction of the surface of wall or ceiling, not to the structure or subsurface of the wall or ceiling. Therefore, a wall constructed of wood but sheathed with an outer layer of gypsum board is permitted to contain boxes set back or recessed not more than ¼ in. Using an opposite example, a wall constructed of metal studs but finished with wood panels requires that contained outlet boxes be mounted flush with the combustible finish.

In walls and ceilings constructed of wood or other combustible surface material, boxes shall be flush with the finished surface or project therefrom.

314.21 Repairing Plaster and Drywall or Plasterboard.

Plaster, drywall, or plasterboard surfaces that are broken or incomplete shall be repaired so there will be no gaps or open spaces greater than 3 mm (⅛ in.) at the edge of the box or fitting.

314.22 Exposed Surface Extensions.

Surface extensions from a flush-mounted box shall be made by mounting and mechanically securing an extension ring over the flush box. Equipment grounding and bonding shall be in accordance with Article 250.

Exception: A surface extension shall be permitted to be made from the cover of a flush-mounted box where the cover is designed so it is unlikely to fall off or be removed if its securing means becomes loose. The wiring method shall be flexible for a length sufficient to permit removal of the cover and provide access to the box interior, and arranged so that any bonding or grounding continuity is independent of the connection between the box and cover.

Exhibit 314.5 shows an example of a flexible surface extension from a flush mounted outlet box. The exception to 314.22, which permits this technique, requires the use of a cover that will not fall off if the securing screws become loose.

314.23 Supports.

Enclosures within the scope of this article shall be supported in accordance with one or more of the provisions in 314.23(A) through (H).

(A) Surface Mounting. An enclosure mounted on a building or other surface shall be rigidly and securely fastened in place. If the surface does not provide rigid and secure

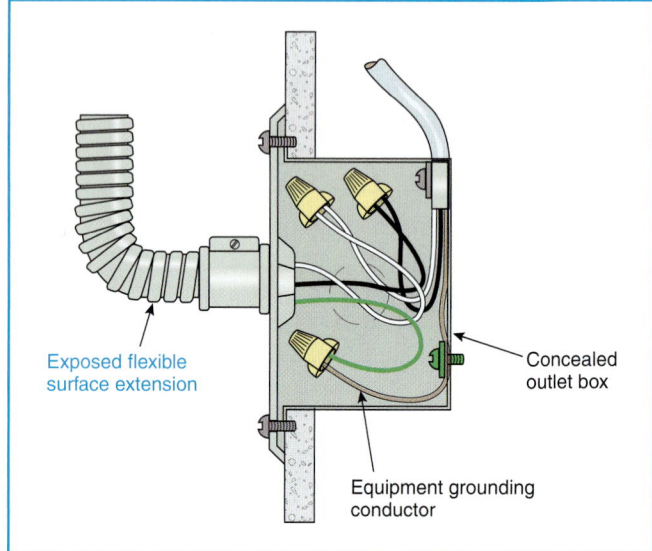

Exhibit 314.5 An example of a flexible surface extension from a flush mounted outlet box.

support, additional support in accordance with other provisions of this section shall be provided.

(B) Structural Mounting. An enclosure supported from a structural member of a building or from grade shall be rigidly supported either directly or by using a metal, polymeric, or wood brace.

(1) Nails and Screws. Nails and screws, where used as a fastening means, shall be attached by using brackets on the outside of the enclosure, or they shall pass through the interior within 6 mm (¼ in.) of the back or ends of the enclosure.

This requirement prevents the nails from interfering with the installation of devices. Permitting nails inside the box within ¼ in. of the ends reduces splitting of the smaller wooden studs used in some frame-type construction. However, splitting sometimes occurs where nails are within ¼ in. of the back of the box.

(2) Braces. Metal braces shall be protected against corrosion and formed from metal that is not less than 0.51 mm (0.020 in.) thick uncoated. Wood braces shall have a cross section not less than nominal 25 mm × 50 mm (1 in. × 2 in.). Wood braces in wet locations shall be treated for the conditions. Polymeric braces shall be identified as being suitable for the use.

(C) Mounting in Finished Surfaces. An enclosure mounted in a finished surface shall be rigidly secured thereto by clamps, anchors, or fittings identified for the application.

Where structural members are lacking or where boxes are cut into existing walls, boxes are permitted to be secured by clamps or anchors. Exhibit 314.6 shows one example of an acceptable mounting method. Shown in the right portion of the exhibit is a device box mounted in an opening in an existing wall by means of a securing bracket that is part of the box.

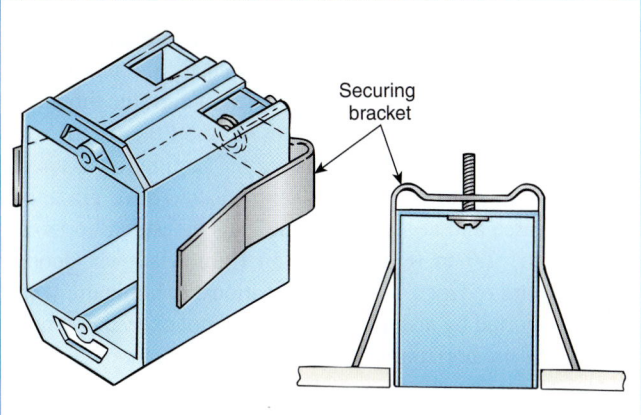

Securing bracket

Exhibit 314.6 One type of device box used for old work.

(D) Suspended Ceilings. An enclosure mounted to structural or supporting elements of a suspended ceiling shall be not more than 1650 cm^3 (100 in.^{3}) in size and shall be securely fastened in place in accordance with either (D)(1) or (D)(2).

(1) Framing Members. An enclosure shall be fastened to the framing members by mechanical means such as bolts, screws, or rivets, or by the use of clips or other securing means identified for use with the type of ceiling framing member(s) and enclosure(s) employed. The framing members shall be adequately supported and securely fastened to each other and to the building structure.

(2) Support Wires. The installation shall comply with the provisions of 300.11(A). The enclosure shall be secured, using methods identified for the purpose, to ceiling support wire(s), including any additional support wire(s) installed for that purpose. Support wire(s) used for enclosure support shall be fastened at each end so as to be taut within the ceiling cavity.

(E) Raceway Supported Enclosure, Without Devices, Luminaires (Fixtures), or Lampholders. An enclosure that does not contain a device(s) other than splicing devices or support a luminaire(s) [fixture(s)], lampholder, or other equipment and is supported by entering raceways shall not exceed 1650 cm^3 (100 in.^{3}) in size. It shall have threaded entries or have hubs identified for the purpose. It shall be

supported by two or more conduits threaded wrenchtight into the enclosure or hubs. Each conduit shall be secured within 900 mm (3 ft) of the enclosure, or within 450 mm (18 in.) of the enclosure if all conduit entries are on the same side.

Boxes are not permitted to be supported by rigid raceways using locknuts and bushings. Enclosures without devices or luminaires, however, are considered to be adequately supported, provided the conduit is connected to the enclosure by threaded hubs and the threaded conduits enter the box on two or more sides and are supported within 3 ft of the enclosure. A box is not permitted to be supported by a single raceway.

Exception: Rigid metal, intermediate metal, or rigid nonmetallic conduit or electrical metallic tubing shall be permitted to support a conduit body of any size, including a conduit body constructed with only one conduit entry, provided the trade size of the conduit body is not larger than the largest trade size of the conduit or electrical metallic tubing.

(F) Raceway Supported Enclosures, with Devices, Luminaires (Fixtures), or Lampholders. An enclosure that contains a device(s) or supports a luminaire(s) [fixture(s)], lampholder, or other equipment and is supported by entering raceways shall not exceed 1650 cm^3 (100 in.^{3}) in size. It shall have threaded entries or have hubs identified for the purpose. It shall be supported by two or more conduits threaded wrenchtight into the enclosure or hubs. Each conduit shall be secured within 450 mm (18 in.) of the enclosure.

The conduit is required to be supported within 18 in. if the enclosure contains devices or luminaires.

Exception No. 1: Rigid metal or intermediate metal conduit shall be permitted to support a conduit body of any size, including a conduit body constructed with only one conduit entry, provided the trade size of the conduit body is not larger than the largest trade size of the conduit.

Exception No. 2: An unbroken length(s) of rigid or intermediate metal conduit shall be permitted to support a box used for luminaire (fixture) or lampholder support, or to support a wiring enclosure that is an integral part of a luminaire (fixture) and used in lieu of a box in accordance with 300.15(B), where all of the following conditions are met.

(a) The conduit is securely fastened at a point so that the length of conduit beyond the last point of conduit support does not exceed 900 mm (3 ft).

(b) The unbroken conduit length before the last point of conduit support is 300 mm (12 in.) or greater, and that portion of the conduit is securely fastened at some point

not less than 300 mm (12 in.) from its last point of support.

(c) *Where accessible to unqualified persons, the luminaire (fixture) or lampholder, measured to its lowest point, is at least 2.5 m (8 ft) above grade or standing area and at least 900 mm (3 ft) measured horizontally to the 2.5 m (8 ft) elevation from windows, doors, porches, fire escapes, or similar locations.*

(d) *A luminaire (fixture) supported by a single conduit does not exceed 300 mm (12 in.) in any direction from the point of conduit entry.*

(e) *The weight supported by any single conduit does not exceed 9 kg (20 lb).*

(f) *At the luminaire (fixture) or lampholder end, the conduit(s) is threaded wrenchtight into the box, conduit body, or integral wiring enclosure, or into hubs identified for the purpose. Where a box or conduit body is used for support, the luminaire (fixture) shall be secured directly to the box or conduit body, or through a threaded conduit nipple not over 75 mm (3 in.) long.*

(G) Enclosures in Concrete or Masonry. An enclosure supported by embedment shall be identified as suitably protected from corrosion and securely embedded in concrete or masonry.

Boxes are permitted to be embedded in masonry or concrete, provided they are rigid and secure. Exhibit 314.7 shows a mud box installed in a concrete ceiling. Additional support is not required.

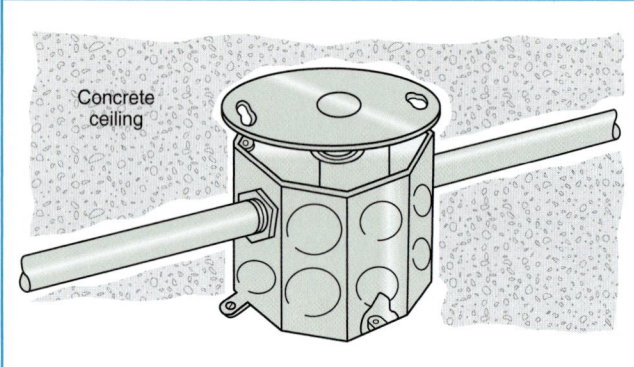

Exhibit 314.7 A mud box installed in a concrete ceiling.

(H) Pendant Boxes. An enclosure supported by a pendant shall comply with 314.23(H)(1) or (2).

(1) Flexible Cord. A box shall be supported from a multiconductor cord or cable in an approved manner that protects

the conductors against strain, such as a strain-relief connector threaded into a box with a hub.

(2) Conduit. A box supporting lampholders or luminaires (lighting fixtures), or wiring enclosures within luminaires (fixtures) used in lieu of boxes in accordance with 300.15(B), shall be supported by rigid or intermediate metal conduit stems. For stems longer than 450 mm (18 in.), the stems shall be connected to the wiring system with flexible fittings suitable for the location. At the luminaire (fixture) end, the conduit(s) shall be threaded wrenchtight into the box or wiring enclosure, or into hubs identified for the purpose.

Where supported by only a single conduit, the threaded joints shall be prevented from loosening by the use of setscrews or other effective means, or the luminaire (fixture), at any point, shall be at least 2.5 m (8 ft) above grade or standing area and at least 900 mm (3 ft) measured horizontally to the 2.5 m (8 ft) elevation from windows, doors, porches, fire escapes, or similar locations. A luminaire (fixture) supported by a single conduit shall not exceed 300 mm (12 in.) in any horizontal direction from the point of conduit entry.

314.24 Depth of Outlet Boxes.

No box shall have an internal depth of less than 12.7 mm (½ in.). Boxes intended to enclose flush devices shall have an internal depth of not less than 23.8 mm (¹⁵⁄₁₆ in.).

The use of a shallow box might become necessary because of old work or existing construction where, for example, very shallow partitions, plumbing pipes, or ductwork is encountered within the partition. The selection of a box used in this type of situation is required to be based on its having sufficient cubic-inch capacity.

314.25 Covers and Canopies.

In completed installations, each box shall have a cover, faceplate, lampholder, or luminaire (fixture) canopy, except where the installation complies with 410.14(B).

(A) Nonmetallic or Metal Covers and Plates. Nonmetallic or metal covers and plates shall be permitted. Where metal covers or plates are used, they shall comply with the grounding requirements of 250.110.

> FPN: For additional grounding requirements, see 410.18(A) for metal luminaire (fixture) canopies, and 404.12 and 406.5(B) for metal faceplates.

(B) Exposed Combustible Wall or Ceiling Finish. Where a luminaire (fixture) canopy or pan is used, any combustible wall or ceiling finish exposed between the edge of the canopy or pan and the outlet box shall be covered with noncombustible material.

(C) Flexible Cord Pendants. Covers of outlet boxes and conduit bodies having holes through which flexible cord pendants pass shall be provided with bushings designed for the purpose or shall have smooth, well-rounded surfaces on which the cords may bear. So-called hard rubber or composition bushings shall not be used.

314.27 Outlet Boxes.

(A) Boxes at Luminaire (Lighting Fixture) Outlets. Boxes used at luminaire (lighting fixture) or lampholder outlets shall be designed for the purpose. At every outlet used exclusively for lighting, the box shall be designed or installed so that a luminaire (lighting fixture) may be attached.

Exception: A wall-mounted luminaire (fixture) weighing not more than 3 kg (6 lb) shall be permitted to be supported on other boxes or plaster rings that are secured to other boxes, provided the luminaire (fixture) or its supporting yoke is secured to the box with no fewer than two No. 6 or larger screws.

Device boxes are designed for the mounting of snap switches, receptacles, and other devices, usually with 6-32 screws (No. 6 screws with 32 threads per inch). Generally, device boxes are not suitable for supporting other than lightweight wall-mounted luminaires. The exception to 314.27(A) permits luminaires such as wall-bracket types or sconces weighing less than 6 lb to be supported by a device box using No. 6 or larger screws. For heavier or ceiling-mounted lighting luminaires, see 410.16, Means of Support. The outlet box is required to provide "adequate" support.

(B) Maximum Luminaire (Fixture) Weight. Outlet boxes or fittings installed as required by 314.23 shall be permitted to support luminaires (lighting fixtures) weighing 23 kg (50 lb) or less. A luminaire (lighting fixture) that weighs more than 23 kg (50 lb) shall be supported independently of the outlet box unless the outlet box is listed for the weight to be supported.

Moved to Chapter 3 in the 2002 *NEC*, this requirement previously appeared in 410.16. Regardless of whether a luminaire is attached to an outlet box or is supported independently of the outlet box, care should be taken to securely fasten the supporting means of the luminaire. Exhibit 314.8 illustrates one method of supporting a fixture in accordance with 314.27(B).

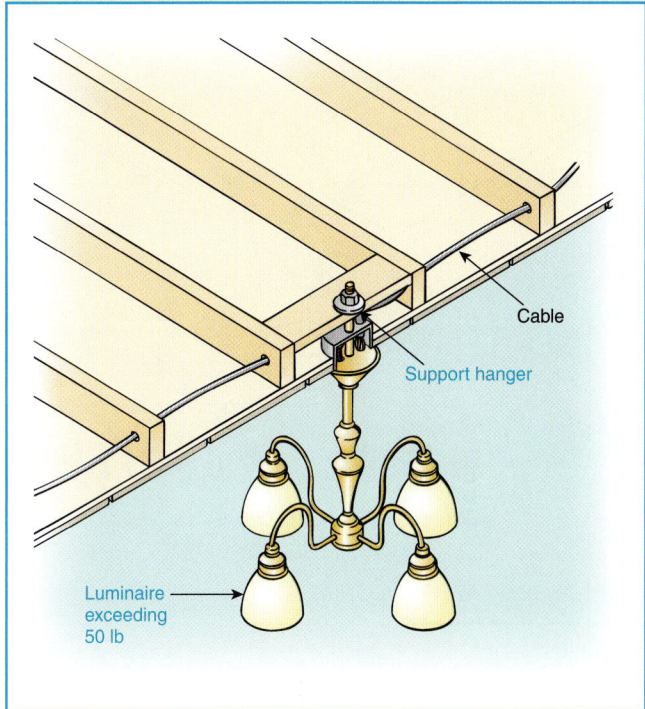

Exhibit 314.8 A wooden brace used to support a heavy luminaire according to 314.27(B).

(C) Floor Boxes. Boxes listed specifically for this application shall be used for receptacles located in the floor.

Exception: Where the authority having jurisdiction judges them free from likely exposure to physical damage, moisture, and dirt, boxes located in elevated floors of show windows and similar locations shall be permitted to be other than those listed for floor applications. Receptacles and covers shall be listed as an assembly for this type of location.

(D) Boxes at Ceiling-Suspended (Paddle) Fan Outlets. Where a box is used as the sole support of a ceiling-suspended (paddle) fan, the box shall be listed for the application and for the weight of the fan to be supported. The installation shall comply with 422.18.

Outlet boxes specifically listed to adequately support ceiling-mounted paddle fans are available, as are several alternative and retrofit methods that can provide suitable support for a paddle fan. One method of supporting a ceiling fan so that the box does not serve as the sole support is shown in Exhibit 314.9. For a detailed view of an outlet box specifically listed to support a paddle fan, see 422.18, the associated commentary, and Exhibit 422.2.

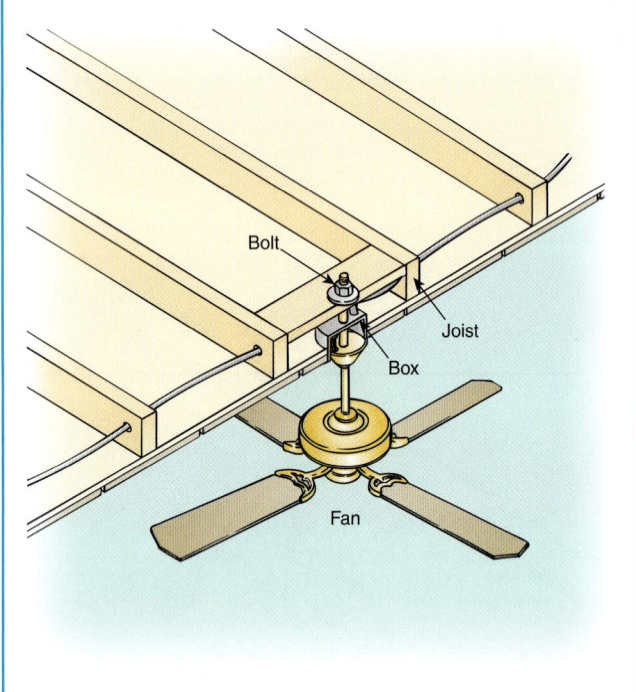

Exhibit 314.9 A ceiling fan supported without depending on the box for sole support.

314.28 Pull and Junction Boxes and Conduit Bodies.

Boxes and conduit bodies used as pull or junction boxes shall comply with 314.28(A) through (D).

Exception: Terminal housings supplied with motors shall comply with the provisions of 430.12.

(A) Minimum Size. For raceways containing conductors of 4 AWG or larger, and for cables containing conductors of 4 AWG or larger, the minimum dimensions of pull or junction boxes installed in a raceway or cable run shall comply with the following. Where an enclosure dimension is to be calculated based on the diameter of entering race-

ways, the diameter shall be the metric designator (trade size) expressed in the units of measurement employed.

(1) Straight Pulls. In straight pulls, the length of the box shall not be less than eight times the metric designator (trade size) of the largest raceway.

Section 314.28(A)(1) applies to minimum dimensions of pull and junction boxes or conduit bodies used with raceways or cables containing conductors 4 AWG or larger. For straight pulls, for example, trade size 2 conduit containing four 4/0 AWG, Type THHW conductors (see Annex C, Table C8) requires a 16-in. long pull box (8×2 in. = 16 in.). It should be understood that although 16 in. is the required minimum length, a longer pull box may be desired for maximum ease in handling this size conductor.

(2) Angle or U Pulls. Where splices or where angle or U pulls are made, the distance between each raceway entry inside the box and the opposite wall of the box shall not be less than six times the metric designator (trade size) of the largest raceway in a row. This distance shall be increased for additional entries by the amount of the sum of the diameters of all other raceway entries in the same row on the same wall of the box. Each row shall be calculated individually, and the single row that provides the maximum distance shall be used.

Exception: Where a raceway or cable entry is in the wall of a box or conduit body opposite a removable cover, the distance from that wall to the cover shall be permitted to comply with the distance required for one wire per terminal in Table 312.6(A).

The distance between raceway entries enclosing the same conductor shall not be less than six times the metric designator (trade size) of the larger raceway.

When transposing cable size into raceway size in 314.28(A)(1) and (A)(2), the minimum metric designator (trade size) raceway required for the number and size of conductors in the cable shall be used.

Exhibits 314.10 and 314.11 provide examples of calculations required by 314.28(A)(2). As the example in Exhibit 314.10 illustrates, where splices, angle pulls, or U pulls are made, the distance between each raceway entry inside the box and the opposite wall of the box must not be less than six times the trade diameter of the largest raceway, plus the distance for additional raceway entries. This additional distance is calculated by adding the diameters of the other raceway entries in one row on the same side of the box. The example in Exhibit 314.11 shows that raceway entries enclosing the

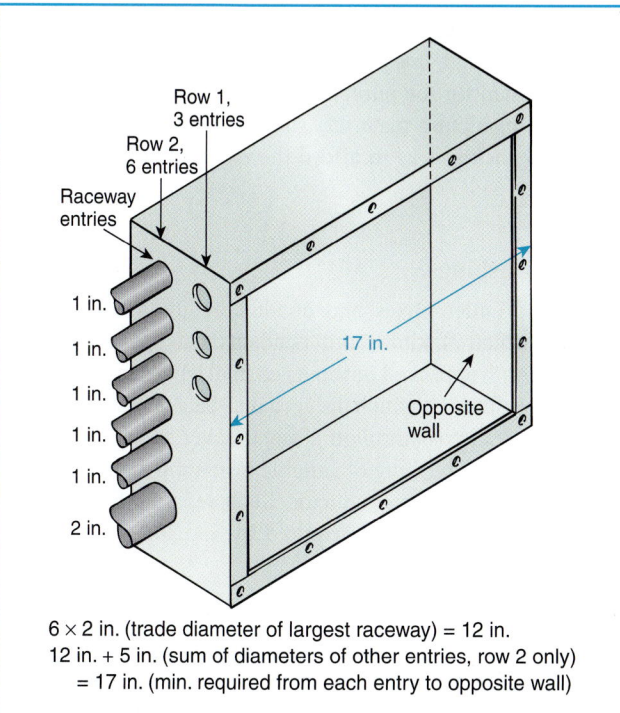

6 × 2 in. (trade diameter of largest raceway) = 12 in.
12 in. + 5 in. (sum of diameters of other entries, row 2 only)
    = 17 in. (min. required from each entry to opposite wall)

Exhibit 314.10 An example showing calculations required by 314.28(A)(2) for splices, angle pulls, or U pulls.

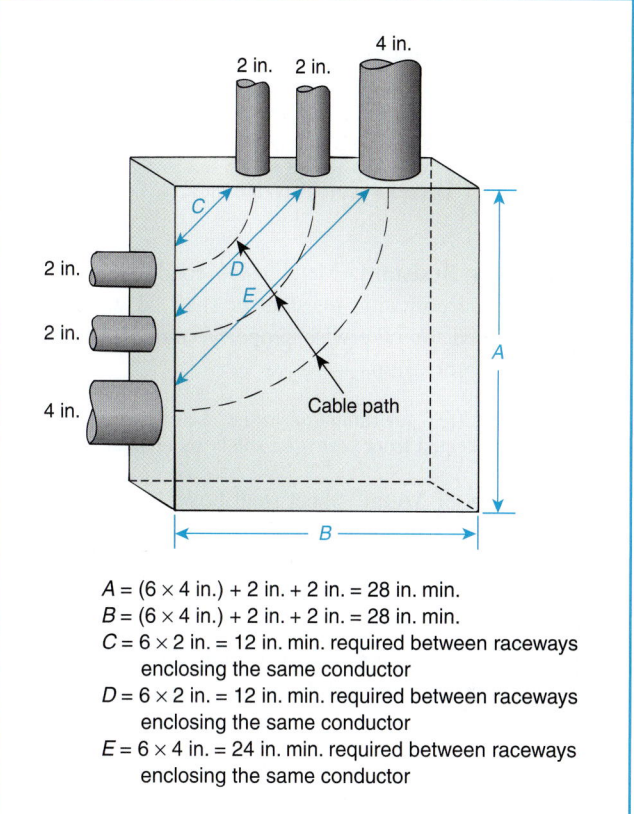

$A = (6 \times 4\ \text{in.}) + 2\ \text{in.} + 2\ \text{in.} = 28\ \text{in. min.}$
$B = (6 \times 4\ \text{in.}) + 2\ \text{in.} + 2\ \text{in.} = 28\ \text{in. min.}$
$C = 6 \times 2\ \text{in.} = 12\ \text{in. min. required between raceways}$
    enclosing the same conductor
$D = 6 \times 2\ \text{in.} = 12\ \text{in. min. required between raceways}$
    enclosing the same conductor
$E = 6 \times 4\ \text{in.} = 24\ \text{in. min. required between raceways}$
    enclosing the same conductor

Exhibit 314.11 An example showing calculations required by 314.28(A)(2) for raceways enclosing the same conductor.

same conductor are required to have a minimum separation between them. The intent is to provide adequate space for the conductor to make the bend.

(3) Smaller Dimensions. Boxes or conduit bodies of dimensions less than those required in 314.28(A)(1) and (A)(2) shall be permitted for installations of combinations of conductors that are less than the maximum conduit or tubing fill (of conduits or tubing being used) permitted by Table 1 of Chapter 9, provided the box or conduit body has been listed for and is permanently marked with the maximum number and maximum size of conductors permitted.

(B) Conductors in Pull or Junction Boxes. In pull boxes or junction boxes having any dimension over 1.8 m (6 ft), all conductors shall be cabled or racked up in an approved manner.

(C) Covers. All pull boxes, junction boxes, and conduit bodies shall be provided with covers compatible with the box or conduit body construction and suitable for the conditions of use. Where metal covers are used, they shall comply with the grounding requirements of 250.110. An extension from the cover of an exposed box shall comply with 314.22, Exception.

(D) Permanent Barriers. Where permanent barriers are installed in a box, each section shall be considered as a separate box.

314.29 Boxes and Conduit Bodies to Be Accessible.

Boxes and conduit bodies shall be installed so that the wiring contained in them can be rendered accessible without removing any part of the building or, in underground circuits, without excavating sidewalks, paving, earth, or other substance that is to be used to establish the finished grade.

Exception: Listed boxes shall be permitted where covered by gravel, light aggregate, or noncohesive granulated soil if their location is effectively identified and accessible for excavation.

Consideration should be given to the accessibility of junction boxes installed on a structural ceiling above a suspended ceiling. A box is permitted to be used at any point for the connection of conduit, tubing, or cable, provided it is not rendered inaccessible. See Article 100 for the definition of *accessible (as applied to wiring methods).* See 300.15 for other requirements for boxes, conduit bodies, or fittings.

III. Construction Specifications

314.40 Metal Boxes, Conduit Bodies, and Fittings.

(A) Corrosion Resistant. Metal boxes, conduit bodies, and fittings shall be corrosion resistant or shall be well-galvanized, enameled, or otherwise properly coated inside and out to prevent corrosion.

> FPN: See 300.6 for limitation in the use of boxes and fittings protected from corrosion solely by enamel.

(B) Thickness of Metal. Sheet steel boxes not over 1650 cm^3 (100 in.^{3}) in size shall be made from steel not less than 1.59 mm (0.0625 in.) thick. The wall of a malleable iron box or conduit body and a die-cast or permanent-mold cast aluminum, brass, bronze, or zinc box or conduit body shall not be less than 2.38 mm ($\frac{3}{32}$ in.) thick. Other cast metal boxes or conduit bodies shall have a wall thickness not less than 3.17 mm ($\frac{1}{8}$ in.).

Exception No. 1: Listed boxes and conduit bodies shown to have equivalent strength and characteristics shall be permitted to be made of thinner or other metals.

Exception No. 2: The walls of listed short radius conduit bodies, as covered in 314.5, shall be permitted to be made of thinner metal.

(C) Metal Boxes Over 1650 cm^3 (100 in.^{3}). Metal boxes over 1650 cm^3 (100 in.^{3}) in size shall be constructed so as to be of ample strength and rigidity. If of sheet steel, the metal thickness shall not be less than 1.35 mm (0.053 in.) uncoated.

(D) Grounding Provisions. A means shall be provided in each metal box for the connection of an equipment grounding conductor. The means shall be permitted to be a tapped hole or equivalent.

For device boxes and other standard outlet boxes, the "means provided" by the box manufacturer is usually in the form of a 10-32 tapped hole marked "GR" or "GRD," or the equivalent, next to the hole. It should be noted, however, that the "means provided" may not necessarily be used.

314.41 Covers.

Metal covers shall be of the same material as the box or conduit body with which they are used, or they shall be lined with firmly attached insulating material that is not less than 0.79 mm ($\frac{1}{32}$ in.) thick, or they shall be listed for the purpose. Metal covers shall be the same thickness as the boxes or conduit body for which they are used, or they shall be listed for the purpose. Covers of porcelain or other approved insulating materials shall be permitted if of such form and thickness as to afford the required protection and strength.

314.42 Bushings.

Covers of outlet boxes and conduit bodies having holes through which flexible cord pendants may pass shall be provided with approved bushings or shall have smooth, well-rounded surfaces on which the cord may bear. Where individual conductors pass through a metal cover, a separate hole equipped with a bushing of suitable insulating material shall be provided for each conductor. Such separate holes shall be connected by a slot as required by 300.20.

314.43 Nonmetallic Boxes.

Provisions for supports or other mounting means for nonmetallic boxes shall be outside of the box, or the box shall be constructed so as to prevent contact between the conductors in the box and the supporting screws.

314.44 Marking.

All boxes and conduit bodies, covers, extension rings, plaster rings, and the like shall be durably and legibly marked with the manufacturer's name or trademark.

IV. Manholes and Other Electric Enclosures Intended for Personnel Entry

314.50 General.

Electric enclosures intended for personnel entry and specifically fabricated for this purpose shall be of sufficient size to provide safe work space about electric equipment with live parts that is likely to require examination, adjustment, servicing, or maintenance while energized. They shall have sufficient size to permit ready installation or withdrawal of the conductors employed without damage to the conductors or to their insulation. They shall comply with the provisions of this part.

Exception: Where electric enclosures covered by Part IV of this article are part of an industrial wiring system operating under conditions of maintenance and supervision that ensure only qualified persons monitor and supervise the system, they shall be permitted to be designed and installed in accordance with appropriate engineering practice. If required by the authority having jurisdiction, design documentation shall be provided.

Part IV of Article 314 covers manholes and other electrical enclosures intended for personnel entry. However, general electrical equipment installation requirements within a manhole or large enclosure are still covered by Article 110.

314.51 Strength.

Manholes, vaults, and their means of access shall be designed under qualified engineering supervision and shall withstand all loads likely to be imposed on the structures.

> FPN: See ANSI C2-1997, *National Electrical Safety Code*, for additional information on the loading that can be expected to bear on underground enclosures.

314.52 Cabling Work Space.

A clear work space not less than 900 mm (3 ft) wide shall be provided where cables are located on both sides, and not less than 750 mm (2½ ft) where cables are only on one side. The vertical headroom shall not be less than 1.8 m (6 ft) unless the opening is within 300 mm (1 ft), measured horizontally, of the adjacent interior side wall of the enclosure.

Exception: A manhole containing only one or more of the following shall be permitted to have one of the horizontal work space dimensions reduced to 600 mm (2 ft) where the other horizontal clear work space is increased so the sum of the two dimensions is not less than 1.8 m (6 ft):

(a) Optical fiber cables as covered in Article 770.
(b) Power-limited fire alarm circuits supplied in accordance with 760.41.
(c) Class 2 or Class 3 remote-control and signaling circuits, or both, supplied in accordance with 725.41.

314.53 Equipment Work Space.

Where electric equipment with live parts that is likely to require examination, adjustment, servicing, or maintenance while energized is installed in a manhole, vault, or other enclosure designed for personnel access, the work space and associated requirements in 110.26 shall be met for installations operating at 600 volts or less. Where the installation is over 600 volts, the work space and associated requirements in 110.34 shall be met. A manhole access cover that weighs over 45 kg (100 lb) shall be considered as meeting the requirements of 110.34(C).

314.54 Bending Space for Conductors.

Bending space for conductors operating at 600 volts or below shall be provided in accordance with the requirements of 314.28(A). Conductors operating over 600 volts shall be provided with bending space in accordance with 314.71(A) and 314.71(B), as applicable. All conductors shall be cabled, racked up, or arranged in an approved manner that provides

ready and safe access for persons to enter for installation and maintenance.

Exception: Where 314.71(B) applies, each row or column of ducts on one wall of the enclosure shall be calculated individually, and the single row or column that provides the maximum distance shall be used.

314.55 Access to Manholes.

(A) Dimensions. Rectangular access openings shall not be less than 650 mm × 550 mm (26 in. × 22 in.). Round access openings in a manhole shall not be less than 650 mm (26 in.) in diameter.

Exception: A manhole that has a fixed ladder that does not obstruct the opening or that contains only one or more of the following shall be permitted to reduce the minimum cover diameter to 600 mm (2 ft):

(a) Optical fiber cables as covered in Article 770.
(b) Power-limited fire alarm circuits supplied in accordance with 760.41.
(c) Class 2 or Class 3 remote-control and signaling circuits, or both, supplied in accordance with 725.41.

(B) Obstructions. Manhole openings shall be free of protrusions that could injure personnel or prevent ready egress.

(C) Location. Manhole openings for personnel shall be located where they are not directly above electric equipment or conductors in the enclosure. Where this is not practicable, either a protective barrier or a fixed ladder shall be provided.

(D) Covers. Covers shall be over 45 kg (100 lb) or otherwise designed to require the use of tools to open. They shall be designed or restrained so they cannot fall into the manhole or protrude sufficiently to contact electrical conductors or equipment within the manhole.

(E) Marking. Manhole covers shall have an identifying mark or logo that prominently indicates their function, such as "electric."

314.56 Access to Vaults and Tunnels.

(A) Location. Access openings for personnel shall be located where they are not directly above electric equipment or conductors in the enclosure. Other openings shall be permitted over equipment to facilitate installation, maintenance, or replacement of equipment.

(B) Locks. In addition to compliance with the requirements of 110.34(C), if applicable, access openings for personnel shall be arranged so that a person on the inside can exit when the access door is locked from the outside, or in the case of normally locking by padlock, the locking arrangement shall be such that the padlock can be closed on the locking system to prevent locking from the outside.

314.57 Ventilation.

Where manholes, tunnels, and vaults have communicating openings into enclosed areas used by the public, ventilation to open air shall be provided wherever practicable.

314.58 Guarding.

Where conductors or equipment, or both, could be contacted by objects falling or being pushed through a ventilating grating, both conductors and live parts shall be protected in accordance with the requirements of 110.27(A)(2) or 110.31(B)(1), depending on the voltage.

314.59 Fixed Ladders.

Fixed ladders shall be corrosion resistant.

V. Pull and Junction Boxes for Use on Systems Over 600 Volts, Nominal

314.70 General.

Where pull and junction boxes are used on systems over 600 volts, the installation shall comply with the provisions of Part V and also with the following general provisions of this article:

(1) In Part I, 314.2, 314.3, and 314.4
(2) In Part II, 314.15; 314.17; 314.20; 314.23(A), (B), or (G); 314.28(B); and 314.29
(3) In Part III, 314.40(A) and (C) and 314.41

314.71 Size of Pull and Junction Boxes.

Pull and junction boxes shall provide adequate space and dimensions for the installation of conductors, and they shall comply with the specific requirements of this section.

Exception: Terminal housings supplied with motors shall comply with the provisions of 430.12.

(A) For Straight Pulls. The length of the box shall not be less than 48 times the outside diameter, over sheath, of the largest shielded or lead-covered conductor or cable entering the box. The length shall not be less than 32 times the outside diameter of the largest nonshielded conductor or cable.

(B) For Angle or U Pulls.

(1) Distance to Opposite Wall. The distance between each cable or conductor entry inside the box and the opposite wall of the box shall not be less than 36 times the outside diameter, over sheath, of the largest cable or conductor. This distance shall be increased for additional entries by the amount of the sum of the outside diameters, over sheath, of

all other cables or conductor entries through the same wall of the box.

Exception No. 1: Where a conductor or cable entry is in the wall of a box opposite a removable cover, the distance from that wall to the cover shall be permitted to be not less than the bending radius for the conductors as provided in 300.34.

Exception No. 2: Where cables are nonshielded and not lead covered, the distance of 36 times the outside diameter shall be permitted to be reduced to 24 times the outside diameter.

(2) Distance Between Entry and Exit. The distance between a cable or conductor entry and its exit from the box shall not be less than 36 times the outside diameter, over sheath, of that cable or conductor.

Exception: Where cables are nonshielded and not lead covered, the distance of 36 times the outside diameter shall be permitted to be reduced to 24 times the outside diameter.

(C) Removable Sides. One or more sides of any pull box shall be removable.

314.72 Construction and Installation Requirements.

(A) Corrosion Protection. Boxes shall be made of material inherently resistant to corrosion or shall be suitably protected, both internally and externally, by enameling, galvanizing, plating, or other means.

(B) Passing Through Partitions. Suitable bushings, shields, or fittings having smooth, rounded edges shall be provided where conductors or cables pass through partitions and at other locations where necessary.

(C) Complete Enclosure. Boxes shall provide a complete enclosure for the contained conductors or cables.

(D) Wiring Is Accessible. Boxes shall be installed so that the wiring is accessible without removing any part of the building. Working space shall be provided in accordance with 110.34.

(E) Suitable Covers. Boxes shall be closed by suitable covers securely fastened in place. Underground box covers that weigh over 45 kg (100 lb) shall be considered meeting this requirement. Covers for boxes shall be permanently marked "DANGER—HIGH VOLTAGE—KEEP OUT." The marking shall be on the outside of the box cover and shall be readily visible. Letters shall be block type and at least 13 mm (½ in.) in height.

(F) Suitable for Expected Handling. Boxes and their covers shall be capable of withstanding the handling to which they may likely be subjected.

ARTICLE 320
Armored Cable: Type AC

Contents

I. General

320.1 Scope.

This article covers the use, installation, and construction specifications for armored cable, Type AC.

Type AC cable (armored cable) is listed by Underwriters Laboratories in sizes 14 AWG through 1 AWG copper and 12 AWG through 1 AWG aluminum or copper-clad aluminum and is rated at 600 volts or less. Exhibit 320.1 shows an example of Type AC cable.

320.2 Definition.

Armored Cable, Type AC. A fabricated assembly of insulated conductors in a flexible metallic enclosure. See 320.100.

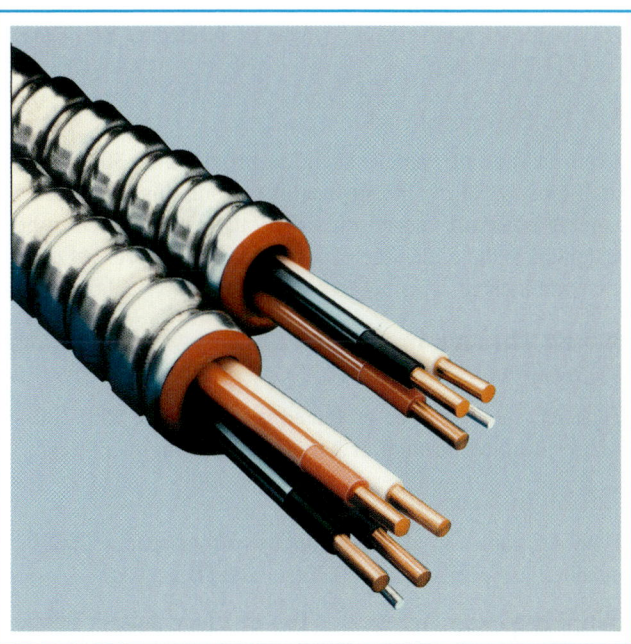

Exhibit 320.1 An example of Type AC cable. (Courtesy of AFC Cable Systems, Inc.)

II. Installation

320.10 Uses Permitted.

Where not subject to physical damage, Type AC cable shall be permitted as follows:

(1) In both exposed and concealed work
(2) In cable trays where identified for such use
(3) In dry locations
(4) Embedded in plaster finish on brick or other masonry, except in damp or wet locations
(5) To be run or fished in the air voids of masonry block or tile walls where such walls are not exposed or subject to excessive moisture or dampness

320.12 Uses Not Permitted.

Type AC cable shall not be used as follows:

(1) In theaters and similar locations, except where permitted in 518.4
(2) In motion picture studios
(3) In hazardous (classified) locations except where permitted in
    a. 501.4(B), Exception
    b. 502.4(B), Exception No. 1
    c. 504.20
(4) Where exposed to corrosive fumes or vapors
(5) In storage battery rooms
(6) In hoistways, or on elevators or escalators, except where permitted in 620.21

(7) In commercial garages where prohibited in 511.4 and 511.7

320.15 Exposed Work.

Exposed runs of cable, except as provided in 300.11(A), shall closely follow the surface of the building finish or of running boards. Exposed runs shall also be permitted to be installed on the underside of joists where supported at each joist and located so as not to be subject to physical damage.

320.17 Through or Parallel to Framing Members.

Type AC cable shall be protected in accordance with 300.4 where installed through or parallel to framing members.

320.23 In Accessible Attics.

Type AC cables in accessible attics or roof spaces shall be installed as specified in 320.23(A) and (B).

(A) Where Run Across the Top of Floor Joists. Where run across the top of floor joists, or within 2.1 m (7 ft) of floor or floor joists across the face of rafters or studding, in attics and roof spaces that are accessible, the cable shall be protected by substantial guard strips that are at least as high as the cable. Where this space is not accessible by permanent stairs or ladders, protection shall only be required within 1.8 m (6 ft) of the nearest edge of the scuttle hole or attic entrance.

In accessible attics, Type AC cable installed across the top of floor joists or within 7 ft of the floor or floor joists across the face of rafters or studs must be protected by guard strips. Where the attic is not accessible by a permanent ladder or stairs, guard strips are required only within 6 ft of the scuttle hole or opening.

(B) Cable Installed Parallel to Framing Members. Where the cable is installed parallel to the sides of rafters, studs, or floor joists, neither guard strips nor running boards shall be required, and the installation shall also comply with 300.4(D).

320.24 Bending Radius.

Bends in Type AC cable shall be made so that the cable will not be damaged. The radius of the curve of the inner edge of any bend shall not be less than five times the diameter of the Type AC cable.

320.30 Securing and Supporting.

Type AC cable shall be secured by staples, cable ties, straps, hangers, or similar fittings designed and installed so as not to damage the cable at intervals not exceeding 1.4 m (4 ft) and within 300 mm (12 in.) of every outlet box, junction box, cabinet, or fitting.

Section 320.30 requires that Type AC cable be secured. Simply draping the cable over air ducts or lower members of bar joists, pipes, and ceiling grid members is not permitted.

(A) Horizontal Runs Through Holes and Notches. In other than vertical runs, cables installed in accordance with 300.4 shall be considered supported and secured where such support does not exceed 1.4-m (4½-ft) intervals and the armored cable is securely fastened in place by an approved means within 300 mm (12 in.) of each box, cabinet, conduit body, or other armored cable termination.

Section 320.30(A) permits Type AC cable, where run horizontally through framing members, to be passed through bored or punched holes in framing members without additional securing, provided the cable is secured within 12 in. of the outlet and the framing members are less than 54 in. apart.

(B) Unsupported Cables. Type AC cable shall be permitted to be unsupported where the cable:

(1) Is fished between access points, where concealed in finished buildings or structures and supporting is impracticable; or
(2) Is not more than 600 mm (2 ft) in length at terminals where flexibility is necessary; or
(3) Is not more than 1.8 m (6 ft) from the last point of support for connections within an accessible ceiling to luminaire(s) [(lighting fixture(s)] or equipment.

(C) Cable Trays. Type AC cable installed in cable trays shall comply with 392.8(B).

320.40 Boxes and Fittings.

At all points where the armor of AC cable terminates, a fitting shall be provided to protect wires from abrasion, unless the design of the outlet boxes or fittings is such as to afford equivalent protection, and, in addition, an insulating bushing or its equivalent protection shall be provided between the conductors and the armor. The connector or clamp by which the Type AC cable is fastened to boxes or cabinets shall be of such design that the insulating bushing or its equivalent will be visible for inspection. Where change is made from Type AC cable to other cable or raceway wiring methods, a box, fitting, or conduit body shall be installed at junction points as required in 300.15.

Armored cable connectors are considered suitable for equipment grounding if installed in accordance with 300.10.

320.80 Ampacity.

The ampacity shall be determined by 310.15.

(A) Thermal Insulation. Armored cable installed in thermal insulation shall have conductors rated at 90°C (194°F). The ampacity of cable installed in these applications shall be that of 60°C (140°F) conductors.

The requirements for armored cable installed in thermal insulation recognize the decrease in heat dissipation capability of cables. Cable marked "ACTH" indicates an armored cable rated 75°C and employing conductors having thermoplastic insulation. Cable marked "ACTHH" indicates an armored cable rated 90°C and employing conductors having thermoplastic insulation. Cable marked "ACHH" indicates armored cable rated 90°C and employing conductors having thermosetting insulation.

(B) Cable Tray. The ampacity of Type AC cable installed in cable tray shall be determined in accordance with 392.11.

III. Construction Specifications

320.100 Construction.

Type AC cable shall have an armor of flexible metal tape and shall have an internal bonding strip of copper or aluminum in intimate contact with the armor for its entire length.

The armor of Type AC cable is recognized as an equipment grounding conductor by 250.118. The required internal bonding strip can be simply cut off at the termination of the armored cable, or it can be bent back on the armor. It is not necessary to connect it to an equipment grounding terminal. It reduces the inductive reactance of the spiral armor and increases the armor's effectiveness as an equipment ground. Many installers use this strip to help prevent the insulating bushing required by 320.40 (the "red head") from falling out during rough wiring.

320.104 Conductors.

Insulated conductors shall be of a type listed in Table 310.13 or those identified for use in this cable. In addition, the conductors shall have an overall moisture-resistant and fire-retardant fibrous covering. For Type ACT, a moisture-resistant fibrous covering shall be required only on the individual conductors.

320.108 Equipment Grounding.

Type AC cable shall provide an adequate path for equipment grounding as required by 250.4(A)(5) or 250.4(B)(4).

320.120 Marking.

The cable shall be marked in accordance with 310.11, except that Type AC shall have ready identification of the manufac-

turer by distinctive external markings on the cable sheath throughout its entire length.

ARTICLE 322
Flat Cable Assemblies: Type FC

Contents

I. General

322.1 Scope.

This article covers the use, installation, and construction specifications for flat cable assemblies, Type FC.

Type FC (flat) cable is an assembly of three or four parallel 10 AWG special stranded copper wires formed integrally with an insulating material web. The cable is marked with the size of the maximum branch circuit to which it may be connected, the cable type designation, manufacturer's identification, maximum working voltage, conductor size, and temperature rating. A marking accompanying the cable on a tag or reel indicates the special metal raceways and specific FC cable fittings with which the cable is intended to be used. Exhibits 322.1 and 322.2 show the basic components of this wiring method.

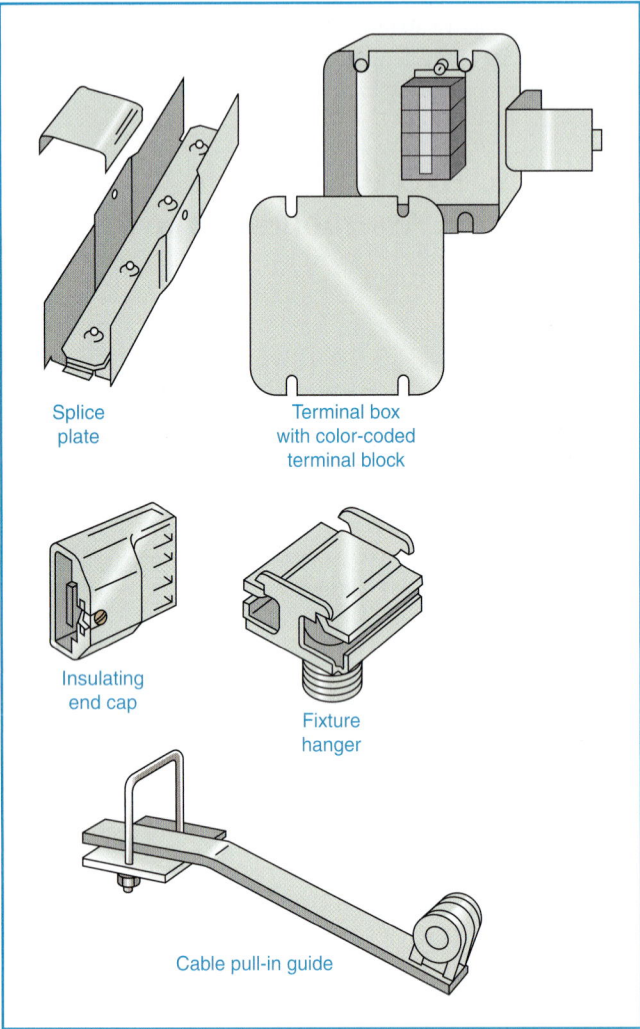

Splice
plate

Terminal box
with color-coded
terminal block

Insulating
end cap

Fixture
hanger

Cable pull-in guide

Exhibit 322.1 Basic components and accessories used for an installation of Type FC cable assembly. (Redrawn from The Wiremold Co.)

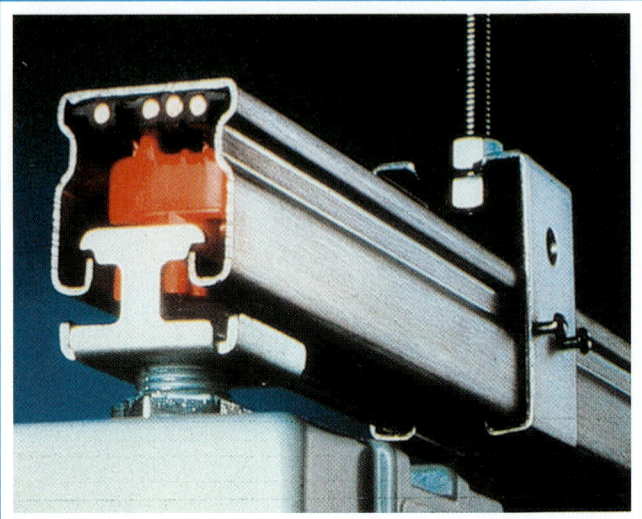

Exhibit 322.2 A fixture hanger used with Type FC cable assembly. (Courtesy of The Wiremold Co.)

322.2 Definition.

Flat Cable Assembly, Type FC. An assembly of parallel conductors formed integrally with an insulating material web specifically designed for field installation in surface metal raceway.

II. Installation

322.10 Uses Permitted.

Flat cable assemblies shall be permitted only as follows:

(1) As branch circuits to supply suitable tap devices for lighting, small appliances, or small power loads. The rating of the branch circuit shall not exceed 30 amperes.
(2) Where installed for exposed work.

(3) In locations where they will not be subjected to physical damage. Where a flat cable assembly is installed less than 2.5 m (8 ft) above the floor or fixed working platform, it shall be protected by a cover identified for the use.
(4) In surface metal raceways identified for the use. The channel portion of the surface metal raceway systems shall be installed as complete systems before the flat cable assemblies are pulled into the raceways.

322.12 Uses Not Permitted.

Flat cable assemblies shall not be used as follows:

(1) Where subject to corrosive vapors unless suitable for the application
(2) In hoistways or on elevators or escalators
(3) In any hazardous (classified) location
(4) Outdoors or in wet or damp locations unless identified for the use

322.30 Securing and Supporting.

The flat cable assemblies shall be supported by means of their special design features, within the surface metal raceways.

The surface metal raceways shall be supported as required for the specific raceway to be installed.

322.40 Boxes and Fittings.

(A) Dead Ends. Each flat cable assembly dead end shall be terminated in an end-cap device identified for the use.

The dead-end fitting for the enclosing surface metal raceway shall be identified for the use.

(B) Luminaire (Fixture) Hangers. Luminaire (fixture) hangers installed with the flat cable assemblies shall be identified for the use.

(C) Fittings. Fittings to be installed with flat cable assemblies shall be designed and installed to prevent physical damage to the cable assemblies.

(D) Extensions. All extensions from flat cable assemblies shall be made by approved wiring methods, within the junction boxes, installed at either end of the flat cable assembly runs.

322.56 Splices and Taps.

(A) Splices. Splices shall be made in listed junction boxes.

(B) Taps. Taps shall be made between any phase conductor and the grounded conductor or any other phase conductor by means of devices and fittings identified for the use. Tap devices shall be rated at not less than 15 amperes, or more than 300 volts to ground, and they shall be color-coded in accordance with the requirements of 322.120(C).

III. Construction

322.100 Construction.

Flat cable assemblies shall consist of two, three, four, or five conductors.

322.104 Conductors.

Flat cable assemblies shall have conductors of 10 AWG special stranded copper wires.

322.112 Insulation.

The entire flat cable assembly shall be formed to provide a suitable insulation covering all the conductors and using one of the materials recognized in Table 310.13 for general branch-circuit wiring.

322.120 Marking.

(A) Temperature Rating. In addition to the provisions of 310.11, Type FC cable shall have the temperature rating durably marked on the surface at intervals not exceeding 600 mm (24 in.).

(B) Identification of Grounded Conductor. The grounded conductor shall be identified throughout its length by means of a distinctive and durable white or gray marking.

> FPN: The color gray may have been used in the past as an ungrounded conductor. Care should be taken when working on existing systems.

(C) Terminal Block Identification. Terminal blocks identified for the use shall have distinctive and durable markings

for color or word coding. The grounded conductor section shall have a white marking or other suitable designation. The next adjacent section of the terminal block shall have a black marking or other suitable designation. The next section shall have a red marking or other suitable designation. The final or outer section, opposite the grounded conductor section of the terminal block, shall have a blue marking or other suitable designation.

ARTICLE 324
Flat Conductor Cable: Type FCC

Contents

I. General

324.1 Scope.

This article covers a field-installed wiring system for branch circuits incorporating Type FCC cable and associated accessories as defined by the article. The wiring system is designed for installation under carpet squares.

The FCC (flat conductor cable) system is designed to provide a completely accessible, flexible power system. As shown in Exhibit 324.1, it also provides an easy method for reworking obsolete wiring systems currently in use in many office facilities. The carpet squares are not permitted to be larger than 36 in. by 36 in. to comply with 324.10(H). This limitation is judged necessary to provide ready access to the cable by lifting a carpet square. It also reduces the likelihood of damage to the cable by an individual cutting through the carpet above the cable with a knife or razor blade and possibly penetrating the top shield of the cable in the process.

Exhibit 324.1 Type FCC cable installed beneath carpet squares. (Courtesy of Tyco Electronics Corp.)

324.2 Definitions.

Bottom Shield. A protective layer that is installed between the floor and Type FCC flat conductor cable to protect the cable from physical damage and may or may not be incorporated as an integral part of the cable.

Cable Connector. A connector designed to join Type FCC cables without using a junction box.

FCC System. A complete wiring system for branch circuits that is designed for installation under carpet squares. The FCC system includes Type FCC cable and associated shielding, connectors, terminators, adapters, boxes, and receptacles.

Insulating End. An insulator designed to electrically insulate the end of a Type FCC cable.

Metal Shield Connections. Means of connection designed to electrically and mechanically connect a metal shield to another metal shield, to a receptacle housing or self-contained device, or to a transition assembly.

Top Shield. A grounded metal shield covering under-carpet components of the FCC system for the purposes of providing protection against physical damage.

Transition Assembly. An assembly to facilitate connection of the FCC system to other wiring systems, incorporating (1) a means of electrical interconnection and (2) a suitable box or covering for providing electrical safety and protection against physical damage.

Type FCC Cable. Three or more flat copper conductors placed edge-to-edge and separated and enclosed within an insulating assembly.

II. Installation

324.10 Uses Permitted.

(A) Branch Circuits. Use of FCC systems shall be permitted both for general-purpose and appliance branch circuits and for individual branch circuits.

(B) Branch-Circuit Ratings.

(1) Voltage. Voltage between ungrounded conductors shall not exceed 300 volts. Voltage between ungrounded conductors and the grounded conductor shall not exceed 150 volts.

(2) Current. General-purpose and appliance branch circuits shall have ratings not exceeding 20 amperes. Individual branch circuits shall have ratings not exceeding 30 amperes.

(C) Floors. Use of FCC systems shall be permitted on hard, sound, smooth, continuous floor surfaces made of concrete, ceramic, or composition flooring, wood, and similar materials.

(D) Walls. Use of FCC systems shall be permitted on wall surfaces in surface metal raceways.

(E) Damp Locations. Use of FCC systems in damp locations shall be permitted.

(F) Heated Floors. Materials used for floors heated in excess of 30°C (86°F) shall be identified as suitable for use at these temperatures.

(G) System Height. Any portion of an FCC system with a height above floor level exceeding 2.3 mm (0.090 in.) shall be tapered or feathered at the edges to floor level.

(H) Coverings. Floor-mounted Type FCC cable, cable connectors, and insulating ends shall be covered with carpet squares not larger than 914 mm (36 in.) square. Those carpet squares that are adhered to the floor shall be attached with release-type adhesives.

(I) Corrosion Resistance. Metal components of the system shall be either corrosion resistant, coated with corrosion-resistant materials, or insulated from contact with corrosive substances.

(J) Metal-Shield Connectors. Metal shields shall be connected to each other and to boxes, receptacle housings, self-contained devices, and transition assemblies using metal-shield connectors.

324.12 Uses Not Permitted.

FCC systems shall not be used:

(1) Outdoors or in wet locations
(2) Where subject to corrosive vapors
(3) In any hazardous (classified) location
(4) In residential, school, and hospital buildings

Section 324.12(4) prohibits Type FCC wiring systems throughout school and hospital buildings, even though parts of these buildings may be office or administrative spaces.

324.18 Crossings.

Crossings of more than two Type FCC cable runs shall not be permitted at any one point. Crossings of a Type FCC cable over or under a flat communications or signal cable shall be permitted. In each case, a grounded layer of metal shielding shall separate the two cables, and crossings of more than two flat cables shall not be permitted at any one point.

324.30 Securing and Supporting.

All FCC system components shall be firmly anchored to the floor or wall using an adhesive or mechanical anchoring system identified for this use. Floors shall be prepared to ensure adherence of the FCC system to the floor until the carpet squares are placed.

324.40 Boxes and Fittings.

(A) Cable Connections and Insulating Ends. All Type FCC cable connections shall use connectors identified for their use, installed such that electrical continuity, insulation, and sealing against dampness and liquid spillage are provided. All bare cable ends shall be insulated and sealed against dampness and liquid spillage using listed insulating ends.

(B) Polarization of Connections. All receptacles and connections shall be constructed and installed so as to maintain proper polarization of the system.

(C) Shields.

(1) Top Shield. A metal top shield shall be installed over all floor-mounted Type FCC cable, connectors, and insulating ends. The top shield shall completely cover all cable runs, corners, connectors, and ends.

(2) Bottom Shield. A bottom shield shall be installed beneath all Type FCC cable, connectors, and insulating ends.

(D) Connection to Other Systems. Power feed, grounding connection, and shield system connection between the FCC system and other wiring systems shall be accomplished in a transition assembly identified for this use.

324.42 Devices.

(A) Receptacles. All receptacles, receptacle housings, and self-contained devices used with the FCC system shall be identified for this use and shall be connected to the Type FCC cable and metal shields. Connection from any grounding conductor of the Type FCC cable shall be made to the shield system at each receptacle.

(B) Receptacles and Housings. Receptacle housings and self-contained devices designed either for floor mounting or for in-wall or on-wall mounting shall be permitted for use with the FCC system. Receptacle housings and self-contained devices shall incorporate means for facilitating entry and termination of Type FCC cable and for electrically connecting the housing or device with the metal shield. Receptacles and self-contained devices shall comply with 406.3. Power and communications outlets installed together in common housing shall be permitted in accordance with 800.52(A)(1)(c), Exception No. 2.

324.56 Splices and Taps.

(A) FCC Systems Alterations. Alterations to FCC systems shall be permitted. New cable connectors shall be used at new connection points to make alterations. It shall be permitted to leave unused cable runs and associated cable connectors in place and energized. All cable ends shall be covered with insulating ends.

(B) Transition Assemblies. All transition assemblies shall be identified for their use. Each assembly shall incorporate means for facilitating entry of the Type FCC cable into the assembly, for connecting the Type FCC cable to grounded conductors, and for electrically connecting the assembly to the metal cable shields and to equipment grounding conductors.

324.60 Grounding.

All metal shields, boxes, receptacle housings, and self-contained devices shall be electrically continuous to the equipment grounding conductor of the supplying branch circuit. All such electrical connections shall be made with connectors identified for this use. The electrical resistivity of such shield system shall not be more than that of one conductor of the Type FCC cable used in the installation.

III. Construction

324.100 Construction.

(A) Type FCC Cable. Type FCC cable shall be listed for use with the FCC system and shall consist of three, four, or five flat copper conductors, one of which shall be an equipment grounding conductor.

Section 324.100 requires all FCC cables to be listed. There was no listing requirement for FCC cable prior to the 1996 *Code*.

(B) Shields.

(1) Materials and Dimensions. All top and bottom shields shall be of designs and materials identified for their use. Top shields shall be metal. Both metallic and nonmetallic materials shall be permitted for bottom shields.

(2) Resistivity. Metal shields shall have cross-sectional areas that provide for electrical resistivity of not more than that of one conductor of the Type FCC cable used in the installation.

324.112 Insulation.

The insulating material of the cable shall be moisture resistant and flame retardant. All insulating materials in the FCC systems shall be identified for their use.

324.120 Markings.

(A) Cable Marking. Type FCC cable shall be clearly and durably marked on both sides at intervals of not more than 610 mm (24 in.) with the information required by 310.11(A) and with the following additional information:

(1) Material of conductors
(2) Maximum temperature rating
(3) Ampacity

(B) Conductor Identification. Conductors shall be clearly and durably identified on both sides throughout their length as specified in 310.12.

ARTICLE 326
Integrated Gas Spacer Cable: Type IGS

Contents

I. General

326.1 Scope.

This article covers the use, installation, and construction specifications for integrated gas spacer cable, Type IGS.

As illustrated in Exhibit 326.1, Type IGS (integrated gas spacer) cable consists of solid aluminum rod conductors, 250-kcmil minimum size. These conductors are insulated with dry kraft paper and are factory installed in a medium-density polyethylene gas pipe, minimum trade size 2, which is then filled with sulfur hexafluoride (SF_6) gas at a pressure of approximately 20 psi.

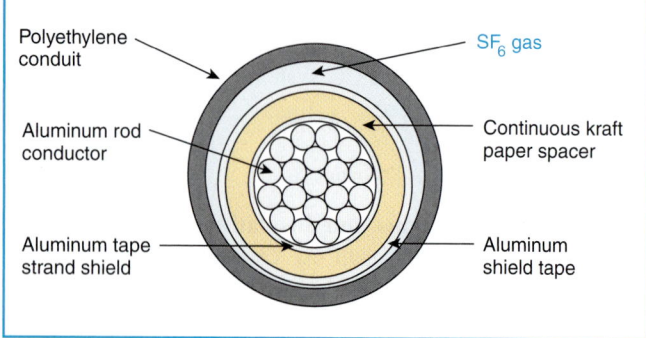

Exhibit 326.1 Cross section of single-conductor, 4750-kcmil Type IGS cable.

326.2 Definition.

Integrated Gas Spacer Cable, Type IGS. A factory assembly of one or more conductors, each individually insulated and enclosed in a loose fit, nonmetallic flexible conduit as an integrated gas spacer cable rated 0 through 600 volts.

II. Installation

326.10 Uses Permitted.

Type IGS cable shall be permitted for use under ground, including direct burial in the earth, as the following:

(1) Service-entrance conductors
(2) Feeder or branch-circuit conductors

326.12 Uses Not Permitted.

Type IGS cable shall not be used as interior wiring or be exposed in contact with buildings.

326.24 Bending Radius.

Where the coilable nonmetallic conduit and cable is bent for installation purposes or is flexed or bent during shipment or installation, the radii of bends measured to the inside of the bend shall not be less than specified in Table 326.24.

Table 326.24 Minimum Radii of Bends

Conduit Size		Minimum Radii	
Metric Designator	Trade Size	mm	in.
53	2	600	24
78	3	900	35
103	4	1150	45

326.26 Bends.

A run of Type IGS cable between pull boxes or terminations shall not contain more than the equivalent of four quarter bends (360 degrees total), including those bends located immediately at the pull box or terminations.

326.40 Fittings.

Terminations and splices for Type IGS cable shall be identified as a type that is suitable for maintaining the gas pressure within the conduit. A valve and cap shall be provided for each length of the cable and conduit to check the gas pressure or to inject gas into the conduit.

326.80 Ampacity.

The ampacity of Type IGS cable shall not exceed the values shown in Table 326.80.

Table 326.80 Ampacity of Type IGS Cable

Size (kcmil)	Amperes	Size (kcmil)	Amperes
250	119	2500	376
500	168	3000	412
750	206	3250	429
1000	238	3500	445
1250	266	3750	461
1500	292	4000	476
1750	344	4250	491
2000	336	4500	505
2250	357	4750	519

III. Construction Specifications

326.104 Conductors.

The conductors shall be solid aluminum rods, laid parallel, consisting of one to nineteen 12.7 mm (½ in.) diameter rods. The minimum conductor size shall be 250 kcmil, and the maximum size shall be 4750 kcmil.

326.112 Insulation.

The insulation shall be dry kraft paper tapes and a pressurized sulfur hexafluoride gas (SF6), both approved for electrical use. The nominal gas pressure shall be 138 kPa gauge (20 pounds per square inch gauge). The thickness of the paper spacer shall be as specified in Table 326.112.

Table 326.112 Paper Spacer Thickness

Size (kcmil)	Thickness	
	mm	in.
250–1000	1.02	0.040
1250–4750	1.52	0.060

326.116 Conduit.

The conduit shall be a medium density polyethylene identified as suitable for use with natural gas rated pipe in metric designator 53, 78, or 103 (trade size 2, 3, or 4). The percent fill dimensions for the conduit are shown in Table 326.116.

Table 326.116 Conduit Dimensions

Conduit Size		Actual Outside Diameter		Actual Inside Diameter	
Metric Designator	Trade Size	mm	in.	mm	in.
53	2	60	2.375	49.46	1.947
78	3	89	3.500	73.30	2.886
103	4	114	4.500	94.23	3.710

The size of the conduit permitted for each conductor size shall be calculated for a percent fill not to exceed those found in Table 1, Chapter 9.

326.120 Marking.

The cable shall be marked in accordance with 310.11(A), 310.11(B)(1), and 310.11(D).

ARTICLE 328
Medium Voltage Cable: Type MV

Contents

I. General
    328.1 Scope
    328.2 Definition
II. Installation
    328.10 Uses Permitted
    328.12 Uses Not Permitted
    328.80 Ampacity
III. Construction Specifications
    328.100 Construction
    328.120 Marking

I. General

328.1 Scope.

This article covers the use, installation, and construction specifications for medium voltage cable, Type MV.

Type MV (medium voltage) cables are rated 2001 to 35,000 volts. Cables rated 2001 to 8000 volts may be shielded or nonshielded. All insulated conductors 8001 volts and higher have electrostatic shielding. When nonshielded cables are used, they must comply with the exception to 310.6. If these cables are installed in underground conduits, they must be encased in 3 in. of concrete to comply with 300.50(A)(2).

328.2 Definition.

Medium Voltage Cable, Type MV. A single or multiconductor solid dielectric insulated cable rated 2001 volts or higher.

II. Installation

328.10 Uses Permitted.

Type MV cables shall be permitted for use on power systems rated up to 35,000 volts, nominal, as follows:

(1) In wet or dry locations
(2) In raceways

(3) In cable trays as specified in 392.3(B)(1)
(4) Direct buried in accordance with 300.50
(5) In messenger-supported wiring

328.12 Uses Not Permitted.

Type MV cable shall not be used unless identified for the use as follows:

Cables intended for installation in cable trays in accordance with Article 392 are marked "For CT Use" or "For Use in Cable Trays."

(1) Where exposed to direct sunlight
(2) In cable trays

328.80 Ampacity.

The ampacity of Type MV cable shall be determined in accordance with 310.60. The ampacity of Type MV cable installed in cable tray shall be determined in accordance with 392.13.

III. Construction Specifications

328.100 Construction.

Type MV cables shall have copper, aluminum, or copper-clad aluminum conductors and shall be constructed in accordance with Article 310.

Cables with aluminum conductors are marked with the word *Aluminum* or the letters *AL*.

328.120 Marking.

Medium voltage cable shall be marked as required in 310.11.

Cables are marked with their conductor size, voltage rating, and insulation level (100 percent or 133 percent).

ARTICLE 330
Metal-Clad Cable: Type MC

Contents

I. General
    330.1 Scope
    330.2 Definition
II. Installation
    330.10 Uses Permitted
      (A) General Uses
      (B) Specific Uses

I. General

330.1 Scope.

This article covers the use, installation, and construction specifications of metal-clad cable, Type MC.

The basic standard to investigate products in this category is UL 1569, *Metal-Clad Cable*. Summary information regarding listed metal-clad cable may be found in the UL *General Information Directory* under category PJAZ.

330.2 Definition.

Metal Clad Cable, Type MC. A factory assembly of one or more insulated circuit conductors with or without optical fiber members enclosed in an armor of interlocking metal tape, or a smooth or corrugated metallic sheath.

Type MC cable is rated for use up to 2000 volts and listed in sizes 18 AWG and larger for copper and 12 AWG and larger for aluminum or copper-clad aluminum, and it employs thermoset or thermoplastic insulated conductors. Composite electrical MC and optical fiber cables are permitted by 770.5(C) and are marked "MC-OF." See Exhibit 770.2 for an example.

Type MC cable must be marked with the maximum rated voltage, the proper insulation type letter or letters, and the AWG size or circular mil area. This marking may be on a marker tape located within the cable and running its complete length, or, if the metallic covering is of smooth construction that permits surface marking, the MC cable may be durably marked on the outer covering at intervals not exceeding 24 in. For MC cable with an outer nonmetallic covering, the marking is permitted on the surface of the nonmetallic jacket. See 310.11 for marking requirements.

Type MC (metal-clad) cable is available in three designs: interlocked metal tape, corrugated metal tube, and smooth metal tube. A nonmetallic jacket may be provided over the metal sheath. Cables suitable for use in cable trays, direct sunlight, or direct burial applications are so marked. Type MC cable includes Type CS (copper sheath) and Type ALS (aluminum sheath). Exhibit 330.1 shows some examples of Type MC cable.

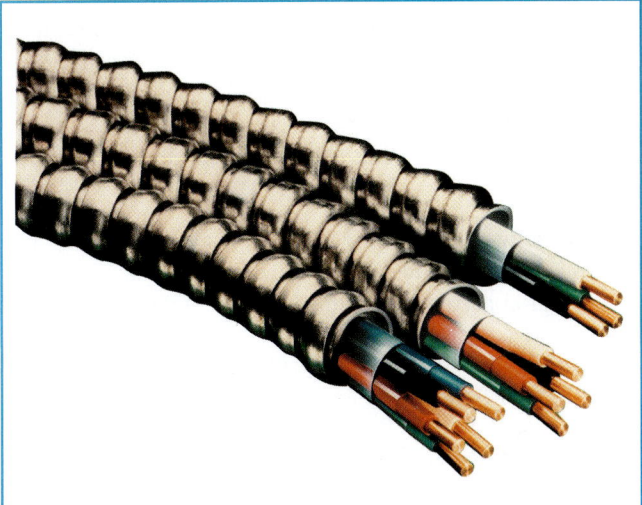

Exhibit 330.1 Examples of Type MC cable. (Courtesy of AFC Cable Systems, Inc.)

II. Installation

330.10 Uses Permitted.

(A) General Uses. Where not subject to physical damage, Type MC cables shall be permitted as follows:

(1) For services, feeders, and branch circuits
(2) For power, lighting, control, and signal circuits
(3) Indoors or outdoors
(4) Where exposed or concealed
(5) Direct buried where identified for such use
(6) In cable tray
(7) In any raceway
(8) As open runs of cable
(9) As aerial cable on a messenger

(10) In hazardous (classified) locations as permitted in Articles 501, 502, 503, 504, and 505

(11) In dry locations and embedded in plaster finish on brick or other masonry except in damp or wet locations

(12) In wet locations where any of the following conditions are met:

   a. The metallic covering is impervious to moisture.
   b. A lead sheath or moisture-impervious jacket is provided under the metal covering.
   c. The insulated conductors under the metallic covering are listed for use in wet locations.

(13) Where single-conductor cables are used, all phase conductors and, where used, the neutral conductor shall be grouped together to minimize induced voltage on the sheath.

> Single-conductor Type MC is now permitted to be used according to 330.10(A)(13). This single-conductor wiring method is new to the 2002 *NEC*. Installation practice for single-conductor wiring methods dictates close circuit conductor spacing not only to minimize induced voltage on the sheath, but also to minimize overall circuit impedance.

(B) Specific Uses. Type MC cable shall be installed in compliance with Articles 300, 490, 725, and 770.52 as applicable and in accordance with 330.10(B)(1) through (B)(4).

(1) Cable Tray. Type MC cable installed in cable tray shall comply with Article 392.

(2) Direct Buried. Direct-buried cable shall comply with 300.5 or 300.50, as appropriate.

(3) Installed as Service-Entrance Cable. Type MC cable installed as service-entrance cable shall comply with Article 230.

(4) Installed Outside of Buildings or as Aerial Cable. Type MC cable installed outside of buildings or as aerial cable shall comply with Article 225 and Article 396.

330.12 Uses Not Permitted.

Type MC cable shall not be used where exposed to the following destructive corrosive conditions, unless the metallic sheath is suitable for the conditions or is protected by material suitable for the conditions:

(1) Direct burial in the earth
(2) In concrete
(3) Where exposed to cinder fills, strong chlorides, caustic alkalis, or vapors of chlorine or of hydrochloric acids

330.17 Through or Parallel to Framing Members.

Type MC cable shall be protected in accordance with 300.4 where installed through or parallel to framing members.

330.23 In Accessible Attics.

The installation of Type MC cable in accessible attics or roof spaces shall also comply with 320.23.

> In accessible attics, Type MC cable installed across the top of floor joists or within 7 ft of the floor or floor joists across the face of rafters or studs must be protected by guard strips. Where the attic is not accessible by a permanent ladder or stairs, guard strips are required only within 6 ft of the scuttle hole or opening.

330.24 Bending Radius.

Bends in Type MC cable shall be made so that the cable will not be damaged. The radius of the curve of the inner edge of any bend shall be not less than shown in 330.24(A) through (C).

(A) Smooth Sheath.

(1) Ten times the external diameter of the metallic sheath for cable not more than 19 mm (¾ in.) in external diameter

(2) Twelve times the external diameter of the metallic sheath for cable more than 19 mm (¾ in.) but not more than 38 mm (1½ in.) in external diameter

(3) Fifteen times the external diameter of the metallic sheath for cable more than 38 mm (1½ in.) in external diameter

(B) Interlocked-Type Armor or Corrugated Sheath. Seven times the external diameter of the metallic sheath.

(C) Shielded Conductors. Twelve times the overall diameter of one of the individual conductors or seven times the overall diameter of the multiconductor cable, whichever is greater.

> The minimum bending radius of 12 times the outside diameter (OD) of a single shielded conductor [330.24(A)(2)] is consistent with ICEA (formerly IPCEA) requirements and good engineering practice. The same minimum on a shielded multiconductor cable, however, would be excessive.

330.30 Securing and Supporting.

Type MC cable shall be supported and secured at intervals not exceeding 1.8 m (6 ft).

(A) Horizontal Runs Through Holes and Notches. In other than vertical runs, cables installed in accordance with 300.4 shall be considered supported and secured where such support does not exceed 1.8-m (6-ft) intervals.

Type MC cable that runs horizontally through framing members (spaced less than 6 ft apart) and passes through bored or punched holes in framing members without additional securing is considered supported by the framing members. Cable ties are not required as the cable passes through these members. However, the MC cable must be secured (fastened in place) within 12 in. of the outlet box.

Section 330.30(A) contains the general support requirements for Type MC cable—that is, supported and secured at least every 6 ft. According to 330.30(C), MC cable containing four or fewer conductors of size 10 AWG or less are required to be secured within 12 in. from every box, cabinet, or fitting. Both requirements are illustrated in Exhibit 330.2.

(B) Unsupported Cables. Type MC cable shall be permitted to be unsupported where the cable:

(1) Is fished between access points, where concealed in finished buildings or structures and supporting is impracticable

(2) Is not more than 1.8 m (6 ft) from the last point of support for connections within an accessible ceiling to luminaire(s) [lighting fixture(s)] or equipment

Section 330.30(B) permits Type MC cable to be fished, as shown in Exhibit 330.3.

(C) At Terminations. Cables containing four or fewer conductors, sized no larger than 10 AWG, shall be secured within 300 mm (12 in.) of every box, cabinet, fitting, or other cable termination.

330.31 Single Conductors.

Where single-conductor cables with a nonferrous armor or sheath are used, the installation shall comply with 300.20.

330.40 Boxes and Fitting.

Fittings used for connecting Type MC cable to boxes, cabinets, or other equipment shall be listed and identified for such use.

Connectors should be selected in accordance with the size and type of cable for which they are designated. Bronze connectors are intended for use only with cable employing

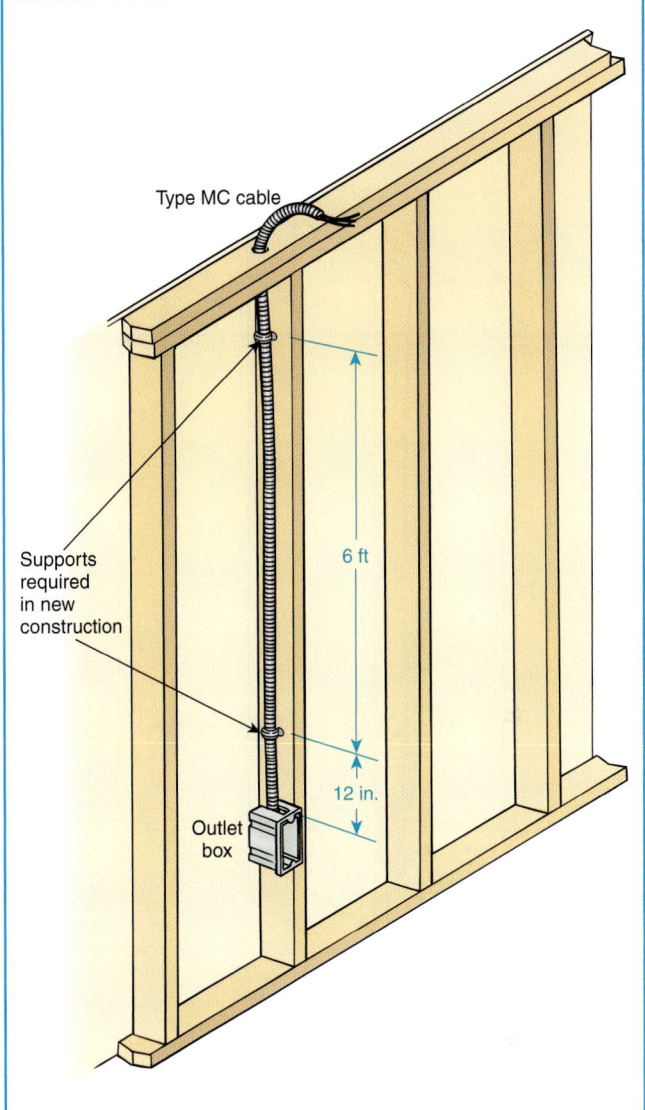

Exhibit 330.2 An application of 330.30(A) and (C), showing Type MC cable supported and secured at intervals not exceeding 6 ft and within 12 in. of the box.

corrugated copper armor. Some Type AC cable connectors are also acceptable for use with Type MC cable when specifically indicated on the device or the shipping carton.

330.80 Ampacity.

The ampacity of Type MC cable shall be determined in accordance with 310.15 or 310.60 for 14 AWG and larger conductors and in accordance with Table 402.5 for 18 AWG and 16 AWG conductors. The installation shall not exceed the temperature ratings of terminations and equipment.

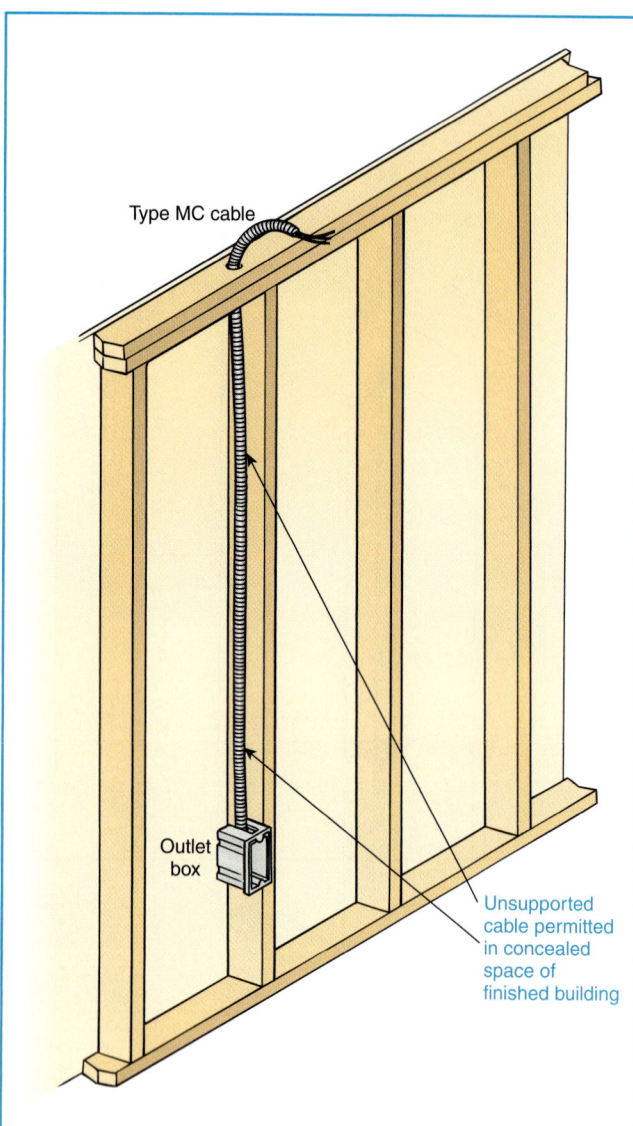

Type MC cable

Outlet box

Unsupported cable permitted in concealed space of finished building

Exhibit 330.3 An application of 330.30(B), which permits Type MC cable to be fished in walls, floors, or ceilings.

(A) Type MC Cable Installed in Cable Tray. The ampacities for Type MC cable installed in cable tray shall be determined in accordance with 392.11 and 392.13.

(B) Single Type MC Conductors Grouped Together. Where single Type MC conductors are grouped together in a triangular or square configuration and installed on a messenger or as open runs with a maintained free airspace of not less than 2.15 times one conductor diameter (2.15 × O.D.) of the largest conductor contained within the configuration and adjacent conductor configurations or cables, the ampacity of the conductors shall not exceed the allowable ampacities of:

(1) Table 310.20 for conductors rated 0 through 2000 volts
(2) Table 310.67 and Table 310.68 for conductors rated over 2000 volts

III. Construction Specifications

330.104 Conductors.

The conductors shall be of copper, aluminum, or copper-clad aluminum, solid or stranded. The minimum conductor size shall be 18 AWG copper and 12 AWG aluminum or copper-clad aluminum.

330.108 Equipment Grounding.

Type MC cable shall provide an adequate path for equipment grounding as required by Article 250.

The armor of interlocking Type MC cable is not recognized by UL as the sole means of providing an equipment grounding circuit but may be used to supplement the internal grounding conductor.

If an individual equipment grounding conductor is installed with a multiconductor Type MC cable installation, the separate equipment grounding conductor must be an integral part of the multiconductor Type MC cable. The size of an equipment grounding conductor is determined according to 250.122.

The following explanation applies only to installations of single-conductor Type MC cable. If single Type MC conductors are installed as open runs or as messenger-supported wiring, the smooth or corrugated metallic sheath on each conductor must have sufficient cross-sectional area to comply with Table 250.122 or concentric conductors may be provided over the conductor under the metallic sheath. Either the metallic sheath or the combination of the concentric conductors in parallel with the metallic sheath may be used to provide the required equipment grounding path. For single Type MC conductors with an interlocking metal tape armor installed as open runs or as messenger-supported wiring, concentric conductors must be provided over the conductor under the metallic sheath. The total cross-sectional area of the concentric conductors must comply with Table 250.122.

If single Type MC conductors are installed in cable tray, the cable tray, an equipment grounding conductor(s) within the cable tray, and/or the equipment grounding provided with the single conductor may be used individually or in any parallel combination to provide the equipment grounding path required by 392.3(C) and 250.118.

If single Type MC conductors are installed under ground in a trench, a separate equipment grounding conductor may be installed in close proximity to the circuit conductors within the same trench. The separate equipment grounding conductor may be used alone or in parallel with the equipment grounding provided with the single conductor cables to provide the equipment grounding path required by 300.5(I) and 250.118.

If single Type MC conductors are installed in parallel, the equipment grounding path provided with each conductor must be sized on the basis of the ampere rating of the overcurrent device protecting the circuit conductors according to 250.122(F). A smaller equipment grounding conductor may be permitted only if ground-fault protection of equipment is installed in accordance with 250.122(F)(2).

330.112 Insulation.

The insulated conductors shall comply with 330.112(A) or (B).

(A) 600 Volts. Insulated conductors in sizes 18 AWG and 16 AWG shall be of a type listed in Table 402.3, with a maximum operating temperature not less than 90°C (194°F) and as permitted by 725.27. Conductors larger than 16 AWG shall be of a type listed in Table 310.13 or of a type identified for use in Type MC cable.

(B) Over 600 Volts. Insulated conductors shall be of a type listed in Table 310.61 through Table 310.64.

330.116 Sheath.

The metallic covering shall be one of the following types: smooth metallic sheath, corrugated metallic sheath, interlocking metal tape armor. The metallic sheath shall be continuous and close fitting. A nonmagnetic sheath or armor shall be used on single conductor Type MC. Supplemental protection of an outer covering of corrosion-resistant material shall be permitted and shall be required where such protection is needed. The sheath shall not be used as a current-carrying conductor.

FPN: See 300.6 for protection against corrosion.

ARTICLE 332
Mineral-Insulated, Metal-Sheathed Cable: Type MI

Contents

I. General

332.1 Scope.

This article covers the use, installation, and construction specifications for mineral-insulated, metal-sheathed cable, Type MI.

332.2 Definition.

Mineral-Insulated, Metal-Sheathed Cable, Type MI. A factory assembly of one or more conductors insulated with a highly compressed refractory mineral insulation and enclosed in a liquidtight and gastight continuous copper or alloy steel sheath.

Type MI mineral-insulated, metal-sheathed cable consists of one or more solid copper conductors insulated with highly compressed magnesium oxide and enclosed in a continuous copper or alloy steel (e.g., stainless steel) sheath with or without a nonmetallic jacket. It is manufactured in size 16 AWG to 500 kcmil single conductor; 16 AWG to 4 AWG, two and three conductor; 16 AWG to 6 AWG, four conductor; and 16 AWG to 10 AWG, seven conductor. The cable is rated 600 volts.

II. Installation

332.10 Uses Permitted.

Type MI cable shall be permitted as follows:

(1) For services, feeders, and branch circuits
(2) For power, lighting, control, and signal circuits

Type MI cable is also constructed with a 300-volt rating for signal circuit applications.

(3) In dry, wet, or continuously moist locations
(4) Indoors or outdoors
(5) Where exposed or concealed
(6) Embedded in plaster, concrete, fill, or other masonry, whether above or below grade
(7) In any hazardous (classified) location
(8) Where exposed to oil and gasoline
(9) Where exposed to corrosive conditions not deteriorating to its sheath
(10) In underground runs where suitably protected against physical damage and corrosive conditions

Type MI cable is suitable for all power and control circuits up to 600 volts and may be used for services, feeders, and branch circuits in exposed and concealed work. It is permitted to be installed in dry and wet locations; in plaster, masonry, concrete, or fill; or under ground. It may be installed where exposed to weather, continuous moisture, oil, or gasoline and in any hazardous (classified) location or in other conditions not having a deteriorating effect on the metal sheath.

332.12 Uses Not Permitted.

Type MI cable shall not be used where exposed to conditions that are destructive and corrosive to the metallic sheath unless additionally protected by materials suitable for the conditions.

In lieu of an overall continuous copper sheath, Type MI cable is available with an overall continuous stainless steel sheath. Type MI cable is also available with an optional overall nonmetallic jacket.

332.17 Through or Parallel to Framing Members.

Type MI cable shall be protected in accordance with 300.4 where installed through or parallel to framing members.

332.24 Bending Radius.

Bends in Type MI cable shall be made so that the cable will not be damaged. The radius of the inner edge of any bend shall not be less than shown as follows:

(1) Five times the external diameter of the metallic sheath for cable not more than 19 mm (¾ in.) in external diameter

(2) Ten times the external diameter of the metallic sheath for cable greater than 19 mm (¾ in.) but not more than 25 mm (1 in.) in external diameter

The minimum bending radius is intended to prevent mechanical damage to the conductor insulation or the sheath that could result in cracking or a hot spot at the point of damage, or both.

As illustrated in Exhibit 332.1, for cables with an external diameter (OD) not greater than ¾ in., the minimum bending radius *(R)* is 5 times the cable OD. For cables greater than ¾ in. but not greater than 1 in. in diameter, the minimum bending radius is 10 times the cable OD.

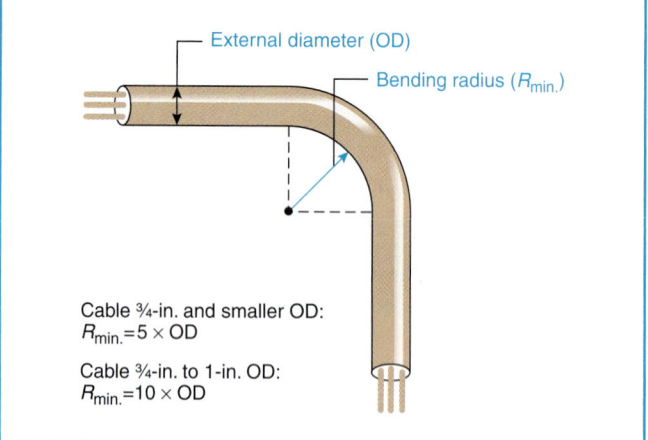

External diameter (OD)
Bending radius ($R_{min.}$)

Cable ¾-in. and smaller OD:
$R_{min.} = 5 \times OD$

Cable ¾-in. to 1-in. OD:
$R_{min.} = 10 \times OD$

Exhibit 332.1 An illustration of 332.24, for bends in Type MI cable.

332.30 Securing and Supporting.

Type MI cable shall be supported securely at intervals not exceeding 1.8 m (6 ft) by straps, staples, hangers, or similar fittings designed and installed so as not to damage the cable.

(A) Horizontal Runs Through Holes and Notches. In other than vertical runs, cables installed in accordance with 300.4 shall be considered supported and secured where such support does not exceed 1.8-m (6-ft) intervals.

(B) Unsupported Cable. Type MI cable shall be permitted to be unsupported where the cable is fished.

(C) Cable Trays. Type MI cable installed in cable trays shall comply with 392.8(B).

332.31 Single Conductors.

Where single-conductor cables are used, all phase conductors and, where used, the neutral conductor shall be grouped together to minimize induced voltage on the sheath.

Section 332.31 permits Type MI cables to be used as single conductors. The larger sizes of Type MI cable are available only as single-conductor cables. Because single conductors in a metal sheath can result in induced voltage on the sheath, this section requires all conductors of the circuit to be grouped together to minimize the voltage on the sheath.

Where single conductors enter a ferrous metal enclosure, inductive heating can occur due to hysteresis loss caused by the magnetic flux occuring in ferrous metals and I^2R losses from the currents induced by the conductor. To minimize this magnetic heating of enclosures, 300.20 requires additional measures, including cutting slots in the metal between the individual holes for each conductor connector. Cable manufacturers offer nonferrous connecting plates that accept individual threaded connections of all circuit conductors, thereby eliminating circulating currents and fully complying with 300.20.

332.40 Boxes and Fittings.

(A) Fittings. Fittings used for connecting Type MI cable to boxes, cabinets, or other equipment shall be identified for such use.

(B) Terminal Seals. Where Type MI cable terminates, an end seal fitting shall be installed immediately after stripping to prevent the entrance of moisture into the insulation. The conductors extending beyond the sheath shall be individually provided with an insulating material.

Terminations specifically investigated for use with this cable are listed as "mineral-insulated cable fittings." Fittings for use on Type MI cable are suitable for use at a maximum operating temperature of 90°C in dry locations and 60°C in wet locations. As shown in Exhibit 332.2, a complete box connector consists of a connector body and a screw-on potting fitting that may be used separately as an end fitting for change to open wiring. The screw-on potting fitting is to be assembled with a special tool and consists of a screw-on pot, insulating cap, insulating sleeving, anchoring bead, and sealing compound.

332.80 Ampacity.

The ampacity of Type MI cable shall be determined in accordance with 310.15. The conductor temperature at the end seal fitting shall not exceed the temperature rating of the listed end seal fitting, and the installation shall not exceed the temperature ratings of terminations or equipment.

(A) Type MI Cable Installed in Cable Tray. The ampacities for Type MI cable installed in cable tray shall be determined in accordance with 392.11.

(B) Single Type MI Conductors Grouped Together. Where single Type MI conductors are grouped together in a triangular or square configuration, as required by 332.31, and installed on a messenger or as open runs with a maintained free air space of not less than 2.15 times one conductor diameter (2.15 × O.D.) of the largest conductor contained within the configuration and adjacent conductor configurations or cables, the ampacity of the conductors shall not exceed the allowable ampacities of Table 310.17.

Determining Type MI cable ampacities according to Tables 310.16 or 310.17 is by far the most common approach. However, where MI cables are terminated at electrical equipment such as circuit breakers, distribution switchgear, transfer switches, and the like, it is important to understand the temperature limitations of electrical equipment terminals and to coordinate those temperature limitations with the ampacity of the MI cables. As stated in both the UL *General Information Directory* and in 110.14(C)(1) of the 2002 *NEC*, unless equipment is listed and marked otherwise, conductor ampacities used in determining equipment terminations must be based on Table 310.16 as modified by 310.15(B)(1) through (6).

III. Construction Specifications

332.104 Conductors.

Type MI cable conductors shall be of solid copper, nickel, or nickel-coated copper with a resistance corresponding to standard AWG and kcmil sizes.

332.108 Equipment Grounding.

Where the outer sheath is made of copper, it shall provide an adequate path for equipment grounding purposes. Where made of steel, an equipment grounding conductor shall be provided.

The copper sheath of Type MI cable is permitted as an equipment grounding conductor according to 250.118(10), but an alloy steel outer sheath of Type MI cable is not. According to the product standard, for Type MI cables with an alloy steel outer sheath, one of the conductors (within the cable assembly) is to be used for equipment grounding. This product standard statement is in concert with the requirement of 300.3 prohibiting the installations of single conductors unless they are part of a wiring method of Chapter 3 of the *Code*. The size of an equipment grounding conductor is determined according to 250.122. If an individual equipment grounding conductor is installed with a multiconductor Type MI cable installation, the separate equipment grounding conductor must be a integral part of the multiconductor Type MI cable.

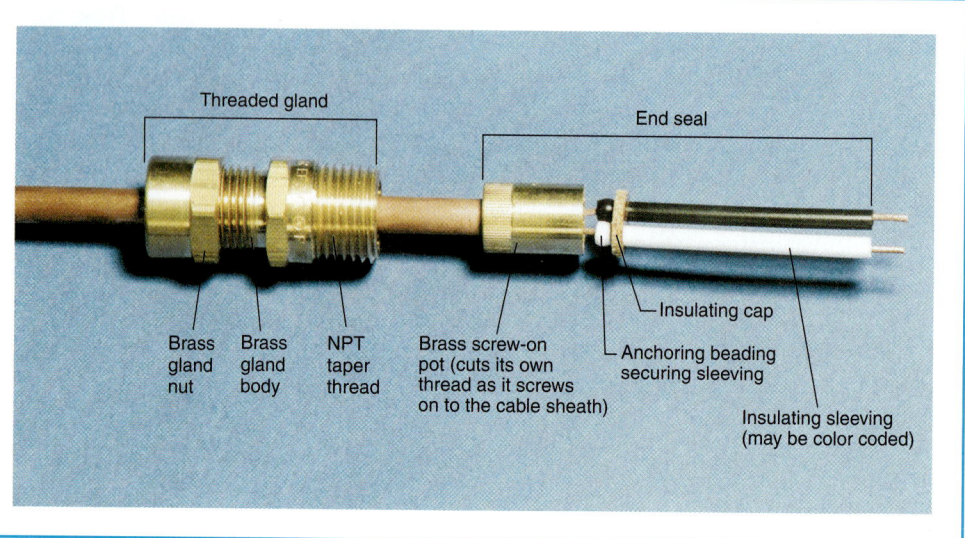

Exhibit 332.2 A Type MI cable fitting used for terminating cable to an enclosure, a box, or directly to equipment. (Courtesy of Pyrotenax Cables Ltd.)

The following explanation applies only to installations of single-conductor Type MI cable. If single Type MI conductors are installed as open runs or as messenger-supported wiring, the copper sheath on each conductor must have sufficient cross-sectional area to comply with Table 250.122.

If single Type MI conductors are installed in cable tray, the cable tray, an equipment grounding conductor(s) within the cable tray, and/or the equipment grounding provided with the single conductor may be used individually or in any parallel combination to provide the equipment grounding path according to the requirements of 392.3(C) and 250.118.

If single Type MI conductors are installed under ground in a trench, a separate equipment grounding conductor may be installed in close proximity to the circuit conductors within the same trench. The separate equipment grounding conductor may be used alone or in parallel with the equipment grounding provided with the single conductor to provide the equipment grounding path required by 300.5(I) and 250.118.

If single Type MI conductors are installed in parallel, the equipment grounding path provided with each conductor must be sized on the basis of the ampere rating of the overcurrent device protecting the circuit conductors in accordance with 250.122(F). A smaller equipment grounding conductor may be permitted only if ground-fault protection of equipment is installed in accordance with 250.122(F)(2).

332.112 Insulation.

The conductor insulation in Type MI cable shall be a highly compressed refractory mineral that provides proper spacing for all conductors.

332.116 Sheath.

The outer sheath shall be of a continuous construction to provide mechanical protection and moisture seal.

ARTICLE 334
Nonmetallic-Sheathed Cable: Types NM, NMC, and NMS

Contents

I. General

334.1 Scope.

This article covers the use, installation, and construction specifications of nonmetallic-sheathed cable.

334.2 Definition.

Nonmetallic-Sheathed Cable. A factory assembly of two or more insulated conductors having an outer sheath of non-metallic material.

Types NM, NMC, and NMS nonmetallic-sheathed cable may be used for either exposed or concealed wiring. Where exposed, the cable should not be subject to physical damage. It was first recognized in the 1928 *NEC* as a substitute for concealed knob-and-tube wiring (Article 394) and open wiring on insulators (Article 398). The original advantages of nonmetallic-sheathed cable over knob-and-tube wiring were that the outer sheath provides continuous protection in addition to the insulation applied to the conductors; the cable is easily fished in partitions of finished buildings; no insulating supports are required; and only one hole need be bored that can accommodate more than one cable passing through a wood cross member.

334.6 Listed.

Type NM, Type NMC, and Type NMS cables shall be listed.

New in the 2002 *Code*, all nonmetallic sheathed cables covered by Article 334 must now be a listed product. UL 719, *Standard for Nonmetallic-Sheathed Cables*, requires a

construction and performance evaluation, including testing relating to flammability, dielectric voltage-withstand, unwinding at low temperatures, pulling through joists, conductor pullout, crushing, and abrasion.

II. Installation

334.10 Uses Permitted.

Type NM, Type NMC, and Type NMS cables shall be permitted to be used in the following:

(1) One- and two-family dwellings.
(2) Multifamily dwellings permitted to be of Types III, IV, and V construction except as prohibited in 334.12.
(3) Other structures permitted to be of Types III, IV, and V construction except as prohibited in 334.12. Cables shall be concealed within walls, floors, or ceilings that provide a thermal barrier of material that has at least a 15-minute finish rating as identified in listings of fire-rated assemblies.

FPN No. 1: Building constructions are defined in NFPA 220-1999, *Standard on Types of Building Construction*, or the applicable building code, or both.
FPN No. 2: See Annex E for determination of building types [NFPA 220, Table 3-1].

A well-established means of codifying fire protection and fire safety requirements is to classify buildings by types of construction, based on materials used for the structural elements and the degree of fire resistance afforded by each element. The five fundamental construction types used by the model building codes are Type I (fire resistive), Type II (noncombustible), Type III (combination of combustible and noncombustible), Type IV (heavy timber), and Type V (wood frame). Types I and II basically require all structural elements to be noncombustible, whereas Types III, IV, and V allow some or all of the structural elements to be combustible (wood).

    The selection of building construction types is regulated by the building code, based on the occupancy, height, and area of the building. The local code official or the architect for a building project can be consulted to determine the minimum allowable (permitted) construction type for the building under consideration. When a building of a selected height (in feet or stories above grade) and area is permitted to be built of combustible construction (i.e., Types III, IV, or V), the installation of nonmetallic sheathed cable is permitted. The common areas (corridors) and incidental and subordinate uses (laundry rooms, lounge rooms, etc.) that serve a multifamily dwelling occupancy are also considered part of the multifamily occupancy, thereby allowing the use of nonmetallic sheathed cable in those areas.

If a building is to be of noncombustible construction (i.e., Type I or II) by the owner's choice, even though the building code would permit combustible construction, the building is allowed to be wired with nonmetallic sheathed cable. In such an instance, nonmetallic sheathed cable may be installed in the noncombustible building because the *Code* would have permitted the building to be of combustible construction.

Annex E provides charts and other explanatory information to assist the user in understanding and categorizing the exact types of construction under consideration. A table to cross reference building types to the various building code types of construction is provided in Annex E also.

(4) Cable trays, where the cables are identified for the use.

> FPN: See 310.10 for temperature limitation of conductors.

(A) Type NM. Type NM cable shall be permitted as follows:

(1) For both exposed and concealed work in normally dry locations except as prohibited in 334.10(3).
(2) To be installed or fished in air voids in masonry block or tile walls

For concealed work, nonmetallic-sheathed cable should be installed where it is protected from physical damage often caused by nails or screws. Where practical, care should be taken to avoid areas where trim, door and window casings, baseboards, moldings, and so on, are likely to be nailed. See 300.4 for details on protection against physical damage.

(B) Type NMC. Type NMC cable shall be permitted as follows:

(1) For both exposed and concealed work in dry, moist, damp, or corrosive locations, except as prohibited in 334.10(3)
(2) In outside and inside walls of masonry block or tile
(3) In a shallow chase in masonry, concrete, or adobe protected against nails or screws by a steel plate at least 1.59 mm (1/16 in.) thick and covered with plaster, adobe, or similar finish

Type NMC (corrosion-resistant) cable is required for installation in dairy barns and similar farm buildings (see Article 547), where cable will be exposed to fumes, vapors, or liquids such as ammonia and barnyard acids. Under such circumstances, ordinary types of nonmetallic-sheathed cable have in some cases deteriorated rapidly due to ammonia fumes or the growth of fungus or mold.

In addition to insulated conductors, nonmetallic-sheathed cable may have an insulated or bare conductor for equipment grounding purposes only. See 250.119 and Table 250.122 for identification requirements and minimum conductor sizes.

(C) Type NMS. Type NMS cable shall be permitted as follows:

(1) For both exposed and concealed work in normally dry locations except as prohibited in 334.10(3)
(2) To be installed or fished in air voids in masonry block or tile walls
(3) To be used as permitted in Article 780

The construction of this hybrid cable, Type NMS, is fully described in 334.116(C) and 780.6(A). Type NMS nonmetallic-sheathed cable is intended for use with "smart house" circuits as permitted in Article 780.

334.12 Uses Not Permitted.

(A) Types NM, NMC, and NMS. Types NM, NMC, and NMS cables shall not be used as follows:

(1) As open runs in dropped or suspended ceilings in other than one- and two-family and multifamily dwellings.

Revised for the 2002 *Code*, 334.12(A)(1) prohibits any nonmetallic sheathed cables installed as "open runs" in the space above accessible hung ceilings. This change does not affect dwelling-type occupancies. Open runs of cable, as used in this requirement, may be defined as cables installed as "open" or "not protected" installations that remain accessible after construction.

For example, cables installed above a dropped sheet rock ceiling or dropped sheet rock soffit would not be considered "open" runs of cable, provided the area above the ceiling is not accessible (does not have removable tiles or does not contain an access panel). Very often, hung or dropped ceilings are accessible: therefore cables installed above these types of ceilings would be considered "open runs of cables" if the cables do not have additional protection.

Additionally, a simple change to an architectural finish schedule during construction could change what is an acceptable wiring method. For example, if a corridor ceiling in an occupancy (other than a dwelling type) called for a painted gypsum board ceiling and the finish schedule changed the ceiling construction to a 2 ft × 2 ft accessible tile ceiling, the wiring method would no longer be permitted to be nonmetallic sheathed cable unless the nonmetallic sheathed cable was installed using additional protection. Examples of additional protection are found in 334.15(B).

(2) As service-entrance cable.
(3) In commercial garages having hazardous (classified) locations as defined in 511.3.
(4) In theaters and similar locations, except where permitted in 518.4.
(5) In motion picture studios.
(6) In storage battery rooms.
(7) In hoistways or on elevators or escalators.
(8) Embedded in poured cement, concrete, or aggregate.
(9) In hazardous (classified) locations, except where permitted in the following:

   a. 501.4(B), Exception
   b. 502.4(B), Exception No. 1
   c. 504.20

(10) Types NM and NMS. Types NM and NMS cable shall not be used as follows:

   a. Where exposed to corrosive fumes or vapors
   b. Where embedded in masonry, concrete, adobe, fill, or plaster
   c. In a shallow chase in masonry, concrete, or adobe and covered with plaster, adobe, or similar finish
   d. Where exposed or subject to excessive moisture or dampness

334.15 Exposed Work.

In exposed work, except as provided in 300.11(A), the cable shall be installed as specified in 334.15(A) through (C).

(A) To Follow Surface. The cable shall closely follow the surface of the building finish or of running boards.

(B) Protection from Physical Damage. The cable shall be protected from physical damage where necessary by conduit, electrical metallic tubing, Schedule 80 PVC rigid nonmetallic conduit, pipe, guard strips, listed surface metal or nonmetallic raceway, or other means. Where passing through a floor, the cable shall be enclosed in rigid metal conduit, intermediate metal conduit, electrical metallic tubing, Schedule 80 PVC rigid nonmetallic conduit, listed surface metal or nonmetallic raceway, or other metal pipe extending at least 150 mm (6 in.) above the floor.

(C) In Unfinished Basements. Where the cable is run at angles with joists in unfinished basements, it shall be permissible to secure cables not smaller than two 6 AWG or three 8 AWG conductors directly to the lower edges of the joists. Smaller cables shall be run either through bored holes in joists or on running boards.

As illustrated in Exhibit 334.1, nonmetallic-sheathed cables installed in an unfinished basement can be run through joists and attached to the side or face of joists or beams and running boards. Section 300.4(D) requires cables that are run parallel to framing members be installed at least 1¼ in. from the nearest edge of studs, joists, or rafters.

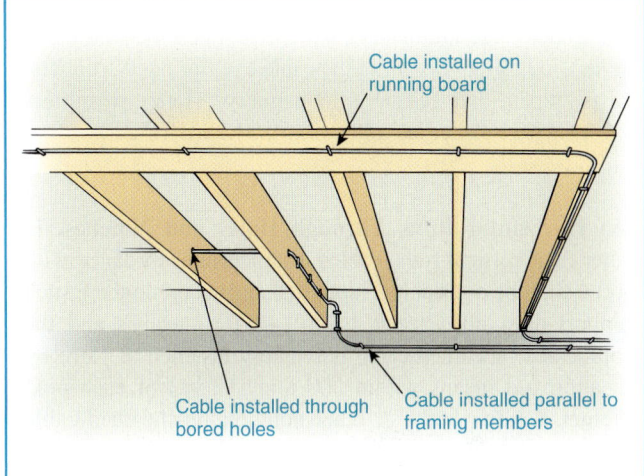

Exhibit 334.1 Nonmetallic sheathed cables installed in an unfinished basement.

334.17 Through or Parallel to Framing Members.

Types NM, NMC, or NMS cable shall be protected in accordance with 300.4 where installed through or parallel to framing members. Grommets used as required in 300.4(B)(1) shall remain in place and be listed for the purpose of cable protection.

In accordance with 300.4(B)(1), where cable passes through factory- or field-punched holes in metal studs or similar members, it is required to be protected by listed bushings or listed grommets covering all metal edges and securely fastened in the opening before being installed. See the commentary following 300.4(B)(1) for further information regarding physical protection of NM cables.

334.23 In Accessible Attics.

The installation of cable in accessible attics or roof spaces shall also comply with 320.23.

334.24 Bending Radius.

Bends in Types NM, NMC, and NMS cable shall be made so that the cable will not be damaged. The radius of the curve of the inner edge of any bend during or after installation shall not be less than five times the diameter of the cable.

334.30 Securing and Supporting.

Nonmetallic-sheathed cable shall be secured by staples, cable ties, straps, hangers, or similar fittings designed and installed so as not to damage the cable at intervals not exceeding 1.4 m (4½ ft) and within 300 mm (12 in.) of every cabinet, box, or fitting. Flat cables shall not be stapled on edge.

The general requirement of 334.30 requires that the cable be secured. Simply draping the cable over air ducts, timbers, joists, pipes, and ceiling grid members is not permitted, except where fished as allowed in 334.30(B)(1).

(A) Horizontal Runs Through Holes and Notches. In other than vertical runs, cables installed in accordance with 300.4 shall be considered supported and secured where such support does not exceed 1.4-m (4½-ft) intervals and the nonmetallic-sheathed cable is securely fastened in place by an approved means within 300 mm (12 in.) of each box, cabinet, conduit body, or other nonmetallic-sheathed cable termination.

> FPN: See 314.17(C) for support where nonmetallic boxes are used.

Nonmetallic sheathed cable that runs horizontally through framing members (spaced less than 54 in. apart) and passes through bored or punched holes in framing members without additional securing is considered supported by the framing members. Cable ties are not required as the cable passes through these members. However, the nonmetallic sheathed cable must be secured (fastened in place) within 12 in. of the outlet box. Where the cable terminates at a nonmetallic outlet box that does not contain a cable clamping device, the cable may be secured (fastened in place) within 8 in. of the outlet box, according to 314.17(C), Exception.

This section also prohibits two-conductor nonmetallic sheathed cable (or other flat configurations) from being stapled on edge. The intent is to prohibit the cable from being installed with its short dimension against a wood joist. When stapled in this manner, two cables are usually placed side by side under the staple. If the staple is driven too far into the stud, damage to the insulation and conductors can occur. See 300.4(C) for support requirements of cables through spaces behind panels designed to allow access.

(B) Unsupported Cables. Nonmetallic-sheathed cable shall be permitted to be unsupported where the cable:

(1) Is fished between access points, where concealed in finished buildings or finished panels for prefabricated buildings and supporting is impracticable

(2) Is not more than 1.4 m (4 ft) from the last point of support for connections within an accessible ceiling to luminaire(s) [lighting fixture(s)] or equipment

Section 330.30(B)(2) permits short, unsupported lengths of nonmetallic sheathed cable for luminaire and equipment connections.

(C) Wiring Device Without a Separate Outlet Box. A wiring device identified for the use, without a separate outlet box, incorporating an integral cable clamp shall be permitted where the cable is secured in place at intervals not exceeding 1.4 m (4½ ft) and within 300 mm (12 in.) from the wiring device wall opening, and there shall be at least a 300 mm (12 in.) loop of unbroken cable or 150 mm (6 in.) of a cable end available on the interior side of the finished wall to permit replacement.

334.40 Boxes and Fittings.

(A) Boxes of Insulating Material. Nonmetallic outlet boxes shall be permitted as provided in 314.3.

Nonmetallic boxes and nonmetallic wiring systems are recommended for use in some corrosive atmospheres. See 314.3, 314.17(C), and Article 547 for details.

(B) Devices of Insulating Material. Switch, outlet, and tap devices of insulating material shall be permitted to be used without boxes in exposed cable wiring and for rewiring in existing buildings where the cable is concealed and fished. Openings in such devices shall form a close fit around the outer covering of the cable, and the device shall fully enclose the part of the cable from which any part of the covering has been removed. Where connections to conductors are by binding-screw terminals, there shall be available as many terminals as conductors.

(C) Devices with Integral Enclosures. Wiring devices with integral enclosures identified for such use shall be permitted as provided in 300.15(E).

334.80 Ampacity.

The ampacity of Types NM, NMC, and NMS cable shall be determined in accordance with 310.15. The ampacity shall be in accordance with the 60°C (140°F) conductor temperature rating. The 90°C (194°F) rating shall be permitted to be used for ampacity derating purposes, provided the final derated ampacity does not exceed that for a 60°C (140°F) rated conductor. The ampacity of Types NM, NMC, and NMS cable installed in cable tray shall be determined in accordance with 392.11.

Section 310.15(B)(2)(a) states in part: "or where single conductors or multiconductor cables are stacked or bundled longer than 600 mm (24 in.) without maintaining spacing and are not installed in raceways, the allowable ampacity of each conductor shall be reduced as shown in Table 310.15(B)(2)(a)." Failure to comply with the appropriate adjustment ampacity derating called for by this table, where nonmetallic sheathed cables may be stacked or bundled, can lead to overheating of conductors.

III. Construction Specifications

334.100 Construction.

The outer cable sheath of nonmetallic-sheathed cable shall be a nonmetallic material.

334.104 Conductors.

The insulated power conductors shall be sizes 14 AWG through 2 AWG with copper conductors or sizes 12 AWG through 2 AWG with aluminum or copper-clad aluminum conductors. The signaling conductors shall comply with 780.5.

334.108 Equipment Grounding.

In addition to the insulated conductors, the cable shall be permitted to have an insulated or bare conductor for equipment grounding purposes only. Where provided, the grounding conductor shall be sized in accordance with Article 250.

334.112 Insulation.

The insulated power conductors shall be one of the types listed in Table 310.13 that is suitable for branch circuit wiring or one that is identified for use in these cables. Conductor insulation shall be rated at 90°C (194°F).

> FPN: Types NM, NMC, and NMS cable identified by the markings NM-B, NMC-B, and NMS-B meet this requirement.

334.116 Sheath.

The outer sheath of nonmetallic-sheathed cable shall comply with 334.116(A), (B), and (C).

(A) Type NM. The overall covering shall be flame retardant and moisture resistant.

(B) Type NMC. The overall covering shall be flame retardant, moisture resistant, fungus resistant, and corrosion resistant.

(C) Type NMS. The overall covering shall be flame retardant and moisture resistant. The sheath shall be applied so as to separate the power conductors from the communications and signaling conductors. The signaling conductors

shall be permitted to be shielded. An optional outer jacket shall be permitted.

> FPN: For composite optical cable, see 770.5 and 770.52.

ARTICLE 336
Power and Control Tray Cable: Type TC

Contents

I. General

336.1 Scope.

This article covers the use, installation, and construction specifications for power and control tray cable, Type TC.

The basic standard to investigate products in this category is UL 1277, *Electrical Power and Control Tray Cables with Optional Optical Fiber-Cables Members*. Summary information regarding listed power and control tray cable may be found in the UL *General Information Directory* under category QPOR.

336.2 Definition.

Power and Control Tray Cable, Type TC. A factory assembly of two or more insulated conductors, with or without associated bare or covered grounding conductors, under a nonmetallic jacket, for installation in cable trays, in raceways, or where supported by a messenger wire.

II. Installation

336.10 Uses Permitted.

Type TC tray cable shall be permitted to be used in the following:

(1) For power, lighting, control, and signal circuits.
(2) In cable trays, or in raceways, or where supported in outdoor locations by a messenger wire.
(3) In cable trays in hazardous (classified) locations as permitted in Articles 392, 501, 502, 504, and 505 in industrial establishments where the conditions of maintenance and supervision ensure that only qualified persons service the installation.
(4) For Class I circuits as permitted in Article 725.
(5) For non–power-limited fire alarm circuits if conductors comply with the requirements of 760.27.
(6) In industrial establishments where the conditions of maintenance and supervision ensure that only qualified persons service the installation, and where the cable is continuously supported and protected against physical damage using mechanical protection, such as struts, angles, or channel, Type TC tray cable that complies with the crush and impact requirements of Type MC cable and is identified for such use shall be permitted between a cable tray and the utilization equipment or device. The cable shall be secured at intervals not exceeding 1.8 m (6 ft). Equipment grounding for the utilization equipment shall be provided by an equipment grounding conductor within the cable. In cables containing conductors size 6 AWG or smaller, the equipment grounding conductor shall be provided within the cable or, at the time of installation, one or more insulated conductors shall be permanently identified as an equipment grounding conductor in accordance with 250.119(B).
(7) Where installed in wet locations, Type TC cable shall also be resistant to moisture and corrosive agents.

> FPN: See 310.10 for temperature limitation of conductors.

According to 336.10(3), Type TC cable is permitted to be installed in a hazardous location only if that location is in an industrial establishment where conditions of maintenance and supervision ensure that only qualified persons will service the installation. The overall jacket on Type TC cable is a "gas/vapor-tight continuous sheath" in the sense discussed in 501.5(D) and 501.5(E). However, Type TC cable is not investigated for transmission of gases or vapors through the core; thus, when these cables are used in hazardous (classified) locations, they may need to be sealed according to 501.5(D) and (E).

Section 336.10(5) permits Type TC tray cable to be used for non-power-limited fire alarm circuits. According to 760.27, the cable must be listed and the conductor material must be copper. Aluminum and copper-clad aluminum conductors are not permitted for fire alarm circuits.

According to 336.10(6), specific types of TC tray cable used in qualifying occupancies are permitted to extend from a cable tray to a piece of equipment without the use of conduit. A restriction of 50 ft previously applied to this type of installation. For the 2002 *Code*, the 50-ft restriction has been removed, allowing virtually any length of cable to be used for the circuit extension from the cable tray to the equipment served. According to the UL *General Information Directory*, a statement in category QPOR indicates that "cables suitable for use as open wiring between cable tray and the utilization equipment according to [336.10(6)] are surface marked 'open wiring'."

336.12 Uses Not Permitted.

Type TC tray cable shall not be used in the following:

(1) Installed where it will be exposed to physical damage
(2) Installed as open cable on brackets or cleats, except as permitted in 340.10(6)
(3) Used where exposed to direct rays of the sun, unless identified as sunlight resistant
(4) Direct buried, unless identified for such use

336.24 Bending Radius.

Bends in Type TC cable shall be made so as not to damage the cable. For Type TC cable without metal shielding, the minimum bending radius shall be as follows:

(1) Four times the overall diameter for cables 25 mm (1 in.) or less in diameter
(2) Five times the overall diameter for cables larger than 25 mm (1 in.) but not more than 50 mm (2 in.) in diameter
(3) Six times the overall diameter for cables larger than 50 mm (2 in.) in diameter

Type TC cables with metallic shielding shall have a minimum bending radius of not less than 12 times the cable overall diameter.

336.80 Ampacity.

The ampacity of Type TC tray cable shall be determined in accordance with 392.11 for 14 AWG and larger conductors, in accordance with 402.5 for 18 AWG through 16 AWG conductors where installed in cable tray, and in accordance with 310.15 where installed in a raceway or as messenger supported wiring.

III. Construction Specifications

336.100 Construction.

A metallic sheath or armor as defined in 330.116 shall not be permitted either under or over the nonmetallic jacket. Metallic shield(s) shall be permitted over groups of conductors, under the outer jacket, or both.

Type TC cables may contain one or more metal shields but do not have a metal sheath or armor. Electrical cables with a metal sheath or armor are covered in either Article 320 as armored cable, Type AC, or Article 330 as metal-clad cable, Type MC.

336.104 Conductors.

The insulated conductors of Type TC tray cable shall be in sizes 18 AWG through 1000 kcmil copper and sizes 12 AWG through 1000 kcmil aluminum or copper-clad aluminum. Insulated conductors of sizes 14 AWG and larger copper and sizes 12 AWG and larger aluminum or copper-clad aluminum shall be one of the types listed in Table 310.13 or Table 310.62 that is suitable for branch circuit and feeder circuits or one that is identified for such use.

(A) Fire Alarm Systems. Where used for fire alarm systems, conductors shall also be in accordance with 760.27.

(B) Thermocouple Circuits. Conductors in Type TC cables used for thermocouple circuits in accordance with Article 725 shall also be permitted to be any of the materials used for thermocouple extension wire.

(C) Class I Circuit Conductors. Insulated conductors of 18 AWG and 16 AWG copper shall also be in accordance with 725.27.

336.116 Jacket.

The outer jacket shall be a flame-retardant, nonmetallic material.

336.120 Marking.

There shall be no voltage marking on a Type TC cable employing thermocouple extension wire.

ARTICLE 338
Service-Entrance Cable: Types SE and USE

Contents

I. General

338.1 Scope.

This article covers the use, installation, and construction specifications of service-entrance cable.

According to the 2001 UL *Electrical Construction Materials Directory,* category TXKT (service cable) and category TYLZ (service-entrance cable rated 600 volts) are listed in sizes 14 AWG and larger for copper and 12 AWG and larger for aluminum or copper-clad aluminum. Type SE cable contains Types RHW, RHW–2, XHHW, XHHW-2, THWN, or THWN-2 conductors. Type USE cable contains conductors with insulation equivalent to RHW or XHHW. Type USE-2 contains insulation equivalent to RHW-2 or XHHW-2 and is rated 90°C wet or dry.

The type designation of the conductors may be marked on the surface of the cable. When used, this marking indicates the temperature rating for the cable corresponding to the temperature rating of the conductors. When this marking does not appear, the temperature rating of the cable is 75°C. The cables are designated as Type SE, Type USE or USE-2, and submersible water pump cable.

Type SE, cable for aboveground installation. Both the individual insulated conductors and the outer jacket or finish of Type SE are suitable for use where exposed to sun.

Type USE or USE-2, cable for underground installation, including burial directly in the earth. Cable in sizes 4/0 AWG and smaller and having all conductors insulated is suitable for all of the underground uses for which Type UF cable is permitted by the *Code.* Types USE and USE-2 are not suitable for use in premises or above ground except to terminate at the service equipment or metering equipment. Both the insulation and the outer covering, when used on single and multiconductor Types USE and USE-2, are suitable for use where exposed to sun.

Submersible water pump cable. This type indicates a multiconductor cable in which two, three, or four single-conductor, Type USE or USE-2 cables are provided in a flat or twisted assembly. The cable is listed in sizes 14 AWG to 4/0 AWG inclusive for copper and 12 AWG to 4/0 AWG inclusive for aluminum or copper-clad aluminum. The cable is tag-marked "For Use Within the Well Casing for Wiring Deep-Well Water Pumps Where the Cable Is Not Subject to Repetitive Handling Caused by Frequent Servicing of the Pump Units." The insulation may also be surface-marked "Pump Cable." The cable may be directly buried in the earth in conjunction with this use. The equipment grounding conductor, where required, is permitted to be bare.

338.2 Definitions.

Service-Entrance Cable. A single conductor or multiconductor assembly provided with or without an overall covering, primarily used for services, and of the following types:

Type SE. Service-entrance cable having a flame-retardant, moisture-resistant covering.

Type USE. Service-entrance cable, identified for underground use, having a moisture-resistant covering, but not required to have a flame-retardant covering.

II. Installation

338.10 Uses Permitted.

(A) Service-Entrance Conductors. Service-entrance cable used as service-entrance conductors shall be installed as required by Article 230.

Type USE used for service laterals shall be permitted to emerge from the ground outside at terminations in meter bases or other enclosures where protected in accordance with 300.5(D).

(B) Branch Circuits or Feeders.

(1) Grounded Conductor Insulated. Type SE service-entrance cables shall be permitted in wiring systems where all of the circuit conductors of the cable are of the rubber-covered or thermoplastic type.

Branch circuits using service-entrance cable as a wiring method are permitted only if all circuit conductors within the cable are fully insulated according to 310.13. The equipment grounding conductor is the only conductor permitted to be bare within service-entrance cable used for branch circuits.

(2) Grounded Conductor Not Insulated. Type SE service-entrance cable shall be permitted for use where the insulated conductors are used for circuit wiring and the uninsulated conductor is used only for equipment grounding purposes.

Exception: Uninsulated conductors shall be permitted as a grounded conductor in accordance with 250.140.

Service-entrance cable containing a bare grounded (neutral) conductor is not permitted for new installations where it is used as a branch circuit to supply appliances such as ranges, wall-mounted ovens, counter-mounted cooking units, or clothes dryers. This exception permits a bare neutral service

entrance cable for existing installations only and is coordinated with 250.140 and 250.142 of the 2002 *Code.*

(3) Temperature Limitations. Type SE service-entrance cable used to supply appliances shall not be subject to conductor temperatures in excess of the temperature specified for the type of insulation involved.

(4) Installation Methods for Branch Circuits and Feeders.

(a) Interior Installations. In addition to the provisions of this article, Type SE service-entrance cable used for interior wiring shall comply with the installation requirements of Parts I and II of Article 334, excluding 334.80.

> FPN: See 310.10 for temperature limitation of conductors.

(b) Exterior Installations. In addition to the provisions of this article, service-entrance cable used for feeders or branch circuits, where installed as exterior wiring, shall be installed as required by Article 225. The cable shall be supported in accordance with 334.30, unless used as messenger-supported wiring as allowed by Article 396.

Type USE cable shall be installed outside in accordance with the provisions of Article 340. Where Type USE cable emerges from the ground at terminations, it shall be protected in accordance with 300.5(D).

Multiconductor service-entrance cable shall be permitted to be installed as messenger-supported wiring in accordance with Articles 225 and 396.

In accordance with 338.2, cables marked only as "Type USE service-entrance cable" are not required to have a flame-retardant covering.

338.24 Bending Radius.

Bends in Types USE and SE cable shall be made so that the cable will not be damaged. The radius of the curve of the inner edge of any bend, during or after installation, shall not be less than five times the diameter of the cable.

III. Construction

338.100 Construction.

Cabled, single-conductor, Type USE constructions recognized for underground use shall be permitted to have a bare copper conductor cabled with the assembly. Type USE single, parallel, or cabled conductor assemblies recognized for underground use shall be permitted to have a bare copper concentric conductor applied. These constructions shall not require an outer overall covering.

> FPN: See 230.41, Exception, item (b), for directly buried, uninsulated service-entrance conductors.

Type SE or USE cable containing two or more conductors shall be permitted to have one conductor uninsulated.

338.120 Marking.

Service-entrance cable shall be marked as required in 310.11. Cable with the neutral conductor smaller than the ungrounded conductors shall be so marked.

ARTICLE 340
Underground Feeder and Branch-Circuit Cable: Type UF

Contents

I. General

340.1 Scope.

This article covers the use, installation, and construction specifications for underground feeder and branch-circuit cable, Type UF.

340.2 Definition.

Underground Feeder and Branch-Circuit Cable, Type UF. A listed factory assembly of one or more insulated conductors with an integral or an overall covering of nonmetallic material suitable for direct burial in the earth.

II. Installation

340.10 Uses Permitted.

Type UF cable shall be permitted as follows:

(1) For use underground, including direct burial in the earth. For underground requirements, see 300.5.
(2) As single-conductor cables. Where installed as single-conductor cables, all conductors of the feeder grounded conductor or branch circuit, including the grounded con-

ductor and equipment grounding conductor, if any, shall be installed in accordance with 300.3.
(3) For wiring in wet, dry, or corrosive locations under the recognized wiring methods of this *Code*.
(4) Installed as nonmetallic-sheathed cable. Where so installed, the installation and conductor requirements shall comply with the provisions of Article 334 and shall be of the multiconductor type.

Where UF cable is installed as nonmetallic sheathed cable, the ampacity of Type UF cable is determined according to 334.80. For Type UF cable used for interior wiring, see the installation requirements and the associated commentary within Parts I and II of Article 334.

(5) For solar photovoltaic systems in accordance with 690.31.
(6) As single-conductor cables as the nonheating leads for heating cables as provided in 424.43.
(7) Supported by cable trays. Type UF cable supported by cable trays shall be of the multiconductor type.

FPN: See 310.10 for temperature limitation of conductors.

340.12 Uses Not Permitted.

Type UF cable shall not be used as follows:

(1) As service-entrance cable
(2) In commercial garages
(3) In theaters and similar locations
(4) In motion picture studios
(5) In storage battery rooms
(6) In hoistways, or on elevators or escalators
(7) In hazardous (classified) locations
(8) Embedded in poured cement, concrete, or aggregate, except where embedded in plaster as nonheating leads where permitted in 424.43
(9) Where exposed to direct rays of the sun, unless identified as sunlight resistant

Type UF cable suitable for exposure to the direct rays of the sun is indicated by tag marking and marking on the cable surface with the designation "Sunlight Resistant."

(10) Where subject to physical damage

This physical protection requirement ensures that Type UF cable, as it emerges from under ground, will be protected from physical damage.

(11) As overhead cable, except where installed as messenger-supported wiring in accordance with Article 396

340.24 Bending Radius.

Bends in Type UF cable shall be made so that the cable shall not be damaged. The radius of the curve of the inner edge of any bend shall not be less than five times the diameter of the cable.

340.80 Ampacity.

The ampacity of Type UF cable shall be that of 60°C (140°F) conductors in accordance with 310.15.

III. Construction Specifications

340.104 Conductors.

The conductors shall be sizes 14 AWG copper or 12 AWG aluminum or copper-clad aluminum through 4/0 AWG.

340.108 Equipment Grounding.

In addition to the insulated conductors, the cable shall be permitted to have an insulated or bare conductor for equipment grounding purposes only.

340.112 Insulation.

The conductors of Type UF shall be one of the moisture-resistant types listed in Table 310.13 that is suitable for branch-circuit wiring or one that is identified for such use.

340.116 Sheath.

The overall covering shall be flame retardant; moisture, fungus, and corrosion resistant; and suitable for direct burial in the earth.

ARTICLE 342
Intermediate Metal Conduit:
Type IMC

Contents

I. General

342.1 Scope.

This article covers the use, installation, and construction specifications for intermediate metal conduit (IMC) and associated fittings.

342.2 Definition.

Intermediate Metal Conduit (IMC). A steel threadable raceway of circular cross section designed for the physical protection and routing of conductors and cables and for use as an equipment grounding conductor when installed with its integral or associated coupling and appropriate fittings.

Intermediate metal conduit (IMC) is a thinner-walled rigid metal conduit that is satisfactory for uses in all locations where rigid metal conduit is permitted to be used. Also, threaded fittings, couplings, connectors, and so on, are interchangeable for either IMC or rigid metal conduit (RMC). Threadless fittings for IMC are suitable only for the type of conduit indicated by the marking on the carton.

342.6 Listing Requirements.

IMC, factory elbows and couplings, and associated fittings shall be listed.

II. Installation

342.10 Uses Permitted.

(A) All Atmospheric Conditions and Occupancies. Use of IMC shall be permitted under all atmospheric conditions and occupancies.

(B) Corrosion Environments. IMC, elbows, couplings, and fittings shall be permitted to be installed in concrete, in direct contact with the earth, or in areas subject to severe corrosive influences where protected by corrosion protection and judged suitable for the condition.

(C) Cinder Fill. IMC shall be permitted to be installed in or under cinder fill where subject to permanent moisture where protected on all sides by a layer of noncinder concrete not less than 50 mm (2 in.) thick; where the conduit is not less than 450 mm (18 in.) under the fill; or where protected by corrosion protection and judged suitable for the condition.

(D) Wet Locations. All supports, bolts, straps, screws, and so forth, shall be of corrosion-resistant materials or protected against corrosion by corrosion-resistant materials.

FPN: See 300.6 for protection against corrosion.

Galvanized IMC installed in concrete does not require supplementary corrosion protection. Similarly, galvanized IMC installed in contact with soil does not generally require supplementary corrosion protection. As a guide in the absence of experience with the corrosive effects of soil in a specific location, soils producing severe corrosive effects are generally characterized by low resistivity, less than 2000 ohm-cm. Wherever ferrous metal conduit runs directly from concrete encasement to soil burial, the metal in contact with the soil can be severely corroded.

342.14 Dissimilar Metals.

Where practicable, dissimilar metals in contact anywhere in the system shall be avoided to eliminate the possibility of galvanic action.

Aluminum fittings and enclosures shall be permitted to be used with IMC.

342.20 Size.

(A) Minimum. IMC smaller than metric designator 16 (trade size ½) shall not be used.

(B) Maximum. IMC larger than metric designator 103 (trade size 4) shall not be used.

FPN: See 300.1(C) for the metric designators and trade sizes. These are for identification purposes only and do not relate to actual dimensions.

Table 300.1(C) identifies a distinct metric designator for each circular raceway trade size. For further explanation of metric designators, see 90.9 and the commentary following Table 300.1(C).

342.22 Number of Conductors.

The number of conductors shall not exceed that permitted by the percentage fill specified in Table 1, Chapter 9.

Cables shall be permitted to be installed where such use is permitted by the respective cable articles. The number of cables shall not exceed the allowable percentage fill specified in Table 1, Chapter 9.

Table 1 of Chapter 9 specifies the maximum fill percentage of a conduit or tubing. Table 4 provides the usable area within the selected conduit or tubing, and Table 5 provides the required area for each conductor. Examples using these tables to calculate a conduit or tubing size are provided following Chapter 9, Table 1, Notes to Tables, Note 6.

If the conductors are of the same wire size, the tables of Annex C can be used instead of doing the calculations. Annex C, which contains 12 sets of tables, very accurately indicates the maximum number of conductors permitted in a conduit or tubing. Examples using this annex to select a conduit or tubing size are provided following the introduction in Annex C.

To select the proper trade size intermediate metal conduit, the section entitled "Article 342—Intermediate Metal Conduit (IMC)" in Table 4 of Chapter 9 should be followed. Annex C Tables C4 and C4A for intermediate metal conduit are also permissible.

342.24 Bends—How Made.

Bends of IMC shall be made so that the conduit will not be damaged and so that the internal diameter of the conduit will not be effectively reduced. The radius of the curve of any field bend to the centerline of the conduit shall not be less than indicated in Table 344.24.

The term *field bend* means any bend or offset made by installers, using proper tools and equipment, during the installation of conduit systems.

342.26 Bends—Number in One Run.

There shall not be more than the equivalent of four quarter bends (360 degrees total) between pull points, for example, conduit bodies and boxes.

See the commentary following 344.26 for the rationale behind limiting the number of bends.

342.28 Reaming and Threading.

All cut ends shall be reamed or otherwise finished to remove rough edges. Where conduit is threaded in the field, a standard cutting die with a taper of 1 in 16 (¾ in. taper per foot) shall be used.

> FPN: See ANSI/ASME B.1.20.1-1983, *Standard for Pipe Threads, General Purpose (Inch).*

Conduit is cut using a saw or a roll cutter (pipe cutter). Care should be taken to ensure a straight cut, given that crooked threads will result from a die not started on the pipe squarely. After the cut is made, the conduit must be reamed. Proper reaming removes any and all burrs from the interior of the cut conduit so that, as wires and cables are pulled through the conduit, no chaffing of the insulation or cable jacket can occur. Finally, the conduit is threaded. The number of threads is important, because cutting too many threads prevents a conduit from being made up properly. If a threaded ring gauge is not available, cut the same number of threads on the conduit as are present on the factory (threaded) end of the conduit.

342.30 Securing and Supporting.

IMC shall be installed as a complete system as provided in Article 300 and shall be securely fastened in place and supported in accordance with 342.30(A) and (B).

(A) Securely Fastened. Each IMC shall be securely fastened within 900 mm (3 ft) of each outlet box, junction box, device box, cabinet, conduit body, or other conduit termination. Fastening shall be permitted to be increased to a distance of 1.5 m (5 ft) where structural members do not readily permit fastening within 900 mm (3 ft). Where approved, conduit shall not be required to be securely fastened within 900 mm (3 ft) of the service head for above-the-roof termination of a mast.

As illustrated in Exhibit 342.1, intermediate metal conduit (IMC) is required to be securely fastened at least every 10 ft. Fastening is also required within 3 ft of outlet boxes, junction boxes, cabinets, and conduit bodies. However, where structural support members do not permit fastening within 3 ft, the support may be located up to 5 ft away.

(B) Supports. IMC shall be supported in accordance with one of the following:

(1) Conduit shall be supported at intervals not exceeding 3 m (10 ft).
(2) The distance between supports for straight runs of conduit shall be permitted in accordance with Table 344.30(B)(2), provided the conduit is made up with threaded couplings and such supports prevent transmission of stresses to termination where conduit is deflected between supports.
(3) Exposed vertical risers from industrial machinery or fixed equipment shall be permitted to be supported at intervals not exceeding 6 m (20 ft), if the conduit is made up with threaded couplings, the conduit is firmly supported at the top and bottom of the riser, and no other means of intermediate support is readily available.
(4) Horizontal runs of IMC supported by openings through framing members at intervals not exceeding 3 m (10 ft) and securely fastened within 900 mm (3 ft) of termination points shall be permitted.

Section 342.30(B)(4) permits lengths of intermediate metal conduit (IMC) to be supported (but not necessarily secured) by framing members at 10-ft intervals, provided the IMC is secured and supported at least 3 ft from the box or enclosure. Installations where the IMC is installed through the bar joists is just one example and is shown in Exhibit 342.2.

342.42 Couplings and Connectors.

(A) Threadless. Threadless couplings and connectors used with conduit shall be made tight. Where buried in masonry or concrete, they shall be the concretetight type. Where installed in wet locations, they shall be the raintight type.

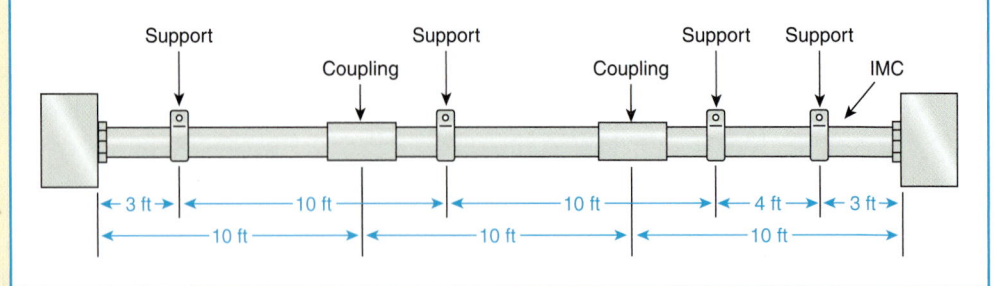

Exhibit 342.1 Minimum fastening requirements for intermediate metal conduit according to 342.30(A).

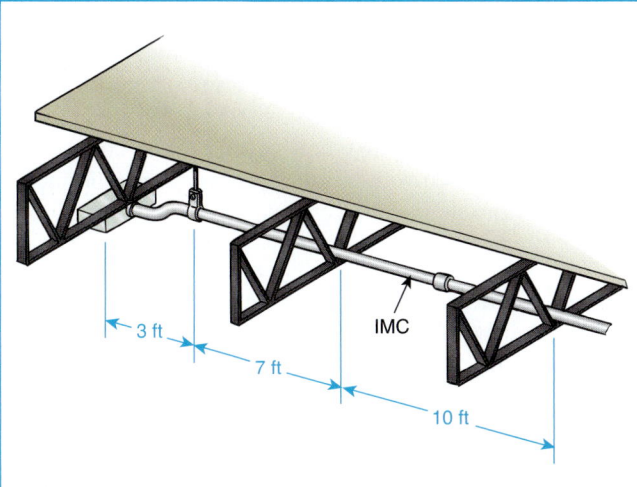

Threadless couplings and connectors shall not be used on threaded conduit ends unless listed for the purpose.

(B) Running Threads. Running threads shall not be used on conduit for connection at couplings.

See the commentary following 344.42(B) for examples of threadless fittings.

342.46 Bushings.

Where a conduit enters a box, fitting, or other enclosure, a bushing shall be provided to protect the wire from abrasion unless the design of the box, fitting, or enclosure is such as to afford equivalent protection.

> FPN: See 300.4(F) for the protection of conductors 4 AWG and larger at bushings.

342.56 Splices and Taps.

Splices and taps shall be made in accordance with 300.15.

342.60 Grounding.

IMC shall be permitted as an equipment grounding conductor.

III. Construction Specifications

342.120 Marking.

Each length shall be clearly and durably marked at least every 1.5 m (5 ft) with the letters IMC. Each length shall be marked as required in 110.21.

342.130 Standard Lengths.

The standard length of IMC shall be 3.05 m (10 ft), including an attached coupling, and each end shall be threaded. Longer or shorter lengths with or without coupling and threaded or unthreaded shall be permitted.

ARTICLE 344
Rigid Metal Conduit: Type RMC

I. General

344.1 Scope.

This article covers the use, installation, and construction specifications for rigid metal conduit (RMC) and associated fittings.

344.2 Definition.

Rigid Metal Conduit (RMC). A threadable raceway of circular cross section designed for the physical protection and routing of conductors and cables and for use as an equipment grounding conductor when installed with its integral or associated coupling and appropriate fittings. RMC is generally made of steel (ferrous) with protective coatings or aluminum (nonferrous). Special use types are silicon bronze and stainless steel.

344.6 Listing Requirements.

RMC, factory elbows and couplings, and associated fittings shall be listed.

II. Installation

344.10 Uses Permitted.

(A) All Atmospheric Conditions and Occupancies. Use of RMC shall be permitted under all atmospheric conditions and occupancies. Ferrous raceways and fittings protected from corrosion solely by enamel shall be permitted only indoors and in occupancies not subject to severe corrosive influences.

Section 344.10(A) makes it clear that aluminum rigid conduit can be used with steel fittings and enclosures, as can aluminum fittings and enclosures with steel rigid conduit. Tests have shown that the galvanic corrosion at steel and aluminum interfaces is minor compared to the natural corrosion on the combination of steel and steel or aluminum and aluminum.

(B) Corrosion Environments. RMC, elbows, couplings, and fittings shall be permitted to be installed in concrete, in direct contact with the earth, or in areas subject to severe corrosive influences where protected by corrosion protection and judged suitable for the condition.

Section 344.10(B) indicates the permitted uses for listed ferrous and nonferrous conduit, including their use in concrete, in direct contact with the earth, and in corrosive areas. The fine print note to 344.10(D) references 300.6 for additional information on protection against corrosion and specific types of corrosion-resistant materials.

The authority having jurisdiction for enforcing this *Code* should be consulted for approval of corrosion-resistant materials or for requirements prior to the installation of nonferrous metal (aluminum) conduit in concrete, since chloride additives in the concrete mix may cause corrosion.

(C) Cinder Fill. RMC shall be permitted to be installed in or under cinder fill where subject to permanent moisture where protected on all sides by a layer of noncinder concrete not less than 50 mm (2 in.) thick; where the conduit is not less than 450 mm (18 in.) under the fill; or where protected by corrosion protection and judged suitable for the condition.

Although cinder fill is not commonly used in modern construction, it is still encountered at older building sites. Care should be taken to ensure the proper installation of rigid metal conduit as permitted by this section. Where cinders have been used as fill, they may contain sulfur, and where they have combined with moisture, sulfuric acid is formed, which can corrode metal raceways.

(D) Wet Locations. All supports, bolts, straps, screws, and so forth, shall be of corrosion-resistant materials or protected against corrosion by corrosion-resistant materials.

FPN: See 300.6 for protection against corrosion.

344.14 Dissimilar Metals.

Where practicable, dissimilar metals in contact anywhere in the system shall be avoided to eliminate the possibility of galvanic action. Aluminum fittings and enclosures shall be permitted to be used with steel RMC, and steel fittings and enclosures shall be permitted to be used with aluminum RMC where not subject to severe corrosive influences.

344.20 Size.

(A) Minimum. RMC smaller than metric designator 16 (trade size ½) shall not be used.

Exception: For enclosing the leads of motors as permitted in 430.145(B).

(B) Maximum. RMC larger than metric designator 155 (trade size 6) shall not be used.

FPN: See 300.1(C) for the metric designators and trade sizes. These are for identification purposes only and do not relate to actual dimensions.

Table 300.1(C) identifies a distinct metric designator for each circular raceway trade size. For further explanation of metric designators, see 90.9 and the commentary following Table 300.1(C).

344.22 Number of Conductors.

The number of conductors or cables shall not exceed that permitted by the percentage fill specified in Table 1, Chapter 9.

Cables shall be permitted to be installed where such use is permitted by the respective cable articles. The number of cables shall not exceed the allowable percentage fill specified in Table 1, Chapter 9.

Table 1 of Chapter 9 specifies the maximum fill percentage of a conduit or tubing. Table 4 provides the usable area within the selected conduit or tubing, and Table 5 provides the required area for each conductor. Examples using these tables to calculate a conduit or tubing size are provided following Chapter 9, Table 1, Notes to Tables, Note 6.

If the conductors are of the same wire size, Annex C can be used instead of doing the calculations. Annex C, through 12 sets of tables, very accurately indicates the maximum number of conductors permitted in a conduit or tubing. Examples using this annex to select a conduit or tubing size are provided following the introduction in Annex C.

To select the proper trade size of rigid metal conduit, the section entitled "Article 344—Rigid Metal Conduit (RMC)" in Table 4 of Chapter 9 should be followed. Annex C Tables C8 and C8A for rigid metal conduit are also permissible.

344.24 Bends—How Made.

Bends of RMC shall be made so that the conduit is not damaged and the internal diameter of the conduit is not effectively reduced. The radius of the curve of any field bend to the centerline of the conduit shall not be less than indicated in Table 344.24.

Table 344.24 Radius of Conduit Bends

Conduit Size		One Shot and Full Shoe Benders		Other Bends	
Metric Designator	Trade Size	mm	in.	mm	in.
16	½	101.6	4	101.6	4
21	¾	114.3	4½	127	5
27	1	146.05	5¾	152.4	6
35	1¼	184.15	7¼	203.2	8
41	1½	209.55	8¼	254	10
53	2	241.3	9½	304.8	12
63	2½	266.7	10½	381	15
78	3	330.2	13	457.2	18
91	3½	381	15	533.4	21
103	4	406.4	16	609.6	24
129	5	609.6	24	762	30
155	6	762	30	914.4	36

The term *field bend* means any bend or offset made by installers, using proper tools and equipment, during the installation of conduit systems.

344.26 Bends—Number in One Run.

There shall not be more than the equivalent of four quarter bends (360 degrees total) between pull points, for example, conduit bodies and boxes.

Limiting the number of bends in a conduit run reduces pulling tension on conductors and helps ensure easy insertion or removal of conductors during later phases of construction, when the conduit may be permanently enclosed by the finish of the building. Adjustments during that time are often impossible. The *Code* does not limit the pull points to conduit bodies and boxes; these are only examples of pull points.

344.28 Reaming and Threading.

All cut ends shall be reamed or otherwise finished to remove rough edges. Where conduit is threaded in the field, a standard cutting die with a 1 in 16 taper (¾-in. taper per foot) shall be used.

> FPN: See ANSI/ASME B.1.20.1-1983, *Standard for Pipe Threads, General Purpose (Inch).*

Conduit is cut using a saw or a roll cutter (pipe cutter). Care should be taken to ensure a straight cut because crooked threads will result from a die not started on the pipe squarely. After the cut is made, the conduit must be reamed. Proper reaming removes any and all burrs from the interior of the cut conduit so that as wires and cables are pulled through the conduit, no chaffing of the insulation or cable jacket can occur. Finally, the conduit is threaded. The number of threads is important. To determine the correct number of threads for a conduit end, cut the same number of threads on the conduit as are present on the factory (threaded) end of the conduit. Where excessive threads are cut on the conduit and threaded couplings are installed, the conduit will butt within the coupling, resulting in a weak mechanical joint and poor grounding continuity.

344.30 Securing and Supporting.

RMC shall be installed as a complete system as provided in Article 300 and shall be securely fastened in place and supported in accordance with 344.30(A) and (B).

(A) Securely Fastened. RMC shall be securely fastened within 900 mm (3 ft) of each outlet box, junction box, device box, cabinet, conduit body, or other conduit termination. Fastening shall be permitted to be increased to a distance of 1.5 m (5 ft) where structural members do not readily permit fastening within 900 mm (3 ft). Where approved, conduit shall not be required to be securely fastened within 900 mm (3 ft) of the service head for above-the-roof termination of a mast.

As illustrated in Exhibit 344.1, rigid metal conduit is required to be securely fastened at least every 10 ft. Secure fastening is also required within 3 ft of outlet boxes, junction boxes,

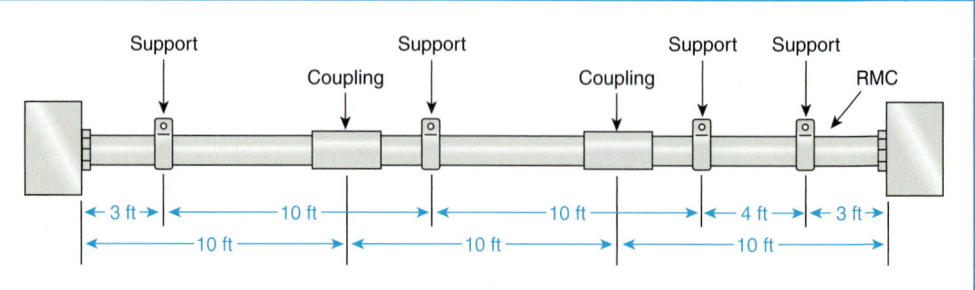

cabinets, and conduit bodies. However, where structural support members do not permit fastening within 3 ft, secure fastening may be located up to 5 ft away.

(B) Supports. RMC shall be supported in accordance with one of the following.

(1) Conduit shall be supported at intervals not exceeding 3 m (10 ft).

(2) The distance between supports for straight runs of conduit shall be permitted in accordance with Table 346.30(B)(2), provided the conduit is made up with threaded couplings, and such supports prevent transmission of stresses to termination where conduit is deflected between supports.

(3) Exposed vertical risers from industrial machinery or fixed equipment shall be permitted to be supported at intervals not exceeding 6 m (20 ft), if the conduit is made up with threaded couplings, the conduit is firmly supported at the top and bottom of the riser, and no other means of intermediate support is readily available.

(4) Horizontal runs of RMC supported by openings through framing members at intervals not exceeding 3 m (10 ft) and securely fastened within 900 mm (3 ft) of termination points shall be permitted.

Section 344.30(B)(4) permits lengths of rigid metal conduit to be supported (but not necessarily secured) by framing members at 10-ft intervals, provided the rigid metal conduit is secured and supported at least 3 ft from the box or enclosure. Installations where the rigid metal conduit is installed through the bar joists is just one example and is shown in Exhibit 344.2.

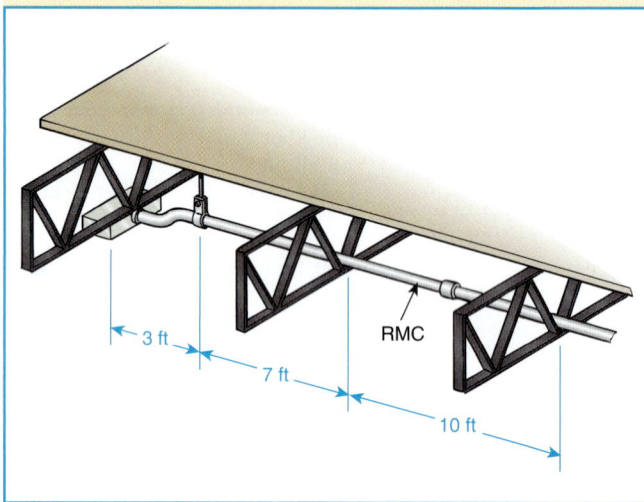

Exhibit 344.2 An example of rigid metal conduit supported by framing members and securely fastened at the 3-ft distance from the box required by 344.30(B)(4).

344.42 Couplings and Connectors.

(A) Threadless. Threadless couplings and connectors used with conduit shall be made tight. Where buried in masonry or concrete, they shall be the concretetight type. Where installed in wet locations, they shall be the raintight type. Threadless couplings and connectors shall not be used on threaded conduit ends unless listed for the purpose.

(B) Running Threads. Running threads shall not be used on conduit for connection at couplings.

Table 344.30(B)(2) Supports for Rigid Metal Conduit

Conduit Size		Maximum Distance Between Rigid Metal Conduit Supports	
Metric Designator	Trade Size	m	ft
16–21	½–¾	3.0	10
27	1	3.7	12
35–41	1¼–1½	4.3	14
53–63	2–2½	4.9	16
78 and larger	3 and larger	6.1	20

Exhibit 344.3 illustrates a threadless connection integral to a conduit body, FS box, and so on. This type of connection may be separate from the conduit body or box as an individual fitting of the compression type (raintight) suitable for wet locations or it may be of the set-screw type.

Threadless fittings are not intended for use over threads because the fitting will not seat properly. The threaded end of the conduit should be cut off and reamed before installation.

Exhibit 344.4 illustrates a three-piece coupling (the electrical equivalent of a pipe union), which is used to join two lengths of conduit where it is impossible to turn either length, such as in underground or concrete slab construction. Another fitting for joining conduit is a bolted split coupling. Running threads are not permitted to join two conduits.

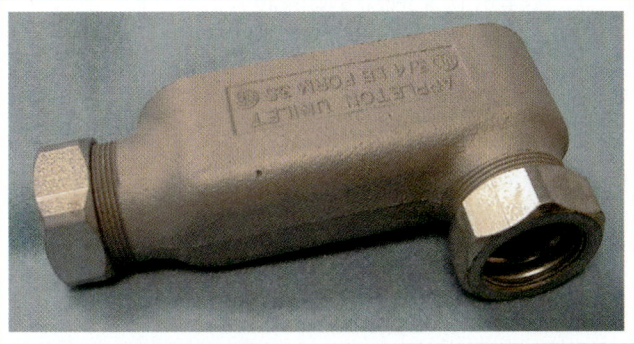

Exhibit 344.3 Conduit body with a threadless connector. (Courtesy of Appleton Electric Co., EGS Electrical Group)

Exhibit 344.4 A three-piece-type (union-type) coupling. (Courtesy of Appleton Electric Co., EGS Electrical Group)

344.46 Bushings.

Where a conduit enters a box, fitting, or other enclosure, a bushing shall be provided to protect the wire from abrasion unless the design of the box, fitting, or enclosure is such as to afford equivalent protection.

FPN: See 300.4(F) for the protection of conductors sizes 4 AWG and larger at bushings.

344.56 Splices and Taps.

Splices and taps shall be made in accordance with 300.15.

344.60 Grounding.

RMC shall be permitted as an equipment grounding conductor.

III. Construction Specifications

344.120 Marking.

Each length shall be clearly and durably identified in every 3 m (10 ft) as required in the first sentence of 110.21. Nonferrous conduit of corrosion-resistant material shall have suitable markings.

344.130 Standard Lengths.

The standard length of RMC shall be 3.05 m (10 ft), including an attached coupling, and each end shall be threaded. Longer or shorter lengths with or without coupling and threaded or unthreaded shall be permitted.

ARTICLE 348
Flexible Metal Conduit: Type FMC

Contents

I. General

348.1 Scope.

This article covers the use, installation, and construction specifications for flexible metal conduit (FMC) and associated fittings.

348.2 Definition.

Flexible Metal Conduit (FMC). A raceway of circular cross section made of helically wound, formed, interlocked metal strip.

348.6 Listing Requirements.

FMC and associated fittings shall be listed.

All flexible metal conduit as well as all flexible metal conduit fittings are required to be listed.

II. Installation

348.10 Uses Permitted.

FMC shall be permitted to be used in exposed and concealed locations.

348.12 Uses Not Permitted.

FMC shall not be used in the following:

(1) In wet locations unless the conductors are approved for the specific conditions and the installation is such that liquid is not likely to enter raceways or enclosures to which the conduit is connected

Listed flexible metal conduit (FMC) is permitted for use in wet locations, provided the completed installation prevents water from entering enclosures or other raceways to which the conduit is connected. Also, for this application, the conductors must be suitable for wet locations. Listed flexible metal conduit ½ in. and larger may be installed in unlimited lengths where an equipment grounding conductor is installed with the circuit conductors. See 250.118(5) and (6) as well as 348.60 for specific requirements related to its use as an equipment grounding conductor.

(2) In hoistways, other than as permitted in 620.21(A)(1)
(3) In storage battery rooms
(4) In any hazardous (classified) location other than as permitted in 501.4(B) and 504.20

(5) Where exposed to materials having a deteriorating effect on the installed conductors, such as oil or gasoline
(6) Underground or embedded in poured concrete or aggregate
(7) Where subject to physical damage

348.20 Size.

(A) Minimum. FMC less than metric designator 16 (trade size ½) shall not be used unless permitted in 348.20(A)(1) through (5) for metric designator 12 (trade size ⅜).

(1) For enclosing the leads of motors as permitted in 430.145(B)
(2) In lengths not in excess of 1.8 m (6 ft) for any of the following uses:
   a. For utilization equipment
   b. As part of a listed assembly
   c. For tap connections to luminaires (lighting fixtures) as permitted in 410.67(C)

Section 348.20(A)(2) makes it clear that ⅜-in. flexible metal conduit is permitted to be used as the manufactured or field-installed metal raceway (1½ ft to 6 ft in length) to enclose tap conductors between the outlet box and the terminal housing of recessed luminaires. Flexible metal conduit is also permitted to be used as a 6-ft fixture whip from an outlet box to a luminaire.

(3) For manufactured wiring systems as permitted in 604.6(A)

Section 604.6(A) permits a smaller minimum size for manufactured wiring systems because the conductors are not as prone to physical damage when assembled under factory-controlled conditions.

(4) In hoistways as permitted in 620.21(A)(1)
(5) As part of a listed assembly to connect wired luminaire (fixture) sections as permitted in 410.77(C)

(B) Maximum. FMC larger than metric designator 103 (trade size 4) shall not be used.

FPN: See 300.1(C) for the metric designators and trade sizes. These are for identification purposes only and do not relate to actual dimensions.

348.22 Number of Conductors.

The number of conductors shall not exceed that permitted by the percentage fill specified in Table 1, Chapter 9, or as permitted in Table 348.22 for metric designator 12 (trade size ⅜).

Table 348.22 Maximum Number of Insulated Conductors in Metric Designator 12 (Trade Size ⅜) Flexible Metal Conduit*

Size (AWG)	Types RFH-2, SF-2		Types TF, XHHW, TW		Types TFN, THHN, THWN		Types FEP, FEBP, PF, PGF	
	Fittings Inside Conduit	Fittings Outside Conduit	Fittings Inside Conduit	Fittings Outside Conduit	Fittings Inside Conduit	Fittings Outside Conduit	Fittings Inside Conduit	Fittings Outside Conduit
18	2	3	3	5	5	8	5	8
16	1	2	3	4	4	6	4	6
14	1	2	2	3	3	4	3	4
12	—	—	1	2	2	3	2	3
10	—	—	1	1	1	1	1	2

*In addition, one covered or bare equipment grounding conductor of the same size shall be permitted.

Table 1 of Chapter 9 specifies the maximum fill percentage of a conduit or tubing. Table 4 provides the usable area within the selected conduit or tubing, and Table 5 provides the required area for each of the conductors. Examples using these tables to calculate a conduit or tubing size are provided following Chapter 9, Table 1, Notes to Tables, Note 6.

If the conductors are of the same wire size, the tables of Annex C may be used instead of doing the calculations. Annex C, with 12 sets of tables, very accurately indicates the maximum number of conductors permitted in a conduit or tubing. Examples using this annex to select a conduit or tubing size are provided following the introduction in Annex C.

To select the proper trade size of flexible metal conduit, the section entitled "Flexible Metal Conduit" in Table 4 of Chapter 9 should be followed. Annex C Tables C3 and C3A for flexible metal conduit are also permissible.

Cables shall be permitted to be installed where such use is permitted by the respective cable articles. The number of cables shall not exceed the allowable percentage fill specified in Table 1, Chapter 9.

348.24 Bends—How Made.

Bends in conduit shall be made so that the conduit is not damaged and the internal diameter of the conduit is not effectively reduced. Bends shall be permitted to be made manually without auxiliary equipment. The radius of the curve to the centerline of any bend shall not be less than shown in Table 344.24 using the column "Other Bends."

348.26 Bends—Number in One Run.

There shall not be more than the equivalent of four quarter bends (360 degrees total) between pull points, for example, conduit bodies and boxes.

As with other raceways, a run of flexible metal conduit installed between boxes, conduit bodies, and so on, is not permitted to contain more than the equivalent of four quarter bends (360 degrees total). Proper shaping and support of this flexible wiring method will ensure that conductors can be easily installed or withdrawn at any time.

348.28 Trimming.

All cut ends shall be trimmed or otherwise finished to remove rough edges, except where fittings that thread into the convolutions are used.

348.30 Securing and Supporting.

FMC shall be securely fastened in place and supported in accordance with 348.30(A) and (B).

(A) Securely Fastened. FMC shall be securely fastened in place by an approved means within 300 mm (12 in.) of each box, cabinet, conduit body, or other conduit termination and shall be supported and secured at intervals not to exceed 1.4 m (4½ ft).

Exception No. 1: Where FMC is fished.

Exception No. 2: Lengths not exceeding 900 mm (3 ft) at terminals where flexibility is required.

Exception No. 3: Lengths not exceeding 1.8 m (6 ft) from a luminaire (fixture) terminal connection for tap connections to luminaires (light fixtures) as permitted in 410.67(C).

(B) Supports. Horizontal runs of flexible metal conduit FMC supported by openings through framing members at intervals not greater than 1.4 m (4½ ft) and securely fastened within 300 mm (12 in.) of termination points shall be permitted.

348.42 Couplings and Connectors.

Angle connectors shall not be used for concealed raceway installations.

348.56 Splices and Taps.

Splices and taps shall be made in accordance with 300.15.

348.60 Grounding and Bonding.

Where used to connect equipment where flexibility is required, an equipment grounding conductor shall be installed.

Where required or installed, equipment grounding conductors shall be installed in accordance with 250.134(B).

Where required or installed, equipment bonding jumpers shall be installed in accordance with 250.102.

According to the product standard UL 1, *Flexible Metal Electrical Conduit*, FMC longer than 6 ft has not been judged to be suitable for grounding purposes. The general rules for permitting or not permitting flexible metal conduit for grounding purposes are found in 250.118(5) and (6). One specific exception is where FMC is used for flexibility. As stated in 348.60, an additional equipment grounding conductor is always required where FMC is used for flexibility. Examples of such installations include using flexible metal conduit to minimize the transmission of vibration from equipment such as motors or to provide flexibility for floodlights, spotlights, or other equipment that requires adjustment.

Another specific exception is the requirement for a bonding jumper where FMC is used in hazardous (classified) locations. See 501.16(B), 502.16(B), and 503.16(B) for details on types of equipment grounding conductors. In addition, 250.102(E) permits the routing of equipment bonding jumpers on the outside of the raceway in lengths that are no longer than 6 ft and bonded at each end.

According to 250.118(6), where the length of the total ground-fault return path exceeds 6 ft or the circuit overcurrent protection exceeds 20 amperes, a separate equipment grounding conductor must be installed with the circuit conductors. The sketch on the right in Exhibit 348.1 shows an acceptable application of flexible metal conduit where the total length of any ground return path is limited to 6 ft. The sketch on the bottom shows an application that is unacceptable because the grounding return path for luminaire 2 exceeds the permitted maximum of 6 ft to the box.

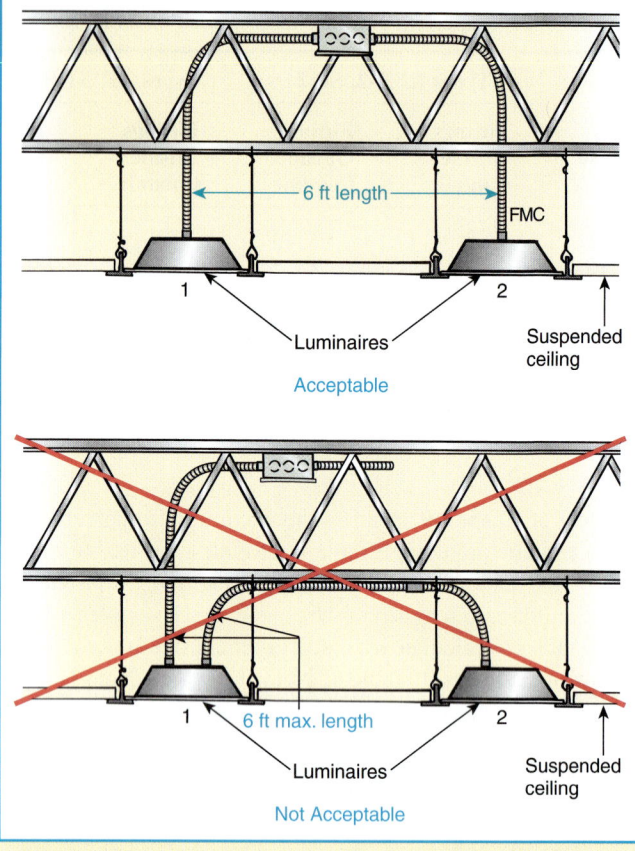

Exhibit 348.1 An example of acceptable and unacceptable applications of flexible metal conduit without separate equipment grounding conductors used as a fixture whip in accordance with 250.118(6)(c).

ARTICLE 350
Liquidtight Flexible Metal Conduit: Type LFMC

Contents

I. General

350.1 Scope.

This article covers the use, installation, and construction specifications for liquidtight flexible metal conduit (LFMC) and associated fittings.

350.2 Definition.

Liquidtight Flexible Metal Conduit (LFMC). A raceway of circular cross section having an outer liquidtight, nonmetallic, sunlight-resistant jacket over an inner flexible metal core with associated couplings, connectors, and fittings for the installation of electric conductors.

350.6 Listing Requirements.

LFMC and associated fittings shall be listed.

II. Installation

350.10 Uses Permitted.

LFMC shall be permitted to be used in exposed or concealed locations as follows:

(1) Where conditions of installation, operation, or maintenance require flexibility or protection from liquids, vapors, or solids
(2) As permitted by 501.4(B), 502.4, 503.3, and 504.20 and in other hazardous (classified) locations where specifically approved, and by 553.7(B)
(3) For direct burial where listed and marked for the purpose

Liquidtight flexible metal conduit (LFMC) is intended for use in wet locations or where exposed to mineral oil, both at a maximum temperature of 140°F. LFMC is not intended for use where exposed to gasoline or similar light petroleum solvents unless so marked on the product. LFMC is required to be a listed product. If properly marked for the application, liquidtight flexible metal conduit is permitted for direct burial in the earth. Note that the requirements of 300.5 are applicable to liquidtight flexible metal conduit if installed underground. LFMC is on the permitted list of wiring methods allowed for services (230.43), provided the length does not exceed 6 ft and an equipment bonding jumper is installed in accordance with 250.102. Liquidtight flexible metal conduit may be installed in unlimited lengths, provided it meets the other requirements of this article and a separate equipment grounding conductor is installed with the circuit conductors.

Liquidtight flexible nonmetallic conduit is also used extensively in the machine tool and related industries. See 17-8 in NFPA 79, *Electrical Standard for Industrial Machinery,* for the uses permitted on an industrial machine.

350.12 Uses Not Permitted.

LFMC shall not be used as follows:

(1) Where subject to physical damage
(2) Where any combination of ambient and conductor temperature produces an operating temperature in excess of that for which the material is approved

350.20 Size.

(A) Minimum. LFMC smaller than metric designator 16 (trade size ½) shall not be used.

Exception: LFMC of metric designator 12 (trade size ⅜) shall be permitted as covered in 348.20(A).

(B) Maximum. The maximum size of LFMC shall be metric designator 103 (trade size 4).

> FPN: See 300.1(C) for the metric designators and trade sizes. These are for identification purposes only and do not relate to actual dimensions.

350.22 Number of Conductors or Cables.

(A) Metric Designators 16 through 103 (Trade Sizes ½ through 4). The number of conductors shall not exceed that permitted by the percentage fill specified in Table 1, Chapter 9.

Cables shall be permitted to be installed where such use is permitted by the respective cable articles. The number of cables shall not exceed the allowable percentage fill specified in Table 1, Chapter 9.

(B) Metric Designator 12 (Trade Size ⅜). The number of conductors shall not exceed that permitted in Table 348.22, "Fittings Outside Conduit" columns.

Table 1 of Chapter 9 specifies the maximum percent fill of a conduit or tubing. Table 4 provides the usable area within the selected conduit or tubing, and Table 5 provides the required area for each of the conductors. Examples using these tables to calculate a conduit or tubing size are provided following Chapter 9, Table 1, Notes to Tables, Note 6.

If the conductors are of the same wire size, the tables of Annex C may be used instead of doing the calculations. Annex C, with 12 sets of tables, very accurately indicates the maximum number of conductors permitted in a conduit or tubing. Examples using this annex to select a conduit or tubing size are provided following the introduction in Annex C.

To select the proper trade size of liquidtight flexible metal conduit, the section entitled "Article 350—Liquidtight Flexible Metal Conduit (LFMC)" in Table 4 of Chapter 9 should be followed. Annex C Tables C7 and C7A for liquidtight flexible metal conduit are also permissible.

350.24 Bends—How Made.

Bends in conduit shall be made so that the conduit will not be damaged and the internal diameter of the conduit will not be effectively reduced. Bends shall be permitted to be made manually without auxiliary equipment. The radius of the curve to the centerline of any bend shall not be less than shown in Table 344.24 using the column "Other Bends."

350.26 Bends—Number in One Run.

There shall not be more than the equivalent of four quarter bends (360 degrees total) between pull points, for example, conduit bodies and boxes.

350.30 Securing and Supporting.

LFMC shall be securely fastened in place and supported in accordance with 350.30(A) and (B).

(A) Securely Fastened. LFMC shall be securely fastened in place by an approved means within 300 mm (12 in.) of each box, cabinet, conduit body, or other conduit termination and shall be supported and secured at intervals not to exceed 1.4 m (4½ ft).

Exception No. 1: Where LFMC is fished.

Exception No. 2: Lengths not exceeding 900 mm (3 ft) at terminals where flexibility is necessary.

Exception No. 3: Lengths not exceeding 1.8 m (6 ft) from a luminaire (fixture) terminal connection for tap conductors to luminaires (lighting fixtures), as permitted in 410.67(C).

(B) Supports. Horizontal runs of LFMC supported by openings through framing members at intervals not greater than 1.4 m (4½ ft) and securely fastened within 300 mm (12 in.) of termination points shall be permitted.

350.42 Couplings and Connectors.

Angle connectors shall not be used for concealed raceway installations.

350.56 Splices and Taps.

Splices and taps shall be made in accordance with 300.15.

350.60 Grounding and Bonding.

Where used to connect equipment where flexibility is required, an equipment grounding conductor shall be installed.

Where required or installed, equipment grounding conductors shall be installed in accordance with 250.134(B).

Where required or installed, equipment bonding jumpers shall be installed in accordance with 250.102.

FPN: See 501.16(B), 502.16(B), and 503.16(B) for types of equipment grounding conductors.

Where liquidtight flexible metal conduit is installed in hazardous (classified) locations, a bonding jumper may be required. There are some exceptions, however, as indicated in the exceptions to 501.16(B), 502.16(B), and 503.16(B).

III. Construction Specifications

350.120 Marking.

LFMC shall be marked according to 110.21. The trade size and other information required by the listing shall also be marked on the conduit. Conduit suitable for direct burial shall be so marked.

ARTICLE 352
Rigid Nonmetallic Conduit: Type RNC

Contents

I. General

352.1 Scope.

This article covers the use, installation, and construction specifications for rigid nonmetallic conduit (RNC) and associated fittings.

352.2 Definition.

Rigid Nonmetallic Conduit (RNC). A nonmetallic raceway of circular cross section, with integral or associated couplings, connectors, and fittings for the installation of electrical conductors.

The 2001 UL *General Information Directory* (White Book) describes three types of rigid nonmetallic conduit recognized for use in the *NEC*, as described in the following extract.

1. Rigid nonmetallic Schedule 40 and Schedule 80 PVC conduit (DZYR)
2. Rigid nonmetallic underground conduit, plastic (EAZX)
3. Rigid nonmetallic fiberglass conduit (DZKT)

Rigid Nonmetallic Schedule 40 and Schedule 80 PVC Conduit (DZYR)

Rigid nonmetallic Schedule 40 PVC conduit is suitable for underground use by direct burial or encasement in concrete. Unless marked "Underground Use Only" or equivalent wording, Schedule 40 conduit is also suitable for above ground use indoors or outdoors exposed to sunlight and weather where not subject to physical damage.

Schedule 80 conduit has a reduced cross-sectional area available for wiring space and is suitable for use wherever Schedule 40 conduit may be used. The marking "Schedule 80" identifies the conduit as suitable for use where exposed to physical damage and for installations on poles to comply with Section 352.10(F) and 352.12(C).

Unless marked for higher temperature, rigid nonmetallic conduit is intended for use with wires rated 75°C or less, including where it is encased in concrete within buildings and where ambient temperature is 50°C or less. Where encased in concrete in trenches outside buildings, it is suitable for use with wires rated 90°C or less.

Listed PVC conduit is inherently resistant to atmosphere containing common industrial corrosive agents and will also withstand vapors or mist of caustic, pickling acids, plating bath, and hydrofluoric and chromic acids.

PVC conduit, elbows, and bends (including couplings) that have been investigated for direct exposure to other reagents may be identified by the designation "Reagent Resistant" printed on the surface of the product. Further information on reagent resistance may be found in the 2001 UL *Electrical Construction Materials Directory* and in UL 651, "Schedule 40 and 80 Rigid PVC Conduit."

PVC conduit is designed for connection to couplings, fittings, and boxes by the use of a suitable solvent-type cement. Instructions supplied by the solvent-type cement manufacturer describe the method of assembly and precautions to be followed.

Rigid Nonmetallic Underground Conduit, Plastic (EAZX)

Plastic types of rigid nonmetallic conduit, for use only when installed underground, may be (1) polyvinyl chloride (PVC) Type A or Type EB, or (2) high-density polyethylene (HDPE) Schedule 40. The various conduit types differ in their inside and outside diameters.

The conduit is intended for underground use under the following conditions, as indicated on the listing mark: (1) when laid with its entire length in concrete (Type A); (2) when laid with its entire length in concrete in any location (Type EB); and (3) direct burial with or without being encased in concrete (HDPE Schedule 40). The conduit is intended for use in ambient temperatures of 50°C or less. Unless marked otherwise, Type A and HDPE Schedule 40 conduit are intended for use with wires rated 75°C or less. Type EB conduit and Type A conduit encased in concrete in trenches outside buildings may be used with wires rated 90°C or less. HDPE Schedule 40 conduit, when directly buried or encased in concrete, may be used with wires rated 90°C or less.

Where conduit emerges from underground installation, the wiring method shall be of a type recognized for the purpose.

PVC conduit is designed for joining with PVC couplings by the use of a solvent-type cement. HDPE conduit is designed for joining by threaded couplings, drive-on couplings, or a butt fusing process. Instructions supplied by the solvent-type cement manufacturer describe the method of assembly and precautions to be followed.

Reinforced Thermosetting Resin Conduit (DZKT)

[formally referred to as Rigid Nonmetallic Fiberglass Conduit and sometimes Fiberglass Reinforced Epoxy Conduit (FRE) conduit.]

Reinforced thermosetting resin conduit (RTRC) marked "Below Ground" or "Type BG" has been evaluated for underground use only—for direct burial, with or without encasement in concrete.

RTRC conduit marked "Above Ground" or AG" has been evaluated for use aboveground, underground, and for direct burial with or without encasement in concrete. This conduit has been evaluated for concealed or exposed work where not subject to physical damage.

Reinforced thermosetting resin conduit has been evaluated for use with wires rated 90°C or less.

Reinforced thermosetting resin conduit is listed in sizes ½ to 6 in. in IPS, ID, RTRC 40 and RTRC 80 dimensions, as marked on the product. Listing includes straight conduit, elbows, bends, and other fittings, unless otherwise noted.

Reinforced thermosetting resin conduit, elbows, bends, and other fittings, which have been investigated for direct exposure to reagents, are identified by the designation "Reagent Resistant" and are marked to indicate the specific reagents.

Reinforced thermosetting resin conduit is designed for connection to couplings, fittings, and boxes by use of a suitable epoxy-type cement or drive-on bell and spigot. Instructions supplied by the epoxy-type cement manufacturer describe the method of assembly and precautions to be followed.

For use of Schedule 80, see 300.5(D), 551.80(B), and 300.50(B).

352.6 Listing Requirements.

RNC, factory elbows, and associated fittings shall be listed.

II. Installation

352.10 Uses Permitted.

The use of RNC shall be permitted under the following conditions.

> FPN: Extreme cold may cause some nonmetallic conduits to become brittle and therefore more susceptible to damage from physical contact.

(A) Concealed. In walls, floors, and ceilings.

(B) Corrosive Influences. In locations subject to severe corrosive influences as covered in 300.6 and where subject to chemicals for which the materials are specifically approved.

(C) Cinders. In cinder fill.

(D) Wet Locations. In portions of dairies, laundries, canneries, or other wet locations and in locations where walls are frequently washed, the entire conduit system including boxes and fittings used therewith shall be installed and equipped so as to prevent water from entering the conduit. All supports, bolts, straps, screws, and so forth, shall be of

corrosion-resistant materials or be protected against corrosion by approved corrosion-resistant materials.

(E) Dry and Damp Locations. In dry and damp locations not prohibited by 352.12.

(F) Exposed. For exposed work where not subject to physical damage if identified for such use.

(G) Underground Installations. For underground installations, see 300.5 and 300.50. Conduits listed for the purpose shall be permitted to be installed underground in continuous lengths from a reel.

(H) Support of Conduit Bodies. Rigid nonmetallic conduit shall be permitted to support nonmetallic conduit bodies not larger than the largest trade size of an entering raceway. The conduit bodies shall not contain devices or support luminaires (fixtures) or other equipment.

352.12 Uses Not Permitted.

RNC shall not be used in the following locations.

(A) Hazardous (Classified) Locations.

(1) In hazardous (classified) locations, except as permitted in 503.3(A), 504.20, 514.8, and 515.8

(2) In Class I, Division 2 locations, except as permitted in 501.4(B), Exception

(B) Support of Luminaires (Fixtures). For the support of luminaires (fixtures) or other equipment not described in 352.10(H).

(C) Physical Damage. Where subject to physical damage unless identified for such use.

(D) Ambient Temperatures. Where subject to ambient temperatures in excess of 50°C (122°F) unless listed otherwise.

(E) Insulation Temperature Limitations. For conductors whose insulation temperature limitations would exceed those for which the conduit is listed.

(F) Theaters and Similar Locations. In theaters and similar locations, except as provided in Articles 518 and 520.

Nonmetallic conduits are not permitted to be installed in ducts, plenums, and other air-handling spaces. See 300.22, which limits the use of materials in ducts, plenums, and other air-handling spaces that may contribute smoke and products of combustion during a fire.

Additionally, rigid nonmetallic conduit is not permitted in places of assembly or theaters except as permitted in 518.4 and 520.5. See these sections for specific details.

352.20 Size.

(A) Minimum. RNC smaller than metric designator 16 (trade size ½) shall not be used.

(B) Maximum. RNC larger than metric designator 155 (trade size 6) shall not be used.

> FPN: The trade sizes and metric designators are for identification purposes only and do not relate to actual dimensions. See 300.1(C).

352.22 Number of Conductors.

The number of conductors shall not exceed that permitted by the percentage fill specified in Table 1, Chapter 9.

Cables shall be permitted to be installed where such use is permitted by the respective cable articles. The number of cables shall not exceed the allowable percentage fill specified in Table 1, Chapter 9.

Table 1 of Chapter 9 specifies the maximum percent fill of a conduit or tubing. Table 4 provides the usable area within the selected conduit or tubing, and Table 5 provides the required area for each of the conductors. Examples using these tables to calculate a conduit or tubing size are provided following Chapter 9, Table 1, Notes to Tables, Note 6.

If the conductors are of the same wire size, the tables of Annex C may be used instead of doing the calculations. Annex C, which contains 12 sets of tables, very accurately indicates the maximum number of conductors permitted in a conduit or tubing. Examples using this annex to select a conduit or tubing size are provided following the introduction in Annex C.

To permit selection of the proper trade size rigid nonmetallic conduit, Table 4 of Chapter 9 contains four separate subtables, one for each type of rigid nonmetallic conduit. The appropriate table for the given type of rigid nonmetallic conduit should be followed. Annex C Tables C9 and C9A through C12 and C12A are also permissible, provided the appropriate table for the given type of rigid nonmetallic conduit is used.

Where schedule 80 RNC is used, notice that the cross-sectional area available for wires is considerably less than that of other raceways of the same trade size due to the extra thick wall of a schedule 80 conduit.

352.24 Bends—How Made.

Bends shall be made so that the conduit will not be damaged and the internal diameter of the conduit will not be effectively reduced. Field bends shall be made only with bending equipment identified for the purpose. The radius of the curve to the centerline of such bends shall not be less than shown in Table 344.24, column "Other Bends."

The installation of rigid metal conduit in runs of PVC conduit installed under ground is covered in 300.5(D) and the associated commentary. The term *field bend* means any bend or offset made by installers, using proper tools and equipment, during the installation of conduit systems.

352.26 Bends—Number in One Run.

There shall not be more than the equivalent of four quarter bends (360 degrees total) between pull points, for example, conduit bodies and boxes.

Limiting the number of bends in a conduit run reduces pulling tension on conductors and helps ensure easy insertion or removal of conductors during later phases of construction, when the conduit may be permanently enclosed by the finish of the building. Adjustments during that time are often impossible. The *Code* does not limit the pull points to conduit bodies and boxes; these are only examples of pull points.

352.28 Trimming.

All cut ends shall be trimmed inside and outside to remove rough edges.

352.30 Securing and Supporting.

RNC shall be installed as a complete system as provided in 300.18 and shall be fastened so that movement from thermal expansion or contraction is permitted. RNC shall be securely fastened and supported in accordance with 352.30(A) and (B).

The requirements of 352.30 are fairly stringent because they are based on ambient temperatures higher than normally encountered and use horizontal-support tests only. Expansion can cause damage to the raceway or its supports. Expansion fittings, therefore, should be used, and the supports must allow expansion/contraction cycles without damage. See the commentary on expansion fittings following 352.44 for details.

(A) Securely Fastened. RNC shall be securely fastened within 900 mm (3 ft) of each outlet box, junction box, device box, conduit body, or other conduit termination. Conduit listed for securing at other than 900 mm (3 ft) shall be permitted to be installed in accordance with the listing.

(B) Supports. RNC shall be supported as required in Table 352.30(B). Conduit listed for support at spacings other than as shown in Table 352.30(B) shall be permitted to be installed in accordance with the listing. Horizontal runs of RNC supported by openings through framing members at intervals not exceeding those in Table 352.30(B) and securely fastened within 900 mm (3 ft) of termination points shall be permitted.

Table 352.30(B) Support of Rigid Nonmetallic Conduit (RNC)

Conduit Size		Maximum Spacing Between Supports	
Metric Designator	Trade Size	mm or m	ft
16–27	½–1	900 mm	3
35–53	1¼–2	1.5 m	5
63–78	2½–3	1.8 m	6
91–129	3½–5	2.1 m	7
155	6	2.5 m	8

352.44 Expansion Fittings.

Expansion fittings for RNC shall be provided to compensate for thermal expansion and contraction where the length change, in accordance with Table 352.44(A) or (B), is expected to be 6 mm (¼ in.) or greater in a straight run between securely mounted items such as boxes, cabinets, elbows, or other conduit terminations.

Expansion fittings are generally provided in exposed runs of rigid nonmetallic conduit where (1) the run is long, (2) the run is subjected to large temperature variations during or after installation, or (3) expansion and contraction measures are provided for the building or other structures. Rigid nonmetallic conduit exhibits a considerably greater change in length per degree change in temperature than do metal raceway systems.

In some parts of the United States and other countries, outdoor temperature variations of over 100°F are common. According to Table 352.44(A), a 100-ft run of PVC rigid nonmetallic conduit will change 4.1 in. in length if the temperature change is 100°F.

The normal expansion range of most larger sizes of rigid nonmetallic conduit expansion couplings is generally 6 in. Information concerning installation and application of this type of coupling may be obtained from manufacturers' instructions.

Expansion fittings are seldom used under ground, where temperatures are relatively constant. If rigid nonmetallic conduit is buried or covered immediately, expansion and contraction are not a problem.

352.46 Bushings.

Where a conduit enters a box, fitting, or other enclosure, a bushing or adapter shall be provided to protect the wire from

Table 352.44(A) Expansion Characteristics of PVC Rigid Nonmetallic Conduit Coefficient of Thermal Expansion = 6.084 × 10^{-5} mm/mm/°C (3.38 × 10^{-5} in./in./°F)

Temperature Change (°C)	Length Change of PVC Conduit (mm/m)	Temperature Change (°F)	Length Change of PVC Conduit (in./100 ft)	Temperature Change (°F)	Length Change of PVC Conduit (in./100 ft)
5	0.30	5	0.20	105	4.26
10	0.61	10	0.41	110	4.46
15	0.91	15	0.61	115	4.66
20	1.22	20	0.81	120	4.87
25	1.52	25	1.01	125	5.07
30	1.83	30	1.22	130	5.27
35	2.13	35	1.42	135	5.48
40	2.43	40	1.62	140	5.68
45	2.74	45	1.83	145	5.88
50	3.04	50	2.03	150	6.08
55	3.35	55	2.23	155	6.29
60	3.65	60	2.43	160	6.49
65	3.95	65	2.64	165	6.69
70	4.26	70	2.84	170	6.90
75	4.56	75	3.04	175	7.10
80	4.87	80	3.24	180	7.30
85	5.17	85	3.45	185	7.50
90	5.48	90	3.65	190	7.71
95	5.78	95	3.85	195	7.91
100	6.08	100	4.06	200	8.11

Table 352.44(B) Expansion Characteristics of Reinforced Thermosetting Resin Conduit (RTRC) Coefficient of Thermal Expansion = 2.7 × 10⁻⁵ mm/mm/°C (1.5 × 10⁻⁵ in./in./°F)

Temperature Change (°C)	Length Change of RTRC Conduit (mm/m)	Temperature Change (°F)	Length Change of RTRC Conduit (in./100 ft)	Temperature Change (°F)	Length Change of RTRC Conduit (in./100 ft)
5	0.14	5	0.09	105	1.89
10	0.27	10	0.18	110	1.98
15	0.41	15	0.27	115	2.07
20	0.54	20	0.36	120	2.16
25	0.68	25	0.45	125	2.25
30	0.81	30	0.54	130	2.34
35	0.95	35	0.63	135	2.43
40	1.08	40	0.72	140	2.52
45	1.22	45	0.81	145	2.61
50	1.35	50	0.90	150	2.70
55	1.49	55	0.99	155	2.79
60	1.62	60	1.08	160	2.88
65	1.76	65	1.17	165	2.97
70	1.89	70	1.26	170	3.06
75	2.03	75	1.35	175	3.15
80	2.16	80	1.44	180	3.24
85	2.30	85	1.53	185	3.33
90	2.43	90	1.62	190	3.42
95	2.57	95	1.71	195	3.51
100	2.70	100	1.80	200	3.60

abrasion unless the box, fitting, or enclosure design provides equivalent protection.

FPN: See 300.4(F) for the protection of conductors 4 AWG and larger at bushings.

352.48 Joints.

All joints between lengths of conduit, and between conduit and couplings, fittings, and boxes, shall be made by an approved method.

352.56 Splices and Taps.

Splices and taps shall be made in accordance with 300.15.

352.60 Grounding.

Where equipment grounding is required by Article 250, a separate equipment grounding conductor shall be installed in the conduit.

Exception No. 1: As permitted in 250.134(B), Exception No. 2, for dc circuits and 250.134(B), Exception No. 1, for separately run equipment grounding conductors.

Exception No. 2: Where the grounded conductor is used to ground equipment as permitted in 250.142.

III. Construction Specifications

352.100 Construction.

RNC and fittings shall be composed of suitable nonmetallic material that is resistant to moisture and chemical atmospheres. For use above ground, it shall also be flame retardant, resistant to impact and crushing, resistant to distortion from heat under conditions likely to be encountered in service, and resistant to low temperature and sunlight effects. For use underground, the material shall be acceptably resistant to moisture and corrosive agents and shall be of sufficient strength to withstand abuse, such as by impact and crushing, in handling and during installation. Where intended for direct burial, without encasement in concrete, the material shall also be capable of withstanding continued loading that is likely to be encountered after installation.

352.120 Marking.

Each length of RNC shall be clearly and durably marked at least every 3 m (10 ft) as required in the first sentence of 110.21. The type of material shall also be included in the marking unless it is visually identifiable. For conduit recognized for use above ground, these markings shall be permanent. For conduit limited to underground use only, these markings shall be sufficiently durable to remain legible until the material is installed. Conduit shall be permitted to be surface marked to indicate special characteristics of the material.

> FPN: Examples of these markings include but are not limited to "limited smoke" and "sunlight resistant."

ARTICLE 354
Nonmetallic Underground Conduit with Conductors: Type NUCC

Contents

I. General

354.1 Scope.

This article covers the use, installation, and construction specifications for nonmetallic underground conduit with conductors (NUCC).

354.2 Definition.

Nonmetallic Underground Conduit with Conductors (NUCC). A factory assembly of conductors or cables inside a nonmetallic, smooth wall conduit with a circular cross section.

Nonmetallic underground conduit with conductors (preassembled conductors in conduit) has been used by electric utilities for outdoor lighting for several years. It is supplied in continuous lengths on coils or reels or in cartons. It consists of nonmetallic conduit with the conductors preinstalled by the manufacturer. The product is designed to allow conductors to be removed and reinserted; therefore, maintenance is an issue.

354.6 Listing Requirements.

NUCC and associated fittings shall be listed.

II. Installation

354.10 Uses Permitted.

The use of NUCC and fittings shall be permitted in the following:

(1) For direct burial underground installation (For minimum cover requirements, see Table 300.5 and Table 300.50 under Rigid Nonmetallic Conduit.)
(2) Encased or embedded in concrete
(3) In cinder fill
(4) In underground locations subject to severe corrosive influences as covered in 300.6 and where subject to chemicals for which the assembly is specifically approved

354.12 Uses Not Permitted.

NUCC shall not be used in the following:

(1) In exposed locations
(2) Inside buildings

Exception: The conductor or the cable portion of the assembly, where suitable, shall be permitted to extend within the building for termination purposes in accordance with 300.3.

(3) In hazardous (classified) locations except as permitted by 503.3(A), 504.20, 514.8, and 515.8, and in Class I, Division 2 locations as permitted in 501.4(B)(3)

354.20 Size.

(A) Minimum. NUCC smaller than metric designator 16 (trade size ½) shall not be used.

(B) Maximum. NUCC larger than metric designator 103 (trade size 4) shall not be used.

> FPN: See 300.1(C) for the metric designators and trade sizes. These are for identification purposes only and do not relate to actual dimensions.

354.22 Number of Conductors.

The number of conductors or cables shall not exceed that permitted by the percentage fill in Table 1, Chapter 9.

354.24 Bends—How Made.

Bends shall be manually made so that the conduit will not be damaged and the internal diameter of the conduit will not be effectively reduced. The radius of the curve of the centerline of such bends shall not be less than shown in Table 354.24.

Table 354.24 Minimum Bending Radius for Nonmetallic Underground Conduit with Conductors (NUCC)

Conduit Size		Minimum Bending Radius	
Metric Designator	Trade Size	mm	in.
16	½	250	10
21	¾	300	12
27	1	350	14
35	1¼	450	18
41	1½	500	20
53	2	650	26
63	2½	900	36
78	3	1200	48
103	4	1500	60

354.26 Bends—Number in One Run.

There shall not be more than the equivalent of four quarter bends (360 degrees total) between termination points.

354.28 Trimming.

For termination, the conduit shall be trimmed away from the conductors or cables using an approved method that will not damage the conductor or cable insulation or jacket. All conduit ends shall be trimmed inside and out to remove rough edges.

354.46 Bushings.

Where the NUCC enters a box, fitting, or other enclosure, a bushing or adapter shall be provided to protect the conductor or cable from abrasion unless the design of the box, fitting, or enclosure provides equivalent protection.

FPN: See 300.4(F) for the protection of conductors size 4 AWG or larger.

354.48 Joints.

All joints between conduit, fittings, and boxes shall be made by an approved method.

354.50 Conductor Terminations.

All terminations between the conductors or cables and equipment shall be made by an approved method for that type of conductor or cable.

354.56 Splices and Taps.

Splices and taps shall be made in junction boxes or other enclosures.

354.60 Grounding.

Where equipment grounding is required by Article 250, an assembly containing a separate equipment grounding conductor shall be used.

III. Construction Specifications

354.100 Construction.

(A) General. NUCC is an assembly that is provided in continuous lengths shipped in a coil, reel, or carton.

(B) Nonmetallic Underground Conduit. The nonmetallic underground conduit shall be listed and composed of a material that is resistant to moisture and corrosive agents. It shall also be capable of being supplied on reels without damage or distortion and shall be of sufficient strength to withstand abuse, such as impact or crushing, in handling and during installation without damage to conduit or conductors.

(C) Conductors and Cables. Conductors and cables used in NUCC shall be listed and shall comply with 310.8(C). Conductors of different systems shall be installed in accordance with 300.3(C).

(D) Conductor Fill. The maximum number of conductors or cables in NUCC shall not exceed that permitted by the percentage fill in Table 1, Chapter 9.

354.120 Marking.

NUCC shall be clearly and durably marked at least every 3.05 m (10 ft) as required by 110.21. The type of conduit material shall also be included in the marking.

Identification of conductors or cables used in the assembly shall be provided on a tag attached to each end of the assembly or to the side of a reel. Enclosed conductors or cables shall be marked in accordance with 310.11.

ARTICLE 356
Liquidtight Flexible Nonmetallic Conduit: Type LFNC

Contents

I. General

356.1 Scope.

This article covers the use, installation, and construction specifications for liquidtight flexible nonmetallic conduit (LFNC) and associated fittings.

356.2 Definition.

Liquidtight Flexible Nonmetallic Conduit (LFNC). A raceway of circular cross section of various types as follows:

(1) A smooth seamless inner core and cover bonded together and having one or more reinforcement layers between the core and covers, designated as Type LFNC-A

(2) A smooth inner surface with integral reinforcement within the conduit wall, designated as Type LFNC-B

(3) A corrugated internal and external surface without integral reinforcement within the conduit wall, designated as LFNC-C.

LFNC is flame resistant and with fittings and is approved for the installation of electrical conductors.

FPN: FNMC is an alternative designation for LFNC.

356.6 Listing Requirements.

LFNC and associated fittings shall be listed.

II. Installation

356.10 Uses Permitted.

LFNC shall be permitted to be used in exposed or concealed locations for the following purposes:

FPN: Extreme cold may cause some types of nonmetallic conduits to become brittle and therefore more susceptible to damage from physical contact.

(1) Where flexibility is required for installation, operation, or maintenance

(2) Where protection of the contained conductors is required from vapors, liquids, or solids

(3) For outdoor locations where listed and marked as suitable for the purpose

(4) For direct burial where listed and marked for the purpose

(5) Type LFNC-B shall be permitted to be installed in lengths longer than 1.8 m (6 ft) where secured in accordance with 356.30

(6) Type LFNC-B as a listed manufactured prewired assembly, metric designator 16 through 27 (trade size through 1) conduit

Prewired Type LFNC-B is a listed assembly where the conductors are required to be installed at the manufacturing facility where controlled conditions prevent damage to the conductor insulation. Special cutting tools are required to be used when cutting prewired Type LFNC-B to prevent nicking the conductor installation. This prewired assembly is shown in Exhibit 356.1.

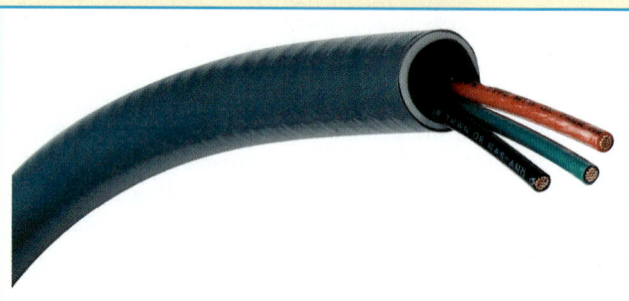

Exhibit 356.1 Listed manufactured prewired assembly of liquidtight flexible nonmetallic conduit, Type B. (Courtesy of Carlon®, Lamson & Sessions)

356.12 Uses Not Permitted.

LFNC shall not be used as follows:

(1) Where subject to physical damage

(2) Where any combination of ambient and conductor temperatures is in excess of that for which the LFNC is approved

(3) In lengths longer than 1.8 m (6 ft), except as permitted by 356.10(5) or where a longer length is approved as essential for a required degree of flexibility

(4) Where voltage of the contained conductors is in excess of 600 volts, nominal

356.20 Size.

(A) Minimum. LFNC smaller than metric designator 16 (trade size ½) shall not be used unless permitted in 356.20(A)(1) through (3) for metric designator 12 (trade size ⅜).

(1) For enclosing the leads of motors as permitted in 430.145(B)
(2) In lengths not exceeding 1.8 m (6 ft) as part of a listed assembly for tap connections to luminaires (lighting fixtures) as required in 410.67(C), or for utilization equipment
(3) For electric sign conductors in accordance with 600.32(A)

(B) Maximum. LFNC larger than metric designator 103 (trade size 4) shall not be used.

> FPN: See 300.1(C) for the metric designators and trade sizes. These are for identification purposes only and do not relate to actual dimensions.

356.22 Number of Conductors.

The number of conductors shall not exceed that permitted by the percentage fill specified in Table 1, Chapter 9.

Cables shall be permitted to be installed where such use is permitted by the respective cable articles. The number of cables shall not exceed the allowable percentage fill specified in Table 1, Chapter 9.

Table 1 of Chapter 9 specifies the maximum fill percentage of a conduit or tubing. Table 4 provides the usable area within the selected conduit or tubing, and Table 5 provides the required area for each of the conductors. Examples using these tables to calculate a conduit or tubing size are provided following Chapter 9, Table 1, Notes to Tables, Note 6.

If the conductors are of the same wire size, the tables of Annex C may be used instead of doing the calculations. Annex C, with 12 sets of tables, very accurately indicates the maximum number of conductors permitted in a conduit or tubing. Examples using this annex to select a conduit or tubing size are provided following the introduction in Annex C.

To permit selection of the proper trade size of liquidtight flexible nonmetallic conduit, Table 4 of Chapter 9 contains two sections entitled "Article 356—Liquidtight Flexible Nonmetallic Conduit (LFNC-B)" and "Article 356—Liquidtight Flexible Nonmetallic Conduit (LFNC-A)." The construction differences between these two raceway types are explained in 356.2.. Annex C Tables C5 and C5A or C6 and C6A for liquidtight flexible nonmetallic conduit (LFNC-A or LFNC-B) are also permissible.

356.24 Bends—How Made.

Bends in conduit shall be made so that the conduit is not damaged and the internal diameter of the conduit is not effectively reduced. Bends shall be permitted to be made manually without auxiliary equipment. The radius of the curve to the centerline of any bend shall not be less than shown in Table 344.24 using the column "Other Bends."

356.26 Bends—Number in One Run.

There shall not be more than the equivalent of four quarter bends (360 degrees total) between pull points, for example, conduit bodies and boxes.

356.28 Trimming.

All cut ends of conduit shall be trimmed inside and outside to remove rough edges.

356.30 Securing and Supporting.

Type LFNC-B shall be securely fastened and supported in accordance with one of the following:

(1) The conduit shall be securely fastened at intervals not exceeding 900 mm (3 ft) and within 300 mm (12 in.) on each side of every outlet box, junction box, cabinet, or fitting.
(2) Securing or supporting of the conduit shall not be required where it is fished, installed in lengths not exceeding 900 mm (3 ft) at terminals where flexibility is required, or installed in lengths not exceeding 1.8 m (6 ft) from a luminaire (fixture) terminal connection for tap conductors to luminaires (lighting fixtures) permitted in 410.67(C).
(3) Horizontal runs of LFNC supported by openings through framing members at intervals not exceeding 900 mm (3 ft) and securely fastened within 300 mm (12 in.) of termination points shall be permitted.

356.42 Couplings and Connectors.

Angle connectors shall not be used for concealed raceway installations.

356.56 Splices and Taps.

Splices and taps shall be made in accordance with 300.15.

356.60 Grounding and Bonding.

Where used to connect equipment where flexibility is required, an equipment grounding conductor shall be installed.

Where required or installed, equipment grounding conductors shall be installed in accordance with 250.134(B).

Where required or installed, equipment bonding jumpers shall be installed in accordance with 250.102.

III. Construction Specifications

356.100 Construction.

LFNC-B as a prewired manufactured assembly shall be provided in continuous lengths capable of being shipped in a coil, reel, or carton without damage.

356.120 Marking.

LFNC shall be marked at least every 600 mm (2 ft) in accordance with 110.21. The marking shall include a type designation in accordance with 356.2 and the trade size. Conduit that is intended for outdoor use or direct burial shall be marked.

The type, size, and quantity of conductors used in prewired manufactured assemblies shall be identified by means of a printed tag or label attached to each end of the manufactured assembly and either the carton, coil, or reel. The enclosed conductors shall be marked in accordance with 310.11.

ARTICLE 358
Electrical Metallic Tubing:
Type EMT

Contents

I. General

358.1 Scope.

This article covers the use, installation, and construction specifications for electrical metallic tubing (EMT) and associated fittings.

358.2 Definition.

Electrical Metallic Tubing (EMT). An unthreaded thin-wall raceway of circular cross section designed for the physical protection and routing of conductors and cables and for use as an equipment grounding conductor when installed utilizing appropriate fittings. EMT is generally made of steel (ferrous) with protective coatings or aluminum (nonferrous).

358.6 Listing Requirements.

EMT, factory elbows, and associated fittings shall be listed.

II. Installation

358.10 Uses Permitted.

(A) Exposed and Concealed. The use of EMT shall be permitted for both exposed and concealed work.

(B) Corrosion Protection. Ferrous or nonferrous EMT, elbows, couplings, and fittings shall be permitted to be installed in concrete, in direct contact with the earth, or in areas subject to severe corrosive influences where protected by corrosion protection and judged suitable for the condition.

(C) Wet Locations. All supports, bolts, straps, screws, and so forth shall be of corrosion-resistant materials or protected against corrosion by corrosion-resistant materials.

FPN: See 300.6 for protection against corrosion.

According to the 2001 UL *General Information for Electrical Equipment Directory* (White Book), category FJMX, galvanized steel electrical metallic tubing (EMT) installed in concrete, on grade or above, generally requires no supplementary corrosion protection. Galvanized steel electrical metallic tubing in concrete slab below grade level may require supplementary corrosion protection. In general, galvanized steel EMT in contact with soil requires supplementary corrosion protection. Where galvanized steel EMT without supplementary corrosion protection extends directly from concrete encasement to soil burial, severe corrosive effects are likely to occur on the metal in contact with the soil.

358.12 Uses Not Permitted.

EMT shall not be used under the following conditions:

(1) Where, during installation or afterward, it will be subject to severe physical damage

(2) Where protected from corrosion solely by enamel

(3) In cinder concrete or cinder fill where subject to permanent moisture unless protected on all sides by a layer of noncinder concrete at least 50 mm (2 in.) thick or unless the tubing is at least 450 mm (18 in.) under the fill

(4) In any hazardous (classified) location except as permitted by 502.4, 503.3, and 504.20

(5) For the support of luminaires (fixtures) or other equipment except conduit bodies no larger than the largest trade size of the tubing

(6) Where practicable, dissimilar metals in contact anywhere in the system shall be avoided to eliminate the possibility of galvanic action

Exception: Aluminum fittings and enclosures shall be permitted to be used with steel EMT where not subject to severe corrosive influences.

358.20 Size.

(A) Minimum. EMT smaller than metric designator 16 (trade size ½) shall not be used.

Exception: For enclosing the leads of motors as permitted in 430.145(B).

(B) Maximum. The maximum size of EMT shall be metric designator 103 (trade size 4).

> FPN: See 300.1(C) for the metric designators and trade sizes. These are for identification purposes only and do not relate to actual dimensions.

358.22 Number of Conductors.

The number of conductors shall not exceed that permitted by the percentage fill specified in Table 1, Chapter 9.

Cables shall be permitted to be installed where such use is permitted by the respective cable articles. The number of cables shall not exceed the allowable percentage fill specified in Table 1, Chapter 9.

Table 1 of Chapter 9 specifies the maximum fill percentage of a conduit or tubing. Table 4 provides the usable area within the selected conduit or tubing, and Table 5 provides the required area for each conductor. Examples using these tables to calculate a conduit or tubing size are provided following Chapter 9, Table 1, Notes to Tables, Note 6.

If the conductors are of the same wire size, instead of doing the calculations, the tables of Annex C may be used. Annex C, with 12 sets of tables, very accurately indicates the maximum number of conductors permitted in a conduit or tubing. Examples using this annex to select a conduit or tubing size are provided following the introduction in Annex C.

To select the proper trade size of electrical metallic tubing, the section entitled "Article 358—Electrical Metallic Tubing (EMT)" in Table 4 of Chapter 9 should be followed. Annex C Tables C1 and C1A for electrical metallic tubing are also permissible.

358.24 Bends—How Made.

Bends shall be made so that the tubing is not damaged and the internal diameter of the tubing is not effectively reduced. The radius of the curve of any field bend to the centerline of the conduit shall not be less than shown in Table 344.24 for one-shot and full shoe benders.

358.26 Bends—Number in One Run.

There shall not be more than the equivalent of four quarter bends (360 degrees total) between pull points, for example, conduit bodies and boxes.

358.28 Reaming and Threading.

(A) Reaming. All cut ends of EMT shall be reamed or otherwise finished to remove rough edges.

In addition to a reamer, a half-round file has proved practical for removing rough edges. The steel handle of a pair of pump pliers, the nose of side-cutting pliers, or an electrician's knife can be effective on the smaller sizes of EMT as well.

(B) Threading. EMT shall not be threaded.

Exception: EMT with factory threaded integral couplings complying with 358.100.

358.30 Securing and Supporting.

EMT shall be installed as a complete system as provided in Article 300 and shall be securely fastened in place and supported in accordance with 358.30(A) and (B).

(A) Securely Fastened. EMT shall be securely fastened in place at least every 3 m (10 ft). In addition, each EMT run between termination points shall be securely fastened within 900 mm (3 ft) of each outlet box, junction box, device box, cabinet, conduit body, or other tubing termination.

"Securely fastened in place" means the EMT must be supported and secured at the prescribed intervals, as illustrated in Exhibit 358.1.

Exception No. 1: Fastening of unbroken lengths shall be permitted to be increased to a distance of 1.5 m (5 ft) where structural members do not readily permit fastening within 900 mm (3 ft).

Exhibit 358.1 Minimum requirements for securely fastening electrical metallic tubing (EMT) unless an exception applies.

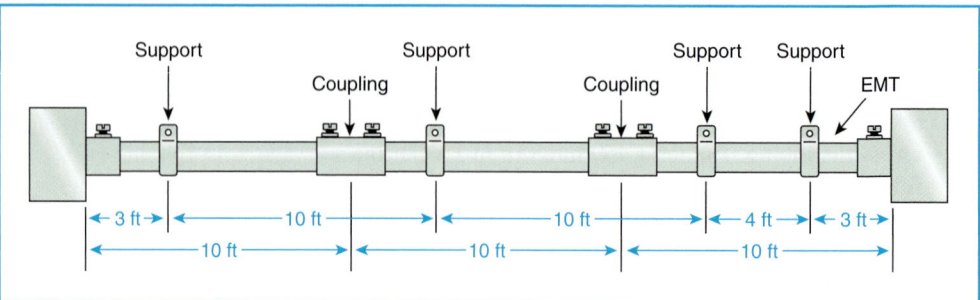

As illustrated in Exhibit 358.2, Exception No. 1 permits boxes secured to ceiling or roof support structural members that are spaced not more than 5 ft apart to serve as support for runs of EMT perpendicular to the axis of the ceiling or roof support members.

Exception No. 2: For concealed work in finished buildings or prefinished wall panels where such securing is impracticable, unbroken lengths (without coupling) of EMT shall be permitted to be fished.

(B) Supports. Horizontal runs of EMT supported by openings through framing members at intervals not greater than 3 m (10 ft) and securely fastened within 900 mm (3 ft) of termination points shall be permitted.

See the commentary and example following 342.30(B)(4), which apply to horizontal runs of intermediate metal conduit.

358.42 Couplings and Connectors.

Couplings and connectors used with EMT shall be made up tight. Where buried in masonry or concrete, they shall be concretetight type. Where installed in wet locations, they shall be of the raintight type.

Fittings have been tested for use only with steel EMT unless there is specific marking on the device or carton to indicate the fittings are suitable for use with aluminum or other material.

According to 358.6, only listed fittings are permitted to be used with EMT. According to UL 797, *Electrical Metallic Tubing*, listed fittings that are suitable for use in poured concrete or where exposed to rain are so indicated on the fitting or carton. The term *raintight* or the equivalent on the carton indicates suitability for use where directly exposed to rain. The term *concretetight* or equivalent on the carton indicates suitability for use in poured concrete. See 225.22 and 230.54(A) for raintight requirements as applied to raceways on exterior surfaces of buildings and to service raceways.

Indentor-type fittings are for use with metallic-coated electrical metallic tubing only and require a special tool supplied by the manufacturer for proper installation. Diametrically opposed indentor-type tools require two sets of inden-

Exhibit 358.2 An example of an EMT installation complying with 358.30(A), Exception No. 1.

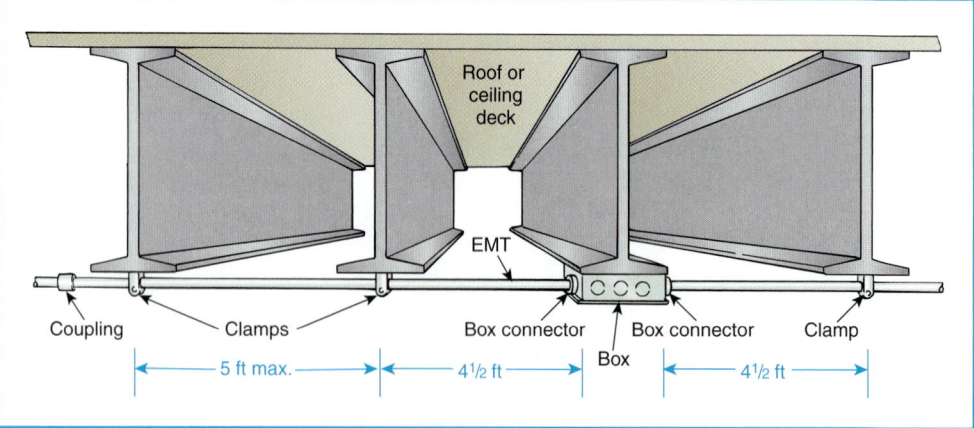

tations nominally 90 degrees apart. Triple-indent tools require one set of indentations.

358.56 Splices and Taps.

Splices and taps shall be made in accordance with 300.15.

358.60 Grounding.

EMT shall be permitted as an equipment grounding conductor.

III. Construction Specifications

358.100 Construction.

Factory-threaded integral couplings shall be permitted. Where EMT with a threaded integral coupling is used, threads for both the tubing and coupling shall be factory-made. The coupling and EMT threads shall be designed so as to prevent bending of the tubing at any part of the thread.

358.120 Marking.

EMT shall be clearly and durably marked at least every 3 m (10 ft) as required in the first sentence of 110.21.

ARTICLE 360
Flexible Metallic Tubing: Type FMT

Contents

I. General

360.1 Scope.

This article covers the use, installation, and construction specifications for flexible metallic tubing (FMT) and associated fittings.

Flexible metallic tubing is a type of raceway used for certain specific applications, particularly under the requirements of 300.22(B) and (C) for wiring in ducts, plenums, and other air-handling spaces. Flexible metallic tubing is very flexible and is rarely affected by vibration or other movement. It is an effective barrier to gases and products of combustion if installed with matching listed fittings and is of adequate mechanical strength for use where not exposed to physical damage.

Flexible metallic tubing not greater than 6 ft in length is suitable for use as a raceway for branch-circuit tap conductors that supply luminaires where outlet boxes are located not less than 1 ft from the luminaire.

360.2 Definition.

Flexible Metallic Tubing (FMT). A raceway that is circular in cross section, flexible, metallic, and liquidtight without a nonmetallic jacket.

360.6 Listing Requirements.

FMT and associated fittings shall be listed.

II. Installation

360.10 Uses Permitted.

FMT shall be permitted to be used for branch circuits as follows:

(1) In dry locations
(2) Where concealed
(3) In accessible locations
(4) For system voltages of 1000 volts maximum

A common application of flexible metallic tubing is as a branch-circuit wiring method for equipment or luminaires mounted on or above suspended ceilings. The 1000-volt limitation prohibits the use of flexible metallic tubing for the secondary circuits of sign ballasts, sign transformers, electronic sign power supplies, or oil burner ignition transformers unless these circuits are less than 1000 volts.

360.12 Uses Not Permitted.

FMT shall not be used as follows:

(1) In hoistways
(2) In storage battery rooms

(3) In hazardous (classified) locations unless otherwise permitted under other articles in this *Code*

(4) Under ground for direct earth burial, or embedded in poured concrete or aggregate

(5) Where subject to physical damage

(6) In lengths over 1.8 m (6 ft)

Unlike flexible metal conduit or liquidtight flexible metal conduit, flexible metallic tubing is limited in use to 6-ft lengths.

360.20 Size.

(A) Minimum. FMT smaller than metric designator 16 (trade size ½) shall not be used.

Exception No. 1: FMT of metric designator 12 (trade size ⅜) shall be permitted to be installed in accordance with 300.22(B) and (C).

Exception No. 2: FMT of metric designator 12 (trade size ⅜) shall be permitted in lengths not in excess of 1.8 m (6 ft) as part of an approved assembly or for luminaires (lighting fixtures). See 410.67(C).

(B) Maximum. The maximum size of FMT shall be metric designator 21 (trade size ¾).

> FPN: See 300.1(C) for the metric designators and trade sizes. These are for identification purposes only and do not relate to actual dimensions.

360.22 Number of Conductors.

(A) FMT—Metric Designators 16 and 21 (Trade Sizes ½ and ¾). The number of conductors in metric designators 16 (trade size ½) and 21 (trade size ¾) shall not exceed that permitted by the percentage fill specified in Table 1, Chapter 9.

Cables shall be permitted to be installed where such use is permitted by the respective cable articles. The number of cables shall not exceed the allowable percentage fill specified in Table 1, Chapter 9.

Table 1 of Chapter 9 specifies the maximum fill percentage of a conduit or tubing. Table 4 provides the usable area within the selected conduit or tubing, and Table 5 provides the required area for each of the conductors. Examples using these tables to calculate a conduit or tubing size are provided following Chapter 9, Table 1, Notes to Tables, Note 6.

If the conductors are of the same wire size, the tables of Annex C may be used instead of doing the calculations. Annex C, with 12 sets of tables, very accurately indicates the maximum number of conductors permitted in a conduit or tubing. Examples using this annex to select a conduit or

tubing size are provided following the introduction in Annex C.

To select the proper trade size of flexible metallic tubing, the section entitled "348—Flexible Metal Conduit (FMC)" in Table 4 of Chapter 9 or the manufacturer's instruction should be followed. Annex C Tables C3 and C3A for flexible metal conduit for sizes ½ in. and ¾ in. are also permissible.

(B) FMT—Metric Designator 12 (Trade Size ⅜). The number of conductors in metric designator 12 (trade size ⅜) shall not exceed that permitted in Table 348.22.

360.24 Bends.

(A) Infrequent Flexing Use. Where FMT may be infrequently flexed in service after installation, the radii of bends measured to the inside of the bend shall not be less than specified in Table 360.24(A).

Table 360.24(A) Minimum Radii for Flexing Use

Metric Designator	Trade Size	Minimum Radii for Flexing Use	
		mm	in.
12	⅜	25.4	10
16	½	317.5	12½
21	¾	444.5	17½

(B) Fixed Bends. Where FMT is bent for installation purposes and is not flexed or bent as required by use after installation, the radii of bends measured to the inside of the bend shall not be less than specified in Table 360.24(B).

Table 360.24(B) Minimum Radii for Fixed Bends

Metric Designator	Trade Size	Minimum Radii for Fixed Bends	
		mm	in.
12	⅜	88.9	3½
16	½	101.6	4
21	¾	127.0	5

360.40 Boxes and Fittings.

Fittings shall effectively close any openings in the connection.

360.56 Splices and Taps.

Splices and taps shall be made in accordance with 300.15.

360.60 Grounding.

FMT shall be permitted as an equipment grounding conductor where installed in accordance with 250.118(8).

III. Construction Specifications

360.120 Marking.

FMT shall be marked according to 110.21.

ARTICLE 362
Electrical Nonmetallic Tubing: Type ENT

Contents

I. General

362.1 Scope.

This article covers the use, installation, and construction specifications for electrical nonmetallic tubing (ENT) and associated fittings.

362.2 Definition.

Electrical Nonmetallic Tubing (ENT). A nonmetallic pliable corrugated raceway of circular cross section with integral or associated couplings, connectors, and fittings for the installation of electric conductors. ENT is composed of a material that is resistant to moisture and chemical atmospheres and is flame retardant.

A pliable raceway is a raceway that can be bent by hand with a reasonable force, but without other assistance.

Electrical nonmetallic tubing (ENT) is made of the same material (PVC) used for rigid nonmetallic conduit (Article 362) suitable for aboveground use. The outside diameters of ENT (½-in. through 2-in. trade sizes only) are such that standard couplings and other fittings for rigid PVC conduit can be used.

Because of the corrugations, the raceway can be bent by hand and has some degree of flexibility. ENT is not intended for use where flexibility is necessary, as at motor terminations to prevent transmission of noise and vibration, or for connection of adjustable luminaires or moving parts. ENT is suitable for the installation of conductors having a temperature rating as indicated on the product. The maximum allowable ambient temperature is 122°F. Exhibit 362.1 shows an example of electrical nonmetallic tubing.

Exhibit 362.1 An example of electrical nonmetallic tubing (ENT). (Courtesy of Carlon®, Lamson & Sessions)

362.6 Listing Requirements.

ENT and associated fittings shall be listed.

II. Installation

362.10 Uses Permitted.

For the purpose of this article, the first floor of a building shall be that floor that has 50 percent or more of the exterior wall surface area level with or above finished grade. One additional level that is the first level and not designed for human habitation and used only for vehicle parking, storage, or similar use shall be permitted. The use of ENT and fittings shall be permitted in the following:

(1) In any building not exceeding three floors above grade

Electrical nonmetallic tubing (ENT) is permitted to be installed, either concealed or exposed. Where exposed and subject to physical damage, ENT is required to be protected and is limited to use in buildings not exceeding three floors above grade, as defined in 362.10(A). [See the definition of *exposed (as applied to wiring methods)* in Article 100.] Where concealed or above a suspended ceiling, ENT is permitted to be installed within walls, floors, or ceilings in buildings of three floors or less without the need for fire-rated construction. The three-floor limitation is based on the likelihood that only a small quantity of ENT will be exposed to fire and that the occupants will have adequate time to exit the building before the products of combustion make the building untenable. Exhibit 362.2 illustrates permitted uses of ENT in a building of three floors or less.

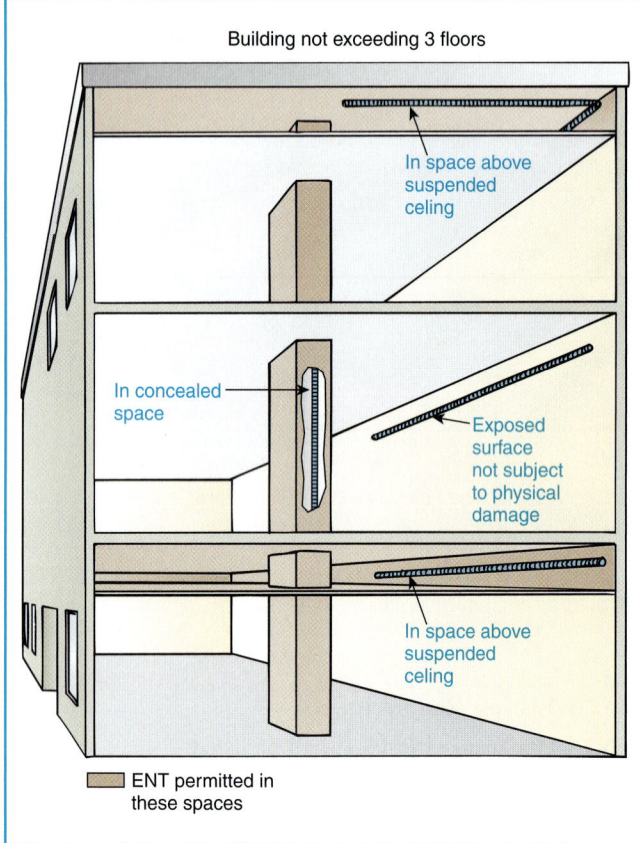

Building not exceeding 3 floors

In space above suspended ceiling

In concealed space

Exposed surface not subject to physical damage

In space above suspended ceiling

ENT permitted in these spaces

Exhibit 362.2 Examples of permitted uses of ENT in a building not exceeding three floors.

   a. For exposed work, where not prohibited by 362.12
   b. Concealed within walls, floors, and ceilings

(2) In any building exceeding three floors above grade, ENT shall be concealed within walls, floors, and ceilings where the walls, floors, and ceilings provide a thermal barrier of material that has at least a 15-minute finish

rating as identified in listings of fire-rated assemblies. The 15-minute-finish-rated thermal barrier shall be permitted to be used for combustible or noncombustible walls, floors, and ceilings.

Exception: Where a fire sprinkler system(s) is installed in accordance with NFPA 13-1999, Standard for the Installation of Sprinkler Systems, on all floors, ENT is permitted to be used within walls, floors, and ceilings, exposed or concealed, in buildings exceeding three floors above grade.

Electrical nonmetallic tubing (ENT) is permitted to be installed within the walls, floors, or ceilings of a building of any height where the walls, floors, or ceilings provide a thermal barrier of material that has at least a 15-minute finish rating. It is not permitted or intended that ENT be used exposed in the first three floors of a building that exceeds three floors except as permitted in 362.10(5). Where installed in a building over three floors, ENT must be installed behind the 15-minute thermal barrier on all floors. Exhibit 362.3 illustrates permitted uses of ENT in a building exceeding three floors. An addition to the 2002 *Code*, the new exception concerning a fire sprinkler system negates the requirement for a 15-minute finish rating.

FPN: A finish rating is established for assemblies containing combustible (wood) supports. The finish rating is defined as the time at which the wood stud or wood joist reaches an average temperature rise of 121°C (250°F) or an individual temperature of 163°C (325°F) as measured on the plane of the wood nearest the fire. A finish rating is not intended to represent a rating for a membrane ceiling.

Interior finish is generally considered to consist of those materials or combinations of materials that form the exposed interior surface of walls and ceilings in a building. Common interior finish materials include plaster, gypsum wallboard, wood, plywood paneling, fibrous ceiling tiles, and a variety of wall coverings. Ordinary paint, wallpaper, or other similar wall coverings not exceeding $^1/_{28}$ in. in thickness are generally considered incidental to interior finish, except where the authority having jurisdiction deems them a hazard. For more information regarding classification of interior finish material, refer to 10.2.1 of NFPA *101, Life Safety Code.*

   The finish rating of a wall or ceiling finish material is the time required for the unexposed surface of the finish membrane to reach an average temperature rise of 250°F above ambient or an individual temperature rise at any one point not exceeding 325°F when the assembly is tested in accordance with NFPA 251, *Standard Methods of Tests of Fire Endurance of Building Construction and Materials* (also known as ANSI/UL 263 or ASTM E119).

   The finish rating of wall and ceiling finish materials

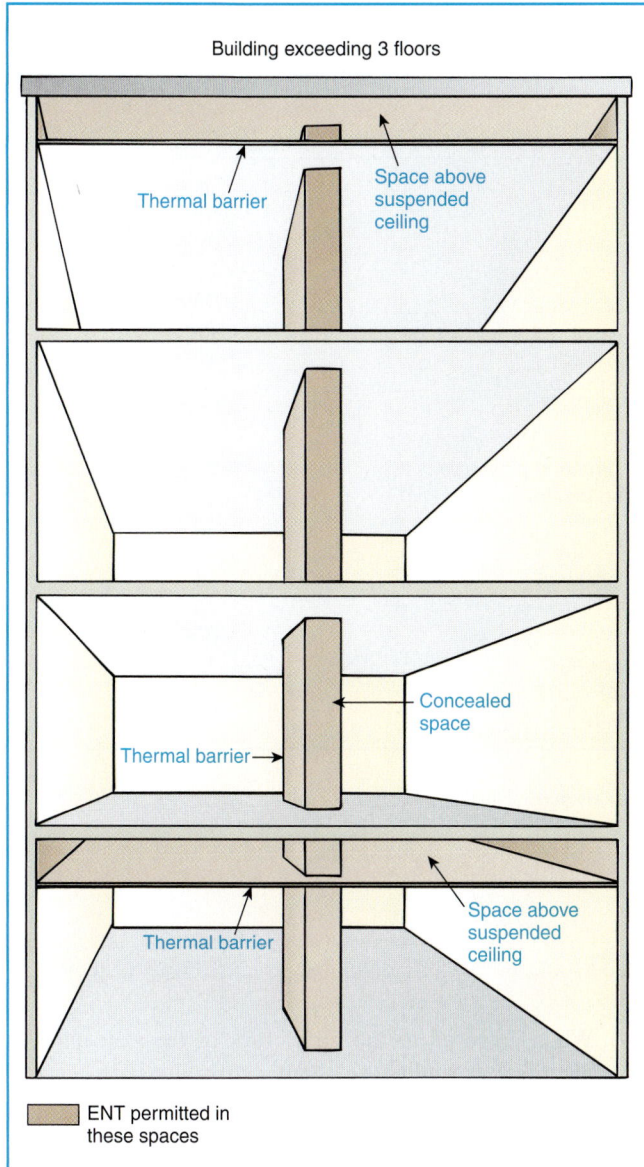

Building exceeding 3 floors

Thermal barrier

Space above suspended ceiling

Concealed space

Thermal barrier

Space above suspended ceiling

Thermal barrier

ENT permitted in these spaces

Exhibit 362.3 Examples of permitted uses of ENT in a building exceeding three floors.

tested and rated by UL as part of wall and ceiling assemblies can be found in the UL *Fire Resistance Directory,* immediately following the assembly rating and just below the design number. Only assemblies containing combustible support members, however, have published finish ratings. Obviously, it is not the intent to limit ENT to constructions consisting of combustible support members. This section is intended to provide a 15-minute thermal barrier as a minimum threshold of acceptability.

Table 362.1, reproduced from the NFPA *Fire Protection Handbook,* (Chapter 7, Table 7-4O, page 7-68), provides ratings for common finish materials. If the finish rating

concealing the ENT is unknown or is less than 15 minutes, the ENT can still be used if the installation meets the criteria in 362.10, including the three-floor limitation, where required, and the installation is not prohibited by 362.12. For finish materials not tested and rated in the *UL Fire Resistance Directory,* use Table 362.1 of the commentary.

(3) In locations subject to severe corrosive influences as covered in 300.6 and where subject to chemicals for which the materials are specifically approved.
(4) In concealed, dry, and damp locations not prohibited by 362.12.
(5) Above suspended ceilings where the suspended ceilings provide a thermal barrier of material that has at least a 15-minute finish rating as identified in listings of fire-rated assemblies, except as permitted in 362.10(1)(a).

Exception: Where a fire sprinkler system(s) is installed in accordance with NFPA 13–1999, Standard for the Installation of Sprinkler Systems–1999, on all floors, ENT is permitted to be used within walls, floors, and ceilings, exposed or concealed, in buildings exceeding three floors above grade.

(6) Encased in poured concrete, or embedded in a concrete slab on grade where ENT is placed on sand or approved screenings, provided fittings identified for this purpose are used for connections.
(7) For wet locations indoors as permitted in this section or in a concrete slab on or below grade, with fittings listed for the purpose.
(8) Metric designator 16 through 27 (trade size ½ through 1) as listed manufactured prewired assembly.

FPN: Extreme cold may cause some types of nonmetallic conduits to become brittle and therefore more susceptible to damage from physical contact.

Prewired ENT is a listed assembly whose conductors must be installed at the manufacturing facility, where controlled conditions will prevent damage to the conductor insulation. Special tools are required when cutting prewired ENT to prevent "nicking" the conductor installation. A prewired assembly is shown in Exhibit 362.4.

362.12 Uses Not Permitted.

ENT shall not be used in the following:

(1) In hazardous (classified) locations, except as permitted by 504.20 and 505.15(A)(1)
(2) For the support of luminaires (fixtures) and other equipment
(3) Where subject to ambient temperatures in excess of 50°C (122°F) unless listed otherwise
(4) For conductors whose insulation temperature limitations would exceed those for which the tubing is listed

Table 362.1 Various Finishes Over Wood Framing, One Side (Combustible) with Exposure on Finish Side

Finish Material	Fire Resistance Rating[1] (minutes)
Fiberboard, ½ in. thick	5
Fiberboard, flameproofed, ½ in. thick	10
Fiberboard, ½ in. thick, with ½-in. 1:2, 1:2 gypsum-sand plaster	15
Gypsum wallboard, ⅜ in. thick	10
Gypsum wallboard, ½ in. thick	15
Gypsum wallboard, ⅝ in. thick	20
Gypsum wallboards, laminated, two ⅜ in. thick	28
Gypsum wallboards, laminated, one ⅜ in. plus one ½ in. thick	37
Gypsum wallboards, laminated, two ½ in. thick	47
Gypsum wallboards, laminated, two ⅝ in. thick	60
Gypsum lath, plain or indented, ⅜ in. thick, with ½-in. 1:2, 1:2 gypsum-sand plaster	20
Gypsum lath, perforated, ⅜ in. thick, with ½-in. 1:2, 1:2 gypsum-sand plaster	30
Gypsum-sand plaster, 1:2, 1:3, ½ in. thick, on wood lath	15
Lime-sand plaster, 1:5, 1:7.5, ½ in. thick, on wood lath	15
Gypsum-sand plaster, 1:2, 1:2, ¾ in. thick, on metal lath (no paper backing)	15
Neat gypsum plaster, ¾ in. thick on metal lath (no paper backing)[2]	15
Neat gypsum plaster, 1 in. thick on metal lath (no paper backing)[2]	35
Lime-sand plaster, 1:5, 1:7.5, ¾ in. thick, on metal lath (no paper backing)	10
Portland cement plaster, ¾ in. thick, on metal lath (no paper backing)	10
Gypsum-sand plaster, 1:2, 1:3, ¾ in. thick, on paper-backed metal lath	20

Note: For SI units, 1 in. = 25.4 mm.
[1]From National Bureau of Standards, now known as the National Institute for Standards and Technology, BMS-92.
[2]Unsanded wood-fiber plaster.

Exhibit 362.4 Listed manufactured prewired ENT assembly. (Courtesy of Carlon®, Lamson & Sessions)

(5) For direct earth burial
(6) Where the voltage is over 600 volts
(7) In exposed locations, except as permitted by 362.10(1), 362.10(5), and 362.10(7)
(8) In theaters and similar locations, except as provided in Articles 518 and 520

See 518.4 and 520.5 for details of permitted wiring methods.

(9) Where exposed to the direct rays of the sun, unless identified as sunlight resistant
(10) Where subject to physical damage

362.20 Size.

(A) Minimum. ENT smaller than metric designator 16 (trade size ½) shall not be used.

(B) Maximum. ENT larger than metric designator 53 (trade size 2) shall not be used.

> FPN: See 300.1(C) for the metric designators and trade sizes. These are for identification purposes only and do not relate to actual dimensions.

362.22 Number of Conductors.

The number of conductors shall not exceed that permitted by the percentage fill in Table 1, Chapter 9.

Cables shall be permitted to be installed where such use is permitted by the respective cable articles. The number of cables shall not exceed the allowable percentage fill specified in Table 1, Chapter 9.

Table 1 of Chapter 9 specifies the maximum fill percentage of a conduit or tubing. Table 4 provides the usable area within the selected conduit or tubing, and Table 5 provides the required area for each conductor. Examples using these tables to calculate a conduit or tubing size are provided following Chapter 9, Table 1, Notes to Tables, Note 6.

If the conductors are of the same wire size, Annex C may be consulted. Annex C, which contains 12 sets of tables, very accurately indicates the maximum number of conductors permitted in a conduit or tubing. Examples using this annex to select a conduit or tubing size are provided following the introduction in Annex C.

To select the proper trade size electrical nonmetallic tubing, the section entitled "348—Electrical Nonmetallic Tubing (ENT)" in Table 4 of Chapter 9 should be followed. Annex C Tables C2 and C2A for electrical nonmetallic tubing are also permissible.

362.24 Bends—How Made.

Bends shall be made so that the tubing will not be damaged and that the internal diameter of the tubing will not be effectively reduced. Bends shall be permitted to be made manually without auxiliary equipment, and the radius of the curve to the centerline of such bends shall not be less than shown in Table 344.24 using the column "Other Bends."

362.26 Bends—Number in One Run.

There shall not be more than the equivalent of four quarter bends (360 degrees total) between pull points, for example, conduit bodies and boxes.

362.28 Trimming.

All cut ends shall be trimmed inside and outside to remove rough edges.

362.30 Securing and Supporting.

ENT shall be installed as a complete system as provided in Article 300 and shall be securely fastened in place and supported in accordance with 362.30(A) and (B).

(A) Securely Fastened. ENT shall be securely fastened at intervals not exceeding 900 mm (3 ft). In addition, ENT shall be securely fastened in place within 900 mm (3 ft) of each outlet box, device box, junction box, cabinet, or fitting where it terminates.

Exception: Lengths not exceeding a distance of 1.8 m (6 ft) from a luminaire (fixture) terminal connection for tap connections to lighting luminaires (fixtures) shall be permitted without being secured.

As illustrated in Exhibit 362.5, where ENT is run on the surface of framing members, it is required to be fastened to

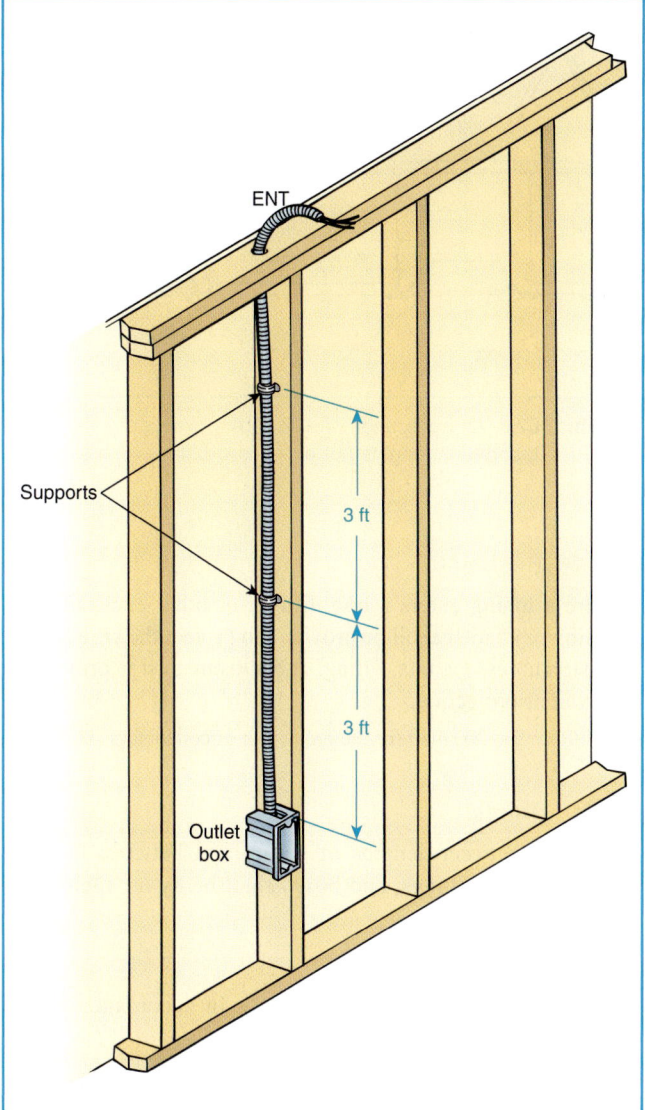

Exhibit 362.5 An application of 362.30(A), showing ENT supported every 3 ft and within 3 ft of the box.

the framing member every 3 ft and within 3 ft of every box. See 300.4(D) for provisions on protection against physical damage.

As illustrated in Exhibit 362.6, ENT is permitted to be used as fixture whip without support for lengths not exceeding 6 ft. See 410.67(C) for details on tap conductor wiring.

(B) Supports. Horizontal runs of ENT supported by openings in framing members at intervals not exceeding 900 mm (3 ft) and securely fastened within 900 mm (3 ft) of termination points shall be permitted.

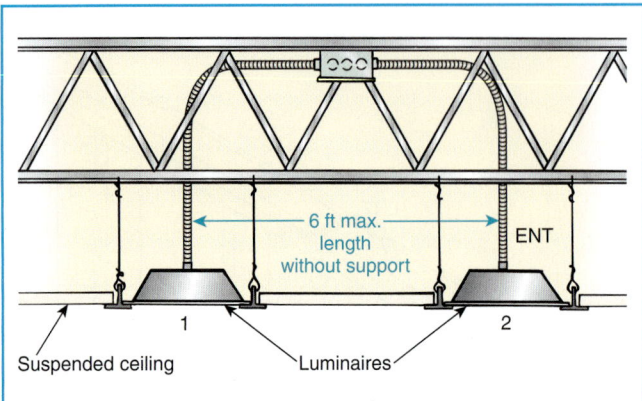

6 ft max. length without support

ENT

1   2

Suspended ceiling   Luminaires

Exhibit 362.6 An application of 362.30(A), Exception, showing ENT unsupported in lengths not exceeding 6 ft.

362.46 Bushings.

Where a tubing enters a box, fitting, or other enclosure, a bushing or adapter shall be provided to protect the wire from abrasion unless the box, fitting, or enclosure design provides equivalent protection.

> FPN: See 300.4(F) for the protection of conductors size 4 AWG or larger.

362.48 Joints.

All joints between lengths of tubing and between tubing and couplings, fittings, and boxes shall be by an approved method.

362.56 Splices and Taps.

Splices and taps shall be made only in accordance with 300.15.

> FPN: See Article 314 for rules on the installation and use of boxes and conduit bodies.

362.60 Grounding.

Where equipment grounding is required by Article 250, a separate equipment grounding conductor shall be installed in the raceway.

III. Construction Specifications

362.100 Construction.

ENT shall be made of material that does not exceed the ignitibility, flammability, smoke generation, and toxicity characteristics of rigid (nonplasticized) polyvinyl chloride.

ENT, as a prewired manufactured assembly, shall be provided in continuous lengths capable of being shipped in a coil, reel, or carton without damage.

362.120 Marking.

ENT shall be clearly and durably marked at least every 3 m (10 ft) as required in the first sentence of 110.21. The type of material shall also be included in the marking. Marking for

limited smoke shall be permitted on the tubing that has limited smoke-producing characteristics.

The type, size, and quantity of conductors used in pre-wired manufactured assemblies shall be identified by means of a printed tag or label attached to each end of the manufactured assembly and either the carton, coil, or reel. The enclosed conductors shall be marked in accordance with 310.11.

ARTICLE 366
Auxiliary Gutters

Contents

366.1 Scope.

This article covers the use, installation and construction requirements of metal auxiliary gutters and nonmetallic auxiliary gutters and associated fittings.

366.2 Use.

Auxiliary gutters shall be permitted to supplement wiring spaces at meter centers, distribution centers, switchboards, and similar points of wiring systems and may enclose conductors or busbars but shall not be used to enclose switches, overcurrent devices, appliances, or other similar equipment.

Auxiliary gutter sections and associated fittings are identical to those of wireways, and if listed by UL, bear the UL listing mark "Listed Wireway" or "Auxiliary Gutter." They differ only in their intended use. See the commentary following 376.2 for a comparative discussion. Gutters (and wireways) are required to be constructed and installed to ensure adequate electrical and mechanical continuity of the complete system per 250.118(14).

Auxiliary gutters installed in wet locations are required to be suitable for such locations. See 620.35 for less restrictive requirements where auxiliary gutters are used for elevators, dumbwaiters, escalators, and moving walks.

366.3 Extension Beyond Equipment.

An auxiliary gutter shall not extend a greater distance than 9 m (30 ft) beyond the equipment that it supplements.

Exception: As permitted in 620.35 for elevators, an auxiliary gutter shall be permitted to extend a distance greater than 9 m (30 ft) beyond the equipment that it supplements.

FPN: For wireways, see Articles 376 and 378. For busways, see Article 368.

366.4 Supports.

(A) Sheet Metal Auxiliary Gutters. Sheet metal auxiliary gutters shall be supported throughout their entire length at intervals not exceeding 1.5 m (5 ft).

(B) Nonmetallic Auxiliary Gutters. Nonmetallic auxiliary gutters shall be supported at intervals not to exceed 900 mm (3 ft) and at each end or joint, unless listed for other support intervals. In no case shall the distance between supports exceed 3 m (10 ft).

366.5 Covers.

Covers shall be securely fastened to the gutter.

366.6 Number of Conductors.

(A) Sheet Metal Auxiliary Gutters. The sum of the cross-sectional areas of all contained conductors at any cross section of a sheet metal auxiliary gutter shall not exceed 20 percent of the interior cross-sectional area of the sheet metal auxiliary gutter. The derating factors in 310.15(B)(2)(a) shall be applied only where the number of current-carrying conductors, including neutral conductors classified as current-carrying under the provisions of 310.15(B)(4), exceeds 30. Conductors for signaling circuits or controller conductors between a motor and its starter and used only for starting duty shall not be considered as current-carrying conductors.

(B) Nonmetallic Auxiliary Gutters. The sum of cross-sectional areas of all contained conductors at any cross section of the nonmetallic auxiliary gutter shall not exceed 20 percent of the interior cross-sectional area of the nonmetallic auxiliary gutter.

Section 366.6 calls out the requirements for metal and non-metallic auxiliary gutters. A common requirement for both types of auxiliary gutters is that all of the contained conductors must not exceed 20 percent fill of the interior cross-sectional area of the gutter. The dimensions of insulated conductors, found in Tables 5 and 5A of Chapter 9, may be used to compute the size of auxiliary gutters.

Where sheet metal auxiliary gutters contain 30 or fewer current-carrying conductors, the correction factors in 310.15(B)(2) do not apply. However, if more than 30 conductors are installed in a sheet metal auxiliary gutter, the ampacity adjustment factors of 310.15(B)(2) apply and there is no limit on the number of current-carrying conductors up to the 20 percent fill.

The requirements for nonmetallic auxiliary gutters limit the cross-sectional area of all conductors to 20 percent. There is no 30-conductor allowance. The derating factors specified in 310.15(B)(2) must be applied.

See the example for calculating the size of a wireway in the commentary following 376.22. This calculation method is also applicable to auxiliary gutters.

No limit is placed on the size of conductors that may be installed in an auxiliary gutter; however, see 366.7 for ampacity limitations of bare copper or aluminum busbars enclosed in gutters.

366.7 Ampacity of Conductors.

(A) Sheet Metal Auxiliary Gutters. Where the number of current-carrying conductors contained in the sheet metal auxiliary gutter is 30 or less, the correction factors specified in 310.15(B)(2)(a) shall not apply. The current carried continuously in bare copper bars in sheet metal auxiliary gutters shall not exceed 1.55 amperes/mm² (1000 amperes/in.²) of cross section of the conductor. For aluminum bars, the current carried continuously shall not exceed 1.09 amperes/mm² (700 amperes/in.²) of cross section of the conductor.

(B) Nonmetallic Auxiliary Gutters. The derating factors specified in 310.15(B)(2)(a) shall be applicable to the current-carrying conductors in the nonmetallic auxiliary gutter.

366.8 Clearance of Bare Live Parts.

Bare conductors shall be securely and rigidly supported so that the minimum clearance between bare current-carrying metal parts of different potential mounted on the same surface will not be less than 50 mm (2 in.), nor less than 25 mm (1 in.) for parts that are held free in the air. A clearance not less than 25 mm (1 in.) shall be secured between bare current-carrying metal parts and any metal surface. Adequate provisions shall be made for the expansion and contraction of busbars.

366.9 Splices and Taps.

Splices and taps shall comply with 366.9(A) through (D).

(A) Within Gutters. Splices or taps shall be permitted within gutters where they are accessible by means of removable covers or doors. The conductors, including splices and taps, shall not fill the gutter to more than 75 percent of its area.

(B) Bare Conductors. Taps from bare conductors shall leave the gutter opposite their terminal connections, and conductors shall not be brought in contact with uninsulated current-carrying parts of different potential.

(C) Suitably Identified. All taps shall be suitably identified at the gutter as to the circuit or equipment that they supply.

(D) Overcurrent Protection. Tap connections from conductors in auxiliary gutters shall be provided with overcurrent protection as required in 240.21.

To prevent abrasion of the conductor insulation, suitable bushings, shields, and so on, must be provided where conductors pass around bends or between gutters and cabinets and other locations. Sections 366.9(C) and (D) require all taps from gutters to be identified (as to circuits or equipment) and be protected with overcurrent devices per 240.21.

366.10 Construction and Installation.

Auxiliary gutters shall comply with 366.10(A) through (F).

(A) Electrical and Mechanical Continuity. Gutters shall be constructed and installed so that adequate electrical and mechanical continuity of the complete system is secured.

(B) Substantial Construction. Gutters shall be of substantial construction and shall provide a complete enclosure for the contained conductors. All surfaces, both interior and exterior, shall be suitably protected from corrosion. Corner joints shall be made tight, and where the assembly is held together by rivets, bolts, or screws, such fasteners shall be spaced not more than 300 mm (12 in.) apart.

(C) Smooth Rounded Edges. Suitable bushings, shields, or fittings having smooth, rounded edges shall be provided where conductors pass between gutters, through partitions, around bends, between gutters and cabinets or junction boxes, and at other locations where necessary to prevent abrasion of the insulation of the conductors.

(D) Deflected Insulated Conductors. Where insulated conductors are deflected within an auxiliary gutter, either at the ends or where conduits, fittings, or other raceways or cables enter or leave the gutter, or where the direction of the gutter is deflected greater than 30 degrees, dimensions corresponding to 312.6 shall apply.

Also, conductors are required to be shaped or formed in a permanent manner so that they are not in contact with bare busbars within the gutter.

(E) Indoor and Outdoor Use.

(1) Sheet Metal Auxiliary Gutters. Sheet metal auxiliary gutters installed in wet locations shall be suitable for such locations.

(2) Nonmetallic Auxiliary Gutters.

(a) Nonmetallic auxiliary gutters installed outdoors shall comply with the following:

(1) Be listed and marked as suitable for exposure to sunlight
(2) Be listed and marked as suitable for use in wet locations
(3) Be listed for the maximum ambient temperature of the installation, and marked for the installed conductor insulation temperature rating
(4) Have expansion fittings installed where the expected length change due to expansion and contraction due to temperature change is more than 6 mm (0.25 in.)

(b) Nonmetallic auxiliary gutters installed indoors shall comply with the following:

(1) Be listed for the maximum ambient temperature of the installation and marked for the installed conductor insulation temperature rating
(2) Have expansion fittings installed where expected length change, due to expansion and contraction due to temperature change, is more than 6 mm (0.25 in.)

FPN: Extreme cold may cause nonmetallic auxiliary gutter to become brittle and therefore more susceptible to damage from physical contact.

This section provides requirements for both indoor and outdoor installations. Nonmetallic gutters must have expansion fittings where temperature changes are expected to change gutter length more than ¼ in. See the fine print note following 378.44 regarding expansion characteristic of PVC rigid nonmetallic conduit and PVC nonmetallic wireway.

(F) Grounding. Grounding shall be in accordance with the provisions of Article 250.

ARTICLE 368
Busways

Contents

I. General Requirements

368.1 Scope.

This article covers service-entrance, feeder, and branch-circuit busways and associated fittings.

368.2 Definition.

Busway. A grounded metal enclosure containing factory-mounted, bare or insulated conductors, which are usually copper or aluminum bars, rods, or tubes.

> FPN: For cablebus, refer to Article 370.

According to the 2001 UL *Electrical Construction Materials Directory*, category CWFT, busways and short-run busways are provided with metal enclosures. These enclosures, and in some cases an additional ground bus, are intended for use as equipment grounding conductors. Some busways are not intended for use ahead of service equipment and are marked with the maximum rating of overcurrent protection to be used on the supply side of the busway. Busways that have been investigated to determine suitability for installation in a specified position, for use in vertical runs, for support at intervals greater than 5 ft, or for outdoor use are so marked. This marking is on or contiguous with the nameplate incorporating the manufacturer's name and electrical rating. A busway or fitting containing a vapor seal is so marked, but unless marked otherwise, the busway or fitting has not been investigated for passage through a fire wall.

Short-run busways are marked to limit the run to 30 ft or less, and no more than 10 ft vertically. They are intended primarily to feed switchboards. Except for transformer stubs, short-run busways are not intended to have intermediate taps. Short-run busways are not ventilated and may be marked for outdoor use.

Busways and associated fittings marked "Short Circuit Current Rating(s) Maximum rms Symmetrical Amps _____ Volts _____" have been investigated for the rating indicated.

Busways that are intended to supply and support industrial and commercial luminaires are classified as "Lighting Busway" and are so marked. Trolley busway is marked "Trolley Busway" and is additionally marked "Lighting Busway" if intended to supply and support industrial and commercial luminaires. Busway with provision for insertion of plug-in devices at any point along its length and intended for general use is classified as "Continuous Plug-in Busway" and is so marked. The marking is contiguous with the marking of the manufacturer's name and the electrical rating.

Busway marked "Lighting Busway" and protected by overcurrent devices rated in excess of 20 amperes is intended for use only with luminaires employing heavy-duty lampholders unless additional overcurrent protection is provided for the luminaire in accordance with this *Code*.

Trolley busway should be installed out of the reach of people, or it should be otherwise installed to prevent accidental contact with exposed conductors.

368.4 Use.

(A) Uses Permitted. Busways shall be permitted to be installed where they are located as follows:

(1) Located in the open and are visible, except as permitted in 368.6, or

(2) Installed behind access panels, provided the busways are totally enclosed, of nonventilating-type construction, and installed so that the joints between sections and at fittings are accessible for maintenance purposes. Where installed behind access panels, means of access shall be provided, and the following conditions shall be met:

   a. The space behind the access panels shall not be used for air-handling purposes, or
   b. Where the space behind the access panels is used for environmental air, other than ducts and plenums, there shall be no provisions for plug-in connections, and the conductors shall be insulated.

Unless busways are mounted in the open and are visible, the installation must comply with 368.4(A)(2). Busways are commonly used as feeders and are mounted horizontally in industrial buildings or mounted vertically in high-rise buildings. See Exhibit 100.2 for an example of a busway mounted above a hung ceiling.

(B) Uses Not Permitted. Busways shall not be installed as follows:

(1) Where subject to severe physical damage or corrosive vapors
(2) In hoistways
(3) In any hazardous (classified) location, unless specifically approved for such use

   FPN: See 501.4(B).

(4) Outdoors or in wet or damp locations unless identified for such use

Lighting busway and trolley busway shall not be installed less than 2.5 m (8 ft) above the floor or working platform unless provided with a cover identified for the purpose.

368.5 Support.

Busways shall be securely supported at intervals not exceeding 1.5 m (5 ft) unless otherwise designed and marked.

368.6 Through Walls and Floors.

(A) Walls. Unbroken lengths of busway shall be permitted to be extended through dry walls.

(B) Floors. Floor penetrations shall comply with (1) and (2):

(1) Busways shall be permitted to be extended vertically through dry floors if totally enclosed (unventilated) where passing through and for a minimum distance of

1.8 m (6 ft) above the floor to provide adequate protection from physical damage.

(2) In other than industrial establishments, where a vertical riser penetrates two or more dry floors, a minimum 100 mm (4 in.) high curb shall be installed around all floor openings for riser busways to prevent liquids from entering the opening. The curb shall be installed within 300 mm (12 in.) of the floor opening. Electrical equipment shall be located so that it will not be damaged by liquids that are retained by the curb.

   FPN: See 300.21 for information concerning the spread of fire or products of combustion.

A busway or fitting containing a vapor seal is so marked, but, unless marked otherwise, the busway or fitting has not been investigated for passage through a fire-rated wall. The requirements of 300.21 are most important in order to confine a fire and the products of combustion at their origin.

Revised for the 1999 *Code,* this section requires that a curb be placed around a busway if the busway penetrates two or more dry floors. Experience has shown that if liquid spills occur on upper floors of normally dry buildings, the spilled liquid often flows to the busway floor penetration and down the vertical rise of the busway. Spills can cause extensive damage to the busway and the building's electrical system. The addition of a 4-in. curb encircling the busway can help eliminate this potentially dangerous situation.

368.7 Dead Ends.

A dead end of a busway shall be closed.

368.8 Branches from Busways.

Branches from busways shall be permitted to be made in accordance with 368.8(A), (B), and (C).

(A) General. Branches from busways shall be made in accordance with Articles 320, 330, 332, 342, 344, 348, 350, 352, 356, 358, 362, 368, 384, 386, and 388. Where a separate equipment grounding conductor is used, connection of the equipment grounding conductor to the busway shall comply with 250.8 and 250.12.

(B) Cord and Cable Assemblies. Suitable cord and cable assemblies approved for extra-hard usage or hard usage and listed bus drop cable shall be permitted as branches from busways for the connection of portable equipment or the connection of stationary equipment to facilitate their interchange in accordance with 400.7 and 400.8 and the following conditions:

(1) The cord or cable shall be attached to the building by an approved means.

(2) The length of the cord or cable from a busway plug-in device to a suitable tension take-up support device shall not exceed 1.8 m (6 ft).

Exhibit 368.1 shows an example of a cable or cord branch from a busway installed according to the provisions of 368.8(B)(2).

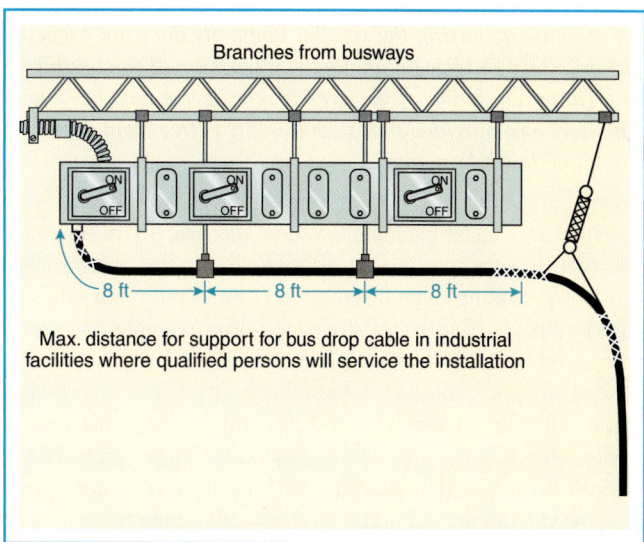

Max. distance for support for bus drop cable in industrial facilities where qualified persons will service the installation

Exhibit 368.2 An example of an installation permitted only in industrial occupancies with other restrictions according to 368.8(B)(2), Exception.

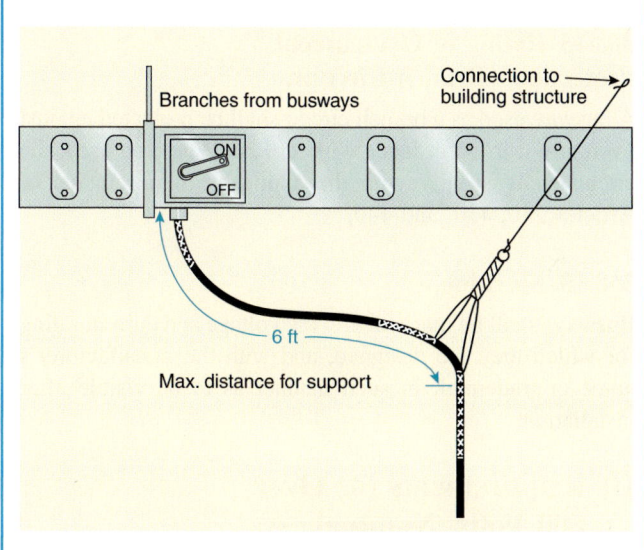

Exhibit 368.1 An example of a cable or cord branch from a busway installed according to 368.8(B).

Exception: In industrial establishments only, where the conditions of maintenance and supervision ensure that only qualified persons service the installation, lengths exceeding 1.8 m (6 ft) shall be permitted between the busway plug-in device and the tension take-up support device where the cord or cable is supported at intervals not exceeding 2.5 m (8 ft).

Section 400.9 specifically prohibits the installation of spliced cords. Exhibit 368.2 shows an example of a cable or cord branch from a busway installed according to 368.8(B)(2), Exception.

(3) The cord or cable shall be installed as a vertical riser from the tension take-up support device to the equipment served.
(4) Strain relief cable grips shall be provided for the cord or cable at the busway plug-in device and equipment terminations.

(C) Branches from Trolley-Type Busways. Suitable cord and cable assemblies approved for extra-hard usage or hard usage and listed bus drop cable shall be permitted as branches

from trolley-type busways for the connection of movable equipment in accordance with 400.7 and 400.8.

368.9 Overcurrent Protection.

Overcurrent protection shall be provided in accordance with 368.10 through 368.13.

368.10 Rating of Overcurrent Protection—Feeders.

A busway shall be protected against overcurrent in accordance with the allowable current rating of the busway.

The rated ampacity of a busway is based on the allowable temperature rise of the conductors and can be determined in the field only by reference to the nameplate data. The requirements of 240.4(B) and 240.4(C) are applicable for busways.

Exception No. 1: The applicable provisions of 240.4 shall be permitted.

Exception No. 2: Where used as transformer secondary ties, the provisions of 450.6(A)(3) shall be permitted.

368.11 Reduction in Ampacity Size of Busway.

Overcurrent protection shall be required where busways are reduced in ampacity.

Exception: For industrial establishments only, omission of overcurrent protection shall be permitted at points where busways are reduced in ampacity, provided that the length of the busway having the smaller ampacity does not exceed 15 m (50 ft) and has an ampacity at least equal to one-third the rating or setting of the overcurrent device next back on the line, and provided that such busway is free from contact with combustible material.

In industrial establishments, where the size of a smaller busway is kept within the specified limits, the additional cost of providing overcurrent protection at the point where the size is changed is not warranted. For example, busway protected by a 1200-ampere overcurrent device may be reduced in size, provided the smaller busway has a current rating of 400 amperes ($^1/_3$ of 1200 amperes) and does not extend more than 50 ft. In this case, overcurrent protection would be required if the smaller busway were rated less than 400 amperes (e.g., 200 amperes, 300 amperes).

368.12 Feeder or Branch Circuits.

Where a busway is used as a feeder, devices or plug-in connections for tapping off feeder or branch circuits from the busway shall contain the overcurrent devices required for the protection of the feeder or branch circuits. The plug-in device shall consist of an externally operable circuit breaker or an externally operable fusible switch. Where such devices are mounted out of reach and contain disconnecting means, suitable means such as ropes, chains, or sticks shall be provided for operating the disconnecting means from the floor.

Externally operated fused switches and circuit breakers that are plugged into busways and mounted out of reach are to be considered accessible if operated by means such as ropes, chains, or hooksticks. Exhibit 368.3 shows a 10-ft section of feeder busway. See Exhibit 100.1 for other example illustrations.

Exhibit 368.3 A 10-ft section of feeder busway.

Exception No. 1: As permitted in 240.21.

Exception No. 2: For fixed or semifixed luminaires (lighting fixtures), where the branch-circuit overcurrent device is part of the luminaire (fixture) cord plug on cord-connected luminaires (fixtures).

Exception No. 3: Where luminaires (fixtures) without cords are plugged directly into the busway and the overcurrent device is mounted on the luminaire (fixture).

368.13 Rating of Overcurrent Protection—Branch Circuits.

A busway used as a branch circuit shall be protected against overcurrent in accordance with 210.20. Where so used, the circuit shall comply with the applicable requirements of Articles 210, 430, and 440.

368.15 Marking.

Busways shall be marked with the voltage and current rating for which they are designed, and with the manufacturer's name or trademark in such manner as to be visible after installation.

II. Requirements for Over 600 Volts, Nominal

368.21 Identification.

Each bus run shall be provided with a permanent nameplate on which the following information shall be provided:

(1) Rated voltage
(2) Rated continuous current; if bus is forced-cooled, both the normal forced-cooled rating and the self-cooled (not forced-cooled) rating for the same temperature rise shall be given
(3) Rated frequency
(4) Rated impulse withstand voltage
(5) Rated 60-Hz withstand voltage (dry)
(6) Rated momentary current
(7) Manufacturer's name or trademark

> FPN: See ANSI C37.23-1987 (R1991), *Guide for Metal-Enclosed Bus and Calculating Losses in Isolated-Phase Bus,* for construction and testing requirements for metal-enclosed buses.

368.22 Grounding.

Metal-enclosed bus shall be grounded in accordance with Article 250.

368.23 Adjacent and Supporting Structures.

Metal-enclosed busways shall be installed so that temperature rise from induced circulating currents in any adjacent metallic parts will not be hazardous to personnel or constitute a fire hazard.

368.24 Neutral.

Neutral bus, where required, shall be sized to carry all neutral load current, including harmonic currents, and shall have adequate momentary and short-circuit rating consistent with system requirements.

368.25 Barriers and Seals.

Bus runs that have sections located both inside and outside of buildings shall have a vapor seal at the building wall to prevent interchange of air between indoor and outdoor sections.

Exception: Vapor seals shall not be required in forced-cooled bus.

Fire barriers shall be provided where fire walls, floors, or ceilings are penetrated.

FPN: See 300.21 for information concerning the spread of fire or products of combustion.

368.26 Drain Facilities.

Drain plugs, filter drains, or similar methods shall be provided to remove condensed moisture from low points in bus run.

368.27 Ventilated Bus Enclosures.

Ventilated bus enclosures shall be installed in accordance with Article 110, Part III, and 490.24.

368.28 Terminations and Connections.

Where bus enclosures terminate at machines cooled by flammable gas, seal-off bushings, baffles, or other means shall be provided to prevent accumulation of flammable gas in the bus enclosures.

Flexible or expansion connections shall be provided in long, straight runs of bus to allow for temperature expansion or contraction, or where the bus run crosses building vibration insulation joints.

All conductor termination and connection hardware shall be accessible for installation, connection, and maintenance.

368.29 Switches.

Switching devices or disconnecting links provided in the bus run shall have the same momentary rating as the bus. Disconnecting links shall be plainly marked to be removable only when bus is de-energized. Switching devices that are not load-break shall be interlocked to prevent operation under load, and disconnecting link enclosures shall be interlocked to prevent access to energized parts.

368.30 Wiring 600 Volts or Less, Nominal.

Secondary control devices and wiring that are provided as part of the metal-enclosed bus run shall be insulated by fire-retardant barriers from all primary circuit elements with the exception of short lengths of wire, such as at instrument transformer terminals.

ARTICLE 370
Cablebus

Contents

370.1 Scope.

This article covers the use and installation requirements of cablebus and associated fittings.

370.2 Definition.

Cablebus. An assembly of insulated conductors with fittings and conductor terminations in a completely enclosed, ventilated protective metal housing. Cablebus is ordinarily assembled at the point of installation from the components furnished or specified by the manufacturer in accordance with instructions for the specific job. This assembly is designed to carry fault current and to withstand the magnetic forces of such current.

As shown in Exhibit 370.1, cablebus consists of a metal structure or framework installed in a manner similar to that for a cable tray support system. Insulated conductors, 1/0 AWG or larger, are field installed within the framework on special insulating blocks at specified intervals to provide controlled spacing between conductors. To completely enclose the conductors, a ventilated top cover is attached to the framework.

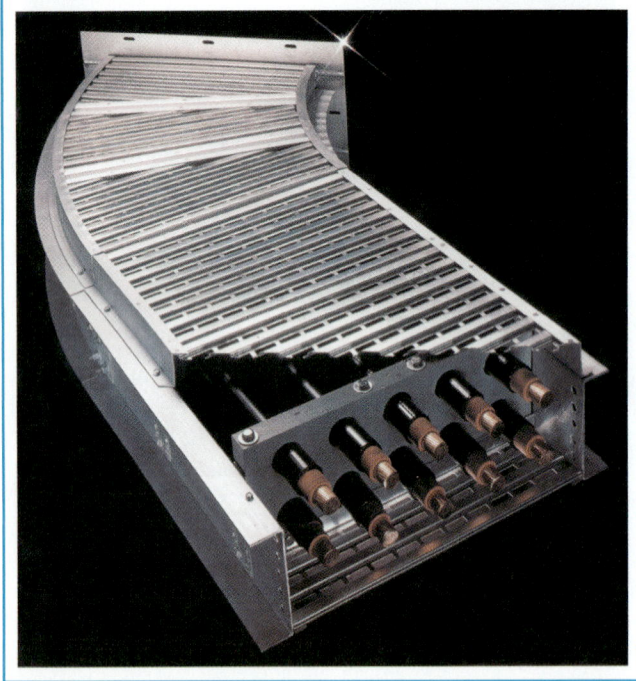

Exhibit 370.1 A section of cablebus with conductors in place and the ventilated top cover ready to be attached to the busway frame. (Courtesy of MPHusky Corp.)

370.3 Use.

Approved cablebus shall be permitted at any voltage or current for which spaced conductors are rated and shall be installed for exposed work only, except as permitted in 370.6. Cablebus installed outdoors or in corrosive, wet, or damp locations shall be identified for such use. Cablebus shall not be installed in hoistways or hazardous (classified) locations unless specifically approved for such use. Cablebus shall be permitted to be used for branch circuits, feeders, and services.

Cablebus framework, where bonded as required by Article 250, shall be permitted as the equipment grounding conductor for branch circuits and feeders.

370.4 Conductors.

(A) Types of Conductors. The current-carrying conductors in cablebus shall have an insulation rating of 75°C

(167°F) or higher of an approved type and suitable for the application in accordance with Articles 310 and 490.

(B) Ampacity of Conductors. The ampacity of conductors in cablebus shall be in accordance with Table 310.17 and Table 310.19, or with Table 310.69 and Table 310.70 for installations over 600 volts.

(C) Size and Number of Conductors. The size and number of conductors shall be that for which the cablebus is designed, and in no case smaller than 1/0 AWG.

(D) Conductor Supports. The insulated conductors shall be supported on blocks or other mounting means designed for the purpose.

The individual conductors in a cablebus shall be supported at intervals not greater than 900 mm (3 ft) for horizontal runs and 450 mm (1½ ft) for vertical runs. Vertical and horizontal spacing between supported conductors shall not be less than one conductor diameter at the points of support.

370.5 Overcurrent Protection.

Cablebus shall be protected against overcurrent in accordance with the allowable ampacity of the cablebus conductors in accordance with 240.4.

Exception: Overcurrent protection shall be permitted in accordance with 240.100 and 240.101 for over 600 volts, nominal.

370.6 Support and Extension Through Walls and Floors.

(A) Support. Cablebus shall be securely supported at intervals not exceeding 3.7 m (12 ft).

Exception: Where spans longer than 3.7 m (12 ft) are required, the structure shall be specifically designed for the required span length.

(B) Transversely Routed. Cablebus shall be permitted to extend transversely through partitions or walls, other than fire walls, provided the section within the wall is continuous, protected against physical damage, and unventilated.

(C) Through Dry Floors and Platforms. Except where firestops are required, cablebus shall be permitted to extend vertically through dry floors and platforms, provided the cablebus is totally enclosed at the point where it passes through the floor or platform and for a distance of 1.8 m (6 ft) above the floor or platform.

(D) Through Floors and Platforms in Wet Locations. Except where firestops are required, cablebus shall be permitted to extend vertically through floors and platforms in wet locations where (1) there are curbs or other suitable means to prevent waterflow through the floor or platform opening, and (2) where the cablebus is totally enclosed at the point where it passes through the floor or platform and for a distance of 1.8 m (6 ft) above the floor or platform.

370.7 Fittings.

A cablebus system shall include approved fittings for the following:

(1) Changes in horizontal or vertical direction of the run
(2) Dead ends
(3) Terminations in or on connected apparatus or equipment or the enclosures for such equipment
(4) Additional physical protection where required, such as guards where subject to severe physical damage

370.8 Conductor Terminations.

Approved terminating means shall be used for connections to cablebus conductors.

370.9 Grounding.

A cablebus installation shall be grounded and bonded in accordance with Article 250, excluding 250.86, Exception No. 2.

370.10 Marking.

Each section of cablebus shall be marked with the manufacturer's name or trade designation and the maximum diameter, number, voltage rating, and ampacity of the conductors to be installed. Markings shall be located so as to be visible after installation.

ARTICLE 372
Cellular Concrete Floor Raceways

Contents

372.1 Scope.

This article covers cellular concrete floor raceways, the hollow spaces in floors constructed of precast cellular concrete slabs, together with suitable metal fittings designed to provide access to the floor cells.

Cellular concrete floor raceways are a form of floor deck construction commonly used in high-rise office buildings. This construction method is very similar in design, application, and adaptation to cellular metal floor raceways. Basically, this wiring method consists of floor cells (that are part of the structural floor system), header ducts laid at right angles to the cells and used to carry conductors from cabinets to cells, and junction boxes.

372.2 Definitions.

Cell. A single, enclosed tubular space in a floor made of precast cellular concrete slabs, the direction of the cell being parallel to the direction of the floor member.

Header. Transverse metal raceways for electric conductors, providing access to predetermined cells of a precast cellular concrete floor, thereby permitting the installation of electric conductors from a distribution center to the floor cells.

372.3 Other Articles.

Cellular concrete floor raceways shall comply with the applicable provisions of Article 300.

372.4 Uses Not Permitted.

Conductors shall not be installed in precast cellular concrete floor raceways as follows:

(1) Where subject to corrosive vapor
(2) In any hazardous (classified) locations except as permitted by 504.20, and in Class I, Division 2 locations as permitted in 501.4(B)(3)
(3) In commercial garages, other than for supplying ceiling outlets or extensions to the area below the floor but not above

> FPN: See 300.8 for installation of conductors with other systems.

Section 300.8 prohibits the installation of electric conductors in raceways or cable trays containing any pipes, tubes, or other means for carrying steam, water, air, gas, or drainage, or for any service other than electrical.

372.5 Header.

The header shall be installed in a straight line at right angles to the cells. The header shall be mechanically secured to the top of the precast cellular concrete floor. The end joints shall be closed by a metal closure fitting and sealed against the entrance of concrete. The header shall be electrically continuous throughout its entire length and shall be electrically bonded to the enclosure of the distribution center.

372.6 Connection to Cabinets and Other Enclosures.

Connections from headers to cabinets and other enclosures shall be made by means of listed metal raceways and listed fittings.

372.7 Junction Boxes.

Junction boxes shall be leveled to the floor grade and sealed against the free entrance of water or concrete. Junction boxes shall be of metal and shall be mechanically and electrically continuous with the header.

372.8 Markers.

A suitable number of markers shall be installed for the future location of cells.

372.9 Inserts.

Inserts shall be leveled and sealed against the entrance of concrete. Inserts shall be of metal and shall be fitted with grounded-type receptacles. A grounding conductor shall connect the insert receptacles to a positive ground connection provided on the header. Where cutting through the cell wall for setting inserts or other purposes (such as providing access openings between header and cells), chips and other dirt shall not be allowed to remain in the raceway, and the tool used shall be designed so as to prevent the tool from entering the cell and damaging the conductors.

372.10 Size of Conductors.

No conductor larger than 1/0 AWG shall be installed, except by special permission.

372.11 Maximum Number of Conductors.

The combined cross-sectional area of all conductors or cables shall not exceed 40 percent of the cross-sectional area of the cell or header.

372.12 Splices and Taps.

Splices and taps shall be made only in header access units or junction boxes.

For the purposes of this section, so-called loop wiring (continuous unbroken conductor connecting the individual outlets) shall not be considered to be a splice or tap.

372.13 Discontinued Outlets.

When an outlet is abandoned, discontinued, or removed, the sections of circuit conductors supplying the outlet shall be removed from the raceway. No splices or reinsulated conductors, such as would be the case of abandoned outlets on loop wiring, shall be allowed in raceways.

ARTICLE 374
Cellular Metal Floor Raceways

Contents

374.1 Scope.

This article covers the use and installation requirements for cellular metal floor raceways.

Cellular metal floor raceways are a form of metal floor deck construction designed for use in steel-frame buildings and consisting of sheet metal formed into shapes that are combined to form cells or raceways. The cells extend across the building and, depending on the structural strength required, can have various shapes and sizes.

374.2 Definitions.

Cellular Metal Floor Raceway. The hollow spaces of cellular metal floors, together with suitable fittings, that may be approved as enclosures for electric conductors.

Cell. A single, enclosed tubular space in a cellular metal floor member, the axis of the cell being parallel to the axis of the metal floor member.

Header. A transverse raceway for electric conductors, providing access to predetermined cells of a cellular metal floor, thereby permitting the installation of electric conductors from a distribution center to the cells.

374.3 Uses Not Permitted.

Conductors shall not be installed in cellular metal floor raceways as follows:

(1) Where subject to corrosive vapor
(2) In any hazardous (classified) location except as permitted by 504.20, and in Class I, Division 2 locations as permitted in 501.4(B)(3)
(3) In commercial garages, other than for supplying ceiling outlets or extensions to the area below the floor but not above

> FPN: See 300.8 for installation of conductors with other systems.

Section 300.8 prohibits the installation of electric conductors in raceways or cable trays containing any pipes, tubes, or other carriers of steam, water, air, gas, or drainage or for any service other than electrical.

I. Installation

374.4 Size of Conductors.

No conductor larger than 1/0 AWG shall be installed, except by special permission.

374.5 Maximum Number of Conductors in Raceway.

The combined cross-sectional area of all conductors or cables shall not exceed 40 percent of the interior cross-sectional area of the cell or header.

Connections to the cells are made by means of headers extending across the cells and connecting only to those cells that are to be used as raceways for the conductors. Two or three separate headers, connecting to different sets of cells, may be used for different systems, such as light and power, signaling, and communications systems.

374.6 Splices and Taps.

Splices and taps shall be made only in header access units or junction boxes.

For the purposes of this section, so-called loop wiring (continuous unbroken conductor connecting the individual outlets) shall not be considered to be a splice or tap.

374.7 Discontinued Outlets.

When an outlet is abandoned, discontinued, or removed, the sections of circuit conductors supplying the outlet shall be removed from the raceway. No splices or reinsulated conductors, such as would be the case with abandoned outlets on loop wiring, shall be allowed in raceways.

374.8 Markers.

A suitable number of markers shall be installed for locating cells in the future.

Markers are brass flat-head screws set into the top side of the cells and adjusted so that their heads are flush with the floor finish and are exposed in order to aid in the location of cells for future installations.

374.9 Junction Boxes.

Junction boxes shall be leveled to the floor grade and sealed against the free entrance of water or concrete. Junction boxes used with these raceways shall be of metal and shall be electrically continuous with the raceway.

Connections to wall outlets are to be made with metal raceways unless there are provisions for equipment grounding termination, as required by 374.11. Installation instructions are supplied by the manufacturer for use by the general contractor, erector, electrical contractor, inspector, and others concerned with the installation.

374.10 Inserts.

Inserts shall be leveled to the floor grade and sealed against the entrance of concrete. Inserts shall be of metal and shall be electrically continuous with the raceway. In cutting through the cell wall and setting inserts, chips and other dirt shall not be allowed to remain in the raceway, and tools shall be used that are designed to prevent the tool from entering the cell and damaging the conductors.

374.11 Connection to Cabinets and Extensions from Cells.

Connections between raceways and distribution centers and wall outlets shall be made by means of flexible metal conduit where not installed in concrete, rigid metal conduit, intermediate metal conduit, electrical metallic tubing, or approved fittings. Where there are provisions for the termination of an equipment grounding conductor, nonmetallic conduit, electrical nonmetallic tubing, or liquidtight flexible nonmetallic conduit where not installed in concrete shall be permitted.

II. Construction Specifications

374.12 General.

Cellular metal floor raceways shall be constructed so that adequate electrical and mechanical continuity of the complete system will be secured. They shall provide a complete enclosure for the conductors. The interior surfaces shall be free from burrs and sharp edges, and surfaces over which conductors are drawn shall be smooth. Suitable bushings or fittings having smooth rounded edges shall be provided where conductors pass.

ARTICLE 376
Metal Wireways

Contents

I. General

376.1 Scope.

This article covers the use, installation, and construction specifications for metal wireways and associated fittings.

376.2 Definition.

Metal Wireways. Sheet metal troughs with hinged or removable covers for housing and protecting electric wires and cable and in which conductors are laid in place after the wireway has been installed as a complete system.

Wireways are sheet-metal enclosures equipped with hinged or removable covers and are manufactured in 1-ft to 10-ft lengths and various widths and depths. Couplings, elbows, end plates, and accessories such as T and X fittings are available. Unlike auxiliary gutters, which are not permitted to extend more than 30 ft from the equipment they supplement, wireways may be run throughout an entire area.

II. Installation

376.10 Uses Permitted.

The use of metal wireways shall be permitted in the following:

(1) For exposed work
(2) In concealed spaces as permitted in 376.10(4)
(3) In hazardous (classified) locations as permitted by 501.4(B) for Class I, Division 2 locations; 502.4(B) for Class II, Division 2 locations; and 504.20 for intrinsically safe wiring. Where installed in wet locations, wireways shall be listed for the purpose.
(4) As extensions to pass transversely through walls if the length passing through the wall is unbroken. Access to the conductors shall be maintained on both sides of the wall.

376.12 Uses Not Permitted.

Metal wireways shall not be used in the following:

(1) Where subject to severe physical damage
(2) Where subject to severe corrosive environments

376.21 Size of Conductors.

No conductor larger than that for which the wireway is designed shall be installed in any wireway.

376.22 Number of Conductors.

The sum of the cross-sectional areas of all contained conductors at any cross section of a wireway shall not exceed 20 percent of the interior cross-sectional area of the wireway. The derating factors in 310.15(B)(2)(a) shall be applied only where the number of current-carrying conductors, including neutral conductors classified as current-carrying under the provisions of 310.15(B)(4), exceeds 30. Conductors for signaling circuits or controller conductors between a motor and its starter and used only for starting duty shall not be considered as current-carrying conductors.

The main requirement of this section is that the total of the cross-sectional areas of all conductors must not exceed 20 percent of the interior cross-sectional area of the wireway. If the quantity of conductors does not exceed 30, the adjustment factors of 310.15(B)(2) do not apply. Where the quantity of conductors does exceed 30, however, the adjustment factors of 310.15(B)(2) do apply. The following example uses a wireway with only 26 current-carrying conductors; therefore, the adjustment factors of 310.15(B)(2) do not apply.

Example

A wireway contains 26 conductors: ten 3/0 AWG XHHW-2, three 6 AWG THWN, three 8 AWG THHN, and ten 12

AWG THHN. Find the minimum standard size wireway required by Article 376.

Solution

Conductor Type and Size	Quantity	Individual Area* (in.²)	Total Area (in.²)
3/0 AWG, XHHW-2	10 ×	0.2642	= 2.6420
6 AWG, THWN	3 ×	0.0507	= 0.1521
8 AWG, THHN	3 ×	0.0366	= 0.1098
12 AWG, THHN	10 ×	0.0133	= 0.1330
Total area occupied by conductors			= 3.0369
Minimum wireway area required:	3.0369 ÷ 20% fill		= 15.1845
Minimum size square wireway required:	15.1845 = 3.9 in. × 3.9 in. or 4 in. × 4 in. wireway		

*Individual area dimensions of conductors are from Chapter 9, Table 5.

376.23 Insulated Conductors.

Insulated conductors installed in a metallic wireway shall comply with 376.23(A) and (B).

(A) Deflected Insulated Conductors. Where insulated conductors are deflected within a metallic wireway, either at the ends or where conduits, fittings, or other raceways or cables enter or leave the metallic wireway, or where the direction of the metallic wireway is deflected greater than 30 degrees, dimensions corresponding to 312.6(A) shall apply.

The intent of 376.23(A) is to provide adequate space for bending conductors without damage to the insulation.

(B) Metallic Wireways Used as Pull Boxes. Where insulated conductors 4 AWG or larger are pulled through a wireway, the distance between raceway and cable entries enclosing the same conductor shall not be less than that required in 314.28(A)(1) for straight pulls and 314.28(A)(2) for angle pulls.

Section 376.23(B) was added to the 2002 *Code* to ensure adequate space for conductors in other raceways as they enter a wireway. Where wireways are used as pull boxes, the same minimum dimension requirements associated with raceway entries of pull boxes apply.

376.30 Securing and Supporting.

Metal wireways shall be supported in accordance with 376.30(A) and (B).

(A) Horizontal Support. Wireways shall be supported where run horizontally at each end and at intervals not to exceed 1.5 m (5 ft) or for individual lengths longer than 1.5 m (5 ft) at each end or joint, unless listed for other support intervals. The distance between supports shall not exceed 3 m (10 ft).

(B) Vertical Support. Vertical runs of wireways shall be securely supported at intervals not exceeding 4.5 m (15 ft) and shall not have more than one joint between supports. Adjoining wireway sections shall be securely fastened together to provide a rigid joint.

376.56 Splices and Taps.

Splices and taps shall be permitted within a wireway provided they are accessible. The conductors, including splices and taps, shall not fill the wireway to more than 75 percent of its area at that point.

Conductors in wireways are accessible through hinged or removable covers. Circuits, taps, or splices may be added or altered if necessary. See 376.22 and the associated example in the commentary regarding the number of conductors permitted.

376.58 Dead Ends.

Dead ends of metal wireways shall be closed.

376.70 Extensions from Metal Wireways.

Extensions from wireways shall be made with cord pendants installed in accordance with 400.10 or any wiring method in Chapter 3 that includes a means for equipment grounding. Where a separate equipment grounding conductor is employed, connection of the equipment grounding conductors in the wiring method to the wireway shall comply with 250.8 and 250.12.

Extensions from wireways using metal raceways, metal-sheathed cables, and nonmetallic-sheathed cables are made through knockouts provided on the wireway or field punched. Rigid nonmetallic conduit, electrical nonmetallic tubing, and liquidtight flexible nonmetallic conduit may also be used. Cables and nonmetallic raceways as well as the wireway must include a means for ensuring an effective continuation of the equipment grounding conductor. Sections of wireways, including accessory fittings (elbows, endplates, flanges, etc.), are bolted together, assuring a rigid mechanical and electrical connection.

III. Construction Specifications

376.120 Marking.

Metal wireways shall be marked so that their manufacturer's name or trademark will be visible after installation.

ARTICLE 378
Nonmetallic Wireways

Contents

I. General

378.1 Scope.

This article covers the use, installation, and construction specifications for nonmetallic wireways and associated fittings.

378.2 Definition.

Nonmetallic Wireways. Flame retardant, nonmetallic troughs with removable covers for housing and protecting electric wires and cables in which conductors are laid in place after the wireway has been installed as a complete system.

Nonmetallic wireways are troughs with removable covers in which conductors are laid after the wireway has been installed as a complete system. The wireway must be installed on the surface of the structure, not concealed. It is allowed to pass through walls, provided that an unbroken length passes through the wall.

378.3 Other Articles.

Installations of nonmetallic wireways shall comply with the applicable provisions of Article 300.

378.6 Listing Requirements.

Nonmetallic wireways and associated fittings shall be listed.

II. Installation

378.10 Uses Permitted.

The use of nonmetallic wireways shall be permitted in the following:

(1) Only for exposed work, except as permitted in 378.10(4).
(2) Where subject to corrosive environments where identified for the use.
(3) In wet locations where listed for the purpose.

> FPN: Extreme cold may cause nonmetallic wireways to become brittle and therefore more susceptible to damage from physical contact.

(4) As extensions to pass transversely through walls if the length passing through the wall is unbroken. Access to the conductors shall be maintained on both sides of the wall.

378.12 Uses Not Permitted.

Nonmetallic wireways shall not be used in the following:

(1) Where subject to physical damage
(2) In any hazardous (classified) location, except as permitted in 504.20
(3) Where exposed to sunlight unless listed and marked as suitable for the purpose
(4) Where subject to ambient temperatures other than those for which nonmetallic wireway is listed

(5) For conductors whose insulation temperature limitations would exceed those for which the nonmetallic wireway is listed

378.21 Size of Conductors.

No conductor larger than that for which the nonmetallic wireway is designed shall be installed in any nonmetallic wireway.

378.22 Number of Conductors.

The sum of cross-sectional areas of all contained conductors at any cross section of the nonmetallic wireway shall not exceed 20 percent of the interior cross-sectional area of the nonmetallic wireway. Conductors for signaling circuits or controller conductors between a motor and its starter and used only for starting duty shall not be considered as current-carrying conductors.

The derating factors specified in 310.15(B)(2)(a) shall be applicable to the current-carrying conductors up to and including the 20 percent fill specified above.

378.23 Insulated Conductors.

Insulated conductors installed in a nonmetallic wireway shall comply with 378.23(A) and (B).

(A) Deflected Insulated Conductors. Where insulated conductors are deflected within a nonmetallic wireway, either at the ends or where conduits, fittings, or other raceways or cables enter or leave the nonmetallic wireway, or where the direction of the nonmetallic wireway is deflected greater than 30 degrees, dimensions corresponding to 312.6(A) shall apply.

(B) Nonmetallic Wireways Used as Pull Boxes. Where insulated conductors 4 AWG or larger are pulled through a wireway, the distance between raceway and cable entries enclosing the same conductor shall not be less than that required in 314.28(A)(1) for straight pulls and in 314.28(A)(2) for angle pulls.

378.30 Securing and Supporting.

Nonmetallic wireway shall be supported in accordance with 378.30(A) and (B).

(A) Horizontal Support. Nonmetallic wireways shall be supported where run horizontally at intervals not to exceed 900 mm (3 ft), and at each end or joint, unless listed for other support intervals. In no case shall the distance between supports exceed 3 m (10 ft).

(B) Vertical Support. Vertical runs of nonmetallic wireway shall be securely supported at intervals not exceeding 1.2 m (4 ft), unless listed for other support intervals, and shall not have more than one joint between supports. Adjoining nonmetallic wireway sections shall be securely fastened together to provide a rigid joint.

378.44 Expansion Fittings.

Expansion fittings for nonmetallic wireway shall be provided to compensate for thermal expansion and contraction where the length change is expected to be 6 mm (0.25 in.) or greater in a straight run.

> FPN: See Table 352.44(A) for expansion characteristics of PVC rigid nonmetallic conduit. The expansion characteristics of PVC nonmetallic wireway are identical.

378.56 Splices and Taps.

Splices and taps shall be permitted within a nonmetallic wireway, provided they are accessible. The conductors, including splices and taps, shall not fill the nonmetallic wireway to more than 75 percent of its area at that point.

378.58 Dead Ends.

Dead ends of nonmetallic wireway shall be closed using listed fittings.

378.60 Grounding.

Where equipment grounding is required by Article 250, a separate equipment grounding conductor shall be installed in the nonmetallic wireway. A separate equipment grounding conductor shall not be required where the grounded conductor is used to ground equipment as permitted in 250.142.

378.70 Extensions from Nonmetallic Wireways.

Extensions from nonmetallic wireway shall be made with cord pendants or any wiring method of Chapter 3. A separate equipment grounding conductor shall be installed in, or an equipment grounding connection shall be made to, any of the wiring methods used for the extension.

III. Construction Specifications

378.120 Marking.

Nonmetallic wireways shall be marked so that the manufacturer's name or trademark and interior cross-sectional area in square inches shall be visible after installation. Marking for limited smoke shall be permitted on the nonmetallic wireways that have limited smoke-producing characteristics.

ARTICLE 380
Multioutlet Assembly

Contents

380.1 Scope.

This article covers the use and installation requirements for multioutlet assemblies.

Multioutlet assemblies are metal and nonmetallic raceways that are usually surface mounted and designed to contain branch-circuit conductors and receptacles. Exhibit 380.1 provides an illustration of a multioutlet assembly. Receptacles may be spaced at desired intervals and may be assembled at the factory or in the field. See 220.3(B)(8) and Exhibit 220.4 for load calculations.

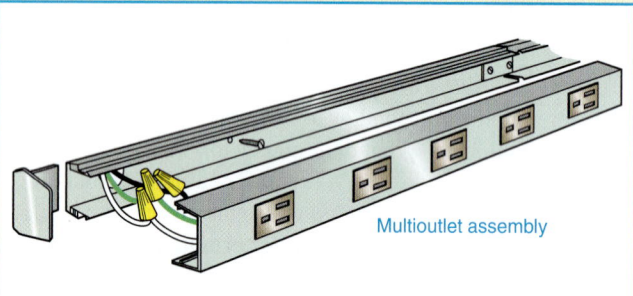

Exhibit 380.1 A typical multioutlet assembly shown in assembly form.

380.2 Use.

(A) Permitted. The use of a multioutlet assembly shall be permitted in dry locations.

(B) Not Permitted. A multioutlet assembly shall not be installed as follows:

(1) Where concealed, except that it shall be permissible to surround the back and sides of a metal multioutlet assembly by the building finish or recess a nonmetallic multioutlet assembly in a baseboard

(2) Where subject to severe physical damage

(3) Where the voltage is 300 volts or more between conductors unless the assembly is of metal having a thickness of not less than 1.02 mm (0.040 in.)

(4) Where subject to corrosive vapors

(5) In hoistways

(6) In any hazardous (classified) locations except Class I, Division 2 locations as permitted in 501.4(B)(3)

380.3 Metal Multioutlet Assembly Through Dry Partitions.

It shall be permissible to extend a metal multioutlet assembly through (not run within) dry partitions if arrangements are made for removing the cap or cover on all exposed portions and no outlet is located within the partitions.

ARTICLE 382
Nonmetallic Extensions

Contents

I. General

382.1 Scope.

This article covers the use, installation, and construction specifications for nonmetallic extensions.

382.2 Definition.

Nonmetallic Extension. An assembly of two insulated conductors within a nonmetallic jacket or an extruded thermoplastic covering. The classification includes surface extensions intended for mounting directly on the surface of walls or ceilings.

II. Installation

382.10 Uses Permitted.

Nonmetallic extensions shall be permitted only where all the conditions in 382.10(A), (B), and (C) are met.

(A) From an Existing Outlet. The extension is from an existing outlet on a 15- or 20-ampere branch circuit.

(B) Exposed and in a Dry Location. The extension is run exposed and in a dry location.

(C) Residential or Offices. For nonmetallic surface extensions mounted directly on the surface of walls or ceilings, the building is occupied for residential or office purposes and does not exceed three floors above grade.

> FPN No. 1: See 310.10 for temperature limitation of conductors.
> FPN No. 2: See 362.10 for definition of *first floor*.

382.12 Uses Not Permitted.

Nonmetallic extensions shall not be used as follows:

(1) In unfinished basements, attics, or roof spaces
(2) Where the voltage between conductors exceeds 150 volts for nonmetallic surface extension and 300 volts for aerial cable
(3) Where subject to corrosive vapors
(4) Where run through a floor or partition, or outside the room in which it originates

382.15 Exposed.

One or more extensions shall be permitted to be run in any direction from an existing outlet, but not on the floor or within 50 mm (2 in.) from the floor.

382.26 Bends.

A bend that reduces the normal spacing between the conductors shall be covered with a cap to protect the assembly from physical damage.

382.30 Securing and Supporting.

Nonmetallic surface extensions shall be secured in place by approved means at intervals not exceeding 200 mm (8 in.), with an allowance for 300 mm (12 in.) to the first fastening where the connection to the supplying outlet is by means of an attachment plug. There shall be at least one fastening between each two adjacent outlets supplied. An extension shall be attached to only woodwork or plaster finish and shall not be in contact with any metal work or other conductive material other than with metal plates on receptacles.

382.40 Boxes and Fittings.

Each run shall terminate in a fitting that covers the end of the assembly. All fittings and devices shall be of a type identified for the use.

382.56 Splices and Taps.

Extensions shall consist of a continuous unbroken length of the assembly, without splices, and without exposed conductors between fittings. Taps shall be permitted where approved fittings completely covering the tap connections are used. Aerial cable and its tap connectors shall be provided with an approved means for polarization. Receptacle-type tap connectors shall be of the locking type.

ARTICLE 384
Strut-Type Channel Raceway

Contents

I. General

384.1 Scope.

This article covers the use, installation, and construction specifications of strut-type channel raceway.

384.2 Definition.

Strut-Type Channel Raceway. A metallic raceway that is intended to be mounted to the surface of or suspended from a structure, with associated accessories for the installation of electrical conductors.

384.6 Listing Requirements.

Strut-type channel raceways, closure strips, and accessories shall be listed and identified for such use.

II. Installation

384.10 Uses Permitted.

The use of strut-type channel raceways shall be permitted in the following:

(1) Where exposed.
(2) In dry locations.
(3) In locations subject to corrosive vapors where protected by finishes judged suitable for the condition.
(4) Where the voltage is 600 volts or less.
(5) As power poles.
(6) In Class I, Division 2 hazardous (classified) locations as permitted in 501.4(B)(3).
(7) As extensions of unbroken lengths through walls, partitions, and floors where closure strips are removable from either side and the portion within the wall, partition, or floor remains covered.
(8) Ferrous channel raceways and fittings protected from corrosion solely by enamel shall be permitted only indoors.

The installation shown in Exhibit 384.1 is typical of how a strut-type channel raceway can be used.

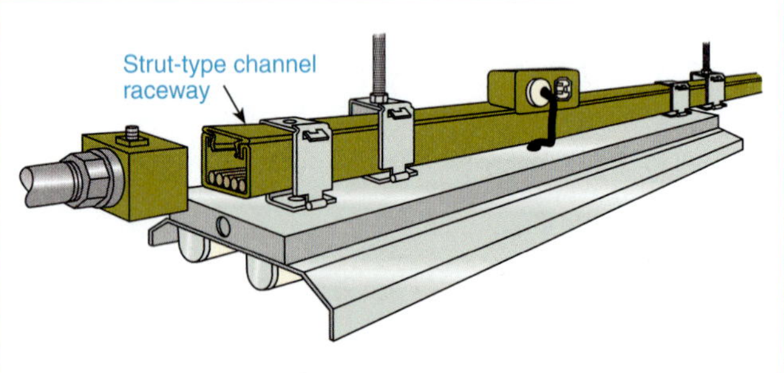

Exhibit 384.1 An example of a strut-type channel raceway using accessories to support and supply power to luminaires. (Redrawn from Allied Tube & Conduit, a Tyco International Co.)

384.12 Uses Not Permitted.

Strut type channel raceways shall not be used as follows:

(1) Where concealed.
(2) Ferrous channel raceways and fittings protected from corrosion solely by enamel shall not be permitted where subject to severe corrosive influences.

384.21 Size of Conductors.

No conductor larger than that for which the raceway is listed shall be installed in strut-type channel raceways.

384.22 Number of Conductors.

The number of conductors permitted in strut-type channel raceways shall not exceed the percentage fill using Table 384.22 and applicable outside diameter (O.D.) dimensions of specific types and sizes of wire given in the tables in Chapter 9.

Table 384.22 Channel Size and Inside Diameter Area

Size Channel	Area		40% Area*		25% Area**	
	in.^{2}	mm^2	in.^{2}	mm^2	in.^{2}	mm^2
1⅝ × 13⁄16	0.887	572	0.355	229	0.222	143
1⅝ × 1	1.151	743	0.460	297	0.288	186
1⅝ × 1⅜	1.677	1076	0.671	433	0.419	270
1⅝ × 1⅝	2.028	1308	0.811	523	0.507	327
1⅝ × 2 7⁄16	3.169	2045	1.267	817	0.792	511
1⅝ × 3¼	4.308	2780	1.723	1112	1.077	695
1½ × ¾	0.849	548	0.340	219	0.212	137
1½ × 1½	1.828	1179	0.731	472	0.457	295
1½ × 1⅞	2.301	1485	0.920	594	0.575	371
1½ × 3	3.854	2487	1.542	995	0.964	622

*Raceways with external joiners shall use a 40 percent wire fill calculation to determine the number of conductors permitted.
**Raceways with internal joiners shall use a 25 percent wire fill calculation to determine the number of conductors permitted.

The adjustment factors of Table 310.15(B)(2) are applicable to strut-type channel raceways because Table 384.22 does not contain any raceways with cross-sectional areas greater than 4 in.^{2}

Example

Calculate the maximum quantity of 10 AWG Type THWN-2 copper conductors permitted in a normal 1½ in. × 1½ in. strut-type channel raceway whose joiners are mounted internally. (Generally, ordinary-duty strut-type channel raceway couplings or joiners are of the internal type, and heavy-duty strut raceway couplings are of the external type.)

Solution

STEP 1. Because the strut-type channel raceway joiners are mounted on the internal surface of the raceway, Note 2 of Table 384.22 requires the use of the "25% Area" column for the maximum usable internal area of the raceway. According to Table 384.22, and using the 25% Area column, the usable internal area of the raceway for a 1½ in. × 1½ in. strut-type channel is 0.457 in.^{2} According to Chapter 9,

Table 5, a 10 AWG, Type THWN-2 copper conductor has an area of 0.0211 in.^{2}

STEP 2. Using the formula for wire fill,

$$n = \frac{ca}{wa}$$

where:

n = number of wires

ca = channel area (in.^{2})

wa = wire area (in.^{2})

and substituting the table values,

$$n = \frac{0.457 \text{ in.}^2}{0.0211 \text{ in.}^2} = 21.66$$

or not more than twenty-one 10 AWG Type THWN-2 copper conductors. However, the adjustment factors of Table 310.15(B)(2)(a) are applicable where a raceway contains more than three current-carrying conductors.

The derating factors of 310.15(B)(2)(a) shall not apply to conductors installed in strut-type channel raceways where all of the following conditions are met:

(1) The cross-sectional area of the raceway exceeds 2500 mm^2 (4 in.^{2}).
(2) The current-carrying conductors do not exceed 30 in number.
(3) The sum of the cross-sectional areas of all contained conductors does not exceed 20 percent of the interior cross-sectional area of the strut-type channel raceways. Formula for wire fill:

$$n = \frac{ca}{wa}$$

where:

n = number of wires

ca = channel area in square inches

wa = wire area

384.30 Securing and Supporting.

(A) Surface Mount. A surface mount strut-type channel raceway shall be secured to the mounting surface with retention straps external to the channel at intervals not exceeding 3 m (10 ft) and within 900 mm (3 ft) of each outlet box, cabinet, junction box, or other channel raceway termination.

(B) Suspension Mount. Strut-type channel raceways shall be permitted to be suspension mounted in air with approved appropriate methods designed for the purpose at intervals not to exceed 3 m (10 ft) and within 900 mm (3 ft) of channel raceway terminations and ends.

384.56 Splices and Taps.

Splices and taps shall be permitted in raceways that are accessible after installation by having a removable cover. The conductors, including splices and taps, shall not fill the raceway to more than 75 percent of its area at that point. All splices and taps shall be made by approved methods.

384.60 Grounding.

Strut-type channel raceway enclosures providing a transition to or from other wiring methods shall have a means for connecting an equipment grounding conductor. Strut-type channel raceways shall be permitted as an equipment grounding conductor in accordance with 250.118(14). Where a snap-fit metal cover for strut-type channel raceways is used to achieve electrical continuity in accordance with the listing, this cover shall not be permitted as the means for providing electrical continuity for a receptacle mounted in the cover.

III. Construction Specifications

384.100 Construction.

Strut-type channel raceways and their accessories shall be of a construction that distinguishes them from other raceways. Raceways and their elbows, couplings, and other fittings shall be designed so that the sections can be electrically and mechanically coupled together and installed without subjecting the wires to abrasion. They shall comply with 384.100(A), (B), and (C).

(A) Material. Raceways and accessories shall be formed of steel, stainless steel, or aluminum.

(B) Corrosion Protection. Steel raceways and accessories shall be protected against corrosion by galvanizing or an organic coating.

> FPN: Enamel and PVC coatings are examples of organic coatings that provide corrosion protection.

(C) Cover. Covers of strut-type channel raceway shall be either metallic or nonmetallic.

384.120 Marking.

Each length of strut-type channel raceways shall be clearly and durably identified as required in the first sentence of 110.21.

ARTICLE 386
Surface Metal Raceways

Contents

I. General

386.1 Scope.

This article covers the use, installation, and construction specifications for surface metal raceways and associated fittings.

386.2 Definition.

Surface Metal Raceway. A metallic raceway that is intended to be mounted to the surface of a structure, with associated couplings, connectors, boxes, and fittings for the installation of electrical conductors.

386.6 Listing Requirements.

Surface metal raceway and associated fittings shall be listed.

II. Installation

386.10 Uses Permitted.

The use of surface metal raceways shall be permitted in the following:

(1) In dry locations.
(2) In Class I, Division 2 hazardous (classified) locations as permitted in 501.4(B)(3).
(3) Under raised floors, as permitted in 645.5(D)(2).
(4) Extension through walls and floors. Surface metal raceway shall be permitted to pass transversely through dry walls, dry partitions, and dry floors if the length passing through is unbroken. Access to the conductors shall be maintained on both sides of the wall, partition, or floor.

The installation shown in Exhibit 386.1 is just one of many ways a surface metal raceway can be used.

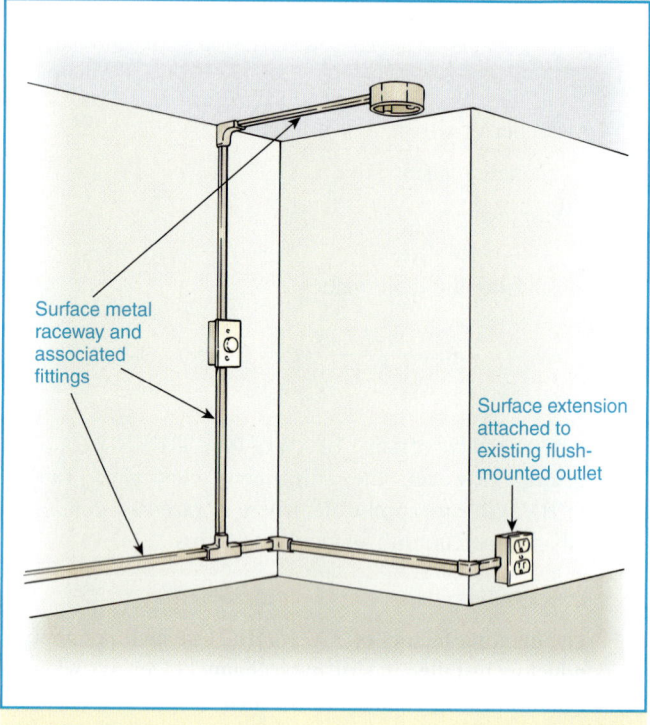

Surface metal raceway and associated fittings

Surface extension attached to existing flush-mounted outlet

Exhibit 386.1 An example of a surface metal raceway extending from an existing receptacle outlet.

386.12 Uses Not Permitted.

Surface metal raceways shall not be used in the following:

(1) Where subject to severe physical damage, unless otherwise approved
(2) Where the voltage is 300 volts or more between conductors, unless the metal has a thickness of not less than 1.02 mm (0.040 in.) nominal
(3) Where subject to corrosive vapors
(4) In hoistways
(5) Where concealed, except as permitted in 386.10

386.21 Size of Conductors.

No conductor larger than that for which the raceway is designed shall be installed in surface metal raceway.

386.22 Number of Conductors or Cables.

The number of conductors or cables installed in surface metal raceway shall not be greater than the number for which the raceway is designed. Cables shall be permitted to be installed where such use is permitted by the respective cable articles.

The derating factors of 310.15(B)(2)(a) shall not apply

to conductors installed in surface metal raceways where all of the following conditions are met:

(1) The cross-sectional area of the raceway exceeds 2500 mm² (4 in.²)
(2) The current-carrying conductors do not exceed 30 in number
(3) The sum of the cross-sectional areas of all contained conductors does not exceed 20 percent of the interior cross-sectional area of the surface metal raceway

The number, type, and sizes of conductors permitted to be installed in a listed surface metal raceway are marked on the raceway or on the package in which it is shipped. Typically, this information is available in detail in the manufacturer's catalog. Exhibit 386.2 provides conductor fill information for specific surface metal raceways.

386.56 Splices and Taps.

Splices and taps shall be permitted in surface metal raceways having a removable cover that is accessible after installation. The conductors, including splices and taps, shall not fill the raceway to more than 75 percent of its area at that point. Splices and taps in surface metal raceways without removable covers shall be made only in junction boxes. All splices and taps shall be made by approved methods.

Taps of Type FC cable installed in surface metal raceway shall be made in accordance with 322.56(B).

386.60 Grounding.

Surface metal raceway enclosures providing a transition from other wiring methods shall have a means for connecting an equipment grounding conductor.

As the example in Exhibit 386.3 shows, where a surface metal raceway is supplied by Type MC or NM cable, a means (e.g., grounding terminal screw or lug) must be available at the surface metal raceway for terminating the equipment grounding conductor provided within the cable.

386.70 Combination Raceways.

When combination surface metal raceways are used both for signaling and for lighting and power circuits, the different systems shall be run in separate compartments identified by sharply contrasting colors of the interior finish, and the same relative position of compartments shall be maintained throughout the premises.

III. Construction Specifications

386.100 Construction.

Surface metal raceways shall be of such construction as will distinguish them from other raceways. Surface metal

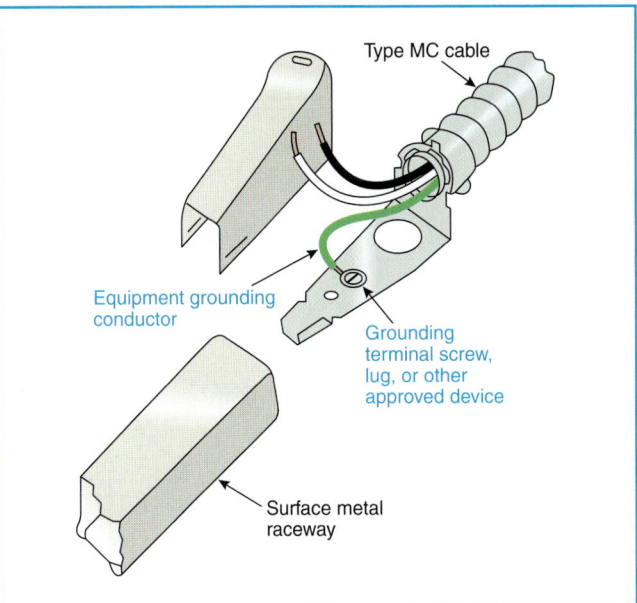

Exhibit 386.3 An example of providing a means for terminating an equipment grounding conductor at a surface metal raceway.

raceways and their elbows, couplings, and similar fittings shall be designed so that the sections can be electrically and mechanically coupled together and installed without subjecting the wires to abrasion.

Where covers and accessories of nonmetallic materials are used on surface metal raceways, they shall be identified for such use.

<h1 style="text-align:center">ARTICLE 388
Surface Nonmetallic Raceways</h1>

Contents

Type of Raceway	Wire Size Gauge No.	Number of Wires: Types RHH, RHW	Type THW	Type TW	Types THHN, THWN
No. 200 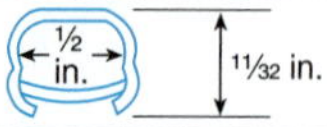 ½ in.; 11/32 in.	12	—	2	3	3
	14	—	2	3	5
No. 500 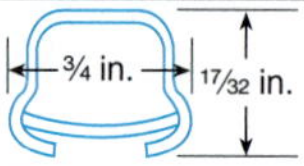 ¾ in.; 17/32 in.	8	—	—	2	2
	10	2	2	3	4
	12	2	3	4	7
	14	2	4	6	9
No. 700 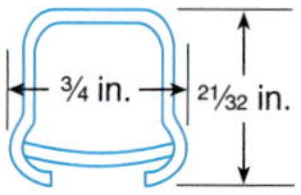 ¾ in.; 21/32 in.	6	—	—	—	2
	8	—	2	2	3
	10	2	3	4	5
	12	2	4	6	8
	14	3	5	7	11
No. 1500 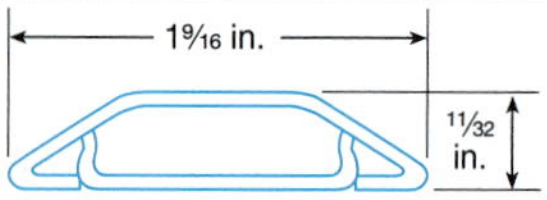 1 9/16 in.; 11/32 in.	6	—	—	—	2
	8	—	—	2	3
	10	2	3	4	5
	12	2	3	5	7
	14	2	4	6	10
No. 2000[a] 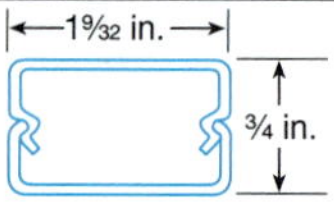 1 9/32 in.; ¾ in.	12	—	—	7	7
	14	—	—	7	7
No. 2100[a] 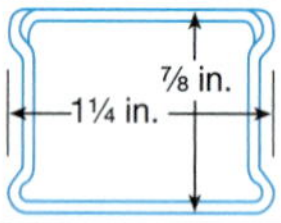 ⅞ in.; 1¼ in.	6	2	4 6	4 8	6
	8	4	10	14	10
	10	7	13	19	17
	12	8	15	24	28
	14	10			37
No. 2200[a] 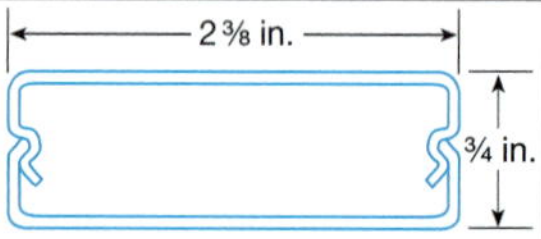 2⅜ in.; ¾ in.	6	5 8	7	3[b] 7	11
	8	13	11	7[b] 14	19
	10	15	19	10[b] 26	32
	12	18	23	10[b] 34	51
	14		29	10[b] 44	69
No. 2600 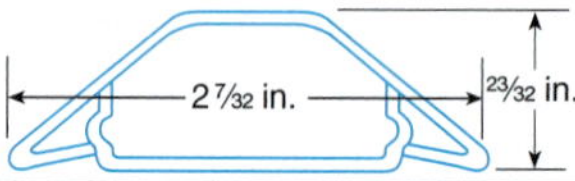 2 7/32 in.; 23/32 in.	6	2	3	3 7	5 9
	8	4	5	12	15
	10	6	9	16	24
	12	7	11	21	33
	14	9	14		
G-3000 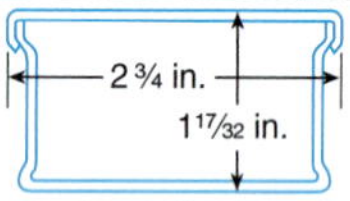 2¾ in.; 1 17/32 in.	6	4[b] 11	6[b] 19	6[b] 17	6[b] 27
	8	6[b] 18	8[b] 26	8[b] 34	8[b] 44
	10	10[b] 30	10[b] 45	10[b] 62	10[b] 76
	12	14[b] 36	18[b] 55	18[b] 81	18[b] 119
	14	16[b] 42	26[b] 67	26[b] 103	26[b] 160
G-4000, with divider 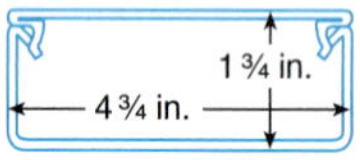 1¾ in.; 4¾ in.	2	— 7 8	— 10	— 10	— 12
	3	— 9	— 11	— 11	— 15
	4	— 12	— 13	— 13	— 17
	6	4[b] 19	4[b] 18	4[b] 18	7[b] 28
	8	7[b] 32	7[b] 28	7[b] 36	8[b] 47
	10	11[b] 39	11[b] 48	11[b] 66	15[b] 81
	12	15[b] 45	15[b] 59	15[b] 86	24[b] 128
	14	17[b]	17[b] 72	17[b] 110	32[b] 171
G-4000, without divider 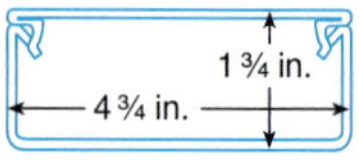 1¾ in.; 4¾ in.	2	— 14	— 20	— 20	— 25
	3	— 16	— 23	— 23	— 30
	4	— 18	— 27	— 27	— 35
	6	8[b] 24	8[b] 36	8[b] 36	10[b] 57
	8	10[b] 39	10[b] 57	10[b] 78	15[b] 94
	10	15[b] 65	15[b] 96	12[b] 133	18[b] 163
	12	21[b] 78	21[b] 119	16[b] 174	34[b] 256
	14	21[b] 91	21[b] 145	17[b] 222	34[b] 344
G-6000 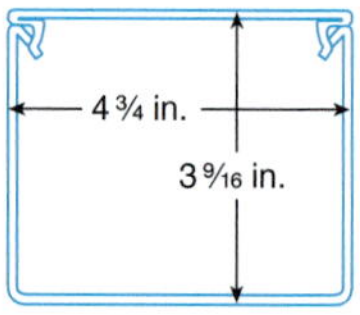 4¾ in.; 3 9/16 in.	2/0	10[b] 17	12[b] 22	12[b] 22	15[b] 27
	1/0	11[b] 20	14[b] 26	14[b] 26	18[b] 33
	1 2	12[b] 23	17[b] 31	17[b] 31	21[b] 39
	3 4	16[b] 30	23[b] 43	23[b] 43	29[b] 53
	6 8	19[b] 34	27[b] 50	27[b] 50	34[b] 63
	10	21[b] 39	32[b] 58	32[b] 58	40[b] 74
	12	27[b] 51	42[b] 77	42[b] 77	66[b] 122
	14	40[b] 74	57[b] 106	73[b] 134	92[b] 169
		75[b] 137	111[b] 203	154[b] 282	187[b] 343
		90[b] 164	137[b] 252	200[b] 386	295[b] 540
		105[b] 193	167[b] 307	255[b] 469	396[b] 726

[a]Figures for Nos. 2000, 2100, 2200, G-3000, G-4000, and G-6000 are without receptacles, except where noted.
[b]With receptacles.

Exhibit 386.2 Conductor fill table for various surface metal raceways. (Redrawn from The Wiremold Co.)

I. General

388.1 Scope.

This article covers the use, installation, and construction specifications for surface nonmetallic raceways and associated fittings.

388.2 Definition.

Surface Nonmetallic Raceway. A nonmetallic raceway that is intended to be mounted to the surface of a structure, with associated couplings, connectors, boxes, and fittings for the installation of electrical conductors.

388.6 Listing Requirements.

Surface nonmetallic raceway and associated fittings shall be listed.

II. Installation

388.10 Uses Permitted.

Surface nonmetallic raceway shall be permitted as follows:

(1) The use of surface nonmetallic raceways shall be permitted in dry locations.
(2) Extension through walls and floors shall be permitted. Surface nonmetallic raceway shall be permitted to pass transversely through dry walls, dry partitions, and dry floors if the length passing through is unbroken. Access to the conductors shall be maintained on both sides of the wall, partition, or floor.

The installations shown in Exhibits 388.1 and 388.2 are typical of how a surface nonmetallic raceway can be used.

388.12 Uses Not Permitted.

Surface nonmetallic raceways shall not be used in the following:

(1) Where concealed, except as permitted in 388.10(2)
(2) Where subject to severe physical damage
(3) Where the voltage is 300 volts or more between conductors, unless listed for higher voltage
(4) In hoistways
(5) In any hazardous (classified) location except Class I, Division 2 locations as permitted in 501.4(B)(3)
(6) Where subject to ambient temperatures exceeding those for which the nonmetallic raceway is listed
(7) For conductors whose insulation temperature limitations would exceed those for which the nonmetallic raceway is listed

388.21 Size of Conductors.

No conductor larger than that for which the raceway is designed shall be installed in surface nonmetallic raceway.

Exhibit 388.1 An example of a surface nonmetallic raceway extending from an existing receptacle outlet. (Courtesy of The Wiremold Co.)

Exhibit 388.2 An example of a surface nonmetallic raceway supplying a speed control switch and a paddle fan outlet. (Courtesy of The Wiremold Co.)

388.22 Number of Conductors or Cables.

The number of conductors or cables installed in surface nonmetallic raceway shall not be greater than the number for which the raceway is designed. Cables shall be permitted to be installed where such use is permitted by the respective cable articles.

388.56 Splices and Taps.

Splices and taps shall be permitted in surface nonmetallic raceways having a removable cover that is accessible after installation. The conductors, including splices and taps, shall not fill the raceway to more than 75 percent of its area at that point. Splices and taps in surface nonmetallic raceways without removable covers shall be made only in junction boxes. All splices and taps shall be made by approved methods.

388.60 Grounding.

Where equipment grounding is required by Article 250, a separate equipment grounding conductor shall be installed in the raceway.

388.70 Combination Raceways.

When combination surface nonmetallic raceways are used both for signaling and for lighting and power circuits, the different systems shall be run in separate compartments identified by sharply contrasting colors of the interior finish.

III. Construction Specifications

388.100 Construction.

Surface nonmetallic raceways shall be of such construction as will distinguish them from other raceways. Surface nonmetallic raceways and their elbows, couplings, and similar fittings shall be designed so that the sections can be mechanically coupled together and installed without subjecting the wires to abrasion.

Surface nonmetallic raceways and fittings are made of suitable nonmetallic material that is resistant to moisture and chemical atmospheres. It shall also be flame retardant, resistant to impact and crushing, resistant to distortion from heat under conditions likely to be encountered in service, and resistant to low-temperature effects.

388.120 Marking.

Surface nonmetallic raceways that have limited smoke-producing characteristics shall be permitted to be so identified.

ARTICLE 390
Underfloor Raceways

Contents

390.1 Scope.

This article covers the use and installation requirements for underfloor raceways.

390.2 Use.

An underfloor raceway is a practical means of bringing light, power, and signal and communications systems to desks, work benches, or tables that are not located adjacent to wall space. This wiring method offers flexibility in layout where used with movable partitions and is commonly used in large retail stores and office buildings to supply power at any desired location.

Underfloor raceways are permitted beneath the surface of concrete, wood, or other flooring material. The wiring method between cabinets, raceway junction boxes, and outlet boxes may be rigid metal conduit, intermediate metal conduit, rigid nonmetallic conduit, liquidtight flexible nonmetallic conduit, electrical nonmetallic tubing, or electrical metallic tubing. Flexible metal conduit may be used where not installed in concrete.

(A) Permitted. The installation of underfloor raceways shall be permitted beneath the surface of concrete or other flooring material or in office occupancies where laid flush with the concrete floor and covered with linoleum or equivalent floor covering.

(B) Not Permitted. Underfloor raceways shall not be installed (1) where subject to corrosive vapors or (2) in any hazardous (classified) locations, except as permitted by 504.20 and in Class I, Division 2 locations as permitted in 501.4(B)(3). Unless made of a material judged suitable for the condition or unless corrosion protection approved for the condition is provided, ferrous or nonferrous metal underfloor raceways, junction boxes, and fittings shall not be installed in concrete or in areas subject to severe corrosive influences.

390.3 Covering.

Raceway coverings shall comply with 390.3(A) through (D).

(A) Raceways Not Over 100 mm (4 in.) Wide. Half-round and flat-top raceways not over 100 mm (4 in.) in width shall have not less than 20 mm (¾ in.) of concrete or wood above the raceway.

Exception: As permitted in 390.3(C) and (D) for flat-top raceways.

As Exhibit 390.1 illustrates, a ¾-in. concrete or wood covering is required over underfloor raceways not over 4 in. wide, except for trench-type and other raceways that are flush with concrete.

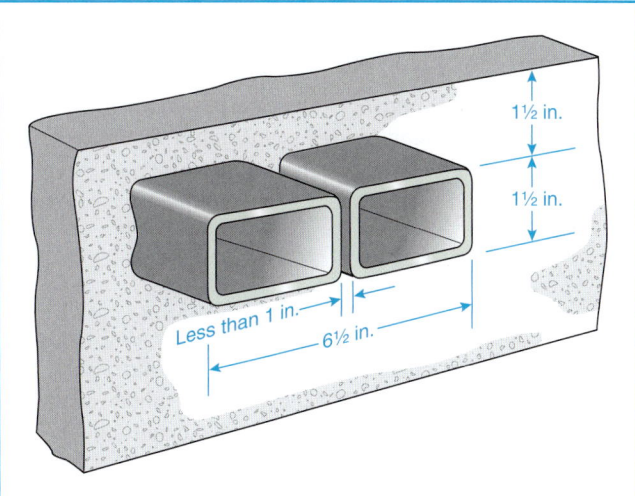

Exhibit 390.2 Two side-by-side underfloor raceways over 4 in. installed in compliance with 390.3(B). (Redrawn from Walker Systems, a Wiremold Co.)

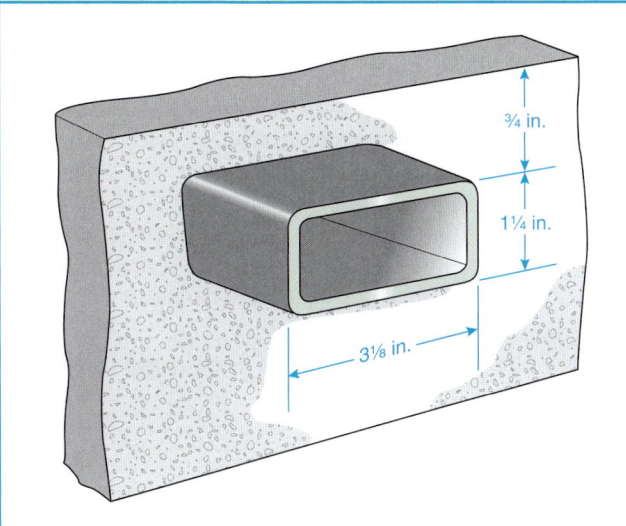

Exhibit 390.1 An underfloor raceway not over 4 in. installed in compliance with 390.3(A). (Redrawn from Walker Systems, a Wiremold Co.)

(B) Raceways Over 100 mm (4 in.) Wide But Not Over 200 mm (8 in.) Wide. Flat-top raceways over 100 mm (4 in.) but not over 200 mm (8 in.) wide with a minimum of 25 mm (1 in.) spacing between raceways shall be covered with concrete to a depth of not less than 25 mm (1 in.). Raceways spaced less than 25 mm (1 in.) apart shall be covered with concrete to a depth of 38 mm (1½ in.).

As Exhibit 390.2 illustrates, flat-top underfloor raceways over 4 in. wide and spaced less than 1 in. apart must be covered with at least 1½ in. of concrete.

(C) Trench-Type Raceways Flush with Concrete. Trench-type flush raceways with removable covers shall be permitted to be laid flush with the floor surface. Such ap-

proved raceways shall be designed so that the cover plates provide adequate mechanical protection and rigidity equivalent to junction box covers.

Approved flush-type underfloor raceways may be installed flush with the floor surface, provided they have covers at least equal to those of junction box covers.

(D) Other Raceways Flush with Concrete. In office occupancies, approved metal flat-top raceways, if not over 100 mm (4 in.) in width, shall be permitted to be laid flush with the concrete floor surface, provided they are covered with substantial linoleum that is not less than 1.6 mm (¹⁄₁₆ in.) thick or with equivalent floor covering. Where more than one and not more than three single raceways are each installed flush with the concrete, they shall be contiguous with each other and joined to form a rigid assembly.

390.4 Size of Conductors.

No conductor larger than that for which the raceway is designed shall be installed in underfloor raceways.

390.5 Maximum Number of Conductors in Raceway.

The combined cross-sectional area of all conductors or cables shall not exceed 40 percent of the interior cross-sectional area of the raceway.

390.6 Splices and Taps.

Splices and taps shall be made only in junction boxes.

For the purposes of this section, so-called loop wiring

(continuous, unbroken conductor connecting the individual outlets) shall not be considered to be a splice or tap.

Exception: Splices and taps shall be permitted in trench-type flush raceway having a removable cover that is accessible after installation. The conductors, including splices and taps, shall not fill more than 75 percent of the raceway area at that point.

390.7 Discontinued Outlets.

When an outlet is abandoned, discontinued, or removed, the sections of circuit conductors supplying the outlet shall be removed from the raceway. No splices or reinsulated conductors, such as would be the case with abandoned outlets on loop wiring, shall be allowed in raceways.

Loop wiring (continuous, unbroken conductors) is recognized where it runs from the underfloor raceway up to the terminals of attached receptacles, back into the raceway, and then on to the next device, as illustrated in Exhibit 390.3. When an outlet is removed, the sections of conductors supplying the outlet must be removed from the raceway as well. As is the case with abandoned outlets on loop wiring, reinsulated or spliced conductors are not allowed in raceways, except in trench-type raceways as covered in the exception to 390.6.

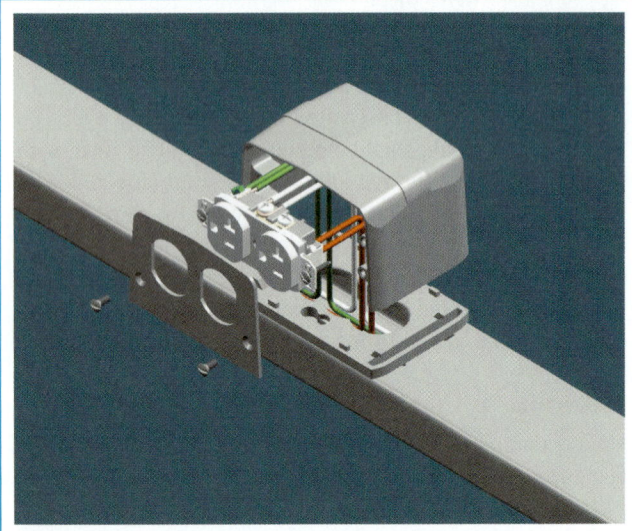

Exhibit 390.3 A receptacle outlet supplied from an underfloor raceway by the loop method of wiring. (Courtesy of Walker Systems, a Wiremold Co.)

390.8 Laid in Straight Lines.

Underfloor raceways shall be laid so that a straight line from the center of one junction box to the center of the next junction box coincides with the centerline of the raceway

system. Raceways shall be firmly held in place to prevent disturbing this alignment during construction.

390.9 Markers at Ends.

A suitable marker shall be installed at or near each end of each straight run of raceways to locate the last insert.

390.10 Dead Ends.

Dead ends of raceways shall be closed.

390.13 Junction Boxes.

Junction boxes shall be leveled to the floor grade and sealed to prevent the free entrance of water or concrete. Junction boxes used with metal raceways shall be metal and shall be electrically continuous with the raceways.

390.14 Inserts.

Inserts shall be leveled and sealed to prevent the entrance of concrete. Inserts used with metal raceways shall be metal and shall be electrically continuous with the raceway. Inserts set in or on fiber raceways before the floor is laid shall be mechanically secured to the raceway. Inserts set in fiber raceways after the floor is laid shall be screwed into the raceway. When cutting through the raceway wall and setting inserts, chips and other dirt shall not be allowed to remain in the raceway, and tools shall be used that are designed so as to prevent the tool from entering the raceway and damaging conductors that may be in place.

390.15 Connections to Cabinets and Wall Outlets.

Connections from underfloor raceways to distribution centers and wall outlets shall be made by approved fittings or by any of the wiring methods in Chapter 3, where installed in accordance with the provisions of the respective articles.

ARTICLE 392
Cable Trays

Contents

392.5 Construction Specifications
   (A) Strength and Rigidity
   (B) Smooth Edges
   (C) Corrosion Protection
   (D) Side Rails
   (E) Fittings
   (F) Nonmetallic Cable Tray
392.6 Installation
   (A) Complete System
   (B) Completed Before Installation
   (C) Supports
   (D) Covers
   (E) Multiconductor Cables Rated 600 Volts or Less
   (F) Cables Rated Over 600 Volts
   (G) Through Partitions and Walls
   (H) Exposed and Accessible
   (I) Adequate Access
   (J) Raceways, Cables, Boxes, and Conduit Bodies Supported from Cable Tray Systems
392.7 Grounding
   (A) Metallic Cable Trays
   (B) Steel or Aluminum Cable Tray Systems
392.8 Cable Installation
   (A) Cable Splices
   (B) Fastened Securely
   (C) Bushed Conduit and Tubing
   (D) Connected in Parallel
   (E) Single Conductors
392.9 Number of Multiconductor Cables, Rated 2000 Volts or Less, in Cable Trays
   (A) Any Mixture of Cables
   (B) Multiconductor Control and/or Signal Cables Only
   (C) Solid Bottom Cable Trays Containing Any Mixture
   (D) Solid Bottom Cable Tray—Multiconductor Control and/or Signal Cables Only
   (E) Ventilated Channel Cable Trays
   (F) Solid Channel Cable Trays
392.10 Number of Single-Conductor Cables, Rated 2000 Volts or Less, in Cable Trays
   (A) Ladder or Ventilated Trough Cable Trays
   (B) Ventilated Channel Cable Trays
392.11 Ampacity of Cables, Rated 2000 Volts or Less, in Cable Trays
   (A) Multiconductor Cables
   (B) Single-Conductor Cables
392.12 Number of Type MV and Type MC Cables (2001 Volts or Over) in Cable Trays
392.13 Ampacity of Type MV and Type MC Cables (2001 Volts or Over) in Cable Trays
   (A) Multiconductor Cables (2001 Volts or Over)
   (B) Single-Conductor Cables (2001 Volts or Over)

392.1 Scope.

This article covers cable tray systems, including ladder, ventilated trough, ventilated channel, solid bottom, and other similar structures.

Cable trays are mechanical support systems. Cable trays are not raceways. See the definition of *raceway* in Article 100.

FPN: For further information on cable trays, see NEMA–VE 1, 1998-*Metal Cable Tray Systems*; NEMA–VE 2-1996, *Metal Cable Tray Installation Guidelines*; and NEMA–FG-1998, *Nonmetallic Cable Tray Systems*.

392.2 Definition.

Cable Tray System. A unit or assembly of units or sections and associated fittings forming a structural system used to securely fasten or support cables and raceways.

392.3 Uses Permitted.

Cable tray shall be permitted to be used as a support system for services, feeders, branch circuits, communications circuits, control circuits, and signaling circuits. Cable tray installations shall not be limited to industrial establishments. Where exposed to direct rays of the sun, insulated conductors and jacketed cables shall be identified as being sunlight resistant. Cable trays and their associated fittings shall be identified for the intended use.

Cable tray installations are typically an industrial-type wiring method. However, cable tray installations have never been restricted to industrial installations. Cable tray installations are being applied more in commercial installations than ever before, especially as a wire-and-cable management system for telecommunications/data installations.

(A) Wiring Methods. The wiring methods in Table 392.3(A) shall be permitted to be installed in cable tray systems under the conditions described in their respective articles and sections.

Section 392.3(A) identifies the raceways and many of the cable types that may be supported in commercial and industrial cable tray installations. Cable tray is rarely used as a major raceway support system. For raceway support systems, the versatility of strut systems exceeds that of cable tray support systems.

Section 392.3(A) does not identify *all* the specific cable types that may be installed in commercial and industrial cable tray systems. According to Table 392.3(A), other

Table 392.3(A) Wiring Methods

Wiring Method	Article	Section
Armored cable	320	
Communication raceways	800	
Electrical metallic tubing	358	
Electrical nonmetallic tubing	362	
Fire alarm cables	760	
Flexible metal conduit	348	
Flexible metallic tubing	360	
Instrumentation tray cable	727	
Intermediate metal conduit	342	
Liquidtight flexible metal conduit	350	
Liquidtight flexible nonmetallic conduit	356	
Metal-clad cable	330	
Mineral-insulated, metal-sheathed cable	332	
Multiconductor service-entrance cable	338	
Multiconductor underground feeder and branch-circuit cable	340	
Multipurpose and communications cables	800	
Nonmetallic-sheathed cable	334	
Power and control tray cable	336	
Power-limited tray cable		725.61(C) and 725.71(F)
Optical fiber cables	770	
Optical fiber raceways	770	
Other factory-assembled, multiconductor control, signal, or power cables that are specifically approved for installation in cable trays		
Rigid metal conduit	344	
Rigid nonmetallic conduit	352	

factory assembled, multiconductor control, signal, and power cables that are specifically approved for installation in cable trays are permitted as well.

(B) In Industrial Establishments. The wiring methods in Table 392.3(A) shall be permitted to be used in any industrial establishment under the conditions described in their respective articles. In industrial establishments only, where conditions of maintenance and supervision ensure that only qualified persons service the installed cable tray system, any of the cables in 392.3(B)(1) and (2) shall be permitted to be installed in ladder, ventilated trough, solid bottom, or ventilated channel cable trays.

Section 392.3(B) permits single-conductor cables (rated 0 to 2000 volts) and Type MV cables to be installed in ladder, ventilated-trough, or ventilated-channel cable trays, provided the installation is located in a qualifying industrial facility. Single conductors and Type MV cable installations are not permitted in solid-bottom cable trays.

(1) Single Conductors. Single-conductor cables shall be permitted to be installed in accordance with the following:

(a) Single-conductor cable shall be 1/0 AWG or larger and shall be of a type listed and marked on the surface for use in cable trays. Where 1/0 AWG through 4/0 AWG single-conductor cables are installed in ladder cable tray, the maximum allowable rung spacing for the ladder cable tray shall be 230 mm (9 in.).

(b) Welding cables shall comply with the provisions of Article 630, Part IV.

Cable trays used to support welding cables are required to be dedicated for welding cable installation. See 630.42 for installation details.

(c) Single conductors used as equipment grounding conductors shall be insulated, covered, or bare, and they shall be 4 AWG or larger.

(2) Medium Voltage. Single- and multiconductor medium voltage cables shall be Type MV cable (Article 328). Single conductors shall be installed in accordance with 392.3(B)(1).

(C) Equipment Grounding Conductors. Metallic cable trays shall be permitted to be used as equipment grounding conductors where continuous maintenance and supervision ensure that qualified persons service the installed cable tray system and the cable tray complies with provisions of 392.7.

Section 392.3(C) expanded the optional use of the cable tray as the equipment grounding conductor. No longer is this practice limited to qualifying industrial installations. To qualify as an equipment grounding conductor, the cable tray system must meet all four requirements of 392.7(B).

(D) Hazardous (Classified) Locations. Cable trays in hazardous (classified) locations shall contain only the cable types permitted in 501.4, 502.4, 503.3, 504.20, and 505.15.

(E) Nonmetallic Cable Tray. In addition to the uses permitted elsewhere in Article 392, nonmetallic cable tray shall be permitted in corrosive areas and in areas requiring voltage isolation.

Fiberglass cable trays are often used to support cables in corrosive environments or in electrolytic cell rooms where voltage isolation is required. See Article 668, Electrolytic Cells.

392.4 Uses Not Permitted.

Cable tray systems shall not be used in hoistways or where subject to severe physical damage. Cable tray systems shall not be used in environmental airspaces, except as permitted in 300.22, to support wiring methods recognized for use in such spaces.

Section 300.22(C) specifically limits the types of wiring methods that may be used within other spaces used for environmental air. Metallic cable trays may be used within these spaces to support only the recognized wiring methods permitted in these spaces. The cable tray types may be ladder, ventilated trough, ventilated channel, or solid bottom. Metal cable trays are not the limiting factor; rather, the cable or wiring method is the limiting factor.

392.5 Construction Specifications.

(A) Strength and Rigidity. Cable trays shall have suitable strength and rigidity to provide adequate support for all contained wiring.

(B) Smooth Edges. Cable trays shall not have sharp edges, burrs, or projections that could damage the insulation or jackets of the wiring.

(C) Corrosion Protection. Cable tray systems shall be corrosion resistant. If made of ferrous material, the system shall be protected from corrosion as required by 300.6.

(D) Side Rails. Cable trays shall have side rails or equivalent structural members.

(E) Fittings. Cable trays shall include fittings or other suitable means for changes in direction and elevation of runs.

(F) Nonmetallic Cable Tray. Nonmetallic cable trays shall be made of flame-retardant material.

392.6 Installation.

(A) Complete System. Cable trays shall be installed as a complete system. Field bends or modifications shall be made so that the electrical continuity of the cable tray system and support for the cables is maintained. Cable tray systems shall be permitted to have mechanically discontinuous segments between cable tray runs or between cable tray runs and equipment. The system shall provide for the support of the cables in accordance with their corresponding articles.

Where cable trays support individual conductors and where the conductors pass from one cable tray to another, or from a cable tray to raceway(s) or from a cable tray to equipment where the conductors are terminated, the distance between cable trays or between the cable tray and the raceway(s) or the equipment shall not exceed 1.8 m (6 ft). The conductors shall be secured to the cable tray(s) at the transition, and they shall be protected, by guarding or by location, from physical damage.

A bonding jumper sized in accordance with 250.102 shall connect the two sections of cable tray, or the cable tray and the raceway or equipment. Bonding shall be in accordance with 250.96.

Runs of cable tray are not required to be totally mechanically continuous from the equipment source to the equipment termination. Breaks in the mechanical continuity of cable tray systems are permitted and often occur at tees, crossovers, elevation changes, fire stops, or for thermal contraction and expansion. Also, cable tray systems are not required to be mechanically connected to the equipment they serve.

The 6-ft distance limit applies to mechanically discontinuous cable tray segments for individual conductors but not to trays containing multiconductor cables. For further information regarding multiconductor Type TC tray cable used with discontinuous cable tray, refer to 336.10(6).

Most important, especially for discontinuous cable tray segments, is the bonding of the entire cable tray system. According to 250.96, properly sized and installed bonding conductors must be installed across any mechanical discontinuities in the cable tray system and across any space between the cable tray and the conductor termination equipment enclosure or its equipment ground bus.

Of course, the cables installed within cable tray systems must always be supported to the minimum requirements of the applicable article. This requirement either limits the gap distance in cable tray runs and between the cable tray and the equipment enclosures or requires intermediate cable supports at the appropriate distances in place of the cable tray.

(B) Completed Before Installation. Each run of cable tray shall be completed before the installation of cables.

(C) Supports. Supports shall be provided to prevent stress on cables where they enter raceways or other enclosures from cable tray systems.

Cable trays shall be supported at intervals in accordance with the installation instructions.

(D) Covers. In portions of runs where additional protection is required, covers or enclosures providing the required protection shall be of a material that is compatible with the cable tray.

(E) Multiconductor Cables Rated 600 Volts or Less. Multiconductor cables rated 600 volts or less shall be permitted to be installed in the same cable tray.

(F) Cables Rated Over 600 Volts. Cables rated over 600 volts and those rated 600 volts or less installed in the same cable tray shall comply with either of the following:

(1) The cables rated over 600 volts are Type MC.
(2) The cables rated over 600 volts are separated from the cables rated 600 volts or less by a solid fixed barrier of a material compatible with the cable tray.

(G) Through Partitions and Walls. Cable trays shall be permitted to extend transversely through partitions and walls or vertically through platforms and floors in wet or dry locations where the installations, complete with installed cables, are made in accordance with the requirements of 300.21.

(H) Exposed and Accessible. Cable trays shall be exposed and accessible except as permitted by 392.6(G).

(I) Adequate Access. Sufficient space shall be provided and maintained about cable trays to permit adequate access for installing and maintaining the cables.

(J) Raceways, Cables, Boxes, and Conduit Bodies Supported from Cable Tray Systems. In industrial facilities where conditions of maintenance and supervision ensure that only qualified persons service the installation and where the cable tray systems are designed and installed to support the load, such systems shall be permitted to support raceways and cables, and boxes and conduit bodies covered in 314.1. For raceways terminating at the tray, a listed cable tray clamp or adapter shall be used to securely fasten the raceway to the cable tray system. Additional supporting and securing of the raceway shall be in accordance with the requirements of the appropriate raceway article.

For raceways or cables running parallel to and attached to the bottom or side of a cable tray system, fastening and supporting shall be in accordance with the requirements of the appropriate raceway or cable article.

For boxes and conduit bodies attached to the bottom or side of a cable tray system, fastening and supporting shall be in accordance with the requirements of 314.23.

Section 392.6(J) permits conduit and cable termination supports as well as outlet boxes supported solely by the cable tray in qualifying industrial facilities only. These items are not permitted to be supported solely by the cable tray in commercial installations.

For commercial installations (and nonqualifying industrial facilities), conduits must be supported within 3 ft of the cable tray or within 5 ft if structural members do not permit fastening within 3 ft of the cable tray. Cables connecting to equipment outside the cable tray system must be supported according to their respective article. For example, Type MC cable in the larger sizes is required to be supported outside a cable tray system at intervals not exceeding 6 ft, according to 330.30.

392.7 Grounding.

(A) Metallic Cable Trays. Metallic cable trays that support electrical conductors shall be grounded as required for conductor enclosures in Article 250.

Section 392.7(A), together with 250.96, requires all cable tray systems that support electrical conductors (whether mechanically continuous or with isolated segments) to be electrically continuous and effectively bonded and grounded. This requirement applies whether or not the cable tray is used as an equipment grounding conductor.

(B) Steel or Aluminum Cable Tray Systems. Steel or aluminum cable tray systems shall be permitted to be used as equipment grounding conductors, provided that all the following requirements are met:

(1) The cable tray sections and fittings shall be identified for grounding purposes.
(2) The minimum cross-sectional area of cable trays shall conform to the requirements in Table 392.7(B).
(3) All cable tray sections and fittings shall be legibly and durably marked to show the cross-sectional area of metal in channel cable trays, or cable trays of one-piece construction, and the total cross-sectional area of both side rails for ladder or trough cable trays.
(4) Cable tray sections, fittings, and connected raceways shall be bonded in accordance with 250.96 using bolted mechanical connectors or bonding jumpers sized and installed in accordance with 250.102.

For cable tray systems in commercial occupancies, designers are afforded the option to specify multiconductor cables without equipment grounding conductors (EGCs) and to use the cable tray system as the required equipment grounding conductor, provided the cable tray system meets the requirements of 392.7(A) and 392.7(B). Exhibit 392.1 shows an example of the grounding and bonding of multiconductor cables in cable trays with conduit runs to power equipment.

For cable tray systems in industrial establishments that qualify, the designer is also afforded the option to specify single-conductor cables without a cable equipment ground-

Table 392.7(B) Metal Area Requirements for Cable Trays Used as Equipment Grounding Conductor

Maximum Fuse Ampere Rating, Circuit Breaker Ampere Trip Setting, or Circuit Breaker Protective Relay Ampere Trip Setting for Ground-Fault Protection of Any Cable Circuit in the Cable Tray System	Minimum Cross-Sectional Area of Metal[a]			
	Steel Cable Trays		Aluminum Cable Trays	
	mm²	in.²	mm²	in.²
60	129	0.20	129	0.20
100	258	0.40	129	0.20
200	451.5	0.70	129	0.20
400	645	1.00	258	0.40
600	967.5	1.50[b]	258	0.40
1000	—	—	387	0.60
1200	—	—	645	1.00
1600	—	—	967.5	1.50
2000	—	—	1290	2.00[b]

[a]Total cross-sectional area of both side rails for ladder or trough cable trays; or the minimum cross-sectional area of metal in channel cable trays or cable trays of one-piece construction.

[b]Steel cable trays shall not be used as equipment grounding conductors for circuits with ground-fault protection above 600 amperes. Aluminum cable trays shall not be used as equipment grounding conductors for circuits with ground-fault protection above 2000 amperes.

ing conductor and to use the cable tray as the required equipment grounding conductor. Again, this option is available only if the cable tray system meets the requirements of 392.7(A) and (B).

392.8 Cable Installation.

(A) Cable Splices. Cable splices made and insulated by approved methods shall be permitted to be located within a cable tray, provided they are accessible and do not project above the side rails.

(B) Fastened Securely. In other than horizontal runs, the cables shall be fastened securely to transverse members of the cable trays.

Other fastening requirements are found in 392.6(A) and (C).

(C) Bushed Conduit and Tubing. A box shall not be required where cables or conductors are installed in bushed conduit and tubing used for support or for protection against physical damage.

(D) Connected in Parallel. Where single conductor cables comprising each phase or neutral of a circuit are connected

in parallel as permitted in 310.4, the conductors shall be installed in groups consisting of not more than one conductor per phase or neutral to prevent current unbalance in the paralleled conductors due to inductive reactance.

Single conductors shall be securely bound in circuit groups to prevent excessive movement due to fault-current magnetic forces unless single conductors are cabled together, such as triplexed assemblies.

The binding or otherwise grouping of 3-phase circuits is good engineering practice. If properly done, it results in the phase reactances of the conductors being balanced, which reduces the voltage unbalance between the phases of the 3-phase circuit.

(E) Single Conductors. Where any of the single conductors installed in ladder or ventilated trough cable trays are 1/0 through 4/0 AWG, all single conductors shall be installed in a single layer. Conductors that are bound together to comprise each circuit group shall be permitted to be installed in other than a single layer.

392.9 Number of Multiconductor Cables, Rated 2000 Volts or Less, in Cable Trays.

The number of multiconductor cables, rated 2000 volts or less, permitted in a single cable tray shall not exceed the requirements of this section. The conductor sizes herein apply to both aluminum and copper conductors.

(A) Any Mixture of Cables. Where ladder or ventilated trough cable trays contain multiconductor power or lighting cables, or any mixture of multiconductor power, lighting, control, and signal cables, the maximum number of cables shall conform to the following:

(1) Where all of the cables are 4/0 AWG or larger, the sum of the diameters of all cables shall not exceed the cable tray width, and the cables shall be installed in a single layer.

(2) Where all of the cables are smaller than 4/0 AWG, the sum of the cross-sectional areas of all cables shall not exceed the maximum allowable cable fill area in Column 1 of Table 392.9 for the appropriate cable tray width.

(3) Where 4/0 AWG or larger cables are installed in the same cable tray with cables smaller than 4/0 AWG, the sum of the cross-sectional areas of all cables smaller than 4/0 AWG shall not exceed the maximum allowable fill area resulting from the computation in Column 2 of Table 392.9 for the appropriate cable tray width. The 4/0 AWG and larger cables shall be installed in a single layer, and no other cables shall be placed on them.

(B) Multiconductor Control and/or Signal Cables Only. Where a ladder or ventilated trough cable tray having a

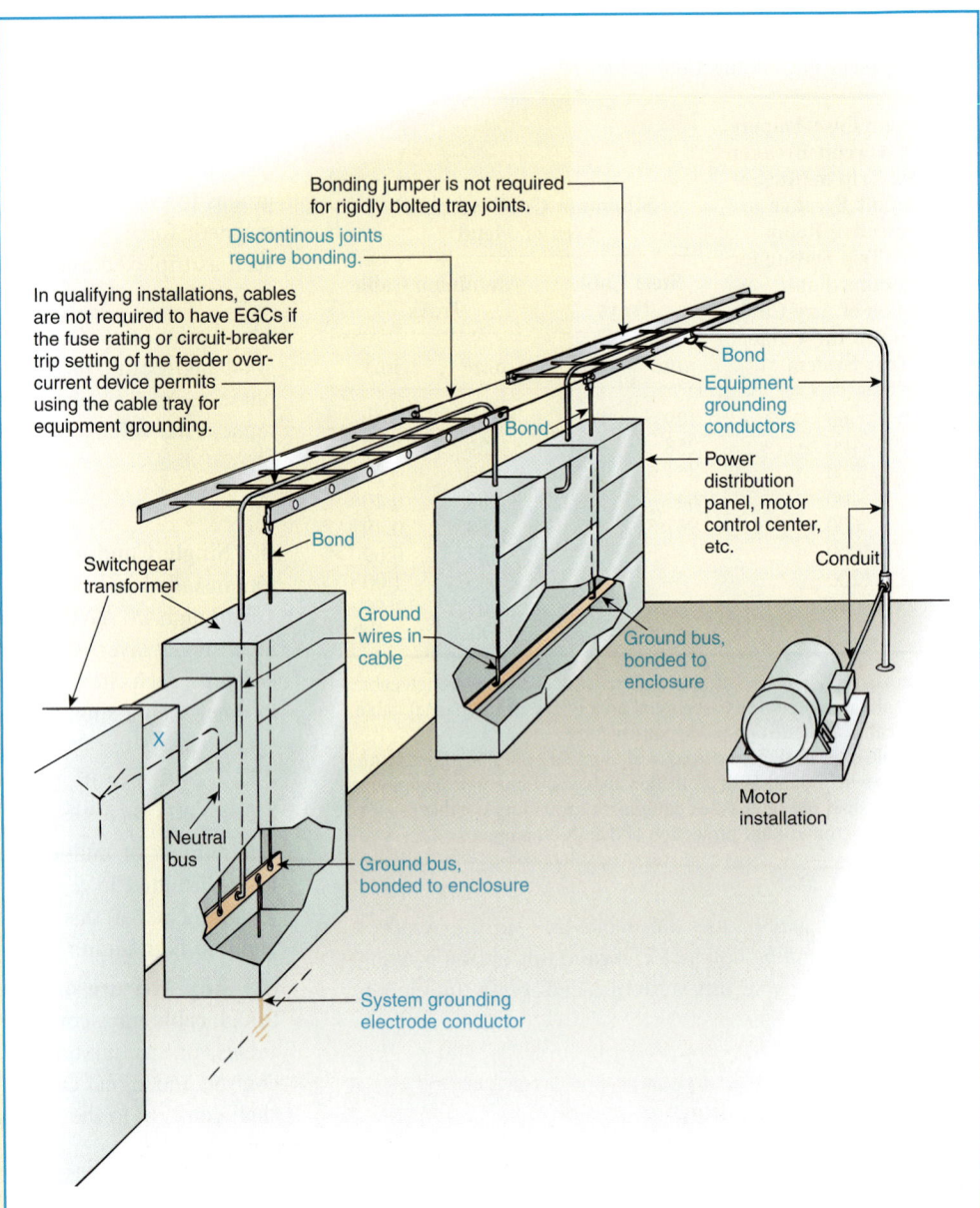

usable inside depth of 150 mm (6 in.) or less contains multi-conductor control and/or signal cables only, the sum of the cross-sectional areas of all cables at any cross section shall not exceed 50 percent of the interior cross-sectional area of the cable tray. A depth of 150 mm (6 in.) shall be used to compute the allowable interior cross-sectional area of any cable tray that has a usable inside depth of more than 150 mm (6 in.).

(C) Solid Bottom Cable Trays Containing Any Mixture. Where solid bottom cable trays contain multiconductor power or lighting cables, or any mixture of multiconductor power, lighting, control, and signal cables, the maximum number of cables shall conform to the following:

(1) Where all of the cables are 4/0 AWG or larger, the sum of the diameters of all cables shall not exceed 90 percent of the cable tray width, and the cables shall be installed in a single layer.

(2) Where all of the cables are smaller than 4/0 AWG, the sum of the cross-sectional areas of all cables shall not exceed the maximum allowable cable fill area in Column 3 of Table 392.9 for the appropriate cable tray width.

(3) Where 4/0 AWG or larger cables are installed in the same cable tray with cables smaller than 4/0 AWG, the sum of the cross-sectional areas of all cables smaller than 4/0 AWG shall not exceed the maximum allowable fill area resulting from the computation in Column 4 of Table 392.9 for the appropriate cable tray width. The

Table 392.9 Allowable Cable Fill Area for Multiconductor Cables in Ladder, Ventilated Trough, or Solid Bottom Cable Trays for Cables Rated 2000 Volts or Less

Inside Width of Cable Tray		Maximum Allowable Fill Area for Multiconductor Cables							
		Ladder or Ventilated Trough Cable Trays, 392.9(A)				Solid Bottom Cable Trays, 392.9(C)			
		Column 1 Applicable for 392.9(A)(2) Only		Column 2[a] Applicable for 392.9(A)(3) Only		Column 3 Applicable for 392.9(C)(2) Only		Column 4[a] Applicable for 392.9(C)(3) Only	
mm	in.	mm²	in.²	mm²	in.²	mm²	in.²	mm²	in.²
150	6.0	4,500	7.0	4,500–(1.2 Sd)[b]	7–(1.2 Sd)[b]	3,500	5.5	3,500–Sd[b]	5.5–Sd[b]
225	9.0	6,800	10.5	6,800–(1.2 Sd)	10.5–(1.2 Sd)	5,100	8.0	5,100–Sd	8.0–Sd
300	12.0	9,000	14.0	9,000–(1.2 Sd)	14–(1.2 Sd)	7,100	11.0	7,100–Sd	11.0–Sd
450	18.0	13,500	21.0	13,500–(1.2 Sd)	21–(1.2 Sd)	10,600	16.5	10,600–Sd	16.5–Sd
600	24.0	18,000	28.0	18,000–(1.2 Sd)	28–(1.2 Sd)	14,200	22.0	14,200–Sd	22.0–Sd
750	30.0	22,500	35.0	22,500–(1.2 Sd)	35–(1.2 Sd)	17,700	27.5	17,700–Sd	27.5–Sd
900	36.0	27,000	42.0	27,000–(1.2 Sd)	42–(1.2 Sd)	21,300	33.0	21,300–Sd	33.0–Sd

[a]The maximum allowable fill areas in Columns 2 and 4 shall be computed. For example, the maximum allowable fill in mm² for a 150-mm wide cable tray in Column 2 shall be 4500 minus (1.2 multiplied by Sd) [the maximum allowable fill, in square inches, for a 6-in. wide cable tray in Column 2 shall be 7 minus (1.2 multiplied by Sd)].
[b]The term Sd in Columns 2 and 4 is equal to the sum of the diameters, in mm, of all cables 107.2 mm (in inches, of all 4/0 AWG) and larger multiconductor cables in the same cable tray with smaller cables.

4/0 AWG and larger cables shall be installed in a single layer, and no other cables shall be placed on them.

(D) Solid Bottom Cable Tray—Multiconductor Control and/or Signal Cables Only. Where a solid bottom cable tray having a usable inside depth of 150 mm (6 in.) or less contains multiconductor control and/or signal cables only, the sum of the cross-sectional areas of all cables at any cross section shall not exceed 40 percent of the interior cross-sectional area of the cable tray. A depth of 150 mm (6 in.) shall be used to compute the allowable interior cross-sectional area of any cable tray that has a usable inside depth of more than 150 mm (6 in.).

(E) Ventilated Channel Cable Trays. Where ventilated channel cable trays contain multiconductor cables of any type, the following shall apply:

(1) Where only one multiconductor cable is installed, the cross-sectional area shall not exceed the value specified in Column 1 of Table 392.9(E).

(2) Where more than one multiconductor cable is installed, the sum of the cross-sectional area of all cables shall not exceed the value specified in Column 2 of Table 392.9(E).

Cable trays may be ladder, ventilated trough, ventilated channel, or solid bottom of various widths. The depth of a cable tray is a structural consideration. The depth (up to 6 in.) is related to fill only where the cable tray contains signal and control cables or if the cable tray contains large cable splices.

Table 392.9(E) Allowable Cable Fill Area for Multiconductor Cables in Ventilated Channel Cable Trays for Cables Rated 2000 Volts or Less

Inside Width of Cable Tray		Maximum Allowable Fill Area for Multiconductor Cables			
		Column 1 One Cable		Column 2 More Than One Cable	
mm	in.	mm²	in.²	mm²	in.²
75	3	1500	2.3	850	1.3
100	4	2900	4.5	1600	2.5
150	6	4500	7.0	2450	3.8

Heat dissipation is generally not a problem for signal and control cables. Inspectors or contractors, therefore, are not expected to compute the various combinations of cable tray fill in the field. Installation handbooks for cable tray applications are available from various cable tray manufacturers.

(F) Solid Channel Cable Trays. Where solid channel cable trays contain multiconductor cables of any type, the following shall apply:

(1) Where only one multiconductor cable is installed, the cross-sectional area of the cable shall not exceed the value specified in Column 1 of Table 392.9(F).

(2) Where more than one multiconductor cable is installed, the sum of the cross-sectional area of all cable shall

Table 392.9(F) Allowable Cable Fill Area for Multiconductor Cables in Solid Channel Cable Trays for Cables Rated 2000 Volts or Less

Inside Width of Cable Tray		Column 1 One Cable		Column 2 More Than One Cable	
mm	in.	mm²	in.²	mm²	in.²
50	2	850	1.3	500	0.8
75	3	1300	2.0	700	1.1
100	4	2400	3.7	1400	2.1
150	6	3600	5.5	2100	3.2

not exceed the value specified in Column 2 of Table 392.9(F).

392.10 Number of Single-Conductor Cables, Rated 2000 Volts or Less, in Cable Trays.

The number of single-conductor cables, rated 2000 volts or less, permitted in a single cable tray section shall not exceed the requirements of this section. The single conductors, or conductor assemblies, shall be evenly distributed across the cable tray. The conductor sizes herein apply to both aluminum and copper conductors.

(A) Ladder or Ventilated Trough Cable Trays. Where ladder or ventilated trough cable trays contain single-conductor cables, the maximum number of single conductors shall conform to the following:

(1) Where all of the cables are 1000 kcmil or larger, the sum of the diameters of all single-conductor cables shall not exceed the cable tray width.

(2) Where all of the cables are from 250 kcmil up to 1000 kcmil, the sum of the cross-sectional areas of all single-conductor cables shall not exceed the maximum allowable cable fill area in Column 1 of Table 392.10(A) for the appropriate cable tray width.

(3) Where 1000 kcmil or larger single-conductor cables are installed in the same cable tray with single-conductor cables smaller than 1000 kcmil, the sum of the cross-sectional areas of all cables smaller than 1000 kcmil shall not exceed the maximum allowable fill area resulting from the computation in Column 2 of Table 392.10(A) for the appropriate cable tray width.

(4) Where any of the single conductor cables are 1/0 through 4/0 AWG, the sum of the diameters of all single conductor cables shall not exceed the cable tray width.

(B) Ventilated Channel Cable Trays. Where 50 mm (2 in.), 75 mm (3 in.), 100 mm (4 in.), or 150 mm (6 in.) wide ventilated channel cable trays contain single-conductor cables, the sum of the diameters of all single conductors shall not exceed the inside width of the channel.

392.11 Ampacity of Cables, Rated 2000 Volts or Less, in Cable Trays.

(A) Multiconductor Cables. The allowable ampacity of multiconductor cables, nominally rated 2000 volts or less, installed according to the requirements of 392.9 shall be as

Table 392.10(A) Allowable Cable Fill Area for Single-Conductor Cables in Ladder or Ventilated Trough Cable Trays for Cables Rated 2000 Volts or Less

Inside Width of Cable Tray		Maximum Allowable Fill Area for Single-Conductor Cables in Ladder or Ventilated Trough Cable Trays			
		Column 1 Applicable for 392.10(A)(2) Only		Column 2[a] Applicable for 392.10(A)(3) Only	
mm	in.	mm²	in.²	mm²	in.²
150	6	4,200	6.5	4,200 – (1.1 Sd)[b]	6.5 – (1.1 Sd)[b]
225	9	6,100	9.5	6,100 – (1.1 Sd)	9.5 – (1.1 Sd)
300	12	8,400	13.0	8,400 – (1.1 Sd)	13.0 – (1.1 Sd)
450	18	12,600	19.5	12,600 – (1.1 Sd)	19.5 – (1.1 Sd)
600	24	16,800	26.0	16,800 – (1.1 Sd)	26.0 – (1.1 Sd)
750	30	21,000	32.5	21,000 – (1.1 Sd)	32.5 – (1.1 Sd)
900	36	25,200	39.0	25,200 – (1.1 Sd)	39.0 – (1.1 Sd)

[a]The maximum allowable fill areas in Column 2 shall be computed. For example, the maximum allowable fill, in mm² for a 150 mm wide cable tray in Column 2 shall be 4192.5 minus (1.1 multiplied by Sd) [the maximum allowable fill, in square inches, for a 6-in. wide cable tray in Column 2 shall be 6.5 minus (1.1 multiplied by Sd)].

[b]The term *Sd* in Column 2 is equal to the sum of the diameters, in mm, of all cables 507 mm² (in inches, of all 1000 kcmil) and larger single-conductor cables in the same ladder or ventilated trough cable tray with small cables.

given in Tables 310.16 and 310.18, subject to the provisions of (1), (2), (3), and 310.15(A)(2).

(1) The derating factors of 310.15(B)(2)(a) shall apply only to multiconductor cables with more than three current-carrying conductors. Derating shall be limited to the number of current-carrying conductors in the cable and not to the number of conductors in the cable tray.

(2) Where cable trays are continuously covered for more than 1.8 m (6 ft) with solid unventilated covers, not over 95 percent of the allowable ampacities of Tables 310.16 and 310.18 shall be permitted for multiconductor cables.

(3) Where multiconductor cables are installed in a single layer in uncovered trays, with a maintained spacing of not less than one cable diameter between cables, the ampacity shall not exceed the allowable ambient temperature-corrected ampacities of multiconductor cables, with not more than three insulated conductors rated 0 through 2000 volts in free air, in accordance with 310.15(C).

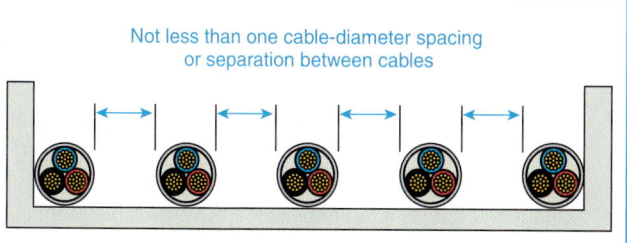

Not less than one cable-diameter spacing or separation between cables

FPN: See Table B.310.3.

(B) Single-Conductor Cables. The allowable ampacity of single-conductor cables shall be as permitted by 310.15(A)(2). The derating factors of 310.15(B)(2)(a) shall not apply to the ampacity of cables in cable trays. The ampacity of single-conductor cables, or single conductors cabled together (triplexed, quadruplexed, etc.), nominally rated 2000 volts or less shall comply with the following:

(1) Where installed according to the requirements of 392.10, the ampacities for 600 kcmil and larger single-

conductor cables in uncovered cable trays shall not exceed 75 percent of the allowable ampacities in Table 310.17 and Table 310.19. Where cable trays are continuously covered for more than 1.8 m (6 ft) with solid unventilated covers, the ampacities for 600 kcmil and larger cables shall not exceed 70 percent of the allowable ampacities in Table 310.17 and Table 310.19.

(2) Where installed according to the requirements of 392.10, the ampacities for 1/0 AWG through 500 kcmil single-conductor cables in uncovered cable trays shall not exceed 65 percent of the allowable ampacities in Table 310.17 and Table 310.19. Where cable trays are continuously covered for more than 1.8 m (6 ft) with solid unventilated covers, the ampacities for 1/0 AWG through 500 kcmil cables shall not exceed 60 percent of the allowable ampacities in Table 310.17 and Table 310.19.

(3) Where single conductors are installed in a single layer in uncovered cable trays, with a maintained space of not less than one cable diameter between individual conductors, the ampacity of 1/0 AWG and larger cables shall not exceed the allowable ampacities in Table 310.17 and Table 310.19.

(4) Where single conductors are installed in a triangular or square configuration in uncovered cable trays, with a maintained free airspace of not less than 2.15 times one conductor diameter (2.15 × O.D.) of the largest conductor contained within the configuration and adjacent conductor configurations or cables, the ampacity of 1/0 AWG and larger cables shall not exceed the allowable ampacities of two or three single insulated conductors rated 0 through 2000 volts supported on a messenger in accordance with 310.15(B).

FPN: See Table 310.20.

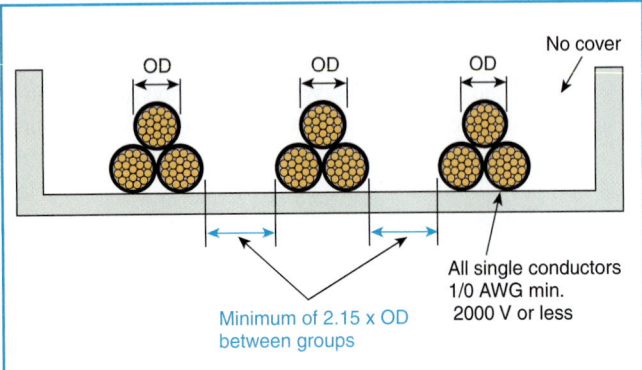

Exhibit 392.3 An illustration of 392.11(B)(4), for three single conductors installed in a triangular configuration with spacing between groups of not less than 2.15 times the conductor diameter (ampacities to be determined from Table 310.20).

tion and are terminated at circuit breakers, distribution switchgear, and similar electrical equipment, it is important to understand the temperature limitations of the electrical equipment terminals and to coordinate those temperature limitations with the ampacity of the single conductor cables. As stated in both the UL *General Information Directory* and in 110.14(C)(1), unless the equipment is listed and marked otherwise, conductor ampacities used in determining equipment terminations must be based on Table 310.16 as modified by 310.15(B)(1) through (6).

392.12 Number of Type MV and Type MC Cables (2001 Volts or Over) in Cable Trays.

The number of cables rated 2001 volts or over permitted in a single cable tray shall not exceed the requirements of this section.

The sum of the diameters of single-conductor and multiconductor cables shall not exceed the cable tray width, and the cables shall be installed in a single layer. Where single conductor cables are triplexed, quadruplexed, or bound together in circuit groups, the sum of the diameters of the single conductors shall not exceed the cable tray width, and these groups shall be installed in single layer arrangement.

392.13 Ampacity of Type MV and Type MC Cables (2001 Volts or Over) in Cable Trays.

The ampacity of cables, rated 2001 volts, nominal, or over, installed according to 392.12 shall not exceed the requirements of this section.

(A) Multiconductor Cables (2001 Volts or Over). The allowable ampacity of multiconductor cables shall be as given in Table 310.75 and Table 310.76, subject to the following provisions:

(1) Where cable trays are continuously covered for more than 1.8 m (6 ft) with solid unventilated covers, not more than 95 percent of the allowable ampacities of Table 310.75 and Table 310.76 shall be permitted for multiconductor cables.

(2) Where multiconductor cables are installed in a single layer in uncovered cable trays, with maintained spacing of not less than one cable diameter between cables, the ampacity shall not exceed the allowable ampacities of Table 310.71 and Table 310.72.

(B) Single-Conductor Cables (2001 Volts or Over). The ampacity of single-conductor cables, or single conductors cabled together (triplexed, quadruplexed, etc.), shall comply with the following:

(1) The ampacities for 1/0 AWG and larger single-conductor cables in uncovered cable trays shall not exceed 75 percent of the allowable ampacities in Table 310.69 and Table 310.70. Where the cable trays are covered for more than 1.8 m (6 ft) with solid unventilated covers, the ampacities for 1/0 AWG and larger single-conductor cables shall not exceed 70 percent of the allowable ampacities in Table 310.69 and Table 310.70.

(2) Where single-conductor cables are installed in a single layer in uncovered cable trays, with a maintained space of not less than one cable diameter between individual conductors, the ampacity of 1/0 AWG and larger cables shall not exceed the allowable ampacities in Table 310.69 and Table 310.70.

(3) Where single conductors are installed in a triangular or square configuration in uncovered cable trays, with a maintained free air space of not less than 2.15 times the diameter (2.15 × O.D.) of the largest conductor contained within the configuration and adjacent conductor configurations or cables, the ampacity of 1/0 AWG and larger cables shall not exceed the allowable ampacities in Table 310.67 and Table 310.68.

ARTICLE 394
Concealed Knob-and-Tube Wiring

Contents

I. General

394.1 Scope.

This article covers the use, installation, and construction specifications of concealed knob-and-tube wiring.

394.2. Definition.

Concealed Knob-and-Tube Wiring. A wiring method using knobs, tubes, and flexible nonmetallic tubing for the protection and support of single insulated conductors.

Open wiring on insulators (Article 398) is required to be exposed, whereas knob-and-tube wiring is allowed to be concealed. Conductors used for knob-and-tube work may be of any general-use type specified by Article 310.

II. Installation

394.10 Uses Permitted.

Concealed knob-and-tube wiring shall be permitted to be installed in the hollow spaces of walls and ceilings or in unfinished attics and roof spaces as provided in 394.23 only as follows:

(1) For extensions of existing installations
(2) Elsewhere by special permission

Concealed knob-and-tube wiring is permitted to be installed only for extensions of existing installations or where special permission is granted by the authority having jurisdiction of enforcement of the *Code*. See definition of *special permission* in Article 100.

394.12 Uses Not Permitted.

Concealed knob-and-tube wiring shall not be used in the following:

(1) Commercial garages
(2) Theaters and similar locations
(3) Motion picture studios
(4) Hazardous (classified) locations
(5) Hollow spaces of walls, ceilings, and attics where such spaces are insulated by loose, rolled, or foamed-in-place insulating material that envelops the conductors

Concealed knob-and-tube wiring is designed for use in hollow spaces of walls, ceilings, and attics and utilizes the free air in such spaces for heat dissipation. Weatherization of hollow spaces by blown-in, foamed-in, or rolled insulation prevents the dissipation of heat into the free air space. This will result in higher conductor temperature, which could cause insulation breakdown and possible ignition of the insulation.

394.17 Through or Parallel to Framing Members.

Conductors shall comply with 398.17 where passing through holes in structural members. Where passing through wood cross members in plastered partitions, conductors shall be protected by noncombustible, nonabsorbent, insulating tubes extending not less than 75 mm (3 in.) beyond the wood member.

The provision for insulated tubes where knob-and-tube wiring passes through wood cross members in plastered partitions is intended to protect the wire from contact with plaster that is likely to accumulate on horizontal wood members.

394.19 Clearances.

(A) General. A clearance of not less than 75 mm (3 in.) shall be maintained between conductors and a clearance of not less than 25 mm (1 in.) between the conductor and the surface over which it passes.

(B) Limited Conductor Space. Where space is too limited to provide these minimum clearances, such as at meters, panelboards, outlets, and switch points, the individual conductors shall be enclosed in flexible nonmetallic tubing, which shall be continuous in length between the last support and the enclosure or terminal point.

(C) Clearance from Piping, Exposed Conductors, and So Forth. Conductors shall comply with 398.19 for clearances from other exposed conductors, piping, and so forth.

394.23 In Accessible Attics.

Conductors in unfinished attics and roof spaces shall comply with 394.23(A) or (B).

FPN: See 310.10 for temperature limitation of conductors.

(A) Accessible by Stairway or Permanent Ladder. Conductors shall be installed along the side of or through bored holes in floor joists, studs, or rafters. Where run through bored holes, conductors in the joists and in studs or rafters to a height of not less than 2.1 m (7 ft) above the floor or floor joists shall be protected by substantial running boards extending not less than 25 mm (1 in.) on each side of the conductors. Running boards shall be securely fastened in place. Running boards and guard strips shall not be required where conductors are installed along the sides of joists, studs, or rafters.

Exhibit 394.1 illustrates the "running board" method of protecting open-type conductors where they are installed at a height less than 7 ft above the floor or floor joists in an accessible attic. This method is applied in attics that are accessible by stairways or permanent ladders and where such spaces are generally used for storage.

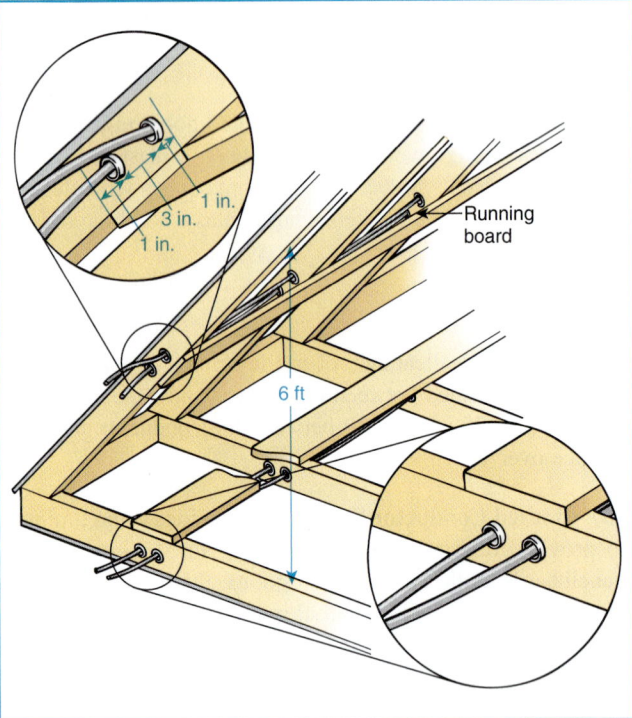

Exhibit 394.1 Open wiring in an accessible attic protected by running boards, as required by 394.23(A).

(B) Not Accessible by Stairway or Permanent Ladder. Conductors shall be installed along the sides of or through bored holes in floor joists, studs, or rafters.

Exception: In buildings completed before the wiring is installed, attic and roof spaces that are not accessible by stairway or permanent ladder and have headroom at all points less than 900 mm (3 ft), the wiring shall be permitted to be installed on the edges of rafters or joists facing the attic or roof space.

394.30 Securing and Supporting.

(A) Supporting. Conductors shall be rigidly supported on noncombustible, nonabsorbent insulating materials and shall not contact any other objects. Supports shall be installed as follows:

(1) Within 150 mm (6 in.) of each side of each tap or splice, and
(2) At intervals not exceeding 1.4 m (4½ ft).

Where it is impracticable to provide supports, conductors shall be permitted to be fished through hollow spaces in dry locations, provided each conductor is individually enclosed in flexible nonmetallic tubing that is in continuous lengths between supports, between boxes, or between a support and a box.

(B) Securing. Where solid knobs are used, conductors shall be securely tied thereto by tie wires having insulation equivalent to that of the conductor.

394.42 Devices.

Switches shall comply with 404.4 and 404.10(B).

394.56 Splices and Taps.

Splices shall be soldered unless approved splicing devices are used. In-line or strain splices shall not be used.

III. Construction Specifications

394.104 Conductors.

Conductors shall be of a type specified by Article 310.

ARTICLE 396
Messenger Supported Wiring

Contents

I. General

396.1 Scope.

This article covers the use, installation, and construction specifications for messenger supported wiring.

Messenger supported wiring systems have been manufactured and successfully used in industrial installations for many years. They have also been used for many years as service drops by utilities for commercial and residential installations. See references to messenger supported wiring in 225.6(A)(1) and 225.6(B).

396.2 Definition.

Messenger Supported Wiring. An exposed wiring support system using a messenger wire to support insulated conductors by any one of the following:

(1) A messenger with rings and saddles for conductor support
(2) A messenger with a field-installed lashing material for conductor support
(3) Factory-assembled aerial cable
(4) Multiplex cables utilizing a bare conductor, factory assembled and twisted with one or more insulated conductors, such as duplex, triplex, or quadruplex type of construction

II. Installation

396.10 Uses Permitted.

(A) Cable Types. The cable types in Table 396.10(A) shall be permitted to be installed in messenger supported wiring under the conditions described in the article or section referenced for each.

(B) In Industrial Establishments. In industrial establishments only, where conditions of maintenance and supervision ensure that only qualified persons service the installed messenger supported wiring, the following shall be permitted:

Table 396.10(A) Cable Types

Cable Type	Section	Article
Metal-clad cable		330
Mineral-insulated, metal-sheathed cable		332
Multiconductor service-entrance cable		338
Multiconductor underground feeder and branch-circuit cable		340
Other factory-assembled, multiconductor control, signal, or power cables that are identified for the use		
Power and control tray cable		336
Power-limited tray cable	725.61(C) and 725.71(E)	

(1) Any of the conductor types shown in Table 310.13 or Table 310.62
(2) MV cable

Where exposed to weather, conductors shall be listed for use in wet locations. Where exposed to direct rays of the sun, conductors or cables shall be sunlight resistant.

Some of the triplex and quadruplex cable used by utilities as service-drop cable does not use conductors recognized in Table 310.13 and does not meet the requirements of Article 310. Such triplex and quadruplex cable would, therefore, be acceptable only where approved by the authority having jurisdiction.

See 310.15(B) and Table 310.20 for two or three single-insulated conductors supported on a messenger. See 310.15(C) and Annex B, Table B.310.3, for ampacities of conductors for other cable types.

(C) Hazardous (Classified) Locations. Messenger supported wiring shall be permitted to be used in hazardous (classified) locations where the contained cables are permitted for such use in 501.4, 502.4, 503.3, and 504.20.

396.12 Uses Not Permitted.

Messenger supported wiring shall not be used in hoistways or where subject to physical damage.

396.30 Messenger Support.

The messenger shall be supported at dead ends and at intermediate locations so as to eliminate tension on the conductors. The conductors shall not be permitted to come into contact with the messenger supports or any structural members, walls, or pipes.

396.56 Conductor Splices and Taps.

Conductor splices and taps made and insulated by approved methods shall be permitted in messenger supported wiring.

396.60 Grounding.

The messenger shall be grounded as required by 250.80 and 250.86 for enclosure grounding.

ARTICLE 398
Open Wiring on Insulators

Contents

I. General

398.1 Scope.

This article covers the use, installation, and construction specifications of open wiring on insulators.

398.2 Definition.

Open Wiring on Insulators. An exposed wiring method using cleats, knobs, tubes, and flexible tubing for the protec-tion and support of single insulated conductors run in or on buildings.

Open wiring on insulators is an exposed wiring method that is not permitted to be concealed by the structure or finish of the building. It is permitted indoors or outdoors, in dry or wet locations, and where subject to corrosive vapors, provided the insulation choice from Table 310.13 is suitable for use in a corrosive environment.

This method of wiring is no longer permitted for tempo-rary lighting and power circuits on construction sites but is permitted for lighting and power circuits in agricultural buildings [see 347.5(A)]. It may also be used for services (see 230.43).

II. Installation

398.10 Uses Permitted.

Open wiring on insulators shall be permitted only for indus-trial or agricultural establishments on systems of 600 volts, nominal, or less, as follows:

(1) Indoors or outdoors
(2) In wet or dry locations
(3) Where subject to corrosive vapors
(4) For services

See Tables 310.17 and 310.19 for ampacities of conductors.

398.12 Uses Not Permitted.

Open wiring on insulators shall not be installed where con-cealed by the building structure.

398.15 Exposed Work.

(A) Dry Locations. In dry locations, where not exposed to severe physical damage, conductors shall be permitted to be separately enclosed in flexible nonmetallic tubing. The tubing shall be in continuous lengths not exceeding 4.5 m (15 ft) and secured to the surface by straps at intervals not exceeding 1.4 m (4½ ft).

(B) Entering Spaces Subject to Dampness, Wetness, or Corrosive Vapors. Conductors entering or leaving loca-tions subject to dampness, wetness, or corrosive vapors shall have drip loops formed on them and shall then pass upward and inward from the outside of the buildings, or from the damp, wet, or corrosive location, through noncombustible, nonabsorbent insulating tubes.

FPN: See 230.52 for individual conductors entering buildings or other structures.

(C) Exposed to Physical Damage. Conductors within 2.1 m (7 ft) from the floor shall be considered exposed to physical damage. Where open conductors cross ceiling joists and wall studs and are exposed to physical damage, they shall be protected by one of the following methods:

(1) Guard strips not less than 25 mm (1 in.) nominal in thickness and at least as high as the insulating supports, placed on each side of and close to the wiring.

(2) A substantial running board at least 13 mm (½ in.) thick in back of the conductors with side protections. Running boards shall extend at least 25 mm (1 in.) outside the conductors, but not more than 50 mm (2 in.), and the protecting sides shall be at least 50 mm (2 in.) high and at least 25 mm (1 in.) nominal in thickness.

(3) Boxing made in accordance with (C)(1) or (C)(2) and furnished with a cover kept at least 25 mm (1 in.) away from the conductors within. Where protecting vertical conductors on side walls, the boxing shall be closed at the top and the holes through which the conductors pass shall be bushed.

(4) Rigid metal conduit, intermediate metal conduit, rigid nonmetallic conduit, or electrical metallic tubing, in which case the rules of Articles 342, 344, 352, or 358 shall apply; or by metal piping, in which case the conductors shall be encased in continuous lengths of approved flexible tubing.

398.17 Through or Parallel to Framing Members.

Open conductors shall be separated from contact with walls, floors, wood cross members, or partitions through which they pass by tubes or bushings of noncombustible, nonabsorbent insulating material. Where the bushing is shorter than the hole, a waterproof sleeve of noninductive material shall be inserted in the hole and an insulating bushing slipped into the sleeve at each end in such a manner as to keep the conductors absolutely out of contact with the sleeve. Each conductor shall be carried through a separate tube or sleeve.

> FPN: See 310.10 for temperature limitation of conductors.

398.19 Clearances.

Open conductors shall be separated at least 50 mm (2 in.) from metal raceways, piping, or other conducting material, and from any exposed lighting, power, or signaling conductor, or shall be separated therefrom by a continuous and firmly fixed nonconductor in addition to the insulation of the conductor. Where any insulating tube is used, it shall be secured at the ends. Where practicable, conductors shall pass over rather than under any piping subject to leakage or accumulations of moisture.

The provision for additional protective insulation on open wiring is to prevent contact with metal piping, metal objects, or exposed conductors of other circuits.

398.23 In Accessible Attics.

Conductors in unfinished attics and roof spaces shall comply with 398.23(A) or (B).

(A) Accessible by Stairway or Permanent Ladder. Conductors shall be installed along the side of or through bored holes in floor joists, studs, or rafters. Where run through bored holes, conductors in the joists and in studs or rafters to a height of not less than 2.1 m (7 ft) above the floor or floor joists shall be protected by substantial running boards extending not less than 25 mm (1 in.) on each side of the conductors. Running boards shall be securely fastened in place. Running boards and guard strips shall not be required for conductors installed along the sides of joists, studs, or rafters.

(B) Not Accessible by Stairway or Permanent Ladder. Conductors shall be installed along the sides of or through bored holes in floor joists, studs, or rafters.

Exception: In buildings completed before the wiring is installed, attic and roof spaces that are not accessible by stairway or permanent ladder and have headroom at all points less than 900 mm (3 ft), the wiring shall be permitted to be installed on the edges of rafters or joists facing the attic or roof space.

398.30 Securing and Supporting.

(A) Conductor Sizes Smaller Than 8 AWG. Conductors smaller than 8 AWG shall be rigidly supported on noncombustible, nonabsorbent insulating materials and shall not contact any other objects. Supports shall be installed as follows:

(1) Within 150 mm (6 in.) from a tap or splice

(2) Within 300 mm (12 in.) of a dead-end connection to a lampholder or receptacle

(3) At intervals not exceeding 1.4 m (4½ ft) and at closer intervals sufficient to provide adequate support where likely to be disturbed

(B) Conductor Sizes 8 AWG and Larger. Supports for conductors 8 AWG or larger installed across open spaces shall be permitted up to 4.5 m (15 ft) apart if noncombustible, nonabsorbent insulating spacers are used at least every 1.4 m (4½ ft) to maintain at least 65 mm (2½ in.) between conductors.

Where not likely to be disturbed in buildings of mill construction, 8 AWG and larger conductors shall be permitted to be run across open spaces if supported from each wood cross member on approved insulators maintaining 150 mm (6 in.) between conductors.

Mill construction is generally considered to be a building where the floors and ceilings are supported by wood timbers or beams or wood cross members spaced approximately 15 ft apart. This type of construction is sometimes referred to as plank-on-timber construction. Section 398.30(B) permits 8 AWG and larger conductors to span this distance where the ceilings are high and free of obstructions and the conductors are unlikely to come into contact with other objects.

(C) Industrial Establishments. In industrial establishments only, where conditions of maintenance and supervision ensure that only qualified persons service the system, conductors of sizes 250 kcmil and larger shall be permitted to be run across open spaces where supported at intervals up to 9.0 m (30 ft) apart.

It is common practice in industrial buildings to install open feeders on insulators, which are mounted on the bottom of roof trusses at every bay location. Many bays are more than 15 ft wide. Therefore, this section permits size 250 kcmil and larger conductors to be supported at 30-ft intervals where it is assured that qualified persons will service the system.

In addition to the ease and economy of installation or alteration of open wiring, it is to be noted that, by close spacing of conductors, the reactance of a circuit is reduced; hence, the voltage drop is reduced.

(D) Mounting of Conductor Supports. Where nails are used to mount knobs, they shall not be smaller than tenpenny. Where screws are used to mount knobs, or where nails or screws are used to mount cleats, they shall be of a length sufficient to penetrate the wood to a depth equal to at least one-half the height of the knob and the full thickness of the cleat. Cushion washers shall be used with nails.

(E) Tie Wires. 8 AWG or larger conductors supported on solid knobs shall be securely tied thereto by tie wires having an insulation equivalent to that of the conductor.

398.42 Devices.

Surface-type snap switches shall be mounted in accordance with 404.10(A), and boxes shall not be required. Other type switches shall be installed in accordance with 404.4.

III. Construction Specifications

398.104 Conductors.

Conductors shall be of a type specified by Article 310.

ARTICLE 400
Flexible Cords and Cables

Contents

I. General

400.1 Scope.

This article covers general requirements, applications, and construction specifications for flexible cords and flexible cables.

Flexible cords and cables, because of the nature of their use, are not considered to be wiring methods. Wiring methods are covered in Chapter 3 of the *Code*. Careful study of 400.7, *Uses Permitted,* and 400.8, *Uses Not Permitted*, is required before choosing flexible cords or cables for a specific application.

400.2 Other Articles.

Flexible cords and flexible cables shall comply with this article and with the applicable provisions of other articles of this *Code*.

400.3 Suitability.

Flexible cords and cables and their associated fittings shall be suitable for the conditions of use and location.

400.4 Types.

Flexible cords and flexible cables shall conform to the description in Table 400.4. Types of flexible cords and flexible cables other than those listed in the table shall be the subject of special investigation.

Type G-CG cable was added to Table 400.4 in the 1999 *Code*. This cable is similar to Type G cable, except that it incorporates an insulated ground-check (GC) conductor. The ground-check conductor is used as part of a low-voltage circuit that monitors the grounding conductor continuity. Notes 11 and 12 to Table 400.4 coordinate these cable types with Article 625, Electric Vehicle Charging System.

400.5 Ampacities for Flexible Cords and Cables.

Table 400.5(A) provides the allowable ampacities, and Table 400.5(B) provides the ampacities for flexible cords and cables with not more than three current-carrying conductors. These tables shall be used in conjunction with applicable end-use product standards to ensure selection of the proper size and type. If the number of current-carrying conductors exceeds three, the allowable ampacity or the ampacity of each conductor shall be reduced from the 3-conductor rating as shown in Table 400.5.

Ultimate Insulation Temperature. In no case shall conductors be associated together in such a way with respect to the kind of circuit, the wiring method used, or the number of conductors such that the limiting temperature of the conductors is exceeded.

A neutral conductor that carries only the unbalanced current from other conductors of the same circuit shall not be required to meet the requirements of a current-carrying conductor.

Table 400.4 Flexible Cords and Cables (See 400.4.)

Trade Name	Type Letter	Voltage	AWG or kcmil	Number of Conductors	Insulation	Nominal Insulation Thickness[1]			Braid on Each Conductor	Outer Covering	Use		
						AWG or kcmil	mm	mils					
Lamp cord	C	300 600	18–16 14–10	2 or more	Thermoset or thermo-plastic	18–16 14–10	0.76 1.14	30 45	Cotton	None	Pendant or portable	Dry locations	Not hard usage
Elevator cable	E See Note 5. See Note 9. See Note 10.	300 or 600	20–2	2 or more	Thermoset	20–16 14–12 12–10 8–2	0.51 0.76 1.14 1.52	20 30 45 60	Cotton	Three cotton, Outer one flame-retardant & moisture-resistant. See Note 3.	Elevator lighting and control	Unclassified locations	
						20–16 14–12 12–10 8–2	0.51 0.76 1.14 1.52	20 30 45 60	Flexible nylon jacket				
Elevator cable	EO See Note 5. See Note 10.	300 or 600	20–2	2 or more	Thermoset	20–16 14–12 12–10 8–2	0.51 0.76 1.14 1.52	20 30 45 60	Cotton	Outer one Three cotton, flame-retardant & moisture-resistant. See Note 3.	Elevator lighting and control	Unclassified locations	
										One cotton and a neoprene jacket. See Note 3.		Hazardous (classified) locations	
Elevator cable	ET See Note 5. See Note 10.	300 or 600	20–2	2 or more	Thermo-plastic	20–16 14–12 12–10 8–2	0.51 0.76 1.14 1.52	20 30 45 60	Rayon	Three cotton or equivalent. Outer one flame-retardant & moisture-resistant. See Note 3.	Unclassified locations		
	ETLB See Note 5. See Note 10.	300 or 600							None				
	ETP See Note 5. See Note 10.	300 or 600							Rayon	Thermoplastic	Hazardous (classified) locations		
	ETT See Note 5. See Note 10.	300 or 600							None	One cotton or equivalent and a thermoplastic jacket			
Portable power cable	G	2000	12–500	2–6 plus grounding conductor(s)	Thermoset	12–2 1–4/0 250–500	1.52 2.03 2.41	60 80 95		Oil-resistant thermoset	Portable and extra hard usage		
	G-GC	2000	12–500	3–6 plus grounding conductors and 1 ground check conductor	Thermoset	12–2 1–4/0 250–500	1.52 2.03 2.41	60 80 95		Oil-resistant thermoset			
Heater cord	HPD	300	18–12	2, 3, or 4	Thermoset	18–16 14–12	0.38 0.76	15 30	None	Cotton or rayon	Portable heaters	Dry locations	Not hard usage
Parallel heater cord	HPN See Note 6.	300	18–12	2 or 3	Oil-resistant thermoset	18–16 14–12	1.14 1.52 2.41	45 60 95	None	Oil-resistant thermoset	Portable	Damp locations	Not hard usage
Thermoset jacketed heater cords	HSJ	300	18–12	2, 3, or 4	Thermoset	18–16	0.76	30	None	Cotton and Thermoset	Portable or portable heater	Damp locations	Hard usage
												Damp and wet locations	

Table 400.4 *Continued*

Trade Name	Type Letter	Voltage	AWG or kcmil	Number of Conductors	Insulation	Nominal Insulation Thickness[1]			Braid on Each Conductor	Outer Covering	Use		
						AWG or kcmil	mm	mils					
	HSJO	300	18–12		Oil-resistant thermoset	14–12	1.14	45		Cotton and oil-resistant thermoset			
	HSJOO	300	18–12										
Non-integral parallel cords	NISP-1 See Note 6.	300	20–18	2 or 3	Thermoset	20–18	0.38	15	None	Thermoset	Pendant or portable	Damp locations	Not hard usage
	NISP-2 See Note 6.	300	18–16			18–16	0.76	30					
	NISPE-1 See Note 6.	300	20–18		Thermoplastic elastomer	20–18	0.38	15		Thermoplastic elastomer			
	NISPE-2 See Note 6.	300	18–16			18–16	0.76	30					
	NISPT-1 See Note 6.	300	20–18		Thermoplastic	20–18	0.38	15		Thermoplastic			
	NISPT-2 See Note 6.	300	18–16			18–16	0.76	30					
Twisted portable cord	PD	300 600	18–16 14–10	2 or more	Thermoset or thermoplastic	18–16 14–10	0.76 1.14	30 45	Cotton	Cotton or rayon	Pendant or portable	Dry locations	Not hard usage
Portable power cable	PPE	2000	12–500	1–6 plus optional grounding conductor(s)	Thermoplastic elastomer	12–2 1–4/0 250–500	1.52 2.03 2.41	60 80 95		Oil-resistant thermoplastic elastomer	Portable, extra hard usage		
Hard service cord	S See Note 4.	600	18–12	2 or more	Thermoset	18–16 14–10 8–2	0.76 1.14 1.52	30 45 60	None	Thermoset	Pendant or portable	Damp locations	Extra hard usage
Flexible stage and lighting power cable	SC	600	8–250	1 or more		8–2 1–4/0 250	1.52 2.03 2.41	60 80 95		Thermoset[2]	Portable, extra hard usage		
	SCE	600			Thermoplastic elastomer					Thermoplastic elastomer[2]			
	SCT	600			Thermoplastic					Thermoplastic[2]			
Hard service cord	SE See Note 4.	600	18–2	2 or more	Thermoplastic elastomer	18–16 14–10 8–2	0.76 1.14 1.52	30 45 60	None	Thermoplastic elastomer	Pendant or portable	Damp locations	Extra hard usage
	SEW See Note 4. See Note 13.	600										Damp and wet locations	
	SEO See Note 4.	600								Oil-resistant thermoplastic elastomer		Damp locations	
	SEOW See Note 4. See Note 13.	600										Damp and wet locations	
	SEOO See Note 4.	600			Oil-resistant thermoplastic elastomer							Damp locations	
	SEOOW See Note 4. See Note 13.	600										Damp and wet locations	
Junior hard service cord	SJ	300	18–10	2–6	Thermoset	18–12	0.76	30	None	Thermoset	Pendant or portable	Damp locations	Hard usage

Table 400.4 *Continued*

Trade Name	Type Letter	Voltage	AWG or kcmil	Number of Conductors	Insulation	Nominal Insulation Thickness[1]			Braid on Each Conductor	Outer Covering		Use	
						AWG or kcmil	mm	mils					
	SJE	300			Thermo-plastic elastomer					Thermoplastic elastomer		Damp locations	
	SJEW See Note 13.	300										Damp and wet locations	
	SJEO	300								Oil-resistant thermoplastic elastomer		Damp locations	
	SJEOW See Note 13.	300										Damp and wet locations	
	SJEOO	300			Oil-resistant thermo-plastic elastomer							Damp locations	
	SJEOOW See Note 13.	300										Damp and wet locations	
	SJO	300			Thermoset					Oil-resistant thermoset		Damp locations	
	SJOW See Note 13.	300										Damp and wet locations	
	SJOO	300			Oil-resistant thermoset							Damp locations	
	SJOOW See Note 13.	300										Damp and wet locations	
	SJT	300			Thermo-plastic	10	1.14	45		Thermoplastic		Damp locations	
	SJTW See Note 13.	300										Damp and wet locations	
	SJTO	300			Thermo-plastic	18–12	0.76	30		Oil-resistant thermoplastic		Damp locations	
	SJTOW See Note 13.	300										Damp and wet locations	
	SJTOO	300			Oil-resistant thermo-plastic							Damp locations	
	SJTOOW See Note 13.	300										Damp and wet locations	
Hard service cord	SO See Note 4.	600	18–2	2 or more	Thermoset	18–16	0.76	30		Oil-resistant thermoset	Pendant or portable	Damp locations	Extra hard usage
	SOW See Note 4. See Note 13.	600										Damp and wet locations	
	SOO See Note 4.	600			Oil-resistant thermoset	14–10 8–2	1.14 1.52	45 60				Damp locations	
	SOOW See Note 4. See Note 13.	600										Damp and wet locations	

Table 400.4 *Continued*

Trade Name	Type Letter	Voltage	AWG or kcmil	Number of Conductors	Insulation	Nominal Insulation Thickness[1]			Braid on Each Conductor	Outer Covering	Use		
						AWG or kcmil	mm	mils					
All thermoset parallel cord	SP-1 See Note 6.	300	20–18	2 or 3	Thermoset	20–18	0.76	30	None	None	Pendant or portable	Damp locations	Not hard usage
	SP-2 See Note 6.	300	18–16			18-16	1.14	45					
	SP-3 See Note 6.	300	18–10			18–16 14 12 10	1.52 2.03 2.41 2.80	60 80 95 110			Refrigerators, room air conditioners, and as permitted in 422.16(B)		
All elastomer (thermoplastic) parallel cord	SPE-1 See Note 6.	300	20–18	2 or 3	Thermoplastic elastomer	20–18	0.76	30	None	None	Pendant or portable	Damp locations	Not hard usage
	SPE-2 See Note 6.	300	18–16			18–16	1.14	45					
	SPE-3 See Note 6.	300	18–10			18–16 14 12 10	1.52 2.03 2.41 2.80	60 80 95 110			Refrigerators, room air conditioners, and as permitted in 422.16(B)		
All plastic parallel cord	SPT-1 See Note 6.	300	20–18	2 or 3	Thermoplastic	20–18	0.76	30	None	None	Pendant or portable	Damp locations	Not hard usage
	SPT-1W See Note 6. See Note 13.	300										Damp and wet locations	
	SPT-2 See Note 6.	300	18–16			18–16	1.14	45				Damp locations	
	SPT-2W See Note 6. See Note 13.	300										Damp and wet locations	
	SPT-3 See Note 6.	300	18–10			18–16 14 12 10	1.52 2.03 2.41 2.80	60 80 95 110			Refrigerators, room air conditioners, and as permitted in 422.16(B)	Damp locations	Not hard usage
Range, dryer cable	SRD	300	10–4	3 or 4	Thermoset	10–4	1.14	45	None	Thermoset	Portable	Damp locations	Ranges, dryers
	SRDE	300	10–4	3 or 4	Thermoplastic elastomer				None	Thermoplastic elastomer			
	SRDT	300	10–4	3 or 4	Thermoplastic				None	Thermoplastic			
Hard service cord	ST See Note 4.	600	18–2	2 or more	Thermoplastic	18–16 14–10 8–2	0.76 1.14 1.52	30 45 60	None	Thermoplastic	Pendant or portable	Damp locations	Extra hard usage
	STW See Note 4. See Note 13.	600										Damp and wet locations	
	STO See Note 4.	600								Oil-resistant thermoplastic		Damp locations	

Table 400.4 *Continued*

Trade Name	Type Letter	Voltage	AWG or kcmil	Number of Conductors	Insulation	Nominal Insulation Thickness[1]			Braid on Each Conductor	Outer Covering	Use		
						AWG or kcmil	mm	mils					
	STOW See Note 4. See Note 13.	600			Thermo-plastic					Oil-resistant thermoset		Damp and wet locations	
	STOO See Note 4.	600			Oil-resistant thermo-plastic							Damp locations	
												Damp and wet locations	
	STOOW See Note 4. See Note 13.	600											
Vacuum cleaner cord	SV See Note 6.	300	18–16	2 or 3	Thermoset	18–16	0.38	15	None	Thermoset	Pendant or portable	Damp locations	Not hard usage
	SVE See Note 6.	300			Thermo-plastic elastomer					Thermoplastic elastomer			
	SVEO See Note 6.	300								Oil-resistant thermoplastic elastomer			
	SVEOO See Note 6.	300			Oil-resistant thermo-plastic elastomer								
	SVO	300			Thermoset					Oil-resistant thermoset			
	SVOO	300			Oil-resistant thermoset					Oil-resistant thermoset			
	SVT See Note 6.	300			Thermo-plastic					Thermoplastic			
	SVTO See Note 6.	300			Thermo-plastic					Oil-resistant thermoplastic			
	SVTOO	300			Oil-resistant thermo-plastic								
Parallel tinsel cord	TPT See Note 2.	300	27	2	Thermo-plastic	27	0.76	30	None	Thermoplastic	Attached to an appliance	Damp locations	Not hard usage
Jacketed tinsel cord	TST See Note 2.	300	27	2	Thermo-plastic	27	0.38	15	None	Thermoplastic	Attached to an appliance	Damp locations	Not hard usage
Portable power-cable	W	2000	12–500 501–1000	1–6 1	Thermoset	12–2 1–4/0 250–500 501–1000	1.52 2.03 2.41 2.80	60 80 95 110		Oil-resistant thermoset	Portable, extra hard usage		
Electric vehicle cable	EV	600	18–500 See Note 11.	2 or more plus grounding conduc-tor(s), plus optional hybrid data, signal communi-cations, and optical fiber cables	Thermoset with optional nylon See Note 12.	18–16 14–10 8–2 1–4/0 250–500	0.76 (0.51) 1.14 (0.76) 1.52 (1.14) 2.03 (1.52) 2.41 (1.90)	30 (20) 45 (30) 60 (45) 80 (60) 95 (75) See Note 12.	Optional	Thermoset	Electric vehicle charging	Wet locations	Extra hard usage
	EVJ	300	18–12 See Note 11.			18–12	0.76 (0.51)	30 (20) See Note 12.					Hard usage

Table 400.4 *Continued*

Trade Name	Type Letter	Voltage	AWG or kcmil	Number of Conductors	Insulation	Nominal Insulation Thickness[1]			Braid on Each Conductor	Outer Covering	Use		
						AWG or kcmil	mm	mils					
	EVE	600	18–500 See Note 11.	2 or more plus grounding conductor(s), plus optional hybrid data, signal communications, and optical fiber cables	Thermoplastic elastomer with optional nylon See Note 12.	18–16 14–10 8–2 1–4/0 250–500	0.76 (0.51) 1.14 (0.76) 1.52 (1.14) 2.03 (1.52) 2.41 (1.90)	30 (20) 45 (30) 60 (45) 80 (60) 95 (75) See Note 12.		Thermoplastic elastomer			Extra hard usage
	EVJE	300	18–12 See Note 11.			18–12	0.76 (0.51)	30 (20) See Note 12.					Hard usage
	EVT	600	18–500 See Note 11.	2 or more plus grounding conductor(s), plus optional hybrid data, signal communications, and optical fiber cables	Thermoplastic with optional nylon See Note 12.	18–16 14–10 8–2 1–4/0 250–500	0.76 (0.51) 1.14 (0.76) 1.52 (1.14) 2.03 (1.52) 2.41 (1.90)	30 (20) 45 (30) 60 (45) 80 (60) 95 (75) See Note 12.	Optional	Thermoplastic	Electric vehicle charging	Wet locations	Extra hard usage
	EVJT	300	18–12 See Note 11.			18–12	0.76 (0.51)	30 (20) See Note 12.					Hard usage

*See Note 8.
**The required outer covering on some single conductor cables may be integral with the insulation.

Notes:

1. All types listed in Table 400.4 shall have individual conductors twisted together except for Types HPN, SP-1, SP-2, SP-3, SPE-1, SPE-2, SPE-3, SPT-1, SPT-2, SPT-3, TPT, NISP-1, NISP-2, NISPT-1, NISPT-2, NISPE-1, NISPE-2, and three-conductor parallel versions of SRD, SRDE, and SRDT.

2. Types TPT and TST shall be permitted in lengths not exceeding 2.5 m (8 ft) where attached directly, or by means of a special type of plug, to a portable appliance rated at 50 watts or less and of such nature that extreme flexibility of the cord is essential.

3. Rubber-filled or varnished cambric tapes shall be permitted as a substitute for the inner braids.

4. Types G, G-GC, S, SC, SCE, SCT, SE, SEO, SEOO, SO, SOO, ST, STO, STOO, PPE, and W shall be permitted for use on theater stages, in garages, and elsewhere where flexible cords are permitted by this *Code*.

5. Elevator traveling cables for operating control and signal circuits shall contain nonmetallic fillers as necessary to maintain concentricity. Cables shall have steel supporting members as required for suspension by 620.41. In locations subject to excessive moisture or corrosive vapors or gases, supporting members of other materials shall be permitted. Where steel supporting members are used, they shall run straight through the center of the cable assembly and shall not be cabled with the copper strands of any conductor.

    In addition to conductors used for control and signaling circuits, Types E, EO, ET, ETLB, ETP, and ETT elevator cables shall be permitted to incorporate in the construction, one or more 20 AWG telephone conductor pairs, one or more coaxial cables, or one or more optical fibers. The 20 AWG conductor pairs shall be permitted to be covered with suitable shielding for telephone, audio, or higher frequency communications circuits; the coaxial cables consist of a center conductor, insulation, and shield for use in video or other radio frequency communications circuits. The optical fiber shall be suitably covered with flame-retardant thermoplastic. The insulation of the conductors shall be rubber or thermoplastic of thickness not less than specified for the other conductors of the particular type of cable. Metallic shields shall have their own protective covering. Where used, these components shall be permitted to be incorporated in any layer of the cable assembly but shall not run straight through the center.

6. The third conductor in these cables shall be used for equipment grounding purpose only. The insulation of the grounding conductor for Types SPE-1, SPE-2, SPE-3, SPT-1, SPT-2, SPT-3, NISPT-1, NISPT-2, NISPE-1, and NISPE-2 shall be permitted to be thermoset polymer.

7. The individual conductors of all cords, except those of heat-resistant cords, shall have a thermoset or thermoplastic insulation, except that the equipment grounding conductor where used shall be in accordance with 400.23(B).

8. Where the voltage between any two conductors exceeds 300, but does not exceed 600, flexible cord of 10 AWG and smaller shall have thermoset or thermoplastic insulation on the individual conductors at least 1.14 mm (45 mils) in thickness, unless Type S, SE, SEO, SEOO, SO, SOO, ST, STO, or STOO cord is used.

9. Insulations and outer coverings that meet the requirements as flame retardant, limited smoke, and are so listed, shall be permitted to be marked for limited smoke after the code type designation.

10. Elevator cables in sizes 20 AWG through 14 AWG are rated 300 volts, and sizes 10 through 2 are rated 600 volts. 12 AWG is rated 300 volts with a 0.76-mm (30-mil) insulation thickness and 600 volts with a 1.14-mm (45-mil) insulation thickness.

11. Conductor size for Types EV, EVJ, EVE, EVJE, EVT, and EVJT cables apply to nonpower-limited circuits only. Conductors of power-limited (data, signal, or communications) circuits may extend beyond the stated AWG size range. All conductors shall be insulated for the same cable voltage rating.

12. Insulation thickness for Types EV, EVJ, EVEJE, EVT, and EVJT cables of nylon construction is indicated in parentheses.

13. Cords that comply with the requirements for outdoor cords and are so listed shall be permitted to be designated as weather and water resistant with the suffix "W" after the code type designation. Cords with the "W" suffix are suitable for use in wet locations.

Table 400.5(A) Allowable Ampacity for Flexible Cords and Cables [Based on Ambient Temperature of 30°C (86°F). See 400.13 and Table 400.4.]

Size (AWG)	Thermoplastic Types TPT, TST	Thermoset Types C, E, EO, PD, S, SJ, SJO, SJOW, SJOO, SJOOW, SO, SOW, SOO, SOOW, SP-1, SP-2, SP-3, SRD, SV, SVO, SVOO / Thermoplastic Types ET, ETLB, ETP, ETT, SE, SEW, SEO, SEOW, SEOOW, SJE, SJEW, SJEO, SJEOW, SJEOOW, SJT, SJTW, SJTO, SJTOW, SJTOO, SJTOOW, SPE-1, SPE-2, SPE-3, SPT-1, SPT-1W, SPT-2, SPT-2W, SPT-3, ST, SRDE, SRDT, STO, STOW, STOO, STOOW, SVE, SVEO, SVT, SVTO, SVTOO		Types HPD, HPN, HSJ, HSJO, HSJOO
		A+	B+	
27*	0.5	—	—	—
20	—	5**	***	—
18	—	7	10	10
17	—	—	12	13
16	—	10	13	15
15	—	—	—	17
14	—	15	18	20
12	—	20	25	30
10	—	25	30	35
8	—	35	40	—
6	—	45	55	—
4	—	60	70	—
2	—	80	95	—

*Tinsel cord.
**Elevator cables only.
***7 amperes for elevator cables only; 2 amperes for other types.
+The allowable currents under subheading A apply to 3-conductor cords and other multiconductor cords connected to utilization equipment so that only 3 conductors are current-carrying. The allowable currents under subheading B apply to 2-conductor cords and other multiconductor cords connected to utilization equipment so that only 2 conductors are current carrying.

In a 3-wire circuit consisting of two phase wires and the neutral of a 4-wire, 3-phase, wye-connected system, a common conductor carries approximately the same current as the line-to-neutral currents of the other conductors and shall be considered to be a current-carrying conductor.

On a 4-wire, 3-phase, wye circuit where the major portion of the load consists of nonlinear loads, there are harmonic currents present in the neutral conductor and the neutral shall be considered to be a current-carrying conductor.

An equipment grounding conductor shall not be considered a current-carrying conductor.

Where a single conductor is used for both equipment grounding and to carry unbalanced current from other conductors, as provided for in 250.140 for electric ranges and electric clothes dryers, it shall not be considered as a current-carrying conductor.

Exception: For other loading conditions, adjustment factors shall be permitted to be calculated under 310.15(C).

> FPN: See Annex B, Table B.310.11, for adjustment factors for more than three current-carrying conductors in a raceway or cable with load diversity.

400.6 Markings.

(A) Standard Markings. Flexible cords and cables shall be marked by means of a printed tag attached to the coil reel or carton. The tag shall contain the information required in 310.11(A). Types S, SC, SCE, SCT, SE, SEO, SEOO, SJ, SJE, SJEO, SJEOO, SJO, SJT, SJTO, SJTOO, SO, SOO, ST, STO, STOO, SEW, SEOW, SEOOW, SJEW, SJEOW, SJEOOW, SJOW, SJTW, SJTOW, SJTOOW, SOW, SOOW, STW, STOW, and STOOW flexible cords and G, G-GC, PPE, and W flexible cables shall be durably marked on the surface at intervals not exceeding 610 mm (24 in.) with the type designation, size, and number of conductors.

(B) Optional Markings. Flexible cords and cable types listed in Table 400.4 shall be permitted to be surface marked to indicate special characteristics of the cable materials. These markings include, but are not limited to, markings for limited smoke, sunlight resistance, and so forth.

The UL *Electrical Construction Materials Directory,* under the category Flexible Cord (ZJCZ), indicates that additional markings may include the following:

"Water Resistant" indicates the cord is suitable for immersion in water.

"For Mobile Home Use," "For Recreational Vehicle Use," or "For Mobile Home and Recreational Vehicle Use" followed by the current rating in amperes indicates suitability for use in mobile homes or recreational vehicles.

"Outdoor" or "W-A" indicates suitability for use outdoors. The minimum temperature rating for these cords is −40°C, unless otherwise marked on the cord.

"W" indicates suitability for use outdoors and for immersion in water. The low temperature rating for these cords is −40°C, unless otherwise marked on the cord with optional ratings of −50, −60, or −70°C. The

Table 400.5(B) Ampacity of Cable Types SC, SCE, SCT, PPE, G, G-GC, and W. [Based on Ambient Temperature of 30°C (86°F). See Table 400.4.] Temperature Rating of Cable.

Size (AWG or kcmil)	60°C (140°F)			75°C (167°F)			90°C (194°F)		
	D[1]	E[2]	F[3]	D[1]	E[2]	F[3]	D[1]	E[2]	F[3]
12	—	31	26	—	37	31	—	42	35
10	—	44	37	—	52	43	—	59	49
8	60	55	48	70	65	57	80	74	65
6	80	72	63	95	88	77	105	99	87
4	105	96	84	125	115	101	140	130	114
3	120	113	99	145	135	118	165	152	133
2	140	128	112	170	152	133	190	174	152
1	165	150	131	195	178	156	220	202	177
1/0	195	173	151	230	207	181	260	234	205
2/0	225	199	174	265	238	208	300	271	237
3/0	260	230	201	310	275	241	350	313	274
4/0	300	265	232	360	317	277	405	361	316
250	340	296	259	405	354	310	455	402	352
300	375	330	289	445	395	346	505	449	393
350	420	363	318	505	435	381	570	495	433
400	455	392	343	545	469	410	615	535	468
500	515	448	392	620	537	470	700	613	536
600	575	—	—	690	—	—	780	—	—
700	630	—	—	755	—	—	855	—	—
750	655	—	—	785	—	—	885	—	—
800	680	—	—	815	—	—	920	—	—
900	730	—	—	870	—	—	985	—	—
1000	780	—	—	935	—	—	1055	—	—

[1]The ampacities under subheading D shall be permitted for single-conductor Types SC, SCE, SCT, PPE, and W cable only where the individual conductors are not installed in raceways and are not in physical contact with each other except in lengths not to exceed 600 mm (24 in.) where passing through the wall of an enclosure.

[2]The ampacities under subheading E apply to two-conductor cables and other multiconductor cables connected to utilization equipment so that only two conductors are current carrying.

[3]The ampacities under subheading F apply to three-conductor cables and other multiconductor cables connected to utilization equipment so that only three conductors are current carrying.

Table 400.5 Adjustment Factors for More Than Three Current-Carrying Conductors in a Flexible Cord or Cable

Number of Conductors	Percent of Value in Tables 400.5(A) and 400.5(B)
4 – 6	80
7 – 9	70
10 – 20	50
21 – 30	45
31 – 40	40
41 and above	35

low temperature ratings are determined by means of a bend test (not a suppleness test) at the given temperature.

"VW-1" indicates that the cord complies with a vertical flame test.

Cords that have been evaluated for leakage currents between the circuit conductor and the grounding conductor and between the circuit conductor and the outer surface of the jacket may have the leakage current values marked on the cable jacket.

400.7 Uses Permitted.

(A) Uses. Flexible cords and cables shall be used only for the following:

(1) Pendants
(2) Wiring of luminaires (fixtures)
(3) Connection of portable lamps, portable and mobile signs, or appliances
(4) Elevator cables
(5) Wiring of cranes and hoists
(6) Connection of utilization equipment to facilitate frequent interchange
(7) Prevention of the transmission of noise or vibration
(8) Appliances where the fastening means and mechanical connections are specifically designed to permit ready

removal for maintenance and repair, and the appliance is intended or identified for flexible cord connection

(9) Data processing cables as permitted by 645.5
(10) Connection of moving parts
(11) Temporary wiring as permitted in 527.4(B) and 527.4(C)

(B) Attachment Plugs. Where used as permitted in 400.7(A)(3), (A)(6), and (A)(8), each flexible cord shall be equipped with an attachment plug and shall be energized from a receptacle outlet.

Exception: As permitted in 368.8.

Flexible cords are permitted to be hard-wired into a junction box if the cord is used for the following:

1. Luminaires and fixtures mentioned in 400.7(A)
2. Supplies to pendant pushbutton stations for cranes
3. Portable lamp (droplight) connections

400.8 Uses Not Permitted.

Unless specifically permitted in 400.7, flexible cords and cables shall not be used for the following:

(1) As a substitute for the fixed wiring of a structure
(2) Where run through holes in walls, structural ceilings, suspended ceilings, dropped ceilings, or floors
(3) Where run through doorways, windows, or similar openings
(4) Where attached to building surfaces

Exception: Flexible cord and cable shall be permitted to be attached to building surfaces in accordance with the provisions of 368.8.

(5) Where concealed by walls, floors, or ceilings or located above suspended or dropped ceilings
(6) Where installed in raceways, except as otherwise permitted in this *Code*

The flexible cords and cables referred to in Article 400 are not limited to use with portable equipment. They may not be used, however, as a substitute for the fixed wiring of a structure or where concealed behind building walls, floors, or ceilings (including structural, suspended, or dropped-type ceilings). See 240.5 and 527.4(B) and (C) for the uses of multiconductor flexible cords for feeder and branch-circuit installations, and for overcurrent protection requirements for flexible cord. See 410.30 for cord-connected luminaires.

400.9 Splices.

Flexible cord shall be used only in continuous lengths without splice or tap where initially installed in applications

permitted by 400.7(A). The repair of hard-service cord and junior hard-service cord (see Trade Name column in Table 400.4) 14 AWG and larger shall be permitted if conductors are spliced in accordance with 110.14(B) and the completed splice retains the insulation, outer sheath properties, and usage characteristics of the cord being spliced.

The requirements of 400.9 are intended to ensure that flexible cords and cables first installed under any of the uses permitted in 400.7(A)(1)–(11) are in their original or near original condition. Damage to a cord can occur under the sometime extreme conditions of use to which it is subjected. The provisions of this section permit repair of a cord in such a manner that the cord will retain its original operating and use integrity. However, if the repaired cord is reused or reinstalled at a new location, the in-line repair is no longer permitted and the cord can be used only in lengths that do not contain a splice.

400.10 Pull at Joints and Terminals.

Flexible cords and cables shall be connected to devices and to fittings so that tension is not transmitted to joints or terminals.

Exception: Listed portable single pole devices that are intended to accommodate such tension at their terminals shall be permitted to be used with single-conductor flexible cable.

FPN: Some methods of preventing pull on a cord from being transmitted to joints or terminals are knotting the cord, winding with tape, and fittings designed for the purpose.

400.11 In Show Windows and Show Cases.

Flexible cords used in show windows and show cases shall be Type S, SE, SEO, SEOO, SJ, SJE, SJEO, SJEOO, SJO, SJOO, SJT, SJTO, SJTOO, SO, SOO, ST, STO, STOO, SEW, SEOW, SEOOW, SJEW, SJEOW, SJEOOW, SJOW, SJOOW, SJTW, SJTOW, SJTOOW, SOW, SOOW, STW, STOW, or STOOW.

Exception No. 1: For the wiring of chain-supported luminaires (lighting fixtures).

Exception No. 2: As supply cords for portable lamps and other merchandise being displayed or exhibited.

Flexible cords listed for hard usage or extra-hard usage should be used in show windows and show cases because such cords may come in contact with combustible materials, such as fabrics, or paper products usually present at these locations and because they are exposed to wear and tear from continual housekeeping and display changes. Flexible cords used in show windows and show cases should be maintained in good condition.

400.12 Minimum Size.

The individual conductors of a flexible cord or cable shall not be smaller than the sizes in Table 400.4.

Exception: The size of the insulated ground-check conductor of Type G-GC cables shall be not smaller than 10 AWG.

Added in the 1999 *Code*, the exception to 400.12 correlates with cable Type G-GC, which was added to Table 400.4 in 1999. This cable is similar to Type G, except that it also incorporates an insulated ground-check (GC) conductor. The ground-check conductor is used as part of a low-voltage circuit that monitors the grounding conductor continuity.

400.13 Overcurrent Protection.

Flexible cords not smaller than 18 AWG, and tinsel cords or cords having equivalent characteristics of smaller size approved for use with specific appliances, shall be considered as protected against overcurrent by the overcurrent devices described in 240.5.

400.14 Protection from Damage.

Flexible cords and cables shall be protected by bushings or fittings where passing through holes in covers, outlet boxes, or similar enclosures.

A variety of bushings and fittings are available for protecting flexible cords and cables, both insulated and noninsulated. Some bushings or fittings include pull-relief means, as required in 400.10. Many insulating bushings are listed by Underwriters Laboratories Inc. in the following product categories:

1. Conduit fittings (bushings and fittings for use on the ends of conduit in boxes and gutters)
2. Insulating devices and materials
3. Bushings (for the protection of cords where they pass through walls or barriers of metal)
4. Outlet bushings and fittings (for use on the ends of conduit, EMT, or armored cable, where a change to open wiring is made)

II. Construction Specifications

400.20 Labels.

Flexible cords shall be examined and tested at the factory and labeled before shipment.

See the definition of *labeled* in Article 100.

400.21 Nominal Insulation Thickness.

The nominal thickness of insulation for conductors of flexible cords and cables shall not be less than specified in Table 400.4.

Exception: The nominal insulation thickness for the ground-check conductors of Type G-GC cables shall not be less than 1.14 mm (45 mils) for 8 AWG and not less than 0.76 mm (30 mils) for 10 AWG.

400.22 Grounded-Conductor Identification.

One conductor of flexible cords that is intended to be used as a grounded circuit conductor shall have a continuous marker that readily distinguishes it from the other conductor or conductors. The identification shall consist of one of the methods indicated in 400.22(A) through (F).

(A) Colored Braid. A braid finished to show a white or gray color and the braid on the other conductor or conductors finished to show a readily distinguishable solid color or colors.

(B) Tracer in Braid. A tracer in a braid of any color contrasting with that of the braid and no tracer in the braid of the other conductor or conductors. No tracer shall be used in the braid of any conductor of a flexible cord that contains a conductor having a braid finished to show white or gray.

Exception: In the case of Types C and PD and cords having the braids on the individual conductors finished to show white or gray. In such cords, the identifying marker shall be permitted to consist of the solid white or gray finish on one conductor, provided there is a colored tracer in the braid of each other conductor.

(C) Colored Insulation. A white or gray insulation on one conductor and insulation of a readily distinguishable color or colors on the other conductor or conductors for cords having no braids on the individual conductors.

For jacketed cords furnished with appliances, one conductor having its insulation colored light blue, with the other conductors having their insulation of a readily distinguishable color other than white or gray.

Exception: Cords that have insulation on the individual conductors integral with the jacket.

The insulation shall be permitted to be covered with an outer finish to provide the desired color.

(D) Colored Separator. A white or gray separator on one conductor and a separator of a readily distinguishable solid color on the other conductor or conductors of cords having insulation on the individual conductors integral with the jacket.

In 400.22(A), (B), (C), and (D), the permission to use "natural gray" colored insulation as a means to identify the grounded conductor of a flexible cord has been revised by deleting the word *natural*. Identification of grounded conductors in flexible cords and cables can now be accomplished through the use of a white or gray colored braid, a white or gray colored tracer in the braid, white or gray colored insulation, or a white or gray colored separator. This revision correlates with the deletion of the term *natural* in other *Code* requirements covering the identification of grounded conductors. In existing installations where a gray colored braid, tracer, or conductor insulation is encountered, caution should be exercised because gray may have been used as means to identify ungrounded conductors.

(E) Tinned Conductors. One conductor having the individual strands tinned and the other conductor or conductors having the individual strands untinned for cords having insulation on the individual conductors integral with the jacket.

(F) Surface Marking. One or more stripes, ridges, or grooves located on the exterior of the cord so as to identify one conductor for cords having insulation on the individual conductors integral with the jacket.

400.23 Equipment Grounding Conductor Identification.

A conductor intended to be used as an equipment grounding conductor shall have a continuous identifying marker readily distinguishing it from the other conductor or conductors. Conductors having a continuous green color or a continuous green color with one or more yellow stripes shall not be used for other than equipment grounding purposes. The identifying marker shall consist of one of the methods in 400.23(A) or (B).

(A) Colored Braid. A braid finished to show a continuous green color or a continuous green color with one or more yellow stripes.

(B) Colored Insulation or Covering. For cords having no braids on the individual conductors, an insulation of a continuous green color or a continuous green color with one or more yellow stripes.

400.24 Attachment Plugs.

Where a flexible cord is provided with an equipment grounding conductor and equipped with an attachment plug, the attachment plug shall comply with 250.138(A) and (B).

III. Portable Cables Over 600 Volts, Nominal

400.30 Scope.

This part applies to multiconductor portable cables used to connect mobile equipment and machinery.

400.31 Construction.

(A) Conductors. The conductors shall be 8 AWG copper or larger and shall employ flexible stranding.

Exception: The size of the insulated ground-check conductor of Type G-GC cables shall be not smaller than 10 AWG.

(B) Shields. Cables operated at over 2000 volts shall be shielded. Shielding shall be for the purpose of confining the voltage stresses to the insulation.

(C) Equipment Grounding Conductor(s). An equipment grounding conductor(s) shall be provided. The total area shall not be less than that of the size of the equipment grounding conductor required in 250.122.

400.32 Shielding.

All shields shall be grounded.

400.33 Grounding.

Grounding conductors shall be connected in accordance with Part V of Article 250.

400.34 Minimum Bending Radii.

The minimum bending radii for portable cables during installation and handling in service shall be adequate to prevent damage to the cable.

400.35 Fittings.

Connectors used to connect lengths of cable in a run shall be of a type that lock firmly together. Provisions shall be made to prevent opening or closing these connectors while energized. Suitable means shall be used to eliminate tension at connectors and terminations.

400.36 Splices and Terminations.

Portable cables shall not contain splices unless the splices are of the permanent molded, vulcanized types in accordance with 110.14(B). Terminations on portable cables rated over 600 volts, nominal, shall be accessible only to authorized and qualified personnel.

ARTICLE 402
Fixture Wires

Contents

402.1 Scope.

This article covers general requirements and construction specifications for fixture wires.

402.2 Other Articles.

Fixture wires shall comply with this article and also with the applicable provisions of other articles of this *Code*.

> FPN: For application in luminaires (lighting fixtures), see Article 410.

402.3 Types.

Fixture wires shall be of a type listed in Table 402.3, and they shall comply with all requirements of that table. The fixture wires listed in Table 402.3 are all suitable for service at 600 volts, nominal, unless otherwise specified.

> FPN: Thermoplastic insulation may stiffen at temperatures colder than −10°C (+14°F), requiring that care be exercised during installation at such temperatures. Thermoplastic insulation may also be deformed at normal temperatures where subjected to pressure, requiring that care be exercised during installation and at points of support.

402.5 Allowable Ampacities for Fixture Wires.

The allowable ampacity of fixture wire shall be as specified in Table 402.5.

No conductor shall be used under such conditions that its operating temperature exceeds the temperature specified in Table 402.3 for the type of insulation involved.

> FPN: See 310.10 for temperature limitation of conductors.

402.6 Minimum Size.

Fixture wires shall not be smaller than 18 AWG.

402.7 Number of Conductors in Conduit or Tubing.

The number of fixture wires permitted in a single conduit or tubing shall not exceed the percentage fill specified in Table 1, Chapter 9.

Table 1 of Chapter 9 specifies the maximum percent fill of a conduit or tubing. Table 4 of Chapter 9 provides the usable area within the selected conduit or tubing, and Table 5 of Chapter 9 provides the required area for each of the conductors. Examples using these tables to calculate a conduit or tubing size are provided following Chapter 9, Table 1, Note 6. The following examples show how to determine the minimum size conduit where the conductors are different sizes and how to select the minimum size conduit directly from the tables in Annex C where the conductors are all the same size.

Example 1

A remote ballast installation requires a single flexible metal conduit to contain fourteen 16 AWG TFFN fixture wires and three 12 AWG THHN conductors. What size flexible metal conduit will be required?

Solution

The solution is found by using Table 1 of Chapter 9 and the accompanying Note 6 following Table 1. Table 1 sets the maximum percentage of conduit and tubing fill based on the internal cross-sectional area of the raceway in question. Note 6 states, "For [calculating] combinations of conductors of different sizes, use Tables 5 and 5A for dimensions of conductors and Table 4 for the applicable conduit or tubing dimensions."

STEP 1. Using Table 1, look up the maximum percent of cross section of conduit permitted for conductors. Table 1 sets the limit of conductor fill for over two conductors at 40 percent of the total cross-sectional area of the raceway.

STEP 2. Look up the individual conductor cross-sectional areas in Chapter 9, Table 5.

$$16 \text{ AWG TFFN} = 0.0072 \text{ in.}^2$$
$$12 \text{ AWG THHN} = 0.0133 \text{ in.}^2$$

STEP 3. Calculate the total area occupied by the wires as follows:

$$16 \text{ AWG TFFN} \times 14 \times 0.0072 = 0.1008 \text{ in.}^2$$
$$12 \text{ AWG TFFN} \times 3 \times 0.0133 = \underline{0.0399 \text{ in.}^2}$$
$$\text{Total area} = 0.1407 \text{ in.}^2$$

STEP 4. Using the 40 percent column of Table 4, look up the appropriate flexible metal conduit size based on 40 percent fill and a total conductor area fill of 0.1407 in.^{2} Because 0.1407 in.^{2} is greater than 0.127 and less than 0.213, select trade size ¾ flexible metal conduit.

If the conductors in a raceway or tubing are all of the same wire size, the Annex C tables may be used instead of doing the calculations. The following example uses Annex C tables to determine electrical metallic tubing size.

Table 402.3 Fixture Wires

Name	Type Letter	Insulation	Thickness of Insulation			Outer Covering	Maximum Operating Temperature	Application Provisions
			AWG	mm	mils			
Heat-resistant rubber-covered fixture wire — flexible stranding	FFH-2	Heat-resistant rubber Cross-linked synthetic polymer	18–16 18–16	0.76 0.76	30 30	Nonmetallic covering	75°C 167°F	Fixture wiring
ECTFE — solid or 7-strand	HF	Ethylene chlorotrifluoroethylene	18–14	0.38	15	None	150°C 302°F	Fixture wiring
ECTFE — flexible stranding	HFF	Ethylene chlorotrifluoroethylene	18–14	0.38	15	None	150°C 302°F	Fixture wiring
Tape insulated fixture wire — solid or 7-strand	KF-1 KF-2	Aromatic polyimide tape Aromatic polyimide tape	18–10 18–10	0.14 0.21	5.5 8.4	None None	200°C 392°F 200°C 392°F	Fixture wiring — limited to 300 volts Fixture wiring
Tape insulated fixture wire — flexible stranding	KFF-1 KFF-2	Aromatic polyimide tape Aromatic polyimide tape	18–10 18–10	0.14 0.21	5.5 8.4	None None	200°C 392°F 200°C 392°F	Fixture wiring — limited to 300 volts Fixture wiring
Perfluoroalkoxy — solid or 7-strand (nickel or nickel-coated copper)	PAF	Perfluoroalkoxy	18–14	0.51	20	None	250°C 482°F	Fixture wiring (nickel or nickel-coated copper)
Perfluoroalkoxy — flexible stranding	PAFF	Perfluoroalkoxy	18–14	0.51	20	None	150°C 302°F	Fixture wiring
Fluorinated ethylene propylene fixture wire — solid or 7-strand	PF	Fluorinated ethylene propylene	18–14	0.51	20	None	200°C 392°F	Fixture wiring
Fluorinated ethylene propylene fixture wire — flexible stranding	PFF	Fluorinated ethylene propylene	18–14	0.51	20	None	150°C 302°F	Fixture wiring
Fluorinated ethylene propylene fixture wire — solid or 7-strand	PGF	Fluorinated ethylene propylene	18–14	0.36	14	Glass braid	200°C 392°F	Fixture wiring
Fluorinated ethylene propylene fixture wire — flexible stranding	PGFF	Fluorinated ethylene propylene	18–14	0.36	14	Glass braid	150°C 302°F	Fixture wiring
Extruded polytetrafluoroethylene — solid or 7-strand (nickel or nickel-coated copper)	PTF	Extruded polytetrafluoroethylene	18–14	0.51	20	None	250°C 482°F	Fixture wiring (nickel or nickel-coated copper)

Table 402.3 *Continued*

Name	Type Letter	Insulation	Thickness of Insulation			Outer Covering	Maximum Operating Temperature	Application Provisions
			AWG	mm	mils			
Extruded polytetrafluoroethylene — flexible stranding 26-36 (AWG silver or nickel-coated copper)	PTFF	Extruded polytetrafluoroethylene	18–14	0.51	20	None	150°C 302°F	Fixture wiring (silver or nickel-coated copper)
Heat-resistant rubber-covered fixture wire — solid or 7-strand	RFH-1	Heat-resistant rubber	18	0.38	15	Nonmetallic covering	75°C 167°F	Fixture wiring — limited to 300 volts
	RFH-2	Heat-resistant rubber Cross-linked synthetic polymer	18–16	0.76	30	None or non-metallic covering	75°C 167°F	Fixture wiring
Heat-resistant cross-linked synthetic polymer-insulated fixture wire — solid or stranded	RFHH-2*	Cross-linked synthetic polymer	18–16	0.76	30	None or non-metallic covering	90°C 194°F	Fixture wiring — multi-conductor cable
	RFHH-3*		18–16	1.14	45			
Silicone insulated fixture wire — solid or 7-strand	SF-1	Silicone rubber	18	0.38	15	Nonmetallic covering	200°C 392°F	Fixture wiring — limited to 300 volts
	SF-2	Silicone rubber	18–12 10	0.76 1.14	30 45	Nonmetallic covering	200°C 392°F	Fixture wiring
Silicone insulated fixture wire — flexible stranding	SFF-1	Silicone rubber	18	0.38	15	Nonmetallic covering	150°C 302°F	Fixture wiring — limited to 300 volts
	SFF-2	Silicone rubber	18–12 10	0.76 1.14	30 45	Nonmetallic covering	150°C 302°F	Fixture wiring
Thermoplastic covered fixture wire — solid or 7-strand	TF*	Thermoplastic	18–16	0.76	30	None	60°C 140°F	Fixture wiring
Thermoplastic covered fixture wire — flexible stranding	TFF*	Thermoplastic	18–16	0.76	30	None	60°C 140°F	Fixture wiring
Heat-resistant thermoplastic covered fixture wire — solid or 7-strand	TFN*	Thermoplastic	18–16	0.38	15	Nylon-jacketed or equivalent	90°C 194°F	Fixture wiring
Heat-resistant thermoplastic covered fixture wire — flexible stranded	TFFN*	Thermoplastic	18–16	0.38	15	Nylon-jacketed or equivalent	90°C 194°F	Fixture wiring
Cross-linked polyolefin insulated fixture wire — solid or 7-strand	XF*	Cross-linked polyolefin	18–14 12–10	0.76 1.14	30 45	None	150°C 302°F	Fixture wiring — limited to 300 volts

Table 402.3 *Continued*

Name	Type Letter	Insulation	Thickness of Insulation			Outer Covering	Maximum Operating Temperature	Application Provisions
			AWG	mm	mils			
Cross-linked polyolefin insulated fixture wire — flexible stranded	XFF*	Cross-linked polyolefin	18–14 12–10	0.76 1.14	30 45	None	150°C 302°F	Fixture wiring — limited to 300 volts
Modified ETFE — solid or 7-strand	ZF	Modified ethylene tetrafluoro-ethylene	18–14	0.38	15	None	150°C 302°F	Fixture wiring
Flexible stranding	ZFF	Modified ethylene tetrafluoro-ethylene	18–14	0.38	15	None	150°C 302°F	Fixture wiring
High temp. modified ETFE — solid or 7-strand	ZHF	Modified ethylene tetrafluoro-ethylene	18–14	0.38	15	None	200°C 392°F	Fixture wiring

*Insulations and outer coverings that meet the requirements of flame retardant, limited smoke, and are so listed shall be permitted to be marked for limited smoke after the *Code* type designation.

Table 402.5 Allowable Ampacity for Fixture Wires

Size (AWG)	Allowable Ampacity
18	6
16	8
14	17
12	23
10	28

Example 2

A fire alarm system requires a riser to contain thirty-six 16 AWG TFFN conductors. What size electrical metallic tubing will be required?

Solution

First, in Annex C, Table C1, find TFFN insulation in the first column. Next, find 16 AWG in the second column. Proceed across the table until the desired number of conductors is equal to or less than the number shown in the table for the respective conduit and tubing sizes. Using this method, a 1-in. EMT is required.

402.8 Grounded Conductor Identification.

One conductor of fixture wires that is intended to be used as a grounded conductor shall be identified by means of stripes or by the means described in 400.22(A) through (E).

402.9 Marking.

(A) Method of Marking. Thermoplastic insulated fixture wire shall be durably marked on the surface at intervals not exceeding 610 mm (24 in.). All other fixture wire shall be marked by means of a printed tag attached to the coil, reel, or carton.

(B) Optional Marking. Fixture wire types listed in Table 402.3 shall be permitted to be surface marked to indicate special characteristics of the cable materials. These markings include, but are not limited to, markings for limited smoke, sunlight resistance, and so forth.

402.10 Uses Permitted.

Fixture wires shall be permitted (1) for installation in luminaires (lighting fixtures) and in similar equipment where enclosed or protected and not subject to bending or twisting in use, or (2) for connecting luminaires (lighting fixtures) to the branch-circuit conductors supplying the luminaires (fixtures).

402.11 Uses Not Permitted.

Fixture wires shall not be used as branch-circuit conductors.

402.12 Overcurrent Protection.

Overcurrent protection for fixture wires shall be as specified in 240.5.

ARTICLE 404
Switches

Contents

I. Installation

404.1 Scope.

The provisions of this article shall apply to all switches, switching devices, and circuit breakers where used as switches.

As part of the Chapter 3 reorganization in the 2002 *Code*, several articles have been relocated to new chapters. Article 404 was formerly Article 380. Chapter 3 contains articles with requirements on wiring methods, and Chapter 4 contains requirements on equipment for general use. For a complete matrix on the reorganization of Chapter 3, see Annex F.

404.2 Switch Connections.

(A) Three-Way and Four-Way Switches. Three-way and four-way switches shall be wired so that all switching is done only in the ungrounded circuit conductor. Where in metal raceways or metal-armored cables, wiring between switches and outlets shall be in accordance with 300.20(A).

Exception: Switch loops shall not require a grounded conductor.

The *NEC* does not specifically prohibit the use of *two* 2-conductor nonmetallic-sheathed cables instead of a *single* 3-conductor cable for wiring 3-way and 4-way switches. However, using two 2-conductor cables could easily result in a violation of 300.20 if metal boxes are used and the cables enter the box through separate knockouts. Also, use of the same clamp or section of a clamp for both cables would, in most cases, be in violation of 110.3(B), because clamps have been tested for only one cable per clamp or section of clamp.

The grounded conductor is not needed in a switch loop [see 300.20(A)] because the ungrounded conductor both enters and leaves the enclosure in the same cable or raceway, thus avoiding inductive heating.

(B) Grounded Conductors. Switches or circuit breakers shall not disconnect the grounded conductor of a circuit.

Exception: A switch or circuit breaker shall be permitted to disconnect a grounded circuit conductor where all circuit conductors are disconnected simultaneously, or where the device is arranged so that the grounded conductor cannot be disconnected until all the ungrounded conductors of the circuit have been disconnected.

404.3 Enclosure.

(A) General. Switches and circuit breakers shall be of the externally operable type mounted in an enclosure listed for the intended use. The minimum wire-bending space at terminals and minimum gutter space provided in switch enclosures shall be as required in 312.6.

Exception No. 1: Pendant- and surface-type snap switches and knife switches mounted on an open-face switchboard or panelboard shall be permitted without enclosures.

Exception No. 2: Switches and circuit breakers installed in accordance with 110.27(A)(1), (2), (3), or (4) shall be permitted without enclosures.

Exception No. 2 to 404.3 recognizes the variety of means allowed by 110.27(A) for guarding of live parts. Switches and circuit breakers guarded by these means are permitted without enclosures.

(B) Used as a Raceway. Enclosures shall not be used as junction boxes, auxiliary gutters, or raceways for conductors feeding through or tapping off to other switches or overcurrent devices, unless the enclosure complies with 312.8.

404.4 Wet Locations.

A switch or circuit breaker in a wet location or outside of a building shall be enclosed in a weatherproof enclosure or cabinet that shall comply with 312.2(A). Switches shall not be installed within wet locations in tub or shower spaces unless installed as part of a listed tub or shower assembly.

404.5 Time Switches, Flashers, and Similar Devices.

Time switches, flashers, and similar devices shall be of the enclosed type or shall be mounted in cabinets or boxes or equipment enclosures. Energized parts shall be barriered to prevent operator exposure when making manual adjustments or switching.

Exception: Devices mounted so they are accessible only to qualified persons shall be permitted without barriers, provided they are located within an enclosure such that any energized parts within 152 mm (6.0 in.) of the manual adjustment or switch are covered by suitable barriers.

404.6 Position and Connection of Switches.

(A) Single-Throw Knife Switches. Single-throw knife switches shall be placed so that gravity will not tend to close

them. Single-throw knife switches, approved for use in the inverted position, shall be provided with a locking device that ensures that the blades remain in the open position when so set.

(B) Double-Throw Knife Switches. Double-throw knife switches shall be permitted to be mounted so that the throw is either vertical or horizontal. Where the throw is vertical, a locking device shall be provided to hold the blades in the open position when so set.

(C) Connection of Switches. Single-throw knife switches and switches with butt contacts shall be connected so that their blades are de-energized when the switch is in the open position. Bolted pressure contact switches shall have barriers that prevent inadvertent contact with energized blades. Single-throw knife switches, bolted pressure contact switches, molded case switches, switches with butt contacts, and circuit breakers used as switches shall be connected so that the terminals supplying the load are de-energized when the switch is in the open position.

New for the 2002 *NEC*, bolted pressure switches that have energized blades when open, such as bottom feed designs, must be provided with barriers or a means to guard against inadvertent contact with the energized blades. This requirement is intended to provide protection against accidental contact with live parts in those cases where personnel are working on energized equipment.

Exception: The blades and terminals supplying the load of a switch shall be permitted to be energized when the switch is in the open position where the switch is connected to circuits or equipment inherently capable of providing a backfeed source of power. For such installations, a permanent sign shall be installed on the switch enclosure or immediately adjacent to open switches with the following words or equivalent: WARNING—LOAD SIDE TERMINALS MAY BE ENERGIZED BY BACKFEED.

Batteries, generators, and double-ended switchboard ties are typical backfeed sources. These sources may cause the load side of the switch or circuit breaker to be energized when it is in the open position, a condition inherent to the circuitry.

404.7 Indicating.

General-use and motor-circuit switches, circuit breakers, and molded case switches, where mounted in an enclosure as described in 404.3, shall clearly indicate whether they are in the open (off) or closed (on) position.

Where these switch or circuit breaker handles are operated vertically rather than rotationally or horizontally, the up position of the handle shall be the (on) position.

Exception: Vertically operated double-throw switches shall be permitted to be in the closed (on) position with the handle in either the up or down position.

404.8 Accessibility and Grouping.

(A) Location. All switches and circuit breakers used as switches shall be located so that they may be operated from a readily accessible place. They shall be installed so that the center of the grip of the operating handle of the switch or circuit breaker, when in its highest position, is not more than 2.0 m (6 ft 7 in.) above the floor or working platform.

Exception No. 1: On busway installations, fused switches and circuit breakers shall be permitted to be located at the same level as the busway. Suitable means shall be provided to operate the handle of the device from the floor.

Exception No. 2: Switches and circuit breakers installed adjacent to motors, appliances, or other equipment that they supply shall be permitted to be located higher than specified in the foregoing and to be accessible by portable means.

Exception No. 3: Hookstick operable isolating switches shall be permitted at greater heights.

(B) Voltage Between Adjacent Devices. A snap switch shall not be grouped or ganged in enclosures with other snap switches, receptacles, or similar devices, unless they are arranged so that the voltage between adjacent devices does not exceed 300 volts, or unless they are installed in enclosures equipped with permanently installed barriers between adjacent devices.

Barriers are required between switches that are ganged in a box and used to control 277-volt lighting on 480Y/277-V systems where two or more phase conductors enter the box. Permanent barriers would be required between devices fed from two different phases of this system because the voltage between the phase conductors would be 480 volts, nominal, and would exceed the 300-volt limit. Barriers are required even if one device space is left empty because the two remaining devices fed from different phase conductors would still be adjacent to each other. This requirement now applies to switches ganged together with any wiring device where the voltage between adjacent conductors exceeds 300 volts.

404.9 Provisions for General-Use Snap Switches.

(A) Faceplates. Faceplates provided for snap switches mounted in boxes and other enclosures shall be installed so

as to completely cover the opening and, where the switch is flush mounted, seat against the finished surface.

(B) Grounding. Snap switches, including dimmer and similar control switches, shall be effectively grounded and shall provide a means to ground metal faceplates, whether or not a metal faceplate is installed. Snap switches shall be considered effectively grounded if either of the following conditions is met.

(1) The switch is mounted with metal screws to a metal box or to a nonmetallic box with integral means for grounding devices.
(2) An equipment grounding conductor or equipment bonding jumper is connected to an equipment grounding termination of the snap switch.

Exception to (B): Where no grounding means exists within the snap-switch enclosure or where the wiring method does not include or provide an equipment ground, a snap switch without a grounding connection shall be permitted for replacement purposes only. A snap switch wired under the provisions of this exception and located within reach of earth, grade conducting floors, or other conducting surfaces shall be provided with a faceplate of nonconducting, noncombustible material.

The provisions of 404.9(B) specify that switching devices, including snap switches, dimmers, and similar control devices, must be grounded. Although the non–current-carrying metal parts of these devices are typically not subject to contact by personnel, there is concern about the use of metal faceplates, which do pose a shock hazard if they become energized. Therefore, the grounded switch must provide a means for grounding the metal faceplate.

The requirements in (1) or (2) of 404.9(B) describe the provisions to satisfy the main requirement. Switch plates in existing installations attached to switches in boxes without an equipment grounding conductor must be made of insulating material. See Exhibit 404.1, following the commentary in 404.12, for an example of the typical method by which a metal faceplate is grounded.

(C) Construction. Metal faceplates shall be of ferrous metal not less than 0.76 mm (0.030 in.) in thickness or of nonferrous metal not less than 1.02 mm (0.040 in.) in thickness. Faceplates of insulating material shall be noncombustible and not less than 2.54 mm (0.010 in.) in thickness, but they shall be permitted to be less than 2.54 mm (0.010 in.) in thickness if formed or reinforced to provide adequate mechanical strength.

404.10 Mounting of Snap Switches.

(A) Surface-Type. Snap switches used with open wiring on insulators shall be mounted on insulating material that separates the conductors at least 13 mm (½ in.) from the surface wired over.

(B) Box Mounted. Flush-type snap switches mounted in boxes that are set back of the wall surface as permitted in 314.20 shall be installed so that the extension plaster ears are seated against the surface of the wall. Flush-type snap switches mounted in boxes that are flush with the wall surface or project from it shall be installed so that the mounting yoke or strap of the switch is seated against the box.

Cooperation is necessary among the building trades (carpenters, drywall installers, plasterers, and so on) in order for electricians to properly set device boxes flush with the finish surface, thereby ensuring a secure seating of the switch yoke and permitting the maximum projection of switch handles through the installed switch plate.

404.11 Circuit Breakers as Switches.

A hand-operable circuit breaker equipped with a lever or handle, or a power-operated circuit breaker capable of being opened by hand in the event of a power failure, shall be permitted to serve as a switch if it has the required number of poles.

FPN: See the provisions contained in 240.81 and 240.83.

Circuit breakers that are capable of being hand operated must clearly indicate whether they are in the open (off) or closed (on) position. See 404.7 for details on handle positions. See 240.83(D) for SWD and HID marking for circuit breakers used as switches for 120-volt and 277-volt fluorescent and high-intensity discharge lighting circuits.

404.12 Grounding of Enclosures.

Metal enclosures for switches or circuit breakers shall be grounded as specified in Article 250. Where nonmetallic enclosures are used with metal raceways or metal-armored cables, provision shall be made for grounding continuity.

Except as covered in 404.9(B), Exception, nonmetallic boxes for switches shall be installed with a wiring method that provides or includes an equipment ground.

Exhibit 404.1 illustrates how the effective grounding of a metal faceplate can be accomplished by connecting the equipment grounding conductor provided with the wiring method to a grounding terminal on a metal yoke or strap.

404.13 Knife Switches.

(A) Isolating Switches. Knife switches rated at over 1200 amperes at 250 volts or less, and at over 600 amperes at

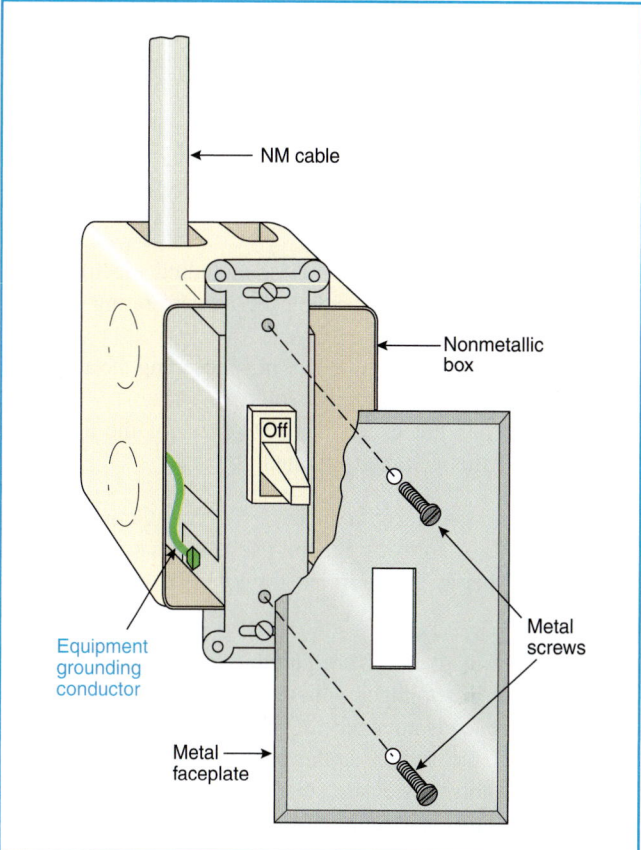

Exhibit 404.1 Grounding of a metal faceplate.

251 to 600 volts, shall be used only as isolating switches and shall not be opened under load.

(B) To Interrupt Currents. To interrupt currents over 1200 amperes at 250 volts, nominal, or less, or over 600 amperes at 251 to 600 volts, nominal, a circuit breaker or a switch of special design listed for such purpose shall be used.

(C) General-Use Switches. Knife switches of ratings less than specified in 404.13(A) and (B) shall be considered general-use switches.

FPN: See definition of *General-Use Switch* in Article 100.

(D) Motor-Circuit Switches. Motor-circuit switches shall be permitted to be of the knife-switch type.

FPN: See definition of a *Motor-Circuit Switch* in Article 100.

404.14 Rating and Use of Snap Switches.

Snap switches shall be used within their ratings and as indicated in 404.14(A) through (D).

FPN No. 1: For switches on signs and outline lighting, see 600.6.

FPN No. 2: For switches controlling motors, see 430.83, 430.109, and 430.110.

(A) Alternating Current General-Use Snap Switch. A form of general-use snap switch suitable only for use on ac circuits for controlling the following:

(1) Resistive and inductive loads, including electric-discharge lamps, not exceeding the ampere rating of the switch at the voltage involved

(2) Tungsten-filament lamp loads not exceeding the ampere rating of the switch at 120 volts

(3) Motor loads not exceeding 80 percent of the ampere rating of the switch at its rated voltage

(B) Alternating-Current or Direct-Current General-Use Snap Switch. A form of general-use snap switch suitable for use on either ac or dc circuits for controlling the following:

(1) Resistive loads not exceeding the ampere rating of the switch at the voltage applied.

(2) Inductive loads not exceeding 50 percent of the ampere rating of the switch at the applied voltage. Switches rated in horsepower are suitable for controlling motor loads within their rating at the voltage applied.

(3) Tungsten-filament lamp loads not exceeding the ampere rating of the switch at the applied voltage if T-rated.

(C) CO/ALR Snap Switches. Snap switches rated 20 amperes or less directly connected to aluminum conductors shall be listed and marked CO/ALR.

(D) Alternating-Current Specific-Use Snap Switches Rated for 347 Volts. Snap switches rated 347 volts ac shall be listed and shall be used only for controlling the following.

(1) Noninductive Loads. Noninductive loads other than tungsten-filament lamps not exceeding the ampere and voltage ratings of the switch.

(2) Inductive Loads. Inductive loads not exceeding the ampere and voltage ratings of the switch. Where particular load characteristics or limitations are specified as a condition of the listing, those restrictions shall be observed regardless of the ampere rating of the load.

The ampere rating of the switch shall not be less than 15 amperes at a voltage rating of 347 volts ac. Flush-type snap switches rated 347 volts ac shall not be readily interchangeable in box mounting with switches identified in 404.14(A) and (B).

Although they are not commonly used in the United States, 600Y/347-volt systems are permitted by the *Code*. In accor-

dance with 210.6 and 225.7(D), these systems can be used to supply installations of outdoor lighting. For the purposes of controlling lighting circuits on these systems, 404.14(D)(2) permits a relatively new type of ac specific-use snap switch that is 347-volt rated. These switches, unless specifically restricted, are permitted to be used on circuits of a lower voltage, such as 277- and 120-volt circuits.

(E) Dimmer Switches. General-use dimmer switches shall be used only to control permanently installed incandescent luminaires (lighting fixtures) unless listed for the control of other loads and installed accordingly.

New for the 2002 *NEC*, general-use dimmers are not permitted to control receptacles or cord-and-plug-connected table and floor lamps. Section 404.14(E) does not apply to commercial dimmers or theater dimmers that can be used for fluorescent lighting and portable lighting. If a dimmer that has been evaluated only for the control of incandescent luminaries is used, the potential for connecting incompatible equipment such as a cord-and-plug-connected motor-operated appliance or a portable fluorescent lamp is increased by using the dimmer to control a receptacle(s).

II. Construction Specifications

404.15 Marking.

(A) Ratings. Switches shall be marked with the current, voltage, and, if horsepower rated, the maximum rating for which they are designed.

(B) Off Indication. Where in the off position, a switching device with a marked OFF position shall completely disconnect all ungrounded conductors to the load it controls.

404.16 600-Volt Knife Switches.

Auxiliary contacts of a renewable or quick-break type or the equivalent shall be provided on all knife switches rated 600 volts and designed for use in breaking current over 200 amperes.

404.17 Fused Switches.

A fused switch shall not have fuses in parallel except as permitted in 240.8.

404.18 Wire-Bending Space.

The wire-bending space required by 404.3 shall meet Table 312.6(B) spacings to the enclosure wall opposite the line and load terminals.

ARTICLE 406
Receptacles, Cord Connectors, and Attachment Plugs (Caps)

Contents

406.1 Scope.

This article covers the rating, type, and installation of receptacles, cord connectors, and attachment plugs (cord caps).

To consolidate the requirements for receptacles, cord connectors, and attachment caps into one location dedicated to these devices, Article 406 was added to the 2002 *NEC*. Article 406 is comprised of information, taken from Article 410, Part L, 410-56, 410-57, and 410-58, and 210.7(d) from the 1999 *NEC*, that has been reorganized and editorially rewritten for the 2002 *NEC*.

406.2 Receptacle Rating and Type.

(A) Receptacles. Receptacles shall be listed for the purpose and marked with the manufacturer's name or identification and voltage and ampere ratings.

(B) Rating. Receptacles and cord connectors shall be rated not less than 15 amperes, 125 volts, or 15 amperes, 250 volts, and shall be of a type not suitable for use as lampholders.

> FPN: See 210.21(B) for receptacle ratings where installed on branch circuits.

(C) Receptacles for Aluminum Conductors. Receptacles rated 20 amperes or less and designed for the direct connection of aluminum conductors shall be marked CO/ALR.

Section 406.2(C) requires that 15- and 20-ampere receptacles directly connected to aluminum conductors be suitable for such use. If the receptacle is not of the CO/ALR type, it can be connected with a copper pigtail to an aluminum branch-circuit conductor only if the wire connector is suitable for such a connection and is marked with the letters AL and CU. The commentary following 110.14(B) further explains the suitability of wire connectors used to join copper and aluminum conductors.

(D) Isolated Ground Receptacles. Receptacles incorporating an isolated grounding connection intended for the reduction of electrical noise (electromagnetic interference) as permitted in 250.146(D) shall be identified by an orange triangle located on the face of the receptacle.

(1) Receptacles so identified shall be used only with grounding conductors that are isolated in accordance with 250.146(D).

(2) Isolated ground receptacles installed in nonmetallic boxes shall be covered with a nonmetallic faceplate.

Exception: Where an isolated ground receptacle is installed in a nonmetallic box, a metal faceplate shall be permitted if the box contains a feature or accessory that permits the effective grounding of the faceplate.

406.3 General Installation Requirements.

Receptacle outlets shall be located in branch circuits in accordance with Part III of Article 210. General installation requirements shall be in accordance with 406.3(A) through (F).

(A) Grounding Type. Receptacles installed on 15- and 20-ampere branch circuits shall be of the grounding type. Grounding-type receptacles shall be installed only on circuits of the voltage class and current for which they are rated, except as provided in Tables 210.21(B)(2) and (B)(3).

Exception: Nongrounding-type receptacles installed in accordance with 406.3(D).

(B) To Be Grounded. Receptacles and cord connectors that have grounding contacts shall have those contacts effectively grounded.

Exception No. 1: Receptacles mounted on portable and vehicle-mounted generators in accordance with 250.34.

Exception No. 2: Replacement receptacles as permitted by 406.3(D).

(C) Methods of Grounding. The grounding contacts of receptacles and cord connectors shall be grounded by connection to the equipment grounding conductor of the circuit supplying the receptacle or cord connector.

> FPN: For installation requirements for the reduction of electrical noise, see 250.146(D).

The branch-circuit wiring method shall include or provide an equipment-grounding conductor to which the grounding contacts of the receptacle or cord connector shall be connected.

> FPN No. 1: 250.118 describes acceptable grounding means.
> FPN No. 2: For extensions of existing branch circuits, see 250.130.

(D) Replacements. Replacement of receptacles shall comply with 406.3(D)(1), (2), and (3) as applicable.

(1) Grounding-Type Receptacles. Where a grounding means exists in the receptacle enclosure or a grounding conductor is installed in accordance with 250.130(C), grounding-type receptacles shall be used and shall be connected to the grounding conductor in accordance with 406.3(C) or 250.130(C).

(2) Ground-Fault Circuit Interrupters. Ground-fault circuit-interrupter protected receptacles shall be provided where replacements are made at receptacle outlets that are required to be so protected elsewhere in this *Code*.

(3) Nongrounding-Type Receptacles. Where grounding means does not exist in the receptacle enclosure, the installation shall comply with (a), (b), or (c).

(a) A nongrounding-type receptacle(s) shall be permitted to be replaced with another nongrounding-type receptacle(s).

(b) A nongrounding-type receptacle(s) shall be permitted to be replaced with a ground-fault circuit interrupter-type of receptacle(s). These receptacles shall be marked "No Equipment Ground." An equipment grounding conductor shall not be connected from the ground-fault circuit-interrupter-type receptacle to any outlet supplied from the ground-fault circuit-interrupter receptacle.

(c) A nongrounding-type receptacle(s) shall be permitted to be replaced with a grounding-type receptacle(s) where supplied through a ground-fault circuit interrupter. Grounding-type receptacles supplied through the ground-fault circuit interrupter shall be marked "GFCI Protected" and "No Equipment Ground." An equipment grounding conductor shall not be connected between the grounding-type receptacles.

(E) Cord-and-Plug-Connected Equipment. The installation of grounding-type receptacles shall not be used as a requirement that all cord-and-plug-connected equipment be of the grounded type.

> FPN: See 250.114 for types of cord-and-plug-connected equipment to be grounded.

(F) Noninterchangeable Types. Receptacles connected to circuits that have different voltages, frequencies, or types of current (ac or dc) on the same premises shall be of such design that the attachment plugs used on these circuits are not interchangeable.

406.4 Receptacle Mounting.

Receptacles shall be mounted in boxes or assemblies designed for the purpose, and such boxes or assemblies shall be securely fastened in place.

(A) Boxes That Are Set Back. Receptacles mounted in boxes that are set back of the wall surface, as permitted in 314.20, shall be installed so that the mounting yoke or strap of the receptacle is held rigidly at the surface of the wall.

(B) Boxes That Are Flush. Receptacles mounted in boxes that are flush with the wall surface or project therefrom shall be installed so that the mounting yoke or strap of the receptacle is held rigidly against the box or raised box cover.

In order to comply with 406.4(B), the outlet box used to enclose a receptacle must be rigidly and securely supported according to 314.23(B) or (C). In addition, mounting outlet boxes with the proper setback, according to 314.20, requires the cooperation of other construction trades (drywall installers, plasterers, and carpenters) and the building designers.

The intent of 406.4 (A), (B), and (C) is to allow attachment plugs to be inserted or removed without moving the receptacle. Additionally, by restricting movement of the receptacle, effective grounding continuity can be maintained for contact devices or receptacle yokes where the box is installed flush with the wall surface or where it projects therefrom. The proper installation of receptacles helps ensure that attachment plugs can be fully inserted, thus providing a better contact.

(C) Receptacles Mounted on Covers. Receptacles mounted to and supported by a cover shall be held rigidly against the cover by more than one screw or shall be a device assembly or box cover listed and identified for securing by a single screw.

Receptacles mounted on raised covers, such as the receptacle illustrated in Exhibit 406.1, are not permitted to be secured by a single screw unless listed and identified for the use.

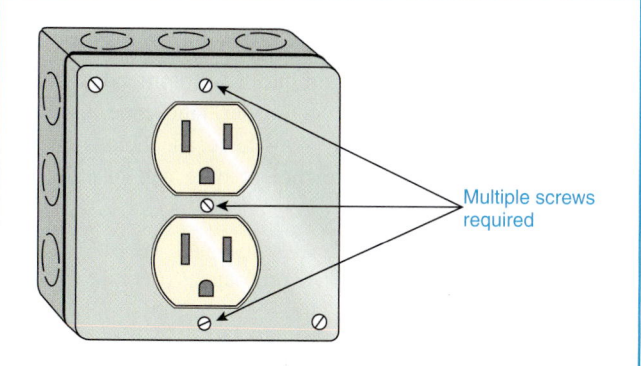

Exhibit 406.1 A receptacle mounted on a raised cover.

(D) Position of Receptacle Faces. After installation, receptacle faces shall be flush with or project from faceplates of insulating material and shall project a minimum of 0.4 mm (0.015 in.) from metal faceplates.

The reason for requiring receptacles to project from metal faceplates is to prevent faults between the blades of attachment plugs and metal faceplates. The proper mounting of faceplates ensures that attachment plugs can be fully inserted, thus providing a better contact. The *NEC* does not specify the position (blades up or blades down) of a common vertically mounted 15- or 20-ampere duplex receptacle. Al-

though many drawings within this handbook show the slots for blades up such as Exhibit 406.1, the receptacle may be installed with the slots for blades down. Receptacles can also be installed horizontally as well as vertically. Refer to 406.8(B) for information on receptacles installed in wet locations.

(E) Receptacles in Countertops and Similar Work Surfaces in Dwelling Units. Receptacles shall not be installed in a face-up position in countertops or similar work surfaces.

(F) Exposed Terminals. Receptacles shall be enclosed so that live wiring terminals are not exposed to contact.

406.5 Receptacle Faceplates (Cover Plates).

Receptacle faceplates shall be installed so as to completely cover the opening and seat against the mounting surface.

(A) Metal faceplates shall be of ferrous metal not less than 0.76 mm (0.030 in.) in thickness or of nonferrous metal not less than 1.02 mm (0.040 in.) in thickness.

(B) Metal faceplates shall be grounded.

Section 406.5(B) requires that metal receptacle faceplates be grounded. Generally, this requirement is easily met by grounding the metal box. However, isolated ground receptacles installed in nonmetallic boxes are problematic because grounding the receptacle in this case does not ground the faceplate. Section 406.2(D)(2) contains two solutions concerning the receptacle faceplate. First, the general solution is to use only nonmetallic faceplates. Second, the exception to 406.2(D)(2) allows a nonmetallic box manufacturer to add a feature or accessory to accomplish effective grounding of a metal faceplate.

(C) Faceplates of insulating material shall be noncombustible and not less than 2.54 mm (0.10 in.) in thickness but shall be permitted to be less than 2.54 mm (0.10 in.) in thickness if formed or reinforced to provide adequate mechanical strength.

406.6 Attachment Plugs.

All attachment plugs and cord connectors shall be listed for the purpose and marked with the manufacturer's name or identification and voltage and ampere ratings.

(A) Attachment plugs and cord connectors shall be constructed so that there are no exposed current-carrying parts except the prongs, blades, or pins. The cover for wire terminations shall be a part that is essential for the operation of an attachment plug or connector (dead-front construction).

(B) Attachment plugs shall be installed so that their prongs, blades, or pins are not energized unless inserted into an energized receptacle. No receptacle shall be installed so as to require an energized attachment plug as its source of supply.

The design requirements found in 406.6(B) (referred to as dead-front construction) have minimized the occurrence of electrical faults between metal plates and attachment plugs with terminal screws exposed on the face of the plug.

The requirements stated in 406.6(B) were originally found within product information only. However, as an aid to the inspection community, these requirements are now clearly stated in the *NEC*. A live attachment plug cap can be a most dangerous situation. Attachment plug caps should never be installed so as to allow the blades to be energized without being plugged into a device.

(C) Attachment Plug Ejector Mechanisms. Attachment plug ejector mechanisms shall not adversely affect engagement of the blades of the attachment plug with the contacts of the receptacle.

Section 406.6(C) permits a device that reduces the likelihood of damage to the cord when the cord is pulled to remove the plug. This device is designed for use by persons with mobility or visual impairment.

406.7 Noninterchangeability.

Receptacles, cord connectors, and attachment plugs shall be constructed so that receptacle or cord connectors do not accept an attachment plug with a different voltage or current rating from that for which the device is intended. However, a 20-ampere T-slot receptacle or cord connector shall be permitted to accept a 15-ampere attachment plug of the same voltage rating. Non–grounding-type receptacles and connectors shall not accept grounding-type attachment plugs.

For information on receptacle and attachment cap configurations, see Exhibits 406.2 and 406.3 which are based on ANSI C73 standard configuration charts.

406.8 Receptacles in Damp or Wet Locations.

(A) Damp Locations. A receptacle installed outdoors in a location protected from the weather or in other damp locations shall have an enclosure for the receptacle that is weatherproof when the receptacle is covered (attachment plug cap not inserted and receptacle covers closed).

An installation suitable for wet locations shall also be considered suitable for damp locations.

A receptacle shall be considered to be in a location protected from the weather where located under roofed open porches, canopies, marquees, and the like, and will not be subjected to a beating rain or water runoff.

(B) Wet Locations.

(1) 15- and 20-Ampere Outdoor Receptacles. 15- and 20-ampere, 125- and 250-volt receptacles installed outdoors in a wet location shall have an enclosure that is weatherproof whether or not the attachment plug cap is inserted.

To ensure the weatherproof integrity of the cord-and-plug connection to a receptacle located in an outdoor wet location, 406.8(B)(1) requires receptacle covers that provide a weatherproof enclosure at all times. The requirement for this type of cover is not contingent on the anticipated use of the receptacle. This requirement applies to all 15- and 20-ampere, 125- and 250-volt receptacles that are installed in outdoor wet locations, including those receptacle outlets at dwelling units specified by 210.52(E). Exhibits 406.4 and 406.5 are examples of the type of receptacle enclosure required by 406.8(B)(1).

(2) Other Receptacles. All other receptacles installed in a wet location shall comply with (a) or (b):

(a) A receptacle installed in a wet location where the product intended to be plugged into it is not attended while in use (e.g., sprinkler system controller, landscape lighting, holiday lights, and so forth) shall have an enclosure that is weatherproof with the attachment plug cap inserted or removed.

Section 406.8(B)(2)(a) applies to receptacles other than those rated 15 and 20 amperes, 125- and 250-volt, that supply cord-and-plug-connected equipment likely to be used outdoors or in a wet location for long periods of time. A portable pump motor is an example of such equipment. Receptacles for this application should remain weatherproof while they are in use.

(b) A receptacle installed in a wet location where the product intended to be plugged into it will be attended while in use (e.g., portable tools, and so forth) shall have an enclosure that is weatherproof when the attachment plug is removed.

Section 406.8(B)(2)(b) applies to receptacles other than those rated 15 and 20 amperes, 125- and 250-volt, that supply cord-and-plug-connected portable tools or other portable equipment likely to be used outdoors for a specific purpose and then removed.

DESCRIPTION		NEMA NUMBER	15 AMPERE		20 AMPERE		30 AMPERE		50 AMPERE		60 AMPERE	
			RECEPTACLE	PLUG	RECEPTACLE	PLUG	RECEPTACLE	PLUG	RECEPTACLE	PLUG	RECEPTACLE	PLUG
2-POLE 2-WIRE	125V	1	1-15R	1-15P POLARIZED / NON POLAR.		2-20P MATES WITH 5-20R		1-30P MATES WITH 5-30R				
	250V	2		2-15P MATES WITH 6-15R	2-20R	2-20P	2-30R	2-30P				
	277V AC	3										
	600V	4										
2-POLE 3-WIRE GROUNDING	125V	5	5-15R	5-15P	5-20R	5-20P	5-30R	5-30P	5-50R	5-50P		
	125V	5ALT			5ALT-20R							
	250V	6	6-15R	6-15P	6-20R	6-20P	6-30R	6-30P	6-50R	6-50P		
	250V	6ALT			6ALT-20R							
	277V AC	7	7-15R	7-15P	7-20R	7-20P	7-30R	7-30P	7-50R	7-50P		
	347V AC	24	24-15R	24-15P	24-20R	24-20P	24-30R	24-30P	24-50R	24-50P		
	480V AC	8										
	600V AC	9										
3-POLE 3-WIRE	125/250V	10			10-20R	10-20P	10-30R	10-30P	10-50R	10-50P		
	3 Ø 250V	11	11-15R	11-15P	11-20R	11-20P	11-30R	11-30P	11-50R	11-50P		
	3 Ø 480V	12										
	3 Ø 600V	13										
3-POLE 4-WIRE GROUNDING	125/250V	14	14-15R	14-15P	14-20R	14-20P	14-30R	14-30P	14-50R	14-50P	14-60R	14-60P
	3 Ø 250V	15	15-15R	15-15P	15-20R	15-20P	15-30R	15-30P	15-50R	15-50P	15-60R	15-60P
	3 Ø 480V	16										
	3 Ø 600V	17										
4-POLE 4-WIRE	3 Ø Y 120/280V	18	18-15R	18-15P	18-20R	18-20P	18-30R	18-30P	18-50R	18-50P	18-60R	18-60P
	3 Ø Y 277/480V	19										
	3 Ø Y 347/600V	20										
4-POLE 5-WIRE GROUNDING	3 Ø Y 120/208V	21										
	3 Ø Y 277/480V	22										
	3 Ø Y 347/600V	23										

Note: Blank spaces reserved for future configurations.

Exhibit 406.2 Configuration chart for general-purpose locking plugs and receptacles. (Reproduced from *Wiring, Devices—Dimensional Requirements*, NEMA WD 6-1997)

Exhibit 406.3 Configuration chart for specific-purpose locking plugs and receptacles. (Reproduced from *Wiring Devices—Dimensional Requirements*, NEMA WD 6-1997)

Exhibit 406.4 A single-gang weatherproof cover suitable for use in wet locations. (Courtesy of L. E. Mason Co.)

Exhibit 406.5 A two-gang weatherproof cover suitable for use in outdoor or indoor wet locations. (Courtesy of Carlon®/Lamson & Sessions)

(C) Bathtub and Shower Space. A receptacle shall not be installed within a bathtub or shower space.

Section 406.8(C) prohibits the installation of receptacles inside bathtub and shower spaces, even if the receptacles are installed in a weatherproof enclosure. Prohibiting this installation helps prevent the use of shavers, radios, hair dryers, and so on, in these areas. The unprotected line side of GFCI-protected receptacles installed in bathtub and shower spaces could possibly become wet and therefore create a shock hazard by energizing surrounding wet surfaces.

(D) Protection for Floor Receptacles. Standpipes of floor receptacles shall allow floor-cleaning equipment to be operated without damage to receptacles.

(E) Flush Mounting with Faceplate. The enclosure for a receptacle installed in an outlet box flush-mounted on a wall surface shall be made weatherproof by means of a weatherproof faceplate assembly that provides a watertight connection between the plate and the wall surface.

406.9 Grounding-Type Receptacles, Adapters, Cord Connectors, and Attachment Plugs.

(A) Grounding Poles. Grounding-type receptacles, cord connectors, and attachment plugs shall be provided with one fixed grounding pole in addition to the circuit poles. The grounding contacting pole of grounding-type plug-in ground-fault circuit interrupters shall be permitted to be of the movable, self-restoring type on circuits operating at not over 150 volts between any two conductors or any conductor and ground.

(B) Grounding-Pole Identification. Grounding-type receptacles, adapters, cord connections, and attachment plugs shall have a means for connection of a grounding conductor to the grounding pole.

A terminal for connection to the grounding pole shall be designated by one of the following:

(1) A green-colored hexagonal-headed or -shaped terminal screw or nut, not readily removable.
(2) A green-colored pressure wire connector body (a wire barrel).
(3) A similar green-colored connection device, in the case of adapters. The grounding terminal of a grounding adapter shall be a green-colored rigid ear, lug, or similar device. The grounding connection shall be designed so that it cannot make contact with current-carrying parts of the receptacle, adapter, or attachment plug. The adapter shall be polarized.

Section 406.9(B)(3) requires the grounding terminal of an adapter to be a green-colored ear, lug, or similar device, thereby prohibiting use of an adapter with an attached pigtail grounding wire, which had been used for many years.

(4) If the terminal for the equipment grounding conductor is not visible, the conductor entrance hole shall be marked with the word *green* or *ground*, the letters G or GR, or the grounding symbol, as shown in Figure 406.9(B)(4), or otherwise identified by a distinctive green color. If the terminal for the equipment grounding conductor is readily removable, the area adjacent to the terminal shall be similarly marked.

Figure 406.9(B)(4) Grounding symbol.

(C) Grounding Terminal Use. A grounding terminal or grounding-type device shall not be used for purposes other than grounding.

(D) Grounding-Pole Requirements. Grounding-type attachment plugs and mating cord connectors and receptacles shall be designed so that the grounding connection is made before the current-carrying connections. Grounding-type devices shall be designed so grounding poles of attachment plugs cannot be brought into contact with current-carrying parts of receptacles or cord connectors.

The grounding blade of the attachment plug cap of most grounding-type combinations is longer than the circuit conductor blades and used to ensure a "make-first, break-last" grounding connection. In some non-ANSI-approved pin-and-sleeve type connections, the grounding contact of the receptacle is closer to the face of the receptacle than it is to other contacts, serving the same purpose.

(E) Use. Grounding-type attachment plugs shall be used only with a cord having an equipment grounding conductor.

FPN: See 200.10(B) for identification of grounded conductor terminals.

406.10 Connecting Receptacle Grounding Terminal to Box.

The connection of the receptacle grounding terminal shall comply with 250.146.

ARTICLE 408
Switchboards and Panelboards

Contents

I. General

408.1 Scope.

This article covers the following:

(1) All switchboards, panelboards, and distribution boards installed for the control of light and power circuits

(2) Battery-charging panels supplied from light or power circuits

See the definitions of *panelboard* and *switchboard* in Article 100.

408.2 Other Articles.

Switches, circuit breakers, and overcurrent devices used on switchboards, panelboards, and distribution boards, and their enclosures, shall comply with this article and also with the requirements of Articles 240, 250, 312, 314, 404, and other articles that apply. Switchboards and panelboards in hazardous (classified) locations shall comply with the requirements of Articles 500 through 517.

408.3 Support and Arrangement of Busbars and Conductors.

(A) Conductors and Busbars on a Switchboard or Panelboard. Conductors and busbars on a switchboard or panelboard shall comply with 408.3(A)(1), (2), and (3) as applicable.

(1) Location. Conductors and busbars shall be located so as to be free from physical damage and shall be held firmly in place.

(2) Service Switchboards. Barriers shall be placed in all service switchboards such that no uninsulated, ungrounded service busbar or service terminal is exposed to inadvertent contact by persons or maintenance equipment while servicing load terminations.

Where it can be demonstrated that it is unfeasible to disconnect or de-energize the service conductors supplying a service switchboard, qualified electricians may be required to work on these switchboards with the load terminals de-energized but with the service bus energized. Barriers are required in service switchboards to provide physical separa-

tion (adequate distance or obstacle) between load terminals and the service busbars and terminals, thus providing some measure of safety against inadvertent contact with line-energized parts during maintenance and installation of new feeders or branch circuits. In most multisection switchboards, barriers are not required, because the line-side conductors and busbars are not in the same switchboard sections that contain the load terminals. It must be clearly understood that de-energizing the load side of a switchboard by operation of the disconnecting means does not de-energize the ungrounded service conductors. Every effort should be made to make arrangements to completely disconnect power from the equipment before performing any work inside. For complete disconnection that is not feasible, the installer should become familiar with NFPA 70E, *Electrical Safety Requirements for Employee Workplaces.* This industry-recognized document provides the guidance for protective equipment and appropriate work rules that must be followed for working on energized equipment.

(3) Same Vertical Section. Other than the required interconnections and control wiring, only those conductors that are intended for termination in a vertical section of a switchboard shall be located in that section.

Exception: Conductors shall be permitted to travel horizontally through vertical sections of switchboards where such conductors are isolated from busbars by a barrier.

The exception to 408.3(A)(3) permits conductors to travel horizontally through vertical sections of a switchboard where barriers are provided to isolate the conductors from the busbars. Horizontal travel of conductors through more than one section of a multisection switchboard is necessary where a raceway or cable entry is made into a switchboard section other than the one at which the conductors are terminated.

(B) Overheating and Inductive Effects. The arrangement of busbars and conductors shall be such as to avoid overheating due to inductive effects.

(C) Used as Service Equipment. Each switchboard or panelboard, if used as service equipment, shall be provided with a main bonding jumper sized in accordance with 250.28(D) or the equivalent placed within the panelboard or one of the sections of the switchboard for connecting the grounded service conductor on its supply side to the switchboard or panelboard frame. All sections of a switchboard shall be bonded together using an equipment grounding conductor sized in accordance with Table 250.122.

Exception: Switchboards and panelboards used as service equipment on high-impedance grounded-neutral systems in accordance with 250.36 shall not be required to be provided with a main bonding jumper.

(D) Terminals. In switchboards and panelboards, load terminals for field wiring, including grounded circuit conductor load terminals and connections to the ground bus for load equipment grounding conductors, shall be located so that it is not necessary to reach across or beyond an uninsulated ungrounded line bus in order to make connections.

(E) Phase Arrangement. The phase arrangement on 3-phase buses shall be A, B, C from front to back, top to bottom, or left to right, as viewed from the front of the switchboard or panelboard. The B phase shall be that phase having the higher voltage to ground on 3-phase, 4-wire, delta-connected systems. Other busbar arrangements shall be permitted for additions to existing installations and shall be marked.

The high leg is common on a 240/120-volt, 3-phase, 4-wire delta system. It is typically designated as "B phase." Section 110.15 requires the high-leg marking to be the color orange or other similar effective means of identification. Electricians should always test each phase to ground with suitable equipment in order to know exactly where this high leg is located in the system.

The exception to 408.3(E) permits the phase leg having the higher voltage to ground to be located at the right-hand position (C phase), making it unnecessary to transpose the panelboard or switchboard busbar arrangement ahead of and beyond a metering compartment. The exception recognizes the fact that metering compartments have been standardized with the high leg at the right position (C phase) rather than in the center on B phase.

See also 110.15, 215.8, and 230.56 for further information on identifying conductors with the higher voltage to ground. Other busbar arrangements for making additions to existing installations are permitted by 408.3(E).

Exception: Equipment within the same single section or multisection switchboard or panelboard as the meter on 3-phase, 4-wire, delta-connected systems shall be permitted to have the same phase configuration as the metering equipment.

> FPN: See 110.15 for requirements on marking the busbar or phase conductor having the higher voltage to ground where supplied from a 4-wire, delta-connected system.

(F) Minimum Wire-Bending Space. The minimum wire-bending space at terminals and minimum gutter space provided in panelboards and switchboards shall be as required in 312.6.

Section 408.3(F) requires that installations in the field comply with 312.6. See also the commentary following 408.35, which covers the size of the enclosure.

408.4 Circuit Directory.

All circuits and circuit modifications shall be legibly identified as to purpose or use on a circuit directory located on the face or inside of the panel door in the case of a panelboard, and at each switch on a switchboard.

The requirement to provide an up-to-date, accurate, and legible circuit directory has been revised to specifically apply to all equipment covered in Article 408, not to panelboards only. The circuit directory is an important feature for the safe operation of an electrical system under normal and emergency conditions. The purpose of an accurate and legible circuit directory in these types of equipment is to provide clear identification of circuit breakers and switches that may need to be operated by service personnel or those who need to operate a switch or circuit breaker in an emergency. This requirement is specific to switchboards and panelboards; however, the identification requirements of 110.22 apply to all disconnecting means.

II. Switchboards

408.5 Location of Switchboards.

Switchboards that have any exposed live parts shall be located in permanently dry locations and then only where under competent supervision and accessible only to qualified persons. Switchboards shall be located so that the probability of damage from equipment or processes is reduced to a minimum.

408.6 Switchboards in Damp or Wet Locations.

Switchboards in damp or wet locations shall be installed to comply with 312.2(A).

408.7 Location Relative to Easily Ignitible Material.

Switchboards shall be placed so as to reduce to a minimum the probability of communicating fire to adjacent combustible materials. Where installed over a combustible floor, suitable protection thereto shall be provided.

One way to comply with the requirement of 408.7 is to form and attach a piece of sheet steel or other suitable noncombustible material to the floor under the electrical equipment.

408.8 Clearances.

(A) From Ceiling. For other than a totally enclosed switchboard, a space not less than 900 mm (3 ft) shall be provided between the top of the switchboard and any combustible ceiling, unless a noncombustible shield is provided between the switchboard and the ceiling.

(B) Around Switchboards. Clearances around switchboards shall comply with the provisions of 110.26.

Sufficient access and working space are required to permit safe operation and maintenance of swtichboards. Table 110.26(A)(1) indicates minimum working clearances from 0 to 600 volts, and Table 110.34(A) is used for voltages over 600 volts.

408.9 Conductor Insulation.

An insulated conductor used within a switchboard shall be listed, shall be flame retardant, and shall be rated not less than the voltage applied to it and not less than the voltage applied to other conductors or busbars with which it may come in contact.

408.10 Clearance for Conductors Entering Bus Enclosures.

Where conduits or other raceways enter a switchboard, floor-standing panelboard, or similar enclosure at the bottom, sufficient space shall be provided to permit installation of conductors in the enclosure. The wiring space shall not be less than shown in Table 408.10 where the conduit or raceways enter or leave the enclosure below the busbars, their supports, or other obstructions. The conduit or raceways, including their end fittings, shall not rise more than 75 mm (3 in.) above the bottom of the enclosure.

Table 408.10 Clearance for Conductors Entering Bus Enclosures

Conductor	Minimum Spacing Between Bottom of Enclosure and Busbars, Their Supports, or Other Obstructions	
	mm	in.
Insulated busbars, their supports, or other obstructions	200	8
Noninsulated busbars	250	10

Section 408.10 should be carefully considered where installing underground conduit or raceways that terminate in the bottom of an open switchboard. For example, larger sizes of conduit used for service laterals or feeders and extending more than 3 in. above the bottom of the enclosure are difficult to shorten. On the other hand, conduits or raceways should

not be installed flush with the finished floor under switchboards that are located on the outside of buildings or in other locations where water could enter the raceways.

408.12 Grounding of Instruments, Relays, Meters, and Instrument Transformers on Switchboards.

Instruments, relays, meters, and instrument transformers located on switchboards shall be grounded as specified in 250.170 through 250.178.

III. Panelboards

408.13 General.

All panelboards shall have a rating not less than the minimum feeder capacity required for the load computed in accordance with Article 220. Panelboards shall be durably marked by the manufacturer with the voltage and the current rating and the number of phases for which they are designed and with the manufacturer's name or trademark in such a manner so as to be visible after installation, without disturbing the interior parts or wiring.

FPN: See 110.22 for additional requirements.

Some panelboards are suitable for use as service equipment and are so marked.

Listed panelboards are used with copper conductors, unless marked to indicate which terminals are suitable for use with aluminum conductors. Such marking must be independent of any marking on terminal connectors and must appear on a wiring diagram or other readily visible location. If all terminals are suitable for use with aluminum conductors as well as with copper conductors, the panelboard will be marked "Use Copper or Aluminum Wire." A panelboard using terminals or main or branch-circuit units individually marked AL-CU will be marked as noted above (Use Copper or Aluminum Wire) or will be marked "Use Copper Wire Only." The latter marking indicates that wiring space or other factors make the panelboard unsuitable for aluminum conductors. [See 110.14(C).]

Panelboards to which units (circuit breakers, switches, etc.) may be added in the field are marked with the name or trademark of the manufacturer and the catalog number or equivalent of those units intended for installation in the field.

Unless the panelboard is marked to indicate otherwise, the termination provisions are based on the use of 60°C (140°F) ampacities for wire sizes 14 AWG through 1 AWG and 75°C (167°F) ampacities for wire sizes 1/0 AWG and larger.

408.14 Classification of Panelboards.

Panelboards shall be classified for the purposes of this article as either lighting and appliance branch-circuit panelboards or power panelboards, based on their content. A lighting and appliance branch circuit is a branch circuit that has a connection to the neutral of the panelboard and that has overcurrent protection of 30 amperes or less in one or more conductors.

(A) Lighting and Appliance Branch-Circuit Panelboard. A lighting and appliance branch-circuit panelboard is one having more than 10 percent of its overcurrent devices protecting lighting and appliance branch circuits.

A lighting and appliance branch-circuit panelboard is one in which more than 10 percent of the installed overcurrent devices are rated 15, 20, or 30 amperes and supply circuits with a neutral conductor. For example, a 24-position (space for 24 full-size circuit breakers), 120/240-volt, residential panelboard contains 21 overcurrent devices. If three or more (10% of 21 = 2.1) of those overcurrent devices supply 15-, 20-, or 30-ampere circuits with a neutral conductor, the panelboard is considered to be a lighting and appliance branch-circuit panelboard. However, if 12 two-pole breakers for electric heat were installed in the panelboard, there would be no circuits with neutral connections, and this panelboard therefore would be a power panelboard and would be subject to the overcurrent protection requirements of 408.16(B).

(B) Power Panelboard. A power panelboard is one having 10 percent or fewer of its overcurrent devices protecting lighting and appliance branch circuits.

A power panelboard is a panelboard that has 10 percent or less of the installed overcurrent devices supplying lighting and appliance branch circuits. Any panelboard that is not classified as a lighting and appliance branch-circuit panelboard is a power panelboard. A typical power panelboard could be located near the service and be designed to supply facility feeder circuits. The feeders from a power panelboard could supply other utilization equipment or other panelboards designed to supply either feeders or branch circuits.

408.15 Number of Overcurrent Devices on One Panelboard.

Not more than 42 overcurrent devices (other than those provided for in the mains) of a lighting and appliance branch-circuit panelboard shall be installed in any one cabinet or cutout box.

A lighting and appliance branch-circuit panelboard shall be provided with physical means to prevent the installation of more overcurrent devices than that number for which the panelboard was designed, rated, and approved.

For the purposes of this article, a 2-pole circuit breaker shall be considered two overcurrent devices; a 3-pole circuit breaker shall be considered three overcurrent devices.

"Class CTL" is the Underwriters Laboratories Inc. designation for the *Code* requirement for circuit limitation within a lighting and appliance branch-circuit panelboard and means "circuit limiting."

Class CTL panelboards incorporate physical features that, in conjunction with the physical size, configuration, or other means provided in Class CTL circuit breakers, fuseholders, or fusible switches, are designed to prevent the installation of more overcurrent protective poles than the number for which the panelboard is designed and rated.

It should be noted that switchboards, unlike panelboards, are not limited to 42 overcurrent devices or 42 switches or devices.

408.16 Overcurrent Protection.

(A) Lighting and Appliance Branch-Circuit Panelboard Individually Protected. Each lighting and appliance branch-circuit panelboard shall be individually protected on the supply side by not more than two main circuit breakers or two sets of fuses having a combined rating not greater than that of the panelboard.

Exception No. 1: Individual protection for a lighting and appliance panelboard shall not be required if the panelboard feeder has overcurrent protection not greater than the rating of the panelboard.

Main overcurrent protection may be an integral part of a panelboard or may be located remote from the panelboard. See also the commentary following 408.15. Exhibit 408.1 shows a panelboard with a 200-ampere main circuit breaker. Exhibit 408.2 illustrates overcurrent protection for the panelboard feeders having a rating not greater than the rating of the panelboard.

Exception No. 2: For existing installations, individual protection for lighting and appliance branch-circuit panelboards shall not be required where such panelboards are used as service equipment in supplying an individual residential occupancy.

The phrase "for existing installations" in Exception No. 2 refers to the existing panelboard. It is not intended that a split-bus panelboard used in an individual residential occupancy be replaced if a circuit is added to the existing panelboard. It does mean, however, that for the installation of new panelboards in new or existing residential occupancies, ampere (*left*) and a 150-ampere (*right*) panelboard. The left panel has two 100-ampere main breakers installed as disconnecting means, and 200-ampere main lugs. The right is a split-bus panel with 150-ampere main lugs and six main breaker disconnecting means. The 150-ampere panelboard is suitable for use as service equipment only if it is not a lighting and appliance panelboard or if it presently exists in an individual residential occupancy.

(B) Power Panelboard Protection. In addition to the requirements of 408.13, a power panelboard with supply conductors that include a neutral and having more than 10 percent of its overcurrent devices protecting branch circuits rated 30 amperes or less shall be protected by an overcurrent protective device having a rating not greater than that of the panelboard. The overcurrent protective device shall be located within or at any point on the supply side of the panelboard.

When a power panelboard is required to have overcurrent protection, such protection can be provided by an overcurrent device in the panelboard or by an overcurrent device protecting the conductors that supply the panelboard. In either case, the rating of the overcurrent device is not permitted to exceed the rating of the panelboard. For example, a feeder protected by a 450-ampere overcurrent device supplies a panelboard with a 600-ampere rating. Because the panelboard is obviously large enough to supply the calculated load and the overcurrent device protecting the feeder does not exceed the rating of the panelboard, an individual overcurrent device in the panelboard is not required. Note that where the panelboard is used as service equipment, the exception to 408.16(B) allows the panelboard overcurrent protection to be multiple devices in accordance with the "six-disconnect rule" in 230.71.

Exhibit 408.1 A panelboard with main circuit breaker disconnect, suitable for use as service equipment. (Courtesy of Square D Co.)

a split-bus six-disconnect panelboard (with more than two circuit breakers or sets of fuses protecting the panelboard) is not permitted for the service equipment.

An individual residential occupancy could be a dwelling unit in a multifamily dwelling where the panelboard is used as service equipment. See the definition of *dwelling unit* in Article 100.

Exhibit 408.3 shows the split-bus circuitry for a 200-

Exhibit 408.2 An arrangement of three individual lighting and appliance branch-circuit panelboards with main overcurrent protection remote from the panelboards.

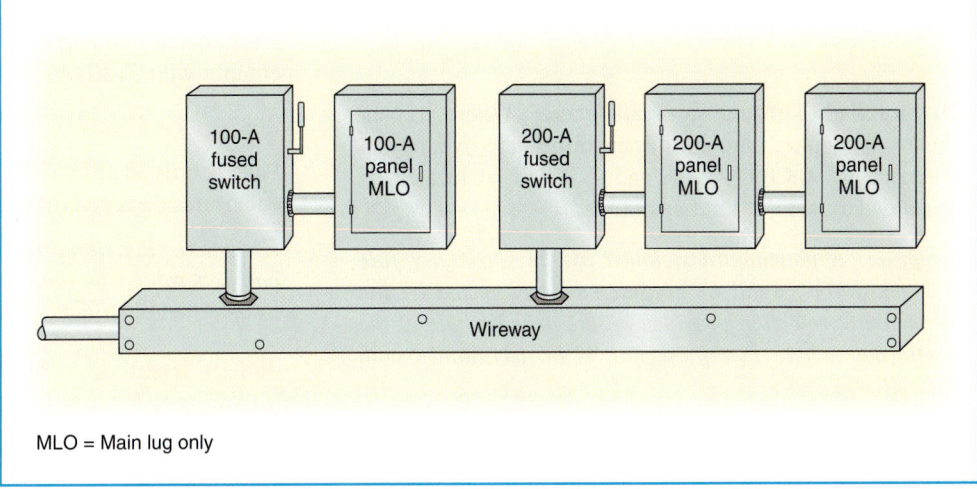

MLO = Main lug only

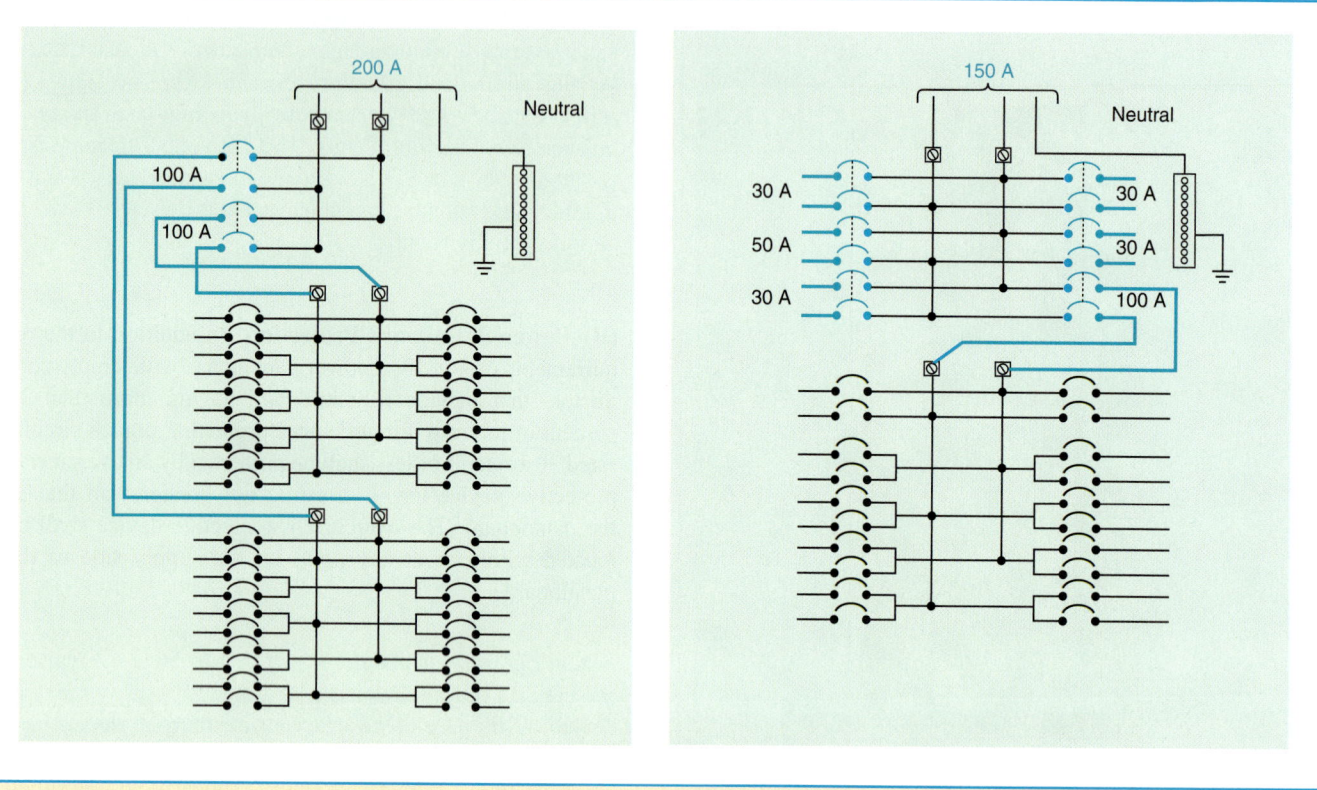

Figure 408.3 Circuitry for a 200-ampere (left) and a 150-ampere (right) split-bus panelboard.

Exception: This individual protection shall not be required for a power panelboard used as service equipment with multiple disconnecting means in accordance with 230.71.

(C) Snap Switches Rated at 30 Amperes or Less. Panelboards equipped with snap switches rated at 30 amperes or less shall have overcurrent protection not in excess of 200 amperes.

The requirement of 408.16(C) is limited to snap switches; it does not apply to panelboards equipped with circuit breakers.

(D) Supplied Through a Transformer. Where a panelboard is supplied through a transformer, the overcurrent protection in 408.16(A), (B), and (C) shall be located on the secondary side of the transformer.

Exception: A panelboard supplied by the secondary side of a transformer shall be considered as protected by the overcurrent protection provided on the primary side of the transformer where that protection is in accordance with 240.21(C)(1).

(E) Delta Breakers. A 3-phase disconnect or overcurrent device shall not be connected to the bus of any panelboard

that has less than 3-phase buses. Delta breakers shall not be installed in panelboards.

(F) Back-Fed Devices. Plug-in-type overcurrent protection devices or plug-in type-main lug assemblies that are backfed and used to terminate field-installed ungrounded supply conductors shall be secured in place by an additional fastener that requires other than a pull to release the device from the mounting means on the panel.

408.17 Panelboards in Damp or Wet Locations.

Panelboards in damp or wet locations shall be installed to comply with 312.2(A).

408.18 Enclosure.

Panelboards shall be mounted in cabinets, cutout boxes, or enclosures designed for the purpose and shall be dead-front.

Exception: Panelboards other than of the dead-front, externally operable type shall be permitted where accessible only to qualified persons.

408.19 Relative Arrangement of Switches and Fuses.

In panelboards, fuses of any type shall be installed on the load side of any switches.

Exception: Fuses installed as part of service equipment in accordance with the provisions of 230.94 shall be permitted on the line side of the service switch.

Sections 230.82 and 230.94 permit the service switch on either the supply side or load side of fuses such as cable limiters and other current-limiting devices. Where fuses of panelboards are accessible to other than qualified persons, such as occupants of a multifamily dwelling, 240.40 requires that disconnecting means be located on the supply side of all fuses in circuits of over 150 volts to ground and in cartridge-type fuses in circuits of any voltage. Thus, when the disconnect switch is opened, the fuses are de-energized, and danger from shock is reduced.

408.20 Grounding of Panelboards.

Panelboard cabinets and panelboard frames, if of metal, shall be in physical contact with each other and shall be grounded. Where the panelboard is used with nonmetallic raceway or cable or where separate grounding conductors are provided, a terminal bar for the grounding conductors shall be secured inside the cabinet. The terminal bar shall be bonded to the cabinet and panelboard frame, if of metal; otherwise it shall be connected to the grounding conductor that is run with the conductors feeding the panelboard.

A separate equipment grounding conductor terminal bar must be installed and bonded to the panelboard for the termination of feeder and branch-circuit equipment grounding conductors. Where installed within service equipment, this terminal is bonded to the neutral terminal bar. Any other connection between the equipment grounding terminal bar and the neutral bar, other than allowed in 250.32, is not permitted. If this downstream connection occurs, current flow in the neutral or grounded conductor would take parallel paths through the equipment grounding conductors (the raceway, the building structure, or earth, for example) back to the service equipment. Normal load currents flowing on the equipment grounding conductors could create a shock hazard. Exposed metal parts of equipment could have a potential difference of several volts created by the load current on the grounding conductors. Another safety hazard created by this effect, where subpanels are used, is arcing or loose connections at connectors and raceway fittings, for example, creating a potential fire hazard. Exhibit 408.4 illustrates the connection of the grounded conductor (neutral bar) to the metallic service equipment enclosure via the main bonding jumper.

Exception: Where an isolated equipment grounding conductor is provided as permitted by 250.146(D), the insulated equipment grounding conductor that is run with the circuit

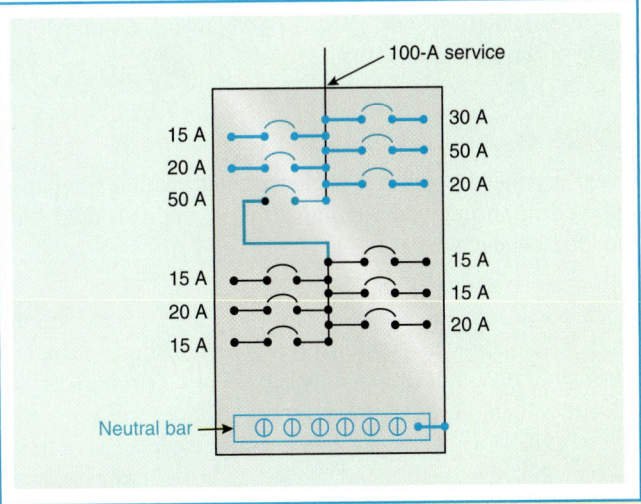

Exhibit 408.4 A split-bus lighting and appliance branch-circuit panelboard supplying an individual residential occupancy.

conductors shall be permitted to pass through the panelboard without being connected to the panelboard's equipment grounding terminal bar.

Grounding conductors shall not be connected to a terminal bar provided for grounded conductors (may be a neutral) unless the bar is identified for the purpose and is located where interconnection between equipment grounding conductors and grounded circuit conductors is permitted or required by Article 250.

The grounding of electronic equipment as well as overall power quality is of concern to the electrical industry. Sensitive electronic equipment used in industrial and commercial power systems may fail to perform properly if electrical noise is present in the equipment grounding conductor.

An isolated equipment grounding terminal is permitted if required for the reduction of electrical noise on the grounding circuit. This equipment grounding terminal must be grounded by an insulated equipment grounding conductor that is run with the circuit conductors. The isolated equipment grounding conductor is also permitted to pass through one or more panelboards (without connection to the panelboard grounding terminal), but it is very important that the equipment grounding conductor terminate directly at the applicable separately derived system or service grounding terminal. If the isolated equipment grounding conductor is run in a separate building, however, 250.146(D) requires the isolated equipment grounding conductor to terminate at a panelboard within the same building.

A connection to only a separate grounding electrode that places the earth in the fault return path may prevent sufficient current for opening overcurrent protection when a ground fault occurs. See the commentary following 250.146(D) FPN and 250.54.

408.21 Grounded Conductor Terminations.

Each grounded conductor shall terminate within the panelboard in an individual terminal that is not also used for another conductor.

In accordance with their listing and the requirement of 110.14(A), conductor terminations are suitable for a single conductor unless the terminal is marked or otherwise identified as suitable for more than one conductor. This requirement applies only to the termination of grounded conductors in panelboards. The use of a single termination point within a panelboard to connect more than one grounded conductor or to connect a grounded conductor and an equipment grounding conductor can be problematic when it is necessary to isolate a particular grounded conductor for testing purposes. For example, if the grounded conductors of two branch circuits were terminated at a single connection point and it was necessary to isolate one branch circuit for the purposes of troubleshooting, the fact that the circuit not being tested remains energized can create an unsafe working condition for service personnel disconnecting the grounded conductor of the circuit that is being tested. There are panelboard instructions that permit the use of a single conductor termination for more than one equipment grounding conductor. See 408.20 for the requirements on panelboard terminations for grounded and equipment grounding conductors.

Exception: Grounded conductors of circuits with parallel conductors shall be permitted to terminate in a single terminal if the terminal is identified for connection of more than one conductor.

IV. Construction Specifications

408.30 Panels.

The panels of switchboards shall be made of moisture-resistant, noncombustible material.

408.31 Busbars.

Insulated or bare busbars shall be rigidly mounted.

408.32 Protection of Instrument Circuits.

Instruments, pilot lights, potential transformers, and other switchboard devices with potential coils shall be supplied by a circuit that is protected by standard overcurrent devices rated 15 amperes or less.

Exception No. 1: Overcurrent devices rated more than 15 amperes shall be permitted where the interruption of the

circuit could create a hazard. Short-circuit protection shall be provided.

Exception No. 2: For ratings of 2 amperes or less, special types of enclosed fuses shall be permitted.

408.33 Component Parts.

Switches, fuses, and fuseholders used on panelboards shall comply with the applicable requirements of Articles 240 and 404.

408.35 Wire-Bending Space in Panelboards.

The enclosure for a panelboard shall have the top and bottom wire-bending space sized in accordance with Table 312.6(B) for the largest conductor entering or leaving the enclosure. Side wire-bending space shall be in accordance with Table 312.6(A) for the largest conductor to be terminated in that space.

Exception No. 1: Either the top or bottom wire-bending space shall be permitted to be sized in accordance with Table 312.6(A) for a lighting and appliance branch-circuit panelboard rated 225 amperes or less.

Exception No. 2: Either the top or bottom wire-bending space for any panelboard shall be permitted to be sized in accordance with Table 312.6(A) where at least one side wire-bending space is sized in accordance with Table 312.6(B) for the largest conductor to be terminated in any side wire-bending space.

Exception No. 3: The top and bottom wire-bending space shall be permitted to be sized in accordance with Table 312.6(A) spacings if the panelboard is designed and constructed for wiring using only one single 90 degree bend for each conductor, including the grounded circuit conductor, and the wiring diagram shows and specifies the method of wiring that shall be used.

Exception No. 4: Either the top or the bottom wire-bending space, but not both, shall be permitted to be sized in accordance with Table 312.6(A) where there are no conductors terminated in that space.

Section 408.35 covers the size of the enclosure for a panelboard. Using Exhibit 312.1 as a reference (see 312.6), the general rule calls for wire-bending spaces T_1 and T_4 to be in accordance with Table 312.6(B) for size M conductors (assuming these are the largest conductors entering the enclosure). Side wire-bending space T_2 must be in accordance with Table 312.6(A) for the wire size to be used with the largest-rated unit facing that side space, and T_3 must be similarly sized for the largest-rated unit facing the right side of the enclosure.

Exception No. 1 to 408.35 permits either T_1 or T_4 (not both) to be reduced to the space required by Table 312.6(A) for size M conductors for a panelboard rated 225 amperes or less.

Exception No. 2 to 408.35 permits either T_1 or T_4 (not both) to be reduced to the space required by Table 312.6(A) for size M conductors for any panelboard. Exception No. 2 is valid where either T_2 or T_3 (or both) is sized in accordance with Table 312.6(B) for the largest conductor to be terminated in either the left or right side spaces. Under the construction rules of 408.35, a panelboard enclosure might not be of adequate size for all manner of wiring, therefore 312.6 must be considered when wiring is planned.

Exception No. 3 to 408.35 permits both the top and bottom wire-bending space to be reduced as noted. A single 90 degree bend, meaning one and only one 90 degree bend, must be present for the ungrounded conductors. A grounded conductor is permitted to be wired straight in if spacing is provided per Table 312.6(B) for the grounded conductor.

Exception No. 4 to 408.35 permits a reduction to the Table 312.6(A) spacing for the top or bottom space where no terminals face that space. In this case, the space is a gutter space, and measurement is on a line perpendicular to the wall of the enclosure and to the closest barrier post or side of a switch, fuse, or circuit breaker unit that is, or may be, installed.

408.36 Minimum Spacings.

The distance between bare metal parts, busbars, and so forth shall not be less than specified in Table 408.36.

Table 408.36 Minimum Spacings Between Bare Metal Parts

Voltage	Opposite Polarity Where Mounted on the Same Surface		Opposite Polarity Where Held Free in Air		Live Parts to Ground*	
	mm	in.	mm	in.	mm	in.
Not over 125 volts, nominal	19.1	¾	12.7	½	12.7	½
Not over 250 volts, nominal	31.8	1¼	19.1	¾	12.7	½
Not over 600 volts, nominal	50.8	2	25.4	1	25.4	1

*For spacing between live parts and doors of cabinets, see 312.11(A)(1), (2), and (3).

Where close proximity does not cause excessive heating, parts of the same polarity at switches, enclosed fuses, and so forth shall be permitted to be placed as close together as convenience in handling will allow.

Exception: The distance shall be permitted to be less than that specified in Table 408.36 at circuit breakers and switches and in listed components installed in switchboards and panelboards.

ARTICLE 410
Luminaires (Lighting Fixtures), Lampholders, and Lamps

Contents

I. General

410.1 Scope.

This article covers luminaires (lighting fixtures), lampholders, pendants, incandescent filament lamps, arc lamps, electric-discharge lamps, the wiring and equipment forming part of such lamps, luminaires (fixtures), and lighting installations.

The identification and installation requirements for receptacles and receptacle covers that were located in Article 410, Part L in previous editions of the *Code* have been relocated into new Article 406, Receptacles, Cord Connectors, and Attachment Plugs (Caps). This new article consolidates receptacle requirements that were formerly located in Articles 210 and 410 into an article specifically on receptacles, cord connectors, and attachment caps. See the commentary following 406.1.

410.2 Application of Other Articles.

Equipment for use in hazardous (classified) locations shall conform to Articles 500 through 517. Lighting systems operating at 30 volts or less shall conform to Article 411. Arc lamps used in theaters shall comply with 520.61, and arc lamps used in projection machines shall comply with 540.20. Arc lamps used on constant-current systems shall comply with the general requirements of Article 490.

410.3 Live Parts.

Luminaires (fixtures), lampholders, and lamps shall have no live parts normally exposed to contact. Exposed accessible terminals in lampholders and switches shall not be installed in metal luminaire (fixture) canopies or in open bases of portable table or floor lamps.

Exception: Cleat-type lampholders located at least 2.5 m (8 ft) above the floor shall be permitted to have exposed terminals.

II. Luminaire (Fixture) Locations

410.4 Luminaires (Fixtures) in Specific Locations.

A pamphlet entitled *Luminaires Marking Guide*, available from Underwriters Laboratories Inc., was developed to help the authority having jurisdiction quickly determine whether common types of UL-listed fluorescent, high-intensity discharge, and incandescent fixtures are installed correctly.

(A) Wet and Damp Locations. Luminaires (fixtures) installed in wet or damp locations shall be installed so that water cannot enter or accumulate in wiring compartments, lampholders, or other electrical parts. All luminaires (fixtures) installed in wet locations shall be marked, "Suitable for Wet Locations." All luminaires (fixtures) installed in damp locations shall be marked, "Suitable for Wet Locations" or "Suitable for Damp Locations."

Where luminaires are exposed to the weather or subject to water saturation, they must be of a type marked "Suitable for Wet Locations." Construction, design, and installation of these luminaires prevents the entrance of rain, snow, ice, and dust. Outdoor parks and parking lots, outdoor recreational areas (tennis, golf, baseball, etc.), car wash areas, and building exteriors are examples of wet locations.

Locations protected from the weather and not subject to water saturation but still exposed to moisture, such as the following, may be considered damp locations:

1. The underside of store or gasoline station canopies or theater marquees
2. Some cold-storage warehouses
3. Some agricultural buildings
4. Some basements
5. Roofed open porches and carports

Luminaires used in these locations must be marked "Suitable for Damp Locations." See the definitions of *location, damp location, dry location,* and *wet location* in Article 100.

(B) Corrosive Locations. Luminaires (fixtures) installed in corrosive locations shall be of a type suitable for such locations.

(C) In Ducts or Hoods. Luminaires (fixtures) shall be permitted to be installed in commercial cooking hoods where all of the following conditions are met:

(1) The luminaire (fixture) shall be identified for use within commercial cooking hoods and installed so that the temperature limits of the materials used are not exceeded.
(2) The luminaire (fixture) shall be constructed so that all exhaust vapors, grease, oil, or cooking vapors are excluded from the lamp and wiring compartment. Diffusers shall be resistant to thermal shock.
(3) Parts of the luminaire (fixture) exposed within the hood shall be corrosion resistant or protected against corrosion, and the surface shall be smooth so as not to collect deposits and to facilitate cleaning.
(4) Wiring methods and materials supplying the luminaire(s) [fixture(s)] shall not be exposed within the cooking hood.

FPN: See 110.11 for conductors and equipment exposed to deteriorating agents.

The requirement in 410.4(C)(4) was initially taken from NFPA 96, *Standard for Ventilation Control and Fire Protection of Commercial Cooking Operations.* NFPA 96 provides the minimum fire safety requirements (preventive and operative) related to the design, installation, operation, inspection, and maintenance of all public and private cooking operations, except single-family residential dwellings. This coverage includes, but is not limited to, all cooking equipment, exhaust hoods, grease removal devices, exhaust ductwork, exhaust fans, dampers, fire-extinguishing equipment, and all other auxiliary or ancillary components or systems that are involved in the capture, containment, and control of grease-laden cooking effluent.

Also, NFPA 96 is intended to include residential cooking equipment where used for purposes other than residential family use, such as employee kitchens or break areas and church and meeting hall kitchens, regardless of frequency of use.

Grease may cause the deterioration of conductor insulation, resulting in short circuits or ground faults in wiring, hence the requirement prohibiting wiring methods and materials (raceways, cables, lampholders) within ducts or hoods. Conventional enclosed and gasketed-type luminaires located in the path of travel of exhaust products are not permitted because a fire could result from the high temperatures on grease-coated glass bowls or globes enclosing the lamps. Recessed or surface gasketed-type luminaires intended for location within hoods must be identified as suitable for the specific purpose and should be installed with the required clearances maintained. Note that wiring systems, including rigid metal conduit, are not permitted to be run exposed within the cooking hood.

For further information, refer to UL 710, *Safety Standard for Exhaust Hoods for Commercial Cooking Equipment.*

(D) Bathtub and Shower Areas. No parts of cord-connected luminaires (fixtures), hanging luminaires (fixtures), lighting track, pendants, or ceiling-suspended (paddle) fans shall be located within a zone measured 900 mm (3 ft) horizontally and 2.5 m (8 ft) vertically from the top of the bathtub rim or shower stall threshold. This zone is all encompassing and includes the zone directly over the tub or shower stall.

The intent of 410.4(D) is to keep cord-connected, hanging, or pendant luminaires and suspended fans out of the reach of an individual standing on a bathtub rim. This list of prohibited items recognizes that the same risk of electric shock is present for each one.

Exhibit 410.1 illustrates the restricted zone in which the specified luminaires, lighting track, and paddle fans are prohibited. This requirement applies to hydromassage bathtubs, as defined in 680.2, as well as other bathtub types and shower areas. See 680.43 for installation requirements for spas and hot tubs (as defined in 680.2) installed indoors.

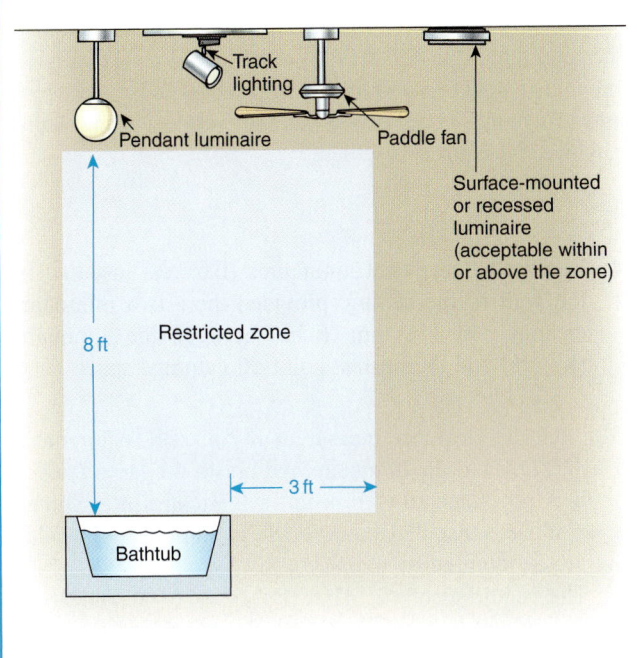

Exhibit 410.1 Luminaires, lighting track, and suspended (paddle) fan located near a bathtub.

410.5 Luminaires (Fixtures) Near Combustible Material.

Luminaires (fixtures) shall be constructed, installed, or equipped with shades or guards so that combustible material is not subjected to temperatures in excess of 90°C (194°F).

Nearly every fire requires an initial heat source, an initial fuel source, and an action that brings them together. The requirements of 410.5, 410.6, 410.7, and 410.8 regulate only the placement of the heat source. It is important to remember that successful fire prevention is most likely to come about if the initial heat source and initial fuel source are treated with due care. Tests have shown that hot particles from broken incandescent lamps can ignite combustibles below the lamps.

410.6 Luminaires (Fixtures) Over Combustible Material.

Lampholders installed over highly combustible material shall be of the unswitched type. Unless an individual switch

is provided for each luminaire (fixture), lampholders shall be located at least 2.5 m (8 ft) above the floor or shall be located or guarded so that the lamps cannot be readily removed or damaged.

Section 410.6 refers to pendants and fixed lighting equipment installed above highly combustible material. If the lamp cannot be located out of reach, the requirement can be met by equipping the lamp with a suitable guard. Section 410.6 does not apply to portable lamps.

410.7 Luminaires (Fixtures) in Show Windows.

Chain-supported luminaires (fixtures) used in a show window shall be permitted to be externally wired. No other externally wired luminaires (fixtures) shall be used.

410.8 Luminaires (Fixtures) in Clothes Closets.

(A) Definition.

Storage Space. The volume bounded by the sides and back closet walls and planes extending from the closet floor vertically to a height of 1.8 m (6 ft) or the highest clothes-hanging rod and parallel to the walls at a horizontal distance of 600 mm (24 in.) from the sides and back of the closet walls, respectively, and continuing vertically to the closet ceiling parallel to the walls at a horizontal distance of 300 mm (12 in.) or the width of the shelf, whichever is greater; for a closet that permits access to both sides of a hanging rod, this space includes the volume below the highest rod extending 300 mm (12 in.) on either side of the rod on a plane horizontal to the floor extending the entire length of the rod.

> FPN: See Figure 410.8.

The 24-in. rule is intended to cover the clothes-hanging space, even if no clothes-hanging rod is installed. If a clothes-hanging rod is installed, the space extends from the floor to the top of the highest rod. If no clothes-hanging rod is installed, the space extends from the floor to a height of 6 ft.

In addition to the space in which clothing will be hung from the closet pole or rod, this requirement also establishes a 12-in. wide shelf space to cover those installations where shelving is not in place at the time of fixture installation. If shelving is installed and the shelves are wider than 12 in., the greater width must be applied in establishing this space.

The storage space for closets that permit access to both sides of the clothes-hanging rod is based on a horizontal plane extending 12 in. from both sides of the rod, from the rod down to the floor. This equates to the 24-in. space

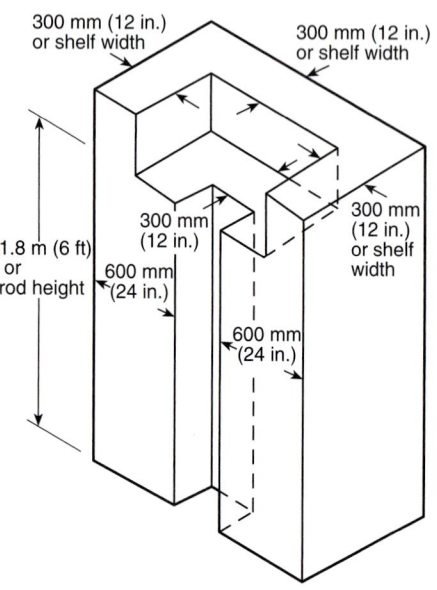

Figure 410.8 Closet storage space.

(B) Luminaire (Fixture) Types Permitted. Listed luminaires (fixtures) of the following types shall be permitted to be installed in a closet:

(1) A surface-mounted or recessed incandescent luminaire (fixture) with a completely enclosed lamp
(2) A surface-mounted or recessed fluorescent luminaire (fixture)

(C) Luminaire (Fixture) Types Not Permitted. Incandescent luminaires (fixtures) with open or partially enclosed lamps and pendant luminaires (fixtures) or lampholders shall not be permitted.

See the commentary following 410.8(D)(3).

(D) Location. Luminaires (fixtures) in clothes closets shall be permitted to be installed as follows:

(1) Surface-mounted incandescent luminaires (fixtures) installed on the wall above the door or on the ceiling, provided there is a minimum clearance of 300 mm (12 in.) between the luminaire (fixture) and the nearest point of a storage space
(2) Surface-mounted fluorescent luminaires (fixtures) installed on the wall above the door or on the ceiling,

provided there is a minimum clearance of 150 mm (6 in.) between the luminaire (fixture) and the nearest point of a storage space
(3) Recessed incandescent luminaires (fixtures) with a completely enclosed lamp installed in the wall or the ceiling, provided there is a minimum clearance of 150 mm (6 in.) between the luminaire (fixture) and the nearest point of a storage space

The requirement in 410.8(D)(3) results from tests that have shown that a hot filament falling from a broken incandescent lamp can ignite combustible material below the luminaire in which the lamp is installed.

(4) Recessed fluorescent luminaires (fixtures) installed in the wall or the ceiling, provided there is a minimum clearance of 150 mm (6 in.) between the luminaire (fixture) and the nearest point of a storage space

Note that the clearance measurement for each requirement in 410.8(D) is to the luminaire and not to the lamp itself.

It is not mandatory to install a luminaire in a clothes closet; if one is installed, however, the conditions for installation are as required by 410.8(D).

The requirements of 410.8(D) apply to incandescent and fluorescent lighting in clothes closets of various kinds of occupancies. The requirement is intended to prevent hot lamps or parts of broken lamps from coming in contact with boxes, cartons, blankets, and the like, stored on shelves, and clothing hung in closets.

410.9 Space for Cove Lighting.

Coves shall have adequate space and shall be located so that lamps and equipment can be properly installed and maintained.

Adequate space is necessary for easy access for relamping luminaires or replacing lampholders, ballasts, and so on; adequate space also improves ventilation.

III. Provisions at Luminaire (Fixture) Outlet Boxes, Canopies, and Pans

410.10 Space for Conductors.

Canopies and outlet boxes taken together shall provide adequate space so that luminaire (fixture) conductors and their connecting devices can be properly installed.

410.11 Temperature Limit of Conductors in Outlet Boxes.

Luminaires (fixtures) shall be of such construction or installed so that the conductors in outlet boxes shall not be subjected to temperatures greater than that for which the conductors are rated.

Branch-circuit wiring, other than 2-wire or multiwire branch circuits supplying power to luminaires (fixtures) connected together, shall not be passed through an outlet box that is an integral part of a luminaire (fixture) unless the luminaire (fixture) is identified for through-wiring.

> FPN: See 410.32 for wiring supplying power to fixtures connected together.

Branch-circuit conductors run to a lighting outlet box are not permitted to be subjected to higher temperatures than those temperatures for which they are rated. For example, conductors are rated 75°C and are to supply a ceiling outlet box for the connection of a surface-mounted luminaire or attached outlet box of a recessed luminaire. The design and installation of the luminaire should be such that the heat of the lamps does not subject the conductors to a greater temperature than 75°C. These types of luminaire are listed by Underwriters Laboratories Inc. based on a heat-contributing factor of the supply conductors of not more than the maximum permitted lamp wattage of the luminaire.

Exhibit 410.2 illustrates recessed luminaires listed by Underwriters Laboratories Inc. for one set of supply conductors, and Exhibit 410.3 illustrates luminaires listed for a feed-through installation.

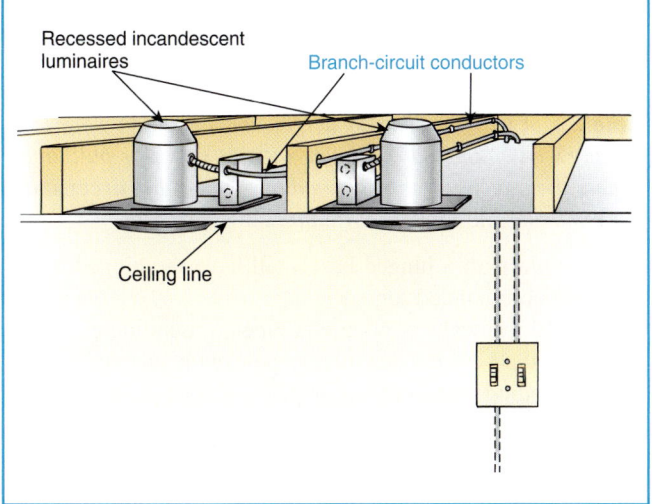

Exhibit 410.2 Recessed luminaires designed for branch-circuit conductors terminating at each luminaire (no feed-through).

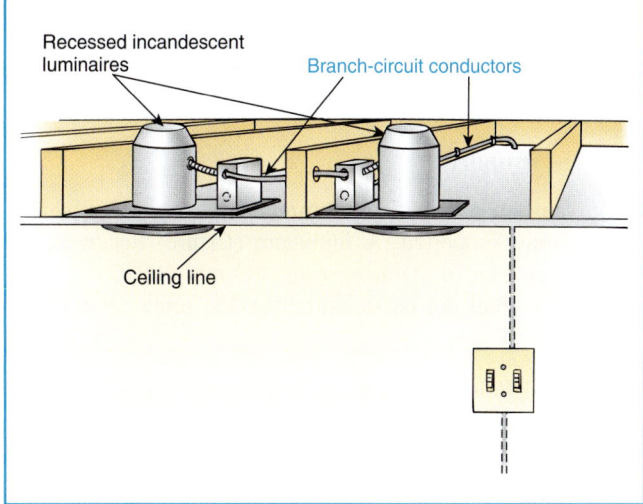

Exhibit 410.3 Recessed luminaires designed for feed-through branch-circuit conductors.

410.12 Outlet Boxes to Be Covered.

In a completed installation, each outlet box shall be provided with a cover unless covered by means of a luminaire (fixture) canopy, lampholder, receptacle, or similar device.

410.13 Covering of Combustible Material at Outlet Boxes.

Any combustible wall or ceiling finish exposed between the edge of a luminaire (fixture) canopy or pan and an outlet box shall be covered with noncombustible material.

Luminaires must be designed and installed not only to prevent overheating of conductors but also to prevent overheating of adjacent combustible wall or ceiling finishes. Hence, it is required that any combustible finish between the edge of a luminaire canopy and an outlet box be covered with a noncombustible material or luminaire accessory. See 314.20 for the requirements covering combustible finishes.

Where luminaires are not directly mounted on outlet boxes, suitable outlet box covers are required.

410.14 Connection of Electric-Discharge Luminaires (Lighting Fixtures).

(A) Independent of the Outlet Box. Electric-discharge luminaires (lighting fixtures) supported independently of the outlet box shall be connected to the branch circuit through metal raceway, nonmetallic raceway, Type MC cable, Type AC cable, Type MI cable, nonmetallic sheathed cable, or by flexible cord as permitted in 410.30(B) or (C).

(B) Access to Boxes. Electric-discharge luminaires (lighting fixtures) surface mounted over concealed outlet, pull, or junction boxes shall be installed with suitable openings in back of the fixture to provide access to the boxes.

IV. Luminaire (Fixture) Supports

410.15 Supports.

(A) General. Luminaires (fixtures) and lampholders shall be securely supported. A luminaire (fixture) that weighs more than 3 kg (6 lb) or exceeds 400 mm (16 in.) in any dimension shall not be supported by the screw shell of a lampholder.

(B) Metal Poles Supporting Luminaires (Lighting Fixtures). Metal poles shall be permitted to be used to support luminaires (lighting fixtures) and as a raceway to enclose supply conductors, provided the following conditions are met:

(1) A metal pole shall have a handhole not less than 50 mm × 100 mm (2 in. × 4 in.) with a raintight cover to provide access to the supply terminations within the pole or pole base.

Exception No. 1: No handhole shall be required in a pole 2.5 m (8 ft) or less in height above grade where the supply wiring method continues without splice or pull point, and where the interior of the pole and any splices are accessible by removing the luminaire (fixture).

Exception No. 1 to 410.15(B)(1) typically applies to both landscape (bollard-type) lighting and pole lights at residential dwellings.

Exception No. 2: No handhole shall be required in a metal pole 6.0 m (20 ft) or less in height above grade that is provided with a hinged base.

Exception No. 2 to 410.15(B)(1) recognizes metal light poles that do not have a handhole, but instead use a hinged-base pole to permit access to splices made in the pole base. The height of the pole is limited to 20 ft. Both parts of the pole must be bonded in accordance with 250.96 (systems operating at 250 volts or less) or 250.97 (circuits operating at over 250 volts), depending on the voltage of the system.

Exhibit 410.4 illustrates a metal light pole with a hinged baseplate that meets the requirements of Exception No. 2 to 410.15(B)(1). A handhole is not necessary because the pole can be tilted to allow access to terminations in the base.

(2) Where raceway risers or cable is not installed within the pole, a threaded fitting or nipple shall be brazed or welded to the pole opposite the handhole for the supply connection.

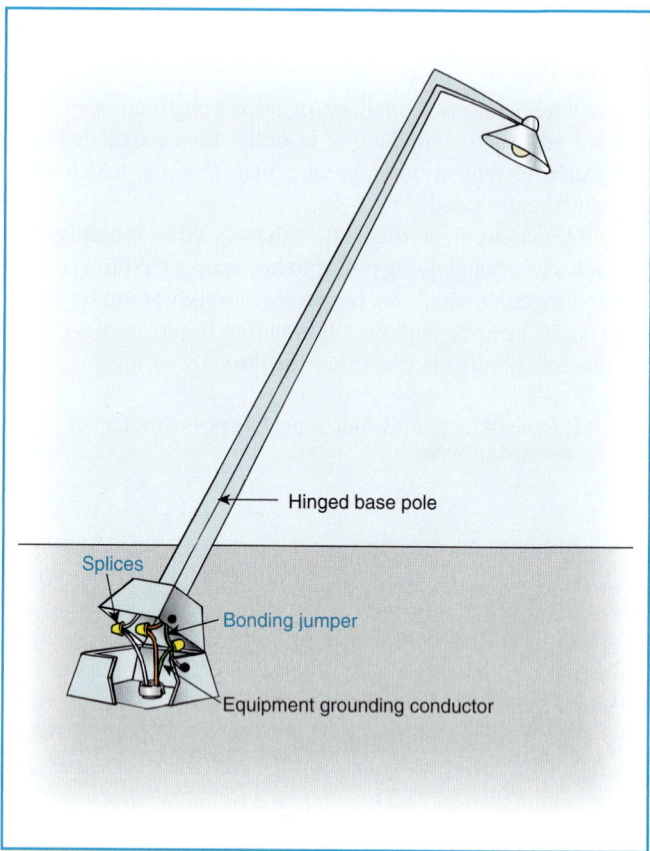

Exhibit 410.4 A hinged-base metal pole supporting a luminaire.

(3) A metal pole shall be provided with a grounding terminal.

   a. A pole with a handhole shall have the grounding terminal accessible from the handhole.

   b. A pole with a hinged base shall have the grounding terminal accessible within the base.

Exception: No grounding terminal shall be required in a pole 2.5 m (8 ft) or less in height above grade where the supply wiring method continues without splice or pull, and where the interior of the pole and any splices are accessible by removing the luminaire (fixture).

(4) A pole with a hinged base shall have the hinged base and pole bonded together.

(5) Metal raceways or other equipment grounding conductors shall be bonded to the pole with an equipment grounding conductor recognized by 250.118 and sized in accordance with 250.122.

(6) Conductors in vertical metal poles used as raceway shall be supported as provided in 300.19.

In 410.15(B) metal poles are permitted to be used as a raceway. If the pole is used as a raceway, the user is reminded to maintain the required separation between power wiring and any communications, signaling, and power-limited circuits that may also be installed within the pole.

For example, where a light pole supports a luminaire and a security camera, the security camera signaling and power-limited wiring may be installed within the pole cavity. Because the pole contains open circuit conductors (e.g., THW or XHHW conductors) to supply the luminaire already, the separation requirements can be fulfilled by enclosing the camera conductors within a flexible raceway and installing that raceway within the pole. Of course, the use of a cable or cord assembly for the lighting circuit conductors remains an optional choice for the user. The provisions of 410.15(B)(6) remind the user that the conductors are to be supported in accordance with 300.19, the same as a vertical raceway.

Section 410.15(B) is not intended to necessarily require the placement of a raceway for communications cables on the exterior of a lighting pole.

410.16 Means of Support.

(A) Outlet Boxes. Outlet boxes or fittings installed as required by 314.23 shall be permitted to support luminaires (fixtures).

Regardless of whether a luminaire is attached to an outlet box or is supported independently of the outlet box, care should be taken to securely fasten the outlet box or support the independent rod or hanger in order to ensure that the luminaire is securely mounted in place. The luminaire may be securely mounted to the box; however, if the box is not secured, it becomes the weak link in the luminaire support.

(B) Inspection. Luminaires (fixtures) shall be installed so that the connections between the luminaire (fixture) conductors and the circuit conductors can be inspected without requiring the disconnection of any part of the wiring unless the luminaires (fixtures) are connected by attachment plugs and receptacles.

(C) Suspended Ceilings. Framing members of suspended ceiling systems used to support luminaires (fixtures) shall be securely fastened to each other and shall be securely attached to the building structure at appropriate intervals. Luminaires (fixtures) shall be securely fastened to the ceiling framing member by mechanical means such as bolts, screws, or rivets. Listed clips identified for use with the type of ceiling framing member(s) and luminaire(s) [fixture(s)] shall also be permitted.

Section 410.16(C) provides requirements for luminaire installations where the luminaires are supported by a sus-

pended or "hung" ceiling. Where luminaires are supported independent of a suspended ceiling, 410.16(C) does not apply.

Section 410.16(C) requires that if clips are used to support luminaires to the framing members of a suspended ceiling, those clips must be of a type listed for the application. However, the use of listed clips for luminaire support does not complete the requirements of this section. Additionally, the ceiling framing members must be securely attached to each other and to the building structure. For the support of wiring that is located within the cavity of floor–ceiling assemblies, see 300.11(A).

(D) Luminaire (Fixture) Studs. Luminaire (fixture) studs that are not a part of outlet boxes, hickeys, tripods, and crowfeet shall be made of steel, malleable iron, or other material suitable for the application.

(E) Insulating Joints. Insulating joints that are not designed to be mounted with screws or bolts shall have an exterior metal casing, insulated from both screw connections.

(F) Raceway Fittings. Raceway fittings used to support a luminaire(s) [lighting fixture(s)] shall be capable of supporting the weight of the complete fixture assembly and lamp(s).

(G) Busways. Luminaires (fixtures) shall be permitted to be connected to busways in accordance with 368.12.

(H) Trees. Outdoor luminaires (lighting fixtures) and associated equipment shall be permitted to be supported by trees.

> FPN No. 1: See 225.26 for restrictions for support of overhead conductors.

The support of overhead conductor spans on trees is prohibited by 225.26.

> FPN No. 2: See 300.5(D) for protection of conductors.

Section 300.5(D) requires buried conductors and cables to be protected from physical damage by the use of raceways from a specified point below grade to a point at least 8 ft above finish grade.

V. Grounding

410.17 General.

Luminaires (fixtures) and lighting equipment shall be grounded as required in Article 250 and Part V of this article.

410.18 Exposed Luminaire (Fixture) Parts.

(A) Exposed Conductive Parts. Exposed metal parts shall be grounded or insulated from ground and other conducting surfaces or be inaccessible to unqualified personnel. Lamp tie wires, mounting screws, clips, and decorative bands on glass spaced at least 38 mm (1½ in.) from lamp terminals shall not be required to be grounded.

(B) Made of Insulating Material. Luminaires (fixtures) directly wired or attached to outlets supplied by a wiring method that does not provide a ready means for grounding shall be made of insulating material and shall have no exposed conductive parts.

Exception: Replacement luminaires (fixtures) shall be permitted to connect an equipment grounding conductor from the outlet in compliance with 250.130(C). The luminaire (fixture) shall then be grounded in accordance wtih 410.18(A).

The exception to 410.18(B) provides a method by which a luminaire with exposed conductive parts can be installed at an outlet where the wiring method is not an equipment grounding conductor per 250.118 or does not provide an equipment grounding conductor. In older installations where luminaires are replaced, the requirement to ground exposed metal parts of the luminaire is not negated simply because there is no means of grounding provided by the existing wiring system. The means allowed by the exception is the same as is permitted for receptacles installed at outlets where no grounding means exists. A single grounding conductor can be run independently of the circuit conductors, from the outlet to a point on the wiring system where an effective grounding connection can be made. The acceptable termination points for this separate grounding conductor are specified by 250.130(C).

410.20 Equipment Grounding Conductor Attachment.

Luminaires (fixtures) with exposed metal parts shall be provided with a means for connecting an equipment grounding conductor for such luminaires (fixtures).

410.21 Methods of Grounding.

Luminaires (fixtures) and equipment shall be considered grounded where mechanically connected to an equipment grounding conductor as specified in 250.118 and sized in accordance with 250.122.

VI. Wiring of Luminaires (Fixtures)

410.22 Luminaire (Fixture) Wiring—General.

Wiring on or within fixtures shall be neatly arranged and shall not be exposed to physical damage. Excess wiring shall be avoided. Conductors shall be arranged so that they are not subjected to temperatures above those for which they are rated.

410.23 Polarization of Luminaires (Fixtures).

Luminaires (fixtures) shall be wired so that the screw shells of lampholders are connected to the same luminaire (fixture) or circuit conductor or terminal. The grounded conductor, where connected to a screw-shell lampholder, shall be connected to the screw shell.

410.24 Conductor Insulation.

Luminaires (fixtures) shall be wired with conductors having insulation suitable for the environmental conditions, current, voltage, and temperature to which the conductors will be subjected.

> FPN: For ampacity of luminaire (fixture) wire, maximum operating temperature, voltage limitations, minimum wire size, and so forth, see Article 402.

410.27 Pendant Conductors for Incandescent Filament Lamps.

(A) Support. Pendant lampholders with permanently attached leads, where used for other than festoon wiring, shall be hung from separate stranded rubber-covered conductors that are soldered directly to the circuit conductors but supported independently thereof.

(B) Size. Unless part of listed decorative lighting assemblies, pendant conductors shall not be smaller than 14 AWG for mogul-base or medium-base screw-shell lampholders or smaller than 18 AWG for intermediate or candelabra-base lampholders.

(C) Twisted or Cabled. Pendant conductors longer than 900 mm (3 ft) shall be twisted together where not cabled in a listed assembly.

410.28 Protection of Conductors and Insulation.

(A) Properly Secured. Conductors shall be secured in a manner that does not tend to cut or abrade the insulation.

(B) Protection Through Metal. Conductor insulation shall be protected from abrasion where it passes through metal.

(C) Luminaire (Fixture) Stems. Splices and taps shall not be located within luminaire (fixture) arms or stems.

(D) Splices and Taps. No unnecessary splices or taps shall be made within or on a luminaire (fixture).

> FPN: For approved means of making connections, see 110.14.

(E) Stranding. Stranded conductors shall be used for wiring on luminaire (fixture) chains and on other movable or flexible parts.

(F) Tension. Conductors shall be arranged so that the weight of the luminaire (fixture) or movable parts does not put tension on the conductors.

410.29 Cord-Connected Showcases.

Individual showcases, other than fixed, shall be permitted to be connected by flexible cord to permanently installed receptacles, and groups of not more than six such showcases shall be permitted to be coupled together by flexible cord and separable locking-type connectors with one of the group connected by flexible cord to a permanently installed receptacle.

The installation shall comply with 410.29(A) through (E).

(A) Cord Requirements. Flexible cord shall be of the hard-service type, having conductors not smaller than the branch-circuit conductors, having ampacity at least equal to the branch-circuit overcurrent device, and having an equipment grounding conductor.

> FPN: See Table 250.122 for size of equipment grounding conductor.

(B) Receptacles, Connectors, and Attachment Plugs. Receptacles, connectors, and attachment plugs shall be of a listed grounding type rated 15 or 20 amperes.

(C) Support. Flexible cords shall be secured to the undersides of showcases so that

(1) Wiring is not exposed to mechanical damage.
(2) A separation between cases not in excess of 50 mm (2 in.), or more than 300 mm (12 in.) between the first case and the supply receptacle, is ensured.
(3) The free lead at the end of a group of showcases has a female fitting not extending beyond the case.

(D) No Other Equipment. Equipment other than showcases shall not be electrically connected to showcases.

(E) Secondary Circuit(s). Where showcases are cord-connected, the secondary circuit(s) of each electric-discharge lighting ballast shall be limited to one showcase.

410.30 Cord-Connected Lampholders and Luminaires (Fixtures).

(A) Lampholders. Where a metal lampholder is attached to a flexible cord, the inlet shall be equipped with an insulating bushing that, if threaded, is not smaller than metric designator 12 (trade size ⅜) pipe size. The cord hole shall be of a size appropriate for the cord, and all burrs and fins

shall be removed in order to provide a smooth bearing surface for the cord.

Bushing having holes 7 mm (⁹⁄₃₂ in.) in diameter shall be permitted for use with plain pendant cord and holes 11 mm (¹³⁄₃₂ in.) in diameter with reinforced cord.

> Metal lampholders (brass- and aluminum-shell type) used with flexible-cord pendants should be equipped with smooth and permanently secured insulating bushings. Nonmetallic-type lampholders do not require a bushing because the material and design afford equivalent protection.

(B) Adjustable Luminaires (Fixtures). Luminaires (fixtures) that require adjusting or aiming after installation shall not be required to be equipped with an attachment plug or cord connector, provided the exposed cord is of the hard-usage or extra-hard-usage type and is not longer than that required for maximum adjustment. The cord shall not be subject to strain or physical damage.

(C) Electric-Discharge Luminaires (Fixtures).

(1) A listed luminaire (fixture) or a listed assembly shall be permitted to be cord connected if the following conditions apply:

(1) The luminaire (fixture) is located directly below the outlet box or busway.
(2) The flexible cord meets all the following:

    a. Is visible for its entire length outside the luminaire (fixture)
    b. Is not subject to strain or physical damage
    c. Is terminated in a grounding-type attachment plug cap or busway plug or has a luminaire (fixture) assembly with a strain relief and canopy

> Section 410.30(C)(1) applies to listed cord-and-plug-connected electric-discharge luminaires, such as the luminaire illustrated in Exhibit 410.5, and to electric-discharge luminaire assemblies. The supply cord is not permitted to penetrate a suspended ceiling, because the cord is required to be continuously visible along its entire length.
>
> Supply cords cannot be used as a supporting means, and the luminaires are suspended directly below the outlet boxes supplying each luminaire. Electric-discharge luminaires are not permitted to be supplied by cord if they are installed in lift-out-type suspended ceilings. If the electric-discharge luminaire is suspended below the lift-out-type ceiling, the cord is not permitted to penetrate the ceiling. Section 400.8 further explains the uses not permitted for cords. According to 368.12, Exception No. 3, electric-discharge luminaires are permitted to be connected to busways by cords

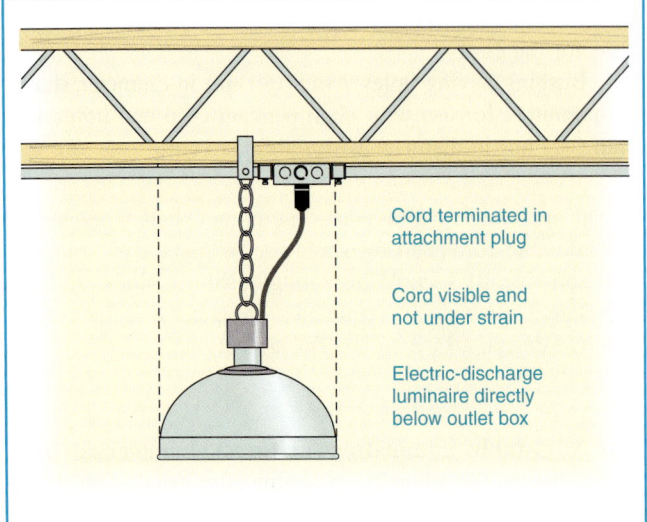

Cord terminated in attachment plug

Cord visible and not under strain

Electric-discharge luminaire directly below outlet box

Exhibit 410.5 A listed cord-and-plug-connected electric-discharge luminaire.

plugged directly into the busway or suspended from the busway.

(2) Electric-discharge luminaires (lighting fixtures) provided with mogul-base, screw-shell lampholders shall be permitted to be connected to branch circuits of 50 amperes or less by cords complying with 240.5. Receptacles and attachment plugs shall be permitted to be of lower ampere rating than the branch circuit but not less than 125 percent of the luminaire (fixture) full-load current.

(3) Electric-discharge luminaires (lighting fixtures) equipped with a flanged surface inlet shall be permitted to be supplied by cord pendants equipped with cord connectors. Inlets and connectors shall be permitted to be of lower ampere rating than the branch circuit but not less than 125 percent of the luminaire (fixture) load current.

410.31 Luminaires (Fixtures) as Raceways.

Luminaires (fixtures) shall not be used as a raceway for circuit conductors unless listed and marked for use as a raceway.

According to the UL *Luminaire Marking Guide*, luminaires listed for use as raceways are marked "Suitable for Use as a Raceway" and also "Maximum of ___ °C permitted in raceway." Without this marking, a row of luminaires connected end to end could not be used as a raceway for circuit conductors other than the two-wire or multiwire circuit supplying those luminaires. Luminaires identified for use as a

raceway have been evaluated for the heat contribution caused by additional current-carrying conductors.

410.32 Wiring Supplying Luminaires (Fixtures) Connected Together.

Luminaires (fixtures) designed for end-to-end connection to form a continuous assembly, or luminaires (fixtures) connected together by recognized wiring methods, shall be permitted to contain the conductors of a 2-wire branch circuit, or one multiwire branch circuit, supplying the connected luminaires (fixtures) and need not be listed as a raceway. One additional 2-wire branch circuit separately supplying one or more of the connected luminaires (fixtures) shall also be permitted.

The provisions of 410.32 facilitate convenient switching and supply circuit arrangements for a physically continuous row of luminaires or a row that is made continuous via the wiring method. A single two-wire or a single multiwire branch circuit supplying the luminaires is permitted to be run through the continuous row(s), and the luminaires are not required to be listed for use as a raceway. An additional two-wire branch circuit is permitted to be run through these luminaires. This circuit may supply only luminaires in the connected row(s) and is commonly employed to switch night lighting as an energy conservation method.

FPN: See Article 100 for the definition of *multiwire branch circuit.*

410.33 Branch Circuit Conductors and Ballasts.

Branch-circuit conductors within 75 mm (3 in.) of a ballast shall have an insulation temperature rating not lower than 90°C (194°F) unless supplying a luminaire (fixture) listed and marked as suitable for a different insulation temperature.

Temperature ratings, along with other insulated conductor specifications, are found in Table 310.13. Note the 90°C rating that may be applied to Type THW conductors for special application in electric-discharge lighting equipment.

VII. Construction of Luminaires (Fixtures)

410.34 Combustible Shades and Enclosures.

Adequate airspace shall be provided between lamps and shades or other enclosures of combustible material.

410.35 Luminaire (Fixture) Rating.

(A) Marking. All luminaires (fixtures) shall be marked with the maximum lamp wattage or electrical rating, manufacturer's name, trademark, or other suitable means of identification. A luminaire (fixture) requiring supply wire rated higher than 60°C (140°F) shall be marked in letters not smaller than 6 mm (¼ in.) high, prominently displayed on the luminaire (fixture) and shipping carton or equivalent.

(B) Electrical Rating. The electrical rating shall include the voltage and frequency and shall indicate the current rating of the unit, including the ballast, transformer, or autotransformer.

410.36 Design and Material.

Luminaires (fixtures) shall be constructed of metal, wood, or other material suitable for the application and shall be designed and assembled so as to secure requisite mechanical strength and rigidity. Wiring compartments, including their entrances, shall be such that conductors may be drawn in and withdrawn without physical damage.

410.37 Nonmetallic Luminaires (Fixtures).

When luminaire (fixture) wiring compartments are constructed from combustible material, armored or lead-covered conductors with suitable fittings shall be used or the wiring compartment shall be lined with metal.

410.38 Mechanical Strength.

(A) Tubing for Arms. Tubing used for arms and stems where provided with cut threads shall not be less than 1.02 mm (0.040 in.) in thickness and where provided with rolled (pressed) threads shall not be less than 0.64 mm (0.025 in.) in thickness. Arms and other parts shall be fastened to prevent turning.

(B) Metal Canopies. Metal canopies supporting lampholders, shades, and so forth exceeding 4 kg (8 lb), or incorporating attachment-plug receptacles, shall not be less than 0.51 mm (0.020 in.) in thickness. Other canopies shall not be less than 0.41 mm (0.016 in.) if made of steel and not less than 0.51 mm (0.020 in.) if of other metals.

(C) Canopy Switches. Pull-type canopy switches shall not be inserted in the rims of metal canopies that are less than 0.64 mm (0.025 in.) in thickness unless the rims are reinforced by the turning of a bead or the equivalent. Pull-type canopy switches, whether mounted in the rims or elsewhere in sheet metal canopies, shall not be located more than 90 mm (3½ in.) from the center of the canopy. Double set-screws, double canopy rings, a screw ring, or equal method shall be used where the canopy supports a pull-type switch or pendant receptacle.

The thickness requirements in the preceding paragraph shall apply to measurements made on finished (formed) canopies.

410.39 Wiring Space.

Bodies of luminaires (fixtures), including portable lamps, shall provide ample space for splices and taps and for the installation of devices, if any. Splice compartments shall be of nonabsorbent, noncombustible material.

410.42 Portable Lamps.

(A) General. Portable lamps shall be wired with flexible cord recognized by 400.4 and an attachment plug of the polarized or grounding type. Where used with Edison-base lampholders, the grounded conductor shall be identified and attached to the screw shell and the identified blade of the attachment plug.

(B) Portable Handlamps. In addition to the provisions of 410.42(A), portable handlamps shall comply with the following.

(1) Metal shell, paper-lined lampholders shall not be used.
(2) Handlamps shall be equipped with a handle of molded composition or other insulating material.
(3) Handlamps shall be equipped with a substantial guard attached to the lampholder or handle.
(4) Metallic guards shall be grounded by means of an equipment grounding conductor run with circuit conductors within the power-supply cord.
(5) Portable handlamps shall not be required to be grounded where supplied through an isolating transformer with an ungrounded secondary of not over 50 volts.

See Exhibit 410.6 for an example of a portable handlamp (drop light) meeting the requirements of 410.42.

410.44 Cord Bushings.

A bushing or the equivalent shall be provided where flexible cord enters the base or stem of a portable lamp. The bushing shall be of insulating material unless a jacketed type of cord is used.

410.45 Tests.

All wiring shall be free from short circuits and grounds and shall be tested for these defects prior to being connected to the circuit.

410.46 Live Parts.

Exposed live parts within porcelain luminaires (fixtures) shall be suitably recessed and located so as to make it improbable that wires come in contact with them. There shall be a spacing of at least 13 mm (½ in.) between live parts and the mounting plane of the luminaire (fixture).

Exhibit 410.6 A portable handlamp with a grounded metal guard and reflector and a swivel-type hook that permits the lamp to be adjusted in different positions. (Courtesy of Daniel Woodhead Co.)

VIII. Installation of Lampholders

410.47 Screw-Shell Type.

Lampholders of the screw-shell type shall be installed for use as lampholders only. Where supplied by a circuit having a grounded conductor, the grounded conductor shall be connected to the screw shell.

Many years ago it was common practice to install screw shell lampholders with screw shell adapters in baseboards and walls to connect cord-connected appliances and lighting equipment. This now-prohibited practice permitted exposed live parts to be contacted by persons when the adapters were removed. See 406.2(B) for permitted uses of receptacles.

410.48 Double-Pole Switched Lampholders.

Where supplied by the ungrounded conductors of a circuit, the switching device of lampholders of the switched type shall simultaneously disconnect both conductors of the circuit.

Single-pole switching may be used to interrupt the ungrounded conductor of a 2-wire circuit having one conductor grounded. The grounded conductor must be connected to the screw shell of the lampholder.

Where a 2-wire circuit is derived from the two ungrounded conductors of a multiwire circuit (3- or 4-wire system) or from the two ungrounded conductors of a 2-wire circuit (3-wire system) and is used with switched lampholders, the switching device is required to be double-pole and to simultaneously disconnect both ungrounded conductors of the circuit. See 410.52 for information on the construction of switched lampholders.

410.49 Lampholders in Wet or Damp Locations.

Lampholders installed in wet or damp locations shall be of the weatherproof type.

IX. Construction of Lampholders

410.50 Insulation.

The outer metal shell and the cap shall be lined with insulating material that prevents the shell and cap from becoming a part of the circuit. The lining shall not extend beyond the metal shell more than 3 mm (⅛ in.) but shall prevent any current-carrying part of the lamp base from being exposed when a lamp is in the lampholding device.

410.52 Switched Lampholders.

Switched lampholders shall be of such construction that the switching mechanism interrupts the electrical connection to the center contact. The switching mechanism shall also be permitted to interrupt the electrical connection to the screw shell if the connection to the center contact is simultaneously interrupted.

X. Lamps and Auxiliary Equipment

410.53 Bases, Incandescent Lamps.

An incandescent lamp for general use on lighting branch circuits shall not be equipped with a medium base if rated over 300 watts, or with a mogul base if rated over 1500 watts. Special bases or other devices shall be used for over 1500 watts.

410.54 Electric-Discharge Lamp Auxiliary Equipment.

(A) Enclosures. Auxiliary equipment for electric-discharge lamps shall be enclosed in noncombustible cases and treated as sources of heat.

The UL *General Information Electrical Equipment Directory* contains two categories for ballasts under Electric Discharge Lamp Control Equipment (FKOT): fluorescent ballasts (FKVS) and HID (high intensity discharge) ballasts (FLCR).

Fluorescent ballast enclosures are categorized by UL as open, indoor, outdoor (both Type 1 and Type 2), and weatherproof. HID ballasts are categorized the same, except there is no open-type ballast for HID ballasts.

Open Type. Open core and coil constructions (i.e., ballasts without complete metal enclosures) are intended for use within suitable enclosures.

Indoor Ballasts. Indoor ballasts are suitable for use in an indoor, dry location only.

Outdoor Ballasts. Type 1 outdoor ballasts are suitable for use in (1) outdoor equipment, (2) fixtures intended for wet or damp locations, or (3) an outdoor sign if the ballasts are within an overall electrical enclosure. Ballasts of this type are marked "Type 1 Outdoor" or "Type 1." Type 2 outdoor ballasts are suitable for use in (1) outdoor equipment, (2) fixtures intended for wet or damp locations, or (3) an outdoor sign if the ballasts, in addition to their own enclosure, are within an overall enclosure. Ballasts of this type are marked "Type 2 Outdoor" or "Type 2."

Weatherproof Ballasts. Weatherproof ballasts are suitable for use where completely exposed to the weather without an additional enclosure and are marked "Weatherproof" or "WP."

(B) Switching. Where supplied by the ungrounded conductors of a circuit, the switching device of auxiliary equipment shall simultaneously disconnect all conductors.

XI. Special Provisions for Flush and Recessed Luminaires (Fixtures)

Formal Interpretation 81-6

Reference: Article 410, Part XI

Question: Is it intended that fixtures installed in suspended ceilings be subject to the requirements of Part XI of Article 410?

Answer: Yes.

Issue Edition: 1981

Reference: 410, Part XI

Issue Date: October 1982

410.64 General.

Luminaires (fixtures) installed in recessed cavities in walls or ceilings shall comply with 410.65 through 410.72.

410.65 Temperature.

(A) Combustible Material. Luminaires (fixtures) shall be installed so that adjacent combustible material will not be subjected to temperatures in excess of 90°C (194°F).

(B) Fire-Resistant Construction. Where a luminaire (fixture) is recessed in fire-resistant material in a building of fire-resistant construction, a temperature higher than 90°C (194°F) but not higher than 150°C (302°F) shall be considered acceptable if the luminaire (fixture) is plainly marked that it is listed for that service.

(C) Recessed Incandescent Luminaires (Fixtures). Incandescent luminaires (fixtures) shall have thermal protection and shall be identified as thermally protected.

Because many recessed incandescent luminaires are suitable for a wide variety of lamp sizes and types and finish trims, the temperature close to the lamp can vary widely. Therefore, many manufacturers have chosen to locate thermal protectors away from the source of heat, such as in the outlet box, and to design the protector so that it will detect a change in temperature resulting from the addition of thermal insulation around the luminaire. This design prevents nuisance tripping of the protector (as a result of changing lamp wattage, for example) but still provides protection against overheating arising from thermal insulation around a recessed luminaire not designed for such use.

Exception No. 1: Thermal protection shall not be required in a recessed luminaire (fixture) identified for use and installed in poured concrete.

Exception No. 2: Thermal protection shall not be required in a recessed luminaire (fixture) whose design, construction, and thermal performance characteristics are equivalent to a thermally protected luminaire (fixture) and are identified as inherently protected.

Recessed incandescent luminaires without thermal protection are not permitted unless they are listed and identified as providing equivalent temperature protection by construction design.

Exhibit 410.7 illustrates a listed Type IC recessed luminaire installed in direct contact with thermal insulation. Thermal protection is provided to deactivate the lamp should the luminaire be mislamped so that it overheats.

410.66 Clearance and Installation.

(A) Clearance.

(1) Non-Type IC. A recessed luminaire (fixture) that is not identified for contact with insulation shall have all recessed

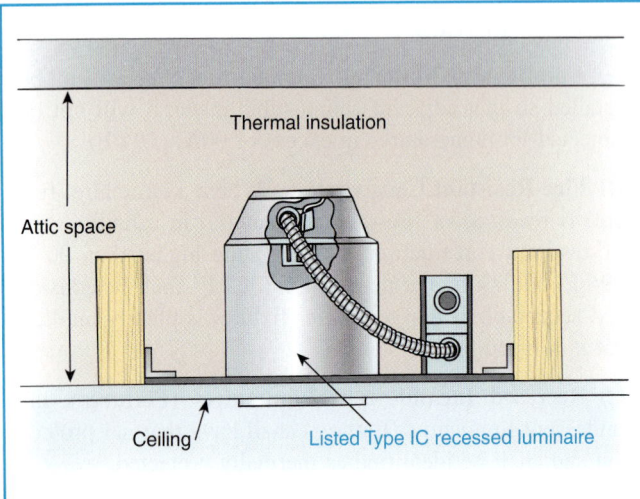

Exhibit 410.7 A listed Type IC recessed luminaire suitable for use in insulated ceilings installed in direct contact with thermal insulation. (Redrawn from Thomas Industries Inc.)

parts spaced not less than 13 mm (½ in.) from combustible materials. The points of support and the trim finishing off the opening in the ceiling or wall surface shall be permitted to be in contact with combustible materials.

(2) Type IC. A recessed luminaire (fixture) that is identified for contact with insulation, Type IC, shall be permitted to be in contact with combustible materials at recessed parts, points of support, and portions passing through or finishing off the opening in the building structure.

(B) Installation. Thermal insulation shall not be installed above a recessed luminaire (fixture) or within 75 mm (3 in.) of the recessed luminaire's (fixture's) enclosure, wiring compartment, or ballast unless it is identified for contact with insulation, Type IC.

410.67 Wiring.

(A) General. Conductors that have insulation suitable for the temperature encountered shall be used.

(B) Circuit Conductors. Branch-circuit conductors that have an insulation suitable for the temperature encountered shall be permitted to terminate in the luminaire (fixture).

(C) Tap Conductors. Tap conductors of a type suitable for the temperature encountered shall be permitted to run from the luminaire (fixture) terminal connection to an outlet box placed at least 300 mm (1 ft) from the luminaire (fixture). Such tap conductors shall be in suitable raceway or Type AC or MC cable of at least 450 mm (18 in.) but not more than 1.8 m (6 ft) in length.

XII. Construction of Flush and Recessed Luminaires (Fixtures)

410.68 Temperature.

Luminaires (fixtures) shall be constructed so that adjacent combustible material is not subject to temperatures in excess of 90°C (194°F).

410.70 Lamp Wattage Marking.

Incandescent lamp luminaires (fixtures) shall be marked to indicate the maximum allowable wattage of lamps. The markings shall be permanently installed, in letters at least 6 mm (¼ in.) high, and shall be located where visible during relamping.

410.71 Solder Prohibited.

No solder shall be used in the construction of a luminaire (fixture) box.

410.72 Lampholders.

Lampholders of the screw-shell type shall be of porcelain or other suitable insulating materials. Where used, cements shall be of the high-heat type.

XIII. Special Provisions for Electric-Discharge Lighting Systems of 1000 Volts or Less

410.73 General.

(A) Open-Circuit Voltage of 1000 Volts or Less. Equipment for use with electric-discharge lighting systems and designed for an open-circuit voltage of 1000 volts or less shall be of a type intended for such service.

(B) Considered as Energized. The terminals of an electric-discharge lamp shall be considered as energized where any lamp terminal is connected to a circuit of over 300 volts.

(C) Transformers of the Oil-Filled Type. Transformers of the oil-filled type shall not be used.

(D) Additional Requirements. In addition to complying with the general requirements for luminaires (lighting fixtures), such equipment shall comply with Part XIII of this article.

(E) Thermal Protection—Fluorescent Luminaires (Fixtures).

(1) Integral Thermal Protection. The ballast of a fluorescent luminaire (fixture) installed indoors shall have integral thermal protection. Replacement ballasts shall also have thermal protection integral with the ballast.

(2) Simple Reactance Ballasts. A simple reactance ballast in a fluorescent luminaire (fixture) with straight tubular lamps shall not be required to be thermally protected.

(3) Exit Fixtures. A ballast in a fluorescent exit luminaire (fixture) shall not have thermal protection.

(4) Emergency Egress Luminaires (Fixtures). A ballast in a fluorescent luminaire (fixture) that is used for egress lighting and energized only during an emergency shall not have thermal protection.

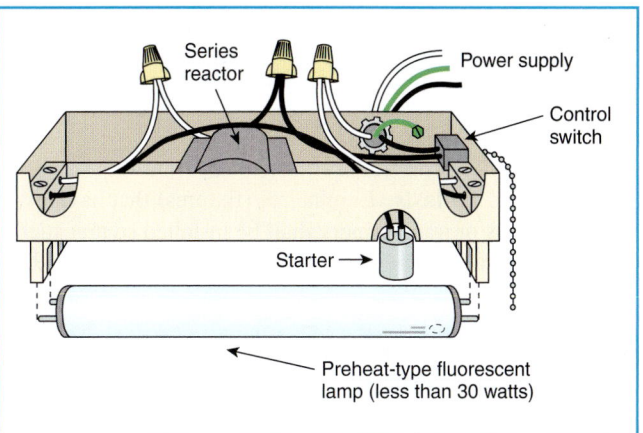

Exhibit 410.8 The circuitry for a simple reactance-type ballast for fluorescent lighting.

Thermal protection that is integral with the ballast is required for fluorescent luminaires installed indoors. Thermally protected ballasts are required as replacements for nonthermally protected ballasts in older fixtures. Thermally protected fluorescent lamp ballasts intended for use in accordance with 410.73(E) are marked "Class P."

Because different Class P ballasts have different heating characteristics, the heating characteristics should be considered when selecting replacements for nonthermally protected ballasts. This type of ballast protection is set to open the circuit at a predetermined temperature, to prevent abnormal ballast heat buildup caused by a fault in one or more of the ballast components or by some lampholder or wiring fault.

Section 410.73(E)(3) exempts exit sign fixtures from the thermal protection requirement because overheating during high ambient conditions could cause the thermal protection to operate. This action could impair evacuation during a fire.

Section 410.73(E)(4) exempts egress lighting from the thermal protection requirement for the same reason that exit signs are exempt. However, this exemption applies to egress lighting that is energized only during the emergency condition.

Exhibit 410.8 illustrates a reactance-type ballast used in series with a 30-watt or less preheat-type fluorescent lamp. This type of ballast does not require thermal protection, and the luminaire may be equipped with automatic-type starters (such as used with medicine cabinet luminaires) or a manual momentary contact starter (such as used with desk lamps and some small under-cabinet luminaires).

(F) High-Intensity Discharge Luminaires (Fixtures).

(1) Recessed. Recessed high-intensity luminaires (fixtures) designed to be installed in wall or ceiling cavities shall have thermal protection and be identified as thermally protected.

(2) Inherently Protected. Thermal protection shall not be required in a recessed high-intensity luminaire (fixture)

whose design, construction, and thermal performance characteristics are equivalent to a thermally protected luminaire (fixture) and are identified as inherently protected.

(3) Installed in Poured Concrete. Thermal protection shall not be required in a recessed high-intensity discharge luminaire (fixture) identified for use and installed in poured concrete.

(4) Recessed Remote Ballasts. A recessed remote ballast for a high-intensity discharge luminaire (fixture) shall have thermal protection that is integral with the ballast and be identified as thermally protected.

410.74 Direct-Current Equipment.

Luminaires (fixtures) installed on dc circuits shall be equipped with auxiliary equipment and resistors designed for dc operation. The luminaires (fixtures) shall be marked for dc operation.

410.75 Open-Circuit Voltage Exceeding 300 Volts.

Equipment having an open-circuit voltage exceeding 300 volts shall not be installed in dwelling occupancies unless such equipment is designed so that there will be no exposed live parts when lamps are being inserted, are in place, or are being removed.

Luminaires intended for use in nondwelling occupancies are so marked. This marking usually indicates that the luminaire has maintenance features that are considered beyond the capabilities of the ordinary homeowner or the luminaire involves voltages in excess of those permitted by this *Code* for dwelling occupancies. For other references to voltage limitations within dwelling units, see 210.6 and 410.80(B).

410.76 Luminaire (Fixture) Mounting.

(A) Exposed Ballasts. Luminaires (fixtures) that have exposed ballasts or transformers shall be installed so that such ballasts or transformers will not be in contact with combustible material.

(B) Combustible Low-Density Cellulose Fiberboard. Where a surface-mounted luminaire (fixture) containing a ballast is to be installed on combustible low-density cellulose fiberboard, it shall be listed for this condition or shall be spaced not less than 38 mm (1½ in.) from the surface of the fiberboard. Where such luminaires (fixtures) are partially or wholly recessed, the provisions of 410.64 through 410.72 shall apply.

> FPN: Combustible low-density cellulose fiberboard includes sheets, panels, and tiles that have a density of 320 kg/m^3 (20 lb/ft^3) or less and that are formed of bonded plant fiber material but does not include solid or laminated wood or fiberboard that has a density in excess of 320 kg/m^3 (20 lb/ft^3) or is a material that has been integrally treated with fire-retarding chemicals to the degree that the flame spread in any plane of the material will not exceed 25, determined in accordance with tests for surface burning characteristics of building materials. See ANSI/ASTM E84-1997, *Test Method for Surface Burning Characteristics of Building Materials.*

Fluorescent lamp luminaires intended for mounting on combustible low-density cellulose fiberboard ceilings have been evaluated with thermal insulation above the ceiling in the vicinity of the luminaire and are marked "Suitable for Surface Mounting on Combustible Low-Density Cellulose Fiberboard."

Fluorescent lamp luminaires not so marked may be directly mounted against a ceiling surface constructed of a material other than *combustible low-density fiberboard* or may be spaced not less than 1½ in. from the surface of the low-density fiberboard.

410.77 Equipment Not Integral with Luminaire (Fixture).

(A) Metal Cabinets. Auxiliary equipment, including reactors, capacitors, resistors, and similar equipment, where not installed as part of a luminaire (lighting fixture) assembly, shall be enclosed in accessible, permanently installed metal cabinets.

(B) Separate Mounting. Separately mounted ballasts that are intended for direct connection to a wiring system shall not be required to be separately enclosed.

(C) Wired Luminaire (Fixture) Sections. Wired luminaire (fixture) sections are paired, with a ballast(s) supplying a lamp or lamps in both. For interconnection between paired units, it shall be permissible to use metric designator 12 (trade size ⅜) flexible metal conduit in lengths not exceeding 7.5 m (25 ft), in conformance with Article 348. Luminaire (fixture) wire operating at line voltage, supplying only the ballast(s) of one of the paired luminaires (fixtures), shall be permitted in the same raceway as the lamp supply wires of the paired luminaires (fixtures).

Wired luminaire sections are shipped in pairs and marked for use in pairs. Each individual unit includes lamps in odd-numbered quantities (1 or 3 is most common), with the odd lamp in each luminaire supplied by a two-lamp ballast located in one luminaire of the pair. Two-lamp ballasts are more energy efficient than single-lamp or three-lamp ballasts.

410.78 Autotransformers.

An autotransformer that is used to raise the voltage to more than 300 volts, as part of a ballast for supplying lighting units, shall be supplied only by a grounded system.

410.79 Switches.

Snap switches shall comply with 404.14.

XIV. Special Provisions for Electric-Discharge Lighting Systems of More Than 1000 Volts

Sections 410.80 through 410.87 apply to interior electric-discharge neon tube-type lighting that contains neon, helium, or argon gas, with or without mercury, at low vapor pressure; long-length fluorescent tube lighting requiring more than 1000 volts; and cold-cathode fluorescent lamp installations arranged to operate with several tubes in series.

410.80 General.

(A) Listing. Electric-discharge lighting systems with an open-circuit voltage exceeding 1000 volts shall be listed and installed in conformance with that listing.

(B) Dwelling Occupancies. Equipment that has an open-circuit voltage exceeding 1000 volts shall not be installed in or on dwelling occupancies.

Section 410.80(B) specifically prohibits electric-discharge lighting, such as neon tube, high-intensity discharge, or fluorescent lighting, that has an open-circuit voltage greater than 1000 volts to be installed in dwelling occupancies. Such lighting systems are often used as decorative lighting as well as for outline lighting and signs. Where used as outline lighting or signs in nonresidential occupancies, see Article 600.

(C) Live Parts. The terminal of an electric-discharge lamp shall be considered as a live part.

(D) Additional Requirements. In addition to complying with the general requirements for luminaires (lighting fixtures), such equipment shall comply with Part XIV of this article.

> FPN: For signs and outline lighting, see Article 600.

410.81 Control.

(A) Disconnection. Luminaires (fixtures) or lamp installation shall be controlled either singly or in groups by an externally operable switch or circuit breaker that opens all ungrounded primary conductors.

(B) Within Sight or Locked Type. The switch or circuit breaker shall be located within sight from the luminaires (fixtures) or lamps, or it shall be permitted elsewhere if it is provided with a means for locking in the open position.

The requirement in 410.81(B) is intended to help protect the service person from the disconnecting means being turned on or closed while the equipment is being serviced.

410.82 Lamp Terminals and Lampholders.

Parts that must be removed for lamp replacement shall be hinged or held captive. Lamps or lampholders shall be designed so that there are no exposed live parts when lamps are being inserted or removed.

410.83 Transformers.

(A) Type. Transformers shall be enclosed, identified for the use, and listed.

(B) Voltage. The secondary-circuit voltage shall not exceed 15,000 volts, nominal, under any load condition. The voltage to ground of any ouput terminals of the secondary circuit shall not exceed 7500 volts, under any load conditions.

(C) Rating. Transformers shall have a secondary short-circuit current rating of not more than 150 mA if the open-circuit voltage is over 7500 volts, and not more than 300 mA if the open-circuit voltage rating is 7500 volts or less.

(D) Secondary Connections. Secondary circuit outputs shall not be connected in parallel or in series.

410.84 Transformer Locations.

(A) Accessible. Transformers shall be accessible after installation.

(B) Secondary Conductors. Transformers shall be installed as near to the lamps as practicable to keep the secondary conductors as short as possible.

(C) Adjacent to Combustible Materials. Transformers shall be located so that adjacent combustible materials are not subjected to temperatures in excess of 90°C (194°F).

410.85 Exposure to Damage.

Lamps shall not be located where normally exposed to physical damage.

410.86 Marking.

Each luminaire (fixture) or each secondary circuit of tubing having an open-circuit voltage of over 1000 volts shall have a clearly legible marking in letters not less than 6 mm (¼ in.) high reading "Caution . . . volts." The voltage indicated shall be the rated open-circuit voltage.

410.87 Switches.

Snap switches shall comply with 404.4.

Refer to 600.32(A) for wiring methods for electric-discharge neon tube-type lighting.

XV. Lighting Track

410.100 Definition.

Lighting Track. A manufactured assembly designed to support and energize luminaires (lighting fixtures) that are capable of being readily repositioned on the track. Its length may be altered by the addition or subtraction of sections of track.

410.101 Installation.

(A) Lighting Track. Lighting track shall be permanently installed and permanently connected to a branch circuit. Only lighting track fittings shall be installed on lighting track. Lighting track fittings shall not be equipped with general-purpose receptacles.

A lighting track fitting differs from a fitting as defined in Article 100 in that it usually performs both an electrical and a mechanical function. Such assemblies are not intended to be used for locating convenience receptacles or as an alternative for required receptacle outlets such as those required in 210.62 for show windows. Lighting track can be removed and relocated and therefore is not a substitute for required outlets.

(B) Connected Load. The connected load on lighting track shall not exceed the rating of the track. Lighting track shall be supplied by a branch circuit having a rating not more than that of the track.

Section 220.12(B) addresses track lighting loads.

The volt-ampere (VA) load for 2 ft of track is 150 volt-amperes because a value of 150 VA is more consistent with standard lamp values for 2 ft of track. It should be understood that 220.12(B) is not intended to limit the number of feet of track on a single branch circuit nor is it intended to limit the number of fixtures on an individual track. Rather, 220.12(B) is intended to be used for load calculations of feeders and services.

Example

Suppose a lighting plan shows 62.5 linear ft of single circuit lighting track for a small department store featuring clothing. Because the actual track luminaires are owner supplied, neither the quantity of track luminaires nor the lamp size is specified. What is the minimum calculated load associated with the lighting track that must be added to the service or feeder supplying this store?

Solution

According to 220.12(B), the minimum calculated load to be added to the service or feeder supplying this track light installation is determined as follows:

62.5 ft ÷ 2 ft = 31.25 ft   31.25 must be rounded
up to 32

32 × 150 VA = 4800 VA

The minimum load that must be added to the service or feeder is 4800 VA.

It is important to note that the branch circuits supplying this installation are covered in 410.101(B). For the lighting track branch-circuit load, the maximum load on the track cannot exceed the rating of the branch circuit supplying the track. Also, the track must be supplied by a branch circuit having a rating not exceeding the rating of the track. The track length does not enter into the branch-circuit calculation.

(C) Locations Not Permitted. Lighting track shall not be installed in the following locations:

(1) Where likely to be subjected to physical damage
(2) In wet or damp locations
(3) Where subject to corrosive vapors
(4) In storage battery rooms
(5) In hazardous (classified) locations
(6) Where concealed
(7) Where extended through walls or partitions
(8) Less than 1.5 m (5 ft) above the finished floor except where protected from physical damage or track operating at less than 30 volts rms open-circuit voltage
(9) Within the zone measured 900 mm (3 ft) horizontally and 2.5 m (8 ft) vertically from the top of the bathtub rim

Low-voltage lighting track operating at less than 30 volts is permitted to be installed less than 5 ft above the floor.

(D) Support. Fittings identified for use on lighting track shall be designed specifically for the track on which they are to be installed. They shall be securely fastened to the track, shall maintain polarization and grounding, and shall be designed to be suspended directly from the track.

410.103 Heavy-Duty Lighting Track.

Heavy-duty lighting track is lighting track identified for use exceeding 20 amperes. Each fitting attached to a heavy-duty lighting track shall have individual overcurrent protection.

410.104 Fastening.

Lighting track shall be securely mounted so that each fastening will be suitable for supporting the maximum weight of luminaires (fixtures) that can be installed. Unless identified for supports at greater intervals, a single section 1.2 m (4 ft) or shorter in length shall have two supports, and, where installed in a continuous row, each individual section of not more than 1.2 m (4 ft) in length shall have one additional support.

410.105 Construction Requirements.

(A) Construction. The housing for the lighting track system shall be of substantial construction to maintain rigidity. The conductors shall be installed within the track housing, permitting insertion of a luminaire (fixture), and designed to prevent tampering and accidental contact with live parts. Components of lighting track systems of different voltages shall not be interchangeable. The track conductors shall be a minimum 12 AWG or equal and shall be copper. The track system ends shall be insulated and capped.

(B) Grounding. Lighting track shall be grounded in accordance with Article 250, and the track sections shall be securely coupled to maintain continuity of the circuitry, polarization, and grounding throughout.

ARTICLE 411
Lighting Systems Operating at 30 Volts or Less

Contents

411.1 Scope.

This article covers lighting systems operating at 30 volts or less and their associated components.

Article 411 addresses low-voltage lighting systems. Article 411 is intended to cover low-voltage interior lighting and low-voltage exterior (landscape) lighting installations. It covers systems operating at 30 volts rms or 42.4 volts peak or less, having a maximum output of 600 volt-amperes.

411.2 Definition.

Lighting Systems Operating at 30 Volts or Less. A lighting system consisting of an isolating power supply operating at 30 volts (42.4 volts peak) or less, under any load condition, with one or more secondary circuits, each limited to 25 amperes maximum, supplying luminaires (lighting fixtures) and associated equipment identified for the use.

411.3 Listing Required.

Lighting systems operating at 30 volts or less shall be listed for the purpose.

411.4 Locations Not Permitted.

Lighting systems operating at 30 volts or less shall not be installed (1) where concealed or extended through a building wall, unless using a wiring method specified in Chapter 3, or (2) within 3.0 m (10 ft) of pools, spas, fountains, or similar locations, except as permitted by Article 680.

The installation requirements of 411.4 recognize that shock and fire hazards still exist, even with low-voltage systems.

411.5 Secondary Circuits.

(A) Grounding. Secondary circuits shall not be grounded.

(B) Isolation. The secondary circuit shall be insulated from the branch circuit by an isolating transformer.

(C) Bare Conductors. Exposed bare conductors and current-carrying parts shall be permitted. Bare conductors shall not be installed less than 2.1 m (7 ft) above the finished floor, unless specifically listed for a lower installation height.

411.6 Branch Circuit.

Lighting systems operating at 30 volts or less shall be supplied from a maximum 20-ampere branch circuit.

411.7 Hazardous (Classified) Locations.

Where installed in hazardous (classified) locations, these systems shall conform with Articles 500 through 517 in addition to this article.

ARTICLE 422
Appliances

Contents

I. General

422.1 Scope.

This article covers electric appliances used in any occupancy.

Article 422 covers electric appliances that may be used in a dwelling unit or in commercial and industrial locations. It also covers appliances that may be fastened in place or be cord-and-plug-connected, such as air-conditioning units, dishwashers, heating appliances, water heaters, infrared heating lamps, and so on. See 422.3 for the requirements of other articles. Also see Article 100 for the definition of *appliance*.

422.3 Other Articles.

Appliances for use in hazardous (classified) locations shall comply with Articles 500 through 517.

The requirements of Article 430 shall apply to the installation of motor-operated appliances, and the requirements of Article 440 shall apply to the installation of appliances containing a hermetic refrigerant motor-compressor(s), except as specifically amended in this article.

422.4 Live Parts.

Appliances shall have no live parts normally exposed to contact other than those parts functioning as open-resistance heating elements, such as the heating element of a toaster, which are necessarily exposed.

II. Installation

422.10 Branch-Circuit Rating.

This section specifies the ratings of branch circuits capable of carrying appliance current without overheating under the conditions specified.

Conductors that form integral parts of appliances are tested as part of the listing or labeling process.

(A) Individual Circuits. The rating of an individual branch circuit shall not be less than the marked rating of the appliance or the marked rating of an appliance having combined loads as provided in 422.62.

The rating of an individual branch circuit for motor-operated appliances not having a marked rating shall be in accordance with Part II of Article 430.

The branch-circuit rating for an appliance that is continuously loaded, other than a motor-operated appliance, shall not be less than 125 percent of the marked rating, or not less than 100 percent of the marked rating if the branch-circuit device and its assembly are listed for continuous loading at 100 percent of its rating.

Branch circuits for household cooking appliances shall be permitted to be in accordance with Table 220.19.

(B) Circuits Supplying Two or More Loads. For branch circuits supplying appliance and other loads, the rating shall be determined in accordance with 210.23.

422.11 Overcurrent Protection.

Appliances shall be protected against overcurrent in accordance with 422.11(A) through (G) and 422.10.

(A) Branch-Circuit Overcurrent Protection. Branch circuits shall be protected in accordance with 240.4.

If a protective device rating is marked on an appliance, the branch-circuit overcurrent device rating shall not exceed the protective device rating marked on the appliance.

If a labeled or listed appliance is provided with installation instructions from the manufacturer, the branch-circuit size is not permitted to be less than the minimum size stated in the installation instructions. See 110.3(B) and its related commentary regarding the installation and use of listed or labeled equipment.

(B) Household-Type Appliance with Surface Heating Elements. A household-type appliance with surface heating elements having a maximum demand of more than 60 amperes computed in accordance with Table 220.19 shall have its power supply subdivided into two or more circuits, each of which shall be provided with overcurrent protection rated at not over 50 amperes.

(C) Infrared Lamp Commercial and Industrial Heating Appliances. Infrared lamp commercial and industrial heating appliances shall have overcurrent protection not exceeding 50 amperes.

(D) Open-Coil or Exposed Sheathed-Coil Types of Surface Heating Elements in Commercial-Type Heating Appliances. Open-coil or exposed sheathed-coil types of surface heating elements in commercial-type heating appli-

ances shall be protected by overcurrent protective devices rated at not over 50 amperes.

(E) Single Nonmotor-Operated Appliance. If the branch circuit supplies a single non–motor-operated appliance, the rating of overcurrent protection shall

(1) Not exceed that marked on the appliance;
(2) If the overcurrent protection rating is not marked and the appliance is rated 13.3 amperes or less, not exceed 20 amperes; or
(3) If the overcurrent protection rating is not marked and the appliance is rated over 13.3 amperes, not exceed 150 percent of the appliance rated current. Where 150 percent of the appliance rating does not correspond to a standard overcurrent device ampere rating, the next higher standard rating shall be permitted.

(F) Electric Heating Appliances Employing Resistance-Type Heating Elements Rated More Than 48 Amperes.

(1) Electric Heating Appliances. Electric heating appliances employing resistance-type heating elements rated more than 48 amperes, other than household appliances with surface heating elements covered by 422.11(B), and commercial-type heating appliances covered by 422.11(D), shall have the heating elements subdivided. Each subdivided load shall not exceed 48 amperes and shall be protected at not more than 60 amperes.

These supplementary overcurrent protective devices shall be (1) factory-installed within or on the heater enclosure or provided as a separate assembly by the heater manufacturer; (2) accessible; and (3) suitable for branch-circuit protection.

The main conductors supplying these overcurrent protective devices shall be considered branch-circuit conductors.

(2) Commercial Kitchen and Cooking Appliances. Commercial kitchen and cooking appliances using sheathed-type heating elements not covered in 422.11(D) shall be permitted to be subdivided into circuits not exceeding 120 amperes and protected at not more than 150 amperes where one of the following is met:

(1) Elements are integral with and enclosed within a cooking surface.
(2) Elements are completely contained within an enclosure identified as suitable for this use.
(3) Elements are contained within an ASME-rated and stamped vessel.

(3) Water Heaters and Steam Boilers. Water heaters and steam boilers employing resistance-type immersion electric heating elements contained in an ASME-rated and stamped vessel or listed instantaneous water heaters shall be permitted to be subdivided into circuits not exceeding 120 amperes and protected at not more than 150 amperes.

(G) Motor-Operated Appliances. Motors of motor-operated appliances shall be provided with overload protection in accordance with Part III of Article 430. Hermetic refrigerant motor-compressors in air-conditioning or refrigerating equipment shall be provided with overload protection in accordance with Part VI of Article 440. Where appliance overcurrent protective devices that are separate from the appliance are required, data for selection of these devices shall be marked on the appliance. The minimum marking shall be that specified in 430.7 and 440.4.

422.12 Central Heating Equipment.

Central heating equipment other than fixed electric space-heating equipment shall be supplied by an individual branch circuit.

Exception: Auxiliary equipment, such as a pump, valve, humidifier, or electrostatic air cleaner directly associated with the heating equipment, shall be permitted to be connected to the same branch circuit.

The exception to 422.12 permits the electric motors, ignition systems, controls, and so on, of fossil-fuel–fired central heating equipment to be connected to the same individual branch circuit, as defined in Article 100 under *branch circuit, individual.*

422.13 Storage-Type Water Heaters.

A branch circuit supplying a fixed storage-type water heater that has a capacity of 450 L (120 gal) or less shall have a rating not less than 125 percent of the nameplate rating of the water heater.

FPN: For branch-circuit rating, see 422.10.

422.14 Infrared Lamp Industrial Heating Appliances.

Infrared industrial heating appliance lampholders shall be permitted to be connected to any of the branch circuits in Article 210 and, in industrial occupancies, shall be permitted to be operated in series on circuits of over 150 volts to ground, provided the voltage rating of the lampholders is not less than the circuit voltage.

Each section, panel, or strip carrying a number of infrared lampholders (including the internal wiring of such section, panel, or strip) shall be considered an appliance. The terminal connection block of each such assembly shall be considered an individual outlet.

422.15 Central Vacuum Outlet Assemblies.

(A) Listed central vacuum outlet assemblies shall be permitted to be connected to a branch circuit in accordance with 210.23(A).

(B) The ampacity of the connecting conductors shall not be less than the ampacity of the branch circuit conductors to which they are connected.

(C) An equipment grounding conductor shall be used where the central vacuum outlet assembly has accessible non–current-carrying metal parts.

Section 422.15 permits listed central vacuum outlet devices to be connected to the ordinary 15- or 20-ampere general-purpose branch circuits that may be located in the same area as the vacuum outlet is installed. Starting and stopping of the central vacuum system is achieved by a Class 2 control circuit that originates at the main unit of the central vacuum system. The circuit is switched at each outlet by the insertion or removal of the matching vacuum hose in the outlet.

422.16 Flexible Cords.

(A) General. Flexible cord shall be permitted (1) for the connection of appliances to facilitate their frequent interchange or to prevent the transmission of noise or vibration or (2) to facilitate the removal or disconnection of appliances that are fastened in place, where the fastening means and mechanical connections are specifically designed to permit ready removal for maintenance or repair and the appliance is intended or identified for flexible cord connection.

It should be understood that a cord-connected appliance is required to be specifically designed, mechanically and electrically, to be readily removable for maintenance and repair.

(B) Specific Appliances.

(1) Electrically Operated Kitchen Waste Disposers. Electrically operated kitchen waste disposers shall be permitted to be cord-and-plug connected with a flexible cord identified as suitable for the purpose in the installation instructions of the appliance manufacturer, where all of the following conditions are met.

(1) The flexible cord shall be terminated with a grounding type attachment plug.

Exception: A listed kitchen waste disposer distinctly marked to identify it as protected by a system of double insulation, or its equivalent, shall not be required to be terminated with a grounding-type attachment plug.

(2) The length of the cord shall not be less than 450 mm (18 in.) and not over 900 mm (36 in.).

(3) Receptacles shall be located to avoid physical damage to the flexible cord.

(4) The receptacle shall be accessible.

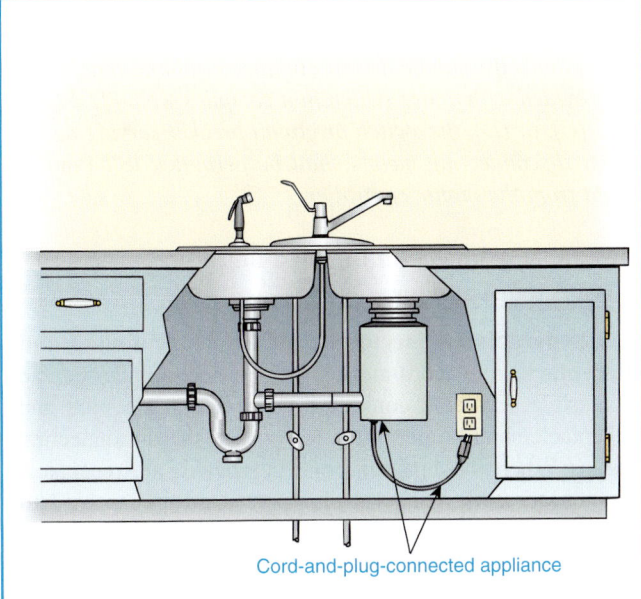

Cord-and-plug-connected appliance

Exhibit 422.1 A cord-and-plug-connected kitchen waste disposer.

(2) Built-in Dishwashers and Trash Compactors. Built-in dishwashers and trash compactors shall be permitted to be cord-and-plug connected with a flexible cord identified as suitable for the purpose in the installation instructions of the appliance manufacturer where all of the following conditions are met.

(1) The flexible cord shall be terminated with a grounding-type attachment plug.

Exception: A listed dishwasher or trash compactor distinctly marked to identify it as protected by a system of double insulation, or its equivalent, shall not be required to be terminated with a grounding-type attachment plug.

(2) The length of the cord shall be 0.9 m to 1.2 m (3 ft to 4 ft) measured from the face of the attachment plug to the plane of the rear of the appliance.

(3) Receptacles shall be located to avoid physical damage to the flexible cord.

(4) The receptacle shall be located in the space occupied by the appliance or adjacent thereto.

(5) The receptacle shall be accessible.

(3) Wall-Mounted Ovens and Counter-Mounted Cooking Units. Wall-mounted ovens and counter-mounted cooking units complete with provisions for mounting and for making electrical connections shall be permitted to be permanently connected or, only for ease in servicing or for installation, cord-and-plug connected.

A separable connector or a plug and receptacle combination in the supply line to an oven or cooking unit shall be approved for the temperature of the space in which it is located.

422.17 Protection of Combustible Material.

Each electrically heated appliance that is intended by size, weight, and service to be located in a fixed position shall be placed so as to provide ample protection between the appliance and adjacent combustible material.

422.18 Support of Ceiling-Suspended (Paddle) Fans.

(A) Ceiling-Suspended (Paddle) Fans 16 kg (35 lb) or Less. Ceiling-suspended (paddle) fans that do not exceed 16 kg (35 lb) in weight, with or without accessories, shall be permitted to be supported by outlet boxes identified for such use and supported in accordance with 314.23 and 314.27.

(B) Ceiling-Suspended (Paddle) Fans Exceeding 16 kg (35 lb). Ceiling-suspended (paddle) fans exceeding 16 kg (35 lb) in weight, with or without accessories, shall be supported independently of the outlet box. See 314.23.

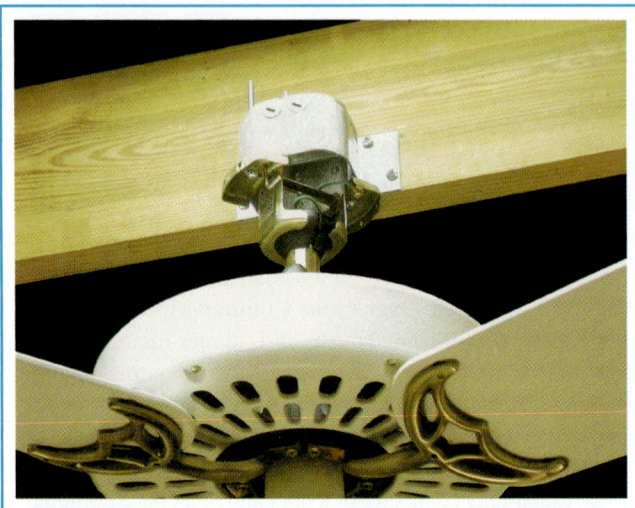

Exhibit 422.2 Supporting a ceiling-suspended (paddle) fan (35 lb or less) with a box identified for such use. (Courtesy of Raco, Inc.)

place. The exception to 422.18(B) allows a box or box system identified or listed for the application to provide the sole support for this heavier class of fan, up to 70 lb.

Exception: Listed outlet boxes or outlet box systems that are identified for the purpose shall be permitted to support ceiling-suspended fans, with or without accessories, that weigh 32 kg (70 lb) or less.

422.20 Other Installation Methods.

Appliances employing methods of installation other than covered by this article shall be permitted to be used only by special permission.

III. Disconnecting Means

422.30 General.

A means shall be provided to disconnect each appliance from all ungrounded conductors in accordance with the following sections of Part III. If an appliance is supplied by more than one source, the disconnecting means shall be grouped and identified.

422.31 Disconnection of Permanently Connected Appliances.

(A) Rated at Not Over 300 Volt-Amperes or ⅛ Horse-power. For permanently connected appliances rated at not over 300 volt-amperes or ⅛ hp, the branch-circuit overcurrent device shall be permitted to serve as the disconnecting means.

(B) Appliances Rated Over 300 Volt-Amperes or ⅛ Horsepower. For permanently connected appliances rated over 300 volt-amperes or ⅛ hp, the branch-circuit switch or circuit breaker shall be permitted to serve as the disconnecting means where the switch or circuit breaker is within sight from the appliance or is capable of being locked in the open position.

> FPN: For appliances employing unit switches, see 422.34.

422.32 Disconnecting Means for Motor-Driven Appliance.

If a switch or circuit breaker serves as the disconnecting means for a permanently connected motor-driven appliance of more than ⅛ hp, it shall be located within sight from the motor controller and shall comply with Part IX of Article 430.

Exception: If a motor-driven appliance of more than ⅛ hp is provided with a unit switch that complies with 422.34(A), (B), (C), or (D), the switch or circuit breaker serving as the other disconnecting means shall be permitted to be out of sight from the motor controller.

422.33 Disconnection of Cord-and-Plug-Connected Appliances.

(A) Separable Connector or an Attachment Plug and Receptacle. For cord-and-plug-connected appliances, an accessible separable connector or an accessible plug and receptacle shall be permitted to serve as the disconnecting means. Where the separable connector or plug and receptacle are not accessible, cord-and-plug-connected appliances shall be provided with disconnecting means in accordance with 422.31.

(B) Connection at the Rear Base of a Range. For cord-and-plug-connected household electric ranges, an attachment plug and receptacle connection at the rear base of a range, if it is accessible from the front by removal of a drawer, shall be considered as meeting the intent of 422.33(A).

(C) Rating. The rating of a receptacle or of a separable connector shall not be less than the rating of any appliance connected thereto.

Exception: Demand factors authorized elsewhere in this Code shall be permitted to be applied to the rating of a receptacle or of a separable connector.

422.34 Unit Switch(es) as Disconnecting Means.

A unit switch(es) with a marked-off position that is a part of an appliance and disconnects all ungrounded conductors shall be permitted as the disconnecting means required by this article where other means for disconnection are provided in the following types of occupancies.

(A) Multifamily Dwellings. In multifamily dwellings, the other disconnecting means shall be within the dwelling unit, or on the same floor as the dwelling unit in which the appliance is installed, and shall be permitted to control lamps and other appliances.

(B) Two-Family Dwellings. In two-family dwellings, the other disconnecting means shall be permitted either inside or outside of the dwelling unit in which the appliance is installed. In this case, an individual switch or circuit breaker for the dwelling unit shall be permitted and shall also be permitted to control lamps and other appliances.

(C) One-Family Dwellings. In one-family dwellings, the service disconnecting means shall be permitted to be the other disconnecting means.

(D) Other Occupancies. In other occupancies, the branch-circuit switch or circuit breaker, where readily accessible for servicing of the appliance, shall be permitted as the other disconnecting means.

422.35 Switch and Circuit Breaker to Be Indicating.

Switches and circuit breakers used as disconnecting means shall be of the indicating type.

IV. Construction

422.40 Polarity in Cord-and-Plug-Connected Appliances.

If the appliance is provided with a manually operated, line-connected, single-pole switch for appliance on–off operation, an Edison-base lampholder, or a 15- or 20-ampere receptacle, the attachment plug shall be of the polarized or grounding type.

A 2-wire, nonpolarized attachment plug shall be permitted to be used on a listed double-insulated shaver.

FPN: For polarity of Edison-base lampholders, see 410.42(A).

422.41 Cord-and-Plug-Connected Appliances Subject to Immersion.

Cord-and-plug-connected portable, freestanding hydromassage units and hand-held hair dryers shall be constructed to provide protection for personnel against electrocution when immersed while in the "on" or "off" position.

Although receptacles in bathrooms of dwelling units have been required to be protected by ground-fault circuit interrupters since the 1975 edition of the *Code,* many receptacles in existing bathrooms are not so protected. Cord-and-plug-connected appliances such as hand-held hair dryers, curling irons, and so on, which can and have accidentally fallen into bathtubs, causing fatalities, are required to be provided with some form of protective device that is part of the appliance. Three types of protectors comply with this requirement:

1. Appliance-leakage circuit interrupters (ALCIs)
2. Immersion-detector circuit interrupters (IDCIs)
3. Ground-fault circuit interrupters (GFCIs)

ALCIs de-energize the supply to the appliance when leakage current exceeds a predetermined value. IDCIs de-energize the supply when a liquid causes a conductive path between a live part and a sensor, and GFCIs de-energize the supply when the current to ground exceeds a predetermined value.

422.42 Signals for Heated Appliances.

In other than dwelling-type occupancies, each electrically heated appliance or group of appliances intended to be applied to combustible material shall be provided with a signal or an integral temperature-limiting device.

A common way to provide a signal light for electrically heated appliances in commercial or industrial locations is to use a red light connected to and within sight of the appliance that indicates that the appliance is energized and operating.

No signal is required for an electrically heated appliance provided with an integral thermostat or its equivalent that limits the temperature to which the appliance can heat.

422.43 Flexible Cords.

(A) Heater Cords. All cord-and-plug-connected smoothing irons and electrically heated appliances that are rated at more than 50 watts and produce temperatures in excess of 121°C (250°F) on surfaces with which the cord is likely to be in contact shall be provided with one of the types of approved heater cords listed in Table 400.4.

(B) Other Heating Appliances. All other cord-and-plug-connected electrically heated appliances shall be connected with one of the approved types of cord listed in Table 400.4, selected in accordance with the usage specified in that table.

422.44 Cord-and-Plug-Connected Immersion Heaters.

Electric heaters of the cord-and-plug-connected immersion type shall be constructed and installed so that current-carrying parts are effectively insulated from electrical contact with the substance in which they are immersed.

422.45 Stands for Cord-and-Plug-Connected Appliances.

Each smoothing iron and other cord-and-plug-connected electrically heated appliance intended to be applied to combustible material shall be equipped with an approved stand, which shall be permitted to be a separate piece of equipment or a part of the appliance.

422.46 Flatirons.

Electrically heated smoothing irons shall be equipped with an identified temperature-limiting means.

422.47 Water Heater Controls.

All storage or instantaneous-type water heaters shall be equipped with a temperature-limiting means in addition to its control thermostat to disconnect all ungrounded conductors. Such means shall be as follows:

(1) Installed to sense maximum water temperature; and
(2) Either a trip-free, manually reset type or a type having a replacement element. Such water heaters shall be marked to require the installation of a temperature and pressure relief valve.

Exception No. 1: Storage water heaters that are identified as being suitable for use with supply water temperature of 82°C (180°F) or above and a capacity of 60 kW or above.

Exception No. 2: Instantaneous-type water heaters that are identified as being suitable for such use, with a capacity of 4 L (1 gal) or less.

> FPN: See ANSI Z21.22-1999/CSA 4.4-M99, *Relief Valves for Hot Water Supply Systems.*

422.48 Infrared Lamp Industrial Heating Appliances.

(A) 300 Watts or Less. Infrared heating lamps rated at 300 watts or less shall be permitted with lampholders of the medium-base, unswitched porcelain type or other types identified as suitable for use with infrared heating lamps rated 300 watts or less.

(B) Over 300 Watts. Screw-shell lampholders shall not be used with infrared lamps rated over 300 watts, unless the lampholders are identified as being suitable for use with infrared heating lamps rated over 300 watts.

Infrared (heat) radiation lamps are tungsten-filament incandescent lamps similar in appearance to lighting lamps; however, they are designed to operate at a lower temperature, thus transferring more heat radiation and less light intensity. Infrared lamps are used for a variety of heating and drying purposes in industrial locations.

422.49 High-Pressure Spray Washers.

All single-phase cord-and-plug-connected high-pressure spray washing machines rated at 250 volts or less shall be provided with factory-installed ground-fault circuit-interrupter protection for personnel. The ground-fault circuit interrupter shall be an integral part of the attachment plug or shall be located in the supply cord within 300 mm (12 in.) of the attachment plug.

High-pressure spray washers may be used without a ground-fault circuit interrupter as part of the supply cord if the washers are rated 3-phase or are over 250 volts.

422.50 Cord-and-Plug-Connected Pipe Heating Assemblies.

Cord-and-plug-connected pipe heating assemblies intended to prevent freezing of piping shall be listed.

The *listing* requirement was added as a result of data submitted that substantiated numerous fires initiated by heat tapes. Additional requirements for ground-fault protection equipment are found in 427.22.

V. Marking

422.60 Nameplate.

(A) Nameplate Marking. Each electric appliance shall be provided with a nameplate giving the identifying name and the rating in volts and amperes, or in volts and watts. If the appliance is to be used on a specific frequency or frequencies, it shall be so marked.

Where motor overload protection external to the appliance is required, the appliance shall be so marked.

> FPN: See 422.11 for overcurrent protection requirements.

(B) To Be Visible. Marking shall be located so as to be visible or easily accessible after installation.

422.61 Marking of Heating Elements.

All heating elements that are rated over one ampere, replaceable in the field, and a part of an appliance shall be legibly marked with the ratings in volts and amperes, or in volts and watts, or with the manufacturer's part number.

422.62 Appliances Consisting of Motors and Other Loads.

(A) Nameplate Horsepower Markings. Where a motor-operated appliance nameplate includes a horsepower rating, that rating shall not be less than the horsepower rating on the motor nameplate. Where an appliance consists of multiple motors, or one or more motors and other loads, the nameplate value shall not be less than the equivalent horsepower of the combined loads, calculated in accordance with 430.110(C)(1).

(B) Additional Nameplate Markings. Appliances, other than those factory-equipped with cords and attachment plugs and with nameplates in compliance with 422.60, shall be marked in accordance with 422.62(B)(1) or (2).

(1) Marking. In addition to the marking required in 422.60, the marking on an appliance consisting of a motor with other load(s) or motors with or without other load(s) shall specify the minimum supply circuit conductor ampacity and the maximum rating of the circuit overcurrent protective device. This requirement shall not apply to an appliance with a nameplate in compliance with 422.60 where both the minimum supply circuit conductor ampacity and maximum rating of the circuit overcurrent protective device are not more than 15 amperes.

(2) Alternate Marking Method. An alternative marking method shall be permitted to specify the rating of the largest motor in volts and amperes, and the additional load(s) in volts and amperes, or volts and watts in addition to the marking required in 422.60. The ampere rating of a motor ⅛ horsepower or less or a nonmotor load 1 ampere or less shall be permitted to be omitted unless such loads constitute the principal load.

ARTICLE 424
Fixed Electric Space-Heating Equipment

Contents

I. General

424.1 Scope.

This article covers fixed electric equipment used for space heating. For the purpose of this article, heating equipment shall include heating cable, unit heaters, boilers, central systems, or other approved fixed electric space-heating equipment. This article shall not apply to process heating and room air conditioning.

424.2 Other Articles.

All requirements of this *Code* shall apply where applicable. Fixed electric space-heating equipment for use in hazardous (classified) locations shall comply with Articles 500 through 517. Fixed electric space-heating equipment incorporating a hermetic refrigerant motor-compressor shall also comply with Article 440.

424.3 Branch Circuits.

(A) Branch-Circuit Requirements. Individual branch circuits shall be permitted to supply any size fixed electric space-heating equipment.

Branch circuits supplying two or more outlets for fixed electric space-heating equipment shall be rated 15, 20, 25, or 30 amperes. In other than residential occupancies, fixed infrared heating equipment shall be permitted to be supplied from branch circuits rated not over 50 amperes.

(B) Branch-Circuit Sizing. The ampacity of the branch-circuit conductors and the rating or setting of overcurrent protective devices supplying fixed electric space-heating equipment consisting of resistance elements with or without a motor shall not be less than 125 percent of the total load of the motors and the heaters. The rating or setting of overcurrent protective devices shall be permitted in accordance with 240.4(B). A contactor, thermostat, relay, or similar device, listed for continuous operation at 100 percent of its rating, shall be permitted to supply its full-rated load as provided in 210.19(A), Exception.

The size of the branch-circuit conductors and overcurrent protective devices supplying fixed electric space-heating equipment, including a hermetic refrigerant motor-compressor with or without resistance units, shall be computed in accordance with 440.34 and 440.35. The provisions of this section shall not apply to conductors that form an integral part of approved fixed electric space-heating equipment.

The sizing of branch-circuit conductors supplying fixed, electric space-heating equipment at 125 percent of the total load of the heaters (and motors) is predicated on the need to protect overcurrent devices, particularly in panelboards, from overheating during periods of prolonged operation. See 408.16(D). Since the branch-circuit conductors and overcurrent devices are required to be rated at 125 percent of the total load, an additional increase of 125 percent is not required for continuous loads.

The requirement in 424.3(B) is phrased to eliminate any question about whether the heating equipment is a "continuous load" as defined in Article 100.

II. Installation

424.9 General.

All fixed electric space-heating equipment shall be installed in an approved manner.

Permanently installed electric baseboard heaters equipped with factory-installed receptacle outlets, or outlets provided as a separate listed assembly, shall be permitted in lieu of a receptacle outlet(s) that is required by 210.50(B).

Such receptacle outlets shall not be connected to the heater circuits.

FPN: Listed baseboard heaters include instructions that may not permit their installation below receptacle outlets.

The second paragraph of 424.9 restates the permission granted in the second paragraph of 210.52, that is, it allows factory-installed receptacle outlets in electric baseboard heaters to satisfy the spacing requirements for receptacle outlets in dwelling units according to 210.52(A).

Heating equipment and systems often have special installation instructions for spacings, types of supply wires, or special control equipment, which must be considered in determining the suitability of the installation.

424.10 Special Permission.

Fixed electric space-heating equipment and systems installed by methods other than covered by this article shall be permitted only by special permission.

424.11 Supply Conductors.

Fixed electric space-heating equipment requiring supply conductors with over 60°C insulation shall be clearly and permanently marked. This marking shall be plainly visible after installation and shall be permitted to be adjacent to the field connection box.

Fixed, electric space-heating equipment may require supply conductors with a temperature rating greater than 60°C.

424.12 Locations.

(A) Exposed to Physical Damage. Where subject to physical damage, fixed electric space-heating equipment shall be protected in an approved manner.

(B) Damp or Wet Locations. Heaters and related equipment installed in damp or wet locations shall be approved for such locations and shall be constructed and installed so that water or other liquids cannot enter or accumulate in or on wired sections, electrical components, or ductwork.

FPN No. 1: See 110.11 for equipment exposed to deteriorating agents.
FPN No. 2: See 680.27(C) for pool deck areas.

424.13 Spacing from Combustible Materials.

Fixed electric space-heating equipment shall be installed to provide the required spacing between the equipment and adjacent combustible material, unless it has been found to be acceptable where installed in direct contact with combustible material.

III. Control and Protection of Fixed Electric Space-Heating Equipment

424.19 Disconnecting Means.

Means shall be provided to disconnect the heater, motor controller(s), and supplementary overcurrent protective device(s) of all fixed electric space-heating equipment from all ungrounded conductors. Where heating equipment is supplied by more than one source, the disconnecting means shall be grouped and marked.

(A) Heating Equipment with Supplementary Overcurrent Protection. The disconnecting means for fixed electric space-heating equipment with supplementary overcurrent protection shall be within sight from the supplementary overcurrent protective device(s), on the supply side of these devices, if fuses, and, in addition, shall comply with either 424.19(A)(1) or (2).

(1) Heater Containing No Motor Rated Over ⅛ Horsepower. The above disconnecting means or unit switches complying with 424.19(C) shall be permitted to serve as the required disconnecting means for both the motor controller(s) and heater under either item (1) or (2):

(1) The disconnecting means provided is also within sight from the motor controller(s) and the heater.
(2) The disconnecting means provided shall be capable of being locked in the open position.

(2) Heater Containing a Motor(s) Rated Over ⅛ Horsepower. The above disconnecting means shall be permitted to serve as the required disconnecting means for both the motor controller(s) and heater by one of the means specified in items (1) through (4):

(1) Where the disconnecting means is also in sight from the motor controller(s) and the heater.
(2) Where the disconnecting means is not within sight from the heater, a separate disconnecting means shall be installed, or the disconnecting means shall be capable of being locked in the open position, or unit switches complying with 424.19(C) shall be permitted.
(3) Where the disconnecting means is not within sight from the motor controller location, a disconnecting means complying with 430.102 shall be provided.
(4) Where the motor is not in sight from the motor controller location, 430.102(B) shall apply.

(B) Heating Equipment Without Supplementary Overcurrent Protection.

(1) Without Motor or with Motor Not Over ⅛ Horsepower. For fixed electric space-heating equipment without a motor rated over ⅛ hp, the branch-circuit switch or circuit breaker shall be permitted to serve as the disconnecting means where the switch or circuit breaker is within sight from the heater or is capable of being locked in the open position.

(2) Over ⅛ Horsepower. For motor-driven electric space-heating equipment with a motor rated over ⅛ hp, a disconnecting means shall be located within sight from the motor controller or shall be permitted to comply with the requirements in 424.19(A)(2).

(C) Unit Switch(es) as Disconnecting Means. A unit switch(es) with a marked "off" position that is part of a fixed heater and disconnects all ungrounded conductors shall be permitted as the disconnecting means required by this article where other means for disconnection are provided in the types of occupancies in 424.19(C)(1) through (C)(4).

Section 424.19(C) permits a unit switch to serve as the disconnecting means, provided it has a marked "off" position and disconnects all ungrounded conductors and that other means are also provided in accordance with paragraphs 424.19(C) (1), (2), (3), and (4). Such *other means* are not required to be capable of being locked in the open position as required by 424.19(B)(1).

See 424.20 for thermostatically controlled switching devices.

(1) Multifamily Dwellings. In multifamily dwellings, the other disconnecting means shall be within the dwelling unit, or on the same floor as the dwelling unit in which the fixed heater is installed, and shall also be permitted to control lamps and appliances.

(2) Two-Family Dwellings. In two-family dwellings, the other disconnecting means shall be permitted either inside or outside of the dwelling unit in which the fixed heater is installed. In this case, an individual switch or circuit breaker for the dwelling unit shall be permitted and shall also be permitted to control lamps and appliances.

(3) One-Family Dwellings. In one-family dwellings, the service disconnecting means shall be permitted to be the other disconnecting means.

(4) Other Occupancies. In other occupancies, the branch-circuit switch or circuit breaker, where readily accessible for servicing of the fixed heater, shall be permitted as the other disconnecting means.

424.20 Thermostatically Controlled Switching Devices.

(A) Serving as Both Controllers and Disconnecting Means. Thermostatically controlled switching devices and combination thermostats and manually controlled switches shall be permitted to serve as both controllers and disconnecting means, provided all of the following conditions are met:

(1) Provided with a marked "off" position
(2) Directly open all ungrounded conductors when manually placed in the "off" position
(3) Designed so that the circuit cannot be energized automatically after the device has been manually placed in the "off" position
(4) Located as specified in 424.19

(B) Thermostats That Do Not Directly Interrupt All Ungrounded Conductors. Thermostats that do not directly interrupt all ungrounded conductors and thermostats that operate remote-control circuits shall not be required to meet the requirements of 424.20(A). These devices shall not be permitted as the disconnecting means.

424.21 Switch and Circuit Breaker to Be Indicating.

Switches and circuit breakers used as disconnecting means shall be of the indicating type.

424.22 Overcurrent Protection.

(A) Branch-Circuit Devices. Electric space-heating equipment, other than such motor-operated equipment as required by Articles 430 and 440 to have additional overcurrent protection, shall be permitted to be protected against overcurrent where supplied by one of the branch circuits in Article 210.

(B) Resistance Elements. Resistance-type heating elements in electric space-heating equipment shall be protected at not more than 60 amperes. Equipment rated more than 48 amperes and employing such elements shall have the heating elements subdivided, and each subdivided load shall not exceed 48 amperes. Where a subdivided load is less than 48 amperes, the rating of the supplementary overcurrent protective device shall comply with 424.3(B). A boiler employing resistance-type immersion heating elements contained in an ASME rated and stamped vessel shall be permitted to comply with 424.72(A).

The reason for subdividing the overcurrent protection is to minimize the amount of damaging energy released into the heating elements during a short circuit, thereby reducing the risk of fire. In addition to safety, a second benefit may be partial continuity of service.

When a short circuit occurs, large amounts of damaging energy are released. This damage comes in the form of both heat and magnetic energy. By limiting the size of the overcurrent device protecting the individual heating elements, the damaging short-circuit energy released at the element is greatly reduced, thereby greatly reducing the risk of fire.

Historically, it has been stated that the subdivision size of 60 amperes was originally selected to use the maximum fuseholder size of 60 amperes while maintaining up to a 48-ampere heating element size (48 amperes $\times$ 125% = 60 amperes).

The following example makes a strong case for the 60-ampere subdivision requirement when a short circuit occurs at the element level. Although this example uses fuses, the same case can be made using circuit breakers.

Example

A 200 kW, 3-phase, 480-volt resistance-type unit heater has a phase current of 240.6 amperes. If the present *Code* rule for subdivision [424.22(B)] was not observed and this load were to be protected by just one device, the selected overcurrent device could be sized by multiplying 240.6 amperes times 125 percent and selecting a maximum overcurrent protective device sized 350 amperes. If the subdivision requirement were followed, the heater would probably contain six separately protected internal circuits limited in size to 60 amperes.

Solution

By using the UL *Electrical Construction Equipment Directory* (Green Book), the energy let-through of a 350-ampere fuse can be compared to the energy let-through of a 60-ampere fuse. In the fuse section (JCQR), the let-through energy, approximated by the current squared and then multiplied by the time, or I^2t, is provided for various fuse classes (UL). For this example, a 600-volt, 60-ampere Class T fuse could have a let-through, I^2t, as high as 30,000 ampere squared seconds. But, a 600-volt, 350-ampere Class T fuse could have a let-through, I^2t, as high as 1,100,000 ampere squared seconds. That means the 350-ampere fuse could let through 36.67 times as much damaging energy as the 60-ampere fuse during a short circuit. The difference in energy let-through between these two overcurrent devices (the single 350-ampere device and the group of 60-ampere devices) is significant enough to make the difference between replacing a single element or replacing a good portion of the entire system.

This example illustrates that subdivision of a circuit greatly reduces the risk of fire.

(C) Overcurrent Protective Devices. The supplementary overcurrent protective devices for the subdivided loads specified in 424.22(B) shall be (1) factory-installed within or on the heater enclosure or supplied for use with the heater as a separate assembly by the heater manufacturer; (2) accessible, but shall not be required to be readily accessible; and (3) suitable for branch-circuit protection.

> FPN: See 240.10.

Where cartridge fuses are used to provide this overcurrent protection, a single disconnecting means shall be permitted to be used for the several subdivided loads.

> FPN No. 1: For supplementary overcurrent protection, see 240.10.
>
> FPN No. 2: For disconnecting means for cartridge fuses in circuits of any voltage, see 240.40.

Where supplementary overcurrent protection is required, the heating equipment manufacturer is required to furnish the necessary overcurrent protective devices.

(D) Branch-Circuit Conductors. The conductors supplying the supplementary overcurrent protective devices shall be considered branch-circuit conductors.

Where the heaters are rated 50 kW or more, the conductors supplying the supplementary overcurrent protective devices specified in 424.22(C) shall be permitted to be sized at not less than 100 percent of the nameplate rating of the heater, provided all of the following conditions are met:

(1) The heater is marked with a minimum conductor size.
(2) The conductors are not smaller than the marked minimum size.
(3) A temperature-actuated device controls the cyclic operation of the equipment.

(E) Conductors for Subdivided Loads. Field-wired conductors between the heater and the supplementary overcurrent protective devices shall be sized at not less than 125 percent of the load served. The supplementary overcurrent protective devices specified in 424.22(C) shall protect these conductors in accordance with 240.4.

Where the heaters are rated 50 kW or more, the ampacity of field-wired conductors between the heater and the supplementary overcurrent protective devices shall be permitted to be not less than 100 percent of the load of their respective subdivided circuits, provided all of the following conditions are met:

(1) The heater is marked with a minimum conductor size.
(2) The conductors are not smaller than the marked minimum size.
(3) A temperature-activated device controls the cyclic operation of the equipment.

IV. Marking of Heating Equipment

424.28 Nameplate.

(A) Marking Required. Each unit of fixed electric space-heating equipment shall be provided with a nameplate giving the identifying name and the normal rating in volts and watts or in volts and amperes.

Electric space-heating equipment intended for use on alternating current only or direct current only shall be marked to so indicate. The marking of equipment consisting of motors over ⅛ hp and other loads shall specify the rating of the motor in volts, amperes, and frequency, and the heating load in volts and watts or in volts and amperes.

(B) Location. This nameplate shall be located so as to be visible or easily accessible after installation.

424.29 Marking of Heating Elements.

All heating elements that are replaceable in the field and are a part of an electric heater shall be legibly marked with the ratings in volts and watts or in volts and amperes.

V. Electric Space-Heating Cables

424.34 Heating Cable Construction.

Heating cables shall be furnished complete with factory-assembled nonheating leads at least 2.1 m (7 ft) in length.

424.35 Marking of Heating Cables.

Each unit shall be marked with the identifying name or identification symbol, catalog number, and ratings in volts and watts or in volts and amperes.

Each unit length of heating cable shall have a permanent legible marking on each nonheating lead located within 75 mm (3 in.) of the terminal end. The lead wire shall have the following color identification to indicate the circuit voltage on which it is to be used:

(1) 120 volt, nominal—yellow
(2) 208 volt, nominal—blue
(3) 240 volt, nominal—red
(4) 277 volt, nominal—brown
(5) 480 volt, nominal—orange

424.36 Clearances of Wiring in Ceilings.

Wiring located above heated ceilings shall be spaced not less than 50 mm (2 in.) above the heated ceiling and shall be considered as operating at an ambient temperature of 50°C (122°F). The ampacity of conductors shall be computed on the basis of the correction factors shown in the 0–2000 volt ampacity tables of Article 310. If this wiring is located above thermal insulation having a minimum thickness of 50 mm (2 in.), the wiring shall not require correction for temperature.

424.37 Location of Branch-Circuit and Feeder Wiring in Exterior Walls.

Wiring methods shall comply with Article 300 and 310.10.

424.38 Area Restrictions.

(A) Shall Not Extend Beyond the Room or Area. Heating cables shall not extend beyond the room or area in which they originate.

(B) Uses Prohibited. Heating cables shall not be installed in the following:

(1) In closets
(2) Over walls
(3) Over partitions that extend to the ceiling, unless they are isolated single runs of embedded cable
(4) Over cabinets whose clearance from the ceiling is less than the minimum horizontal dimension of the cabinet to the nearest cabinet edge that is open to the room or area

(C) In Closet Ceilings as Low-Temperature Heat Sources to Control Relative Humidity. The provisions of 424.38(B) shall not prevent the use of cable in closet ceilings as low-temperature heat sources to control relative humidity, provided they are used only in those portions of the ceiling that are unobstructed to the floor by shelves or other permanent luminaires (fixtures).

424.39 Clearance from Other Objects and Openings.

Heating elements of cables shall be separated at least 200 mm (8 in.) from the edge of outlet boxes and junction boxes that are to be used for mounting surface luminaires (lighting fixtures). A clearance of not less than 50 mm (2 in.) shall be provided from recessed luminaires (fixtures) and their trims, ventilating openings, and other such openings in room surfaces. Sufficient area shall be provided to ensure that no heating cable will be covered by any surface-mounted units.

424.40 Splices.

Embedded cables shall be spliced only where necessary and only by approved means, and in no case shall the length of the heating cable be altered.

424.41 Installation of Heating Cables on Dry Board, in Plaster, and on Concrete Ceilings.

(A) In Walls. Cables shall not be installed in walls unless it is necessary for an isolated single run of cable to be installed down a vertical surface to reach a dropped ceiling.

(B) Adjacent Runs. Adjacent runs of cable not exceeding 9 watts/m (2¾ watts/ft) shall not be installed less than 38 mm (1½ in.) on centers.

(C) Surfaces to Be Applied. Heating cables shall be applied only to gypsum board, plaster lath, or other fire-resistant material. With metal lath or other electrically conductive surfaces, a coat of plaster shall be applied to completely separate the metal lath or conductive surface from the cable.

> FPN: See also 424.41(F).

(D) Splices. All heating cables, the splice between the heating cable and nonheating leads, and 75-mm (3-in.) minimum of the nonheating lead at the splice shall be embedded in plaster or dry board in the same manner as the heating cable.

(E) Ceiling Surface. The entire ceiling surface shall have a finish of thermally noninsulating sand plaster that has a nominal thickness of 13 mm (½ in.), or other noninsulating material identified as suitable for this use and applied according to specified thickness and directions.

(F) Secured. Cables shall be secured by means of approved stapling, tape, plaster, nonmetallic spreaders, or other approved means either at intervals not exceeding 400 mm (16 in.) or at intervals not exceeding 1.8 m (6 ft) for cables identified for such use. Staples or metal fasteners that straddle the cable shall not be used with metal lath or other electrically conductive surfaces.

(G) Dry Board Installations. In dry board installations, the entire ceiling below the heating cable shall be covered with gypsum board not exceeding 13 mm (½ in.) thickness. The void between the upper layer of gypsum board, plaster lath, or other fire-resistant material and the surface layer of gypsum board shall be completely filled with thermally conductive, nonshrinking plaster or other approved material or equivalent thermal conductivity.

(H) Free from Contact with Conductive Surfaces. Cables shall be kept free from contact with metal or other electrically conductive surfaces.

(I) Joists. In dry board applications, cable shall be installed parallel to the joist, leaving a clear space centered under the joist of 65 mm (2½ in.) (width) between centers of adjacent runs of cable. A surface layer of gypsum board shall be mounted so that the nails or other fasteners do not pierce the heating cable.

(J) Crossing Joists. Cables shall cross joists only at the ends of the room unless the cable is required to cross joists elsewhere in order to satisfy the manufacturer's instructions that the installer avoid placing the cable too close to ceiling penetrations and luminaires (lighting fixtures).

424.42 Finished Ceilings.

Finished ceilings shall not be covered with decorative panels or beams constructed of materials that have thermal insulating properties, such as wood, fiber, or plastic. Finished ceilings shall be permitted to be covered with paint, wallpaper, or other approved surface finishes.

424.43 Installation of Nonheating Leads of Cables.

(A) Free Nonheating Leads. Free nonheating leads of cables shall be installed in accordance with approved wiring methods from the junction box to a location within the ceiling. Such installations shall be permitted to be single conductors in approved raceways, single or multiconductor Type UF, Type NMC, Type MI, or other approved conductors.

(B) Leads in Junction Box. Not less than 150 mm (6 in.) of free nonheating lead shall be within the junction box. The marking of the leads shall be visible in the junction box.

(C) Excess Leads. Excess leads of heating cables shall not be cut but shall be secured to the underside of the ceiling and embedded in plaster or other approved material, leaving only a length sufficient to reach the junction box with not less than 150 mm (6 in.) of free lead within the box.

424.44 Installation of Cables in Concrete or Poured Masonry Floors.

(A) Watts per Linear Foot. Constant wattage heating cables shall not exceed 54 watts/linear meter (16½ watts/linear foot) of cable.

(B) Spacing Between Adjacent Runs. The spacing between adjacent runs of cable shall not be less than 25 mm (1 in.) on centers.

(C) Secured in Place. Cables shall be secured in place by nonmetallic frames or spreaders or other approved means while the concrete or other finish is applied.

Cables shall not be installed where they bridge expansion joints unless protected from expansion and contraction.

(D) Spacings Between Heating Cable and Metal Embedded in the Floor. Spacings shall be maintained between the heating cable and metal embedded in the floor, unless the cable is a grounded metal-clad cable.

(E) Leads Protected. Leads shall be protected where they leave the floor by rigid metal conduit, intermediate metal conduit, rigid nonmetallic conduit, electrical metallic tubing, or by other approved means.

(F) Bushings or Approved Fittings. Bushings or approved fittings shall be used where the leads emerge within the floor slab.

(G) Ground-Fault Circuit-Interrupter Protection for Heated Floors of Bathrooms, and in Hydromassage Bathtub, Spa, and Hot Tub Locations. Ground-fault circuit-interrupter protection for personnel shall be provided for electrically heated floors in bathrooms, and in hydromassage bathtub, spa, and hot tub locations.

Section 424.44(G) requires the use of GFCI protection where cables are installed in concrete or poured masonry floors, thereby reducing shock hazards to persons with bare feet in these areas. For the 2002 *Code*, the reference to conductive floor coverings was removed to clarify the requirement for GFCI protection in all of the areas identified in 424.44(G), regardless of the type of floor covering over the concrete or poured masonry.

424.45 Inspection and Tests.

Cable installations shall be made with due care to prevent damage to the cable assembly and shall be inspected and approved before cables are covered or concealed.

VI. Duct Heaters

424.57 General.

Part VI shall apply to any heater mounted in the airstream of a forced-air system where the air-moving unit is not provided as an integral part of the equipment.

424.58 Identification.

Heaters installed in an air duct shall be identified as suitable for the installation.

424.59 Airflow.

Means shall be provided to ensure uniform and adequate airflow over the face of the heater in accordance with the manufacturer's instructions.

> FPN: Heaters installed within 1.2 m (4 ft) of the outlet of an air-moving device, heat pump, air conditioner, elbows, baffle plates, or other obstructions in ductwork may require turning vanes, pressure plates, or other devices on the inlet side of the duct heater to ensure an even distribution of air over the face of the heater.

424.60 Elevated Inlet Temperature.

Duct heaters intended for use with elevated inlet air temperature shall be identified as suitable for use at the elevated temperatures.

424.61 Installation of Duct Heaters with Heat Pumps and Air Conditioners.

Heat pumps and air conditioners having duct heaters closer than 1.2 m (4 ft) to the heat pump or air conditioner shall have both the duct heater and heat pump or air conditioner identified as suitable for such installation and so marked.

424.62 Condensation.

Duct heaters used with air conditioners or other air-cooling equipment that could result in condensation of moisture shall be identified as suitable for use with air conditioners.

424.63 Fan Circuit Interlock.

Means shall be provided to ensure that the fan circuit is energized when any heater circuit is energized. However, time- or temperature-controlled delay in energizing the fan motor shall be permitted.

424.64 Limit Controls.

Each duct heater shall be provided with an approved, integral, automatic-reset temperature-limiting control or controllers to de-energize the circuit or circuits.

In addition, an integral independent supplementary control or controllers shall be provided in each duct heater that disconnects a sufficient number of conductors to interrupt current flow. This device shall be manually resettable or replaceable.

424.65 Location of Disconnecting Means.

Duct heater controller equipment shall be either accessible with the disconnecting means installed at or within sight from the controller or as permitted by 424.19(A).

424.66 Installation.

Duct heaters shall be installed in accordance with the manufacturer's instructions in such a manner that operation will not create a hazard to persons or property. Furthermore, duct heaters shall be located with respect to building construction and other equipment so as to permit access to the heater. Sufficient clearance shall be maintained to permit replacement of controls and heating elements and for adjusting and cleaning of controls and other parts requiring such attention. See 110.26.

> FPN: For additional installation information, see NFPA 90A-1999, *Standard for the Installation of Air Conditioning and Ventilating Systems*, and NFPA 90B-1999, *Standard for the Installation of Warm Air Heating and Air Conditioning Systems*.

VII. Resistance-Type Boilers

424.70 Scope.

The provisions in Part VII of this article shall apply to boilers employing resistance-type heating elements. Electrode-type boilers shall not be considered as employing resistance-type heating elements. See Part VIII of this article.

424.71 Identification.

Resistance-type boilers shall be identified as suitable for the installation.

424.72 Overcurrent Protection.

(A) Boiler Employing Resistance-Type Immersion Heating Elements in an ASME Rated and Stamped Vessel. A boiler employing resistance-type immersion heating elements contained in an ASME rated and stamped vessel shall have the heating elements protected at not more than 150 amperes. Such a boiler rated more than 120 amperes shall have the heating elements subdivided into loads not exceeding 120 amperes.

Where a subdivided load is less than 120 amperes, the rating of the overcurrent protective device shall comply with 424.3(B).

(B) Boiler Employing Resistance-Type Heating Elements Rated More Than 48 Amperes and Not Contained in an ASME Rated and Stamped Vessel. A boiler employing resistance-type heating elements not contained in an ASME rated and stamped vessel shall have the heating elements protected at not more than 60 amperes. Such a boiler rated more than 48 amperes shall have the heating elements subdivided into loads not exceeding 48 amperes.

Where a subdivided load is less than 48 amperes, the rating of the overcurrent protective device shall comply with 424.3(B).

> See the commentary following 424.22(B) for an explanation of the subdivision requirement.

(C) Supplementary Overcurrent Protective Devices. The supplementary overcurrent protective devices for the subdivided loads as required by 424.72(A) and (B) shall be as follows:

(1) Factory-installed within or on the boiler enclosure or provided as a separate assembly by the boiler manufacturer
(2) Accessible, but need not be readily accessible
(3) Suitable for branch-circuit protection

Where cartridge fuses are used to provide this overcurrent protection, a single disconnecting means shall be permitted for the several subdivided circuits. See 240.40.

(D) Conductors Supplying Supplementary Overcurrent Protective Devices. The conductors supplying these supplementary overcurrent protective devices shall be considered branch-circuit conductors.

Where the heaters are rated 50 kW or more, the conductors supplying the overcurrent protective device specified in

424.72(C) shall be permitted to be sized at not less than 100 percent of the nameplate rating of the heater, provided all of the following conditions are met:

(1) The heater is marked with a minimum conductor size.
(2) The conductors are not smaller than the marked minimum size.
(3) A temperature- or pressure-actuated device controls the cyclic operation of the equipment.

(E) Conductors for Subdivided Loads. Field-wired conductors between the heater and the supplementary overcurrent protective devices shall be sized at not less than 125 percent of the load served. The supplementary overcurrent protective devices specified in 424.72(C) shall protect these conductors in accordance with 240.4.

Where the heaters are rated 50 kW or more, the ampacity of field-wired conductors between the heater and the supplementary overcurrent protective devices shall be permitted to be not less than 100 percent of the load of their respective subdivided circuits, provided all of the following conditions are met:

(1) The heater is marked with a minimum conductor size.
(2) The conductors are not smaller than the marked minimum size.
(3) A temperature-activated device controls the cyclic operation of the equipment.

424.73 Overtemperature Limit Control.

Each boiler designed so that in normal operation there is no change in state of the heat transfer medium shall be equipped with a temperature-sensitive limiting means. It shall be installed to limit maximum liquid temperature and shall directly or indirectly disconnect all ungrounded conductors to the heating elements. Such means shall be in addition to a temperature regulating system and other devices protecting the tank against excessive pressure.

424.74 Overpressure Limit Control.

Each boiler designed so that in normal operation there is a change in state of the heat transfer medium from liquid to vapor shall be equipped with a pressure-sensitive limiting means. It shall be installed to limit maximum pressure and shall directly or indirectly disconnect all ungrounded conductors to the heating elements. Such means shall be in addition to a pressure regulating system and other devices protecting the tank against excessive pressure.

VIII. Electrode-Type Boilers

424.80 Scope.

The provisions in Part VIII of this article shall apply to boilers for operation at 600 volts, nominal, or less, in which

heat is generated by the passage of current between electrodes through the liquid being heated.

FPN: For over 600 volts, see Part V of Article 490.

424.81 Identification.

Electrode-type boilers shall be identified as suitable for the installation.

424.82 Branch-Circuit Requirements.

The size of branch-circuit conductors and overcurrent protective devices shall be calculated on the basis of 125 percent of the total load (motors not included). A contactor, relay, or other device, approved for continuous operation at 100 percent of its rating, shall be permitted to supply its full-rated load. See 210.19(A), Exception. The provisions of this section shall not apply to conductors that form an integral part of an approved boiler.

Where an electrode boiler is rated 50 kW or more, the conductors supplying the boiler electrode(s) shall be permitted to be sized at not less than 100 percent of the nameplate rating of the electrode boiler, provided all the following conditions are met:

(1) The electrode boiler is marked with a minimum conductor size.
(2) The conductors are not smaller than the marked minimum size.
(3) A temperature- or pressure-actuated device controls the cyclic operation of the equipment.

424.83 Overtemperature Limit Control.

Each boiler designed so that in normal operation there is no change in state of the heat transfer medium shall be equipped with a temperature-sensitive limiting means. It shall be installed to limit maximum liquid temperature and shall directly or indirectly interrupt all current flow through the electrodes. Such means shall be in addition to the temperature regulating system and other devices protecting the tank against excessive pressure.

424.84 Overpressure Limit Control.

Each boiler designed so that in normal operation there is a change in state of the heat transfer medium from liquid to vapor shall be equipped with a pressure-sensitive limiting means. It shall be installed to limit maximum pressure and shall directly or indirectly interrupt all current flow through the electrodes. Such means shall be in addition to a pressure regulating system and other devices protecting the tank against excessive pressure.

424.85 Grounding.

For those boilers designed such that fault currents do not pass through the pressure vessel, and the pressure vessel is electrically isolated from the electrodes, all exposed non–current-carrying metal parts, including the pressure vessel, supply, and return connecting piping, shall be grounded in accordance with Article 250.

For all other designs, the pressure vessel containing the electrodes shall be isolated and electrically insulated from ground.

424.86 Markings.

All electrode-type boilers shall be marked to show the following:

(1) The manufacturer's name
(2) The normal rating in volts, amperes, and kilowatts
(3) The electrical supply required specifying frequency, number of phases, and number of wires
(4) The marking "Electrode-Type Boiler"
(5) A warning marking, "All Power Supplies Shall Be Disconnected Before Servicing, Including Servicing the Pressure Vessel"

The nameplate shall be located so as to be visible after installation.

IX. Electric Radiant Heating Panels and Heating Panel Sets

424.90 Scope.

The provisions of Part IX of this article shall apply to radiant heating panels and heating panel sets.

424.91 Definitions.

Heating Panel. A complete assembly provided with a junction box or a length of flexible conduit for connection to a branch circuit.

Heating Panel Set. A rigid or nonrigid assembly provided with nonheating leads or a terminal junction assembly identified as being suitable for connection to a wiring system.

424.92 Markings.

(A) Markings shall be permanent and in a location that is visible prior to application of panel finish.

(B) Each unit shall be identified as suitable for the installation.

(C) Each unit shall be marked with the identifying name or identification symbol, catalog number, and rating in volts and watts or in volts and amperes.

(D) The manufacturers of heating panels or heating panel sets shall provide marking labels that indicate that the space-heating installation incorporates heating panels or heating

panel sets and instructions that the labels shall be affixed to the panelboards to identify which branch circuits supply the circuits to those space-heating installations. If the heating panels and heating panel set installations are visible and distinguishable after installation, the labels shall not be required to be provided and affixed to the panelboards.

424.93 Installation.

(A) General.

(1) Manufacturer's Instructions. Heating panels and heating panel sets shall be installed in accordance with the manufacturer's instructions.

(2) Locations Not Permitted. The heating portion shall not

(1) Be installed in or behind surfaces where subject to physical damage.
(2) Be run through or above walls, partitions, cupboards, or similar portions of structures that extend to the ceiling.
(3) Be run in or through thermal insulation, but shall be permitted to be in contact with the surface of thermal insulation.

(3) Separation from Outlets for Luminaires (Lighting Fixtures). Edges of panels and panel sets shall be separated by not less than 200 mm (8 in.) from the edges of any outlet boxes and junction boxes that are to be used for mounting surface luminaires (lighting fixtures). A clearance of not less than 50 mm (2 in.) shall be provided from recessed luminaires (fixtures) and their trims, ventilating openings, and other such openings in room surfaces, unless the heating panels and panel sets are listed and marked for lesser clearances, in which case they shall be permitted to be installed at the marked clearances. Sufficient area shall be provided to ensure that no heating panel or heating panel set is to be covered by any surface-mounted units.

(4) Surfaces Covering Heating Panels. After the heating panels or heating panel sets are installed and inspected, it shall be permitted to install a surface that has been identified by the manufacturer's instructions as being suitable for the installation. The surface shall be secured so that the nails or other fastenings do not pierce the heating panels or heating panel sets.

(5) Surface Coverings. Surfaces permitted by 424.93(A)(4) shall be permitted to be covered with paint, wallpaper, or other approved surfaces identified in the manufacturer's instructions as being suitable.

(B) Heating Panel Sets.

(1) Mounting Location. Heating panel sets shall be permitted to be secured to the lower face of joists or mounted in between joists, headers, or nailing strips.

(2) Parallel to Joists or Nailing Strips. Heating panel sets shall be installed parallel to joists or nailing strips.

(3) Installation of Nails, Staples, or Other Fasteners. Nailing or stapling of heating panel sets shall be done only through the unheated portions provided for this purpose. Heating panel sets shall not be cut through or nailed through any point closer than 6 mm (¼ in.) to the element. Nails, staples, or other fasteners shall not be used where they penetrate current-carrying parts.

(4) Installed as Complete Unit. Heating panel sets shall be installed as complete units unless identified as suitable for field cutting in an approved manner.

424.94 Clearances of Wiring in Ceilings.

Wiring located above heated ceilings shall be spaced not less than 50 mm (2 in.) above the heated ceiling and shall be considered as operating at an ambient of 50°C (122°F). The ampacity shall be computed on the basis of the correction factors given in the 0–2000 volt ampacity tables of Article 310. If this wiring is located above thermal insulations having a minimum thickness of 50 mm (2 in.), the wiring shall not require correction for temperature.

424.95 Location of Branch-Circuit and Feeder Wiring in Walls.

(A) Exterior Walls. Wiring methods shall comply with Article 300 and 310.10.

(B) Interior Walls. Any wiring behind heating panels or heating panel sets located in interior walls or partitions shall be considered as operating at an ambient temperature of 40°C (104°F), and the ampacity shall be computed on the basis of the correction factors given in the 0–2000 volt ampacity tables of Article 310.

424.96 Connection to Branch-Circuit Conductors.

(A) General. Heating panels or heating panel sets assembled together in the field to form a heating installation in one room or area shall be connected in accordance with the manufacturer's instructions.

(B) Heating Panels. Heating panels shall be connected to branch-circuit wiring by an approved wiring method.

(C) Heating Panel Sets.

(1) Connection to Branch Circuit Wiring. Heating panel sets shall be connected to branch-circuit wiring by a method identified as being suitable for the purpose.

(2) Panel Sets with Terminal Junction Assembly. A heating panel set provided with terminal junction assembly shall

be permitted to have the nonheating leads attached at the time of installation in accordance with the manufacturer's instructions.

424.97 Nonheating Leads.

Excess nonheating leads of heating panels or heating panel sets shall be permitted to be cut to the required length. They shall meet the installation requirements of the wiring method employed in accordance with 424.96. Nonheating leads shall be an integral part of a heating panel and a heating panel set and shall not be subjected to the ampacity requirements of 424.3(B) for branch circuits.

424.98 Installation in Concrete or Poured Masonry.

(A) Maximum Heated Area. Heating panels or heating panel sets shall not exceed 355 watts/m^2 (33 watts/ft^2) of heated area.

(B) Secured in Place and Identified as Suitable. Heating panels or heating panel sets shall be secured in place by means specified in the manufacturer's instructions and identified as suitable for the installation.

(C) Expansion Joints. Heating panels or heating panel sets shall not be installed where they bridge expansion joints unless provision is made for expansion and contraction.

(D) Spacings. Spacings shall be maintained between heating panels or heating panel sets and metal embedded in the floor. Grounded metal-clad heating panels shall be permitted to be in contact with metal embedded in the floor.

(E) Protection of Leads. Leads shall be protected where they leave the floor by rigid metal conduit, intermediate metal conduit, rigid nonmetallic conduit, or electrical metallic tubing, or by other approved means.

(F) Bushings or Fittings Required. Bushings or approved fittings shall be used where the leads emerge within the floor slabs.

424.99 Installation Under Floor Covering.

(A) Identification. Heating panels or heating panel sets for installation under floor covering shall be identified as suitable for installation under floor covering.

(B) Maximum Heated Area. Heating panels or panel sets, installed under floor covering, shall not exceed 160 watts/m^2 (15 watts/ft^2) of heated area.

(C) Installation. Listed heating panels or panel sets, if installed under floor covering, shall be installed on floor surfaces that are smooth and flat in accordance with the manufacturer's instructions and shall also comply with 424.99(C)(1) through (C)(5).

(1) Expansion Joints. Heating panels or heating panel sets shall not be installed where they bridge expansion joints unless protected from expansion and contraction.

(2) Connection to Conductors. Heating panels and heating panel sets shall be connected to branch-circuit and supply wiring by wiring methods recognized in Chapter 3.

(3) Anchoring. Heating panels and heating panel sets shall be firmly anchored to the floor using an adhesive or anchoring system identified for this use.

(4) Coverings. After heating panels or heating panel sets are installed and inspected, they shall be permitted to be covered by a floor covering that has been identified by the manufacturer as being suitable for the installation. The covering shall be secured to the heating panel or heating panel sets with release-type adhesives or by means identified for this use.

(5) Fault Protection. A device to open all ungrounded conductors supplying the heating panels or heating panel sets, provided by the manufacturer, shall function when a low- or high-resistance line-to-line, line-to-grounded conductor, or line-to-ground fault occurs, such as the result of a penetration of the element or element assembly.

> FPN: An integral grounding shield may be required to provide this protection.

A system using conductive-film heating elements is an example of a heating system that could be installed under a floor covering.

ARTICLE 426
Fixed Outdoor Electric Deicing and Snow-Melting Equipment

Contents

I. General

426.1 Scope.

The requirements of this article shall apply to electrically energized heating systems and the installation of these systems.

(A) Embedded. Embedded in driveways, walks, steps, and other areas.

(B) Exposed. Exposed on drainage systems, bridge structures, roofs, and other structures.

Article 426 includes requirements for resistance heating elements, impedance heating systems, or skin-effect heating systems used for deicing and snow melting. These systems are defined in 426.2.

426.2 Definitions.

For the purpose of this article:

Heating System. A complete system consisting of components such as heating elements, fastening devices, nonheating circuit wiring, leads, temperature controllers, safety signs, junction boxes, raceways, and fittings.

Impedance Heating System. A system in which heat is generated in a pipe or rod, or combination of pipes and rods, by causing current to flow through the pipe or rod by direct connection to an ac voltage source from a dual-winding transformer. The pipe or rod shall be permitted to be embedded in the surface to be heated, or constitute the exposed components to be heated.

Resistance Heating Element. A specific separate element to generate heat that is embedded in or fastened to the surface to be heated.

> FPN: Tubular heaters, strip heaters, heating cable, heating tape, and heating panels are examples of resistance heaters.

Skin-Effect Heating System. A system in which heat is generated on the inner surface of a ferromagnetic envelope embedded in or fastened to the surface to be heated.

> FPN: Typically, an electrically insulated conductor is routed through and connected to the envelope at the other end. The envelope and the electrically insulated

conductor are connected to an ac voltage source from a dual-winding transformer.

426.3 Application of Other Articles.

All requirements of this *Code* shall apply except as specifically amended in this article. Cord-and-plug-connected fixed outdoor electric deicing and snow-melting equipment intended for specific use and identified as suitable for this use shall be installed according to Article 422. Fixed outdoor electric deicing and snow-melting equipment for use in hazardous (classified) locations shall comply with Articles 500 through 516.

426.4 Branch-Circuit Sizing.

The ampacity of branch-circuit conductors and the rating or setting of overcurrent protective devices supplying fixed outdoor electric deicing and snow-melting equipment shall not be less than 125 percent of the total load of the heaters. The rating or setting of overcurrent protective devices shall be permitted in accordance with 240.4(B).

II. Installation

426.10 General.

Equipment for outdoor electric deicing and snow melting shall be identified as being suitable for the following:

(1) The chemical, thermal, and physical environment

(2) Installation in accordance with the manufacturer's drawings and instructions

426.11 Use.

Electrical heating equipment shall be installed in such a manner as to be afforded protection from physical damage.

Underwriters Laboratories Inc. requires that manufacturers of UL listed mat or cable deicing and snow-melting equipment provide specific installation instructions for these products. These instructions supplement the requirements contained in Article 426. For example, if the equipment can only be installed in concrete that is double poured (poured in two parts), the installation instructions will specifically require that installation technique. Where the instructions do not specifically require that a double-pour installation process be used, it is acceptable to use either a single- or double-pour method of installation. See 110.3(B) regarding the installation and use of listed or labeled equipment.

426.12 Thermal Protection.

External surfaces of outdoor electric deicing and snow-melting equipment that operate at temperatures exceeding 60°C (140°F) shall be physically guarded, isolated, or thermally insulated to protect against contact by personnel in the area.

426.13 Identification.

The presence of outdoor electric deicing and snow-melting equipment shall be evident by the posting of appropriate caution signs or markings where clearly visible.

426.14 Special Permission.

Fixed outdoor deicing and snow-melting equipment employing methods of construction or installation other than covered by this article shall be permitted only by special permission.

See the definition of *special permission* in Article 100.

III. Resistance Heating Elements

426.20 Embedded Deicing and Snow-Melting Equipment.

(A) Watt Density. Panels or units shall not exceed 1300 watts/m^2 (120 watts/ft^2) of heated area.

(B) Spacing. The spacing between adjacent cable runs is dependent upon the rating of the cable and shall be not less than 25 mm (1 in.) on centers.

(C) Cover. Units, panels, or cables shall be installed as follows:

(1) On a substantial asphalt or masonry base at least 50 mm (2 in.) thick and have at least 38 mm (1½ in.) of asphalt or masonry applied over the units, panels, or cables; or

(2) They shall be permitted to be installed over other approved bases and embedded within 90 mm (3½ in.) of masonry or asphalt but not less than 38 mm (1½ in.) from the top surface; or

(3) Equipment that has been specially investigated for other forms of installation shall be installed only in the manner for which it has been investigated.

(D) Secured. Cables, units, and panels shall be secured in place by frames or spreaders or other approved means while the masonry or asphalt finish is applied.

(E) Expansion and Contraction. Cables, units, and panels shall not be installed where they bridge expansion joints unless provision is made for expansion and contraction.

426.21 Exposed Deicing and Snow-Melting Equipment.

(A) Secured. Heating element assemblies shall be secured to the surface being heated by approved means.

(B) Overtemperature. Where the heating element is not in direct contact with the surface being heated, the design

of the heater assembly shall be such that its temperature limitations shall not be exceeded.

(C) Expansion and Contraction. Heating elements and assemblies shall not be installed where they bridge expansion joints unless provision is made for expansion and contraction.

(D) Flexural Capability. Where installed on flexible structures, the heating elements and assemblies shall have a flexural capability that is compatible with the structure.

426.22 Installation of Nonheating Leads for Embedded Equipment.

(A) Grounding Sheath or Braid. Nonheating leads having a grounding sheath or braid shall be permitted to be embedded in the masonry or asphalt in the same manner as the heating cable without additional physical protection.

(B) Raceways. All but 25 mm to 150 mm (1 in. to 6 in.) of nonheating leads of Type TW and other approved types not having a grounding sheath shall be enclosed in a rigid conduit, electrical metallic tubing, intermediate metal conduit, or other raceways within asphalt or masonry; and the distance from the factory splice to raceway shall not be less than 25 mm (1 in.) or more than 150 mm (6 in.).

(C) Bushings. Insulating bushings shall be used in the asphalt or masonry where leads enter conduit or tubing.

See 300.4(F) and the associated commentary for further information regarding insulating bushings.

(D) Expansion and Contraction. Leads shall be protected in expansion joints and where they emerge from masonry or asphalt by rigid conduit, electrical metallic tubing, intermediate metal conduit, other raceways, or other approved means.

(E) Leads in Junction Boxes. Not less than 150 mm (6 in.) of free nonheating lead shall be within the junction box.

426.23 Installation of Nonheating Leads for Exposed Equipment.

(A) Nonheating Leads. Power supply nonheating leads (cold leads) for resistance elements shall be suitable for the temperature encountered. Preassembled nonheating leads on approved heaters shall be permitted to be shortened if the markings specified in 426.25 are retained. Not less than 150 mm (6 in.) of nonheating leads shall be provided within the junction box.

(B) Protection. Nonheating power supply leads shall be enclosed in a rigid conduit, intermediate metal conduit, electrical metallic tubing, or other approved means.

426.24 Electrical Connection.

(A) Heating Element Connections. Electrical connections, other than factory connections of heating elements to nonheating elements embedded in masonry or asphalt or on exposed surfaces, shall be made with insulated connectors identified for the use.

(B) Circuit Connections. Splices and terminations at the end of the nonheating leads, other than the heating element end, shall be installed in a box or fitting in accordance with 110.14 and 300.15.

426.25 Marking.

Each factory-assembled heating unit shall be legibly marked within 75 mm (3 in.) of each end of the nonheating leads with the permanent identification symbol, catalog number, and ratings in volts and watts or in volts and amperes.

426.26 Corrosion Protection.

Ferrous and nonferrous metal raceways, cable armor, cable sheaths, boxes, fittings, supports, and support hardware shall be permitted to be installed in concrete or in direct contact with the earth, or in areas subject to severe corrosive influences, where made of material suitable for the condition, or where provided with corrosion protection identified as suitable for the condition.

426.27 Grounding Braid or Sheath.

Grounding means, such as copper braid, metal sheath, or other approved means, shall be provided as part of the heated section of the cable, panel, or unit.

426.28 Equipment Protection.

Ground-fault protection of equipment shall be provided for fixed outdoor electric deicing and snow-melting equipment, except for equipment that employs mineral-insulated, metal-sheathed cable embedded in a noncombustible medium.

Section 426.28 was revised for the 1999 *Code*. Section 426.28 states that ground-fault protection of equipment is to be provided for fixed outdoor electric deicing and snow-melting equipment. Rather than protecting the entire branch circuit, the ground-fault protection requirement is focused on protecting just the equipment itself. This affords the manufacturer and the user an option of providing both circuit

and equipment protection, or just the required equipment protection. This required protection for fixed outdoor deicing and snow-melting equipment may be accomplished by using circuit breakers equipped with ground-fault equipment protection (GFEP) or an integral device supplied as part of the deicing or snow-melting equipment that is sensitive to leakage currents in the magnitude of 6 milliamperes to 50 milliamperes [referred to by UL as ground-fault equipment protection circuit interrupters (GFEPCI)]. These protection devices, if applied properly, will substantially reduce the risk of a fire being started by low-level electrical arcing.

It is important to understand that this required equipment protection is not the same as a ground-fault circuit interrupter used for personal protection that trips at 5 milliamperes (±1 milliampere).

For further information regarding ground-fault equipment protection used to comply with 426.28 and 427.22, refer to the UL *General Information for Electrical Equipment Directory* (White Book), category KCZI.

IV. Impedance Heating

426.30 Personnel Protection.

Exposed elements of impedance heating systems shall be physically guarded, isolated, or thermally insulated with a weatherproof jacket to protect against contact by personnel in the area.

426.31 Isolation Transformer.

A dual-winding transformer with a grounded shield between the primary and secondary windings shall be used to isolate the distribution system from the heating system.

426.32 Voltage Limitations.

Unless protected by ground-fault circuit-interrupter protection for personnel, the secondary winding of the isolation transformer connected to the impedance heating elements shall not have an output voltage greater than 30 volts ac.

Where ground-fault circuit-interrupter protection for personnel is provided, the voltage shall be permitted to be greater than 30 but not more than 80 volts.

426.33 Induced Currents.

All current-carrying components shall be installed in accordance with 300.20.

426.34 Grounding.

An impedance heating system that is operating at a voltage greater than 30 but not more than 80 shall be grounded at a designated point(s).

V. Skin-Effect Heating

426.40 Conductor Ampacity.

The current through the electrically insulated conductor inside the ferromagnetic envelope shall be permitted to exceed the ampacity values shown in Article 310, provided it is identified as suitable for this use.

426.41 Pull Boxes.

Where pull boxes are used, they shall be accessible without excavation by location in suitable vaults or above grade. Outdoor pull boxes shall be of watertight construction.

426.42 Single Conductor in Enclosure.

The provisions of 300.20 shall not apply to the installation of a single conductor in a ferromagnetic envelope (metal enclosure).

426.43 Corrosion Protection.

Ferromagnetic envelopes, ferrous or nonferrous metal raceways, boxes, fittings, supports, and support hardware shall be permitted to be installed in concrete or in direct contact with the earth, or in areas subjected to severe corrosive influences, where made of material suitable for the condition, or where provided with corrosion protection identified as suitable for the condition. Corrosion protection shall maintain the original wall thickness of the ferromagnetic envelope.

426.44 Grounding.

The ferromagnetic envelope shall be grounded at both ends; and, in addition, it shall be permitted to be grounded at intermediate points as required by its design.

The provisions of 250.30 shall not apply to the installation of skin-effect heating systems.

FPN: For grounding methods, see Article 250.

VI. Control and Protection

426.50 Disconnecting Means.

(A) Disconnection. All fixed outdoor deicing and snow-melting equipment shall be provided with a means for disconnection from all ungrounded conductors. Where readily accessible to the user of the equipment, the branch-circuit switch or circuit breaker shall be permitted to serve as the disconnecting means. Switches used as the disconnecting means shall be of the indicating type.

(B) Cord-and-Plug-Connected Equipment. The factory-installed attachment plug of cord-and-plug-connected equipment rated 20 amperes or less and 150 volts or less to ground shall be permitted to be the disconnecting means.

426.51 Controllers.

(A) Temperature Controller with "Off" Position. Temperature controlled switching devices that indicate an "off" position and that interrupt line current shall open all ungrounded conductors when the control device is in the "off" position. These devices shall not be permitted to serve as the disconnecting means unless provided with a positive lockout in the "off" position.

(B) Temperature Controller Without "Off" Position. Temperature controlled switching devices that do not have an "off" position shall not be required to open all ungrounded conductors and shall not be permitted to serve as the disconnecting means.

(C) Remote Temperature Controller. Remote controlled temperature-actuated devices shall not be required to meet the requirements of 426.51(A). These devices shall not be permitted to serve as the disconnecting means.

(D) Combined Switching Devices. Switching devices consisting of combined temperature-actuated devices and manually controlled switches that serve both as the controller and the disconnecting means shall comply with all of the following conditions:

(1) Open all ungrounded conductors when manually placed in the "off" position
(2) Be so designed that the circuit cannot be energized automatically if the device has been manually placed in the "off" position
(3) Be provided with a positive lockout in the "off" position

426.52 Overcurrent Protection.

Fixed outdoor electric deicing and snow-melting equipment shall be permitted to be protected against overcurrent where supplied by a branch circuit as specified in 426.4.

426.54 Cord-and-Plug-Connected Deicing and Snow-Melting Equipment.

Cord-and-plug-connected deicing and snow-melting equipment shall be listed.

According to the UL *General Information for Electrical Equipment Directory* (White Book), category KOBQ, UL listed deicing and snow-melting equipment is provided with means for permanent wiring connection, except the equipment rated 20 amperes or less and 150 volts or less to ground may be of cord-and-plug-connected construction. See the definition of *listed* in Article 100.

ARTICLE 427
Fixed Electric Heating Equipment for Pipelines and Vessels

Contents

I. General

427.1 Scope.

The requirements of this article shall apply to electrically energized heating systems and the installation of these systems used with pipelines or vessels or both.

Article 427 includes requirements for impedance heating, induction heating, and skin-effect heating, in addition to resistance heating elements. Definitions of the various systems are provided in 427.2.

427.2 Definitions.

Impedance Heating System. A system in which heat is generated in a pipeline or vessel wall by causing current to flow through the pipeline or vessel wall by direct connection to an ac voltage source from a dual-winding transformer.

Induction Heating System. A system in which heat is generated in a pipeline or vessel wall by inducing current and hysteresis effect in the pipeline or vessel wall from an external isolated ac field source.

Integrated Heating System. A complete system consisting of components such as pipelines, vessels, heating elements, heat transfer medium, thermal insulation, moisture barrier, nonheating leads, temperature controllers, safety signs, junction boxes, raceways, and fittings.

Pipeline. A length of pipe including pumps, valves, flanges, control devices, strainers, and/or similar equipment for conveying fluids.

Resistance Heating Element. A specific separate element to generate heat that is applied to the pipeline or vessel externally or internally.

> FPN: Tubular heaters, strip heaters, heating cable, heating tape, heating blankets, and immersion heaters are examples of resistance heaters.

Skin-Effect Heating System. A system in which heat is generated on the inner surface of a ferromagnetic envelope attached to a pipeline or vessel, or both.

> FPN: Typically, an electrically insulated conductor is routed through and connected to the envelope at the

other end. The envelope and the electrically insulated conductor are connected to an ac voltage source from a dual-winding transformer.

Vessel. A container such as a barrel, drum, or tank for holding fluids or other material.

427.3 Application of Other Articles.

All requirements of this *Code* shall apply except as specifically amended in this article. Cord-connected pipe heating assemblies intended for specific use and identified as suitable for this use shall be installed according to Article 422. Fixed electric pipeline and vessel heating equipment for use in hazardous (classified) locations shall comply with Articles 500 through 516.

427.4 Branch-Circuit Sizing.

The ampacity of branch-circuit conductors and the rating or setting of overcurrent protective devices that supply fixed electric heating equipment for pipelines and vessels shall be not less than 125 percent of the total load of the heaters. The rating or setting of overcurrent protective devices shall be permitted in accordance with 240.4(B).

II. Installation

427.10 General.

Equipment for pipeline and vessel electrical heating shall be identified as being suitable for (1) the chemical, thermal, and physical environment and (2) installation in accordance with the manufacturer's drawings and instructions.

427.11 Use.

Electrical heating equipment shall be installed in such a manner as to be afforded protection from physical damage.

427.12 Thermal Protection.

External surfaces of pipeline and vessel heating equipment that operate at temperatures exceeding 60°C (140°F) shall be physically guarded, isolated, or thermally insulated to protect against contact by personnel in the area.

427.13 Identification.

The presence of electrically heated pipelines, vessels, or both, shall be evident by the posting of appropriate caution signs or markings at frequent intervals along the pipeline or vessel.

III. Resistance Heating Elements

427.14 Secured.

Heating element assemblies shall be secured to the surface being heated by means other than the thermal insulation.

427.15 Not in Direct Contact.

Where the heating element is not in direct contact with the pipeline or vessel being heated, means shall be provided to prevent overtemperature of the heating element unless the design of the heater assembly is such that its temperature limitations will not be exceeded.

427.16 Expansion and Contraction.

Heating elements and assemblies shall not be installed where they bridge expansion joints unless provisions are made for expansion and contraction.

427.17 Flexural Capability.

Where installed on flexible pipelines, the heating elements and assemblies shall have a flexural capability that is compatible with the pipeline.

427.18 Power Supply Leads.

(A) Nonheating Leads. Power supply nonheating leads (cold leads) for resistance elements shall be suitable for the temperature encountered. Preassembled nonheating leads on approved heaters shall be permitted to be shortened if the markings specified in 427.20 are retained. Not less than 150 mm (6 in.) of nonheating leads shall be provided within the junction box.

(B) Power Supply Leads Protection. Nonheating power supply leads shall be protected where they emerge from electrically heated pipeline or vessel heating units by rigid metal conduit, intermediate metal conduit, electrical metallic tubing, or other raceways identified as suitable for the application.

(C) Interconnecting Leads. Interconnecting nonheating leads connecting portions of the heating system shall be permitted to be covered by thermal insulation in the same manner as the heaters.

427.19 Electrical Connections.

(A) Nonheating Interconnections. Nonheating interconnections, where required under thermal insulation, shall be made with insulated connectors identified as suitable for this use.

(B) Circuit Connections. Splices and terminations outside the thermal insulation shall be installed in a box or fitting in accordance with 110.14 and 300.15.

427.20 Marking.

Each factory-assembled heating unit shall be legibly marked within 75 mm (3 in.) of each end of the nonheating leads with the permanent identification symbol, catalog number, and ratings in volts and watts or in volts and amperes.

427.22 Equipment Protection.

Ground-fault protection of equipment shall be provided for electric heat tracing and heating panels. This requirement shall not apply in industrial establishments where there is alarm indication of ground faults and

(1) Conditions of maintenance and supervision ensure that only qualified persons service the installed systems.
(2) Continued circuit operation is necessary for safe operation of equipment or processes.

Section 427.22 states that electric heat tracing and heating panels must have ground-fault protection for equipment except in industrial establishments where an alarm indication of a ground fault is provided and the two conditions of 427.22 are met.

Rather than protecting the entire branch circuit, the ground-fault protection requirement is focused on protecting just the equipment itself. This affords the manufacturer and the user an option of providing both circuit and equipment protection, or just the required equipment protection. This required protection may be accomplished by using circuit breakers equipped with ground-fault equipment protection (GFEP) or an integral device supplied as part of the pipeline or vessel heating equipment that is sensitive to leakage currents in the magnitude of 6 milliamperes to 50 milliamperes [referred to by UL as ground-fault equipment protection circuit-interrupters (GFEPCI)]. These protection devices, if applied properly, will substantially reduce the risk of a fire being started by low-level electrical arcing.

It is important to understand that this required equipment protection is not the same as a ground-fault circuit interrupter used for personal protection that trips at 5 milliamperes (±1 milliampere).

For further information regarding ground-fault equipment protection used to comply with 426.28 and 427.22, refer to the UL *General Information for Electrical Equipment Directory* (White Book), category KCZI.

427.23 Grounded Conductive Covering.

Electric heating equipment shall be listed and have a grounded conductive covering in accordance with 427.23(A) or (B). The conductive covering shall provide an effective ground path for equipment protection.

The requirement in 427.23 was changed for the 2002 *Code* and requires a listed grounded conductive covering on all heaters. This grounded conductive covering is intended to provide a ground fault current path in order to trip circuit or ground-fault protective devices, thus reducing the potential for fire and electric shock. It also provides added mechanical protection of the heating cable or panel.

(A) Heating Wires or Cables. Heating wires or cables shall have a grounded conductive covering that surrounds the heating element and bus wires, if any, and their electrical insulation.

(B) Heating Panels. Heating panels shall have a grounded conductive covering over the heating element and its electrical insulation on the side opposite the side attached to the surface to be heated.

IV. Impedance Heating

427.25 Personnel Protection.

All accessible external surfaces of the pipeline, vessel, or both, being heated shall be physically guarded, isolated, or thermally insulated (with a weatherproof jacket for outside installations) to protect against contact by personnel in the area.

427.26 Isolation Transformer.

A dual-winding transformer with a grounded shield between the primary and secondary windings shall be used to isolate the distribution system from the heating system.

427.27 Voltage Limitations.

Unless protected by ground-fault circuit-interrupter protection for personnel, the secondary winding of the isolation transformer connected to the pipeline or vessel being heated shall not have an output voltage greater than 30 volts ac.

Where ground-fault circuit-interrupter protection for personnel is provided, the voltage shall be permitted to be greater than 30 but not more than 80 volts.

427.28 Induced Currents.

All current-carrying components shall be installed in accordance with 300.20.

427.29 Grounding.

The pipeline, vessel, or both, being heated that is operating at a voltage greater than 30 but not more than 80 shall be grounded at designated points.

427.30 Secondary Conductor Sizing.

The ampacity of the conductors connected to the secondary of the transformer shall be at least 100 percent of the total load of the heater.

V. Induction Heating

427.35 Scope.

This part covers the installation of line frequency induction heating equipment and accessories for pipelines and vessels.

FPN: See Article 665 for other applications.

427.36 Personnel Protection.

Induction coils that operate or may operate at a voltage greater than 30 volts ac shall be enclosed in a nonmetallic or split metallic enclosure, isolated, or made inaccessible by location to protect personnel in the area.

427.37 Induced Current.

Induction coils shall be prevented from inducing circulating currents in surrounding metallic equipment, supports, or structures by shielding, isolation, or insulation of the current paths. Stray current paths shall be bonded to prevent arcing.

VI. Skin-Effect Heating

427.45 Conductor Ampacity.

The ampacity of the electrically insulated conductor inside the ferromagnetic envelope shall be permitted to exceed the values given in Article 310, provided it is identified as suitable for this use.

427.46 Pull Boxes.

Pull boxes for pulling the electrically insulated conductor in the ferromagnetic envelope shall be permitted to be buried under the thermal insulation, provided their locations are indicated by permanent markings on the insulation jacket surface and on drawings. For outdoor installations, pull boxes shall be of watertight construction.

427.47 Single Conductor in Enclosure.

The provisions of 300.20 shall not apply to the installation of a single conductor in a ferromagnetic envelope (metal enclosure).

427.48 Grounding.

The ferromagnetic envelope shall be grounded at both ends, and, in addition, it shall be permitted to be grounded at intermediate points as required by its design. The ferromagnetic envelope shall be bonded at all joints to ensure electrical continuity.

The provisions of 250.30 shall not apply to the installation of skin-effect heating systems.

FPN: See Article 250 for grounding methods.

VII. Control and Protection

427.55 Disconnecting Means.

(A) Switch or Circuit Breaker. Means shall be provided to disconnect all fixed electric pipeline or vessel heating equipment from all ungrounded conductors. The branch-circuit switch or circuit breaker, where readily accessible to the user of the equipment, shall be permitted to serve as the disconnecting means. The disconnecting means shall be of the indicating type and shall be provided with a positive lockout in the "off" position.

(B) Cord-and-Plug-Connected Equipment. The factory-installed attachment plug of cord-and-plug-connected equipment rated 20 amperes or less and 150 volts or less to ground shall be permitted to be the disconnecting means.

427.56 Controls.

(A) Temperature Control with "Off" Position. Temperature controlled switching devices that indicate an "off" position and that interrupt line current shall open all ungrounded conductors when the control device is in this "off" position. These devices shall not be permitted to serve as the disconnecting means unless provided with a positive lockout in the "off" position.

(B) Temperature Control Without "Off" Position. Temperature controlled switching devices that do not have an "off" position shall not be required to open all ungrounded conductors and shall not be permitted to serve as the disconnecting means.

(C) Remote Temperature Controller. Remote controlled temperature-actuated devices shall not be required to meet the requirements of 427.56(A) and (B). These devices shall not be permitted to serve as the disconnecting means.

(D) Combined Switching Devices. Switching devices consisting of combined temperature-actuated devices and manually controlled switches that serve both as the controllers and the disconnecting means shall comply with all the following conditions:

(1) Open all ungrounded conductors when manually placed in the "off" position
(2) Be designed so that the circuit cannot be energized automatically if the device has been manually placed in the "off" position
(3) Be provided with a positive lockout in the "off" position

427.57 Overcurrent Protection.

Heating equipment shall be considered as protected against overcurrent where supplied by a branch circuit as specified in 427.4.

ARTICLE 430
Motors, Motor Circuits, and Controllers

Contents

I. General

Most electrical equipment is rated in volt-amperes (VA) or watt input. Basic to the understanding of Article 430 is the fact that motors have traditionally been rated in horsepower output. Circuits supplying motors are sized according to the input to the motor (input equals output plus losses of the motor). The losses are not the type of information found on the nameplate of a motor. Tables 430.147 through 430.151(B) contain very accurate industry-wide input ampere ratings for motors.

However, some motors are available with their output ratings expressed in watts and kilowatts. (One horsepower equals approximately 746 watts.) It is important to understand that circuits that supply motors not rated in horsepower must still be sized according to the input of the motor, rated in amperes. Sizing circuits based solely on kilowatt output results in seriously undersized conductors and the improper application of overcurrent devices. See 430.6 for ampacity and motor rating determination.

430.1 Scope.

This article covers motors, motor branch-circuit and feeder conductors and their protection, motor overload protection, motor control circuits, motor controllers, and motor control centers.

FPN No. 1: Installation requirements for motor control centers are covered in 110.26(F). Air conditioning and refrigerating equipment are covered in Article 440.

FPN No. 2: Figure 430.1 is for information only.

Figure 430.1 in the *Code* text is intended to assist the user in following the provisions of Article 430. The requirements for motors in Article 430 are broken down into 13 individual parts. Many of these parts are pictorially identified in *Code* Figure 430.1 as they pertain to the installation of motors.

General, 430.1 through 430.18	Part I
Motor Circuit Conductors, 430.21 through 430.29	Part II
Motor and Branch-Circuit Overload Protection, 430.31 through 430.44	Part III
Motor Branch-Circuit Short-Circuit and Ground-Fault Protection, 430.51 through 430.58	Part IV
Motor Feeder Short-Circuit and Ground-Fault Protection, 430.61 through 430.63	Part V
Motor Control Circuits, 430.71 through 430.74	Part VI
Motor Controllers, 430.81 through 430.91	Part VII
Motor Control Centers, 430.92 through 430.98	Part VIII
Disconnecting Means, 430.101 through 430.113	Part IX
Over 600 Volts, Nominal, 430.121 through 430.127	Part X
Protection of Live Parts—All Voltages, 430.131 through 430.133	Part XI
Grounding—All Voltages, 430.141 through 430.145	Part XII
Tables, Tables 430.147 through 430.151(B)	Part XIII

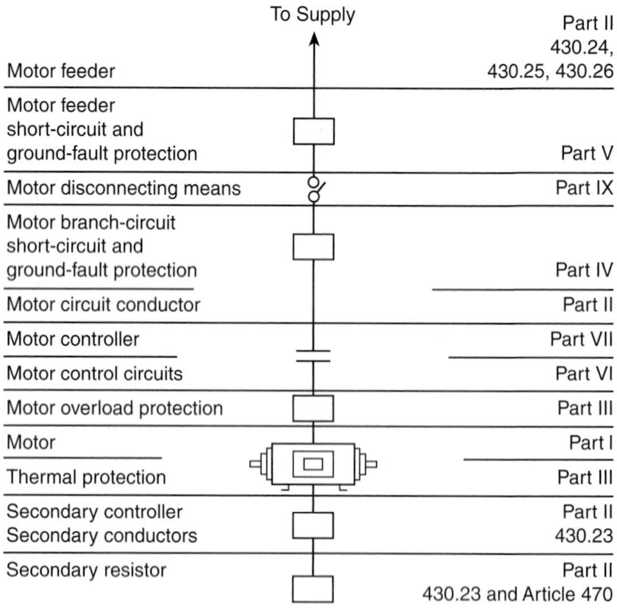

Figure 430.1 Article 430 contents.

430.2 Adjustable-Speed Drive Systems.

The incoming branch circuit or feeder to power conversion equipment included as a part of an adjustable-speed drive system shall be based on the rated input to the power conversion equipment. Where the power conversion equipment is marked to indicate that overload protection is included, additional overload protection shall not be required.

The disconnecting means shall be permitted to be in the incoming line to the conversion equipment and shall have a rating not less than 115 percent of the rated input current of the conversion unit.

FPN: Electrical resonance can result from the interaction of the nonsinusoidal currents from this type of load with power factor correction capacitors.

The operating characteristics of the adjustable-speed drive systems may create harmonic currents that could excite a resonance condition in the circuit if capacitors are improperly applied for power factor correction. Power factor correction with harmonic filtering can safely be applied to this type of load through proper design. For specific details of the harmonic effects related to capacitors, the capacitor manufacturer should be consulted.

430.3 Part-Winding Motors.

A part-winding start induction or synchronous motor is one that is arranged for starting by first energizing part of its primary (armature) winding and, subsequently, energizing the remainder of this winding in one or more steps. A standard part-winding start induction motor is arranged so that one-half of its primary winding can be energized initially, and, subsequently, the remaining half can be energized, both halves then carrying equal current. A hermetic refrigerant compressor motor shall not be considered a standard part-winding start induction motor.

Where separate overload devices are used with a standard part-winding start induction motor, each half of the motor winding shall be individually protected in accordance with 430.32 and 430.37 with a trip current one-half that specified.

Each motor-winding connection shall have branch-circuit short-circuit and ground-fault protection rated at not more than one-half that specified by 430.52.

Exception: A short-circuit and ground-fault protective device shall be permitted for both windings if the device will allow the motor to start. Where time-delay (dual-element) fuses are used, they shall be permitted to have a rating not exceeding 150 percent of the motor full-load current.

430.5 Other Articles.

Motors and controllers shall also comply with the applicable provisions of Table 430.5.

430.6 Ampacity and Motor Rating Determination.

The size of conductors supplying equipment covered by Article 430 shall be selected from the allowable ampacity tables in accordance with 310.15(B) or shall be calculated in accordance with 310.15(C). Where flexible cord is used, the size of the conductor shall be selected in accordance with 400.5. The required ampacity and motor ratings shall be determined as specified in 430.6(A), (B), and (C).

(A) General Motor Applications. For general motor applications, current ratings shall be determined based on (1) and (2).

Table 430.5 Other Articles

Equipment/Occupancy	Article	Section
Air-conditioning and refrigerating equipment	440	
Capacitors		460.8, 460.9
Commercial garages; aircraft hangars; motor fuel dispensing facilities; bulk storage plants; spray application, dipping, and coating processes; and inhalation anesthetizing locations	511, 513, 514, 515, 516, and 517 Part IV	
Cranes and hoists	610	
Electrically driven or controlled irrigation machines	675	
Elevators, dumbwaiters, escalators, moving walks, wheelchair lifts, and stairway chair lifts	620	
Fire pumps	695	
Hazardous (classified) locations	500–503 and 505	
Industrial machinery	670	
Motion picture projectors		540.11 and 540.20
Motion picture and television studios and similar locations	530	
Resistors and reactors	470	
Theaters, audience areas of motion picture and television studios, and similar locations		520.48
Transformers and transformer vaults	450	

(1) Table Values. The values given in Table 430.147, Table 430.148, Table 430.149, and Table 430.150, including notes, shall be used to determine the ampacity of conductors or ampere ratings of switches, branch-circuit short-circuit and ground-fault protection, instead of the actual current rating marked on the motor nameplate. Where a motor is marked in amperes, but not horsepower, the horsepower rating shall be assumed to be that corresponding to the value given in Table 430.147, Table 430.148, Table 430.149, and Table 430.150, interpolated if necessary.

Exception No. 1: Multispeed motors shall be in accordance with 430.22(A) and 430.52.

Exception No. 2: For equipment that employs a shaded-pole or permanent-split capacitor-type fan or blower motor that is marked with the motor type, the full load current for such motor marked on the nameplate of the equipment in which the fan or blower motor is employed shall be used instead of the horsepower rating to determine the ampacity or rating of the disconnecting means, the branch-circuit conductors, the controller, the branch-circuit short-circuit and ground-fault protection, and the separate overload protection. This marking on the equipment nameplate shall not be less than the current marked on the fan or blower motor nameplate.

Exception No. 3: For a listed motor-operated appliance that is marked with both motor horsepower and full-load current, the motor full-load current marked on the nameplate of the appliance shall be used instead of the horsepower rating on the appliance nameplate to determine the ampacity or rating of the disconnecting means, the branch-circuit conductors, the controller, the branch-circuit short-circuit and ground-fault protection, and any separate overload protection.

Exception No. 3 to 430.6(A)(1) is intended to resolve confusion that may result when motor-operated appliances are labeled with both horsepower and ampere ratings. The nameplate current rating in amperes is closely evaluated by the testing laboratories and is considered more accurate than the marked horsepower rating for motor-operated appliances.

(2) Nameplate Values. Separate motor overload protection shall be based on the motor nameplate current rating.

For general motor applications other than ac adjustable voltage motors or torque motors, the ampacity of motor branch-circuit conductors, branch-circuit and ground-fault protection, and ampere rating of the motor disconnecting means are determined by the ampere values listed in Tables 430.147 through 430.150. The ampere values are based on the horsepower rating and nominal voltage listed on the motor nameplate.

The ampere rating provided on the motor nameplate is used to size the overload protective devices intended to protect the motor, motor control apparatus, and motor branch-circuit conductors.

(B) Torque Motors. For torque motors, the rated current shall be locked-rotor current, and this nameplate current shall be used to determine the ampacity of the branch-circuit conductors covered in 430.22 and 430.24, the ampere rating of the motor overload protection, and the ampere rating of motor branch-circuit short-circuit and ground-fault protection in accordance with 430.52(B).

FPN: For motor controllers and disconnecting means, see 430.83(D) and 430.110.

(C) Alternating-Current Adjustable Voltage Motors. For motors used in alternating-current, adjustable voltage,

variable torque drive systems, the ampacity of conductors, or ampere ratings of switches, branch-circuit short-circuit and ground-fault protection, and so forth, shall be based on the maximum operating current marked on the motor or control nameplate, or both. If the maximum operating current does not appear on the nameplate, the ampacity determination shall be based on 150 percent of the values given in Tables 430.149 and 430.150.

430.7 Marking on Motors and Multimotor Equipment.

(A) Usual Motor Applications. A motor shall be marked with the following information.

(1) Manufacturer's name.
(2) Rated volts and full-load amperes. For a multispeed motor, full-load amperes for each speed, except shaded-pole and permanent-split capacitor motors where amperes are required only for maximum speed.
(3) Rated frequency and number of phases if an ac motor.
(4) Rated full-load speed.
(5) Rated temperature rise or the insulation system class and rated ambient temperature.
(6) Time rating. The time rating shall be 5, 15, 30, or 60 minutes, or continuous.
(7) Rated horsepower if ⅛ hp or more. For a multispeed motor ⅛ hp or more, rated horsepower for each speed, except shaded-pole and permanent-split capacitor motors ⅛ hp or more where rated horsepower is required only for maximum speed. Motors of arc welders are not required to be marked with the horsepower rating.
(8) Code letter or locked-rotor amperes if an alternating-current motor rated ½ hp or more. On polyphase wound-rotor motors, the code letter shall be omitted.

FPN: See 430.7(B).

(9) Design letter for design B, C, D, or E motors.

FPN: Motor design letter definitions are found in ANSI/NEMA MG 1-1993, *Motors and Generators, Part 1, Definitions,* and in IEEE 100-1996, *Standard Dictionary of Electrical and Electronic Terms.*

Design letters indicate a motor's speed/torque characteristic curve and are not to be confused with code letters. For technical accuracy, code letters should be referred to as "locked-rotor indicating code letters," which are explained in 430.7(B). Design letters reflect characteristics inherent in motor design, such as locked-rotor current, slip at rated load, and locked-rotor and breakdown torque.

Design E motors have significant differences from the older and very common Design B motors. For example, Design E motors are very energy efficient (lower I^2R losses) and have high locked-rotor currents [see Table 430.151(B)].

Some Design E motors may even have higher full-load currents than their equivalent Design B motors. It is imperative that these electrical differences be understood before a Design E motor is used for an initial installation or as a replacement.

Where existing motors are being considered for replacement with Design E motors, the existing motor circuits must be evaluated for minimum *Code* compliance in areas such as disconnect rating, controller rating, short-circuit and ground-fault protection, and overload protection, as well as standard design compliance issues such as starting voltage-drop or power quality impact on the facility. Of course, it is prudent, where energy conservation is concerned, to also upgrade the mechanical portion of the system, since the electric motor does not reflect all the losses of a typical mechanical system.

Because Design E motors differ greatly, various sections in Article 430 have been adjusted to safely permit the installation and proper use of Design E motors for this edition as well as in previous editions of the *NEC*.

(10) Secondary volts and full-load amperes if a wound-rotor induction motor.

(11) Field current and voltage for dc excited synchronous motors.

(12) Winding—straight shunt, stabilized shunt, compound, or series, if a dc motor. Fractional horsepower dc motors 175 mm (7 in.) or less in diameter shall not be required to be marked.

(13) A motor provided with a thermal protector complying with 430.32(A)(2) or (B)(2) shall be marked "Thermally Protected." Thermally protected motors rated 100 watts or less and complying with 430.32(B)(2) shall be permitted to use the abbreviated marking "T.P."

(14) A motor complying with 430.32(B)(4) shall be marked "Impedance Protected." Impedance protected motors rated 100 watts or less and complying with 430.32(B)(4) shall be permitted to use the abbreviated marking "Z.P."

(B) Locked-Rotor Indicating Code Letters. Code letters marked on motor nameplates to show motor input with locked rotor shall be in accordance with Table 430.7(B).

The code letter indicating motor input with locked rotor shall be in an individual block on the nameplate, properly designated.

(1) Multispeed Motors. Multispeed motors shall be marked with the code letter designating the locked-rotor kilovolt-ampere (kVA) per horsepower for the highest speed at which the motor can be started.

Exception: Constant horsepower multispeed motors shall be marked with the code letter giving the highest locked-rotor kilovolt-ampere (kVA) per horsepower.

Table 430.7(B) Locked-Rotor Indicating Code Letters

Code Letter	Kilovolt-Amperes per Horsepower with Locked Rotor
A	0–3.14
B	3.15–3.54
C	3.55–3.99
D	4.0–4.49
E	4.5–4.99
F	5.0–5.59
G	5.6–6.29
H	6.3–7.09
J	7.1–7.99
K	8.0–8.99
L	9.0–9.99
M	10.0–11.19
N	11.2–12.49
P	12.5–13.99
R	14.0–15.99
S	16.0–17.99
T	18.0–19.99
U	20.0–22.39
V	22.4 and up

(2) Single-Speed Motors. Single-speed motors starting on wye connection and running on delta connections shall be marked with a code letter corresponding to the locked-rotor kilovolt-ampere (kVA) per horsepower for the wye connection.

(3) Dual-Voltage Motors. Dual-voltage motors that have a different locked-rotor kilovolt-ampere (kVA) per horsepower on the two voltages shall be marked with the code letter for the voltage giving the highest locked-rotor kilovolt-ampere (kVA) per horsepower.

(4) 50/60 Hz Motors. Motors with 50- and 60-Hz ratings shall be marked with a code letter designating the locked-rotor kilovolt-ampere (kVA) per horsepower on 60 Hz.

(5) Part Winding Motors. Part-winding start motors shall be marked with a code letter designating the locked-rotor kilovolt-ampere (kVA) per horsepower that is based on the locked-rotor current for the full winding of the motor.

The following example shows how to determine the locked-rotor current for a specific motor using the values from Table 430.7(B).

Example

Using Table 430.7(B), find the maximum locked-rotor current for a 20-hp, 460-volt, 3-phase motor with a nameplate kilovolt-ampere code letter "G."

Solution

Table 430.7(B) lists a range of values for "G." The maximum value in the range is 6.29, or 6.29 kilovolt-amperes per horsepower. Use the following formula to find the maximum locked-rotor current.

$$\frac{\text{Locked-rotor}}{\text{amperes}} = \text{motor hp}$$
$$\times \text{maximum code letter value } \frac{\text{kVA}}{\text{hp}}$$
$$\times \frac{1000}{\text{volts} \times 1.73}$$

Substitute as follows:

$$\frac{\text{Locked-rotor}}{\text{amperes}} = 20 \text{ hp} \times 6.29 \frac{\text{kVA}}{\text{hp}} \times \frac{1000}{460 \text{ volts} \times 1.73}$$
$$= 158 \text{ amperes}$$

Therefore, the maximum locked-rotor current for a 20-hp, 460-volt motor with a code letter "G" is 158 amperes.

(C) Torque Motors. Torque motors are rated for operation at standstill and shall be marked in accordance with 430.7(A), except that locked-rotor torque shall replace horsepower.

(D) Multimotor and Combination-Load Equipment.

(1) Factory-Wired. Multimotor and combination-load equipment shall be provided with a visible nameplate marked with the manufacturer's name, the rating in volts, frequency, number of phases, minimum supply circuit conductor ampacity, and the maximum ampere rating of the circuit short-circuit and ground-fault protective device. The conductor ampacity shall be computed in accordance with 430.24 and counting all of the motors and other loads that will be operated at the same time. The short-circuit and ground-fault protective device rating shall not exceed the value computed in accordance with 430.53. Multimotor equipment for use on two or more circuits shall be marked with the preceding information for each circuit.

The nameplate marking for the maximum ampere rating of the branch-circuit short-circuit and ground-fault protective device may limit the type of protective device to a fuse by stipulating "fuse" without reference to a circuit breaker. This means that the circuit to the equipment must be protected by fuses, such as by a fused disconnect switch. The fused switch may be supplied from a circuit breaker in a panelboard.

(2) Not Factory-Wired. Where the equipment is not factory-wired and the individual nameplates of motors and other loads are visible after assembly of the equipment, the individual nameplates shall be permitted to serve as the required marking.

430.8 Marking on Controllers.

A controller shall be marked with the manufacturer's name or identification, the voltage, the current or horsepower rating, and such other necessary data to properly indicate the motors for which it is suitable. A controller that includes motor overload protection suitable for group motor application shall be marked with the motor overload protection and the maximum branch-circuit short-circuit and ground-fault protection for such applications.

Combination controllers that employ adjustable instantaneous trip circuit breakers shall be clearly marked to indicate the ampere settings of the adjustable trip element.

Where a controller is built-in as an integral part of a motor or of a motor-generator set, individual marking of the controller shall not be required if the necessary data are on the nameplate. For controllers that are an integral part of equipment approved as a unit, the above marking shall be permitted on the equipment nameplate.

FPN: See 110.10 for information on circuit impedance and other characteristics.

430.9 Terminals.

(A) Markings. Terminals of motors and controllers shall be suitably marked or colored where necessary to indicate the proper connections.

(B) Conductors. Motor controllers and terminals of control circuit devices shall be connected with copper conductors unless identified for use with a different conductor.

Terminals for motor controllers are tested, designed, and listed using copper conductors, unless marked for other conductors. Section 430.9(B) highlights this limitation while permitting other conductors to be used if they are determined to be suitable and are identified as such.

(C) Torque Requirements. Control circuit devices with screw-type pressure terminals used with 14 AWG or smaller copper conductors shall be torqued to a minimum of 0.8 N•m (7 lb-in.) unless identified for a different torque value.

Proper torque is essential for safe and reliable connections. Section 430.9(C) enhances safety by providing a minimum torque value for screw-type pressure terminals. See the commentary following 110.14(B) for more information on electrical connections.

430.10 Wiring Space in Enclosures.

(A) General. Enclosures for motor controllers and disconnecting means shall not be used as junction boxes, auxiliary gutters, or raceways for conductors feeding through or tapping off to the other apparatus unless designs are employed that provide adequate space for this purpose.

FPN: See 312.8 for switch and overcurrent-device enclosures.

During the planning stages of a motor(s) installation, location and proper working spaces (as required by 110.26 and 110.34) for motor controllers and disconnects should be considered, including provisions for the use of auxiliary gutters or junction boxes, to ensure space for conductors feeding through or tapping off to other apparatus. For switch and overcurrent device enclosures, see 312.8 and the commentary that follows.

(B) Wire-Bending Space in Enclosures. Minimum wire-bending space within the enclosures for motor controllers shall be in accordance with Table 430.10(B) where measured in a straight line from the end of the lug or wire connector (in the direction the wire leaves the terminal) to the wall or barrier. Where alternate wire termination means are substituted for that supplied by the manufacturer of the controller, they shall be of a type identified by the manufacturer for use with the controller and shall not reduce the minimum wire-bending space.

Table 430.10(B) Minimum Wire-Bending Space at the Terminals of Enclosed Motor Controllers

Size of Wire (AWG or kcmil)	Wires per Terminal*			
	1		2	
	mm	in.	mm	in.
14–10	Not specified		—	—
8–6	38	1½	—	—
4–3	50	2	—	—
2	65	2½	—	—
1	75	3	—	—
1/0	125	5	125	5
2/0	150	6	150	6
3/0–4/0	175	7	175	7
250	200	8	200	8
300	250	10	250	10
350–500	300	12	300	12
600–700	350	14	400	16
750–900	450	18	475	19

*Where provision for three or more wires per terminal exists, the minimum wire-bending space shall be in accordance with the requirements of Article 312.

Exhibit 430.1 illustrates application of the wiring bending space requirements of either 430.10(B) or 312.6(B) within an enclosure for a motor controller.

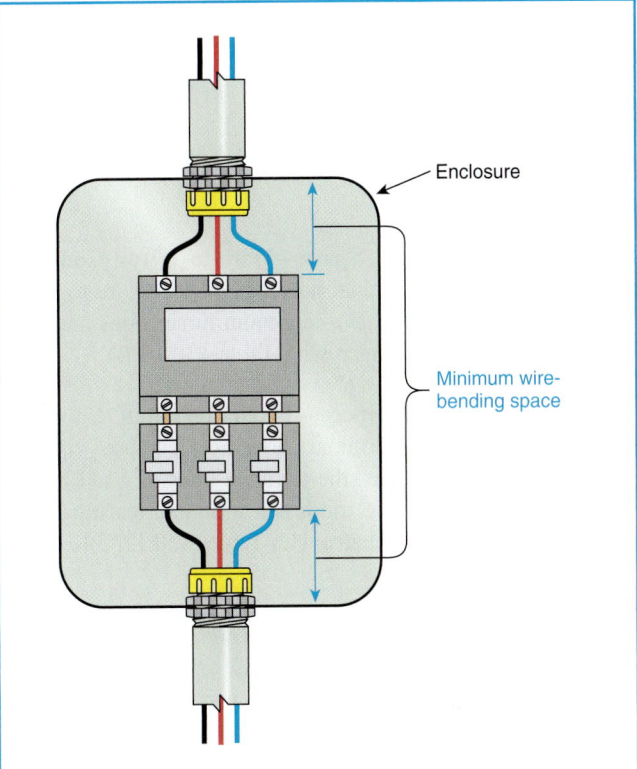

Exhibit 430.1 Wire-bending space in enclosures for motor controllers.

430.11 Protection Against Liquids.

Suitable guards or enclosures shall be provided to protect exposed current-carrying parts of motors and the insulation of motor leads where installed directly under equipment, or in other locations where dripping or spraying oil, water, or other injurious liquid may occur, unless the motor is designed for the existing conditions.

Exposed current-carrying parts and insulated leads of motors should be suitably protected from liquids (dripping or spraying oil or other lubricants, water, or excessive moisture). The presence of liquids may cause deterioration and insulation breakdown.

430.12 Motor Terminal Housings.

(A) Material. Where motors are provided with terminal housings, the housings shall be of metal and of substantial construction.

Exception: In other than hazardous (classified) locations, substantial, nonmetallic, nonburning housings shall be permitted, provided an internal grounding means between the motor frame and the equipment grounding connection is incorporated within the housing.

Nonmetallic terminal housings are permitted on motors of any size, provided the housing material has been determined to be nonburning, with, for example, a flammability rating of at least 94-5V in accordance with UL 746C, *Polymeric Materials—Use in Electrical Equipment Evaluations.*

(B) Dimensions and Space—Wire-to-Wire Connections. Where these terminal housings enclose wire-to-wire connections, they shall have minimum dimensions and usable volumes in accordance with Table 430.12(B).

(C) Dimensions and Space—Fixed Terminal Connections. Where these terminal housings enclose rigidly mounted motor terminals, the terminal housing shall be of sufficient size to provide minimum terminal spacings and usable volumes in accordance with Tables 430.12(C)(1) and 430.12(C)(2).

(D) Large Wire or Factory Connections. For motors with larger ratings, greater number of leads, or larger wire sizes, or where motors are installed as a part of factory-wired equipment, without additional connection being required at the motor terminal housing during equipment installation, the terminal housing shall be of ample size to make connections, but the foregoing provisions for the volumes of terminal housings shall not be considered applicable.

(E) Equipment Grounding Connections. A means for attachment of an equipment grounding conductor termination in accordance with 250.8 shall be provided at motor terminal housings for wire-to-wire connections or fixed terminal connections. The means for such connections shall be permitted to be located either inside or outside the motor terminal housing.

Exception: Where a motor is installed as a part of factory-wired equipment that is required to be grounded and without additional connection being required at the motor terminal housing during equipment installation, a separate means for motor grounding at the motor terminal housing shall not be required.

430.13 Bushing.

Where wires pass through an opening in an enclosure, conduit box, or barrier, a bushing shall be used to protect the conductors from the edges of openings having sharp edges. The bushing shall have smooth, well-rounded surfaces where it may be in contact with the conductors. If used where oils, greases, or other contaminants may be present, the bushing shall be made of material not deleteriously affected.

FPN: For conductors exposed to deteriorating agents, see 310.9.

Table 430.12(B) Terminal Housings — Wire-to-Wire Connections Motors 275 mm (11 in.) in Diameter or Less

Horsepower	Cover Opening Minimum Dimension		Usable Volume Minimum	
	mm	in.	cm³	in.³
1 and smaller[a]	41	1⅝	170	10.5
1½, 2, and 3[b]	45	1¾	275	16.8
5 and 7½	50	2	365	22.4
10 and 15	65	2½	595	36.4

Motors Over 275 mm (11 in.) in Diameter — Alternating-Current Motors

Maximum Full Load Current for 3-Phase Motors with Maximum of 12 Leads (Amperes)	Terminal Box Cover Opening Minimum Dimension		Usable Volume Minimum		Typical Maximum Horsepower 3-Phase	
	mm	in.	cm³	in.³	230 Volt	460 Volt
45	65	2.5	595	36.4	15	30
70	84	3.3	1,265	77	25	50
110	100	4.0	2,295	140	40	75
160	125	5.0	4,135	252	60	125
250	150	6.0	7,380	450	100	200
400	175	7.0	13,775	840	150	300
600	200	8.0	25,255	1540	250	500

Direct-Current Motors

Maximum Full-Load Current for Motors with Maximum of 6 Leads (Amperes)	Terminal Box Minimum Dimensions		Usable Volume Minimum	
	mm	in.	cm³	in.³
68	65	2.5	425	26
105	84	3.3	900	55
165	100	4.0	1,640	100
240	125	5.0	2,950	180
375	150	6.0	5,410	330
600	175	7.0	9,840	600
900	200	8.0	18,040	1,100

Note: Auxiliary leads for such items as brakes, thermostats, space heaters, and exciting fields shall be permitted to be neglected if their current-carrying area does not exceed 25 percent of the current-carrying area of the machine power leads. [a]For motors rated 1 hp and smaller and with the terminal housing partially or wholly integral with the frame or end shield, the volume of the terminal housing shall not be less than 18.0 cm³ (1.1 in.³) per wire-to-wire connection. The minimum cover opening dimension is not specified.
[b]For motors rated 1½, 2, and 3 hp and with the terminal housing partially or wholly integral with the frame or end shield, the volume of the terminal housing shall not be less than 23.0 cm³ (1.4 in.³) per wire-to-wire connection. The minimum cover opening dimension is not specified.

Table 430.12(C)(1) Terminal Spacings — Fixed Terminals

	Minimum Spacing			
	Between Line Terminals		Between Line Terminals and Other Uninsulated Metal Parts	
Nominal Volts	mm	in.	mm	in.
240 or less	6	¼	6	¼
Over 250 – 600	10	⅜	10	⅜

Table 430.12(C)(2) Usable Volumes — Fixed Terminals

Power-Supply Conductor Size (AWG)	Minimum Usable Volume per Power-Supply Conductor	
	cm³	in.³
14	16	1
12 and 10	20	1¼
8 and 6	37	2¼

430.14 Location of Motors.

(A) Ventilation and Maintenance. Motors shall be located so that adequate ventilation is provided and so that maintenance, such as lubrication of bearings and replacing of brushes, can be readily accomplished.

Exception: Ventilation shall not be required for submersible types of motors.

(B) Open Motors. Open motors that have commutators or collector rings shall be located or protected so that sparks cannot reach adjacent combustible material.

Exception: Installation of these motors on wooden floors or supports shall be permitted.

430.16 Exposure to Dust Accumulations.

In locations where dust or flying material collects on or in motors in such quantities as to seriously interfere with the ventilation or cooling of motors and thereby cause dangerous temperatures, suitable types of enclosed motors that do not overheat under the prevailing conditions shall be used.

> FPN: Especially severe conditions may require the use of enclosed pipe-ventilated motors, or enclosure in separate dusttight rooms, properly ventilated from a source of clean air.

For motors exposed to combustible dust or readily ignitible flying material, see the requirements of 502.8 and 502.9 (Class II, Divisions 1 and 2) and 503.6 and 503.7 (Class III, Divisions 1 and 2). For classification of locations, see 500.5(C) (Class II locations) and 500.(D) (Class III locations).

430.17 Highest Rated or Smallest Rated Motor.

In determining compliance with 430.24, 430.53(B), and 430.53(C), the highest rated or smallest rated motor shall be based on the rated full-load current as selected from Tables 430.147, 430.148, 430.149, and 430.150.

430.18 Nominal Voltage of Rectifier Systems.

The nominal value of the ac voltage being rectified shall be used to determine the voltage of a rectifier derived system.

Exception: The nominal dc voltage of the rectifier shall be used if it exceeds the peak value of the ac voltage being rectified.

II. Motor Circuit Conductors

430.21 General.

Part II specifies ampacities of conductors that are capable of carrying the motor current without overheating under the conditions specified.

The provisions of Part II shall not apply to motor circuits rated over 600 volts, nominal.

The provisions of Articles 250, 300, and 310 shall not apply to conductors that form an integral part of equipment, such as motors, motor controllers, motor control centers, or other factory-assembled control equipment.

> FPN No. 1: See 300.1(B) and 310.1 for similar requirements.
> FPN No. 2: See 110.14(C) and 430.9(B) for equipment device terminal requirements.
> FPN No. 3: For over 600 volts, nominal, see Part XI.

430.22 Single Motor.

(A) General. Branch-circuit conductors that supply a single motor used in a continuous duty application shall have an ampacity of not less than 125 percent of the motor's full-load current rating as determined by 430.6(A)(1).

Exception No. 1: For dc motors operating from a rectified single-phase power supply, the conductors between the field wiring terminals of the rectifier and the motor shall have an ampacity of not less than the following percent of the motor full-load current rating:

(a) Where a rectifier bridge of the single-phase half-wave type is used, 190 percent.

(b) Where a rectifier bridge of the single-phase full-wave type is used, 150 percent.

Exception No. 2: Circuit conductors supplying power conversion equipment included as part of an adjustable-speed drive system shall have an ampacity not less than 125 percent of the rated input to the power conversion equipment.

Section 430.22 was editorially rewritten for the 2002 *Code*. Section 430.22 describes the branch-circuit requirements for single motor installations. Generally, the branch circuit that serves a continuous duty motor must be sized at 125 percent of the motor full-load current or greater. The provision for a conductor with an ampacity of at least 125 percent of the motor full-load current rating does not constitute a conductor derating; rather, it is based on the need to provide for a sustained running current that is greater than the rated full-load current and for protection of the conductors by the motor overload protective device set above the motor full-load current rating.

The ampacity of the motor branch-circuit conductors is based on the full-load current rating values provided in Tables 430.147 through 430.150. Motor nameplate full-load current is not to be used to size branch-circuit conductors.

Exhibit 430.2 illustrates each motor on an individual branch circuit with branch-circuit short-circuit and ground-fault protective devices and disconnecting means in one location, and controllers and overload protection at the motor locations.

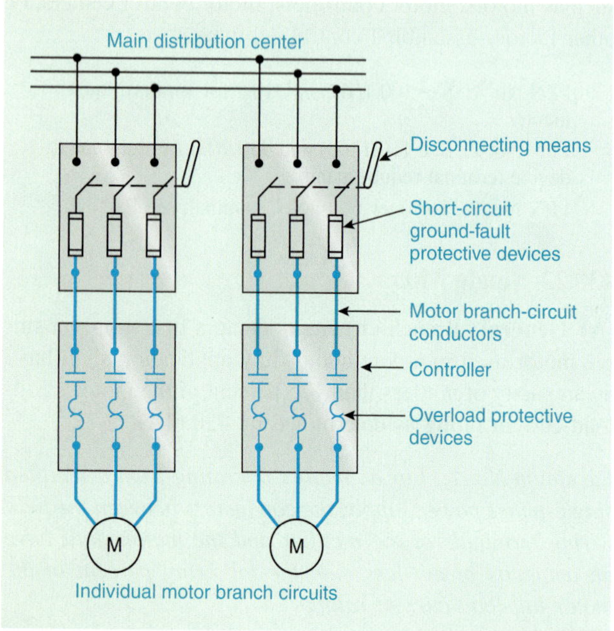

Exhibit 430.3 also illustrates each motor on an individual branch circuit, but, unlike Exhibit 430.2, the branch circuits are tapped from a feeder at a convenient location, such as a junction box or wireway, or from open wiring. The tap conductors are required to terminate in a branch-circuit protective device located not more than 25 ft from where the taps are connected to the feeder, in accordance with 430.28. Also see 430.28, Exception, which permits a 100-ft tap under some conditions in high-bay manufacturing facilities.

If motors are connected to a 15- or 20-ampere branch circuit that also supplies lighting or other appliance loads, as illustrated in Exhibit 430.4, the provisions of Articles 210 and 430 apply. Motors rated less than 1 hp may be connected to these circuits, and they must be provided with overload protective devices unless the motors are not permanently installed, are started manually, and are within sight from the controller location. For additional information on the installation of motors (1 hp or less), see 430.32(B) and (C) and 430.53(A).

Exhibit 430.5 illustrates the following essential parts of a motor branch circuit:

1. Branch-circuit conductors
2. Disconnecting means

Exhibit 430.2 A main distribution center supplying individual branch circuits to each motor (branch-circuit short-circuit and ground-fault protective devices).

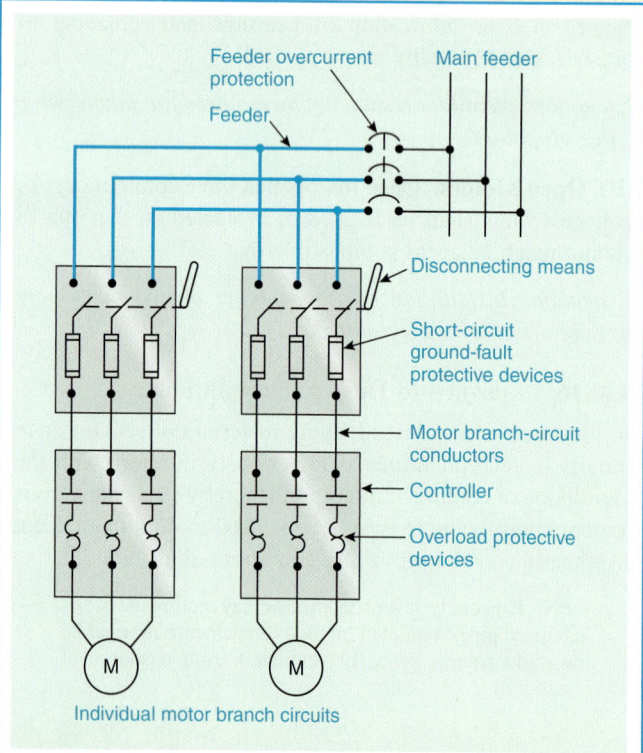

Exhibit 430.3 A feeder supplying individual branch circuits to each motor.

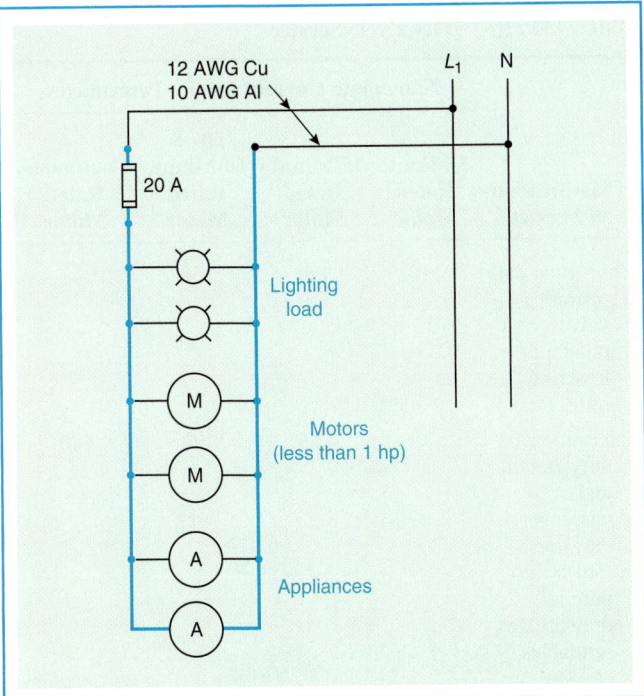

Exhibit 430.4 A 20-ampere branch circuit supplying lighting, small motors, and appliances.

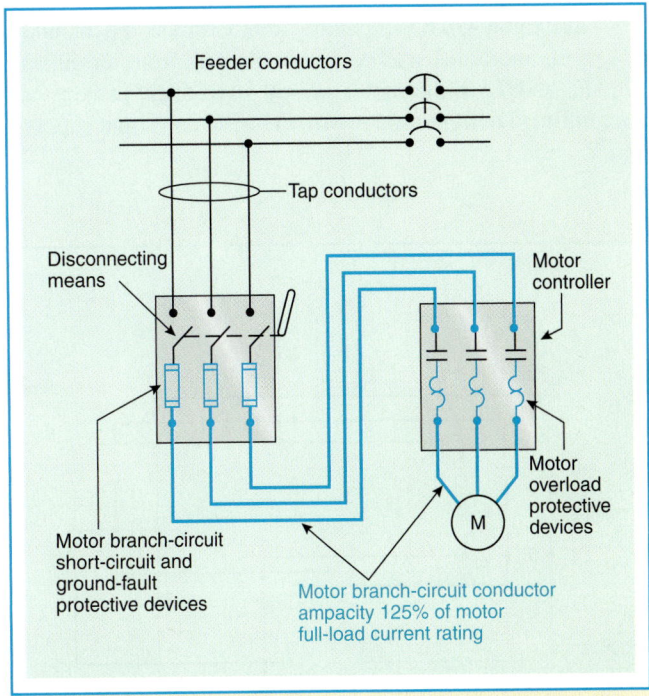

Exhibit 430.5 A motor branch circuit showing the essential parts.

3. Branch-circuit short-circuit and ground-fault protective devices
4. Motor-controller
5. Motor overload protective devices

The branch-circuit short-circuit and ground-fault protective device may be a fuse or a circuit breaker and must be capable of carrying the starting current of the motor without opening the circuit. See Table 430.52.

In general, every motor must be provided with overload protective devices intended to protect the motor windings, motor-control apparatus, and motor branch-circuit conductors against excessive heating due to motor overloads and failure to start. Overload in equipment is defined as operation in excess of normal full-load rating, which, when it persists for a sufficient length of time, will cause damage or dangerous overheating. Overload in a motor includes a stalled rotor but does not include fault currents due to short circuits or ground faults. See 430.44 for conditions where providing automatic opening of a motor circuit due to overload may be objectionable.

(B) Multispeed Motor. For a multispeed motor, the selection of branch-circuit conductors on the line side of the controller shall be based on the highest of the full-load current ratings shown on the motor nameplate. The selection

of branch-circuit conductors between the controller and the motor shall be based on the current rating of the winding(s) that the conductors energize.

(C) Wye-Start, Delta-Run Motor. For a wye-start, delta-run connected motor, the selection of branch-circuit conductors on the line side of the controller shall be based on the motor full-load current. The selection of conductors between the controller and the motor shall be based on 58 percent of the motor full-load current.

A wye-start, delta-run winding configuration is a method of providing reduced-voltage starting for a polyphase induction motor. This method requires a specific type of motor controller and a delta-wired motor with all leads brought out to the terminal box. This method of starting finds wide application in certain compressors used for air conditioning and where the driven machinery is allowed to start unloaded. During starting, the windings are configured in a wye configuration. The wye-start configuration results in a reduced starting voltage of a mathematical ratio of $1/\sqrt{3} = 0.5774$, or 58 percent of the full line voltage, which results in approximately 58 percent starting current and about one-third of the normal starting torque. Once the motor attains speed, the windings are reconfigured to run as delta, giving full line voltage to the individual windings, which allows the motor to have full torque capability.

In Exhibit 430.6, conductors from terminals T_1, T_2, and T_3 to the motor, as well as the conductors from terminals T_4, T_5, and T_6 to the motor, are all sized at 58 percent of the full-load current used to size the conductors that supply L_1, L_2, and L_3.

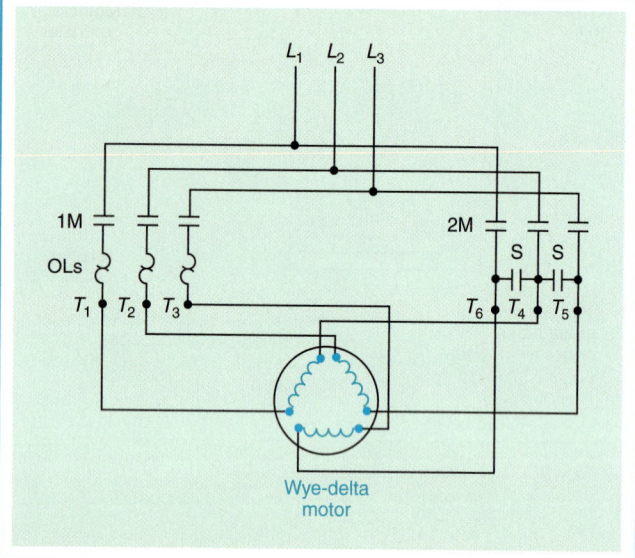

Exhibit 430.6 An elementary wiring diagram of a typical wye-start, delta-run motor and controller. During START, contacts 1M and S are closed and contacts 2M are open. During RUN, contacts 1M and 2M are closed and contacts S are open. (Redrawn from Square D Co.)

(D) Part-Winding Motor. For a part-winding connected motor, the selection of branch-circuit conductors on the line side of the controller shall be based on the motor full-load current. The selection of conductors between the controller and the motor shall be based on 50 percent of the motor full-load current.

(E) Other Than Continuous Duty. Conductors for a motor used in a short-time, intermittent, periodic, or varying duty application shall have an ampacity of not less than the percentage of the motor nameplate current rating shown in Table 430.22(E), unless the authority having jurisdiction grants special permission for conductors of lower ampacity.

Most motor applications are continuous duty, meaning they operate at a constant load for an indefinitely long time. For motors that are not continuous duty, the motor nameplate currents and Table 430.22(E) are used to determine the branch circuit ampacity. A motor is considered to be for continuous duty unless the nature of the apparatus it drives

Table 430.22(E) Duty-Cycle Service

Classification of Service	Nameplate Current Rating Percentages			
	5-Minute Rated Motor	15-Minute Rated Motor	30- & 60-Minute Rated Motor	Continuous Rated Motor
Short-time duty operating valves, raising or lowering rolls, etc.	110	120	150	—
Intermittent duty freight and passenger elevators, tool heads, pumps, drawbridges, turntables, etc. (for arc welders, see 630.11)	85	85	90	140
Periodic duty rolls, ore- and coal-handling machines, etc.	85	90	95	140
Varying duty	110	120	150	200

Note: Any motor application shall be considered as continuous duty unless the nature of the apparatus it drives is such that the motor will not operate continuously with load under any condition of use.

is such that the motor cannot operate continuously with load under any condition of use. Conductors for a motor used for short-time, intermittent, periodic, or varying duty are required to have an ampacity in accordance with Table 430.22(E). Branch-circuit conductors for a motor with a rated horsepower used for 5-minute short-time duty service are permitted to be sized smaller than for the same motor with a 60-minute rating, due to the cooling intervals between operating periods. The terms *continuous duty, intermittent duty, periodic duty, short-time duty*, and *varying duty* are defined in Article 100.

(F) Separate Terminal Enclosure. The conductors between a stationary motor rated 1 hp or less and the separate terminal enclosure permitted in 430.145(B) shall be permitted to be smaller than 14 AWG but not smaller than 18 AWG, provided they have an ampacity as specified in 430.22(A).

430.23 Wound-Rotor Secondary.

(A) Continuous Duty. For continuous duty, the conductors connecting the secondary of a wound-rotor ac motor to its controller shall have an ampacity not less than 125 percent of the full-load secondary current of the motor.

(B) Other Than Continuous Duty. For other than continuous duty, these conductors shall have an ampacity, in percent of full-load secondary current, not less than that specified in Table 430.22(E).

(C) Resistor Separate from Controller. Where the secondary resistor is separate from the controller, the ampacity of the conductors between controller and resistor shall not be less than that shown in Table 430.23(C).

Table 430.23(C) Secondary Conductor

Resistor Duty Classification	Ampacity of Conductor in Percent of Full-Load Secondary Current
Light starting duty	35
Heavy starting duty	45
Extra-heavy starting duty	55
Light intermittent duty	65
Medium intermittent duty	75
Heavy intermittent duty	85
Continuous duty	110

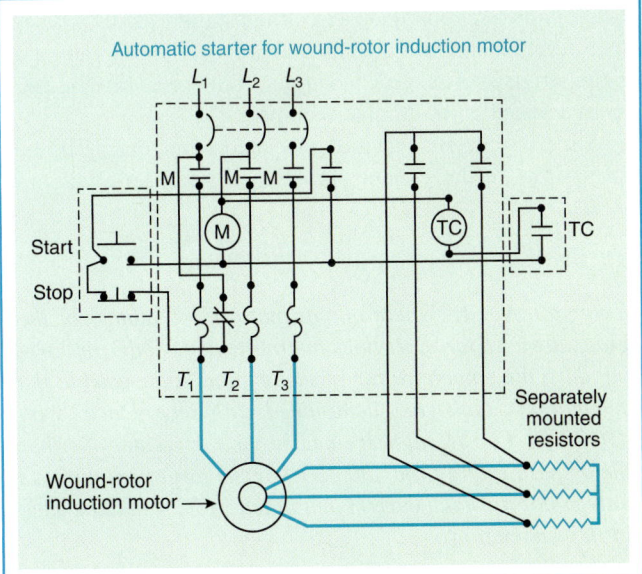

Exhibit 430.7 A branch circuit to a wound-rotor induction motor showing a controller and separate bank of resistors for motor starting and speed regulation.

Before the advent of adjustable speed drives, wound-rotor ac motors were generally used where speed control was desired; where high starting torque for a rapid, smooth acceleration to full load was required; for frequent starting; and for low starting current. Wound-rotor ac motors are also known as slip-ring motors because three slip rings are mounted on the shaft, and brushes in contact with the slip rings are connected to field-installed external resistance units and a controller, as shown in Exhibit 430.7. The resistors are a part of the rotor circuit, and the full value of the external resistance is in the circuit when the motor starts. This value is gradually reduced until the motor attains the desired speed.

Controllers used for speed regulation, usually a dial-type or a drum-type switch, are basically for two types of loads: constant torque (machine loads) and variable torque (fan loads).

The ampacities of the conductors between the controller and the resistance units are the allowable percentages of Table 430.23(C) for the resistor duty classification.

430.24 Several Motors or a Motor(s) and Other Load(s).

Conductors supplying several motors, or a motor(s) and other load(s), shall have an ampacity not less than 125 percent of the full-load current rating of the highest rated motor plus the sum of the full-load current ratings of all the other motors in the group, as determined by 430.6(A), plus the ampacity required for the other loads.

FPN: See Annex D, Example No. D8.

Where feeders serve motors and/or other electrical loads, the highest rating or setting of the feeder short-circuit and ground-fault protective devices for the minimum-size feeder conductor permitted by 430.24 is specified in 430.62.

Where the selection of a feeder protective device of higher rating or setting is based on the simultaneous starting of two or more motors, the size of the feeder conductors is required to be increased accordingly.

These requirements, and those of 430.62 for the short-circuit and ground-fault protection of power feeders, are based on the principle that the power feeder conductors should be sized to have an ampacity equal to 125 percent of the full-load current of the largest motor plus the full-load currents of all other motors and all other loads supplied by the feeder.

Except where two or more motors may be started simultaneously, the heaviest load that a power feeder will ever be required to carry occurs when the largest motor is started and all the other motors supplied by the same feeder are running and delivering their full-rated horsepower.

Where the conductors are branch-circuit conductors to multimotor equipment, 430.53 specifies the maximum rating of the branch-circuit short-circuit and ground-fault protective device and 430.7(D)(1) requires the maximum ampere rating of the short-circuit and ground-fault protective device to be marked on multimotor equipment.

See 430.62(B) and the associated commentary if the size of the feeder conductors is larger than the minimum size.

Exception No. 1: Where one or more of the motors of the group are used for short-time, intermittent, periodic, or varying duty, the ampere rating of such motors to be used in the summation shall be determined in accordance with 430.22(E). For the highest rated motor, the greater of either the ampere rating from 430.22(E) or the largest continuous duty motor full-load current multiplied by 1.25 shall be used in the summation.

Exception No. 2: The ampacity of conductors supplying motor-operated fixed electric space-heating equipment shall conform with 424.3(B).

Exception No. 3: Where the circuitry is interlocked so as to prevent operation of selected motors or other loads at the same time, the conductor ampacity shall be permitted to be based on the summation of the currents of the motors and other loads to be operated at the same time that results in the highest total current.

430.25 Multimotor and Combination-Load Equipment.

The ampacity of the conductors supplying multimotor and combination-load equipment shall not be less than the minimum circuit ampacity marked on the equipment in accordance with 430.7(D). Where the equipment is not factory-wired and the individual nameplates are visible in accordance with 430.7(D)(2), the conductor ampacity shall be determined in accordance with 430.24.

When computing the load for the minimum allowable conductor size for a combination lighting (or lighting and appliance) load and motor load, the capacity for the lighting load is determined in accordance with Article 220 (and other applicable articles and sections, for example, Article 422, for appliances, etc.) and the motor load is determined in accordance with 430.22 (single motor) or 430.24 (two or more motors). The lighting load and the motor load are added together to determine the minimum conductor ampacity.

430.26 Feeder Demand Factor.

Where reduced heating of the conductors results from motors operating on duty-cycle, intermittently, or from all motors not operating at one time, the authority having jurisdiction

may grant permission for feeder conductors to have an ampacity less than specified in 430.24, provided the conductors have sufficient ampacity for the maximum load determined in accordance with the sizes and number of motors supplied and the character of their loads and duties.

The authority having jurisdiction may grant permission to allow a demand factor of less than 100 percent if operational procedures, production demands, or the nature of the work is such that not all the motors are running at one time. Engineering study or evaluation of motor operation may provide information that will allow a demand factor of less than 100 percent.

430.27 Capacitors with Motors.

Where capacitors are installed in motor circuits, conductors shall comply with 460.8 and 460.9.

430.28 Feeder Taps.

Feeder tap conductors shall have an ampacity not less than that required by Part II, shall terminate in a branch-circuit protective device, and, in addition, shall meet one of the following requirements:

(1) Be enclosed either by an enclosed controller or by a raceway, be not more than 3.0 m (10 ft) in length, and, for field installation, be protected by an overcurrent device on the line side of the tap conductor, the rating or setting of which shall not exceed 1000 percent of the tap conductor ampacity

(2) Have an ampacity of at least one-third that of the feeder conductors, be suitably protected from physical damage or enclosed in a raceway, and be not more than 7.5 m (25 ft) in length

(3) Have the same ampacity as the feeder conductors

Exception: Feeder taps over 7.5 m (25 ft) long. In high-bay manufacturing buildings [over 11 m (35 ft) high at walls], where conditions of maintenance and supervision ensure that only qualified persons service the systems, conductors tapped to a feeder shall be permitted to be not over 7.5 m (25 ft) long horizontally and not over 30.0 m (100 ft) in total length where all of the following conditions are met:

(a) The ampacity of the tap conductors is not less than one-third that of the feeder conductors.

(b) The tap conductors terminate with a single circuit breaker or a single set of fuses conforming with (1) Part IV, where the load-side conductors are a branch circuit, or (2) Part V, where the load-side conductors are a feeder.

(c) The tap conductors are suitably protected from physical damage and are installed in raceways.

(d) The tap conductors are continuous from end-to-end and contain no splices.

(e) The tap conductors shall be 6 AWG copper or 4 AWG aluminum or larger.

(f) The tap conductors shall not penetrate walls, floors, or ceilings.

(g) The tap shall not be made less than 9.0 m (30 ft) from the floor.

Section 430.28 contains three basic requirements for feeder taps that supply motor circuits.

First, the tap conductor from the feeder to the motor overcurrent device must be sized according to Part II of Article 430. For a single motor load, the tap conductors are sized the same as the motor branch-circuit conductors, that is, according to 430.22. Section 430.22 requires that motor branch-circuit conductors be sized at least 125 percent of the full-load current value for the motor given in Tables 430.147 through 430.150. The table value, rather than the nameplate value, is the full-load current used for conductor sizing according to 430.6(A).

Second, the tap conductors must terminate in a set of fuses or a circuit breaker, thus limiting the load on the tap conductors. It is important to point out that reduced size tap conductors are protected from overload by the terminal overcurrent device, but protected from short-circuit (and ground-fault) only from the feeder overcurrent device.

Third, where the tap conductor ampacity is less than the ampacity of the feeder, the tap conductor installation must meet the additional requirements associated with their tap conductor distance limits, that is, 10 ft, 25 ft, or by exception, 100 ft.

The requirements for tap conductors that supply motor loads are somewhat similar to the basic tap requirements found in 240.21. For example, where tap conductors supply a motor load and do not exceed 10 ft, the tap conductors must be sized for the load, terminate in a set of fuses or a circuit breaker, be enclosed by a controller or a raceway, and be protected by a feeder overcurrent device not exceeding 10 times the tap conductor ampacity.

Where the tap conductors supply a motor load and do not exceed 25 ft, the tap conductors must be sized for the load, terminate in a set of fuses or a circuit breaker, be protected from physical damage or be enclosed in a raceway, and have an ampacity at least one-third that of the feeder conductor.

In a high-bay manufacturing building, feeder taps up to 100 ft long are conditionally permitted under 430.28, Exception.

Example

A 15-hp, 230-volt, 3-phase, NEMA Design B, squirrel cage induction motor with a service factor of 1.15 is to be supplied by a tap from a 250-kcmil feeder. Assuming three conductors in an individual raceway, all Type THWN copper, and no ambient correction factor, the feeder has an ampacity of 255 amperes (from Table 310.16, 75°C column). Where the tap conductors are not over 25 ft long (see Exhibit 430.8), 4 AWG conductors with an ampacity of 85 amperes are permitted ($\frac{1}{3} \times 255$ amperes = 85 amperes).

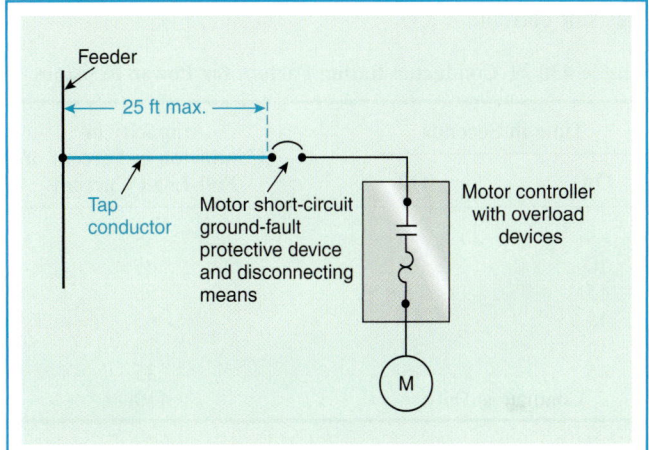

Exhibit 430.8 Protective devices (branch-circuit short-circuit and ground-fault) for a motor branch circuit located not more than 25 ft from the point where the conductors are tapped to a feeder.

Solution

Based on 430.6(A) and Table 430.150, the full-load current of the 15-hp motor is 42 amperes. According to 430.52(C)(1), the motor branch-circuit short-circuit and ground-fault protective device cannot exceed the values given in Table 430.52. The maximum time-delay fuse value is 42 × 1.75 = 73.5 amperes. The maximum inverse time circuit breaker value is 42 × 2.50 = 105 amperes. Section 430.52(C)(1), Exception No. 1, allows the next higher standard size—80 and 110 amperes, respectively. A higher size, based on Exception No. 2, is allowed if the 80- or 110-ampere size is not adequate to start the motor. Based on 430.32, the motor overload protective devices (heaters) are required to be set at a value not greater than 125 percent of the full-load ampere rating marked on the motor nameplate. A higher-sized motor overload protective device setting of up to 140 percent may be used according to the permissive rules set forth in 430.32(C). Regardless of the exact setting, with the motor overload protection set at approximately 50 amperes, the 4 AWG THWN copper motor branch-circuit tap conductors are well protected from overload.

For additional information concerning taps supplying motor circuits for group installations, see 430.53(D) and the associated commentary.

430.29 Constant Voltage Direct-Current Motors—Power Resistors.

Conductors connecting the motor controller to separately mounted power accelerating and dynamic braking resistors in the armature circuit shall have an ampacity not less than the value calculated from Table 430.29 using motor full-load current. If an armature shunt resistor is used, the power accelerating resistor conductor ampacity shall be calculated using the total of motor full-load current and armature shunt resistor current.

Table 430.29 Conductor Rating Factors for Power Resistors

| Time in Seconds | | Ampacity of Conductor in Percent of Full-Load Current |
On	Off	
5	75	35
10	70	45
15	75	55
15	45	65
15	30	75
15	15	85
Continuous Duty		110

Armature shunt resistor conductors shall have an ampacity of not less than that calculated from Table 430.29 using rated shunt resistor current as full-load current.

III. Motor and Branch-Circuit Overload Protection

430.31 General.

Part III specifies overload devices intended to protect motors, motor-control apparatus, and motor branch-circuit conductors against excessive heating due to motor overloads and failure to start.

Overload in electrical apparatus is an operating overcurrent that, when it persists for a sufficient length of time, would cause damage or dangerous overheating of the apparatus. It does not include short circuits or ground faults.

These provisions shall not be interpreted as requiring overload protection where it might introduce additional or increased hazards, as in the case of fire pumps.

FPN: For protection of fire pump supply conductors, see 695.6.

The provisions of Part III shall not apply to motor circuits rated over 600 volts, nominal.

FPN No. 1: For over 600 volts, nominal, see Part X.
FPN No. 2: See Annex D, Example No. D8.

Section 430.31 sets the general guidelines for motor overload protection. The purpose of this protection is to guard the motor against abnormal operating conditions such as failure to start from a locked rotor, single-phase condition or too much friction on the driven load. The overload protection also guards against excessive heating in the motor caused by an overload condition or from a loss of phase condition. Adequately applied overload protection should protect the motor from any overload condition prior to damage occurring in the motor. Overload protection is not designed or may not be capable of breaking short-circuit current or ground-fault current. Overload protection may not be installed where it could cause increased hazards as identified for fire pumps.

430.32 Continuous-Duty Motors.

(A) More Than 1 Horsepower. Each continuous-duty motor rated more than 1 hp shall be protected against overload by one of the means in 430.32(A)(1) through (A)(4).

The basic premise behind 430.32(A) through (E) is that the operation of a motor in excess of its normal full-load rating for a prolonged period of time causes damage or dangerous overheating that may start a fire. Overload protection is intended to protect the motor and the system components from damaging overload currents.

A continuous-duty motor with a marked service factor of 1.15 or greater or with a marked temperature rise of 40°C or less can carry a 25 percent overload for an extended period without damage to the motor. Other similar types of motors are those with a service factor of less than 1.15 or those with a marked temperature rise of greater than 40°C that are incapable of withstanding a prolonged overload, where the motor overload protective device opens the circuit if the motor continues to draw 115 percent of its rated full-load current.

(1) Separate Overload Device. A separate overload device that is responsive to motor current. This device shall be selected to trip or shall be rated at no more than the following percent of the motor nameplate full-load current rating:

Motors with a marked service factor 1.15 or greater	125%
Motors with a marked temperature rise 40°C or less	125%
All other motors	115%

Modification of this value shall be permitted as provided in 430.32(C). For a multispeed motor, each winding connection shall be considered separately.

Where a separate motor overload device is connected so that it does not carry the total current designated on the motor nameplate, such as for wye-delta starting, the proper percentage of nameplate current applying to the selection or setting of the overload device shall be clearly designated on the equipment, or the manufacturer's selection table shall take this into account.

> FPN: Where power factor correction capacitors are installed on the load side of the motor overload device, see 460.9.

Motors are required to be protected from overloads. To protect a motor from an overload, the motor nameplate full-load current is used to select the overload protection rather than the full-load current values from Tables 430.47 through 430.150 that are used to design the feeder and branch circuit wiring.

(2) Thermal Protector. A thermal protector integral with the motor, approved for use with the motor it protects on the basis that it will prevent dangerous overheating of the motor due to overload and failure to start. The ultimate trip current of a thermally protected motor shall not exceed the following percentage of motor full-load current given in Tables 430.148, 430.149, and 430.150:

Motor full-load current 9 amperes or less	170%
Motor full-load current from 9.1 to, and including, 20 amperes	156%
Motor full-load current greater than 20 amperes	140%

If the motor current-interrupting device is separate from the motor and its control circuit is operated by a protective device integral with the motor, it shall be arranged so that the opening of the control circuit will result in interruption of current to the motor.

(3) Integral with Motor. A protective device integral with a motor that will protect the motor against damage due to failure to start shall be permitted if the motor is part of an approved assembly that does not normally subject the motor to overloads.

(4) Larger Than 1500 Horsepower. For motors larger than 1500 hp, a protective device having embedded temperature detectors that cause current to the motor to be interrupted when the motor attains a temperature rise greater than marked on the nameplate in an ambient temperature of 40°C.

Continuous-duty-rated motors of more than 1 hp can be protected against overload conditions by any one of the following four methods in accordance with 430.32(A):

1. An overload device located in the motor controller, such as a bimetallic element or eutectic material
2. A thermal protector located in the motor that senses excessive current or temperature

3. A protective device in the motor, if the motor is part of an assembly that does not normally subject the motor to overloads
4. For motors larger than 1500 hp, a temperature-sensitive device embedded in the motor windings that will de-energize the motor

(B) One Horsepower or Less, Automatically Started. Any motor of 1 hp or less that is started automatically shall be protected against overload by one of the following means.

(1) Separate Overload Device. A separate overload device that is responsive to motor current. This device shall be selected to trip or shall be rated at not more than the following percentage of the motor nameplate full-load current rating:

Motors with a marked service factor 1.15 or greater	125%
Motors with a marked temperature rise 40°C or less	125%
All other motors	115%

For a multispeed motor, each winding connection shall be considered separately. Modification of this value shall be permitted as provided in 430.32(C).

(2) Thermal Protector. A thermal protector integral with the motor, approved for use with the motor that it protects on the basis that it will prevent dangerous overheating of the motor due to overload and failure to start. Where the motor current-interrupting device is separate from the motor and its control circuit is operated by a protective device integral with the motor, it shall be arranged so that the opening of the control circuit results in interruption of current to the motor.

A thermal protector located inside the motor housing, as shown in Exhibit 430.9, is connected in series with the motor winding. This protective device commonly consists of a set of normally closed contacts attached to a bimetallic disk through which the circuit is normally closed. The thermal protector heating coil (in series with the motor winding) causes the disk to heat rapidly. The heat-actuated disk snaps the contacts open to protect the motor windings from overheating due to failure to start, a sudden heavy overload, or a prolonged overload.

After the circuit opens and the motor has cooled to a normal temperature, the contacts automatically close and restart the motor. In some cases, this may not be desirable. For such applications, the protective device is designed so that it must be returned to the closed position by a manually controlled reset, as required by 430.43. For larger motors (usually over 1 hp), a similar device is used. This device, upon abnormal overload, acts as a control-circuit switch and

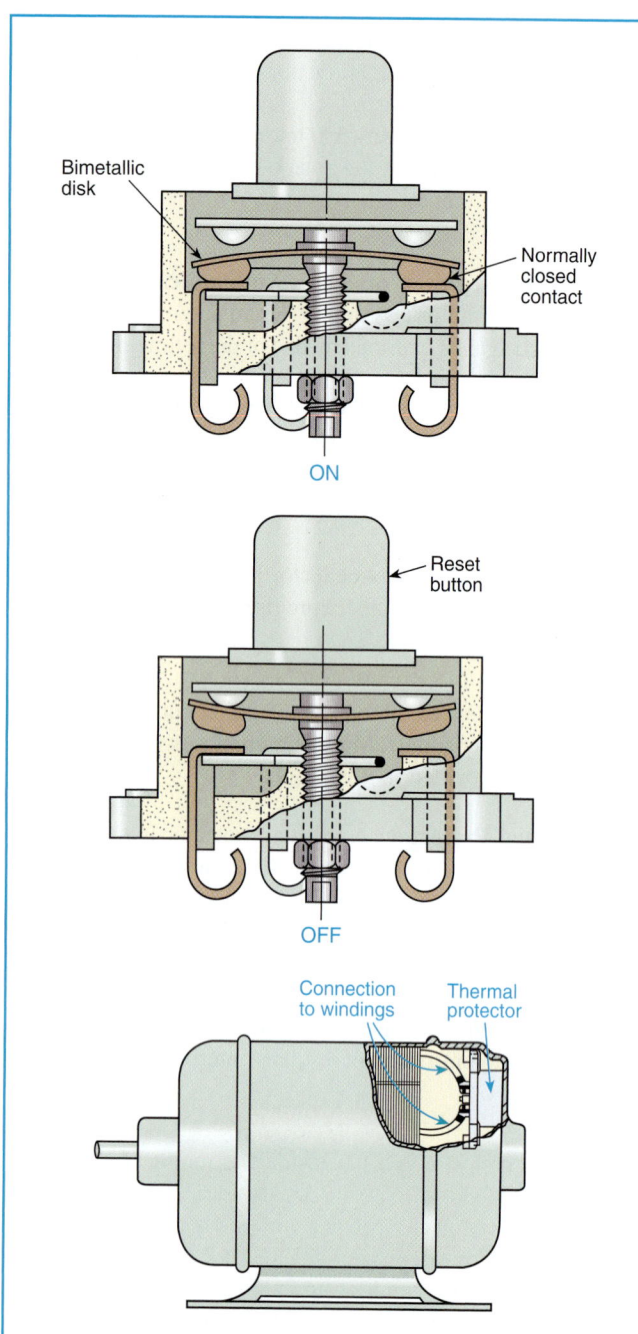

Exhibit 430.9 A thermal protector for a motor, in which a heat-sensitive snap-action disk opens contacts and protects the motor against dangerous overheating. This device is integrally mounted within the motor. (Redrawn from Texas Instruments, Inc.)

operates the control circuit of a motor current-interrupting device, usually a motor contactor or starter, located separately from the motor. A thermal protector and circuit-interrupting device should be approved for use with the motor it protects and is required to open the circuit on an overcurrent, as specified in 430.32(A)(2).

(3) Integral with Motor. A protective device integral with a motor that protects the motor against damage due to failure to start shall be permitted (1) if the motor is part of an approved assembly that does not subject the motor to overloads, or (2) if the assembly is also equipped with other safety controls (such as the safety combustion controls on a domestic oil burner) that protect the motor against damage due to failure to start. Where the assembly has safety controls that protect the motor, it shall be so indicated on the nameplate of the assembly where it will be visible after installation.

(4) Impedance-Protected. In case the impedance of the motor windings is sufficient to prevent overheating due to failure to start, the motor shall be permitted to be protected as specified in 430.32(D)(1) for manually started motors if the motor is part of an approved assembly in which the motor will limit itself so that it will not be dangerously overheated.

> FPN: Many ac motors of less than $\frac{1}{20}$ hp, such as clock motors, series motors, and so forth, and also some larger motors such as torque motors, come within this classification. It does not include split-phase motors having automatic switches that disconnect the starting windings.

(C) Selection of Overload Relay. Where the sensing element or setting of the overload relay selected in accordance with 430.32(A)(1) and 430.32(B)(1) is not sufficient to start the motor or to carry the load, higher size sensing elements or incremental settings shall be permitted to be used, provided the trip current of the overload relay does not exceed the following percentage of motor nameplate full-load current rating:

Motors with marked service factor 1.15 or greater	140%
Motors with a marked temperature rise 40°C or less	140%
All other motors	130%

If not shunted during the starting period of the motor as provided in 430.35, the overload device shall have sufficient time delay to permit the motor to start and accelerate its load.

> FPN: A Class 20 or 30 overload relay will provide a longer motor acceleration time than a Class 10 or 20, respectively. Use of a higher class overload relay may preclude the need for selection of a higher trip current.

(D) One Horsepower or Less, Nonautomatically Started.

(1) Within Sight from Controller. Each continuous-duty motor rated at 1 hp or less that is not permanently installed, is nonautomatically started, and is within sight from the controller location shall be permitted to be protected against overload by the branch-circuit short-circuit and ground-fault protective device. This branch-circuit protective device shall not be larger than that specified in Part IV of Article 430.

Exception: Any such motor shall be permitted on a nominal 120-volt branch circuit protected at not over 20 amperes.

Motors rated 1 hp or less that are not permanently installed and not automatically started, such as for bench grinders, drill presses, and portable electric tools, are not required to have overload protection and may be protected by the branch-circuit short-circuit fuse or circuit breaker. This type of equipment is usually attended by the operator, who can immediately shut off power to the motor should it overheat and start smoking.

(2) Not Within Sight from Controller. Any such motor that is not in sight from the controller location shall be protected as specified in 430.32(B). Any motor rated at 1 hp or less that is permanently installed shall be protected in accordance with 430.32(B).

(E) Wound-Rotor Secondaries. The secondary circuits of wound-rotor ac motors, including conductors, controllers, resistors, and so forth, shall be permitted to be protected against overload by the motor-overload device.

430.33 Intermittent and Similar Duty.

A motor used for a condition of service that is inherently short-time, intermittent, periodic, or varying duty, as illustrated by Table 430.22(E), shall be permitted to be protected against overload by the branch-circuit short-circuit and ground-fault protective device, provided the protective device rating or setting does not exceed that specified in Table 430.52.

Any motor application shall be considered to be for continuous duty unless the nature of the apparatus it drives is such that the motor cannot operate continuously with load under any condition of use.

If a motor is selected for duty-cycle service (short-time, intermittent, periodic, or varying), it can be assumed that the motor will not operate continuously, due to the nature of the apparatus or machinery it drives. Therefore, prolonged overloads are rare unless mechanical failure in the driven apparatus stalls the motor; in this case, however, the branch-circuit protective device would open the circuit. The omission of overload protective devices for such motors is based on the type of duty and not on the time rating of the motor.

430.35 Shunting During Starting Period.

(A) Nonautomatically Started. For a nonautomatically started motor, the overload protection shall be permitted to be shunted or cut out of the circuit during the starting period of the motor if the device by which the overload protection

is shunted or cut out cannot be left in the starting position and if fuses or inverse time circuit breakers rated or set at not over 400 percent of the full-load current of the motor are located in the circuit so as to be operative during the starting period of the motor.

(B) Automatically Started. The motor overload protection shall not be shunted or cut out during the starting period if the motor is automatically started.

Exception: The motor overload protection shall be permitted to be shunted or cut out during the starting period on an automatically started motor where

(a) *The motor starting period exceeds the time delay of available motor overload protective devices, and*
(b) *Listed means are provided to*
 (1) *Sense motor rotation and to automatically prevent the shunting or cutout in the event that the motor fails to start, and*
 (2) *Limit the time of overload protection shunting or cutout to less than the locked rotor time rating of the protected motor, and*
 (3) *Provide for shutdown and manual restart if motor running condition is not reached.*

If not shunted during the starting period of the motor, the overload device must have sufficient time delay to start and accelerate its load; whereas, if shunting is employed, the overload protection is permitted to be bypassed only during the starting period of the motor. See Exhibit 430.10, which illustrates this type of bypass during motor starting. When the switch is thrown momentarily in one direction (starting position), the overload protective fuses are shunted or cut out of the circuit. The switch is then thrown in the opposite direction (running position) and must be designed so that it cannot be left in the starting position.

If fuses are used as overload protection, they may be shunted or cut out of the circuit during the starting period by a device (in this case a double-throw switch) designed so that it cannot be left in the starting position. Therefore, during the starting period, the motor is protected only by the branch-circuit fuses that are always rated within the limits of 430.35(A). If there are no branch-circuit fuses, as permitted by 430.53, then a starter (shunting) device is not allowed during the starting period unless the feeder protection is within the limits of 430.35(B) (not over 400 percent of the full-load motor current).

430.36 Fuses—In Which Conductor.

Where fuses are used for motor overload protection, a fuse shall be inserted in each ungrounded conductor and also in the grounded conductor if the supply system is 3-wire, 3-phase ac with one conductor grounded.

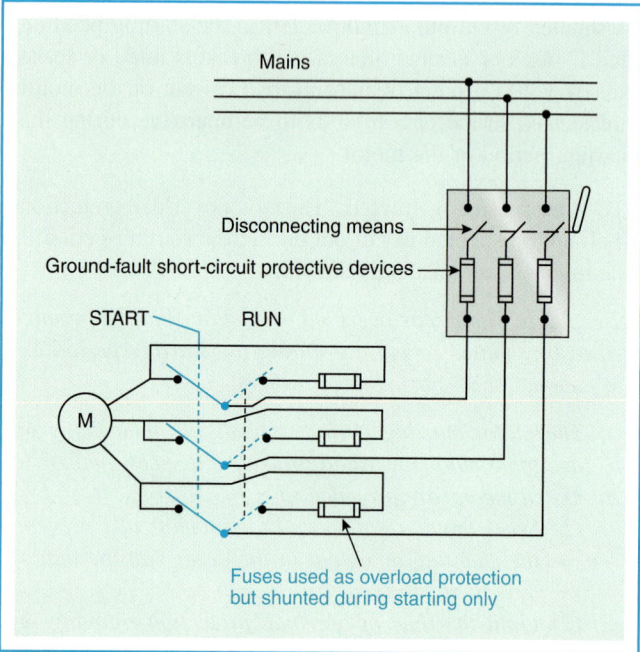

Exhibit 430.10 Arrangement for across-the-line, or full-voltage, starting of a motor.

Table 430.37 Overload Units

Kind of Motor	Supply System	Number and Location of Overload Units, Such as Trip Coils or Relays
1-phase ac or dc	2-wire, 1-phase ac or dc ungrounded	1 in either conductor
1-phase ac or dc	2-wire, 1-phase ac or dc, one conductor grounded	1 in ungrounded conductor
1-phase ac or dc	3-wire, 1-phase ac or dc, grounded neutral	1 in either ungrounded conductor
1-phase ac	Any 3-phase	1 in ungrounded conductor
2-phase ac	3-wire, 2-phase ac, ungrounded	2, one in each phase
2-phase ac	3-wire, 2-phase ac, one conductor grounded	2 in ungrounded conductors
2-phase ac	4-wire, 2-phase ac, grounded or ungrounded	2, one per phase in ungrounded conductors
2-phase ac	Grounded neutral or 5-wire, 2-phase ac, ungrounded	2, one per phase in any ungrounded phase wire
3-phase ac	Any 3-phase	3, one in each phase*

All 3-phase motors, except those protected by other approved means, must be provided with three overload units, one in each phase. Examples of those motors protected by other means include specially designed or integral-type detectors, with or without supplementary external protective devices. See 430.36 for instances in which fuses used as overloads are required even in the grounded conductor.

Exception: An overload unit in each phase shall not be required where overload protection is provided by other approved means.

430.37 Devices Other Than Fuses—In Which Conductor.

Where devices other than fuses are used for motor overload protection, Table 430.37 shall govern the minimum allowable number and location of overload units such as trip coils or relays.

430.38 Number of Conductors Opened by Overload Device.

Motor overload devices, other than fuses or thermal protectors, shall simultaneously open a sufficient number of ungrounded conductors to interrupt current flow to the motor.

430.39 Motor Controller as Overload Protection.

A motor controller shall also be permitted to serve as an overload device if the number of overload units complies with Table 430.37 and if these units are operative in both the starting and running position in the case of a dc motor, and in the running position in the case of an ac motor.

For the purpose of Article 430, a controller may be a switch, a circuit breaker, a contactor, or any other device that starts and stops a motor by making and breaking the motor circuit current. The controller must be capable of interrupting the stalled-rotor current of the motor and must have a horsepower rating that is not lower than the horsepower rating of the motor. Motor controllers are covered in Part VII of Article 430.

Dual-element fuses can be sized to provide motor overload protection (see 430.36). Automatically operated contactors or circuit breakers (with trip units) are governed by the requirements of 430.37 and 430.38 where these devices are used to provide overload protection.

430.40 Overload Relays.

Overload relays and other devices for motor overload protection that are not capable of opening short circuits or ground faults shall be protected by fuses or circuit breakers with ratings or settings in accordance with 430.52 or by a motor short-circuit protector in accordance with 430.52.

Some overload devices are marked with a maximum short-circuit and ground-fault protective device rating or setting. This rating sets the limit on the maximum fuse or breaker size that may be upstream from the overload device. The rating also notifies the user that coordination between the overload device and the short-circuit and ground-fault device is required, which is most often the case for group motor installation.

Exception: Where approved for group installation and marked to indicate the maximum size of fuse or inverse time circuit breaker by which they must be protected, the overload devices shall be protected in accordance with this marking.

> FPN: For instantaneous trip circuit breakers or motor short-circuit protectors, see 430.52.

430.42 Motors on General-Purpose Branch Circuits.

Overload protection for motors used on general-purpose branch circuits as permitted in Article 210 shall be provided as specified in 430.42(A), (B), (C), or (D).

(A) Not Over 1 Horsepower. One or more motors without individual overload protection shall be permitted to be connected to a general-purpose branch circuit only where the installation complies with the limiting conditions specified in 430.32(B) and (D) and 430.53(A)(1) and (A)(2).

(B) Over 1 Horsepower. Motors of ratings larger than specified in 430.53(A) shall be permitted to be connected to general-purpose branch circuits only where each motor is protected by overload protection selected to protect the motor as specified in 430.32. Both the controller and the motor overload device shall be approved for group installation with the short-circuit and ground-fault protective device selected in accordance with 430.53.

(C) Cord-and-Plug Connected. Where a motor is connected to a branch circuit by means of an attachment plug and receptacle and individual overload protection is omitted as provided in 430.42(A), the rating of the attachment plug

and receptacle shall not exceed 15 amperes at 125 volts or 250 volts. Where individual overload protection is required as provided in 430.42(B) for a motor or motor-operated appliance that is attached to the branch circuit through an attachment plug and receptacle, the overload device shall be an integral part of the motor or of the appliance. The rating of the attachment plug and receptacle shall determine the rating of the circuit to which the motor may be connected, as provided in Article 210.

(D) Time Delay. The branch-circuit short-circuit and ground-fault protective device protecting a circuit to which a motor or motor-operated appliance is connected shall have sufficient time delay to permit the motor to start and accelerate its load.

430.43 Automatic Restarting.

A motor overload device that can restart a motor automatically after overload tripping shall not be installed unless approved for use with the motor it protects. A motor overload device that can restart a motor automatically after overload tripping shall not be installed if automatic restarting of the motor can result in injury to persons.

An integral motor overload protective device may be of the type that, after tripping and sufficiently cooling, will automatically restart the motor, or it may be of the type that, after tripping, is closed by use of a manually operated reset button. See the commentary following 430.32(B)(2).

430.44 Orderly Shutdown.

If immediate automatic shutdown of a motor by a motor overload protective device(s) would introduce additional or increased hazard(s) to a person(s) and continued motor operation is necessary for safe shutdown of equipment or process, a motor overload sensing device(s) conforming with the provisions of Part III of this article shall be permitted to be connected to a supervised alarm instead of causing immediate interruption of the motor circuit, so that corrective action or an orderly shutdown can be initiated.

IV. Motor Branch-Circuit Short-Circuit and Ground-Fault Protection

430.51 General.

Part IV specifies devices intended to protect the motor branch-circuit conductors, the motor control apparatus, and the motors against overcurrent due to short circuits or grounds. These rules add to or amend the provisions of Article 240. The devices specified in Part IV do not include the types of devices required by 210.8, 230.95, and 527.6.

The provisions of Part IV shall not apply to motor circuits rated over 600 volts, nominal.

FPN No. 1: For over 600 volts, nominal, see Part X.
FPN No. 2: See Annex D, Example D8.

430.52 Rating or Setting for Individual Motor Circuit.

(A) General. The motor branch-circuit short-circuit and ground-fault protective device shall comply with 430.52(B) and either 430.52(C) or (D), as applicable.

Section 430.52(A) establishes the maximum allowable ratings or settings of devices (fuses or circuit breakers) acceptable for motor branch-circuit short-circuit and ground-fault protection and states that these devices are expected to carry the starting current of the motor and provide short-circuit and ground-fault protection. For certain exceptions to the maximum rating or setting of these motor branch-circuit protective devices, as specified in Table 430.52, see 430.52, 430.53, and 430.54.

Section 430.6 requires that if the current rating of a motor is used to determine the ampacity of conductors or ampere ratings of switches, branch-circuit overcurrent devices, and so on, the values given in Tables 430.147 through 430.150 (including notes) must be used instead of the actual motor nameplate current rating. Separate motor overload protection must be based on the motor nameplate current rating.

Exhibit 430.5 illustrates a typical motor circuit in which the branch-circuit short-circuit and ground-fault protective fuse or circuit breaker rating must carry the starting current and may be sized 150 to 300 percent of the motor full-load current (depending on the type of motor, but excluding Design E). Note that it is not necessary to size the branch-circuit conductors to the percentages (150 to 300) permitted for the branch-circuit short-circuit and ground-fault protective devices.

The rules for short-circuit and ground-fault protection are specific for particular situations. A *short circuit* is a fault between two conductors or between phases. A *ground fault* is a fault between an ungrounded conductor and ground. During a short-circuit or phase-to-ground condition, the extremely high current causes the protective fuses or circuit breakers to open the circuit. Excess current flow caused by an overload condition passes through the overload protective device at the motor controller, thereby causing the device to open the control circuit or motor circuit conductors. Branch-circuit conductors with an ampacity of 125 percent (not 150 to 300 percent) of the motor full-load current are reasonably protected by motor-protective devices set to operate at nearly the same current as the ampacity of the conductors. Branch-circuit short-circuit and ground-fault protective devices will open the circuit under short-circuit conditions and thereby provide short-circuit and ground-fault protection for both

the motor and overload protective device; however, the overload protective device is not intended to open short circuits or ground faults.

The selected rating or setting of the branch-circuit short-circuit and ground-fault protective device should be as low as possible for maximum protection. However, if the rating or setting specified in Table 430.52 or permitted by 430.52(C)(1), Exception No. 1, is not sufficient for the starting current of the motor, a higher rating or setting is allowed per 430.52(C)(1), Exception No. 2. For example, a higher rating would be allowed for a motor under severe starting conditions in which the motor and its driven machinery required an extended period of time to reach the desired speed.

(B) All Motors. The motor branch-circuit short-circuit and ground-fault protective device shall be capable of carrying the starting current of the motor.

(C) Rating or Setting.

(1) In Accordance with Table 430.52. A protective device that has a rating or setting not exceeding the value calculated according to the values given in Table 430.52 shall be used.

Exception No. 1: Where the values for branch-circuit short-circuit and ground-fault protective devices determined by Table 430.52 do not correspond to the standard sizes or ratings of fuses, nonadjustable circuit breakers, thermal protective devices, or possible settings of adjustable circuit breakers, a higher size, rating, or possible setting that does not exceed the next higher standard ampere rating shall be permitted.

Exception No. 2: Where the rating specified in Table 430.52, as modified by Exception No. 1, is not sufficient for the starting current of the motor:

(a) *The rating of a nontime-delay fuse not exceeding 600 amperes or a time-delay Class CC fuse shall be permitted to be increased but shall in no case exceed 400 percent of the full-load current.*

(b) *The rating of a time-delay (dual-element) fuse shall be permitted to be increased but shall in no case exceed 225 percent of the full-load current.*

(c) *The rating of an inverse time circuit breaker shall be permitted to be increased but shall in no case exceed 400 percent for full-load currents of 100 amperes or less or 300 percent for full-load currents greater than 100 amperes.*

(d) *The rating of a fuse of 601–6000 ampere classification shall be permitted to be increased but shall in no case exceed 300 percent of the full-load current.*

FPN: See Annex D, Example D8, and Figure 430.1.

Table 430.52 Maximum Rating or Setting of Motor Branch-Circuit Short-Circuit and Ground-Fault Protective Devices

Type of Motor	Percentage of Full-Load Current			
	Nontime Delay Fuse[1]	Dual Element (Time-Delay) Fuse[1]	Instantaneous Trip Breaker	Inverse Time Breaker[2]
Single-phase motors	300	175	800	250
AC polyphase motors other than wound-rotor				
Squirrel cage — other than Design E or Design B energy efficient	300	175	800	250
Design E or Design B energy efficient	300	175	1100	250
Synchronous[3]	300	175	800	250
Wound rotor	150	150	800	150
Direct current (constant voltage)	150	150	250	150

Note: For certain exceptions to the values specified, see 430.54.
[1]The values in the Nontime Delay Fuse column apply to Time-Delay Class CC fuses.
[2]The values given in the last column also cover the ratings of nonadjustable inverse time types of circuit breakers that may be modified as in 430.52(C), Exception No. 1 and No. 2.
[3]Synchronous motors of the low-torque, low-speed type (usually 450 rpm or lower), such as are used to drive reciprocating compressors, pumps, and so forth, that start unloaded, do not require a fuse rating or circuit-breaker setting in excess of 200 percent of full-load current.

Although Class CC fuses are rated as time delay, they are permitted to be sized according to the requirements of nontime-delay rated fuses because they are so fast acting. Examples of Class CC fuses are shown in Exhibit 430.11.

(2) Overload Relay Table. Where maximum branch-circuit short-circuit and ground-fault protective device ratings are shown in the manufacturer's overload relay table for use with a motor controller or are otherwise marked on the equipment, they shall not be exceeded even if higher values are allowed as shown above.

(3) Instantaneous Trip Circuit Breaker. An instantaneous trip circuit breaker shall be used only if adjustable and if part of a listed combination motor controller having coordinated motor overload and short-circuit and ground-

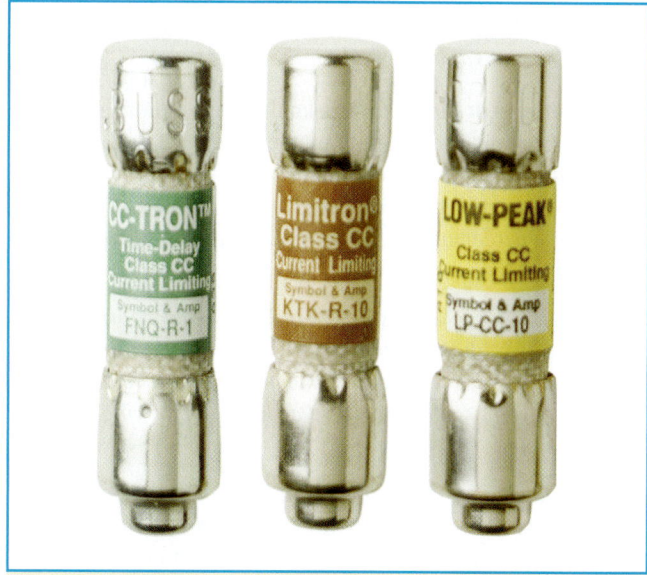

Exhibit 430.11 Class CC fuses. (Courtesy of Bussmann Division, Cooper Industries)

fault protection in each conductor, and the setting is adjusted to no more than the value specified in Table 430.52.

FPN: For the purpose of this article, instantaneous-trip circuit breakers may include a damping means to accommodate a transient motor inrush current without nuisance tripping of the circuit breaker.

Exception No. 1: Where the setting specified in Table 430.52 is not sufficient for the starting current of the motor, the setting of an instantaneous trip circuit breaker shall be permitted to be increased but shall in no case exceed 1300 percent of the motor full-load current for other than Design E motors or Design B energy efficient motors and no more than 1700 percent of full-load motor current for Design E motors or Design B energy efficient motors. Trip settings above 800 percent for other than Design E motors or Design B energy efficient motors and above 1100 percent for Design E motors or Design B energy efficient motors shall be permitted where the need has been demonstrated by engineering evaluation. In such cases, it shall not be necessary to first apply an instantaneous-trip circuit breaker at 800 percent or 1100 percent.

FPN: For additional information on the requirements for a motor to be classified "energy efficient," see NEMA Standards Publication No. MG1-1993, Revision, *Motors and Generators,* Part 12.59.

Exception No. 2: Where the motor full-load current is 8 amperes or less, the setting of the instantaneous-trip circuit breaker with a continuous current rating of 15 amperes or less in a listed combination motor controller that provides coordinated motor branch-circuit overload and short-circuit and ground-fault protection shall be permitted to be increased to the value marked on the controller.

(4) Multispeed Motor. For a multispeed motor, a single short-circuit and ground-fault protective device shall be permitted for two or more windings of the motor, provided the rating of the protective device does not exceed the above applicable percentage of the nameplate rating of the smallest winding protected.

Exception: For a multispeed motor, a single short-circuit and ground-fault protective device shall be permitted to be used and sized according to the full-load current of the highest current winding, where all of the following conditions are met:

(a) Each winding is equipped with individual overload protection sized according to its full-load current.
(b) The branch-circuit conductors supplying each winding are sized according to the full-load current of the highest full-load current winding.
(c) The controller for each winding has a horsepower rating not less than that required for the winding having the highest horsepower rating.

(5) Power Electronic Devices. Suitable fuses shall be permitted in lieu of devices listed in Table 430.52 for power electronic devices in a solid state motor controller system, provided that the marking for replacement fuses is provided adjacent to the fuses.

(6) Self-Protected Combination Controller. A listed self-protected combination controller shall be permitted in lieu of the devices specified in Table 430.52. Adjustable instantaneous-trip settings shall not exceed 1300 percent of full-load motor current for other than Design E motors or Design B energy efficient motors and not more than 1700 percent of full-load motor current for Design E motors or Design B energy efficient motors.

Section 430.52 recognizes a new product referred to as a "self-protected combination controller." This unit combines the functions of short-circuit protection, disconnect, controller, and overload protection into one single unit. Only listed units are permitted, therefore they must be applied within their ratings.

For the 2002 *Code*, former Table 430.152 was relocated and renumbered Table 430.52.

(7) Motor Short-Circuit Protector. A motor short-circuit protector shall be permitted in lieu of devices listed in Table 430.52 if the motor short-circuit protector is part of a listed combination motor controller having coordinated motor overload protection and short-circuit and ground-fault protection in each conductor and it will open the circuit at

currents exceeding 1300 percent of motor full-load current for other than Design E motors or Design B energy efficient motors and 1700 percent of motor full-load motor current for Design E motors or Design B energy efficient motors.

(D) Torque Motors. Torque motor branch circuits shall be protected at the motor nameplate current rating in accordance with 240.4(B).

430.53 Several Motors or Loads on One Branch Circuit.

Two or more motors or one or more motors and other loads shall be permitted to be connected to the same branch circuit under conditions specified in 430.53(D) and in 430.53(A), (B), or (C).

(A) Not Over 1 Horsepower. Several motors, each not exceeding 1 hp in rating, shall be permitted on a nominal 120-volt branch circuit protected at not over 20 amperes or a branch circuit of 600 volts, nominal, or less, protected at not over 15 amperes, if all of the following conditions are met:

(1) The full-load rating of each motor does not exceed 6 amperes.
(2) The rating of the branch-circuit short-circuit and ground-fault protective device marked on any of the controllers is not exceeded.
(3) Individual overload protection conforms to 430.32.

Two or more motors or one or more motors and other loads may be connected to the same 120-volt, 15- or 20-ampere, single-phase lighting circuit as long as each motor is rated not more than 1 hp, the full-load rating of each motor does not exceed 6 amperes, and the rating of the branch-circuit protective device is not exceeded.

The requirements for overload protection, as provided in 430.32, must be applied in all cases, regardless of the number (one or more) of motors or the type of branch circuit.

(B) If Smallest Rated Motor Protected. If the branch-circuit short-circuit and ground-fault protective device is selected not to exceed that allowed by 430.52 for the smallest rated motor, two or more motors or one or more motors and other load(s), with each motor having individual overload protection, shall be permitted to be connected to a branch circuit where it can be determined that the branch-circuit short-circuit and ground-fault protective device will not open under the most severe normal conditions of service that might be encountered.

(C) Other Group Installations. Two or more motors of any rating or one or more motors and other load(s), with each motor having individual overload protection, shall be permitted to be connected to one branch circuit where the motor controller(s) and overload device(s) are (1) installed as a listed factory assembly and the motor branch-circuit short-circuit and ground-fault protective device either is provided as part of the assembly or is specified by a marking on the assembly, or (2) the motor branch-circuit short-circuit and ground-fault protective device, the motor controller(s), and overload device(s) are field-installed as separate assemblies listed for such use and provided with manufacturers' instructions for use with each other, and (3) all of the following conditions are complied with:

(1) Each motor overload device is listed for group installation with a specified maximum rating of fuse, inverse time circuit breaker, or both.

(2) Each motor controller is listed for group installation with a specified maximum rating of fuse, circuit breaker, or both.

(3) Each circuit breaker is one of the inverse time type and listed for group installation.

(4) The branch circuit shall be protected by fuses or inverse time circuit breakers having a rating not exceeding that specified in 430.52 for the highest rated motor connected to the branch circuit plus an amount equal to the sum of the full-load current ratings of all other motors and the ratings of other loads connected to the circuit. Where this calculation results in a rating less than the ampacity of the supply conductors, it shall be permitted to increase the maximum rating of the fuses or circuit breaker to a value not exceeding that permitted by 240.4(B).

(5) The branch-circuit fuses or inverse time circuit breakers are not larger than allowed by 430.40 for the overload relay protecting the smallest rated motor of the group.

FPN: See 110.10 for circuit impedance and other characteristics.

Section 110.10 addresses characteristics of components such as impedance and short-circuit current ratings. Devices with the same ampere rating may have significantly different short-circuit current ratings. Proper selection of components includes consideration of the characteristics of all components so that a fault will not cause unacceptable damage.

(D) Single Motor Taps. For group installations described above, the conductors of any tap supplying a single motor shall not be required to have an individual branch-circuit short-circuit and ground-fault protective device, provided they comply with one of the following:

(1) No conductor to the motor shall have an ampacity less than that of the branch-circuit conductors.

(2) No conductor to the motor shall have an ampacity less than one-third that of the branch-circuit conductors, with a minimum in accordance with 430.22, the conductors to the motor overload device being not more than 7.5 m (25 ft) long and being protected from physical damage.

(3) Conductors from the branch-circuit short-circuit and ground-fault protective device to a listed manual motor controller additionally marked "Suitable for Tap Conductor Protection in Group Installations" shall be permitted to have an ampacity not less than ¹⁄₁₀ the rating or setting of the branch-circuit short-circuit and ground-fault protective device. The conductors from the controller to the motor shall have an ampacity in accordance with 430.22. The conductors from the branch-circuit short-circuit and ground-fault protective device to the controller shall (1) be suitably protected from physical damage and enclosed either by an enclosed controller or by a raceway and shall be not more than 3 m (10 ft) long or (2) shall have an ampacity not less than that of the branch circuit conductors.

For group motor applications covered in 430.53(C), the provisions of 430.53(D)(3) add a third alternative for tapping a branch circuit to supply a single motor. The conditions for applying this tap rule are similar to those in 430.28 covering motor supply conductors tapped to a feeder. The short-circuit ground-fault device on the line side of the tap conductors is protecting more than one set of conductors that supply individual motors, thus eliminating the need for an individual short-circuit ground-fault device for each set of conductors that supply a motor. This approach requires that the tap conductors meet certain size, physical protection, length, and termination conditions. The tap conductors always have to meet the conductor size requirements of 430.22 and must have an ampacity not less than ¹⁄₁₀ the rating of the upstream short-circuit ground-fault protective device. Where the conductors are sized according to this provision, their length cannot exceed 10 ft, and they have to be enclosed in a raceway or by the motor controller. If the conductor ampacity is not less than the rating of the upstream short-circuit ground-fault protective device, the length of the conductor is not limited. The tap conductors are permitted to terminate in a listed manual motor controller that is marked "Suitable for Tap Conductor Protection in Group Installations." This controller provides an instantaneous trip mechanism, motor overload protection, and provisions for disconnecting the motor.

Exhibit 430.12 illustrates main branch-circuit conductors supplying a motor that is part of a group installation. The tap conductors have an ampacity equal to the ampacity of the main branch-circuit conductors; therefore, branch-circuit short-circuit and ground-fault protective devices, fuses, or circuit breakers for the conductors in the tap are

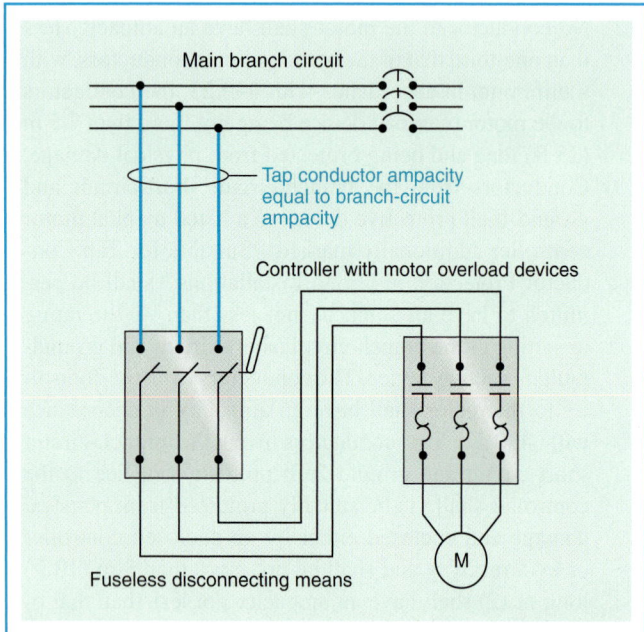

Exhibit 430.12 An example of the permissible omission of motor branch-circuit protective devices for tap conductors that have the same ampacity as the main conductors.

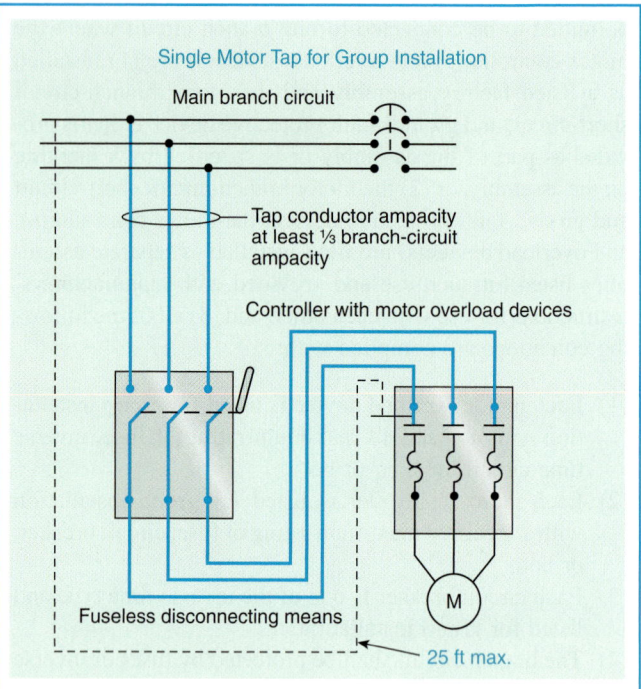

Exhibit 430.13 An example of the permissible omission of motor branch-circuit protective devices for tap conductors that have at least one-third the ampacity of the main conductors, are not over 25 ft long, and are protected from physical damage.

not required at the point of connection of the tap conductors to the main conductors, provided that the motor controller and motor overload protective device are listed for group installation with the size of the main branch-circuit short-circuit and ground-fault protective device used.

Exhibit 430.13 also illustrates main branch-circuit conductors supplying a motor that is part of a group installation. Here, the tap conductors have an ampacity at least one-third the ampacity of the main branch-circuit conductors, are not more than 25 ft in length, and are suitably protected from physical damage. The motor controller and motor overload protective device must be listed for group installation with the size of the main branch-circuit short-circuit and ground-fault protective device used.

In both examples, the main branch-circuit fuses or circuit breakers would operate in the event of a short circuit, and the overload protective device would operate to protect the motor and tap conductors under overload conditions.

The tap conductors should never be of a smaller size and ampacity than the branch-circuit conductors required by 430.22. That is, a tap conductor (25 ft or less) may be one-third the ampacity of the main branch-circuit conductor to which it is connected; however, this ampacity must be equal to or larger than 125 percent of the motor's full-load current rating (see 430.22).

Example

A branch circuit sized at 2/0 AWG THW copper typically has an ampacity of 175 amperes (see Table 310.16). A tap

conductor (25 ft or less) would normally be permitted to be sized at 6 AWG THW copper (65 amperes). But one-third of 175 amperes is 58 amperes, and the motor circuit conductors must have an ampacity of at least 85 amperes. If a 25-hp, 230-volt, 3-phase squirrel-cage motor is to be supplied from this branch circuit, a 6 AWG tap conductor would not meet the requirements of 430.22. That is, 125 percent of the full-load current of the motor (68 amperes from Table 430.150) is 85 amperes (1.25 × 68 amperes = 85 amperes). Therefore, the branch-circuit tap conductors are not permitted to be smaller than 4 AWG THW copper, with a normal ampacity of 85 amperes (see Table 310.16). Note that the ampacities in Table 310.16 are reduced for ambient temperatures above 30°C and for more than three conductors in the raceway or cable.

430.54 Multimotor and Combination-Load Equipment.

The rating of the branch-circuit short-circuit and ground-fault protective device for multimotor and combination-load equipment shall not exceed the rating marked on the equipment in accordance with 430.7(D).

430.55 Combined Overcurrent Protection.

Motor branch-circuit short-circuit and ground-fault protection and motor overload protection shall be permitted to be combined in a single protective device where the rating or setting of the device provides the overload protection specified in 430.32.

Fuses and circuit breakers are not permitted to be sized as overload protection according to the values of 430.32(C). Rather fuses are only permitted to be sized as overload protection according to the values found in 430.32(A)(1), (B)(1), and (D)(1).

Either a circuit breaker with inverse time characteristics or a dual-element (time-delay) fuse may serve as both motor overload protection and also as the branch-circuit short-circuit and ground-fault protection.

One-time, time-delay dual-element and Type S dual-element fuses and adapters are available with up to a 30-ampere rating. Type S fuses are designed to prevent oversize fusing. See 240.50 through 240.54 for more information about these fuses and adapters.

Exhibits 430.14 and 430.15 are examples of time-delay, cartridge-type dual-element fuses that are able to withstand the normal motor starting current if sized at or near the motor full-load rating but that open when subjected to prolonged overload or blow quickly during a short circuit or ground fault. The dual-element characteristics are the thermal cutout element, which permits harmless high-inrush currents to flow for short periods (but which would open the circuit during a prolonged period), and the fuse link element, which has current-limiting ability for short-circuit currents (and which would blow quickly). Dual-element fuses may be used in larger sizes to provide only short-circuit and ground-fault protection.

430.56 Branch-Circuit Protective Devices—In Which Conductor.

Branch-circuit protective devices shall comply with the provisions of 240.20.

430.57 Size of Fuseholder.

Where fuses are used for motor branch-circuit short-circuit and ground-fault protection, the fuseholders shall not be of a smaller size than required to accommodate the fuses specified by Table 430.52.

Exception: Where fuses having time delay appropriate for the starting characteristics of the motor are used, it shall be permitted to use fuseholders sized to fit the fuses that are used.

The use of dual-element (time-delay) fuses makes it possible to use smaller fuses, thereby providing better protection because of the smaller fuses' lower ratings. Dual-element fuses also save in installation cost by allowing smaller-size switches and panels, and they allow for easier arrangement of equipment where space is at a premium at motor control centers.

430.58 Rating of Circuit Breaker.

A circuit breaker for motor branch-circuit short-circuit and ground-fault protection shall have a current rating in accordance with 430.52 and 430.110.

V. Motor Feeder Short-Circuit and Ground-Fault Protection

430.61 General.

Part V specifies protective devices intended to protect feeder conductors supplying motors against overcurrents due to short circuits or grounds.

FPN: See Annex D, Example D8.

430.62 Rating or Setting—Motor Load.

(A) Specific Load. A feeder supplying a specific fixed motor load(s) and consisting of conductor sizes based on

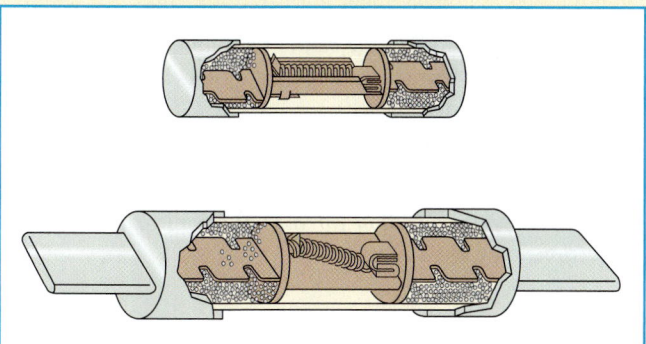

Exhibit 430.14 A Fusetron cartridge-type fuse. (Redrawn from Bussmann Division, Cooper Industries)

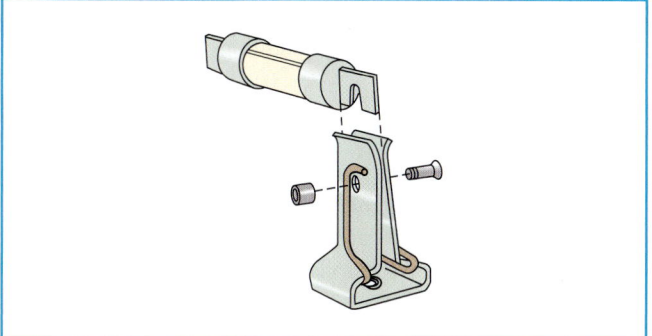

Exhibit 430.15 A Class R dual-element fuse with physical rejection feature to prevent interchangeability. (Redrawn from Bussmann Division, Cooper Industries)

430.24 shall be provided with a protective device having a rating or setting not greater than the largest rating or setting of the branch-circuit short-circuit and ground-fault protective device for any motor supplied by the feeder [based on the maximum permitted value for the specific type of a protective device in accordance with 430.52, or 440.22(A) for hermetic refrigerant motor-compressors], plus the sum of the full-load currents of the other motors of the group.

Where the same rating or setting of the branch-circuit short-circuit and ground-fault protective device is used on two or more of the branch circuits supplied by the feeder, one of the protective devices shall be considered the largest for the above calculations.

The rating of a motor feeder short-circuit ground-fault protective device is determined by adding the rating of the largest branch-circuit short-circuit ground-fault protective device for any motor supplied by the feeder to the sum of the full-load currents of all of the other motors supplied by that feeder. The largest branch-circuit short-circuit ground-fault protective device is based on 430.52 and Table 430.52. The largest rating can be based on either of the exceptions to 430.52. For the purposes of sizing the feeder protective device, it is assumed that the same type of protective device is being used for the feeder and the branch circuits. This is necessary if the feeder protective device and the largest branch-circuit protective device are different types, for example, one is a fuse and the other is a circuit breaker.

Section 430.62(A) recognizes the lower setting for motor overload devices that is required for hermetic refrigerant motor-compressors.

Exception No. 1: Where one or more instantaneous trip circuit breakers or motor short-circuit protectors are used for motor branch-circuit short-circuit and ground-fault protection as permitted in 430.52(C), the procedure provided above for determining the maximum rating of the feeder protective device shall apply with the following provision: For the purpose of the calculation, each instantaneous trip circuit breaker or motor short-circuit protector shall be assumed to have a rating not exceeding the maximum percentage of motor full-load current permitted by Table 430.52 for the type of feeder protective device employed.

Exception No. 2: Where the feeder overcurrent protective device also provides overcurrent protection for a motor control center, the provisions of 430.94 shall apply.

FPN: See Annex D, Example D8.

(B) Other Installations. Where feeder conductors have an ampacity greater than required by 430.24, the rating or setting of the feeder overcurrent protective device shall be permitted to be based on the ampacity of the feeder conductors.

Section 430.62(B) explains how to size a motor feeder that is larger than the minimum size required by the *Code*. If the motor feeder conductors are sized larger than the minimum required, then the size of the overcurrent device for the feeder is based on the size of the feeder conductors.

Exception No. 2 correlates the requirements of 430.62(B) for determining feeder short-circuit ground-fault protection with the requirements of 430.94 covering overcurrent protection for motor-control centers. Where the motor feeder short-circuit ground-fault protective device is also the overcurrent protective device for a motor control center, its rating cannot exceed that allowed for protecting the common power bus of the motor control center.

430.63 Rating or Setting—Power and Lighting Loads.

Where a feeder supplies a motor load and, in addition, a lighting or a lighting and appliance load, the feeder protective device shall have a rating sufficient to carry the lighting or lighting and appliance load, plus the following:

(1) For a single motor, the rating permitted by 430.52
(2) For a single hermetic refrigerant motor-compressor, the rating permitted by 440.22
(3) For two or more motors, the rating permitted by 430.62

Exception: Where the feeder overcurrent device provides the overcurrent protection for a motor control center, the provisions of 430.94 shall apply.

See commentary on 430.62(B) Exception No. 2.

VI. Motor Control Circuits

430.71 General.

Part VI contains modifications of the general requirements and applies to the particular conditions of motor control circuits.

FPN: See 430.9(B) for equipment device terminal requirements.

Definition: Motor Control Circuit. The circuit of a control apparatus or system that carries the electric signals directing the performance of the controller but does not carry the main power current.

430.72 Overcurrent Protection.

(A) General. A motor control circuit tapped from the load side of a motor branch-circuit short-circuit and ground-fault protective device(s) and functioning to control the motor(s) connected to that branch circuit shall be protected against overcurrent in accordance with 430.72. Such a tapped control circuit shall not be considered to be a branch circuit and shall be permitted to be protected by either a supplementary or branch-circuit overcurrent protective device(s). A motor control circuit other than such a tapped control circuit shall be protected against overcurrent in accordance with 725.23 or the notes to Table 11(A) and (B) in Chapter 9, as applicable.

(B) Conductor Protection. The overcurrent protection for conductors shall be provided as specified in 430.72(B)(1) or (B)(2).

Exception No. 1: Where the opening of the control circuit would create a hazard as, for example, the control circuit of a fire pump motor, and the like, conductors of control circuits shall require only short-circuit and ground-fault protection and shall be permitted to be protected by the motor branch-circuit short-circuit and ground-fault protective device(s).

Exception No. 2: Conductors supplied by the secondary side of a single-phase transformer having only a two-wire (single-voltage) secondary shall be permitted to be protected by overcurrent protection provided on the primary (supply) side of the transformer, provided this protection does not exceed the value determined by multiplying the appropriate maximum rating of the overcurrent device for the secondary conductor from Table 430.72(B) by the secondary-to-primary voltage ratio. Transformer secondary conductors (other than two-wire) shall not be considered to be protected by the primary overcurrent protection.

(1) Separate Overcurrent Protection. Where the motor branch-circuit short-circuit and ground-fault protective device does not provide protection in accordance with 430.72(B)(2), separate overcurrent protection shall be provided. The overcurrent protection shall not exceed the values specified in Column A of Table 430.72(B).

(2) Branch-Circuit Overcurrent Protective Device. Conductors shall be permitted to be protected by the motor branch-circuit short-circuit and ground-fault protective device and shall require only short-circuit and ground-fault protection. Where the conductors do not extend beyond the motor control equipment enclosure, the rating of the protective device(s) shall not exceed the value specified in Column B of Table 430.72(B). Where the conductors extend beyond the motor control equipment enclosure, the rating of the protective device(s) shall not exceed the value specified in Column C of Table 430.72(B).

(C) Control Circuit Transformer. Where a motor control circuit transformer is provided, the transformer shall be protected in accordance with 430.72(C)(1), (2), (3), (4), or (5).

Exception: Overcurrent protection shall be omitted where the opening of the control circuit would create a hazard as, for example, the control circuit of a fire pump motor and the like.

(1) Compliance with Article 725. Where the transformer supplies a Class 1 power-limited circuit, Class 2, or Class 3 remote-control circuit conforming with the requirements of Article 725, the protection shall comply with Article 725.

(2) Compliance with Article 450. Protection shall be permitted to be provided in accordance with 450.3.

Table 430.72(B) Maximum Rating of Overcurrent Protective Device in Amperes

| | Column A Separate Protection Provided | | Protection Provided by Motor Branch-Circuit Protective Device(s) | | | |
| | | | Column B Conductors Within Enclosure | | Column C Conductors Extend Beyond Enclosure | |
Control Circuit Conductor Size (AWG)	Copper	Aluminum or Copper-Clad Aluminum	Copper	Aluminum or Copper-Clad Aluminum	Copper	Aluminum or Copper-Clad Aluminum
18	7	—	25	—	7	—
16	10	—	40	—	10	—
14	(Note 1)	—	100	—	45	—
12	(Note 1)	(Note 1)	120	100	60	45
10	(Note 1)	(Note 1)	160	140	90	75
Larger than 10	(Note 1)	(Note 1)	(Note 2)	(Note 2)	(Note 3)	(Note 3)

Notes:
1. Value specified in 310.15 as applicable.
2. 400 percent of value specified in Table 310.17 for 60°C conductors.
3. 300 percent of value specified in Table 310.16 for 60°C conductors.

(3) Less Than 50 Volt-Amperes. Control circuit transformers rated less than 50 volt-amperes (VA) and that are an integral part of the motor controller and located within the motor controller enclosure shall be permitted to be protected by primary overcurrent devices, impedance limiting means, or other inherent protective means.

(4) Primary Less Than 2 Amperes. Where the control circuit transformer rated primary current is less than 2 amperes, an overcurrent device rated or set at not more than 500 percent of the rated primary current shall be permitted in the primary circuit.

(5) Other Means. Protection shall be permitted to be provided by other approved means.

Motor control circuits are allowed to receive their power from either the load side of the motor short-circuit and ground-fault protective device or from a separate source, such as a panelboard.

Motor control circuits that receive their power from a separate source are protected against overcurrent in accordance with 725.23 for Class 1 circuits. Conductor sizes 14 AWG and larger are protected according to their ampacity listed in Tables 310.16 through 310.20. Conductor sizes 16 and 18 AWG must be protected at not more than 10 and 7 amperes, respectively, as specified in Table 430.72(B).

If a motor control circuit is tapped from the load side of the motor branch-circuit short-circuit and ground-fault protective device, the size of the tapped conductor and rating of the overcurrent device are based on whether the conductor stays within the motor control enclosure or leaves it. The load on a motor control circuit is similar to a motor branch-circuit load in that there is a predetermined connected load. There is also an initial high inrush of current, until the armature of the relay is seated and the current decreases to a steady state. Therefore, the overcurrent protection is similar to the short-circuit and ground-fault protection provided for a motor and is allowed to be greater than the ampacity of the control circuit conductor.

430.73 Mechanical Protection of Conductor.

Where damage to a motor control circuit would constitute a hazard, all conductors of such a remote motor control circuit that are outside the control device itself shall be installed in a raceway or be otherwise suitably protected from physical damage.

Where one side of the motor control circuit is grounded, the motor control circuit shall be arranged so that an accidental ground in the control circuit remote from the motor controller will (1) not start the motor and (2) not bypass

manually operated shutdown devices or automatic safety shutdown devices.

If damage to the motor control circuit conductors would constitute a fire or accident hazard, then physical protection of the motor control circuit conductors is extremely important. If damage to the control circuit conductors could result in an accidental ground fault or short circuit, causing the device to operate or rendering the device inoperative (either condition constituting a hazard to persons or property), conductors must be installed in a raceway. Where boilers or furnaces are equipped with an automatic safety control device, damage to the conductors of the low-voltage control circuit (for example, a thermostat) does not constitute a hazard (see Article 725, Part III).

The second paragraph of 430.73 requires that if one side of the motor control circuit is grounded, the circuit must be arranged so that an accidental ground in the remote control device will not start the motor. For example, see the control wiring illustrated in Exhibit 430.16. If the control circuit is a 120-volt, single-phase circuit derived from a 208-volt, 3-phase wye system supplying the motor, one side of the control circuit will be the grounded neutral. If the start button of the motor control circuit is in the grounded neutral, a ground fault on the coil side of the start button can short-circuit the start circuit and start the motor. The same condition exists if the ground fault is in the wiring rather than in the control device itself. This hazardous condition can be alleviated by locating the start button in the ungrounded side for the control circuit as shown in Exhibit 430.17.

Combinations of ground faults in motor and motor control circuits can also result in inadvertent motor starting. If the circuit is ungrounded, the first fault may go undetected. One solution is to use double-pole control devices, with one pole in each of the two control lines.

430.74 Disconnection.

(A) General. Motor control circuits shall be arranged so that they will be disconnected from all sources of supply when the disconnecting means is in the open position. The disconnecting means shall be permitted to consist of two or more separate devices, one of which disconnects the motor and the controller from the source(s) of power supply for the motor, and the other(s), the motor control circuit(s) from its power supply. Where separate devices are used, they shall be located immediately adjacent to each other.

Exception No. 1: Where more than 12 motor control circuit conductors are required to be disconnected, the disconnecting means shall be permitted to be located other than immediately adjacent to each other where all of the following conditions are complied with.

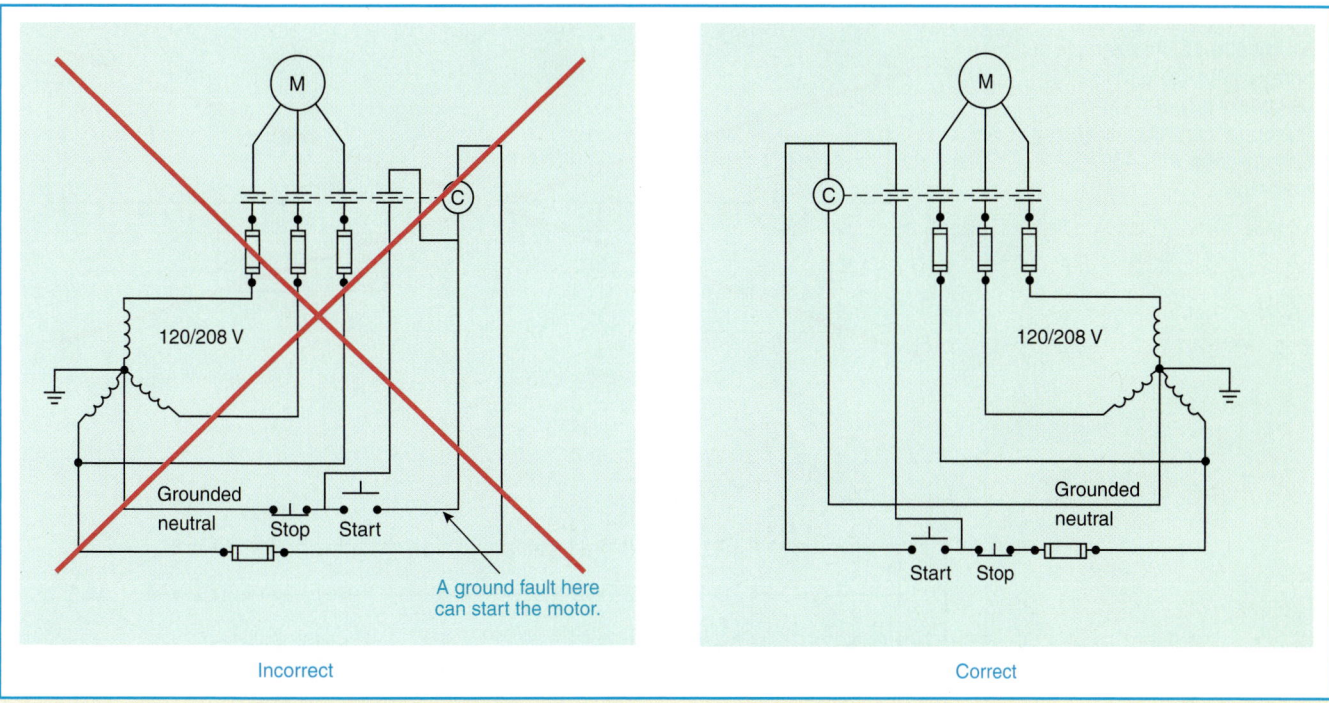

Incorrect   Correct

Exhibit 430.16 An example of control wiring in violation of 430.73, paragraph 2 (incorrect, left), and in compliance with 430.73, paragraph 2 (correct, right). (For simplification, motor overload elements and disconnecting means are not shown.)

(a) Access to energized parts is limited to qualified persons in accordance with Part XI of this article.

(b) A warning sign is permanently located on the outside of each equipment enclosure door or cover permitting access to the live parts in the motor control circuit(s), warning that motor control circuit disconnecting means are remotely located and specifying the location and identification of each disconnect. Where energized parts are not in an equipment enclosure as permitted by 430.132 and 430.133, an additional warning sign(s) shall be located where visible to persons who may be working in the area of the energized parts.

Exception No. 2: The motor control circuit disconnecting means shall be permitted to be remote from the motor controller power supply disconnecting means where the opening of one or more motor control circuit disconnect means may result in potentially unsafe conditions for personnel or property and the conditions of items (a) and (b) of Exception No. 1 are complied with.

(B) Control Transformer in Controller Enclosure. Where a transformer or other device is used to obtain a reduced voltage for the motor control circuit and is located in the controller enclosure, such transformer or other device

shall be connected to the load side of the disconnecting means for the motor control circuit.

VII. Motor Controllers

430.81 General.

Part VII is intended to require suitable controllers for all motors.

(A) Definition. For the definition of *Controller*, see Article 100. For the purpose of this article, a controller is any switch or device that is normally used to start and stop a motor by making and breaking the motor circuit current.

(B) Stationary Motor of ⅛ Horsepower or Less. For a stationary motor rated at ⅛ hp or less that is normally left running and is constructed so that it cannot be damaged by overload or failure to start, such as clock motors and the like, the branch-circuit protective device shall be permitted to serve as the controller.

(C) Portable Motor of ⅓ Horsepower or Less. For a portable motor rated at ⅓ hp or less, the controller shall be permitted to be an attachment plug and receptacle.

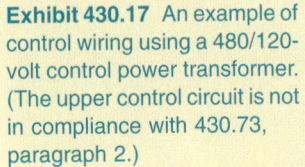

Exhibit 430.17 An example of control wiring using a 480/120-volt control power transformer. (The upper control circuit is not in compliance with 430.73, paragraph 2.)

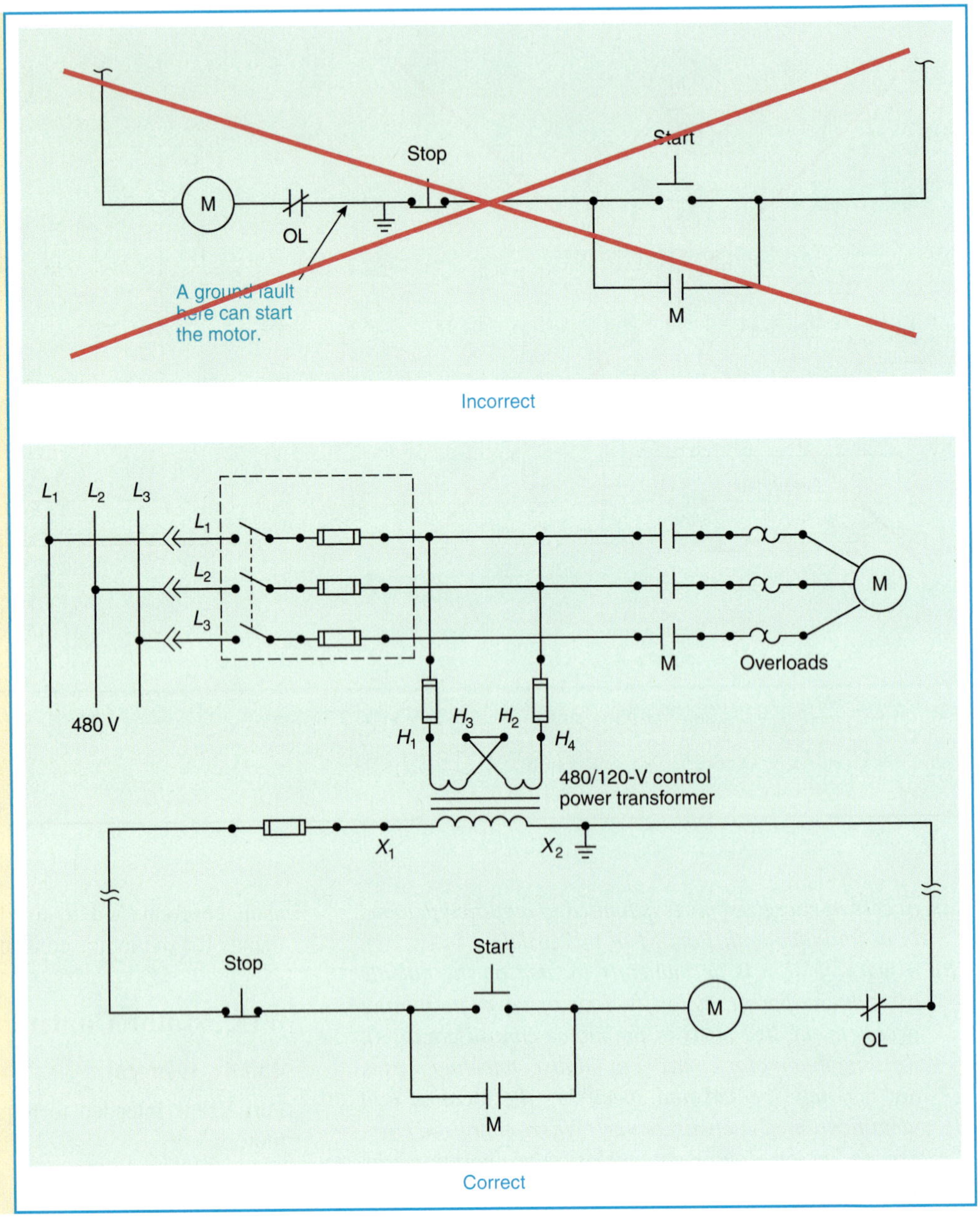

Stop

Start

M

OL

A ground fault here can start the motor.

Incorrect

L_1 L_2 L_3

L_1

L_2

L_3

480 V

H_1 H_3 H_2 H_4

480/120-V control power transformer

X_1 X_2

M   Overloads

M

Stop

Start

M

M

OL

Correct

430.82 Controller Design.

(A) Starting and Stopping. Each controller shall be capable of starting and stopping the motor it controls and shall be capable of interrupting the locked-rotor current of the motor.

(B) Autotransformer. An autotransformer starter shall provide an "off" position, a running position, and at least one starting position. It shall be designed so that it cannot rest in the starting position or in any position that will render the overload device in the circuit inoperative.

(C) Rheostats. Rheostats shall be in compliance with the following:

(1) Motor-starting rheostats shall be designed so that the contact arm cannot be left on intermediate segments. The point or plate on which the arm rests when in the starting position shall have no electrical connection with the resistor.

(2) Motor-starting rheostats for dc motors operated from a constant voltage supply shall be equipped with automatic devices that will interrupt the supply before the speed of the motor has fallen to less than one-third its normal rate.

430.83 Ratings.

The controller shall have a rating as specified in 430.83(A), unless otherwise permitted in 430.83(B) or (C), or as specified in 430.83(D), under the conditions specified.

(A) General.

(1) Horsepower Ratings. Controllers, other than inverse time circuit breakers and molded case switches, shall have horsepower ratings at the application voltage not lower than the horsepower rating of the motor. A controller for a Design E motor rated more than 2 hp shall (1) be marked as rated for use with a Design E motor or (2) have a horsepower rating not less than 1.4 times the rating of a motor rated 3 through 100 hp or not less than 1.3 times the rating of a motor rated over 100 hp.

(2) Circuit Breaker. A branch-circuit inverse time circuit breaker rated in amperes shall be permitted as a controller for all motors, including Design E. Where this circuit breaker is also used for overload protection, it shall conform to the appropriate provisions of this article governing overload protection.

(3) Molded Case Switch. A molded case switch rated in amperes shall be permitted as a controller for all motors, including Design E.

A molded case switch has the same frame appearance as a molded case circuit breaker and is designed to fit in circuit breaker enclosures. However, where the device is marked with only a short-circuit current withstand rating, this rating indicates that the switch does not provide overcurrent protection. Those fused molded case switches that do provide overcurrent protection are marked with a short-circuit current interrupting rating. Both fused and unfused molded case switches can be used in motor circuits. Molded case switches are permitted as motor disconnecting means per 430.109. These switches are rated only in amperes and, where used in a motor circuit, must be sized at 115 percent of the motor full-load current rating.

(B) Small Motors. Devices as specified in 430.81(B) and (C) shall be permitted as a controller.

(C) Stationary Motors of 2 Horsepower or Less. For stationary motors rated at 2 hp or less and 300 volts or less, the controller shall be permitted to be either of the following:

(1) A general-use switch having an ampere rating not less than twice the full-load current rating of the motor
(2) On ac circuits, a general-use snap switch suitable only for use on ac (not general-use ac–dc snap switches) where the motor full-load current rating is not more than 80 percent of the ampere rating of the switch

(D) Torque Motors. For torque motors, the controller shall have a continuous-duty, full-load current rating not less than the nameplate current rating of the motor. For a motor controller rated in horsepower but not marked with the foregoing current rating, the equivalent current rating shall be determined from the horsepower rating by using Tables 430.147, 430.148, 430.149, or 430.150.

(E) Voltage Rating. A controller with a straight voltage rating, for example, 240 volts or 480 volts, shall be permitted to be applied in a circuit in which the nominal voltage between any two conductors does not exceed the controller's voltage rating. A controller with a slash rating, for example, 120/240 volts or 480/277 volts, shall only be applied in a circuit in which the nominal voltage to ground from any conductor does not exceed the lower of the two values of the controller's voltage rating and the nominal voltage between any two conductors does not exceed the higher value of the controller's voltage rating.

Section 430.83(E) requires that controllers identified with slash voltage ratings, such as 120/240-volt and 480/277-volt grounded systems, are allowed to be used only on electrical systems in which the nominal voltage to ground does not exceed the lower voltage rating of the controller, and the nominal voltage between any two phases of the electrical system is not greater than the higher value of the controller voltage rating.

430.84 Need Not Open All Conductors.

The controller shall not be required to open all conductors to the motor.

Exception: Where the controller serves also as a disconnecting means, it shall open all ungrounded conductors to the motor as provided in 430.111.

A controller that does not also serve as a disconnecting means must open only as many motor circuit conductors as may be necessary to stop the motor, that is, one conductor for a dc or single-phase motor circuit, two conductors for a 3-phase motor circuit, and three conductors for a 2-phase motor circuit.

430.85 In Grounded Conductors.

One pole of the controller shall be permitted to be placed in a permanently grounded conductor, provided the controller is designed so that the pole in the grounded conductor cannot be opened without simultaneously opening all conductors of the circuit.

Generally, one conductor of a 120-volt circuit is grounded, and a single-pole device must be connected in the ungrounded conductor to serve as a controller. A 2-pole controller is permitted for such a circuit, where both conductors (grounded and ungrounded) are opened simultaneously. The same requirement can be applied to other circuits, such as 240-volt, 3-wire circuits with one conductor grounded.

430.87 Number of Motors Served by Each Controller.

Each motor shall be provided with an individual controller.

Exception: For motors rated 600 volts or less, a single controller rated at not less than the equivalent horsepower, as determined in accordance with 430.110(C)(1), of all the motors in the group shall be permitted to serve the group under any of the following conditions:

(a) *Where a number of motors drive several parts of a single machine or piece of apparatus, such as metal and woodworking machines, cranes, hoists, and similar apparatus*

(b) *Where a group of motors is under the protection of one overcurrent device as permitted in 430.53(A)*

(c) *Where a group of motors is located in a single room within sight from the controller location*

The conditions stated in the exception to 430.87 are similar to those specified in the exception to 430.112, which permit the use of a single disconnecting means for a group of motors.

430.88 Adjustable-Speed Motors.

Adjustable-speed motors that are controlled by means of field regulation shall be equipped and connected so that they cannot be started under a weakened field.

Exception: Starting under a weakened field shall be permitted where the motor is designed for such starting.

The torque and speed of a motor depend on the amount of current passing through the armature. This current is a function of shunt field strength and rpm of the armature. A reduction of the shunt field magnetic flux causes a reduction of the counterelectromotive force in the armature, resulting in an increase in armature current, thereby increasing torque and thus increasing speed.

Because of excessive armature starting currents, field regulated, adjustable-speed motors are not permitted to be started under a weakened field condition unless some means is provided to limit the speed to within safe limits.

430.89 Speed Limitation.

Machines of the following types shall be provided with speed-limiting devices or other speed-limiting means:

(1) Separately excited dc motors
(2) Series motors
(3) Motor-generators and converters that can be driven at excessive speed from the dc end, as by a reversal of current or decrease in load

Exception: Separate speed-limiting devices or means shall not be required under either of the following conditions:

(a) *Where the inherent characteristics of the machines, the system, or the load and the mechanical connection thereto are such as to safely limit the speed*

(b) *Where the machine is always under the manual control of a qualified operator*

It is still fairly common for dc motors to be used where speed control is essential, such as in the case of electric railways and elevators, where a smooth start, controlled acceleration, and a smooth stop are necessary.

If the load is removed from a series motor when it is running, the speed of the motor will increase until it is dangerously high. To produce the necessary counterelectromotive force with a weakened field, the armature must turn correspondingly faster. Series motors are commonly used as gear-drive traction motors of electric locomotives and, thus, are continuously loaded.

Separately excited dc motors, series motors, and motor (compound-wound dc) generators and (synchronous) converters must be provided with speed-limiting devices (note exceptions), such as a centrifugal device on the shaft of the machine or a remotely located overspeed device, which may be set at a predetermined speed to operate a set of contacts and thereby trip a circuit breaker and de-energize the machine.

430.90 Combination Fuseholder and Switch as Controller.

The rating of a combination fuseholder and switch used as a motor controller shall be such that the fuseholder will accommodate the size of the fuse specified in Part III of this article for motor overload protection.

Exception: Where fuses having time delay appropriate for the starting characteristics of the motor are used, fuseholders of smaller size than specified in Part III of this article shall be permitted.

Time-delay (dual-element) fuses can commonly be used for both motor overload and branch-circuit short-circuit and ground-fault protection and can be sized in accordance with 430.32. See also 430.36, 430.55, and 430.57 for other requirements regarding fuses and fuseholders.

or contamination that may occur within the enclosure or enter via the conduit or unsealed openings. These internal conditions shall require special consideration by the installer and user.

430.91 Motor Controller Enclosure Types.

Table 430.91 provides the basis for selecting enclosures for use in specific locations other than hazardous (classified) locations. The enclosures are not intended to protect against conditions such as condensation, icing, corrosion,

See the commentary following the definition of *enclosure* in Article 100. Enclosure type numbers are described in greater detail in industry standards such as ANSI/NEMA ICS6, NEMA Pub. 250-179, UL 508, and the controller manufacturer's literature. For other than general-use Type

Table 430.91 Motor Controller Enclosure Selection

For Outdoor Use										
Provides a Degree of Protection Against the Following Environmental Conditions	**Enclosure Type Number[1]**									
	3	**3R**	**3S**	**3X**	**3RX**	**3SX**	**4**	**4X**	**6**	**6P**
Incidental contact with the enclosed equipment	X	X	X	X	X	X	X	X	X	X
Rain, snow, and sleet	X	X	X	X	X	X	X	X	X	X
Sleet[2]	—	—	X	—	—	X	—	—	—	—
Windblown dust	X	—	X	X	—	X	X	X	X	X
Hosedown	—	—	—	—	—	—	X	X	X	X
Corrosive agents	—	—	—	X	X	X	—	X	—	X
Temporary submersion	—	—	—	—	—	—	—	—	X	X
Prolonged submersion	—	—	—	—	—	—	—	—	—	X

For Indoor Use										
Provides a Degree of Protection Against the Following Environmental Conditions	**Enclosure Type Number[1]**									
	1	**2**	**4**	**4X**	**5**	**6**	**6P**	**12**	**12K**	**13**
Incidental contact with the enclosed equipment	X	X	X	X	X	X	X	X	X	X
Falling dirt	X	X	X	X	X	X	X	X	X	X
Falling liquids and light splashing	—	X	X	X	X	X	X	X	X	X
Circulating dust, lint, fibers, and flyings	—	—	X	X	—	X	X	X	X	X
Settling airborne dust, lint, fibers, and flyings	—	—	X	X	X	X	X	X	X	X
Hosedown and splashing water	—	—	X	X	—	X	X	—	—	—
Oil and coolant seepage	—	—	—	—	—	—	—	X	X	X
Oil or coolant spraying and splashing	—	—	—	—	—	—	—	—	—	X
Corrosive agents	—	—	—	X	—	—	X	—	—	—
Temporary submersion	—	—	—	—	—	X	X	—	—	—
Prolonged submersion	—	—	—	—	—	—	X	—	—	—

[1]Enclosure type number shall be marked on the motor controller enclosure.
[2]Mechanism shall be operable when ice covered.

1 enclosures, the type number must be marked on the motor-controller enclosure.

VIII. Motor Control Centers

430.92 General.

Part VIII covers motor control centers installed for the control of motors, lighting, and power circuits.

Part VIII of Article 430 provides requirements for the installation of motor control centers. These requirements cover subjects that include overcurrent protection use as service equipment, grounding, and construction. In addition to Part VIII, 430.1, FPN No. 1, refers to installation requirements for motor control centers contained in 110.26(F). The requirements of 110.26(F) specify dedicated space for a motor control center and physical protection from mechanical systems that might leak or otherwise adversely impact a motor control center.

Motor control centers are made up of a number of motor starters, controls, and disconnect switches assembled in one large enclosure. Motor control centers are allowed to be used as service equipment if provided with a single main disconnecting means. A second service disconnecting means, however, is permitted in the motor control center if it is provided to serve other loads.

Access and working space clearances as well as dedicated space requirements of 110.26 are applicable to motor control centers.

430.94 Overcurrent Protection.

Motor control centers shall be provided with overcurrent protection in accordance with Parts I, II, and IX of Article 240. The ampere rating or setting of the overcurrent protective device shall not exceed the rating of the common power bus. This protection shall be provided by (1) an overcurrent protective device located ahead of the motor control center or (2) a main overcurrent protective device located within the motor control center.

The overcurrent protection cannot exceed the rating of the common power bus of a motor control center. It is permitted to use an overcurrent protective device with a rating less than the common power bus, provided it is of sufficient size to carry the load determined in accordance with Part II of Article 430.

430.95 Service-Entrance Equipment.

Where used as service equipment, each motor control center shall be provided with a single main disconnecting means to disconnect all ungrounded service conductors.

Exception: A second service disconnect shall be permitted to supply additional equipment.

Where a grounded conductor is provided, the motor control center shall be provided with a main bonding jumper, sized in accordance with 250.28(D), within one of the sections for connecting the grounded conductor, on its supply side, to the motor control center equipment ground bus.

Exception: High-impedance grounded neutral systems shall be permitted to be connected as provided in 250.36.

430.96 Grounding.

Multisection motor control centers shall be bonded together with an equipment grounding conductor or an equivalent grounding bus sized in accordance with Table 250.122. Equipment grounding conductors shall terminate on this grounding bus or to a grounding termination point provided in a single-section motor control center.

430.97 Busbars and Conductors.

(A) Support and Arrangement. Busbars shall be protected from physical damage and be held firmly in place. Other than for required interconnections and control wiring, only those conductors that are intended for termination in a vertical section shall be located in that section.

Exception: Conductors shall be permitted to travel horizontally through vertical sections where such conductors are isolated from the busbars by a barrier.

(B) Phase Arrangement. The phase arrangement on 3-phase horizontal common power and vertical buses shall be A, B, C from front to back, top to bottom, or left to right, as viewed from the front of the motor control center. The B phase shall be that phase having the higher voltage to ground on 3-phase, 4-wire, delta-connected systems. Other busbar arrangements shall be permitted for additions to existing installations and shall be marked.

Exception: Rear-mounted units connected to a vertical bus that is common to front-mounted units shall be permitted to have a C, B, A phase arrangement where properly identified.

(C) Minimum Wire-Bending Space. The minimum wire-bending space at the motor control center terminals and minimum gutter space shall be as required in Article 312.

(D) Spacings. Spacings between motor control center bus terminals and other bare metal parts shall not be less than specified in Table 430.97.

(E) Barriers. Barriers shall be placed in all service-entrance motor control centers to isolate service busbars and terminals from the remainder of the motor control center.

Table 430.97 Minimum Spacing Between Bare Metal Parts

Nominal Voltage	Opposite Polarity Where Mounted on the Same Surface		Opposite Polarity Where Held Free in Air		Live Parts to Ground	
	mm	in.	mm	in.	mm	in.
Not over 125 volts, nominal	19.1	¾	12.7	½	12.7	½
Not over 250 volts, nominal	31.8	1¼	19.1	¾	12.7	½
Not over 600 volts, nominal	50.8	2	25.4	1	25.4	1

430.98 Marking.

(A) Motor Control Centers. Motor control centers shall be marked according to 110.21, and such marking shall be plainly visible after installation. Marking shall also include common power bus current rating and motor control center short-circuit rating.

(B) Motor Control Units. Motor control units in a motor control center shall comply with 430.8.

IX. Disconnecting Means

430.101 General.

Part IX is intended to require disconnecting means capable of disconnecting motors and controllers from the circuit.

> FPN No. 1: See Figure 430.1.
> FPN No. 2: See 110.22 for identification of disconnecting means.

430.102 Location.

(A) Controller. An individual disconnecting means shall be provided for each controller and shall disconnect the controller. The disconnecting means shall be located in sight from the controller location.

Exception No. 1: For motor circuits over 600 volts, nominal, a controller disconnecting means capable of being locked in the open position shall be permitted to be out of sight of the controller, provided the controller is marked with a warning label giving the location of the disconnecting means.

Exception No. 2: A single disconnecting means shall be permitted for a group of coordinated controllers that drive several parts of a single machine or piece of apparatus. The disconnecting means shall be located in sight from the controllers, and both the disconnecting means and the controllers shall be located in sight from the machine or apparatus.

(B) Motor. A disconnecting means shall be located in sight from the motor location and the driven machinery location. The disconnecting means required in accordance with

430.102(A) shall be permitted to serve as the disconnecting means for the motor if it is located in sight from the motor location and the driven machinery location.

Exception: The disconnecting means shall not be required to be in sight from the motor and the driven machinery location under either condition (a) or (b), provided the disconnecting means required in accordance with 430.102(A) is individually capable of being locked in the open position. The provision for locking or adding a lock to the disconnecting means shall be permanently installed on or at the switch or circuit breaker used as the disconnecting means.

(a) *Where such a location of the disconnecting means is impracticable or introduces additional or increased hazards to persons or property*

(b) *In industrial installations, with written safety procedures, where conditions of maintenance and supervision ensure that only qualified persons service the equipment*

> FPN No. 1: Some examples of increased or additional hazards include, but are not limited to, motors rated in excess of 100 hp, multimotor equipment, submersible motors, motors associated with variable frequency drives, and motors located in hazardous (classified) locations.
> FPN No. 2: For information on lockout/tagout procedures, see NFPA 70E-2000, *Standard for Electrical Safety Requirements for Employee Workplaces.*

The main rules of 430.102(A) and (B) require that the disconnecting means be in sight from the controller, the motor location, and the driven machinery location. For motors over 600 volts, the controller disconnecting means may be out of sight of the controller, as illustrated in Exhibit 430.18, provided that the controller has a warning label indicating the location and identification of the disconnecting means, which must be capable of being locked in the open position.

A single disconnecting means may be located adjacent to a group of coordinated controllers, as illustrated in Exhibit 430.19, where the controllers are mounted on a multimotor continuous process machine.

The exception to 430.102(B) was revised for the 2002 *Code.* The disconnecting means may only be out of sight

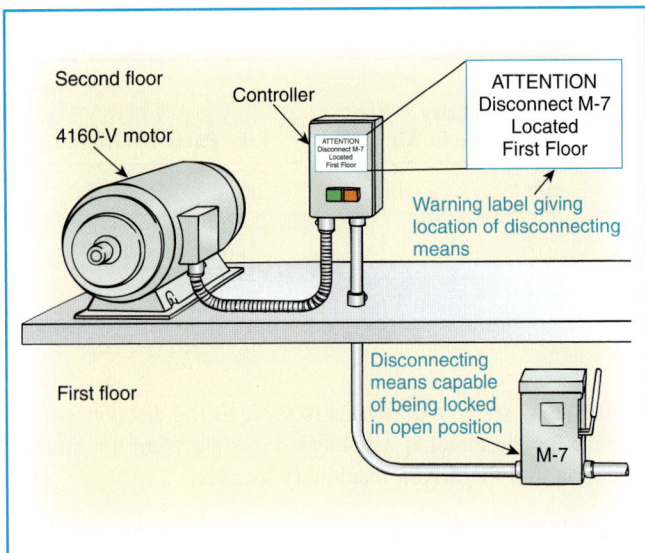

Exhibit 430.18 A motor installation over 600 volts, where the motor controller is not located within sight of its disconnecting means.

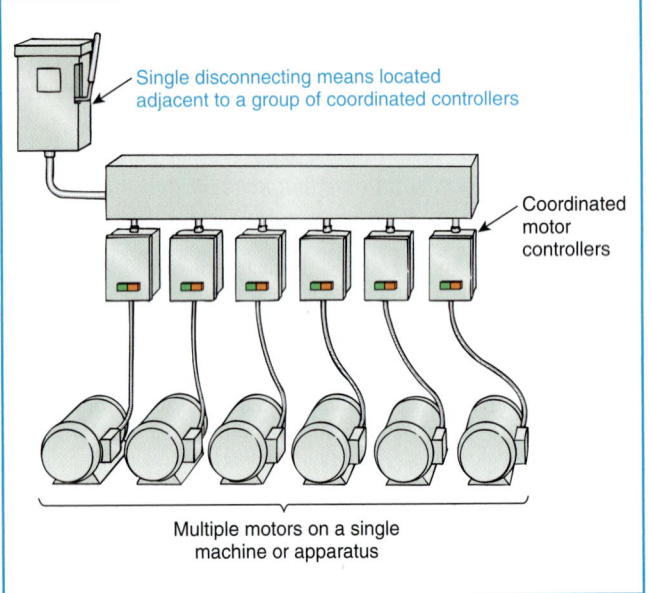

Exhibit 430.19 A single disconnecting means located adjacent to a group of coordinated controllers mounted on a multimotor continuous process machine.

of the motor, as illustrated in Exhibit 430.20, if the disconnecting means complying with 430.102(A) is individually capable of being locked in the open position and meets the criteria of either (a) or (b) in the exception. If locating the disconnecting means close to the motor location and driven machinery is impracticable due to the type of machinery, the type of facility, the lack of space for locating large

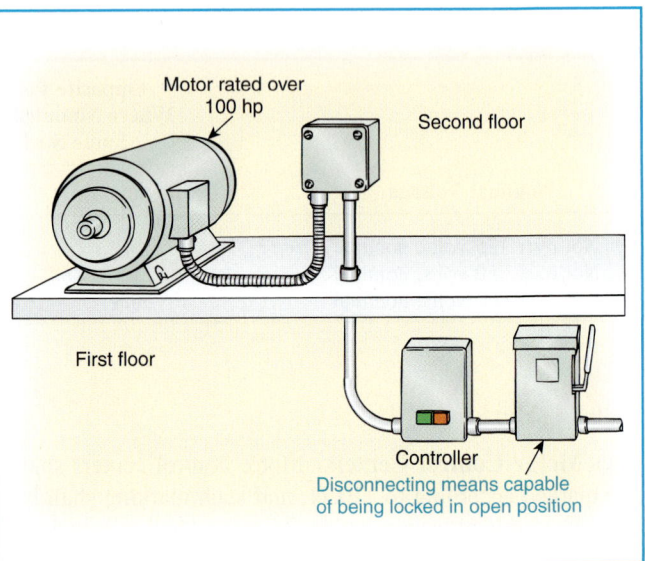

Exhibit 430.20 A controller disconnecting means that is out of sight of the motor.

equipment such as disconnecting means rated over 600 volts, or any increased hazard to persons or property, then the disconnecting means is permitted to be located remotely. Industrial facilities that comply with OSHA, CFR 1910.147, *The Control of Hazardous Energy* (lockout/tagout), are permitted to have the disconnecting means located remotely.

Section 430.102 clearly requires that individual disconnect switches or circuit breakers must be capable of being locked in the open position. Disconnect switches or circuit breakers that are only located behind the locked door of a panelboard or located within locked rooms do not comply with the requirements of 430.102.

Fine Print Note No. 2 points out an important consideration and reference standard for employee safety in the workplace. *Part II, Chapter 5,* of NFPA 70E, *Standard for Electrical Safety Requirements for Employee Workplaces,* requires in part that, "All electrical circuit conductors and circuit parts shall not be considered to be in an electrically safe condition until all sources of energy are removed, the disconnecting means is under lockout/tagout, the absence of voltage is verified by an approved voltage testing device, etc." Further, it states that "Lockout/tagout requirements shall apply to fixed permanently installed equipment, temporarily installed equipment, and to portable equipment." The principles and procedures set forth in NFPA 70E establish strict work rules requiring locking-off and tagging-out of disconnect switches.

430.103 Operation.

The disconnecting means shall open all ungrounded supply conductors and shall be designed so that no pole can be operated independently. The disconnecting means shall be permitted in the same enclosure with the controller.

FPN: See 430.113 for equipment receiving energy from more than one source.

The *Code* requires that a switch, circuit breaker, or other device serve as a disconnecting means for both the controller and the motor, thereby providing safety during maintenance and inspection shutdown periods. The disconnecting means also disconnects the controller; therefore, it cannot be a part of the controller. However, separate disconnects and controllers may be mounted on the same panel or be contained in the same enclosure, such as combination fused-switch, magnetic-starter units.

Depending on the size of the motor and other conditions, the type of disconnecting means required may be a motor circuit switch, a circuit breaker, a general-use switch, an isolating switch, an attachment plug and receptacle, or a branch-circuit short-circuit and ground-fault protective device, as specified in 430.109.

If a motor is stalled or under heavy overload and the motor controller fails to properly open the circuit, the disconnecting means, which must be rated to interrupt locked-rotor current, can be used to open the circuit. For motors larger than 100 hp ac or 40 hp dc, the disconnecting means is, in accordance with 430.109(E), permitted to be a general-use or an isolating switch where plainly marked "Do not operate under load."

430.104 To Be Indicating.

The disconnecting means shall plainly indicate whether it is in the open (off) or closed (on) position.

430.105 Grounded Conductors.

One pole of the disconnecting means shall be permitted to disconnect a permanently grounded conductor, provided the disconnecting means is designed so that the pole in the grounded conductor cannot be opened without simultaneously disconnecting all conductors of the circuit.

430.107 Readily Accessible.

At least one of the disconnecting means shall be readily accessible.

430.108 Every Switch.

Every disconnecting means in the motor circuit between the point of attachment to the feeder and the point of connection to the motor shall comply with the requirements of 430.109 and 430.110.

Exhibit 430.21 represents an example of a listed disconnect switch rated in horsepower meeting the requirements of 430.109 and 430.110.

Exhibit 430.21 Heavy-duty safety switches, UL-listed for use on systems up to 200,000 amperes, fault current rms symmetrical with Class J or Class R fuses installed. (Courtesy of Square D Co.)

430.109 Type.

The disconnecting means shall be a type specified in 430.109(A), unless otherwise permitted in 430.109(B) through (G), under the conditions specified.

(A) General.

(1) Motor Circuit Switch. A listed motor-circuit switch rated in horsepower. For Design E motors rated greater than 2 hp, the motor circuit switch shall be either (1) marked as rated for use with Design E motors or (2) have a horsepower rating not less than 1.4 times the rating of a motor rated 3–100 hp, or not less than 1.3 times the rating of a motor rated over 100 hp.

(2) Molded Case Circuit Breaker. A listed molded case circuit breaker.

(3) Molded Case Switch. A listed molded case switch.

(4) Instantaneous Trip Circuit Breaker. An instantaneous trip circuit breaker that is part of a listed combination motor controller.

(5) Self-Protected Combination Controller. Listed self-protected combination controller.

(6) Manual Motor Controller. Listed manual motor controllers additionally marked "Suitable as Motor Disconnect" shall be permitted as a disconnecting means where installed between the final motor branch-circuit short-circuit and ground-fault protective device and the motor.

(B) Stationary Motors of ⅛ Horsepower or Less. For stationary motors of ⅛ hp or less, the branch-circuit overcurrent device shall be permitted to serve as the disconnecting means.

(C) Stationary Motors of 2 Horsepower or Less. For stationary motors rated at 2 hp or less and 300 volts or less, the disconnecting means shall be permitted to be one of the devices specified in (1), (2), or (3):

(1) A general-use switch having an ampere rating not less than twice the full-load current rating of the motor
(2) On ac circuits, a general-use snap switch suitable only for use on ac (not general-use ac–dc snap switches) where the motor full-load current rating is not more than 80 percent of the ampere rating of the switch
(3) A listed manual motor controller having a horsepower rating not less than the rating of the motor and marked "Suitable as Motor Disconnect"

(D) Autotransformer-Type Controlled Motors. For motors of over 2 hp to and including 100 hp, the separate disconnecting means required for a motor with an autotransformer-type controller shall be permitted to be a general-use switch where all of the following provisions are met:

(1) The motor drives a generator that is provided with overload protection.
(2) The controller is capable of interrupting the locked-rotor current of the motors, is provided with a no voltage release, and is provided with running overload protection not exceeding 125 percent of the motor full-load current rating.
(3) Separate fuses or an inverse time circuit breaker rated or set at not more than 150 percent of the motor full-load current are provided in the motor branch circuit.

(E) Isolating Switches. For stationary motors rated at more than 40 hp dc or 100 hp ac, the disconnecting means shall be permitted to be a general-use or isolating switch where plainly marked "Do not operate under load."

(F) Cord-and-Plug-Connected Motors. For a cord-and-plug-connected motor, a horsepower-rated attachment plug and receptacle having ratings no less than the motor ratings shall be permitted to serve as the disconnecting means for other than a Design E motor and for a Design E motor rated 2 hp or less. For a Design E motor rated more than 2 hp, an attachment plug and receptacle used as the disconnecting means shall have a horsepower rating not less than 1.4 times the motor rating. A horsepower-rated attachment plug and receptacle shall not be required for a cord-and-plug-connected appliance in accordance with 422.32, a room air conditioner in accordance with 440.63, or a portable motor rated ⅓ hp or less.

Section 430.109 requires the disconnecting means to be a motor circuit switch, a circuit breaker, or a molded-case switch (nonautomatic circuit interrupter). A motor circuit switch is a horsepower-rated switch capable of interrupting the maximum overload current of a motor (see the definition of *switch, motor-circuit* in Article 100). A molded-case switch (nonautomatic circuit interrupter) is a circuit-breaker-like device without the overcurrent element and automatic-trip mechanism. It is rated in amperes and is suitable for use as a motor circuit disconnect based on its ampere rating, as is a circuit breaker. The disconnecting means must be listed.

Exhibits 430.22, 430.23, 430.24, and 430.25 illustrate various methods of providing motor disconnecting means as permitted by 430.109(B), (C), (E), and (F), respectively.

Section 430.109(A) recognizes that Design E motors have a high starting (locked-rotor) current. Disconnect switches used with Design E motors over 2 hp must be identified for use with these design motors, or the size of the disconnect switch must be adjusted to compensate for this high starting current.

Where horsepower-rated fused switches are required, it should be noted that marking within the enclosure usually permits a dual horsepower rating. The standard horsepower

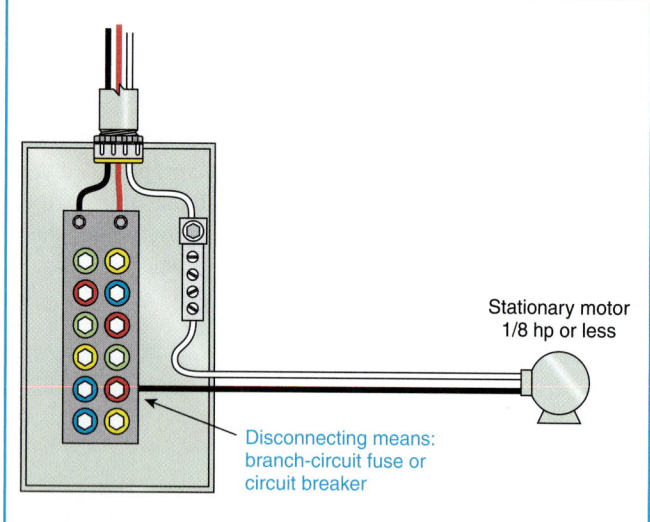

Stationary motor
1/8 hp or less

Disconnecting means:
branch-circuit fuse or
circuit breaker

Exhibit 430.22 A branch-circuit overcurrent device serving as the disconnecting means for a stationary motor of ⅛ hp or less according to 430.109(B).

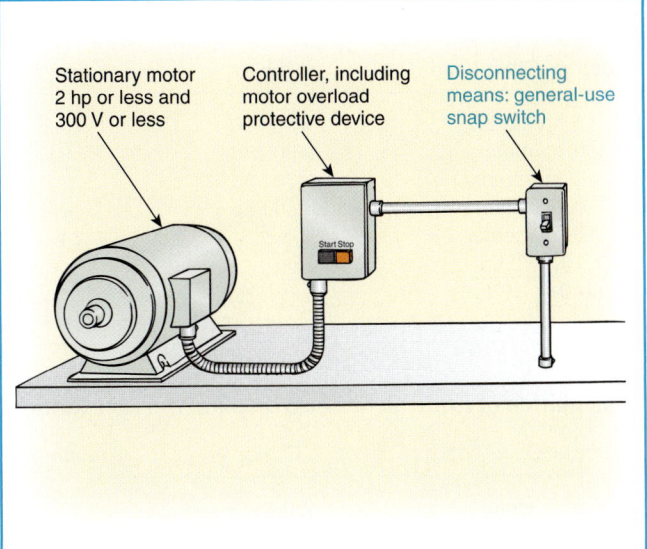

Stationary motor 2 hp or less and 300 V or less

Controller, including motor overload protective device

Disconnecting means: general-use snap switch

Start Stop

Exhibit 430.23 A general-use snap switch having an ampere rating not less than twice the motor full-load current serving as the disconnecting means for a stationary motor rated at 2 hp or less and at 300 volts or less according to 430.109(C).

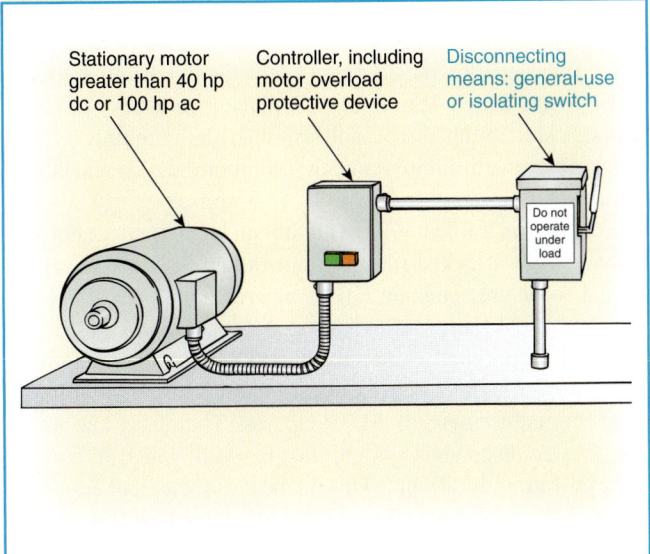

Stationary motor greater than 40 hp dc or 100 hp ac

Controller, including motor overload protective device

Disconnecting means: general-use or isolating switch

Do not operate under load

Exhibit 430.24 A general-use or an isolation switch serving as the disconnecting means for a stationary motor rated at more than 40 hp dc or 100 hp ac according to 430.109(E).

rating is based on the largest non-time-delay (non-dual-element) fuse rating that can be used in the switch and that will permit the motor to start. The maximum horsepower rating is based on the largest rated time-delay (dual-element) fuse that can be used in the switch and that will permit the motor to start. Thus, where time-delay fuses are used, smaller-size switches and fuseholders can be used (see 430.57, Exception).

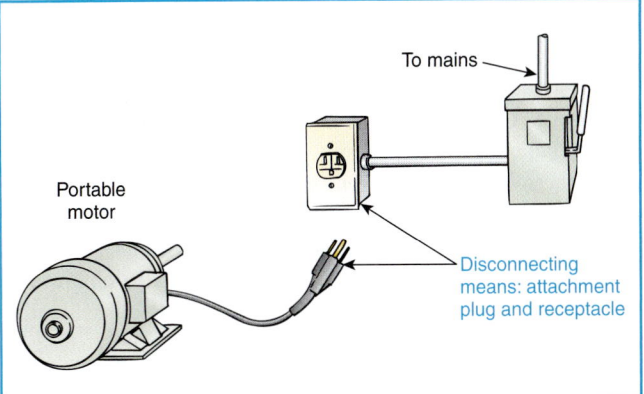

To mains

Portable motor

Disconnecting means: attachment plug and receptacle

Exhibit 430.25 An attachment plug and receptacle of the proper rating serving as the disconnecting means for a certain cord-and-plug-connected motor according to 430.109(F).

(G) Torque Motors. For torque motors, the disconnecting means shall be permitted to be a general-use switch.

430.110 Ampere Rating and Interrupting Capacity.

(A) General. The disconnecting means for motor circuits rated 600 volts, nominal, or less shall have an ampere rating of at least 115 percent of the full-load current rating of the motor.

Exception: A listed nonfused motor-circuit switch having a horsepower rating equal to or greater than the motor horsepower shall be permitted to have an ampere rating less than 115 percent of the full-load current rating of the motor.

(B) For Torque Motors. Disconnecting means for a torque motor shall have an ampere rating of at least 115 percent of the motor nameplate current.

(C) For Combination Loads. Where two or more motors are used together or where one or more motors are used in combination with other loads, such as resistance heaters, and where the combined load may be simultaneous on a single disconnecting means, the ampere and horsepower ratings of the combined load shall be determined as follows.

(1) Horsepower Rating. The rating of the disconnecting means shall be determined from the sum of all currents, including resistance loads, at the full-load condition and also at the locked-rotor condition. The combined full-load current and the combined locked-rotor current so obtained shall be considered as a single motor for the purpose of this requirement as follows.

The full-load current equivalent to the horsepower rating of each motor shall be selected from Tables 430.147, 430.148, 430.149, or 430.150. These full-load currents shall be added to the rating in amperes of other loads to obtain an equivalent full-load current for the combined load.

The locked-rotor current equivalent to the horsepower rating of each motor shall be selected from Table 430.151(A) or Table 430.151(B). The locked-rotor currents shall be added to the rating in amperes of other loads to obtain an equivalent locked-rotor current for the combined load. Where two or more motors or other loads cannot be started simultaneously, the largest sum of locked rotor currents of a motor or group of motors that can be started simultaneously and the full load currents of other concurrent loads shall be permitted to be used to determine the equivalent locked-rotor current for the simultaneous combined loads.

Exception: Where part of the concurrent load is resistance load, and where the disconnecting means is a switch rated in horsepower and amperes, the switch used shall be permitted to have a horsepower rating that is not less than the combined load of the motor(s), if the ampere rating of the switch is not less than the locked-rotor current of the motor(s) plus the resistance load.

(2) Ampere Rating. The ampere rating of the disconnecting means shall not be less than 115 percent of the sum of all currents at the full-load condition determined in accordance with 430.110(C)(1).

Exception: A listed nonfused motor-circuit switch having a horsepower rating equal to or greater than the equivalent horsepower of the combined loads, determined in accordance with 430.110(C)(1), shall be permitted to have an ampere rating less than 115 percent of the sum of all currents at the full-load condition.

(3) Small Motors. For small motors not covered by Tables 430.147, 430.148, 430.149, or 430.150, the locked-rotor current shall be assumed to be six times the full-load current.

A general-use switch, circuit breaker, molded-case switch (nonautomatic circuit interrupter), or attachment plug and receptacle used as a disconnecting means must have an ampere rating of not less than 115 percent of the motor full-load current.

Following is an example of how to determine the size of a disconnect for a combination load.

Example

Assume the load consists of one 5-hp, one 3-hp, and two ½-hp motors, plus a 10-kilowatt heater, all rated 240 volts, 3-phase. All motors are Design B motors. The full-load and locked-rotor current equivalents are calculated, using the appropriate tables, as follows:

Calculation of Equivalent Full-Load Current Rating

Motor or Other Load	Table	Amperes
5-hp motor	430.150	15.2
3-hp motor	430.150	9.6
½-hp motor	430.150	2.2
½-hp motor	430.150	2.2
10-kW heater		24.1
$\dfrac{10 \times 1000}{(240 \times 1.732)}$		
	Total	53.3

Calculation of Equivalent Locked-Rotor Current Rating

Motor or Other Load	Table	Amperes
5-hp motor	430.151B	92
3-hp motor	430.151B	64
½-hp motor	430.151B	20
½-hp motor	430.151B	20
10-kW heater		24.1
$\dfrac{10 \times 1000}{(240 \times 1.732)}$		
	Total	220.1

Solution

The disconnecting means for a motor must have an ampere rating that is not less than 115 percent and also have a horsepower rating that is not less than the combined load. Therefore, the minimum ampere rating of the disconnecting means is 61.3 amperes (1.15 × 53.3 amperes = 61.3 amperes). Using Table 430.151B to obtain an equivalent horsepower from a locked-rotor current rating, the closest value equal to or greater than 220.1 amperes under the 230-volt column is 232 amperes, which equates to 15 hp. A 15-hp general-purpose or heavy-duty switch satisfies the minimum horsepower requirement but fails to satisfy the minimum current requirement of 61.3 amperes. Therefore, the next larger size disconnect switch must be used, which in horsepower rating is 20 hp. This is because the locked-rotor current for a 20-hp, 3-phase, 230-volt motor is 290 amperes (over the needed 220.1 amperes) and the full-load current is 54 amperes (over the full-load rating of 53.3 amperes).

Listed circuit breakers and molded-case switches are tested under overload conditions at six times their rating, to cover motor circuit applications, and are suitable for use as a motor disconnecting means.

430.111 Switch or Circuit Breaker as Both Controller and Disconnecting Means.

A switch or circuit breaker shall be permitted to be used as both the controller and disconnecting means if it complies with 430.111(A) and is one of the types specified in 430.111(B).

(A) General. The switch or circuit breaker complies with the requirements for controllers specified in 430.83, opens all ungrounded conductors to the motor, and is protected by an overcurrent device in each ungrounded conductor (which shall be permitted to be the branch-circuit fuses). The overcurrent device protecting the controller shall be permitted to be part of the controller assembly or shall be permitted to be separate. An autotransformer-type controller shall be provided with a separate disconnecting means.

(B) Type. The device shall be one of the types specified in 430.111(B)(1), (2), or (3).

(1) Air-Break Switch. An air-break switch, operable directly by applying the hand to a lever or handle.

(2) Inverse Time Circuit Breaker. An inverse time circuit breaker operable directly by applying the hand to a lever or handle. The circuit breaker shall be permitted to be both power and manually operable.

(3) Oil Switch. An oil switch used on a circuit whose rating does not exceed 600 volts or 100 amperes, or by special permission on a circuit exceeding this capacity where under expert supervision. The oil switch shall be permitted to be both power and manually operable.

If used as a controller, a switch or circuit breaker must meet all the requirements for controllers and be protected by branch-circuit short-circuit and ground-fault protective devices (fuses or a circuit breaker), which ensure that all ungrounded conductors will be opened.

If the controller consists of a manually operable air-break switch, an inverse time circuit breaker, or a 100-ampere maximum oil switch (higher rating by special permission), the controller is considered a satisfactory disconnecting means. It is the intent of 430.11 to permit omission of an additional device to serve as a disconnecting means.

Note that a separate disconnecting means must be provided if the controller is of the autotransformer or compensator type. (This switch may be combined in the same enclosure with a motor overload protective device.)

430.112 Motors Served by Single Disconnecting Means.

Each motor shall be provided with an individual disconnecting means.

Exception: A single disconnecting means shall be permitted to serve a group of motors under any one of the conditions of (a), (b), and (c). The single disconnecting means shall be rated in accordance with 430.110(C).

(a) Where a number of motors drive several parts of a single machine or piece of apparatus, such as metal and woodworking machines, cranes, and hoists.

(b) Where a group of motors is under the protection of one set of branch-circuit protective devices as permitted by 430.53(A).

(c) Where a group of motors is in a single room within sight from the location of the disconnecting means.

The exception after 430.112 permits a single disconnecting means to serve a group of motors. The disconnecting means must have a rating equal to the sum of the horsepower or current of each motor in the group. If the sum is over 2 hp, a motor circuit switch (horsepower-rated) must be used; thus, for five 2-hp motors, the disconnecting means should be a motor circuit switch rated at not less than 10 hp.

Part (a) of the exception to 430.112 indicates that a single disconnecting means may be used where a number of motors drive several parts of a single machine, such as cranes (see 610.31, 610.32, and 610.33), metal or woodworking machines, steel rolling mill machinery, and so on. The single disconnecting means for multimotor machinery provides a positive means of simultaneously de-energizing all motor branch circuits, including remote control circuits, interlocking circuits, limit-switch circuits, and operator control stations.

Part (b) of the exception to 430.112 refers to 430.53(A), which permits a group of motors under the protection of the same branch-circuit device, provided the device is rated not more than 20 amperes at 125 volts or 15 amperes at more than 125 volts but not more than 600 volts. The motors must be rated 1 hp or less, and the full-load current for each motor is not permitted to exceed 6 amperes. A single disconnecting means is both practical and economical for a group of such small motors.

Part (c) of the exception to 430.112 covers the common situation in which a group of motors is located in one room, such as a pump room, compressor room, mixer room, and so on. It is therefore possible to design the layout of a single disconnecting means with an unobstructed view (not more than 50 ft) from each motor.

These conditions for an individual disconnecting means are similar to those specified in 430.87, which permits the use of a single controller for a group of motors.

430.113 Energy from More Than One Source.

Motor and motor-operated equipment receiving electrical energy from more than one source shall be provided with disconnecting means from each source of electrical energy immediately adjacent to the equipment served. Each source shall be permitted to have a separate disconnecting means. Where multiple disconnecting means are provided, a permanent warning sign shall be provided on or adjacent to each disconnecting means.

Exception No. 1: Where a motor receives electrical energy from more than one source, the disconnecting means for the main power supply to the motor shall not be required to be immediately adjacent to the motor, provided the controller disconnecting means is capable of being locked in the open position.

Exception No. 2: A separate disconnecting means shall not be required for a Class 2 remote-control circuit conforming with Article 725, rated not more than 30 volts, and that is isolated and ungrounded.

Certain motors may require multiple separate sources of power to operate properly. For example, large synchronous motors commonly receive electrical energy from more than one source. Section 430.113 could apply as well to control circuits that supply power to sensors mounted within or otherwise attached to the motor or the driven machine. Exception No. 2 to 430.113 relieves the user from the disconnect requirement only for Class 2 circuits.

New for the 2002 *Code*, where there are multiple disconnecting means for the individual sources, a permanent warning sign is required to warn the user that other power sources are present.

X. Over 600 Volts, Nominal

430.121 General.

Part X recognizes the additional hazard due to the use of higher voltages. It adds to or amends the other provisions of this article.

430.122 Marking on Controllers.

In addition to the marking required by 430.8, a controller shall be marked with the control voltage.

430.123 Conductor Enclosures Adjacent to Motors.

Flexible metal conduit or liquidtight flexible metal conduit not exceeding 1.8 m (6 ft) in length shall be permitted to be employed for raceway connection to a motor terminal enclosure.

430.124 Size of Conductors.

Conductors supplying motors shall have an ampacity not less than the current at which the motor overload protective device(s) is selected to trip.

430.125 Motor-Circuit Overcurrent Protection.

(A) General. Each motor circuit shall include coordinated protection to automatically interrupt overload and fault currents in the motor, the motor-circuit conductors, and the motor control apparatus.

Exception: Where a motor is vital to operation of the plant and the motor should operate to failure if necessary to prevent a greater hazard to persons, the sensing device(s) shall be permitted to be connected to a supervised annunciator or alarm instead of interrupting the motor circuit.

(B) Overload Protection.

(1) Type of Overload Device. Each motor shall be protected against dangerous heating due to motor overloads and failure to start by a thermal protector integral with the motor or external current-sensing devices, or both.

(2) Wound-Rotor AC Motors. The secondary circuits of wound-rotor ac motors including conductors, controllers, and resistors rated for the application shall be considered as protected against overcurrent by the motor overload protection means.

(3) Operation. Operation of the overload interrupting device shall simultaneously disconnect all ungrounded conductors.

(4) Automatic Reset. Overload sensing devices shall not automatically reset after trip unless resetting of the overload sensing device does not cause automatic restarting of the motor or there is no hazard to persons created by automatic restarting of the motor and its connected machinery.

(C) Fault-Current Protection.

(1) Type of Protection. Fault-current protection shall be provided in each motor circuit by one of the following means.

(a) A circuit breaker of suitable type and rating arranged so that it can be serviced without hazard. The circuit breaker shall simultaneously disconnect all ungrounded conductors. The circuit breaker shall be permitted to sense the fault current by means of integral or external sensing elements.

(b) Fuses of a suitable type and rating placed in each ungrounded conductor. Fuses shall be used with suitable disconnecting means, or they shall be of a type that can also serve as the disconnecting means. They shall be arranged so that they cannot be serviced while they are energized.

(2) Reclosing. Fault-current interrupting devices shall not automatically reclose the circuit.

Exception: Automatic reclosing of a circuit shall be permitted where the circuit is exposed to transient faults and where such automatic reclosing does not create a hazard to persons.

(3) Combination Protection. Overload protection and fault-current protection shall be permitted to be provided by the same device.

430.126 Rating of Motor Control Apparatus.

The ultimate trip current of overcurrent (overload) relays or other motor-protective devices used shall not exceed 115 percent of the controller's continuous current rating. Where the motor branch-circuit disconnecting means is separate from the controller, the disconnecting means current rating shall not be less than the ultimate trip setting of the overcurrent relays in the circuit.

430.127 Disconnecting Means.

The controller disconnecting means shall be capable of being locked in the open position.

XI. Protection of Live Parts— All Voltages

430.131 General.

Part XI specifies that live parts shall be protected in a manner judged adequate for the hazard involved.

430.132 Where Required.

Exposed live parts of motors and controllers operating at 50 volts or more between terminals shall be guarded against accidental contact by enclosure or by location as follows:

(1) By installation in a room or enclosure that is accessible only to qualified persons
(2) By installation on a suitable balcony, gallery, or platform, elevated and arranged so as to exclude unqualified persons
(3) By elevation 2.5 m (8 ft) or more above the floor

Exception: Live parts of motors operating at more than 50 volts between terminals shall not require additional guarding for stationary motors that have commutators, collectors, and brush rigging located inside of motor-end brackets and not conductively connected to supply circuits operating at more than 150 volts to ground.

430.133 Guards for Attendants.

Where live parts of motors or controllers operating at over 150 volts to ground are guarded against accidental contact only by location as specified in 430.132, and where adjustment or other attendance may be necessary during the operation of the apparatus, suitable insulating mats or platforms shall be provided so that the attendant cannot readily touch live parts unless standing on the mats or platforms.

FPN: For working space, see 110.26 and 110.34.

XII. Grounding—All Voltages

430.141 General.

Part XII specifies the grounding of exposed non–current-carrying metal parts, likely to become energized, of motor and controller frames to prevent a voltage above ground in the event of accidental contact between energized parts and frames. Insulation, isolation, or guarding are suitable alternatives to grounding of motors under certain conditions.

430.142 Stationary Motors.

The frames of stationary motors shall be grounded under any of the following conditions:

Any motor in a wet location and subject to contact by personnel constitutes a serious hazard and, unless it is isolated, elevated, or guarded from reach, must be grounded.

Stationary motors are usually supplied by wiring that is enclosed in metal raceways, flexible metal conduit, or cables with metal sheaths. When effectively attached to the motor junction box or frame, the metal raceway or cable armor serves as the equipment grounding conductor. See 250.118 for more information on types of equipment grounding conductors.

(1) Where supplied by metal-enclosed wiring
(2) Where in a wet location and not isolated or guarded
(3) If in a hazardous (classified) location as covered in Articles 500 through 517
(4) If the motor operates with any terminal at over 150 volts to ground

Where the frame of the motor is not grounded, it shall be permanently and effectively insulated from the ground.

430.143 Portable Motors.

The frames of portable motors that operate at over 150 volts to ground shall be guarded or grounded.

FPN No. 1: See 250.114(4) for grounding of portable appliances in other than residential occupancies.
FPN No. 2: See 250.119(B) for color of equipment grounding conductor.

430.144 Controllers.

Controller enclosures shall be grounded regardless of voltage. Controller enclosures shall have means for attachment of an equipment grounding conductor termination in accordance with 250.8.

Exception: Enclosures attached to ungrounded portable equipment shall not be required to be grounded.

430.145 Method of Grounding.

Where required, grounding shall be done in the manner specified in Part V of Article 250.

(A) Grounding Through Terminal Housings. Where the wiring to fixed motors is metal-enclosed cable or in metal raceways, junction boxes to house motor terminals shall be provided, and the armor of the cable or the metal raceways shall be connected to them in the manner specified in Article 250.

> FPN: See 430.12(E) for equipment grounding connection means required at motor terminal housings.

(B) Separation of Junction Box from Motor. The junction box required by 430.145(A) shall be permitted to be separated from the motor by not more than 1.8 m (6 ft), provided the leads to the motor are Type AC cable or armored cord or are stranded leads enclosed in liquidtight flexible metal conduit, flexible metal conduit, intermediate metal conduit, rigid metal conduit, or electrical metallic tubing not smaller than metric designator 12 (trade size ⅜), the armor or raceway being connected both to the motor and to the box.

Liquidtight flexible nonmetallic conduit and rigid nonmetallic conduit shall be permitted to enclose the leads to the motor, provided the leads are stranded and the required equipment grounding conductor is connected to both the motor and to the box.

Where stranded leads are used, protected as specified above, each strand within the conductor shall be not larger than 10 AWG and shall comply with other requirements of this *Code* for conductors to be used in raceways.

(C) Grounding of Controller-Mounted Devices. Instrument transformer secondaries and exposed non–current-carrying metal or other conductive parts or cases of instrument transformers, meters, instruments, and relays shall be grounded as specified in 250.170 through 250.178.

Most motors are subject to vibration, and good wiring practice requires that, in nearly all cases, the wiring to motors that are fixed be installed with a short section (not more than 6 ft) of liquidtight flexible metal or nonmetallic conduit or of flexible metal conduit to the motor terminal housing. Such use of flexible conduit requires an equipment grounding conductor.

XIII. Tables

Tables 430.147 through 430.150 accurately reflect the typical and most used 4-pole and 2-pole induction motors in use.

Table 430.147 Full-Load Current in Amperes, Direct-Current Motors
The following values of full-load currents[*] are for motors running at base speed.

Horse-power	Armature Voltage Rating[*]					
	90 Volts	120 Volts	180 Volts	240 Volts	500 Volts	550 Volts
¼	4.0	3.1	2.0	1.6	—	—
⅓	5.2	4.1	2.6	2.0	—	—
½	6.8	5.4	3.4	2.7	—	—
¾	9.6	7.6	4.8	3.8	—	—
1	12.2	9.5	6.1	4.7	—	—
1½	—	13.2	8.3	6.6	—	—
2	—	17	10.8	8.5	—	—
3	—	25	16	12.2	—	—
5	—	40	27	20	—	—
7½	—	58	—	29	13.6	12.2
10	—	76	—	38	18	16
15	—	—	—	55	27	24
20	—	—	—	72	34	31
25	—	—	—	89	43	38
30	—	—	—	106	51	46
40	—	—	—	140	67	61
50	—	—	—	173	83	75
60	—	—	—	206	99	90
75	—	—	—	255	123	111
100	—	—	—	341	164	148
125	—	—	—	425	205	185
150	—	—	—	506	246	222
200	—	—	—	675	330	294

*These are average dc quantities.

Table 430.148 Full-Load Currents in Amperes, Single-Phase Alternating-Current Motors
The following values of full-load currents are for motors running at usual speeds and motors with normal torque characteristics. Motors built for especially low speeds or high torques may have higher full-load currents, and multispeed motors will have full-load current varying with speed, in which case the nameplate current ratings shall be used.

The voltages listed are rated motor voltages. The currents listed shall be permitted for system voltage ranges of 110 to 120 and 220 to 240 volts.

Horsepower	115 Volts	200 Volts	208 Volts	230 Volts
⅙	4.4	2.5	2.4	2.2
¼	5.8	3.3	3.2	2.9
⅓	7.2	4.1	4.0	3.6
½	9.8	5.6	5.4	4.9
¾	13.8	7.9	7.6	6.9
1	16	9.2	8.8	8.0
1½	20	11.5	11.0	10
2	24	13.8	13.2	12
3	34	19.6	18.7	17
5	56	32.2	30.8	28
7½	80	46.0	44.0	40
10	100	57.5	55.0	50

Table 430.149 Full-Load Current, Two-Phase Alternating-Current Motors (4-Wire)

The following values of full-load current are for motors running at speeds usual for belted motors and motors with normal torque characteristics. Motors built for especially low speeds or high torques may require more running current, and multispeed motors will have full-load current varying with speed, in which case the nameplate current rating shall be used. Current in the common conductor of a 2-phase, 3-wire system will be 1.41 times the value given.

The voltages listed are rated motor voltages. The currents listed shall be permitted for system voltage ranges of 110 to 120, 220 to 240, 440 to 480, and 550 to 600 volts.

	Induction-Type Squirrel Cage and Wound Rotor (Amperes)				
Horsepower	115 Volts	230 Volts	460 Volts	575 Volts	2300 Volts
½	4.0	2.0	1.0	0.8	—
¾	4.8	2.4	1.2	1.0	—
1	6.4	3.2	1.6	1.3	—
1½	9.0	4.5	2.3	1.8	—
2	11.8	5.9	3.0	2.4	—
3	—	8.3	4.2	3.3	—
5	—	13.2	6.6	5.3	—
7½	—	19	9.0	8.0	—

Table 430.149 Continued

	Induction-Type Squirrel Cage and Wound Rotor (Amperes)				
Horsepower	115 Volts	230 Volts	460 Volts	575 Volts	2300 Volts
10	—	24	12	10	—
15	—	36	18	14	—
20	—	47	23	19	—
25	—	59	29	24	—
30	—	69	35	28	—
40	—	90	45	36	—
50	—	113	56	45	—
60	—	133	67	53	14
75	—	166	83	66	18
100	—	218	109	87	23
125	—	270	135	108	28
150	—	312	156	125	32
200	—	416	208	167	43

Table 430.150 Full-Load Current, Three-Phase Alternating-Current Motors

The following values of full-load currents are typical for motors running at speeds usual for belted motors and motors with normal torque characteristics.

Motors built for low speeds (1200 rpm or less) or high torques may require more running current, and multispeed motors will have full-load current varying with speed. In these cases, the nameplate current rating shall be used.

The voltages listed are rated motor voltages. The currents listed shall be permitted for system voltage ranges of 110 to 120, 220 to 240, 440 to 480, and 550 to 600 volts.

	Induction-Type Squirrel Cage and Wound Rotor (Amperes)							Synchronous-Type Unity Power Factor* (Amperes)			
Horsepower	115 Volts	200 Volts	208 Volts	230 Volts	460 Volts	575 Volts	2300 Volts	230 Volts	460 Volts	575 Volts	2300 Volts
½	4.4	2.5	2.4	2.2	1.1	0.9	—	—	—	—	—
¾	6.4	3.7	3.5	3.2	1.6	1.3	—	—	—	—	—
1	8.4	4.8	4.6	4.2	2.1	1.7	—	—	—	—	—
1½	12.0	6.9	6.6	6.0	3.0	2.4	—	—	—	—	—
2	13.6	7.8	7.5	6.8	3.4	2.7	—	—	—	—	—
3	—	11.0	10.6	9.6	4.8	3.9	—	—	—	—	—
5	—	17.5	16.7	15.2	7.6	6.1	—	—	—	—	—
7½	—	25.3	24.2	22	11	9	—	—	—	—	—
10	—	32.2	30.8	28	14	11	—	—	—	—	—
15	—	48.3	46.2	42	21	17	—	—	—	—	—
20	—	62.1	59.4	54	27	22	—	—	—	—	—
25	—	78.2	74.8	68	34	27	—	53	26	21	—
30	—	92	88	80	40	32	—	63	32	26	—
40	—	120	114	104	52	41	—	83	41	33	—
50	—	150	143	130	65	52	—	104	52	42	—
60	—	177	169	154	77	62	16	123	61	49	12
75	—	221	211	192	96	77	20	155	78	62	15
100	—	285	273	248	124	99	26	202	101	81	20
125	—	359	343	312	156	125	31	253	126	101	25
150	—	414	396	360	180	144	37	302	151	121	30
200		552	528	480	240	192	49	400	201	161	40
250	—	—	—	—	302	242	60	—	—	—	—
300	—	—	—	—	361	289	72	—	—	—	—
350	—	—	—	—	414	336	83	—	—	—	—
400	—	—	—	—	477	382	95	—	—	—	—
450	—	—	—	—	515	412	103	—	—	—	—
500	—	—	—	—	590	472	118	—	—	—	—

*For 90 and 80 percent power factor, the figures shall be multiplied by 1.1 and 1.25, respectively.

Table 430.151(A) Conversion Table of Single-Phase Locked-Rotor Currents for Selection of Disconnecting Means and Controllers as Determined from Horsepower and Voltage Rating
For use only with 430.110, 440.12, 440.41 and 455.8(C).

Rated Horsepower	Maximum Locked-Rotor Current in Amperes, Single Phase			Rated Horsepower	Maximum Locked-Rotor Current in Amperes, Single Phase		
	115 Volts	208 Volts	230 Volts		115 Volts	208 Volts	230 Volts
½	58.8	32.5	29.4	3	204	113	102
¾	82.8	45.8	41.4	5	336	186	168
1	96	53	48	7½	480	265	240
1½	120	66	60	10	600	332	300
2	144	80	72				

Table 430.151(B) Conversion Table of Polyphase Design B, C, D, and E Maximum Locked-Rotor Currents for Selection of Disconnecting Means and Controllers as Determined from Horsepower and Voltage Rating and Design Letter
For use only with 430.110, 440.12[*], 440.41[*] and 455.8(C).

Rated Horsepower	Maximum Motor Locked-Rotor Current in Amperes, Two- and Three-Phase, Design B, C, D, and E											
	115 Volts		200 Volts		208 Volts		230 Volts		460 Volts		575 Volts	
	B, C, D	E	B, C, D	E	B, C, D	E	B, C, D	E	B, C, D	E	B, C, D	E
½	40	40	23	23	22.1	22.1	20	20	10	10	8	8
¾	50	50	28.8	28.8	27.6	27.6	25	25	12.5	12.5	10	10
1	60	60	34.5	34.5	33	33	30	30	15	15	12	12
1½	80	80	46	46	44	44	40	40	20	20	16	16
2	100	100	57.5	57.5	55	55	50	50	25	25	20	20
3	—	—	73.6	84	71	81	64	73	32	36.5	25.6	29.2
5	—	—	105.8	140	102	135	92	122	46	61	36.8	48.8
7½	—	—	146	210	140	202	127	183	63.5	91.5	50.8	73.2
10	—	—	186.3	259	179	249	162	225	81	113	64.8	90
15	—	—	267	388	257	373	232	337	116	169	93	135
20	—	—	334	516	321	497	290	449	145	225	116	180
25	—	—	420	646	404	621	365	562	183	281	146	225
30	—	—	500	775	481	745	435	674	218	337	174	270
40	—	—	667	948	641	911	580	824	290	412	232	330
50	—	—	834	1185	802	1139	725	1030	363	515	290	412
60	—	—	1001	1421	962	1367	870	1236	435	618	348	494
75	—	—	1248	1777	1200	1708	1085	1545	543	773	434	618
100	—	—	1668	2154	1603	2071	1450	1873	725	937	580	749
125	—	—	2087	2692	2007	2589	1815	2341	908	1171	726	936
150	—	—	2496	3230	2400	3106	2170	2809	1085	1405	868	1124
200	—	—	3335	4307	3207	4141	2900	3745	1450	1873	1160	1498
250	—	—	—	—	—	—	—	—	1825	2344	1460	1875
300	—	—	—	—	—	—	—	—	2200	2809	1760	2247
350	—	—	—	—	—	—	—	—	2550	3277	2040	2622
400	—	—	—	—	—	—	—	—	2900	3745	2320	2996
450	—	—	—	—	—	—	—	—	3250	4214	2600	3371
500	—	—	—	—	—	—	—	—	3625	4682	2900	3746

*In determining compliance with 440.12 and 440.41, the values in the B, C, D columns shall be used.

ARTICLE 440
Air-Conditioning and Refrigerating Equipment

Contents

I. General

440.1 Scope.

The provisions of this article apply to electric motor-driven air-conditioning and refrigerating equipment and to the branch circuits and controllers for such equipment. It provides for the special considerations necessary for circuits supplying hermetic refrigerant motor-compressors and for any air-conditioning or refrigerating equipment that is supplied from a branch circuit that supplies a hermetic refrigerant motor-compressor.

440.2 Definitions.

Branch-Circuit Selection Current. The value in amperes to be used instead of the rated-load current in determining the ratings of motor branch-circuit conductors, disconnecting means, controllers, and branch-circuit short-circuit and ground-fault protective devices wherever the running overload protective device permits a sustained current greater than the specified percentage of the rated-load current. The value of branch-circuit selection current will always be equal to or greater than the marked rated-load current.

Hermetic Refrigerant Motor-Compressor. A combination consisting of a compressor and motor, both of which are enclosed in the same housing, with no external shaft or shaft seals, the motor operating in the refrigerant.

Leakage Current Detection and Interruption (LCDI) Protection. A device provided in a power supply cord or cord set that senses leakage current flowing between or from the cord conductors and interrupts the circuit at a predetermined level of leakage current.

As indicated by the definition, power supply cords or cord sets with this type of protection automatically interrupt the circuit where leakage current between or from the conductors exceeds a predetermined level. Opening of the circuit is accomplished through the use of electronic switching or by "air-break" contacts. The circuit remains open until the cause of the leakage current is eliminated or the protection device is manually reset. Leakage current detection and interrupter protection is one of the protection methods for the power supply cord or cord set of a room air conditioner specified in 440.65.

Rated-Load Current. The rated-load current for a hermetic refrigerant motor-compressor is the current resulting when the motor-compressor is operated at the rated load, rated voltage, and rated frequency of the equipment it serves.

440.3 Other Articles.

(A) Article 430. These provisions are in addition to, or amendatory of, the provisions of Article 430 and other articles in this *Code*, which apply except as modified in this article.

(B) Articles 422, 424, or 430. The rules of Articles 422, 424, or 430, as applicable, shall apply to air-conditioning and refrigerating equipment that does not incorporate a hermetic refrigerant motor-compressor. This equipment includes devices that employ refrigeration compressors driven by conventional motors, furnaces with air-conditioning evaporator coils installed, fan-coil units, remote forced air-cooled condensers, remote commercial refrigerators, and so forth.

(C) Article 422. Equipment such as room air conditioners, household refrigerators and freezers, drinking water coolers,

and beverage dispensers shall be considered appliances, and the provisions of Article 422 shall also apply.

(D) Other Applicable Articles. Hermetic refrigerant motor-compressors, circuits, controllers, and equipment shall also comply with the applicable provisions of Table 440.3(D).

Table 440.3(D) Other Articles

Equipment/Occupancy	Article	Section
Capacitors		460.9
Commercial garages, aircraft hangars, motor fuel dispensing facilities, bulk storage plants, spray application, dipping, and coating processes, and inhalation anesthetizing locations	511, 513, 514, 515, 516, and 517 Part IV	
Hazardous (classified) locations	500–503 and 505	
Motion picture and television studios and similar locations	530	
Resistors and reactors	470	

Article 440 provides special considerations necessary for circuits supplying *hermetic* refrigerant motor-compressors and is in addition to or amendatory of the provisions of Article 430 and other applicable articles. However, many requirements, such as disconnecting means, controllers, single or group installations, and sizing of conductors, are the same as or similar to those applied in Article 430.

Article 440 does not apply unless a hermetic refrigerant motor-compressor is supplied. Article 440 must be applied in conjunction with Article 430.

Note the terms *rated-load current* and *branch-circuit selection current,* defined in 440.2. When a branch-circuit selection current is marked on a nameplate, it must be used instead of the rated-load current to determine the size of the disconnecting means, the controller, the motor branch-circuit conductors, and the overcurrent protective devices for the branch-circuit conductors and the motor. The value of branch-circuit selection current will always be greater than the marked rated-load current.

440.4 Marking on Hermetic Refrigerant Motor-Compressors and Equipment.

(A) Hermetic Refrigerant Motor-Compressor Nameplate. A hermetic refrigerant motor-compressor shall be provided with a nameplate that shall indicate the manufacturer's name, trademark, or symbol; identifying designation; phase; voltage; and frequency. The rated-load current in amperes of the motor-compressor shall be marked by the equipment manufacturer on either or both the motor-compressor nameplate and the nameplate of the equipment in which the motor-compressor is used. The locked-rotor current of each single-phase motor-compressor having a rated-load current of more than 9 amperes at 115 volts, or more than 4.5 amperes at 230 volts, and each polyphase motor-compressor shall be marked on the motor-compressor nameplate. Where a thermal protector complying with 440.52(A)(2) and (B)(2) is used, the motor-compressor nameplate or the equipment nameplate shall be marked with the words "thermally protected." Where a protective system complying with 440.52(A)(4) and (B)(4) is used and is furnished with the equipment, the equipment nameplate shall be marked with the words, "thermally protected system." Where a protective system complying with 440.52(A)(4) and (B)(4) is specified, the equipment nameplate shall be appropriately marked.

(B) Multimotor and Combination-Load Equipment. Multimotor and combination-load equipment shall be provided with a visible nameplate marked with the maker's name, the rating in volts, frequency and number of phases, minimum supply circuit conductor ampacity, and the maximum rating of the branch-circuit short-circuit and ground-fault protective device. The ampacity shall be calculated by using Part IV and counting all the motors and other loads that will be operated at the same time. The branch-circuit short-circuit and ground-fault protective device rating shall not exceed the value calculated by using Part III. Multimotor or combination-load equipment for use on two or more circuits shall be marked with the above information for each circuit.

Exception No. 1: Multimotor and combination-load equipment that is suitable under the provisions of this article for connection to a single 15- or 20-ampere, 120-volt, or a 15-ampere, 208- or 240-volt, single-phase branch circuit shall be permitted to be marked as a single load.

Exception No. 2: The minimum supply circuit conductor ampacity and the maximum rating of the branch-circuit short-circuit and ground-fault protective device shall not be required to be marked on a room air conditioner conforming with 440.62(A).

(C) Branch-Circuit Selection Current. A hermetic refrigerant motor-compressor, or equipment containing such a compressor, having a protection system that is approved for use with the motor-compressor that it protects and that permits continuous current in excess of the specified percentage of nameplate rated-load current given in 440.52(B)(2) or (B)(4) shall also be marked with a branch-circuit selection current that complies with 440.52(B)(2) or (B)(4). This marking shall be provided by the equipment manufacturer and shall be on the nameplate(s) where the rated-load current(s) appears.

440.5 Marking on Controllers.

A controller shall be marked with the manufacturer's name, trademark, or symbol; identifying designation; voltage; phase; full-load and locked-rotor current (or horsepower) rating; and such other data as may be needed to properly indicate the motor-compressor for which it is suitable.

440.6 Ampacity and Rating.

The size of conductors for equipment covered by this article shall be selected from Table 310.16 through Table 310.19 or calculated in accordance with 310.15 as applicable. The required ampacity of conductors and rating of equipment shall be determined according to 440.6(A) and (B).

(A) Hermetic Refrigerant Motor-Compressor. For a hermetic refrigerant motor-compressor, the rated-load current marked on the nameplate of the equipment in which the motor-compressor is employed shall be used in determining the rating or ampacity of the disconnecting means, the branch-circuit conductors, the controller, the branch-circuit short-circuit and ground-fault protection, and the separate motor overload protection. Where no rated-load current is shown on the equipment nameplate, the rated-load current shown on the compressor nameplate shall be used.

Exception No. 1: Where so marked, the branch-circuit selection current shall be used instead of the rated-load current to determine the rating or ampacity of the disconnecting means, the branch-circuit conductors, the controller, and the branch-circuit short-circuit and ground-fault protection.

Exception No. 2: For cord-and-plug-connected equipment, the nameplate marking shall be used in accordance with 440.22(B), Exception No. 2.

> FPN: For disconnecting means and controllers, see 440.12 and 440.41.

(B) Multimotor Equipment. For multimotor equipment employing a shaded-pole or permanent split-capacitor-type fan or blower motor, the full-load current for such motor marked on the nameplate of the equipment in which the fan or blower motor is employed shall be used instead of the horsepower rating to determine the ampacity or rating of the disconnecting means, the branch-circuit conductors, the controller, the branch-circuit short-circuit and ground-fault protection, and the separate overload protection. This marking on the equipment nameplate shall not be less than the current marked on the fan or blower motor nameplate.

440.7 Highest Rated (Largest) Motor.

In determining compliance with this article and with 430.24, 430.53(B) and (C), and 430.62(A), the highest rated (largest) motor shall be considered to be the motor that has the highest rated-load current. Where two or more motors have the same highest rated-load current, only one of them shall be considered as the highest rated (largest) motor. For other than hermetic refrigerant motor-compressors, and fan or blower motors as covered in 440.6(B), the full-load current used to determine the highest rated motor shall be the equivalent value corresponding to the motor horsepower rating selected from Tables 430.148, 430.149, or 430.150.

Exception: Where so marked, the branch-circuit selection current shall be used instead of the rated-load current in determining the highest rated (largest) motor-compressor.

440.8 Single Machine.

An air-conditioning or refrigerating system shall be considered to be a single machine under the provisions of 430.87, Exception, and 430.112, Exception. The motors shall be permitted to be located remotely from each other.

II. Disconnecting Means

440.11 General.

The provisions of Part II are intended to require disconnecting means capable of disconnecting air-conditioning and refrigerating equipment, including motor-compressors and controllers from the circuit conductors.

440.12 Rating and Interrupting Capacity.

(A) Hermetic Refrigerant Motor-Compressor. A disconnecting means serving a hermetic refrigerant motor-compressor shall be selected on the basis of the nameplate rated-load current or branch-circuit selection current, whichever is greater, and locked-rotor current, respectively, of the motor-compressor as follows.

(1) Ampere Rating. The ampere rating shall be at least 115 percent of the nameplate rated-load current or branch-circuit selection current, whichever is greater.

Exception: A listed nonfused motor circuit switch having a horsepower rating not less than the equivalent horsepower determined in accordance with 440.12(A)(2) shall be permitted to have an ampere rating less than 115 percent of the specified current.

(2) Equivalent Horsepower. To determine the equivalent horsepower in complying with the requirements of 430.109, the horsepower rating shall be selected from Tables 430.148, 430.149, or 430.150 corresponding to the rated-load current or branch-circuit selection current, whichever is greater, and also the horsepower rating from Table 430.151(A) or Table 430.151(B) corresponding to the locked-rotor current. In case the nameplate rated-load current or branch-circuit selection current and locked-rotor current do not correspond to the currents shown in Tables 430.148, 430.149, 430.150, 430.151(A), or 430.151(B), the horsepower rating corres-

ponding to the next higher value shall be selected. In case different horsepower ratings are obtained when applying these tables, a horsepower rating at least equal to the larger of the values obtained shall be selected.

(B) Combination Loads. Where the combined load of two or more hermetic refrigerant motor-compressors or one or more hermetic refrigerant motor-compressor with other motors or loads may be simultaneous on a single disconnecting means, the rating for the disconnecting means shall be determined in accordance with 440.12(B)(1) and (B)(2).

(1) Horsepower Rating. The horsepower rating of the disconnecting means shall be determined from the sum of all currents, including resistance loads, at the rated-load condition and also at the locked-rotor condition. The combined rated-load current and the combined locked-rotor current so obtained shall be considered as a single motor for the purpose of this requirement as follows:

(a) The full-load current equivalent to the horsepower rating of each motor, other than a hermetic refrigerant motor-compressor, and fan or blower motors as covered in 440.6(B) shall be selected from Tables 430.148, 430.149, or 430.150. These full-load currents shall be added to the motor-compressor rated-load current(s) or branch-circuit selection current(s), whichever is greater, and to the rating in amperes of other loads to obtain an equivalent full-load current for the combined load.

(b) The locked-rotor current equivalent to the horsepower rating of each motor, other than a hermetic refrigerant motor-compressor, shall be selected from Table 430.151(A) or Table 430.151(B), and, for fan and blower motors of the shaded-pole or permanent split-capacitor type marked with the locked-rotor current, the marked value shall be used. The locked-rotor currents shall be added to the motor-compressor locked-rotor current(s) and to the rating in amperes of other loads to obtain an equivalent locked-rotor current for the combined load. Where two or more motors or other loads such as resistance heaters, or both, cannot be started simultaneously, appropriate combinations of locked-rotor and rated-load current or branch-circuit selection current, whichever is greater, shall be an acceptable means of determining the equivalent locked-rotor current for the simultaneous combined load.

Exception: Where part of the concurrent load is a resistance load and the disconnecting means is a switch rated in horsepower and amperes, the switch used shall be permitted to have a horsepower rating not less than the combined load to the motor-compressor(s) and other motor(s) at the locked-rotor condition, if the ampere rating of the switch is not less than this locked-rotor load plus the resistance load.

(2) Full-Load Current Equivalent. The ampere rating of the disconnecting means shall be at least 115 percent of the sum of all currents at the rated-load condition determined in accordance with 440.12(B)(1).

Exception: A listed nonfused motor circuit switch having a horsepower rating not less than the equivalent horsepower determined by 440.12(B)(1) shall be permitted to have an ampere rating less than 115 percent of the sum of all currents.

(C) Small Motor-Compressors. For small motor-compressors not having the locked-rotor current marked on the nameplate, or for small motors not covered by Tables 430.147, 430.148, 430.149, or 430.150, the locked-rotor current shall be assumed to be six times the rated-load current.

(D) Every Switch. Every disconnecting means in the refrigerant motor-compressor circuit between the point of attachment to the feeder and the point of connection to the refrigerant motor-compressor shall comply with the requirements of 440.12.

(E) Disconnecting Means Rated in Excess of 100 Horsepower. Where the rated-load or locked-rotor current as determined above would indicate a disconnecting means rated in excess of 100 hp, the provisions of 430.109(E) shall apply.

440.13 Cord-Connected Equipment.

For cord-connected equipment such as room air conditioners, household refrigerators and freezers, drinking water coolers, and beverage dispensers, a separable connector or an attachment plug and receptacle shall be permitted to serve as the disconnecting means.

FPN: For room air conditioners, see 440.63.

440.14 Location.

Disconnecting means shall be located within sight from and readily accessible from the air-conditioning or refrigerating equipment. The disconnecting means shall be permitted to be installed on or within the air-conditioning or refrigerating equipment.

The disconnecting means shall not be located on panels that are designed to allow access to the air-conditioning or refrigeration equipment.

Exception No. 1: Where the disconnecting means provided in accordance with 430.102(A) is capable of being locked in the open position, and the refrigerating or air-conditioning equipment is essential to an industrial process in a facility where the conditions of maintenance and the supervision ensure that only qualified persons service the equipment, a disconnecting means within sight from the equipment shall not be required.

Exception No. 1 accommodates special conditions associated with process refrigeration equipment. Typically, this equipment is very large, so rated disconnects may not be available. Additionally, this equipment may be in hazardous locations, and locating disconnecting means within sight of the motor may introduce additional hazards.

Exception No. 2: Where an attachment plug and receptacle serve as the disconnecting means in accordance with 440.13, their location shall be accessible but shall not be required to be readily accessible.

FPN: See Parts VII and IX of Article 430 for additional requirements.

The reference to Parts VII and IX of Article 430 in the fine print note is intended to call attention to the additional disconnect location requirements in 430.102, 430.107, and 430.113. Because 440.3(A) makes the requirements in Article 440 in addition to or amendatory of the provisions of Article 430, the requirement of 440.14 mandates that the equipment disconnecting means be within sight from and readily accessible from the equipment, even if there is also a remote disconnect capable of being locked in the open position under the provision of 430.102(B), Exception.

This special requirement for air-conditioning and refrigeration equipment covered by Article 440 is more stringent than the provisions in Article 430 to provide protection for service personnel working on equipment located in attics, on roofs, or outside in a remote location where it is difficult to gain access to a remote lockable disconnect. See 440.14, Exception No. 1.

III. Branch-Circuit Short-Circuit and Ground-Fault Protection

440.21 General.

The provisions of Part III specify devices intended to protect the branch-circuit conductors, control apparatus, and motors in circuits supplying hermetic refrigerant motor-compressors against overcurrent due to short circuits and grounds. They are in addition to or amendatory of the provisions of Article 240.

Where an air conditioner is listed by a qualified electrical testing laboratory with a nameplate that reads "maximum fuse size," the listing restricts the use of this unit to fuse protection only and does not cover its use with circuit breakers. If the air conditioner has been evaluated for both fuses and ordinary circuit breakers, or both fuses and HACR-type circuit breakers, it may be so marked. UL-listed circuit breakers that have been found suitable for use with heating, air-conditioning, and refrigeration equipment comprising multimotor or combination loads are marked "Listed HACR Type." It is the intent of 110.3(B) to require that the manufacturer's installation specifications be closely followed and that any restriction of the listing be applied to the installation of the equipment, in order to comply with the *Code.*

The UL *Electrical Appliance and Utilization Equipment Directory* states the following under Air-Conditioners, Central Cooling (ACAV): "This marked protective device rating is the maximum for which the equipment has been investigated and found acceptable. Where the marking specifies fuses, or 'HACR Type' circuit breakers, the circuit is intended to be protected only by the type of protective device specified." Exhibit 440.1 illustrates three wiring configurations where the equipment is under fuse protection only.

The UL *Electrical Appliance and Utilization Equipment Directory* further indicates that "in units employing two or more motors or a motor(s) and other loads operating from a single supply circuit, the motor overload protective devices (including thermal protectors for motors) and other factory-installed motor circuit components and wiring are investigated on the basis of compliance with the motor branch-circuit short-circuit and ground-fault protection requirements of 430.53(C), Other Group Installations. Such multimotor and combination load equipment is to be connected only to a circuit protected by fuses or a circuit breaker with a rating that does not exceed the value marked on the data plate of the equipment."

Current-limiting overcurrent devices, which may reduce the amount of fault current to which the equipment will be subjected, can be installed in the branch circuit supplying the equipment. See 240.2 for the definition of *current-limiting*

device overcurrent protective device and commentary explaining short-circuit damage. This is important particularly for the larger commercial and industrial installations. See also 110.3(B) and 110.10 and associated commentary regarding the installation and use of listed or labeled equipment and the selection of overcurrent protective devices (such as fuses and circuit breakers).

440.22 Application and Selection.

(A) Rating or Setting for Individual Motor-Compressor. The motor-compressor branch-circuit short-circuit and ground-fault protective device shall be capable of carrying the starting current of the motor. A protective device having a rating or setting not exceeding 175 percent of the motor-compressor rated-load current or branch-circuit selection current, whichever is greater, shall be permitted, provided that, where the protection specified is not sufficient for the starting current of the motor, the rating or setting shall be permitted to be increased but shall not exceed 225 percent of the motor rated-load current or branch-circuit selection current, whichever is greater.

Exception: The rating of the branch-circuit short-circuit and ground-fault protective device shall not be required to be less than 15 amperes.

(B) Rating or Setting for Equipment. The equipment branch-circuit short-circuit and ground-fault protective device shall be capable of carrying the starting current of the equipment. Where the hermetic refrigerant motor-compressor is the only load on the circuit, the protection shall conform with 440.22(A). Where the equipment incorporates more than one hermetic refrigerant motor-compressor or a

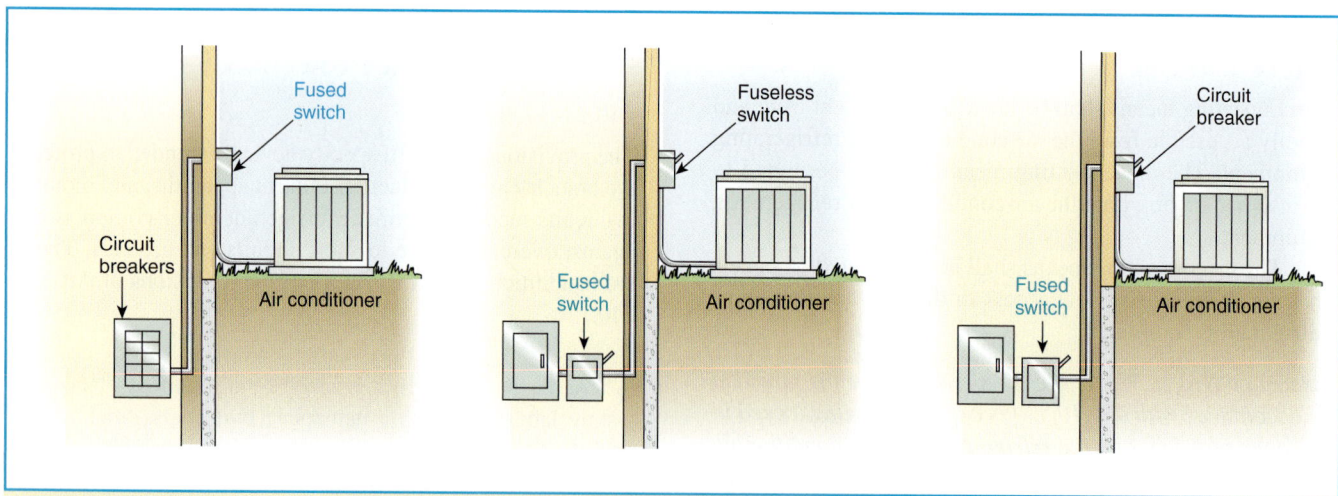

Exhibit 440.1 Three correct alternate wiring configurations satisfying a nameplate that specifies fuses, thus restricting the equipment to protection by fuses only. (Note that the fuse rating cannot exceed the maximum fuse size specified on the air-conditioner nameplate.)

hermetic refrigerant motor-compressor and other motors or other loads, the equipment short-circuit and ground-fault protection shall conform with 430.53 and the following.

(1) Motor-Compressor Largest Load. Where a hermetic refrigerant motor-compressor is the largest load connected to the circuit, the rating or setting of the branch-circuit short-circuit and ground-fault protective device shall not exceed the value specified in 440.22(A) for the largest motor-compressor plus the sum of the rated-load current or branch-circuit selection current, whichever is greater, of the other motor-compressor(s) and the ratings of the other loads supplied.

(2) Motor-Compressor Not Largest Load. Where a hermetic refrigerant motor-compressor is not the largest load connected to the circuit, the rating or setting of the branch-circuit short-circuit and ground-fault protective device shall not exceed a value equal to the sum of the rated-load current or branch-circuit selection current, whichever is greater, rating(s) for the motor-compressor(s) plus the value specified in 430.53(C)(4) where other motor loads are supplied, or the value specified in 240.4 where only nonmotor loads are supplied in addition to the motor-compressor(s).

Exception No. 1: Equipment that starts and operates on a 15- or 20-ampere 120-volt, or 15-ampere 208- or 240-volt single-phase branch circuit, shall be permitted to be protected by the 15- or 20-ampere overcurrent device protecting the branch circuit, but if the maximum branch-circuit short-circuit and ground-fault protective device rating marked on the equipment is less than these values, the circuit protective device shall not exceed the value marked on the equipment nameplate.

Exception No. 2: The nameplate marking of cord-and-plug-connected equipment rated not greater than 250 volts, single-phase, such as household refrigerators and freezers, drinking water coolers, and beverage dispensers, shall be used in determining the branch-circuit requirements, and each unit shall be considered as a single motor unless the nameplate is marked otherwise.

(C) Protective Device Rating Not to Exceed the Manufacturer's Values. Where maximum protective device ratings shown on a manufacturer's overload relay table for use with a motor controller are less than the rating or setting selected in accordance with 440.22(A) and (B), the protective device rating shall not exceed the manufacturer's values marked on the equipment.

IV. Branch-Circuit Conductors

440.31 General.

The provisions of Part IV and Article 310 specify ampacities of conductors required to carry the motor current without

overheating under the conditions specified, except as modified in 440.6(A), Exception No. 1.

The provisions of these articles shall not apply to integral conductors of motors, motor controllers and the like, or to conductors that form an integral part of approved equipment.

FPN: See 300.1(B) and 310.1 for similar requirements.

440.32 Single Motor-Compressor.

Branch-circuit conductors supplying a single motor-compressor shall have an ampacity not less than 125 percent of either the motor-compressor rated-load current or the branch-circuit selection current, whichever is greater.

For a wye-start, delta-run connected motor-compressor, the selection of branch-circuit conductors between the controller and the motor-compressor shall be permitted to be based on 58 percent of either the motor-compressor rated-load current or the branch-circuit selection current, whichever is greater.

The use of wye-start, delta-run hermetic motor compressors is a means of reducing starting current in large air-conditioning equipment. Starting the motor using a wye-configured winding results in a phase voltage that is 1/1.732 or 58 percent of the line voltage. This reduction is phase voltage results in a directly proportional reduction in phase current. In the starting configuration, the phase current and the line current are equal.

The second paragraph of 440.32 permits the two sets of conductors between the controller and the compressor motor to be sized at 58 percent of the larger of either the motor-compressor rated-load current or the branch circuit selection current. The conductors on the line side of the controller are sized at not less than 125 percent of either the motor-compressor rated-load current or the branch circuit selection current, whichever is larger. For more information on wye-start, delta-run motors, see the commentary on 430.32(C).

440.33 Motor-Compressor(s) With or Without Additional Motor Loads.

Conductors supplying one or more motor-compressor(s) with or without an additional load(s) shall have an ampacity not less than the sum of the rated-load or branch-circuit selection current ratings, whichever is larger, of all the motor-compressors plus the full-load currents of the other motors, plus 25 percent of the highest motor or motor-compressor rating in the group.

Exception No. 1: Where the circuitry is interlocked so as to prevent the starting and running of a second motor-compressor or group of motor-compressors, the conductor size shall be determined from the largest motor-compressor or group of motor-compressors that is to be operated at a given time.

Exception No. 2: The branch circuit conductors for room air conditioners shall be in accordance with Part VII of Article 440.

Branch circuits for listed air-conditioning and refrigeration equipment that have a nameplate marked with the branch-circuit conductor size and branch-circuit short-circuit protective device size are not required to have the branch-circuit conductors sized in accordance with 440.33. The testing laboratory standard includes the 25 percent increase for the largest motor or compressor in the group plus the other nonmotor or noncompressor load; therefore, the actual nameplate full-load amperes for the complete assembly can be used to size the branch-circuit conductors.

440.34 Combination Load.

Conductors supplying a motor-compressor load in addition to a lighting or appliance load as computed from Article 220 and other applicable articles shall have an ampacity sufficient for the lighting or appliance load plus the required ampacity for the motor-compressor load determined in accordance with 440.33, or, for a single motor-compressor, in accordance with 440.32.

Exception: Where the circuitry is interlocked so as to prevent simultaneous operation of the motor-compressor(s) and all other loads connected, the conductor size shall be determined from the largest size required for the motor-compressor(s) and other loads to be operated at a given time.

440.35 Multimotor and Combination-Load Equipment.

The ampacity of the conductors supplying multimotor and combination-load equipment shall not be less than the minimum circuit ampacity marked on the equipment in accordance with 440.4(B).

V. Controllers for Motor-Compressors

440.41 Rating.

(A) Motor-Compressor Controller. A motor-compressor controller shall have both a continuous-duty full-load current rating and a locked-rotor current rating not less than the nameplate rated-load current or branch-circuit selection current, whichever is greater, and locked-rotor current, respectively, of the compressor. In case the motor controller is rated in horsepower but is without one or both of the foregoing current ratings, equivalent currents shall be determined from the ratings as follows. Tables 430.148, 430.149, and

430.150 shall be used to determine the equivalent full-load current rating. Table 430.151(A) and Table 430.151(B) shall be used to determine the equivalent locked-rotor current ratings.

(B) Controller Serving More Than One Load. A controller serving more than one motor-compressor or a motor-compressor and other loads shall have a continuous-duty full-load current rating and a locked-rotor current rating not less than the combined load as determined in accordance with 440.12(B).

VI. Motor-Compressor and Branch-Circuit Overload Protection

440.51 General.

The provisions of Part VI specify devices intended to protect the motor-compressor, the motor-control apparatus, and the branch-circuit conductors against excessive heating due to motor overload and failure to start.

> FPN: See 240.4(G) for application of Parts III and VI of Article 440.

440.52 Application and Selection.

(A) Protection of Motor-Compressor. Each motor-compressor shall be protected against overload and failure to start by one of the following means:

(1) A separate overload relay that is responsive to motor-compressor current. This device shall be selected to trip at not more than 140 percent of the motor-compressor rated-load current.

(2) A thermal protector integral with the motor-compressor, approved for use with the motor-compressor that it protects on the basis that it will prevent dangerous overheating of the motor-compressor due to overload and failure to start. If the current-interrupting device is separate from the motor-compressor and its control circuit is operated by a protective device integral with the motor-compressor, it shall be arranged so that the opening of the control circuit will result in interruption of current to the motor-compressor.

(3) A fuse or inverse time circuit breaker responsive to motor current, which shall also be permitted to serve as the branch-circuit short-circuit and ground-fault protective device. This device shall be rated at not more than 125 percent of the motor-compressor rated-load current. It shall have sufficient time delay to permit the motor-compressor to start and accelerate its load. The equipment or the motor-compressor shall be marked with this maximum branch-circuit fuse or inverse time circuit breaker rating.

(4) A protective system, furnished or specified and approved for use with the motor-compressor that it protects

on the basis that it will prevent dangerous overheating of the motor-compressor due to overload and failure to start. If the current-interrupting device is separate from the motor-compressor and its control circuit is operated by a protective device that is not integral with the current-interrupting device, it shall be arranged so that the opening of the control circuit will result in interruption of current to the motor-compressor.

(B) Protection of Motor-Compressor Control Apparatus and Branch-Circuit Conductors. The motor-compressor controller(s), the disconnecting means, and the branch-circuit conductors shall be protected against overcurrent due to motor overload and failure to start by one of the following means, which shall be permitted to be the same device or system protecting the motor-compressor in accordance with 440.52(A):

Exception: Overload protection of motor-compressors and equipment on 15- and 20-ampere, single-phase, branch circuits shall be permitted to be in accordance with 440.54 and 440.55.

(1) An overload relay selected in accordance with 440.52(A)(1)
(2) A thermal protector applied in accordance with 440.52(A)(2), that will not permit a continuous current in excess of 156 percent of the marked rated-load current or branch-circuit selection current
(3) A fuse or inverse time circuit breaker selected in accordance with 440.52(A)(3)
(4) A protective system, in accordance with 440.52(A)(4), that will not permit a continuous current in excess of 156 percent of the marked rated-load current or branch-circuit selection current

440.53 Overload Relays.

Overload relays and other devices for motor overload protection that are not capable of opening short circuits shall be protected by fuses or inverse time circuit breakers with ratings or settings in accordance with Part III unless approved for group installation or for part-winding motors and marked to indicate the maximum size of fuse or inverse time circuit breaker by which they shall be protected.

Exception: The fuse or inverse time circuit breaker size marking shall be permitted on the nameplate of approved equipment in which the overload relay or other overload device is used.

440.54 Motor-Compressors and Equipment on 15- or 20-Ampere Branch Circuits—Not Cord-and-Attachment-Plug-Connected.

Overload protection for motor-compressors and equipment used on 15- or 20-ampere 120-volt, or 15-ampere 208- or 240-volt single-phase branch circuits as permitted in Article 210 shall be permitted as indicated in 440.54(A) and (B).

(A) Overload Protection. The motor-compressor shall be provided with overload protection selected as specified in 440.52(A). Both the controller and motor overload protective device shall be approved for installation with the short-circuit and ground-fault protective device for the branch circuit to which the equipment is connected.

(B) Time Delay. The short-circuit and ground-fault protective device protecting the branch circuit shall have sufficient time delay to permit the motor-compressor and other motors to start and accelerate their loads.

440.55 Cord-and-Attachment-Plug-Connected Motor-Compressors and Equipment on 15- or 20-Ampere Branch Circuits.

Overload protection for motor-compressors and equipment that are cord-and-attachment-plug-connected and used on 15- or 20-ampere 120-volt, or 15-ampere 208- or 240-volt single-phase branch circuits as permitted in Article 210 shall be permitted as indicated in 440.55(A), (B), and (C).

(A) Overload Protection. The motor-compressor shall be provided with overload protection as specified in 440.52(A). Both the controller and the motor overload protective device shall be approved for installation with the short-circuit and ground-fault protective device for the branch circuit to which the equipment is connected.

(B) Attachment Plug and Receptacle Rating. The rating of the attachment plug and receptacle shall not exceed 20 amperes at 125 volts or 15 amperes at 250 volts.

(C) Time Delay. The short-circuit and ground-fault protective device protecting the branch circuit shall have sufficient time delay to permit the motor-compressor and other motors to start and accelerate their loads.

VII. Provisions for Room Air Conditioners

440.60 General.

The provisions of Part VII shall apply to electrically energized room air conditioners that control temperature and humidity. For the purpose of Part VII, a room air conditioner (with or without provisions for heating) shall be considered as an ac appliance of the air-cooled window, console, or in-wall type that is installed in the conditioned room and that incorporates a hermetic refrigerant motor-compressor(s). The provisions of Part VII cover equipment rated not over 250 volts, single phase, and such equipment shall be permitted to be cord-and-attachment-plug-connected.

A room air conditioner that is rated three phase or rated

over 250 volts shall be directly connected to a wiring method recognized in Chapter 3, and provisions of Part VII shall not apply.

440.61 Grounding.

Room air conditioners shall be grounded in accordance with 250.110, 250.112, and 250.114.

440.62 Branch-Circuit Requirements.

(A) Room Air Conditioner as a Single Motor Unit. A room air conditioner shall be considered as a single motor unit in determining its branch-circuit requirements where all the following conditions are met:

(1) It is cord-and-attachment-plug-connected.
(2) Its rating is not more than 40 amperes and 250 volts, single phase.
(3) Total rated-load current is shown on the room air-conditioner nameplate rather than individual motor currents.
(4) The rating of the branch-circuit short-circuit and ground-fault protective device does not exceed the ampacity of the branch-circuit conductors or the rating of the receptacle, whichever is less.

(B) Where No Other Loads Are Supplied. The total marked rating of a cord-and-attachment-plug-connected room air conditioner shall not exceed 80 percent of the rating of a branch circuit where no other loads are supplied.

(C) Where Lighting Units or Other Appliances Are Also Supplied. The total marked rating of a cord-and-attachment-plug-connected room air conditioner shall not exceed 50 percent of the rating of a branch circuit where lighting outlets, other appliances, or general-use receptacles are also supplied. Where the circuitry is interlocked to prevent simultaneous operation of the room air conditioner and energization of other outlets on the same branch circuit, a cord-and-attachment-plug-connected room air conditioner shall not exceed 80 percent of the branch-circuit rating.

440.63 Disconnecting Means.

An attachment plug and receptacle shall be permitted to serve as the disconnecting means for a single-phase room air conditioner rated 250 volts or less if (1) the manual controls on the room air conditioner are readily accessible and located within 1.8 m (6 ft) of the floor or (2) an approved manually operable switch is installed in a readily accessible location within sight from the room air conditioner.

440.64 Supply Cords.

Where a flexible cord is used to supply a room air conditioner, the length of such cord shall not exceed 3.0 m (10 ft) for a nominal, 120-volt rating or 1.8 m (6 ft) for a nominal, 208- or 240-volt rating.

440.65 Leakage Current Detection and Interruption (LCDI) and Arc Fault Circuit Interrupter (AFCI).

Single-phase cord-and-plug-connected room air conditioners shall be provided with factory-installed LCDI or AFCI protection. The LCDI or AFCI protection shall be an integral part of the attachment plug or be located in the power supply cord within 300 mm (12 in.) of the attachment plug.

Generally, portable room air conditioners are used only on a seasonal basis and are removed and stored at the end of the cooling season. During the life of a room air conditioner, the installation and removal occurs many times, and there is an increased likelihood of a damaged cord as a result of the unit being set on the cord or pushed against it. To provide enhanced protection against fires initiated by damaged supply cords, all single-phase cord-and-plug-connected room air conditioners are required to be equipped with either leakage current detection and interruption protection or arc-fault circuit interrupter protection. The protective device is required to be installed in the appliance cord and must be within 12 in. of the attachment plug.

ARTICLE 445
Generators

Contents

445.1 Scope.

This article covers the installation of generators.

445.3 Other Articles.

Generators and their associated wiring and equipment shall also comply with the applicable provisions of Articles 695, 700, 701, 702, and 705.

445.10 Location.

Generators shall be of a type suitable for the locations in which they are installed. They shall also meet the requirements for motors in 430.14. Generators installed in hazardous (classified) locations as described in Articles 500 through 503 and Article 505, or in other locations as described in Articles 510 through 517, and in Articles 518, 520, 525, 530, 665, and 695 shall also comply with the applicable provisions of those articles.

445.11 Marking.

Each generator shall be provided with a nameplate giving the manufacturer's name, the rated frequency, power factor, number of phases if of alternating current, the rating in kilowatts or kilovolt amperes, the normal volts and amperes corresponding to the rating, rated revolutions per minute, insulation system class and rated ambient temperature or rated temperature rise, and time rating.

445.12 Overcurrent Protection.

(A) Constant-Voltage Generators. Constant-voltage generators, except ac generator exciters, shall be protected from overloads by inherent design, circuit breakers, fuses, or other acceptable overcurrent protective means suitable for the conditions of use.

(B) Two-Wire Generators. Two-wire, dc generators shall be permitted to have overcurrent protection in one conductor only if the overcurrent device is actuated by the entire current generated other than the current in the shunt field. The overcurrent device shall not open the shunt field.

(C) 65 Volts or Less. Generators operating at 65 volts or less and driven by individual motors shall be considered as protected by the overcurrent device protecting the motor if these devices will operate when the generators are delivering not more than 150 percent of their full-load rated current.

(D) Balancer Sets. Two-wire, dc generators used in conjunction with balancer sets to obtain neutrals for 3-wire systems shall be equipped with overcurrent devices that will disconnect the 3-wire system in case of excessive unbalancing of voltages or currents.

(E) Three-Wire, Direct-Current Generators. Three-wire, dc generators, whether compound or shunt wound, shall be equipped with overcurrent devices, one in each armature lead, and connected so as to be actuated by the entire current from the armature. Such overcurrent devices shall consist either of a double-pole, double-coil circuit breaker or of a 4-pole circuit breaker connected in the main and equalizer leads and tripped by two overcurrent devices, one in each armature lead. Such protective devices shall be interlocked so that no one pole can be opened without simultaneously disconnecting both leads of the armature from the system.

Exception to (A) through (E): Where deemed by the authority having jurisdiction, a generator is vital to the operation of an electrical system and the generator should operate to failure to prevent a greater hazard to persons. The overload sensing device(s) shall be permitted to be connected to an annunciator or alarm supervised by authorized personnel instead of interrupting the generator circuit.

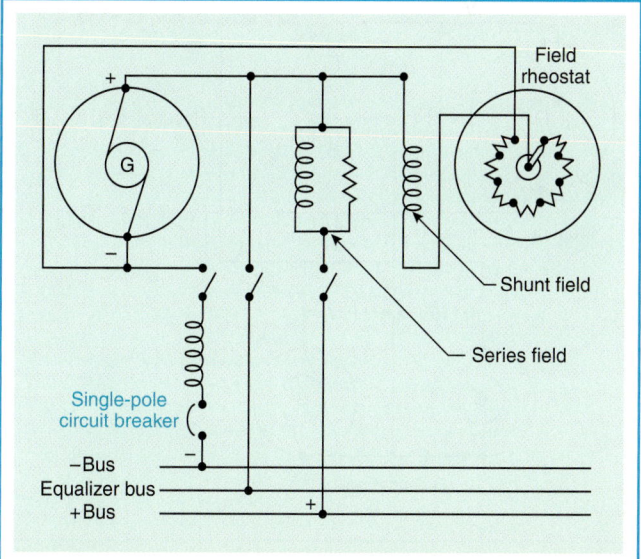

Exhibit 445.1 A 2-wire dc generator protected by a single-pole circuit breaker.

the series fields in parallel so as to maintain equal output voltage for each generator. The current could divide at the positive terminal, some flowing through the series field and positive lead and some flowing through the equalizer lead. The entire current generated flows through the negative lead; therefore, the fuse or circuit breaker (or at least the operating coil of the circuit breaker) must be placed in the negative lead. Overcurrent devices must be connected so as to be actuated by the entire armature output current.

An overcurrent device should not be placed in the shunt field circuit because, if the circuit were to open when the field was at full strength, an extremely high voltage would be induced that could damage the field winding insulation and the generator.

Section 445.12(C) indicates that generators operating at 65 volts or less are to be thought of as protected by the overcurrent devices that also protect the drive motor, provided these devices operate when the generator delivers 150 percent of its full-load rated current.

Exhibit 445.2 illustrates a two-pole circuit breaker with one pole connected in each lead of the main generator and with the operating coil properly designed to be connected in the neutral lead from the balancer, and arranged so as to be operated by either of the A coils or by the B coil. Each of the two generators used as a balancer set carries approximately half the unbalanced load and, thus, is always smaller than the main generator. During an excessive imbalance of the load, the balancer set would be overloaded, with no

overload on the main generator; hence, a double-pole circuit breaker is connected (as noted) to guard against this condition.

Note that the authority having jurisdiction may judge that having the generator operate to failure is preferable to providing automatic means to shut it down, which, in many cases, could present a greater hazard to personnel. An overload sensing device(s) would be permitted to be connected to an annunciator or an alarm (instead of interrupting the generator) and allow operating personnel to shut down load-side equipment in a safe and orderly fashion.

445.13 Ampacity of Conductors.

The ampacity of the conductors from the generator terminals to the first distribution device(s) containing overcurrent protection shall not be less than 115 percent of the nameplate current rating of the generator. It shall be permitted to size the neutral conductors in accordance with 220.22. Conductors that must carry ground-fault currents shall not be smaller than required by 250.24(B). Neutral conductors of dc generators that must carry ground-fault currents shall not be smaller than the minimum required size of the largest conductor.

Exception: Where the design and operation of the generator prevent overloading, the ampacity of the conductors shall not be less than 100 percent of the nameplate current rating of the generator.

445.14 Protection of Live Parts.

Live parts of generators operated at more than 50 volts to ground shall not be exposed to accidental contact where accessible to unqualified persons.

445.15 Guards for Attendants.

Where necessary for the safety of attendants, the requirements of 430.133 shall apply.

445.16 Bushings.

Where wires pass through an opening in an enclosure, conduit box, or barrier, a bushing shall be used to protect the conductors from the edges of an opening having sharp edges. The bushing shall have smooth, well-rounded surfaces where it may be in contact with the conductors. If used where oils, grease, or other contaminants may be present, the bushing shall be made of a material not deleteriously affected.

445.17 Generator Terminal Housings.

Generator terminal housings shall comply with 430.12. Where a horsepower rating is required to determine the required minimum size of the generator terminal housing, the full-load current of the generator shall be compared with comparable motors in Tables 430.147 through 430.150. The

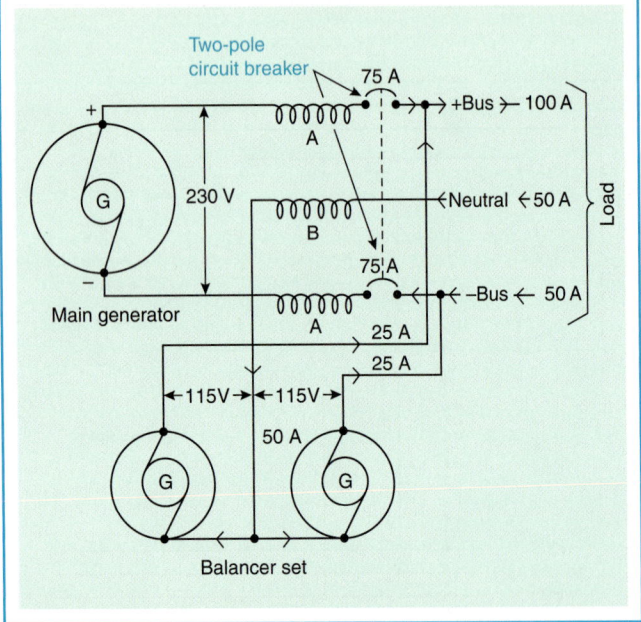

Exhibit 445.2 A two-pole circuit breaker (one pole connected in each lead of the main generator) with the opening coil connected in the neutral of the balancer set.

higher horsepower rating of Tables 430.147 and 430.150 shall be used whenever the generator selection is between two ratings.

445.18 Disconnecting Means Required for Generators.

Generators shall be equipped with a disconnect by means of which the generator and all protective devices and control apparatus are able to be disconnected entirely from the circuits supplied by the generator except where:

(1) The driving means for the generator can be readily shut down; and
(2) The generator is not arranged to operate in parallel with another generator or other source of voltage.

Added for the 1999 *Code*, 445.18 requires that generators be equipped with a disconnect switch or circuit breaker unless the prime mover can be readily shut down and the generator is not operating in parallel with another generator or source of power.

ARTICLE 450
Transformers and Transformer Vaults (Including Secondary Ties)

Contents

450.1 Scope.

This article covers the installation of all transformers.

Exception No. 1: Current transformers.

See 110.23 for the requirement on energized current transformers that are not in use.

Exception No. 2: Dry-type transformers that constitute a component part of other apparatus and comply with the requirements for such apparatus.

Exception No. 3: Transformers that are an integral part of an X-ray, high-frequency, or electrostatic-coating apparatus.

Exception No. 4: Transformers used with Class 2 and Class 3 circuits that comply with Article 725.

Exception No. 5: Transformers for sign and outline lighting that comply with Article 600.

Exception No. 6: Transformers for electric-discharge lighting that comply with Article 410.

Exception No. 7: Transformers used for power-limited fire alarm circuits that comply with Part III of Article 760.

Exception No. 8: Transformers used for research, development, or testing, where effective arrangements are provided to safeguard persons from contacting energized parts.

This article covers the installation of transformers dedicated to supplying power to a fire pump installation as modified by Article 695.

This article also covers the installation of transformers in hazardous (classified) locations as modified by Articles 501 through 504.

I. General Provisions

450.2 Definition.

For the purpose of this article, the following definition shall apply.

Transformer. An individual transformer, single- or polyphase, identified by a single nameplate, unless otherwise indicated in this article.

450.3 Overcurrent Protection.

Overcurrent protection of transformers shall comply with 450.3(A), (B), or (C). As used in this section, the word *transformer* shall mean a transformer or polyphase bank of two or more single-phase transformers operating as a unit.

Section 450.3 was reorganized for the 1999 *Code* in order to present transformer protection requirements in a simpler and more user friendly table format. The fundamental protection requirements were not changed, only the presentation of the requirements is different. Instead of 11 paragraphs, 10 exceptions, and 2 tables, 450.3 contains two basic tables with notes and one exception. Remember, a note to a table is part of the requirements of the table, unlike the information

contained in fine print notes (FPNs), which per 90.5(C) are explanatory in nature and thus are not enforceable.

> FPN No. 1: See 240.4, 240.21, 240.100, and 240.101 for overcurrent protection of conductors.

The requirements for protection of transformer secondaries in Article 450 apply only to the protection of transformers, not to the protection of conductors. Article 240 applies only to the protection of conductors, not to the protection of transformers. It is possible that the overcurrent protection required by Article 450 also satisfies the requirements in Article 240 for protection of the conductors, and vice versa, but it is also possible that they will not.

> FPN No. 2: Nonlinear loads can increase heat in a transformer without operating its overcurrent protective device.

The increased heating effects of nonlinear load currents must be taken into account when determining the load on a transformer. There are several methods for dealing with the heating effects of nonlinear loads, including derating equipment, oversizing equipment, increasing insulation ratings, installing thermal protection systems, and using K-factor transformers. The optimum method for dealing with transformer overheating will vary, depending on several technical and economic factors, and should be considered during the design phase of the electrical system.

(A) Transformers Over 600 Volts, Nominal. Overcurrent protection shall be provided in accordance with Table 450.3(A).

For Note 1, for standard ratings of circuit breakers and fuses, see 240.6.

For Note 2, overcurrent protection to protect the secondary of a transformer is allowed to consist of not more than six sets of fuses or six circuit breakers, under the following conditions.

1. They must be grouped in one location.
2. The sum of the overcurrent devices must not exceed the maximum value of a single device.
3. Where a combination of fuses and circuit breakers is used, the maximum value will be that of a single set of fuses.

See Exhibits 450.1 and 450.2 for an application of Note 2. This note also appears in Table 450.3(B) and applies to transformers rated 600 volts and less as well.

For Note 3, a supervised location is where maintenance of the equipment is performed by personnel familiar with

Table 450.3(A) Maximum Rating or Setting of Overcurrent Protection for Transformers Over 600 Volts (as a Percentage of Transformer-Rated Current)

| Location Limitations | Transformer Rated Impedance | Primary Protection Over 600 Volts | | Secondary Protection (See Note 2.) | | |
| | | | | Over 600 Volts | | 600 Volts or Below |
		Circuit Breaker (See Note 4.)	Fuse Rating	Circuit Breaker (See Note 4.)	Fuse Rating	Circuit Breaker or Fuse Rating
Any Location	Not more than 6%	600% (See Note 1.)	300% (See Note 1.)	300% (See Note 1.)	250% (See Note 1.)	125% (See Note 1.)
	More than 6% and not more than 10%	400% (See Note 1.)	300% (See Note 1.)	250% (See Note 1.)	225% (See Note 1.)	125% (See Note 1.)
Supervised Locations Only (See Note 3.)	Any	300% (See Note 1.)	250% (See Note 1.)	Not required	Not required	Not required
	Not more than 6%	600%	300%	300% (See Note 5.)	250% (See Note 5.)	250% (See Note 5.)
	More than 6% and not more than 10%	400%	300%	250% (See Note 5.)	225% (See Note 5.)	250% (See Note 5.)

Notes:

1. Where the required fuse rating or circuit breaker setting does not correspond to a standard rating or setting, a higher rating or setting that does not exceed the next higher standard rating or setting shall be permitted.

2. Where secondary overcurrent protection is required, the secondary overcurrent device shall be permitted to consist of not more than six circuit breakers or six sets of fuses grouped in one location. Where multiple overcurrent devices are utilized, the total of all the device ratings shall not exceed the allowed value of a single overcurrent device. If both circuit breakers and fuses are used as the overcurrent device, the total of the device ratings shall not exceed that allowed for fuses.

3. A supervised location is a location where conditions of maintenance and supervision ensure that only qualified persons monitor and service the transformer installation.

4. Electronically actuated fuses that may be set to open at a specific current shall be set in accordance with settings for circuit breakers.

5. A transformer equipped with a coordinated thermal overload protection by the manufacturer shall be permitted to have separate secondary protection omitted.

proper operation of the equipment and aware of the hazards associated with it.

For Note 4, *electronically actuated fuse* is defined in Article 100, Part II, Over 600 Volts, Nominal.

The ratings or settings obtained from Table 450.3(A) are based on the type of protective device (fuse, electronic fuse, or circuit breaker), transformer rated current and impedance, and primary and secondary voltages. According to Table 450.3(A), the maximum ratings or settings of an overcurrent protective device for transformers rated over 600 volts is separated into two broad categories: *any location* (or unsupervised) and *supervised locations only*.

The first category for over 600-volt transformers is not limited by location and it is referred to as *any location*. The maximum ratings or settings for overcurrent devices permitted in the *any location* row are applicable for all unsupervised locations. An *any location* transformer installation must be provided with overcurrent protection in both the primary and secondary circuit. Of course, the user may select the *any location* row even if the location qualifies as a *supervised locations only* because the requirements for the *any location* row are the same or exceed the *supervised locations only* row. See Exhibit 450.3 for an example of an installation using circuit breakers on the primary and the secondary for an over 600-volt transformer with 6 percent impedance.

The second category for over 600-volt transformers is *supervised locations only*. The maximum ratings or settings for overcurrent devices permitted in the *supervised locations only* rows are strictly limited to the conditions explained in Note 3 of Table 450.3(A). If the location of a transformer does not qualify as a *supervised* location, then it is necessary to select the general rows of *any location* in Table 450.3(A). Notice that the installation shown in Exhibit 450.3 fulfills

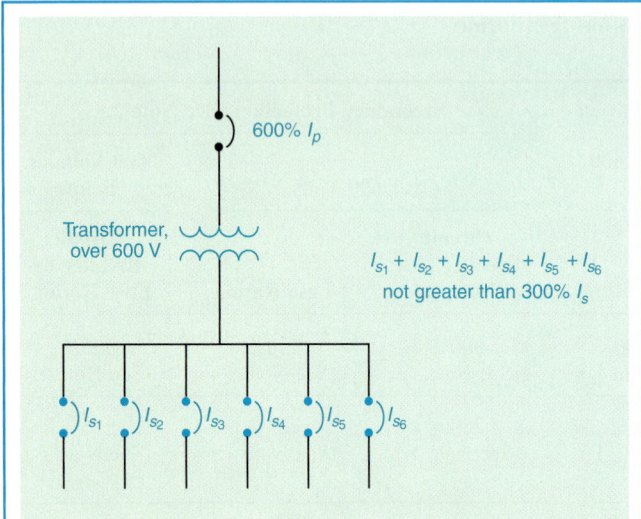

Exhibit 450.1 A transformer rated over 600 volts with a secondary rated over 600 volts, with secondary protection consisting of six circuit breakers. The sum of the ratings of the circuit breakers is not permitted to exceed 300 percent of the rated secondary current.

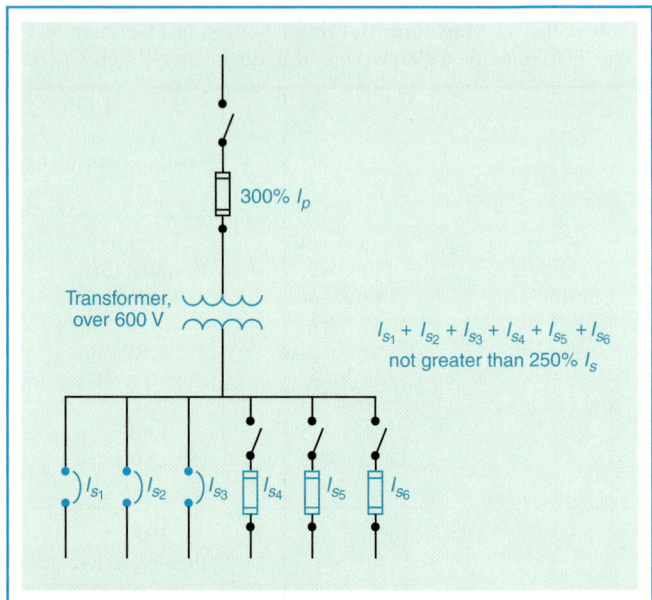

Exhibit 450.2 A transformer rated over 600 volts with a secondary rated over 600 volts, with secondary protection consisting of fuses and circuit breakers. The sum of the ratings of all the overcurrent devices is not permitted to exceed the rating permitted for fuses.

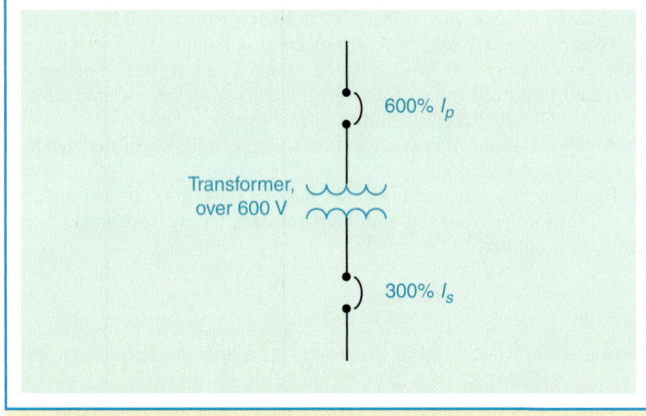

Exhibit 450.3 A transformer with 6 percent impedance and rated over 600 volts using circuit-breaker protection for both the primary and the secondary. For the transformer in this figure, both the primary and the secondary voltages are over 600 volts.

the requirements of both *any location* and *supervised locations only.*

Article 240, Overcurrent Protection, contains Part VIII, Supervised Industrial Installations, which contains many revised overcurrent protection requirements for feeders and feeder taps associated with transformers. Also, requirements for the overcurrent protection of transformer secondary conductors are found in 240.4(F).

(B) Transformers 600 Volts, Nominal, or Less. Overcurrent protection shall be provided in accordance with Table 450.3(B).

The ratings or settings of the overcurrent protective device obtained from Table 450.3(B) are based on the transformer-rated current and whether or not secondary protection is provided. According to Table 450.3(B), the maximum ratings or settings of overcurrent protective devices for transformers rated 600 volts and less are separated into two categories: *primary only protection* and *primary and secondary protection.*

According to Table 450.3(B), transformers with currents of 9 amperes or more must be protected by either of two methods. Method 1 requires primary protection only and is set at not more than 125 percent of the primary side rating. Method 1 does not require secondary side overcurrent protection. Method 2 requires secondary side overcurrent protection to be set at not more than 125 percent, provided the primary side overcurrent protection is set at not more than

250 percent of the primary side rating. Although not required, following either protection method will free the user from any further protection requirements of Table 450.3(B). According to this table, smaller transformers have protection requirements that are less restrictive. For overcurrent protection of motor control circuit transformers, see 430.72.

An example of *primary only protection* is shown in Exhibit 450.4. An example of *primary and secondary protection* is shown in Exhibit 450.5.

Table 450.3(B) Maximum Rating or Setting of Overcurrent Protection for Transformers 600 Volts and Less (as a Percentage of Transformer-Rated Current)

Protection Method	Primary Protection			Secondary Protection (See Note 2.)	
	Currents of 9 Amperes or More	Currents Less Than 9 Amperes	Currents Less Than 2 Amperes	Currents of 9 Amperes or More	Currents Less Than 9 Amperes
Primary only protection	125% (See Note 1.)	167%	300%	Not required	Not required
Primary and secondary protection	250% (See Note 3.)	250% (See Note 3.)	250% (See Note 3.)	125% (See Note 1.)	167%

Notes:

1. Where 125 percent of this current does not correspond to a standard rating of a fuse or nonadjustable circuit breaker, a higher rating that does not exceed the next higher standard rating shall be permitted.

2. Where secondary overcurrent protection is required, the secondary overcurrent device shall be permitted to consist of not more than six circuit breakers or six sets of fuses grouped in one location. Where multiple overcurrent devices are utilized, the total of all the device ratings shall not exceed the allowed value of a single overcurrent device. If both breakers and fuses are utilized as the overcurrent device, the total of the device ratings shall not exceed that allowed for fuses.

3. A transformer equipped with coordinated thermal overload protection by the manufacturer and arranged to interrupt the primary current shall be permitted to have primary overcurrent protection rated or set at a current value that is not more than six times the rated current of the transformer for transformers having not more than 6 percent impedance and not more than four times the rated current of the transformer for transformers having more than 6 percent but not more than 10 percent impedance.

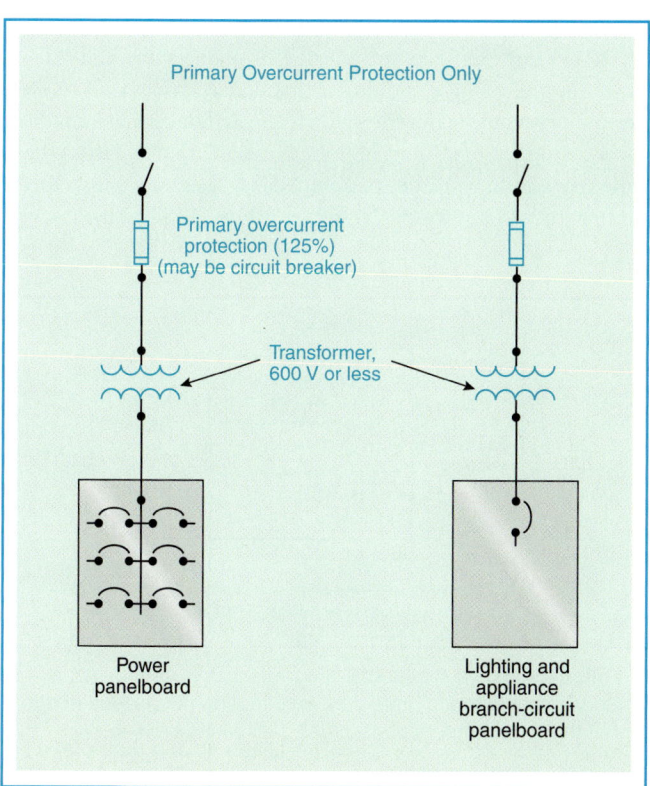

Exhibit 450.4 Two transformers (with currents of 9 amperes or more) rated 600 volts or less with only primary overcurrent protection according to Table 450.3(B).

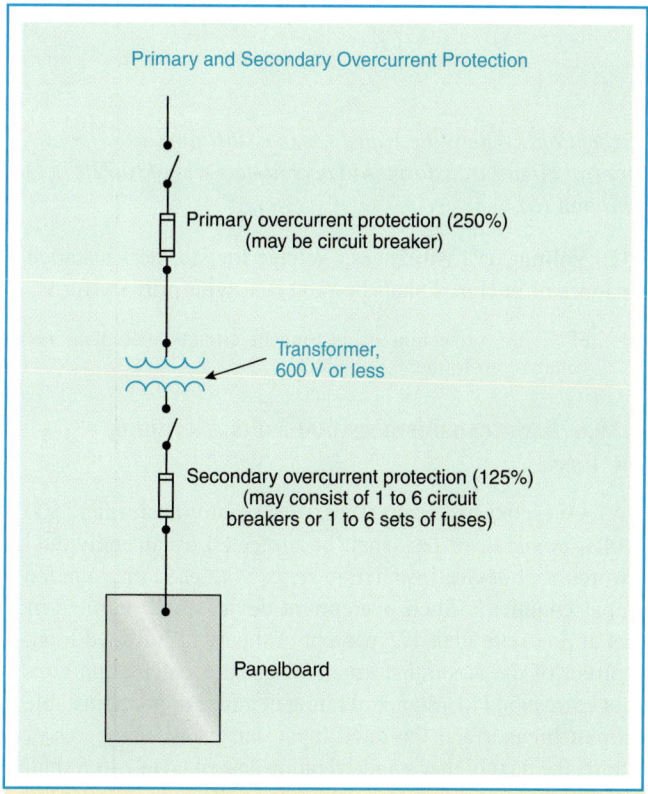

Exhibit 450.5 A transformer (9 amperes or more) rated 600 volts or less and protected by a combination of primary and secondary overcurrent protection, according to Table 450.3(B).

Questions frequently arise as to whether the overcurrent protection required for transformers, as specified in 450.3, provides satisfactory protection for the primary and secondary conductors. Where polyphase transformers are involved, primary and secondary conductors will usually not be properly protected. The rules in 450.3 are intended to protect the transformer alone. The primary overcurrent device provides short-circuit protection for the primary conductors and a degree of overload protection for the transformer, and secondary overcurrent devices prevent the transformer and secondary conductors from being overloaded. The transformer is considered the point of supply, and the conductors it supplies must be protected in accordance with their ampacity.

Section 240.4(F) permits the secondary circuit conductors from a transformer to be protected by overcurrent devices in the primary circuit conductors of the transformer only in two special cases. The first case is a transformer with a 2-wire primary and a 2-wire secondary, provided the transformer primary is protected in accordance with 450.3. The second case is a 3-phase, delta-delta-connected transformer having a 3-wire, single-voltage secondary, provided its primary is protected in accordance with 450.3. In cases where the primary feeder to the transformer incorporates overcurrent protective devices rated (or set) at a level not to exceed those prescribed herein, it is not necessary to duplicate them at the transformer.

Exception: Where the transformer is installed as a motor-control circuit transformer in accordance with 430.72(C)(1) through (5).

(C) Voltage Transformers. Voltage transformers installed indoors or enclosed shall be protected with primary fuses.

> FPN: For protection of instrument circuits including voltage transformers, see 408.32.

450.4 Autotransformers 600 Volts, Nominal, or Less.

(A) Overcurrent Protection. Each autotransformer 600 volts, nominal, or less shall be protected by an individual overcurrent device installed in series with each ungrounded input conductor. Such overcurrent device shall be rated or set at not more than 125 percent of the rated full-load input current of the autotransformer. Where this calculation does not correspond to a standard rating of a fuse or nonadjustable circuit breaker and the rated input current is 9 amperes or more, the next higher standard rating described in 240.6 shall be permitted. An overcurrent device shall not be installed in series with the shunt winding (the winding common to both the input and the output circuits) of the autotransformer between Points A and B as shown in Figure 450.4.

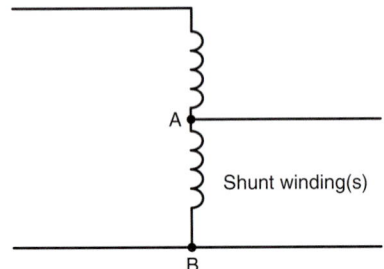

Figure 450.4 Autotransformer.

Exception: Where the rated input current of the autotransformer is less than 9 amperes, an overcurrent device rated or set at not more than 167 percent of the input current shall be permitted.

Because of the voltage feedback problem that may occur, an overcurrent device is not permitted between points A and B in *Code* Figure 450.4, Autotransformer.

Exhibit 450.6 provides an example of overcurrent protection for an autotransformer. It shows a two-winding, single-phase transformer connected to boost a 208-volt supply to 240 volts. The autotransformer is provided with a two-pole disconnect switch with both overcurrent devices (OC-1a and OC-1b) located on the supply side of the autotransformer. If an overcurrent device were located between points A and B and this overcurrent device opened, the full 208-volt supply voltage would be applied across the 32-volt secondary winding in series with the load. Under these conditions, a higher-than-normal voltage would appear across the primary winding. If the load impedance were very low, this voltage could approach $208/32 \times 208 = 1352$ volts.

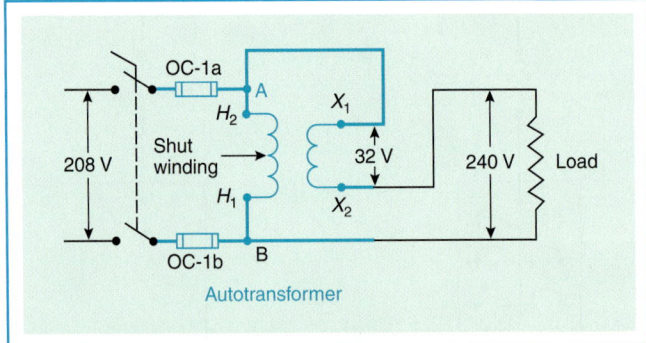

Exhibit 450.6 A disconnect switch with overcurrent devices properly connected to protect an autotransformer and located to meet the requirements of 450.4(A), last sentence.

(B) Transformer Field-Connected as an Autotransformer. A transformer field-connected as an autotransformer shall be identified for use at elevated voltage.

The requirement in 450.4(B) is necessary because of the dielectric voltage withstand test requirements applied to transformers. The test is conducted at 2500 volts for windings rated 250 volts or less, and 4000 volts for higher-rated windings. A transformer intended for buck or boost operation would require that the test for the low-voltage winding be based on the sum of the primary and secondary voltage ratings.

FPN: For information on permitted uses of autotransformers, see 210.9 and 215.11.

450.5 Grounding Autotransformers.

Grounding autotransformers covered in this section are zigzag or T-connected transformers connected to 3-phase, 3-wire ungrounded systems for the purpose of creating a 3-phase, 4-wire distribution system or providing a neutral reference for grounding purposes. Such transformers shall have a continuous per-phase current rating and a continuous neutral current rating.

FPN: The phase current in a grounding autotransformer is one-third the neutral current.

(A) Three-Phase, 4-Wire System. A grounding autotransformer used to create a 3-phase, 4-wire distribution system from a 3-phase, 3-wire ungrounded system shall conform to 450.5(A)(1) through (A)(4).

(1) Connections. The transformer shall be directly connected to the ungrounded phase conductors and shall not be switched or provided with overcurrent protection that is independent of the main switch and common-trip overcurrent protection for the 3-phase, 4-wire system.

(2) Overcurrent Protection. An overcurrent sensing device shall be provided that will cause the main switch or common-trip overcurrent protection referred to in 450.5 (A)(1) to open if the load on the autotransformer reaches or exceeds 125 percent of its continuous current per-phase or neutral rating. Delayed tripping for temporary overcurrents sensed at the autotransformer overcurrent device shall be permitted for the purpose of allowing proper operation of branch or feeder protective devices on the 4-wire system.

(3) Transformer Fault Sensing. A fault-sensing system that causes the opening of a main switch or common-trip overcurrent device for the 3-phase, 4-wire system shall be provided to guard against single-phasing or internal faults.

FPN: This can be accomplished by the use of two subtractive-connected donut-type current transformers installed to sense and signal when an unbalance occurs in the line current to the autotransformer of 50 percent or more of rated current.

(4) Rating. The autotransformer shall have a continuous neutral-current rating that is sufficient to handle the maximum possible neutral unbalanced load current of the 4-wire system.

Exhibit 450.7 shows the proper method of protecting a grounding autotransformer where it is used to provide a neutral for a 3-phase system where necessary to supply a group of single-phase, line-to-neutral loads. Separate overcurrent protection is not provided for the autotransformer, because there will be no control of the system line-to-neutral voltages if the autotransformer becomes disconnected. Consequently, simultaneous interruption of the power supply to all the line-to-neutral loads is necessary whenever the grounding autotransformer is switched off.

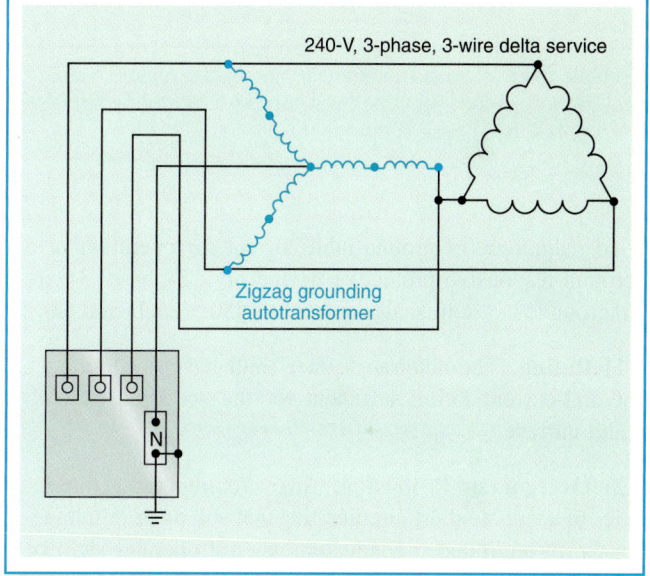

240-V, 3-phase, 3-wire delta service

Zigzag grounding autotransformer

N

Exhibit 450.7 A zigzag autotransformer used to create a 3-phase, 4-wire distribution system or to provide a neutral reference for grounding purposes.

The donut-type current transformers CT-1, CT-2, and CT-3 shown in Exhibit 450.8 must be arranged to trip the circuit breaker located upstream of both the autotransformer and line-to-neutral connected loads, to satisfy the requirements of 450.5(A)(1). All three relays are intended to trip the main breaker if the current in any phase or the neutral conductor exceeds 125 percent of the rated current, as specified in 450.5(A)(2). The current transformers CT-2 and CT-3 are also differentially connected, to protect against an internal failure of the autotransformer, as required by 450.5(A)(3).

(B) Ground Reference for Fault Protection Devices. A grounding autotransformer used to make available a speci-

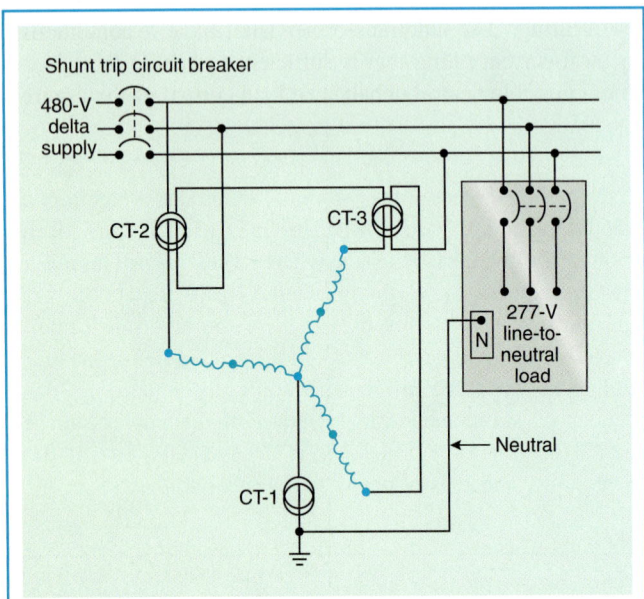

Exhibit 450.8 A zigzag autotransformer used to establish a neutral connection for a 480Y/277-volt, 3-phase ungrounded system to supply single-phase line-to-neutral loads.

fied magnitude of ground-fault current for operation of a ground-responsive protective device on a 3-phase, 3-wire ungrounded system shall conform to 450.5(B)(1) and (2).

(1) Rating. The autotransformer shall have a continuous neutral-current rating sufficient for the specified ground-fault current.

(2) Overcurrent Protection. An overcurrent protective device of adequate short-circuit rating that will open simultaneously all ungrounded conductors when it operates shall be applied in the grounding autotransformer branch circuit and shall be rated or set at a current not exceeding 125 percent of the autotransformer continuous per-phase current rating or 42 percent of the continuous-current rating of any series connected devices in the autotransformer neutral connection. Delayed tripping for temporary overcurrents to permit the proper operation of ground-responsive tripping devices on the main system shall be permitted but shall not exceed values that would be more than the short-time current rating of the grounding autotransformer or any series connected devices in the neutral connection thereto.

Exhibit 450.9 shows the proper method of protecting a grounding autotransformer where it is used as a ground reference for fault protection devices. The overcurrent protective device is to have a rating (or setting) not in excess of 125 percent of the rated phase current of the autotrans-

former (42 percent of the neutral current rating) and not more than 42 percent of the continuous current rating of the neutral grounding resistor or other current-carrying device in the neutral connection, as specified in 450.5(B)(2).

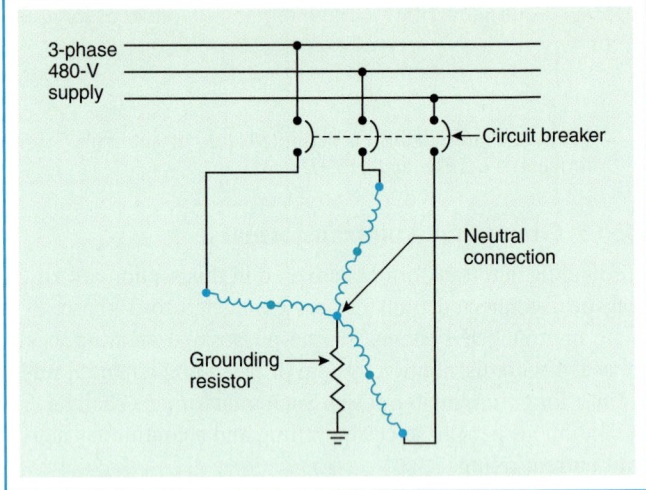

Exhibit 450.9 A zigzag autotransformer used to establish a reference ground-fault current for fault-protective-device operation or for damping-transitory-overvoltage surges.

(C) Ground Reference for Damping Transitory Overvoltages. A grounding autotransformer used to limit transitory overvoltages shall be of suitable rating and connected in accordance with 450.5(A)(1).

For installations involving a high-resistance grounding package, the functional performance of the installation parallels that described in 450.5(B), differing only in that the magnitude of available ground-fault current would likely be a lower value. It would be appropriate to employ the connections displayed in Figure 450.9 and conform with the overcurrent protection requirements prescribed in 450.5(B)(2).

With any of the grounding autotransformer applications covered by 450.5(A), (B), or (C), it is important to emphasize the use of a ganged 3-pole switching interrupter for connecting and disconnecting the autotransformer, in order to accomplish simultaneous connection (and disconnection) of the three line terminals. If, at any time, one or two of the line connections to the autotransformer should open, which could occur if the protective devices were single pole, the grounding autotransformer would cease to function in the desired fashion and would act as a high-inductive-reactance connection between the electrical system and ground. The latter connection is prone to create high-value transitory overvoltages, line-to-ground.

450.6 Secondary Ties.

A secondary tie is a circuit operating at 600 volts, nominal, or less between phases that connects two power sources or power supply points, such as the secondaries of two transformers. The tie shall be permitted to consist of one or more conductors per phase.

As used in this section, the word *transformer* means a transformer or a bank of transformers operating as a unit.

(A) Tie Circuits. Tie circuits shall be provided with overcurrent protection at each end as required in Article 240.

Under the conditions described in 450.6(A)(1) and 450.6(A)(2), the overcurrent protection shall be permitted to be in accordance with 450.6(A)(3).

(1) Loads at Transformer Supply Points Only. Where all loads are connected at the transformer supply points at each end of the tie and overcurrent protection is not provided in accordance with Article 240, the rated ampacity of the tie shall not be less than 67 percent of the rated secondary current of the largest transformer connected to the secondary tie system.

(2) Loads Connected Between Transformer Supply Points. Where load is connected to the tie at any point between transformer supply points and overcurrent protection is not provided in accordance with Article 240, the rated ampacity of the tie shall not be less than 100 percent of the rated secondary current of the largest transformer connected to the secondary tie system.

Exception: As otherwise provided in 450.6(A)(4).

(3) Tie Circuit Protection. Under the conditions described in 450.6(A)(1) and (A)(2), both ends of each tie conductor shall be equipped with a protective device that opens at a predetermined temperature of the tie conductor under short-circuit conditions. This protection shall consist of one of the following: (1) a fusible link cable connector, terminal, or lug, commonly known as a limiter, each being of a size corresponding with that of the conductor and of construction and characteristics according to the operating voltage and the type of insulation on the tie conductors or (2) automatic circuit breakers actuated by devices having comparable current-time characteristics.

(4) Interconnection of Phase Conductors Between Transformer Supply Points. Where the tie consists of more than one conductor per phase, the conductors of each phase shall comply with one of the following provisions.

(a) Interconnected. The conductors shall be interconnected in order to establish a load supply point, and the protection specified in 450.6(A)(3) shall be provided in each tie conductor at this point.

(b) Not Interconnected. The loads shall be connected to one or more individual conductors of a paralleled conductor tie without interconnecting the conductors of each phase and without the protection specified in 450.6(A)(3) at load connection points. Where this is done, the tie conductors of each phase shall have a combined capacity of not less than 133 percent of the rated secondary current of the largest transformer connected to the secondary tie system, the total load of such taps shall not exceed the rated secondary current of the largest transformer, and the loads shall be equally divided on each phase and on the individual conductors of each phase as far as practicable.

(5) Tie Circuit Control. Where the operating voltage exceeds 150 volts to ground, secondary ties provided with limiters shall have a switch at each end that, when open, de-energizes the associated tie conductors and limiters. The current rating of the switch shall not be less than the rated current of the conductors connected to the switch. It shall be capable of opening its rated current, and it shall be constructed so that it will not open under the magnetic forces resulting from short-circuit current.

(B) Overcurrent Protection for Secondary Connections. Where secondary ties are used, an overcurrent device rated or set at not more than 250 percent of the rated secondary current of the transformers shall be provided in the secondary connections of each transformer. In addition, an automatic circuit breaker actuated by a reverse-current relay set to open the circuit at not more than the rated secondary current of the transformer shall be provided in the secondary connection of each transformer.

The requirements of 450.6 apply specifically to network systems for power distribution commonly employed where the load density is high and reliability of service is important. Such a system is illustrated in Exhibit 450.10. This type of distribution system introduces a variety of problems not encountered in the more common radial-type distribution system and must be designed by experienced electrical engineers. Exhibit 450.10 shows a typical 3-phase network system for an industrial plant fed by two primary feeders, preferably from separate substations, energized at any standard voltage up to 34,500 volts. Each of the transformers is supplied by the two primary feeders, which are arranged by means of a double-throw switch at the transformer so that the transformer may be supplied by either feeder.

Each of the network transformers is rated in the range of 300 to 1000 kilovolt-amperes (kVA) and is required to be protected as illustrated in Exhibit 450.11. The primary and secondary protection is in accordance with 450.3, but an additional protective device must also be provided on the

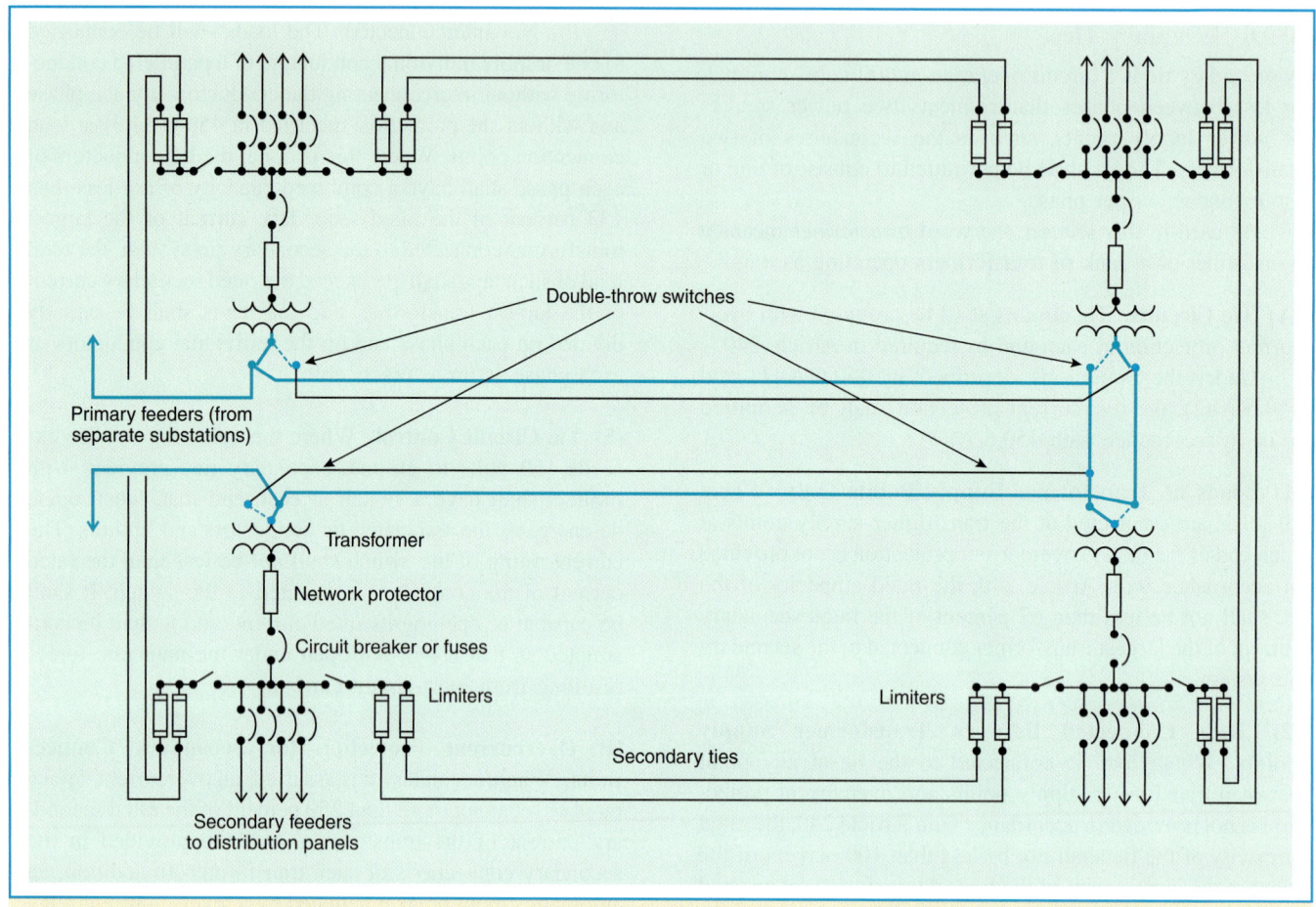

Exhibit 450.10 A typical 3-phase network system for an industrial plant fed by two primary feeders.

secondary side. This protective device is known as a *network protector*, which consists of a circuit breaker and a reverse-current relay. The network protector operates on reverse current to prevent power from being fed back into the transformer through the secondary ties should a fault occur in the transformer or a primary feeder. The reverse-current relay is set to trip the circuit breaker at a current value not more than the rated secondary current of the transformer. The relay is not designed to trip the circuit breaker in the event of an overload on the secondary of the transformer.

The secondary ties shown in Exhibit 450.10 must be protected at each end with an overcurrent device, in accordance with 450.6(A)(3). The overcurrent device most commonly provided for this purpose is a special type of fuse known as a *current limiter*, which is illustrated in Exhibit 450.12. This high-interrupting-capacity device is designed to provide short-circuit protection only for the secondary ties, which will open safely before temperatures damaging to the cable insulation are reached. See 240.2 for the defini-

tion of *current-limiting overcurrent protective device* and its accompanying commentary. The secondary ties form a closed loop equipped with switching devices so that any part of the loop may be isolated when repairs are needed or a current limiter must be replaced.

450.7 Parallel Operation.

Transformers shall be permitted to be operated in parallel and switched as a unit provided the overcurrent protection for each transformer meets the requirements of 450.3(A) for primary and secondary protective devices over 600 volts or 450.3(B) for primary and secondary protective devices 600 volts or less.

Parallel operation of transformers that are not switched as a unit can present dangerous backfeed situations for workers performing electrical maintenance. Appropriate lockout/tag-

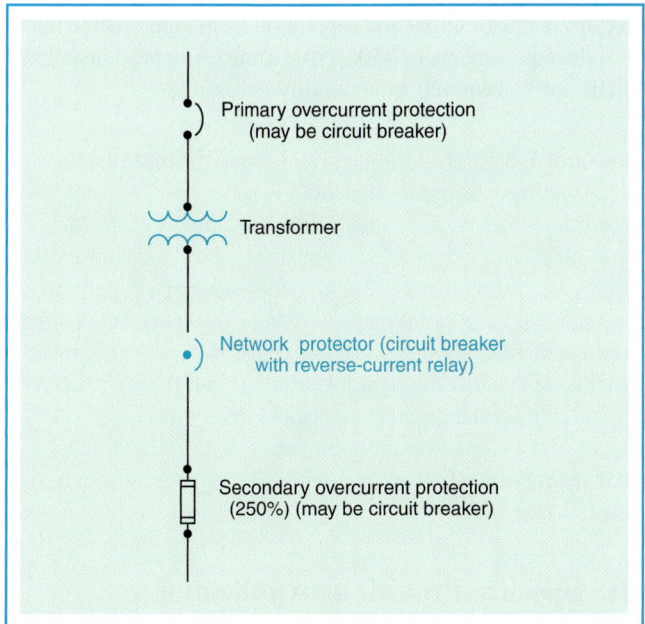

Exhibit 450.11 Primary and secondary overcurrent protection for a transformer in a network system, showing a network protector (an automatic circuit breaker actuated by a reverse-current relay).

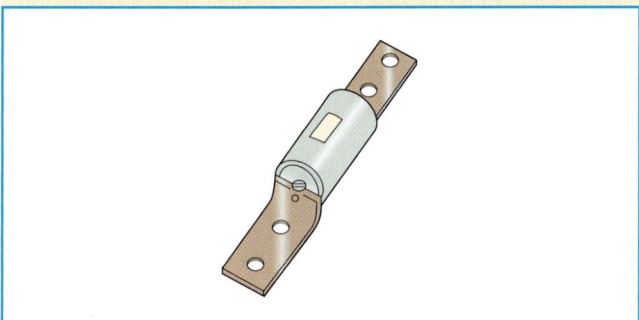

Exhibit 450.12 A current limiter (a special type of high-interrupting-capacity fuse).

out procedures must be implemented during maintenance of electrical equipment operated or connected in parallel. See NFPA 70E, *Standard for Electrical Safety Requirements for Employee Workplaces,* for safety-related work practices and appropriate lockout/tagout procedures.

450.8 Guarding.

Transformers shall be guarded as specified in 450.8(A) through (D).

(A) Mechanical Protection. Appropriate provisions shall be made to minimize the possibility of damage to transformers from external causes where the transformers are exposed to physical damage.

One method of providing mechanical protection is to strategically place bollards around the transformer. This practice provides a degree of protection from vehicles.

(B) Case or Enclosure. Dry-type transformers shall be provided with a noncombustible moisture-resistant case or enclosure that provides protection against the accidental insertion of foreign objects.

(C) Exposed Energized Parts. Switches or other equipment operating at 600 volts, nominal, or less and serving only equipment within a transformer enclosure shall be permitted to be installed in the transformer enclosure if accessible to qualified persons only. All energized parts shall be guarded in accordance with 110.27 and 110.34.

(D) Voltage Warning. The operating voltage of exposed live parts of transformer installations shall be indicated by signs or visible markings on the equipment or structures.

450.9 Ventilation.

The ventilation shall be adequate to dispose of the transformer full-load losses without creating a temperature rise that is in excess of the transformer rating.

> FPN No. 1: See ANSI/IEEE C57.12.00-1993, *General Requirements for Liquid-Immersed Distribution, Power, and Regulating Transformers,* and ANSI/IEEE C57.12.01-1989, *General Requirements for Dry-Type Distribution and Power Transformers.*
>
> FPN No. 2: Additional losses may occur in some transformers where nonsinusoidal currents are present, resulting in increased heat in the transformer above its rating. See ANSI/IEEE C57.110-1993, *Recommended Practice for Establishing Transformer Capability When Supplying Nonsinusoidal Load Currents,* where transformers are utilized with nonlinear loads.

Transformers with ventilating openings shall be installed so that the ventilating openings are not blocked by walls or other obstructions. The required clearances shall be clearly marked on the transformer.

Section 450.9 is intended to clarify that transformers are not permitted to be installed directly against walls or other obstructions that block openings for ventilation and that the required clearances should be clearly marked on the transformer (see 450.11).

Fine Print Note No. 2 of 450.9 warns of increased heating of transformers. See the commentary following 450.3, FPN No. 2, and the commentary following 310.15(B)(4) for additional information concerning nonlinear loads.

450.10 Grounding.

Exposed non–current-carrying metal parts of transformer installations, including fences, guards, and so forth, shall be grounded where required under the conditions and in the manner specified for electric equipment and other exposed metal parts in Article 250.

450.11 Marking.

Each transformer shall be provided with a nameplate giving the name of the manufacturer, rated kilovolt-amperes, frequency, primary and secondary voltage, impedance of transformers 25 kVA and larger, required clearances for transformers with ventilating openings, and the amount and kind of insulating liquid where used. In addition, the nameplate of each dry-type transformer shall include the temperature class for the insulation system.

> The information given on a transformer nameplate is necessary to determine whether special precautions must be used pertaining to clearances for ventilation, overcurrent protection, or liquid confinement.

450.12 Terminal Wiring Space.

The minimum wire-bending space at fixed, 600-volt and below terminals of transformer line and load connections shall be as required in 312.6. Wiring space for pigtail connections shall conform to Table 314.16(B).

> The requirement in 450.12 ensures adequate wire bending space at fixed terminals of transformer line and load connections rated 600 volts or less, as this is a point of maximum mechanical and electrical stress on the conductor insulation.

450.13 Accessibility.

All transformers and transformer vaults shall be readily accessible to qualified personnel for inspection and maintenance or shall meet the requirements of 450.13(A) or (B).

> Transformers are not accessible if wiring methods or other equipment obstruct the access of a worker or prevent removal of the covers for inspection or maintenance. Practical clearance considerations required for removal and replacement of the transformer are also important.

(A) Open Installations. Dry-type transformers 600 volts, nominal, or less, located in the open on walls, columns, or structures, shall not be required to be readily accessible.

(B) Hollow Space Installations. Dry-type transformers 600 volts, nominal, or less and not exceeding 50 kVA shall be permitted in hollow spaces of buildings not permanently closed in by structure, provided they meet the ventilation

requirements of 450.9 and separation from combustible materials requirements of 450.21(A). Transformers so installed shall not be required to be readily accessible.

> Section 450.13(B) continues to permit the installation of dry-type transformers rated 600 volts or less and not exceeding 50 kVA in hollow spaces of hung ceiling areas, provided these spaces are fire resistant, ventilated, and accessible. According to 300.22(C)(2), transformers are permitted to be installed in hollow spaces where the space is used for environmental air provided the transformer is in a metal enclosure (ventilated or nonventilated) and the transformer is suitable for the ambient air temperature within the hollow space. Of course, the requirement of 450.13(B) applies to transformer installations in "other spaces used for environmental air."

II. Specific Provisions Applicable to Different Types of Transformers

450.21 Dry-Type Transformers Installed Indoors.

(A) Not Over 112½ kVA. Dry-type transformers installed indoors and rated 112½ kVA or less shall have a separation of at least 305 mm (12 in.) from combustible material unless separated from the combustible material by a fire-resistant, heat-insulated barrier.

Exception: This rule shall not apply to transformers rated for 600 volts, nominal, or less that are completely enclosed, with or without ventilating openings.

(B) Over 112½ kVA. Individual dry-type transformers of more than 112½ kVA rating shall be installed in a transformer room of fire-resistant construction. Unless specified otherwise in this article, the term *fire resistant* means a construction having a minimum fire rating of 1 hour.

Exception No. 1: Transformers with Class 155 or higher insulation systems and separated from combustible material by a fire-resistant, heat-insulating barrier or by not less than 1.83 m (6 ft) horizontally and 3.7 m (12 ft) vertically.

Exception No. 2: Transformers with Class 155 or higher insulation systems and completely enclosed except for ventilating openings.

> Dry-type transformers with a Class 155 or higher insulation system rating are not required to be installed in transformer rooms or vaults if space separation or a fire-resistant heat-insulating barrier is provided. Although these units are designed for higher operating temperatures, the need for a transformer vault is mitigated by the fire-resistant characteristics of high-temperature insulations.

The two exceptions to 450.21(B) were revised for the 1999 *Code* by eliminating the reference to the specific temperature rating (80°C rise or higher rating) and substituting the class insulation system (Class 155 or higher.) The transformer class insulation system provides a more complete reference than simply using the permitted temperature rise. Further information on specific transformer class insulation systems may be found in UL 1561, *Dry-Type General Purpose and Power Transformers.*

FPN: See ANSI/ASTM E119-1995, *Method for Fire Tests of Building Construction and Materials*, and NFPA 251-1999, *Standard Methods of Tests of Fire Endurance of Building Construction and Materials.*

(C) Over 35,000 Volts. Dry-type transformers rated over 35,000 volts shall be installed in a vault complying with Part III of this article.

Dry-type transformers depend on the surrounding air for adequate ventilation and, where rated 112½ kilovolt-amperes or less, are not required to be installed in a fire-resistant transformer room but must comply with 450.9.

Dry-type transformers, or gas-filled or less-flammable liquid-insulated transformers (see 450.23), installed indoors with a primary voltage of not more than 35,000 volts are commonly used because a transformer vault is not required.

For the same reason, askarel-filled transformers have been extensively used indoors in the past. Askarel, which contains a polychlorinated biphenyl (PCB), is no longer being manufactured. Acceptable substitutes that comply with 450.23 are readily available.

Exhibit 450.13 shows a dry-type transformer with the outside casing in place and with the latest core and coil design for a typical dry-type power transformer rated at 1000 kilovolt-amperes, 13,800 volts to 480 volts, 3-phase, 60 Hz. This transformer has a high-voltage and low-voltage flange for connection to switchgear and a high-voltage, 2-position (double-throw), 3-pole-load air-break switch that may be attached to the case and arranged as a selector switch for the connection of the transformer primary to either of two feeder sources.

Dry-type transformers rated 112½ kilovolt-amperes or less require 12 in. of separation from combustible material or separation by fire-resistant barriers. Transformers rated less than 600 volts and completely enclosed, except for ventilating openings, are exempt from this requirement unless the manufacturer's installation instructions specify clearance distances. Noncombustible insulations used in transformers, such as mica, porcelain, and glass, which can withstand high temperatures, have permitted the application of larger dry-type transformers. Combustible materials, however, such as varnishes, may have been used with those

Exhibit 450.13 A dry-type transformer with a core and coil design rated at 1000 kVA, 13,800 volts to 480 volts, 3-phase, 60 Hz. (Courtesy of Square D Co.)

insulations, and, under short-circuit conditions, flames can escape from the transformer enclosure. Transformers rated over 112½ kilovolt-amperes must be located in fire-resistant transformer rooms or vaults unless either of the exceptions to 450.21(B) apply.

450.22 Dry-Type Transformers Installed Outdoors.

Dry-type transformers installed outdoors shall have a weatherproof enclosure.

Transformers exceeding 112½ kVA shall not be located within 305 mm (12 in.) of combustible materials of buildings unless the transformer has Class 155 insulation systems or higher and is completely enclosed except for ventilating openings.

See the commentary following 450.21(B), Exception No. 2, for an explanation of the change from specific temperature rating (80°C rise or higher rating) to the class insulation system (Class 155 or higher).

450.23 Less-Flammable Liquid-Insulated Transformers.

Transformers insulated with listed less-flammable liquids that have a fire point of not less than 300°C shall be permitted to be installed in accordance with 450.23(A) or (B).

(A) Indoor Installations. Indoor installations shall be permitted in accordance with one of the following:

(1) In Type I or Type II buildings, in areas where all of the following requirements are met:

    a. The transformer is rated 35,000 volts or less.
    b. No combustible materials are stored.
    c. A liquid confinement area is provided.
    d. The installation complies with all restrictions provided for in the listing of the liquid.

(2) With an automatic fire extinguishing system and a liquid confinement area, provided the transformer is rated 35,000 volts or less

(3) In accordance with 450.26

(B) Outdoor Installations. Less-flammable liquid-filled transformers shall be permitted to be installed outdoors, attached to, adjacent to, or on the roof of buildings, where installed in accordance with (1) or (2):

(1) For Type I and Type II buildings, the installation shall comply with all restrictions provided for in the listing of the liquid.

    FPN: Installations adjacent to combustible material, fire escapes, or door and window openings may require additional safeguards such as those listed in 450.27.

(2) In accordance with 450.27.

    FPN No. 1: As used in this section, *Type I and Type II buildings* refers to Type I and Type II building construction as defined in NFPA 220-1999, *Standard on Types of Building Construction. Combustible materials* refers to those materials not classified as noncombustible or limited-combustible as defined in NFPA 220-1999, *Standard on Types of Building Construction.*

NFPA 220, *Standard on Types of Building Construction,* defines Type I building construction as "that type in which the structural members, including walls, columns, beams, girders, trusses, arches, floors, and roofs, are of approved noncombustible or limited-combustible materials and have fire resistance ratings not less than those specified in Table 3-1."

Type II building construction is defined in NFPA 220 as "that type not qualifying as Type I construction in which the structural members, including walls, columns, beams, girders, trusses, arches, floors, and roofs, are of approved

noncombustible or limited-combustible materials and shall have fire resistance ratings not less than those specified in Table 3-1." Table 3-1 is reprinted here as Table 450.1.

Table 450.1 Fire Resistance Ratings (in Hours) for Type I and Type II Construction*

	Type I		Type II	
Exterior Bearing Walls —				
Supporting more than one floor, columns, or other bearing walls	4	3	2	1
Supporting one floor only	4	3	2	1
Supporting a roof only	4	3	1	1
Interior Bearing Walls —				
Supporting more than one floor, columns, or other bearing walls	4	3	2	1
Supporting one floor only	3	2	2	1
Supporting roofs only	3	2	1	1
Columns —				
Supporting more than one floor, columns, or other bearing walls	4	3	2	1
Supporting one floor only	3	2	2	1
Supporting roofs only	3	2	1	1
Beams, Girders, Trusses & Arches —				
Supporting more than one floor, columns, or other bearing walls	4	3	2	1
Supporting one floor only	3	2	2	1
Supporting roofs only	3	2	1	1
Floor Construction	3	2	2	1
Roof Construction	2	1½	1	1
Exterior Nonbearing Walls	0	0	0	0

*For further information, see NFPA 220, *Standard on Types of Building Construction.*

Source: Reprinted from NFPA 220, *Standard on Types of Building Construction,* 1999 edition.

FPN No. 2: See definition of *Listed* in Article 100.

Two listing agencies, Factory Mutual Research and Underwriters Laboratories Inc., list less-flammable liquids for transformers. These liquids have a fire point of at least 300°C. Exhibit 450.14 shows an example of a liquid-insulated transformer.

The Factory Mutual Research listing is based on the use of a Factory Mutual Research approved less-flammable

Exhibit 450.14 A liquid-insulated transformer filled with a listed less-flammable liquid having fire point of at least 300°C. (Courtesy of Square D Co.)

fluid in a transformer tank that meets certain criteria. Pressure-relief devices must be provided. Factory Mutual Research also recommends the use of enhanced electrical protection. Spacing from adjacent combustibles must be provided, based on the fluid capacity of the transformer tank, as illustrated in Exhibit 450.15. In the event of a leak, the

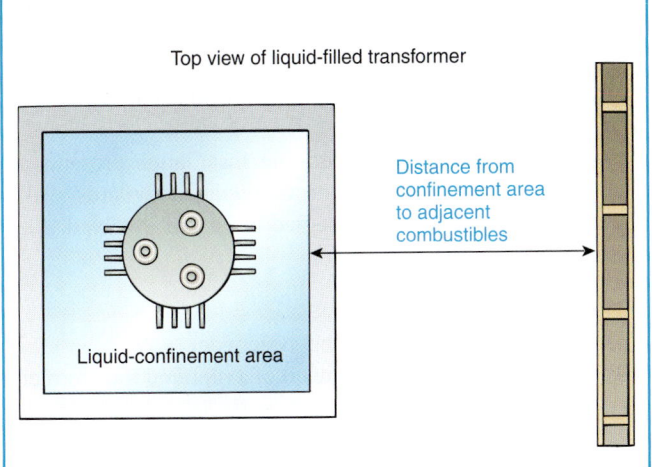

Exhibit 450.15 A transformer tank containing a Factory Mutual Research listed less-flammable fluid, where the spacing from adjacent combustibles to the liquid confinement area is based on the capacity of the tank.

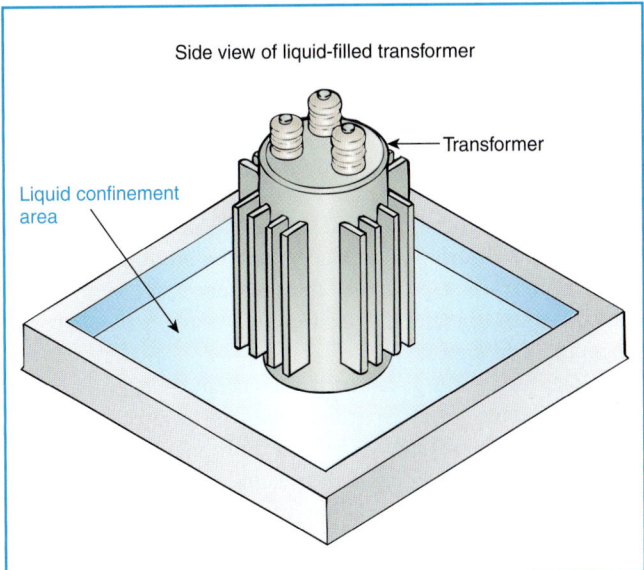

Exhibit 450.16 Side view of a transformer tank containing a Factory Mutual Research listed less-flammable fluid, where the volume of the confinement area is based on the capacity of the tank.

liquid confinement area is intended to prevent transformer dielectric fluid from spreading beyond the vicinity of the transformer, as illustrated in Exhibit 450.16. Further information on applications may be found in the Factory Mutual Loss Prevention Data Sheet 5-4/14-8.

The Underwriters Laboratories Inc. listing is based on UL requirements that no tank rupture or noted fluid leakage occur during low- and high-current arcing fault tests. Further information may be obtained from the UL *Gas and Oil Equipment Directory*, under Transformer Fluids (EOVK), or from the manufacturer.

450.24 Nonflammable Fluid-Insulated Transformers.

Transformers insulated with a dielectric fluid identified as nonflammable shall be permitted to be installed indoors or outdoors. Such transformers installed indoors and rated over 35,000 volts shall be installed in a vault. Such transformers installed indoors shall be furnished with a liquid confinement area and a pressure-relief vent. The transformers shall be furnished with a means for absorbing any gases generated by arcing inside the tank, or the pressure-relief vent shall be connected to a chimney or flue that will carry such gases to an environmentally safe area.

FPN: Safety may be increased if fire hazard analyses are performed for such transformer installations.

For the purposes of this section, a nonflammable dielectric fluid is one that does not have a flash point or fire point and is not flammable in air.

Section 450.24 requires a liquid confinement area and a pressure-relief vent. The liquid confinement area is intended to limit the extent of a spill if the tank leaks or ruptures. If a means for absorbing gases generated by arcing within the transformer is not provided, the pressure-relief vent must be connected to a chimney or flue that vents to an environmentally safe area.

The need for a gas absorption system or a chimney or flue that vents to an environmentally safe area is due to concerns about products generated during arcing. The high arc temperatures may cause the insulating medium to break down, resulting in the evolution of toxic or corrosive compounds.

450.25 Askarel-Insulated Transformers Installed Indoors.

Askarel-insulated transformers installed indoors and rated over 25 kVA shall be furnished with a pressure-relief vent. Where installed in a poorly ventilated place, they shall be furnished with a means for absorbing any gases generated by arcing inside the case, or the pressure-relief vent shall be connected to a chimney or flue that carries such gases outside the building. Askarel-insulated transformers rated over 35,000 volts shall be installed in a vault.

Askarel-insulated transformers are no longer manufactured. The information in the *Code* is for reference and for the modification of existing askarel-insulated installations.

450.26 Oil-Insulated Transformers Installed Indoors.

Oil-insulated transformers installed indoors shall be installed in a vault constructed as specified in Part III of this article.

Exception No. 1: Where the total capacity does not exceed 112½ kVA, the vault specified in Part III of this article shall be permitted to be constructed of reinforced concrete that is not less than 100 mm (4 in.) thick.

Exception No. 2: Where the nominal voltage does not exceed 600, a vault shall not be required if suitable arrangements are made to prevent a transformer oil fire from igniting other materials and the total capacity in one location does not exceed 10 kVA in a section of the building classified as combustible or 75 kVA where the surrounding structure is classified as fire-resistant construction.

Exception No. 3: Electric furnace transformers that have a total rating not exceeding 75 kVA shall be permitted to be

installed without a vault in a building or room of fire-resistant construction, provided suitable arrangements are made to prevent a transformer oil fire from spreading to other combustible material.

Exception No. 4: A transformer that has a total rating not exceeding 75 kVA and a supply voltage of 600 volts or less that is an integral part of charged-particle-accelerating equipment shall be permitted to be installed without a vault in a building or room of noncombustible or fire-resistant construction, provided suitable arrangements are made to prevent a transformer oil fire from spreading to other combustible material.

Exception No. 5: Transformers shall be permitted to be installed in a detached building that does not comply with Part III of this article if neither the building nor its contents present a fire hazard to any other building or property, and if the building is used only in supplying electric service and the interior is accessible only to qualified persons.

Exception No. 6: Oil-insulated transformers shall be permitted to be used without a vault in portable and mobile surface mining equipment (such as electric excavators) if each of the following conditions is met:

(a) Provision is made for draining leaking fluid to the ground.
(b) Safe egress is provided for personnel.
(c) A minimum 6-mm (¼-in.) steel barrier is provided for personnel protection.

450.27 Oil-Insulated Transformers Installed Outdoors.

Combustible material, combustible buildings, and parts of buildings, fire escapes, and door and window openings shall be safeguarded from fires originating in oil-insulated transformers installed on roofs, attached to or adjacent to a building or combustible material.

In cases where the transformer installation presents a fire hazard, one or more of the following safeguards shall be applied according to the degree of hazard involved:

(1) Space separations
(2) Fire-resistant barriers
(3) Automatic fire suppression systems
(4) Enclosures that confine the oil of a ruptured transformer tank

Oil enclosures shall be permitted to consist of fire-resistant dikes, curbed areas or basins, or trenches filled with coarse, crushed stone. Oil enclosures shall be provided with trapped drains where the exposure and the quantity of oil involved are such that removal of oil is important.

FPN: For additional information on transformers installed on poles or structures or under ground, see ANSI C2-1997, *National Electrical Safety Code.*

450.28 Modification of Transformers.

When modifications are made to a transformer in an existing installation that change the type of the transformer with respect to Part II of this article, such transformer shall be marked to show the type of insulating liquid installed, and the modified transformer installation shall comply with the applicable requirements for that type of transformer.

Askarel-insulated transformers are permitted to be modified by replacing the askarel with either oil or a less-flammable liquid. Where such a modification takes place, the completed installation must have the same degree of safety as a new installation. For example, replacement of askarel with oil in an indoor installation without a vault may not be acceptable (see 450.26 and its exceptions). The same is true if the replacement liquid is a less-flammable liquid (see 450.23).

III. Transformer Vaults

450.41 Location.

Vaults shall be located where they can be ventilated to the outside air without using flues or ducts wherever such an arrangement is practicable.

450.42 Walls, Roofs, and Floors.

The walls and roofs of vaults shall be constructed of materials that have adequate structural strength for the conditions with a minimum fire resistance of 3 hours. The floors of vaults in contact with the earth shall be of concrete that is not less than 100 mm (4 in.) thick, but where the vault is constructed with a vacant space or other stories below it, the floor shall have adequate structural strength for the load imposed thereon and a minimum fire resistance of 3 hours. For the purposes of this section, studs and wallboard construction shall not be acceptable.

Exception: Where transformers are protected with automatic sprinkler, water spray, carbon dioxide, or halon, construction of 1-hour rating shall be permitted.

FPN No. 1: For additional information, see ANSI/ASTM E119-1995, *Method for Fire Tests of Building Construction and Materials*, and NFPA 251-1999, *Standard Methods of Tests of Fire Endurance of Building Construction and Materials.*
FPN No. 2: A typical 3-hour construction is 150 mm (6 in.) thick reinforced concrete.

Vaults are intended primarily as passive fire protection. The need for vaults is dictated by the combustibility of the dielec-

tric media and the size of the transformer. Transformers insulated with mineral oil have the greatest need for passive protection, to prevent the spread of burning oil to other combustible materials.

Although it may be possible to construct a 3-hour-rated wall using studs and wallboard, this construction method is not permitted for transformer vaults. A reduction in fire-resistance rating from 3 hours to 1 hour is permitted for vaults equipped with an automatic fire suppression system.

There is less need for a vault around a dry-type transformer of less than 35,000 volts. If the transformer has adequate clearance from combustible construction and storage (see 450.21), a fire would be confined to the transformer.

Askarel is no longer manufactured as a transformer insulating fluid. Askarel-insulated transformers of less than 35,000 volts do not require vaults, because askarel is considered a noncombustible fluid. Transformers with a listed less-flammable liquid insulation may be installed without a vault, as permitted in 450.23. See the commentary following 450.23(B)(2), which relates to Type I and Type II building construction.

450.43 Doorways.

Vault doorways shall be protected in accordance with 450.43(A), (B), and (C).

(A) Type of Door. Each doorway leading into a vault from the building interior shall be provided with a tight-fitting door that has a minimum fire rating of 3 hours. The authority having jurisdiction shall be permitted to require such a door for an exterior wall opening where conditions warrant.

Exception: Where transformers are protected with automatic sprinkler, water spray, carbon dioxide, or halon, construction of 1-hour rating shall be permitted.

FPN: For additional information, see NFPA 80-1999, *Standard for Fire Doors and Fire Windows.*

(B) Sills. A door sill or curb that is of sufficient height to confine the oil from the largest transformer within the vault shall be provided, and in no case shall the height be less than 100 mm (4 in.).

(C) Locks. Doors shall be equipped with locks, and doors shall be kept locked, access being allowed only to qualified persons. Personnel doors shall swing out and be equipped with panic bars, pressure plates, or other devices that are normally latched but open under simple pressure.

Section 450.43 prohibits the use of conventional rotation-type door knobs on transformer vault doors. It is believed that an injured worker attempting to escape from a transformer vault may not be able to operate a rotating-type door knob but would be able to operate panic-type door hardware.

450.45 Ventilation Openings.

Where required by 450.9, openings for ventilation shall be provided in accordance with 450.45(A) through (F).

(A) Location. Ventilation openings shall be located as far as possible from doors, windows, fire escapes, and combustible material.

(B) Arrangement. A vault ventilated by natural circulation of air shall be permitted to have roughly half of the total area of openings required for ventilation in one or more openings near the floor and the remainder in one or more openings in the roof or in the sidewalls near the roof, or all of the area required for ventilation shall be permitted in one or more openings in or near the roof.

(C) Size. For a vault ventilated by natural circulation of air to an outdoor area, the combined net area of all ventilating openings, after deducting the area occupied by screens, gratings, or louvers, shall not be less than 1900 mm^2 (3 in.^{2}) per kVA of transformer capacity in service, and in no case shall the net area be less than 0.1 m^2 (1 ft^2) for any capacity under 50 kVA.

(D) Covering. Ventilation openings shall be covered with durable gratings, screens, or louvers, according to the treatment required in order to avoid unsafe conditions.

(E) Dampers. All ventilation openings to the indoors shall be provided with automatic closing fire dampers that operate in response to a vault fire. Such dampers shall possess a standard fire rating of not less than 1½ hours.

> FPN: See ANSI/UL 555-1995, *Standard for Fire Dampers.*

(F) Ducts. Ventilating ducts shall be constructed of fire-resistant material.

450.46 Drainage.

Where practicable, vaults containing more than 100 kVA transformer capacity shall be provided with a drain or other means that will carry off any accumulation of oil or water in the vault unless local conditions make this impractical. The floor shall be pitched to the drain where provided.

450.47 Water Pipes and Accessories.

Any pipe or duct system foreign to the electrical installation shall not enter or pass through a transformer vault. Piping or other facilities provided for vault fire protection, or for transformer cooling, shall not be considered foreign to the electrical installation.

Section 450.47 permits automatic sprinkler protection for transformer vaults. Piping or ductwork for cooling of the transformer is also permitted to be installed in a transformer vault. No other piping or ductwork is permitted to enter or pass through a transformer vault.

450.48 Storage in Vaults.

Materials shall not be stored in transformer vaults.

ARTICLE 455
Phase Converters

Contents

I. General

I. General

455.1 Scope.

This article covers the installation and use of phase converters.

A *phase converter* is an electrical device that converts single-phase electrical power to 3-phase, for the operation of equipment that normally operates from a 3-phase electrical supply. Phase converters are of two types: *static*, with no moving parts, and *rotary*, with an internal rotor that must be rotating before a load is applied (see 455.2 for definitions of *rotary-phase converter* and *static-phase converter*).

Phase converters are most commonly used to supply 3-phase motor loads in locations where only single-phase power is available from the local utility. Electrical installations on farms and in other remote or rural areas are examples of such locations. Although their most common loads are motors, phase converters are increasingly used to supply such loads as cellular telephone and other communication transmitter sites.

455.2 Definitions.

Manufactured Phase. The manufactured or derived phase originates at the phase converter and is not solidly connected to either of the single-phase input conductors.

Phase Converter. An electrical device that converts single-phase power to 3-phase electrical power.

> FPN: Phase converters have characteristics that modify the starting torque and locked-rotor current of motors served, and consideration is required in selecting a phase converter for a specific load.

Rotary-Phase Converter. A device that consists of a rotary transformer and capacitor panel(s) that permits the operation of 3-phase loads from a single-phase supply.

Static-Phase Converter. A device without rotating parts, sized for a given 3-phase load to permit operation from a single-phase supply.

455.3 Other Articles.

All applicable requirements of this *Code* shall apply to phase converters except as amended by this article.

455.4 Marking.

Each phase converter shall be provided with a permanent nameplate indicating the following:

(1) Manufacturer's name
(2) Rated input and output voltages
(3) Frequency
(4) Rated single-phase input full-load amperes
(5) Rated minimum and maximum single load in kilovolt-amperes (kVA) or horsepower
(6) Maximum total load in kilovolt-amperes (kVA) or horsepower
(7) For a rotary-phase converter, 3-phase amperes at full load

455.5 Equipment Grounding Connection.

A means for attachment of an equipment grounding conductor termination in accordance with 250.8 shall be provided.

455.6 Conductors.

(A) Ampacity. The ampacity of the single-phase supply conductors shall be determined by 455.6(A)(1) or (A)(2).

> FPN: Single-phase conductors sized to prevent a voltage drop not exceeding 3 percent from the source of supply to the phase converter may help ensure proper starting and operation of motor loads.

(1) Variable Loads. Where the loads to be supplied are variable, the conductor ampacity shall not be less than 125 percent of the phase converter nameplate single-phase input full-load amperes.

(2) Fixed Loads. Where the phase converter supplies specific fixed loads, and the conductor ampacity is less than 125 percent of the phase converter nameplate single-phase input full-load amperes, the conductors shall have an ampacity not less than 250 percent of the sum of the full-load, 3-phase current rating of the motors and other loads served where the input and output voltages of the phase converter are identical. Where the input and output voltages of the phase converter are different, the current as determined by this section shall be multiplied by the ratio of output to input voltage.

(B) Manufactured Phase Marking. The manufactured phase conductors shall be identified in all accessible locations with a distinctive marking. The marking shall be consistent throughout the system and premises.

455.7 Overcurrent Protection.

The single-phase supply conductors and phase converter shall be protected from overcurrent by 455.7(A) or (B). Where the required fuse rating or circuit breaker setting does not correspond to a standard rating or setting, the next higher standard rating or setting shall be permitted.

(A) Variable Loads. Where the loads to be supplied are variable, overcurrent protection shall be set at not more than 125 percent of the phase converter nameplate single-phase input full-load amperes.

(B) Fixed Loads. Where the phase converter supplies specific fixed loads and the conductors are sized in accordance with 455.6(A)(2), the conductors shall be protected in accordance with their ampacity. The overcurrent protection determined from this section shall not exceed 125 percent of the phase converter nameplate single-phase input amperes.

455.8 Disconnecting Means.

Means shall be provided to disconnect simultaneously all ungrounded single-phase supply conductors to the phase converter.

(A) Location. The disconnecting means shall be readily accessible and located in sight from the phase converter.

(B) Type. The disconnecting means shall be a switch rated in horsepower, a circuit breaker, or a molded-case switch. Where only nonmotor loads are served, an ampere-rated switch shall be permitted.

(C) Rating. The ampere rating of the disconnecting means shall not be less than 115 percent of the rated maximum single-phase input full-load amperes or, for specific fixed loads, shall be permitted to be selected from 455.7(C)(1) or (C)(2).

(1) Current Rated Disconnect. The disconnecting means shall be a circuit breaker or molded-case switch with an ampere rating not less than 250 percent of the sum of the following:

(1) Full-load, 3-phase current ratings of the motors
(2) Other loads served

(2) Horsepower Rated Disconnect. The disconnecting means shall be a switch with a horsepower rating. The equivalent locked rotor current of the horsepower rating of the switch shall not be less than 200 percent of the sum of the following:

(1) Nonmotor loads
(2) The 3-phase, locked-rotor current of the largest motor as determined from Table 430.151(B)
(3) The full-load current of all other 3-phase motors operating at the same time

(D) Voltage Ratios. The calculations in 455.8(C) shall apply directly where the input and output voltages of the phase converter are identical. Where the input and output voltages of the phase converter are different, the current shall be multiplied by the ratio of the output to input voltage.

455.9 Connection of Single-Phase Loads.

Where single-phase loads are connected on the load side of a phase converter, they shall not be connected to the manufactured phase.

455.10 Terminal Housings.

A terminal housing shall be provided on a phase converter, and the terminal housing shall be in accordance with the provisions of 430.12.

II. Specific Provisions Applicable to Different Types of Phase Converters

455.20 Disconnecting Means.

The single-phase disconnecting means for the input of a static phase converter shall be permitted to serve as the disconnecting means for the phase converter and a single load if the load is within sight of the disconnecting means.

455.21 Start-Up.

Power to the utilization equipment shall not be supplied until the rotary-phase converter has been started.

455.22 Power Interruption.

Utilization equipment supplied by a rotary-phase converter shall be controlled in such a manner that power to the equipment will be disconnected in the event of a power interruption.

> FPN: Magnetic motor starters, magnetic contactors, and similar devices, with manual or time delay restarting for the load, provide restarting after power interruption.

455.23 Capacitors.

Capacitors that are not an integral part of the rotary-phase conversion system but are installed for a motor load shall be connected to the line side of that motor overload protective device.

ARTICLE 460
Capacitors

Contents

460.1 Scope.

This article covers the installation of capacitors on electric circuits.

Surge capacitors or capacitors included as a component part of other apparatus and conforming with the requirements of such apparatus are excluded from these requirements.

This article also covers the installation of capacitors in hazardous (classified) locations as modified by Articles 501 through 503.

460.2 Enclosing and Guarding.

(A) Containing More Than 11 L (3 gal) of Flammable Liquid. Capacitors containing more than 11 L (3 gal) of flammable liquid shall be enclosed in vaults or outdoor fenced enclosures complying with Article 110, Part III. This limit shall apply to any single unit in an installation of capacitors.

(B) Accidental Contact. Where capacitors are accessible to unauthorized and unqualified persons, they shall be enclosed, located, or guarded so that persons cannot come into accidental contact or bring conducting materials into accidental contact with exposed energized parts, terminals, or buses associated with them. However, no additional guarding is required for enclosures accessible only to authorized and qualified persons.

Means are required to drain off the stored charge in a capacitor after the supply circuit has been opened. Otherwise, a person servicing the equipment could receive a severe shock, or damage may occur to the equipment.

Exhibit 460.1, diagram (a), shows a method in which capacitors are connected in a motor circuit so that they may be switched with the motor. In this arrangement, the stored charge will drain off through the windings when the circuit is opened. Diagram (b) shows another arrangement in which the capacitor is connected to the line side of the motor starter contacts. An automatic discharge device and a separate disconnecting means are required.

As shown in Exhibit 460.2, capacitors are often quipped with built-in resistors to drain off the stored charge, although this type of capacitor is not needed where connected as shown in Exhibit 460.1, diagram (a).

I. 600 Volts, Nominal, and Under

460.6 Discharge of Stored Energy.

Capacitors shall be provided with a means of discharging stored energy.

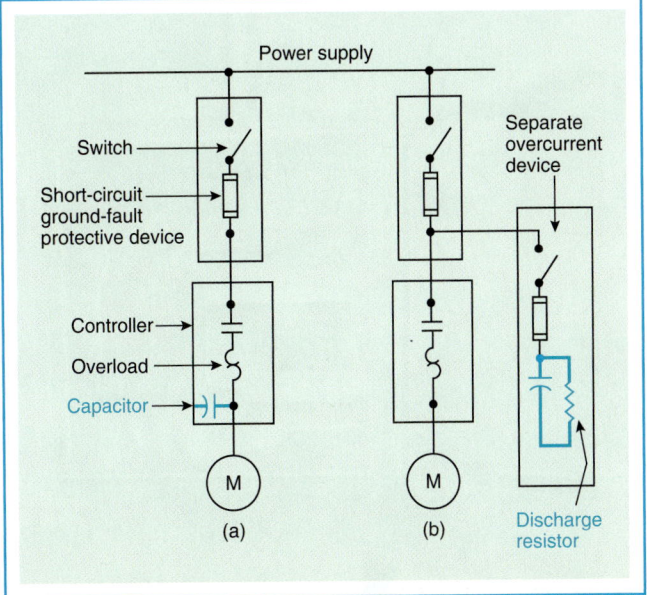

Exhibit 460.1 Methods of connecting capacitors in induction motor circuit for power factor correction.

(A) Time of Discharge. The residual voltage of a capacitor shall be reduced to 50 volts, nominal, or less within 1 minute after the capacitor is disconnected from the source of supply.

(B) Means of Discharge. The discharge circuit shall be either permanently connected to the terminals of the capacitor or capacitor bank or provided with automatic means of connecting it to the terminals of the capacitor bank on removal of voltage from the line. Manual means of switching or connecting the discharge circuit shall not be used.

460.8 Conductors.

(A) Ampacity. The ampacity of capacitor circuit conductors shall not be less than 135 percent of the rated current of the capacitor. The ampacity of conductors that connect a capacitor to the terminals of a motor or to motor circuit conductors shall not be less than one-third the ampacity of the motor circuit conductors and in no case less than 135 percent of the rated current of the capacitor.

(B) Overcurrent Protection. An overcurrent device shall be provided in each ungrounded conductor for each capacitor bank. The rating or setting of the overcurrent device shall be as low as practicable.

Except as permitted in the exception to 460.8(B), it is intended that the overcurrent device be separate from the overcurrent device protecting any other equipment or conductor. See Exhibit 460.1, diagrams (a) and (b).

Exhibit 460.2 Power factor correction capacitors with discharge resistors.

Exception: A separate overcurrent device shall not be required for a capacitor connected on the load side of a motor overload protective device.

(C) Disconnecting Means. A disconnecting means shall be provided in each ungrounded conductor for each capacitor bank and shall meet the following requirements:

(1) The disconnecting means shall open all ungrounded conductors simultaneously.

(2) The disconnecting means shall be permitted to disconnect the capacitor from the line as a regular operating procedure.

(3) The rating of the disconnecting means shall not be less than 135 percent of the rated current of the capacitor.

Exception: A separate disconnecting means shall not be required where a capacitor is connected on the load side of a motor controller.

Capacitors are rated in kilovars, which is abbreviated "kVAr" and stands for reactive kilovolt-amperes. Sometimes capacitors are rated in "kVAc," which stands for kilovolt amperes capacitive. Both ratings are synonymous. The kVAr rating shows how many reactive kilovolt-amperes the capacitor will supply in order to cancel out the reactive kilovolt-amperes caused by inductance. For example, a 20-kVAr capacitor will cancel out 20 kilovolt-amperes of inductive reactive kilovolt-amperes.

The basic capacitor is 3-phase and delta-connected internally, but single-phase and 2-phase units are also available. The capacitors are constructed with built-in fuses for short-circuit protection and discharge resistors that reduce the voltage to a 50-volt crest or less when disconnected from the power supply. This will occur within 1 minute on 600-volt units and within 5 minutes on 2400- and 4160-volt units.

The capacitor circuit conductors and disconnecting means must have an ampacity not less than 135 percent of the rated current of the capacitor. The reason is that all capacitors are manufactured with a tolerance of zero percent to 15 percent, so a 100-kVAr capacitor may actually draw a current equivalent to a 115-kVAr capacitor. In addition, the current drawn by a capacitor varies directly with the line voltage, and any variation in the line voltage from a pure sine wave form causes the capacitor to draw an increased current. Considering these several factors, the increased current can amount to 135 percent of the rated current of the capacitor.

The current corresponding to the kVAr rating of a 3-phase capacitor, I_c, is computed from the following formula:

$$I_c = \frac{kVAr \times 1000}{1.73 \text{ volts}}$$

The ampacity of the conductors and the switching device is then determined by multiplying I_c by 1.35. The most effective power factor correction is obtained where the individual capacitors are connected directly to the terminals of the motors, transformers, and other inductive machinery.

Where capacitors are connected together and operated as a unit, no complicated calculations are needed to determine the proper size capacitor to use. Capacitor manufacturers publish tables in which the required capacitor value is obtained by referring to the speed and horsepower of the motor. These values will improve the motor power factor to approximately 95 percent. To improve a plant power factor, capacitor manufacturers also publish tables to assist in calculating the total kVAr rating of capacitors required to improve the power factor to any desired value.

Care should be given to using capacitors where harmonic-producing loads are present. Adding capacitors to an electrical system can place the system in a harmonic resonance condition. The harmonic loads can excite the electrical system at the harmonic resonance frequency and cause overcurrent and overvoltage conditions. If capacitors are to be placed on electrical systems with harmonic loads, an engineering study that evaluates the size and placement of capacitors and the reactive impedance and load of the system should be conducted. Capacitors may need a reactor placed in series with them to help de-tune the electrical system from a harmonic resonance condition.

460.9 Rating or Setting of Motor Overload Device.

Where a motor installation includes a capacitor connected on the load side of the motor overload device, the rating or setting of the motor overload device shall be based on the improved power factor of the motor circuit.

The effect of the capacitor shall be disregarded in determining the motor circuit conductor rating in accordance with 430.22.

Where a capacitor is connected on the load side of the overload relays, as shown in Exhibit 460.1, diagram (a), consideration must be given when selecting the rating or setting of the motor overload device because the line current will be reduced due to an improved power factor. A value lower than indicated in 430.32 should be used for proper protection of the motor.

460.10 Grounding.

Capacitor cases shall be grounded in accordance with Article 250.

Exception: Capacitor cases shall not be grounded where the capacitor units are supported on a structure designed to operate at other than ground potential.

460.12 Marking.

Each capacitor shall be provided with a nameplate giving the name of the manufacturer, rated voltage, frequency, kilovar or amperes, number of phases, and, if filled with a combustible liquid, the volume of liquid. Where filled with a nonflammable liquid, the nameplate shall so state. The nameplate shall also indicate whether a capacitor has a discharge device inside the case.

II. Over 600 Volts, Nominal

460.24 Switching.

(A) Load Current. Group-operated switches shall be used for capacitor switching and shall be capable of the following:

(1) Carrying continuously not less than 135 percent of the rated current of the capacitor installation
(2) Interrupting the maximum continuous load current of each capacitor, capacitor bank, or capacitor installation that will be switched as a unit
(3) Withstanding the maximum inrush current, including contributions from adjacent capacitor installations
(4) Carrying currents due to faults on capacitor side of switch

(B) Isolation.

(1) General. A means shall be installed to isolate from all sources of voltage each capacitor, capacitor bank, or capacitor installation that will be removed from service as a unit. The isolating means shall provide a visible gap in the electrical circuit adequate for the operating voltage.

(2) Isolating or Disconnecting Switches with No Interrupting Rating. Isolating or disconnecting switches (with no interrupting rating) shall be interlocked with the load-interrupting device or shall be provided with prominently displayed caution signs in accordance with 490.22 to prevent switching load current.

(C) Additional Requirements for Series Capacitors. The proper switching sequence shall be ensured by use of one of the following:

(1) Mechanically sequenced isolating and bypass switches
(2) Interlocks
(3) Switching procedure prominently displayed at the switching location

460.25 Overcurrent Protection.

(A) Provided to Detect and Interrupt Fault Current. A means shall be provided to detect and interrupt fault current likely to cause dangerous pressure within an individual capacitor.

(B) Single Pole or Multipole Devices. Single-pole or multipole devices shall be permitted for this purpose.

(C) Protected Individually or in Groups. Capacitors shall be permitted to be protected individually or in groups.

(D) Protective Devices Rated or Adjusted. Protective devices for capacitors or capacitor equipment shall be rated or adjusted to operate within the limits of the safe zone for individual capacitors. If the protective devices are rated or adjusted to operate within the limits for Zone 1 or Zone 2, the capacitors shall be enclosed or isolated.

In no event shall the rating or adjustment of the protective devices exceed the maximum limit of Zone 2.

FPN: For definitions of *Safe Zone, Zone 1,* and *Zone 2,* see ANSI/IEEE 18-1992, *Shunt Power Capacitors.*

The reference to Zones 1 and 2 of ANSI/IEEE 18-1992 in the FPN pertains to the performance of the capacitors under fault conditions. If a fault current exceeds the limit established for Zone 2, the capacitor tank may burst.

460.26 Identification.

Each capacitor shall be provided with a permanent nameplate giving the manufacturer's name, rated voltage, frequency, kilovar or amperes, number of phases, and the volume of liquid identified as flammable, if such is the case.

460.27 Grounding.

Capacitor neutrals and cases, if grounded, shall be grounded in accordance with Article 250.

Exception: Where the capacitor units are supported on a structure that is designed to operate at other than ground potential.

460.28 Means for Discharge.

(A) Means to Reduce the Residual Voltage. A means shall be provided to reduce the residual voltage of a capacitor to 50 volts or less within 5 minutes after the capacitor is disconnected from the source of supply.

(B) Connection to Terminals. A discharge circuit shall be either permanently connected to the terminals of the capacitor or provided with automatic means of connecting it to the terminals of the capacitor bank after disconnection of the capacitor from the source of supply. The windings of motors, transformers, or other equipment directly connected to capacitors without a switch or overcurrent device interposed shall meet the requirements of 460.28(A).

ARTICLE 470
Resistors and Reactors

Contents

I. 600 Volts, Nominal, and Under

470.1 Scope.

This article covers the installation of separate resistors and reactors on electric circuits.

Resistors are made in many sizes and shapes and for different purposes. They may be wire or ribbon wound, form wound, edgewise wound, cast grid, punched steel grid, or box resistors. They may be mounted in the open or in ventilated metal boxes or cabinets, depending on their use and location. Because they give off heat, resistors must be guarded and located at safe distances from combustible materials. Where mounted on switchboards or installed in control panels, they are not required to have additional guards.

Reactors are installed in a circuit to introduce inductance for motor starting, combined with a capacitor to make a filter, controlling the current, and paralleling transformers. Current-limiting reactors are installed to limit the amount of current that can flow in a circuit when a short circuit occurs. Reactors can be divided into two classes: those with iron cores and those with no magnetic materials in the windings. Either type may be air cooled or oil immersed.

Mechanical stresses exist between adjacent air-core reactors due to their external fields, and the manufacturer's recommendations should be followed in spacing and bracing units and fastening supporting insulators.

Saturable reactors may be used for theater dimming [see 520.25(A) and commentary]. These reactors have, in addition to the ac winding, an auxiliary winding connected line-to-line or line-to-ground, in order to neutralize charging current and prevent a voltage rise. Those reactors used on high-voltage systems may be oil immersed.

Exception: Resistors and reactors that are component parts of other apparatus.

This article also covers the installation of resistors and reactors in hazardous (classified) locations as modified by Articles 501 through 504.

470.2 Location.

Resistors and reactors shall not be placed where exposed to physical damage.

470.3 Space Separation.

A thermal barrier shall be required if the space between the resistors and reactors and any combustible material is less than 305 mm (12 in.).

470.4 Conductor Insulation.

Insulated conductors used for connections between resistance elements and controllers shall be suitable for an operating temperature of not less than 90°C (194°F).

Exception: Other conductor insulations shall be permitted for motor starting service.

II. Over 600 Volts, Nominal

470.18 General.

(A) Protected Against Physical Damage. Resistors and reactors shall be protected against physical damage.

(B) Isolated by Enclosure or Elevation. Resistors and reactors shall be isolated by enclosure or elevation to protect personnel from accidental contact with energized parts.

(C) Combustible Materials. Resistors and reactors shall not be installed in close enough proximity to combustible materials to constitute a fire hazard and shall have a clearance of not less than 305 mm (12 in.) from combustible materials.

(D) Clearances. Clearances from resistors and reactors to grounded surfaces shall be adequate for the voltage involved.

FPN: See Article 490.

(E) Temperature Rise from Induced Circulating Currents. Metallic enclosures of reactors and adjacent metal parts shall be installed so that the temperature rise from induced circulating currents is not hazardous to personnel or does not constitute a fire hazard.

470.19 Grounding.

Resistor and reactor cases or enclosures shall be grounded in accordance with Article 250.

Exception: Resistor or reactor cases or enclosures supported on a structure designed to operate at other than ground potential shall not be grounded.

470.20 Oil-Filled Reactors.

Installation of oil-filled reactors, in addition to the above requirements, shall comply with applicable requirements of Article 450.

ARTICLE 480
Storage Batteries

Contents

480.1 Scope.

The provisions of this article shall apply to all stationary installations of storage batteries.

There are two general types of storage cells: the *lead-acid* type and the *alkali* (nickel-cadmium) type. Basically, a lead-acid cell consists of a positive plate, usually lead peroxide (a semisolid compound) mounted on a framework or grid for support, and a negative plate, made of sponge lead mounted on a grid. Grids are generally made of a lead alloy, such as lead-calcium, lead-antimony, or lead-selenium. The electrolyte is sulfuric acid and distilled water.

Lead-acid cells may be of the vented or sealed (valve-regulated) type. Under normal charging conditions, the vented type will liberate gases, hydrogen at the negative plate and oxygen at the positive plate. The valve-regulated type provides a means to recombine this gas, thus minimizing emissions from the cell.

In the alkali, or nickel-cadmium battery, the principal active material in the positive plate is nickelous hydroxide; in the negative plate, it is cadmium hydroxide. The electrolyte is potassium hydroxide (an alkali).

In stationary installations, nickel-cadmium cells are generally of the vented type and will liberate hydrogen and oxygen during normal charging. Hermetically sealed nickel-cadmium cells are sometimes used, but they require special charging equipment to prevent gas emissions.

480.2 Definitions.

Nominal Battery Voltage. The voltage computed on the basis of 2 volts per cell for the lead-acid type and 1.2 volts per cell for the alkali type.

Sealed Cell or Battery. A sealed cell or battery is one that has no provision for the addition of water or electrolyte or for external measurement of electrolyte specific gravity. The individual cells shall be permitted to contain a venting arrangement as described in 480.10(B).

Storage Battery. A battery comprised of one or more rechargeable cells of the lead-acid, nickel-cadmium, or other rechargeable electrochemical types.

480.3 Wiring and Equipment Supplied from Batteries.

Wiring and equipment supplied from storage batteries shall be subject to the requirements of this *Code* applying to wiring and equipment operating at the same voltage, unless otherwise permitted by 480.4.

480.4 Overcurrent Protection for Prime Movers.

Overcurrent protection shall not be required for conductors from a battery rated less than 50 volts if the battery provides power for starting, ignition, or control of prime movers. Section 300.3 shall not apply to these conductors.

The overcurrent protection requirements of Article 240 do not apply to field-installed conductors used for starting, ignition, or control of prime movers, provided that the supply source for these conductors is a battery rated less than 50 volts. In addition, the requirement to use single conductors only in conjunction with a Chapter 3 wiring method is not applicable to the battery-powered conductors. For example, if it were necessary at a generator location to extend the conductors from the battery to the prime mover starting solenoid, these conductors would not be required to have overcurrent protection and could be run as open, single conductors.

480.5 Grounding.

The requirements of Article 250 shall apply.

480.6 Insulation of Batteries Not Over 250 Volts.

This section shall apply to storage batteries having cells connected so as to operate at a nominal battery voltage of not over 250 volts.

(A) Vented Lead-Acid Batteries. Cells and multicompartment batteries with covers sealed to containers of nonconductive, heat-resistant material shall not require additional insulating support.

(B) Vented Alkaline-Type Batteries. Cells with covers sealed to jars of nonconductive, heat-resistant material shall require no additional insulation support. Cells in jars of conductive material shall be installed in trays of nonconductive material with not more than 20 cells (24 volts, nominal) in the series circuit in any one tray.

(C) Rubber Jars. Cells in rubber or composition containers shall require no additional insulating support where the total nominal voltage of all cells in series does not exceed 150 volts. Where the total voltage exceeds 150 volts, batteries shall be sectionalized into groups of 150 volts or less, and each group shall have the individual cells installed in trays or on racks.

(D) Sealed Cells or Batteries. Sealed cells and multicompartment sealed batteries constructed of nonconductive, heat-resistant material shall not require additional insulating support. Batteries constructed of a conducting container shall have insulating support if a voltage is present between the container and ground.

480.7 Insulation of Batteries of Over 250 Volts.

The provisions of 480.6 shall apply to storage batteries having the cells connected so as to operate at a nominal voltage exceeding 250 volts, and, in addition, the provisions of this section shall also apply to such batteries. Cells shall be installed in groups having a total nominal voltage of not over 250 volts. Insulation, which can be air, shall be provided between groups and shall have a minimum separation between live battery parts of opposite polarity of 50 mm (2 in.) for battery voltages not exceeding 600 volts.

480.8 Racks and Trays.

Racks and trays shall comply with 480.8(A) and (B).

(A) Racks. Racks, as required in this article, are rigid frames designed to support cells or trays. They shall be substantial and be made of one of the following:

(1) Metal, treated so as to be resistant to deteriorating action by the electrolyte and provided with nonconducting members directly supporting the cells or with continuous insulating material other than paint on conducting members

(2) Other construction such as fiberglass or other suitable nonconductive materials

(B) Trays. Trays are frames, such as crates or shallow boxes usually of wood or other nonconductive material, constructed or treated so as to be resistant to deteriorating action by the electrolyte.

480.9 Battery Locations.

Battery locations shall conform to 480.9(A), (B) and (C).

(A) Ventilation. Provisions shall be made for sufficient diffusion and ventilation of the gases from the battery to prevent the accumulation of an explosive mixture.

Compliance with 480.9(A) is necessary to prevent classification of a battery location as a hazardous (classified) location, in accordance with Article 500.

It is not the intent of 480.9(A) to mandate mechanical ventilation. Hydrogen disperses rapidly and requires little air movement to prevent accumulation. Unrestricted natural air movement in the vicinity of the battery, together with normal air changes for occupied spaces or heat removal, will normally be sufficient. If the space is confined, mechanical ventilation may be required in the vicinity of the battery.

Hydrogen is lighter than air and will tend to concentrate at ceiling level, so some form of ventilation should be provided at the upper portion of the structure. Ventilation can be a fan, roof ridge vent, or louvered area.

Although valve-regulated batteries are often referred to as "sealed," they actually emit very small quantities of hydrogen gas under normal operation, and are capable of liberating large quantities of explosive gases if overcharged. These batteries therefore require the same amount of ventilation as their vented counterparts.

(B) Live Parts. Guarding of live parts shall comply with 110.27.

Batteries should be located in clean, dry rooms. Batteries must be arranged to provide sufficient work space for inspection and maintenance. Provisions must also be made for adequate ventilation, in order to prevent an accumulation of an explosive mixture of the gases from the batteries.

The fumes given off by storage batteries are very corrosive; therefore, wiring and its insulation must be of a type that will withstand corrosive action (see 310.9). Special precautions are necessary to ensure that all metalwork (metal raceways, metal racks, etc.) is designed or treated so as to be corrosion resistant. Manufacturers sometimes suggest that aluminum or plastic conduit be used to withstand the corrosive battery fumes, or, if steel conduit is used, that it be zinc coated and corrosion protected with a coating of an asphaltum-type paint (see 300.6).

Overcharging heats a battery and causes gassing and loss of water. A battery should not be allowed to reach temperatures over 110°F (43.3°C), because heat causes a shedding of active materials from the plates that will eventually form a sediment buildup in the bottom of the case and short circuit the plates and the cell. Because mixtures of oxygen and hydrogen are highly explosive, flame or sparks should never be allowed near a cell, especially if the filler cap is removed.

(C) Working Space. Working space about the battery systems shall comply with 110.26. Working clearance shall be measured from the edge of the battery rack.

480.10 Vents.

(A) Vented Cells. Each vented cell shall be equipped with a flame arrester that is designed to prevent destruction of the cell due to ignition of gases within the cell by an external spark or flame under normal operating conditions.

(B) Sealed Cells. Sealed battery or cells shall be equipped with a pressure-release vent to prevent excessive accumulation of gas pressure, or the battery or cell shall be designed to prevent scatter of cell parts in event of a cell explosion.

ARTICLE 490
Equipment, Over 600 Volts, Nominal

Contents

I. General

490.1 Scope.

This article covers the general requirements for equipment operating at more than 600 volts, nominal.

FPN No. 1: See NFPA 70E-2000, *Standard for Electrical Safety Requirements for Employee Workplaces*, for electrical safety requirements for employee workplaces.

FPN No. 2: For further information on hazard signs and labels, see ANSI Z535-4, *Product Signs and Safety Labels*.

490.2 Definition.

High Voltage. For the purposes of this article, more than 600 volts, nominal.

490.3 Oil-Filled Equipment.

Installation of electrical equipment, other than transformers covered in Article 450, containing more than 38 L (10 gal) of flammable oil per unit shall meet the requirements of Parts II and III of Article 450.

II. Equipment—Specific Provisions

490.21 Circuit-Interrupting Devices.

(A) Circuit Breakers.

(1) Location.

(a) Circuit breakers installed indoors shall be mounted either in metal-enclosed units or fire-resistant cell-mounted units, or they shall be permitted to be open-mounted in locations accessible to qualified persons only.

(b) Circuit breakers used to control oil-filled transformers shall either be located outside the transformer vault or be capable of operation from outside the vault.

(c) Oil circuit breakers shall be arranged or located so that adjacent readily combustible structures or materials are safeguarded in an approved manner.

(2) Operating Characteristics. Circuit breakers shall have the following equipment or operating characteristics:

(1) An accessible mechanical or other approved means for manual tripping, independent of control power.
(2) Be release free (trip free).
(3) If capable of being opened or closed manually while energized, the main contacts shall operate independently of the speed of the manual operation.
(4) A mechanical position indicator at the circuit breaker to show the open or closed position of the main contacts.
(5) A means of indicating the open and closed position of the breaker at the point(s) from which they may be operated.

(3) Nameplate. A circuit breaker shall have a permanent and legible nameplate showing manufacturer's name or trademark, manufacturer's type or identification number, continuous current rating, interrupting rating in megavolt-amperes (MVA) or amperes, and maximum voltage rating. Modification of a circuit breaker affecting its rating(s) shall be accompanied by an appropriate change of nameplate information.

(4) Rating. Circuit breakers shall have the following ratings:

(1) The continuous current rating of a circuit breaker shall not be less than the maximum continuous current through the circuit breaker.

(2) The interrupting rating of a circuit breaker shall not be less than the maximum fault current the circuit breaker will be required to interrupt, including contributions from all connected sources of energy.

(3) The closing rating of a circuit breaker shall not be less than the maximum asymmetrical fault current into which the circuit breaker can be closed.

(4) The momentary rating of a circuit breaker shall not be less than the maximum asymmetrical fault current at the point of installation.

(5) The rated maximum voltage of a circuit breaker shall not be less than the maximum circuit voltage.

(B) ... ... Fuseholders.

... are used to protect conductors and ... be placed in each ungrounded con-... shall be permitted to be used in ... load if both fuses have identical ... stalled in an identified common ... connections that will divide the ... wer fuses of the vented type shall not be ... under ground, or in metal enclosures unless id ... ed for the use.

(2) Interrupting Rating. The interrupting rating of power fuses shall not be less than the maximum fault current the fuse will be required to interrupt, including contributions from all connected sources of energy.

(3) Voltage Rating. The maximum voltage rating of power fuses shall not be less than the maximum circuit voltage. Fuses having a minimum recommended operating voltage shall not be applied below this voltage.

(4) Identification of Fuse Mountings and Fuse Units. Fuse mountings and fuse units shall have permanent and legible nameplates showing the manufacturer's type or designation, continuous current rating, interrupting current rating, and maximum voltage rating.

(5) Fuses. Fuses that expel flame in opening the circuit shall be designed or arranged so that they function properly without hazard to persons or property.

(6) Fuseholders. Fuseholders shall be designed or installed so that they are de-energized while a fuse is being replaced.

Exception: Fuses and fuseholders designed to permit fuse replacement by qualified persons using equipment designed for the purpose without de-energizing the fuseholder shall be permitted.

(7) High-Voltage Fuses. Metal-enclosed switchgear and substations that utilize high-voltage fuses shall be provided with a gang-operated disconnecting switch. Isolation of the fuses from the circuit shall be provided by either connecting a switch between the source and the fuses or providing roll-out switch and fuse-type construction. The switch shall be of the load-interrupter type, unless mechanically or electrically interlocked with a load-interrupting device arranged to reduce the load to the interrupting capability of the switch.

Exception: More than one switch shall be permitted as the disconnecting means for one set of fuses where the switches are installed to provide connection to more than one set of supply conductors. The switches shall be mechanically or electrically interlocked to permit access to the fuses only when all switches are open. A conspicuous sign shall be placed at the fuses identifying the presence of more than one source.

(C) Distribution Cutouts and Fuse Links—Expulsion Type.

(1) Installation. Cutouts shall be located so that they may be readily and safely operated and re-fused, and so that the exhaust of the fuses does not endanger persons. Distribution cutouts shall not be used indoors, underground, or in metal enclosures.

(2) Operation. Where fused cutouts are not suitable to interrupt the circuit manually while carrying full load, an approved means shall be installed to interrupt the entire load. Unless the fused cutouts are interlocked with the switch to prevent opening of the cutouts under load, a conspicuous sign shall be placed at such cutouts identifying that they shall not be operated under load.

(3) Interrupting Rating. The interrupting rating of distribution cutouts shall not be less than the maximum fault current the cutout is required to interrupt, including contributions from all connected sources of energy.

(4) Voltage Rating. The maximum voltage rating of cutouts shall not be less than the maximum circuit voltage.

(5) Identification. Distribution cutouts shall have on their body, door, or fuse tube a permanent and legible nameplate or identification showing the manufacturer's type or designation, continuous current rating, maximum voltage rating, and interrupting rating.

(6) Fuse Links. Fuse links shall have a permanent and legible identification showing continuous current rating and type.

(7) Structure Mounted Outdoors. The height of cutouts mounted outdoors on structures shall provide safe clearance between lowest energized parts (open or closed position) and standing surfaces, in accordance with 110.34(E).

(D) Oil-Filled Cutouts.

(1) Continuous Current Rating. The continuous current rating of oil-filled cutouts shall not be less than the maximum continuous current through the cutout.

(2) Interrupting Rating. The interrupting rating of oil-filled cutouts shall not be less than the maximum fault current the oil-filled cutout is required to interrupt, including contributions from all connected sources of energy.

(3) Voltage Rating. The maximum voltage rating of oil-filled cutouts shall not be less than the maximum circuit voltage.

(4) Fault Closing Rating. Oil-filled cutouts shall have a fault closing rating not less than the maximum asymmetrical fault current that can occur at the cutout location, unless suitable interlocks or operating procedures preclude the possibility of closing into a fault.

(5) Identification. Oil-filled cutouts shall have a permanent and legible nameplate showing the rated continuous current, rated maximum voltage, and rated interrupting current.

(6) Fuse Links. Fuse links shall have a permanent and legible identification showing the rated continuous current.

(7) Location. Cutouts shall be located so that they are readily and safely accessible for re-fusing, with the top of the cutout not over 1.5 m (5 ft) above the floor or platform.

(8) Enclosure. Suitable barriers or enclosures shall be provided to prevent contact with nonshielded cables or energized parts of oil-filled cutouts.

(E) Load Interrupters. Load-interrupter switches shall be permitted if suitable fuses or circuit breakers are used in conjunction with these devices to interrupt fault currents. Where these devices are used in combination, they shall be coordinated electrically so that they will safely withstand the effects of closing, carrying, or interrupting all possible currents up to the assigned maximum short-circuit rating.

Where more than one switch is installed with interconnected load terminals to provide for alternate connection to different supply conductors, each switch shall be provided with a conspicuous sign identifying this hazard.

(1) Continuous Current Rating. The continuous current rating of interrupter switches shall equal or exceed the maximum continuous current at the point of installation.

(2) Voltage Rating. The maximum voltage rating of interrupter switches shall equal or exceed the maximum circuit voltage.

(3) Identification. Interrupter switches shall have a permanent and legible nameplate including the following information: manufacturer's type or designation, continuous current rating, interrupting current rating, fault closing rating, maximum voltage rating.

(4) Switching of Conductors. The switching mechanism shall be arranged to be operated from a location where the operator is not exposed to energized parts and shall be arranged to open all ungrounded conductors of the circuit simultaneously with one operation. Switches shall be arranged to be locked in the open position. Metal-enclosed switches shall be operable from outside the enclosure.

(5) Stored Energy for Opening. The stored-energy operator shall be permitted to be left in the uncharged position after the switch has been closed if a single movement of the operating handle charges the operator and opens the switch.

(6) Supply Terminals. The supply terminals of fused interrupter switches shall be installed at the top of the switch enclosure, or, if the terminals are located elsewhere, the equipment shall have barriers installed so as to prevent persons from accidentally contacting energized parts or dropping tools or fuses into energized parts.

See Exhibits 490.1 and 490.2 for an example of a fused interrupter switch and the fuseholder components.

490.22 Isolating Means.

Means shall be provided to completely isolate an item of equipment. The use of isolating switches shall not be required where there are other ways of de-energizing the equipment for inspection and repairs, such as draw-out-type metal-enclosed switchgear units and removable truck panels.

Isolating switches not interlocked with an approved circuit-interrupting device shall be provided with a sign warning against opening them under load.

A fuseholder and fuse, designed for the purpose, shall be permitted as an isolating switch.

490.23 Voltage Regulators.

Proper switching sequence for regulators shall be ensured by use of one of the following:

(1) Mechanically sequenced regulator bypass switch(es)
(2) Mechanical interlocks

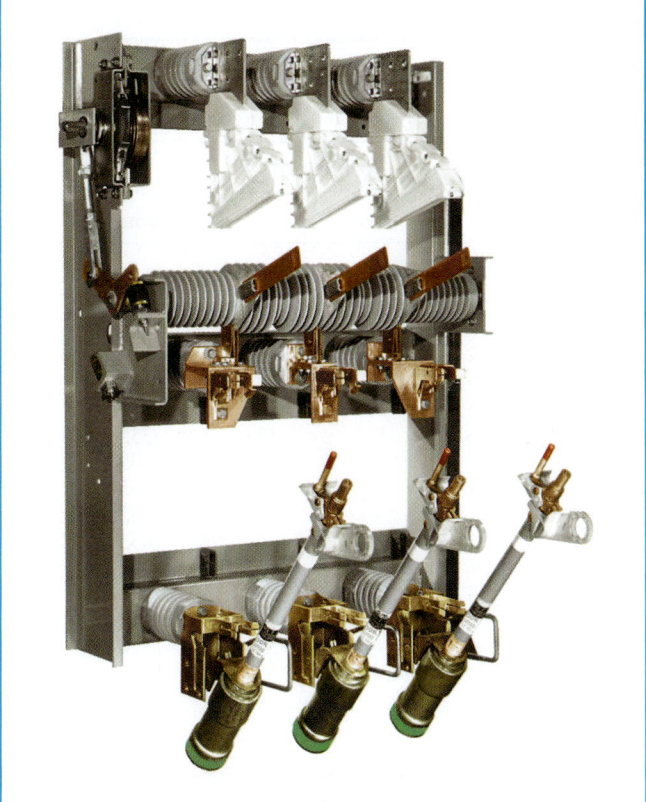

Exhibit 490.1 Group-operated interrupter-switch and powerfuse combination rated at 13.8 kV, 600 amperes continuous and interrupting, 40,000 amperes momentary, 40,000 amperes fault closing. (Courtesy of S&C Electric Co.)

Exhibit 490.2 Components of the indoor solid-material (SM) power fuseholder (boric-acid arc-extinguishing type) with a 14.4 kV, 400E-ampere maximum, 40,000-ampere rms asymmetrical interrupting rating. Shown here are the spring and cable assembly, refill unit, holder, and snuffler. (Courtesy of S&C Electric Co.)

(3) Switching procedure prominently displayed at the switching location

490.24 Minimum Space Separation.

In field-fabricated installations, the minimum air separation between bare live conductors and between such conductors and adjacent grounded surfaces shall not be less than the values given in Table 490.24. These values shall not apply to interior portions or exterior terminals of equipment designed, manufactured, and tested in accordance with accepted national standards.

III. Equipment—Metal-Enclosed Power Switchgear and Industrial Control Assemblies

490.30 General.

This part covers assemblies of metal-enclosed power switchgear and industrial control, including but not limited to switches, interrupting devices and their control, metering, protection and regulating equipment, where an integral part of the assembly, with associated interconnections and supporting structures. This part also includes metal-enclosed power switchgear assemblies that form a part of unit substations, power centers, or similar equipment.

For the control and protection of feeders leaving a substation, Exhibit 490.3 illustrates a typical example of modern, metal-enclosed switchgear. This industrial unit substation includes a high-voltage disconnect switch, transformer, and low-voltage switchgear with a fully functioning ground-fault relay protection system.

Indicator instruments, such as voltmeters, ammeters, wattmeters, and protective relays, may be mounted on the panel doors as desired. This switchgear affords a high degree of safety because all live parts are metal-enclosed, and interlocks are provided for safe operation.

An example of a high-voltage pad-mounted transformer and enclosure that may contain primary and secondary switches or circuit breakers is shown in Exhibit 490.4.

Table 490.24 Minimum Clearance of Live Parts*

Nominal Voltage Rating (kV)	Impulse Withstand, B.I.L (kV)		Minimum Clearance of Live Parts							
			Phase-to-Phase				Phase-to-Ground			
			Indoors		Outdoors		Indoors		Outdoors	
	Indoors	Outdoors	mm	in.	mm	in.	mm	in.	mm	in.
2.4–4.16	60	95	115	4.5	180	7	80	3.0	155	6
7.2	75	95	140	5.5	180	7	105	4.0	155	6
13.8	95	110	195	7.5	305	12	130	5.0	180	7
14.4	110	110	230	9.0	305	12	170	6.5	180	7
23	125	150	270	10.5	385	15	190	7.5	255	10
34.5	150	150	320	12.5	385	15	245	9.5	255	10
	200	200	460	18.0	460	18	335	13.0	335	13
46	—	200	—	—	460	18	—	—	335	13
	—	250	—	—	535	21	—	—	435	17
69	—	250	—	—	535	21	—	—	435	17
	—	350	—	—	790	31	—	—	635	25
115	—	550	—	—	1350	53	—	—	1070	42
138	—	550	—	—	1350	53	—	—	1070	42
	—	650	—	—	1605	63	—	—	1270	50
161	—	650	—	—	1605	63	—	—	1270	50
	—	750	—	—	1830	72	—	—	1475	58
230	—	750	—	—	1830	72	—	—	1475	58
	—	900	—	—	2265	89	—	—	1805	71
	—	1050	—	—	2670	105	—	—	2110	83

*The values given are the minimum clearance for rigid parts and bare conductors under favorable service conditions. They shall be increased for conductor movement or under unfavorable service conditions or wherever space limitations permit. The selection of the associated impulse withstand voltage for a particular system voltage is determined by the characteristics of the surge protective equipment.

Exhibit 490.3 An assembly of metal-enclosed switchgear. (Courtesy of Square D Co.)

490.31 Arrangement of Devices in Assemblies.

Arrangement of devices in assemblies shall be such that individual components can safely perform their intended function without adversely affecting the safe operation of other components in the assembly.

Exhibit 490.4 A 300-kVA, 15-kV pad-mounted transformer integral unit containing a primary hookstick-operated switch with a limited number of secondary breakers or switches. (Courtesy of Square D Co.)

490.32 Guarding of High-Voltage Energized Parts Within a Compartment.

Where access for other than visual inspection is required to a compartment that contains energized high-voltage parts, barriers shall be provided to prevent accidental contact by persons, tools, or other equipment with energized parts. Exposed live parts shall only be permitted in compartments accessible to qualified persons. Fuses and fuseholders designed to enable future replacement without de-energizing the fuse holder shall only be permitted for use by qualified persons.

490.33 Guarding of Low-Voltage Energized Parts Within a Compartment.

Energized bare parts mounted on doors shall be guarded where the door must be opened for maintenance of equipment or removal of draw-out equipment.

490.34 Clearance for Cable Conductors Entering Enclosure.

The unobstructed space opposite terminals or opposite raceways or cables entering a switchgear or control assembly shall be adequate for the type of conductor and method of termination.

490.35 Accessibility of Energized Parts.

(A) High-Voltage Equipment. Doors that would provide unqualified persons access to high-voltage energized parts shall be locked.

(B) Low-Voltage Control Equipment. Low-voltage control equipment, relays, motors, and the like shall not be installed in compartments with exposed high-voltage energized parts or high-voltage wiring unless either of the following conditions is met:

(1) The access means is interlocked with the high-voltage switch or disconnecting means to prevent the access means from being opened or removed.

(2) The high-voltage switch or disconnecting means is in the isolating position.

(C) High-Voltage Instruments or Control Transformers and Space Heaters. High-voltage instrument or control transformers and space heaters shall be permitted to be installed in the high-voltage compartment without access restrictions beyond those that apply to the high-voltage compartment generally.

490.36 Grounding.

Frames of switchgear and control assemblies shall be grounded.

490.37 Grounding of Devices.

Devices with metal cases or frames, or both, such as instruments, relays, meters, and instrument and control transformers, located in or on switchgear or control, shall have the frame or case grounded.

490.38 Door Stops and Cover Plates.

External hinged doors or covers shall be provided with stops to hold them in the open position. Cover plates intended to be removed for inspection of energized parts or wiring shall be equipped with lifting handles and shall not exceed 1.1 m^2 (12 ft^2) in area or 27 kg (60 lb) in weight, unless they are hinged and bolted or locked.

490.39 Gas Discharge from Interrupting Devices.

Gas discharged during operating of interrupting devices shall be directed so as not to endanger personnel.

490.40 Inspection Windows.

Windows intended for inspection of disconnecting switches or other devices shall be of suitable transparent material.

490.41 Location of Devices.

(A) Control and Instrument Transfer Switch Handles or Push Buttons. Control and instrument transfer switch handles or push buttons other than those covered in 490.41(B) shall be in a readily accessible location at an elevation of not over 2.0 m (78 in.).

Exception: Operating handles requiring more than 23 kg (50 lb) of force shall be located no higher than 1.7 m (66 in.) in either the open or closed position.

(B) Infrequently Operated Devices. Operating handles for infrequently operated devices, such as drawout fuses, fused potential or control transformers and their primary disconnects, and bus transfer switches, shall be permitted to be located where they are safely operable and serviceable from a portable platform.

490.42 Interlocks—Interrupter Switches.

Interrupter switches equipped with stored energy mechanisms shall have mechanical interlocks to prevent access to the switch compartment unless the stored energy mechanism is in the discharged or blocked position.

490.43 Stored Energy for Opening.

The stored energy operator shall be permitted to be left in the uncharged position after the switch has been closed if a single movement of the operating handle charges the operator and opens the switch.

490.44 Fused Interrupter Switches.

(A) Supply Terminals. The supply terminals of fused interrupter switches shall be installed at the top of the switch enclosure or, if the terminals are located elsewhere, the equipment shall have barriers installed so as to prevent persons from accidentally contacting energized parts or dropping tools or fuses into energized parts.

(B) Backfeed. Where fuses can be energized by backfeed, a sign shall be placed on the enclosure door identifying this hazard.

(C) Switching Mechanism. The switching mechanism shall be arranged to be operated from a location outside the enclosure where the operator is not exposed to energized parts and shall be arranged to open all ungrounded conductors of the circuit simultaneously with one operation. Switches shall be capable of being locked in the open position.

490.45 Circuit Breakers—Interlocks.

(A) Circuit Breakers. Circuit breakers equipped with stored energy mechanisms shall be designed to prevent the release of the stored energy unless the mechanism has been fully charged.

(B) Mechanical Interlocks. Mechanical interlocks shall be provided in the housing to prevent the complete withdrawal of the circuit breaker from the housing when the stored energy mechanism is in the fully charged position, unless a suitable device is provided to block the closing function of the circuit breaker before complete withdrawal.

IV. Mobile and Portable Equipment

490.51 General.

(A) Covered. The provisions of this part shall apply to installations and use of high-voltage power distribution and utilization equipment that is portable, mobile, or both, such as substations and switch houses mounted on skids, trailers, or cars; mobile shovels; draglines; cranes; hoists; drills; dredges; compressors; pumps; conveyors; underground excavators; and the like.

(B) Other Requirements. The requirements of this part shall be additional to, or amendatory of, those prescribed in Articles 100 through 725 of this *Code*. Special attention shall be paid to Article 250.

(C) Protection. Adequate enclosures, guarding, or both, shall be provided to protect portable and mobile equipment from physical damage.

(D) Disconnecting Means. Disconnecting means shall be installed for mobile and portable high-voltage equipment according to the requirements of Part VIII of Article 230 and shall disconnect all ungrounded conductors.

490.52 Overcurrent Protection.

Motors driving single or multiple dc generators supplying a system operating on a cyclic load basis do not require overload protection, provided that the thermal rating of the ac drive motor cannot be exceeded under any operating condition. The branch-circuit protective device(s) shall provide short-circuit and locked-rotor protection and shall be permitted to be external to the equipment.

490.53 Enclosures.

All energized switching and control parts shall be enclosed in effectively grounded metal cabinets or enclosures. These cabinets or enclosures shall be marked "DANGER—HIGH VOLTAGE—KEEP OUT" and shall be locked so that only authorized and qualified persons can enter. Circuit breakers and protective equipment shall have the operating means projecting through the metal cabinet or enclosure so these units can be reset without opening locked doors. With doors closed, reasonable safe access for normal operation of these units shall be provided.

490.54 Collector Rings.

The collector ring assemblies on revolving-type machines (shovels, draglines, etc.) shall be guarded to prevent accidental contact with energized parts by personnel on or off the machine.

490.55 Power Cable Connections to Mobile Machines.

A metallic enclosure shall be provided on the mobile machine for enclosing the terminals of the power cable. The enclosure shall include provisions for a solid connection for the ground wire(s) terminal to effectively ground the machine frame. Ungrounded conductors shall be attached to insulators or be terminated in approved high-voltage cable couplers (which include ground wire connectors) of proper voltage and ampere rating. The method of cable termination used shall prevent any strain or pull on the cable from stressing the electrical connections. The enclosure shall have provision for locking so only authorized and qualified persons may open it and shall be marked

DANGER—HIGH VOLTAGE—KEEP OUT.

490.56 High-Voltage Portable Cable for Main Power Supply.

Flexible high-voltage cable supplying power to portable or mobile equipment shall comply with Article 250 and Article 400, Part III.

V. Electrode-Type Boilers

490.70 General.

The provisions of this part shall apply to boilers operating over 600 volts, nominal, in which heat is generated by the passage of current between electrodes through the liquid being heated.

490.71 Electric Supply System.

Electrode-type boilers shall be supplied only from a 3-phase, 4-wire solidly grounded wye system, or from isolating transformers arranged to provide such a system. Control circuit voltages shall not exceed 150 volts, shall be supplied from a grounded system, and shall have the controls in the ungrounded conductor.

490.72 Branch-Circuit Requirements.

(A) Rating. Each boiler shall be supplied from an individual branch circuit rated not less than 100 percent of the total load.

(B) Common-Trip Fault-Interrupting Device. The circuit shall be protected by a 3-phase, common-trip fault-interrupting device, which shall be permitted to automatically reclose the circuit upon removal of an overload condition but shall not reclose after a fault condition.

(C) Phase-Fault Protection. Phase-fault protection shall be provided in each phase, consisting of a separate phase-overcurrent relay connected to a separate current transformer in the phase.

(D) Ground Current Detection. Means shall be provided for detection of the sum of the neutral and ground currents and shall trip the circuit-interrupting device if the sum of those currents exceeds the greater of 5 amperes or 7½ percent of the boiler full-load current for 10 seconds or exceeds an instantaneous value of 25 percent of the boiler full-load current.

(E) Grounded Neutral Conductor. The grounded neutral conductor shall be as follows:

(1) Connected to the pressure vessel containing the electrodes
(2) Insulated for not less than 600 volts
(3) Have not less than the ampacity of the largest ungrounded branch-circuit conductor
(4) Installed with the ungrounded conductors in the same raceway, cable, or cable tray, or, where installed as open conductors, in close proximity to the ungrounded conductors
(5) Not used for any other circuit

490.73 Pressure and Temperature Limit Control.

Each boiler shall be equipped with a means to limit the maximum temperature, pressure, or both, by directly or indirectly interrupting all current flow through the electrodes. Such means shall be in addition to the temperature, pressure, or both, regulating systems and pressure relief or safety valves.

490.74 Grounding.

All exposed non–current-carrying metal parts of the boiler and associated exposed grounded structures or equipment shall be bonded to the pressure vessel or to the neutral conductor to which the vessel is connected in accordance with 250.102, except the ampacity of the bonding jumper shall not be less than the ampacity of the neutral conductor.

ARTICLE 500
Hazardous (Classified) Locations, Classes I, II, and III, Divisions 1 and 2

Contents

FPN: Rules that are followed by a reference in brackets contain text that has been extracted from NFPA 497, *Recommended Practice for the Classification of Flammable Liquids, Gases, or Vapors and of Hazardous (Classified) Locations for Electrical Installations in Chemical Process Areas*, 1997 edition, and NFPA 499, *Recommended Practice for the Classification of Combustible Dusts and of Hazardous (Classified) Locations for Electrical Installation in Chemical Process Areas*, 1997 edition. Only editorial changes were made to the extracted text to make it consistent with this *Code*.

500.1 Scope—Articles 500 Through 504.

Articles 500 through 504 cover the requirements for electrical and electronic equipment and wiring for all voltages in Class I, Divisions 1 and 2; Class II, Divisions 1 and 2; and Class III, Divisions 1 and 2 locations where fire or explosion hazards may exist due to flammable gases or vapors, flammable liquids, combustible dust, or ignitible fibers or flyings.

> FPN: For the requirements for electrical and electronic equipment and wiring for all voltages in Class I, Zone 0, Zone 1, and Zone 2 hazardous (classified) locations where fire or explosion hazards may exist due to flammable gases or vapors or flammable liquids, refer to Article 505.

Article 500 is limited to locations classified as Class I, II, or III, Division 1 or 2. Article 505 contains the requirements for using the Class I, Zone 0, Zone 1, and Zone 2 method of area classification in accordance with International Electrotechnical Commission (IEC) 60079-10, *Electrical Apparatus for Explosive Gas Atmospheres*. See the commentary following 505.1.

The *Code* does not classify areas where the manufacture, transportation, storage, and use of explosive materials, such as ammunition, dynamite, and blasting powder, occur. Areas where such materials are present are not considered hazardous (classified) locations in accordance with Article 500. However, many organizations responsible for the safety of such areas require equipment and wiring methods suitable for hazardous locations as part of many safety precautions, even though the equipment and wiring have not been investigated for such locations. The logic is that hazardous location equipment and wiring methods are safer than ordinary location equipment and wiring methods. Further information on these locations can be found in NFPA 495, *Explosive Materials Code*.

The *National Electrical Code* does not classify specific Class I, Class II, and Class III locations. NFPA technical committees with the experience and expertise of working with flammable liquids, gases, vapors, dusts, and flyings that are inherent to a process or may be present under abnormal conditions of operation, determine the parameters, distances, and degrees of hazard associated with classified locations. Some of this information has been extracted from other NFPA documents and is included in Articles 511 through 517 of the *Code*.

Articles 500 through 505 cover the requirements for electrical installations in locations that are classified as hazardous locations due to the materials handled, processed, or stored in those locations. Some of the most common materials encountered in hazardous (classified) locations are flammable liquids. A flammable liquid is one that has a flash point below 100°F. A flammable or combustible liquid must be at its flash point for an explosion to occur. No. 1-D diesel fuel oil and kerosene have flash points higher than 100°F and therefore do not emit flammable vapors unless heated above their flash points. They are considered combustible liquids.

For information on hazardous locations in general, including background on the classification of areas, equipment protection systems, ignition sources, static electricity and lightning, and requirements that apply outside the United States, see *Electrical Installations in Hazardous Locations,* by Peter J. Schram and Mark W. Earley. This book is available from NFPA.

500.2 Definitions.

For purposes of Articles 500 through 504 and Articles 510 through 516, the following definitions apply.

Associated Nonincendive Field Wiring Apparatus. Apparatus in which the circuits are not necessarily nonincendive themselves but that affect the energy in nonincendive field wiring circuits and are relied upon to maintain nonincendive energy levels. Associated nonincendive field wiring apparatus may be either of the following:

(1) Electrical apparatus that has an alternative type of protection for use in the appropriate hazardous (classified) location
(2) Electrical apparatus not so protected that shall not be used in a hazardous (classified) location

FPN: Associated nonincendive field wiring apparatus has designated associated nonincendive field wiring apparatus connections for nonincendive field wiring apparatus and may also have connections for other electrical apparatus.

Combustible Gas Detection System. A protection technique utilizing stationary gas detectors in industrial establishments.

Control Drawing. A drawing or other document provided by the manufacturer of the intrinsically safe or associated apparatus, or of the nonincendive field wiring apparatus or associated nonincendive field wiring apparatus, that details the allowed interconnections between the intrinsically safe and associated apparatus or between the nonincendive field wiring apparatus or associated nonincendive field wiring apparatus.

Dust-Ignitionproof. Equipment enclosed in a manner that excludes dusts and does not permit arcs, sparks, or heat otherwise generated or liberated inside of the enclosure to cause ignition of exterior accumulations or atmospheric suspensions of a specified dust on or in the vicinity of the enclosure.

FPN: For further information on dust-ignitionproof enclosures, see Type 9 enclosure in ANSI/NEMA 250-

1991, *Enclosures for Electrical Equipment*, and ANSI/UL 1203-1994, *Explosionproof and Dust-Ignitionproof Electrical Equipment for Hazardous (Classified) Locations.*

Dusttight. Enclosures constructed so that dust will not enter under specified test conditions.

FPN: See ANSI/ISA 12.12.01-2000, *Nonincendive Electrical Equipment for Use in Class I and II, Division 2, and Class III, Divisions 1 and 2 Hazardous (Classified) Locations* , and UL 1604-1994, *Electrical Equipment for Use in Class I and II, Division 2 and Class III Hazardous (Classified) Locations.*

Electrical and Electronic Equipment. Materials, fittings, devices, appliances, and the like that are part of, or in connection with, an electrical installation.

FPN: Portable or transportable equipment having self-contained power supplies, such as battery-operated equipment, could potentially become an ignition source in hazardous (classified) locations.

Explosionproof Apparatus. Apparatus enclosed in a case that is capable of withstanding an explosion of a specified gas or vapor that may occur within it and of preventing the ignition of a specified gas or vapor surrounding the enclosure by sparks, flashes, or explosion of the gas or vapor within, and that operates at such an external temperature that a surrounding flammable atmosphere will not be ignited thereby.

FPN: For further information, see ANSI/UL 1203-1994, *Explosion-Proof and Dust-Ignition-Proof Electrical Equipment for Use in Hazardous (Classified) Locations.*

Hermetically Sealed. Equipment sealed against the entrance of an external atmosphere where the seal is made by fusion, for example, soldering, brazing, welding, or the fusion of glass to metal.

FPN: For further information, see ANSI/ISA 12.12.01-2000, *Nonincendive Electrical Equipment for Use in Class I and II, Division 2, and Class III, Division 1 and 2 Hazardous (Classified) Locations.*

Nonincendive Circuit. A circuit, other than field wiring, in which any arc or thermal effect produced under intended operating conditions of the equipment is not capable, under specified test conditions, of igniting the flammable gas–air, vapor–air, or dust–air mixture.

FPN: Conditions are described in ANSI/ISA 12.12.01-2000, *Nonincendive Electrical Equipment for Use in Class I and II, Division 2, and Class III, Divisions 1 and 2 Hazardous (Classified) Locations.*

Nonincendive Component. A component having contacts for making or breaking an incendive circuit and the contacting mechanism is constructed so that the component is incapable of igniting the specified flammable gas–air or vapor–air mixture. The housing of a nonincendive component is not intended to exclude the flammable atmosphere or contain an explosion.

> FPN: For further information, see UL 1604-1994, *Electrical Equipment for Use in Class I and II, Division 2, and Class III Hazardous (Classified) Locations.*

Nonincendive Equipment. Equipment having electrical/electronic circuitry that is incapable, under normal operating conditions, of causing ignition of a specified flammable gas–air, vapor–air, or dust–air mixture due to arcing or thermal means.

> FPN: For further information, see ANSI/ISA 12.12.01-2000, *Nonincendive Electrical Equipment for Use in Class I and II, Division 2, and Class III, Divisions 1 and 2 Hazardous (Classified) Locations.*

Nonincendive Field Wiring. Wiring that enters or leaves an equipment enclosure and, under normal operating conditions of the equipment, is not capable, due to arcing or thermal effects, of igniting the flammable gas–air, vapor–air, or dust–air mixture. Normal operation includes opening, shorting, or grounding the field wiring.

Nonincendive Field Wiring Apparatus. Apparatus intended to be connected to nonincendive field wiring.

> FPN: For further information see ANSI/ISA 12.12.01-2000, *Nonincendive Electrical Equipment for Use in Class I and II, Division 2, and Class III, Divisions 1 and 2 Hazardous (Classified) Locations.*

Oil Immersion. Electrical equipment immersed in a protective liquid in such a way that an explosive atmosphere that may be above the liquid or outside the enclosure cannot be ignited.

> FPN: For further information, see ANSI/UL 698-1995, *Industrial Control Equipment for Use in Hazardous (Classified) Locations.*

Purged and Pressurized. The process of supplying an enclosure with a protective gas at a sufficient flow and positive pressure to reduce the concentration of any flammable gas or vapor initially present to an acceptable level.

> FPN: For further information, see ANSI/ NFPA 496-1998, *Purged and Pressurized Enclosures for Electrical Equipment.*

Unclassified Locations. Locations determined to be neither Class I, Division 1; Class I, Division 2; Class I, Zone 0; Class I, Zone 1; Class I, Zone 2; Class II, Division 1; Class II, Division 2; Class III, Division 1; Class III, Division 2; or any combination thereof.

500.3 Other Articles.

Except as modified in Articles 500 through 504, all other applicable rules contained in this *Code* shall apply to electri-

cal equipment and wiring installed in hazardous (classified) locations.

The first four chapters of the *Code* cover general installation requirements for all electrical equipment and wiring (see 90.3). This means that materials and equipment must be suitable for environmental conditions such as rain, snow, ice, altitude, and heat; deteriorating effects on the conductors and equipment; and interrupting rating sufficient for the nominal circuit voltage and available fault current, just to name a few. The requirements in Articles 500 through 505 amend or modify the general rules to ensure the integrity of the electrical installation and to minimize the possibility of the electrical equipment being an ignition source in the volatile or potentially volatile environments covered in these articles. An example of how these articles modify the general requirements is found in the wiring method requirements of 501.4, 502.4, and 503.3. These sections of the *Code* limit the wiring methods that can be used in Class I, Class II, and Class III locations in order to provide the highest degree of protection against physical damage. The installation of any wiring methods allowed by Articles 500 through 505 must be in accordance with the Chapter 3 article that governs that particular wiring method and with any modifications that occur in Articles 500 through 517.

500.4 General.

(A) Documentation. All areas designated as hazardous (classified) locations shall be properly documented. This documentation shall be available to those authorized to design, install, inspect, maintain, or operate electrical equipment at the location.

One type of documentation consists of area classification drawings. This type of documentation provides the necessary information for installers, service personnel, and authorities having jurisdiction to ensure that any electrical equipment installed or maintained in those classified areas is of the proper type. See the fine print note to 505.4(A).

(B) Reference Standards. Important information relating to topics covered in Chapter 5 may be found in other publications.

The NFPA and ANSI standards referenced in Articles 500 through 517 are essential for proper application of these articles. The following NFPA codes, standards, and recommended practices include information on hazardous (classified) locations and the extent of hazardous (classified) locations in specific occupancies or industries.

NFPA 30, *Flammable and Combustible Liquids Code*

NFPA 30A, *Code for Motor Fuel Dispensing Facilities and Repair Garages*

NFPA 32, *Standard for Drycleaning Plants*

NFPA 33, *Standard for Spray Application Using Flammable or Combustible Materials*

NFPA 34, *Standard for Dipping and Coating Processes Using Flammable or Combustible Liquids*

NFPA 35, *Standard for the Manufacture of Organic Coatings*

NFPA 36, *Standard for Solvent Extraction Plants*

NFPA 45, *Standard on Fire Protection for Laboratories Using Chemicals*

NFPA 50A, *Standard for Gaseous Hydrogen Systems at Consumer Sites*

NFPA 50B, *Standard for Liquefied Hydrogen Systems at Consumer Sites*

NFPA 51, *Standard for the Design and Installation of Oxygen-Fuel Gas Systems for Welding, Cutting, and Allied Processes*

NFPA 51A, *Standard for Acetylene Cylinder Charging Plants*

NFPA 52, *Compressed Natural Gas (CNG) Vehicular Fuel Systems Code*

NFPA 54, *Natural Fuel Gas Code*

NFPA 58, *Liquefied Petroleum Gas Code*

NFPA 59, *Utility LP-Gas Plant Code*

NFPA 59A, *Standard for the Production, Storage, and Handling of Liquefied Natural Gas (LNG)*

NFPA 61, *Standard for the Prevention of Fires and Dust Explosions in Agricultural and Food Products Facilities*

NFPA 85, *Boiler and Combustion Systems Hazards Code*

NFPA 88A, *Standard for Parking Structures*

NFPA 88B, *Standard for Repair Garages*

NFPA 99, *Standard for Health Care Facilities*

NFPA 407, *Standard for Aircraft Fuel Servicing*

NFPA 409, *Standard on Aircraft Hangars*

NFPA 480, *Standard for the Storage, Handling, and Processing of Magnesium Solids and Powders*

NFPA 481, *Standard for the Production, Processing, Handling, and Storage of Titanium*

NFPA 495, *Explosive Materials Code*

NFPA 496, *Standard for Purged and Pressurized Enclosures for Electrical Equipment*

NFPA 497, *Recommended Practice for the Classification of Flammable Liquids, Gases, or Vapors and of Hazardous (Classified) Locations for Electrical Installations in Chemical Process Areas*

NFPA 499, *Recommended Practice for the Classification of Combustible Dusts and of Hazardous (Classified) Locations for Electrical Installations in Chemical Process Areas*

NFPA 651, *Standard for the Machining and Finishing of Aluminum and the Production and Handling of Aluminum Powders*

NFPA 654, *Standard for the Prevention of Fire and Dust Explosions from the Manufacturing, Processing, and Handling of Combustible Particulate Solids*

NFPA 655, *Standard for Prevention of Sulfur Fires and Explosions*

FPN No. 1: It is important that the authority having jurisdiction be familiar with recorded industrial experience as well as with the standards of the National Fire Protection Association (NFPA), the American Petroleum Institute (API), and the Instrumentation, Systems, and Automation Society (ISA) that may be of use in the classification of various locations, the determination of adequate ventilation, and the protection against static electricity and lightning hazards.

FPN No. 2: For further information on the classification of locations, see NFPA 30-2000, *Flammable and Combustible Liquids Code*; NFPA 32-2000, *Standard for Drycleaning Plants*; NFPA 33-2000, *Standard for Spray Application Using Flammable or Combustible Materials*; NFPA 34-2000, *Standard for Dipping and Coating Processes Using Flammable or Combustible Liquids*; NFPA 35-1999, *Standard for the Manufacture of Organic Coatings*; NFPA 36-2001, *Standard for Solvent Extraction Plants*; NFPA 45-2000, *Standard on Fire Protection for Laboratories Using Chemicals*; NFPA 50A-1999, *Standard for Gaseous Hydrogen Systems at Consumer Sites*; NFPA 50B-1999, *Standard for Liquefied Hydrogen Systems at Consumer Sites*; NFPA 58-2001, *Liquefied Petroleum Gas Code*; NFPA 59-2001, *Utility LP-Gas Plant Code*; NFPA 497-1997, *Recommended Practice for the Classification of Flammable Liquids, Gases, or Vapors and of Hazardous (Classified) Locations for Electrical Installations in Chemical Process Areas*; NFPA 499-1997, *Recommended Practice for the Classification of Combustible Dusts and of Hazardous (Classified) Locations for Electrical Installations in Chemical Process Areas*; NFPA 820-1999, *Standard for Fire Protection in Wastewater Treatment and Collection Facilities*; ANSI/API RP500-1997, *Recommended Practice for Classification of Locations of Electrical Installations at Petroleum Facilities Classified as Class I, Division 1 and Division 2*; ISA 12.10-1988, *Area Classification In Hazardous (Classified) Dust Locations*.

FPN No. 3: For further information on protection against static electricity and lightning hazards in hazardous (classified) locations, see NFPA 77-2000, *Recommended Practice on Static Electricity*; NFPA 780-1997, *Standard for the Installation of Lightning Protection Systems*; and API RP 2003-1998, *Protection Against Ignitions Arising Out of Static Lightning and Stray Currents.*

FPN No. 4: For further information on ventilation, see NFPA 30-2000, *Flammable and Combustible Liquids Code*; and API RP 500-1997, *Recommended Practice for Classification of Locations for Electrical Installations at Petroleum Facilities Classified as Class I, Division 1 and Division 2.*

FPN No. 5: For further information on electrical systems for hazardous (classified) locations on offshore oil- and gas-producing platforms, see ANSI/API RP 14F-1999, *Recommended Practice for Design and Installation of Electrical Systems for Fixed and Floating Offshore Petroleum Facilities for Unclassified and Class I, Division 1 and Division 2 Locations.*

500.5 Classifications of Locations.

(A) Classifications of Locations. Locations shall be classified depending on the properties of the flammable vapors, liquids, or gases, or combustible dusts or fibers that may be present, and the likelihood that a flammable or combustible concentration or quantity is present. Where pyrophoric materials are the only materials used or handled, these locations shall not be classified. Each room, section, or area shall be considered individually in determining its classification.

Pyrophoric materials ignite spontaneously upon contact with air. The use of electrical equipment that is suitable for a hazardous (classified) location will not prevent ignition of pyrophoric materials. The process containment system should be designed to prevent contact between pyrophoric material and air.

FPN: Through the exercise of ingenuity in the layout of electrical installations for hazardous (classified) locations, it is frequently possible to locate much of the equipment in a reduced level of classification or in an unclassified location and, thus, to reduce the amount of special equipment required.

Rooms and areas containing ammonia refrigeration systems that are equipped with adequate mechanical ventilation may be classified as "unclassified" locations.

FPN: For further information regarding classification and ventilation of areas involving ammonia, see ANSI/ASHRAE 15-1994, *Safety Code for Mechanical Refrigeration*, and ANSI/CGA G2.1-1989, *Safety Requirements for the Storage and Handling of Anhydrous Ammonia.*

(B) Class I Locations. Class I locations are those in which flammable gases or vapors are or may be present in the air in quantities sufficient to produce explosive or ignitible mixtures. Class I locations shall include those specified in 500.5(B)(1) and (B)(2).

(1) Class I, Division 1. A Class I, Division 1 location is a location

(1) In which ignitible concentrations of flammable gases or vapors can exist under normal operating conditions, or

(2) In which ignitible concentrations of such gases or vapors may exist frequently because of repair or maintenance operations or because of leakage, or

(3) In which breakdown or faulty operation of equipment or processes might release ignitible concentrations of flammable gases or vapors and might also cause simultaneous failure of electrical equipment in such a way as to directly cause the electrical equipment to become a source of ignition.

FPN No. 1: This classification usually includes the following locations:

(1) Where volatile flammable liquids or liquefied flammable gases are transferred from one container to another

(2) Interiors of spray booths and areas in the vicinity of spraying and painting operations where volatile flammable solvents are used

(3) Locations containing open tanks or vats of volatile flammable liquids

(4) Drying rooms or compartments for the evaporation of flammable solvents

(5) Locations containing fat- and oil-extraction equipment using volatile flammable solvents

(6) Portions of cleaning and dyeing plants where flammable liquids are used

(7) Gas generator rooms and other portions of gas manufacturing plants where flammable gas may escape

(8) Inadequately ventilated pump rooms for flammable gas or for volatile flammable liquids

(9) The interiors of refrigerators and freezers in which volatile flammable materials are stored in open, lightly stoppered, or easily ruptured containers

(10) All other locations where ignitible concentrations of flammable vapors or gases are likely to occur in the course of normal operations

FPN No. 2: In some Division 1 locations, ignitible concentrations of flammable gases or vapors may be present continuously or for long periods of time. Examples include the following:

(1) The inside of inadequately vented enclosures containing instruments normally venting flammable gases or vapors to the interior of the enclosure

(2) The inside of vented tanks containing volatile flammable liquids

(3) The area between the inner and outer roof sections of a floating roof tank containing volatile flammable fluids

(4) Inadequately ventilated areas within spraying or coating operations using volatile flammable fluids

(5) The interior of an exhaust duct that is used to vent ignitible concentrations of gases or vapors

Experience has demonstrated the prudence of avoiding the installation of instrumentation or other electric equipment in these particular areas altogether or where it cannot be avoided because it is essential to the process and other locations are not feasible [see 500.5(A), FPN] using electric equipment or instrumentation approved for the specific application or consisting of intrinsically safe systems as described in Article 504.

Fine print note No. 2 describes locations that are defined in 500.5(B)(1) as Class I, Division 1 locations. These locations are Class I, Zone 0 locations in accordance with 505.5(B)(1). Where classified as a Zone 0 location, only intrinsically safe equipment and wiring suitable for Zone 0 locations is permitted to be used.

(2) Class I, Division 2. A Class I, Division 2 location is a location

(1) In which volatile flammable liquids or flammable gases are handled, processed, or used, but in which the liquids, vapors, or gases will normally be confined within closed containers or closed systems from which they can escape only in case of accidental rupture or breakdown of such containers or systems or in case of abnormal operation of equipment, or

(2) In which ignitible concentrations of gases or vapors are normally prevented by positive mechanical ventilation, and which might become hazardous through failure or abnormal operation of the ventilating equipment, or

(3) That is adjacent to a Class I, Division 1 location, and to which ignitible concentrations of gases or vapors might occasionally be communicated unless such communication is prevented by adequate positive-pressure ventilation from a source of clean air and effective safeguards against ventilation failure are provided.

FPN No. 1: This classification usually includes locations where volatile flammable liquids or flammable gases or vapors are used but that, in the judgment of the authority having jurisdiction, would become hazardous only in case of an accident or of some unusual operating condition. The quantity of flammable material that might escape in case of accident, the adequacy of ventilating equipment, the total area involved, and the record of the industry or business with respect to explosions or fires are all factors that merit consideration in determining the classification and extent of each location.

FPN No. 2: Piping without valves, checks, meters, and similar devices would not ordinarily introduce a hazardous condition even though used for flammable liquids or gases. Depending on factors such as the quantity and size of the containers and ventilation, locations used for the storage of flammable liquids or liquefied or compressed gases in sealed containers may be considered either hazardous (classified) or unclassified locations. See NFPA 30-2000, *Flammable and Combustible Liquids Code*, and NFPA 58-2001, *Liquefied Petroleum Gas Code*.

(C) Class II Locations. Class II locations are those that are hazardous because of the presence of combustible dust. Class II locations shall include those specified in 500.5(C)(1) and (C)(2).

(1) Class II, Division 1. A Class II, Division 1 location is a location

(1) In which combustible dust is in the air under normal operating conditions in quantities sufficient to produce explosive or ignitible mixtures, or

(2) Where mechanical failure or abnormal operation of machinery or equipment might cause such explosive or ignitible mixtures to be produced, and might also provide a source of ignition through simultaneous failure of electric equipment, through operation of protection devices, or from other causes, or

(3) In which combustible dusts of an electrically conductive nature may be present in hazardous quantities.

FPN: Combustible dusts that are electrically nonconductive include dusts produced in the handling and processing of grain and grain products, pulverized sugar and cocoa, dried egg and milk powders, pulverized spices, starch and pastes, potato and wood-flour, oil meal from beans and seed, dried hay, and other organic materials that may produce combustible dusts when processed or handled. Only Group E dusts are considered to be electrically conductive for classification purposes. Dusts containing magnesium or aluminum are particularly hazardous, and the use of extreme precaution is necessary to avoid ignition and explosion.

(2) Class II, Division 2. A Class II, Division 2 location is a location

(1) Where combustible dust is not normally in the air in quantities sufficient to produce explosive or ignitible mixtures, and dust accumulations are normally insufficient to interfere with the normal operation of electrical equipment or other apparatus, but combustible dust may be in suspension in the air as a result of infrequent malfunctioning of handling or processing equipment and

(2) Where combustible dust accumulations on, in, or in the vicinity of the electrical equipment may be sufficient to interfere with the safe dissipation of heat from electrical equipment or may be ignitible by abnormal operation or failure of electrical equipment.

FPN No. 1: The quantity of combustible dust that may be present and the adequacy of dust removal systems are factors that merit consideration in determining the classification and may result in an unclassified area.

FPN No. 2: Where products such as seed are handled in a manner that produces low quantities of dust, the amount of dust deposited may not warrant classification.

(D) Class III Locations. Class III locations are those that are hazardous because of the presence of easily ignitible fibers or flyings, but in which such fibers or flyings are not likely to be in suspension in the air in quantities sufficient to produce ignitible mixtures. Class III locations shall include those specified in 500.5(D)(1) and (D)(2).

(1) Class III, Division 1. A Class III, Division 1 location is a location in which easily ignitible fibers or materials producing combustible flyings are handled, manufactured, or used.

> FPN No. 1: Such locations usually include some parts of rayon, cotton, and other textile mills; combustible fiber manufacturing and processing plants; cotton gins and cotton-seed mills; flax-processing plants; clothing manufacturing plants; woodworking plants; and establishments and industries involving similar hazardous processes or conditions.

> FPN No. 2: Easily ignitible fibers and flyings include rayon, cotton (including cotton linters and cotton waste), sisal or henequen, istle, jute, hemp, tow, cocoa fiber, oakum, baled waste kapok, Spanish moss, excelsior, and other materials of similar nature.

(2) Class III, Division 2. A Class III, Division 2 location is a location in which easily ignitible fibers are stored or handled other than in the process of manufacture.

Sections 500.5(B), (C), and (D) describe three classes of hazardous (classified) locations based on the type of material involved. Within each class there are varying degrees of hazard, so each class is subdivided into two divisions. The classification by division is based on the likelihood the material will be present. The requirements for Division 1 of each class are more stringent than those for Division 2.

The materials in the three classes are defined as follows: Class I, flammable gases or vapors; Class II, combustible dust; and Class III, combustible fibers or flyings.

When a given location is classified as hazardous, it should be easy to determine which of the three classes it belongs to (it may belong in more than one class). Common sense and good judgment must prevail in classifying an area that is likely to become hazardous and in determining those portions of the premises to be classified Division 1 or Division 2. However, if different types of material exist in a process, such as flammable liquids, gases, or vapors, and combustible dust, the area must be classified as both a Class I and Class II location. If a location is classified due to two different hazards, such as a dust hazard and a flammable liquids hazard, protection must be provided for both hazards. Equipment that is approved for a Class I location may not be suitable for a Class II location and vice versa.

500.6 Material Groups.

For purposes of testing, approval, and area classification, various air mixtures (not oxygen-enriched) shall be grouped in accordance with 500.6(A) and 500.6(B).

Oxygen enrichment can drastically change the explosion characteristics of materials. It lowers the minimum ignition energies, increases explosion pressures, and can reduce the maximum experimental safe gap, rendering both intrinsically safe and explosionproof equipment unsafe unless the equipment has been tested for the conditions involved.

Exception: Equipment identified for a specific gas, vapor, or dust.

> FPN: This grouping is based on the characteristics of the materials. Facilities are available for testing and identifying equipment for use in the various atmospheric groups.

Determining the proper group classification for flammable gases and vapors involves evaluating explosion pressures and maximum safe clearances between parts of a clamped joint under several conditions, and comparison of the values obtained with those obtained for presently classified materials under the same test conditions. Although some work has been done on the classification of flammable materials on the basis of chemical structure, this method is not sufficiently refined or accurate enough to ensure proper classification of all flammable materials.

For additional information on the rationale for classification, reference may be made to the following:

1. *Rationale for Classification of Combustible Gases, Vapors, and Dusts with Reference to the National Electrical Code*, Publication NMAB 353-6, 1982, a report of the Committee on Evaluation of Industrial Hazards, The National Materials Advisory Board, Commission on Engineering and Technical Systems, National Research Council. This publication is available from the National Technical Information Service (NTIS), Springfield, VA 22151.

2. *An Investigation of Fifteen Flammable Gases or Vapors with Respect to Explosion-Proof Electrical Equipment*, Bulletin of Research No. 58 by Underwriters Laboratories Inc., August 1969 (also Bulletin of Research Nos. 58A and 58B, which supplement No. 58). These publications are available from Underwriters Laboratories Inc., Publications Stock, Northbrook, IL 60062.

3. *Electrical Installations in Hazardous Locations*, Chapter 2, Peter J. Schram and Mark W. Earley. This book is available from NFPA.

(A) Class I Group Classifications. Class I groups shall be according to 500.6(A)(1) through (A)(4).

> FPN No. 1: FPN Nos. 2 and 3 apply to 500.6(A).
> FPN No. 2: The explosion characteristics of air mixtures of gases or vapors vary with the specific material in-

volved. For Class I locations, Groups A, B, C, and D, the classification involves determinations of maximum explosion pressure and maximum safe clearance between parts of a clamped joint in an enclosure. It is necessary, therefore, that equipment be identified not only for class but also for the specific group of the gas or vapor that will be present.

FPN No. 3: Certain chemical atmospheres may have characteristics that require safeguards beyond those required for any of the Class I groups. Carbon disulfide is one of these chemicals because of its low ignition temperature [100°C (212°F)] and the small joint clearance permitted to arrest its flame.

(1) Group A. Acetylene. [NFPA 497, 1-3]

(2) Group B. Flammable gas, flammable liquid–produced vapor, or combustible liquid–produced vapor mixed with air that may burn or explode, having either a maximum experimental safe gap (MESG) value less than or equal to 0.45 mm or a minimum igniting current ratio (MIC ratio) less than or equal to 0.40. [NFPA 497, 1-3]

FPN: A typical Class I, Group B material is hydrogen.

Exception No. 1: Group D equipment shall be permitted to be used for atmospheres containing butadiene, provided all conduit runs into explosionproof equipment are provided with explosionproof seals installed within 450 mm (18 in.) of the enclosure.

Exception No. 2: Group C equipment shall be permitted to be used for atmospheres containing allyl glycidyl ether, n-butyl glycidyl ether, ethylene oxide, propylene oxide, and acrolein, provided all conduit runs into explosionproof equipment are provided with explosionproof seals installed within 450 mm (18 in.) of the enclosure.

The specific materials identified in Exception Nos. 1 and 2 produce high pressures because of pressure piling in unsealed conduits. If all conduits are sealed, however, this problem is eliminated.

(3) Group C. Flammable gas, flammable liquid–produced vapor, or combustible liquid–produced vapor mixed with air that may burn or explode, having either a maximum experimental safe gap (MESG) value greater than 0.45 mm and less than or equal to 0.75 mm, or a minimum igniting current ratio (MIC ratio) greater than 0.40 and less than or equal to 0.80. [NFPA 497, 1-3]

FPN: A typical Class I, Group C material is ethylene.

(4) Group D. Flammable gas, flammable liquid–produced vapor, or combustible liquid–produced vapor mixed with air that may burn or explode, having either a maximum experimental safe gap (MESG) value greater than 0.75 mm

or a minimum igniting current ratio (MIC ratio) greater than 0.80. [NFPA 497, 1-3]

FPN No. 1: A typical Class I, Group D material is propane.

FPN No. 2: For classification of areas involving ammonia atmospheres, see ANSI/ASHRAE 15-1994, *Safety Code for Mechanical Refrigeration*, and ANSI/CGA G2.1-1989, *Safety Requirements for the Storage and Handling of Anhydrous Ammonia*.

Class I flammable gases or vapors are separated into four different atmospheric groups—A, B, C, and D. The *Code* requirements for Class I locations do not vary for different kinds of gas or vapor contained in the atmosphere, except in those cases where seals may be used in all conduits to change the group classification. See Exception Nos. 1 and 2 to 500.6(A)(2) for materials such as butadiene and ethylene oxide. It is necessary to select equipment designed for use in the particular group involved. The reason for designating the groups this way is that explosive mixtures have different igniting current ratios and maximum safe clearances between parts of a joint in an enclosure.

Underwriters Laboratories Inc. and Factory Mutual Research Corp. list or approve electrical equipment suitable for use in all groups of Class I locations. Further information is available from the UL *Hazardous Locations Equipment Directory* and the FM *Approval Guide*. It should be noted (from the UL Directory) that "only those products bearing the appropriate listing mark and the company's name, trade name, trademark, or other recognized identification should be considered as covered by UL's Listing and Follow-Up Service." Other testing laboratories may also list equipment for hazardous locations for use in the United States. The acceptance of the listing agency is the responsibility of the authority having jurisdiction.

Several testing laboratories outside the United States also provide listing, approval, or certification of equipment for use in hazardous locations. However, they may not be testing and investigating the equipment for use in hazardous locations, as defined in Article 500. Some international laboratories certify equipment for installation where the classification is according to the IEC classification scheme covered in Article 505.

Table 5.1 is an alphabetical listing of selected combustible materials with their group classification and relevant physical properties shown. All of the materials included in this table have been evaluated for the purpose of designating the appropriate gas group. This information is used to properly select electrical equipment for use in Class I locations. The combustible materials in Table 5.1 are classified into four Class I, division groups—A, B, C, and D—or three Class I, zone groups—IIC, IIB, and IIA—depending on their properties. Table 5.1 is extracted from NFPA 497,

Table 5.1 Selected Chemicals

Chemical	CAS No.	NEC Group	Type[6]	Flash Point (°C)	AIT (°C)	%LFL	%UFL	Vapor Density (Air=1)	Vapor Pressure[7] (mm Hg)	Class I Zone Group[8]	MIE (mJ) TR	MIC Ratio	MESG (mm)
Acetaldehyde	75-07-0	C*	I	−38	175	4.0	60.0	1.5	874.9	IIA	0.37	0.98	0.92
Acetic Acid	64-19-7	D*	II	43	464	4.0	19.9	2.1	15.6	IIA		2.67	1.76
Acetic Acid-Tert.-Butyl Ester	540-88-5	D	II			1.7	9.8	4.0	40.6				
Acetic Anhydride	108-24-7	D	II	54	316	2.7	10.3	3.5	4.9				
Acetone	67-64-1	D*	I		465	2.5	12.8	2.0	230.7	IIA	1.15	1.00	1.02
Acetone Cyanohydrin	75-86-5	D	IIIA	74	688	2.2	12.0	2.9	0.3				
Acetonitrile	75-05-8	D	I	6	524	3.0	16.0	1.4	91.1	IIA			1.50
Acetylene	74-86-2	A*	GAS		305	2.5	99.9	0.9	36600	IIC	0.017	0.28	0.25
Acrolein (Inhibited)	107-02-8	B(C)*	I		235	2.8	31.0	1.9	274.1	IIB	0.13		
Acrylic Acid	79-10-7	D	II	54	438	2.4	8.0	2.5	4.3				
Acrylonitrile	107-13-1	D*	I	−26	481	3.0	17.0	1.8	108.5	IIB	0.16	0.78	0.87
Adiponitrile	111-69-3	D	IIIA	93	550			1.0	0.002				
Allyl Alcohol	107-18-6	C*	I	22	378	2.5	18.0	2.0	25.4				0.84
Allyl Chloride	107-05-1	D	I	-32	485	2.9	11.1	2.6	366			1.33	1.17
Allyl Glycidyl Ether	106-92-3	B(C)[1]	II		57			3.9					
Alpha-Methyl Styrene	98-83-9	D	II		574	0.8	11.0	4.1	2.7				
n-Amyl Acetate	628-63-7	D	I	25	360	1.1	7.5	4.5	4.2				1.02
sec-Amyl Acetate	626-38-0	D	I	23		1.1	7.5	4.5		IIA			
Ammonia	7664-41-7	D*[2]	I		498	15.0	28.0	0.6	7498.0	IIA	680	6.85	3.17
Aniline	62-53-3	D	IIIA	70	615	1.3	11.0	3.2	0.7	IIA			
Benzene	71-43-2	D*	I	−11	498	1.2	7.8	2.8	94.8	IIA	0.20	1.00	0.99
Benzyl Chloride	98-87-3	D	IIIA		585	1.1		4.4	0.5				
Bromopropyne	106-96-7	D	I	10	324	3.0							
n-Butane	3583-47-9	D*5	GAS		288	1.9	8.5	2.0			0.25	0.94	1.07
1,3-Butadiene	106-99-0	B(D)*[1]	GAS	−76	420	2	12	1.9		IIB	0.13	0.76	0.79
1-Butanol	71-36-3	D*	I	36	343	1.4	11.2	2.6	7.0	IIA			0.91
2-Butanol	71-36-5	D*	I	36	405	1.7	9.8	2.6		IIA			
Butylamine	109-73-9	D	GAS	−12	312	1.7	9.8	2.5	92.9			1.13	
Butylene	25167-67-3	D	I		385	1.6	10.0	1.9	2214.6				
n-Butyraldehyde	123-72-8	C*	I	−12	218	1.9	12.5	2.5	112.2				0.92
n-Butyl Acetate	123-86-4	D*	I	22	421	1.7	7.6	4.0	11.5	IIA		1.08	1.04
sec-Butyl Acetate	105-46-4	D	II	−8		1.7	9.8	4.0	22.2				
tert.-Butyl Acetate	540-88-5	D	II			1.7	9.8	4.0	40.6				
n-Butyl Acrylate (Inhibited)	141-32-2	D	II	49	293	1.7	9.9	4.4	5.5				
n-Butyl Glycidyl Ether	2426-08-6	B(C)[1]	II										
n-Butyl Formal	110-62-3	C	IIIA						34.3				
Butyl Mercaptan	109-79-5	C	I	2				3.1	46.4				
Butyl-2-Propenoate	141-32-2	D	II	49		1.7	9.9	4.4	5.5				
para tert.-Butyl Toluene	98-51-1	D	IIIA										
n-Butyric Acid	107-92-6	D	IIIA	72	443	2.0	10.0	3.0	0.8				
Carbon Disulfide	75-15-0	*3	I	−30	90	1.3	50.0	2.6	358.8	IIC	0.009	0.39	0.20
Carbon Monoxide	630-08-0	C*	GAS	609	700	12.5	74.0	0.97		IIA			
Chloroacetaldehyde	107-20-0	C	IIIA	88					63.1				
Chlorobenzene	108-90-7	D	I	29	593	1.3	9.6	3.9	11.9				

Table 5.1 *Continued*

Chemical	CAS No.	NEC Group	Type[6]	Flash Point (°C)	AIT (°C)	%LFL	%UFL	Vapor Density (Air=1)	Vapor Pressure[7] (mm Hg)	Class I Zone Group[3]	MIE (mJ) TR	MIC Ratio	MESG (mm)
1-Chloro-1-Nitropropane	2425-66-3	C	IIIA										
Chloroprene	126-99-8	D	GAS	−20		4.0	20.0	3.0					
Cresol	1319-77-3	D	IIIA	81	559	1.1		3.7					
Crotonaldehyde	4170-30-3	C*	I	13	232	2.1	15.5	2.4	33.1	IIB			0.81
Cumene	98-82-8	D	I	36	424	0.9	6.5	4.1	4.6	IIA			
Cyclohexane	110-82-7	D	I	−17	245	1.3	8.0	2.9	98.8	IIA	0.22	1.0	0.94
Cyclohexanol	108-93-0	D	IIIA	68	300			3.5	0.7	IIA			
Cyclohexanone	108-94-1	D	II	44	245	1.1	9.4	3.4	4.3	IIA			0.98
Cyclohexene	110-83-8	D	I	−6	244	1.2		2.8	89.4			0.97	
Cyclopropane	75-19-4	D*	I		503	2.4	10.4	1.5	5430	IIB	0.17	0.84	0.91
p-Cymene	99-87-6	D	II	47	436	0.7	5.6	4.6	1.5	IIA			
Decene	872-05-9	D	II		235			4.8	1.7				
n-Decaldehyde	112-31-2	C	IIIA						0.09				
n-Decanol	112-30-1	D	IIIA	82	288			5.3	0.008				
Decyl Alcohol	112-30-1	D	IIIA	82	288			5.3	0.008				
Diacetone Alcohol	123-42-2	D	IIIA	64	603	1.8	6.9	4.0	1.4				
Di-Isobutylene	25167-70-8	D*	I	2	391	0.8	4.8	3.8			0.96		
Di-Isobutyl Ketone	108-83-8	D	II	60	396	0.8	7.1	4.9	1.7				
o-Dichlorobenzene	955-50-1	D	IIIA	66	647	2.2	9.2	5.1		IIA			
1,4-Dichloro-2,3 Epoxybutane	3583-47-9	D*	I			1.9	8.5	2.0			0.25	0.98	1.07
1,1-Dichloroethane	1300-21-6	D	I		438	6.2	16.0	3.4	227				1.82
1,2-Dichloroethylene	156-59-2	D	I	97	460	5.6	12.8	3.4	204	IIA			
1,1-Dichloro-1-Nitroethane	594-72-9	C	IIIA	76				5.0					
1,3-Dichloropropene	10061-02-6	D	I	35		5.3	14.5	3.8					
Dicyclopentadiene	77-73-6	C	I	32	503				2.8				0.91
Diethylamine	109-87-9	C*	I	−28	312	1.8	10.1	2.5		IIA			1.15
Diethylaminoethanol	100-37-8	C	IIIA	60	320			4.0	1.6	IIA			
Diethyl Benzene	25340-17-4	D	II	57	395			4.6					
Diethyl Ether	60-29-7	C*	I	12	160	1.9	36.0	2.6	38.2	IIB	0.19	0.88	0.83
Diethylene Glycol Monobutyl Ether	112-34-5	C	IIIA	78	228	0.9	24.6	5.6	0.02				
Diethylene Glycol Monomethyl Ether	111-77-3	C	IIIA	93	241				0.2				
n-n-Dimethyl Aniline	121-69-7	C	IIIA	63	371	1.0		4.2	0.7				
Dimethyl Formamide	68-12-2	D	II	58	455	2.2	15.2	2.5	4.1				1.08
Dimethyl Sulfate	77-78-1	D	IIIA	83	188			4.4	0.7				
Dimethylamine	124-40-3	C	GAS		400	2.8	14.4	1.6		IIA			
2,2-Dimethylbutane	75-83-2	D[5]	I	−48	405				319.3				
2,3-Dimethylbutane		D[5]	I		396								
3,3-Dimethylheptane	1071-26-7	D[5]	I		325				10.8				
2,3-Dimethylhexane	31394-54-4	D[5]	I		438								
2,3-Dimethylpentane	107-83-5	D[5]	I		335				211.7				
Di-N-Propylamine	142-84-7	C	I	17	299				27.1				
1,4-Dioxane	123-91-1	C*	I	12	180	2.0	22.0	3.0	38.2	IIB	0.19		0.70
Dipentene	138-86-3	D	II	45	237	0.7	6.1	4.7					1.18

Table 5.1 *Continued*

Chemical	CAS No.	NEC Group	Type[6]	Flash Point (°C)	AIT (°C)	%LFL	%UFL	Vapor Density (Air = 1)	Vapor Pressure[7] (mm Hg)	Class I Zone Group[3]	MIE (mJ) TR	MIC Ratio	MESG (mm)
Dipropylene Glycol Methyl Ether	34590-94-8	C	IIIA	85		1.1	3.0	5.1	0.5				
Diisopropylamine	108-18-9	C	GAS	−6	316	1.1	7.1	3.5					
Dodecene	6842-15-5	D	IIIA	100	255								
Epichlorohydrin	3132-64-7	C*	I	33	411	3.8	21.0	3.2	13.0				
Ethane	74-84-0	D*	GAS	−29	472	3.0	12.5	1.0		IIA	0.24	0.82	0.91
Ethanol	64-17-5	D*	I	13	363	3.3	19.0	1.6	59.5	IIA		0.88	0.89
Ethylamine	75-04-7	D*	I	−18	385	3.5	14.0	1.6	1048		2.4		
Ethylene	74-85-1	C*	GAS	0	450	2.7	36.0	1.0		IIB	0.070	0.53	0.65
Ethylenediamine	107-15-3	D*	I	33	385	2.5	12.0	2.1	12.5				
Ethylenimine	151-56-4	C*	I	−11	320	3.3	54.8	1.5	211		0.48		
Ethylene Chlorohydrin	107-07-3	D	IIIA	59	425	4.9	15.9	2.8	7.2				
Ethylene Dichloride	107-06-2	D*	I	13	413	6.2	16.0	3.4	79.7				
Ethylene Glycol Monoethyl Ether Acetate	111-15-9	C	II	47	379	1.7		4.7	2.3			0.53	0.97
Ethylene Glycol Monobutyl Ether Acetate	112-07-2	C	IIIA		340	0.9	8.5		0.9				
Ethylene Glycol Monobutyl Ether	111-76-2	C	IIIA		238	1.1	12.7	4.1	1.0				
Ethylene Glycol Monoethyl Ether	110-80-5	C	II		235	1.7	15.6	3.0	5.4				0.84
Ethylene Glycol Monomethyl Ether	109-86-4	D	II		285	1.8	14.0	2.6	9.2				0.85
Ethylene Oxide	75-21-8	B(C)*1	I	−20	429	3.0	99.9	1.5	1314	IIB	0.065	0.47	0.59
2-Ethylhexaldehyde	123-05-7	C	II	52	191	0.8	7.2	4.4	1.9				
2-Ethylhexanol	104-76-7	D	IIIA	81		0.9	9.7	4.5	0.2				
2-Ethylhexyl Acrylate	103-09-3	D	IIIA	88	252				0.3				
Ethyl Acetate	141-78-6	D*	I	−4	427	2.0	11.5	3.0	93.2		0.46		0.99
Ethyl Acrylate (Inhibited)	140-88-5	D*	I	9	372	1.4	14.0	3.5	37.5	IIA			0.86
Ethyl Alcohol	64-17-5	D*	I	13	363	3.3	19.0	1.6	59.5				0.89
Ethyl Sec − Amyl Ketone	541-85-5	D	II	59									
Ethyl Benzene	100-41-4	D	I	21	432	0.8	6.7	3.7	9.6				
Ethyl Butanol	97-95-0	D	II	57		1.2	7.7	3.5	1.5				
Ethyl Butyl Ketone	106-35-4	D	II	46				4.0	3.6				
Ethyl Chloride	75-00-3	D	GAS	−50	519	3.8	15.4	2.2					
Ethyl Ether	60-29-7	C*	I	−45	160	1.9	36.0	2.6	538		0.19	0.88	0.84
Ethyl Formate	109-94-4	D	GAS	−20	455	2.8	16.0	2.6		IIA			0.94
Ethyl Mercaptan	75-08-1	C*	I	−18	300	2.8	18.0	2.1	527.4			0.90	0.90
n-Ethyl Morpholine	100-74-3	C	I	32				4.0					
2-Ethyl-3-Propyl Acrolein	645-62-5	C	IIIA	68				4.4					
Ethyl Silicate	78-10-4	D	II					7.2					
Formaldehyde (Gas)	50-00-0	B	GAS	60	429	7.0	73.0	1.0					0.57
Formic Acid	64-18-6	D	II	50	434	18.0	57.0	1.6	42.7				1.86

Table 5.1 *Continued*

Chemical	CAS No.	NEC Group	Type[6]	Flash Point (°C)	AIT (°C)	%LFL	%UFL	Vapor Density (Air = 1)	Vapor Pressure[7] (mm Hg)	Class I Zone Group[3]	MIE (mJ) TR	MIC Ratio	MESG (mm)
Fuel Oil 1	8008-20-6	D	II	72	210	0.7	5.0						0.94
Furfural	98-01-1	C	IIIA	60	316	2.1	19.3	3.3	2.3				
Furfuryl Alcohol	98-00-0	C	IIIA	75	490	1.8	16.3	3.4	0.6				
Gasoline	8006-61-9	D*	I	-46	280	1.4	7.6	3.0					
n-Heptane	142-82-5	D*	I	-4	204	1.0	6.7	3.5	45.5	IIA	0.24	0.88	0.91
n-Heptene	81624-04-6	D[5]	I	-1	204			3.4					0.97
n-Hexane	110-54-3	D*[5]	I	-23	225	1.1	7.5	3.0	152	IIA	0.24	0.88	0.93
Hexanol	111-27-3	D	IIIA	63				3.5	0.8	IIA			0.98
2-Hexanone	591-78-6	D	I	35	424	1.2	8.0	3.5	10.6				
Hexene	592-41-6	D	I	-26	245	1.2	6.9		186				
sec-Hexyl Acetate	108-84-9	D	II	45				5.0					
Hydrazine	302-01-2	C	II	38	23		98.0	1.1	14.4				
Hydrogen	1333-74-0	B*	GAS		520	4.0	75.0	0.1		IIC	0.019	0.25	0.28
Hydrogen Cyanide	74-90-8	C*	GAS	-18	538	5.6	40.0	0.9		IIB			0.80
Hydrogen Selenide	7783-07-5	C	I						7793				
Hydrogen Sulfide	7783-06-4	C*	GAS		260	4.0	44.0	1.2			0.068		0.90
Isoamyl Acetate	123-92-2	D	I	25	360	1.0	7.5	4.5	6.1				
Isoamyl Alcohol	123-51-3	D	II	43	350	1.2	9.0	3.0	3.2				1.02
Isobutane	75-28-5	D[5]	GAS		460	1.8	8.4	2.0					
Isobutyl Acetate	110-19-0	D*	I	18	421	2.4	10.5	4.0	17.8				
Isobutyl Acrylate	106-63-8	D	I		427			4.4	7.1				
Isobutyl Alcohol	78-83-1	D*	I	-40	416	1.2	10.9	2.5	10.5			0.92	0.98
Isobutyraldehyde	78-84-2	C	GAS	-40	196	1.6	10.6	2.5					
Isodecaldehyde	112-31-2	C	IIIA					5.4	0.09				
Isohexane	107-83-5	D[5]			264				211.7			1.00	
Isopentane	78-78-4	D[5]			420				688.6				
Isooctyl Aldehyde	123-05-7	C	II		197				1.9				
Isophorone	78-59-1	D		84	460	0.8	3.8	4.8	0.4				
Isoprene	78-79-5	D*	I	-54	220	1.5	8.9	2.4	550.6				
Isopropyl Acetate	108-21-4	D	I		460	1.8	8.0	3.5	60.4				
Isopropyl Ether	108-20-3	D*	I	-28	443	1.4	7.9	3.5	148.7			1.14	0.94
Isopropyl Glycidyl Ether	4016-14-2	C	I										
Isopropylamine	75-31-0	D	GAS	-26	402	2.3	10.4	2.0			2.0		
Kerosene	8008-20-6	D	II	72	210	0.7	5.0			IIA			
Liquefied Petroleum Gas	68476-85-7	D	I		405								
Mesityl Oxide	141-97-9	D*	I	31	344	1.4	7.2	3.4	47.6				
Methane	74-82-8	D*	GAS	-223	630	5.0	15.0	0.6		IIA	0.28	1.00	1.12
Methanol	67-56-1	D*	I	12	385	6.0	36.0	1.1	126.3	IIA	0.14	0.82	0.92
Methyl Acetate	79-20-9	D	GAS	-10	454	3.1	16.0	2.6		IIB		1.08	0.99
Methyl Acrylate	96-33-3	D	GAS	-3	468	2.8	25.0	3.0				0.98	0.85
Methyl Alcohol	67-56-1	D*	I		385	6.0	36.0	1.1	126.3				0.91
Methyl Amyl Alcohol	108-11-2	D	II	41		1.0	5.5	3.5	5.3				1.01
Methyl Chloride	74-87-3	D	GAS	-46	632	8.1	17.4	1.7					1.00
Methyl Ether	115-10-6	C*	GAS	-41	350	3.4	27.0	1.6				0.85	0.84
Methyl Ethyl Ketone	78-93-3	D*	I	-6	404	1.4	11.4	2.5	92.4		0.53	0.92	0.84
Methyl Formal	534-15-6	C*	I	1	238			3.1					

Table 5.1 *Continued*

Chemical	CAS No.	*NEC* Group	Type[6]	Flash Point (°C)	AIT (°C)	%LFL	%UFL	Vapor Density (Air = 1)	Vapor Pressure[7] (mm Hg)	Class I Zone Group[3]	MIE (mJ)	MIC Ratio	MESG (mm)
Methyl Formate	107-31-3	D	GAS	-19	449	4.5	23.0	2.1					0.94
2-Methylhexane	31394-54-4	D[5]	I		280								
Methyl Isobutyl Ketone	141-79-7	D*	I	31	440	1.2	8.0	3.5	11				
Methyl Isocyanate	624-83-9	D	GAS	-15	534	5.3	26.0	2.0					
Methyl Mercaptan	74-93-1	C	GAS	-18		3.9	21.8	1.7					
Methyl Methacrylate	80-62-6	D	I	10	422	1.7	8.2	3.6	37.2	IIA			0.95
Methyl N-Amyl Ketone	110-43-0	D	II	49	393	1.1	7.9	3.9	3.8				
Methyl Tertiary Butyl Ether	1634-04-4	D	I	-80	435	1.6	8.4	0.2	250.1				
2-Methyloctane	3221-61-2				220				6.3				
2-Methylpropane	75-28-5	D[5]	I		460				2639				
Methyl-1-Propanol	78-83-1	D*	I	-40	416	1.2	10.9	2.5	10.1				0.98
Methyl-2-Propanol	75-65-0	D*	I	10	360	2.4	8.0	2.6	42.2				
2-Methyl-5-Ethyl Pyridine	104-90-5	D		74		1.1	6.6	4.2					
Methylacetylene	74-99-7	C*	I			1.7		1.4	4306		0.11		
Methylacetylene-Propadiene	27846-30-6	C	I										0.74
Methylal	109-87-5	C	I	-18	237	1.6	17.6	2.6	398				
Methylamine	74-89-5	D	GAS		430	4.9	20.7	1.0		IIA			1.10
2-Methylbutane	78-78-4	D[5]		-56	420	1.4	8.3	2.6	688.6				
Methylcyclohexane	208-87-2	D	I	-4	250	1.2	6.7	3.4			0.27		
Methylcyclohexanol	25630-42-3	D		68	296			3.9					
2-Methylcyclohexanone	583-60-8	D	II					3.9					.
2-Methylheptane		D[5]			420								
3-Methylhexane	589-34-4	D[5]			280				61.5				
3-Methylpentane	94-14-0	D[5]			278								
2-Methylpropane	75-28-5	D[5]	I		460				2639				
2-Methyl-1-Propanol	78-83-1	D*	I	-40	223	1.2	10.9	2.5	10.5				
2-Methyl-2-Propanol	75-65-0	D*	I		478	2.4	8.0	2.6	42.2				
2-Methyloctane	2216-32-2	D[5]			220								
3-Methyloctane	2216-33-3	D[5]			220				6.3				
4-Methyloctane	2216-34-4	D[5]			225				6.8				
Monoethanolamine	141-43-5	D		85	410			2.1	0.4	IIA			
Monoiso-propanolamine	78-96-6	D		77	374			2.6	1.1				
Monomethyl Aniline	100-61-8	C			482				0.5				
Monomethyl Hydrazine	60-34-4	C	I	23	194	2.5	92.0	1.6					
Morpholine	110-91-8	C*	II	35	310	1.4	11.2	3.0	10.1				0.95
Naphtha (Coal Tar)	8030-30-6	D	II	42	277					IIA			
Naphtha (Petroleum)	8030-30-6	D*[4]	I	42	288	1.1	5.9	2.5		IIA			
Neopentane	463-82-1	D[5]		-65	450	1.4	8.3	2.6	1286				
Nitrobenzene	98-95-3	D		88	482	1.8		4.3	0.3				0.94
Nitroethane	79-24-3	C	I	28	414	3.4		2.6	20.7	IIA			0.87
Nitromethane	75-52-5	C	I	35	418	7.3		2.1	36.1	IIA		0.92	1.17
1-Nitropropane	108-03-2	C	I	34	421	2.2		3.1	10.1				0.84

Table 5.1 *Continued*

Chemical	CAS No.	NEC Group	Type[6]	Flash Point (°C)	AIT (°C)	%LFL	%UFL	Vapor Density (Air=1)	Vapor Pressure[7] (mm Hg)	Class I Zone Group[3]	MIE (mJ) TR	MIC Ratio	MESG (mm)
2-Nitropropane	79-46-9	C*	I	28	428	2.6	11.0	3.1	17.1				
n-Nonane	111-84-2	D[5]	I	31	205	0.8	2.9	4.4	4.4	IIA			
Nonene	27214-95-8	D	I			0.8		4.4					
Nonyl Alcohol	143-08-8	D				0.8	6.1	5.0	0.02	IIA			
n-Octane	111-65-9	D*[5]	I	13	206	1.0	6.5	3.9	14.0	IIA			0.94
Octene	25377-83-7	D	I	8	230	0.9		3.9					
n-Octyl Alcohol	111-87-5	D						4.5	0.08	IIA			1.05
n-Pentane	109-66-0	D*[5]	I	-40	243	1.5	7.8	2.5	513		0.28	0.97	0.93
1-Pentanol	71-41-0	D*	I	33	300	1.2	10.0	3.0	2.5	IIA			
2-Pentanone	107-87-9	D	I	7	452	1.5	8.2	3.0	35.6				0.99
1-Pentene	109-67-1	D	I	-18	275	1.5	8.7	2.4	639.7				
2-Pentene	109-68-2	D	I	-18				2.4					
2-Pentyl Acetate	626-38-0	D	I	23		1.1	7.5	4.5					
Phenylhydrazine	100-63-0	D		89				3.7	0.03				
Process Gas > 30% H_2	1333-74-0	B**	GAS		520	4.0	75.0	0.1			0.019	0.45	
Propane	74-98-6	D*	GAS	-104	450	2.1	9.5	1.6		IIA	0.25	0.82	0.97
1-Propanol	71-23-8	D*	I	15	413	2.2	13.7	2.1	20.7	IIA			0.89
2-Propanol	67-63-0	D*	I	12	399	2.0	12.7	2.1	45.4		0.65		1.00
Propiolactone	57-57-8	D				2.9		2.5	2.2				
Propionaldehyde	123-38-6	C	I	-9	207	2.6	17.0	2.0	318.5				
Propionic Acid	79-09-4	D	II	54	466	2.9	12.1	2.5	3.7				
Propionic Anhydride	123-62-6	D		74	285	1.3	9.5	4.5	1.4				
n-Propyl Acetate	109-60-4	D	I	14	450	1.7	8.0	3.5	33.4				1.05
n-Propyl Ether	111-43-3	C*	I	21	215	1.3	7.0	3.5	62.3				
Propyl Nitrate	627-13-4	B*	I	20	175	2.0	100.0						
Propylene	115-07-1	D*	GAS	-108	455	2.0	11.1	1.5			0.28		0.91
Propylene Dichloride	78-87-5	D	I	16	557	3.4	14.5	3.9	51.7				1.32
Propylene Oxide	75-56-9	B(C)*[1]	I	-37	449	2.3	36.0	2.0	534.4		0.13		0.70
Pyridine	110-86-1	D*	I	20	482	1.8	12.4	2.7	20.8	IIA			
Styrene	100-42-5	D*	I	31	490	0.9	6.8	3.6	6.1	IIA		1.21	
Tetrahydrofuran	109-99-9	C*	I	-14	321	2.0	11.8	2.5	161.6	IIB	0.54		0.87
Tetrahydro-naphthalene	119-64-2	D	IIIA		385	0.8	5.0	4.6	0.4				
Tetramethyl Lead	75-74-1	C	II	38				9.2					
Toluene	108-88-3	D*	I	4	480	1.1	7.1	3.1	28.53	IIA	0.24		
n-Tridecene	2437-56-1	D	IIIA			0.6		6.4	593.4				
Triethylamine	121-44-8	C*	I	-9	249	1.2	8.0	3.5	68.5	IIA	0.75		
Triethylbenzene	25340-18-5	D		83			56.0	5.6					
2,2,3-Trimethylbutane		D[5]			442								
2,2,4-Trimethylbutane		D[5]			407								
2,2,3-Trimethylpentane		D[5]			396								
2,2,4-Trimethylpentane		D[5]			415								
2,3,3-Trimethylpentane		D[5]			425								
Tripropylamine	102-69-2	D	II	41				4.9	1.5				1.13

Table 5.1 *Continued*

Chemical	CAS No.	NEC Group	Type[6]	Flash Point (°C)	AIT (°C)	%LFL	%UFL	Vapor Density (Air = 1)	Vapor Pressure[7] (mm Hg)	Class I Zone Group[3]	MIE (mJ) TR	MIC Ratio	MESG (mm)
Turpentine	8006-64-2	D	I	35	253	0.8			4.8				
n-Undecene	28761-27-5	D	IIIA			0.7		5.5					
Unsymmetrical Dimethyl Hydrazine	57-14-7	C*	I	-15	249	2.0	95.0	1.9					0.85
Valeraldehyde	110-62-3	C	I	280	222			3.0	34.3				
Vinyl Acetate	108-05-4	D*	I	-6	402	2.6	13.4	3.0	113.4	IIA	0.70		0.94
Vinyl Chloride	75-01-4	D*	GAS	-78	472	3.6	33.0	2.2					0.96
Vinyl Toluene	25013-15-4	D		52	494	0.8	11.0	4.1					
Vinylidene Chloride	75-35-4	D	I		570	6.5	15.5	3.4	599.4				3.91
Xylene	1330-20-7	D*	I	25	464	0.9	7.0	3.7		IIA	0.2		
Xylidine	121-69-7	C	IIIA	63	371	1.0		4.2	0.7				

* Material has been classified by test.
** Fuel and process gas mixtures found by test not to present hazards similar to those of hydrogen, may be grouped based on the test results.
Notes:
1. If explosionproof equipment is isolated by sealing all conduits 1/2 in. or larger, in accordance with 501.5(A) of NFPA 70, *National Electrical Code,* equipment for the group classification shown in parentheses is permitted.
2. For classification of areas involving ammonia, see *Safety Code for Mechanical Refrigeration,* ANSI/ASHRAE 15, and *Safety Requirements for the Storage and Handling of Anhydrous Ammonia,* ANSI/CGA G2.1.
3. Certain chemicals may have characteristics that require safeguards beyond those required for any of the above groups. Carbon disulfide is one of these chemicals because of its low autoignition temperature and the small joint clearance necessary to arrest its flame propagation.
4. Petroleum naphtha is a saturated hydrocarbon mixture whose boiling range is 68°F to 275°F (20°C to 135°C). It is also known as benzine, ligroin, petroleum ether, and naphtha.
5. Commercial grades of aliphatic hydrocarbon solvents are mixtures of several isomers of the same chemical formula (or molecular weight). The autoignition temperatures of the individual isomers are significantly different. The electrical equipment should be suitable for the AIT of the solvent mixture.
6. Type is used to designate if the material is a gas, flammable liquid, or combustible liquid.
7. Vapor pressure reflected in units of mm Hg at 77°F (25°C) unless stated otherwise.
8. Class I, Zone Groups are based upon "Electrical apparatus for explosive gas atmospheres — Part 20: Data for flammable gases and vapors, relating to the use of electrical apparatus, IEC 79-20 (1996)."
Source: Table 2-1 in NFPA 497, *Recommended Practice for the Classification of Flammable Liquids, Gases, or Vapors and of Hazardous (Classified) Locations for Electrical Installations in Chemical Process Areas,* 1997 edition.

Recommended Practice for the Classification of Flammable Liquids, Gases, or Vapors and of Hazardous (Classified) Locations for Electrical Installations in Chemical Process Areas. For the definitions of terms used in this table, refer to the original document, NFPA 497.

Publication NMAB 353-4, *Classifications of Dusts Relative to Electrical Equipment in Class II Hazardous Locations,* published in 1982 by the Committee on Evaluation of Industrial Hazards, National Materials Advisory Board, Commission on Engineering and Technical Systems, National Research Council, and National Academy of Sciences, includes a description of the test methods used to determine the ignition temperatures and electrical resistivity of combustible dusts. The publication is available from the National Technical Information Service (NTIS), Springfield, VA 22151.

(B) Class II Group Classifications. Class II groups shall be according to 500.6(B)(1) through (B)(3).

(1) Group E. Atmospheres containing combustible metal dusts, including aluminum, magnesium, and their commercial alloys, or other combustible dusts whose particle size, abrasiveness, and conductivity present similar hazards in the use of electrical equipment. [NFPA 499, 1-3]

FPN: Certain metal dusts may have characteristics that require safeguards beyond those required for atmospheres containing the dusts of aluminum, magnesium, and their commercial alloys. For example, zirconium, thorium, and uranium dusts have extremely low ignition temperatures [as low as 20°C (68°F)] and minimum ignition energies lower than any material classified in any of the Class I or Class II groups.

(2) Group F. Atmospheres containing combustible carbonaceous dusts that have more than 8 percent total entrapped volatiles (see ASTM D 3175-89, *Standard Test Method for Volatile Material in the Analysis Sample for Coal and Coke*, for coal and coke dusts) or that have been sensitized by other materials so that they present an explosion hazard. Coal, carbon black, charcoal, and coke dusts are examples of carbonaceous dusts. [NFPA 499, 1-3]

(3) Group G. Atmospheres containing combustible dusts not included in Group E or F, including flour, grain, wood, plastic, and chemicals.

> FPN No. 1: For additional information on group classification of Class II materials, see NFPA 499-1997, *Recommended Practice for the Classification of Combustible Dusts and of Hazardous (Classified) Locations for Electrical Installations in Chemical Process Areas*.
>
> FPN No. 2: The explosion characteristics of air mixtures of dust vary with the materials involved. For Class II

locations, Groups E, F, and G, the classification involves the tightness of the joints of assembly and shaft openings to prevent the entrance of dust in the dust-ignitionproof enclosure, the blanketing effect of layers of dust on the equipment that may cause overheating, and the ignition temperature of the dust. It is necessary, therefore, that equipment be identified not only for the class, but also for the specific group of dust that will be present.

> FPN No. 3: Certain dusts may require additional precautions due to chemical phenomena that can result in the generation of ignitible gases. See ANSI C2-1997, *National Electrical Safety Code*, Section 127A, Coal Handling Areas.

Table 5.2 is an alphabetical listing of selected combustible materials with their group classification and relevant physical properties shown. All of the materials included in this

Table 5.2 Selected Combustible Materials

Chemical Name	CAS No.	*NEC* Group	Code	Layer or Cloud Ignition Temp °C
Acetal, Linear		G	NL	440
Acetoacet-p-phenetidide	122-82-7	G	NL	560
Acetoacetanilide	102-01-2	G	M	440
Acetylamino-t-nitrothiazole		G		450
Acrylamide Polymer		G		240
Acrylonitrile Polymer		G		460
Acrylonitrile-Vinyl Chloride-Vinylidene Chloride Copolymer (70-20-10)		G		210
Acrylonitrile-Vinyl Pyridine Copolymer		G		240
Adipic Acid	124-04-9	G	M	550
Alfalfa Meal		G		200
Alkyl Ketone Dimer Sizing Compound		G		160
Allyl Alcohol Derivative (CR-39)		G	NL	500
Almond Shell		G		200
Aluminum, A422 Flake	7429-90-5	E		320
Aluminum, Atomized Collector Fines		E	CL	550
Aluminum—cobalt alloy (60-40)		E		570
Aluminum—copper alloy (50-50)		E		830
Aluminum—lithium alloy (15% Li)		E		400
Aluminum—magnesium alloy (Dowmetal)		E	CL	430
Aluminum—nickel alloy (58-42)		E		540
Aluminum—silicon alloy (12% Si)		E	NL	670
Amino-5-nitrothiazole	121-66-4	G		460
Anthranilic Acid	118-92-3	G	M	580
Apricot Pit		G		230
Aryl-nitrosomethylamide		G	NL	490
Asphalt	8052-42-4	F		510
Aspirin [acetol (2)]	50-78-2	G	M	660
Azelaic Acid	109-31-9	G	M	610
Azo-bis-butyronitrile	78-67-1	G		350
Benzethonium Chloride		G	CL	380

Table 5.2 *Continued*

Chemical Name	CAS No.	*NEC* Group	Code	Layer or Cloud Ignition Temp °C
Benzoic Acid	65-85-0	G	M	440
Benzotriazole	95-14-7	G	M	440
Beta-naphthalene-axo-dimethylaniline		G		175
Bis(2-hydroxy-5-chlorophenyl) Methane	97-23-4	G	NL	570
Bisphenol-A	80-05-7	G	M	570
Boron, Commercial Amorphous (85% B)	7440-42-8	E		400
Calcium Silicide		E		540
Carbon Black (More Than 8% Total Entrapped Volatiles)		F		
Carboxymethyl Cellulose	9000-11-7	G		290
Carboxypolymethylene		G	NL	520
Cashew Oil, Phenolic, Hard		G		180
Cellulose		G		260
Cellulose Acetate		G		340
Cellulose Acetate Butyrate		G	NL	370
Cellulose Triacetate		G	NL	430
Charcoal (Activated)	64365-11-3	F		180
Charcoal (More Than 8% Total Entrapped Volatiles)		F		
Cherry Pit		G		220
Chlorinated Phenol		G	NL	570
Chlorinated Polyether Alcohol		G		460
Chloroacetoacetanilide	101-92-8	G	M	640
Chromium (97%) Electrolytic, Milled	7440-47-3	E		400
Cinnamon		G		230
Citrus Peel		G		270
Coal, Kentucky Bituminous		F		180
Coal, Pittsburgh Experimental		F		170
Coal, Wyoming		F		
Cocoa Bean Shell		G		370
Cocoa, Natural, 19% Fat		G		240
Coconut Shell		G		220
Coke (More Than 8% Total Entrapped Volatiles)		F		
Cork		G		210
Corn		G		250
Corn Dextrine		G		370
Corncob Grit		G		240
Cornstarch, Commercial		G		330
Cornstarch, Modified		G		200
Cottonseed Meal		G		200
Coumarone-Indene, Hard		G	NL	520
Crag No. 974	533-74-4	G	CL	310
Cube Root, South America	83-79-4	G		230
Di-alphacumyl Peroxide, 40-60 on CA	80-43-3	G		180
Diallyl Phthalate	131-17-9	G	M	480
Dicyclopentadiene Dioxide		G	NL	420
Dieldrin (20%)	60-57-1	G	NL	550
Dihydroacetic Acid		G	NL	430
Dimethyl Isophthalate	1459-93-4	G	M	580
Dimethyl Terephthalate	120-61-6	G	M	570
Dinitro-o-toluamide	148-01-6	G	NL	500
Dinitrobenzoic Acid		G	NL	460

Table 5.2 *Continued*

Chemical Name	CAS No.	*NEC* Group	Code	Layer or Cloud Ignition Temp °C
Diphenyl	92-52-4	G	M	630
Ditertiary-butyl-paracresol	128-37-0	G	NL	420
Dithane m-45	8018-01-7	G		180
Epoxy		G	NL	540
Epoxy-bisphenol A		G	NL	510
Ethyl Cellulose		G	CL	320
Ethyl Hydroxyethyl Cellulose		G	NL	390
Ethylene Oxide Polymer		G	NL	350
Ethylene-maleic Anhydride Copolymer		G	NL	540
Ferbam™	14484-64-1	G		150
Ferromanganese, Medium Carbon	12604-53-4	E		290
Ferrosilicon (88% Si, 9% Fe)	8049-17-0	E		800
Ferrotitanium (19% Ti, 74.1% Fe, 0.06% C)		E	CL	380
Flax Shive		G		230
Fumaric Acid	110-17-8	G	M	520
Garlic, Dehydrated		G	NL	360
Gilsonite	12002-43-6	F		500
Green Base Harmon Dye		G		175
Guar Seed		G	NL	500
Gulasonic Acid, Diacetone		G	NL	420
Gum, Arabic		G		260
Gum, Karaya		G		240
Gum, Manila		G	CL	360
Gum, Tragacanth	9000-65-1	G		260
Hemp Hurd		G		220
Hexamethylene Tetramine	100-97-0	G	S	410
Hydroxyethyl Cellulose		G	NL	410
Iron, 98% H_2 Reduced		E		290
Iron, 99% Carbonyl	13463-40-6	E		310
Isotoic Anhydride		G	NL	700
L-sorbose		G	M	370
Lignin, Hydrolized, Wood-type, Fine		G	NL	450
Lignite, California		F		180
Lycopodium		G		190
Malt Barley		G		250
Manganese	7439-96-5	E		240
Magnesium, Grade B, Milled		E		430
Manganese Vancide		G		120
Mannitol	69-65-8	G	M	460
Methacrylic Acid Polymer		G		290
Methionine (l-methionine)	63-68-3	G		360
Methyl Cellulose		G		340
Methyl Methacrylate Polymer	9011-14-7	G	NL	440
Methyl Methacrylate-ethyl Acrylate		G	NL	440
Methyl Methacrylate-styrene-butadiene		G	NL	480
Milk, Skimmed		G		200
N,N-Dimethylthio-formamide		G		230
Nitropyridone	100703-82-0	G	M	430
Nitrosamine		G	NL	270
Nylon Polymer	63428-84-2	G		430

Table 5.2 *Continued*

Chemical Name	CAS No.	*NEC* Group	Code	Layer or Cloud Ignition Temp °C
Para-oxy-benzaldehyde	123-08-0	G	CL	380
Paraphenylene Diamine	106-50-3	G	M	620
Paratertiary Butyl Benzoic Acid	98-73-7	G	M	560
Pea Flour		G		260
Peach Pit Shell		G		210
Peanut Hull		G		210
Peat, Sphagnum	94114-14-4	G		240
Pecan Nut Shell	8002-03-7	G		210
Pectin	5328-37-0	G		200
Pentaerythritol	115-77-5	G	M	400
Petrin Acrylate Monomer	7659-34-9	G	NL	220
Petroleum Coke (More Than 8% Total Entrapped Volatiles)		F		
Petroleum Resin	64742-16-1	G		500
Phenol Formaldehyde	9003-35-4	G	NL	580
Phenol Formaldehyde, Polyalkylene-p	9003-35-4	G		290
Phenol Furfural	26338-61-4	G		310
Phenylbetanaphthylamine	135-88-6	G	NL	680
Phthalic Anydride	85-44-9	G	M	650
Phthalimide	85-41-6	G	M	630
Pitch, Coal Tar	65996-93-2	F	NL	710
Pitch, Petroleum	68187-58-6	F	NL	630
Polycarbonate		G	NL	710
Polyethylene, High Pressure Process	9002-88-4	G		380
Polyethylene, Low Pressure Process	9002-88-4	G	NL	420
Polyethylene Terephthalate	25038-59-9	G	NL	500
Polyethylene Wax	68441-04-8	G	NL	400
Polypropylene (no antioxidant)	9003-07-0	G	NL	420
Polystyrene Latex	9003-53-6	G		500
Polystyrene Molding Compound	9003-53-6	G	NL	560
Polyurethane Foam, Fire Retardant	9009-54-5	G		390
Polyurethane Foam, No Fire Retardant	9009-54-5	G		440
Polyvinyl Acetate	9003-20-7	G	NL	550
Polyvinyl Acetate/Alcohol	9002-89-5	G		440
Polyvinyl Butyral	63148-65-2	G		390
Polyvinyl Chloride-dioctyl Phthalate		G	NL	320
Potato Starch, Dextrinated	9005-25-8	G	NL	440
Pyrethrum	8003-34-7	G		210
Rayon (Viscose) Flock	61788-77-0	G		250
Red Dye Intermediate		G		175
Rice		G		220
Rice Bran		G	NL	490
Rice Hull		G		220
Rosin, DK	8050-09-7	G	NL	390
Rubber, Crude, Hard	9006-04-6	G	NL	350
Rubber, Synthetic, Hard (33% S)	64706-29-2	G	NL	320
Safflower Meal		G		210
Salicylanilide	87-17-2	G	M	610
Sevin	63-25-2	G		140
Shale, Oil	68308-34-9	F		
Shellac	9000-59-3	G	NL	400

Table 5.2 *Continued*

Chemical Name	CAS No.	*NEC* Group	Code	Layer or Cloud Ignition Temp °C
Sodium Resinate	61790-51-0	G		220
Sorbic Acid (Copper Sorbate or Potash)	110-44-1	G		460
Soy Flour	68513-95-1	G		190
Soy Protein	9010-10-0	G		260
Stearic Acid, Aluminum Salt	637-12-7	G		300
Stearic Acid, Zinc Salt	557-05-1	G	M	510
Styrene Modified Polyester-Glass Fiber	100-42-5	G		360
Styrene-acrylonitrile (70-30)	9003-54-7	G	NL	500
Styrene-butadiene Latex (>75% styrene)	903-55-8	G	NL	440
Styrene-maleic Anhydride Copolymer	9011-13-6	G	CL	470
Sucrose	57-50-1	G	CL	350
Sugar, Powdered	57-50-1	G	CL	370
Sulfur	7704-34-9	G		220
Tantalum	7440-25-7	E		300
Terephthalic Acid	100-21-0	G	NL	680
Thorium, 1.2% O_2	7440-29-1	E	CL	280
Tin, 96%, Atomized, (2% Pb)	7440-31-5	E		430
Titanium, 99% Ti	7440-32-6	E	CL	330
Titanium Hydride (95% Ti, 3.8% H_2)	7704-98-5	E	CL	480
Trithiobisdimethylthio-formamide		G		230
Tung, Kernels, Oil-free	8001-20-5	G		240
Urea Formaldehyde Molding Compound	9011-05-6	G	NL	460
Urea Formaldehyde-phenol Formaldehyde	25104-55-6	G		240
Vanadium, 86.4%	7440-62-2	E		490
Vinyl Chloride-acrylonitrile Copolymer	9003-00-3	G		470
Vinyl Toluene-acrylonitrile Butadiene	76404-69-8	G	NL	530
Violet 200 Dye		G		175
Vitamin B1, Mononitrate	59-43-8	G	NL	360
Vitamin C	50-81-7	G		280
Walnut Shell, Black		G		220
Wheat		G		220
Wheat Flour	130498-22-5	G		360
Wheat Gluten, Gum	100684-25-1	G	NL	520
Wheat Starch		G	NL	380
Wheat Straw		G		220
Wood Flour		G		260
Woodbark, Ground		G		250
Yeast, Torula	68602-94-8	G		260
Zirconium Hydride	7704-99-6	E		270
Zirconium		E	CL	330

Notes: 1. Normally, the minimum ignition temperature of a layer of a specific dust is lower than the minimum ignition temperature of a cloud of that dust. Since this is not universally true, the lower of the two minimum ignition temperatures is listed. If no symbol appears between the two temperature columns, then the layer ignition temperature is shown. "CL" means the cloud ignition temperature is shown. "NL" means that no layer ignition temperature is available, and the cloud ignition temperature is shown. "M" signifies that the dust layer melts before it ignites; the cloud ignition temperature is shown. "S" signifies that the dust layer sublimes before it ignites; the cloud ignition temperature is shown.
2. Certain metal dusts may have characteristics that require safeguards beyond those required for atmospheres containing the dusts of aluminum, magnesium, and their commercial alloys. For example, zirconium, thorium, and uranium dusts have extremely low ignition temperatures [as low as 68°F (20°C)] and minimum ignition energies lower than any material classified in any of the Class I or Class II groups.
Source: Table 2-2 in NFPA 497, Recommended Practice for the Classification of Flammable Liquids, Gases, or Vapors and of Hazardous (Classified) Locations for Electrical Installations in Chemical Process Areas, 1997 edition.

table have been evaluated for the purpose of designating the appropriate dust group. This information is used to properly select electrical equipment for use in Class II locations. Combustible dusts are classified into three Class II, division groups—E, F, and G—depending on their properties. Table 5.2 is extracted from NFPA 499, *Recommended Practice for the Classification of Combustible Dusts and of Hazardous (Classified) Locations for Electrical Installations in Chemical Process Areas.* For the definitions of terms used in the table, refer to the original document, NFPA 499.

As in Class I locations, equipment must be approved not only for the class but also for the specific group. It is important that, in addition to the proper selection of equipment, high standards of installation be maintained for subsequent additions or alterations.

500.7 Protection Techniques.

Section 500.7(A) through (L) shall be acceptable protection techniques for electrical and electronic equipment in hazardous (classified) locations.

(A) Explosionproof Apparatus. This protection technique shall be permitted for equipment in Class I, Division 1 or 2 locations.

(B) Dust Ignitionproof. This protection technique shall be permitted for equipment in Class II, Division 1 or 2 locations.

(C) Dusttight. This protection technique shall be permitted for equipment in Class II, Division 2 or Class III, Division 1 or 2 locations.

(D) Purged and Pressurized. This protection technique shall be permitted for equipment in any hazardous (classified) location for which it is identified.

NFPA 496, *Standard for Purged and Pressurized Enclosures for Electrical Equipment,* covers purged and pressurized enclosures for electrical equipment in Class I and Class II hazardous (classified) locations.

In Class I locations, purged and pressurized enclosures are used to eliminate or reduce, within the enclosure, a Class I hazardous (classified) location classification, as defined in Article 500 of the *Code.* Purged and pressurized enclosures make it possible for equipment that is not otherwise acceptable for hazardous (classified) locations to be used in these locations, in accordance with the *Code.*

Purging is the process of supplying an enclosure with a protective gas at a sufficient flow and positive pressure to reduce the concentration of any flammable gas or vapor initially present to an acceptable level. The types of pressurizing are as follows:

1. Type X pressurizing reduces the classification within a protected enclosure from Division 1 to nonclassified.

2. Type Y pressurizing reduces the classification within a protected enclosure from Division 1 to Division 2.
3. Type Z pressurizing reduces the classification within a protected enclosure from Division 2 to nonclassified.

In Class II hazardous (classified) locations, pressurized enclosures prevent the entrance of dusts into an enclosure. Pressurized enclosures make it possible for equipment that is not otherwise acceptable for hazardous (classified) locations to be used in these locations, in accordance with the *Code. Pressurization*, for the purposes of NFPA 496, is the process of supplying an enclosure with a protective gas, with or without continuous flow, at sufficient pressure to prevent the entrance of a flammable gas or vapor, a combustible dust, or an ignitible fiber.

It should be noted that an atmosphere that is made hazardous by combustible dust inside an enclosure cannot be reduced to a safe level by supplying a flow of protective gas in the same manner as with gases or vapors. The enclosure must be opened, and the dust must be removed. Visual inspection can determine if the dust has been removed. Positive pressure then prevents dust from entering a clean enclosure.

Field-installed devices and equipment, such as pushbutton controls and pilot lights, are permitted in purged and pressurized enclosures, provided the conductor terminals are within the purged or pressurized atmosphere. Section 5.3.4.1 of NFPA 30, *Flammable and Combustible Liquids Code,* and Section 4.6 of API RP500 provide guidelines on what is considered to be adequate ventilation.

(E) Intrinsic Safety. This protection technique shall be permitted for equipment in Class I, Division 1 or 2; or Class II, Division 1 or 2; or Class III, Division 1 or 2 locations. The provisions of Articles 501 through 503 and Articles 510 through 516 shall not be considered applicable to such installations, except as required by Article 504, and installation of intrinsically safe apparatus and wiring shall be in accordance with the requirements of Article 504.

(F) Nonincendive Circuit. This protection technique shall be permitted for equipment in Class I, Division 2; Class II, Division 2; or Class III, Division 1 or 2 locations.

(G) Nonincendive Equipment. This protection technique shall be permitted for equipment in Class I, Division 2; Class II, Division 2; or Class III, Division 1 or 2 locations.

(H) Nonincendive Component. This protection technique shall be permitted for equipment in Class I, Division 2; Class II, Division 2; or Class III, Division 1 or 2 locations.

(I) Oil Immersion. This protection technique shall be permitted for current-interrupting contacts in Class I, Division 2 locations as described in 501.6(B)(1)(2).

(J) Hermetically Sealed. This protection technique shall be permitted for equipment in Class I, Division 2; Class II, Division 2; or Class III, Division 1 or 2 locations.

(K) Combustible Gas Detection System. A combustible gas detection system shall be permitted as a means of protection in industrial establishments with restricted public access and where the conditions of maintenance and supervision ensure that only qualified persons service the installation. Gas detection equipment shall be listed for detection of the specific gas or vapor to be encountered. Where such a system is installed, equipment specified in 500.7(K)(1), (2), or (3) shall be permitted.

(1) Inadequate Ventilation. In a Class I, Division 1 location that is so classified due to inadequate ventilation, electrical equipment suitable for Class I, Division 2 locations shall be permitted.

(2) Interior of a Building. In a building located in, or with an opening into, a Class I, Division 2 location where the interior does not contain a source of flammable gas or vapor, electrical equipment for unclassified locations shall be permitted.

(3) Interior of a Control Panel. In the interior of a control panel containing instrumentation utilizing or measuring flammable liquids, gases, or vapors, electrical equipment suitable for Class I, Division 2 locations shall be permitted.

> FPN No. 1: For further information, see ANSI/ISA-12.13.01, *Performance Requirements, Combustible Gas Detectors.*
>
> FPN No. 2: For further information., see ANSI/API RP 500, *Recommended Practice for Classification of Locations for Electrical Installations at Petroleum Facilities Classified as Class I, Division I or Division 2.*
>
> FPN No. 3: For further information, see ISA-RP12.13.02, *Installation, Operation, and Maintenance of Combustible Gas Detection Instruments.*

(L) Other Protection Techniques. Other protection techniques used in equipment identified for use in hazardous (classified) locations.

Some listed equipment employs unique protection techniques or a combination of protection techniques, such as listed attachment plugs for use in hazardous locations and listed battery-operated flashlights and lanterns.

500.8 Equipment.

Articles 500 through 504 require equipment construction and installation that ensure safe performance under conditions of proper use and maintenance.

> FPN No. 1: It is important that inspection authorities and users exercise more than ordinary care with regard to installation and maintenance.
>
> FPN No. 2: Low ambient conditions require special consideration. Explosionproof or dust-ignitionproof equipment may not be suitable for use at temperatures lower than $-25°C$ ($-13°F$) unless they are identified for low-temperature service. However, at low ambient temperatures, flammable concentrations of vapors may not exist in a location classified Class I, Division 1 at normal ambient temperature.

At low ambient temperatures, such as those encountered in the Arctic, explosion pressures increase at very low temperatures. The strengths of materials change, and the explosion pressure within the explosionproof equipment may increase beyond the safe operating strength of the material. In addition, some sealing materials for sealing fittings may become brittle. However, the extent of the hazardous (classified) location may also change under low ambient conditions. The material may be used in a location where the temperature range is so low that no vapors are produced based on the flash point of the material involved.

(A) Approval for Class and Properties.

(1) Equipment shall be identified not only for the class of location but also for the explosive, combustible, or ignitible properties of the specific gas, vapor, dust, fiber, or flyings that will be present. In addition, Class I equipment shall not have any exposed surface that operates at a temperature in excess of the ignition temperature of the specific gas or vapor. Class II equipment shall not have an external temperature higher than that specified in 500.8(C)(2). Class III equipment shall not exceed the maximum surface temperatures specified in 503.1.

> FPN: Luminaires (lighting fixtures) and other heat-producing apparatus, switches, circuit breakers, and plugs and receptacles are potential sources of ignition and are investigated for suitability in classified locations. Such types of equipment, as well as cable terminations for entry into explosionproof enclosures, are available as listed for Class I, Division 2 locations. Fixed wiring, however, may utilize wiring methods that are not evaluated with respect to classified locations. Wiring products such as cable, raceways, boxes, and fittings, therefore, are not marked as being suitable for Class I, Division 2 locations. Also see Exception No. 3 to 500.8(B).

Suitability of identified equipment shall be determined by any of the following:

(1) Equipment listing or labeling
(2) Evidence of equipment evaluation from a qualified testing laboratory or inspection agency concerned with product evaluation

(3) Evidence acceptable to the authority having jurisdiction such as a manufacturer's self-evaluation or an owner's engineering judgment.

(2) Equipment that has been identified for a Division 1 location shall be permitted in a Division 2 location of the same class and group.

(3) Where specifically permitted in Articles 501 through 503, general-purpose equipment or equipment in general-purpose enclosures shall be permitted to be installed in Division 2 locations if the equipment does not constitute a source of ignition under normal operating conditions.

(4) Equipment, regardless of the classification of the location in which it is installed, that depends on a single compression seal, diaphragm, or tube to prevent flammable or combustible fluids from entering the equipment shall be identified for a Class I, Division 2 location. Equipment installed in a Class I, Division 1 location shall be identified for the Class I, Division 1 location.

> FPN: See 501.5(F)(3) for additional requirements.

(5) Unless otherwise specified, normal operating conditions for motors shall be assumed to be rated full-load steady conditions.

It is not intended that locked-rotor or other motor overload conditions, such as single phasing, be considered when evaluating motor operating temperatures (internal and external) in Class I, Division 2 locations. However, such abnormal load conditions must be considered when evaluating the external temperatures of explosionproof motors for Class I, Division 1 locations and motors such as dust-ignitionproof motors for Class II, Division 1 locations.

(6) Where flammable gases or combustible dusts are or may be present at the same time, the simultaneous presence of both shall be considered when determining the safe operating temperature of the electrical equipment.

> FPN: The characteristics of various atmospheric mixtures of gases, vapors, and dusts depend on the specific material involved.

A coal-handling facility is an example of a location where methane gas and coal dust may be present at the same time.

(B) Marking. Equipment shall be marked to show the class, group, and operating temperature or temperature class referenced to a 40°C ambient.

The marked operating temperature or temperature range is normally referenced to a 40°C (104°F) ambient. Unless the equipment is provided with thermally actuated sensors that limit the temperature to that marked on the equipment, operation in ambient temperatures higher than 40°C (104°F) will increase the operating temperature of the equipment. Many explosionproof and dust-ignitionproof motors are equipped with thermal protectors. In like manner, operation in ambient temperatures lower than 40°C (104°F) will usually reduce the operating temperature.

Exception No. 1: Equipment of the non–heat-producing type, such as junction boxes, conduit, and fittings, and equipment of the heat-producing type having a maximum temperature not more than 100°C (212°F) shall not be required to have a marked operating temperature or temperature class.

Exception No. 2: Fixed luminaires (lighting fixtures) marked for use in Class I, Division 2 or Class II, Division 2 locations only shall not be required to be marked to indicate the group.

Exception No. 3: Fixed general-purpose equipment in Class I locations, other than fixed luminaires (lighting fixtures), that is acceptable for use in Class I, Division 2 locations shall not be required to be marked with the class, group, division, or operating temperature.

A squirrel-cage induction motor without brushes, switching mechanisms, or similar arc-producing devices is an example of fixed general-purpose equipment. See 501.8(B) and its associated commentary for more information on motors in Class I, Division 2 locations.

Exception No. 4: Fixed dusttight equipment other than fixed luminaires (lighting fixtures) that is acceptable for use in Class II, Division 2 and Class III locations shall not be required to be marked with the class, group, division, or operating temperature.

Exception No. 5: Electric equipment suitable for ambient temperatures exceeding 40°C (104°F) shall be marked with both the maximum ambient temperature and the operating temperature or temperature class at that ambient temperature.

> FPN: Equipment not marked to indicate a division, or marked "Division 1" or "Div. 1," is suitable for both Division 1 and 2 locations. Equipment marked "Division 2" or "Div. 2" is suitable for Division 2 locations only.

The temperature class, if provided, shall be indicated using the temperature class (T Codes) shown in Table

500.8(B). The temperature class (T Code) marked on equipment nameplates shall be in accordance with Table 500.8(B). Equipment for Class I and Class II shall be marked with the maximum safe operating temperature, as determined by simultaneous exposure to the combinations of Class I and Class II conditions.

Table 500.8(B) Classification of Maximum Surface Temperature

Maximum Temperature		Temperature Class (T Code)
C°	F°	
450	842	T1
300	572	T2
280	536	T2A
260	500	T2B
230	446	T2C
215	419	T2D
200	392	T3
180	356	T3A
165	329	T3B
160	320	T3C
135	275	T4
120	248	T4A
100	212	T5
85	185	T6

FPN: Since there is no consistent relationship between explosion properties and ignition temperature, the two are independent requirements.

(C) Temperature.

(1) Class I Temperature. The temperature marking specified in 500.8(B) shall not exceed the ignition temperature of the specific gas or vapor to be encountered.

The ignition temperature of a solid, liquid, or gaseous substance is the minimum temperature required to initiate or cause self-sustained combustion independent of the heating or heated element. The ignition temperature and the flash point are unrelated properties, except that the flash point is always lower than the ignition temperature.

FPN: For information regarding ignition temperatures of gases and vapors, see NFPA 497-1997, *Recommended Practice for the Classification of Flammable Liquids, Gases, or Vapors, and of Hazardous (Classified) Locations for Electrical Installations in Chemical Process Areas.*

(2) Class II Temperature. The temperature marking specified in 500.8(B) shall be less than the ignition temperature of the specific dust to be encountered. For organic dusts that may dehydrate or carbonize, the temperature marking shall

not exceed the lower of either the ignition temperature or 165°C (329°F).

FPN: See NFPA 499-1997, *Recommended Practice for the Classification of Combustible Dusts and of Hazardous (Classified) Locations for Electrical Installations in Chemical Process Areas*, for minimum ignition temperatures of specific dusts.

The ignition temperature for which equipment was approved prior to this requirement shall be assumed to be as shown in Table 500.8(C)(2).

Table 500.8(C)(2) Class II Temperatures

Class II Group	Equipment Not Subject to Overloading		Equipment (Such as Motors or Power Transformers) That May Be Overloaded			
			Normal Operation		Abnormal Operation	
	°C	°F	°C	°F	°C	°F
E	200	392	200	392	200	392
F	200	392	150	302	200	392
G	165	329	120	248	165	329

(D) Threading. All threaded conduit or fittings referred to herein shall be threaded with a National (American) Standard Pipe Taper (NPT) standard conduit cutting die that provides a taper of 1 in 16 (¾-in. taper per foot). Such conduit shall be made wrenchtight to prevent sparking when fault current flows through the conduit system and to ensure the explosionproof or flameproof integrity of the conduit system where applicable. Equipment provided with threaded entries for field wiring connections shall be installed in accordance with 500.8(D)(1) or (D)(2).

(1) Equipment Provided with Threaded Entries for NPT Threaded Conduit or Fittings. For equipment provided with threaded entries for NPT threaded conduit or fittings, listed conduit, conduit fittings, or cable fittings shall be used.

FPN: Thread form specifications for NPT threads are located in ANSI/ASME B1.20.1-1983, *Pipe Threads, General Purpose (Inch).*

(2) Equipment Provided with Threaded Entries for Metric Threaded Conduit or Fittings. For equipment with metric threaded entries, such entries shall be identified as being metric, or listed adapters to permit connection to conduit or NPT-threaded fittings shall be provided with the equipment. Adapters shall be used for connection to conduit or NPT-threaded fittings. Listed cable fittings that have metric threads shall be permitted to be used.

FPN: Threading specifications for metric threaded entries are located in ISO 965/1-1980, *Metric Screw Threads*, and ISO 965/3-1980, *Metric Screw Threads*.

All conduit joints must be made up wrenchtight to prevent arcing between the conduit and the coupling, fitting, or enclosure of the conduit under ground-fault conditions. The use of a bonding jumper in lieu of a wrenchtight connection is not permitted. The integrity of the ground-fault current path is critical in hazardous locations in order to prevent ignition-capable arcing or sparking.

The information on metric threads in 500.8(D) has been included to allow for safe electrical and mechanical connections where the enclosure has metric threads and the raceway or cable has NPT threads. Equipment with metric threaded entries must be identified or provided with suitable adapters that permit the connection of conduit and fittings that are NPT threaded.

(E) Fiber Optic Cable Assembly. Where a fiber optic cable assembly contains conductors that are capable of carrying current, the fiber optic cable assembly shall be installed in accordance with the requirements of Articles 500, 501, 502, or 503, as applicable.

The requirements of Articles 500, 501, 502, or 503 apply, even if the conductor is grounded.

500.9 Specific Occupancies.

Articles 510 through 517 cover garages, aircraft hangars, motor fuel dispensing facilities, bulk storage plants, spray application, dipping and coating processes, and health care facilities.

ARTICLE 501
Class I Locations

Contents

501.1 General.

The general rules of this *Code* shall apply to the electric wiring and equipment in locations classified as Class I in 500.5. Equipment listed and marked in accordance with 505.9(C)(2) for use in Class I, Zone 0, 1, or 2 locations shall be permitted in Class I, Division 2 locations for the same gas and with a suitable temperature class.

Exception: As modified by this article.

The most common Class I locations are those areas involved in the handling or processing of volatile flammable liquids such as gasoline, naphtha, benzene, diethyl ether, and acetone, or flammable gases such as hydrogen, methane, and propane.

Where ignitible concentrations (concentrations within the flammable or explosive limits) of flammable gases or vapors are present, atmospheres exist that are explosive when ignited by an arc, a spark, or high temperature. NFPA 497, *Recommended Practice for the Classification of Flammable Liquids, Gases, or Vapors and of Hazardous (Classified) Locations for Electrical Installations in Chemical Process Areas,* includes information on the explosive limits of flammable liquids and gases.

All electrical equipment that may cause ignition-capable arcs or sparks should be kept out of Class I locations where practicable. If this is not practicable, such apparatus must be approved for the purpose and installed properly. The arc produced at the contacts of listed or labeled intrinsically safe equipment is not ignition-capable because the energy available is insufficient to cause ignition.

Hermetic sealing of all electrical equipment is impractical, because equipment such as motors, conventional switches, and circuit breakers has movable parts that must be operated through the enclosing case; that is, the lever of a switch or the shaft of a motor must have sufficient clearance to operate freely. In addition, in many cases, it is necessary to have access to the inside of enclosures for installation, servicing, or alterations.

It is practically impossible to make threaded conduit joints gastight. The conduit system and apparatus enclosure "breathe" due to temperature changes, and any flammable gases or vapors in the room may slowly enter the conduit or enclosure, creating an explosive mixture. Should an arc occur, an explosion could take place.

When an explosion occurs within the enclosure or conduit system, the burning mixture or hot gases must be sufficiently confined within the system to prevent ignition of any explosive mixture that might be present in the area outside the enclosures or conduit system. An apparatus enclosure must be designed with sufficient strength to withstand the maximum pressure generated by an internal explosion in order to prevent rupture and the release of burning or hot gases. Enclosures have been designed to withstand such internal explosions. The ability to withstand an internal explosion is one criterion by which explosionproof enclosures are evaluated.

During an explosion within an enclosure, gases escape through any paths or openings that exist, but the gases are sufficiently cooled if they are carried out through an opening that is long in proportion to its width; that is, the spiral path of at least five fully engaged threads of a screw-on type junction box cover, as illustrated in Exhibit 501.1. This principle is also applied in the design of explosionproof enclosures for apparatus in which a wide machined flange on the body of the enclosure and a similar machined flange on the cover are provided, as illustrated in Exhibit 501.2. These machined flanges are ground so that when the cover

is seated in place, the clearance between the two surfaces at no point exceeds, for example, 0.0015 in. If an explosion occurs within the enclosure, escaping gas travels a considerable distance through a very small opening. The gas therefore is cooled sufficiently when it enters and mixes with the surrounding atmosphere, thus preventing ignition of the external explosive mixture.

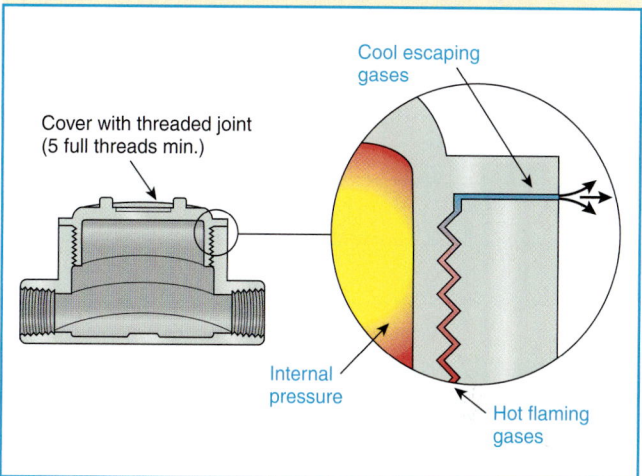

Exhibit 501.1 Hot gases are cooled as they pass through the threads of a screw-type cover of an explosionproof junction box.

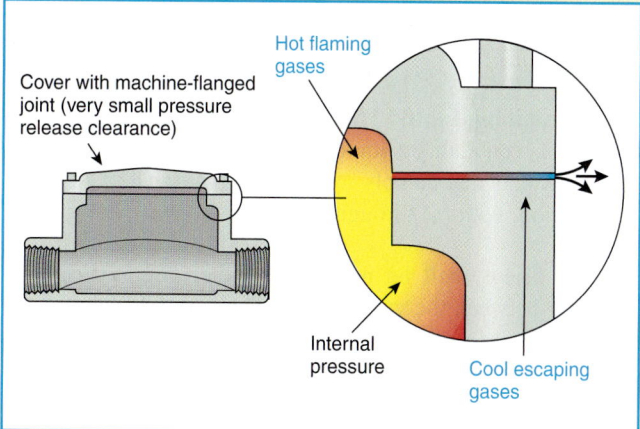

Exhibit 501.2 Hot gases are cooled as they pass across a machine-flanged joint. The clearance between the machined surfaces is kept very small.

The clearance between flat surfaces may increase somewhat under explosion conditions because the internal pressures created by the explosion tend to force the surfaces apart, as shown in Exhibit 501.3. The amount of increase in the joint clearance depends on the stiffness of the enclosure parts; the size, strength, and spacing of the bolts; and the explosion pressure. Simply measuring the joint width and clearance when there are no internal pressures does not indicate the actual clearances under the dynamic conditions of an explosion. Explosion tests are usually needed to demonstrate the acceptability of the design.

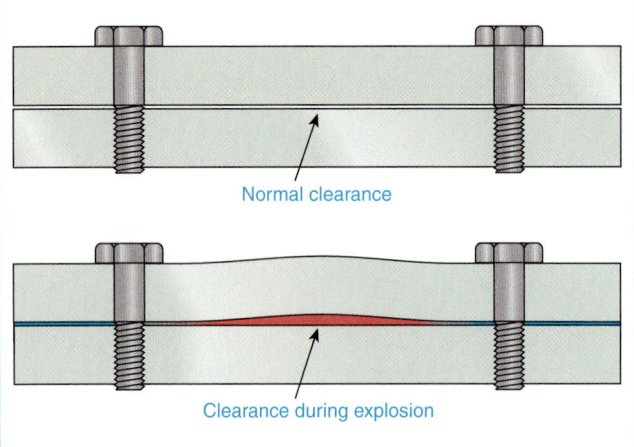

Exhibit 501.3 Effect of internal explosion (bottom) on cover-to-body joint clearance in an explosionproof enclosure. (Redrawn from Underwriters Laboratories Inc.)

501.2 Transformers and Capacitors.

(A) Class I, Division 1. In Class I, Division 1 locations, transformers and capacitors shall comply with 501.2(A)(1) and (A)(2).

(1) Containing Liquid That Will Burn. Transformers and capacitors containing a liquid that will burn shall be installed only in vaults that comply with 450.41 through 450.48 and, in addition, with (a) through (d).

(a) There shall be no door or other communicating opening between the vault and the Division 1 location.

(b) Ample ventilation shall be provided for the continuous removal of flammable gases or vapors.

(c) Vent openings or ducts shall lead to a safe location outside of buildings.

(d) Vent ducts and openings shall be of sufficient area to relieve explosion pressures within the vault, and all portions of vent ducts within the buildings shall be of reinforced concrete construction.

(2) Not Containing Liquid That Will Burn. Transformers and capacitors that do not contain a liquid that will burn shall be installed in vaults complying with 501.2(A)(1) or be approved for Class I locations.

(B) Class I, Division 2. In Class I, Division 2 locations, transformers and capacitors shall comply with 450.21 through 450.27.

501.3 Meters, Instruments, and Relays.

(A) Class I, Division 1. In Class I, Division 1 locations, meters, instruments, and relays, including kilowatt-hour meters, instrument transformers, resistors, rectifiers, and thermionic tubes, shall be provided with enclosures identified for Class I, Division 1 locations. Enclosures for Class I, Division 1 locations include explosionproof enclosures and purged and pressurized enclosures.

> FPN: See NFPA 496-1998, *Standard for Purged and Pressurized Enclosures for Electrical Equipment.*

See the commentary on purged and pressurized enclosures for electrical equipment in hazardous (classified) locations following 500.7(D) and the commentary on explosionproof enclosures following the exception to 501.1.

(B) Class I, Division 2. In Class I, Division 2 locations, meters, instruments, and relays shall comply with 501.3(B)(1) through (B)(6).

(1) Contacts. Switches, circuit breakers, and make-and-break contacts of pushbuttons, relays, alarm bells, and horns shall have enclosures identified for Class I, Division 1 locations in accordance with 501.3(A).

Exception: General-purpose enclosures shall be permitted if current-interrupting contacts are

(a) Immersed in oil, or
(b) Enclosed within a chamber that is hermetically sealed against the entrance of gases or vapors, or

Generally speaking, there are several types of hermetic seals, including fusion seals such as the glass-to-metal seals in mercury-tube switches and some reed switches, welded seals, soldered seals, and seals made with gaskets. Seals of the glass-to-metal-fusion type are usually the most reliable. Soft soldered seals can be relatively porous, and their effectiveness is highly dependent on workmanship. Although gasketed seals can be very effective, depending on the gasket material used, gasket materials can be damaged and can deteriorate rapidly if exposed to atmospheres that contain solvent vapors. Gasketed enclosures may be considered hermetically sealed under some conditions; however, in accordance with the 500.2 definition of *hermetically sealed*, such enclosures cannot be used to satisfy those requirements in which hermetic sealing is recognized as a protection technique.

(c) In nonincendive circuits, or

(d) Part of a listed nonincendive component.

(2) Resistors and Similar Equipment. Resistors, resistance devices, thermionic tubes, rectifiers, and similar equipment that are used in or in connection with meters, instruments, and relays shall comply with 501.3(A).

Exception: General-purpose-type enclosures shall be permitted if such equipment is without make-and-break or sliding contacts [other than as provided in 501.3(B)(1)] and if the maximum operating temperature of any exposed surface will not exceed 80 percent of the ignition temperature in degrees Celsius of the gas or vapor involved or has been tested and found incapable of igniting the gas or vapor. This exception shall not apply to thermionic tubes.

The intent of the phrase "or has been tested and found incapable of igniting the gas or vapor" is to permit the use of listed equipment with operating temperatures higher than 80 percent of the ignition temperature. If the equipment has been tested, the safety factor inherent in this 80 percent rule is not needed. The system of temperature measurement must be specified, as 80 percent of a temperature in degrees Celsius is not the same temperature as 80 percent of that temperature in degrees Fahrenheit.

The last sentence of the exception to 501.3(B)(2) concerns ionization of the air from thermionic tubes, such as cathode-ray tubes.

(3) Without Make-or-Break Contacts. Transformer windings, impedance coils, solenoids, and other windings that do not incorporate sliding or make-or-break contacts shall be provided with enclosures. General-purpose-type enclosures shall be permitted.

(4) General-Purpose Assemblies. Where an assembly is made up of components for which general-purpose enclosures are acceptable as provided in 501.3(B)(1), (B)(2), and (B)(3), a single general-purpose enclosure shall be acceptable for the assembly. Where such an assembly includes any of the equipment described in 501.3(B)(2), the maximum

obtainable surface temperature of any component of the assembly shall be clearly and permanently indicated on the outside of the enclosure. Alternatively, equipment shall be permitted to be marked to indicate the temperature class for which it is suitable, using the temperature class (T Code) of Table 500.8(B).

(5) Fuses. Where general-purpose enclosures are permitted in 501.3(B)(1), (B)(2), (B)(3), and (B)(4), fuses for overcurrent protection of instrument circuits not subject to overloading in normal use shall be permitted to be mounted in general-purpose enclosures if each such fuse is preceded by a switch complying with 501.3(B)(1).

(6) Connections. To facilitate replacements, process control instruments shall be permitted to be connected through flexible cord, attachment plug, and receptacle, provided all of the following conditions apply:

(1) A switch complying with 501.3(B)(1) is provided so that the attachment plug is not depended on to interrupt current.
(2) The current does not exceed 3 amperes at 120 volts, nominal.
(3) The power-supply cord does not exceed 900 mm (3 ft), is of a type listed for extra-hard usage or for hard usage if protected by location, and is supplied through an attachment plug and receptacle of the locking and grounding type.
(4) Only necessary receptacles are provided.
(5) The receptacle carries a label warning against unplugging under load.

501.4 Wiring Methods.

Wiring methods shall comply with 501.4(A) or (B).

(A) Class I, Division 1.

(1) General. In Class I, Division 1 locations, the wiring methods in (a) through (d) shall be permitted.

(a) Threaded rigid metal conduit or threaded steel intermediate metal conduit. Threaded joints shall be made up with at least five threads fully engaged.

Rigid metal conduit and intermediate metal conduit must be threaded with an NPT standard conduit cutting die that provides a ¾-in. taper per foot, and five full threads must be engaged.

The *Code* recognizes electrical equipment with metric threaded entries [see 500.8(D)(2)]. Equipment with metric threaded entries must be identified to indicate that metric threads are provided or be provided with listed adapters to allow the connection of NPT-threaded conduit or fittings to the equipment. Each joint must be made up wrenchtight at couplings and unions, threaded hubs of junction boxes, device boxes, conduit bodies, and so on.

Exception: Rigid nonmetallic conduit complying with Article 352 shall be permitted where encased in a concrete envelope a minimum of 50 mm (2 in.) thick and provided with not less than 600 mm (24 in.) of cover measured from the top of the conduit to grade. The concrete encasement shall be permitted to be omitted where subject to the provisions of 511.4, Exception; 514.8, Exception No. 2; and 515.8(A). Threaded rigid metal conduit or threaded steel intermediate metal conduit shall be used for the last 600 mm (24 in.) of the underground run to emergence or to the point of connection to the aboveground raceway. An equipment grounding conductor shall be included to provide for electrical continuity of the raceway system and for grounding of non–current-carrying metal parts.

The exception to 501.4(A)(1)(a) permits the use of rigid nonmetallic conduit in some underground installations. If rigid nonmetallic conduit is used for underground wiring, threaded rigid metal conduit or threaded steel intermediate metal conduit must be used for the last 2 ft of the underground run to the point of emergence or to the point of connection to the aboveground raceway. The rigid nonmetallic conduit, including rigid nonmetallic conduit elbows and fittings, must be located not less than 2 ft below grade. The conduit must also be encased in not less than 2 in. of concrete.

The requirements covering the use of rigid nonmetallic conduit in 511.4(A)(1) Exception, 514.8 Exception No. 2, and 515.8 do not require concrete encasement. These provisions are for specific occupancies where there has been considerable experience with underground nonmetallic conduit. The exception to 501.4(A)(1)(a) applies to other occupancies with underground Class I, Division 1 locations.

If rigid nonmetallic conduit is used, an equipment grounding conductor must be included and must be bonded to the metal raceways that extend from the underground rigid nonmetallic conduit.

(b) Type MI cable with termination fittings listed for the location. Type MI cable shall be installed and supported in a manner to avoid tensile stress at the termination fittings.

The requirement in 501.4(A)(1)(b) now specifies that termination fittings used with Type MI cable must be listed for use in Class I, Division 1 hazardous (classified) locations. In previous editions of the *Code*, termination fittings used with Type MI cable were required to be *approved*. (See the

definition of *approved* in Article 100.) This change means that MI cable fittings must be evaluated in accordance with an appropriate product standard or be tested for the specific use. A fitting that is approved must be acceptable to the authority having jurisdiction, who, in many cases, requires the use of listed equipment as a basis for approval. However, the term *approved* does not mandate product evaluation or testing. The requirement that fittings used with MI cables be specifically listed for use in the particular hazardous (classified) location class and group involved provides a more objective basis for selecting the proper fitting. Type MI cable fittings, as shown in Exhibit 501.4, have a clamp-type joint that must be investigated to determine that it is explosionproof. Type MI cable fittings not investigated for use in hazardous locations may not be explosionproof. Type MI cable fittings that are suitable for nonhazardous locations may not be suitable for Class I, Division 1 hazardous (classified) locations.

Exhibit 501.5 shows an explosionproof junction box with two hubs and a threaded opening for the screw-type cover. Unused openings must be effectively closed by inserting threaded metal plugs that engage at least five full threads and afford protection equivalent to that of the wall of the box.

(c) In industrial establishments with restricted public access, where the conditions of maintenance and supervision ensure that only qualified persons service the installation, Type MC-HL cable, listed for use in Class I, Division 1 locations, with a gas/vaportight continuous corrugated metallic sheath, an overall jacket of suitable polymeric material, separate grounding conductors in accordance with 250.122, and provided with termination fittings listed for the application.

FPN: See 330.10 and 330.12 for restrictions on use of Type MC cable.

Due to the potential for physical damage to Type MC cable, its use is limited to cable that is listed specifically for use in Class I, Division 1 locations and installed at facilities that have full-time, qualified maintenance personnel. This special-use cable is identified as Type MC-HL. Qualified maintenance personnel are those who, in the course of regular maintenance procedures, would notice whether cables were damaged, understand the associated hazards, and are able to de-energize the circuit to repair the installation.

(d) In industrial establishments with restricted public access, where the conditions of maintenance and supervision ensure that only qualified persons service the installation,

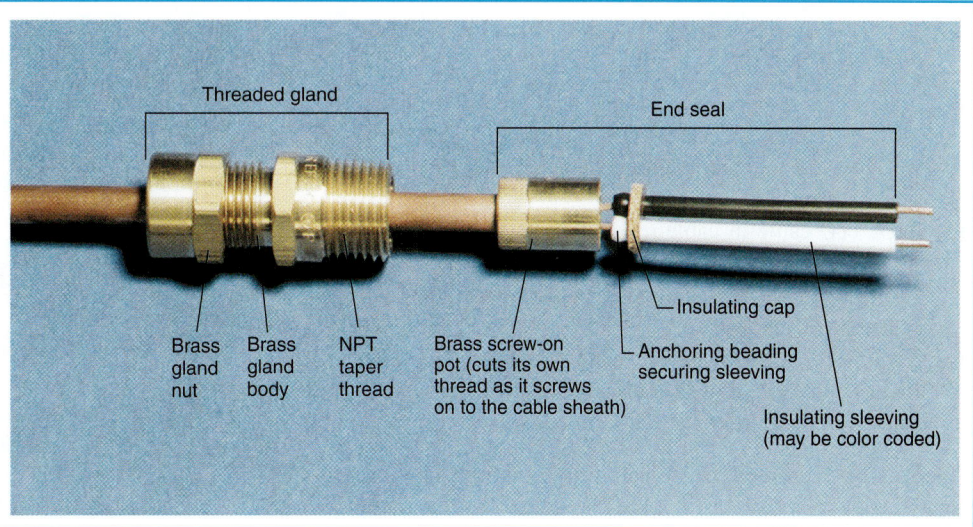

Threaded gland

End seal

Brass gland nut

Brass gland body

NPT taper thread

Brass screw-on pot (cuts its own thread as it screws on to the cable sheath)

Insulating cap

Anchoring beading securing sleeving

Insulating sleeving (may be color coded)

Exhibit 501.5 An explosionproof junction box with a screw-type cover. (Courtesy of O.Z/Gedney Co.)

Type ITC-HL cable, listed for use in Class I, Division 1 locations, with a gas/vaportight continuous corrugated metallic sheath, an overall jacket of suitable polymeric material and provided with termination fittings listed for the application.

Type ITC-HL cable has a gas/vaportight continuous corrugated metallic sheath and a polymeric jacket and is listed for use in Class I, Division 1 locations. Type ITC-HL cable is permitted to be used in industrial establishments that have

specific conditions of operation. The conditions under which Type ITC cable can be used in Class I, Division 1 locations parallel those in 501.4(A)(1)(c) for Type MC-HL cable.

(2) Flexible Connections. Where necessary to employ flexible connections, as at motor terminals, flexible fittings listed for Class I, Division 1 locations or flexible cord in accordance with the provisions of 501.11 shall be permitted.

Flexible connection fittings are available in lengths up to 3 ft for use in Class I, Division 1 locations. A flexible connection fitting consists of a deeply corrugated bronze tube with an internal nonmetallic tubular protective liner and an outer cover of braided fine bronze wires. A threaded fitting is securely attached to each end of the flexible tube. This type of flexible fitting is commonly used at motor connections, can withstand continuous vibration for long periods, is explosionproof, and affords maximum protection to any enclosed conductors.

(3) Boxes, Fittings, and Joints. All boxes, fittings, and joints shall be approved for Class I, Division 1.

(B) Class I, Division 2.

(1) General. In Class I, Division 2 locations, the following wiring methods shall be permitted:

(1) All wiring methods permitted in Article 501.4(A).
(2) Threaded rigid metal conduit, threaded steel intermediate metal conduit.

(3) Enclosed gasketed busways, enclosed gasketed wireways.

(4) Type PLTC cable in accordance with the provisions of Article 725, or in cable tray systems. PLTC shall be installed in a manner to avoid tensile stress at the termination fittings.

(5) Type ITC cable in cable trays, in raceways, supported by messenger wire, afforded mechanical protection and run as open wiring, or directly buried where the cable is listed for this use.

(6) Type MI, MC, MV, or TC cable with termination fittings, or in cable tray systems and installed in a manner to avoid tensile stress at the termination fittings.

(2) Flexible Connections. Where provision must be made for limited flexibility, flexible metal fittings, flexible metal conduit with listed fittings, liquidtight flexible metal conduit with listed fittings, liquidtight flexible nonmetallic conduit with listed fittings, or flexible cord listed for extra-hard usage and provided with listed bushed fittings shall be used. An additional conductor for grounding shall be included in the flexible cord.

> FPN: See 501.16(B) for grounding requirements where flexible conduit is used.

(3) Nonincendive Field Wiring. Nonincendive field wiring shall be permitted using any of the wiring methods permitted for unclassified locations. Nonincendive field wiring systems shall be installed in accordance with the control drawing(s). Simple apparatus, not shown on the control drawing, shall be permitted in a nonincendive field wiring circuit, provided the simple apparatus does not interconnect the nonincendive field wiring circuit to any other circuit.

> FPN: Simple apparatus is defined in 504.2.

Separate nonincendive field wiring circuits shall be installed in accordance with one of the following:

(1) In separate cables
(2) In multiconductor cables where the conductors of each circuit are within a grounded metal shield
(3) In multiconductor cables, where the conductors of each circuit have insulation with a minimum thickness of 0.25 mm (0.01 in.)

The installation of nonincendive field wiring is covered in 501.4(B)(3). See the commentary following 501.3(B)(1), Exception (c) and the definitions of *nonincendive circuit* and *nonincendive field wiring* in 500.2. Many low-voltage, low-energy circuits are of the nonincendive type. However, a Class 2 circuit, as defined in Article 725, is not necessarily nonincendive. Testing laboratories, such as Factory Mutual Research Corp. and Underwriters Laboratories Inc., list many types of equipment that have nonincendive circuits intended for connection of nonincendive field wiring. This equipment is evaluated for use in one or more of the Class I gas or vapor groups and is permitted for use only in Division 2 locations. Some common telephone circuits and thermocouple circuits are also nonincendive.

(4) Boxes, Fittings, and Joints. Boxes, fittings, and joints shall not be required to be explosionproof except as required by 501.3(B)(1), 501.6(B)(1), and 501.14(B).

In Class I, Division 2 locations, boxes, fittings, and joints are not required to be explosionproof at lighting outlets or at enclosures containing no arcing devices, such as solenoids and control transformers, if the maximum operating temperature of any exposed surface does not exceed 80 percent of the ignition temperature in degrees Celsius. Where general-purpose enclosures are permitted by 501.4(B)(4), rigid or intermediate metal conduit may be used with locknuts and bushings. However, a bonding jumper with proper fittings or bonding-type locknuts is required to be used between the enclosure and the raceway to ensure adequate bonding from the hazardous area to the point of grounding at the service equipment or separately derived system. See 501.16(A) for grounding and bonding requirements.

Where limited flexibility is necessary and approved fittings are required for use with flexible metal conduit, liquidtight flexible conduit, and extra-hard-usage flexible cord, the fittings are not required to be specifically approved for Class I locations. Also, where flexible conduit or liquidtight flexible conduit is used, internal or external bonding jumpers with proper fittings must be provided, in accordance with 501.16(B), unless liquidtight flexible conduit is installed under the conditions described in 501.16(B), Exception.

Section 501.4(B)(1) permits a variety of cable types, cable tray systems in accordance with 392.3(D), enclosed gasketed wireways, and enclosed gasketed busways. The cable and cable fittings, cable trays, wireways, and busways are not required to be specifically listed or labeled for Class I, Division 2 locations. For example, if Type ITC or MC cable is used, neither the cable nor the fittings need to be listed for use in hazardous (classified) locations. Type AC cable is not a permitted wiring method in 501.4(B) because of concern about arcing between convolutions during ground-fault conditions.

Any wiring method suitable for ordinary locations may be used for nonincendive field wiring. See 501.4(B)(3).

501.5 Sealing and Drainage.

Seals in conduit and cable systems shall comply with 501.5(A) through (F). Sealing compound shall be used in

Type MI cable termination fittings to exclude moisture and other fluids from the cable insulation.

> FPN No. 1: Seals are provided in conduit and cable systems to minimize the passage of gases and vapors and prevent the passage of flames from one portion of the electrical installation to another through the conduit. Such communication through Type MI cable is inherently prevented by construction of the cable. Unless specifically designed and tested for the purpose, conduit and cable seals are not intended to prevent the passage of liquids, gases, or vapors at a continuous pressure differential across the seal. Even at differences in pressure across the seal equivalent to a few inches of water, there may be a slow passage of gas or vapor through a seal and through conductors passing through the seal. See 501.5(E)(2). Temperature extremes and highly corrosive liquids and vapors can affect the ability of seals to perform their intended function. See 501.5(C)(2).
>
> FPN No. 2: Gas or vapor leakage and propagation of flames may occur through the interstices between the strands of standard stranded conductors larger than 2 AWG. Special conductor constructions, for example, compacted strands or sealing of the individual strands, are means of reducing leakage and preventing the propagation of flames.

The sealing compound used in conduit seal fittings is somewhat porous, so that gases, particularly those under slight pressure and those with small molecules such as hydrogen, can pass slowly through the sealing compound. Also, the seal is around the insulation on the conductor, and gases can be transmitted slowly through the air spaces (the interstices) between strands of stranded conductors. The cable core does not include the interstices of the conductor strands. See the commentary following 501.5(E)(2), FPN No. 2.

Experience has shown, however, that under normal conditions for smaller conductors, and with only normal atmospheric pressure differentials across the seal, the passage of gas through a seal is not sufficient to result in a hazard. For larger conductors, however, gas or vapor leakage and flame propagation may occur through the interstices between the strands, of stranded conductors. Special conductor constructions, such as compacted strands or sealing individual strands, may reduce leakage and prevent flame propagation.

Sealing fittings should be used only with the sealing compound or compounds recommended by the fitting manufacturer. Different sealing compounds have different rates of expansion and contraction that may affect their performance within a given fitting. Sealing compound must be used as soon as possible on Type MI cable terminations to exclude moisture from cable insulation.

The use of Teflon tapes or joint compounds on conduit threads may weaken the seal fitting and interrupt the equipment grounding path. Cracks have developed in fittings during hydrostatic testing in which these materials were used.

(A) Conduit Seals, Class I, Division 1. In Class I, Division 1 locations, conduit seals shall be located in accordance with 501.5(A)(1) through (A)(4).

(1) Entering Enclosures. In each conduit entry into an explosionproof enclosure where either

(1) The enclosure contains apparatus, such as switches, circuit breakers, fuses, relays, or resistors, that may produce arcs, sparks, or high temperatures that are considered to be an ignition source in normal operation, or
(2) The entry is metric designator 53 (trade size 2) or larger and the enclosure contains terminals, splices, or taps.

For the purposes of this section, high temperatures shall be considered to be any temperatures exceeding 80 percent of the autoignition temperature in degrees Celsius of the gas or vapor involved.

Exception to 501.5(A)(1)(1): Seals shall not be required for conduit entering an enclosure where such switches, circuit breakers, fuses, relays, or resistors are

(a) *Enclosed within a chamber hermetically sealed against the entrance of gases or vapors, or*
(b) *Immersed in oil in accordance with 501.6(B)(1)(2), or*
(c) *Enclosed within a factory-sealed explosionproof chamber located within the enclosure, identified for the location, and marked "factory sealed" or equivalent, unless the enclosure entry is metric designator 53 (trade size 2) or larger.*
(d) *In nonincendive circuits.*

Factory-sealed enclosures shall not be considered to serve as a seal for another adjacent explosionproof enclosure that is required to have a conduit seal.

Conduit seals shall be installed within 450 mm (18 in.) from the enclosure. Only explosionproof unions, couplings, reducers, elbows, capped elbows, and conduit bodies similar to L, T, and Cross types that are not larger than the trade size of the conduit shall be permitted between the sealing fitting and the explosionproof enclosure.

(2) Pressurized Enclosures. In each conduit entry into a pressurized enclosure where the conduit is not pressurized as part of the protection system. Conduit seals shall be installed within 450 mm (18 in.) from the pressurized enclosure.

> FPN No. 1: Installing the seal as close as possible to the enclosure will reduce problems with purging the dead airspace in the pressurized conduit.
>
> FPN No. 2: For further information, see NFPA 496-1998, *Standard for Purged and Pressurized Enclosures for Electrical Equipment.*

(3) Two or More Explosionproof Enclosures. Where two or more explosionproof enclosures for which conduit seals are required under 501.5(A)(1) are connected by nipples or by runs of conduit not more than 900 mm (36 in.) long, a single conduit seal in each such nipple connection or run of conduit shall be considered sufficient if located not more than 450 mm (18 in.) from either enclosure.

An example of 501.5(A) requirements for the location of conduit seals in Class I, Division 1 locations is illustrated in Exhibit 501.6. In the example shown in Exhibit 501.6, two seals are required so that the run of conduit between Enclosure No. 1 and Enclosure No. 2 is sealed.

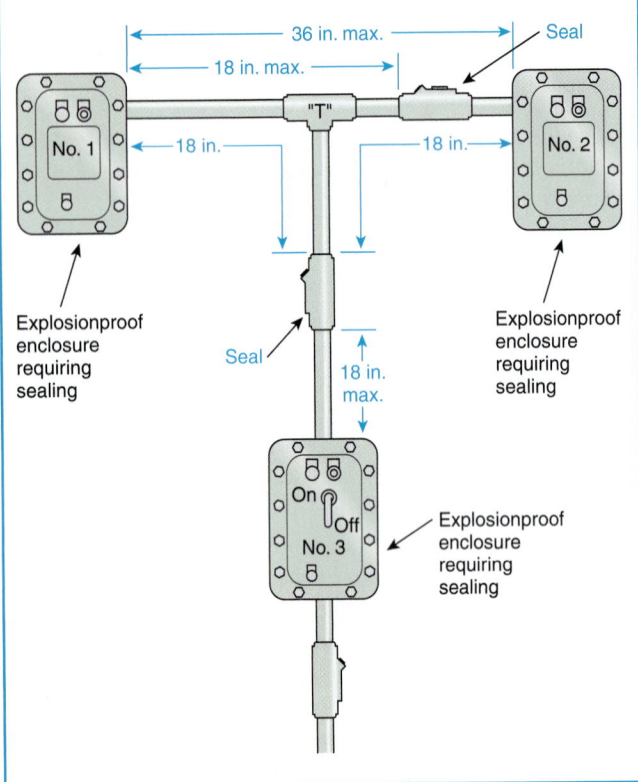

Exhibit 501.6 Two seals are required so that the run of conduit between Enclosure No. 1 and Enclosure No. 2 is sealed. Even if Enclosure No. 3 were not required to be sealed, the vertical seal in the vertical run of conduit to Enclosure No. 3 would be required to be sealed within 18 in. of Enclosure No. 1, because the vertical conduit run to the "T" fitting is a conduit run to Enclosure No. 1.

(4) Class I, Division 1 Boundary. In each conduit run leaving a Class I, Division 1 location. The sealing fitting shall be permitted on either side of the boundary of such location within 3.05 m (10 ft) of the boundary and shall be designed and installed so as to minimize the amount of gas

or vapor within the Division 1 portion of the conduit from being communicated to the conduit beyond the seal. Except for listed explosionproof reducers at the conduit seal, there shall be no union, coupling, box, or fitting between the conduit seal and the point at which the conduit leaves the Division 1 location.

Exception No. 1: Metal conduit that contains no unions, couplings, boxes, or fittings and passes completely through a Class I, Division 1 location with no fittings less than 300 mm (12 in.) beyond each boundary shall not require a conduit seal if the termination points of the unbroken conduit are in unclassified locations.

Exception No. 2: For underground conduit installed in accordance with 300.5 where the boundary is beneath the ground, the sealing fitting shall be permitted to be installed after the conduit leaves the ground, but there shall be no union, coupling, box, or fitting, other than listed explosionproof reducers at the sealing fitting, in the conduit between the sealing fitting and the point at which the conduit leaves the ground.

(B) Conduit Seals, Class I, Division 2. In Class I, Division 2 locations, conduit seals shall be located in accordance with 501.5(B)(1) and (B)(2).

(1) Entering Enclosures. For connections to enclosures that are required to be explosionproof, a conduit seal shall be provided in accordance with 501.5(A)(1)(1) and (A)(3). All portions of the conduit run or nipple between the seal and such enclosure shall comply with 501.4(A).

(2) Class I, Division 2 Boundary. In each conduit run passing from a Class I, Division 2 location into an unclassified location. The sealing fitting shall be permitted on either side of the boundary of such location within 3.05 m (10 ft) of the boundary and shall be designed and installed so as to minimize the amount of gas or vapor within the Division 2 portion of the conduit from being communicated to the conduit beyond the seal. Rigid metal conduit or threaded steel intermediate metal conduit shall be used between the sealing fitting and the point at which the conduit leaves the Division 2 location, and a threaded connection shall be used at the sealing fitting. Except for listed explosionproof reducers at the conduit seal, there shall be no union, coupling, box, or fitting between the conduit seal and the point at which the conduit leaves the Division 2 location.

Exception No. 1: Metal conduit that contains no unions, couplings, boxes, or fittings and passes completely through a Class I, Division 2 location with no fittings less than 300 mm (12 in.) beyond each boundary shall not be required to be sealed if the termination points of the unbroken conduit are in unclassified locations.

Exception No. 2: Conduit systems terminating at an unclassified location where a wiring method transition is made to cable tray, cablebus, ventilated busway, Type MI cable, or open wiring shall not be required to be sealed where passing from the Class I, Division 2 location into the unclassified location. The unclassified location shall be outdoors or, if the conduit system is all in one room, it shall be permitted to be indoors. The conduits shall not terminate at an enclosure containing an ignition source in normal operation.

Exception No. 3: Conduit systems passing from an enclosure or room that is unclassified as a result of pressurization into a Class I, Division 2 location shall not require a seal at the boundary.

> FPN: For further information, refer to NFPA 496-1998, *Standard for Purged and Pressurized Enclosures for Electrical Equipment.*

Exception No. 4: Segments of aboveground conduit systems shall not be required to be sealed where passing from a Class I, Division 2 location into an unclassified location if the following conditions are met:

(a) No part of the conduit system segment passes through a Class I, Division 1 location where the conduit contains unions, couplings, boxes, or fittings within 300 mm (12 in.) of the Class I, Division 1 location; and

(b) The conduit system segment is located entirely in outdoor locations; and

(c) The conduit system segment is not directly connected to canned pumps, process or service connections for flow, pressure, or analysis measurement, and so forth, that depend on a single compression seal, diaphragm, or tube to prevent flammable or combustible fluids from entering the conduit system; and

(d) The conduit system segment contains only threaded metal conduit, unions, couplings, conduit bodies, and fittings in the unclassified location; and

(e) The conduit system segment is sealed at its entry to each enclosure or fitting housing terminals, splices, or taps in Class I, Division 2 locations

(C) Class I, Divisions 1 and 2. Where required, seals in Class I, Division 1 and 2 locations shall comply with 501.5(C)(1) through (C)(6).

(1) Fittings. Enclosures for connections or equipment shall be provided with an integral means for sealing, or sealing fittings listed for the location shall be used. Sealing fittings shall be listed for use with one or more specific compounds and shall be accessible.

(2) Compound. The compound shall provide a seal against passage of gas or vapors through the seal fitting, shall not be affected by the surrounding atmosphere or liquids, and shall not have a melting point of less than 93°C (200°F).

(3) Thickness of Compounds. In a completed seal, the minimum thickness of the sealing compound shall not be less than the trade size of the sealing fitting and, in no case, less than 16 mm (⅝ in.).

Exception: Listed cable sealing fittings shall not be required to have a minimum thickness equal to the trade size of the fitting.

(4) Splices and Taps. Splices and taps shall not be made in fittings intended only for sealing with compound, nor shall other fittings in which splices or taps are made be filled with compound.

(5) Assemblies. In an assembly where equipment that may produce arcs, sparks, or high temperatures is located in a compartment separate from the compartment containing splices or taps, and an integral seal is provided where conductors pass from one compartment to the other, the entire assembly shall be identified for the location. Seals in conduit connections to the compartment containing splices or taps shall be provided in Class I, Division 1 locations where required by 501.5(A)(1)(2).

(6) Conductor Fill. The cross-sectional area of the conductors permitted in a seal shall not exceed 25 percent of the cross-sectional area of a rigid metal conduit of the same trade size unless it is specifically identified for a higher percentage of fill.

(D) Cable Seals, Class I, Division 1. In Class I, Division 1 locations, cable seals shall be located according to 501.5(D)(1) through (D)(3).

(1) At Terminations. Cable shall be sealed at all terminations. The sealing fitting shall comply with 501.5(C). Multiconductor Type MC-HL cables with a gas/vaportight continuous corrugated metallic sheath and an overall jacket of suitable polymeric material shall be sealed with a listed fitting after removing the jacket and any other covering so that the sealing compound surrounds each individual insulated conductor in such a manner as to minimize the passage of gases and vapors.

> In accordance with 501.4(A)(1)(c), Type MC-HL cable is permitted as a wiring method in Class I, Division 1 areas. Type MC-HL cable has a continuous corrugated metallic sheath that is gas/vaportight and an outer nonmetallic material that enables it to be installed in wet locations. Type MC-HL cables are available with ratings up to 35,000 volts. The cable is specifically listed for use in Class I, Division 1 locations.
>
> The provisions of 501.5(D) contain the sealing requirements for cables installed in Class I, Division 1 locations,

which differ from the requirements for sealing conduits. Conduits entering explosionproof enclosures must be sealed if the enclosure contains equipment that produces arcs, sparks, or high temperatures or if the conduit entering the enclosure is trade size 2 or larger. In accordance with 501.5(D)(1), cables must be sealed at all terminations, irrespective of the type of equipment contained in the enclosure or the diameter of the cable. Exhibit 501.7 shows an example of a cable sealing fitting for Type MC-HL cable.

Exhibit 501.7 An explosionproof sealing fitting for Type MC-HL cable. (Courtesy of Cooper Crouse-Hinds)

Exception: Shielded cables and twisted pair cables shall not require the removal of the shielding material or separation of the twisted pairs, provided the termination is by an approved means to minimize the entrance of gases or vapors and prevent propagation of flame into the cable core.

Shielded cables and twisted pair cables, commonly used for signaling and instrumentation circuits, are permitted by the exception to 501.5(D) to be sealed without removal of the outer sheathing or the separation of the twisted conductors. The need to provide a suitable seal while not adversely impacting the operational performance of these cables is accomplished through this exception.

(2) Cables Capable of Transmitting Gases or Vapors. Cables in conduit with a gas/vaportight continuous sheath capable of transmitting gases or vapors through the cable core shall be sealed in the Division 1 location after removing the jacket and any other coverings so that the sealing compound will surround each individual insulated conductor and the outer jacket.

Exception: Multiconductor cables with a gas/vaportight continuous sheath capable of transmitting gases or vapors through the cable core shall be permitted to be considered as a single conductor by sealing the cable in the conduit within 450 mm (18 in.) of the enclosure and the cable end within the enclosure by an approved means to minimize the entrance of gases or vapors and prevent the propagation of flame into the cable core, or by other approved methods. For shielded cables and twisted pair cables, it shall not be required to remove the shielding material or separate the twisted pair.

The intent of the exception to 501.5(D)(2) is to permit flat computer cables, coaxial cables, and twisted pair cables to be treated as single conductors if installed in conduit, because separating the individual conductors or removing the outer jacket (of a coaxial cable or a twisted pair, for example) is impractical and can destroy the electrical properties of the cable.

In addition to the cable seal, the end of the cable within the enclosure must also be sealed.

(3) Cables Incapable of Transmitting Gases or Vapors. Each multiconductor cable in conduit shall be considered as a single conductor if the cable is incapable of transmitting gases or vapors through the cable core. These cables shall be sealed in accordance with 501.5(A).

(E) Cable Seals, Class I, Division 2. In Class I, Division 2 locations, cable seals shall be located in accordance with 501.5(E)(1) through (E)(4).

(1) Terminations. Cables entering enclosures that are required to be explosionproof shall be sealed at the point of entrance. The sealing fitting shall comply with 501.5(B)(1). Multiconductor cables with a gas/vaportight continuous sheath capable of transmitting gases or vapors through the cable core shall be sealed in a listed fitting in the Division 2 location after removing the jacket and any other coverings so that the sealing compound surrounds each individual insulated conductor in such a manner as to minimize the passage of gases and vapors. Multiconductor cables in conduit shall be sealed as described in 501.5(D).

Exception No. 1: Cables passing from an enclosure or room that is unclassified as a result of Type Z pressurization into a Class I, Division 2 location shall not require a seal at the boundary.

If cables are run from a Type Z pressurized room or enclosure into a Class I, Division 2 location, Exception No. 1 to 501.5(E)(1) allows the cables to be installed without a sealing fitting at the boundary. This correlates with a similar allowance for conduit systems found in 501.5(B)(2), Exception No. 3.

Exception No. 2: Shielded cables and twisted pair cables shall not require the removal of the shielding material or separation of the twisted pairs, provided the termination is by an approved means to minimize the entrance of gases or vapors and prevent propagation of flame into the cable core.

(2) Cables That Do Not Transmit Gases or Vapors. Cables that have a gas/vaportight continuous sheath and do not transmit gases or vapors through the cable core in excess of the quantity permitted for seal fittings shall not be required to be sealed except as required in 501.5(E). The minimum length of such cable run shall not be less than that length that limits gas or vapor flow through the cable core to the rate permitted for seal fittings [200 cm^3/hr (0.007 ft^3/hr) of air at a pressure of 1500 pascals (6 in. of water)].

The ability of a cable to transmit gases or vapors through the core (primarily between insulated conductors) depends not only on how tightly packed the conductors are within the outer sheaths, and the location and composition of fillers, but also on how the cable has been handled and the geometry of the cable run. If there is any question as to whether or not the cable run is capable of transmitting gases or vapors through the core, a sealing fitting should be installed. See the commentary following 501.5, FPN No. 2.

FPN No. 1: See ANSI/UL 886-1994, *Outlet Boxes and Fittings for Use in Hazardous (Classified) Locations.*
FPN No. 2: The cable core does not include the interstices of the conductor strands.

When performing a leak rate test, the ends of each individual conductor in the cable are sealed to prevent migration of gases or vapors between the individual strands of wire. Sealing can be achieved by dipping the cable end in hot wax. The rate of flow through the filler between the insulated conductors can now be accurately measured, excluding any leakage through the conductor strands. If this is done, the wax should be removed before making the connections and placing the system in service.

(3) Cables Capable of Transmitting Gases or Vapors. Cables with a gas/vaportight continuous sheath capable of transmitting gases or vapors through the cable core shall not be required to be sealed except as required in 501.5(E), unless the cable is attached to process equipment or devices that may cause a pressure in excess of 1500 pascals (6 in. of water) to be exerted at a cable end, in which case a seal, barrier, or other means shall be provided to prevent migration of flammables into an unclassified location.

Exception: Cables with an unbroken gas/vaportight continuous sheath shall be permitted to pass through a Class I, Division 2 location without seals.

(4) Cables Without Gas/Vaportight Sheath. Cables that do not have gas/vaportight continuous sheath shall be sealed at the boundary of the Division 2 and unclassified location in such a manner as to minimize the passage of gases or vapors into an unclassified location.

Table 5.3 summarizes the sealing requirements of 501.5.

(F) Drainage.

(1) Control Equipment. Where there is a probability that liquid or other condensed vapor may be trapped within enclosures for control equipment or at any point in the raceway system, approved means shall be provided to prevent accumulation or to permit periodic draining of such liquid or condensed vapor.

(2) Motors and Generators. Where the authority having jurisdiction judges that there is a probability that liquid or condensed vapor may accumulate within motors or generators, joints and conduit systems shall be arranged to minimize the entrance of liquid. If means to prevent accumulation or to permit periodic draining are judged necessary, such means shall be provided at the time of manufacture and shall be considered an integral part of the machine.

(3) Canned Pumps, Process, or Service Connections, etc. For canned pumps, process, or service connections for flow, pressure, or analysis measurement, and so forth, that depend on a single compression seal, diaphragm, or tube to prevent flammable or combustible fluids from entering the

Table 5.3 Conduit and Cable Sealing Requirements

Classification	Application	Location of Seal
Conduit Seals Class I, Division 1	Switch enclosure	In conduit run within 18 in. of enclosure
	Circuit breaker enclosure Fuse enclosure Relay enclosure Resistor enclosure Arcing or sparking apparatus High-temperature apparatus	
	Explosionproof enclosure containing arcing or sparking contacts that are hermetically sealed against gas or vapor entry	In conduit runs of 1½ in. and smaller, no seal is required. If conduit is larger than 1½ in., in conduit run within 18 in. of enclosure
	Explosionproof enclosure containing arcing or sparking contacts that are immersed in oil, in accordance with 501.6(B)(1)(2)	
	Enclosure containing terminals, splices, or taps fitting containing terminals, splices, or taps	In conduit runs smaller than trade size 2, no seal is required. If conduit is 2 in. or larger, in conduit run within 18 in. of enclosure
	Two explosionproof enclosures with a conduit run between them of 36 in. or less	In conduit run within 18 in. of each enclosure. Permitted to use a single seal in each run as long as the seal is within 18 in. of each enclosure
	Two explosionproof enclosures with a conduit run between them greater than 36 in.	In conduit run within 18 in. of each enclosure
	Conduit run leaving Division 1 location	On either side of boundary. No unions, couplings, boxes, or fittings (other than explosionproof reducers) permitted between the seal fitting and the point where the conduit leaves the Division 1 location.
	Metal conduit containing no unions, couplings, boxes, or fittings that passes completely through a Class I, Division 1 location, with no fittings less than 12 in. beyond each boundary	Not required to be sealed if the termination points of the unbroken conduit are in unclassified locations.
Class I, Division 2	Enclosure required to be explosionproof	Seal as required for similar equipment in Division 1 location
	Conduit run leaving Division 2 location	On either side of boundary. No unions, couplings, boxes, or fittings (other than explosionproof reducers) permitted between the seal fitting and the point where the conduit leaves the Division 2 location.
	Metal conduit containing no unions, couplings, boxes, or fittings that passes completely through a Division 2 location with no fittings less than 12 in. beyond each boundary	Not required to be sealed if the termination points of the unbroken conduit are in unclassified locations.
	Conduit systems terminating at an unclassified location where a wiring method transition is made to cable tray, cablebus, ventilated busway, Type MI cable, or open wiring	Not required to be sealed if passing from the Class I, Division 2 location into an outdoor unclassified location or an indoor location if the conduit system is all in one room. The conduits do not terminate at an enclosure containing an ignition source in normal operation.
Cable Seals Class I, Division 1	Enclosure with integral seal	Conduit seal fitting not required
	Multiconductor Type MC-HL cables with a gas/vaportight continuous corrugated metallic sheath and an overall jacket of suitable polymeric material	Seal at all terminations with an approved fitting after removing the jacket and any other covering, so that the sealing compound surrounds each individual insulated conductor.
	Cables in conduit with a gas/vaportight continuous sheath capable of transmitting gases or vapors through the cable core	Seal in the Division 1 location after removing the jacket and any other coverings, so that the sealing compound surrounds each individual insulated conductor and the outer jacket.

Table 5.3 *Continued*

Classification	Application	Location of Seal
	Multiconductor cables with a gas/vaportight continuous sheath capable of transmitting gases or vapors through the cable core	Permitted to be considered a single conductor by sealing the cable in the conduit within 18 in. of the enclosure and the cable end within the enclosure by an approved means, to minimize the entrance of gases or vapors and prevent the propagation of flame into the cable core, or by other approved methods.
	For shielded cables and twisted pair cables	Removal of the shielding material or separation of the twisted pair is not required. Sealing the cable in the conduit within 18 in. of the enclosure and the cable end within the enclosure by an approved means, to minimize the entrance of gases or vapors and prevent the propagation of flame into the cable core, or by other approved methods.
	Each multiconductor cable in conduit if the cable is incapable of transmitting gases or vapors through the cable core	Considered a single conductor. These cables are sealed in accordance with Section 501.5(A)
Class I, Division 2	Cables entering enclosures that are required to be approved for Class I locations	Sealed at the point of entrance
	Multiconductor cables in conduit	Sealed in accordance with the requirements for Division 1 locations
	Multiconductor cables with a gas/vaportight continuous sheath capable of transmitting gases or vapors through the cable core	Sealed in an approved fitting in the Division 2 location after removing the jacket and any other coverings, so that the sealing compound surrounds each individual insulated conductor.
	Cables with a gas/vaportight continuous sheath that will not transmit gases or vapors through the cable core in excess of the quantity permitted for seal fittings. The minimum length of such cable run is not less than that length that limits gas or vapor flow through the cable core to the rate permitted for seal fittings (0.007 cu ft per hour of air at a pressure of 6 in. of water).	Not required to be sealed unless entering an enclosure that is required to be approved for Class I locations.
	Cables with a gas/vaportight continuous sheath capable of transmitting gases or vapors through the cable core	Not required to be sealed unless entering an enclosure that is required to be approved for Class I locations or unless the cable is attached to process equipment or devices that may cause a pressure in excess of 6 in. of water to be exerted at a cable end, in which case a seal, barrier, or other means is provided to prevent migration of flammables into an unclassified area.
	Cables with an unbroken gas/vaportight continuous sheath that pass through a Class I, Division 2 location	No seal required
	Cables that do not have a gas/vaportight continuous sheath	Sealed at the boundary of the Division 2 and unclassified location in such a manner as to minimize the passage of gases or vapors into an unclassified location.

electrical raceway or cable system capable of transmitting fluids, an additional approved seal, barrier, or other means shall be provided to prevent the flammable or combustible fluid from entering the raceway or cable system capable of transmitting fluids beyond the additional devices or means,

if the primary seal fails. The additional approved seal or barrier and the interconnecting enclosure shall meet the temperature and pressure conditions to which they will be subjected upon failure of the primary seal, unless other approved means are provided to accomplish the purpose above. Drains, vents, or other devices shall be provided so that primary seal leakage will be obvious.

Canned pumps and other process equipment that operate above atmospheric pressure are provided with a primary seal where the electrical conductors enter the pump or equipment containing flammable liquids or gases under pressure. Sealant for sealing fittings is not designed to withstand high pressure or extremely low temperatures, which might be encountered in canned pump installations. Therefore, a second seal or barrier is required to prevent fluid from entering the electrical conduit or cable system. In addition to this seal or barrier, a drain, vent, or other similar device that indicates failure of the primary seal must be provided. This redundant protection system may be achieved by installing a vented junction (box) enclosure within the classified area where the conductors terminate on busbars. Terminating the conductors in this manner allows any fluid that escapes through the primary seal and that has traveled through the stranding of the conductors to vent at the terminations. The circuit continues through the vented enclosure at normal atmospheric pressure to another set of conductors that also must be sealed with a sealing fitting if they travel into a different classified area.

FPN: See also the fine print notes to 501.5.

A seal in a conduit prevents an explosion from traveling through the conduit to another enclosure and minimizes the passage of gases or vapors from a hazardous (classified) location to a nonhazardous location. If the conduit enters an enclosure that contains arcing or high-temperature equipment, a sealing fitting must be placed within 18 in. of the enclosure it isolates; conduit bodies ("L," "T," etc.), couplings, unions, and elbows are the only enclosures or fittings permitted between the seal and the enclosure. Exhibit 501.6 illustrates the placement of conduit seals. See Exhibit 501.8 for an approved type of union. If two enclosures are spaced not more than 36 in. apart, a single seal may be placed between two connecting nipples if the seal is located not more than 18 in. from either enclosure.

In each 2-in. or larger conduit, a sealing fitting must be placed within 18 in. of the conduit entrance to any explosionproof enclosure, regardless of whether the enclosure contains arcing or sparking equipment or if it contains only splices, taps, or terminals.

A sealing fitting is also required at the point where the conduit leaves a Division 1 location or passes from a Division 2 location to a nonhazardous location. The sealing fitting is permitted on either side of the boundary, and no union, coupling, box, or similar fitting is permitted between

Exhibit 501.8 An explosionproof union. (Courtesy of Thomas & Betts Corp.)

the seal and the boundary. However, approved explosionproof reducers are permitted to be installed in conduit seals.

It is preferable to locate the sealing fitting on the nonhazardous side of the boundary. A sealing fitting located as such serves two purposes: it completes the explosionproof wiring method, and it completes the explosionproof enclosure system. Note that a ½-in. conduit connected to an explosionproof box that contains only splices, even in a Division 1 location, is not required to be sealed within 18 in. of the box. The sealing fitting at the boundary of the Division 1 location serves to complete the explosionproof system. The sealing fitting at the boundary also prevents the conduit system from serving as a pipe to transmit flammable mixtures from a Division 1 or Division 2 location to a nonhazardous location.

In Class I, Division 2 locations, a seal is required in each conduit entering an enclosure that is required to be explosionproof, in order to complete the explosionproof enclosure.

The maximum permitted fill for the conduit is 40 percent; the maximum permitted fill for most conduit seals is 25 percent. If the conduit fill exceeds 25 percent of the cross-sectional area of the sealing fitting, a larger trade size seal may be required. Reducers are allowed for connection of a larger trade size sealing fitting to the conduit.

Exhibit 501.9 illustrates the proper sealing of a fitting. A dam must be provided to prevent the sealing material, while still in the liquid state, from running out of the fitting. All conductors must be separated to permit the sealing material to run between them. The sealing compound must have a minimum thickness of not less than the trade size of the conduit, and in no case less than ⅝ in. Conduit fittings for

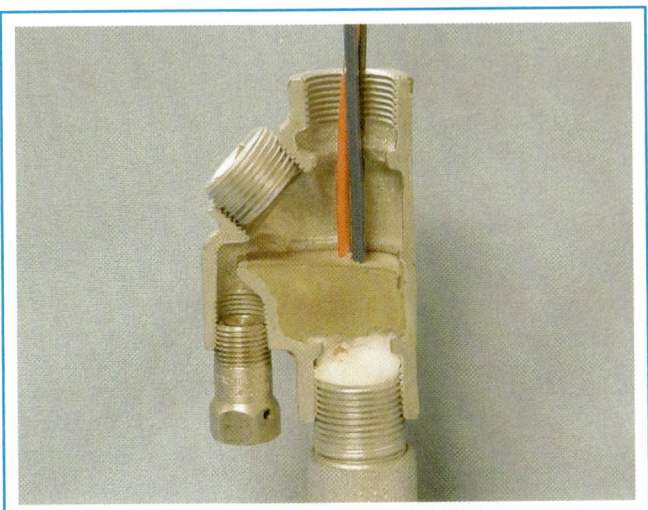

Exhibit 501.9 A sealing fitting placed in a run of conduit to minimize the passage of gases from one portion of the electrical installation to another. (Courtesy of Appleton Electric Co., EGS Electrical Group)

Exhibit 501.10 A sealing fitting with an automatic drain plug. (Courtesy of Appleton Electric Co., EGS Electrical Group)

Exhibit 501.11 A combination breather-drainage fitting. (Courtesy of Appleton Electric Co., EGS Electrical Group)

sealing are to be used only with sealing compound that is supplied with the fitting and specified by the manufacturer in instructions furnished with the fitting.

Unless the additional seal or barrier, described in 501.5(F)(3), and interconnecting enclosures meet the performance requirements of the primary seal, the application of pressure or exposure to extreme temperatures must be prevented at the additional seal or barrier so that the process fluid will not enter the conduit system if the primary seal fails. If the process fluid is a gas or can become a gas under ordinary atmospheric conditions (liquefied natural gas, for example), the drain mentioned in 501.5(F)(3) should be a vent. See the commentary following 501.5(F)(3).

The necessary sealing may be accomplished by a sealing fitting and compound. To eliminate the time-consuming task of field-poured seals, a factory-sealed device with the seal designed into the device is permissible. A wide selection of factory-sealed devices are available for a variety of installations in hazardous (classified) locations. For example, explosionproof motors are normally factory sealed and therefore require no additional seal. Factory-sealed devices are usually marked as such. If a conduit terminates in a motor, however, and if the conduit is 2 in. or larger, a seal must be placed within 18 in. of the motor terminal housing.

Exhibit 501.10 shows a sealing fitting designed for use in a vertical run of conduit to provide drainage for any condensation of moisture trapped by the seal above the enclosure. Any accumulation of water runs down over the surface of the sealing compound, flowing through an explosionproof drain.

Exhibit 501.11 shows a combination drain and breather fitting. This type of fitting is specifically designed to serve as a water drain and air vent while providing positive explosionproof protection. The fitting permits the escape of accumulated water through its drain, and the breather allows the continuous circulation of air, preventing condensation of any moisture that may be present. Individual drain or breather fittings are also available. It is good practice to consider the installation of drain, breather, or combination fittings to guard against water accumulation, which can cause future insulation failures, even though prevalent conditions may not indicate a need.

Exhibit 501.12 illustrates a Class I, Division 1 location using threaded rigid metal conduit or threaded intermediate metal conduit and explosionproof fittings and equipment, including motors, motor controllers, push-button stations, lighting outlets, and junction boxes. The enclosures for the disconnecting means and motor controller for the motor (right portion of the drawing) are placed in a nonhazardous location and are thus not required to be explosionproof.

In Exhibit 501.12, each of the three conduits is sealed on the nonhazardous side before passing into the hazardous (classified) location. The pigtail leads of both motors are factory sealed at the motor-terminal housing, and, unless the size of the flexible fitting entering the motor-terminal housing is trade size 2 or larger, no other seals are needed at this point. Because the push-button control station and the motor controller and disconnect (left portion of the drawing) are considered arc-producing devices, conduits are sealed within 18 in. of the entrance to these enclosures. Seals are required even though the contacts may be immersed in oil.

In Exhibit 501.12, a seal is provided within 18 in. of the switch controlling the lighting. The design of the luminaire, as required by ANSI/UL 844, is such that the explosionproof chamber for the wiring must be separated or sealed

from the lamp compartment; hence, a separate seal is not required adjacent to luminaires that comply with ANSI/UL 844. The luminaire is suspended on a conduit stem threaded into the cover of an explosionproof ceiling box. See 501.9 for luminaire requirements.

501.6 Switches, Circuit Breakers, Motor Controllers, and Fuses.

(A) Class I, Division 1. In Class I, Division 1 locations, switches, circuit breakers, motor controllers, and fuses, including pushbuttons, relays, and similar devices, shall be provided with enclosures, and the enclosure in each case, together with the enclosed apparatus, shall be identified as a complete assembly for use in Class I locations.

(B) Class I, Division 2. Switches, circuit breakers, motor controllers, and fuses in Class I, Division 2 locations shall comply with 501.6(B)(1) through (B)(4).

(1) Type Required. Circuit breakers, motor controllers, and switches intended to interrupt current in the normal performance of the function for which they are installed shall be provided with enclosures identified for Class I, Division 1 locations in accordance with 501.3(A), unless general-purpose enclosures are provided and any of the following apply:

Exhibit 501.12 A Class I, Division 1 location where threaded metal conduits, sealing fittings, explosionproof fittings, and equipment for power and lights are used.

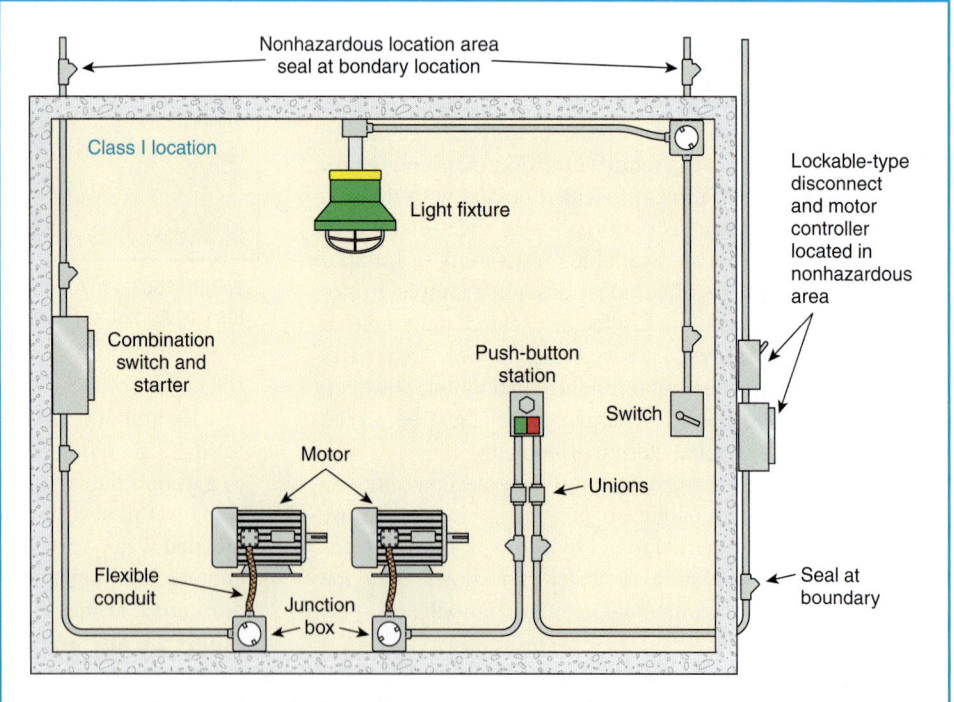

(1) The interruption of current occurs within a chamber hermetically sealed against the entrance of gases and vapors.

(2) The current make-and-break contacts are oil-immersed and of the general-purpose type having a 50-mm (2-in.) minimum immersion for power contacts and a 25-mm (1-in.) minimum immersion for control contacts.

(3) The interruption of current occurs within a factory-sealed explosionproof chamber.

(4) The device is a solid state, switching control without contacts, where the surface temperature does not exceed 80 percent of the ignition temperature in degrees Celsius of the gas or vapor involved.

(2) Isolating Switches. Fused or unfused disconnect and isolating switches for transformers or capacitor banks that are not intended to interrupt current in the normal performance of the function for which they are installed shall be permitted to be installed in general-purpose enclosures.

(3) Fuses. For the protection of motors, appliances, and lamps, other than as provided in 501.6(B)(4), standard plug or cartridge fuses shall be permitted, provided they are placed within enclosures identified for the location; or fuses shall be permitted if they are within general-purpose enclosures, and if they are of a type in which the operating element is immersed in oil or other approved liquid, or the operating element is enclosed within a chamber hermetically sealed against the entrance of gases and vapors, or the fuse is a nonindicating, filled, current-limiting type.

(4) Fuses Internal to Luminaires (Lighting Fixtures). Listed cartridge fuses shall be permitted as supplementary protection within luminaires (lighting fixtures).

Exhibit 501.13 shows an explosionproof panelboard that consists of an assembly of branch-circuit devices enclosed in a cast metal explosionproof housing. Explosionproof panelboards are provided with bolted access covers and threaded conduit-entry hubs designed to withstand the force of an internal explosion.

Exhibit 501.14 shows a cylindrical-type (spin-top) combination motor controller, motor control starter, and circuit breaker in an explosionproof enclosure. The top and bottom covers are threaded on for quick removal for installation and servicing. Exhibit 501.15 shows the same type of equipment in a rectangular enclosure with a hinged, bolted-on cover. These types of housings are designed to accommodate a wide range of manually or magnetically operated across-the-line types of motor starters in a variety of ratings.

Exhibit 501.16 illustrates a standard toggle switch in an explosionproof enclosure.

Exhibit 501.13 An explosionproof panelboard. (Courtesy of Appleton Electric Co., EGS Electrical Group)

In Class I, Division 2 locations, it is assumed that fuses or circuit breakers will seldom open the circuit where used to protect feeders or branch circuits supplying lamps in fixed positions only. Division 2 locations are not normally hazardous but may become so [see 500.5(B)(2)(3)], and because it is unlikely that the fuse or circuit breaker in such a circuit will operate simultaneously with the occurrence of an explosive mixture inside the enclosure, general-purpose enclosures are permitted for such overcurrent devices.

Fuses used for supplementary ballast protection in high-intensity-discharge and outdoor fluorescent fixtures are permitted by 501.6(B)(4).

501.7 Control Transformers and Resistors.

Transformers, impedance coils, and resistors used as, or in conjunction with, control equipment for motors, generators, and appliances shall comply with 501.7(A) and (B).

(A) Class I, Division 1. In Class I, Division 1 locations, transformers, impedance coils, and resistors, together with any switching mechanism associated with them, shall be

Exhibit 501.14 An explosionproof enclosure for a motor control starter and circuit breaker. (Courtesy of Appleton Electric Co., EGS Electrical Group)

Exhibit 501.15 A magnetic motor starter for use in a Class I, Group D location. Note the number of securing bolts and the width of the flange. (Courtesy of O.Z/Gedney Co.)

provided with enclosures identified for Class I, Division 1 locations in accordance with 501.3(A).

(B) Class I, Division 2. In Class I, Division 2 locations, control transformers and resistors shall comply with 501.7(B)(1) through (B)(3).

(1) Switching Mechanisms. Switching mechanisms used in conjunction with transformers, impedance coils, and resistors shall comply with 501.6(B).

(2) Coils and Windings. Enclosures for windings of transformers, solenoids, or impedance coils shall be permitted to be of the general-purpose type.

(3) Resistors. Resistors shall be provided with enclosures; and the assembly shall be identified for Class I locations, unless resistance is nonvariable and maximum operating temperature, in degrees Celsius, will not exceed 80 percent of the ignition temperature of the gas or vapor involved or

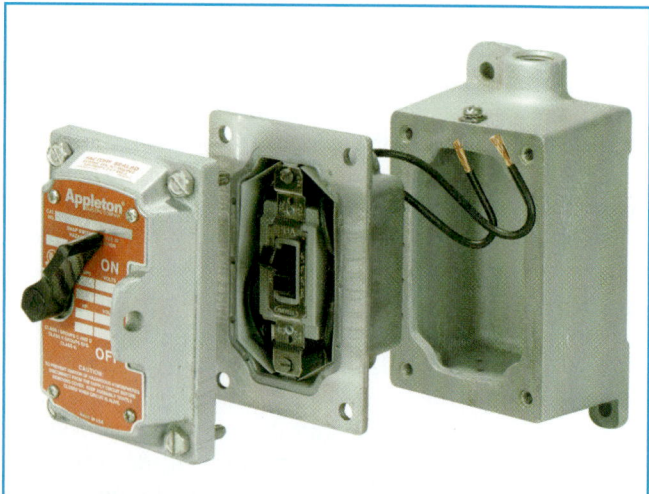

Exhibit 501.16 A standard toggle switch in an explosionproof enclosure. (Courtesy of Appleton Electric Co.)

has been tested and found incapable of igniting the gas or vapor.

501.8 Motors and Generators.

(A) Class I, Division 1. In Class I, Division 1 locations, motors, generators, and other rotating electric machinery shall be as follows:

(1) Identified for Class I, Division 1 locations; or
(2) Of the totally enclosed type supplied with positive-pressure ventilation from a source of clean air with discharge to a safe area, so arranged to prevent energizing of the machine until ventilation has been established and the enclosure has been purged with at least 10 volumes of air, and also arranged to automatically de-energize the equipment when the air supply fails; or
(3) Of the totally enclosed inert gas-filled type supplied with a suitable reliable source of inert gas for pressuring the enclosure, with devices provided to ensure a positive pressure in the enclosure and arranged to automatically de-energize the equipment when the gas supply fails; or
(4) Of a type designed to be submerged in a liquid that is flammable only when vaporized and mixed with air, or in a gas or vapor at a pressure greater than atmospheric and that is flammable only when mixed with air; and the machine is arranged so to prevent energizing it until it has been purged with the liquid or gas to exclude air, and also arranged to automatically de-energize the equipment when the supply of liquid or gas or vapor fails or the pressure is reduced to atmospheric.

The intent of 501.8(A)(4) is to permit nonexplosionproof motors to be submerged in liquefied natural gas (LNG), liquefied petroleum gas (LP-Gas), and so on. The provisions of 501.8(A)(4) do not permit nonexplosionproof motors under water, such as in wet pits, unless the motors are provided with some other system of explosion protection, for example, if they are purged and pressurized per NFPA 496, *Standard for Purged and Pressurized Enclosures for Electrical Equipment.*

The ASTM test procedure is used to determine the ignition temperature of some flammable and combustible liquids.

Totally enclosed motors of the types specified in 501.8(A)(2) or (A)(3) shall have no external surface with an operating temperature in degrees Celsius in excess of 80 percent of the ignition temperature of the gas or vapor involved. Appropriate devices shall be provided to detect and automatically de-energize the motor or provide an adequate alarm if there is any increase in temperature of the motor

beyond designed limits. Auxiliary equipment shall be of a type identified for the location in which it is installed.

FPN: See D 2155-69, ASTM Test Procedure.

(B) Class I, Division 2. In Class I, Division 2 locations, motors, generators, and other rotating electric machinery in which are employed sliding contacts, centrifugal or other types of switching mechanism (including motor overcurrent, overloading, and overtemperature devices), or integral resistance devices, either while starting or while running, shall be identified for Class I, Division 1 locations, unless such sliding contacts, switching mechanisms, and resistance devices are provided with enclosures identified for Class I, Division 2 locations in accordance with 501.3(B). The exposed surface of space heaters used to prevent condensation of moisture during shutdown periods shall not exceed 80 percent of the ignition temperature in degrees Celsius of the gas or vapor involved when operated at rated voltage, and the maximum surface temperature [based on a 40°C (104°F) ambient] shall be permanently marked on a visible nameplate mounted on the motor. Otherwise, space heaters shall be identified for Class I, Division 2 locations. In Class I, Division 2 locations, the installation of open or nonexplosionproof enclosed motors, such as squirrel-cage induction motors without brushes, switching mechanisms, or similar arc-producing devices that are not identified for use in a Class I, Division 2 location, shall be permitted.

It is intended that the phrase "other rotating electric machinery" include electric brakes. Listed and labeled electric brakes are available for Class I, Division 1, Group C and D locations.

Many motor heaters are de-energized automatically when the motor is running. However, the heater ratings are usually low when compared with the normal heat generated during motor operation. Unless otherwise indicated on the motor wiring diagram or in instructions provided with the motor, there is no need to de-energize the heater except to save energy. Note the requirement for marking the heater temperature on the motor or that the heaters be identified for the location.

FPN No. 1: It is important to consider the temperature of internal and external surfaces that may be exposed to the flammable atmosphere.

FPN No. 2: It is important to consider the risk of ignition due to currents arcing across discontinuities and overheating of parts in multisection enclosures of large motors and generators. Such motors and generators may need equipotential bonding jumpers across joints in the enclosure and from enclosure to ground. Where the presence of ignitible gases or vapors is suspected, clean-air purging may be needed immediately prior to and during start-up periods.

Exhibit 501.17 shows a totally enclosed fan-cooled motor listed for use in explosive atmospheres. The main frame and end-bells are designed with sufficient strength to withstand an internal explosion. Flames or hot gases are cooled while escaping because of the wide metal-to-metal joints between the frame and end-bells and the long, close-tolerance clearance provided for the free turn of the shaft. Air circulation outside the motor is maintained by a nonsparking (aluminum, bronze, or non-static-generating-type plastic) fan on the end opposite the shaft end of the motor. A sheet metal housing surrounds this fan to reduce the likelihood of an individual or object coming into contact with the moving blades and to direct the flow of air. An internal fan on the shaft, as shown in Exhibit 501.18, circulates air around the windings.

Motors that have arc- or spark-producing devices, such as commutators, internal switches, or other control devices, must be explosionproof. General-purpose squirrel-cage induction motors without arc- or spark-producing devices may be used in Division 2 locations.

Some open-type motors are permitted in Class I, Division 2 locations. Motor types used where flammable gases or vapors with very low ignition temperatures may be present should be selected with great care. Modern motors with high-temperature insulation systems, such as Class H [180°C (356°F)], may operate close to or above the ignition temperature of the flammable mixture.

Exhibit 501.18 View showing internal fan of motor in Exhibit 501.17.

Exhibit 501.17 Terminal housing of a motor listed for use in specific hazardous locations. Note integral sealing of the motor. (Courtesy of General Electric Co.)

501.9 Luminaires (Lighting Fixtures).

Luminaires (lighting fixtures) shall comply with 501.9(A) or (B).

(A) Class I, Division 1. In Class I, Division 1 locations, luminaires (lighting fixtures) shall comply with 501.9(A)(1) through (A)(4).

(1) Luminaires (Lighting Fixtures). Each luminaire (lighting fixture) shall be identified as a complete assembly for the Class I, Division 1 location and shall be clearly marked to indicate the maximum wattage of lamps for which it is identified. Luminaires (lighting fixtures) intended for portable use shall be specifically listed as a complete assembly for that use.

(2) Physical Damage. Each luminaire (lighting fixture) shall be protected against physical damage by a suitable guard or by location.

(3) Pendant Luminaires (Lighting Fixtures). Pendant luminaires (lighting fixtures) shall be suspended by and supplied through threaded rigid metal conduit stems or threaded steel intermediate conduit stems, and threaded joints shall be provided with set-screws or other effective means to prevent loosening. For stems longer than 300 mm (12 in.), permanent and effective bracing against lateral displacement shall be provided at a level not more than 300 mm (12 in.) above the lower end of the stem, or flexibility in the form of a fitting or flexible connector identified for the Class I, Division 1 location shall be provided not more than 300 mm (12 in.) from the point of attachment to the supporting box or fitting.

(4) Supports. Boxes, box assemblies, or fittings used for the support of luminaires (lighting fixtures) shall be identified for Class I locations.

(B) Class I, Division 2. In Class I, Division 2 locations, luminaires (lighting fixtures) shall comply with 501.9(B)(1) through (B)(5).

(1) Portable Lighting Equipment. Portable lighting equipment shall comply with 501.9(A)(1).

Exception: Where portable lighting equipment is mounted on movable stands and is connected by flexible cords, as covered in 501.11, it shall be permitted, where mounted in any position, if it conforms to 501.9(B)(2).

(2) Fixed Luminaires (Lighting Fixtures). Luminaires (lighting fixtures) for fixed lighting shall be protected from physical damage by suitable guards or by location. Where there is danger that falling sparks or hot metal from lamps or fixtures might ignite localized concentrations of flammable vapors or gases, suitable enclosures or other effective protective means shall be provided. Where lamps are of a size or type that may, under normal operating conditions, reach surface temperatures exceeding 80 percent of the ignition temperature in degrees Celsius of the gas or vapor involved, fixtures shall comply with 501.9(A)(1) or shall be of a type that has been tested in order to determine the marked operating temperature or temperature class (T Code).

(3) Pendant Luminaires (Fixtures). Pendant luminaires (lighting fixtures) shall be suspended by threaded rigid metal conduit stems, threaded steel intermediate metal conduit stems, or other approved means. For rigid stems longer than 300 mm (12 in.), permanent and effective bracing against lateral displacement shall be provided at a level not more than 300 mm (12 in.) above the lower end of the stem, or flexibility in the form of an identified fitting or flexible connector shall be provided not more than 300 mm (12 in.) from the point of attachment to the supporting box or fitting.

(4) Switches. Switches that are a part of an assembled fixture or of an individual lampholder shall comply with 501.6(B)(1).

(5) Starting Equipment. Starting and control equipment for electric-discharge lamps shall comply with 501.7(B).

Exception: A thermal protector potted into a thermally protected fluorescent lamp ballast if the luminaire (lighting fixture) is identified for the location.

Exhibit 501.19 shows a typical luminaire for Class I, Group C and D locations. The outlet boxes have an internally threaded opening designed to receive the cover. A pendant fixture is attached to the cover by threaded rigid metal con-

duit or threaded intermediate metal conduit. To prevent loosening from vibration or lamp changing, threaded joints must be provided with setscrews. The setscrews should not interrupt the explosionproof joint. Rigid metal conduit or intermediate metal conduit stems longer than 12 in. require effective bracing or a flexible fitting approved for the purpose, placed not more than 12 in. from the point of attachment to the supporting box, cover, or fitting.

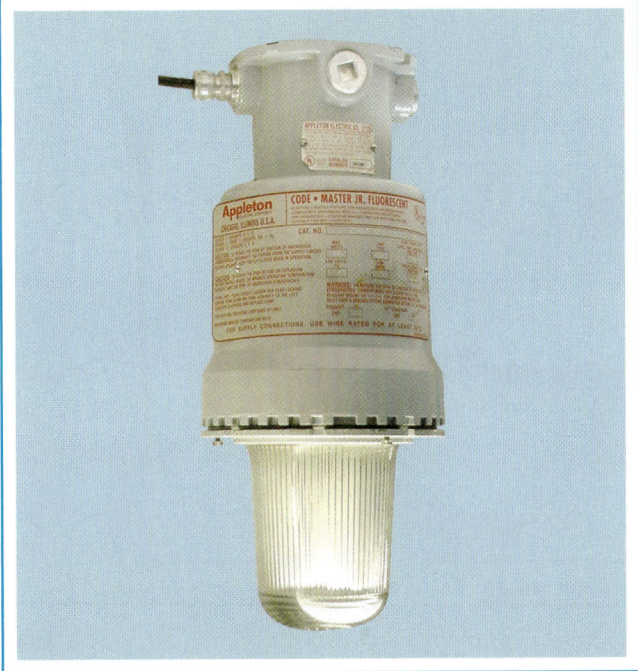

Exhibit 501.19 A typical lighting fixture for use in Class I, Group C and D locations. (Courtesy of Appleton Electric Co.)

A globe holder is threaded onto the body of the fixture housing and supports a heavy glass globe, guard, and reflector. It is available in sizes suitable for lamps from 40 watts through 500 watts. In designing any hazardous (classified) location lighting system, operating temperatures must be considered. Therefore, if the area is Class I, Division 1, fixtures approved for this location that are properly marked must be used. Generally, enclosed and gasketed fixtures (previously called vaportight fixtures) without guards, if breakage is unlikely, or fixtures approved for Class I, Division 2 locations, are required in Division 2 locations. Fixtures listed by Underwriters Laboratories Inc. for use in any of the groups under Class I, either Division 1 or 2 locations, or both, are designed to operate without igniting surrounding flammable gas or vapor atmospheres and are marked with the operating temperature or temperature class, as shown in Table 500.8(B).

Exhibit 501.20 shows an explosionproof hand lamp. Lamp compartments must be sealed from the terminal compartment. Provisions must be made for the connection of 3-conductor (one must be a grounding conductor) flexible, extra-hard-usage cord. See 501.11.

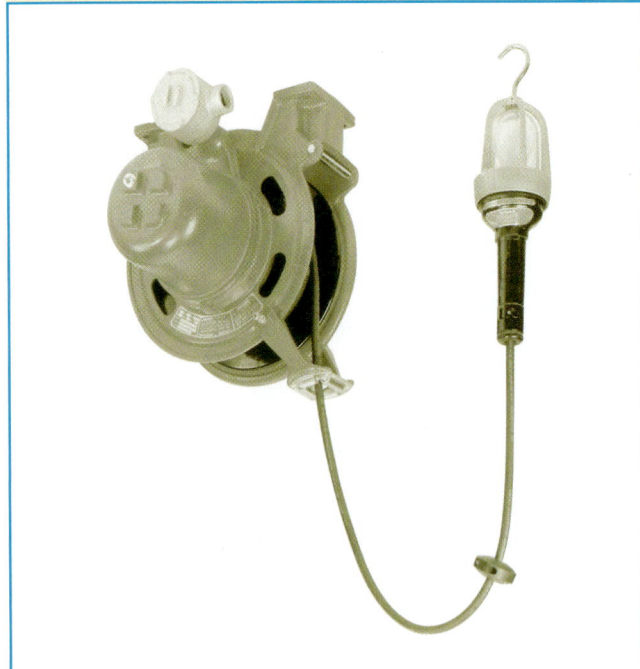

Exhibit 501.20 An explosionproof hand lamp for use in Class I locations. (Courtesy of Appleton Electric Co.)

501.10 Utilization Equipment.

(A) Class I, Division 1. In Class I, Division 1 locations, all utilization equipment shall be identified for Class I, Division 1 locations.

(B) Class I, Division 2. In Class I, Division 2 locations, all utilization equipment shall comply with 501.10(B)(1) through (B)(3).

(1) Heaters. Electrically heated utilization equipment shall conform with either item (1) or (2):

(1) The heater shall not exceed 80 percent of the ignition temperature in degrees Celsius of the gas or vapor involved on any surface that is exposed to the gas or vapor when continuously energized at the maximum rated ambient temperature. If a temperature controller is not provided, these conditions shall apply when the heater is operated at 120 percent of rated voltage.

Exception No. 1: For motor-mounted anticondensation space heaters, see 501.8(B).

Exception No. 2: Where current-limiting device is applied to the circuit serving the heater to limit the current in the heater to a value less than that required to raise the heater surface temperature to 80 percent of the ignition temperature.

(2) The heater shall be identified for Class I, Division 1 locations.

Exception: Electrical resistance heat tracing identified for Class I, Division 2 locations.

(2) Motors. Motors of motor-driven utilization equipment shall comply with 501.8(B).

(3) Switches, Circuit Breakers, and Fuses. Switches, circuit breakers, and fuses shall comply with 501.6(B).

The requirements for utilization equipment in Class I locations are virtually identical for Division 1 and 2 locations, except for heaters. Electric pipe heat-tracing systems listed for Class I, Division 2 locations and complying with 501.10(B)(1)(2), Exception, are available.

501.11 Flexible Cords, Class I, Divisions 1 and 2.

A flexible cord shall be permitted for connection between portable lighting equipment or other portable utilization equipment and the fixed portion of their supply circuit. Flexible cord shall also be permitted for that portion of the circuit where the fixed wiring methods of 501.4(A) cannot provide the necessary degree of movement for fixed and mobile electrical utilization equipment, in an industrial establishment where conditions of maintenance and engineering supervision ensure that only qualified persons install and service the installation, and the flexible cord is protected by location or by a suitable guard from damage. The length of the flexible cord shall be continuous. Where flexible cords are used, the cords shall be as follows:

(1) Of a type listed for extra-hard usage
(2) Contain, in addition to the conductors of the circuit, a grounding conductor complying with 400.23
(3) Connected to terminals or to supply conductors in an approved manner
(4) Supported by clamps or by other suitable means in such a manner that there is no tension on the terminal connections
(5) Provided with suitable seals where the flexible cord enters boxes, fittings, or enclosures of the explosionproof type

Exception: As provided in 501.3(B)(6) and 501.4(B).

Electric submersible pumps with means for removal without entering the wet-pit shall be considered portable utilization equipment. The extension of the flexible cord within a suitable raceway between the wet-pit and the power source shall be permitted. Electric mixers intended for travel into and out of open-type mixing tanks or vats shall be considered portable utilization equipment.

FPN: See 501.13 for flexible cords exposed to liquids having a deleterious effect on the conductor insulation.

The second paragraph of 501.11 recognizes a wet-pit type of installation that is finding increasing acceptance for use in waste-water systems. Section 501.11 permits a flexible cord to be run in raceway to a junction box outside the wet pit. See the commentary following 501.8(A) for more information on motors installed in wet pits.

Due to the operating conditions associated with electric mixers used in mixing tanks, the mixers are considered portable utilization equipment and may be wired with flexible cord.

501.12 Receptacles and Attachment Plugs, Class I, Divisions 1 and 2.

Receptacles and attachment plugs shall be of the type providing for connection to the grounding conductor of a flexible cord and shall be identified for the location.

Exception: As provided in 501.3(B)(6).

Exhibit 501.21 shows an explosionproof receptacle and attachment plug with an interlocking switch. The design of this device is such that when the switch is in the on position, the plug cannot be removed. Also, the switch cannot be placed in the on position when the plug has been removed; that is, the plug cannot be inserted or removed unless the switch is in the off position. The receptacle is factory sealed, with a provision for threaded-conduit entry to the switch compartment. The plug is to be used with Type S or equivalent extra-hard-service flexible cord having a grounding conductor.

Exhibit 501.22 shows a 30-ampere, 4-pole receptacle and attachment plug assembly that is suitable for use without a switch. The design is such that the mating parts of the receptacle and plug are enclosed in a chamber that seals the arc and, by delayed-action construction, prevents complete removal of the plug until the arc or hot metal has cooled. The receptacle is factory sealed, and the attachment plug is designed for use with a 4-conductor cord (3-conductor, 3-phase circuit with one grounding conductor) or a 3-conductor cord (two circuit conductors and one grounding conductor).

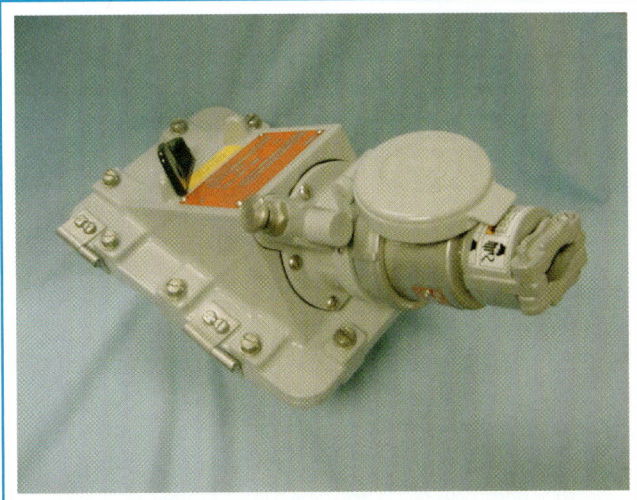

Exhibit 501.21 A receptacle and attachment plug of the explosionproof type with an interlocking switch. The switch must be in the off position before the attachment plug can be removed. (Courtesy of Appleton Electric Co., EGS Electrical Group)

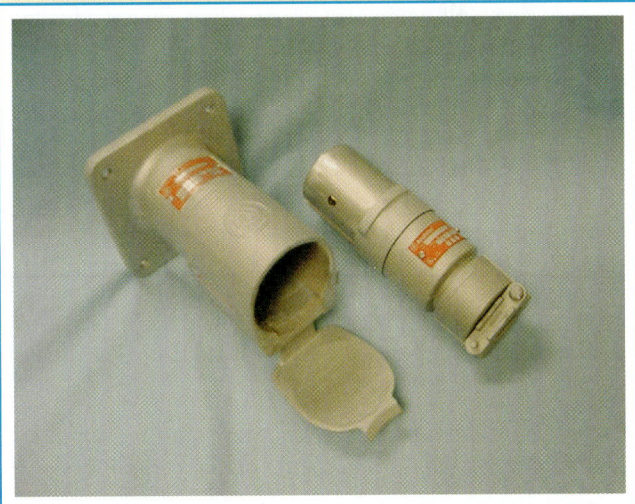

Exhibit 501.22 A 4-pole (delayed action) explosionproof receptacle and attachment plug suitable for use without a switch. (Courtesy of Appleton Electric Co., EGS Electrical Group)

501.13 Conductor Insulation, Class I, Divisions 1 and 2.

Where condensed vapors or liquids may collect on, or come in contact with, the insulation on conductors, such insulation shall be of a type identified for use under such conditions; or the insulation shall be protected by a sheath of lead or by other approved means.

Nylon-jacketed conductors, such as Type THWN, that are suitable for use where exposed to gasoline have gained widespread acceptance because of their ease in handling and application as well as for economic reasons.

Underwriters Laboratories Inc.'s *Electrical Construction Materials Directory* states the following under "Wires, Thermoplastic-Insulated."

THWN—wire that is suitable for exposure to mineral oil, and to liquid gasoline and gasoline vapors at ordinary ambient temperature, is marked "Gasoline and Oil Resistant I" if suitable for exposure to mineral oil at 60°C or "Gasoline and Oil Resistant II" if the compound is suitable for exposure to mineral oil at 75°C. Gasoline-resistant wire has been tested at 23°C when immersed in gasoline. It is considered inherently resistant to gasoline vapors within the limits of the temperature rating of the wire type.

501.14 Signaling, Alarm, Remote-Control, and Communications Systems.

(A) Class I, Division 1. In Class I, Division 1 locations, all apparatus and equipment of signaling, alarm, remote-control, and communications systems, regardless of voltage, shall be identified for Class I, Division 1 locations, and all wiring shall comply with 501.4(A) and 501.5(A) and (C).

(B) Class I, Division 2. In Class I, Division 2 locations, signaling, alarm, remote-control, and communications systems shall comply with 501.14(B)(1) through (B)(4).

(1) Contacts. Switches, circuit breakers, and make-and-break contacts of pushbuttons, relays, alarm bells, and horns shall have enclosures identified for Class I, Division 1 locations in accordance with 501.3(A).

Exception: General-purpose enclosures shall be permitted if current-interrupting contacts are one of the following:

(a) *Immersed in oil, or*
(b) *Enclosed within a chamber hermetically sealed against the entrance of gases or vapors, or*
(c) *In nonincendive circuits, or*

See the commentary following 501.3(B)(1), Exception (c), for information on nonincendive circuits.

(d) *Part of a listed nonincendive component*

(2) Resistors and Similar Equipment. Resistors, resistance devices, thermionic tubes, rectifiers, and similar equipment shall comply with 501.3(B)(2).

(3) Protectors. Enclosures shall be provided for lightning protective devices and for fuses. Such enclosures shall be permitted to be of the general-purpose type.

(4) Wiring and Sealing. All wiring shall comply with 501.4(B) and 501.5(B) and (C).

Audible signaling devices, such as bells, sirens, and horns, other than electronic types, usually involve make-and-break contacts that are capable of producing a spark of sufficient energy to cause ignition of a hazardous atmospheric mixture. If used in Class I locations, therefore, this type of equipment, as shown in Exhibit 501.23, must be contained in explosionproof or purged and pressurized enclosures, wiring methods must comply with 501.4, and sealing fittings must be provided in accordance with 501.5.

Exhibit 501.23 An audible signaling device for use in hazardous (classified) locations. (Courtesy of Cooper Crouse-Hinds)

Explosionproof devices or explosionproof enclosures may prove more practical than oil-immersed contacts because maintaining the condition and level of the oil can be a problem. Hermetically sealed enclosures, such as float-operated mercury-tube switches, are available for some applications. Electronic signal devices without make-and-break contacts usually do not require explosionproof enclosures in Division 2 locations.

501.15 Live Parts, Class I, Divisions 1 and 2.

There shall be no exposed live parts.

Exposed live parts are prohibited because contact with the circuit could produce sparks that could cause an explosion in a hazardous (classified) location.

501.16 Grounding, Class I, Divisions 1 and 2.

Wiring and equipment in Class I, Division 1 and 2 locations shall be grounded as specified in Article 250 and with the requirements in 501.16(A) and (B).

(A) Bonding. The locknut-bushing and double-locknut types of contacts shall not be depended on for bonding purposes, but bonding jumpers with proper fittings or other approved means of bonding shall be used. Such means of bonding shall apply to all intervening raceways, fittings, boxes, enclosures, and so forth between Class I locations and the point of grounding for service equipment or point of grounding of a separately derived system.

Exception: The specific bonding means shall only be required to the nearest point where the grounded circuit conductor and the grounding electrode are connected together on the line side of the building or structure disconnecting means as specified in 250.32(A), (B), and (C), provided the branch-circuit overcurrent protection is located on the load side of the disconnecting means.

The exception to 501.16(A) covers the grounding and bonding requirements that are specific to hazardous (classified) locations where the installation occurs at a multibuilding or multistructure setting. If the service equipment and the electrical equipment operating in the hazardous (classified) location are not located in the same building or structure, it is not necessary to apply the bonding requirement of 501.16(A) from the hazardous location back to the service equipment. It is only necessary to apply the bonding requirement from the hazardous location back to the grounding electrode on the line side of the building or structure disconnecting means. This connection must be ahead of the branch circuits that are on the load side of the disconnecting means for the building or structure.

FPN: See 250.100 for additional bonding requirements in hazardous (classified) locations.

(B) Types of Equipment Grounding Conductors. Where flexible metal conduit or liquidtight flexible metal conduit is used as permitted in 501.4(B) and is to be relied on to complete a sole equipment grounding path, it shall be installed with internal or external bonding jumpers in parallel with each conduit and complying with 250.102.

Special consideration is necessary in the grounding and bonding of exposed non-current-carrying metal parts of

equipment, such as the frames or metal exteriors of motors, fixed or portable lamps, luminaires, enclosures, and raceways, to ensure permanent and effective mechanical and electrical connections in order to prevent the possibility of arcs or sparks caused by ineffective or poor grounding methods. One example of an external bonding jumper is shown in Exhibit 501.24.

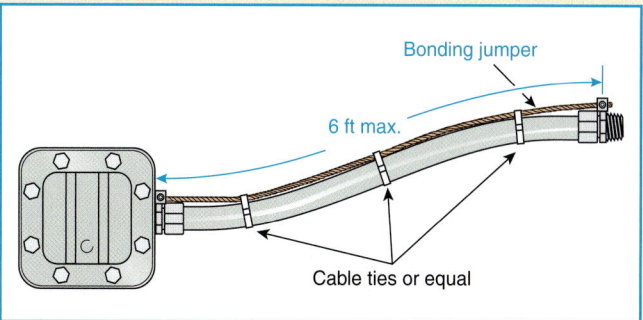

Exhibit 501.24 A fitting for the connection of an external bonding jumper used with liquidtight flexible metal conduit.

To be effective, proper grounding and bonding applies to all interconnected raceways, fittings, enclosures, and so on, between hazardous (classified) locations and the point of grounding for service equipment or the point of grounding of building disconnecting means that supplies the branch-circuit overcurrent protection. If conduit is used in hazardous (classified) locations, it is preferable that threaded connections also be employed in the nonhazardous location.

Exception: In Class I, Division 2 locations, the bonding jumper shall be permitted to be deleted where all the following conditions are met.

(a) Listed liquidtight flexible metal conduit 1.8 m (6 ft) or less in length, with fittings listed for grounding, is used.

(b) Overcurrent protection in the circuit is limited to 10 amperes or less.

(c) The load is not a power utilization load.

501.17 Surge Protection.

(A) Class I, Division 1. Surge arresters, including their installation and connection, shall comply with Article 280. The surge arresters and capacitors shall be installed in enclosures identified for Class I, Division 1 locations. Surge-protective capacitors shall be of a type designed for specific duty.

(B) Class I, Division 2. Surge arresters shall be nonarcing, such as metal-oxide varistor (MOV), sealed type, and surge-protective capacitors shall be of a type designed for specific duty. Installation and connection shall comply with Article 280. Enclosures shall be permitted to be of the general-purpose type. Surge protection of types other than described above shall be installed in enclosures identified for Class I, Division 1 locations.

Some surge arresters, such as older style lightning arresters, are spark-producing devices. Others, such as solid-state types, are not. Surge arresters should be connected to the service conductors outside the building and be bonded to the service-entrance raceway system. For services less than 1000 volts, the arrester grounding conductor is connected as provided in Article 280, Part III.

Where the service voltage is less than 600 volts, the supply system is a secondary system. Therefore, the grounded service conductor should always be bonded to the equipment grounding conductor, as required by Article 250.

In Class I, Division 1 locations, all surge arresters must be installed in explosionproof or purged enclosures. In Class I, Division 2 locations, only the spark-producing types of surge arresters require such protection. Surge arresters can also be installed in oil-filled enclosures or have the arcing or sparking contacts enclosed in hermetically sealed chambers. Nonsparking-type surge arresters need no special enclosure in a Class I, Division 2 location.

501.18 Multiwire Branch Circuits.

In a Class I, Division 1 location, a multiwire branch circuit shall not be permitted.

Exception: Where the disconnect device(s) for the circuit opens all ungrounded conductors of the multiwire circuit simultaneously.

The requirement in 501.18 does not permit multiwire branch circuits in Class I hazardous locations unless the disconnecting device opens all the ungrounded conductors of the multiwire branch circuit simultaneously. See the definition of *branch circuit, multiwire* in Article 100. The requirement in 501.18 addresses the inherent hazard of working on a multiwire branch circuit in which all of the ungrounded conductors are not opened by the disconnecting means. Although the ungrounded conductor supplying equipment that is being serviced may be disconnected, the common grounded conductor in this circuit configuration is still a current-carrying conductor in a circuit with an ungrounded conductor(s) that is not disconnected. In a Division 1 or 2 location, the opening of the grounded conductor in this scenario could result in an ignition-capable arc.

ARTICLE 502
Class II Locations

Contents

502.1 General.

The general rules of this *Code* shall apply to the electric wiring and equipment in locations classified as Class II locations in 500.5(C).

Two different types of dust environments typically warrant a Class II, Division 1 area classification. The first is where a cloud of combustible dust is likely to be present continuously or intermittently under normal operating conditions as a result of repair or maintenance operations or leakage. The other environment is one in which a dust layer is likely to accumulate to a depth of greater than ⅛ in. on major horizontal surfaces over a defined period of time, generally 24 hours. The size of the dust particle is the primary factor in determining whether the area should be classified as Class II or Class III. Combustible dust, as defined in NFPA 499, *Recommended Practice for the Classification of Combustible Dusts and of Hazardous (Classified) Locations for Electrical Installations in Chemical Process Areas,* is any finely divided solid material 420 microns (0.0165 in.) or less in diameter (i.e., material passing through a U.S. No. 40 standard sieve) that presents a fire or explosion hazard when dispersed.

Exception: As modified by this article.

Equipment installed in Class II locations shall be able to function at full rating without developing surface temperatures high enough to cause excessive dehydration or gradual carbonization of any organic dust deposits that may occur.

FPN: Dust that is carbonized or excessively dry is highly susceptible to spontaneous ignition.

Explosionproof equipment and wiring shall not be required and shall not be acceptable in Class II locations unless identified for such locations. Where Class II, Group E dusts are present in hazardous quantities, there are only Division 1 locations.

Class II, Division 1 and 2 locations are defined in 500.5(C) as hazardous due to the presence of combustible dust. These locations are separated into three groups: Group E, Group F, and Group G [see 500.6(B)].

It should be noted that equipment suitable for one class and group is not necessarily suitable for any other class and group. Class I equipment is not necessarily better for, or even suitable for, a Class II location, because the hazard contemplated in the equipment design is different. Class II equipment is designed to prevent the ignition of layers of dust, which may increase the operating temperature of the equipment. Class I equipment is not designed for dust layering unless it is also designed and approved for Class II locations. To protect against explosions in hazardous (classified) locations, all electrical equipment exposed to the haz-

ardous atmosphere must be suitable for such locations. Grain dust, for example, ignites at a temperature lower than that of most flammable vapors. Motors listed for use in Class I locations may not have dust shields on the bearings to prevent entrance of dust into the bearing race, thereby causing overheating of the bearing and resulting in ignition of dust on the motor.

One or more of the following four hazards may be present in a Class II location:

1. An explosive mixture of air and dust in suspension
2. Accumulation of dust that acts as a thermal blanket and interferes with the safe dissipation of heat from electrical equipment
3. Accumulation of electrically conductive dust lodged between terminals that have a difference of potential, thereby causing tracking and glowing hot particles, short-circuits, or ground faults that may ignite dust accumulated in the vicinity
4. Deposits of dust that could be ignited by arcs or sparks

In the layout of electrical installations for hazardous (classified) locations, it is preferable to locate service equipment, switchboards, panelboards, and much of the electrical equipment in less hazardous areas, usually in a separate room. The use of pressurized rooms, as described in NFPA 496, *Standard for Purged and Pressurized Enclosures for Electrical Equipment,* is a common method of protecting panelboards and switchboards in grain elevators and similar locations.

502.2 Transformers and Capacitors.

(A) Class II, Division 1. In Class II, Division 1 locations, transformers and capacitors shall comply with 502.2(A)(1) through (A)(3).

(1) Containing Liquid That Will Burn. Transformers and capacitors containing a liquid that will burn shall be installed only in vaults complying with 450.41 through 450.48, and, in addition, (a), (b), and (c) shall apply.

(a) Doors or other openings communicating with the Division 1 location shall have self-closing fire doors on both sides of the wall, and the doors shall be carefully fitted and provided with suitable seals (such as weather stripping) to minimize the entrance of dust into the vault.

(b) Vent openings and ducts shall communicate only with the outside air.

(c) Suitable pressure-relief openings communicating with the outside air shall be provided.

(2) Not Containing Liquid That Will Burn. Transformers and capacitors that do not contain a liquid that will burn shall be installed in vaults complying with 450.41 through 450.48 or be identified as a complete assembly, including terminal connections for Class II locations.

(3) Metal Dusts. No transformer or capacitor shall be installed in a location where dust from magnesium, aluminum, aluminum bronze powders, or other metals of similarly hazardous characteristics may be present.

(B) Class II, Division 2. In Class II, Division 2 locations, transformers and capacitors shall comply with 502.2(B)(1) through (B)(3).

(1) Containing Liquid That Will Burn. Transformers and capacitors containing a liquid that will burn shall be installed in vaults that comply with 450.41 through 450.48.

(2) Containing Askarel. Transformers containing askarel and rated in excess of 25 kVA shall be as follows:

(1) Provided with pressure-relief vents
(2) Provided with a means for absorbing any gases generated by arcing inside the case, or the pressure-relief vents shall be connected to a chimney or flue that will carry such gases outside the building
(3) Have an airspace of not less than 150 mm (6 in.) between the transformer cases and any adjacent combustible material

(3) Dry-Type Transformers. Dry-type transformers shall be installed in vaults or shall have their windings and terminal connections enclosed in tight metal housings without ventilating or other openings and shall operate at not over 600 volts, nominal.

Where it is necessary to install a transformer, it may be possible to use a small, low-voltage, dusttight (without ventilating openings) dry-type transformer, but transformers that have a primary voltage rating of over 600 volts must be either less flammable liquid-insulated or installed in a vault. In almost all cases, transformers can be remotely located from dust atmospheres.

Capacitors used for power-factor correction of individual motors are of sealed construction but, if installed in Class II, Division 1 locations, they must also be identified as a complete assembly, including dusttight terminal enclosures. The only special requirement for capacitors in Division 2 locations is that they are not permitted to contain oil or any other liquid that burns, otherwise they are to be installed in vaults.

502.4 Wiring Methods.

Wiring methods shall comply with 502.4(A) or (B).

(A) Class II, Division 1.

(1) General. In Class II, Division 1 locations, the wiring methods in (a) through (e) shall be permitted.

(a) Threaded rigid metal conduit, or threaded steel intermediate metal conduit.

(b) Type MI cable with termination fittings listed for the location. Type MI cable shall be installed and supported in a manner to avoid tensile stress at the termination fittings.

(c) In industrial establishments with limited public access, where the conditions of maintenance and supervision ensure that only qualified persons service the installation, Type MC cable, listed for use in Class II, Division 1 locations, with a gas/vaportight continuous corrugated metallic sheath, an overall jacket of suitable polymeric material, separate grounding conductors in accordance with 250.122, and provided with termination fittings listed for the application, shall be permitted.

(d) Fittings and boxes shall be provided with threaded bosses for connection to conduit or cable terminations and shall be dusttight. Fittings and boxes in which taps, joints, or terminal connections are made, or that are used in Group E locations, shall be identified for Class II locations.

In Class II, Division 1 locations, boxes or fittings, such as conduit "L," "T," or "C" fittings, that contain splices, taps, or terminations must be of the threaded type with close-fitting covers. There can be no holes in the box or fitting, such as mounting holes, that would allow dust to enter or sparks or burning material to be ejected from the enclosure. Where used as stated above, the equipment must be approved for Class II locations, usually of the dust-ignitionproof or pressurized type.

Boxes or fittings of the type described in the preceding paragraph that do not enclose splices, taps, or terminations must be dusttight, with threaded bosses for connection of conduit or cable connectors.

(e) Where necessary to employ flexible connections, dusttight flexible connectors, liquidtight flexible metal conduit with listed fittings, liquidtight flexible nonmetallic conduit with listed fittings, or flexible cord listed for extra-hard usage and provided with bushed fittings shall be used. Where flexible cords are used, they shall comply with 502.12. Where flexible connections are subject to oil or other corrosive conditions, the insulation of the conductors shall be of a type listed for the condition or shall be protected by means of a suitable sheath.

FPN: See 502.16(B) for grounding requirements where flexible conduit is used.

Due to the potential for physical damage to Type MC cable, its use in Class II, Division 1 locations is limited to cable that is listed specifically for use in Class II, Division 1 locations and installed at facilities that have full-time, qualified maintenance personnel. Qualified maintenance personnel are those who, in the course of regular maintenance procedures, would notice if cables were damaged, understand the associated hazards, and be able to de-energize the circuit to repair the installation.

(B) Class II, Division 2.

(1) General. In Class II, Division 2 locations, the following wiring methods shall be permitted:

(1) All wiring methods permitted in 502.4(A).
(2) Rigid metal conduit, intermediate metal conduit, electrical metallic tubing, dusttight wireways.
(3) Type MC or MI cable with listed termination fittings.
(4) Type PLTC in cable trays.
(5) Type ITC in cable trays.
(6) Type MC, MI, or TC cable installed in ladder, ventilated trough, or ventilated channel cable trays in a single layer, with a space not less than the larger cable diameter between the two adjacent cables, shall be the wiring method employed.

Exception: Type MC cable listed for use in Class II, Division 1 locations shall be permitted to be installed without the above required spacings.

(2) Flexible Connections. Where provision must be made for flexibility, 502.4(A)(1)(e) shall apply.

Where it is necessary to use flexible connections, liquidtight flexible conduit or extra-hard-usage flexible cord is permitted. Another method would be to use a flexible fitting, as described in the commentary on 501.4(A)(2). Where liquidtight flexible conduit is used, a bonding jumper (internal or external) must be provided around such conduit. See 502.16(B). An additional conductor for grounding must be provided where flexible cord is used.

(3) Nonincendive Field Wiring. Nonincendive field wiring shall be permitted using any of the wiring methods permitted for unclassified locations. Nonincendive field wiring systems shall be installed in accordance with the control drawing(s). Simple apparatus, not shown on the control drawing, shall be permitted in a nonincendive field wiring circuit, provided the simple apparatus does not interconnect the nonincendive field wiring circuit to any other circuit.

FPN: Simple apparatus is defined in 504.2.

Separate nonincendive field wiring circuits shall be as follows:

(1) In separate cables, or
(2) In multiconductor cables where the conductors of each circuit are within a grounded metal shield, or
(3) In multiconductor cables where the conductors of each circuit have insulation with a minimum thickness of 0.25 mm (0.01 in.)

(4) Boxes and Fittings. All boxes and fittings shall be dusttight.

In Division 1 locations, boxes containing taps, joints, or terminal connections must be dust-ignitionproof and must be provided with threaded hubs, as shown in Exhibit 502.1. Threaded hubs also provide adequate bonding in Division 2 locations (see Exhibit 502.1.)

Exhibit 502.1 also shows a close-fitting cover, which is required for Class II locations. Standard pressed-steel boxes are permitted as long as they do not contain taps, joints, or terminal connections, and a bonding jumper is provided around the box.

Exhibit 502.1 Junction box with threaded hubs, suitable for use in Class II, Group E hazardous atmospheres. (Courtesy of Appleton Electric Co.)

502.5 Sealing, Class II, Divisions 1 and 2.

Where a raceway provides communication between an enclosure that is required to be dust-ignitionproof and one that is not, suitable means shall be provided to prevent the entrance of dust into the dust-ignitionproof enclosure through the raceway. One of the following means shall be permitted:

(1) A permanent and effective seal
(2) A horizontal raceway not less than 3.05 m (10 ft) long, or
(3) A vertical raceway not less than 1.5 m (5 ft) long and extending downward from the dust-ignitionproof enclosure

Where a raceway provides communication between an enclosure that is required to be dust-ignitionproof and an enclosure in an unclassified location, seals shall not be required.

Sealing fittings shall be accessible.

Seals shall not be required to be explosionproof.

FPN: Electrical sealing putty is a method of sealing.

The requirements of 502.5 provide three suitable ways to prevent dust from entering dust-ignitionproof enclosures through the raceway. A seal in the raceway, a horizontal separation between enclosures of not less than 10 ft, or a vertical separation between enclosures of not less than 5 ft in a raceway extending downward from a dust-ignitionproof enclosure are methods of sealing in Class II locations and are shown in Exhibit 502.2. The requirement to provide a seal applies if a raceway connects an enclosure that is required to be dust-ignitionproof to one that is also required to be dust-ignitionproof. If a raceway extends from a dust-ignitionproof enclosure to an unclassified location, it is not necessary to provide a seal in that raceway. Sealing fittings designed for

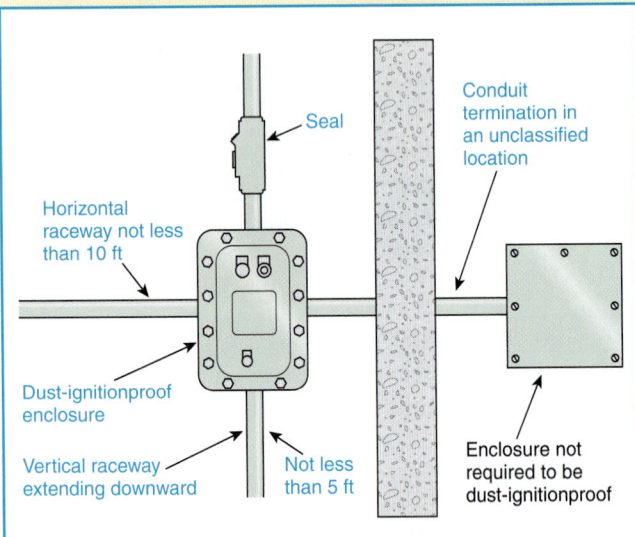

Exhibit 502.2 Three methods for preventing dust from entering a dust-ignitionproof enclosure through the raceway.

use in Class I locations are acceptable. However, because the Class I location pressure-piling considerations are not inherent to Class II locations, conduit seals are not required to be explosionproof. Conduit seals are expected only to prevent the migration of dust into dust-ignitionproof enclosures. No sealing method is needed in the special, but not unusual, situation in which dust cannot enter the raceway in the hazardous (classified) location.

502.6 Switches, Circuit Breakers, Motor Controllers, and Fuses.

(A) Class II, Division 1. In Class II, Division 1 locations, switches, circuit breakers, motor controllers, and fuses shall comply with 502.6(A)(1) through (A)(3).

(1) Type Required. Switches, circuit breakers, motor controllers, and fuses, including pushbuttons, relays, and similar devices that are intended to interrupt current during normal operation or that are installed where combustible dusts of an electrically conductive nature may be present, shall be provided with identified dust-ignitionproof enclosures.

(2) Isolating Switches. Disconnecting and isolating switches containing no fuses and not intended to interrupt current and not installed where dusts may be of an electrically conductive nature shall be provided with tight metal enclosures that shall be designed to minimize the entrance of dust and that shall (1) be equipped with telescoping or close-fitting covers or with other effective means to prevent the escape of sparks or burning material and (2) have no openings (such as holes for attachment screws) through which, after installation, sparks or burning material might escape or through which exterior accumulations of dust or adjacent combustible material might be ignited.

(3) Metal Dusts. In locations where dust from magnesium, aluminum, aluminum bronze powders, or other metals of similarly hazardous characteristics may be present, fuses, switches, motor controllers, and circuit breakers shall have enclosures identified for such locations.

(B) Class II, Division 2. In Class II, Division 2 locations, enclosures for fuses, switches, circuit breakers, and motor controllers, including pushbuttons, relays, and similar devices, shall be dusttight.

The electrical equipment required in Class II locations is different from that required for Class I locations. Dust-ignitionproof enclosures for Class II locations are not required to be explosionproof. However, explosionproof equipment is allowed to be used in Class II locations if the equipment is dual rated and identified as suitable for the Class II division and group. Explosionproof enclosures are not necessarily dust-ignitionproof.

Explosionproof enclosures, where used in an environment that is only a Class II location, are not required to be sealed within 18 in. of the enclosure to complete the explosionproof assembly, as required in a Class I environment. However, explosionproof enclosures must be provided with seals to prevent the entrance of dust into the dust-ignitionproof enclosure where a raceway provides a means for dust to enter the system, such as in a conduit run between a dust-ignitionproof enclosure and a general-purpose junction box.

Where equipment is located in a Class II, Division 1 location and is likely to produce arcs or sparks during normal operation, it must be installed in a dust-ignitionproof or pressurized enclosure. In addition, heat-generating equipment, such as control transformers, solenoids, impedance coils, resistors, and any associated overcurrent devices or switching mechanisms, must have dust-ignitionproof enclosures or pressurized enclosures installed in accordance with NFPA 496, *Standard for Purged and Pressurized Enclosures for Electrical Equipment.* Caution should be used where metal dusts, such as magnesium, aluminum, or other metals with similar hazardous characteristics, may be present. Enclosures must be approved specifically for that environment (suitable for Class II, Division 1, Group E).

Exhibit 502.3 shows a pushbutton station with pilot light that is suitable for both Class I and Class II hazardous (classified) locations. Exhibit 502.4 shows a panelboard that is suitable for use in Class II locations only. Many, but not

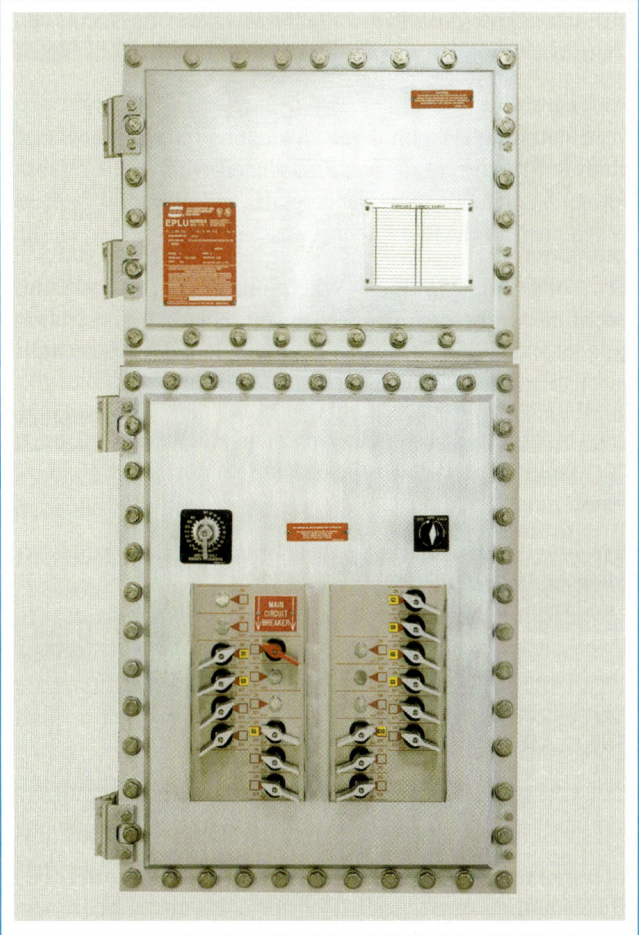

Exhibit 502.4 A dust-ignitionproof panelboard for use in Class II, Group E, F, and G locations. (Courtesy of Cooper Crouse-Hinds)

all, of the switches or circuit breakers approved for Class I, Division 1 locations are also approved for Class II locations. It is important to always look for the listing and identification of the hazardous (classified) locations for which the listing has been given.

502.7 Control Transformers and Resistors.

(A) Class II, Division 1. In Class II, Division 1 locations, control transformers, solenoids, impedance coils, resistors, and any overcurrent devices or switching mechanisms associated with them shall have dust-ignitionproof enclosures identified for Class II locations. No control transformer, impedance coil, or resistor shall be installed in a location where dust from magnesium, aluminum, aluminum bronze powders, or other metals of similarly hazardous characteristics may be present unless provided with an enclosure identified for the specific location.

Exhibit 502.3 An explosionproof and dusttight pushbutton control station suitable for use in Class I, Group C and D, and Class II, Group E, F, and G locations. (Courtesy of Appleton Electric Co., EGS Electrical Group)

(B) Class II, Division 2. In Class II, Division 2 locations, transformers and resistors shall comply with 502.7(B)(1) through (B)(3).

(1) Switching Mechanisms. Switching mechanisms (including overcurrent devices) associated with control transformers, solenoids, impedance coils, and resistors shall be provided with dusttight enclosures.

(2) Coils and Windings. Where not located in the same enclosure with switching mechanisms, control transformers, solenoids, and impedance coils shall be provided with tight metal housings without ventilating openings.

(3) Resistors. Resistors and resistance devices shall have dust-ignitionproof enclosures identified for Class II locations.

Exception: Where the maximum normal operating temperature of the resistor will not exceed 120°C (248°F), nonadjustable resistors or resistors that are part of an automatically timed starting sequence shall be permitted to have enclosures complying with 502.7(B)(2).

502.8 Motors and Generators.

(A) Class II, Division 1. In Class II, Division 1 locations, motors, generators, and other rotating electrical machinery shall be

(1) Identified for Class II, Division 1 locations, or
(2) Totally enclosed pipe-ventilated, meeting temperature limitations in 502.1.

It is intended that the phrase "other rotating electrical machinery" include electric brakes. Listed and labeled electric brakes are available for Class II, Group E, F, and G locations.

Although some explosionproof motors for Class I, Division 1 locations are dust-ignitionproof and approved for both Class I and II locations, this is not true of all motors. The marking should always be checked to be sure the motor is designed and tested for the Class II location involved. If control wiring to the motor is necessary, according to the motor installation instructions, the control circuit must be installed and connected properly. Most motors for Class II locations require internal thermal protection to comply with the temperature limitations in 500.8(C)(2). Integral horsepower Class II motors may require both power and control circuit wiring from the motor controller to the motor.

(B) Class II, Division 2. In Class II, Division 2 locations, motors, generators, and other rotating electrical equipment shall be totally enclosed nonventilated, totally enclosed pipe-ventilated, totally enclosed water-air-cooled, totally enclosed fan-cooled or dust-ignitionproof for which maximum full-load external temperature shall be in accordance with 500.8(C)(2) for normal operation when operating in free air (not dust blanketed) and shall have no external openings.

Exception: If the authority having jurisdiction believes accumulations of nonconductive, nonabrasive dust will be moderate and if machines can be easily reached for routine cleaning and maintenance, the following shall be permitted to be installed:

(a) *Standard open-type machines without sliding contacts, centrifugal or other types of switching mechanism (including motor overcurrent, overloading, and overtemperature devices), or integral resistance devices*
(b) *Standard open-type machines with such contacts, switching mechanisms, or resistance devices enclosed within dusttight housings without ventilating or other openings*
(c) *Self-cleaning textile motors of the squirrel-cage type*

The requirements in 502.8(B) permit all types of totally enclosed motors in Class II, Division 2 locations if the external surface temperatures, without a dust blanket, do not exceed the temperatures indicated under the maximum full-load (normal operation) conditions in 500.8(C)(2). Totally enclosed fan-cooled (TEFC) motors are specifically mentioned. The motor should be examined carefully to be sure there are no external openings, even though the motor may be marked TEFC.

Totally enclosed motors with no special provision for cooling may be used in Class II, Division 2 locations, but to deliver the same horsepower, they must be considerably larger than an open-type, fan-cooled, or pipe-ventilated motor.

502.9 Ventilating Piping.

Ventilating pipes for motors, generators, or other rotating electric machinery, or for enclosures for electric equipment, shall be of metal not less than 0.53 mm (0.021 in.) in thickness or of equally substantial noncombustible material and shall comply with all of the following:

(1) Lead directly to a source of clean air outside of buildings
(2) Be screened at the outer ends to prevent the entrance of small animals or birds
(3) Be protected against physical damage and against rusting or other corrosive influences

Ventilating pipes shall also comply with 502.9(A) and (B).

(A) Class II, Division 1. In Class II, Division 1 locations, ventilating pipes, including their connections to motors or to the dust-ignitionproof enclosures for other equipment, shall be dusttight throughout their length. For metal pipes, seams and joints shall comply with one of the following:

(1) Be riveted and soldered
(2) Be bolted and soldered
(3) Be welded
(4) Be rendered dusttight by some other equally effective means

(B) Class II, Division 2. In Class II, Division 2 locations, ventilating pipes and their connections shall be sufficiently tight to prevent the entrance of appreciable quantities of dust into the ventilated equipment or enclosure and to prevent the escape of sparks, flame, or burning material that might ignite dust accumulations or combustible material in the vicinity. For metal pipes, lock seams and riveted or welded joints shall be permitted; and tight-fitting slip joints shall be permitted where some flexibility is necessary, as at connections to motors.

502.10 Utilization Equipment.

(A) Class II, Division 1. In Class II, Division 1 locations, all utilization equipment shall be identified for Class II locations. Where dust from magnesium, aluminum, aluminum bronze powders, or other metals of similarly hazardous characteristics may be present, such equipment shall be identified for the specific location.

(B) Class II, Division 2. In Class II, Division 2 locations, all utilization equipment shall comply with 502.10(B)(1) through (B)(4).

(1) Heaters. Electrically heated utilization equipment shall be identified for Class II locations.

Exception: Metal-enclosed radiant heating panel equipment shall be dusttight and marked in accordance with 500.8(B).

(2) Motors. Motors of motor-driven utilization equipment shall comply with 502.8(B).

(3) Switches, Circuit Breakers, and Fuses. Enclosures for switches, circuit breakers, and fuses shall be dusttight.

(4) Transformers, Solenoids, Impedance Coils, and Resistors. Transformers, solenoids, impedance coils, and resistors shall comply with 502.7(B).

502.11 Luminaires (Lighting Fixtures).

Luminaires (lighting fixtures) shall comply with 502.11(A) and (B).

(A) Class II, Division 1. In Class II, Division 1 locations, luminaires (lighting fixtures) for fixed and portable lighting shall comply with 502.11(A)(1) through (A)(4).

(1) Fixtures. Each luminaire (fixture) shall be identified for Class II locations and shall be clearly marked to indicate the maximum wattage of the lamp for which it is designed. In locations where dust from magnesium, aluminum, aluminum bronze powders, or other metals of similarly hazardous characteristics may be present, luminaires (fixtures) for fixed or portable lighting and all auxiliary equipment shall be identified for the specific location.

(2) Physical Damage. Each luminaire (fixture) shall be protected against physical damage by a suitable guard or by location.

(3) Pendant Luminaires (Fixtures). Pendant luminaires (fixtures) shall be suspended by threaded rigid metal conduit stems, threaded steel intermediate metal conduit stems, by chains with approved fittings, or by other approved means. For rigid stems longer than 300 mm (12 in.), permanent and effective bracing against lateral displacement shall be provided at a level not more than 300 mm (12 in.) above the lower end of the stem, or flexibility in the form of a fitting or a flexible connector listed for the location shall be provided not more than 300 mm (12 in.) from the point of attachment to the supporting box or fitting. Threaded joints shall be provided with set-screws or other effective means to prevent loosening. Where wiring between an outlet box or fitting and a pendant luminaire (fixture) is not enclosed in conduit, flexible cord listed for hard usage shall be used, and suitable seals shall be provided where the cord enters the luminaire (fixture) and the outlet box or fitting. Flexible cord shall not serve as the supporting means for a fixture.

(4) Supports. Boxes, box assemblies, or fittings used for the support of luminaires (lighting fixtures) shall be identified for Class II locations.

(B) Class II, Division 2. In Class II, Division 2 locations, luminaires (lighting fixtures) shall comply with 502.11(B)(1) through (B)(5).

(1) Portable Lighting Equipment. Portable lighting equipment shall be identified for Class II locations. They shall be clearly marked to indicate the maximum wattage of lamps for which they are designed.

(2) Fixed Lighting. Luminaires (lighting fixtures) for fixed lighting, where not of a type identified for Class II locations, shall provide enclosures for lamps and lampholders that shall be designed to minimize the deposit of dust on lamps and to prevent the escape of sparks, burning material, or hot metal. Each fixture shall be clearly marked to indicate the maximum wattage of the lamp that shall be permitted without exceeding an exposed surface temperature in accordance with 500.8(C)(2) under normal conditions of use.

(3) Physical Damage. Luminaires (lighting fixtures) for fixed lighting shall be protected from physical damage by suitable guards or by location.

(4) Pendant Luminaires (Fixtures). Pendant luminaires (fixtures) shall be suspended by threaded rigid metal conduit stems, threaded steel intermediate metal conduit stems, by chains with approved fittings, or by other approved means. For rigid stems longer than 300 mm (12 in.), permanent and effective bracing against lateral displacement shall be provided at a level not more than 300 mm (12 in.) above the lower end of the stem, or flexibility in the form of an identified fitting or a flexible connector shall be provided not more than 300 mm (12 in.) from the point of attachment to the supporting box or fitting. Where wiring between an outlet box or fitting and a pendant luminaire (fixture) is not enclosed in conduit, flexible cord listed for hard usage shall be used. Flexible cord shall not serve as the supporting means for a fixture.

(5) Electric-Discharge Lamps. Starting and control equipment for electric-discharge lamps shall comply with the requirements of 502.7(B).

Exhibit 502.5 shows a listed fixture suitable for use in Class II, Group E, F, and G locations. Luminaires, fixed or portable, and auxiliary equipment (such as ballasts) must be approved for use in Group E atmospheres if metal dusts are present.

Other than the requirement that the fixture be marked to indicate maximum lamp wattage, the only requirement for fixtures in Division 2 locations is that lamps be enclosed in suitable globes to minimize dust deposits on the lamps and prevent the escape of sparks or burning material. Metal guards must be provided, unless globe breakage is unlikely.

Flexible cord of the hard-usage type is permitted with approved sealed connections for the wiring of chain-suspended or hook-and-eye-suspended fixtures. Flexible cords are not intended to be used as cord pendants or drop cords.

The portable hand lamp shown in Exhibit 501.20 is approved as a complete assembly for use in Class I locations and also in any Class II, Group F or G location.

502.12 Flexible Cords—Class II, Divisions 1 and 2.

Flexible cords used in Class II locations shall comply with all of the following:

(1) Be of a type listed for extra-hard usage

Exception: Flexible cord listed for hard usage as permitted by 502.11(A)(3) and (B)(4).

(2) Contain, in addition to the conductors of the circuit, a grounding conductor complying with 400.23

Exhibit 502.5 A typical luminaire for use in Class II, Division 1 locations. (Courtesy of Cooper Crouse-Hinds)

(3) Be connected to terminals or to supply conductors in an approved manner
(4) Be supported by clamps or by other suitable means in such a manner that there will be no tension on the terminal connections
(5) Be provided with suitable seals to prevent the entrance of dust where the flexible cord enters boxes or fittings that are required to be dust-ignitionproof

502.13 Receptacles and Attachment Plugs.

(A) Class II, Division 1. In Class II, Division 1 locations, receptacles and attachment plugs shall be of the type providing for connection to the grounding conductor of the flexible cord and shall be identified for Class II locations.

(B) Class II, Division 2. In Class II, Division 2 locations, receptacles and attachment plugs shall be of the type that provides for connection to the grounding conductor of the flexible cord and shall be designed so that connection to the supply circuit cannot be made or broken while live parts are exposed.

502.14 Signaling, Alarm, Remote-Control, and Communications Systems; and Meters, Instruments, and Relays.

FPN: See Article 800 for rules governing the installation of communications circuits.

(A) Class II, Division 1. In Class II, Division 1 locations, signaling, alarm, remote-control, and communications systems; and meters, instruments, and relays shall comply with 502.14(A)(1) through (A)(6).

(1) Wiring Methods. The wiring method shall comply with 502.4(A).

(2) Contacts. Switches, circuit breakers, relays, contactors, fuses and current-breaking contacts for bells, horns, howlers, sirens, and other devices in which sparks or arcs may be produced shall be provided with enclosures identified for a Class II location.

Exception: Where current-breaking contacts are immersed in oil or where the interruption of current occurs within a chamber sealed against the entrance of dust, enclosures shall be permitted to be of the general-purpose type.

(3) Resistors and Similar Equipment. Resistors, transformers, choke coils, rectifiers, thermionic tubes, and other heat-generating equipment shall be provided with enclosures identified for Class II locations.

Exception: Where resistors or similar equipment are immersed in oil or enclosed in a chamber sealed against the entrance of dust, enclosures shall be permitted to be of the general-purpose type.

(4) Rotating Machinery. Motors, generators, and other rotating electric machinery shall comply with 502.8(A).

(5) Combustible, Electrically Conductive Dusts. Where dusts are of a combustible, electrically conductive nature, all wiring and equipment shall be identified for Class II locations.

(6) Metal Dusts. Where dust from magnesium, aluminum, aluminum bronze powders, or other metals of similarly hazardous characteristics may be present, all apparatus and equipment shall be identified for the specific conditions.

(B) Class II, Division 2. In Class II, Division 2 locations, signaling, alarm, remote-control, and communications systems; and meters, instruments, and relays shall comply with the following.

(1) Contacts. Enclosures shall comply with 502.14(A), (2) or contacts shall have tight metal enclosures designed to minimize the entrance of dust and shall have telescoping or tight-fitting covers and no openings through which, after installation, sparks or burning material might escape.

Exception: In nonincendive circuits, enclosures shall be permitted to be of the general-purpose type.

(2) Transformers and Similar Equipment. The windings and terminal connections of transformers, choke coils, and similar equipment shall be provided with tight metal enclosures without ventilating openings.

(3) Resistors and Similar Equipment. Resistors, resistance devices, thermionic tubes, rectifiers, and similar equipment shall comply with 502.14(A)(3).

Exception: Enclosures for thermionic tubes, nonadjustable resistors, or rectifiers for which maximum operating temperature will not exceed 120°C (248°F) shall be permitted to be of the general-purpose type.

(4) Rotating Machinery. Motors, generators, and other rotating electric machinery shall comply with 502.8(B).

(5) Wiring Methods. The wiring method shall comply with 502.4(B).

502.15 Live Parts, Class II, Divisions 1 and 2.

Live parts shall not be exposed.

502.16 Grounding, Class II, Divisions 1 and 2.

Wiring and equipment in Class II, Divisions 1 and 2 locations shall be grounded as specified in Article 250 and with the requirements in 502.16(A) and (B).

(A) Bonding. The locknut-bushing and double-locknut types of contact shall not be depended on for bonding purposes, but bonding jumpers with proper fittings or other approved means of bonding shall be used. Such means of bonding shall apply to all intervening raceways, fittings, boxes, enclosures, and so forth, between Class II locations and the point of grounding for service equipment or point of grounding of a separately derived system.

Exception: The specific bonding means shall only be required to the nearest point where the grounded circuit conductor and the grounding electrode conductor are connected together on the line side of the building or structure disconnecting means as specified in 250.32(A), (B), and (C), if the branch-circuit overcurrent protection is located on the load side of the disconnecting means.

FPN: See 250.100 for additional bonding requirements in hazardous (classified) locations.

(B) Types of Equipment Grounding Conductors. Where flexible conduit is used as permitted in 502.4, it shall be installed with internal or external bonding jumpers in parallel with each conduit and complying with 250.102.

Exception: In Class II, Division 2 locations, the bonding jumper shall be permitted to be deleted where all the following conditions are met:

(a) Listed liquidtight flexible metal conduit 1.8 m (6 ft) or less in length, with fittings listed for grounding, is used.

(b) Overcurrent protection in the circuit is limited to 10 amperes or less.

(c) The load is not a power utilization load.

In Class II locations, a raceway connected to an enclosure via a double locknut connection or a single locknut and bushing connection is not an acceptable method of bonding. Bonding jumpers or other approved means with proper fittings are required for the interconnection of all raceways, junction boxes, fittings, enclosures, and so on, installed between the hazardous area and the grounding electrode conductor connection point at the service equipment, the grounding electrode connection at a feeder or branch circuit disconnecting means at separate buildings, or the grounding electrode connection to the source of a separately derived system. If installed outside the raceway or enclosure, the grounding conductor must not exceed 6 ft and must be routed with the raceway or enclosure. See 250.102(E) for equipment bonding jumper installation requirements.

502.17 Surge Protection—Class II, Divisions 1 and 2.

Surge arresters, including their installation and connection, shall comply with Article 280. In addition, surge arresters, if installed in a Class II, Division 1 location, shall be in suitable enclosures. Surge-protective capacitors shall be of a type designed for specific duty.

502.18 Multiwire Branch Circuits.

In a Class II, Division 1 location, a multiwire branch circuit shall not be permitted.

Exception: Where the disconnect device(s) for the circuit opens all ungrounded conductors of the multiwire circuit simultaneously.

ARTICLE 503
Class III Locations

Contents

503.1 General.

The general rules of this *Code* shall apply to electric wiring and equipment in locations classified as Class III locations in 500.5(D).

Exception: As modified by this article.

Equipment installed in Class III locations shall be able to function at full rating without developing surface temperatures high enough to cause excessive dehydration or gradual carbonization of accumulated fibers or flyings. Organic material that is carbonized or excessively dry is highly susceptible to spontaneous ignition. The maximum surface temperatures under operating conditions shall not exceed 165°C (329°F) for equipment that is not subject to overloading, and 120°C (248°F) for equipment (such as motors or power transformers) that may be overloaded.

FPN: For electric trucks, see NFPA 505-1999, *Fire Safety Standard for Powered Industrial Trucks Including Type Designations, Areas of Use, Conversions, Maintenance, and Operation.*

Class III locations usually include textile mills that process cotton, rayon, and so on, where easily ignitible fibers or combustible flyings are present in the manufacturing process. Sawmills and other woodworking plants, where sawdust, wood shavings, and combustible fibers or flyings are present, may also become hazardous. If wood flour (dust) is present, the location is a Class II, Group G location, not a Class III location.

Fibers or flyings are hazardous not only because they are easily ignited, but also because flames spread through them quickly. Such fires travel with a rapidity approaching an explosion and are commonly called flash fires.

Class III, Division 1 applies to locations where material is handled, manufactured, or used. Division 2 applies to locations where material is stored or handled but where no manufacturing processes are performed. Unlike Class I locations (Groups A, B, C, and D) and Class II locations (Groups E, F, and G), there are no material group designations in Class III locations.

503.2 Transformers and Capacitors—Class III, Divisions 1 and 2.

Transformers and capacitors shall comply with 502.2(B).

503.3 Wiring Methods.

Wiring methods shall comply with 503.3(A) or (B).

(A) Class III, Division 1. In Class III, Division 1 locations, the wiring method shall be rigid metal conduit, rigid nonmetallic conduit, intermediate metal conduit, electrical metallic tubing, dusttight wireways, or Type MC or MI cable with listed termination fittings.

(1) Boxes and Fittings. All boxes and fittings shall be dusttight.

(2) Flexible Connections. Where necessary to employ flexible connections, dusttight flexible connectors, liquidtight flexible metal conduit with listed fittings, liquidtight flexible nonmetallic conduit with listed fittings, or flexible cord in conformance with 503.10 shall be used.

FPN: See 503.16(B) for grounding requirements where flexible conduit is used.

(B) Class III, Division 2. In Class III, Division 2 locations, the wiring method shall comply with 503.3(A).

Exception: In sections, compartments, or areas used solely for storage and containing no machinery, open wiring on insulators shall be permitted where installed in accordance with Article 398, but only on condition that protection as required by 398.15(C) be provided where conductors are not run in roof spaces and are well out of reach of sources of physical damage.

503.4 Switches, Circuit Breakers, Motor Controllers, and Fuses—Class III, Divisions 1 and 2.

Switches, circuit breakers, motor controllers, and fuses, including pushbuttons, relays, and similar devices, shall be provided with dusttight enclosures.

See the definition of *dusttight* in Article 100.

503.5 Control Transformers and Resistors—Class III, Divisions 1 and 2.

Transformers, impedance coils, and resistors used as or in conjunction with control equipment for motors, generators, and appliances shall be provided with dusttight enclosures complying with the temperature limitations in 503.1.

See the definition of *dusttight* in Article 100.

503.6 Motors and Generators—Class III, Divisions 1 and 2.

In Class III, Divisions 1 and 2 locations, motors, generators, and other rotating machinery shall be totally enclosed nonventilated, totally enclosed pipe ventilated, or totally enclosed fan cooled.

Exception: In locations where, in the judgment of the authority having jurisdiction, only moderate accumulations of lint or flyings are likely to collect on, in, or in the vicinity of a rotating electric machine and where such machine is readily accessible for routine cleaning and maintenance, one of the following shall be permitted:

(a) Self-cleaning textile motors of the squirrel-cage type
(b) Standard open-type machines without sliding contacts, centrifugal or other types of switching mechanisms, including motor overload devices
(c) Standard open-type machines having such contacts, switching mechanisms, or resistance devices enclosed within tight housings without ventilating or other openings

It is intended that the phrase "other rotating machinery" in 503.6 include electric brakes. Listed and labeled electric brakes are available for Class II, Group G locations, and, according to the Underwriters Laboratories Inc. *Hazardous Location Equipment Directory*, these brakes are suitable for Class III locations.

503.7 Ventilating Piping—Class III, Divisions 1 and 2.

Ventilating pipes for motors, generators, or other rotating electric machinery, or for enclosures for electric equipment, shall be of metal not less than 0.53 mm (0.021 in.) in thickness, or of equally substantial noncombustible material, and shall comply with the following:

(1) Lead directly to a source of clean air outside of buildings
(2) Be screened at the outer ends to prevent the entrance of small animals or birds
(3) Be protected against physical damage and against rusting or other corrosive influences

Ventilating pipes shall be sufficiently tight, including their connections, to prevent the entrance of appreciable quantities of fibers or flyings into the ventilated equipment or enclosure and to prevent the escape of sparks, flame, or burning material that might ignite accumulations of fibers or flyings or combustible material in the vicinity. For metal pipes, lock seams and riveted or welded joints shall be permitted; and tight-fitting slip joints shall be permitted where some flexibility is necessary, as at connections to motors.

503.8 Utilization Equipment—Class III, Divisions 1 and 2.

(A) Heaters. Electrically heated utilization equipment shall be identified for Class III locations.

(B) Motors. Motors of motor-driven utilization equipment shall comply with 503.6.

(C) Switches, Circuit Breakers, Motor Controllers, and Fuses. Switches, circuit breakers, motor controllers, and fuses shall comply with 503.4.

503.9 Luminaires (Lighting Fixtures)—Class III, Divisions 1 and 2.

(A) Fixed Lighting. Luminaires (lighting fixtures) for fixed lighting shall provide enclosures for lamps and lampholders that are designed to minimize entrance of fibers and flyings and to prevent the escape of sparks, burning material, or hot metal. Each luminaire (fixture) shall be clearly marked to show the maximum wattage of the lamps that shall be permitted without exceeding an exposed surface temperature of 165°C (329°F) under normal conditions of use.

(B) Physical Damage. A luminaire (fixture) that may be exposed to physical damage shall be protected by a suitable guard.

(C) Pendant Luminaires (Fixtures). Pendant luminaires (fixtures) shall be suspended by stems of threaded rigid metal conduit, threaded intermediate metal conduit, threaded metal tubing of equivalent thickness, or by chains with approved fittings. For stems longer than 300 mm (12 in.), permanent and effective bracing against lateral displacement shall be provided at a level not more than 300 mm (12 in.) above the lower end of the stem, or flexibility in the form of an identified fitting or a flexible connector shall be provided not more than 300 mm (12 in.) from the point of attachment to the supporting box or fitting.

(D) Portable Lighting Equipment. Portable lighting equipment shall be equipped with handles and protected with substantial guards. Lampholders shall be of the unswitched type with no provision for receiving attachment plugs. There shall be no exposed current-carrying metal parts, and all exposed non–current-carrying metal parts shall be grounded. In all other respects, portable lighting equipment shall comply with 503.9(A).

503.10 Flexible Cords—Class III, Divisions 1 and 2.

Flexible cords shall comply with the following:

(1) Be of a type listed for extra-hard usage
(2) Contain, in addition to the conductors of the circuit, a grounding conductor complying with 400.23
(3) Be connected to terminals or to supply conductors in an approved manner
(4) Be supported by clamps or other suitable means in such a manner that there will be no tension on the terminal connections
(5) Be provided with suitable means to prevent the entrance of fibers or flyings where the cord enters boxes or fittings

503.11 Receptacles and Attachment Plugs—Class III, Divisions 1 and 2.

Receptacles and attachment plugs shall be of the grounding type and shall be designed so as to minimize the accumulation or the entry of fibers or flyings, and shall prevent the escape of sparks or molten particles.

Exception: In locations where, in the judgment of the authority having jurisdiction, only moderate accumulations of lint or flyings will be likely to collect in the vicinity of a receptacle, and where such receptacle is readily accessible for routine cleaning, general-purpose grounding-type receptacles mounted so as to minimize the entry of fibers or flyings shall be permitted.

503.12 Signaling, Alarm, Remote-Control, and Local Loudspeaker Intercommunications Systems—Class III, Divisions 1 and 2.

Signaling, alarm, remote-control, and local loudspeaker intercommunications systems shall comply with the requirements of Article 503 regarding wiring methods, switches, transformers, resistors, motors, luminaires (lighting fixtures), and related components.

503.13 Electric Cranes, Hoists, and Similar Equipment—Class III, Divisions 1 and 2.

Where installed for operation over combustible fibers or accumulations of flyings, traveling cranes and hoists for material handling, traveling cleaners for textile machinery, and similar equipment shall comply with 503.13(A) through (D).

(A) Power Supply. Power supply to contact conductors shall be isolated from all other systems and shall be equipped with an acceptable ground detector that gives an alarm and automatically de-energizes the contact conductors in case of a fault to ground or gives a visual and audible alarm as long as power is supplied to the contact conductors and the ground fault remains.

(B) Contact Conductors. Contact conductors shall be located or guarded so as to be inaccessible to other than authorized persons and shall be protected against accidental contact with foreign objects.

(C) Current Collectors. Current collectors shall be arranged or guarded so as to confine normal sparking and prevent escape of sparks or hot particles. To reduce sparking, two or more separate surfaces of contact shall be provided for each contact conductor. Reliable means shall be provided to keep contact conductors and current collectors free of accumulations of lint or flyings.

(D) Control Equipment. Control equipment shall comply with 503.4 and 503.5.

In a Class III location, cranes that are installed over accumulations of fibers or flyings and equipped with rolling or sliding collectors that make contact with bare conductors introduce two hazards.

The first hazard results from any arcing between a conductor and a collector rail igniting combustible fibers or lint that has accumulated on or near the bare conductor. This hazard may be prevented by maintaining the proper alignment of the bare conductor, by using a collector designed so that proper contact is always maintained, and by using guards or shields to confine hot metal particles that result from arcing.

The second hazard occurs if enough moisture is present and fibers and flyings accumulating on the insulating supports of the bare conductors form a conductive path between the conductors or from one conductor to ground, permitting enough current to flow to ignite the fibers. If the system is ungrounded, a current flow to ground is unlikely to start a fire. A suitable recording ground detector will sound an alarm and automatically de-energize contact conductors when the insulation resistance is lowered by an accumulation of fibers on the insulators or in case of a fault to ground. A ground-fault indicator is permitted that will maintain an alarm until the system is de-energized or the ground fault is cleared.

503.14 Storage Battery Charging Equipment—Class III, Divisions 1 and 2.

Storage battery charging equipment shall be located in separate rooms built or lined with substantial noncombustible materials. The rooms shall be constructed to prevent the entrance of ignitible amounts of flyings or lint and shall be well ventilated.

503.15 Live Parts—Class III, Divisions 1 and 2.

Live parts shall not be exposed.

Exception: As provided in 503.13.

503.16 Grounding—Class III, Divisions 1 and 2.

Wiring and equipment in Class III, Divisions 1 and 2 locations shall be grounded as specified in Article 250 and with the following additional requirements in 503.16(A) and (B).

(A) Bonding. The locknut-bushing and double-locknut types of contacts shall not be depended on for bonding purposes, but bonding jumpers with proper fittings or other approved means of bonding shall be used. Such means of bonding shall apply to all intervening raceways, fittings, boxes, enclosures, and so forth, between Class III locations and the point of grounding for service equipment or point of grounding of a separately derived system.

Exception: The specific bonding means shall only be required to the nearest point where the grounded circuit conductor and the grounding electrode conductor are connected together on the line side of the building or structure disconnecting means as specified in 250.32(A), (B), and (C), if the branch-circuit overcurrent protection is located on the load side of the disconnecting means.

FPN: See 250.100 for additional bonding requirements in hazardous (classified) locations.

(B) Types of Equipment Grounding Conductors. Where flexible conduit is used as permitted in 503.3, it shall be installed with internal or external bonding jumpers in parallel with each conduit and complying with 250.102.

Exception: In Class III, Division 1 and 2 locations, the bonding jumper shall be permitted to be deleted where all the following conditions are met:

(a) *Listed liquidtight flexible metal 1.8 m (6 ft) or less in length, with fittings listed for grounding, is used.*
(b) *Overcurrent protection in the circuit is limited to 10 amperes or less.*
(c) *The load is not a power utilization load.*

ARTICLE 504
Intrinsically Safe Systems

Contents

504.1 Scope.

This article covers the installation of intrinsically safe (I.S.) apparatus, wiring, and systems for Class I, II, and III locations.

> FPN: For further information, see ANSI/ISA RP 12.6-1995, *Wiring Practices for Hazardous (Classified) Locations Instrumentation—Part 1: Intrinsic Safety.*

Article 504 was first included in the 1990 *NEC*. Previously, the installation requirements for intrinsically safe systems were in ANSI/ISA RP 12.6.

504.2 Definitions.

The standard used in the United States for construction and performance requirements for intrinsically safe systems is ANSI/UL 913, *Intrinsically Safe Apparatus and Associated Apparatus for Use in Class I, II, and III, Division 1 Hazardous (Classified) Locations.* ANSI/UL 913 is similar to standards in other countries; all standards for intrinsically safe equipment are based on the IEC (International Electrotechnical Commission) standard. The *NEC* offers the choice of designating hazardous (classified) locations into two divisions (1 and 2) or three zones (0, 1, and 2). ANSI/UL 913 requirements, however, are based on the IEC Zone 0 requirements, which are the most stringent. Equipment certified by a testing laboratory for Zone 1 would not necessarily meet ANSI/UL 913 requirements for Division 1.

Associated Apparatus. Apparatus in which the circuits are not necessarily intrinsically safe themselves but that affect the energy in the intrinsically safe circuits and are relied on to maintain intrinsic safety. Associated apparatus may be either of the following:

(1) Electrical apparatus that has an alternative-type protection for use in the appropriate hazardous (classified) location, or
(2) Electrical apparatus not so protected that shall not be used within a hazardous (classified) location

> FPN No. 1: Associated apparatus has identified intrinsically safe connections for intrinsically safe apparatus and also may have connections for nonintrinsically safe apparatus.
>
> FPN No. 2: An example of associated apparatus is an intrinsic safety barrier, which is a network designed to limit the energy (voltage and current) available to the protected circuit in the hazardous (classified) location, under specified fault conditions.

For an illustration of an intrinsic safety barrier, see Exhibit 504.1.

Control Drawing. See definition in 500.2.

Different Intrinsically Safe Circuits. Intrinsically safe circuits in which the possible interconnections have not been evaluated and identified as intrinsically safe.

Intrinsically Safe Apparatus. Apparatus in which all the circuits are intrinsically safe.

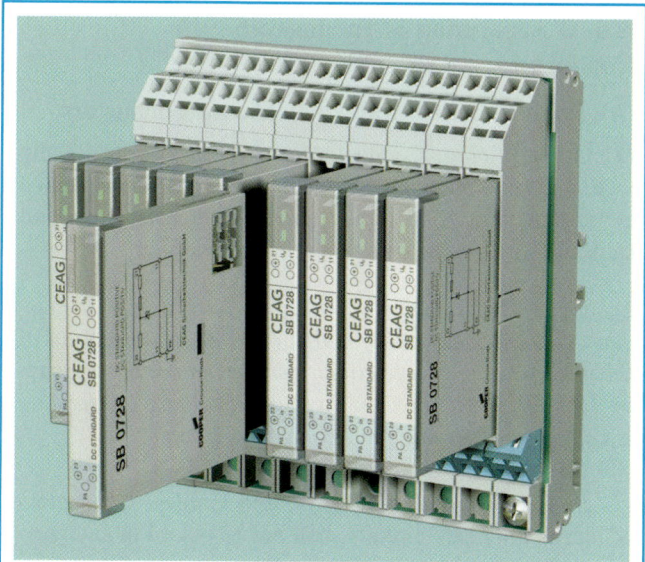

Exhibit 504.1 A typical intrinsic safety barrier that limits energy to the hazardous location. (Courtesy of Cooper Crouse-Hinds)

Intrinsically Safe Circuit. A circuit in which any spark or thermal effect is incapable of causing ignition of a mixture of flammable or combustible material in air under prescribed test conditions.

FPN: Test conditions are described in ANSI/UL 913-1997, *Standard for Safety, Intrinsically Safe Apparatus and Associated Apparatus for Use in Class I, II, and III, Division 1, Hazardous (Classified) Locations.*

Due to its physical and electrical characteristics, an intrinsically safe circuit does not develop sufficient electrical energy (millijoules) in an arc or spark to cause ignition or sufficient thermal energy resulting from an overload condition to cause the temperature of the installed circuit to exceed the ignition temperature of a specified gas or vapor under normal or abnormal operating conditions.

An abnormal condition may be due to accidental damage, failure of electrical components, excessive voltage, or improper adjustment or maintenance of the equipment.

Intrinsically Safe System. An assembly of interconnected intrinsically safe apparatus, associated apparatus, and interconnecting cables in that those parts of the system that may be used in hazardous (classified) locations are intrinsically safe circuits.

Although low-energy devices, such as thermocouples, crystal strain, or pressure transducers, generate millivolts and currents in the microampere range, they are not necessarily

intrinsically safe. Low-energy devices are normally connected to amplifiers and power supplies that are connected to 120-volt or higher circuits. Should a fault occur within the amplifier or power supply, or a voltage surge occur in the electrical supply system, high-energy arcing, sparking, or overheating of the low-energy portion of the circuit could occur.

FPN: An intrinsically safe system may include more than one intrinsically safe circuit.

Simple Apparatus. An electrical component or combination of components of simple construction with well-defined electrical parameters that does not generate more than 1.5 volts, 100 milliamps, and 25 milliwatts, or a passive component that does not dissipate more than 1.3 watts and is compatible with the intrinsic safety of the circuit in which it is used.

FPN: The following apparatus are examples of simple apparatus:

(a) Passive components; for example, switches, junction boxes, resistance temperature devices, and simple semiconductor devices such as LEDs
(b) Sources of generated energy, for example, thermocouples and photocells, which do not generate more than 1.5 V, 100 mA, and 25 mW

The definition of *simple apparatus* clarifies the use of the term in 504.4, Exception, and 504.10, Exception. The intent is to permit the use of apparatus that stores little or no energy without requiring the apparatus to be listed or to comply with the control drawing. See the fine print note following the definition of *simple apparatus* for examples of the apparatus contemplated.

504.3 Application of Other Articles.

Except as modified by this article, all applicable articles of this *Code* shall apply.

Because intrinsically safe wiring must be low-energy wiring in order to be intrinsically safe, the wiring itself is most likely to be Class 2, in accordance with 725.41, or, in a fire-protective signaling system, power-limited in accordance with 760.41. See Article 725 or 760, as appropriate, for the requirements for such wiring. The installation may also fall under the scope of Article 800. The intrinsically safe apparatus and associated apparatus, on the other hand, may be supplied by ordinary power circuits, in which case other *Code* rules may apply. It is common for intrinsically safe apparatus or associated apparatus supplied by a power circuit to be located in a hazardous (classified) location, with the apparatus protected by one of the protection systems required by Articles 500 through 503 or 505, that is, explosionproof, dust-ignitionproof, purged, and pressurized. In this case, Articles 500 through 503 and 505 also apply. Also, intrinsically safe systems are not exempt from the grounding and bonding requirements of 501.16, 502.16, 503.16, and 505.25.

504.4 Equipment.

All intrinsically safe apparatus and associated apparatus shall be listed.

Exception: Simple apparatus, as described on the control drawing, shall not be required to be listed.

504.10 Equipment Installation.

(A) Control Drawing. Intrinsically safe apparatus, associated apparatus, and other equipment shall be installed in accordance with the control drawing(s).

Exception: A simple apparatus that does not interconnect intrinsically safe circuits.

> FPN: The control drawing identification is marked on the apparatus.

The control drawing may put limitations on cables and on the separation of circuits within an intrinsically safe system. The control drawing also illustrates what is permitted to be connected in the system. Compliance with all the provisions in the control drawing is essential if intrinsic safety is to be maintained. The investigation of the equipment by third-party testing laboratories is based on installation in accordance with the control drawing.

(B) Location. Intrinsically safe apparatus shall be permitted to be installed in any hazardous (classified) locations for which it has been identified. General-purpose enclosures shall be permitted for intrinsically safe apparatus.

Associated apparatus shall be permitted to be installed in any hazardous (classified) location for which it has been identified or, if protected by other means, permitted by Articles 501 through 503 and Article 505.

504.20 Wiring Methods.

Intrinsically safe apparatus and wiring shall be permitted to be installed using any of the wiring methods suitable for unclassified locations, including Chapter 7 and Chapter 8. Sealing shall be as provided in 504.70, and separation shall be as provided in 504.30.

See the commentary following 504.3 for more information on the types of circuits typically used as part of an intrinsically safe system.

504.30 Separation of Intrinsically Safe Conductors.

(A) From Nonintrinsically Safe Circuit Conductors.

(1) Open Wiring. Conductors and cables of intrinsically safe circuits not in raceways or cable trays shall be separated at least 50 mm (2 in.) and secured from conductors and cables of any nonintrinsically safe circuits.

Exception: Where either (1) all of the intrinsically safe circuit conductors are in Type MI or MC cables or (2) all of the nonintrinsically safe circuit conductors are in raceways or Type MI or MC cables where the sheathing or cladding is capable of carrying fault current to ground.

(2) In Raceways, Cable Trays, and Cables. Conductors of intrinsically safe circuits shall not be placed in any raceway, cable tray, or cable with conductors of any nonintrinsically safe circuit.

Exception No. 1: Where conductors of intrinsically safe circuits are separated from conductors of nonintrinsically safe circuits by a distance of at least 50 mm (2 in.) and secured, or by a grounded metal partition or an approved insulating partition.

> FPN: No. 20 gauge sheet metal partitions 0.91 mm (0.0359 in.) or thicker are generally considered acceptable.

Exception No. 2: Where either (1) all of the intrinsically safe circuit conductors or (2) all of the nonintrinsically safe circuit conductors are in grounded metal-sheathed or metal-clad cables where the sheathing or cladding is capable of carrying fault current to ground.

> FPN: Cables meeting the requirements of Articles 330 and 332 are typical of those considered acceptable.

Type MI cable with a copper sheath or Type MC cable of smooth metallic sheath or corrugated metallic sheath construction meet the conditions prescribed in Exception No. 2. The metallic sheath of interlocked-tape Type MC cable (as opposed to the smooth or corrugated continuous outer sheath) is generally not investigated for suitability as an equipment grounding conductor and, therefore, may not be capable of carrying a fault current to ground as required by Exception No. 2.

(3) Within Enclosures.

(a) Conductors of intrinsically safe circuits shall be separated at least 50 mm (2 in.) from conductors of any nonintrinsically safe circuits, or as specified in 504.30(A)(2).

(b) All conductors shall be secured so that any conductor that might come loose from a terminal cannot come in contact with another terminal.

> FPN No. 1: The use of separate wiring compartments for the intrinsically safe and nonintrinsically safe terminals is the preferred method of complying with this requirement.
> FPN No. 2: Physical barriers such as grounded metal partitions or approved insulating partitions or approved restricted access wiring ducts separated from other such ducts by at least 19 mm (¾ in.) can be used to help ensure the required separation of the wiring.

The intent of 504.30(A) is to prevent the intrusion of unsafe energy into the intrinsically safe system as a result of a wiring fault. Because low-voltage, low-energy, but nonintrinsically safe circuits are permitted by Articles 725 and 800 to be minimally insulated and may not be protected by being installed in raceways or cables, particularly in nonhazardous locations, it is essential that nonintrinsically safe circuits and intrinsically safe circuits be physically and electrically separated.

(B) From Different Intrinsically Safe Circuit Conductors. Different intrinsically safe circuits shall be in separate cables or shall be separated from each other by one of the following means:

(1) The conductors of each circuit are within a grounded metal shield.
(2) The conductors of each circuit have insulation with a minimum thickness of 0.25 mm (0.01 in.).

Exception: Unless otherwise identified.

504.50 Grounding.

(A) Intrinsically Safe Apparatus, Associated Apparatus, and Raceways. Intrinsically safe apparatus, associated apparatus, cable shields, enclosures, and raceways, if of metal, shall be grounded.

> FPN: Supplementary bonding to the grounding electrode may be needed for some associated apparatus, for example, zener diode barriers, if specified in the control drawing. See ANSI/ISA RP 12.6-1995, *Wiring Practices for Hazardous (Classified) Locations Instrumentation Part 1: Intrinsic Safety.*

It is very important to maintain a low-impedance path to ground for zener diode barrier systems because such systems shunt fault currents to ground.

Exhibit 504.2 illustrates a common type of zener diode barrier system. F_1 limits the duration of the current in Z_1

and Z_2 to the power rating of the zener diodes under fault conditions, so that the diodes need not be excessively large. Although the diodes themselves can be considered subject to open-circuit fault, they are redundant components, and either one alone provides the necessary protection in the event of a first fault, that is, high voltage across input terminals 1 and 2. Resistor R_2 is the current-limiting resistor and has been investigated as a protective component not subject to short-circuit fault. Resistor R_1 is used primarily to permit testing to determine that the diodes are intact. Terminals 2 and 4 and the ends of the two diodes are connected to a ground bus that, in turn, is connected to a ground system to which all grounds in the intrinsically safe system are connected. A very low impedance (1 ohm or less is usually recommended) is necessary so that the voltage level on the ground bus will not be raised to an unsafe level under high-current fault conditions.

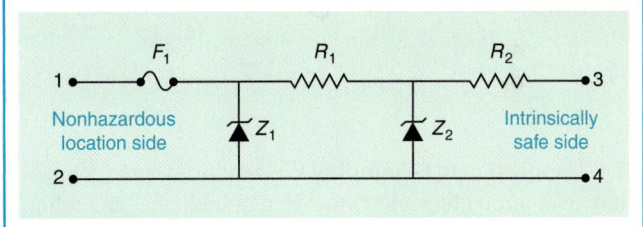

Exhibit 504.2 Fused zener diode barrier.

A shunt diode barrier system is commonly tested with 250 volts ac across terminals 1 and 2, representing a fault in associated apparatus on the nonhazardous location side of the barrier, which imposes a voltage up to 250 volts ac on these terminals. Diode Z_1 conducts at its rated voltage, typically 14 or 15 volts for a barrier designed for a 12-volt (normal operation) system, thus limiting the output voltage of the circuit to 14 or 15 volts. The voltage at which this diode conducts is designed to be higher than the rated input voltage (under no-fault conditions) at terminals 1 and 2. The fuse is selected so that it will open before the power rating of Z_1 is exceeded. Diode Z_2 usually conducts within one or two volts of Z_1 and serves as a backup to Z_1 in the event Z_1 fails in an open-circuit condition for any reason.

Resistor R_2 limits the current in the intrinsically safe circuit to the required value at the output voltage of Z_1 and Z_2. Therefore, even though up to 250 volts ac may be applied to the input of the circuit (terminals 1 and 2) as a result of a fault in the equipment or circuit connected to terminals 1 and 2, the output at terminals 3 and 4 cannot exceed the voltage and current permitted by the diodes and resistor.

Shunt diode barriers normally have maximum inductance and capacitance ratings on the intrinsically safe side because, even though both the voltage and current are limited at terminals 3 and 4, too much inductance in the circuit connected to terminals 3 and 4 could result in the release of an ignition-capable spark when the circuit is opened. In a like manner, too much capacitance between terminals 3 and 4 could result in the release of an ignition-capable spark if there were a short-circuit between wiring connected to terminals 3 and 4 or between wiring connected to terminal 3 and ground.

Wiring always has inductance and capacitance associated with it, depending on the spacing between conductors, size of conductors, and length of conductors. It is necessary, therefore, to limit the length of conductors connected to terminals 3 and 4, just as it is necessary to limit the inductance and capacitance of connected equipment. The length limitation is usually on the order of thousands of feet. The manufacturer's instructions provide the information on installation limitations.

By adjusting the values of its components, the barrier can be designed for a variety of uses, including different types of instrument systems and thermocouples. Barriers such as described are available commercially from a number of manufacturers.

(B) Connection to Grounding Electrodes. Where connection to a grounding electrode is required, the grounding electrode shall be as specified in 250.52(A)(1), (2), (3), and (4) and shall comply with 250.30(A)(3). Section 250.52(A)(5), (6) and (7) shall not be used if electrodes specified in 250.52(A)(1), (2), (3) or (4) are available.

The electrodes specified in 250.52(A)(1), (2), (3) and (4) include metal underground water pipes, the metal frames of building, concrete-encased electrodes, and ground rings. These electrodes usually provide lower resistance grounds than electrodes, such as ground rods and plate electrodes, covered in 250.52(A)(5), and (6).

(C) Shields. Where shielded conductors or cables are used, shields shall be grounded.

Exception: Where a shield is part of an intrinsically safe circuit.

504.60 Bonding.

(A) Hazardous Locations. In hazardous (classified) locations, intrinsically safe apparatus shall be bonded in the hazardous (classified) location in accordance with 250.100.

(B) Unclassified. In unclassified or nonhazardous locations, where metal raceways are used for intrinsically safe system wiring in hazardous (classified) locations, associated apparatus shall be bonded in accordance with 501.16(A), 502.16(A), 503.16(A), or 505.25, as applicable.

504.70 Sealing.

Conduits and cables that are required to be sealed by 501.5 and 502.5 shall be sealed to minimize the passage of gases, vapors, or dusts. Such seals shall not be required to be explosionproof or flameproof.

Exception: Seals shall not be required for enclosures that contain only intrinsically safe apparatus, except as required by 501.5(F)(3).

504.80 Identification.

Labels required by this section shall be suitable for the environment where they are installed with consideration given to exposure to chemicals and sunlight.

(A) Terminals. Intrinsically safe circuits shall be identified at terminal and junction locations in a manner that will prevent unintentional interference with the circuits during testing and servicing.

(B) Wiring. Raceways, cable trays, and other wiring methods for intrinsically safe system wiring shall be identified with permanently affixed labels with the wording "Intrinsic Safety Wiring" or equivalent. The labels shall be located so as to be visible after installation and placed so that they may be readily traced through the entire length of the installation. Intrinsic safety circuit labels shall appear in every section of the wiring system that is separated by enclosures, walls, partitions, or floors. Spacing between labels shall not be more than 7.5 m (25 ft).

Exception: Circuits run underground shall be permitted to be identified where they become accessible after emergence from the ground.

> FPN No. 1: Wiring methods permitted in unclassified locations may be used for intrinsically safe systems in hazardous (classified) locations. Without labels to identify the application of the wiring, enforcement authorities cannot determine that an installation is in compliance with the *Code.*
> FPN No. 2: In unclassified locations, the identification is necessary to ensure that nonintrinsically safe wire will not be inadvertently added to existing raceways at a later date.

(C) Color Coding. Color coding shall be permitted to identify intrinsically safe conductors where they are colored light blue and where no other conductors colored light blue are used. Likewise, color coding shall be permitted to identify raceways, cable trays, and junction boxes where they are colored light blue and contain only intrinsically safe wiring.

ARTICLE 505
Class I, Zone 0, 1, and 2 Locations

Contents

FPN: Rules that are followed by a reference in brackets contain text that has been extracted from NFPA 497-1997, *Recommended Practice for the Classification of Flammable Liquids, Gases, or Vapors and of Hazardous (Classified) Locations for Electrical Installations in Chemical Process Areas*. Only editorial changes were made to the extracted text to make it consistent with this *Code*.

505.1 Scope.

This article covers the requirements for the zone classification system as an alternative to the division classification system covered in Article 500 for electrical and electronic equipment and wiring for all voltages in Class I, Zone 0, Zone 1, and Zone 2 hazardous (classified) locations where fire or explosion hazards may exist due to flammable gases, vapors, or liquids.

FPN: For the requirements for electrical and electronic equipment and wiring for all voltages in Class I, Division 1 or Division 2; Class II, Division 1 or Division 2; and Class III, Division 1 or Division 2 hazardous (classified) locations where fire or explosion hazards may exist due to flammable gases or vapors, flammable liquids, or combustible dusts or fibers, refer to Articles 500 through 504.

New requirements that cover wiring methods, sealing, and flexible cord use have been added to Article 505 for the 2002 *Code* so that the Zone 0, Zone 1, and Zone 2 classification concept can be used as a set of requirements parallel to those in Articles 500 and 501 for installations in which the Division 1 and 2 concept is used. The zone classification concept offers an alternative method of classifying Class I hazardous locations. The zone classification method is based on the standards for area classification used by the International Electrotechnical Commission (IEC).

The IEC classification scheme includes underground mines. In the United States, mines are under the jurisdiction of the Mine Safety and Health Administration (MSHA) and are outside of the scope of the *Code*.

505.2 Definitions.

For purposes of this article, the following definitions apply.

Combustible Gas Detection System. A protection technique utilizing stationary gas detectors in industrial establishments.

Electrical and Electronic Equipment. Materials, fittings, devices, appliances, and the like that are part of, or in connection with, an electrical installation.

> FPN: Portable or transportable equipment having self-contained power supplies, such as battery-operated equipment, could potentially become an ignition source in hazardous (classified) locations.

Encapsulation "m." Type of protection where electrical parts that could ignite an explosive atmosphere by either sparking or heating are enclosed in a compound in such a way that this explosive atmosphere cannot be ignited.

> FPN: See ISA 12.23.01-1998, *Electrical Apparatus for Use in Class I, Zone 1 Hazardous (Classified) Locations, Type of Protection—Encapsulation "m"*; IEC 60079-18-1992, *Electrical Apparatus for Explosive Gas Atmospheres—Part 18: Encapsulation "m"* ; and ANSI/UL 2279-1997(Part 18), *Electrical Equipment for Use in Class I, Zone 0, 1, and 2 Hazardous (Classified) Locations*.

Flameproof "d." Type of protection where the enclosure will withstand an internal explosion of a flammable mixture that has penetrated into the interior, without suffering damage and without causing ignition, through any joints or structural openings in the enclosure, of an external explosive gas atmosphere consisting of one or more of the gases or vapors for which it is designed.

> FPN: See ISA 12.22.01-1998, *Electrical Apparatus for Use in Class I, Zone 1 and 2 Hazardous (Classified) Locations, Type of Protection—Flameproof "d"*; IEC 60079-1-2000, *Electrical Apparatus for Explosive Gas Atmospheres, Part 1—Construction and Verification Test of Flameproof Enclosures of Electrical Apparatus*; ANSI/ UL 2279-1997 (Part 1), *Electrical Equipment for Use in Class I, Zone 0, 1, and 2 Hazardous (Classified) Locations*.

Increased Safety "e." Type of protection applied to electrical equipment that does not produce arcs or sparks in normal service and under specified abnormal conditions, in which additional measures are applied so as to give increased security against the possibility of excessive temperatures and of the occurrence of arcs and sparks.

> FPN: See ISA—12.16.01-1998, *Electrical Apparatus for Use in Class I, Zone 1 Hazardous (Classified) Locations, Type of Protection—Increased Safety "e"*; IEC 60079-7-1990, *Electrical Apparatus for Explosive Gas Atmospheres—Part 7: Increased Safety "e"*, Amendment No. 1 (1991) and Amendment No. 2 (1993); and ANSI/ UL 2279-1997(Part 7), *Electrical Equipment for Use in*

Class I, Zone 0, 1, and 2 Hazardous (Classified) Locations.

Intrinsic Safety "i." Type of protection where any spark or thermal effect is incapable of causing ignition of a mixture of flammable or combustible material in air under prescribed test conditions.

> FPN No. 1: See ANSI/UL 913-1997, *Intrinsically Safe Apparatus and Associated Apparatus for Use in Class I, II, and III, Hazardous Locations*; ISA —12.02.01-1999, *Electrical Apparatus for Use in Class I, Zones 0, 1 and 2 Hazardous (Classified) Locations—Intrinsic Safety "i"*; IEC 60079-11-1999, *Electrical Apparatus for Explosive Gas Atmospheres—Part 11: Intrinsic Safety "i"*; and ANSI/UL 2279-1997 (Part 11), *Electrical Equipment for Use in Class I, Zone 0, 1, and 2 Hazardous (Classified) Locations*.
>
> FPN No. 2: Intrinsic safety is designated type of protection "ia" for use in Zone 0 locations. Intrinsic safety is designated type of protection "ib" for use in Zone 1 locations.
>
> FPN No. 3: Intrinsically safe associated apparatus, designated by [ia] or [ib], is connected to intrinsically safe apparatus ("ia" or "ib," respectively) but is located outside the hazardous (classified) location unless also protected by another type of protection (such as flameproof).

Oil Immersion "o." Type of protection where electrical equipment is immersed in a protective liquid in such a way that an explosive atmosphere that may be above the liquid or outside the enclosure cannot be ignited.

> FPN: See ISA 12.26.01-1998, *Electrical Apparatus for Use in Class I, Zone 1 Hazardous (Classified) Locations, Type of Protection—Oil-Immersion "o"*; IEC 60079-6-1995, *Electrical Apparatus for Explosive Gas Atmospheres*, Part 6—Oil-Immersion "o" ; and ANSI/UL 2279-1997 (Part 6), *Electrical Equipment for Use in Class I, Zone 0, 1, and 2 Hazardous (Classified) Locations*.

Powder Filling "q." Type of protection where electrical parts capable of igniting an explosive atmosphere are fixed in position and completely surrounded by filling material (glass or quartz powder) to prevent the ignition of an external explosive atmosphere.

> FPN: See ISA-12.25.01-1996, *Electrical Apparatus for Use in Class I, Zone 1 Hazardous (Classified) Locations Type of Protection—Powder Filling "q"*; IEC 60079-5-1996, *Electrical Apparatus for Explosive Gas Atmospheres—Part 5: Powder Filling, Type of Protection "q"*; and ANSI/UL 2279-1997 (Part 5), *Electrical Equipment for Use in Class I, Zone 0, 1, and 2 Hazardous (Classified) Locations*.

Purged and Pressurized. Type of protection for electrical equipment that uses the technique of guarding against the ingress of the external atmosphere, which may be explosive, into an enclosure by maintaining a protective gas therein at a pressure above that of the external atmosphere.

FPN No. 1: See NFPA 496-1998, *Standard for Purged and Pressurized Enclosures for Electrical Equipment.*

FPN No. 2: See IEC 60079-2-2000, *Electrical Apparatus for Explosive Gas Atmospheres—Part 2: Electrical Apparatus, Type of Protection "p"*; and IEC 60079-13-1982, *Electrical Apparatus for Explosive Gas Atmospheres—Part 13: Construction and Use of Rooms or Buildings Protected by Pressurization.*

Type of Protection "n." Type of protection where electrical equipment, in normal operation, is not capable of igniting a surrounding explosive gas atmosphere and a fault capable of causing ignition is not likely to occur.

FPN: See IEC 60079-15-2000, *Electrical Apparatus for Explosive Gas Atmospheres, Part 15—Electrical Apparatus with Type of Protection "n"*; and ANSI/UL 2279-1997 (Part 15), *Electrical Equipment for Use in Class I, Zone 0, 1, and 2 Hazardous (Classified) Locations.*

Unclassified Locations. Locations determined to be neither Class I, Division 1; Class I, Division 2; Class I, Zone 0; Class I, Zone 1; Class I, Zone 2; Class II, Division 1; Class II, Division 2; Class III, Division 1; Class III, Division 2; or any combination thereof.

505.3 Other Articles.

All other applicable rules contained in this *Code* shall apply to electrical equipment and wiring installed in hazardous (classified) locations.

Exception: As modified by Article 504 and this article.

See the commentary following 500.3 for more information on other applicable *Code* requirements that apply to electrical equipment and wiring.

505.4 General.

(A) Documentation for Industrial Occupancies. All areas in industrial occupancies designated as hazardous (classified) locations shall be properly documented. This documentation shall be available to those authorized to design, install, inspect, maintain, or operate electrical equipment at the location.

FPN: For examples of area classification drawings, see ANSI/API RP 505-1997, *Recommended Practice for Classification of Locations for Electrical Installations at Petroleum Facilities Classified as Class I, Zone 0, Zone 1, or Zone 2*; ISA RP12.24.01-1998, *Recommended Practice for Classification of Locations for Electrical Installations Classified as Class I, Zone 0, Zone 1, or Zone 2*; IEC 60079-10-1995, *Electrical Apparatus for Explosive Gas Atmospheres, Classification of Hazardous Areas*; and *Model Code of Safe Practice in the Petroleum Industry, Part 15: Area Classification Code for Petroleum Installations*, IP 15, The Institute of Petroleum, London.

(B) Reference Standards. Important information relating to topics covered in Chapter 5 may be found in other publications.

FPN No. 1: It is important that the authority having jurisdiction be familiar with recorded industrial experience as well as with standards of the National Fire Protection Association (NFPA), the American Petroleum Institute (API), Instrumentation, Systems, and Automation Society (ISA), and the International Electrotechnical Commission (IEC) that may be of use in the classification of various locations, the determination of adequate ventilation, and the protection against static electricity and lightning hazards.

FPN No. 2: For further information on the classification of locations, see ANSI/API RP 505-1997, *Recommended Practice for Classification of Locations for Electrical Installations at Petroleum Facilities Classified as Class I, Zone 0, Zone 1, or Zone 2*; ISA RP12.24.01-1998, *Recommended Practice for Classification of Locations for Electrical Installations Classified as Class I, Zone 0, Zone 1, or Zone 2*; IEC 60079-10-1995, *Electrical Apparatus for Explosive Gas Atmospheres, Classification of Hazardous Areas*; and *Model Code of Safe Practice in the Petroleum Industry, Part 15: Area Classification Code for Petroleum Installations*, IP 15, The Institute of Petroleum, London.

FPN No. 3: For further information on protection against static electricity and lightning hazards in hazardous (classified) locations, see NFPA 77-2000, *Recommended Practice on Static Electricity*; NFPA 780-1997, *Standard for the Installation of Lightning Protection Systems*; and API RP 2003-1998, *Protection Against Ignitions Arising Out of Static Lightning and Stray Currents.*

FPN No. 4: For further information on ventilation, see NFPA 30-2000, *Flammable and Combustible Liquids Code*, and ANSI/API RP 505-1997, *Recommended Practice for Classification of Locations for Electrical Installations at Petroleum Facilities Classified as Class I, Zone 0, Zone 1, or Zone 2.*

FPN No. 5: For further information on electrical systems for hazardous (classified) locations on offshore oil and gas producing platforms, see ANSI/API RP 14FZ-2000, *Recommended Practice for Design and Installation of Electrical Systems for Fixed and Floating Offshore Petroleum Facilities for Unclassified and Class I, Zone 0, Zone 1, and Zone 2 Locations.*

FPN No. 6: For further information on the installation of electrical equipment in hazardous (classified) locations in general, see IEC 60079-14-1996, *Electrical Apparatus for Explosive Gas Atmospheres—Part 14: Electrical Installations in Explosive Gas Atmospheres (Other Than Mines)*, and IEC 60079-16-1990, *Electrical Apparatus for Explosive Gas Atmospheres—Part 16: Artificial Ventilation for the Protection of Analyzer(s) Houses.*

FPN No. 7: For further information on application of electrical equipment in hazardous (classified) locations in general, see ISA 12.00.01-1999, *Electrical Apparatus for Use in Class I, Zones 0 and 1, Hazardous (Classified) Locations: General Requirements*; ISA 12.01.01-1999, *Definitions and Information Pertaining to Electrical Apparatus in Hazardous (Classified) Locations*; and ANSI/UL 2279-1997 (Part 0), *Electrical Equipment for Use in*

Class I, Zone 0, 1, and 2 Hazardous (Classified) Locations.

505.5 Classifications of Locations.

(A) Classification of Locations. Locations shall be classified depending on the properties of the flammable vapors, liquids, or gases that may be present and the likelihood that a flammable or combustible concentration or quantity is present. Where pyrophoric materials are the only materials used or handled, these locations shall not be classified. Each room, section, or area shall be considered individually in determining its classification.

FPN No. 1: See 505.7 for restrictions on area classification.

FPN No. 2: Through the exercise of ingenuity in the layout of electrical installations for hazardous (classified) locations, it is frequently possible to locate much of the equipment in reduced level of classification or in an unclassified location and, thus, to reduce the amount of special equipment required.

Rooms and areas containing ammonia refrigeration systems that are equipped with adequate mechanical ventilation may be classified as "unclassified" locations.

FPN: For further information regarding classification and ventilation of areas involving ammonia, see ANSI/ASHRAE 15-1994, *Safety Code for Mechanical Refrigeration*; and ANSI/CGA G2.1-1989 (14-39), *Safety Requirements for the Storage and Handling of Anhydrous Ammonia*.

(B) Class I, Zone 0, 1, and 2 Locations. Class I, Zone 0, 1, and 2 locations are those in which flammable gases or vapors are or may be present in the air in quantities sufficient to produce explosive or ignitible mixtures. Class I, Zone 0, 1, and 2 locations shall include those specified in 505(B)(1), (B)(2), and (B)(3).

(1) Class I, Zone 0. A Class I, Zone 0 location is a location in which

(1) Ignitible concentrations of flammable gases or vapors are present continuously, or

(2) Ignitible concentrations of flammable gases or vapors are present for long periods of time.

FPN No. 1: As a guide in determining when flammable gases or vapors are present continuously or for long periods of time, refer to ANSI/API RP 505-1997, *Recommended Practice for Classification of Locations for Electrical Installations of Petroleum Facilities Classified as Class I, Zone 0, Zone 1 or Zone 2*; ISA 12.24.01-1998, *Recommended Practice for Classification of Locations for Electrical Installations Classified as Class I, Zone 0, Zone 1, or Zone 2*; IEC 60079-10-1995, *Electrical Apparatus for Explosive Gas Atmospheres, Classifications of Hazardous Areas*; and *Area Classification Code*

for Petroleum Installations, Model Code, Part 15, Institute of Petroleum.

FPN No. 2: This classification includes locations inside vented tanks or vessels that contain volatile flammable liquids; inside inadequately vented spraying or coating enclosures, where volatile flammable solvents are used; between the inner and outer roof sections of a floating roof tank containing volatile flammable liquids; inside open vessels, tanks and pits containing volatile flammable liquids; the interior of an exhaust duct that is used to vent ignitible concentrations of gases or vapors; and inside inadequately ventilated enclosures that contain normally venting instruments utilizing or analyzing flammable fluids and venting to the inside of the enclosures.

FPN No. 3: It is not good practice to install electrical equipment in Zone 0 locations except when the equipment is essential to the process or when other locations are not feasible. [See 505.5(A) FPN No. 2.] If it is necessary to install electrical systems in a Zone 0 location, it is good practice to install intrinsically safe systems as described by Article 504.

(2) Class I, Zone 1. A Class I, Zone 1 location is a location

(1) In which ignitible concentrations of flammable gases or vapors are likely to exist under normal operating conditions; or

(2) In which ignitible concentrations of flammable gases or vapors may exist frequently because of repair or maintenance operations or because of leakage; or

(3) In which equipment is operated or processes are carried on, of such a nature that equipment breakdown or faulty operations could result in the release of ignitible concentrations of flammable gases or vapors and also cause simultaneous failure of electrical equipment in a mode to cause the electrical equipment to become a source of ignition; or

(4) That is adjacent to a Class I, Zone 0 location from which ignitible concentrations of vapors could be communicated, unless communication is prevented by adequate positive pressure ventilation from a source of clean air and effective safeguards against ventilation failure are provided.

FPN No. 1: Normal operation is considered the situation when plant equipment is operating within its design parameters. Minor releases of flammable material may be part of normal operations. Minor releases include the releases from mechanical packings on pumps. Failures that involve repair or shutdown (such as the breakdown of pump seals and flange gaskets, and spillage caused by accidents) are not considered normal operation.

FPN No. 2: This classification usually includes locations where volatile flammable liquids or liquefied flammable gases are transferred from one container to another. In areas in the vicinity of spraying and painting operations where flammable solvents are used; adequately ventilated drying rooms or compartments for evaporation of flammable solvents; adequately ventilated locations containing fat and oil extraction equipment using volatile flammable solvents; portions of cleaning and dyeing plants where volatile flammable liquids are used; adequately ventilated gas generator rooms and other portions of gas manufacturing plants where flammable gas may escape; inadequately ventilated pump rooms for flammable gas or for volatile flammable liquids; the interiors of refrigerators and freezers in which volatile flammable materials are stored in the open, lightly stoppered, or in easily ruptured containers; and other locations where ignitible concentrations of flammable vapors or gases are likely to occur in the course of normal operation but not classified Zone 0.

(3) Class I, Zone 2. A Class I, Zone 2 location is a location

(1) In which ignitible concentrations of flammable gases or vapors are not likely to occur in normal operation and, if they do occur, will exist only for a short period; or

(2) In which volatile flammable liquids, flammable gases, or flammable vapors are handled, processed, or used but in which the liquids, gases, or vapors normally are confined within closed containers of closed systems from which they can escape, only as a result of accidental rupture or breakdown of the containers or system, or as a result of the abnormal operation of the equipment with which the liquids or gases are handled, processed, or used; or

(3) In which ignitible concentrations of flammable gases or vapors normally are prevented by positive mechanical ventilation but which may become hazardous as a result of failure or abnormal operation of the ventilation equipment; or

(4) That is adjacent to a Class I, Zone 1 location, from which ignitible concentrations of flammable gases or vapors could be communicated, unless such communication is prevented by adequate positive-pressure ventilation from a source of clean air and effective safeguards against ventilation failure are provided.

FPN: The Zone 2 classification usually includes locations where volatile flammable liquids or flammable gases or vapors are used but which would become hazardous only in case of an accident or of some unusual operating condition.

505.6 Material Groups.

For purposes of testing, approval, and area classification, various air mixtures (not oxygen enriched) shall be grouped as required in 505.6(A), (B), and (C).

FPN: Group I is intended for use in describing atmospheres that contain firedamp (a mixture of gases, composed mostly of methane, found underground, usually in mines). This *Code* does not apply to installations underground in mines. See 90.2(B).

Group II shall be subdivided into IIC, IIB, and IIA, as noted in 505.6(A), (B), and (C), according to the nature of

the gas or vapor, for protection techniques "d," "ia," "ib," "[ia]," and "[ib]," and, where applicable, "n" and "o."

FPN No. 1: The gas and vapor subdivision as described above is based on the maximum experimental safe gap (MESG), minimum igniting current (MIC), or both. Test equipment for determining the MESG is described in IEC 60079-1A-1975, Amendment No. 1 (1993), *Construction and Verification Tests of Flameproof Enclosures of Electrical Apparatus*; and *UL Technical Report No. 58* (1993). The test equipment for determining MIC is described in IEC 60079-11-1999, *Electrical Apparatus for Explosive Gas Atmospheres—Part 11: Intrinsic Safety "i"*. The classification of gases or vapors according to their maximum experimental safe gaps and minimum igniting currents is described in IEC 60079-12-1978, *Classification of Mixtures of Gases or Vapours with Air According to Their Maximum Experimental Safe Gaps and Minimum Igniting Currents*.

FPN No. 2: Verification of electrical equipment utilizing protection techniques "e," "m," "p," and "q," due to design technique, does not require tests involving MESG or MIC. Therefore, Group II is not required to be subdivided for these protection techniques.

FPN No. 3: It is necessary that the meanings of the different equipment markings and Group II classifications be carefully observed to avoid confusion with Class I, Divisions 1 and 2, Groups A, B, C, and D.

In the zone classification system, the gas or vapor group order is the approximate inverse of the gas or vapor groups specified in Article 500. For example, Group IIC includes Article 500, Groups A and B. Determination of a gas or vapor for the purposes of grouping includes the evaluation of the maximum safe experimental gap ratio as well as minimum igniting current ratio. Although the maximum safe experimental gap for Group A is less than that for Group B in some circumstances, the minimum igniting current ratio is less for hydrogen (Group B) than it is for acetylene (Group A). This difference has been accounted for in ANSI/UL 913, *Intrinsically Safe Apparatus and Associated Apparatus for Use in Class I, II, III, Division 1, Hazardous (Classified) Locations,* because it is a consideration for the evaluation of intrinsically safe apparatus.

Class I, Zone 0, 1, and 2, groups shall be as follows:

(A) Group IIC. Atmospheres containing acetylene, hydrogen, or flammable gas, flammable liquid-produced vapor, or combustible liquid-produced vapor mixed with air that may burn or explode, having either a maximum experimental safe gap (MESG) value less than or equal to 0.50 mm or minimum igniting current ratio (MIC ratio) less than or equal to 0.45. [NFPA 497, 1-3]

FPN: Group IIC is equivalent to a combination of Class I, Group A, and Class I, Group B, as described in 500.6(A)(1) and 500.6(A)(2).

(B) Group IIB. Atmospheres containing acetaldehyde, ethylene, or flammable gas, flammable liquid-produced vapor, or combustible liquid-produced vapor mixed with air that may burn or explode, having either maximum experimental safe gap (MESG) values greater than 0.50 mm and less than or equal to 0.90 mm or minimum igniting current ratio (MIC ratio) greater than 0.45 and less than or equal to 0.80. [NFPA 497, 1-3]

> FPN: Group IIB is equivalent to Class I, Group C, as described in 500.6(A)(3).

(C) Group IIA. Atmospheres containing acetone, ammonia, ethyl alcohol, gasoline, methane, propane, or flammable gas, flammable liquid-produced vapor, or combustible liquid-produced vapor mixed with air that may burn or explode, having either a maximum experiment safe gap (MESG) value greater than 0.90 mm or minimum igniting current ratio (MIC ratio) greater than 0.80. [NFPA 497, 1-3]

> FPN: Group IIA is equivalent to Class I, Group D as described in 500.6(A)(4).

505.7 Special Precaution.

Article 505 requires equipment construction and installation that ensures safe performance under conditions of proper use and maintenance.

> FPN No. 1: It is important that inspection authorities and users exercise more than ordinary care with regard to the installation and maintenance of electrical equipment in hazardous (classified) locations.
>
> FPN No. 2: Low ambient conditions require special consideration. Electrical equipment depending on the protection techniques described by 505.8(A) may not be suitable for use at temperatures lower than $-20°C$ ($-4°F$) unless they are identified for use at lower temperatures. However, at low ambient temperatures, flammable concentrations of vapors may not exist in a location classified Class I, Zones 0, 1, or 2 at normal ambient temperature.

(A) Supervision of Work. Classification of areas and selection of equipment and wiring methods shall be under the supervision of a qualified Registered Professional Engineer.

> Note that 505.7(A) requires area classification, wiring, and equipment selection to be under the supervision of a qualified Registered Professional Engineer for installations in Class I, Zone 0, 1, and 2 locations.

(B) Dual Classification. In instances of areas within the same facility classified separately, Class I, Zone 2 locations shall be permitted to abut, but not overlap, Class I, Division 2 locations. Class I, Zone 0 or Zone 1 locations shall not abut Class I, Division 1 or Division 2 locations.

> An installation is permitted to be designed using either the classification scheme of Article 500 or the classification scheme of Article 505. Both schemes cannot be used for classifying the same area. In areas within the same facility, Class I, Zone 2 locations are allowed to be adjacent to and share the same border, but not overlap Class I, Division 2 locations. However, Class I, Zone 0 or Zone 1 locations are not allowed to be adjacent to and share the same border with Class I, Division 1 or Division 2 locations.

(C) Reclassification Permitted. A Class I, Division 1 or Division 2 location shall be permitted to be reclassified as a Class I, Zone 0, Zone 1, or Zone 2 location, provided all of the space that is classified because of a single flammable gas or vapor source is reclassified under the requirements of this article.

(D) Solid Obstacles. Flameproof equipment with flanged joints shall not be installed such that the flange openings are closer than the distances shown in Table 505.7(D) to any solid obstacle that is not a part of the equipment (such as steelworks, walls, weather guards, mounting brackets, pipes, or other electrical equipment) unless the equipment is listed for a smaller distance of separation.

Table 505.7(D) Minimum Distance of Obstructions from Flameproof "d" Flange Openings

Gas Group	Minimum Distance	
	mm	**in.**
IIC	40	$1\,37/64$
IIB	30	$1\,3/16$
IIA	10	$25/64$

505.8 Protection Techniques.

Acceptable protection techniques for electrical and electronic equipment in hazardous (classified) locations shall be as described in 505.8(A) through (I).

> FPN: For additional information, see ISA 12.00.01-1999, *Electrical Apparatus for Use in Class I, Zones 0 and 1 Hazardous (Classified) Locations, General Requirements*; ISA 12.01.01-1999, *Definitions and Information Pertaining to Electrical Apparatus in Hazardous (Classified) Locations*; ANSI/UL 2279, 1997, *Electrical Equipment for Use in Class I, Zone 0, 1, and 2 Hazardous (Classified) Locations*; and IEC 60079-0-1998, *Electrical Apparatus for Explosive Gas Atmospheres—Part 0: General Requirements*.

> Where the area is classified in accordance with the zone method, electrical and electronic equipment may be protected by the following methods:

1. In Class I, Zone 1, equipment approved as flameproof "d," which is very similar to explosionproof equipment in the U.S.
2. Purged and pressurized equipment for Class I, Zone 1 or Zone 2 locations for which it is approved
3. Intrinsic safety techniques for the Class I, Zone 0 or Zone 1 for which the technique is listed
4. In Class I, Zone 2, equipment approved as type "n," which under normal operation is not capable of igniting a surrounding explosive gas atmosphere and is not likely to create a fault that is capable of causing ignition
5. Oil immersion "o" technique for Class I, Zone 1, where the equipment or part of the equipment is immersed in a protective liquid so that explosive gases above or outside the enclosure cannot be ignited
6. A type of protection technique of increased safety "e" approved for Class I, Zone 1, in which the electrical equipment involved does not produce arcs or sparks or excessive temperature under normal operation, as well as under abnormal conditions, under specified conditions of operation
7. Encapsulation "m" technique in which the arcing, sparking, or hot parts are completely surrounded in a compound in such a way that an explosive gas or vapor cannot be ignited in a Class I, Zone 1 area
8. A technique using powder filling "q" in which the arcing, sparking, or hot parts of the equipment are surrounded by a material such as glass or quartz powder to prevent ignition of an external explosive atmosphere in a Class I, Zone 1 area

The letter in quotation marks following the protection method is the letter used in the marking on the equipment. See 505.9(C) for marking requirements.

Of the many protection techniques, intrinsic safety, flameproof, and increased safety are the most common for Zone 1 locations.

(A) Flameproof "d". This protection technique shall be permitted for equipment in Class I, Zone 1 or Zone 2 locations.

Flameproof protection is commonly combined with increased safety protection. For example, motor control and other switching contacts are commonly protected by flameproof enclosures with the field wiring terminals protected in a separate but attached enclosure by increased safety. The conductors between the enclosures are protected by flameproof feed-through insulators. The equipment shown in Exhibit 505.1 is an example of where this combination of protection techniques is employed.

Exhibit 505.1 Typical control stations with the combination of flameproof and increased safety types of protection suitable for use in Class I, Zone 1 areas. (Courtesy of Cooper Crouse-Hinds)

(B) Purged and Pressurized. This protection technique shall be permitted for equipment in those Class I, Zone 1 or Zone 2 locations for which it is identified.

(C) Intrinsic Safety. This protection technique shall be permitted for apparatus and associated apparatus in Class I, Zone 0, Zone 1, or Zone 2 locations for which it is listed.

The identifying letter for intrinsic safety is "i" followed by either "a" or "b," identifying whether the equipment is suitable for Zone 0 (ia) or Zone 1 (ib). The associated apparatus is identified by the same letters in brackets, that is, [ia] or [ib].

(D) Type of Protection "n." This protection technique shall be permitted for equipment in Class I, Zone 2 locations. Type of protection "n" is further subdivided into nA, nC, and nR.

FPN: See Table 505.9(C)(2)(4) for the descriptions of subdivisions for type of protection "n."

(E) Oil Immersion "o." This protection technique shall be permitted for equipment in Class I, Zone 1 or Zone 2 locations.

(F) Increased Safety "e." This protection technique shall be permitted for equipment in Class I, Zone 1 or Zone 2 locations.

The increased safety protection technique is commonly used for motors and generators (see 505.22) and fluorescent luminaires. It is also used for terminal boxes.

(G) Encapsulation "m." This protection technique shall be permitted for equipment in Class I, Zone 1 or Zone 2 locations.

(H) Powder Filling "q." This protection technique shall be permitted for equipment in Class I, Zone 1 or Zone 2 locations.

(I) Combustible Gas Detection System. A combustible gas detection system shall be permitted as a means of protection in industrial establishments with restricted public access and where the conditions of maintenance and supervision ensure that only qualified persons service the installation. Gas detection equipment shall be listed for detection of the specific gas or vapor to be encountered. Where such a system is installed, equipment specified in 505.8(I)(1), (2), or (3) shall be permitted.

(1) Inadequate Ventilation. In a Class I, Zone 1 location that is so classified due to inadequate ventilation, electrical equipment suitable for Class I, Zone 2 locations shall be permitted.

(2) Interior of a Building. In a building located in, or with an opening into, a Class I, Zone 2 location where the interior does not contain a source of flammable gas or vapor, electrical equipment for unclassified locations shall be permitted.

(3) Interior of a Control Panel. In the interior of a control panel containing instrumentation utilizing or measuring flammable liquids, gases, or vapors, electrical equipment suitable for Class I, Zone 2 locations shall be permitted.

> FPN No. 1: For further information, see ANSI/ISA-12.13.01, *Performance Requirements, Combustible Gas Detectors.*
>
> FPN No. 2: For further information, see ANSI/API RP 500, *Recommended Practice for Classification of Locations for Electrical Installations at Petroleum Facilities Classified as Class I, Division 1 or Division 2.*
>
> FPN No. 3: For further information, see ISA-RP12.13.02, *Installation, Operation, and Maintenance of Combustible Gas Detection Instruments.*

505.9 Suitability of Equipment.

(A) Suitability. Suitability of identified equipment shall be determined by one of the following:

(1) Equipment listing or labeling
(2) Evidence of equipment evaluation from a qualified testing laboratory or inspection agency concerned with product evaluation
(3) Evidence acceptable to the authority having jurisdiction such as a manufacturer's self-evaluation or an owner's engineering judgment

(B) Listing.

(1) Equipment that is listed for a Zone 0 location shall be permitted in a Zone 1 or Zone 2 location of the same gas or vapor. Equipment that is listed for a Zone 1 location shall be permitted in a Zone 2 location of the same gas or vapor.
(2) Equipment shall be permitted to be listed for a specific gas or vapor, specific mixtures of gases or vapors, or any specific combination of gases or vapors.

> FPN: One common example is equipment marked for "IIB. + H2."

(C) Marking. Equipment shall be marked in accordance with 505.9(C)(1) or (2).

(1) Division Equipment. Equipment identified for Class I, Division 1 or Class I, Division 2 shall, in addition to being marked in accordance with 500.8(B), be permitted to be marked with the following:

(1) Class I, Zone 1 or Class I, Zone 2 (as applicable), and
(2) Applicable gas classification group(s) in accordance with Table 505.9(C), and
(3) Temperature classification in accordance with 505.9(D)(1).

Table 505.9(C) Gas Classification Groups

Gas Group	Comment
IIC	See 505.6(A)(1)
IIB	See 505.6(A)(2)
IIA	See 505.6(A)(3)

(2) Zone Equipment. Equipment meeting one or more of the protection techniques described in 505.8 shall be marked with the following in the order shown:

The symbol AEx identifies the equipment as meeting American standards. In European Common Market countries, the symbol is EEx. In the IEC standards on which American and European standards are based, the symbol is Ex.

(1) Class
(2) Zone
(3) Symbol "AEx"
(4) Protection technique(s) in accordance with Table 505.9(C)(2)(4)
(5) Applicable gas classification group(s) in accordance with Table 505.9(C)
(6) Temperature classification in accordance with 505.9(D)(1).

Table 505.9(C)(2)(4) Types of Protection Designation

Designation	Technique	Zone*
d	Flameproof enclosure	1
e	Increased safety	1
ia	Intrinsic safety	0
ib	Intrinsic safety	1
[ia]	Intrinsically safe associated apparatus	Unclassified
[ib]	Intrinsically safe associated apparatus	Unclassified
m	Encapsulation	1
nA	Nonsparking equipment	2
nC	Sparking equipment in which the contacts are suitably protected other than by restricted breathing enclosure	2
nR	Restricted breathing enclosure	2
o	Oil immersion	1
p	Purged and pressurized	1 or 2
q	Powder filled	1

*Does not address use where a combination of techniques is used.

Exception: Intrinsically safe associated apparatus shall be required to be marked only with (3), (4), and (5)

Electrical equipment of types of protection "e," "m," "p," or "q," shall be marked Group II. Electrical equipment of types of protection "d," "ia," "ib," "[ia]," or "[ib]" shall be marked Group IIA, IIB, or IIC, or for a specific gas or vapor. Electrical equipment of types of protection "n" shall be marked Group II unless it contains enclosed-break devices, nonincendive components, or energy-limited equipment or circuits, in which case it shall be marked Group IIA, IIB, or IIC, or a specific gas or vapor. Electrical equipment of other types of protection shall be marked Group II unless the type of protection utilized by the equipment requires that it shall be marked Group IIA, IIB, or IIC, or a specific gas or vapor.

> FPN: An example of such a required marking is "Class I, Zone 0, AEx ia IIC T6." An explanation of the marking that is required is shown in FPN Figure 505.9(C)(2).

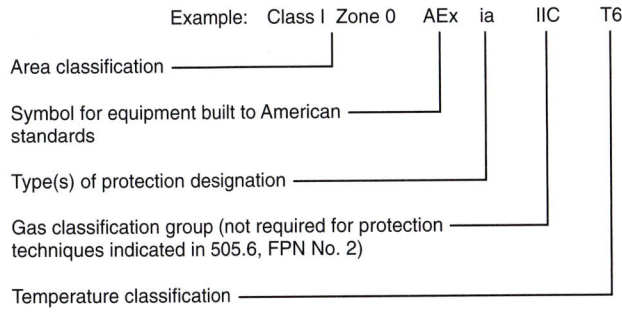

Example: Class I Zone 0   AEx ia   IIC   T6

Area classification ————————

Symbol for equipment built to American standards ————————

Type(s) of protection designation ————————

Gas classification group (not required for protection techniques indicated in 505.6, FPN No. 2) ————————

Temperature classification ————————

FPN Figure 505.9(C)(2) Zone equipment marking.

(D) Class I Temperature. The temperature marking specified below shall not exceed the ignition temperature of the specific gas or vapor to be encountered.

> FPN: For information regarding ignition temperatures of gases and vapors, see NFPA 497-1997, *Recommended Practice for the Classification of Flammable Liquids, Gases, or Vapors and of Hazardous (Classified) Locations for Electrical Installations in Chemical Process Areas*; and IEC 60079-20-1996, *Electrical Apparatus for Explosive Gas Atmospheres, Data for Flammable Gases and Vapours, Relating to the Use of Electrical Apparatus*.

(1) Temperature Classifications. Equipment shall be marked to show the operating temperature or temperature class referenced to a 40°C (104°F) ambient. The temperature class, if provided, shall be indicated using the temperature class (T Code) shown in Table 505.9(D)(1).

Table 505.9(D)(1) Classification of Maximum Surface Temperature for Group II Electrical Equipment

Temperature Class (T Code)	Maximum Surface Temperature (°C)
T1	≤450
T2	≤300
T3	≤200
T4	≤135
T5	≤100
T6	≤85

Electrical equipment designed for use in the ambient temperature range between −20°C and +40°C shall require no additional ambient temperature marking.

Electrical equipment that is designed for use in a range of ambient temperatures other than −20°C and +40°C is considered to be special; and the ambient temperature range shall then be marked on the equipment, including either the symbol "Ta" or "Tamb" together with the special range of ambient temperatures. As an example, such a marking might be "−30°C ≤ Ta ≤ + 40°C."

Electrical equipment suitable for ambient temperatures exceeding 40°C (104°F) shall be marked with both the maximum ambient temperature and the operating temperature or temperature class at that ambient temperature.

Exception No. 1: Equipment of the non–heat-producing type, such as conduit fittings, and equipment of the heat-producing type having a maximum temperature of not more than 100°C (212°F) shall not be required to have a marked operating temperature or temperature class.

Exception No. 2: Equipment identified for Class I, Division 1 or Division 2 locations as permitted by 505.20(B) and (C) shall be permitted to be marked in accordance with 500.8(B) and Table 500.8(B).

(E) Threading. All threaded conduit referred to herein shall be threaded with a National (American) Standard Pipe Taper (NPT) standard conduit cutting die that provides a taper of 1 in 16 (¾-in. taper per foot). Such conduit shall be made wrenchtight to prevent sparking when fault current flows through the conduit system, and to ensure the explosionproof or flameproof integrity of the conduit system where applicable. Threaded joints shall be made up with at least five threads fully engaged for entries into flameproof or explosionproof equipment.

Equipment provided with threaded entries for field wiring connections shall be installed in accordance with 505.9(D)(1) or (2).

(1) Equipment Provided with Threaded Entries for NPT Threaded Conduit or Fittings. For equipment provided with threaded entries for NPT threaded conduit or fittings, listed conduit fittings, or cable fittings shall be used.

> FPN: Thread form specifications for NPT threads are located in ANSI/ASME B1.20.1-1983, *Pipe Threads, General Purpose (Inch).*

(2) Equipment Provided with Threaded Entries for Metric Threaded Conduit or Fittings. For equipment with metric threaded entries, such entries shall be identified as being metric, or listed adapters to permit connection to conduit or NPT-threaded fittings shall be provided with the equipment. Adapters shall be used for connection to conduit or NPT-threaded fittings. Listed cable fittings that have metric threads shall be permitted to be used.

> FPN: Threading specifications for metric threaded entries are located in ISO 965/1-1980, *Metric Screw Threads*; and ISO 965/3-1980, *Metric Screw Threads.*

See the commentary on 500.8(D)(2) for information on metric threads. Exhibit 505.2 is an example of an adapter that provides a means of connecting conduit or fitting with NPT threads to an "increased safety" enclosure that has metric threads.

505.15 Wiring Methods.

Wiring methods shall maintain the integrity of protection techniques and shall comply with 505.15(A), (B), or (C).

(A) Class I, Zone 0. In Class I, Zone 0 locations, only intrinsically safe wiring methods in accordance with Article 504 shall be permitted.

> FPN: Article 504 only includes protection technique "ia."

The provision in 505.15(A) is one of the most significant differences between the zone and division area classification requirements. The degree of hazard within a Zone 0 area is

Exhibit 505.2 *A typical hub providing an NPT threaded entry for conduit or cable into an increased safety enclosure. (Courtesy of Cooper Crouse-Hinds)*

considered so severe that all wiring is this area must be intrinsically safe. In general, only instrumentation and signaling circuits installed in accordance with Article 504 can be used in a Zone 0 area. Explosionproof power utilization equipment, such as motors and luminaires, is not permitted in Class I, Zone 0 locations.

(B) Class I, Zone 1.

(1) General. In Class I, Zone 1 locations, the wiring methods in (a) through (e) shall be permitted.

(a) In industrial establishments with restricted public access, where the conditions of maintenance and supervision ensure that only qualified persons service the installation, and where the cable is not subject to physical damage, Type MC-HL cable listed for use in Class I, Zone 1 or Division 1 locations, with a gas/vaportight continuous corrugated metallic sheath, an overall jacket of suitable polymeric material, separate grounding conductors in accordance with 250.122, and provided with termination fittings listed for the application.

> FPN: See 330.10 and 330.12 for restrictions on use of Type MC cable.

(b) In industrial establishments with restricted public access, where the conditions of maintenance and supervision ensure that only qualified persons service the installation, and where the cable is not subject to physical damage, Type ITC-HL cable, listed for use in Class I, Zone 1 or Division 1 locations, with a gas/vaportight continuous corrugated metallic sheath, an overall jacket of suitable polymeric material and provided with termination fittings listed for the application.

(c) Type MI cable with termination fittings listed for Class I, Zone 1 or Division 1 locations. Type MI cable shall be installed and supported in a manner to avoid tensile stress at the termination fittings.

(d) Threaded rigid metal conduit, or threaded steel intermediate metal conduit.

(e) Rigid nonmetallic conduit complying with Article 352 shall be permitted where encased in a concrete envelope a minimum of 50 mm (2 in.) thick and provided with not less than 600 mm (24 in.) of cover measured from the top of the conduit to grade. Threaded rigid metal conduit or threaded steel intermediate metal conduit shall be used for the last 600 mm (24 in.) of the underground run to emergence or to the point of connection to the aboveground raceway. An equipment grounding conductor shall be included to provide for electrical continuity of the raceway system and for grounding of non–current-carrying metal parts.

(2) Flexible Connections. Where necessary to employ flexible connections, flexible fittings listed for Class I, Zone 1 or Division 1 locations or flexible cord in accordance with the provisions of 505.17 shall be permitted.

(C) Class I, Zone 2.

(1) General. In Class I, Zone 2 locations, the wiring methods in (a) through (g) shall be permitted.

(a) All wiring methods permitted by 505.15(B).

(b) Types MI, MC, MV, or TC cable with termination fittings, or in cable tray systems and installed in a manner to avoid tensile stress at the termination fittings.

(c) Type ITC cable in cable trays, in raceways, supported by messenger wire, where afforded mechanical protection and run as open wiring, or directly buried where the cable is listed for this use.

(d) Type PLTC cable in accordance with the provisions of Article 725, or in cable tray systems. PLTC shall be installed in a manner to avoid tensile stress at the termination fittings.

(e) Enclosed gasketed busways, enclosed gasketed wireways.

(f) Threaded rigid metal conduit, threaded steel intermediate metal conduit.

(g) Nonincendive field wiring shall be permitted using any of the wiring methods permitted for unclassified locations. Nonincendive field wiring systems shall be installed in accordance with the control drawing(s). Simple apparatus, not shown on the control drawing, shall be permitted in a nonincendive field wiring circuit, provided the simple apparatus does not interconnect the nonincendive field wiring circuit to any other circuit.

FPN: Simple apparatus is defined in 504.2.

Separate nonincendive field wiring circuits shall be installed as follows:

(1) In separate cables, or
(2) In multiconductor cables where the conductors of each circuit are within a grounded metal shield, or
(3) In multiconductor cables where the conductors of each circuit have insulation with a minimum thickness of 0.25 mm (0.01 in.)

(2) Flexible Connections. Where provision must be made for limited flexibility, flexible metal fittings, flexible metal conduit with listed fittings, liquidtight flexible metal conduit with listed fittings, liquidtight flexible nonmetallic conduit with listed fittings, or flexible cord in accordance with the provisions of 505.17 shall be permitted.

FPN: See 505.25(B) for grounding requirements where flexible conduit is used.

505.16 Sealing and Drainage.

Seals in conduit and cable systems shall comply with 505.16(A) through (E). Sealing compound shall be used in Type MI cable termination fittings to exclude moisture and other fluids from the cable insulation.

FPN No. 1: Seals are provided in conduit and cable systems to minimize the passage of gases and vapors and prevent the passage of flames from one portion of the electrical installation to another through the conduit. Such communication through Type MI cable is inherently prevented by construction of the cable. Unless specifically designed and tested for the purpose, conduit and cable seals are not intended to prevent the passage of liquids, gases, or vapors at a continuous pressure differential across the seal. Even at differences in pressure across the seal equivalent to a few inches of water, there may be a slow passage of gas or vapor through a seal, and through conductors passing through the seal. See 505.16(C)(2)(b). Temperature extremes and highly corrosive liquids and vapors can affect the ability of seals to perform their intended function. See 505.16(D)(2).

FPN No. 2: Gas or vapor leakage and propagation of flames may occur through the interstices between the strands of standard stranded conductors larger than 2

AWG. Special conductor constructions, for example, compacted strands or sealing of the individual strands, are means of reducing leakage and preventing the propagation of flames.

(A) Zone 0. In Class I, Zone 0 locations, seals shall be located according to 505.16(A)(1), (A)(2), and (A)(3).

(1) Conduit Seals. Seals shall be provided within 3.05 m (10 ft) of where a conduit leaves a Zone 0 location. There shall be no unions, couplings, boxes, or fittings, except listed reducers at the seal, in the conduit run between the seal and the point at which the conduit leaves the location.

Exception: A rigid unbroken conduit that passes completely through the Zone 0 location with no fittings less than 300 mm (12 in.) beyond each boundary shall not be required to be sealed if the termination points of the unbroken conduit are in unclassified locations.

(2) Cable Seals. Seals shall be provided on cables at the first point of termination after entry into the Zone 0 location.

(3) Not Required to Be Explosionproof or Flameproof. Seals shall not be required to be explosionproof or flameproof.

(B) Zone 1. In Class I, Zone 1 locations, seals shall be located in accordance with 505.16(B)(1) through (B)(8).

(1) Type of Protection "d" or "e" Enclosures. Conduit seals shall be provided for each conduit entering enclosures having type of protection "d" or "e."

Exception: Where the enclosure having type of protection "d" is marked to indicate that a seal is not required.

(2) Explosionproof Equipment. Conduit seals shall be provided for each conduit entering explosionproof equipment according to (a), (b), and (c).

(a) In each conduit entry into an explosionproof enclosure where either (1) the enclosure contains apparatus, such as switches, circuit breakers, fuses, relays, or resistors, that may produce arcs, sparks, or high temperatures that are considered to be an ignition source in normal operation, or (2) the entry is metric designator 53 (trade size 2) or larger and the enclosure contains terminals, splices, or taps. For the purposes of this section, high temperatures shall be considered to be any temperatures exceeding 80 percent of the autoignition temperature in degrees Celsius of the gas or vapor involved.

Exception: Conduit entering an enclosure where such switches, circuit breakers, fuses, relays, or resistors are

(a) *Enclosed within a chamber hermetically sealed against the entrance of gases or vapors, or*

(b) *Immersed in oil, or*

(c) *Enclosed within a factory-sealed explosionproof chamber located within the enclosure, identified for the location, and marked "factory sealed" or equivalent, unless the entry is metric designator 53 (trade size 2) or larger. Factory-sealed enclosures shall not be considered to serve as a seal for another adjacent explosionproof enclosure that is required to have a conduit seal.*

(b) Conduit seals shall be installed within 450 mm (18 in.) from the enclosure. Only explosionproof unions, couplings, reducers, elbows, capped elbows, and conduit bodies similar to L, T, and Cross types that are not larger than the trade size of the conduit shall be permitted between the sealing fitting and the explosionproof enclosure.

(c) Where two or more explosionproof enclosures for which conduit seals are required under 505.16(B)(2) are connected by nipples or by runs of conduit not more than 900 mm (36 in.) long, a single conduit seal in each such nipple connection or run of conduit shall be considered sufficient if located not more than 450 mm (18 in.) from either enclosure.

(3) Pressurized Enclosures. Conduit seals shall be provided in each conduit entry into a pressurized enclosure where the conduit is not pressurized as part of the protection system. Conduit seals shall be installed within 450 mm (18 in.) from the pressurized enclosure.

> FPN No. 1: Installing the seal as close as possible to the enclosure reduces problems with purging the dead airspace in the pressurized conduit.
>
> FPN No. 2: For further information, see NFPA 496-1998, *Standard for Purged and Pressurized Enclosures for Electrical Equipment.*

(4) Class I, Zone 1 Boundary. Conduit seals shall be provided in each conduit run leaving a Class I, Zone 1 location. The sealing fitting shall be permitted on either side of the boundary of such location within 3.05 m (10 ft) of the boundary and shall be designed and installed so as to minimize the amount of gas or vapor within the Zone 1 portion of the conduit from being communicated to the conduit beyond the seal. Except for listed explosionproof reducers at the conduit seal, there shall be no union, coupling, box, or fitting between the conduit seal and the point at which the conduit leaves the Zone 1 location.

Exception: Metal conduit containing no unions, couplings, boxes, or fittings and passing completely through a Class I, Zone 1 location with no fittings less than 300 mm (12 in.) beyond each boundary shall not require a conduit seal if the termination points of the unbroken conduit are in unclassified locations.

(5) Cables Capable of Transmitting Gases or Vapors. Conduits containing cables with a gas/vaportight continuous sheath capable of transmitting gases or vapors through the cable core shall be sealed in the Zone 1 location after removing the jacket and any other coverings so that the sealing compound surrounds each individual insulated conductor and the outer jacket.

Exception: Multiconductor cables with a gas/vaportight continuous sheath capable of transmitting gases or vapors through the cable core shall be permitted to be considered as a single conductor by sealing the cable in the conduit within 450 mm (18 in.) of the enclosure and the cable end within the enclosure by an approved means to minimize the entrance of gases or vapors and prevent the propagation of flame into the cable core, or by other approved methods. For shielded cables and twisted pair cables, it shall not be required to remove the shielding material or separate the twisted pair.

(6) Cables Incapable of Transmitting Gases or Vapors. Each multiconductor cable in conduit shall be considered as a single conductor if the cable is incapable of transmitting gases or vapors through the cable core. These cables shall be sealed in accordance with 505.16(D).

(7) Cables Entering Enclosures. Cable seals shall be provided for each cable entering flameproof or explosionproof enclosures. The seal shall comply with 505.16(D).

(8) Class I, Zone 1 Boundary. Cables shall be sealed at the point at which they leave the Zone 1 location.

Exception: Where cable is sealed at the termination point.

(C) Zone 2. In Class I, Zone 2 locations, seals shall be located in accordance with 505.16(C)(1) and (C)(2).

(1) Conduit Seals. Conduit seals shall be located in accordance with (a) and (b).

(a) For connections to enclosures that are required to be flameproof or explosionproof, a conduit seal shall be provided in accordance with 505.16(B)(1) and 505.16(B)(2). All portions of the conduit run or nipple between the seal and such enclosure shall comply with 505.16(B).

(b) In each conduit run passing from a Class I, Zone 2 location into an unclassified location. The sealing fitting shall be permitted on either side of the boundary of such location within 3.05 m (10 ft) of the boundary and shall be designed and installed so as to minimize the amount of gas or vapor within the Zone 2 portion of the conduit from being communicated to the conduit beyond the seal. Rigid metal conduit or threaded steel intermediate metal conduit shall be used between the sealing fitting and the point at which the conduit leaves the Zone 2 location, and a threaded connection shall be used at the sealing fitting. Except for listed explosionproof reducers at the conduit seal, there shall be

no union, coupling, box, or fitting between the conduit seal and the point at which the conduit leaves the Zone 2 location.

Exception No. 1: Metal conduit containing no unions, couplings, boxes, or fittings and passing completely through a Class I, Zone 2 location with no fittings less than 300 mm (12 in.) beyond each boundary shall not be required to be sealed if the termination points of the unbroken conduit are in unclassified locations.

Exception No. 2: Conduit systems terminating at an unclassified location where a wiring method transition is made to cable tray, cablebus, ventilated busway, Type MI cable, or open wiring shall not be required to be sealed where passing from the Class I, Zone 2 location into the unclassified location. The unclassified location shall be outdoors or, if the conduit system is all in one room, it shall be permitted to be indoors. The conduits shall not terminate at an enclosure containing an ignition source in normal operation.

Exception No. 3: Conduit systems passing from an enclosure or room that is unclassified as a result of pressurization into a Class I, Zone 2 location shall not require a seal at the boundary.

> FPN: For further information, refer to NFPA 496-1998, *Standard for Purged and Pressurized Enclosures for Electrical Equipment.*

Exception No. 4: Segments of aboveground conduit systems shall not be required to be sealed where passing from a Class I, Zone 2 location into an unclassified location if all the following conditions are met:

(a) *No part of the conduit system segment passes through a Class I, Zone 0 or Class I, Zone 1 location where the conduit contains unions, couplings, boxes, or fittings within 300 mm (12 in.) of the Class I, Zone 0 or Class I, Zone 1 location.*

(b) *The conduit system segment is located entirely in outdoor locations.*

(c) *The conduit system segment is not directly connected to canned pumps, process or service connections for flow, pressure, or analysis measurement, and so forth, that depend on a single compression seal, diaphragm, or tube to prevent flammable or combustible fluids from entering the conduit system.*

(d) *The conduit system segment contains only threaded metal conduit, unions, couplings, conduit bodies, and fittings in the unclassified location.*

(e) *The conduit system segment is sealed at its entry to each enclosure or fitting housing terminals, splices, or taps in Class I, Zone 2 locations.*

(2) Cable Seals. Cable seals shall be located in accordance with (a), (b), and (c).

(a) Explosionproof and Flameproof Enclosures. Cables entering enclosures required to be flameproof or explosionproof shall be sealed at the point of entrance. The seal shall comply with 505.16(D). Multiconductor cables with a gas/vaportight continuous sheath capable of transmitting gases or vapors through the cable core shall be sealed in the Zone 2 location after removing the jacket and any other coverings so that the sealing compound surrounds each individual insulated conductor in such a manner as to minimize the passage of gases and vapors. Multiconductor cables in conduit shall be sealed as described in 505.16(B)(4).

Exception No. 1: Cables passing from an enclosure or room that is unclassified as a result of Type Z pressurization into a Class I, Zone 2 location shall not require a seal at the boundary.

Exception No. 2: Shielded cables and twisted pair cables shall not require the removal of the shielding material or separation of the twisted pairs, provided the termination is by an approved means to minimize the entrance of gases or vapors and prevent propagation of flame into the cable core.

(b) Cables That Will Not Transmit Gases or Vapors. Cables with a gas/vaportight continuous sheath and that will not transmit gases or vapors through the cable core in excess of the quantity permitted for seal fittings shall not be required to be sealed except as required in 505.16(C)(2)(a). The minimum length of such cable run shall not be less than the length that limits gas or vapor flow through the cable core to the rate permitted for seal fittings [200 cm^3/hr (0.007 ft^3/hr) of air at a pressure of 1500 pascals (6 in. of water)].

> FPN No. 1: See ANSI/UL 886-1994, *Outlet Boxes and Fittings for Use in Hazardous (Classified) Locations.*
> FPN No. 2: The cable core does not include the interstices of the conductor strands.

(c) Cables Capable of Transmitting Gases or Vapors. Cables with a gas/vaportight continuous sheath capable of transmitting gases or vapors through the cable core shall not be required to be sealed except as required in 505.16(C)(2)(a), unless the cable is attached to process equipment or devices that may cause a pressure in excess of 1500 pascals (6 in. of water) to be exerted at a cable end, in which case a seal, barrier, or other means shall be provided to prevent migration of flammables into an unclassified area.

Exception: Cables with an unbroken gas/vaportight continuous sheath shall be permitted to pass through a Class I, Zone 2 location without seals.

(d) Cables Without Gas/Vaportight Continuous Sheath. Cables that do not have gas/vaportight continuous sheath shall be sealed at the boundary of the Zone 2 and unclassified location in such a manner as to minimize the passage of gases or vapors into an unclassified location.

> FPN: The cable sheath may be either metal or a nonmetallic material.

(D) Class I, Zones 0, 1, and 2. Where required, seals in Class I, Zones 0, 1, and 2 locations shall comply with 505.16(D)(1) through (D)(5).

(1) Fittings. Enclosures for connections or equipment shall be provided with an integral means for sealing, or sealing fittings listed for the location shall be used. Sealing fittings shall be listed for use with one or more specific compounds and shall be accessible.

(2) Compound. The compound shall provide a seal against passage of gas or vapors through the seal fitting, shall not be affected by the surrounding atmosphere or liquids, and shall not have a melting point less than 93°C (200°F).

(3) Thickness of Compounds. In a completed seal, the minimum thickness of the sealing compound shall not be less than the trade size of the sealing fitting and, in no case, less than 16 mm (⅝ in.).

Exception: Listed cable sealing fittings shall not be required to have a minimum thickness equal to the trade size of the fitting.

(4) Splices and Taps. Splices and taps shall not be made in fittings intended only for sealing with compound, nor shall other fittings in which splices or taps are made be filled with compound.

(5) Conductor Fill. The cross-sectional area of the conductors permitted in a seal shall not exceed 25 percent of the cross-sectional area of a rigid metal conduit of the same trade size unless it is specifically listed for a higher percentage of fill.

(E) Drainage.

(1) Control Equipment. Where there is a probability that liquid or other condensed vapor may be trapped within enclosures for control equipment or at any point in the raceway system, approved means shall be provided to prevent accumulation or to permit periodic draining of such liquid or condensed vapor.

(2) Motors and Generators. Where the authority having jurisdiction judges that there is a probability that liquid or condensed vapor may accumulate within motors or generators, joints and conduit systems shall be arranged to minimize entrance of liquid. If means to prevent accumulation or to permit periodic draining are judged necessary, such means shall be provided at the time of manufacture and shall be considered an integral part of the machine.

(3) Canned Pumps, Process or Service Connections, and So Forth. For canned pumps, process or service connections for flow, pressure, or analysis measurement, and so forth, that depend upon a single compression seal, diaphragm, or tube to prevent flammable or combustible fluids from entering the electrical conduit system, an additional approved seal, barrier, or other means shall be provided to prevent the flammable or combustible fluid from entering the conduit system beyond the additional devices or means if the primary seal fails.

The additional approved seal or barrier and the interconnecting enclosure shall meet the temperature and pressure conditions to which they will be subjected upon failure of the primary seal, unless other approved means are provided to accomplish the purpose in the preceding paragraph.

Drains, vents, or other devices shall be provided so that primary seal leakage is obvious.

505.17 Flexible Cords, Class I, Zones 1 and 2.

A flexible cord shall be permitted for connection between portable lighting equipment or other portable utilization equipment and the fixed portion of their supply circuit. Flexible cord shall also be permitted for that portion of the circuit where the fixed wiring methods of 505.15(B) cannot provide the necessary degree of movement for fixed and mobile electrical utilization equipment, in an industrial establishment where conditions of maintenance and engineering supervision ensure that only qualified persons install and service the installation, and the flexible cord is protected by location or by a suitable guard from damage. The length of the flexible cord shall be continuous. Where flexible cords are used, the cords shall be as follows:

(1) Of a type listed for extra-hard usage;
(2) Contain, in addition to the conductors of the circuit, a grounding conductor complying with 400.23;
(3) Connected to terminals or to supply conductors in an approved manner;
(4) Be supported by clamps or by other suitable means in such a manner that there will be no tension on the terminal connections; and
(5) Be provided with listed seals where the flexible cord enters boxes, fittings, or enclosures that are required to be explosionproof or flameproof.

Exception: As provided in 505.16.

Electric submersible pumps with means for removal without entering the wet-pit shall be considered portable utilization equipment. The extension of the flexible cord within a suitable raceway between the wet-pit and the power source shall be permitted.

Electric mixers intended for travel into and out of open-type mixing tanks or vats shall be considered portable utilization equipment.

FPN: See 505.18 for flexible cords exposed to liquids having a deleterious effect on the conductor insulation.

505.18 Conductors and Conductor Insulation.

(A) Conductors. For type of protection "e," field wiring conductors shall be copper.

(B) Conductor Insulation. Where condensed vapors or liquids may collect on, or come in contact with, the insulation on conductors, such insulation shall be of a type identified for use under such conditions or the insulation shall be protected by a sheath of lead or by other approved means.

505.19 Live Parts.

There shall be no exposed live parts.

505.20 Equipment Requirements.

(A) Zone 0. In Class I, Zone 0 locations, only equipment specifically listed and marked as suitable for the location shall be permitted.

Exception: Intrinsically safe apparatus listed for use in Class I, Division 1 locations for the same gas, or as permitted by 505.9(B)(2), and with a suitable temperature class shall be permitted.

The exception to 505.20(A) results from the fact that ANSI/UL 913, the standard used to evaluate intrinsically safe systems for Class I, Division 1 locations, is based on the IEC requirements for intrinsically safe equipment for Class I, Zone 0 locations.

(B) Zone 1. In Class I, Zone 1 locations, only equipment specifically listed and marked as suitable for the location shall be permitted.

Exception No. 1: Equipment identified for use in Class I, Division 1 or listed for use in Class I, Zone 0 locations for the same gas, or as permitted by 505.9(B)(2), and with a suitable temperature class shall be permitted.

Exception No. 2: Equipment identified for Class I, Zone 1, or Zone 2 type of protection "p" shall be permitted.

(C) Zone 2. In Class I, Zone 2 locations, only equipment specifically listed and marked as suitable for the location shall be permitted.

Exception No. 1: Equipment listed for use in Class I, Zone 0 or Zone 1 locations for the same gas, or as permitted by 505.9(B)(2), and with a suitable temperature class, shall be permitted.

Exception No. 2: Equipment identified for Class I, Zone 1 or Zone 2 type of protection "p" shall be permitted.

Exception No. 3: Equipment identified for use in Class I, Division 1 or Division 2 locations for the same gas, or as permitted by 505.9(B)(2), and with a suitable temperature class shall be permitted.

Exception No. 4: In Class I, Zone 2 locations, the installation of open or nonexplosionproof or nonflameproof enclosed motors, such as squirrel-cage induction motors without brushes, switching mechanisms, or similar arc-producing devices that are not identified for use in a Class I, Zone 2 location shall be permitted.

FPN No. 1: It is important to consider the temperature of internal and external surfaces that may be exposed to the flammable atmosphere.

FPN No. 2: It is important to consider the risk of ignition due to currents arcing across discontinuities and overheating of parts in multisection enclosures of large motors and generators. Such motors and generators may need equipotential bonding jumpers across joints in the enclosure and from enclosure to ground. Where the presence of ignitible gases or vapors is suspected, clean air purging may be needed immediately prior to and during start-up periods.

See the commentary following 501.8(B), FPN No. 2 for more information on electric motors installed in hazardous (classified) locations.

(D) Manufacturer's Instructions. Electrical equipment installed in hazardous (classified) locations shall be installed in accordance with the instructions (if any) provided by the manufacturer.

505.21 Multiwire Branch Circuits.

In a Class I, Zone 1 location, a multiwire branch circuit shall not be permitted.

Exception: Where the disconnect device(s) for the circuit opens all ungrounded conductors of the multiwire circuit simultaneously.

505.22 Increased Safety "e" Motors and Generators.

In Class I, Zone 1 locations, Increased Safety "e" motors and generators of all voltage ratings shall be listed for Class I, Zone 1 locations, and shall comply with the following:

(1) Motors shall be marked with the current ratio, I_A/I_N, and time, t_E.
(2) Motors shall have controllers marked with the model or identification number, output rating (horsepower or kilowatt), full-load amperes, starting current ratio (I_A/I_N), and time (t_E) of the motors that they are intended to protect; the controller marking shall also include

the specific overload protection type (and setting, if applicable) that is listed with the motor or generator.

(3) Connections shall be made with the specific terminals listed with the motor or generator.
(4) Terminal housings shall be permitted to be of substantial, nonmetallic, nonburning material, provided an internal grounding means between the motor frame and the equipment grounding connection is incorporated within the housing.
(5) The provisions of Part III of Article 430 shall apply regardless of the voltage rating of the motor.
(6) The motors shall be protected against overload by a separate overload device that is responsive to motor current. This device shall be selected to trip or shall be rated in accordance with the listing of the motor and its overload protection.
(7) Sections 430.32(C) and 430.44 shall not apply to such motors.
(8) The motor overload protection shall not be shunted or cut out during the starting period.

505.25 Grounding and Bonding.

Grounding and bonding shall comply with Article 250 and the requirements in 505.25(A) and (B).

(A) Bonding. The locknut-bushing and double-locknut types of contacts shall not be depended on for bonding purposes, but bonding jumpers with proper fittings or other approved means of bonding shall be used. Such means of bonding shall apply to all intervening raceways, fittings, boxes, enclosures, and so forth, between Class I locations and the point of grounding for service equipment or point of grounding of a separately derived system.

Exception: The specific bonding means shall only be required to the nearest point where the grounded circuit conductor and the grounding electrode are connected together on the line side of the building or structure disconnecting means as specified in 250.32(A), (B), and (C), provided the branch-circuit overcurrent protection is located on the load side of the disconnecting means.

FPN: See 250.100 for additional bonding requirements in hazardous (classified) locations.

(B) Types of Equipment Grounding Conductors. Where flexible metal conduit or liquidtight flexible metal conduit is used as permitted in 505.15(C) and is to be relied on to complete a sole equipment grounding path, it shall be installed with internal or external bonding jumpers in parallel with each conduit and complying with 250.102.

Exception: In Class I, Zone 2 locations, the bonding jumper shall be permitted to be deleted where all the following conditions are met:

(a) *Listed liquidtight flexible metal conduit 1.8 m (6 ft) or less in length, with fittings listed for grounding, is used.*

(b) *Overcurrent protection in the circuit is limited to 10 amperes or less.*

(c) *The load is not a power utilization load.*

ARTICLE 510
Hazardous (Classified) Locations—Specific

Contents

510.1 Scope
510.2 General

510.1 Scope.

Articles 511 through 517 cover occupancies or parts of occupancies that are or may be hazardous because of atmospheric concentrations of flammable liquids, gases, or vapors, or because of deposits or accumulations of materials that may be readily ignitible.

510.2 General.

The general rules of this *Code* and the provisions of Articles 500 through 504 shall apply to electric wiring and equipment in occupancies within the scope of Articles 511 through 517, except as such rules are modified in Articles 511 through 517. Where unusual conditions exist in a specific occupancy, the authority having jurisdiction shall judge with respect to the application of specific rules.

Some of the requirements contained in Articles 511 through 517 have been extracted from other NFPA codes and standards. For example, Tables 514.3(B)(1), 514.3(B)(2), and 515.3 are extracted from NFPA 30A, *Code for Motor Fuel Dispensing Facilities and Repair Garages,* and NFPA 30, *Flammable and Combustible Liquids Code.* The tables were developed by the NFPA Technical Committees responsible for those documents. The documents were developed through the same process as the *NEC*; however, the National Electrical Code Committee is not directly responsible for the technical content of extracted material.

NFPA publishes a number of standards and recommended practices that provide requirements or guidance on the classification of hazardous locations in specific occupancies. Information and copies of standards may be obtained from NFPA, 1 Batterymarch Park, P.O. Box 9101, Quincy, MA 02269-9101.

ARTICLE 511
Commercial Garages, Repair and Storage

Contents

FPN: Rules that are followed by a reference in brackets contain text that has been extracted from NFPA 88B-1997, *Standard for Repair Garages.* Only editorial changes were made to the extracted text to make it consistent with this *Code.*

511.1 Scope.

These occupancies shall include locations used for service and repair operations in connection with self-propelled vehicles (including, but not limited to, passenger automobiles, buses, trucks, and tractors) in which volatile flammable liquids or flammable gases are used for fuel or power.

Article 100 defines *garage* as a building or portion of a building in which one or more self-propelled vehicles can be kept for use, sale, storage, rental, repair, exhibition, or demonstration purposes. Article 511 applies to commercial garages in which the primary operation is the service and repair of self-propelled vehicles that use flammable gases or liquids for fuel. The commercial garages covered by Article 511 include automotive service centers; repair garages for commercial vehicles, such as trucks and tractors;

and service garages for fleet vehicles, such as buses, cars, and trucks.

The requirements of Article 511 are intended to mitigate the potential for an ignition-capable arc or spark from electrical wiring or equipment used in or above hazardous (classified) locations. Additionally, there are requirements for personnel protection in occupancies that are frequently wet or damp in which service personnel are subject to contact with large grounded surfaces, such as concrete slabs in direct contact with the earth. The increasing number of service operations in which minor repairs, such as oil changes, occur is covered under the requirements of this article plus the underfloor area classification requirements of Article 514. See Exception No. 2 to 511.3(B)(3) and its associated commentary.

Parking, storage, and similar occupancies are not required to be classified, provided that any repair that occurs is minor and does not involve the use of electrical equipment. In accordance with NFPA 88A, *Standard for Parking Structures,* a mechanical ventilating system that is capable of continuously providing a ventilation rate of one cubic foot per minute for each square foot of floor area is required for all enclosed, basement and underground parking garages.

Operations that involve open flames or electric arcs, including fusion gas welding and electric welding, are covered by NFPA 88B, *Standard for Repair Garages,* and must be restricted to areas specifically provided for such purposes.

Approved suspended unit heaters may be used in commercial garages, provided they are located not less than 8 ft above the floor and are installed in accordance with the conditions of their approval. This requirement also comes from NFPA 88B.

For requirements covering locations used for the storage or repair of boats, see Article 555 of this *Code* and NFPA 303, *Fire Protection Standard for Marinas and Boatyards.*

511.3 Classifications of Locations.

(A) Unclassified Locations. Parking garages used for parking or storage and where no repair work is done except exchange of parts and routine maintenance requiring no use of electrical equipment, open flame, welding, or the use of volatile flammable liquids are not classified.

The storage, handling, or dispensing into motor vehicles of alcohol-based windshield washer fluid in areas used for the service and repair operations of the vehicles shall not cause such areas to be classified as hazardous (classified) locations.

FPN No. 1: For further information, see NFPA 88A-1998, *Standard for Parking Structures* and NFPA 88B-1997, *Standard for Repair Garages.*

FPN No. 2: For further information, see 8.3.5 of NFPA 30A, *Code for Motor Fuel Dispensing Facilities and Repair Garages.*

(B) Classified Locations. Classification shall be in accordance with Article 500. Areas in which flammable fuel is transferred to vehicle fuel tanks shall also conform to Article 514.

(1) Up to a Level of 450 mm (18 in.) Above the Floor. For each floor, the entire area up to a level of 450 mm (18 in.) above the floor shall be considered to be a Class I, Division 2 location.

Exception: Where the enforcing agency determines that there is mechanical ventilation providing a minimum of four air changes per hour or one cubic foot per minute of exchanged air for each square foot of floor area. Ventilation shall provide for air exchange across the entire floor area within 0.3 m (12 in.) of the floor.

(2) Within 450 mm (18 in.) of the Ceiling. Where compressed natural gas (CNG) vehicles are repaired or stored, the area within 450 mm (18 in.) of the ceiling shall be classified as Class I, Division 2, except where ventilation of at least four air changes per hour is provided. [NFPA 88B, 3-1.1]

(3) Any Pit or Depression Below Floor Level. Any pit or depression below floor level shall be considered to be a Class I, Division 1 location and shall extend up to said floor level.

Exception No. 1: Any pit or depression in which six air changes per hour are exhausted at the floor level of the pit shall be permitted to be judged by the enforcing agency to be a Class I, Division 2 location.

Exception No. 2: Lubrication and service rooms without dispensing shall be classified in accordance with Table 514.3(B)(1).

Exception No. 2 to 511.3(B)(3) provides guidance for determining the classification of facilities that primarily offer oil and filter change and lubrication-type service, not the dispensing of fuel. If the lower level work area of a lubritorium is provided with exhaust ventilation at a rate specified in Table 514.3(B)(1) under "Lubrication or Service Room—Without Dispensing," the lower level is not a hazardous (classified) location. Table 514.3(B)(1) is extracted from NFPA 30A, *Code for Motor Fuel Dispensing Facilities and Repair Garages.*

(4) Areas Adjacent to Defined Locations or with Positive-Pressure Ventilation. Areas adjacent to defined locations in which flammable vapors are not likely to be released, such as stock rooms, switchboard rooms, and other similar locations, shall not be classified where mechanically ventilated at a rate of four or more air changes per hour or where effectively cut off by walls or partitions.

(5) Adjacent Areas by Special Permission. Adjacent areas that by reason of ventilation, air pressure differentials, or physical spacing are such that, in the opinion of the authority enforcing this *Code*, no ignition hazard exists, shall be unclassified.

511.4 Wiring and Equipment in Class I Locations.

(A) Wiring Located in Class I Locations. Within Class I locations as classified in 511.3, wiring shall conform to applicable provisions of Article 501.

(1) Raceways. Raceways embedded in a masonry wall or buried beneath a floor shall be considered to be within the Class I location above the floor if any connections or extensions lead into or through such areas.

The requirement in 511.4(A)(1) applies to raceways located in walls and below floors of the Class I locations in commercial garages. A raceway in a masonry wall or buried beneath a floor is not permitted to have any connections or extensions leading into or through a Class I, Division 1 or 2 location if it is to be considered located in a nonhazardous location and not subject to the provisions of Article 501. See Exhibit 511.1.

Article 501 applies to the raceway if any part of it is not embedded in the wall or if the wall is not masonry.

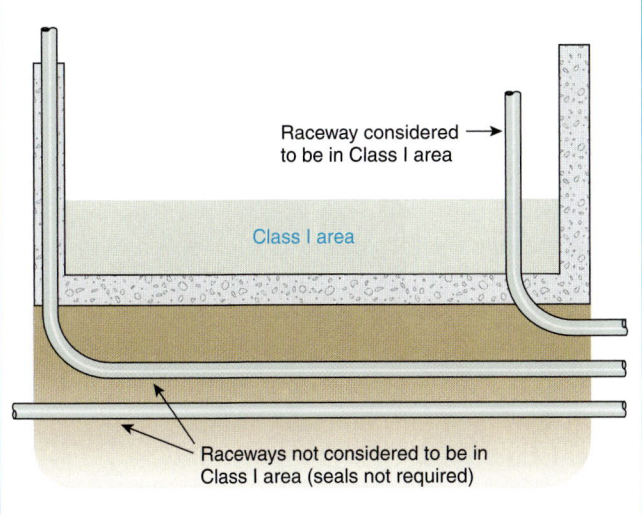

Raceway considered to be in Class I area

Class I area

Raceways not considered to be in Class I area (seals not required)

Exhibit 511.1 Raceways embedded in a masonry wall or buried beneath the floor are not considered located in the Class I location only under conditions where there are no extensions into the Class I location.

Article 501 applies to the raceway if any part of it is not buried beneath the floor. The floor material is not specified.

Exception: Rigid nonmetallic conduit that complies with Article 352 shall be permitted where buried under not less than 600 mm (24 in.) of cover. Where rigid nonmetallic conduit is used, threaded rigid metal conduit or threaded steel intermediate metal conduit shall be used for the last 600 mm (24 in.) of the underground run to emergence or to the point of connection to the aboveground raceway and an equipment grounding conductor shall be included to provide electrical continuity of the raceway system and for grounding of non–current-carrying metal parts.

The exception to 511.4(A)(1) permits underground installations of rigid nonmetallic conduit at commercial garages. Underground conduit in similar facilities has been damaged by corrosion. The conditions under which the rigid nonmetallic conduit can be used are similar to those contained in the exception to 501.4(A)(1)(a) except that concrete encasement is not required for the commercial garage application. If nonmetallic raceways are used, the bonding and grounding requirements of 501.16 apply.

(B) Equipment Located in Class I Locations. Within Class I locations as defined in 511.3, equipment shall conform to applicable provisions of Article 501.

(1) Fuel-Dispensing Units. Where fuel-dispensing units (other than liquid petroleum gas, which is prohibited) are located within buildings, the requirements of Article 514 shall govern.

See Figure 514.3 in the *Code* and Exhibit 514.1 for information on classified areas in the vicinity of dispensing units.

Where mechanical ventilation is provided in the dispensing area, the control shall be interlocked so that the dispenser cannot operate without ventilation as prescribed in 500.5(B)(2).

(2) Portable Lighting Equipment. Portable lighting equipment shall be equipped with handle, lampholder, hook, and substantial guard attached to the lampholder or handle. All exterior surfaces that might come in contact with battery terminals, wiring terminals, or other objects shall be of nonconducting material or shall be effectively protected with insulation. Lampholders shall be of an unswitched type and shall not provide means for plug-in of attachment plugs. The outer shell shall be of molded composition or other suitable material. Unless the lamp and its cord are supported or arranged in such a manner that they cannot be used in the locations classified in 511.3, they shall be of a type identified for Class I, Division 1 locations.

Often, a reel-type portable handlamp is used for supplemental lighting during vehicle servicing. See Exhibit 511.2 for an illustration of a cord reel assembly.

Where there is no transfer of flammable liquids to vehicle fuel tanks, but there is use of electrical diagnostic equipment, electrically operated tools or machinery, or open flames such as used for welding or cutting, the area must be classified in accordance with 511.3. See Exhibit 511.3. Fuel dispensing units located within the garage are governed by the requirements of Article 514.

The Class I, Division 2 location above grade within a commercial garage extends 18 in. above floor level, unless the authority having jurisdiction determines otherwise because mechanical ventilation provides at least four air changes per hour.

The Class I, Division 1 location below grade extends from the floor of the pit or depression to floor level, unless the authority having jurisdiction permits the pit or depression to be classified as Class I, Division 2 because ventilation providing at least six air changes per hour exhausts air at the floor level of the pit or depression. Areas suitably cut off and areas adjacent to unclassified, ventilated garages are not classified as hazardous.

Exhibit 511.2 Cord reel. This cord reel, which is part of a portable lamp assembly, must be arranged so that the lamp cannot be used in a Class I location. Otherwise, the lamp must be an explosionproof type approved for Class I, Division 1 hazardous locations. (Courtesy of Appleton Electric Co., EGS Electrical Group)

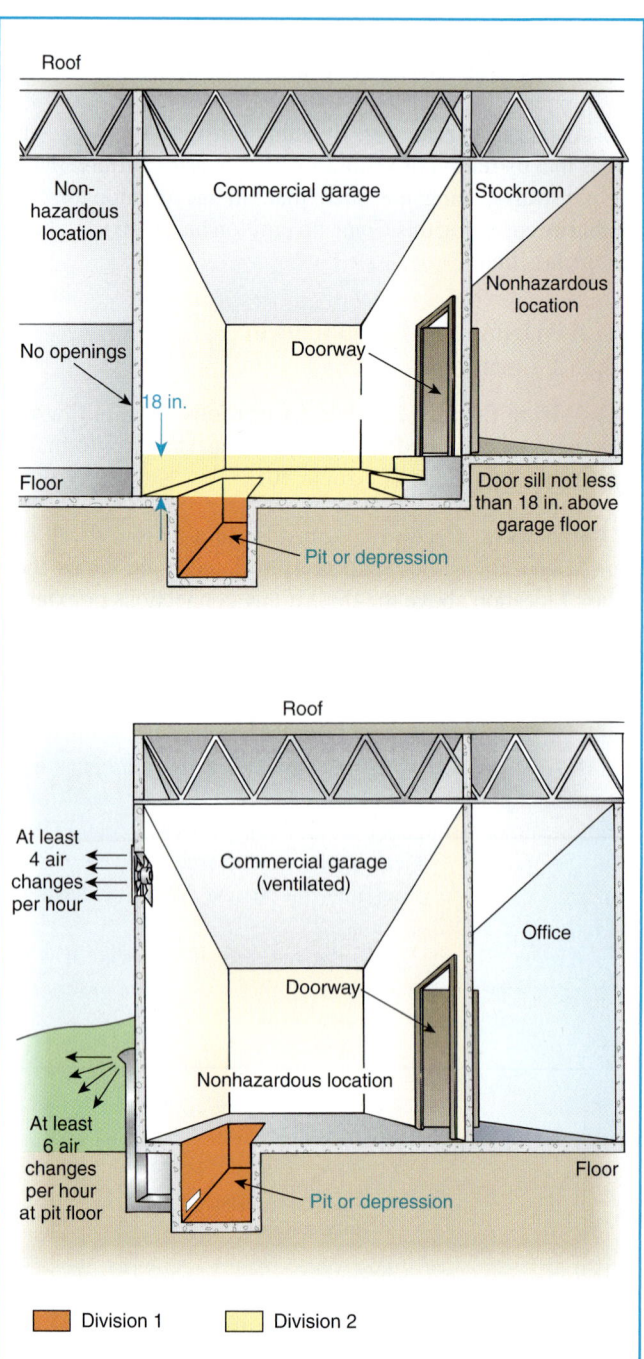

Exhibit 511.3 Classification of locations in commercial garages.

511.7 Wiring and Equipment Installed Above Class I Locations.

For the installation of electrical wiring in areas above those designated by 511.3 as Class I locations, 511.7 specifies a number of raceway and cable systems. The integrity of the wiring system above the classified location is necessary to ensure that an ignition capable arc or spark does not migrate into the classified location. Wiring installed above unclassified areas can be selected from the methods covered in Chapter 3, provided the respective article covering that wiring method does not contain any restrictions that would limit its use in commercial garages.

(A) Wiring in Spaces Above Class I Locations.

(1) Fixed Wiring Above Class I Locations. All fixed wiring above Class I locations shall be in metal raceways, rigid nonmetallic conduit, electrical nonmetallic tubing, flexible metal conduit, liquidtight flexible metal conduit, or liquidtight flexible nonmetallic conduit or shall be Type MC, MI, manufactured wiring systems, or PLTC cable in accordance with Article 725, or Type TC cable or Type ITC cable in accordance with Article 727. Cellular metal floor raceways or cellular concrete floor raceways shall be permitted to be used only for supplying ceiling outlets or extensions to the area below the floor, but such raceways shall have no connections leading into or through any Class I location above the floor.

(2) Pendant. For pendants, flexible cord suitable for the type of service and listed for hard usage shall be used.

(B) Electrical Equipment Installed Above Class I Locations.

(1) Fixed Electrical Equipment. Electrical equipment in a fixed position shall be located above the level of any defined Class I location or shall be identified for the location.

(a) Arcing Equipment. Equipment that is less than 3.7 m (12 ft) above the floor level and that may produce arcs, sparks, or particles of hot metal, such as cutouts, switches, charging panels, generators, motors, or other equipment (excluding receptacles, lamps, and lampholders) having make-and-break or sliding contacts, shall be of the totally enclosed type or constructed so as to prevent the escape of sparks or hot metal particles.

(b) Fixed Lighting. Lamps and lampholders for fixed lighting that is located over lanes through which vehicles are commonly driven or that may otherwise be exposed to physical damage shall be located not less than 3.7 m (12 ft) above floor level, unless of the totally enclosed type or constructed so as to prevent escape of sparks or hot metal particles.

511.9 Sealing.

Seals conforming to the requirements of 501.5 and 501.5(B)(2) shall be provided and shall apply to horizontal as well as vertical boundaries of the defined Class I locations.

Seals are required if any part of the raceway is in, or passes through, a Class I, Division 2 location. See the commentary on seals following 501.5(F)(3,) FPN.

511.10 Special Equipment.

(A) Battery Charging Equipment. Battery chargers and their control equipment, and batteries being charged, shall not be located within locations classified in 511.3.

(B) Electric Vehicle Charging Equipment.

(1) General. All electrical equipment and wiring shall be installed in accordance with Article 625, except as noted in 511.10(B)(2) and (B)(3). Flexible cords shall be of a type identified for extra-hard usage.

(2) Connector Location. No connector shall be located within a Class I location as defined in 511.3.

(3) Plug Connections to Vehicles. Where the cord is suspended from overhead, it shall be arranged so that the lowest point of sag is at least 150 mm (6 in.) above the floor. Where an automatic arrangement is provided to pull both cord and plug beyond the range of physical damage, no additional connector shall be required in the cable or at the outlet.

511.12 Ground-Fault Circuit-Interrupter Protection for Personnel.

All 125-volt, single-phase, 15- and 20-ampere receptacles installed in areas where electrical diagnostic equipment, electrical hand tools, or portable lighting equipment are to be used shall have ground-fault circuit-interrupter protection for personnel.

Ground-fault circuit interrupters (GFCIs) intended to protect personnel from shock hazards are designed to trip when a ground-fault current of 5 milliamperes (plus or minus 1 mA) or greater is detected. The GFCI necessary to comply with the requirement in 511.12 may be either a receptacle-type or a circuit-breaker-type. This requirement applies to receptacles supplying specific types of utilization equipment that will be in use by repair personnel in environments where the floor surface (typically, concrete slabs with direct or indirect earth contact) and the possibility of dampness or even standing water increases the potential for electric shock.

See the definition of *ground-fault circuit interrupter* in Article 100 and see Exhibits 210.11 and 210.12.

511.16 Grounded and Grounding Requirements.

(A) General Grounding Requirements. All metal raceways, the metal armor or metallic sheath on cables, and all non–current-carrying metal parts of fixed or portable electrical equipment, regardless of voltage, shall be grounded as provided in Article 250.

(B) Supplying Circuits with Grounded and Grounding Conductors in Class I Locations. Grounding in Class I locations shall comply with 501.16.

(1) Circuits Supplying Portable Equipment or Pendants. Where a circuit supplies portables or pendants and includes a grounded conductor as provided in Article 200, receptacles, attachment plugs, connectors, and similar devices shall be of the grounding type, and the grounded conductor of the flexible cord shall be connected to the screw shell of any lampholder or to the grounded terminal of any utilization equipment supplied.

(2) Approved Means. Approved means shall be provided for maintaining continuity of the grounding conductor between the fixed wiring system and the non–current-carrying metal portions of pendant luminaires (fixtures), portable lamps, and portable utilization equipment.

ARTICLE 513
Aircraft Hangars

Contents

513.1 Scope.

This article shall apply to buildings or structures in any part of which aircraft containing Class I (flammable) liquids or Class II (combustible) liquids whose temperatures are above their flash points are housed or stored and in which aircraft might undergo service, repairs, or alterations. It shall not apply to locations used exclusively for aircraft that have never contained fuel or unfueled aircraft.

The scope of Article 513 clarifies that it is not necessary to classify areas in which the only fuel contained in the aircraft is a Class II combustible liquid, unless the fuel is going to be used or stored above its flash point. A Class II liquid has a closed-cup flash point at or above 100°F (37.8°C). See the definition of *combustible liquid* in NFPA 30, *Flammable and Combustible Liquids Code*. Some aviation fuel, such as that used in jet engines, is a Class II combustible liquid. An aircraft manufacturing plant in which the aircraft under construction have never contained fuel is an example of a facility that is not covered by the requirements of Article 513.

For further information, see NFPA 409, *Standard on Aircraft Hangars*.

> FPN No. 1: For definitions of aircraft hangar and unfueled aircraft, see NFPA 409-1995, *Standard on Aircraft Hangars*.
>
> FPN No. 2: For further information on fuel classification see NFPA 30-2000, *Flammable and Combustible Liquids Code*.

513.2 Definitions.

For the purpose of this article, the following definitions shall apply.

The definitions in 513.2 apply only to the portable and mobile equipment covered in Article 513.

Mobile Equipment. Equipment with electric components suitable to be moved only with mechanical aids or is provided with wheels for movement by person(s) or powered devices.

Portable Equipment. Equipment with electric components suitable to be moved by a single person without mechanical aids.

513.3 Classification of Locations.

(A) Below Floor Level. Any pit or depression below the level of the hangar floor shall be classified as a Class I, Division 1 or Zone 1 location that shall extend up to said floor level.

(B) Areas Not Cut Off or Ventilated. The entire area of the hangar, including any adjacent and communicating areas not suitably cut off from the hangar, shall be classified as a Class I, Division 2 or Zone 2 location up to a level 450 mm (18 in.) above the floor.

(C) Vicinity of Aircraft. The area within 1.5 m (5 ft) horizontally from aircraft power plants or aircraft fuel tanks shall be classified as a Class I, Division 2 or Zone 2 location that shall extend upward from the floor to a level 1.5 m (5 ft) above the upper surface of wings and of engine enclosures.

In order to properly classify the area in accordance with 513.3(C), it is necessary to obtain information on the aircraft parking patterns, the types of aircraft, and the operations to be performed in the hangar. See Exhibit 513.1 for area classification in aircraft hangars.

Consideration of future changes in aircraft types and locations is appropriate to avoid the need for costly wiring and equipment alterations as a result of changes in the area classification.

(D) Areas Suitably Cut Off and Ventilated. Adjacent areas in which flammable liquids or vapors are not likely to be released, such as stock rooms, electrical control rooms, and other similar locations, shall not be classified where adequately ventilated and where effectively cut off from the hangar itself by walls or partitions.

513.4 Wiring and Equipment in Class I Locations.

(A) General. All wiring and equipment that is or may be installed or operated within any of the Class I locations

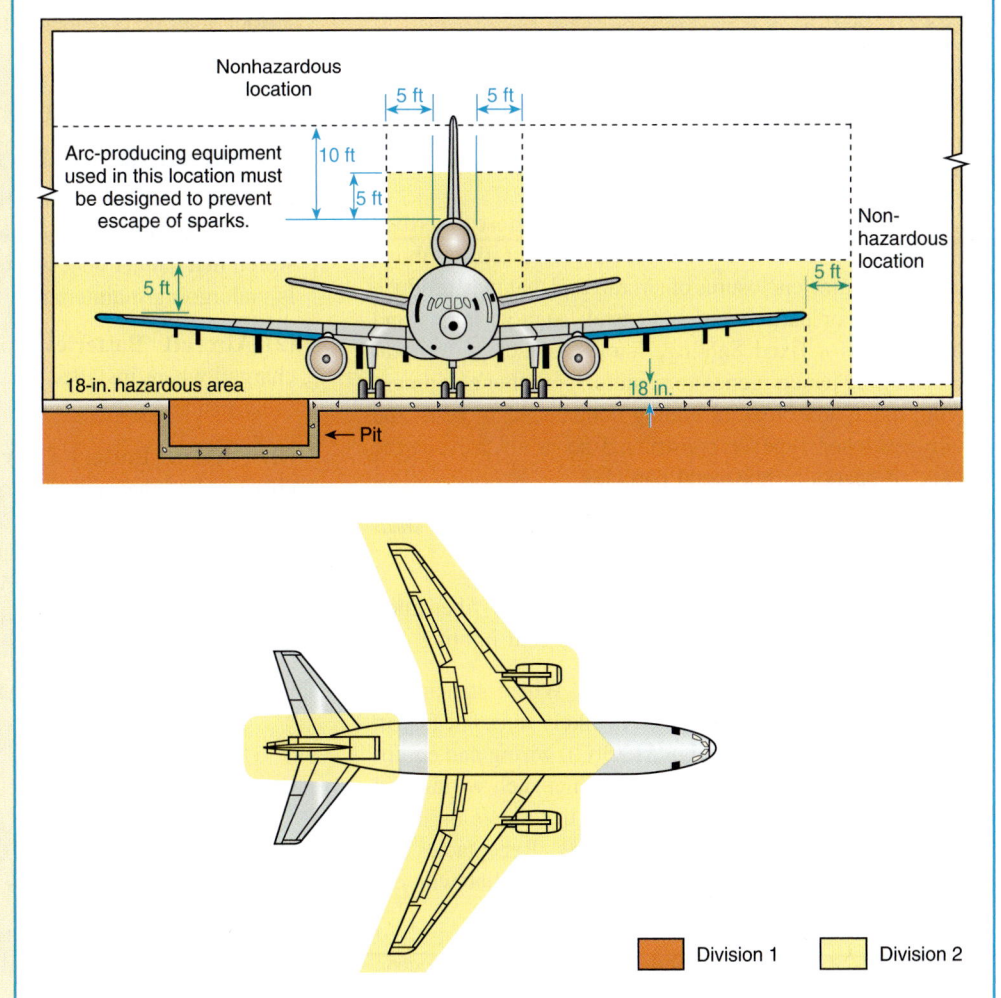

Exhibit 513.1 Area classification in aircraft hangars.

defined in 513.3 shall comply with the applicable provisions of Article 501 or Article 505 for the division or zone in which they are used.

Attachment plugs and receptacles in Class I locations shall be identified for Class I locations or shall be designed so that they cannot be energized while the connections are being made or broken.

(B) Stanchions, Rostrums, and Docks. Electric wiring, outlets, and equipment (including lamps) on or attached to stanchions, rostrums, or docks that are located or likely to be located in a Class I location, as defined in 513.3(C), shall comply with the applicable provisions of Article 501 or Article 505 for the division or zone in which they are used.

513.7 Wiring and Equipment Not Installed in Class I Locations.

(A) Fixed Wiring. All fixed wiring in a hangar but not installed in a Class I location as classified in 513.3 shall be installed in metal raceways or shall be Type MI, TC, or MC cable.

Exception: Wiring in unclassified locations, as classified in 513.3(D), shall be of a type recognized in Chapter 3.

(B) Pendants. For pendants, flexible cord suitable for the type of service and identified for hard usage or extra-hard usage shall be used. Each such cord shall include a separate equipment grounding conductor.

(C) Arcing Equipment. In locations above those described in 513.3, equipment that is less than 3.0 m (10 ft) above wings and engine enclosures of aircraft and that may produce arcs, sparks, or particles of hot metal, such as lamps and lampholders for fixed lighting, cutouts, switches, receptacles, charging panels, generators, motors, or other equipment having make-and-break or sliding contacts, shall be of the totally enclosed type or constructed so as to prevent the escape of sparks or hot metal particles.

Exception: Equipment in areas described in 513.3(D) shall be permitted to be of the general-purpose type.

(D) Lampholders. Lampholders of metal-shell, fiber-lined types shall not be used for fixed incandescent lighting.

(E) Stanchions, Rostrums, or Docks. Where stanchions, rostrums, or docks are not located or likely to be located in a Class I location, as defined in 513.3(C), wiring and equipment shall comply with 513.7, except that such wiring and equipment not more than 457 mm (18 in.) above the floor in any position shall comply with 513.4(B). Receptacles and attachment plugs shall be of a locking type that will not readily disconnect.

(F) Mobile Stanchions. Mobile stanchions with electric equipment complying with 513.7(E) shall carry at least one

permanently affixed warning sign with the following words or equivalent:

> WARNING
> KEEP 5 FT CLEAR OF AIRCRAFT ENGINES AND
> FUEL TANK AREAS

or

> WARNING
> KEEP 1.5 METERS CLEAR OF AIRCRAFT ENGINES
> AND FUEL TANK AREAS

513.8 Wiring and Equipment Embedded, Under Slab, or Under Ground.

All wiring installed in or under the hangar floor shall comply with the requirements for Class I, Division 1 locations. Where such wiring is located in vaults, pits, or ducts, adequate drainage shall be provided.

513.9 Sealing.

Seals shall be provided in accordance with 501.5 and 505.16. Sealing requirements specified shall apply to horizontal as well as to vertical boundaries of the defined Class I locations. Raceways embedded in a concrete floor or buried beneath a floor shall be considered to be within the Class I location above the floor.

513.10 Special Equipment.

(A) Aircraft Electrical Systems.

(1) De-energizing Aircraft Electrical Systems. Aircraft electrical systems shall be de-energized when the aircraft is stored in a hangar and, whenever possible, while the aircraft is undergoing maintenance.

(2) Aircraft Batteries. Aircraft batteries shall not be charged where installed in an aircraft located inside or partially inside a hangar.

(B) Aircraft Battery Charging and Equipment. Battery chargers and their control equipment shall not be located or operated within any of the Class I locations defined in 513.3 and shall preferably be located in a separate building or in an area such as defined in 513.3(D). Mobile chargers shall carry at least one permanently affixed warning sign with the following words or equivalent:

> WARNING
> KEEP 5 FT CLEAR OF AIRCRAFT ENGINES AND
> FUEL TANK AREAS

or

> WARNING
> KEEP 1.5 METERS CLEAR OF AIRCRAFT ENGINES
> AND FUEL TANK AREAS

Tables, racks, trays, and wiring shall not be located within a Class I location and, in addition, shall comply with Article 480.

(C) External Power Sources for Energizing Aircraft.

(1) Not Less Than 450 mm (18 in.) Above Floor. Aircraft energizers shall be designed and mounted so that all electric equipment and fixed wiring will be at least 450 mm (18 in.) above floor level and shall not be operated in a Class I location as defined in 513.3(C).

(2) Marking for Mobile Units. Mobile energizers shall carry at least one permanently affixed warning sign with the following words or equivalent:

WARNING
KEEP 5 FT CLEAR OF AIRCRAFT ENGINES AND
FUEL TANK AREAS.

or

WARNING
KEEP 1.5 METERS CLEAR OF AIRCRAFT ENGINES
AND FUEL TANK AREAS

(3) Cords. Flexible cords for aircraft energizers and ground support equipment shall be identified for the type of service and extra-hard usage and shall include an equipment grounding conductor.

(D) Mobile Servicing Equipment with Electric Components.

(1) General. Mobile servicing equipment (such as vacuum cleaners, air compressors, air movers, etc.) having electric wiring and equipment not suitable for Class I, Division 2 or Zone 2 locations shall be designed and mounted so that all such fixed wiring and equipment will be at least 450 mm (18 in.) above the floor. Such mobile equipment shall not be operated within the Class I location defined in 513.3(C) and shall carry at least one permanently affixed warning sign with the following words or equivalent:

WARNING
KEEP 5 FT CLEAR OF AIRCRAFT ENGINES AND
FUEL TANK AREAS

or

WARNING
KEEP 1.5 METERS CLEAR OF AIRCRAFT ENGINES
AND FUEL TANK AREAS

(2) Cords and Connectors. Flexible cords for mobile equipment shall be suitable for the type of service and identified for extra-hard usage and shall include an equipment grounding conductor. Attachment plugs and receptacles shall be identified for the location in which they are installed and shall provide for connection of the equipment grounding conductor.

(3) Restricted Use. Equipment that is not identified as suitable for Class I, Division 2 locations shall not be operated in locations where maintenance operations likely to release flammable liquids or vapors are in progress.

(E) Portable Equipment.

(1) Portable Lighting Equipment. Portable lighting equipment that is used within a hangar shall be identified for the location in which they are used. For portable lamps, flexible cord suitable for the type of service and identified for extra-hard usage shall be used. Each such cord shall include a separate equipment grounding conductor.

(2) Portable Utilization Equipment. Portable utilization equipment that is or may be used within a hangar shall be of a type suitable for use in Class I, Division 2 or Zone 2 locations. For portable utilization equipment flexible cord suitable for the type of service and approved for extra-hard usage shall be used. Each such cord shall include a separate equipment grounding conductor.

513.16 Grounded and Grounding Requirements.

(A) General Grounding Requirements. All metal raceways, the metal armor or metallic sheath on cables, and all non–current-carrying metal parts of fixed or portable electrical equipment, regardless of voltage, shall be grounded as provided in Article 250. Grounding in Class I locations shall comply with 501.16 for Class I, Division 1 and 2 locations and 505.25 for Class I, Zone 0, 1, and 2 locations.

(B) Supplying Circuits with Grounded and Grounding Conductors in Class I Locations.

(1) Circuits Supplying Portable Equipment or Pendants. Where a circuit supplies portables or pendants and includes a grounded conductor as provided in Article 200, receptacles, attachment plugs, connectors, and similar devices shall be of the grounding type, and the grounded conductor of the flexible cord shall be connected to the screw shell of any lampholder or to the grounded terminal of any utilization equipment supplied.

(2) Approved Means. Approved means shall be provided for maintaining continuity of the grounding conductor between the fixed wiring system and the non–current-carrying metal portions of pendant luminaires (fixtures), portable lamps, and portable utilization equipment.

ARTICLE 514
Motor Fuel Dispensing Facilities

Contents

FPN: Rules that are followed by a reference in brackets contain text that has been extracted from NFPA 30A-2000, *Code for Motor Fuel Dispensing Facilities and Repair Garages*. Only editorial changes were made to the extracted text to make it consistent with this *Code*.

514.1 Scope.

These occupancies shall include locations where gasoline or other volatile flammable liquids or liquefied flammable gases are transferred to the fuel tanks (including auxiliary fuel tanks) of self-propelled vehicles or approved containers.

The title of Article 514 has been revised to encompass all locations where volatile flammable liquids or gases are dispensed into the fuel tanks of self-propelled vehicles or other approved fuel tanks. The phrase "approved containers" covers portable gasoline containers and is also intended to apply to dispensing locations for liquefied petroleum gas (LPG), including those locations that do not serve self-propelled vehicles. The popularity of outdoor cooking appliances using liquefied petroleum gas has resulted in a marked increase in portable container dispensing sites. Electrical area classification for these sites is specified in Table 514.3(B)(2).

FPN: For further information regarding safeguards for motor fuel dispensing facilities, see NFPA 30A-2000, *Code for Motor Fuel Dispensing Facilities and Repair Garages*.

514.2 Definition.

Motor Fuel Dispensing Facility. A location where gasoline or other volatile flammable liquids or liquefied flammable gases are transferred to the fuel tanks (including auxiliary fuel tanks) of self-propelled vehicles or approved containers.

FPN: Refer to Articles 510 and 511 with respect to electric wiring and equipment for other areas used as lubritori-

ums, service rooms, repair rooms, offices, salesrooms, compressor rooms, and similar locations.

514.3 Classification of Locations.

(A) Unclassified Locations. Where the authority having jurisdiction can satisfactorily determine that flammable liquids having a flash point below 38°C (100°F), such as gasoline, will not be handled, such location shall not be required to be classified.

(B) Classified Locations.

(1) Class I Locations. Table 514.3(B)(1) shall be applied where Class I liquids are stored, handled, or dispensed and shall be used to delineate and classify motor fuel dispensing facilities and commercial garages as defined in Article 511. Table 515.3 shall be used for the purpose of delineating and classifying aboveground tanks. A Class I location shall not extend beyond an unpierced wall, roof, or other solid partition. [NFPA 30A, 8.1, 8.3]

(2) Compressed Natural Gas, Liquefied Natural Gas, and Liquefied Petroleum Gas Areas. Table 514.3(B)(2) shall be used to delineate and classify areas where compressed natural gas (CNG), liquefied natural gas (LNG), or liquefied petroleum gas (LPG) are stored, handled, or dispensed. Where CNG or LNG dispensers are installed beneath a canopy or enclosure, either the canopy or enclosure shall be designed to prevent accumulation or entrapment of ignitible vapors, or all electrical equipment installed beneath the canopy or enclosure shall be suitable for Class I, Division 2 hazardous (classified) locations. Dispensing devices for liquefied petroleum gas shall be located not less than 1.5 m (5 ft) from any dispensing device for Class I liquids. [NFPA 30A, 12.1, 12.4, 12.5]

FPN No. 1: For information on area classification where liquefied petroleum gases are dispensed, see NFPA 58-2001, *Liquefied Petroleum Gas Code*.

FPN No. 2: For information on classified areas pertaining to LP-Gas systems other than residential or commercial, see NFPA 58-2001, *Liquefied Petroleum Gas Code* and NFPA 59-2001, *Utility LP-Gas Plant Code*.

FPN No. 3: See 555.21 for gasoline dispensing stations in marinas and boatyards.
[NFPA 30A, Table 12.6]

Tables 514.3(B)(1) and 514.3(B)(2) are extracted from NFPA 30A, *Code for Motor Fuel Dispensing Facilities and Repair Garages*. See the commentary following 511.3(B)(3), Exception No. 2, for information regarding the classification of facilities that primarily offer lubrication-type service. See Exhibit 514.1 for an illustration of the Class I location around overhead motor fuel dispensing units.

The use of aboveground tanks with dispensing equip-

Table 514.3(B)(1) Class I Locations — Motor Fuel Dispensing Facilities and Commercial Garages

Location	Class I, Group D Division	Extent of Classified Location
Underground Tank		
Fill opening	1	Any pit, box, or space below grade level, any part of which is within the Division 1 or Division 2, Zone 1 or Zone 2 classified location
	2	Up to 450 mm (18 in.) above grade level within a horizontal radius of 3.0 m (10 ft) from a loose fill connection and within a horizontal radius of 1.5 m (5 ft) from a tight fill connection
Vent — discharging upward	1	Within 900 mm (3 ft) of open end of vent, extending in all directions
	2	Space between 900 mm (3 ft) and 1.5 m (5 ft) of open end of vent, extending in all directions
Dispensing Device[1,4] (except overhead type)[2]		
Pits	1	Any pit, box, or space below grade level, any part of which is within the Division 1 or Division 2, Zone 1 or Zone 2 classified location
Dispenser		FPN: Space classification inside the dispenser enclosure is covered in ANSI/UL 87-1995, *Power Operated Dispensing Devices for Petroleum Products.*
	2	Within 450 mm (18 in.) horizontally in all directions extending to grade from the dispenser enclosure or that portion of the dispenser enclosure containing liquid-handling components FPN: Space classification inside the dispenser enclosure is covered in ANSI/UL 87-1995, *Power Operated Dispensing Devices for Petroleum Products.*
Outdoor	2	Up to 450 mm (18 in.) above grade level within 6.0 m (20 ft) horizontally of any edge of enclosure.
Indoor with mechanical ventilation	2	Up to 450 mm (18 in.) above grade or floor level within 6.0 m (20 ft) horizontally of any edge of enclosure
with gravity ventilation	2	Up to 450 mm (18 in.) above grade or floor level within 7.5 m (25 ft) horizontally of any edge of enclosure
Dispensing Device[4]		
Overhead type[2]	1	The space within the dispenser enclosure, and all electrical equipment integral with the dispensing hose or nozzle
	2	A space extending 450 mm (18 in.) horizontally in all directions beyond the enclosure and extending to grade
	2	Up to 450 mm (18 in.) above grade level within 6.0 m (20 ft) horizontally measured from a point vertically below the edge of any dispenser enclosure
Remote Pump — Outdoor	1	Any pit, box, or space below grade level if any part is within a horizontal distance of 3.0 m (10 ft) from any edge of pump
	2	Within 900 mm (3 ft) of any edge of pump, extending in all directions. Also up to 450 mm (18 in.) above grade level within 3.0 m (10 ft) horizontally from any edge of pump
Remote Pump — Indoor	1	Entire space within any pit
	2	Within 1.5 m (5 ft) of any edge of pump, extending in all directions. Also up to 900 mm (3 ft) above grade level within 7.5 m (25 ft) horizontally from any edge of pump
Lubrication or Service Room — with Dispensing	1	Any pit within any unventilated space
	2	Any pit with ventilation
	2	Space up to 450 mm (18 in.) above floor or grade level and 900 mm (3 ft) horizontally from a lubrication pit
Dispenser for Class I liquids	2	Within 900 mm (3 ft) of any fill or dispensing point, extending in all directions
Lubrication or Service Room — Without Dispensing	2	Entire area within any pit used for lubrication or similar services where Class I liquids may be released
	2	Area up to 450 mm (18 in.) above any such pit and extending a distance of 900 mm (3 ft) horizontally from any edge of the pit
	2	Entire unventilated area within any pit, belowgrade area, or subfloor area
	2	Area up to 450 mm (18 in.) above any such unventilated pit, belowgrade work area, or subfloor work area and extending a distance of 900 mm (3 ft) horizontally from the edge of any such pit, belowgrade work area, or subfloor work area
	Unclassified	Any pit, belowgrade work area, or subfloor work area that is provided with exhaust ventilation at a rate of not less than 0.3 m³/min/m² (1 cfm/ft²) of floor area at all times that the building is occupied or when vehicles are parked in or over this area and where exhaust air is taken from a point within 300 mm (12 in.) of the floor of the pit, belowgrade work area, or subfloor work area

Table 514.3(B)(1) *Continued*

Location	Class I, Group D Division	Extent of Classified Location
Special Enclosure Inside Building[3]	1	Entire enclosure
Sales, Storage, and Rest Rooms	Unclassified	If there is any opening to these rooms within the extent of a Division 1 location, the entire room shall be classified as Division 1
Vapor Processing Systems Pits	1	Any pit, box, or space below grade level, any part of which is within a Division 1 or Division 2 classified location or that houses any equipment used to transport or process vapors
Vapor Processing Equipment Located Within Protective Enclosures FPN: See 10.1.7 of NFPA 30A-2000, *Code for Motor Fuel Dispensing Facilities and Repair Garages.*	2	Within any protective enclosure housing vapor processing equipment
Vapor Processing Equipment Not Within Protective Enclosures (excluding piping and combustion devices)	2	The space within 450 mm (18 in.) in all directions of equipment containing flammable vapor or liquid extending to grade level. Up to 450 mm (18 in.) above grade level within 3.0 m (10 ft) horizontally of the vapor processing equipment
Equipment Enclosures	1	Any space within the enclosure where vapor or liquid is present under normal operating conditions
Vacuum-Assist Blowers	2	The space within 450 mm (18 in.) in all directions extending to grade level. Up to 450 mm (18 in.) above grade level within 3.0 m (10 ft) horizontally

[1]Refer to Figure 514.3 for an illustration of classified location around dispensing devices.
[2]Ceiling mounted hose reel.
[3]FPN: See 4.3.9 of NFPA 30A-2000, *Code for Motor Fuel Dispensing Facilities and Repair Garages.*
[4]FPN: Area classification inside the dispenser enclosure is covered in ANSI/UL 87-1995, *Power-Operated Dispensing Devices for Petroleum Products.*
[NFPA 30A, Table 8.3.1]

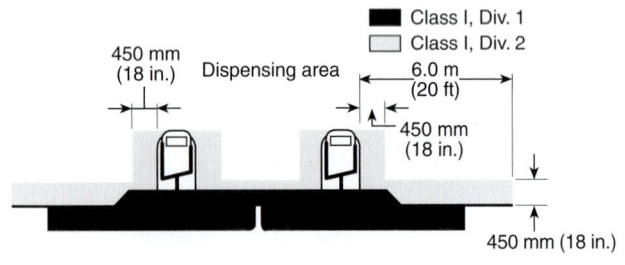

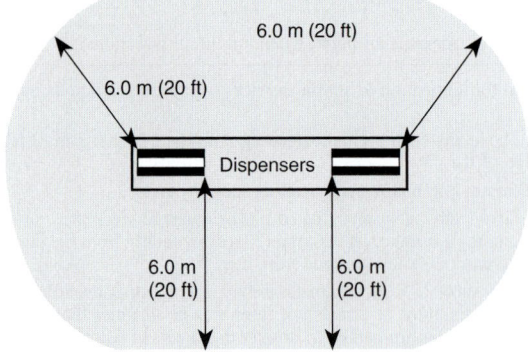

Figure 514.3 Classified locations adjacent to dispensers as detailed in Table 514.3(B)(1). [NFPA 30A, Figure 8.3]

Table 514.3(B)(2) Electrical Equipment Classified Areas for Dispensing Devices

| Dispensing Device | Extent of Classified Area | |
	Class I, Division 1	Class I, Division 2
Compressed Natural Gas	Entire space within the dispenser enclosure	1.5 m (5 ft) in all directions from dispenser enclosure
Liquefied Natural Gas	Entire space within the dispenser enclosure and 1.5 m (5 ft) in all directions from the dispenser enclosure	From 1.5 m to 3.0 m (5 ft to 10 ft) in all directions from the dispenser enclosure
Liquefied Petroleum Gas	Entire space within the dispenser enclosure; 450 mm (18 in.) from the exterior surface of the dispenser enclosure to an elevation of 1.2 m (4 ft) above the base of the dispenser; the entire pit or open space beneath the dispenser and within 6.0 m (20 ft) horizontally from any edge of the dispenser when the pit or trench is not mechanically ventilated.	Up to 450 mm (18 in.) aboveground and within 6.0 m (20 ft) horizontally from any edge of the dispenser enclosure, including pits or trenches within this area when provided with adequate mechanical ventilation

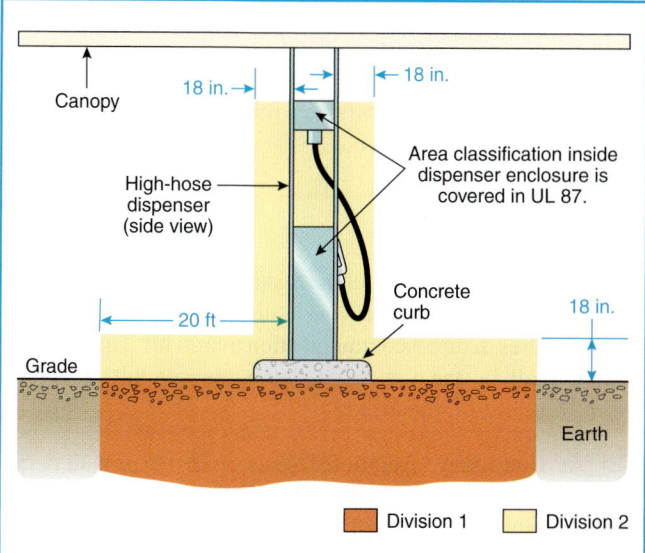

Exhibit 514.1 Extent of Class I location around overhead motor fuel dispensing units, in accordance with Table 514.3(B)(1).

ment is an acceptable alternative when underground tanks are impractical. Aboveground tanks with dispensing equipment are frequently used at fleet motor fuel-dispensing facilities. Hazardous area classification for aboveground tank installations is performed in accordance with Table 515.3.

514.4 Wiring and Equipment Installed in Class I Locations.

All electrical equipment and wiring installed in Class I locations as classified in 514.3 shall comply with the applicable provisions of Article 501.

Exception: As permitted in 514.8.

FPN: For special requirements for conductor insulation, see 501.13.

The wiring methods and equipment required in Article 501 must be used within the Class I areas at a motor fuel dispensing facility. An explosionproof outlet box of the type frequently used in gasoline dispensing units is shown in Exhibit 514.2. The branch-circuit conductors for dispenser power, lighting, or both, connect to the internal wiring of the dispenser in the explosionproof outlet box. For gasoline- and oil-resistant insulated conductors, see the commentary following 501.13.

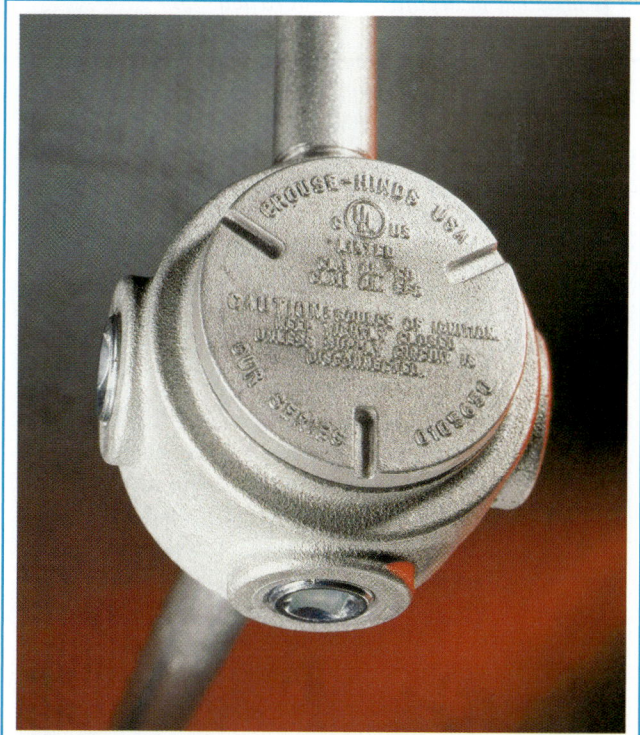

Exhibit 514.2 A typical explosionproof outlet box used in dispenser applications. (Courtesy of Cooper Crouse-Hinds)

514.7 Wiring and Equipment Above Class I Locations.

Wiring and equipment above the Class I locations as classified in 514.3 shall comply with 511.7.

514.8 Underground Wiring.

Underground wiring shall be installed in threaded rigid metal conduit or threaded steel intermediate metal conduit. Any portion of electrical wiring or equipment that is below the surface of a Class I, Division 1, or a Class I, Division 2 location [as classified in Table 514.3(B)(1) and Table 514.3(B)(2)] shall be considered to be in a Class I, Division 1, location that shall extend at least to the point of emergence above grade. Refer to Table 300.5.

Experience has shown that the fuel spilled in the vicinity of gasoline pumps in service stations tends to accumulate underground, where it can enter electrical conduits and accumulate in voids. Therefore, 514.8 classifies the space below those surface areas that are subject to fuel spills as Class I, Division 1 locations. Tables 514.3(B)(1) and 514.3(B)(2) define the extent of the aboveground Class I, Divisions 1 and 2 locations.

Exception No. 1: Type MI cable shall be permitted where it is installed in accordance with Article 332.

Exception No. 2: Rigid nonmetallic conduit complying with Article 352 shall be permitted where buried under not less than 600 mm (2 ft) of cover. Where rigid nonmetallic conduit is used, threaded rigid metal conduit or threaded steel intermediate metal conduit shall be used for the last 600 mm (2 ft) of the underground run to emergence or to the point of connection to the aboveground raceway, and an equipment grounding conductor shall be included to provide electrical continuity of the raceway system and for grounding of non–current-carrying metal parts.

Exception No. 2 to 514.8 makes it clear that if rigid nonmetallic conduit is used for underground wiring, threaded rigid metal conduit or threaded steel intermediate metal conduit must be used for the last 2 ft of the underground run to the point of emergence or to the point of connection to the aboveground raceway. The rigid nonmetallic conduit, including rigid nonmetallic conduit elbows and fittings, must be located not less than 2 ft below grade, as shown in Exhibit 514.3.

If rigid nonmetallic conduit is used, an equipment grounding conductor must be included and must be bonded to the explosionproof raceway system inside the dispenser. This is accomplished by terminating the equipment ground-

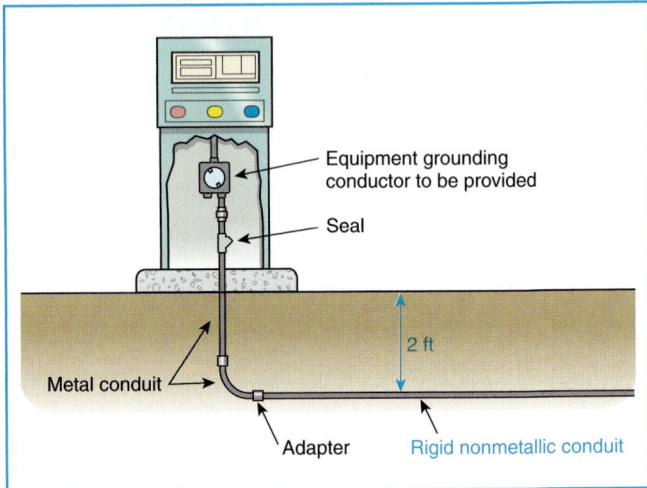

Exhibit 514.3 Use of rigid nonmetallic conduit in accordance with 514.8, Exception No. 2.

ing conductor on the ground screw (or other means) provided in the dispenser junction box.

514.9 Sealing.

(A) At Dispenser. A listed seal shall be provided in each conduit run entering or leaving a dispenser or any cavities or enclosures in direct communication therewith. The sealing fitting shall be the first fitting after the conduit emerges from the earth or concrete.

(B) At Boundary. Additional seals shall be provided in accordance with 501.5. Sections 501.5(A)(4) and (B)(2) shall apply to horizontal as well as to vertical boundaries of the defined Class I locations.

Sealing fittings are required in all conduits leaving a Class I location. All conduits passing under the boundaries of the hazardous (classified) locations (20-ft radius from dispenser) or the tank fill-pipe (10-ft radius from a loose-fill connection and 5-ft radius from a tight-fill connection) are considered in a Class I, Division 1 location (see 514.8), and the seal is to be the first fitting at the point of emergence. A seal must be provided in each conduit run entering or leaving a dispenser. Therefore, even though a conduit runs from dispenser to dispenser and does not leave the hazardous (classified) location, a seal is necessary where leaving, and again where entering, the dispenser. Panelboards are generally located in a room classified as a nonhazardous location; however, any conduit coming from the dispenser or passing under the hazardous (classified) location boundaries from the dispenser or tank fill-pipe would require a seal at the panelboard location to minimize the likelihood of gas migration into the remote location. If the panelboard is located in the lube or repair room, all conduits emerging into the 18-in. hazardous (classified) location would require seals. See Exhibits 514.4 and 514.5.

514.11 Circuit Disconnects.

(A) General. Each circuit leading to or through dispensing equipment, including equipment for remote pumping systems, shall be provided with a clearly identified and readily accessible switch or other acceptable means, located remote from the dispensing devices, to disconnect simultaneously from the source of supply, all conductors of the circuits, including the grounded conductor, if any.

Single-pole breakers utilizing handle ties shall not be permitted.

The disconnecting means required by 514.11(A) must be clearly marked and readily accessible. The disconnecting means also must be remote from the dispensing device, so that if an emergency occurs that requires rapid shutdown of the dispensing equipment, the person operating the disconnecting means will not be exposed to the hazard.

It is important to note that all conductors of a circuit, including the grounded conductor, that may be present within a dispensing device must be provided with a switch or

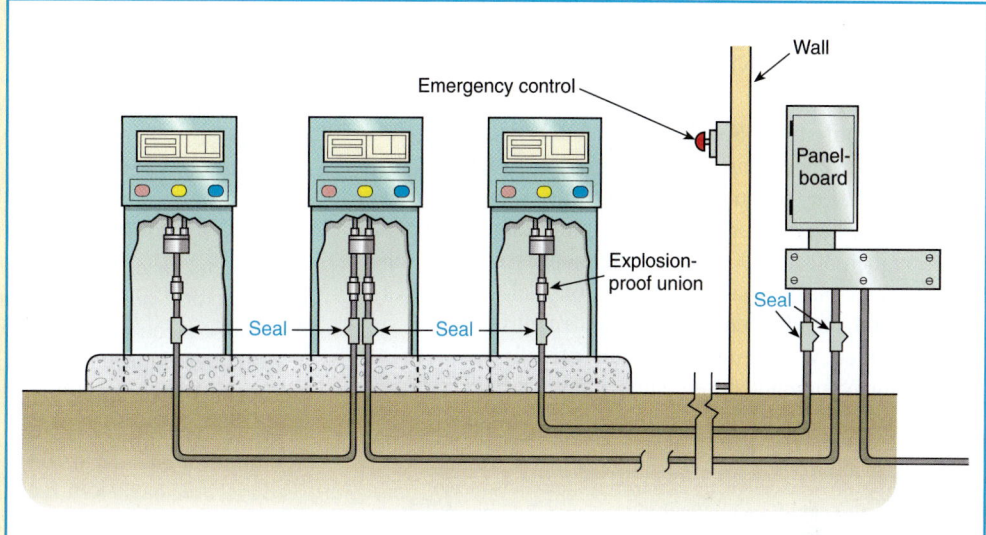

Exhibit 514.4 A gasoline dispenser installation indicating locations for sealing fittings. Emergency controls are required for self-service stations.

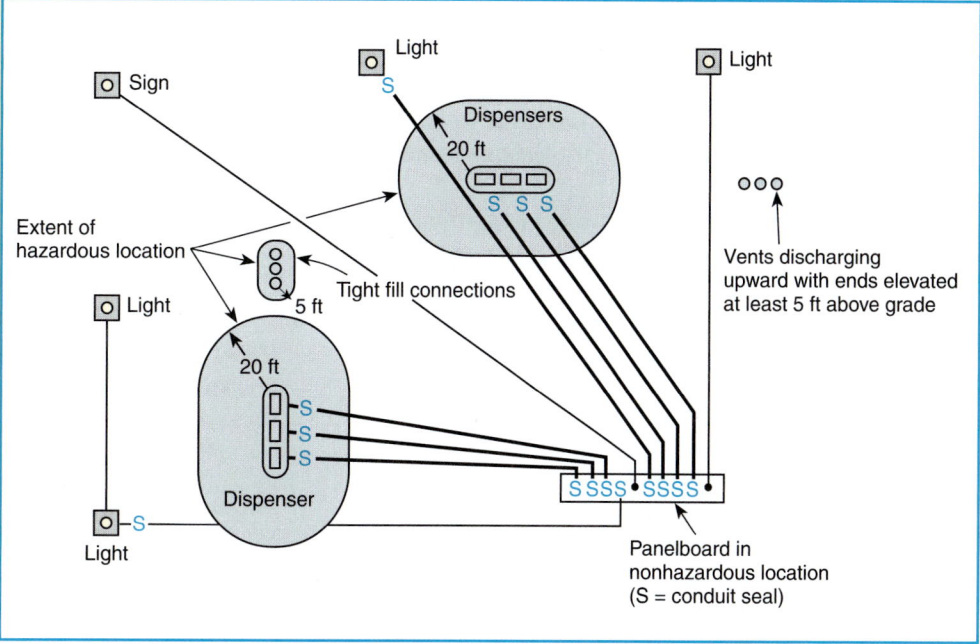

Exhibit 514.5 Seals are required at points marked "S." Seals are not required at the sign and two of the lights because conduit runs do not pass through a hazardous location.

special-type circuit breaker that simultaneously disconnects all conductors. Handle ties on single-pole circuit breakers are not permitted. The intent is that no energized conductors are in the dispenser vicinity during maintenance or alteration. Considering possible accidental reversal of the polarities of conductors at panelboards, the grounded conductor must be able to be switched to the open or off position. Grounded conductors may be present in old-style pump motors, or they may pass through a dispenser as part of a circuit for the dispensing island lighting.

(B) Attended Self-Service Motor Fuel Dispensing Facilities. Emergency controls as specified in 514.11(A) shall be installed at a location acceptable to the authority having jurisdiction, but controls shall not be more than 30 m (100 ft) from dispensers. [NFPA 30A, 6.7.1]

(C) Unattended Self-Service Motor Fuel Dispensing Facilities. Emergency controls as specified in 514.11(A) shall be installed at a location acceptable to the authority having jurisdiction, but the control shall be more than 6 m (20 ft) but less than 30 m (100 ft) from the dispensers. Additional emergency controls shall be installed on each group of dispensers or the outdoor equipment used to control the dispensers. Emergency controls shall shut off all power to all dispensing equipment at the station. Controls shall be manually reset only in a manner approved by the authority having jurisdiction. [NFPA 30A, 6.7.2]

> FPN: For additional information, see 6.7.1 and 6.7.2 of NFPA 30A-2000, *Code For Motor Fuel Dispensing Facilities and Repair Garages.*

Because a fire or large gasoline spill at the dispensing island may make it impossible for a person to approach and shut off the flow of gasoline by operating a disconnecting means located at the dispensing island, Section 6.7 of NFPA 30A, *Code for Motor Fuel Dispensing Facilities and Repair Garages,* requires an easily accessible and clearly identified emergency power shutoff to be provided at a location remote from the dispensing device. The requirements in 514.11(B) and 514.11(C) are extracted from NFPA 30A and provide the maximum and minimum distances for the location of the emergency control for attended and unattended self-service dispensing facilities.

The term *clearly identified* means that a sign must be posted indicating where the shutoff switch is located. This emergency power shutoff must be readily accessible and not blocked by the storage of such things as tires, cases of lubricating oil, or merchandise on display. All dispensing facility operators as well as responding fire fighters should know the location of the emergency power shutoff.

514.13 Provisions for Maintenance and Service of Dispensing Equipment.

Each dispensing device shall be provided with a means to remove all external voltage sources, including feedback, during periods of maintenance and service of the dispensing equipment. The location of this means shall be permitted to be other than inside or adjacent to the dispensing device.

This requirement is intended to enhance the level of safety for personnel servicing dispensing equipment. As more sophisticated control circuitry is integrated into dispensing equipment, simply shutting off the main power source to the dispenser or remote pump does not necessarily ensure that the equipment being worked on has been isolated from all sources of voltage. To ensure that the equipment is completely isolated from all voltage sources, a means must be provided to remove all external voltage sources from each dispensing device, including sources that may backfeed into the dispenser.

514.16 Grounding.

All metal raceways, the metal armor or metallic sheath on cables, and all non–current-carrying metal parts of fixed portable electrical equipment, regardless of voltage, shall be grounded as provided in Article 250. Grounding in Class I locations shall comply with 501.16.

ARTICLE 515
Bulk Storage Plants

Contents

> FPN: Rules that are followed by a reference in brackets contain text that has been extracted from NFPA 30-2000,

Flammable and Combustible Liquids Code. Only editorial changes were made to the extracted text to make it consistent with this *Code.*

515.1 Scope.

This article covers a property or portion of a property where flammable liquids are received by tank vessel, pipelines, tank car, or tank vehicle and are stored or blended in bulk for the purpose of distributing such liquids by tank vessel, pipeline, tank car, tank vehicle, portable tank, or container.

515.2 Definition.

Bulk Storage Plant. That portion of a property where flammable liquids are received by tank vessel, pipelines, tank car, or tank vehicle, and are stored or blended in bulk for the purpose of distributing such liquids by tank vessel, pipelines, tank car, tank vehicle, portable tank, or container.

> FPN: For further information, see NFPA 30-2000, *Flammable and Combustible Liquids Code.*

515.3 Class I Locations.

Table 515.3 shall be applied where Class I liquids are stored, handled, or dispensed and shall be used to delineate and classify bulk storage plants. The class location shall not extend beyond a floor, wall, roof, or other solid partition that has no communicating openings. [NFPA 30, 5.9.5.1, 5.9.5.3]

Table 515.3 is essentially the same as Table 6.2.2 in NFPA 30, *Flammable and Combustible Liquids Code.* Section 5-3.4.5 of NFPA 30, as referenced in the first two items in column 1 of Table 515.3, states:

5-3.4.5 Equipment such as dispensing stations, open centrifuges, plate and frame filters, and open vacuum filters used in a building and the ventilation of the building shall be designed to limit flammable vapor-air mixtures under normal operating conditions to the interior of equipment and to not more than 5 ft

Table 515.3 Electrical Area Classifications

Location	NEC Class I Division	Zone	Extent of Classified Area
Indoor equipment installed in accordance with 5.3 of NFPA 30 where flammable vapor–air mixtures can exist under normal operation	1	0	The entire area associated with such equipment where flammable gases or vapors are present continuously or for long periods of time
	1	1	Area within 1.5 m (5 ft) of any edge of such equipment, extending in all directions
	2	2	Area between 1.5 m and 2.5 m (5 ft and 8 ft) of any edge of such equipment, extending in all directions; also, space up to 900 mm (3 ft) above floor or grade level within 1.5 m to 7.5 m (5 ft to 25 ft) horizontally from any edge of such equipment[1]
Outdoor equipment of the type covered in 5.3 of NFPA 30 where flammable vapor–air mixtures may exist under normal operation	1	0	The entire area associated with such equipment where flammable gases or vapors are present continuously or for long periods of time
	1	1	Area within 900 mm (3 ft) of any edge of such equipment, extending in all directions
	2	2	Area between 900 mm (3 ft) and 2.5 m (8 ft) of any edge of such equipment, extending in all directions; also, space up to 900 mm (3 ft) above floor or grade level within 900 mm to 3.0 m (3 ft to 10 ft) horizontally from any edge of such equipment
Tank storage installations inside buildings	1	1	All equipment located below grade level
	2	2	Any equipment located at or above grade level
Tank – aboveground	1	0	Inside fixed roof tank
	1	1	Area inside dike where dike height is greater than the distance from the tank to the dike for more than 50 percent of the tank circumference
Shell, ends, or roof and dike area	2	2	Within 3.0 m (10 ft) from shell, ends, or roof of tank; also, area inside dike to level of top of tank

Table 515.3 *Continued*

Location	NEC Class I Division	Zone	Extent of Classified Area
Vent	1	0	Area inside of vent piping or opening
	1	1	Within 1.5 m (5 ft) of open end of vent, extending in all directions
	2	2	Area between 1.5 m and 3.0 m (5 ft and 10 ft) from open end of vent, extending in all directions
Floating roof with fixed outer roof	1	0	Area between the floating and fixed roof sections and within the shell
Floating roof with no fixed outer roof	1	1	Area above the floating roof and within the shell
Underground tank fill opening	1	1	Any pit, or space below grade level, if any part is within a Division 1 or 2, or Zone 1 or 2, classified location
	2	2	Up to 450 mm (18 in.) above grade level within a horizontal radius of 3.0 m (10 ft) from a loose fill connection, and within a horizontal radius of 1.5 m (5 ft) from a tight fill connection
Vent – discharging upward	1	0	Area inside of vent piping or opening
	1	1	Within 900 mm (3 ft) of open end of vent, extending in all directions
	2	2	Area between 900 mm and 1.5 m (3 ft and 5 ft) of open end of vent, extending in all directions
Drum and container filling – outdoors or indoors	1	0	Area inside the drum or container
	1	1	Within 900 mm (3 ft) of vent and fill openings, extending in all directions
	2	2	Area between 900 mm and 1.5 m (3 ft and 5 ft) from vent or fill opening, extending in all directions; also, up to 450 mm (18 in.) above floor or grade level within a horizontal radius of 3.0 m (10 ft) from vent or fill opening
Pumps, bleeders, withdrawal fittings,			
Indoors	2	2	Within 1.5 m (5 ft) of any edge of such devices, extending in all directions; also, up to 900 mm (3 ft) above floor or grade level within 7.5 m (25 ft) horizontally from any edge of such devices
Outdoors	2	2	Within 900 mm (3 ft) of any edge of such devices, extending in all directions. Also, up to 450 mm (18 in.) above grade level within 3.0 m (10 ft) horizontally from any edge of such devices
Pits and sumps			
Without mechanical ventilation	1	1	Entire area within a pit or sump if any part is within a Division 1 or 2, or Zone 1 or 2, classified location
With adequate mechanical ventilation	2	2	Entire area within a pit or sump if any part is within a Division 1 or 2, or Zone 1 or 2, classified location
Containing valves, fittings, or piping, and not within a Division 1 or 2, or Zone 1 or 2, classified location	2	2	Entire pit or sump
Drainage ditches, separators, impounding basins			
Outdoors	2	2	Area up to 450 mm (18 in.) above ditch, separator, or basin; also, area up to 450 mm (18 in.) above grade within 4.5 m (15 ft) horizontally from any edge
Indoors			Same classified area as pits

Table 515.3 *Continued*

Location	NEC Class I Division	Zone	Extent of Classified Area
Tank vehicle and tank car[2] loading through open dome	1	0	Area inside of the tank
	1	1	Within 900 mm (3 ft) of edge of dome, extending in all directions
	2	2	Area between 900 mm and 4.5 m (3 ft and 15 ft) from edge of dome, extending in all directions
Loading through bottom connections with atmospheric venting	1	0	Area inside of the tank
	1	1	Within 900 mm (3 ft) of point of venting to atmosphere, extending in all directions
	2	2	Area between 900 mm and 4.5 m (3 ft and 15 ft) from point of venting to atmosphere, extending in all directions; also, up to 450 mm (18 in.) above grade within a horizontal radius of 3.0 m (10 ft) from point of loading connection
Office and rest rooms	Ordinary		If there is any opening to these rooms within the extent of an indoor classified location, the room shall be classified the same as if the wall, curb, or partition did not exist.
Loading through closed dome with atmospheric venting	1	1	Within 900 mm (3 ft) of open end of vent, extending in all directions
	2	2	Area between 900 mm and 4.5 m (3 ft and 15 ft) from open end of vent, extending in all directions; also, within 900 mm (3 ft) of edge of dome, extending in all directions
Loading through closed dome with vapor control	2	2	Within 900 mm (3 ft) of point of connection of both fill and vapor lines extending in all directions
Bottom loading with vapor control or any bottom unloading	2	2	Within 900 mm (3 ft) of point of connections, extending in all directions; also up to 450 mm (18 in.) above grade within a horizontal radius of 3.0 m (10 ft) from point of connections
Storage and repair garage for tank vehicles	1	1	All pits or spaces below floor level
	2	2	Area up to 450 mm (18 in.) above floor or grade level for entire storage or repair garage
Garages for other than tank vehicles	Ordinary		If there is any opening to these rooms within the extent of an outdoor classified location, the entire room shall be classified the same as the area classification at the point of the opening.
Outdoor drum storage	Ordinary		
Inside rooms or storage lockers used for the storage of Class I liquids	2	2	Entire room
Indoor warehousing where there is no flammable liquid transfer	Ordinary		If there is any opening to these rooms within the extent of an indoor classified location, the room shall be classified the same as if the wall, curb, or partition did not exist.
Piers and wharves			See Figure 515.3.

[1]The release of Class I liquids may generate vapors to the extent that the entire building, and possibly an area surrounding it, should be considered a Class I, Division 2 or Zone 2 location.

[2]When classifying extent of area, consideration shall be given to fact that tank cars or tank vehicles may be spotted at varying points. Therefore, the extremities of the loading or unloading positions shall be used. [NFPA 30, Table 5-9.5.3]

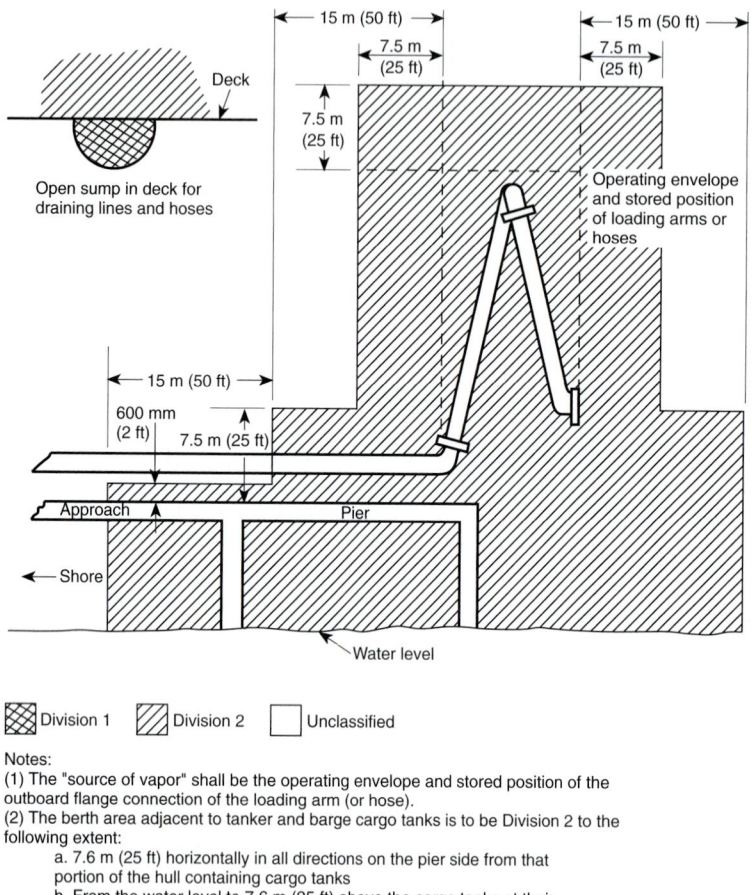

Deck

Open sump in deck for
draining lines and hoses

⟵ 15 m (50 ft) ⟶   ⟵ 15 m (50 ft) ⟶
⟵ 7.5 m
(25 ft) ⟶   ⟵ 7.5 m
(25 ft) ⟶

7.5 m
(25 ft)

Operating envelope
and stored position
of loading arms or
hoses

⟵ 15 m (50 ft) ⟶

600 mm
(2 ft)

7.5 m (25 ft)

Approach   Pier

⟵ Shore

⟵ Water level

▨ Division 1   ▨ Division 2   ☐ Unclassified

Notes:
(1) The "source of vapor" shall be the operating envelope and stored position of the
outboard flange connection of the loading arm (or hose).
(2) The berth area adjacent to tanker and barge cargo tanks is to be Division 2 to the
following extent:
    a. 7.6 m (25 ft) horizontally in all directions on the pier side from that
    portion of the hull containing cargo tanks
    b. From the water level to 7.6 m (25 ft) above the cargo tanks at their
    highest position
(3) Additional locations may have to be classified as required by the presence of other
sources of flammable liquids on the berth, by Coast Guard, or other regulations.

Figure 515.3 Marine terminal handling flammable liquids. [NFPA 30, Figure 5.7.16]

(1.5 m) from equipment that exposes Class I liquids
to the air.

Exhibits 515.1 through 515.7 illustrate the hazardous (classified) locations associated with several types of flammable
liquid containers and operations.

FPN No. 1: The area classifications listed in Table 515.3
are based on the premise that the installation meets the
applicable requirements of NFPA 30-2000, *Flammable
and Combustible Liquids Code*, Chapter 5, in all respects.
Should this not be the case, the authority having jurisdiction has the authority to classify the extent of the classified space.
FPN No. 2: See 555.21 for gasoline dispensing stations
in marinas and boatyards.

515.4 Wiring and Equipment Located in Class I Locations.

All electrical wiring and equipment within the Class I locations defined in 515.2 shall comply with the applicable pro-

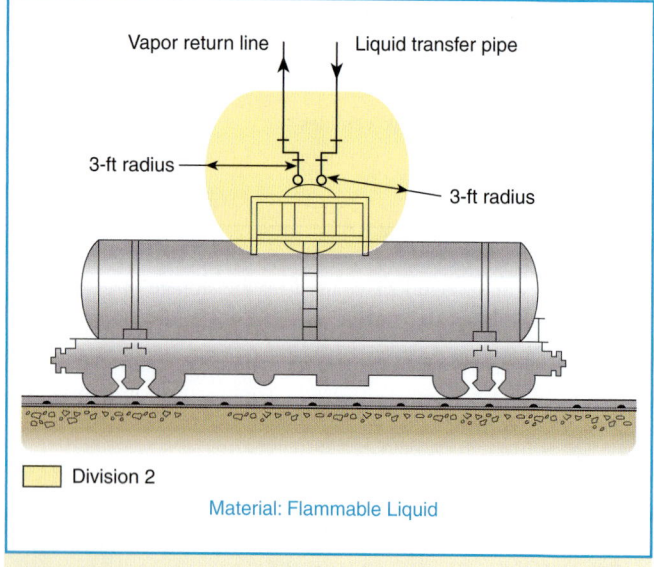

Vapor return line   Liquid transfer pipe

3-ft radius

3-ft radius

☐ Division 2

Material: Flammable Liquid

Exhibit 515.1 Tank car/tank truck loading and unloading via
closed system. Transfer through dome only.

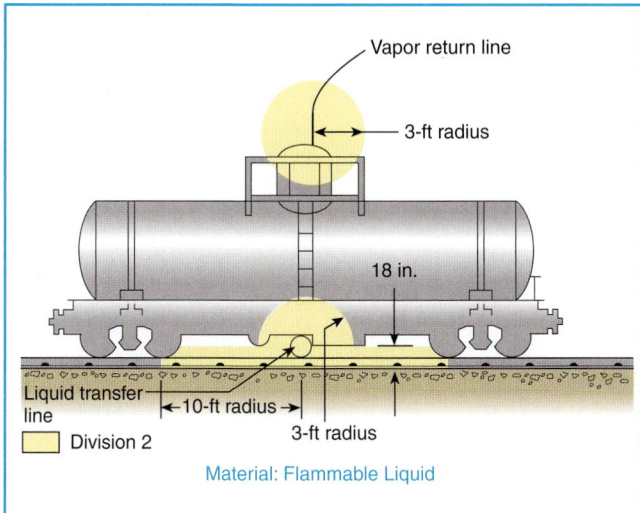

Exhibit 515.2 Tank car/tank truck loading and unloading via closed system. Bottom product transfer only.

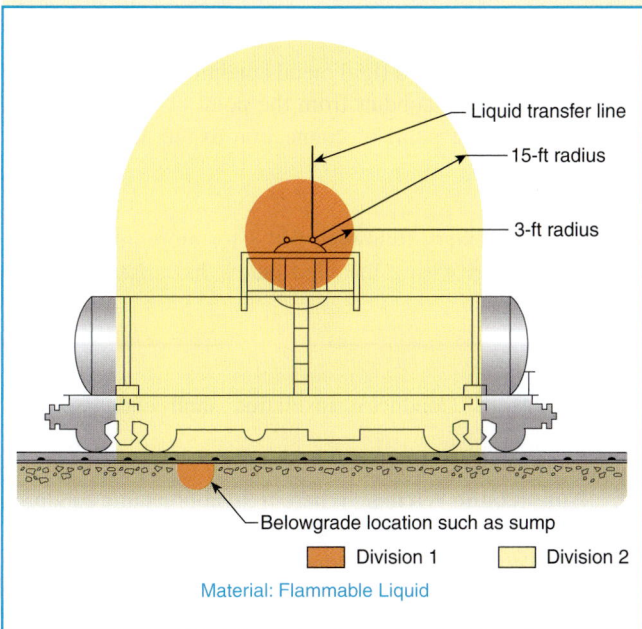

Exhibit 515.3 Tank car/tank truck loading and unloading via open system. Top or bottom product transfer.

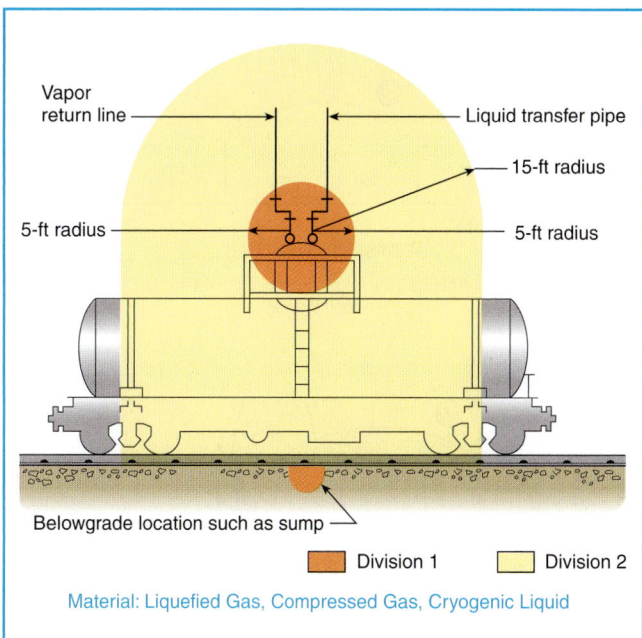

Exhibit 515.4 Tank car/tank truck loading and unloading via closed system. Transfer through dome only.

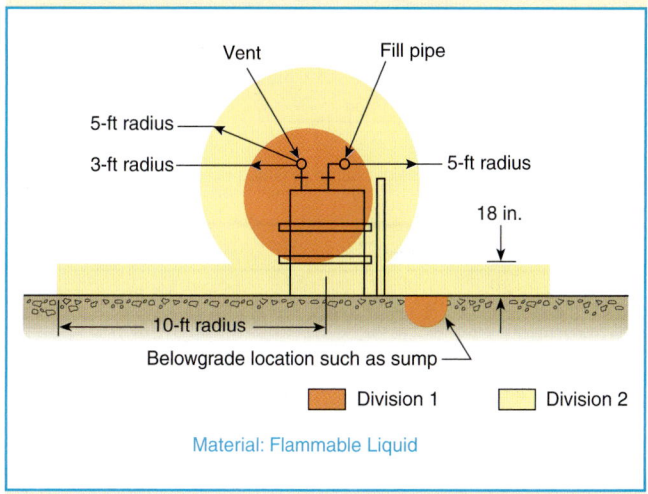

Exhibit 515.5 Drum filling station, outdoors or indoors, with adequate ventilation.

visions of Article 501 or Article 505 for the division or zone in which they are used.

Exception: As permitted in 515.8.

515.7 Wiring and Equipment Above Class I Locations.

(A) Fixed Wiring. All fixed wiring above Class I locations shall be in metal raceways or PVC Schedule 80 rigid nonme-

tallic conduit, or equivalent, or be Type MI, TC, or MC cable.

(B) Fixed Equipment. Fixed equipment that may produce arcs, sparks, or particles of hot metal, such as lamps and lampholders for fixed lighting, cutouts, switches, receptacles, motors, or other equipment having make-and-break or sliding contacts, shall be of the totally enclosed type or be constructed so as to prevent the escape of sparks or hot metal particles.

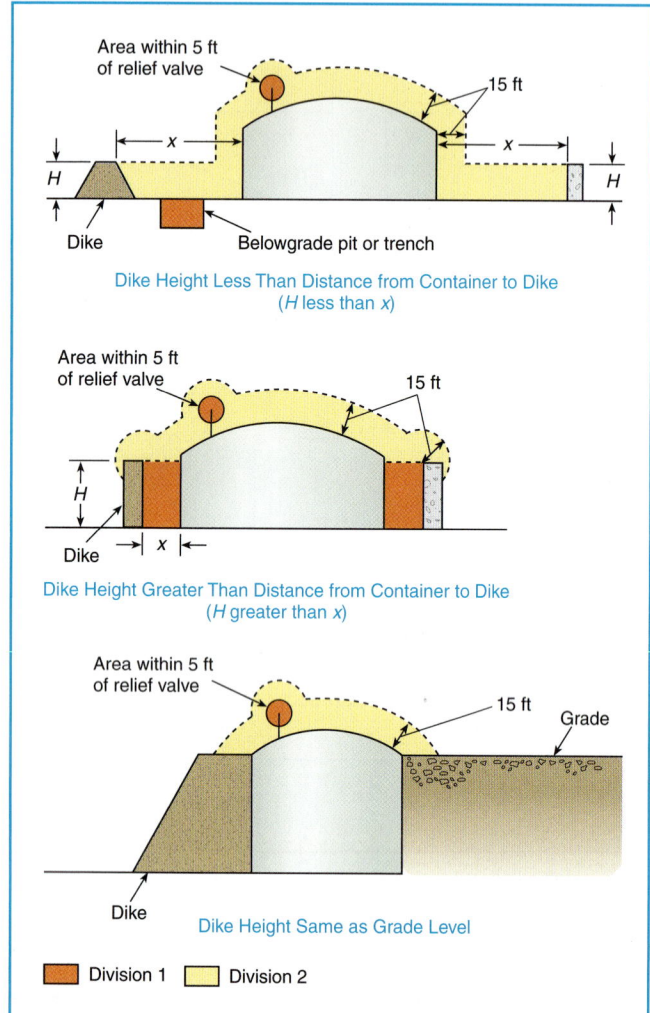

Exhibit 515.6 Storage tanks for cryogenic liquids. [From NFPA 59A, Standard for the Production, Storage, and Handling of Liquefied Natural Gas (LNG)]

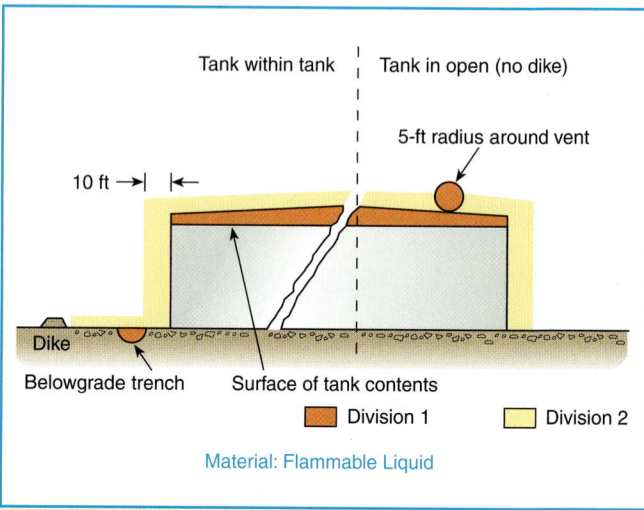

Exhibit 515.7 Storage tanks, outdoors at grade. (From API Recommended Practice 500A)

(C) Portable Lamps or Other Utilization Equipment. Portable lamps or other utilization equipment and their flexible cords shall comply with the provisions of Article 501 or Article 505 for the class of location above which they are connected or used.

515.8 Underground Wiring.

(A) Wiring Method. Underground wiring shall be installed in threaded rigid metal conduit or threaded steel intermediate metal conduit or, where buried under not less than 600 mm (2 ft) of cover, shall be permitted in rigid nonmetallic conduit or a listed cable. Where rigid nonmetallic conduit is used, threaded rigid metal conduit or threaded steel intermediate metal conduit shall be used for the last 600 mm (2 ft) of the conduit run to emergence or to the point of connection to the aboveground raceway. Where cable is used, it shall

be enclosed in threaded rigid metal conduit or threaded steel intermediate metal conduit from the point of lowest buried cable level to the point of connection to the aboveground raceway.

See the commentary following 514.8 for more information on underground wiring installed below hazardous (classified) locations.

(B) Insulation. Conductor insulation shall comply with 501.13.

(C) Nonmetallic Wiring. Where rigid nonmetallic conduit or cable with a nonmetallic sheath is used, an equipment grounding conductor shall be included to provide for electrical continuity of the raceway system and for grounding of non–current-carrying metal parts.

515.9 Sealing.

Sealing requirements shall apply to horizontal as well as to vertical boundaries of the defined Class I locations. Buried raceways and cables under defined Class I locations shall be considered to be within a Class I, Division 1 or Zone 1 location.

515.10 Special Equipment—Gasoline Dispensers.

Where gasoline or other volatile flammable liquids or liquefied flammable gases are dispensed at bulk stations, the applicable provisions of Article 514 shall apply.

515.16 Grounding.

All metal raceways, the metal armor or metallic sheath on cables, and all non–current-carrying metal parts of fixed or portable electrical equipment, regardless of voltage, shall be grounded as provided in Article 250. Grounding in Class I locations shall comply with 501.16 for Class I, Division 1 and 2 locations and Article 505.25 for Class I, Zone 0, 1, and 2 locations.

> FPN: For information on grounding for static protection, see 5.6.3.4 and 5.6.3.5 of NFPA 30-2000, *Flammable and Combustible Liquids Code*.

ARTICLE 516
Spray Application, Dipping, and Coating Processes

Contents

> FPN: Rules that are followed by a reference in brackets contain text that has been extracted from NFPA 33-2000, *Standard for Spray Application Using Flammable and Combustible Materials*, or NFPA 34-2000, *Standard for Dipping and Coating Processes Using Flammable or Combustible Liquids*. Only editorial changes were made to the extracted text to make it consistent with this *Code*.

516.1 Scope.

This article covers the regular or frequent application of flammable liquids, combustible liquids, and combustible powders by spray operations and the application of flammable liquids, or combustible liquids at temperatures above their flashpoint, by dipping, coating, or other means.

> FPN: For further information regarding safeguards for these processes, such as fire protection, posting of warning signs, and maintenance, see NFPA 33-2000, *Standard for Spray Application Using Flammable and Combustible Materials*, and NFPA 34-2000, *Standard for Dipping and Coating Processes Using Flammable or Combustible Liquids*. For additional information regarding ventilation, see NFPA 91-1999, *Standard for Exhaust Systems for Air Conveying of Vapors, Gases, Mists, and Noncombustible Particulate Solids*.

516.2 Definitions.

For the purpose of this article, the following definitions shall apply.

Spray Area. Normally locations outside of buildings or localized operations within a larger room or space. Such are normally provided with some local vapor extraction/ventilation system. In automated operations, the area limits shall be the maximum area in the direct path of spray operations. In manual operations, the area limits shall be the maximum area of spray when aimed at 180 degrees to the application surface. [NFPA 33, 1.6, Definitions]

Spray Booth. An enclosure or insert within a larger room used for spray/coating/dipping applications. A spray booth may be fully enclosed or have open front or face and may include separate conveyor entrance and exit. The spray booth is provided with a dedicated ventilation exhaust but may draw supply air from the larger room or have a dedicated air supply. [NFPA 33, 1.6, Definitions]

Spray Room. A purposefully enclosed room built for spray/coating/dipping applications provided with dedicated ventilation supply and exhaust. Normally the room is configured to house the item to be painted, providing reasonable access around the item/process. Depending on the size of the item being painted, such rooms may actually be the entire building or the major portion thereof. [NFPA 33, 1.6, Definitions]

516.3 Classification of Locations.

Classification is based on dangerous quantities of flammable vapors, combustible mists, residues, dusts, or deposits.

(A) Class I or Class II, Division 1 Locations. The following spaces shall be considered Class I or Class II, Division 1 locations, as applicable:

(1) The interior of spray booths and rooms except as specifically provided in 516.3(D).
(2) The interior of exhaust ducts.
(3) Any area in the direct path of spray operations.
(4) For dipping and coating operations, all space within

a 1.5-m (5-ft) radial distance from the vapor sources extending from these surfaces to the floor. The vapor source shall be the liquid exposed in the process and the drainboard, and any dipped or coated object from which it is possible to measure vapor concentrations exceeding 25 percent of the lower flammable limit at a distance of 300 mm (1 ft), in any direction, from the object.

(5) Sumps, pits, or belowgrade channels within 7.5 m (25 ft) horizontally of a vapor source. If the sump, pit, or channel extends beyond 7.5 m (25 ft) from the vapor source, it shall be provided with a vapor stop or it shall be classified as Class I, Division 1 for its entire length.

(6) The interior of any enclosed dipping or coating process or apparatus. [NFPA 33, 1.6 Definitions; NFPA 34, 1.6, Definitions, 4.2.1, 4.2.2, 4.3.1]

(B) Class I or Class II, Division 2 Locations. The following spaces shall be considered Class I or Class II, Division 2 as applicable.

(1) Open Spraying. For open spraying, all space outside of but within 6 m (20 ft) horizontally and 3 m (10 ft) vertically of the Class I, Division 1 location as defined in 516.3(A), and not separated from it by partitions. See Figure 516.3(B)(1).

(2) Closed-Top, Open-Face, and Open-Front Spraying. If spray application operations are conducted within a closed-top, open-face, or open-front booth or room, any electrical wiring or utilization equipment located outside of the booth or room but within the boundaries designated as Division 2 in Figure 516.3(B)(2) shall be suitable for Class I, Division 2 or Class II, Division 2 locations, whichever is applicable. The Class I, Division 2 or Class II, Division 2 locations shown in Figure 516.3(B)(2) shall extend from the edges of the open face or open front of the booth or room in accordance with the following:

(a) If the exhaust ventilation system is interlocked with the spray application equipment, then the Division 2 location shall extend 1.5 m (5 ft) horizontally and 900 mm (3 ft) vertically from the open face or open front of the booth or room, as shown in Figure 516.3(B)(2), top.

(b) If the exhaust ventilation system is not interlocked with the spray application equipment, then the Division 2 location shall extend 3 m (10 ft) horizontally and 900 mm (3 ft) vertically from the open face or open front of the booth or room, as shown in Figure 516.3(B)(2), bottom.

For the purposes of this subsection, *interlocked* shall mean that the spray application equipment cannot be operated unless the exhaust ventilation system is operating and functioning properly and spray application is automatically stopped if the exhaust ventilation system fails.

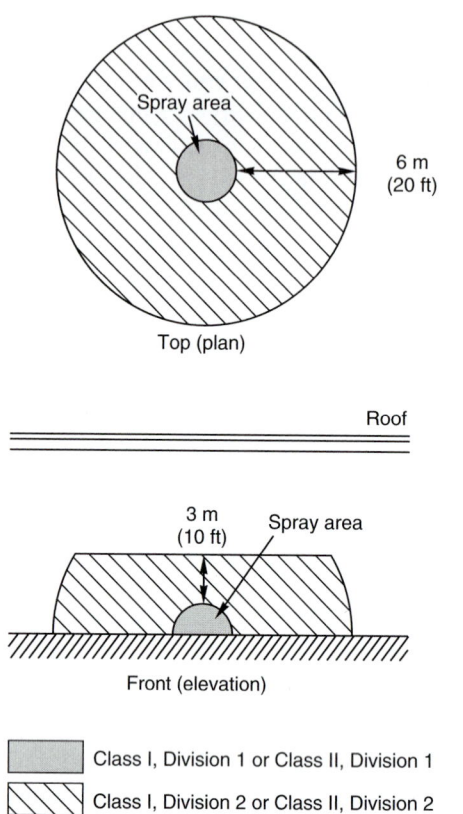

Figure 516.3(B)(1) Electrical area classification for open spray areas. [NFPA 33, Figure 4.3.1]

(3) Open-Top Spraying. For spraying operations conducted within an open top spray booth, the space 900 mm (3 ft) vertically above the booth and within 900 mm (3 ft) of other booth openings shall be considered Class I or Class II, Division 2.

(4) Enclosed Booths and Rooms. For spraying operations confined to an enclosed spray booth or room, the space within 900 mm (3 ft) in all directions from any openings shall be considered Class I or Class II, Division 2 as shown in Figure 516.3(B)(4).

(5) Dip Tanks and Drain Boards—Surrounding Space. For dip tanks and drain boards, the 914-mm (3-ft) space surrounding the Class I, Division 1 location as defined in 516.3(A)(4) and as shown in Figure 516.3(B)(5).

(6) Dip Tanks and Drain Boards—Space Above Floor. For dip tanks and drain boards, the space 900 mm (3 ft) above the floor and extending 6 m (20 ft) horizontally in all directions from the Class I, Division 1 location.

Exception: This space shall not be required to be considered a hazardous (classified) location where the vapor source area is 0.46 m² (5 ft²) or less, and where the contents of

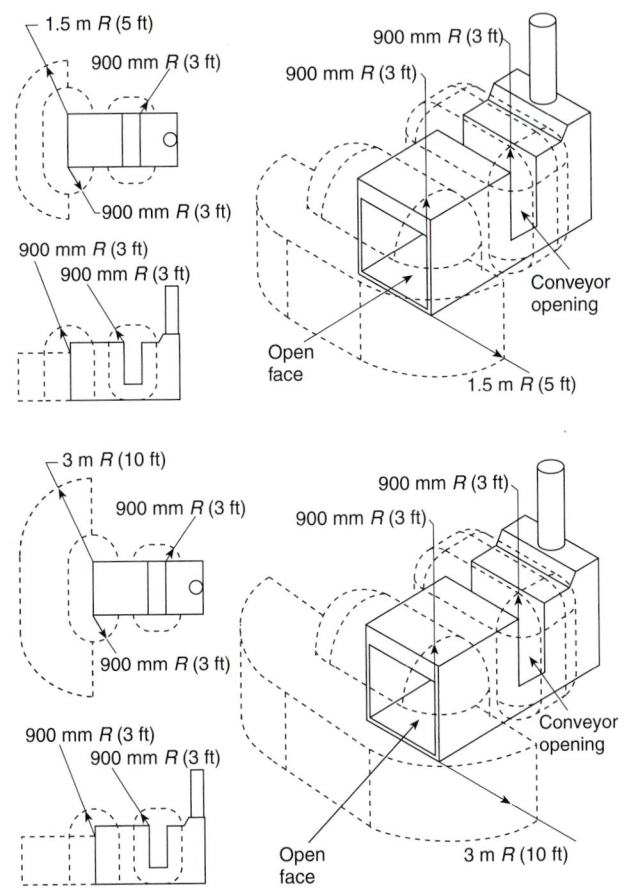

Figure 516.3(B)(2) Class I or Class II, Division 2 locations adjacent to a closed top, open face, or open front spray booth or room. [NFPA 33, Figures 4.3.2(a) and 4.3.2(b)]

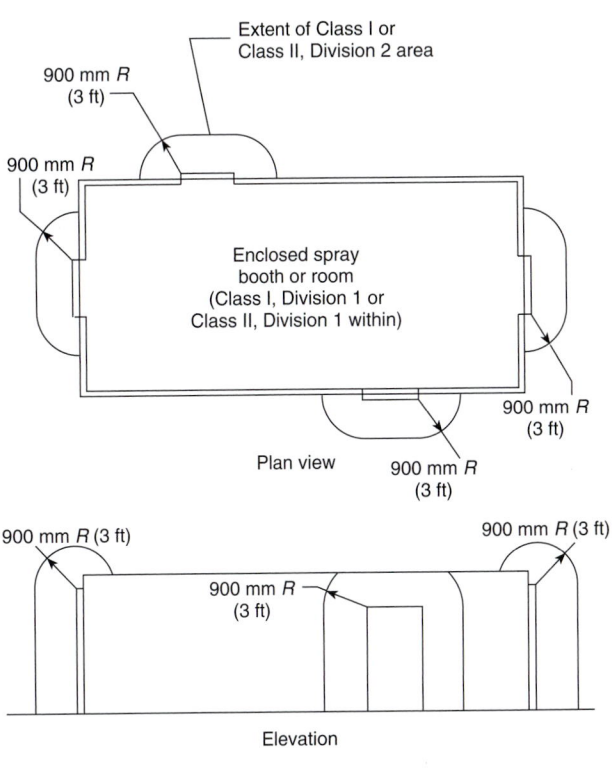

Figure 516.3(B)(4) Class I (or Class II), Division 2 locations adjacent to an enclosed spray booth or spray room. [NFPA 33, Figure 4.3.4]

the open tank trough or container do not exceed 19 L (5 gal). In addition, the vapor concentration during operation and shutdown periods shall not exceed 25 percent of the lower flammable limit outside the Class I location specified in 516.3(A)(4). [NFPA 33, 4.3.1, 4.3.2, 4.3.3, 4.3.4; NFPA 34, 4.2.3, 4.2.4]

(C) Enclosed Coating and Dipping Operations. The space adjacent to an enclosed dipping or coating process or apparatus shall be considered unclassified.

Exception: The space within 900 mm (3 ft) in all directions from any opening in the enclosures shall be classified as Class I, Division 2. [NFPA 34, 4.3.2]

(D) Adjacent Locations. Adjacent locations that are cut off from the defined Class I or Class II locations by tight partitions without communicating openings, and within which flammable vapors or combustible powders are not likely to be released, shall be unclassified.

(E) Unclassified Locations. Locations using drying, curing, or fusion apparatus and provided with positive mechanical ventilation adequate to prevent accumulation of flammable concentrations of vapors, and provided with effective interlocks to de-energize all electrical equipment (other than equipment identified for Class I locations) in case the ventilating equipment is inoperative, shall be permitted to be unclassified where the authority having jurisdiction so judges.

FPN: For further information regarding safeguards, see NFPA 86-1999, *Standard for Ovens and Furnaces.*

516.4 Wiring and Equipment in Class I Locations.

(A) Wiring and Equipment—Vapors. All electric wiring and equipment within the Class I location (containing vapor only—not residues) defined in 516.3 shall comply with the applicable provisions of Article 501.

(B) Wiring and Equipment—Vapors and Residues. Unless specifically listed for locations containing deposits of dangerous quantities of flammable or combustible vapors, mists, residues, dusts, or deposits (as applicable), there shall be no electrical equipment in any spray area as herein defined whereon deposits of combustible residue may readily accu-

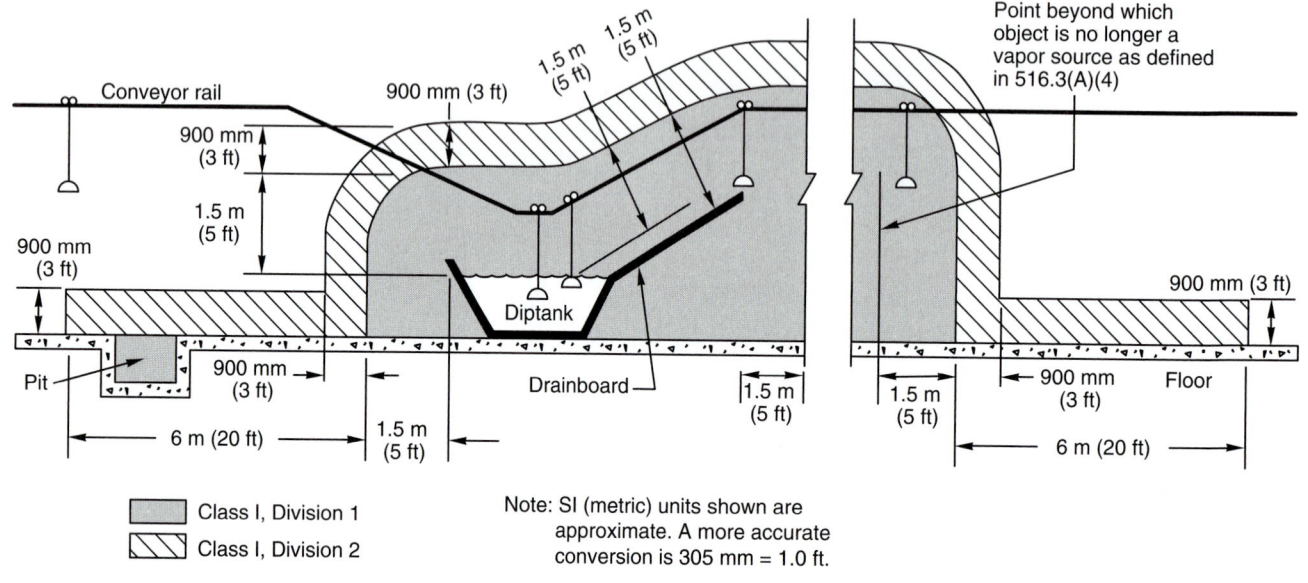

Figure 516.3(B)(5) Electrical area classification for open processes without vapor containment or ventilation. [NFPA 34, Figure 4.2(a)]

mulate, except wiring in rigid metal conduit, intermediate metal conduit, Type MI cable, or in metal boxes or fittings containing no taps, splices, or terminal connections. [NFPA 33, 4.2]

Only electrical equipment that is specifically listed for the location is permitted where deposits of combustible residue may accumulate, as stated in 516.4(B). The latter part of the requirement in 516.4(B), however, permits wiring in rigid metal conduit, intermediate metal conduit, Type MI cable, or metal boxes or fittings without splices, taps, or terminal connections.

(C) Illumination. Illumination of readily ignitible areas through panels of glass or other transparent or translucent material shall be permitted only if it complies with the following:

(1) Fixed lighting units are used as the source of illumination.
(2) The panel effectively isolates the Class I location from the area in which the lighting unit is located.
(3) The lighting unit is identified for its specific location.
(4) The panel is of a material or is protected so that breakage is unlikely.
(5) The arrangement is such that normal accumulations of hazardous residue on the surface of the panel will not be raised to a dangerous temperature by radiation or conduction from the source of illumination.

Underwriters Laboratories Inc.'s *Hazardous Locations Equipment Directory* indicates that listed hazardous location fixtures suitable for use in paint spray booths are evaluated for deposits of readily combustible paint residues on the side of the fixture that forms part of the interior ceiling or wall surface of the spray booth.

(D) Portable Equipment. Portable electric lamps or other utilization equipment shall not be used in a spray area during spray operations.

Exception No. 1: Where portable electric lamps are required for operations in spaces not readily illuminated by fixed lighting within the spraying area, they shall be of the type identified for Class I, Division 1 locations where readily ignitable residues may be present.

Exception No. 2: Where portable electric drying apparatus is used in automobile refinishing spray booths and the following requirements are met.

(a) The apparatus and its electrical connections are not located within the spray enclosure during spray operations.
(b) Electrical equipment within 450 mm (18 in.) of the floor is identified for Class I, Division 2 locations.
(c) All metallic parts of the drying apparatus are electrically bonded and grounded.
(d) Interlocks are provided to prevent the operation of spray equipment while drying apparatus is within the spray enclosure, to allow for a 3-minute purge of the enclosure

before energizing the drying apparatus and to shut off drying apparatus on failure of ventilation system. [NFPA 33, 4.7]

(E) Electrostatic Equipment. Electrostatic spraying or detearing equipment shall be installed and used only as provided in 516.10.

> FPN: For further information, see NFPA 33-2000, *Standard for Spray Application Using Flammable or Combustible Materials.*

516.7 Wiring and Equipment Not Within Class I and II Locations.

(A) Wiring. All fixed wiring above the Class I and II locations shall be in metal raceways, rigid nonmetallic conduit, or electrical nonmetallic tubing, or shall be Type MI, TC, or MC cable. Cellular metal floor raceways shall be permitted only for supplying ceiling outlets or extensions to the area below the floor of a Class I or II location, but such raceways shall have no connections leading into or through the Class I or II location above the floor unless suitable seals are provided.

(B) Equipment. Equipment that may produce arcs, sparks, or particles of hot metal, such as lamps and lampholders for fixed lighting, cutouts, switches, receptacles, motors, or other equipment having make-and-break or sliding contacts, where installed above a Class I or II location or above a location where freshly finished goods are handled, shall be of the totally enclosed type or be constructed so as to prevent the escape of sparks or hot metal particles.

516.10 Special Equipment.

(A) Fixed Electrostatic Equipment. This section shall apply to any equipment using electrostatically charged elements for the atomization, charging, and/or precipitation of hazardous materials for coatings on articles or for other similar purposes in which the charging or atomizing device is attached to a mechanical support or manipulator. This shall include robotic devices. This section shall not apply to devices that are held or manipulated by hand. Where robot or programming procedures involve manual manipulation of the robot arm while spraying with the high voltage on, the provisions of 516.10(B) shall apply. The installation of electrostatic spraying equipment shall comply with 516.10(A)(1) through (A)(10). Spray equipment shall be listed. All automatic electrostatic equipment systems shall comply with 516.4(A)(1) through (A)(9).

(1) Power and Control Equipment. Transformers, high-voltage supplies, control apparatus, and all other electric portions of the equipment shall be installed outside of the Class I location as defined in 516.3 or be of a type identified for the location.

Exception: High-voltage grids, electrodes, electrostatic atomizing heads, and their connections shall be permitted within the Class I location.

(2) Electrostatic Equipment. Electrodes and electrostatic atomizing heads shall be adequately supported in permanent locations and shall be effectively insulated from ground. Electrodes and electrostatic atomizing heads that are permanently attached to their bases, supports, reciprocators, or robots shall be deemed to comply with this section.

(3) High-Voltage Leads. High-voltage leads shall be properly insulated and protected from mechanical damage or exposure to destructive chemicals. Any exposed element at high voltage shall be effectively and permanently supported on suitable insulators and shall be effectively guarded against accidental contact or grounding.

(4) Support of Goods. Goods being coated using this process shall be supported on conveyors or hangers. The conveyors or hangers shall be arranged (1) to ensure that the parts being coated are electrically connected to ground with a resistance of 1 megohm or less and (2) to prevent parts from swinging.

(5) Automatic Controls. Electrostatic apparatus shall be equipped with automatic means that will rapidly de-energize the high-voltage elements under any of the following conditions:

(1) Stoppage of ventilating fans or failure of ventilating equipment from any cause
(2) Stoppage of the conveyor carrying goods through the high-voltage field unless stoppage is required by the spray process
(3) Occurrence of excessive current leakage at any point in the high-voltage system
(4) De-energizing the primary voltage input to the power supply

(6) Grounding. All electrically conductive objects in the spray area, except those objects required by the process to be at high voltage, shall be adequately grounded. This requirement shall apply to paint containers, wash cans, guards, hose connectors, brackets, and any other electrically conductive objects or devices in the area.

(7) Isolation. Safeguards such as adequate booths, fencing, railings, interlocks, or other means shall be placed about the equipment or incorporated therein so that they, either by their location, character, or both, ensure that a safe separation of the process is maintained.

(8) Signs. Signs shall be conspicuously posted to convey the following:

(1) Designate the process zone as dangerous with regard to fire and accident

(2) Identify the grounding requirements for all electrically conductive objects in the spray area

(3) Restrict access to qualified personnel only

(9) Insulators. All insulators shall be kept clean and dry.

(10) Other Than Nonincendive Equipment. Spray equipment that cannot be classified as nonincendive shall comply with (a) and (b).

(a) Conveyors or hangers shall be arranged so as to maintain a safe distance of at least twice the sparking distance between goods being painted and electrodes, electrostatic atomizing heads, or charged conductors. Warnings defining this safe distance shall be posted.

(b) The equipment shall provide an automatic means of rapidly de-energizing the high-voltage elements in the event the distance between the goods being painted and the electrodes or electrostatic atomizing heads falls below that specified in (a). [NFPA 33, Chapter 9]

(B) Electrostatic Hand-Spraying Equipment. This section shall apply to any equipment using electrostatically charged elements for the atomization, charging, and/or precipitation of materials for coatings on articles, or for other similar purposes in which the atomizing device is hand-held or manipulated during the spraying operation. Electrostatic hand-spraying equipment and devices used in connection with paint-spraying operations shall be of listed types and shall comply with 516.10(B)(1) through (B)(5).

(1) General. The high-voltage circuits shall be designed so as not to produce a spark of sufficient intensity to ignite the most readily ignitible of those vapor–air mixtures likely to be encountered, or result in appreciable shock hazard upon coming in contact with a grounded object under all normal operating conditions. The electrostatically charged exposed elements of the handgun shall be capable of being energized only by an actuator that also controls the coating material supply.

(2) Power Equipment. Transformers, power packs, control apparatus, and all other electric portions of the equipment shall be located outside of the Class I location or be identified for the location.

Exception: The handgun itself and its connections to the power supply shall be permitted within the Class I location.

(3) Handle. The handle of the spraying gun shall be electrically connected to ground by a metallic connection and be constructed so that the operator in normal operating position is in intimate electrical contact with the grounded handle to prevent buildup of a static charge on the operator's body.

Signs indicating the necessity for grounding other persons entering the spray area shall be conspicuously posted.

(4) Electrostatic Equipment. All electrically conductive objects in the spraying area shall be adequately grounded. This requirement shall apply to paint containers, wash cans, and any other electrical conductive objects or devices in the area. The equipment shall carry a prominent, permanently installed warning regarding the necessity for this grounding feature.

(5) Support of Objects. Objects being painted shall be maintained in metallic contact with the conveyor or other grounded support. Hooks shall be regularly cleaned to ensure adequate grounding of 1 megohm or less. Areas of contact shall be sharp points or knife edges where possible. Points of support of the object shall be concealed from random spray where feasible; and, where the objects being sprayed are supported from a conveyor, the point of attachment to the conveyor shall be located so as to not collect spray material during normal operation. [NFPA 33, Chapter 10]

(C) Powder Coating. This section shall apply to processes in which combustible dry powders are applied. The hazards associated with combustible dusts are present in such a process to a degree, depending on the chemical composition of the material, particle size, shape, and distribution.

(1) Electric Equipment and Sources of Ignition. Electric equipment and other sources of ignition shall comply with the requirements of Article 502. Portable electric lamps and other utilization equipment shall not be used within a Class II location during operation of the finishing processes. Where such lamps or utilization equipment are used during cleaning or repairing operations, they shall be of a type identified for Class II, Division 1 locations, and all exposed metal parts shall be effectively grounded.

Exception: Where portable electric lamps are required for operations in spaces not readily illuminated by fixed lighting within the spraying area, they shall be of the type listed for Class II, Division 1 locations where readily ignitible residues may be present.

(2) Fixed Electrostatic Spraying Equipment. The provisions of 516.10(A) and 516.10(C)(1) shall apply to fixed electrostatic spraying equipment.

(3) Electrostatic Hand-Spraying Equipment. The provisions of 516.10(B) and 516.10(C)(1) shall apply to electrostatic hand-spraying equipment.

(4) Electrostatic Fluidized Beds. Electrostatic fluidized beds and associated equipment shall be of identified types. The high-voltage circuits shall be designed so that any discharge produced when the charging electrodes of the bed are approached or contacted by a grounded object shall not be of sufficient intensity to ignite any powder–air mixture likely to be encountered or to result in an appreciable shock hazard.

(a) Transformers, power packs, control apparatus, and all other electric portions of the equipment shall be located outside the powder-coating area or shall otherwise comply with the requirements of 516.10(C)(1).

Exception: The charging electrodes and their connections to the power supply shall be permitted within the powder-coating area.

(b) All electrically conductive objects within the powder-coating area shall be adequately grounded. The powder-coating equipment shall carry a prominent, permanently installed warning regarding the necessity for grounding these objects.

(c) Objects being coated shall be maintained in electrical contact (less than 1 megohm) with the conveyor or other support in order to ensure proper grounding. Hangers shall be regularly cleaned to ensure effective electrical contact. Areas of electrical contact shall be sharp points or knife edges where possible.

(d) The electric equipment and compressed air supplies shall be interlocked with a ventilation system so that the equipment cannot be operated unless the ventilating fans are in operation. [NFPA 33, Chapter 13]

516.16 Grounding.

All metal raceways, the metal armors or metallic sheath on cables, and all non–current-carrying metal parts of fixed or portable electrical equipment, regardless of voltage, shall be grounded as provided in Article 250. Grounding in Class I and Class II locations shall comply with 501.16 and 502.16, respectively.

NFPA 33, *Standard for Spray Application Using Flammable or Combustible Materials,* covers the spray application of flammable or combustible materials by means of compressed air atomization, airless or hydraulic atomization, or electrostatic application methods, or any other means in continuous or intermittent processes. NFPA 33 also covers the application of combustible powders applied by powder spray guns, electrostatic powder spray guns, and fluidized bed application method or electrostatic fluidized bed application method. NFPA 33 also contains requirements for the maintenance of safe conditions as well as personal safety.

The proper maintenance and operation of processes and process areas where flammable and combustible materials are handled and applied is critical with respect to the protec-

tion of life and property from fire or explosion. An analysis of actual experience in industry demonstrates that the largest fire losses and frequency of fires have occurred where the proper codes and standards have not been used or applied properly.

Notes on Electrical Installations. The safety of life and property from fire or explosion as a result of spray applications of flammable and combustible materials, such as paints, finishes, and adhesives, depends on the arrangement and operation of a particular installation. The principal hazards of spray application operations originate from flammable or combustible liquids or powders and their vapors or mists, as well as from highly combustible residues or powders.

Properly constructed spray booths, with adequate mechanical ventilation, may be used to discharge vapors or powder to a safe location and reduce the possibility of an explosion. In like manner, the accumulation of overspray residues, many of which are not only highly combustible but also subject to spontaneous ignition, can be controlled.

The elimination of all sources of ignition in those areas where flammable or combustible liquids, vapors, mists, or combustible residues are present, together with constant supervision and maintenance, is essential to the safe operation of spraying.

The human element necessitates careful consideration of the location of the operation and the installation of extinguishing equipment to reduce the possibility of fire spreading to other property and to minimize the probability of damage to other property.

No open flames or spark-producing equipment should be in any area where, because of inadequate ventilation, explosive vapor–air mixtures or mists are present. Equally important, no open flames or spark-producing equipment should be located where highly combustible spray residues will be deposited on them. Because some residues may be ignited at very low temperatures, additional consideration must be given to operating temperatures of equipment subject to residue deposits. Many deposits may be ignited at temperatures produced by low-pressure steam pipes or incandescent light globes, even fixtures of the explosionproof type.

It should be noted that electrical equipment is generally not permitted inside any spray booth, in the exhaust duct from a spray booth, in the entrained air of an exhaust system from a spraying operation, or in the direct path of spray, unless such equipment is specifically listed for both readily ignitible deposits and flammable vapor.

The determination of the extent of hazardous areas involved in spray application requires an understanding of the multiple hazards of flammable vapors, mists, powders, and highly combustible deposits applied at each individual location.

Where electrical equipment is installed in locations not subject to deposits of combustible residues but, due to inadequate ventilation, is subject to explosive concentrations of flammable vapors or mists, only approved explosionproof or other types of equipment approved for Class I, Division 1 locations (for example, purged, pressurized, or intrinsically safe equipment or systems) are permitted.

Where spray areas contain dangerous quantities of flammable or combustible vapors, mists, residues, dusts, or deposits under normal operation, the adjacent unpartitioned areas, which are safe under normal operating conditions but that may become dangerous due to accident or careless operation, should be given consideration. Equipment known to produce sparks or flames under normal operating conditions should not be installed in these adjacent unpartitioned areas.

Where spraying operations are confined to adequately ventilated spray booths or rooms, there should be no deposits of combustible residues or dangerous concentrations of flammable vapors, mists, or dusts outside the spray booth under normal operating conditions. In the interest of safety, however, unless separated by partitions, an area within a certain distance [see 516.3(B)] of the Class I (or Class II), Division 1 spraying area, depending on the arrangement, is classified as Division 2; that is, the area should contain no equipment that produces ignition-capable sparks under normal operation. Furthermore, within this distance, electric lamps must be enclosed to prevent hot particles from falling on freshly painted stock or other readily ignitible material and, if subject to physical damage, must be properly guarded. See 516.7(B).

Even though it is contemplated that areas adjacent to spray booths (particularly where coating-material stocks are located) will be provided with ventilation sufficient to prevent the presence of flammable vapors or deposits, it is nevertheless advisable that electric lamps be totally enclosed to prevent hot particles from falling in any area where there may be freshly painted stock, accidentally spilled flammable or combustible materials, readily ignitible refuse, or flammable or combustible liquid containers that have been left open accidentally. See 516.7(B).

Where electric lamps are in areas subject to atmospheres of flammable vapor, lamps should be replaced while electricity is off; otherwise there may be a spark from this source.

Sufficient lighting for coating operations, booth cleaning, and booth repair work should be provided at the time the equipment is installed in order to avoid the use of temporary or emergency lamps connected to ordinary extension cords in this area. See 516.4(D). A satisfactory and practical method of lighting is the use of ¼-in.-thick wired or tempered glass panels in the top or sides of the spray booth, with electrical luminaires located outside the booth, avoiding the direct path of the spray. See 516.4(C).

To prevent sparks from the accumulation of static elec-

tricity, all electrically conductive objects, including metal parts of spray booths, exhaust ducts, piping systems conveying flammable or combustible liquids or paint, solvent tanks, and canisters, should be properly grounded. See Section 4-5 of NFPA 33 and Section 9.3 of NFPA 77, *Recommended Practice on Static Electricity.*

ARTICLE 517
Health Care Facilities

Contents

FPN: Rules that are followed by a reference in brackets contain text that has been extracted from NFPA 99-1999, *Standard for Health Care Facilities*. Only editorial changes were made to the extracted text to make it consistent with this *Code*.

I. General

517.1 Scope.

The provisions of this article shall apply to electrical construction and installation criteria in health care facilities that provide services to human beings.

The requirements in Parts II and III not only apply to single-function buildings but are also intended to be individually applied to their respective forms of occupancy within a multifunction building (e.g., a doctor's examining room located within a limited care facility would be required to meet the provisions of 517.10).

FPN: For information concerning performance, maintenance, and testing criteria, refer to the appropriate health care facilities documents.

The requirements of Article 517 are intended to apply to all types of health care facilities. The requirements for each type of health care facility are nevertheless intended to be applied in a very specific manner. For example, in a suite of doctors' offices within an office building, a doctor's business office would be treated as an ordinary occupancy and would be required to meet only the applicable portions of other parts of this *Code*. However, the examining rooms attached to the doctor's business office would be required to meet the provisions of Part II and 517.45 in Article 517.

The scope of Article 517 also includes health care facili-

ties that may be mobile or supply very limited outpatient services. However, the scope does not include animal hospitals or veterinary offices.

Other standards referenced in Article 517 are NFPA 99, *Standard for HealthCare Facilities;* NFPA *101®, Life Safety Code®;* and NFPA 20, *Standard for the Installation of Stationary Pumps for Fire Protection.*

517.2 Definitions.

Alternate Power Source. One or more generator sets, or battery systems where permitted, intended to provide power during the interruption of the normal electrical services or the public utility electrical service intended to provide power during interruption of service normally provided by the generating facilities on the premises.

As stated in 517.30(B)(6), alternate power sources are permitted to serve essential electrical systems of contiguous or multiple building facilities.

Ambulatory Health Care Facility. A building or part thereof used to provide services or treatment to four or more patients at the same time and meeting either (1) or (2).

Ambulatory health care facilities, such as outpatient surgery centers, freestanding emergency medical centers, and hemodialysis units, are subject to the requirements of Part II and 517.45. The definition of *ambulatory health care facility* in Article 517 correlates with the definition of the same term in NFPA 99, *Standard for Health Care Facilities.*

(1) Those facilities that provide, on an outpatient basis, treatment for patients that would render them incapable of taking action for self-preservation under emergency conditions without assistance from others, such as hemodialysis units or freestanding emergency medical units.
(2) Those facilities that provide, on an outpatient basis, surgical treatment requiring general anesthesia.

Anesthetizing Location. Any area of a facility that has been designated to be used for the administration of any flammable or nonflammable inhalation anesthetic agent in the course of examination or treatment, including the use of such agents for relative analgesia.

The definition of *anesthetizing location* recognizes that in an emergency it may be necessary to administer an anesthetic almost anywhere in a health care facility. However, only those areas in a health care facility that are set aside specifically for the induction of anesthetics are required to meet the provisions of Part IV of Article 517. This definition and the provisions of Part IV are not intended to apply to the administering of analgesic or local anesthetics, such as might be used in minor medical or dental procedures.

The definition of *anesthetizing location* applies to health care facilities where inhalation anesthetics are used for *relative analgesia*. The term *relative analgesia* (see definition of this term in 517.2) is sometimes referred to as "conscious sedation" and is a state of sedation in which the perception of pain is partially blocked and the patient does not lose consciousness. Oral surgeons often use this form of anesthesia. For guidance on flammable anesthetizing locations, see Annex 2 of NFPA 99, *Standard for Health Care Facilities*.

Critical Branch. A subsystem of the emergency system consisting of feeders and branch circuits supplying energy to task illumination, special power circuits, and selected receptacles serving areas and functions related to patient care, and which are connected to alternate power sources by one or more transfer switches during interruption of the normal power source.

Electrical Life-Support Equipment. Electrically powered equipment whose continuous operation is necessary to maintain a patient's life.

Emergency System. A system of circuits and equipment intended to supply alternate power to a limited number of prescribed functions vital to the protection of life and safety.

The requirements for sizing a generator and a feeder capacity for the essential electrical system in a health care facility are found in 517.30(D). Emergency systems in other occupancies are installed primarily for life safety. The emergency systems in hospitals are for life safety systems as well as critical patient care systems.

Equipment System. A system of circuits and equipment arranged for delayed, automatic, or manual connection to the alternate power source and that serves primarily 3-phase power equipment.

Essential Electrical System. A system comprised of alternate sources of power and all connected distribution systems and ancillary equipment, designed to ensure continuity of electrical power to designated areas and functions of a health care facility during disruption of normal power sources, and also designed to minimize disruption within the internal wiring system.

Exposed Conductive Surfaces. Those surfaces that are capable of carrying electric current and that are unprotected, unenclosed, or unguarded, permitting personal contact.

Paint, anodizing, and similar coatings are not considered suitable insulation, unless they are listed for such use.

Fault Hazard Current. *See* Hazard Current.

Flammable Anesthetics. Gases or vapors, such as fluroxene, cyclopropane, divinyl ether, ethyl chloride, ethyl ether, and ethylene, which may form flammable or explosive mixtures with air, oxygen, or reducing gases such as nitrous oxide.

Flammable Anesthetizing Location. Any area of the facility that has been designated to be used for the administration of any flammable inhalation anesthetic agents in the normal course of examination or treatment.

Hazard Current. For a given set of connections in an isolated power system, the total current that would flow through a low impedance if it were connected between either isolated conductor and ground.

Fault Hazard Current. The hazard current of a given isolated system with all devices connected except the line isolation monitor.

Monitor Hazard Current. The hazard current of the line isolation monitor alone.

Total Hazard Current. The hazard current of a given isolated system with all devices, including the line isolation monitor, connected.

Health Care Facilities. Buildings or portions of buildings in which medical, dental, psychiatric, nursing, obstetrical, or surgical care are provided. Health care facilities include, but are not limited to, hospitals, nursing homes, limited care facilities, clinics, medical and dental offices, and ambulatory care centers, whether permanent or movable.

Hospital. A building or part thereof used for the medical, psychiatric, obstetrical, or surgical care, on a 24-hour basis, of four or more inpatients. *Hospital*, wherever used in this *Code*, shall include general hospitals, mental hospitals, tuberculosis hospitals, children's hospitals, and any such facilities providing inpatient care.

Isolated Power System. A system comprising an isolating transformer or its equivalent, a line isolation monitor, and its ungrounded circuit conductors.

Isolation Transformer. A transformer of the multiple-winding type, with the primary and secondary windings physically separated, which inductively couples its secondary winding to the grounded feeder systems that energize its primary winding.

Life Safety Branch. A subsystem of the emergency system consisting of feeders and branch circuits, meeting the requirements of Article 700 and intended to provide adequate power needs to ensure safety to patients and personnel, and which are automatically connected to alternate power sources during interruption of the normal power source.

Limited Care Facility. A building or part thereof used on a 24-hour basis for the housing of four or more persons who are incapable of self-preservation because of age, physical limitation due to accident or illness, or mental limitations, such as mental retardation/developmental disability, mental illness, or chemical dependency.

Line Isolation Monitor. A test instrument designed to continually check the balanced and unbalanced impedance from each line of an isolated circuit to ground and equipped with a built-in test circuit to exercise the alarm without adding to the leakage current hazard.

Monitor Hazard Current. *See* Hazard Current.

Nurses' Stations. Areas intended to provide a center of nursing activity for a group of nurses serving bed patients, where the patient calls are received, nurses are dispatched, nurses' notes written, inpatient charts prepared, and medications prepared for distribution to patients. Where such activities are carried on in more than one location within a nursing unit, all such separate areas are considered a part of the nurses' station.

Nursing Home. A building or part thereof used for the lodging, boarding, and nursing care, on a 24-hour basis, of four or more persons who, because of mental or physical incapacity, may be unable to provide for their own needs and safety without the assistance of another person. *Nursing home,* wherever used in this *Code,* shall include nursing and convalescent homes, skilled nursing facilities, intermediate care facilities, and infirmaries of homes for the aged.

Patient Bed Location. The location of an inpatient sleeping bed; or the bed or procedure table used in a critical patient care area.

Patient Care Area. Any portion of a health care facility wherein patients are intended to be examined or treated. Areas of a health care facility in which patient care is administered are classified as general care areas or critical care areas, either of which may be classified as a wet location. The governing body of the facility designates these areas in accordance with the type of patient care anticipated and with the following definitions of the area classification.

> FPN: Business offices, corridors, lounges, day rooms, dining rooms, or similar areas typically are not classified as patient care areas.

General Care Areas. Patient bedrooms, examining rooms, treatment rooms, clinics, and similar areas in which it is intended that the patient will come in contact with ordinary appliances such as a nurse call system, electrical beds, examining lamps, telephone, and entertainment devices. In such areas, it may also be intended that patients be connected to

electromedical devices (such as heating pads, electrocardiographs, drainage pumps, monitors, otoscopes, ophthalmoscopes, intravenous lines, etc.).

Critical Care Areas. Those special care units, intensive care units, coronary care units, angiography laboratories, cardiac catheterization laboratories, delivery rooms, operating rooms, and similar areas in which patients are intended to be subjected to invasive procedures and connected to line-operated, electromedical devices.

Wet Locations. Those patient care areas that are normally subject to wet conditions while patients are present. These include standing fluids on the floor or drenching of the work area, either of which condition is intimate to the patient or staff. Routine housekeeping procedures and incidental spillage of liquids do not define a wet location.

> The definition of *patient care area* applies to hospitals as well as patient care areas in outpatient facilities. A patient bed location in a nursing home can be considered a patient care area if a person is examined or treated in that location. However, it excludes such areas as laundry rooms, boiler rooms, and utility areas, which, although routinely wet, are not patient care areas. The governing body of the health care facility may elect to include such areas as hydrotherapy areas, dialysis laboratories, and certain wet laboratories under this definition. Lavatories or bathrooms within a health care facility are not intended to be classified as wet locations. For infection control purposes, many patient and treatment areas have a sink for hand washing, which is not intended to be a wet location either.

Patient Equipment Grounding Point. A jack or terminal bus that serves as the collection point for redundant grounding of electric appliances serving a patient vicinity or for grounding other items in order to eliminate electromagnetic interference problems.

Patient Vicinity. In an area in which patients are normally cared for, the *patient vicinity* is the space with surfaces likely to be contacted by the patient or an attendant who can touch the patient. Typically in a patient room, this encloses a space within the room not less than 1.8 m (6 ft) beyond the perimeter of the bed in its nominal location, and extending vertically not less than 2.3 m (7½ ft) above the floor.

> The patient vicinity area is limited to patient beds in their normal position, that is, the position of the bed as called for in the architect's plans, rather than the temporary position of the bed subject to movement by housekeeping staff or the convenience of the medical staff.

Psychiatric Hospital. A building used exclusively for the psychiatric care, on a 24-hour basis, of four or more inpatients.

Reference Grounding Point. The ground bus of the panelboard or isolated power system panel supplying the patient care area.

Relative Analgesia. A state of sedation and partial block of pain perception produced in a patient by the inhalation of concentrations of nitrous oxide insufficient to produce loss of consciousness (conscious sedation).

Selected Receptacles. A minimum number of electric receptacles to accommodate appliances ordinarily required for local tasks or likely to be used in patient care emergencies.

Task Illumination. Provision for the minimum lighting required to carry out necessary tasks in the described areas, including safe access to supplies and equipment, and access to exits.

Therapeutic High Frequency Diathermy Equipment. Therapeutic high-frequency diathermy equipment is therapeutic induction and dielectric heating equipment.

Total Hazard Current. *See* Hazard Current.

X-Ray Installations, Long-Time Rating. A rating based on an operating interval of 5 minutes or longer.

X-Ray Installations, Mobile. X-ray equipment mounted on a permanent base with wheels, casters, or a combination of both to facilitate moving the equipment while completely assembled.

X-Ray Installations, Momentary Rating. A rating based on an operating interval that does not exceed 5 seconds.

X-Ray Installations, Portable. X-ray equipment designed to be hand carried.

X-Ray Installations, Transportable. X-ray equipment to be installed in a vehicle or that may be readily disassembled for transport in a vehicle.

II. Wiring and Protection

Formal Interpretation 99-1
Reference: Article 517, Part II

Question: Does Part II of Article 517 of the *NEC* apply to patient sleeping rooms of nursing homes or limited care

facilities where patient care activities do not involve the use of electrical or electronic life support systems; or invasive procedures where patients are electrically connected to line connected electromedical devices?

Answer: No

Issue Edition: 1999

Reference: Article 517

Issue Date: August 1, 2000

Effective Date: August 21, 2000

517.10 Applicability.

(A) Part II shall apply to patient care areas of all health care facilities.

(B) Part II shall not apply to the following:

(1) Business offices, corridors, waiting rooms, and the like in clinics, medical and dental offices, and outpatient facilities

(2) Areas of nursing homes and limited care facilities wired in accordance with Chapters 1 through 4 of this *Code* where these areas are used exclusively as patient sleeping rooms

FPN: See NFPA *101®-2000, Life Safety Code®.*

In accordance with the conditions specified in 517.10(B)(2), areas in nursing homes that are designated as patient sleeping rooms are not considered to be patient care areas. This is often the case in nursing homes where the resident requires assistance to attend to their personal needs and safety, but does not require special medical care. However, areas of nursing homes in which it is intended that patients be examined or treated can be considered as patient bedrooms or examining rooms and are classified in the broader category as patient care areas. See the definition of *patient care area* in 517.2.

See the Formal Interpretation on Article 517, Part II applicability.

517.11 General Installation—Construction Criteria.

It is the purpose of this article to specify the installation criteria and wiring methods that minimize electrical hazards by the maintenance of adequately low potential differences only between exposed conductive surfaces that are likely to become energized and could be contacted by a patient.

FPN: In a health care facility, it is difficult to prevent the occurrence of a conductive or capacitive path from the patient's body to some grounded object, because that path may be established accidentally or through instrumentation directly connected to the patient. Other electrically conductive surfaces that may make an additional contact with the patient, or instruments that may be connected to the patient, then become possible sources of electric currents that can traverse the patient's body. The hazard is increased as more apparatus is associated with the patient, and, therefore, more intensive precautions are needed. Control of electric shock hazard requires the limitation of electric current that might flow in an electric circuit involving the patient's body by raising the resistance of the conductive circuit that includes the patient, or by insulating exposed surfaces that might become energized, in addition to reducing the potential difference that can appear between exposed conductive surfaces in the patient vicinity, or by combinations of these methods. A special problem is presented by the patient with an externalized direct conductive path to the heart muscle. The patient may be electrocuted at current levels so low that additional protection in the design of appliances, insulation of the catheter, and control of medical practice is required.

This fine print note recognizes the possibility of increased sensitivity to electric shock by patients whose body resistance may be compromised either accidentally or by a necessary medical procedure. For example, incontinence or the insertion of a catheter may render a patient much more vulnerable to the effects of an electric current. Therefore, it is essential that those responsible for the design, installation, and maintenance of the electrical system in patient care areas be well acquainted with at least the rudiments of the hazard as explained in this note.

Since the original recognition of this hazard in the 1971 *Code*, continued clinical evaluation of the problem has provided a better understanding of the extent of the hazard, bringing about the changes in both value and wiring methods now found in the *Code*.

The *Code* assigns responsibility for the designation of the types of patient care areas to the governing body of the health care facility. Both the design and inspection of a patient care area must, therefore, be based on the governing body's designation rather than the superficial appearance of the area.

517.12 Wiring Methods.

Except as modified in this article, wiring methods shall comply with the applicable requirements of Chapters 1 through 4 of this *Code*.

517.13 Grounding of Receptacles and Fixed Electric Equipment in Patient Care Areas.

Wiring in patient care areas shall comply with 517.13(A) and (B).

(A) Wiring Methods. All branch circuits serving patient care areas shall be provided with a ground path for fault

current by installation in a metal raceway system, or a cable armor or sheath assembly. The metal raceway system, or cable armor, or sheath assembly, shall itself qualify as an equipment grounding return path in accordance with 250.118. Type AC, Type MC, Type MI cables shall have an outer metal armor or sheath that is identified as an acceptable grounding return path.

The wiring method provisions of 517.13(A) apply to the branch circuits in areas used for patient care and are not limited to patient rooms. Additional areas, such as examining rooms, therapy areas, recreational areas, solaria, and certain patient corridors, are also included. The branch circuit wiring method used in these areas is one component of a two-part redundant grounding scheme unique to patient care areas. The second component in this approach is the separate insulated copper conductor required by 517.13(B). Therefore, the metal raceway or metal cable armor or sheath must qualify as an equipment grounding conductor in accordance with 250.118, independent of any separate wire-type equipment grounding conductor. Metal-sheathed cable assemblies are not permitted for emergency circuits in the patient vicinity because 517.30(C)(3) requires such wiring to be protected by installation in metal raceways.

(B) Insulated Equipment Grounding Conductor. In an area used for patient care, the grounding terminals of all receptacles and all non–current-carrying conductive surfaces of fixed electric equipment likely to become energized that are subject to personal contact, operating at over 100 volts, shall be grounded by an insulated copper conductor. The grounding conductor shall be sized in accordance with Table 250.122 and installed in metal raceways or metal-clad cables with the branch-circuit conductors supplying these receptacles or fixed equipment.

Exception No. 1: Metal faceplates shall be permitted to be grounded by means of a metal mounting screw(s) securing the faceplate to a grounded outlet box or grounded wiring device.

Exception No. 2: Luminaires (light fixtures) more than 2.3 m (7½ ft) above the floor and switches located outside of the patient vicinity shall not be required to be grounded by an insulated equipment grounding conductor.

The requirements in 517.13(B) cover the second component of the redundant grounding approach. An insulated, copper, equipment grounding conductor sized in accordance with 250.122 must be installed with the branch-circuit conductors in the wiring method that meets the provisions of 517.13(A). The conductor can be either solid or stranded. It is not required to run a separate, insulated equipment grounding conductor to the branch-circuit panelboard where the feeder wiring method is recognized as an equipment grounding conductor per 250.18, or the feeder wiring method contains an equipment grounding conductor.

The grounding requirements of 517.13 for patient care areas are not limited to hospitals. They are also required for patient care areas in other health care facilities, such as nursing homes, clinics, medical and dental offices, and so on.

Exception No. 1 to 517.13(B) permits metal faceplates to be grounded by means of the metal mounting screws rather than by having a separate equipment grounding conductor or bonding jumper run to the metal plate. See 404.9(B), which requires switches and their metal faceplates to be effectively grounded.

Exception No. 2 to 517.13(B) exempts luminaires mounted 7½ ft above the floor and switches located outside of the patient vicinity from redundant grounding requirements because it is unlikely that the patient or attendants will contact these items and the patient at the same time. The patient vicinity space consists of a volume 6 ft horizontally in all directions from the bed and up to a height of 7½ ft above the floor.

517.14 Panelboard Bonding.

The equipment grounding terminal buses of the normal and essential branch-circuit panelboards serving the same individual patient vicinity shall be bonded together with an insulated continuous copper conductor not smaller than 10 AWG. Where more than two panels serve the same location, this conductor shall be continuous from panel to panel but shall be permitted to be broken in order to terminate on the ground bus in each panel.

517.16 Receptacles with Insulated Grounding Terminals.

Receptacles with insulated grounding terminals, as permitted in 250.146(D), shall be identified; such identification shall be visible after installation.

FPN: Caution is important in specifying such a system with receptacles having insulated grounding terminals, since the grounding impedance is controlled only by the grounding conductors and does not benefit functionally from any parallel grounding paths.

The requirement in 517.16 prevents indiscriminate use of isolated ground receptacles. Isolated ground receptacles are required to be identified by an orange triangle located on the face of the receptacle. Older isolated ground-type receptacles may be identified differently.

517.17 Ground-Fault Protection.

(A) Feeders. Where ground-fault protection is provided for operation of the service disconnecting means or feeder disconnecting means as specified by 230.95 or 215.10, an additional step of ground-fault protection shall be provided in the next level of feeder disconnecting means downstream toward the load. Such protection shall consist of overcurrent devices and current transformers or other equivalent protective equipment that shall cause the feeder disconnecting means to open.

The additional levels of ground-fault protection shall not be installed as follows:

(1) On the load side of an essential electrical system transfer switch
(2) Between the on-site generating unit(s) described in 517.35(B) and the essential electrical system transfer switch(es)
(3) On electrical systems that are not solidly grounded wye systems with greater than 150 volts to ground but not exceeding 600 volts phase-to-phase

(B) Selectivity. Ground-fault protection for operation of the service and feeder disconnecting means shall be fully selective such that the feeder device and not the service device shall open on ground faults on the load side of the feeder device. A six-cycle minimum separation between the service and feeder ground-fault tripping bands shall be provided. Operating time of the disconnecting devices shall be considered in selecting the time spread between these two bands to achieve 100 percent selectivity.

FPN: See 230.95, fine print note, for transfer of alternate source where ground-fault protection is applied.

Wherever ground-fault protection (GFP) of equipment is applied to the service providing power to a health care facility, whether by design or by reason of the requirements of 215.10 or 230.95, an additional level of ground-fault protection is required downstream. Under this rule, ground-fault protection must be applied to every feeder, and additional ground-fault protective devices may be applied farther downstream at the option of the governing body of the health care facility. With proper coordination, this additional ground-fault protection is intended to limit a ground fault to a single feeder and thereby prevent a total outage of the entire health care system. Coordination includes consideration of the trip setting, the time setting, and the time required for operation (opening time) of each level of the ground-fault protection system.

However, where a health care installation, such as a doctor's office, is a part of a larger general-use facility (for example, a business office building) that has service ground-fault protection in accordance with 230.95, it is not intended that ground-fault protection be required on the health care facility (doctor's office) feeder because it will be the only feeder in that complex required to have the second level ground-fault protection. Insofar as the doctor's office in a multiple use type building is concerned, no additional protection is achieved by having the second level of ground-fault protection.

It is not intended that ground-fault protection be installed between the on-site generator and the transfer switch or on the load side of the essential electrical system transfer switch.

FPN No. 3 to 230.95(C), calls attention to problems that may arise when ground-fault-protected systems are transferred to another supply system.

(C) Testing. When equipment ground-fault protection is first installed, each level shall be performance tested to ensure compliance with 517.17(B).

See 230.95(C) and its commentary for more information on performance testing.

517.18 General Care Areas.

(A) Patient Bed Location. Each patient bed location shall be supplied by at least two branch circuits, one from the emergency system and one from the normal system. All branch circuits from the normal system shall originate in the same panelboard.

Exception No. 1: Branch circuits serving only special-purpose outlets or receptacles, such as portable X-ray outlets, shall not be required to be served from the same distribution panel or panels.

Exception No. 2: Requirements of 517.18(A) shall not apply to patient bed locations in clinics, medical and dental offices, and outpatient facilities; psychiatric, substance abuse, and rehabilitation hospitals; sleeping rooms of nursing homes and limited care facilities meeting the requirements of 517.10(B)(2).

Exception No. 3: A general care patient bed location served from two separate transfer switches on the emergency system shall not be required to have circuits from the normal system.

Patient bed locations in general care areas are prohibited from deriving *all* their branch circuits from the emergency system. At least one branch circuit for each patient bed location must originate in a normal system panelboard. This is a reflection of the requirements in 517.33.

Exception No. 3 to 517.18 allows the two required

branch circuits for the general care area both to be supplied by the emergency system, provided they are supplied by two separate transfer switches. A normal branch circuit is not required in this case. Two emergency branch circuits have a higher reliability than one normal and one emergency branch circuit.

(B) Patient Bed Location Receptacles. Each patient bed location shall be provided with a minimum of four receptacles. They shall be permitted to be of the single or duplex types or a combination of both. All receptacles, whether four or more, shall be listed "hospital grade" and so identified. Each receptacle shall be grounded by means of an insulated copper conductor sized in accordance with Table 250.122.

Exception No. 1: Requirements of 517.18(B) shall not apply to psychiatric, substance abuse, and rehabilitation hospitals meeting the requirements of 517.10(B)(2).

Exception No. 2: Psychiatric security rooms shall not be required to have receptacle outlets installed in the room.

> FPN: It is not intended that there be a total, immediate replacement of existing non-hospital grade receptacles. It is intended, however, that non-hospital grade receptacles be replaced with hospital grade receptacles upon modification of use, renovation, or as existing receptacles need replacement.

Since the 1990 *Code,* the provisions of 517.18(B) have required hospital-grade receptacles in general care patient bed locations. See the commentary following 517.19(B)(2) for more information.

(C) Pediatric Locations. Receptacles located within the patient care areas of pediatric wards, rooms, or areas shall be listed tamper resistant or shall employ a listed tamper resistant cover.

The receptacle safeguarding requirement of 517.18(C) has been revised to cover all receptacles installed in the patient care areas of pediatric locations. This safeguarding can be achieved through the use of either listed tamper resistant receptacles or listed tamper resistant covers. The use of locking covers over ordinary receptacles does not meet this requirement.

517.19 Critical Care Areas.

(A) Patient Bed Location Branch Circuits. Each patient bed location shall be supplied by at least two branch circuits, one or more from the emergency system and one or more circuits from the normal system. At least one branch circuit from the emergency system shall supply an outlet(s) only at that bed location. All branch circuits from the normal system shall be from a single panelboard. Emergency system receptacles shall be identified and shall also indicate the panelboard and circuit number supplying them.

Exception No. 1: Branch circuits serving only special-purpose receptacles or equipment in critical care areas shall be permitted to be served by other panelboards.

Exception No. 2: Critical care locations served from two separate transfer switches on the emergency system shall not be required to have circuits from the normal system.

Exception No. 2 to 517.19(A) covers the special case in which two separate transfer switches supply a single patient care area. Branch circuits supplied from two separate transfer switches provide the same level of redundancy as required by the main requirement.

The requirements in 517.19(A) are similar to those in 517.18(A). See the commentary following 517.18(A), Exception No. 3.

(B) Patient Bed Location Receptacles.

(1) Minimum Number and Supply. Each patient bed location shall be provided with a minimum of six receptacles, at least one of which shall be connected to either of the following:

(1) The normal system branch circuit required in 517.19(A)
(2) An emergency system branch circuit supplied by a different transfer switch than the other receptacles at the same location

(2) Receptacle Requirements. The receptacles required in 517.19(B)(1) shall be permitted to be of the single or duplex types or a combination of both. All receptacles, whether six or more, shall be listed "hospital grade" and so identified. Each receptacle shall be grounded to the reference grounding point by means of an insulated copper equipment grounding conductor.

The number and type of branch circuits required for patient bed locations in critical care areas is covered in 517.19(A). The number of receptacles and the type of circuit supplying them is covered in 517.19(B). Each patient bed location must be provided with at least six receptacles. A single receptacle counts as one receptacle and a duplex receptacle counts as two receptacles, therefore, three duplex receptacles meet the requirement.

Each patient bed location must be supplied by at least two branch circuits, one from the *normal* panel and one from the emergency panel, as shown in Exhibit 517.1. The normal circuits must be supplied from the same panel

(L-1). The emergency circuits are permitted to be supplied from different panels (EML-1 and EML-2). However, the emergency branch circuit to patient bed location A cannot supply emergency receptacles for patient bed location B. The patient bed location receptacles can also be supplied by two different emergency circuits, instead of one emergency and one normal, provided the emergency circuits are supplied from two different emergency transfer switches.

Receptacles may be of the single or duplex type, provided they are listed hospital-grade type and are identified as such. A typical method of marking hospital-grade receptacles is by a green dot on the face of the receptacle, as shown in Exhibit 517.2. The emergency system receptacles must be marked to indicate the panelboard and circuit number supplying them. The requirements for the number and type of branch circuits in critical care areas are intended to ensure that critical care patients will not be without electrical power regardless of whether the equipment, the branch circuits, or the normal system itself is at fault.

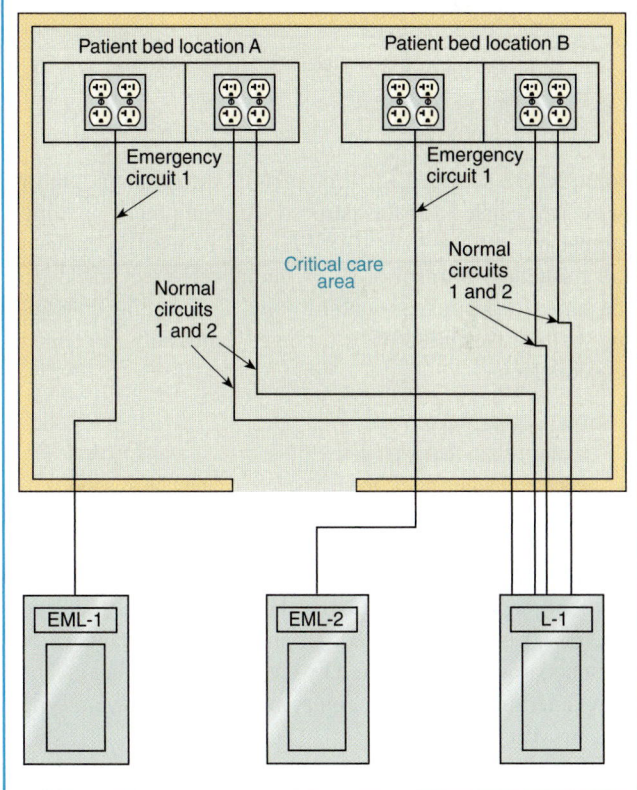

Exhibit 517.1 Examples of normal and emergency circuits supplying patient bed locations in a critical care area.

(C) Patient Vicinity Grounding and Bonding (Optional). A patient vicinity shall be permitted to have a patient equipment grounding point. The patient equipment grounding

Exhibit 517.2 A tamper-resistant hospital-grade receptacle, identified by a green dot on its face, that fulfills the requirements of 517.18(B) and 517.18(C). (Courtesy of Pass & Seymour/Legrand®)

point, where supplied, shall be permitted to contain one or more jacks listed for the purpose. An equipment bonding jumper not smaller than 10 AWG shall be used to connect the grounding terminal of all grounding-type receptacles to the patient equipment grounding point. The bonding conductor shall be permitted to be arranged centrically or looped as convenient.

> FPN: Where there is no patient equipment grounding point, it is important that the distance between the reference grounding point and the patient vicinity be as short as possible to minimize any potential differences.

(D) Panelboard Grounding. Where a grounded electrical distribution system is used and metal feeder raceway or Type MC or MI cable is installed, grounding of a panelboard or switchboard shall be ensured by one of the following means at each termination or junction point of the raceway or Type MC or MI cable:

(1) A grounding bushing and a continuous copper bonding jumper, sized in accordance with 250.122, with the bonding jumper connected to the junction enclosure or the ground bus of the panel
(2) Connection of feeder raceways or Type MC or MI cable to threaded hubs or bosses on terminating enclosures
(3) Other approved devices such as bonding-type locknuts or bushings

(E) Additional Protective Techniques in Critical Care Areas (Optional). Isolated power systems shall be permitted to be used for critical care areas, and, if used, the isolated power system equipment shall be listed for the purpose and the system designed and installed so that it meets the provisions of and is in accordance with 517.160.

Exception: The audible and visual indicators of the line isolation monitor shall be permitted to be located at the nursing station for the area being served.

(F) Isolated Power System Grounding. Where an isolated ungrounded power source is used and limits the first-fault current to a low magnitude, the grounding conductor associated with the secondary circuit shall be permitted to be run outside of the enclosure of the power conductors in the same circuit.

> FPN: Although it is permitted to run the grounding conductor outside of the conduit, it is safer to run it with the power conductors to provide better protection in case of a second ground fault.

Installing the conductor inside the raceway with the conductors delivering the fault current reduces the impedance of the grounding path.

(G) Special-Purpose Receptacle Grounding. The equipment grounding conductor for special-purpose receptacles, such as the operation of mobile X-ray equipment, shall be extended to the reference grounding points of branch circuits for all locations likely to be served from such receptacles. Where such a circuit is served from an isolated ungrounded system, the grounding conductor shall not be required to be run with the power conductors; however, the equipment grounding terminal of the special-purpose receptacle shall be connected to the reference grounding point.

517.20 Wet Locations.

(A) Receptacles and Fixed Equipment. All receptacles and fixed equipment within the area of the wet location shall have ground-fault circuit-interrupter protection for personnel if interruption of power under fault conditions can be tolerated, or be served by an isolated power system if such interruption cannot be tolerated.

Exception: Branch circuits supplying only listed, fixed, therapeutic and diagnostic equipment shall be permitted to be supplied from a normal grounded service, single- or 3-phase system, provided that

(a) Wiring for grounded and isolated circuits does not occupy the same raceway, and
(b) All conductive surfaces of the equipment are grounded.

(B) Isolated Power Systems. Where an isolated power system is utilized, the equipment shall be listed for the purpose and installed so that it meets the provisions of and is in accordance with 517.160.

> FPN: For requirements for installation of therapeutic pools and tubs, see Part VI of Article 680.

In areas that are designated patient care wet locations by the governing body of the facility, ground-fault circuit-interrupter protection is required for the protection of receptacles and fixed equipment if a circuit interruption can be tolerated. Otherwise, an isolated power system is required. See the commentary following the definition of *patient care area* in 517.2.

517.21 Ground-Fault Circuit-Interrupter Protection for Personnel.

Ground-fault circuit-interrupter protection for personnel shall not be required for receptacles installed in those critical care areas where the toilet and basin are installed within the patient room.

In critical care areas, the patients are bedridden. The bathroom accommodations in the critical care area for patients are not the same as those in other patient areas. Because of the unique use of the bathroom facility, only those receptacles in the specific location are exempt from the ground-fault circuit-interrupter requirement. The provisions of 517.21 do not exempt the requirements for GFCI-protected receptacles in other bathrooms for patients, staff, or the public, as required by 210.8(B)(1).

III. Essential Electrical System

517.25 Scope.

The essential electrical system for these facilities shall comprise a system capable of supplying a limited amount of lighting and power service, which is considered essential for life safety and orderly cessation of procedures during the time normal electrical service is interrupted for any reason. This includes clinics, medical and dental offices, outpatient facilities, nursing homes, limited care facilities, hospitals, and other health care facilities serving patients.

> FPN: For information as to the need for an essential electrical system, see NFPA 99-1999, *Standard for Health Care Facilities.*

517.30 Essential Electrical Systems for Hospitals.

(A) Applicability. The requirements of Part III, 517.30 through 517.35, shall apply to hospitals where an essential electrical system is required.

> FPN No. 1: For performance, maintenance, and testing requirements of essential electrical systems in hospitals, see NFPA 99-1999, *Standard for Health Care Facilities.* For installation of centrifugal fire pumps, see NFPA 20-1999, *Standard for the Installation of Stationary Fire Pumps for Fire Protection.*
>
> FPN No. 2: For additional information, see NFPA 99-1999, *Standard for Health Care Facilities.*

(B) General.

(1) Separate Systems. Essential electrical systems for hospitals shall be comprised of two separate systems capable of supplying a limited amount of lighting and power service, which is considered essential for life safety and effective hospital operation during the time the normal electrical service is interrupted for any reason. These two systems shall be the emergency system and the equipment system.

(2) Emergency Systems. The emergency system shall be limited to circuits essential to life safety and critical patient care. These are designated the life safety branch and the critical branch.

(3) Equipment System. The equipment system shall supply major electrical equipment necessary for patient care and basic hospital operation.

(4) Transfer Switches. The number of transfer switches to be used shall be based on reliability, design, and load considerations. Each branch of the emergency system and each equipment system shall have one or more transfer switches. One transfer switch shall be permitted to serve one or more branches or systems in a facility with a maximum demand on the essential electrical system of 150 kVA.

> FPN No. 1: See NFPA 99-1999, *Standard for Health Care Facilities*: 3.4.3.2, Transfer Switch Operation Type I; 3.4.2.1.4, Automatic Transfer Switch Features; and 3.4.2.1.6, Nonautomatic Transfer Device Features.
>
> FPN No. 2: See FPN Figure 517.30, No. 1.
>
> FPN No. 3: See FPN Figure 517.30, No. 2.

FPN Figures 517.30, No. 1 and 517.30, No. 2 illustrate possible electrical system connections for hospitals. For a small electrical system having a maximum demand on the essential electrical system of 150 kVA, see FPN Figure 517.30, No. 2. A small load can be served by a single transfer switch that can handle the loads associated with both the emergency system and the equipment system. This, of course, is based on the assumption that the transfer switch has sufficient capacity to handle the combined loads and that the alternate source of power is sufficiently large to withstand the impact of the simultaneous transfer of both systems in the event of a normal power loss. For further explanation of loads permitted on an emergency system, see NFPA 99, *Standard for Health Care Facilities,* 3-4.2.2.

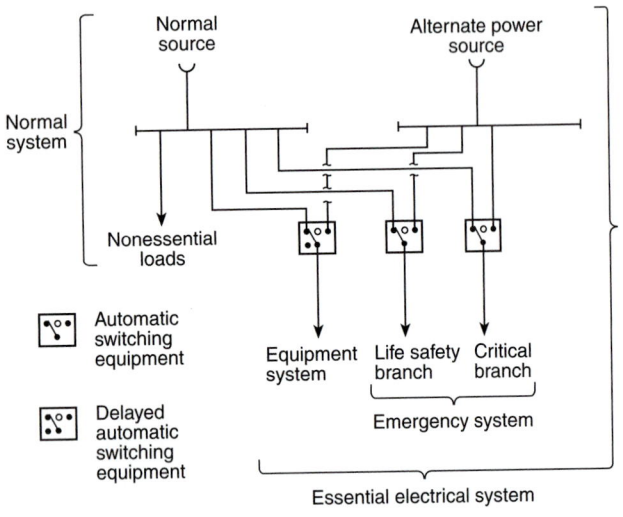

FPN Figure 517.30, No. 1 Hospital—minimum requirement for transfer switch arrangement.

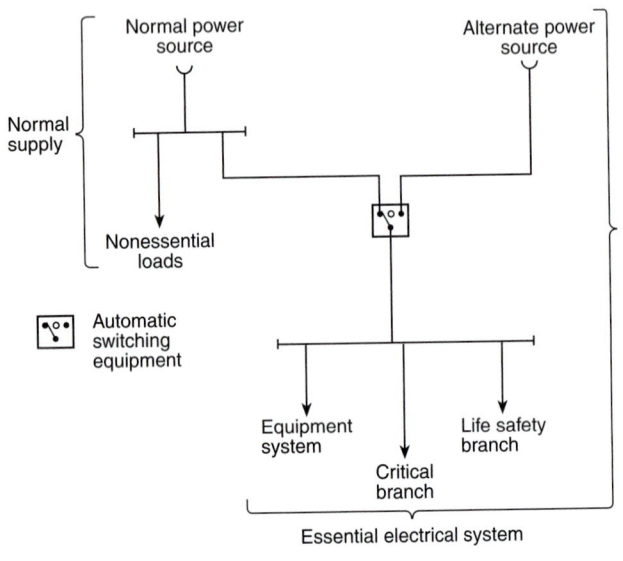

FPN Figure 517.30, No. 2 Hospital—minimum requirement (150 kVA or less) for transfer switch arrangement.

(5) Other Loads. Loads served by the generating equipment not specifically named in Article 517 shall be served by their own transfer switches such that these loads:

(1) Shall not be transferred if the transfer will overload the generating equipment.
(2) Shall be automatically shed upon generating equipment overloading.

(6) Contiguous Facilities. Hospital power sources and alternate power sources shall be permitted to serve the essential electrical systems of contiguous or same site facilities. [NFPA 99, 3.4.2.2.1, 12.3.3.2]

The requirement in 517.30(B)(6) correlates with 12-3.3.2 of NFPA 99, *Standard for Health Care Facilities.*

(C) Wiring Requirements.

(1) Separation from Other Circuits. The life safety branch and critical branch of the emergency system shall be kept entirely independent of all other wiring and equipment and shall not enter the same raceways, boxes, or cabinets with each other or other wiring.

Wiring of the life safety branch and the critical branch shall be permitted to occupy the same raceways, boxes, or cabinets of other circuits not part of the branch where such wiring is as follows:

(1) In transfer equipment enclosures, or
(2) In exit or emergency luminaires (lighting fixtures) supplied from two sources, or
(3) In a common junction box attached to exit or emergency luminaires (lighting fixtures) supplied from two sources, or
(4) For two or more emergency circuits supplied from the same branch

The wiring of the equipment system shall be permitted to occupy the same raceways, boxes, or cabinets of other circuits that are not part of the emergency system.

Multiwire branch circuits supplied by the same panelboard are permitted by 517.30(C)(1)(4).

(2) Isolated Power Systems. Where isolated power systems are installed in any of the areas in 517.33(A)(1) and (A)(2), each system shall be supplied by an individual circuit serving no other load.

(3) Mechanical Protection of the Emergency System. The wiring of the emergency system of a hospital shall be mechanically protected by installation in nonflexible metal raceways, or shall be wired with Type MI cable.

The wiring of emergency systems in hospitals requires additional protection not normally required in other occupancies. Only metal raceways of the nonflexible type and Type MI cable are permitted as a wiring method for hospital emergency systems. (Emergency system circuits can include services, feeders, and branch circuits.) As described in 517.31, the emergency system in a hospital consists of the life safety branch and the critical branch. There are five exceptions to the requirement in 517.30(C)(3).

Exception No. 1: Flexible power cords of appliances, or other utilization equipment, connected to the emergency system shall not be required to be enclosed in raceways.

Exception No. 2: Secondary circuits of transformer-powered communications or signaling systems shall not be required to be enclosed in raceways unless otherwise specified by Chapters 7 or 8.

The branch-circuit conductors that supply the power supplies of equipment, such as fire alarm panels, public address system amplifiers, and nurses' call must be installed in nonflexible metal raceways or Type MI cable.

Exception No. 2 exempts the secondary conductors of limited energy systems, such as nurse call, telephone, and alarm circuits, from being run in metal raceways, provided they comply with their applicable articles elsewhere in the *Code*. Although this requirement allows substantial latitude in the wiring method, it should be noted that the restrictions of 300.22 (ducts, plenums, and other air-handling spaces) apply, unless cables specifically listed for use in these environments are used. See the requirements for these cables in Articles 725, 760, 770, 800, and 820.

Exception No. 3: Schedule 80 rigid nonmetallic conduit shall be permitted if the branch circuits do not serve patient care areas and it is not prohibited elsewhere in this Code.

Exception No. 4: Where encased in not less than 50 mm (2 in.) of concrete, Schedule 40 rigid nonmetallic conduit or electrical nonmetallic tubing shall be permitted if the branch circuits do not serve patient care areas.

Exception No. 5: Flexible metal raceways and cable assemblies shall be permitted to be used in listed prefabricated medical headwalls, listed office furnishings, or where necessary for flexible connection to equipment.

Exception No. 5 recognizes that flexible connections are sometimes required for equipment and, for practical reasons, permits their limited use.

FPN: See 517.13 for additional grounding requirements in patient care areas.

(D) Capacity of Systems. The essential electrical system shall have adequate capacity to meet the demand for the operation of all functions and equipment to be served by each system and branch.

Feeders shall be sized in accordance with Articles 215 and 220. The generator set(s) shall have sufficient capacity and proper rating to meet the demand produced by the load of the essential electrical system(s) at any given time.

Demand calculations for sizing of the generator set(s) shall be based on the following:

(1) Prudent demand factors and historical data, or
(2) Connected load, or
(3) Feeder calculation procedures described in Article 220, or
(4) Any combination of the above

The sizing requirements in 700.5 and 701.6 shall not apply to hospital generator set(s).

The intent of 517.30(D) is to permit the sizing of generators based on actual demand likely to be produced by the connected load of the system at any one time. This method of calculation facilitates practical sizing of generators in health care facilities and helps eliminate prime mover operational problems associated with lightly loaded generators.

(E) Receptacle Identification. The cover plates for the electrical receptacles or the electrical receptacles themselves supplied from the emergency system shall have a distinctive color or marking so as to be readily identifiable. [NFPA 99, 3.4.2.2.4(b)2]

The 1999 *Code* required identification of cover plates or the receptacles themselves where supplied by the critical branch of the emergency system. This identification requirement now applies to all receptacles supplied by the emergency system, thus it is now required to also identify those receptacles supplied by the life safety branch.

517.31 Emergency System.

Those functions of patient care depending on lighting or appliances that are connected to the emergency system shall be divided into two mandatory branches: the life safety branch and the critical branch, described in 517.32 and 517.33.

The branches of the emergency system shall be installed and connected to the alternate power source so that all functions specified herein for the emergency system shall be automatically restored to operation within 10 seconds after interruption of the normal source. [NFPA 99, 3.4.2.2.2(a), 3.5.2.2.2]

517.32 Life Safety Branch.

No function other than those listed in 517.32(A) through (G) shall be connected to the life safety branch. The life safety branch of the emergency system shall supply power for the following lighting, receptacles, and equipment.

(A) Illumination of Means of Egress. Illumination of means of egress, such as lighting required for corridors, passageways, stairways, and landings at exit doors, and all necessary ways of approach to exits. Switching arrangements to transfer patient corridor lighting in hospitals from general illumination circuits to night illumination circuits shall be permitted, provided only one of two circuits can be selected and both circuits cannot be extinguished at the same time.

FPN: See NFPA *101-2000*, *Life Safety Code*, 5.8 and 5.9.

(B) Exit Signs. Exit signs and exit directional signs.

FPN: See NFPA *101-2000*, *Life Safety Code*, 5.10.

(C) Alarm and Alerting Systems. Alarm and alerting systems including the following:

(1) Fire alarms

FPN: See NFPA *101-2000*, *Life Safety Code*, 7.6 and 12.3.4.

(2) Alarms required for systems used for the piping of nonflammable medical gases

FPN: See NFPA 99-1999, *Standard for Health Care Facilities*, 12.3.4.1.

(D) Communications Systems. Hospital communications systems, where used for issuing instructions during emergency conditions.

(E) Generator Set Location. Task illumination battery charger for emergency battery-powered lighting unit(s) and selected receptacles at the generator set location.

(F) Elevators. Elevator cab lighting, control, communications, and signal systems.

(G) Automatic Doors. Automatically operated doors used for building egress. [NFPA 99, 3.4.2.2.2(b)]

517.33 Critical Branch.

(A) Task Illumination and Selected Receptacles. The critical branch of the emergency system shall supply power for task illumination, fixed equipment, selected receptacles, and special power circuits serving the following areas and functions related to patient care:

(1) Critical care areas that utilize anesthetizing gases—task illumination, selected receptacles, and fixed equipment
(2) The isolated power systems in special environments
(3) Patient care areas—task illumination and selected receptacles in the following:

a. Infant nurseries
b. Medication preparation areas
c. Pharmacy dispensing areas
d. Selected acute nursing areas
e. Psychiatric bed areas (omit receptacles)
f. Ward treatment rooms

g. Nurses' stations (unless adequately lighted by corridor luminaires)

(4) Additional specialized patient care task illumination and receptacles, where needed
(5) Nurse call systems
(6) Blood, bone, and tissue banks
(7) Telephone equipment rooms and closets
(8) Task illumination, selected receptacles, and selected power circuits for the following:

a. General care beds (at least one duplex receptacle per patient bedroom)
b. Angiographic labs
c. Cardiac catheterization labs
d. Coronary care units
e. Hemodialysis rooms or areas
f. Emergency room treatment areas (selected)
g. Human physiology labs
h. Intensive care units
i. Postoperative recovery rooms (selected)

(9) Additional task illumination, receptacles, and selected power circuits needed for effective hospital operation. Single-phase fractional horsepower motors shall be permitted to be connected to the critical branch. [NFPA 99, 3.4.2.2.2(c)]

The critical branch is intended to serve a limited number of receptacles and locations, to reduce the load, and to minimize the chances of a fault condition. Receptacles in general patient care area corridors are permitted on the critical branch, but they must be identified in some manner (color-coded or labeled) as part of the emergency system in accordance with 517.30(E).

(B) Subdivision of the Critical Branch. It shall be permitted to subdivide the critical branch into two or more branches.

FPN: It is important to analyze the consequences of supplying an area with only critical care branch power when failure occurs between the area and the transfer switch. Some proportion of normal and critical power or critical power from separate transfer switches may be appropriate.

517.34 Equipment System Connection to Alternate Power Source.

The equipment system shall be installed and connected to the alternate power source such that the equipment described in 517.34(A) is automatically restored to operation at appropriate time-lag intervals following the energizing of the emergency system. Its arrangement shall also provide for the subsequent connection of equipment described in 517.34(B). [NFPA 99, 3.4.2.2.3(b)]

Exception: For essential electrical systems under 150 kVA, deletion of the time-lag intervals feature for delayed automatic connection to the equipment system shall be permitted.

(A) Equipment for Delayed Automatic Connection. The following equipment shall be arranged for delayed automatic connection to the alternate power source.

(1) Central suction systems serving medical and surgical functions, including controls. Such suction systems shall be permitted on the critical branch.

(2) Sump pumps and other equipment required to operate for the safety of major apparatus, including associated control systems and alarms.

(3) Compressed air systems serving medical and surgical functions, including controls. Such air systems shall be permitted on the critical branch.

(4) Smoke control and stair pressurization systems, or both.

(5) Kitchen hood supply or exhaust systems, or both, if required to operate during a fire in or under the hood. [NFPA 99, 3.4.2.2.3(d)]

Exception: Sequential delayed automatic connection to the alternate power source to prevent overloading the generator shall be permitted where engineering studies indicate it is necessary.

(B) Equipment for Delayed Automatic or Manual Connection. The following equipment shall be arranged for either delayed automatic or manual connection to the alternate power source:

(1) Heating equipment to provide heating for operating, delivery, labor, recovery, intensive care, coronary care, nurseries, infection/isolation rooms, emergency treatment spaces, and general patient rooms and pressure maintenance (jockey or make-up) pump(s) for water-based fire protection systems.

Exception: Heating of general patient rooms and infection/isolation rooms during disruption of the normal source shall not be required under any of the following conditions:

(a) The outside design temperature is higher than −6.7°C (20°F).

(b) The outside design temperature is lower than −6.7°C (20°F), and where a selected room(s) is provided for the needs of all confined patients, only such room(s) need be heated.

(c) The facility is served by a dual source of normal power.

FPN No. 1: The design temperature is based on the 97½ percent design value as shown in Chapter 24 of the ASHRAE *Handbook of Fundamentals* (1997).

FPN No. 2: For a description of a dual source of normal power, see 517.35(C), FPN.

In some areas, it is common practice to install individual room heating/air conditioners rather than have a central heating/air-conditioning plant. If these individual units are electrically powered, it may not be practical to apply their high-demand load to the generator. If the governing body of the nursing home has full-time skilled attendants who can move people to one room(s) that will be heated when this smaller load is picked up by the generator, the intent of the *Code* is satisfied. The provisions for limited heating during emergency conditions are based on consideration of outside design temperature.

(2) An elevator(s) selected to provide service to patient, surgical, obstetrical, and ground floors during interruption of normal power. In instances where interruption of normal power would result in other elevators stopping between floors, throw-over facilities shall be provided to allow the temporary operation of any elevator for the release of patients or other persons who may be confined between floors.

(3) Supply, return, and exhaust ventilating systems for airborne infectious/isolation rooms, protective environment rooms, exhaust fans for laboratory fume hoods, nuclear medicine areas where radioactive material is used, ethylene oxide evacuation and anesthesia evacuation. Where delayed automatic connection is not appropriate, such ventilation systems shall be permitted to be placed on the critical branch. [NFPA 99, 3.4.2.2.3(e)(4)]

(4) Hyperbaric facilities.

(5) Hypobaric facilities.

(6) Automatically operated doors.

(7) Minimal electrically heated autoclaving equipment shall be permitted to be arranged for either automatic or manual connection to the alternate source.

(8) Controls for equipment listed in 517.34.

(9) Other selected equipment shall be permitted to be served by the equipment system. [NFPA 99, 3.4.2.2.3(e)]

517.35 Sources of Power.

(A) Two Independent Sources of Power. Essential electrical systems shall have a minimum of two independent sources of power: a normal source generally supplying the entire electrical system and one or more alternate sources for use when the normal source is interrupted. [NFPA 99, 3.4.1.1.2]

(B) Alternate Source of Power. The alternate source of power shall be one of the following:

(1) Generator(s) driven by some form of prime mover(s) and located on the premises

(2) Another generating unit(s) where the normal source consists of a generating unit(s) located on the premises

(3) An external utility service when the normal source consists of a generating unit(s) located on the premises

(C) Location of Essential Electrical System Components. Careful consideration shall be given to the location of the spaces housing the components of the essential electrical system to minimize interruptions caused by natural forces common to the area (e.g., storms, floods, earthquakes, or hazards created by adjoining structures or activities). Consideration shall also be given to the possible interruption of normal electrical services resulting from similar causes as well as possible disruption of normal electrical service due to internal wiring and equipment failures.

> FPN: Facilities in which the normal source of power is supplied by two or more separate central station-fed services experience greater than normal electrical service reliability than those with only a single feed. Such a dual source of normal power consists of two or more electrical services fed from separate generator sets or a utility distribution network that has multiple power input sources and is arranged to provide mechanical and electrical separation so that a fault between the facility and the generating sources is not likely to cause an interruption of more than one of the facility service feeders.

517.40 Essential Electrical Systems for Nursing Homes and Limited Care Facilities.

(A) Applicability. The requirements of Part III, 517.40(C) through 517.44, shall apply to nursing homes and limited care facilities.

Exception: The requirements of Part III, 517.40(C) through 517.44, shall not apply to freestanding buildings used as nursing homes and limited care facilities, provided that the following apply:

(a) *Admitting and discharge policies are maintained that preclude the provision of care for any patient or resident who may need to be sustained by electrical life-support equipment.*

(b) *No surgical treatment requiring general anesthesia is offered.*

(c) *An automatic battery-operated system(s) or equipment is provided that shall be effective for at least 1½ hours and is otherwise in accordance with 700.12 and that shall be capable of supplying lighting for exit lights, exit corridors, stairways, nursing stations, medical preparation areas, boiler rooms, and communications areas. This system shall also supply power to operate all alarm systems. [NFPA 99, 16.3.3.2 Exception, 17.3.3.2 Exception]*

NFPA 99, *Standard for Health Care Facilities,* recognizes two classes of nursing homes or limited care facilities. For the smaller, less complex facility, only a minimum alternate lighting and alarm service need be furnished.

At nursing homes or limited care facilities where patients are sustained by electrical life-support equipment or inpatient hospital care is provided, the requirements of 517.41 through 517.44 apply. The branches of the emergency system for this class of occupancy bear identical titles to their counterparts for hospital-type occupancies.

> FPN: See NFPA *101-2000, Life Safety Code.*

(B) Inpatient Hospital Care Facilities. Nursing homes and limited care facilities that provide inpatient hospital care shall comply with the requirements of Part III, 517.30 through 517.35.

Regardless of the name given to the facility, the type of electrical system depends on the type of patient care provided. If such care is clearly inpatient hospital care, a hospital-type electrical system must be installed. The type of care that can be provided at a nursing home or limited care facility is generally regulated through the administrative agency that licenses the facility.

(C) Facilities Contiguous or Located on the Same Site with Hospitals. Nursing homes and limited care facilities that are contiguous or located on the same site with a hospital shall be permitted to have their essential electrical systems supplied by that of the hospital.

> FPN: For performance, maintenance, and testing requirements of essential electrical systems in nursing homes and limited care facilities, see NFPA 99-1999, *Standard for Health Care Facilities.*

If a nursing home or limited care facility shares essentially the same building with a hospital, the nursing home is not required to have its own essential electrical system as long as it derives its power supply from the hospital. It should be noted, however, that this rule applies only to the electrical supply and does not permit the sharing of transfer devices and the like.

517.41 Essential Electrical Systems.

(A) General. Essential electrical systems for nursing homes and limited care facilities shall be comprised of two separate branches capable of supplying a limited amount of lighting and power service, which is considered essential for the protection of life safety and effective operation of the institution during the time normal electrical service is interrupted for any reason. These two separate branches shall be the life safety branch and the critical branch. [NFPA 99, 3-5.2.2.1]

(B) Transfer Switches. The number of transfer switches to be used shall be based on reliability, design, and load considerations. Each branch of the essential electrical system shall be served by one or more transfer switches. One transfer switch shall be permitted to serve one or more branches or systems in a facility with a maximum demand on the essential electrical system of 150 kVa. [NFPA 99, 3.5.2.2.1]

> FPN No. 1: See NFPA 99-1999, *Standard for Health Care Facilities*, 3.5.3.2, Transfer Switch Operation Type II; 3.4.2.1.4, Automatic Transfer Switch Features; and 3.4.2.1.6, Nonautomatic Transfer Device Features.
>
> FPN No. 2: See FPN Figure 517.41, No. 1.
>
> FPN No. 3: See FPN Figure 517.41, No. 2.

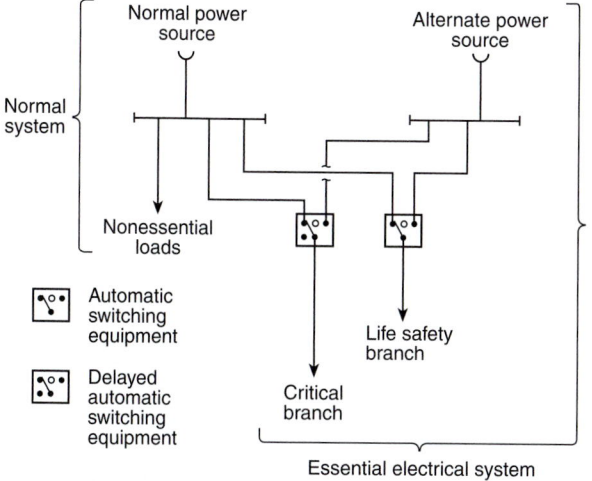

FPN Figure 517.41, No. 1 Nursing home and limited health care facilities—minimum requirement for transfer switch arrangement.

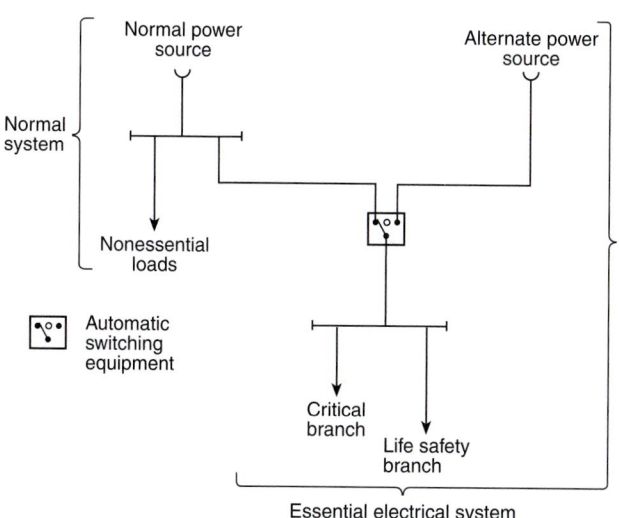

FPN Figure 517.41, No. 2 Nursing home and limited health care facilities—minimum requirement (150 kVa or less) for transfer switch arrangement.

(C) Capacity of System. The essential electrical system shall have adequate capacity to meet the demand for the operation of all functions and equipment to be served by each branch at one time.

(D) Separation from Other Circuits. The life safety branch shall be kept entirely independent of all other wiring and equipment and shall not enter the same raceways, boxes, or cabinets with other wiring except as follows:

(1) In transfer switches
(2) In exit or emergency luminaires (lighting fixtures) supplied from two sources, or
(3) In a common junction box attached to exit or emergency luminaires (lighting fixtures) supplied from two sources

The wiring of the critical branch shall be permitted to occupy the same raceways, boxes, or cabinets of other circuits that are not part of the life safety branch.

(E) Receptacle Identification. The cover plates for the electrical receptacles or the electrical receptacles themselves supplied from the emergency system shall have a distinctive color or marking so as to be readily identifiable. [NFPA 99, 3-5.2.2.4]

517.42 Automatic Connection to Life Safety Branch.

The life safety branch shall be installed and connected to the alternate source of power so that all functions specified herein shall be automatically restored to operation within 10 seconds after the interruption of the normal source. No functions other than those listed in 517.42(A) through (G) shall be connected to the life safety branch. The life safety branch shall supply power for the following lighting, receptacles, and equipment.

> FPN: The life safety branch is called the emergency system in NFPA 99-1999, *Standard for Health Care Facilities.*

(A) Illumination of Means of Egress. Illumination of means of egress as is necessary for corridors, passageways, stairways, landings, and exit doors and all ways of approach to exits. Switching arrangement to transfer patient corridor lighting from general illumination circuits shall be permitted, providing only one of two circuits can be selected and both circuits cannot be extinguished at the same time.

> FPN: See NFPA *101*-2000, *Life Safety Code*, 5.8 and 5.9.

(B) Exit Signs. Exit signs and exit directional signs.

> FPN: See NFPA *101*-2000, *Life Safety Code*, 5.10.

(C) Alarm and Alerting Systems. Alarm and alerting systems, including the following:

(1) Fire alarms

> FPN: See NFPA *101-2000, Life Safety Code,,* 7.6 and 12.3.4.

(2) Alarms required for systems used for the piping of nonflammable medical gases

> FPN: See NFPA 99-1999, *Standard for Health Care Facilities,* 16.3.4.1.

(D) Communications Systems. Communications systems, where used for issuing instructions during emergency conditions.

(E) Dining and Recreation Areas. Sufficient lighting in dining and recreation areas to provide illumination to exit ways.

(F) Generator Set Location. Task illumination and selected receptacles in the generator set location.

(G) Elevators. Elevator cab lighting, control, communications, and signal systems. [NFPA 99, 3.5.2.2.2, 3.5.3.1]

517.43 Connection to Critical Branch.

The critical branch shall be installed and connected to the alternate power source so that the equipment listed in 517.43(A) shall be automatically restored to operation at appropriate time-lag intervals following the restoration of the life safety branch to operation. Its arrangement shall also provide for the additional connection of equipment listed in 517.43(B) by either delayed automatic or manual operation.

Exception: For essential electrical systems under 150 kVA, deletion of the time-lag intervals feature for delayed automatic connection to the equipment system shall be permitted.

(A) Delayed Automatic Connection. The following equipment shall be connected to the critical branch and shall be arranged for delayed automatic connection to the alternate power source:

(1) Patient care areas—task illumination and selected receptacles in the following:

   a. Medication preparation areas

   b. Pharmacy dispensing areas

   c. Nurses' stations (unless adequately lighted by corridor luminaires)

(2) Sump pumps and other equipment required to operate for the safety of major apparatus and associated control systems and alarms

(3) Smoke control and stair pressurization systems

(4) Kitchen hood supply and/or exhaust systems, if required to operate during a fire in or under the hood

(5) Supply, return and exhaust ventilating systems for airborne infectious isolation rooms

(B) Delayed Automatic or Manual Connection. The following equipment shall be connected to the critical branch and shall be arranged for either delayed automatic or manual connection to the alternate power source.

(1) Heating equipment to provide heating for patient rooms.

Exception: Heating of general patient rooms during disruption of the normal source shall not be required under any of the following conditions:

(a) *The outside design temperature is higher than −6.7°C (20°F), or*

(b) *The outside design temperature is lower than −6.7°C (20°F) and where a selected room(s) is provided for the needs of all confined patients, only such room(s) need be heated.*

(c) *The facility is served by a dual source of normal power as described in 517.44(C), FPN.*

> FPN: The outside design temperature is based on the 97½ percent design values as shown in Chapter 24 of the ASHRAE *Handbook of Fundamentals* (1997).

(2) Elevator service—in instances where disruption of power would result in elevators stopping between floors, throw-over facilities shall be provided to allow the temporary operation of any elevator for the release of passengers. For elevator cab lighting, control, and signal system requirements, see 517.42(G).

(3) Additional illumination, receptacles, and equipment shall be permitted to be connected only to the critical branch. [NFPA 99, 3.5.2.2.3]

517.44 Sources of Power.

(A) Two Independent Sources of Power. Essential electrical systems shall have a minimum of two independent sources of power: a normal source generally supplying the entire electrical system and one or more alternate sources for use when the normal source is interrupted. [NFPA 99, 3.4.1.1.2, 3.5.1]

(B) Alternate Source of Power. The alternate source of power shall be a generator(s) driven by some form of prime mover(s) and located on the premises.

Exception No. 1: Where the normal source consists of generating units on the premises, the alternate source shall be either another generator set or an external utility service.

Exception No. 2: Nursing homes or limited care facilities meeting the requirements of 517.40(A), Exception, shall be permitted to use a battery system or self-contained battery integral with the equipment. [NFPA 99, 3.4.1.1.3, 3.5.1, 16.3.3.2.1, 17.3.3.2.1]

(C) Location of Essential Electrical System Components. Careful consideration shall be given to the location of the spaces housing the components of the essential electrical system to minimize interruptions caused by natural forces common to the area (e.g., storms, floods, earthquakes, or hazards created by adjoining structures or activities). Consideration shall also be given to the possible interruption of normal electrical services resulting from similar causes as well as possible disruption of normal electrical service due to internal wiring and equipment failures.

FPN: Facilities in which the normal source of power is supplied by two or more separate central station-fed services experience greater than normal electrical service reliability than those with only a single feed. Such a dual source of normal power consists of two or more electrical services fed from separate generator sets or a utility distribution network that has multiple power input sources and is arranged to provide mechanical and electrical separation so that a fault between the facility and the generating sources will not likely cause an interruption of more than one of the facility service feeders.

517.45 Essential Electrical Systems for Other Health Care Facilities.

(A) Essential Electrical Distribution. The essential electrical distribution system shall be a battery or generator system. [NFPA 99, 13.3.3.2]

(B) Electrical Life Support Equipment. Where electrical life support equipment is required, the essential electrical distribution system shall be as described in 517.30 through 517.35. [NFPA 99, 13.3.3.2.1]

(C) Critical Care Areas. Where critical care areas are present, the essential electrical distribution system shall be as described in 517.30 through 517.35. [NFPA 99, 13.3.3.2.2]

(D) Power Systems. Battery systems shall be installed in accordance with the requirements of Article 700, and generator systems shall be as described in 517.30 through 517.35.

The provisions of 517.45 and NFPA 99, *Standard for Health Care Facilities,* provide the essential electrical system requirements for health care facilities other than hospitals, nursing homes, and limited care facilities. These facilities include medical and dental offices and ambulatory health care facilities. Depending on the type and level of patient care, these facilities may require an alternate source of power since the patients are treated in essentially the same manner as in hospitals, even though the facility may not be called a hospital.

IV. Inhalation Anesthetizing Locations

FPN: For further information regarding safeguards for anesthetizing locations, see NFPA 99-1999, *Standard for Health Care Facilities.*

517.60 Anesthetizing Location Classification.

FPN: If either of the anesthetizing locations in 517.60(A) or (B) is designated a wet location, refer to 517.20.

(A) Hazardous (Classified) Location.

(1) Use Location. In a location where flammable anesthetics are employed, the entire area shall be considered to be a Class I, Division 1 location that shall extend upward to a level 1.52 m (5 ft) above the floor. The remaining volume up to the structural ceiling is considered to be above a hazardous (classified) location. [NFPA 99, Annex 2, 2.1, 2.2]

(2) Storage Location. Any room or location in which flammable anesthetics or volatile flammable disinfecting agents are stored shall be considered to be a Class I, Division 1 location from floor to ceiling.

Past editions of NFPA 99, *Standard for Health Care Facilities*, contained requirements for flammable anesthetizing locations throughout the standard. Now, all of these requirements are located in Annex 2 of NFPA 99. The reason for this change is explained in the second note of Annex 2 in NFPA 99 and reads as follows:

The text of this annex is a compilation of requirements included in previous editions of NFPA 99 on safety practices for facilities that used flammable inhalation anesthetics. This material is being retained in this annex by the Technical Committee on Anesthesia Services for the following reasons: 1) the Committee is aware that some countries outside the United States still use this type of anesthetics, and rely on the safety measures herein; and 2) while the Committee is unaware of any medical schools in the U.S. still teaching the proper use of flammable anesthetics or any health care facilities in the U.S. using flammable anesthetics, retaining this material will serve as a reminder of the precautions that would be necessary should the use of this type of anesthetics be re-instituted.

In 517.60, anesthetizing locations are designated as hazardous (classified) locations, where flammable or nonflammable anesthetics may be interchangeably employed [517.60(A)], or as other-than-hazardous (classified) locations, where only nonflammable anesthetics are used [517.60(B)]. In the case of the flammable anesthetizing location, the entire volume of the room, extending upward from a level 5 ft above the floor to the surface of the structural ceiling of the room and including the space between a drop ceiling and the structural ceiling, is considered to be above a hazardous (classified) location.

(B) Other-Than-Hazardous (Classified) Location. Any inhalation anesthetizing location designated for the exclusive use of nonflammable anesthetizing agents shall be considered to be an other-than-hazardous (classified) location.

517.61 Wiring and Equipment.

(A) Within Hazardous (Classified) Anesthetizing Locations.

(1) Isolation. Except as permitted in 517.160, each power circuit within, or partially within, a flammable anesthetizing location as referred to in 517.60 shall be isolated from any distribution system by the use of an isolated power system. [NFPA 99, Annex 2, 2.6.3.2]

(2) Design and Installation. Isolated power system equipment shall be listed for the purpose and the system designed and installed so that it meets the provisions and is in accordance with Part VII.

(3) Equipment Operating at More Than 10 Volts. In hazardous (classified) locations referred to in 517.60, all fixed wiring and equipment and all portable equipment, including lamps and other utilization equipment, operating at more than 10 volts between conductors shall comply with the requirements of 501.1 through 501.15 and 501.16(A) and (B) for Class I, Division 1 locations. All such equipment shall be specifically approved for the hazardous atmospheres involved. [NFPA 99, Annex 2, 2.2.1, 2.4.5, 2.4.6, 2.4.7]

(4) Extent of Location. Where a box, fitting, or enclosure is partially, but not entirely, within a hazardous (classified) location(s), the hazardous (classified) location(s) shall be considered to be extended to include the entire box, fitting, or enclosure.

(5) Receptacles and Attachment Plugs. Receptacles and attachment plugs in a hazardous (classified) location(s) shall be listed for use in Class I, Group C hazardous (classified) locations and shall have provision for the connection of a grounding conductor.

(6) Flexible Cord Type. Flexible cords used in hazardous (classified) locations for connection to portable utilization equipment, including lamps operating at more than 8 volts between conductors, shall be of a type approved for extra-hard usage in accordance with Table 400.4 and shall include an additional conductor for grounding.

(7) Flexible Cord Storage. A storage device for the flexible cord shall be provided and shall not subject the cord to bending at a radius of less than 75 mm (3 in.).

(B) Above Hazardous (Classified) Anesthetizing Locations.

(1) Wiring Methods. Wiring above a hazardous (classified) location referred to in 517.60 shall be installed in rigid metal conduit, electrical metallic tubing, intermediate metal conduit, Type MI cable, or Type MC cable that employs a continuous, gas/vaportight metal sheath.

(2) Equipment Enclosure. Installed equipment that may produce arcs, sparks, or particles of hot metal, such as lamps and lampholders for fixed lighting, cutouts, switches, generators, motors, or other equipment having make-and-break or sliding contacts, shall be of the totally enclosed type or be constructed so as to prevent escape of sparks or hot metal particles.

Exception: Wall-mounted receptacles installed above the hazardous (classified) location in flammable anesthetizing locations shall not be required to be totally enclosed or have openings guarded or screened to prevent dispersion of particles.

(3) Luminaires (Lighting Fixtures). Surgical and other luminaires (lighting fixtures) shall conform to 501.9(B).

Exception No. 1: The surface temperature limitations set forth in 501.9(B)(2) shall not apply.

Exception No. 2: Integral or pendant switches that are located above and cannot be lowered into the hazardous (classified) location(s) shall not be required to be explosionproof.

(4) Seals. Approved seals shall be provided in conformance with 501.5, and 501.5(A)(4) shall apply to horizontal as well as to vertical boundaries of the defined hazardous (classified) locations.

(5) Receptacles and Attachment Plugs. Receptacles and attachment plugs located above hazardous (classified) anesthetizing locations shall be listed for hospital use for services of prescribed voltage, frequency, rating, and number of conductors with provision for the connection of the grounding conductor. This requirement shall apply to attachment plugs and receptacles of the 2-pole, 3-wire grounding type for single-phase, 120-volt, nominal, ac service.

See the commentary following 517.19(B)(2) regarding receptacles listed for hospital use.

(6) 250-Volt Receptacles and Attachment Plugs Rated 50 and 60 Amperes. Receptacles and attachment plugs rated 250 volts, for connection of 50-ampere and 60-ampere ac medical equipment for use above hazardous (classified) locations shall be arranged so that the 60-ampere receptacle will accept either the 50-ampere or the 60-ampere plug.

Fifty-ampere receptacles shall be designed so as not to accept the 60-ampere attachment plug. The attachment plugs shall be of the 2-pole, 3-wire design with a third contact connecting to the insulated (green or green with yellow stripe) equipment grounding conductor of the electrical system.

(C) Other-Than-Hazardous (Classified) Anesthetizing Locations.

(1) Wiring Methods. Wiring serving other-than-hazardous (classified) locations, as defined in 517.60, shall be installed in a metal raceway system or cable assembly. The metal raceway system or cable armor or sheath assembly shall qualify as an equipment grounding return path in accordance with 250.118. Type MC and Type MI cable shall have an outer metal armor or sheath that is identified as an acceptable grounding return path.

Exception: Pendant receptacle constructions that employ at least Type SJO or equivalent flexible cords suspended not less than 1.8 m (6 ft) from the floor shall not be required to be installed in a metal raceway or cable assembly.

(2) Receptacles and Attachment Plugs. Receptacles and attachment plugs installed and used in other-than-hazardous (classified) locations shall be listed for hospital use for services of prescribed voltage, frequency, rating, and number of conductors with provision for connection of the grounding conductor. This requirement shall apply to 2-pole, 3-wire grounding type for single-phase, 120-, 208-, or 240-volt, nominal, ac service.

See the commentary following 517.19(B)(2) regarding receptacles listed for hospital use.

(3) 250-Volt Receptacles and Attachment Plugs Rated 50 and 60 Amperes. Receptacles and attachment plugs rated 250 volts, for connection of 50-ampere and 60-ampere ac medical equipment for use in other-than-hazardous (classified) locations shall be arranged so that the 60-ampere receptacle will accept either the 50-ampere or the 60-ampere plug. Fifty-ampere receptacles shall be designed so as not to accept the 60-ampere attachment plug. The attachment plugs shall be of the 2-pole, 3-wire design with a third contact connecting to the insulated (green or green with yellow stripe) equipment grounding conductor of the electrical system.

517.62 Grounding.

In any anesthetizing area, all metal raceways and metal-sheathed cables and all non–current-carrying conductive portions of fixed electric equipment shall be grounded. Grounding in Class I locations shall comply with 501.16.

Exception: Equipment operating at not more than 10 volts between conductors shall not be required to be grounded.

The grounding requirements for anesthetizing locations apply only to metal raceways, metal-sheathed cables, and electrical equipment. Carts, tables, and other nonelectrical items are not required to be grounded. In flammable anesthetizing locations, however, portable carts and tables usually have a resistance to ground of not over 1,000,000 ohms, through the use of conductive tires and wheels and conductive flooring, to avoid the buildup of static electrical charges. See NFPA 99, *Standard for Health Care Facilities*, Annex 2, 2-6.3.10, for electrostatic safeguards.

517.63 Grounded Power Systems in Anesthetizing Locations.

(A) Battery-Powered Emergency Lighting Units. One or more battery-powered emergency lighting units shall be provided in accordance with 700.12(E).

The failure of the emergency circuit feeder that supplies the operating room will ordinarily plunge the room into darkness. The requirement to install at least one battery-operated emergency lighting unit results in immediate illumination upon loss of power that helps mitigate the potentially dangerous impact of total power interruption.

(B) Branch-Circuit Wiring. Branch circuits supplying only listed, fixed, therapeutic and diagnostic equipment, permanently installed above the hazardous (classified) location and in other-than-hazardous (classified) locations, shall be permitted to be supplied from a normal grounded service, single- or three-phase system, provided the following apply:

(1) Wiring for grounded and isolated circuits does not occupy the same raceway or cable.
(2) All conductive surfaces of the equipment are grounded.
(3) Equipment (except enclosed X-ray tubes and the leads to the tubes) are located at least 2.5 m (8 ft) above the floor or outside the anesthetizing location.
(4) Switches for the grounded branch circuit are located outside the hazardous (classified) location.

Exception: Sections 517.63(B)(3) and (B)(4) shall not apply in other-than-hazardous (classified) locations.

(C) Fixed Lighting Branch Circuits. Branch circuits supplying only fixed lighting shall be permitted to be supplied by a normal grounded service, provided the following apply:

(1) Such luminaires (fixtures) are located at least 2.5 m (8 ft) above the floor.

(2) All conductive surfaces of luminaires (fixtures) are grounded.

(3) Wiring for circuits supplying power to luminaires (fixtures) does not occupy the same raceway or cable for circuits supplying isolated power.

(4) Switches are wall-mounted and located above hazardous (classified) locations.

Exception: Sections 517.63(C)(1) and (C)(4) shall not apply in other-than-hazardous (classified) locations.

(D) Remote-Control Stations. Wall-mounted remote-control stations for remote-control switches operating at 24 volts or less shall be permitted to be installed in any anesthetizing location.

(E) Location of Isolated Power Systems. An isolated power center listed for the purpose and its grounded primary feeder shall be permitted to be located in an anesthetizing location, provided it is installed above a hazardous (classified) location or in an other-than-hazardous (classified) location.

(F) Circuits in Anesthetizing Locations. Except as permitted above, each power circuit within, or partially within, a flammable anesthetizing location as referred to in 517.60 shall be isolated from any distribution system supplying other-than-anesthetizing locations.

517.64 Low-Voltage Equipment and Instruments.

(A) Equipment Requirements. Low-voltage equipment that is frequently in contact with the bodies of persons or has exposed current-carrying elements shall be as follows:

(1) Operate on an electrical potential of 10 volts or less, or

(2) Approved as intrinsically safe or double-insulated equipment, or

(3) Moisture resistant

(B) Power Supplies. Power shall be supplied to low-voltage equipment from the following:

(1) An individual portable isolating transformer (autotransformers shall not be used) connected to an isolated power circuit receptacle by means of an appropriate cord and attachment plug, or

(2) A common low-voltage isolating transformer installed in an other-than-hazardous (classified) location, or

(3) Individual dry-cell batteries, or

(4) Common batteries made up of storage cells located in an other-than-hazardous (classified) location

(C) Isolated Circuits. Isolating-type transformers for supplying low-voltage circuits shall have the following:

(1) Approved means for insulating the secondary circuit from the primary circuit, and

(2) The core and case grounded

(D) Controls. Resistance or impedance devices shall be permitted to control low-voltage equipment but shall not be used to limit the maximum available voltage to the equipment.

(E) Battery-Powered Appliances. Battery-powered appliances shall not be capable of being charged while in operation unless their charging circuitry incorporates an integral isolating-type transformer.

(F) Receptacles or Attachment Plugs. Any receptacle or attachment plug used on low-voltage circuits shall be of a type that does not permit interchangeable connection with circuits of higher voltage.

> FPN: Any interruption of the circuit, even circuits as low as 10 volts, either by any switch or loose or defective connections anywhere in the circuit, may produce a spark that is sufficient to ignite flammable anesthetic agents. See 7.5.1.2.3 of NFPA 99-1999, *Standard for Health Care Facilities.*

V. X-Ray Installations

Nothing in this part shall be construed as specifying safeguards against the useful beam or stray X-ray radiation.

> FPN No. 1: Radiation safety and performance requirements of several classes of X-ray equipment are regulated under Public Law 90-602 and are enforced by the Department of Health and Human Services.
>
> FPN No. 2: In addition, information on radiation protection by the National Council on Radiation Protection and Measurements is published as *Reports of the National Council on Radiation Protection and Measurement.* These reports are obtainable from NCRP Publications, P.O. Box 30175, Washington, DC 20014.

517.71 Connection to Supply Circuit.

(A) Fixed and Stationary Equipment. Fixed and stationary X-ray equipment shall be connected to the power supply by means of a wiring method that meets the general requirements of this *Code.*

Exception: Equipment properly supplied by a branch circuit rated at not over 30 amperes shall be permitted to be supplied through a suitable attachment plug and hard-service cable or cord.

(B) Portable, Mobile, and Transportable Equipment. Individual branch circuits shall not be required for portable, mobile, and transportable medical X-ray equipment requiring a capacity of not over 60 amperes.

(C) Over 600-Volt Supply. Circuits and equipment operated on a supply circuit of over 600 volts shall comply with Article 490.

517.72 Disconnecting Means.

(A) Capacity. A disconnecting means of adequate capacity for at least 50 percent of the input required for the momentary rating or 100 percent of the input required for the long-time rating of the X-ray equipment, whichever is greater, shall be provided in the supply circuit.

(B) Location. The disconnecting means shall be operable from a location readily accessible from the X-ray control.

(C) Portable Equipment. For equipment connected to a 120-volt branch circuit of 30 amperes or less, a grounding-type attachment plug and receptacle of proper rating shall be permitted to serve as a disconnecting means.

517.73 Rating of Supply Conductors and Overcurrent Protection.

(A) Diagnostic Equipment.

(1) Branch Circuits. The ampacity of supply branch-circuit conductors and the current rating of overcurrent protective devices shall not be less than 50 percent of the momentary rating or 100 percent of the long-time rating, whichever is greater.

(2) Feeders. The ampacity of supply feeders and the current rating of overcurrent protective devices supplying two or more branch circuits supplying X-ray units shall not be less than 50 percent of the momentary demand rating of the largest unit plus 25 percent of the momentary demand rating of the next largest unit plus 10 percent of the momentary demand rating of each additional unit. Where simultaneous biplane examinations are undertaken with the X-ray units, the supply conductors and overcurrent protective devices shall be 100 percent of the momentary demand rating of each X-ray unit.

> FPN: The minimum conductor size for branch and feeder circuits is also governed by voltage regulation requirements. For a specific installation, the manufacturer usually specifies minimum distribution transformer and conductor sizes, rating of disconnecting means, and overcurrent protection.

(B) Therapeutic Equipment. The ampacity of conductors and rating of overcurrent protective devices shall not be less than 100 percent of the current rating of medical X-ray therapy equipment.

> FPN: The ampacity of the branch-circuit conductors and the ratings of disconnecting means and overcurrent protection for X-ray equipment are usually designated by the manufacturer for the specific installation.

517.74 Control Circuit Conductors.

(A) Number of Conductors in Raceway. The number of control circuit conductors installed in a raceway shall be determined in accordance with 300.17.

(B) Minimum Size of Conductors. Size 18 AWG or 16 AWG fixture wires as specified in 725.27 and flexible cords shall be permitted for the control and operating circuits of X-ray and auxiliary equipment where protected by not larger than 20-ampere overcurrent devices.

517.75 Equipment Installations.

All equipment for new X-ray installations and all used or reconditioned X-ray equipment moved to and reinstalled at a new location shall be of an approved type.

517.76 Transformers and Capacitors.

Transformers and capacitors that are part of X-ray equipment shall not be required to comply with Articles 450 and 460.

Capacitors shall be mounted within enclosures of insulating material or grounded metal.

517.77 Installation of High-Tension X-Ray Cables.

Cables with grounded shields connecting X-ray tubes and image intensifiers shall be permitted to be installed in cable trays or cable troughs along with X-ray equipment control and power supply conductors without the need for barriers to separate the wiring.

517.78 Guarding and Grounding.

(A) High-Voltage Parts. All high-voltage parts, including X-ray tubes, shall be mounted within grounded enclosures. Air, oil, gas, or other suitable insulating media shall be used to insulate the high-voltage from the grounded enclosure. The connection from the high-voltage equipment to X-ray tubes and other high-voltage components shall be made with high-voltage shielded cables.

(B) Low-Voltage Cables. Low-voltage cables connecting to oil-filled units that are not completely sealed, such as transformers, condensers, oil coolers, and high-voltage switches, shall have insulation of the oil-resistant type.

(C) Noncurrent–Carrying Metal Parts. Noncurrent-carrying metal parts of X-ray and associated equipment (controls, tables, X-ray tube supports, transformer tanks, shielded cables, X-ray tube heads, etc.) shall be grounded in the manner specified in Article 250, as modified by 517.13(A) and (B).

VI. Communications, Signaling Systems, Data Systems, Fire Alarm Systems, and Systems Less Than 120 Volts, Nominal

517.80 Patient Care Areas.

Equivalent insulation and isolation to that required for the electrical distribution systems in patient care areas shall be provided for communications, signaling systems, data system circuits, fire alarm systems, and systems less than 120 volts, nominal.

> FPN: An acceptable alternate means of providing isolation for patient/nurse call systems is by the use of nonelectrified signaling, communications, or control devices held by the patient or within reach of the patient.

517.81 Other-Than-Patient-Care Areas.

In other-than-patient-care areas, installations shall be in accordance with the appropriate provisions of Articles 640, 725, 760, and 800.

517.82 Signal Transmission Between Appliances.

(A) General. Permanently installed signal cabling from an appliance in a patient location to remote appliances shall employ a signal transmission system that prevents hazardous grounding interconnection of the appliances.

> FPN: See 517.13(A) for additional grounding requirements in patient care areas.

(B) Common Signal Grounding Wire. Common signal grounding wires (i.e., the chassis ground for single-ended transmission) shall be permitted to be used between appliances all located within the patient vicinity, provided the appliances are served from the same reference grounding point.

VII. Isolated Power Systems

517.160 Isolated Power Systems.

(A) Installations.

(1) Isolated Power Circuits. Each isolated power circuit shall be controlled by a switch that has a disconnecting pole in each isolated circuit conductor to simultaneously disconnect all power. Such isolation shall be accomplished by means of one or more transformers having no electrical connection between primary and secondary windings, by means of motor generator sets, or by means of suitably isolated batteries.

(2) Circuit Characteristics. Circuits supplying primaries of isolating transformers shall operate at not more than 600 volts between conductors and shall be provided with proper overcurrent protection. The secondary voltage of such transformers shall not exceed 600 volts between conductors of each circuit. All circuits supplied from such secondaries shall be ungrounded and shall have an approved overcurrent device of proper ratings in each conductor. Circuits supplied directly from batteries or from motor generator sets shall be ungrounded and shall be protected against overcurrent in the same manner as transformer-fed secondary circuits. If an electrostatic shield is present, it shall be connected to the reference grounding point. [NFPA 99, 3.3.2.2.1]

(3) Equipment Location. The isolating transformers, motor generator sets, batteries and battery chargers, and associated primary or secondary overcurrent devices shall not be installed in hazardous (classified) locations. The isolated secondary circuit wiring extending into a hazardous anesthetizing location shall be installed in accordance with 501.4.

(4) Isolation Transformers. An isolation transformer shall not serve more than one operating room except as covered in (a) and (b).

For purposes of this section, anesthetic induction rooms are considered part of the operating room or rooms served by the induction rooms.

(a) Induction Rooms. Where an induction room serves more than one operating room, the isolated circuits of the induction room shall be permitted to be supplied from the isolation transformer of any one of the operating rooms served by that induction room.

(b) Higher Voltages. Isolation transformers shall be permitted to serve single receptacles in several patient areas where the following apply:

(1) The receptacles are reserved for supplying power to equipment requiring 150 volts or higher, such as portable X-ray units.
(2) The receptacles and mating plugs are not interchangeable with the receptacles on the local isolated power system. [NFPA 99, 12.4.1.2.6(d), 12.4.1.2.6(e)]

(5) Conductor Identification. The isolated circuit conductors shall be identified as follows:

(1) Isolated Conductor No. 1—Orange
(2) Isolated Conductor No. 2—Brown

For 3-phase systems, the third conductor shall be identified as yellow. Where isolated circuit conductors supply

125-volt, single-phase, 15- and 20-ampere receptacles, the orange conductor(s) shall be connected to the terminal(s) on the receptacles that are identified in accordance with 200.10(B) for connection to the grounded circuit conductor.

(6) Wire-Pulling Compounds. Wire-pulling compounds that increase the dielectric constant shall not be used on the secondary conductors of the isolated power supply.

> FPN No. 1: It is desirable to limit the size of the isolation transformer to 10 kVA or less and to use conductor insulation with low leakage to meet impedance requirements.
>
> FPN No. 2: Minimizing the length of branch-circuit conductors and using conductor insulations with a dielectric constant less than 3.5 and insulation resistance constant greater than 6100 megohm-meters (20,000 megohm-ft) at 16°C (60°F) reduces leakage from line to ground, reducing the hazard current.

See Exhibit 517.3 for an example of a hospital isolated power system panel.

Exhibit 517.3 An example of a hospital isolated power system panel with built-in isolation transformer, line isolation monitor, load center, and grounded busbar. (Courtesy of Square D Co.)

(B) Line Isolation Monitor.

(1) Characteristics. In addition to the usual control and overcurrent protective devices, each isolated power system shall be provided with a continually operating line isolation monitor that indicates total hazard current. The monitor shall be designed so that a green signal lamp, conspicuously visible to persons in each area served by the isolated power system, remains lighted when the system is adequately isolated from ground. An adjacent red signal lamp and an audible warning signal (remote if desired) shall be energized when the total hazard current (consisting of possible resistive and capacitive leakage currents) from either isolated conductor to ground reaches a threshold value of 5 mA under nominal line voltage conditions. The line monitor shall not alarm for a fault hazard of less than 3.7 mA or for a total hazard current of less than 5 mA.

Exception: A system shall be permitted to be designed to operate at a lower threshold value of total hazard current. A line isolation monitor for such a system shall be permitted to be approved with the provision that the fault hazard current shall be permitted to be reduced but not to less than 35 percent of the corresponding threshold value of the total hazard current, and the monitor hazard current is to be correspondingly reduced to not more than 50 percent of the alarm threshold value of the total hazard current.

(2) Impedance. The line isolation monitor shall be designed to have sufficient internal impedance such that, when properly connected to the isolated system, the maximum internal current that can flow through the line isolation monitor, when any point of the isolated system is grounded, shall be 1 mA.

Exception: The line isolation monitor shall be permitted to be of the low-impedance type such that the current through the line isolation monitor, when any point of the isolated system is grounded, will not exceed twice the alarm threshold value for a period not exceeding 5 milliseconds.

> FPN: Reduction of the monitor hazard current, provided this reduction results in an increased "not alarm" threshold value for the fault hazard current, will increase circuit capacity.

(3) Ammeter. An ammeter calibrated in the total hazard current of the system (contribution of the fault hazard current plus monitor hazard current) shall be mounted in a plainly visible place on the line isolation monitor with the "alarm on" zone at approximately the center of the scale.

Exception: The line isolation monitor shall be permitted to be a composite unit, with a sensing section cabled to a separate display panel section on which the alarm or test functions are located.

> FPN: It is desirable to locate the ammeter so that it is conspicuously visible to persons in the anesthetizing location.

ARTICLE 518
Places of Assembly

Contents

518.1 Scope.

This article covers all buildings or portions of buildings or structures designed or intended for the assembly of 100 or more persons.

Article 518 applies to places of assembly designed or intended for 100 or more persons. Article 518 would apply, for example, to a church chapel or auditorium for occupancy of 100 or more persons, its capacity determined by the methods for occupancy population capacity in accordance with NFPA *101, Life Safety Code.* Article 518 does not apply to a supermarket, even though a supermarket may contain 100 or more persons, because a supermarket is not specifically designed or intended for the assembly of persons, nor is it considered to be an auditorium. Article 518 does not apply to an office building or a school building, even though such buildings, as a rule, are designed for occupancy by 100 or more persons. Article 518 does, however, apply to assembly halls, restaurants, and so on, within office or school buildings if these parts of the building are designed or intended for the assembly of 100 or more persons.

The following information for determining new assembly occupancy capacity is extracted from NFPA *101, Life Safety Code:*

12.1.7 Occupant Load.
12.1.7.1 The occupant load, in number of persons for whom means of egress and other provisions are required, shall be determined on the basis of the occupant load factors of Table 7.3.1.2 [shown here as commentary Table 5.4] that are characteristic of the use of the space or shall be determined as the maximum probable population of the space under consideration, whichever is greater. In areas not in excess of 10,000 ft^2 (930 m^2), the occupant load shall not exceed one person in 5 ft^2 (0.46 m^2); in areas in excess of 10,000 ft^2 (930 m^2), the occupant load shall not exceed one person in 7 ft^2 (0.65 m^2).

Table 5.4 Occupant Load Factor

Use	ft^2† (per person)	m^2† (per person)
Assembly Use		
Concentrated use, without fixed seating	7 net	0.65 net
Less concentrated use, without fixed seating	15 net	1.4 net
Bench-type seating	1 person/18 linear in.	1 person/45.7 linear cm
Fixed seating	Number of fixed seats	Number of fixed seats
Waiting spaces	*See 12.1.7.2 and 13.1.7.2.*	*See 12.1.7.2 and 13.1.7.2.*
Kitchens	100	9.3
Library stack areas	100	9.3
Library reading rooms	50 net	4.6 net
Swimming pools	50 – of water surface	4.6 – of water surface
Swimming pool decks	30	2.8
Exercise rooms with equipment	50	4.6
Exercise rooms without equipment	15	1.4
Stages	15 net	1.4 net

Table 5.4 *Continued*

Use	ft²† (per person)	m²† (per person)
Lighting and access catwalks, galleries, gridirons	100 net	9.3 net
Casinos and similar gaming areas	11	1
Skating rinks	50	4.6
Educational Use		
Classrooms	20 net	1.9 net
Shops, laboratories, vocational rooms	50 net	4.6 net
Day-Care Use	35 net	3.3 net
Health Care Use		
Inpatient treatment departments	240	22.3
Sleeping departments	120	11.1
Detention and Correctional Use	120	11.1
Residential Use		
Hotels and dormitories	200	18.6
Apartment buildings	200	18.6
Board and care, large	200	18.6
Industrial Use		
General and high hazard industrial	100	9.3
Special purpose industrial	NA‡	NA‡
Business Use	100	9.3
Storage Use (other than mercantile storerooms)	NA‡	NA‡
Mercantile Use		
Sales area on street floor§◇	30	2.8
Sales area on two or more street floors◇	40	3.7
Sales area on floor below street floor◇	30	2.8
Sales area on floors above street floor◇	60	5.6
Floors or portions of floors used only for offices	*See business use.*	*See business use.*
Floors or portions of floors used only for storage, receiving, and shipping, and not open to general public	300	27.9
Covered mall buildings	Per factors applicable to use of space#	Per factors applicable to use of space#

†All factors expressed in gross area unless marked "net."

‡Not applicable. The occupant load shall be not less than the maximum probable number of occupants present at any time.

§For the purpose of determining occupant load in mercantile occupancies where, due to differences in grade of streets on different sides, two or more floors directly accessible from streets (not including alleys or similar back streets) exist, each such floor shall be considered a street floor. The occupant load factor shall be one person for each 40 ft² (3.7 m²) of gross floor area of sales space.

◇In mercantile occupancies with no street floor, as defined in 3.3.196, but with access directly from the street by stairs or escalators, the principal floor at the point of entrance to the mercantile occupancy shall be considered the street floor.

#The portions of the covered mall, where considered a pedestrian way and not used as gross leasable area, shall not be assessed an occupant load based on Table 7.3.1.2. However, means of egress from a covered mall pedestrian way shall be provided for an occupant load determined by dividing the gross leasable area of the covered mall building (not including anchor stores) by the appropriate lowest whole number occupant load factor from Figure 7.3.1.2.

Each individual tenant space shall have means of egress to the outside or to the covered mall based on occupant loads figured by using the appropriate occupant load factor from Table 7.3.1.2.

Each individual anchor store shall have means of egress independent of the covered mall.

Source: Table 7.3.1.2 of NFPA *101, Life Safety Code.*

7.3.1 Occupant Load.

7.3.1.1 The total capacity of the means of egress for any story, balcony, tier, or other occupied space shall be sufficient for the occupant load thereof.

7.3.1.2* The occupant load in any building or portion thereof shall be not less than the number of persons determined by dividing the floor area assigned to that use by the occupant load factor for that use as specified in Table 7.3.1.2. Where both gross and net area figures are given for the same occupancy, calculations shall be made by applying the gross area figure to the gross area of the portion of the building devoted to the use for which the gross area figure is specified and by applying the net area figure to the net area of the use for which the net area figure is specified.

518.2 General Classifications.

(A) Examples. Places of assembly shall include but not be limited to the following:

Armories	Courtrooms
Assembly halls	Dance halls
Auditoriums	Dining facilities
Auditoriums within	Exhibition halls
Business establishments	Gymnasiums
Mercantile establishments	Mortuary chapels
Other occupancies	Multipurpose rooms
Schools	Museums
Bowling lanes	Places of awaiting transportation
Church chapels	Pool rooms
Club rooms	Restaurants
Conference rooms	Skating rinks

(B) Multiple Occupancies. Occupancy of any room or space for assembly purposes by less than 100 persons in a building of other occupancy, and incidental to such other occupancy, shall be classified as part of the other occupancy and subject to the provisions applicable thereto.

(C) Theatrical Areas. Where any such building structure, or portion thereof, contains a projection booth or stage platform or area for the presentation of theatrical or musical productions, either fixed or portable, the wiring for that area, including associated audience seating areas, and all equipment that is used in the referenced area, and portable equipment and wiring for use in the production that will not be connected to permanently installed wiring, shall comply with Article 520.

See the commentary following 520.1.

FPN: For methods of determining population capacity, see local building code or, in its absence, NFPA *101-2000, Life Safety Code.*

518.3 Other Articles.

(A) Hazardous (Classified) Areas. Electrical installations in hazardous (classified) areas located in places of assembly shall comply with Article 500.

(B) Temporary Wiring. In exhibition halls used for display booths, as in trade shows, the temporary wiring shall be installed in accordance with Article 527. Flexible cables and cords approved for hard or extra-hard usage shall be permitted to be laid on floors where protected from contact by the general public. The ground-fault circuit-interrupter requirements of 527.6 shall not apply.

A treadle, such as the one shown in Exhibit 518.1, is an example of a protection technique used to protect cords from abuse in areas where the cords are laid across pedestrian ways. Treadles, for example, may protect temporary wiring used in exhibition halls.

Exhibit 518.1 A treadle used to protect cords. (Courtesy of Daniel Woodhead Co.)

Exception: Where conditions of supervision and maintenance ensure that only qualified persons will service the installation, flexible cords or cables identified in Table 400.4 for hard usage or extra-hard usage shall be permitted in cable trays used only for temporary wiring. All cords or cables shall be installed in a single layer. A permanent sign shall be attached to the cable tray at intervals not to exceed 7.5 m (25 ft). The sign shall read

    CABLE TRAY FOR TEMPORARY WIRING ONLY

(C) Emergency Systems. Control of emergency systems shall comply with Article 700.

518.4 Wiring Methods.

(A) General. The fixed wiring methods shall be metal raceways, flexible metal raceways, nonmetallic raceways encased in not less than 50 mm (2 in.) of concrete, Type MI, MC, or AC cable containing an insulated equipment grounding conductor sized in accordance with Table 250.122.

Exception: Fixed wiring methods shall be as provided in

(a) *Audio signal processing, amplification, and reproduction equipment—Article 640*
(b) *Communications circuits—Article 800*
(c) *Class 2 and Class 3 remote-control and signaling circuits—Article 725*
(d) *Fire alarm circuits—Article 760*

(B) Nonrated Construction. Nonmetallic-sheathed cable, Type AC cable, electrical nonmetallic tubing, and rigid nonmetallic conduit shall be permitted to be installed in those buildings or portions thereof that are not required to be of fire-rated construction by the applicable building code.

FPN: Fire-rated construction is the fire-resistive classification used in building codes.

(C) Spaces with Finish Rating. Electrical nonmetallic tubing and rigid nonmetallic conduit shall be permitted to be installed in club rooms, college and university classrooms, conference and meeting rooms in hotels or motels, courtrooms, drinking establishments, dining facilities, restaurants, mortuary chapels, museums, passenger stations and terminals of air, surface, underground, and marine public transportation facilities, libraries, and places of religious worship where the following apply:

(1) The electrical nonmetallic tubing or rigid nonmetallic conduit is installed concealed within walls, floors, and ceilings where the walls, floors, and ceilings provide a thermal barrier of material that has at least a 15-minute finish rating as identified in listings of fire-rated assemblies.
(2) The electrical nonmetallic tubing or rigid nonmetallic conduit is installed above suspended ceilings where the suspended ceilings provide a thermal barrier of material that has at least a 15-minute finish rating as identified in listings of fire-rated assemblies.

Electrical nonmetallic tubing and rigid nonmetallic conduit are not recognized for use in other space used for environmental air in accordance with 300.22(C).

FPN: A finish rating is established for assemblies containing combustible (wood) supports. The finish rating is defined as the time at which the wood stud or wood joist

reaches an average temperature rise of 121°C (250°F) or an individual temperature rise of 163°C (325°F) as measured on the plane of the wood nearest the fire. A finish rating is not intended to represent a rating for a membrane ceiling.

Referring to Exhibit 518.2, the wash rooms and office area of this single-story facility are not places of assembly, as defined in 518.2, and therefore require no special wiring methods. Ordinary wiring methods can be used on the inside surface of the storage area walls and on or in the partitions between storage areas, as these also are not places of assembly. Inside any hollow spaces of the fire-rated storage area walls, however, the main requirements of 518.4 apply, because the serving corridors are part of the places of assembly as a result of this particular building design. If the hollow spaces of fire-rated walls or ceiling also provide a 15-minute finish rating and are not other spaces for environmental air, as described in 300.22(C), then electrical nonmetallic tubing as well as rigid nonmetallic conduit would be permitted for specifically described occupancies.

In like manner, wiring in ceilings or floors that is required to be of fire-rated construction in a place of assembly as defined in 518.2 must also comply with 518.4, except as

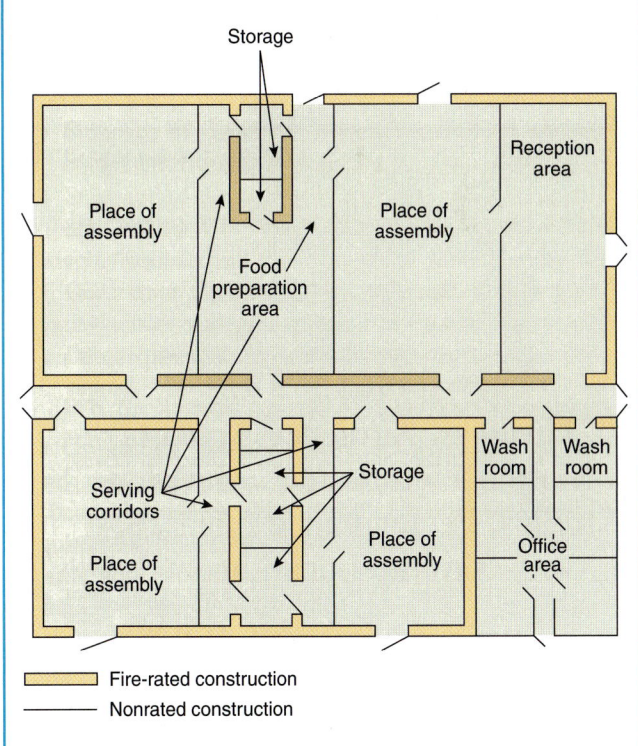

Exhibit 518.2 The walls represented by the wide gold lines are those required by the local building code to be of fire-rated construction. The thin black lines represent walls not required by the local building code to be of fire-rated construction.

noted in 518.4(A), Exception.

The intent is that, within a place of assembly, as defined in 518.2, 518.4(A) and the exception to 518.4(A) apply to any wall, floor, or ceiling. The provisions of 518.4(B) apply to those portions of the building and those places of assembly not required to be fire rated. The permission to use electrical nonmetallic tubing and rigid nonmetallic conduit as specified in 518.4(C), applies only to the specific occupancies described, provided these wiring methods are installed concealed behind a surface that has a 15-minute finish rating.

Places of assembly frequently require emergency wiring, particularly for emergency illumination and exit lighting. The requirements of 700.9(D) contain special fire protection requirements for emergency circuits in assembly occupancies with an occupant capacity of 1000 or more.

518.5 Supply.

Portable switchboards and portable power distribution equipment shall be supplied only from listed power outlets of sufficient voltage and ampere rating. Such power outlets shall be protected by overcurrent devices. Such overcurrent devices and power outlets shall not be accessible to the general public. Provisions for connection of an equipment grounding conductor shall be provided. The neutral of feeders supplying solid-state, 3-phase, 4-wire dimmer systems shall be considered a current-carrying conductor.

In accordance with 518.5, portable switchboards and portable power distribution equipment must be supplied only from listed power outlets rated for the voltage and current they are used for. See Exhibit 518.3. The overcurrent devices and power outlets used to supply such equipment must be located so as not to be accessible to the general public.

ARTICLE 520
Theaters, Audience Areas of Motion Picture and Television Studios, Performance Areas, and Similar Locations

Contents

Exhibit 518.3 A listed power outlet for connection of portable switchboards in a place of assembly. (Courtesy of Union Connector Co, Inc.)

I. General

520.1 Scope.

This article covers all buildings or that part of a building or structure, indoor or outdoor, designed or used for presentation, dramatic, musical, motion picture projection, or similar purposes and to specific audience seating areas within motion picture or television studios.

The special requirements of Article 520 apply only to that part of a building used as a theater or for a similar purpose and do not necessarily apply to the entire building. For example, the requirements of Article 520 would apply to an auditorium in a school building used for dramatic or other performances. The special requirements of this chapter apply to the stage, auditorium, dressing rooms, and main corridors leading to the auditorium, but not to other parts of the building that are not involved in the use of the auditorium for performances or entertainment. The theater space may be a traditional theater, where the audience sits in the auditorium (house) facing the proscenium arch and views the performance on the stage on the other side of the arch, or other spaces, such as a simple stage platform, either indoors or outdoors, with seats on three or four sides facing the platform.

The audience areas of motion picture and television studios (as defined and covered in Article 530) are also covered by the requirements of Article 520.

520.2 Definitions.

Border Light. A permanently installed overhead strip light.

Breakout Assembly. An adapter used to connect a multipole connector containing two or more branch circuits to multiple individual branch-circuit connectors.

Bundled. Cables or conductors that are physically tied, wrapped, taped, or otherwise periodically bound together.

Connector Strip. A metal wireway containing pendant or flush receptacles.

Drop Box. A box containing pendant- or flush-mounted receptacles attached to a multiconductor cable via strain relief or a multipole connector.

Footlight. A border light installed on or in the stage.

Grouped. Cables or conductors positioned adjacent to one another but not in continuous contact with each other.

Performance Area. The stage and audience seating area associated with a temporary stage structure, whether indoors or outdoors, constructed of scaffolding, truss, platforms, or similar devices, that is used for the presentation of theatrical or musical productions or for public presentations.

Portable Equipment. Equipment fed with portable cords or cables intended to be moved from one place to another.

Portable Power Distribution Unit. A power distribution box containing receptacles and overcurrent devices.

Proscenium. The wall and arch that separates the stage from the auditorium (house).

Stand Lamp (Work Light). A portable stand that contains a general-purpose luminaire (lighting fixture) or lampholder with guard for the purpose of providing general illumination on the stage or in the auditorium.

Strip Light. A luminaire (lighting fixture) with multiple lamps arranged in a row.

Two-Fer. An adapter cable containing one male plug and two female cord connectors used to connect two loads to one branch circuit.

A two-fer, as shown in Exhibit 520.1, consists of two cord connectors on separate cords connected to a single supply cord.

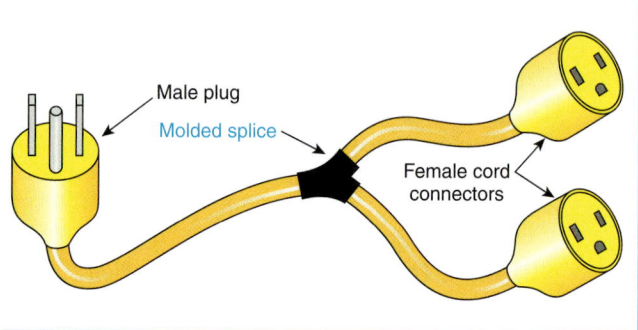

Male plug
Molded splice
Female cord connectors

Exhibit 520.1 A two-fer.

520.3 Motion Picture Projectors.

Motion picture equipment and its installation and use shall comply with Article 540.

520.4 Audio Signal Processing, Amplification, and Reproduction Equipment.

Audio signal processing, amplification, and reproduction equipment and its installation shall comply with Article 640.

520.5 Wiring Methods.

(A) General. The fixed wiring method shall be metal raceways, nonmetallic raceways encased in at least 50 mm (2 in.) of concrete, Type MI cable, MC cable, or AC cable containing an insulated equipment grounding conductor sized in accordance with Table 250.122.

Exception: Fixed wiring methods shall be as provided in Article 640 for audio signal processing, amplification, and reproduction equipment, in Article 800 for communication circuits, in Article 725 for Class 2 and Class 3 remote-control and signaling circuits, and in Article 760 for fire alarm circuits.

(B) Portable Equipment. The wiring for portable switchboards, stage set lighting, stage effects, and other wiring not fixed as to location shall be permitted with approved flexible cords and cables as provided elsewhere in Article 520. Fastening such cables and cords by uninsulated staples or nailing shall not be permitted.

(C) Nonrated Construction. Nonmetallic-sheathed cable, Type AC cable, electrical nonmetallic tubing, and rigid nonmetallic conduit shall be permitted to be installed in those buildings or portions thereof that are not required to be of fire-rated construction by the applicable building code.

Theaters and similar buildings are usually required to be of fire-rated construction, as determined by applicable building codes; therefore, the fixed wiring methods are limited. See 518.4 for the requirements on wiring methods.

The exception to the requirements for metal-enclosed, or concrete- enclosed, fixed wiring permits the installation of communications circuits, Class 2 and Class 3 remote-control and signaling circuits, sound-reproduction wiring, and fire alarm circuits using wiring methods from the respective articles covering these systems in Chapters 7 and 8. Where portability, flexibility, and adjustments are necessary for portable switchboards, stage lighting, and special effects, suitable cords and cables are permitted. In accordance with 520.5(C), Type NM cable, Type AC cable, ENT, and RNC are permitted as the wiring method in buildings or portions of buildings that are not required to be of fire-rated construction. In this application, Type AC cable is not required to contain an insulated equipment grounding conductor.

520.6 Number of Conductors in Raceway.

The number of conductors permitted in any metal conduit, rigid nonmetallic conduit as permitted in this article, or electrical metallic tubing for border or stage pocket circuits or for remote-control conductors shall not exceed the percentage fill shown in Table 1 of Chapter 9. Where contained within an auxiliary gutter or a wireway, the sum of the cross-sectional areas of all contained conductors at any cross section shall not exceed 20 percent of the interior cross-sectional area of the auxiliary gutter or wireway. The 30-conductor limitation of 366.6 and 376.22 shall not apply.

520.7 Enclosing and Guarding Live Parts.

Live parts shall be enclosed or guarded to prevent accidental contact by persons and objects. All switches shall be of the externally operable type. Dimmers, including rheostats, shall be placed in cases or cabinets that enclose all live parts.

520.8 Emergency Systems.

Control of emergency systems shall comply with Article 700.

520.9 Branch Circuits.

A branch circuit of any size supplying one or more receptacles shall be permitted to supply stage set lighting. The voltage rating of the receptacles shall not be less than the circuit voltage. Receptacle ampere ratings and branch-circuit conductor ampacity shall not be less than the branch-circuit overcurrent device ampere rating. Table 210.21(B)(2) shall not apply.

The stage set lighting and associated equipment, such as stage effects, both fixed and portable, must be as flexible as possible. Connectors are often used for different purposes and are therefore marked on a show-by-show basis as to the voltage, current, and type of current actually employed. The provisions of 520.9 only require that connectors be rated sufficiently for the parameters involved, thus permitting connectors with voltage and current ratings higher than the branch-circuit rating to be used.

The intent of 520.9 is to exclude the occupancies referenced in Article 520 from all the general requirements relating to connector rating and branch-circuit loading found elsewhere in the *Code*, such as in Table 210.21(B)(2). The requirements of 520.9 modify several other sections, such as 210.23(C) and (D), which would disallow 40-ampere and larger branch circuits from serving 5000-watt and larger portable stage lighting equipment found in the theater.

Stage set lighting is usually planned in advance, and the loads on each receptacle are known. Loads are not casually connected, as they might be at a typical general-use wall receptacle. Care is taken to ensure that circuits are not overloaded, thereby avoiding nuisance tripping during a performance.

520.10 Portable Equipment.

Portable stage and studio lighting equipment and portable power distribution equipment shall be permitted for temporary use outdoors, provided the equipment is supervised by qualified personnel while energized and barriered from the general public.

In accordance with 520.10, portable indoor stage or studio equipment that is not marked suitable for wet or damp locations is permitted to be used temporarily in outdoor locations. If rain occurs, this equipment is typically de-energized, and a protective cover is installed before it is re-energized. At the end of the day, this equipment is either de-energized and protected or dismantled and stored.

II. Fixed Stage Switchboards

520.21 Dead Front.

Stage switchboards shall be of the dead-front type and shall comply with Part IV of Article 384 unless approved based on suitability as a stage switchboard as determined by a qualified testing laboratory and recognized test standards and principles.

Early stage switchboards were vertical marble or slate slabs mounted on the stage near the proscenium wall, with exposed knife switches and fuseholders mounted on them and with exposed resistance-type dimmer plates across the top. The "dead front," "guarded back," and "metal hood" requirements of the *Code* are intended to provide the operator with some sort of protection from shock, and the heat-producing equipment with some sort of protection from flammable curtains and scenery likely to be above and around the equipment. For these reasons, modern switchboards are totally enclosed.

520.22 Guarding Back of Switchboard.

Stage switchboards having exposed live parts on the back of such boards shall be enclosed by the building walls, wire mesh grills, or by other approved methods. The entrance to this enclosure shall be by means of a self-closing door.

520.23 Control and Overcurrent Protection of Receptacle Circuits.

Means shall be provided at a stage-lighting switchboard to which load circuits are connected for overcurrent protection of stage-lighting branch circuits, including branch circuits supplying stage and auditorium receptacles used for cord- and plug-connected stage equipment. Where the stage switchboard contains dimmers to control nonstage lighting, the locating of the overcurrent protective devices for these branch circuits at the stage switchboard shall be permitted.

The purpose of 520.23 is to ensure that the overcurrent protection devices are readily accessible to stage personnel during the presentation.

520.24 Metal Hood.

A stage switchboard that is not completely enclosed dead-front and dead-rear or recessed into a wall shall be provided with a metal hood extending the full length of the board to protect all equipment on the board from falling objects.

Because stages are usually crowded and a great deal of flammable material is often present, a stage switchboard is not permitted to have exposed live parts on its front. Moreover, the space at the rear of a stage switchboard must be guarded in order to prevent entrance or contact by unqualified and unauthorized persons. One accepted method of accomplishing this is by enclosing the space between the rear of the switchboard and the wall in a sheet-steel housing with a door at one end.

520.25 Dimmers.

Dimmers shall comply with 520.25(A) through (D).

(A) Disconnection and Overcurrent Protection. Where dimmers are installed in ungrounded conductors, each dim-

mer shall have overcurrent protection not greater than 125 percent of the dimmer rating and shall be disconnected from all ungrounded conductors when the master or individual switch or circuit breaker supplying such dimmer is in the open position.

A modern, high-density digital dimmer rack typically contains one dimmer (usually of 20-, 50-, or 100-ampere capacity) for each branch circuit connected to it. The rack is usually serviced by a 3-phase, 4-wire-plus-ground feeder, which is distributed via buses to all dimmers in the rack. Typical dimmer racks contain between 12 and 96 dimmers and may have total power capacities of up to 288 kW. In large theatrical systems, many racks may be bused together. A central control electronics module drives multiple dimmers in the rack. A digital data link may connect the dimmer rack to the remotely located computer control console. Exhibit 520.2 shows a high-density digital SCR dimmer switchboard, and Exhibit 520.3 shows its schematic diagram.

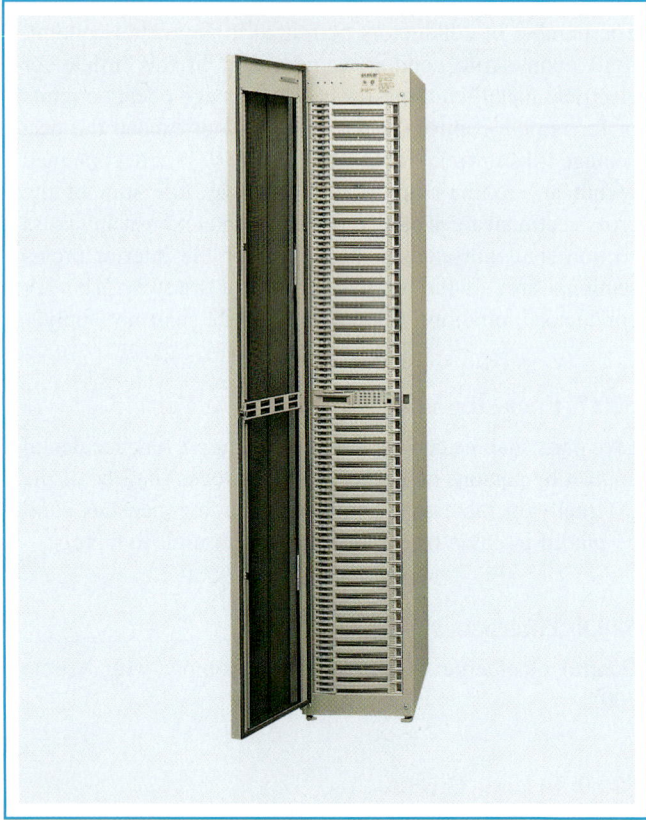

Exhibit 520.2 A typical high-density digital SCR dimmer switchboard. (Courtesy of Electronic Theatre Controls, Inc.)

(B) Resistance- or Reactor-Type Dimmers. Resistance- or series reactor-type dimmers shall be permitted to be placed

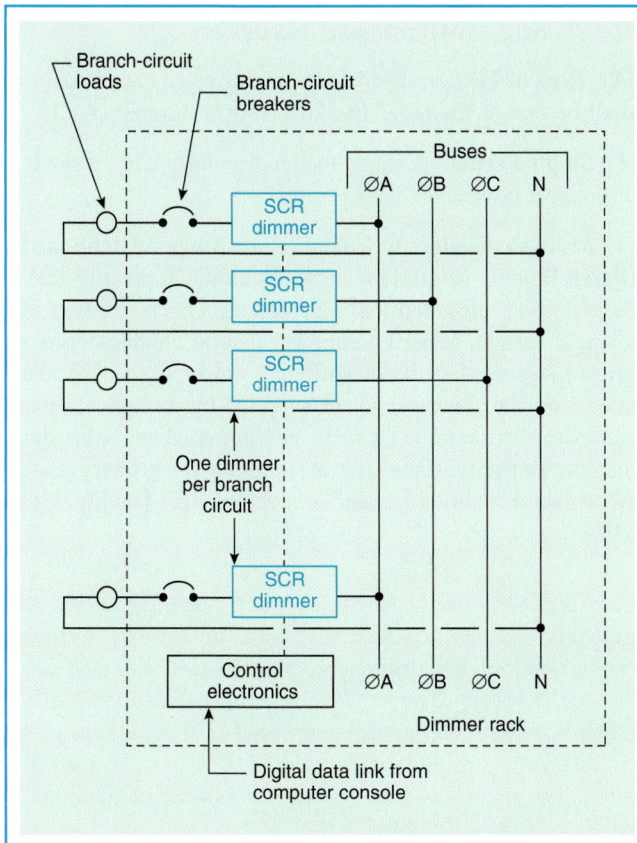

in either the grounded or the ungrounded conductor of the circuit. Where designed to open either the supply circuit to the dimmer or the circuit controlled by it, the dimmer shall then comply with 404.1. Resistance- or reactor-type dimmers placed in the grounded neutral conductor of the circuit shall not open the circuit.

(C) Autotransformer-Type Dimmers. The circuit supplying an autotransformer-type dimmer shall not exceed 150 volts between conductors. The grounded conductor shall be common to the input and output circuits.

> FPN: See 210.9 for circuits derived from autotransformers.

Circuits supplying autotransformer-type dimmers are not permitted to exceed 150 volts between conductors. Any desired voltage may be applied to the lamps, from full-line voltage to voltage so low that the lamps provide no illumination, by means of a movable contact tap. Typical connections for an autotransformer-type dimmer are shown in Exhibit 520.4 This type of dimmer produces very little

heat and operates at high efficiency. Its dimming effect, within its maximum rating, is independent of the wattage of the load. Autotransformer-type dimmers are currently seldom used. See the commentary that follows 470.4, which discusses saturable reactors that are sometimes used for stage dimmers.

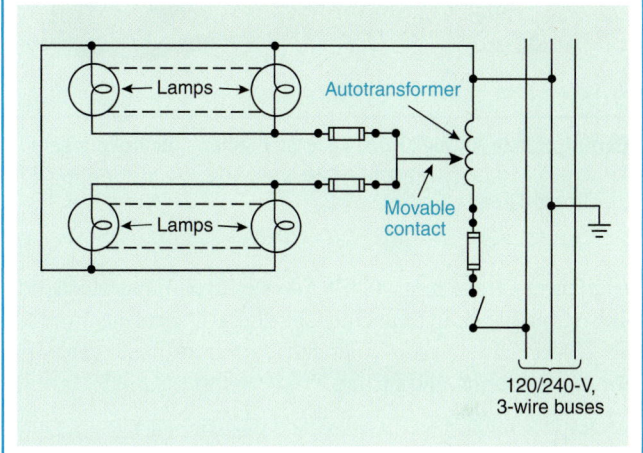

(D) Solid-State-Type Dimmers. The circuit supplying a solid-state dimmer shall not exceed 150 volts between conductors unless the dimmer is listed specifically for higher voltage operation. Where a grounded conductor supplies a dimmer, it shall be common to the input and output circuits. Dimmer chassis shall be connected to the equipment grounding conductor.

Modern stage switchboards are usually of the remote-control type. The switchboard is operated from a remote console, typically a computer system such as the one shown in Exhibit 520.5. The switchboard or dimmer rack is normally located offstage in a dimmer room, where proper climate control can be furnished and noise from the rack cooling fans will not interfere with the performance onstage. Branch circuits are usually connected to the dimmer rack on a "dimmer per circuit" basis. A digital control cable connects the computer and the dimmer rack, allowing the operator to be positioned on stage or in the auditorium for easy viewing of the performance.

A front view of a typical high-density digital SCR dimmer rack is shown in Exhibit 520.2. A schematic for this type of dimmer rack is shown in Exhibit 520.3. Dimmers for individual circuits are contained in dual plug-in dimmer modules. These modules also contain circuit breakers for

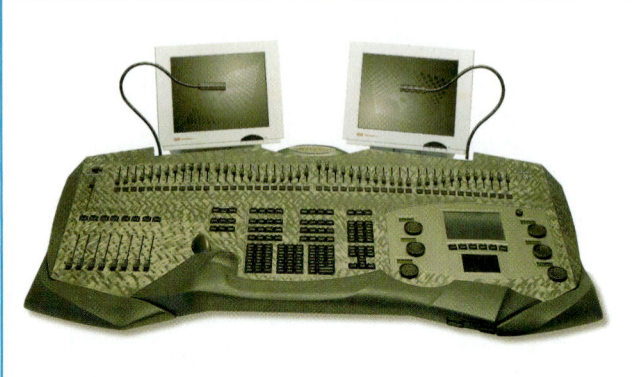

Exhibit 520.5 An electronic computer lighting control console for remotely controlling solid-state-type dimmers. (Courtesy of Electronic Theatre Controls, Inc.)

overcurrent protection and filter chokes to eliminate acoustic noise from the lamp filaments. The digital control electronics are contained in a plug-in module with front-panel controls for configuration and testing.

520.26 Type of Switchboard.

A stage switchboard shall be either one or a combination of the types specified in 520.26(A), (B), and (C).

(A) Manual. Dimmers and switches are operated by handles mechanically linked to the control devices.

Manual-type switchboards usually contain resistance-type or autotransformer-type dimmers. Exhibit 520.4 is a schematic of a manual autotransformer-type dimmer.

(B) Remotely Controlled. Devices are operated electrically from a pilot-type control console or panel. Pilot control panels either shall be part of the switchboard or shall be permitted to be at another location.

(C) Intermediate. A stage switchboard with circuit interconnections is a secondary switchboard (patch panel) or panelboard remote to the primary stage switchboard. It shall contain overcurrent protection. Where the required branch-circuit overcurrent protection is provided in the dimmer panel, it shall be permitted to be omitted from the intermediate switchboard.

An intermediate stage switchboard, usually called a patch panel, is located between the dimmer switchboard and the branch circuits. Its purpose is to either break down larger dimmer circuits to smaller branch circuits or to select branch circuits to be controlled by a dimmer, or both.

520.27 Stage Switchboard Feeders.

(A) Type of Feeder. Feeders supplying stage switchboards shall be one of the types in 520.27(A)(1) through (A)(3).

(1) Single Feeder. A single feeder disconnected by a single disconnect device.

(2) Multiple Feeders to Intermediate Stage Switchboard (Patch Panel). Multiple feeders of unlimited quantity shall be permitted, provided that all multiple feeders are part of a single system. Where combined, neutral conductors in a given raceway shall be of sufficient ampacity to carry the maximum unbalanced current supplied by multiple feeder conductors in the same raceway, but they need not be greater than the ampacity of the neutral supplying the primary stage switchboard. Parallel neutral conductors shall comply with 310.4.

The feeders to patch panels are often many dimmer-controlled circuits at 100 amperes or less, single phase, so they can be distributed to different combinations of the same size or smaller branch circuits. This type of installation usually requires a common neutral, and because of the quantity of circuits, many installations require several parallel neutrals running in several raceways. Generally, these parallel neutrals are sized as follows:

1. Size the common neutral to the feeder of the primary switchboard.
2. Split this neutral into multiple parallel conductors, one per raceway.
3. Equally divide, per phase, and size each ungrounded conductor of the many single-phase circuits among the raceways.

In no case is it acceptable to install the ungrounded conductors in one raceway and the common neutral in another.

(3) Separate Feeders to Single Primary Stage Switchboard (Dimmer Bank). Installations with separate feeders to a single primary stage switchboard shall have a disconnecting means for each feeder. The primary stage switchboard shall have a permanent and obvious label stating the number and location of disconnecting means. If the disconnecting means are located in more than one distribution switchboard, the primary stage switchboard shall be provided with barriers to correspond with these multiple locations.

Larger primary stage switchboards usually consist of several sections, often called dimmer racks, which form a dimmer bank. See Exhibit 520.2. These dimmer racks may be fed separately or be bused together to accept one or more feeder circuits. If an intermediate stage switchboard is connected to a primary stage switchboard, a single large feeder usually supplies the primary stage switchboard, because the intermediate stage switchboard patches only the ungrounded conductors and requires a common neutral. Modern theaters do not use intermediate stage switchboards, and dimmer banks may have one or several feeders.

(B) Neutral. The neutral of feeders supplying solid-state, 3-phase, 4-wire dimming systems shall be considered a current-carrying conductor.

(C) Supply Capacity. For the purposes of computing supply capacity to switchboards, it shall be permissible to consider the maximum load that the switchboard is intended to control in a given installation, provided that the following apply:

(1) All feeders supplying the switchboard shall be protected by an overcurrent device with a rating not greater than the ampacity of the feeder.
(2) The opening of the overcurrent device shall not affect the proper operation of the egress or emergency lighting systems.

> FPN: For computation of stage switchboard feeder loads, see 220.10.

The feeder for single, primary stage switchboards is sized in accordance with the maximum load the switchboard is intended to control for a specific location. The feeder(s) must be protected by an overcurrent device that has a rating not greater than the feeder ampacity. Operation of the overcurrent device is not allowed to have any effect on egress or emergency lighting systems. The neutral of feeders supplying solid-state, 3-phase, 4-wire dimming systems will carry third-harmonic currents that are present even under balanced load conditions.

III. Fixed Stage Equipment Other Than Switchboards

520.41 Circuit Loads.

(A) Circuits Rated 20 Amperes or Less. Footlights, border lights, and proscenium sidelights shall be arranged so that no branch circuit supplying such equipment carries a load exceeding 20 amperes.

(B) Circuits Rated Greater Than 20 Amperes. Where heavy-duty lampholders only are used, such circuits shall be permitted to comply with Article 210 for circuits supplying heavy-duty lampholders.

In accordance with 210.23(B) and (C), 30-, 40-, or 50-ampere branch circuits are permitted if heavy-duty lampholders, such as medium- or mogul-base Edison screw shell types, are used for fixed lighting.

520.42 Conductor Insulation.

Foot, border, proscenium, or portable strip lights and connector strips shall be wired with conductors that have insulation suitable for the temperature at which the conductors are operated, but not less than 125°C (257°F). The ampacity of the 125°C (257°F) conductors shall be that of 60°C (140°F) conductors. All drops from connector strips shall be 90°C (194°F) wire sized to the ampacity of 60°C (140°F) cords and cables with no more than 150 mm (6 in.) of conductor extending into the connector strip. Section 310.15(B)(2)(a) shall not apply.

> FPN: See Table 310.13 for conductor types.

The 125°C (257°F) minimum temperature rating is based on the heat from the lamps raising the ambient temperature in which the wiring is located. Drops from connector strips are usually flexible cord. Although the 90°C-rated cord is also in the higher ambient, it is not in sufficient contact with other circuits that might also heat it. The derating factors of 310.15(B)(2)(a) are judged unnecessary because the conductors are not all energized at one time, are not often energized at full intensity (dimmed), and are not energized continuously.

520.43 Footlights.

(A) Metal Trough Construction. Where metal trough construction is employed for footlights, the trough containing the circuit conductors shall be made of sheet metal not lighter than 0.81 mm (0.032 in.) and treated to prevent oxidation. Lampholder terminals shall be kept at least 13 mm (½ in.) from the metal of the trough. The circuit conductors shall be soldered to the lampholder terminals.

(B) Other-Than-Metal Trough Construction. Where the metal trough construction specified in 520.43(A) is not used, footlights shall consist of individual outlets with lampholders wired with rigid metal conduit, intermediate metal conduit, or flexible metal conduit, Type MC cable, or mineral-insulated, metal-sheathed cable. The circuit conductors shall be soldered to the lampholder terminals.

(C) Disappearing Footlights. Disappearing footlights shall be arranged so that the current supply is automatically disconnected when the footlights are replaced in the storage recesses designed for them.

The footlights described in 520.43(A) and (B) are generally obsolete units that were built in the field. Modern footlights are compartmentalized, factory-wired assemblies for field installation, as shown in Exhibit 520.6. Footlight assemblies may be permanently exposed or be of the disappearing type. Disappearing footlights are arranged to automatically disconnect the current supply when the footlights are in the closed position, thereby preventing heat entrapment that could cause a fire. Disconnection is accomplished by mercury switches in the terminal compartment.

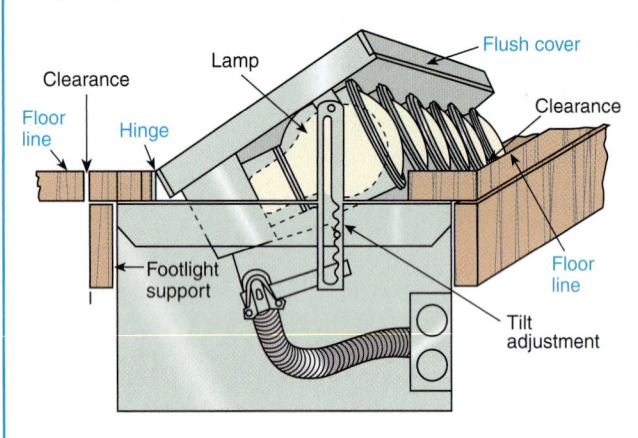

Exhibit 520.6 Disappearing footlights.

520.44 Borders and Proscenium Sidelights.

(A) General. Borders and proscenium sidelights shall be as follows:

(1) Constructed as specified in 520.43
(2) Suitably stayed and supported
(3) Designed so that the flanges of the reflectors or other adequate guards protect the lamps from mechanical damage and from accidental contact with scenery or other combustible material

Exhibit 520.7 shows a modern border light installed over a stage. Exhibit 520.8 is a cross-sectional view that illustrates construction details. This particular border light is designed for 200-watt lamps. To obtain the highest illumination efficiency, each lamp is provided with its own reflector. Fitted to each reflector is a glass roundel available in any color. Commonly, lampholders are wired alternately on three or four circuits. A splice box is provided on top of the housing for enclosing connections between the cable supplying the border light and the border light's internal wiring, which consists of wiring from the splice box to the lamp sockets in a trough extending the length of the border.

Exhibit 520.7 A suspended border light assembly for installation over a stage. (Redrawn from Kliegl Bros.)

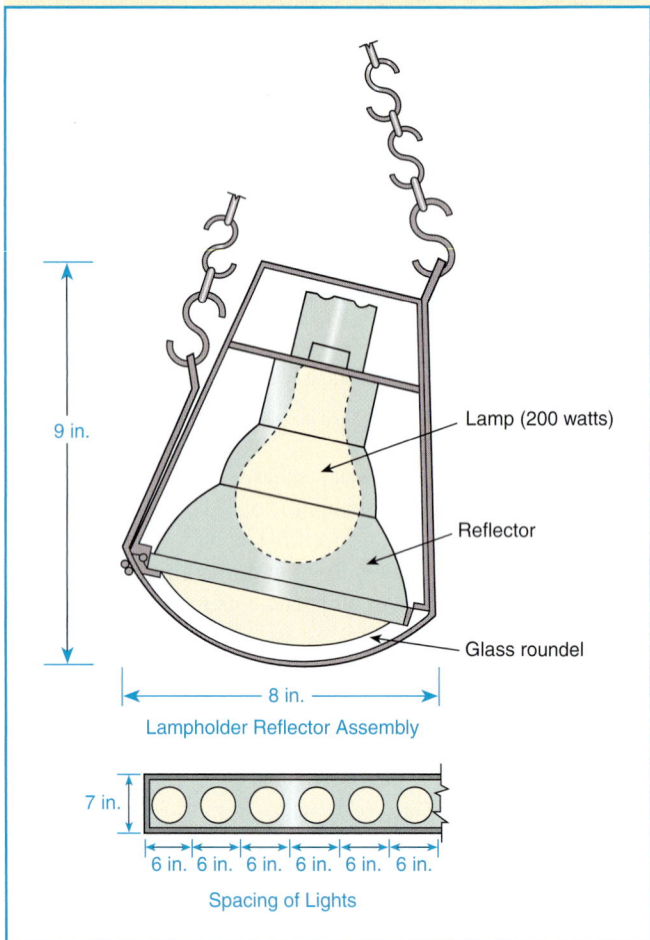

Exhibit 520.8 A cross-sectional view of a typical light in the border light assembly shown in Exhibit 520.7. (Redrawn from Kliegl Bros.)

(B) Cords and Cables for Border Lights.

(1) General. Cords and cables for supply to border lights shall be listed for extra-hard usage. The cords and cables shall be suitably supported. Such cords and cables shall be employed only where flexible conductors are necessary. Ampacity of the conductors shall be as provided in 400.5.

To facilitate height adjustment for cleaning and lamp replacement, border lights are usually supported by steel cables, as shown in Exhibit 520.9. Therefore, the circuit conductors supplying the border lights must be carried to the border light in a flexible cable. Each of these flexible cables usually contains many circuits; however, its overall size is limited by its ability to travel up and down without getting tangled.

(2) Cords and Cables Not in Contact with Heat-Producing Equipment. Listed multiconductor extra-hard-usage-type cords and cables not in direct contact with equipment containing heat-producing elements shall be permitted to have their ampacity determined by Table 520.44. Maximum load current in any conductor with an ampacity determined by Table 520.44 shall not exceed the values in Table 520.44.

The provisions of 520.44(B)(2) permit extra-hard-usage cords not in direct contact with heat-producing equipment to have their ampacity determined by Table 520.44 instead of 400.5.

Table 520.44 is based on a minimum 50 percent diversity factor. It includes the fact that not all circuits are on at the same time, not all circuits are at full intensity (dimmed), and not all circuits are on for a long period of time. If the load diversity does not follow this pattern, such as border lights that are all left on at full intensity to light the stage for rehearsal, lecture, or classroom purposes, this table must not be used.

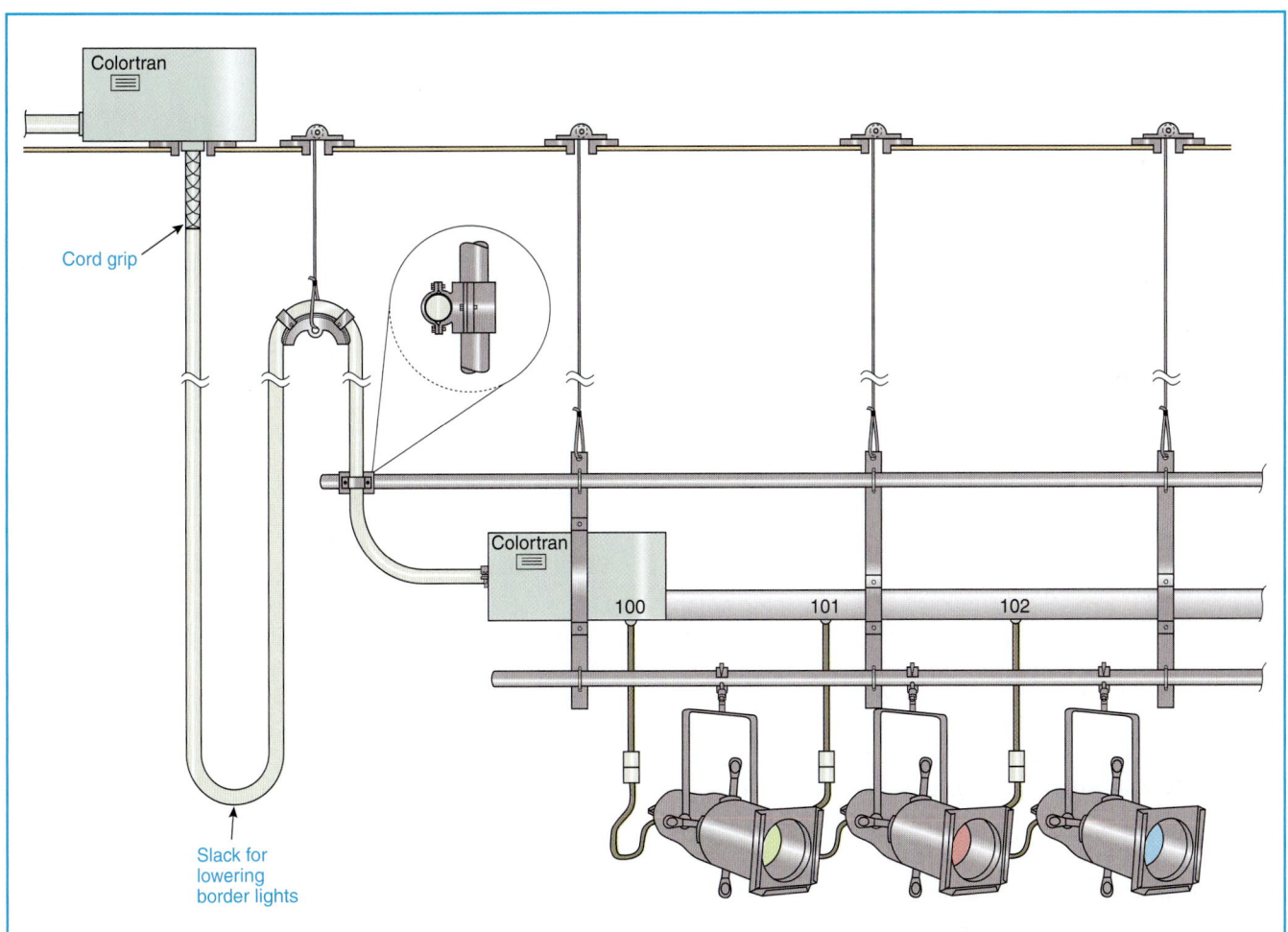

Exhibit 520.9 A suspended connector strip with border light cable attached. (Redrawn from Colortran, Inc.)

Table 520.44 Ampacity of Listed Extra-Hard-Usage Cords and Cables with Temperature Ratings of 75°C (167°F) and 90°C (194°F)* [Based on Ambient Temperature of 30°C (86°F)]

Size (AWG)	Temperature Rating of Cords and Cables		Maximum Rating of Overcurrent Device
	75°C (167°F)	90°C (194°F)	
14	24	28	15
12	32	35	20
10	41	47	25
8	57	65	35
6	77	87	45
4	101	114	60
2	133	152	80

*Ampacity shown is the ampacity for multiconductor cords and cables where only three copper conductors are current-carrying as described in 400.5. If the number of current-carrying conductors in a cord or cable exceeds three and the load diversity factor is a minimum of 50 percent, the ampacity of each conductor shall be reduced as shown in the following table.

Number of Conductors	Percent of Ampacity
4–6	80
7–24	70
25–42	60
43 and above	50

Note: Ultimate insulation temperature. In no case shall conductors be associated together in such a way with respect to the kind of circuit, the wiring method used, or the number of conductors such that the temperature limit of the conductors is exceeded.

A neutral conductor that carries only the unbalanced current from other conductors of the same circuit need not be considered as a current-carrying conductor.

In a 3-wire circuit consisting of two phase wires and the neutral of a 4-wire, 3-phase, wye-connected system, a common conductor carries approximately the same current as the line-to-neutral currents of the other conductors and shall be considered to be a current-carrying conductor.

On a 4-wire, 3-phase, wye circuit where the major portion of the load consists of nonlinear loads such as electric-discharge lighting, electronic computer/data processing, or similar equipment, there are harmonic currents present in the neutral conductor, and the neutral shall be considered to be a current-carrying conductor.

520.45 Receptacles.

Receptacles for electrical equipment on stages shall be rated in amperes. Conductors supplying receptacles shall be in accordance with Articles 310 and 400.

520.46 Connector Strips, Drop Boxes, Floor Pockets, and Other Outlet Enclosures.

Receptacles for the connection of portable stage-lighting equipment shall be pendant or mounted in suitable pockets or enclosures and shall comply with 520.45. Supply cables for connector strips and drop boxes shall be as specified in 520.44(B).

Exhibit 520.9 shows a hanging connector strip with its associated border light cable. Border lights are hung and supplied in a similar manner. Exhibits 520.9, 520.10, and 520.11 illustrate different types of connections for portable stage lighting equipment.

Exhibit 520.10 A 4-gang, 4-receptacle pin-plug outlet box designed for flush mounting. (Courtesy of Electronic Theatre Controls, Inc.)

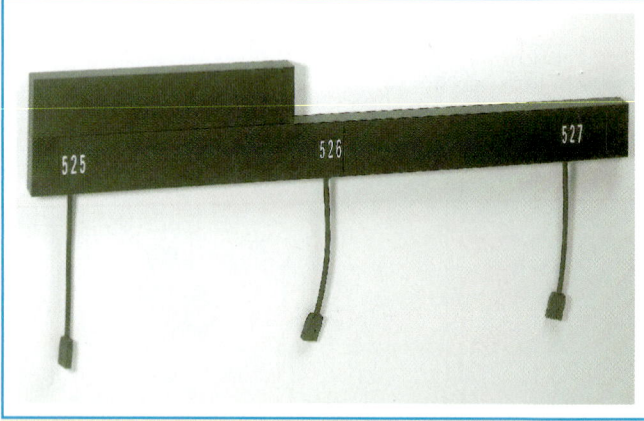

Exhibit 520.11 A typical three-circuit connector strip designed for wall or pipe mounting. (Courtesy of Electronic Theatre Controls, Inc.)

520.47 Backstage Lamps (Bare Bulbs).

Lamps (bare bulbs) installed in backstage and ancillary areas where they can come in contact with scenery shall be located and guarded so as to be free from physical damage and shall provide an air space of not less than 50 mm (2 in.) between such lamps and any combustible material.

Exception: Decorative lamps installed in scenery shall not be considered to be backstage lamps for the purpose of this section.

520.48 Curtain Machines.

Curtain machines shall be listed.

520.49 Smoke Ventilator Control.

Where stage smoke ventilators are released by an electrical device, the circuit operating the device shall be normally closed and shall be controlled by at least two externally operable switches, one switch being placed at a readily accessible location on stage and the other where designated by the authority having jurisdiction. The device shall be designed for the full voltage of the circuit to which it is connected, no resistance being inserted. The device shall be located in the loft above the scenery and shall be enclosed in a suitable metal box having a tight, self-closing door.

In addition to the smoke ventilators being controlled from two externally operable switches at different locations, the design of a normally closed circuit ensures that the smoke ventilators will operate when the circuit opens for any reason, such as a circuit breaker tripping or a fuse blowing.

IV. Portable Switchboards on Stage

520.50 Road Show Connection Panel (A Type of Patch Panel).

A panel designed to allow for road show connection of portable stage switchboards to fixed lighting outlets by means of permanently installed supplementary circuits. The panel, supplementary circuits, and outlets shall comply with 520.50(A) through (D).

Also known as a road show interconnect or intercept panel, a road show connection panel is designed to connect the load side of a portable switchboard to the fixed building branch circuits and associated outlets. It may also provide for the fixed branch circuits to be connected to a fixed switchboard when the portable switchboard is not installed.

(A) Load Circuits. Circuits shall terminate in grounding-type polarized inlets of current and voltage rating that match the fixed-load receptacle.

The grounding-type polarized inlets may be flush or pendant. The fixed-load receptacle is on the other end of the branch circuit that emanates from the panel.

(B) Circuit Transfer. Circuits that are transferred between fixed and portable switchboards shall have all circuit conductors transferred simultaneously.

In accordance with 520.50(B), simultaneous transfer of all conductors of the circuit, including any grounded conductors, is required.

(C) Overcurrent Protection. The supply devices of these supplementary circuits shall be protected by branch-circuit overcurrent protective devices. The individual supplementary circuit, within the road show connection panel and theater shall be protected by branch-circuit overcurrent protective devices of suitable ampacity installed within the road show connection panel.

The branch-circuit overcurrent protection should normally be in the switchboard but because some older units do not have this protection, backup overcurrent protection is provided by 520.50(C).

(D) Enclosure. Panel construction shall be in accordance with Article 408.

520.51 Supply.

Portable switchboards shall be supplied only from power outlets of sufficient voltage and ampere rating. Such power outlets shall include only externally operable, enclosed fused switches or circuit breakers mounted on stage or at the permanent switchboard in locations readily accessible from the stage floor. Provisions for connection of an equipment grounding conductor shall be provided. The neutral of feeders supplying solid-state, 3-phase, 4-wire dimmer systems shall be considered a current-carrying conductor.

Power outlets, known in the entertainment industry as company switches or bull switches, are the point in the wiring system where portable feeder cables connect to the fixed building wiring. They may be as simple as an overcurrent-protected multipole receptacle designed to accept the supply cable described in 520.53(P), Exception, or they may be multiple sets of parallel single-conductor feeder cables. These single-conductor feeder cables, as described in 520.53(H), may be terminated via single-pole separable connectors, as described in 520.53(K), or directly to busbars, fused disconnect switches, or circuit breakers with wire connectors (lugs).

520.52 Overcurrent Protection.

Circuits from portable switchboards directly supplying equipment containing incandescent lamps of not over 300 watts shall be protected by overcurrent protective devices having a rating or setting of not over 20 amperes. Circuits for lampholders over 300 watts shall be permitted where overcurrent protection complies with Article 210.

520.53 Construction and Feeders.

Portable switchboards and feeders for use on stages shall comply with 520.53(A) through (P).

See Exhibit 520.12 for an example of a portable switchboard.

Exhibit 520.12 A large, portable SCR dimmer switchboard (rolling rack). (Courtesy of Electronic Theatre Controls, Inc.)

(A) Enclosure. Portable switchboards shall be placed within an enclosure of substantial construction, which shall be permitted to be arranged so that the enclosure is open during operation. Enclosures of wood shall be completely lined with sheet metal of not less than 0.51 mm (0.020 in.) and shall be well galvanized, enameled, or otherwise properly coated to prevent corrosion or be of a corrosion-resistant material.

(B) Energized Parts. There shall not be exposed energized parts within the enclosure.

(C) Switches and Circuit Breakers. All switches and circuit breakers shall be of the externally operable, enclosed type.

(D) Circuit Protection. Overcurrent devices shall be provided in each ungrounded conductor of every circuit supplied through the switchboard. Enclosures shall be provided for all overcurrent devices in addition to the switchboard enclosure.

(E) Dimmers. The terminals of dimmers shall be provided with enclosures, and dimmer faceplates shall be arranged so that accidental contact cannot be readily made with the faceplate contacts.

(F) Interior Conductors.

(1) Type. All conductors other than busbars within the switchboard enclosure shall be stranded. Conductors shall be approved for an operating temperature at least equal to the approved operating temperature of the dimming devices used in the switchboard and in no case less than the following:

(1) Resistance-type dimmers—200°C (392°F); or
(2) Reactor-type, autotransformer, and solid-state dimmers—125°C (257°F).

All control wiring shall comply with Article 725.

(2) Protection. Each conductor shall have an ampacity not less than the rating of the circuit breaker, switch, or fuse that it supplies. Circuit interrupting and bus bracing shall be in accordance with 110.9 and 110.10. The short-circuit current rating shall be marked on the switchboard.

Conductors shall be enclosed in metal wireways or shall be securely fastened in position and shall be bushed where they pass through metal.

(G) Pilot Light. A pilot light shall be provided within the enclosure and shall be connected to the circuit supplying the board so that the opening of the master switch does not cut off the supply to the lamp. This lamp shall be on an individual branch circuit having overcurrent protection rated or set at not over 15 amperes.

The requirement of 520.53(G) applies only to switchboards with a main disconnect, if provided, on the switchboard. The pilot light serves as a warning at the switchboard that indicates the presence of power before the main disconnect is activated.

(H) Supply Conductors.

(1) General. The supply to a portable switchboard shall be by means of listed extra-hard usage cords or cables. The supply cords or cable shall terminate within the switchboard enclosure, in an externally operable fused master switch or circuit breaker or in a connector assembly identified for the purpose. The supply cords or cable (and connector assembly) shall have sufficient ampacity to carry the total load connected to the switchboard and shall be protected by overcurrent devices.

(2) Single-Conductor Cables. Single-conductor portable supply cable sets shall not be smaller than 2 AWG conductors. The equipment grounding conductor shall not be smaller than 6 AWG conductor. Single-conductor grounded neutral cables for a supply shall be sized as per 520.53(O)(2). Where single conductors are paralleled for increased ampacity, the paralleled conductors shall be of the same length and size. Single-conductor supply cables shall be grouped together but not bundled. The equipment grounding conductor shall be permitted to be of a different type, provided it meets the other requirements of this section, and it shall be permitted to be reduced in size as permitted by 250.122. Grounded (neutral) and equipment grounding conductors shall be identified in accordance with 200.6, 250.119, and 310.12. Grounded conductors shall be permitted to be identified by marking at least the first 150 mm (6 in.) from both ends of each length of conductor with white or gray. Equipment grounding conductors shall be permitted to be identified by marking at least the first 150 mm (6 in.) from both ends of each length of conductor with green or green with yellow stripes. Where more than one nominal voltage exists within the same premises, each ungrounded conductor shall be identified by system.

(3) Supply Conductors Not Over 3.0 m (10 ft) Long. Where supply conductors do not exceed 3.0 m (10 ft) in length between supply and switchboard or supply and a subsequent overcurrent device, the supply conductors shall be permitted to be reduced in size where all of the following conditions are met:

(1) The ampacity of the supply conductors shall be at least one-quarter of the ampacity of the supply overcurrent protection device.
(2) The supply conductors shall terminate in a single overcurrent protection device that will limit the load to the ampacity of the supply conductors. This single overcurrent device shall be permitted to supply additional overcurrent devices on its load side.

(3) The supply conductors shall not penetrate walls, floors, or ceilings or be run through doors or traffic areas. The supply conductors shall be adequately protected from physical damage.
(4) The supply conductors shall be suitably terminated in an approved manner.
(5) Conductors shall be continuous without splices or connectors.
(6) Conductors shall not be bundled.
(7) Conductors shall be supported above the floor in an approved manner.

(4) Supply Conductors Not Over 6.0 m (20 ft) Long. Where supply conductors do not exceed 6.0 m (20 ft) in length between supply and switchboard or supply and a subsequent overcurrent protection device, the supply conductors shall be permitted to be reduced in size where all of the following conditions are met:

(1) The ampacity of the supply conductors shall be at least one-half of the ampacity of the supply overcurrent protection device.
(2) The supply conductors shall terminate in a single overcurrent protection device that limits the load to the ampacity of the supply conductors. This single overcurrent device shall be permitted to supply additional overcurrent devices on its load side.
(3) The supply conductors shall not penetrate walls, floors, or ceilings or be run through doors or traffic areas. The supply conductors shall be adequately protected from physical damage.
(4) The supply conductors shall be suitably terminated in an approved manner.
(5) The supply conductors shall be supported in an approved manner at least 2.1 m (7 ft) above the floor except at terminations.
(6) The supply conductors shall not be bundled.
(7) Tap conductors shall be in unbroken lengths.

devices, either fixed or portable, must be provided for each of the smaller switchboards.

The requirement that the conductors not be bundled is so that Column D of Table 400.5(B) can be employed. If the conductors were bundled, Column F and all applicable derating factors would apply. Most devices used in the theater to terminate single-conductor cables are rated for use at 90°C ampacity. However, if single-conductor cables are terminated directly to a circuit breaker or fused switch, a 75°C ampacity or lower would most likely apply.

(5) Supply Conductors Not Reduced in Size. Supply conductors not reduced in size under provisions of 520.53(H)(3) or 520.53(H)(4) shall be permitted to pass through holes in walls specifically designed for the purpose. If penetration is through the fire-resistant–rated wall, it shall be in accordance with 300.21.

(I) Cable Arrangement. Cables shall be protected by bushings where they pass through enclosures and shall be arranged so that tension on the cable is not transmitted to the connections. Where power conductors pass through metal, the requirements of 300.20 shall apply.

Tension on the connections is removed by using conventional strain relief devices, or often by lashing the cable to the enclosure with rope.

(J) Number of Supply Interconnections. Where connectors are used in a supply conductor, there shall be a maximum number of three interconnections (mated connector pairs) where the total length from supply to switchboard does not exceed 30 m (100 ft). In cases where the total length from supply to switchboard exceeds 30 m (100 ft), one additional interconnection shall be permitted for each additional 30 m (100 ft) of supply conductor.

The addition of excessive numbers of interconnections could jeopardize the mechanical and electrical integrity of the supply conductors.

(K) Single-Pole Separable Connectors. Where single-pole portable cable connectors are used, they shall be listed and of the locking type. Sections 400.10, 406.6, and 406.7 shall not apply to listed single-pole separable connectors and single-conductor cable assemblies utilizing listed single-pole separable connectors. Where paralleled sets of current-carrying, single-pole separable connectors are provided as input devices, they shall be prominently labeled with a warning indicating the presence of internal parallel connections.

The use of single-pole separable connectors shall comply with at least one of the following conditions:

(1) Connection and disconnection of connectors are only possible where the supply connectors are interlocked to the source and it is not possible to connect or disconnect connectors when the supply is energized.

(2) Line connectors are of the listed sequential-interlocking type so that load connectors shall be connected in the following sequence:

    a. Equipment grounding conductor connection

    b. Grounded circuit conductor connection, if provided

    c. Ungrounded conductor connection, and that disconnection shall be in the reverse order

(3) A caution notice shall be provided adjacent to the line connectors indicating that plug connection shall be in the following order:

    a. Equipment grounding conductor connectors

    b. Grounded circuit conductor connectors, if provided

    c. Ungrounded conductor connectors, and that disconnection shall be in the reverse order

The requirements in 520.53(K) provide for a special type of connection device suitable for connecting single-conductor feeder cables. The connection device must be listed and of the locking type, reducing the likelihood of its separating while under load. The connectors must be used in sets, because they are only single-pole types. It is important that the grounding conductor be connected first and disconnected last, and that the grounded conductor be connected next-to-first and disconnected next-to-last. The connector sets must be arranged so as to reduce the likelihood that the connections will be made in the incorrect order, in accordance with one of the following methods.

1. Provide a scheme whereby the main disconnect cannot be energized until all conductors are connected.
2. Provide a scheme whereby the connectors are precluded from being connected in any order other than the proper one.
3. Provide a scheme whereby the individual connectors, free of any special electromechanical intervention, are marked with instructions to the user regarding proper connection.

Single-pole separable connectors are quick-connect feeder splicing and terminating devices, not attachment plugs or receptacles. They are designed to be sized, terminated, and inspected by a qualified person before being energized, and are to be guarded from accidental disconnection before being de-energized.

(L) Protection of Supply Conductors and Connectors. All supply conductors and connectors shall be protected against physical damage by an approved means. This protection shall not be required to be raceways.

Rubber mats and commercially available rubber bridges are often used for the protection of supply conductors and connectors.

(M) Flanged Surface Inlets. Flanged surface inlets (recessed plugs) that are used to accept the power shall be rated in amperes.

(N) Terminals. Terminals to which stage cables are connected shall be located so as to permit convenient access to the terminals.

The requirement in 520.53(N) facilitates the field connection and disconnection of the large feeder cables as the show travels from place to place.

(O) Neutral.

(1) Neutral Terminal. In portable switchboard equipment designed for use with 3-phase, 4-wire with ground supply, the supply neutral terminal, its associated busbar, or equivalent wiring, or both, shall have an ampacity equal to at least twice the ampacity of the largest ungrounded supply terminal.

Exception: Where portable switchboard equipment is specifically constructed and identified to be internally converted in the field, in an approved manner, from use with a balanced 3-phase, 4-wire with ground supply to a balanced single-phase, 3-wire with ground supply, the supply neutral terminal and its associated busbar, equivalent wiring, or both, shall have an ampacity equal to at least that of the largest ungrounded single-phase supply terminal.

The requirement in 520.53(O)(1) requires careful study because overlapping concepts are involved. If a 3-phase, 4-wire switchboard of any kind is brought into a space that has only single-phase, 3-wire service, the switchboard will most likely be connected with two phases to one leg and one phase to the other. This connection could double the current flowing through the neutral, so the neutral must be double size to allow for this possibility. The exception to 520.53(O)(1) provides for a smaller neutral sized for the single-phase feed where a switchboard contains switching devices that can divide the B-phase load equally between the A-phase and C-phase buses for single-phase operation.

Additionally, 3-phase, 4-wire switchboards that contain solid-state dimming devices must, when connected to a 3-phase, 4-wire supply, be connected to that supply with a multiconductor cable sized by counting the neutral as a current-carrying conductor, or with a set of single-conductor cables where the neutral is sized 130 percent greater than the phases.

For example, a 3-phase, 4-wire switchboard containing six 50-ampere SCR dimmers (100 amperes per phase) without a reassignment switching system would have to have a 200-ampere neutral. (A single-phase, 3-wire-only switchboard would not have to meet this special requirement.) This 200 percent rule would cover all the components making up the neutral conductor system inside or permanently attached to the switchboard, so as to allow for a full-size, single-phase, 3-wire feed when two of the 3-phase, 4-wire phase conductors are terminated to one single-phase, 3-wire leg. Note that the 200 percent neutral already covers the derating requirements (125 percent for a multiconductor feeder system and 130 percent for a single-conductor feeder system) when used in the 3-phase mode. If a reassignment system were added, the neutral would be required to be only 150 amperes. Again, when used in the 3-phase mode, the derating factors would be covered.

Note that the double-neutral requirement covers the terminal and associated busbar or wiring. This requirement begins at the main input terminals or busing, main input inlet connector, or attached main input cord-and-plug set and includes all wiring on the load side of that point. Power supply feeders easily detached at the terminals or inlet connector need not adhere to the 200 percent neutral rule because they can easily be sized on a show-by-show basis for the type of supply encountered. These cables must, however, adhere to the requirements of the neutral as a current-carrying conductor, or of the 130 percent single-conductor-cable neutral.

(2) Supply Neutral. The power supply conductors for portable switchboards shall be sized considering the neutral as a current-carrying conductor. Where single-conductor feeder cables, not installed in raceways, are used on multiphase circuits, the grounded neutral conductor shall have an ampacity of at least 130 percent of the ungrounded circuit conductors feeding the portable switchboard.

(P) Qualified Personnel. The routing of portable supply conductors, the making and breaking of supply connectors and other supply connections, and the energization and de-energization of supply services shall be performed by qualified personnel, and portable switchboards shall be so marked, indicating this requirement in a permanent and conspicuous manner.

Exception: A portable switchboard shall be permitted to be connected to a permanently installed supply receptacle by other than qualified personnel, provided that the supply receptacle is protected for its rated ampacity by an overcurrent device of not greater than 150 amperes, and where the receptacle, interconnection, and switchboard further

(a) *Employ listed multipole connectors suitable for the purpose for every supply interconnection, and*

(b) *Prevent access to all supply connections by the general public, and*

(c) *Employ listed extra-hard usage multiconductor cords or cables with an ampacity suitable for the type of load and not less than the ampere rating of the connectors.*

The intent of 520.53(P) is to divide the acceptable practices in what are most likely to be professional and professional-grade educational venues from those in amateur or amateur-grade educational venues. The basic requirements allow for such things as single-conductor feeder systems, feeders sized for the current-connected load, tap rules, and so on, and require the services of a qualified person. The exception to 520.53(P) provides for a conventional feeder system suitable for use by an untrained person.

V. Portable Stage Equipment Other Than Switchboards

520.61 Arc Lamps.

Arc lamps, including enclosed arc lamps and associated ballasts, shall be listed. Interconnecting cord sets and interconnecting cords and cables shall be extra-hard usage type and listed.

520.62 Portable Power Distribution Units.

Portable power distribution units shall comply with 520.62(A) through (E).

(A) Enclosure. The construction shall be such that no current-carrying part will be exposed.

(B) Receptacles and Overcurrent Protection. Receptacles shall comply with 520.45 and shall have branch-circuit overcurrent protection in the box. Fuses and circuit breakers shall be protected against physical damage. Cords or cables supplying pendant receptacles shall be listed for extra-hard usage.

(C) Busbars and Terminals. Busbars shall have an ampacity equal to the sum of the ampere ratings of all the circuits connected to the busbar. Lugs shall be provided for the connection of the master cable.

(D) Flanged Surface Inlets. Flanged surface inlets (recessed plugs) that are used to accept the power shall be rated in amperes.

(E) Cable Arrangement. Cables shall be adequately protected where they pass through enclosures and be arranged so that tension on the cable is not transmitted to the terminations.

520.63 Bracket Fixture Wiring.

(A) Bracket Wiring. Brackets for use on scenery shall be wired internally, and the fixture stem shall be carried through to the back of the scenery where a bushing shall be placed on the end of the stem. Externally wired brackets or other fixtures shall be permitted where wired with cords designed for hard usage that extend through scenery and without joint or splice in canopy of fixture back and terminate in an approved-type stage connector located, where practical, within 450 mm (18 in.) of the fixture.

(B) Mounting. Fixtures shall be securely fastened in place.

520.64 Portable Strips.

Portable strips shall be constructed in accordance with the requirements for border lights and proscenium sidelights in 520.44(A). The supply cable shall be protected by bushings where it passes through metal and shall be arranged so that tension on the cable will not be transmitted to the connections.

> FPN No. 1: See 520.42 for wiring of portable strips.
> FPN No. 2: See 520.68(A)(3) for insulation types required on single conductors.

520.65 Festoons.

Joints in festoon wiring shall be staggered. Lamps enclosed in lanterns or similar devices of combustible material shall be equipped with guards.

Festoon lighting is defined in Article 100. Joints in festoon wiring must be staggered and properly insulated. This arrangement ensures that connections will not be opposite one another, which could cause sparking due to improper insulation or unraveling of insulation, which, in turn, could ignite lanterns or other combustible material enclosing lamps. Where lampholders have terminals of a type that puncture the conductor insulation and make contact with the conductors, stranded conductors should be used.

520.66 Special Effects.

Electrical devices used for simulating lightning, waterfalls, and the like shall be constructed and located so that flames, sparks, or hot particles cannot come in contact with combustible material.

520.67 Multipole Branch-Circuit Cable Connectors.

Multipole branch-circuit cable connectors, male and female, for flexible conductors shall be constructed so that tension on the cord or cable is not transmitted to the connections. The female half shall be attached to the load end of the power supply cord or cable. The connector shall be rated in amperes and designed so that differently rated devices cannot be connected together; however, a 20-ampere T-slot receptacle shall be permitted to accept a 15-ampere attachment plug of the same voltage rating. Alternating-current multipole connectors shall be polarized and comply with 406.6 and 406.9.

FPN: See 400.10 for pull at terminals.

520.68 Conductors for Portables.

(A) Conductor Type.

(1) General. Flexible conductors, including cable extensions, used to supply portable stage equipment shall be listed extra-hard usage cords or cables.

(2) Stand Lamps. Reinforced cord shall be permitted to supply stand lamps where the cord is not subject to severe physical damage and is protected by an overcurrent device rated at not over 20 amperes.

See 520.2 for the definition of *stand lamp* (work light).

(3) High-Temperature Applications. A special assembly of conductors in sleeving not longer than 1.0 m (3.3 ft) shall be permitted to be employed in lieu of flexible cord if the individual wires are stranded and rated not less than 125°C (257°F) and the outer sleeve is glass fiber with a wall thickness of at least 0.635 mm (0.025 in.).

Portable stage equipment requiring flexible supply conductors with a higher temperature rating where one end is permanently attached to the equipment shall be permitted to employ alternate, suitable conductors as determined by a qualified testing laboratory and recognized test standards.

The requirements of 520.68(A)(3) cover the connection of high-temperature equipment including stage lighting fixtures, which often do operate at elevated temperatures. High-temperature (150°C to 250°C) extra-hard-usage cords are, in general, not available. Less than extra-hard-usage cords are limited to 3.3 ft in length to reduce the likelihood that they could be placed on the floor or other area where they might be damaged by traffic or moving scenery.

(4) Breakouts. Listed, hard usage (junior hard service) cords shall be permitted in breakout assemblies where all of the following conditions are met:

(1) The cords are utilized to connect between a single multipole connector containing two or more branch circuits and multiple 2-pole, 3-wire connectors.
(2) The longest cord in the breakout assembly does not exceed 6.0 m (20 ft).
(3) The breakout assembly is protected from physical damage by attachment over its entire length to a pipe, truss, tower, scaffold, or other substantial support structure.
(4) All branch circuits feeding the breakout assembly are protected by overcurrent devices rated at not over 20 amperes.

The provisions of 520.68(A)(4) apply to multiconductor cable assemblies with multipole connectors that contain more than one branch circuit. The breakout assembly is a multipole connector with several pendant receptacles connected to it, separating the multiple branch circuits into individual branch circuits. It is also possible to use a similar arrangement of pendant plugs to form a breaking assembly on the other end of the multiconductor cable.

(B) Conductor Ampacity. The ampacity of conductors shall be as given in 400.5, except multiconductor, listed, extra-hard usage portable cords that are not in direct contact with equipment containing heat-producing elements shall be permitted to have their ampacity determined by Table 520.44. Maximum load current in any conductor with an ampacity determined by Table 520.44 shall not exceed the values in Table 520.44.

In accordance with 520.68(B), portable, multicircuit, multiconductor cable is permitted to be sized in accordance with Table 520.44, similar to the method used for border light cable. If portable, multicircuit, multiconductor cable is located horizontally directly above heat-producing equipment, in lieu of a connector strip, it should be spaced sufficiently above that equipment to avoid the elevated temperatures or should be sized in accordance with 400.5.

Exception: Where alternate conductors are allowed in 520.68(A)(3), their ampacity shall be as given in the appropriate table in this Code for the types of conductors employed.

520.69 Adapters.

Adapters, two-fers, and other single- and multiple-circuit outlet devices shall comply with 520.69(A), (B), and (C).

(A) No Reduction in Current Rating. Each receptacle and its corresponding cable shall have the same current and voltage rating as the plug supplying it. It shall not be utilized in a stage circuit with a greater current rating.

(B) Connectors. All connectors shall be wired in accordance with 520.67.

Adapters are available where cords and connector bodies of one ampacity are connected to a plug of a larger rating. For example, a 12 AWG conductor with an ampacity of 20 amperes could be connected to a 100-ampere circuit. An overload could result in a fire because the circuit breaker or fuse would not provide adequate protection. The plug and receptacle must be of the same rating, in accordance with 520.69(B).

(C) Conductor Type. Conductors for adapters and two-fers shall be listed, extra-hard usage or listed, hard usage (junior hard service) cord. Hard usage (junior hard service) cord shall be restricted in overall length to 1.0 m (3.3 ft).

VI. Dressing Rooms

520.71 Pendant Lampholders.

Pendant lampholders shall not be installed in dressing rooms.

520.72 Lamp Guards.

All exposed incandescent lamps in dressing rooms, where less than 2.5 m (8 ft) from the floor, shall be equipped with open-end guards riveted to the outlet box cover or otherwise sealed or locked in place.

Because of the many types of flammable materials present in dressing rooms, such as costumes and wigs, pendant lampholders are not permitted. Lamps must be provided with suitable open-end guards that permit relamping and are not easily removed. This makes it difficult to circumvent the guard's intended purpose of preventing contact between the lamps and flammable material.

520.73 Switches Required.

All lights and any receptacles adjacent to the mirror(s) and above the dressing table counter(s) installed in dressing rooms shall be controlled by wall switches installed in the dressing room(s). Each switch controlling receptacles adjacent to the mirror(s) and above the dressing table counter(s) shall be provided with a pilot light located outside the dressing room, adjacent to the door to indicate when the receptacles are energized. Other outlets installed in the dressing room shall not be required to be switched.

The requirement in 520.73 only addresses receptacles located adjacent to the mirror and on the countertop. The receptacles located elsewhere in the room are not subject to the disconnect and pilot light requirements of 520.73. The purpose of the switching requirement is to make sure that all coffee pots, curling irons, hair dryers, and other similar countertop appliances can be readily disconnected at the end of the performance.

VII. Grounding

520.81 Grounding.

All metal raceways and metal-sheathed cables shall be grounded. The metal frames and enclosures of all equipment, including border lights and portable luminaires (lighting fixtures), shall be grounded. Grounding, where used, shall be in accordance with Article 250.

ARTICLE 525
Carnivals, Circuses, Fairs, and Similar Events

Contents

I. General Requirements

525.1 Scope.

This article covers the installation of portable wiring and equipment for carnivals, circuses, fairs, and similar functions, including wiring in or on all structures.

Article 525 addresses the installation of portable wiring and equipment for temporary attractions, such as carnivals, circuses, and fairs. Article 525 is intended to apply to all wiring in or on portable structures, whereas Articles 518 and 520 apply to permanent structures. Installations for portable equipment used at carnivals, circuses, fairs, and the like, were formerly covered under Article 527, Temporary Installations. Article 525 was developed because the requirements for temporary installations contained in Article 527 apply more to construction sites than to events such as fairs and carnivals that are open to the general public. Additionally, Article 525 significantly expands the requirements and scope for installing electrical equipment at these locations.

525.3 Other Articles.

(A) Portable Wiring and Equipment. Wherever the requirements of other articles of this *Code* and Article 525 differ, the requirements of Article 525 shall apply to the portable wiring and equipment.

(B) Permanent Structures. Articles 518 and 520 shall apply to wiring in permanent structures.

(C) Audio Signal Processing, Amplification, and Reproduction Equipment. Article 640 shall apply to the wiring and installation of audio signal processing, amplification, and reproduction equipment.

(D) Attractions Utilizing Pools, Fountains, and Similar Installations with Contained Volumes of Water. This equipment shall be installed to comply with the applicable requirements of Article 680.

525.5 Overhead Conductor Clearances.

(A) Vertical Clearances. Conductors shall have a vertical clearance to ground in accordance with 225.18. These clearances shall apply only to wiring installed outside of tents and concessions.

(B) Clearance to Rides and Attractions. Amusement rides and amusement attractions shall be maintained not less than 4.5 m (15 ft) in any direction from overhead conductors operating at 600 volts or less, except for the conductors supplying the amusement ride or attraction. Amusement rides or attractions shall not be located under or within 4.5 m (15 ft) horizontally of conductors operating in excess of 600 volts.

525.6 Protection of Electrical Equipment.

Electrical equipment and wiring methods in or on rides, concessions, or other units shall be provided with mechanical protection where such equipment or wiring methods are subject to physical damage.

II. Power Sources

525.10 Separately Derived Systems.

(A) Generators. Generators shall comply with the requirements of Article 445.

(B) Transformers. Transformers shall comply with the applicable requirements of 240.4(A), (B)(3), and (C); 250.30; and Article 450.

525.11 Services.

Services shall be installed in accordance with the applicable requirements of Article 230 and, in addition, shall comply with 525.11(A) and (B).

(A) Guarding. Service equipment shall not be installed in a location that is accessible to unqualified persons, unless the equipment is lockable.

Service equipment must be installed in accordance with Article 230 and must be lockable where accessible to unqualified persons. At fairgrounds, carnivals, and similar events, there is significant pedestrian traffic throughout the site, including through those areas where electrical equipment is located. This requirement helps safeguard the general public from accidentally coming in contact with energized service equipment.

(B) Mounting and Location. Service equipment shall be mounted on a solid backing and be installed so as to be protected from the weather, unless of weatherproof construction.

III. Wiring Methods

525.20 Wiring Methods.

(A) Type. Where flexible cords or cables are used, they shall be listed for extra hard usage. Where flexible cords or cables are used and are not subject to physical damage, they shall be permitted to be listed for hard usage. Where used outdoors, flexible cords and cables shall also be listed for wet locations and shall be sunlight resistant. Extra-hard usage flexible cords or cables shall be permitted for use as permanent wiring on portable amusement rides and attractions where not subject to physical damage.

(B) Single-Conductor. Single-conductor cable shall be permitted only in sizes 2 AWG or larger.

(C) Open Conductors. Open conductors are prohibited except as part of a listed assembly or festoon lighting installed in accordance with Article 225.

(D) Splices. Flexible cords or cables shall be continuous without splice or tap between boxes or fittings.

(E) Cord Connectors. Cord connectors shall not be laid on the ground unless listed for wet locations. Connectors and cable connections shall not be placed in audience traffic paths or within areas accessible to the public unless guarded.

(F) Support. Wiring for an amusement ride, attraction, tent, or similar structure shall not be supported by any other ride or structure unless specifically designed for the purpose.

(G) Protection. Flexible cords or cables accessible to the public shall be arranged to minimize the tripping hazard and shall be permitted to be covered with nonconductive matting, provided that the matting does not constitute a greater tripping hazard than the uncovered cables. It shall be permitted to bury cables. The requirements of 300.5 shall not apply.

(H) Boxes and Fittings. A box or fitting shall be installed at each connection point, outlet, switchpoint, or junction point.

525.21 Rides, Tents and Concessions.

(A) Disconnecting Means. Each ride and concession shall be provided with a fused disconnect switch or circuit breaker located within sight and within 1.8 m (6 ft) of the operator's station. The disconnecting means shall be readily accessible to the operator, including when the ride is in operation. Where accessible to unqualified persons, the enclosure for the switch or circuit breaker shall be of the lockable type. A shunt trip device that opens the fused disconnect or circuit breaker when a switch located in the ride operator's console is closed shall be a permissible method of opening the circuit.

(B) Portable Wiring Inside Tents and Concessions. Electrical wiring for lighting, where installed inside of tents and concessions, shall be securely installed and, where subject to physical damage, shall be provided with mechanical protection. All lamps for general illumination shall be protected from accidental breakage by a suitable fixture or lampholder with a guard.

525.22 Portable Distribution or Termination Boxes.

Portable distribution or termination boxes shall comply with 525.22(A) through (D).

(A) Construction. Boxes shall be designed so that no live parts are exposed to accidental contact. Where installed outdoors, the box shall be of weatherproof construction and mounted so that the bottom of the enclosure is not less than 150 mm (6 in.) above the ground.

Portable distribution or termination equipment must be mounted so that the bottom of the enclosure is at least 6 in. above the ground. This prevents excessive moisture from entering the equipment and allows for proper radius of bend on conductors entering and exiting the equipment from below.

(B) Busbars and Terminals. Busbars shall have an ampere rating not less than the overcurrent device supplying the feeder supplying the box. Where conductors terminate directly on busbars, busbar connectors shall be provided.

(C) Receptacles and Overcurrent Protection. Receptacles shall have overcurrent protection installed within the box. The overcurrent protection shall not exceed the ampere rating of the receptacle, except as permitted in Article 430 for motor loads.

(D) Single-Pole Connectors. Where single-pole connectors are used, they shall comply with 530.22.

525.23 Ground-Fault Circuit-Interrupter (GFCI) Protection for Personnel.

(A) General-Use 15- and 20-Ampere, 125-Volt Receptacles. All 125-volt, single-phase, 15- and 20-ampere receptacle outlets that are in use by personnel shall have listed ground-fault circuit-interrupter protection for personnel. The ground-fault circuit interrupter shall be permitted to be an integral part of the attachment plug or located in the power-supply cord, within 300 mm (12 in.) of the attachment plug. For the purposes of this section, listed cord sets incorporating ground-fault circuit-interrupter protection for personnel shall be permitted. Egress lighting shall not be connected to the load side terminals of a ground-fault circuit-interrupter receptacle.

(B) Appliance Receptacles. Receptacles supplying items, such as cooking and refrigeration equipment, that are incompatible with ground-fault circuit-interrupter devices shall not be required to have ground-fault circuit-interrupter protection.

(C) Other Receptacles. Other receptacle outlets not covered in 525.23(A) or (B) shall be permitted to have ground-fault circuit-interrupter protection for personnel, or a written procedure shall be continuously enforced at the site by one or more designated persons to ensure the safety of equipment grounding conductors for all cord sets and receptacles, as described in 527.6(B)(2).

IV. Grounding and Bonding

525.30 Equipment Bonding.

The following equipment connected to the same source shall be bonded:

(1) Metal raceways and metal-sheathed cable
(2) Metal enclosures of electric equipment
(3) Metal frames and metal parts of rides, concessions, tents, trailers, trucks, or other equipment that contain or support electrical equipment

525.31 Equipment Grounding.

All equipment requiring grounding shall be grounded by an equipment grounding conductor of a type and size recognized by 250.118 and installed in accordance with Article 250. The equipment grounding conductor shall be bonded to the system grounded conductor at the service disconnecting means or, in the case of a separately derived system such as a generator, at the generator or first disconnecting means supplied by the generator. The grounded circuit conductor shall not be connected to the equipment grounding conductor on the load side of the service disconnecting means or on the load side of a separately derived system disconnecting means.

525.32 Grounding Conductor Continuity Assurance.

The continuity of the grounding conductor system used to reduce electrical shock hazards as required by 250.114, 250.138, 406.3(C), and 527.4(D) shall be verified each time that portable electrical equipment is connected.

The transient nature of amusements and, in some cases, the entire electrical distribution system associated with fairs, carnivals, and circuses increases the possibility that continuity of the equipment grounding conductor system could be interrupted. The verification of the grounding system continuity helps ensure the safety of workers and the general public who may come in contact with exposed non–current-carrying surfaces of electrical equipment or equipment that is electrically powered. The verification of the grounding system continuity is required each time that portable equipment is reconnected.

ARTICLE 527
Temporary Installations

Contents

527.1 Scope.

The provisions of this article apply to temporary electrical power and lighting installations.

527.2 All Wiring Installations.

(A) Other Articles. Except as specifically modified in this article, all other requirements of this *Code* for permanent wiring shall apply to temporary wiring installations.

Temporary installations of electrical equipment must be installed in accordance with all applicable permanent installation requirements except as modified by the rules in this article. For example, the requirements of 300.15 specify that a box or other enclosure must be used where splices are made. This rule is amended by 527.4(G), which, for construction sites, permits splices to be made in multiconductor cords and cables without the use of a box.

(B) Approval. Temporary wiring methods shall be acceptable only if approved based on the conditions of use and any special requirements of the temporary installation.

The provisions of 527.2(B) require that all temporary wiring methods be approved based on criteria such as (1) length of time in service, (2) severity of physical abuse, (3) exposure to weather, and (4) other special requirements. Special requirements may range from tunnel construction projects and tent cities constructed after a natural disaster to flammable hazardous material reclamation projects.

527.3 Time Constraints.

(A) During the Period of Construction. Temporary electrical power and lighting installations shall be permitted during the period of construction, remodeling, maintenance, repair, or demolition of buildings, structures, equipment, or similar activities.

(B) 90 Days. Temporary electrical power and lighting installations shall be permitted for a period not to exceed 90 days for holiday decorative lighting and similar purposes.

Note that the 90-day time limit in 527.3(B) applies only to temporary electrical installations associated with holiday displays. Construction and emergency and test temporary wiring installations are not bound by this time limit.

(C) Emergencies and Tests. Temporary electrical power and lighting installations shall be permitted during emergencies and for tests, experiments, and developmental work.

(D) Removal. Temporary wiring shall be removed immediately upon completion of construction or purpose for which the wiring was installed.

Due to the modifications permitted by Article 527, temporary wiring installations may not meet all of the requirements for a permanent installation. Therefore, all temporary wiring must be not only disconnected but it also must be removed from the building, structure, or other location of installation.

527.4 General.

(A) Services. Services shall be installed in conformance with Article 230.

(B) Feeders. Feeders shall be protected as provided in Article 240. They shall originate in an approved distribution center. Conductors shall be permitted within cable assemblies or within multiconductor cords or cables of a type identified in Table 400.4 for hard usage or extra-hard usage. For the purpose of this section, Type NM and Type NMC cables shall be permitted to be used in any dwelling, building, or structure without any height limitation.

Temporary feeders are permitted to be (1) cable assemblies, (2) multiconductor cords, or (3) single-conductor cords. Cords used as feeders must be identified for hard or extra-hard usage according to Table 400.4. Individual conductors, as described in Table 310.13, are not permitted as open conductors but, rather, must be part of a cable assembly or used in a raceway system. Open or individual conductor feeders are permitted only during emergencies or tests.

All temporary wiring methods must be approved by the authority having jurisdiction. [See 527.2(B).]

Exception: Single insulated conductors shall be permitted where installed for the purpose(s) specified in 527.3(C), where accessible only to qualified persons.

(C) Branch Circuits. All branch circuits shall originate in an approved power outlet or panelboard. Conductors shall be permitted within cable assemblies or within multiconductor cord or cable of a type identified in Table 400.4 for hard usage or extra-hard usage. All conductors shall be protected as provided in Article 240. For the purposes of this section, Type NM and Type NMC cables shall be permitted to be used in any dwelling, building, or structure without any height limitation.

The basic requirement for safety in 527.4(C) is that temporary wiring be located and installed so that it will not be physically damaged. In accordance with 527.2(A), temporary wiring must be installed in accordance with the appropriate Chapter 3 article for the wiring method employed (unless modified in Article 527).

Note that hard-usage or extra-hard-usage extension cords are permitted to be laid on the floor.

Exception: Branch circuits installed for the purposes specified in 527.3(B) or (C) shall be permitted to be run as single insulated conductors. Where the wiring is installed in accordance with 527.3(B), the voltage to ground shall not exceed 150 volts, the wiring shall not be subject to physical damage, and the conductors shall be supported on insulators at intervals of not more than 3.0 m (10 ft); or, for festoon lighting, the conductors shall be arranged so that excessive strain is not transmitted to the lampholders.

(D) Receptacles. All receptacles shall be of the grounding type. Unless installed in a continuous grounded metal raceway or metal-covered cable, all branch circuits shall contain a separate equipment grounding conductor, and all receptacles shall be electrically connected to the equipment grounding conductors. Receptacles on construction sites shall not be installed on branch circuits that supply temporary lighting. Receptacles shall not be connected to the same ungrounded conductor of multiwire circuits that supply temporary lighting.

The intent of the branch-circuit provisions in 527.4(D) is to require separate ungrounded conductors for lighting and receptacle loads so that the activation of a fuse, circuit breaker, or ground-fault circuit interrupter, due to a fault or equipment overload, will not de-energize the lighting circuit.

(E) Disconnecting Means. Suitable disconnecting switches or plug connectors shall be installed to permit the disconnection of all ungrounded conductors of each temporary circuit. Multiwire branch circuits shall be provided with a means to disconnect simultaneously all ungrounded conductors at the power outlet or panelboard where the branch circuit originated. Approved handle ties shall be permitted.

(F) Lamp Protection. All lamps for general illumination shall be protected from accidental contact or breakage by a suitable fixture or lampholder with a guard.

Brass shell, paper-lined sockets, or other metal-cased sockets shall not be used unless the shell is grounded.

(G) Splices. On construction sites, a box shall not be required for splices or junction connections where the circuit conductors are multiconductor cord or cable assemblies,

provided that the equipment grounding continuity is maintained with or without the box. See 110.14(B) and 400.9. A box, conduit body, or terminal fitting having a separately bushed hole for each conductor shall be used wherever a change is made to a conduit or tubing system or a metal-sheathed cable system.

(H) Protection from Accidental Damage. Flexible cords and cables shall be protected from accidental damage. Sharp corners and projections shall be avoided. Where passing through doorways or other pinch points, protection shall be provided to avoid damage.

Unlike the requirement in 400.8, flexible cords and cables, because of the nature of their use, are permitted to pass through doorways, in accordance with 527.4(H).

(I) Termination(s) at Devices. Flexible cords and cables entering enclosures containing devices requiring termination shall be secured to the box with fittings designed for the purpose.

(J) Support. Cable assemblies and flexible cords and cables shall be supported in place at intervals that ensure that they will be protected from physical damage. Support shall be in the form of staples, cable ties, straps, or similar type fittings installed so as not to cause damage. Vegetation shall not be used for support of overhead spans of branch circuits or feeders.

Per 527.4(J), temporary wiring methods do not have to be supported in accordance with the permanent installation requirements (from Chapter 3) for the particular wiring method. It should be noted the temporary wiring must be removed upon completion of construction and adequate support is needed only to minimize the possibility of damage to the wiring method during its temporary period of use. It is not permitted to use vegetation as a support structure for overhead spans of branch-circuit and feeder conductors.

527.6 Ground-Fault Protection for Personnel.

Ground-fault protection for personnel for all temporary wiring installations shall be provided to comply with 527.6(A) and (B). This section shall apply only to temporary wiring installations used to supply temporary power to equipment used by personnel during construction, remodeling, maintenance, repair, or demolition of buildings, structures, equipment, or similar activities.

(A) Receptacle Outlets. All 125-volt, single-phase, 15-, 20-, and 30-ampere receptacle outlets that are not a part of the permanent wiring of the building or structure and that are in use by personnel shall have ground-fault circuit interrupter protection for personnel. If a receptacle(s) is installed or exists as part of the permanent wiring of the building or structure and is used for temporary electric power, ground-fault circuit-interrupter protection for personnel shall be provided. For the purposes of this section, cord sets or devices incorporating listed ground-fault circuit interrupter protection for personnel identified for portable use shall be permitted.

Exception: In industrial establishments only, where conditions of maintenance and supervision ensure that only qualified personnel are involved, an assured equipment grounding conductor program as specified in 527.6(B)(2) shall be permitted for only those receptacle outlets used to supply equipment that would create a greater hazard if power was interrupted or having a design that is not compatible with GFCI protection.

(B) Use of Other Outlets. Receptacles other than 125-volt, single-phase, 15-, 20-, and 30-ampere receptacles shall have protection in accordance with (1) or, the assured equipment grounding conductor program in accordance with (2).

(1) GFCI Protection. Ground-fault circuit interrupter protection for personnel.

(2) Assured Equipment Grounding Conductor Program. A written assured equipment grounding conductor program continuously enforced at the site by one or more designated persons to ensure that equipment grounding conductors for all cord sets, receptacles that are not a part of the permanent wiring of the building or structure, and equipment connected by cord and plug are installed and maintained in accordance with the applicable requirements of 250.114, 250.138, 406.3(C), and 527.4(D).

(a) The following tests shall be performed on all cord sets, receptacles that are not part of the permanent wiring of the building or structure, and cord- and plug-connected equipment required to be grounded:

(1) All equipment grounding conductors shall be tested for continuity and shall be electrically continuous.
(2) Each receptacle and attachment plug shall be tested for correct attachment of the equipment grounding conductor. The equipment grounding conductor shall be connected to its proper terminal.
(3) All required tests shall be performed as follows:

    a. Before first use on site

    b. When there is evidence of damage

    c. Before equipment is returned to service following any repairs

    d. At intervals not exceeding 3 months

(b) The tests required in item (2)(a) shall be recorded and made available to the authority having jurisdiction.

Due to the more severe environmental conditions often encountered by personnel using temporary wiring while performing activities such as construction, remodeling, maintenance, repair, and demolition, there is generally an elevated exposure to electrical shock or electrocution hazards. The requirement of 527.6(A) for GFCI protection of all temporarily installed, 125-volt, single-phase, 15-, 20-, and 30-ampere receptacles is intended to protect personnel using these receptacles from shock hazards that may be encountered during construction and maintenance activities.

The exception to 527.6(A) is limited in scope and application. The exception applies only to those industrial occupancies in which qualified persons will be using 125-volt, single-phase, 15-, 20-, and 30-ampere receptacles. Additionally, the nature of the equipment being supplied by these receptacles either has to be of such importance that the hazard of power interruption outweighs the benefits of GFCI protection or the equipment has been demonstrated to be incompatible with the proper operation of GFCI protective devices. In those instances where the conditions specified by the exception are present, the use of the assured equipment grounding conductor program specified in 527.6(B)(2) is permitted. An electrically operated air supply for personnel working in toxic environments is an example of where the loss of power is the greater hazard. Some electrically operated testing equipment has proven to be incompatible with GFCI protection.

Receptacle configurations, other than the 125-volt, single-phase, 15-, 20- and 30-ampere types, must be GFCI protected or installed and maintained in accordance with the assured equipment grounding conductor program of 527.6(B)(2).

According to OSHA 29 CFR 1926.404(b)(1)(iii),

The employer shall establish and implement an assured equipment grounding conductor program on construction sites covering all cord sets, receptacles which are not a part of the building or structure, and equipment connected by cord and plug which are available for use or used by employees. This program shall comply with the following minimum requirements:

(A) A written description of the program, including the specific procedures adopted by the employer, shall be available at the jobsite for inspection and copying by the Assistant Secretary and any affected employee.

(B) The employer shall designate one or more competent persons . . . .

These OSHA requirements are very similar to the present *NEC* requirements for an assured grounding program.

GFCI protection for construction or maintenance personnel using receptacles that are part of the permanent wiring and are not GFCI protected may be provided by using cord sets or listed portable GFCIs identified for portable use. An example of a GFCI cord set that is identified for portable use is shown in Exhibit 527.1.

Exhibits 527.1 through 527.4 show some examples of ways to implement the temporary wiring requirements of 527.6.

Exhibit 527.3 A watertight plug and connector used to prevent tripping of GFCI protective devices in wet or damp weather. (Courtesy of Hubbell, Inc.)

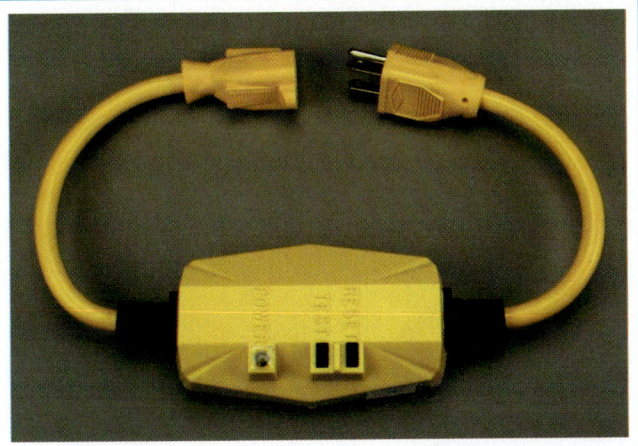

Exhibit 527.1 A raintight GFCI with open neutral protection that is designed for use on the line end of a flexible cord. (Courtesy of Pass & Seymour/Legrand®)

Exhibit 527.4 A 15-ampere duplex receptacle with integral GFCI that also protects downstream loads. (Courtesy of Pass & Seymour/Legrand®)

527.7 Guarding.

For wiring over 600 volts, nominal, suitable fencing, barriers, or other effective means shall be provided to limit access only to authorized and qualified personnel.

Exhibit 527.2 A temporary power outlet unit commonly used on construction sites with a variety of configurations, including GFCI protection. (Courtesy of Hubbell, Inc.)

ARTICLE 530
Motion Picture and Television Studios and Similar Locations

Contents

I. General
    530.1 Scope
    530.2 Definitions

I. General

530.1 Scope.

The requirements of this article shall apply to television studios and motion picture studios using either film or electronic cameras, except as provided in 520.1, and exchanges, factories, laboratories, stages, or a portion of the building in which film or tape more than 22 mm (⅞ in.) in width is exposed, developed, printed, cut, edited, rewound, repaired, or stored.

> FPN: For methods of protecting against cellulose nitrate film hazards, see NFPA 40-1997, *Standard for the Storage and Handling of Cellulose Nitrate Motion Picture Film.*

The requirements for motion picture studios and television studios are virtually the same and are intended to apply only to those locations presenting special hazards, that is, temporary structures constructed of wood or other combustible material. Otherwise, the conditions are similar to theater stages. Therefore, the applicable provisions of Article 520 should be observed, such as those for stages and dressing rooms.

530.2 Definitions.

Alternating-Current Power Distribution Box (Alternating-Current Plugging Box, Scatter Box). An ac distribution center or box that contains one or more grounding-type polarized receptacles that may contain overcurrent protection devices.

Bull Switch. An externally operated wall-mounted safety switch that may or may not contain overcurrent protection and is designed for the connection of portable cables and cords.

Location (Shooting Location). A place outside a motion picture studio where a production or part of it is filmed or recorded.

Location Board (Deuce Board). Portable equipment containing a lighting contactor or contactors and overcurrent protection designed for remote control of stage lighting.

Motion Picture Studio (Lot). A building or group of buildings and other structures designed, constructed, or permanently altered for use by the entertainment industry for the purpose of motion picture or television production.

Portable Equipment. Equipment intended to be moved from one place to another.

Plugging Box. A dc device consisting of one or more 2-pole, 2-wire, nonpolarized, nongrounding-type receptacles intended to be used on dc circuits only.

Single-Pole Separable Connector. A device that is installed at the ends of portable, flexible, single-conductor cable that is used to establish connection or disconnection between two cables or one cable and a single-pole, panel-mounted separable connector.

Spider (Cable Splicing Block). A device that contains busbars that are insulated from each other for the purpose of splicing or distributing power to portable cables and cords that are terminated with single-pole busbar connectors.

Stage Effect (Special Effect). An electrical or electromechanical piece of equipment used to simulate a distinctive visual or audible effect such as wind machines, lightning simulators, sunset projectors, and the like.

Stage Property. An article or object used as a visual element in a motion picture or television production, except painted backgrounds (scenery) and costumes.

Stage Set. A specific area set up with temporary scenery and properties designed and arranged for a particular scene in a motion picture or television production.

Stand Lamp (Work Light). A portable stand that contains a general-purpose luminaire (lighting fixture) or lampholder with guard for the purpose of providing general illumination in the studio or stage.

Television Studio or Motion Picture Stage (Sound Stage). A building or portion of a building usually insulated from the outside noise and natural light for use by the entertainment industry for the purpose of motion picture, television, or commercial production.

530.6 Portable Equipment.

Portable stage and studio lighting equipment and portable power distribution equipment shall be permitted for temporary use outdoors if the equipment is supervised by qualified personnel while energized and barriered from the general public.

See the commentary following 520.10 for more information on portable equipment.

II. Stage or Set

530.11 Permanent Wiring.

The permanent wiring shall be Type MC cable, Type AC cable containing an insulated equipment grounding conduc-

tor sized in accordance with Table 250.122, Type MI cable, or in approved raceways.

Exception: Communications circuits; audio signal processing, amplification, and reproduction circuits; Class 1, Class 2, and Class 3 remote-control or signaling circuits and power-limited fire alarm circuits shall be permitted to be wired in accordance with Articles 640, 725, 760, and 800.

530.12 Portable Wiring.

(A) Stage Set Wiring. The wiring for stage set lighting and other supply wiring not fixed as to location shall be done with listed hard usage flexible cords and cables. Where subject to physical damage, such wiring shall be listed extra-hard usage flexible cords and cables. Splices or taps in cables shall be permitted if the total connected load does not exceed the maximum ampacity of the cable.

(B) Stage Effects and Electrical Equipment Used as Stage Properties. The wiring for stage effects and electrical equipment used as stage properties shall be permitted to be wired with single- or multiconductor listed flexible cords or cables if the conductors are protected from physical damage and secured to the scenery by approved cable ties or by insulated staples. Splices or taps shall be permitted where such are made with listed devices and the circuit is protected at not more than 20 amperes.

(C) Other Electrical Equipment. Cords and cables other than extra-hard usage, where supplied as a part of a listed assembly, shall be permitted.

530.13 Stage Lighting and Effects Control.

Switches used for studio stage set lighting and effects (on the stages and lots and on location) shall be of the externally operable type. Where contactors are used as the disconnecting means for fuses, an individual externally operable switch, such as a tumbler switch, for the control of each contactor shall be located at a distance of not more than 1.8 m (6 ft) from the contactor, in addition to remote-control switches. A single externally operable switch shall be permitted to simultaneously disconnect all the contactors on any one location board, where located at a distance of not more than 1.8 m (6 ft) from the location board.

530.14 Plugging Boxes.

Each receptacle of dc plugging boxes shall be rated at not less than 30 amperes.

530.15 Enclosing and Guarding Live Parts.

(A) Live Parts. Live parts shall be enclosed or guarded to prevent accidental contact by persons and objects.

(B) Switches. All switches shall be of the externally operable type.

(C) Rheostats. Rheostats shall be placed in approved cases or cabinets that enclose all live parts, having only the operating handles exposed.

(D) Current-Carrying Parts. Current-carrying parts of bull switches, location boards, spiders, and plugging boxes shall be enclosed, guarded, or located so that persons cannot accidentally come into contact with them or bring conductive material into contact with them.

530.16 Portable Lamps.

Portable lamps and work lights shall be equipped with flexible cords, composition or metal-sheathed porcelain sockets, and substantial guards.

Exception: Portable lamps used as properties in a motion picture set or television stage set, on a studio stage or lot, or on location shall not be considered to be portable lamps for the purpose of this section.

530.17 Portable Arc Lamps.

(A) Portable Carbon Arc Lamps. Portable carbon arc lamps shall be substantially constructed. The arc shall be provided with an enclosure designed to retain sparks and carbons and to prevent persons or materials from coming into contact with the arc or bare live parts. The enclosures shall be ventilated. All switches shall be of the externally operable type.

(B) Portable Noncarbon Arc Electric-Discharge Lamps. Portable noncarbon arc lamps, including enclosed arc lamps, and associated ballasts shall be listed. Interconnecting cord sets and interconnecting cords and cables shall be extra-hard usage type and listed.

530.18 Overcurrent Protection—General.

Automatic overcurrent protective devices (circuit breakers or fuses) for motion picture studio stage set lighting and the stage cables for such stage set lighting shall be as given in 530.18(A) through (G). The maximum ampacity allowed on a given conductor, cable, or cord size shall be as given in the applicable tables of Articles 310 and 400.

(A) Stage Cables. Stage cables for stage set lighting shall be protected by means of overcurrent devices set at not more than 400 percent of the ampacity given in the applicable tables of Articles 310 and 400.

(B) Feeders. In buildings used primarily for motion picture production, the feeders from the substations to the stages shall be protected by means of overcurrent devices (generally located in the substation) having a suitable ampere rating. The overcurrent devices shall be permitted to be multipole or single-pole gang operated. No pole shall be required in the neutral conductor. The overcurrent device setting for each feeder shall not exceed 400 percent of the ampacity of the feeder, as given in the applicable tables of Article 310.

An overcurrent device setting of up to 400 percent is permitted if the loads are of short duration. In accordance with 530.18(B), the use of short-term ratings where the equipment operates for 20 minutes or less is permitted. A longer period of operation may pose a fire hazard.

(C) Cable Protection. Cables shall be protected by bushings where they pass through enclosures and shall be arranged so that tension on the cable is not transmitted to the connections. Where power conductors pass through metal, the requirements of 300.20 shall apply.

Portable feeder cables shall be permitted to temporarily penetrate fire-rated walls, floors, or ceilings provided that all of the following apply:

(1) The opening is of noncombustible material.
(2) When in use, the penetration is sealed with a temporary seal of a listed firestop material.
(3) When not in use, the opening shall be capped with a material of equivalent fire rating.

(D) Location Boards. Overcurrent protection (fuses or circuit breakers) shall be provided at the location boards. Fuses in the location boards shall have an ampere rating of not over 400 percent of the ampacity of the cables between the location boards and the plugging boxes.

(E) Plugging Boxes. Cables and cords supplied through plugging boxes shall be of copper. Cables and cords smaller than 8 AWG shall be attached to the plugging box by means of a plug containing two cartridge fuses or a 2-pole circuit breaker. The rating of the fuses or the setting of the circuit breaker shall be not over 400 percent of the rated ampacity of the cables or cords as given in the applicable tables of Articles 310 and 400. Plugging boxes shall not be permitted on ac systems.

(F) Alternating-Current Power Distribution Boxes. Alternating-current power distribution boxes used on sound stages and shooting locations shall contain connection receptacles of a polarized, grounding type.

(G) Lighting. Work lights, stand lamps, and luminaires (fixtures) rated 1000 watts or less and connected to dc plugging boxes shall be by means of plugs containing two cartridge fuses not larger than 20 amperes, or they shall be permitted to be connected to special outlets on circuits protected by fuses or circuit breakers rated at not over 20 amperes. Plug fuses shall not be used unless they are on the load side of the fuse or circuit breakers on the location boards.

530.19 Sizing of Feeder Conductors for Television Studio Sets.

(A) General. It shall be permissible to apply the demand factors listed in Table 530.19(A) to that portion of the maximum possible connected load for studio or stage set lighting for all permanently installed feeders between substations and stages and to all permanently installed feeders between the main stage switchboard and stage distribution centers or location boards.

Table 530.19(A) Demand Factors for Stage Set Lighting

Portion of Stage Set Lighting Load to Which Demand Factor Applied (volt-amperes)	Feeder Demand Factor
First 50,000 or less at	100%
From 50,001 to 100,000 at	75%
From 100,001 to 200,000 at	60%
Remaining over 200,000 at	50%

(B) Portable Feeders. A demand factor of 50 percent of maximum possible connected load shall be permitted for all portable feeders.

530.20 Grounding.

Type MC cable, Type MI cable, metal raceways, and all non–current-carrying metal parts of appliances, devices, and equipment shall be grounded as specified in Article 250. This shall not apply to pendant and portable lamps, to stage lighting and stage sound equipment, or to other portable and special stage equipment operating at not over 150 volts dc to ground.

530.21 Plugs and Receptacles.

(A) Rating. Plugs and receptacles shall be rated in amperes. The voltage rating of the plugs and receptacles shall not be less than the circuit voltage. Plug and receptacle ampere ratings for ac circuits shall not be less than the feeder or branch-circuit overcurrent device ampere rating. Table 210.21(B)(2) shall not apply.

(B) Interchangeability. Plugs and receptacles used in portable professional motion picture and television equipment shall be permitted to be interchangeable for ac or dc use on the same premises provided they are listed for ac/dc use and

marked in a suitable manner to identify the system to which they are connected.

530.22 Single-Pole Separable Connectors.

(A) General. Where ac single-pole portable cable connectors are used, they shall be listed and of the locking type. Sections 400.10, 406.6, and 406.7 shall not apply to listed single-pole separable connections and single-conductor cable assemblies utilizing listed single-pole separable connectors. Where paralleled sets of current-carrying single-pole separable connectors are provided as input devices, they shall be prominently labeled with a warning indicating the presence of internal parallel connections. The use of single-pole separable connectors shall comply with at least one of the following conditions:

(1) Connection and disconnection of connectors are only possible where the supply connectors are interlocked to the source and it is not possible to connect or disconnect connectors when the supply is energized.
(2) Line connectors are of the listed sequential-interlocking type so that load connectors shall be connected in the following sequence:
   a. Equipment grounding conductor connection
   b. Grounded circuit conductor connection, if provided
   c. Ungrounded conductor connection, and that disconnection shall be in the reverse order
(3) A caution notice shall be provided adjacent to the line connectors, indicating that plug connection shall be in the following order:
   a. Equipment grounding conductor connectors
   b. Grounded circuit-conductor connectors, if provided
   c. Ungrounded conductor connectors, and that disconnection shall be in the reverse order

(B) Interchangeability. Single-pole separable connectors used in portable professional motion picture and television equipment shall be permitted to be interchangeable for ac or dc use or for different current ratings on the same premises, provided they are listed for ac/dc use and marked in a suitable manner to identify the system to which they are connected.

530.23 Branch Circuits.

A branch circuit of any size supplying one or more receptacles shall be permitted to supply stage set lighting loads.

III. Dressing Rooms

530.31 Dressing Rooms.

Fixed wiring in dressing rooms shall be installed in accordance with the wiring methods covered in Chapter 3. Wiring for portable dressing rooms shall be approved.

IV. Viewing, Cutting, and Patching Tables

530.41 Lamps at Tables.

Only composition or metal-sheathed, porcelain, keyless lampholders equipped with suitable means to guard lamps from physical damage and from film and film scrap shall be used at patching, viewing, and cutting tables.

V. Cellulose Nitrate Film Storage Vaults

530.51 Lamps in Cellulose Nitrate Film Storage Vaults.

Lamps in cellulose nitrate film storage vaults shall be installed in rigid fixtures of the glass-enclosed and gasketed type. Lamps shall be controlled by a switch having a pole in each ungrounded conductor. This switch shall be located outside of the vault and provided with a pilot light to indicate whether the switch is on or off. This switch shall disconnect from all sources of supply all ungrounded conductors terminating in any outlet in the vault.

530.52 Electrical Equipment in Cellulose Nitrate Film Storage Vaults.

Except as permitted in 530.51, no receptacles, outlets, heaters, portable lights, or other portable electric equipment shall be located in cellulose nitrate film storage vaults. Electric motors shall be permitted, provided they are listed for the application and comply with Article 500, Class I, Division 2.

VI. Substations

530.61 Substations.

Wiring and equipment of over 600 volts, nominal, shall comply with Article 490.

530.62 Portable Substations.

Wiring and equipment in portable substations shall conform to the sections applying to installations in permanently fixed substations, but, due to the limited space available, the working spaces shall be permitted to be reduced, provided that the equipment shall be arranged so that the operator can work safely and so that other persons in the vicinity cannot accidentally come into contact with current-carrying parts or bring conducting objects into contact with them while they are energized.

530.63 Overcurrent Protection of Direct-Current Generators.

Three-wire generators shall have overcurrent protection in accordance with 445.12(E).

530.64 Direct-Current Switchboards.

(A) General. Switchboards of not over 250 volts dc between conductors, where located in substations or switchboard rooms accessible to qualified persons only, shall not be required to be dead-front.

(B) Circuit Breaker Frames. Frames of dc circuit breakers installed on switchboards shall not be required to be grounded.

ARTICLE 540
Motion Picture Projection Rooms

I. General

540.1 Scope.

The provisions of this article apply to motion picture projection rooms, motion picture projectors, and associated equipment of the professional and nonprofessional types using incandescent, carbon arc, xenon, or other light source equipment that develops hazardous gases, dust, or radiation.

FPN: For further information, see NFPA 40-1997, *Standard for the Storage and Handling of Cellulose Nitrate Motion Picture Film.*

Hazardous (classified) locations, as defined in Article 500, do not include a motion picture projection room, even though some of the older types of film, such as cellulose nitrate film (rarely used now), are highly flammable. In comparison, cellulose acetate film, called safety film, is in wide use today. Because film is not volatile at ordinary temperatures and no flammable gases are present, the wiring installation is not required to be suitable for hazardous (classified) locations, as defined in Article 500, but should be installed with special care to protect against the hazards of fire.

540.2 Definitions.

Nonprofessional Projector. Nonprofessional projectors are those types other than as described in 540.2.

Professional Projector. A type of projector using 35- or 70-mm film that has a minimum width of 35 mm (1⅜ in.) and has on each edge 212 perforations per meter (5.4 perforations per inch), or a type using carbon arc, xenon, or other light source equipment that develops hazardous gases, dust, or radiation.

II. Equipment and Projectors of the Professional Type

540.10 Motion Picture Projection Room Required.

Every professional-type projector shall be located within a projection room. Every projection room shall be of permanent construction, approved for the type of building in which the projection room is located. All projection ports, spotlight ports, viewing ports, and similar openings shall be provided with glass or other approved material so as to completely close the opening. Such rooms shall not be considered as hazardous (classified) locations as defined in Article 500.

> FPN: For further information on protecting openings in projection rooms handling cellulose nitrate motion picture film, see NFPA *101*-2000, *Life Safety Code*.

540.11 Location of Associated Electrical Equipment.

(A) Motor Generator Sets, Transformers, Rectifiers, Rheostats, and Similar Equipment. Motor generator sets, transformers, rectifiers, rheostats, and similar equipment for the supply or control of current to projection or spotlight equipment shall, where nitrate film is used, be located in a separate room. Where placed in the projection room, they shall be located or guarded so that arcs or sparks cannot come in contact with film, and the commutator end or ends of motor generator sets shall comply with one of the conditions in 540.11(A)(1) through (A)(6).

(1) Types. Be of the totally enclosed, enclosed fan-cooled, or enclosed pipe-ventilated type.

(2) Separate Rooms or Housings. Be enclosed in separate rooms or housings built of noncombustible material constructed so as to exclude flyings or lint, and properly ventilated from a source of clean air.

(3) Solid Metal Covers. Have the brush or sliding-contact end of motor-generator enclosed by solid metal covers.

(4) Tight Metal Housings. Have brushes or sliding contacts enclosed in substantial, tight metal housings.

(5) Upper and Lower Half Enclosures. Have the upper half of the brush or sliding-contact end of the motor-generator enclosed by a wire screen or perforated metal and the lower half enclosed by solid metal covers.

(6) Wire Screens or Perforated Metal. Have wire screens or perforated metal placed at the commutator of brush ends. No dimension of any opening in the wire screen or perforated metal shall exceed 1.27 mm (0.05 in.), regardless of the shape of the opening and of the material used.

(B) Switches, Overcurrent Devices, or Other Equipment. Switches, overcurrent devices, or other equipment not normally required or used for projectors, sound reproduction, flood or other special effect lamps, or other equipment shall not be installed in projection rooms.

Exception No. 1: In projection rooms approved for use only with cellulose acetate (safety) film, the installation of appurtenant electrical equipment used in conjunction with the operation of the projection equipment and the control of lights, curtains, and audio equipment, and so forth, shall be permitted. In such projection rooms, a sign reading "Safety Film Only Permitted in This Room" shall be posted on the outside of each projection room door and within the projection room itself in a conspicuous location.

Exception No. 2: Remote-control switches for the control of auditorium lights or switches for the control of motors operating curtains and masking of the motion picture screen shall be permitted to be installed in projection rooms.

(C) Emergency Systems. Control of emergency systems shall comply with Article 700.

The plan of a projection room of a motion picture theater is illustrated in Exhibit 540.1. This plan shows one stereopticon, or "effect machine" (L), two spot machines (S), and three motion picture projectors (P) that are supplied from the dc panelboard.

A dc arc lamp is the light source in each of the six machines. The dc supply is furnished by two motor-generator sets, which are usually installed in soundproof areas to avoid interfering with the sound-reproducing equipment and are controlled from the general control panel in the projection room. Two 500-kcmil feeder cables are run from each generator to the dc panelboard.

A branch circuit consisting of two, 2/0 AWG cables runs from the dc panelboard to each projector (P) and to each spot machine (S). One of the two branch-circuit conductors runs directly to a projector (P); the other passes through an auxiliary gutter to the bank of resistors in the rheostat room and then to the projector. The resistors are equipped with short-circuiting switches, so that the total resistance in series with each arc may be preset to a desired value.

Since the stereopticon or effect machine (L) contains two arc lamps, two circuits with 1 AWG conductors are routed to this machine.

The provisions of 540.13 require that the conductors supplying outlets for arc and xenon lamps of the professional type are not to be smaller than 8 AWG and must be of sufficient size for the lamps employed. In each case, therefore, the maximum current drawn by the lamps should be determined. In this example, with the arc lamps sized for a large picture, the arc in each projector draws nearly 150 amperes.

Four outlets, in addition to the main outlet for supplying the arc, are located at each projector for the following auxiliary circuits:

1. Outlet C supplies a small incandescent lamp inside each lamphouse and/or projector.
2. Outlet G is for the 8 AWG equipment grounding conductor, which is connected to each projector frame and to a metal water pipe.
3. Outlet F supplies a foot switch that controls a solenoid-operated shutter behind each lens, for changing from one projector to another.
4. Outlet M supplies the motor used to operate each projector.

Two exhaust fans and two duct systems, one exhausting from the ceiling of the projection room and one connected to the arc lamp housing of each machine, provide ventilation.

540.12 Work Space.

Each motion picture projector, floodlight, spotlight, or similar equipment shall have clear working space not less than 750 mm (30 in.) wide on each side and at the rear thereof.

Exception: One such space shall be permitted between adjacent pieces of equipment.

Exhibit 540.1 A typical layout of a projection room, including associated generator supplied equipment. Modern projectors contain rectifiers as an integral part of their equipment, thereby eliminating generators and other associated equipment.

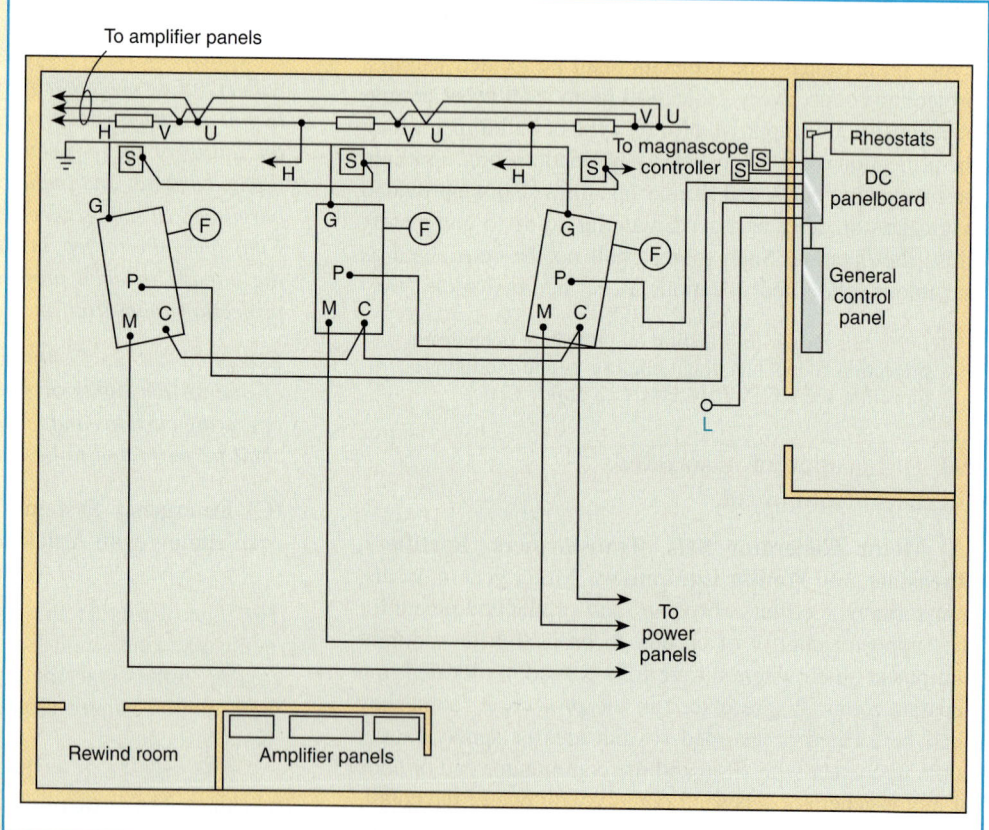

540.13 Conductor Size.

Conductors supplying outlets for arc and xenon projectors of the professional type shall not be smaller than 8 AWG and shall be of sufficient size for the projector employed. Conductors for incandescent-type projectors shall conform to normal wiring standards as provided in 210.24.

540.14 Conductors on Lamps and Hot Equipment.

Insulated conductors having a rated operating temperature of not less than 200°C (392°F) shall be used on all lamps or other equipment where the ambient temperature at the conductors as installed will exceed 50°C (122°F).

540.15 Flexible Cords.

Cords approved for hard usage, as provided in Table 400.4, shall be used on portable equipment.

540.20 Approval.

Projectors and enclosures for arc, xenon and incandescent lamps and rectifiers, transformers, rheostats, and similar equipment shall be listed.

540.21 Marking.

Projectors and other equipment shall be marked with the manufacturer's name or trademark and with the voltage and current for which they are designed in accordance with 110.21.

III. Nonprofessional Projectors

540.31 Motion Picture Projection Room Not Required.

Projectors of the nonprofessional or miniature type, where employing cellulose acetate (safety) film, shall be permitted to be operated without a projection room.

540.32 Approval.

Projection equipment shall be listed.

IV. Audio Signal Processing, Amplification, and Reproduction Equipment

540.50 Audio Signal Processing, Amplification, and Reproduction Equipment.

Audio signal processing, amplification, and reproduction equipment shall be installed as provided in Article 640.

ARTICLE 545
Manufactured Buildings

Contents

545.1 Scope.

This article covers requirements for a manufactured building and building components as herein defined.

The term *manufactured building* is defined in 545.3. In respect to this construction method being used for dwelling units, it is important to make the distinction between manufactured buildings covered in Article 545 and *manufactured homes* as covered and defined in Article 550. The most distinguishing feature between the two types of structures is how they are placed on the building site. Manufactured homes are built on a chassis and installed on-site with or without a permanent foundation. Manufactured buildings are generally constructed within a factory or assembly plant and then transported to the building site. They are not built on a chassis and are designed to be installed on a permanent foundation.

In addition, the organizations responsible for construction standards of these units differ. In the case of manufactured homes, the U.S. Department of Housing and Urban Development CFR-Part 3280, *Manufactured Home Construction and Safety Standards,* contains construction requirements for manufactured homes. Manufactured homes bear a nameplate documenting that the unit has been constructed in accordance with the federal standard. In accordance with federal law, this identifying mark is universally recognized throughout the United States.

545.2 Other Articles.

Wherever the requirements of other articles of this *Code* and Article 545 differ, the requirements of Article 545 shall apply.

545.3 Definitions.

Building Component. Any subsystem, subassembly, or other system designed for use in or integral with or as part of a structure, which can include structural, electrical, mechanical, plumbing, and fire protection systems, and other systems affecting health and safety.

Building System. Plans, specifications, and documentation for a system of manufactured building or for a type or a system of building components, which can include structural, electrical, mechanical, plumbing, and fire protection systems, and other systems affecting health and safety, and including such variations thereof as are specifically permitted by regulation, and which variations are submitted as part of the building system or amendment thereto.

Closed Construction. Any building, building component, assembly, or system manufactured in such a manner that all concealed parts of processes of manufacture cannot be inspected before installation at the building site without disassembly, damage, or destruction.

Manufactured Building. Any building that is of closed construction and is made or assembled in manufacturing facilities on or off the building site for installation, or for assembly and installation on the building site, other than manufactured homes, mobile homes, park trailers, or recreational vehicles.

545.4 Wiring Methods.

(A) Methods Permitted. All raceway and cable wiring methods included in this *Code* and such other wiring systems specifically intended and listed for use in manufactured buildings shall be permitted with listed fittings and with fittings listed and identified for manufactured buildings.

(B) Securing Cables. In closed construction, cables shall be permitted to be secured only at cabinets, boxes, or fittings where 10 AWG or smaller conductors are used and protection against physical damage is provided.

545.5 Supply Conductors.

Provisions shall be made to route the service-entrance, service-lateral, feeder, or branch-circuit supply to the service or building disconnecting means conductors.

545.6 Installation of Service-Entrance Conductors.

Service-entrance conductors shall be installed after erection at the building site.

Exception: Where point of attachment is known prior to manufacture.

545.7 Service Equipment.

Service equipment shall be installed in accordance with 230.70.

545.8 Protection of Conductors and Equipment.

Protection shall be provided for exposed conductors and equipment during processes of manufacturing, packaging, in transit, and erection at the building site.

545.9 Boxes.

(A) Other Dimensions. Boxes of dimensions other than those required in Table 314.16(A) shall be permitted to be installed where tested, identified, and listed to applicable standards.

(B) Not Over 1650 cm³ (100 in.³). Any box not over 1650 cm³ (100 in.³) in size, intended for mounting in closed construction, shall be affixed with anchors or clamps so as to provide a rigid and secure installation.

545.10 Receptacle or Switch with Integral Enclosure.

A receptacle or switch with integral enclosure and mounting means, where tested, identified, and listed to applicable standards, shall be permitted to be installed.

See the commentary following 300.15(E) for additional discussion about wiring devices with integral enclosures.

545.11 Bonding and Grounding.

Prewired panels and building components shall provide for the bonding, or bonding and grounding, of all exposed metals likely to become energized, in accordance with Article 250, Parts V, VI, and VII.

545.12 Grounding Electrode Conductor.

Provisions shall be made to route a grounding electrode conductor from the service, feeder, or branch-circuit supply to the point of attachment to the grounding electrode.

545.13 Component Interconnections.

Fittings and connectors that are intended to be concealed at the time of on-site assembly, where tested, identified, and listed to applicable standards, shall be permitted for on-site interconnection of modules or other building components. Such fittings and connectors shall be equal to the wiring method employed in insulation, temperature rise, and fault-current withstand and shall be capable of enduring the vibration and minor relative motions occurring in the components of manufactured building.

Structural components or modules are usually constructed in manufacturing facilities and then transported over the road to a building site for complete assembly of a structure, such as a dwelling unit, motel, or office building. At the on-site location, approved wiring methods are used to interconnect two or more modules. Exhibit 545.1 shows a type of nonmetallic-sheathed cable connector permitted for such interconnections.

Exhibit 545.1 A type of nonmetallic-sheathed cable connector used for interconnecting modules in a manufacturing building. (Courtesy of Pass & Seymour/Legrand®)

ARTICLE 547
Agricultural Buildings

Contents

547.1 Scope.

The provisions of this article shall apply to the following agricultural buildings or that part of a building or adjacent areas of similar or like nature as specified in 547.1(A) and (B).

(A) Excessive Dust and Dust with Water. Agricultural buildings where excessive dust and dust with water may accumulate, including all areas of poultry, livestock, and fish confinement systems, where litter dust or feed dust, including mineral feed particles, may accumulate.

(B) Corrosive Atmosphere. Agricultural buildings where a corrosive atmosphere exists. Such buildings include areas where the following conditions exist:

(1) Poultry and animal excrement may cause corrosive vapors.

(2) Corrosive particles may combine with water.

(3) The area is damp and wet by reason of periodic washing for cleaning and sanitizing with water and cleansing agents.

(4) Similar conditions exist.

Article 547 applies not only to buildings but also to adjacent areas of similar or like nature. The requirements in Article 547 address the severe environmental conditions that regularly exist on agricultural premises. Damp and wet conditions, dust from feed and litter, and corrosive agents from livestock excrement are all present in these settings as part of normal operating conditions. Grounding and bonding requirements unique to agricultural settings are necessary due to the sensitivity of livestock to differences in potential between surfaces that they are in direct contact with. The wet or damp concrete common to animal confinement areas enhances this sensitivity.

547.2 Definitions.

Distribution Point. An electrical supply point from which service drops, service laterals, feeders, or branch circuits to agricultural buildings, associated farm dwelling(s), and associated buildings under single management are supplied.

> FPN No. 1: Distribution points are also known as the center yard pole, meterpole or the common distribution point.
> FPN No. 2: The service point as defined in Article 100 is typically at the distribution point.

Equipotential Plane. An area where wire mesh or other conductive elements are embedded in or placed under concrete, bonded to all metal structures and fixed nonelectrical equipment that may become energized, and connected to the electrical grounding system to prevent a difference in voltage from developing within the plane.

547.3 Other Articles.

For agricultural buildings not having conditions as specified in 547.1, the electrical installations shall be made in accordance with the applicable articles in this *Code*.

547.4 Surface Temperatures.

Electrical equipment or devices installed in accordance with the provisions of this article shall be installed in a manner such that they will function at full rating without developing surface temperatures in excess of the specified normal safe operating range of the equipment or device.

547.5 Wiring Methods.

(A) Wiring Systems. Types UF, NMC, copper SE cables, jacketed Type MC cable, rigid nonmetallic conduit, liquidtight flexible nonmetallic conduit, or other cables or raceways suitable for the location, with approved termination fittings, shall be the wiring methods employed. Article 398 and Article 502 wiring methods shall be permitted for areas described in 547.1(A).

> FPN: See 300.7 and 352.44 for installation of raceway systems exposed to widely different temperatures.

(B) Mounting. All cables shall be secured within 200 mm (8 in.) of each cabinet, box, or fitting. The 6-mm (¼-in.) airspace required for nonmetallic boxes, fittings, conduit, and cables in 300.6(C) shall not be required in buildings covered by this article.

Cables must be secured within 8 in. of cabinets, boxes, or fittings installed in agricultural buildings. This distance is less than that required for cables in other types of occupancies. The requirement for a ¼-in. airspace in 300.6(C) is judged unnecessary in agricultural buildings, provided nonmetallic wiring methods are used. Decreasing the support spacing requirements coupled with eliminating the ¼-in. airspace requirement reduces the potential for mechanical damage to cable-type wiring methods. Locating the wiring methods directly on the interior surface of the building allows a sealant to be placed along the wiring method to facilitate cleaning. See also 300.6(C), Exception.

(C) Equipment Enclosures, Boxes, Conduit Bodies, and Fittings.

(1) Excessive Dust. Equipment enclosures, boxes, conduit bodies, and fittings installed in areas of buildings where excessive dust may be present shall be designed to minimize the entrance of dust and shall have no openings (such as holes for attachment screws) through which dust could enter the enclosure.

(2) Damp or Wet Locations. In damp or wet locations, equipment enclosures, boxes, conduit bodies, and fittings shall be placed or equipped so as to prevent moisture from entering or accumulating within the enclosure, box, conduit body, or fitting. In wet locations, including normally dry or damp locations where surfaces are periodically washed or sprayed with water, boxes, conduit bodies, and fittings shall be listed for use in wet locations and equipment enclosures shall be weatherproof.

(3) Corrosive Atmosphere. Where wet dust, excessive moisture, corrosive gases or vapors, or other corrosive conditions may be present, equipment enclosures, boxes, conduit bodies, and fittings shall have corrosion resistance properties suitable for the conditions.

> FPN No. 1: See Table 430.91 for appropriate enclosure type designations.
>
> FPN No. 2: Aluminum and magnetic ferrous materials may corrode in agricultural environments.

Gasketed weatherproof enclosures may not provide adequate ventilation for certain types of sensitive equipment, such as electronic equipment. In some cases, it may be necessary to provide ventilation by using other types of enclosures. In accordance with 547.5(C), appropriate enclosures for the conditions encountered are required. This includes consideration of both the type of enclosure and the proper materials used in the construction of the enclosure.

(D) Flexible Connections. Where necessary to employ flexible connections, dusttight flexible connectors, liquidtight flexible conduit, or flexible cord listed and identified for hard usage shall be used. All connectors and fittings used shall be listed and identified for the purpose.

(E) Physical Protection. All electrical wiring and equipment subject to physical damage shall be protected.

(F) Separate Equipment Grounding Conductor. Non–current-carrying metal parts of equipment, raceways, and other enclosures, where required to be grounded, shall be grounded by a copper equipment grounding conductor installed between the equipment and the building disconnecting means. If installed underground, the equipment grounding conductor shall be insulated or covered.

The requirements in 547.5(F) improve the longevity of equipment grounding conductors installed above ground and under ground in the highly corrosive locations that are typical of many farm buildings.

(G) Receptacles. All 125-volt, single-phase, 15- and 20-ampere general-purpose receptacles installed in the following locations shall have ground-fault circuit-interrupter protection for personnel:

(1) In areas having an equipotential plane
(2) Outdoors
(3) Damp or wet locations

547.6 Switches, Receptacles, Circuit Breakers, Controllers, and Fuses.

Switches, including pushbuttons, relays, and similar devices, receptacles, circuit breakers, controllers, and fuses, shall be provided with enclosures as specified in 547.5(C).

547.7 Motors.

Motors and other rotating electrical machinery shall be totally enclosed or designed so as to minimize the entrance of dust, moisture, or corrosive particles.

547.8 Luminaires (Lighting Fixtures).

Luminaires (lighting fixtures) shall comply with the following.

(A) Minimize the Entrance of Dust. Luminaires (lighting fixtures) shall be installed to minimize the entrance of dust, foreign matter, moisture, and corrosive material.

(B) Exposed to Physical Damage. Any luminaire (lighting fixture) that may be exposed to physical damage shall be protected by a suitable guard.

(C) Exposed to Water. A luminaire (fixture) that may be exposed to water from condensation, building cleansing water, or solution shall be watertight.

547.9 Electrical Supply to Building or Structures from a Distribution Point.

(A) Site-Isolating Device. A disconnecting means shall be installed at the distribution point where two or more agricultural buildings, structures, associated farm dwelling(s), or other buildings are supplied from the distribution point. For the purposes of applying the requirements of this section, this disconnecting means shall be classified as a site-isolating device and shall have provisions for bonding the grounding electrode conductor to the grounded conductor.

(1) Purpose. The disconnecting means shall simultaneously interrupt all ungrounded conductors for the purposes of isolation, system maintenance, emergency disconnection, or connection of optional standby systems.

(2) Series Disconnects. An additional disconnecting means shall not be required where the serving utility provides a disconnecting means as part of their service requirements and this disconnecting means is accessible to the user and meets the requirements of this section.

(3) Rating. The disconnecting means shall be rated for the calculated load as determined by Part IV of Article 220.

(4) Overcurrent. The disconnecting means shall not be required to contain overload protection.

(5) Accessibility. Where not readily accessible, the disconnecting means shall be capable of operation from a readily accessible point.

(6) Grounding. The grounded conductor of the system shall be connected to a grounding electrode through a grounding electrode conductor at the disconnecting means.

(B) Electrical Supply. The buildings or structures shall be permitted to be supplied by either 547.9(B)(1) or (B)(2).

(1) Building(s) or Structure(s). Where the disconnecting means and overcurrent protection are located at the buildings or structures, the supply conductors shall be sized in accordance with Part IV of Article 220 and installed in accordance with the requirements of Part II of Article 225.

For each building or structure, the conditions in either (a) or (b) shall be permitted.

(a) The grounded circuit conductor shall be permitted to be connected to the building disconnecting means and to the grounding electrode system of that building or structure where all the requirements of 250.32(B)(2) are met.

(b) A separate equipment grounding conductor shall be run with the supply conductors to the building(s) or structure(s) and the following conditions shall be met:

(1) The equipment grounding conductor is the same size as the largest supply conductor, if of the same material, or is adjusted in size in accordance with the equivalent size columns of Table 250.122 if of different materials.
(2) The equipment grounding conductor is bonded to the grounded circuit conductor at the disconnecting means enclosure at the distribution point or at the source of a separately derived system.
(3) A grounding electrode system is provided in accordance with Part III of Article 250 and connected to the equipment grounding conductor at the building(s) or structure(s) disconnecting means.
(4) The grounded circuit conductor is not connected to a grounding electrode or to any equipment grounding conductor on the load side of the distribution point.

(2) Disconnecting Means and Overcurrent Protection at the Distribution Point. Where the disconnecting means and overcurrent protection for each set of feeder conductors are located at the distribution point, feeders to building(s) or structure(s) shall meet the requirements of 250.32 and Article 225, Parts I and II.

> FPN: Methods to reduce neutral-to-earth voltages in livestock facilities include supplying buildings or structures with 4-wire, single-phase services, sizing of 3-wire service conductors to limit voltage drop to 2 percent, and connecting loads line-to-line.

(C) Underground Equipment Grounding Conductors. Where livestock is housed, any portion of the equipment grounding conductor run underground to the building or structure shall be insulated or covered copper.

The requirements in 547.9 cover the installation of conductors that originate from an electrical distribution point and supply agricultural buildings. The term *distribution point* is defined in 547.2 as a supply point for service conductors, feeders, or branch circuits supplying agricultural premises.

Many agricultural sites consist of multiple buildings that are directly related to the agricultural operation or support the operation, as is the case with dwelling units. A distribution point, sometimes referred to as the center yard pole, is often used as a means of centrally locating the origin of the electrical distribution system that supplies multiple buildings at an agricultural site. A means to disconnect all ungrounded conductors run to the buildings and structures that are supplied from the distribution point is required. This site-isolating device provides a means to disconnect all power to the buildings at the agricultural site from a single location. This is useful in the event of an emergency, for the purposes of maintenance, or for connection to a stand-by power source. A site-isolating device provided by an electric utility is permitted, provided that it meets the requirements of 547.9. If the supply system includes a grounded conductor, it must be connected to a grounding electrode system at the site-isolating device.

The intent of referring to this disconnecting means as the site-isolation device is so that it is not considered the service disconnecting means. This site-isolation device is not required to have provisions for overcurrent protection, nor is overcurrent protection for the load side conductors required to be located immediately adjacent to this device. Based on the requirements in 547.9(A) and (B) the location of the service disconnecting means is on the load side of the site-isolation switch and is located either at the distribution point or at the building or structure supplied.

A site-isolating device can be considered as the service disconnecting means if it has provisions for overcurrent protection, is identified as suitable for use as service equipment, is grounded in accordance with 250.24, and meets all other applicable requirements for service disconnecting means. In this case, all the load side conductors are feeders or branch circuits.

If the site-isolating device does not provide overcurrent protection for the conductors run to the individual buildings or structures, the overcurrent protection is permitted to be located either at the distribution point or at the building or structure supplied. Grounding on the load side of the site-isolating device must be in accordance with 250.32.

In Exhibit 547.1, the site-isolation device is located at the distribution point. A set of service conductors is run to each of the three structures on the premises and, at each of the buildings, a disconnecting means and overcurrent protection is installed. A grounding electrode system is required at the distribution point and a grounding electrode conductor connection to the supply system grounded conductor must be made at the site-isolation device. A grounding electrode system is also required at each of the buildings. At these locations, a grounding electrode conductor connection to the system grounded conductor is permitted, provided that such a connection does not create a parallel path for neutral current between the distribution point and the structure disconnecting means. If a parallel path for neutral current is created through multiple grounding connections, 547.9(B)(1)(b) requires the installation of a separate equipment grounding conductor between the distribution point and the disconnecting means at each building.

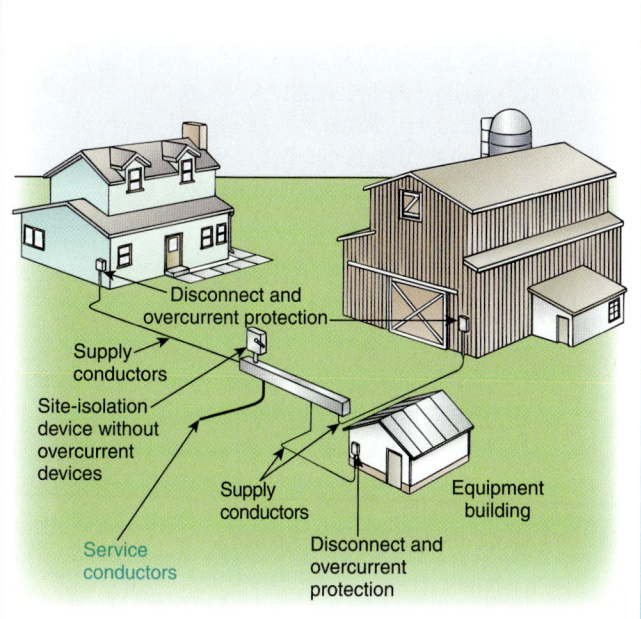

Exhibit 547.1 Site-isolation device located at the distribution point with service conductors run to each building. A disconnecting means is installed at each building.

The equipment grounding conductor must be the same size as the largest supply conductor or if it is of different conductor material than the supply conductors, it must be adjusted in size based on Table 250.122. Any portion of the equipment grounding conductor installed underground must be insulated or covered copper. This provision only applies to conductors that are run to buildings in which livestock is housed. The intent of this requirement is to reduce leakage current in those areas where livestock are kept, as the prevention of stray voltage at agricultural premises is of the utmost importance.

In Exhibit 547.2, the site-isolation device and the feeder disconnecting means and overcurrent protection are located at the distribution point. A set of feeder conductors is run to each of the three structures on the premises. At the distri-

bution point, a grounding electrode system is required and, in accordance with 250.32, a grounding electrode system is required at each of the buildings. In addition, a disconnecting means at each of the structures is required per 225.31 and 225.32. This distribution arrangement for agricultural sites is covered in 547.9(B)(2).

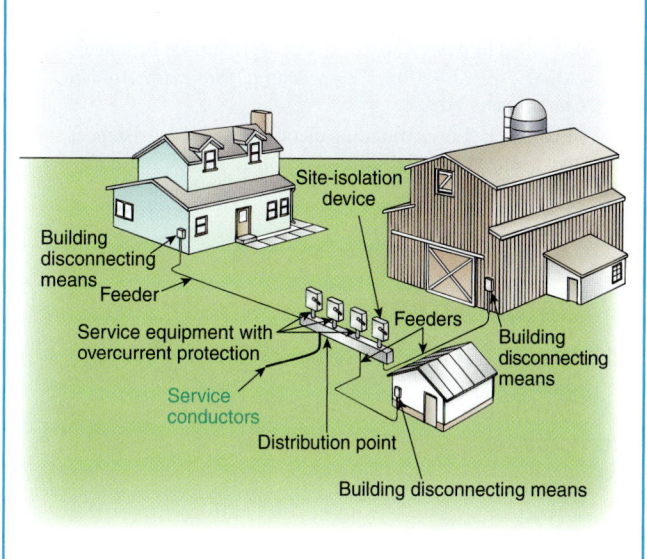

Exhibit 547.2 Site-isolation device and feeder disconnecting means located at the distribution point with feeders run to each building.

547.10 Equipotential Planes and Bonding of Equipotential Planes.

For the purposes of this section, the term *livestock* shall not include poultry.

(A) Areas Requiring Equipotential Planes. Equipotential planes shall be installed in all concrete floor confinement areas of livestock buildings that contain metallic equipment that is accessible to animals and likely to become energized. Outdoor confinement areas, such as feedlots, shall have equipotential planes installed around metallic equipment that is accessible to animals and likely to become energized. The equipotential plane shall encompass the area around the equipment where the animal stands while accessing the equipment.

(B) Areas Not Requiring Equipotential Planes. Equipotential planes shall not be required in dirt confinement areas containing metallic equipment that is accessible to animals and likely to become energized. All circuits providing electric power to equipment that is accessible to animals in dirt confinement areas shall have GFCI protection.

(C) Bonding. Equipotential planes shall be bonded to the electrical grounding system. The bonding conductor shall be copper, insulated, covered or bare, and not smaller and 8 AWG. The means of bonding to wire mesh or conductive elements shall be by pressure connectors or clamps of brass, copper, copper alloy, or an equally substantial approved means. Slatted floors that are supported by structures that are a part of an equipotential plane shall not require bonding.

> FPN No. 1: Methods to establish equipotential planes are described in American Society of Agricultural Engineers (ASAE) EP473-2001, *Equipotential Planes in Animal Containment Areas.*
>
> FPN No. 2: Low grounding electrode system resistances may reduce potential differences in livestock facilities.

ARTICLE 550
Mobile Homes, Manufactured Homes, and Mobile Home Parks

Contents

I. General

550.1 Scope.

The provisions of this article cover the electrical conductors and equipment installed within or on mobile and manufactured homes, the conductors that connect mobile and manufactured homes to a supply of electricity, and the installation of electrical wiring, luminaires (fixtures), equipment, and appurtenances related to electrical installations within a mobile home park up to the mobile home service-entrance conductors or, if none, the mobile home service equipment.

> FPN: For additional information on manufactured housing see NFPA 501-1999, *Standard on Manufactured Housing*, and Part 3280, *Manufactured Home Construction and Safety Standards*, of the Federal Department of Housing and Urban Development.

The *Federal Mobile Home Construction and Safety Standard,* issued by the Federal Housing and Urban Development Administration (HUD), has incorporated many of the provisions of Article 550 of the *NEC.* The federal standard contains the requirements for electrical systems, conductors, and equipment installed within or on mobile homes and the conductors that connect mobile homes to a supply of electricity. Mobile homes are defined as manufactured homes in the HUD regulations. For the purposes of this *Code,* and unless otherwise indicated, the term *mobile home* includes manufactured homes.

    The regulations pertaining to electrical systems are located in the *Code of Federal Regulations,* Title 24, Sections 3280.801 through 3280.816. They require that new manufactured homes comply with the federal standard. In some cases, HUD has delegated the enforcement of this standard to state

and private inspection agencies and qualified testing laboratories. The service equipment and feeders installed at the mobile or manufactured home site is covered by the requirements in Part III of this article.

    See Article 545 for the requirements covering electrical systems in manufactured buildings.

550.2 Definitions.

Appliance, Fixed. An appliance that is fastened or otherwise secured at a specific location.

Appliance, Portable. An appliance that is actually moved or can easily be moved from one place to another in normal use.

> FPN: For the purpose of this article, the following major appliances, other than built-in, are considered portable if cord connected: refrigerators, range equipment, clothes washers, dishwashers without booster heaters, or other similar appliances.

Appliance, Stationary. An appliance that is not easily moved from one place to another in normal use.

Distribution Panelboard. See definition of panelboard in Article 100.

Feeder Assembly. The overhead or under-chassis feeder conductors, including the grounding conductor, together with the necessary fittings and equipment or a power-supply cord listed for mobile home use, designed for the purpose of delivering energy from the source of electrical supply to the distribution panelboard within the mobile home.

Laundry Area. An area containing or designed to contain a laundry tray, clothes washer, or a clothes dryer.

Manufactured Home. A structure, transportable in one or more sections, that is 2.5 m (8 body ft) or more in width or 12 m (40 body ft) or more in length in the traveling mode or, when erected on site, is 30 m² (320 ft²) or more; which is built on a chassis and designed to be used as a dwelling, with or without a permanent foundation, when connected to the required utilities, including the plumbing, heating, air conditioning, and electrical systems contained therein. Calculations used to determine the number of square meters (square feet) in a structure will be based on the structure's exterior dimensions, measured at the largest horizontal projections when erected on site. These dimensions include all expandable rooms, cabinets, and other projections containing interior space, but do not include inside bay windows.

    For the purpose of this *Code* and unless otherwise indicated, the term *mobile home* includes manufactured homes.

> FPN No. 1: See the applicable building code for definition of the term *permanent foundation.*
>
> FPN No. 2: See Part 3280, *Manufactured Home Construction and Safety Standards,* of the Federal Depart-

ment of Housing and Urban Development, for additional information on the definition.

Mobile Home. A factory-assembled structure or structures transportable in one or more sections that is built on a permanent chassis and designed to be used as a dwelling without a permanent foundation where connected to the required utilities and that includes the plumbing, heating, air-conditioning, and electric systems contained therein.

For the purpose of this *Code* and unless otherwise indicated, the term *mobile home* includes manufactured homes.

Mobile Home Accessory Building or Structure. Any awning, cabana, ramada, storage cabinet, carport, fence, windbreak, or porch established for the use of the occupant of the mobile home on a mobile home lot.

Mobile home is the original term covering a structure that is built on a chassis, designed to be transportable and intended for installation on a site with or without a permanent foundation. The *Code* has covered these units and their associated site equipment since the 1965 edition. Manufactured homes (not to be confused with manufactured buildings, covered in Article 545) are also covered by Article 550 and, for the purposes of this article, are considered mobile homes. The term *manufactured home* is used in the federal standard that covers their construction. The requirements in Article 550 treat mobile and manufactured homes the same unless specifically stated otherwise. An example of where a distinction is made between the two is found in 550.23, which contains provisions exclusive to manufactured homes that cover mounting of service equipment on the structure.

The requirements contained in Article 550, Part III cover the installation of service equipment and feeders at mobile and manufactured home sites. It is intended that mobile and manufactured homes constructed in accordance with the requirements of Article 550, Parts I and II (or under the HUD Section 3280 regulations) be installed at their sites in accordance with the requirements of Part III. For more information on the distinction between manufactured homes and manufactured buildings, see the commentary following 545.1.

Mobile Home Lot. A designated portion of a mobile home park designed for the accommodation of one mobile home and its accessory buildings or structures for the exclusive use of its occupants.

Mobile Home Park. A contiguous parcel of land that is used for the accommodation of occupied mobile homes.

Mobile Home Service Equipment. The equipment containing the disconnecting means, overcurrent protective de-

vices, and receptacles or other means for connecting a mobile home feeder assembly.

Park Electrical Wiring Systems. All of the electrical wiring, luminaires (fixtures), equipment, and appurtenances related to electrical installations within a mobile home park, including the mobile home service equipment.

550.3 Other Articles.

Wherever the requirements of other articles of this *Code* and Article 550 differ, the requirements of Article 550 shall apply.

550.4 General Requirements.

(A) Mobile Home Not Intended as a Dwelling Unit. A mobile home not intended as a dwelling unit—for example, those equipped for sleeping purposes only, contractor's on-site offices, construction job dormitories, mobile studio dressing rooms, banks, clinics, mobile stores, or intended for the display or demonstration of merchandise or machinery—shall not be required to meet the provisions of this article pertaining to the number or capacity of circuits required. It shall, however, meet all other applicable requirements of this article if provided with an electrical installation intended to be energized from a 120-volt or 120/240-volt ac power supply system. Where different voltage is required by either design or available power supply system, adjustment shall be made in accordance with other articles and sections for the voltage used.

(B) In Other Than Mobile Home Parks. Mobile homes installed in other than mobile home parks shall comply with the provisions of this article.

(C) Connection to Wiring System. The provisions of this article shall apply to mobile homes intended for connection to a wiring system rated 120/240 volts, nominal, 3-wire ac, with grounded neutral.

(D) Listed or Labeled. All electrical materials, devices, appliances, fittings, and other equipment shall be listed or labeled by a qualified testing agency and shall be connected in an approved manner when installed.

II. Mobile and Manufactured Homes

550.10 Power Supply.

(A) Feeder. The power supply to the mobile home shall be a feeder assembly consisting of not more than one listed 50-ampere mobile home power-supply cord with an integrally molded or securely attached plug cap or a permanently installed feeder.

Exception No. 1: A mobile home that is factory equipped with gas or oil-fired central heating equipment and cooking appliances shall be permitted to be provided with a listed mobile home power-supply cord rated 40 amperes.

Exception No. 2: Manufactured homes constructed in accordance with 550.32(B).

Exception No. 2 modifies the requirement of 550.10(A) only for manufactured homes. The installation of service equipment is permitted in or on a manufactured home per 550.23(B).

(B) Power-Supply Cord. If the mobile home has a power-supply cord, it shall be permanently attached to the distribution panelboard or to a junction box permanently connected to the distribution panelboard, with the free end terminating in an attachment plug cap.

Cords with adapters and pigtail ends, extension cords, and similar items shall not be attached to, or shipped with, a mobile home.

A suitable clamp or the equivalent shall be provided at the distribution panelboard knockout to afford strain relief for the cord to prevent strain from being transmitted to the terminals when the power-supply cord is handled in its intended manner.

The cord shall be a listed type with four conductors, one of which shall be identified by a continuous green color or a continuous green color with one or more yellow stripes for use as the grounding conductor.

(C) Attachment Plug Cap. The attachment plug cap shall be a 3-pole, 4-wire, grounding type, rated 50 amperes, 125/250 volts with a configuration as shown in Figure 550.10(C) and intended for use with the 50-ampere, 125/250-volt receptacle configuration shown in Figure 550.10(C). It shall be listed, by itself or as part of a power-supply cord assembly, for the purpose and shall be molded to or installed on the flexible cord so that it is secured tightly to the cord at the point where the cord enters the attachment plug cap. If a right-angle cap is used, the configuration shall be oriented so that the grounding member is farthest from the cord.

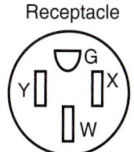 

125/250-V, 50-A, 3-pole, 4-wire, grounding type

Figure 550.10(C) 50-ampere, 125/250-volt receptacle and attachment plug cap configurations, 3-pole, 4-wire, grounding-types, used for mobile home supply cords and mobile home parks.

FPN: Complete details of the 50-ampere plug and receptacle configuration can be found in the National Electrical Manufacturers Association *Standard for Dimensions of Attachment Plugs and Receptacles*, ANSI/NEMA WD 6-1989, Figure 14-50.

(D) Overall Length of a Power-Supply Cord. The overall length of a power-supply cord, measured from the end of the cord, including bared leads, to the face of the attachment plug cap shall not be less than 6.4 m (21 ft) and shall not exceed 11 m (36½ ft). The length of the cord from the face of the attachment plug cap to the point where the cord enters the mobile home shall not be less than 6.0 m (20 ft).

(E) Marking. The power-supply cord shall bear the following marking:

FOR USE WITH MOBILE HOMES—40 AMPERES.

or

FOR USE WITH MOBILE HOMES—50 AMPERES.

(F) Point of Entrance. The point of entrance of the feeder assembly to the mobile home shall be in the exterior wall, floor, or roof.

(G) Protected. Where the cord passes through walls or floors, it shall be protected by means of conduits and bushings or equivalent. The cord shall be permitted to be installed within the mobile home walls, provided a continuous raceway having a maximum size of 32 mm (1¼ in.) is installed from the branch-circuit panelboard to the underside of the mobile home floor.

(H) Protection Against Corrosion and Mechanical Damage. Permanent provisions shall be made for the protection of the attachment plug cap of the power-supply cord and any connector cord assembly or receptacle against corrosion and mechanical damage if such devices are in an exterior location while the mobile home is in transit.

(I) Mast Weatherhead or Raceway. Where the calculated load exceeds 50 amperes or where a permanent feeder is used, the supply shall be by means of either of the following:

(1) One mast weatherhead installation, installed in accordance with Article 230, containing four continuous, insulated, color-coded feeder conductors, one of which shall be an equipment grounding conductor

(2) A metal raceway or rigid nonmetallic conduit from the disconnecting means in the mobile home to the underside of the mobile home, with provisions for the attachment to a suitable junction box or fitting to the raceway on the underside of the mobile home [with or without conductors as in 550.10(I)(1)]. The manufacturer shall provide written installation instructions stating the

proper feeder conductor sizes for the raceway and the size of the junction box to be used.

New mobile homes are generally larger and have more electrically powered equipment than most of the earlier units, which in most cases results in a feeder calculation that exceeds 50 amperes. Cord-and-plug connection is permitted for units in which the calculated load does not exceed 50 amperes. If the calculated load exceeds 50 amperes, the *Code* requires a permanently connected feeder, as covered in 550.10(I) and 550.33. Local requirements must be checked for the approved method of installing overhead and underground feeder assemblies.

A raceway is required from the distribution panelboard in the mobile home to the underside of the mobile home. Typically, the feeder conductors in this raceway are installed when the mobile home is located at its site. The raceway provides a means to install the feeder conductors to the mobile home panelboard without having to damage the interior finish. The feeder assembly must be comprised of four continuous, insulated, color-coded conductors, as indicated in 550.10(I)(1) and 550.33(A).

550.11 Disconnecting Means and Branch-Circuit Protective Equipment.

The branch-circuit equipment shall be permitted to be combined with the disconnecting means as a single assembly. Such a combination shall be permitted to be designated as a distribution panelboard. If a fused distribution panelboard is used, the maximum fuse size for the mains shall be plainly marked with lettering at least 6 mm (¼ in.) high and visible when fuses are changed.

Where plug fuses and fuseholders are used, they shall be tamper-resistant Type S, enclosed in dead-front fuse panelboards. Electrical distribution panelboards containing circuit breakers shall also be dead-front type.

> FPN: See 110.22 concerning identification of each disconnecting means and each service, feeder, or branch circuit at the point where it originated and the type marking needed.

(A) Disconnecting Means. A single disconnecting means shall be provided in each mobile home consisting of a circuit breaker, or a switch and fuses and its accessories installed in a readily accessible location near the point of entrance of the supply cord or conductors into the mobile home. The main circuit breakers or fuses shall be plainly marked "Main." This equipment shall contain a solderless type of grounding connector or bar for the purposes of grounding, with sufficient terminals for all grounding conductors. The neutral bar termination of the grounded circuit conductors shall be insulated in accordance with 550.16(A). The disconnecting equipment shall have a rating suitable for the con-

nected load. The distribution equipment, either circuit breaker or fused type, shall be located a minimum of 600 mm (24 in.) from the bottom of such equipment to the floor level of the mobile home.

> FPN: See 550.20(B) for information on disconnecting means for branch circuits designed to energize heating or air-conditioning equipment, or both, located outside the mobile home, other than room air conditioners.

A distribution panelboard shall be rated not less than 50 amperes and employ a 2-pole circuit breaker rated 40 amperes for a 40-ampere supply cord, or 50 amperes for a 50-ampere supply cord. A distribution panelboard employing a disconnect switch and fuses shall be rated 60 amperes and shall employ a single 2-pole, 60-ampere fuseholder with 40- or 50-ampere main fuses for 40- or 50-ampere supply cords, respectively. The outside of the distribution panelboard shall be plainly marked with the fuse size.

The distribution panelboard shall be located in an accessible location but shall not be located in a bathroom or a clothes closet. A clear working space at least 750 mm (30 in.) wide and 750 mm (30 in.) in front of the distribution panelboard shall be provided. This space shall extend from the floor to the top of the distribution panelboard.

(B) Branch-Circuit Protective Equipment. Branch-circuit distribution equipment shall be installed in each mobile home and shall include overcurrent protection for each branch circuit consisting of either circuit breakers or fuses.

The branch-circuit overcurrent devices shall be rated as follows:

(1) Not more than the circuit conductors; and
(2) Not more than 150 percent of the rating of a single appliance rated 13.3 amperes or more that is supplied by an individual branch circuit; but
(3) Not more than the overcurrent protection size and of the type marked on the air conditioner or other motor-operated appliance.

(C) Two-Pole Circuit Breakers. Where circuit breakers are provided for branch-circuit protection, 240-volt circuits shall be protected by a 2-pole common or companion trip, or handle-tied paired circuit breakers.

(D) Electrical Nameplates. A metal nameplate on the outside adjacent to the feeder assembly entrance shall read:

THIS CONNECTION FOR 120/240-VOLT, 3-POLE, 4-WIRE, 60-HERTZ, _____ AMPERE SUPPLY

The correct ampere rating shall be marked in the blank space.

Exception: For manufactured homes, the manufacturer shall provide in its written installation instructions or in the data plate the minimum ampere rating of the feeder assembly or, where provided, the service entrance conductors intended

for connection to the manufactured home. The rating provided shall not be less than the minimum load calculated in accordance with 550.18.

550.12 Branch Circuits.

The number of branch circuits required shall be determined in accordance with 550.12(A) through (E).

(A) Lighting. Based on 33 volt-amperes/m^2 (3 VA/ft^2) times outside dimensions of the mobile home (coupler excluded) divided by 120 volts to determine the number of 15- or 20-ampere lighting area circuits, for example,

$$\frac{3 \times \text{length} \times \text{width}}{120 \times 15 \text{ (or 20)}}$$

(B) Small Appliances. For the small-appliance load in kitchens, pantries, dining rooms, and breakfast rooms, two or more 20-ampere appliance branch circuits, in addition to the number of branch circuits required by other parts of this section, shall be provided for all receptacle outlets required by 550.13(D) in these rooms. Such circuits shall have no other outlets.

Exception No. 1: A receptacle installed solely for the electrical supply to and support of an electric clock in any of the rooms specified in (B).

Exception No. 2: Receptacles installed to provide power for supplemental equipment and lighting on gas-fired ranges, ovens, or counter-mounted cooking units.

Countertop receptacle outlets installed in the kitchen shall be supplied by not less than two small-appliance branch circuits, either or both of which shall be permitted to supply receptacle outlets in the kitchen and other rooms specified above.

(C) Laundry Area. Where a laundry area is provided, a 20-ampere branch circuit shall be provided to supply the laundry receptacle outlet(s).

(D) General Appliances. (Including furnace, water heater, range, and central or room air conditioner, etc.). There shall be one or more circuits of adequate rating in accordance with the following:

FPN: For central air conditioning, see Article 440.

(1) The ampere rating of fixed appliances shall not be over 50 percent of the circuit rating if lighting outlets (receptacles, other than kitchen, dining area, and laundry, considered as lighting outlets) are on the same circuit.

(2) For fixed appliances on a circuit without lighting outlets, the sum of rated amperes shall not exceed the branch-circuit rating. Motor loads or other continuous duty loads shall not exceed 80 percent of the branch-circuit rating.

(3) The rating of a single cord-and-plug-connected appliance on a circuit having no other outlets shall not exceed 80 percent of the circuit rating.

(4) The rating of a range branch circuit shall be based on the range demand as specified for ranges in 550.18(B)(5).

(E) Bathrooms. Bathroom receptacle outlets shall be supplied by at least one 20-ampere branch circuit. Such circuits shall have no other outlets other than as provided for in 550.13(E)(2).

550.13 Receptacle Outlets.

(A) Grounding-Type Receptacle Outlets. All receptacle outlets shall comply with the following:

(1) Be of grounding type
(2) Be installed according to 406.3
(3) Except where supplying specific appliances, be 15- or 20-ampere, 125-volt, either single or duplex, and accept parallel-blade attachment plugs

(B) Ground-Fault Circuit Interrupters (GFCI). All 125-volt, single-phase, 15- and 20-ampere receptacle outlets installed outdoors, in compartments accessible from outside the unit, or in bathrooms, including receptacles in luminaires (light fixtures), shall have GFCI protection for personnel. GFCI protection for personnel shall be provided for receptacle outlets serving countertops in kitchens, and receptacle outlets located within 1.8 m (6 ft) of a wet bar sink.

Exception: Receptacles installed for appliances in dedicated spaces, such as for dishwashers, disposals, refrigerators, freezers, and laundry equipment.

Feeders supplying branch circuits shall be permitted to be protected by a ground-fault circuit-interrupter in lieu of the provision for such interrupters specified herein.

(C) Cord-Connected Fixed Appliance. A grounding-type receptacle outlet shall be provided for each cord-connected fixed appliance installed.

(D) Receptacle Outlets Required. Except in the bath, closet, and hall areas, receptacle outlets shall be installed at wall spaces 600 mm (2 ft) wide or more so that no point along the floor line is more than 1.8 m (6 ft) measured horizontally from an outlet in that space. In addition, a receptacle outlet shall be installed in the following locations:

(1) Over or adjacent to countertops in the kitchen [at least one on each side of the sink if countertops are on each side and are 300 mm (12 in.) or over in width].

(2) Adjacent to the refrigerator and freestanding gas-range space. A duplex receptacle shall be permitted to serve as the outlet for a countertop and a refrigerator.

(3) At countertop spaces for built-in vanities.

(4) At countertop spaces under wall-mounted cabinets.

(5) In the wall at the nearest point to where a bar-type counter attaches to the wall.

(6) In the wall at the nearest point to where a fixed room divider attaches to the wall.

(7) In laundry areas within 1.8 m (6 ft) of the intended location of the laundry appliance(s).

(8) At least one receptacle outlet located outdoors and accessible at grade level and not more than 2.0 m (6½ ft) above grade. A receptacle outlet located in a compartment accessible from the outside of the unit shall be considered an outdoor receptacle.

(9) At least one receptacle outlet shall be installed in bathrooms within 900 mm (36 in.) of the outside edge of each basin. The receptacle outlet shall be located above or adjacent to the basin location. This receptacle shall be in addition to any receptacle that is a part of a luminaire (fixture) or appliance. The receptacle shall not be enclosed within a bathroom cabinet or vanity.

(E) Pipe Heating Cable(s) Outlet. For the connection of pipe heating cable(s), a receptacle outlet shall be located on the underside of the unit as follows:

(1) Within 600 mm (2 ft) of the cold water inlet.

(2) Connected to an interior branch circuit, other than a small appliance branch circuit. It shall be permitted to use a bathroom receptacle circuit for this purpose.

(3) On a circuit where all of the outlets are on the load side of the ground-fault circuit-interrupter.

(4) This outlet shall not be considered as the receptacle required by 550.13(D)(8).

The provisions of 550.13(E) require a receptacle outlet on the underside of mobile homes to supply cord-and-plug-connected pipe heating cables (sometimes referred to as heat tape outlets). The receptacle must be GFCI protected and connected to a branch circuit that serves the interior of the mobile home. All of the outlets supplied by this branch circuit must be on the load (downstream) side of a ground-fault circuit interrupter. The purpose of arranging the supply circuit to the pipe heating cable outlet in this manner is to allow supervision of the power supply to, and GFCI protection of, this outlet from the interior of the mobile home. If the overcurrent protective device or GFCI device opens, the occupants of the mobile home are more likely to notice it than if the outlet were supplied by a dedicated circuit and the GFCI device were located at the outlet.

(F) Receptacle Outlets Not Permitted. Receptacle outlets shall not be permitted in the following locations:

(1) Receptacle outlets shall not be installed in or within reach [750 mm (30 in.)] of a shower or bathtub space.

(2) A receptacle shall not be installed in a face-up position in any countertop.

(3) Receptacle outlets shall not be installed above electric baseboard heaters, unless provided for in the listing or manufacturer's instructions.

(G) Receptacle Outlets Not Required. Receptacle outlets shall not be located in the following locations:

(1) In the wall space occupied by built-in kitchen or wardrobe cabinets

(2) In the wall space behind doors that can be opened fully against a wall surface

(3) In room dividers of the lattice type that are less than 2.5 m (8 ft) long, not solid, and within 150 mm (6 in.) of the floor

(4) In the wall space afforded by bar-type counters

550.14 Luminaires (Fixtures) and Appliances.

(A) Fasten Appliances in Transit. Means shall be provided to securely fasten appliances when the mobile home is in transit. (See 550.16 for provisions on grounding.)

(B) Accessibility. Every appliance shall be accessible for inspection, service, repair, or replacement without removal of permanent construction.

(C) Pendants. Listed pendant-type luminaires (fixtures) or pendant cords shall be permitted.

(D) Bathtub and Shower Luminaires (Fixtures). Where a luminaire (lighting fixture) is installed over a bathtub or in a shower stall, it shall be of the enclosed and gasketed type listed for wet locations.

550.15 Wiring Methods and Materials.

Except as specifically limited in this section, the wiring methods and materials included in this *Code* shall be used in mobile homes. Aluminum conductors, aluminum alloy conductors, and aluminum core conductors such as copper-clad aluminum shall not be acceptable for use as branch-circuit wiring.

(A) Nonmetallic Boxes. Nonmetallic boxes shall be permitted only with nonmetallic cable or nonmetallic raceways.

(B) Nonmetallic Cable Protection. Nonmetallic cable located 380 mm (15 in.) or less above the floor, if exposed, shall be protected from physical damage by covering boards, guard strips, or raceways. Cable likely to be damaged by stowage shall be so protected in all cases.

(C) Metal-Covered and Nonmetallic Cable Protection. Metal-covered and nonmetallic cables shall be permitted to pass through the centers of the wide side of 2 by 4 studs. However, they shall be protected where they pass through

2 by 2 studs or at other studs or frames where the cable or armor would be less than 32 mm (1¼ in.) from the inside or outside surface of the studs where the wall covering materials are in contact with the studs. Steel plates on each side of the cable, or a tube, with not less than 1.35 mm (0.053 in.) wall thickness shall be required to protect the cable. These plates or tubes shall be securely held in place.

(D) Metal Faceplates. Where metal faceplates are used, they shall be effectively grounded.

(E) Installation Requirements. If a range, clothes dryer, or similar appliance is connected by metal-covered cable or flexible metal conduit, a length of not less than 900 mm (3 ft) of free cable or conduit shall be provided to permit moving the appliance. The cable or flexible metal conduit shall be secured to the wall. Type NM or Type SE cable shall not be used to connect a range or dryer. This shall not prohibit the use of Type NM or Type SE cable between the branch-circuit overcurrent-protective device and a junction box or range or dryer receptacle.

(F) Raceways. Where rigid metal conduit or intermediate metal conduit is terminated at an enclosure with a locknut and bushing connection, two locknuts shall be provided, one inside and one outside of the enclosure. Rigid nonmetallic conduit, electrical nonmetallic tubing, or surface raceway shall be permitted. All cut ends of conduit and tubing shall be reamed or otherwise finished to remove rough edges.

(G) Switches. Switches shall be rated as follows:

(1) For lighting circuits, switches shall be rated not less than 10 amperes, 120 to 125 volts, and in no case less than the connected load.
(2) For motors or other loads, switches shall have ampere or horsepower ratings, or both, adequate for loads controlled. (An ac general-use snap switch shall be permitted to control a motor 2 hp or less with full-load current not over 80 percent of the switch ampere rating.)

(H) Under-Chassis Wiring (Exposed to Weather). Where outdoor or under-chassis line-voltage (120 volts, nominal, or higher) wiring is exposed to moisture or physical damage, it shall be protected by rigid metal conduit or intermediate metal conduit. The conductors shall be suitable for wet locations.

Exception: Electrical metallic tubing or rigid nonmetallic conduit shall be permitted where closely routed against frames and equipment enclosures.

(I) Boxes, Fittings, and Cabinets. Boxes, fittings, and cabinets shall be securely fastened in place and shall be supported from a structural member of the home, either directly or by using a substantial brace.

Exception: Snap-in-type boxes. Boxes provided with special wall or ceiling brackets and wiring devices with integral enclosures that securely fasten to walls or ceilings and are identified for the use shall be permitted without support from a structural member or brace. The testing and approval shall include the wall and ceiling construction systems for which the boxes and devices are intended to be used.

(J) Appliance Terminal Connections. Appliances having branch-circuit terminal connections that operate at temperatures higher than 60°C (140°F) shall have circuit conductors as described in the following:

(1) Branch-circuit conductors having an insulation suitable for the temperature encountered shall be permitted to be run directly to the appliance.
(2) Conductors having an insulation suitable for the temperature encountered shall be run from the appliance terminal connection to a readily accessible outlet box placed at least 300 mm (1 ft) from the appliance. These conductors shall be in a suitable raceway or Type AC or MC cable of at least 450 mm (18 in.) but not more than 1.8 m (6 ft) in length.

(K) Component Interconnections. Fittings and connectors that are intended to be concealed at the time of assembly shall be listed and identified for the interconnection of building components. Such fittings and connectors shall be equal to the wiring method employed in insulation, temperature rise, and fault-current withstanding and shall be capable of enduring the vibration and shock occurring in mobile home transportation.

FPN: See 550.19 for interconnection of multiple section units.

550.16 Grounding.

Grounding of both electrical and nonelectrical metal parts in a mobile home shall be through connection to a grounding bus in the mobile home distribution panelboard. The grounding bus shall be grounded through the green-colored insulated conductor in the supply cord or the feeder wiring to the service ground in the service-entrance equipment located adjacent to the mobile home location. Neither the frame of the mobile home nor the frame of any appliance shall be connected to the grounded circuit conductor (neutral) in the mobile home. Where service equipment is installed in or on a manufactured home as permitted in 550.32(B), the neutral conductors and the ground bus shall be permitted to be connected in the distribution panel.

(A) Grounded (Neutral) Conductor.

(1) Insulated. The grounded circuit conductor (neutral) shall be insulated from the grounding conductors and from equipment enclosures and other grounded parts. The grounded (neutral) circuit terminals in the distribution panelboard and in ranges, clothes dryers, counter-mounted cooking units, and wall-mounted ovens shall be insulated from the equipment enclosure. Bonding screws, straps, or buses in the distribution panelboard or in appliances shall be removed and discarded. Where service equipment is installed in or on a manufactured home as permitted in 550.32(B), the neutral conductors and the ground bus shall be permitted to be connected in the distribution panel.

(2) Connections of Ranges and Clothes Dryers. Connections of ranges and clothes dryers with 120/240-volt, 3-wire ratings shall be made with 4-conductor cord and 3-pole, 4-wire, grounding-type plugs or by Type AC cable, Type MC cable, or conductors enclosed in flexible metal conduit.

The provisions of 550.33(A) require that the feeder assembly for a mobile home consist of a listed cord or four color-coded insulated conductors, one of which is the grounded conductor (white) and one of which is used for grounding purposes (green). Thus, the grounded and grounding conductors are kept independent of each other and are connected only at the service equipment (at the point of connection of the grounding electrode conductor). Grounding of both electrical and nonelectrical metal parts, including the frame of the mobile home or the frame of any appliance, is accomplished by connection to the equipment grounding bus [never to the grounded conductor (neutral bus)]. The purpose of this requirement is to prevent incidental contact between the grounded conductor and non–current-carrying metal parts of electrical equipment. Without the separation of the grounded and grounding conductors, this contact could result in the metal structure or metal sheathing of the mobile home becoming a parallel path for neutral current.

Bonding screws, straps, or buses, which bond the grounded (neutral) circuit conductors to the non–current-carrying metal parts in the mobile home panelboard or to the metal frame of an appliance (ranges, clothes dryers), must not be installed or, in the case of ranges and clothes dryers, they must be removed.

(B) Equipment Grounding Means.

(1) Supply Cord or Permanent Feeder. The green-colored insulated grounding wire in the supply cord or permanent feeder wiring shall be connected to the grounding bus in the distribution panelboard or disconnecting means.

(2) Electrical System. In the electrical system, all exposed metal parts, enclosures, frames, lamp fixture canopies, and so forth shall be effectively bonded to the grounding terminal or enclosure of the distribution panelboard.

(3) Cord-Connected Appliances. Cord-connected appliances, such as washing machines, clothes dryers, and refrigerators, and the electrical system of gas ranges and so forth, shall be grounded by means of a cord with grounding conductor and grounding-type attachment plug.

(C) Bonding of Non–Current-Carrying Metal Parts.

(1) Exposed Non–Current-Carrying Metal Parts. All exposed non–current-carrying metal parts that may become energized shall be effectively bonded to the grounding terminal or enclosure of the distribution panelboard. A bonding conductor shall be connected between the distribution panelboard and accessible terminal on the chassis.

(2) Grounding Terminals. Grounding terminals shall be of the solderless type and listed as pressure-terminal connectors recognized for the wire size used. The bonding conductor shall be solid or stranded, insulated or bare, and shall be 8 AWG copper minimum, or equivalent. The bonding conductor shall be routed so as not to be exposed to physical damage.

(3) Metallic Piping and Ducts. Metallic gas, water, and waste pipes and metallic air-circulating ducts shall be considered bonded if they are connected to the terminal on the chassis [see 550.16(C)(1)] by clamps, solderless connectors, or by suitable grounding-type straps.

(4) Metallic Roof and Exterior Coverings. Any metallic roof and exterior covering shall be considered bonded if the following conditions are met:

(1) The metal panels overlap one another and are securely attached to the wood or metal frame parts by metallic fasteners.
(2) The lower panel of the metallic exterior covering is secured by metallic fasteners at a cross member of the chassis by two metal straps per mobile home unit or section at opposite ends.

The bonding strap material shall be a minimum of 100 mm (4 in.) in width of material equivalent to the skin or a material of equal or better electrical conductivity. The straps shall be fastened with paint-penetrating fittings such as screws and starwashers or equivalent.

550.17 Testing.

(A) Dielectric Strength Test. The wiring of each mobile home shall be subjected to a 1-minute, 900-volt, dielectric strength test (with all switches closed) between live parts (including neutral) and the mobile home ground. Alternatively, the test shall be permitted to be performed at 1080 volts for 1 second. This test shall be performed after branch circuits are complete and after luminaires (fixtures) or appliances are installed.

Exception: Listed luminaires (fixtures) or appliances shall not be required to withstand the dielectric strength test.

(B) Continuity and Operational Tests and Polarity Checks. Each mobile home shall be subjected to all of the following:

(1) An electrical continuity test to ensure that all exposed electrically conductive parts are properly bonded

(2) An electrical operational test to demonstrate that all equipment, except water heaters and electric furnaces, is connected and in working order

(3) Electrical polarity checks of permanently wired equipment and receptacle outlets to determine that connections have been properly made

550.18 Calculations.

The following method shall be employed in computing the supply-cord and distribution-panelboard load for each feeder assembly for each mobile home in lieu of the procedure shown in Article 220 and shall be based on a 3-wire, 120/240-volt supply with 120-volt loads balanced between the two legs of the 3-wire system.

(A) Lighting, Small Appliance, and Laundry Load.

(1) Lighting Volt-Amperes. Length times width of mobile home floor (outside dimensions) times 33 volt-amperes/m² (3 VA/ft²), for example, length × width × 3 = lighting volt-amperes.

(2) Small Appliance Volt-Amperes. Number of circuits times 1500 volt-amperes for each 20-ampere appliance receptacle circuit (see definition of Appliance, Portable with note in 550.2), for example, number of circuits × 1500 = small appliance volt-amperes.

(3) Laundry Area Circuit Volt-Amperes. 1500 volt-amperes.

(4) Total Volt-Amperes. Lighting volt-amperes plus small appliance volt-amperes plus laundry area volt-amperes equals total volt-amperes.

(5) Net Volt-Amperes. First 3000 total volt-amperes at 100 percent plus remainder at 35 percent equals volt-amperes to be divided by 240 volts to obtain current (amperes) per leg.

(B) Total Load for Determining Power Supply. Total load for determining power supply is the sum of the following:

(1) Lighting and small appliance load as calculated in 550.18(A)(5).

(2) Nameplate amperes for motors and heater loads (exhaust fans, air conditioners, electric, gas, or oil heating). Omit

smaller of the heating and cooling loads, except include blower motor if used as air-conditioner evaporator motor. Where an air conditioner is not installed and a 40-ampere power-supply cord is provided, allow 15 amperes per leg for air conditioning.

(3) Twenty-five percent of current of largest motor in (2).

(4) Total of nameplate amperes for waste disposer, dishwasher, water heater, clothes dryer, wall-mounted oven, cooking units. Where the number of these appliances exceeds three, use 75 percent of total.

(5) Derive amperes for freestanding range (as distinguished from separate ovens and cooking units) by dividing the following values by 240 volts:

Nameplate Rating (watts)	Use (volt-amperes)
0 – 10,000	80 percent of rating
Over 10,000 – 12,500	8,000
Over 12,500 – 13,500	8,400
Over 13,500 – 14,500	8,800
Over 14,500 – 15,500	9,200
Over 15,500 – 16,500	9,600
Over 16,500 – 17,500	10,000

(6) If outlets or circuits are provided for other than factory-installed appliances, include the anticipated load.

FPN: Refer to Annex D, Example D11, for an illustration of the application of this calculation.

(C) Optional Method of Calculation for Lighting and Appliance Load. The optional method for calculating lighting and appliance load shown in 220.30 shall be permitted.

550.19 Interconnection of Multiple-Section Mobile or Manufactured Home Units.

(A) Wiring Methods. Approved and listed fixed-type wiring methods shall be used to join portions of a circuit that must be electrically joined and are located in adjacent sections after the home is installed on its support foundation. The circuit's junction shall be accessible for disassembly when the home is prepared for relocation.

FPN: See 550.15(K) for component interconnections.

(B) Disconnecting Means. Expandable or multi-unit manufactured homes not having permanently installed feeders that are to be moved from one location to another shall be permitted to have disconnecting means with branch-circuit protective equipment in each unit when so located that after assembly or joining together of units, the requirements of 550.10 will be met.

550.20 Outdoor Outlets, Luminaires (Fixtures), Air-Cooling Equipment, and So Forth.

(A) Listed for Outdoor Use. Outdoor luminaires (fixtures) and equipment shall be listed for outdoor use. Outdoor receptacle or convenience outlets shall be of a gasketed-cover type for use in wet locations. Where located on the underside of the home or located under roof extensions or similarly protected locations, outdoor luminaires (fixtures) and equipment shall be listed for use in damp locations.

(B) Outside Heating Equipment, Air-Conditioning Equipment, or Both. A mobile home provided with a branch circuit designed to energize outside heating equipment, air-conditioning equipment, or both, located outside the mobile home, other than room air conditioners, shall have such branch-circuit conductors terminate in a listed outlet box, or disconnecting means, located on the outside of the mobile home. A label shall be permanently affixed adjacent to the outlet box and shall contain the following information:

THIS CONNECTION IS FOR HEATING AND/OR AIR-CONDITIONING EQUIPMENT. THE BRANCH CIRCUIT IS RATED AT NOT MORE THAN _____ AMPERES, AT _____ VOLTS, 60-HERTZ, _____ CONDUCTOR AMPACITY. A DISCONNECTING MEANS SHALL BE LOCATED WITHIN SIGHT OF THE EQUIPMENT.

The correct voltage and ampere rating shall be given. The tag shall be not less than 0.51 mm (0.020 in.) thick etched brass, stainless steel, anodized or alclad aluminum, or equivalent. The tag shall not be less than 75 mm by 45 mm (3 in. by 1¾ in.) minimum size.

550.25 Arc-Fault Circuit-Interrupter Protection.

(A) Definition. Arc-fault circuit interrupters are defined in 210.12(A).

(B) Bedrooms of Mobile Homes and Manufactured Homes. All branch circuits that supply 125-volt, single-phase, 15- and 20-ampere outlets installed in bedrooms of mobile homes and manufactured homes shall be protected by arc-fault circuit interrupter(s).

The requirement for arc-fault circuit-interrupter protection in mobile and manufactured homes is similar to that found in 210.12 for site-built dwelling units. Arc-fault circuit-interrupter protection must be provided for the branch circuits supplying 125-volt, 15- and 20-ampere outlets in bedrooms. The branch circuits covered by this requirement are all those that fall within the voltage and current ratings

specified and supply lighting outlets, receptacle outlets, and other power outlets to which devices or utilization equipment are connected.

III. Services and Feeders

550.30 Distribution System.

The mobile home park secondary electrical distribution system to mobile home lots shall be single-phase, 120/240 volts, nominal. For the purpose of Part III, where the park service exceeds 240 volts, nominal, transformers and secondary distribution panelboards shall be treated as services.

Mobile homes are intended for connection to a wiring system nominally rated 120/240 volts, 3-wire ac, with a grounded neutral; therefore, distribution systems at mobile home parks must supply 120/240 volts to the mobile home lot. Because appliances and other equipment are usually installed during the manufacturing process of mobile homes and are rated 120/240 volts, a 120/208-volt supply derived from a 4-wire, 120/208-volt wye system is unsuitable.

550.31 Allowable Demand Factors.

Park electrical wiring systems shall be calculated (at 120/240 volts) on the larger of the following:

(1) 16,000 volt-amperes for each mobile home lot
(2) The load calculated in accordance with 550.18 for the largest typical mobile home that each lot will accept

It shall be permissible to compute the feeder or service load in accordance with Table 550.31. No demand factor shall be allowed for any other load, except as provided in this *Code*.

Table 550.31 Demand Factors for Services and Feeders

Number of Mobile Homes	Demand Factor (percent)
1	100
2	55
3	44
4	39
5	33
6	29
7–9	28
10–12	27
13–15	26
16–21	25
22–40	24
41–60	23
61 and over	22

Service and feeder conductors to a mobile home in compliance with 310.15(B)(6) shall be permitted.

> In accordance with 550.31, park electrical wiring systems must be calculated on the basis of the larger of (1) not less than 16,000 volt-amperes (at 120/240 volts) for each mobile home lot, or (2) the calculated load of the largest typical mobile home the lot will accommodate. However, the ampacity of the feeder-circuit conductors to each mobile home lot cannot be less than 100 amperes (at 120/240 volts), per 550.33(B).

550.32 Service Equipment.

(A) Mobile Home Service Equipment. The mobile home service equipment shall be located adjacent to the mobile home and not mounted in or on the mobile home. The service equipment shall be located in sight from and not more than 9.0 m (30 ft) from the exterior wall of the mobile home it serves. The service equipment shall be permitted to be located elsewhere on the premises, provided that a disconnecting means suitable for service equipment is located in sight from and not more than 9.0 m (30 ft) from the exterior wall of the mobile home it serves. Grounding at the disconnecting means shall be in accordance with 250.32.

> Mobile home service equipment must be located in sight of the mobile home, but the equipment can be up to 30 ft from any point on the exterior wall of the mobile home. This requirement recognizes the use of feeder raceways that are external to the mobile home. Service equipment may be located more than 30 ft from the mobile home if an additional disconnecting means is located within 30 ft of the mobile home and grounding and bonding of this additional disconnecting means is performed in accordance with the provisions of 250.32. In a mobile home park, this arrangement facilitates locating service equipment at one or more centralized locations that are not within the required 30-ft proximity to the mobile home. Feeders are installed from this service equipment to the properly located mobile home site disconnecting means.

(B) Manufactured Home Service Equipment. The manufactured home service equipment shall be permitted to be installed in or on a manufactured home, provided that all of the following conditions are met:

(1) The manufacturer shall include in its written installation instructions information indicating that the home shall be secured in place by an anchoring system or installed on and secured to a permanent foundation.

(2) The installation of the service equipment shall comply with Article 230.

(3) Means shall be provided for the connection of a grounding electrode conductor to the service equipment and routing it outside the structure.

(4) Bonding and grounding of the service shall be in accordance with Article 250.

(5) The manufacturer shall include in its written installation instructions one method of grounding the service equipment at the installation site. The instructions shall clearly state that other methods of grounding are found in Article 250.

(6) The minimum size grounding electrode conductor shall be specified in the instructions.

(7) A red warning label shall be mounted on or adjacent to the service equipment. The label shall state the following:

WARNING
DO NOT PROVIDE ELECTRICAL POWER UNTIL THE GROUNDING ELECTRODE(S) IS INSTALLED AND CONNECTED (SEE INSTALLATION INSTRUCTIONS).

Where the service equipment is not installed in or on the unit, the installation shall comply with the other provisions of this section.

> The provisions of 550.32(B) specify the conditions required in order to install the service equipment in or on a manufactured home. The concern over the unit being moved off-site (intentionally or unintentionally) without the ability to disconnect the electrical supply is addressed in condition (1). A manufactured home with a service in or on the unit must be anchored in place or secured to a permanent foundation.
>
> Other conditions specified by 550.32(B) cover the need to provide proper grounding and bonding conductors, systems, and connections and to install the service equipment in accordance with the applicable requirements in Article 230. The provisions of 550.32(B) only apply to manufactured homes as defined in 550.2.

(C) Rating. Mobile home service equipment shall be rated at not less than 100 amperes at 120/240 volts, and provisions shall be made for connecting a mobile home feeder assembly by a permanent wiring method. Power outlets used as mobile home service equipment shall also be permitted to contain receptacles rated up to 50 amperes with appropriate overcurrent protection. Fifty-ampere receptacles shall conform to the configuration shown in Figure 550.10(C).

> FPN: Complete details of the 50-ampere plug and receptacle configuration can be found in ANSI/NEMA WD 6-1989, National Electrical Manufacturers Association *Standard for Wiring Devices—Dimensional Requirements*, Figure 14-50.

(D) Additional Outside Electrical Equipment. Means for connecting a mobile home accessory building or structure or additional electrical equipment located outside a mobile home by a fixed wiring method shall be provided in either the mobile home service equipment or the local external disconnecting means permitted in 550.32(A).

(E) Additional Receptacles. Additional receptacles shall be permitted for connection of electrical equipment located outside the mobile home, and all such 125-volt, single-phase, 15- and 20-ampere receptacles shall be protected by a listed ground-fault circuit interrupter.

(F) Mounting Height. Outdoor mobile home disconnecting means shall be installed so the bottom of the enclosure containing the disconnecting means is not less than 600 mm (2 ft) above finished grade or working platform. The disconnecting means shall be installed so that the center of the grip of the operating handle, when in the highest position, is not more than 2.0 m (6 ft 7 in.) above the finished grade or working platform.

(G) Marking. Where a 125/250-volt receptacle is used in mobile home service equipment, the service equipment shall be marked as follows:

TURN DISCONNECTING SWITCH OR
CIRCUIT BREAKER OFF BEFORE
INSERTING OR REMOVING PLUG.
PLUG MUST BE FULLY INSERTED
OR REMOVED.

The marking shall be located on the service equipment adjacent to the receptacle outlet.

550.33 Feeder.

(A) Feeder Conductors. Feeder conductors shall consist of either a listed cord, factory installed in accordance with 550.10(B), or a permanently installed feeder consisting of four, insulated, color-coded conductors that shall be identified by the factory or field marking of the conductors in compliance with 310.12. Equipment grounding conductors shall not be identified by stripping the insulation.

Exception: Where a feeder is installed between service equipment and a disconnecting means as covered in 550.32(A), it shall be permitted to omit the equipment grounding conductor where the grounded circuit conductor is grounded at the disconnecting means as required in 250.32(B).

(B) Adequate Feeder Capacity. Mobile home and manufactured home lot feeder circuit conductors shall have adequate capacity for the loads supplied and shall be rated at not less than 100 amperes at 120/240 volts.

ARTICLE 551
Recreational Vehicles and Recreational Vehicle Parks

Contents

I. General

551.1 Scope.

The provisions of this article cover the electrical conductors and equipment installed within or on recreational vehicles, the conductors that connect recreational vehicles to a supply of electricity, and the installation of equipment and devices related to electrical installations within a recreational vehicle park.

Laws in many states require a factory inspection by either a governmental or private inspection agency. The requirements of such laws follow closely the requirements of NFPA 1192, *Standard on Recreational Vehicles*. Section 1-5 in NFPA 1192 specifies that electrical installations in recreational vehicles must comply with Part A (Part I in the 2002 *NEC*) of Article 551 and other applicable sections of the *Code*. This reference to Part A of Article 551 is based on a previous edition of the *NEC*. Parts II through VI of Article 551 in this edition of the *NEC* were formerly located in Part A.

551.2 Definitions.

(See Article 100 for additional definitions.)

Air-Conditioning or Comfort-Cooling Equipment. All of that equipment intended or installed for the purpose of processing the treatment of air so as to control simultaneously its temperature, humidity, cleanliness, and distribution to meet the requirements of the conditioned space.

Appliance, Fixed. An appliance that is fastened or otherwise secured at a specific location.

Appliance, Portable. An appliance that is actually moved or can easily be moved from one place to another in normal use.

> FPN: For the purpose of this article, the following major appliances, other than built-in, are considered portable if cord connected: refrigerators, range equipment, clothes washers, dishwashers without booster heaters, or other similar appliances.

Appliance, Stationary. An appliance that is not easily moved from one place to another in normal use.

Camping Trailer. A vehicular portable unit mounted on wheels and constructed with collapsible partial side walls that fold for towing by another vehicle and unfold at the campsite to provide temporary living quarters for recreational, camping, or travel use. (*See* Recreational Vehicle.)

Converter. A device that changes electrical energy from one form to another, as from alternating current to direct current.

Dead Front (as applied to switches, circuit breakers, switchboards, and distribution panelboards). Designed, constructed, and installed so that no current-carrying parts are normally exposed on the front.

Disconnecting Means. The necessary equipment usually consisting of a circuit breaker or switch and fuses, and their accessories, located near the point of entrance of supply conductors in a recreational vehicle and intended to constitute the means of cutoff for the supply to that recreational vehicle.

Distribution Panelboard. A single panel or group of panel units designed for assembly in the form of a single panel, including buses, and with or without switches and/or automatic overcurrent-protective devices for the control of light, heat, or power circuits of small individual as well as aggregate capacity; designed to be placed in a cabinet or cutout box placed in or against a wall or partition and accessible only from the front.

Frame. Chassis rail and any welded addition thereto of metal thickness of 1.35 mm (0.053 in.) or greater.

Low Voltage. An electromotive force rated 24 volts, nominal, or less, supplied from a transformer, converter, or battery.

Motor Home. A vehicular unit designed to provide temporary living quarters for recreational, camping, or travel use built on or permanently attached to a self-propelled motor vehicle chassis or on a chassis cab or van that is an integral part of the completed vehicle. (*See* Recreational Vehicle.)

Power-Supply Assembly. The conductors, including ungrounded, grounded, and equipment grounding conductors, the connectors, attachment plug caps, and all other fittings, grommets, or devices installed for the purpose of delivering energy from the source of electrical supply to the distribution panel within the recreational vehicle.

Recreational Vehicle. A vehicular-type unit primarily designed as temporary living quarters for recreational, camping, or travel use, which either has its own motive power or is mounted on or drawn by another vehicle. The basic entities are travel trailer, camping trailer, truck camper, and motor home.

Recreational Vehicle Park. A plot of land upon which two or more recreational vehicle sites are located, established, or maintained for occupancy by recreational vehicles of the general public as temporary living quarters for recreation or vacation purposes.

Recreational Vehicle Site. A plot of ground within a recreational vehicle park set aside for the accommodation of a recreational vehicle on a temporary basis. It can be used as either a recreational vehicle site or as a camping unit site.

Recreational Vehicle Site Feeder Circuit Conductors. The conductors from the park service equipment to the recreational vehicle site supply equipment.

Recreational Vehicle Site Supply Equipment. The necessary equipment, usually a power outlet, consisting of a circuit breaker or switch and fuse and their accessories, located near the point of entrance of supply conductors to a recreational vehicle site and intended to constitute the disconnecting means for the supply to that site.

Recreational Vehicle Stand. That area of a recreational vehicle site intended for the placement of a recreational vehicle.

Transformer. A device that, when used, raises or lowers the voltage of alternating current of the original source.

Travel Trailer. A vehicular unit, mounted on wheels, designed to provide temporary living quarters for recreational, camping, or travel use, of such size or weight as not to require special highway movement permits when towed by a motorized vehicle, and of gross trailer area less than 30 m^2 (320 ft^2). (*See* Recreational Vehicle.)

Truck Camper. A portable unit constructed to provide temporary living quarters for recreational, travel, or camping use, consisting of a roof, floor, and sides, designed to be loaded onto and unloaded from the bed of a pick-up truck. (*See* Recreational Vehicle.)

551.3 Other Articles.

Wherever the requirements of other articles of this *Code* and Article 551 differ, the requirements of Article 551 shall apply.

551.4 General Requirements.

(A) Not Covered. A recreational vehicle not used for the purposes as defined in 551.2 shall not be required to meet the provisions of Part I pertaining to the number or capacity of circuits required. It shall, however, meet all other applicable requirements of this article if the recreational vehicle is provided with an electrical installation intended to be energized from a 120- or 120/240-volt, nominal, ac power-supply system.

(B) Systems. This article covers battery and other low-voltage power systems (24 volts or less), combination electrical systems, generator installations, and 120- or 120/240-volt, nominal, systems.

II. Low-Voltage Systems

551.10 Low-Voltage Systems.

(A) Low-Voltage Circuits. Low-voltage circuits furnished and installed by the recreational vehicle manufacturer, other than automotive vehicle circuits or extensions thereof, are subject to this *Code*. Circuits supplying lights subject to federal or state regulations shall comply with applicable government regulations and this *Code*.

The wiring referred to in 551.10(A) is the low-voltage wiring within the recreational vehicle that would be used in place of 120-volt ac supplies and does not refer to the wiring of the automotive vehicle circuits (e.g., circuits used for ignition, headlights, or other vehicle functions).

(B) Low-Voltage Wiring.

(1) Material. Copper conductors shall be used for low-voltage circuits.

Exception: Metal chassis or frame shall be permitted as the return path to the source of supply.

It is not the intent of 551.10(B)(1) to permit the sidewalls or the roof to serve as the ground return path. See the definition of *frame* in 551.2.

(2) Conductor Types. Conductors shall conform to the requirements for Type GXL, HDT, SGT, SGR, or Type SXL or shall have insulation in accordance with Table 310.13 or the equivalent. Conductor sizes 6 AWG through 18 AWG or SAE shall be listed. Single-wire, low-voltage conductors shall be of the stranded type.

> FPN: See SAE Standard J1128-1995 for Types GXL, HDT, and SXL, and SAE Standard J1127-1995 for Types SGT and SGR.

(3) Marking. All insulated low-voltage conductors shall be surface marked at intervals not greater than 1.2 m (4 ft) as follows:

(1) Listed conductors shall be marked as required by the listing agency.
(2) SAE conductors shall be marked with the name or logo of the manufacturer, specification designation, and wire gauge.
(3) Other conductors shall be marked with the name or logo of the manufacturer, temperature rating, wire gauge, conductor material, and insulation thickness.

(4) Insulation Rating. Conductors shall have a minimum insulation rating of 90°C (194°F) for interior installations and 125°C (257°F) for all engine compartment wiring or any under-chassis installations where conductors are located less than 450 mm (18 in.) from any component of an internal combustion engine exhaust system.

The provisions of 551.10(B)(4) address the location of electrical equipment that is adjacent to hot components of engines and exhaust systems. Its requirements correlate with ANSI/RVIA 12V, *Standard for Low Voltage Systems in Conversion Vehicles.*

(C) Low-Voltage Wiring Methods.

(1) Physical Protection. Conductors shall be protected against physical damage and shall be secured. Where insulated conductors are clamped to the structure, the conductor insulation shall be supplemented by an additional wrap or layer of equivalent material, except that jacketed cables shall not be required to be so protected. Wiring shall be routed away from sharp edges, moving parts, or heat sources.

(2) Splices. Conductors shall be spliced or joined with splicing devices that provide a secure connection or by brazing, welding, or soldering with a fusible metal or alloy. Soldered splices shall first be spliced or joined so as to be mechanically and electrically secure without solder and then soldered. All splices, joints, and free ends of conductors shall be covered with an insulation equivalent to that on the conductors.

(3) Separation. Battery and dc circuits shall be physically separated by at least a 13-mm (½-in.) gap or other approved means from circuits of a different power source. Acceptable methods shall be by clamping, routing, or equivalent means

that ensure permanent total separation. Where circuits of different power sources cross, the external jacket of the nonmetallic-sheathed cables shall be deemed adequate separation.

(4) Ground Connections. Ground connections to the chassis or frame shall be made in an accessible location and shall be mechanically secure. Ground connections shall be by means of copper conductors and copper or copper-alloy terminals of the solderless type identified for the size of wire used. The surface on which ground terminals make contact shall be cleaned and be free from oxide or paint or shall be electrically connected through the use of a cadmium, tin, or zinc-plated internal/external-toothed lockwasher or locking terminals. Ground terminal attaching screws, rivets or bolts, nuts, and lockwashers shall be cadmium, tin, or zinc-plated except rivets shall be permitted to be unanodized aluminum where attaching to aluminum structures.

The chassis-grounding terminal of the battery shall be bonded to the vehicle chassis with a minimum 8 AWG copper conductor. In the event the power lead from the battery exceeds 8 AWG, then the bonding conductor shall be of an equal size.

Section 551.10(C)(4) requires that the chassis-grounding terminal of the battery be bonded to the vehicle chassis in a mechanically secure manner and be placed in an accessible location using a minimum 8 AWG copper conductor. The ac equipment grounding conductor of the appliance may not have sufficient ampacity to safely conduct the dc fault current. This will necessitate installation of the battery bonding conductor. Some recreational vehicles already have one side of the battery circuit bonded to the frame by an 8 AWG or larger copper conductor.

(D) Battery Installations. Storage batteries subject to the provisions of this *Code* shall be securely attached to the vehicle and installed in an area vaportight to the interior and ventilated directly to the exterior of the vehicle. Where batteries are installed in a compartment, the compartment shall be ventilated with openings having a minimum area of 1100 mm² (1.7 in.²) at both the top and at the bottom. Where compartment doors are equipped for ventilation, the openings shall be within 50 mm (2 in.) of the top and bottom. Batteries shall not be installed in a compartment containing spark- or flame-producing equipment, except that they shall be permitted to be installed in the engine generator compartment if the only charging source is from the engine generator.

(E) Overcurrent Protection.

(1) Rating. Low-voltage circuit wiring shall be protected by overcurrent protective devices rated not in excess of the ampacity of copper conductors, in accordance with Table 551.10(E)(1).

Table 551.10(E)(1) Low-Voltage Overcurrent Protection

Wire Size (AWG)	Ampacity	Wire Type
18	6	Stranded only
16	8	Stranded only
14	15	Stranded or solid
12	20	Stranded or solid
10	30	Stranded or solid

(2) Type. Circuit breakers or fuses shall be of an approved type, including automotive types. Fuseholders shall be clearly marked with maximum fuse size, and both circuit breakers and fuses shall be protected against shorting and physical damage by a cover or equivalent means.

> FPN: For further information, see ANSI/SAE J554-1987, *Standard for Electric Fuses (Cartridge Type)*; SAE J1284-1988, *Standard for Blade Type Electric Fuses,*; and UL 275-1993, *Standard for Automotive Glass Tube Fuses.*

The requirement for protection of fuseholders by a cover or equivalent means is intended to reduce the possibility of the low-voltage system shorting to ground.

(3) Appliances. DC appliances, such as pumps, compressors, heater blowers, and similar motor-driven appliances, shall be installed in accordance with the manufacturer's instructions.

Motors that are controlled by automatic switching or by latching-type manual switches shall be protected in accordance with 430.32(B).

(4) Location. The overcurrent protective device shall be installed in an accessible location on the vehicle within 450 mm (18 in.) of the point where the power supply connects to the vehicle circuits. If located outside the recreational vehicle, the device shall be protected against weather and physical damage.

Exception: External low-voltage supply shall be permitted to have the overcurrent protective device within 450 mm (18 in.) after entering the vehicle or after leaving a metal raceway.

(F) Switches. Switches shall have a dc rating not less than the connected load.

(G) Luminaires (Lighting Fixtures). All low-voltage interior luminaires (lighting fixtures) rated more than 4 watts, employing lamps rated more than 1.2 watts, shall be listed.

Ceiling luminaires (fixtures) in camping trailers shall be automatically de-energized by an interlock when folding down the trailer, or it shall be physically impossible to fold down the trailer unless the ceiling luminaire(s) (fixtures) are disconnected.

(H) Cigarette Lighter Receptacles. Twelve-volt receptacles that will accept and energize cigarette lighters shall be installed in a noncombustible outlet box, or the assembly shall be identified by the manufacturer of the product as thermally protected.

Twelve-volt systems for running and signal lights, similar to those used in a conventional automobile, are covered in 551.10, 551.20, and 551.30. In many recreational vehicles, 12-volt systems are also used for interior lighting and other small loads. The 12-volt system is often supplied from an on-board battery or through a transfer switch from a 120/12-volt transformer in conjunction with a full-wave rectifier.

III. Combination Electrical Systems

551.20 Combination Electrical Systems.

(A) General. Vehicle wiring suitable for connection to a battery or dc supply source shall be permitted to be connected to a 120-volt source, provided the entire wiring system and equipment are rated and installed in full conformity with Parts I, III, IV, V, and VI requirements of this article covering 120-volt electrical systems. Circuits fed from ac transformers shall not supply dc appliances.

(B) Voltage Converters (120-Volt Alternating Current to Low-Voltage Direct Current). The 120-volt ac side of the voltage converter shall be wired in full conformity with Parts I, III, IV, V, and VI requirements of this article for 120-volt electrical systems.

Exception: Converters supplied as an integral part of a listed appliance shall not be subject to the above.

All converters and transformers shall be listed for use in recreation vehicles and designed or equipped to provide over-temperature protection. To determine the converter rating, the following formula shall be applied to the total connected load, including average battery charging rate, of all 12-volt equipment:

The first 20 amperes of load at 100 percent, plus

The second 20 amperes of load at 50 percent, plus

All load above 40 amperes at 25 percent

Exception: A low-voltage appliance that is controlled by a momentary switch (normally open) that has no means for holding in the closed position or refrigerators with a 120-

volt function shall not be considered as a connected load when determining the required converter rating. Momentarily energized appliances shall be limited to those used to prepare the vehicle for occupancy or travel.

(C) Bonding Voltage Converter Enclosures. The non–current-carrying metal enclosure of the voltage converter shall be bonded to the frame of the vehicle with a minimum 8 AWG copper conductor. The voltage converter shall be provided with a separate chassis bonding conductor that shall not be used as a current-carrying conductor.

The intent of 551.20(C) is to reduce the possibility of damage to the power supply cord by large dc fault currents that may find their way back to the vehicle frame or battery through the ac grounding conductor of the converter. Metal enclosures of listed converters are provided with an external pressure terminal connector for this purpose. See the commentary following 551.10(C)(4).

(D) Dual-Voltage Fixtures, Including Luminaires or Appliances. Fixtures, including luminaires, or appliances having both 120-volt and low-voltage connections shall be listed for dual voltage.

In the dual-voltage fixtures described in 551.20(D), barriers are used to separate the 120-volt and the 12-volt wiring connections.

(E) Autotransformers. Autotransformers shall not be used.

(F) Receptacles and Plug Caps. Where a recreational vehicle is equipped with a 120-volt or 120/240-volt ac system, a low-voltage system, or both, receptacles and plug caps of the low-voltage system shall differ in configuration from those of the 120- or 120/240-volt system. Where a vehicle equipped with a battery or other low-voltage system has an external connection for low-voltage power, the connector shall have a configuration that will not accept 120-volt power.

IV. Other Power Sources

551.30 Generator Installations.

(A) Mounting. Generators shall be mounted in such a manner as to be effectively bonded to the recreational vehicle chassis.

(B) Generator Protection. Equipment shall be installed to ensure that the current-carrying conductors from the engine generator and from an outside source are not connected to a vehicle circuit at the same time.

Receptacles used as disconnecting means shall be accessible (as applied to wiring methods) and capable of interrupting their rated current without hazard to the operator.

(C) Installation of Storage Batteries and Generators. Storage batteries and internal-combustion-driven generator units (subject to the provisions of this *Code*) shall be secured in place to avoid displacement from vibration and road shock.

(D) Ventilation of Generator Compartments. Compartments accommodating internal-combustion-driven generator units shall be provided with ventilation in accordance with instructions provided by the manufacturer of the generator unit.

FPN: For generator compartment construction requirements, see NFPA 1192-1999, *Standard on Recreational Vehicles*.

(E) Supply Conductors. The supply conductors from the engine generator to the first termination on the vehicle shall be of the stranded type and be installed in listed flexible conduit or listed liquidtight flexible conduit. The point of first termination shall be in one of the following:

(1) Panelboard
(2) Junction box with a blank cover
(3) Junction box with a receptacle
(4) Enclosed transfer switch
(5) Receptacle assembly listed in conjunction with the generator

The panelboard or junction box with a receptacle shall be installed within the vehicle's interior and within 450 mm (18 in.) of the compartment wall but not inside the compartment. If the generator is below the floor level and not in a compartment, the panelboard or junction box with receptacle shall be installed within the vehicle interior within 450 mm (18 in.) of the point of entry into the vehicle. A junction box with a blank cover shall be mounted on the compartment wall and shall be permitted inside or outside the compartment. A receptacle assembly listed in conjunction with the generator shall be mounted in accordance with its listing. If the generator is below floor level and not in a compartment, the junction box with blank cover shall be mounted either to any part of the generator supporting structure (but not to the generator) or to the vehicle floor within 450 mm (18 in.) of any point directly above the generator on either the inside or outside of the floor surface. Overcurrent protection in accordance with 240.4 shall be provided for supply conductors as an integral part of a listed generator or shall be located within 450 mm (18 in.) of their point of entry into the vehicle.

551.31 Multiple Supply Source.

(A) Multiple Supply Sources. Where a multiple supply system consisting of an alternate power source and a power-supply cord is installed, the feeder from the alternate power source shall be protected by an overcurrent-protective device. Installation shall be in accordance with 551.30(A) and (B) and 551.40.

(B) Calculation of Loads. Calculation of loads shall be in accordance with 551.42.

(C) Multiple Supply Sources Capacity. The multiple supply sources shall not be required to be of the same capacity.

(D) Alternate Power Sources Exceeding 30 Amperes. If an alternate power source exceeds 30 amperes, 120 volts, nominal, it shall be permissible to wire it as a 120-volt, nominal, system or a 120/240-volt, nominal, system, provided an overcurrent-protective device of the proper rating is installed in the feeder.

(E) Power-Supply Assembly Not Less Than 30 Amperes. The external power-supply assembly shall be permitted to be less than the calculated load but not less than 30 amperes and shall have overcurrent protection not greater than the capacity of the external power-supply assembly.

551.32 Other Sources.

Other sources of ac power, such as inverters or motor generators, shall be listed for use in recreational vehicles and shall be installed in accordance with the terms of the listing. Other sources of ac power shall be wired in full conformity with the requirements in Parts I, III, IV, V, and VI of this article covering 120-volt electrical systems.

551.33 Alternate Source Restriction.

Transfer equipment, if not integral with the listed power source, shall be installed to ensure that the current-carrying conductors from other sources of ac power and from an outside source are not connected to the vehicle circuit at the same time.

V. Nominal 120-Volt or 120/240-Volt Systems

551.40 120-Volt or 120/240-Volt, Nominal, Systems.

(A) General Requirements. The electrical equipment and material of recreational vehicles indicated for connection to a wiring system rated 120 volts, nominal, 2-wire with ground, or a wiring system rated 120/240 volts, nominal, 3-wire with ground, shall be listed and installed in accordance with the requirements of Parts I, III, IV, V, and VI of this article.

(B) Materials and Equipment. Electrical materials, devices, appliances, fittings, and other equipment installed in, intended for use in, or attached to the recreational vehicle shall be listed. All products shall be used only in the manner in which they have been tested and found suitable for the intended use.

(C) Ground-Fault Circuit-Interrupter Protection. The internal wiring of a recreational vehicle having only one 15- or 20-ampere branch circuit as permitted in 551.42(A) and (B) shall have ground-fault circuit-interrupter protection for personnel. The ground-fault circuit interrupter shall be installed at the point where the power supply assembly terminates within the recreational vehicle. Where a separable cord set is not employed, the ground-fault circuit interrupter shall be permitted to be an integral part of the attachment plug of the power supply assembly. The ground-fault circuit interrupter shall provide protection also under the conditions of an open grounded circuit conductor, interchanged circuit conductors, or both.

551.41 Receptacle Outlets Required.

(A) Spacing. Receptacle outlets shall be installed at wall spaces 600 mm (2 ft) wide or more so that no point along the floor line is more than 1.8 m (6 ft), measured horizontally, from an outlet in that space.

Exception No. 1: Bath and hall areas.

Exception No. 2: Wall spaces occupied by kitchen cabinets, wardrobe cabinets, built-in furniture, behind doors that may open fully against a wall surface, or similar facilities.

(B) Location. Receptacle outlets shall be installed as follows:

(1) Adjacent to countertops in the kitchen [at least one on each side of the sink if countertops are on each side and are 300 mm (12 in.) or over in width]
(2) Adjacent to the refrigerator and gas range space, except where a gas-fired refrigerator or cooking appliance, requiring no external electrical connection, is factory installed
(3) Adjacent to countertop spaces of 300 mm (12 in.) or more in width that cannot be reached from a receptacle required in 551.41(B)(1) by a cord of 1.8 m (6 ft) without crossing a traffic area, cooking appliance, or sink

(C) Ground-Fault Circuit-Interrupter Protection. Where provided, each 125-volt, single-phase, 15- or 20-ampere receptacle outlet shall have ground-fault circuit-interrupter protection for personnel in the following locations:

(1) Adjacent to a bathroom lavatory

(2) Where the receptacles are installed to serve the countertop surfaces and are within 1.8 m (6 ft) of any lavatory or sink

Exception No. 1: Receptacles installed for appliances in dedicated spaces, such as for dishwashers, disposals, refrigerators, freezers, and laundry equipment.

Exception No. 2: Single receptacles for interior connections of expandable room sections.

Exception No. 3: De-energized receptacles that are within 1.8 m (6 ft) of any sink or lavatory due to the retraction of the expandable room section.

(3) In the area occupied by a toilet, shower, tub, or any combination thereof

(4) On the exterior of the vehicle

Exception: Receptacles that are located inside of an access panel that is installed on the exterior of the vehicle to supply power for an installed appliance shall not be required to have ground-fault circuit-interrupter protection.

The receptacle outlet shall be permitted in a listed luminaire (lighting fixture). A receptacle outlet shall not be installed in a tub or combination tub–shower compartment.

In accordance with 551.41(C)(4), a bathroom receptacle is permitted to be mounted in the side of a lavatory cabinet where installation of a receptacle outlet is not possible in a wall that does not provide the necessary depth. The receptacle must be GFCI protected in accordance with 551.41(C)(1). Receptacles of any type are not permitted in a tub or tub–shower compartment.

(D) Face-Up Position. A receptacle shall not be installed in a face-up position in any countertop or similar horizontal surfaces within the living area.

551.42 Branch Circuits Required.

Each recreational vehicle containing a 120-volt electrical system shall contain one of the following.

(A) One 15-Ampere Circuit. One 15-ampere circuit to supply lights, receptacle outlets, and fixed appliances. Such recreational vehicles shall be equipped with one 15-ampere switch and fuse or one 15-ampere circuit breaker.

(B) One 20-Ampere Circuit. One 20-ampere circuit to supply lights, receptacle outlets, and fixed appliances. Such recreational vehicles shall be equipped with one 20-ampere switch and fuse or one 20-ampere circuit breaker.

(C) Two to Five 15- or 20-Ampere Circuits. A maximum of five 15- or 20-ampere circuits to supply lights, receptacle outlets, and fixed appliances shall be permitted. Such recreational vehicles shall be equipped with a distribution panelboard rated at 120 volts maximum with a 30-ampere rated main power supply assembly. Not more than two 120-volt thermostatically controlled appliances (i.e., air conditioner and water heater) shall be installed in such systems unless appliance isolation switching, energy management systems, or similar methods are used.

Exception: Additional 15- or 20-ampere circuits shall be permitted where a listed energy management system rated at 30-ampere maximum is employed within the system.

> FPN: See 210.23(A) for permissible loads. See 551.45(C) for main disconnect and overcurrent protection requirements.

(D) More Than Five Circuits Without a Listed Energy Management System. A 50-ampere, 120/240-volt power-supply assembly shall be used where six or more circuits are employed. The load distribution shall ensure a reasonable current balance between phases.

Experience and field data indicates that supply calculations are not necessary and, in some cases, such calculations have resulted in oversized power supplies. In recreational vehicles with six or more circuits, the power supply assembly shall have a minimum rating of 50 amperes. Reasonable balancing of the electrical load between phases is required.

551.43 Branch-Circuit Protection.

(A) Rating. The branch-circuit overcurrent devices shall be rated as follows:

(1) Not more than the circuit conductors, and

(2) Not more than 150 percent of the rating of a single appliance rated 13.3 amperes or more and supplied by an individual branch circuit, but

(3) Not more than the overcurrent protection size marked on an air conditioner or other motor-operated appliances

(B) Protection for Smaller Conductors. A 20-ampere fuse or circuit breaker shall be permitted for protection for fixtures, including a luminaire leads, cords, or small appliances, and 14 AWG tap conductors, not over 1.8 m (6 ft) long for recessed luminaires (lighting fixtures).

(C) Fifteen-Ampere Receptacle Considered Protected by 20 Amperes. If more than one receptacle or load is on a branch circuit, a 15-ampere receptacle shall be permitted to be protected by a 20-ampere fuse or circuit breaker.

551.44 Power-Supply Assembly.

Each recreational vehicle shall have only one of the following main power-supply assemblies.

(A) Fifteen-Ampere Main Power-Supply Assembly. Recreational vehicles wired in accordance with 551.42(A) shall use a listed 15-ampere or larger main power-supply assembly.

(B) Twenty-Ampere Main Power-Supply Assembly. Recreational vehicles wired in accordance with 551.42(B) shall use a listed 20-ampere or larger main power-supply assembly.

(C) Thirty-Ampere Main Power-Supply Assembly. Recreational vehicles wired in accordance with 551.42(C) shall use a listed 30-ampere or larger main power-supply assembly.

(D) Fifty-Ampere Power-Supply Assembly. Recreational vehicles wired in accordance with 551.42(D) shall use a listed 50-ampere, 120/240-volt main power-supply assembly.

551.45 Distribution Panelboard.

(A) Listed and Appropriately Rated. A listed and appropriately rated distribution panelboard or other equipment specifically listed for the purpose shall be used. The grounded conductor termination bar shall be insulated from the enclosure as provided in 551.54(C). An equipment grounding terminal bar shall be attached inside the metal enclosure of the panelboard.

(B) Location. The distribution panelboard shall be installed in a readily accessible location. Working clearance for the panelboard shall be not less than 600 mm (24 in.) wide and 750 mm (30 in.) deep.

Exception No. 1: Where the panelboard cover is exposed to the inside aisle space, one of the working clearance dimensions shall be permitted to be reduced to a minimum of 550 mm (22 in.). A panelboard is considered exposed where the panelboard cover is within 50 mm (2 in.) of the aisle's finished surface.

Exception No. 2: Compartment doors used for access to a generator shall be permitted to be equipped with a locking system.

(C) Dead-Front Type. The distribution panelboard shall be of the dead-front type and shall consist of one or more circuit breakers or Type S fuseholders. A main disconnecting means shall be provided where fuses are used or where more than two circuit breakers are employed. A main overcurrent protective device not exceeding the power-supply assembly

rating shall be provided where more than two branch circuits are employed.

551.46 Means for Connecting to Power Supply.

(A) Assembly. The power-supply assembly or assemblies shall be factory supplied or factory installed and be of one of the types specified herein.

(1) Separable. Where a separable power-supply assembly consisting of a cord with a female connector and molded attachment plug cap is provided, the vehicle shall be equipped with a permanently mounted, flanged surface inlet (male, recessed-type motor-base attachment plug) wired directly to the distribution panelboard by an approved wiring method. The attachment plug cap shall be of a listed type.

(2) Permanently Connected. Each power-supply assembly shall be connected directly to the terminals of the distribution panelboard or conductors within a junction box and provided with means to prevent strain from being transmitted to the terminals. The ampacity of the conductors between each junction box and the terminals of each distribution panelboard shall be at least equal to the ampacity of the power-supply cord. The supply end of the assembly shall be equipped with an attachment plug of the type described in 551.46(C). Where the cord passes through the walls or floors, it shall be protected by means of conduit and bushings or equivalent. The cord assembly shall have permanent provisions for protection against corrosion and mechanical damage while the vehicle is in transit.

(B) Cord. The cord exposed usable length shall be measured from the point of entrance to the recreational vehicle or the face of the flanged surface inlet (motor-base attachment plug) to the face of the attachment plug at the supply end.

The cord exposed usable length, measured to the point of entry on the vehicle exterior, shall be a minimum of 7.5 m (25 ft) where the point of entrance is at the side of the vehicle or shall be a minimum 9.0 m (30 ft) where the point of entrance is at the rear of the vehicle.

Where the cord entrance into the vehicle is more than 900 mm (3 ft) above the ground, the minimum cord lengths above shall be increased by the vertical distance of the cord entrance heights above 900 mm (3 ft).

FPN: See 551.46(E).

(C) Attachment Plugs.

(1) Units with One 15-Ampere Branch Circuit. Recreational vehicles having only one 15-ampere branch circuit as permitted by 551.42(A) shall have an attachment plug that shall be 2-pole, 3-wire grounding type, rated 15 amperes, 125 volts, conforming to the configuration shown in Figure 551.46(C).

Receptacles   **Caps**

20-A, 125-V,
2-pole, 3-wire,
grounding type

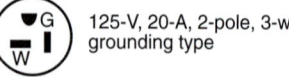

125-V, 20-A, 2-pole, 3-wire,
grounding type

125-V, 15-A, 2-pole, 3-wire,
grounding type

30-A, 125-V, 2-pole, 3-wire, grounding type

50-A, 125/250-V, 3-pole, 4-wire, grounding type

Figure 551.46(C) Configurations for grounding-type receptacles and attachment plug caps used for recreational vehicle supply cords and recreational vehicle lots.

FPN: Complete details of this configuration can be found in National Electrical Manufacturers Association's ANSI/NEMA WD 6-1989, *Standard for Dimensions of Attachment Plugs and Receptacle*, Figure 5.15.

(2) Units with One 20-Ampere Branch Circuit. Recreational vehicles having only one 20-ampere branch circuit as permitted in 551.42(B) shall have an attachment plug that shall be 2-pole, 3-wire grounding type, rated 20 amperes, 125 volts, conforming to the configuration shown in Figure 551.46(C).

FPN: Complete details of this configuration can be found in ANSI/NEMA WD 6-1989, National Electrical Manufacturers Association's *Standard for Dimensions of Attachment Plugs and Receptacles*, Figure 5.20.

(3) Units with Two to Five 15- or 20-Ampere Branch Circuits. Recreational vehicles wired in accordance with 551.42(C) shall have an attachment plug that shall be 2-pole, 3-wire grounding type, rated 30 amperes, 125 volts, conforming to the configuration shown in Figure 551.46(C), intended for use with units rated at 30 amperes, 125 volts.

FPN: Complete details of this configuration can be found in ANSI/NEMA WD 6-1989, National Electrical Manufacturers Association's *Standard for Dimensions of Attachment Plugs and Receptacles*, Figure TT.

(4) Units with 50-Ampere Power Supply Assembly. Recreational vehicles having a power-supply assembly rated 50 amperes as permitted by 551.42(D) shall have a 3-pole, 4-wire grounding-type attachment plug rated 50 amperes, 125/250 volts, conforming to the configuration shown in Figure 551.46(C).

FPN: Complete details of this configuration can be found in ANSI/NEMA WD 6-1989, National Electrical Manufacturers Association's *Standard for Dimensions of Attachment Plugs and Receptacles*, Figure 14.50.

The 20- and 50-ampere receptacle and plug configurations used for recreational vehicles appear in Exhibit 406.2 (NEMA configuration chart); however, the 30-ampere plug and receptacle configurations are unique to recreational vehicles and are described in UL 498, *Standard for Safety, Attachment Plugs and Receptacles.*

(D) Labeling at Electrical Entrance. Each recreational vehicle shall have permanently affixed to the exterior skin, at or near the point of entrance of the power-supply cord(s), a label 75 mm × 45 mm (3 in. × 1¾ in.) minimum size, made of etched, metal-stamped, or embossed brass, stainless steel, or anodized or alclad aluminum not less than 0.51 mm (0.020 in.) thick, or other suitable material [e.g., 0.13 mm (0.005 in.) thick plastic laminate] that reads, as appropriate, either

THIS CONNECTION IS FOR 110–125-VOLT AC, 60 HZ, _____ AMPERE SUPPLY.

or

THIS CONNECTION IS FOR 120/240-VOLT AC, 3-POLE, 4-WIRE, 60 HZ, _____ AMPERE SUPPLY.

The correct ampere rating shall be marked in the blank space.

(E) Location. The point of entrance of a power-supply assembly shall be located within 4.5 m (15 ft) of the rear, on the left (road) side or at the rear, left of the longitudinal center of the vehicle, within 450 mm (18 in.) of the outside wall.

Exception No. 1: A recreational vehicle equipped with only a listed flexible drain system or a side-vent drain system shall be permitted to have the electrical point of entrance located on either side, provided the drain(s) for the plumbing system is (are) located on the same side.

Exception No. 2: A recreational vehicle shall be permitted to have the electrical point of entrance located more than 4.5 m (15 ft) from the rear. Where this occurs, the distance beyond the 4.5-m (15-ft) dimension shall be added to the cord's minimum length as specified in 551.46(B).

551.47 Wiring Methods.

(A) Wiring Systems. Cables and raceways installed in accordance with Articles 320, 322, 330 through 340, 342 through 362, 386, and 388 shall be permitted in accordance with their applicable article, except as otherwise specified in this article. An equipment grounding means shall be provided in accordance with 250.118.

See the commentary following 348.60 for information on the use of flexible metal conduit for equipment grounding.

(B) Conduit and Tubing. Where rigid metal conduit or intermediate metal conduit is terminated at an enclosure with a locknut and bushing connection, two locknuts shall be provided, one inside and one outside of the enclosure. All cut ends of conduit and tubing shall be reamed or otherwise finished to remove rough edges.

See the commentary following 358.28(A) and 300.4(F) for information on protecting conductor insulation against abrasion at conduit and tubing terminations.

(C) Nonmetallic Boxes. Nonmetallic boxes shall be acceptable only with nonmetallic-sheathed cable or nonmetallic raceways.

(D) Boxes. In walls and ceilings constructed of wood or other combustible material, boxes and fittings shall be flush with the finished surface or project therefrom.

(E) Mounting. Wall and ceiling boxes shall be mounted in accordance with Article 314.

Exception No. 1: Snap-in-type boxes or boxes provided with special wall or ceiling brackets that securely fasten boxes in walls or ceilings shall be permitted.

Exception No. 2: A wooden plate providing a 38-mm (1½-in.) minimum width backing around the box and of a thickness of 13 mm (½ in.) or greater (actual) attached directly to the wall panel shall be considered as approved means for mounting outlet boxes.

Exception No. 2 to 551.47(E) permits mounting outlet boxes by screws to a wooden plate that is secured directly to the back of the wall panel. This wooden plate must be not less than ½ in. thick and must extend at least 1½ in. around the box. This requirement recognizes the special construction of recreational vehicle walls, which often makes it quite difficult or impossible to attach an outlet box to a structural member, as required by 314.23(B).

(F) Raceway and Cable Continuity. Raceways and cable sheaths shall be continuous between boxes and other enclosures.

(G) Protected. Metal-clad, Type AC, or nonmetallic-sheathed cables and electrical nonmetallic tubing shall be permitted to pass through the centers of the wide side of 2 by 4 wood studs. However, they shall be protected where they pass through 2 by 2 wood studs or at other wood studs or frames where the cable or tubing would be less than 32 mm (1¼ in.) from the inside or outside surface. Steel plates on each side of the cable or tubing or a steel tube, with not less than 1.35 mm (0.053 in.) wall thickness, shall be installed to protect the cable or tubing. These plates or tubes shall be securely held in place. Where nonmetallic-sheathed cables pass through punched, cut, or drilled slots or holes in metal members, the cable shall be protected by bushings or grommets securely fastened in the opening prior to installation of the cable.

(H) Bends. No bend shall have a radius of less than five times the cable diameter.

(I) Cable Supports. Where connected with cable connectors or clamps, cables shall be supported within 300 mm (12 in.) of outlet boxes, distribution panelboards, and splice boxes on appliances. Supports shall be provided every 1.4 m (4½ ft) at other places.

(J) Nonmetallic Box Without Cable Clamps. Nonmetallic-sheathed cables shall be supported within 200 mm (8 in.) of a nonmetallic outlet box without cable clamps. Where wiring devices with integral enclosures are employed with a loop of extra cable to permit future replacement of the device, the cable loop shall be considered as an integral portion of the device.

(K) Physical Damage. Where subject to physical damage, exposed nonmetallic cable shall be protected by covering boards, guard strips, raceways, or other means.

(L) Metal Faceplates. Metal faceplates shall be of ferrous metal not less than 0.76 mm (0.030 in.) in thickness or of nonferrous metal not less than 1.0 mm (0.040 in.) in thickness. Nonmetallic faceplates shall be listed.

(M) Metal Faceplates Effectively Grounded. Where metal faceplates are used, they shall be effectively grounded.

(N) Moisture or Physical Damage. Where outdoor or under-chassis wiring is 120 volts, nominal, or over and is exposed to moisture or physical damage, the wiring shall be protected by rigid metal conduit, by intermediate metal conduit, or by electrical metallic tubing or rigid nonmetallic conduit that is closely routed against frames and equipment enclosures or other raceway or cable identified for the application.

(O) Component Interconnections. Fittings and connectors that are intended to be concealed at the time of assembly shall be listed and identified for the interconnection of building components. Such fittings and connectors shall be equal to the wiring method employed in insulation, temperature rise, and fault-current withstanding and shall be capable of enduring the vibration and shock occurring in recreational vehicles.

(P) Method of Connecting Expandable Units. The method of connecting expandable units to the main body of the vehicle shall comply with the following as applicable:

(1) That portion of a branch circuit that is installed in an expandable unit shall be permitted to be connected to the portion of the branch circuit in the main body of the vehicle by means of a flexible cord or attachment plug and cord listed for hard usage. The cord and its connections shall conform to all provisions of Article 400 and shall be considered as a permitted use under 400.7. Where the attachment plug and cord are located within the vehicle's interior, use of plastic thermoset or elastomer parallel cord Type SPT-3, SP-3, or SPE shall be permitted.

(2) If the receptacle provided for connection of the cord to the main circuit is located on the outside of the vehicle, it shall be protected with a ground-fault circuit interrupter for personnel and be listed for wet locations. A cord located on the outside of a vehicle shall be identified for outdoor use.

(3) Unless removable or stored within the vehicle interior, the cord assembly shall have permanent provisions for protection against corrosion and mechanical damage while the vehicle is in transit.

(4) If an attachment plug and cord is used, it shall be installed so as not to permit exposed live attachment plug pins.

(Q) Prewiring for Air-Conditioning Installation. Prewiring installed for the purpose of facilitating future air-conditioning installation shall conform to the applicable portions of this article and the following:

(1) An overcurrent protective device with a rating compatible with the circuit conductors shall be installed in the distribution panelboard and wiring connections completed.

(2) The load end of the circuit shall terminate in a junction box with a blank cover or a device listed for the purpose. Where a junction box with a blank cover is used, the free ends of the conductors shall be adequately capped or taped.

(3) A label conforming to 551.46(D) shall be placed on or adjacent to the junction box and shall read

AIR-CONDITIONING CIRCUIT. THIS CONNECTION IS FOR AIR CONDITIONERS RATED 110–125-VOLT AC, 60 HZ, _____ AMPERES MAXIMUM. DO NOT EXCEED CIRCUIT RATING.

An ampere rating, not to exceed 80 percent of the circuit rating, shall be legibly marked in the blank space.

(4) The circuit shall serve no other purpose.

(R) Prewiring for Generator Installation. Prewiring installed for the purpose of facilitating future generator installation shall conform to the other applicable portions of this article and the following:

(1) Circuit conductors shall be appropriately sized in relation to the anticipated load and shall be protected by an overcurrent device in accordance with their ampacities.

(2) Where junction boxes are utilized at either of the circuit originating or terminus points, free ends of the conductors shall be adequately capped or taped.

(3) Where devices such as receptacle outlet, transfer switch, and so forth, are installed, the installation shall be complete, including circuit conductor connections. All devices shall be listed and appropriately rated.

(4) A label conforming to 551.46(D) shall be placed on the cover of each junction box containing incomplete circuitry and shall read, as appropriate, either

GENERATOR CIRCUIT. THIS CONNECTION IS FOR GENERATORS RATED 110–125-VOLT AC, 60 HZ, _____ AMPERES MAXIMUM.

or

GENERATOR CIRCUIT. THIS CONNECTION IS FOR GENERATORS RATED 120/240-VOLT AC, 60 HZ, _____ AMPERES MAXIMUM.

The correct ampere rating shall be legibly marked in the blank space.

551.48 Conductors and Boxes.

The maximum number of conductors permitted in boxes shall be in accordance with 314.16.

551.49 Grounded Conductors.

The identification of grounded conductors shall be in accordance with 200.6.

551.50 Connection of Terminals and Splices.

Conductor splices and connections at terminals shall be in accordance with 110.14.

551.51 Switches.

Switches shall be rated in accordance with 551.51(A) and (B).

(A) Lighting Circuits. For lighting circuits, switches shall be rated not less than 10 amperes, 120–125 volts and in no case less than the connected load.

(B) Motors or Other Loads. For motors or other loads, switches shall have ampere or horsepower ratings, or both, adequate for loads controlled. (An ac general-use snap switch shall be permitted to control a motor 2 hp or less with full-load current not over 80 percent of the switch ampere rating.)

551.52 Receptacles.

All receptacle outlets shall be of the grounding type and installed in accordance with 406.3 and 210.21.

551.53 Luminaires (Lighting Fixtures).

(A) General. Any combustible wall or ceiling finish exposed between the edge of a luminaire (fixture) canopy, or pan and the outlet box, shall be covered with noncombustible material or a material identified for the purpose.

(B) Shower Luminaires (Fixtures). If a luminaire (lighting fixture) is provided over a bathtub or in a shower stall, it shall be of the enclosed and gasketed type and listed for the type of installation, and it shall be ground-fault circuit-interrupter protected.

Many shower luminaires have a metal base and, due to low ceilings in recreational vehicles, may be easily reached by most persons under a shower or standing in a bathtub. In accordance with 551.53(B), the installation of luminaires that are listed for wet locations and have ground-fault circuit-interrupter protection is permitted above a tub or shower enclosure.

The switch for shower luminaires (lighting fixtures) and exhaust fans, located over a tub or in a shower stall, shall be located outside the tub or shower space.

(C) Outdoor Outlets, Luminaires (Fixtures), Air-Cooling Equipment, and So On. Outdoor luminaires (fixtures) and other equipment shall be listed for outdoor use.

551.54 Grounding.

(See also 551.56 on bonding of non–current-carrying metal parts.)

(A) Power-Supply Grounding. The grounding conductor in the supply cord or feeder shall be connected to the grounding bus or other approved grounding means in the distribution panelboard.

(B) Distribution Panelboard. The distribution panelboard shall have a grounding bus with sufficient terminals for all grounding conductors or other approved grounding means.

(C) Insulated Neutral. The grounded circuit conductor (neutral) shall be insulated from the equipment grounding conductors and from equipment enclosures and other grounded parts. The grounded (neutral) circuit terminals in the distribution panelboard and in ranges, clothes dryers, counter-mounted cooking units, and wall-mounted ovens shall be insulated from the equipment enclosure. Bonding screws, straps, or buses in the distribution panelboard or in appliances shall be removed and discarded. Connection of electric ranges and electric clothes dryers utilizing a grounded (neutral) conductor, if cord-connected, shall be made with 4-conductor cord and 3-pole, 4-wire grounding-type plug caps and receptacles.

551.55 Interior Equipment Grounding.

(A) Exposed Metal Parts. In the electrical system, all exposed metal parts, enclosures, frames, luminaire (lighting fixture) canopies, and so forth, shall be effectively bonded to the grounding terminals or enclosure of the distribution panelboard.

(B) Equipment Grounding and Bonding Conductors. Bare wires, insulated wire with an outer finish that is green or green with one or more yellow stripes shall be used for equipment grounding or bonding conductors only.

(C) Grounding of Electrical Equipment. Where grounding of electrical equipment is specified, it shall be permitted as follows:

(1) Connection of metal raceway (conduit or electrical metallic tubing), the sheath of Type MC and Type MI cable where the sheath is identified for grounding, or the armor of Type AC cable to metal enclosures.
(2) A connection between the one or more equipment grounding conductors and a metal box by means of a grounding screw, which shall be used for no other purpose, or a listed grounding device.
(3) The equipment grounding conductor in nonmetallic-sheathed cable shall be permitted to be secured under a screw threaded into the luminaire (fixture) canopy other than a mounting screw or cover screw, or attached to a listed grounding means (plate) in a nonmetallic outlet box for luminaire (fixture) mounting. [Grounding means shall also be permitted for luminaire (fixture) attachment screws.]

(D) Grounding Connection in Nonmetallic Box. A connection between the one or more grounding conductors brought into a nonmetallic outlet box shall be so arranged that a connection can be made to any fitting or device in that box that requires grounding.

(E) Grounding Continuity. Where more than one equipment grounding or bonding conductor of a branch circuit enters a box, all such conductors shall be in good electrical contact with each other, and the arrangement shall be such that the disconnection or removal of a receptacle, luminaire (fixture), or other device fed from the box will not interfere with or interrupt the grounding continuity.

(F) Cord-Connected Appliances. Cord-connected appliances, such as washing machines, clothes dryers, refrigerators, and the electrical system of gas ranges, and so forth, shall be grounded by means of an approved cord with equipment grounding conductor and grounding-type attachment plug.

551.56 Bonding of Non–Current-Carrying Metal Parts.

(A) Required Bonding. All exposed non–current-carrying metal parts that may become energized shall be effectively bonded to the grounding terminal or enclosure of the distribution panelboard.

(B) Bonding Chassis. A bonding conductor shall be connected between any distribution panelboard and an accessible terminal on the chassis. Aluminum or copper-clad aluminum conductors shall not be used for bonding if such conductors or their terminals are exposed to corrosive elements.

Exception: Any recreational vehicle that employs a unitized metal chassis-frame construction to which the distribution panelboard is securely fastened with a bolt(s) and nut(s) or by welding or riveting shall be considered to be bonded.

(C) Bonding Conductor Requirements. Grounding terminals shall be of the solderless type and listed as pressure terminal connectors recognized for the wire size used. The bonding conductor shall be solid or stranded, insulated or bare, and shall be 8 AWG copper minimum, or equal.

(D) Metallic Roof and Exterior Bonding. The metal roof and exterior covering shall be considered bonded where

(1) The metal panels overlap one another and are securely attached to the wood or metal frame parts by metal fasteners, and
(2) The lower panel of the metal exterior covering is secured by metal fasteners at each cross member of the chassis, or the lower panel is bonded to the chassis by a metal strap.

(E) Gas, Water, and Waste Pipe Bonding. The gas, water, and waste pipes shall be considered grounded if they are bonded to the chassis.

(F) Furnace and Metal Air Duct Bonding. Furnace and metal circulating air ducts shall be bonded.

551.57 Appliance Accessibility and Fastening.

Every appliance shall be accessible for inspection, service, repair, and replacement without removal of permanent construction. Means shall be provided to securely fasten appliances in place when the recreational vehicle is in transit.

VI. Factory Tests

551.60 Factory Tests (Electrical).

Each recreational vehicle shall be subjected to the following tests.

(A) Circuits of 120 Volts or 120/240 Volts. Each recreational vehicle designed with a 120-volt or a 120/240-volt electrical system shall withstand the applied potential without electrical breakdown of a 1-minute, 900-volt dielectric strength test, or a 1-second, 1080-volt dielectric strength test, with all switches closed, between ungrounded and grounded conductors and the recreational vehicle ground. During the test, all switches and other controls shall be in the "on" position. Fixtures, including luminaires and permanently installed appliances shall not be required to withstand this test. The test shall be performed after branch circuits are complete prior to energizing the system and again after all outer coverings and cabinetry have been secured.

Each recreational vehicle shall be subjected to all of the following:

(1) A continuity test to ensure that all metal parts are properly bonded
(2) Operational tests to demonstrate that all equipment is properly connected and in working order
(3) Polarity checks to determine that connections have been properly made

(B) Low-Voltage Circuits. An operational test of all low-voltage circuits shall be conducted to demonstrate that all equipment is connected and in electrical working order. This test shall be performed in the final stages of production after all outer coverings and cabinetry have been secured.

VII. Recreational Vehicle Parks

551.71 Type Receptacles Provided.

Every recreational vehicle site with electrical supply shall be equipped with at least one 20-ampere, 125-volt receptacle. A minimum of 5 percent of all recreational vehicle sites, with electrical supply, shall each be equipped with a 50-ampere, 125/250-volt receptacle conforming to the configuration as identified in Figure 551.46(C). These electrical supplies shall be permitted to include additional receptacles that have configurations in accordance with 551.81. A minimum of 70 percent of all recreational vehicle sites with electrical supply shall each be equipped with a 30-ampere, 125-volt receptacle conforming to Figure 551.46(C). This supply shall be permitted to include additional receptacle configurations conforming to 551.81. The remainder of all recreational vehicle sites with electrical supply shall be equipped with one or more of the receptacle configurations conforming to 551.81. Dedicated tent sites with a 15- or 20-ampere electrical supply shall be permitted to be excluded when determining the percentage of recreational vehicle sites with 30- or 50-ampere receptacles.

Additional receptacles shall be permitted for the connection of electrical equipment outside the recreational vehicle within the recreational vehicle park.

All 125-volt, single-phase, 15- and 20-ampere receptacles shall have listed ground-fault circuit-interrupter protection for personnel.

In accordance with 551.71, at least one 20-ampere, 125-volt receptacle must be installed at each recreational vehicle campsite. Existing recreational vehicle campgrounds may have some sites that are equipped with 30-ampere receptacles only. Adapter plugs or "cheater" cords are often used to connect a recreational vehicle with a 20-ampere supply cord to a 30-ampere receptacle outlet. This practice does not provide adequate overload protection for the cord or the connected load. The requirement in 551.71 ensures the availability of the properly rated receptacle.

Some newer recreational vehicles have a 50-ampere, 120/240-volt supply installed, and 551.71 contains provisions that require a limited number of RV sites with 50-ampere, 125/250-volt receptacles. Receptacle configurations are shown in Figure 551.46(C) and receptacle ratings are described in 551.81.

551.72 Distribution System.

Receptacles rated at 50 amperes shall be supplied from a branch circuit of the voltage class and rating of the receptacle. Other recreational vehicle sites with 125-volt, 20- and 30-ampere receptacles shall be permitted to be derived from any grounded distribution system that supplies 120-volt single-phase power. The neutral conductors shall not be reduced in size below the size of the ungrounded conductors for the site distribution. The neutral conductors shall be permitted to be reduced in size below the minimum required size of the ungrounded conductors for 240-volt, line-to-line, permanently connected loads only.

The reason for requiring single-phase, 120/240-volt, 3-wire systems to the recreational vehicle site is that the recreational vehicles are designed for connection to such systems. If the recreational vehicle park supplies another type of system to

the recreational vehicle site (for example, two-phase conductors and a neutral of a 3-phase, 4-wire, 208Y/120-volt system), a hazard could result due to a supply of less than 240 volts, nominal, to motors, for example, or attempts to adapt the recreational vehicle to a different supply system than the one for which it was designed.

551.73 Calculated Load.

(A) **Basis of Calculations.** Electrical service and feeders shall be calculated on the basis of not less than 9600 volt-amperes per site equipped with 50-ampere, 120/240-volt supply facilities; 3600 volt-amperes per site equipped with both 20-ampere and 30-ampere supply facilities; 2400 volt-amperes per site equipped with only 20-ampere supply facilities; and 600 volt-amperes per site equipped with only 20-ampere supply facilities that are dedicated to tent sites. The demand factors set forth in Table 551.73 shall be the minimum allowable demand factors that shall be permitted in calculating load for service and feeders. Where the electrical supply for a recreational vehicle site has more than one receptacle, the calculated load shall only be computed for the highest rated receptacle.

Table 551.73 Demand Factors for Site Feeders and Service-Entrance Conductors for Park Sites

Number of Recreational Vehicle Sites	Demand Factor (percent)
1	100
2	90
3	80
4	75
5	65
6	60
7–9	55
10–12	50
13–15	48
16–18	47
19–21	45
22–24	43
25–35	42
36 plus	41

Dedicated tent sites supplied with electricity are not intended to accommodate recreational vehicles; therefore the calculated load for these sites can be smaller.

(B) **Transformers and Secondary Distribution Panelboards.** For the purpose of this *Code*, where the park service exceeds 240 volts, transformers and secondary distribution panelboards shall be treated as services.

(C) Demand Factors. The demand factor for a given number of sites shall apply to all sites indicated. For example, 20 sites calculated at 45 percent of 3600 volt-amperes results in a permissible demand of 1620 volt-amperes per site or a total of 32,400 volt-amperes for 20 sites.

> FPN: These demand factors may be inadequate in areas of extreme hot or cold temperature with loaded circuits for heating or air conditioning.

(D) Feeder-Circuit Capacity. Recreational vehicle site feeder-circuit conductors shall have adequate ampacity for the loads supplied and shall be rated at not less than 30 amperes. The grounded conductors shall have the same ampacity as the ungrounded conductors.

> FPN: Due to the long circuit lengths typical in most recreational vehicle parks, feeder conductor sizes found in the ampacity tables of Article 310 may be inadequate to maintain the voltage regulation suggested in the fine print note to 210.19. Total circuit voltage drop is a sum of the voltage drops of each serial circuit segment, where the load for each segment is calculated using the load that segment sees and the demand factors of 551.73(A).

Loads for other amenities such as, but not limited to, service buildings, recreational buildings, and swimming pools shall be sized separately and then be added to the value calculated for the recreational vehicle sites where they are all supplied by one service.

551.74 Overcurrent Protection.

Overcurrent protection shall be provided in accordance with Article 240.

551.75 Grounding.

All electrical equipment and installations in recreational vehicle parks shall be grounded as required by Article 250.

551.76 Grounding—Recreational Vehicle Site Supply Equipment.

(A) Exposed Non–Current-Carrying Metal Parts. Exposed non–current-carrying metal parts of fixed equipment, metal boxes, cabinets, and fittings that are not electrically connected to grounded equipment shall be grounded by a continuous equipment grounding conductor run with the circuit conductors from the service equipment or from the transformer of a secondary distribution system. Equipment grounding conductors shall be sized in accordance with 250.122 and shall be permitted to be spliced by listed means.

The arrangement of equipment grounding connections shall be such that the disconnection or removal of a receptacle or other device will not interfere with, or interrupt, the grounding continuity.

(B) Secondary Distribution System. Each secondary distribution system shall be grounded at the transformer.

(C) Neutral Conductor Not to Be Used as an Equipment Ground. The neutral conductor shall not be used as an equipment ground for recreational vehicles or equipment within the recreational vehicle park.

(D) No Connection on the Load Side. No connection to a grounding electrode shall be made to the neutral conductor on the load side of the service disconnecting means or transformer distribution panelboard.

551.77 Recreational Vehicle Site Supply Equipment.

(A) Location. Where provided on back-in sites, the recreational vehicle site electrical supply equipment shall be located on the left (road) side of the parked vehicle, on a line that is 1.5 m to 2.1 m (5 ft to 7 ft) from the left edge (driver's side of the parked RV) of the stand and shall be located at any point on this line from the rear of the stand to 4.5 m (15 ft) forward of the rear of the stand.

For pull-through sites, the electrical supply equipment shall be permitted to be located at any point along the line that is 1.5 m to 2.1 m (5 ft to 7 ft) from the left edge (driver's side of the parked RV) from 4.9 m (16 ft) forward of the rear of the stand to the center point between the two roads that gives access to and egress from the pull-through sites.

The left edge (driver's side of the parked RV) of the stand shall be marked.

> The requirements of 551.77(A) are intended to accommodate vehicles towing boats and the like. The location of the site supply equipment permitted for pull-through sites will reduce the use of extension cords.

(B) Disconnecting Means. A disconnecting switch or circuit breaker shall be provided in the site supply equipment for disconnecting the power supply to the recreational vehicle.

(C) Access. All site supply equipment shall be accessible by an unobstructed entrance or passageway not less than 600 mm (2 ft) wide and 2.0 m (6 ft 6 in.) high.

(D) Mounting Height. Site supply equipment shall be located not less than 600 mm (2 ft) or more than 2.0 m (6 ft 6 in.) above the ground.

(E) Working Space. Sufficient space shall be provided and maintained about all electrical equipment to permit ready and safe operation, in accordance with 110.26.

(F) Marking. Where the site supply equipment contains a 125/250-volt receptacle, the equipment shall be marked as follows: "Turn disconnecting switch or circuit breaker off before inserting or removing plug. Plug must be fully inserted or removed." The marking shall be located on the equipment adjacent to the receptacle outlet.

Partially engaged attachment plugs may result in intermittent neutral (grounded conductor) contact. Loss of the neutral could momentarily apply the line-to-line voltage (240 volts) across 125-volt equipment, causing malfunction or damage.

551.78 Protection of Outdoor Equipment.

(A) Wet Locations. All switches, circuit breakers, receptacles, control equipment, and metering devices located in wet locations or outside of a building shall be rainproof equipment.

(B) Meters. If secondary meters are installed, meter sockets without meters installed shall be blanked off with an approved blanking plate.

551.79 Clearance for Overhead Conductors.

Open conductors of not over 600 volts, nominal, shall have a vertical clearance of not less than 5.5 m (18 ft) and a horizontal clearance of not less than 900 mm (3 ft) in all areas subject to recreational vehicle movement. In all other areas, clearances shall conform to 225.18 and 225.19.

FPN: For clearances of conductors over 600 volts, nominal, see ANSI C2-1997, *National Electrical Safety Code*.

551.80 Underground Service, Feeder, Branch-Circuit, and Recreational Vehicle Site Feeder-Circuit Conductors.

(A) General. All direct-burial conductors, including the equipment grounding conductor if of aluminum, shall be insulated and identified for the use. All conductors shall be continuous from equipment to equipment. All splices and taps shall be made in approved junction boxes or by use of material listed and identified for the purpose.

(B) Protection Against Physical Damage. Direct-buried conductors and cables entering or leaving a trench shall be protected by rigid metal conduit, intermediate metal conduit, electrical metallic tubing with supplementary corrosion protection, rigid nonmetallic conduit, liquidtight flexible nonmetallic conduit, liquidtight flexible metal conduit, or other approved raceways or enclosures. Where subject to physical damage, the conductors or cables shall be protected by rigid metal conduit, intermediate metal conduit, or Schedule 80 rigid nonmetallic conduit. All such protection shall extend at least 450 mm (18 in.) into the trench from finished grade.

FPN: See 300.5 and Article 340 for conductors or Type UF cable used underground or in direct burial in earth.

551.81 Receptacles.

A receptacle to supply electric power to a recreational vehicle shall be one of the configurations shown in Figure 551.46(C) in the following ratings.

(1) 50-ampere—125/250-volt, 50-ampere, 3-pole, 4-wire grounding type for 120/240-volt systems
(2) 30-ampere—125-volt, 30-ampere, 2-pole, 3-wire grounding type for 120-volt systems
(3) 20-ampere—125-volt, 20-ampere, 2-pole, 3-wire grounding type for 120-volt systems

FPN: Complete details of these configurations can be found in ANSI/NEMA WD 6-1989, National Electrical Manufacturers Association's *Standard for Dimensions of Attachment Plugs and Receptacles*, Figures 14-50, TT, and 5-20.

ARTICLE 552
Park Trailers

Contents

I. General

552.1 Scope.

The provisions of this article cover the electrical conductors and equipment installed within or on park trailers not covered fully under Articles 550 and 551.

The scope of Article 552 covers park trailers that have a single chassis and wheels, not exceeding 400 ft^2 (set up), and that are not used as permanent residences. Additionally, Article 552 does not apply to units that meet the definition of park trailer but are used for commercial purposes.

It is not uncommon for park trailers to be equipped with electrical loads similar to those used in mobile homes. It is also not uncommon for the park trailer to be located in the same park trailer community for several years without relocation.

Park trailers are somewhat similar to mobile homes and recreational vehicles, and many requirements in Article 552 are the same or similar to those contained within Articles

550 and 551. Article 552, therefore, is similar in structure to Articles 550 and 551.

552.2 Definition.

(See Articles 100, 550, and 551 for additional definitions.)

Park Trailer. A unit that is built on a single chassis mounted on wheels and has a gross trailer area not exceeding 37 m² (400 ft²) in the set-up mode.

552.3 Other Articles.

Wherever the requirements of other articles of this *Code* and Article 552 differ, the requirements of Article 552 shall apply.

552.4 General Requirements.

A park trailer as specified in 552.2 is intended for seasonal use. It is not intended as a permanent dwelling unit or for commercial uses such as banks, clinics, offices, or similar.

II. Low-Voltage Systems

552.10 Low-Voltage Systems.

(A) Low-Voltage Circuits. Low-voltage circuits furnished and installed by the park trailer manufacturer, other than those related to braking, are subject to this *Code*. Circuits supplying lights subject to federal or state regulations shall comply with applicable government regulations and this *Code*.

In accordance with 552.10(A), the requirements of Article 552, Part II apply to the low-voltage wiring within the park trailer that would be used in place of 120-volt ac supplies. These requirements do not apply to braking circuits.

(B) Low-Voltage Wiring.

(1) Material. Copper conductors shall be used for low-voltage circuits.

Exception: A metal chassis or frame shall be permitted as the return path to the source of supply.

It is not the intent of 552.10(B)(1) to permit the sidewalls or the roof of a park trailer to serve as the ground return path. See the definition of *frame* in 551.2.

(2) Conductor Types. Conductors shall conform to the requirements for Type GXL, HDT, SGT, SGR, or Type SXL or shall have insulation in accordance with Table 310.13 or the equivalent. Conductor sizes 6 AWG through 18 AWG

or SAE shall be listed. Single-wire, low-voltage conductors shall be of the stranded type.

> FPN: See SAE Standard J1128-1995 for Types GXL, HDT, and SXL and SAE Standard J1127-1995 for Types SGT and SGR.

(3) Marking. All insulated low-voltage conductors shall be surface marked at intervals not greater than 1.2 m (4 ft) as follows:

(1) Listed conductors shall be marked as required by the listing agency.
(2) SAE conductors shall be marked with the name or logo of the manufacturer, specification designation, and wire gauge.
(3) Other conductors shall be marked with the name or logo of the manufacturer, temperature rating, wire gauge, conductor material, and insulation thickness.

(C) Low-Voltage Wiring Methods.

(1) Physical Protection. Conductors shall be protected against physical damage and shall be secured. Where insulated conductors are clamped to the structure, the conductor insulation shall be supplemented by an additional wrap or layer of equivalent material, except that jacketed cables shall not be required to be so protected. Wiring shall be routed away from sharp edges, moving parts, or heat sources.

(2) Splices. Conductors shall be spliced or joined with splicing devices that provide a secure connection or by brazing, welding, or soldering with a fusible metal or alloy. Soldered splices shall first be spliced or joined to be mechanically and electrically secure without solder and then soldered. All splices, joints, and free ends of conductors shall be covered with an insulation equivalent to that on the conductors.

(3) Separation. Battery and other low-voltage circuits shall be physically separated by at least a 13-mm (½-in.) gap or other approved means from circuits of a different power source. Acceptable methods shall be by clamping, routing, or equivalent means that ensure permanent total separation. Where circuits of different power sources cross, the external jacket of the nonmetallic-sheathed cables shall be deemed adequate separation.

(4) Ground Connections. Ground connections to the chassis or frame shall be made in an accessible location and shall be mechanically secure. Ground connections shall be by means of copper conductors and copper or copper-alloy terminals of the solderless type identified for the size of wire used. The surface on which ground terminals make contact shall be cleaned and be free from oxide or paint or shall be electrically connected through the use of a cadmium, tin, or zinc-plated internal/external-toothed lockwasher or locking terminals. Ground terminal attaching screws, rivets or bolts, nuts, and lockwashers shall be cadmium, tin, or zinc-plated except rivets shall be permitted to be unanodized aluminum where attaching to aluminum structures.

The chassis-grounding terminal of the battery shall be bonded to the unit chassis with a minimum 8 AWG copper conductor. In the event the power lead from the battery exceeds 8 AWG, the bonding conductor shall be of an equal size.

The provisions of 552.10(C)(4) require that the chassis-grounding terminal of the battery be bonded to the vehicle chassis in a mechanically secure manner and be placed in an accessible location using a minimum 8 AWG copper conductor. This minimizes the possibility of low-voltage circuit-fault currents passing through the ac panelboard bonding conductor and the equipment grounding conductor of the combination ac/dc appliance and subsequently passing through the negative dc conductor feeding the appliance that also may be bonded to the external metal cover of the appliance. The ac equipment grounding conductor of the appliance may not have sufficient ampacity to safely conduct the dc fault current. This will necessitate installation of the battery bonding conductor. Some recreational vehicles already have one side of the battery circuit bonded to the frame by an 8 AWG or larger copper conductor.

(D) Battery Installations. Storage batteries subject to the provisions of this *Code* shall be securely attached to the unit and installed in an area vaportight to the interior and ventilated directly to the exterior of the unit. Where batteries are installed in a compartment, the compartment shall be ventilated with openings having a minimum area of 1100 mm² (1.7 in.²) at both the top and at the bottom. Where compartment doors are equipped for ventilation, the openings shall be within 50 mm (2 in.) of the top and bottom. Batteries shall not be installed in a compartment containing spark- or flame-producing equipment.

(E) Overcurrent Protection.

(1) Rating. Low-voltage circuit wiring shall be protected by overcurrent protective devices rated not in excess of the ampacity of copper conductors, in accordance with Table 552.10(E)(1).

Table 552.10(E)(1) Low-Voltage Overcurrent Protection

Wire Size (AWG)	Ampacity	Wire Type
18	6	Stranded only
16	8	Stranded only
14	15	Stranded or solid
12	20	Stranded or solid
10	30	Stranded or solid

(2) Type. Circuit breakers or fuses shall be of an approved type, including automotive types. Fuseholders shall be clearly marked with maximum fuse size and shall be protected against shorting and physical damage by a cover or equivalent means.

> FPN: For further information, see ANSI/SAE J554-1987, *Standard for Electric Fuses (Cartridge Type)*; SAE J1284-1988, *Standard for Blade Type Electric Fuses*; and UL 275-1993, *Standard for Automotive Glass Tube Fuses*.

The requirement for protection of fuseholders by a cover or equivalent means is intended to reduce the possibility of the low-voltage system shorting to ground.

(3) Appliances. Higher-current-consuming dc appliances such as pumps, compressors, heater blowers, and similar motor-driven appliances shall be installed in accordance with the manufacturer's instructions.

Motors that are controlled by automatic switching or by latching-type manual switches shall be protected in accordance with 430.32(B).

(4) Location. The overcurrent protective device shall be installed in an accessible location on the unit within 450 mm (18 in.) of the point where the power supply connects to the unit circuits. If located outside the park trailer, the device shall be protected against weather and physical damage.

Exception: External low-voltage supply shall be permitted to have the overcurrent protective device within 450 mm (18 in.) after entering the unit or after leaving a metal raceway.

(F) Switches. Switches shall have a dc rating not less than the connected load.

(G) Luminaires (Lighting Fixtures). All low-voltage interior luminaires (lighting fixtures) rated more than 4 watts, employing lamps rated more than 1.2 watts, shall be listed.

Twelve-volt systems for running and signal lights, similar to those used in a conventional automobile, are covered in 552.10 and 552.20. In some park trailers, 12-volt systems are also used for interior lighting and other small loads. The 12-volt system is often supplied from an on-board battery or through a transfer switch from a 120/12-volt transformer in conjunction with a full-wave rectifier.

III. Combination Electrical Systems

552.20 Combination Electrical Systems.

(A) General. Unit wiring suitable for connection to a battery or other low-voltage supply source shall be permitted to be connected to a 120-volt source, provided that the entire wiring system and equipment are rated and installed in full conformity with Parts I, III, IV, and V requirements of this article covering 120-volt electrical systems. Circuits fed from ac transformers shall not supply dc appliances.

(B) Voltage Converters (120-Volt Alternating Current to Low-Voltage Direct Current). The 120-volt ac side of the voltage converter shall be wired in full conformity with Parts I, III, IV, and V requirements of this article for 120-volt electrical systems.

Exception: Converters supplied as an integral part of a listed appliance shall not be subject to the above.

All converters and transformers shall be listed for use in recreation units and designed or equipped to provide over-temperature protection. To determine the converter rating, the following formula shall be applied to the total connected load, including average battery charging rate, of all 12-volt equipment:

The first 20 amperes of load at 100 percent; plus

The second 20 amperes of load at 50 percent; plus

All load above 40 amperes at 25 percent

Exception: A low-voltage appliance that is controlled by a momentary switch (normally open) that has no means for holding in the closed position shall not be considered as a connected load when determining the required converter rating. Momentarily energized appliances shall be limited to those used to prepare the unit for occupancy or travel.

(C) Bonding Voltage Converter Enclosures. The non–current-carrying metal enclosure of the voltage converter shall be bonded to the frame of the unit with a 8 AWG copper conductor minimum. The grounding conductor for the battery and the metal enclosure shall be permitted to be the same conductor.

(D) Dual-Voltage Fixtures Including Luminaires or Appliances. Fixtures, including luminaires, or appliances having both 120-volt and low-voltage connections shall be listed for dual voltage.

In the dual-voltage fixtures described in 552.20(D), barriers are used to separate the 120-volt and the 12-volt wiring connections.

(E) Autotransformers. Autotransformers shall not be used.

(F) Receptacles and Plug Caps. Where a park trailer is equipped with a 120-volt or 120/240-volt ac system, a low-voltage system, or both, receptacles and plug caps of the

low-voltage system shall differ in configuration from those of the 120-volt or 120/240-volt system. Where a unit equipped with a battery or dc system has an external connection for low-voltage power, the connector shall have a configuration that will not accept 120-volt power.

IV. Nominal 120-Volt or 120/240-Volt Systems

552.40 120-Volt or 120/240-Volt, Nominal, Systems.

(A) General Requirements. The electrical equipment and material of park trailers indicated for connection to a wiring system rated 120 volts, nominal, 2-wire with ground, or a wiring system rated 120/240 volts, nominal, 3-wire with ground, shall be listed and installed in accordance with the requirements of Parts I, III, IV, and V of this article.

(B) Materials and Equipment. Electrical materials, devices, appliances, fittings, and other equipment installed, intended for use in, or attached to the park trailer shall be listed. All products shall be used only in the manner in which they have been tested and found suitable for the intended use.

552.41 Receptacle Outlets Required.

(A) Spacing. Receptacle outlets shall be installed at wall spaces 600 mm (2 ft) wide or more so that no point along the floor line is more than 1.8 m (6 ft), measured horizontally, from an outlet in that space.

Exception No. 1: Bath and hall areas.

Exception No. 2: Wall spaces occupied by kitchen cabinets, wardrobe cabinets, built-in furniture; behind doors that may open fully against a wall surface; or similar facilities.

(B) Location. Receptacle outlets shall be installed as follows:

(1) Adjacent to countertops in the kitchen [at least one on each side of the sink if countertops are on each side and are 300 mm (12 in.) or over in width]
(2) Adjacent to the refrigerator and gas range space, except where a gas-fired refrigerator or cooking appliance, requiring no external electrical connection, is factory-installed
(3) Adjacent to countertop spaces of 300 mm (12 in.) or more in width that cannot be reached from a receptacle required in 552.41(B)(1) by a cord of 1.8 m (6 ft) without crossing a traffic area, cooking appliance, or sink

(C) Ground-Fault Circuit-Interrupter Protection. Where provided, each 125-volt, single-phase, 15- or 20-ampere receptacle outlet shall have ground-fault circuit-interrupter protection for personnel in the following locations:

(1) Adjacent to a bathroom lavatory

(2) Within 1.8 m (6 ft) of any lavatory or sink

Exception: Receptacles installed for appliances in dedicated spaces, such as for dishwashers, disposals, refrigerators, freezers, and laundry equipment.

(3) In the area occupied by a toilet, shower, tub, or any combination thereof

(4) On the exterior of the unit

Exception: Receptacles that are located inside of an access panel that is installed on the exterior of the unit to supply power for an installed appliance shall not be required to have ground-fault circuit-interrupter protection.

The receptacle outlet shall be permitted in a listed luminaire (lighting fixture). A receptacle outlet shall not be installed in a tub or combination tub–shower compartment.

In accordance with 552.41(C)(4), a bathroom receptacle is permitted to be mounted in the side of a lavatory cabinet where installation of a receptacle outlet is not possible in a wall that does not provide the necessary depth. The receptacle must be GFCI protected in accordance with 551.42(C)(1). Receptacles of any type are not permitted in a tub or tub–shower compartment.

(D) Pipe Heating Cable Outlet. Where a pipe heating cable outlet is installed, the outlet shall be as follows:

(1) Located within 600 mm (2 ft) of the cold water inlet

(2) Connected to an interior branch circuit, other than a small appliance branch circuit

(3) On a circuit where all of the outlets are on the load side of the ground-fault circuit-interrupter protection for personnel

(4) Mounted on the underside of the park trailer and shall not be considered to be the outdoor receptacle outlet required in 552.41(E)

(E) Outdoor Receptacle Outlets. At least one receptacle outlet shall be installed outdoors. A receptacle outlet located in a compartment accessible from the outside of the park trailer shall be considered an outdoor receptacle. Outdoor receptacle outlets shall be protected as required in 552.41(C)(4).

(F) Receptacle Outlets Not Permitted.

(1) Shower or Bathtub Space. Receptacle outlets shall not be installed in or within reach [750 mm (30 in.)] of a shower or bathtub space.

(2) Face-Up Position. A receptacle shall not be installed in a face-up position in any countertop.

552.43 Power Supply.

(A) Feeder. The power supply to the park trailer shall be a feeder assembly consisting of not more than one listed 30-ampere or 50-ampere park trailer power-supply cord with an integrally molded or securely attached cap, or a permanently installed feeder.

(B) Power-Supply Cord. If the park trailer has a power-supply cord, it shall be permanently attached to the distribution panelboard or to a junction box permanently connected to the distribution panelboard, with the free end terminating in a molded-on attachment plug cap.

Cords with adapters and pigtail ends, extension cords, and similar items shall not be attached to, or shipped with, a park trailer.

A suitable clamp or the equivalent shall be provided at the distribution panelboard knockout to afford strain relief for the cord to prevent strain from being transmitted to the terminals when the power-supply cord is handled in its intended manner.

The cord shall be a listed type with 3-wire, 120-volt or 4-wire, 120/240-volt conductors, one of which shall be identified by a continuous green color or a continuous green color with one or more yellow stripes for use as the grounding conductor.

(C) Mast Weatherhead or Raceway. Where the calculated load exceeds 50 amperes or where a permanent feeder is used, the supply shall be by means of one of the following:

(1) One mast weatherhead installation, installed in accordance with Article 230, containing four continuous, insulated, color-coded feeder conductors, one of which shall be an equipment grounding conductor

(2) A metal raceway, rigid nonmetallic conduit, or liquidtight flexible nonmetallic conduit from the disconnecting means in the park trailer to the underside of the park trailer, with provisions for the attachment to a suitable junction box or fitting to the raceway on the underside of the park trailer [with or without conductors as in 550.10(I)(1)]

552.44 Cord.

(A) Permanently Connected. Each power-supply assembly shall be factory supplied or factory installed and connected directly to the terminals of the distribution panelboard or conductors within a junction box and provided with means to prevent strain from being transmitted to the terminals. The ampacity of the conductors between each junction box and the terminals of each distribution panelboard shall be at least equal to the ampacity of the power-supply cord. The supply end of the assembly shall be equipped with an attachment plug of the type described in 552.44(C). Where the cord passes through the walls or floors, it shall be protected by means of conduit and bushings or equivalent. The cord assembly shall have permanent provisions for protection against corrosion and mechanical damage while the unit is in transit.

(B) Cord Length. The cord-exposed usable length shall be measured from the point of entrance to the park trailer or the face of the flanged surface inlet (motor-base attachment plug) to the face of the attachment plug at the supply end.

The cord-exposed usable length, measured to the point of entry on the unit exterior, shall be a minimum of 7.0 m (23 ft) where the point of entrance is at the side of the unit, or shall be a minimum 8.5 m (28 ft) where the point of entrance is at the rear of the unit. The maximum length shall not exceed 11 m (36½ ft).

Where the cord entrance into the unit is more than 900 mm (3 ft) above the ground, the minimum cord lengths above shall be increased by the vertical distance of the cord entrance heights above 900 mm (3 ft).

(C) Attachment Plugs.

(1) Units with Two to Five 15- or 20-Ampere Branch Circuits. Park trailers wired in accordance with 552.46(A) shall have an attachment plug that shall be 2-pole, 3-wire grounding-type, rated 30 amperes, 125 volts, conforming to the configuration shown in Figure 552.44(C) intended for use with units rated at 30 amperes, 125 volts.

> FPN: Complete details of this configuration can be found in ANSI/NEMA WD 6-1989, National Electrical Manufacturers Association's *Standard for Dimensions of Attachment Plugs and Receptacles*, Figure TT.

(2) Units with 50-Ampere Power Supply Assembly. Park trailers having a power-supply assembly rated 50 amperes as permitted by 552.43(B) shall have a 3-pole, 4-wire grounding-type attachment plug rated 50 amperes, 125/250 volts, conforming to the configuration shown in Figure 552.44(C).

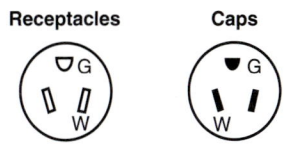

30-A, 125-V, 2-pole, 3-wire, grounding type

50-A, 125/250-V, 3-pole, 4-wire, grounding type

Figure 552.44(C) Attachment cap and receptacle configurations.

> FPN: Complete details of this configuration can be found in ANSI/NEMA WD 6-1989, National Electrical Manufacturers Association *Standard for Dimensions of Attachment Plugs and Receptacles*, Figure 14-50.

The 50-ampere receptacle and plug configurations used for park trailers appear in Exhibit 406.2 (NEMA configuration chart); however, the 30-ampere plug and receptacle configurations are unique to recreational vehicles and are described in UL 498, *Standard for Safety, Attachment Plugs and Receptacles.*

(D) Labeling at Electrical Entrance. Each park trailer shall have permanently affixed to the exterior skin, at or near the point of entrance of the power-supply assembly, a label 75 mm × 45 mm (3 in. × 1¾ in.) minimum size, made of etched, metal-stamped, or embossed brass, stainless steel, or anodized or alclad aluminum not less than 0.51 mm (0.020 in.) thick, or other suitable material [e.g., 0.13 mm (0.005 in.) thick plastic laminate], that reads, as appropriate, either

THIS CONNECTION IS FOR 110–125-VOLT AC, 60 HZ, 30 AMPERE SUPPLY

or

THIS CONNECTION IS FOR 120/240 VOLT AC, 3-POLE, 4-WIRE, 60 HZ, _____ AMPERE SUPPLY.

The correct ampere rating shall be marked in the blank space.

(E) Location. The point of entrance of a power-supply assembly shall be located within 4.5 m (15 ft) of the rear, on the left (road) side or at the rear, left of the longitudinal center of the unit, within 450 mm (18 in.) of the outside wall.

Exception: A park trailer shall be permitted to have the electrical point of entrance located more than 4.5 m (15 ft) from the rear. Where this occurs, the distance beyond the 4.5-m (15-ft) dimension shall be added to the cord's minimum length as specified in 551.46(B).

552.45 Distribution Panelboard.

(A) Listed and Appropriately Rated. A listed and appropriately rated distribution panelboard or other equipment specifically listed for the purpose shall be used. The grounded conductor termination bar shall be insulated from the enclosure as provided in 552.55(C). An equipment grounding terminal bar shall be attached inside the metal enclosure of the panelboard.

(B) Location. The distribution panelboard shall be installed in a readily accessible location. Working clearance for the panelboard shall be not less than 600 mm (24 in.) wide and 750 mm (30 in.) deep.

Exception: Where the panelboard cover is exposed to the inside aisle space, one of the working clearance dimensions shall be permitted to be reduced to a minimum of 550 mm (22 in.). A panelboard shall be considered exposed where the panelboard cover is within 50 mm (2 in.) of the aisle's finished surface.

(C) Dead-Front Type. The distribution panelboard shall be of the dead-front type. A main disconnecting means shall be provided where fuses are used or where more than two circuit breakers are employed. A main overcurrent protective device not exceeding the power-supply assembly rating shall be provided where more than two branch circuits are employed.

552.46 Branch Circuits.

Branch circuits shall be determined in accordance with 552.46(A) and (B).

(A) Two to Five 15- or 20-Ampere Circuits. Two to five 15- or 20-ampere circuits to supply lights, receptacle outlets, and fixed appliances shall be permitted. Such park trailers shall be equipped with a distribution panelboard rated at 120 volts maximum with a 30-ampere rated main power supply assembly. Not more than two 120-volt thermostatically controlled appliances (e.g., air conditioner and water heater) shall be installed in such systems unless appliance isolation switching, energy management systems, or similar methods are used.

Exception: Additional 15- or 20-ampere circuits shall be permitted where a listed energy management system rated at 30 amperes maximum is employed within the system.

(B) More Than Five Circuits. Where more than five circuits are needed, they shall be determined in accordance with 552.46(B)(1), (B)(2), and (B)(3).

(1) Lighting. Based on 33 volt-amperes/m^2 (3 VA/ft^2) multiplied by the outside dimensions of the park trailer (coupler excluded) divided by 120 volts to determine the number of 15- or 20-ampere lighting area circuits, for example,

$$\frac{3 \times \text{length} \times \text{width}}{120 \times 15 \text{ (or 20)}} = \text{No. of 15 (or 20) ampere circuits}$$

The lighting circuits shall be permitted to serve built-in gas ovens with electric service only for lights, clocks or timers, or listed cord-connected garbage disposal units.

(2) Small Appliances. Small appliance branch circuits shall be installed in accordance with 210.11(C)(1).

(3) General Appliances. (including furnace, water heater, space heater, range, and central or room air conditioner, etc.) An individual branch circuit shall be permitted to supply any load for which it is rated. There shall be one or more circuits of adequate rating in accordance with (a) through (d).

FPN No. 1: For the laundry branch circuit, see 210.11(C)(2).
FPN No. 2: For central air conditioning, see Article 440.

(a) The total rating of fixed appliances shall not exceed 50 percent of the circuit rating if lighting outlets, general-use receptacles, or both, are also supplied.

(b) For fixed appliances with a motor(s) larger than ⅛ horsepower the total computed load shall be based on 125 percent of the largest motor plus the sum of the other loads. Where a branch circuit supplies continuous load(s) or any combination of continuous and noncontinuous loads, the branch-circuit conductor size shall be in accordance with 210.19(A).

(c) The rating of a single cord- and plug-connected appliance supplied by other than an individual branch circuit shall not exceed 80 percent of the circuit rating.

(d) The rating of a range branch circuit shall be based on the range demand as specified for ranges in 552.47(B)(5).

552.47 Calculations.

The following method shall be employed in computing the supply-cord and distribution-panelboard load for each feeder assembly for each park trailer in lieu of the procedure shown in Article 220 and shall be based on a 3-wire, 120/240-volt supply with 120-volt loads balanced between the two phases of the 3-wire system.

(A) Lighting and Small Appliance Load. Lighting Volt-Amperes: Length times width of park trailer floor (outside dimensions) times 33 volt-amperes/m^2 (3 VA/ft^2), for example,

Length × width × 3 = lighting volt-amperes

Small Appliance Volt-Amperes: Number of circuits times 1500 volt-amperes for each 20-ampere appliance receptacle circuit (see definition of *Appliance, Portable* with note) including 1500 volt-amperes for laundry circuit, for example,

No. of circuits × 1500 = small appliance volt-amperes

Total: Lighting volt-amperes plus small appliance volt-amperes = total volt-amperes

First 3000 total volt-amperes at 100 percent plus remainder at 35 percent = volt-amperes to be divided by 240 volts to obtain current (amperes) per leg

(B) Total Load for Determining Power Supply. Total load for determining power supply is the sum of the following:

(1) Lighting and small appliance load as calculated in 552.47(A).
(2) Nameplate amperes for motors and heater loads (exhaust fans, air conditioners, electric, gas, or oil heating). Omit smaller of the heating and cooling loads, except include blower motor if used as air-conditioner evaporator motor. Where an air conditioner is not installed and a 50-ampere power-supply cord is provided, allow 15 amperes per phase for air conditioning.
(3) Twenty-five percent of current of largest motor in (2).
(4) Total of nameplate amperes for disposal, dishwasher, water heater, clothes dryer, wall-mounted oven, cooking units. Where the number of these appliances exceeds three, use 75 percent of total.
(5) Derive amperes for freestanding range (as distinguished from separate ovens and cooking units) by dividing the following values by 240 volts:

Nameplate Rating (watts)	Use (volt-amperes)
0–10,000	80 percent of rating
Over 10,000–12,500	8,000
Over 12,500–13,500	8,400
Over 13,500–14,500	8,800
Over 14,500–15,500	9,200
Over 15,500–16,500	9,600
Over 16,500–17,500	10,000

(6) If outlets or circuits are provided for other than factory-installed appliances, include the anticipated load.

FPN: Refer to Annex D, Example D12, for an illustration of the application of this calculation.

(C) Optional Method of Calculation for Lighting and Appliance Load. For park trailers, the optional method for calculating lighting and appliance load shown in 220.30 shall be permitted.

552.48 Wiring Methods.

(A) Wiring Systems. Cables and raceways installed in accordance with Articles 320, 322, 330 through 340, 342 through 362, Article 386, and Article 388 shall be permitted in accordance with their applicable article, except as otherwise specified in this article. An equipment grounding means shall be provided in accordance with 250.118.

See the commentary following 348.60 for information on the use of flexible metal conduit for equipment grounding.

(B) Conduit and Tubing. Where rigid metal conduit or intermediate metal conduit is terminated at an enclosure with

a locknut and bushing connection, two locknuts shall be provided, one inside and one outside of the enclosure. All cut ends of conduit and tubing shall be reamed or otherwise finished to remove rough edges.

See the commentary following 358.28(A) and 300.4(F) for information on protecting conductor insulation against abrasion at conduit and tubing terminations.

(C) Nonmetallic Boxes. Nonmetallic boxes shall be acceptable only with nonmetallic-sheathed cable or nonmetallic raceways.

(D) Boxes. In walls and ceilings constructed of wood or other combustible material, boxes and fittings shall be flush with the finished surface or project therefrom.

(E) Mounting. Wall and ceiling boxes shall be mounted in accordance with Article 314.

Exception No. 1: Snap-in-type boxes or boxes provided with special wall or ceiling brackets that securely fasten boxes in walls or ceilings shall be permitted.

Exception No. 2: A wooden plate providing a 38-mm (1½-in.) minimum width backing around the box and of a thickness of 13 mm (½ in.) or greater (actual) attached directly to the wall panel shall be considered as approved means for mounting outlet boxes.

Exception No. 2 to 552.48(E) permits the mounting of outlet boxes by screws to a wooden plate that is secured directly to the back of the wall panel. This wooden plate must be not less than ½ in. thick and must extend at least 1½ in. around the box. This requirement recognizes the special construction of recreational vehicle walls, which often makes it quite difficult or impossible to attach an outlet box to a structural member, as required by 314.23(B).

(F) Sheath Armor. The sheath of nonmetallic-sheathed cable, metal-clad cable, and Type AC cable shall be continuous between outlet boxes and other enclosures.

(G) Protected. Metal-clad, Type AC, or nonmetallic-sheathed cables and electrical nonmetallic tubing shall be permitted to pass through the centers of the wide side of 2 by 4 wood studs. However, they shall be protected where they pass through 2 by 2 wood studs or at other wood studs or frames where the cable or tubing would be less than 32 mm (1¼ in.) from the inside or outside surface. Steel plates on each side of the cable or tubing, or a steel tube, with not less than 1.35 mm (0.053 in.) wall thickness, shall be installed to protect the cable or tubing. These plates or tubes shall be securely held in place. Where nonmetallic-sheathed cables pass through punched, cut, or drilled slots or holes in metal members, the cable shall be protected by bushings or grommets securely fastened in the opening prior to installation of the cable.

(H) Cable Supports. Where connected with cable connectors or clamps, cables shall be supported within 300 mm (12 in.) of outlet boxes, distribution panelboards, and splice boxes on appliances. Supports shall be provided every 1.4 m (4½ ft) at other places.

(I) Nonmetallic Box Without Cable Clamps. Nonmetallic-sheathed cables shall be supported within 200 mm (8 in.) of a nonmetallic outlet box without cable clamps.

Exception: Where wiring devices with integral enclosures are employed with a loop of extra cable to permit future replacement of the device, the cable loop shall be considered as an integral portion of the device.

(J) Physical Damage. Where subject to physical damage, exposed nonmetallic cable shall be protected by covering boards, guard strips, raceways, or other means.

(K) Metal Faceplates. Metal faceplates shall be of ferrous metal not less than 0.76 mm (0.030 in.) in thickness or of nonferrous metal not less than 1.0 mm (0.040 in.) in thickness. Nonmetallic faceplates shall be listed.

(L) Metal Faceplates Effectively Grounded. Where metal faceplates are used, they shall be effectively grounded.

(M) Moisture or Physical Damage. Where outdoor or under-chassis wiring is 120 volts, nominal, or over and is exposed to moisture or physical damage, the wiring shall be protected by rigid metal conduit, by intermediate metal conduit, or by electrical metallic tubing or rigid nonmetallic conduit that is closely routed against frames and equipment enclosures or other raceway or cable identified for the application.

(N) Component Interconnections. Fittings and connectors that are intended to be concealed at the time of assembly shall be listed and identified for the interconnection of building components. Such fittings and connectors shall be equal to the wiring method employed in insulation, temperature rise, and fault-current withstanding, and shall be capable of enduring the vibration and shock occurring in park trailers.

(O) Method of Connecting Expandable Units. The method of connecting expandable units to the main body of the vehicle shall comply with the following as applicable:

(1) That portion of a branch circuit that is installed in an expandable unit shall be permitted to be connected to the branch circuit in the main body of the vehicle by means of a flexible cord or attachment plug and cord listed for hard usage. The cord and its connections shall conform to all provisions of Article 400 and shall be considered as a permitted use under 400.7.

(2) If the receptacle provided for connection of the cord to the main circuit is located on the outside of the unit, it shall be protected with a ground-fault circuit interrupter for personnel and be listed for wet locations. A cord located on the outside of a unit shall be identified for outdoor use.

(3) Unless removable or stored within the unit interior, the cord assembly shall have permanent provisions for protection against corrosion and mechanical damage while the unit is in transit.

(4) If an attachment plug and cord is used, it shall be installed so as not to permit exposed live attachment plug pins.

(P) Prewiring for Air-Conditioning Installation. Prewiring installed for the purpose of facilitating future air-conditioning installation shall conform to the applicable portions of this article and the following:

(1) An overcurrent protective device with a rating compatible with the circuit conductors shall be installed in the distribution panelboard and wiring connections completed.

(2) The load end of the circuit shall terminate in a junction box with a blank cover or a device listed for the purpose. Where a junction box with a blank cover is used, the free ends of the conductors shall be adequately capped or taped.

(3) A label conforming to 552.44(D) shall be placed on or adjacent to the junction box and shall read

AIR-CONDITIONING CIRCUIT. THIS CONNECTION IS FOR AIR CONDITIONERS RATED 110–125-VOLT AC, 60 HZ, _____ AMPERES MAXIMUM. DO NOT EXCEED CIRCUIT RATING.

An ampere rating, not to exceed 80 percent of the circuit rating, shall be legibly marked in the blank space.

(4) The circuit shall serve no other purpose.

552.49 Maximum Number of Conductors in Boxes.

The maximum number of conductors permitted in boxes shall be in accordance with 314.16.

552.50 Grounded Conductors.

The identification of grounded conductors shall be in accordance with 200.6.

552.51 Connection of Terminals and Splices.

Conductor splices and connections at terminals shall be in accordance with 110.14.

552.52 Switches.

Switches shall be rated as follows.

(A) Lighting Circuits. For lighting circuits, switches shall be rated not less than 10 amperes, 120/125 volts and in no case less than the connected load.

(B) Motors or Other Loads. For motors or other loads, switches shall have ampere or horsepower ratings, or both, adequate for loads controlled. (An ac general-use snap switch shall be permitted to control a motor 2 hp or less with full-load current not over 80 percent of the switch ampere rating.)

552.53 Receptacles.

All receptacle outlets shall be of the grounding type and installed in accordance with 210.21 and 406.3.

552.54 Luminaires (Lighting Fixtures).

(A) General. Any combustible wall or ceiling finish exposed between the edge of a luminaire (fixture) canopy or pan and the outlet box shall be covered with noncombustible material or a material identified for the purpose.

(B) Shower Luminaires (Fixtures). If a luminaire (lighting fixture) is provided over a bathtub or in a shower stall, it shall be of the enclosed and gasketed type and listed for the type of installation, and it shall be ground-fault circuit-interrupter protected.

Many shower luminaires have a metal base and, due to low ceilings in recreational vehicles, may be easily reached by most persons under a shower or standing in a bathtub. In accordance with 552.54(B), the installation of luminaires that are listed for wet locations and have ground-fault circuit-interrupter protection is permitted above a tub or shower enclosure.

The switch for shower luminaires (lighting fixtures) and exhaust fans, located over a tub or in a shower stall, shall be located outside the tub or shower space.

(C) Outdoor Outlets, Luminaires (Fixtures), Air-Cooling Equipment, and So On. Outdoor luminaires (fixtures) and other equipment shall be listed for outdoor use.

552.55 Grounding.

(See also 552.57 on bonding of non–current-carrying metal parts.)

(A) Power-Supply Grounding. The grounding conductor in the supply cord or feeder shall be connected to the grounding bus or other approved grounding means in the distribution panelboard.

(B) Distribution Panelboard. The distribution panelboard shall have a grounding bus with sufficient terminals for all grounding conductors or other approved grounding means.

(C) Insulated Neutral. The grounded circuit conductor (neutral) shall be insulated from the equipment grounding conductors and from equipment enclosures and other grounded parts. The grounded (neutral) circuit terminals in the distribution panelboard and in ranges, clothes dryers, counter-mounted cooking units, and wall-mounted ovens shall be insulated from the equipment enclosure. Bonding screws, straps, or buses in the distribution panelboard or in appliances shall be removed and discarded. Connection of electric ranges and electric clothes dryers utilizing a grounded (neutral) conductor, if cord-connected, shall be made with 4-conductor cord and 3-pole, 4-wire, grounding-type plug caps and receptacles.

552.56 Interior Equipment Grounding.

(A) Exposed Metal Parts. In the electrical system, all exposed metal parts, enclosures, frames, luminaire (lighting fixture) canopies, and so forth, shall be effectively bonded to the grounding terminals or enclosure of the distribution panelboard.

(B) Equipment Grounding Conductors. Bare wires, green-colored wires, or green wires with a yellow stripe(s) shall be used for equipment grounding conductors only.

(C) Grounding of Electrical Equipment. Where grounding of electrical equipment is specified, it shall be permitted as follows:

(1) Connection of metal raceway (conduit or electrical metallic tubing), the sheath of Type MC and Type MI cable where the sheath is identified for grounding, or the armor of Type AC cable to metal enclosures.
(2) A connection between the one or more equipment grounding conductors and a metal box by means of a grounding screw, which shall be used for no other purpose, or a listed grounding device.
(3) The equipment grounding conductor in nonmetallic-sheathed cable shall be permitted to be secured under a screw threaded into the luminaire (fixture) canopy other than a mounting screw or cover screw or attached to a listed grounding means (plate) in a nonmetallic outlet box for luminaire (fixture) mounting [grounding means shall also be permitted for luminaire (fixture) attachment screws].

(D) Grounding Connection in Nonmetallic Box. A connection between the one or more grounding conductors brought into a nonmetallic outlet box shall be arranged so that a connection can be made to any fitting or device in that box that requires grounding.

(E) Grounding Continuity. Where more than one equipment grounding conductor of a branch circuit enters a box, all such conductors shall be in good electrical contact with each other, and the arrangement shall be such that the disconnection or removal of a receptacle, fixture, including a luminaire, or other device fed from the box will not interfere with or interrupt the grounding continuity.

(F) Cord-Connected Appliances. Cord-connected appliances, such as washing machines, clothes dryers, refrigerators, and the electrical system of gas ranges, and so on, shall be grounded by means of an approved cord with equipment grounding conductor and grounding-type attachment plug.

552.57 Bonding of Non–Current-Carrying Metal Parts.

(A) Required Bonding. All exposed non–current-carrying metal parts that may become energized shall be effectively bonded to the grounding terminal or enclosure of the distribution panelboard.

(B) Bonding Chassis. A bonding conductor shall be connected between any distribution panelboard and an accessible terminal on the chassis. Aluminum or copper-clad aluminum conductors shall not be used for bonding if such conductors or their terminals are exposed to corrosive elements.

Exception: Any park trailer that employs a unitized metal chassis-frame construction to which the distribution panelboard is securely fastened with a bolt(s) and nut(s) or by welding or riveting shall be considered to be bonded.

(C) Bonding Conductor Requirements. Grounding terminals shall be of the solderless type and listed as pressure terminal connectors recognized for the wire size used. The bonding conductor shall be solid or stranded, insulated or bare, and shall be 8 AWG copper minimum or equivalent.

(D) Metallic Roof and Exterior Bonding. The metal roof and exterior covering shall be considered bonded where

(1) The metal panels overlap one another and are securely attached to the wood or metal frame parts by metal fasteners, and

(2) The lower panel of the metal exterior covering is secured by metal fasteners at each cross member of the chassis, or the lower panel is bonded to the chassis by a metal strap.

(E) Gas, Water, and Waste Pipe Bonding. The gas, water, and waste pipes shall be considered grounded if they are bonded to the chassis.

(F) Furnace and Metal Air Duct Bonding. Furnace and metal circulating air ducts shall be bonded.

552.58 Appliance Accessibility and Fastening.

Every appliance shall be accessible for inspection, service, repair, and replacement without removal of permanent construction. Means shall be provided to securely fasten appliances in place when the park trailer is in transit.

552.59 Outdoor Outlets, Fixtures, Including Luminaires, Air-Cooling Equipment, and So On.

(A) Listed for Outdoor Use. Outdoor fixtures, including luminaires, and equipment shall be listed for outdoor use. Outdoor receptacle or convenience outlets shall be of a gasketed-cover type for use in wet locations.

(B) Outside Heating Equipment, Air-Conditioning Equipment, or Both. A park trailer provided with a branch circuit designed to energize outside heating equipment or air-conditioning equipment, or both, located outside the park trailer, other than room air conditioners, shall have such branch-circuit conductors terminate in a listed outlet box or disconnecting means located on the outside of the park trailer. A label shall be permanently affixed adjacent to the outlet box and shall contain the following information:

> THIS CONNECTION IS FOR HEATING AND/OR AIR-CONDITIONING EQUIPMENT. THE BRANCH CIRCUIT IS RATED AT NOT MORE THAN _____ AMPERES, AT _____ VOLTS, 60-Hz, _____ CONDUCTOR AMPACITY. A DISCONNECTING MEANS SHALL BE LOCATED WITHIN SIGHT OF THE EQUIPMENT.

The correct voltage and ampere rating shall be given. The tag shall not be less than 0.51 mm (0.020 in.) thick etched brass, stainless steel, anodized or alclad aluminum, or equivalent. The tag shall not be less than 75 mm × 45 mm (3 in. × 1¾ in.) minimum size.

V. Factory Tests

552.60 Factory Tests (Electrical).

Each park trailer shall be subjected to the following tests.

(A) Circuits of 120 Volts or 120/240 Volts. Each park trailer designed with a 120-volt or a 120/240-volt electrical system shall withstand the applied potential without electrical breakdown of a 1-minute, 900-volt dielectric strength test, or a 1-second, 1080-volt dielectric strength test, with all switches closed, between ungrounded and grounded conductors and the park trailer ground. During the test, all switches and other controls shall be in the on position. Fixtures, including luminaires, and permanently installed appliances shall not be required to withstand this test.

Each park trailer shall be subjected to the following:

(1) A continuity test to ensure that all metal parts are properly bonded
(2) Operational tests to demonstrate that all equipment is properly connected and in working order
(3) Polarity checks to determine that connections have been properly made
(4) Receptacles requiring GFCI protection shall be tested for correct function by the use of a GFCI testing device

(B) Low-Voltage Circuits. Low-voltage circuit conductors in each park trailer shall withstand the applied potential without electrical breakdown of a 1-minute, 500-volt or a 1-second, 600-volt dielectric strength test. The potential shall be applied between ungrounded and grounded conductors.

The test shall be permitted on running light circuits before the lights are installed, provided the unit's outer covering and interior cabinetry have been secured. The braking circuit shall be permitted to be tested before being connected to the brakes, provided the wiring has been completely secured.

ARTICLE 553
Floating Buildings

Contents

I. General

553.1 Scope.

This article covers wiring, services, feeders, and grounding for floating buildings.

553.2 Definition.

Floating Building. A building unit as defined in Article 100 that floats on water, is moored in a permanent location, and has a premises wiring system served through connection by permanent wiring to an electricity supply system not located on the premises.

553.3 Application of Other Articles.

Wiring for floating buildings shall comply with the applicable provisions of other articles of this *Code*, except as modified by this article.

II. Services and Feeders

553.4 Location of Service Equipment.

The service equipment for a floating building shall be located adjacent to, but not in or on, the building.

The intent of 553.4 is to ensure that supply conductors to a floating building can be disconnected in an emergency, such as during a storm, when the floating dwelling unit has to be moved quickly.

Overcurrent protection for supply conductors is provided, since these conductors may develop excessive leakage where located underwater or where a short circuit or ground-fault occurs.

553.5 Service Conductors.

One set of service conductors shall be permitted to serve more than one set of service equipment.

553.6 Feeder Conductors.

Each floating building shall be supplied by a single set of feeder conductors from its service equipment.

Exception: Where the floating building has multiple occupancy, each occupant shall be permitted to be supplied by a single set of feeder conductors extended from the occupant's service equipment to the occupant's panelboard.

553.7 Installation of Services and Feeders.

(A) Flexibility. Flexibility of the wiring system shall be maintained between floating buildings and the supply conductors. All wiring shall be installed so that motion of the water surface and changes in the water level will not result in unsafe conditions.

(B) Wiring Methods. Liquidtight flexible metal conduit or liquidtight flexible nonmetallic conduit with approved fittings shall be permitted for feeders and where flexible connections are required for services. Extra-hard usage portable power cable listed for both wet locations and sunlight resistance shall be permitted for a feeder to a floating building where flexibility is required. Other raceways suitable for the location shall be permitted to be installed where flexibility is not required.

FPN: See 555.1 and 555.13.

Type W cables included in Table 400.4 may or may not be listed for wet locations. Liquidtight flexible nonmetallic conduit with approved fittings is permitted where flexible connections are required.

III. Grounding

553.8 General Requirements.

Grounding of both electrical and nonelectrical parts in a floating building shall be through connection to a grounding bus in the building panelboard. The grounding bus shall be grounded through a green-colored insulated equipment grounding conductor run with the feeder conductors and connected to a grounding terminal in the service equipment. The grounding terminal in the service equipment shall be grounded by connection through an insulated grounding electrode conductor to a grounding electrode on shore.

553.9 Insulated Neutral.

The grounded circuit conductor (neutral) shall be an insulated conductor identified in conformance with 200.6. The neutral conductor shall be connected to the equipment grounding terminal in the service equipment, and, except for that connection, it shall be insulated from the equipment grounding conductors, equipment enclosures, and all other grounded parts. The neutral circuit terminals in the panelboard and in ranges, clothes dryers, counter-mounted cooking units, and the like shall be insulated from the enclosures.

553.10 Equipment Grounding.

(A) Electrical Systems. All enclosures and exposed metal parts of electrical systems shall be bonded to the grounding bus.

(B) Cord-Connected Appliances. Where required to be grounded, cord-connected appliances shall be grounded by means of an equipment grounding conductor in the cord and a grounding-type attachment plug.

553.11 Bonding of Non–Current-Carrying Metal Parts.

All metal parts in contact with the water, all metal piping, and all non–current-carrying metal parts that may become energized shall be bonded to the grounding bus in the panelboard.

ARTICLE 555
Marinas and Boatyards

Contents

555.1 Scope.

This article covers the installation of wiring and equipment in the areas comprising fixed or floating piers, wharfs, docks, and other areas in marinas, boatyards, boat basins, boathouses, yacht clubs, boat condominiums, docking facilities associated with residential condominiums, any multiple docking facility, or similar occupancies, and facilities that are used, or intended for use, for the purpose of repair, berthing, launching, storage, or fueling of small craft and the moorage of floating buildings.

Private, noncommercial docking facilities constructed or occupied for the use of the owner or residents of the associated single-family dwelling are not covered by this article.

> FPN: See NFPA 303-2000, *Fire Protection Standard for Marinas and Boatyards*, for additional information.

The requirements of Article 555 apply to public and private docking, storage, repair, and fueling facilities for small craft. The term *small craft* is not defined, however, based on the scope of NFPA 303, *Fire Protection Standard for Marinas and Boatyards*, the term includes recreational and commercial boats, yachts, and other craft that do not exceed 300 gross tons. For facilities that serve larger craft and ships, see NFPA 307, *Standard for the Construction and Fire Protection of Marine Terminals, Piers, and Wharves*. See Article 553 for requirements for floating buildings, including floating dwelling units.

The requirements of Article 555 apply to stand-alone boathouses except for those constructed and used in association with a single-family dwelling. See 210.8(A)(8) for the GFCI protection requirement of 125-volt, single-phase, 15- and 20- ampere receptacles installed in or on boathouses located at a single-family dwelling. Electrical installations on docks and piers located at a single-family dwelling are not subject to the requirements of Article 555, however, all applicable requirements in Chapters 1 through 4 for these outdoor, wet locations, including the GFCI requirements of 210.8(A)(3) for outdoor receptacles, are applicable.

555.2 Definitions.

Electrical Datum Plane. The electrical datum plane is defined as follows:

Throughout Article 555, the physical location of electrical equipment is referenced to the electrical datum plane. This term is used as a horizontal benchmark on land and on floating piers. The definition of *electrical datum plane* encompasses areas subject to tidal movement and areas in which the water level is impacted only by conditions such as climate (rain or snow fall) or by human intervention (the opening or closing of dams and floodgates). In either case, the term covers the normal highest water level, such as astronomical high tides. The term does not cover extremes due to natural or manmade disasters.

(1) In land areas subject to tidal fluctuation, the electrical datum plane is a horizontal plane 606 mm (2 ft) above the highest tide level for the area occurring under normal circumstances, that is, highest high tide.

(2) In land areas not subject to tidal fluctuation, the electrical datum plane is a horizontal plane 606 mm (2 ft)

above the highest water level for the area occurring under normal circumstances.

(3) The electrical datum plane for floating piers and landing stages that are (a) installed to permit rise and fall response to water level, without lateral movement, and (b) that are so equipped that they can rise to the datum plane established for (1) or (2), is a horizontal plane 762 mm (30 in.) above the water level at the floating pier or landing stage and a minimum of 305 mm (12 in.) above the level of the deck.

Marine Power Outlet. An enclosed assembly that can include receptacles, circuit breakers, fused switches, fuses, watt-hour meter(s), and monitoring means approved for marine use.

555.4 Distribution System.

Yard and pier distribution systems shall not exceed 600 volts phase to phase.

555.5 Transformers.

Transformers and enclosures shall be specifically approved for the intended location. The bottom of enclosures for transformers shall not be located below the electrical datum plane.

555.7 Location of Service Equipment.

The service equipment for floating docks or marinas shall be located adjacent to, but not on or in, the floating structure.

The requirement covering service equipment location in 555.7 is similar to that in 553.4 for service equipment supplying floating buildings.

555.9 Electrical Connections.

All electrical connections shall be located at least 305 mm (12 in.) above the deck of a floating pier. All electrical connections shall be located at least 305 mm (12 in.) above the deck of a fixed pier but not below the electrical datum plane.

555.10 Electrical Equipment Enclosures.

(A) Securing and Supporting. Electrical equipment enclosures installed on piers above deck level shall be securely and substantially supported by structural members, independent of any conduit connected to them. If enclosures are not attached to mounting surfaces by means of external ears or lugs, the internal screw heads shall be sealed to prevent seepage of water through mounting holes.

(B) Location. Electrical equipment enclosures on piers shall be located so as not to interfere with mooring lines.

555.11 Circuit Breakers, Switches, Panelboards, and Marine Power Outlets.

Circuit breakers and switches installed in gasketed enclosures shall be arranged to permit required manual operation without exposing the interior of the enclosure. All such enclosures shall be arranged with a weep hole to discharge condensation.

555.12 Load Calculations for Service and Feeder Conductors.

General lighting and other loads shall be calculated in accordance with Article 220, and, in addition, the load for each service and/or feeder circuit supplying receptacles that provide shore power for boats shall be calculated using the demand factors shown in Table 555.12. These calculations shall be permitted to be modified as indicated in notes (1) and (2).

Table 555.12 Demand Factors

Number of Receptacles	Sum of the Rating of the Receptacles (percent)
1–4	100
5–8	90
9–14	80
15–30	70
31–40	60
41–50	50
51–70	40
71-plus	30

Notes:
1. Where shore power accommodations provide two receptacles specifically for an individual boat slip and these receptacles have different voltages (for example, one 30 ampere, 125 volt and one 50 ampere, 125/250 volt), only the receptacle with the larger kilowatt demand shall be required to be calculated.
2. If the facility being installed includes individual kilowatt-hour submeters for each slip and is being calculated using the criteria listed in Table 555.12, the total demand amperes may be multiplied by 0.9 to achieve the final demand amperes.

> FPN: These demand factors may be inadequate in areas of extreme hot or cold temperatures with loaded circuits for heating, air-conditioning, or refrigerating equipment.

555.13 Wiring Methods and Installation.

(A) Wiring Methods.

(1) General. Wiring methods of Chapter 3 shall be permitted where identified for use in wet locations.

(2) Portable Power Cables. Extra-hard usage portable power cables rated not less than 167°F (75°C), 600 volts, listed for both wet locations and sunlight resistance, having an outer jacket rated to be resistant to temperature extremes, oil, gasoline, ozone, abrasion, acids, and chemicals shall be permitted as follows:

(1) As permanent wiring on the underside of piers (floating or fixed)
(2) Where flexibility is necessary as on piers composed of floating sections

Table 400.4 identifies Types G, PPE, and W as portable power cables suitable for extra-hard usage. The requirements in 555.13(A)(2) for cable construction are necessary due to the cable's exposure to extremes in weather conditions and to operational hazards such as oil and gasoline spills. The use of these cables on floating piers and docks provides the necessary degree of flexibility to compensate for tidal and wave action.

(3) Temporary Wiring. Temporary wiring, except as permitted by Article 527, shall not be used to supply power to boats.

(B) Installation.

(1) Overhead Wiring. Overhead wiring shall be installed to avoid possible contact with masts and other parts of boats being moved in the yard.

Conductors and cables shall be routed to avoid wiring closer than 6.0 m (20 ft) from the outer edge or any portion of the yard that can be used for moving vessels or stepping or unstepping masts.

(2) Outside Branch Circuits and Feeders. Outside branch circuits and feeders shall comply with Article 225 except that clearances for overhead wiring in portions of the yard other than those described in 555.13(B)(1) shall not be less than 5.49 m (18 ft) above grade.

(3) Wiring Over and Under Navigable Water. Wiring over and under navigable water shall be subject to approval by the authority having jurisdiction.

> FPN: See NFPA 303-2000, *Fire Protection Standard for Marinas and Boatyards*, for warning sign requirements.

Some federal and local agencies, such as the Army Corps of Engineers, Coast Guard, or local harbormasters, have specific authority over navigable waterways. Therefore, approval of any proposed installation over or under such a waterway should be obtained from the appropriate authority.

(4) Portable Power Cables.

(a) Where portable power cables are permitted by 555.13(A)(2), the installation shall comply with the following:

(1) Cables shall be properly supported.
(2) Cables shall be located on the underside of the pier.
(3) Cables shall be securely fastened by nonmetallic clips to structural members other than the deck planking.

(4) Cables shall not be installed where subject to physical damage.

(5) Where cables pass through structural members, they shall be protected against chafing by a permanently installed oversized sleeve of nonmetallic material.

(b) Where portable power cables are used as permitted in 555.13(A)(2)(2), there shall be an approved junction box of corrosion-resistant construction with permanently installed terminal blocks on each pier section to which the feeder and feeder extensions are to be connected. Metal junction boxes and their covers, and metal screws and parts that are exposed externally to the boxes, shall be of corrosion-resistant materials or protected by material resistant to corrosion.

(5) Protection. Rigid metal or nonmetallic conduit suitable for the location shall be installed to protect wiring above decks of piers and landing stages and below the enclosure that it serves. The conduit shall be connected to the enclosure by full standard threads. The use of special fittings of nonmetallic material to provide a threaded connection into enclosures on rigid nonmetallic conduit, employing joint design as recommended by the conduit manufacturer for attachment of the fitting to the conduit shall be acceptable, provided the equipment and method of attachment are approved and the assembly meets the requirements of installation in damp or wet locations as applicable.

555.15 Grounding.

Wiring and equipment within the scope of this article shall be grounded as specified in Article 250 and with the following additional requirements.

(A) Equipment to Be Grounded. The following items shall be connected to an equipment grounding conductor run with the circuit conductors in the same raceway, cable, or trench:

(1) Metal boxes, metal cabinets, and all other metal enclosures

(2) Metal frames of utilization equipment

(3) Grounding terminals of grounding-type receptacles

(B) Type of Equipment Grounding Conductor. The equipment grounding conductor shall be an insulated copper conductor with a continuous outer finish that is either green or green with one or more yellow stripes. The equipment grounding conductor of Type MI cable shall be permitted to be identified at terminations. For conductors larger than 6 AWG, or where multiconductor cables are used, re-identification of conductors as allowed in 250.119(A)(2) and (A)(3) or 250.119(B)(2) and (B)(3) shall be permitted.

The provisions of 555.15(B) require an insulated equipment grounding conductor that ensures a high integrity path for ground-fault current. Because of corrosive conditions present in marinas and boatyards, metal raceways are not permitted to serve as the sole equipment grounding conductor.

(C) Size of Equipment Grounding Conductor. The insulated copper equipment grounding conductor shall be sized in accordance with 250.122 but not smaller than 12 AWG.

(D) Branch-Circuit Equipment Grounding Conductor. The insulated equipment grounding conductor for branch circuits shall terminate at a grounding terminal in a remote panelboard or the grounding terminal in the main service equipment.

(E) Feeder Equipment Grounding Conductors. Where a feeder supplies a remote panelboard, an insulated equipment grounding conductor shall extend from a grounding terminal in the service equipment to a grounding terminal in the remote panelboard.

555.17 Disconnecting Means for Shore Power Connection(s).

Disconnecting means shall be provided to isolate each boat from its supply connection(s).

(A) Type. The disconnecting means shall be permitted to consist of a circuit breaker, switch, or both, and shall be properly identified as to which receptacle it controls.

(B) Location. The disconnecting means shall be readily accessible, located not more than 762 mm (30 in.) from the receptacle it controls, and shall be located in the supply circuit ahead of the receptacle. Circuit breakers or switches located in marine power outlets complying with this section shall be permitted as the disconnecting means.

555.19 Receptacles.

Receptacles shall be mounted not less than 305 mm (12 in.) above the deck surface of the pier and not below the electrical datum plane on a fixed pier.

In accordance with 555.19, the location of enclosures for receptacles on fixed and floating piers is based on the electrical datum plane defined in 555.2. For floating piers, the datum plane is 12 in. above the deck of the pier. The purpose of this requirement is to prevent submersion of receptacle enclosures.

The requirements for enclosures in 555.19(A)(1) address their exposure to the severe weather (wind-driven rain) and environmental conditions (splashing from breaking waves or wakes) frequently encountered at marine locations.

(A) Shore Power Receptacles.

(1) Enclosures. Receptacles intended to supply shore power to boats shall be housed in marine power outlets listed as marina power outlets or listed for set locations, or shall be installed in listed enclosures protected from the weather or in listed weatherproof enclosures. The integrity of the assembly shall not be affected when the receptacles are in use with any type of booted or nonbooted attachment plug/cap inserted.

(2) Strain Relief. Means shall be provided where necessary to reduce the strain on the plug and receptacle caused by the weight and catenary angle of the shore power cord.

(3) Branch Circuits. Each single receptacle that supplies shore power to boats shall be supplied from a marine power outlet or panelboard by an individual branch circuit of the voltage class and rating corresponding to the rating of the receptacle.

> FPN: Supplying receptacles at voltages other than the voltages marked on the receptacle may cause overheating or malfunctioning of connected equipment, for example, supplying single-phase, 120/240-volt, 3-wire loads from a 208Y/120-volt, 3-wire source.

Each single receptacle that supplies shore power to boats must be supplied from an individual branch circuit. The requirement for shore power receptacles to be supplied by individual branch circuits can be met through the use of multiwire branch circuits derived from single-phase, 3-wire systems, or from 3-phase, 4-wire systems. Although the ungrounded conductors of a multiwire branch circuit share the same grounded (neutral) conductor, this configuration can be considered multiple branch circuits in accordance with 210.4(A). See the commentary following 300.13(B) regarding device removal for multiwire branch circuits.

Locking- and grounding-type receptacles and attachment caps must ensure proper connections to prevent unintentional disconnection of on-board equipment, such as bilge pumps, refrigerators, and so on.

(4) Ratings. Receptacles that provide shore power for boats shall be rated not less than 30 amperes and shall be single outlet type.

> FPN: For locking- and grounding-type receptacles for auxiliary power to boats, see NFPA 303-2000, *Fire Protection Standard for Marinas and Boatyards.*

(a) Receptacles rated not less than 30 amperes or more than 50 amperes shall be of the locking and grounding type.

> FPN: For various configurations and ratings of locking and grounding-type receptacles and caps, see ANSI/

NEMA 18WD 6-1989, National Electrical Manufacturers Association's *Standard for Dimensions of Attachment Plugs and Receptacles.*

(b) Receptacles rated for 60 amperes or 100 amperes shall be of the pin and sleeve type.

> FPN: For various configurations and ratings of pin and sleeve receptacles, see ANSI/UL 1686, *UL Standard for Safety Pin and Sleeve Configurations.*

Single locking- and grounding-type receptacles are required for providing shore power to boats. Exhibit 555.1 and Exhibit 406.3 together illustrate a complete chart of grounding-type locking plug and receptacle configurations. Exhibit 555.2 shows a pin-and sleeve-type receptacle configurations.

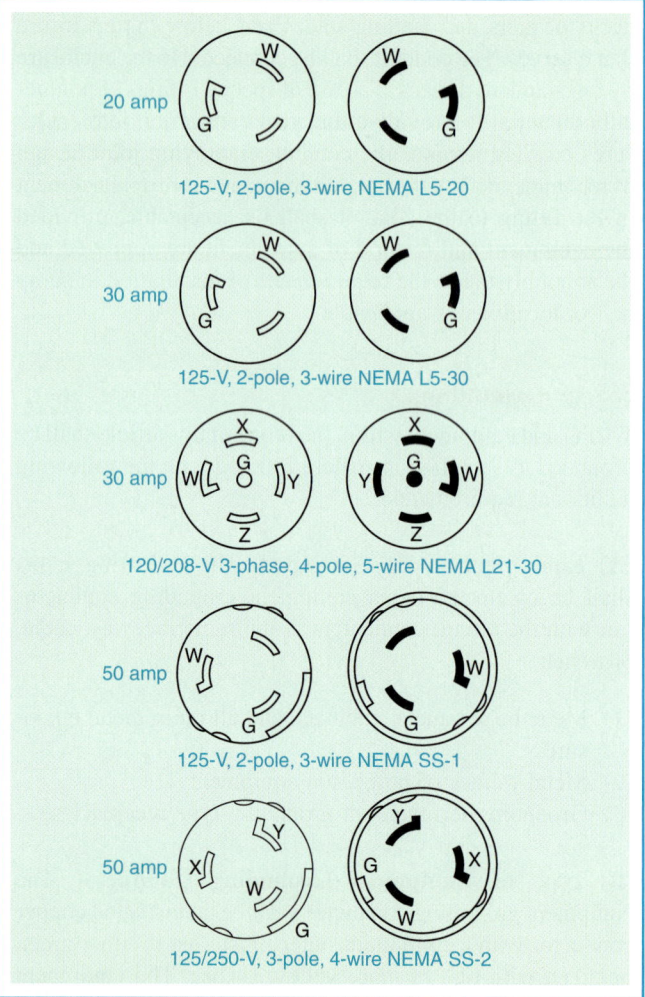

20 amp
125-V, 2-pole, 3-wire NEMA L5-20

30 amp
125-V, 2-pole, 3-wire NEMA L5-30

30 amp
120/208-V 3-phase, 4-pole, 5-wire NEMA L21-30

50 amp
125-V, 2-pole, 3-wire NEMA SS-1

50 amp
125/250-V, 3-pole, 4-wire NEMA SS-2

Exhibit 555.1 Typical configurations for single locking- and grounding-type receptacles and attachment plug caps used to provide shore power for boats in marinas and boatyards. These configurations are 30 amperes to 50 amperes.

Exhibit 555.2 Typical configurations for safety pin-and-sleeve-type receptacles, plugs, connectors, and power inlets used to provide shore power for boats in marinas and boatyards. These configurations are 60 amperes or 100 amperes.

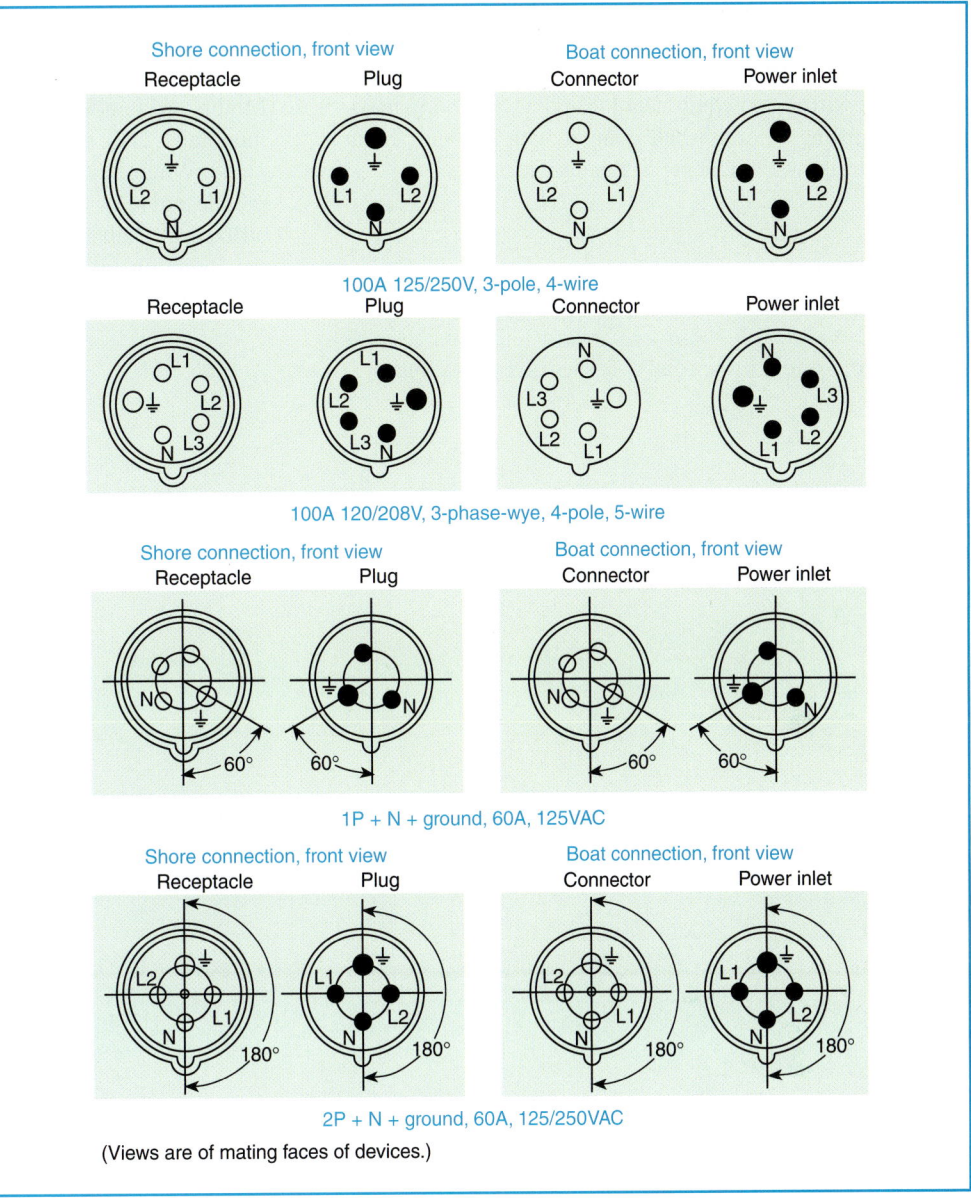

(Views are of mating faces of devices.)

(B) Other Than Shore Power.

(1) Ground-Fault Circuit-Interrupter (GFCI) Protection for Personnel. Fifteen- and 20-ampere, single-phase, 125-volt receptacles installed outdoors, in boathouses, in buildings used for storage, maintenance, or repair where portable electrical hand tools, electrical diagnostic equipment, or portable lighting equipment are to be used shall be provided with GFCI protection for personnel. Receptacles in other locations shall be protected in accordance with 210.8(B).

Fifteen- and 20-ampere, single-phase, 125-volt receptacles, other than those supplying shore power to boats and used for maintenance or other purposes at piers, wharves, and so on, may be of the general-purpose, nonlocking type and must be protected by ground-fault circuit interrupters. See Exhibit 210.8.

(2) Marking. Receptacles other than those supplying shore power to boats shall be permitted to be housed in marine power outlets with the receptacles that provide shore power to boats, provided they are marked to clearly indicate that they are not to be used to supply power to boats.

555.21 Gasoline Dispensing Stations— Hazardous (Classified) Locations.

Electrical wiring and equipment located at or serving gasoline dispensing stations shall comply with Article 514 in addition to the requirements of this article.

In addition to the requirements in Article 514, Section 6-3 in NFPA 303, *Fire Protection Standard for Marinas and Boatyards,* and NFPA 30A, *Code for Motor Fuel Dispensing Facilities and Repair Garages,* contain requirements pertaining to gasoline dispensing facilities and operations.

555.23 Marine Hoists, Railways, Cranes, and Monorails.

Motors and controls for marine hoists, railways, cranes, and monorails shall not be located below the electrical datum plane. Where it is necessary to provide electric power to a mobile crane or hoist in the yard, and a trailing cable is utilized, it shall be a listed portable power cable rated for the conditions of use and be provided with an outer jacket of distinctive color for safety.

ARTICLE 600
Electric Signs and Outline Lighting

Contents

I. General

600.1 Scope.

This article covers the installation of conductors and equipment for electric signs and outline lighting as defined in Article 100.

> FPN: As defined in Article 100, electric signs and outline lighting include all products and installations utilizing neon tubing, such as signs, decorative elements, skeleton tubing, or art forms.

A common misunderstanding arises regarding which requirements apply to signs, those in Article 410 or those in Article 600. Section 90.3 specifies that the general requirements in Chapters 1 through 4 be applied to the electrical installations covered in Chapters 5, 6, and 7. Requirements that are unique to specific occupancies, equipment, or conditions and modify the requirements of Chapters 1 through 4 are found in Chapters 5, 6, or 7. Certain requirements in Article 410 that apply to the types of equipment and supply systems used in conjunction with electric sign installations must be followed unless they are modified or amended by Article 600. An example of Article 600 modifying Article 410 is seen in 600.23(B). This requirement covers transformer and electronic power supply secondary-circuit ground-fault protection. Such protection is required only for systems that supply equipment covered in Article 600.

The requirements in Article 600 apply to electric signs of the fixed, stationary, or portable self-contained type. They may have letters or symbols that provide illumination, or they may be illuminated from a source other than a letter or symbol.

by special permission (i.e., the written consent of the authority having jurisdiction). A listed sign consists of the transformer, channel letter, plastic face, glass tube supports, raceways, glass cups or insulating boots if provided, and disconnecting means when provided by the manufacturer. A listed sign section usually consists of a channel letter, plastic face, glass tube supports, and metal-enclosed electrode receptacles. Sign sections or entire signs are listed for installation indoors or outdoors.

Skeleton tubing that is not listed is permitted by 600.3(A) to be field installed, provided that the installation is made in accordance with the applicable requirements of the *Code*. In 600.3(B), the installation of nonlisted outline lighting that is made up of listed lighting fixtures is permitted, provided that it is installed in accordance with Chapter 3.

Large signs are often transported in several parts from the factory to the location where they are assembled. It is at this time that an inspection authority should be present to ensure that the components are assembled in conformity with their listing.

600.2 Definitions.

Electric-Discharge Lighting. Systems of illumination utilizing fluorescent lamps, high-intensity discharge (HID) lamps, or neon tubing.

Neon Tubing. Electric-discharge tubing manufactured into shapes that form letters, parts of letters, skeleton tubing, outline lighting, other decorative elements, or art forms, and filled with various inert gases.

Sign Body. A portion of a sign that may provide protection from the weather but is not an electrical enclosure.

Skeleton Tubing. Neon tubing that is itself the sign or outline lighting and not attached to an enclosure or sign body.

600.3 Listing.

Electric signs and outline lighting — fixed, mobile, or portable — shall be listed and installed in conformance with that listing, unless otherwise approved by special permission.

(A) Field Installed Skeleton Tubing. Field installed skeleton tubing shall not be required to be listed where installed in conformance with this *Code*.

(B) Outline Lighting. Outline lighting shall not be required to be listed as a system when it consists of listed luminaires (lighting fixtures) wired in accordance with Chapter 3.

In 600.3 there are requirements specifying that any electric sign, including outline lighting, be listed unless it is approved

600.4 Markings.

(A) Signs and Outline Lighting Systems. Signs and outline lighting systems shall be marked with the manufacturer's name, trademark, or other means of identification; and input voltage and current rating.

(B) With Incandescent Lamp Holders. Signs and outline lighting systems with incandescent lamp holders shall be marked to indicate the maximum allowable wattage of lamps. The markings shall be permanently installed, in letters at least 6 mm (¼ in.) high, and shall be located where visible during relamping.

600.5 Branch Circuits.

(A) Required Branch Circuit. Each commercial building and each commercial occupancy accessible to pedestrians shall be provided with at least one outlet in an accessible location at each entrance to each tenant space for sign or outline lighting system use. The outlet(s) shall be supplied by a branch circuit rated at least 20 amperes that supplies no other load. Service hallways or corridors shall not be considered accessible to pedestrians.

The use of electric signs and outline lighting to attract the attention of consumers is used extensively by commercial enterprises. The need for an accessible provision to facilitate the connection of an electric sign at commercial buildings is covered by this requirement. A 20-ampere outlet on a circuit dedicated to the purpose of supplying an electric sign(s) is required to be installed at the entrance to single-occupant commercial buildings and at the entrance to each occupancy of multiple-occupant commercial buildings, for example, shopping malls. This requirement is not contingent on whether an electric sign will be installed at the time an occupant moves into a commercial space, since it is not uncommon for an electric sign to be installed after the space is occupied or when a new occupant moves into an existing space. The minimum load for the sign outlet is covered in 220.3(B)(6).

(B) Rating. Branch circuits that supply signs shall be rated as follows.

(1) Incandescent and Fluorescent. Branch circuits that supply signs and outline lighting systems containing incandescent and fluorescent forms of illumination shall be rated not to exceed 20 amperes.

(2) Neon. Branch circuits that supply neon tubing installations shall not be rated in excess of 30 amperes.

Branch circuits that supply electric signs are limited by 600.5(B)(1) and (B)(2) to ratings of 20 and 30 amperes, respectively. Large signs often have load requirements that exceed the ratings permitted by 600.5(B). It is typical for signs having large electrical loads to be supplied by a feeder that in turn supplies branch circuits operating within the parameters specified by this requirement. In some cases, particularly for signs installed along highways, a utility service dedicated to the sign is provided. The rating of the feeder is not limited by this requirement.

(C) Wiring Methods. Wiring methods used to supply signs shall comply with 600.5(C)(1), (C)(2), and (C)(3).

(1) Supply. The wiring method used to supply signs and outline lighting systems shall terminate within a sign, an outline lighting system enclosure, a suitable box, or a conduit body.

(2) Enclosures as Pull Boxes. Signs and transformer enclosures shall be permitted to be used as pull or junction boxes for conductors supplying other adjacent signs, outline lighting systems, or floodlights that are part of a sign and shall be permitted to contain both branch and secondary circuit conductors.

(3) Metal Poles. Metal poles used to support signs shall be permitted to enclose supply conductors, provided the poles and conductors are installed in accordance with 410.15(B).

600.6 Disconnects.

Each sign and outline lighting system, or feeder circuit or branch circuit supplying a sign or outline lighting system,

shall be controlled by an externally operable switch or circuit breaker that will open all ungrounded conductors. Signs and outline lighting systems located within fountains shall have the disconnect located in accordance with 680.12.

Exception No. 1: A disconnecting means shall not be required for an exit directional sign located within a building.

Exception No. 2: A disconnecting means shall not be required for cord-connected signs with an attachment plug.

(A) Location.

(1) Within Sight of the Sign. The disconnecting means shall be within sight of the sign or outline lighting system that it controls. Where the disconnecting means is out of the line of sight from any section that may be energized, the disconnecting means shall be capable of being locked in the open position.

The requirement of 600.6(A)(1) covers sign installations where the branch circuit or feeder is run directly to the sign. Each branch circuit or feeder supplying a sign must have an externally operable switch or circuit breaker to open the ungrounded conductors. Two options are permitted for locating the sign disconnecting means. The disconnecting means is required either to be located within sight of the sign or to be equipped with the provision to lock it in the open position. Exhibit 600.1 depicts a sign with two supply circuits. These circuits could be feeders or branch circuits. Each circuit is provided with an externally operable switch that is located within sight of the sign. Exhibit 600.2 illustrates three compliant alternatives. The supply circuit discon-

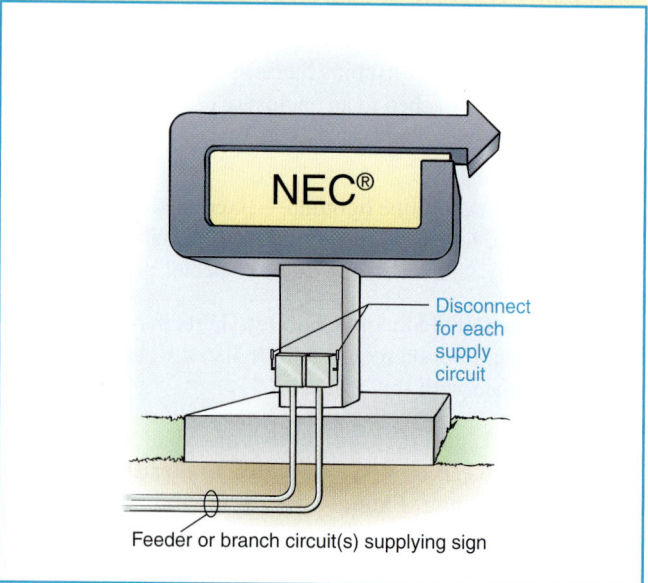

Exhibit 600.1 Supply circuit-disconnecting means located at or on an electric sign.

Exhibit 600.2 Three acceptable methods of providing disconnecting means for electric signs.

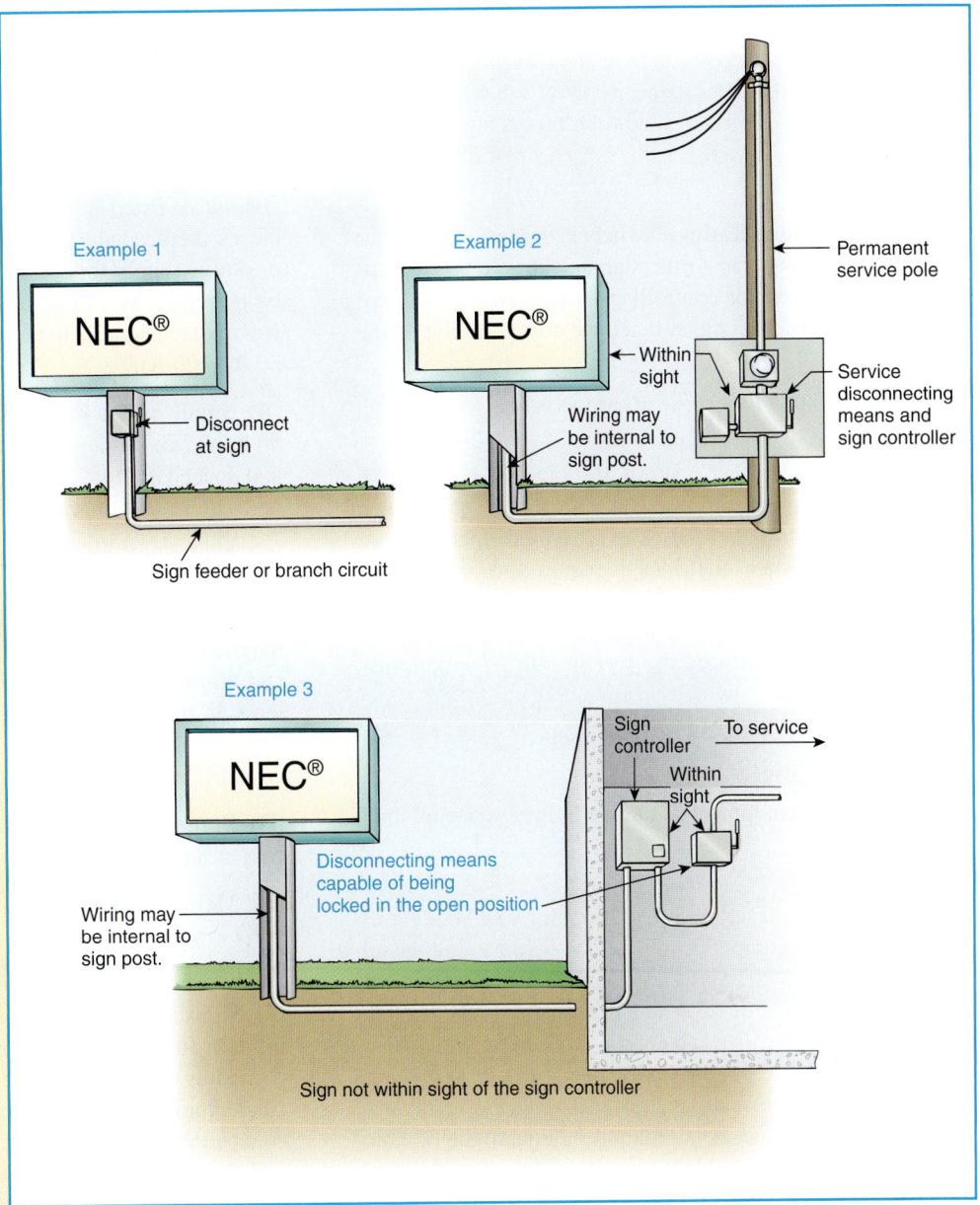

Example 1

NEC®

Disconnect at sign

Sign feeder or branch circuit

Example 2

NEC®

Within sight

Wiring may be internal to sign post.

Permanent service pole

Service disconnecting means and sign controller

Example 3

NEC®

Wiring may be internal to sign post.

Disconnecting means capable of being locked in the open position

Sign controller

To service

Within sight

Sign not within sight of the sign controller

necting means shown in Example 1 is externally operable and located at and within sight of the sign. The disconnecting means in Example 2 is externally operable, and its location, though not at or on the sign, is acceptable because it meets the definition of *within sight*. Where the disconnecting means is not located within sight of the sign, as shown in Example 3, it is required to be located within sight of the controller and must be capable of being locked in the open position.

(2) Within Sight of the Controller. The following shall apply for signs or outline lighting systems operated by electronic or electromechanical controllers located external to the sign or outline lighting system:

(1) The disconnecting means shall be permitted to be located within sight of the controller or in the same enclosure with the controller.
(2) The disconnecting means shall disconnect the sign or outline lighting system and the controller from all ungrounded supply conductors.
(3) The disconnecting means shall be designed so that no pole can be operated independently and shall be capable of being locked in the open position.

For signs or outline lighting systems operated by mechanical or electromechanical controllers located external to the sign, the disconnecting means is required to be located within sight of or in the same enclosure as the controller and must be capable of being locked in the open position. This requirement enhances safe working conditions for persons servicing the controller or the sign.

(B) Control Switch Rating. Switches, flashers, and similar devices controlling transformers and electronic power supplies shall be rated for controlling inductive loads or have a current rating not less than twice the current rating of the transformer.

> FPN: See 404.14 for rating of snap switches.

A switching device that controls the primary circuit of a transformer supplying a luminous gas tube is subject to a highly inductive load that causes severe arcing of its contacts. Therefore, the switch or flasher is required to be rated for the inductive load (a general-use ac snap switch used in accordance with 404.14 is permitted) or must have a current rating that is at least twice the current rating of the transformer it controls.

600.7 Grounding.

Signs and metal equipment of outline lighting systems shall be grounded.

The requirements of 600.7 cover grounding and bonding of signs and metal equipment of outline lighting. All metal parts larger than 2 in. are to be bonded. It is common in the sign industry to use flexible metal raceways to enclose the conductors supplied from the secondary circuit of a transformer or electronic power supply. In addition to providing protection from physical damage, enclosing the secondary conductors in flexible metal conduit or liquidtight flexible metal conduit is permitted as the bonding means for non–current-carrying metal parts of electric signs. Where used for bonding in the secondary circuit, the total length of flexible metal conduit in the secondary circuit is not permitted to exceed 100 ft.

Where rigid or flexible nonmetallic conduit is used on secondary wiring of transformers and power supplies, copper bonding conductors must be located external to the nonmetallic raceway. Bonding conductor spacing is required to be at least 1½ in. for circuit frequencies of 100 Hz or less, and 1¾ in. for circuit frequencies over 100 Hz. These raceways normally contain only one conductor, which is connected to one side of the neon tube. When rigid nonmetallic conduit or liquidtight flexible nonmetallic conduit is used and any sign parts are required to be bonded, the bonding conductor(s) must be run outside of and be separated from the nonmetallic conduit. Installing bonding conductors inside

the nonmetallic conduit with secondary power supply conductors is not permitted, since it could increase the chance of failure of the conductor or nonmetallic tubing.

(A) Flexible Metal Conduit Length. Listed flexible metal conduit or listed liquidtight flexible metal conduit that encloses the secondary circuit conductor from a transformer or power supply for use with electric discharge tubing shall be permitted as a bonding means if the total accumulative length of the conduit in the secondary circuit does not exceed 30 m (100 ft).

(B) Small Metal Parts. Small metal parts not exceeding 50 mm (2 in.) in any dimension, not likely to be energized, and spaced at least 19 mm (¾ in.) from neon tubing shall not require bonding.

(C) Nonmetallic Conduit. Where listed nonmetallic conduit is used to enclose the secondary circuit conductor from a transformer or power supply and a bonding conductor is required, the bonding conductor shall be installed separate and remote from the nonmetallic conduit and be spaced at least 38 mm (1½ in.) from the conduit when the circuit is operated at 100 Hz or less or 45 mm (1¾ in.) when the circuit is operated at over 100 Hz.

(D) Bonding Conductors. Bonding conductors shall be copper and not smaller than 14 AWG.

(E) Metal Building Parts. Metal parts of a building shall not be permitted as a secondary return conductor or an equipment grounding conductor.

(F) Signs in Fountains. Signs or outline lighting installed inside a fountain shall have all metal parts and equipment grounding conductors bonded to the equipment grounding conductor for the fountain recirculating system. The bonding connection shall be as near as practicable to the fountain and shall be permitted to be made to metal piping systems that are bonded in accordance with 680.53.

> FPN: Refer to 600.32(J) for restrictions on length of high-voltage secondary conductors.

600.8 Enclosures.

Live parts other than lamps and neon tubing shall be enclosed.

Exception: A transformer or electronic power supply provided with an integral enclosure, including a primary and secondary circuit splice enclosure, shall not be required to be provided with an additional enclosure.

(A) Strength. Enclosures shall have ample structural strength and rigidity.

(B) Material. Sign and outline lighting system enclosures shall be constructed of metal or shall be listed.

(C) Minimum Thickness of Enclosure Metal. Sheet copper or aluminum shall be at least 0.51 mm (0.020 in.) thick. Sheet steel shall be at least 0.41 mm (0.016 in.) thick.

(D) Protection of Metal. Metal parts of equipment shall be protected from corrosion.

600.9 Location.

(A) Vehicles. Sign or outline lighting system equipment shall be at least 4.3 m (14 ft) above areas accessible to vehicles unless protected from physical damage.

(B) Pedestrians. Neon tubing, other than dry-location portable signs, accessible to pedestrians shall be protected from physical damage.

(C) Adjacent to Combustible Materials. Signs and outline lighting systems shall be installed so that adjacent combustible materials are not subjected to temperatures in excess of 90°C (194°F).

The spacing between wood or other combustible materials and an incandescent or HID lamp or lampholder shall not be less than 50 mm (2 in.).

(D) Wet Location. Signs and outline lighting system equipment for wet location use, other than listed watertight type, shall be weatherproof and have drain holes, as necessary, in accordance with the following:

(1) Drain holes shall not be larger than 13 mm (½ in.) or smaller than 6 mm (¼ in.).
(2) Every low point or isolated section of the equipment shall have at least one drain hole.
(3) Drain holes shall be positioned such that there will be no external obstructions.

600.10 Portable or Mobile Signs.

Portable or mobile electric signs, such as those mounted on trailers, are subject to the requirements of 600.10. These requirements address the safety concerns associated with equipment that is frequently moved and that may be used in damp or wet environments.

(A) Support. Portable or mobile signs shall be adequately supported and readily movable without the use of tools.

(B) Attachment Plug. An attachment plug shall be provided for each portable or mobile sign.

(C) Wet or Damp Location. Portable or mobile signs in wet or damp locations shall comply with 600.10(C)(1) and (C)(2).

(1) Cords. All cords shall be junior hard service or hard service types as designated in Table 400.4 and have an equipment grounding conductor.

(2) Ground-Fault Circuit Interrupter. Portable or mobile signs shall be provided with factory-installed ground-fault circuit-interrupter protection for personnel. The ground-fault circuit interrupter shall be an integral part of the attachment plug or shall be located in the power-supply cord within 300 mm (12 in.) of the attachment plug.

It is intended that a ground-fault circuit interrupter identified for use with portable electric signs be provided with open neutral protection so that the ground-fault circuit interrupter opens the load circuit when the neutral conductor supplying the sign and ground-fault circuit interrupter is interrupted. These protective devices are required to be original equipment installed by the manufacturer and can be located in-line with the supply cord as shown in Exhibit 600.3 or an attachment plug with an integrated GFCI device can be used.

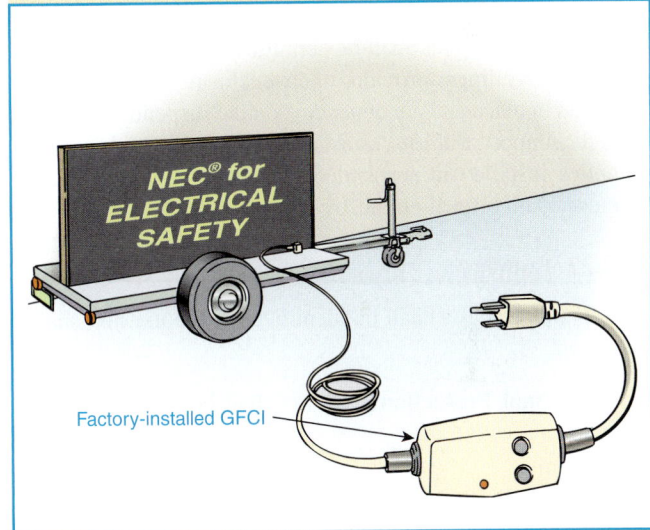

Exhibit 600.3 A factory-installed GFCI device located in the power supply cord within 12 in. of the attachment plug.

(D) Dry Location. Portable or mobile signs in dry locations shall meet the following:

(1) Cords shall be SP-2, SPE-2, SPT-2, or heavier, as designated in Table 400.4.
(2) The cord shall not exceed 4.5 m (15 ft) in length.

600.21 Ballasts, Transformers, and Electronic Power Supplies.

(A) Accessibility. Ballasts, transformers, and electronic power supplies shall be located where accessible and shall be securely fastened in place.

(B) Location. Ballasts, transformers, and electronic power supplies shall be installed as near to the lamps or neon tubing as practicable to keep the secondary conductors as short as possible.

(C) Wet Location. Ballasts, transformers, and electronic power supplies used in wet locations shall be of the weatherproof type or be of the outdoor type and protected from the weather by placement in a sign body or separate enclosure.

(D) Working Space. A working space at least 900 mm (3 ft) high, 900 mm (3 ft) wide, by 900 mm (3 ft) deep shall be provided at each ballast, transformer, and electronic power supply or its enclosure where not installed in a sign.

(E) Attic and Soffit Locations. Ballasts, transformers, and electronic power supplies shall be permitted to be located in attics and soffits, provided there is an access door at least 900 mm by 600 mm (3 ft by 2 ft) and a passageway of at least 900 mm (3 ft) high by 600 mm (2 ft) wide with a suitable permanent walkway at least 300 mm (12 in.) wide extending from the point of entry to each component.

(F) Suspended Ceilings. Ballasts, transformers, and electronic power supplies shall be permitted to be located above suspended ceilings, provided their enclosures are securely fastened in place and not dependent on the suspended ceiling grid for support. Ballasts, transformers, and electronic power supplies installed in suspended ceilings shall not be connected to the branch circuit by flexible cord.

600.22 Ballasts.

(A) Type. Ballasts shall be identified for the use and shall be listed.

(B) Thermal Protection. Ballasts shall be thermally protected.

600.23 Transformers and Electronic Power Supplies.

(A) Type. Transformers and electronic power supplies shall be identified for the use and shall be listed.

(B) Secondary-Circuit Ground-Fault Protection. Transformers and electronic power supplies other than the following shall have secondary-circuit ground-fault protection:

(1) Transformers with isolated ungrounded secondaries and with a maximum open circuit voltage of 7500 volts or less

(2) Transformers with integral porcelain or glass secondary housing for the neon tubing and requiring no field wiring of the secondary circuit

(C) Voltage. Secondary-circuit voltage shall not exceed 15,000 volts, nominal, under any load condition. The voltage

to ground of any output terminals of the secondary circuit shall not exceed 7500 volts, under any load condition.

(D) Rating. Transformers and electronic power supplies shall have a secondary-circuit current rating of not more than 300 mA.

(E) Secondary Connections. Secondary circuit outputs shall not be connected in parallel or in series.

(F) Marking. A transformer or power supply shall be marked to indicate that it has secondary-circuit ground-fault protection.

II. Field-Installed Skeleton Tubing

600.30 Applicability.

Part II of this article shall apply only to field-installed skeleton tubing. These requirements are in addition to the requirements of Part I.

Field-installed skeleton tubing often involves the use of electrically isolated metal components, such as tube supports, fasteners, or decorative channel over the tubing. These metallic components should be located so as to minimize the possibility of their becoming energized.

600.31 Neon Secondary-Circuit Conductors, 1000 Volts or Less, Nominal.

(A) Wiring Method. Conductors shall be installed using any wiring method included in Chapter 3 suitable for the conditions.

(B) Insulation and Size. Conductors shall be insulated, listed for the purpose, and not smaller than 18 AWG.

(C) Number of Conductors in Raceway. The number of conductors in a raceway shall be in accordance with Table 1 of Chapter 9.

(D) Installation. Conductors shall be installed so they are not subject to physical damage.

(E) Protection of Leads. Bushings shall be used to protect wires passing through an opening in metal.

600.32 Neon Secondary Circuit Conductors, Over 1000 Volts, Nominal.

(A) Wiring Methods.

(1) Installation. Conductors shall be installed on insulators, in rigid metal conduit, intermediate metal conduit, rigid nonmetallic conduit, liquidtight flexible nonmetallic conduit, flexible metal conduit, liquidtight flexible metal conduit, electrical metallic tubing, metal enclosures, or other equipment listed for the purpose and shall be installed in accordance with the requirements of Chapter 3.

(2) Number of Conductors. Conduit or tubing shall contain only one conductor.

(3) Size. Conduit or tubing shall be a minimum of metric designator 16 (trade size ½).

(4) Spacing from Ground. Other than at the location of connection to a metal enclosure or sign body, nonmetallic conduit or flexible nonmetallic conduit shall be spaced no less than 38 mm (1½ in.) from grounded or bonded parts when the conduit contains a conductor operating at 100 Hz or less and shall be spaced no less than 45 mm (1¾ in.) from grounded or bonded parts when the conduit contains a conductor operating at more than 100 Hz.

Locating GTO cable in close proximity to a grounded surface will result in damaging stress to the cable insulation. The requirements in this section provide for minimum separation between nonmetallic raceways containing neon secondary circuits and grounded or bonded metal parts.

(5) Metal Building Parts. Metal parts of a building shall not be permitted as a secondary return conductor or an equipment grounding conductor.

See the commentary following 600.7.

(B) Insulation and Size. Conductors shall be insulated, listed as Gas Tube Sign and Ignition Cable Type GTO, rated for 5, 10, or 15 kV, not smaller than 18 AWG, and have a minimum temperature rating of 105°C (221°F).

The 2002 *Code* revised the requirement for the type of conductors permitted in neon secondary circuits. The requirement previously specified that the use of conductors listed for the purpose be installed in the secondary circuit. The revised text requires the use of cable listed as gas tube sign and ignition cable, Type GTO, rated at 5, 10, or 15 kV. Annex A identifies the product standard for this cable as UL 814, *Gas-Tube-Sign and Ignition Cable.*

(C) Installation. Conductors shall be installed so they are not subject to physical damage.

(D) Bends in Conductors. Sharp bends in insulated conductors shall be avoided.

(E) Spacing. Secondary conductors shall be separated from each other and from all objects other than insulators or neon

tubing by a spacing of not less than 38 mm (1½ in.). GTO cable installed in metal conduit or tubing requires no spacing between the cable insulation and the conduit or tubing.

(F) Insulators and Bushings. Insulators and bushings for conductors shall be listed for the purpose.

(G) Conductors in Raceways.

(1) Damp or Wet Locations. In damp or wet locations, the insulation on all conductors shall extend not less than 100 mm (4 in.) beyond the metal conduit or tubing.

(2) Dry Locations. In dry locations, the insulation on all conductors shall extend not less than 65 mm (2½ in.) beyond the metal conduit or tubing.

(H) Between Neon Tubing and Midpoint Return. Conductors shall be permitted to run between the ends of neon tubing or to the secondary circuit midpoint return of transformers or electronic power supplies listed for the purpose and provided with terminals or leads at the midpoint.

(I) Dwelling Occupancies. Equipment having an open circuit voltage exceeding 1000 volts shall not be installed in or on dwelling occupancies.

(J) Length of Secondary Circuit Conductors.

(1) Secondary Conductor to the First Electrode. The length of secondary circuit conductors from a high-voltage terminal or lead of a transformer or electronic power supply to the first neon tube electrode shall not exceed the following:

(1) 6 m (20 ft) where installed in metal conduit or tubing
(2) 15 m (50 ft) where installed in nonmetallic conduit

(2) Other Secondary Circuit Conductors. All other sections of secondary circuit conductor in a neon tube circuit shall be as short as practicable.

600.41 Neon Tubing.

(A) Design. The length and design of the tubing shall not cause a continuous overcurrent beyond the design loading of the transformer or electronic power supply.

(B) Support. Tubing shall be supported by listed tube supports.

(C) Spacing. A spacing of not less than 6 mm (¼ in.) shall be maintained between the tubing and the nearest surface, other than its support.

Electric discharge tubing is required to be of such length and design that it will not cause a continuous overvoltage on the transformer. A tube that is too long or too small in diameter increases the impedance of the load and thus stresses the transformer insulation. Gas tube sign transformers are designed to operate at or near short-circuit current. Generally, the primary voltage of the transformers is 120 volts, and proper installation and maintenance of transformers and high-voltage secondary conductors minimize the possibility of injury or fire. Precautions should be taken to ensure that secondary conductors are properly terminated to the tube electrodes and that these connections are protected from contact by unauthorized persons or contact with any flammable or combustible material. Broken tubes should be replaced or de-energized.

600.42 Electrode Connections.

(A) Accessibility. Terminals of the electrode shall not be accessible to unqualified persons.

(B) Electrode Connections. Connections shall be made by use of a connection device, twisting of the wires together, or use of an electrode receptacle. Connections shall be electrically and mechanically secure and shall be in an enclosure listed for the purpose.

(C) Support. The neon tubing and conductor shall be supported not more than 150 mm (6 in.) from the electrode connection.

(D) Receptacles. Electrode receptacles shall be listed for the purpose.

(E) Bushings. Where electrodes penetrate an enclosure, bushings listed for the purpose shall be used unless receptacles are provided.

(F) Wet Locations. A listed cap shall be used to close the opening between neon tubing and a receptacle where the receptacle penetrates a building. Where a bushing or neon tubing penetrates a building, the opening between neon tubing and the bushing shall be sealed.

(G) Electrode Enclosures. Electrode enclosures shall be listed for the purpose.

ARTICLE 604
Manufactured Wiring Systems

Contents

604.1 Scope.

The provisions of this article apply to field-installed wiring using off-site manufactured subassemblies for branch circuits, remote-control circuits, signaling circuits, and communications circuits in accessible areas.

604.2 Definition.

Manufactured Wiring System. A system containing component parts that are assembled in the process of manufacture and cannot be inspected at the building site without damage or destruction to the assembly.

604.3 Other Articles.

Except as modified by the requirements of this article, all other applicable articles of this *Code* shall apply.

604.4 Uses Permitted.

The manufactured wiring systems shall be permitted in accessible and dry locations and in plenums and spaces used for environmental air, where listed for this application and installed in accordance with 300.22.

Article 604 covers manufactured wiring systems, which are typically comprised of Type AC or Type MC cables and connection devices with molded plugs and receptacles. The connection devices used with these systems greatly reduce installation time and facilitate relocation of equipment such as luminaires. These systems are used extensively for the installation of branch circuit and tap conductors supplying luminaires in open and suspended ceiling construction.

For further information, see Article 100 for the definition of *accessible (as applied to wiring methods).*

Exception No. 1: In concealed spaces, one end of tapped cable shall be permitted to extend into hollow walls for direct termination at switch and outlet points.

Exception No. 2: For use in outdoor locations where listed for the purpose.

Exception No. 2 permits the manufacture and installation of manufactured wiring systems listed for use in outdoor locations.

604.5 Uses Not Permitted.

Manufactured wiring system types shall not be permitted where limited by the applicable article in Chapter 3 for the wiring method used in its construction.

604.6 Construction.

(A) Types.

(1) Cables. Cable shall be listed armored cable or metal-clad cable containing nominal 600-volt 10 or 12 AWG copper-insulated conductors with a bare or insulated copper equipment grounding conductor equivalent in size to the ungrounded conductor.

Other cables as listed in 725.61, 800.50, 820.50, and 830.5 shall be permitted in manufactured wiring systems for wiring of equipment within the scope of their respective articles.

(2) Conduits. Conduit shall be listed flexible metal conduit or listed liquidtight flexible conduit containing nominal 600-volt 10 or 12 AWG copper-insulated conductors with a bare or insulated copper equipment grounding conductor equivalent in size to the ungrounded conductor.

Exception No. 1 to (1) and (2): A luminaire (fixture) tap, maximum 1.8 m (6 ft) long, intended for connection to a single luminaire (fixture) shall be permitted to contain conductors smaller than 12 AWG but not smaller than 18 AWG.

Exception No. 1 permits 6-ft lengths of the wiring methods covered in 604.6(A)(1) and 604.6(A)(2) to contain conductors smaller than 12 AWG but not smaller than 18 AWG for use as tap conductors to supply a single luminaire.

Exception No. 2 to (1) and (2): Conductors smaller than 12 AWG shall be permitted for remote-control, signaling, or communications circuits. The assembly shall be listed for the purpose.

(3) Flexible Cord. Flexible cord suitable for hard usage, with minimum 12 AWG conductors, shall be permitted as part of a listed factory-made assembly not exceeding 1.8 m (6 ft) in length when making a transition between components of a manufactured wiring system and utilization equipment not permanently secured to the building structure. The cord shall be visible for its entire length and shall not be subject to strain or physical damage.

This provision facilitates a transition between manufactured wiring systems and utilization equipment found in display cases, merchandise racks, temporary work stations, and the like. This transition is limited, however, to hard-usage cord not over 6 ft in length in order to minimize damage, as illustrated in Exhibit 604.1.

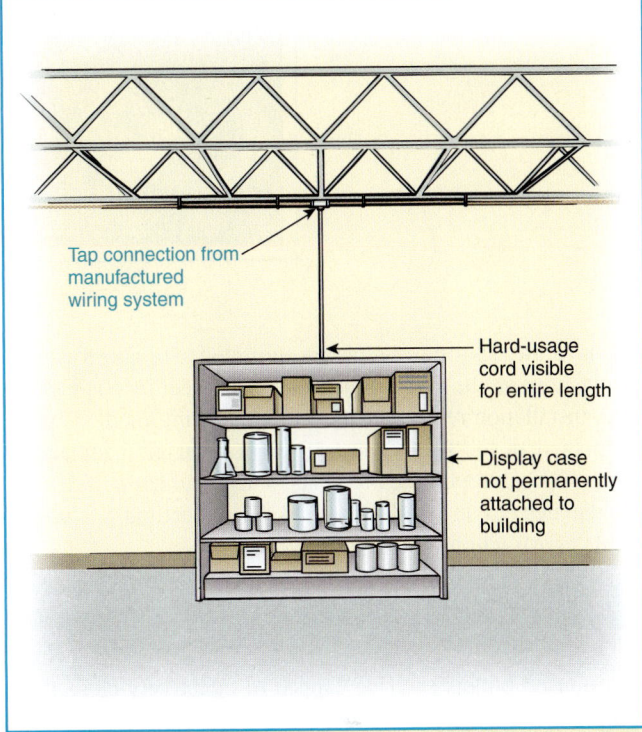

Tap connection from manufactured wiring system

Hard-usage cord visible for entire length

Display case not permanently attached to building

Exhibit 604.1 Transition wiring between a manufactured wiring system and utilization equipment.

(B) Marking. Each section shall be marked to identify the type of cable, flexible cord, or conduit.

(C) Receptacles and Connectors. Receptacles and connectors shall be of the locking type, uniquely polarized and identified for the purpose, and shall be part of a listed assembly for the appropriate system.

Examples of polarized receptacles and connectors are shown in Exhibit 604.2.

(D) Other Component Parts. Other component parts shall be listed for the appropriate system.

(E) Support. Manufactured wiring systems shall be supported in accordance with the applicable cable or conduit article for the cable or conduit type employed.

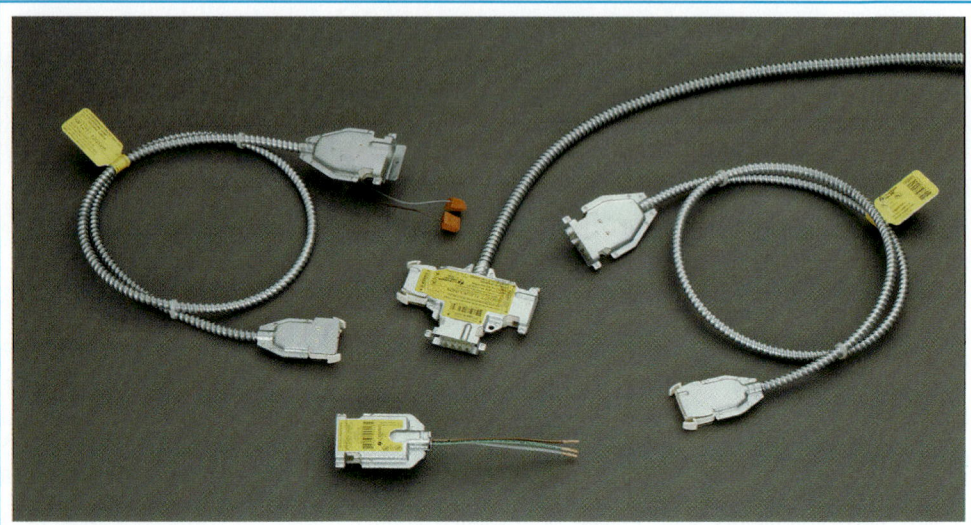

Exhibit 604.2 Polarized receptacles and connectors of a manufactured wiring system. (Courtesy of Lithonia Lighting, Reloc Wiring Systems)

Manufactured wiring systems are permitted to be constructed using any of the wiring methods described in 604.6(A). The installation requirements for these wiring methods are contained in the respective articles of Chapter 3 for each method. The support requirements for a manufactured wiring system depend on which Chapter 3 wiring method was used in its construction.

604.7 Unused Outlets.

All unused outlets shall be capped to effectively close the connector openings.

ARTICLE 605
Office Furnishings (Consisting of Lighting Accessories and Wired Partitions)

Contents

605.1 Scope.

This article covers electrical equipment, lighting accessories, and wiring systems used to connect, or contained within, or installed on relocatable wired partitions.

605.2 General.

Wiring systems shall be identified as suitable for providing power for lighting accessories and appliances in wired partitions. These partitions shall not extend from floor to ceiling.

Exception: Where permitted by the authority having jurisdiction, these relocatable wired partitions shall be permitted to extend to the ceiling but shall not penetrate the ceiling.

(A) Use. These assemblies shall be installed and used only as provided for by this article.

(B) Other Articles. Except as modified by the requirements of this article, all other articles of this *Code* shall apply.

(C) Hazardous (Classified) Locations. Where used in hazardous (classified) locations, these assemblies shall conform with Articles 500 through 517 in addition to this article.

605.3 Wireways.

All conductors and connections shall be contained within wiring channels of metal or other material identified as suitable for the conditions of use. Wiring channels shall be free of projections or other conditions that may damage conductor insulation.

A wiring channel that is separate from the channel containing the branch circuits for light and power may be provided within the system components for the routing of communications, signaling, and fiber optic cables.

605.4 Partition Interconnections.

The electrical connection between partitions shall be a flexible assembly identified for use with wired partitions or shall be permitted to be installed using flexible cord, provided all the following conditions are met:

(1) The cord is extra-hard usage type with 12 AWG or larger conductors, with an insulated grounding conductor.
(2) The partitions are mechanically contiguous.
(3) The cord is not longer than necessary for maximum positioning of the partitions but is in no case to exceed 600 mm (2 ft).
(4) The cord is terminated at an attachment plug and cord connector with strain relief.

605.5 Lighting Accessories.

Lighting equipment listed and identified for use with wired partitions shall comply with 605.5(A), (B), and (C).

(A) Support. A means for secure attachment or support shall be provided.

(B) Connection. Where cord and plug connection is provided, the cord length shall be suitable for the intended application but shall not exceed 2.7 m (9 ft) in length. The cord shall not be smaller than 18 AWG, shall contain an equipment grounding conductor, and shall be of the hard usage type. Connection by other means shall be identified as suitable for the condition of use.

(C) Receptacle Outlet. Convenience receptacles shall not be permitted in lighting accessories.

605.6 Fixed-Type Partitions.

Wired partitions that are fixed (secured to building surfaces) shall be permanently connected to the building electrical system by one of the wiring methods of Chapter 3.

605.7 Freestanding-Type Partitions.

Partitions of the freestanding type (not fixed) shall be permitted to be permanently connected to the building electrical system by one of the wiring methods of Chapter 3.

605.8 Freestanding-Type Partitions, Cord-and-Plug-Connected.

Individual partitions of the freestanding type, or groups of individual partitions that are electrically connected, are mechanically contiguous, and do not exceed 9.0 m (30 ft) when assembled, shall be permitted to be connected to the building electrical system by a single flexible cord and plug, provided all of the conditions of 605.8(A) through (D) are met.

(A) Flexible Power-Supply Cord. The flexible power-supply cord shall be extra-hard usage type with 12 AWG or larger conductors with an insulated equipment grounding conductor and not exceeding 600 mm (2 ft) in length.

(B) Receptacle Supplying Power. The receptacle(s) supplying power shall be on a separate circuit serving only panels and no other loads and shall be located not more than 300 mm (12 in.) from the partition that is connected to it.

(C) Receptacle Outlets, Maximum. Individual partitions or groups of interconnected individual partitions shall not contain more than thirteen 15-ampere, 125-volt receptacle outlets.

(D) Multiwire Circuits, Not Permitted. Individual partitions or groups of interconnected individual partitions shall not contain multiwire circuits.

FPN: See 210.4 for circuits supplying partitions in 605.6 and 605.7.

Partitions (other than the cord-and-plug-connected type described in 605.8) are permitted to contain multiwire branch circuits.

ARTICLE 610
Cranes and Hoists

Contents

I. General

610.1 Scope.

This article covers the installation of electrical equipment and wiring used in connection with cranes, monorail hoists, hoists, and all runways.

> FPN: For further information, see ANSI B-30, *Safety Code for Cranes, Derricks, Hoists, Jacks, and Slings.*

610.2 Special Requirements for Particular Locations.

(A) Hazardous (Classified) Locations. All equipment that operates in a hazardous (classified) location shall conform to Article 500.

(1) Class I Locations. Equipment used in locations that are hazardous because of the presence of flammable gases or vapors shall conform to Article 501.

(2) Class II Locations. Equipment used in locations that are hazardous because of combustible dust shall conform to Article 502.

(3) Class III Locations. Equipment used in locations that are hazardous because of the presence of easily ignitible fibers or flyings shall conform to Article 503.

See the commentary following 503.13(D).

(B) Combustible Materials. Where a crane, hoist, or monorail hoist operates over readily combustible material, the resistors shall be located as permitted in the following:

(1) A well-ventilated cabinet composed of noncombustible material constructed so that it does not emit flames or molten metal

(2) A cage or cab constructed of noncombustible material that encloses the sides of the cage or cab from the floor

to a point at least 150 mm (6 in.) above the top of the resistors

(C) Electrolytic Cell Lines. See 668.32.

Special precautions are necessary on electrolytic cell lines to prevent the introduction of exposed grounded parts, as described in 668.32.

II. Wiring

610.11 Wiring Method.

Conductors shall be enclosed in raceways or be Type AC cable with insulated grounding conductor, Type MC cable, or Type MI cable unless otherwise permitted or required in 610.11(A) through (E).

Type AC cable with an insulated grounding conductor is permitted as an acceptable wiring method. The insulated equipment grounding conductor in Type AC cable ensures a reliable grounding path.

(A) Contact Conductor. Contact conductors are not required to be enclosed in raceways.

(B) Open Conductors. Short lengths of open conductors at resistors, collectors, and other equipment are not required to be enclosed in raceways.

(C) Flexible Connections to Motors and Similar Equipment. Where flexible connections are necessary, flexible stranded conductors shall be used. Conductors shall be in flexible metal conduit, liquidtight flexible metal conduit, liquidtight flexible nonmetallic conduit, multiconductor cable, or an approved nonmetallic flexible raceway.

The use of short lengths of open wiring on cranes and hoists is permitted by 610.11(B). Short runs of open conductors facilitate connection to resistors, collectors, and similar equipment. Each conductor is required to be provided with a separately bushed hole in boxes, as well as in cable and raceway fittings used where the transition to open wiring is made. In addition to other types of raceways and cables, 610.11(C) permits flexible metal conduit, liquidtight flexible metal conduit, liquidtight flexible nonmetallic conduit, multiconductor cable, or an approved nonmetallic enclosure (tubing) where flexibility is necessary.

(D) Pushbutton Stations Multiconductor Cable. Where multiconductor cable is used with a suspended pushbutton station, the station shall be supported in some satisfactory manner that protects the electric conductors against strain.

Exhibit 610.1 shows an example of suitable strain relief for a cord that supports a control pushbutton station.

Exhibit 610.1 A suitable strain relief grip for a cord-suspended pushbutton station. (Courtesy of Hubbell Inc., Kellems Division)

(E) Flexibility to Moving Parts. Where flexibility is required for power or control to moving parts, a cord suitable for the purpose shall be permitted provided the following apply:

(1) Suitable strain relief and protection from physical damage is provided.
(2) In Class I, Division 2 locations, the cord is approved for extra-hard usage.

610.12 Raceway or Cable Terminal Fittings.

Conductors leaving raceways or cables shall comply with either of 610.12(A) or (B).

(A) Separately Bushed Hole. A box or terminal fitting that has a separately bushed hole for each conductor shall be used wherever a change is made from a raceway or cable to open wiring. A fitting used for this purpose shall not contain taps or splices and shall not be used at luminaire (fixture) outlets.

(B) Bushing in Lieu of a Box. A bushing shall be permitted to be used in lieu of a box at the end of a rigid metal conduit, intermediate metal conduit, or electrical metallic tubing where the raceway terminates at unenclosed controls or similar equipment, including contact conductors, collectors, resistors, brakes, power-circuit limit switches, and dc split-frame motors.

610.13 Types of Conductors.

Conductors shall comply with Table 310.13 unless otherwise permitted in 610.13(A) through (D).

(A) Exposed to External Heat or Connected to Resistors. A conductor(s) exposed to external heat or connected to resistors shall have a flame-resistant outer covering or be covered with flame-resistant tape individually or as a group.

(B) Contact Conductors. Contact conductors along runways, crane bridges, and monorails shall be permitted to be bare and shall be copper, aluminum, steel, or other alloys or combinations thereof in the form of hard drawn wire, tees, angles, tee rails, or other stiff shapes.

(C) Flexibility. Where flexibility is required, flexible cord or cable shall be permitted to be used and, where necessary, cable reels or take-up devices shall be used.

(D) Class 1, Class 2, and Class 3 Circuits. Conductors for Class 1, Class 2, and Class 3 remote-control, signaling, and power-limited circuits, installed in accordance with Article 725, shall be permitted.

> Added in the 1999 *Code*, 610.13(D) references Article 725 for Class 1, Class 2, and Class 3 control circuit conductors. This cross-reference is necessary because 90.3 does not allow Article 725 to modify Article 610. The change allows the methods of wiring control and signaling circuits permitted in Article 725 to be used for crane and hoist control circuits.

610.14 Rating and Size of Conductors.

(A) Ampacity. The allowable ampacities of conductors shall be as shown in Table 610.14(A).

> FPN: For the ampacities of conductors between controllers and resistors, see 430.23.

(B) Secondary Resistor Conductors. Where the secondary resistor is separate from the controller, the minimum size of the conductors between controller and resistor shall be calculated by multiplying the motor secondary current by the appropriate factor from Table 610.14(B) and selecting a wire from Table 610.14(A).

(C) Minimum Size. Conductors external to motors and controls shall not be smaller than 16 AWG unless otherwise permitted in (1) and (2).

(1) 18 AWG wire in multiconductor cord shall be permitted for control circuits at not over 7 amperes.

(2) Wires not smaller than 20 AWG shall be permitted for electronic circuits.

(D) Contact Conductors. Contact wires shall have an ampacity not less than that required by Table 610.14(A) for 75°C (167°F) wire, and in no case shall they be smaller than as shown in Table 610.14(D).

(E) Calculation of Motor Load.

(1) Single Motor. For one motor, 100 percent of motor nameplate full-load ampere rating shall be used.

(2) Multiple Motors on Single Crane or Hoist. For multiple motors on a single crane or hoist, the minimum ampacity of the power supply conductors shall be the nameplate full-load ampere rating of the largest motor or group of motors for any single crane motion, plus 50 percent of the nameplate full-load ampere rating of the next largest motor or group of motors, using that column of Table 610.14(A) that applies to the longest time-rated motor.

(3) Multiple Cranes or Hoists on a Common Conductor System. For multiple cranes, hoists, or both, supplied by a common conductor system, compute the motor minimum ampacity for each crane as defined in 610.14(E), add them together, and multiply the sum by the appropriate demand factor from Table 610.14(E).

(F) Other Loads. Additional loads, such as heating, lighting, and air conditioning, shall be provided for by application of the appropriate sections of this *Code*.

(G) Nameplate. Each crane, monorail, or hoist shall be provided with a visible nameplate marked with the manufacturer's name, the rating in volts, frequency, number of phases, and circuit amperes as calculated in 610.14(E) and (F).

610.15 Common Return.

Where a crane or hoist is operated by more than one motor, a common-return conductor of proper ampacity shall be permitted.

III. Contact Conductors

610.21 Installation of Contact Conductors.

Contact conductors shall comply with 610.21(A) through (H).

(A) Locating or Guarding Contact Conductors. Runway contact conductors shall be guarded, and bridge contact conductors shall be located or guarded in such a manner that persons cannot inadvertently touch energized current-carrying parts.

Table 610.14(A) Ampacities of Insulated Copper Conductors Used with Short-Time Rated Crane and Hoist Motors. Based on Ambient Temperature of 30°C (86°F). Up to Four Conductors in Raceway or Cable.[1] Up to 3 ac[2] or 4 dc[1] Conductors in Raceway or Cable

Maximum Operating Temperature	75°C (167°F)		90°C (194°F)		125°C (257°F)		Maximum Operating Temperature
	Types MTW, RHW, THW, THWN, XHHW, USE, ZW		Types TA, TBS, SA, SIS, PFA, FEP, FEPB, RHH, THHN, XHHW, Z, ZW		Types FEP, FEPB, PFA, PFAH, SA, TFE, Z, ZW		
Size (AWG or kcmil)	60 Min	30 Min	60 Min	30 Min	60 Min	30 Min	Size (AWG or kcmil)
16	10	12	—	—	—	—	16
14	25	26	31	32	38	40	14
12	30	33	36	40	45	50	12
10	40	43	49	52	60	65	10
8	55	60	63	69	73	80	8
6	76	86	83	94	101	119	6
5	85	95	95	106	115	134	5
4	100	117	111	130	133	157	4
3	120	141	131	153	153	183	3
2	137	160	148	173	178	214	2
1	143	175	158	192	210	253	1
1/0	190	233	211	259	253	304	1/0
2/0	222	267	245	294	303	369	2/0
3/0	280	341	305	372	370	452	3/0
4/0	300	369	319	399	451	555	4/0
250	364	420	400	461	510	635	250
300	455	582	497	636	587	737	300
350	486	646	542	716	663	837	350
400	538	688	593	760	742	941	400
450	600	765	660	836	818	1042	450
500	660	847	726	914	896	1143	500

AMPACITY CORRECTION FACTORS

Ambient Temperature (°C)	For ambient temperatures other than 30°C (86°F), multiply the ampacities shown above by the appropriate factor shown below.						Ambient Temperature (°F)
21–25	1.05	1.05	1.04	1.04	1.02	1.02	70–77
26–30	1.00	1.00	1.00	1.00	1.00	1.00	79–86
31–35	0.94	0.94	0.96	0.96	0.97	0.97	88–95
36–40	0.88	0.88	0.91	0.91	0.95	0.95	97–104
41–45	0.82	0.82	0.87	0.87	0.92	0.92	106–113
46–50	0.75	0.75	0.82	0.82	0.89	0.89	115–122
51–55	0.67	0.67	0.76	0.76	0.86	0.86	124–131
56–60	0.58	0.58	0.71	0.71	0.83	0.83	133–140
61–70	0.33	0.33	0.58	0.58	0.76	0.76	142–158
71–80	—	—	0.41	0.41	0.69	0.69	160–176
81–90	—	—	—	—	0.61	0.61	177–194
91–100	—	—	—	—	0.51	0.51	195–212
101–120	—	—	—	—	0.40	0.40	213–248

Note: Other insulations shown in Table 310.13 and approved for the temperature and location shall be permitted to be substituted for those shown in Table 610.14(A). The allowable ampacities of conductors used with 15-minute motors shall be the 30-minute ratings increased by 12 percent.

[1] For 5 to 8 simultaneously energized power conductors in raceway or cable, the ampacity of each power conductor shall be reduced to a value of 80 percent of that shown in the table.

[2] For 4 to 6 simultaneously energized 125°C (257°F) ac power conductors in raceway or cable, the ampacity of each power conductor shall be reduced to a value of 80 percent of that shown in the table.

Table 610.14(B) Secondary Conductor Rating Factors

Time in Seconds		Ampacity of Wire in Percent of Full-Load Secondary Current
On	**Off**	
5	75	35
10	70	45
15	75	55
15	45	65
15	30	75
15	15	85
Continuous Duty		110

Table 610.14(D) Contact Conductor Supports

Distance Between End Strain Insulators or Clamp-Type Intermediate Supports	Size of Wire (AWG)
Less than 9.0 m (30 ft)	6
9.0 m–18 m (30 ft–60 ft)	4
Over 18 m (60 ft)	2

Table 610.14(E) Demand Factors

Number of Cranes or Hoists	Demand Factor
2	0.95
3	0.91
4	0.87
5	0.84
6	0.81
7	0.78

(B) Contact Wires. Wires that are used as contact conductors shall be secured at the ends by means of approved strain insulators and shall be mounted on approved insulators so that the extreme limit of displacement of the wire does not bring the latter within less than 38 mm (1½ in.) from the surface wired over.

(C) Supports Along Runways. Main contact conductors carried along runways shall be supported on insulating supports placed at intervals not exceeding 6.0 m (20 ft) unless otherwise permitted in 610.21(F).

Such conductors shall be separated not less than 150 mm (6 in.), other than for monorail hoists where a spacing of not less than 75 mm (3 in.) shall be permitted. Where necessary, intervals between insulating supports shall be permitted to be increased up to 12 m (40 ft), the separation between conductors being increased proportionally.

(D) Supports on Bridges. Bridge wire contact conductors shall be kept at least 65 mm (2½ in.) apart, and, where the span exceeds 25 m (80 ft), insulating saddles shall be placed at intervals not exceeding 15 m (50 ft).

(E) Supports for Rigid Conductors. Conductors along runways and crane bridges, that are of the rigid type specified in 610.13(B) and not contained within an approved enclosed assembly, shall be carried on insulating supports spaced at intervals of not more than 80 times the vertical dimension of the conductor, but in no case greater than 4.5 m (15 ft), and spaced apart sufficiently to give a clear electrical separation of conductors or adjacent collectors of not less than 25 mm (1 in.).

(F) Track as Circuit Conductor. Monorail, tram rail, or crane runway tracks shall be permitted as a conductor of current for one phase of a 3-phase, ac system furnishing power to the carrier, crane, or trolley, provided all of the following conditions are met:

(1) The conductors supplying the other two phases of the power supply are insulated.
(2) The power for all phases is obtained from an insulating transformer.
(3) The voltage does not exceed 300 volts.
(4) The rail serving as a conductor is effectively grounded at the transformer and also shall be permitted to be grounded by the fittings used for the suspension or attachment of the rail to a building or structure.

Crane runway tracks are permitted as a current-carrying conductor where part of a 3-phase system that is furnishing power to the crane. Exhibit 610.2 illustrates a 3-phase, isolated-delta secondary with one phase grounded (a corner-grounded delta secondary system) at the transformer. The track is also permitted to be grounded through the metal supporting means attached to the metal frame of a building.

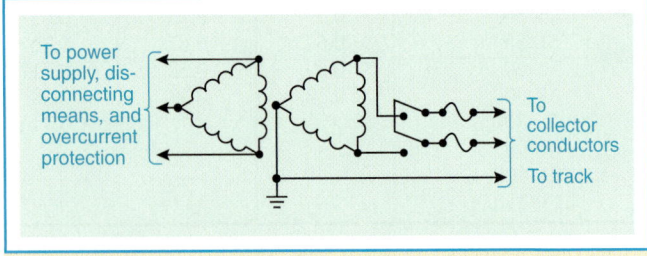

Exhibit 610.2 Three-phase, delta isolating transformer.

(G) Electrical Continuity of Contact Conductors. All sections of contact conductors shall be mechanically joined to provide a continuous electrical connection.

(H) Not to Supply Other Equipment. Contact conductors shall not be used as feeders for any equipment other than the crane(s) or hoist(s) that they are primarily designed to serve.

610.22 Collectors.

Collectors shall be designed so as to reduce to a minimum sparking between them and the contact conductor; and, where operated in rooms used for the storage of easily ignitible combustible fibers and materials, they shall comply with 503.13.

IV. Disconnecting Means

610.31 Runway Conductor Disconnecting Means.

A disconnecting means that has a continuous ampere rating not less than that computed in 610.14(E) and (F) shall be provided between the runway contact conductors and the power supply. Such disconnecting means shall consist of a motor-circuit switch, circuit breaker, or molded case switch. This disconnecting means shall be as follows:

(1) Readily accessible and operable from the ground or floor level
(2) Capable of being locked in the open position
(3) Open all ungrounded conductors simultaneously
(4) Placed within view of the runway contact conductors

610.32 Disconnecting Means for Cranes and Monorail Hoists.

A motor-circuit switch, molded-case switch, or circuit breaker shall be provided in the leads from the runway contact conductors or other power supply on all cranes and monorail hoists. The disconnecting means shall be capable of being locked in the open position.

Where a monorail hoist or hand-propelled crane bridge installation meets all of the following, the disconnecting means shall be permitted to be omitted:

(1) The unit is controlled from the ground or floor level.
(2) The unit is within view of the power supply disconnecting means.
(3) No fixed work platform has been provided for servicing the unit.

Where the disconnecting means is not readily accessible from the crane or monorail hoist operating station, means shall be provided at the operating station to open the power circuit to all motors of the crane or monorail hoist.

Many crane installations are not arranged so that the unit is within view of the power supply disconnecting means.

Therefore, a disconnecting means (lock-open type) must be provided in the contact conductors. However, personnel should be aware that when one crane is being serviced, another unit on the same system could remain energized and be run into the person performing maintenance on the locked-out unit.

610.33 Rating of Disconnecting Means.

The continuous ampere rating of the switch or circuit breaker required by 610.32 shall not be less than 50 percent of the combined short-time ampere rating of the motors or less than 75 percent of the sum of the short-time ampere rating of the motors required for any single motion.

V. Overcurrent Protection

610.41 Feeders, Runway Conductors.

(A) Single Feeder. The runway supply conductors and main contact conductors of a crane or monorail shall be protected by an overcurrent device(s) that shall not be greater than the largest rating or setting of any branch-circuit protective device plus the sum of the nameplate ratings of all the other loads with application of the demand factors from Table 610.14(E).

(B) More Than One Feeder Circuit. Where more than one feeder circuit is installed to supply runway conductors, each feeder circuit shall be sized and protected in compliance with 610.41(A).

Multiple feeders are sometimes used to supply long runway conductors in order to minimize voltage drops on the runway conductors.

610.42 Branch-Circuit Short-Circuit and Ground-Fault Protection.

Branch circuits shall be protected in accordance with 610.42(A). Branch-circuit taps, where made, shall comply with 610.42(B).

(A) Fuse or Circuit Breaker Rating. Crane, hoist, and monorail hoist motor branch circuits shall be protected by fuses or inverse-time circuit breakers that have a rating in accordance with Table 430.52. Where two or more motors operate a single motion, the sum of their nameplate current ratings shall be considered as that of a single motor.

(B) Taps.

(1) Multiple Motors. Where two or more motors are connected to the same branch circuit, each tap conductor to an individual motor shall have an ampacity not less than one-third that of the branch circuit. Each motor shall be protected from overload according to 610.43.

(2) Control Circuits. Where taps to control circuits originate on the load side of a branch-circuit protective device, each tap and piece of equipment shall be protected in accordance with 430.72.

(3) Brake Coils. Taps without separate overcurrent protection shall be permitted to brake coils.

610.43 Overload Protection.

(A) Motor and Branch-Circuit Overload Protection. Each motor, motor controller, and branch-circuit conductor shall be protected from overload by one of the following means:

(1) A single motor shall be considered as protected where the branch-circuit overcurrent device meets the rating requirements of 610.42.
(2) Overload relay elements in each ungrounded circuit conductor, with all relay elements protected from short circuit by the branch-circuit protection.
(3) Thermal sensing devices, sensitive to motor temperature or to temperature and current, that are thermally in contact with the motor winding(s). A hoist or trolley shall be considered to be protected if the sensing device is connected in the hoist's upper limit switch circuit so as to prevent further hoisting during an overload condition of either motor.

(B) Manually Controlled Motor. If the motor is manually controlled, with spring return controls, the overload protective device shall not be required to protect the motor against stalled rotor conditions.

(C) Multimotor. Where two or more motors drive a single trolley, truck, or bridge and are controlled as a unit and protected by a single set of overload devices with a rating equal to the sum of their rated full-load currents, a hoist or trolley shall be considered to be protected if the sensing device is connected in the hoist's upper limit switch circuit so as to prevent further hoisting during an overtemperature condition of either motor.

(D) Hoists and Monorail Hoists. Hoists and monorail hoists and their trolleys that are not used as part of an overhead traveling crane shall not require individual motor overload protection, provided the largest motor does not exceed 7½ hp and all motors are under manual control of the operator.

VI. Control

610.51 Separate Controllers.

Each motor shall be provided with an individual controller unless otherwise permitted in 610.51(A) or (B).

(A) Motions with More Than One Motor. Where two or more motors drive a single hoist, carriage, truck, or bridge, they shall be permitted to be controlled by a single controller.

(B) Multiple Motion Controller. One controller shall be permitted to be switched between motors, under the following conditions:

(1) The controller has a horsepower rating that is not lower than the horsepower rating of the largest motor.
(2) Only one motor is operated at one time.

610.53 Overcurrent Protection.

Conductors of control circuits shall be protected against overcurrent. Control circuits shall be considered as protected by overcurrent devices that are rated or set at not more than 300 percent of the ampacity of the control conductors, unless otherwise permitted in 610.53(A) or (B).

(A) Taps to Control Transformers. Taps to control transformers shall be considered as protected where the secondary circuit is protected by a device rated or set at not more than 200 percent of the rated secondary current of the transformer and not more than 200 percent of the ampacity of the control circuit conductors.

(B) Continuity of Power. Where the opening of the control circuit would create a hazard, as for example, the control circuit of a hot metal crane, the control circuit conductors shall be considered as being properly protected by the branch-circuit overcurrent devices.

610.55 Limit Switch.

A limit switch or other device shall be provided to prevent the load block from passing the safe upper limit of travel of all hoisting mechanisms.

610.57 Clearance.

The dimension of the working space in the direction of access to live parts that are likely to require examination, adjustment, servicing, or maintenance while energized shall be a minimum of 750 mm (2½ ft). Where controls are enclosed in cabinets, the door(s) shall either open at least 90 degrees or be removable.

VII. Grounding

610.61 Grounding.

All exposed non–current-carrying metal parts of cranes, monorail hoists, hoists, and accessories, including pendant controls, shall be metallically joined together into a continuous electrical conductor so that the entire crane or hoist will be grounded in accordance with Article 250. Moving parts, other than removable accessories or attachments, that have metal-to-metal bearing surfaces shall be considered to be electrically connected to each other through the bearing surfaces for grounding purposes. The trolley frame and bridge frame shall be considered as electrically grounded through the bridge and trolley wheels and their respective tracks unless local conditions, such as paint or other insulating material, prevent reliable metal-to-metal contact. In this case, a separate bonding conductor shall be provided.

It is not intended that the trolley frame or bridge frame serve as the equipment grounding conductor for electrical equipment (such as motors, motor controllers, lighting fixtures, and transformers) on a crane. The equipment grounding conductors that are run with the circuit conductors are required to be one of the types described in 250.118. However, 610.61 does not require the use of a separate bar of runway contact conductor as an equipment grounding conductor to serve the mobile portion of the crane, unless the metal-to-metal contact through the bridge and trolley wheels and their respective tracks does not provide a reliable and effective equipment grounding path.

ARTICLE 620
Elevators, Dumbwaiters, Escalators, Moving Walks, Wheelchair Lifts, and Stairway Chair Lifts

Contents

I. General

620.1 Scope.

This article covers the installation of electrical equipment and wiring used in connection with elevators, dumbwaiters, escalators, moving walks, wheelchair lifts, and stairway chair lifts.

Since the 1996 edition of the *Code,* there has been a continuing effort to revise Article 620 in order to harmonize elevator requirements throughout North America and to keep pace with modern advances in elevator technology. These changes continue to reflect time-honored safety concerns and more accurately address the use of modern equipment.

FPN No. 1: For further information, see ASME/ANSI A17.1-1996, *Safety Code for Elevators and Escalators.*

FPN No. 2: For further information, see ASME/ANSI A17.5-1996 (CSA B44.1-1996), *Elevator and Escalator Electrical Equipment Certification Standard.*

Fine print note No. 1 to 620.1 points out the existence of the widely recognized elevator code, ANSI/ASME A17.1-1998, *Safety Code for Elevators and Escalators.* ASME A17.1 contains many construction and maintenance requirements, from required machine room lighting to seismic considerations.

620.2 Definitions.

Control System. The overall system governing the starting, stopping, direction of motion, acceleration, speed, and retardation of the moving member.

Controller, Motion. The electric device(s) for that part of the control system that governs the acceleration, speed, retardation, and stopping of the moving member.

Controller, Motor. The operative units of the control system comprised of the starter device(s) and power conversion equipment used to drive an electric motor, or the pumping unit used to power hydraulic control equipment.

Controller, Operation. The electric device(s) for that part of the control system that initiates the starting, stopping, and direction of motion in response to a signal from an operating device.

Operating Device. The car switch, push buttons, key or toggle switch(s), or other devices used to activate the operation controller.

Signal Equipment. Includes audible and visual equipment such as chimes, gongs, lights, and displays that convey information to the user.

FPN No. 1: The motor controller, motion controller, and operation controller may be located in a single enclosure or a combination of enclosures.

FPN No. 2: FPN Figure 620.2 is for information only.

FPN Figure 620.2 illustrates a typical elevator control system. The figure does not suggest that every elevator control system should have these exact components.

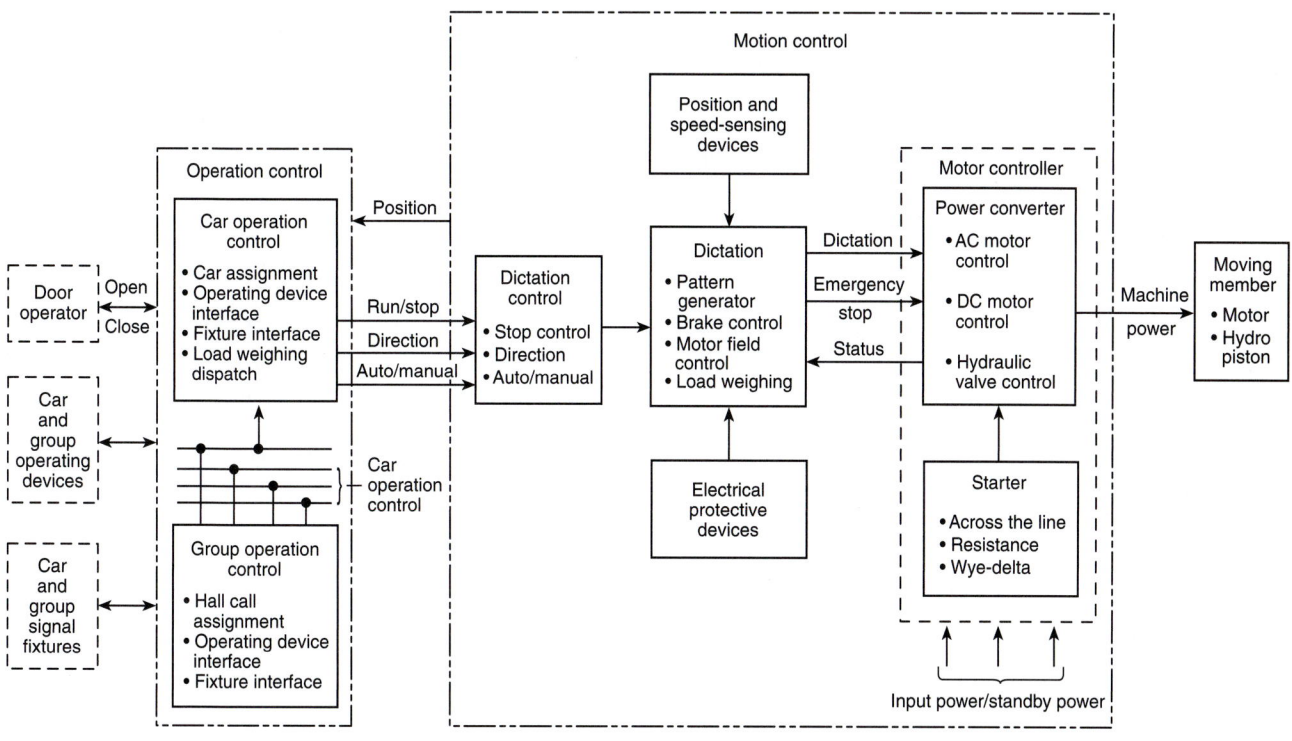

FPN Figure 620.2 Control system.

The definitions in 620.2 and FPN Figure 620.2 separate the control system into its functional parts. Article 620 uses these definitions to address and specify the proper safety concerns of each of these controllers: motor, motion, and operation.

620.3 Voltage Limitations.

The supply voltage shall not exceed 300 volts between conductors unless otherwise permitted in 620.3(A) through (C).

(A) Power Circuits. Branch circuits to door operator controllers and door motors and branch circuits and feeders to motor controllers, driving machine motors, machine brakes, and motor-generator sets shall not have a circuit voltage in excess of 600 volts. Internal voltages of power conversion and functionally associated equipment, including the interconnecting wiring, shall be permitted to have higher voltages, provided that all such equipment and wiring shall be listed for the higher voltages. Where the voltage exceeds 600 volts, warning labels or signs that read "DANGER — HIGH VOLTAGE" shall be attached to the equipment and shall be plainly visible.

(B) Lighting Circuits. Lighting circuits shall comply with the requirements of Article 410.

(C) Heating and Air-Conditioning Circuits. Branch circuits for heating and air-conditioning equipment located on the elevator car shall not have a circuit voltage in excess of 600 volts.

The voltage limitations for power, lighting, and heating and air-conditioning circuits are described in 620.3. Voltage limitations for power conversion units and functionally associated equipment have been replaced with a warning sign requirement for voltages above 600 volts.

620.4 Live Parts Enclosed.

All live parts of electrical apparatus in the hoistways, at the landings, in or on the cars of elevators and dumbwaiters, in the wellways or the landings of escalators or moving walks, or in the runways and machinery spaces of wheelchair lifts and stairway chair lifts shall be enclosed to protect against accidental contact.

FPN: See 110.27 for guarding of live parts (600 volts, nominal, or less).

620.5 Working Clearances.

Working space shall be provided about controllers, disconnecting means, and other electrical equipment. The minimum working space shall not be less than that specified in 110.26(A).

Where conditions of maintenance and supervision ensure that only qualified persons examine, adjust, service, and maintain the equipment, the clearance requirements of 110.26(A) shall be waived as permitted in 620.5(A) through (D).

(A) Flexible Connections to Equipment. Electrical equipment in (1) through (4) shall be permitted to be provided with flexible leads to all external connections so that it can be repositioned to meet the clear working space requirements of 110.26(A).

Due to the physical constraints of the locations where this equipment is typically installed and the necessity of performing diagnostic work on it while it is energized, 620.5(A) permits flexible leads on equipment so that it can be relocated to meet the working clearance requirements of 110.26(A).

(1) Controllers and disconnecting means for dumbwaiters, escalators, moving walks, wheelchair lifts, and stairway chair lifts installed in the same space with the driving machine
(2) Controllers and disconnecting means for elevators installed in the hoistway or on the car
(3) Controllers for door operators
(4) Other electrical equipment installed in the hoistway or on the car

(B) Guards. Live parts of the electrical equipment are suitably guarded, isolated, or insulated, and the equipment can be examined, adjusted, serviced, or maintained while energized without removal of this protection.

> FPN: See definition of *Exposed* in Article 100.

(C) Examination, Adjusting, and Servicing. Electrical equipment is not required to be examined, adjusted, serviced, or maintained while energized.

(D) Low Voltage. Uninsulated parts are at a voltage not greater than 30 volts rms, 42 volts peak, or 60 volts dc.

II. Conductors

620.11 Insulation of Conductors.

The insulation of conductors shall comply with 620.11(A) through (D).

> FPN: One method of determining that conductors are flame retardant is by testing the conductors to the VW-1 (Vertical-Wire) Flame Test in ANSI/UL 1581-1991, *Reference Standard for Electrical Wires, Cables, and Flexible Cords.*

(A) Hoistway Door Interlock Wiring. The conductors to the hoistway door interlocks from the hoistway riser shall be flame retardant and suitable for a temperature of not less than 200°C (392°F). Conductors shall be Type SF or equivalent.

(B) Traveling Cables. Traveling cables used as flexible connections between the elevator or dumbwaiter car or counterweight and the raceway shall be of the types of elevator cable listed in Table 400.4 or other approved types.

(C) Other Wiring. All conductors in raceways shall have flame-retardant insulation.

Conductors shall be Type MTW, TF, TFF, TFN, TFFN, THHN, THW, THWN, TW, XHHW, hoistway cable, or any other conductor with insulation designated as flame retardant. Shielded conductors shall be permitted if such conductors are insulated for the maximum nominal circuit voltage applied to any conductor within the cable or raceway system.

(D) Insulation. All conductors shall have an insulation voltage rating equal to at least the maximum nominal circuit voltage applied to any conductor within the enclosure, cable, or raceway. Insulations and outer coverings that are marked for limited smoke and are so listed shall be permitted.

Hoistway door interlock wiring is required to be suitable for 200C (392°F). See Table 310.13, Conductor Application and Insulations. See also Table 310.18.

See Table 400.4 for approved types of elevator cables for use in hazardous (classified) and nonhazardous locations. See also Note 5 to Table 400.4. A characteristic equally important with respect to safety is the need to prevent twisting of cables as they rise and fall with the elevator or dumbwaiter.

620.12 Minimum Size of Conductors.

The minimum size of conductors, other than conductors that form an integral part of control equipment, shall be in accordance with 620.12(A) and (B).

(A) Traveling Cables.

(1) Lighting Circuits. For lighting circuits: 14 AWG copper; 20 AWG copper or larger conductors shall be permitted in parallel, provided the ampacity is equivalent to at least that of 14 AWG copper.

(2) Other Circuits. For other circuits, 20 AWG copper.

(B) Other Wiring. 24 AWG copper; smaller size listed conductors shall be permitted.

In general, the requirements of 310.4 provide the conditions under which conductors can be installed in parallel for power and lighting circuits. One of those conditions stipulates that the minimum size for parallel conductors is 1/0 AWG. In high-rise structures, the length of the elevator traveling cables is problematic with respect to maintaining an acceptable level of voltage drop for equipment on or within the car. To require compliance with the general rules for parallel conductors would result in exceedingly large traveling cables. This section amends the general requirements for parallel conductors and permits 20 AWG and larger conductors to be installed in parallel for lighting circuits, provided that the combined ampacity of the paralleled conductors is not less than that of a 14 AWG conductor. This provision is unique to Article 620 and is an example of the structure of the *Code* as set forth in 90.3, in that a requirement in Chapter 6 modifies a general requirement from Chapter 3.

The extensive use of electronics with correspondingly lower currents permits the use of smaller wire sizes. The use of conductors smaller than 24 AWG is permitted by 620.12(B) where they are listed for the purpose. One application may be the shielded cables interconnecting various microprocessors in an elevator distributive system. All conductors should, of course, have the necessary strength and durability for the conditions to which they will be exposed.

620.13 Feeder and Branch-Circuit Conductors.

Conductors shall have an ampacity in accordance with 620.13(A) through (D). With generator field control, the conductor ampacity shall be based on the nameplate current rating of the driving motor of the motor-generator set that supplies power to the elevator motor.

> FPN No. 1: The heating of conductors depends on root-mean-square current values, which, with generator field control, are reflected by the nameplate current rating of the motor-generator driving motor rather than by the rating of the elevator motor, which represents actual but short-time and intermittent full-load current values.
>
> FPN No. 2: See Figure 620.13.

Code Figure 620.13 depicts the appropriate reference for each part of an elevator circuit.

(A) Conductors Supplying Single Motor. Conductors supplying a single motor shall have an ampacity not less than the percentage of motor nameplate current determined from 430.22(A) and (E).

> FPN: Elevator motor currents, or those of similar functions, may exceed the nameplate value, but since they are inherently intermittent duty and the heating of the motor and conductors is dependent on the root-mean-square (rms) current value, conductors are sized for duty cycle service as shown in Table 430.22(E).

(B) Conductors Supplying a Single Motor Controller. Conductors supplying a single motor controller shall

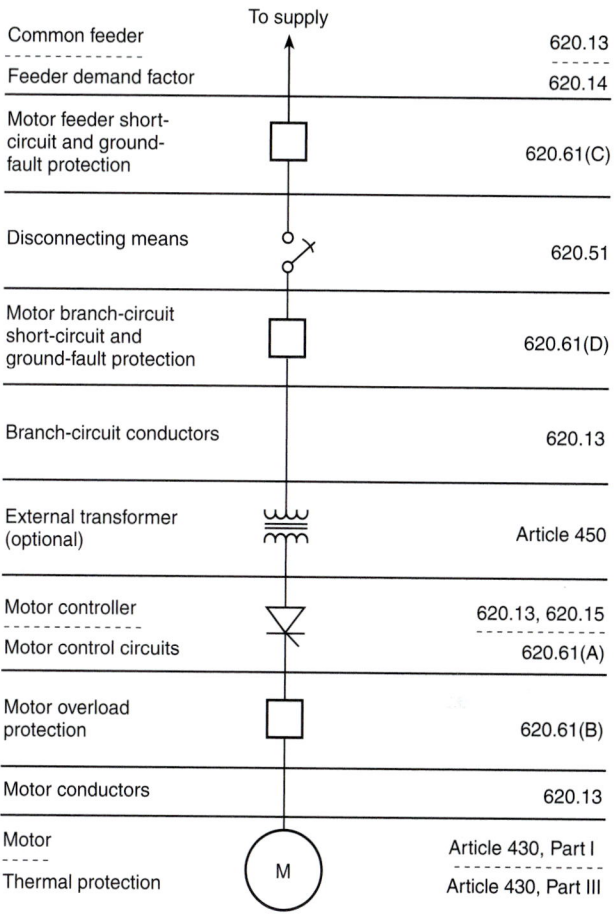

Figure 620.13 Single-line diagram.

have an ampacity not less than the motor controller nameplate current rating, plus all other connected loads.

> FPN: Motor controller nameplate current rating may be derived based on the rms value of the motor current using an intermittent duty cycle and other control system loads, if applicable.

(C) Conductors Supplying a Single Power Transformer. Conductors supplying a single power transformer shall have an ampacity not less than the nameplate current rating of the power transformer plus all other connected loads.

> FPN No. 1: The nameplate current rating of a power transformer supplying a motor controller reflects the nameplate current rating of the motor controller at line voltage (transformer primary).
>
> FPN No. 2: See Annex D, Example No. D10.

(D) Conductors Supplying More Than One Motor, Motor Controller, or Power Transformer. Conductors supplying more than one motor, motor controller, or power transformer shall have an ampacity not less than the sum of the nameplate current ratings of the equipment plus all other connected loads. The ampere ratings of motors to be used in the summation shall be determined from Table 430.22(E), and 430.24 and 430.24, Exception No. 1.

> FPN: See Annex D, Example Nos. D9 and D10.

620.14 Feeder Demand Factor.

Feeder conductors of less ampacity than required by 620.13 shall be permitted subject to the requirements of Table 620.14.

> FPN: Demand factors are based on 50 percent duty cycle (i.e., half time on and half time off).

620.15 Motor Controller Rating.

The motor controller rating shall comply with 430.83. The rating shall be permitted to be less than the nominal rating of the elevator motor, when the controller inherently limits the available power to the motor and is marked as power limited.

The inherent power-limiting ability of certain adjustable-speed drive controllers is the basis for permitting the controller to have a lower current or horsepower rating than that of the motor. To use a controller in this manner, the manufacturer's marking must indicate that it is power limiting.

> FPN: For controller markings, see 430.8.

III. Wiring

620.21 Wiring Methods.

Conductors and optical fibers located in hoistways, in escalator and moving walk wellways, in wheelchair lifts, stairway chair lift runways, machinery spaces, control spaces, in or on cars, in machine rooms and control rooms, not including the traveling cables connecting the car or counterweight and hoistway wiring, shall be installed in rigid metal conduit,

Table 620.14 Feeder Demand Factors for Elevators

Number of Elevators on a Single Feeder	Demand Factor
1	1.00
2	0.95
3	0.90
4	0.85
5	0.82
6	0.79
7	0.77
8	0.75
9	0.73
10 or more	0.72

intermediate metal conduit, electrical metallic tubing, rigid nonmetallic conduit, or wireways, or shall be Type MC, MI, or AC cable unless otherwise permitted in 620.21(A) through (C).

(A) Elevators.

(1) Hoistways.

(a) Flexible metal conduit, liquidtight flexible metal conduit, or liquidtight flexible nonmetallic conduit shall be permitted in hoistways between risers and limit switches, interlocks, operating buttons, and similar devices.

(b) Cables used in Class 2 power-limited circuits shall be permitted to be installed between risers and signal equipment and operating devices, provided the cables are supported and protected from physical damage and are of a jacketed and flame-retardant type.

(c) Flexible cords and cables that are components of listed equipment and used in circuits operating at 30 volts rms or less or 42 volts dc or less shall be permitted in lengths not to exceed 1.8 m (6 ft), provided the cords and cables are supported and protected from physical damage and are of a jacketed and flame-retardant type.

Limited (6 ft) lengths of flexible cord and cable are permitted to be used on elevator cars, where the cord or cable is part of listed equipment such as transducers (position, velocity, direction) and the circuit is limited to 30 volts rms or 42 volts dc.

(d) Flexible metal conduit, liquidtight flexible metal conduit, liquidtight flexible nonmetallic conduit or flexible cords and cables, or conductors grouped together and taped or corded that are part of listed equipment, a driving machine, or a driving machine brake shall be permitted in the hoistway, in lengths not to exceed 1.8 m (6 ft), without being installed in a raceway and where located to be protected from physical damage and are of a flame-retardant type.

(2) Cars.

(a) Flexible metal conduit, liquidtight flexible metal conduit, or liquidtight flexible nonmetallic conduit of metric designator 12 (trade size ⅜), or larger, not exceeding 1.8 m (6 ft) in length, shall be permitted on cars where located so as to be free from oil and if securely fastened in place.

Exception: Liquidtight flexible nonmetallic conduit of metric designator 12 (trade size ⅜), or larger, as defined by 356.2, shall be permitted in lengths in excess of 1.8 m (6 ft).

(b) Hard-service cords and junior hard-service cords that conform to the requirements of Article 400 (Table 400.4) shall be permitted as flexible connections between the fixed wiring on the car and devices on the car doors or gates. Hard-service cords only shall be permitted as flexible connections for the top-of-car operating device or the car-top work light. Devices or luminaires (fixtures) shall be grounded by means of an equipment grounding conductor run with the circuit conductors. Cables with smaller conductors and other types and thicknesses of insulation and jackets shall be permitted as flexible connections between the fixed wiring on the car and devices on the car doors or gates, if listed for this use.

(c) Flexible cords and cables that are components of listed equipment and used in circuits operating at 30 volts rms or less or 42 volts dc or less shall be permitted in lengths not to exceed 1.8 m (6 ft), provided the cords and cables are supported and protected from physical damage and are of a jacketed and flame-retardant type.

(d) Flexible metal conduit, liquidtight flexible metal conduit, liquidtight flexible nonmetallic conduit or flexible cords and cables, or conductors grouped together and taped or corded that are part of listed equipment, a driving machine, or a driving machine brake shall be permitted on the car assembly, in lengths not to exceed 1.8 m (6 ft) without being installed in a raceway and where located to be protected from physical damage and are of a flame-retardant type.

> The requirements of 620.21(A)(2)(d) describe the permitted wiring methods where a driving machine or driving machine brake is located on the car. In addition to flexible metal and nonmetallic conduits, the use of single conductors that are taped or corded together is permitted. This use includes rack and pinion or screw column drives located on cars. See also 620.71(B).

(3) Within Machine Rooms, Control Rooms, and Machinery Spaces and Control Spaces.

(a) Flexible metal conduit, liquidtight flexible metal conduit, or liquidtight flexible nonmetallic conduit of metric designator 12 (trade size ⅜), or larger, not exceeding 1.8 m (6 ft) in length, shall be permitted between control panels and machine motors, machine brakes, motor-generator sets, disconnecting means, and pumping unit motors and valves.

Exception: Liquidtight flexible nonmetallic conduit metric designator 12 (trade size ⅜) or larger, as defined in 356.2(2), shall be permitted to be installed in lengths in excess of 1.8 m (6 ft).

(b) Where motor-generators, machine motors, or pumping unit motors and valves are located adjacent to or underneath control equipment and are provided with extra-length

terminal leads not exceeding 1.8 m (6 ft) in length, such leads shall be permitted to be extended to connect directly to controller terminal studs without regard to the carrying-capacity requirements of Articles 430 and 445. Auxiliary gutters shall be permitted in machine and control rooms between controllers, starters, and similar apparatus.

(c) Flexible cords and cables that are components of listed equipment and used in circuits operating at 30 volts rms or less or 42 volts dc or less shall be permitted in lengths not to exceed 1.8 m (6 ft), provided the cords and cables are supported and protected from physical damage and are of a jacketed and flame-retardant type.

(d) On existing or listed equipment, conductors shall also be permitted to be grouped together and taped or corded without being installed in a raceway. Such cable groups shall be supported at intervals not over 900 mm (3 ft) and located so as to be protected from physical damage.

(4) Counterweight. Flexible metal conduit, liquidtight flexible metal conduit, liquidtight flexible nonmetallic conduit or flexible cords and cables, or conductors grouped together and taped or corded that are part of listed equipment, a driving machine, or a driving machine brake shall be permitted on the counterweight assembly, in lengths not to exceed 1.8 m (6 ft) without being installed in a raceway and where located to be protected from physical damage and are of a flame-retardant type.

(B) Escalators.

(1) Wiring Methods. Flexible metal conduit, liquidtight flexible metal conduit, or liquidtight flexible nonmetallic conduit shall be permitted in escalator and moving walk wellways. Flexible metal conduit or liquidtight flexible conduit of metric designator 12 (trade size ⅜) shall be permitted in lengths not in excess of 1.8 m (6 ft).

Exception: Metric designator 12 (trade size ⅜), nominal, or larger liquidtight flexible nonmetallic conduit, as defined in 356.2(2), shall be permitted to be installed in lengths in excess of 1.8 m (6 ft).

(2) Class 2 Circuit Cables. Cables used in Class 2 power-limited circuits shall be permitted to be installed within escalators and moving walkways, provided the cables are supported and protected from physical damage and are of a jacketed and flame-retardant type.

(3) Flexible Cords. Hard-service cords that conform to the requirements of Article 400 (Table 400.4) shall be permitted as flexible connections on escalators and moving walk control panels and disconnecting means where the entire control panel and disconnecting means are arranged for removal from machine spaces as permitted in 620.5.

(C) Wheelchair Lifts and Stairway Chair Lift Raceways.

(1) Wiring Methods. Flexible metal conduit or liquidtight flexible metal conduit shall be permitted in wheelchair lifts and stairway chair lift runways and machinery spaces. Flexible metal conduit or liquidtight flexible conduit of metric designator 12 (trade size ⅜) shall be permitted in lengths not in excess of 1.8 m (6 ft).

Exception: Metric designator 12 (trade size ⅜) or larger liquidtight flexible nonmetallic conduit, as defined in 356.2(2), shall be permitted to be installed in lengths in excess of 1.8 m (6 ft).

(2) Class 2 Circuit Cables. Cables used in Class 2 power-limited circuits shall be permitted to be installed within wheelchair lifts and stairway chair lift runways and machinery spaces, provided the cables are supported and protected from physical damage and are of a jacketed and flame-retardant type.

620.22 Branch Circuits for Car Lighting, Receptacle(s), Ventilation, Heating, and Air Conditioning.

(A) Car Light Source. A separate branch circuit shall supply the car lights, receptacle(s), auxiliary lighting power source, and ventilation on each elevator car. The overcurrent device protecting the branch circuit shall be located in the elevator machine room or control room/machinery space or control space.

(B) Air-Conditioning and Heating Source. A dedicated branch circuit shall supply the air-conditioning and heating units on each elevator car. The overcurrent device protecting the branch circuit shall be located in the elevator machine room or control room/machinery space or control space.

The requirements in 620.22(A) and 620.22(B) specify that the overcurrent devices protecting the branch circuits for elevator car lighting and heating/air conditioning are to be located in the elevator machine room or control room or in the elevator machinery or control space. This requirement facilitates ease of maintenance and troubleshooting and correlates with other North American code requirements. See Exhibit 620.1.

620.23 Branch Circuits for Machine Room or Control Room/Machinery Space or Control Space Lighting and Receptacle(s).

(A) Separate Branch Circuit. A separate branch circuit shall supply the machine room or control room/machinery space or control space lighting and receptacle(s).

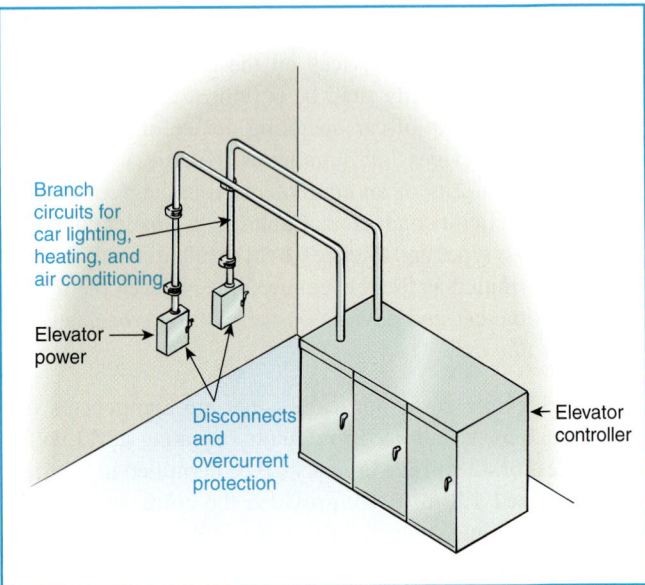

Exhibit 620.1 Disconnecting means and overcurrent devices for elevator power, car lighting, and car heating/air-conditioning branch circuits in the elevator machine room.

Required lighting shall not be connected to the load side of a ground-fault circuit interrupter.

Luminaires are not permitted to be connected to the load side of a GFCI device of any kind, because the machine room lighting could be de-energized during certain fault conditions.

(B) Lighting Switch. The machine room or control room/machinery space or control space lighting switch shall be located at the point of entry.

(C) Duplex Receptacle. At least one 125-volt, single-phase, duplex receptacle shall be provided in each machine room or control room and machinery space or control space.

FPN: See ANSI/ASME A17.1-1996, *Safety Code for Elevators and Escalators*, for illumination levels.

620.24 Branch Circuit for Hoistway Pit Lighting and Receptacle(s).

(A) Separate Branch Circuit. A separate branch circuit shall supply the hoistway pit lighting and receptacle(s).

Required lighting shall not be connected to the load side of a ground-fault circuit interrupter.

(B) Lighting Switch. The lighting switch shall be located so as to be readily accessible from the pit access door.

(C) Duplex Receptacle. At least one 125-volt, single-phase, duplex receptacle shall be provided in the hoistway pit.

FPN: See ANSI/ASME A17.1-1996, *Safety Code for Elevators and Escalators*, for illumination levels.

The receptacles required by 620.23 and 620.24 are also required to be provided with ground-fault circuit-interrupter protection for personnel (see 620.85). Luminaires are not permitted to be connected to the load side of GFCI devices. See also the commentary following 620.23(A).

ANSI/ASME A17.1-1996, *Safety Code for Elevators and Escalators,* requires a minimum of 5 ft-candles (54 lux) at the pit floor and requires luminaires to be externally guarded to prevent accidental breakage.

Luminaires in pits should be mounted so that the car or counterweight does not strike them when on fully compressed buffers. It is recommended that luminaires be mounted in the corners of the pit.

620.25 Branch Circuits for Other Utilization Equipment.

(A) Additional Branch Circuits. Additional branch circuit(s) shall supply utilization equipment not identified in 620.22, 620.23, and 620.24. Other utilization equipment shall be restricted to that equipment identified in 620.1.

(B) Overcurrent Devices. The overcurrent devices protecting the branch circuit(s) shall be located in the elevator machinery room or control room/machinery space or control space.

IV. Installation of Conductors

620.32 Metal Wireways and Nonmetallic Wireways.

The sum of the cross-sectional area of the individual conductors in a wireway shall not be more than 50 percent of the interior cross-sectional area of the wireway.

Vertical runs of wireways shall be securely supported at intervals not exceeding 4.5 m (15 ft) and shall have not more than one joint between supports. Adjoining wireway sections shall be securely fastened together to provide a rigid joint.

620.33 Number of Conductors in Raceways.

The sum of the cross-sectional area of the individual conductors in raceways shall not exceed 40 percent of the interior cross-sectional area of the raceway, except as permitted in 620.32 for wireways.

620.34 Supports.

Supports for cables or raceways in a hoistway or in an escalator or moving walk wellway or wheelchair lift and stairway chair lift runway shall be securely fastened to the guide rail; escalator or moving walk truss; or to the hoistway, wellway, or runway construction.

620.35 Auxiliary Gutters.

Auxiliary gutters shall not be subject to the restrictions of 366.3 as to length or of 366.6 as to number of conductors.

620.36 Different Systems in One Raceway or Traveling Cable.

Optical fiber cables and conductors for operating devices, operation and motion control, power, signaling, fire alarm, lighting, heating, and air-conditioning circuits of 600 volts or less shall be permitted to be run in the same traveling cable or raceway system if all conductors are insulated for the maximum voltage applied to any conductor within the cables or raceway system and if all live parts of the equipment are insulated from ground for this maximum voltage. Such a traveling cable or raceway shall also be permitted to include shielded conductors and/or one or more coaxial cables, if such conductors are insulated for the maximum voltage applied to any conductor within the cable or raceway system. Conductors shall be permitted to be covered with suitable shielding for telephone, audio, video, or higher frequency communications circuits.

With the use of greater numbers of individual cables and the use of much longer cables in tall buildings, there is a possibility of intertwisting cable loops. To eliminate the practice of tying a cable to the traveling cable, one elevator cable or raceway is permitted to enclose optical fiber cables and all the conductors for power, control, lighting, video, fire alarm, and communications circuits if all conductors are insulated for the maximum voltage applied to any conductor within the cable or raceway and all live parts are also insulated from ground for the maximum voltage present.

The revision of 620.36 in the 1999 *NEC* clarified that power-limited and non–power-limited fire alarm conductors are permitted in the same raceway or traveling cable as power and other types of signaling conductors. All power, signaling, and fire alarm conductors are to be insulated for the maximum voltage applied to any conductor in the raceway or cable.

620.37 Wiring in Hoistways, Machine Rooms, Control Rooms, Machinery Spaces, and Control Spaces.

(A) Uses Permitted. Only such electric wiring, raceways, and cables used directly in connection with the elevator or dumbwaiter, including wiring for signals, for communication with the car, for lighting, heating, air conditioning, and ventilating the elevator car, for fire detecting systems, for pit sump pumps, and for heating, lighting, and ventilating the hoistway, shall be permitted inside the hoistway, machine rooms, control rooms, machinery spaces, and control spaces.

(B) Lightning Protection. Bonding of elevator rails (car and/or counterweight) to a lightning protection system grounding down conductor(s) shall be permitted. The lightning protection system grounding down conductor(s) shall not be located within the hoistway. Elevator rails or other hoistway equipment shall not be used as the grounding down conductor for lightning protection systems.

Elevator hoistways are often convenient locations to run wiring from the basement to the roof. However, 620.37(B) prohibits equipment and wiring that is not associated with the elevator from being installed in elevator machine rooms and hoistways. Only electrical equipment and wiring used directly in connection with the elevator may be installed inside the hoistway and machine room.

Where a lightning protection system is provided, and where a lightning protection system grounding "down" conductor(s) located outside the hoistway is within a critical horizontal distance of the elevator rails, bonding of the rails to the lightning protection system grounding "down" conductor(s) is required by NFPA 780, *Standard for the Installation of Lightning Protection Systems*. Bonding prevents a dangerous side flash between the lightning protection system grounding "down" conductor(s) and the elevator rails. A lightning strike on the building air terminal will be conducted through the lightning protection system grounding "down" conductor(s), and, if the elevator rails are not at the same potential as the lightning protection system grounding "down" conductor(s), a side flash may occur. Generally, "down" conductors are installed vertically near the perimeter of the structure, so this requirement may have application to "outside" elevators.

FPN: See 250.106 for bonding requirements. For further information, see NFPA 780-1997, *Standard for the Installation of Lightning Protection Systems*.

(C) Main Feeders. Main feeders for supplying power to elevators and dumbwaiters shall be installed outside the hoistway unless as follows:

(1) By special permission, feeders for elevators shall be permitted within an existing hoistway if no conductors are spliced within the hoistway.
(2) Feeders shall be permitted inside the hoistway for elevators with driving machine motors located in the hoistway or on the car or counterweight.

620.38 Electrical Equipment in Garages and Similar Occupancies.

Electrical equipment and wiring used for elevators, dumbwaiters, escalators, moving walks, and wheelchair lifts and stairway chair lifts in garages shall comply with the requirements of Article 511.

FPN: Garages used for parking or storage and where no repair work is done in accordance with 511.3 are not classified.

V. Traveling Cables

620.41 Suspension of Traveling Cables.

Traveling cables shall be suspended at the car and hoistways' ends, or counterweight end where applicable, so as to reduce the strain on the individual copper conductors to a minimum.

Traveling cables shall be supported by one of the following means:

(1) By its steel supporting member(s)
(2) By looping the cables around supports for unsupported lengths less than 30 m (100 ft)
(3) By suspending from the supports by a means that automatically tightens around the cable when tension is increased for unsupported lengths up to 60 m (200 ft)

FPN: Unsupported length for the hoistway suspension means is that length of cable as measured from the point of suspension in the hoistway to the bottom of the loop, with the elevator car located at the bottom landing. Unsupported length for the car suspension means is that length of cable as measured from the point of suspension on the car to the bottom of the loop, with the elevator car located at the top landing.

Traveling cables between fixed suspension points are not required to be installed in a raceway. If the fixed suspension point is on top of the car, the cable might be exposed on the side of the car. Suitable guards are necessary to protect these cables from damage. If the suspension point is under the car, the cables might be run up the side of the car to the car top junction box. If these runs are over 6 ft, the cables are required to be run in a raceway. Refer to 620.44 and Exhibit 620.2.

620.42 Hazardous (Classified) Locations.

In hazardous (classified) locations, traveling cables shall be of a type approved for hazardous (classified) locations and shall comply with 501.11, 502.12, or 503.10, as applicable.

620.43 Location of and Protection for Cables.

Traveling cable supports shall be located so as to reduce to a minimum the possibility of damage due to the cables coming in contact with the hoistway construction or equip-

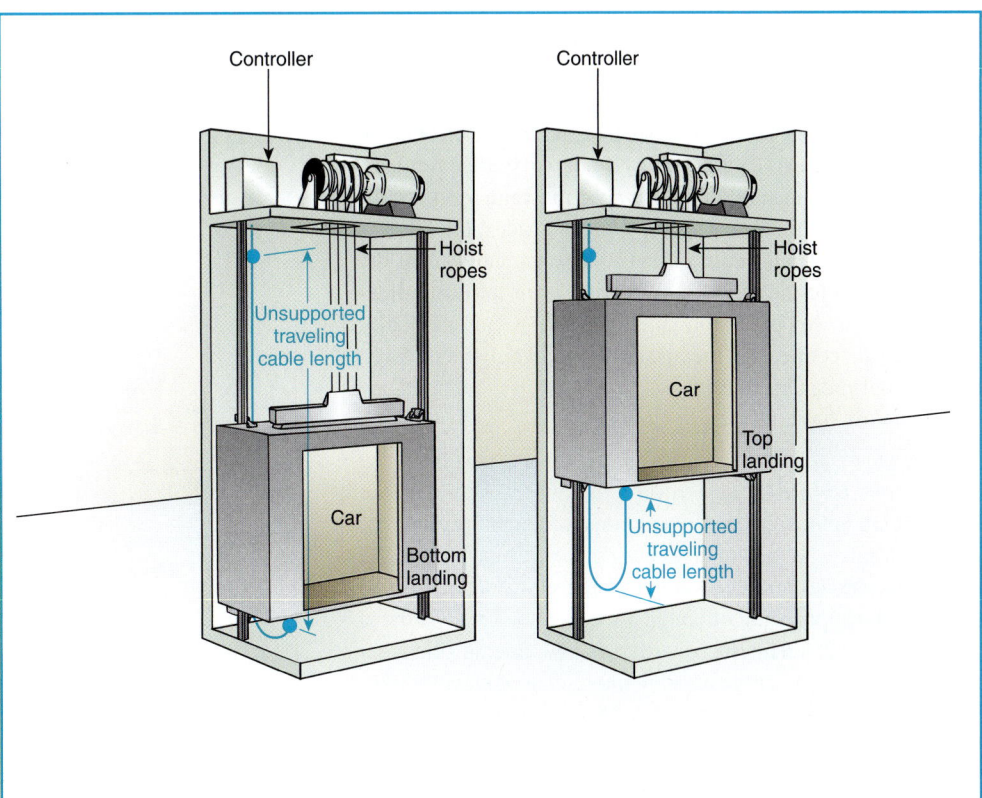

Exhibit 620.2 Unsupported lengths of traveling cable.

ment in the hoistway. Where necessary, suitable guards shall be provided to protect the cables against damage.

620.44 Installation of Traveling Cables.

Traveling cable shall be permitted to be run without the use of a raceway for a distance not exceeding 1.8 m (6 ft) in length as measured from the first point of support on the elevator car or hoistway wall, or counterweight where applicable, provided the conductors are grouped together and taped or corded, or in the original sheath.

Traveling cables shall be permitted to be continued to elevator controller enclosures and to elevator car and machine room, control room, machinery space, and control space connections, as fixed wiring, provided they are suitably supported and protected from physical damage.

VI. Disconnecting Means and Control

620.51 Disconnecting Means.

A single means for disconnecting all ungrounded main power supply conductors for each unit shall be provided and be designed so that no pole can be operated independently. Where multiple driving machines are connected to a single elevator, escalator, moving walk, or pumping unit, there shall be one disconnecting means to disconnect the motor(s) and control valve operating magnets.

The disconnecting means for the main power supply

conductors shall not disconnect the branch circuit required in 620.22, 620.23, and 620.24.

The branch circuits that supply elevator car lighting, receptacles, ventilation, air conditioning, and heating are required to be independent of the control portion of the elevator. In addition, the branch circuits supplying hoistway pit lighting and receptacles, and machine room or control room lights and receptacles, are not permitted to be disconnected by the main elevator power disconnect. This requirement provides for passenger safety and comfort and for the safety of elevator maintenance personnel during an inadvertent or emergency shutdown of the main power circuit to the elevator.

(A) Type. The disconnecting means shall be an enclosed externally operable fused motor circuit switch or circuit breaker capable of being locked in the open position. The disconnecting means shall be a listed device.

FPN: For additional information, see ASME/ANSI A17.1-1996, *Safety Code for Elevators and Escalators.*

Exception: Where an individual branch circuit supplies a wheelchair lift, the disconnecting means required by 620.51(C)(4) shall be permitted to comply with 430.109(C).

This disconnecting means shall be listed and shall be capable of being locked in the open position.

(B) Operation. No provision shall be made to open or close this disconnecting means from any other part of the premises. If sprinklers are installed in hoistways, machine rooms, control rooms, machinery spaces, or control spaces, the disconnecting means shall be permitted to automatically open the power supply to the affected elevator(s) prior to the application of water. No provision shall be made to automatically close this disconnecting means. Power shall only be restored by manual means.

ANSI/ASME A17.1-1998, *Safety Code for Elevators and Escalators,* Rule 102.2(c), requires that where sprinklers are installed in hoistways, machine rooms, or machinery spaces, a means must be provided to automatically disconnect the main line power supply to the affected elevator(s) upon or prior to the application of water. Water on elevator electrical equipment can result in hazards such as uncontrolled car movement (wet machine brakes), movement of elevator with open doors (water on safety circuits bypassing car and/or hoistway door interlocks), and shock hazards.

Automatic disconnection of the main line power supply is not required by ANSI/ASME A17.1-1998, *Safety Code for Elevators and Escalators,* where hoistways and machine rooms are not sprinklered. NFPA 13, *Standard for the Installation of Sprinkler Systems,* provides requirements for the installation of sprinklers in machine rooms, hoistways, and pits.

Elevator shutdown is generally accomplished through the use of heat detectors located near sprinkler heads. These heat detectors are designed to actuate before the sprinkler heads discharge water and generate an alarm signal. An output control relay powered by the fire alarm system then provides a monitored output to the main line disconnecting means control circuit, which activates the shunt trip. This practice ensures that all components have secondary power and are monitored for integrity, as required by *NFPA 72®, National Fire Alarm Code®.* Stand-alone heat detectors connected directly to the elevator disconnecting means control circuit are not monitored for integrity, have no secondary power supply, and are not permitted by *NFPA 72.*

Elevator shutdown can occur even if the car is not at a landing. However, to avoid trapping occupants in the car(s), it is highly desirable to recall the car(s) to the designated landing prior to disconnecting the main line power. Most fires produce detectable quantities of smoke before there is sufficient heat to activate a sprinkler head. Therefore, ANSI/ASME A17.1-1998, *Safety Code for Elevators and Escalators,* Rule 102.2(c), requires smoke detectors to be installed in hoistways that are sprinklered for the purposes of recalling the elevator car(s) before the main line power is discon-

nected. See 3-9.4 of *NFPA 72, National Fire Alarm Code,* for additional requirements relating to the fire alarm system and elevator shutdown.

Exhibit 620.3 illustrates a typical method of supervising control power using a fire alarm system. Loss of control power would produce a supervisory signal at the fire alarm control unit that would then be investigated.

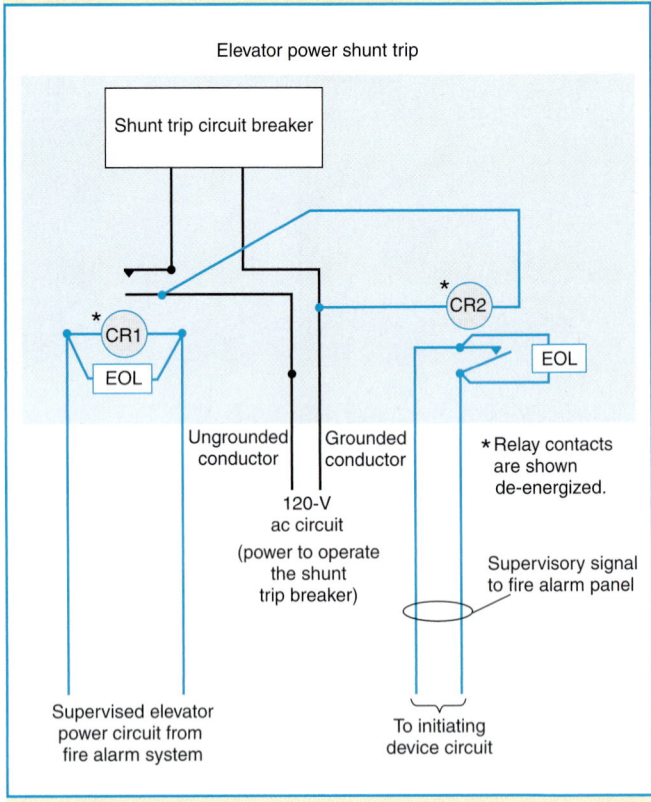

Exhibit 620.3 Typical method of control power supervision using a fire alarm control unit.

FPN: To reduce hazards associated with water on live elevator electrical equipment.

(C) Location. The disconnecting means shall be located where it is readily accessible to qualified persons.

(1) On Elevators Without Generator Field Control. On elevators without generator field control, the disconnecting means shall be located within sight of the motor controller. Driving machines or motion and operation controllers not within sight of the disconnecting means shall be provided with a manually operated switch installed in the control circuit to prevent starting. The manually operated switch(es) shall be installed adjacent to this equipment.

Where the driving machine of an electric elevator or the hydraulic machine of a hydraulic elevator is located in a remote machine room or remote machinery space, a single means for disconnecting all ungrounded main power supply conductors shall be provided and be capable of being locked in the open position.

See Exhibit 620.4 for disconnecting means for driving machines or motion and operation controllers not within sight of the main line disconnecting means.

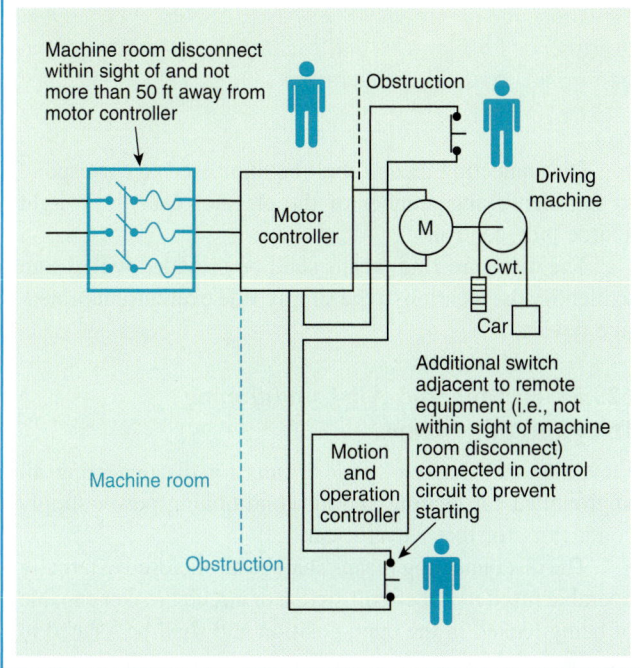

Exhibit 620.4 Disconnecting means for driving machines or motion and operation controllers not within sight of the main line disconnecting means. (Redrawn from ASME)

(2) On Elevators with Generator Field Control. On elevators with generator field control, the disconnecting means shall be located within sight of the motor controller for the driving motor of the motor-generator set. Driving machines, motor-generator sets, or motion and operation controllers not within sight of the disconnecting means shall be provided with a manually operated switch installed in the control circuit to prevent starting. The manually operated switch(es) shall be installed adjacent to this equipment.

Where the driving machine or the motor-generator set is located in a remote machine room or remote machinery space, a single means for disconnecting all ungrounded main power supply conductors shall be provided and be capable of being locked in the open position.

See Exhibits 620.5 and 620.6 for examples of disconnecting means for a motor-generator set and for driving machines in remote locations.

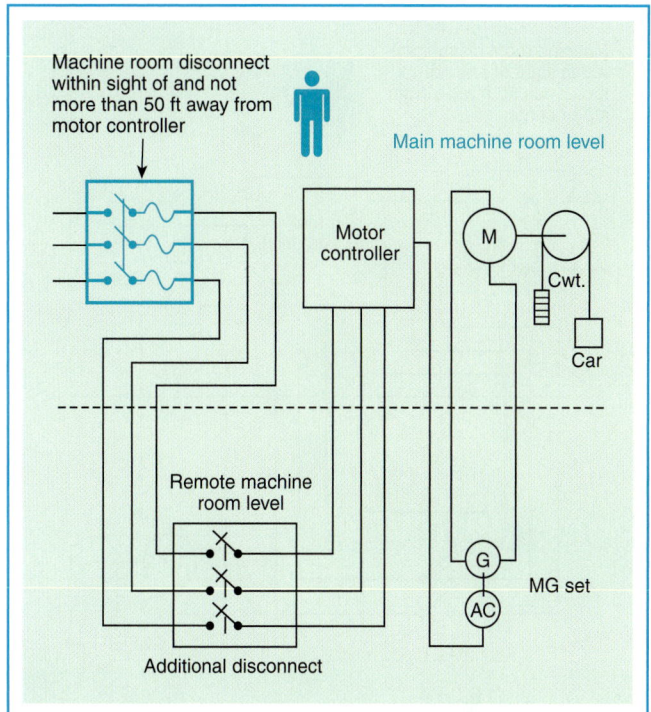

Exhibit 620.5 Disconnecting means for a motor-generator (MG) set in a remote location. (Redrawn from ASME)

(3) On Escalators and Moving Walks. On escalators and moving walks, the disconnecting means shall be installed in the space where the controller is located.

(4) On Wheelchair Lifts and Stairway Chair Lifts. On wheelchair lifts and stairway chair lifts, the disconnecting means shall be located within sight of the motor controller.

(D) Identification and Signs. Where there is more than one driving machine in a machine room, the disconnecting means shall be numbered to correspond to the identifying number of the driving machine that they control.

The disconnecting means shall be provided with a sign to identify the location of the supply side overcurrent protective device.

Sign requirements for the location of supply-side overcurrent devices assist the elevator mechanic in troubleshooting during a power loss.

620.52 Power from More Than One Source.

(A) Single-Car and Multicar Installations. On single-car and multicar installations, equipment receiving electrical power from more than one source shall be provided with a disconnecting means for each source of electrical power. The disconnecting means shall be within sight of the equipment served.

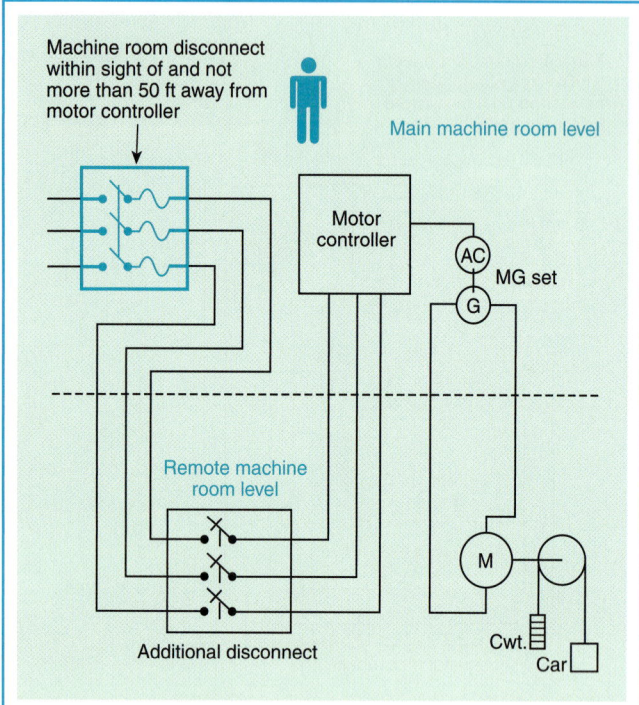

Exhibit 620.6 Disconnecting means for driving machines in a remote location. (Redrawn from ASME)

(B) Warning Sign for Multiple Disconnecting Means.
Where multiple disconnecting means are used and parts of the controllers remain energized from a source other than the one disconnected, a warning sign shall be mounted on or next to the disconnecting means. The sign shall be clearly legible and shall read

WARNING
PARTS OF THE CONTROLLER ARE NOT
DE-ENERGIZED BY THIS SWITCH.

(C) Interconnection Multicar Controllers. Where interconnections between controllers are necessary for the operation of the system on multicar installations that remain energized from a source other than the one disconnected, a warning sign in accordance with 620.52(B) shall be mounted on or next to the disconnecting means.

620.53 Car Light, Receptacle(s), and Ventilation Disconnecting Means.

Elevators shall have a single means for disconnecting all ungrounded car light, receptacle(s), and ventilation power-supply conductors for that elevator car.

The disconnecting means shall be an enclosed externally operable fused motor circuit switch or circuit breaker capable of being locked in the open position and shall be located in the machine room or control room for that elevator car. Where there is no machine room or control room, the discon-

necting means shall be located in the same space as the disconnecting means required by 620.51.

This requirement specifies the location of the disconnecting means for lighting, receptacle, and ventilation branch circuits associated with elevators that do not have a machine room. This type of installation includes those designs using drive systems located on the car, on the counterweight, or in the hoistway. Such designs include screw drive or linear induction motor drives. See ANSI/ASME A17.1-1998, *Safety Code for Elevators and Escalators,* for more information on this type of arrangement.

Disconnecting means shall be numbered to correspond to the identifying number of the elevator car whose light source they control.

The disconnecting means shall be provided with a sign to identify the location of the supply side overcurrent protective device.

620.54 Heating and Air-Conditioning Disconnecting Means.

Elevators shall have a single means for disconnecting all ungrounded car heating and air-conditioning power-supply conductors for that elevator car.

The disconnecting means shall be an enclosed externally operable fused motor circuit switch or circuit breaker capable of being locked in the open position and shall be located in the machine room or control room for that elevator car. Where there is no machine room or control room, the disconnecting means shall be located in the same space as the disconnecting means required by 620.51.

Where there is equipment for more than one elevator car in the machine room, the disconnecting means shall be numbered to correspond to the identifying number of the elevator car whose heating and air-conditioning source they control.

The disconnecting means shall be provided with a sign to identify the location of the supply side overcurrent protective device.

620.55 Utilization Equipment Disconnecting Means.

Each branch circuit for other utilization equipment shall have a single means for disconnecting all ungrounded conductors. The disconnecting means shall be capable of being locked in the open position and shall be located in the machine room or control room/machine space or control space. Where there is more than one branch circuit for other utilization equipment, the disconnecting means shall be numbered to correspond to the identifying number of the equipment served. The disconnecting means shall be provided with a sign to identify the location of the supply side overcurrent protective device.

VII. Overcurrent Protection

620.61 Overcurrent Protection.

Overcurrent protection shall be provided in accordance with 620.61(A) through (D).

(A) Operating Devices and Control and Signaling Circuits. Operating devices and control and signaling circuits shall be protected against overcurrent in accordance with the requirements of 725.23 and 725.24.

Class 2 power-limited circuits shall be protected against overcurrent in accordance with the requirements of Chapter 9, Notes to Tables 11(A) and 11(B).

(B) Overload Protection for Motors.

(1) Duty Rating on Elevator, Dumbwaiter, and Motor-Generator Sets Driving Motors. Duty on elevator and dumbwaiter driving machine motors and driving motors of motor-generators used with generator field control shall be rated as intermittent. Such motors shall be protected against overload in accordance with 430.33.

(2) Duty Rating on Escalator Motors. Duty on escalator and moving walk driving machine motors shall be rated as continuous. Such motors shall be protected against overload in accordance with 430.32.

(3) Overload Protection. Escalator and moving walk driving machine motors and driving motors of motor-generator sets shall be protected against running overload as provided in Table 430.37.

(4) Duty Rating and Overload Protection on Wheelchair and Stairway Chair Lift Motors. Duty on wheelchair lift and stairway chair lift driving machine motors shall be rated as intermittent. Such motors shall be protected against overload in accordance with 430.33.

> FPN: For further information, see 430.44 for orderly shutdown.

(C) Motor Feeder Short-Circuit and Ground-Fault Protection. Motor feeder short-circuit and ground-fault protection shall be as required in Article 430, Part V.

(D) Motor Branch-Circuit Short-Circuit and Ground-Fault Protection. Motor branch-circuit short-circuit and ground-fault protection shall be as required in Article 430, Part IV.

620.62 Selective Coordination.

Where more than one driving machine disconnecting means is supplied by a single feeder, the overcurrent protective devices in each disconnecting means shall be selectively coordinated with any other supply side overcurrent protective devices.

Coordination of the overcurrent protective devices is important. For example, if a building contains three elevators and a fault occurs in the circuit conductors to one of the elevators, only the overcurrent device ahead of that faulted circuit should open. Coordination leaves the remaining two elevators in operation. This is especially important because elevators are commonly used to carry fire fighters and equipment closer to the fire during fire-fighting operations.

Where the overcurrent devices in the elevator room do not have proper coordination with the upstream feeder overcurrent device, there is increased potential for interruption of power to all three elevators.

For selective coordination of overcurrent protective devices, the manufacturer's time-current curves, let-through and withstand capacity data, and unlatching times data must be used for sizing or setting overcurrent devices. See Exhibit 620.7.

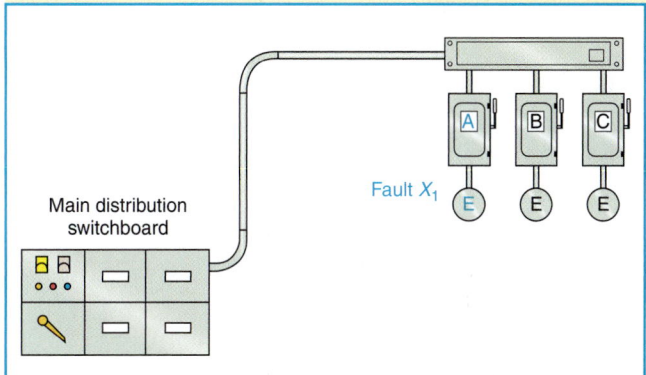

Exhibit 620.7 Example of selective coordination, where only overcurrent devices in switch A should open for a fault at X_1. Feeder overcurrent devices in the main distribution panel should not open, so that two elevators can remain in use. See 620.51(A) for the requirement regarding power supply disconnecting means.

VIII. Machine Rooms, Control Rooms, Machinery Spaces, and Control Spaces

620.71 Guarding Equipment.

Elevator, dumbwaiter, escalator, and moving walk driving machines; motor-generator sets; motor controllers; and disconnecting means shall be installed in a room or space set aside for that purpose unless otherwise permitted in 620.71(A) or (B). The room or space shall be secured against unauthorized access.

(A) Motor Controllers. Motor controllers shall be permitted outside the spaces herein specified, provided they are in enclosures with doors or removable panels that are capable of being locked in the closed position and the disconnecting means is located adjacent to or is an integral part of the motor controller. Motor controller enclosures for escalator or moving walks shall be permitted in the balustrade on the side located away from the moving steps or moving treadway. If the disconnecting means is an integral part of the motor controller, it shall be operable without opening the enclosure.

(B) Driving Machines. Elevators with driving machines located on the car, on the counterweight, or in the hoistway, and driving machines for dumbwaiters, wheelchair lifts, and stairway lifts shall be permitted outside the spaces herein specified.

IX. Grounding

620.81 Metal Raceways Attached to Cars.

Metal raceways, Type MC cable, Type MI cable, or Type AC cable attached to elevator cars shall be bonded to grounded metal parts of the car that they contact.

620.82 Electric Elevators.

For electric elevators, the frames of all motors, elevator machines, controllers, and the metal enclosures for all electrical equipment in or on the car or in the hoistway shall be grounded in accordance with Article 250.

620.83 Nonelectric Elevators.

For elevators other than electric having any electric conductors attached to the car, the metal frame of the car, where normally accessible to persons, shall be grounded in accordance with Article 250.

620.84 Escalators, Moving Walks, Wheelchair Lifts, and Stairway Chair Lifts.

Escalators, moving walks, wheelchair lifts, and stairway chair lifts shall comply with Article 250.

620.85 Ground-Fault Circuit-Interrupter Protection for Personnel.

Each 125-volt, single-phase, 15- and 20-ampere receptacle installed in pits, in hoistways, on elevator car tops, and in escalator and moving walk wellways shall be of the ground-fault circuit-interrupter type.

All 125-volt, single-phase, 15- and 20-ampere receptacles installed in machine rooms and machinery spaces shall have ground-fault circuit-interrupter protection for personnel.

A single receptacle supplying a permanently installed sump pump shall not require ground-fault circuit-interrupter protection.

The GFCI requirements of 620.85 are intended to reduce the shock hazard to maintenance personnel who service elevator equipment using portable hand tools and temporary lighting.

The first paragraph of 620.85 requires a GFCI-type receptacle for each 15- and 20-ampere receptacle installed in pits, on elevator car tops, and in escalator and moving walk wellways. This requirement is based on the premise that the reset pushbutton for a tripped GFCI receptacle should be within easy reach of an elevator mechanic working in confined spaces.

The second paragraph of 620.85 requires that all 15- and 20-ampere receptacles installed in machine rooms and machinery spaces have ground-fault circuit-interrupter protection for personnel. This protection can be afforded by either a GFCI-type circuit breaker or a GFCI-type receptacle because machine spaces usually do not cause access hazards for service personnel.

X. Emergency and Standby Power Systems

620.91 Emergency and Standby Power Systems.

An elevator(s) shall be permitted to be powered by an emergency or standby power system.

> FPN: See ASME/ANSI A17.1-1996, Rule 211.2, and CAN/CSA-B44-1994, Clause 3.12.13, for additional information.

(A) Regenerative Power. For elevator systems that regenerate power back into the power source that is unable to absorb the regenerative power under overhauling elevator load conditions, a means shall be provided to absorb this power.

(B) Other Building Loads. Other building loads, such as power and lighting, shall be permitted as the energy absorption means required in 620.91(A), provided that such loads are automatically connected to the emergency or standby power system operating the elevators and are large enough to absorb the elevator regenerative power.

(C) Disconnecting Means. The disconnecting means required by 620.51 shall disconnect the elevator from both the emergency or standby power system and the normal power system.

Where an additional power source is connected to the load side of the disconnecting means, the disconnecting means required in 620.51 shall be provided with an auxiliary contact that is positively opened mechanically, and the opening shall not be solely dependent on springs. This contact shall cause the additional power source to be disconnected from its load when the disconnecting means is in the open position.

ARTICLE 625
Electric Vehicle Charging System

Contents

I. General

A variety of street- and highway-worthy electric and combination electric/fossil fuel vehicles are becoming available to consumers. New and proposed legislation in several regions around the United States calls for increasing deployment of electric vehicles as a way to reduce air pollution. In California, the Air Resources Board has mandated that 10 percent of all vehicles produced by major manufacturers and available for sale in that state must be "zero polluting" by 2003. Other states have adopted similar requirements. In addition, the Clean Air Act Amendments of 1990 and the National Energy Policy Act of 1992 have requirements for public and private purchases of clean fuel vehicles and alternatively fueled vehicles, respectively. Electric vehicles fulfill both of those requirements. It is apparent that electric vehicle charging will be occurring in all occupancies, including residential, commercial, retail, and public sites.

Article 625 sets forth installation safety requirements for typical hard-wired conductive connections of battery charging equipment, as well as the safety concerns of the new "smart" inductive coupling connections of battery charging equipment. In particular, this article covers the wiring methods, equipment construction, control and protection, and equipment locations for automotive-type vehicle charging equipment. Throughout Article 625, the intent is to prevent the users of electric equipment associated with the vehicle charging system from being exposed to energized live parts and to provide for a safe vehicle charging environment.

625.1 Scope.

The provisions of this article cover the electrical conductors and equipment external to an electric vehicle that connect an electric vehicle to a supply of electricity by conductive or inductive means, and the installation of equipment and devices related to electric vehicle charging.

The scope of Article 625 is intended to cover all electrical wiring and equipment installed between the service point and the skin of the "automotive-type" electric vehicle. Automotive-type electric vehicles are emphasized because they are much different from other electric vehicles commonly used today. Most existing electric vehicles are off-road types, such as industrial forklifts, hoists, lifts, transports, golf carts, and airport personnel trams. The charging requirements and other exterior electrical connections are usually serviced and maintained by trained mechanics or technicians. The *NEC* has adequate provisions to allow the authority having jurisdiction to make interpretations that provide the safety levels needed for these installations.

Article 625 specifically excludes off-road vehicles, to

avoid conflict with existing articles. Motorcycles are not covered by Article 625 because motorcycles typically have smaller propulsion systems that operate at lower voltages, 12 to 24 volts dc versus 100 to 350 volts dc for electric automotive vehicles. Typically, motorcycles are charged from standard 120-volt, 15-ampere receptacles due to lower battery capacity. GFCI protection is not mandatory for charging electric motorcycles. However, 210.8(A)(2) and (A)(3) require GFCI protection of receptacles in the locations where an electric motorcycle would typically be charged.

FPN: For industrial trucks, see NFPA 505-1999, *Fire Safety Standard for Powered Industrial Trucks Including Type Designations, Areas of Use, Conversions, Maintenance, and Operation.*

625.2 Definitions.

Several of the definitions in 625.2 correlate with such industry standards as Society of Automotive Engineers (SAE) J1772 and J1773 and Underwriters Laboratories (UL) 2331.

Electric Vehicle. An automotive-type vehicle for highway use, such as passenger automobiles, buses, trucks, vans, and the like, primarily powered by an electric motor that draws current from a rechargeable storage battery, fuel cell, photovoltaic array, or other source of electric current. For the purpose of this article, electric motorcycles and similar type vehicles and off-road self-propelled electric vehicles, such as industrial trucks, hoists, lifts, transports, golf carts, airline ground support equipment, tractors, boats, and the like, are not included.

The primary difference between electric vehicles as defined in Article 625 and electric vehicles presently covered by other sections in the *NEC* is in their road and highway worthiness. The automotive electric vehicles under consideration are comparable in performance and function to the conventional automobiles and light trucks in use today. See Exhibit 625.1 for an example of such an electric vehicle. The automotive electric vehicles under development must be capable of complying with the Federal Motor Vehicle Safety Standards and other Department of Transportation, National Highway Traffic Safety Administration, and U.S. Environmental Protection Agency requirements. Special-purpose or limited-capability electric vehicles, such as lift trucks and golf carts, are not covered by Article 625.

Electric Vehicle Connector. A device that, by insertion into an electric vehicle inlet, establishes an electrical connection to the electric vehicle for the purpose of charging and

Exhibit 625.1 The Ford "Th!nk City" electric vehicle. (Courtesy of Ford Motor Co.)

information exchange. This device is part of the electric vehicle coupler.

Electric Vehicle Coupler. A mating electric vehicle inlet and electric vehicle connector set.

Electric Vehicle Inlet. The device on the electric vehicle into which the electric vehicle connector is inserted for charging and information exchange. This device is part of the electric vehicle coupler. For the purposes of this *Code,* the electric vehicle inlet is considered to be part of the electric vehicle and not part of the electric vehicle supply equipment.

Electric Vehicle Nonvented Storage Battery. A hermetically sealed battery comprised of one or more rechargeable electrochemical cells that has no provision for release of excessive gas pressure, or for the addition of water or electrolyte, or for external measurements of electrolyte specific gravity.

Electric Vehicle Supply Equipment. The conductors, including the ungrounded, grounded, and equipment grounding conductors and the electric vehicle connectors, attachment plugs, and all other fittings, devices, power outlets, or apparatus installed specifically for the purpose of delivering energy from the premises wiring to the electric vehicle.

Electric vehicle supply equipment, as illustrated in Exhibit 625.2, comprises the components between the skin of the electric vehicle and the premises wiring, including any flexible cable, disconnecting means, enclosures, power outlet, and electric vehicle connector. The definition of *electric vehicle* includes all off-vehicle charging equipment and does not include charging equipment installed on the vehicle.

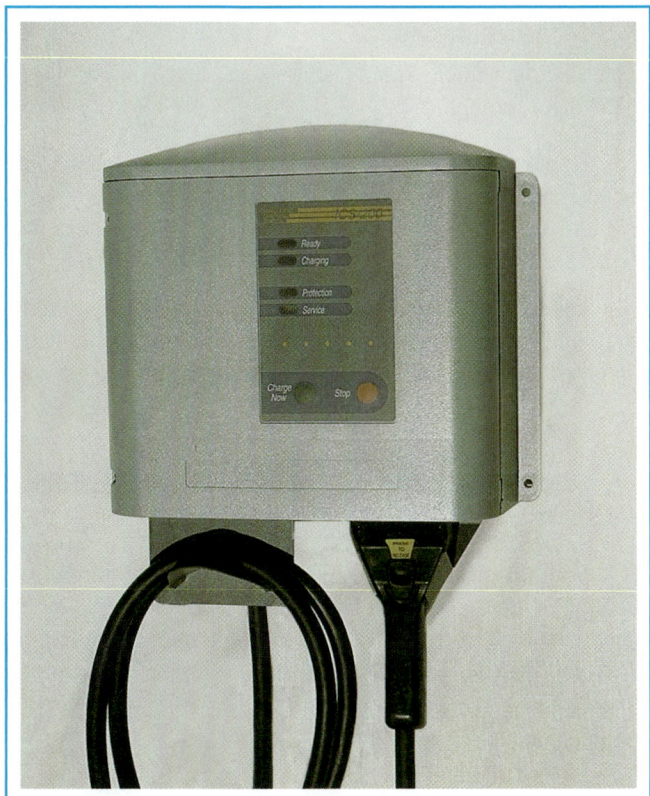

Exhibit 625.2 Example of electric vehicle supply equipment. (Courtesy of Ford Motor Co.)

Personnel Protection System. A system of personnel protection devices and constructional features that when used together provide protection against electric shock of personnel.

625.3 Other Articles.

Wherever the requirements of other articles of this *Code* and Article 625 differ, the requirements of Article 625 shall apply.

625.4 Voltages.

Unless other voltages are specified, the nominal ac system voltages of 120, 120/240, 208Y/120, 240, 480Y/277, 480, 600Y/347, and 600 volts shall be used to supply equipment covered by this article.

625.5 Listed or Labeled.

All electrical materials, devices, fittings, and associated equipment shall be listed or labeled.

II. Wiring Methods

625.9 Electric Vehicle Coupler.

The electric vehicle coupler shall comply with 625.9(A) through (F).

The electric vehicle connector is the device that inserts into the electric vehicle inlet (charge port) of the vehicle. The electric vehicle inlet is not a premises wiring receptacle or an attachment cap. An electric vehicle coupler is the mating set of the electric vehicle connector and electric vehicle inlet. A requirement was added for the 1999 *Code* to require the coupler to be noninterchangeable to prevent equipment damage or personal injury. Couplers for the electric vehicle charging equipment are not permitted to be standard NEMA-configuration wiring devices.

(A) Polarization. The electric vehicle coupler shall be polarized unless part of a system identified and listed as suitable for the purpose.

(B) Noninterchangeability. The electric vehicle coupler shall have a configuration that is noninterchangeable with wiring devices in other electrical systems. Nongrounding-type electric vehicle couplers shall not be interchangeable with grounding-type electric vehicle couplers.

(C) Construction and Installation. The electric vehicle coupler shall be constructed and installed so as to guard against inadvertent contact by persons with parts made live from the electric vehicle supply equipment or the electric vehicle battery.

The requirements for coupler construction in 625.9(C) provide a safe interface component for persons connecting the vehicle to or disconnecting the vehicle from the charging system. This type of activity generally is performed daily by persons who typically do not have any knowledge of the equipment operation and its associated hazards.

(D) Unintentional Disconnection. The electric vehicle coupler shall be provided with a positive means to prevent unintentional disconnection.

(E) Grounding Pole. The electric vehicle coupler shall be provided with a grounding pole, unless part of a system identified and listed as suitable for the purpose in accordance with Article 250.

(F) Grounding Pole Requirements. If a grounding pole is provided, the electric vehicle coupler shall be designed so that the grounding pole connection is the first to make and the last to break contact.

III. Equipment Construction

625.13 Electric Vehicle Supply Equipment.

Electric vehicle supply equipment rated at 125 volts, single phase, 15 or 20 amperes or part of a system identified and listed as suitable for the purpose and meeting the requirements of 625.18, 625.19, and 625.29 shall be permitted to be cord-and-plug connected. All other electric vehicle supply equipment shall be permanently connected and fastened in place. This equipment shall have no exposed live parts.

> Some manufacturers produce 125-volt, single-phase, 15- or 20-ampere portable charging units for convenience charging. These charging units may be stored in the vehicle. However, 625.13 makes it clear that nonportable equipment must be mounted and permanently wired. This equipment may be physically attached to the wall, floor, or ceiling. The provision for no exposed live parts is a safety concern for the general public.

625.14 Rating.

Electric vehicle supply equipment shall have sufficient rating to supply the load served. For the purposes of this article, electric vehicle charging loads shall be considered to be continuous loads.

> Considering both near-term and long-term requirements for electric vehicle (EV) charging, three methods have been identified for recommended development. Referred to as Level 1, Level 2, and Level 3 EV charging, they cover the range of power levels anticipated for charging EVs.
>
> *Level 1.* This method, which allows broad access to charge an EV, permits plugging into a common, grounded 120-volt electrical receptacle (NEMA 5-15R or 5-20R). The maximum load on this receptacle is 12 amperes or 1.4 kVA. The minimum circuit and overcurrent rating for this connection is 15 amperes for a 15-ampere receptacle and 20 amperes for a 20-ampere receptacle.
>
> *Level 2.* This is the primary and preferred method of EV charging at both private and public facilities. It requires special equipment and connection to an electric power supply dedicated to EV charging. The voltage of this connection is either 240 volts or 208 volts. The maximum load is 32 amperes (7.7 kVA at 240 volts or 6.7 kVA at 208 volts). The minimum circuit and overcurrent rating for this connection is 40 amperes (32 × 1.25 = 40 amperes). Electric vehicles are treated as continuous loads. See 625.21 for sizing overcurrent protection devices.
>
> *Level 3.* The EV equivalent of a commercial gasoline dispensing station, this high-speed, high-power method charges an EV in about the same time it takes to refuel a conventional vehicle. Because of individual supply requirements and available source voltages, exact voltage and load specifications for Level 3 charging have not been defined as in Level 1 and Level 2. These power requirements are specified by the equipment manufacturer.

625.15 Markings.

The electric vehicle supply equipment shall comply with 625.15(A) through (C).

(A) General. All electric vehicle supply equipment shall be marked by the manufacturer as follows:

FOR USE WITH ELECTRIC VEHICLES.

(B) Ventilation Not Required. Where marking is required by 625.29(C), the electric vehicle supply equipment shall be clearly marked by the manufacturer as follows:

VENTILATION NOT REQUIRED.

The marking shall be located so as to be clearly visible after installation.

(C) Ventilation Required. Where marking is required by 625.29(D), the electric vehicle supply equipment shall be clearly marked by the manufacturer "Ventilation Required." The marking shall be located so as to be clearly visible after installation.

625.16 Means of Coupling.

The means of coupling to the electric vehicle shall be either conductive or inductive. Attachment plugs, electric vehicle connectors, and electric vehicle inlets shall be listed or labeled for the purpose.

625.17 Cable.

The electric vehicle supply equipment cable shall be Type EV, EVJ, EVE, EVJE, EVT, or EVJT flexible cable as specified in Article 400 and Table 400.4. Ampacities shall be as specified in Table 400.5(A) for 10 AWG and smaller and Table 400.5(B) for 8 AWG and larger. The overall length of the cable shall not exceed 7.5 m (25 ft) unless equipped with a cable management system that is listed as suitable for the purpose. Other cable types and assemblies listed as being suitable for the purpose, including optional hybrid communications, signal, and optical fiber cables, shall be permitted.

> The 25-ft cable length is established by adding the 15-ft car length to the 7-ft car width, plus 3 ft to the power outlet securement point. This limits excessive cable lengths that may be exposed to damage. To use a single electric vehicle charging system for multiple electric vehicles, the 2002 *Code* permits cable lengths in excess of 25 ft where a listed cable management system is installed. For commercial parking areas, this change allows flexibility in site planning and meeting any legislated requirements that may be in place on the number of charging spaces that must be provided.

625.18 Interlock.

Electric vehicle supply equipment shall be provided with an interlock that de-energizes the electric vehicle connector and its cable whenever the electric connector is uncoupled from the electric vehicle. An interlock shall not be required for portable cord-and-plug-connected electric vehicle supply equipment intended for connection to receptacle outlets rated at 125 volts, single phase, 15 and 20 amperes.

To reduce shock hazard, a pilot or communications interlock establishes power through the electric vehicle supply equipment. Loss of the pilot or communications circuit locks out power, isolating possible hazardous situations in the electric vehicle supply equipment. See 625.29(D) for mechanical ventilation interlock requirements.

For ventilation interlock, see 625.29(D)(3) for 125-volt receptacles intended to charge electric vehicles.

625.19 Automatic De-Energization of Cable.

The electric vehicle supply equipment or the cable-connector combination of the equipment shall be provided with an automatic means to de-energize the cable conductors and electric vehicle connector upon exposure to strain that could result in either cable rupture or separation of the cable from the electric connector and exposure of live parts. Automatic means to de-energize the cable conductors and electric vehicle connector shall not be required for portable cord-and-plug-connected electric vehicle supply equipment intended for connection to receptacle outlets rated at 125 volts, single phase, 15 and 20 amperes.

IV. Control and Protection

625.21 Overcurrent Protection.

Overcurrent protection for feeders and branch circuits supplying electric vehicle supply equipment shall be sized for continuous duty and shall have a rating of not less than 125 percent of the maximum load of the electric vehicle supply equipment. Where noncontinuous loads are supplied from the same feeder or branch circuit, the overcurrent device shall have a rating of not less than the sum of the noncontinuous loads plus 125 percent of the continuous loads.

625.22 Personnel Protection System.

The electric vehicle supply equipment shall have a listed system of protection against electric shock of personnel. The personnel protection system shall be composed of listed personnel protection devices and constructional features. Where cord-and-plug-connected electric vehicle supply equipment is used, the interrupting device of a listed person-

nel protection system shall be provided and shall be an integral part of the attachment plug or shall be located in the power supply cable not more than 300 mm (12 in.) from the attachment plug.

The personnel protection system may consist of one or more components that provide protection against electric shock for different portions of the electric vehicle supply equipment circuitry, which may be operating at frequencies other than 50/60 Hz, at direct current potentials, and/or voltages above 150 volts to ground.

Standard GFCI devices do not provide the range of protection needed for the various types of charging systems being developed. Devices or methods that may be used include basic insulation, double insulation, grounding monitors, insulation monitors with interrupters, and leakage current monitors. Many combinations and variations of these devices can be used to provide the personnel protection required. For systems operating above 150 volts to ground, the protective system may include monitoring systems to ensure that proper grounding is provided and maintained during charging.

625.23 Disconnecting Means.

For electric vehicle supply equipment rated more than 60 amperes or more than 150 volts to ground, the disconnecting means shall be provided and installed in a readily accessible location. The disconnecting means shall be capable of being locked in the open position.

625.25 Loss of Primary Source.

Means shall be provided such that, upon loss of voltage from the utility or other electric system(s), energy cannot be backfed through the electric vehicle supply equipment to the premises wiring system. The electric vehicle shall not be permitted to serve as a standby power supply.

V. Electric Vehicle Supply Equipment Locations

625.28 Hazardous (Classified) Locations.

Where electric vehicle supply equipment or wiring is installed in a hazardous (classified) location, the requirements of Articles 500 through 516 shall apply.

The installation of EV charging equipment is permitted in hazardous locations where the installation is made in accordance with the requirements of Chapter 5. The increased use of electric vehicles makes this provision necessary in order to cover installations at commercial repair garages and combination gasoline/EV charging stations (see 511.10).

625.29 Indoor Sites.

Indoor sites shall include, but not be limited to, integral, attached, and detached residential garages; enclosed and underground parking structures; repair and nonrepair commercial garages; and agricultural buildings.

(A) Location. The electric vehicle supply equipment shall be located to permit direct connection to the electric vehicle.

(B) Height. Unless specifically listed for the purpose and location, the coupling means of the electric vehicle supply equipment shall be stored or located at a height of not less than 450 mm (18 in.) and not more than 1.2 m (4 ft) above the floor level.

(C) Ventilation Not Required. Where electric vehicle nonvented storage batteries are used or where the electric vehicle supply equipment is listed or labeled as suitable for charging electric vehicles indoors without ventilation and marked in accordance with 625.15(B), mechanical ventilation shall not be required.

Major auto manufacturers are taking the necessary steps to make electric vehicle systems safe. Most batteries used in electric vehicles manufactured by major automakers will not emit hydrogen gas in quantities that could cause an explosion. Preventive measures such as mechanical or passive ventilation are not required, because the electric vehicle batteries and charging systems are designed to prevent or limit the emission of hydrogen during charging. The Society of Automotive Engineers (SAE) has developed a recommended practice, J-1718, *Measurement of Hydrogen Gas Emission from Battery-Powered Passenger Cars and Light Trucks During Battery Charging,* that can be used to assess suitability for indoor charging. This standard includes provisions for tests during normal charging operations and potential equipment failure modes. See Exhibit 625.3 for an illustration of a garage without ventilation. In this application, when the electric vehicle is connected to the charging equipment, a signal is received at the electric vehicle charging equipment. The signal indicates that the electric vehicle either is equipped with a nonvented storage battery or is listed or labeled as suitable to be charged indoors. Failure to receive a verification signal from the vehicle will prevent initiation of the charging operation. The electric vehicle supply equipment is required to be marked in accordance with 625.15(B).

(D) Ventilation Required. Where the electric vehicle supply equipment is listed or labeled as suitable for charging electric vehicles that require ventilation for indoor charging and marked in accordance with 625.15(C), mechanical ventilation, such as a fan, shall be provided. The ventilation shall include both supply and exhaust equipment and shall be permanently installed and located to intake from, and vent directly to, the outdoors. Positive pressure ventilation systems shall be permitted only in buildings or areas that have been specifically designed and approved for that application. Mechanical ventilation requirements shall be determined by one of the methods specified in 625.29(D)(1) through (D)(4).

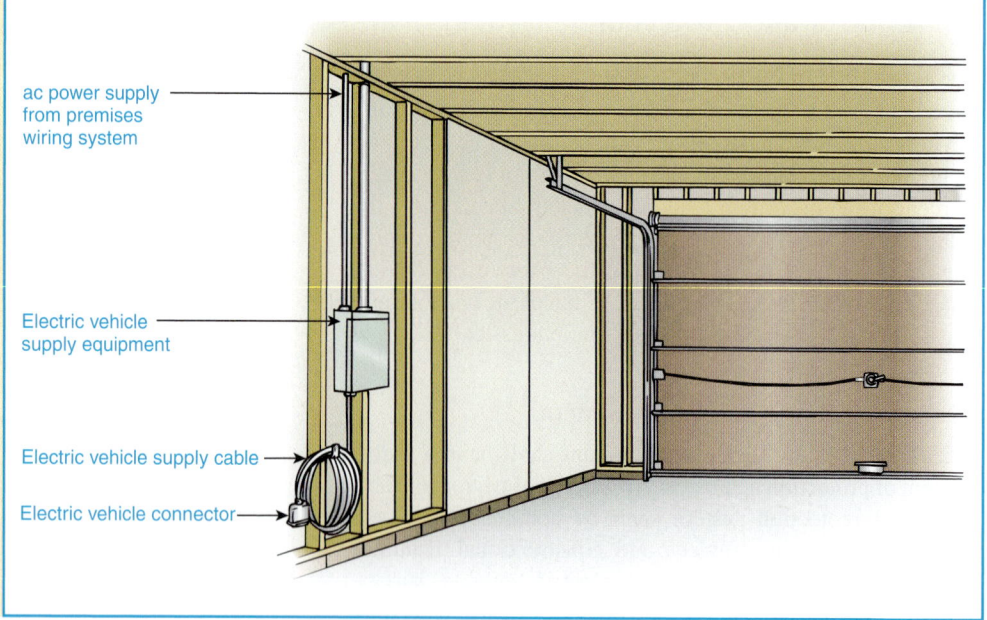

Exhibit 625.3 Garage interior with electric vehicle supply equipment where ventilation is not required.

ac power supply from premises wiring system

Electric vehicle supply equipment

Electric vehicle supply cable

Electric vehicle connector

(1) Table Values. For supply voltages and currents specified in Table 625.29(D)(1) or Table 625.29(D)(2), the minimum ventilation requirements shall be as specified in Table 625.29(D)(1) or Table 625.29(D)(2) for each of the total number of electric vehicles that can be charged at one time.

(2) Other Values. For supply voltages and currents other than specified in Table 625.29(D)(1) or Table 625.29(D)(2), the minimum ventilation requirements shall be calculated by means of the following general formulas as applicable.

(1) Single phase:

Ventilation $_{\text{single phase}}$ in cubic meters per minute $(\text{m}^3/\text{min}) =$

$$\frac{(\text{volts})(\text{amperes})}{1718}$$

Ventilation $_{\text{single phase}}$ in cubic feet per minute (cfm) $=$

$$\frac{(\text{volts})(\text{amperes})}{48.7}$$

(2) Three phase:

Ventilation $_{\text{three phase}}$ in cubic meters per minute $(\text{m}^3/\text{min}) =$

$$\frac{1.732(\text{volts})(\text{amperes})}{1718}$$

Ventilation$_{\text{three phase}}$ in cubic feet per minute (cfm) $=$

$$\frac{1.732(\text{volts})(\text{amperes})}{48.7}$$

(3) Engineered Systems. For an electric vehicle supply equipment ventilation system designed by a person qualified to perform such calculations as an integral part of a building's total ventilation system, the minimum ventilation requirements shall be permitted to be determined per calculations specified in the engineering study.

(4) Supply Circuits. The supply circuit to the mechanical ventilation equipment shall be electrically interlocked with the electric vehicle supply equipment and shall remain energized during the entire electric vehicle charging cycle. Electric vehicle supply equipment shall be marked in accordance with 625.15. Electric vehicle supply equipment receptacles rated at 125 volts, single phase, 15 and 20 amperes shall be marked in accordance with 625.15(C) and shall be switched, and the mechanical ventilation system shall be electrically interlocked through the switch supply power to the receptacle.

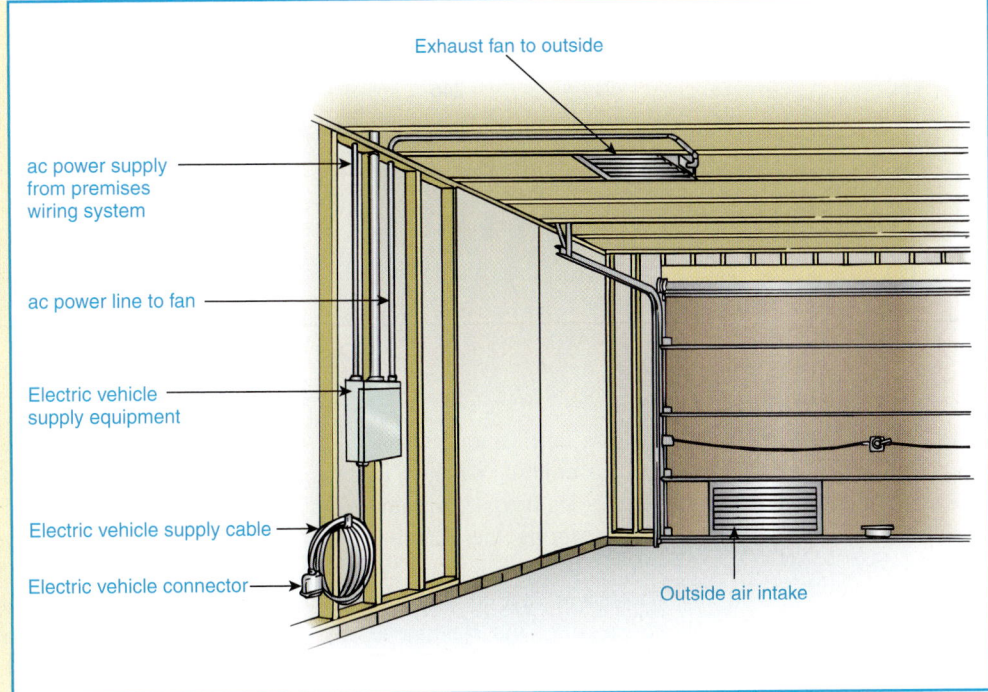

Exhibit 625.4 Garage interior with electric vehicle supply equipment and interlocked ventilation.

Exhaust fan to outside

ac power supply from premises wiring system

ac power line to fan

Electric vehicle supply equipment

Electric vehicle supply cable

Electric vehicle connector

Outside air intake

Table 625.29(D)(1) Minimum Ventilation Required in Cubic Meters per Minute (m³/min) for Each of the Total Number of Electric Vehicles That Can Be Charged at One Time

Branch-Circuit Ampere Rating	Branch-Circuit Voltage						
	Single Phase			3 Phase			
	120 V	208 V	240 V or 120/240 V	208 V or 208 Y/120 V	240 V	480 V or 480 Y/277 V	600 V or 600 Y/347 V
15	1.1	1.8	2.1	—	—	—	—
20	1.4	2.4	2.8	4.2	4.8	9.7	12
30	2.1	3.6	4.2	6.3	7.2	15	18
40	2.8	4.8	5.6	8.4	9.7	19	24
50	3.5	6.1	7.0	10	12	24	30
60	4.2	7.3	8.4	13	15	29	36
100	7.0	12	14	21	24	48	60
150	—	—	—	31	36	73	91
200	—	—	—	42	48	97	120
250	—	—	—	52	60	120	150
300	—	—	—	63	73	145	180
350	—	—	—	73	85	170	210
400	—	—	—	84	97	195	240

Table 625.29(D)(2) Minimum Ventilation Required in Cubic Feet per Minute (cfm) for Each of the Total Number of Electric Vehicles That Can Be Charged at One Time

Branch-Circuit Ampere Rating	Branch-Circuit Voltage						
	Single Phase			3 Phase			
	120 V	208 V	240 V or 120/240 V	208 V or 208 Y/120 V	240 V	480 V or 480 Y/277 V	600 V or 600 Y/347 V
15	37	64	74	—	—	—	—
20	49	85	99	148	171	342	427
30	74	128	148	222	256	512	641
40	99	171	197	296	342	683	854
50	123	214	246	370	427	854	1066
60	148	256	296	444	512	1025	1281
100	246	427	493	740	854	1708	2135
150	—	—	—	1110	1281	2562	3203
200	—	—	—	1480	1708	3416	4270
250	—	—	—	1850	2135	4270	5338
300	—	—	—	2221	2562	5125	6406
350	—	—	—	2591	2989	5979	7473
400	—	—	—	2961	3416	6832	8541

The intent of 625.29(D) is to ensure sufficient diffusion and dilution of hydrogen gas from gas-emitting batteries to prevent a hazardous condition. Certain batteries used in some electric vehicles emit hydrogen gas during the charging process.

Hydrogen is a colorless, odorless, tasteless, nontoxic flammable gas. At atmospheric pressure, the flammable range for hydrogen is 4 to 75 percent by volume in air.

NFPA 69, *Standard on Explosion Prevention Systems,* establishes requirements to ensure safety with flammable mixtures. The provisions of 3.3, Design and Operating Requirements, of NFPA 69 specify that combustible gas concentrations be restricted to 25 percent of the lower flammable limit. This design criterion provides a safety margin for personnel working with atmospheres containing hydrogen. Safety is accomplished by keeping the concentration of hydrogen below 25 percent of the lower flammability limit, or 1 percent (25 percent × 4 percent = 1 percent) hydrogen by volume in air, that is, below 10,000 ppm hydrogen.

A ventilation system for a typical residential-type garage includes both supply and mechanical exhaust equipment and is permanently installed. The equipment is located in the space such that it takes in air from outdoors to the space, circulates the air through the space, and exhausts the air directly to the outdoors. Typically, the equipment includes a passive vent for intake on one side of the enclosed space and an exhaust fan vented to the outside on the other side of the space. In enclosed commercial garages and other structures, additional ventilation is not required if the exhaust, as required by the building code for carbon monoxide or other purposes, is greater than the quantity listed in the table. Other engineered electric vehicle ventilation systems are allowed, provided they are designed properly. The electric vehicle charging area is permitted to be ventilated by the building ventilation system. The ventilation system and the charging system must be interlocked to prevent charging if the ventilation is not operating, as shown in Exhibit 625.5. This charging arrangement can be used with electric vehicles equipped with a self-contained charging system in which activation of the charging system does not depend on a signal from the electric vehicle. A manually operated switch controls the receptacle used to supply the vehicle charging system, and it is also interlocked with the power supply to the ventilation fan. This arrangement ensures that the

ventilation fan is operating whenever the vehicle charging receptacle is energized. A qualified person must perform the calculation of the ventilation requirements.

625.30 Outdoor Sites.

Outdoor sites shall include but not be limited to residential carports and driveways, curbside, open parking structures, parking lots, and commercial charging facilities.

Where the operation is conducted in outdoor or open locations, the off-gassing of hydrogen resulting from battery charging does not pose the same risk of creating an ignitible environment compared to indoor locations. The lighter-than-air hydrogen readily diffuses into the atmosphere. In addition to driveways and parking lots, structures with adequate natural ventilation, such as carports and open parking structures, do not require mechanical ventilation. NFPA 88A, *Standard for Parking Structures*, provides a quantifiable definition of the term *open parking structure*.

(A) Location. The electric vehicle supply equipment shall be located to permit direct connection to the electric vehicle.

(B) Height. Unless specifically listed for the purpose and location, the coupling means of electric vehicle supply equipment shall be stored or located at a height of not less than

Exhibit 625.5 An example of ventilation equipment electrically interlocked with the electric vehicle charging equipment receptacle.

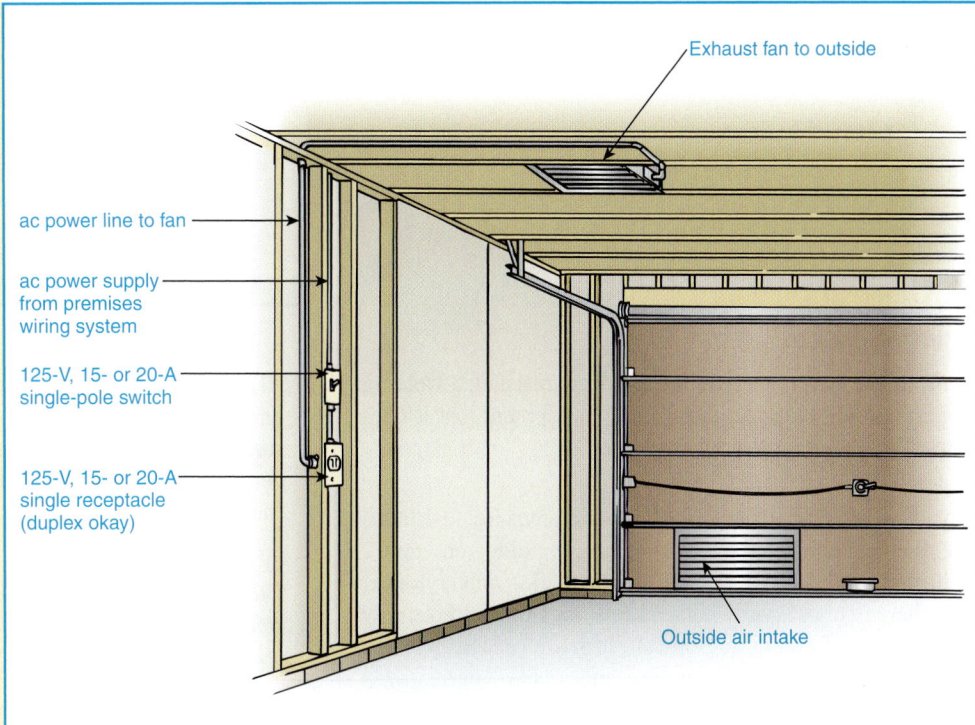

ac power line to fan

ac power supply from premises wiring system

125-V, 15- or 20-A single-pole switch

125-V, 15- or 20-A single receptacle (duplex okay)

Exhaust fan to outside

Outside air intake

600 mm (24 in.) and not more than 1.2 m (4 ft) above the parking surface.

ARTICLE 630
Electric Welders

Contents

I. General

630.1 Scope.

This article covers electric arc welding, resistance welding apparatus, and other similar welding equipment that is connected to an electric supply system.

The two general types of electric welding are resistance welding and arc welding. Resistance welding, or "spot" welding, is the process of joining, or electrically fusing together, two or more metal sheets or parts without any preparation of stock. The metal parts are placed between two electrodes or welding points, and a heavy current at a low voltage is passed through the electrodes. The metal parts

offer resistance to the flow of current such that they heat to a molten state, and a weld is made.

Arc welding is the butting of two metal parts to be welded and then the striking of an arc at this joint with a metal electrode (a flux-coated wire rod). The electrode itself is melted and supplies the extra metal necessary for joining the metal parts.

A transformer supplies current for one ac arc welder, and a generator or rectifier supplies current for one or more dc arc welders.

II. Arc Welders

630.11 Ampacity of Supply Conductors.

The ampacity of conductors for arc welders shall be in accordance with 630.11(A) and (B).

(A) **Individual Welders.** The ampacity of the supply conductors shall not be less than the I_{1eff} value on the rating plate. Alternatively, if the I_{1eff} is not given, the ampacity of the supply conductors shall not be less than the current value determined by multiplying the rated primary current in amperes given on the welder rating plate and the factor shown in Table 630.11(A) based on the duty cycle of the welder.

Table 630.11(A) Duty Cycle Multiplication Factors for Arc Welders

	Multiplier for Arc Welders	
Duty Cycle	Nonmotor Generator	Motor Generator
100	1.00	1.00
90	0.95	0.96
80	0.89	0.91
70	0.84	0.86
60	0.78	0.81
50	0.71	0.75
40	0.63	0.69
30	0.55	0.62
20 or less	0.45	0.55

(B) **Group of Welders.** Conductor ampacity shall be based on the individual currents determined in 630.11(A) as the sum of 100 percent of the two largest welders, plus 85 percent of the third largest welder, plus 70 percent of the fourth largest welder, plus 60 percent of all remaining welders.

Exception: Percentage values lower than those given in (B) shall be permitted in cases where the work is such that a high-operating duty cycle for individual welders is impossible.

FPN: Duty cycle considers welder loading based on the use to be made of each welder and the number of welders supplied by the conductors that will be in use at the same time. The load value used for each welder considers both the magnitude and the duration of the load while the welder is in use.

Even under high-production conditions, the loads on transformer arc welders are considered intermittent; therefore, it is permissible to reduce the ampacity of feeder conductors supplying several transformers (three or more) to the allowable percentages permitted in 630.11(B). It is obvious that intermittent transformer arc welder loads would be considerably less than a continuous load equal to the sum of the full-load current ratings of all the transformers. See also 630.31(B). The ampacity of conductors supplying welders is based on the I_{1eff} rating on the welder rating plate. If the I_{1eff} rating is not available, the size of supply conductors to welders may be calculated. The calculation is done by selecting the appropriate factor from 630.11(A) based on the type of welder and duty cycle of the welder. The selected factor is then multiplied by the primary current rating from the welder rating plate to determine the minimum ampacity of the supply conductors.

Obsolete terms and ratings such as *nameplate* and *1-hour duty cycle* were removed from 630.11 in the 1999 *NEC*. The terms I_{1eff} and I_{1max} were added to the marking requirements. A new FPN includes a formula for I_{1eff}, which is preferred over the value derived from using the factors in 630.11. The calculated value is still allowed. I_{1max} is basically the same as the rated primary current.

630.12 Overcurrent Protection.

Overcurrent protection for arc welders shall be as provided in 630.12(A) and (B). Where the values as determined by this section do not correspond with the standard ampere ratings provided in 240.6 or the rating or setting specified results in unnecessary opening of the overcurrent device, the next higher standard rating or setting shall be permitted.

(A) For Welders. Each welder shall have overcurrent protection rated or set at not more than 200 percent of I_{1max}. Alternatively, if the I_{1max} is not given, the overcurrent protection shall be rated or set at not more than 200 percent of the rated primary current of the welder.

An overcurrent device shall not be required for a welder that has supply conductors protected by an overcurrent device rated or set at not more than 200 percent of I_{1max} or the rated primary current of the welder.

If the supply conductors for a welder are protected by an overcurrent device rated or set at not more than 200 percent of I_{1max} or rated primary current of the welder, a separate overcurrent device shall not be required.

(B) For Conductors. Conductors that supply one or more welders shall be protected by an overcurrent device rated or set at not more than 200 percent of the conductor ampacity.

Some arc-welding machines have a welding range involving an excess secondary-current output capacity beyond that indicated by the secondary rating marked on the machines. This excess capacity (generally not more than 150 percent of the marked output capacity) is usually supplied by means of one or more secondary taps in addition to the tap(s) intended for normal output current; the higher currents thus available are intended to provide for heavier welding work, including the use of larger-sized electrodes. This excess capacity is somewhat analogous to the inherent overload capacity of motors and transformers. However, the use of this excess current capacity and the overloading of welding machines, except for relatively short periods of time, could be hazardous and should be undertaken with caution.

FPN: I_{1max} is the maximum value of the rated supply current at maximum rated output. I_{1eff} is the maximum value of the effective supply current, calculated from the rated supply current (I_1), the corresponding duty cycle (duty factor) (X), and the supply current at no-load (I_0) by the following formula:

$$I_{1eff} = \sqrt{I_1^2 X + I_0^2(1 - X)}$$

630.13 Disconnecting Means.

A disconnecting means shall be provided in the supply circuit for each arc welder that is not equipped with a disconnect mounted as an integral part of the welder.

The disconnecting means shall be a switch or circuit breaker, and its rating shall not be less than that necessary to accommodate overcurrent protection as specified under 630.12.

630.14 Marking.

A rating plate shall be provided for arc welders giving the following information:

(1) Name of manufacturer
(2) Frequency
(3) Number of phases
(4) Primary voltage
(5) I_{1max} and I_{1eff}, or rated primary current
(6) Maximum open-circuit voltage
(7) Rated secondary current and
(8) Basis of rating, such as the duty cycle

630.15 Grounding of Welder Secondary Circuit.

The secondary circuit conductors of an arc welder, consisting of the electrode conductor and the work conductor, shall not be considered as premises wiring for the purpose of applying Article 250.

In theory and in accordance with the *NEC* definition of *separately derived system,* the secondary circuit of an arc welder could be viewed as such a system. However, the intended operation of a welder is to create a high-current circuit between the electrode and the work surface. In the normal operation of an ac power distribution system, such an event would be considered a fault, and the operation of an overcurrent device to open the circuit and clear the fault is a fundamental concept of Articles 240 and 250. In the case of an arc welder, the opening of an overcurrent device is not intended unless the welding operation significantly exceeds the operating parameters of the welder. Grounding of a welder secondary terminal has the potential to cause excessive and potentially degrading parallel currents on power system equipment grounding conductors.

This new requirement clarifies that for the purposes of Article 250, specifically the requirements covering grounding of separately derived systems, the secondary circuit of a welder is not treated as premises wiring and is not required to be grounded as such. This new wording modifies Article 250 for the purposes of electric welder secondary circuits and thereby removes any potential conflict where grounding in the welder secondary circuit occurs at the work object.

FPN: Connecting welder secondary circuits to grounded objects can create parallel paths and can cause objectionable current over equipment grounding conductors.

III. Resistance Welders

630.31 Ampacity of Supply Conductors.

The ampacity of the supply conductors for resistance welders necessary to limit the voltage drop to a value permissible for the satisfactory performance of the welder is usually greater than that required to prevent overheating as described in 630.31(A) and (B).

(A) Individual Welders. The rated ampacity for conductors for individual welders shall comply with the following:

(1) The ampacity of the supply conductors for a welder that may be operated at different times at different values of primary current or duty cycle shall not be less than 70 percent of the rated primary current for seam and automatically fed welders and 50 percent of the rated primary current for manually operated nonautomatic welders.

(2) The ampacity of the supply conductors for a welder wired for a specific operation for which the actual primary current and duty cycle are known and remain unchanged shall not be less than the product of the actual primary current and the multiplier specified in Table 630.31(A) for the duty cycle at which the welder will be operated.

Table 630.31(A)(2) Duty Cycle Multiplication Factors for Resistance Welders

Duty Cycle (percent)	Multiplier
50	0.71
40	0.63
30	0.55
25	0.50
20	0.45
15	0.39
10	0.32
7.5	0.27
5 or less	0.22

(B) Groups of Welders. The ampacity of conductors that supply two or more welders shall not be less than the sum of the value obtained in accordance with 630.31(A) for the largest welder supplied and 60 percent of the values obtained for all the other welders supplied.

FPN: **Explanation of Terms**

(1) The *rated primary current* is the rated kilovolt-amperes (kVA) multiplied by 1000 and divided by the rated primary voltage, using values given on the nameplate.
(2) The *actual primary current* is the current drawn from the supply circuit during each welder operation at the particular heat tap and control setting used.
(3) The *duty cycle* is the percentage of the time during which the welder is loaded. For instance, a spot welder supplied by a 60-Hz system (216,000 cycles per hour) making four hundred 15-cycle welds per hour would have a duty cycle of 2.8 percent (400 multiplied by 15, divided by 216,000, multiplied by 100). A seam welder operating 2 cycles "on" and 2 cycles "off" would have a duty cycle of 50 percent.

The ampacity of supply conductors for a welder that is not wired for a specific function (i.e., one operated at varying intervals for different applications, such as dissimilar metals or thicknesses) is permitted to be 70 percent of the rated primary current for automatically fed welders and 50 percent of the rated primary current for manually operated welders. The rated primary current can be determined using the following equation with the values given on the welder nameplate:

$$\text{Rated primary current} = \frac{\text{welder kVA} \times 1000}{\text{rated primary voltage}}$$

Where the actual primary current and the duty cycle are known, such as for a welder wired for a specific operation, the ampacity of the supply conductors is not permitted to be less than the product of the actual primary current (current drawn during weld operation) and the multiplier, as provided in 630.31(A)(2), for the duty cycle at which the welder will be operated. For example, a spot welder is specifically set to perform 300 welds per hour on a 60-Hz system. Each weld draws current for 16 cycles. During the 1-hour period, the welder draws current for 4800 cycles (300 × 16). There are 216,000 cycles per hour (60 × 60 × 60). The duty cycle is calculated as follows:

$$\frac{4800}{216,000} \times 100\% = 2.2\% \quad \text{(duty cycle)}$$

For a seam welder that draws current for 3 cycles and is off for 4 cycles during every 7-cycle period, the duty cycle is calculated as follows:

$$\frac{3}{7} \times 100\% = 42.9\% \quad \text{(duty cycle)}$$

An instrument capable of measuring current impulses for 3 cycles ($\frac{1}{20}$ second), as shown in the preceding example, is required to measure the actual primary current. The duty cycle is set for a specific operation by adjusting the controller for the welder. For the sizing of supply conductors, voltage drop should be limited to a value permissible for the satisfactory performance of the welder.

630.32 Overcurrent Protection.

Overcurrent protection for resistance welders shall be as provided in 630.32(A) and (B). Where the values as determined by this section do not correspond with the standard ampere ratings provided in 240.6 or the rating or setting specified results in unnecessary opening of the overcurrent device, a higher rating or setting that does not exceed the next higher standard ampere rating shall be permitted.

(A) For Welders. Each welder shall have an overcurrent device rated or set at not more than 300 percent of the rated primary current of the welder. If the supply conductors for a welder are protected by an overcurrent device rated or set at not more than 200 percent of the rated primary current of the welder, a separate overcurrent device shall not be required.

(B) For Conductors. Conductors that supply one or more welders shall be protected by an overcurrent device rated or set at not more than 300 percent of the conductor rating.

630.33 Disconnecting Means.

A switch or circuit breaker shall be provided by which each resistance welder and its control equipment can be disconnected from the supply circuit. The ampere rating of this disconnecting means shall not be less than the supply conductor ampacity determined in accordance with 630.31. The supply circuit switch shall be permitted as the welder

disconnecting means where the circuit supplies only one welder.

630.34 Marking.

A nameplate shall be provided for each resistance welder, giving the following information:

(1) Name of manufacturer
(2) Frequency
(3) Primary voltage
(4) Rated kilovolt-amperes (kVA) at 50 percent duty cycle
(5) Maximum and minimum open-circuit secondary voltage
(6) Short-circuit secondary current at maximum secondary voltage
(7) Specified throat and gap setting

IV. Welding Cable

630.41 Conductors.

Insulation of conductors intended for use in the secondary circuit of electric welders shall be flame retardant.

630.42 Installation.

Cables shall be permitted to be installed in a dedicated cable tray as provided in 630.42(A), (B), and (C).

(A) Cable Support. The cable tray shall provide support at not greater than 150-mm (6-in.) intervals.

(B) Spread of Fire and Products of Combustion. The installation shall comply with 300.21.

(C) Signs. A permanent sign shall be attached to the cable tray at intervals not greater than 6.0 m (20 ft). The sign shall read as follows:

CABLE TRAY
FOR WELDING CABLES ONLY

ARTICLE 640
Audio Signal Processing, Amplification, and Reproduction Equipment

Contents

I. General

640.1 Scope.

This article covers equipment and wiring for audio signal generation, recording, processing, amplification and reproduction; distribution of sound; public address; speech input systems; temporary audio system installations; and electronic organs or other electronic musical instruments. This also includes audio systems subject to Article 517, Part VI, and Articles 518, 520, 525, and 530.

Equipment covered by Article 640 includes amplifiers; public address (PA) systems and centralized sound systems used in schools, factories, businesses, stadiums, and similar locations; intercommunications devices and systems; and devices used for recording or reproducing voice or music. The scope remains limited to equipment whose main function is the processing, distribution, amplification, and reproduction of audio frequency bandwidth signals. This does not preclude equipment that uses radio frequency or other forms of transmission between equipment components, such as wireless microphone systems.

Electronic organs are synthesizers, and synthesizers also generate audio signals. For the sake of clarity, electronic organs are still uniquely cited in the scope, and electronic musical instruments have been added to cover all other forms of electronic tone generation. Electronic musical instruments create an electronic signal as their sole or primary output and require amplification and reproduction equipment to be audible.

FPN No. 1: Examples of permanently installed distributed audio system locations include, but are not limited to, restaurant, hotel, business office, commercial and retail sales environments, churches, and schools. Both portable and permanently installed equipment locations include, but are not limited to, residences, auditoriums, theaters, stadiums, and movie and television studios.

Temporary installations include, but are not limited to, auditoriums, theaters, stadiums (which use both temporary and permanently installed systems), and outdoor events such as fairs, festivals, circuses, public events, and concerts.

FPN No. 2: Fire and burglary alarm signaling devices are specifically not encompassed by this article.

640.2 Definitions.

For purposes of this article, the following definitions apply.

Abandoned Audio Distribution Cable. Installed audio distribution cable that is not terminated at equipment and not identified for future use with a tag.

Audio Amplifier or Pre-Amplifier. Electronic equipment that increases the current or voltage, or both, potential of an audio signal intended for use by another piece of audio equipment. *Amplifier* is the term used to denote an audio amplifier within this article.

Audio Autotransformer. A transformer with a single winding and multiple taps intended for use with an amplifier loudspeaker signal output.

Audio Signal Processing Equipment. Electrically operated equipment that produces or processes, or both, electronic signals that, when appropriately amplified and reproduced by a loudspeaker, produce an acoustic signal within the range of normal human hearing (typically 20-20 kHz). Within this article, the terms *equipment* and *audio equipment* are assumed to be equivalent to audio signal processing equipment.

> FPN: This equipment includes, but is not limited to, loudspeakers; headphones; pre-amplifiers; microphones and their power supplies; mixers; MIDI (musical instrument digital interface) equipment or other digital control systems; equalizers, compressors, and other audio signal processing equipment; audio media recording and playback equipment, including turntables, tape decks and disk players (audio and multimedia), synthesizers, tone generators, and electronic organs. Electronic organs and synthesizers may have integral or separate amplification and loudspeakers. With the exception of amplifier outputs, virtually all such equipment is used to process signals (utilizing analog or digital techniques) that have nonhazardous levels of voltage or current potential.

The definition of audio signal processing equipment clarifies the limits of signal processing (frequency bandwidth), which falls under Article 640. The FPN enumerates the breadth of equipment considered to fall within the defined scope of Article 640. It is intended to provide a sufficiently broad list of current technology equipment to assist in determining the applicability of Article 640 to the equipment under review.

"MIDI (musical instrument digital interface) equipment or other digital control systems" is mentioned specifically

because, while MIDI or similar digital control signals may issue from an electronic musical instrument, such signals also can be obtained from a computer that is appropriately configured to perform similar controlling functions.

Audio System. Within this article, the totality of all equipment and interconnecting wiring used to fabricate a fully functional audio signal processing, amplification, and reproduction system.

Audio Transformer. A transformer with two or more electrically isolated windings and multiple taps intended for use with an amplifier loudspeaker signal output.

The definition of audio transformer is included to clearly state that such transformers are intended only for use with audio signals, not light and power.

Equipment Rack. A framework for the support, enclosure, or both, of equipment. May be portable or stationary. See ANSI/EIA/310-D-1992, *Cabinets, Racks, Panels and Associated Equipment*.

ANSI/EIA/310-D-1992, Cabinets, Racks, Panels and Associated Equipment, defines commercial equipment racks. However, custom-fabricated racks that utilize equipment mounting hole patterns that generally comply with this standard might not be constructed of steel. Within Article 640, both equipment rack and rack are used to refer to equipment enclosures that are conceptually similar in intended use to those defined by the ANSI/EIA standard.

Loudspeaker. Equipment that converts an ac electric signal into an acoustic signal. The term *speaker* is commonly used to mean *loudspeaker*.

Exhibit 640.1 is an example of a surface-mounted loudspeaker.

Maximum Output Power. The maximum output power delivered by an amplifier into its rated load as determined under specified test conditions. This may exceed the manufacturer's rated output power for the same amplifier.

Mixer. Equipment used to combine and level match a multiplicity of electronic signals, such as from microphones, electronic instruments, and recorded audio.

Typical peak signal operating voltages for such equipment vary from a few millivolts for microphones to 2 to 4 volts for disk players. A mixer's purpose is to balance these inputs to provide (typically) a 1-volt peak output signal to an amplifier.

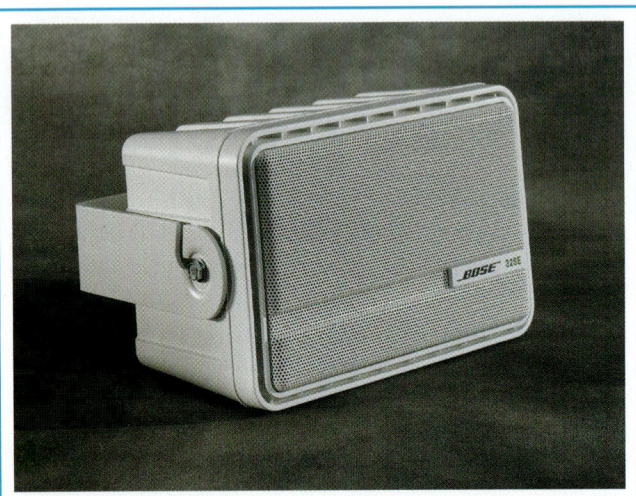

Exhibit 640.1 Surface-mounted loudspeaker for indoor or outdoor use. (Courtesy of Bose Corp.)

Mixer–Amplifier. Equipment that combines the functions of a mixer and amplifier within a single enclosure.

Portable Equipment. Equipment fed with portable cords or cables intended to be moved from one place to another.

Powered Loudspeaker. Equipment that consists of a loudspeaker and amplifier within the same enclosure. Other signal processing may also be included.

Rated Load Impedance. The amplifier manufacturer's stated or marked speaker impedance into which an amplifier will deliver its rated output power. 2Ω, 4Ω, and 8Ω are typical ratings.

Rated Output Power. The amplifier manufacturer's stated or marked output power capability into its rated load.

Rated Output Voltage. For audio amplifiers of the constant-voltage type, the nominal output voltage when the amplifier is delivering full rated power. Rated output voltage is used for determining approximate acoustic output in distributed speaker systems that typically employ impedance matching transformers. Typical ratings are 25 volts, 70.7 volts, and 100 volts.

Technical Power System. An electrical distribution system with grounding in accordance with 250.146(D), where the equipment grounding conductor is isolated from the premises grounded conductor except at a single grounded termination point within a branch-circuit panelboard, at the originating (main breaker) branch-circuit panelboard, or at the premises grounding electrode.

The terms *technical power* and *technical ground* are commonly used by audio/video technicians and electricians to

designate a wiring system that is in compliance with 250.146(D). Including the definition of *technical power system* in Article 640 is intended to broaden the scope of this term to include the commonly employed distribution systems fabricated in compliance with 250.146(D).

Temporary Equipment. Portable wiring and equipment intended for use with events of a transient or temporary nature where all equipment is presumed to be removed at the conclusion of the event.

Temporary equipment may be used in facilities of a permanent or temporary nature or with no facilities other than a source of electrical power. Locations would include both indoor and outdoor areas such as athletic facilities, halls, auditoriums and concert shells, athletic fields, beaches, or other places designated for public assembly.

640.3 Locations and Other Articles.

Circuits and equipment shall comply with 640.3(A) through (L), as applicable.

(A) Spread of Fire or Products of Combustion. The accessible portion of abandoned audio distribution cables shall not be permitted to remain. See 300.21.

(B) Ducts, Plenums, and Other Air-Handling Spaces. See 300.22 for circuits and equipment installed in ducts or plenums or other space used for environmental air.

> FPN: NFPA 90A-1999, *Standard for the Installation of Air Conditioning and Ventilation Systems*, 2-3.10.2(a), Exception No. 3, permits loudspeakers, loudspeaker assemblies, and their accessories listed in accordance with UL 2043-1996, *Fire Test for Heat and Visible Smoke Release for Discrete Products and Their Accessories Installed in Air-Handling Spaces*, to be installed in other spaces used for environmental air (ceiling cavity plenums).

(C) Cable Trays. Cable trays shall be used in accordance with Article 392.

> FPN: See 725.61(C) for the use of Class 2, Class 3, and Type PLTC cable in cable trays.

(D) Hazardous (Classified) Locations. Equipment used in hazardous (classified) locations shall comply with the applicable requirements of Chapter 5.

(E) Places of Assembly. Equipment used in places of assembly shall comply with Article 518.

Places of assembly, as described in 518.2, are some of the most common locations for the installation of both distributed audio systems (e.g., background music) and centralized systems (permanently installed sound reinforcement systems for meeting rooms, auditoriums, gymnasiums, and the like).

(F) Theaters, Audience Areas of Motion Picture and Television Studios, and Similar Locations. Equipment used in theaters, audience areas of motion picture and television studios, and similar locations shall comply with Article 520.

(G) Carnivals, Circuses, Fairs, and Similar Events. Equipment used in carnivals, circuses, fairs, and similar events shall comply with Article 525.

(H) Motion Picture and Television Studios. Equipment used in motion picture and television studios shall comply with Article 530.

(I) Swimming Pools, Fountains, and Similar Locations. Audio equipment used in or near swimming pools, fountains, and similar locations shall comply with Article 680.

The underwater installation of audio equipment is found in 680.27. The acceptable placement, wiring, and use of equipment used near (rather than immersed within) bodies of water, both natural and artificial, are covered in 640.10. Article 680 does not cover natural bodies of water, such as lakes, rivers, and streams.

(J) Combination Systems. Where the authority having jurisdiction permits audio systems for paging or music, or both, to be combined with fire alarm systems, the wiring shall comply with Article 760.

In addition to alarm tones, fire alarm systems frequently use loudspeakers for verbal announcements. All such systems need to comply with Article 760 and *NFPA 72, National Fire Alarm Code*. Audio systems that use a paging or background music system are permitted to be used as part of a fire alarm warning system, but they require compliance with Article 760. The installation of fire alarm systems is governed by the *National Fire Alarm Code*. Refer to NFPA *101®, Life Safety Code®*, for multipurpose systems.

FPN: For installation requirements for such combination systems, refer to NFPA 72-1999, *National Fire Alarm Code®, and NFPA 101®-2000, Life Safety Code®*.

(K) Antennas. Equipment used in audio systems that contain an audio or video tuner and an antenna input shall comply with Article 810. Wiring other than antenna wiring that connects such equipment to other audio equipment shall comply with this article.

The term *receiver* is commonly used in the consumer market to mean an amplifier combined with a radio tuner (typically AM/FM) and other signal processing and/or switching functions. Except for the tuner function and the antenna input, the signal processing functions are the same as those provided by equipment in Article 640. Article 810, Part II, covers the antenna installation for such equipment, and 810.3 references Article 640 as appropriate for wiring requirements (other than for the antenna).

(L) Generators. Generators shall be installed in accordance with 445.2 through 445.10. Grounding of portable and vehicle-mounted generators shall be in accordance with 250.34.

640.4 Protection of Electrical Equipment.

Amplifiers, loudspeakers, and other equipment shall be so located or protected so as to guard against environmental exposure or physical damage, such as might result in fire, shock, or personal hazard.

640.5 Access to Electrical Equipment Behind Panels Designed to Allow Access.

Access to equipment shall not be denied by an accumulation of wires and cables that prevents removal of panels, including suspended ceiling panels.

640.6 Mechanical Execution of Work.

Equipment and cabling shall be installed in a neat and workmanlike manner. Cables installed exposed on the surface of ceiling and sidewalls shall be supported by the structural components of the building structure in such a manner that the cable will not be damaged by normal building use. Such cables shall be attached to structural components by straps, staples, hangers, or similar fittings designed and installed so as not to damage the cable. The installation shall also conform with 300.4(D).

640.7 Grounding.

(A) General. Wireways and auxiliary gutters shall be grounded and bonded in accordance with the requirements of Article 250. Where the wireway or auxiliary gutter does not contain power-supply wires, the equipment grounding conductor shall not be required to be larger than 14 AWG copper or its equivalent. Where the wireway or auxiliary gutter contains power-supply wires, the equipment grounding conductor shall not be smaller than specified in 250.122.

(B) Separately Derived Systems with 60 Volts to Ground. Grounding of separately derived systems with 60 volts to ground shall be in accordance with 647.6.

(C) Isolated Ground Receptacles. Isolated grounding-type receptacles shall be permitted as described in 250.146(D), and for the implementation of other technical power systems in compliance with Article 250. For separately derived systems with 60 volts to ground, the branch-circuit equipment grounding conductor shall be terminated as required in 647.6(B).

The reference to 647.6 in 640.7(B) provides guidance for grounding of separately derived systems operating at 60 volts to ground. The requirements for these systems, formerly located in Part G of Article 530 in the 1999 *Code,* have been relocated in the 2002 *Code* to new Article 647, Sensitive Electronic Equipment. One of the most significant impacts of this relocation is that these systems can now be used in other than motion picture and television studios. This separately derived system is used as a technique for the reduction of electromagnetic noise in audio systems and in video systems.

The intent of 640.7(C) is to clarify the proper use of isolated ground receptacles when used with technical power systems of the type that are separately derived systems with 60 volts to ground.

FPN: See 406.2(D) for grounding-type receptacles and required identification.

640.8 Grouping of Conductors.

Insulated conductors of different systems grouped or bundled so as to be in close physical contact with each other in the same raceway or other enclosure, or in portable cords or cables, shall comply with 300.3(C)(1).

640.9 Wiring Methods.

(A) Wiring to and Between Audio Equipment.

(1) Power Wiring. Wiring and equipment from source of power to and between devices connected to the premises wiring systems shall comply with the requirements of Chapters 1 through 4, except as modified by this article.

(2) Separately Derived Power Systems. Separately derived systems shall comply with the applicable articles of this *Code*, except as modified by this article. Separately derived systems with 60 volts to ground shall be permitted for use in audio system installations as specified in Article 647.

(3) Other Wiring. All wiring not connected to the premises wiring system or to a wiring system separately derived from the premises wiring system shall comply with Article 725.

(B) Auxiliary Power Supply Wiring. Equipment that has a separate input for an auxiliary power supply shall be wired in compliance with Article 725. Battery installation shall be in accordance with Article 480.

FPN No. 1: This section does not apply to the use of uninterruptible power supply (*ups*) equipment, or other sources of supply, that are intended to act as a direct replacement for the primary circuit power source and are connected to the primary circuit input.

Audio equipment with a separate input for an auxiliary power supply is typically used for emergency paging or fire alarm systems. These auxiliary power supply inputs typically range from 12 to 48 volts dc. Article 480 adequately covers installation and overcurrent protection of battery circuits of this type. The term *auxiliary* is used to indicate that the equipment is also capable of being powered by the premises wiring system through an independent input connector, cord, or cable.

FPN No. 1 clarifies that, while the equipment can be powered by a replacement source for the premises wiring system, such a source (a UPS or a standby generator) is not considered to be supplying the equipment auxiliary power supply unless it is directly connected to the auxiliary power supply input and supplying a dc voltage.

FPN No. 2: Refer to NFPA 72-1999, *National Fire Alarm Code*, where equipment is used for a fire alarm system.

(C) Output Wiring and Listing of Amplifiers. Amplifiers with output circuits carrying audio program signals shall be permitted to employ Class 1, Class 2, or Class 3 wiring where the amplifier is listed and marked for use with the specific class of wiring method. Such listing shall ensure the energy output is equivalent to the shock and fire risk of the same class as stated in Article 725. Overcurrent protection shall be provided and shall be permitted to be inherent in the amplifier.

Audio amplifier output circuits wired using Class 1 wiring methods shall be considered equivalent to Class 1 circuits and be installed in accordance with 725.25, where applicable.

Audio amplifier output circuits wired using Class 2 or Class 3 wiring methods shall be considered equivalent to Class 2 or Class 3 circuits respectively. They shall use conductors insulated at not less than the requirements of 725.71, and shall be installed in accordance with 725.54 and 725.61.

FPN No. 1: ANSI/UL 1711-1994, *Amplifiers for Fire Protective Signaling Systems*, contains requirements for the listing of amplifiers used for fire alarm systems in compliance with NFPA 72-1999, *National Fire Alarm Code*.

FPN No. 2: Examples of requirements for listing amplifiers used in residential, commercial, and professional use are found in ANSI/UL 813-1996, *Commercial Audio Equipment*, ANSI/UL 1419-1997, *Professional Video and Audio Equipment*, ANSI/UL 1492-1996, *Audio-*

Video Products and Accessories, or ANSI/UL 6500-1996, *Audio/Video and Musical Instrument Apparatus for Household, Commercial, and Similar Use.*

(D) Use of Audio Transformers and Autotransformers. Audio transformers and autotransformers shall only be used for audio signals in a manner so as not to exceed the manufacturer's stated input or output voltage, impedance, or power limitations. The input or output wires of an audio transformer or autotransformer shall be allowed to connect directly to the amplifier or loudspeaker terminals. No electrical terminal or lead shall be required to be grounded or bonded.

Audio transformers and autotransformers are commonly used between the amplifier output and the loudspeaker input for the following reasons:

1. At the output of the amplifier: to change the amplifier's operating voltage to match the design impedance of the loudspeaker
2. At the loudspeaker, where the inherently low voice coil impedance is raised: to match the output voltage of the amplifier (or autotransformer)
3. Between the amplifier output and loudspeaker input: as an attenuating device (volume control)

Audio autotransformers are similar only in concept to autotransformers used for light and power, but they are not designed for such use. Audio transformers are commonly used to provide electrical isolation of the speakers from the signal source. Either type of audio transformer (two winding or autotransformer) is frequently referred to as an "impedance matching transformer." The last sentence of 640.9(D) specifically addresses the fact that electrical terminals are not to be treated in the same manner as transformers used for light and power might be (e.g., grounding the common terminal of an autotransformer). Some amplifier outputs are deliberately isolated from equipment ground, in which case such a connection could damage the amplifier and violate the manufacturer's recommended use. The frame of the transformer may or may not require bonding, depending on the manufacturer's installation instructions.

640.10 Audio Systems Near Bodies of Water.

Audio systems near bodies of water, either natural or artificial, shall be subject to the restrictions specified in 640.10(A) and (B).

Exception: This section does not include audio systems intended for use on boats, yachts, or other forms of land or water transportation used near bodies of water, whether or not supplied by branch-circuit power.

FPN: See 680.27(A) for installation of underwater audio equipment.

(A) Equipment Supplied by Branch-Circuit Power. Audio system equipment supplied by branch-circuit power shall not be placed laterally within 1.5 m (5 ft) of the inside wall of a pool, spa, hot tub, or fountain, or within 1.5 m (5 ft) of the prevailing or tidal high water mark. The equipment shall be provided with branch-circuit power protected by a ground-fault circuit interrupter where required by other articles.

Article 680 is limited by its scope to the use of underwater loudspeakers. This particular application for audio equipment is unique in construction and wiring to pools and is appropriate for Article 680.

Other locations where audio equipment might be used "near bodies of water" are not addressed in Article 680. The purpose of 640.10(A), in conjunction with the other requirements of Article 640, is to address locations where audio equipment is used near bodies of water.

The exception to 640.10(A) excludes sound systems on vehicles used in or near water, such as amphibious vehicles and boats of all sizes.

The term *prevailing or tidal high water mark* recognizes that the edges of natural bodies of water can advance or recede. This phrase makes clear that such changes can be anticipated.

The requirement for use of a ground-fault circuit interrupter is placed here to ensure that personnel are properly protected. Where the equipment (specifically an amplifier or a receiver) is installed in an electrical cabinet or a room not near a body of water, the requirement for GFCI protection does not apply, unless required by other sections of the *Code.*

(B) Equipment Not Supplied by Branch-Circuit Power. Audio system equipment powered by a listed Class 2 power supply or by the output of an amplifier listed as permitting the use of Class 2 wiring shall only be restricted in placement by the manufacturer's recommendations.

FPN: Placement of the power supply or amplifier, if supplied by branch-circuit power, is still subject to 640.10(A).

II. Permanent Audio System Installations

Permanent audio systems are characterized by fixed locations for the wiring, signal processing equipment, and reproduction equipment. Wiring is attached to the building structure and is frequently concealed. Speakers in commercial, hospital, school, and restaurant areas are commonly recessed into ceiling or wall surfaces, or they are mounted to structure surfaces using brackets, usually beyond the reach of a standing person.

640.21 Use of Flexible Cords and Cables.

(A) Between Equipment and Branch-Circuit Power. Power supply cords for audio equipment shall be suitable for the use and shall be permitted to be used where the interchange, maintenance, or repair of such equipment is facilitated through the use of a power supply cord.

(B) Between Loudspeakers and Amplifiers or Between Loudspeakers. Cables used to connect loudspeakers to each other or to an amplifier shall comply with Article 725. Other listed cable types and assemblies, including optional hybrid communications, signal, and optical fiber cables, shall be permitted.

Some loudspeakers are specifically identified as being for outdoor use. The requirements of 110.11 specify that electrical equipment and conductors be identified for use in the operating environment. This general requirement applies to equipment and conductors used in conjunction with the installation of audio equipment. Exhibit 640.2 shows a loudspeaker identified for outdoor use and located partially in ground. The conductors supplying this outdoor speaker must also be identified for the environment.

Exhibit 640.2 Loudspeaker for outdoor use aboveground or (as shown) partially in ground. (Courtesy of Bose Corp.)

(C) Between Equipment. Cables used for the distribution of audio signals between equipment shall comply with Article 725. Other listed cable types and assemblies, including optional hybrid communications, signal, and optical fiber cables, shall be permitted. Other cable types and assemblies specified by the equipment manufacturer as acceptable for the use shall be permitted in accordance with 110.3(B).

(D) Between Equipment and Power Supplies Other Than Branch-Circuit Power. The following power supplies, other than branch-circuit power supplies, shall be installed and wired between equipment in accordance with the requirements of this *Code* for the voltage and power delivered:

(1) Storage batteries
(2) Transformers
(3) Transformer rectifiers
(4) Other ac or dc power supplies

> FPN: For some equipment, these sources such as in items (1) and (2) serve as the only source of power. These could, in turn, be supplied with intermittent or continuous branch-circuit power.

(E) Between Equipment Racks and Premises Wiring System. Flexible cords and cables shall be permitted for the electrical connection of permanently installed equipment racks to the premises wiring system to facilitate access to equipment or for the purpose of isolating the technical power system of the rack from the premises ground. Connection shall be made either using approved plugs and receptacles or by direct connection within an approved enclosure. Flexible cords and cables shall not be subjected to physical manipulation or abuse while the rack is in use.

640.22 Wiring of Equipment Racks and Enclosures.

Metal equipment racks and enclosures shall be grounded. Bonding shall not be required if the rack is connected to a technical power ground.

Equipment racks shall be wired in a neat and workmanlike manner. Wires, cables, structural components, or other equipment shall not be placed in such a manner as to prevent reasonable access to equipment power switches and resettable or replaceable circuit overcurrent protection devices.

Supply cords or cables, if used, shall terminate within the equipment rack enclosure in an identified connector assembly. The supply cords or cable (and connector assembly, if used) shall have sufficient ampacity to carry the total load connected to the equipment rack and shall be protected by overcurrent devices.

640.23 Conduit or Tubing.

(A) Number of Conductors. The number of conductors permitted in a single conduit or tubing shall not exceed the percentage fill specified in Table 1, Chapter 9.

(B) Nonmetallic Conduit or Tubing and Insulating Bushings. The use of nonmetallic conduit or tubing and insulating bushings shall be permitted where a technical power system is employed and shall comply with applicable articles.

640.24 Wireways, Gutters, and Auxiliary Gutters.

The use of metallic and nonmetallic wireways, gutters, and auxiliary gutters shall be permitted for use with audio signal conductors and shall comply with applicable articles with respect to permitted locations, construction, and fill.

640.25 Loudspeaker Installation in Fire Resistance-Rated Partitions, Walls, and Ceilings.

Loudspeakers installed in a fire resistance-rated partition, wall, or ceiling shall be listed for the purpose or installed in an enclosure or recess that maintains the fire resistance rating.

> FPN: Fire-rated construction is the fire-resistive classification used in building codes. One method of determining fire rating is testing in accordance with NFPA 251-1999, *Standard Methods of Tests of Fire Endurance of Building Construction and Materials*.

Revised for the 1999 *NEC*, 640.25 requires an enclosure that maintains the requisite fire resistance rating of the wall or ceiling in which a flush-mount loudspeaker is installed. Listed enclosures are available for this purpose; however, prior to their availability, site-built enclosures were installed with the approval of the authority having jurisdiction and have been used as a method to maintain the fire-resistance rating of the wall or ceiling.

The reference to NFPA 251, *Standard Methods of Tests of Fire Endurance of Building Construction and Materials*, is intended to clarify a typical classification method for a component of the ceiling or for components installed therein.

III. Portable and Temporary Audio System Installations

Portable and temporary audio system installations are characterized by the portable nature of the signal processing equipment and the reproduction equipment. While the equipment may not differ fundamentally from that used in permanent installations, the enclosures that serve as portable equipment racks provide both transit protection and mechanical protection while the equipment is in use. Such enclosures may accommodate one or multiple pieces of equipment and may be of metal, wood, plastic, or reinforced plastic construction. The nonmetal construction enclosures frequently do not comply with ANSI/EIA/310-D-1992, *Cabinets, Racks, Panels and Associated Equipment*, except for the mounting attachment points for the equipment.

Three conditions distinguish a portable installation from a temporary installation. The first is the nature of use

and possession. For example, a theater might employ portable equipment that is moved during the course of a performance but is never used for a nontheater event. The second condition is the transitory nature of the event. A performer could use his or her personal portable equipment for a performance in a theater, but because the presence of the equipment within the theater terminates with the end of the performance, the equipment's use qualifies as temporary. Finally, an event that is considered temporary will obviously use temporary audio systems. Examples include fairs, circuses, outdoor concerts, and folk festivals. Part III of Article 640 covers portable and temporary installations, the differences being primarily ones of intent and, for outdoor temporary use, adequate protection of the equipment from environmental hazards.

640.41 Multipole Branch-Circuit Cable Connectors.

Multipole branch-circuit cable connectors, male and female, for power supply cords and cables shall be constructed so that tension on the cord or cable is not transmitted to the connections. The female half shall be attached to the load end of the power supply cord or cable. The connector shall be rated in amperes and designed so that differently rated devices cannot be connected together. Alternating-current multipole connectors shall be polarized and comply with 406.6(A) and (B) and 406.9. Alternating-current or direct-current multipole connectors utilized for connection between loudspeakers and amplifiers, or between loudspeakers, shall not be compatible with nonlocking 15- or 20-ampere rated connectors intended for branch-circuit power or with connectors rated 250 volts or greater and of either the locking or nonlocking type. Signal cabling not intended for such loudspeaker and amplifier interconnection shall not be permitted to be compatible with multipole branch-circuit cable connectors of any accepted configuration.

> FPN: See 400.10 for pull at terminals.

640.42 Use of Flexible Cords and Cables.

(A) Between Equipment and Branch-Circuit Power. Power supply cords for audio equipment shall be listed and shall be permitted to be used where the interchange, maintenance, or repair of such equipment is facilitated through the use of a power supply cord.

(B) Between Loudspeakers and Amplifiers, or Between Loudspeakers. Flexible cords and cables used to connect loudspeakers to each other or to an amplifier shall comply with Article 400 and Article 725, respectively. Cords and cables listed for portable use, either hard or extra-hard usage as defined by Article 400, shall also be permitted. Other listed cable types and assemblies, including optional hybrid communications, signal, and optical fiber cables, shall be permitted.

(C) Between Equipment and/or Between Equipment Racks. Flexible cords and cables used for the distribution of audio signals between equipment shall comply with Article 400 and Article 725, respectively. Cords and cables listed for portable use, either hard or extra-hard service as defined by Article 400, shall also be permitted. Other listed cable types and assemblies, including optional hybrid communications, signal, and optical fiber cables, shall be permitted.

(D) Between Equipment, Equipment Racks, and Power Supplies Other Than Branch-Circuit Power. Wiring between the following power supplies, other than branch-circuit power supplies, shall be installed, connected, or wired in accordance with the requirements of this *Code* for the voltage and power required:

(1) Storage batteries
(2) Transformers
(3) Transformer rectifiers
(4) Other ac or dc power supplies

(E) Between Equipment Racks and Branch-Circuit Power. The supply to a portable equipment rack shall be by means of listed extra-hard usage cords or cables, as defined in Table 400.4. For outdoor portable or temporary use, the cords or cables shall be further listed as being suitable for wet locations and sunlight resistant. Sections 520.5, 520.10, and 525.3 shall apply as appropriate when the following conditions exist:

(1) Where equipment racks include audio and lighting and/or power equipment
(2) When using or constructing cable extensions, adapters, and breakout assemblies

640.43 Wiring of Equipment Racks.

Equipment racks fabricated of metal shall be grounded. Nonmetallic racks with covers (if provided) removed shall not allow access to Class 1, Class 3, or primary circuit power without the removal of covers over terminals or the use of tools.

Equipment racks shall be wired in a neat and workmanlike manner. Wires, cables, structural components, or other equipment shall not be placed in such a manner as to prevent reasonable access to equipment power switches and resettable or replaceable circuit overcurrent protection devices.

Wiring that exits the equipment rack for connection to other equipment or to a power supply shall be relieved of strain or otherwise suitably terminated such that a pull on the flexible cord or cable shall not increase the risk of damage to the cable or connected equipment such as to cause an unreasonable risk of fire or electric shock.

640.44 Environmental Protection of Equipment.

Temporary outdoor, unsheltered placement or use of portable equipment not listed for the purpose shall be permitted only where appropriate protection of such equipment from adverse weather conditions is provided to prevent risk of fire or electrical shock. Where the system is intended to remain operable during adverse weather, arrangements shall be made for maintaining operation and ventilation of heat dissipating equipment.

Most audio equipment used in temporary audio systems is not listed for use in an outdoor environment. The use of such equipment with an appropriate enclosure or other means to protect the equipment from anticipated adverse weather conditions is permitted by 640.44.

640.45 Protection of Wiring.

Where accessible to the public, flexible cords and cables laid or run on the ground or on the floor shall be covered with approved nonconductive mats. Cables and mats shall be arranged so as not to present a tripping hazard.

640.46 Equipment Access.

Equipment likely to present a risk of fire, electrical shock, or physical injury to the public shall be protected by barriers or supervised by qualified personnel so as to prevent public access.

ARTICLE 645
Information Technology
Equipment

Contents

645.1 Scope.

This article covers equipment, power-supply wiring, equipment interconnecting wiring, and grounding of information technology equipment and systems, including terminal units, in an information technology equipment room.

The term *information technology equipment* replaced other terms that describe computer-based business, personal, and industrial equipment. This terminology is also used by UL 1950, as well as international standards, as a more inclusive term for the equipment addressed by Article 645.

FPN: For further information, see NFPA 75-1999, *Standard for the Protection of Electronic Computer/Data Processing Equipment*.

645.2 Special Requirements for Information Technology Equipment Room.

This article shall apply, provided all the following conditions are met:

(1) Disconnecting means complying with 645.10 are provided.
(2) A separate heating/ventilating/air-conditioning (HVAC) system is provided that is dedicated for information technology equipment use and is separated from other areas of occupancy. Any HVAC system that serves other occupancies shall be permitted to also serve the information technology equipment room if fire/smoke dampers are provided at the point of penetration of the room boundary. Such dampers shall operate on activation of smoke detectors and also by operation of the disconnecting means required by 645.10.

FPN: For further information, see NFPA 75-1999, *Standard for the Protection of Electronic Computer/Data Processing Equipment*, Chapter 8, 8-1, 8-1.1, 8-1.2, and 8-1.3.

(3) Listed information technology equipment is installed.
(4) The room is occupied only by those personnel needed for the maintenance and functional operation of the installed information technology equipment.
(5) The room is separated from other occupancies by fire-resistant-rated walls, floors, and ceilings with protected openings.

FPN: For further information on room construction requirements, see NFPA 75-1999, *Standard for the Protection of Electronic Computer/Data Processing Equipment*, Chapter 3.

The requirements in Article 645 are based on the assumption that the room complies with NFPA 75, *Standard for the Protection of Electronic Computer/Data Processing Equipment*.

The requirements in Article 645 amend the first four chapters of the *Code*. This article has less stringent requirements for electrical installations in rooms that comply with NFPA 75, provided that the five prerequisites in 645.2 are satisfied.

Article 645 applies only to equipment and wiring located within the information technology equipment room. An information technology equipment room is an enclosed area, with one or more means of entry, that contains computer-based business and industrial equipment. It is designed to comply with the special construction and fire protection provisions of NFPA 75, as well as 645.2 of the *Code*.

Small terminals, such as remote telephone terminal units, remote data terminals, personal computers, and cash registers in stores and supermarkets, are not covered by Article 645.

645.5 Supply Circuits and Interconnecting Cables.

(A) Branch-Circuit Conductors. The branch-circuit conductors supplying one or more units of a data processing system shall have an ampacity not less than 125 percent of the total connected load.

(B) Cord-and-Plug Connections. The data processing system shall be permitted to be connected to a branch circuit by any of the following means listed for the purpose:

(1) Flexible cord and attachment plug cap not to exceed 4.5 m (15 ft).
(2) Cord set assembly. Where run on the surface of the floor, they shall be protected against physical damage.

(C) Interconnecting Cables. Separate data processing units shall be permitted to be interconnected by means of cables and cable assemblies listed for the purpose. Where run on the surface of the floor, they shall be protected against physical damage.

(D) Under Raised Floors. Power cables, communications cables, connecting cables, interconnecting cables, and receptacles associated with the information technology equipment shall be permitted under a raised floor, provided the following conditions are met.

(1) The raised floor is of suitable construction, and the area under the floor is accessible.
(2) The branch-circuit supply conductors to receptacles or field-wired equipment are in rigid metal conduit, rigid nonmetallic conduit, intermediate metal conduit, electrical metallic tubing, electrical nonmetallic tubing, metal wireway, nonmetallic wireway, surface metal raceway with metal cover, nonmetallic surface raceway, flexible metal conduit, liquidtight flexible metal conduit, or liq-

uidtight flexible nonmetallic conduit, Type MI cable, Type MC cable, or Type AC cable. These supply conductors shall be installed in accordance with the requirements of 300.11.

Branch-circuit conductors installed under the raised floor of an information technology equipment room using any of the wiring methods listed in 645.5(D)(2) are required to conform to the specific article for the wiring method used. In addition, Article 300 applies, except where modified by Article 645. For example, 300.11 requires raceways, cables, and boxes to be securely fastened in place, even though they are installed below a raised floor.

(3)   Ventilation in the underfloor area is used for the information equipment room only. The ventilation system shall be so arranged, with approved smoke detection devices, that upon the detection of fire or products of combustion in the underfloor space the circulation of air will cease.

This requirement has been revised for the 2002 *Code*. The underfloor area is required to be provided with smoke detection device(s). Upon detection of smoke, the circulation of air in the underfloor area must be interrupted. The most common method of interrupting air circulation is to open the circuit that supplies power to the air circulation fan. In addition to causing cessation of air circulation in the underfloor area, the smoke detectors may provide other fire protection functions as part of a complete building fire alarm system.

(4)   Openings in raised floors for cables protect cables against abrasions and minimize the entrance of debris beneath the floor.
(5)   Cables, other than those covered in (2) and those complying with (a), (b), and (c), shall be listed as Type DP cable having adequate fire-resistant characteristics suitable for use under raised floors of an information technology equipment room.

  (a)   Interconnecting cables enclosed in a raceway.
  (b)   Interconnecting cables listed with equipment manufactured prior to July 1, 1994, being installed with that equipment.
  (c)   Cable type designations Type TC (Article 336); Types CL2, CL3, and PLTC (Article 725); Type ITC (Article 727); Types NPLF and FPL (Article 760); Types OFC and OFN (Article 770); Types CM and MP (Article 800); and Type CATV (Article 820). These designations shall be permitted to have an additional letter P or R or G. Green, with one or more yellow stripes, insulated single conductor

cables, 4 AWG and larger, marked "for use in cable trays" or "for CT use" shall be permitted for equipment grounding.

FPN: One method of defining fire resistance is by establishing that the cables do not spread fire to the top of the tray in the "Vertical Tray Flame Test" referenced in ANSI/UL 1581-1991, *Standard for Electrical Wires, Cables, and Flexible Cords*. Another method of defining fire resistance is for the damage (char length) not to exceed 1.5 m (4 ft 11 in.) when performing the CSA "Vertical Flame Test — Cables in Cable Trays," as described in CSA C22.2 No. 0.3-M-1985, *Test Methods for Electrical Wires and Cables*.

Interconnecting cables used under raised floors (other than branch-circuit conductors) are required by 645.5(D)(5) to be listed as Type DP cables. Cables listed as part of equipment manufactured before the effective date of July 1, 1994, were not required to be listed. Cables in raceways are also exempt. Cables that pass ANSI/UL 1581-1991, *Vertical Tray Flame Test*, or CSA C22.2 No. 0.3-M-1992, *Vertical Flame Test — Cables in Cable Trays,* (where not more than 4 ft of cable is damaged during the CSA test), are permitted to be installed under raised floors of computer rooms. Type DP cables that satisfy these tests are also permitted under raised floors.

(6)   Abandoned cables shall not be permitted to remain unless contained in metal raceways.

(E) Securing in Place. Power cables; communications cables; connecting cables; interconnecting cables; and associated boxes, connectors, plugs, and receptacles that are listed as part of, or for, information technology equipment shall not be required to be secured in place.

645.6 Cables Not in Information Technology Equipment Room.

Cables extending beyond the information technology equipment room shall be subject to the applicable requirements of this *Code*.

FPN: For signaling circuits, refer to Article 725; for fiber optic circuits, refer to Article 770; and for communications circuits, refer to Article 800. For fire alarm systems, refer to Article 760.

645.7 Penetrations.

Penetrations of the fire-resistant room boundary shall be in accordance with 300.21.

645.10 Disconnecting Means.

A means shall be provided to disconnect power to all electronic equipment in the information technology equipment room. There shall also be a similar means to disconnect the power to all dedicated HVAC systems serving the room and cause all required fire/smoke dampers to close. The control for these disconnecting means shall be grouped and identified and shall be readily accessible at the principal exit doors. A single means to control both the electronic equipment and HVAC systems shall be permitted. Where a pushbutton is used as a means to disconnect power, pushing the button in shall disconnect the power.

In 645.10, two separate disconnecting means are required, but a single control, such as one pushbutton, is permitted to electrically operate both disconnecting means. The disconnecting means is required to disconnect the conductors of each circuit from their supply source and close all required fire/smoke dampers. (See the definition of *disconnecting means* in Article 100.) The disconnecting means is permitted to be remote-controlled switching devices, such as relays, with pushbutton stations at the principal exit doors. The 2002 *Code* specifies that the actuation of the emergency pushbutton(s) be accomplished by pushing the button in, rather than pulling it out. The requirement recognizes that in an emergency situation the intuitive reaction to operating the control is to push, not pull, the button.

The requirements of 645.10 and those of 645.7 for sealing penetrations are intended to minimize the passage of smoke or fire to other parts of the building.

Exception: Installations qualifying under the provisions of Article 685.

645.11 Uninterruptible Power Supplies (UPS).

Unless otherwise permitted in (1) or (2), UPS systems installed within the information technology room, and their supply and output circuits, shall comply with 645.10. The disconnecting means shall also disconnect the battery from its load.

(1) Installations qualifying under the provisions of Article 685
(2) Power sources capable of supplying 750 volt-amperes or less derived either from UPS equipment or from battery circuits integral to electronic equipment

645.15 Grounding.

All exposed non–current-carrying metal parts of an information technology system shall be grounded in accordance with Article 250 or shall be double insulated. Power systems derived within listed information technology equipment that supply information technology systems through receptacles or cable assemblies supplied as part of this equipment shall not be considered separately derived for the purpose of

applying 250.20(D). Where signal reference structures are installed, they shall be bonded to the equipment grounding system provided for the information technology equipment.

The last sentence of 645.15 was added for the 1999 *Code* to recognize that properly bonded high-frequency signal reference structures provide additional safety measures.

FPN No. 1: The bonding and grounding requirements in the product standards governing this listed equipment ensure that it complies with Article 250.
FPN No. 2: Where isolated grounding-type receptacles are used, see 250.146(D) and 406.2(D).

645.16 Marking.

Each unit of an information technology system supplied by a branch circuit shall be provided with a manufacturer's nameplate, which shall also include the input power requirements for voltage, frequency, and maximum rated load in amperes.

ARTICLE 647
Sensitive Electronic Equipment

Contents

647.1 Scope.

This article covers the installation and wiring of separately derived systems operating at 120 volts line-to-line and 60 volts to ground for sensitive electronic equipment.

The use of this type of supply system as a means to reduce objectionable noise and its adverse effect on the performance of electronic audio and video equipment has been recognized since the 1996 edition of the *Code*. Until the 2002 *NEC*, the use of separately derived, three-wire, 120-volt line-to-line, 60-volt-to-ground technical power systems was limited to motion picture and television studios. This new article expands the permitted use of this type of supply system to all commercial and industrial applications where sensitive audio/video or similar electronic equipment is used. This system can be used only under the close supervision of qualified individuals.

Unlike electrical distribution systems that supply lighting and appliance branch circuits, the supply systems covered by Article 647 are subject to mandatory voltage-drop requirements. The voltage-drop requirements are needed to ensure the operation of overcurrent devices to protect conductors and equipment supplied by these systems. Because the use of standard overcurrent devices and distribution equipment with higher voltage ratings is permitted, the impedance in circuits supplied by these systems under fault conditions is a primary concern, hence the mandatory voltage-drop requirement.

647.3 General.

Use of a separately derived 120-volt single-phase 3-wire system with 60 volts on each of two ungrounded conductors to a grounded neutral conductor shall be permitted for the purpose of reducing objectionable noise in senstive electronic equipment locations provided that the following conditions apply:

(1) The system is installed only in commercial or industrial occupancies.
(2) The system's use is restricted to areas under close supervision by qualified personnel.
(3) All of the requirements in 647.4 through 647.8 are met.

647.4 Wiring Methods.

(A) Panelboards and Overcurrent Protection. Use of standard single-phase panelboards and distribution equipment with a higher voltage rating shall be permitted. The system shall be clearly marked on the face of the panel or on the inside of the panel doors. Common-trip two-pole circuit breakers that are identified for operation at the system voltage shall be provided for both ungrounded conductors in all feeders and branch circuits.

(B) Junction Boxes. All junction box covers shall be clearly marked to indicate the distribution panel and the system voltage.

(C) Color Coding. All feeders and branch-circuit conductors installed under this section shall be identified as to system at all splices and terminations by color, marking, tagging, or equally effective means. The means of identification shall be posted at each branch-circuit panelboard and at the disconnecting means for the building.

(D) Voltage Drop. The voltage drop on any branch circuit shall not exceed 1.5 percent. The combined voltage drop of feeder and branch-circuit conductors shall not exceed 2.5 percent.

(1) Fixed Equipment. The voltage drop on branch circuits supplying equipment connected using wiring methods in Chapter 3 shall not exceed 1.5 percent. The combined voltage drop of feeder and branch-circuit conductors shall not exceed 2.5 percent.

(2) Cord-Connected Equipment. The voltage drop on branch circuits supplying receptacles shall not exceed 1 percent. For the purposes of making this calculation, the load connected to the receptacle outlet shall be considered to be 50 percent of the branch-circuit rating. The combined voltage drop of feeder and branch-circuit conductors shall not exceed 2.0 percent.

> FPN: The purpose of this provision is to limit voltage drop to 1.5 percent where portable cords may be used as a means of connecting equipment.

647.5 Three-Phase Systems.

Where 3-phase power is supplied, a separately derived 6-phase "wye" system with 60 volts to ground installed under this article shall be configured as three separately derived 120-volt single-phase systems having a combined total of no more than six main disconnects.

647.6 Grounding.

(A) General. The system shall be grounded as provided in 250.30 as a separately derived single-phase 3-wire system.

(B) Grounding Conductors Required. Permanently wired utilization equipment and receptacles shall be grounded by means of an equipment grounding conductor run with the circuit conductors to an equipment grounding bus prominently marked "Technical Equipment Ground" in the originating branch-circuit panelboard. The grounding bus shall be connected to the grounded conductor on the line side of the separately derived system's disconnecting means. The grounding conductor shall not be smaller than that specified in Table 250.122 and run with the feeder conductors. The technical equipment grounding bus need not be bonded to the panelboard enclosure. Other grounding methods authorized elsewhere in this *Code* shall be permitted where the impedance of the grounding return path does not exceed the impedance of equipment grounding conductors sized and installed in accordance with this article.

FPN No. 1: See 250.122 for equipment grounding conductor sizing requirements where circuit conductors are adjusted in size to compensate for voltage drop.

FPN No. 2: These requirements limit the impedance of the ground fault path where only 60 volts apply to a fault condition instead of the usual 120 volts.

647.7 Receptacles.

(A) General. Where receptacles are used as a means of connecting equipment, the following conditions shall be met:

(1) All 15- and 20-ampere receptacles shall be GFCI protected.

(2) All outlet strips, adapters, receptacle covers, and faceplates shall be marked with the following words or equivalent:

> WARNING — TECHNICAL POWER
> Do not connect to lighting equipment.
> For electronic equipment use only.
> 60/120 V. 1φac
> GFCI protected

(3) A 125-volt, single-phase, 15- or 20-ampere-rated receptacle outlet having one of its current-carrying poles connected to a grounded circuit conductor shall be located within 1.8 m (6 ft) of all permanently installed 15- or 20-ampere-rated 60/120-volt technical power-system receptacles.

(4) All 125-volt receptacles used for 60/120-volt technical power shall have a unique configuration and be identified for use with this class of system. All 125-volt, single-phase, 15- or 20-ampere-rated receptacle outlets and attachment plugs that are identified for use with grounded circuit conductors shall be permitted in machine rooms, control rooms, equipment rooms, equipment racks, and other similar locations that are restricted to use by qualified personnel.

(B) Isolated Ground Receptacles. Isolated ground receptacles shall be permitted as described in 250.146(D); however, the branch circuit equipment grounding conductor shall be terminated as required in 647.6(B).

647.8 Lighting Equipment.

Lighting equipment installed under this article for the purpose of reducing electrical noise originating from lighting equipment shall meet the conditions of 647.8(A) through (C).

(A) Disconnecting Means. All luminaires (lighting fixtures) connected to separately derived systems operating at 60 volts to ground and associated control equipment, if provided, shall have a disconnecting means that simultaneously opens all ungrounded conductors. The disconnecting means shall be located within sight of the luminaire (lighting fixture) or be capable of being locked in the open position.

(B) Luminaires (Lighting Fixtures). All luminaires (lighting fixtures) shall be permanently installed and listed for connection to a separately derived system at 120 volts line-to-line and 60 volts to ground.

(C) Screw-shell. Luminaires installed under this section shall not have an exposed lamp screw-shell.

ARTICLE 650
Pipe Organs

Contents

650.1 Scope.

This article covers those electrical circuits and parts of electrically operated pipe organs that are employed for the control of the sounding apparatus and keyboards.

650.2 Other Articles.

Electronic organs shall comply with the appropriate provisions of Article 640.

650.3 Source of Energy.

The source of power shall be a transformer-type rectifier, the dc potential of which shall not exceed 30 volts dc.

650.4 Grounding.

The rectifier shall be grounded according to the provisions in 250.112(B).

650.5 Conductors.

Conductors shall comply with 650.5(A) through (D).

(A) Size. Conductors shall be not less than 28 AWG for electronic signal circuits and not less than 26 AWG for electromagnetic valve supply and the like. A main common-return conductor in the electromagnetic supply shall not be less than 14 AWG.

Note that 14 AWG wire is required only for a main common return conductor in the electromagnetic supply.

(B) Insulation. Conductors shall have thermoplastic or thermosetting insulation.

(C) Conductors to Be Cabled. Except for the common-return conductor and conductors inside the organ proper, the organ sections and the organ console conductors shall be cabled. The common-return conductors shall be permitted under an additional covering enclosing both cable and return conductor, or they shall be permitted as a separate conductor and shall be permitted to be in contact with the cable.

(D) Cable Covering. Each cable shall be provided with an outer covering, either overall or on each of any subassemblies of grouped conductors. Tape shall be permitted in place of a covering. Where not installed in metal raceway, the covering shall be resistant to flame spread or the cable or each cable subassembly shall be covered with a closely wound listed fireproof tape.

> FPN: One method of determining that cable is resistant to flame spread is by testing the cable to the VW-1 (vertical-wire) flame test in the ANSI/UL 1581-1991, *Reference Standard for Electrical Wires, Cables and Flexible Cords.*

650.6 Installation of Conductors.

Cables shall be securely fastened in place and shall be permitted to be attached directly to the organ structure without insulating supports. Cables shall not be placed in contact with other conductors.

> Insulating supports are not required; however, measures should be taken to prevent contact between the pipe organ cables and conductors of other systems.

650.7 Overcurrent Protection.

Circuits shall be so arranged that 26 AWG and 28 AWG conductors shall be protected by an overcurrent device rated at not more than 6 amperes. Other conductor sizes shall be protected in accordance with their ampacity. A common return conductor shall not require overcurrent protection.

ARTICLE 660
X-Ray Equipment

Contents

I. General

660.1 Scope.

This article covers all X-ray equipment operating at any frequency or voltage for industrial or other nonmedical or nondental use.

> FPN: See Article 517, Part V, for X-ray installations in health care facilities.

Nothing in this article shall be construed as specifying safeguards against the useful beam or stray X-ray radiation.

> FPN No. 1: Radiation safety and performance requirements of several classes of X-ray equipment are regulated under Public Law 90-602 and are enforced by the Department of Health and Human Services.

> FPN No. 2: In addition, information on radiation protection by the National Council on Radiation Protection and Measurements is published as *Reports of the National Council on Radiation Protection and Measurement.* These reports can be obtained from NCRP Publications, 7910 Woodmont Ave., Suite 1016, Bethesda, MD 20814.

660.2 Definitions.

Long-Time Rating. A rating based on an operating interval of 5 minutes or longer.

Mobile. X-ray equipment mounted on a permanent base with wheels and/or casters for moving while completely assembled.

Momentary Rating. A rating based on an operating interval that does not exceed 5 seconds.

Portable. X-ray equipment designed to be hand-carried.

Transportable. X-ray equipment that is to be installed in a vehicle or that may be readily disassembled for transport in a vehicle.

660.3 Hazardous (Classified) Locations.

Unless approved for the location, X-ray and related equipment shall not be installed or operated in hazardous (classified) locations.

FPN: See Article 517, Part IV.

X-ray equipment in industrial establishments or similar locations is commonly used for inspecting a process or product. This method permits nondestructive testing without dismantling or applying stress to detect cracks, flaws, or structural defects. Welded joints are frequently inspected with X-ray equipment to detect hidden defects that can cause failure under stress.

Among the industrial applications of X-rays, the most common is radiography, in which shadow pictures of the object are produced on photographic film. The type and thickness of the material involved govern the voltage to be employed, which can range from a few thousand volts (kilovolts) to millions of volts (megavolts). It is possible to X-ray metal objects that are 20 in. thick.

Fluoroscopy is another X-ray technique used for industrial or commercial applications. Fluoroscopy is similar to radiography, but it operates at a much lower voltage range (less than 250 kilovolts). Instead of producing a film, it projects a shadow picture on a screen, similar to those used for security checks of luggage at airport terminals. Fluoroscopy is capable of detecting minute flaws or defects.

660.4 Connection to Supply Circuit.

(A) Fixed and Stationary Equipment. Fixed and stationary X-ray equipment shall be connected to the power supply by means of a wiring method meeting the general requirements of this *Code*. Equipment properly supplied by a branch circuit rated at not over 30 amperes shall be permitted to be supplied through a suitable attachment plug cap and hard-service cable or cord.

(B) Portable, Mobile, and Transportable Equipment. Individual branch circuits shall not be required for portable, mobile, and transportable X-ray equipment requiring a capacity of not over 60 amperes. Portable and mobile types of X-ray equipment of any capacity shall be supplied through a suitable hard-service cable or cord. Transportable X-ray equipment of any capacity shall be permitted to be connected to its power supply by suitable connections and hard-service cable or cord.

(C) Over 600 Volts, Nominal. Circuits and equipment operated at more than 600 volts, nominal, shall comply with Article 490.

660.5 Disconnecting Means.

A disconnecting means of adequate capacity for at least 50 percent of the input required for the momentary rating or 100 percent of the input required for the long-time rating of the X-ray equipment, whichever is greater, shall be provided in the supply circuit. The disconnecting means shall be operable from a location readily accessible from the X-ray control. For equipment connected to a 120-volt, nominal, branch circuit of 30 amperes or less, a grounding-type attachment plug cap and receptacle of proper rating shall be permitted to serve as a disconnecting means.

660.6 Rating of Supply Conductors and Overcurrent Protection.

(A) Branch-Circuit Conductors. The ampacity of supply branch-circuit conductors and the overcurrent protective devices shall not be less than 50 percent of the momentary rating or 100 percent of the long-time rating, whichever is greater.

(B) Feeder Conductors. The rated ampacity of conductors and overcurrent devices of a feeder for two or more branch circuits supplying X-ray units shall not be less than 100 percent of the momentary demand rating [as determined by 660.6(A)] of the two largest X-ray apparatus plus 20 percent of the momentary ratings of other X-ray apparatus.

FPN: The minimum conductor size for branch and feeder circuits is also governed by voltage regulation requirements. For a specific installation, the manufacturer usually specifies minimum distribution transformer and conductor sizes, rating of disconnect means, and overcurrent protection.

660.7 Wiring Terminals.

X-ray equipment not provided with a permanently attached cord or cord set shall be provided with suitable wiring terminals or leads for the connection of power-supply conductors of the size required by the rating of the branch circuit for the equipment.

660.8 Number of Conductors in Raceway.

The number of control circuit conductors installed in a raceway shall be determined in accordance with 300.17.

660.9 Minimum Size of Conductors.

Size 18 AWG or 16 AWG fixture wires, as specified in 725.27, and flexible cords shall be permitted for the control and operating circuits of X-ray and auxiliary equipment where protected by not larger than 20-ampere overcurrent devices.

660.10 Equipment Installations.

All equipment for new X-ray installations and all used or reconditioned X-ray equipment moved to and reinstalled at a new location shall be of an approved type.

II. Control

660.20 Fixed and Stationary Equipment.

(A) Separate Control Device. A separate control device, in addition to the disconnecting means, shall be incorporated in the X-ray control supply or in the primary circuit to the high-voltage transformer. This device shall be a part of the X-ray equipment but shall be permitted in a separate enclosure immediately adjacent to the X-ray control unit.

(B) Protective Device. A protective device, which shall be permitted to be incorporated into the separate control device, shall be provided to control the load resulting from failures in the high-voltage circuit.

660.21 Portable and Mobile Equipment.

Portable and mobile equipment shall comply with 660.20, but the manually controlled device shall be located in or on the equipment.

660.23 Industrial and Commercial Laboratory Equipment.

(A) Radiographic and Fluoroscopic Types. All radiographic- and fluoroscopic-type equipment shall be effectively enclosed or shall have interlocks that de-energize the equipment automatically to prevent ready access to live current-carrying parts.

(B) Diffraction and Irradiation Types. Diffraction- and irradiation-type equipment or installations not effectively enclosed or provided with interlocks to prevent access to live current-carrying parts during operation shall be provided with a positive means to indicate when they are energized. The indicator shall be a pilot light, readable meter deflection, or equivalent means.

660.24 Independent Control.

Where more than one piece of equipment is operated from the same high-voltage circuit, each piece or each group of equipment as a unit shall be provided with a high-voltage switch or equivalent disconnecting means. This disconnecting means shall be constructed, enclosed, or located so as to avoid contact by persons with its live parts.

A control device provides means for initiating and terminating X-ray exposures and automatically times their duration.

III. Transformers and Capacitors

660.35 General.

Transformers and capacitors that are part of an X-ray equipment shall not be required to comply with Articles 450 and 460.

High-ratio step-up transformers that are an integral part of an X-ray are not required to comply with Article 450 and are generally used to provide the high voltage necessary for X-ray tubes. Because there is a lesser degree of fire hazard due to the low primary voltage, X-ray transformers are not required to be installed in fire-resistant vaults.

660.36 Capacitors.

Capacitors shall be mounted within enclosures of insulating material or grounded metal.

IV. Guarding and Grounding

660.47 General.

(A) High-Voltage Parts. All high-voltage parts, including X-ray tubes, shall be mounted within grounded enclosures. Air, oil, gas, or other suitable insulating media shall be used to insulate the high voltage from the grounded enclosure. The connection from the high-voltage equipment to X-ray tubes and other high-voltage components shall be made with high-voltage shielded cables.

(B) Low-Voltage Cables. Low-voltage cables connecting to oil-filled units that are not completely sealed, such as transformers, condensers, oil coolers, and high-voltage switches, shall have insulation of the oil-resistant type.

Grounded enclosures are required to be provided for all high-voltage X-ray equipment, including X-ray tubes. High-voltage shielded cables are required to be used to connect high-voltage equipment to X-ray tubes, and the shield is required to be grounded, as specified in 660.48.

660.48 Grounding.

Non–current-carrying metal parts of X-ray and associated equipment (controls, tables, X-ray tube supports, transformer tanks, shielded cables, X-ray tube heads, and so forth) shall be grounded in the manner specified in Article 250. Portable and mobile equipment shall be provided with an approved grounding-type attachment plug cap.

Exception: Battery-operated equipment.

ARTICLE 665
Induction and Dielectric Heating Equipment

Contents

I. General

665.1 Scope.

This article covers the construction and installation of dielectric heating, induction heating, induction melting, and induction welding equipment and accessories for industrial and scientific applications. Medical or dental applications, appliances, or line frequency pipeline and vessel heating are not covered in this article.

> FPN: See Article 427, Part V, for line frequency induction heating of pipelines and vessels.

To prevent spurious radiation caused by induction and dielectric heating equipment and to ensure that the frequency spectrum is utilized equitably, the Federal Communications Commission (FCC) has established rules that govern the use of this type of industrial heating equipment operating above 10 kHz [*Code of Federal Regulations (CFR),* Title 47, Part 18].

665.2 Definitions.

Converting Device. That part of the heating equipment that converts input mechanical or electrical energy to the voltage, current, and frequency suitable for the heating applicator. A converting device shall consist of equipment using mains frequency, all static multipliers, oscillator-type units using vacuum tubes, inverters using solid state devices, or motor generator equipment.

Dielectric Heating. Heating of a nominally insulating material due to its own dielectric losses when the material is placed in a varying electric field.

Heating Equipment. As used in this article, any equipment that is used for heating purposes and whose heat is generated by induction or dielectric methods.

Heating Equipment Applicator. The heating equipment applicator is the device used to transfer energy between the output circuit and the object or mass to be heated.

Induction Heating, Melting, and Welding. The heating, melting, or welding of a nominally conductive material due to its own I^2R losses when the material is placed in a varying electromagnetic field.

Induction and dielectric heating are used for ovens, furnaces, and industrial equipment where pieces of material are heated by a rapidly alternating magnetic or electric field. For further information on electric heating systems using an induction heater or a dielectric heater in ovens and furnaces, see NFPA 86, *Standard for Ovens and Furnaces,* and NFPA 86D, *Standard for Industrial Furnaces Using Vacuum as an Atmosphere.*

THEORY OF OPERATION—SOLID-STATE CONVERTER POWER CIRCUIT

The solid-state power converter consists of three sections: the rectifier section, the inverter section, and the output section, which includes the load coil and is usually located outside the power supply. Exhibit 665.1 is an example of an enclosed power supply for an induction heating process. There are two basic types of inverters, voltage fed (Exhibit 665.2) and current fed (Exhibit 665.3).

The rectifier section converts 3-phase, line frequency

Exhibit 665.1 Enclosed power supply for an induction heating process. (Courtesy of Tocco Division, Park Ohio Industries)

voltage to direct current. The input voltage can be any desired voltage—480 volts, 575 volts, and 1150 volts are some typical voltages. The dc output is filtered with either a large

choke (current-fed inverter) or a filter capacitor (voltage-fed inverter). The output of the rectifier section supplies energy for the inverter section. The inverter section converts the energy stored in the magnetic field of the choke (current-fed inverter) or in the electric field of the filter capacitor (voltage-fed inverter) to a variable frequency for the output circuit. The variable output frequency controls the power delivered to the load.

The output section consists of a capacitor in parallel (current fed) or in series (voltage fed) with the coil. Because the output capacitance and coil inductance have a resonant frequency, as the output frequency approaches this resonant frequency, the power to the load approaches the maximum output. At minimum frequency, the output power is very low.

INDUCTION HEATING

Induction heating occurs when an electrically conductive material (the load) is placed in a varying magnetic field. The magnetic field is generated by a coil (inductor) around or adjacent to the workpiece to be heated. The varying magnetic field induces current in the electrically conductive load. Heat in the load is generated by the resulting I^2R losses in the load. Induction heating can be further subdivided into heating, melting, and welding.

Induction heating raises the temperature of the load to some temperature below its melting point, usually for the purposes of hardening, tempering, annealing, forging, extruding, or rolling. Frequencies used for induction heating range from about 50 Hz to 500 kHz. Power levels range from 5 kW to 42,000 kW.

Exhibit 665.2 Schematic of a half-bridge, voltage-fed inverter. (Redrawn from Inductotherm Corp.)

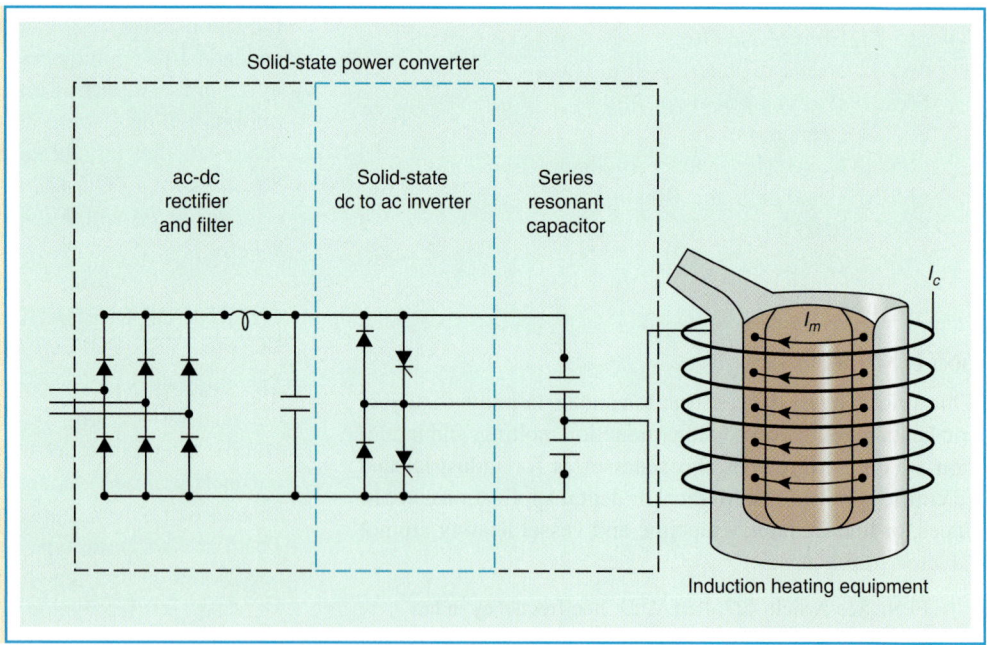

Exhibit 665.3 Schematic of a full-bridge, current-fed inverter. (Redrawn from Radyne Corp.)

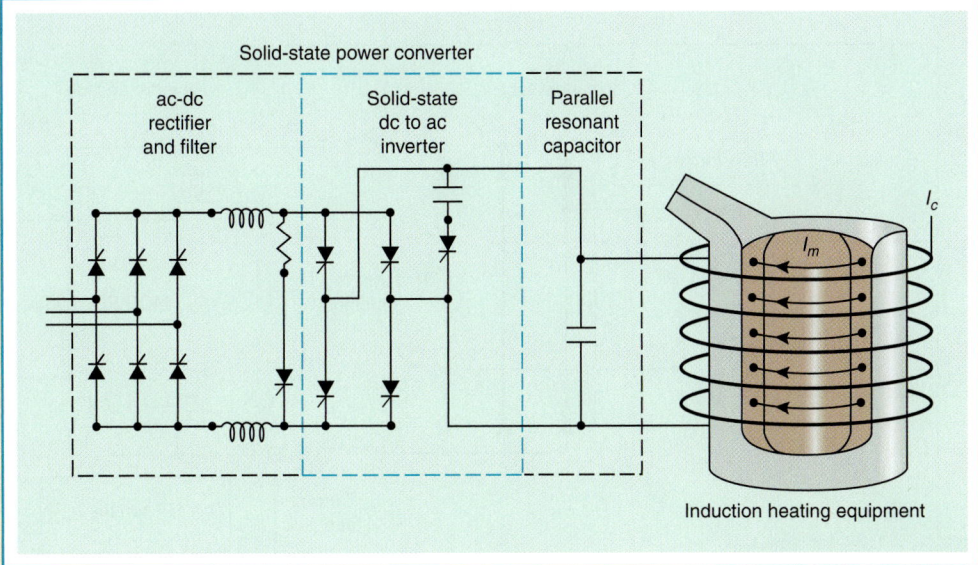

Solid-state power converter

ac-dc rectifier and filter · Solid-state dc to ac inverter · Parallel resonant capacitor

Induction heating equipment

Induction melting raises the temperature of the load to some temperature above its melting point, so the molten material can be alloyed, homogenized, and/or poured. Frequencies used for induction melting range from about 50 Hz to 10 kHz. Power levels range from 5 kW to 16,500 kW.

Induction welding is primarily used in the manufacture of welded pipe and tubing. In this process, a high-frequency current is passed through an induction coil in the proximity of the conducting metal surfaces to be joined. By proper arrangement of the coil and by the addition of ferrite elements to control the currents induced in the surfaces, selected portions are heated nearly instantaneously to the forging temperature. The surfaces are then joined under pressure to produce a forge weld. Frequencies used for induction welding range from about 100 kHz to 800 kHz. Power levels range from 20 kW to 1000 kW. Induction heating (or melting or welding) of magnetic loads, such as iron or carbon steel, must contend with the change in magnetic permeability when the temperature of the load material passes its Curie temperature (at which point the load changes from the magnetic to the nonmagnetic state). The solid-state converter varies its output frequency to maintain constant output power through the entire range of temperatures of the load. See Exhibit 665.4.

DIELECTRIC HEATING

Dielectric heating equipment is similar to induction heating equipment, except that it is used to heat nonmetallic materials as opposed to metals. Typical applications include the drying of textiles after dyeing, drying of water-based coatings on paper, preheating of wood fibers for the MDF (medium-density fiberboard) industry, welding of plastic materials, food processing, and many other diverse applications.

At radio frequencies, the material to be heated forms a lossy dielectric when placed between metal capacitor plates connected across the output of the generator. A high-frequency alternating electric field is created between these electrode plates. The molecules in the dielectric field are made to vibrate, causing dissipation of energy through the material and frictional heating of the dielectric material.

At higher (microwave) frequencies, the process is similar, but the generator is coupled to a resonant cavity into which the dielectric material is placed.

The frequency of operation of dielectric heating equipment is considerably higher than for induction heating. By international agreement, these machines operate at the assigned radio frequencies of 13.56 MHz, 27.12 MHz, and 40.68 MHz or at microwave frequencies of 915 MHz and 2450 MHz.

All industrial dielectric heating equipment should meet appropriate U.S. federal regulations for RF emissions (CFR Title 47, Part 18) and occupational safety (OSHA 29CFR B1910.97).

The majority of installed machines use vacuum tube generators, and powers range from 0.5 kW to 1 MW. Solid state generators also have been installed, although powers have been limited to 5 kW or less. Exhibit 665.5 illustrates the components of a vacuum tube generator in a dielectric heating process.

665.3 Other Articles.

Unless specifically amended by this article, wiring from the source of power to the heating equipment shall comply with Chapters 1 through 4.

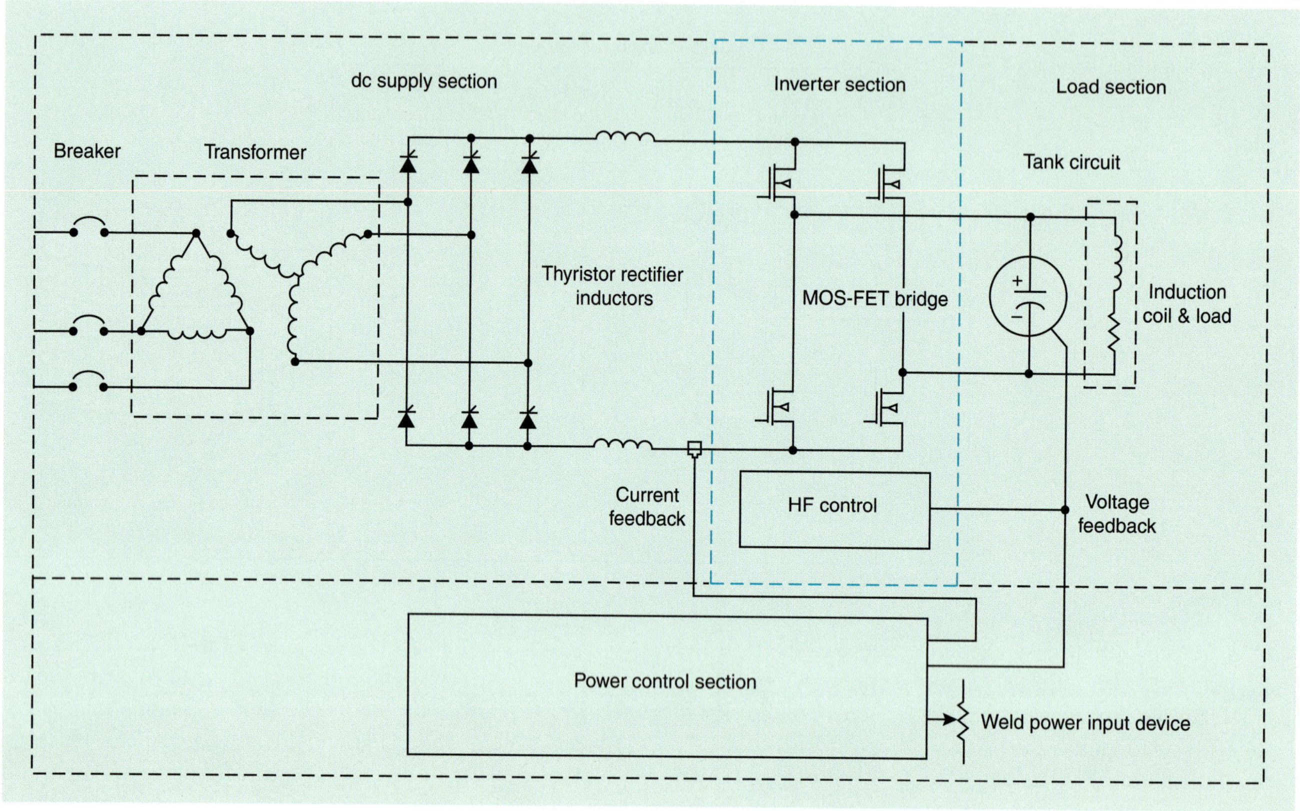

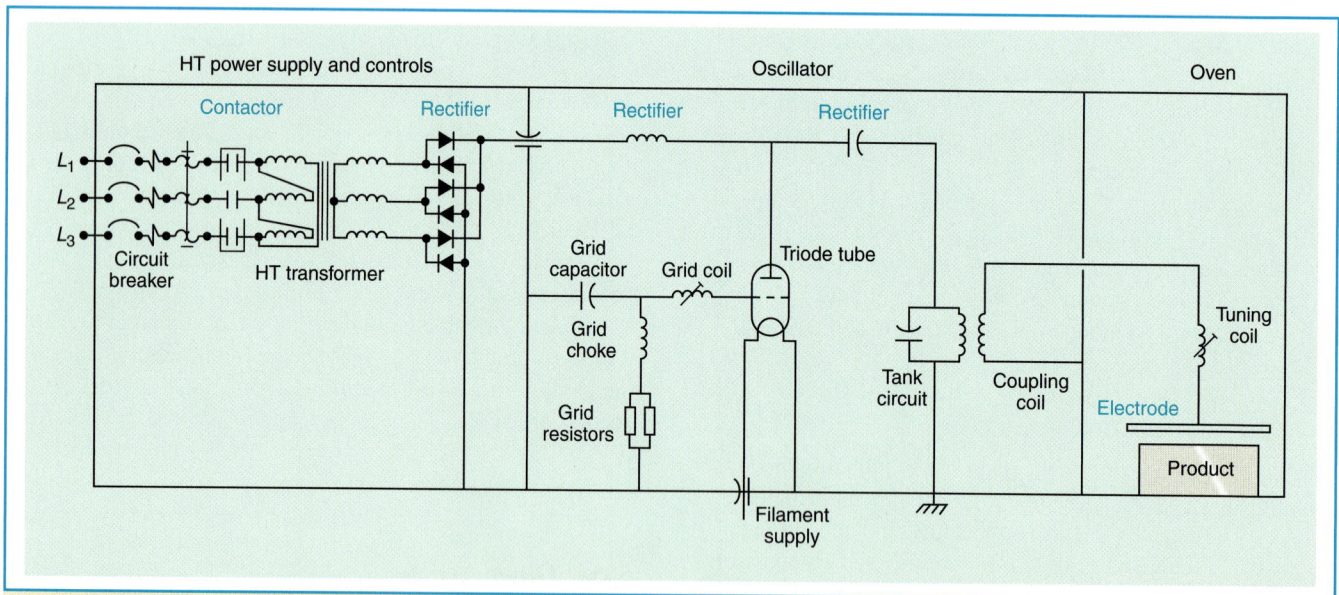

Exhibit 665.4 Simplified diagram of a solid-state inverter used for induction welding. (Redrawn from Thermatool Corp.)

Exhibit 665.5 Simplified diagram of the components of a vacuum tube generator used for dielectric heating. (Redrawn from Strayfield Ltd.)

665.4 Hazardous (Classified) Locations.

Heating equipment shall not be installed in hazardous (classified) locations as defined in Article 500 unless the equipment and wiring are designed and approved for the hazardous (classified) locations.

665.5 Output Circuit.

The output circuit shall include all output components external to the converting device, including contactors, switches, bus bars, and other conductors. The current flow from the output circuit to ground under operating and ground-fault conditions shall be limited to a value that does not cause 50 volts or more to ground to appear on any accessible part of the heating equipment and its load. The output circuit shall be permitted to be isolated from ground.

If the load (the object or mass being heated) accidentally comes in contact with the output coil, a voltage to ground will appear on the load, depending on the various impedances to ground of the coil and the load. If the voltage on the load is limited to less than 50 volts, guarding per 110.27(A) is not required. If the coil is isolated from ground and the load is grounded through an impedance that is low (less than 1 percent) relative to the coil impedance to ground, the voltage of the load to ground will be low no matter where the load contacts the coil.

In induction melting furnaces, an additional reason for isolating the coil from ground is to limit the fault current when a coil does go to ground. Limiting the fault current is intended to prevent severe damage to the water-cooled coil, resulting in a water leak and the potential for a water molten-metal explosion. If water is trapped under molten metal, the rapid transfer of heat to the water causes the water to turn almost instantly into steam. The resulting 1600-to-1 expansion of the steam results in the ejection of molten metal from the furnace.

665.7 Remote Control.

(A) Multiple Control Points. Where multiple control points are used for applicator energization, a means shall be provided and interlocked so that the applicator can be energized from only one control point at a time. A means for de-energizing the applicator shall be provided at each control point.

(B) Foot Switches. Switches operated by foot pressure shall be provided with a shield over the contact button to avoid accidental closing of a foot switch.

665.10 Ampacity of Supply Conductors.

The ampacity of supply conductors shall be determined by 665.10(A) or (B).

(A) Nameplate Rating. The ampacity of conductors supplying one or more pieces of equipment shall be not less than the sum of the nameplate ratings for the largest group of machines capable of simultaneous operation, plus 100 percent of the standby currents of the remaining machines. Where standby currents are not given on the nameplate, the nameplate rating shall be used as the standby current.

(B) Motor-Generator Equipment. The ampacity of supply conductors for motor generator equipment shall be determined in accordance with Article 430, Part II.

665.11 Overcurrent Protection.

Overcurrent protection for the heating equipment shall be provided as specified in Article 240. This overcurrent protection shall be permitted to be provided separately or as a part of the equipment.

665.12 Disconnecting Means.

A readily accessible disconnecting means shall be provided to disconnect each heating equipment from its supply circuit. The disconnecting means shall be located within sight from the controller or be capable of being locked in the open position. The rating of this disconnecting means shall not be less than the nameplate rating of the heating equipment. Motor-generator equipment shall comply with Article 430, Part IX. The supply circuit disconnecting means shall be permitted to serve as the heating equipment disconnecting means where only one heating equipment is supplied.

II. Guarding, Grounding, and Labeling

665.19 Component Interconnection.

The interconnection components required for a complete heating equipment installation shall be guarded.

665.20 Enclosures.

The converting device (excluding the component interconnections) shall be completely contained within an enclosure(s) of noncombustible material.

665.21 Control Panels.

All control panels shall be of dead-front construction.

665.22 Access to Internal Equipment.

Access doors or detachable access panels shall be employed for internal access to heating equipment. Access doors to internal compartments containing equipment employing voltages from 150 volts to 1000 volts ac or dc shall be capable of being locked closed or shall be interlocked to prevent the supply circuit from being energized while the door(s) is open. Access doors to internal compartments containing equipment employing voltages exceeding 1000 volts ac or dc shall be provided with a disconnecting means equipped with mechanical lockouts to prevent access while the heating equipment is energized, or the access doors shall be capable of being locked closed and interlocked to prevent the supply circuit from being energized while the door(s) is open. Detachable panels not normally used for access to such parts shall be fastened in a manner that will make them inconvenient to remove.

665.23 Warning Labels or Signs.

Warning labels or signs that read "DANGER – HIGH VOLTAGE – KEEP OUT" shall be attached to the equipment and shall be plainly visible where persons might come in contact with energized parts when doors are open or closed or when panels are removed from compartments containing over 150 volts ac or dc.

665.24 Capacitors.

The time and means of discharge shall be in accordance with 460.6 for capacitors rated 600 volts, nominal, and under. The time and means of discharge shall be in accordance with 460.28 for capacitors rated over 600 volts, nominal. Capacitor internal pressure switches connected to a circuit-interruptor device shall be permitted for capacitor overcurrent protection.

It should be noted that it is necessary to provide enhanced protection against rupture of capacitor cases when capacitors are operated at the higher frequencies used for induction and dielectric heating. A high-resistance fault condition can cause case pressure to build up inside the capacitor over a very short time. Capacitor internal pressure switches are the preferred method to detect this type of failure.

Consider a 5000-kVAR, 2500-V, 300-Hz capacitor. Nominal current is 2000 amperes. A "high-resistance" fault of 10 ohms will result in 250 amperes of resistive current, or a total capacitor current of 2016 amperes rms. This small increase in rms current will not result in the opening of an overcurrent device even though 625 kW of thermal energy is being generated inside the capacitor, which was designed to dissipate about 1.5 kW of losses.

665.25 Dielectric Heating Applicator Shielding.

Protective cages or adequate shielding shall be used to guard dielectric heating applicators. Interlock switches shall be used on all hinged access doors, sliding panels, or other easy means of access to the applicator. All interlock switches shall be connected in such a manner as to remove all power from the applicator when any one of the access doors or panels is open.

665.26 Grounding and Bonding.

Grounding or inter-unit bonding, or both, shall be used wherever required for circuit operation, for limiting to a safe value

radio frequency voltages between all exposed non–current-carrying parts of the equipment and earth ground, between all equipment parts and surrounding objects, and between such objects and earth ground. Such grounding and bonding shall be installed in accordance with Article 250, Parts II and V.

Because of stray currents flowing between units of the equipment or to the ground, bonding presents special problems at radio frequencies. Special bonding requirements are particularly needed at dielectric heating frequencies (100 to 200 MHz) because of the differences in radio frequency potential that can exist between the equipment and surrounding metal units or other units of the installation. Satisfactory bonding can be accomplished by placing all units of the equipment on a flooring or base consisting of a copper or aluminum sheet and then thoroughly bonding by soldering, welding, or bolting.

Such special bonding holds the radio frequency resistance and reactance between units to a minimum, and any stray circulating currents flowing through the bonding will not cause a dangerous voltage drop.

It is necessary to protect the operator from high radio frequency potentials by shielding at dielectric heating frequencies. Interference with radio communications systems at such high frequencies can be eliminated by totally enclosing all components in a shielding of copper or aluminum.

FPN: Under certain conditions, contact between the object being heated and the applicator results in an unsafe condition, such as eruption of heated materials. This unsafe condition may be prevented by grounding of the object being heated and ground detection.

665.27 Marking.

Each heating equipment shall be provided with a nameplate giving the manufacturer's name and model identification and the following input data: line volts, frequency, number of phases, maximum current, full-load kilovolt-amperes (kVA), and full-load power factor. Additional data shall be permitted.

ARTICLE 668
Electrolytic Cells

Contents

668.1 Scope.

The provisions of this article apply to the installation of the electrical components and accessory equipment of electrolytic cells, electrolytic cell lines, and process power supply for the production of aluminum, cadmium, chlorine, copper, fluorine, hydrogen peroxide, magnesium, sodium, sodium chlorate, and zinc.

Not covered by this article are cells used as a source of electric energy and for electroplating processes and cells used for the production of hydrogen.

> FPN No. 1: In general, any cell line or group of cell lines operated as a unit for the production of a particular metal, gas, or chemical compound may differ from any other cell lines producing the same product because of variations in the particular raw materials used, output capacity, use of proprietary methods or process practices, or other modifying factors to the extent that detailed *Code* requirements become overly restrictive and do not accomplish the stated purpose of this *Code*.

> FPN No. 2: For further information, see IEEE 463-1993, *Standard for Electrical Safety Practices in Electrolytic Cell Line Working Zones*.

Within a cell line working zone, both an electrolytic cell line and its dc process power supply circuit are treated as an individual machine supplied from a single source, even though they might cover acres of space, have a load current in excess of 400,000 amperes dc, or have a circuit voltage in excess of 1000 volts dc. The cell line process current passes through each cell in a series connection, and the load current cannot be subdivided, as it can in the heating circuit of a resistance-type electric furnace. Because a cell line is supplied by its individual dc rectifier system, the rectifier or the entire cell line circuit is de-energized by removing its source of primary power.

In some electrolytic cell systems, the terminal voltage of the process supply can be appreciable. The voltage to ground of exposed live parts from one end of a cell line to the other is variable between the limits of the terminal voltage. Hence, operating and maintenance personnel and their tools are required to be insulated from ground. See Exhibit 668.1 for an example of a pot room in an aluminum reduction plant.

668.2 Definitions.

Cell Line. An assembly of electrically interconnected electrolytic cells supplied by a source of direct-current power.

Cell Line Attachments and Auxiliary Equipment. As applied to this article, a term that includes, but is not limited to, auxiliary tanks; process piping; ductwork; structural supports; exposed cell line conductors; conduits and other raceways; pumps, positioning equipment, and cell cutout or bypass electrical devices. Auxiliary equipment includes tools, welding machines, crucibles, and other portable equipment used for operation and maintenance within the electrolytic cell line working zone.

In the cell line working zone, auxiliary equipment includes the exposed conductive surfaces of ungrounded cranes and crane-mounted cell-servicing equipment.

Exhibit 668.1 A typical pot room in an aluminum reduction plant. (Courtesy of Alcoa Inc.)

Electrically Connected. A connection capable of carrying current as distinguished from connection through electromagnetic induction.

Electrolytic Cell. A tank or vat in which electrochemical reactions are caused by applying electrical energy for the purpose of refining or producing usable materials.

Electrolytic Cell Line Working Zone. The space envelope wherein operation or maintenance is normally performed on or in the vicinity of exposed energized surfaces of electrolytic cell lines or their attachments.

668.3 Other Articles.

(A) Lighting, Ventilating, Material Handling. Chapters 1 through 4 shall apply to services, feeders, branch circuits, and apparatus for supplying lighting, ventilating, material handling, and the like that are outside the electrolytic cell line working zone.

(B) Systems Not Electrically Connected. Those elements of a cell line power-supply system that are not electrically connected to the cell supply system, such as the primary winding of a two-winding transformer, the motor of a motor-generator set, feeders, branch circuits, disconnecting means, motor controllers, and overload protective equipment, shall be required to comply with all applicable provisions of this *Code*.

(C) Electrolytic Cell Lines. Electrolytic cell lines shall comply with the provisions of Chapters 1, 2, 3, and 4 except as amended in 668.3(C)(1), (C)(2), (C)(3), or (C)(4).

(1) Conductors. The electrolytic cell line conductors shall not be required to comply with the provisions of Articles 110, 210, 215, 220, and 225. See 668.11.

(2) Overcurrent Protection. Overcurrent protection of electrolytic cell dc process power circuits shall not be required to comply with the requirements of Article 240.

(3) Grounding. Equipment located or used within the electrolytic cell line working zone or associated with the cell line dc power circuits shall not be required to comply with the provisions of Article 250.

(4) Working Zone. The electrolytic cells, cell line attachments, and the wiring of auxiliary equipments and devices within the cell line working zone shall not be required to comply with the provisions of Articles 110, 210, 215, 220, and 225. See 668.30.

> FPN: See 668.15 for equipment, apparatus, and structural component grounding.

668.10 Cell Line Working Zone.

(A) Area Covered. The space envelope of the cell line working zone shall encompass spaces that meet any of the following conditions:

(1) Is within 2.5 m (96 in.) above energized surfaces of electrolytic cell lines or their energized attachments.
(2) Is below energized surfaces of electrolytic cell lines or their energized attachments, provided the headroom in the space beneath is less than 2.5 m (96 in.).

(3) Is within 1.0 m (42 in.) horizontally from energized surfaces of electrolytic cell lines or their energized attachments or from the space envelope described in 668.10(A)(1) or (A)(2).

(B) Area Not Covered. The cell line working zone shall not be required to extend through or beyond walls, floors, roofs, partitions, barriers, or the like.

668.11 Direct-Current Cell Line Process Power Supply.

(A) Not Grounded. The dc cell line process power-supply conductors shall not be required to be grounded.

(B) Metal Enclosures Grounded. All metal enclosures of dc cell line process power-supply apparatus operating at a power-supply potential between terminals of over 50 volts shall be grounded as follows:

(1) Through protective relaying equipment, or
(2) By a minimum 2/0 AWG copper grounding conductor or a conductor of equal or greater conductance

(C) Grounding Requirements. The grounding connections required by 668.11(B) shall be installed in accordance with 250.8, 250.10, 250.12, 250.68, and 250.70.

668.12 Cell Line Conductors.

(A) Insulation and Material. Cell line conductors shall be either bare, covered, or insulated and of copper, aluminum, copper-clad aluminum, steel, or other suitable material.

(B) Size. Cell line conductors shall be of such cross-sectional area that the temperature rise under maximum load conditions and at maximum ambient shall not exceed the safe operating temperature of the conductor insulation or the material of the conductor supports.

(C) Connections. Cell line conductors shall be joined by bolted, welded, clamped, or compression connectors.

668.13 Disconnecting Means.

(A) More Than One Process Power Supply. Where more than one dc cell line process power supply serves the same cell line, a disconnecting means shall be provided on the cell line circuit side of each power supply to disconnect it from the cell line circuit.

(B) Removable Links or Conductors. Removable links or removable conductors shall be permitted to be used as the disconnecting means.

668.14 Shunting Means.

(A) Partial or Total Shunting. Partial or total shunting of cell line circuit current around one or more cells shall be permitted.

(B) Shunting One or More Cells. The conductors, switches, or combination of conductors and switches used for shunting one or more cells shall comply with the applicable requirements of 668.12.

668.15 Grounding.

For equipment, apparatus, and structural components that are required to be grounded by provisions of Article 668, the provisions of Article 250 shall apply, except a water pipe electrode shall not be required to be used. Any electrode or combination of electrodes described in 250.52 shall be permitted.

668.20 Portable Electrical Equipment.

(A) Portable Electrical Equipment Not to Be Grounded. The frames and enclosures of portable electrical equipment used within the cell line working zone shall not be grounded.

Exception No. 1: Where the cell line voltage does not exceed 200 volts dc, these frames and enclosures shall be permitted to be grounded.

Exception No. 2: These frames and enclosures shall be permitted to be grounded where guarded.

(B) Isolating Transformers. Electrically powered, hand-held, cord-connected portable equipment with ungrounded frames or enclosures used within the cell line working zone shall be connected to receptacle circuits that have only ungrounded conductors such as a branch circuit supplied by an isolating transformer with an ungrounded secondary.

(C) Marking. Ungrounded portable electrical equipment shall be distinctively marked and shall employ plugs and receptacles of a configuration that prevents connection of this equipment to grounding receptacles and that prevents inadvertent interchange of ungrounded and grounded portable electrical equipments.

668.21 Power Supply Circuits and Receptacles for Portable Electrical Equipment.

(A) Isolated Circuits. Circuits supplying power to ungrounded receptacles for hand-held, cord-connected equipments shall be electrically isolated from any distribution system supplying areas other than the cell line working zone and shall be ungrounded. Power for these circuits shall be supplied through isolating transformers. Primaries of such transformers shall operate at not more than 600 volts between conductors and shall be provided with proper overcurrent protection. The secondary voltage of such transformers shall not exceed 300 volts between conductors, and all circuits supplied from such secondaries shall be ungrounded and shall have an approved overcurrent device of proper rating in each conductor.

(B) Noninterchangeability. Receptacles and their mating plugs for ungrounded equipment shall not have provision for a grounding conductor and shall be of a configuration that prevents their use for equipment required to be grounded.

(C) Marking. Receptacles on circuits supplied by an isolating transformer with an ungrounded secondary shall be a distinctive configuration, distinctively marked, and shall not be used in any other location in the plant.

668.30 Fixed and Portable Electrical Equipment.

(A) Electrical Equipment Not Required to Be Grounded. Alternating-current systems supplying fixed and portable electrical equipments within the cell line working zone shall not be required to be grounded.

(B) Exposed Conductive Surfaces Not Required to Be Grounded. Exposed conductive surfaces, such as electrical equipment housings, cabinets, boxes, motors, raceways, and the like, that are within the cell line working zone shall not be required to be grounded.

(C) Wiring Methods. Auxiliary electrical equipment such as motors, transducers, sensors, control devices, and alarms, mounted on an electrolytic cell or other energized surface, shall be connected to premises wiring systems by any of the following means:

(1) Multiconductor hard usage cord.
(2) Wire or cable in suitable raceways or metal or nonmetallic cable trays. If metal conduit, cable tray, armored cable, or similar metallic systems are used, they shall be installed with insulating breaks such that they do not cause a potentially hazardous electrical condition.

(D) Circuit Protection. Circuit protection shall not be required for control and instrumentation that are totally within the cell line working zone.

(E) Bonding. Bonding of fixed electrical equipment to the energized conductive surfaces of the cell line, its attachments, or auxiliaries shall be permitted. Where fixed electrical equipment is mounted on an energized conductive surface, it shall be bonded to that surface.

668.31 Auxiliary Nonelectric Connections.

Auxiliary nonelectric connections, such as air hoses, water hoses, and the like, to an electrolytic cell, its attachments, or auxiliary equipments shall not have continuous conductive reinforcing wire, armor, braids, and the like. Hoses shall be of a nonconductive material.

668.32 Cranes and Hoists.

(A) Conductive Surfaces to Be Insulated from Ground. The conductive surfaces of cranes and hoists that enter the cell line working zone shall not be required to be grounded. The portion of an overhead crane or hoist that contacts an energized electrolytic cell or energized attachments shall be insulated from ground.

(B) Hazardous Electrical Conditions. Remote crane or hoist controls that could introduce hazardous electrical conditions into the cell line working zone shall employ one or more of the following systems:

(1) Isolated and ungrounded control circuit in accordance with 668.21(A)
(2) Nonconductive rope operator
(3) Pendant pushbutton with nonconductive supporting means and having nonconductive surfaces or ungrounded exposed conductive surfaces
(4) Radio

668.40 Enclosures.

General-purpose electrical equipment enclosures shall be permitted where a natural draft ventilation system prevents the accumulation of gases.

ARTICLE 669
Electroplating

Contents

669.1 Scope.

The provisions of this article apply to the installation of the electrical components and accessory equipment that supply the power and controls for electroplating, anodizing, electropolishing, and electrostripping. For purposes of this article, the term *electroplating* shall be used to identify any or all of these processes.

> Because of the extremely high currents and low voltages normally involved, conventional wiring methods cannot be used in electroplating, anodizing, electropolishing, and electrostripping processes. Note the permission in 669.6(A) and (B) to use bare conductors, even in systems exceeding 50 volts dc. Some systems in the aluminum anodizing process have potentials up to 240 volts. Warning signs are required to be posted to indicate the presence of bare conductors.

669.2 Other Articles.

Except as modified by this article, wiring and equipment used for electroplating processes shall comply with the applicable requirements of Chapters 1 through 4.

669.3 General.

Equipment for use in electroplating processes shall be identified for such service.

669.5 Branch-Circuit Conductors.

Branch-circuit conductors supplying one or more units of equipment shall have an ampacity of not less than 125 percent of the total connected load. The ampacities for busbars shall be in accordance with 366.7.

669.6 Wiring Methods.

Conductors connecting the electrolyte tank equipment to the conversion equipment shall be in accordance with 669.6(A) and (B).

(A) Systems Not Exceeding 50 Volts Direct Current. Insulated conductors shall be permitted to be run without insulated support, provided they are protected from physical damage. Bare copper or aluminum conductors shall be permitted where supported on insulators.

(B) Systems Exceeding 50 Volts Direct Current. Insulated conductors shall be permitted to be run on insulated supports, provided they are protected from physical damage. Bare copper or aluminum conductors shall be permitted where supported on insulators and guarded against accidental contact up to the point of termination in accordance with 110.27.

669.7 Warning Signs.

Warning signs shall be posted to indicate the presence of bare conductors.

669.8 Disconnecting Means.

(A) More Than One Power Supply. Where more than one power supply serves the same dc system, a disconnecting means shall be provided on the dc side of each power supply.

(B) Removable Links or Conductors. Removable links or removable conductors shall be permitted to be used as the disconnecting means.

669.9 Overcurrent Protection.

Direct-current conductors shall be protected from overcurrent by one or more of the following:

(1) Fuses or circuit breakers
(2) A current-sensing device that operates a disconnecting means
(3) Other approved means

ARTICLE 670
Industrial Machinery

> **Contents**
>
> 670.1 Scope
> 670.2 Definitions
> 670.3 Machine Nameplate Data
>   (A) Permanent Nameplate
>   (B) Overcurrent Protection
> 670.4 Supply Conductors and Overcurrent Protection
>   (A) Size
>   (B) Overcurrent Protection
> 670.5 Clearance

670.1 Scope.

This article covers the definition of, the nameplate data for, and the size and overcurrent protection of supply conductors to industrial machinery.

> FPN: For further information, see NFPA 79-1997, *Electrical Standard for Industrial Machinery.*

670.2 Definitions.

> Exhibit 670.1 shows an example of an industrial machine. The scope of Article 670, described in 670.1, and the definitions contained in 670.2 permit the inclusion of other types of industrial machines without the need to continuously modify the scope of Article 670 and NFPA 79, *Electrical Standard for Industrial Machinery.* Also, the scope and defi-

Exhibit 670.1 All-electric injection molding machine with PC-based control. (Courtesy of Cincinnati Milacron)

nitions are more in harmony with NFPA 79 and IEC publication 60204-1.

Industrial Machinery (Machine). A power-driven machine (or a group of machines working together in a coordinated manner), not portable by hand while working, that is used to process material by cutting; forming; pressure; electrical, thermal, or optical techniques; lamination; or a combination of these processes. It can include associated equipment used to transfer material or tooling, including fixtures, to assemble/disassemble, to inspect or test, or to package. [The associated electrical equipment, including the logic controller(s) and associated software or logic together with the machine actuators and sensors, are considered as part of the industrial machine.]

Industrial Manufacturing System. A systematic array of one or more industrial machines that is not portable by hand and includes any associated material handling, manipulating, gauging, measuring, or inspection equipment.

670.3 Machine Nameplate Data.

(A) Permanent Nameplate. A permanent nameplate that lists supply voltage, phase, frequency, full-load current, the maximum ampere rating of the short-circuit and ground-fault protective device, ampere rating of largest motor or load, short-circuit interrupting rating of the machine overcurrent-protective device, if furnished, and diagram number shall be attached to the control equipment enclosure or machine where plainly visible after installation.

The full-load current shown on the nameplate shall not be less than the sum of the full-load currents required for all motors and other equipment that may be in operation at the same time under normal conditions of use. Where unusual type loads, duty cycles, and so forth require oversized conductors or permit reduced-size conductors, the required

capacity shall be included in the marked "full-load current." Where more than one incoming supply circuit is to be provided, the nameplate shall state the above information for each circuit.

The second paragraph of 670.3(A) has been revised to recognize that the operating characteristics of an industrial machine may permit the use of a feeder demand factor, as covered in 430.26. An industrial machine containing motors that are sized for high torque, but that in normal operation run at close to no-load current values, is an example of where it may be appropriate to reduce the full-load current marking on the machine nameplate.

(B) Overcurrent Protection. Where overcurrent protection is provided in accordance with 670.4(B), the machine shall be marked "overcurrent protection provided at machine supply terminals."

670.4 Supply Conductors and Overcurrent Protection.

(A) Size. The size of the supply conductor shall be such as to have an ampacity not less than 125 percent of the full-load current rating of all resistance heating loads plus 125 percent of the full-load current rating of the highest rated motor plus the sum of the full-load current ratings of all other connected motors and apparatus based on their duty cycle that may be in operation at the same time.

In conjunction with the revision to 670.3, the requirements for determining the minimum ampacity of a supply circuit conductor for an industrial machine now specifically reference the duty cycle of motors and apparatus as a consideration. Depending on the operating characteristics of the motor, the duty cycle of the apparatus might not always result in reduction of the supply conductor ampacity. Where motors are used in other than a continuous-duty mode of operation, Table 430.22(E) provides percentages by which the full-load current (FLC) of a given motor is increased or decreased for the purpose of sizing motor circuit conductors. A motor that is loaded continuously under any conditions of use is regarded to be a continuous-duty application.

FPN: See the 0–2000-volt ampacity tables of Article 310 for ampacity of conductors rated 600 volts and below.

(B) Overcurrent Protection. A machine shall be considered as an individual unit and therefore shall be provided with a disconnecting means. The disconnecting means shall be permitted to be supplied by branch circuits protected by either fuses or circuit breakers. The disconnecting means shall not be required to incorporate overcurrent protection. Where furnished as part of the machine, overcurrent protection shall consist of a single circuit breaker or set of fuses, the machine shall bear the marking required in 670.3, and the supply conductors shall be considered either as feeders or taps as covered by 240.21.

The rating or setting of the overcurrent protective device for the circuit supplying the machine shall not be greater than the sum of the largest rating or setting of the branch-circuit short-circuit and ground-fault protective device provided with the machine, plus 125 percent of the full-load current rating of all resistance heating loads, plus the sum of the full-load currents of all other motors and apparatus that could be in operation at the same time.

NFPA 79, *Electrical Standard for Industrial Machinery,* states, in part:

The operating handle of the disconnecting means shall be readily accessible. . . . The center of the grip of the operating handle of the disconnecting means, when in its highest position, shall not be more than 6 ft 7 in. (2 m) above the floor. A permanent operating platform, readily accessible by means of a permanent stair or ladder, shall be considered as the floor for the purpose of this requirement. . . . The operating handle shall be capable of being locked only in the open (off) position. . . . When the control enclosure door is closed, the operating handle shall positively indicate whether the disconnecting means is in the open (off) or closed position. [NFPA 79-1997, 7.10]

Exception: Where one or more instantaneous trip circuit breakers or motor short-circuit protectors are used for motor branch-circuit short-circuit and ground-fault protection as

permitted by 430.52(C), the procedure specified above for determining the maximum rating of the protective device for the circuit supplying the machine shall apply with the following provision: For the purpose of the calculation, each instantaneous trip circuit breaker or motor short-circuit protector shall be assumed to have a rating not exceeding the maximum percentage of motor full-load current permitted by Table 430.52 for the type of machine supply circuit protective device employed.

Where no branch-circuit short-circuit and ground-fault protective device is provided with the machine, the rating or setting of the overcurrent protective device shall be based on 430.52 and 430.53, as applicable.

A commonly asked question is, "When does the *NEC* apply to wiring for machinery defined in Article 670?" The equipment and wiring of industrial machinery, for which different component parts may be purchased and assembled at the location of use, must be installed in accordance with the applicable articles in the *NEC*.

Machinery assembled by the manufacturer, in accordance with NFPA 79, *Electrical Standard for Industrial Machinery*, then disassembled for shipping and reassembled at its place of use, comes only under Article 670 and any *NEC* sections referenced therein. In this case, the machinery is treated as a package unit. The nameplate provides the necessary information to size the branch-circuit conductors, disconnecting means, and overcurrent protection. The computation of motor and nonmotor loads is reflected on the nameplate as full-load amperes, and no further calculation is necessary.

670.5 Clearance.

Where the conditions of maintenance and supervision ensure that only qualified persons service the installation, the dimensions of the working space in the direction of access to live parts operating at not over 150 volts line-to-line or line-to-ground that are likely to require examination, adjustment, servicing, or maintenance while energized shall be a minimum of 750 mm (2½ ft). Where controls are enclosed in cabinets, the door(s) shall open at least 90 degrees or be removable.

Exception: Where the enclosure requires a tool to open, and where only diagnostic and troubleshooting testing is involved on live parts, the clearances shall be permitted to be less than 750 mm (2½ ft).

It is essential that *only qualified persons* perform the service or maintenance on energized live parts. See the definitions of *energized* and *live parts* in Article 100.

ARTICLE 675
Electrically Driven or Controlled Irrigation Machines

Contents

I. General

675.1 Scope.

The provisions of this article apply to electrically driven or controlled irrigation machines, and to the branch circuits and controllers for such equipment.

Electric pump motors used to supply water to irrigation machines are governed by the general requirements of the *Code*, not by Article 675.

675.2 Definitions.

Center Pivot Irrigation Machines. A multimotored irrigation machine that revolves around a central pivot and employs alignment switches or similar devices to control individual motors.

Collector Rings. An assembly of slip rings for transferring electrical energy from a stationary to a rotating member.

Irrigation Machines. An electrically driven or controlled machine, with one or more motors, not hand portable, and used primarily to transport and distribute water for agricultural purposes.

675.3 Other Articles.

These provisions are in addition to, or amendatory of, the provisions of Article 430 and other articles in this *Code* that apply except as modified in this article.

The requirements of Article 675 apply to special equipment for a particular condition; they supplement or modify the general rules. See 90.3, Code Arrangement.

675.4 Irrigation Cable.

(A) Construction. The cable used to interconnect enclosures on the structure of an irrigation machine shall be an assembly of stranded, insulated conductors with nonhygroscopic and nonwicking filler in a core of moisture- and flame-resistant nonmetallic material overlaid with a metallic covering and jacketed with a moisture-, corrosion-, and sunlight-resistant nonmetallic material.

The conductor insulation shall be of a type listed in Table 310.13 for an operating temperature of 75°C (167°F) and for use in wet locations. The core insulating material thickness shall not be less than 0.76 mm (30 mils), and the metallic overlay thickness shall not be less than 0.20 mm (8 mils). The jacketing material thickness shall not be less than 1.27 mm (50 mils).

A composite of power, control, and grounding conductors in the cable shall be permitted.

(B) Alternate Wiring Methods. Other cables listed for the purpose.

(C) Supports. Irrigation cable shall be secured by straps, hangers, or similar fittings identified for the purpose and installed as not to damage the cable. Cable shall be supported at intervals not exceeding 1.2 m (4 ft).

(D) Fittings. Fittings shall be used at all points where irrigation cable terminates. The fittings shall be designed for use with the cable and shall be suitable for the conditions of service.

675.5 More Than Three Conductors in a Raceway or Cable.

The signal and control conductors of a raceway or cable shall not be counted for the purpose of derating the conductors as required in 310.15(B)(2)(a).

675.6 Marking on Main Control Panel.

The main control panel shall be provided with a nameplate that shall give the following information:

(1) The manufacturer's name, the rated voltage, the phase, and the frequency
(2) The current rating of the machine
(3) The rating of the main disconnecting means and size of overcurrent protection required

675.7 Equivalent Current Ratings.

Where intermittent duty is not involved, the provisions of Article 430 shall be used for determining ratings for controllers, disconnecting means, conductors, and the like. Where irrigation machines have inherent intermittent duty, the determinations of equivalent current ratings in 675.7(A) and (B) shall be used.

(A) Continuous-Current Rating. The equivalent continuous-current rating for the selection of branch-circuit conductors and overcurrent protection shall be equal to 125 percent of the motor nameplate full-load current rating of the largest motor plus a quantity equal to the sum of each of the motor nameplate full-load current ratings of all remaining motors on the circuit multiplied by the maximum percent duty cycle at which they can continuously operate.

(B) Locked-Rotor Current. The equivalent locked-rotor current rating shall be equal to the numerical sum of the locked-rotor current of the two largest motors plus 100 percent of the sum of the motor nameplate full-load current ratings of all the remaining motors on the circuit.

675.8 Disconnecting Means.

(A) Main Controller. A controller that is used to start and stop the complete machine shall meet all of the following requirements:

(1) An equivalent continuous current rating not less than specified in 675.7(A) or 675.22(A)
(2) A horsepower rating not less than the value from Tables 430.151(A) and (B), based on the equivalent locked-rotor current specified in 675.7(B) or 675.22(B)

(B) Main Disconnecting Means. The main disconnecting means for the machine shall provide overcurrent protection, shall be at the point of connection of electrical power to the machine or shall be visible and not more than 15 m (50 ft) from the machine, and shall be readily accessible and capable of being locked in the open position. This disconnecting means shall have a horsepower and current rating not less than required for the main controller.

In accordance with 675.8(B), the main disconnecting means is permitted to be up to 50 ft from the machine but must be readily accessible and capable of being locked in the open position. This eliminates one set of overcurrent protective devices and one disconnecting means where the circuit originates at the motor control panel for the irrigation pump and the panel is located within 50 ft of the center pivot machine. It also alleviates some potential problems with machines designed to be towed to a second site.

Exception: Circuit breakers without marked horsepower ratings shall be permitted in accordance with 430.109.

(C) Disconnecting Means for Individual Motors and Controllers. A disconnecting means shall be provided to simultaneously disconnect all ungrounded conductors for each motor and controller and shall be located as required by Article 430, Part IX. The disconnecting means shall not be required to be readily accessible.

Article 430, Part IX, provides for safety during maintenance and inspection shutdown periods. See the commentary following 430.103.

675.9 Branch-Circuit Conductors.

The branch-circuit conductors shall have an ampacity not less than specified in 675.7(A) or 675.22(A).

675.10 Several Motors on One Branch Circuit.

The requirements of 430.53 provide for motor branch-circuit short-circuit and ground-fault protection where several motors are supplied by one branch circuit. A combination of these requirements, which are modified for this special equipment application, is found in 675.10.

(A) Protection Required. Several motors, each not exceeding 2 hp rating, shall be permitted to be used on an irrigation machine circuit protected at not more than 30 amperes at 600 volts, nominal, or less, provided all of the following conditions are met:

(1) The full-load rating of any motor in the circuit shall not exceed 6 amperes.

(2) Each motor in the circuit shall have individual overload protection in accordance with 430.32.

(3) Taps to individual motors shall not be smaller than 14 AWG copper and not more than 7.5 m (25 ft) in length.

(B) Individual Protection Not Required. Individual branch-circuit short-circuit protection for motors and motor controllers shall not be required where the requirements of 675.10(A) are met.

675.11 Collector Rings.

(A) Transmitting Current for Power Purposes. Collector rings shall have a current rating not less than 125 percent of the full-load current of the largest device served plus the full-load current of all other devices served, or as determined from 675.7(A) or 675.22(A).

(B) Control and Signal Purposes. Collector rings for control and signal purposes shall have a current rating not less than 125 percent of the full-load current of the largest device served plus the full-load current of all other devices served.

(C) Grounding. The collector ring used for grounding shall have a current rating of not less than that sized in accordance with 675.11(A).

(D) Protection. Collector rings shall be protected from the expected environment and from accidental contact by means of a suitable enclosure.

675.12 Grounding.

The following equipment shall be grounded:

(1) All electrical equipment on the irrigation machine
(2) All electrical equipment associated with the irrigation machine
(3) Metal junction boxes and enclosures
(4) Control panels or control equipment that supply or control electrical equipment to the irrigation machine

Exception: Grounding shall not be required on machines where all of the following provisions are met:

(a) The machine is electrically controlled but not electrically driven.

(b) The control voltage is 30 volts or less.

(c) The control or signal circuits are current limited as specified in Chapter 9, Tables 11(A) and 11(B).

675.13 Methods of Grounding.

Machines that require grounding shall have a non–current-carrying equipment grounding conductor provided as an integral part of each cord, cable, or raceway. This grounding conductor shall be sized not less than the largest supply conductor in each cord, cable, or raceway. Feeder circuits

supplying power to irrigation machines shall have an equipment grounding conductor sized according to Table 250.122.

675.14 Bonding.

Where electrical grounding is required on an irrigation machine, the metallic structure of the machine, metallic conduit, or metallic sheath of cable shall be bonded to the grounding conductor. Metal-to-metal contact with a part that is bonded to the grounding conductor and the non–current-carrying parts of the machine shall be considered as an acceptable bonding path.

675.15 Lightning Protection.

If an irrigation machine has a stationary point, a grounding electrode system in accordance with Article 250, Part III, shall be connected to the machine at the stationary point for lightning protection.

Where the electrical power supply to irrigation machine equipment is a service, the requirements of Article 250 for grounding the system and equipment are applicable. Due to the physical location of irrigation equipment, the most likely grounding electrode of the types covered in 250.52 is a driven ground rod or ground plate. Consideration should be given to the requirements of 250.60 and NFPA 780, *Standard for the Installation of Lightning Protection Systems,* in areas where lightning protection is critical. A common electrode system is not permitted to be used for the dual function of grounding the electric service and grounding the lightning protection system. The separate electrode systems are required to be bonded together.

675.16 Energy from More Than One Source.

Equipment within an enclosure receiving electrical energy from more than one source shall not be required to have a disconnecting means for the additional source, provided that its voltage is 30 volts or less and it meets the requirements of Part III of Article 725.

675.17 Connectors.

External plugs and connectors on the equipment shall be of the weatherproof type.

Unless provided solely for the connection of circuits meeting the requirements of Part III of Article 725, external plugs and connectors shall be constructed as specified in 250.124(A).

II. Center Pivot Irrigation Machines

675.21 General.

The provisions of Part II are intended to cover additional special requirements that are peculiar to center pivot irrigation machines. See 675.2 for definition of *Center Pivot Irrigation Machines.*

675.22 Equivalent Current Ratings.

In order to establish ratings of controllers, disconnecting means, conductors, and the like, for the inherent intermittent duty of center pivot irrigation machines, the determinations in 675.22(A) and (B) shall be used.

The ratings of electrical components of any circuit should be selected so as to avoid extensive damage to the equipment during a short circuit or ground fault. Requirements for establishing ratings of components of special equipment for inherent intermittent duty are covered in 675.22. Also see the commentary following 110.10 and 430.52.

(A) Continuous-Current Rating. The equivalent continuous-current rating for the selection of branch-circuit conductors and branch-circuit devices shall be equal to 125 percent of the motor nameplate full-load current rating of the largest motor plus 60 percent of the sum of the motor nameplate full-load current ratings of all remaining motors on the circuit.

(B) Locked-Rotor Current. The equivalent locked-rotor current rating shall be equal to the numerical sum of two times the locked-rotor current of the largest motor plus 80 percent of the sum of the motor nameplate full-load current ratings of all the remaining motors on the circuit.

ARTICLE 680
Swimming Pools, Fountains, and Similar Installations

Contents

I. General

680.1 Scope.

The provisions of this article apply to the construction and installation of electrical wiring for and equipment in or adjacent to all swimming, wading, therapeutic, and decorative pools, fountains, hot tubs, spas, and hydromassage bathtubs, whether permanently installed or storable, and to metallic auxiliary equipment, such as pumps, filters, and similar equipment. The term *body of water* used throughout Part I applies to all bodies of water covered in this scope unless otherwise amended.

Article 680 applies to decorative pools and fountains; swimming, wading, and wave pools; therapeutic tubs and tanks; hot tubs; spas; hydromassage bathtubs; and similar installations. The installations covered by this article can be indoors or outdoors, permanent or storable, and may or may not be directly supplied by electrical circuits of any nature.

   Studies conducted by Underwriters Laboratories, various manufacturers, and others indicate that a person in a swimming pool can receive a severe electric shock by reaching over and touching the energized casing of a faulty appliance — such as a radio or a hair dryer — as the person's body establishes a conductive path through the water to earth. Also, a person not in contact with a faulty appliance or any grounded object can receive an electric shock and be rendered immobile by a potential gradient in the water itself. Accordingly, the requirements of Article 680 covering effective bonding and grounding, installation of receptacles and luminaires, use of ground-fault circuit interrupters, modified wiring methods, and so on, apply not only to the installation of the pool but also to installations and equipment adjacent to or associated with the pool.

680.2 Definitions.

Cord-and-Plug-Connected Lighting Assembly. A lighting assembly consisting of a luminaire (lighting fixture) intended for installation in the wall of a spa, hot tub, or storable pool, and a cord-and-plug-connected transformer.

Dry-Niche Luminaire (Lighting Fixture). A luminaire (lighting fixture) intended for installation in the wall of a pool or fountain in a niche that is sealed against the entry of pool water.

Equipment, Fixed. Equipment that is fastened or otherwise secured at a specific location.

Equipment, Portable. Equipment that is actually moved or can easily be moved from one place to another in normal use.

Equipment, Stationary. Equipment that is not easily moved from one place to another in normal use.

Forming Shell. A structure designed to support a wet-niche luminaire (lighting fixture) assembly and intended for mounting in a pool or fountain structure.

Fountain. Fountains, ornamental pools, display pools, and reflection pools. The definition does not include drinking fountains.

Hydromassage Bathtub. A permanently installed bathtub equipped with a recirculating piping system, pump, and associated equipment. It is designed so it can accept, circulate, and discharge water upon each use.

See the commentary following Part VII, Hydromassage Bathtubs.

Maximum Water Level. The highest level that water can reach before it spills out.

No-Niche Luminaire (Lighting Fixture). A luminaire (lighting fixture) intended for installation above or below the water without a niche.

Packaged Spa or Hot Tub Equipment Assembly. A factory-fabricated unit consisting of water-circulating, heating, and control equipment mounted on a common base, intended to operate a spa or hot tub. Equipment can include pumps, air blowers, heaters, lights, controls, sanitizer generators, and so forth.

The definition of *packaged spa or hot tub equipment assembly* clarifies which assemblies are subject to the requirements of 680.44.

Packaged Therapeutic Tub or Hydrotherapeutic Tank Equipment Assembly. A factory-fabricated unit consisting of water-circulating, heating, and control equipment mounted on a common base, intended to operate a therapeu-

tic tub or hydrotherapeutic tank. Equipment can include pumps, air blowers, heaters, lights, controls, sanitizer generators, and so forth.

Permanently Installed Decorative Fountains and Reflection Pools. Those that are constructed in the ground, on the ground, or in a building in such a manner that the fountain cannot be readily disassembled for storage, whether or not served by electrical circuits of any nature. These units are primarily constructed for their aesthetic value and are not intended for swimming or wading.

Permanently Installed Swimming, Wading, and Therapeutic Pools. Those that are constructed in the ground or partially in the ground, and all others capable of holding water in a depth greater than 1.0 m (42 in.), and all pools installed inside of a building, regardless of water depth, whether or not served by electrical circuits of any nature.

See the commentary following Part VI, Pools and Tubs for Therapeutic Use.

Pool. Manufactured or field-constructed equipment designed to contain water on a permanent or semipermanent basis and used for swimming, wading, or other purposes.

Pool Cover, Electrically Operated. Motor-driven equipment designed to cover and uncover the water surface of a pool by means of a flexible sheet or rigid frame.

The requirements for electrically operated pool covers are found in 680.27(B).

Self-Contained Spa or Hot Tub. Factory-fabricated unit consisting of a spa or hot tub vessel with all water-circulating, heating, and control equipment integral to the unit. Equipment can include pumps, air blowers, heaters, lights, controls, sanitizer generators, and so forth.

The definition of *self-contained spa or hot tub* clarifies which assemblies are subject to the requirements of 680.44.

Self-Contained Therapeutic Tubs or Hydrotherapeutic Tanks. A factory-fabricated unit consisting of a therapeutic tub or hydrotherapeutic tank with all water-circulating, heating, and control equipment integral to the unit. Equipment may include pumps, air blowers, heaters, light controls, sanitizer generators, and so forth.

Spa or Hot Tub. A hydromassage pool, or tub for recreational or therapeutic use, not located in health care facilities, designed for immersion of users, and usually having a filter, heater, and motor-driven blower. It may be installed indoors or outdoors, on the ground or supporting structure, or in the ground or supporting structure. Generally, a spa or hot tub is not designed or intended to have its contents drained or discharged after each use.

See the commentary following Part IV, Spas and Hot Tubs.

Storable Swimming or Wading Pool. Those that are constructed on or above the ground and are capable of holding water to a maximum depth of 1.0 m (42 in.), or a pool with nonmetallic, molded polymeric walls or inflatable fabric walls regardless of dimension.

See the commentary following 680.30 and Exhibit 680.16.

Originally, storable pools were not specifically addressed in the *NEC*. Article 680 was written to provide guidance relative to permanent, in-ground pools and their unique construction requirements because of the unusual earth-water-electricity-human body environment created in the finished product. The conductivity of moist concrete or metal walls buried in the ground, the incorporation of large masses of reinforcing steel, and the inclusion of stainless-steel handrails and diving-board stands as well as 120-volt lights in the pool structure all called for the strict wiring, bonding, and grounding requirements of Article 680.

Storable pools, on the other hand, are intended to be temporary structures, without the need for special wiring or modification to the pool site. They are usually sold as a complete package, consisting of the pool walls, vinyl liner, plumbing kit, and pump/filter device. A storable pool is often disassembled and stored during the winter months. Regional preferences, weather patterns, economic considerations, and design characteristics of the pool are all factors influencing this action. The original Article 680 definition of a storable pool was "One that is so constructed that it may be readily disassembled for storage and reassembled to its original integrity."

Part III of Article 680 was created to address the special equipment specifications of storable pools, and Underwriters Laboratories developed testing and labeling criteria for listing the pump/filter units designed especially for these pools. This equipment has the following notable aspects:

1. It must have an approved system of double insulation or the equivalent.
2. It is permitted to have a flexible cord equipped with a parallel-blade, grounding-type attachment plug for electrical connection.
3. It must have a grounding conductor included in the flexible cord.
4. The flexible cord is not limited to 3 ft, as required in 680.7, and is specified by UL to be not less than 25 ft long. This length was chosen to discourage the use of extension cords.

The UL labeling requirement for these listed units includes the wording "Do Not Use with Permanently Installed Pools." In some cases, consumers and members of the swimming pool industry, however, have found it desirable to use these pump/filter units on any aboveground or on-ground pool, regardless of the pool's dimensions or "storability."

Storable pools are supplied as two distinct types. One type is intended to be disassembled at the end of each swimming season. The second type, by the nature of its construction, can be disassembled, but manufacturers recommend leaving it assembled. The pools in the latter category frequently require special modification to and preparation of the pool site, making them impractical to disassemble. Draining these pools, especially the larger ones, increases the likelihood of costly damage caused by shrinkage of the vinyl liner material.

The main factor differentiating the two types of pools is wall height. Generally, pools, other than the inflatable type, intended to be disassembled at season's end have wall heights of 42 in. or less, while those not intended for disassembly have wall heights of 48 in. or more. The surface area of the pools is not a factor. Inflatable pools are treated as storable pools regardless of their wall height.

Through-Wall Lighting Assembly. A lighting assembly intended for installation above grade, on or through the wall of a pool, consisting of two interconnected groups of components separated by the pool wall.

Wet-Niche Luminaire (Lighting Fixture). A luminaire (lighting fixture) intended for installation in a forming shell mounted in a pool or fountain structure where the luminaire (fixture) will be completely surrounded by water.

680.3 Other Articles.

Except as modified by this article, wiring and equipment in or adjacent to pools and fountains shall comply with other applicable provisions of this *Code*, including those provisions identified in Table 680.3.

Table 680.3 Other Articles

Topic	Section or Article
Wiring	Chapters 1–4
Junction box support	314.23
Rigid nonmetallic conduit	352.12
Audio Equipment	Article 640, Parts I and II
Adjacent to pools and fountains	640.10
Underwater speakers*	

*Underwater loudspeakers shall be installed in accordance with 680.27(A).

Note that 314.23(D) specifies the requirements for the support of threaded boxes that do not contain devices and that 352.12 does not permit luminaires or most other electrical equipment to be supported by rigid nonmetallic conduit. Exhibit 680.1 shows a properly supported junction box for a wet-niche fixture. Also see the commentary following 314.23(D).

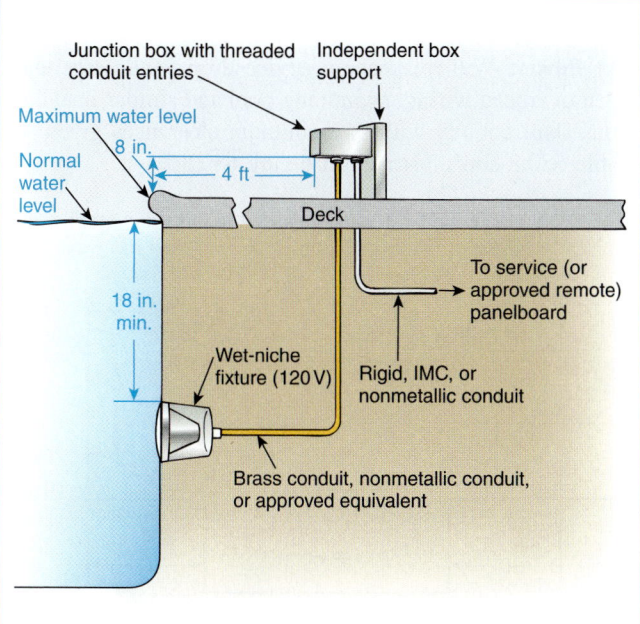

Exhibit 680.1 Wet-niche fixture installation with junction box supported above pool deck.

680.4 Approval of Equipment.

All electrical equipment installed in the water, walls, or decks of pools, fountains, and similar installations shall comply with the provisions of this article.

680.5 Ground-Fault Circuit Interrupters.

Ground-fault circuit interrupters (GFCIs) shall be self-contained units, circuit-breaker or receptacle types, or other listed types.

See the definition of *ground-fault circuit interrupter* in Article 100.

A ground-fault circuit interrupter is intended to be used only in a circuit that has a solidly grounded conductor; however, an equipment grounding conductor is not necessary in order for the GFCI to function. A Class A GFCI trips where the current to ground has a value in the range of 4 through 6 mA; it is suitable for use in swimming pool circuits. It should be noted, however, that circuits supplying

pool equipment that were installed before local adoption of the 1965 edition of the *Code* may have sufficient leakage current to cause a Class A GFCI to trip. A Class B GFCI trips if the current to ground exceeds 20 mA; it is suitable for use only with underwater swimming pool lighting fixtures installed before the local adoption of the 1965 *Code*.

680.6 Grounding.

Electrical equipment shall be grounded in accordance with Parts V, VI, and VII of Article 250 and connected by wiring methods of Chapter 3, except as modified by this article. The following equipment shall be grounded:

(1) Through-wall lighting assemblies and underwater luminaires (lighting fixtures), other than those low-voltage systems listed for the application without a grounding conductor
(2) All electrical equipment located within 1.5 m (5 ft) of the inside wall of the specified body of water
(3) All electrical equipment associated with the recirculating system of the specified body of water
(4) Junction boxes
(5) Transformer enclosures
(6) Ground-fault circuit interrupters
(7) Panelboards that are not part of the service equipment and that supply any electrical equipment associated with the specified body of water

Electrical equipment other than underwater lighting fixtures and pool-associated motors is required to be connected by the wiring methods of Chapter 3 and grounded in accordance with Article 250. For example, an outdoor receptacle installed to meet the requirements of 680.22(A)(3) is permitted to be wired with Type UF cable containing an insulated or bare conductor for equipment grounding purposes. Circuits for pools may be derived from an existing remote panelboard supplied by an approved cable assembly, as specified in 680.25(A), Exception. The requirements of 680.6 permit Type UF cable to be used for the receptacle required by 680.22(A)(3) and for some pool-related equipment, but circuit conductors for underwater lighting fixtures are required to be run in raceways. Circuit conductors for pool-associated motors other than flexible cord, as permitted by 680.7, are required to be installed in raceways except in the interior of one-family dwelling units, where any raceway or cable assembly permitted by Chapter 3 is acceptable if the equipment grounding conductor is at least 12 AWG copper and is enclosed by the wiring method.

Equipment grounding requirements are contained in 680.6, 680.21(A)(1), 680.23(F), and 680.25(B). These requirements specify that equipment grounding conductors be connected to non–current-carrying metal parts of the specified equipment. These equipment grounding conductors are required to be run with the circuit conductors in rigid metal conduit, intermediate conduit, listed MC cable (for motors only), or rigid nonmetallic conduit (electrical metallic tubing is permitted in or on buildings, and electrical nonmetallic tubing is permitted inside buildings), and they must be terminated at the grounding terminal bus of the service panelboard, the source of the separately derived system, or the subpanel. This equipment grounding conductor provides a path of low impedance that limits the voltage to ground and facilitates operation of the circuit overcurrent protective device(s). The equipment grounding conductor is required to be an insulated copper conductor not smaller than 12 AWG.

The requirements of 680.6, 680.21(A)(1), 680.23(F)(2), and 680.25(B) are in addition to the bonding requirements in 680.26. The intent of the bonding requirements is to establish an equipotential plane to limit the voltage between all non–current-carrying parts of electrical and nonelectrical equipment in the pool area.

Bonding conductors may be insulated, covered, or bare and are required to be 8 AWG solid copper or larger. They may be direct buried, and, if connected to metal parts of the pool structure or metal parts of electrical equipment, they may be externally clamped or attached and are not required to be accessible. All these parts form a common bonding grid that establishes an equipotential grounding system, and they do not have to be run to the equipment grounding terminals of panelboards or service equipment.

680.7 Cord-and-Plug-Connected Equipment.

Fixed or stationary equipment other than an underwater luminaire (lighting fixture) for a permanently installed pool shall be permitted to be connected with a flexible cord to facilitate the removal or disconnection for maintenance or repair.

(A) Length. For other than storable pools, the flexible cord shall not exceed 900 mm (3 ft) in length.

(B) Equipment Grounding. The flexible cord shall have a copper equipment grounding conductor sized in accordance with 250.122 but not smaller than 12 AWG. The cord shall terminate in a grounding-type attachment plug.

(C) Construction. The equipment grounding conductors shall be connected to a fixed metal part of the assembly. The removable part shall be mounted on or bonded to the fixed metal part.

In some climates, it is preferable to disconnect and remove a permanent pool's filter pump during cold-weather months. A 3-ft cord is permitted, to facilitate the removal of fixed

or stationary equipment for maintenance and storage. The 3-ft cord limitation does not apply to cord-and-plug-connected filter pumps used with storable-type pools (covered in Part III of Article 680), since these pumps are neither fixed nor stationary. Listed filter pumps for use with storable pools are considered portable and are permitted to be equipped with cords longer than 3 ft.

680.8 Overhead Conductor Clearances.

(A) Power. With respect to service drop conductors and open overhead wiring, swimming pool and similar installations shall comply with the minimum clearances given in Table 680.8 and illustrated in Figure 680.8.

FPN: Open overhead wiring as used in this article typically refers to conductor(s) not in an enclosed raceway.

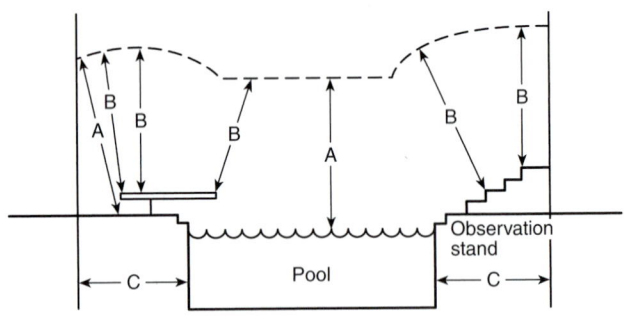

Figure 680.8 Clearances from pool structures.

(B) Communications Systems. Communication, radio, and television coaxial cables within the scope of Articles 800 through 820 shall be permitted at a height of not less than 3.0 m (10 ft) above swimming and wading pools, diving structures, and observation stands, towers, or platforms.

(C) Network-Powered Broadband Communications Systems. The minimum clearances for overhead network-powered broadband communications systems conductors from pools or fountains shall comply with the provisions in Table 680.8 for conductors operating at 0 to 750 volts to ground.

Service drop conductors, conductors of network-powered broadband communications systems, and aerial feeders and branch circuits are permitted to be located above a swimming pool and associated pool structures where provided with the clearances specified in Table 680.8. Overhead conductors of communications systems are required to comply with 680.8(B). These clearances consider such factors as the use of skimmers with aluminum handles and provide sufficient separation between the conductors and the pool. In some instances, locating a swimming pool below electric conduc-

Table 680.8 Overhead Conductor Clearances

Clearance Parameters	Insulated Cables, 0–750 Volts to Ground, Supported on and Cabled Together with an Effectively Grounded Bare Messenger or Effectively Grounded Neutral Conductor		All Other Conductors Voltage to Ground			
			0 through 15 kV		Over 15 through 50 kV	
	m	ft	m	ft	m	ft
A. Clearance in any direction to the water level, edge of water surface, base of diving platform, or permanently anchored raft	6.9	22.5	7.5	25	8.0	27
B. Clearance in any direction to the observation stand, tower, or diving platform	4.4	14.5	5.2	17	5.5	18
C. Horizontal limit of clearance measured from inside wall of the pool	This limit shall extend to the outer edge of the structures listed in A and B of this table but not less than 3 m (10 ft).					

tors is unavoidable, for example, on a building lot with limited area or an existing lot where the electric supply lines are already in place. The clearances for conductors from pools and pool structures were increased in the 1999 *Code*. These changes harmonize the *NEC* with ANSI C2, *National Electrical Safety Code* (NESC).

680.9 Electric Pool Water Heaters.

All electric pool water heaters shall have the heating elements subdivided into loads not exceeding 48 amperes and protected at not over 60 amperes. The ampacity of the branch-circuit conductors and the rating or setting of overcurrent protective devices shall not be less than 125 percent of the total nameplate-rated load.

680.10 Underground Wiring Location.

Underground wiring shall not be permitted under the pool or within the area extending 1.5 m (5 ft) horizontally from the inside wall of the pool unless this wiring is necessary to supply pool equipment permitted by this article. Where space limitations prevent wiring from being routed a distance 1.5 m (5 ft) or more from the pool, such wiring shall be permitted where installed in rigid metal conduit, intermediate metal conduit, or a nonmetallic raceway system. All metal conduit shall be corrosion resistant and suitable for the location. The minimum burial depth shall be as given in Table 680.10.

Table 680.10 Minimum Burial Depths

Wiring Method	Minimum Burial	
	mm	in.
Rigid metal conduit	150	6
Intermediate metal conduit	150	6
Nonmetallic raceways listed for direct burial without concrete encasement	450	18
Other approved raceways*	450	18

*Raceways approved for burial only where concrete encased shall require a concrete envelope not less than 50 mm (2 in.) thick.

680.11 Equipment Rooms and Pits.

Electric equipment shall not be installed in rooms or pits that do not have drainage that adequately prevents water accumulation during normal operation or filter maintenance.

680.12 Maintenance Disconnecting Means.

One or more means to disconnect all ungrounded conductors shall be provided for all utilization equipment other than lighting. Each means shall be accessible and within sight from its equipment.

A disconnecting means is required to be installed within sight of the pool, spa, and hot tub equipment to allow service personnel to disconnect the power while servicing these units. This requirement ensures that a disconnect is available to workers servicing pool, spa, and hot tub equipment such as motors, heaters, and control panels. This requirement has been revised for the 2002 *NEC* to clarify that lighting equipment installed in swimming pools is not subject to this requirement. See Exhibit 680.2.

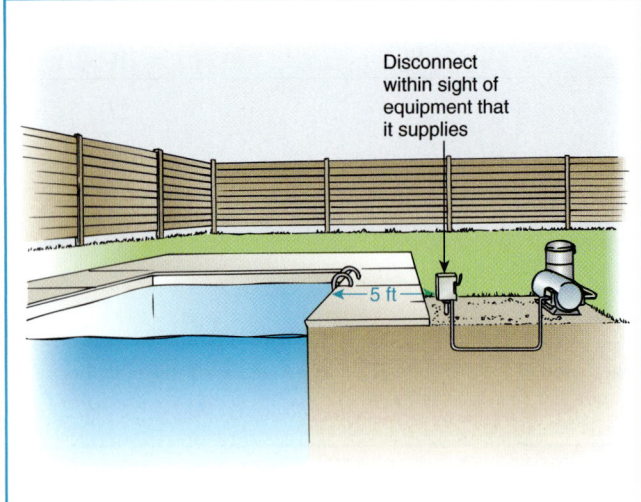

Exhibit 680.2 Pool equipment disconnect required. The disconnect for pool equipment must be located within sight of the pool equipment and at least 5 ft from the pool.

II. Permanently Installed Pools

680.20 General.

Electrical installations at permanently installed pools shall comply with the provisions of Part I and Part II of this article.

680.21 Motors.

(A) Wiring Methods.

(1) General. The branch circuits for pool-associated motors shall be installed in rigid metal conduit, intermediate metal conduit, rigid nonmetallic conduit, or Type MC cable listed for the location. Other wiring methods and materials shall be permitted in specific locations or applications as covered in this section. Any wiring method employed shall contain a copper equipment grounding conductor sized in accordance with 250.122 but not smaller than 12 AWG.

Type MC cable listed for the application is permitted as a wiring method for swimming pool motor circuits. This listing requirement ensures that the MC cable is suitable for the environmental and installation conditions that are typically encountered with swimming pool pump motors. Type MC

cables that are listed for installation in direct sunlight or direct burial are marked to indicate suitability for these applications.

(2) On or Within Buildings. Where installed on or within buildings, electrical metallic tubing shall be permitted.

(3) Flexible Connections. Where necessary to employ flexible connections at or adjacent to the motor, liquidtight flexible metal or nonmetallic conduit with approved fittings shall be permitted.

(4) One-Family Dwellings. In the interior of one-family dwellings, or in the interior of accessory buildings associated with a one-family dwelling, any of the wiring methods recognized in Chapter 3 of this *Code* shall be permitted that comply with the provisions of this paragraph. Where run in a raceway, the equipment grounding conductor shall be insulated. Where run in a cable assembly, the equipment grounding conductor shall be permitted to be uninsulated, but it shall be enclosed within the outer sheath of the cable assembly.

(5) Cord-and-Plug Connections. Pool-associated motors shall be permitted to employ cord-and-plug connections. The flexible cord shall not exceed 900 mm (3 ft) in length. The flexible cord shall include an equipment grounding conductor sized in accordance with 250.122 and shall terminate in a grounding-type attachment plug.

(B) Double Insulated Pool Pumps. A listed cord-and-plug-connected pool pump incorporating an approved system of double insulation that provides a means for grounding only the internal and nonaccessible, non–current-carrying metal parts of the pump shall be connected to any wiring method recognized in Chapter 3 that is suitable for the location.

Cord-and-plug-connected double-insulated swimming pool filter pumps have been used with permanently installed aboveground pools and some storable pools, regardless of the pool's size, for many years without any known field-related problems. The internal metal parts of a swimming pool filter pump incorporating a system of double insulation are grounded; however, they are not required to be incorporated into the bonding system required by 680.26(B), since the act of bonding compromises the double-insulation system.

680.22 Area Lighting, Receptacles, and Equipment.

(A) Receptacles.

(1) Circulation and Sanitation System, Location. Receptacles that provide power for water-pump motors or for other loads directly related to the circulation and sanitation system shall be located at least 3.0 m (10 ft) from the inside walls of the pool, or not less than 1.5 m (5 ft) from the inside walls of the pool if they meet all of the following conditions:

(1) Consist of single receptacles
(2) Employ a locking configuration
(3) Are of the grounding type
(4) Have GFCI protection

(2) Other Receptacles, Location. Other receptacles shall be not less than 3.0 m (10 ft) from the inside walls of a pool.

(3) Dwelling Unit(s). Where a permanently installed pool is installed at a dwelling unit(s), no fewer than one 125-volt 15- or 20-ampere receptacle on a general-purpose branch circuit shall be located not less than 3.0 m (10 ft) from and not more than 6.0 m (20 ft) from the inside wall of the pool. This receptacle shall be located not more than 2.0 m (6 ft 6 in.) above the floor, platform, or grade level serving the pool.

(4) Restricted Space. Where a pool is within 3.0 m (10 ft) of a dwelling and the dimensions of the lot preclude meeting the required clearances, not more than one receptacle outlet shall be permitted if not less than 1.5 m (5 ft) measured horizontally from the inside wall of the pool.

(5) GFCI Protection. All 125-volt receptacles located within 6.0 m (20 ft) of the inside walls of a pool or fountain shall be protected by a ground-fault circuit interrupter. Receptacles that supply pool pump motors and that are rated 15 or 20 amperes, 120 volt through 240 volts, single phase, shall be provided with GFCI protection.

All single-phase, 15- and 20-ampere, 120 through 240 receptacles that supply swimming pool pump motors are required to have GFCI protection. While this requirement applied only to installations at other than dwellings in the 1999 *Code*, the 2002 *Code* has been revised to require GFCI protection of these receptacles for all swimming pool installations. It should be noted that this requirement applies to these receptacles regardless of their proximity to the swimming pool, and it applies only to cord-and-plug-connected pump motors.

(6) Measurements. In determining the dimensions in this section addressing receptacle spacings, the distance to be measured shall be the shortest path the supply cord of an appliance connected to the receptacle would follow without piercing a floor, wall, ceiling, doorway with hinged or sliding door, window opening, or other effective permanent barrier.

The requirements of 680.22(A) apply to receptacles located near a permanently installed pool or fountain. They do not apply to direct-connected equipment. Permission is given in 680.22(A)(1) to allow a single locking- and grounding-type receptacle to supply a recirculation pump motor where the receptacle is located not less than 5 ft from the inside walls of the pool or fountain and is protected by a GFCI.

As required by 680.22(A)(3), each permanently installed pool in a residential setting is required to have at least one receptacle, which must be located at least 10 ft from the pool and not more than 20 ft from the pool. The intent of this requirement is to permit ordinary appliances to be safely plugged in and used near the pool but to avoid the need for extension cords in the vicinity of the pool. The 10-ft minimum dimension was chosen so that an appliance with a 6-ft cord could not be accidentally knocked into the pool.

The 2002 *NEC* contains a new provision [680.22(A)(5)] that covers receptacle outlet installation at dwelling units where the spatial constraints prevent locating the required receptacle 10 ft or more from the inside walls of the pool. Where this condition exists, one GFCI-protected receptacle is permitted to be located closer than 10 ft but not less than 5 ft from the inside walls of the pool.

Ground-fault circuit-interrupter protection of all 125-volt receptacles located within 20 ft of a pool or fountain is required by 680.22(A)(5). This rule applies to pools located outdoors or indoors, permanently installed or storable, and for residential or commercial use. Since people within 20 ft of a pool are normally subjected to dampness and moisture, the GFCI requirements within the 20-ft space is warranted.

Examples of receptacles meeting the requirements of 680.22(A) are shown in Exhibits 680.3 and 680.4. Exhibit 680.5 illustrates that the determination of the minimum distance for receptacles from a pool does not include receptacles within a structure. The receptacles within the structure are permitted to be less than 10 ft from the pool. Where this installation is at a dwelling unit, it is necessary to provide at least one receptacle between 10 ft and 20 ft from the inside walls of the pool. This location precludes having to run the cord of an appliance used on the pool deck through a doorway.

(B) Luminaires (Lighting Fixtures), Lighting Outlets, and Ceiling-Suspended (Paddle) Fans.

(1) New Outdoor Installation Clearances. In outdoor pool areas, luminaires (lighting fixtures), lighting outlets, and ceiling-suspended (paddle) fans installed above the pool or the area extending 1.5 m (5 ft) horizontally from the inside walls of the pool shall be installed at a height not less than 3.7 m (12 ft) above the maximum water level of the pool.

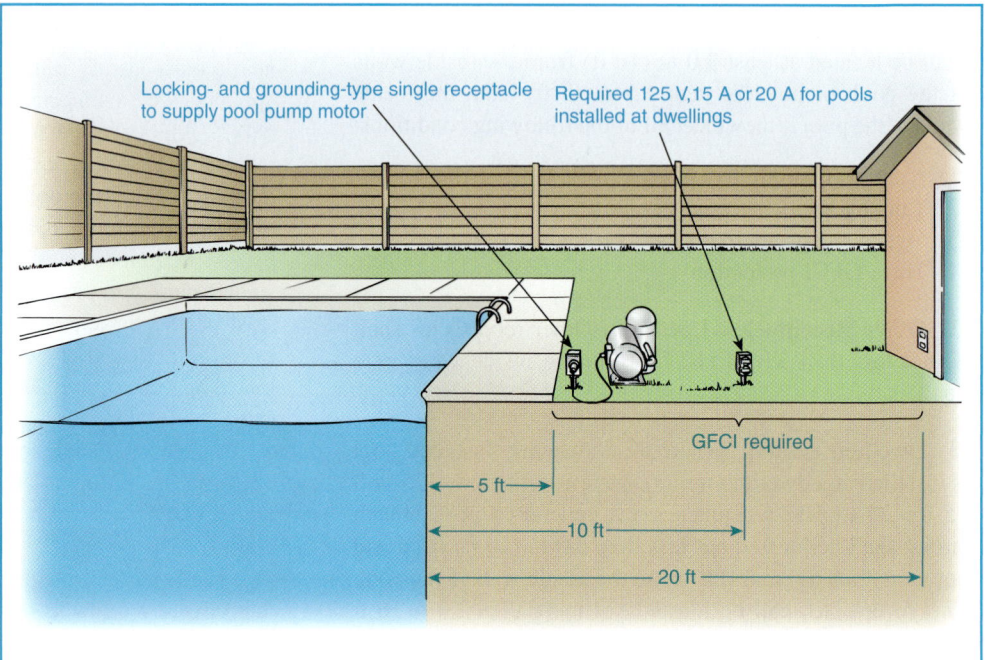

Exhibit 680.3 An example of a receptacle installed according to 680.22(A). For permanently installed pools at a dwelling unit(s), it is mandatory to install a 125-volt receptacle between 10 ft and 20 ft from the inside wall of the pool.

Locking- and grounding-type single receptacle to supply pool pump motor

Required 125 V,15 A or 20 A for pools installed at dwellings

GFCI required

5 ft

10 ft

20 ft

(2) Indoor Clearances. For installations in indoor pool areas, the clearances shall be the same as for outdoor areas unless modified as provided in this paragraph. If the branch circuit supplying the equipment is protected by a ground-fault circuit interrupter, the following equipment shall be permitted at a height not less than 2.3 m (7 ft 6 in.) above the maximum pool water level:

(1) Totally enclosed luminaires (fixtures)
(2) Ceiling-suspended (paddle) fans identified for use beneath ceiling structures such as provided on porches or patios

(3) Existing Installations. Existing luminaires (lighting fixtures) and lighting outlets located less than 1.5 m (5 ft) measured horizontally from the inside walls of a pool shall be not less than 1.5 m (5 ft) above the surface of the maximum water level, shall be rigidly attached to the existing structure, and shall be protected by a ground-fault circuit interrupter.

(4) GFCI Protection in Adjacent Areas. Luminaires (lighting fixtures), lighting outlets, and ceiling-suspended (paddle) fans installed in the area extending between 1.5 m (5 ft) and 3.0 m (10 ft) horizontally from the inside walls of a pool shall be protected by a ground-fault circuit interrupter unless installed not less than 1.5 m (5 ft) above the maximum water level and rigidly attached to the structure adjacent to or enclosing the pool.

(5) Cord-and-Plug-Connected Luminaires (Lighting Fixtures). Cord-and-plug-connected luminaires (lighting fixtures) shall comply with the requirements of 680.7 where

installed within 4.9 m (16 ft) of any point on the water surface, measured radially.

See Exhibit 680.6 for diagrams that clarify the limitations applicable to certain zones surrounding outdoor and indoor pools.

(C) Switching Devices. Switching devices shall be located at least 1.5 m (5 ft) horizontally from the inside walls of a pool unless separated from the pool by a solid fence, wall, or other permanent barrier. Alternatively, a switch that is listed as being acceptable for use within 1.5 m (5 ft) shall be permitted.

Panelboards, time clocks, pool light switches, and so on, where located not less than 5 ft horizontally from the inside walls of a pool without a solid fence, wall, or other permanent barrier, must be out of reach of persons who are in the pool, thereby preventing contact and possible shock hazards.

680.23 Underwater Luminaires (Lighting Fixtures).

This section covers all luminaires (lighting fixtures) installed below the normal water level of the pool.

(A) General.

(1) Luminaire (Fixture) Design, Normal Operation. The design of an underwater luminaire (lighting fixture) supplied from a branch circuit either directly or by way of a trans-

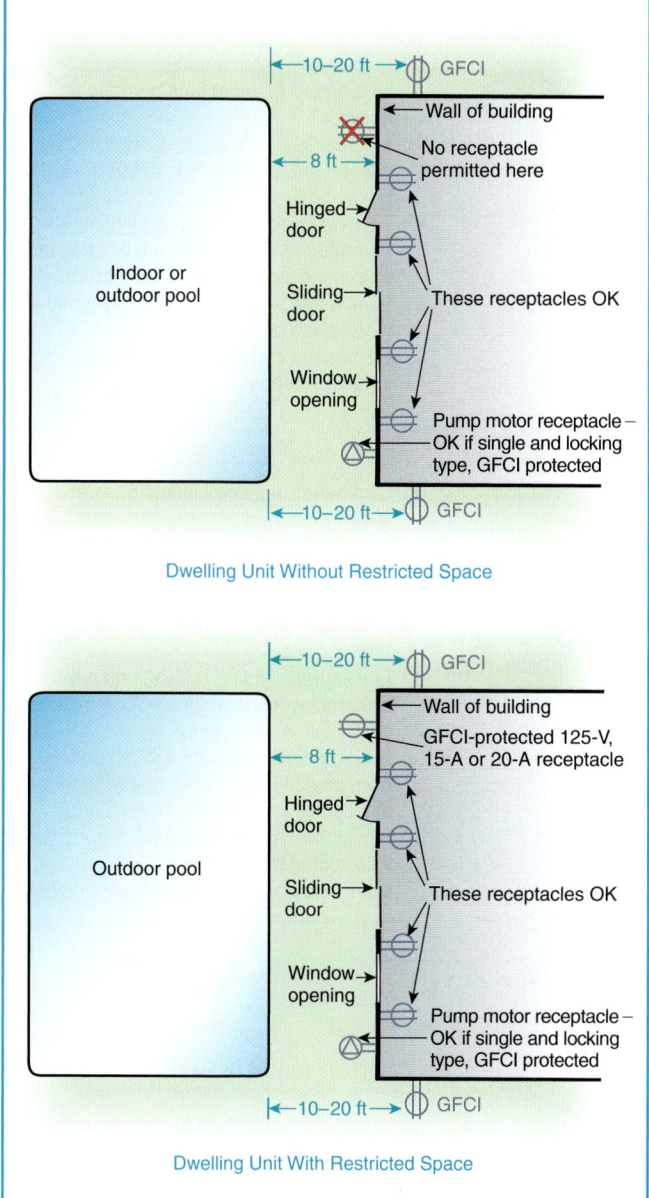

Exhibit 680.4 Acceptable receptacle locations within 20 ft of a permanently installed swimming pool.

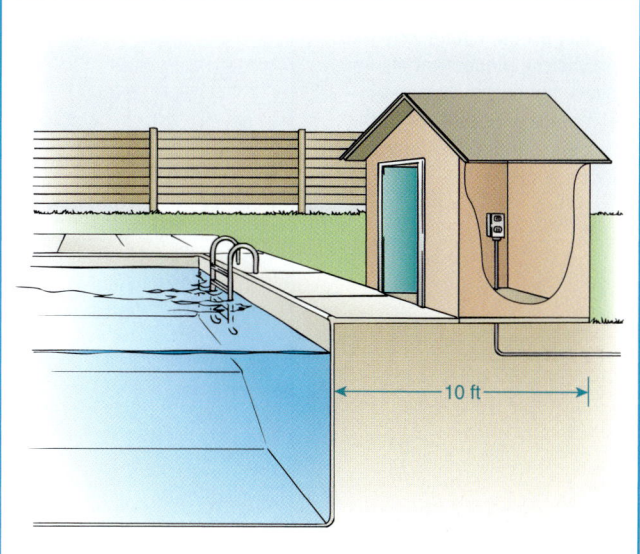

Exhibit 680.5 Permitted receptacle location less than 10 ft from the inside wall of a permanently installed pool. The minimum distance required by 680.22(A) does not apply to a receptacle located in a structure.

Branch-circuit conductors for dry-niche fixtures are required to be installed in approved rigid metal conduit, intermediate metal conduit, or rigid nonmetallic conduit from the fixture to a panelboard or the service equipment. Branch-circuit conductors for wet-niche fixtures leaving the pool junction box are required to be enclosed in rigid metal conduit, intermediate metal conduit, liquidtight flexible nonmetallic conduit, or rigid nonmetallic conduit, except where located in or on buildings, where the conductors are permitted to be installed in electrical metallic tubing or electrical nonmetallic tubing. Unlike wet-niche fixtures, a junction box is not required for dry-niche fixtures. If one is used, it is not required to be elevated or located as specified in 680.24(A)(2). (See Exhibits 680.1 and 680.7.)

(2) Transformers. Transformers used for the supply of underwater luminaires (fixtures), together with the transformer enclosure, shall be listed for the purpose. The transformer shall be an isolated winding type with an ungrounded secondary that has a grounded metal barrier between the primary and secondary windings.

Unless marked otherwise, UL-listed swimming pool and spa transformers are not suitable for connection to a conduit that extends directly to an underwater pool light forming shell. Swimming pool and spa transformers are not permitted to be used outdoors unless marked "For Outdoor Use" or in an equivalent manner that signifies that they have been found acceptable for both outdoor and indoor use. See 110.3(B).

former meeting the requirements of this section shall be such that, where the luminaire (fixture) is properly installed without a ground-fault circuit interrupter, there is no shock hazard with any likely combination of fault conditions during normal use (not relamping).

Dry-niche, no-niche, or wet-niche underwater luminaires operating at more than 15 volts require ground-fault circuit-interrupter protection. See the commentary following 680.5.

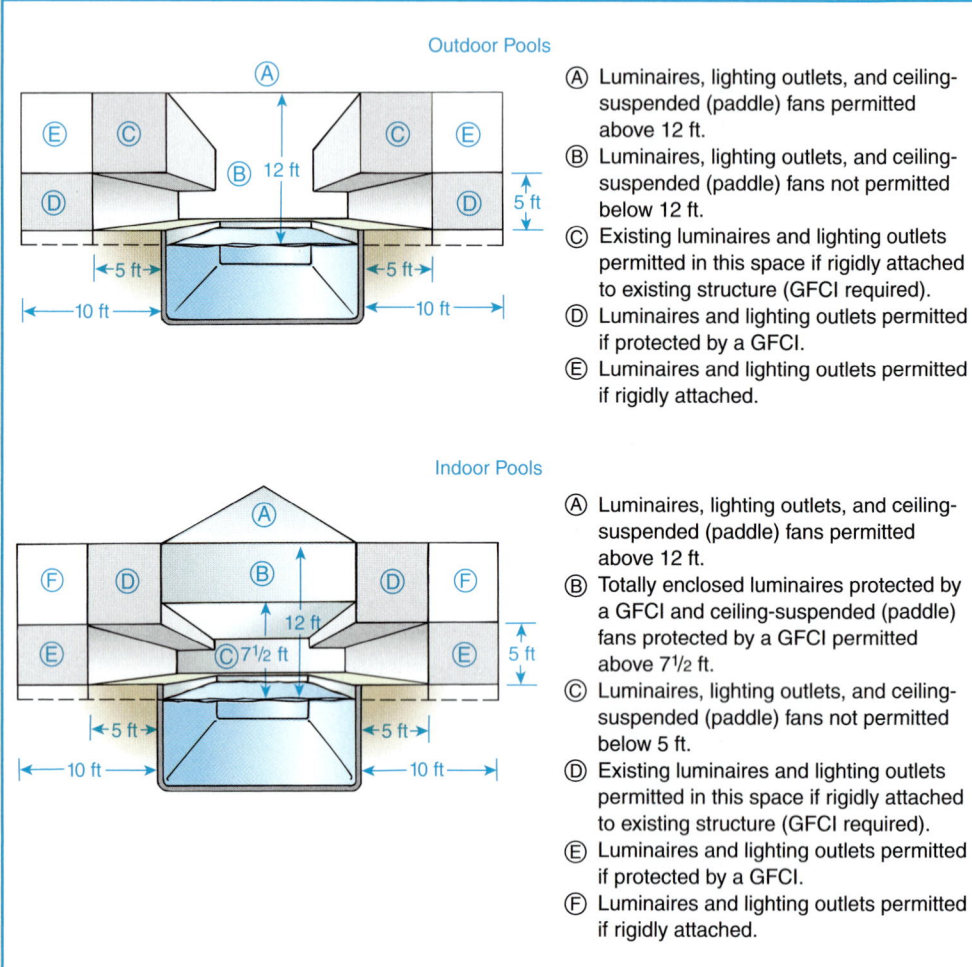

Exhibit 680.6 Limitations that apply to the placement of luminaires, lighting outlets, and ceiling-suspended fans in the area surrounding outdoor and indoor pools.

Outdoor Pools

Ⓐ Luminaires, lighting outlets, and ceiling-suspended (paddle) fans permitted above 12 ft.

Ⓑ Luminaires, lighting outlets, and ceiling-suspended (paddle) fans not permitted below 12 ft.

Ⓒ Existing luminaires and lighting outlets permitted in this space if rigidly attached to existing structure (GFCI required).

Ⓓ Luminaires and lighting outlets permitted if protected by a GFCI.

Ⓔ Luminaires and lighting outlets permitted if rigidly attached.

Indoor Pools

Ⓐ Luminaires, lighting outlets, and ceiling-suspended (paddle) fans permitted above 12 ft.

Ⓑ Totally enclosed luminaires protected by a GFCI and ceiling-suspended (paddle) fans protected by a GFCI permitted above 7½ ft.

Ⓒ Luminaires, lighting outlets, and ceiling-suspended (paddle) fans not permitted below 5 ft.

Ⓓ Existing luminaires and lighting outlets permitted in this space if rigidly attached to existing structure (GFCI required).

Ⓔ Luminaires and lighting outlets permitted if protected by a GFCI.

Ⓕ Luminaires and lighting outlets permitted if rigidly attached.

(3) GFCI Protection, Relamping. A ground-fault circuit interrupter shall be installed in the branch circuit supplying luminaires (fixtures) operating at more than 15 volts, so that there is no shock hazard during relamping. The installation of the ground-fault circuit interrupter shall be such that there is no shock hazard with any likely fault-condition combination that involves a person in a conductive path from any ungrounded part of the branch circuit or the luminaire (fixture) to ground.

(4) Voltage Limitation. No luminaires (lighting fixtures) shall be installed for operation on supply circuits over 150 volts between conductors.

(5) Location, Wall-Mounted Luminaires (Fixtures). Luminaires (lighting fixtures) mounted in walls shall be installed with the top of the luminaire (fixture) lens not less than 450 mm (18 in.) below the normal water level of the pool, unless the luminaire (lighting fixture) is listed and identified for use at lesser depths. No luminaire (fixture)

shall be installed less than 100 mm (4 in.) below the normal water level of the pool.

The reason for the 18-in. minimum submergence requirement is to reduce the likelihood that a person in the water and hanging onto the side of the pool directly over the fixture will have his or her chest in line with the fixture. This section covers fixtures that have been investigated and found acceptable for use where a person's chest may be directly in front of the fixture. The highest level of leakage current in the pool coming from a wet-niche fixture with a broken lens and bulb is found directly in front of the fixture.

(6) Bottom-Mounted Luminaires (Fixtures). A luminaire (lighting fixture) facing upward shall have the lens adequately guarded to prevent contact by any person.

(7) Dependence on Submersion. Luminaires (fixtures) that depend on submersion for safe operation shall be inherently protected against the hazards of overheating when not submerged.

(8) Compliance. Compliance with these requirements shall be obtained by the use of a listed underwater luminaire (lighting fixture) and by installation of a listed ground-fault circuit interrupter in the branch circuit or a listed transformer for luminaires (fixtures) operating at not more than 15 volts.

(B) Wet-Niche Luminaires (Fixtures).

(1) Forming Shells. Forming shells shall be installed for the mounting of all wet-niche underwater luminaires (fixtures) and shall be equipped with provisions for conduit entries. Metal parts of the luminaire (fixture) and forming shell in contact with the pool water shall be of brass or other approved corrosion-resistant metal. All forming shells used with nonmetallic conduit systems, other than those that are part of a listed low-voltage lighting system not requiring grounding, shall include provisions for terminating an 8 AWG copper conductor.

(2) Wiring Extending Directly to the Forming Shell. Conduit shall be installed from the forming shell to a suitable junction box or other enclosure located as provided in 680.24. Conduit shall be rigid metal, intermediate metal, liquidtight flexible nonmetallic, or rigid nonmetallic.

(a) Metal Conduit. Metal conduit shall be approved and shall be of brass or other approved corrosion-resistant metal.

(b) Nonmetallic Conduit. Where a nonmetallic conduit is used, an 8 AWG insulated solid or stranded copper equipment grounding conductor shall be installed in this conduit unless a listed low-voltage lighting system not requiring grounding is used. The equipment grounding conductor shall be terminated in the forming shell, junction box or transformer enclosure, or ground-fault circuit-interrupter enclosure. The termination of the 8 AWG equipment grounding conductor in the forming shell shall be covered with, or encapsulated in, a listed potting compound to protect the connection from the possible deteriorating effect of pool water.

(3) Equipment Grounding Provisions for Cords. Wet-niche luminaires (lighting fixtures) that are supplied by a flexible cord or cable shall have all exposed non–current-carrying metal parts grounded by an insulated copper equipment grounding conductor that is an integral part of the cord or cable. This grounding conductor shall be connected to a grounding terminal in the supply junction box, transformer enclosure, or other enclosure. The grounding conductor shall not be smaller than the supply conductors and not smaller than 16 AWG.

(4) Luminaire (Fixture) Grounding Terminations. The end of the flexible-cord jacket and the flexible-cord conductor terminations within a luminaire (fixture) shall be covered with, or encapsulated in, a suitable potting compound to prevent the entry of water into the luminaire (fixture) through the cord or its conductors. In addition, the grounding connection within a luminaire (fixture) shall be similarly treated to protect such connection from the deteriorating effect of pool water in the event of water entry into the luminaire (fixture).

(5) Luminaire (Fixture) Bonding. The luminaire (fixture) shall be bonded to and secured to the forming shell by a positive locking device that ensures a low-resistance contact and requires a tool to remove the luminaire (fixture) from the forming shell. Bonding shall not be required for luminaires (fixtures) that are listed for the application and have no non–current-carrying metal parts.

(C) Dry-Niche Luminaires (Fixtures).

(1) Construction. A dry-niche luminaire (lighting fixture) shall be provided with a provision for drainage of water and a means for accommodating one equipment grounding conductor for each conduit entry.

(2) Junction Box. A junction box shall not be required but, if used, shall not be required to be elevated or located as specified in 680.24(A)(2) if the luminaire (fixture) is specifically identified for the purpose.

(D) No-Niche Luminaires (Fixtures). A no-niche luminaire (fixture) shall meet the construction requirements of 680.23(B)(3) and be installed in accordance with the requirements of 680.23(B). Where connection to a forming shell is specified, the connection shall be to the mounting bracket.

(E) Through-Wall Lighting Assembly. A through-wall lighting assembly shall be equipped with a threaded entry or hub, or a nonmetallic hub listed for the purpose, for the purpose of accommodating the termination of the supply conduit. A through-wall lighting assembly shall meet the construction requirements of 680.23(B)(3) and be installed in accordance with the requirements of 680.23. Where connection to a forming shell is specified, the connection shall be to the conduit termination point.

(F) Branch-Circuit Wiring.

(1) Wiring Methods. Branch-circuit wiring on the supply side of enclosures and junction boxes connected to conduits run to wet-niche and no-niche luminaires (fixtures), and the field wiring compartments of dry-niche luminaires (fixtures), shall be installed using rigid metal conduit, intermediate metal conduit, liquidtight flexible nonmetallic conduit, or rigid nonmetallic conduit. Where installed on buildings, electrical metallic tubing shall be permitted, and where installed within buildings, electrical nonmetallic tubing or electrical metallic tubing shall be permitted.

Exception: Where connecting to transformers for pool lights, liquidtight flexible metal conduit or liquidtight flexible nonmetallic conduit shall be permitted. The length shall not exceed 1.8 m (6 ft) for any one length or exceed 3.0 m (10 ft) in total length used. Liquidtight flexible nonmetallic conduit, Type B (LFNC-B), shall be permitted in lengths longer than 1.8 m (6 ft).

(2) Equipment Grounding. Through-wall lighting assemblies, wet-niche, dry-niche, or no-niche luminaires (lighting fixtures) shall be connected to an insulated copper equipment grounding conductor installed with the circuit conductors. The equipment grounding conductor shall be installed without joint or splice except as permitted in (a) and (b). The equipment grounding conductor shall be sized in accordance with Table 250.122 but shall not be smaller than 12 AWG.

Exception: An equipment grounding conductor between the wiring chamber of the secondary winding of a transformer and a junction box shall be sized in accordance with the overcurrent device in this circuit.

(a) If more than one underwater luminaire (lighting fixture) is supplied by the same branch circuit, the equipment grounding conductor, installed between the junction boxes, transformer enclosures, or other enclosures in the supply circuit to wet-niche luminaires (fixtures), or between the field-wiring compartments of dry-niche luminaires (fix-

tures), shall be permitted to be terminated on grounding terminals.

(b) If the underwater luminaire (lighting fixture) is supplied from a transformer, ground-fault circuit interrupter, clock-operated switch, or a manual snap switch that is located between the panelboard and a junction box connected to the conduit that extends directly to the underwater luminaire (lighting fixture), the equipment grounding conductor shall be permitted to terminate on grounding terminals on the transformer, ground-fault circuit interrupter, clock-operated switch enclosure, or an outlet box used to enclose a snap switch.

See the commentary following 680.23(A)(2).

(3) Conductors. Conductors on the load side of a ground-fault circuit interrupter or of a transformer, used to comply with the provisions of 680.23(A)(8), shall not occupy raceways, boxes, or enclosures containing other conductors unless one of the following conditions applies:

(1) The other conductors are protected by ground-fault circuit interrupters.
(2) The other conductors are grounding conductors.
(3) The other conductors are supply conductors to a feed-through type ground-fault circuit interrupter.
(4) Ground-fault circuit interrupters shall be permitted in a panelboard that contains circuits protected by other than ground-fault circuit interrupters.

680.24 Junction Boxes and Enclosures for Transformers or Ground-Fault Circuit Interrupters.

(A) Junction Boxes. A junction box connected to a conduit that extends directly to a forming shell or mounting bracket of a no-niche luminaire (fixture) shall meet the requirements of this section.

(1) Construction. The junction box shall be listed and labeled for the purpose and shall comply with the following conditions:

(1) Be equipped with threaded entries or hubs or a nonmetallic hub listed for the purpose
(2) Be comprised of copper, brass, suitable plastic, or other approved corrosion-resistant material
(3) Be provided with electrical continuity between every connected metal conduit and the grounding terminals by means of copper, brass, or other approved corrosion-resistant metal that is integral with the box

(2) Installation. Where the luminaire (fixture) operates over 15 volts, the junction box location shall comply with

(a) and (b). Where the luminaire (fixture) operates at less than 15 volts, the junction box location shall be permitted to comply with (c).

(a) Vertical Spacing. The junction box shall be located not less than 100 mm (4 in.), measured from the inside of the bottom of the box, above the ground level, or pool deck, or not less than 200 mm (8 in.) above the maximum pool water level, whichever provides the greater elevation.

(b) Horizontal Spacing. The junction box shall be located not less than 1.2 m (4 ft) from the inside wall of the pool, unless separated from the pool by a solid fence, wall, or other permanent barrier.

(c) Flush Deck Box. If used on a lighting system operating at 15 volts or less, a flush deck box shall be permitted if both of the following apply:

(1) An approved potting compound is used to fill the box to prevent the entrance of moisture.
(2) The flush deck box is located not less than 1.2 m (4 ft) from the inside wall of the pool.

(B) Other Enclosures. An enclosure for a transformer, ground-fault circuit interrupter, or a similar device connected to a conduit that extends directly to a forming shell or mounting bracket of a no-niche luminaire (fixture) shall meet the requirements of this section.

(1) Construction. The enclosure shall be listed and labeled for the purpose and meet the following requirements:

(1) Equipped with threaded entries or hubs or a nonmetallic hub listed for the purpose
(2) Comprised of copper, brass, suitable plastic, or other approved corrosion-resistant material
(3) Provided with an approved seal, such as duct seal at the conduit connection, that prevents circulation of air between the conduit and the enclosures
(4) Provided with electrical continuity between every connected metal conduit and the grounding terminals by means of copper, brass, or other approved corrosion-resistant metal that is integral with the box

(2) Installation.

(a) Vertical Spacing. The enclosure shall be located not less than 100 mm (4 in.), measured from the inside of the bottom of the box, above the ground level, or pool deck, or not less than 200 mm (8 in.) above the maximum pool water level, whichever provides the greater elevation.

(b) Horizontal Spacing. The enclosure shall be located not less than 1.2 m (4 ft) from the inside wall of the pool, unless separated from the pool by a solid fence, wall, or other permanent barrier.

(C) Protection. Junction boxes and enclosures mounted above the grade of the finished walkway around the pool shall not be located in the walkway unless afforded additional protection, such as by location under diving boards, adjacent to fixed structures, and the like.

(D) Grounding Terminals. Junction boxes, transformer enclosures, and ground-fault circuit-interrupter enclosures connected to a conduit that extends directly to a forming shell or mounting bracket of a no-niche luminaire (fixture) shall be provided with a number of grounding terminals that shall be no fewer than one more than the number of conduit entries.

(E) Strain Relief. The termination of a flexible cord of an underwater luminaire (lighting fixture) within a junction box, transformer enclosure, ground-fault circuit interrupter, or other enclosure shall be provided with a strain relief.

(F) Grounding. The junction box, transformer enclosure, or other enclosure in the supply circuit to a wet-niche or no-niche luminaire (lighting fixture) and the field-wiring chamber of a dry-niche luminaire (lighting fixture) shall be grounded to the equipment grounding terminal of the panelboard. This terminal shall be directly connected to the panelboard enclosure.

The requirements in 680.24(A) through (F) cover the construction and installation of boxes and enclosures associated with underwater luminaires. Boxes and enclosures used for the supply wiring to wet-niche and no-niche underwater luminaires must be listed for the purpose by a recognized testing laboratory. The provisions of 680.24(D) ensure the availability of integral grounding terminals necessary for the grounding and bonding of underwater luminaires. A box that is listed, but not specifically for use with swimming pools, does not provide the correct number of integral grounding and bonding terminals. The number of grounding terminals in a box or enclosure is required to be one more than the number of conduit entries for which the box is designed. In an installation where nonmetallic conduit is the wiring method between the wet-niche forming shell and the deck (junction) box, two equipment grounding conductors in that conduit must be terminated in the junction box. The first equipment grounding conductor is covered in 680.23(B)(2)(b). The use of nonmetallic conduit requires the installation of an insulated, copper equipment grounding conductor in that section of conduit between the deck box and the wet-niche forming shell. This conductor can be solid or stranded and must not be smaller than 8 AWG. The function of this conductor is twofold. It permanently bonds all non–current-carrying metal surfaces of the forming shell to any non–current-carrying parts of the deck box and to the equipment grounding conductor of the circuit that supplies the wet-niche luminaire. Additionally, this conductor will serve as the path for ground-fault current in the event of a ground fault when the wet-niche luminaire is removed from the forming shell, as is typically done during relamping. Damage to the wet-niche luminaire supply cord could result in this ground-fault scenario.

The second equipment grounding conductor is the one contained in the flexible cord supplying the wet-niche luminaire. In accordance with 680.23(B)(3), this conductor is required to be insulated, copper, and sized no smaller than the circuit conductors within the cord, but not smaller than 16 AWG.

In addition to the two equipment grounding conductors contained in the section of nonmetallic conduit between the forming shell and the deck box, the wiring method from the deck box to the power source is also required to contain a separate equipment grounding conductor. This equipment grounding conductor is required by 680.23(F)(2) and must be insulated, copper, and not smaller than 12 AWG. The grounding terminals within the deck (junction) box are used to terminate and bond together all of these equipment grounding conductors.

Exhibit 680.7 illustrates an installation of a forming shell for a wet-niche luminaire and a flush junction (deck) box. (See Exhibit 680.1 for surface deck boxes.)

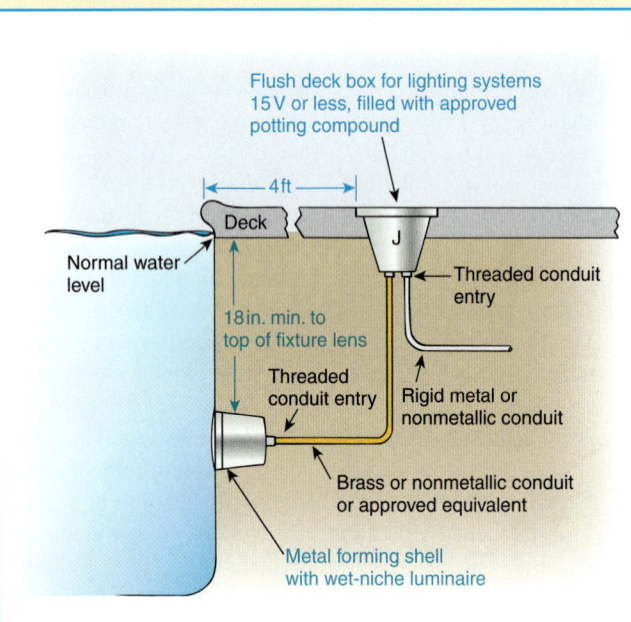

Flush deck box for lighting systems 15 V or less, filled with approved potting compound

4 ft

Deck

Normal water level

18 in. min. to top of fixture lens

Threaded conduit entry

Threaded conduit entry

Rigid metal or nonmetallic conduit

Brass or nonmetallic conduit or approved equivalent

Metal forming shell with wet-niche luminaire

Exhibit 680.7 A flush junction (deck) box and a forming shell for a wet-niche luminaire installed according to 680.24(A)(2).

680.25 Feeders.

These provisions shall apply to any feeder on the supply side of panelboards supplying branch circuits for pool equipment covered in Part II of this article and on the load side of the service equipment or the source of a separately derived system.

(A) Wiring Methods. Feeders shall be installed in rigid metal conduit, intermediate metal conduit, liquidtight flexible nonmetallic conduit, or rigid nonmetallic conduit. Electrical metallic tubing shall be permitted where installed on or within a building, and electrical nonmetallic tubing shall be permitted where installed within a building.

Exception: An existing feeder between an existing remote panelboard and service equipment shall be permitted to run in flexible metal conduit or an approved cable assembly that includes an equipment grounding conductor within its outer sheath. The equipment grounding conductor shall comply with 250.24(A)(5).

(B) Grounding. An equipment grounding conductor shall be installed with the feeder conductors between the grounding terminal of the pool equipment panelboard and the grounding terminal of the applicable service equipment or source of a separately derived system. For other than (1) existing feeders covered in 680.25(A), Exception or (2) feeders to separate buildings that do not utilize an insulated equipment grounding conductor in accordance with 680.25(B)(2), this equipment grounding conductor shall be insulated.

(1) Size. This conductor shall be sized in accordance with 250.122 but not smaller than 12 AWG. On separately derived systems, this conductor shall be sized in accordance with Table 250.66 but not smaller than 8 AWG.

(2) Separate Buildings. A feeder to a separate building shall be permitted to supply swimming pool equipment branch circuits, or feeders supplying swimming pool equipment branch circuits, if the grounding arrangements in the separate building meet the requirements in 250.32. Where installed, a separate equipment grounding conductor shall be an insulated conductor.

The insulated equipment grounding conductor can be aluminum or copper and is required to be installed in a raceway. It should be understood that for an existing remote panelboard, the 680.25(A) Exception permits an approved cable assembly with an insulated or covered aluminum or copper equipment grounding conductor. See Exhibit 680.8.

Swimming pool equipment supplied by a separately derived system is covered in 680.25(B). Where a remote panelboard supplying a pool is supplied by a separately derived system, the rules covering the grounding conductor apply only to the feeder between the separately derived system and the panelboard, not all the way back to the service, which might be high voltage.

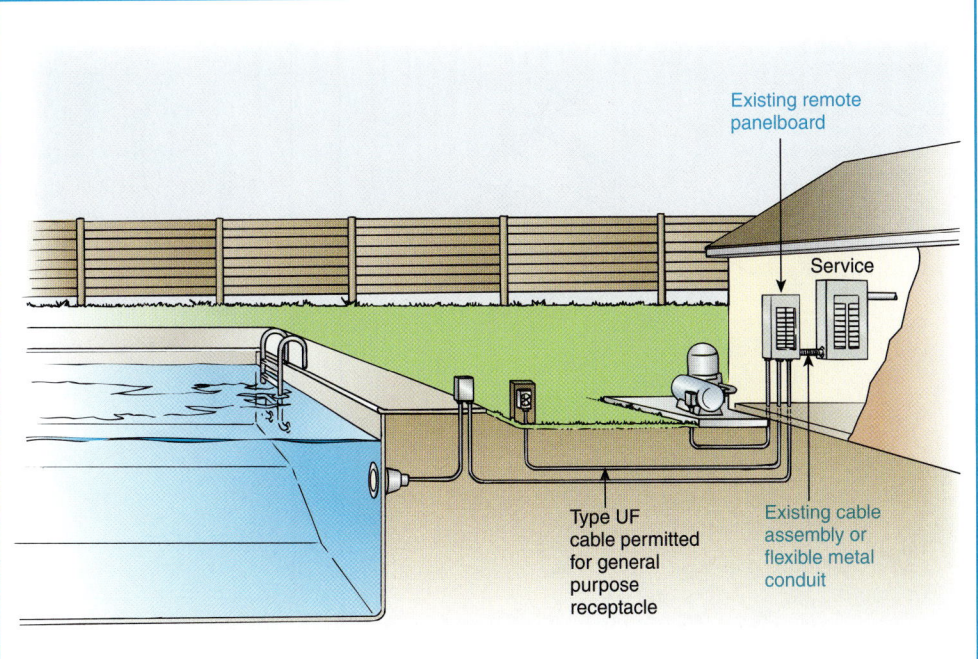

Exhibit 680.8 An existing remote panelboard supplying new pool equipment. A raceway is not required for this application if the existing feeder wiring method contains an insulated or covered equipment grounding conductor.

The general rule in 680.25(B) requires an equipment grounding conductor to be installed between a panelboard serving swimming pool equipment and the service or the source of a separately derived system. Added in the 1999 *Code*, 680.25(B)(2) allows pool equipment to be supplied from a remote panelboard in a separate building where an equipment grounding conductor is not installed with the feeder circuit conductors run from the service (or derived system) to the panelboard. In this case, grounding at the separate building must meet the requirements of 250.32. See Exhibit 680.9.

680.26 Bonding.

(A) Performance. The bonding required by this section shall be installed to eliminate voltage gradients in the pool area as prescribed.

> FPN: This section does not require that the 8 AWG or larger solid copper bonding conductor be extended or attached to any remote panelboard, service equipment, or any electrode.

The primary purpose of bonding is to ensure that voltage gradients in the pool area are eliminated. The fine print note explains that the 8 AWG conductor's only function is to eliminate the voltage gradient in the pool area. It is not required to provide a path for fault current that may occur as a result of electrical equipment failure. See Exhibit 680.10.

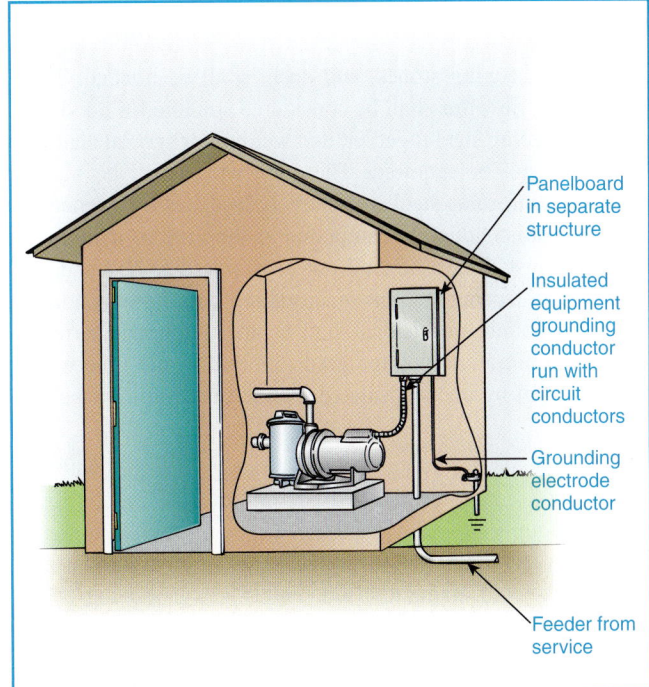

Exhibit 680.9 Grounding requirements per 680.25(B)(2) for remote panelboard and swimming pool equipment located in a structure remote from the service equipment.

(B) Bonded Parts. The parts specified in 680.26(B)(1) through (B)(5) shall be bonded together.

(1) Metallic Structural Components. All metallic parts of the pool structure, including the reinforcing metal of the

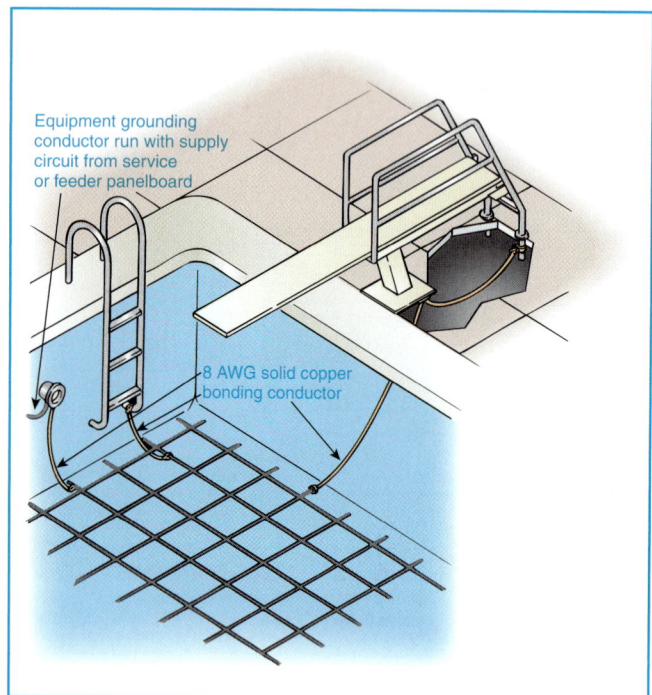

Exhibit 680.10 Bonding in a swimming pool.

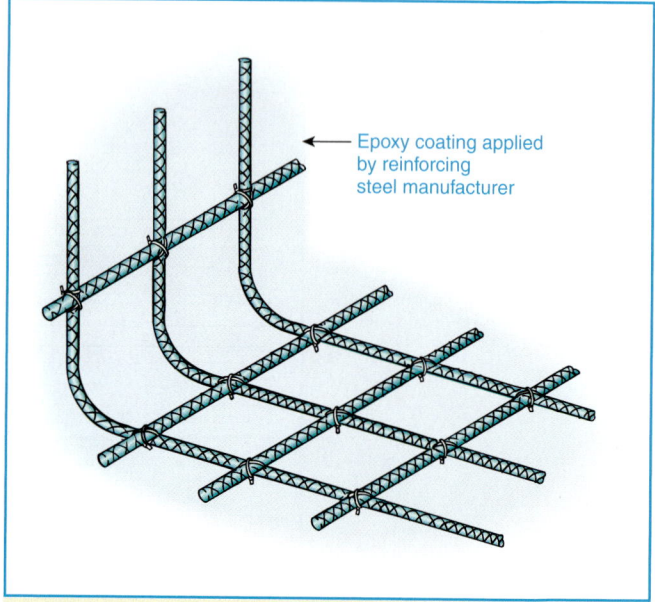

Exhibit 680.11 Epoxy-coated rebar, which does not require bonding.

pool shell, coping stones, and deck, shall be bonded. The usual steel tie wires shall be considered suitable for bonding the reinforcing steel together, and welding or special clamping shall not be required. These tie wires shall be made tight. If reinforcing steel is effectively insulated by an encapsulating nonconductive compound at the time of manufacture and installation, it shall not be required to be bonded. Where reinforcing steel is encapsulated with a nonconductive compound, provisions shall be made for an alternate means to eliminate voltage gradients that would otherwise be provided by unencapsulated, bonded reinforcing steel.

Encapsulated reinforcing steel might not provide the conductivity necessary to establish the required common bonding grid. A common bonding grid will not be formed if the steel is effectively encapsulated by a listed compound during installation and manufacturing. Therefore, a bonding connection is not required for this type of application. See Exhibit 680.11.

In Exhibit 680.12, the structural reinforcing steel serves as the common bonding grid to which all metal appurtenances associated with the pool are connected. Safety-rope hooks are not required to be bonded, as specified in 680.26(B)(3). The flush deck box meets the provisions of 680.24(A).

(2) Underwater Lighting. All forming shells and mounting brackets of no-niche luminaires (fixtures) shall be

bonded unless a listed low-voltage lighting system with nonmetallic forming shells not requiring bonding is used.

(3) Metal Fittings. All metal fittings within or attached to the pool structure shall be bonded. Isolated parts that are not over 100 mm (4 in.) in any dimension and do not penetrate into the pool structure more than 25 mm (1 in.) shall not require bonding.

(4) Electrical Equipment. Metal parts of electrical equipment associated with the pool water circulating system, including pump motors and metal parts of equipment associated with pool covers, including electric motors, shall be bonded. Metal parts of listed equipment incorporating an approved system of double insulation and providing a means for grounding internal nonaccessible, non–current-carrying metal parts shall not be bonded.

Where a double-insulated water-pump motor is installed under the provisions of this rule, a solid 8 AWG copper conductor that is of sufficient length to make a bonding connection to a replacement motor shall be extended from the bonding grid to an accessible point in the motor vicinity. Where there is no connection between the swimming pool bonding grid and the equipment grounding system for the premises, this bonding conductor shall be connected to the equipment grounding conductor of the motor circuit.

(5) Metal Wiring Methods and Equipment. Metal-sheathed cables and raceways, metal piping, and all fixed metal parts except those separated from the pool by a perma-

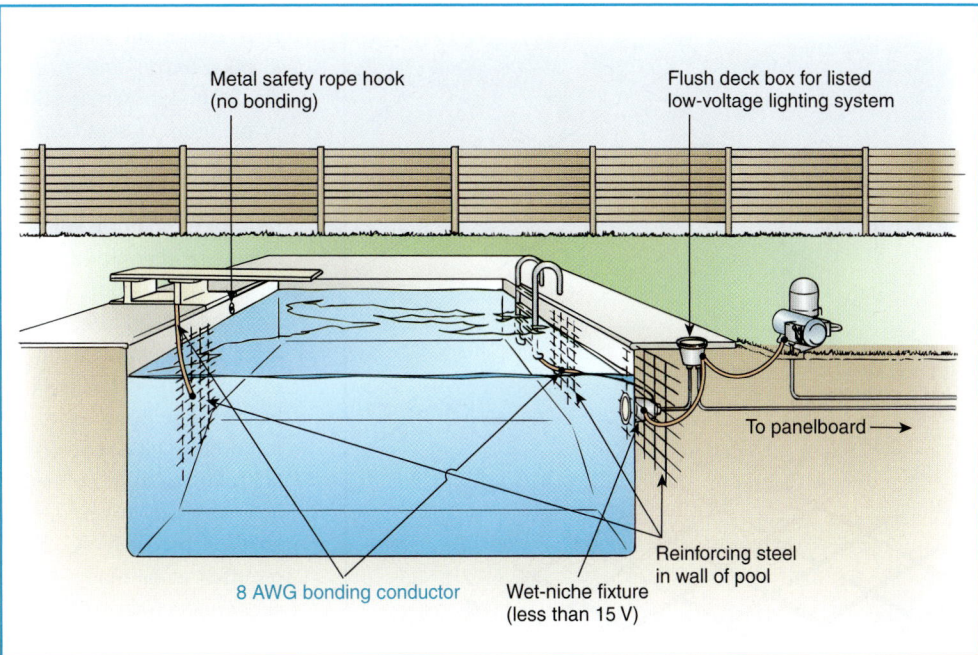

Exhibit 680.12 A poured-concrete pool with structural reinforcing steel that serves as the common bonding grid.

nent barrier shall be bonded that are within the following distances of the pool:

Examples of fixed metal parts bound by this requirement include metal fences, metal awnings, metal door frames, and metal window frames.

(1) Within 1.5 m (5 ft) horizontally of the inside walls of the pool
(2) Within 3.7 m (12 ft) measured vertically above the maximum water level of the pool, or any observation stands, towers, or platforms, or any diving structures

(C) Common Bonding Grid. The parts specified in 680.26(B) shall be connected to a common bonding grid with a solid copper conductor, insulated, covered, or bare, not smaller than 8 AWG. Connection shall be made by exothermic welding or by pressure connectors or clamps that are labeled as being suitable for the purpose and are of stainless steel, brass, copper, or copper alloy. The common bonding grid shall be permitted to be any of the following:

Exothermic welding is permitted as a means of connecting to a common bonding grid. Pressure connectors and clamps specifically listed for the purpose are also acceptable connection methods. Connections in pool areas must be suitable for wet conditions and high levels of chlorine. See Exhibit 680.13 for an illustration of bonding connections.

(1) The structural reinforcing steel of a concrete pool where the reinforcing rods are bonded together by the usual steel tie wires or the equivalent
(2) The wall of a bolted or welded metal pool
(3) A solid copper conductor, insulated, covered, or bare, not smaller than 8 AWG
(4) Rigid metal conduit or intermediate metal conduit of brass or other identified corrosion-resistant metal conduit

Brass rigid metal conduit, brass intermediate metal conduit, and other corrosion-resistant metal conduits are an effective means to provide a common bonding grid, particularly where they are installed aboveground and in accessible locations. See Exhibit 680.14.

The requirement in 250.8 specifies that the connection be made by exothermic welding, listed clamps, connectors, or other listed means. If a pool is built with structural reinforcing steel (rebar) encapsulated by a nonconductive coating, such as epoxy or plastic, the rebar is not required to be bonded to the other metal parts of the pool system. See the commentary following 680.26(B)(1).

(D) Connections. Where structural reinforcing steel or the walls of bolted or welded metal pool structures are used as a common bonding grid for nonelectrical parts, the connections shall be made in accordance with 250.8.

(E) Pool Water Heaters. For pool water heaters rated at more than 50 amperes that have specific instructions regard-

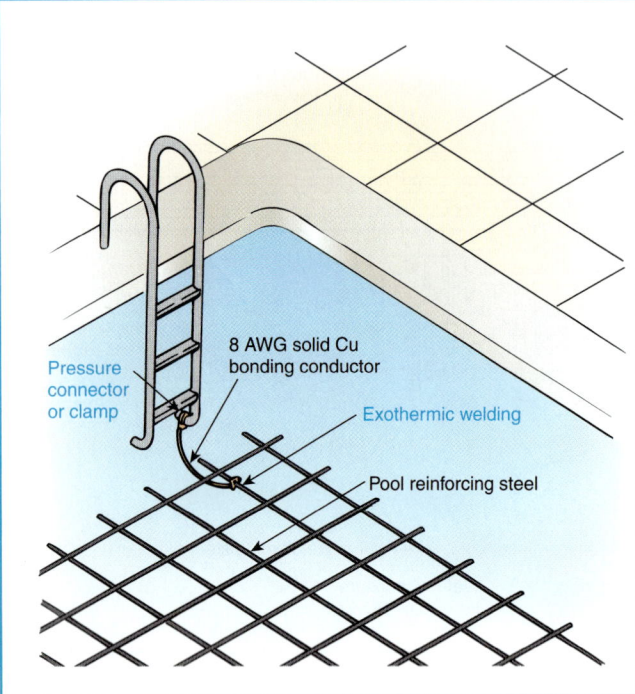

Exhibit 680.13 Bonding connections in a swimming pool.

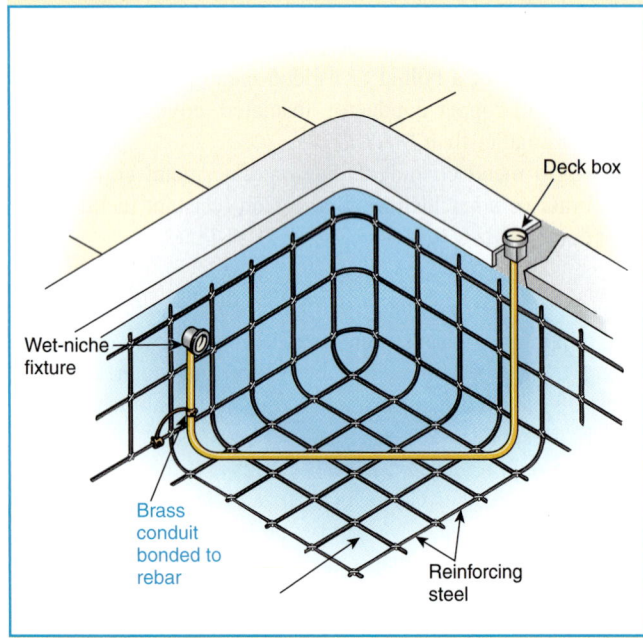

Exhibit 680.14 Corrosion-resistant conduit used as part of the common bonding grid.

It is important to understand the difference between the terms *bonding* and *grounding* as they apply to Article 680. As defined in Article 100, *bonding* is "the permanent joining of metallic parts to form an electrically conductive path that ensures electrical continuity and the capacity to conduct safely any current likely to be imposed." The metal parts required to be bonded per 680.26(B) include all metal parts of electrical equipment associated with the water-circulating system of the pool, all metal parts of the pool structure, and all fixed metal parts, which include conduit and piping, metal door frames, and metal window frames, within 5 ft of the inside walls of the pool and not separated by a permanent barrier. The bonding of these parts does not mean they are required to be connected to each other; rather, it means they are required to be connected to a common bonding grid with an insulated, covered, or bare solid copper conductor not smaller than 8 AWG. See Exhibit 680.15. Connections are required to be made by exothermic welds or by listed pressure connectors, clamps, or other listed means, in accordance with 250.8.

The reason for connecting metal parts (ladders, handrails, water-circulating equipment, forming shells, diving boards, etc.) to a common bonding grid (pool reinforcing steel, pool metal wall, or an 8 AWG solid conductor) is to ensure that all such metal parts will be at the same electrical potential. This grid reduces any possible shock hazard created by stray currents in the ground or piping connected to the swimming pool. Stray currents can also exist in nonmetallic piping because of the low resistivity of chlorinated water.

Because corrosion is normally associated with the wet conditions of swimming pool areas, wiring and connections should be checked periodically, especially bonding connections between the 8 AWG copper conductor and, for instance, an aluminum (or other dissimilar metal) ladder.

680.27 Specialized Pool Equipment.

(A) Underwater Audio Equipment. All underwater audio equipment shall be identified for the purpose.

(1) Speakers. Each speaker shall be mounted in an approved metal forming shell, the front of which is enclosed by a captive metal screen, or equivalent, that is bonded to and secured to the forming shell by a positive locking device that ensures a low-resistance contact and requires a tool to open for installation or servicing of the speaker. The forming shell shall be installed in a recess in the wall or floor of the pool.

(2) Wiring Methods. Rigid metal conduit or intermediate metal conduit of brass or other identified corrosion-resistant metal, liquidtight flexible nonmetallic conduit (LFNC-B), or rigid nonmetallic conduit shall extend from the forming shell to a listed junction box or other enclosure as provided

ing bonding and grounding, only those parts designated to be bonded shall be bonded, and only those parts designated to be grounded shall be grounded.

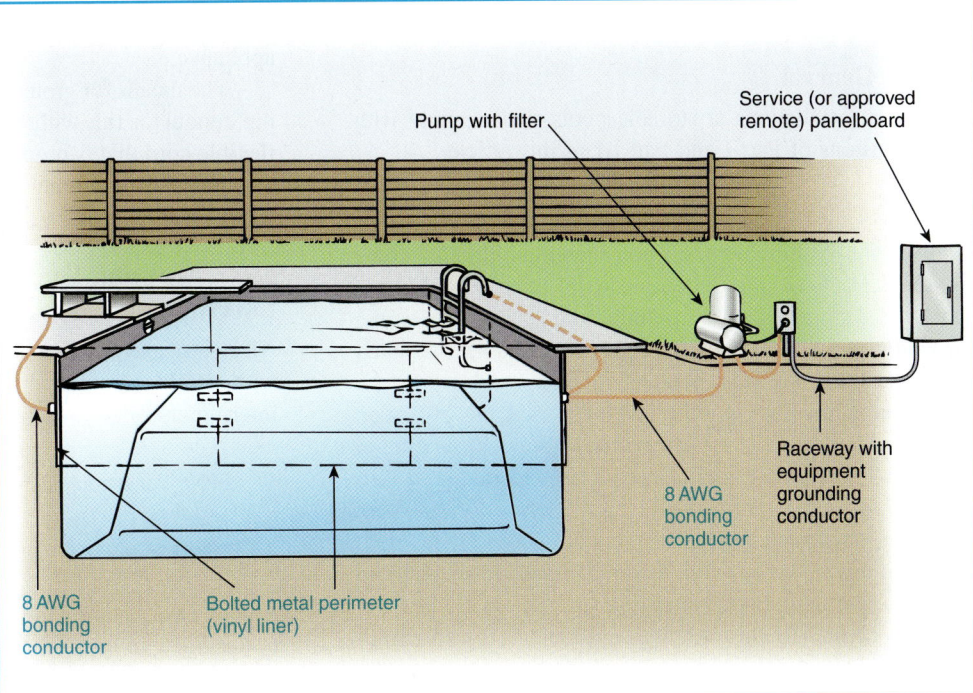

Exhibit 680.15 A metal-perimeter (e.g., steel or aluminum) pool with bolted or welded sections. The metal perimeter serves as the common bonding grid to which the metal ladder, metal diving board, and pump motor are connected.

Pump with filter

Service (or approved remote) panelboard

Raceway with equipment grounding conductor

8 AWG bonding conductor

8 AWG bonding conductor

Bolted metal perimeter (vinyl liner)

in 680.24. Where rigid nonmetallic conduit or liquidtight flexible nonmetallic conduit is used, an 8 AWG insulated solid or stranded copper equipment grounding conductor shall be installed in this conduit. The equipment grounding conductor shall be terminated in the forming shell and the junction box. The termination of the 8 AWG equipment grounding conductor in the forming shell shall be covered with, or encapsulated in, a listed potting compound to protect such connection from the possible deteriorating effect of pool water.

(3) Forming Shell and Metal Screen. The forming shell and metal screen shall be of brass or other approved corrosion-resistant metal. All forming shells shall include provisions for terminating an 8 AWG copper conductor.

(B) Electrically Operated Pool Covers.

(1) Motors and Controllers. The electric motors, controllers, and wiring shall be located not less than 1.5 m (5 ft) from the inside wall of the pool unless separated from the pool by a wall, cover, or other permanent barrier. Electric motors installed below grade level shall be of the totally enclosed type. The device that controls the operation of the motor for an electrically operated pool cover shall be located so that the operator has full view of the pool.

> FPN No. 1: For cabinets installed in damp and wet locations, see 312.2(A).
> FPN No. 2: For switches or circuit breakers installed in wet locations, see 404.4.
> FPN No. 3: For protection against liquids, see 430.11.

(2) Protection. The electric motor and controller shall be connected to a circuit protected by a ground-fault circuit interrupter.

(C) Deck Area Heating. These provisions of this section shall apply to all pool deck areas, including a covered pool, where electrically operated comfort heating units are installed within 6.0 m (20 ft) of the inside wall of the pool.

(1) Unit Heaters. Unit heaters shall be rigidly mounted to the structure and shall be of the totally enclosed or guarded types. Unit heaters shall not be mounted over the pool or within the area extending 1.5 m (5 ft) horizontally from the inside walls of a pool.

(2) Permanently Wired Radiant Heaters. Radiant electric heaters shall be suitably guarded and securely fastened to their mounting device(s). Heaters shall not be installed over a pool or within the area extending 1.5 m (5 ft) horizontally from the inside walls of the pool and shall be mounted at least 3.7 m (12 ft) vertically above the pool deck unless otherwise approved.

(3) Radiant Heating Cables Not Permitted. Radiant heating cables embedded in or below the deck shall not be permitted.

> Only unit heaters and permanently connected radiant heaters are permitted in the area that extends 5 ft to 20 ft horizontally from the inside walls of a pool. Radiant heat cables embedded in the deck are not permitted.

III. Storable Pools

680.30 General.

Electrical installations at storable pools shall comply with the provisions of Part I and Part III of this article.

Storable pools can be readily disassembled. The maximum wall height for storable pools other than the inflatable type is 42 in. Pools of any dimension with inflatable walls are considered storable. See the definition of *storable swimming or wading pool* in 680.2. This type of pool and its associated equipment do not require bonding conductors. However, the filter pump must be double insulated, and the provision of grounding means consisting of an equipment grounding conductor that is an integral part of the flexible cord also is required. There are portable filter pumps for use with storable pools listed by Underwriters Laboratories. All electrical equipment used with a storable pool is required to have ground-fault circuit-interrupter protection for personnel. Exhibit 680.16 illustrates the requirements for a storable-type pool.

680.31 Pumps.

A cord-connected pool filter pump shall incorporate an approved system of double insulation or its equivalent and shall be provided with means for grounding only the internal

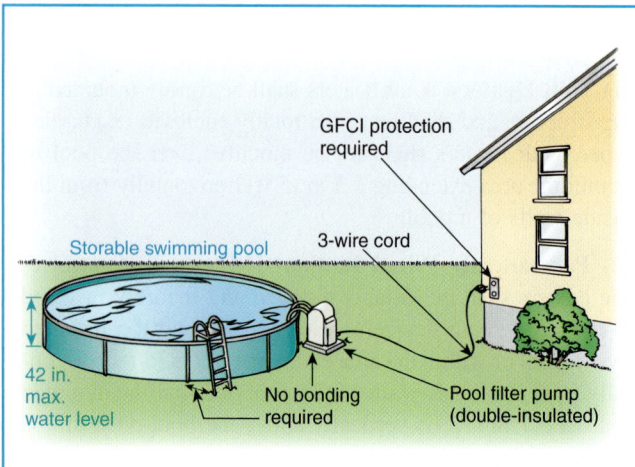

Exhibit 680.16 The requirements for a storable-type pool. Metal appurtenances are not required to be bonded. The 3-wire cord may be longer than 3 ft (listed filter pumps are equipped with cords 25 ft long). The receptacle shown can be a GFCI-type receptacle, a receptacle supplied through a GFCI-type receptacle, or a receptacle protected by a GFCI-type circuit breaker. The maximum wall height of 42 in. does not apply to storable swimming pools.

and nonaccessible non–current-carrying metal parts of the appliance.

The means for grounding shall be an equipment grounding conductor run with the power-supply conductors in the flexible cord that is properly terminated in a grounding-type attachment plug having a fixed grounding contact member.

680.32 Ground-Fault Circuit Interrupters Required.

All electrical equipment, including power-supply cords, used with storable pools shall be protected by ground-fault circuit interrupters.

> FPN: For flexible cord usage, see 400.4.

680.33 Luminaires (Lighting Fixtures).

An underwater luminaire (lighting fixture), if installed, shall be installed in or on the wall of the storable pool. It shall comply with one of the following two provisions.

(A) 15 Volts or Less. A luminaire (lighting fixture) shall be part of a cord-and-plug-connected lighting assembly. This assembly shall be listed as an assembly for the purpose and have the following construction features:

(1) No exposed metal parts
(2) A luminaire (fixture) lamp that operates at 15 volts or less
(3) An impact-resistant polymeric lens, luminaire (fixture) body, and transformer enclosure
(4) A transformer meeting the requirements of 680.23(A)(2) with a primary rating not over 150 volts

(B) Over 15 Volts But Not Over 150 Volts. A lighting assembly without a transformer and with the luminaire (fixture) lamp(s) operating at not over 150 volts shall be permitted to be cord-and-plug connected where the assembly is listed as an assembly for the purpose. The installation shall comply with 680.23(A)(5), and the assembly shall have the following construction features:

(1) No exposed metal parts
(2) An impact-resistant polymeric lens and luminaire (fixture) body
(3) A ground-fault circuit interrupter with open neutral protection as an integral part of the assembly
(4) The luminaire (fixture) lamp permanently connected to the ground-fault circuit interrupter with open-neutral protection
(5) Compliance with the requirements of 680.23(A)

Switch labeled as emergency shutoff

5 ft min. – 50 ft max. and within sight

Spa or hot tub

This requirement permits lighting fixtures to be installed in or on storable pools. These cord-and-plug-connected fixtures are required to be listed as an assembly.

IV. Spas and Hot Tubs

680.40 General.

Electrical installations at spas and hot tubs shall comply with the provisions of Part I and Part IV of this article.

680.41 Emergency Switch for Spas and Hot Tubs.

A clearly labeled emergency shutoff or control switch for the purpose of stopping the motor(s) that provide power to the recirculation system and jet system shall be installed at a point readily accessible to the users and not less than 1.5 m (5 ft) away, adjacent to, and within sight of the spa or hot tub. This requirement shall not apply to single-family dwellings.

The provisions of 680.41 require a local disconnecting device for spas and hot tubs that is capable of being used in an emergency. This requirement was added to address entrapment hazards associated with spas and hot tubs. The definitive publication on this issue, *Guideline for Entrapment Hazards: Making Pools and Spas Safer* (Pub. No. 363), is available from the U.S. Consumer Product Safety Commission, Washington, DC 20207.

The emergency shutoff switch must be installed within sight of and at least 5 ft from the spa or hot tub and must be clearly labeled "Emergency Shutoff." See Exhibit 680.17 for an illustration of the switch location. The shutoff switch can be either a line-operated device or a remote-control circuit that causes the pump circuit to open. This requirement does not apply to one-family dwellings.

680.42 Outdoor Installations.

A spa or hot tub installed outdoors shall comply with the provisions of Parts I and II of this article, except as permitted in 680.42(A) and (B), that would otherwise apply to pools installed outdoors.

(A) Flexible Connections. Listed packaged spa or hot tub equipment assemblies or self-contained spas or hot tubs utilizing a factory-installed or assembled control panel or panelboard shall be permitted to use flexible connections as covered in 680.42(A)(1) and (A)(2):

(1) Flexible Conduit. Liquidtight flexible metal conduit or liquidtight flexible nonmetallic conduit shall be permitted in lengths of not more than 1.8 m (6 ft).

The use of liquidtight flexible metal or nonmetallic conduit is permitted by 680.41(A)(1). This modifies the requirements for wiring methods in 680.25(A).

(2) Cord-and-Plug Connections. Cord-and-plug connections with a cord not longer than 4.6 m (15 ft) shall be permitted where protected by a ground-fault circuit interrupter.

(B) Bonding. Bonding by metal-to-metal mounting on a common frame or base shall be permitted. The metal bands or hoops used to secure wooden staves shall not be required to be bonded as required in 680.26.

(C) Interior Wiring to Outdoor Installations. In the interior of a one-family dwelling or in the interior of another building or structure associated with a one-family dwelling, any of the wiring methods recognized in Chapter 3 of this *Code* that contain a copper equipment grounding conductor that is insulated or enclosed within the outer sheath of the wiring method and not smaller than 12 AWG shall be permitted to be used for the connection to motor, heating, and control loads that are part of a self-contained spa or hot tub, or a packaged spa or hot tub equipment assembly. Wiring to an underwater light shall comply with 680.23 or 680.33.

680.43 Indoor Installations.

A spa or hot tub installed indoors shall comply with the provisions of Parts I and II of this article except as modified by this section, and shall be connected by the wiring methods of Chapter 3.

Exception: Listed spa and hot tub packaged units rated 20 amperes or less shall be permitted to be cord-and-plug connected to facilitate the removal or disconnection of the unit for maintenance and repair.

(A) Receptacles. At least one 125-volt, 15- or 20-ampere receptacle on a general-purpose branch circuit shall be located not less than of 1.5 m (5 ft) from and not exceeding 3.0 m (10 ft) from the inside wall of the spa or hot tub.

(1) Location. Receptacles shall be located at least 1.5 m (5 ft) measured horizontally from the inside walls of the spa or hot tub.

(2) Protection, General. Receptacles rated 125 volts and 30 amperes or less and located within 3.0 m (10 ft) of the inside walls of a spa or hot tub shall be protected by a ground-fault circuit interrupter.

(3) Protection, Spa or Hot Tub Supply Receptacle. Receptacles that provide power for a spa or hot tub shall be ground-fault circuit-interrupter protected.

(4) Measurements. In determining the dimensions in this section addressing receptacle spacings, the distance to be measured shall be the shortest path the supply cord of an appliance connected to the receptacle would follow without piercing a floor, wall, ceiling, doorway with hinged or sliding door, window opening, or other effective permanent barrier.

(B) Installation of Luminaires (Lighting Fixtures), Lighting Outlets, and Ceiling-Suspended (Paddle) Fans.

(1) Elevation. Luminaires (lighting fixtures), except as covered in 680.43(B)(2), lighting outlets, and ceiling-suspended (paddle) fans located over the spa or hot tub or within 1.5 m (5 ft) from the inside walls of the spa or hot tub shall comply with the clearances specified in (a), (b), and (c) above the maximum water level.

(a) Without GFCI. Where no GFCI protection is provided, the mounting height shall be not less than 3.7 m (12 ft).

(b) With GFCI. Where GFCI protection is provided, the mounting height shall be permitted to be not less than 2.3 m (7 ft 6 in.).

(c) Below 2.3 m (7 ft 6 in.). Luminaires (lighting fixtures) meeting the requirements of item (1) or (2) and protected by a ground-fault circuit interrupter shall be permitted to be installed less than 2.3 m (7 ft 6 in.) over a spa or hot tub.

(1) Recessed luminaires (fixtures) with a glass or plastic lens, nonmetallic or electrically isolated metal trim, and suitable for use in damp locations
(2) Surface-mounted luminaires (fixtures) with a glass or plastic globe, a nonmetallic body, or a metallic body isolated from contact, and suitable for use in damp locations

(2) Underwater Applications. Underwater luminaires (lighting fixtures) shall comply with the provisions of 680.23 or 680.33.

(C) Wall Switches. Switches shall be located at least 1.5 m (5 ft), measured horizontally, from the inside walls of the spa or hot tub.

Receptacles, wall switches, and electrical devices and controls not associated with a spa or hot tub are required to be located at least 5 ft from the inside wall of the spa or hot tub. Receptacles within 10 ft are required to be protected by a GFCI. Receptacles supplying power to a spa or hot tub are also required to be protected by a GFCI unless the unit is a listed package unit with integral GFCI protection.

Lighting fixtures, lighting outlets, and ceiling-suspended (paddle) fans located less than 12 ft over a spa or hot tub and within 5 ft horizontally from the inside walls of the spa or hot tub are required to be protected by a GFCI.

(D) Bonding. The following parts shall be bonded together:

(1) All metal fittings within or attached to the spa or hot tub structure

(2) Metal parts of electrical equipment associated with the spa or hot tub water circulating system, including pump motors

(3) Metal conduit and metal piping that are within 1.5 m (5 ft) of the inside walls of the spa or hot tub and that are not separated from the spa or hot tub by a permanent barrier

(4) All metal surfaces that are within 1.5 m (5 ft) of the inside walls of the spa or hot tub and that are not separated from the spa or hot tub area by a permanent barrier

Exception: Small conductive surfaces not likely to become energized, such as air and water jets and drain fittings, where not connected to metallic piping, towel bars, mirror frames, and similar nonelectrical equipment, shall not be required to be bonded.

(5) Electrical devices and controls that are not associated with the spas or hot tubs and that are located not less than 1.5 m (5 ft) from such units; otherwise they shall be bonded to the spa or hot tub system

Bonding and grounding requirements are similar to those in Parts I and II of Article 680, except that metal-to-metal mounting on a common frame or base is an acceptable bonding method.

Small conductive surfaces such as air and water jets, drain fittings, and towel bars are not required to be bonded. See 680.43(D)(4), Exception.

Listed packaged units are permitted to be cord connected.

(E) Methods of Bonding. All metal parts associated with the spa or hot tub shall be bonded by any of the following methods:

(1) The interconnection of threaded metal piping and fittings
(2) Metal-to-metal mounting on a common frame or base
(3) The provisions of a copper bonding jumper, insulated, covered, or bare, not smaller than 8 AWG solid

(F) Grounding. The following equipment shall be grounded:

(1) All electric equipment located within 1.5 m (5 ft) of the inside wall of the spa or hot tub
(2) All electric equipment associated with the circulating system of the spa or hot tub

(G) Underwater Audio Equipment. Underwater audio equipment shall comply with the provisions of Part II of this article.

680.44 Protection.

Except as otherwise provided in this section, the outlet(s) that supplies a self-contained spa or hot tub, a packaged spa or hot tub equipment assembly, or a field-assembled spa or hot tub shall be protected by a ground-fault circuit interrupter.

The requirements of 680.44 specify that field-assembled spas and hot tubs with heater loads of 50 amperes or less are to be GFCI protected. Spas and hot tubs utilizing voltages over 250 volts or 3-phase power are not required to have GFCI protection because GFCI devices are not available in all voltage, amperage, and phasing arrangements. Combination spa-pool or hot tub-pool arrangements are not required to have GFCI protection if they share a common bonding grid.

(A) Listed Units. If so marked, a listed self-contained unit or listed packaged equipment assembly that includes integral ground-fault circuit-interrupter protection for all electrical parts within the unit or assembly (pumps, air blowers, heaters, lights, controls, sanitizer generators, wiring, and so forth) shall be permitted without additional GFCI protection.

(B) Other Units. A field assembled spa or hot tub rated 3 phase or rated over 250 volts or with a heater load of more than 50 amperes shall not require the supply to be protected by a ground-fault circuit interrupter.

(C) Combination Pool and Spa or Hot Tub. A combination pool/hot tub or spa assembly commonly bonded need not be protected by a ground-fault circuit interrupter.

FPN: See 680.2 for definitions of *self-contained spa or hot tub* and for *packaged spa or hot tub equipment assembly.*

V. Fountains

Part V applies to permanently installed decorative fountains and reflecting pools in the ground, partially in the ground, or in a building. These units are primarily for aesthetic value and are not intended for swimming or wading.

Part V does not cover installations in natural lakes, rivers, or ponds. However, it may be used in conjunction with the rest of the *Code* where electrical equipment is installed in a natural body of water.

680.50 General.

The provisions of Part I and Part V of this article shall apply to all permanently installed fountains as defined in 680.2. Fountains that have water common to a pool shall additionally comply with the requirements in Part II of this article.

Part V does not cover self-contained, portable fountains not larger than 1.5 m (5 ft) in any dimension. Portable fountains shall comply with Parts II and III of Article 422.

680.51 Luminaires (Lighting Fixtures), Submersible Pumps, and Other Submersible Equipment.

(A) Ground-Fault Circuit Interrupter. Fountain equipment, unless listed for operation at 15 volts or less and supplied by a transformer that complies with 680.23(A)(2), shall be protected by a ground-fault circuit interrupter.

(B) Operating Voltage. No luminaires (lighting fixtures) shall be installed for operation on supply circuits over 150 volts between conductors. Submersible pumps and other submersible equipment shall operate at 300 volts or less between conductors.

(C) Luminaire (Lighting Fixture) Lenses. Luminaires (lighting fixtures) shall be installed with the top of the luminaire (fixture) lens below the normal water level of the fountain unless listed for above-water locations. A luminaire (lighting fixture) facing upward shall have the lens adequately guarded to prevent contact by any person.

(D) Overheating Protection. Electrical equipment that depends on submersion for safe operation shall be protected against overheating by a low-water cutoff or other approved means when not submerged.

(E) Wiring. Equipment shall be equipped with provisions for threaded conduit entries or be provided with a suitable flexible cord. The maximum length of exposed cord in the fountain shall be limited to 3.0 m (10 ft). Cords extending beyond the fountain perimeter shall be enclosed in approved wiring enclosures. Metal parts of equipment in contact with water shall be of brass or other approved corrosion-resistant metal.

(F) Servicing. All equipment shall be removable from the water for relamping or normal maintenance. Luminaires (fixtures) shall not be permanently embedded into the fountain structure such that the water level must be reduced or the fountain drained for relamping, maintenance, or inspection.

(G) Stability. Equipment shall be inherently stable or be securely fastened in place.

680.52 Junction Boxes and Other Enclosures.

(A) General. Junction boxes and other enclosures used for other than underwater installation shall comply with 680.24.

(B) Underwater Junction Boxes and Other Underwater Enclosures. Junction boxes and other underwater enclosures shall meet the requirements of 680.52(B)(1) and (B)(2).

(1) Construction.

(a) Underwater enclosures shall be equipped with provisions for threaded conduit entries or compression glands or seals for cord entry.

(b) Underwater enclosures shall be submersible, and made of copper, brass, or other approved corrosion-resistant material.

(2) Installation. Underwater enclosure installations shall comply with (a) and (b).

(a) Underwater enclosures shall be filled with an approved potting compound to prevent the entry of moisture.

(b) Underwater enclosures shall be firmly attached to the supports or directly to the fountain surface and bonded as required. Where the junction box is supported only by the conduit, the conduit shall be of copper, brass, or other approved corrosion-resistant metal. Where the box is fed by nonmetallic conduit, it shall have additional supports and fasteners of copper, brass, or other approved corrosion-resistant material.

FPN: See 314.23 for support of enclosures.

680.53 Bonding.

All metal piping systems associated with the fountain shall be bonded to the equipment grounding conductor of the branch circuit supplying the fountain.

FPN: See 250.122 for sizing of these conductors.

680.54 Grounding.

The following equipment shall be grounded:

(1) All electrical equipment located within the fountain or within 1.5 m (5 ft) of the inside wall of the fountain
(2) All electrical equipment associated with the recirculating system of the fountain
(3) Panelboards that are not part of the service equipment and that supply any electrical equipment associated with the fountain

680.55 Methods of Grounding.

(A) Applied Provisions. The provisions of 680.21(A), 680.23(B)(3), 680.23(F)(1) and (2), 680.24(F), and 680.25 shall apply.

(B) Supplied by a Flexible Cord. Electrical equipment that is supplied by a flexible cord shall have all exposed non–current-carrying metal parts grounded by an insulated copper equipment grounding conductor that is an integral part of this cord. The grounding conductor shall be connected to a grounding terminal in the supply junction box, transformer enclosure, or other enclosure.

680.56 Cord-and-Plug-Connected Equipment.

(A) Ground-Fault Circuit Interrupter. All electrical equipment, including power-supply cords, shall be protected by ground-fault circuit interrupters.

(B) Cord Type. Flexible cord immersed in or exposed to water shall be of a type for extra-hard usage, as designated in Table 400.4 and shall be a listed type with a "W" suffix.

(C) Sealing. The end of the flexible cord jacket and the flexible cord conductor termination within equipment shall be covered with, or encapsulated in, a suitable potting compound to prevent the entry of water into the equipment through the cord or its conductors. In addition, the ground connection within equipment shall be similarly treated to protect such connections from the deteriorating effect of water that may enter into the equipment.

(D) Terminations. Connections with flexible cord shall be permanent, except that grounding-type attachment plugs and receptacles shall be permitted to facilitate removal or disconnection for maintenance, repair, or storage of fixed or stationary equipment not located in any water-containing part of a fountain.

680.57 Signs.

(A) General. This section covers electric signs installed within or adjacent to fountains.

(B) Ground-Fault Circuit-Interrupter Protection for Personnel. All circuits supplying the sign shall have ground-fault circuit-interrupter protection for personnel.

(C) Location.

(1) Fixed or Stationary. A fixed or stationary electric sign installed within a fountain shall be not less than 1.5 m (5 ft) inside the fountain measured from the outside edges of the fountain.

(2) Portable. A portable electric sign shall not be placed within a pool or fountain or within 1.5 m (5 ft) measured horizontally from the inside walls of the fountain.

(D) Disconnect. A sign shall have a local disconnecting means in accordance with 600.6 and 680.12.

(E) Bonding and Grounding. A sign shall be grounded and bonded in accordance with 600.7.

The use of electric signs in fountains has become increasingly popular. The requirements for electric signs in fountains were added for the 1999 *Code*. Signs were previously addressed by Article 600, and fountains were addressed by Article 680. In the 2002 *Code*, 680.57 addresses them together. Electric signs in fountains are required to have GFCI protection for personnel. This protection may be provided in the feeder or branch circuit. To prevent contact by persons around the fountain, the sign must be at least 5 ft from the edge of the fountain (see Exhibit 680.18). Disconnecting and bonding requirements in Article 600 apply, and grounding must be provided in accordance with Article 250.

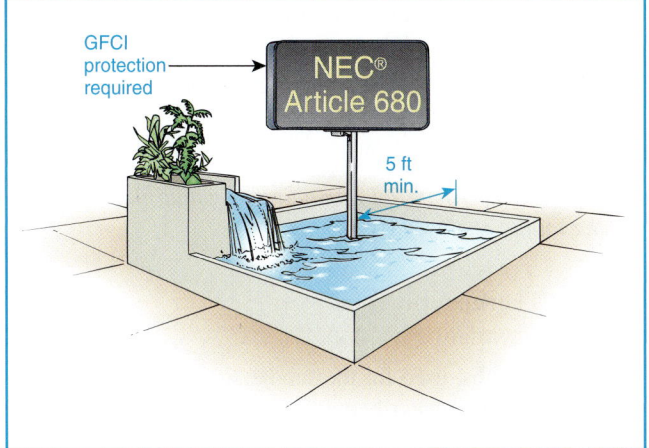

Exhibit 680.18 Electric sign located in a fountain as described in 680.57.

VI. Pools and Tubs for Therapeutic Use

Part VI recognizes therapeutic equipment in other locations, such as athletic training rooms and health care facilities, where conditions of use are the same or similar. Portable therapeutic appliances, which are covered by the provisions of Article 422, are required to provide protection from electrocution while in the on or off position. The device used is an immersion-detection circuit interrupter (IDCI).

Permanently installed therapeutic pools that cannot be readily disassembled are required to comply with Parts I and II of Article 680. The limitations regarding lighting fixtures over and around a swimming pool do not apply to therapeutic pools and tubs. The lighting fixtures in the tub area are required to be totally enclosed. Therapeutic tubs that cannot easily be moved are subject to the same basic requirements.

Bonding and grounding requirements are similar to those in Parts I and II of Article 680, except that metal-to-metal mounting on a common frame or base is permitted. Where equipment is connected by a flexible cord, the equipment grounding conductor is required to be connected to a fixed metal part of the assembly.

680.60 General.

The provisions of Part I and Part VI of this article shall apply to pools and tubs for therapeutic use in health care facilities, gymnasiums, athletic training rooms, and similar areas. Portable therapeutic appliances shall comply with Parts II and III of Article 422.

FPN: See 517.2 for definition of health care facilities.

680.61 Permanently Installed Therapeutic Pools.

Therapeutic pools that are constructed in the ground, on the ground, or in a building in such a manner that the pool cannot be readily disassembled shall comply with Parts I and II of this article.

Exception: The limitations of 680.22(B)(1) through (B)(4) shall not apply where all luminaires (lighting fixtures) are of the totally enclosed type.

680.62 Therapeutic Tubs (Hydrotherapeutic Tanks).

Therapeutic tubs, used for the submersion and treatment of patients, that are not easily moved from one place to another in normal use or that are fastened or otherwise secured at a specific location, including associated piping systems, shall conform to this part.

(A) Protection. Except as otherwise provided in this section, the outlet(s) that supplies a self-contained therapeutic tub or hydrotherapeutic tank, a packaged therapeutic tub or hydrotherapeutic tank, or a field-assembled therapeutic tub or hydrotherapeutic tank shall be protected by a ground-fault circuit interrupter.

(1) Listed Units. If so marked, a listed self-contained unit or listed packaged equipment assembly that includes integral ground-fault circuit-interrupter protection for all electrical parts within the unit or assembly (pumps, air blowers, heaters, lights, controls, sanitizer generators, wiring, and so forth) shall be permitted without additional GFCI protection.

(2) Other Units. A therapeutic tub or hydrotherapeutic tank rated 3 phase or rated over 250 volts or with a heater load of more than 50 amperes shall not require the supply to be protected by a ground-fault circuit interrupter.

The requirements in 680.62(A)(2) for large therapeutic tanks and therapeutic tubs are similar to the requirements for large hot tubs and spas. Large field-assembled therapeutic tubs are not required to have GFCI protection in their electrical supply when the heater load is over 50 amperes.

(B) Bonding. The following parts shall be bonded together:

(1) All metal fittings within or attached to the tub structure
(2) Metal parts of electrical equipment associated with the tub water circulating system, including pump motors
(3) Metal-sheathed cables and raceways and metal piping that are within 1.5 m (5 ft) of the inside walls of the tub and not separated from the tub by a permanent barrier
(4) All metal surfaces that are within 1.5 m (5 ft) of the inside walls of the tub and not separated from the tub area by a permanent barrier
(5) Electrical devices and controls that are not associated with the therapeutic tubs and that are not located a minimum of 1.5 m (5 ft) from such units

(C) Methods of Bonding. All metal parts required to be bonded by this section shall be bonded by any of the following methods:

(1) The interconnection of threaded metal piping and fittings
(2) Metal-to-metal mounting on a common frame or base
(3) Connections by suitable metal clamps
(4) By the provisions of a solid copper bonding jumper, insulated, covered, or bare, not smaller than 8 AWG

(D) Grounding.

(1) Fixed or Stationary Equipment. The equipment specified in (a) and (b) shall be grounded.

(a) Location. All electrical equipment located within 1.5 m (5 ft) of the inside wall of the tub shall be grounded.
(b) Circulation System. All electrical equipment associated with the circulating system of the tub shall be grounded.

(2) Portable Equipment. Portable therapeutic appliances shall meet the grounding requirements in 250.114.

(E) Receptacles. All receptacles within 1.5 m (5 ft) of a therapeutic tub shall be protected by a ground-fault circuit interrupter.

(F) Luminaires (Lighting Fixtures). All luminaires (lighting fixtures) used in therapeutic tub areas shall be of the totally enclosed type.

VII. Hydromassage Bathtubs

680.70 General.

Hydromassage bathtubs as defined in 680.2 shall comply with Part VII of this article. They shall not be required to comply with other parts of this article.

680.71 Protection.

Hydromassage bathtubs and their associated electrical components shall be protected by a ground-fault circuit interrupter. All 125-volt, single-phase receptacles not exceeding 30 amperes and located within 1.5 m (5 ft) measured horizontally of the inside walls of a hydromassage tub shall be protected by a ground-fault circuit interrupter(s).

Hydromassage bathtubs (see definition in 680.2) are required to be protected by a GFCI. In addition, all 125-volt, single-phase, 15-, 20-, and 30-ampere receptacles within 5 ft of the inside wall of the hydromassage bathtub are required to be GFCI protected. Hydromassage bathtubs are treated the same as ordinary bathtubs in regard to the installation of luminaires, switches, and other electrical equipment. See 410.4(D) for special requirements relating to cord-connected fixtures, hanging fixtures, and pendants near bathtubs. Also see 210.8(A)(1) and 210.8(B)(1) for requirements for GFCI protection of bathroom receptacles.

680.72 Other Electrical Equipment.

Luminaires (lighting fixtures), switches, receptacles, and other electrical equipment located in the same room, and not directly associated with a hydromassage bathtub, shall be installed in accordance with the requirements of Chapters 1 through 4 in this *Code* covering the installation of that equipment in bathrooms.

680.73 Accessibility.

Hydromassage bathtub electrical equipment shall be accessible without damaging the building structure or building finish.

This section requires access to electrical equipment associated with the hydromassage tub. Building codes and plumbing codes might not require access to this equipment. This requirement is intended to ensure that the electrical equipment associated with hydromassage bathtubs can be accessed for maintenance and repair without damaging the finish or structure of the building. The access may be either an integral part of the tub or one that is provided in the finish that encloses the tub. See Exhibit 680.19.

680.74 Bonding.

All metal piping systems, metal parts of electrical equipment, and pump motors associated with the hydromassage tub shall be bonded together using a copper bonding jumper, insulated, covered, or bare, not smaller than 8 AWG solid. Metal parts of listed equipment incorporating an approved system of double insulation and providing a means for

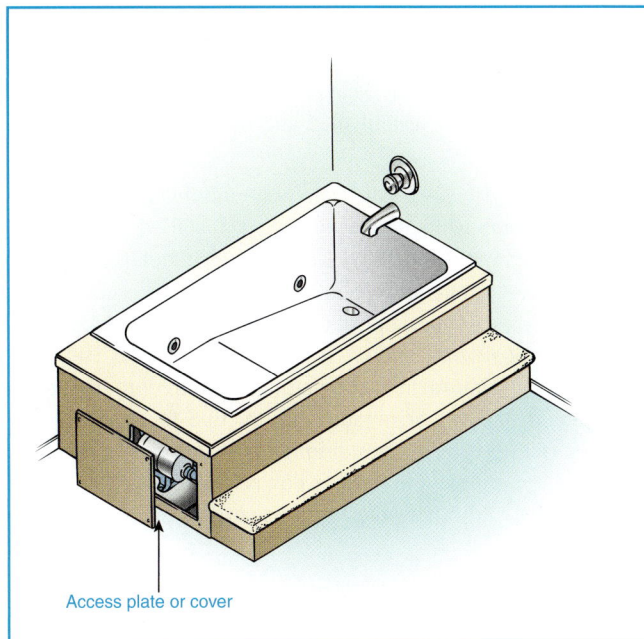

Access plate or cover

Exhibit 680.19 Access plate for hydromassage tub electrical equipment, located as described in 680.73.

grounding internal nonaccessible, non–current-carrying metal parts shall not be bonded.

ARTICLE 685
Integrated Electrical Systems

Contents

I. General
    685.1 Scope
    685.2 Application of Other Articles
II. Orderly Shutdown
    685.10 Location of Overcurrent Devices in
        or on Premises
    685.12 Direct-Current System Grounding
    685.14 Ungrounded Control Circuits

I. General

685.1 Scope.

This article covers integrated electrical systems, other than unit equipment, in which orderly shutdown is necessary to ensure safe operation. An *integrated electrical system* as used in this article is a unitized segment of an industrial wiring system where all of the following conditions are met:

(1) An orderly shutdown is required to minimize personnel hazard and equipment damage.

(2) The conditions of maintenance and supervision ensure that qualified persons service the system.

(3) Effective safeguards, acceptable to the authority having jurisdiction, are established and maintained.

> The integrated electrical systems commonly used in large industrial establishments are designed, installed, and operated by engineering workforces. The control equipment, including overcurrent devices, is located so as to be accessible to qualified personnel, but that location might not meet the definition of *readily accessible* as given in Article 100.
>
> Orderly shutdown is sometimes required to prevent equipment damage or personal injury due to sudden loss of electrical power to the equipment. Orderly shutdown is commonly employed in nuclear power-generating facilities, paper mills, and other areas with hazardous processes.

685.2 Application of Other Articles.

The articles/sections in Table 685.2 apply to particular cases of installation of conductors and equipment, where there are orderly shutdown requirements that are in addition to those of this article or are modifications of them.

Table 685.2 Application of Other Articles

Conductor/Equipment	Section
More than one building or other structure	225, Part II
Ground-fault protection of equipment	230.95, Exception No. 1
Protection of conductors	240.4
Electrical system coordination	240.12
Ground-fault protection of equipment	240.13(1)
Grounding ac systems of 50 volts to 1000 volts	250.21
Equipment protection	427.22
Orderly shutdown	430.44
Disconnection	430.74, Exception Nos. 1 and 2
Disconnecting means in sight from controller	430.102(A), Exception No. 2
Energy from more than one source	430.113, Exception Nos. 1 and 2
Disconnecting means	645.10, Exception
Uninterruptible power supplies (UPS)	645.11(1)
Point of connection	705.12(A)

II. Orderly Shutdown

685.10 Location of Overcurrent Devices in or on Premises.

Location of overcurrent devices that are critical to integrated electrical systems shall be permitted to be accessible, with

mounting heights permitted to ensure security from operation by nonqualified personnel.

685.12 Direct-Current System Grounding.

Two-wire dc circuits shall be permitted to be ungrounded.

685.14 Ungrounded Control Circuits.

Where operational continuity is required, control circuits of 150 volts or less from separately derived systems shall be permitted to be ungrounded.

ARTICLE 690
Solar Photovoltaic Systems

Contents

I. General

690.1 Scope.

The provisions of this article apply to solar photovoltaic electrical energy systems, including the array circuit(s), inverter(s), and controller(s) for such systems. [See Figures 690.1(A) and (B).] Solar photovoltaic systems covered by this article may be interactive with other electrical power production sources or stand-alone, with or without electrical energy storage such as batteries. These systems may have ac or dc output for utilization.

In accordance with 705.3, Article 705 is also applicable to a solar photovoltaic system that is interconnected with another power production source. The other power source may be an electric utility or an on-site generating system, such as a wind turbine or a hydroelectric generator.

Exhibit 690.1 shows a custom-designed home with a solar photovoltaic electrical system.

Typical solar photovoltaic systems are illustrated in Exhibits 690.2 through 690.5. Other circuit arrangements are permissible.

690.2 Definitions.

Alternating-Current Module (Alternating-Current Photovoltaic Module). A complete, environmentally protected unit consisting of solar cells, optics, inverter, and other com-

ponents, exclusive of tracker, designed to generate ac power when exposed to sunlight.

An alternating-current photovoltaic (ACPV) module consists of a single mechanical unit. Because there is no accessible, field-installed direct-current wiring in this single unit, the dc photovoltaic source circuit requirements in this *Code* are not applicable to the dc wiring in an ACPV module.

Array. A mechanically integrated assembly of modules or panels with a support structure and foundation, tracker, and other components, as required, to form a direct-current power-producing unit.

The building blocks of an array are illustrated in Exhibit 690.6.

Bipolar Photovoltaic Array. A photovoltaic array that has two outputs, each having opposite polarity to a common reference point or center tap.

Blocking Diode. A diode used to block reverse flow of current into a photovoltaic source circuit.

Blocking diodes are not required by this *Code,* although the instructions or labels supplied with the photovoltaic module may require them.

Charge Controller. Equipment that controls dc voltage or dc current, or both, used to charge a battery.

Diversion Charge Controller. Equipment that regulates the charging process of a battery by diverting power from energy storage to direct-current or alternating-current loads or to an interconnected utility service.

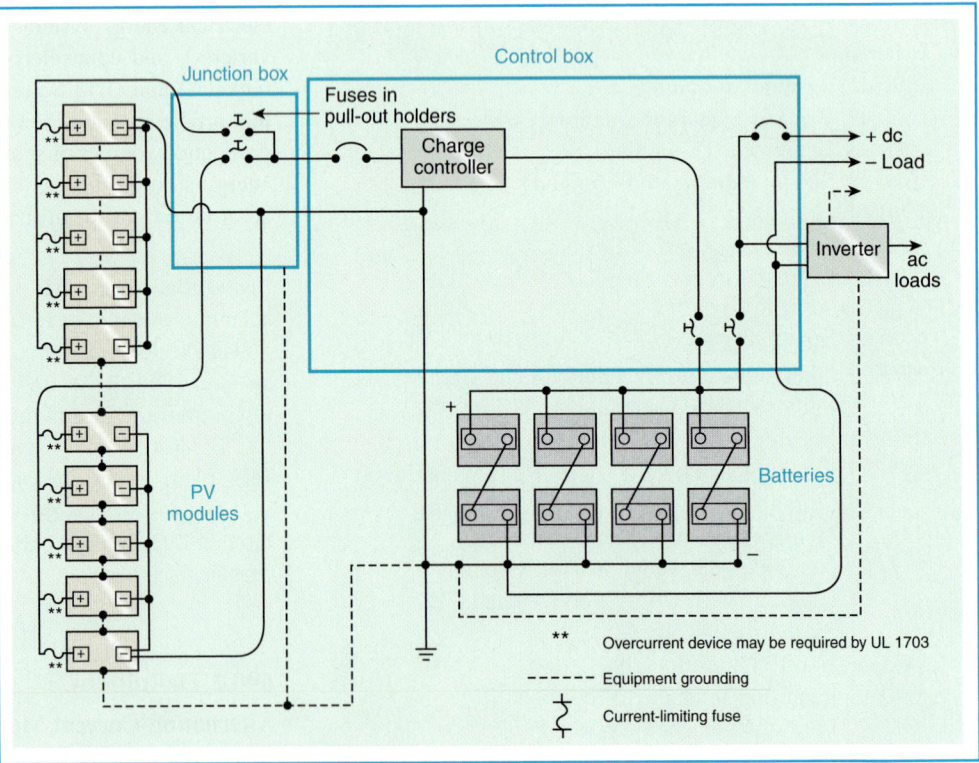

Exhibit 690.2 Simplified circuit diagram of a small residential stand-alone system.

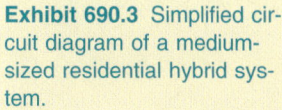

Exhibit 690.3 Simplified circuit diagram of a medium-sized residential hybrid system.

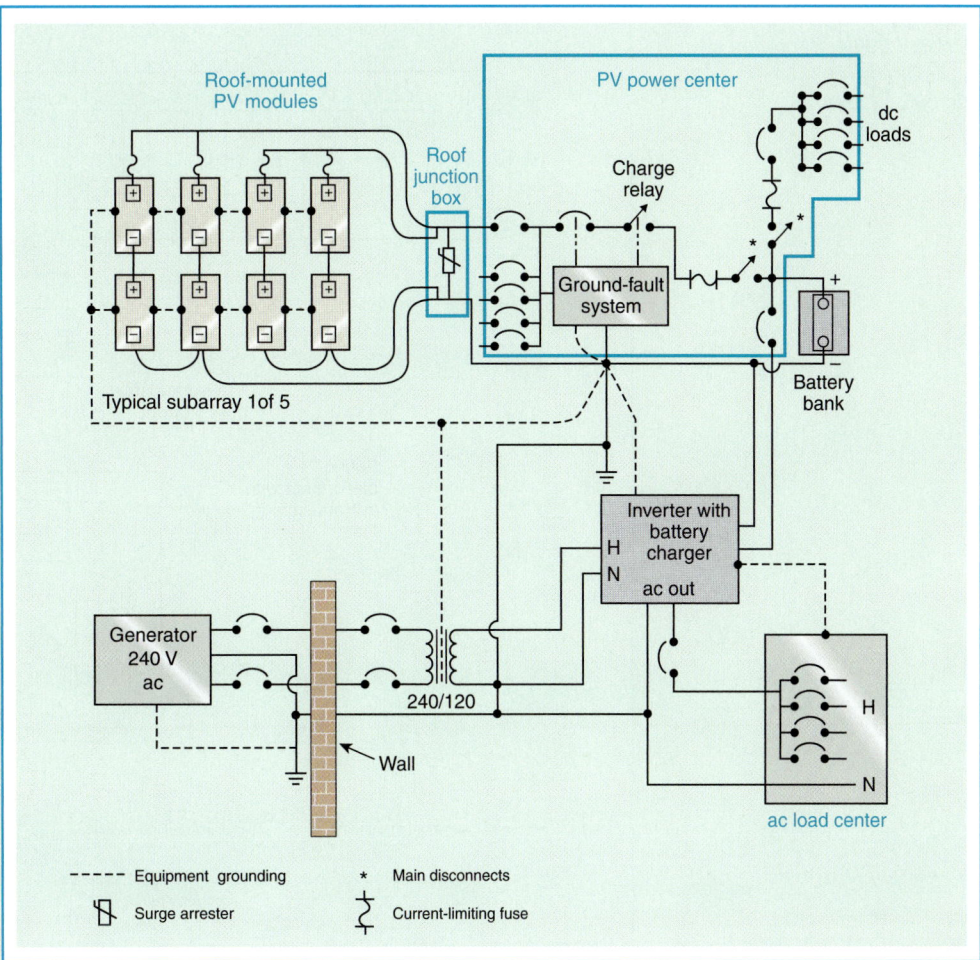

Electrical Production and Distribution Network. A power production, distribution, and utilization system, such as a utility system and connected loads, that is external to and not controlled by the photovoltaic power system.

Hybrid System. A system comprised of multiple power sources. These power sources may include photovoltaic, wind, micro-hydro generators, engine-driven generators, and others, but do not include electrical production and distribution network systems. Energy storage systems, such as batteries, do not constitute a power source for the purpose of this definition.

Interactive System. A solar photovoltaic system that operates in parallel with and may deliver power to an electrical production and distribution network. For the purpose of this definition, an energy storage subsystem of a solar photovoltaic system, such as a battery, is not another electrical production source.

Inverter. Equipment that is used to change voltage level or waveform, or both, of electrical energy. Commonly, an

inverter [also known as a power conditioning unit (PCU) or power conversion system (PCS)] is a device that changes dc input to an ac output. Inverters may also function as battery chargers that use alternating current from another source and convert it into direct current for charging batteries.

Inverter Input Circuit. Conductors between the inverter and the battery in stand-alone systems or the conductors between the inverter and the photovoltaic output circuits for electrical production and distribution network.

Inverter Output Circuit. Conductors between the inverter and an ac load center for stand-alone systems or the conductors between the inverter and the service equipment or another electric power production source, such as a utility, for electrical production and distribution network.

Module. A complete, environmentally protected unit consisting of solar cells, optics, and other components, exclusive of tracker, designed to generate dc power when exposed to sunlight.

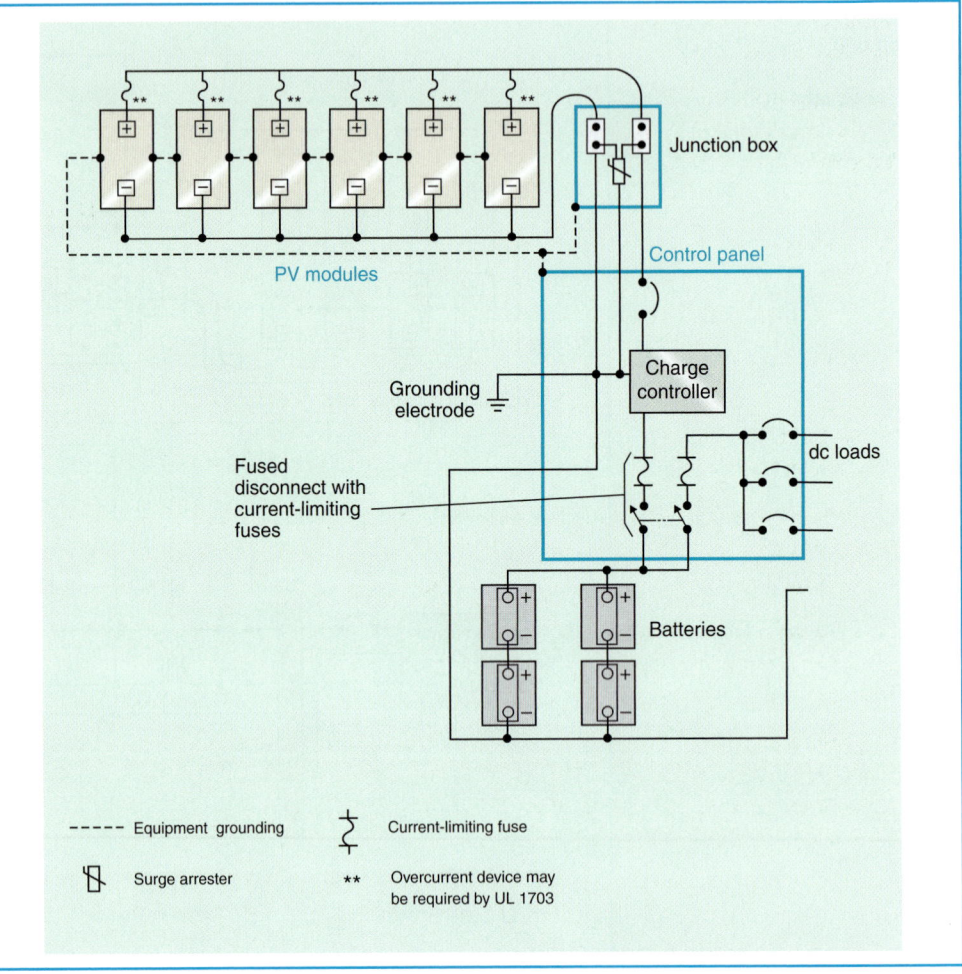

Exhibit 690.4 Simplified circuit diagram of a remote-cabin dc-only system.

Junction box

Control panel

PV modules

Grounding electrode

Charge controller

dc loads

Fused disconnect with current-limiting fuses

Batteries

- - - - - Equipment grounding

Surge arrester

Current-limiting fuse

** Overcurrent device may be required by UL 1703

Panel. A collection of modules mechanically fastened together, wired, and designed to provide a field-installable unit.

Photovoltaic Output Circuit. Circuit conductors between the photovoltaic source circuit(s) and the inverter or dc utilization equipment.

Photovoltaic Power Source. An array or aggregate of arrays that generates dc power at system voltage and current.

Photovoltaic Source Circuit. Circuits between modules and from modules to the common connection point(s) of the dc system.

Photovoltaic Systems Voltage. The direct current (dc) voltage of any photovoltaic source or photovoltaic output circuit. For bipolar or multiwire installations, the photovoltaic systems voltage is the highest voltage between any two dc conductors.

Solar Cell. The basic photovoltaic device that generates electricity when exposed to light.

Solar Photovoltaic System. The total components and subsystems that, in combination, convert solar energy into electrical energy suitable for connection to a utilization load.

Stand-Alone System. A solar photovoltaic system that supplies power independently of an electrical production and distribution network.

The simplified circuit diagrams in Exhibits 690.2 through 690.5 demonstrate the use of various components in a photovoltaic system. Specific requirements for overcurrent protection, disconnecting means, and grounding are covered in other sections of Article 690 and should not be assumed based on these diagrams. Instructions for or labels on the photovoltaic module might require additional overcurrent devices that may not be shown.

690.3 Other Articles.

Wherever the requirements of other articles of this *Code* and Article 690 differ, the requirements of Article 690 shall

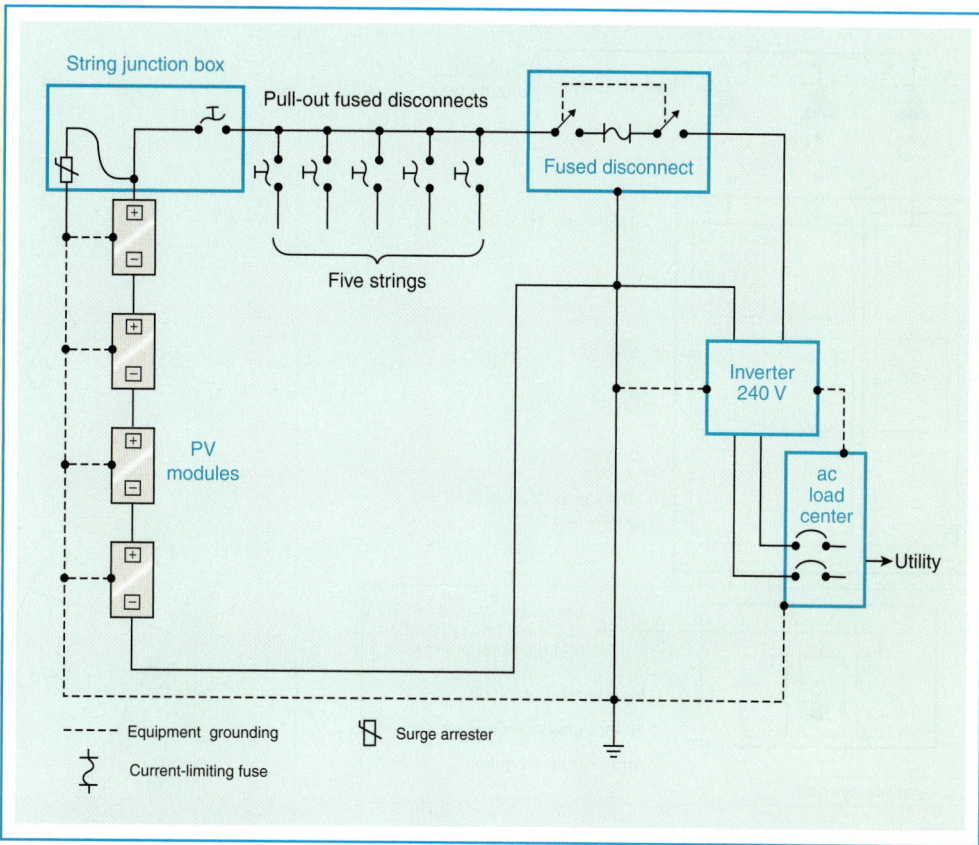

Exhibit 690.5 Simplified circuit diagram of a rooftop grid-connected system.

String junction box

Pull-out fused disconnects

Fused disconnect

Five strings

Inverter 240 V

PV modules

ac load center

Utility

- - - - - Equipment grounding   Surge arrester

Current-limiting fuse

apply and, if the system is operated in parallel with a primary source(s) of electricity, the requirements in 705.14, 705.16, 705.32, and 705.43 shall apply.

690.4 Installation.

(A) Solar Photovoltaic System. A solar photovoltaic system shall be permitted to supply a building or other structure in addition to any service(s) of another electricity supply system(s).

(B) Conductors of Different Systems. Photovoltaic source circuits and photovoltaic output circuits shall not be contained in the same raceway, cable tray, cable, outlet box, junction box, or similar fitting as feeders or branch circuits of other systems, unless the conductors of the different systems are separated by a partition or are connected together.

For example, 690.4(B) requires the conductors of a yard light located near a roof-mounted photovoltaic (PV) array to be in a separate raceway or cable from the conductors of PV source circuits or PV output circuits.

(C) Module Connection Arrangement. The connections to a module or panel shall be arranged so that removal of a module or panel from a photovoltaic source circuit does

not interrupt a grounded conductor to another photovoltaic source circuit. Sets of modules interconnected as systems rated at 50 volts or less, with or without blocking diodes, and having a single overcurrent device shall be considered as a single-source circuit. Supplementary overcurrent devices used for the exclusive protection of the photovoltaic modules are not considered as overcurrent devices for the purpose of this section.

In general, 690.4(C) requires that a jumper be installed between a module terminal or lead and the connection point to the grounded photovoltaic source circuit conductor. That way, a module can be removed without interrupting the grounded conductor to other photovoltaic source circuits. If interrupted, such conductors, although identified as grounded, would be operating at the system potential with respect to ground, and a shock hazard could result. The reverse current protection on nearly all PV modules (as indicated by the fuse requirement labeled on the back of each module) generally dictates that each module or string of modules has a series overcurrent device and becomes a source circuit.

(D) Equipment. Inverters or motor generators shall be identified for use in solar photovoltaic systems.

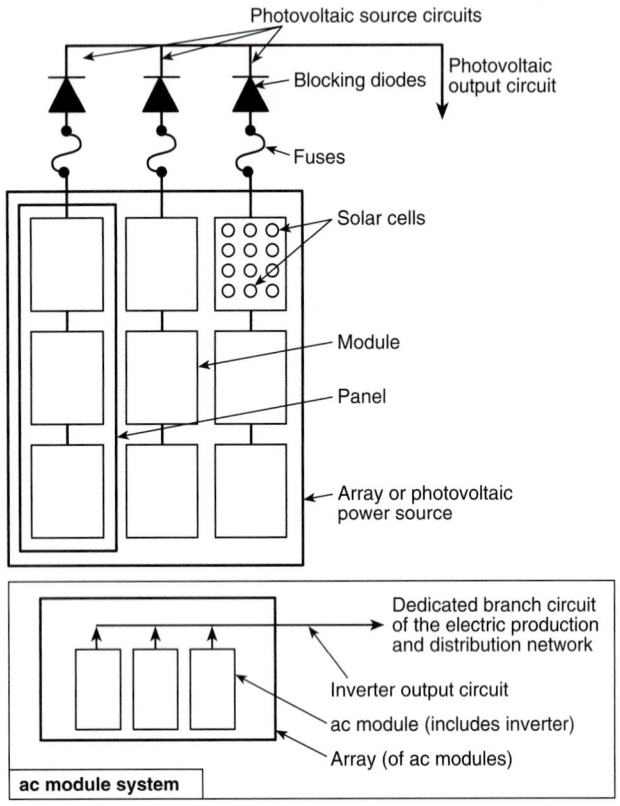

Figure 690.1(A) Identification of solar photovoltaic system components.

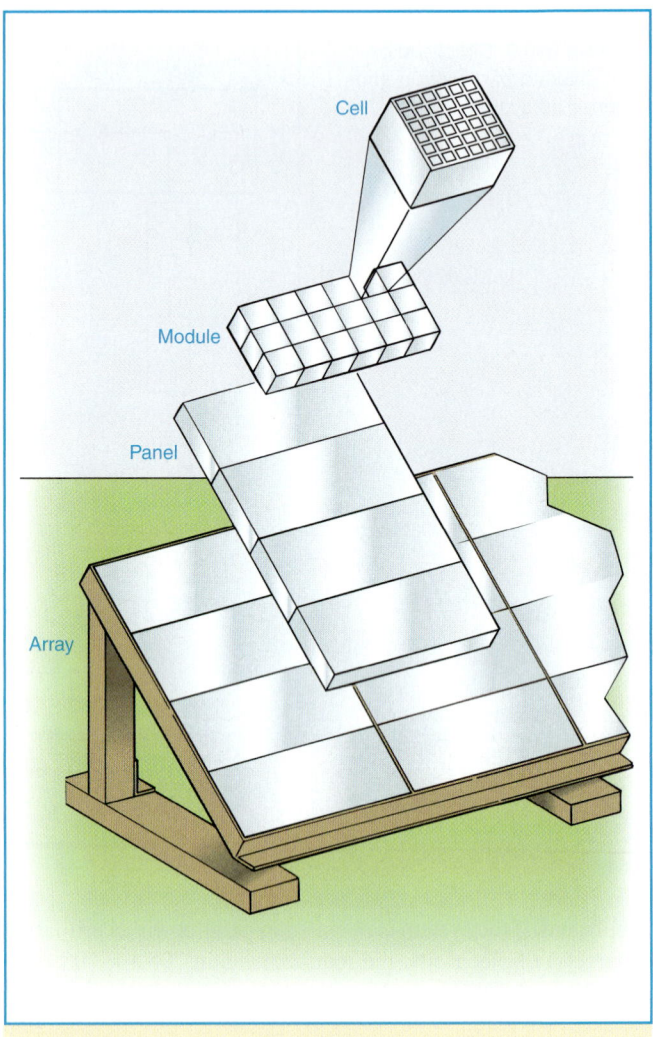

690.5 Ground-Fault Protection.

Roof-mounted dc photovoltaic arrays located on dwellings shall be provided with dc ground-fault protection to reduce fire hazards.

Ground-fault detection and interruption for solar photovoltaic systems should not be confused with the requirements for ground-fault circuit-interrupter (GFCI) protection, as defined in Article 100. A GFCI is intended for the protection of personnel in single-phase ac systems. It functions to open the ungrounded conductor when a 5-mA fault current is detected. In contrast, 690.5 is intended to prevent fires in dc photovoltaic circuits due to ground faults.

(A) Ground-Fault Detection and Interruption. The ground-fault protection device or system shall be capable of detecting a ground fault, interrupting the flow of fault current, and providing an indication of the fault.

(B) Disconnection of Conductors. The ungrounded conductors of the faulted source circuit shall be automatically disconnected. If the grounded conductors of the faulted source circuit are disconnected to comply with the requirements of 690.5(A), all conductors of the faulted source circuit shall be opened automatically and simultaneously. Opening the grounded conductor of the array or opening the faulted sections of the array shall be permitted to interrupt the ground-fault current path.

(C) Labels and Markings. Labels and markings shall be applied near the ground-fault indicator at a visible location, stating that, if a ground fault is indicated, the normally grounded conductors may be energized and ungrounded.

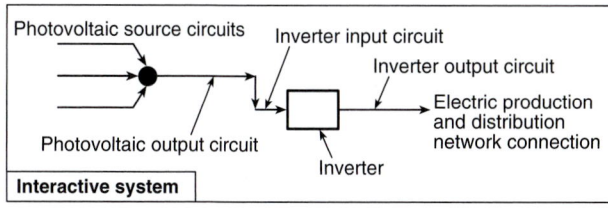

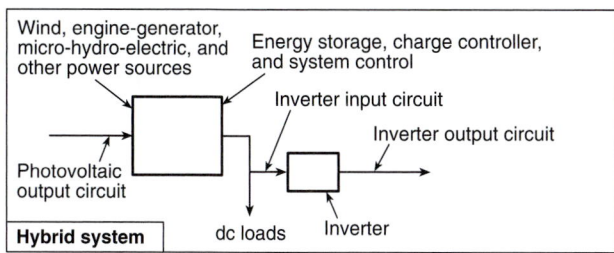

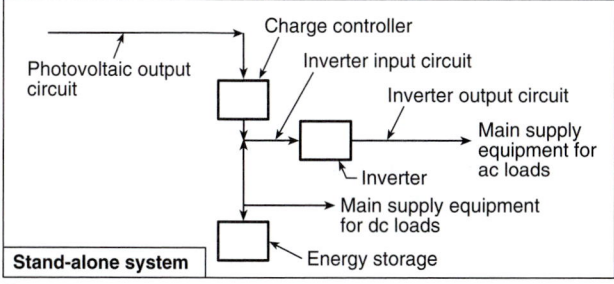

Notes:
1. These diagrams are intended to be a means of identification for photovoltaic system components, circuits, and connections.
2. Disconnecting means and overcurrent protection required by Article 690 are not shown.
3. System grounding and equipment grounding are not shown. See Article 690, Part V.
4. Custom designs occur in each configuration and some components are optional.

Figure 690.1(B) Identification of solar photovoltaic system components in common system configurations.

> Many types of ground-fault detection and interruption equipment break the negative-to-ground bond to interrupt the fault currents. The now ungrounded PV negative conductor will generally be at open-circuit voltage below the ground reference (e.g., −400 volts).

690.6 Alternating-Current Modules.

(A) Photovoltaic Source Circuits. The requirements of Article 690 pertaining to photovoltaic source circuits shall not apply to ac modules. The photovoltaic source circuit, conductors, and inverters shall be considered as internal wiring of an ac module.

(B) Inverter Output Circuit. The output of an ac module shall be considered an inverter output circuit.

(C) Disconnecting Means. A single disconnecting means, in accordance with 690.15 and 690.17, shall be permitted

for the combined ac output of one or more ac modules. Additionally, each ac module in a multiple ac-module system shall be provided with a connector, bolted, or terminal-type disconnecting means.

> Alternating-current photovoltaic modules, as utility-interactive devices, are designed to produce ac power only when they are connected to an external source of ac power at the correct voltage and frequency. A single disconnecting means will remove the external source and turn off all ac photovoltaic modules connected to that disconnecting device.

(D) Ground-Fault Detection. Alternating-current-module systems shall be permitted to use a single detection device to detect only ac ground faults and to disable the array by removing ac power to the ac module(s).

> The permissive language of 690.6(D) for ac photovoltaic modules replaces the requirements of 690.5 that apply only to conventional dc photovoltaic modules. As in 690.5, this is a fire prevention device and is not intended to be a shock prevention device. Existing GFCI and equipment GFP devices are generally not listed for backfeeding or suitable for meeting this requirement.

(E) Overcurrent Protection. The output circuits of ac modules shall be permitted to have overcurrent protection and conductor sizing in accordance with 240.5(B)(2).

II. Circuit Requirements

690.7 Maximum Voltage.

(A) Maximum Photovoltaic System Voltage. In a dc photovoltaic source circuit or output circuit, the maximum photovoltaic system voltage for that circuit shall be computed as the sum of the rated open-circuit voltage of the series-connected photovoltaic modules corrected for the lowest expected ambient temperature. For crystalline and multicrystalline silicon modules, the rated open-circuit voltage shall be multiplied by the correction factor provided in Table 690.7. This voltage shall be used to determine the voltage rating of cables, disconnects, overcurrent devices, and other equipment. Where the lowest expected ambient temperature is below −40°C (−40°F), or where other than crystalline or multicrystalline silicon photovoltaic modules are used, the system voltage adjustment shall be made in accordance with the manufacturer's instructions.

> A photovoltaic source is not a constant-voltage source, and the difference between the rated operating voltage determined under controlled laboratory conditions and the open-circuit voltage under field-installed conditions can be signifi-

Table 690.7 Voltage Correction Factors for Crystalline and Multicrystalline Silicon Modules

Ambient Temperature (°C)	Correction Factors for Ambient Temperatures Below 25°C (77°F) (Multiply the rated open-circuit voltage by the appropriate correction factor shown below.)	Ambient Temperature (°F)
25 to 10	1.06	77 to 50
9 to 0	1.10	49 to 32
−1 to −10	1.13	31 to 14
−11 to −20	1.17	13 to −4
−21 to −40	1.25	−5 to −40

cant. Consequently, the higher-rated open-circuit voltage must be used to select circuit components with proper voltage ratings.

The voltage potential (both open circuit and operating) of a photovoltaic power source increases with decreasing temperature. The installer should note the temperature conditions under which the photovoltaic device was rated. If the anticipated lowest daytime temperature at the installation site is lower than the rating condition (25°C), Table 690.7 should be used to adjust the maximum open-circuit voltage of the crystalline system before selecting conductors, overcurrent devices, and switchgear.

Previous to the 1999 edition of this *Code,* the temperature adjustment requirement was included in the instructions provided with listed photovoltaic modules and was established at a fixed 125 percent, regardless of temperature.

Bipolar photovoltaic systems (with positive and negative voltages) must use the sum of the absolute values of the open-circuit voltages to determine the rated open-circuit system voltage. For example, a system with open-circuit voltages of plus and minus 480 volts would have a system open-circuit voltage of 960 volts. This voltage should be multiplied by a temperature-dependent factor from Table 690.7, yielding a system design voltage of up to 1200 volts. This system design voltage should be used in selecting cables and other equipment. Also see the definition of *photovoltaic system voltage* in 690.2. Certain methods of connecting bipolar PV arrays meeting the requirements of 690.7(E) may have different requirements for calculating the maximum system voltage.

(B) Direct-Current Utilization Circuits. The voltage of dc utilization circuits shall conform with 210.6.

The requirements of 690.7(B) cover installations where the photovoltaic output is connected to direct-current utilization circuits.

(C) Photovoltaic Source and Output Circuits. In one- and two-family dwellings, photovoltaic source circuits and

photovoltaic output circuits that do not include lampholders, fixtures, or receptacles shall be permitted to have a maximum system voltage up to 600 volts. Other installations with a maximum system voltage over 600 volts shall comply with Article 690, Part I.

Photovoltaic direct-current circuits in buildings are permanently connected using wiring systems recognized by this *Code.* Requirements for protecting unqualified persons from contact with these circuits are included in 690.7(B) and (D). Unqualified persons are not likely to service equipment in these circuits due to its complexity. There is a significant difference between the rated open-circuit voltage and the operating voltage in photovoltaic direct-current circuits. For the photovoltaic system to perform its intended function, rated direct-current open-circuit voltages of up to 600 volts may be present.

(D) Circuits Over 150 Volts to Ground. In one- and two-family dwellings, live parts in photovoltaic source circuits and photovoltaic output circuits over 150 volts to ground shall not be accessible to other than qualified persons while energized.

> FPN: See 110.27 for guarding of live parts, and 210.6 for voltage to ground and between conductors.

Where direct-current circuitry over 150 volts to ground is present in one- and two-family dwellings, additional protection for unqualified persons may be needed. Protection may be in the form of conduit, a closed cabinet, or an enclosure that requires the use of tools to open it and that permits entry only by qualified persons.

(E) Bipolar Source and Output Circuits. For 2-wire circuits connected to bipolar systems, the maximum system voltage shall be the highest voltage between the conductors of the 2-wire circuit if all of the following conditions apply:

(1) One conductor of each circuit is solidly grounded.
(2) Each circuit is connected to a separate subarray.
(3) The equipment is clearly marked with a label as follows: Warning—Bipolar Photovoltaic Array. Disconnection of neutral or grounded conductors may result in overvoltage on array or inverter.

690.8 Circuit Sizing and Current.

(A) Computation of Maximum Circuit Current. The maximum current for the specific circuit shall be computed in accordance with 690.8(A)(1) through (A)(4).

(1) Photovoltaic Source Circuit Currents. The maximum current shall be the sum of parallel module rated short-circuit currents multiplied by 125 percent.

The use of the array short-circuit current allows for proper sizing of conductors to handle the current generated during extended periods of operation under a short-circuit current operating point.

The 125 percent factor mentioned in 690.8(A)(1) is required because photovoltaic modules, photovoltaic source circuits, and photovoltaic output circuits can deliver output currents higher than the rated short-circuit currents for more than 3 hours near solar noon. Prior to the 1999 edition of this *Code,* this requirement was specified in the instructions provided with each listed module. This 125 percent requirement is in addition to the 125 percent factor required by 690.8(B).

Photovoltaic modules in hot climates operate at temperatures of 60°C to 80°C due to solar heating. Conductors with insulation types rated at least 90°C should be used, and these conductors should have the ampacity corrected in accordance with Table 310.16 or Table 310.17.

(2) Photovoltaic Output Circuit Currents. The maximum current shall be the sum of parallel source circuit maximum currents as calculated in 690.8(A)(1).

(3) Inverter Output Circuit Current. The maximum current shall be the inverter continuous output current rating.

Both stand-alone and utility-interactive inverters are power-limited devices. Output circuits connected to these devices are sized on the continuous rated outputs of these devices and are not based on load calculations.

(4) Stand-Alone Inverter Input Circuit Current. The maximum current shall be the stand-alone continuous inverter input current rating when the inverter is producing rated power at the lowest input voltage.

Stand-alone inverters are constant-output-voltage devices. As the input battery voltage decreases, the input battery current increases to maintain a constant ac output. The input current for such inverters is calculated by taking the rated full-power output of the inverter in watts and dividing it by the lowest operating battery voltage and then by the rated efficiency of the inverter. For example, the input current for a 4000-watt, 24-volt inverter that is 85 percent efficient at 22 volts can be calculated as follows:

$$\text{Ampere input} = \frac{\text{watt output}}{\text{voltage} \times \text{efficiency}}$$
$$= \frac{4000 \text{ W}}{22 \text{ V} \times 0.85}$$
$$= 214 \text{ A}$$

Ripple currents might be present in the dc-input circuits of single-phase, stand-alone inverters. These ripple currents might cause nuisance operation of overcurrent devices at continuous high inverter outputs. In such cases, the measured maximum rms (root mean square) value of the input current, which will be greater than the average current calculated here, should be used to determine conductor sizes and overcurrent device ratings.

(B) Ampacity and Overcurrent Device Ratings. Photovoltaic system currents shall be considered continuous.

(1) Sizing of Conductors and Overcurrent Devices. The circuit conductors and overcurrent devices shall be sized to carry not less than 125 percent of the maximum currents as computed in 690.8(A). The rating or setting of overcurrent devices shall be permitted in accordance with 240.4(B) and (C).

Exception: Circuits containing an assembly, together with its overcurrent device(s), that is listed for continuous operation at 100 percent of its rating shall be permitted to be utilized at 100 percent of its rating.

The exception to 690.8(B)(1) permits use at the full rating of assemblies, such as panelboards, incorporating overcurrent devices listed for continuous operation at 100 percent of the rating.

(2) Internal Current Limitation. Overcurrent protection for photovoltaic output circuits with devices that internally limit the current from the photovoltaic output circuit shall be permitted to be rated at less than the value computed in 690.8(B)(1). This reduced rating shall be at least 125 percent of the limited current value. Photovoltaic output circuit conductors shall be sized in accordance with 690.8(B)(1).

Exception: An overcurrent device in an assembly listed for continuous operation at 100 percent of its rating shall be permitted to be utilized at 100 percent of its rating.

(C) Systems with Multiple Direct-Current Voltages. For a photovoltaic power source that has multiple output circuit voltages and employs a common-return conductor, the ampacity of the common-return conductor shall not be less than the sum of the ampere ratings of the overcurrent devices of the individual output circuits.

(D) Sizing of Module Interconnection Conductors. Where a single overcurrent device is used to protect a set of two or more parallel-connected module circuits, the ampacity of each of the module interconnection conductors shall not be less than the sum of the rating of the single fuse plus 125 percent of the short-circuit current from the other parallel-connected modules.

Normally, labeling or module instructions require reverse overcurrent protection for each module or string of modules. In some cases, modules with low rated short-circuit currents and high values of the required series protective fuse may allow the use of one overcurrent device to provide reverse current protection for two modules or strings of modules and overcurrent protection for the conductors. The PV module manufacturer should be contacted for specific information.

690.9 Overcurrent Protection.

(A) Circuits and Equipment. Photovoltaic source circuit, photovoltaic output circuit, inverter output circuit, and storage battery circuit conductors and equipment shall be protected in accordance with the requirements of Article 240. Circuits connected to more than one electrical source shall have overcurrent devices located so as to provide overcurrent protection from all sources.

Exception: An overcurrent device shall not be required for circuit conductors sized in accordance with 690.8(B) and located where one of the following apply:

(a) *There are no external sources such as parallel-connected source circuits, batteries, or backfeed from inverters.*

(b) *The short-circuit currents from all sources do not exceed the ampacity of the conductors.*

> FPN: Possible backfeed of current from any source of supply, including a supply through an inverter into the photovoltaic output circuit and photovoltaic source circuits, is a consideration in determining whether adequate overcurrent protection from all sources is provided for conductors and modules.

In the circuits illustrated in Exhibits 690.2, 690.3, 690.4, 690.5, and 690.7, the photovoltaic source circuit overcurrent devices are required to be rated so that the source circuit conductors are protected in accordance with Article 240 and the overcurrent device ratings do not exceed the maximum overcurrent device rating marked on the modules. Possible backfeed from the other photovoltaic source circuits, other supply sources through the inverter, and storage battery circuits, if any, have to be considered.

Diodes (possibly required by the module manufacturer for performance reasons) can lose their blocking ability because of overtemperature conditions or internal breakdown. Therefore, overcurrent protection has to be considered with a condition of shorted blocking diodes if they are used in the circuit.

At the inverter or battery/charge controller end of the

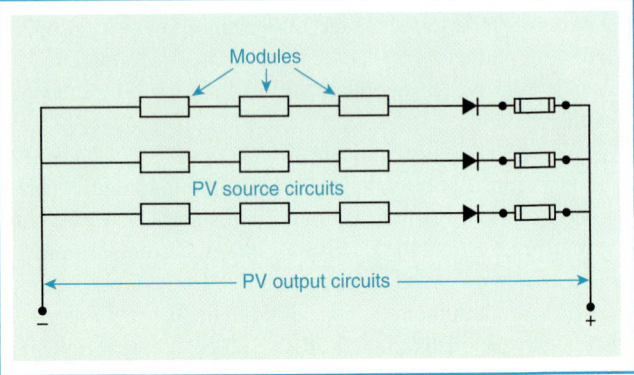

Exhibit 690.7 Application of blocking diodes and source circuit overcurrent devices.

photovoltaic output circuit, the need for overcurrent protection has to be considered with respect to the maximum backfeed fault current available from other sources.

(B) Power Transformers. Overcurrent protection for a transformer with a source(s) on each side shall be provided in accordance with 450.3 by considering first one side of the transformer, then the other side of the transformer, as the primary.

Exception: A power transformer with a current rating on the side connected toward the photovoltaic power source not less than the short-circuit output current rating of the inverter shall be permitted without overcurrent protection from that source.

(C) Photovoltaic Source Circuits. Branch-circuit or supplementary-type overcurrent devices shall be permitted to provide overcurrent protection in photovoltaic source circuits. The overcurrent devices shall be accessible but shall not be required to be readily accessible.

Standard values of supplementary overcurrent devices allowed by this section shall be in one ampere size increments starting at one ampere up to and including 15 amperes. Higher standard values above 15 amperes for supplementary overcurrent devices shall be based on the standard sizes provided in 240.6(A).

If the overcurrent protection of photovoltaic source circuits is considered as supplementary overcurrent protection, use of overcurrent devices with ratings other than those suitable for branch-circuit protection is permitted. The use of such devices permits module protection closer to the specified ratings required on the labels attached to listed modules. It is anticipated that only qualified service personnel will replace or reset overcurrent devices in photovoltaic source circuits. Consequently, ready access to the user need not be provided. These supplementary overcurrent devices must be listed for dc operation and have appropriate voltage and current ratings.

(D) Direct-Current Rating. Overcurrent devices, either fuses or circuit breakers, used in any dc portion of a photovoltaic power system shall be listed for use in dc circuits and shall have the appropriate voltage, current, and interrupt ratings.

Direct-current faults are considerably harder to interrupt than alternating-current faults. Overcurrent devices marked or listed only for ac use should not be used in dc circuits. Automotive- and marine-type fuses, although listed for use in automotive dc systems, are not suitable for use in electrical power systems meeting the requirements of this *Code*.

(E) Series Overcurrent Protection. In series-connected strings of two or more modules, a single overcurrent protection device shall be permitted.

690.10 Stand-Alone Systems.

The premises wiring system shall be adequate to meet the requirements of this *Code* for a similar installation connected to a service. The wiring on the supply side of the building or structure disconnecting means shall comply with this *Code* except as modified by 690.10(A), (B), and (C).

(A) Inverter Output. The ac inverter output from a stand-alone system shall be permitted to supply ac power to the building or structure disconnecting means at current levels below the rating of that disconnecting means.

A stand-alone residential or commercial photovoltaic installation may have an alternating-current output and be connected to a building wired in full compliance with all articles of this *Code*. Even though such an installation may have service-entrance equipment rated at 100 or 200 amperes at 120/240 volts, there is no requirement that the photovoltaic source provide either the rated full current or the dual voltages of the service equipment. While safety requirements dictate full compliance with the ac wiring sections of this *Code*, a photovoltaic installation is usually designed so that the actual ac demands on the system are sized to the output rating of the photovoltaic system.

(B) Sizing and Protection. The circuit conductors between the inverter output and the building or structure disconnecting means shall be sized based on the output rating of the inverter. These conductors shall be protected from overcurrents in accordance with Article 240. The overcurrent protection shall be located at the output of the inverter.

(C) Single 120-Volt Supply. The inverter output of a stand-alone solar photovoltaic system shall be permitted to supply 120 volts to single-phase, 3-wire, 120/240-volt service equipment or distribution panels where there are no 240-volt outlets and where there are no multiwire branch circuits. In all installations, the rating of the overcurrent device connected to the output of the inverter shall be less than the rating of the neutral bus in the service equipment. This equipment shall be marked with the following words or equivalent:

Multiwire branch circuits are common in single- and two-family dwelling units. When connected to a normal 120/240-volt ac service, the currents in the neutral conductors of these multiwire branch circuits (typically 14-3 AWG) cancel or are, at most, no larger than the rating of the branch-circuit overcurrent device. When these electrical systems are connected to a single 120-volt photovoltaic power system inverter by paralleling the two ungrounded conductors in the service entrance load center, the currents in the neutral conductor for each multiwire branch circuit add and do not cancel. The currents in the neutral conductor may be as high as twice the rating of the branch-circuit overcurrent device. Neutral conductor overloading is possible.

WARNING
SINGLE 120-VOLT SUPPLY. DO NOT CONNECT
MULTIWIRE BRANCH CIRCUITS!

III. Disconnecting Means

690.13 All Conductors.

Means shall be provided to disconnect all current-carrying conductors of a photovoltaic power source from all other conductors in a building or other structure. Where a circuit grounding connection is not designed to be automatically interrupted as part of the ground-fault protection system required by 690.5, a switch or circuit breaker used as a disconnecting means shall not have a pole in the grounded conductor.

FPN: The grounded conductor may have a bolted or terminal disconnecting means to allow maintenance or troubleshooting by qualified personnel.

690.14 Additional Provisions.

Photovoltaic disconnecting means shall comply with 690.14(A) through (C).

(A) Disconnecting Means. The disconnecting means shall not be required to be suitable as service equipment and shall be rated in accordance with 690.17.

A disconnecting means rated in accordance with 690.17 may be used, but it is not required to be marked suitable for use as service equipment.

(B) Equipment. Equipment such as photovoltaic source circuit isolating switches, overcurrent devices, and blocking diodes shall be permitted on the photovoltaic side of the photovoltaic disconnecting means.

In general, equipment that needs servicing must be disconnected from sources of supply. In a solar photovoltaic system, however, some equipment, as indicated, is permitted to be located on the photovoltaic power source side of the disconnecting means. See Exhibit 690.8. Servicing the exempted equipment might require disabling all or portions of the array, as explained in the commentary following 690.18.

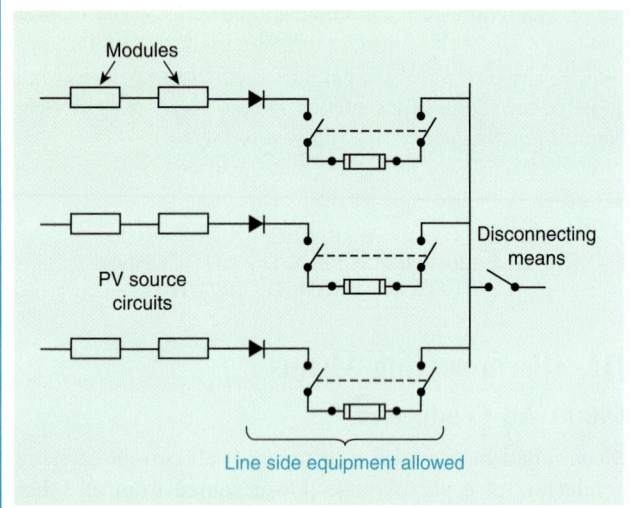

Exhibit 690.8 Equipment permitted on the photovoltaic (PV) power source side of the photovoltaic power source disconnecting means.

There is no intent or requirement to have a disconnecting means located in each photovoltaic source circuit or located physically at each photovoltaic module location. Unlike load circuits (e.g., rooftop air conditioners), photovoltaic source-circuit conductors may be energized at any time from the photovoltaic modules. A centrally located disconnect meeting the requirements of 690.14(C)(1) near the inverter or batteries serves to disconnect the photovoltaic source circuits from the other portions of the electrical power system.

(C) Requirements for Disconnecting Means. Means shall be provided to disconnect all conductors in a building or other structure from the photovoltaic system conductors.

(1) Location. The photovoltaic disconnecting means shall be installed at a readily accessible location either outside of a building or structure or inside nearest the point of entrance of the system conductors.

The photovoltaic system disconnecting means shall not be installed in bathrooms.

These requirements will generally prohibit long runs of PV source and output circuits inside a building before reaching the required PV disconnect. A short conductor run through a wall at the point of first penetration to reach a disconnect mounted inside the building is allowed.

(2) Marking. Each photovoltaic system disconnecting means shall be permanently marked to identify it as a photovoltaic system disconnect.

(3) Suitable for Use. Each photovoltaic system disconnecting means shall be suitable for the prevailing conditions. Equipment installed in hazardous (classified) locations shall comply with the requirements of Articles 500 through 517.

(4) Maximum Number of Disconnects. The photovoltaic system disconnecting means shall consist of not more than six switches or six circuit breakers mounted in a single enclosure, in a group of separate enclosures, or in or on a switchboard.

(5) Grouping. The photovoltaic system disconnecting means shall be grouped with other disconnecting means for the system to comply with 690.14(C)(4). A photovoltaic disconnecting means shall not be required at the photovoltaic module or array location.

Complex PV systems may have multiple sources of power, including the utility, the PV array, a backup generator, and a wind system. No more than six disconnects for all sources of power to the building are allowed, and they should be grouped together.

690.15 Disconnection of Photovoltaic Equipment.

Means shall be provided to disconnect equipment, such as inverters, batteries, charge controllers, and the like, from all ungrounded conductors of all sources. If the equipment is energized from more than one source, the disconnecting means shall be grouped and identified.

A single disconnecting means in accordance with 690.17 shall be permitted for the combined ac output of one or more inverters or ac modules in an interactive system.

690.16 Fuses.

Disconnecting means shall be provided to disconnect a fuse from all sources of supply if the fuse is energized from both directions and is accessible to other than qualified persons. Such a fuse in a photovoltaic source circuit shall be capable of being disconnected independently of fuses in other photovoltaic source circuits.

Switches, pullouts, or similar devices that have suitable ratings may serve as means to disconnect fuses from all sources of supply.

690.17 Switch or Circuit Breaker.

The disconnecting means for ungrounded conductors shall consist of a manually operable switch(es) or circuit breaker(s) complying with all of the following requirements:

(1) Located where readily accessible
(2) Externally operable without exposing the operator to contact with live parts
(3) Plainly indicating whether in the open or closed position
(4) Having an interrupting rating sufficient for the nominal circuit voltage and the current that is available at the line terminals of the equipment

Where all terminals of the disconnecting means may be energized in the open position, a warning sign shall be mounted on or adjacent to the disconnecting means. The sign shall be clearly legible and have the following words or equivalent:

WARNING.
ELECTRIC SHOCK HAZARD.
DO NOT TOUCH TERMINALS.
TERMINALS ON BOTH THE LINE AND
LOAD SIDES MAY BE ENERGIZED
IN THE OPEN POSITION.

Exception No. 1: A disconnecting means located on the dc side shall be permitted to have an interrupting rating less than the current-carrying rating where the system is designed so that the dc switch cannot be opened under load.

Exception No. 2: A connector shall be permitted to be used as an ac or a dc disconnecting means provided that it complies with the requirements of 690.33 and is listed and identified for the use.

690.18 Installation and Service of an Array.

Open circuiting, short circuiting, or opaque covering shall be used to disable an array or portions of an array for installation and service.

FPN: Photovoltaic modules are energized while exposed to light. Installation, replacement, or servicing of array components while a module(s) is irradiated may expose persons to electric shock.

To prevent contact by personnel with energized parts during installation, servicing, or other procedures, a number of methods can be used to disable an array or portions of an array. One method, used infrequently because of the expense in time and materials, is to cover all or portions of an array with an opaque material. Care must be taken that all of the area to be covered is shielded from light.

Another method divides the array into nonhazardous segments, which can be accomplished by switches or connectors. Also see 690.33.

Short circuiting all or portions of an array by means of switches or plug-in connectors in conjunction with bypass diodes can also provide the necessary disablement. (Bypass diodes are incorporated in photovoltaic power sources for performance purposes.)

IV. Wiring Methods

690.31 Methods Permitted.

(A) Wiring Systems. All raceway and cable wiring methods included in this *Code* and other wiring systems and fittings specifically intended and identified for use on photovoltaic arrays shall be permitted. Where wiring devices with integral enclosures are used, sufficient length of cable shall be provided to facilitate replacement.

(B) Single Conductor Cable. Types SE, UF, USE, and USE-2 single-conductor cable shall be permitted in photovoltaic source circuits where installed in the same manner as a Type UF multiconductor cable in accordance with Article 339. Where exposed to sunlight, Type UF cable identified as sunlight-resistant shall be used.

Some photovoltaic modules are designed for a direct series connection by having terminations located at both ends. To accommodate such a direct series connection without the waste of one or more conductors in a multiconductor cable, use of a single-conductor Type USE, SE, UF, or USE-2 cable is permitted in photovoltaic source circuits. The Article 340 reference to installation as a multiconductor cable permits the single-conductor cable to be routed separately, not necessarily with the other conductors of a circuit.

Because long-time constants in dc circuits may result in improper operation of overcurrent devices, it is suggested that, wherever possible, both positive and negative conductors of each circuit and the equipment-grounding conductor be routed as closely together as possible to minimize the circuit time constant. Because photovoltaic modules may operate at high temperatures and are installed in outdoor, exposed locations, the use of high-temperature, wet-rated conductors such as USE-2 is suggested. UF conductors might not have the necessary temperature ratings required.

(C) Flexible Cords and Cables. Flexible cords and cables, where used to connect the moving parts of tracking PV modules, shall comply with Article 400 and shall be of a type identified as a hard service cord or portable power cable; they shall be suitable for extra-hard usage, listed for outdoor use, water resistant, and sunlight resistant. Allowable ampacities shall be in accordance with 400.5. For ambient temperatures exceeding 30°C (86°F), the ampacities shall be derated by the appropriate factors given in Table 690.31(C).

(D) Small-Conductor Cables. Single-conductor cables listed for outdoor use that are sunlight resistant and moisture resistant in sizes 16 AWG and 18 AWG shall be permitted for module interconnections where such cables meet the ampacity requirements of 690.8. Section 310.15 shall be used to determine the cable ampacity and temperature derating factors.

> Because these smaller cables are not normally marked with standard *Code*-recognized markings (e.g., USE, UF, SE), the photovoltaic module manufacturer or installer should provide information establishing that these cables have the necessary sunlight and moisture resistance and are suitable for exposed, outdoor use.
>
> In accordance with 200.6(A), grounded conductors smaller than 6 AWG that are used in photovoltaic source circuits are permitted to be marked with a white marking.

690.32 Component Interconnections.

Fittings and connectors that are intended to be concealed at the time of on-site assembly, where listed for such use, shall be permitted for on-site interconnection of modules or other array components. Such fittings and connectors shall be equal to the wiring method employed in insulation, temperature rise, and fault-current withstand, and shall be capable of resisting the effects of the environment in which they are used.

690.33 Connectors.

The connectors permitted by Article 690 shall comply with 690.33(A) through (E).

(A) Configuration. The connectors shall be polarized and shall have a configuration that is noninterchangeable with receptacles in other electrical systems on the premises.

(B) Guarding. The connectors shall be constructed and installed so as to guard against inadvertent contact with live parts by persons.

(C) Type. The connectors shall be of the latching or locking type.

(D) Grounding Member. The grounding member shall be the first to make and the last to break contact with the mating connector.

(E) Interruption of Circuit. The connectors shall be capable of interrupting the circuit current without hazard to the operator.

690.34 Access to Boxes.

Junction, pull, and outlet boxes located behind modules or panels shall be installed so that the wiring contained in them can be rendered accessible directly or by displacement of a module(s) or panel(s) secured by removable fasteners and connected by a flexible wiring system.

V. Grounding

690.41 System Grounding.

For a photovoltaic power source, one conductor of a two-wire system with a system voltage over 50 volts and the reference (center tap) conductor of a bipolar system shall be solidly grounded or shall use other methods that accomplish equivalent system protection in accordance with 250.4(A) and that utilize equipment listed and identified for the use.

Table 690.31(C) Correction Factors

Ambient Temperature (°C)	Temperature Rating of Conductor				Ambient Temperature (°F)
	60°C (140°F)	75°C (167°F)	90°C (194°F)	105°C (221°F)	
30	1.00	1.00	1.00	1.00	86
31–35	0.91	0.94	0.96	0.97	87–95
36–40	0.82	0.88	0.91	0.93	96–104
41–45	0.71	0.82	0.87	0.89	105–113
46–50	0.58	0.75	0.82	0.86	114–122
51–55	0.41	0.67	0.76	0.82	123–131
56–60	—	0.58	0.71	0.77	132–140
61–70	—	0.33	0.58	0.68	141–158
71–80	—	—	0.41	0.58	159–176

Low-voltage systems that are not grounded must have over-current protection in each of the ungrounded conductors, as required by 240.21.

Other methods that employ available equipment may be used to achieve objectives contained in 250.2(A), thereby providing protection for the photovoltaic power source circuits equivalent to solid grounding.

690.42 Point of System Grounding Connection.

The dc circuit grounding connection shall be made at any single point on the photovoltaic output circuit.

> FPN: Locating the grounding connection point as close as practicable to the photovoltaic source better protects the system from voltage surges due to lightning.

If other than solid grounding is utilized, as permitted by 690.41, the connections should be made in accordance with the markings found on the equipment or in the installation instructions.

Stand-alone photovoltaic systems might require the grounding connection point to be located close to the high-current conductors associated with the battery and the inverter.

Photovoltaic systems on the roofs of dwellings that require ground-fault protection of equipment (see 690.5) might require that the single-point grounding connection be made inside the ground-fault protection equipment or inside the utility-interactive inverter. Connections should be made in accordance with markings on the equipment or in the installation instructions.

690.43 Equipment Grounding.

Exposed non–current-carrying metal parts of module frames, equipment, and conductor enclosures shall be grounded in accordance with 250.134 or 250.136(A) regardless of voltage.

Equipment grounding is required even in low-voltage (12- and 24-volt) systems not otherwise required to have a system ground. A grounding electrode must be added to an ungrounded system to accommodate the equipment grounds.

To maintain the shortest electrical time constant in each dc circuit, the equipment-grounding conductor should be routed as close as possible to the circuit conductors. This will facilitate the operation of overcurrent devices.

690.45 Size of Equipment Grounding Conductor.

Where not protected by the ground-fault protection equipment required by 690.5, the equipment-grounding conductor

for photovoltaic source and photovoltaic output circuits shall be sized for 125 percent of the photovoltaic-originated short-circuit currents in that circuit. Where protected by the ground-fault protection equipment required by 690.5, the equipment-grounding conductors for photovoltaic source and photovoltaic output circuits shall be sized in accordance with 250.122.

In systems where a 690.5 ground-fault protection device is not used, and the circuit conductors are oversized for voltage drop, the requirements of 250.122(B) must be followed.

690.47 Grounding Electrode System.

The ac and dc grounding electrode conductors for a photovoltaic system, where both are present, should connect to a common grounding electrode or grounding electrode system. It is suggested that only one grounding electrode conductor from the dc system and one grounding electrode conductor from the ac system be connected to the grounding electrode and that neither of these conductors be associated with an equipment-grounding system. If separate ac and dc grounding electrodes are used, they should be bonded together, either directly or through the equipment-grounding system.

(A) Alternating-Current Systems. If installing an ac system, a grounding electrode system shall be provided in accordance with 250.50 through 250.60. The grounding electrode conductor shall be installed in accordance with 250.64.

(B) Direct-Current Systems. If installing a dc system, a grounding electrode system shall be provided in accordance with 250.166 for grounded systems or 250.169 for ungrounded systems. The grounding electrode conductor shall be installed in accordance with 250.64.

VI. Marking

690.51 Modules.

Modules shall be marked with identification of terminals or leads as to polarity, maximum overcurrent device rating for module protection, and with the following ratings:

(1) Open-circuit voltage
(2) Operating voltage
(3) Maximum permissible system voltage
(4) Operating current
(5) Short-circuit current
(6) Maximum power

690.52 Alternating-Current Photovoltaic Modules.

Alternating-current modules shall be marked with identification of terminals or leads and with identification of the following ratings:

(1) Nominal operating ac voltage
(2) Nominal operating ac frequency
(3) Maximum ac power
(4) Maximum ac current
(5) Maximum overcurrent device rating for ac module protection

690.53 Photovoltaic Power Source.

A marking, specifying the photovoltaic power source rated as follows, shall be provided by the installer at the site at an accessible location at the disconnecting means for the photovoltaic power source:

(1) Operating current
(2) Operating voltage
(3) Maximum system voltage
(4) Short-circuit current

> FPN: Reflecting systems used for irradiance enhancement may result in increased levels of output current and power.

After installation of photovoltaic arrays, it may be difficult to determine the system's rated voltage and current. These ratings, along with the open-circuit voltage and short-circuit current, are necessary to size the remainder of the system components, as specified elsewhere in Article 690.

Generally, the marking described in 690.53 is required to be provided by the installer. The rated values for the photovoltaic power source can be calculated by adding voltage ratings of series-connected modules and adding current ratings of parallel-connected modules or photovoltaic source circuits.

With respect to the fine print note, a deliberate increase in the level of irradiance by reflectors or the like can cause the power source to operate at levels above those recommended by the manufacturer. See 110.3.

690.54 Interactive System Point of Interconnection.

All interactive system(s) points of interconnection with other sources shall be marked at an accessible location at the disconnecting means as a power source with the maximum ac output operating current and the operating ac voltage.

690.55 Photovoltaic Power Systems Employing Energy Storage.

Photovoltaic power systems employing energy storage shall also be marked with the maximum operating voltage, including any equalization voltage and the polarity of the grounded circuit conductor.

690.56 Identification of Power Sources.

(A) Facilities with Stand-Alone Systems. Any structure or building with a photovoltaic power system that is not connected to a utility service source and is a stand-alone system shall have a permanent plaque or directory installed on the exterior of the building or structure at a readily visible location acceptable to the authority having jurisdiction. The plaque or directory shall indicate the location of system disconnecting means and that the structure contains a stand-alone electrical power system.

(B) Facilities with Utility Services and PV Systems. Buildings or structures with both utility service and a photovoltaic system shall have a permanent plaque or directory providing the location of the service disconnecting means and the photovoltaic system disconnecting means, if not located at the same location.

VII. Connection to Other Sources

690.60 Identified Interactive Equipment.

Only inverters and ac modules listed and identified as interactive shall be permitted in interactive systems.

690.61 Loss of Interactive System Power.

An inverter or an ac module in an interactive solar photovoltaic system shall automatically de-energize its output to the connected electrical production and distribution network upon loss of voltage in that system and shall remain in that state until the electrical production and distribution network voltage has been restored.

A normally interactive solar photovoltaic system shall be permitted to operate as a stand-alone system to supply loads that have been disconnected from electrical production and distribution network sources.

The requirement of 690.61 prevents energizing of otherwise de-energized system conductors or output conductors of other off-site sources (e.g., an electrical utility) and is intended to prevent electric shock. This feature normally is provided as part of the power-conditioning unit.

690.62 Ampacity of Neutral Conductor.

If a single-phase, 2-wire inverter output is connected to the neutral and one ungrounded conductor (only) of a 3-wire system or of a 3-phase, 4-wire wye-connected system, the maximum load connected between the neutral and any one ungrounded conductor plus the inverter output rating shall not exceed the ampacity of the neutral conductor.

690.63 Unbalanced Interconnections.

(A) Single Phase. Single-phase inverters for photovoltaic systems and ac modules in interactive solar photovoltaic systems shall not be connected to 3-phase power systems unless the interconnected system is designed so that significant unbalanced voltages cannot result.

(B) Three Phase. Three-phase inverters and 3-phase ac modules in interactive systems shall have all phases automatically de-energized upon loss of, or unbalanced, voltage in one or more phases unless the interconnected system is designed so that significant unbalanced voltages will not result.

690.64 Point of Connection.

The output of a photovoltaic power source shall be connected as specified in 690.64(A) or (B).

(A) Supply Side. A photovoltaic power source shall be permitted to be connected to the supply side of the service disconnecting means as permitted in 230.82(5).

(B) Load Side. A photovoltaic power source shall be permitted to be connected to the load side of the service disconnecting means of the other source(s) at any distribution equipment on the premises, provided that all of the following conditions are met:

(1) Each source interconnection shall be made at a dedicated circuit breaker or fusible disconnecting means.
(2) The sum of the ampere ratings of overcurrent devices in circuits supplying power to a busbar or conductor shall not exceed the rating of the busbar or conductor.

Exception: For a dwelling unit, the sum of the ampere ratings of the overcurrent devices shall not exceed 120 percent of the rating of the busbar or conductor.

(3) The interconnection point shall be on the line side of all ground-fault protection equipment.

Exception: Connection shall be permitted to be made to the load side of ground-fault protection, provided that there is ground-fault protection for equipment from all ground-fault current sources.

(4) Equipment containing overcurrent devices in circuits supplying power to a busbar or conductor shall be marked to indicate the presence of all sources.

Exception: Equipment with power supplied from a single point of connection.

(5) Equipment such as circuit breakers, if backfed, shall be identified for such operation.

VIII. Storage Batteries

690.71 Installation.

(A) General. Storage batteries in a solar photovoltaic system shall be installed in accordance with the provisions

of Article 480. The interconnected battery cells shall be considered grounded where the photovoltaic power source is installed in accordance with 690.41.

> Batteries in photovoltaic power systems are usually grounded when the photovoltaic power system is grounded in accordance with Article 690, Part VI.

(B) Dwellings.

(1) Operating Voltage. Storage batteries for dwellings shall have the cells connected so as to operate at less than 50 volts, nominal.

Exception: Where live parts are not accessible during routine battery maintenance, a battery system voltage in accordance with 690.7 shall be permitted.

(2) Guarding of Live Parts. Live parts of battery systems for dwellings shall be guarded to prevent accidental contact by persons or objects, regardless of voltage or battery type.

> FPN: Batteries in solar photovoltaic systems are subject to extensive charge–discharge cycles and typically require frequent maintenance, such as checking electrolyte and cleaning connections.

> At any voltage, a primary safety concern in battery systems is that a fault (e.g., a metal tool dropped onto a terminal) might cause a fire or an explosion. *Guarded,* as defined in Article 100, describes the best method to reduce this hazard.

(C) Current Limiting. A listed, current-limiting, overcurrent device shall be installed in each circuit adjacent to the batteries where the available short-circuit current from a battery or battery bank exceeds the interrupting or withstand ratings of other equipment in that circuit. The installation of current-limiting fuses shall comply with 690.16.

> Large banks of storage batteries can deliver significant amounts of short-circuit current. Current-limiting overcurrent devices should be used if necessary.

(D) Battery Nonconductive Cases and Conductive Racks. Flooded, vented, lead-acid batteries with more than twenty-four 2-volt cells connected in series (48 volts, nominal) shall not use conductive cases or shall not be installed in conductive cases. Conductive racks used to support the nonconductive cases shall be permitted where no rack material is located within 150 mm (6 in.) of the tops of the nonconductive cases.

This requirement shall not apply to any type of valve-regulated lead-acid (VRLA) battery or any other types of sealed batteries that may require steel cases for proper operation.

(E) Disconnection of Series Battery Circuits. Battery circuits subject to field servicing, where more than twenty-four 2-volt cells are connected in series (48 volts, nominal), shall have provisions to disconnect the series-connected strings into segments of 24 cells or less for maintenance by qualified persons. Non–load-break bolted or plug-in disconnects shall be permitted.

(F) Battery Maintenance Disconnecting Means. Battery installations, where there are more than twenty-four 2-volt cells connected in series (48 volts, nominal), shall have a disconnecting means, accessible only to qualified persons, that disconnects the grounded circuit conductor(s) in the battery electrical system for maintenance. This disconnecting means shall not disconnect the grounded circuit conductor(s) for the remainder of the photovoltaic electrical system. A non–load-break-rated switch shall be permitted to be used as the disconnecting means.

(G) Battery Systems of More Than 48 Volts. On photovoltaic systems where the battery system consists of more than twenty-four 2-volt cells connected in series (more than 48 volts, nominal), the battery system shall be permitted to operate with ungrounded conductors, provided the conditions in 690.71(G)(1) through (G)(4) are met:

(1) The photovoltaic array source and output circuits shall comply with 690.41.
(2) The dc and ac load circuits shall be solidly grounded.
(3) All main ungrounded battery input/output circuit conductors shall be provided with switched disconnects and overcurrent protection.
(4) A ground-fault detector and indicator shall be installed to monitor for ground faults in the battery bank.

690.72 Charge Control.

(A) General. Equipment shall be provided to control the charging process of the battery. Charge control shall not be required where the design of the photovoltaic source circuit is matched to the voltage rating and charge current requirements of the interconnected battery cells and the maximum charging current multiplied by 1 hour is less than 3 percent of the rated battery capacity expressed in ampere-hours or as recommended by the battery manufacturer.

All adjusting means for control of the charging process shall be accessible only to qualified persons.

FPN: Certain battery types such as valve-regulated lead acid or nickel cadmium can experience thermal failure when overcharged.

(B) Diversion Charge Controller.

(1) Sole Means of Regulating Charging. A photovoltaic power system employing a diversion charge controller as the sole means of regulating the charging of a battery shall be equipped with a second independent means to prevent overcharging of the battery.

(2) Circuits with Direct Current Diversion Charge Controller and Diversion Load. Circuits containing a dc diversion charge controller and a dc diversion load shall comply with the following:

(1) The current rating of the diversion load shall be rated at least 150 percent of the current rating of the diversion charge controller.
(2) The conductor ampacity and the rating of the overcurrent device for this circuit shall be at least 150 percent of the maximum current rating of the diversion charge controller.

(3) Circuits Using Inverters. The requirements in 690.72(B)(2) shall not apply to ac or dc circuits using inverters to control the battery charging process by feeding power into the utility system. These circuits, used in several modes, shall be sized and protected as required in 690.8.

690.74 Battery Interconnections.

Flexible cables, as identified in Article 400, in sizes 2/0 AWG and larger shall be permitted within the battery enclosure from battery terminals to a nearby junction box where they shall be connected to an approved wiring method. Flexible battery cables shall also be permitted between batteries and cells within the battery enclosure. Such cables shall be listed for hard service use and identified as moisture resistant.

be distorted. The use of flexible cables (see Article 400) reduces such distortions. Listed cables with the appropriate physical and chemical resistant properties should be used. Welding and "battery" cables are not allowed or described in the *NEC* for this use. Flexible "building wire"-type cables (Chapter 3) are also available and suitable for this use.

IX. Systems Over 600 Volts

690.80 General.

Solar photovoltaic systems with a maximum system voltage over 600 volts dc shall comply with Article 490 and other requirements applicable to installations rated over 600 volts.

690.85 Definitions.

For the purposes of Part IX of this article, the voltages used to determine cable and equipment ratings are as follows.

Battery Circuits. In battery circuits, the highest voltage experienced under charging or equalizing conditions.

Photovoltaic Circuits. In dc photovoltaic source circuits and photovoltaic output circuits, the maximum system voltage.

ARTICLE 692
Fuel Cell Systems

Contents

I. General

692.1 Scope.

This article identifies the requirements for the installation of fuel cell power systems, which may be stand-alone or interactive with other electrical power production sources and may be with or without electrical energy storage such as batteries.

The rising demand for electrical power has led to the development of power sources that are viable alternatives to or can be interconnected with electric utility distribution systems. Article 692 is a new article that covers the installation of on-premises electrical supply systems where the power is derived from an emerging technology—fuel cells.

The principle of operation is that direct current is generated through a chemical reaction in which fuel such as natural gas or LP-Gas is consumed. As opposed to internal combustion prime movers, the consumption of the fuel gas is via an electrochemical process rather than a combustion process. A power inverter converts the dc to ac. The installation requirements of this new article allow power derived from fuel cells to be safely delivered into residential and light commercial occupancies as the sole source of electrical power or as an integrated source with a utility or other power source.

692.2 Definitions.

Fuel Cell. An electrochemical system that consumes fuel to produce an electrical current. The main chemical reaction used in a fuel cell for producing electrical power is not combustion. There may, however, be sources of combustion used within the overall fuel cell system such as reformers/fuel processors.

Fuel Cell System. The complete aggregate of equipment used to convert chemical fuel into usable electricity. A fuel cell system typically consists of a reformer, stack, power inverter, and auxiliary equipment.

Interactive System. A fuel cell system that operates in parallel with and may deliver power to an electrical production and distribution network. For the purpose of this definition, an energy storage subsystem of a fuel cell system, such as a battery, is not another electrical production source.

Maximum System Voltage. The highest fuel cell inverter output voltage between any ungrounded conductors present at accessible output terminals.

Output Circuit. The conductors used to connect the fuel cell system to its electrical point of delivery. In the case of sites that have series- or parallel-connected multiple units, the term *output circuit* also refers to the conductors used to electrically interconnect the fuel cell system(s).

Point of Common Coupling. The point at which the power production and distribution network and the customer interface occurs in an interactive system. Typically, this is the load side of the power network meter.

Stand-Alone System. A fuel cell system that supplies power independently of an electrical production and distribution network.

692.3 Other Articles.

Wherever the requirements of other articles of this *Code* and Article 692 differ, the requirements of Article 692 shall apply, and, if the system is operated in parallel with a primary source(s) of electricity, the requirements in 705.14, 705.16, 705.32, and 705.43 shall apply.

692.4 Installation.

(A) Fuel Cell System. A fuel cell system shall be permitted to supply a building or other structure in addition to any service(s) of another electricity supply system(s).

(B) Identification. A permanent plaque or directory, denoting all electrical power sources on or in the premises, shall be installed at each service equipment location.

692.6 Listing Requirement.

The fuel cell system shall be evaluated and listed for its intended application prior to installation.

Because the fuel cell system typically will be a component of the premises wiring system, to facilitate its implementation into commercial use, 692.6 requires that the system be evaluated and listed for its intended use.

II. Circuit Requirements

692.8 Circuit Sizing and Current.

(A) Nameplate Rated Circuit Current. The nameplate(s) rated circuit current shall be the rated current indicated on the fuel cell nameplate(s).

(B) Conductor Ampacity and Overcurrent Device Ratings. The ampacity of the feeder circuit conductors from the fuel cell system(s) to the premises wiring system shall not be less than the greater of (1) nameplate(s) rated circuit current or (2) the rating of the fuel cell system(s) overcurrent protective device(s).

(C) Ampacity of Grounded or Neutral Conductor. If interactive single-phase, 2-wire fuel cell output(s) is connected to the grounded or neutral conductor and a single ungrounded conductor of a 3-wire system or of a 3-phase, 4-wire wye-connected system, the maximum unbalanced neutral load current plus the fuel cell system(s) output rating shall not exceed the ampacity of the grounded or neutral conductor.

692.9 Overcurrent Protection.

(A) Circuits and Equipment. If the fuel cell system is provided with overcurrent protection sufficient to protect the circuit conductors that supply the load, additional circuit overcurrent devices shall not be required. Equipment and conductors connected to more than one electrical source shall be protected.

(B) Accessibility. Overcurrent devices shall be readily accessible.

692.10 Stand-Alone Systems.

The premises wiring system shall meet the requirements of this *Code* except as modified by 692.10(A), (B), and (C).

(A) Fuel Cell System Output. The fuel cell system output from a stand-alone system shall be permitted to supply ac power to the building or structure disconnecting means at current levels below the rating of that disconnecting means.

(B) Sizing and Protection. The circuit conductors between the fuel cell system(s) output and the building or structure disconnecting means shall be sized based on the output rating of the fuel cell system(s). These conductors shall be protected from overcurrents in accordance with 240.4. The overcurrent protection shall be located at the output of the fuel cell system(s).

(C) Single 120-Volt Nominal Supply. The inverter output of a stand-alone fuel cell system shall be permitted to supply 120 volts, nominal, to single-phase, 3-wire 120/240-volt service equipment or distribution panels where there are no 240-volt loads and where there are no multiwire branch circuits. In all installations, the rating of the overcurrent device connected to the output of the fuel cell system(s) shall be less than the rating of the service equipment. This equipment shall be marked as follows:

WARNING
SINGLE 120-VOLT SUPPLY. DO NOT CONNECT
MULTIWIRE BRANCH CIRCUITS!

III. Disconnecting Means

692.13 All Conductors.

Means shall be provided to disconnect all current-carrying conductors of a fuel cell system power source from all other conductors in a building or other structure.

692.14 Provisions.

The provisions of 225.31 and 225.33 through 225.40 shall apply to the fuel cell source disconnecting means. The disconnecting means shall not be required to be suitable as service equipment and shall be rated in accordance with 692.17.

692.17 Switch or Circuit Breaker.

The disconnecting means for ungrounded conductors shall consist of readily accessible, manually operable switch(es) or circuit breaker(s).

Where all terminals of the disconnecting means may be energized in the open position, a warning sign shall be mounted on or adjacent to the disconnecting means. The sign shall be clearly legible and shall have the following words or equivalent:

DANGER
ELECTRIC SHOCK HAZARD.
DO NOT TOUCH TERMINALS.
TERMINALS ON BOTH THE LINE AND
LOAD SIDES MAY BE ENERGIZED
IN THE OPEN POSITION.

IV. Wiring Methods

692.31 Wiring Systems.

All raceway and cable wiring methods included in Chapter 3 of this *Code* and other wiring systems and fittings specifically intended and identified for use with fuel cell systems shall be permitted. Where wiring devices with integral enclosures are used, sufficient length of cable shall be provided to facilitate replacement.

V. Grounding

692.41 System Grounding.

For a fuel cell system output circuit, one conductor of a 2-wire system rated over 50 volts and a neutral conductor of a 3-wire system shall be solidly grounded by either 692.41(A) or (B).

(A) Stand-Alone Systems. Grounding and bonding shall be in accordance with 250.30.

(B) Other Than Stand-Alone Systems.

(1) Two-Wire Systems. One conductor shall be terminated at the grounded circuit conductor terminal of the premises wiring system.

(2) Three-Wire Systems. The neutral conductor shall be terminated at the grounded circuit conductor terminal of the premises wiring system.

692.44 Equipment Grounding Conductor.

A separate equipment grounding conductor shall be installed.

692.45 Size of Equipment Grounding Conductor.

The equipment grounding conductor shall be sized in accordance with 250.122.

692.47 Grounding Electrode System.

Any supplementary grounding electrode(s) required by the manufacturer shall be connected to the equipment grounding conductor specified in 250.118.

VI. Marking

692.53 Fuel Cell Power Sources.

A marking specifying the fuel cell system, output voltage, output power rating, and continuous output current rating shall be provided at the disconnecting means for the fuel cell power source at an accessible location on the site.

692.54 Fuel Shut-Off.

The location of the manual fuel shut-off valve shall be at the location of the primary disconnecting means of the building or circuits supplied.

692.56 Stored Energy.

A fuel cell system that stores electrical energy shall require the following warning sign, or equivalent, at the location of the service disconnecting means of the premises:

WARNING
FUEL CELL POWER SYSTEM CONTAINS
ELECTRICAL ENERGY STORAGE DEVICES.

VII. Connection to Other Circuits

692.59 Transfer Switch.

A transfer switch shall be required in non–grid-interactive systems that use utility grid backup. The transfer switch shall maintain isolation between the electrical production and distribution network and the fuel cell system. The transfer switch shall be permitted to be located externally or internally to the fuel cell system unit. When the utility service conductors of the structure are connected to the transfer switch, the switch shall comply with Article 230, Part V.

692.60 Identified Interactive Equipment.

Only fuel cell systems listed and identified as interactive shall be permitted in interactive systems.

692.61 Output Characteristics.

The output of a fuel cell system operating in parallel with an electric supply system shall be compatible with the voltage, wave shape, and frequency of the system to which it is connected.

FPN: The term *compatible* does not necessarily mean matching the primary source wave shape.

692.62 Loss of Interactive System Power.

The fuel cell system shall be provided with a means of detecting when the electrical production and distribution network has become de-energized and shall not feed the electrical production and distribution network side of the point of common coupling during this condition. The fuel cell system shall remain in that state until the electrical production and distribution network voltage has been restored.

A normally interactive fuel cell system shall be permitted to operate as a stand-alone system to supply loads that have been disconnected from electrical production and distribution network sources.

692.64 Unbalanced Interconnections.

(A) Single Phase. Single-phase interactive fuel cell systems shall not be connected to a 3-phase power system unless the interactive system is designed so that significant unbalanced voltages cannot result.

(B) Three Phase. Three-phase interactive fuel cell systems shall have all phases automatically de-energized upon loss of voltage, or unbalance of voltage in one or more phases, unless the interactive system is designed so that significant unbalanced voltages will not result.

692.65 Point of Connection.

The output of a fuel cell system power source shall be connected as specified in 692.65(A) or (B).

(A) Supply Side. A fuel cell system power source shall be permitted to be connected to the supply side of the service disconnecting means as permitted in 230.82(5).

(B) Load Side. A fuel cell system power source shall be permitted to be connected to the load side of the service disconnecting means of the other source(s) at any distribution equipment on the premises, provided that all of the following conditions are met:

(1) Each source interconnection shall be made at a dedicated circuit breaker or fusible disconnecting means.
(2) The sum of the ampere ratings of overcurrent devices in circuits supplying power to a busbar or conductor shall not exceed the rating of the busbar or conductor.

Exception: For a dwelling unit, the sum of the ampere ratings of the overcurrent devices shall not exceed 120 percent of the rating of the busbar or conductor.

(3) The interconnection point shall be on the line side of all ground-fault protection equipment.
(4) Equipment containing overcurrent devices in circuits supplying power to a busbar or conductor shall be marked to indicate the presence of all sources.
(5) Equipment such as circuit breakers, if backfed, shall be identified for such operation.
(6) The circuit breaker on the dedicated output of a utility-interactive inverter shall be positioned in the distribution equipment at the opposite (load) end from the input feeder connection or main circuit location. A permanent warning label shall be applied to the distribution equipment with the following, or equivalent:

WARNING.
FUEL CELL POWER SYSTEM OUTPUT. DO NOT
RELOCATE THIS CIRCUIT BREAKER.

VIII. Outputs Over 600 Volts

692.80 General.

Fuel cell systems with a maximum output voltage over 600 volts ac shall comply with the requirements of other articles applicable to such installations.

ARTICLE 695
Fire Pumps

Contents

FPN: Rules that are followed by a reference in brackets contain text that has been extracted from NFPA 20-1999, *Standard for the Installation of Stationary Pumps for Fire Protection.* Only editorial changes were made to the extracted text to make it consistent with this *Code.*

695.1 Scope.

(A) Covered. This article covers the installation of the following:

(1) Electric power sources and interconnecting circuits
(2) Switching and control equipment dedicated to fire pump drivers

(B) Not Covered. This article does not cover the following:

(1) The performance, maintenance, and acceptance testing of the fire pump system, and the internal wiring of the components of the system
(2) Pressure maintenance (jockey or makeup) pumps

FPN: See NFPA 20-1999, *Standard for the Installation of Stationary Pumps for Fire Protection,* for further information.

Article 695 has been extensively revised since its first inclusion in the *Code* in 1996. These revisions involve some important changes, for example, provisions to allow for power sources that do not directly originate from a utility. These revisions also harmonize the *NEC* and NFPA 20, *Standard for the Installation of Stationary Pumps for Fire Protection.* Another major issue that has been addressed is that of providing power for electric fire pumps installed in multibuilding campus-style arrangements such as those found in institutional and industrial settings.

Testing and performance criteria for a fire pump system remain within the domain of NFPA 20.

Article 695 does not apply to pumps used to supply sprinkler systems in one- and two-family dwellings. NFPA 13D, *Standard for the Installation of Sprinkler Systems in One- and Two-Family Dwellings and Manufactured Homes,* does not require the use of a listed pump; thus, neither NFPA 20 nor Article 695 is applicable.

Although jockey pumps are not covered by Article 695, they are permitted to be connected to the fire pump feeder.

695.2 Definitions.

Fault Tolerant External Control Conductors. Those control conductors entering and/or leaving the fire pump controller enclosure, which if broken, disconnected, or shorted will not prevent the controller from starting the fire pump from all other internal or external means and may cause the controller to start the pump under these conditions.

On-Site Power Production Facility. The normal supply of electric power for the site that is expected to be constantly producing power.

On-Site Standby Generator. A facility producing electric power on site as the alternate supply of electric power. It differs from an on-site power production facility in that it is not constantly producing power.

695.3 Power Source(s) for Electric Motor-Driven Fire Pumps.

Electric motor-driven fire pumps shall have a reliable source of power.

The power source for an electric motor-driven fire pump must be reliable and have adequate capacity to carry the locked-rotor currents of the fire pump motor and accessory equipment. These two main requirements ensure that the fire pump will operate in the event of a fire without being accidentally disconnected and that the fire pump will continue to operate until the fire is extinguished, the fire pump is purposely shut down, or the pump itself is destroyed.

One or more of the following three basic types of power sources are permitted. If only one source is used, that source must be capable of supplying locked-rotor current to the fire pump and accessory equipment. Power may be supplied by the following:

1. A separate utility service or connection ahead of the main disconnecting means
2. An on-site power production system
3. Multiple sources (two or more feeders) on a multibuilding campus where utility or power produced on site is not reasonably available and this method of power supply is approved by the authority having jurisdiction

If none of these options is individually capable of providing reliable power with adequate capacity, a combination (two or more) of these sources or a combination of one or more of these sources with an on-site standby generator must be used.

(A) Individual Sources. Where reliable, and where capable of carrying indefinitely the sum of the locked-rotor current of the fire pump motor(s) and the pressure maintenance pump motor(s) and the full-load current of the associated fire pump accessory equipment when connected to this power supply, the power source for an electric motor-driven fire pump shall be one or more of the following.

(1) Electric Utility Service Connection. A fire pump shall be permitted to be supplied by a separate service, or by a tap located ahead of and not within the same cabinet, enclosure, or vertical switchboard section as the service discon-

necting means. The connection shall be located and arranged so as to minimize the possibility of damage by fire from within the premises and from exposing hazards. A tap ahead of the service disconnecting means shall comply with 230.82(4). The service equipment shall comply with the labeling requirements in 230.2 and the location requirements in 230.72(B).

(2) On-Site Power Production Facility. A fire pump shall be permitted to be supplied by an on-site power production facility. The source facility shall be located and protected to minimize the possibility of damage by fire. [NFPA 20, 6.2.1, 6.2.2, 6.2.4.4]

Examples of power sources are illustrated in Exhibits 695.1 and 695.2.

The determination of whether the serving electric utility is a reliable source of power is an issue for the authority having jurisdiction. Appendix A of NFPA 20, *Standard for the Installation of Stationary Pumps for Fire Protection*, provides guidance for evaluating the reliability of the power supply.

Performance requirements for an alternative source of electric power can be found in NFPA 110, *Standard for Emergency and Standby Power Systems*.

Most important in 695.3(A)(2) is the requirement that a fire in one source not affect the reliability of another source. For example, if an emergency feeder were physically routed above the normal switchboard and a fire occurred in the normal switchboard, the reliability of the emergency feeder during the fire would be questionable. There are many other examples for such interruption scenarios. The requirements for fire protection of the fire pump supply conductors are located in 695.6(A) and (B).

(B) Multiple Sources. Where reliable power cannot be obtained from a source described in 695.3(A), power shall be supplied from an approved combination of two or more of either of such sources, or from an approved combination of feeders constituting two or more power sources as covered in 695.3(B)(2), or from an approved combination of one or more of such power sources in combination with an on-site standby generator complying with 695.3(B)(1) and (B)(3).

(1) Generator Capacity. An on-site generator(s) used to comply with this section shall be of sufficient capacity to allow normal starting and running of the motor(s) driving the fire pump(s) while supplying all other simultaneously operated load. Automatic shedding of one or more optional standby loads in order to comply with this capacity requirement shall be permitted. A tap ahead of the on-site generator disconnecting means shall not be required. The requirements of 430.113 shall not apply.

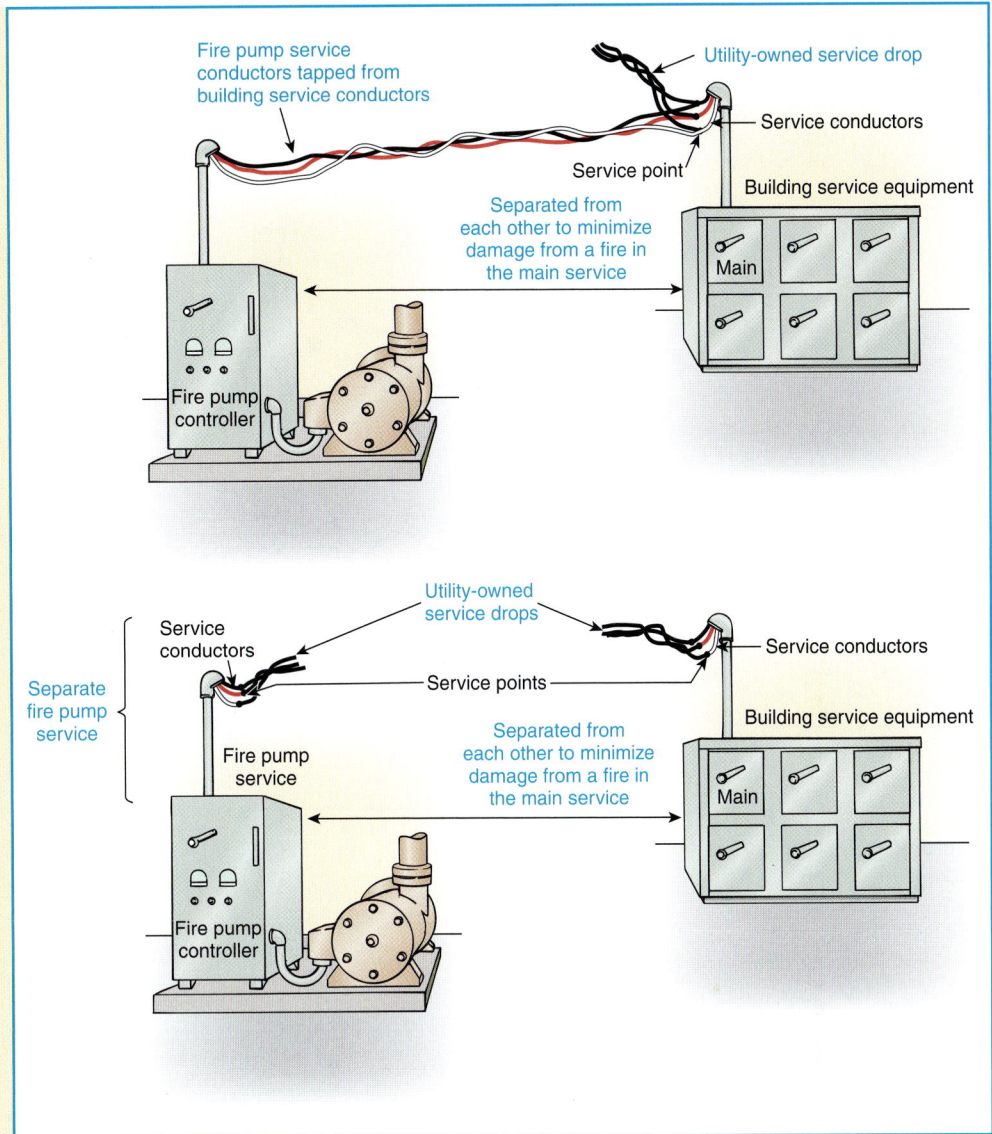

Exhibit 695.1 Two permitted configurations for connecting to electric utility-owned service drops.

Where the alternative source of power is an on-site generator, the alternative source disconnecting means and the alternative source overcurrent protective device(s) for the electric-drive fire pump are not required to be sized for locked-rotor current of the fire pump motor(s). Rather, the circuit components of the alternative source are permitted to be sized according to Article 430, provided they are "selected or set to allow instantaneous pickup and running of the fire pump load." See 445.12(A).

(2) Feeder Sources. This section applies to multibuilding campus-style complexes with fire pumps at one or more buildings. Where sources in 695.3(A) are not practicable, and with the approval of the authority having jurisdiction, two or more feeder sources shall be permitted as one power source or as more than one power source where such feeders are connected to or derived from separate utility services. The connection(s), overcurrent protective device(s), and disconnecting means for such feeders shall meet the requirements of 695.4(B).

The requirements of 695.3(B)(2) were an important revision to the 1999 *Code.* This change permits the use of feeder sources for campus-style applications. Chapter 6 of NFPA 20, *Standard for the Installation of Stationary Pumps for Fire Protection,* permits the use of a reliable feeder to supply a fire pump if it is acceptable to the authority having jurisdiction. See Exhibit 695.3. In NFPA 20, the use of a reliable feeder as a power supply for a fire pump is based on performance criteria, such as the following:

Exhibit 695.2 On-site power production facility as a power source for a fire pump installation.

On-site power production facility located where it is protected from fire, such as outdoors or in a room of fire-rated construction

Generator

Fire pump controller

Fire pump supply conductors

Exhibit 695.3 Multiple feeder sources for campus-style application, as described in 695.3(B)(2).

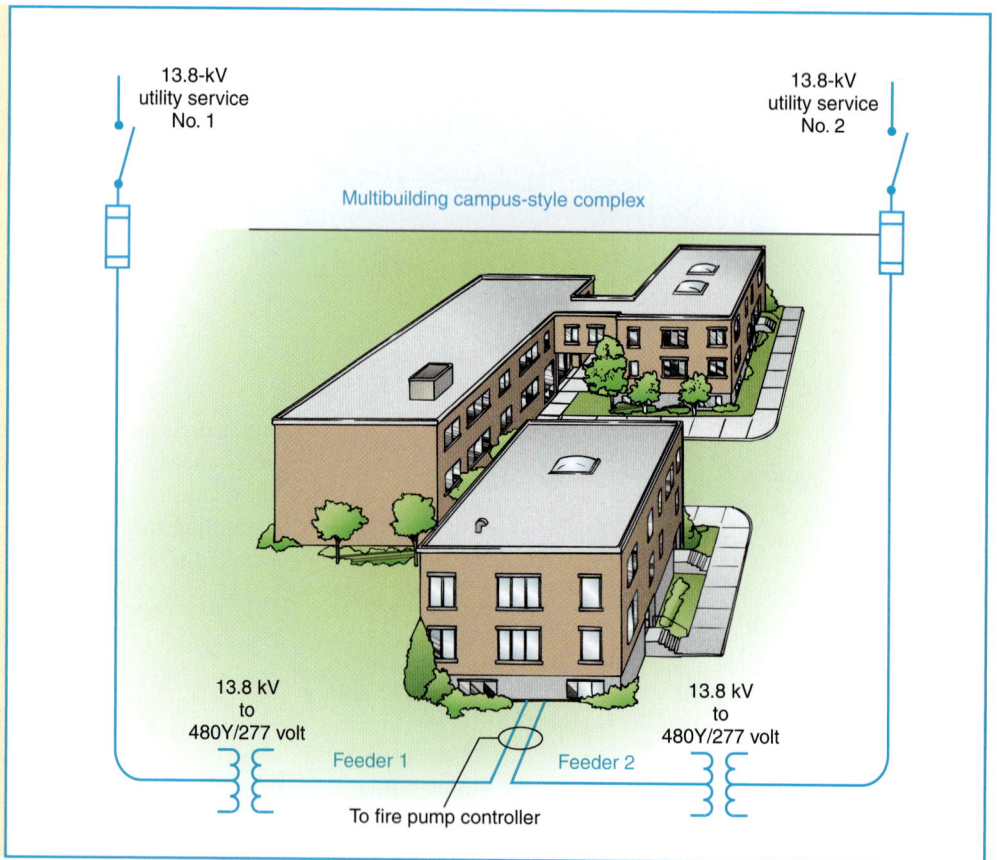

13.8-kV utility service No. 1

13.8-kV utility service No. 2

Multibuilding campus-style complex

13.8 kV to 480Y/277 volt

13.8 kV to 480Y/277 volt

Feeder 1

Feeder 2

To fire pump controller

1. Redundant power supply features
2. Special requirements for generators
3. Special requirements for transfer switches
4. Generator overcurrent protection

(3) Arrangement. The power sources shall be arranged so that a fire at one source will not cause an interruption at the other source. [NFPA 20, 6.2.3, 6.2.4.1, 6.2.4.3, 6.6.1]

695.4 Continuity of Power.

Circuits that supply electric motor-driven fire pumps shall be supervised from inadvertent disconnection as covered in 695.4(A) or (B).

(A) Direct Connection. The supply conductors shall directly connect the power source to either a listed fire pump controller or listed combination fire pump controller and power transfer switch.

(B) Supervised Connection. A single disconnecting means and associated overcurrent protective device(s) shall be permitted to be installed between a remote power source and one of the following:

(1) A listed fire pump controller
(2) A listed fire pump power transfer switch
(3) A listed combination fire pump controller and power transfer switch

For systems installed under the provisions of 695.3(B)(2) only, such additional disconnecting means and associated overcurrent protective device(s) shall be permitted as required to comply with other provisions of this *Code*. Overcurrent protective devices between an on-site standby generator and a fire pump controller shall be selected and sized according to 430.62 to provide short-circuit protection only. All disconnecting devices and overcurrent protective devices that are unique to the fire pump loads shall comply with 695.4(B)(1) through (B)(4).

(1) Overcurrent Device Selection. The overcurrent protective device(s) shall be selected or set to carry indefinitely the sum of the locked-rotor current of the fire pump motor(s) and the pressure maintenance pump motor(s) and the full-load current of the associated fire pump accessory equipment when connected to this power supply.

(2) Disconnecting Means. The disconnecting means shall comply with all of the following:

(1) Be identified as suitable for use as service equipment
(2) Be lockable in the closed position
(3) Be located sufficiently remote from other building or other fire pump source disconnecting means such that

inadvertent contemporaneous operation would be unlikely

(3) Disconnect Marking. The disconnecting means shall be marked "Fire Pump Disconnecting Means." The letters shall be at least 25 mm (1 in.) in height, and they shall be visible without opening enclosure doors or covers.

(4) Controller Marking. A placard shall be placed adjacent to the fire pump controller, stating the location of this disconnecting means and the location of the key (if the disconnecting means is locked).

Typically, the supply conductors of the power source are directly connected, without a disconnecting means or overcurrent protection, to the fire pump controller or combination fire pump controller and power transfer switch. However, a single disconnecting means is allowed, provided all the conditions in 695.4(B), Supervised Connection, are met.

In addition to the fire pump controller, where a separate service disconnect with overcurrent protection is provided for the fire pump circuit, the disconnecting means will be as follows:

1. The overcurrent device is sized to carry the locked-rotor currents of all the fire pump motors.
2. The fire pump disconnection means is located away from the other service disconnecting means. This disconnecting means is rated as service equipment and is lockable in the on position.
3. The disconnect and controller is marked and placarded as described.
4. The circuit is supervised in the closed position, in accordance with 695.4(B)(5).

Specifically, the size of the overcurrent protective device is the sum of the locked-rotor currents of all the permitted motors plus the sum of any other fire pump auxiliary loads.

Example

A fusible service disconnect switch supplies power to a 100-hp, 460-volt, 3-phase fire pump and to a 1½-hp, 460-volt, 3-phase jockey pump. The fire pump feeder circuit will be installed in a raceway between the disconnecting means and fire pump controller. The raceway is considered outside of the building per 230.6. Using the requirements of Article 695, determine the sizes of the disconnecting means and overcurrent protection device. Also determine the minimum ampacity of the feeder conductors.

Solution

STEP 1. Determine the minimum ratings of the disconnecting means and the overcurrent protective device. According to the motor nameplates, the locked-rotor current

(LRC) is 725 amperes for the 100-hp motor and 20 amperes for the 1½-hp motor. If the locked-rotor amperes are not on the nameplates, the locked-rotor currents found in Table 430.151(B) must be used. Calculate the size by summing the locked-rotor currents of both motors and then going to the next larger standard-size overcurrent device, as follows:

$$100\text{-hp, 3-phase LRC} = 725 \text{ A}$$
$$1\frac{1}{2}\text{-hp, 3-phase LRC} = \underline{\quad 20 \text{ A}}$$
$$\text{Total LRC} = 745 \text{ A}$$

The next larger standard-size disconnect switch and overcurrent device is 800 amperes. An adjustable-trip circuit breaker of 750 amperes is also permitted, because it, too, will carry the locked-rotor current indefinitely.

STEP 2. Determine the minimum ampacity for the fire pump feeder conductor. Even though the disconnect switch, fuse, and circuit breakers are sized according to locked-rotor currents, the feeder conductors to the fire pump and associated equipment are required to have an ampacity not less than 125 percent of the full-load current (FLC) rating of the fire pump motor(s) and pressure maintenance pump motor(s), plus 100 percent of associated accessory equipment. Using the same motors as above, calculate the size of the feeder to the fire pump controller as follows, using 430.6(A)(1) and Table 430.150 for the full-load currents of the motors:

100-hp, 3-phase FLC

$$(124 \text{ A} \times 1.25) = 155.00 \text{ A}$$

1½-hp, 3-phase FLC

$$(3 \text{ A} \times 1.25) = \underline{\quad 3.75 \text{ A}}$$

$$\text{Total FLC} = 158.75 \text{ A} \quad \text{or} \quad 159 \text{ A}$$

Thus, the minimum ampacity for the feeder conductors is 159 amperes. Using the 75°C column from Table 310.16, a 2/0 copper conductor is the minimum size required, per 110.14(C)(1)(b).

(5) Supervision. The disconnecting means shall be supervised in the closed position by one of the following methods:

(1) Central station, proprietary, or remote station signal device
(2) Local signaling service that causes the sounding of an audible signal at a constantly attended point
(3) Locking the disconnecting means in the closed position
(4) Sealing of disconnecting means and approved weekly recorded inspections when the disconnecting means are located within fenced enclosures or in buildings under the control of the owner [NFPA 20, 6.3.2.2.1, 6.3.2.2.2, 6.3.2.2.3]

Supervision of the disconnecting means by a local (protected premises) fire alarm system, central station, proprietary supervising station, or remote supervising station requires a connection to the premises fire alarm system. The connection is generally made through the use of dry contacts in the fire pump controller that are connected to a fire alarm system initiating device circuit. This circuit is programmed to generate a supervisory signal at the fire alarm control unit on loss of voltage to the fire pump controller. A supervisory signal indicates that the suppression system is "off-normal." Connection of this device must cause a supervisory signal and may not cause a trouble or alarm signal. For more information, see *NFPA 72, National Fire Alarm Code.*

695.5 Transformers.

Where the service or system voltage is different from the utilization voltage of the fire pump motor, transformer(s) protected by disconnecting means and overcurrent protective devices shall be permitted to be installed between the system supply and the fire pump controller in accordance with 695.5(A) and (B), or (C). Only transformers covered in 695.5(C) shall be permitted to supply loads not directly associated with the fire pump system.

(A) Size. Where a transformer supplies an electric motor-driven fire pump, it shall be rated at a minimum of 125 percent of the sum of the fire pump motor(s) and pressure maintenance pump(s) motor loads, and 100 percent of the associated fire pump accessory equipment supplied by the transformer.

(B) Overcurrent Protection. The primary overcurrent protective device(s) shall be selected or set to carry indefinitely the sum of the locked-rotor current of the fire pump motor(s) and the pressure maintenance pump motor(s) and the full-load current of the associated fire pump accessory equipment when connected to this power supply. Secondary overcurrent protection shall not be permitted.

See Exhibit 695.4.

Dedicated transformer and overcurrent protection sizing can be broken down into three basic requirements. Generally stated, they are as follows:

1. The transformer must be sized to at least 125 percent of the sum of the loads.
2. The transformer primary overcurrent device must be at least a specified minimum size.
3. The transformer secondary must not contain any overcurrent devices whatsoever.

Example

A 4160/480-volt, 3-phase, dedicated transformer supplies power to a 100-hp, 460-volt, 3-phase, code letter G fire

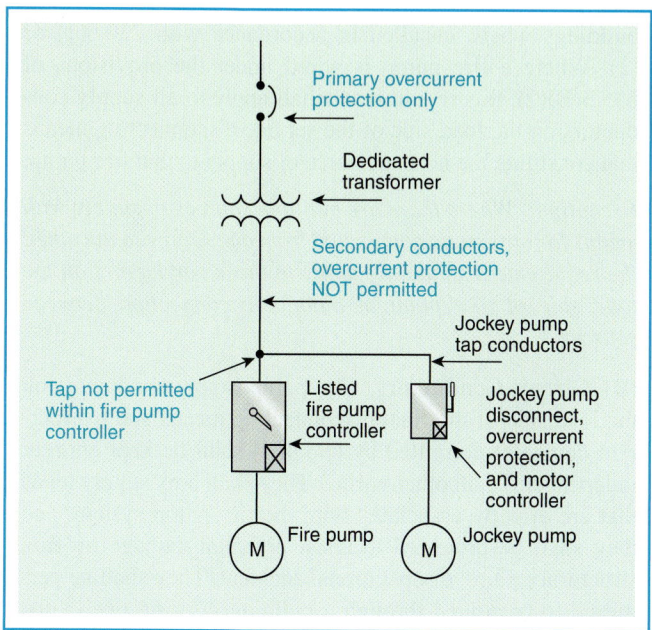

Exhibit 695.4 The overcurrent device in the primary of a transformer supplying a fire pump installation. The device is required to be sized to carry the locked-rotor current motor(s) and associated fire pump accessory equipment indefinitely.

pump and to a 1½-hp, 460-volt, 3-phase, code letter H jockey pump. Using the requirements of Article 695, determine the sizes of the dedicated transformer and its primary overcurrent protection.

Solution

STEP 1. Determine the minimum standard-size transformer. According to 695.5(A), to determine the minimum current value to use in the 3-phase power calculation, add the full-load currents of the fire pump motor(s) and jockey pump motor(s), then increase the total to 125 percent. The result of this calculation provides the minimum kVA rating of a transformer dedicated to a fire pump installation. Any transformer with a kVA rating equal to or greater than the value determined by this calculation is acceptable. First, add the full-load currents of the two motors, using the full-load current (FLC) values from Table 430.150, as follows:

$$
\begin{aligned}
\text{100-hp, 3-phase FLC} &= 124 \text{ A} \\
\text{1½-hp, 3-phase FLC} &= \underline{\ \ 3 \text{ A}} \\
\text{Total FLC} &= 127 \text{ A}
\end{aligned}
$$

Now, increase the sum of the fire pump motor and the jockey pump motor to 125 percent, as follows:

$$127 \text{ A} \times 1.25 = 158.75 \text{ A}$$

Then, size the transformer as follows:

$$\text{Transformer kVA} = \frac{\text{volts} \times \text{amperes} \times \sqrt{3}}{1000}$$

$$\text{Transformer kVA} = \frac{480 \times 158.75 \times \sqrt{3}}{1000}$$

$$= 131.98 \text{ kVA}$$

Thus, the minimum-size transformer permitted by 695.5(A) is 131.98 kVA. The next larger standard-size transformer available is 150 kVA, but any larger size is permitted.

STEP 2. Calculate the minimum-size primary overcurrent protection device permitted for this transformer. According to 695.5(B), the minimum primary overcurrent protection device must allow the transformer secondary to supply the locked-rotor current to the fire pump and, in this case, the jockey pump. The locked-rotor current (LRC) of each motor must be individually calculated if it is not available on the motor nameplate. In this example, however, we are assuming that only the kVA code letters are available. According to 430.7(B) and using the maximum values for the individual code letters per Table 430.7(B), calculate the maximum locked-rotor currents, as follows.

For the 100-hp motor, code letter G:

$$\text{LRC} = \text{motor hp} \times \text{max. code letter value}$$

$$\times \frac{1000}{\text{motor voltage} \times \text{3-phase factor}}$$

$$\text{LRC} = 100 \text{ hp} \times 6.29 \frac{\text{kVA}}{\text{hp}} \times \frac{1000}{460 \times \sqrt{3}} = 789.49 \text{ A}$$

For the 1½-hp motor, code letter H (using the same formula):

$$\text{LRC} = 1\tfrac{1}{2}\text{-hp} \times 7.09 \frac{\text{kVA}}{\text{hp}} \times \frac{1000}{460 \times \sqrt{3}} = 13.35 \text{ A}$$

For the total LRC:

$$
\begin{aligned}
\text{100-hp LRC} &= 789.49 \text{ A} \\
\text{1½-hp LRC} &= \underline{\ \ 13.35 \text{ A}} \\
\text{Total LRC} &= 802.84 \text{ A} \quad \text{or} \quad 803 \text{ A}
\end{aligned}
$$

Now, calculate the equivalent locked-rotor current on the primary side of the transformer, based on the calculated locked-rotor current of the secondary of the transformer, as follows:

$$\text{LRC}_{\text{primary}} = \frac{\text{secondary voltage}}{\text{primary voltage}} \times \text{LRC}_{\text{secondary}}$$

$$= \frac{480 \text{ V}}{4160 \text{ V}} \times 803 \text{ A}$$

$$= 92.65 \text{ A} \quad \text{or} \quad 93 \text{ A}$$

This value of 93 amperes represents the secondary locked-rotor current reflected to the primary side of the transformer. Because this value is the absolute smallest overcurrent protective device permitted, the next larger standard size, according to 240.6, is 100 amperes. Thus, the minimum standard-size overcurrent protective device is 100 amperes.

Conclusion: The calculation for a 4160/480-volt, 3-phase transformer supplying a 100-hp fire pump and a 1½-hp jockey pump, both at 460 volts, 3 phase, can be summarized as follows:

1. The smallest standard-size transformer that is permitted is 150 kVA.
2. The smallest standard-size overcurrent protective device permitted on the primary of the transformer is 100 amperes.
3. A secondary overcurrent protective device is not permitted.

(C) Feeder Source. Where a feeder source is provided in accordance with 695.3(B)(2), transformers supplying the fire pump system shall be permitted to supply other loads. All other loads shall be calculated in accordance with Article 220, including demand factors as applicable.

(1) Size. Transformers shall be rated at a minimum of 125 percent of the sum of the fire pump motor(s) and pressure maintenance pump(s) motor loads, and 100 percent of the remaining load supplied by the transformer.

(2) Overcurrent Protection. The transformer size, the feeder size, and the overcurrent protective device(s) shall be coordinated such that overcurrent protection is provided for the transformer in accordance with 450.3 and for the feeder in accordance with 215.3, and such that the overcurrent protective device(s) is selected or set to carry indefinitely the sum of the locked-rotor current of the fire pump motor(s), the pressure maintenance pump motor(s), the full-load current of the associated fire pump accessory equipment, and 100 percent of the remaining loads supplied by the transformer.

695.6 Power Wiring.

Power circuits and wiring methods shall comply with the requirements in 695.6(A) through (G), and as permitted in 230.90(A), Exception No. 4; 230.94, Exception No. 4; 230.95, Exception No. 2; 240.13; 230.208; 240.4(A); and 430.31.

(A) Service Conductors. Supply conductors shall be physically routed outside a building(s) and shall be installed as service entrance conductors in accordance with Article 230. Where supply conductors cannot be physically routed outside buildings, they shall be permitted to be routed through

buildings where installed in accordance with 230.6(1) or (2). Where a fire pump is wired under the provisions of 695.3(B)(2), this requirement shall apply to all supply conductors on the load side of the service disconnecting means that constitute the normal source of supply to that fire pump.

Exception: Where there are multiple sources of supply with means for automatic connection from one source to the other, the requirement shall only apply to those conductors on the load side of that point of automatic connection between sources.

(B) Circuit Conductors. Fire pump supply conductors on the load side of the final disconnecting means and overcurrent device(s) permitted by 695.4(B) shall be kept entirely independent of all other wiring. They shall only supply loads that are directly associated with the fire pump system, and they shall be protected to resist potential damage by fire, structural failure, or operational accident. They shall be permitted to be routed through a building(s) using one of the following methods:

It is important to understand the difference between a 1-hour fire rating of an electrical circuit, such as a conduit with wires, and a 1-hour fire-resistance rating of a structural member, such as a wall. Simply stated, at the end of a 1-hour fire test on an electrical conduit with wires, the circuit must function electrically (no short circuits, grounds, or opens are permitted). The circuit and its insulation must be intact and electrically functioning. A wall subjected to a 1-hour fire-resistance test must only prevent a fire from passing through or past the wall, without regard to damage to the wall. All fire ratings and fire-resistance ratings are based on the assumption that the structural supports for the assembly are not impaired by the effects of the fire.

The UL 2001 *Fire Resistance Directory, Volume 2,* describes three categories of products that can be used in the fire protection of electrical circuits for fire pumps: Various Electrical Circuit Protective Systems (FHIT), Electrical Circuit Protective Materials (FHIY), and Fire Resistive Cables (FHJR). (The four-letter codes in parentheses are the UL product category guide designations.) For information on electrical circuit protective systems, see UL Subject 1724, *Fire Tests for Electrical Circuit Protective Systems.*

(1) Be encased in a minimum 50 mm (2 in.) of concrete
(2) Be within an enclosed construction dedicated to the fire pump circuit(s) and having a minimum of a 1-hour fire resistive rating
(3) Be a listed electrical circuit protective system with a minimum 1-hour fire rating

Exception: The supply conductors located in the electrical equipment room where they originate and in the fire pump room shall not be required to have the minimum 1-hour fire separation or fire resistance rating, unless otherwise required by 700.9(D) of this Code.

(C) Conductor Size.

(1) Fire Pump Motors and Other Equipment. Conductors supplying a fire pump motor(s), pressure maintenance pumps, and associated fire pump accessory equipment shall have a rating not less than 125 percent of the sum of the fire pump motor(s) and pressure maintenance motor(s) full-load current(s), and 100 percent of the associated fire pump accessory equipment.

(2) Fire Pump Motors Only. Conductors supplying only a fire pump motor(s) shall have a rating not less than 125 percent of the fire pump motor(s) full-load current(s).

(D) Overload Protection. Power circuits shall not have automatic protection against overloads. Except as provided in 695.5(C)(2), branch-circuit and feeder conductors shall be protected against short circuit only. Where a tap is made to supply a fire pump, and the tap wiring is run in accordance with 230.6, the applicable distance and size restrictions in 240.21 shall not apply.

The requirements for ground-fault protection are not permitted to apply to the fire pump power wiring. [Ground-fault protection (GFP) is equipment protection and should not be confused with GFCI protection for personnel.] See 230.95, Exception No. 2, and 240.13(3).

Exception No. 1: Conductors between storage batteries and the engine shall not require overcurrent protection or disconnecting means.

Exception No. 2: For on-site standby generator(s) that produce continuous currents in excess of 225 percent of the full-load amperes of the fire pump motor, the conductors between the on-site generator(s) and the combination fire pump transfer switch controller or separately mounted transfer switch shall be installed in accordance with 695.6(B) or protected in accordance with 430.52.

The protection provided shall be in accordance with the short-circuit current rating of the combination fire pump transfer switch controller or separately mounted transfer switch.

(E) Pump Wiring. All wiring from the controllers to the pump motors shall be in rigid metal conduit, intermediate metal conduit, liquidtight flexible metal conduit, or liquidtight flexible nonmetallic conduit Type LFNC-B, or Type MI cable.

The requirement of 695.6(E) does not apply to light switches, convenience receptacles, telephone outlets, fire detectors, and similar equipment located in the fire pump room.

(F) Junction Points. Where wire connectors are used in the fire pump circuit, the connectors shall be listed. A fire pump controller or fire pump power transfer switch, where provided, shall not be used as a junction box to supply other equipment, including a pressure maintenance (jockey) pump(s). A fire pump controller and fire pump power transfer switch, where provided, shall not serve any load other than the fire pump for which it is intended.

(G) Mechanical Protection. All wiring from engine controllers and batteries shall be protected against physical damage and shall be installed in accordance with the controller and engine manufacturer's instructions.

695.7 Voltage Drop.

The voltage at the controller line terminals shall not drop more than 15 percent below normal (controller-rated voltage) under motor starting conditions. The voltage at the motor terminals shall not drop more than 5 percent below the voltage rating of the motor when the motor is operating at 115 percent of the full-load current rating of the motor.

Exception: This limitation shall not apply for emergency run mechanical starting. [NFPA 20, 6.4]

695.10 Listed Equipment.

Diesel engine fire pump controllers, electric fire pump controllers, electric motors, fire pump power transfer switches, foam pump controllers, and limited service controllers shall be listed for fire pump service. [NFPA 20, 6.5.1.1, 7.1.2.1, 9.1.1.1]

This requirement parallels those provisions in NFPA 20, *Standard for the Installation of Stationary Pumps for Fire Protection,* that require components used in fire pump systems to be listed.

695.12 Equipment Location.

(A) Controllers and Transfer Switches. Electric motor-driven fire pump controllers and power transfer switches shall be located as close as practicable to the motors that they control and shall be within sight of the motors.

(B) Engine-Drive Controllers. Engine-drive fire pump controllers shall be located as close as is practical to the engines that they control and shall be within sight of the engines.

(C) Storage Batteries. Storage batteries for diesel engine drives shall be rack supported above the floor, secured against displacement, and located where they will not be subject to physical damage, flooding with water, excessive temperature, or excessive vibration.

(D) Energized Equipment. All energized equipment parts shall be located at least 300 mm (12 in.) above the floor level.

(E) Protection Against Pump Water. Fire pump controllers and power transfer switches shall be located or protected so that they will not be damaged by water escaping from pumps or pump connections.

(F) Mounting. All fire pump control equipment shall be mounted in a substantial manner on noncombustible supporting structures.

Neither NFPA 20, *Standard for the Installation of Stationary Pumps for Fire Protection,* nor this *Code* mandates a dedicated room for the fire pump. However, NFPA 20 does specify a suitable space for this equipment.

The phrase "as close as practicable" may require that additional space be available to achieve the minimum maintenance working space set forth in 110.26.

Generally, fire pump controllers are housed in substantial enclosures suitable to protect the contents against limited amounts of falling water and dirt. In addition, all energized parts in the enclosure must be mounted at least 12 in. above the floor. Typically, the floor space for this area is equipped with a floor drain.

The requirement of 695.12(F) does not permit fire pump control equipment to be mounted on combustible backboards (such as plywood).

695.14 Control Wiring.

(A) Control Circuit Failures. External control circuits that extend outside the fire pump room shall be arranged so that failure of any external circuit (open or short circuit) shall not prevent the operation of a pump(s) from all other internal or external means. Breakage, disconnecting, shorting of the wires, or loss of power to these circuits could cause continuous running of the fire pump but shall not prevent the controller(s) from starting the fire pump(s) due to causes other than these external control circuits. All control conductors within

the fire pump room that are not fault tolerant shall be protected against physical damage. [NFPA 20, 7.5.2.5]

(B) Sensor Functioning. No undervoltage, phase-loss, frequency-sensitive, or other sensor(s) shall be installed that automatically or manually prohibit actuation of the motor contactor. [NFPA 20, 7.4.5.6]

Exception: A phase loss sensor(s) shall be permitted only as a part of a listed fire pump controller.

(C) Remote Device(s). No remote device(s) shall be installed that will prevent automatic operation of the transfer switch. [NFPA 20, 7.8.1.3]

(D) Engine-Drive Control Wiring. All wiring between the controller and the diesel engine shall be stranded and sized to continuously carry the charging or control currents as required by the controller manufacturer. Such wiring shall be protected against physical damage. Controller manufacturer's specifications for distance and wire size shall be followed. [NFPA 20, 9.3.5.1]

(E) Electric Fire Pump Control Wiring Methods. All electric motor-driven fire pump control wiring shall be in rigid metal conduit, intermediate metal conduit, liquidtight flexible metal conduit, liquidtight flexible nonmetallic conduit Type B (LFNC-B), or Type MI cable.

The wiring methods described in 695.14(E) apply only to the control wiring for electric motor-driven fire pumps. They do not apply to the control wiring for engine-driven fire pumps, because there are no similar requirements in NFPA 20, *Standard for the Installation of Stationary Pumps for Fire Protection.*

(F) Generator Control Wiring Methods. Control conductors installed between the fire pump power transfer switch and the standby generator supplying the fire pump during normal power loss shall be kept entirely independent of all other wiring. They shall be protected to resist potential damage by fire or structural failure. They shall be permitted to be routed through a building(s) encased in 50 mm (2 in.) of concrete or within enclosed construction dedicated to the fire pump circuits and having a minimum 1-hour fire resistance rating, or circuit protective systems with a minimum of 1-hour fire resistance. The installation shall comply with any restrictions provided in the listing of the electrical circuit protective system used.

Special Conditions

ARTICLE 700
Emergency Systems

Contents

I. General

700.1 Scope.

The provisions of this article apply to the electrical safety of the installation, operation, and maintenance of emergency systems consisting of circuits and equipment intended to supply, distribute, and control electricity for illumination, power, or both, to required facilities when the normal electrical supply or system is interrupted.

Emergency systems are those systems legally required and classed as emergency by municipal, state, federal, or other codes, or by any governmental agency having jurisdiction. These systems are intended to automatically supply illumination, power, or both, to designated areas and equipment in the event of failure of the normal supply or in the event of accident to elements of a system intended to supply, distribute, and control power and illumination essential for safety to human life.

Emergency systems are designed and installed to maintain a specific degree of illumination or provide power for essential equipment if the normal power supply fails. Examples of essential equipment include fire pumps and operating room and life-support equipment in hospitals.

Article 700 applies to the installation of emergency systems that are essential for safety to human life and are legally required by municipal, state, federal, or other codes or by a governmental agency having jurisdiction.

Article 700 does not dictate whether emergency systems are required or where emergency or exit lights should be located. These determinations may be made by using NFPA 101®, Life Safety Code®.

If authorities determine that emergency lighting, including the proper placement of exit signs, is required for safe egress from various classes of buildings or parts of buildings, then corridors, passageways, stairways, lobbies, and so on must also be sufficiently illuminated.

FPN No. 1: For further information regarding wiring and installation of emergency systems in health care facilities, see Article 517.

FPN No. 2: For further information regarding performance and maintenance of emergency systems in health care facilities, see NFPA 99-1999, *Standard for Health Care Facilities*.

FPN No. 3: Emergency systems are generally installed in places of assembly where artificial illumination is required for safe exiting and for panic control in buildings subject to occupancy by large numbers of persons, such as hotels, theaters, sports arenas, health care facilities, and similar institutions. Emergency systems may also provide power for such functions as ventilation where essential to maintain life, fire detection and alarm systems, elevators, fire pumps, public safety communications systems, industrial processes where current interruption would produce serious life safety or health hazards, and similar functions.

FPN No. 4: For specification of locations where emergency lighting is considered essential to life safety, see NFPA *101*®-2000, *Life Safety Code*®.

FPN No. 5: For further information regarding performance of emergency and standby power systems, see NFPA 110-1999, *Standard for Emergency and Standby Power Systems.*

700.2 Application of Other Articles.

Except as modified by this article, all applicable articles of this *Code* shall apply.

700.3 Equipment Approval.

All equipment shall be approved for use on emergency systems.

700.4 Tests and Maintenance.

(A) Conduct or Witness Test. The authority having jurisdiction shall conduct or witness a test of the complete system upon installation and periodically afterward.

(B) Tested Periodically. Systems shall be tested periodically on a schedule acceptable to the authority having jurisdiction to ensure the systems are maintained in proper operating condition.

(C) Battery Systems Maintenance. Where battery systems or unit equipments are involved, including batteries used for starting, control, or ignition in auxiliary engines, the authority having jurisdiction shall require periodic maintenance.

(D) Written Record. A written record shall be kept of such tests and maintenance.

(E) Testing Under Load. Means for testing all emergency lighting and power systems during maximum anticipated load conditions shall be provided.

> FPN: For testing and maintenance procedures of emergency power supply systems (EPSSs), see NFPA 110-1999, *Standard for Emergency and Standby Power Systems.*

Emergency system testing may be broken down into two general categories — acceptance testing and operational testing. Section 700.4 requires both types of testing as well as written records of both types of testing and maintenance.

Acceptance testing is performed after the emergency system has been installed but before the system is used. Acceptance testing ensures that the emergency system meets or exceeds the original installation specification. Portable load banks may be used for the acceptance testing of the system to maximum design load.

Operational testing, which is performed during the life of the system, ensures that the emergency system remains functional and that maintenance has been performed adequately. One method of operational testing is running the generating system to power the load of the facility. Generally, actual emergency system loads are smaller than the design capacity of the emergency generator system. Actual peak loads of the emergency system should be kept as part of the written record.

Further information on tests and maintenance may be found in *NFPA 72*®, *National Fire Alarm Code*®; NFPA 99, *Standard for Health Care Facilities;* NFPA *101*®, *Life Safety Code*®; NFPA 110, *Standard for Emergency and Standby Power Systems;* and NFPA 111, *Standard on Stored Electrical Energy Emergency and Standby Power Systems.*

700.5 Capacity.

(A) Capacity and Rating. An emergency system shall have adequate capacity and rating for all loads to be operated simultaneously. The emergency system equipment shall be suitable for the maximum available fault current at its terminals.

The emergency system must be designed with adequate capacity and rating to safely carry the entire load connected to the emergency system at one time. It must be capable of restarting emergency loads, such as motors, that may have stopped, and it must be suitable for the available fault current. Using devices that limit the available fault current is one method of achieving suitability.

(B) Selective Load Pickup, Load Shedding, and Peak Load Shaving. The alternate power source shall be permitted to supply emergency, legally required standby, and optional standby system loads where automatic selective load pickup and load shedding is provided as needed to ensure adequate power to (1) the emergency circuits, (2) the legally required standby circuits, and (3) the optional standby circuits, in that order of priority. The alternate power source shall be permitted to be used for peak load shaving, provided the above conditions are met.

Peak load-shaving operation shall be permitted for satisfying the test requirement of 700.4(B), provided all other conditions of 700.4 are met.

A portable or temporary alternate source shall be available whenever the emergency generator is out of service for major maintenance or repair.

Section 700.5(B) permits a generator to serve more than one level of emergency, legally required standby, or other operational standby system loads. Section 700.5(B) also permits the use of a generator for peak load shaving, supplying backup power, and other uses. However, assurance is required that priority loads will be properly and reliably served. To provide the necessary assurance, such systems must be maintained and periodically tested.

If a generator is used for peak load shaving or in a cogeneration system, major downtime for maintenance of the generator must be anticipated. On the other hand, using the emergency generator on a regular basis for nonemergency loads provides assurance that the emergency generator will supply emergency power when it is needed. The requirement for a portable or temporary alternate source is to provide emergency power when the generator set is out of service for a long time. A major maintenance or repair procedure is one that keeps the generator set out of service for more than a few hours.

700.6 Transfer Equipment.

(A) General. Transfer equipment, including automatic transfer switches, shall be automatic, identified for emergency use, and approved by the authority having jurisdiction. Transfer equipment shall be designed and installed to prevent the inadvertent interconnection of normal and emergency sources of supply in any operation of the transfer equipment. Transfer equipment and electric power production systems installed to permit operation in parallel with the normal source shall meet the requirements of Article 705.

New in the 2002 *NEC*, the transfer equipment described in 700.6(A) is permitted to allow parallel operation of the generation equipment with the normal source as long as the requirements of Article 705 are met. Traditional automatic transfer switches are not designed to permit parallel operation of generation equipment and the normal source. Therefore, traditional automatic transfer switches need not comply with Article 705. However, certain automatic transfer switch configurations are intentionally designed to briefly (a few cycles) parallel the generation equipment with the normal source upon load transfer from generator to normal source. This load transfer can occur with minimal disturbance or effect on the load. Transfer switches that employ this type of paralleling must comply with Article 705.

If continuous parallel operation of generation equipment and the source is desired, paralleling switchgear or paralleling equipment with appropriate protection is required. (See Article 705.)

(B) Bypass Isolation Switches. Means shall be permitted to bypass and isolate the transfer equipment. Where bypass isolation switches are used, inadvertent parallel operation shall be avoided.

(C) Automatic Transfer Switches. Automatic transfer switches shall be electrically operated and mechanically held.

Section 700.6(C) was added in the 1999 *Code* to ensure that relay contacts would be mechanically held in the event of coil failure. This change correlates the *NEC* with the requirements found in NFPA 110, *Standard for Emergency and Standby Power Systems*.

(D) Use. Transfer equipment shall supply only emergency loads.

Although the alternate power source is permitted to supply emergency loads as well as other loads, the transfer switch used for the emergency system is strictly limited to emergency loads only, that is, loads classified as emergency in accordance with 700.1. Other loads, such as legally required standby loads or optional standby loads (covered by Articles 701 and 702), are not permitted to be supplied from the emergency system transfer switch. If a single generator is used to supply both emergency and nonemergency loads, then multiple transfer switches are required.

700.7 Signals.

Audible and visual signal devices shall be provided, where practicable, for the purpose described in 700.7(A) through (D).

(A) Derangement. To indicate derangement of the emergency source.

(B) Carrying Load. To indicate that the battery is carrying load.

(C) Not Functioning. To indicate that the battery charger is not functioning.

The major causes of emergency equipment failure are improper testing or lack of testing, inadequate maintenance, and the failure of attendants to see the signals that indicate malfunctioning battery or battery-charging equipment. One way to minimize equipment failure is to install signal devices to annunciate trouble where attendants or other personnel familiar with the operation of the emergency equipment can see or hear them. In theaters, assembly halls, or similar locations, audible signal bells or horns that annunciate the functions specified in 700.7 should be located where their sounding will not cause panic.

Battery-operated unit equipment generally has a test switch to simulate a failure of the normal system and an indicating light that glows brightly while charging and dims when ready. Transparent cases for lead-acid batteries allow easy viewing of electrolyte levels.

A storage battery system is normally capable of delivering 12 volts, 24 volts, 32 volts, or 120 volts and consists of monitoring and distribution cabinets and a console with battery and charger. A storage battery system generally includes audio, visual, and remote signal devices and a test switch, and it may include a trouble bell and silence switch.

(D) Ground Fault. To indicate a ground fault in solidly grounded wye emergency systems of more than 150 volts to ground and circuit-protective devices rated 1000 amperes or more. The sensor for the ground-fault signal devices shall be located at, or ahead of, the main system disconnecting means for the emergency source, and the maximum setting of the signal devices shall be for a ground-fault current of 1200 amperes. Instructions on the course of action to be taken in event of indicated ground fault shall be located at or near the sensor location.

> FPN: For signals for generator sets, see NFPA 110-1999, *Standard for Emergency and Standby Power Systems.*

Although 700.26 indicates that ground-fault protection of equipment is not required on the alternate source for emergency systems, ground faults can occur on such systems, and they can result in equipment burndown. Because of the emergency nature of such systems, automatic disconnect in the event of a ground fault is inappropriate. Detection of such a fault, however, is desirable so that the condition can be corrected.

700.8 Signs.

(A) Emergency Sources. A sign shall be placed at the service entrance equipment indicating type and location of on-site emergency power sources.

Exception: A sign shall not be required for individual unit equipment as specified in 700.12(E).

(B) Grounding. Where the grounded circuit conductor connected to the emergency source is connected to a grounding electrode conductor at a location remote from the emer-

gency source, there shall be a sign at the grounding location that shall identify all emergency and normal sources connected at that location.

Section 700.8(B) requires a sign at the grounding location if the emergency source is a separately derived system and is connected to a grounding electrode conductor at a location that is remote from the emergency source.

II. Circuit Wiring

700.9 Wiring, Emergency System.

(A) Identification. All boxes and enclosures (including transfer switches, generators, and power panels) for emergency circuits shall be permanently marked so they will be readily identified as a component of an emergency circuit or system.

The marking may be by color code, the words "emergency system," or any other type of identification that identifies the box or enclosure as a component of the emergency system.

(B) Wiring. Wiring of two or more emergency circuits supplied from the same source shall be permitted in the same raceway, cable, box, or cabinet. Wiring from an emergency source or emergency source distribution overcurrent protection to emergency loads shall be kept entirely independent of all other wiring and equipment, unless otherwise permitted in (1) through (4):

(1) Wiring from the normal power source located in transfer equipment enclosures
(2) Wiring supplied from two sources in exit or emergency luminaires (lighting fixtures)
(3) Wiring from two sources in a common junction box, attached to exit or emergency luminaires (lighting fixtures)
(4) Wiring within a common junction box attached to unit equipment, containing only the branch circuit supplying the unit equipment and the emergency circuit supplied by the unit equipment

Emergency circuit wiring is not permitted to enter the same raceway, cable, box, or cabinet with the regular or normal wiring of the building concerned. Wiring for the emergency circuits must be completely independent of all other wiring and equipment. This practice ensures that any fault on the normal wiring circuits will not affect the performance of the emergency wiring or equipment.

To effect an immediate transfer from one system to the other, both the normal source and the emergency source must be present within a transfer switch enclosure per 700.9(B)(1).

Sections 700.9(B)(2) and (3) permit the use of two-lamp exit or two-lamp emergency fixtures where one lamp is connected to the normal supply and one lamp is connected to the alternate supply. Both lamps may be illuminated as part of the regular lighting operation.

Wiring on the load side of a transfer switch serves as both the emergency circuit wiring and the normal circuit wiring. It is not intended that two sets of wiring supply emergency loads from the transfer switch to the emergency load distribution panel, as shown in Exhibits 700.3 and 700.4, or from these emergency distribution panels to the emergency loads.

(C) Wiring Design and Location. Emergency wiring circuits shall be designed and located so as to minimize the hazards that might cause failure due to flooding, fire, icing, vandalism, and other adverse conditions.

The purpose of 700.9(C) is to minimize the likelihood of impairment of the emergency system due to flood, fire, icing, vandalism, and other adverse conditions. The same requirement applies to sources of power covered in 700.12.

(D) Fire Protection. Emergency systems shall meet the following additional requirements in assembly occupancies for not less than 1000 persons or in buildings above 23 m (75 ft) in height with any of the following occupancy classes: assembly, educational, residential, detention and correctional, business, and mercantile.

(1) Feeder-Circuit Wiring. Feeder-circuit wiring shall meet one of the following conditions:

(1) Be installed with buildings that are fully protected by an approved automatic fire suppression system
(2) Be a listed electrical circuit protective system with a minimum 1-hour fire rating
(3) Be protected by a listed thermal barrier system for electrical system components
(4) Be protected by a fire-rated assembly listed to achieve a minimum fire rating of 1 hour
(5) Be embedded in not less than 50 mm (2 in.) of concrete
(6) Be a cable listed to maintain circuit integrity for not less than 1 hour when installed in accordance with the listing requirements

(2) Feeder-Circuit Equipment. Equipment for feeder circuits (including transfer switches, transformers, and panelboards) shall be located either in spaces fully protected by approved automatic fire suppression systems (including sprinklers, carbon dioxide systems) or in spaces with a 1-hour fire resistance rating.

FPN: For the definition of *occupancy class*, see 4.1 of NFPA *101*-2000, *Life Safety Code*.

The proper operation of emergency electrical systems is critical for densely populated occupancies and for high-rise occupancies. Therefore, fire protection requirements for both emergency system feeder circuits and equipment ensure the integrity as well as the performance of the emergency electrical system. If feeders and equipment are located within buildings that are fully protected by an approved fire suppression system, then no further fire protection techniques are generally required.

Sprinkler systems are the most common fire suppression systems, and they are covered in NFPA 13, *Standard for the Installation of Sprinkler Systems*. Buildings that are fully protected by automatic sprinkler systems meet the requirements of 700.9(D). Additional fire suppression systems are included in the following standards:

1. NFPA 11, *Standard for Low-Expansion Foam*
2. NFPA 12, *Standard on Carbon Dioxide Extinguishing Systems*
3. NFPA 12A, *Standard on Halon 1301 Fire Extinguishing Systems*
4. NFPA 15, *Standard for Water Spray Fixed Systems for Fire Protection*
5. NFPA 17, *Standard for Dry Chemical Extinguishing Systems*
6. NFPA 2001, *Standard on Clean Agent Fire Extinguishing Systems*

If feeders and equipment are not located within buildings that are fully protected by an approved fire suppression system, other methods and protection techniques are available to comply with the fire protection requirements of 700.9(D)(2). These additional fire protection methods and techniques include the following:

1. *Listed Electrical Circuit Protective Systems*. These systems are described in the UL *Building Materials Directory*. The four-letter code (shown in parentheses) following each of the category headings in the directory is the UL product category guide designation. Examples of these systems include electrical circuit protective systems (FHIT), electrical circuit protective materials (FHIY), and fire resistive cables (FHJR).
2. *Listed Thermal Barrier Systems*. These systems are described in the UL *Building Materials Directory* as thermal barrier systems (XCLF). Examples of the thermal barrier protection technique include batts and blankets (XCLR), packing material (XCMD), and preformed

mineral and fiber units (XCMK) wrapped or otherwise formed over the conduit to achieve a predetermined fire rating.

3. *Fire-Rated Assembly.* These systems are described in the *UL Fire Resistance Directory,* Volumes 1 and 2. The assemblies found in Volume 1 include hourly ratings for beams, floors, roofs, columns, and walls and partitions. Volume 2 of this directory includes hourly ratings for joint systems and through-penetration firestop systems. All fire ratings and fire resistance ratings are based on the assumption that the structural supports for the assembly are not impaired by the fire.

4. *Embedded in Concrete.* Embedding a conduit in concrete is most effective when implemented during original construction. This method has been successful for many years in protecting premises from service conductors. According to 230.6, conductors embedded in not less than 2 in. of concrete are considered outside of the building.

5. *Cables Listed to Maintain Circuit Integrity.* Circuit integrity cables are classified by the UL *Building Materials Directory* under the existing product category of fire resistive cables (FHJR).

It is important to understand the difference between a 1-hour fire rating of an electrical cable and a 1-hour fire resistance rating of a structural member, such as a wall. Simply stated, at the end of a 1-hour fire rating test on an electrical cable, the circuit and its insulation must be intact and electrically functioning. (No short circuits, grounds, or opens are permitted.) However, a wall subjected to a 1-hour fire resistance test must only prevent a fire from passing through or past the wall, without regard to damage to the wall.

III. Sources of Power

700.12 General Requirements.

Current supply shall be such that, in the event of failure of the normal supply to, or within, the building or group of buildings concerned, emergency lighting, emergency power, or both shall be available within the time required for the application but not to exceed 10 seconds. The supply system for emergency purposes, in addition to the normal services to the building and meeting the general requirements of this section, shall be one or more of the types of systems described in 700.12(A) through (D). Unit equipment in accordance with 700.12(E) shall satisfy the applicable requirements of this article.

In selecting an emergency source of power, consideration shall be given to the occupancy and the type of service to be rendered, whether of minimum duration, as for evacuation of a theater, or longer duration, as for supplying emergency power and lighting due to an indefinite period of current failure from trouble either inside or outside the building.

Equipment shall be designed and located to minimize the hazards that might cause complete failure due to flooding, fires, icing, and vandalism.

Equipment for sources of power as described in 700.12(A) through (D) where located within assembly occupancies for greater than 1000 persons or in buildings above 23 m (75 ft) in height with any of the following occupancy classes — assembly, educational, residential, detention and correctional, business, and mercantile — shall be installed either in spaces fully protected by approved automatic fire suppression systems (sprinklers, carbon dioxide systems, and so forth), or in spaces with a 1-hour fire rating.

> FPN No. 1: For the definition of occupancy class, see 4.1 of NFPA *101*-2000, *Life Safety Code.*
>
> FPN No. 2: Assignment of degree of reliability of the recognized emergency supply system depends on the careful evaluation of the variables at each particular installation.

(A) Storage Battery. Storage batteries used as a source of power for emergency systems shall be of suitable rating and capacity to supply and maintain the total load for a period of 1½ hours minimum, without the voltage applied to the load falling below 87½ percent of normal.

Batteries, whether of the acid or alkali type, shall be designed and constructed to meet the requirements of emergency service and shall be compatible with the charger for that particular installation.

For a sealed battery, the container shall not be required to be transparent. However, for the lead acid battery that requires water additions, transparent or translucent jars shall be furnished. Automotive-type batteries shall not be used.

An automatic battery charging means shall be provided.

(B) Generator Set.

(1) Prime Mover-Driven. For a generator set driven by a prime mover acceptable to the authority having jurisdiction and sized in accordance with 700.5, means shall be provided for automatically starting the prime mover on failure of the normal service and for automatic transfer and operation of all required electrical circuits. A time-delay feature permitting a 15-minute setting shall be provided to avoid retransfer in case of short-time reestablishment of the normal source.

(2) Internal Combustion as Prime Movers. Where internal combustion engines are used as the prime mover, an onsite fuel supply shall be provided with an on-premise fuel supply sufficient for not less than 2 hours' full-demand operation of the system. Where power is needed for the operation of the fuel transfer pumps to deliver fuel to a generator set day tank, this pump shall be connected to the emergency power system.

(3) Dual Supplies. Prime movers shall not be solely dependent on a public utility gas system for their fuel supply or municipal water supply for their cooling systems. Means shall be provided for automatically transferring from one fuel supply to another where dual fuel supplies are used.

Exception: Where acceptable to the authority having jurisdiction, the use of other than on-site fuels shall be permitted where there is a low probability of a simultaneous failure of both the off-site fuel delivery system and power from the outside electrical utility company.

(4) Battery Power and Dampers. Where a storage battery is used for control or signal power or as the means of starting the prime mover, it shall be suitable for the purpose and shall be equipped with an automatic charging means independent of the generator set. Where the battery charger is required for the operation of the generator set, it shall be connected to the emergency system. Where power is required for the operation of dampers used to ventilate the generator set, the dampers shall be connected to the emergency system.

(5) Auxiliary Power Supply. Generator sets that require more than 10 seconds to develop power shall be permitted if an auxiliary power supply energizes the emergency system until the generator can pick up the load.

(6) Outdoor Generator Sets. Where an outdoor housed generator set is equipped with a readily accessible discon-

necting means located within sight of the building or structure supplied, an additional disconnecting means shall not be required where ungrounded conductors pass through the building or structure.

(C) Uninterruptible Power Supplies. Uninterruptible power supplies used to provide power for emergency systems shall comply with the applicable provisions of 700.12(A) and (B).

(D) Separate Service. Where acceptable to the authority having jurisdiction as suitable for use as an emergency source, a second service shall be permitted. This service shall be in accordance with Article 230, with separate service drop or lateral, widely separated electrically and physically from the normal service to minimize the possibility of simultaneous interruption of supply.

(E) Unit Equipment. Individual unit equipment for emergency illumination shall consist of the following:

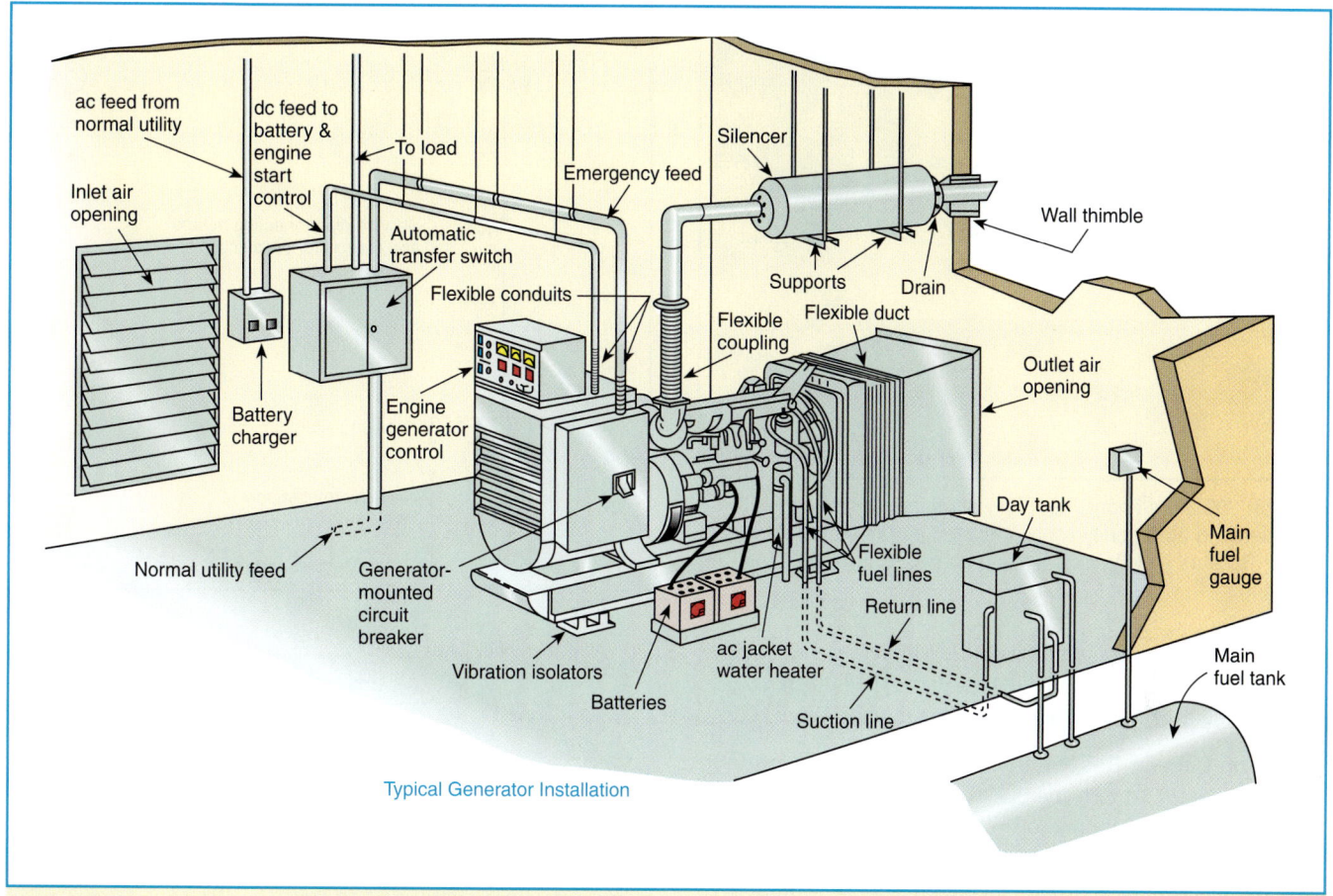

Typical Generator Installation

Exhibit 700.1 A typical generator installation supplying standby power in ratings from 55 kW to 930 kW, 60 Hz. (Redrawn from Caterpillar)

or diesel engines or in isolated areas where maintenance or refueling could be a problem.

Some types of drivers, particularly large ones, may take longer than 10 seconds to accelerate and develop generator voltage. Gas and steam turbines and large internal-combustion engines may have prolonged starting times. Depending on the specific loads, short-time supply could be provided by an uninterruptible power supply; a generator shared with other loads; or a generator with limited emergency supply, such as an expander, a steam turbine, or a waste heat system.

3. Uninterruptible power supplies (UPS), which generally include a rectifier, storage battery, and inverter to ac. Uninterruptible power supplies may be very complex systems with redundant components and high-speed solid-state switching. It is common practice to include an automatic bypass for UPS malfunction to permit maintenance.

4. The use of a separate service, which requires a judgment by the authority having jurisdiction. Such judgment should be based on the nature of the emergency loads and the expected reliability of the other available sources.

5. Unit equipment wired with a flexible cord (not longer than 3 ft) and attachment plug cap. Unit equipment must be permanently fixed in place, usually by mounting screws that are accessible only from within the unit. One or more lamps may be mounted on or remote from the unit. The unit should be located where it can be readily checked or tested for proper performance. See Exhibit 700.2.

Unit equipment is intended to provide illumination for the area where it is installed. For instance, if a unit is located in a corridor, it must be connected to the branch circuit supplying the normal corridor lights (on the line side of any switching arrangements). If normal power fails, the unit would automatically energize the unit lamps, restoring illumination to the corridor. A separate circuit is not permitted for unit equipment [except as noted in the exception to 700.12(E)] because, if applied to the above example, failure

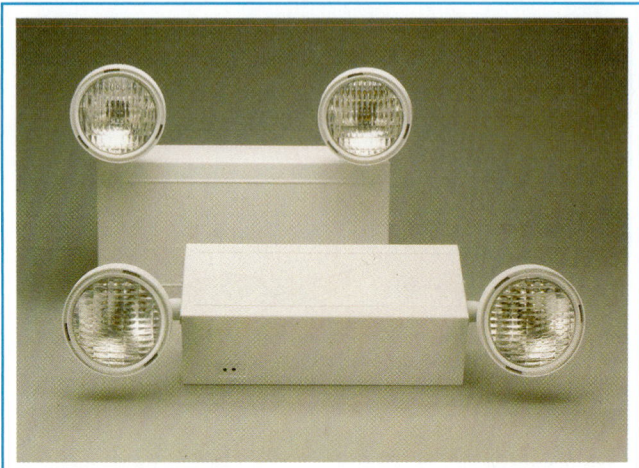

Exhibit 700.2 Self-contained, fully automatic unit equipment for operating emergency lighting located on the unit or for remotely located exit signs or lighting heads. (Courtesy of Dual-Lite, Inc.)

of the normal corridor circuit would not affect the unit equipment, and the corridor would remain dark. The branch circuit feeding the unit must be identified at the panelboard.

Notes on General Requirements for Emergency Lighting Systems. At least two sources of power must be provided — one normal supply and one or more emergency systems described in 700.12. The sources may be (1) two services, one normal supply and one emergency supply (preferably from separate utility stations), (2) one normal service and a storage battery (or unit equipment) system, or (3) one normal service and a generator set. (See Exhibits 700.3 and 700.4.)

A transfer means (or throw-over switch) must be provided to energize the emergency equipment from the alternate supply when the normal source of supply is interrupted.

If a separate service is used, both may operate normally, but equipment for emergency lighting and power must be arranged to be energized from either service.

If the alternate or emergency source of supply is a storage battery or generator set, then the single emergency system is usually operated on the normal service, and the battery (or batteries) or generator operates only if the normal service fails. A generator may be used, however, for peak load shaving and the like in accordance with 700.5.

Two or more separate and complete systems may provide power for emergency lighting, but means must be provided for energizing one system if the other one fails.

It should be noted that provisions for disconnecting means and overcurrent protection (see Exhibits 700.3 and 700.4) are to be provided for emergency systems as required by Article 230.

(1) A rechargeable battery
(2) A battery charging means

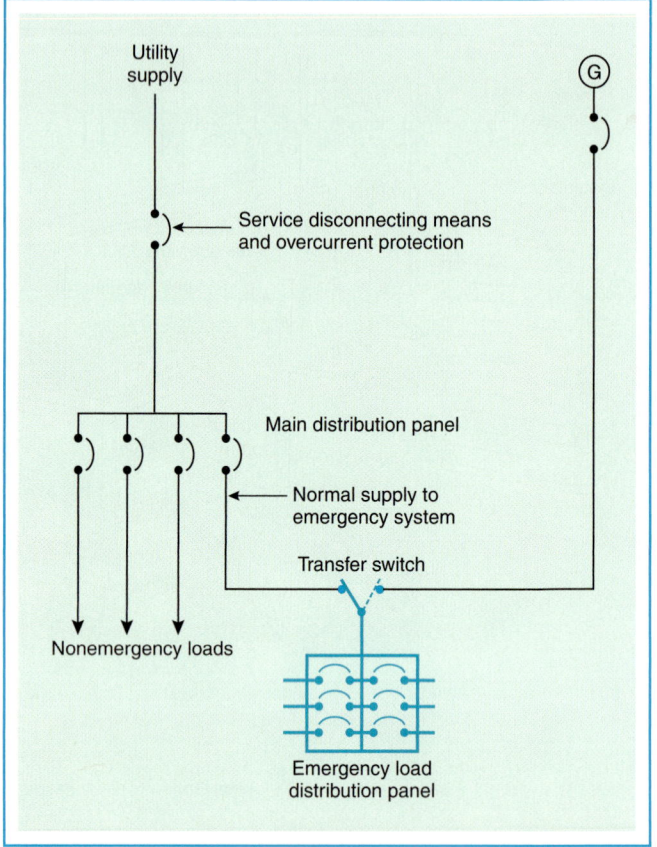

Exhibit 700.3 Emergency load arranged to be supplied from a generator, as permitted by 700.12(B).

(3) Provisions for one or more lamps mounted on the equipment, or shall be permitted to have terminals for remote lamps, or both
(4) A relaying device arranged to energize the lamps automatically upon failure of the supply to the unit equipment

The batteries shall be of suitable rating and capacity to supply and maintain at not less than 87½ percent of the nominal battery voltage for the total lamp load associated with the unit for a period of at least 1½ hours, or the unit equipment shall supply and maintain not less than 60 percent of the initial emergency illumination for a period of at least 1½ hours. Storage batteries, whether of the acid or alkali type, shall be designed and constructed to meet the requirements of emergency service.

Unit equipment shall be permanently fixed in place (i.e., not portable) and shall have all wiring to each unit installed in accordance with the requirements of any of the wiring methods in Chapter 3. Flexible cord-and-plug connection shall be permitted, provided that the cord does not exceed 900 mm (3 ft) in length. The branch circuit feeding the unit equipment shall be the same branch circuit as that serving the normal lighting in the area and connected ahead of any

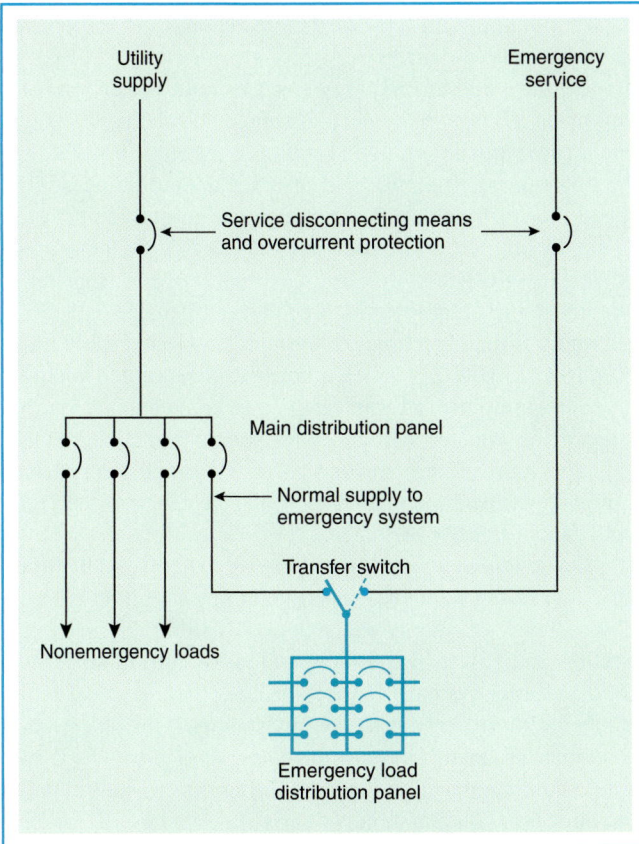

Utility supply

Emergency service

Service disconnecting means and overcurrent protection

Main distribution panel

Normal supply to emergency system

Transfer switch

Nonemergency loads

Emergency load distribution panel

Exhibit 700.4 Emergency load arranged to be supplied from two widely separated services, as permitted by 700.12(D). When one service fails, the emergency load will be transferred to the other service.

local switches. The branch circuit that feeds unit equipment shall be clearly identified at the distribution panel. Emergency luminaires (illumination fixtures) that obtain power from a unit equipment and are not part of the unit equipment shall be wired to the unit equipment as required by 700.9 and by one of the wiring methods of Chapter 3.

Exception: In a separate and uninterrupted area supplied by a minimum of three normal lighting circuits, a separate branch circuit for unit equipment shall be permitted if it originates from the same panelboard as that of the normal lighting circuits and is provided with a lock-on feature.

IV. Emergency System Circuits for Lighting and Power

700.15 Loads on Emergency Branch Circuits.

No appliances and no lamps, other than those specified as required for emergency use, shall be supplied by emergency lighting circuits.

700.16 Emergency Illumination.

Emergency illumination shall include all required means of egress lighting, illuminated exit signs, and all other lights specified as necessary to provide required illumination.

Emergency lighting systems shall be designed and installed so that the failure of any individual lighting element, such as the burning out of a light bulb, cannot leave in total darkness any space that requires emergency illumination.

Where high-intensity discharge lighting such as high- and low-pressure sodium, mercury vapor, and metal halide is used as the sole source of normal illumination, the emergency lighting system shall be required to operate until normal illumination has been restored.

Exception: Alternative means that ensure emergency lighting illumination level is maintained shall be permitted.

High-intensity discharge (HID) fixtures take some time to start once they are energized. Therefore, if HID fixtures are the sole source of normal illumination in an area, the *Code* requires that the emergency lighting system operate not only until the normal system is returned to service but also until the HID fixtures provide illumination. This may require a timing circuit, photoelectric monitoring system, or the equivalent.

700.17 Circuits for Emergency Lighting.

Branch circuits that supply emergency lighting shall be installed to provide service from a source complying with 700.12 when the normal supply for lighting is interrupted. Such installations shall provide either one of the following:

(1) An emergency lighting supply, independent of the general lighting supply, with provisions for automatically transferring the emergency lights upon the event of failure of the general lighting system supply
(2) Two or more separate and complete systems with independent power supply, each system providing sufficient current for emergency lighting purposes

Unless both systems are used for regular lighting purposes and are both kept lighted, means shall be provided for automatically energizing either system upon failure of the other. Either or both systems shall be permitted to be a part of the general lighting system of the protected occupancy if circuits supplying lights for emergency illumination are installed in accordance with other sections of this article.

GENERAL CONSIDERATIONS FOR TRANSFER SWITCHES

Automatic transfer switches of double-throw construction are used primarily for emergency and standby power generation systems rated 600 volts or less. These transfer switches do not normally incorporate overcurrent protection and are designed and applied in accordance with the *Code,* particularly Articles 700 and 701. Automatic transfer switches are available in ratings from 30 to 3000 amperes. For reliability, most automatic transfer switches rated above 100 amperes are mechanically held and electrically operated from the power source to which the load is to be transferred.

An automatic transfer switch is usually located in the main or secondary distribution bus that feeds the branch circuits. Because of its location in the system, the capabilities that must be designed into the transfer switch are unique and extensive as compared with the design requirements for other branch-circuit and feeder devices. For example, special consideration should be given to the following characteristics of an automatic transfer:

1. Its ability to close against high inrush currents
2. Its ability to carry full-rated current continuously from normal and emergency sources
3. Its ability to withstand fault currents
4. Its ability to interrupt six times the full-load current

In addition to considering each of the preceding characteristics individually, it is also necessary to consider the effect each has on the other.

In arrangements that provide protection against failure of the utility service, consideration should also be given to the following:

1. An open circuit within the building area on the load side of the incoming service
2. Overload or fault condition
3. Electrical or mechanical failure of the electric power distribution system within the building

It is, therefore, desirable to locate transfer switches close to the load and to keep the operation of the transfer switches independent of overcurrent protection. It is often desirable to use multiple transfer switches of lower current rating located near the load rather than one large transfer switch at the point of incoming service.

LOCATION OF OVERCURRENT DEVICES

The location of overcurrent devices for both normal and emergency power is covered by 240.21 and is not affected by the installation of an automatic transfer switch. Transfer switches should be rated for continuous duty and have low contact temperature rise.

SOLID NEUTRAL ON ALTERNATING-CURRENT AND DIRECT-CURRENT SYSTEMS

If automatic ac-to-ac transfer switches are used, then solid neutrals can be used with the grounding connections, as required in 250.24. If multiple grounding creates objection-able ground current, then corrective action, as specified in 250.6(B), must be made.

Section 230.95 requires ground-fault protection of equipment. Because the normal source and the emergency source are typically grounded at their locations, the multiple neutral-to-ground connections usually require some additional means or devices to ensure proper ground-fault sensing by the ground-fault protection device. Additional means or devices are generally required because the usual alterations used to stop objectionable current, per 250.6(B), do not apply if the objectionable current is a ground-fault current. [See 250.6(C).] Rather, solutions often include adding an overlapping neutral transfer pole or conventional fourth pole to the transfer switch. Other solutions include using isolation transformers and special ground-fault circuits or using the service ground to also ground the generator neutral with 3-pole transfer switches.

On ac-to-dc automatic transfer switches, a solid neutral tie between the ac and dc neutrals is not permitted where both sources of supply are exterior distribution systems. Section 250.164, which addresses the location of grounds for dc exterior systems, clearly specifies that the dc system can be grounded only at the supply station. If the dc system is an interior isolated system, such as a storage battery, then solid neutral connection between the ac system neutral and the dc source is acceptable.

On an ac-to-dc automatic transfer switch where the neutral must be switched, the size of the neutral switching pole must be considered. A 4-pole, double-throw switch must be used where a 3-phase, 4-wire normal source and a 2-wire dc emergency source are transferred. A 4-pole, double-throw transfer switch is required because the neutral is switched. In this instance, one pole of the dc emergency source carries three times the current of the other poles.

CLOSE DIFFERENTIAL VOLTAGE SUPERVISION OF NORMAL SOURCE

Most often, the normal source is an electric utility company whose power is transmitted many miles to the point of utilization. The automatic transfer switch control panel at the utility company continuously monitors the voltage of all phases. (Because utility frequency is, for all practical purposes, constant, only the voltage needs to be monitored.) For single-phase power systems, the line-to-line voltage is monitored. For 3-phase power systems, all three line-to-line voltages should be monitored to provide full-phase protection.

Monitoring protects against operation at reduced voltage, such as during brownouts, which can damage loads such as motors. Because the voltage sensitivity of loads varies, the pickup (acceptable) voltage setting and dropout (unacceptable) voltage setting of the monitors should be adjustable. The typical range of adjustment for the pickup is 85 percent to 100 percent of nominal, while the dropout setting, which is a function of the pickup setting, is 75 to 98 percent of the pickup selected. Typical settings for most loads are 95 percent of nominal for pickup and 85 percent of nominal for dropout (90 percent of pickup). Consideration must be given to voltage supervision at closer differential for many installations where the load circuits are critical to voltage.

Electronic equipment load is frequently voltage critical. Installations that use such equipment include patient care equipment in health care facilities, X-ray equipment, television stations, cable TV centers, microwave communications, telephone communications, computers, computer-operated equipment, computer centers, and similar applications.

Polyphase motors operating at low load have a tendency to single phase, despite the loss of voltage in one phase, leading to burnout of the motor. A close differential of voltage supervision should be applied to automatic transfer switches for motor installations of the polyphase type. Differential voltage relays with a close adjustment of 2 percent for transfer and retransfer values aid in the detection of phase outages and provide protection from single phasing.

AUTOMATIC TRANSFER SWITCHES WITH EMERGENCY SOURCE ON AUTOMATICALLY STARTED POWER PLANT

In installations that require an emergency power source, the normal source is usually a utility power line, and the emergency source is an automatically started engine generator set that starts when the normal source fails. To ensure maximum reliability, a minimum installation should be arranged to do the following:

1. Initiate engine starting of the power plant from a contact on the automatic transfer switch control panel.
2. Sustain connection of load circuits to the normal source during the starting period to provide utilization of any existing service from the normal source.
3. Measure output voltage and frequency of emergency source through the use of a voltage-frequency-sensitive monitor and effect transfer of the load circuits to the power plant only when both voltage and frequency of the power plant are approximately normal. Sensing of the emergency source need only be single phase, because most applications involve an on-site engine generator with a relatively short line run to the automatic transfer switch. In addition to monitoring voltage, the emergency source's frequency should be monitored. Unlike the utility power, the engine generator frequency can vary during start-up. Frequency monitoring will avoid overloading the engine generator while it is starting and can thus prevent stalling the engine. Combined frequency and voltage monitoring will prevent loads

from being transferred to an engine generator set with an unacceptable output.
4. Provide visual signal and auxiliary contact for remote indication when the power plant is feeding the load.

TIME-DELAY DEVICES ON AUTOMATIC TRANSFER SWITCHES

Time delay controls are essential to the operation of the automatic transfer switch. To avoid unnecessary starting and transfer to the alternate supply, a nominal 1-second time delay, adjustable up to 6 seconds, can override momentary interruptions and temporary reductions in normal source voltage but still allow starting and transfer if the reduction or outage is sustained. Local electric utilities can provide circuit protection schemes and timing information. Because momentary outages may last 2 seconds, nominal, a correctly set time delay should prevent the transfer switch from operating (and prevent generator starting) during utility automatic circuit protection operation. However, the time delay should be set fast enough to effectively operate the transfer switch and provide backup power for long-term outages.

The advantages of a time delay are realized in all types of automatic transfer installations. In standby plant installations, the reduced number of false starts is especially important to minimize wear on the starting gear, battery, and associated equipment. This delay is generally set at 1 second but may be set higher if reclosers or circuit breakers on the utility power lines take longer to operate or if momentary power dips exceed 1 second. If longer delay settings are used, care must be taken to ensure that sufficient time remains to meet 10-second power restoration requirements. The authority having jurisdiction may determine that a longer-term power failure does not occur until the utility automatic protective devices fail to restore power to the facility. For example, the 10-second power restoration requirements would become effective after the 2-second, nominal, recloser or circuit breaker cycle.

Once the load is transferred to the alternate source, another timer delays retransfer to the normal source until that source has time to stabilize. See the commentary following the exception to 700.16. This timer is required by 700.12(B)(1) and is controlled by the preferred source voltage monitors. The timer is adjustable from 0 to 30 minutes and is normally set at 30 minutes. Another important function of this retransfer timer is to allow an engine generator to operate under load for a preselected minimum time to ensure continued good performance of the set and its starting system. This delay should be automatically nullified if the alternate source fails and the normal source is available, as determined by the voltage monitors.

Engine generator manufacturers often recommend a cool-down period for their sets that allows them to run unloaded after the load is retransferred to the normal source.

A third time delay, usually 5 minutes, is provided for this purpose. Running an unloaded engine for more than 5 minutes is neither necessary nor recommended because it can cause deterioration in engine performance.

If more than one automatic transfer switch is connected to the same engine generator, it is sometimes recommended that transfer of the loads be purposely sequenced to the alternate source. Using such a sequencing scheme can reduce starting kVA capacity requirements of the generator. A fourth timer, adjustable from 0 to 5 minutes, will delay transfer to the emergency supply source for this and other similar requirements.

700.18 Circuits for Emergency Power.

For branch circuits that supply equipment classed as emergency, there shall be an emergency supply source to which the load will be transferred automatically upon the failure of the normal supply.

V. Control—Emergency Lighting Circuits

700.20 Switch Requirements.

The switch or switches installed in emergency lighting circuits shall be arranged so that only authorized persons have control of emergency lighting.

Exception No. 1: Where two or more single-throw switches are connected in parallel to control a single circuit, at least one of these switches shall be accessible only to authorized persons.

Exception No. 2: Additional switches that act only to put emergency lights into operation but not disconnect them shall be permissible.

Switches connected in series or 3- and 4-way switches shall not be used.

700.21 Switch Location.

All manual switches for controlling emergency circuits shall be in locations convenient to authorized persons responsible for their actuation. In places of assembly, such as theaters, a switch for controlling emergency lighting systems shall be located in the lobby or at a place conveniently accessible thereto.

In no case shall a control switch for emergency lighting in a theater, motion-picture theater, or place of assembly be placed in a motion-picture projection booth or on a stage or platform.

Exception: Where multiple switches are provided, one such switch shall be permitted in such locations where arranged so that it can energize the circuit only but cannot de-energize the circuit.

700.22 Exterior Lights.

Those lights on the exterior of a building that are not required for illumination when there is sufficient daylight shall be permitted to be controlled by an automatic light-actuated device.

VI. Overcurrent Protection

700.25 Accessibility.

The branch-circuit overcurrent devices in emergency circuits shall be accessible to authorized persons only.

> FPN: Fuses and circuit breakers for emergency circuit overcurrent protection, where coordinated to ensure selective clearing of fault currents, increase overall reliability of the system.

700.26 Ground-Fault Protection of Equipment.

The alternate source for emergency systems shall not be required to have ground-fault protection of equipment with automatic disconnecting means. Ground-fault indication of the emergency source shall be provided per 700.7(D).

ARTICLE 701
Legally Required Standby Systems

Contents

I. General

701.1 Scope.

The provisions of this article apply to the electrical safety of the installation, operation, and maintenance of legally required standby systems consisting of circuits and equipment intended to supply, distribute, and control electricity to required facilities for illumination or power, or both, when the normal electrical supply or system is interrupted.

The systems covered by this article consist only of those that are permanently installed in their entirety, including the power source.

FPN No. 1: For additional information, see NFPA 99-1999, *Standard for Health Care Facilities*.

FPN No. 2: For further information regarding performance of emergency and standby power systems, see NFPA 110-1999, *Standard for Emergency and Standby Power Systems*.

FPN No. 3: For further information, see ANSI/IEEE 446-1995, *Recommended Practice for Emergency and Standby Power Systems for Industrial and Commercial Applications*.

Legally required standby systems are intended to provide electric power to aid in fire fighting, rescue operations, control of health hazards, and similar operations. In comparison, emergency systems (see Article 700) are those systems essential for safety to life. Optional standby systems (see Article 702) are those in which failure can cause physical discomfort, serious interruption of an industrial process, damage to process equipment, or disruption of business, for example.

The requirements for legally required standby systems are much the same as for emergency systems. There are, however, some differences. When normal power is lost, legally required systems must be able to supply standby power in 60 seconds or less, instead of the 10 seconds or less required of emergency systems. Wiring for legally required standby systems may occupy the same raceways, cables, boxes, and cabinets as other general wiring. Wiring for emergency systems must be kept entirely independent of other wiring. Legally required standby systems take second priority to emergency systems if they are involved in sharing an alternate supply and/or load shedding or peak shaving schemes.

701.2 Definition.

Legally Required Standby Systems. Those systems required and so classed as legally required standby by municipal, state, federal, or other codes or by any governmental agency having jurisdiction. These systems are intended to automatically supply power to selected loads (other than those classed as emergency systems) in the event of failure of the normal source.

FPN: Legally required standby systems are typically installed to serve loads, such as heating and refrigeration systems, communications systems, ventilation and smoke removal systems, sewage disposal, lighting systems, and industrial processes, that, when stopped during any interruption of the normal electrical supply, could create hazards or hamper rescue or fire-fighting operations.

701.3 Application of Other Articles.

Except as modified by this article, all applicable articles of this *Code* shall apply.

701.4 Equipment Approval.

All equipment shall be approved for the intended use.

701.5 Tests and Maintenance for Legally Required Standby Systems.

(A) **Conduct or Witness Test.** The authority having jurisdiction shall conduct or witness a test of the complete system upon installation.

(B) **Tested Periodically.** Systems shall be tested periodically on a schedule and in a manner acceptable to the authority having jurisdiction to ensure the systems are maintained in proper operating condition.

(C) **Battery Systems Maintenance.** Where batteries are used for control, starting, or ignition of prime movers, the authority having jurisdiction shall require periodic maintenance.

(D) **Written Record.** A written record shall be kept on such tests and maintenance.

(E) **Testing Under Load.** Means for testing legally required standby systems under load shall be provided.

FPN: For testing and maintenance procedures of emergency power supply systems (EPSSs), see NFPA 110-1999, *Standard for Emergency and Standby Power Systems*.

701.6 Capacity and Rating.

A legally required standby system shall have adequate capacity and rating for the supply of all equipment intended to be operated at one time. Legally required standby system equipment shall be suitable for the maximum available fault current at its terminals.

The alternate power source shall be permitted to supply legally required standby and optional standby system loads where automatic selective load pickup and load shedding is provided as needed to ensure adequate power to the legally required standby circuits.

701.7 Transfer Equipment.

(A) General. Transfer equipment, including automatic transfer switches, shall be automatic and identified for standby use and approved by the authority having jurisdiction. Transfer equipment shall be designed and installed to prevent the inadvertent interconnection of normal and alternate sources of supply in any operation of the transfer equipment. Transfer equipment and electric power production systems installed to permit operation in parallel with the normal source shall meet the requirements of Article 705.

Revised for the 2002 *NEC*, 701.7(A) now permits transfer equipment to allow parallel operation of the generation equipment with the normal source as long as the requirements of Article 705 are met. Traditional automatic transfer switches are not designed to permit parallel operation of generation equipment and the normal source. Therefore, traditional automatic transfer switches need not comply with Article 705. However, certain automatic transfer switch configurations are intentionally designed to briefly (a few cycles) parallel the generation equipment with the normal source upon load transfer from generator to normal source. This load transfer can occur with minimal disturbance or effect on the load. Transfer switches that employ this type of paralleling must comply with Article 705.

(B) Bypass Isolation Switches. Means to bypass and isolate the transfer switch equipment shall be permitted. Where bypass isolation switches are used, inadvertent parallel operation shall be avoided.

(C) Automatic Transfer Switches. Automatic transfer switches shall be electrically operated and mechanically held.

The intent of 701.7(C) is to ensure that relay contacts will be mechanically held in the event of coil failure. This requirement also correlates the *NEC* with NFPA 110, *Standard for Emergency and Standby Power Systems*.

701.8 Signals.

Audible and visual signal devices shall be provided, where practicable, for the purposes described in 701.8(A), (B), and (C).

(A) Derangement. To indicate derangement of the standby source.

(B) Carrying Load. To indicate that the standby source is carrying load.

(C) Not Functioning. To indicate that the battery charger is not functioning.

FPN: For signals for generator sets, see NFPA 110-1999, *Standard for Emergency and Standby Power Systems*.

701.9 Signs.

(A) Mandated Standby. A sign shall be placed at the service entrance indicating type and location of on-site legally required standby power sources.

Exception: A sign shall not be required for individual unit equipment as specified in 701.11(F).

(B) Grounding. Where the grounded circuit conductor connected to the legally required standby power source is connected to a grounding electrode conductor at a location remote from the legally required standby power source, there shall be a sign at the grounding location that shall identify all legally required standby power and normal sources connected at that location.

II. Circuit Wiring

701.10 Wiring Legally Required Standby Systems.

The legally required standby system wiring shall be permitted to occupy the same raceways, cables, boxes, and cabinets with other general wiring.

III. Sources of Power

701.11 Legally Required Standby Systems.

Current supply shall be such that, in the event of failure of the normal supply to, or within, the building or group of buildings concerned, legally required standby power will be available within the time required for the application but not to exceed 60 seconds. The supply system for legally required standby purposes, in addition to the normal services to the building, shall be permitted to comprise one or more of the types of systems described in 701.11(A) through (E). Unit equipment in accordance with 701.11(F) shall satisfy the applicable requirements of this article.

In selecting a legally required standby source of power, consideration shall be given to the type of service to be rendered, whether of short-time duration or long duration.

Consideration shall be given to the location or design, or both, of all equipment to minimize the hazards that might cause complete failure due to floods, fires, icing, and vandalism.

> FPN: Assignment of degree of reliability of the recognized legally required standby supply system depends on the careful evaluation of the variables at each particular installation.

(A) Storage Battery. A storage battery shall be of suitable rating and capacity to supply and maintain at not less than 87½ percent of system voltage the total load of the circuits supplying legally required standby power for a period of at least 1½ hours.

Batteries, whether of the acid or alkali type, shall be designed and constructed to meet the service requirements of emergency service and shall be compatible with the charger for that particular installation.

For a sealed battery, the container shall not be required to be transparent. However, for the lead acid battery that requires water additions, transparent or translucent jars shall be furnished. Automotive-type batteries shall not be used.

An automatic battery charging means shall be provided.

(B) Generator Set.

(1) Prime Mover-Driven. For a generator set driven by a prime mover acceptable to the authority having jurisdiction and sized in accordance with 701.6, means shall be provided for automatically starting the prime mover upon failure of the normal service and for automatic transfer and operation of all required electrical circuits. A time-delay feature permitting a 15-minute setting shall be provided to avoid re-transfer in case of short-time re-establishment of the normal source.

(2) Internal Combustion Engines as Prime Mover. Where internal combustion engines are used as the prime mover, an on-site fuel supply shall be provided with an on-premise fuel supply sufficient for not less than 2 hours' full-demand operation of the system.

(3) Dual Fuel Supplies. Prime movers shall not be solely dependent on a public utility gas system for their fuel supply or municipal water supply for their cooling systems. Means shall be provided for automatically transferring one fuel supply to another where dual fuel supplies are used.

Exception: Where acceptable to the authority having jurisdiction, the use of other than on-site fuels shall be permitted where there is a low probability of a simultaneous failure of both the off-site fuel delivery system and power from the outside electrical utility company.

(4) Battery Power. Where a storage battery is used for control or signal power, or as the means of starting the prime mover, it shall be suitable for the purpose and shall be equipped with an automatic charging means independent of the generator set.

(5) Outdoor Generator Sets. Where an outdoor housed generator set is equipped with a readily accessible disconnecting means located within sight of the building or structure supplied, an additional disconnecting means shall not be required where ungrounded conductors pass through the building or structure.

Section 701.11(B)(5) was added to the 2002 *NEC* to clarify the requirements for the disconnecting means for an outdoor generator set. The disconnecting means on the generator can be used as disconnecting means required in 225.31, provided the disconnecting means is readily accessible and is within sight of the building. See the definition of the terms *accessible, readily,* and *in sight from* in Article 100.

(C) Uninterruptible Power Supplies. Uninterruptible power supplies used to provide power for legally required standby systems shall comply with the applicable provisions of 701.11(A) and (B).

(D) Separate Service. Where acceptable to the authority having jurisdiction, a second service shall be permitted. This service shall be in accordance with Article 230, with separate service drop or lateral widely separated electrically and physically from the normal service to minimize the possibility of simultaneous interruption of supply.

(E) Connection Ahead of Service Disconnecting Means. Where acceptable to the authority having jurisdiction, connections located ahead of and not within the same cabinet, enclosure, or vertical switchboard section as the service disconnecting means shall be permitted. The legally required standby service shall be sufficiently separated from the normal main service disconnecting means to prevent simultaneous interruption of supply through an occurrence within the building or groups of buildings served.

> FPN: See 230.82 for equipment permitted on the supply side of a service disconnecting means.

Section 230.82 provides requirements for service equipment. These requirements provide safe interruption of available fault current from the utility.

(F) Unit Equipment. Individual unit equipment for legally required standby illumination shall consist of the following:

(1) A rechargeable battery
(2) A battery charging means
(3) Provisions for one or more lamps mounted on the equipment and shall be permitted to have terminals for remote lamps
(4) A relaying device arranged to energize the lamps automatically upon failure of the supply to the unit equipment

The batteries shall be of suitable rating and capacity to supply and maintain at not less than 87½ percent of the nominal battery voltage for the total lamp load associated with the unit for a period of at least 1½ hours, or the unit equipment shall supply and maintain not less than 60 percent of the initial legally required standby illumination for a period of at least 1½ hours. Storage batteries, whether of the acid or alkali type, shall be designed and constructed to meet the requirements of emergency service.

Unit equipment shall be permanently fixed in place (i.e., not portable) and shall have all wiring to each unit installed in accordance with the requirements of any of the wiring methods in Chapter 3. Flexible cord-and-plug connection shall be permitted, provided that the cord does not exceed 900 mm (3 ft) in length. The branch circuit feeding the unit equipment shall be the same branch circuit as that serving the normal lighting in the area and connected ahead of any local switches. Legally required standby luminaires (illumination fixtures) that obtain power from a unit equipment and are not part of the unit equipment shall be wired to the unit equipment by one of the wiring methods of Chapter 3.

Exception: In a separate and uninterrupted area supplied by a minimum of three normal lighting circuits, a separate branch circuit for unit equipment shall be permitted if it originates from the same panelboard as that of the normal lighting circuits and is provided with a lock-on feature.

IV. Overcurrent Protection

701.15 Accessibility.

The branch-circuit overcurrent devices in legally required standby circuits shall be accessible to authorized persons only.

701.17 Ground-Fault Protection of Equipment.

The alternate source for legally required standby systems shall not be required to have ground-fault protection of equipment.

ARTICLE 702
Optional Standby Systems

Contents

I. General

702.1 Scope.

The provisions of this article apply to the installation and operation of optional standby systems.

The systems covered by this article consist of those that are permanently installed in their entirety, including prime movers, and those that are arranged for a connection to a premises wiring system from a portable alternate power supply.

The scope of Article 702 was revised for the 2002 *NEC* to clarify that it applies not only to permanently installed generators and prime movers, but also to portable alternate power supplies that can be connected to an optional standby system. For example, upon failure of an optional standby generator at a frozen food processing plant, a vehicle-mounted generator can be brought in and connected to the plant's optional standby system, which has provisions for such a connection.

Optional standby systems are those in which failure can cause physical discomfort, serious interruption of an industrial process, damage to process equipment, or disruption of business.

702.2 Definition.

Optional Standby Systems. Those systems intended to protect public or private facilities or property where life safety does not depend on the performance of the system.

Optional standby systems are intended to supply on-site generated power to selected loads either automatically or manually.

> FPN: Optional standby systems are typically installed to provide an alternate source of electric power for such facilities as industrial and commercial buildings, farms, and residences and to serve loads such as heating and refrigeration systems, data processing and communications systems, and industrial processes that, when stopped during any power outage, could cause discomfort, serious interruption of the process, damage to the product or process, or the like.

702.3 Application of Other Articles.

Except as modified by this article, all applicable articles of this *Code* shall apply.

702.4 Equipment Approval.

All equipment shall be approved for the intended use.

702.5 Capacity and Rating.

An optional standby system shall have adequate capacity and rating for the supply of all equipment intended to be operated at one time. Optional standby system equipment shall be suitable for the maximum available fault current at its terminals. The user of the optional standby system shall be permitted to select the load connected to the system.

702.6 Transfer Equipment.

Transfer equipment shall be suitable for the intended use and designed and installed so as to prevent the inadvertent interconnection of normal and alternate sources of supply in any operation of the transfer equipment. Transfer equipment and electric power production systems installed to permit operation in parallel with the normal source shall meet the requirements of Article 705.

Transfer equipment, located on the load side of branch circuit protection, shall be permitted to contain supplementary overcurrent protection having an interrupting rating sufficient for the available fault current that the generator can deliver. The supplementary overcurrent protection devices shall be part of a listed transfer equipment.

Transfer equipment shall be required for all standby systems subject to the provisions of this article and for which an electric-utility supply is either the normal or standby source.

Revised for the 2002 *NEC*, 702.6 now permits transfer equipment to allow parallel operation of the generation equipment with the normal source as long as the requirements of Article 705 are met. Traditional automatic transfer switches are not designed to permit parallel operation of generation equip-

ment and the normal source. Therefore, traditional automatic transfer switches need not comply with Article 705. However, certain automatic transfer switch configurations are intentionally designed to briefly (a few cycles) parallel the generation equipment with the normal source upon load transfer from generator to normal source. This load transfer can occur with minimal disturbance or effect on the load. Transfer switches that employ this type of paralleling must comply with Article 705.

702.7 Signals.

Audible and visual signal devices shall be provided, where practicable, for the following purposes.

(1) Derangement. To indicate derangement of the optional standby source.

(2) Carrying Load. To indicate that the optional standby source is carrying load.

702.8 Signs.

(A) Standby. A sign shall be placed at the service-entrance equipment that indicates the type and location of on-site optional standby power sources. A sign shall not be required for individual unit equipment for standby illumination.

(B) Grounding. Where the grounded circuit conductor connected to the optional standby power source is connected to a grounding electrode conductor at a location remote from the optional standby power source, there shall be a sign at the grounding location that shall identify all optional standby power and normal sources connected at that location.

II. Circuit Wiring

702.9 Wiring Optional Standby Systems.

The optional standby system wiring shall be permitted to occupy the same raceways, cables, boxes, and cabinets with other general wiring.

III. Grounding

702.10 Portable Generator Grounding.

(A) Separately Derived System. Where a portable optional standby source is used as a separately derived system, it shall be grounded to a grounding electrode in accordance with 250.30.

(B) Nonseparately Derived System. Where a portable optional standby source is used as a nonseparately derived system, the equipment grounding conductor shall be bonded to the system grounding electrode.

ARTICLE 705
Interconnected Electric Power Production Sources

Contents

705.1 Scope.

This article covers installation of one or more electric power production sources operating in parallel with a primary source(s) of electricity.

> FPN: Examples of the types of primary sources are a utility supply, on-site electric power source(s), or other sources.

Article 705 sets forth basic safety requirements for the installation of generators and other types of power production sources that are interconnected and operate in parallel as distributed generation. Power sources include any systems that produce electric power. They include not only electric utility sources but also on-premises sources ranging from rotating generators (see Article 445) to solar photovoltaic systems (see Article 690) to fuel cells (see Article 692).

Article 705 addresses the basic safety requirements, specifically related to parallel operation, for the generators and other power sources, the power system that interconnects the power sources, and the equipment that is connected to these systems. The proper application of these systems requires a thorough review of the entire power system.

705.2 Definition.

For purposes of this article, the following definition applies.

Interactive System. An electric power production system that is operating in parallel with and capable of delivering energy to an electric primary source supply system.

705.3 Other Articles.

Interconnected electric power production sources shall comply with this article and also the applicable requirements of the articles in Table 705.3.

Table 705.3 Other Articles

Equipment/System	Article
Generators	445
Emergency systems	700
Legally required standby systems	701
Optional standby systems	702

Exception No. 1: Installation of solar photovoltaic systems operated as interconnected power sources shall be in accordance with Article 690.

Exception No. 2: Installation of fuel cell systems operated as interconnected power sources shall be in accordance with Article 692.

705.10 Directory.

A permanent plaque or directory, denoting all electrical power sources on or in the premises, shall be installed at each service equipment location and at locations of all electric power production sources capable of being interconnected.

Exception: Installations with large numbers of power production sources shall be permitted to be designated by groups.

705.12 Point of Connection.

The outputs of electric power production systems shall be interconnected at the premises service disconnecting means.

(A) Integrated Electric System. The outputs shall be permitted to be interconnected at a point or points elsewhere on the premises where the system qualifies as an integrated electric system and incorporates protective equipment in accordance with all applicable sections of Article 685.

(B) General. The outputs shall be permitted to be interconnected at a point or points elsewhere on the premises where all of the following conditions are met:

(1) The aggregate of nonutility sources of electricity has a capacity in excess of 100 kW, or the service is above 1000 volts.
(2) The conditions of maintenance and supervision ensure that qualified persons service and operate the system.
(3) Safeguards and protective equipment are established and maintained.

The point of interconnection must be at the premises service disconnecting means, which may be difficult to accomplish for systems larger than 100 kW. This requirement is intended to prevent the indiscriminate interconnection of small generators or other sources of power without proper protection against fire and electric shock. (See Exhibit 705.1.) It is important to use disconnect devices (switches, etc.) that are suitable for the purpose.

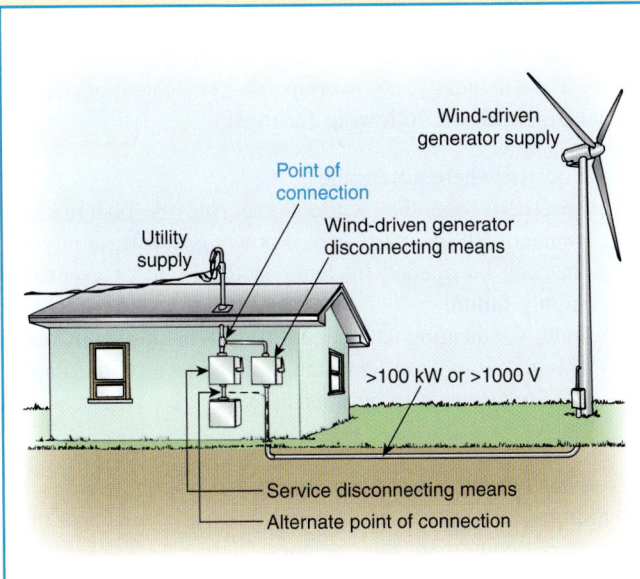

Exhibit 705.1 The point of interconnection, located at the premises service disconnecting means, as required by 705.12.

The requirement specifying "at the premises service disconnecting means" permits connection ahead of the disconnect or on the load side. This practice accommodates the safe work practices of many utilities, which provide a readily accessible disconnect for distributed generation.

Sections 705.12(A) and (B) recognize that generators and other power sources can be safely connected elsewhere on the premises system. These locations include where the premises has an integrated electrical system as set forth in Article 685, where the total generator capacity on premises is greater than 100 kW, and where the service is greater than 1000 volts.

705.14 Output Characteristics.

The output of a generator or other electric power production source operating in parallel with an electric supply system shall be compatible with the voltage, wave shape, and frequency of the system to which it is connected.

> FPN: The term *compatible* does not necessarily mean matching the primary source wave shape.

The level and quality of output power of interconnected sources must be controlled to facilitate proper operation of the interconnected generator or other power production source and the electric system. Control of the generator or power production source should include real power, reactive power, and harmonic content of the output. Section 705.14 states that the interconnected equipment must be compatible with the electric supply system in voltage, wave shape, and frequency.

The output characteristics of a rotating generator are significantly different from those of a solid-state power source. Their compatibility with other sources and with different types of loads will be limited in different ways. Control of the driver speed causes real power (kW) to flow from an induction generator. Control of the prime mover torque causes real power (kW) to flow from a synchronous generator. Control of voltage causes reactive power (kVAr) to flow to or from a synchronous generator. Induction generators have no means to control reactive power (kVAr) flow and continuously draw reactive power. The parallel operation of generators is a complex balance of several variables. These variables are design parameters and are therefore beyond the scope of the *Code*.

Where either the power source or the loads have solid-state equipment, such as inverters, uninterruptible power supplies (UPS), or solid-state variable speed drives, harmonic currents may flow in the system. (See Exhibits 705.2, 705.3, and 705.4.) These multiples of the basic supply frequency (usually 60 Hz) cause additional heating, which may require derating of generators, transformers, cables, and motors. Special generator voltage control systems are required to avoid erratic operation or destruction of control devices. Circuit breakers may require derating if the higher harmonics become significant.

Significant magnitudes of harmonics may be inadvertently matched to system resonance and result in the opening of capacitor fuses, overheating of circuits, and erratic operation of controls. The usual solution is to detune the system by rearrangement or installation of reactors, or both.

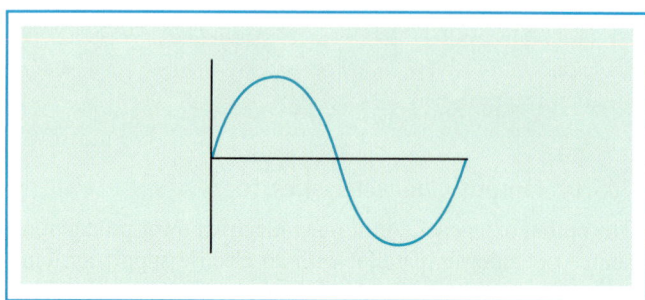

Exhibit 705.2 Typical output wave shape of rotating generator and system wave shape normally encountered with motor, lighting, and heating loads.

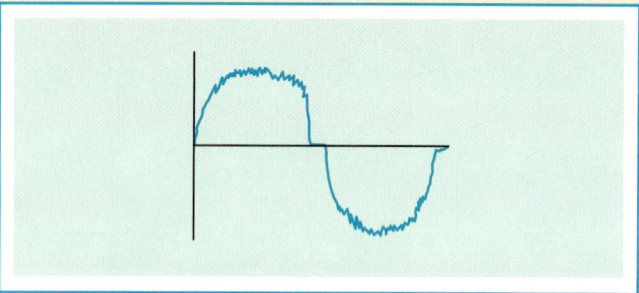

Exhibit 705.3 Typical output wave shape with inverter source. Motors and transformers will be driven by harmonic-rich voltage and may require derating.

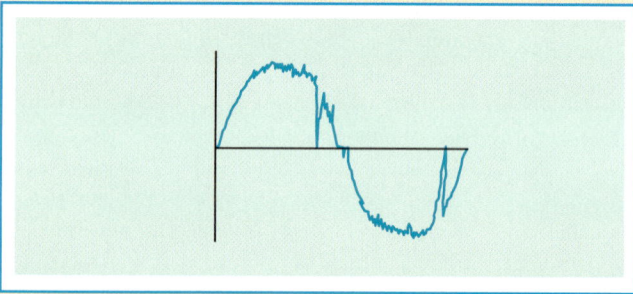

Exhibit 705.4 Wave shape typical of system with variable speed drive, rectifier, elevator, and uninterruptible power to supply loads. Source generator may require derating, and special voltage control may be needed.

705.16 Interrupting and Short-Circuit Current Rating.

Consideration shall be given to the contribution of fault currents from all interconnected power sources for the interrupting and short-circuit current ratings of equipment on interactive systems.

705.20 Disconnecting Means, Sources.

Means shall be provided to disconnect all ungrounded conductors of an electric power production source(s) from all other conductors.

705.21 Disconnecting Means, Equipment.

Means shall be provided to disconnect equipment, such as inverters or transformers associated with a power production source, from all ungrounded conductors of all sources of supply. Equipment intended to be operated and maintained as an integral part of a power production source exceeding 1000 volts shall not be required to have a disconnecting means.

An example of equipment that is covered under the second sentence of 705.21 is a generator designed for 4160 volts and connected to a 13,800-volt system through a transformer. In this example, the transformer and generator are to be operated as a unit; therefore, a disconnecting means is not required between the generator and transformer. A disconnecting means, however, is required between the transformer and the point of connection to the power system. Section 445.18 addresses the general requirements for disconnecting means associated with generators.

705.22 Disconnect Device.

The disconnecting means for ungrounded conductors shall consist of a manually or power operable switch(es) or circuit breaker(s) with the following features:

(1) Located where accessible
(2) Externally operable without exposing the operator to contact with live parts and, if power operable, of a type that can be opened by hand in the event of a power supply failure
(3) Plainly indicating whether in the open or closed position
(4) Having ratings not less than the load to be carried and the fault current to be interrupted

For disconnect equipment energized from both sides, a marking shall be provided to indicate that all contacts of the disconnect equipment may be energized.

FPN No. 1: In parallel generation systems, some equipment, including knife blade switches and fuses, is likely to be energized from both directions. See 240.40.
FPN No. 2: Interconnection to an off-premises primary source could require a visibly verifiable disconnecting device.

The requirements for disconnects in 705.22 are very important. A disconnecting means must serve each generating source. This disconnecting means will be the service-entrance disconnect. Still another disconnecting means may be applied to separate the generating systems.

The basic requirement in 705.22 recognizes the success

of applying switches as well as circuit breakers as the disconnecting means for ungrounded conductors. Most safe work practices on the premises use these disconnect devices.

The disconnect at the service entrance is required for disconnecting the premises wiring system from the utility. The utility safe work practices may also use this disconnect device. Utility safe work practices may require a visibly verifiable disconnect device. For this reason, some utility contracts require that a visible break be provided. The second fine print note to 705.22(4) brings attention to this common utility requirement.

705.30 Overcurrent Protection.

Conductors shall be protected in accordance with Article 240. Equipment and conductors connected to more than one electrical source shall have a sufficient number of overcurrent devices located so as to provide protection from all sources.

(A) Generators. Generators shall be protected in accordance with 445.12.

(B) Solar Photovoltaic Systems. Solar photovoltaic systems shall be protected in accordance with Article 690.

(C) Transformers. Overcurrent protection for a transformer with a source(s) on each side shall be provided in accordance with 450.3 by considering first one side of the transformer, then the other side of the transformer, as the primary.

(D) Fuel Cell Systems. Fuel cell systems shall be protected in accordance with Article 692.

705.32 Ground-Fault Protection.

Where ground-fault protection is used, the output of an interactive system shall be connected to the supply side of the ground-fault protection.

Exception: Connection shall be permitted to be made to the load side of ground-fault protection, provided that there is ground-fault protection for equipment from all ground-fault current sources.

Larger optional standby systems have become very sophisticated and may involve many special relays. The more common switchgear-type relays include under- and over-voltage, under- and over-frequency, voltage-restrained overcurrent, anti-motoring, loss-of-excitation, over-temperature, and shutdown for derangement of the mechanical driver.

Small generator installations cannot justify the cost of switchgear-type relays. Therefore, small generator protection is comprised of more common devices with fewer features. Application guides for relay protection are available in manufacturers' technical literature.

The requirements in Article 705 are only for protection against conditions that may occur because generating sources are being operated in parallel. The installation must also meet the requirements of the referenced articles to provide the basic protection and safeguards for all equipment, whether or not the equipment is involved with more than one source.

705.40 Loss of Primary Source.

Upon loss of primary source, an electric power production source shall be automatically disconnected from all ungrounded conductors of the primary source and shall not be reconnected until the primary source is restored.

FPN No. 1: Risks to personnel and equipment associated with the primary source could occur if an interactive electric power production source can operate as an island. Special detection methods can be required to determine that a primary source supply system outage has occurred and whether there should be automatic disconnection. When the primary source supply system is restored, special detection methods can be required to limit exposure of power production sources to out-of-phase reconnection.

FPN No. 2: Induction-generating equipment on systems with significant capacitance can become self-excited upon loss of primary source and experience severe overvoltage as a result.

When two interconnected power systems separate, they drift out of synchronism. When the utility and the interconnected power system separate, there is a risk of damage to the system if restoration of the utility occurs out of phase. If the timing of the reconnection is random, then violent electromechanical stresses can destroy mechanical components such as gears, couplings, and shafts, and can displace coils. Therefore the premises wiring system or generator must be disconnected from the primary source.

Many technical guides are available from which to select appropriate protective systems and equipment for large systems. A limited choice of low-cost devices is still available for application to small systems. Induction generators are commonly used today. Induction generators have characteristics quite different from synchronous machines. They are more rugged because of the construction of the rotor, and they are less expensive because of their basic design, availability, and type of starting and control equipment. Theoretically, an induction machine can continue to run on an isolated system if a large capacitor bank provides excitation. In reality, an induction machine will probably lose stability and be shut down quickly by one of the protective devices.

705.42 Unbalanced Interconnections.

A 3-phase electric power production source shall be automatically disconnected from all ungrounded conductors of the interconnected systems when one of the phases of that source opens. This requirement shall not be applicable to an electric power production source providing power for an emergency or legally required standby system.

705.43 Synchronous Generators.

Synchronous generators in a parallel system shall be provided with the necessary equipment to establish and maintain a synchronous condition.

705.50 Grounding.

Interconnected electric power production sources shall be grounded in accordance with Article 250.

Exception: For direct-current systems connected through an inverter directly to a grounded service, other methods that accomplish equivalent system protection and that utilize equipment listed and identified for the use shall be permitted.

ARTICLE 720
Circuits and Equipment Operating at Less Than 50 Volts

Contents

720.1 Scope.

This article covers installations operating at less than 50 volts, direct current or alternating current.

720.2 Other Articles.

Installations operating at less than 50 volts, direct current or alternating current as covered in Articles 411, 551, 650, 669, 690, 725, and 760 shall not be required to comply with this article.

Article 411 covers lighting systems operating at 30 volts or less.

720.3 Hazardous (Classified) Locations.

Installations coming within the scope of this article and installed in hazardous (classified) locations shall also comply with the appropriate provisions of Articles 500 through 517.

Low voltage alone does not render a circuit incapable of igniting flammable atmospheres. Ordinary flashlights using two 1½-volt D-cell batteries can become a source of ignition in some hazardous (classified) locations.

720.4 Conductors.

Conductors shall not be smaller than 12 AWG copper or equivalent. Conductors for appliance branch circuits supplying more than one appliance or appliance receptacle shall not be smaller than 10 AWG copper or equivalent.

720.5 Lampholders.

Standard lampholders that have a rating of not less than 660 watts shall be used.

720.6 Receptacle Rating.

Receptacles shall have a rating of not less than 15 amperes.

720.7 Receptacles Required.

Receptacles of not less than 20-ampere rating shall be provided in kitchens, laundries, and other locations where portable appliances are likely to be used.

720.8 Overcurrent Protection.

Overcurrent protection shall comply with Article 240.

720.9 Batteries.

Installations of storage batteries shall comply with Article 480.

720.10 Grounding.

Grounding shall be as provided in Article 250.

720.11 Mechanical Execution of Work.

Circuits operating at less than 50 volts shall be installed in a neat and workmanlike manner. Cables shall be supported by the building structure in such a manner that the cable will not be damaged by normal building use.

Section 720.11 requires cables to be installed in a manner that is consistent with standard industry practice.

ARTICLE 725
Class 1, Class 2, and Class 3 Remote-Control, Signaling, and Power-Limited Circuits

Contents

725.61 Applications of Listed Class 2, Class 3, and PLTC Cables
- (A) Plenum
- (B) Riser
- (C) Cable Trays
- (D) Hazardous (Classified) Locations
- (E) Other Wiring Within Buildings
- (F) Cross-Connect Arrays
- (G) Class 2 and Class 3 Cable Uses and Permitted Substitutions

725.71 Listing and Marking of Class 2, Class 3, and Type PLTC Cables
- (A) Types CL2P and CL3P
- (B) Types CL2R and CL3R
- (C) Types CL2 and CL3
- (D) Types CL2X and CL3X
- (E) Type PLTC
- (F) Class 2 and Class 3 Cable Voltage Ratings
- (G) Class 3 Single Conductors
- (H) Marking

I. General

725.1 Scope.

This article covers remote-control, signaling, and power-limited circuits that are not an integral part of a device or appliance.

> FPN: The circuits described herein are characterized by usage and electrical power limitations that differentiate them from electric light and power circuits; therefore, alternative requirements to those of Chapters 1 through 4 are given with regard to minimum wire sizes, derating factors, overcurrent protection, insulation requirements, and wiring methods and materials.

The scope of Article 725 may include such systems as burglar alarm circuits (see Exhibit 725.1), access control circuits, sound circuits, nurse call circuits, intercom circuits, some computer network systems, some control circuits for lighting dimmer systems, and some low-voltage control circuits that originate from listed appliances or originate from listed computer equipment. Article 760 covers fire alarm circuits and systems. Article 800 covers communications circuits. Article 830 covers network-powered broadband communications circuits.

The installation requirements for the low-voltage wiring of information technology equipment (electronic data processing and computer equipment) located within the confines of a room that is constructed according to the requirements of NFPA 75, *Standard for the Protection of Electronic Computer/Data Processing Equipment,* are not covered by Article 725. Low-voltage wiring within these specially constructed rooms is covered in Article 645.

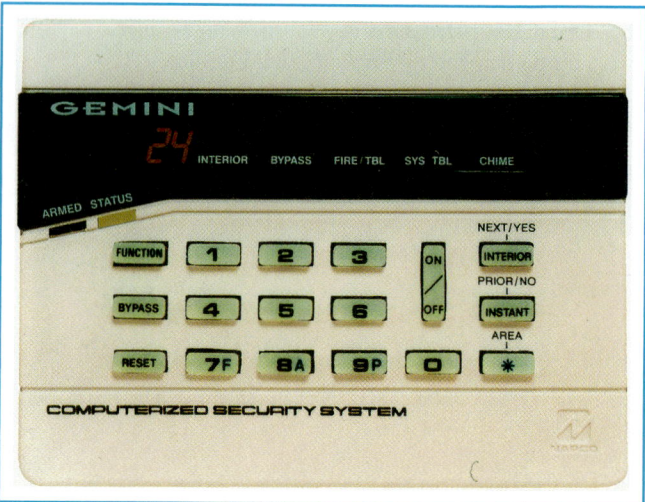

Exhibit 725.1 Typical burglar alarm system keypad. (Courtesy of NAPCO Security Systems, Inc.)

Also, if listed computer equipment is interconnected and all of the interconnected equipment is in close proximity, the wiring is considered an integral part of the equipment and therefore not subject to the requirements of Article 725. If the wiring leaves the group of equipment to connect to other devices in the same room or elsewhere in the building, the wiring is considered "wiring within buildings" and is subject to the requirements of Article 725.

The format for this article is similar to the format for Articles 760, 770, 800, 820, and 830.

General Discussion of Remote-Control, Signaling, and Power-Limited Circuits

The wiring methods required by Chapters 1 through 4 of the *Code* apply to remote-control, signaling, and power-limited circuits, except as amended by Article 725 for specified conditions.

A remote-control, signaling, or power-limited circuit is the portion of the wiring system between the load side of the overcurrent device or the power-limited supply and all connected equipment. The circuit is categorized as Class 1, Class 2, or Class 3.

Class 1 circuits are not permitted to exceed 600 volts. In many cases, Class 1 circuits are extensions of power systems and are subject to the requirements of the power systems, except under the following conditions:

1. Conductors sized 16 AWG and 18 AWG may be used if properly protected against overcurrent (see 725.23).
2. Where damage to the circuit would introduce a hazard, the circuit must be mechanically protected by a suitable means [see 725.8(B)].
3. The adjustment factors of 310.15(B)(2) apply only if such conductors carry a continuous load (see 725.28

for the exact requirements for adjustment factors affecting ampacity).

Class 1 remote-control circuits are commonly used to operate motor controllers in conjunction with moving equipment or mechanical processes, elevators, conveyors, and other such equipment. Class 1 remote-control circuits may also be used as shunt trip circuits for circuit breakers. Class 1 signaling circuits often operate at 120 volts but are not limited to this value.

Conductors and equipment on the supply side of overcurrent protection, transformers, or current-limiting devices of Class 2 and Class 3 circuits must be installed according to the applicable requirements of Chapter 3. Load-side conductors and equipment must comply with Article 725. Class 2 and Class 3 conductors are required to be separated from, and not occupy, the same raceways, cable trays, cables, or enclosures as electric light, power, and Class 1 conductors. Exceptions are noted in 725.55.

Dry-cell batteries are considered Class 2 power supplies, provided the voltage is 30 volts or less and the capacity is equal to or less than that available from series-connected No. 6 carbon-zinc cells. (A No. 6 dry-cell battery is cylindrically shaped with nominal dimensions of 2.5 in. in diameter by 6 in. tall and weighing just over 2 lb. A No. 6 dry-cell battery is about 10 times the volume of the standard D-cell battery commonly used in flashlights.)

Circuits originating from thermocouples are classified Class 2 circuits also. Neither dry-cell batteries nor thermocouples are required to be listed.

725.2 Definitions.

For purposes of this article, the following definitions apply.

Abandoned Class 2, Class 3, and PLTC Cable. Installed Class 2, Class 3, and PLTC cable that is not terminated at equipment and not identified for future use with a tag.

This definition has been added to the 2002 *Code* for use with 725.3(B), which now requires removal of accessible abandoned Class 2, Class 3, and PLTC cables. Abandoned cable increases fire loading unnecessarily, and, where installed in plenums, it can affect airflow. Similar requirements can be found in Articles 640, 645, 725, 760, 770, 820, and 830.

Class 1 Circuit. The portion of the wiring system between the load side of the overcurrent device or power-limited supply and the connected equipment. The voltage and power limitations of the source are in accordance with 725.21.

Class 2 Circuit. The portion of the wiring system between the load side of a Class 2 power source and the connected equipment. Due to its power limitations, a Class 2 circuit considers safety from a fire initiation standpoint and provides acceptable protection from electric shock.

Class 3 Circuit. The portion of the wiring system between the load side of a Class 3 power source and the connected equipment. Due to its power limitations, a Class 3 circuit considers safety from a fire initiation standpoint. Since higher levels of voltage and current than Class 2 are permitted, additional safeguards are specified to provide protection from an electric shock hazard that could be encountered.

725.3 Locations and Other Articles.

Circuits and equipment shall comply with the articles or sections listed in 725.3(A) through (F). Only those sections of Article 300 referenced in this article shall apply to Class 1, Class 2, and Class 3 circuits.

(A) Number and Size of Conductors in Raceway. Section 300.17.

(B) Spread of Fire or Products of Combustion. Section 300.21. The accessible portion of abandoned Class 2, Class 3, and PLTC cables shall not be permitted to remain.

(C) Ducts, Plenums, and Other Air-Handling Spaces. Section 300.22 for Class 1, Class 2, and Class 3 circuits installed in ducts, plenums, or other space used for environmental air. Type CL2P or CL3P cables shall be permitted for Class 2 and Class 3 circuits.

See the commentary following 300.22(B), 300.22(C), and 725.71(A), FPN for more information on wiring in ducts, plenums, and other air-handling spaces.

(D) Hazardous (Classified) Locations. Articles 500 through 516 and Article 517, Part IV, where installed in hazardous (classified) locations.

(E) Cable Trays. Article 392, where installed in cable tray.

(F) Motor Control Circuits. Article 430, Part VI, where tapped from the load side of the motor branch-circuit protective device(s) as specified in 430.72(A).

725.5 Access to Electrical Equipment Behind Panels Designed to Allow Access.

Access to electrical equipment shall not be denied by an accumulation of wires and cables that prevents removal of panels, including suspended ceiling panels.

The excess accumulation of wires and cables can limit access to electrical equipment by preventing the removal of access panels. To service, rearrange, or install electrical equipment, the worker must have an accessible work space for working safely on energized equipment. See 300.11(A), which permits the use of support wires and approved fittings that are independent of the suspended ceiling support wires. (See Exhibit 725.2; also see 725.7.)

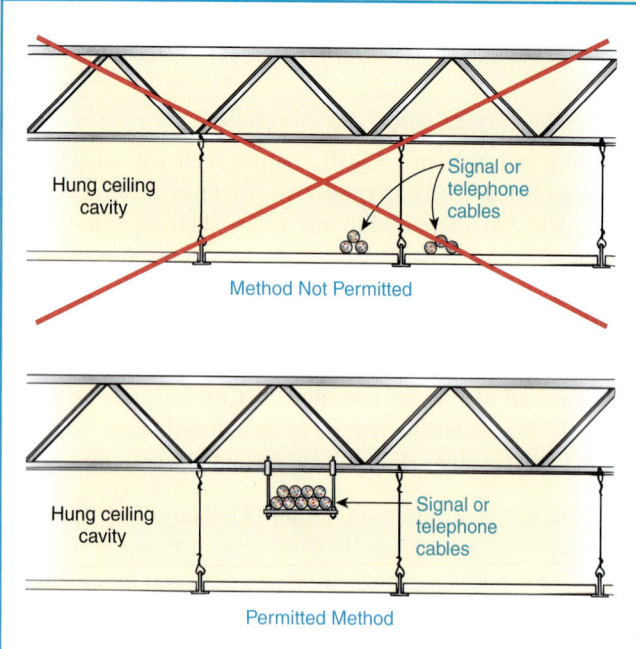

Exhibit 725.2 Installations of wires and cables, which can prevent access to equipment or cables. Correct and incorrect methods are shown.

725.6 Mechanical Execution of Work.

Class 1, Class 2, and Class 3 circuits shall be installed in a neat and workmanlike manner. Cables and conductors installed exposed on the outer surface of ceiling and sidewalls shall be supported by structural components of the building in such a manner that the cable or conductors will not be damaged by normal building use. Such cables shall be attached to structural components by straps, staples, hangers, or similar fittings designed and installed so as not to damage the cable. The installation shall also conform with 300.4(D).

The requirement in 725.6 makes it clear that the *Code* requires cables to be installed in a neat and workmanlike manner. Revised for the 2002 *Code*, this section now provides definitive requirements for workmanship. Cable must now be attached to or supported by the structure by straps, clamps, hangers, and the like. The installation method must not damage the cable. In addition, the location of the cable

should be carefully evaluated to ensure that activities and processes within the building do not cause damage to the cable. [See Exhibit 725.2 and 725.54(D).] The reference to 300.4(D) calls attention to the hazards to which cables are exposed where they are installed on framing members. Such cables are required to be installed in a manner that protects them from nail or screw penetration.

725.8 Safety-Control Equipment.

(A) Remote-Control Circuits. Remote-control circuits for safety-control equipment shall be classified as Class 1 if the failure of the equipment to operate introduces a direct fire or life hazard. Room thermostats, water temperature regulating devices, and similar controls used in conjunction with electrically controlled household heating and air conditioning shall not be considered safety-control equipment.

(B) Physical Protection. Where damage to remote-control circuits of safety control equipment would introduce a hazard, as covered in 725.8(A), all conductors of such remote-control circuits shall be installed in rigid metal conduit, intermediate metal conduit, rigid nonmetallic conduit, electrical metallic tubing, Type MI cable, Type MC cable, or be otherwise suitably protected from physical damage.

The remote-control circuits to safety-control devices may be required to be classified as Class 1 if failure of the safety-control circuit could cause a direct fire or life hazard. One example of a direct fire hazard could be a boiler explosion caused by the failure of the low-water cutoff circuit. See Exhibit 725.3. This is an example of a direct link between failure and the initiation of a hazard, that is, where the failure of the equipment itself causes the hazard.

Generally, fire alarm systems and nurses' call systems do not fit this category. These systems do not have a direct link to the initiation of fire or the initiation of a life hazard but, rather, serve as the reporting or warning link of a hazard initiated by some other (indirect) cause.

725.9 Class 1, Class 2, and Class 3 Circuit Grounding.

Class 1, Class 2, and Class 3 circuits and equipment shall be grounded in accordance with Article 250.

725.10 Class 1, Class 2, and Class 3 Circuit Identification.

Class 1, Class 2, and Class 3 circuits shall be identified at terminal and junction locations, in a manner that prevents unintentional interference with other circuits during testing and servicing.

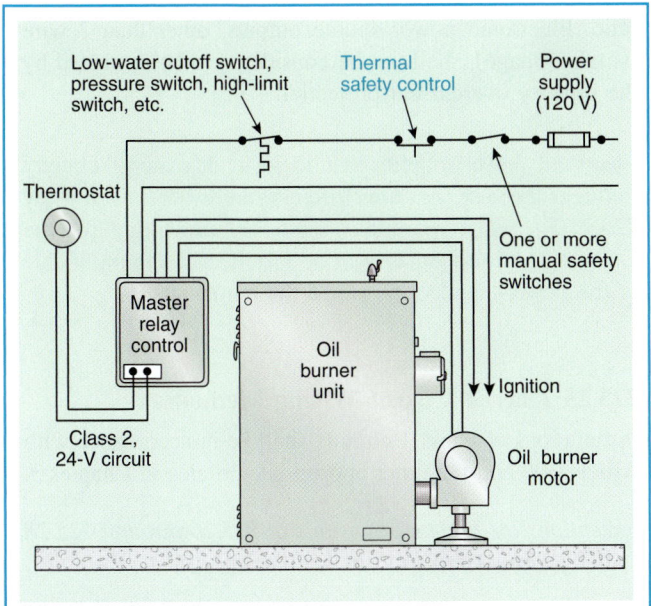

Exhibit 725.3 Typical installation of an automatic oil burner unit for a boiler employing a safety shutdown circuit, according to the requirements of 725.8.

725.15 Class 1, Class 2, and Class 3 Circuit Requirements.

A remote-control, signaling, or power-limited circuit shall comply with the following parts of this article:

(1) Class 1 Circuits, Parts I and II
(2) Class 2 and Class 3 Circuits, Parts I and III

II. Class 1 Circuits

725.21 Class 1 Circuit Classifications and Power Source Requirements.

Class 1 circuits shall be classified as either Class 1 power-limited circuits where they comply with the power limitations of 725.21(A) or as Class 1 remote-control and signaling circuits where they are used for remote control or signaling purposes and comply with the power limitations of 725.21(B).

(A) Class 1 Power-Limited Circuits. These circuits shall be supplied from a source that has a rated output of not more than 30 volts and 1000 volt-amperes.

(1) Class 1 Transformers. Transformers used to supply power-limited Class 1 circuits shall comply with Article 450.

(2) Other Class 1 Power Sources. Power sources other than transformers shall be protected by overcurrent devices

rated at not more than 167 percent of the volt-ampere rating of the source divided by the rated voltage. The overcurrent devices shall not be interchangeable with overcurrent devices of higher ratings. The overcurrent device shall be permitted to be an integral part of the power supply.

To comply with the 1000 volt-ampere limitation of 725.21(A), the maximum output (VA_{max}) of power sources other than transformers shall be limited to 2500 volt-amperes, and the product of the maximum current (I_{max}) and maximum voltage (V_{max}) shall not exceed 10,000 volt-amperes. These ratings shall be determined with any overcurrent-protective device bypassed.

VA_{max} is the maximum volt-ampere output after one minute of operation regardless of load and with overcurrent protection bypassed, if used. Current-limiting impedance shall not be bypassed when determining VA_{max}.

I_{max} is the maximum output current under any noncapacitive load, including short circuit, and with overcurrent protection bypassed, if used. Current-limiting impedance should not be bypassed when determining I_{max}. Where a current-limiting impedance, listed for the purpose or as part of a listed product, is used in combination with a stored energy source, for example, storage battery, to limit the output current, I_{max} limits apply after 5 seconds.

V_{max} is the maximum output voltage regardless of load with rated input applied.

(B) Class 1 Remote-Control and Signaling Circuits. These circuits shall not exceed 600 volts. The power output of the source shall not be required to be limited.

Often, remote-control and signaling circuits that do not meet the requirements of a Class 2 or a Class 3 circuit are classified Class 1 circuits by default. For example, a listed nurses' call system may contain a power supply with an output of 500 watts at 24 volts. Because this power supply obviously exceeds the maximum permitted output of a Class 2 power supply and the output terminals are not marked to indicate this equipment as suitable for a Class 2 power supply, the output circuit wiring is classified as Class 1 and subject to all the Class 1 circuit requirements.

725.23 Class 1 Circuit Overcurrent Protection.

Overcurrent protection for conductors 14 AWG and larger shall be provided in accordance with the conductor ampacity, without applying the derating factors of 310.15 to the ampacity calculation. Overcurrent protection shall not exceed 7 amperes for 18 AWG conductors and 10 amperes for 16 AWG.

Exception: Where other articles of this Code permit or require other overcurrent protection.

FPN: For example, see 430.72 for motors, 610.53 for cranes and hoists, and 517.74(B) and 660.9 for X-ray equipment.

725.24 Class 1 Circuit Overcurrent Device Location.

Overcurrent devices shall be located as specified in 725.24(A) through (E).

(A) Point of Supply. Overcurrent devices shall be located at the point where the conductor to be protected receives its supply.

(B) Feeder Taps. Class 1 circuit conductors shall be permitted to be tapped, without overcurrent protection at the tap, where the overcurrent device protecting the circuit conductor is sized to protect the tap conductor.

(C) Transformer Taps. Class 1 circuit conductors 14 AWG and larger that are tapped from the load side of the overcurrent-protective device(s) of a controlled light and power circuit shall require only short-circuit and ground-fault protection and shall be permitted to be protected by the branch-circuit overcurrent protective device(s) where the rating of the protective device(s) is not more than 300 percent of the ampacity of the Class 1 circuit conductor.

Section 725.24(C) correlates with 240.4(G).

(D) Primary Side of Transformer. Class 1 circuit conductors supplied by the secondary of a single-phase transformer having only a 2-wire (single-voltage) secondary shall be permitted to be protected by overcurrent protection provided on the primary side of the transformer, provided this protection is in accordance with 450.3 and does not exceed the value determined by multiplying the secondary conductor ampacity by the secondary-to-primary transformer voltage ratio. Transformer secondary conductors other than 2 wire shall not be considered to be protected by the primary overcurrent protection.

Section 725.24(D) correlates with 240.4(F) and 430.72(B), Exception No. 2.

(E) Input Side of Electronic Power Source. Class 1 circuit conductors supplied by the output of a single-phase, listed electronic power source, other than a transformer, having only a 2-wire (single voltage) output for connection to Class 1 circuits shall be permitted to be protected by overcurrent protection provided on the input side of the electronic power source, provided this protection does not exceed the value determined by multiplying the Class 1 circuit conductor ampacity by the output-to-input voltage

ratio. Electronic power source outputs, other than 2 wire (single voltage), shall not be considered to be protected by the primary overcurrent protection.

Electronic power supplies that do not supply energy directly through the use of transformers are now covered by 725.24(E), which permits overcurrent protection for the Class 1 circuit conductors to be installed on the input side of the power source rather than the output side.

725.25 Class 1 Circuit Wiring Methods.

Installations of Class 1 circuits shall be in accordance with Article 300 and the other appropriate articles in Chapter 3.

Exception No. 1: The provisions of 725.26 through 725.28 shall be permitted to apply in installations of Class 1 circuits.

Exception No. 2: Methods permitted or required by other articles of this Code shall apply to installations of Class 1 circuits.

725.26 Conductors of Different Circuits in the Same Cable, Cable Tray, Enclosure, or Raceway.

Class 1 circuits shall be permitted to be installed with other circuits as specified in 725.26(A) and (B).

(A) Two or More Class 1 Circuits. Class 1 circuits shall be permitted to occupy the same cable, cable tray, enclosure, or raceway without regard to whether the individual circuits are alternating current or direct current, provided all conductors are insulated for the maximum voltage of any conductor in the cable, cable tray, enclosure, or raceway.

(B) Class 1 Circuits with Power Supply Circuits. Class 1 circuits shall be permitted to be installed with power supply conductors as specified in 725.26(B)(1) through (B)(4).

(1) In a Cable, Enclosure, or Raceway. Class 1 circuits and power supply circuits shall be permitted to occupy the same cable, enclosure, or raceway only where the equipment powered is functionally associated.

(2) In Factory- or Field-Assembled Control Centers. Class 1 circuits and power supply circuits shall be permitted to be installed in factory- or field-assembled control centers.

(3) In a Manhole. Class 1 circuits and power supply circuits shall be permitted to be installed as underground conductors in a manhole in accordance with one of the following:

Section 725.26(B)(3) permits Class 1 power-limited circuit conductors to be installed in manholes with wiring of non-power-limited systems where suitable separation provides the same degree of protection as required for the following:

1. Class 2 and Class 3 power-limited circuits in 725.55
2. Communications circuits in 800.52(A)
3. Radio/television antennas and lead-in conductors in 800.18(A)
4. CATV conductors in 820.52(A)

(1) The power-supply or Class 1 circuit conductors are in a metal-enclosed cable or Type UF cable.
(2) The conductors are permanently separated from the power-supply conductors by a continuous firmly fixed nonconductor, such as flexible tubing, in addition to the insulation on the wire.
(3) The conductors are permanently and effectively separated from the power supply conductors and securely fastened to racks, insulators, or other approved supports.

(4) In cable trays, where the Class 1 circuit conductors and power-supply conductors not functionally associated with them are separated by a solid fixed barrier of a material compatible with the cable tray, or where the power-supply or Class 1 circuit conductors are in a metal-enclosed cable.

725.27 Class 1 Circuit Conductors.

(A) Sizes and Use. Conductors of sizes 18 AWG and 16 AWG shall be permitted to be used, provided they supply loads that do not exceed the ampacities given in 402.5 and are installed in a raceway, an approved enclosure, or a listed cable. Conductors larger than 16 AWG shall not supply loads greater than the ampacities given in 310.15. Flexible cords shall comply with Article 400.

(B) Insulation. Insulation on conductors shall be suitable for 600 volts. Conductors larger than 16 AWG shall comply with Article 310. Conductors in sizes 18 AWG and 16 AWG shall be Type FFH-2, KF-2, KFF-2, PAF, PAFF, PF, PFF, PGF, PGFF, PTF, PTFF, RFH-2, RFHH-2, RFHH-3, SF-2, SFF-2, TF, TFF, TFFN, TFN, ZF, or ZFF. Conductors with other types and thicknesses of insulation shall be permitted if listed for Class 1 circuit use.

Section 725.27 requires all Class 1 circuit conductors to be rated at 600 volts. This effectively requires Class 1 circuits to be wired using the wiring methods found in Chapter 3, or to use conductors specifically listed for Class 1 circuit use.

725.28 Number of Conductors in Cable Trays and Raceway, and Derating.

(A) Class 1 Circuit Conductors. Where only Class 1 circuit conductors are in a raceway, the number of conductors shall be determined in accordance with 300.17. The derating factors given in 310.15(B)(2)(a) shall apply only if such conductors carry continuous loads in excess of 10 percent of the ampacity of each conductor.

(B) Power-Supply Conductors and Class 1 Circuit Conductors. Where power-supply conductors and Class 1 circuit conductors are permitted in a raceway in accordance with 725.26, the number of conductors shall be determined in accordance with 300.17. The derating factors given in 310.15(B)(2)(a) shall apply as follows:

(1) To all conductors where the Class 1 circuit conductors carry continuous loads in excess of 10 percent of the ampacity of each conductor and where the total number of conductors is more than three
(2) To the power-supply conductors only, where the Class 1 circuit conductors do not carry continuous loads in excess of 10 percent of the ampacity of each conductor and where the number of power-supply conductors is more than three

(C) Class 1 Circuit Conductors in Cable Trays. Where Class 1 circuit conductors are installed in cable trays, they shall comply with the provisions of 392.9 through 392.11.

725.29 Circuits Extending Beyond One Building.

Class 1 circuits that extend aerially beyond one building shall also meet the requirements of Article 225.

III. Class 2 and Class 3 Circuits

725.41 Power Sources for Class 2 and Class 3 Circuits.

(A) Power Source. The power source for a Class 2 or a Class 3 circuit shall be as specified in 725.41(A)(1), (2), (3), (4), or (5):

> FPN No. 1: Figure 725.41 illustrates the relationships between Class 2 or Class 3 power sources, their supply, and the Class 2 or Class 3 circuits.
> FPN No. 2: Tables 11(A) and 11(B) in Chapter 9 provide the requirements for listed Class 2 and Class 3 power sources.

(1) A listed Class 2 or Class 3 transformer
(2) A listed Class 2 or Class 3 power supply
(3) Other listed equipment marked to identify the Class 2 or Class 3 power source

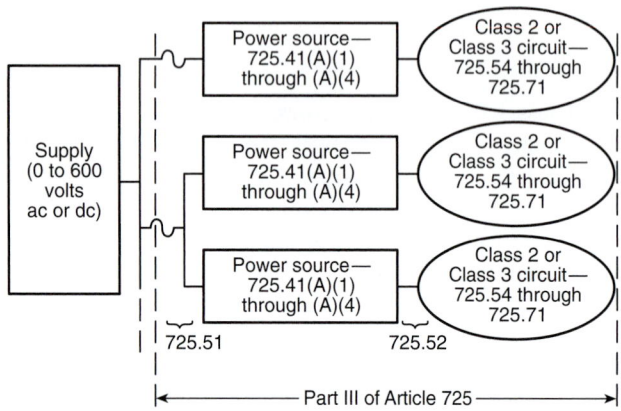

Figure 725.41 Class 2 and Class 3 circuits.

Exception: Thermocouples shall not require listing as a Class 2 power source.

FPN: Examples of other listed equipment are as follows:

(1) A circuit card listed for use as a Class 2 or Class 3 power source where used as part of a listed assembly
(2) A current-limiting impedance, listed for the purpose, or part of a listed product, used in conjunction with a non–power-limited transformer or a stored energy source, for example, storage battery, to limit the output current
(3) A thermocouple

(4) Listed information technology (computer) equipment limited power circuits.

FPN: One way to determine applicable requirements for listing of information technology (computer) equipment is to refer to UL 1950-1995, *Standard for Safety of Information Technology Equipment, Including Electrical Business Equipment.* Typically such circuits are used to interconnect information technology equipment for the purpose of exchanging information (data).

(5) A dry cell battery shall be considered an inherently limited Class 2 power source, provided the voltage is 30 volts or less and the capacity is equal to or less than that available from series connected No. 6 carbon zinc cells.

(B) Interconnection of Power Sources. Class 2 or Class 3 power sources shall not have the output connections paralleled or otherwise interconnected unless listed for such interconnection.

Except for dry-cell batteries and thermocouples, 725.41(B) requires listed power sources for Class 2 and Class 3 circuits.

725.42 Circuit Marking.

The equipment shall be durably marked where plainly visible to indicate each circuit that is a Class 2 or Class 3 circuit.

Section 725.42, new in the 2002 *Code*, requires equipment to be marked to clearly identify Class 2 and Class 3 circuits.

725.51 Wiring Methods on Supply Side of the Class 2 or Class 3 Power Source.

Conductors and equipment on the supply side of the power source shall be installed in accordance with the appropriate requirements of Chapters 1 through 4. Transformers or other devices supplied from electric light or power circuits shall be protected by an overcurrent device rated not over 20 amperes.

Exception: The input leads of a transformer or other power source supplying Class 2 and Class 3 circuits shall be permitted to be smaller than 14 AWG, but not smaller than 18 AWG if they are not over 12 in. (305 mm) long and if they have insulation that complies with 725.27(B).

Listed Class 2 and Class 3 transformers must be protected by an overcurrent device not exceeding 20 amperes, unless the transformers are fed from circuits other than power or lighting.

725.52 Wiring Methods and Materials on Load Side of the Class 2 or Class 3 Power Source.

Class 2 and Class 3 circuits on the load side of the power source shall be permitted to be installed using wiring methods and materials in accordance with either 725.52(A) or (B).

(A) Class 1 Wiring Methods and Materials. Installation shall be in accordance with 725.25.

Exception No. 1: The derating factors that are given in 310.15(B)(2)(a) shall not apply.

Exception No. 2: Class 2 and Class 3 circuits shall be permitted to be reclassified and installed as Class 1 circuits if the Class 2 and Class 3 markings required in 725.42 are eliminated and the entire circuit is installed using the wiring methods and materials in accordance with Part II, Class 1 circuits. Class 2 and Class 3 circuits reclassified and installed as Class 1 circuits shall not be classified as Class 2 or Class 3 circuits, regardless of the continued connection to a Class 2 or Class 3 power source.

(B) Class 2 and Class 3 Wiring Methods. Conductors on the load side of the power source shall be insulated at not less than the requirements of 725.71 and shall be installed in accordance with 725.54 and 725.61.

Exception No. 1: As provided for in 620.21 for elevators and similar equipment.

Exception No. 2: Other wiring methods and materials installed in accordance with the requirements of 725.3 shall be permitted to extend or replace the conductors and cables described in 725.71 and permitted by 725.52(B).

Revised in the 2002 *Code*, 725.52 now clearly permits the following two distinct wiring methods for circuits connected to Class 2 or Class 3 power sources:

- Class 1 wiring methods
- Class 2 and Class 3 wiring methods

Where it is necessary to locate Class 2 or Class 3 circuits inside the same cable or raceway as a Class 1 circuit, 725.52(A), Exception No. 2, permits a Class 2 or Class 3 circuit to be reclassified and installed as Class 1, providing that the Class 2 or Class 3 marking is removed, that overcurrent protection is provided in accordance with 725.23, and that the reclassified circuit maintains separation from other Class 2 and Class 3 circuits in accordance with 725.55.

725.54 Installation of Conductors and Equipment in Cables, Compartments, Cable Trays, Enclosures, Manholes, Outlet Boxes, Device Boxes, and Raceways for Class 2 and Class 3 Circuits.

Conductors and equipment for Class 2 and Class 3 circuits shall be installed in accordance with 725.55 through 725.58.

725.55 Separation from Electric Light, Power, Class 1, Non–Power-Limited Fire Alarm Circuit Conductors, and Medium Power Network-Powered Broadband Communications Cables.

(A) General. Cables and conductors of Class 2 and Class 3 circuits shall not be placed in any cable, cable tray, compartment, enclosure, manhole, outlet box, device box, raceway, or similar fitting with conductors of electric light, power, Class 1, non–power-limited fire alarm circuits, and medium power network-powered broadband communications circuits unless permitted by 725.55(B) through (J).

Section 725.55(A) specifically includes cables of Class 2 and Class 3 circuits. Jackets of listed Class 2 and Class 3 cables do not have sufficient construction specifications to permit them to be installed with electric light, power, Class 1, non–power-limited fire alarm circuits, and medium power network-powered broadband communications cables. Failure of the cable insulation due to a fault could lead to hazardous voltages being imposed on the Class 2 or Class 3 circuit conductors.

(B) Separated by Barriers. Class 2 and Class 3 circuits shall be permitted to be installed together with Class 1, non–power-limited fire alarm and medium power network-powered broadband communications circuits where they are separated by a barrier.

(C) Raceways Within Enclosures. In enclosures, Class 2 and Class 3 circuits shall be permitted to be installed in a raceway to separate them from Class 1, non–power-limited fire alarm and medium power network-powered broadband communications circuits.

(D) Associated Systems Within Enclosures. Class 2 and Class 3 circuit conductors in compartments, enclosures, device boxes, outlet boxes, or similar fittings shall be permitted to be installed with electric light, power, Class 1, non–power-limited fire alarm, and medium power network-powered broadband communications circuits where they are introduced solely to connect the equipment connected to Class 2 and Class 3 circuits, and where (1) or (2) applies:

(1) The electric light, power, Class 1, non–power-limited fire alarm, and medium power network-powered broadband communications circuit conductors are routed to maintain a minimum of 6 mm (0.25 in.) separation from the conductors and cables of Class 2 and Class 3 circuits.
(2) The circuit conductors operate at 150 volts or less to ground and also comply with one of the following:

   a. The Class 2 and Class 3 circuits are installed using Type CL3, CL3R, or CL3P or permitted substitute cables, provided these Class 3 cable conductors extending beyond the jacket are separated by a minimum of 6 mm (0.25 in.) or by a nonconductive sleeve or nonconductive barrier from all other conductors.
   b. The Class 2 and Class 3 circuit conductors are installed as a Class 1 circuit in accordance with 725.21.

(E) Enclosures with Single Opening. Class 2 and Class 3 circuit conductors entering compartments, enclosures, device boxes, outlet boxes, or similar fittings shall be permitted to be installed with Class 1, non–power-limited fire alarm and medium power network-powered broadband communications circuits where they are introduced solely to connect the equipment connected to Class 2 and Class 3 circuits. Where Class 2 and Class 3 circuit conductors must enter an enclosure that is provided with a single opening, they shall be permitted to enter through a single fitting (such as a tee), provided the conductors are separated from the conductors of the other circuits by a continuous and firmly fixed nonconductor, such as flexible tubing.

Power circuit and Class 2 circuit conductors can be permitted in the same motor–starter enclosure, where the Class 2 circuit source is the secondary of a control transformer in the same motor–starter enclosure. In such an installation, the Class 2 conductor insulation is not required to have the same voltage rating as the insulation on the power conductors in the same enclosure.

(F) Manholes. Underground Class 2 and Class 3 circuit conductors in a manhole shall be permitted to be installed with Class 1, non–power-limited fire alarm and medium power network-powered broadband communications circuits where one of the following conditions is met:

(1) The electric light, power, Class 1, non–power-limited fire alarm and medium power network-powered broadband communications circuit conductors are in a metal-enclosed cable or Type UF cable.

(2) The Class 2 and Class 3 circuit conductors are permanently and effectively separated from the conductors of other circuits by a continuous and firmly fixed nonconductor, such as flexible tubing, in addition to the insulation or covering on the wire.

(3) The Class 2 and Class 3 circuit conductors are permanently and effectively separated from conductors of the other circuits and securely fastened to racks, insulators, or other approved supports.

Section 725.55(F) permits the installation of Class 2 and Class 3 power-limited circuit conductors in manholes that have wiring for electric light, power, Class 1, and non–power-limited fire alarm circuit systems, and from medium power network-powered broadband communications cables where suitable separation provides the same degree of protection as required in 725.55(A). [See the commentary following 725.26(B)(3).]

(G) Article 780. Class 2 and Class 3 conductors as permitted by 780.6(A) shall be permitted to be installed in accordance with Article 780.

(H) Cable Trays. Class 2 and Class 3 circuit conductors shall be permitted to be installed in cable trays, where the conductors of the electric light, Class 1, and non–power-limited fire alarm circuits are separated by a solid fixed barrier of a material compatible with the cable tray or where the Class 2 or Class 3 circuits are installed in Type MC cable.

Section 725.55(H) permits mixing of circuits in some cable tray applications because, under these circumstances, jacket damage that results in faults between circuits is very unlikely.

(I) In Hoistways. In hoistways, Class 2 or Class 3 circuit conductors shall be installed in rigid metal conduit, rigid nonmetallic conduit, intermediate metal conduit, liquidtight flexible nonmetallic conduit, or electrical metallic tubing. For elevators or similar equipment, these conductors shall be permitted to be installed as provided in 620.21.

(J) Other Applications. For other applications, conductors of Class 2 and Class 3 circuits shall be separated by at least 50 mm (2 in.) from conductors of any electric light, power, Class 1 non–power-limited fire alarm or medium power network-powered broadband communications circuits unless one of the following conditions is met:

(1) Either (a) all of the electric light, power, Class 1, non–power-limited fire alarm and medium power network-powered broadband communications circuit conductors or (b) all of the Class 2 and Class 3 circuit conductors are in a raceway or in metal-sheathed, metal-clad, non–metallic-sheathed, or Type UF cables.

(2) All of the electric light, power, Class 1 non–power-limited fire alarm, and medium power network-powered broadband communications circuit conductors are permanently separated from all of the Class 2 and Class 3 circuit conductors by a continuous and firmly fixed nonconductor, such as porcelain tubes or flexible tubing, in addition to the insulation on the conductors.

725.56 Installation of Conductors of Different Circuits in the Same Cable, Enclosure, or Raceway.

(A) Two or More Class 2 Circuits. Conductors of two or more Class 2 circuits shall be permitted within the same cable, enclosure, or raceway.

(B) Two or More Class 3 Circuits. Conductors of two or more Class 3 circuits shall be permitted within the same cable, enclosure, or raceway.

(C) Class 2 Circuits with Class 3 Circuits. Conductors of one or more Class 2 circuits shall be permitted within the same cable, enclosure, or raceway with conductors of Class 3 circuits, provided that the insulation of the Class 2 circuit conductors in the cable, enclosure, or raceway is at least that required for Class 3 circuits.

(D) Class 2 and Class 3 Circuits with Communications Circuits.

(1) Classified as Communications Circuits. Class 2 and Class 3 circuit conductors shall be permitted in the same cable with communications circuits, in which case the Class 2 and Class 3 circuits shall be classified as communications circuits and shall be installed in accordance with the requirements of Article 800. The cables shall be listed as communications cables or multipurpose cables.

(2) Composite Cables. Cables constructed of individually listed Class 2, Class 3, and communications cables under a common jacket shall be permitted to be classified as communications cables. The fire resistance rating of the composite cable shall be determined by the performance of the composite cable.

In a typical office environment consisting of a group of computers in a local area network, data wiring is as prevalent as telephone wiring. A common way to minimize the amount of cabling is to run the telephone and data circuits in the same cable. Exhibit 725.4 illustrates such an arrangement. Section 724.56(D) requires that either a communications or multipurpose cable be used for this purpose.

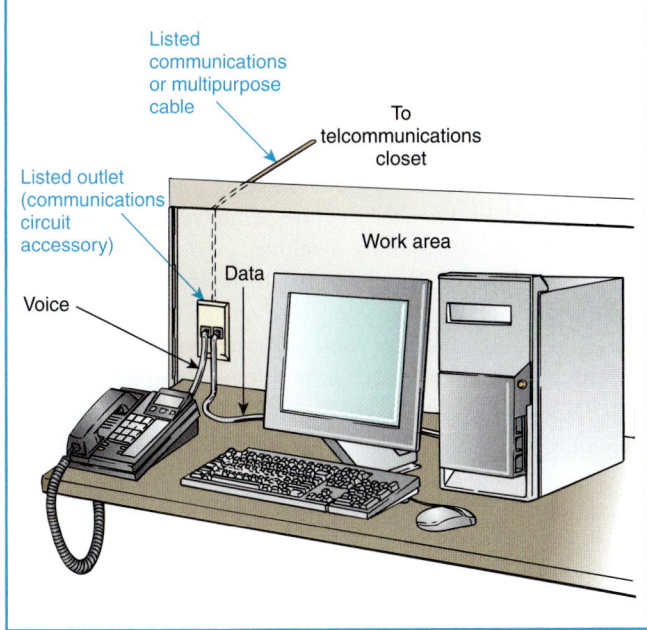

Exhibit 725.4 Telephone and data circuits in the same cable.

(E) Class 2 or Class 3 Cables with Other Circuit Cables. Jacketed cables of Class 2 or Class 3 circuits shall be permitted in the same enclosure or raceway with jacketed cables of any of the following:

(1) Power-limited fire alarm systems in compliance with Article 760
(2) Nonconductive and conductive optical fiber cables in compliance with Article 770
(3) Communications circuits in compliance with Article 800
(4) Community antenna television and radio distribution systems in compliance with Article 820

(5) Low-power, network-powered broadband communications in compliance with Article 830

725.57 Installation of Circuit Conductors Extending Beyond One Building.

Where Class 2 or Class 3 circuit conductors extend beyond one building and are run so as to be subject to accidental contact with electric light or power conductors operating over 300 volts to ground, or are exposed to lightning on interbuilding circuits on the same premises, the requirements of the following shall also apply:

(1) Sections 800.10, 800.12, 800.13, 800.31, 800.32, 800.33, and 800.40 for other than coaxial conductors
(2) Sections 820.10, 820.33, and 820.40 for coaxial conductors

725.58 Support of Conductors.

Class 2 or Class 3 circuit conductors shall not be strapped, taped, or attached by any means to the exterior of any conduit or other raceway as a means of support. These conductors shall be permitted to be installed as permitted by 300.11(B)(2).

See the commentary following 725.6 for more information on the support of conductors.

725.61 Applications of Listed Class 2, Class 3, and PLTC Cables.

Class 2, Class 3, and PLTC cables shall comply with any of the requirements described in 725.61(A) through (F).

Sections 725.61(A) and 725.61(B) have been revised for the 2002 *Code* for use with the definition of *abandoned Class 2, Class 3, and PLTC cable* in 725.2. This section now requires the removal of accessible abandoned cable. Abandoned cable increases fire loading unnecessarily, and, where installed in plenums, it can affect airflow. Similar requirements can be found in Articles 640, 645, 760, 770, 800, 820, and 830. See the definition of *abandoned Class 2, Class 3, and PLTC cable* in 725.2.

(A) Plenum. Cables installed in ducts, plenums, and other spaces used for environmental air shall be Type CL2P or CL3P. Abandoned cables shall not be permitted to remain. Listed wires and cables installed in compliance with 300.22 shall be permitted.

(B) Riser. Cables installed in risers shall be as described in any of (1), (2), or (3):

(1) Cables installed in vertical runs and penetrating more than one floor, or cables installed in vertical runs in a

shaft, shall be Type CL2R or CL3R. Floor penetrations requiring Type CL2R or CL3R shall contain only cables suitable for riser or plenum use. Abandoned cables shall not be permitted to remain.

(2) Other cables as covered in Table 725.61 and other listed wiring methods as covered in Chapter 3 shall be installed in metal raceways or located in a fireproof shaft having firestops at each floor.

(3) Type CL2, CL3, CL2X, and CL3X cables shall be permitted in one- and two-family dwellings.

> FPN: See 300.21 for firestop requirements for floor penetrations.

(C) Cable Trays. Cables installed in cable trays outdoors shall be Type PLTC. Cables installed in cable trays indoors shall be Types PLTC, CL3P, CL3R, CL3, CL2P, CL2R, and CL2.

> FPN: See 800.52(D) for cables permitted in cable trays.

(D) Hazardous (Classified) Locations. Cables installed in hazardous locations shall be as described in 725.61(D)(1) through (D)(4).

(1) Type PLTC. Cables installed in hazardous (classified) locations shall be Type PLTC. Where the use of Type PLTC cable is permitted by 501.4(B), 502.4(B), and 504.20, the cable shall be installed in cable trays, in raceways supported by messenger wire, or otherwise adequately supported and mechanically protected by angles, struts, channels, or other mechanical means. The cable shall be permitted to be directly buried where the cable is listed for this use.

(2) Nonincendive Field Wiring. Wiring for Class 2 circuits as permitted by 501.4(B)(3) shall be permitted.

(3) Thermocouple Circuits. Conductors in Type PLTC cables used for Class 2 thermocouple circuits shall be permitted to be any of the materials used for thermocouple extension wire.

(4) In Industrial Establishments. In industrial establishments where the conditions of maintenance and supervision ensure that only qualified persons service the installation, and where the cable is not subject to physical damage, Type PLTC cable that complies with the crush and impact requirements of Type MC cable and is identified for such use shall be permitted as open wiring between cable tray and utilization equipment in lengths not to exceed 15 m (50 ft). The cable shall be supported and protected against physical damage using mechanical protection such as dedicated struts, angles, or channels. The cable shall be supported and secured at intervals not exceeding 1.75 m (6 ft).

Section 725.61(D)(4) allows limited use of a special Type PLTC cable in open wiring applications between a cable

tray and utilization equipment in industrial establishments. This allowance offers the installer an alternative to Type MC cable. If used in the manner described in 725.61(D)(4), Type PLTC cable must meet the crush and impact requirements of Type MC cable. Type PLTC cable must also be mechanically protected and installed in runs of 50 ft or less.

(E) Other Wiring Within Buildings. Cables installed in building locations other than those covered in 725.61(A) through (D) shall be as described in any of (1) through (6). Abandoned cables in hollow spaces shall not be permitted to remain.

(1) Type CL2 or CL3 shall be permitted.

(2) Type CL2X or CL3X shall be permitted to be installed in a raceway or in accordance with other wiring methods covered in Chapter 3.

(3) Cables shall be permitted to be installed in nonconcealed spaces where the exposed length of cable does not exceed 3 m (10 ft).

(4) Listed Type CL2X cables less than 6 mm (0.25 in.) in diameter and listed Type CL3X cables less than 6 mm (0.25 in.) in diameter shall be permitted to be installed in one- and two-family dwellings.

(5) Listed Type CL2X cables less than 6 mm (0.25 in.) in diameter and listed Type CL3X cables less than 6 mm (0.25 in.) in diameter shall be permitted to be installed in nonconcealed spaces in multifamily dwellings.

(6) Type CMUC undercarpet communications wires and cables shall be permitted to be installed under carpet.

(F) Cross-Connect Arrays. Type CL2 or CL3 conductors or cables shall be used for cross-connect arrays.

Cross-connect arrays and patch panels located within building spaces must be cross wired, jumpered, or interconnected using listed cables.

(G) Class 2 and Class 3 Cable Uses and Permitted Substitutions. The uses and permitted substitutions for Class 2 and Class 3 cables listed in Table 725.61 shall be considered suitable for the purpose and shall be permitted.

> FPN: For information on Types CMP, CMR, CH, and CMX cables, see 800.51.

725.71 Listing and Marking of Class 2, Class 3, and Type PLTC Cables.

Class 2, Class 3, and Type PLTC cables installed as wiring within buildings shall be listed as being resistant to the spread of fire and other criteria in accordance with 725.71(A)

Table 725.61 Cable Uses and Permitted Substitutions

Cable Type	Use	References	Permitted Substitutions
CL3P	Class 3 plenum cable	725.61(A)	CMP
CL2P	Class 2 plenum cable	725.61(A)	CMP, CL3P
CL3R	Class 3 riser cable	725.61(B)	CMP, CL3P, CMR
CL2R	Class 2 riser cable	725.61(B)	CMP, CL3P, CL2P, CMR, CL3R
PLTC	Power-limited tray cable	725.61(C) and (D)	
CL3	Class 3 cable	725.61(B), (E), and (F)	CMP, CL3P, CMR, CL3R, CMG, CM, PLTC
CL2	Class 2 cable	725.61(B), (E), and (F)	CMP, CL3P, CL2P, CMR, CL3R, CL2R, CMG, CM, PLTC, CL3
CL3X	Class 3 cable, limited use	725.61(B) and (E)	CMP, CL3P, CMR, CL3R, CMG, CM, PLTC, CL3, CMX
CL2X	Class 2 cable, limited use	725.61(B) and (E)	CMP, CL3P, CL2P, CMR, CL3R, CL2R, CMG, CM, PLTC, CL3, CL2, CMX, CL3X

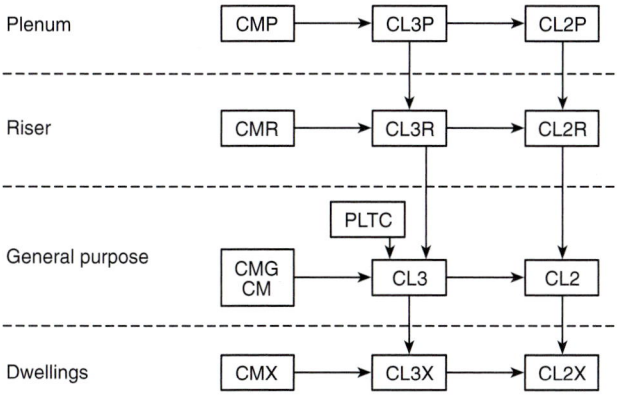

Type CM—Communications wires and cables
Type CL2 and CL3—Class 2 and Class 3 remote-control, signaling, and power-limited cables
Type PLTC—Power-limited tray cable

[A] → [B]   Cable A shall be permitted to be used in place of cable B.

Figure 725.61 Cable substitution hierarchy

through (G) and shall be marked in accordance with 725.71(H).

(A) Types CL2P and CL3P. Types CL2P and CL3P plenum cables shall be listed as being suitable for use in ducts, plenums, and other space used for environmental air and shall also be listed as having adequate fire-resistant and low smoke-producing characteristics.

> FPN: One method of defining *low smoke-producing cable* is by establishing an acceptable value of the smoke produced when tested in accordance with NFPA 262-1999, *Standard Method of Test for Flame Travel and Smoke of Wires and Cables for Use in Air-Handling Spaces,* to a maximum peak optical density of 0.5 and a maximum average optical density of 0.15. Similarly, one method of defining fire-resistant cables is by establishing a maximum allowable flame travel distance of 1.52 m (5 ft) when tested in accordance with the same test.

NFPA 262, *Standard Method of Test for Flame Travel and Smoke of Wires and Cables for Use in Air-Handling Spaces,* is a test method for electrical wires and cables that are to be installed without raceways in plenums and other spaces used for environmental air. NFPA 262 was originally developed as UL 910, which is an adaptation of the Steiner Tunnel Test (NFPA 255/ASTM E84/UL 723).

The test is conducted in a 25-ft-long horizontal duct lined with insulating masonry faced with a row of refractory fire brick. See Exhibit 725.5. One side of the duct is provided with a row of double-paned observation windows that permit monitoring of the progress of the test. One end of the chamber, designated the "fire end," is provided with two gas burners that deliver flames upward to engulf the test specimen. The burners are positioned on each side of the centerline of the furnace so that flame is evenly distributed across the sample. Test specimens are installed in a cable tray that is 23.9 ft long.

Smoke output is measured in a metal vent pipe at the vent end of the furnace, which is opposite the fire end. Where smoke passes through the vent pipe, it passes between a light source and a photocell. The output signal of the photocell is directly proportional to the amount of light received.

The cable specimens are exposed to a fire source with an output of 300,000 Btu/hour for a period of 20 minutes. A graph is plotted, illustrating the flame distance beyond 4½ ft (measured from the end of the flame delivered by the burners) versus the time taken for the duration of the test. This graph is plotted against a flame spread curve representative of a red oak specimen. A graph is also plotted for optical density versus time for the sample. Optical density is calculated as follows:

$$\text{Optical density} = \log_{10} \frac{T_0}{T}$$

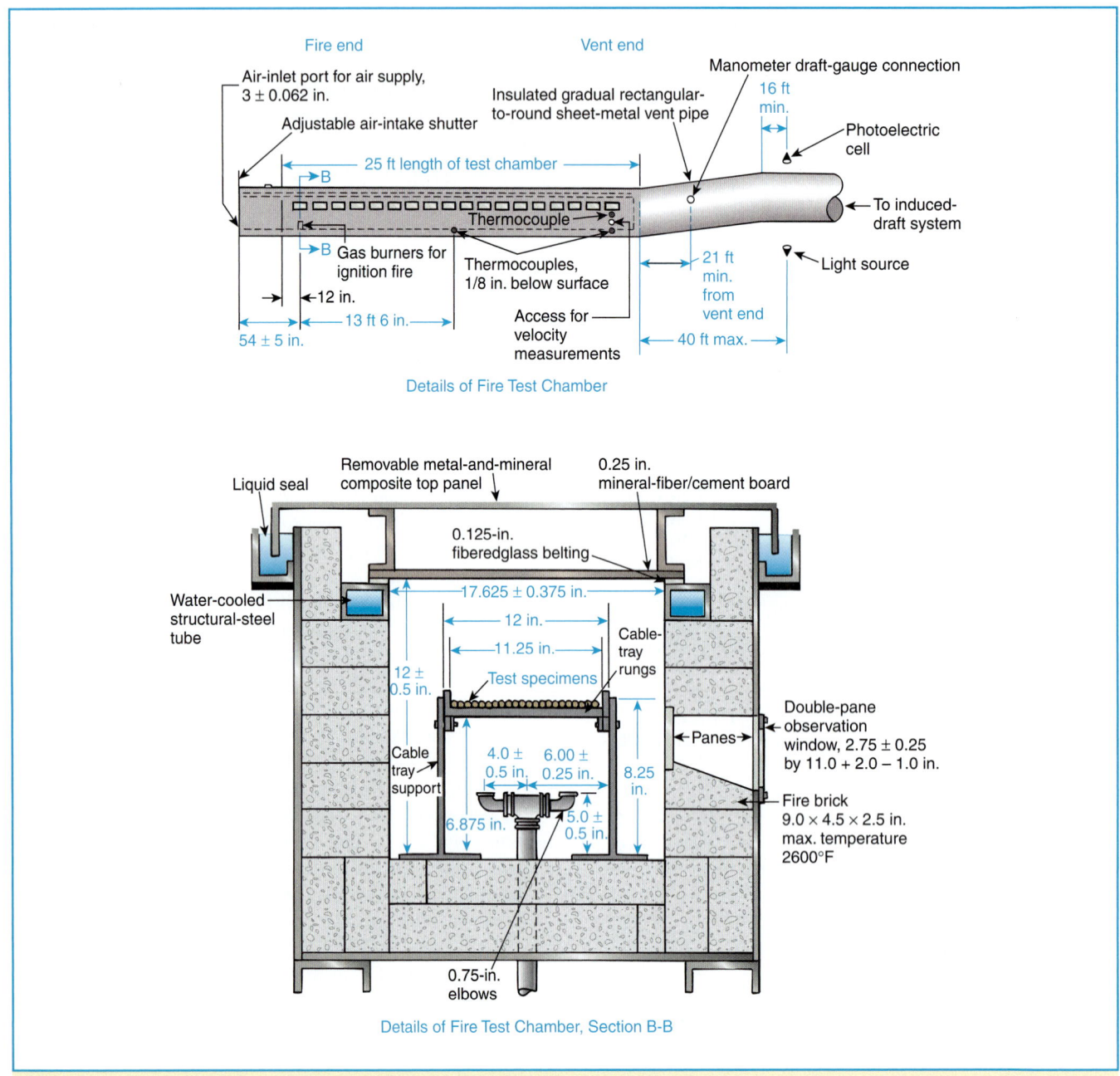

Fire end   Vent end

Air-inlet port for air supply,
3 ± 0.062 in.

Adjustable air-intake shutter

Manometer draft-gauge connection

Insulated gradual rectangular-
to-round sheet-metal vent pipe

16 ft min.

Photoelectric cell

25 ft length of test chamber

B

Thermocouple

To induced-draft system

B Gas burners for
ignition fire

Thermocouples,
1/8 in. below surface

21 ft min. from vent end

Light source

12 in.

13 ft 6 in.

Access for
velocity
measurements

54 ± 5 in.

40 ft max.

Details of Fire Test Chamber

Liquid seal

Removable metal-and-mineral
composite top panel

0.25 in.
mineral-fiber/cement board

0.125-in.
fiberedglass belting

Water-cooled
structural-steel
tube

17.625 ± 0.375 in.

12 in.

11.25 in.

Cable-
tray
rungs

12 ±
0.5 in.

Test specimens

Double-pane
observation
window, 2.75 ± 0.25
by 11.0 + 2.0 − 1.0 in.

Cable
tray
support

4.0 ±
0.5 in.

6.00 ±
0.25 in.

8.25
in.

Panes

Fire brick
9.0 × 4.5 × 2.5 in.
max. temperature
2600°F

6.875 in.

5.0 ±
0.5 in.

0.75-in.
elbows

Details of Fire Test Chamber, Section B-B

Exhibit 725.5 Steiner Tunnel Test chamber used in NFPA 262/UL 910, *Standard Method of Test for Flame Travel and Smoke of Wires and Cables for Use in Air-Handling Spaces.*

where:

T_0 = initial light transmission

T = light transmission during test

NFPA 262 does not list pass/fail criteria. The criteria for acceptance of a given application are given in the appropriate sections of the *Code*. See the fine print notes that follow 725.71(A), 760.31(C), 770.51(A), 800.51(A), and 820.51(A).

A Class 2 or Class 3 cable that has passed the requirements of this test may be used in ducts, plenums, or other air-handling spaces. In addition, such cable may be used anywhere in a building where Class 2 or Class 3 cable is permitted. (See Table 725.61.)

(B) Types CL2R and CL3R. Types CL2R and CL3R riser cables shall be listed as being suitable for use in a vertical run in a shaft or from floor to floor and shall also be listed as having fire-resistant characteristics capable of preventing the carrying of fire from floor to floor.

> FPN: One method of defining fire-resistant characteristics capable of preventing the carrying of fire from floor to floor is that the cables pass the requirements of ANSI/UL 1666-1997, *Test for Flame Propagation Height of Electrical and Optical-Fiber Cable Installed Vertically in Shafts.*

In the fire test covered in UL 1666, cables are arranged in a simulated vertical shaft and subjected to an ignition source. The shaft is a 19-ft-high concrete shaft divided into two compartments at the 12-ft level. There is a 1 ft by 2 ft opening between the two compartments. The ignition source is a burner with a heat output rate of 495,000 Btu/hr. See Exhibit 725.6.

The individual cable lengths are suspended from a support system on the second floor and held in place just below the support system and at the first-floor slot.

The support frame consists of a steel bar located on the second floor above the slot opening. The individual cable lengths are suspended from the support frame by one of the following methods:

1. The cables are draped over the support frame.
2. The cables are arranged in a clamping device that is suspended from the support frame by two hooks.
3. In the case of large-diameter cables, each individual cable is placed in a wire-mesh grip. Each grip is attached to a hook that is suspended from the support frame.

In order to pass, cables must not propagate flame to the top of the 12-ft-high compartment during the 30-minute test.

(C) Types CL2 and CL3. Types CL2 and CL3 cables shall be listed as being suitable for general-purpose use, with the exception of risers, ducts, plenums, and other space used for environmental air and shall also be listed as being resistant to the spread of fire.

> FPN: One method of defining *resistant to the spread of fire* is that the cables do not spread fire to the top of the tray in the vertical tray flame test in ANSI/UL 1581-1991, *Reference Standard for Electrical Wires, Cables and Flexible Cords.*
>
> Another method of defining *resistant to the spread of fire* is for the damage (char length) not to exceed 1.5 m (4 ft 11 in.) when performing the CSA vertical flame test for cables in cable trays, as described in CSA C22.2 No. 0.3-M-1985, *Test Methods for Electrical Wires and Cables.*

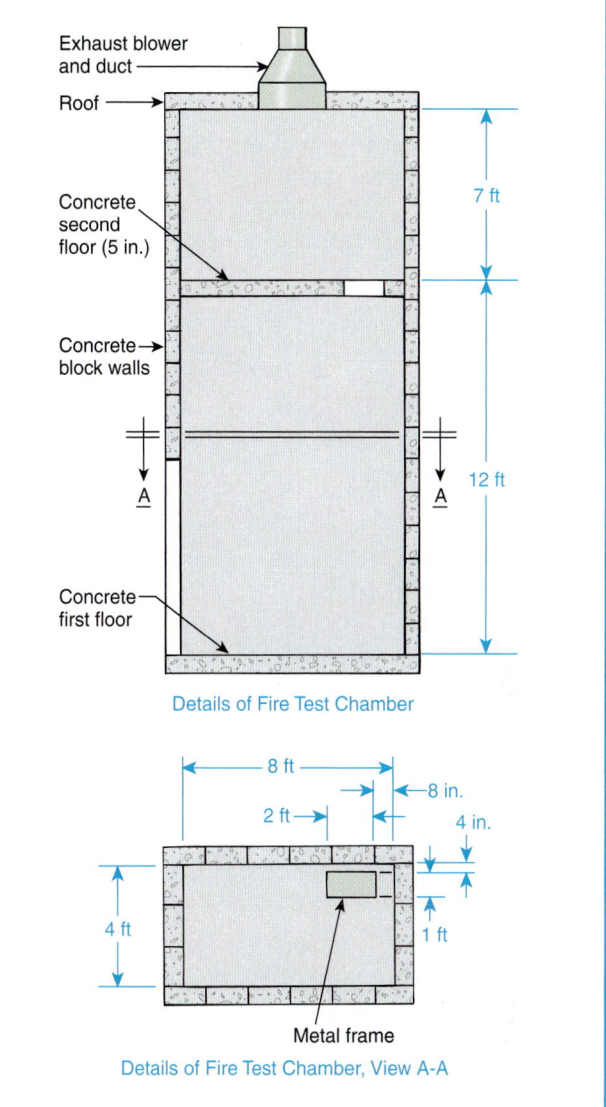

Details of Fire Test Chamber

Details of Fire Test Chamber, View A-A

Exhibit 725.6 Fire test chamber for Test for Flame Propagation Height of Electrical and Optical-Fiber Cable Installed Vertically in Shafts, ANSI/UL 1666-1997.

The UL Vertical Tray Flame Test determines whether cables installed in a ladder-type cable tray will propagate fire from a given exposure. The test flame is supplied by a strip or ribbon-type propane gas burner.

The cables are installed in a steel ladder-type cable tray that is 12 in. wide, 3 in. deep, and 96 in. long. Cables are installed in the tray spaced half a diameter apart, filling 75 percent of the tray width. A burner supplying 70,000 Btu/hr (20.32 kW/hr) is positioned 18 in. above the bottom of the tray, midway between two rungs. The burner flame is applied to the samples for a period of 20 minutes. See

Exhibit 725.7. The samples are considered to have passed the test if flame has not propagated to the top of the cable tray by the test's conclusion. See also the commentary following 725.71(D), FPN.

A cable that has passed this testing may be listed as a Type CL2 or Type CL3 cable.

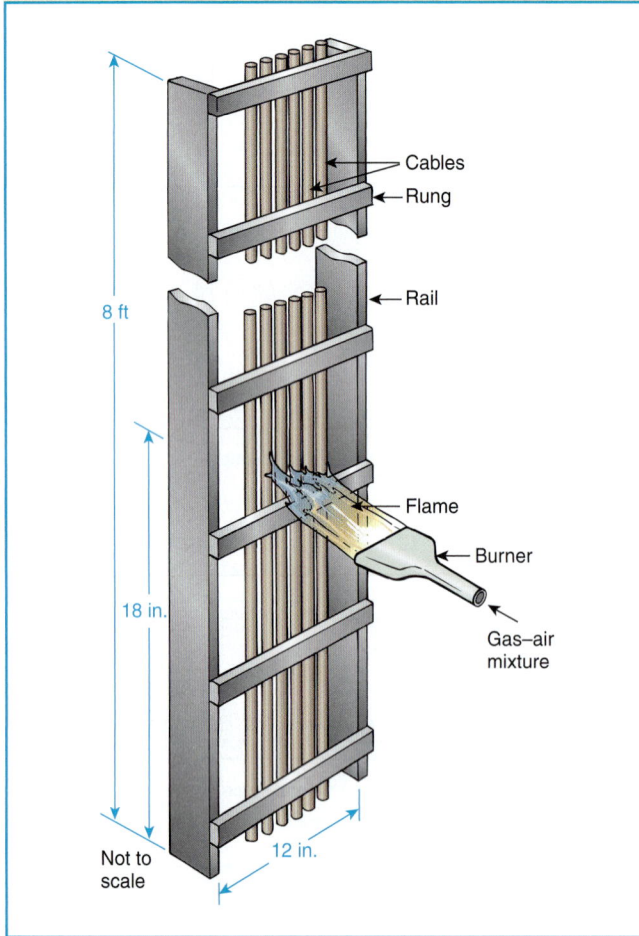

Exhibit 725.7 UL Vertical Tray Flame Test.

(D) Types CL2X and CL3X. Types CL2X and CL3X limited-use cables shall be listed as being suitable for use in dwellings and for use in raceway and shall also be listed as being resistant to flame spread.

> FPN: One method of determining that cable is resistant to flame spread is by testing the cable to the VW-1 (vertical-wire) flame test in ANSI/UL 1581-1991, *Reference Standard for Electrical Wires, Cables and Flexible Cords.*

UL 1581, *Reference Standard for Electrical Wires, Cables and Flexible Cords,* contains basic requirements for conduc-

tors, insulation, jackets, and other coverings, and the methods of sample preparation, specimen selection and conditioning, and measurements and calculations required in the standards for UL 44, *Rubber-Insulated Wires and Cables;* UL 83, *Thermoplastic-Insulated Wires and Cables;* and UL 62, *Flexible Cord and Fixture Wire.* The flame test methods of these standards include the Vertical Wire Flame Test (VW-1) and the Vertical Tray Flame Test. [See the commentary following 725.71(C), FPN.]

The VW-1 test is a small-scale test of the flammability of insulating material. It uses a single specimen, secured in a vertical position in a three-sided metal enclosure that has a layer of untreated surgical cotton on the bottom. See Exhibit 725.8. For thermoplastic- or rubber-insulated wire and cable, the specimen used is an 18-in. insulated 14 AWG copper or 12 AWG aluminum conductor. A small kraft-paper indicator flag is attached to the specimen near its top. A flame is impinged 10 in. below the bottom of the flag for a period of 15 seconds. The flame is then removed for 15 seconds and reapplied for 15 seconds. This procedure is repeated for five 15-second flame applications. The specimen is considered to be capable of propagating fire if

1. More than 25 percent of the indicator flag is burned away or charred,
2. Flaming or glowing particles are emitted that ignite the surgical cotton, or
3. The specimen continues to flame longer than 60 seconds after removal of the burner.

A sample that has passed this test may be listed as Type CL2X or Type CL3X, suitable for use as a limited-use, power-limited cable. This cable may be used in one- and two-family and multifamily dwellings.

(E) Type PLTC. Type PLTC nonmetallic-sheathed, power-limited tray cable shall be listed as being suitable for cable trays and shall consist of a factory assembly of two or more insulated conductors under a nonmetallic jacket. The insulated conductors shall be 22 AWG through 12 AWG. The conductor material shall be copper (solid or stranded). Insulation on conductors shall be suitable for 300 volts. The cable core shall be either (1) two or more parallel conductors, (2) one or more group assemblies of twisted or parallel conductors, or (3) a combination thereof. A metallic shield or a metallized foil shield with drain wire(s) shall be permitted to be applied either over the cable core, over groups of conductors, or both. The cable shall be listed as being resistant to the spread of fire. The outer jacket shall be a sunlight- and moisture-resistant nonmetallic material.

Exception No. 1: Where a smooth metallic sheath, continuous corrugated metallic sheath, or interlocking tape armor

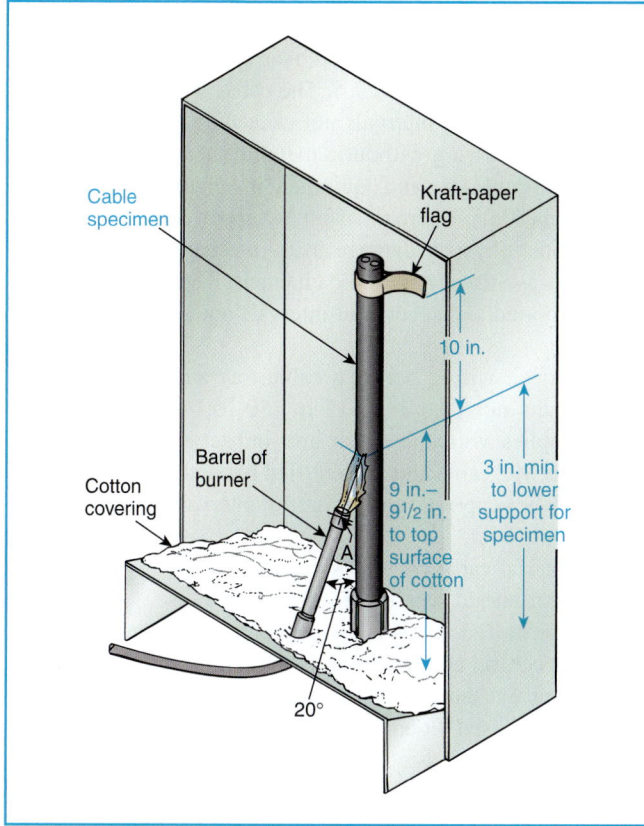

Labels on image: Cable specimen; Kraft-paper flag; 10 in.; Barrel of burner; Cotton covering; 9 in.–9½ in. to top surface of cotton; 3 in. min. to lower support for specimen; A; 20°

Exhibit 725.8 UL Vertical Wire Flame Test.

is applied over the nonmetallic jacket, an overall nonmetallic jacket shall not be required. On metallic-sheathed cable without an overall nonmetallic jacket, the information required in 310.11 shall be located on the nonmetallic jacket under the sheath.

Exception No. 2: Conductors in PLTC cables used for Class 2 thermocouple circuits shall be permitted to be any of the materials used for thermocouple extension wire.

> FPN: One method of defining *resistant to the spread of fire* is that the cables do not spread fire to the top of the tray in the vertical tray flame test in ANSI/UL 1581-1991, *Reference Standard for Electrical Wires, Cables and Flexible Cords.*
> Another method of defining *resistant to the spread of fire* is for the damage (char length) not to exceed 1.5 m (4 ft 11 in.) when performing the CSA vertical flame test for cables in cable trays, as described in CSA C22.2 No. 0.3-M-1985, *Test Methods for Electrical Wires and Cables.*

(F) Class 2 and Class 3 Cable Voltage Ratings. Class 2 cables shall have a voltage rating of not less than 150 volts.

Class 3 cables shall have a voltage rating of not less than 300 volts.

(G) Class 3 Single Conductors. Class 3 single conductors used as other wiring within buildings shall not be smaller than 18 AWG and shall be Type CL3. Conductor types described in 725.27(B) that are also listed as Type CL3 shall be permitted.

> FPN: One method of defining *resistant to the spread of fire* is that the cables do not spread fire to the top of the tray in the vertical tray flame test in ANSI/UL 1581-1991, *Reference Standard for Electrical Wires, Cables and Flexible Cords.*
> Another method of defining *resistant to the spread of fire* is for the damage (char length) not to exceed 1.5 m (4 ft 11 in.) when performing the CSA vertical flame test for cables in cable trays as described in CSA C22.2 No. 0.3-M-1985, *Test Methods for Electrical Wires and Cables.*

> Section 725.71(G) clarifies that Class 3 single conductors must also be resistant to the spread of fire.

(H) Marking. Cables shall be marked in accordance with 310.11(A)(2), (3), (4), and (5) and Table 725.71. Voltage ratings shall not be marked on the cables.

Table 725.71 Cable Markings

Cable Marking	Type	Listing References
CL3P	Class 3 plenum cable	725.71(A), (F), and (H)
CL2P	Class 2 plenum cable	725.71(A) and (H)
CL3R	Class 3 riser cable	725.71(B), (F), and (H)
CL2R	Class 2 riser cable	725.71(B) and (H)
PLTC	Power-limited tray cable	725.71(E) and (H)
CL3	Class 3 cable	725.71(C), (F), and (H)
CL2	Class 2 cable	725.71(C), (F), and (H)
CL3X	Class 3 cable, limited use	725.71(D), (F), and (H)
CL2X	Class 2 cable, limited use	725.71(D), (F), and (H)

> FPN: Voltage markings on cables may be misinterpreted to suggest that the cables may be suitable for Class 1 electric light and power applications.

Exception: Voltage markings shall be permitted where the cable has multiple listings and a voltage marking is required for one or more of the listings.

> FPN: Class 2 and Class 3 cable types are listed in descending order of fire resistance rating, and Class 3 cables are listed above Class 2 cables, because Class 3 cables can substitute for Class 2 cables.

ARTICLE 727
Instrumentation Tray Cable: Type ITC

Contents

727.1 Scope.

This article covers the use, installation, and construction specifications of instrumentation tray cable for application to instrumentation and control circuits operating at 150 volts or less and 5 amperes or less.

Article 727 permits an alternate wiring method for circuits that do not exceed 5 amperes and 150 volts. Instrument tray cable is particularly suited for instrumentation circuits in industrial establishments where qualified persons perform service and maintenance.

727.2 Definition.

Type ITC Instrumentation Tray Cable.   A factory assembly of two or more insulated conductors, with or without a grounding conductor(s), enclosed in a nonmetallic sheath.

727.3 Other Articles.

In addition to the provisions of this article, installation of Type ITC cable shall comply with other applicable articles of this *Code*, such as Articles 240, 250, 300, and 392.

727.4 Uses Permitted.

Type ITC cable shall be permitted to be used as follows in industrial establishments where the conditions of maintenance and supervision ensure that only qualified persons service the installation:

(1) In cable trays.
(2) In raceways.
(3) In hazardous locations as permitted in 501.4, 502.4, 503.3, 504.20, 504.30, 504.80, and 505.15.
(4) As open wiring where enclosed in a smooth metallic sheath, continuous corrugated metallic sheath, or interlocking tape armor applied over the nonmetallic sheath in accordance with 727.6. The cable shall be supported and secured at intervals not exceeding 1.8 m (6 ft).
(5) As open wiring without a metallic sheath or armor between cable tray and equipment in lengths not to exceed 15 m (50 ft), where the cable is supported and protected against physical damage using mechanical protection, such as struts, angles, or channels. The cable shall be supported and secured at intervals not exceeding 1.8 m (6 ft).
(6) As open wiring between cable tray and equipment in lengths not to exceed 15 m (50 ft), where the cable complies with the crush and impact requirements of Type MC cable and is identified for such use. The cable shall be supported and secured at intervals not exceeding 1.8 m (6 ft).
(7) As aerial cable on a messenger.
(8) Direct buried where identified for the use.
(9) Under raised floors in rooms containing industrial process control equipment and rack rooms where arranged to prevent damage to the cable.
(10) Under raised floors in information technology equipment rooms in accordance with 645.5(D)(5)(c).

727.5 Uses Not Permitted.

Type ITC cable shall not be installed on circuits operating at more than 150 volts or more than 5 amperes.

Installation of Type ITC cable with other cables shall be subject to the stated provisions of the specific articles for the other cables. Where the governing articles do not contain stated provisions for installation with Type ITC cable, the installation of Type ITC cable with the other cables shall not be permitted.

Type ITC cable shall not be installed with power, lighting, Class 1, or non–power-limited circuits.

Exception No. 1: Where terminated within equipment or junction boxes and separations are maintained by insulating barriers or other means.

Exception No. 2: Where a metallic sheath or armor is applied over the nonmetallic sheath of the Type ITC cable.

727.6 Construction.

The insulated conductors of Type ITC cable shall be in sizes 22 AWG through 12 AWG. The conductor material shall be copper or thermocouple alloy. Insulation on the conductors shall be rated for 300 volts. Shielding shall be permitted.

The cable shall be listed as being resistant to the spread of fire. The outer jacket shall be sunlight and moisture resistant.

Where a smooth metallic sheath, continuous corrugated metallic sheath, or interlocking tape armor is applied over the nonmetallic sheath, an overall nonmetallic jacket shall not be required.

727.7 Marking.

The cable shall be marked in accordance with 310.11(A)(2), (3), (4), and (5). Voltage ratings shall not be marked on the cable.

727.8 Allowable Ampacity.

The allowable ampacity of the conductors shall be 5 amperes, except for 22 AWG conductors that shall have an allowable ampacity of 3 amperes.

727.9 Overcurrent Protection.

Overcurrent protection shall not exceed 5 amperes for 20 AWG and larger conductors, and 3 amperes for 22 AWG conductors.

727.10 Bends.

Bends in Type ITC cables shall be made so as not to damage the cable.

ARTICLE 760
Fire Alarm Systems

Contents

I. General

760.1 Scope.

This article covers the installation of wiring and equipment of fire alarm systems including all circuits controlled and powered by the fire alarm system.

> FPN No. 1: Fire alarm systems include fire detection and alarm notification, guard's tour, sprinkler waterflow, and sprinkler supervisory systems. Circuits controlled and powered by the fire alarm system include circuits for the control of building systems safety functions, elevator capture, elevator shutdown, door release, smoke doors and damper control, fire doors and damper control and fan shutdown, but only where these circuits are powered by and controlled by the fire alarm system. For further information on the installation and monitoring for integrity requirements for fire alarm systems, refer to the NFPA 72®-1999, *National Fire Alarm Code®*.

> FPN No. 2: Class 1, 2, and 3 circuits are defined in Article 725.

Article 760 covers only those circuits that are powered and controlled by the fire alarm system, including fire safety features such as smoke door control, damper control, fan shutdown, and elevator recall. Circuits powered and controlled by other building systems such as heating, ventilating, and air conditioning (HVAC); security; lighting controls; and time recording are covered by Article 725.

Article 760 covers the wiring between the devices and equipment required by *NFPA 72, National Fire Alarm Code®. NFPA 72* provides the requirements for the selection, installation, performance, use, and testing and maintenance of the fire alarm system components. The provision for whether an occupancy is required to have a fire alarm system is found in NFPA *101® Life Safety Code®,* or other local codes.

Examples of fire alarm devices and equipment are shown in Exhibits 760.1 and 760.2. The installation of these system component items is covered by *NFPA 72,* but the circuit wiring associated with these components must be installed in accordance with the requirements of Article 760.

NFPA 72 requires that all wiring, cable, and equipment be in accordance with NFPA 70, *National Electrical Code,* and specifically with Article 760. Additionally, *NFPA 72* requires all equipment to be listed for its intended purpose.

NFPA 1221, *Standard for the Installation, Maintenance, and Use of Emergency Services Communication Systems,* covers the installation, maintenance, and use of all public fire service communication facilities. These facilities include public reporting, dispatching, telephone, and both two-way and microwave radio systems, all of which fulfill two princi-

Exhibit 760.1 Typical fire alarm control unit.

Exhibit 760.2 Typical spot-type smoke detector.

pal functions: the receipt of fire alarms or other emergency calls from the public and the retransmission of these alarms and emergency calls to fire companies and other appropriate agencies.

The format for Article 760 is similar to that for Articles 725, 770, 800, 820, and 830.

760.2 Definitions.

For purposes of this article, the following definitions apply.

Abandoned Fire Alarm Cable. Installed fire alarm cable that is not terminated at equipment other than a connector and not identified for future use with a tag.

This definition has been added to the 2002 *Code* for use with 760.61, which now requires removal of accessible abandoned fire alarm cable. Abandoned cable increases fire loading unnecessarily, and, where installed in plenums, it can affect airflow. Similar requirements can be found in Articles 640, 645, 725, 770, 800, 820, and 830.

Fire Alarm Circuit. The portion of the wiring system between the load side of the overcurrent device or the power-limited supply and the connected equipment of all circuits powered and controlled by the fire alarm system. Fire alarm circuits are classified as either non–power-limited or power-limited.

Fire Alarm Circuit Integrity (CI) Cable. Cable used in fire alarm systems to ensure continued operation of critical circuits during a specified time under fire conditions.

Non–Power-Limited Fire Alarm Circuit (NPLFA). A fire alarm circuit powered by a source that complies with 760.21 and 760.23.

Power-Limited Fire Alarm Circuit (PLFA). A fire alarm circuit powered by a source that complies with 760.41.

760.3 Locations and Other Articles.

Circuits and equipment shall comply with 760.3(A) through (F). Only those sections of Article 300 referenced in this article shall apply to fire alarm systems.

(A) Spread of Fire or Products of Combustion. Section 300.21. The accessible portion of abandoned fire alarm cables shall not be permitted to remain.

(B) Ducts, Plenums, and Other Air-Handling Spaces. Section 300.22, where installed in ducts or plenums or other spaces used for environmental air.

Exception: As permitted in 760.30(B)(1) and (2) and 760.61(A).

See the commentary following 300.22(B) and (C) for more information on wiring installed in ducts, plenums, or other spaces used for environmental air.

(C) Hazardous (Classified) Locations. Articles 500 through 516 and Article 517, Part IV, where installed in hazardous (classified) locations.

(D) Corrosive, Damp, or Wet Locations. Sections 110.11, 300.6, and 310.9 where installed in corrosive, damp, or wet locations.

Section 760.3(D) requires cables and equipment that are used in wet or damp locations, high ambient temperature areas, or corrosive locations to be identified as suitable for the particular use. Underground installations are considered wet locations.

(E) Building Control Circuits. Article 725 where building control circuits (e.g., elevator capture, fan shutdown) are associated with the fire alarm system.

(F) Optical Fiber Cables. Where optical fiber cables are utilized for fire alarm circuits, the cables shall be installed in accordance with Article 770.

760.5 Access to Electrical Equipment Behind Panels Designed to Allow Access.

Access to electrical equipment shall not be denied by an accumulation of conductors and cables that prevents removal of panels, including suspended ceiling panels.

An excess accumulation of wires and cables can limit access to equipment by preventing the removal of access panels. See Exhibit 760.3.

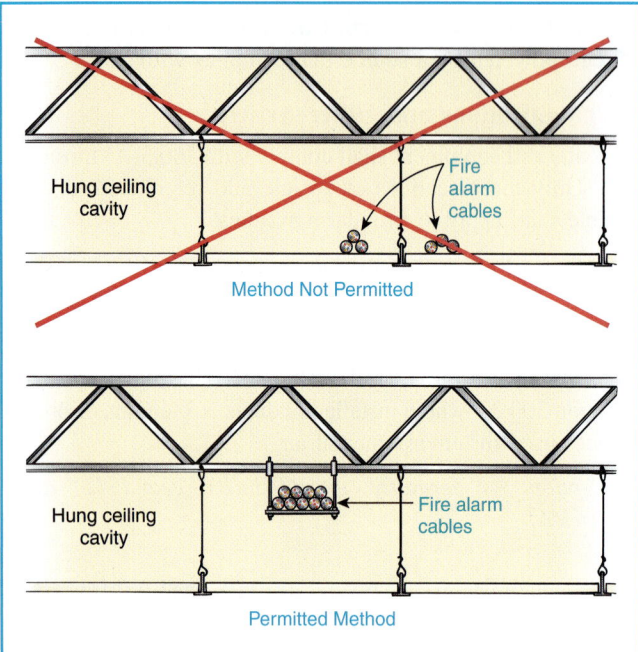

Exhibit 760.3 Installations of conductors and cables, which can prevent access to equipment or cables. Correct and incorrect methods are shown.

760.6 Mechanical Execution of Work.

Fire alarm circuits shall be installed in a neat and workman-like manner. Cables and conductors installed exposed on the surface of ceiling and sidewalls shall be supported by structural components of the building in such a manner that the cable or conductors will not be damaged by normal building use. Such cables shall be attached to structural components by straps, staples, hangers, or similar fittings designed and installed so as not to damage the cable. The installation shall also conform with 300.4(D).

Revised for the 2002 *Code*, this section now provides definitive requirements for workmanship. Cable must now be attached to or supported by the structure by straps, clamps, hangers, and the like. The installation method must not damage the cable. In addition, the location of the cable should be carefully evaluated to ensure that activities and processes within the building do not cause damage to the cable.

The reference to 300.4(D) calls attention to the hazards to which cables are exposed where they are installed on framing members. Such cables are required to be installed in a manner that protects them from nail or screw penetration.

760.7 Fire Alarm Circuits Extending Beyond One Building.

Power-limited fire alarm circuits that extend beyond one building and run outdoors either shall meet the installation requirements of Parts II, III, and IV of Article 800 or shall meet the installation requirements of Part I of Article 300. Non–power-limited fire alarm circuits that extend beyond one building and run outdoors shall meet the installation requirements of Part I of Article 300 and the applicable sections of Part I of Article 225.

760.9 Fire Alarm Circuit and Equipment Grounding.

Fire alarm circuits and equipment shall be grounded in accordance with Article 250.

760.10 Fire Alarm Circuit Identification.

Fire alarm circuits shall be identified at terminal and junction locations, in a manner that will prevent unintentional interference with the signaling circuit during testing and servicing.

760.15 Fire Alarm Circuit Requirements.

Fire alarm circuits shall comply with the following parts of this article.

(A) Non–Power–Limited Fire Alarm (NPLFA) Circuits. See Parts I and II.

(B) Power-Limited Fire Alarm (PLFA) Circuits. See Parts I and III.

Exact power source limitations for power-limited fire alarm circuits used by testing laboratories are found in Chapter 9, Tables 12(A) and 12(B). Table 12(A) covers alternating-current source limitations, and Table 12(B) covers direct-current source limitations.

II. Non–Power-Limited Fire Alarm (NPLFA) Circuits

760.21 NPLFA Circuit Power Source Requirements.

The power source of non–power-limited fire alarm circuits shall comply with Chapters 1 through 4, and the output voltage shall not be more than 600 volts, nominal. These circuits shall not be supplied through ground-fault circuit interrupters.

FPN: See 210.8(A)(5), Exception No. 3 for receptacles in dwelling-unit unfinished basements that supply power for fire alarm systems.

760.23 NPLFA Circuit Overcurrent Protection.

Overcurrent protection for conductors 14 AWG and larger shall be provided in accordance with the conductor ampacity without applying the derating factors of 310.15 to the ampacity calculation. Overcurrent protection shall not exceed 7 amperes for 18 AWG conductors and 10 amperes for 16 AWG conductors.

Exception: Where other articles of this Code permit or require other overcurrent protection.

760.24 NPLFA Circuit Overcurrent Device Location.

Overcurrent devices shall be located at the point where the conductor to be protected receives its supply.

Exception No. 1: Where the overcurrent device protecting the larger conductor also protects the smaller conductor.

Exception No. 2: Transformer secondary conductors. Non–power-limited fire alarm circuit conductors supplied by the secondary of a single-phase transformer that has only a 2-wire (single-voltage) secondary shall be permitted to be protected by overcurrent protection provided by the primary (supply) side of the transformer, provided the protection is in accordance with 450.3 and does not exceed the value determined by multiplying the secondary conductor ampacity by the secondary-to-primary transformer voltage ratio. Transformer secondary conductors other than 2-wire shall not be considered to be protected by the primary overcurrent protection.

Exception No. 3: Electronic power source output conductors. Non–power-limited circuit conductors supplied by the output of a single-phase, listed electronic power source, other than a transformer, having only a 2-wire (single-voltage) output for connection to non–power-limited circuits shall be permitted to be protected by overcurrent protection provided on the input side of the electronic power source, provided this protection does not exceed the value determined by multiplying the non–power-limited circuit conductor ampacity by the output-to-input voltage ratio. Electronic power source outputs, other than 2-wire (single voltage), connected to non–power-limited circuits shall not be considered to be protected by overcurrent protection on the input of the electronic power source.

> FPN: A single-phase, listed electronic power supply whose output supplies a 2-wire (single-voltage) circuit is an example of a non–power-limited power source that meets the requirements of 760.21.

Non–power-limited electronic power supplies that do not supply energy directly through the use of transformers are covered by Exception No. 3 to 760.24, which permits overcurrent protection for the non–power-limited circuit conduc-

tors to be installed on the input side of the electronic power source rather than on the output side for 2-wire circuits only.

760.25 NPLFA Circuit Wiring Methods.

Installation of non–power-limited fire alarm circuits shall be in accordance with 110.3(B), 300.11(A), 300.15, 300.17, and other appropriate articles of Chapter 3.

Exception No. 1: As provided in 760.26 through 760.30.

Exception No. 2: Where other articles of this Code require other methods.

Section 760.25 requires that the appropriate wiring methods in Chapter 3 be used for non–power-limited circuits. However, 760.25, Exception No. 1, permits special non–power-limited cable types to be used in place of Chapter 3 wiring methods.

Section 760.25 requires that devices be mounted in accordance with Chapter 3. Section 300.11(A) requires devices and equipment to be securely mounted. Section 300.15(B) is referenced to require non–power-limited circuit terminations to be made in a box or conduit body. However, 300.15(E) permits devices with integral terminal enclosures and mounting brackets to be used without a box. Devices must be mounted on a box or conduit body where the instructions or listing require the use of a box. Fire alarm system components such as manual fire alarm boxes are frequently tested. Therefore, secure mounting of the back box is necessary to ensure that manual fire alarm device will remain in place. (See Exhibit 760.4.)

Exhibit 760.4 Typical manual fire alarm box.

760.26 Conductors of Different Circuits in Same Cable, Enclosure, or Raceway.

(A) Class 1 with NPLFA Circuits. Class 1 and non–power-limited fire alarm circuits shall be permitted to occupy the same cable, enclosure, or raceway without regard to whether the individual circuits are alternating current or direct current, provided all conductors are insulated for the maximum voltage of any conductor in the enclosure or raceway.

(B) Fire Alarm with Power-Supply Circuits. Power-supply and fire alarm circuit conductors shall be permitted in the same cable, enclosure, or raceway only where connected to the same equipment.

760.27 NPLFA Circuit Conductors.

(A) Sizes and Use. Only copper conductors shall be permitted to be used for fire alarm systems. Size 18 AWG and 16 AWG conductors shall be permitted to be used, provided they supply loads that do not exceed the ampacities given in Table 402.5 and are installed in a raceway, an approved enclosure, or a listed cable. Conductors larger than 16 AWG shall not supply loads greater than the ampacities given in 310.15, as applicable.

The minimum size conductors permitted to be used on non–power-limited fire protective signaling circuits is 18 AWG. The load must not exceed the conductor ampacities specified in Table 402.5.

NFPA 72, National Fire Alarm Code, requires fire alarm device and appliance voltages to be between 85 and 110 percent of nominal rated voltage. Calculations should be made to ensure that all devices or appliances will be operating within these limits at full circuit load. Where future circuit extensions are anticipated, larger conductors should be considered. Some manufacturers specify maximum circuit loop resistances. The equipment specifications should be consulted to ensure that maximum allowable loop resistances are not exceeded.

(B) Insulation. Insulation on conductors shall be suitable for 600 volts. Conductors larger than 16 AWG shall comply with Article 310. Conductors 18 AWG and 16 AWG shall be Type KF-2, KFF-2, PAFF, PTFF, PF, PFF, PGF, PGFF, RFH-2, RFHH-2, RFHH-3, SF-2, SFF-2, TF, TFF, TFN, TFFN, ZF, or ZFF. Conductors with other types and thickness of insulation shall be permitted if listed for non–power-limited fire alarm circuit use.

FPN: For application provisions, see Table 402.3.

(C) Conductor Materials. Conductors shall be solid or stranded copper.

Exception to (B) and (C): Wire Types PAF and PTF shall be permitted only for high-temperature applications between 90°C (194°F) and 250°C (482°F).

760.28 Number of Conductors in Cable Trays and Raceways, and Derating.

(A) NPLFA Circuits and Class 1 Circuits. Where only non–power-limited fire alarm circuit and Class 1 circuit conductors are in a raceway, the number of conductors shall be determined in accordance with 300.17. The derating factors given in 310.15(B)(2)(a) shall apply if such conductors carry continuous load in excess of 10 percent of the ampacity of each conductor.

(B) Power-Supply Conductors and Fire Alarm Circuit Conductors. Where power-supply conductors and fire alarm circuit conductors are permitted in a raceway in accordance with 760.26, the number of conductors shall be determined in accordance with 300.17. The derating factors given in 310.15(B)(2)(a) shall apply as follows:

(1) To all conductors where the fire alarm circuit conductors carry continuous loads in excess of 10 percent of the ampacity of each conductor and where the total number of conductors is more than three
(2) To the power-supply conductors only, where the fire alarm circuit conductors do not carry continuous loads in excess of 10 percent of the ampacity of each conductor and where the number of power-supply conductors is more than three

(C) Cable Trays. Where fire alarm circuit conductors are installed in cable trays, they shall comply with 392.9 through 392.11.

760.30 Multiconductor NPLFA Cables.

Multiconductor non–power-limited fire alarm cables that meet the requirements of 760.31 shall be permitted to be used on fire alarm circuits operating at 150 volts or less and shall be installed in accordance with 760.30(A) and (B).

(A) NPLFA Wiring Method. Multiconductor non–power-limited fire alarm circuit cables shall be installed in accordance with 760.30(A)(1), (A)(2), and (A)(3).

(1) Exposed or Fished in Concealed Spaces. In raceway or exposed on surface of ceiling and sidewalls or fished in concealed spaces. Cable splices or terminations shall be made in listed fittings, boxes, enclosures, fire alarm devices, or utilization equipment. Where installed exposed, cables shall be adequately supported and installed in such a way that maximum protection against physical damage is afforded by building construction such as baseboards, door frames, ledges, and so forth. Where located within 2.1 m (7 ft) of the floor, cables shall be securely fastened in an approved manner at intervals of not more than 450 mm (18 in.).

(2) Passing Through a Floor or Wall. In metal raceway or rigid nonmetallic conduit where passing through a floor or wall to a height of 2.1 m (7 ft) above the floor unless adequate protection can be afforded by building construction such as detailed in 760.30(A)(1) or unless an equivalent solid guard is provided.

(3) In Hoistways. In rigid metal conduit, rigid nonmetallic conduit, intermediate metal conduit, liquidtight flexible non-metallic tubing, or electrical metallic tubing where installed in hoistways.

Exception: As provided for in 620.21 for elevators and similar equipment.

(B) Applications of Listed NPLFA Cables. The use of non–power-limited fire alarm circuit cables shall comply with 760.30(B)(1) through (B)(4).

(1) Ducts and Plenums. Multiconductor non–power-limited fire alarm circuit cables, Types NPLFP, NPLFR, and NPLF, shall not be installed exposed in ducts or plenums.

FPN: See 300.22(B).

Wiring methods for non–power-limited circuits in ducts and plenums must be in accordance with the Chapter 3 wiring methods covered by 300.22(B). It is important to note that cables marked NPLFP may not be installed in plenums. While the "P" designation was used for consistency, the higher possible voltages and currents of non–power-limited fire alarm circuits preclude the use of the listed cables inside plenums.

(2) Other Spaces Used for Environmental Air. Cables installed in other spaces used for environmental air shall be Type NPLFP.

Exception No. 1: Types NPLFR and NPLF cables installed in compliance with 300.22(C).

Exception No. 2: Other wiring methods in accordance with 300.22(C) and conductors in compliance with 760.27(C).

Other spaces used for environmental air are covered by 300.22(C) and the related fine print note. Spaces over suspended ceilings used as an environmental air-handling return are considered by the *Code* as other spaces used for environmental air. Non–power-limited cables used in other spaces used for environmental air must, however, be marked NPLFP. [See 760.31(C).]

(3) Riser. Cables installed in vertical runs and penetrating more than one floor or cables installed in vertical runs in a shaft shall be Type NPLFR. Floor penetrations requiring Type NPLFR shall contain only cables suitable for riser or plenum use.

Exception No. 1: Type NPLF or other cables that are specified in Chapter 3 and are in compliance with 760.27(C) and encased in metal raceway.

Exception No. 2: Type NPLF cables located in a fireproof shaft having firestops at each floor.

FPN: See 300.21 for firestop requirements for floor penetrations.

(4) Other Wiring Within Buildings. Cables installed in building locations other than the locations covered in 760.30(B)(1), (B)(2), and (B)(3) shall be Type NPLF.

Exception No. 1: Chapter 3 wiring methods with conductors in compliance with 760.27(C).

Exception No. 2: Type NPLFP or Type NPLFR cables shall be permitted.

760.31 Listing and Marking of NPLFA Cables.

Non–power-limited fire alarm cables installed as wiring within buildings shall be listed in accordance with 760.31(A) and (B) and as being resistant to the spread of fire in accordance with 760.31(C) through (F), and shall be marked in accordance with 760.31(G).

(A) NPLFA Conductor Materials. Conductors shall be 18 AWG or larger solid or stranded copper.

(B) Insulated Conductors. Insulated conductors shall be suitable for 600 volts. Insulated conductors 14 AWG and larger shall be one of the types listed in Table 310.13 or one that is identified for this use. Insulated conductors 18 AWG and 16 AWG shall be in accordance with 760.27.

(C) Type NPLFP. Type NPLFP non–power-limited fire alarm cable for use in other space used for environmental air shall be listed as being suitable for use in other space used for environmental air as described in 300.22(C) and shall also be listed as having adequate fire-resistant and low smoke-producing characteristics.

FPN: One method of defining low smoke-producing cable is by establishing an acceptable value of the smoke produced when tested in accordance with NFPA 262-1999, *Standard Method of Test for Flame Travel and Smoke of Wires and Cables for Use in Air-Handling Spaces*, to a maximum peak optical density of 0.5 and a maximum average optical density of 0.15. Similarly, one method of defining fire-resistant cables is by establishing a maximum allowable flame travel distance of 1.52 m (5 ft) when tested in accordance with the same test.

For further information on the fire test method for Type NPLFP cables, see the commentary following 725.71(A), FPN. Also see the commentary following 760.30(B)(2), Exception No. 2, which discusses spaces used for environmental air.

(D) Type NPLFR. Type NPLFR non–power-limited fire alarm riser cable shall be listed as being suitable for use in a vertical run in a shaft or from floor to floor and shall also be listed as having fire-resistant characteristics capable of preventing the carrying of fire from floor to floor.

> FPN: One method of defining fire-resistant characteristics capable of preventing the carrying of fire from floor to floor is that the cables pass ANSI/UL 1666-1997, *Test for Flame Propagation Height of Electrical and Optical-Fiber Cables Installed Vertically in Shafts.*

For further information on the fire test method for Type NPLFR cables, see the commentary following 725.71(B), FPN.

(E) Type NPLF. Type NPLF non–power-limited fire alarm cable shall be listed as being suitable for general-purpose fire alarm use, with the exception of risers, ducts, plenums, and other space used for environmental air, and shall also be listed as being resistant to the spread of fire.

> FPN No. 1: One method of defining *resistant to the spread of fire* is that the cables do not spread fire to the top of the tray in the vertical-tray flame test in ANSI/UL 1581-1991, *Reference Standard for Electrical Wires, Cables and Flexible Cords.*
>
> FPN No. 2: Another method of defining *resistant to the spread of fire* is for the damage (char length) not to exceed 1.5 m (4 ft 11 in.) when performing the CSA vertical flame test for cables in cable trays, as described in CSA C22.2 No. 0.3-M-1985, *Test Methods for Electrical Wires and Cables.*

For further information on the fire test method for Type NPLF cables used as other wiring within buildings, see the commentary following 725.71(C), FPN.

(F) Fire Alarm Circuit Integrity (CI) Cable. Cables suitable for use in fire alarm systems to ensure survivability of critical circuits during a specified time under fire conditions shall be listed as circuit integrity (CI) cable. Cables identified in 760.31(C), (D), and (E) that meet the requirements for circuit integrity shall have the additional classification using the suffix "CI" (for example, NPLFP-CI, NPLFR-CI, and NPLF-CI).

> FPN No. 1: This cable may be used for fire alarm circuits to comply with the survivability requirements of NFPA

72-1999, *National Fire Alarm Code®,* 3-4.2.2.2, 3-8.4.1.1.4, and 3-8.4.1.3.3.3(3), that the cable maintain its electrical function during fire conditions for a defined period of time.

> FPN No. 2: One method of defining circuit integrity (CI) cable is by establishing a minimum 2-hour fire resistance rating for the cable when tested in accordance with UL 2196-1995, *Standard for Tests of Fire Resistive Cables.*

Circuit integrity (CI) cable was added in the 1999 *Code* to meet the performance requirements for survivability required by *NFPA 72, National Fire Alarm Code.* This type of cable is designed to retain vital electrical performance during and immediately after fire exposure. Circuit integrity cable, which carries the CI suffix, is considered a 2-hour-rated cable assembly and provides an alternative to fire-rated mineral insulated cable (Type MI).

(G) NPLFA Cable Markings. Multiconductor non–power-limited fire alarm cables shall be marked in accordance with Table 760.31(G). Non–power-limited fire alarm circuit cables shall be permitted to be marked with a maximum usage voltage rating of 150 volts. Cables that are listed for circuit integrity shall be identified with the suffix "CI" as defined in 760.31(F).

Table 760.31(G) NPLFA Cable Markings

Cable Marking	Type	Reference
NPLFP	Non–power-limited fire alarm circuit cable for use in other space used for environmental air	760.31(C) and (G)
NPLFR	Non–power-limited fire alarm circuit riser cable	760.31(D) and (G)
NPLF	Non–power-limited fire alarm circuit cable	760.31(E) and (G)

Note: Cables identified in 760.31(C), (D), and (E) and meeting the requirements for circuit integrity shall have the additional classification using the suffix "CI" (for example, NPLFP-CI, NPLFR-CI, and NPLF-CI).

> FPN: Cable types are listed in descending order of fire resistance rating.

III. Power-Limited Fire Alarm (PLFA) Circuits

760.41 Power Sources for PLFA Circuits.

The power source for a power-limited fire alarm circuit shall be as specified in 760.41(A), (B), or (C). These circuits shall not be supplied through ground-fault circuit interrupters.

FPN No. 1: Tables 12(A) and 12(B) in Chapter 9 provide the listing requirements for power-limited fire alarm circuit sources.

FPN No. 2: See 210.8(A)(5), Exception No. 3, for receptacles in dwelling-unit unfinished basements that supply power for fire alarm systems.

(A) Transformers. A listed PLFA or Class 3 transformer.

(B) Power Supplies. A listed PLFA or Class 3 power supply.

(C) Listed Equipment. Listed equipment marked to identify the PLFA power source.

FPN: Examples of listed equipment are a fire alarm control panel with integral power source; a circuit card listed for use as a PLFA source, where used as part of a listed assembly; a current-limiting impedance, listed for the purpose or part of a listed product, used in conjunction with a non–power-limited transformer or a stored energy source, for example, storage battery, to limit the output current.

760.42 Circuit Marking.

The equipment shall be durably marked where plainly visible to indicate each circuit that is a power-limited fire alarm circuit.

FPN: See 760.52(A), Exception No. 3 where a power-limited circuit is to be reclassified as a non–power-limited circuit.

760.51 Wiring Methods on Supply Side of the PLFA Power Source.

Conductors and equipment on the supply side of the power source shall be installed in accordance with the appropriate requirements of Part II and Chapters 1 through 4. Transformers or other devices supplied from power-supply conductors shall be protected by an overcurrent device rated not over 20 amperes.

Exception: The input leads of a transformer or other power source supplying power-limited fire alarm circuits shall be permitted to be smaller than 14 AWG, but not smaller than 18 AWG, if they are not over 300 mm (12 in.) long and if they have insulation that complies with 760.27(B).

760.52 Wiring Methods and Materials on Load Side of the PLFA Power Source.

Fire alarm circuits on the load side of the power source shall be permitted to be installed using wiring methods and materials in accordance with either 760.52(A) or (B).

Section 760.52 permits individual power-limited circuits to be installed using Chapter 3 wiring methods, using non–power-limited fire alarm circuit wiring methods, or using power-limited circuit wiring methods. If it is desirable

to run power-limited circuits in the same cable or raceway with non–power-limited circuits, the power-limited circuits may be reclassified as permitted by 760.52(A), Exception No. 3. Also note the information contained in the fine print note that follows 760.52(A), Exception No. 3, regarding circuit classification.

(A) NPLFA Wiring Methods and Materials. Installation shall be in accordance with 760.25, and conductors shall be solid or stranded copper.

Section 760.52(A) requires power-limited fire alarm circuits using non–power-limited fire alarm wiring methods and materials to be installed in accordance with 760.25. Section 760.25 requires that non–power-limited devices be mounted in accordance with Chapter 3. Section 300.11(A) requires these devices and equipment to be securely mounted. Section 300.15(B) is also referenced to require non–power-limited circuit terminations to be made in a box or conduit body. However, 300.15(E) permits devices with integral terminal enclosures and mounting brackets to be used without a box. Devices are required to be mounted on a box or conduit body where the instructions or listing indicate the use of a box. Fire alarm system components such as manual fire alarm boxes are frequently tested. Secure mounting is necessary to ensure that they will remain in place.

Exception No. 1: The derating factors given in 310.15(B)(2)(a) shall not apply.

Exception No. 2: Conductors and multiconductor cables described in and installed in accordance with 760.27 and 760.30 shall be permitted.

Exception No. 3: Power-limited circuits shall be permitted to be reclassified and installed as non–power-limited circuits if the power-limited fire alarm circuit markings required by 760.42 are eliminated and the entire circuit is installed using the wiring methods and materials in accordance with Part II, Non–Power-Limited Fire Alarm Circuits.

FPN: Power-limited circuits reclassified and installed as non–power-limited circuits are no longer power-limited circuits, regardless of the continued connection to a power-limited source.

Section 760.52(A) permits any of the wiring methods in Chapter 3 to be used for power-limited circuits. In addition, 760.52(A), Exception No. 3, allows power-limited circuits to be reclassified and installed in accordance with the requirements for non–power-limited circuits. Where installed as non–power-limited circuits, the power-limited marking must be removed from equipment, overcurrent protection must be provided in accordance with 760.23, and reclassified circuits must maintain separation from power-limited circuits, in accordance with 760.26 and 760.54.

(B) PLFA Wiring Methods and Materials. Power-limited fire alarm conductors and cables described in 760.71 shall be installed as detailed in 760.52(B)(1), (2), or (3) of this section. Devices shall be installed in accordance with 110.3(B), 300.11(A), and 300.15.

Section 760.52(B) requires mechanical protection at splices and termination points. Since failure of a circuit often occurs at splices or termination points, this requirement offers more protection and strain relief for these cable connections.

(1) Exposed or Fished in Concealed Spaces. In raceway or exposed on the surface of ceiling and sidewalls or fished in concealed spaces. Cable splices or terminations shall be made in listed fittings, boxes, enclosures, fire alarm devices, or utilization equipment. Where installed exposed, cables shall be adequately supported and installed in such a way that maximum protection against physical damage is afforded by building construction such as baseboards, door frames, ledges, and so forth. Where located within 2.1 m (7 ft) of the floor, cables shall be securely fastened in an approved manner at intervals of not more than 450 mm (18 in.).

(2) Passing Through a Floor or Wall. In metal raceways or rigid nonmetallic conduit where passing through a floor or wall to a height of 2.1 m (7 ft) above the floor, unless adequate protection can be afforded by building construction such as detailed in 760.52(B)(1) or unless an equivalent solid guard is provided.

(3) In Hoistways. In rigid metal conduit, rigid nonmetallic conduit, intermediate metal conduit, or electrical metallic tubing where installed in hoistways.

Exception No. 1: As provided for in 620.21 for elevators and similar equipment.

Exception No. 2: Other wiring methods and materials installed in accordance with the requirements of 760.3 shall be permitted to extend or replace the conductors and cables described in 760.71 and permitted by 760.52(B).

Section 760.52(B)(3), Exception No. 2, permits the mixing of different wiring methods on a single circuit only for circuit extensions, for replacement of conductors and cables, or in certain locations covered by 760.3. Mixing of different wiring methods on a single circuit would be permitted when the circuit passes through a classified area, a wet location,

or a plenum, or when the circuit extends outdoors to other buildings.

760.54 Installation of Conductors and Equipment in Cables, Compartments, Cable Trays, Enclosures, Manholes, Outlet Boxes, Device Boxes, and Raceways for Power-Limited Circuits.

Conductors and equipment for power-limited fire alarm circuits shall be installed in accordance with 760.55 through 760.58.

760.55 Separation from Electric Light, Power, Class 1, NPLFA, and Medium Power Network-Powered Broadband Communications Circuit Conductors.

(A) General. Power-limited fire alarm circuit cables and conductors shall not be placed in any cable, cable tray, compartment, enclosure, manhole, outlet box, device box, raceway, or similar fitting with conductors of electric light, power, Class 1, non–power-limited fire alarm circuits, and medium power network-powered broadband communications circuits unless permitted by 760.55(B) through (G).

Jackets of listed power-limited fire alarm cables do not have sufficient construction specifications to permit them to be installed with electric light, power, Class 1, non–power-limited fire alarm circuits, and medium power network-powered broadband communications cables. Failure of the cable insulation due to a fault could lead to hazardous voltages being imposed on the power-limited fire alarm circuit conductors.

(B) Separated by Barriers. Power-limited fire alarm circuit cables shall be permitted to be installed together with Class 1, non–power-limited fire alarm, and medium power network-powered broadband communications circuits where they are separated by a barrier.

(C) Raceways Within Enclosures. In enclosures, power-limited fire alarm circuits shall be permitted to be installed in a raceway within the enclosure to separate them from Class 1, non–power-limited fire alarm, and medium power network-powered broadband communications circuits.

(D) Associated Systems Within Enclosures. Power-limited fire alarm conductors in compartments, enclosures, device boxes, outlet boxes, or similar fittings shall be permitted to be installed with electric light, power, Class 1, non–power-limited fire alarm, and medium power network-powered broadband communications circuits where they are introduced solely to connect the equipment connected to power-limited fire alarm circuits, and comply with either of the following conditions:

(1) The electric light, power, Class 1, non–power-limited fire alarm, and medium power network-powered broadband communications circuit conductors are routed to maintain a minimum of 6 mm (0.25 in.) separation from the conductors and cables of power-limited fire alarm circuits.

(2) The circuit conductors operate at 150 volts or less to ground and also comply with one of the following:

    a. The fire alarm power-limited circuits are installed using Type FPL, FPLR, FPLP, or permitted substitute cables, provided these power-limited cable conductors extending beyond the jacket are separated by a minimum of 6 mm (0.25 in.) or by a nonconductive sleeve or nonconductive barrier from all other conductors.

    b. The power-limited fire alarm circuit conductors are installed as non–power-limited circuits in accordance with 760.25.

(E) Enclosures with Single Opening. Power-limited fire alarm circuit conductors entering compartments, enclosures, device boxes, outlet boxes, or similar fittings shall be permitted to be installed with electric light, power, Class 1 non–power-limited fire alarm, and medium power network-powered broadband communications circuits where they are introduced solely to connect the equipment connected to power-limited fire alarm circuits or to other circuits controlled by the fire alarm system to which the other conductors in the enclosure are connected. Where power-limited fire alarm circuit conductors must enter an enclosure that is provided with a single opening, they shall be permitted to enter through a single fitting (such as a tee), provided the conductors are separated from the conductors of the other circuits by a continuous and firmly fixed nonconductor, such as flexible tubing.

(F) In Hoistways. In hoistways, power-limited fire alarm circuit conductors shall be installed in rigid metal conduit, rigid nonmetallic conduit, intermediate metal conduit, liquidtight flexible nonmetallic conduit, or electrical metallic tubing. For elevators or similar equipment, these conductors shall be permitted to be installed as provided in 620.21.

(G) Other Applications. For other applications, power-limited fire alarm circuit conductors shall be separated by at least 50 mm (2 in.) from conductors of any electric light, power, Class 1, non–power-limited fire alarm, or medium power network-powered broadband communications circuits unless one of the following conditions is met:

(1) Either (a) all of the electric light, power, Class 1, non–power-limited fire alarm, and medium power net-work-powered broadband communications circuit conductors or (b) all of the power-limited fire alarm circuit conductors are in a raceway or in metal-sheathed, metal-clad, nonmetallic-sheathed, or Type UF cables.

(2) All of the electric light, power, Class 1 non–power-limited fire alarm, and medium power network-powered broadband communications circuit conductors are permanently separated from all of the power-limited fire alarm circuit conductors by a continuous and firmly fixed nonconductor, such as porcelain tubes or flexible tubing, in addition to the insulation on the conductors.

760.56 Installation of Conductors of Different PLFA Circuits, Class 2, Class 3, and Communications Circuits in the Same Cable, Enclosure, or Raceway.

(A) Two or More PLFA Circuits. Cable and conductors of two or more power-limited fire alarm circuits, communications circuits, or Class 3 circuits shall be permitted within the same cable, enclosure, or raceway.

(B) Class 2 Circuits with PLFA Circuits. Conductors of one or more Class 2 circuits shall be permitted within the same cable, enclosure, or raceway with conductors of power-limited fire alarm circuits, provided that the insulation of the Class 2 circuit conductors in the cable, enclosure, or raceway is at least that required by the power-limited fire alarm circuits.

(C) Low-Power Network-Powered Broadband Communications Cables and PLFA Cables. Low-power network-powered broadband communications circuits shall be permitted in the same enclosure or raceway with PLFA cables.

760.57 Support of Conductors.

Power-limited fire alarm circuit conductors shall not be strapped, taped, or attached by any means to the exterior of any conduit or other raceway as a means of support.

See the commentary following 760.6.

760.58 Conductor Size.

Conductors of 26 AWG shall be permitted only where spliced with a connector listed as suitable for 26 AWG to 24 AWG or larger conductors that are terminated on equipment or where the 26 AWG conductors are terminated on equipment listed as suitable for 26 AWG conductors. Single conductors shall not be smaller than 18 AWG.

Due to a signaling method called *multiplexing* used with digitally addressable fire alarm systems, power-limited fire alarm cable may contain circuit conductors as small as 26 AWG. In the past, these small conductors were typically reserved for communications circuits, but due to recent technological advances, they now have application within the fire alarm industry. Of course, these small circuit conductors are only permitted to be used as specified in 760.58 and as permitted by the listing or installation instructions of specific fire alarm equipment.

760.59 Current-Carrying Continuous Line-Type Fire Detectors.

(A) Application. Listed continuous line-type fire detectors, including insulated copper tubing of pneumatically operated detectors, employed for both detection and carrying signaling currents shall be permitted to be used in power-limited circuits.

(B) Installation. Continuous line-type fire detectors shall be installed in accordance with 760.42 through 760.52 and 760.54.

760.61 Applications of Listed PLFA Cables.

PLFA cables shall comply with the requirements described in either 760.61(A), (B), or (C) or where cable substitutions are made as shown in 760.61(D).

Sections 760.61(A) and 760.61(B) have been revised for the 2002 *Code* for use with the definition of *abandoned fire alarm cable* in 760.2. These sections now require the removal of accessible abandoned fire alarm cable. Abandoned cable increases fire loading unnecessarily, and, where installed in plenums, it can affect airflow. Similar requirements can be found in Articles 640, 645, 725, 770, 800, 820, and 830. See the definition of *abandoned fire alarm cable* in 760.2.

(A) Plenum. Cables installed in ducts, plenums, and other spaces used for environmental air shall be Type FPLP. Abandoned cables shall not be permitted to remain. Types FPLP, FPLR, and FPL cables installed in compliance with 300.22 shall be permitted.

(B) Riser. Cables installed in risers shall be as described in either (1), (2), or (3):

(1) Cables installed in vertical runs and penetrating more than one floor, or cables installed in vertical runs in a shaft, shall be Type FPLR. Floor penetrations requiring Type FPLR shall contain only cables suitable for riser or plenum use. Abandoned cables shall not be permitted to remain.

(2) Other cables shall be installed in metal raceways or located in a fireproof shaft having firestops at each floor.

(3) Type FPL cable shall be permitted in one- and two-family dwellings.

   FPN: See 300.21 for firestop requirements for floor penetrations.

(C) Other Wiring Within Buildings. Cables installed in building locations other than those covered in 760.61(A) or (B) shall be as described in either (1), (2), (3), or (4).

(1) Type FPL shall be permitted.
(2) Cables shall be permitted to be installed in raceways.
(3) Cables specified in Chapter 3 and meeting the requirements of 760.71(A) and (B) shall be permitted to be installed in nonconcealed spaces where the exposed length of cable does not exceed 3 m (10 ft).
(4) A portable fire alarm system provided to protect a stage or set when not in use shall be permitted to use wiring methods in accordance with 530.12.

(D) Fire Alarm Cable Uses and Permitted Substitutions. The uses and permitted substitutions for fire alarm cables listed in Table 760.61 shall be considered suitable for the purpose and shall be permitted.

   FPN: For information on multipurpose cables (Types MPP, MPR, MPG, MP) and communications cables (Types CMP, CMR, CMG, CM), see 800.51.

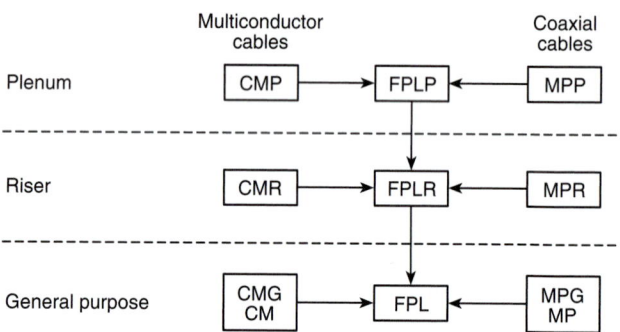

Type CM—Communications wires and cables
Type FPL—Power-limited fire alarm cables
Type MP—Multipurpose cables (coaxial cables only)
[A]→[B] Cable A shall be permitted to be used in place of cable B.
26 AWG minimum

Figure 760.61 Cable substitution hierarchy.

760.71 Listing and Marking of PLFA Cables and Insulated Continuous Line-Type Fire Detectors.

Type FPL cables installed as wiring within buildings shall be listed as being resistant to the spread of fire and other criteria in accordance with 760.71(A) through (H) and shall be marked in accordance with 760.71(I). Insulated continu-

Table 760.61 Cable Uses and Permitted Substitutions

Cable Type	Use	References	Permitted Substitutions	
			Multiconductor	Coaxial
FPLP	Power-limited fire alarm plenum cable	760.61(A)	CMP	MPP
FPLR	Power-limited fire alarm riser cable	760.61(B)	CMP, FPLP, CMR	MPP, MPR
FPL	Power-limited fire alarm cable	760.61(C)	CMP, FPLP, CMR, FPLR, CMG, CM	MPP, MPR, MPG, MP

ous line-type fire detectors shall be listed in accordance with 760.71(J).

(A) Conductor Materials. Conductors shall be solid or stranded copper.

Some line-type fire detectors may not be made exclusively of copper but are listed for the application nevertheless.

(B) Conductor Size. The size of conductors in a multiconductor cable shall not be smaller than 26 AWG. Single conductors shall not be smaller than 18 AWG.

(C) Ratings. The cable shall have a voltage rating of not less than 300 volts.

(D) Type FPLP. Type FPLP power-limited fire alarm plenum cable shall be listed as being suitable for use in ducts, plenums, and other space used for environmental air and shall also be listed as having adequate fire-resistant and low smoke-producing characteristics.

> FPN: One method of defining low smoke-producing cable is by establishing an acceptable value of the smoke produced when tested in accordance with NFPA 262-1999, *Standard Method of Test for Flame Travel and Smoke of Wires and Cables for Use in Air-Handling Spaces*, to a maximum peak optical density of 0.5 and a maximum average optical density of 0.15. Similarly, one method of defining fire-resistant cables is by establishing maximum allowable flame travel distance of 1.52 m (5 ft) when tested in accordance with the same test.

For further information on the fire test method for Type FPLP cables, see the commentary following 725.71(A), FPN.

(E) Type FPLR. Type FPLR power-limited fire alarm riser cable shall be listed as being suitable for use in a vertical run in a shaft or from floor to floor and shall also be listed as having fire-resistant characteristics capable of preventing the carrying of fire from floor to floor.

> FPN: One method of defining fire-resistant characteristics capable of preventing the carrying of fire from floor to floor is that the cables pass the requirements of ANSI/

UL 1666-1997, *Standard Test for Flame Propagation Height of Electrical and Optical-Fiber Cable Installed Vertically in Shafts*.

For further information on the fire test method for Type FPLR cables, see the commentary following 725.71(B) FPN.

(F) Type FPL. Type FPL power-limited fire alarm cable shall be listed as being suitable for general-purpose fire alarm use, with the exception of risers, ducts, plenums, and other spaces used for environmental air and shall also be listed as being resistant to the spread of fire.

> FPN: One method of defining *resistant to the spread of fire* is that the cables do not spread fire to the top of the tray in the vertical-tray flame test in ANSI/UL 1581-1991, *Reference Standard for Electrical Wires, Cables and Flexible Cords*. Another method of defining *resistant to the spread of fire* is for the damage (char length) not to exceed 1.5 m (4 ft 11 in.) when performing the CSA vertical flame test for cables in cable trays, as described in CSA C22.2 No. 0.3-M-1985, *Test Methods for Electrical Wires and Cables*.

For further information on the fire test method for Type FPL cables used as other wiring within buildings, see the commentary following 725.71(C), FPN.

(G) Fire Alarm Circuit Integrity (CI) Cable. Cables suitable for use in fire alarm systems to ensure survivability of critical circuits during a specified time under fire conditions shall be listed as circuit integrity (CI) cable. Cables identified in 760.71(D), (E), and (F) that meet the requirements for circuit integrity shall have the additional classification using the suffix "CI" (for example, FPLP-CI, FPLR-CI, and FPL-CI).

> FPN No. 1: This cable is used for fire alarm circuits as one method of complying with the survivability requirements of NFPA 72-1999, *National Fire Alarm Code*, 3-4.2.2.2, 3-8.4.1.1.4, and 3-8.4.1.3.3.3(3), that the cable maintain its electrical function during fire conditions for a defined period of time.
>
> FPN No. 2: One method of defining circuit integrity (CI) cable is by establishing a minimum 2-hour fire resistance

rating for the cable when tested in accordance with UL 2196-1995, *Standard for Tests of Fire Resistive Cables.*

There are provisions in *NFPA 72, National Fire Alarm Code,* that require continued operation of the fire alarm system, including circuit wiring, under severe conditions such as attack by fire. To provide this integrity, *NFPA 72* recognizes the use of 2-hour fire-rated cable assemblies. FPN No. 2 to 760.71(G) refers to cables tested in accordance with UL 2196 as an example of the type of wiring method that would qualify as circuit integrity (CI) cable. For one such example of CI cable, see Exhibit 760.5.

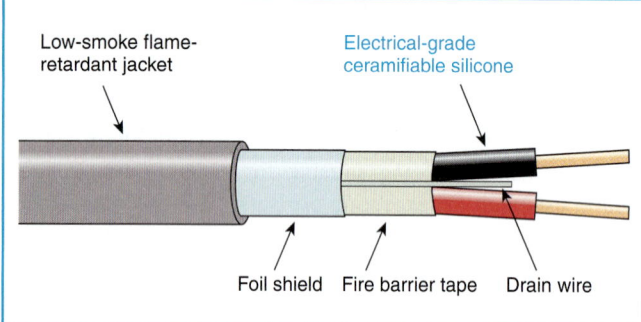

Exhibit 760.5. Circuit integrity cable. (Redrawn from Rockbestos-Suprenant Cable Corp.)

(H) Coaxial Cables. Coaxial cables shall be permitted to use 30 percent conductivity copper-covered steel center conductor wire and shall be listed as Type FPLP, FPLR, or FPL cable.

(I) Cable Marking. The cable shall be marked in accordance with Table 760.71(I). The voltage rating shall not be marked on the cable. Cables that are listed for circuit integrity shall be identified with the suffix CI as defined in 760.71(G).

Table 760.71(I) Cable Markings

Cable Marking	Type	Listing References
FPLP	Power-limited fire alarm plenum cable	760.71(D) and (I)
FPLR	Power-limited fire alarm riser cable	760.71(E) and (I)
FPL	Power-limited fire alarm cable	760.71(F) and (I)

Note: Cables identified in (D), (E), and (F) meeting the requirements for circuit integrity shall have the additional classification using the suffix "CI" (for example, FPLP-CI, FPLR-CI, and FPL-CI).

FPN: Voltage ratings on cables may be misinterpreted to suggest that the cables may be suitable for Class 1, electric light, and power applications.

Exception: Voltage markings shall be permitted where the cable has multiple listings and voltage marking is required for one or more of the listings.

FPN: Cable types are listed in descending order of fire-resistance rating.

(J) Insulated Continuous Line-Type Fire Detectors. Insulated continuous line-type fire detectors shall be rated in accordance with 760.71(C), listed as being resistant to the spread of fire in accordance with 760.71(D) through (F), marked in accordance with 760.71(I), and the jacket compound shall have a high degree of abrasion resistance.

ARTICLE 770
Optical Fiber Cables and Raceways

Contents

I. General

770.1 Scope.

The provisions of this article apply to the installation of optical fiber cables and raceways. This article does not cover the construction of optical fiber cables and raceways.

Article 770 permits the orderly development and usage of optical fiber technology where used in conjunction with electrical conductors for communications, signaling, and control circuits in lieu of metallic conductors. The most common optical fiber cable used in buildings is nonconductive. (See Exhibit 770.1.)

Optical fiber cables may be desirable in some circumstances to transmit data or other communications where electrical noise is a problem, as they are not affected by

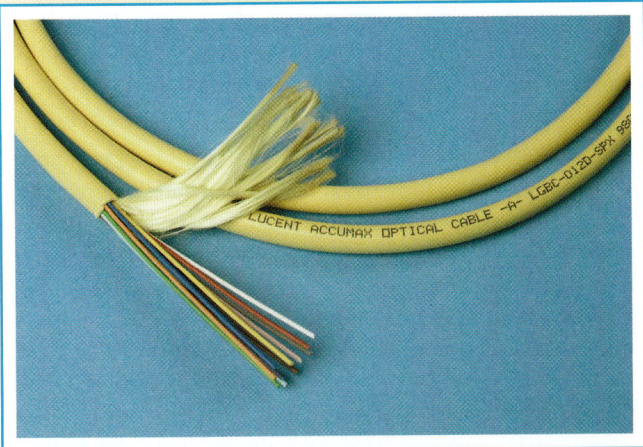

Exhibit 770.1 An example of a nonconductive optical fiber cable.

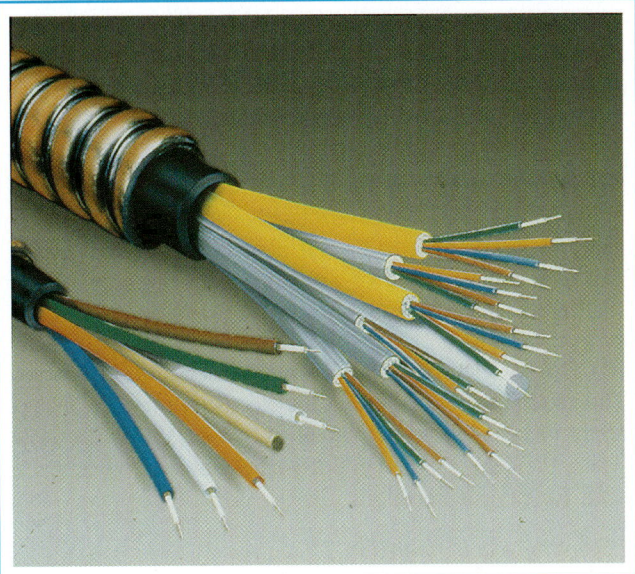

Exhibit 770.2 An example of a composite optical fiber cable. This cable also meets the requirements of Article 330 and is referred to as Type MC cable. (Courtesy of AFC Cable Systems, Inc.)

electrical noise. Optical fiber cables may be nonconductive or they may contain electrical conductors. See Exhibits 770.1 and 770.2.

770.2 Definitions.

Abandoned Optical Fiber Cable. Installed optical fiber cable that is not terminated at equipment other than a connector and not identified for future use with a tag.

This definition has been added to the 2002 *Code* for use with 770.53, which now requires removal of accessible abandoned optical fiber cable. Abandoned cable increases fire loading unnecessarily, and, where installed in plenums, it can affect airflow. Similar requirements can be found in Articles 640, 645, 725, 760, 800, 820, and 830.

Exposed. The circuit is in such a position that, in case of failure of supports and insulation, contact with another circuit may result.

 FPN: See Article 100 for two other definitions of *Exposed*.

Optical Fiber Raceway. A raceway designed for enclosing and routing listed optical fiber cables.

Point of Entrance. The point at which the wire or cable emerges from an external wall, from a concrete floor slab, or from a rigid metal conduit or an intermediate metal conduit grounded to an electrode in accordance with 800.40(B).

770.3 Locations and Other Articles.

Circuits and equipment shall comply with 770.3(A) and (B). Only those sections of Article 300 referenced in this article shall apply to optical fiber cables and raceways.

(A) Spread of Fire or Products of Combustion. The requirements of 300.21 for electrical installations shall also apply to installations of optical fiber cables and raceways. The accessible portion of abandoned optical fiber cables shall not be permitted to remain.

(B) Ducts, Plenums, and Other Air-Handling Spaces. The requirements of 300.22 for electric wiring shall also apply to installations of optical fiber cables and raceways where they are installed in ducts or plenums or other space used for environmental air.

Exception: As permitted in 770.53(A).

See the commentary following 300.22(B) and 300.22(C) for more information on wiring systems installed in ducts, plenums, or other spaces used for environmental air.

770.4 Optical Fiber Cables.

Optical fiber cables transmit light for control, signaling, and communications through an optical fiber.

770.5 Types.

Optical fiber cables can be grouped into three types.

(A) Nonconductive. These cables contain no metallic members and no other electrically conductive materials.

(B) Conductive. These cables contain non–current-carrying conductive members such as metallic strength members, metallic vapor barriers, and metallic armor or sheath.

(C) Composite. These cables contain optical fibers and current-carrying electrical conductors, and shall be permitted to contain non–current-carrying conductive members such as metallic strength members and metallic vapor barriers. Composite optical fiber cables shall be classified as electrical cables in accordance with the type of electrical conductors.

770.6 Raceways for Optical Fiber Cables.

The raceway shall be of a type permitted in Chapter 3 and installed in accordance with Chapter 3.

Exception: Listed nonmetallic optical fiber raceway identified as general-purpose, riser, or plenum optical fiber raceway in accordance with 770.51 and installed in accordance with 362.24 through 362.56, where the requirements applica-

ble to electrical nonmetallic tubing shall apply. Unlisted underground or outside plant construction plastic innerduct shall be terminated at the point of entrance.

Conduit fill requirements apply where the optical fiber is installed in a raceway with electrical conductors.

FPN: For information on listing requirements for optical fiber raceways, see UL 2024, *Standard for Optical Fiber Raceways.*

Where optical fiber cables are installed within the raceway without current-carrying conductors, the raceway fill tables of Chapter 3 and Chapter 9 shall not apply.

Where nonconductive optical fiber cables are installed with electric conductors in a raceway, the raceway fill tables of Chapter 3 and Chapter 9 shall apply.

770.7 Access to Electrical Equipment Behind Panels Designed to Allow Access.

Access to electrical equipment shall not be denied by an accumulation of cables that prevents removal of panels, including suspended ceiling panels.

See the commentary following 725.5 for information on safe access to electrical equipment behind panels.

770.8 Mechanical Execution of Work.

Optical fiber cables shall be installed in a neat and workmanlike manner. Cables installed exposed on the surface of ceiling and sidewalls shall be supported by the structural components of the building structure in such a manner that the cable will not be damaged by normal building use. Such cables shall be attached to structural components by straps, staples, hangers, or similar fittings designed and installed so as not to damage the cable. The installation shall also conform with 300.4(D).

Section 770.8 makes it clear that the *Code* requires these optical fiber cables to be installed in a neat and workmanlike manner. Revised for the 2002 *Code*, this section now provides definitive requirements for workmanship. Cable must now be attached to or supported by the structure by straps, clamps, hangers, and the like. The installation method must not damage the cable. In addition, the location of the cable should be carefully evaluated to ensure that activities and processes within the building do not cause damage to the cable. The reference to 300.4(D) calls attention to the hazard to which cables are exposed where they are installed on

II. Protection

770.33 Grounding of Entrance Cables.

Where exposed to contact with electric light or power conductors, the non–current-carrying metallic members of optical fiber cables entering buildings shall be grounded as close to the point of entrance as practicable or shall be interrupted as close to the point of entrance as practicable by an insulating joint or equivalent device.

III. Cables Within Buildings

770.49 Fire Resistance of Optical Fiber Cables.

Optical fiber cables installed as wiring within buildings shall be listed as being resistant to the spread of fire in accordance with 770.50 and 770.51.

770.50 Listing, Marking, and Installation of Optical Fiber Cables.

Optical fiber cables in a building shall be listed as being suitable for the purpose, and cables shall be marked in accordance with Table 770.50.

Table 770.50 Cable Markings

Cable Marking	Type	Reference
OFNP	Nonconductive optical fiber plenum cable	770.51(A) and 770.53(A)
OFCP	Conductive optical fiber plenum cable	770.51(A) and 770.53(A)
OFNR	Nonconductive optical fiber riser cable	770.51(B) and 770.53(B)
OFCR	Conductive optical fiber riser cable	770.51(B) and 770.53(B)
OFNG	Nonconductive optical fiber general-purpose cable	770.51(C) and 770.53(C)
OFCG	Conductive optical fiber general-purpose cable	770.51(C) and 770.53(C)
OFN	Nonconductive optical fiber general-purpose cable	770.51(D) and 770.53(C)
OFC	Conductive optical fiber general-purpose cable	770.51(D) and 770.53(C)

Exception No. 1: Optical fiber cables shall not be required to be listed and marked where the length of the cable within the building, measured from its point of entrance, does not exceed 15 m (50 ft) and the cable enters the building from the outside and is terminated in an enclosure.

FPN: Splice cases or terminal boxes, both metallic and plastic types, are typically used as enclosures for splicing or terminating optical fiber cables.

Exception No. 2: Conductive optical fiber cable shall not be required to be listed and marked where the cable enters the building from the outside and is run in rigid metal conduit or intermediate metal conduit and such conduits are grounded to an electrode in accordance with 800.40(B).

Exception No. 3: Nonconductive optical fiber cables shall not be required to be listed and marked where the cable enters the building from the outside and is run in raceway installed in compliance with Chapter 3.

FPN No. 1: Cable types are listed in descending order of fire resistance rating. Within each fire resistance rating, nonconductive cable is listed first, since it may substitute for the conductive cable.

FPN No. 2: See the referenced sections for requirements and permitted uses.

770.51 Listing Requirements for Optical Fiber Cables and Raceways.

Optical fiber cables shall be listed in accordance with 770.51(A) through (D), and optical fiber raceways shall be listed in accordance with 770.51(E) through (G).

(A) Types OFNP and OFCP. Types OFNP and OFCP nonconductive and conductive optical fiber plenum cables shall be listed as being suitable for use in ducts, plenums, and other space used for environmental air and shall also be listed as having adequate fire-resistant and low smoke-producing characteristics.

FPN: One method of defining low smoke-producing cables is by establishing an acceptable value of the smoke produced when tested in accordance with NFPA 262-1999, *Standard Method of Test for Flame Travel and Smoke of Wires and Cables for Use in Air-Handling Spaces,* to a maximum peak optical density of 0.5 and a maximum average optical density of 0.15. Similarly, one method of defining fire-resistant cables is by defining maximum allowable flame travel distance of 1.52 m (5 ft) when tested in accordance with the same test.

For further information on the fire test method for optical fiber plenum cables, see the commentary following 725.71(A), FPN.

(B) Types OFNR and OFCR. Types OFNR and OFCR nonconductive and conductive optical fiber riser cables shall be listed as being suitable for use in a vertical run in a shaft or from floor to floor and shall also be listed as having fire-resistant characteristics capable of preventing the carrying of fire from floor to floor.

FPN: One method of defining fire-resistant characteristics capable of preventing the carrying of fire from floor to floor is that the cables pass the requirements of ANSI/UL 1666-1997, *Standard Test for Flame Propagation*

Height of Electrical and Optical-Fiber Cable Installed Vertically in Shafts.

For further information on the fire test method for optical fiber riser cables, see the commentary following 725.71(B) FPN.

(C) Types OFNG and OFCG. Types OFNG and OFCG nonconductive and conductive general-purpose optical fiber cables shall be listed as being suitable for general-purpose use, with the exception of risers and plenums, and shall also be listed as being resistant to the spread of fire.

> FPN: One method of defining *resistance to the spread of fire* is for the damage (char length) not to exceed 1.5 m (4 ft 11 in.) when performing the vertical flame test for cables in cable trays, as described in CSA C22.2 No. 0.3-M-1985, *Test Methods for Electrical Wires and Cables.*

For further information on the fire test method for optical fiber cables used as other wiring within buildings, see the commentary following 725.71(C), FPN.

(D) Types OFN and OFC. Types OFN and OFC nonconductive and conductive optical fiber cables shall be listed as being suitable for general-purpose use, with the exception of risers, plenums, and other spaces used for environmental air, and shall also be listed as being resistant to the spread of fire.

> FPN: One method of defining *resistant to the spread of fire* is that the cables do not spread fire to the top of the tray in the vertical-tray flame test in ANSI/UL 1581-1991, *Reference Standard for Electrical Wires, Cables, and Flexible Cords.*
>
> Another method of defining *resistant to the spread of fire* is for the damage (char length) not to exceed 1.5 m (4 ft 11 in.) when performing the vertical flame test for cables in cable trays, as described in CSA C22.2 No. 0.3-M-1985, *Test Methods for Electrical Wires and Cables.*

(E) Plenum Optical Fiber Raceway. Plenum optical fiber raceways shall be listed as having adequate fire-resistant and low smoke-producing characteristics.

(F) Riser Optical Fiber Raceway. Riser optical fiber raceways shall be listed as having fire-resistant characteristics capable of preventing the carrying of fire from floor to floor.

(G) General-Purpose Optical Fiber Cable Raceway. General-purpose optical fiber cable raceway shall be listed as being resistant to the spread of fire.

The optical fiber raceways covered in 770.51(E), (F), and (G) are listed raceways used in plenum, riser, or general-purpose applications. This listing includes raceways and fittings for installations of nonconductive optical fiber cables in accordance with Article 770. These raceways are not suitable for installation of current-carrying conductors, cords, or cables. Nor are these raceways suitable for installations of hybrid cables that contain optical fiber members and current-carrying conductors.

PLENUM RACEWAY

A raceway marked "plenum" is suitable for use in ducts, plenums, or other spaces used for environmental air in accordance with 770.53(A) where used to enclose optical fiber cables marked OFNP. This plenum raceway exhibits a maximum peak optical density of 0.5, a maximum average optical density of 0.15, and a maximum flame-spread distance of 5 ft when tested in accordance with the *Test for Flame Propagation and Smoke-Density Values for Electrical and Optical-Fiber Cables Used in Spaces Transporting Environmental Air,* UL 910. This raceway is identified by a marking on the surface of the raceway or on a marker tape indicating "plenum." A raceway marked "plenum" is also suitable for installation in risers where used to enclose optical fiber cables marked OFNP or OFNR, and for general-purpose use where used to enclose optical fiber cables marked OFNP, OFNR, OFNG, or OFN.

RISER RACEWAY

A raceway marked "riser" is suitable for installation in risers in accordance with 770.53(B) where used to enclose optical fiber cable marked OFNP or OFNR. This raceway has fire-resistant characteristics capable of preventing the carrying of fire from floor to floor. Riser raceway meets the test requirements of the *Standard Test for Flame Propagation Height of Electrical and Optical-Fiber Cable Installed Vertically in Shafts,* UL 1666. This raceway is identified by a marking on the surface of the raceway or on a marker tape indicating "riser." A raceway marked "riser" is also suitable for general-purpose use when used to enclose optical fiber cable marked OFNP, OFNR, OFNG, or OFN.

GENERAL PURPOSE RACEWAY

A raceway marked "general purpose" is suitable for installation in general-purpose areas in accordance with 770.53(C) where used to enclose optical fiber cable marked OFNP, OFNR, OFNG, or OFN. General-purpose raceway has fire-resistant characteristics that are capable of preventing the spread of fire.

Pliable raceway is raceway that can be bent by hand without the use of tools. The smallest radius of the curve of the inner edge of any bend to which the raceway may be bent without cracking either on the outer surface or internally is not less than 2½ times the outside diameter of the raceway.

770.52 Installation of Optical Fibers and Electrical Conductors.

(A) With Conductors for Electric Light, Power, Class 1, Non–Power-Limited Fire Alarm, or Medium Power Network-Powered Broadband Communications Circuits. Optical fibers shall be permitted within the same composite cable for electric light, power, Class 1, non–power-limited fire alarm, or medium power network-powered broadband communications circuits operating at 600 volts or less only where the functions of the optical fibers and the electrical conductors are associated.

Nonconductive optical fiber cables shall be permitted to occupy the same cable tray or raceway with conductors for electric light, power, Class 1, non–power-limited fire alarm, or medium power network-powered broadband communications circuits operating at 600 volts or less. Conductive optical fiber cables shall not be permitted to occupy the same cable tray or raceway with conductors for electric light, power, Class 1, non–power-limited fire alarm, or medium power network-powered broadband communications circuits.

Composite optical fiber cables containing only current-carrying conductors for electric light, power, Class 1 circuits rated 600 volts or less shall be permitted to occupy the same cabinet, cable tray, outlet box, panel, raceway, or other termination enclosure with conductors for electric light, power, or Class 1 circuits operating at 600 volts or less.

Nonconductive optical fiber cables shall not be permitted to occupy the same cabinet, outlet box, panel, or similar enclosure housing the electrical terminations of an electric light, power, Class 1, non–power-limited fire alarm, or medium power network-powered broadband communications circuit.

Exception No. 1: Occupancy of the same cabinet, outlet box, panel, or similar enclosure shall be permitted where nonconductive optical fiber cable is functionally associated with the electric light, power, Class 1, non–power-limited fire alarm, or medium power network-powered broadband communications circuit.

Exception No. 2: Occupancy of the same cabinet, outlet box, panel, or similar enclosure shall be permitted where nonconductive optical fiber cables are installed in factory- or field-assembled control centers.

Exception No. 3: In industrial establishments only, where conditions of maintenance and supervision ensure that only qualified persons service the installation, nonconductive optical fiber cables shall be permitted with circuits exceeding 600 volts.

Exception No. 4: In industrial establishments only, where conditions of maintenance and supervision ensure that only qualified persons service the installation, composite optical fiber cables shall be permitted to contain current-carrying conductors operating over 600 volts.

Installations in raceway shall comply with 300.17.

(B) With Other Conductors. Optical fibers shall be permitted in the same cable, and conductive and nonconductive optical fiber cables shall be permitted in the same cable tray, enclosure, or raceway with conductors of any of the following:

(1) Class 2 and Class 3 remote-control, signaling, and power-limited circuits in compliance with Article 725
(2) Power-limited fire alarm systems in compliance with Article 760
(3) Communications circuits in compliance with Article 800
(4) Community antenna television and radio distribution systems in compliance with Article 820
(5) Low-power network-powered broadband communications circuits in compliance with Article 830

(C) Grounding. Non–current-carrying conductive members of optical fiber cables shall be grounded in accordance with Article 250.

770.53 Applications of Listed Optical Fiber Cables and Raceways.

Nonconductive and conductive optical fiber cables shall comply with any of the requirements given in 770.53(A) through (E) or where cable substitutions are made as shown in 770.53(F).

Sections 770.53(A) and 770.53(B) have been revised for the 2002 *Code* for use with the definition of *abandoned optical fiber cable* in 770.2. These sections now require the removal of accessible abandoned optical fiber cable. Abandoned cable increases fire loading unnecessarily, and, where installed in plenums, it can affect airflow. Similar requirements can be found in Articles 640, 645, 725, 760, 800, 820, and 830. See the definition of *abandoned communications cable* in 800.2.

(A) Plenum. Cables installed in ducts, plenums, and other spaces used for environmental air shall be Type OFNP or OFCP. Abandoned cables shall not be permitted to remain. Types OFNR, OFCR, OFNG, OFN, OFCG, and OFC cables installed in compliance with 300.22 shall be permitted. Listed plenum optical fiber raceways shall be permitted to be installed in ducts and plenums as described in 300.22(B) and in other spaces used for environmental air as described in 300.22(C). Only types OFNP and OFCP cables shall be permitted to be installed in these raceways.

(B) Riser. Cables installed in risers shall be as described in any of the following:

(1) Cables installed in vertical runs and penetrating more than one floor, or cables installed in vertical runs in a shaft, shall be Type OFNR or OFCR. Floor penetrations requiring Type OFNR or OFCR shall contain only cables suitable for riser or plenum use. Abandoned cables shall not be permitted to remain. Listed riser optical fiber raceways shall be permitted to be installed in vertical riser runs in a shaft from floor to floor. Only Types OFNP, OFCP, OFNR and OFCR cables shall be permitted to be installed in these raceways.

(2) Types OFNG, OFN, OFCG, and OFC cables shall be permitted to be encased in a metal raceway or located in a fireproof shaft having firestops at each floor.

(3) Types OFNG, OFN, OFCG, and OFC cables shall be permitted in one- and two-family dwellings.

> FPN: See 300.21 for firestop requirements for floor penetrations.

(C) Other Wiring Within Buildings. Cables installed in building locations other than the locations covered in 770.53(A) and (B) shall be Type OFNG, OFN, OFCG, or OFC. Such cables shall be permitted to be installed in listed general-purpose optical fiber raceways.

(D) Hazardous (Classified) Locations. Cables installed in hazardous (classified) locations shall be any type indicated in Table 770.53.

Table 770.53 Cable Substitutions

Cable Type	Permitted Substitutions
OFNP	None
OFCP	OFNP
OFNR	OFNP
OFCR	OFNP, OFCP, OFNR
OFNG, OFN	OFNP, OFNR
OFCG, OFC	OFNP, OFCP, OFNR, OFCR, OFNG, OFN

(E) Cable Trays. Optical fiber cables of the types listed in Table 770.50 shall be permitted to be installed in cable trays.

> FPN: It is not the intent to require that these optical fiber cables be listed specifically for use in cable trays.

(F) Cable Substitutions. The substitutions for optical fiber cables listed in Table 770.53 shall be permitted.

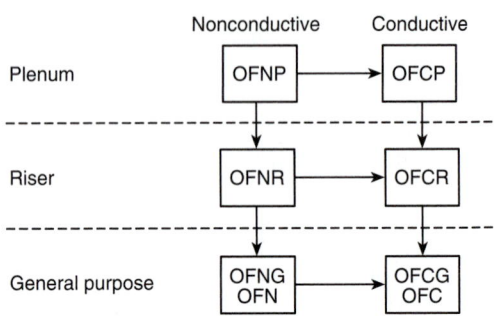

Figure 770.53 Cable substitution hierarchy.

ARTICLE 780
Closed-Loop and Programmed Power Distribution

Contents

780.1 Scope
780.2 General
  (A) Other Articles
  (B) Component Parts
780.3 Control
  (A) Characteristic Electrical Identification Required
  (B) Conditions for De-Energization
  (C) Additional Conditions for De-Energization When an Alternate Source of Power Is Used
  (D) Controller Malfunction
780.5 Power Limitation in Signaling Circuits
780.6 Cables and Conductors
  (A) Hybrid Cable
  (B) Cables and Conductors in the Same Cabinet, Panel, or Box
780.7 Noninterchangeability

780.1 Scope.

The provisions of this article apply to premise power distribution systems jointly controlled by a signaling between the energy controlling equipment and utilization equipment.

Article 780 provides requirements for the "smart house" concept, which involves universal cable terminating in universal outlets.

Buildings wired by conventional methods require separate sets of conductors for different systems. In the smart house, however, multiple conductors for 120-volt ac power, 24-volt dc UPS, telephone, remote-control and signaling, as well as coaxial cable, are combined in a single construction known as *hybrid cabling*.

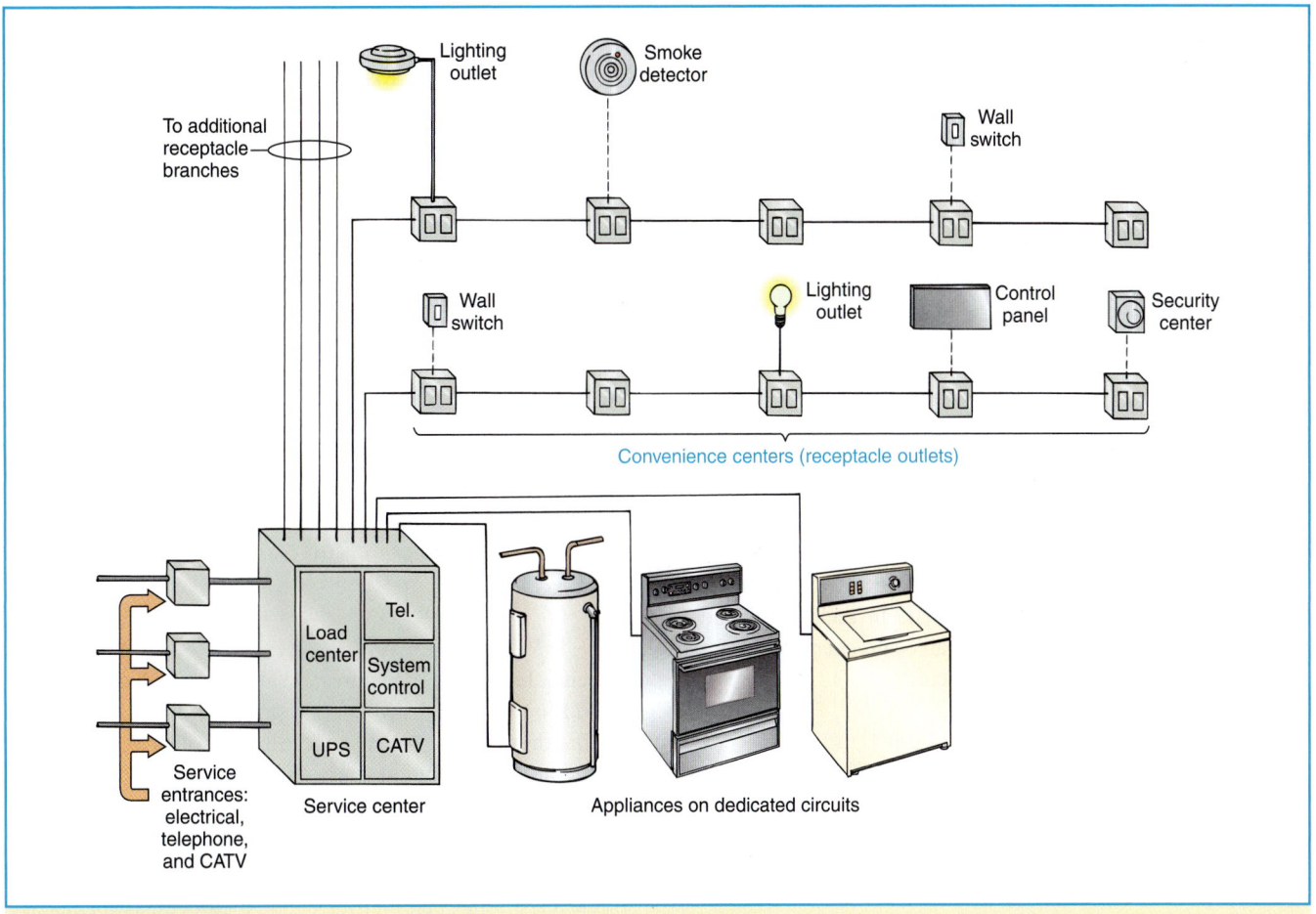

Exhibit 780.1 A typical smart house installation. (Redrawn from Smart House)

Hybrid cabling serves multipurpose receptacle outlets known as *convenience centers,* which are capable of supplying different types of energy and signals to specific appliances or equipment.

The smart house uses an energy safety technique called *closed-loop control* to reduce shock hazard. In conventional wiring, receptacles are energized at all times under normal operating conditions. In the closed-loop configuration, receptacles are not energized until the insertion of an attachment plug generates a characteristic electrical identification.

Exhibit 780.1 illustrates a typical smart house installation. Present smart house technology uses 120/240-volts ac, with 24-volts dc UPS, to maintain system electronics in the event of a transient or utility power outage.

780.2 General.

(A) Other Articles. Except as modified by the requirements of this article, all other applicable articles of this *Code* shall apply.

(B) Component Parts. All equipment and conductors shall be listed and identified.

780.3 Control.

The control equipment and all power switching devices operated by the control equipment shall be listed and identified. The system shall operate in accordance with 780.2(A) through (D).

(A) Characteristic Electrical Identification Required. Outlets shall not be energized unless the utilization equipment first exhibits a characteristic electrical identification.

Receptacles are energized with 120-volt ac power only when electronic circuitry in the convenience center receives this characteristic identification.

(B) Conditions for De-Energization. Outlets shall be de-energized when any of the following conditions occur:

(1) A nominal-operation acknowledgment signal is not being received from the utilization equipment connected to the outlet.

Convenience center receptacles are de-energized when the characteristic electrical identification ceases (when the attachment plug is withdrawn). In addition, appliances with built-in smart house communications chips transmit a continuous nominal-operation signal to the convenience center electronics. If this signal is interrupted, indicating a possible malfunction or safety problem, the receptacle is automatically de-energized.

(2) A ground-fault condition exists.
(3) An overcurrent condition exists.

(C) Additional Conditions for De-Energization When an Alternate Source of Power Is Used. In addition to the requirements in 780.2(B), outlets shall be de-energized when any of the following conditions occur:

(1) The grounded conductor is not properly grounded.
(2) Any ungrounded conductor is not at nominal voltage.

(D) Controller Malfunction. In the event of a controller malfunction, all associated outlets shall be de-energized.

780.5 Power Limitation in Signaling Circuits.

For signaling circuits not exceeding 24 volts, the current required shall not exceed 1 ampere where protected by an overcurrent device or an inherently limited power source.

780.6 Cables and Conductors.

(A) Hybrid Cable. Listed hybrid cable consisting of power, communications, and signaling conductors shall be permitted under a common jacket. The jacket shall be applied so as to separate the power conductors from the communications and signaling conductors. An optional outer jacket shall be permitted to be applied. The individual conductors of a hybrid cable shall conform to the *Code* provisions applicable to their current, voltage, and insulation rating. The signaling conductors shall not be smaller than 24 AWG copper.

(B) Cables and Conductors in the Same Cabinet, Panel, or Box. The power, communications, and signaling conductors of listed hybrid cable are permitted to occupy the same cabinet, panel, or outlet box (or similar enclosure housing the electrical terminations of electric light or power circuits) only if connectors specifically listed for hybrid cable are employed.

780.7 Noninterchangeability.

Receptacles, cord connectors, and attachment plugs used on closed-loop power distribution systems shall be constructed so that they are not interchangeable with other receptacles, cord connectors, and attachment plugs.

Convenience center receptacles are constructed so that they will not accept an attachment plug with a different voltage or current rating than that for which the device is intended. Attachment plugs for use with closed-loop power distribution systems will not fit into conventional receptacles, which ensures that "smart" appliances are not used on other power distribution systems that lack closed-loop control features.

Communications Systems

ARTICLE 800
Communications Circuits

Contents

I. General

800.1 Scope.

This article covers telephone, telegraph (except radio), outside wiring for fire alarm and burglar alarm, and similar central station systems; and telephone systems not connected to a central station system but using similar types of equipment, methods of installation, and maintenance.

> FPN No. 1: For further information for fire alarm, guard tour, sprinkler waterflow, and sprinkler supervisory systems, see Article 760.
>
> FPN No. 2: For installation requirements of optical fiber cables, see Article 770.
>
> FPN No. 3: For installation requirements for network-powered broadband communications circuits, see Article 830.

Section 90.3, Code Arrangement, states that Chapter 8—comprising Articles 800, 810, 820, and 830—covers communications systems and is not subject to the requirements of Chapters 1–7 except where a requirement from these chapters is specifically referenced in Chapter 8. For instance, 800.10(A)(3) references 225.14(D), 800.30(C) references Article 500, and 800.52(C) references 300.22(C).

Although information technology equipment systems are often used for or with communications systems, Article 800 does not cover wiring of this equipment. Instead, Article 645 provides requirements for wiring contained solely within an information technology equipment (computer) room. See 645.2 for a description of the type of information technology equipment room to which Article 645 applies. Article 725 provides requirements for wiring extending beyond a computer room, and Article 760 covers wiring requirements for a fire alarm system.

In some cases, telephone system wiring is also used for data transmission; this use is covered by Article 800. Telephone company central offices are exempt from the requirements of Article 800 by 90.2(B)(4). The format for Article 800 is similar to that for Articles 725, 760, 770, and 820.

Article 830 was added to the 1999 *Code* to cover network-powered broadband communications systems.

800.2 Definitions.

See Article 100. For purposes of this article, the following additional definitions apply.

Abandoned Communications Cable. Installed communications cable that is not terminated at both ends at a connector or other equipment and not identified for future use with a tag.

This definition has been added to the 2002 *Code* for use with 800.52(B), which now requires removal of accessible abandoned communications cable. Abandoned cable increases fire loading unnecessarily, and, where installed in plenums, it can affect airflow. Similar requirements can be found in Articles 640, 645, 725, 760, 770, 820, and 830.

Block. A square or portion of a city, town, or village enclosed by streets and including the alleys so enclosed, but not any street.

Cable. A factory assembly of two or more conductors having an overall covering.

Cable Sheath. A covering over the conductor assembly that may include one or more metallic members, strength members, or jackets.

Exposed. A circuit that is in such a position that, in case of failure of supports and insulation, contact with another circuit may result.

FPN: See Article 100 for two other definitions of *Exposed*.

Point of Entrance. Within a building, the point at which the wire or cable emerges from an external wall, from a concrete floor slab, or from a rigid metal conduit or an intermediate metal conduit grounded to an electrode in accordance with 800.40(B).

Premises. The land and buildings of a user located on the user side of the utility-user network point of demarcation.

Wire. A factory assembly of one or more insulated conductors without an overall covering.

See Article 100 for definitions of *conductor, equipment,* and *raceway.*

800.3 Hybrid Power and Communications Cables.

The provisions of 780.6 shall apply for listed hybrid power and communications cables in closed-loop and programmed power distribution.

FPN: See 800.51(I) for hybrid power and communications cable in other applications.

See 800.51(I) for listing requirements and applications of hybrid power and communications cable in one- and two-family residences for other than closed-loop and programmed power distribution.

800.4 Equipment.

Equipment intended to be electrically connected to a telecommunications network shall be listed for the purpose. Installation of equipment shall also comply with 110.3(B).

FPN: One way to determine applicable requirements is to refer to UL 1950-1993, *Standard for Safety of Information Technology Equipment, Including Electrical Business Equipment,* third edition; UL 1459-1995, *Standard for Safety, Telephone Equipment,* third edition; or UL 1863-1995, *Standard for Safety, Communications Circuit Accessories,* second edition. For information on listing requirements for communications raceways, see UL 2024-1995, *Standard for Optical Fiber Raceways.*

UL 1459 and UL 1863 are two safety standards that contain requirements for determining whether equipment connected to a telecommunications network is suitable for the intended purpose. Listed equipment that is connected to the telecommunications network and evaluated according to other U.S. safety standards is also subject to telecommunications requirements appropriate for the equipment. Examples of this equipment include information technology equipment, audiovideo equipment, and signaling equipment connected to a central station. The appropriate requirements are contained within the applicable safety standard, are extracted from UL 1459 or UL 1863, or both.

Exception: This listing requirement shall not apply to test equipment that is intended for temporary connection to a telecommunications network by qualified persons during the course of installation, maintenance, or repair of telecommunications equipment or systems.

Except for test equipment, all permanently installed electrical components of the communications network are subject to the listing requirements of 800.4.

800.5 Access to Electrical Equipment Behind Panels Designed to Allow Access.

Access to electrical equipment shall not be denied by an accumulation of wires and cables that prevents removal of panels, including suspended ceiling panels.

An excess accumulation of wires and cables can limit access to equipment by preventing the removal of access panels. (See Exhibit 800.1.)

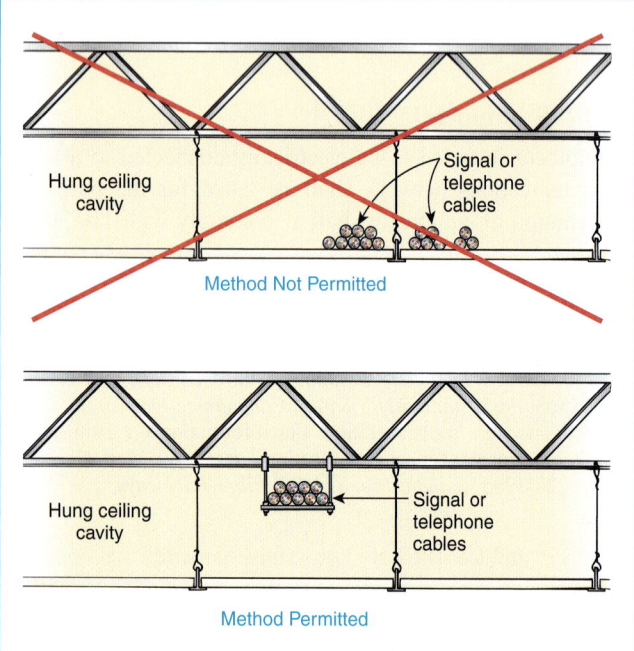

Exhibit 800.1 Installations of conductors and cables, which can prevent access to equipment or cables. Correct and incorrect methods are shown.

800.6 Mechanical Execution of Work.

Communications circuits and equipment shall be installed in a neat and workmanlike manner. Cables installed exposed on the outer surface of ceiling and sidewalls shall be supported by the structural components of the building structure in such a manner that the cable is not damaged by normal building use. Such cables shall be attached to structural components by straps, staples, hangers, or similar fittings designed and installed so as not to damage the cable. The installation shall also conform with 300.4(D).

Revised for the 2002 *Code*, this section now provides definitive requirements for workmanship. Cable must now be attached to or supported by the structure by straps, clamps, hangers, and the like. The installation method must not damage the cable. In addition, the location of the cable should be carefully evaluated to ensure that activities and processes within the building do not cause damage to the cable.

800.8 Hazardous (Classified) Locations.

Communications circuits and equipment installed in a location that is classified in accordance with Article 500 shall comply with the applicable requirements of Chapter 5.

Section 800.8 alerts users that communications circuits installed in locations classified in accordance with Article 500 must conform to the applicable requirements of Chapter 5.

II. Conductors Outside and Entering Buildings

800.10 Overhead Communications Wires and Cables.

Overhead communications wires and cables entering buildings shall comply with 800.10(A) and (B).

(A) On Poles and In-Span. Where communications wires and cables and electric light or power conductors are supported by the same pole or run parallel to each other in-span, the conditions described in 800.10(A)(1) through (A)(4) shall be met.

(1) Relative Location. Where practicable, the communications wires and cables shall be located below the electric light or power conductors.

(2) Attachment to Crossarms. Communications wires and cables shall not be attached to a crossarm that carries electric light or power conductors.

(3) Climbing Space. The climbing space through communications wires and cables shall comply with the requirements of 225.14(D).

(4) Clearance. Supply service drops of 0–750 volts running above and parallel to communications service drops shall have a minimum separation of 300 mm (12 in.) at any point in the span, including the point of and at their attachment to the building, provided the nongrounded conductors are insulated and that a clearance of not less than 1.0 m (40 in.) is maintained between the two services at the pole.

(B) Above Roofs. Communications wires and cables shall have a vertical clearance of not less than 2.5 m (8 ft) from all points of roofs above which they pass.

Exception No. 1: Auxiliary buildings, such as garages and the like.

Exception No. 2: A reduction in clearance above only the overhanging portion of the roof to not less than 450 mm (18 in.) shall be permitted if (a) not more than 1.2 m (4 ft) of communications service-drop conductors pass above the roof overhang and (b) they are terminated at a through- or above-the-roof raceway or approved support.

Exception No. 3: Where the roof has a slope of not less than 100 mm in 300 mm (4 in. in 12 in.), a reduction in clearance to not less than 900 mm (3 ft) shall be permitted.

> FPN: For additional information regarding overhead wires and cables, see ANSI C2-1997, *National Electric Safety Code*, Part 2 Safety Rules For Overhead Lines.

800.11 Underground Circuits Entering Buildings.

Underground communications wires and cables entering buildings shall comply with 800.11(A) and (B).

(A) With Electric Light or Power Conductors. Underground communications wires and cables in a raceway, handhole, or manhole containing electric light, power, Class 1, or non–power-limited fire alarm circuit conductors shall be in a section separated from such conductors by means of brick, concrete, or tile partitions or by means of a suitable barrier.

(B) Underground Block Distribution. Where the entire street circuit is run underground and the circuit within the block is placed so as to be free from the likelihood of accidental contact with electric light or power circuits of over 300 volts to ground, the insulation requirements of 800.12(A) and (C) shall not apply, insulating supports shall not be required for the conductors, and bushings shall not be required where the conductors enter the building.

800.12 Circuits Requiring Primary Protectors.

Circuits that require primary protectors as provided in 800.30 shall comply with 800.12(A), (B), and (C).

(A) Insulation, Wires, and Cables. Communications wires and cables without a metallic shield, running from the last outdoor support to the primary protector, shall be listed as being suitable for the purpose and shall have current-carrying capacity as specified in 800.30(A)(1)(b) or 800.30(A)(1)(c).

(B) On Buildings. Communications wires and cables in accordance with 800.12(A) shall be separated at least 100 mm (4 in.) from electric light or power conductors not in a raceway or cable or be permanently separated from conductors of the other system by a continuous and firmly fixed nonconductor in addition to the insulation on the wires, such as porcelain tubes or flexible tubing. Communications wires and cables in accordance with 800.12(A) exposed to accidental contact with electric light and power conductors operating at over 300 volts to ground and attached to buildings shall be separated from woodwork by being supported on glass, porcelain, or other insulating material.

Exception: Separation from woodwork shall not be required where fuses are omitted as provided for in 800.30(A)(1), or where conductors are used to extend circuits to a building from a cable having a grounded metal sheath.

(C) Entering Buildings. Where a primary protector is installed inside the building, the communications wires and cables shall enter the building either through a noncombustible, nonabsorbent insulating bushing or through a metal raceway. The insulating bushing shall not be required where the entering communications wires and cables (1) are in metal-sheathed cable, (2) pass through masonry, (3) meet the requirements of 800.12(A) and fuses are omitted as provided in 800.30(A)(1), or (4) meet the requirements of 800.12(A) and are used to extend circuits to a building from a cable having a grounded metallic sheath. Raceways or bushings shall slope upward from the outside or, where this cannot be done, drip loops shall be formed in the communications wires and cables immediately before they enter the building.

Raceways shall be equipped with an approved service head. More than one communications wire and cable shall be permitted to enter through a single raceway or bushing. Conduits or other metal raceways located ahead of the primary protector shall be grounded.

800.13 Lightning Conductors.

Where practicable, a separation of at least 1.8 m (6 ft) shall be maintained between communications wires and cables on buildings and lightning conductors.

III. Protection

800.30 Protective Devices.

(A) Application. A listed primary protector shall be provided on each circuit run partly or entirely in aerial wire or aerial cable not confined within a block. Also, a listed primary protector shall be provided on each circuit, aerial or underground, located within the block containing the building served so as to be exposed to accidental contact with electric light or power conductors operating at over 300 volts to ground. In addition, where there exists a lightning exposure, each interbuilding circuit on a premises shall be protected by a listed primary protector at each end of the interbuilding circuit. Installation of primary protectors shall also comply with 110.3(B).

> FPN No. 1: On a circuit not exposed to accidental contact with power conductors, providing a listed primary protector in accordance with this article helps protect against other hazards, such as lightning and above-normal voltages induced by fault currents on power circuits in proximity to the communications circuit.
>
> FPN No. 2: Interbuilding circuits are considered to have a lightning exposure unless one or more of the following conditions exist:
> (1) Circuits in large metropolitan areas where buildings are close together and sufficiently high to intercept lightning.
> (2) Interbuilding cable runs of 42 m (140 ft) or less, directly buried or in underground conduit, where a continuous metallic cable shield or a continuous metallic conduit containing the cable is bonded to each building grounding electrode system.
> (3) Areas having an average of five or fewer thunderstorm days per year and earth resistivity of less than 100 ohm-meters. Such areas are found along the Pacific coast.

Telephone utility companies ordinarily provide primary protectors where telephone lines are exposed to lightning. Installers of private networks that include interbuilding cable should also install primary protectors where cables are exposed to lightning. Generally, cable is considered to be exposed to lightning unless one or more of the conditions in FPN No. 2 exist. A primary protector is required at each end of an interbuilding communications circuit where lightning exposure exists.

(1) Fuseless Primary Protectors. Fuseless-type primary protectors shall be permitted under any of the conditions given in (a) through (e).

(a) Where conductors enter a building through a cable with grounded metallic sheath member(s) and if the conductors in the cable safely fuse on all currents greater than the current-carrying capacity of the primary protector and of the primary protector grounding conductor

(b) Where insulated conductors in accordance with 800.12(A) are used to extend circuits to a building from a cable with an effectively grounded metallic sheath member(s) and if the conductors in the cable or cable stub, or the connections between the insulated conductors and the exposed plant, safely fuse on all currents greater than the current-carrying capacity of the primary protector, or the associated insulated conductors and of the primary protector grounding conductor

(c) Where insulated conductors in accordance with 800.12(A) or (B) are used to extend circuits to a building from other than a cable with a metallic sheath member(s) if (1) the primary protector is listed for this purpose, and (2) the connections of the insulated conductors to the exposed plant or the conductors of the exposed plant safely fuse on all currents greater than the current-carrying capacity of the primary protector, or the associated insulated conductors and of the primary protector grounding conductor

(d) Where insulated conductors in accordance with 800.12(A) are used to extend circuits aerially to a building from an unexposed buried or underground circuit

(e) Where insulated conductors in accordance with 800.12(A) are used to extend circuits to a building from cable with an effectively grounded metallic sheath member(s) and if (1) the combination of the primary protector and insulated conductors is listed for this purpose, and (2) the insulated conductors safely fuse on all currents greater than the current-carrying capacity of the primary protector and of the primary protector grounding conductor

The term *effectively grounded* is defined in Article 100.

(2) Fused Primary Protectors. Where the requirements listed under 800.30(A)(1)(a) through (1)(e) are not met, fused-type primary protectors shall be used. Fused-type primary protectors shall consist of an arrester connected between each line conductor and ground, a fuse in series with each line conductor, and an appropriate mounting arrangement. Primary protector terminals shall be marked to indicate line, instrument, and ground, as applicable.

(B) Location. The primary protector shall be located in, on, or immediately adjacent to the structure or building served and as close as practicable to the point of entrance.

> FPN: See 800.2 for the definition of *point of entrance*.

See Exhibit 800.2 for an example of a primary protector unit typically installed in commercial buildings.

See Exhibit 800.3 for an example of applications of listed communications and multipurpose cable.

Exhibit 800.2 A primary protector unit typically installed in commercial buildings. This is the interface to the outside plant cable.

For purposes of this section, primary protectors located at mobile home service equipment located in sight from and not more than 9.0 m (30 ft) from the exterior wall of the mobile home it serves, or at a mobile home disconnecting means grounded in accordance with 250.32 and located in sight from and not more than 9.0 m (30 ft) from the exterior wall of the mobile home it serves, shall be considered to meet the requirements of this section.

FPN: Selecting a primary protector location to achieve the shortest practicable primary protector grounding conductor helps limit potential differences between communications circuits and other metallic systems.

(C) Hazardous (Classified) Locations. The primary protector shall not be located in any hazardous (classified) location as defined in Article 500 or in the vicinity of easily ignitible material.

Exception: As permitted in 501.14, 502.14, and 503.12.

800.31 Primary Protector Requirements.

The primary protector shall consist of an arrester connected between each line conductor and ground in an appropriate mounting. Primary protector terminals shall be marked to indicate line and ground as applicable.

FPN: One way to determine applicable requirements for a listed primary protector is to refer to ANSI/UL 497-1995, *Standard for Protectors for Paired Conductor Communications Circuits.*

800.32 Secondary Protector Requirements.

Where a secondary protector is installed in series with the indoor communications wire and cable between the primary protector and the equipment, it shall be listed for the purpose. The secondary protector shall provide means to safely limit currents to less than the current-carrying capacity of listed indoor communications wire and cable, listed telephone set line cords, and listed communications terminal equipment having ports for external wire line communications circuits. Any overvoltage protection, arresters, or grounding connection shall be connected on the equipment terminals side of the secondary protector current-limiting means.

FPN No. 1: One way to determine applicable requirements for a listed secondary protector is to refer to UL 497A-1996, *Standard for Secondary Protectors for Communications Circuits.*

FPN No. 2: Secondary protectors on exposed circuits are not intended for use without primary protectors.

800.33 Cable Grounding.

The metallic sheath of communications cables entering buildings shall be grounded as close as practicable to the point of entrance or shall be interrupted as close to the point of entrance as practicable by an insulating joint or equivalent device.

FPN: See 800.2 for the definition of *point of entrance.*

IV. Grounding Methods

800.40 Cable and Primary Protector Grounding.

The metallic member(s) of the cable sheath, where required to be grounded by 800.33, and primary protectors shall be grounded as specified in 800.40(A) through (D).

(A) Grounding Conductor.

(1) Insulation. The grounding conductor shall be insulated and shall be listed as suitable for the purpose.

(2) Material. The grounding conductor shall be copper or other corrosion-resistant conductive material, stranded or solid.

(3) Size. The grounding conductor shall not be smaller than 14 AWG.

Exhibit 800.3 An example of applications of listed communications and multipurpose cables.

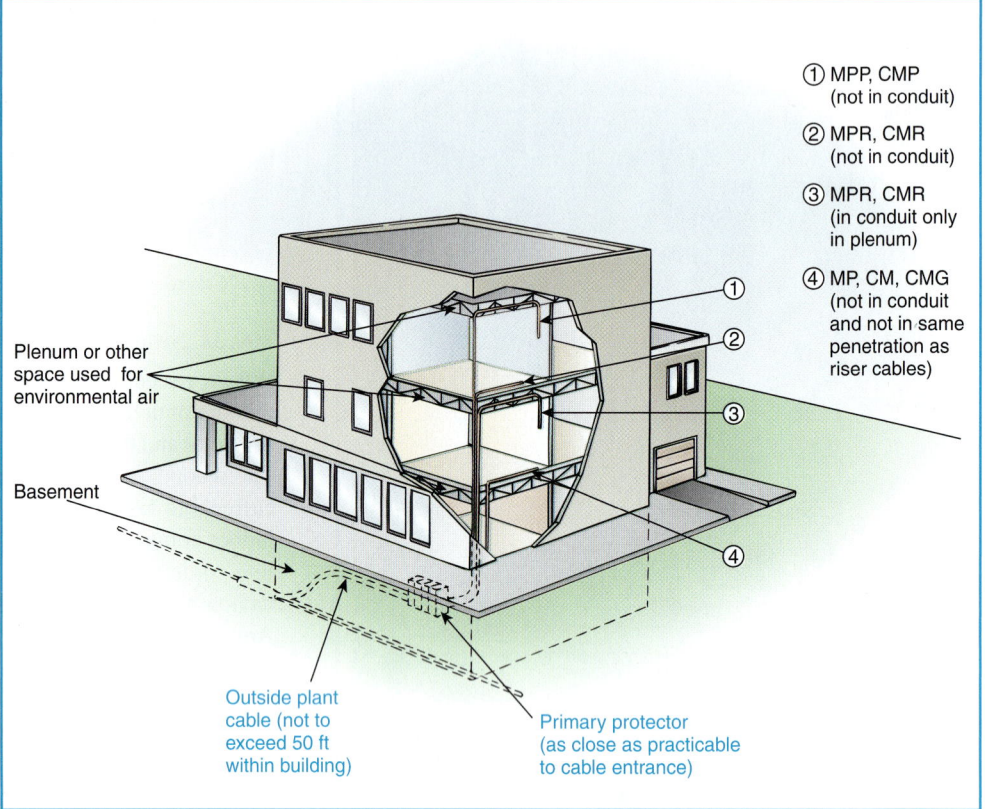

① MPP, CMP
(not in conduit)

② MPR, CMR
(not in conduit)

③ MPR, CMR
(in conduit only
in plenum)

④ MP, CM, CMG
(not in conduit
and not in same
penetration as
riser cables)

Plenum or other
space used for
environmental air

Basement

Outside plant
cable (not to
exceed 50 ft
within building)

Primary protector
(as close as practicable
to cable entrance)

(4) Length. The primary protector grounding conductor shall be as short as practicable. In one- and two-family dwellings, the primary protector grounding conductor shall be as short as practicable, not to exceed 6.0 m (20 ft) in length.

For one- and two-family dwellings, 800.40(A)(4) has been revised by restricting the length of the primary protector grounding conductor to 20 ft. This restriction conductor length reduces the impedance of the grounding conductor and results in a lower potential difference between the communications system conductors and equipment and the electrical conductors and equipment in the building. This will reduce the fire hazard and shock hazard to persons in the event that electric utility power lines come in contact with communication conductors. Section 800.40(D) requires bonding of communications and power grounding electrodes at the same building or structure.

Exception: In one- and two-family dwellings where it is not practicable to achieve an overall maximum primary protector grounding conductor length of 6.0 m (20 ft), a separate communications ground rod meeting the minimum dimensional criteria of 800.40(B)(2)(2) shall be driven, the primary protector shall be grounded to the communications ground rod in accordance with 800.40(C), and the communi-

cations ground rod bonded to the power grounding electrode system in accordance with 800.40(D).

(5) Run in Straight Line. The grounding conductor shall be run to the grounding electrode in as straight a line as practicable.

(6) Physical Damage. Where necessary, the grounding conductor shall be guarded from physical damage. Where the grounding conductor is run in a metal raceway, both ends of the raceway shall be bonded to the grounding conductor or the same terminal or electrode to which the grounding conductor is connected.

(B) Electrode. The grounding conductor shall be connected in accordance with 800.40(B)(1) and (B)(2).

(1) In Buildings or Structures with Grounding Means. To the nearest accessible location on the following:

(1) The building or structure grounding electrode system as covered in 250.50
(2) The grounded interior metal water piping system, within 1.5 m (5 ft) from its point of entrance to the building, as covered in 250.52

See the commentary following 250.52(A)(1) for information on water pipes as grounding electrodes.

(3) The power service accessible means external to enclosures as covered in 250.94

(4) The metallic power service raceway

(5) The service equipment enclosure

(6) The grounding electrode conductor or the grounding electrode conductor metal enclosure

(7) The grounding conductor or the grounding electrode of a building or structure disconnecting means that is grounded to an electrode as covered in 250.32.

For purposes of this section, the mobile home service equipment or the mobile home disconnecting means, as described in 800.30(B), shall be considered accessible.

(2) In Buildings or Structures Without Grounding Means. If the building or structure served has no grounding means, as described in 800.40(B)(1):

(1) To any one of the individual electrodes described in 250.52(A)(1), (2), (3), (4); or

(2) If the building or structure served has no grounding means, as described in 800.40(B)(1) or (B)(2)(1), to an effectively grounded metal structure or to a ground rod or pipe not less than 1.5 m (5 ft) in length and 12.7 mm (½ in.) in diameter, driven, where practicable, into permanently damp earth and separated from lightning conductors as covered in 800.13 and at least 1.8 m (6 ft) from electrodes of other systems. Steam or hot water pipes or air terminal conductors (lightning-rod conductors) shall not be employed as electrodes for protectors.

(C) Electrode Connection. Connections to grounding electrodes shall comply with 250.70. Connectors, clamps, fittings, or lugs used to attach grounding conductors and bonding jumpers to grounding electrodes or to each other that are to be concrete-encased or buried in the earth shall be suitable for its application.

(D) Bonding of Electrodes. A bonding jumper not smaller than 6 AWG copper or equivalent shall be connected between the communications grounding electrode and power grounding electrode system at the building or structure served where separate electrodes are used. Bonding together of all separate electrodes shall be permitted.

Exception: At mobile homes as covered in 800.41.

FPN No. 1: See 250.60 for use of air terminals (lightning rods).

FPN No. 2: Bonding together of all separate electrodes limits potential differences between them and between their associated wiring systems.

800.41 Primary Protector Grounding and Bonding at Mobile Homes.

(A) Grounding. Where there is no mobile home service equipment located in sight from, and not more than 9.0 m

(30 ft) from, the exterior wall of the mobile home it serves, or there is no mobile home disconnecting means grounded in accordance with 250.32 and located within sight from, and not more than 9.0 m (30 ft) from, the exterior wall of the mobile home it serves, the primary protector ground shall be in accordance with 800.40(B)(2).

(B) Bonding. The primary protector grounding terminal or grounding electrode shall be bonded to the metal frame or available grounding terminal of the mobile home with a copper grounding conductor not smaller than 12 AWG under any of the following conditions:

(1) Where there is no mobile home service equipment or disconnecting means as in 800.41(A)

(2) Where the mobile home is supplied by cord and plug

V. Communications Wires and Cables Within Buildings

Data circuits between computers are classified as Class 2 circuits. In a typical office environment consisting of a group of computers connected to a local area network, data wiring is as prevalent as telephone wiring. One common way to minimize the amount of cabling is to run the telephone and data circuits in the same cable, as illustrated in Exhibit 800.4. Section 725.56(D) requires that either a communications cable or a multipurpose cable be used for this purpose.

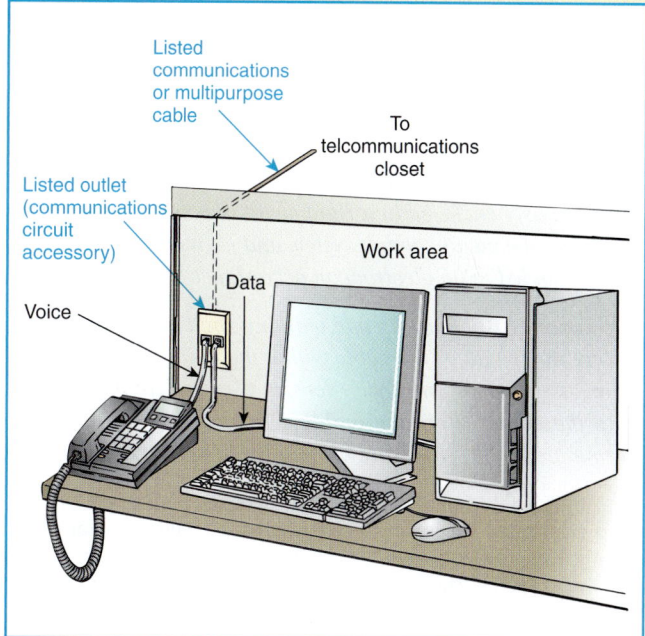

Exhibit 800.4 An example of telephone and data circuits in the same cable.

800.48 Raceways for Communications Wires and Cables.

Where communications wire and cables are installed in a raceway, the raceway shall be either of a type permitted in Chapter 3 and installed in accordance with Chapter 3 or a listed nonmetallic raceway complying with 800.51(J), (K), or (L), as applicable, and installed in accordance with 362.24 through 362.56, where the requirements applicable to electrical nonmetallic tubing apply.

Exception: Conduit fill restrictions shall not apply.

800.49 Fire Resistance of Communications Wires and Cables.

Communications wires and cables installed as wiring within a building shall be listed as being resistant to the spread of fire in accordance with 800.50 and 800.51.

800.50 Listing, Marking, and Installation of Communications Wires and Cables.

Communications wires and cables installed as wiring within buildings shall be listed as being suitable for the purpose and installed in accordance with 800.52. Communications cables and undercarpet communications wires shall be marked in accordance with Table 800.50. The cable voltage rating shall not be marked on the cable or on the undercarpet communications wire.

> FPN: Voltage markings on cables may be misinterpreted to suggest that the cables may be suitable for Class 1, electric light, and power applications.

Exception No. 1: Voltage markings shall be permitted where the cable has multiple listings and voltage marking is required for one or more of the listings.

Exception No. 2: Listing and marking shall not be required where the cable enters the building from the outside and is continuously enclosed in a rigid metal conduit system or an intermediate metal conduit system and such conduit systems are grounded to an electrode in accordance with 800.40(B).

Exception No. 3: Listing and marking shall not be required where the length of the cable within the building, measured from its point of entrance, does not exceed 15 m (50 ft) and the cable enters the building from the outside and is terminated in an enclosure or on a listed primary protector.

> FPN No. 1: Splice cases or terminal boxes, both metallic and plastic types, are typically used as enclosures for splicing or terminating telephone cables.
> FPN No. 2: This exception limits the length of unlisted outside plant cable to 15 m (50 ft), while 800.30(B) requires that the primary protector be located as close as practicable to the point at which the cable enters the building. Therefore, in installations requiring a primary protector, the outside plant cable may not be permitted

Table 800.50 Cable Markings

Cable Marking	Type	Reference
MPP	Multipurpose plenum cable	800.51(G) and 800.53(A)
CMP	Communications plenum cable	800.51(A) and 800.53(A)
MPR	Multipurpose riser cable	800.51(G) and 800.53(B)
CMR	Communications riser cable	800.51(B) and 800.53(B)
MPG	Multipurpose general-purpose cable	800.51(G) and 800.53(D) and (E)(1)
CMG	Communications general-purpose cable	800.51(C) and 800.53(D) and (E)(1)
MP	Multipurpose general-purpose cable	800.51(G) and 800.53(D) and (E)(1)
CM	Communications general-purpose cable	800.51(D) and 800.53(D) and (E)(1)
CMX	Communications cable, limited use	800.51(E) and 800.53(C), (D), and (E)
CMUC	Undercarpet communications wire and cable	800.51(F) and 800.53(F)(6)

to extend 15 m (50 ft) into the building if it is practicable to place the primary protector closer than 15 m (50 ft) to the entrance point.

Exception No. 4: Multipurpose cables shall be considered as being suitable for the purpose and shall be permitted to substitute for communications cables as provided for in 800.53(G).

> FPN No. 1: Cable types are listed in descending order of fire resistance rating, and multipurpose cables are listed above communications cables because multipurpose cables may substitute for communications cables.
> FPN No. 2: See the referenced sections for permitted uses.

800.51 Listing Requirements for Communications Wires and Cables and Communications Raceways.

Communications wires and cables shall have a voltage rating of not less than 300 volts and shall be listed in accordance with 800.51(A) through (J), and communications raceways shall be listed in accordance with 800.51(K) through (L). Conductors in communications cables, other than in a coaxial cable, shall be copper.

> FPN: See 800.4 for listing requirement for equipment.

(A) Type CMP. Type CMP communications plenum cable shall be listed as being suitable for use in ducts, plenums, and other spaces used for environmental air and shall also be listed as having adequate fire-resistant and low smoke-producing characteristics.

> FPN: One method of defining low smoke-producing cables is by establishing an acceptable value of the smoke produced when tested in accordance with NFPA 262-1999, *Standard Method of Test for Flame Travel and Smoke of Wire and Cables for Use in Air-Handling Spaces*, to a maximum peak optical density of 0.5 and a maximum average optical density of 0.15. Similarly, one method of defining fire-resistant cables is by establishing a maximum allowable flame travel distance of 1.52 m (5 ft) when tested in accordance with the same test.

See the commentary following 725.71(A), FPN, for information on a test method for wires and cables to be installed without raceways in plenums and other spaces used for environmental air.

(B) Type CMR. Type CMR communications riser cable shall be listed as being suitable for use in a vertical run in a shaft or from floor to floor and shall also be listed as having fire-resistant characteristics capable of preventing the carrying of fire from floor to floor.

> FPN: One method of defining fire-resistant characteristics capable of preventing the carrying of fire from floor to floor is that the cables pass the requirements of ANSI/UL 1666-1997, *Standard Test for Flame Propagation Height of Electrical and Optical-Fiber Cable Installed Vertically in Shafts*.

See the commentary following 725.71(B), FPN, for information on a test for defining fire-resistant characteristics capable of preventing fire spread from floor to floor.

(C) Type CMG. Type CMG general-purpose communications cable shall be listed as being suitable for general-purpose communications use, with the exception of risers and plenums, and shall also be listed as being resistant to the spread of fire.

> FPN: One method of defining *resistant to the spread of fire* is for the damage (char length) not to exceed 1.5 m (4 ft 11 in.) when performing the vertical flame test for cables in cable trays, as described in CSA C22.2 No. 0.3-M 1985, *Test Methods for Electrical Wires and Cables*.

See the commentary following 725.71(C), FPN, for information on the UL Vertical Tray Flame Test.

(D) Type CM. Type CM communications cable shall be listed as being suitable for general-purpose communications use, with the exception of risers and plenums, and shall also be listed as being resistant to the spread of fire.

> FPN: One method of defining *resistant to the spread of fire* is that the cables do not spread fire to the top of the tray in the vertical-tray flame test in ANSI/UL 1581-1991, *Reference Standard for Electrical Wires, Cables and Flexible Cords*. Another method of defining *resistant to the spread of fire* is for the damage (char length) not to exceed 1.5 m (4 ft 11 in.) when performing the vertical flame test for cables in cable trays, as described in CSA C22.2 No. 0.3-M-1985, *Test Methods for Electrical Wires and Cables*.

See the commentary following 725.71(D), FPN, for information on test methods for determining whether cable is resistant to the spread of fire.

(E) Type CMX. Type CMX limited-use communications cable shall be listed as being suitable for use in dwellings and for use in raceway and shall also be listed as being resistant to flame spread.

> FPN: One method of determining that cable is resistant to flame spread is by testing the cable to the VW-1 (vertical-wire) flame test in ANSI/UL 1581-1991, *Reference Standard for Electrical Wires, Cables and Flexible Cords*.

(F) Type CMUC Undercarpet Wire and Cable. Type CMUC undercarpet communications wire and cable shall be listed as being suitable for undercarpet use and shall also be listed as being resistant to flame spread.

> FPN: One method of determining that cable is resistant to flame spread is by testing the cable to the VW-1 (vertical-wire) flame test in ANSI/UL 1581-1991, *Reference Standard for Electrical Wires, Cables and Flexible Cords*.

(G) Multipurpose (MP) Cables. Until July 1, 2003, cables that meet the requirements for Types CMP, CMR, CMG, and CM and also satisfy the requirements of 760.71(B) for multiconductor cables and 760.71(H) for coaxial cables shall be permitted to be listed and marked as multipurpose cable Types MPP, MPR, MPG, and MP, respectively.

(H) Communications Wires. Communications wires, such as distributing frame wire and jumper wire, shall be listed as being resistant to the spread of fire.

FPN: One method of defining *resistant to the spread of fire* is that the cables do not spread fire to the top of the tray in the vertical-tray flame test in ANSI/UL 1581-1991, *Reference Standard for Electrical Wires, Cables and Flexible Cords*. Another method of defining *resistant to the spread of fire* is for the damage (char length) not to exceed 1.5 m (4 ft 11 in.) when performing the vertical flame test for cables in cable trays, as described in CSA C22.2 No. 0.3-M-1985, *Test Methods for Electrical Wires and Cables*.

(I) Hybrid Power and Communications Cable. Listed hybrid power and communications cable shall be permitted where the power cable is a listed Type NM or NM-B conforming to the provisions of Article 334, and the communications cable is a listed Type CM, the jackets on the listed NM or NM-B and listed CM cables are rated for 600 volts minimum, and the hybrid cable is listed as being resistant to the spread of fire.

FPN: One method of defining *resistant to the spread of fire* is that the cables do not spread fire to the top of the tray in the vertical-tray flame test in ANSI/UL 1581-1991, *Reference Standard for Electrical Wires, Cables and Flexible Cords*. Another method of defining *resistant to the spread of fire* is for the damage (char length) not to exceed 1.5 m (4 ft 11 in.) when performing the vertical flame test for cables in cable trays, as described in CSA C22.2 No. 0.3-M-1985, *Test Methods for Electrical Wires and Cables*.

(J) Plenum Communications Raceways. Plenum communications raceways listed as plenum optical fiber raceways shall be permitted for use in ducts, plenums, and other spaces used for environmental air and shall also be listed as having adequate fire-resistant and low smoke-producing characteristics.

(K) Riser Communications Raceway. Riser communications raceways shall be listed as having adequate fire-resistant characteristics capable of preventing the carrying of fire from floor to floor.

(L) General-Purpose Communications Raceway. General-purpose communications raceways shall be listed as being resistant to the spread of fire.

800.52 Installation of Communications Wires, Cables, and Equipment.

Communications wires and cables from the protector to the equipment or, where no protector is required, communications wires and cables attached to the outside or inside of the building shall comply with 800.52(A) through (E).

Section 800.52 was revised for the 1999 *Code* to include non–power-limited fire alarm circuits covered by Article 760, and network-powered broadband communications circuits covered by Article 830.

(A) Separation from Other Conductors.

(1) In Raceways, Boxes, and Cables.

(a) Other Power-Limited Circuits. Communications cables shall be permitted in the same raceway or enclosure with cables of any of the following:

(1) Class 2 and Class 3 remote-control, signaling, and power-limited circuits in compliance with Article 725
(2) Power-limited fire alarm systems in compliance with Article 760
(3) Nonconductive and conductive optical fiber cables in compliance with Article 770
(4) Community antenna television and radio distribution systems in compliance with Article 820
(5) Low-power network-powered broadband communications circuits in compliance with Article 830

(b) Class 2 and Class 3 Circuits. Class 1 circuits shall not be run in the same cable with communications circuits. Class 2 and Class 3 circuit conductors shall be permitted in the same cable with communications circuits, in which case the Class 2 and Class 3 circuits shall be classified as communications circuits and shall meet the requirements of this article. The cables shall be listed as communications cables or multipurpose cables.

Exception: Cables constructed of individually listed Class 2, Class 3, and communications cables under a common jacket shall not be required to be classified as communications cable. The fire-resistance rating of the composite cable shall be determined by the performance of the composite cable.

(c) Electric Light, Power, Class 1, Non–Power-Limited Fire Alarm, and Medium Power Network-Powered Broadband Communications Circuits in Raceways, Compartments, and Boxes. Communications conductors shall not be placed in any raceway, compartment, outlet box, junction box, or similar fitting with conductors of electric light, power, Class 1, non–power-limited fire alarm or medium power network-powered broadband communications circuits.

Exception No. 1: Where all of the conductors of electric light, power, Class 1, non–power-limited fire alarm, and

medium power network-powered broadband communications circuits are separated from all of the conductors of communications circuits by a barrier.

Exception No. 2: Power conductors in outlet boxes, junction boxes, or similar fittings or compartments where such conductors are introduced solely for power supply to communications equipment. The power circuit conductors shall be routed within the enclosure to maintain a minimum of 6 mm (0.25 in.) separation from the communications circuit conductors.

Exception No. 3: As permitted by 620.36.

(2) Other Applications. Communications wires and cables shall be separated at least 50 mm (2 in.) from conductors of any electric light, power, Class 1, non–power-limited fire alarm, or medium power network-powered broadband communications circuits.

Exception No. 1: Where either (1) all of the conductors of the electric light, power, Class 1, non–power-limited fire alarm, and medium power network-powered broadband communications circuits are in a raceway or in metal-sheathed, metal-clad, nonmetallic-sheathed, Type AC, or Type UF cables, or (2) all of the conductors of communications circuits are encased in raceway.

Exception No. 2: Where the communications wires and cables are permanently separated from the conductors of electric light, power, Class 1, non–power-limited fire alarm, and medium power network-powered broadband communications circuits by a continuous and firmly fixed nonconductor, such as porcelain tubes or flexible tubing, in addition to the insulation on the wire.

(B) Spread of Fire or Products of Combustion. Installations in hollow spaces, vertical shafts, and ventilation or air-handling ducts shall be made so that the possible spread of fire or products of combustion is not substantially increased. Openings around penetrations through fire resistance-rated walls, partitions, floors, or ceilings shall be firestopped using approved methods to maintain the fire resistance rating.

The accessible portion of abandoned communications cables shall not be permitted to remain.

FPN: Directories of electrical construction materials published by qualified testing laboratories contain many listing installation restrictions necessary to maintain the fire-resistive rating of assemblies where penetrations or openings are made.

This section has been revised for the 2002 *Code* for use with the definition of *abandoned communications cable* in 800.2. This section now requires the removal of accessible abandoned communications cable. Abandoned cable increases fire loading unnecessarily, and, where installed in plenums, it can affect airflow. Similar requirements can be found in Articles 640, 645, 725, 760, 770, 820, and 830. See the definition of *abandoned communications cable* in 800.2.

(C) Equipment in Other Space Used for Environmental Air. Section 300.22(C) shall apply.

(D) Cable Trays. Types MPP, MPR, MPG, and MP multipurpose cables and Types CMP, CMR, CMG, and CM communications cables shall be permitted to be installed in cable trays. Communications raceways, as described in 800.51, shall be permitted to be installed in cable trays.

(E) Support of Conductors. Raceways shall be used for their intended purpose. Communications cables or wires shall not be strapped, taped, or attached by any means to the exterior of any conduit or raceway as a means of support.

See 800.5 and 800.6. These sections require that communications cable be supported by the building structure in such a manner that it will not be damaged by ordinary building use.

Exception: Overhead (aerial) spans of communications cables or wires shall be permitted to be attached to the exterior of a raceway-type mast intended for the attachment and support of such conductors.

In some instances, the only way to achieve the proper clearance above roadways, driveways, or structures is by use of a mast. The exception to 800.52(E) permits overhead spans of communications cable to be attached to the exterior of a raceway-type mast only if the mast is installed to support communications cable. Section 230.28 prohibits the attachment of communications cable to a service mast.

800.53 Applications of Listed Communications Wires and Cables and Communications Raceways.

Communications wires and cables shall comply with the requirements of 800.53(A) through (F) or where cable substitutions are made in accordance with 800.53(G).

Note that the length of unlisted outside-plant cable permitted in a building depends on the location of the primary protector, in accordance with 800.30(B) and 800.50, Exception No. 3.

(A) Plenum. Cables installed in ducts, plenums, and other spaces used for environmental air shall be Type CMP. Abandoned cables shall not be permitted to remain. Types CMP, CMR, CMG, CM, and CMX and communications wire installed in compliance with 300.22 shall be permitted. Listed plenum communications raceways shall be permitted to be installed in ducts and plenums as described in 300.22(B) and in other spaces used for environmental air as described in 300.22(C). Only Type CMP cable shall be permitted to be installed in these raceways.

Section 800.53(A) covers listed plenum communications raceways. These raceways provide limited mechanical protection and ease of installation, but they are limited to Type CMP plenum-rated cable if installed in ducts and plenums.

(B) Riser. Cables installed in risers shall comply with 800.53(B)(1), (B)(2), or (B)(3).

Section 800.53(B) covers riser raceways. They provide limited mechanical protection and ease of installation, but they are limited to Type CMP plenum-rated cable or Type CMR riser-rated cable if installed in risers.

(1) Cables in Vertical Runs. Cables installed in vertical runs and penetrating more than one floor, or cables installed in vertical runs in a shaft, shall be Type CMR. Floor penetrations requiring Type CMR shall contain only cables suitable for riser or plenum use. Abandoned cables shall not be permitted to remain. Listed riser communications raceways shall be permitted to be installed in vertical riser runs in a shaft from floor to floor. Only Type CMR and CMP cables shall be permitted to be installed in these raceways.

(2) Metal Raceways or Fireproof Shafts. Listed communications cables shall be encased in a metal raceway or located in a fireproof shaft having firestops at each floor.

(3) One- and Two-Family Dwellings. Type CM and CMX cable shall be permitted in one- and two-family dwellings.

> FPN: See 800.52(B) for firestop requirements for floor penetrations.

(C) Distributing Frames and Cross-Connect Arrays. Listed communications wire and Types CMP, CMR, CMG, and CM communications cables shall be used in distributing frames and cross-connect arrays.

(D) Cable Trays. Types MPP, MPR, MPG, and MP multipurpose cables and Types CMP, CMR, CMG, and CM communications cables shall be permitted to be installed in cable trays.

(E) Other Wiring Within Buildings. Cables installed in building locations other than the locations covered in 800.53(A) through (D) shall be in accordance with 800.53(E)(1) through (E)(6).

(1) General. Cables shall be Type CMG or Type CM. Listed communications general-purpose raceways shall be permitted. Only Types CMG, CM, CMR, or CMP cables

Table 800.53 Cable Uses and Permitted Substitutions

Cable Type	Use	References	Permitted Substitutions
CMP	Communications plenum cable	800.53(A)	MPP
CMR	Communications riser cable	800.53(B)	MPP, CMP, MPR
CMG, CM	Communications general-purpose cable	800.53(E)(1)	MPP, CMP, MPR, CMR, MPG, MP
CMX	Communications cable, limited use	800.53(E)	MPP, CMP, MPR, CMR, MPG, MP, CMG, CM

Note: See Figure 800.53, Cable substitution hierarchy.

Table 800.53 lists the permitted uses of field applications for various cable types.

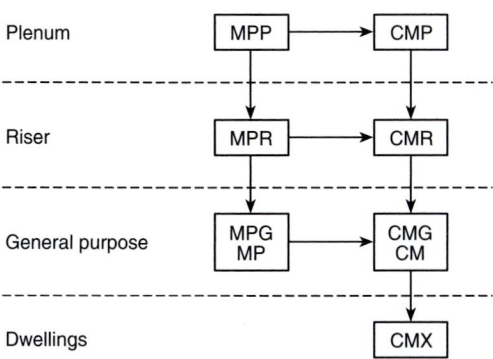

Type CM—Communications cables
Type MP—Multipurpose cable

A → B   Cable A shall be permitted to be used in place of cable B.

Figure 800.53 Cable substitution hierarchy.

shall be permitted to be installed in general-purpose communications raceways.

(2) In Raceways. Listed communications wires that are enclosed in a raceway of a type included in Chapter 3 shall be permitted.

(3) Nonconcealed Spaces. Type CMX communications cable shall be permitted to be installed in nonconcealed spaces where the exposed length of cable does not exceed 3 m (10 ft).

(4) One- and Two-Family Dwellings. Type CMX communications cable less than 6 mm (0.25 in.) in diameter shall be permitted to be installed in one- and two-family dwellings.

(5) Multi-Family Dwellings. Type CMX communications cable less than 6 mm (0.25 in.) in diameter shall be permitted to be installed in nonconcealed spaces in multi-family dwellings.

(6) Under Carpets. Type CMUC undercarpet communications wires and cables shall be permitted to be installed under carpet.

(F) Hybrid Power and Communications Cable. Hybrid power and communications cable listed in accordance with 800.51(J) shall be permitted to be installed in one- and two-family dwellings.

(G) Cable Substitutions. The uses and permitted substitutions for communications cables listed in Table 800.53 shall be considered suitable for the purpose and shall be permitted.

FPN: For information on Types CMP, CMR, CMG, CM, and CMX cables, see 800.51.

ARTICLE 810
Radio and Television Equipment

Contents

Article 810 covers wiring requirements for television and radio receiving equipment, specifically including digital satellite receiving equipment for television signals, and wiring for amateur radio equipment. Chapters 1 through 4 cover wiring for the power supply. Article 640 contains requirements for sound distribution systems. The interior wiring of coaxial cable is covered by Article 820.

810.2 Definitions.

For definitions applicable to this article, see Article 100.

This section was revised and relocated for the 2002 *Code*. In accordance with 90.3, the articles in Chapter 8 are not subject to the requirements contained in the other articles of the *Code* unless specifically referenced. Section 810.2 clearly states that the definitions in Article 100 also apply to Article 810.

810.3 Other Articles.

Wiring from the source of power to and between devices connected to the interior wiring system shall comply with Chapters 1 through 4 other than as modified by Parts I and II of Article 640. Wiring for audio signal processing, amplification, and reproduction equipment shall comply with Article 640. Coaxial cables that connect antennas to equipment shall comply with Article 820.

810.4 Community Television Antenna.

The antenna shall comply with this article. The distribution system shall comply with Article 820.

810.5 Radio Noise Suppressors.

Radio interference eliminators, interference capacitors, or noise suppressors connected to power-supply leads shall be of a listed type. They shall not be exposed to physical damage.

I. General

810.1 Scope.

This article covers antenna systems for radio and television receiving equipment, amateur radio transmitting and receiving equipment, and certain features of transmitter safety. This article covers antennas such as multi-element, vertical rod, and dish, and also covers the wiring and cabling that connects them to equipment. This article does not cover equipment and antennas used for coupling carrier current to power line conductors.

II. Receiving Equipment— Antenna Systems

810.11 Material.

Antennas and lead-in conductors shall be of hard-drawn copper, bronze, aluminum alloy, copper-clad steel, or other high-strength, corrosion-resistant material.

Exception: Soft-drawn or medium-drawn copper shall be permitted for lead-in conductors where the maximum span between points of support is less than 11 m (35 ft).

810.12 Supports.

Outdoor antennas and lead-in conductors shall be securely supported. The antennas or lead-in conductors shall not be attached to the electric service mast. They shall not be attached to poles or similar structures carrying open electric light or power wires or trolley wires of over 250 volts between conductors. Insulators supporting the antenna conductors shall have sufficient mechanical strength to safely support the conductors. Lead-in conductors shall be securely attached to the antennas.

810.13 Avoidance of Contacts with Conductors of Other Systems.

Outdoor antennas and lead-in conductors from an antenna to a building shall not cross over open conductors of electric light or power circuits and shall be kept well away from all such circuits so as to avoid the possibility of accidental contact. Where proximity to open electric light or power service conductors of less than 250 volts between conductors cannot be avoided, the installation shall be such as to provide a clearance of at least 600 mm (2 ft).

Where practicable, antenna conductors shall be installed so as not to cross under open electric light or power conductors.

One of the leading causes of electrical shock and electrocution, according to statistical reports, is the accidental contact of radio, television, and amateur radio transmitting and receiving antennas and equipment with light or power conductors. Extreme caution should therefore be exercised during this type of installation, and periodic visual inspections should be conducted thereafter.

810.14 Splices.

Splices and joints in antenna spans shall be made mechanically secure with approved splicing devices or by such other means as will not appreciably weaken the conductors.

Conductor spans from antennas should be of sufficient size and strength to maintain clearances and avoid possible contact with light or power conductors. Splices and joints should be made with approved connectors or other means that provide sufficient mechanical strength so that conductors are not weakened appreciably, a condition that could cause them to break and come into contact with higher-voltage conductors.

810.15 Grounding.

Masts and metal structures supporting antennas shall be grounded in accordance with 810.21.

810.16 Size of Wire-Strung Antenna—Receiving Station.

(A) Size of Antenna Conductors. Outdoor antenna conductors for receiving stations shall be of a size not less than given in Table 810.16(A).

Table 810.16(A) Size of Receiving Station Outdoor Antenna Conductors

Material	Minimum Size of Conductors (AWG) Where Maximum Open Span Length Is		
	Less Than 11 m (35 ft)	11 m to 45 m (35 ft to 150 ft)	Over 45 m (150 ft)
Aluminum alloy, hard-drawn copper	19	14	12
Copper-clad steel, bronze, or other high-strength material	20	17	14

(B) Self-Supporting Antennas. Outdoor antennas, such as vertical rods, dishes, or dipole structures, shall be of corrosion-resistant materials and of strength suitable to withstand ice and wind loading conditions and shall be located well away from overhead conductors of electric light and power circuits of over 150 volts to ground, so as to avoid the possibility of the antenna or structure falling into or making accidental contact with such circuits.

Section 810.16(B) includes dish-type (parabolic) antennas.

810.17 Size of Lead-in—Receiving Station.

Lead-in conductors from outside antennas for receiving stations shall, for various maximum open span lengths, be of such size as to have a tensile strength at least as great as that of the conductors for antennas as specified in 810.16. Where the lead-in consists of two or more conductors that are twisted together, are enclosed in the same covering, or are concentric, the conductor size shall, for various maximum open span lengths, be such that the tensile strength of the combination is at least as great as that of the conductors for antennas as specified in 810.16.

810.18 Clearances—Receiving Stations.

(A) Outside of Buildings. Lead-in conductors attached to buildings shall be installed so that they cannot swing closer than 600 mm (2 ft) to the conductors of circuits of 250 volts or less between conductors, or 3.0 m (10 ft) to the conductors of circuits of over 250 volts between conductors, except that in the case of circuits not over 150 volts between conductors, where all conductors involved are supported so as to ensure permanent separation, the clearance shall be permitted to be reduced but shall not be less than 100 mm (4 in.). The clearance between lead-in conductors and any conductor forming a part of a lightning rod system shall not be less than 1.8 m (6 ft) unless the bonding referred to in 250.60 is accomplished. Underground conductors shall be separated at least 300 mm (12 in.) from conductors of any light or power circuits or Class 1 circuits.

Exception: Where the electric light or power conductors, Class 1 conductors, or lead-in conductors are installed in raceways or metal cable armor.

(B) Antennas and Lead-ins — Indoors. Indoor antennas and indoor lead-ins shall not be run nearer than 50 mm (2 in.) to conductors of other wiring systems in the premises.

Exception No. 1: Where such other conductors are in metal raceways or cable armor.

Exception No. 2: Where permanently separated from such other conductors by a continuous and firmly fixed nonconductor, such as porcelain tubes or flexible tubing.

(C) In Boxes or Other Enclosures. Indoor antennas and indoor lead-ins shall be permitted to occupy the same box or enclosure with conductors of other wiring systems where separated from such other conductors by an effective permanently installed barrier.

810.19 Electric Supply Circuits Used in Lieu of Antenna—Receiving Stations.

Where an electric supply circuit is used in lieu of an antenna, the device by which the radio receiving set is connected to the supply circuit shall be listed.

The approved connecting device is usually a small, fixed capacitor connecting the antenna terminal of the receiver and one wire of the supply circuit. As is the case with most receivers, the capacitor should be designed for operation at not less than 300 volts. This rating ensures a high degree of safety and minimizes the possibility of a breakdown in the capacitor, thereby avoiding a short circuit to ground through the antenna coil of the set.

810.20 Antenna Discharge Units—Receiving Stations.

(A) Where Required. Each conductor of a lead-in from an outdoor antenna shall be provided with a listed antenna discharge unit.

Exception: Where the lead-in conductors are enclosed in a continuous metallic shield that either is permanently and effectively grounded or is protected by an antenna discharge unit.

(B) Location. Antenna discharge units shall be located outside the building or inside the building between the point of entrance of the lead-in and the radio set or transformers and as near as practicable to the entrance of the conductors to the building. The antenna discharge unit shall not be located near combustible material or in a hazardous (classified) location as defined in Article 500.

An antenna discharge unit (lightning arrester) is not required if the lead-in conductors are enclosed in a continuous metal shield, such as rigid or intermediate metal conduit, electrical metallic tubing, or any metal raceway or metal-shielded cable that is effectively grounded. A lightning discharge will take the path of lower impedance and jump from the lead-in conductors to the metal raceway or shield rather than take the path through the antenna coil of the receiver.

(C) Grounding. The antenna discharge unit shall be grounded in accordance with 810.21.

810.21 Grounding Conductors—Receiving Stations.

Grounding conductors shall comply with 810.21(A) through (J).

(A) Material. The grounding conductor shall be of copper, aluminum, copper-clad steel, bronze, or similar corrosion-resistant material. Aluminum or copper-clad aluminum grounding conductors shall not be used where in direct contact with masonry or the earth or where subject to corrosive conditions. Where used outside, aluminum or copper-clad aluminum shall not be installed within 450 mm (18 in.) of the earth.

(B) Insulation. Insulation on grounding conductors shall not be required.

(C) Supports. The grounding conductors shall be securely fastened in place and shall be permitted to be directly attached to the surface wired over without the use of insulating supports.

Exception: Where proper support cannot be provided, the size of the grounding conductors shall be increased proportionately.

(D) Mechanical Protection. The grounding conductor shall be protected where exposed to physical damage, or the size of the grounding conductors shall be increased proportionately to compensate for the lack of protection. Where the grounding conductor is run in a metal raceway, both ends of the raceway shall be bonded to the grounding conductor or to the same terminal or electrode to which the grounding conductor is connected.

If metal enclosures such as steel conduit are used to enclose the grounding conductor, bonding must be provided at both ends to ensure an adequate low-impedance current path.

(E) Run in Straight Line. The grounding conductor for an antenna mast or antenna discharge unit shall be run in as straight a line as practicable from the mast or discharge unit to the grounding electrode.

(F) Electrode. The grounding conductor shall be connected as follows:

(1) To the nearest accessible location on the following:

    a. The building or structure grounding electrode system as covered in 250.50

    b. The grounded interior metal water piping systems, within 1.52 m (5 ft) from its point of entrance to the building, as covered in 250.52

See the commentary following 250.52(A)(1).

    c. The power service accessible means external to the building, as covered in 250.94

    d. The metallic power service raceway

    e. The service equipment enclosure, or

    f. The grounding electrode conductor or the grounding electrode conductor metal enclosures; or

(2) If the building or structure served has no grounding means, as described in 810.21(F)(1), to any one of the individual electrodes described in 250.52; or

(3) If the building or structure served has no grounding means, as described in 810.21(F)(1) or (F)(2), to an effectively grounded metal structure or to any of the individual electrodes described in 250.52.

(G) Inside or Outside Building. The grounding conductor shall be permitted to be run either inside or outside the building.

(H) Size. The grounding conductor shall not be smaller than 10 AWG copper, 8 AWG aluminum, or 17 AWG copper-clad steel or bronze.

(I) Common Ground. A single grounding conductor shall be permitted for both protective and operating purposes.

(J) Bonding of Electrodes. A bonding jumper not smaller than 6 AWG copper or equivalent shall be connected between

the radio and television equipment grounding electrode and the power grounding electrode system at the building or structure served where separate electrodes are used.

The requirements for grounding are in accordance with Article 250. Antenna masts must be grounded to the same grounding electrode used for the building's electrical system, to ensure that all exposed, non–current-carrying metal parts are at the same potential. In many cases, masts are connected incorrectly to conveniently located vent pipes, metal gutters, or downspouts. Such a connection could create potential differences between lead-in conductors and various metal parts located in or on buildings, resulting in possible shock and fire hazards. An underground gas piping system is not permitted to be used as a grounding electrode.

    Section 810.21(J) clarifies that the bonding requirement applies only to electrodes at the same building or structure. The use of separate radio/television grounding electrodes is not required.

III. Amateur Transmitting and Receiving Stations— Antenna Systems

810.51 Other Sections.

In addition to complying with Part III, antenna systems for amateur transmitting and receiving stations shall also comply with 810.11 through 810.15.

810.52 Size of Antenna.

Antenna conductors for transmitting and receiving stations shall be of a size not less than given in Table 810.52.

Table 810.52 Size of Amateur Station Outdoor Antenna Conductors

Material	Minimum Size of Conductors (AWG) Where Maximum Open Span Length Is	
	Less Than 45 m (150 ft)	Over 45 m (150 ft)
Hard-drawn copper	14	10
Copper-clad steel, bronze, or other high-strength material	14	12

810.53 Size of Lead-in Conductors.

Lead-in conductors for transmitting stations shall, for various maximum span lengths, be of a size at least as great as that of conductors for antennas as specified in 810.52.

810.54 Clearance on Building.

Antenna conductors for transmitting stations, attached to buildings, shall be firmly mounted at least 75 mm (3 in.) clear of the surface of the building on nonabsorbent insulating supports, such as treated pins or brackets equipped with insulators having not less than 75-mm (3-in.) creepage and airgap distances. Lead-in conductors attached to buildings shall also comply with these requirements.

Exception: Where the lead-in conductors are enclosed in a continuous metallic shield that is permanently and effectively grounded, they shall not be required to comply with these requirements. Where grounded, the metallic shield shall also be permitted to be used as a conductor.

Creepage distance is measured from the conductor across the face of the supporting insulator to the building surface. Air gap distance is measured from the conductor (at its closest point) across the air space (not necessarily in a straight line) to the surface of the building. This exception covers coaxial cable with the shield permanently and effectively grounded.

810.55 Entrance to Building.

Except where protected with a continuous metallic shield that is permanently and effectively grounded, lead-in conductors for transmitting stations shall enter buildings by one of the following methods:

(1) Through a rigid, noncombustible, nonabsorbent insulating tube or bushing
(2) Through an opening provided for the purpose in which the entrance conductors are firmly secured so as to provide a clearance of at least 50 mm (2 in.)
(3) Through a drilled window pane

810.56 Protection Against Accidental Contact.

Lead-in conductors to radio transmitters shall be located or installed so as to make accidental contact with them difficult.

810.57 Antenna Discharge Units— Transmitting Stations.

Each conductor of a lead-in for outdoor antennas shall be provided with an antenna discharge unit or other suitable means that drain static charges from the antenna system.

If an antenna discharge unit is not installed at a transmitting station, protection against lightning may be provided by a switch that connects the lead-in to ground during the time the station is not in operation.

Exception No. 1: Where protected by a continuous metallic shield that is permanently and effectively grounded.

Exception No. 2: Where the antenna is permanently and effectively grounded.

810.58 Grounding Conductors—Amateur Transmitting and Receiving Stations.

Grounding conductors shall comply with 810.58(A) through (C).

(A) Other Sections. All grounding conductors for amateur transmitting and receiving stations shall comply with 810.21(A) through (J).

(B) Size of Protective Grounding Conductor. The protective grounding conductor for transmitting stations shall be as large as the lead-in but not smaller than 10 AWG copper, bronze, or copper-clad steel.

(C) Size of Operating Grounding Conductor. The operating grounding conductor for transmitting stations shall not be less than 14 AWG copper or its equivalent.

IV. Interior Installation— Transmitting Stations

810.70 Clearance from Other Conductors.

All conductors inside the building shall be separated at least 100 mm (4 in.) from the conductors of any electric light, power, or signaling circuit.

Exception No. 1: As provided in Article 640.

Exception No. 2: Where separated from other conductors by raceway or some firmly fixed nonconductor, such as porcelain tubes or flexible tubing.

810.71 General.

Transmitters shall comply with 810.71(A) through (C).

(A) Enclosing. The transmitter shall be enclosed in a metal frame or grille, or separated from the operating space by a barrier or other equivalent means, all metallic parts of which are effectively connected to ground.

(B) Grounding of Controls. All external metal handles and controls accessible to the operating personnel shall be effectively grounded.

(C) Interlocks on Doors. All access doors shall be provided with interlocks that disconnect all voltages of over 350 volts between conductors when any access door is opened.

ARTICLE 820
Community Antenna Television and Radio Distribution Systems

Contents

I. General

820.1 Scope.

This article covers coaxial cable distribution of radio frequency signals typically employed in community antenna television (CATV) systems.

Article 820 covers the installation of coaxial cable for the distribution of radio frequency (RF) signals associated with closed-circuit television, cable television, and security television cameras. This article also covers interior coaxial cable for radio and television receiving equipment. Article 830 was added to the 1999 *Code* to cover network-powered broadband system installations.

820.2 Definitions.

See Article 100. For the purposes of this article, the following additional definitions apply.

Abandoned Coaxial Cable. Installed coaxial cable that is not terminated at equipment other than a coaxial connector and not identified for future use with a tag.

This definition is new in the 2002 *Code*. It is used with 820.3(A) and 820.53, which now require removal of accessible abandoned communications cable. Abandoned cable increases fire loading unnecessarily, and, where installed in plenums, it can affect airflow. Similar requirements can be found in Articles 640, 645, 725, 760, 770, 800, and 830.

Exposed. An exposed cable is one that is in such a position that, in case of failure of supports and insulation, contact with another circuit could result.

FPN: See Article 100 for two other definitions of *exposed*.

Point of Entrance. The point within a building at which the cable emerges from an external wall, from a concrete floor slab, or from a rigid metal conduit or an intermediate metal conduit grounded to an electrode in accordance with 820.40(B).

Premises. The land and buildings of a user located on the user side of utility-user network point of demarcation.

820.3 Locations and Other Articles.

Circuits and equipment shall comply with 820.3(A) through (G).

Section 820.3(G) permits Article 830 wiring methods to substitute for those covered in Article 820. The substitution of these wiring methods facilitates an upgrade of Article 820 installations to network-powered broadband applications. It is also important to note that 820.3(H) in the 2002 edition of the *Code* now requires that accessible abandoned CATV cable be removed.

(A) Spread of Fire or Products of Combustion. Section 300.21 shall apply. The accessible portion of abandoned coaxial cables shall not be permitted to remain.

(B) Ducts, Plenums, and Other Air-Handling Spaces. Section 300.22, where installed in ducts or plenums or other spaces used for environmental air, shall apply.

Exception: As permitted in 820.53(A).

(C) Installation and Use. Section 110.3 shall apply.

(D) Installations of Conductive and Nonconductive Optical Fiber Cables. Article 770 shall apply.

(E) Communications Circuits. Article 800 shall apply.

(F) Network-Powered Broadband Communications Systems. Article 830 shall apply.

(G) Alternate Wiring Methods. The wiring methods of Article 830 shall be permitted to substitute for the wiring methods of Article 820.

FPN: Use of Article 830 wiring methods will facilitate the upgrading of Article 820 installations to network-powered broadband applications.

820.4 Energy Limitations.

The coaxial cable shall be permitted to deliver low-energy power to equipment that is directly associated with the radio frequency distribution system if the voltage is not over 60 volts and if the current supply is from a transformer or other device that has energy-limiting characteristics.

820.5 Access to Electrical Equipment Behind Panels Designed to Allow Access.

Access to electrical equipment shall not be denied by an accumulation of wires and cables that prevents removal of panels, including suspended ceiling panels.

The excess accumulation of wires and cabling can limit access to equipment by preventing the removal of access panels. See Exhibit 820.1.

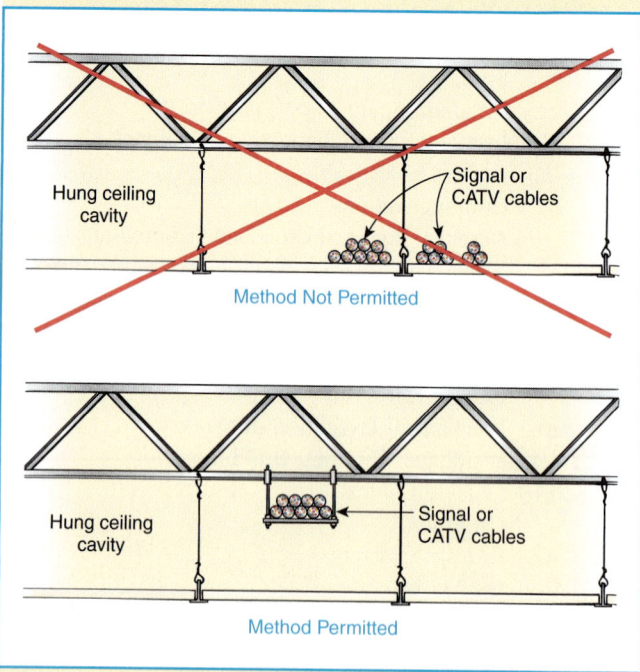

Exhibit 820.1 Installations of conductors and cables, which can prevent access to equipment or cables. Correct and incorrect methods are shown.

820.6 Mechanical Execution of Work.

Community antenna television and radio distribution systems shall be installed in a neat and workmanlike manner. Cables installed exposed on the surface of ceiling and sidewalls shall be supported by the structural components of the building structure in such a manner that the cable is not damaged by normal building use. Such cables shall be attached to structural components by straps, staples, hangers, or similar fittings designed and installed so as not to damage the cable. The installation shall also conform with 300.4(D).

Revised for the 2002 *Code*, this section provides clear requirements for workmanship. Cables are now required to be attached to or supported by the structure by straps, clamps, hangers, and the like. The installation method must not damage the cable. In addition, the location of the cable should be carefully evaluated to ensure that activities and processes within the building do not cause damage to the cable.

II. Cables Outside and Entering Buildings

820.10 Outside Cables.

Coaxial cables, prior to the point of grounding, as defined in 820.33, shall comply with 820.10(A) through (F).

(A) On Poles. Where practicable, conductors on poles shall be located below the electric light, power, Class 1, or non–power-limited fire alarm circuit conductors and shall not be attached to a crossarm that carries electric light or power conductors.

(B) Lead-in Clearance. Lead-in or aerial-drop cables from a pole or other support, including the point of initial attachment to a building or structure, shall be kept away from electric light, power, Class 1, or non–power-limited fire alarm circuit conductors so as to avoid the possibility of accidental contact.

Exception: Where proximity to electric light, power, Class 1, or non–power-limited fire alarm circuit service conductors cannot be avoided, the installation shall be such as to provide clearances of not less than 300 mm (12 in.) from light, power, Class 1, or non–power-limited fire alarm circuit service drops. The clearance requirement shall apply at all points along the drop, and it shall increase to 1.02 m (40 in.) at the pole.

(C) On Masts. Aerial cable shall be permitted to be attached to an above-the-roof raceway mast that does not enclose or support conductors of electric light or power circuits.

(D) Above Roofs. Cables shall have a vertical clearance of not less than 2.5 m (8 ft) from all points of roofs above which they pass.

Exception No. 1: Auxiliary buildings such as garages and the like.

Exception No. 2: A reduction in clearance above only the overhanging portion of the roof to not less than 450 mm (18 in.) shall be permitted if (1) not more than 1.2 m (4 ft) of communications service drop conductors pass above the roof overhang, and (2) they are terminated at a raceway mast or other approved support.

Exception No. 3: Where the roof has a slope of not less than 100 mm (4 in.) in 300 mm (12 in.), a reduction in clearance to not less than 900 mm (3 ft) shall be permitted.

(E) Between Buildings. Cables extending between buildings and also the supports or attachment fixtures shall be acceptable for the purpose and shall have sufficient strength to withstand the loads to which they may be subjected.

Wind and ice loads, which can be excessive, should be considered.

Exception: Where a cable does not have sufficient strength to be self-supporting, it shall be attached to a supporting messenger cable that, together with the attachment fixtures or supports, shall be acceptable for the purpose and shall have sufficient strength to withstand the loads to which they may be subjected.

(F) On Buildings. Where attached to buildings, cables shall be securely fastened in such a manner that they will be separated from other conductors in accordance with 820.10(F)(1), (F)(2), and (F)(3).

(1) Electric Light or Power. The coaxial cable shall have a separation of at least 100 mm (4 in.) from electric light, power, Class 1, or non–power-limited fire alarm circuit conductors not in raceway or cable or be permanently separated from conductors of the other system by a continuous and firmly fixed nonconductor in addition to the insulation on the wires.

(2) Other Communications Systems. Coaxial cable shall be installed so that there will be no unnecessary interference in the maintenance of the separate systems. In no case shall the conductors, cables, messenger strand, or equipment of one system cause abrasion to the conductors, cable, messenger strand, or equipment of any other system.

(3) Lightning Conductors. Where practicable, a separation of at least 1.8 m (6 ft) shall be maintained between any coaxial cable and lightning conductors.

FPN: For additional information regarding overhead wires and cables, see ANSI C2-1997, *National Electric Safety Code*, Part 2, Safety Rules for Overhead Lines.

820.11 Entering Buildings.

(A) Underground Systems. Underground coaxial cables in a duct, pedestal, handhole, or manhole that contains electric light or power conductors or Class 1 circuits shall be in a section permanently separated from such conductors by means of a suitable barrier.

(B) Direct-Buried Cables and Raceways. Direct-buried coaxial cable shall be separated at least 300 mm (12 in.) from conductors of any light or power or Class 1 circuit.

Exception No. 1: Where electric service conductors or coaxial cables are installed in raceways or have metal cable armor.

Exception No. 2: Where electric light or power branch-circuit or feeder conductors or Class 1 circuit conductors are installed in a raceway or in metal-sheathed, metal-clad, or Type UF or Type USE cables; or the coaxial cables have metal cable armor or are installed in a raceway.

III. Protection

820.33 Grounding of Outer Conductive Shield of a Coaxial Cable.

The outer conductive shield of the coaxial cable shall be grounded at the building premises as close to the point of cable entrance or attachment as practicable.

For purposes of this section, grounding located at mobile home service equipment located in sight from, and not more than 9.0 m (30 ft) from, the exterior wall of the mobile home it serves, or at a mobile home disconnecting means grounded in accordance with 250.32 and located in sight from and not more than 9.0 m (30 ft) from the exterior wall of the mobile home it serves, shall be considered to meet the requirements of this section.

> FPN: Selecting a grounding location to achieve the shortest practicable grounding conductor helps limit potential differences between CATV and other metallic systems.

(A) Shield Grounding. Where the outer conductive shield of a coaxial cable is grounded, no other protective devices shall be required.

(B) Shield Protection Devices. Grounding of a coaxial drop cable shield by means of a protective device that does not interrupt the grounding system within the premises shall be permitted.

Section 820.33(B) permits the use of a shield protection device that does not interrupt the grounding system within the premises. The electric utility supply, the CATV system, and the premises wiring are all grounded. When a ground fault occurs at the premises, the current tries to return to its source and follows the multiple paths through the different premises metal water piping systems or earth. This causes current to flow on the CATV shield, whose primary function is to prevent RF leakage out of the cable. The fault current in the cable shield can cause the shield to burn open and also damage the cable insulation. A device that can safely conduct current at 60 Hz and block current at the higher frequencies can be connected between the cable shield and ground, thereby maintaining grounding integrity. An ordinary fuse, for example, would not be suitable.

IV. Grounding Methods

820.40 Cable Grounding.

Where required by 820.33, the shield of the coaxial cable shall be grounded as specified in 820.40(A) through (D).

(A) Grounding Conductor.

(1) Insulation. The grounding conductor shall be insulated and shall be listed as suitable for the purpose.

(2) Material. The grounding conductor shall be copper or other corrosion-resistant conductive material, stranded or solid.

(3) Size. The grounding conductor shall not be smaller than 14 AWG. It shall have a current-carrying capacity approximately equal to that of the outer conductor of the coaxial cable. The grounding conductor shall not be required to exceed 6 AWG.

(4) Length. The grounding conductor shall be as short as practicable. In one- and two-family dwellings, the grounding conductor shall be as short as practicable, not to exceed 6.0 m (20 ft) in length.

This section is new for the 2002 *Code*. The limitation on length will result in a lower impedance, which will limit the potential difference between broadband communications systems and other systems during a lightning strike. Large potential differences between grounding conductors can result in increased damage if a lightning strike were to occur.

Exception: In one- and two-family dwellings where it is not practicable to achieve an overall maximum grounding conductor length of 6.0 m (20 ft), a separate ground as specified in 250.52(A)(5), (6), or (7) shall be used, the grounding conductor shall be grounded to the separate ground in accordance with 250.70, and the separate ground bonded to the power grounding electrode system in accordance with 820.40(D).

(5) Run in Straight Line. The grounding conductor shall be run to the grounding electrode in as straight a line as practicable.

(6) Physical Protection. Where subject to physical damage, the grounding conductor shall be adequately protected. Where the grounding conductor is run in a metal raceway, both ends of the raceway shall be bonded to the grounding conductor or the same terminal or electrode to which the grounding conductor is connected.

(B) Electrode. The grounding conductor shall be connected in accordance with 820.40(B)(1) and (B)(2).

(1) In Buildings or Structures with Grounding Means. To the nearest accessible location on the following:

(1) The building or structure grounding electrode system as covered in 250.50;
(2) The grounded interior metal water piping system, within 1.52 m (5 ft) from its point of entrance to the building, as covered in 250.52;

See commentary following 250.52(A)(1).

(3) The power service accessible means external to enclosures as covered in 250.94;
(4) The metallic power service raceway;
(5) The service equipment enclosure;
(6) The grounding electrode conductor or the grounding electrode conductor metal enclosure; or
(7) The grounding conductor or the grounding electrode of a building or structure disconnecting means that is grounded to an electrode as covered in 250.32.

(2) In Buildings or Structures Without Grounding Means. If the building or structure served has no grounding means, as described in 820.40(B)(1):

(1) To any one of the individual electrodes described in 250.52(A)(1), (2), (3), (4); or,
(2) If the building or structure served has no grounding means, as described in 820.40(B)(1) or (B)(2)(1), to an effectively grounded metal structure or to any one of the individual electrodes described in 250.52(A)(5), (6), and (7).

(C) Electrode Connection. Connections to grounding electrodes shall comply with 250.70.

(D) Bonding of Electrodes. A bonding jumper not smaller than 6 AWG copper or equivalent shall be connected between the antenna systems grounding electrode and the power grounding electrode system at the building or structure served where separate electrodes are used.

Section 820.40(D) requires bonding of CATV and power grounding electrodes at the same building or structure.

Exception: At mobile homes as covered in 820.42.

> FPN No. 1: See 250.60 for use of air terminals (lightning rods).
> FPN No. 2: Bonding together of all separate electrodes limits potential differences between them and between their associated wiring systems.

The most common error made in grounding CATV systems is to connect the coaxial cable sheath to a ground rod driven

by the CATV installer at a convenient location near the point of cable entry to the building, instead of bonding it to the electrical service grounding electrode system, service raceway, or other components that make up the grounding electrode system. A separate grounding electrode is permitted by the *Code* only if the building or structure has none of the grounding means described in 820.40(B)(1) or 820.40(B)(2), which is rare. The *Code* requires that some means that is accessible and external to the service equipment be provided for making the bonding and grounding connection for other systems. One of the following means must be provided:

1. Exposed service raceways
2. Exposed grounding electrode conductor
3. An approved means for the external connection of a conductor (A 6 AWG copper conductor with one end bonded to the service raceway or equipment with about 6 in. exposed is acceptable.)

Proper bonding of the CATV system coaxial cable sheath to the electrical power ground is needed to prevent potential fire and shock hazards. The earth cannot be used as an equipment grounding conductor or bonding conductor because it does not have the low-impedance path required. (See 250.54.)

Both CATV systems and power systems are subject to current surges as a result, for example, of induced voltages from lightning in the vicinity of the usually extensive outside distribution systems. Surges also result from switching operations on power systems. If the grounded conductors and parts of the two systems are not bonded by a low-impedance path, such line surges can raise the potential difference between the two systems to many thousands of volts. This can result in arcing between the two systems, for example, wherever the coaxial cable jacket contacts a grounded part, such as a metal water pipe or metal structural member, inside the building.

If an individual is the interface between the two systems and the bonding has not been done in accordance with the *Code*, the high-voltage surge could result in electric shock. More common, however, is burnout of the television tuner because this part is almost always an interface between the two systems. The tuner is connected to the power system ground through the grounded neutral of the power supply, even if the television itself is not provided with an equipment grounding conductor.

Also see the commentary following 250.92(B), FPN No. 2, and 820.33(B).

820.41 Equipment Grounding.

Unpowered equipment and enclosures or equipment powered by the coaxial cable shall be considered grounded where connected to the metallic cable shield.

820.42 Bonding and Grounding at Mobile Homes.

(A) Grounding. Where there is no mobile home service equipment located in sight from, and not more than 9.0 m (30 ft) from, the exterior wall of the mobile home it serves or there is no mobile home disconnecting means grounded in accordance with 250.32 and located within sight from, and not more than 9.0 m (30 ft) from, the exterior wall of the mobile home it serves, the coaxial cable shield ground, or surge arrester ground, shall be in accordance with 820.40(B)(2).

(B) Bonding. The coaxial cable shield grounding terminal, surge arrester grounding terminal, or grounding electrode shall be bonded to the metal frame or available grounding terminal of the mobile home with a copper grounding conductor not smaller than 12 AWG under any of the following conditions:

(1) Where there is no mobile home service equipment or disconnecting means as in 820.42(A)
(2) Where the mobile home is supplied by cord and plug

V. Cables Within Buildings

820.49 Fire Resistance of CATV Cables.

Coaxial cables installed as wiring within buildings shall be listed as being resistant to the spread of fire in accordance with 820.50 and 820.51.

820.50 Listing, Marking, and Installation of Coaxial Cables.

Coaxial cables in a building shall be listed as being suitable for the purpose, and cables shall be marked in accordance with Table 820.50. The cable voltage rating shall not be marked on the cable.

> FPN: Voltage markings on cables could be misinterpreted to suggest that the cables may be suitable for Class 1, electric light, and power applications.

Table 820.50 Cable Markings

Cable Marking	Type	Reference
CATVP	CATV plenum cable	820.51(A) and 820.53(A)
CATVR	CATV riser cable	820.51(B) and 820.53(B)
CATV	CATV cable	820.51(C) and 820.53(C)
CATVX	CATV cable, limited use	820.51(D) and 820.53(C)

Exception No. 1: Voltage markings shall be permitted where the cable has multiple listings and voltage marking is required for one or more of the listings.

Exception No. 2: Listing and marking shall not be required where the cable enters the building from the outside and is run in rigid metal conduit or intermediate metal conduit, and such conduits are grounded to an electrode in accordance with 820.40(B).

Exception No. 3: Listing and marking shall not be required where the length of the cable within the building, measured from its point of entrance, does not exceed 15 m (50 ft) and the cable enters the building from the outside and is terminated at a grounding block.

> FPN No. 1: Cable types are listed in descending order of fire-resistance rating.
>
> FPN No. 2: See the referenced sections for listing requirements and permitted uses.

820.51 Additional Listing Requirements.

Cables shall be listed in accordance with 820.51(A) through (D).

(A) Type CATVP. Type CATVP community antenna television plenum cable shall be listed as being suitable for use in ducts, plenums, and other spaces used for environmental air and shall also be listed as having adequate fire-resistant and low smoke-producing characteristics.

> FPN: One method of defining low smoke-producing cables is by establishing an acceptable value of the smoke produced when tested in accordance with NFPA 262-1999, *Standard Method for Test for Flame Travel and Smoke of Wire and Cables for Use in Air-Handling Spaces*, to a maximum peak optical density of 0.5 and a maximum average optical density of 0.15. Similarly, one method of defining fire-resistant cables is by establishing maximum allowable flame travel distance of 1.52 m (5 ft) when tested in accordance with the same test.

> See the commentary following 725.71(A), FPN, for more information on a test for wires and cables to be installed in plenums and other spaces used for environmental air.

(B) Type CATVR. Type CATVR community antenna television riser cable shall be listed as being suitable for use in a vertical run in a shaft or from floor to floor and shall also be listed as having fire-resistant characteristics capable of preventing the carrying of fire from floor to floor.

> FPN: One method of defining fire-resistant characteristics capable of preventing the carrying of fire from floor to floor is that the cables pass the requirements of ANSI/UL 1666-1997, *Standard Test for Flame Propagation Height of Electrical and Optical-Fiber Cable Installed Vertically in Shafts*.

See the commentary following 725.71(B), FPN, for information on a test method for defining fire-resistant characteristics capable of preventing fire spread from floor to floor.

(C) Type CATV. Type CATV community antenna television cable shall be listed as being suitable for general-purpose CATV use, with the exception of risers and plenums, and shall also be listed as being resistant to the spread of fire.

> FPN: One method of defining *resistant to the spread of fire* is that the cables do not spread fire to the top of the tray in the vertical-tray flame test in ANSI/UL 1581-1991, *Reference Standard for Electrical Wires, Cables and Flexible Cords.*
>
> Another method of defining *resistant to the spread of fire* is for the damage (char length) not to exceed 1.5 m (4 ft 11 in.) when performing the vertical flame test for cables in cable trays, as described in CSA C22.2 No. 0.3-M-1985, *Test Methods for Electrical Wires and Cables.*

See the commentary following 725.71(C), FPN, for information on the UL Vertical Flame Test.

(D) Type CATVX. Type CATVX limited-use community antenna television cable shall be listed as being suitable for use in dwellings and for use in raceway and shall also be listed as being resistant to flame spread.

> FPN: One method of determining that cable is resistant to flame spread is by testing the cable to the VW-1 (vertical-wire) flame test in ANSI/UL 1581-1991, *Reference Standard for Electrical Wires, Cables and Flexible Cords.*

See the commentary following 725.71(D), FPN, for information on test methods for determining whether cable is resistant to fire spread.

820.52 Installation of Cables and Equipment.

Beyond the point of grounding, as defined in 820.33, the cable installation shall comply with 820.33(A) through (D).

Section 820.52 was revised for the 1999 *Code* to specifically include separation from network-powered broadband communications circuits. Jackets of coaxial cable do not have sufficient construction specifications to permit them to be installed with electric light, power, Class 1, non–power-limited fire alarm circuits, and medium- and high-power network-powered broadband communications cable. Failure

of the cable insulation due to a fault could lead to hazardous voltages being imposed on the Class 2 or Class 3 circuit conductors.

(A) Separation from Other Conductors.

(1) In Raceways and Boxes.

(a) Other Circuits. Coaxial cables shall be permitted in the same raceway or enclosure with jacketed cables of any of the following:

(1) Class 2 and Class 3 remote-control, signaling, and power-limited circuits in compliance with Article 725
(2) Power-limited fire alarm systems in compliance with Article 760
(3) Nonconductive and conductive optical fiber cables in compliance with Article 770
(4) Communications circuits in compliance with Article 800
(5) Low-power network-powered broadband communications circuits in compliance with Article 830

(b) Electric Light, Power, Class 1, Non–Power-Limited Fire Alarm, and Medium Power Network-Powered Broadband Communications Circuits. Coaxial cable shall not be placed in any raceway, compartment, outlet box, junction box, or other enclosures with conductors of electric light, power, Class 1, non–power-limited fire alarm, or medium power network-powered broadband communications circuits.

Exception No. 1: Where all of the conductors of electric light, power, Class 1, non–power-limited fire alarm, and medium power network-powered broadband communications circuits are separated from all of the coaxial cables by a barrier.

Exception No. 2: Power circuit conductors in outlet boxes, junction boxes, or similar fittings or compartments where such conductors are introduced solely for power supply to the coaxial cable system distribution equipment. The power circuit conductors shall be routed within the enclosure to maintain a minimum 6-mm (0.25-in.) separation from coaxial cables.

(2) Other Applications. Coaxial cable shall be separated at least 50 mm (2 in.) from conductors of any electric light, power, Class 1, non–power-limited fire alarm, or medium power network-powered broadband communications circuits.

Exception No. 1: Where either (1) all of the conductors of electric light, power, Class 1, non–power-limited fire alarm, and medium power network-powered broadband communications and circuits are in a raceway, or in metal-sheathed, metal-clad, nonmetallic-sheathed Type AC or Type UF cables, or (2) all of the coaxial cables are encased in raceway.

Exception No. 2: Where the coaxial cables are permanently separated from the conductors of electric light, power, Class 1, non–power-limited fire alarm, and medium power network-powered broadband communications circuits by a continuous and firmly fixed nonconductor, such as porcelain tubes or flexible tubing, in addition to the insulation on the wire.

(B) Equipment in Other Space Used for Environmental Air. Section 300.22(C) shall apply.

(C) Hybrid Power and Coaxial Cabling. The provisions of 780.6 shall apply for listed hybrid power and coaxial cabling in closed-loop and programmed power distribution.

(D) Support of Cables. Raceways shall be used for their intended purpose. Coaxial cables shall not be strapped, taped, or attached by any means to the exterior of any conduit or raceway as a means of support.

Exception: Overhead (aerial) spans of coaxial cables shall be permitted to be attached to the exterior of a raceway-type mast intended for the attachment and support of such cables.

820.53 Applications of Listed CATV Cables.

CATV cables shall comply with the requirements of 820.53(A) through (D) or where cable substitutions are made as shown in Table 820.53.

Table 820.53 Coaxial Cable Uses and Permitted Substitutions

Cable Type	Use	References	Permitted Substitutions
CATVP	Coaxial plenum cable	820.53(A)	CMP
CATVR	Coaxial riser cable	820.53(B)	CATVP, CMP, CMR
CATV	Coaxial general-purpose cable	820.53(D)	CATVP, CMP, CATVR, CMR, CMG, CM
CATVX	Coaxial cable, limited use	820.53(D)	CATVP, CMP, CATVR, CMR, CATV, CMG, CM

Note: See Figure 820.53, Cable substitution hierarchy.

Table 820.53 was extensively revised for the 1999 *Code* to reflect the current field applications for various cable types.

Sections 820.53(A) and (B)(1) have been revised for the 2002 *Code*. Used in conjunction with the definition of *abandoned coaxial cable* in 820.2 and the general requirement in 820.3(A), this section now requires removal of accessible abandoned CATV cable. Abandoned cable increases fire loading unnecessarily, and, where installed in plenums, it can affect airflow. Similar requirements can be found in Articles 640, 645, 725, 760, 770, 800, and 830. See the definition of *abandoned coaxial cable* in 820.2.

FPN: The substitute cables in Table 820.53 are only coaxial-type cables.

(A) Plenum. Cables installed in ducts, plenums, and other spaces used for environmental air shall be Type CATVP. Abandoned cables shall not be permitted to remain. Types CATVP, CATVR, CATV, and CATVX cables installed in compliance with 300.22 shall be permitted.

(B) Riser. Cables installed in risers shall comply with any of the requirements of 820.53(B)(1) through (B)(3).

(1) Cables in Vertical Runs. Cables installed in vertical runs and penetrating more than one floor, or cables installed in vertical runs in a shaft, shall be Type CATVR. Floor penetrations requiring Type CATVR shall contain only cables suitable for riser or plenum use. Abandoned cables shall not be permitted to remain.

(2) Metal Raceways or Fireproof Shafts. Types CATV and CATVX cables shall be permitted to be encased in a metal raceway or located in a fireproof shaft having firestops at each floor.

(3) One- and Two-Family Dwellings. Types CATV and CATVX cables shall be permitted in one- and two-family dwellings.

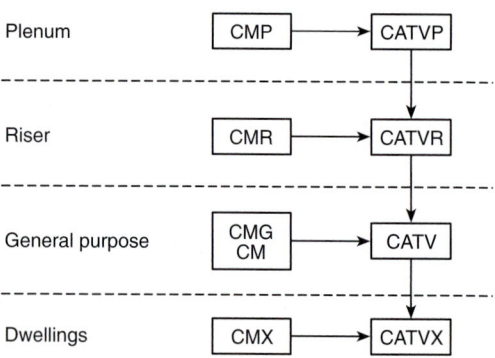

Type CATV—Community antenna television cables
Type CM—Communications cables

A → B Coaxial cable A shall be permitted to be used in place of coaxial cable B.

Figure 820.53 Cable substitution hierarchy.

FPN: See 820.53(A) for the firestop requirements for floor penetrations.

(C) Cable Trays. Cables installed in cable trays shall be Types CATVP, CATVR, and CATV.

(D) Other Wiring Within Buildings. Cables installed in building locations other than the locations covered in 820.53(A) and (B) shall be with any of the requirements in 820.53(D)(1) through (5). Abandoned cables in hollow spaces shall not be permitted to remain.

(1) General. Type CATV shall be permitted.

(2) In Raceways. Type CATVX shall be permitted to be installed in a raceway.

(3) Nonconcealed Spaces. Type CATVX shall be permitted to be installed in nonconcealed spaces where the exposed length of cable does not exceed 3 m (10 ft).

(4) One- and Two-Family Dwellings. Type CATVX cables less than 10 mm (0.375 in.) in diameter shall be permitted to be installed in one- and two-family dwellings.

(5) Multifamily Dwellings. Type CATVX cables less than 10 mm (0.375 in.) in diameter shall be permitted to be installed in multifamily dwellings.

ARTICLE 830
Network-Powered Broadband Communications Systems

Contents

I. General

830.1 Scope.

This article covers network-powered broadband communications systems that provide any combination of voice, audio, video, data, and interactive services through a network interface unit.

> FPN No. 1: A typical basic system configuration includes a cable supplying power and broadband signal to a network interface unit that converts the broadband signal to the component signals. Typical cables are coaxial cable with both broadband signal and power on the center conductor, composite metallic cable with a coaxial member for the broadband signal and a twisted pair for power, and composite optical fiber cable with a pair of conductors for power. Larger systems may also include network components such as amplifiers that require network power.
>
> FPN No. 2: See 90.2(B)(4) for installations of broadband communications systems that are not covered.

Article 830 covers network-powered broadband communications circuits, which provide a wide array of subscriber services, including voice, data (such as internet access), interactive services, and television signals.

830.2 Definitions.

See Article 100. For purposes of this article, the following additional definitions apply.

Abandoned Network-Powered Broadband Communications Cable. Installed network-powered broadband communications cable that is not terminated at equipment other than a connector and not identified for future use with a tag.

This definition is new in the 2002 *Code*, for use with 830.3(A) and 830.55, which now require removal of accessible abandoned broadband communications cable. Abandoned cable increases fire loading unnecessarily, and, where installed in plenums, it can affect airflow. Similar requirements can be found in Articles 640, 645, 725, 760, 770, 800, and 820.

Block. A square or portion of a city, town, or village enclosed by streets, including the alleys so enclosed but not any street.

Exposed to Accidental Contact with Electrical Light or Power Conductors. A circuit in such a position that, in case of failure of supports or insulation, contact with another circuit may result.

Fault Protection Device. An electronic device that is intended for the protection of personnel and functions under fault conditions, such as network-powered broadband communications cable short or open circuit, to limit the current or voltage, or both, for a low-power network-powered broadband communications circuit and provide acceptable protection from electric shock.

Network Interface Unit (NIU). A device that converts a broadband signal into component voice, audio, video, data, and interactive services signals. The NIU provides isolation between the network power and the premises signal circuits. The NIU may also contain primary and secondary protectors.

Network-Powered Broadband Communications Circuit. The circuit extending from the communications utility's serving terminal or tap up to and including the NIU.

> FPN: A typical single-family network-powered communications circuit consists of a communications drop or communications service cable and an NIU and includes the communications utility's serving terminal or tap where it is not under the exclusive control of the communications utility.

Point of Entrance. The point within a building at which the cable emerges from an external wall, from a concrete floor slab, or from a rigid metal conduit or an intermediate metal conduit grounded to an electrode in accordance with 830.40(B).

Premises Wiring. The circuits located on the user side of the network interface unit.

830.3 Locations and Other Articles.

Circuits and equipment shall comply with 830.3(A) through (D).

Article 830 contains provisions for wiring both the inside and outside of buildings. However, other articles cover the wiring derived from the network interface unit (NIU) into the premises. For example, Article 725 covers wiring of Class 2 and Class 3 circuits, Article 760 covers wiring of fire alarm systems, Article 770 covers the installation of optical fiber cable, Article 800 covers telephone wiring, and Article 820 covers coaxial cable installations for television signals. The major difference between Article 820 and Article 830 is the voltage present on the circuit conductors. Article 820 systems are limited to 60 volts, but Article 830 systems are permitted to have ratings as high as 150 volts.

Higher voltages allow systems to power more sophisticated electronics and provide a wider variety of services. (See Exhibit 830.1.)

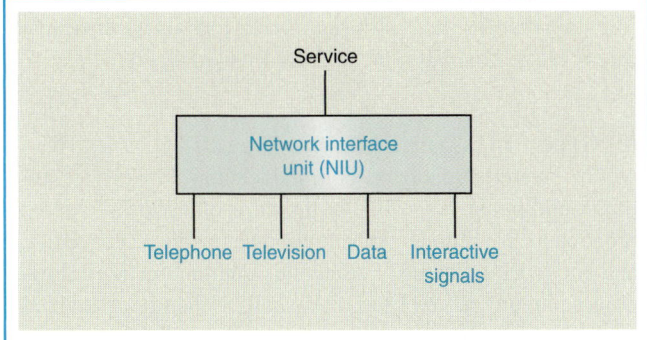

Service

Network interface unit (NIU)

Telephone Television Data Interactive signals

Exhibit 830.1 A network interface unit (NIU) diagram showing derived circuits.

(A) Spread of Fire or Products of Combustion. Section 300.21 shall apply. The accessible portion of abandoned network-powered broadband communications cables shall not be permitted to remain.

(B) Ducts, Plenums, and Other Air-Handling Spaces. Section 300.22 shall apply, where installed in ducts or plenums or other spaces used for environmental air.

Exception: As permitted in 830.55(B).

(C) Installation and Use. Section 110.3(B) shall apply.

(D) Output Circuits. As appropriate for the services provided, the output circuits derived from the network interface unit shall comply with the requirements of the following:

(1) Installations of communications circuits — Article 800
(2) Installations of community antenna television and radio distribution circuits — Article 820

Exception: 830.30(B)(3) shall apply where protection is provided in the output of the NIU.

(3) Installations of optical fiber cables — Article 770
(4) Installations of Class 2 and Class 3 circuits — Article 725
(5) Installations of power-limited fire alarm circuits — Article 760

830.4 Power Limitations.

Network-powered broadband communications systems shall be classified as having low or medium power sources as defined in Table 830.4.

Table 830.4 Limitations for Network-Powered Broadband Communications Systems

Network Power Source	Low	Medium
Circuit voltage, V_{max} (volts)[1]	0–100	0–150
Power limitation, VA_{max} (volt-amperes)[1]	250	250
Current limitation, I_{max} (amperes)[1]	$1000/V_{max}$	$1000/V_{max}$
Maximum power rating (volt-amperes)	100	100
Maximum voltage rating (volts)	100	150
Maximum overcurrent protection (amperes)[2]	$100/V_{max}$	NA

[1] V_{max}, I_{max}, and VA_{max} are determined with the current-limiting impedance in the circuit (not bypassed) as follows:
V_{max}—Maximum system voltage regardless of load with rated input applied.
I_{max}—Maximum system current under any noncapacitive load, including short circuit, and with overcurrent protection bypassed if used. I_{max} limits apply after 1 minute of operation.
VA_{max}—Maximum volt-ampere output after 1 minute of operation regardless of load and overcurrent protection bypassed if used.
[2] Overcurrent protection is not required where the current-limiting device provides equivalent current limitation and the current-limiting device does not reset until power or the load is removed.

High-power network-powered broadband communications circuits are not covered by this *Code*.

830.5 Network-Powered Broadband Communications Equipment and Cables.

Network-powered broadband communications equipment and cables shall be listed as suitable for the purpose.

Exception No. 1: This listing requirement shall not apply to community antenna television and radio distribution system coaxial cables that were installed prior to January 1, 2000, in accordance with Article 820 and are used for low-power network-powered broadband communications circuits. See 830.10.

Exception No. 2: Substitute cables for network-powered broadband communications cables shall be permitted as shown in Table 830.58.

(A) Listing and Marking. Listing and marking of network-powered broadband communications cables shall comply with 830.5(A)(1) or (A)(2).

Article 830 permits three types of listed cable, including coaxial, coaxial with a twisted pair of conductors, and fiber optical with a twisted pair of conductors. In the coaxial configuration, power and signal are carried on the coaxial center conductor. With coaxial and a twisted pair of conductors, the signal is carried on the coaxial center conductor and the power on the twisted pair. The configuration of fiber optic and twisted pair conductors carries the signal on the fiber and the power on the twisted pair.

(1) Type BMU, Type BM, and Type BMR Cables. Network-powered broadband communications medium power underground cable, Type BMU; network-powered broadband communications medium power cable, Type BM; and network-powered broadband communications medium power riser cable, Type BMR, shall be factory-assembled cables consisting of a jacketed coaxial cable, a jacketed combination of coaxial cable and multiple individual conductors, or a jacketed combination of an optical fiber cable and multiple individual conductors. The insulation for the individual conductors shall be rated for 300 volts minimum. Cables intended for outdoor use shall be listed as suitable for the application. Cables shall be marked in accordance with 310.11. Type BMU cables shall be jacketed and listed as being suitable for outdoor underground use. Type BM cables shall be listed as being suitable for general-purpose use, with the exception of risers and plenums, and shall also be listed as being resistant to the spread of fire. Type BMR cables shall be listed as being suitable for use in a vertical run in a shaft or from floor to floor and shall also be listed as having fire-resistant characteristics capable of preventing the carrying of fire from floor to floor.

A rating of 300 volts is necessary for the following reasons:

1. To coordinate with protector installation requirements (i.e., protectors are not required within a block unless the cable is exposed to over 300 volts)
2. To recognize the fact that primary protectors are designed to allow voltages below 300 to pass
3. To accommodate the voltages ordinarily found on a network-powered broadband communications circuit (voltage up to 150 volts rms)

FPN No. 1: One method of defining *resistant to spread of fire* is that the cables do not spread fire to the top of the tray in the vertical tray flame test in ANSI/UL 1581-1991, *Reference Standard for Electrical Wires, Cables and Flexible Cords*. Another method of defining *resistant to the spread of fire* is for the damage (char length) not to exceed 1.5 m (4 ft 11 in.) when performing the CSA vertical flame test for cables in cable trays, as described in CSA C22.2 No. 0.3-M-1985, *Test Methods for Electrical Wires and Cables*.

See the commentary following 725.71(C), FPN, for information on the UL Vertical Tray Flame Test.

FPN No. 2: One method of defining fire-resistant characteristics capable of preventing the carrying of fire from

floor to floor is that the cables pass the requirements of ANSI/UL 1666-1997, *Standard Test for Flame Propagation Height of Electrical and Optical-Fiber Cable Installed Vertically in Shafts*.

See the commentary following 725.71(B), FPN, for information on a test for defining fire-resistant characteristics capable of preventing fire spread from floor to floor.

(2) Type BLU, Type BLX, and Type BLP Cables. Network-powered broadband communications low-power underground cable, Type BLU; limited use network-powered broadband communications low-power cable, Type BLX; and network-powered broadband communications low-power plenum cable, Type BLP, shall be factory assembled cables consisting of a jacketed coaxial cable, a jacketed combination of coaxial cable and multiple individual conductors, or a jacketed combination of an optical fiber cable and multiple individual conductors. The insulation for the individual conductors shall be rated for 300 volts minimum. Cables intended for outdoor use shall be listed as suitable for the application. Cables shall be marked in accordance with 310.11. Type BLU cables shall be jacketed and listed as being suitable for outdoor underground use. Type BLX limited-use cables shall be listed as being suitable for use outside, for use in dwellings, and for use in raceways and shall also be listed as being resistant to flame spread. Type BLP cables shall be listed as being suitable for use in ducts, plenums, and other spaces for environmental air and shall also be listed as having adequate fire-resistant and low smoke-producing characteristics.

FPN No. 1: One method of determining that cable is resistant to flame spread is by testing the cable to VW-1 (vertical-wire) flame test in ANSI/UL 1581-1991, *Reference Standard for Electrical Wires, Cables and Flexible Cords*.

See the commentary following 725.71(D), FPN, for information on test methods for determining the fire resistance of cable.

FPN No. 2: One method of defining low smoke-producing cable is by establishing an acceptable value of the smoke produced when tested in accordance with NFPA 262-1999, *Standard Method of Test for Flame Travel and Smoke of Wires and Cables for Use in Air-Handling Spaces*, to a maximum peak optical density of 0.5 and a maximum average optical density of 0.15. Similarly, one method of defining fire-resistant cables is by establishing maximum allowable flame travel distance of 1.52 m (5 ft) when tested in accordance with the same test.

See the commentary following 725.71(A), FPN, for information on a test method for wires and cables to be installed without raceways in plenums and other spaces used for environmental air.

830.6 Access to Electrical Equipment Behind Panels Designed to Allow Access.

Access to electrical equipment shall not be denied by an accumulation of wires and cables that prevents removal of panels. including suspended ceiling panels.

An excess accumulation of wires and cabling can limit access to equipment by preventing the removal of access panels. (See Exhibit 830.2.)

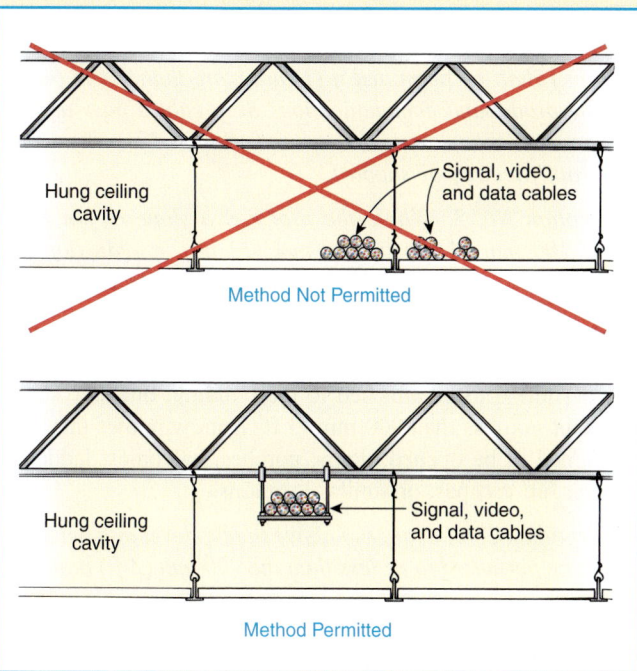

Exhibit 830.2 Installations of conductors and cables, which can prevent access to equipment or cables. Correct and incorrect methods are shown.

830.7 Mechanical Execution of Work.

Network-powered broadband communications circuits and equipment shall be installed in a neat and workmanlike manner. Cables installed exposed on the surface of ceiling and sidewalls shall be supported by the structural components of the building structure in such a manner that the cable is not damaged by normal building use. Such cables shall be attached to structural components by straps, staples, hangers, or similar fittings designed and installed so as not to damage the cable. The installation shall also conform with 300.4(D).

Revised for the 2002 *Code*, this section now provides definitive requirements for workmanship. Cable must now be attached to or supported by the structure by straps, clamps, hangers, and the like. The installation method must not damage the cable. In addition, the location of the cable should be carefully evaluated to ensure that activities and processes within the building do not cause damage to the cable.

830.9 Hazardous (Classified) Locations.

Network-powered broadband communications circuits and equipment installed in a location that is classified in accordance with Article 500 shall comply with the applicable requirements of Chapter 5.

II. Cables Outside and Entering Buildings

830.10 Entrance Cables.

Cables installed outdoors shall be listed as suitable for the application. In addition, network-powered broadband communications cables located outside and entering buildings shall comply with 830.10(A) and (B).

(A) Medium Power Circuits. Medium power network-powered broadband communications circuits located outside and entering buildings shall be installed using Type BMU, Type BM, or Type BMR network-powered broadband communications medium power cables.

(B) Low-Power Circuits. Low-power network-powered broadband communications circuits located outside and entering buildings shall be installed using Type BLU or Type BLX low-power network-powered broadband communications cables. Cables shown in Table 830.58 shall be permitted to substitute.

Exception: Outdoor community antenna television and radio distribution system coaxial cables installed prior to January 1, 2000, and installed in accordance with Article 820, shall be permitted for low-power-type, network-powered broadband communications circuits.

Network-powered broadband communications systems may contain sufficient energy to shock and kill. For that reason, they are subjected to requirements similar to those for other high-powered circuits.

Conductor spans should be of sufficient size and strength to maintain clearances and avoid possible contact with light or power conductors. Splices and joints should be made with approved connectors or other means that provide sufficient mechanical strength so that conductors are not

830.11 Aerial Cables.

Aerial powered broadband communications cables shall comply with 830.11(A) through (I).

> FPN: For additional information regarding overhead wires and cables, see ANSI C2-1997, *National Electric Safety Code*, Part 2, Safety Rules For Overhead Lines.

(A) On Poles. Where practicable, network-powered broadband communications cables on poles shall be located below the electric light, power, Class 1, or non–power-limited fire alarm circuit conductors and shall not be attached to a cross-arm that carries electric light or power conductors.

(B) Climbing Space. The climbing space through network-powered broadband communications cables shall comply with the requirements of 225.14(D).

(C) Lead-in Clearance. Lead-in or aerial-drop network-powered broadband communications cables from a pole or other support, including the point of initial attachment to a building or structure, shall be kept away from electric light, power, Class 1, or non–power-limited fire alarm circuit conductors so as to avoid the possibility of accidental contact.

Exception: Where proximity to electric light, power, Class 1, or non–power-limited fire alarm circuit service conductors cannot be avoided, the installation shall be such as to provide clearances of not less than 300 mm (12 in.) from light, power, Class 1, or non–power-limited fire alarm circuit service drops. The clearance requirement shall apply to all points along the drop, and it shall increase to 1.02 m (40 in.) at the pole.

(D) Clearance from Ground. Overhead spans of network-powered broadband communication cables shall conform to not less than the following:

(1) 2.9 m (9.5 ft)—above finished grade, sidewalks, or from any platform or projection from which they might be reached and accessible to pedestrians only
(2) 3.5 m (11.5 ft)—over residential property and driveways, and those commercial areas not subject to truck traffic
(3) 4.7 m (15.5 ft)—over public streets, alleys, roads, parking areas subject to truck traffic, driveways on other than residential property, and other land traversed by vehicles such as cultivated, grazing, forest, and orchard

> FPN: These clearances have been specifically chosen to correlate with ANSI C2-1997, *National Electrical Safety Code*, Table 232-1, which provides for clearances of

wires, conductors, and cables above ground and roadways, rather than using the clearances referenced in 225.18. Because Article 800 and Article 820 have had no required clearances, the communications industry has used the clearances from the NESC for their installed cable plant.

(E) Over Pools. Clearance of network-powered broadband communications cable in any direction from the water level, edge of pool, base of diving platform, or anchored raft shall comply with those clearances in 680.8.

(F) Above Roofs. Network-powered broadband communications cables shall have a vertical clearance of not less than 2.5 m (8 ft) from all points of roofs above which they pass.

Exception No. 1: Auxiliary buildings such as garages and the like.

Exception No. 2: A reduction in clearance above only the overhanging portion of the roof to not less than 450 mm (18 in.) shall be permitted if (1) not more than 1.2 m (4 ft) of the broadband communications drop cables pass above the roof overhang, and (2) they are terminated at a through-the-roof raceway or support.

Exception No. 3: Where the roof has a slope of not less than 100 mm (4 in.) in 300 mm (12 in.), a reduction in clearance to not less than 900 mm (3 ft) shall be permitted.

(G) Final Spans. Final spans of network-powered broadband communications cables without an outer jacket shall be permitted to be attached to the building, but they shall be kept not less than 900 mm (3 ft) from windows that are designed to be opened, doors, porches, balconies, ladders, stairs, fire escapes, or similar locations.

Exception: Conductors run above the top level of a window shall be permitted to be less than the 900-mm (3-ft) requirement above.

Overhead network-powered broadband communications cables shall not be installed beneath openings through which materials may be moved, such as openings in farm and commercial buildings, and shall not be installed where they will obstruct entrance to these building openings.

(H) Between Buildings. Network-powered broadband communications cables extending between buildings and also the supports or attachment fixtures shall be acceptable for the purpose and shall have sufficient strength to withstand the loads to which they may be subjected.

Exception: Where a network-powered broadband communications cable does not have sufficient strength to be self-supporting, it shall be attached to a supporting messenger cable that, together with the attachment fixtures or supports, shall be acceptable for the purpose and shall have sufficient strength to withstand the loads to which they may be subjected.

(I) On Buildings. Where attached to buildings, network-powered broadband communications cables shall be securely fastened in such a manner that they are separated from other conductors in accordance with 830.11(I)(1) through (I)(4).

(1) Electric Light or Power. The network-powered broadband communications cable shall have a separation of at least 100 mm (4 in.) from electric light, power, Class 1, or non–power-limited fire alarm circuit conductors not in raceway or cable, or be permanently separated from conductors of the other system by a continuous and firmly fixed nonconductor in addition to the insulation on the wires.

(2) Other Communications Systems. Network-powered broadband communications cables shall be installed so that there will be no unnecessary interference in the maintenance of the separate systems. In no case shall the conductors, cables, messenger strand, or equipment of one system cause abrasion to the conductors, cables, messenger strand, or equipment of any other system.

(3) Lightning Conductors. Where practicable, a separation of at least 1.8 m (6 ft) shall be maintained between any network-powered broadband communications cable and lightning conductors.

(4) Protection from Damage. Network-powered broadband communications cables attached to buildings and located within 2.5 m (8 ft) of finished grade shall be protected by enclosures, raceways, or other approved means.

Exception: A low-power network-powered broadband communications circuit that is equipped with a listed fault protection device, appropriate to the network-powered broadband communications cable used, and located on the network side of the network-powered broadband communications cable being protected.

830.12 Underground Circuits Entering Buildings.

(A) Underground Systems. Underground network-powered broadband communications cables in a duct, pedestal, handhole, or manhole that contains electric light, power conductors, non–power-limited fire alarm circuit conductors, or Class 1 circuits shall be in a section permanently separated from such conductors by means of a suitable barrier.

(B) Direct-Buried Cables and Raceways. Direct-buried network-powered broadband communications cables shall be separated at least 300 mm (12 in.) from conductors of

any light, power, non–power-limited fire alarm circuit conductors or Class 1 circuit.

Exception No. 1: Where electric service conductors or network-powered broadband communications cables are installed in raceways or have metal cable armor.

Exception No. 2: Where electric light or power branch-circuit or feeder conductors, non–power-limited fire alarm circuit conductors, or Class 1 circuit conductors are installed in a raceway or in metal-sheathed, metal-clad, or Type UF or Type USE cables; or the network-powered broadband communications cables have metal cable armor or are installed in a raceway.

(C) Mechanical Protection. Direct-buried cable, conduit, or other raceways shall be installed to meet the minimum cover requirements of Table 830.12. In addition, direct-buried cables emerging from the ground shall be protected by enclosures, raceways, or other approved means extending from the minimum cover distance required by Table 830.12 below grade to a point at least 2.5 m (8 ft) above finished grade. In no case shall the protection be required to exceed 450 mm (18 in.) below finished grade. Type BMU and BLU direct-buried cables emerging from the ground shall be installed in rigid metal conduit, intermediate metal conduit, rigid nonmetallic conduit, or other approved means extending from the minimum cover distance required by Table 830.12 below grade to the point of entrance.

Exception: A low-power network-powered broadband communications circuit that is equipped with a listed fault protection device, appropriate to the network-powered broadband communications cable used, and located on the network side of the network-powered broadband communications cable being protected.

(D) Pools. Cables located under the pool or within the area extending 1.5 m (5 ft) horizontally from the inside wall of the pool shall meet those clearances and requirements specified in 680.10.

III. Protection

830.30 Primary Electrical Protection.

(A) Application. Primary electrical protection shall be provided on all network-powered broadband communications conductors that are neither grounded nor interrupted and are run partly or entirely in aerial cable not confined within a block. Also, primary electrical protection shall be provided on all aerial or underground network-powered broadband communications conductors that are neither grounded nor interrupted and are located within the block containing the building served so as to be exposed to lightning or accidental contact with electric light or power conductors operating at over 300 volts to ground.

Table 830.12 Network-Powered Broadband Communications Systems Minimum Cover Requirements (*Cover is the shortest distance measured between a point on the top surface of any direct-buried cable, conduit, or other raceway and the top surface of finished grade, concrete, or similar cover.*)

Location of Wiring Method or Circuit	Direct Burial Cables		Rigid Metal Conduit or Intermediate Metal Conduit		Nonmetallic Raceways Listed for Direct Burial; Without Concrete Encasement or Other Approved Raceways	
	mm	in.	mm	in.	mm	in.
All locations not specified below	450	18	150	6	300	12
In trench below 50-mm (2-in.) thick concrete or equivalent	300	12	150	6	150	6
Under a building (in raceway only)	0	0	0	0	0	0
Under minimum of 100-mm (4-in.) thick concrete exterior slab with no vehicular traffic and the slab extending not less than 150 mm (6 in.) beyond the underground installation	300	12	100	4	100	4
One- and two-family dwelling driveways and outdoor parking areas and used only for dwelling-related purposes	300	12	300	12	300	12

Notes:
1. Raceways approved for burial only where concrete encased shall require a concrete envelope not less than 50 mm (2 in.) thick.
2. Lesser depths shall be permitted where cables rise for terminations or splices or where access is otherwise required.
3. Where solid rock is encountered, all wiring shall be installed in metal or nonmetallic raceway permitted for direct burial. The raceways shall be covered by a minimum of 50 mm (2 in.) of concrete extending down to rock.
4. Low-power network-powered broadband communications circuits using directly buried community antenna television and radio distribution system coaxial cables that were installed outside and entering buildings prior to January 1, 2000, in accordance with Article 820 shall be permitted where buried to a minimum depth of 300 mm (12 in.).

Exception: Where electrical protection is provided on the derived circuit(s) (output side of the NIU) in accordance with 830.30(B)(3).

FPN No. 1: On network-powered broadband communications conductors not exposed to lightning or accidental contact with power conductors, providing primary electrical protection in accordance with this article helps protect against other hazards, such as ground potential rise caused by power fault currents, and above-normal voltages induced by fault currents on power circuits in proximity to the network-powered broadband communications conductors.

FPN No. 2: Network-powered broadband communications circuits are considered to have a lightning exposure unless one or more of the following conditions exist:

(1) Circuits in large metropolitan areas where buildings are close together and sufficiently high to intercept lightning.
(2) Areas having an average of five or fewer thunderstorm days per year and earth resistivity of less than 100 ohm-meters. Such areas are found along the Pacific coast.

Utility companies may provide primary protectors if conductors are exposed to lightning. Generally, cable is not consid-

ered to be exposed to lightning unless one or both of the conditions in FPN No. 2 exist. A primary protector is required at each end of a communications circuit if lightning exposure exists, unless protection is provided on the output side of the network interface unit (NIU).

(1) Fuseless Primary Protectors. Fuseless-type primary protectors shall be permitted where power fault currents on all protected conductors in the cable are safely limited to a value no greater than the current-carrying capacity of the primary protector and of the primary protector grounding conductor.

(2) Fused Primary Protectors. Where the requirements listed in 830.30(A)(1) are not met, fused-type primary protectors shall be used. Fused-type primary protectors shall consist of an arrester connected between each conductor to be protected and ground, a fuse in series with each conductor to be protected, and an appropriate mounting arrangement. Fused primary protector terminals shall be marked to indicate line, instrument, and ground, as applicable.

(B) Location. The location of the primary protector, where required, shall comply with (1), (2), or (3):

(1) A listed primary protector shall be applied on each network-powered broadband communications cable external to and on the network side of the network interface unit.

(2) The primary protection function shall be an integral part of and contained in the network interface unit. The network interface unit shall be listed for the purpose and shall have an external marking indicating that it contains primary electrical protection.

(3) The primary protector(s) shall be provided on the derived circuit(s) (output side of the NIU), and the combination of the NIU and the protector(s) shall be listed for the purpose.

A primary protector, whether provided integrally or external to the network interface unit, shall be located as close as practicable to the point of entrance.

For purposes of this section, a network interface unit and any externally provided primary protectors located at mobile home service equipment located in sight from and not more than 9.0 m (30 ft) from the exterior wall of the mobile home it serves, or at a mobile home disconnecting means grounded in accordance with 250.32 and located in sight from and not more than 9.0 m (30 ft) from the exterior wall of the mobile home it serves, shall be considered to meet the requirements of this section.

FPN: Selecting a network interface unit and primary protector location to achieve the shortest practicable primary protector grounding conductor helps limit potential differences between communications circuits and other metallic systems.

(C) Hazardous (Classified) Locations. The primary protector or equipment providing the primary protection function shall not be located in any hazardous (classified) location as defined in Article 500 or in the vicinity of easily ignitible material.

Exception: As permitted in 501.14, 502.14, and 503.12.

830.33 Grounding or Interruption of Metallic Members of Network-Powered Broadband Communications Cables.

The shields of network-powered broadband communications cables used for communications or powering shall be grounded at the building as close as practicable to the point of entrance or attachment of the NIU. Metallic cable members not used for communications or powering shall be grounded or interrupted by an insulating joint or equivalent device as close as practicable to the point of entrance or attachment of the NIU.

For purposes of this section, grounding or interruption of network-powered broadband communications cable metallic members installed at mobile home service equipment located

in sight from and no more than 9.0 m (30 ft) from the exterior wall of the mobile home it serves, or at a mobile home disconnecting means grounded in accordance with 250.32 and located in sight from and not more than 9.0 m (30 ft) from the exterior wall of the mobile home it serves, shall be considered to meet the requirements of this section.

FPN: Selecting a grounding location to achieve the shortest practicable grounding conductor helps limit potential differences between the network-powered broadband communications circuits and other metallic systems.

IV. Grounding Methods

830.40 Cable, Network Interface Unit, and Primary Protector Grounding.

Network interface units containing protectors, NIUs with metallic enclosures, primary protectors, and the metallic members of the network-powered broadband communications cable that are intended to be grounded shall be grounded as specified in 830.40(A) through (D).

(A) Grounding Conductor.

(1) Insulation. The grounding conductor shall be insulated and shall be listed as suitable for the purpose.

(2) Material. The grounding conductor shall be copper or other corrosion-resistant conductive material, stranded or solid.

(3) Size. The grounding conductor shall not be smaller than 14 AWG and shall have a current-carrying capacity approximately equal to that of the grounded metallic member(s) and protected conductor(s) of the network-powered broadband communications cable. The grounding conductor shall not be required to exceed 6 AWG.

(4) Length. The grounding conductor shall be as short as practicable. In one-family and multifamily dwellings, the grounding conductor shall be as short as permissible, not to exceed 6.0 m (20 ft) in length.

This section is new for the 2002 *Code*. The limitation on length will result in a lower impedance, which will limit the potential difference between broadband communications systems and other systems during a lightning strike. Large potential differences between grounding conductors can result in increased damage if a lightning strike were to occur.

Exception: In one- and two-family dwellings where it is not practicable to achieve an overall maximum grounding conductor length of 6.0 m (20 ft), a separate communications ground rod meeting the minimum dimensional criteria of 830.40(B)(2)(2) shall be driven, and the grounding conductor connected to the communications ground rod in accor-

dance with 830.40(C). The communications ground rod shall be bonded to the power grounding electrode system in accordance with 830.40(D).

(5) Run in Straight Line. The grounding conductor shall be run to the grounding electrode in as straight a line as practicable.

(6) Physical Protection. Where subject to physical damage, the grounding conductor shall be adequately protected. Where the grounding conductor is run in a metal raceway, both ends of the raceway shall be bonded to the grounding conductor or the same terminal or electrode to which the grounding conductor is connected.

(B) Electrode. The grounding conductor shall be connected as follows.

(1) In Buildings or Structures with Grounding Means. To the nearest accessible location on the following:

(1) The building or structure grounding electrode system as covered in 250.50;
(2) The grounded interior metal water piping system, within 1.5 m (5 ft) from its point of entrance to the building, as covered in 250.52;

See the commentary following 250.52(A)(1).

(3) The power service accessible means external to enclosures as covered in 250.94;
(4) The metallic power service raceway;
(5) The service equipment enclosure;
(6) The grounding electrode conductor or the grounding electrode metal enclosure; or
(7) The grounding conductor or the grounding electrode of a building or structure disconnecting means that is grounded to an electrode as covered in 250.32.

For purposes of this section, the mobile home service equipment or the mobile home disconnecting means, as described in 830.33, shall be considered accessible.

(2) In Buildings or Structures Without Grounding Means. If the building or structure served has no grounding means, as described in (B)(1):

(1) To any one of the individual electrodes described in 250.52(A)(1), (2), (3), (4); or
(2) If the building or structure served has no grounding means, as described in 830.40(B)(1) or (B)(2)(1), to an effectively grounded metal structure or to a ground rod or pipe not less than 1.5 m (5 ft) in length and 12.7 mm (½ in.) in diameter, driven, where practicable, into permanently damp earth and separated from lightning con-

ductors as covered in 800.13 and at least 1.8 m (6 ft) from electrodes of other systems. Steam or hot water pipes or lightning-rod conductors shall not be employed as electrodes for protectors, NIUs with integral protection, grounded metallic members, NIUs with metallic enclosures, and other equipment.

(C) Electrode Connection. Connections to grounding electrodes shall comply with 250.70. Connectors, clamps, fittings, or lugs used to attach grounding conductors and bonding jumpers to grounding electrodes or to each other that are to be concrete encased or buried in the earth shall be suitable for its application.

(D) Bonding of Electrodes. A bonding jumper not smaller than 6 AWG copper or equivalent shall be connected between the network-powered broadband communications system grounding electrode and the power grounding electrode system at the building or structure served where separate electrodes are used.

Exception: At mobile homes as covered in 830.42.

> FPN No. 1: See 250.60 for use of lightning rods.
> FPN No. 2: Bonding together of all separate electrodes limits potential differences between them and between their associated wiring systems.

The most common error made in grounding network-powered broadband communications systems is to connect the cable sheath to a ground rod driven by the utility installer at a convenient location near the point of cable entry to the building, instead of bonding it to the electrical service grounding electrode system, service raceway, or other components that make up the grounding electrode system. A separate grounding electrode is permitted by the *Code* only if the building or structure has none of the grounding means described in 820.40(B). The *Code* requires that some means that is accessible and external to the service equipment be provided for making the bonding and grounding connection for other systems. One of the following means must be provided:

1. Exposed service raceways
2. An exposed grounding electrode conductor
3. An approved means for the external connection of a conductor (A 6 AWG copper conductor with one end bonded to the service raceway or equipment with about 6 in. exposed is acceptable.)

Proper bonding of the network-powered broadband communications system cable sheath to the electrical power ground is needed to prevent potential fire and shock hazards.

The earth cannot be used as an equipment grounding conductor or bonding conductor because it does not have the required low-impedance path. (See 250.54.)

Both network-powered broadband communications systems and power systems are subject to current surges as a result, for example, of induced voltages from lightning in the vicinity of the usually extensive outside distribution systems. Surges also result from switching operations on power systems. If the grounded conductors and parts of the two systems are not bonded by a low-impedance path, such line surges can raise the potential difference between the two systems to many thousands of volts. This can result in arcing between the two systems, for example, wherever the coaxial cable jacket contacts a grounded part, such as a metal water pipe or metal structural member, inside the building.

If an individual is the interface between the two systems and the bonding has not been accomplished in accordance with the *Code,* the high-voltage surge could result in electric shock. More common, however, is burnout of a television tuner because this part is almost always an interface between the two systems. The tuner is connected to the power system ground through the grounded neutral of the power supply, even if the television itself is not provided with an equipment grounding conductor.

Also see the commentary following 250.92(B), FPN No. 2, and 820.33(B).

830.42 Bonding and Grounding at Mobile Homes.

(A) Grounding. Where there is no mobile home service equipment located in sight from and not more than 9.0 m (30 ft) from the exterior wall of the mobile home it serves, or there is no mobile home disconnecting means grounded in accordance with 250.32 and located within sight from and not more than 9.0 m (30 ft) from the exterior wall of the mobile home it serves, the network-powered broadband communications cable, network interface unit, and primary protector ground shall be installed in accordance with 830.40(B)(2).

(B) Bonding. The network-powered broadband communications cable grounding terminal, network interface unit grounding terminal, if present, and primary protector grounding terminal shall be bonded together with a copper bonding conductor not smaller than 12 AWG. The network-powered broadband communications cable grounding terminal, network interface unit grounding terminal, primary protector grounding terminal, or the grounding electrode shall be bonded to the metal frame or available grounding terminal of the mobile home with a copper bonding conductor not smaller than 12 AWG under any of the following conditions:

(1) Where there is no mobile home service equipment or disconnecting means as in (A)
(2) Where the mobile home is supplied by cord and plug

V. Wiring Methods Within Buildings

830.54 Medium Power Network-Powered Broadband Communications System Wiring Methods.

Medium power network-powered broadband communications systems shall be installed within buildings using listed Type BM or Type BMR, network-powered broadband communications medium power cables.

(A) Ducts, Plenums, and Other Air-Handling Spaces. Section 300.22 shall apply.

Section 300.22 requires that the appropriate methods of Chapter 3 be used for the installation of medium-power network-powered broadband communications circuits in plenums and other spaces used for environmental air within buildings. There is no plenum-rated cable for medium-power network-powered broadband communications circuits. However, if the wiring downstream of the NIU is covered by other articles, then listed cable may be available.

(B) Riser. Cables installed in vertical runs and penetrating more than one floor, or cables installed in vertical runs in a shaft, shall be Type BMR. Floor penetrations requiring Type BMR shall contain only cables suitable for riser or plenum use.

Exception No. 1: Type BM cables encased in metal raceway or located in a fireproof shaft that has firestops at each floor.

Exception No. 2: Type BM cables in one- and two-family dwellings.

(C) Other Wiring. Cables installed in locations other than the locations covered in 830.54(A) and (B) shall be Type BM.

Exception: Type BMU cable where the cable enters the building from the outside and is run in rigid metal conduit or intermediate metal conduit, and such conduits are grounded to an electrode in accordance with 830.40(B).

830.55 Low-Power Network-Powered Broadband Communications System Wiring Methods.

Low-power network-powered broadband communications systems shall comply with any of the requirements of 830.55(A) through (D).

(A) In Buildings. Low-power network-powered broadband communications systems shall be installed within buildings using listed Type BLX or Type BLP network-powered broadband communications low power cables.

> Section 830.55(A) permits the use of listed Type BLP cable for low-power network-powered broadband communications circuits within plenums and other spaces used for environmental air within buildings.

(B) Ducts, Plenums, and Other Air-Handling Spaces. Cables installed in ducts, plenums, and other spaces used for environmental air shall be Type BLP. Abandoned cables shall not be permitted to remain. Type BLX cable installed in compliance with 300.22 shall be permitted.

> Sections 830.55(B) and (C)(1) have been revised for the 2002 *Code.* Used in conjunction with the definition of *abandoned network-powered broadband communications cable* in 830.2 and the general requirement in 830.3(A), these sections now require removal of accessible abandoned broadband communications cable. Abandoned cable increases fire loading unnecessarily, and, where installed in plenums, it can affect airflow. Similar requirements can be found in Articles 640, 645, 725, 760, 770, 800, and 820. See the definition of *abandoned network-powered broadband communications cable* in 830.2.

(C) Riser. Cables installed in risers shall comply with any of the requirements in 830.55(C)(1), (C)(2), or (C)(3).

(1) Cables in Vertical Runs. Cables installed in vertical runs and penetrating more than one floor, or cables installed in vertical runs in a shaft, shall be Type BLP or BMR. Floor penetrations requiring Type BMR shall contain only cables suitable for riser or plenum use. Abandoned cables shall not be permitted to remain.

(2) Metal Raceways or Fireproof Shafts. Type BLX cables shall be permitted to be encased in a metal raceway or located in a fireproof shaft having firestops at each floor.

(3) One- and Two-Family Dwellings. Type BLX cables less than 10 mm (0.375 in.) in diameter shall be permitted in one- and two-family dwellings.

(D) Other Wiring. Cables installed in locations other than the locations covered in 830.55(A), (B), and (C) shall comply with the requirements of 830.55(D)(1) through (D)(5).

(1) General. Type BLP or BM shall be permitted.

(2) In Raceways. Type BLX shall be permitted to be installed in a raceway.

(3) Type BLU Cable. Type BLU cable entering the building from outside shall be permitted to be run in rigid metal conduit or intermediate metal conduit. Such conduits shall be grounded to an electrode in accordance with 830.40(B).

(4) One- and Two-Family Dwellings. Type BLX cable less than 10 mm (0.375 in.) in diameter shall be permitted to be installed in one- and two-family dwellings.

(5) Type BLX Cable. Type BLX cable entering the building from outside and terminated at a grounding block or a primary protection location shall be permitted to be installed, provided that the length of cable within the building does not exceed 15 m (50 ft).

> FPN: This provision limits the length of Type BLX cable to 15 m (50 ft), while 830.30(B) requires that the primary protector, or NIU with integral protection, be located as close as practicable to the point at which the cable enters the building. Therefore, in installations requiring a primary protector, or NIU with integral protection, Type BLX cable may not be permitted to extend 15 m (50 ft) into the building if it is practicable to place the primary protector closer than 15 m (50 ft) to the entrance point.

830.56 Protection Against Physical Damage.

Section 300.4 shall apply.

830.57 Bends.

Bends in network broadband cable shall be made so as not to damage the cable.

830.58 Installation of Network-Powered Broadband Communications Cables and Equipment.

Cable and equipment installations within buildings shall comply with 830.58(A) through (E), as applicable.

(A) Separation of Conductors.

(1) In Raceways and Enclosures.

(a) Low and Medium Power Network-Powered Broadband Communications Circuit Cables. Low and medium power network-powered broadband communications cables shall be permitted in the same raceway or enclosure.

(b) Low Power Network-Powered Broadband Communications Circuit Cables. Low power network-powered broadband communications cables shall be permitted in the same raceway or enclosure with jacketed cables of any of the following circuits:

(1) Class 2 and Class 3 remote-control, signaling, and power-limited circuits in compliance with Article 725
(2) Power-limited fire alarm systems in compliance with Article 760

(3) Communications circuits in compliance with Article 800

(4) Nonconductive and conductive optical fiber cables in compliance with Article 770

(5) Community antenna television and radio distribution systems in compliance with Article 820

(c) **Medium Power Network-Powered Broadband Communications Circuit Cables.** Medium power network-powered broadband communications cables shall not be permitted in the same raceway or enclosure with conductors of any of the following circuits:

(1) Class 2 and Class 3 remote-control, signaling, and power-limited circuits in compliance with Article 725

(2) Power-limited fire alarm systems in compliance with Article 760

(3) Communications circuits in compliance with Article 800

(4) Conductive optical fiber cables in compliance with Article 770

(5) Community antenna television and radio distribution systems in compliance with Article 820

(d) **Electric Light, Power, Class 1, Non–Powered Broadband Communications Circuit Cables.** Network-powered broadband communications cable shall not be placed in any raceway, compartment, outlet box, junction box, or similar fittings with conductors of electric light, power, Class 1, or non–power-limited fire alarm circuit cables.

Exception No. 1: Where all of the conductors of electric light, power, Class 1, non–power-limited fire alarm circuits are separated from all of the network-powered broadband communications cables by a barrier.

Exception No. 2: Power circuit conductors in outlet boxes, junction boxes, or similar fittings or compartments where such conductors are introduced solely for power supply to the network-powered broadband communications system distribution equipment. The power circuit conductors shall be routed within the enclosure to maintain a minimum 6-mm (0.25-in.) separation from network-powered broadband communications cables.

(2) **Other Applications.** Network-powered broadband communications cable shall be separated at least 50 mm (2 in.) from conductors of any electric light, power, Class 1, and non–power-limited fire alarm circuits.

Exception No. 1: Where either (1) all of the conductors of electric light, power, Class 1, and non–power-limited fire alarm circuits are in a raceway, or in metal-sheathed, metal-clad, nonmetallic-sheathed, Type AC, or Type UF cables, or (2) all of the network-powered broadband communications cables are encased in raceway.

Exception No. 2: Where the network-powered broadband communications cables are permanently separated from the conductors of electric light, power, Class 1, and non–power-limited fire alarm circuits by a continuous and firmly fixed nonconductor, such as porcelain tubes or flexible tubing, in addition to the insulation on the wire.

(B) **Spread of Fire or Products of Combustion.** Installations in hollow spaces, vertical shafts, and ventilation or air-handling ducts shall be made so that the possible spread of fire or products of combustion will not be substantially increased. Openings around penetrations through fire-resistance–rated walls, partitions, floors, or ceilings shall be firestopped using approved methods to maintain the fire resistance rating.

(C) **Equipment in Other Space Used for Environmental Air.** Section 300.22(C) shall apply.

(D) **Support of Conductors.** Raceways shall be used for their intended purpose. Network-powered broadband communications cables shall not be strapped, taped, or attached by any means to the exterior of any conduit or raceway as a means of support.

See the commentary following 300.11(B)(3).

(E) **Cable Substitutions.** The substitutions for network-powered broadband cables listed in Table 830.58 shall be permitted. All cables in Table 830.58, other than network-powered broadband cables, shall be coaxial cables.

Table 830.58 Cable Substitutions

Cable Type	Permitted Cable Substitutions
BM	BMR
BLP	CMP, CL3P
BLX	CMP, CL3P, CMR, CL3R, CMG, CM, CL3, CMX, CL3X, BMR, BM, BLP

Because Chapter 3 of the *Code* was reorganized for the 2002 edition, Table 4 in Chapter 9 was modified to include the new article numbers so the user can more readily locate the appropriate article. In addition, metric designations for standard trade sizes were added. The trade size designations were also changed. Now, the dimension units for trade size designations have been deleted. For example, ½-in. conduit is now designated as trade size ½ conduit. For further information on metrication, see the commentary following 90.9.

Because conduits and tubing from different manufacturers have different internal diameters for the same trade size, the tables in Annex C provide the diameter and the actual area of different raceways for 31, 40, 53, and 100 percent fill. In addition, the format of the tables for dimensions of conductors has been changed to provide greater accuracy.

Annex C contains 12 raceway conductor fill tables, which were expanded to provide specific and accurate information about raceway conductor fill. There are separate tables for both metallic- and nonmetallic-type conduit and tubing raceways and for flexible raceways. Examples of how to use the tables have been included both here and in Annex C.

Table 1 Percent of Cross Section of Conduit and Tubing for Conductors

Number of Conductors	All Conductor Types
1	53
2	31
Over 2	40

FPN No. 1: Table 1 is based on common conditions of proper cabling and alignment of conductors where the length of the pull and the number of bends are within reasonable limits. It should be recognized that, for certain conditions, a larger size conduit or a lesser conduit fill should be considered.

Table 1 sets forth the absolute maximum fill permitted for conduit or tubing. Fine Print Note No. 1 is a warning that, in some cases, underfilling or oversizing of the conduit may be necessary.

FPN No. 2: When pulling three conductors or cables into a raceway, if the ratio of the raceway (inside diameter) to the conductor or cable (outside diameter) is between 2.8 and 3.2, jamming can occur. While jamming can occur when pulling four or more conductors or cables into a raceway, the probability is very low.

Fine Print Note No. 2 alerts the user that conductor jamming may occur during the installation (pulling) of conductors into a conduit even if fill allowances of 40 percent are

observed. During the installation of three conductors or cables into the raceway, it is possible that one conductor may slip between the other two conductors. This is more likely to take place at bends, where the raceway may be slightly oval.

Annex C, Table C1, permits three 8 AWG conductors in trade size ½ EMT. Using this example, ½-in. EMT has an internal diameter (ID) of 0.622 in. (from Table 4), and an 8 AWG conductor has an outside diameter (OD) of 0.216 in. (from Table 5).

While the EMT in a straight run has an internal diameter of 0.622 in., it may not be round at a bend, where one conductor may slip between the other two and cause a jam as the conductors exit the bend. In a straight run, assuming no variation in the EMT's internal diameter or a conductor's outside diameter, one conductor usually cannot slip between the other two, because the total of the outside diameters of the conductors (3×0.216 in. $= 0.648$ in.) is greater than the EMT's internal diameter of 0.622 in. However, at a bend, the major internal diameter of the raceway may increase due to bending, particularly in tubing, to a diameter slightly larger than 0.648 in., permitting the middle conductor to be pulled between the outer two conductors. As the conductors exit the bend and the raceway's internal diameter returns to a normal circle of 0.622 in., the conductors may jam. This can also occur in straight runs where the ratio of the raceway's internal diameter to the conductor's outside diameter approaches 3. The jam ratio is calculated as follows:

$$\text{Jam ratio} = \frac{\text{ID of raceway}}{\text{OD of conductor}} = \frac{0.622}{0.216} = 2.88$$

Industry representatives recommend avoiding a jam ratio of 2.8 to 3.2.

Notes to Tables

(1) See Annex C for the maximum number of conductors and fixture wires, all of the same size (total cross-sectional area including insulation) permitted in trade sizes of the applicable conduit or tubing.

The percentage fill allowed for conduit and tubing was not changed for the 1999 *Code*, and it has not been changed for the 2002 *Code*. Annex C is a useful tool for determining the correct size of conduit or tubing based on the size and number of conductors to be installed in the raceway. See the commentary in Annex C for examples of how to use the tables.

(2) Table 1 applies only to complete conduit or tubing systems and is not intended to apply to sections of conduit or tubing used to protect exposed wiring from physical damage.

The maximum fill requirements do not apply to short sections of conduit or tubing used for the physical protection of conductors and cables. Cables are commonly protected from physical damage by conduit or tubing sleeves sized to enable the cable to be passed through with relative ease without injuring or abrading the protective jacket of the cable. However, a fitting is required on the end(s) of the conduit or tubing to protect the conductors or cables from abrasion. [See 300.15(C).]

(3) Equipment grounding or bonding conductors, where installed, shall be included when calculating conduit or tubing fill. The actual dimensions of the equipment grounding or bonding conductor (insulated or bare) shall be used in the calculation.

All conductors, whether insulated, covered, or bare, occupy space within a raceway. Therefore, all conductors, including equipment grounding conductors, bonding conductors, and grounded conductors, must be counted when the percent fill of a conduit or tubing is being calculated. The only exception to this rule is the addition of an uninsulated grounding conductor permitted within trade size ⅜ flexible metal conduit (see footnote to Table 348.22). The dimensions of bare conductors are given in Table 8.

(4) Where conduit or tubing nipples having a maximum length not to exceed 600 mm (24 in.) are installed between boxes, cabinets, and similar enclosures, the nipples shall be permitted to be filled to 60 percent of their total cross-sectional area, and 310.15(B)(2)(a) adjustment factors need not apply to this condition.
(5) For conductors not included in Chapter 9, such as multiconductor cables, the actual dimensions shall be used.

For conductors not included in Chapter 9, such as high-voltage types, the cross-sectional area can be calculated in the following manner, using the actual dimensions of each conductor:

$$\text{cross-sectional area} = d^2 \text{ cmil}$$

where:

d = outside diameter of a conductor (including insulation) [1 in. = 1000 mil (1 mil = 0.001 in.)]

cmil = circular mil, a unit measure of area equal to $\pi/4$ (3.1416/4 = 0.7854) square mil

Example

Three 15-kV single conductors are to be installed in conduit. The outside diameter of each conductor measures 1⅝ in.

(1⅝ in. = 1.625 in.). What size rigid metal conduit will accommodate the three conductors?

Solution

STEP 1. Find the area within the cross section of the conduit that will be displaced by the three conductors:

$$1.625 \text{ in.} \times 1.625 \text{ in.} \times 0.7854 \times 3 = 6.2218 \text{ in.}^2$$
$$\text{or } 6.222 \text{ in.}^2$$

STEP 2. Determine the size conduit that will accommodate the three conductors. Table 1 allows 40 percent conduit fill for three or more conductors, and Table 4 indicates that 40 percent of trade size 5 rigid metal conduit is 8.085 in.² Thus, trade size 5 rigid metal conduit will accommodate three 15-kV single conductors.

(6) For combinations of conductors of different sizes, use Table 5 and Table 5A for dimensions of conductors and Table 4 for the applicable conduit or tubing dimensions.

The following two examples calculate the minimum trade size conduit or tubing required for mixed conductor sizes.

Example

A 200-ampere feeder is routed in various wiring methods [EMT, PVC (Schedule 40), and rigid metal conduit] from the main switchboard in one building to a second building. The circuit consists of four 4/0 AWG XHHW conductors and one 6 AWG XHHW. Select the proper trade size for the various types of conduit and tubing to be used for this feeder.

Solution

All the raceways for this example require conduit fill to be calculated according to Table 1 of Chapter 9. (See 344.22 for RMC, 352.22 for PVC, and 358.22 for EMT.) Table 1 of Chapter 9 permits a conduit fill to a maximum of 40 percent for over two conductors in a raceway. Note 6 directs the user to Table 5 for the area required for each insulated conductor. Multiplying the quantity of conductors by the area of each conductor results in the total conductor-occupied space within the conduit. Then Note 6 directs the user to Table 4 to select the appropriate trade size conduits or tubing. Table 4 contains the applicable cross-sectional area of conduit and tubing based on conductor-occupied space (40 percent maximum in this example).

STEP 1. Calculate the total area occupied by the wires, using the approximate areas listed in Table 5:

For four 4/0 AWG XHHW:

$$4 \times 0.3197 \text{ in.}^2 = 1.2788 \text{ in.}^2$$

For one 6 AWG XHHW:

$$1 \times 0.0590 \text{ in.}^2 = 0.0590 \text{ in.}^2$$

Total area   1.3378 in.^{2} or   1.338 in.^{2}

STEP 2. Determine the proper trade size EMT, rigid metal conduit, and PVC (Schedule 40) from Table 4. The EMT portion of this feeder will require a minimum trade size 2 EMT. Trade size 2 EMT has 1.342 in.^{2} of available space for over two conductors, and the minimum required space is 1.338 in.^{2}, which is within the trade size 2 EMT limit. Rigid metal conduit also will require a minimum trade size 2, since trade size 2 RMC has 1.363 in.^{2} of available space for over two conductors. PVC (Schedule 40), however, will require a minimum trade size 2½ conduit. Trade size 2 PVC has only 1.316 in.^{2} available space for over two conductors, and the circuit under consideration requires at least 1.338 in.^{2}

Example

Determine the minimum size rigid metal conduit allowed for the following 10 mixed conductor sizes and types:

Quantity	Wire Size and Type	Cross-Sectional Area of Each (from Table 5)	Subtotal Cross-Sectional Area
4	12 THWN	0.0133	0.0532
3	8 TW	0.0437	0.1311
3	6 THW	0.0726	0.2178
		Total cross-sectional area	0.4021

Solution

The "Over 2 Wires" column in Table 4 indicates that 40 percent of a trade size 1¼ rigid metal conduit is 0.610 in.^{2} Therefore, trade size 1¼ is the minimum permitted conduit size for these 10 conductors.

(7) When calculating the maximum number of conductors permitted in a conduit or tubing, all of the same size (total cross-sectional area including insulation), the next higher whole number shall be used to determine the maximum number of conductors permitted when the calculation results in a decimal of 0.8 or larger.

Example

Determine how many 10 AWG THHN conductors are permitted in a trade size 1¼ rigid metal conduit.

Solution

Table 1 permits a 40 percent fill for over two conductors. From Table 4, 40 percent fill for a trade size 1¼ rigid metal conduit is 0.610 in., and, from Table 5, the cross-sectional area of a 10 AWG THHN conductor is 0.0211 in.^{2} The number of conductors permitted is calculated as follows:

$$\frac{0.610 \text{ in.}^2}{0.0211 \text{ in.}^2 \text{ per conductor}} = 28.910 \text{ conductors}$$

Without Note 7, the maximum quantity of conductors could not exceed 28. However, 29 conductors are permitted in trade size 1¼ rigid metal conduit. This is because the decimal 0.910 is more than 0.8, and the quantity of conductors may be increased to the next higher whole number. Annex C, Table C8, verifies that twenty-nine 10 AWG THWN conductors are permitted in a trade size 1¼ rigid metal conduit.

If the decimal portion is less than 0.8, it is dropped.

(8) Where bare conductors are permitted by other sections of this *Code*, the dimensions for bare conductors in Table 8 shall be permitted.

(9) A multiconductor cable of two or more conductors shall be treated as a single conductor for calculating percentage conduit fill area. For cables that have elliptical cross sections, the cross-sectional area calculation shall be based on using the major diameter of the ellipse as a circle diameter.

Table 4 Dimensions and Percent Area of Conduit and Tubing (Areas of Conduit or Tubing for the Combinations of Wires Permitted in Table 1, Chapter 9)

Article 358 — Electrical Metallic Tubing (EMT)

Metric Designator	Trade Size	Nominal Internal Diameter mm	Nominal Internal Diameter in.	Total Area 100% mm²	Total Area 100% in.²	2 Wires 31% mm²	2 Wires 31% in.²	Over 2 Wires 40% mm²	Over 2 Wires 40% in.²	1 Wire 53% mm²	1 Wire 53% in.²	60% mm²	60% in.²
16	½	15.8	0.622	196	0.304	61	0.094	78	0.122	104	0.161	118	0.182
21	¾	20.9	0.824	343	0.533	106	0.165	137	0.213	182	0.283	206	0.320
27	1	26.6	1.049	556	0.864	172	0.268	222	0.346	295	0.458	333	0.519
35	1¼	35.1	1.380	968	1.496	300	0.464	387	0.598	513	0.793	581	0.897
41	1½	40.9	1.610	1314	2.036	407	0.631	526	0.814	696	1.079	788	1.221
53	2	52.5	2.067	2165	3.356	671	1.040	866	1.342	1147	1.778	1299	2.013
63	2½	69.4	2.731	3783	5.858	1173	1.816	1513	2.343	2005	3.105	2270	3.515
78	3	85.2	3.356	5701	8.846	1767	2.742	2280	3.538	3022	4.688	3421	5.307
91	3½	97.4	3.834	7451	11.545	2310	3.579	2980	4.618	3949	6.119	4471	6.927
103	4	110.1	4.334	9521	14.753	2951	4.573	3808	5.901	5046	7.819	5712	8.852

Article 362 — Electrical Nonmetallic Tubing (ENT)

Metric Designator	Trade Size	Nominal Internal Diameter mm	Nominal Internal Diameter in.	Total Area 100% mm²	Total Area 100% in.²	2 Wires 31% mm²	2 Wires 31% in.²	Over 2 Wires 40% mm²	Over 2 Wires 40% in.²	1 Wire 53% mm²	1 Wire 53% in.²	60% mm²	60% in.²
16	½	14.2	0.560	158	0.246	49	0.076	63	0.099	84	0.131	95	0.148
21	¾	19.3	0.760	293	0.454	91	0.141	117	0.181	155	0.240	176	0.272
27	1	25.4	1.000	507	0.785	157	0.243	203	0.314	269	0.416	304	0.471
35	1¼	34.0	1.340	908	1.410	281	0.437	363	0.564	481	0.747	545	0.846
41	1½	39.9	1.570	1250	1.936	388	0.600	500	0.774	663	1.026	750	1.162
53	2	51.3	2.020	2067	3.205	641	0.993	827	1.282	1095	1.699	1240	1.923
63	2½	—	—	—	—	—	—	—	—	—	—	—	—
78	3	—	—	—	—	—	—	—	—	—	—	—	—
91	3½	—	—	—	—	—	—	—	—	—	—	—	—

Article 348 — Flexible Metal Conduit (FMC)

Metric Designator	Trade Size	Nominal Internal Diameter mm	Nominal Internal Diameter in.	Total Area 100% mm²	Total Area 100% in.²	2 Wires 31% mm²	2 Wires 31% in.²	Over 2 Wires 40% mm²	Over 2 Wires 40% in.²	1 Wire 53% mm²	1 Wire 53% in.²	60% mm²	60% in.²
12	⅜	9.7	0.384	74	0.116	23	0.036	30	0.046	39	0.061	44	0.069
16	½	16.1	0.635	204	0.317	63	0.098	81	0.127	108	0.168	122	0.190
21	¾	20.9	0.824	343	0.533	106	0.165	137	0.213	182	0.283	206	0.320
27	1	25.9	1.020	527	0.817	163	0.253	211	0.327	279	0.433	316	0.490
35	1¼	32.4	1.275	824	1.277	256	0.396	330	0.511	437	0.677	495	0.766
41	1½	39.1	1.538	1201	1.858	372	0.576	480	0.743	636	0.985	720	1.115
53	2	51.8	2.040	2107	3.269	653	1.013	843	1.307	1117	1.732	1264	1.961
63	2½	63.5	2.500	3167	4.909	982	1.522	1267	1.963	1678	2.602	1900	2.945
78	3	76.2	3.000	4560	7.069	1414	2.191	1824	2.827	2417	3.746	2736	4.241
91	3½	88.9	3.500	6207	9.621	1924	2.983	2483	3.848	3290	5.099	3724	5.773
103	4	101.6	4.000	8107	12.566	2513	3.896	3243	5.027	4297	6.660	4864	7.540

Table 4 *Continued*

Article 342 — Intermediate Metal Conduit (IMC)

Metric Designator	Trade Size	Nominal Internal Diameter mm	Nominal Internal Diameter in.	Total Area 100% mm²	Total Area 100% in.²	2 Wires 31% mm²	2 Wires 31% in.²	Over 2 Wires 40% mm²	Over 2 Wires 40% in.²	1 Wire 53% mm²	1 Wire 53% in.²	60% mm²	60% in.²
12	⅜	—	—	—	—	—	—	—	—	—	—	—	—
16	½	16.8	0.660	222	0.342	69	0.106	89	0.137	117	0.181	133	0.205
21	¾	21.9	0.864	377	0.586	117	0.182	151	0.235	200	0.311	226	0.352
27	1	28.1	1.105	620	0.959	192	0.297	248	0.384	329	0.508	372	0.575
35	1¼	36.8	1.448	1064	1.647	330	0.510	425	0.659	564	0.873	638	0.988
41	1½	42.7	1.683	1432	2.225	444	0.690	573	0.890	759	1.179	859	1.335
53	2	54.6	2.150	2341	3.630	726	1.125	937	1.452	1241	1.924	1405	2.178
63	2½	64.9	2.557	3308	5.135	1026	1.592	1323	2.054	1753	2.722	1985	3.081
78	3	80.7	3.176	5115	7.922	1586	2.456	2046	3.169	2711	4.199	3069	4.753
91	3½	93.2	3.671	6822	10.584	2115	3.281	2729	4.234	3616	5.610	4093	6.351
103	4	105.4	4.166	8725	13.631	2705	4.226	3490	5.452	4624	7.224	5235	8.179

Article 356— Liquidtight Flexible Nonmetallic Conduit (LFNC-B*)

Metric Designator	Trade Size	Nominal Internal Diameter mm	Nominal Internal Diameter in.	Total Area 100% mm²	Total Area 100% in.²	2 Wires 31% mm²	2 Wires 31% in.²	Over 2 Wires 40% mm²	Over 2 Wires 40% in.²	1 Wire 53% mm²	1 Wire 53% in.²	60% mm²	60% in.²
12	⅜	12.5	0.494	123	0.192	38	0.059	49	0.077	65	0.102	74	0.115
16	½	16.1	0.632	204	0.314	63	0.097	81	0.125	108	0.166	122	0.188
21	¾	21.1	0.830	350	0.541	108	0.168	140	0.216	185	0.287	210	0.325
27	1	26.8	1.054	564	0.873	175	0.270	226	0.349	299	0.462	338	0.524
35	1¼	35.4	1.395	984	1.528	305	0.474	394	0.611	522	0.810	591	0.917
41	1½	40.3	1.588	1276	1.981	395	0.614	510	0.792	676	1.050	765	1.188
53	2	51.6	2.033	2091	3.246	648	1.006	836	1.298	1108	1.720	1255	1.948

*Corresponds to 356.2(2)

Article 356 — Liquidtight Flexible Nonmetallic Conduit (LFNC-A*)

Metric Designator	Trade Size	Nominal Internal Diameter mm	Nominal Internal Diameter in.	Total Area 100% mm²	Total Area 100% in.²	2 Wires 31% mm²	2 Wires 31% in.²	Over 2 Wires 40% mm²	Over 2 Wires 40% in.²	1 Wire 53% mm²	1 Wire 53% in.²	60% mm²	60% in.²
12	⅜	12.6	0.495	125	0.192	39	0.060	50	0.077	66	0.102	75	0.115
16	½	16.0	0.630	201	0.312	62	0.097	80	0.125	107	0.165	121	0.187
21	¾	21.0	0.825	346	0.535	107	0.166	139	0.214	184	0.283	208	0.321
27	1	26.5	1.043	552	0.854	171	0.265	221	0.342	292	0.453	331	0.513
35	1¼	35.1	1.383	968	1.502	300	0.466	387	0.601	513	0.796	581	0.901
41	1½	40.7	1.603	1301	2.018	403	0.626	520	0.807	690	1.070	781	1.211
53	2	52.4	2.063	2157	3.343	669	1.036	863	1.337	1143	1.772	1294	2.006

*Corresponds to 356.2(1)

Table 4 *Continued*

Article 350 — Liquidtight Flexible Metal Conduit (LFMC)

Metric Designator	Trade Size	Nominal Internal Diameter		Total Area 100%		2 Wires 31%		Over 2 Wires 40%		1 Wire 53%		60%	
		mm	in.	mm²	in.²	mm²	in.²	mm²	in.²	mm²	in.²	mm²	in.²
12	3/8	12.5	0.494	123	0.192	38	0.059	49	0.077	65	0.102	74	0.115
16	1/2	16.1	0.632	204	0.314	63	0.097	81	0.125	108	0.166	122	0.188
21	3/4	21.1	0.830	350	0.541	108	0.168	140	0.216	185	0.287	210	0.325
27	1	26.8	1.054	564	0.873	175	0.270	226	0.349	299	0.462	338	0.524
35	1 1/4	35.4	1.395	984	1.528	305	0.474	394	0.611	522	0.810	591	0.917
41	1 1/2	40.3	1.588	1276	1.981	395	0.614	510	0.792	676	1.050	765	1.188
53	2	51.6	2.033	2091	3.246	648	1.006	836	1.298	1108	1.720	1255	1.948
63	2 1/2	63.3	2.493	3147	4.881	976	1.513	1259	1.953	1668	2.587	1888	2.929
78	3	78.4	3.085	4827	7.475	1497	2.317	1931	2.990	2559	3.962	2896	4.485
91	3 1/2	89.4	3.520	6277	9.731	1946	3.017	2511	3.893	3327	5.158	3766	5.839
103	4	102.1	4.020	8187	12.692	2538	3.935	3275	5.077	4339	6.727	4912	7.615
129	5	—	—	—	—	—	—	—	—	—	—	—	—
155	6	—	—	—	—	—	—	—	—	—	—	—	—

Article 344 — Rigid Metal Conduit (RMC)

Metric Designator	Trade Size	Nominal Internal Diameter		Total Area 100%		2 Wires 31%		Over 2 Wires 40%		1 Wire 53%		60%	
		mm	in.	mm²	in.²	mm²	in.²	mm²	in.²	mm²	in.²	mm²	in.²
12	3/8	—	—	—	—	—	—	—	—	—	—	—	—
16	1/2	16.1	0.632	204	0.314	63	0.097	81	0.125	108	0.166	122	0.188
21	3/4	21.2	0.836	353	0.549	109	0.170	141	0.220	187	0.291	212	0.329
27	1	27.0	1.063	573	0.887	177	0.275	229	0.355	303	0.470	344	0.532
35	1 1/4	35.4	1.394	984	1.526	305	0.473	394	0.610	522	0.809	591	0.916
41	1 1/2	41.2	1.624	1333	2.071	413	0.642	533	0.829	707	1.098	800	1.243
53	2	52.9	2.083	2198	3.408	681	1.056	879	1.363	1165	1.806	1319	2.045
63	2 1/2	63.2	2.489	3137	4.866	972	1.508	1255	1.946	1663	2.579	1882	2.919
78	3	78.5	3.090	4840	7.499	1500	2.325	1936	3.000	2565	3.974	2904	4.499
91	3 1/2	90.7	3.570	6461	10.010	2003	3.103	2584	4.004	3424	5.305	3877	6.006
103	4	102.9	4.050	8316	12.882	2578	3.994	3326	5.153	4408	6.828	4990	7.729
129	5	128.9	5.073	13050	20.212	4045	6.266	5220	8.085	6916	10.713	7830	12.127
155	6	154.8	6.093	18821	29.158	5834	9.039	7528	11.663	9975	15.454	11292	17.495

Article 352 — Rigid PVC Conduit (RNC), Schedule 80

Metric Designator	Trade Size	Nominal Internal Diameter		Total Area 100%		2 Wires 31%		Over 2 Wires 40%		1 Wire 53%		60%	
		mm	in.	mm²	in.²	mm²	in.²	mm²	in.²	mm²	in.²	mm²	in.²
12	3/8	—	—	—	—	—	—	—	—	—	—	—	—
16	1/2	13.4	0.526	141	0.217	44	0.067	56	0.087	75	0.115	85	0.130
21	3/4	18.3	0.722	263	0.409	82	0.127	105	0.164	139	0.217	158	0.246
27	1	23.8	0.936	445	0.688	138	0.213	178	0.275	236	0.365	267	0.413
35	1 1/4	31.9	1.255	799	1.237	248	0.383	320	0.495	424	0.656	480	0.742
41	1 1/2	37.5	1.476	1104	1.711	342	0.530	442	0.684	585	0.907	663	1.027
53	2	48.6	1.913	1855	2.874	575	0.891	742	1.150	983	1.523	1113	1.725
63	2 1/2	58.2	2.290	2660	4.119	825	1.277	1064	1.647	1410	2.183	1596	2.471
78	3	72.7	2.864	4151	6.442	1287	1.997	1660	2.577	2200	3.414	2491	3.865
91	3 1/2	84.5	3.326	5608	8.688	1738	2.693	2243	3.475	2972	4.605	3365	5.213
103	4	96.2	3.786	7268	11.258	2253	3.490	2907	4.503	3852	5.967	4361	6.755
129	5	121.1	4.768	11518	17.855	3571	5.535	4607	7.142	6105	9.463	6911	10.713
155	6	145.0	5.709	16513	25.598	5119	7.935	6605	10.239	8752	13.567	9908	15.359

Table 4 *Continued*

Article 352 — Rigid PVC Conduit (RNC), Schedule 40, and HDPE Conduit

Metric Designator	Trade Size	Nominal Internal Diameter		Total Area 100%		2 Wires 31%		Over 2 Wires 40%		1 Wire 53%		60%	
		mm	in.	mm²	in.²	mm²	in.²	mm²	in.²	mm²	in.²	mm²	in.²
12	⅜	—	—	—	—	—	—	—	—	—	—	—	—
16	½	15.3	0.602	184	0.285	57	0.088	74	0.114	97	0.151	110	0.171
21	¾	20.4	0.804	327	0.508	101	0.157	131	0.203	173	0.269	196	0.305
27	1	26.1	1.029	535	0.832	166	0.258	214	0.333	284	0.441	321	0.499
35	1¼	34.5	1.360	935	1.453	290	0.450	374	0.581	495	0.770	561	0.872
41	1½	40.4	1.590	1282	1.986	397	0.616	513	0.794	679	1.052	769	1.191
53	2	52.0	2.047	2124	3.291	658	1.020	849	1.316	1126	1.744	1274	1.975
63	2½	62.1	2.445	3029	4.695	939	1.455	1212	1.878	1605	2.488	1817	2.817
78	3	77.3	3.042	4693	7.268	1455	2.253	1877	2.907	2487	3.852	2816	4.361
91	3½	89.4	3.521	6277	9.737	1946	3.018	2511	3.895	3327	5.161	3766	5.842
103	4	101.5	3.998	8091	12.554	2508	3.892	3237	5.022	4288	6.654	4855	7.532
129	5	127.4	5.016	12748	19.761	3952	6.126	5099	7.904	6756	10.473	7649	11.856
155	6	153.2	6.031	18433	28.567	5714	8.856	7373	11.427	9770	15.141	11060	17.140

Article 352 — Type A, Rigid PVC Conduit (RNC)

Metric Designator	Trade Size	Nominal Internal Diameter		Total Area 100%		2 Wires 31%		Over 2 Wires 40%		1 Wire 53%		60%	
		mm	in.	mm²	in.²	mm²	in.²	mm²	in.²	mm²	in.²	mm²	in.²
16	½	17.8	0.700	249	0.385	77	0.119	100	0.154	132	0.204	149	0.231
21	¾	23.1	0.910	419	0.650	130	0.202	168	0.260	222	0.345	251	0.390
27	1	29.8	1.175	697	1.084	216	0.336	279	0.434	370	0.575	418	0.651
35	1¼	38.1	1.500	1140	1.767	353	0.548	456	0.707	604	0.937	684	1.060
41	1½	43.7	1.720	1500	2.324	465	0.720	600	0.929	795	1.231	900	1.394
53	2	54.7	2.155	2350	3.647	728	1.131	940	1.459	1245	1.933	1410	2.188
63	2½	66.9	2.635	3515	5.453	1090	1.690	1406	2.181	1863	2.890	2109	3.272
78	3	82.0	3.230	5281	8.194	1637	2.540	2112	3.278	2799	4.343	3169	4.916
91	3½	93.7	3.690	6896	10.694	2138	3.315	2758	4.278	3655	5.668	4137	6.416
103	4	106.2	4.180	8858	13.723	2746	4.254	3543	5.489	4695	7.273	5315	8.234
129	5	—	—	—	—	—	—	—	—	—	—	—	—
155	6	—	—	—	—	—	—	—	—	—	—	—	—

Article 352 — Type EB, PVC Conduit (RNC)

Metric Designator	Trade Size	Nominal Internal Diameter		Total Area 100%		2 Wires 31%		Over 2 Wires 40%		1 Wire 53%		60%	
		mm	in.	mm²	in.²	mm²	in.²	mm²	in.²	mm²	in.²	mm²	in.²
16	½	—	—	—	—	—	—	—	—	—	—	—	—
21	¾	—	—	—	—	—	—	—	—	—	—	—	—
27	1	—	—	—	—	—	—	—	—	—	—	—	—
35	1¼	—	—	—	—	—	—	—	—	—	—	—	—
41	1½	—	—	—	—	—	—	—	—	—	—	—	—
53	2	56.4	2.221	2498	3.874	774	1.201	999	1.550	1324	2.053	1499	2.325
63	2½	—	—	—	—	—	—	—	—	—	—	—	—
78	3	84.6	3.330	5621	8.709	1743	2.700	2248	3.484	2979	4.616	3373	5.226
91	3½	96.6	3.804	7329	11.365	2272	3.523	2932	4.546	3884	6.023	4397	6.819
103	4	108.9	4.289	9314	14.448	2887	4.479	3726	5.779	4937	7.657	5589	8.669
129	5	135.0	5.316	14314	22.195	4437	6.881	5726	8.878	7586	11.763	8588	13.317
155	6	160.9	6.336	20333	31.530	6303	9.774	8133	12.612	10776	16.711	12200	18.918

Table 5 Dimensions of Insulated Conductors and Fixture Wires

Type	Size (AWG or kcmil)	Approximate Diameter mm	Approximate Diameter in.	Approximate Area mm²	Approximate Area in.²
Type: FFH-2, RFH-1, RFH-2, RHH*, RHW*, RHW-2*, RHH, RHW, RHW-2, SF-1, SF-2, SFF-1, SFF-2, TF, TFF, THHW, THW, THW-2, TW, XF, XFF					
RFH-2, FFH-2	18	3.454	0.136	9.355	0.0145
	16	3.759	0.148	11.10	0.0172
RHW-2, RHH, RHW	14	4.902	0.193	18.90	0.0293
	12	5.385	0.212	22.77	0.0353
	10	5.994	0.236	28.19	0.0437
	8	8.280	0.326	53.87	0.0835
	6	9.246	0.364	67.16	0.1041
	4	10.46	0.412	86.00	0.1333
	3	11.18	0.440	98.13	0.1521
	2	11.99	0.472	112.9	0.1750
	1	14.78	0.582	171.6	0.2660
	1/0	15.80	0.622	196.1	0.3039
	2/0	16.97	0.668	226.1	0.3505
	3/0	18.29	0.720	262.7	0.4072
	4/0	19.76	0.778	306.7	0.4754
	250	22.73	0.895	405.9	0.6291
	300	24.13	0.950	457.3	0.7088
	350	25.43	1.001	507.7	0.7870
	400	26.62	1.048	556.5	0.8626
	500	28.78	1.133	650.5	1.0082
	600	31.57	1.243	782.9	1.2135
	700	33.38	1.314	874.9	1.3561
	750	34.24	1.348	920.8	1.4272
	800	35.05	1.380	965.0	1.4957
	900	36.68	1.444	1057	1.6377
	1000	38.15	1.502	1143	1.7719
	1250	43.92	1.729	1515	2.3479
	1500	47.04	1.852	1738	2.6938
	1750	49.94	1.966	1959	3.0357
	2000	52.63	2.072	2175	3.3719
SF-2, SFF-2	18	3.073	0.121	7.419	0.0115
	16	3.378	0.133	8.968	0.0139
	14	3.759	0.148	11.10	0.0172
SF-1, SFF-1	18	2.311	0.091	4.194	0.0065
RFH-1, XF, XFF	18	2.692	0.106	5.161	0.0080
TF, TFF, XF, XFF	16	2.997	0.118	7.032	0.0109
TW, XF, XFF, THHW, THW, THW-2	14	3.378	0.133	8.968	0.0139
TW, THHW, THW, THW-2	12	3.861	0.152	11.68	0.0181
	10	4.470	0.176	15.68	0.0243
	8	5.994	0.236	28.19	0.0437
RHH*, RHW*, RHW-2*	14	4.140	0.163	13.48	0.0209
RHH*, RHW*, RHW-2*, XF, XFF	12	4.623	0.182	16.77	0.0260

Table 5 *Continued*

Type	Size (AWG or kcmil)	Approximate Diameter mm	Approximate Diameter in.	Approximate Area mm²	Approximate Area in.²
Type: RHH*, RHW*, RHW-2*, THHN, THHW, THW, THW-2, TFN, TFFN, THWN, THWN-2, XF, XFF					
RHH*, RHW*, RHW-2*, XF, XFF	10	5.232	0.206	21.48	0.0333
RHH*, RHW*, RHW-2*	8	6.756	0.266	35.87	0.0556
TW, THW, THHW, THW-2, RHH*, RHW*, RHW-2*	6	7.722	0.304	46.84	0.0726
	4	8.941	0.352	62.77	0.0973
	3	9.652	0.380	73.16	0.1134
	2	10.46	0.412	86.00	0.1333
	1	12.50	0.492	122.6	0.1901
	1/0	13.51	0.532	143.4	0.2223
	2/0	14.68	0.578	169.3	0.2624
	3/0	16.00	0.630	201.1	0.3117
	4/0	17.48	0.688	239.9	0.3718
	250	19.43	0.765	296.5	0.4596
	300	20.83	0.820	340.7	0.5281
	350	22.12	0.871	384.4	0.5958
	400	23.32	0.918	427.0	0.6619
	500	25.48	1.003	509.7	0.7901
	600	28.27	1.113	627.7	0.9729
	700	30.07	1.184	710.3	1.1010
	750	30.94	1.218	751.7	1.1652
	800	31.75	1.250	791.7	1.2272
	900	33.38	1.314	874.9	1.3561
	1000	34.85	1.372	953.8	1.4784
	1250	39.09	1.539	1200	1.8602
	1500	42.21	1.662	1400	2.1695
	1750	45.11	1.776	1598	2.4773
	2000	47.80	1.882	1795	2.7818
TFN, TFFN	18	2.134	0.084	3.548	0.0055
	16	2.438	0.096	4.645	0.0072
THHN, THWN, THWN-2	14	2.819	0.111	6.258	0.0097
	12	3.302	0.130	8.581	0.0133
	10	4.166	0.164	13.61	0.0211
	8	5.486	0.216	23.61	0.0366
	6	6.452	0.254	32.71	0.0507
	4	8.230	0.324	53.16	0.0824
	3	8.941	0.352	62.77	0.0973
	2	9.754	0.384	74.71	0.1158
	1	11.33	0.446	100.8	0.1562
	1/0	12.34	0.486	119.7	0.1855
	2/0	13.51	0.532	143.4	0.2223
	3/0	14.83	0.584	172.8	0.2679
	4/0	16.31	0.642	208.8	0.3237
	250	18.06	0.711	256.1	0.3970
	300	19.46	0.766	297.3	0.4608

Table 5 *Continued*

Type	Size (AWG or kcmil)	Approximate Diameter		Approximate Area	
		mm	in.	mm²	in.²
Type: FEP, FEPB, PAF, PAFF, PF, PFA, PFAH, PFF, PGF, PGFF, PTF, PTFF, TFE, THHN, THWN, THWN-2, Z, ZF, ZFF					
THHN,	350	20.75	0.817	338.2	0.5242
THWN,	400	21.95	0.864	378.3	0.5863
THWN-2	500	24.10	0.949	456.3	0.7073
	600	26.70	1.051	559.7	0.8676
	700	28.50	1.122	637.9	0.9887
	750	29.36	1.156	677.2	1.0496
	800	30.18	1.188	715.2	1.1085
	900	31.80	1.252	794.3	1.2311
	1000	33.27	1.310	869.5	1.3478
PF, PGFF,	18	2.184	0.086	3.742	0.0058
PGF, PFF, PTF,	16	2.489	0.098	4.839	0.0075
PAF, PTFF,					
PAFF					
PF, PGFF,	14	2.870	0.113	6.452	0.0100
PGF, PFF, PTF,					
PAF, PTFF,					
PAFF, TFE,					
FEP, PFA,					
FEPB, PFAH					
TFE, FEP,	12	3.353	0.132	8.839	0.0137
PFA, FEPB,	10	3.962	0.156	12.32	0.0191
PFAH	8	5.232	0.206	21.48	0.0333
	6	6.198	0.244	30.19	0.0468
	4	7.417	0.292	43.23	0.0670
	3	8.128	0.320	51.87	0.0804
	2	8.941	0.352	62.77	0.0973
TFE, PFAH	1	10.72	0.422	90.26	0.1399
TFE, PFA,	1/0	11.73	0.462	108.1	0.1676
PFAH, Z	2/0	12.90	0.508	130.8	0.2027
	3/0	14.22	0.560	158.9	0.2463
	4/0	15.70	0.618	193.5	0.3000
ZF, ZFF	18	1.930	0.076	2.903	0.0045
	16	2.235	0.088	3.935	0.0061
Z, ZF, ZFF	14	2.616	0.103	5.355	0.0083
Z	12	3.099	0.122	7.548	0.0117
	10	3.962	0.156	12.32	0.0191
	8	4.978	0.196	19.48	0.0302
	6	5.944	0.234	27.74	0.0430
	4	7.163	0.282	40.32	0.0625
	3	8.382	0.330	55.16	0.0855
	2	9.195	0.362	66.39	0.1029
	1	10.21	0.402	81.87	0.1269

Table 5 *Continued*

Type	Size (AWG or kcmil)	Approximate Diameter		Approximate Area	
		mm	in.	mm²	in.²
Type: KF-1, KF-2, KFF-1, KFF-2, XHH, XHHW, XHHW-2, ZW					
XHHW, ZW,	14	3.378	0.133	8.968	0.0139
XHHW-2,	12	3.861	0.152	11.68	0.0181
XHH	10	4.470	0.176	15.68	0.0243
	8	5.994	0.236	28.19	0.0437
	6	6.960	0.274	38.06	0.0590
	4	8.179	0.322	52.52	0.0814
	3	8.890	0.350	62.06	0.0962
	2	9.703	0.382	73.94	0.1146
XHHW,	1	11.23	0.442	98.97	0.1534
XHHW-2,					
XHH	1/0	12.24	0.482	117.7	0.1825
	2/0	13.41	0.528	141.3	0.2190
	3/0	14.73	0.58	170.5	0.2642
	4/0	16.21	0.638	206.3	0.3197
	250	17.91	0.705	251.9	0.3904
	300	19.30	0.76	292.6	0.4536
	350	20.60	0.811	333.3	0.5166
	400	21.79	0.858	373.0	0.5782
	500	23.95	0.943	450.6	0.6984
	600	26.75	1.053	561.9	0.8709
	700	28.55	1.124	640.2	0.9923
	750	29.41	1.158	679.5	1.0532
	800	30.23	1.190	717.5	1.1122
	900	31.85	1.254	796.8	1.2351
	1000	33.32	1.312	872.2	1.3519
	1250	37.57	1.479	1108	1.7180
	1500	40.69	1.602	1300	2.0157
	1750	43.59	1.716	1492	2.3127
	2000	46.28	1.822	1682	2.6073
KF-2, KFF-2	18	1.600	0.063	2.000	0.0031
	16	1.905	0.075	2.839	0.0044
	14	2.286	0.090	4.129	0.0064
	12	2.769	0.109	6.000	0.0093
	10	3.378	0.133	8.968	0.0139
KF-1, KFF-1	18	1.448	0.057	1.677	0.0026
	16	1.753	0.069	2.387	0.0037
	14	2.134	0.084	3.548	0.0055
	12	2.616	0.103	5.355	0.0083
	10	3.226	0.127	8.194	0.0127

*Types RHH, RHW, and RHW-2 without outer covering.

Table 5A Compact Aluminum Building Wire Nominal Dimensions* and Areas

Size (AWG or kcmil)	Bare Conductor			Types THW and THHW				Type THHN				Type XHHW				Size (AWG or kcmil)
	Number of Strands	Diameter		Approximate Diameter		Approximate Area		Approximate Diameter		Approximate Area		Approximate Diameter		Approximate Area		
		mm	in.	mm	in.	mm²	in.²	mm	in.	mm²	in.²	mm	in.	mm²	in.²	
8	7	3.404	0.134	6.477	0.255	32.90	0.0510	—	—	—	—	5.690	0.224	25.42	0.0394	8
6	7	4.293	0.169	7.366	0.290	42.58	0.0660	6.096	0.240	29.16	0.0452	6.604	0.260	34.19	0.0530	6
4	7	5.410	0.213	8.509	0.335	56.84	0.0881	7.747	0.305	47.10	0.0730	7.747	0.305	47.10	0.0730	4
2	7	6.807	0.268	9.906	0.390	77.03	0.1194	9.144	0.360	65.61	0.1017	9.144	0.360	65.61	0.1017	2
1	19	7.595	0.299	11.81	0.465	109.5	0.1698	10.54	0.415	87.23	0.1352	10.54	0.415	87.23	0.1352	1
1/0	19	8.534	0.336	12.70	0.500	126.6	0.1963	11.43	0.450	102.6	0.1590	11.43	0.450	102.6	0.1590	1/0
2/0	19	9.550	0.376	13.84	0.545	150.5	0.2332	12.57	0.495	124.1	0.1924	12.45	0.490	121.6	0.1885	2/0
3/0	19	10.74	0.423	14.99	0.590	176.3	0.2733	13.72	0.540	147.7	0.2290	13.72	0.540	147.7	0.2290	3/0
4/0	19	12.07	0.475	16.38	0.645	210.8	0.3267	15.11	0.595	179.4	0.2780	14.99	0.590	176.3	0.2733	4/0
250	37	13.21	0.520	18.42	0.725	266.3	0.4128	17.02	0.670	227.4	0.3525	16.76	0.660	220.7	0.3421	250
300	37	14.48	0.570	19.69	0.775	304.3	0.4717	18.29	0.720	262.6	0.4071	18.16	0.715	259.0	0.4015	300
350	37	15.65	0.616	20.83	0.820	340.7	0.5281	19.56	0.770	300.4	0.4656	19.30	0.760	292.6	0.4536	350
400	37	16.74	0.659	21.97	0.865	379.1	0.5876	20.70	0.815	336.5	0.5216	20.32	0.800	324.3	0.5026	400
500	37	18.69	0.736	23.88	0.940	447.7	0.6939	22.48	0.885	396.8	0.6151	22.35	0.880	392.4	0.6082	500
600	61	20.65	0.813	26.67	1.050	558.6	0.8659	25.02	0.985	491.6	0.7620	24.89	0.980	486.6	0.7542	600
700	61	22.28	0.877	28.19	1.110	624.3	0.9676	26.67	1.050	558.6	0.8659	26.67	1.050	558.6	0.8659	700
750	61	23.06	0.908	29.21	1.150	670.1	1.0386	27.31	1.075	585.5	0.9076	27.69	1.090	602.0	0.9331	750
1000	61	26.92	1.060	32.64	1.285	836.6	1.2968	31.88	1.255	798.1	1.2370	31.24	1.230	766.6	1.1882	1000

*Dimensions are from industry sources.

Most aluminum building wire in Types THW, THWN/THHN, and XHHW conductors is compact stranded. Table 5A provides appropriate dimensions for these types of wire.

Table 8 Conductor Properties

Size (AWG or kcmil)	Area mm²	Area Circular mils	Stranding Quantity	Stranding Diameter mm	Stranding Diameter in.	Overall Diameter mm	Overall Diameter in.	Overall Area mm²	Overall Area in.²	Copper Uncoated ohm/km	Copper Uncoated ohm/kFT	Copper Coated ohm/km	Copper Coated ohm/kFT	Aluminum ohm/km	Aluminum ohm/kFT
18	0.823	1620	1	—	—	1.02	0.040	0.823	0.001	25.5	7.77	26.5	8.08	42.0	12.8
18	0.823	1620	7	0.39	0.015	1.16	0.046	1.06	0.002	26.1	7.95	27.7	8.45	42.8	13.1
16	1.31	2580	1	—	—	1.29	0.051	1.31	0.002	16.0	4.89	16.7	5.08	26.4	8.05
16	1.31	2580	7	0.49	0.019	1.46	0.058	1.68	0.003	16.4	4.99	17.3	5.29	26.9	8.21
14	2.08	4110	1	—	—	1.63	0.064	2.08	0.003	10.1	3.07	10.4	3.19	16.6	5.06
14	2.08	4110	7	0.62	0.024	1.85	0.073	2.68	0.004	10.3	3.14	10.7	3.26	16.9	5.17
12	3.31	6530	1	—	—	2.05	0.081	3.31	0.005	6.34	1.93	6.57	2.01	10.45	3.18
12	3.31	6530	7	0.78	0.030	2.32	0.092	4.25	0.006	6.50	1.98	6.73	2.05	10.69	3.25
10	5.261	10380	1	—	—	2.588	0.102	5.26	0.008	3.984	1.21	4.148	1.26	6.561	2.00
10	5.261	10380	7	0.98	0.038	2.95	0.116	6.76	0.011	4.070	1.24	4.226	1.29	6.679	2.04
8	8.367	16510	1	—	—	3.264	0.128	8.37	0.013	2.506	0.764	2.579	0.786	4.125	1.26
8	8.367	16510	7	1.23	0.049	3.71	0.146	10.76	0.017	2.551	0.778	2.653	0.809	4.204	1.28
6	13.30	26240	7	1.56	0.061	4.67	0.184	17.09	0.027	1.608	0.491	1.671	0.510	2.652	0.808
4	21.15	41740	7	1.96	0.077	5.89	0.232	27.19	0.042	1.010	0.308	1.053	0.321	1.666	0.508
3	26.67	52620	7	2.20	0.087	6.60	0.260	34.28	0.053	0.802	0.245	0.833	0.254	1.320	0.403
2	33.62	66360	7	2.47	0.097	7.42	0.292	43.23	0.067	0.634	0.194	0.661	0.201	1.045	0.319
1	42.41	83690	19	1.69	0.066	8.43	0.332	55.80	0.087	0.505	0.154	0.524	0.160	0.829	0.253
1/0	53.49	105600	19	1.89	0.074	9.45	0.372	70.41	0.109	0.399	0.122	0.415	0.127	0.660	0.201
2/0	67.43	133100	19	2.13	0.084	10.62	0.418	88.74	0.137	0.3170	0.0967	0.329	0.101	0.523	0.159
3/0	85.01	167800	19	2.39	0.094	11.94	0.470	111.9	0.173	0.2512	0.0766	0.2610	0.0797	0.413	0.126
4/0	107.2	211600	19	2.68	0.106	13.41	0.528	141.1	0.219	0.1996	0.0608	0.2050	0.0626	0.328	0.100
250		—	37	2.09	0.082	14.61	0.575	168	0.260	0.1687	0.0515	0.1753	0.0535	0.2778	0.0847
300		—	37	2.29	0.090	16.00	0.630	201	0.312	0.1409	0.0429	0.1463	0.0446	0.2318	0.0707
350		—	37	2.47	0.097	17.30	0.681	235	0.364	0.1205	0.0367	0.1252	0.0382	0.1984	0.0605
400		—	37	2.64	0.104	18.49	0.728	268	0.416	0.1053	0.0321	0.1084	0.0331	0.1737	0.0529
500		—	37	2.95	0.116	20.65	0.813	336	0.519	0.0845	0.0258	0.0869	0.0265	0.1391	0.0424
600		—	61	2.52	0.099	22.68	0.893	404	0.626	0.0704	0.0214	0.0732	0.0223	0.1159	0.0353
700		—	61	2.72	0.107	24.49	0.964	471	0.730	0.0603	0.0184	0.0622	0.0189	0.0994	0.0303
750		—	61	2.82	0.111	25.35	0.998	505	0.782	0.0563	0.0171	0.0579	0.0176	0.0927	0.0282
800		—	61	2.91	0.114	26.16	1.030	538	0.834	0.0528	0.0161	0.0544	0.0166	0.0868	0.0265
900		—	61	3.09	0.122	27.79	1.094	606	0.940	0.0470	0.0143	0.0481	0.0147	0.0770	0.0235
1000		—	61	3.25	0.128	29.26	1.152	673	1.042	0.0423	0.0129	0.0434	0.0132	0.0695	0.0212
1250		—	91	2.98	0.117	32.74	1.289	842	1.305	0.0338	0.0103	0.0347	0.0106	0.0554	0.0169
1500		—	91	3.26	0.128	35.86	1.412	1011	1.566	0.02814	0.00858	0.02814	0.00883	0.0464	0.0141
1750		—	127	2.98	0.117	38.76	1.526	1180	1.829	0.02410	0.00735	0.02410	0.00756	0.0397	0.0121
2000		—	127	3.19	0.126	41.45	1.632	1349	2.092	0.02109	0.00643	0.02109	0.00662	0.0348	0.0106

Notes:

1. These resistance values are valid **only** for the parameters as given. Using conductors having coated strands, different stranding type, and, especially, other temperatures changes the resistance.

2. Formula for temperature change: $R_2 = R_1 [1 + \alpha (T_2 - 75)]$ where $\alpha_{cu} = 0.00323$, $\alpha_{AL} = 0.00330$ at 75°C.

3. Conductors with compact and compressed stranding have about 9 percent and 3 percent, respectively, smaller bare conductor diameters than those shown. See Table 5A for actual compact cable dimensions.

4. The IACS conductivities used: bare copper = 100%, aluminum = 61%.

5. Class B stranding is listed as well as solid for some sizes. Its overall diameter and area is that of its circumscribing circle.

FPN: The construction information is per NEMA WC8-1992 or ANSI/UL 1581-1998. The resistance is calculated per National Bureau of Standards Handbook 100, dated 1966, and Handbook 109, dated 1972.

Traditionally, wire sizes have been expressed as American Wire Gage (AWG), circular mil (cmil) area, or thousands of circular mil (kcmil) area. Today, wire is available with its cross-sectional area expressed in square millimeters (mm^2) as well.

The 2002 *NEC* specifically requires that insulated conductors be marked with their sizes and that the sizes be expressed in either American Wire Gage or circular mil area. [See 110.6 and 310.11(A)(4).] There are no exceptions to either of these two requirements. Because Article 310 does not specifically prohibit optional marking on insulated conductors, the *Code* permits square millimeter (mm^2) markings on conductors, but only if they are in addition to the required traditional markings of AWG or circular mil area.

According to the American National Standards Institute (ANSI), in SI 10-1997, *Standard for Use of the International System of Units (SI): The Modern Metric System*, to convert circular mils to square meters, you would multiply circular mils by 5.067075×10^{-10}. However, because square millimeters, rather than square meters, is the standard marking for wire size, and because the reciprocal is more appropriate for this conversion, a simpler conversion factor to convert from square millimeters to circular mils (approximately) follows:

$$k = 1973.53 \ \frac{\text{circular mils}}{mm^2}$$

To compare the square millimeter wire gauge to traditional wire sizes, the following example is provided.

Example

What traditional wire size does the size 125 mm^2 approximately represent?

Solution

Circular mil area = wire size (mm^2) $\times$ conversion factor

$$= 125 \ mm^2 \times 1973.53 \ \frac{\text{circular mils}}{mm^2}$$

$$= 246{,}691 \text{ circular mils}$$

$$\text{or} \quad 246.691 \text{ kcmil}$$

Therefore, the 125-mm^2 wire is larger than 4/0 AWG but smaller than a 250-kcmil conductor.

Conclusion If a 125-mm^2 wire is determined to be the minimum or recommended size conductor, it is important to understand that size 250 kcmil would be the only Table 8 conductor with equivalent cross-sectional area. Size 4/0 AWG is simply not enough metal, so 250 kcmil would be the choice for minimum equivalency. However, it is important to note that the 250-kcmil conductor ampacity could not be used for a 125-mm^2 conductor because the metric conductor size is smaller. The 4/0 AWG ampacity can be used, or the ampacity can be calculated under engineering supervision.

Table 9 Alternating-Current Resistance and Reactance for 600-Volt Cables, 3-Phase, 60 Hz, 75°C (167°F) — Three Single Conductors in Conduit

Size (AWG or kcmil)	X_L (Reactance) for All Wires		Alternating-Current Resistance for Uncoated Copper Wires			Alternating-Current Resistance for Aluminum Wires			Effective Z at 0.85 PF for Uncoated Copper Wires			Effective Z at 0.85 PF for Aluminum Wires			Size (AWG or kcmil)
	PVC, Aluminum Conduits	Steel Conduit	PVC Conduit	Aluminum Conduit	Steel Conduit	PVC Conduit	Aluminum Conduit	Steel Conduit	PVC Conduit	Aluminum Conduit	Steel Conduit	PVC Conduit	Aluminum Conduit	Steel Conduit	
14	0.190 0.058	0.240 0.073	10.2 3.1	10.2 3.1	10.2 3.1	— —	— —	— —	8.9 2.7	8.9 2.7	8.9 2.7	— —	— —	— —	14
12	0.177 0.054	0.223 0.068	6.6 2.0	6.6 2.0	6.6 2.0	10.5 3.2	10.5 3.2	10.5 3.2	5.6 1.7	5.6 1.7	5.6 1.7	9.2 2.8	9.2 2.8	9.2 2.8	12
10	0.164 0.050	0.207 0.063	3.9 1.2	3.9 1.2	3.9 1.2	6.6 2.0	6.6 2.0	6.6 2.0	3.6 1.1	3.6 1.1	3.6 1.1	5.9 1.8	5.9 1.8	5.9 1.8	10
8	0.171 0.052	0.213 0.065	2.56 0.78	2.56 0.78	2.56 0.78	4.3 1.3	4.3 1.3	4.3 1.3	2.26 0.69	2.26 0.69	2.30 0.70	3.6 1.1	3.6 1.1	3.6 1.1	8
6	0.167 0.051	0.210 0.064	1.61 0.49	1.61 0.49	1.61 0.49	2.66 0.81	2.66 0.81	2.66 0.81	1.44 0.44	1.48 0.45	1.48 0.45	2.33 0.71	2.36 0.72	2.36 0.72	6
4	0.157 0.048	0.197 0.060	1.02 0.31	1.02 0.31	1.02 0.31	1.67 0.51	1.67 0.51	1.67 0.51	0.95 0.29	0.95 0.29	0.98 0.30	1.51 0.46	1.51 0.46	1.51 0.46	4
3	0.154 0.047	0.194 0.059	0.82 0.25	0.82 0.25	0.82 0.25	1.31 0.40	1.35 0.41	1.31 0.40	0.75 0.23	0.79 0.24	0.79 0.24	1.21 0.37	1.21 0.37	1.21 0.37	3
2	0.148 0.045	0.187 0.057	0.62 0.19	0.66 0.20	0.66 0.20	1.05 0.32	1.05 0.32	1.05 0.32	0.62 0.19	0.62 0.19	0.66 0.20	0.98 0.30	0.98 0.30	0.98 0.30	2
1	0.151 0.046	0.187 0.057	0.49 0.15	0.52 0.16	0.52 0.16	0.82 0.25	0.85 0.26	0.82 0.25	0.52 0.16	0.52 0.16	0.52 0.16	0.79 0.24	0.79 0.24	0.82 0.25	1
1/0	0.144 0.044	0.180 0.055	0.39 0.12	0.43 0.13	0.39 0.12	0.66 0.20	0.69 0.21	0.66 0.20	0.43 0.13	0.43 0.13	0.43 0.13	0.62 0.19	0.66 0.20	0.66 0.20	1/0
2/0	0.141 0.043	0.177 0.054	0.33 0.10	0.33 0.10	0.33 0.10	0.52 0.16	0.52 0.16	0.52 0.16	0.36 0.11	0.36 0.11	0.36 0.11	0.52 0.16	0.52 0.16	0.52 0.16	2/0
3/0	0.138 0.042	0.171 0.052	0.253 0.077	0.269 0.082	0.259 0.079	0.43 0.13	0.43 0.13	0.43 0.13	0.289 0.088	0.302 0.092	0.308 0.094	0.43 0.13	0.43 0.13	0.46 0.14	3/0
4/0	0.135 0.041	0.167 0.051	0.203 0.062	0.220 0.067	0.207 0.063	0.33 0.10	0.36 0.11	0.33 0.10	0.243 0.074	0.256 0.078	0.262 0.080	0.36 0.11	0.36 0.11	0.36 0.11	4/0
250	0.135 0.041	0.171 0.052	0.171 0.052	0.187 0.057	0.177 0.054	0.279 0.085	0.295 0.090	0.282 0.086	0.217 0.066	0.230 0.070	0.240 0.073	0.308 0.094	0.322 0.098	0.33 0.10	250
300	0.135 0.041	0.167 0.051	0.144 0.044	0.161 0.049	0.148 0.045	0.233 0.071	0.249 0.076	0.236 0.072	0.194 0.059	0.207 0.063	0.213 0.065	0.269 0.082	0.282 0.086	0.289 0.088	300
350	0.131 0.040	0.164 0.050	0.125 0.038	0.141 0.043	0.128 0.039	0.200 0.061	0.217 0.066	0.207 0.063	0.174 0.053	0.190 0.058	0.197 0.060	0.240 0.073	0.253 0.077	0.262 0.080	350
400	0.131 0.040	0.161 0.049	0.108 0.033	0.125 0.038	0.115 0.035	0.177 0.054	0.194 0.059	0.180 0.055	0.161 0.049	0.174 0.053	0.184 0.056	0.217 0.066	0.233 0.071	0.240 0.073	400
500	0.128 0.039	0.157 0.048	0.089 0.027	0.105 0.032	0.095 0.029	0.141 0.043	0.157 0.048	0.148 0.045	0.141 0.043	0.157 0.048	0.164 0.050	0.187 0.057	0.200 0.061	0.210 0.064	500
600	0.128 0.039	0.157 0.048	0.075 0.023	0.092 0.028	0.082 0.025	0.118 0.036	0.135 0.041	0.125 0.038	0.131 0.040	0.144 0.044	0.154 0.047	0.167 0.051	0.180 0.055	0.190 0.058	600

Table 9 *Continued*

Size (AWG or kcmil)	X_L (Reactance) for All Wires		Alternating-Current Resistance for Uncoated Copper Wires			Alternating-Current Resistance for Aluminum Wires			Effective Z at 0.85 *PF* for Uncoated Copper Wires			Effective Z at 0.85 *PF* for Aluminum Wires			Size (AWG or kcmil)
	PVC, Aluminum Conduits	Steel Conduit	PVC Conduit	Aluminum Conduit	Steel Conduit	PVC Conduit	Aluminum Conduit	Steel Conduit	PVC Conduit	Aluminum Conduit	Steel Conduit	PVC Conduit	Aluminum Conduit	Steel Conduit	
750	0.125 0.038	0.157 0.048	0.062 0.019	0.079 0.024	0.069 0.021	0.095 0.029	0.112 0.034	0.102 0.031	0.118 0.036	0.131 0.040	0.141 0.043	0.148 0.045	0.161 0.049	0.171 0.052	750
1000	0.121 0.037	0.151 0.046	0.049 0.015	0.062 0.019	0.059 0.018	0.075 0.023	0.089 0.027	0.082 0.025	0.105 0.032	0.118 0.036	0.131 0.040	0.128 0.039	0.138 0.042	0.151 0.046	1000

Notes:

1. These values are based on the following constants: UL-Type RHH wires with Class B stranding, in cradled configuration. Wire conductivities are 100 percent IACS copper and 61 percent IACS aluminum, and aluminum conduit is 45 percent IACS. Capacitive reactance is ignored, since it is negligible at these voltages. These resistance values are valid only at 75°C (167°F) and for the parameters as given, but are representative for 600-volt wire types operating at 60 Hz.

2. *Effective Z* is defined as $R \cos(\theta) + X \sin(\theta)$, where θ is the power factor angle of the circuit. Multiplying current by effective impedance gives a good approximation for line-to-neutral voltage drop. Effective impedance values shown in this table are valid only at 0.85 power factor. For another circuit power factor (*PF*), effective impedance (*Ze*) can be calculated from *R* and *XL* values given in this table as follows: $Ze = R \times PF + X_L \sin[\arccos(PF)]$.

Voltage-drop calculations using the dc-resistance formula are not always accurate for ac circuits, especially for those with a less-than-unity power factor or for those that use conductors larger than 2 AWG. Table 9 allows the *Code* user to perform simple ac voltage-drop calculations. Table 9 was compiled using the Neher-McGrath ac-resistance calculation method, and the values presented are both reliable and conservative. This table contains completed calculations of effective impedance *(Z)* for the average ac circuit with an 85 percent power factor (see Example 1 below). If calculations with a different power factor are necessary, Table 9 also contains the appropriate values of inductive reactance and ac resistance (see Example 2 below).

The basic assumptions and the limitations of Table 9 are as follows:

1. Capacitive reactance is ignored.
2. There are three conductors in a raceway.
3. The calculated voltage-drop values are approximate.
4. For circuits with other parameters, the Neher-McGrath ac-resistance calculation method is used.

Example 1

A feeder has a 100-ampere continuous load. The system source is 240 volts, 3 phase, and the supplying circuit breaker is 125 amperes. The feeder is in a trade size 1¼ aluminum conduit with three 1 AWG THHN copper conductors operating at their maximum temperature rating of 75°C. The circuit length is 150 ft, and the power factor is 85 percent. Using Table 9, determine the approximate voltage drop of this circuit.

Solution

STEP 1. Find the approximate line-to-neutral voltage drop. Using the Table 9 column "Effective *Z* at 0.85 *PF* for Uncoated Copper Wires," select aluminum conduit and size 1 AWG copper wire. Use the given value of 0.16 ohm per 1000 ft in the following formula:

$$\text{Voltage drop}_{\text{(line-to-neutral)}} = \text{table value} \times \frac{\text{circular length}}{1000 \text{ ft}}$$
$$\times \text{ circuit load}$$
$$= 0.16 \text{ ohm} \times \frac{150 \text{ ft}}{1000 \text{ ft}}$$
$$\times 100 \text{ A}$$
$$= 2.40 \text{ V}$$

STEP 2. Find the line-to-line voltage drop.

$$\text{Voltage drop}_{\text{(line-to-line)}} = \text{voltage drop}_{\text{(line-to-neutral)}}$$
$$\times \sqrt{3}$$
$$= 2.40 \text{ V} \times 1.732$$
$$= 4.157 \text{ V}$$

STEP 3. Find the voltage drop expressed as a percentage of the circuit voltage.

$$\text{Percentage voltage drop}_{\text{(line-to-line)}} = \frac{4.157 \text{ V}}{240 \text{ V}} \times 100$$
$$= 1.73\%$$

STEP 4. Find the voltage present at the load end of the circuit.

$$240 \text{ V} - 4.157 \text{ V} = 235.84 \text{ V}$$

Example 2

A 270-ampere continuous load is present on a feeder. The circuit consists of a single 4-in. PVC conduit with three 600-kcmil XHHW/USE aluminum conductors fed from a 480-volt, 3-phase, 3-wire source. The conductors are operating at their maximum rated temperature of 75°C. If the power factor is 0.7 and the circuit length is 250 ft, is the voltage drop excessive?

Solution

STEP 1. Using the Table 9 column "X_L (Reactance) for All Wires," select PVC conduit and the row for size 600 kcmil. A value of 0.039 ohm per 1000 ft is given as this X_L. Next, using the column "Alternating-Current Resistance for Aluminum Wires," select PVC conduit and the row for size 600 kcmil. A value of 0.036 ohm per 1000 ft is given as this R.

STEP 2. Find the angle representing a power factor of 0.7. Using a calculator with trigonometric functions or a trigonometric function table, find the arccosine $(\cos^{-1})\theta$ of 0.7, which is 45.57 degrees. For this example, we will call this angle θ. Continuing to use the table or a calculator, find the sin of 45.57 degrees, which is 0.7141.

STEP 3. Find the impedance (Z_c) corrected to 0.7 power factor (Z_c).

$$
\begin{aligned}
Z_c &= (R \times \cos \theta) + (X_L \times \sin \theta) \\
&= (0.036 \times 0.7) + (0.039 \times 0.7141) \\
&= 0.0252 \times 0.0279 \\
&= 0.0531 \text{ ohm to neutral}
\end{aligned}
$$

STEP 4. As in Example 1, find the approximate line-to-neutral voltage drop.

$$
\begin{aligned}
\text{Voltage drop}_{\text{(line-to-neutral)}} &= Z_c \times \frac{\text{circular length}}{1000 \text{ ft}} \\
&\quad \times \text{ circuit load} \\
&= 0.0530 \times \frac{250 \text{ ft}}{1000 \text{ ft}} \times 270 \text{ A} \\
&= 3.577 \text{ V}
\end{aligned}
$$

STEP 5. Find the approximate line-to-line voltage drop.

$$
\begin{aligned}
\text{Voltage drop}_{\text{(line-to-line)}} &= \text{voltage drop}_{\text{(line-to-neutral)}} \\
&\quad \times \sqrt{3} \\
&= 3.577 \text{ V} \times 1.732 \\
&= 6.196 \text{ V}
\end{aligned}
$$

STEP 6. Find the approximate voltage drop expressed as a percentage of the circuit voltage.

$$
\begin{aligned}
\text{Percentage voltage drop}_{\text{(line-to-line)}} &= \frac{6.196 \text{ V}}{480 \text{ V}} \times 100 \\
&= 1.29\%
\end{aligned}
$$

STEP 7. Find the voltage present at the load end of the circuit.

$$480 \text{ volts} - 6.196 \text{ V} = 473.8 \text{ V}$$

Conclusion According to 210.19(A)(1), FPN No. 4, this voltage drop does not appear to be excessive.

Tables 11(A) and 11(B)

For listing purposes, Table 11(A) and Table 11(B) provide the required power source limitations for Class 2 and Class 3 power sources. Table 11(A) applies for alternating-current sources, and Table 11(B) applies for direct-current sources.

The power for Class 2 and Class 3 circuits shall be either (1) inherently limited, requiring no overcurrent protection, or (2) not inherently limited, requiring a combination of power source and overcurrent protection. Power sources designed for interconnection shall be listed for the purpose.

As part of the listing, the Class 2 or Class 3 power source shall be durably marked where plainly visible to indicate the class of supply and its electrical rating. A Class 2 power source not suitable for wet location use shall be so marked.

Exception: Limited power circuits used by listed information technology equipment.

Overcurrent devices, where required, shall be located at the point where the conductor to be protected receives its supply and shall not be interchangeable with devices of higher ratings. The overcurrent device shall be permitted as an integral part of the power source.

The information in Tables 11(A) and 11(B) was previously in Tables 725-31(a) and (b). Because of the listing requirements for Class 2 and Class 3 power supplies, this information is no longer useful to the average user. It has been retained in Chapter 9 as Tables 11(A) and 11(B) to provide direction for organizations that are properly equipped and qualified to evaluate these products.

Table 11(A) Class 2 and Class 3 Alternating-Current Power Source Limitations

Power Source	Inherently Limited Power Source (Overcurrent Protection Not Required)				Not Inherently Limited Power Source (Overcurrent Protection Required)			
	Class 2			Class 3	Class 2			Class 3
Source voltage V_{max} (volts) (see Note 1)	0 through 20*	Over 20 and through 30*	Over 30 and through 150	Over 30 and through 100	0 through 20*	Over 20 and through 30*	Over 30 and through 100	Over 100 and through 150
Power limitations VA_{max} (volt-amperes) (see Note 1)	—	—	—	—	250 (see Note 3)	250	250	N.A.
Current limitations I_{max} (amperes) (see Note 1)	8.0	8.0	0.005	$150/V_{max}$	$1000/V_{max}$	$1000/V_{max}$	$1000/V_{max}$	1.0
Maximum overcurrent protection (amperes)	—	—	—	—	5.0	$100/V_{max}$	$100/V_{max}$	1.0
Power source maximum nameplate rating — VA (volt-amperes)	$5.0 \times V_{max}$	100	$0.005 \times V_{max}$	100	$5.0 \times V_{max}$	100	100	100
Power source maximum nameplate rating — Current (amperes)	5.0	$100/V_{max}$	0.005	$100/V_{max}$	5.0	$100/V_{max}$	$100/V_{max}$	$100/V_{max}$

*Voltage ranges shown are for sinusoidal ac in indoor locations or where wet contact is not likely to occur. For nonsinusoidal or wet contact conditions, see Note 2.

Table 11(B) Class 2 and Class 3 Direct-Current Power Source Limitations

Power Source	Inherently Limited Power Source (Overcurrent Protection Not Required)					Not Inherently Limited Power Source (Overcurrent Protection Required)			
	Class 2				Class 3	Class 2			Class 3
Source voltage V_{max} (volts) (see Note 1)	0 through 20*	Over 20 and through 30*	Over 30 and through 60*	Over 60 and through 150	Over 60 and through 100	0 through 20*	Over 20 and through 60*	Over 60 and through 100	Over 100 and through 150
Power limitations VA_{max} (volt-amperes) (see Note 1)	—	—	—	—	—	250 (see Note 3)	250	250	N.A.
Current limitations I_{max} (amperes) (see Note 1)	8.0	8.0	$150/V_{max}$	0.005	$150/V_{max}$	$1000/V_{max}$	$1000/V_{max}$	$1000/V_{max}$	1.0
Maximum overcurrent protection (amperes)	—	—	—	—	—	5.0	$100/V_{max}$	$100/V_{max}$	1.0
Power source maximum nameplate rating — VA (volt-amperes)	$5.0 \times V_{max}$	100	100	$0.005 \times V_{max}$	100	$5.0 \times V_{max}$	100	100	100
Power source maximum nameplate rating — Current (amperes)	5.0	$100/V_{max}$	$100/V_{max}$	0.005	$100/V_{max}$	5.0	$100/V_{max}$	$100/V_{max}$	$100/V_{max}$

*Voltage ranges shown are for continuous dc in indoor locations or where wet contact is not likely to occur. For interrupted dc or wet contact conditions, see Note 4.

Notes for Tables 11(A) and 11(B)

1. V_{max}, I_{max}, and VA_{max} are determined with the current-limiting impedance in the circuit (not bypassed) as follows:

V_{max}: Maximum output voltage regardless of load with rated input applied.

I_{max}: Maximum output current under any noncapacitive load, including short circuit, and with overcurrent protection bypassed if used. Where a transformer limits the output current, I_{max} limits apply after 1 minute of operation. Where a current-limiting impedance, listed for the purpose, or as part of a listed product, is used in combination with a nonpower-limited transformer or a stored energy source, e.g., storage battery, to limit the output current, I_{max} limits apply after 5 seconds.

VA_{max}: Maximum volt-ampere output after 1 minute of operation regard-less of load and overcurrent protection bypassed if used.

2. For nonsinusoidal ac, V_{max} shall not be greater than 42.4 volts peak. Where wet contact (immersion not included) is likely to occur, Class 3 wiring methods shall be used or V_{max} shall not be greater than 15 volts for sinusoidal ac and 21.2 volts peak for nonsinusoidal ac.

3. If the power source is a transformer, VA_{max} is 350 or less when V_{max} is 15 or less.

4. For dc interrupted at a rate of 10 to 200 Hz, V_{max} shall not be greater than 24.8 volts peak. Where wet contact (immersion not included) is likely to occur, Class 3 wiring methods shall be used, or V_{max} shall not be greater than 30 volts for continuous dc; 12.4 volts peak for dc that is interrupted at a rate of 10 to 200 Hz.

Tables 12(A) and 12(B)

For listing purposes, Tables 12(A) and 12(B) provide the required power source limitations for power-limited fire alarm sources. Table 12(A) applies for alternating-current sources, and Table 12(B) applies for direct-current sources.

The power for power-limited fire alarm circuits shall be either (1) inherently limited, requiring no overcurrent protection, or (2) not inherently limited, requiring the power to be limited by a combination of power source and overcurrent protection.

As part of the listing, the PLFA power source shall be durably marked where plainly visible to indicate that it is a power-limited fire alarm power source. The overcurrent device, where required, shall be located at the point where the conductor to be protected receives its supply and shall not be interchangeable with devices of higher ratings. The overcurrent device shall be permitted as an integral part of the power source.

The information in Tables 12(A) and 12(B) was previously in Tables 760-21(a) and (b). Because of the listing requirements for these power supplies, this information is no longer useful to the average user. These tables have been moved to Chapter 9 as Tables 12(A) and 12(B) to provide direction for organizations that are properly equipped and qualified to evaluate these products.

Table 12(A) PLFA Alternating-Current Power Source Limitations

Power Source		Inherently Limited Power Source (Overcurrent Protection Not Required)			Not Inherently Limited Power Source (Overcurrent Protection Required)		
Circuit voltage V_{max} (volts) (see Note 1)		0 through 20	Over 20 and through 30	Over 30 and through 100	0 through 20	Over 20 and through 100	Over 100 and through 150
Power limitations VA_{max} (volt-amperes) (see Note 1)		—	—	—	250 (see Note 2)	250	N.A.
Current limitations I_{max} (amperes) (see Note 1)		8.0	8.0	$150/V_{max}$	$1000/V_{max}$	$1000/V_{max}$	1.0
Maximum overcurrent protection (amperes)		—	—	—	5.0	$100/V_{max}$	1.0
Power source maximum nameplate ratings	VA (volt-amperes)	$5.0 \times V_{max}$	100	100	$5.0 \times V_{max}$	100	100
	Current (amperes)	5.0	$100/V_{max}$	$100/V_{max}$	5.0	$100/V_{max}$	$100/V_{max}$

Table 12(B) PLFA Direct-Current Power Source Limitations

Power Source		Inherently Limited Power Source (Overcurrent Protection Not Required)				Not Inherently Limited Power Source (Overcurrent Protection Required)		
Circuit voltage V_{max} (volts) (see Note 1)		0 through 20	Over 20 and through 30	Over 30 and through 100	Over 100 and through 250	0 through 20	Over 20 and through 100	Over 100 and through 150
Power limitations VA_{max} (volt-amperes) (see Note 1)		—	—	—	—	250 (see Note 2)	250	N.A.
Current limitations I_{max} (amperes) (see Note 1)		8.0	8.0	$150/V_{max}$	0.030	$1000/V_{max}$	$1000/V_{max}$	1.0
Maximum overcurrent protection (amperes)		—	—	—	—	5.0	$100/V_{max}$	1.0
Power source maximum nameplate ratings	VA (volt-amperes)	$5.0 \times V_{max}$	100	100	$0.030 \times V_{max}$	$5.0 \times V_{max}$	100	100
	Current (amperes)	5.0	$100/V_{max}$	$100/V_{max}$	0.030	5.0	$100/V_{max}$	$100/V_{max}$

Notes for Tables 12(A) and 12(B)

1. V_{max}, I_{max}, and VA_{max} are determined as follows:

V_{max}: Maximum output voltage regardless of load with rated input applied.

I_{max}: Maximum output current under any noncapacitive load, including short circuit, and with overcurrent protection bypassed if used. Where a transformer limits the output current, I_{max} limits apply after 1 minute of operation. Where a current-limiting impedance, listed for the purpose, is used in combination with a nonpower-limited transformer or a stored energy source, e.g., storage battery, to limit the output current, I_{max} limits apply after 5 seconds.

VA_{max}: Maximum volt-ampere output after 1 minute of operation regardless of load and overcurrent protection bypassed if used. Current limiting impedance shall not be bypassed when determining I_{max} and VA_{max}.

2. If the power source is a transformer, VA_{max} is 350 or less when V_{max} is 15 or less.

Product Safety Standards

Annex A is not a part of the requirements of this NFPA document but is included for informational purposes only.

This informational annex provides a list of product safety standards used for product listing where that listing is required by this *Code*. It is recognized that this list is current at the time of publication but that new standards or modifications to existing standards can occur at any time while this edition of the *Code* is in effect.

This annex does not form a mandatory part of the requirements of this *Code* but is intended only to provide *Code* users with informational guidance about the product characteristics about which *Code* requirements have been based.

Product Standard Name	Product Standard Number
Antenna-Discharge Units	UL 452
Armored Cable	UL 4
Attachment Plugs and Receptacles	UL 498
Audio/Video and Musical Instrument Apparatus for Household, Commercial, and Similar General Use	UL 6500
Audio-Video Products and Accessories	UL 1492
Busways and Associated Fittings	UL 857
Cables — Thermoplastic-Insulated Underground Feeder and Branch-Circuit Cables	UL 493
Cables — Thermoplastic-Insulated Wires and Cables	UL 83
Cables — Thermoset-Insulated Wires and Cables	UL 44
Cables for Non–Power-Limited Fire-Alarm Circuits	UL 1425
Cables for Power-Limited Fire-Alarm Circuits	UL 1424
Cellular Metal Floor Raceways and Fittings	UL 209
Class 2 Power Units	UL 1310
Commercial Audio Equipment	UL 813
Communication Circuit Accessories	UL 1863
Communications Cables	UL 444
Community-Antenna Television Cables	UL 1655
Conduit — Type EB and A Rigid PVC Conduit and HDPE Conduit	UL 651A
Continuous Length High Density Polyethylene Conduit	UL 651B
Control Centers for Changing Message Type Electric Signs	UL 1433
Cord Sets and Power-Supply Cords	UL 817

Product Standard Name	Product Standard Number
Data-Processing Cable	UL 1690
Dead-Front Switchboards	UL 891
Electric Signs	UL 48
Electric Spas, Equipment Assemblies, and Associated Equipment	UL 1563
Electric Water Heaters for Pools and Tubs	UL 1261
Electrical Apparatus for Use in Class I, Zone 1 Hazardous (Classified) Locations Type of Protection — Encapsulation "m"	ISA S12.23.01
Electrical Apparatus for Use in Class I, Zones 0 & 1 Hazardous (Classified) Locations: General Requirements	ISA 12.0.01
Electrical Apparatus for Use in Class I, Zone 1 Hazardous (Classified) Locations: Type of Protection — Increased Safety "e"	ISA S12.16.01
Electrical Apparatus for Use in Class I, Zone 1 Hazardous (Classified) Locations: Type of Protection — Flameproof "d"	ISA S12.22.01
Electrical Apparatus for Use in Class I, Zone 1 Hazardous (Classified) Locations: Type of Protection — Powder Filling "q"	ISA S12.25.01
Electrical Apparatus for Use in Class I, Zone 1 Hazardous (Classified) Locations: Type of Protection — Oil-Immersion "o"	ISA S12.26.01
Electrical Equipment for Use in Class I, Zone 0, 1, and 2 Hazardous (Classified) Locations	UL 2279
Electrical Metallic Tubing	UL 797

Product Standard Name	Product Standard Number	Product Standard Name	Product Standard Number
Electrical Nonmetallic Tubing	UL 1653	Metal-Clad Cables and Cable-Sealing Fittings for Use in Hazardous (Classified) Locations	UL 2225
Electric-Battery-Powered Industrial Trucks	UL 583		
Electrode Receptacles for Gas-Tube Signs	UL 879		
Enclosed and Dead-Front Switches	UL 98	Metallic Outlet Boxes	UL 514A
Enclosures for Electrical Equipment	UL 50	Mobile Home Pipe Heating Cable	UL 1462
Energy Management Equipment	UL 916	Molded-Case Circuit Breakers, Molded-Case Switches, and Circuit-Breaker Enclosures	UL 489
Fire Pump Controllers	UL 218		
Fittings for Cable and Conduit	UL 514B		
Flexible Cord and Fixture Wire	UL 62	Molded-Case Switches	UL 1087
Flexible Metal Conduit	UL 1	Neon Transformers and Power Supplies	UL 2161
Fluorescent Lighting Fixtures	UL 1570	Nonincendive Electrical Equipment for Use in Class I and II, Division 2 and Class III, Divisions 1 and 2 Hazardous (Classified) Locations	ISA S12.12
Fluorescent-Lamp Ballasts	UL 935		
Gas-Tube-Sign and Ignition Cable	UL 814		
General-Use Snap Switches	UL 20		
Ground-Fault Circuit-Interrupters	UL 943		
Ground-Fault Sensing and Relaying Equipment	UL 1053	Nonmetallic Outlet Boxes, Flush-Device Boxes, and Covers	UL 514C
Grounding and Bonding Equipment	UL 467	Nonmetallic Surface Raceways and Fittings	UL 5A
High Intensity Discharge Lighting Fixtures	UL 1572		
High-Intensity-Discharge Lamp Ballasts	UL 1029	Nonmetallic Underground Conduit with Conductors	UL 1990
Incandescent Lighting Fixtures	UL 1571	Office Furnishings	UL 1286
Industrial Battery Chargers	UL 1564	Optical Fiber Cable	UL 1651
Industrial Control Equipment	UL 508	Optical Fiber Cable Raceway	UL 2024
Instrumentation Tray Cable	UL 2250	Panelboards	UL 67
Insulated Wire Connector Systems for Underground Use or in Damp or Wet Locations	UL 486D	Personal Protection Systems for Electric Vehicle Supply Circuits: General Requirements	UL 2231-1
Intermediate Metal Conduit	UL 1242	Personal Protection Systems for Electric Vehicle Supply Circuits: Particular Requirements for Protection Devices for Use in Charging Systems	UL 2231-2
Isolated Power Systems Equipment	UL 1047		
Junction Boxes for Swimming Pool Lighting Fixtures	UL 1241		
Liquid-Tight Flexible Nonmetallic Conduit	UL 1660	Portable Electric Lamps	UL 153
Liquid-Tight Flexible Steel Conduit	UL 360	Potting Compounds for Swimming Pool, Fountain, and Spa Equipment	UL 676A
Low Voltage Landscape Lighting Systems	UL 1838		
Low-Voltage Fuses — Part 1: General Requirements	UL 248-1	Power Outlets	UL 231
		Power Units Other Than Class 2	UL 1012
Low-Voltage Fuses — Part 2: Class C Fuses	UL 248-2	Power-Limited Circuit Cables	UL 13
Low-Voltage Fuses — Part 3: Class CA and CB Fuses	UL 248-3	Professional Video and Audio Equipment	UL 1419
		Protectors for Coaxial Communications Circuits	UL 497C
Low-Voltage Fuses — Part 4: Class CC Fuses	UL 248-4	Protectors for Data Communication and Fire Alarm Circuits	UL 497B
Low-Voltage Fuses — Part 5: Class G Fuses	UL 248-5		
Low-Voltage Fuses — Part 6: Class H Non-Renewable Fuses	UL 248-6	Protectors for Paired Conductor Communications Circuits	UL 497
Low-Voltage Fuses — Part 7: Class H Renewable Fuses	UL 248-7	Radio Receivers, Audio Systems, and Accessories	UL 1270
Low-Voltage Fuses — Part 8: Class J Fuses	UL 248-8	Reference Standard for Electrical Wires, Cables, and Flexible Cords	UL 1581
Low-Voltage Fuses — Part 9: Class K Fuses	UL 248-9		
Low-Voltage Fuses — Part 10: Class L Fuses	UL 249-10	Reinforced Thermosetting Resin Conduit (RTRC) and Fittings	UL 1684
		Residential Pipe Heating Cable	UL 2049
Low-Voltage Fuses — Part 11: Plug Fuses	UL 248-11	Rigid Metal Conduit	UL 6
Low-Voltage Fuses — Part 12: Class R Fuses	UL 248-12	Roof and Gutter De-Icing Cable Units	UL 1588
		Safety of Information Technology Equipment, Including Electrical Business Equipment	UL 1950
Low-Voltage Fuses — Part 15: Class T Fuses	UL 248-15		
Machine-Tool Wires and Cables	UL 1063	Schedule 40 and 80 Rigid PVC Conduit	UL 651
Manufactured Wiring Systems	UL 183	Secondary Protectors for Communications Circuits	UL 497A
Medical and Dental Equipment	UL 544		
Medium-Voltage Power Cables	UL 1072	Service-Entrance Cables	UL 854
Metal-Clad Cables	UL 1569	Smoke Detectors for Fire Protective Signaling Systems	UL 268

Product Standard Name	Product Standard Number	Product Standard Name	Product Standard Number
Specialty Transformers	UL 506	Telephone Equipment	UL 1459
Splicing Wire Connectors	UL 486C	Transfer Switch Equipment	UL 1008
Static Inverters and Charge Controllers for Use in Photovoltaic Power Systems	UL 1741	Transient Voltage Surge Suppressors	UL 1449
		Underfloor Raceways and Fittings	UL 884
Strut-Type Channel Raceways and Fittings	UL 5B	Underwater Lighting Fixtures	UL 676
Surface Metal Raceways and Fittings	UL 5	Vacuum Cleaners, Blower Cleaners, and Household Floor Finishing Machines	UL 1017
Surface Raceways and Fittings for Use with Data, Signal and Control Circuits	UL 5C		
Surge Arresters — Gapped Silicon-Carbide Surge Arresters for AC Power Circuits	IEEE C62.1	Wire Connectors and Soldering Lugs for Use with Copper Conductors	UL 486A
Surge Arresters — Metal-Oxide Surge Arresters for AC Power Circuits	IEEE C62.11	Wire Connectors for Use with Aluminum Conductors	UL 486B
Swimming Pool Pumps, Filters, and Chlorinators	UL 1081	Wireways, Auxiliary Gutters, and Associated Fittings	UL 870

Annex A, which is new in the 2002 *Code*, is a list of product standards for products covered by the *Code*. The specific requirements governing a product are found in the product standard, rather than in the *Code*. As a result of product testing, a laboratory may impose installation requirements on the use of a product. Section 110.3(B) makes those installation requirements a mandatory part of the *Code* requirements. Consult the listing directory of the testing laboratory for additional requirements or limitations of product use.

Application Information for Ampacity Calculation

This annex is not a part of the requirements of this NFPA document but is included for informational purposes only.

B.310.15(B)(1) Formula Application Information. This annex provides application information for ampacities calculated under engineering supervision.

The data in Annex B is based on calculations using the Neher-McGrath formula, which is used to calculate conductor ampacities and which can be found in 310.15(C). See the commentary following 310.15(C).

B.310.15(B)(2) Typical Applications Covered by Tables. Typical ampacities for conductors rated 0 through 2000 volts are shown in Table B.310.1 through Table B.310.10. Underground electrical duct bank configurations, as detailed in Figure B.310.3, Figure B.310.4, and Figure B.310.5, are utilized for conductors rated 0 through 5000 volts. In Figure B.310.2 through Figure B.310.5, where adjacent duct banks are used, a separation of 1.5 m (5 ft) between the centerlines of the closest ducts in each bank or 1.2 m (4 ft) between the extremities of the concrete envelopes is sufficient to prevent derating of the conductors due to mutual heating. These ampacities were calculated as detailed in the basic ampacity paper, AIEE Paper 57-660, *The Calculation of the Temperature Rise and Load Capability of Cable Systems,* by J. H. Neher and M. H. McGrath. For additional information concerning the application of these ampacities, see IEEE/ICEA Standard S-135/P-46-426, *Power Cable Ampacities,* and IEEE Standard 835-1994, *Standard Power Cable Ampacity Tables.*

If other factors remain the same, a soil resistivity higher than 90 will reduce ampacities below those listed in Tables B.310.5, B.310.6, B.310.7, B.310.8, B.310.9, and B.310.10 for underground ampacity. Conversely, a load factor less than 100 percent will increase ampacities if other factors remain the same. See B.310.15(B)(7) for allowable adjustments if the load factor is less than 100 percent. Reduced load factors are used in Figures B.310.3, B.310.4, and B.310.5.

Typical values of thermal resistivity (Rho) are as follows:
Average soil (90 percent of USA)= 90
Concrete = 55
Damp soil (coastal areas, high water table) = 60
Paper insulation = 550
Polyethylene (PE) = 450
Polyvinyl chloride (PVC) = 650
Rubber and rubber-like = 500
Very dry soil (rocky or sandy) = 120

Thermal resistivity, as used in this annex, refers to the heat transfer capability through a substance by conduction. It is the reciprocal of thermal conductivity and is normally expressed in the units °C-cm/watt. For additional information on determining soil thermal resistivity (Rho), see ANSI/IEEE Standard 442-1996, *Guide for Soil Thermal Resistivity Measurements.*

B.310.15(B)(3) Criteria Modifications. Where values of load factor and Rho are known for a particular electrical duct bank installation and they are different from those shown in a specific table or figure, the ampacities shown in the table or figure can be modified by the application of factors derived from the use of Figure B.310.1.

Where two different ampacities apply to adjacent portions of a circuit, the higher ampacity can be used beyond the point of transition, a distance equal to 3 m (10 ft) or 10 percent of the circuit length figured at the higher ampacity, whichever is less.

The information given in B.310.15(B)(3) is the same as that found in the exception to 310.15(A)(2). See the commentary following that exception.

Where the burial depth of direct burial or electrical duct bank circuits are modified from the values shown in a figure or table, ampacities can be modified as shown in (a) and (b) as follows.

(a) Where burial depths are increased in part(s) of an electrical duct run to avoid underground obstructions, no decrease in ampacity of the conductors is needed, provided the total length of parts of the duct run increased in depth to avoid obstructions is less than 25 percent of the total run length.

See Exhibit B.1.

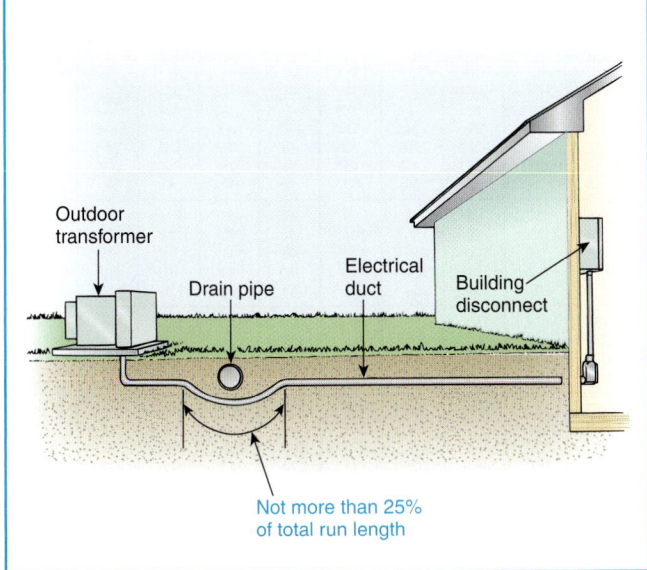

Exhibit B.1 If that portion deeper than 30 in. does not exceed 25 percent of the total run length, no decrease in ampacity is required, even if part of the run is more than 30 in. deep, which is the maximum depth assumed in Figure B.310.2, Note 1, to maintain the accuracy of the tables.

(b) Where burial depths are deeper than shown in a specific underground ampacity table or figure, an ampacity derating factor of 6 percent per increased 300 mm (foot) of depth for all values of Rho can be utilized. No rating change is needed where the burial depth is decreased.

For example, in accordance with Table B.310.7 and Figure B.310.2, the ampacity of six parallel runs of 500-kcmil, Type XHHW copper conductors (Rho = 90, as shown in Detail 3 in Figure B.310.2) is 6 × 273 = 1638 amperes at a depth measuring no more than 30 in. to the top duct in the bank. If the burial depth is 6 ft, the ampacity is 1638 — (3.5 × 0.06 × 1638) = 1294 amperes.

B.310.15(B)(4) Electrical Ducts. The term *electrical duct(s)* is defined in 310.60.

B.310.15(B)(5) Tables B.310.6 and B.310.7.

(a) To obtain the ampacity of cables installed in two electrical ducts in one horizontal row with 190-mm (7.5-in.) center-to-center spacing between electrical ducts, similar to Figure B.310.2, Detail 1, multiply the ampacity shown for one duct in Table B.310.6 and Table B.310.7 by 0.88.

(b) To obtain the ampacity of cables installed in four electrical ducts in one horizontal row with 190-mm (7.5-in.) center-to-center spacing between electrical ducts, similar to Figure B.310.2, Detail 2, multiply the ampacity shown for three electrical ducts in Table B.310.6 and Table B.310.7 by 0.94.

The underground ampacity tables (Tables B.310.5 through B.310.10) are based on the 7.5-in. center-to-center spacing illustrated in Figure B.310.2. Although moving directly buried cables or electrical ducts farther apart will increase ampacities, the effect is surprisingly small. One calculation indicates that two side-by-side electrical ducts buried 30 in. below grade would have to be spaced about 5 ft apart before they could be considered single electrical ducts as shown in Detail 1, Figure B.310.2. Decreasing the burial depth, decreasing the thermal resistivity of the earth or other surrounding medium, and decreasing the load factor each has a much greater effect in increasing ampacity than does increasing the horizontal spacing.

B.310.15(B)(6) Electrical Ducts Used in Figure B.310.2. If spacing between electrical ducts, as shown in Figure B.310.2, is less than specified in Figure B.310.2, where electrical ducts enter equipment enclosures from underground, the ampacity of conductors contained within such electrical ducts need not be reduced.

B.310.15(B)(7) Examples Showing Use of Figure B.310.1 for Electrical Duct Bank Ampacity Modifications. Figure B.310.1 is used for interpolation or extrapolation for values of Rho and load factor for cables installed in electrical ducts. The upper family of curves shows the variation in ampacity and Rho at unity load factor in terms of I_1, the ampacity for Rho = 60, and 50 percent load factor. Each curve is designated for a particular ratio I_2/I_1, where I_2 is the ampacity at Rho = 120 and 100 percent load factor.

The lower family of curves shows the relationship between Rho and load factor that will give substantially the same ampacity as the indicated value of Rho at 100 percent load factor.

As an example, to find the ampacity of a 500 kcmil copper cable circuit for six electrical ducts as shown in Table B.310.5: At the Rho = 60, LF = 50, I_1 = 583; for Rho = 120 and LF = 100, I_2 = 400. The ratio I_2/I_1 = 0.686. Locate Rho = 90 at the bottom of the chart and follow the 90 Rho line to the intersection with 100 percent load factor where the equivalent Rho = 90. Then follow the 90 Rho line to I_2/I_1 ratio of 0.686 where F = 0.74. The desired ampacity = 0.74 × 583 = 431, which agrees with the table for Rho = 90, LF = 100.

To determine the ampacity for the same circuit where Rho = 80 and LF = 75, using Figure B.310.1, the equivalent Rho = 43, F = 0.855, and the desired ampacity = 0.855 × 583 = 498 amperes. Values for using Figure B.310.1 are found in the electrical duct bank ampacity tables of this annex.

Table B.310.1 Ampacities of Two or Three Insulated Conductors, Rated 0 Through 2000 Volts, Within an Overall Covering (Multiconductor Cable), in Raceway in Free Air Based on Ambient Air Temperature of 30°C (86°F)

	Temperature Rating of Conductor. (See Table 310.13.)						
	60°C (140°F)	75°C (167°F)	90°C (194°F)	60°C (140°F)	75°C (167°F)	90°C (194°F)	
	Types TW, UF	Types RHW, THHW, THW, THWN, XHHW, ZW	Types THHN, THHW, THW-2, THWN-2, RHH, RWH-2, USE-2, XHHW, XHHW-2, ZW-2	Type TW	Types RHW, THHW, THW, THWN, XHHW	Types THHN, THHW, THW-2, THWN-2, RHH, RWH-2, USE-2, XHHW, XHHW-2, ZW-2	
Size AWG or kcmil	COPPER			ALUMINUM OR COPPER-CLAD ALUMINUM			Size (AWG or kcmil)
14	16*	18*	21*	—	—	—	14
12	20*	24*	27*	16*	18*	21*	12
10	27*	33*	36*	21*	25*	28*	10
8	36	43	48	28	33	37	8
6	48	58	65	38	45	51	6
4	66	79	89	51	61	69	4
3	76	90	102	59	70	79	3
2	88	105	119	69	83	93	2
1	102	121	137	80	95	106	1
1/0	121	145	163	94	113	127	1/0
2/0	138	166	186	108	129	146	2/0
3/0	158	189	214	124	147	167	3/0
4/0	187	223	253	147	176	197	4/0
250	205	245	276	160	192	217	250
300	234	281	317	185	221	250	300
350	255	305	345	202	242	273	350
400	274	328	371	218	261	295	400
500	315	378	427	254	303	342	500
600	343	413	468	279	335	378	600
700	376	452	514	310	371	420	700
750	387	466	529	321	384	435	750
800	397	479	543	331	397	450	800
900	415	500	570	350	421	477	900
1000	448	542	617	382	460	521	1000

Correction Factors

Ambient Temp. (°C)	For ambient temperatures other than 30°C (86°F), multiply the ampacities shown above by the appropriate factor shown below.						Ambient Temp. (°F)
21–25	1.08	1.05	1.04	1.08	1.05	1.04	70–77
26–30	1.00	1.00	1.00	1.00	1.00	1.00	79–86
31–35	0.91	0.94	0.96	0.91	0.94	0.96	88–95
36–40	0.82	0.88	0.91	0.82	0.88	0.91	97–104
41–45	0.71	0.82	0.87	0.71	0.82	0.87	106–113
46–50	0.58	0.75	0.82	0.58	0.75	0.82	115–122
51–55	0.41	0.67	0.76	0.41	0.67	0.76	124–131
56–60	—	0.58	0.71	—	0.58	0.71	133–140
61–70	—	0.33	0.58	—	0.33	0.58	142–158
71–80	—	—	0.41	—	—	0.41	160–176

*Unless otherwise specifically permitted elsewhere in this *Code*, the overcurrent protection for these conductor types shall not exceed 15 amperes for 14 AWG, 20 amperes for 12 AWG, and 30 amperes for 10 AWG copper; or 15 amperes for 12 AWG and 25 AWG amperes for 10 AWG aluminum and copper-clad aluminum.

Table B.310.3 Ampacities of Multiconductor Cables with Not More Than Three Insulated Conductors, Rated 0 Through 2000 Volts, in Free Air Based on Ambient Air Temperature of 40°C (104°F) (For Types TC, MC, MI, UF, and USE Cables)

	Temperature Rating of Conductor. (See Table 310-13.)								
	60°C (140°F)	75°C (167°F)	85°C (185°F)	90°C (194°F)	60°C (140°F)	75°C (167°F)	85°C (185°F)	90°C (194°F)	
Size AWG or kcmil	COPPER				ALUMINUM OR COPPER-CLAD ALUMINUM				Size AWG or kcmil
18	—	—	—	11*	—	—	—	—	18
16	—	—	—	16*	—	—	—	—	16
14	18*	21*	24*	—	—	—	—	—	14
12	21*	28*	30*	32*	18*	21*	24*	25*	12
10	28*	36*	41*	43*	21*	28*	30*	32*	10
8	39	50	56	59	30	39	44	46	8
6	52	68	75	79	41	53	59	61	6
4	69	89	100	104	54	70	78	81	4
3	81	104	116	121	63	81	91	95	3
2	92	118	132	138	72	92	103	108	2
1	107	138	154	161	84	108	120	126	1
1/0	124	160	178	186	97	125	139	145	1/0
2/0	143	184	206	215	111	144	160	168	2/0
3/0	165	213	238	249	129	166	185	194	3/0
4/0	190	245	274	287	149	192	214	224	4/0
250	212	274	305	320	166	214	239	250	250
300	237	306	341	357	186	240	268	280	300
350	261	337	377	394	205	265	296	309	350
400	281	363	406	425	222	287	317	334	400
500	321	416	465	487	255	330	368	385	500
600	354	459	513	538	284	368	410	429	600
700	387	502	562	589	306	405	462	473	700
750	404	523	586	615	328	424	473	495	750
800	415	539	604	633	339	439	490	513	800
900	438	570	639	670	362	469	514	548	900
1000	461	601	674	707	385	499	558	584	1000

Correction Factors

Ambient Temp. (°C)	For ambient temperatures other than 40°C (104°F), multiply the ampacities shown above by the appropriate factor shown below.								Ambient Temp. (°F)
21–25	1.32	1.20	1.15	1.14	1.32	1.20	1.15	1.14	70–77
26–30	1.22	1.13	1.11	1.10	1.22	1.13	1.11	1.10	79–86
31–35	1.12	1.07	1.05	1.05	1.12	1.07	1.05	1.05	88–95
36–40	1.00	1.00	1.00	1.00	1.00	1.00	1.00	1.00	97–104
41–45	0.87	0.93	0.94	0.95	0.87	0.93	0.94	0.95	106–113
46–50	0.71	0.85	0.88	0.89	0.71	0.85	0.88	0.89	115–122
51–55	0.50	0.76	0.82	0.84	0.50	0.76	0.82	0.84	124–131
56–60	—	0.65	0.75	0.77	—	0.65	0.75	0.77	133–140
61–70	—	0.38	0.58	0.63	—	0.38	0.58	0.63	142–158
71–80	—	—	0.33	0.44	—	—	0.33	0.44	160–176

*Unless otherwise specifically permitted elsewhere in this *Code*, the overcurrent protection for these conductor types shall not exceed 15 amperes for 14 AWG, 20 amperes for 12 AWG, and 30 amperes for 10 AWG copper; or 15 amperes for 12 AWG and 25 amperes for 10 AWG aluminum and copper-clad aluminum.

Table B.310.5 Ampacities of Single Insulated Conductors, Rated 0 through 2000 Volts, in Nonmagnetic Underground Electrical Ducts (One Conductor per Electrical Duct), Based on Ambient Earth Temperature of 20°C (68°F), Electrical Duct Arrangement per Figure B.310.2, Conductor Temperature 75°C (167°F)

Size (kcmil)	3 Electrical Ducts (Fig. B.310.2, Detail 2) Types RHW, THHW, THW, THWN, XHHW, USE — COPPER			6 Electrical Ducts (Fig. B.310.2, Detail 3) Types RHW, THHW, THW, THWN, XHHW, USE — COPPER			9 Electrical Ducts (Fig. B.310.2, Detail 4) Types RHW, THHW, THW, THWN, XHHW, USE — COPPER			3 Electrical Ducts (Fig. B.310.2, Detail 2) Types RHW, THHW, THW, THWN, XHHW, USE — ALUMINUM OR COPPER-CLAD ALUMINUM			6 Electrical Ducts (Fig. B.310.2, Detail 3) Types RHW, THHW, THW, THWN, XHHW, USE — ALUMINUM OR COPPER-CLAD ALUMINUM			9 Electrical Ducts (Fig. B.310.2, Detail 4) Types RHW, THHW, THW, THWN, XHHW, USE — ALUMINUM OR COPPER-CLAD ALUMINUM			Size (kcmil)
	RHO 60 LF 50	RHO 90 LF 100	RHO 120 LF 100	RHO 60 LF 50	RHO 90 LF 100	RHO 120 LF 100	RHO 60 LF 50	RHO 90 LF 100	RHO 120 LF 100	RHO 60 LF 50	RHO 90 LF 100	RHO 120 LF 100	RHO 60 LF 50	RHO 90 LF 100	RHO 120 LF 100	RHO 60 LF 50	RHO 90 LF 100	RHO 120 LF 100	
250	410	344	327	386	295	275	369	270	252	320	269	256	302	230	214	288	211	197	250
350	503	418	396	472	355	330	446	322	299	393	327	310	369	277	258	350	252	235	350
500	624	511	484	583	431	400	545	387	360	489	401	379	457	337	313	430	305	284	500
750	794	640	603	736	534	494	674	469	434	626	505	475	581	421	389	538	375	347	750
1000	936	745	700	864	617	570	776	533	493	744	593	557	687	491	453	629	432	399	1000
1250	1055	832	781	970	686	632	854	581	536	848	668	627	779	551	508	703	478	441	1250
1500	1160	907	849	1063	744	685	918	619	571	941	736	689	863	604	556	767	517	477	1500
1750	1250	970	907	1142	793	729	975	651	599	1026	796	745	937	651	598	823	550	507	1750
2000	1332	1027	959	1213	836	768	1030	683	628	1103	850	794	1005	693	636	877	581	535	2000

Ambient Temp. (°C)	Correction Factors						Ambient Temp. (°F)
6–10	1.09	1.09	1.09	1.09	1.09	1.09	43–50
11–15	1.04	1.04	1.04	1.04	1.04	1.04	52–59
16–20	1.00	1.00	1.00	1.00	1.00	1.00	61–68
21–25	0.95	0.95	0.95	0.95	0.95	0.95	70–77
26–30	0.90	0.90	0.90	0.90	0.90	0.90	79–86

Table B.310.6 Ampacities of Three Insulated Conductors, Rated 0 through 2000 Volts, Within an Overall Covering (Three-Conductor Cable) in Underground Electrical Ducts (One Cable per Electrical Duct) Based on Ambient Earth Temperature of 20°C (68°F), Electrical Duct Arrangement per Figure B.310.2, Conductor Temperature 75°C (167°F)

Size (AWG or kcmil)	1 Electrical Duct (Fig. B.310.2, Detail 1) Types RHW, THHW, THW, THWN, XHHW, USE — COPPER			3 Electrical Ducts (Fig. B.310.2, Detail 2) Types RHW, THHW, THW, THWN, XHHW, USE — COPPER			6 Electrical Ducts (Fig. B.310.2, Detail 3) Types RHW, THHW, THW, THWN, XHHW, USE — COPPER			1 Electrical Duct (Fig. B.310.2, Detail 1) Types RHW, THHW, THW, THWN, XHHW, USE — ALUMINUM OR COPPER-CLAD ALUMINUM			3 Electrical Ducts (Fig. B.310.2, Detail 2) Types RHW, THHW, THW, THWN, XHHW, USE — ALUMINUM OR COPPER-CLAD ALUMINUM			6 Electrical Ducts (Fig. B.310.2, Detail 3) Types RHW, THHW, THW, THWN, XHHW, USE — ALUMINUM OR COPPER-CLAD ALUMINUM			Size (AWG or kcmil)
	RHO 60 LF 50	RHO 90 LF 100	RHO 120 LF 100	RHO 60 LF 50	RHO 90 LF 100	RHO 120 LF 100	RHO 60 LF 50	RHO 90 LF 100	RHO 120 LF 100	RHO 60 LF 50	RHO 90 LF 100	RHO 120 LF 100	RHO 60 LF 50	RHO 90 LF 100	RHO 120 LF 100	RHO 60 LF 50	RHO 90 LF 100	RHO 120 LF 100	
8	58	54	53	56	48	46	53	42	39	45	42	41	43	37	36	41	32	30	8
6	77	71	69	74	63	60	70	54	51	60	55	54	57	49	47	54	42	39	6
4	101	93	91	96	81	77	91	69	65	78	72	71	75	63	60	71	54	51	4
2	132	121	118	126	105	100	119	89	83	103	94	92	98	82	78	92	70	65	2
1	154	140	136	146	121	114	137	102	95	120	109	106	114	94	89	107	79	74	1
1/0	177	160	156	168	137	130	157	116	107	138	125	122	131	107	101	122	90	84	1/0
2/0	203	183	178	192	156	147	179	131	121	158	143	139	150	122	115	140	102	95	2/0
3/0	233	210	204	221	178	158	205	148	137	182	164	159	172	139	131	160	116	107	3/0
4/0	268	240	232	253	202	190	234	168	155	209	187	182	198	158	149	183	131	121	4/0
250	297	265	256	280	222	209	258	184	169	233	207	201	219	174	163	202	144	132	250
350	363	321	310	340	267	250	312	219	202	285	252	244	267	209	196	245	172	158	350
500	444	389	375	414	320	299	377	261	240	352	308	297	328	254	237	299	207	190	500
750	552	478	459	511	388	362	462	314	288	446	386	372	413	314	293	374	254	233	750
1000	628	539	518	579	435	405	522	351	321	521	447	430	480	361	336	433	291	266	1000

Ambient Temp. (°C)	Correction Factors						Ambient Temp. (°F)
6–10	1.09	1.09	1.09	1.09	1.09	1.09	43–50
11–15	1.04	1.04	1.04	1.04	1.04	1.04	52–59
16–20	1.00	1.00	1.00	1.00	1.00	1.00	61–68
21–25	0.95	0.95	0.95	0.95	0.95	0.95	70–77
26–30	0.90	0.90	0.90	0.90	0.90	0.90	79–86

Table B.310.7 Ampacities of Three Single Insulated Conductors, Rated 0 Through 2000 Volts, in Underground Electrical Ducts (Three Conductors per Electrical Duct) Based on Ambient Earth Temperature of 20°C (68°F), Electrical Duct Arrangement per Figure B.310.2, Conductor Temperature 75°C (167°F)

Size (AWG or kcmil)	1 Electrical Duct (Fig. B.310.2, Detail 1) Types RHW, THHW, THW, THWN, XHHW, USE — COPPER			3 Electrical Ducts (Fig. B.310.2, Detail 2) Types RHW, THHW, THW, THWN, XHHW, USE			6 Electrical Ducts (Fig. B.310.2, Detail 3) Types RHW, THHW, THW, THWN, XHHW, USE			1 Electrical Duct (Fig. B.310.2, Detail 1) Types RHW, THHW, THW, THWN, XHHW, USE — ALUMINUM OR COPPER-CLAD ALUMINUM			3 Electrical Ducts (Fig. B.310.2, Detail 2) Types RHW, THHW, THW, THWN, XHHW, USE			6 Electrical Ducts (Fig. B.310.2, Detail 3) Types RHW, THHW, THW, THWN, XHHW, USE			Size (AWG or kcmil)
	RHO 60 LF 50	RHO 90 LF 100	RHO 120 LF 100	RHO 60 LF 50	RHO 90 LF 100	RHO 120 LF 100	RHO 60 LF 50	RHO 90 LF 100	RHO 120 LF 100	RHO 60 LF 50	RHO 90 LF 100	RHO 120 LF 100	RHO 60 LF 50	RHO 90 LF 100	RHO 120 LF 100	RHO 60 LF 50	RHO 90 LF 100	RHO 120 LF 100	
8	63	58	57	61	51	49	57	44	41	49	45	44	47	40	38	45	34	32	8
6	84	77	75	80	67	63	75	56	53	66	60	58	63	52	49	59	44	41	6
4	111	100	98	105	86	81	98	73	67	86	78	76	79	67	63	77	57	52	4
3	129	116	113	122	99	94	113	83	77	101	91	89	83	77	73	84	65	60	3
2	147	132	128	139	112	106	129	93	86	115	103	100	108	87	82	101	73	67	2
1	171	153	148	161	128	121	149	106	98	133	119	115	126	100	94	116	83	77	1
1/0	197	175	169	185	146	137	170	121	111	153	136	132	144	114	107	133	94	87	1/0
2/0	226	200	193	212	166	156	194	136	126	176	156	151	165	130	121	151	106	98	2/0
3/0	260	228	220	243	189	177	222	154	142	203	178	172	189	147	138	173	121	111	3/0
4/0	301	263	253	280	215	201	255	175	161	235	205	198	219	168	157	199	137	126	4/0
250	334	290	279	310	236	220	281	192	176	261	227	218	242	185	172	220	150	137	250
300	373	321	308	344	260	242	310	210	192	293	252	242	272	204	190	245	165	151	300
350	409	351	337	377	283	264	340	228	209	321	276	265	296	222	207	266	179	164	350
400	442	376	361	394	302	280	368	243	223	349	297	284	321	238	220	288	191	174	400
500	503	427	409	460	341	316	412	273	249	397	338	323	364	270	250	326	216	197	500
600	552	468	447	511	371	343	457	296	270	446	373	356	408	296	274	365	236	215	600
700	602	509	486	553	402	371	492	319	291	488	408	389	443	321	297	394	255	232	700
750	632	529	505	574	417	385	509	330	301	508	425	405	461	334	309	409	265	241	750
800	654	544	520	597	428	395	527	338	308	530	439	418	481	344	318	427	273	247	800
900	692	575	549	628	450	415	554	355	323	563	466	444	510	365	337	450	288	261	900
1000	730	605	576	659	472	435	581	372	338	597	494	471	538	385	355	475	304	276	1000

Ambient Temp. (°C)	Correction Factors						Ambient Temp. (°F)
6–10	1.09	1.09	1.09	1.09	1.09	1.09	43–50
11–15	1.04	1.04	1.04	1.04	1.04	1.04	52–59
16–20	1.00	1.00	1.00	1.00	1.00	1.00	61–68
21–25	0.95	0.95	0.95	0.95	0.95	0.95	70–77
26–30	0.90	0.90	0.90	0.90	0.90	0.90	79–86

Table B.310.8 Ampacities of Two or Three Insulated Conductors, Rated 0 Through 2000 Volts, Cabled Within an Overall (Two- or Three-Conductor) Covering, Directly Buried in Earth, Based on Ambient Earth Temperature of 20°C (68°F), Arrangement per Figure B.310.2, 100 Percent Load Factor, Thermal Resistance (Rho) of 90

Size (AWG or kcmil)	1 Cable (Fig. B.310.2, Detail 5) 60°C (140°F) UF	1 Cable (Fig. B.310.2, Detail 5) 75°C (167°F) RHW, THHW, THW, THWN, XHHW, USE	2 Cables (Fig. B.310.2, Detail 6) 60°C (140°F) UF	2 Cables (Fig. B.310.2, Detail 6) 75°C (167°F) RHW, THHW, THW, THWN, XHHW, USE	1 Cable (Fig. B.310.2, Detail 5) 60°C (140°F) UF	1 Cable (Fig. B.310.2, Detail 5) 75°C (167°F) RHW, THHW, THW, THWN, XHHW, USE	2 Cables (Fig. B.310.2, Detail 6) 60°C (140°F) UF	2 Cables (Fig. B.310.2, Detail 6) 75°C (167°F) RHW, THHW, THW, THWN, XHHW, USE	Size (AWG or kcmil)
	COPPER				**ALUMINUM OR COPPER-CLAD ALUMINUM**				
8	64	75	60	70	51	59	47	55	8
6	85	100	81	95	68	75	60	70	6
4	107	125	100	117	83	97	78	91	4
2	137	161	128	150	107	126	110	117	2
1	155	182	145	170	121	142	113	132	1
1/0	177	208	165	193	138	162	129	151	1/0
2/0	201	236	188	220	157	184	146	171	2/0
3/0	229	269	213	250	179	210	166	195	3/0
4/0	259	304	241	282	203	238	188	220	4/0
250	—	333	—	308	—	261	—	241	250
350	—	401	—	370	—	315	—	290	350
500	—	481	—	442	—	381	—	350	500
750	—	585	—	535	—	473	—	433	750
1000	—	657	—	600	—	545	—	497	1000

Ambient Temp. (°C)	Correction Factors								Ambient Temp. (°F)
6–10	1.12	1.09	1.12	1.09	1.12	1.09	1.12	1.09	43–50
11–15	1.06	1.04	1.06	1.04	1.06	1.04	1.06	1.04	52–59
16–20	1.00	1.00	1.00	1.00	1.00	1.00	1.00	1.00	61–68
21–25	0.94	0.95	0.94	0.95	0.94	0.95	0.94	0.95	70–77
26–30	0.87	0.90	0.87	0.90	0.87	0.90	0.87	0.90	79–86

Note: For ampacities of Type UF cable in underground electrical ducts, multiply the ampacities shown in the table by 0.74.

Table B.310.9 Ampacities of Three Triplexed Single Insulated Conductors, Rated 0 Through 2000 Volts, Directly Buried in Earth Based on Ambient Earth Temperature of 20°C (68°F), Arrangement per Figure B.310.2, 100 Percent Load Factor, Thermal Resistance (Rho) of 90

Size (AWG or kcmil)	See Fig. B.310.2, Detail 7 60°C (140°F) UF	See Fig. B.310.2, Detail 7 75°C (167°F) USE	See Fig. B.310.2, Detail 8 60°C (140°F) UF	See Fig. B.310.2, Detail 8 75°C (167°F) USE	See Fig. B.310.2, Detail 7 60°C (140°F) UF	See Fig. B.310.2, Detail 7 75°C (167°F) USE	See Fig. B.310.2, Detail 8 60°C (140°F) UF	See Fig. B.310.2, Detail 8 75°C (167°F) USE	Size (AWG or kcmil)
	COPPER				**ALUMINUM OR COPPER-CLAD ALUMINUM**				
8	72	84	66	77	55	65	51	60	8
6	91	107	84	99	72	84	66	77	6
4	119	139	109	128	92	108	85	100	4
2	153	179	140	164	119	139	109	128	2
1	173	203	159	186	135	158	124	145	1
1/0	197	231	181	212	154	180	141	165	1/0
2/0	223	262	205	240	175	205	159	187	2/0
3/0	254	298	232	272	199	233	181	212	3/0
4/0	289	339	263	308	226	265	206	241	4/0
250	—	370	—	336	—	289	—	263	250
350	—	445	—	403	—	349	—	316	350
500	—	536	—	483	—	424	—	382	500
750	—	654	—	587	—	525	—	471	750
1000	—	744	—	665	—	608	—	544	1000

Ambient Temp. (°C)	Correction Factors								Ambient Temp. (°F)
6–10	1.12	1.09	1.12	1.09	1.12	1.09	1.12	1.09	43–50
11–15	1.06	1.04	1.06	1.04	1.06	1.04	1.06	1.04	52–59
16–20	1.00	1.00	1.00	1.00	1.00	1.00	1.00	1.00	61–68
21–25	0.94	0.95	0.94	0.95	0.94	0.95	0.94	0.95	70–77
26–30	0.87	0.90	0.87	0.90	0.87	0.90	0.87	0.90	79–86

Table B.310.10 Ampacities of Three Single Insulated Conductors, Rated 0 Through 2000 Volts, Directly Buried in Earth Based on Ambient Earth Temperature of 20°C (68°F), Arrangement per Figure B.310.2, 100 Percent Load Factor, Thermal Resistance (Rho) of 90

	See Fig. B.310.2, Detail 9		See Fig. B.310.2, Detail 10		See Fig. B.310.2, Detail 9		See Fig. B.310.2, Detail 10		
	60°C (140°F)	75°C (167°F)	60°C (140°F)	75°C (167°F)	60°C (140°F)	75°C (167°F)	60°C (140°F)	75°C (167°F)	
	TYPES				TYPES				
	UF	USE	UF	USE	UF	USE	UF	USE	
Size (AWG or kcmil)	COPPER				ALUMINUM OR COPPER-CLAD ALUMINUM				Size (AWG or kcmil)
8	84	98	78	92	66	77	61	72	8
6	107	126	101	118	84	98	78	92	6
4	139	163	130	152	108	127	101	118	4
2	178	209	165	194	139	163	129	151	2
1	201	236	187	219	157	184	146	171	1
1/0	230	270	212	249	179	210	165	194	1/0
2/0	261	306	241	283	204	239	188	220	2/0
3/0	297	348	274	321	232	272	213	250	3/0
4/0	336	394	309	362	262	307	241	283	4/0
250	—	429	—	394	—	335	—	308	250
350	—	516	—	474	—	403	—	370	350
500	—	626	—	572	—	490	—	448	500
750	—	767	—	700	—	605	—	552	750
1000	—	887	—	808	—	706	—	642	1000
1250	—	979	—	891	—	787	—	716	1250
1500	—	1063	—	965	—	862	—	783	1500
1750	—	1133	—	1027	—	930	—	843	1750
2000	—	1195	—	1082	—	990	—	897	2000

Ambient Temp. (°C)	Correction Factors								Ambient Temp. (°F)
6–10	1.12	1.09	1.12	1.09	1.12	1.09	1.12	1.09	43–50
11–15	1.06	1.04	1.06	1.04	1.06	1.04	1.06	1.04	52–59
16–20	1.00	1.00	1.00	1.00	1.00	1.00	1.00	1.00	61–68
21–25	0.94	0.95	0.94	0.95	0.94	0.95	0.94	0.95	70–77
26–30	0.87	0.90	0.87	0.90	0.87	0.90	0.87	0.90	79–86

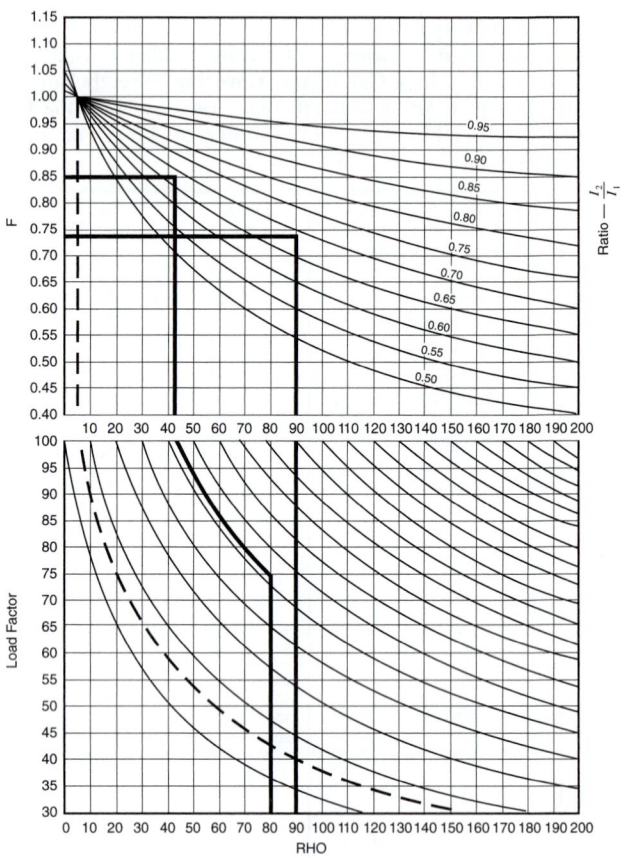

Figure B.310.1 Interpolation chart for cables in a duct bank
I_1 = ampacity for Rho = 60, 50 LF; I_2 = ampacity for Rho = 120, 100 LF (load factor); desired ampacity = F × I_1.

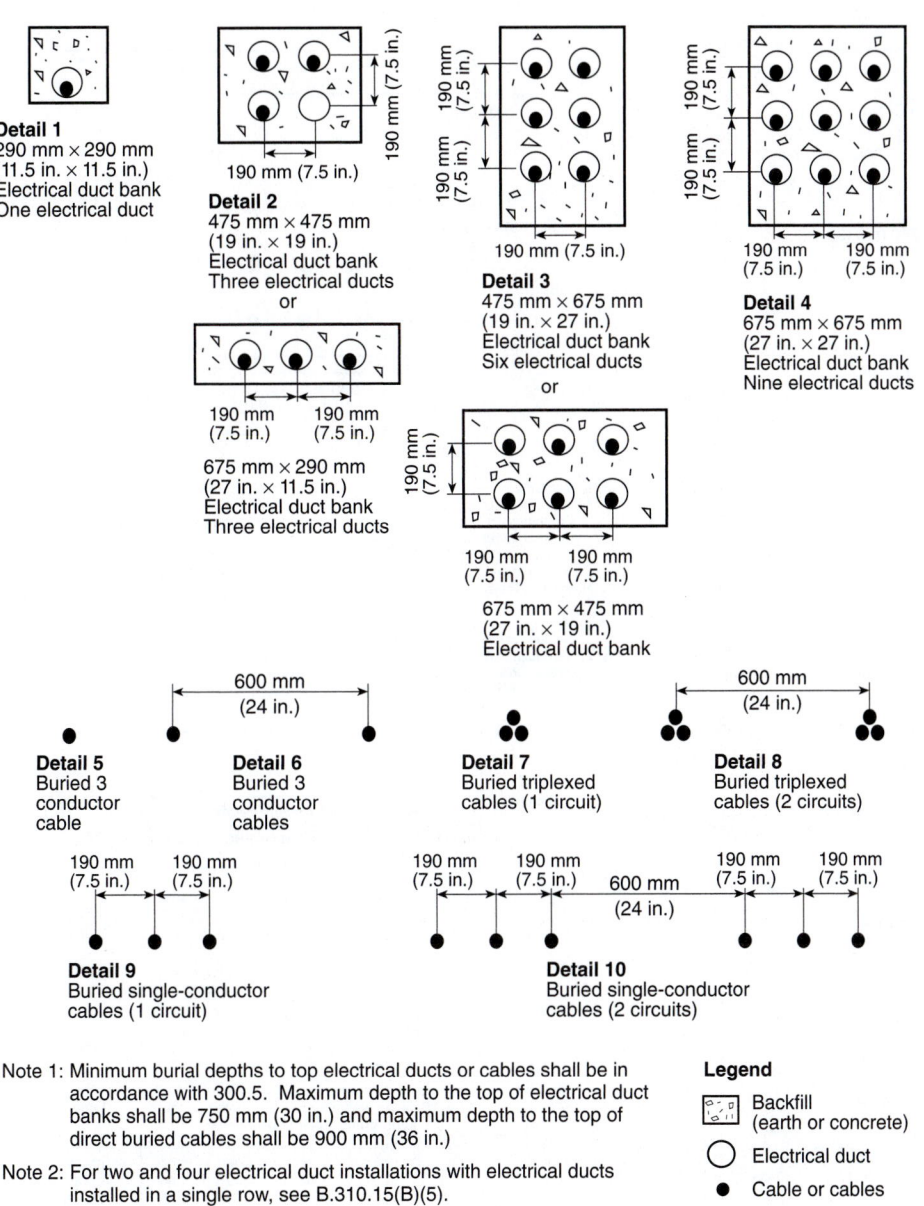

Detail 1
290 mm × 290 mm
(11.5 in. × 11.5 in.)
Electrical duct bank
One electrical duct

Detail 2
475 mm × 475 mm
(19 in. × 19 in.)
Electrical duct bank
Three electrical ducts
or

190 mm (7.5 in.)

675 mm × 290 mm
(27 in. × 11.5 in.)
Electrical duct bank
Three electrical ducts

Detail 3
475 mm × 675 mm
(19 in. × 27 in.)
Electrical duct bank
Six electrical ducts
or

675 mm × 475 mm
(27 in. × 19 in.)
Electrical duct bank

Detail 4
675 mm × 675 mm
(27 in. × 27 in.)
Electrical duct bank
Nine electrical ducts

600 mm (24 in.)

Detail 5
Buried 3
conductor
cable

Detail 6
Buried 3
conductor
cables

Detail 7
Buried triplexed
cables (1 circuit)

600 mm (24 in.)

Detail 8
Buried triplexed
cables (2 circuits)

190 mm (7.5 in.) 190 mm (7.5 in.)

Detail 9
Buried single-conductor
cables (1 circuit)

190 mm (7.5 in.) 190 mm (7.5 in.) 600 mm (24 in.) 190 mm (7.5 in.) 190 mm (7.5 in.)

Detail 10
Buried single-conductor
cables (2 circuits)

Note 1: Minimum burial depths to top electrical ducts or cables shall be in accordance with 300.5. Maximum depth to the top of electrical duct banks shall be 750 mm (30 in.) and maximum depth to the top of direct buried cables shall be 900 mm (36 in.)

Note 2: For two and four electrical duct installations with electrical ducts installed in a single row, see B.310.15(B)(5).

Legend

Backfill (earth or concrete)

○ Electrical duct

● Cable or cables

Figure B.310.2 Cable installation dimensions for use with Table B.310.5 through Table B.310.10.

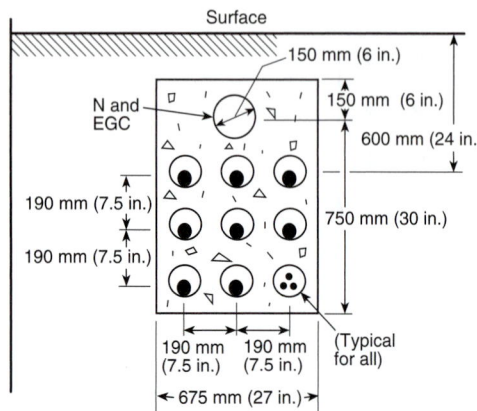

Design Criteria
Neutral and Equipment
  Grounding conductor (EGC)
    Duct = 150 mm (6 in.)
Phase Ducts = 75 to 125 mm (3 to 5 in.)
Conductor Material = Copper
Number of Cables per Duct = 3

Number of Cables per Phase = 9
Rho concrete = Rho Earth – 5

Rho PVC Duct = 650
Rho Cable Insulation = 500
Rho Cable Jacket = 650

Notes:
1. Neutral configuration per 300.5(I), Exception No. 2, for isolated phase installations in nonmagnetic ducts.
2. Phasing is A, B, C in rows or columns. Where magnetic electrical ducts are used, conductors are installed A, B, C per electrical duct with the neutral and all equipment grounding conductors in the same electrical duct. In this case, the 6-in. trade size neutral duct is eliminated.
3. Maximum harmonic loading on the neutral conductor cannot exceed 50 percent of the phase current for the ampacities shown in the table.
4. Metallic shields of Type MV-90 cable shall be grounded at one point only where using A, B, C phasing in rows or columns.

Size kcmil	TYPES RHW, THHW, THW, THWN, XHHW, USE, OR MV-90*			Size kcmil
	Total per Phase Ampere Rating			
	RHO EARTH 60 LF 50	RHO EARTH 90 LF 100	RHO EARTH 120 LF 100	
250	2340 (260A/Cable)	1530 (170A/Cable)	1395 (155A/Cable)	250
350	2790 (310A/Cable)	1800 (200A/Cable)	1665 (185A/Cable)	350
500	3375 (375A/Cable)	2160 (240A/Cable)	1980 (220A/Cable)	350

Ambient Temp. (°C)	For ambient temperatures other than 20°C (68°F), multiply the ampacities shown above by the appropriate factor shown below.					Ambient Temp. (°F)
6–10	1.09	1.09	1.09	1.09	1.09	43–50
11–15	1.04	1.04	1.04	1.04	1.04	52–59
16–20	1.00	1.00	1.00	1.00	1.00	61–68
21–25	0.95	0.95	0.95	0.95	0.95	70–77
26–30	0.90	0.90	0.90	0.90	0.90	79–86

*Limited to 75°C conductor temperature.

FPN Figure B.310.3 Ampacities of single insulated conductors rated 0 through 5000 volts in underground electrical ducts (three conductors per electrical duct), nine single-conductor cables per phase based on ambient earth temperature of 20°C (68°F), conductor temperature 75°C (167°F).

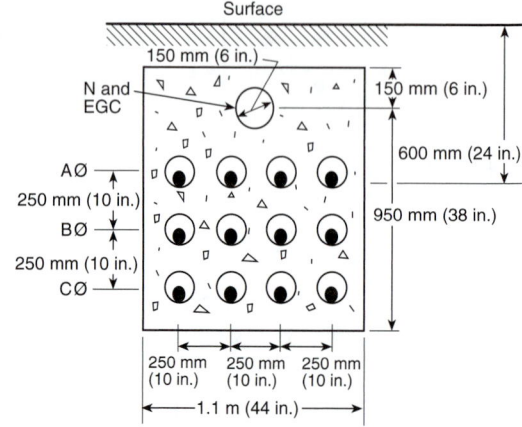

Design Criteria
Neutral and Equipment
  Grounding Conductor (EGC)
    Duct = 150 mm (6 in.)
Phase Ducts = 75 mm (3 in.)
Conductor Material = Copper
Number of Cables per Duct = 1

Number of Cables per Phase = 4
Rho Concrete = Rho Earth – 5
Rho PVC Duct = 650

Rho Cable Insulation = 500
Rho Cable Jacket = 650

Notes:
1. Neutral configuration per 300.5(I), Exception No. 2.
2. Maximum harmonic loading on the neutral conductor cannot exceed 50 percent of the phase current for the ampacities shown in the table.
3. Metallic shields of Type MV-90 cable shall be grounded at one point only.

Size kcmil	TYPES RHW, THHW, THW, THWN, XHHW, USE, OR MV-90*			Size kcmil
	Total per Phase Ampere Rating			
	RHO EARTH 60 LF 50	RHO EARTH 90 LF 100	RHO EARTH 120 LF 100	
750	2820 (705A/Cable)	1860 (465A/Cable)	1680 (420A/Cable)	750
1000	3300 (825A/Cable)	2140 (535A/Cable)	1920 (480A/Cable)	1000
1250	3700 (925A/Cable)	2380 (595A/Cable)	2120 (530A/Cable)	1250
1500	4060 (1015A/Cable)	2580 (645A/Cable)	2300 (575A/Cable)	1500
1750	4360 (1090A/Cable)	2740 (685A/Cable)	2460 (615A/Cable)	1750

Ambient Temp. (°C)	For ambient temperatures other than 20°C (68°F), multiply the ampacities shown above by the appropriate factor shown below.					Ambient Temp. (°F)
6–10	1.09	1.09	1.09	1.09	1.09	43–50
11–15	1.04	1.04	1.04	1.04	1.04	52–59
16–20	1.00	1.00	1.00	1.00	1.00	61–68
21–25	0.95	0.95	0.95	0.95	0.95	70–77
26–30	0.90	0.90	0.90	0.90	0.90	79–86

*Limited to 75°C conductor temperature.

FPN Figure B.310.4 Ampacities of single insulated conductors rated 0 through 5000 volts in nonmagnetic underground electrical ducts (one conductor per electrical duct), four single-conductor cables per phase based on ambient earth temperature of 20°C (68°F), conductor temperature 75°C (167°F).

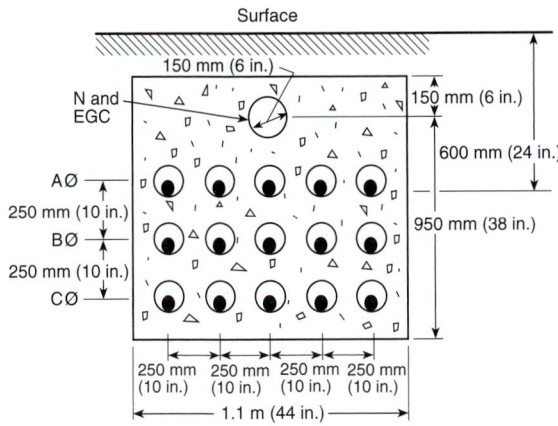

Surface

150 mm (6 in.)

N and
EGC

150 mm (6 in.)

600 mm (24 in.)

AØ

250 mm (10 in.)

BØ

950 mm (38 in.)

250 mm (10 in.)

CØ

250 mm   250 mm   250 mm   250 mm
(10 in.)   (10 in.)   (10 in.)   (10 in.)

1.1 m (44 in.)

Design Criteria
Neutral and Equipment
    Grounding Conductor (EGC)
    Duct = 150 mm (6 in.)
Phase Ducts = 75 mm (3 in.)
Conductor Material = Copper
Number of Cables per Duct = 1

Number of Cables per Phase = 5
Rho Concrete = Rho Earth – 5
Rho PVC Duct = 650

Rho Cable Insulation = 500
Rho Cable Jacket = 650

Notes:
1. Neutral configuration per 300.5(I), Exception No. 2.
2. Maximum harmonic loading on the neutral conductor cannot exceed 50 percent of the
   phase current for the ampacities shown in the table.
3. Metallic shields of Type MV-90 cable shall be grounded at one point only.

Size kcmil	TYPES RHW, THHW, THW, THWN, XHHW, USE, OR MV-90*			Size kcmil
	Total per Phase Ampere Rating			
	RHO EARTH 60 LF 50	RHO EARTH 90 LF 100	RHO EARTH 120 LF 100	
2000	5575 (1115A/Cable)	3375 (675A/Cable)	3000 (600A/Cable)	2000

Ambient Temp. (°C)	For ambient temperatures other than 20°C (68°F), multiply the ampacities shown above by the appropriate factor shown below.					Ambient Temp. (°F)
6–10	1.09	1.09	1.09	1.09	1.09	43–50
11–15	1.04	1.04	1.04	1.04	1.04	52–59
16–20	1.00	1.00	1.00	1.00	1.00	61–68
21–25	0.95	0.95	0.95	0.95	0.95	70–77
26–30	0.90	0.90	0.90	0.90	0.90	79–86

*Limited to 75°C conductor temperature.

FPN Figure B.310.5 Ampacities of single insulated conductors rated 0 through 5000 volts in nonmagnetic underground electrical ducts (one conductor per electrical duct), five single-conductor cables per phase based on ambient earth temperature of 20°C (68°F), conductor temperature 75°C (167°F).

Where the load factor is less than 100 percent and can be verified by measurement or calculation, the ampacity of electrical duct bank installations can be modified as shown. Different values of Rho can be accommodated in the same manner.

Table B.310.11 Adjustment Factors for More Than Three Current-Carrying Conductors in a Raceway or Cable with Load Diversity

Number of Current-Carrying Conductors	Percent of Values in Tables as Adjusted for Ambient Temperature if Necessary
4 – 6	80
7 – 9	70
10 – 24	70*
25 – 42	60*
43 – 85	50*

*These factors include the effects of a load diversity of 50 percent.

FPN: The ampacity limit for the number of current-carrying conductors in 10 through 85 is based on the following formula. For greater than 85 conductors, special calculations are required that are beyond the scope of this table.

$$A_2 = \sqrt{\frac{0.5N}{E}} \times (A^1) \text{ or } A^1, \text{ whichever is less}$$

where:

A_1 = ampacity from Table 310.16; Table 310.18; Table B.310.1; Table B.310.6; and Table B.310.7 multiplied by the appropriate factor from Table B.310.11.

N = total number of conductors used to obtain multiplying factor from Table B.310.11

E = desired number of current-carrying conductors in the raceway or cable

A_2 = ampacity limit for the current-carrying conductors in the raceway or cable

Example 1
Calculate the ampacity limit for twelve 14 AWG THWN current-carrying conductors (75°C) in a raceway that contains 24 conductors.

$$A^2 = \sqrt{\frac{(0.5)(24)}{12}} \times 20(0.7)$$

$$= 14 \text{ amperes (i.e., 50 percent diversity)}$$

Example 2
Calculate the ampacity limit for eighteen 14 AWG THWN current-carrying conductors (75°C) in a raceway that contains 24 conductors.

$$A^2 = \sqrt{\frac{(0.5)(24)}{18}} \times 20(0.7) = 11.5 \text{ amperes}$$

This annex is not a part of the requirements of this NFPA document but is included for informational purposes only.

Table

C1 — Electrical Metallic Tubing (EMT)
C1(A)* — Electrical Metallic Tubing (EMT)
C2 — Electrical Nonmetallic Tubing (ENT)
C2(A)* — Electrical Nonmetallic Tubing (ENT)
C3 — Flexible Metal Conduit (FMC)
C3(A)* — Flexible Metal Conduit (FMC)
C4 — Intermediate Metal Conduit (IMC)
C4(A)* — Intermediate Metal Conduit (IMC)
C5 — Liquidtight Flexible Nonmetallic Conduit (Type LFNC-B)
C5(A)* — Liquidtight Flexible Nonmetallic Conduit (Type LFNC-B)
C6 — Liquidtight Flexible Nonmetallic Conduit (Type LFNC-A)
C6(A)* — Liquidtight Flexible Nonmetallic Conduit (Type LFNC-A)
C7 — Liquidtight Flexible Metal Conduit (LFML)
C7(A)* — Liquidtight Flexible Metal Conduit (LFML)
C8 — Rigid Metal Conduit (RMC)
C8(A)* — Rigid Metal Conduit (RMC)
C9 — Rigid PVC Conduit, Schedule 80
C9(A)* — Rigid PVC Conduit, Schedule 80
C10 — Rigid PVC Conduit, Schedule 40 and HDPE Conduit
C10(A)* — Rigid PVC Conduit, Schedule 40 and HDPE Conduit
C11 — Type A, Rigid PVC Conduit
C11(A)* — Type A, Rigid PVC Conduit
C12 — Type EB, PVC Conduit
C12(A)* — Type EB, PVC Conduit

*Where this table is used in conjunction with Tables C1 through C12, the conductors installed must be of the compact type.

Contents

in Rigid Metal Conduit (RMC) (Based on Table 1, Chapter 9)

Table C9. Maximum Number of Conductors or Fixture Wires in Rigid PVC Conduit, Schedule 80 (Based on Table 1, Chapter 9)

Table C9(A). Maximum Number of Compact Conductors in Rigid PVC Conduit, Schedule 80 (Based on Table 1, Chapter 9)

Table C10. Maximum Number of Conductors or Fixture Wires in Rigid PVC Conduit, Schedule 40 and HDPE Conduit (Based on Table 1, Chapter 9)

Table C10(A). Maximum Number of Compact Conductors in Rigid PVC Conduit, Schedule 40 and HDPE Conduit (Based on Table 1, Chapter 9)

Table C11. Maximum Number of Conductors or Fixture Wires in Type A, Rigid PVC Conduit (Based on Table 1, Chapter 9)

Table C11(A). Maximum Number of Compact Conductors in Type A, Rigid PVC Conduit (Based on Table 1, Chapter 9)

Table C12. Maximum Number of Conductors in Type EB, PVC Conduit (Based on Table 1, Chapter 9)

Table C12(A). Maximum Number of Compact Conductors in Type EB, PVC Conduit (Based on Table 1, Chapter 9)

Annex C is not a part of the requirements of this *Code* and is included only for information. The tables in Annex C provide the user with accurate calculations that are based on Chapter 9, Table 1.

The Annex C tables came about because of the different internal dimensions of the many types of conduit and tubing. Tables 3A, 3B, and 3C in previous editions of the *Code* were no longer being used in calculating conduit fill.

As in past editions of the *Code,* Chapter 9, Table 1, sets forth the percentage fill required, and Tables 4, 5, and 5A list the accurate conduit, tubing, and wire dimensions. The user can calculate the percent fill, as permitted in past editions of the *Code,* or use the tables in Annex C, all of which were generated using Chapter 9, Tables 1, 4, 5, and 5A.

At first glance, the number of tables contained in this annex may appear daunting. However, after closer inspection, the reader will realize that there are only 12 sets of tables, corresponding with the 12 wiring methods.

Each set of tables contains three types of tables. First are the conductors for general wiring, which appear in an order similar to previous Chapter 9, Tables 3A, 3B, and 3C. Next are fixture wires, in an order similar to the previous Chapter 9, Table 2. Finally, compact stranded conductors are listed, similar to previous Chapter 9, Table 5A. Tables that use compact stranding are listed as "A" tables.

To select the correct trade size conduit or tubing, the user should proceed as follows.

STEP 1. Select the appropriate wiring method from Tables C1 through C12 using the following lists of metallic and nonmetallic wiring methods.

Metallic Wiring Methods

Type of Wiring	Appropriate Table
Electrical metallic tubing	Table C1
Flexible metal conduit	Table C3
Intermediate metal conduit	Table C4
Liquidtight flexible metal conduit	Table C7
Rigid metal conduit	Table C8

Nonmetallic Wiring Methods

Type of Wiring	Appropriate Table
Electrical nonmetallic tubing	Table C2
Liquidtight flexible nonmetallic conduit (LFNC-B)	Table C5
Liquidtight flexible nonmetallic conduit (LFNC-A)	Table C6
Rigid PVC conduit, Schedule 80	Table C9
Rigid PVC conduit, Schedule 40 and HDPE conduit	Table C10
Type A, rigid PVC conduit	Table C11
Type EB PVC conduit	Table C12

STEP 2. Choose the appropriate conductors (general wiring conductors, fixture wires, or compact conductors).

STEP 3. Choose the appropriate insulation.

STEP 4. Select the correct trade size conduit or tubing for the given quantity and size of conductors required.

The following examples show how to determine the correct trade size conduit or tubing.

Example 1

A circuit will require ten 10 AWG THWN-2 conductors in an underground conduit across a parking lot for exterior lighting. What size PVC conduit will be required?

Solution

Annex C lists the following types of PVC conduit:

Wiring Method	Table	Trade Size Conduit or Tubing
PVC Schedule 40 and HDPE	Table C10	1
PVC Type A	Table C11	¾
PVC Schedule 80	Table C9	1
PVC Type EB	Table C12	Available only in trade sizes 2 and larger

Example 2

An underground service lateral requires four 600-kcmil XHHW compact-stranded aluminum conductors. What trade size conduit will be required for RMC, IMC, PVC Schedule 40, PVC Schedule 80, and PVC Type EB?

Solution

Annex C lists the following:

Wiring Method	Table	Trade Size Conduit or Tubing
RMC	Table C8A	3
IMC	Table C4A	3
PVC Schedule 40	Table C10A	3
PVC Schedule 80	Table C9A	3½
PVC Type EB	Table C12A	3

Most aluminum building wire in Types THW, THWN/THHN, and XHHW is compact stranded.

Example 3

A 40-hp motor will be supplied by three 4 AWG TW conductors. What size metal conduit or tubing will be required? What size metal flex will be required at the motor termination?

Solution

Annex C lists various types of metal conduit and metal flex as follows:

Wiring Method	Table	Trade Size Conduit or Tubing
EMT	Table C1	1
IMC	Table C4	1
RMC	Table C8	1
FMC	Table C3	1
Liquidtight FMC	Table C7	1

Example 4

A fire alarm system will require the riser to contain twenty-one 16 AWG TFF conductors. What size electrical metallic tubing will be required?

Solution

According to Table C1, trade size 1 EMT is required.

Table C1 Maximum Number of Conductors or Fixture Wires in Electrical Metallic Tubing (EMT) *(Based on Table 1, Chapter 9)*

Type	Conductor Size (AWG kcmil)	16 (½)	21 (¾)	27 (1)	35 (1¼)	41 (1½)	53 (2)	63 (2½)	78 (3)	91 (3½)	103 (4)
RHH,	14	4	7	11	20	27	46	80	120	157	201
RHW,	12	3	6	9	17	23	38	66	100	131	167
RHW-2	10	2	5	8	13	18	30	53	81	105	135
	8	1	2	4	7	9	16	28	42	55	70
	6	1	1	3	5	8	13	22	34	44	56
	4	1	1	2	4	6	10	17	26	34	44
	3	1	1	1	4	5	9	15	23	30	38
	2	1	1	1	3	4	7	13	20	26	33
	1	0	1	1	1	3	5	9	13	17	22
	1/0	0	1	1	1	2	4	7	11	15	19
	2/0	0	1	1	1	2	4	6	10	13	17
	3/0	0	0	1	1	1	3	5	8	11	14
	4/0	0	0	1	1	1	3	5	7	9	12
	250	0	0	0	1	1	1	3	5	7	9
	300	0	0	0	1	1	1	3	5	6	8
	350	0	0	0	1	1	1	3	4	6	7
	400	0	0	0	1	1	1	2	4	5	7
	500	0	0	0	0	1	1	2	3	4	6
	600	0	0	0	0	1	1	1	3	4	5
	700	0	0	0	0	0	1	1	2	3	4
	750	0	0	0	0	0	1	1	2	3	4
	800	0	0	0	0	0	1	1	2	3	4
	900	0	0	0	0	0	1	1	1	3	3
	1000	0	0	0	0	0	1	1	1	2	3
	1250	0	0	0	0	0	0	1	1	1	2
	1500	0	0	0	0	0	0	1	1	1	1
	1750	0	0	0	0	0	0	1	1	1	1
	2000	0	0	0	0	0	0	1	1	1	1
TW,	14	8	15	25	43	58	96	168	254	332	424
THHW,	12	6	11	19	33	45	74	129	195	255	326
THW,	10	5	8	14	24	33	55	96	145	190	243
THW-2	8	2	5	8	13	18	30	53	81	105	135
RHH*,	14	6	10	16	28	39	64	112	169	221	282
RHW*,	12	4	8	13	23	31	51	90	136	177	227
RHW-2*	10	3	6	10	18	24	40	70	106	138	177
	8	1	4	6	10	14	24	42	63	83	106
RHH*,	6	1	3	4	8	11	18	32	48	63	81
RHW*,	4	1	1	3	6	8	13	24	36	47	60
RHW-2*,	3	1	1	3	5	7	12	20	31	40	52
TW, THW,	2	1	1	2	4	6	10	17	26	34	44
THW											
THHW,	1	1	1	1	3	4	7	12	18	24	31
THW-2	1/0	0	1	1	2	3	6	10	16	20	26
	2/0	0	1	1	1	3	5	9	13	17	22
	3/0	0	1	1	1	2	4	7	11	15	19
	4/0	0	0	1	1	1	3	6	9	12	16
	250	0	0	1	1	1	3	5	7	10	13
	300	0	0	1	1	1	2	4	6	8	11
	350	0	0	0	1	1	1	4	6	7	10
	400	0	0	0	1	1	1	3	5	7	9
	500	0	0	0	1	1	1	3	4	6	7
	600	0	0	0	1	1	1	2	3	4	6
	700	0	0	0	0	1	1	1	3	4	5
	750	0	0	0	0	1	1	1	3	4	5
	800	0	0	0	0	1	1	1	3	3	5
	900	0	0	0	0	0	1	1	2	3	4
	1000	0	0	0	0	0	1	1	2	3	4
	1250	0	0	0	0	0	1	1	1	2	3
	1500	0	0	0	0	0	1	1	1	1	2
	1750	0	0	0	0	0	0	1	1	1	2
	2000	0	0	0	0	0	0	1	1	1	1

Table C1 Continued

CONDUCTORS

Type	Conductor Size (AWG kcmil)	Metric Designator (Trade Size)									
		16 (½)	21 (¾)	27 (1)	35 (1¼)	41 (1½)	53 (2)	63 (2½)	78 (3)	91 (3½)	103 (4)
THHN, THWN, THWN-2	14	12	22	35	61	84	138	241	364	476	608
	12	9	16	26	45	61	101	176	266	347	443
	10	5	10	16	28	38	63	111	167	219	279
	8	3	6	9	16	22	36	64	96	126	161
	6	2	4	7	12	16	26	46	69	91	116
	4	1	2	4	7	10	16	28	43	56	71
	3	1	1	3	6	8	13	24	36	47	60
	2	1	1	3	5	7	11	20	30	40	51
	1	1	1	1	4	5	8	15	22	29	37
	1/0	1	1	1	3	4	7	12	19	25	32
	2/0	0	1	1	2	3	6	10	16	20	26
	3/0	0	1	1	1	3	5	8	13	17	22
	4/0	0	1	1	1	2	4	7	11	14	18
	250	0	0	1	1	1	3	6	9	11	15
	300	0	0	1	1	1	3	5	7	10	13
	350	0	0	1	1	1	2	4	6	9	11
	400	0	0	0	1	1	1	4	6	8	10
	500	0	0	0	1	1	1	3	5	6	8
	600	0	0	0	1	1	1	2	4	5	7
	700	0	0	0	1	1	1	2	3	4	6
	750	0	0	0	0	1	1	1	3	4	5
	800	0	0	0	0	1	1	1	3	4	5
	900	0	0	0	0	1	1	1	3	3	4
	1000	0	0	0	0	1	1	1	2	3	4
FEP, FEPB, PFA, PFAH, TFE	14	12	21	34	60	81	134	234	354	462	590
	12	9	15	25	43	59	98	171	258	337	430
	10	6	11	18	31	42	70	122	185	241	309
	8	3	6	10	18	24	40	70	106	138	177
	6	2	4	7	12	17	28	50	75	98	126
	4	1	3	5	9	12	20	35	53	69	88
	3	1	2	4	7	10	16	29	44	57	73
	2	1	1	3	6	8	13	24	36	47	60
PFA, PFAH, TFE	1	1	1	2	4	6	9	16	25	33	42
PFA, PFAH, TFE, Z	1/0	1	1	1	3	5	8	14	21	27	35
	2/0	0	1	1	3	4	6	11	17	22	29
	3/0	0	1	1	2	3	5	9	14	18	24
	4/0	0	1	1	1	2	4	8	11	15	19
Z	14	14	25	41	72	98	161	282	426	556	711
	12	10	18	29	51	69	114	200	302	394	504
	10	6	11	18	31	42	70	122	185	241	309
	8	4	7	11	20	27	44	77	117	153	195
	6	3	5	8	14	19	31	54	82	107	137
	4	1	3	5	9	13	21	37	56	74	94
	3	1	2	4	7	9	15	27	41	54	69
	2	1	1	3	6	8	13	22	34	45	57
	1	1	1	2	4	6	10	18	28	36	46
XHH, XHHW, XHHW-2, ZW	14	8	15	25	43	58	96	168	254	332	424
	12	6	11	19	33	45	74	129	195	255	326
	10	5	8	14	24	33	55	96	145	190	243
	8	2	5	8	13	18	30	53	81	105	135
	6	1	3	6	10	14	22	39	60	78	100
	4	1	2	4	7	10	16	28	43	56	72
	3	1	1	3	6	8	14	24	36	48	61
	2	1	1	3	5	7	11	20	31	40	51
XHH, XHHW, XHHW-2	1	1	1	1	4	5	8	15	23	30	38
	1/0	1	1	1	3	4	7	13	19	25	32
	2/0	0	1	1	2	3	6	10	16	21	27
	3/0	0	1	1	1	3	5	9	13	17	22
	4/0	0	1	1	1	2	4	7	11	14	18
	250	0	0	1	1	1	3	6	9	12	15
	300	0	0	1	1	1	3	5	8	10	13
	350	0	0	1	1	1	2	4	7	9	11
	400	0	0	0	1	1	1	4	6	8	10
	500	0	0	0	1	1	1	3	5	6	8
	600	0	0	0	1	1	1	2	4	5	6
	700	0	0	0	0	1	1	2	3	4	6
	750	0	0	0	0	1	1	1	3	4	5
	800	0	0	0	0	1	1	1	3	4	5
	900	0	0	0	0	1	1	1	3	3	4
	1000	0	0	0	0	0	1	1	2	3	4
	1250	0	0	0	0	0	1	1	1	2	3
	1500	0	0	0	0	0	1	1	1	1	3
	1750	0	0	0	0	0	0	1	1	1	2
	2000	0	0	0	0	0	0	1	1	1	1

Table C1 Continued

FIXTURE WIRES

Type	Conductor Size (AWG/ kcmil)	Metric Designator (Trade Size)					
		16 (½)	21 (¾)	27 (1)	35 (1¼)	41 (1½)	53 (2)
FFH-2, RFH-2, RFHH-3	18	8	14	24	41	56	92
	16	7	12	20	34	47	78
SF-2, SFF-2	18	10	18	30	52	71	116
	16	8	15	25	43	58	96
	14	7	12	20	34	47	78
SF-1, SFF-1	18	18	33	53	92	125	206
RFH-1, RFHH-2, TF, TFF, XF, XFF	18	14	24	39	68	92	152
RFHH-2, TF, TFF, XF, XFF	16	11	19	31	55	74	123
XF, XFF	14	8	15	25	43	58	96
TFN, TFFN	18	22	38	63	108	148	244
	16	17	29	48	83	113	186
PF, PFF, PGF, PGFF, PAF, PTF, PTFF, PAFF	18	21	36	59	103	140	231
	16	16	28	46	79	108	179
	14	12	21	34	60	81	134
ZF, ZFF, ZHF, HF, HFF	18	27	47	77	133	181	298
	16	20	35	56	98	133	220
	14	14	25	41	72	98	161
KF-2, KFF-2	18	39	69	111	193	262	433
	16	27	48	78	136	185	305
	14	19	33	54	93	127	209
	12	13	23	37	64	87	144
	10	8	15	25	43	58	96
KF-1, KFF-1	18	46	82	133	230	313	516
	16	33	57	93	161	220	362
	14	22	38	63	108	148	244
	12	14	25	41	72	98	161
	10	9	16	27	47	64	105
XF, XFF	12	4	8	13	23	31	51
	10	3	6	10	18	24	40

Note; This table is for concentric stranded conductors only. For compact stranded conductors, Table C1(A) should be used.
*Types RHH, RHW, and RHW-2 without outer covering.

Table C1(A) Maximum Number of Compact Conductors in Electrical Metallic Tubing (EMT) (*Based on Table 1, Chapter 9*)

COMPACT CONDUCTORS

Type	Conductor Size (AWG/ kcmil)	16 (½)	21 (¾)	27 (1)	35 (1¼)	41 (1½)	53 (2)	63 (2½)	78 (3)	91 (3½)	103 (4)
THW,	8	2	4	6	11	16	26	46	69	90	115
THW-2,	6	1	3	5	9	12	20	35	53	70	89
THHW	4	1	2	4	6	9	15	26	40	52	67
	2	1	1	3	5	7	11	19	29	38	49
	1	1	1	1	3	4	8	13	21	27	34
	1/0	1	1	1	3	4	7	12	18	23	30
	2/0	0	1	1	2	3	5	10	15	20	25
	3/0	0	1	1	1	3	5	8	13	17	21
	4/0	0	1	1	1	2	4	7	11	14	18
	250	0	0	1	1	1	3	5	8	11	14
	300	0	0	1	1	1	3	5	7	9	12
	350	0	0	1	1	1	2	4	6	8	11
	400	0	0	0	1	1	1	4	6	8	10
	500	0	0	0	1	1	1	3	5	6	8
	600	0	0	0	1	1	1	2	4	5	7
	700	0	0	0	1	1	1	2	3	4	6
	750	0	0	0	0	1	1	1	3	4	5
	1000	0	0	0	0	1	1	1	2	3	4
THHN,	8	—	—	—	—	—	—	—	—	—	—
THWN,	6	2	4	7	13	18	29	52	78	102	130
THWN-2	4	1	3	4	8	11	18	32	48	63	81
	2	1	1	3	6	8	13	23	34	45	58
	1	1	1	2	4	6	10	17	26	34	43
	1/0	1	1	1	3	5	8	14	22	29	37
	2/0	1	1	1	3	4	7	12	18	24	30
	3/0	0	1	1	2	3	6	10	15	20	25
	4/0	0	1	1	1	3	5	8	12	16	21
	250	0	1	1	1	1	4	6	10	13	16
	300	0	0	1	1	1	3	5	8	11	14
	350	0	0	1	1	1	3	5	7	10	12
	400	0	0	1	1	1	2	4	6	9	11
	500	0	0	0	1	1	1	4	5	7	9
	600	0	0	0	1	1	1	3	4	6	7
	700	0	0	0	1	1	1	2	4	5	7
	750	0	0	0	1	1	1	2	4	5	6
	1000	0	0	0	0	1	1	1	3	3	4
XHHW,	8	3	5	8	15	20	34	59	90	117	149
XHHW-2	6	1	4	6	11	15	25	44	66	87	111
	4	1	3	4	8	11	18	32	48	63	81
	2	1	1	3	6	8	13	23	34	45	58
	1	1	1	2	4	6	10	17	26	34	43
	1/0	1	1	1	3	5	8	14	22	29	37
	2/0	1	1	1	3	4	7	12	18	24	31
	3/0	0	1	1	2	3	6	10	15	20	25
	4/0	0	1	1	1	3	5	8	13	17	21
	250	0	1	1	1	2	4	7	10	13	17
	300	0	0	1	1	1	3	6	9	11	14
	350	0	0	1	1	1	3	5	8	10	13
	400	0	0	1	1	1	2	4	7	9	11
	500	0	0	0	1	1	1	4	6	7	9
	600	0	0	0	1	1	1	3	4	6	8
	700	0	0	0	1	1	1	2	4	5	7
	750	0	0	0	1	1	1	2	3	5	6
	1000	0	0	0	0	1	1	1	3	4	5

Definition: *Compact stranding* is the result of a manufacturing process where the standard conductor is compressed to the extent that the interstices (voids between strand wires) are virtually eliminated.

Table C2 Maximum Number of Conductors or Fixture Wires in Electrical Nonmetallic Tubing (ENT) (*Based on Table 1, Chapter 9*)

CONDUCTORS

Type	Conductor Size (AWG/ kcmil)	16 (½)	21 (¾)	27 (1)	35 (1¼)	41 (1½)	53 (2)
RHH, RHW, RHW-2	14	3	6	10	19	26	43
	12	2	5	9	16	22	36
RHH, RHW, RHW-2	10	1	4	7	13	17	29
	8	1	1	3	6	9	15
	6	1	1	3	5	7	12
	4	1	1	2	4	6	9
	3	1	1	1	3	5	8
	2	0	1	1	3	4	7
	1	0	1	1	1	3	5
	1/0	0	0	1	1	2	4
	2/0	0	0	1	1	1	3
	3/0	0	0	1	1	1	3
	4/0	0	0	1	1	1	2
	250	0	0	0	1	1	1
	300	0	0	0	1	1	1
	350	0	0	0	1	1	1
	400	0	0	0	1	1	1
	500	0	0	0	0	1	1
	600	0	0	0	0	1	1
	700	0	0	0	0	0	1
	750	0	0	0	0	0	1
	800	0	0	0	0	0	1
	900	0	0	0	0	0	1
	1000	0	0	0	0	0	1
	1250	0	0	0	0	0	0
	1500	0	0	0	0	0	0
	1750	0	0	0	0	0	0
	2000	0	0	0	0	0	0
TW, THHW, THW, THW-2	14	7	13	22	40	55	92
	12	5	10	17	31	42	71
	10	4	7	13	23	32	52
	8	1	4	7	13	17	29
RHH*, RHW*, RHW-2*	14	4	8	15	27	37	61
RHH*, RHW*, RHW-2*	12	3	7	12	21	29	49
	10	3	5	9	17	23	38
RHH*, RHW*, RHW-2*	8	1	3	5	10	14	23
RHH*, RHW*, RHW-2*, TW, THW, THHW, THW-2	6	1	2	4	7	10	17
	4	1	1	3	5	8	13
	3	1	1	2	5	7	11
	2	1	1	2	4	6	9
	1	0	1	1	3	4	6
	1/0	0	1	1	2	3	5
	2/0	0	1	1	1	3	5
	3/0	0	0	1	1	2	4
	4/0	0	0	1	1	1	3
	250	0	0	1	1	1	2
	300	0	0	0	1	1	2
	350	0	0	0	1	1	1
	400	0	0	0	1	1	1
	500	0	0	0	1	1	1
	600	0	0	0	0	1	1
	700	0	0	0	0	1	1
	750	0	0	0	0	1	1
	800	0	0	0	0	1	1
	900	0	0	0	0	0	1
	1000	0	0	0	0	0	1
	1250	0	0	0	0	0	1
	1500	0	0	0	0	0	0
	1750	0	0	0	0	0	0
	2000	0	0	0	0	0	0

Table C2 Continued

		CONDUCTORS					
		Metric Designator (Trade Size)					
Type	Conductor Size (AWG/kcmil)	16 (½)	21 (¾)	27 (1)	35 (1¼)	41 (1½)	53 (2)
THHN, THWN, THWN-2	14	10	18	32	58	80	132
	12	7	13	23	42	58	96
	10	4	8	15	26	36	60
	8	2	5	8	15	21	35
	6	1	3	6	11	15	25
	4	1	1	4	7	9	15
	3	1	1	3	5	8	13
	2	1	1	2	5	6	11
	1	1	1	1	3	5	8
	1/0	0	1	1	3	4	7
	2/0	0	1	1	2	3	5
	3/0	0	1	1	1	3	4
	4/0	0	0	1	1	2	4
	250	0	0	1	1	1	3
	300	0	0	1	1	1	2
	350	0	0	0	1	1	2
	400	0	0	0	1	1	1
	500	0	0	0	1	1	1
	600	0	0	0	1	1	1
	700	0	0	0	0	1	1
	750	0	0	0	0	1	1
	800	0	0	0	0	1	1
	900	0	0	0	0	1	1
	1000	0	0	0	0	0	1
FEP, FEPB, PFA, PFAH, TFE	14	10	18	31	56	77	128
	12	7	13	23	41	56	93
	10	5	9	16	29	40	67
	8	3	5	9	17	23	38
	6	1	4	6	12	16	27
	4	1	2	4	8	11	19
	3	1	1	4	7	9	16
	2	1	1	3	5	8	13
PFA, PFAH, TFE	1	1	1	1	4	5	9
PFA, PFAH, TFE, Z	1/0	0	1	1	3	4	7
	2/0	0	1	1	2	4	6
	3/0	0	1	1	1	3	5
	4/0	0	1	1	1	2	4
Z	14	12	22	38	68	93	154
	12	8	15	27	48	66	109
	10	5	9	16	29	40	67
	8	3	6	10	18	25	42
	6	1	4	7	13	18	30
	4	1	3	5	9	12	20
	3	1	1	3	6	9	15
	2	1	1	3	5	7	12
	1	1	1	2	4	6	10
XHH, XHHW, XHHW-2, ZW	14	7	13	22	40	55	92
	12	5	10	17	31	42	71
	10	4	7	13	23	32	52
	8	1	4	7	13	17	29
	6	1	3	5	9	13	21
	4	1	1	4	7	9	15
	3	1	1	3	6	8	13
	2	1	1	2	5	6	11
XHH, XHHW, XHHW-2	1	1	1	1	3	5	8
	1/0	0	1	1	3	4	7
	2/0	0	1	1	2	3	6
	3/0	0	1	1	1	3	5
	4/0	0	0	1	1	2	4
	250	0	0	1	1	1	3
	300	0	0	1	1	1	3
	350	0	0	1	1	1	2
	400	0	0	0	1	1	1
	500	0	0	0	1	1	1
	600	0	0	0	1	1	1
	700	0	0	0	0	1	1
	750	0	0	0	0	1	1
	800	0	0	0	0	1	1
	900	0	0	0	0	1	1
	1000	0	0	0	0	0	1
	1250	0	0	0	0	0	1
	1500	0	0	0	0	0	1
	1750	0	0	0	0	0	0
	2000	0	0	0	0	0	0

Table C2 Continued

		FIXTURE WIRES					
		Metric Designator (Trade Size)					
Type	Conductor Size (AWG/kcmil)	16 (½)	21 (¾)	27 (1)	35 (1¼)	41 (1½)	53 (2)
FFH-2, RFH-2, RFHH-3 SF-2, SFF-2	18	6	12	21	39	53	88
	16	5	10	18	32	45	74
	18	8	15	27	49	67	111
	16	7	13	22	40	55	92
	14	5	10	18	32	45	74
SF-1, SFF-1	18	15	28	48	86	119	197
RFH-1, RFHH-2, TF, TFF, XF, XFF	18	11	20	35	64	88	145
RFHH-2, TF, TFF, XF, XFF	16	9	16	29	51	71	117
XF, XFF	14	7	13	22	40	55	92
TFN, TFFN	18	18	33	57	102	141	233
	16	13	25	43	78	107	178
PF, PFF, PGF, PGFF, PAF, PTF, PTFF, PAFF	18	17	31	54	97	133	221
	16	13	24	42	75	103	171
	14	10	18	31	56	77	128
ZF, ZFF, ZHF, HF, HFF	18	22	40	70	125	172	285
	16	16	29	51	92	127	210
	14	12	22	38	68	93	154
KF-2, KFF-2	18	31	58	101	182	250	413
	16	22	41	71	128	176	291
	14	15	28	49	88	121	200
	12	10	19	33	60	83	138
	10	7	13	22	40	55	92
KF-1, KFF-1	18	38	69	121	217	298	493
	16	26	49	85	152	209	346
	14	18	33	57	102	141	233
	12	12	22	38	68	93	154
	10	7	14	24	44	61	101
XF, XFF	12	3	7	12	21	29	49
	10	3	5	9	17	23	38

Note: This table is for concentric stranded conductors only. For compact stranded conductors, Table C2(A) should be used.
*Types RHH, RHW, and RHW-2 without outer covering.

Table C2(A) Maximum Number of Compact Conductors in Electrical Nonmetallic Tubing (ENT) (*Based on Table 1, Chapter 9*)

COMPACT CONDUCTORS

Type	Conductor Size (AWG/kcmil)	Metric Designator (Trade Size)					
		16 (½)	21 (¾)	27 (1)	35 (1¼)	41 (1½)	53 (2)
THW, THW-2, THHW	8	1	3	6	11	15	25
	6	1	2	4	8	11	19
	4	1	1	3	6	8	14
	2	1	1	2	4	6	10
	1	0	1	1	3	4	7
	1/0	0	1	1	3	4	6
	2/0	0	1	1	2	3	5
	3/0	0	1	1	1	3	4
	4/0	0	0	1	1	2	4
	250	0	0	1	1	1	3
	300	0	0	1	1	1	2
	350	0	0	0	1	1	2
	400	0	0	0	1	1	1
	500	0	0	0	1	1	1
	600	0	0	0	1	1	1
	700	0	0	0	0	1	1
	750	0	0	0	0	1	1
	1000	0	0	0	0	0	1
THHN, THWN, THWN-2	8	—	—	—	—	—	—
	6	1	4	7	12	17	28
	4	1	2	4	7	10	17
	2	1	1	3	5	7	12
	1	1	1	2	4	5	9
	1/0	1	1	1	3	5	8
	2/0	0	1	1	3	4	6
	3/0	0	1	1	2	3	5
	4/0	0	1	1	1	2	4
	250	0	0	1	1	1	3
	300	0	0	1	1	1	3
	350	0	0	1	1	1	2
	400	0	0	0	1	1	2
	500	0	0	0	1	1	1
	600	0	0	0	1	1	1
	700	0	0	0	1	1	1
	750	0	0	0	1	1	1
	1000	0	0	0	0	1	1
XHHW, XHHW-2	8	2	4	8	14	19	32
	6	1	3	6	10	14	24
	4	1	2	4	7	10	17
	2	1	1	3	5	7	12
	1	1	1	2	4	5	9
	1/0	1	1	1	3	5	8
	2/0	0	1	1	3	4	7
	3/0	0	1	1	2	3	5
	4/0	0	1	1	1	3	4
	250	0	0	1	1	1	3
	300	0	0	1	1	1	3
	350	0	0	1	1	1	3
	400	0	0	1	1	1	2
	500	0	0	0	1	1	1
	600	0	0	0	1	1	1
	700	0	0	0	1	1	1
	750	0	0	0	1	1	1
	1000	0	0	0	0	1	1

Definition: Compact stranding is the result of a manufacturing process where the standard conductor is compressed to the extent that the interstices (voids between strand wires) are virtually eliminated.

Table C3 Maximum Number of Conductors or Fixture Wires in Flexible Metal Conduit (FMC) (*Based on Table 1, Chapter 9*)

CONDUCTORS

Type	Conductor Size (AWG/kcmil)	Metric Designator (Trade Size)									
		16 (½)	21 (¾)	27 (1)	35 (1¼)	41 (1½)	53 (2)	63 (2½)	78 (3)	91 (3½)	103 (4)
RHH, RHW, RHW-2	14	4	7	11	17	25	44	67	96	131	171
	12	3	6	9	14	21	37	55	80	109	142
RHH, RHW, RHW-2	10	3	5	7	11	17	30	45	64	88	115
	8	1	2	4	6	9	15	23	34	46	60
	6	1	1	3	5	7	12	19	27	37	48
	4	1	1	2	4	5	10	14	21	29	37
	3	1	1	1	3	5	8	13	18	25	33
	2	1	1	1	3	4	7	11	16	22	28
	1	0	1	1	1	2	5	7	10	14	19
	1/0	0	1	1	1	2	4	6	9	12	16
	2/0	0	1	1	1	1	3	5	8	11	14
	3/0	0	0	1	1	1	3	5	7	9	12
	4/0	0	0	1	1	1	2	4	6	8	10
	250	0	0	0	1	1	1	3	4	6	8
	300	0	0	0	1	1	1	2	4	5	7
	350	0	0	0	1	1	1	2	3	5	6
	400	0	0	0	0	1	1	1	3	4	6
	500	0	0	0	0	1	1	1	3	4	5
	600	0	0	0	0	1	1	1	2	3	4
	700	0	0	0	0	0	1	1	1	3	3
	750	0	0	0	0	0	1	1	1	2	3
	800	0	0	0	0	0	1	1	1	2	3
	900	0	0	0	0	0	1	1	1	2	3
	1000	0	0	0	0	0	1	1	1	1	3
	1250	0	0	0	0	0	0	1	1	1	1
	1500	0	0	0	0	0	0	1	1	1	1
	1750	0	0	0	0	0	0	1	1	1	1
	2000	0	0	0	0	0	0	0	1	1	1
TW, THHW, THW, THW-2	14	9	15	23	36	53	94	141	203	277	361
	12	7	11	18	28	41	72	108	156	212	277
	10	5	8	13	21	30	54	81	116	158	207
	8	3	5	7	11	17	30	45	64	88	115
RHH*, RHW*, RHW-2*	14	6	10	15	24	35	62	94	135	184	240
RHH*, RHW*, RHW-2*	12	5	8	12	19	28	50	75	108	148	193
	10	4	6	10	15	22	39	59	85	115	151
RHH*, RHW*, RHW-2*	8	1	4	6	9	13	23	35	51	69	90
RHH*, RHW*, RHW-2*, TW, THW, THHW, THW-2	6	1	3	4	7	10	18	27	39	53	69
	4	1	1	3	5	7	13	20	29	39	51
	3	1	1	3	4	6	11	17	25	34	44
	2	1	1	2	4	5	10	14	21	29	37
	1	1	1	1	2	4	7	10	15	20	26
	1/0	0	1	1	1	3	6	9	12	17	22
	2/0	0	1	1	1	3	5	7	10	14	19
	3/0	0	1	1	1	2	4	6	9	12	16
	4/0	0	0	1	1	1	3	5	7	10	13
	250	0	0	1	1	1	3	4	6	8	11
	300	0	0	1	1	1	2	3	5	7	9
	350	0	0	0	1	1	1	3	4	6	8
	400	0	0	0	1	1	1	3	4	6	7
	500	0	0	0	1	1	1	2	3	5	6
	600	0	0	0	0	1	1	1	3	4	5
	700	0	0	0	0	1	1	1	2	3	4
	750	0	0	0	0	1	1	1	2	3	4
	800	0	0	0	0	1	1	1	1	3	4
	900	0	0	0	0	0	1	1	1	3	3
	1000	0	0	0	0	0	1	1	1	2	3
	1250	0	0	0	0	0	1	1	1	1	2
	1500	0	0	0	0	0	0	1	1	1	1
	1750	0	0	0	0	0	0	1	1	1	1
	2000	0	0	0	0	0	0	1	1	1	1

Table C3 Continued

CONDUCTORS

Type	Conductor Size (AWG/kcmil)	16 (½)	21 (¾)	27 (1)	35 (1¼)	41 (1½)	53 (2)	63 (2½)	78 (3)	91 (3½)	103 (4)
THHN, THWN, THWN-2	14	13	22	33	52	76	134	202	291	396	518
	12	9	16	24	38	56	98	147	212	289	378
	10	6	10	15	24	35	62	93	134	182	238
	8	3	6	9	14	20	35	53	77	105	137
	6	2	4	6	10	14	25	38	55	76	99
	4	1	2	4	6	9	16	24	34	46	61
	3	1	1	3	5	7	13	20	29	39	51
	2	1	1	3	4	6	11	17	24	33	43
	1	1	1	1	3	4	8	12	18	24	32
	1/0	1	1	1	2	4	7	10	15	20	27
	2/0	0	1	1	1	3	6	9	12	17	22
	3/0	0	1	1	1	2	5	7	10	14	18
	4/0	0	1	1	1	1	4	6	8	12	15
	250	0	0	1	1	1	3	5	7	9	12
	300	0	0	1	1	1	3	4	6	8	11
	350	0	0	1	1	1	2	3	5	7	9
	400	0	0	0	1	1	1	3	5	6	8
	500	0	0	0	1	1	1	2	4	5	7
	600	0	0	0	0	1	1	1	3	4	5
	700	0	0	0	0	1	1	1	3	4	5
	750	0	0	0	0	1	1	1	2	3	4
	800	0	0	0	0	1	1	1	2	3	4
	900	0	0	0	0	0	1	1	1	3	4
	1000	0	0	0	0	0	1	1	1	3	3
FEP, FEPB, PFA, PFAH, TFE	14	12	21	32	51	74	130	196	282	385	502
	12	9	15	24	37	54	95	143	206	281	367
	10	6	11	17	26	39	68	103	148	201	263
	8	4	6	10	15	22	39	59	85	115	151
	6	2	4	7	11	16	28	42	60	82	107
	4	1	3	5	7	11	19	29	42	57	75
	3	1	2	4	6	9	16	24	35	48	62
	2	1	1	3	5	7	13	20	29	39	51
PFA, PFAH, TFE	1	1	1	2	3	5	9	14	20	27	36
PFA, PFAH, TFE, Z	1/0	1	1	1	3	4	8	11	17	23	30
	2/0	1	1	1	2	3	6	9	14	19	24
	3/0	0	1	1	1	3	5	8	11	15	20
	4/0	0	1	1	1	2	4	6	9	13	16
Z	14	15	25	39	61	89	157	236	340	463	605
	12	11	18	28	43	63	111	168	241	329	429
	10	6	11	17	26	39	68	103	148	201	263
	8	4	7	11	17	24	43	65	93	127	166
	6	3	5	7	12	17	30	45	65	89	117
	4	1	3	5	8	12	21	31	45	61	80
	3	1	2	4	6	8	15	23	33	45	58
	2	1	1	3	5	7	12	19	27	37	49
	1	1	1	2	4	6	10	15	22	30	39
XHH, XHHW, XHHW-2, ZW	14	9	15	23	36	53	94	141	203	277	361
	12	7	11	18	28	41	72	108	156	212	277
	10	5	8	13	21	30	54	81	116	158	207
	8	3	5	7	11	17	30	45	64	88	115
	6	1	3	5	8	12	22	33	48	65	85
	4	1	2	4	6	9	16	24	34	47	61
	3	1	1	3	5	7	13	20	29	40	52
	2	1	1	3	4	6	11	17	24	33	44
XHH, XHHW, XHHW-2	1	1	1	1	3	5	8	13	18	25	32
	1/0	1	1	1	2	4	7	10	15	21	27
	2/0	0	1	1	2	3	6	9	13	17	23
	3/0	0	1	1	1	3	5	7	10	14	19
	4/0	0	1	1	1	2	4	6	9	12	15
	250	0	0	1	1	1	3	5	7	10	13
	300	0	0	1	1	1	3	4	6	8	11
	350	0	0	1	1	1	2	4	5	7	9
	400	0	0	0	1	1	1	3	5	6	8
	500	0	0	0	1	1	1	3	4	5	7
	600	0	0	0	0	1	1	1	3	4	5
	700	0	0	0	0	1	1	1	3	4	5
	750	0	0	0	0	1	1	1	2	3	4
	800	0	0	0	0	1	1	1	2	3	4
	900	0	0	0	0	0	1	1	1	3	4
	1000	0	0	0	0	0	1	1	1	3	3
	1250	0	0	0	0	0	1	1	1	1	3
	1500	0	0	0	0	0	1	1	1	1	2
	1750	0	0	0	0	0	0	1	1	1	1
	2000	0	0	0	0	0	0	1	1	1	1

*Types RHH, RHW, and RHW-2 without outer covering.

Table C3 Continued

FIXTURE WIRES

Type	Conductor Size (AWG/kcmil)	16 (½)	21 (¾)	27 (1)	35 (1¼)	41 (1½)	53 (2)
FFH-2, RFH-2, RFHH-3	18	8	14	22	35	51	90
	16	7	12	19	29	43	76
SF-2, SFF-2	18	11	18	28	44	64	113
	16	9	15	23	36	53	94
	14	7	12	19	29	43	76
SF-1, SFF-1	18	19	32	50	78	114	201
RFH-1, RFHH-2, TF, TFF, XF, XFF	18	14	24	37	58	84	148
RFHH-2, TF, TFF, XF, XFF	16	11	19	30	47	68	120
XF, XFF	14	9	15	23	36	53	94
TFN, TFFN	18	23	38	59	93	135	237
	16	17	29	45	71	103	181
PF, PFF, PGF, PGFF, PAF, PTF, PTFF, PAFF	18	22	36	56	88	128	225
	16	17	28	43	68	99	174
	14	12	21	32	51	74	130
ZF, ZFF, ZHF, HF, HFF	18	28	47	72	113	165	290
	16	20	35	53	83	121	214
	14	15	25	39	61	89	157
KF-2, KFF-2	18	41	68	105	164	239	421
	16	28	48	74	116	168	297
	14	19	33	51	80	116	204
	12	13	23	35	55	80	140
	10	9	15	23	36	53	94
KF-1, KFF-1	18	48	82	125	196	285	503
	16	34	57	88	138	200	353
	14	23	38	59	93	135	237
	12	15	25	39	61	89	157
	10	10	16	25	40	58	103
XF, XFF	12	5	8	12	19	28	50
	10	4	6	10	15	22	39

Note: This table is for concentric stranded conductors only. For compact stranded conductors, Table C3(A) should be used.

Table C3(A) Maximum Number of Compact Conductors in Flexible Metal Conduit (FMC) (*Based on Table 1, Chapter 9*)

COMPACT CONDUCTORS

Type	Conductor Size (AWG/ kcmil)	16 (½)	21 (¾)	27 (1)	35 (1¼)	41 (1½)	53 (2)	63 (2½)	78 (3)	91 (3½)	103 (4)
THW,	8	2	4	6	10	14	25	38	55	75	98
THHW,	6	1	3	5	7	11	20	29	43	58	76
THW-2	4	1	2	3	5	8	15	22	32	43	57
	2	1	1	2	4	6	11	16	23	32	42
	1	1	1	1	3	4	7	11	16	22	29
	1/0	1	1	1	2	3	6	10	14	19	25
	2/0	0	1	1	1	3	5	8	12	16	21
	3/0	0	1	1	1	2	4	7	10	14	18
	4/0	0	1	1	1	1	4	6	8	11	15
	250	0	0	1	1	1	3	4	7	9	12
	300	0	0	1	1	1	2	4	6	8	10
	350	0	0	1	1	1	2	3	5	7	9
	400	0	0	0	1	1	1	3	5	6	8
	500	0	0	0	1	1	1	3	4	5	7
	600	0	0	0	0	1	1	1	3	4	6
	700	0	0	0	0	1	1	1	3	4	5
	750	0	0	0	0	1	1	1	2	3	5
	1000	0	0	0	0	0	1	1	1	3	4
THHN,	8	—	—	—	—	—	—	—	—	—	—
THWN,	6	3	4	7	11	16	29	43	62	85	111
THWN-2	4	1	3	4	7	10	18	27	38	52	69
	2	1	1	3	5	7	13	19	28	38	49
	1	1	1	2	3	5	9	14	21	28	37
	1/0	1	1	1	3	4	8	12	17	24	31
	2/0	1	1	1	2	4	6	10	14	20	26
	3/0	0	1	1	1	3	5	8	12	17	22
	4/0	0	1	1	1	2	4	7	10	14	18
	250	0	1	1	1	1	3	5	8	11	14
	300	0	0	1	1	1	3	5	7	9	12
	350	0	0	1	1	1	3	4	6	8	10
	400	0	0	1	1	1	2	3	5	7	9
	500	0	0	0	1	1	1	3	4	6	8
	600	0	0	0	1	1	1	2	3	5	6
	700	0	0	0	0	1	1	1	3	4	6
	750	0	0	0	0	1	1	1	3	4	5
	1000	0	0	0	0	0	1	1	1	3	4
XHHW,	8	3	5	8	13	19	33	50	71	97	127
XHHW-2	6	2	4	6	9	14	24	37	53	72	95
	4	1	3	4	7	10	18	27	38	52	69
	2	1	1	3	5	7	13	19	28	38	49
	1	1	1	2	3	5	9	14	21	28	37
	1/0	1	1	1	3	4	8	12	17	24	31
	2/0	1	1	1	2	4	7	10	15	20	26
	3/0	0	1	1	1	3	5	8	12	17	22
	4/0	0	1	1	1	2	4	7	10	14	18
	250	0	1	1	1	1	4	5	8	11	14
	300	0	0	1	1	1	3	5	7	9	12
	350	0	0	1	1	1	3	4	6	8	11
	400	0	0	1	1	1	2	4	5	7	10
	500	0	0	0	1	1	1	3	4	6	8
	600	0	0	0	0	1	1	2	3	5	6
	700	0	0	0	0	1	1	1	3	4	6
	750	0	0	0	0	1	1	1	3	4	5
	1000	0	0	0	0	1	1	1	2	3	4

Definition: *Compact stranding* is the result of a manufacturing process where the standard conductor is compressed to the extent that the interstices (voids between strand wires) are virtually eliminated.

Table C4 Maximum Number of Conductors or Fixture Wires in Intermediate Metal Conduit (IMC) (*Based on Table 1, Chapter 9*)

CONDUCTORS

Type	Conductor Size (AWG/ kcmil)	16 (½)	21 (¾)	27 (1)	35 (1¼)	41 (1½)	53 (2)	63 (2½)	78 (3)	91 (3½)	103 (4)
RHH, RHW,	14	4	8	13	22	30	49	70	108	144	186
RHW-2	12	4	6	11	18	25	41	58	89	120	154
RHH, RHW,	10	3	5	8	15	20	33	47	72	97	124
RHW-2	8	1	3	4	8	10	17	24	38	50	65
	6	1	1	3	6	8	14	19	30	40	52
	4	1	1	3	5	6	11	15	23	31	41
	3	1	1	2	4	6	9	13	21	28	36
	2	1	1	1	3	5	8	11	18	24	31
	1	0	1	1	2	3	5	7	12	16	20
	1/0	0	1	1	1	3	4	6	10	14	18
	2/0	0	1	1	1	2	4	6	9	12	15
	3/0	0	0	1	1	1	3	5	7	10	13
	4/0	0	0	1	1	1	3	4	6	9	11
	250	0	0	1	1	1	1	3	5	6	8
	300	0	0	0	1	1	1	3	4	6	7
	350	0	0	0	1	1	1	2	4	5	7
	400	0	0	0	1	1	1	2	3	5	6
	500	0	0	0	1	1	1	1	3	4	5
	600	0	0	0	0	1	1	1	2	3	4
	700	0	0	0	0	1	1	1	2	3	4
	750	0	0	0	0	1	1	1	1	3	4
	800	0	0	0	0	0	1	1	1	3	3
	900	0	0	0	0	0	1	1	1	2	3
	1000	0	0	0	0	0	1	1	1	2	3
	1250	0	0	0	0	0	1	1	1	1	2
	1500	0	0	0	0	0	1	1	1	1	1
	1750	0	0	0	0	0	0	1	1	1	1
	2000	0	0	0	0	0	0	1	1	1	1
TW, THHW,	14	10	17	27	47	64	104	147	228	304	392
THW,	12	7	13	21	36	49	80	113	175	234	301
THW-2	10	5	9	15	27	36	59	84	130	174	224
	8	3	5	8	15	20	33	47	72	97	124
RHH*,	14	6	11	18	31	42	69	98	151	202	261
RHW*, RHW-2											
RHH*,	12	5	9	14	25	34	56	79	122	163	209
RHW*, RHW-2*	10	4	7	11	19	26	43	61	95	127	163
RHH*, RHW*, RHW-2*	8	2	4	7	12	16	26	37	57	76	98
RHH*,	6	1	3	5	9	12	20	28	43	58	75
RHW*, RHW-2*	4	1	2	4	6	9	15	21	32	43	56
TW, THW,	3	1	1	3	6	8	13	18	28	37	48
THHW,	2	1	1	3	5	6	11	15	23	31	41
THW-2	1	1	1	1	3	4	7	11	16	22	28
	1/0	1	1	1	3	4	6	9	14	19	24
	2/0	0	1	1	2	3	5	8	12	16	20
	3/0	0	1	1	1	3	4	6	10	13	17
	4/0	0	1	1	1	2	4	5	8	11	14
	250	0	0	1	1	1	3	4	7	9	12
	300	0	0	1	1	1	2	4	6	8	10
	350	0	0	1	1	1	2	3	5	7	9
	400	0	0	0	1	1	1	3	4	6	8
	500	0	0	0	1	1	1	2	4	5	7
	600	0	0	0	1	1	1	1	3	4	5
	700	0	0	0	0	1	1	1	3	4	5
	750	0	0	0	0	1	1	1	2	3	4
	800	0	0	0	0	1	1	1	2	3	4
	900	0	0	0	0	1	1	1	2	3	4
	1000	0	0	0	0	0	1	1	1	3	3
	1250	0	0	0	0	0	1	1	1	1	3
	1500	0	0	0	0	0	1	1	1	1	2
	1750	0	0	0	0	0	0	1	1	1	1
	2000	0	0	0	0	0	0	1	1	1	1

Table C4 Continued

CONDUCTORS

Type	Conductor Size (AWG/kcmil)	Metric Designator (Trade Size)									
		16 (½)	21 (¾)	27 (1)	35 (1¼)	41 (1½)	53 (2)	63 (2½)	78 (3)	91 (3½)	103 (4)
THHN, THWN, THWN-2	14	14	24	39	68	91	149	211	326	436	562
	12	10	17	29	49	67	109	154	238	318	410
	10	6	11	18	31	42	68	97	150	200	258
	8	3	6	10	18	24	39	56	86	115	149
	6	2	4	7	13	17	28	40	62	83	107
	4	1	3	4	8	10	17	25	38	51	66
	3	1	2	4	6	9	15	21	32	43	56
	2	1	1	3	5	7	12	17	27	36	47
	1	1	1	2	4	5	9	13	20	27	35
	1/0	1	1	1	3	4	8	11	17	23	29
	2/0	1	1	1	3	4	6	9	14	19	24
	3/0	0	1	1	2	3	5	7	12	16	20
	4/0	0	1	1	1	2	4	6	9	13	17
	250	0	0	1	1	1	3	5	8	10	13
	300	0	0	1	1	1	3	4	7	9	12
	350	0	0	1	1	1	2	4	6	8	10
	400	0	0	1	1	1	2	3	5	7	9
	500	0	0	0	1	1	1	3	4	6	7
	600	0	0	0	1	1	1	2	3	5	6
	700	0	0	0	1	1	1	1	3	4	5
	750	0	0	0	1	1	1	1	3	4	5
	800	0	0	0	0	1	1	1	3	4	5
	900	0	0	0	0	1	1	1	2	3	4
	1000	0	0	0	0	1	1	1	2	3	4
FEP, FEPB, PFA, PFAH, TFE	14	13	23	38	66	89	145	205	317	423	545
	12	10	17	28	48	65	106	150	231	309	398
	10	7	12	20	34	46	76	107	166	221	285
	8	4	7	11	19	26	43	61	95	127	163
	6	3	5	8	14	19	31	44	67	90	116
	4	1	3	5	10	13	21	30	47	63	81
	3	1	3	4	8	11	18	25	39	52	68
	2	1	2	4	6	9	15	21	32	43	56
PFA, PFAH, TFE	1	1	1	2	4	6	10	14	22	30	39
PFA, PFAH, TFE, Z	1/0	1	1	1	4	5	8	12	19	25	32
	2/0	1	1	1	3	4	7	10	15	21	27
	3/0	0	1	1	2	3	6	8	13	17	22
	4/0	0	1	1	1	3	5	7	10	14	18
Z	14	16	28	46	79	107	175	247	381	510	657
	12	11	20	32	56	76	124	175	271	362	466
	10	7	12	20	34	46	76	107	166	221	285
	8	4	7	12	21	29	48	68	105	140	180
	6	3	5	9	15	20	33	47	73	98	127
	4	1	3	6	10	14	23	33	50	67	87
	3	1	2	4	7	10	17	24	37	49	63
	2	1	1	3	6	8	14	20	30	41	53
	1	1	1	3	5	7	11	16	25	33	43
XHH, XHHW, XHHW-2, ZW	14	10	17	27	47	64	104	147	228	304	392
	12	7	13	21	36	49	80	113	175	234	301
	10	5	9	15	27	36	59	84	130	174	224
	8	3	5	8	15	20	33	47	72	97	124
	6	1	4	6	11	15	24	35	53	71	92
	4	1	3	4	8	11	18	25	39	52	67
	3	1	2	4	7	9	15	21	33	44	56
	2	1	1	3	5	7	12	18	27	37	47
XHH, XHHW, XHHW-2	1	1	1	2	4	5	9	13	20	27	35
	1/0	1	1	1	3	5	8	11	17	23	30
	2/0	1	1	1	3	4	6	9	14	19	25
	3/0	0	1	1	2	3	5	7	12	16	20
	4/0	0	1	1	1	2	4	6	10	13	17
	250	0	0	1	1	1	3	5	8	11	14
	300	0	0	1	1	1	3	4	7	9	12
	350	0	0	1	1	1	3	4	6	8	10
	400	0	0	1	1	1	2	3	5	7	9
	500	0	0	0	1	1	1	3	4	6	8
	600	0	0	0	1	1	1	2	3	5	6
	700	0	0	0	1	1	1	1	3	4	5
	750	0	0	0	1	1	1	1	3	4	5
	800	0	0	0	0	1	1	1	3	4	5
	900	0	0	0	0	1	1	1	2	3	4
	1000	0	0	0	0	1	1	1	2	3	4
	1250	0	0	0	0	0	1	1	1	2	3
	1500	0	0	0	0	0	1	1	1	1	2
	1750	0	0	0	0	0	1	1	1	1	2
	2000	0	0	0	0	0	0	1	1	1	1

Table C4 Continued

FIXTURE WIRES

Type	Conductor Size (AWG/kcmil)	Metric Designator (Trade Size)					
		16 (½)	21 (¾)	27 (1)	35 (1¼)	41 (1½)	53 (2)
FHH-2, RFH-2, RFHH-3	18	9	16	26	45	61	100
	16	8	13	22	38	51	84
SF-2, SFF-2	18	12	20	33	57	77	126
	16	10	17	27	47	64	104
	14	8	13	22	38	51	84
SF-1, SFF-1	18	21	36	59	101	137	223
RFH-1, RFHH-2, TF, TFF, XF, XFF	18	15	26	43	75	101	165
RFH-2, TF, TFF, XF, XFF	16	12	21	35	60	81	133
XF, XFF	14	10	17	27	47	64	104
TFN, TFFN	18	25	42	69	119	161	264
	16	19	32	53	91	123	201
PF, PFF, PGF, PGFF, PAF, PTF, PTFF, PAFF	18	23	40	66	113	153	250
	16	18	31	51	87	118	193
	14	13	23	38	66	89	145
ZF, ZFF, ZHF, HF, HFF	18	30	52	85	146	197	322
	16	22	38	63	108	145	238
	14	16	28	46	79	107	175
KF-2, KFF-2	18	44	75	123	212	287	468
	16	31	53	87	149	202	330
	14	21	36	60	103	139	227
	12	14	25	41	70	95	156
	10	10	17	27	47	64	104
KF-1, KFF-1	18	52	90	147	253	342	558
	16	37	63	103	178	240	392
	14	25	42	69	119	161	264
	12	16	28	46	79	107	175
	10	10	18	30	52	70	114
XF, XFF	12	5	9	14	25	34	56
	10	4	7	11	19	26	43

Note: This table is for concentric stranded conductors only. For compact stranded conductors, Table C4(A) should be used.
*Types RHH, RHW, and RHW-2 without outer covering.

Table C4(A) Maximum Number of Compact Conductors in Intermediate Metal Conduit (IMC) (*Based on Table 1, Chapter 9*)

COMPACT CONDUCTORS

Type	Conductor Size (AWG/kcmil)	16 (½)	21 (¾)	27 (1)	35 (1¼)	41 (1½)	53 (2)	63 (2½)	78 (3)	91 (3½)	103 (4)
THW, THW-2, THHW	8	2	4	7	13	17	28	40	62	83	107
	6	1	3	6	10	13	22	31	48	64	82
	4	1	2	4	7	10	16	23	36	48	62
	2	1	1	3	5	7	12	17	26	35	45
	1	1	1	1	4	5	8	12	18	25	32
	1/0	1	1	1	3	4	7	10	16	21	27
	2/0	0	1	1	3	4	6	9	13	18	23
	3/0	0	1	1	2	3	5	7	11	15	20
	4/0	0	1	1	1	2	4	6	9	13	16
	250	0	0	1	1	1	3	5	7	10	13
	300	0	0	1	1	1	3	4	6	9	11
	350	0	0	1	1	1	2	4	6	8	10
	400	0	0	1	1	1	2	3	5	7	9
	500	0	0	0	1	1	1	3	4	6	8
	600	0	0	0	1	1	1	2	3	5	6
	700	0	0	0	1	1	1	1	3	4	5
	750	0	0	0	1	1	1	1	3	4	5
	1000	0	0	0	0	1	1	1	2	3	4
THHN, THWN, THWN-2	8	—	—	—	—	—	—	—	—	—	—
	6	3	5	8	14	19	32	45	70	93	120
	4	1	3	5	9	12	20	28	43	58	74
	2	1	1	3	6	8	14	20	31	41	53
	1	1	1	3	5	6	10	15	23	31	40
	1/0	1	1	2	4	5	9	13	20	26	34
	2/0	1	1	1	3	4	7	10	16	22	28
	3/0	0	1	1	3	4	6	9	14	18	24
	4/0	0	1	1	2	3	5	7	11	15	19
	250	0	1	1	1	2	4	6	9	12	15
	300	0	0	1	1	1	3	5	7	10	13
	350	0	0	1	1	1	3	4	7	9	11
	400	0	0	1	1	1	2	4	6	8	10
	500	0	0	1	1	1	2	3	5	7	9
	600	0	0	0	1	1	1	2	4	5	7
	700	0	0	0	1	1	1	2	3	5	6
	750	0	0	0	1	1	1	1	3	4	6
	1000	0	0	0	0	1	1	1	2	3	4
XHHW, XHHW-2	8	3	6	9	16	22	37	52	80	107	138
	6	2	4	7	12	16	27	38	59	80	103
	4	1	3	5	9	12	20	28	43	58	74
	2	1	1	3	6	8	14	20	31	41	53
	1	1	1	3	5	6	10	15	23	31	40
	1/0	1	1	2	4	5	9	13	20	26	34
	2/0	1	1	1	3	4	7	11	17	22	29
	3/0	0	1	1	3	4	6	9	14	18	24
	4/0	0	1	1	2	3	5	7	11	15	20
	250	0	1	1	1	2	4	6	9	12	16
	300	0	0	1	1	1	3	5	8	10	13
	350	0	0	1	1	1	3	4	7	9	12
	400	0	0	1	1	1	3	4	6	8	11
	500	0	0	1	1	1	2	3	5	7	9
	600	0	0	0	1	1	1	2	4	5	7
	700	0	0	0	1	1	1	2	3	5	6
	750	0	0	0	1	1	1	1	3	4	6
	1000	0	0	0	0	1	1	1	2	3	4

Definition: *Compact stranding* is the result of a manufacturing process where the standard conductor is compressed to the extent that interstices (voids between strand wires) are virtually eliminated.

Table C5 Maximum Number of Conductors or Fixture Wires in Liquidtight Flexible Nonmetallic Conduit (Type LFNC-B*) (*Based on Table 1, Chapter 9*)

CONDUCTORS

Type	Conductor Size (AWG/kcmil)	12 (⅜)	16 (½)	21 (¾)	27 (1)	35 (1¼)	41 (1½)	53 (2)
RHH, RHW, RHW-2	14	2	4	7	12	21	27	44
	12	1	3	6	10	17	22	36
RHH, RHW, RHW-2	10	1	3	5	8	14	18	29
	8	1	1	2	4	7	9	15
	6	1	1	1	3	6	7	12
	4	0	1	1	2	4	6	9
	3	0	1	1	1	4	5	8
	2	0	1	1	1	3	4	7
	1	0	0	1	1	1	3	5
	1/0	0	0	1	1	1	2	4
	2/0	0	0	1	1	1	1	3
	3/0	0	0	0	1	1	1	3
	4/0	0	0	0	1	1	1	2
	250	0	0	0	0	1	1	1
	300	0	0	0	0	1	1	1
	350	0	0	0	0	1	1	1
	400	0	0	0	0	1	1	1
	500	0	0	0	0	1	1	1
	600	0	0	0	0	0	1	1
	700	0	0	0	0	0	0	1
	750	0	0	0	0	0	0	1
	800	0	0	0	0	0	0	1
	900	0	0	0	0	0	0	1
	1000	0	0	0	0	0	0	1
	1250	0	0	0	0	0	0	0
	1500	0	0	0	0	0	0	0
	1750	0	0	0	0	0	0	0
	2000	0	0	0	0	0	0	0
TW, THHW, THW, THW-2	14	5	9	15	25	44	57	93
	12	4	7	12	19	33	43	71
	10	3	5	9	14	25	32	53
	8	1	3	5	8	14	18	29
RHH†, RHW†, RHW-2†	14	3	6	10	16	29	38	62
RHH†, RHW†, RHW-2†	12	3	5	8	13	23	30	50
	10	1	3	6	10	18	23	39
RHH†, RHW†, RHW-2†	8	1	1	4	6	11	14	23
RHH†, RHW†, RHW-2†, TW, THW	6	1	1	3	5	8	11	18
	4	1	1	1	3	6	8	13
	3	1	1	1	3	5	7	11
THHW, THW-2	2	0	1	1	2	4	6	9
	1	0	1	1	1	3	4	7
	1/0	0	0	1	1	2	3	6
	2/0	0	0	1	1	2	3	5
	3/0	0	0	1	1	1	2	4
	4/0	0	0	0	1	1	1	3
	250	0	0	0	1	1	1	3
	300	0	0	0	1	1	1	2
	350	0	0	0	0	1	1	1
	400	0	0	0	0	1	1	1
	500	0	0	0	0	1	1	1
	600	0	0	0	0	1	1	1
	700	0	0	0	0	0	1	1
	750	0	0	0	0	0	1	1
	800	0	0	0	0	0	1	1
	900	0	0	0	0	0	0	1
	1000	0	0	0	0	0	0	1
	1250	0	0	0	0	0	0	1
	1500	0	0	0	0	0	0	0
	1750	0	0	0	0	0	0	0
	2000	0	0	0	0	0	0	0

Table C5 Continued

CONDUCTORS

Type	Conductor Size (AWG/kcmil)	12 (3/8)	16 (1/2)	21 (3/4)	27 (1)	35 (1¼)	41 (1½)	53 (2)
THHN, THWN, THWN-2	14	8	13	22	36	63	81	133
	12	5	9	16	26	46	59	97
	10	3	6	10	16	29	37	61
	8	1	3	6	9	16	21	35
	6	1	2	4	7	12	15	25
	4	1	1	2	4	7	9	15
	3	1	1	1	3	6	8	13
	2	1	1	1	3	5	7	11
	1	0	1	1	1	4	5	8
	1/0	0	1	1	1	3	4	7
	2/0	0	0	1	1	2	3	6
	3/0	0	0	1	1	1	3	5
	4/0	0	0	1	1	1	2	4
	250	0	0	0	1	1	1	3
	300	0	0	0	1	1	1	3
	350	0	0	0	1	1	1	2
	400	0	0	0	0	1	1	1
	500	0	0	0	0	1	1	1
	600	0	0	0	0	1	1	1
	700	0	0	0	0	1	1	1
	750	0	0	0	0	0	1	1
	800	0	0	0	0	0	1	1
	900	0	0	0	0	0	1	1
	1000	0	0	0	0	0	0	1
FEP, FEPB, PFA, PFAH, TFE	14	7	12	21	35	61	79	129
	12	5	9	15	25	44	57	94
	10	4	6	11	18	32	41	68
	8	1	3	6	10	18	23	39
	6	1	2	4	7	13	17	27
	4	1	1	3	5	9	12	19
	3	1	1	2	4	7	10	16
	2	1	1	1	3	6	8	13
PFA, PFAH, TFE	1	0	1	1	2	4	5	9
PFA, PFAH	1/0	0	1	1	1	3	4	7
TFE, Z	2/0	0	1	1	1	3	4	6
	3/0	0	0	1	1	2	3	5
	4/0	0	0	1	1	1	2	4
Z	14	9	15	26	42	73	95	156
	12	6	10	18	30	52	67	111
	10	4	6	11	18	32	41	68
	8	2	4	7	11	20	26	43
	6	1	3	5	8	14	18	30
	4	1	1	3	5	9	12	20
	3	1	1	2	4	7	9	15
	2	0	1	1	3	6	7	12
	1	0	1	1	2	5	6	10
XHH, XHHW, XHHW-2, ZW	14	5	9	15	25	44	57	93
	12	4	7	12	19	33	43	71
	10	3	5	9	14	25	32	53
	8	1	3	5	8	14	18	29
	6	1	1	3	6	10	13	22
	4	1	1	2	4	7	9	16
	3	1	1	1	3	6	8	13
	2	1	1	1	3	5	7	11
XHH, XHHW, XHHW-2	1	0	1	1	1	4	5	8
	1/0	0	1	1	1	3	4	7
	2/0	0	0	1	1	2	3	6
	3/0	0	0	1	1	1	3	5
	4/0	0	0	1	1	1	2	4
	250	0	0	0	1	1	1	3
	300	0	0	0	1	1	1	3
	350	0	0	0	1	1	1	2
	400	0	0	0	0	1	1	1
	500	0	0	0	0	1	1	1
	600	0	0	0	0	1	1	1
	700	0	0	0	0	1	1	1
	750	0	0	0	0	0	1	1
	800	0	0	0	0	0	1	1
	900	0	0	0	0	0	1	1
	1000	0	0	0	0	0	0	1
	1250	0	0	0	0	0	0	1
	1500	0	0	0	0	0	0	1
	1750	0	0	0	0	0	0	0
	2000	0	0	0	0	0	0	0

Table C5 Continued

FIXTURE WIRES

Type	Conductor Size (AWG/kcmil)	12 (3/8)	16 (1/2)	21 (3/4)	27 (1)	35 (1¼)	41 (1½)	53 (2)
FFH-2, RFH-2 SF-2, SFF-2	18	5	8	15	24	42	54	89
	16	4	7	12	20	35	46	75
	18	6	11	19	30	53	69	113
	16	5	9	15	25	44	57	93
	14	4	7	12	20	35	46	75
SF-1, SFF-1	18	11	19	33	53	94	122	199
RFH-1, RFHH-2, TF, TFF, XF, XFF	18	8	14	24	39	69	90	147
RFHH-2, TF, TFF, XF, XFF	16	7	11	20	32	56	72	119
XF, XFF	14	5	9	15	25	44	57	93
TFN, TFFN	18	14	23	39	63	111	144	236
	16	10	17	30	48	85	110	180
PF, PFF, PGF, PGFF, PAF, PTF, PTFF, PAFF	18	13	21	37	60	105	136	223
	16	10	16	29	46	81	105	173
	14	7	12	21	35	61	79	129
HF, HFF, ZF, ZFF, ZHF	18	17	28	48	77	136	176	288
	16	12	20	35	57	100	129	212
	14	9	15	26	42	73	95	156
KF-2, KFF-2	18	24	40	70	112	197	255	418
	16	17	28	49	79	139	180	295
	14	12	19	34	54	95	123	202
	12	8	13	23	37	65	85	139
	10	5	9	15	25	44	57	93
KF-1, KFF-1	18	29	48	83	134	235	304	499
	16	20	34	58	94	165	214	350
	14	14	23	39	63	111	144	236
	12	9	15	26	42	73	95	156
	10	6	10	17	27	48	62	102
XF, XFF	12	3	5	8	13	23	30	50
	10	1	3	6	10	18	23	39

Note: This table is for concentric stranded conductors only. For compact stranded conductors, Table C5(A) should be used.
*Corresponds to 356.2(2).
†Types RHH, RHW, and RHW-2 without outer covering.

Table C5(A) Maximum Number of Compact Conductors in Liquidtight Flexible Nonmetallic Conduit (Type LFNC-B*) (*Based on Table 1, Chapter 9*)

COMPACT CONDUCTORS

Type	Conductor Size (AWG/kcmil)	Metric Designator (Trade Size)						
		12 (³⁄₈)	16 (¹⁄₂)	21 (³⁄₄)	27 (1)	35 (1¼)	41 (1½)	53 (2)
THW, THW-2, THHW	8	1	2	4	7	12	15	25
	6	1	1	3	5	9	12	19
	4	1	1	2	4	7	9	14
	2	1	1	1	3	5	6	11
	1	0	1	1	1	3	4	7
	1/0	0	1	1	1	3	4	6
	2/0	0	0	1	1	2	3	5
	3/0	0	0	1	1	1	3	4
	4/0	0	0	1	1	1	2	4
	250	0	0	0	1	1	1	3
	300	0	0	0	1	1	1	2
	350	0	0	0	1	1	1	2
	400	0	0	0	0	1	1	1
	500	0	0	0	0	1	1	1
	600	0	0	0	0	1	1	1
	700	0	0	0	0	1	1	1
	750	0	0	0	0	0	1	1
	1000	0	0	0	0	0	1	1
THHN, THWN, THWN-2	8	—	—	—	—	—	—	—
	6	1	2	4	7	13	17	28
	4	1	1	3	4	8	11	17
	2	1	1	1	3	6	7	12
	1	0	1	1	2	4	6	9
	1/0	0	1	1	1	4	5	8
	2/0	0	1	1	1	3	4	6
	3/0	0	0	1	1	2	3	5
	4/0	0	0	1	1	1	3	4
	250	0	0	1	1	1	1	3
	300	0	0	0	1	1	1	3
	350	0	0	0	1	1	1	2
	400	0	0	0	1	1	1	2
	500	0	0	0	0	1	1	1
	600	0	0	0	0	1	1	1
	700	0	0	0	0	1	1	1
	750	0	0	0	0	1	1	1
	1000	0	0	0	0	0	1	1
XHHW, XHHW-2	8	1	3	5	9	15	20	33
	6	1	2	4	6	11	15	24
	4	1	1	3	4	8	11	17
	2	1	1	1	3	6	7	12
	1	0	1	1	2	4	6	9
	1/0	0	1	1	1	4	5	8
	2/0	0	1	1	1	3	4	7
	3/0	0	0	1	1	2	3	5
	4/0	0	0	1	1	1	3	4
	250	0	0	1	1	1	1	3
	300	0	0	0	1	1	1	3
	350	0	0	0	1	1	1	3
	400	0	0	0	1	1	1	2
	500	0	0	0	0	1	1	1
	600	0	0	0	0	1	1	1
	700	0	0	0	0	1	1	1
	750	0	0	0	0	1	1	1
	1000	0	0	0	0	0	1	1

*Corresponds to 356.2(2).
Definition: *Compact stranding* is the result of a manufacturing process where the standard conductor is compressed to the extent that the interstices (voids between strand wires) are virtually eliminated.

Table C6 Maximum Number of Conductors or Fixture Wires in Liquidtight Flexible Nonmetallic Conduit (Type LFNC-A*) (*Based on Table 1, Chapter 9*)

CONDUCTORS

Type	Conductor Size (AWG/kcmil)	Metric Designator (Trade Size)						
		12 (³⁄₈)	16 (¹⁄₂)	21 (³⁄₄)	27 (1)	35 (1¼)	41 (1½)	53 (2)
RHH, RHW, RHW-2	14	2	4	7	11	20	27	45
	12	1	3	6	9	17	23	38
	10	1	3	5	8	13	18	30
	8	1	1	2	4	7	9	16
	6	1	1	1	3	5	7	13
	4	0	1	1	2	4	6	10
	3	0	1	1	1	4	5	8
	2	0	1	1	1	3	4	7
	1	0	0	1	1	1	3	5
	1/0	0	0	1	1	1	2	4
	2/0	0	0	1	1	1	1	4
	3/0	0	0	0	1	1	1	3
	4/0	0	0	0	1	1	1	3
	250	0	0	0	0	1	1	1
	300	0	0	0	0	1	1	1
	350	0	0	0	0	1	1	1
	400	0	0	0	0	1	1	1
	500	0	0	0	0	0	1	1
	600	0	0	0	0	0	1	1
	700	0	0	0	0	0	0	1
	750	0	0	0	0	0	0	1
	800	0	0	0	0	0	0	1
	900	0	0	0	0	0	0	1
	1000	0	0	0	0	0	0	1
	1250	0	0	0	0	0	0	0
	1500	0	0	0	0	0	0	0
	1750	0	0	0	0	0	0	0
	2000	0	0	0	0	0	0	0
TW, THHW, THW, THW-2	14	5	9	15	24	43	58	96
	12	4	7	12	19	33	44	74
	10	3	5	9	14	24	33	55
	8	1	3	5	8	13	18	30
RHH†, RHW†, RHW-2†	14	3	6	10	16	28	38	64
	12	3	4	8	13	23	31	51
	10	1	3	6	10	18	24	40
	8	1	1	4	6	10	14	24
RHH†, RHW†, RHW-2†, TW, THW, THHW, THW-2	6	1	1	3	4	8	11	18
	4	1	1	1	3	6	8	13
	3	1	1	1	3	5	7	11
	2	0	1	1	2	4	6	10
	1	0	1	1	1	3	4	7
	1/0	0	0	1	1	2	3	6
	2/0	0	0	1	1	1	3	5
	3/0	0	0	1	1	1	2	4
	4/0	0	0	0	1	1	1	3
	250	0	0	0	1	1	1	3
	300	0	0	0	1	1	1	2
	350	0	0	0	0	1	1	1
	400	0	0	0	0	1	1	1
	500	0	0	0	0	1	1	1
	600	0	0	0	0	1	1	1
	700	0	0	0	0	0	1	1
	750	0	0	0	0	0	1	1
	800	0	0	0	0	0	1	1
	900	0	0	0	0	0	0	1
	1000	0	0	0	0	0	0	1
	1250	0	0	0	0	0	0	1
	1500	0	0	0	0	0	0	1
	1750	0	0	0	0	0	0	0
	2000	0	0	0	0	0	0	0

Table C6 Continued

CONDUCTORS

Type	Conductor Size (AWG/kcmil)	12 (³⁄₈)	16 (¹⁄₂)	21 (³⁄₄)	27 (1)	35 (1¼)	41 (1½)	53 (2)
THHN, THWN, THWN-2	14	8	13	22	35	62	83	137
	12	5	9	16	25	45	60	100
	10	3	6	10	16	28	38	63
	8	1	3	6	9	16	22	36
	6	1	2	4	6	12	16	26
	4	1	1	2	4	7	9	16
	3	1	1	1	3	6	8	13
	2	1	1	1	3	5	7	11
	1	0	1	1	1	4	5	8
	1/0	0	1	1	1	3	4	7
	2/0	0	0	1	1	2	3	6
	3/0	0	0	1	1	1	3	5
	4/0	0	0	1	1	1	2	4
	250	0	0	0	1	1	1	3
	300	0	0	0	1	1	1	3
	350	0	0	0	1	1	1	2
	400	0	0	0	0	1	1	1
	500	0	0	0	0	1	1	1
	600	0	0	0	0	1	1	1
	700	0	0	0	0	1	1	1
	750	0	0	0	0	0	1	1
	800	0	0	0	0	0	1	1
	900	0	0	0	0	0	1	1
	1000	0	0	0	0	0	0	1
FEP, FEPB, PFA, PFAH, TFE	14	7	12	21	34	60	80	133
	12	5	9	15	25	44	59	97
	10	4	6	11	18	31	42	70
	8	1	3	6	10	18	24	40
	6	1	2	4	7	13	17	28
	4	1	1	3	5	9	12	20
	3	1	1	2	4	7	10	16
	2	1	1	1	3	6	8	13
PFA, PFAH, TFE	1	0	1	1	2	4	5	9
PFA, PFAH, TFE, Z	1/0	0	1	1	1	3	5	8
	2/0	0	1	1	1	3	4	6
	3/0	0	0	1	1	2	3	5
	4/0	0	0	1	1	1	2	4
Z	14	9	15	25	41	72	97	161
	12	6	10	18	29	51	69	114
	10	4	6	11	18	31	42	70
	8	2	4	7	11	20	26	44
	6	1	3	5	8	14	18	31
	4	1	1	3	5	9	13	21
	3	1	1	2	4	7	9	15
	2	1	1	1	3	6	8	13
	1	1	1	1	2	4	6	10
XHH, XHHW, XHHW-2, ZW	14	5	9	15	24	43	58	96
	12	4	7	12	19	33	44	74
	10	3	5	9	14	24	33	55
	8	1	3	5	8	13	18	30
	6	1	1	3	5	10	13	22
	4	1	1	2	4	7	10	16
	3	1	1	1	3	6	8	14
	2	1	1	1	3	5	7	11
XHH, XHHW, XHHW-2	1	0	1	1	1	4	5	8
	1/0	0	1	1	1	3	4	7
	2/0	0	0	1	1	2	3	6
	3/0	0	0	1	1	1	3	5
	4/0	0	0	1	1	1	2	4
	250	0	0	0	1	1	1	3
	300	0	0	0	1	1	1	3
	350	0	0	0	1	1	1	2
	400	0	0	0	0	1	1	1
	500	0	0	0	0	1	1	1
	600	0	0	0	0	1	1	1
	700	0	0	0	0	1	1	1
	750	0	0	0	0	0	1	1
	800	0	0	0	0	0	1	1
	900	0	0	0	0	0	1	1
	1000	0	0	0	0	0	0	1
	1250	0	0	0	0	0	0	1
	1500	0	0	0	0	0	0	1
	1750	0	0	0	0	0	0	0
	2000	0	0	0	0	0	0	0

Table C6 Continued

FIXTURE WIRES

Type	Conductor Size (AWG/kcmil)	12 (³⁄₈)	16 (¹⁄₂)	21 (³⁄₄)	27 (1)	35 (1¼)	41 (1½)	53 (2)
FFH-2, RFH-2, RFHH-3	18	5	8	14	23	41	55	92
	16	4	7	12	20	35	47	77
SF-2, SFF-2	18	6	11	18	29	52	70	116
	16	5	9	15	24	43	58	96
	14	4	7	12	20	35	47	77
SF-1, SFF-1	18	12	19	33	52	92	124	205
RFH-1, RFHH-2, TF, TFF, XF, XFF	18	8	14	24	39	68	91	152
RFHH-2, TF, TFF, XF, XFF	16	7	11	19	31	55	74	122
XF, XFF	14	5	9	15	24	43	58	96
TFN, TFFN	18	14	22	39	62	109	146	243
	16	10	17	29	47	83	112	185
PF, PFF, PGF, PGFF, PAF, PTF, PTFF, PAFF	18	13	21	37	59	103	139	230
	16	10	16	28	45	80	107	178
	14	7	12	21	34	60	80	133
HF, HFF, ZF, ZFF, ZHF	18	17	27	47	76	133	179	297
	16	12	20	35	56	98	132	219
	14	9	15	25	41	72	97	161
KF-2, KFF-2	18	25	40	69	110	193	260	431
	16	17	28	48	77	136	183	303
	14	12	19	33	53	94	126	209
	12	8	13	23	36	64	86	143
	10	5	9	15	24	43	58	96
KF-1, KFF-1	18	29	48	82	131	231	310	514
	16	21	33	57	92	162	218	361
	14	14	22	39	62	109	146	243
	12	9	15	25	41	72	97	161
	10	6	10	17	27	47	63	105
XF, XFF	12	3	4	8	13	23	31	51
	10	1	3	6	10	18	24	40

Note: This table is for concentric stranded conductors only. For compact stranded conductors, Table C6(A) should be used.
*Corresponds to 356.2(1).
†Types RHH, RHW, and RHW-2 without outer covering.

Table C6(A) Maximum Number of Compact Conductors in Liquidtight Flexible Nonmetallic Conduit (Type LFNC-A*) *(Based on Table 1, Chapter 9)*

COMPACT CONDUCTORS

Type	Conductor Size (AWG/ kcmil)	Metric Designator (Trade Size)						
		12 (3/8)	16 (1/2)	21 (3/4)	27 (1)	35 (1¼)	41 (1½)	53 (2)
THW, THW-2, THHW	8	1	2	4	6	11	16	26
	6	1	1	3	5	9	12	20
	4	1	1	2	4	7	9	15
	2	1	1	1	3	5	6	11
	1	0	1	1	1	3	4	8
	1/0	0	1	1	1	3	4	7
	2/0	0	0	1	1	2	3	5
	3/0	0	0	1	1	1	3	5
	4/0	0	0	1	1	1	2	4
	250	0	0	0	1	1	1	3
	300	0	0	0	1	1	1	3
	350	0	0	0	1	1	1	2
	400	0	0	0	0	1	1	1
	500	0	0	0	0	1	1	1
	600	0	0	0	0	1	1	1
	700	0	0	0	0	1	1	1
	750	0	0	0	0	0	1	1
	1000	0	0	0	0	0	1	1
THHN, THWN, THWN-2	8	—	—	—	—	—	—	—
	6	1	2	4	7	13	18	29
	4	1	1	3	4	8	11	18
	2	1	1	1	3	6	8	13
	1	0	1	1	2	4	6	10
	1/0	0	1	1	1	3	5	8
	2/0	0	1	1	1	3	4	7
	3/0	0	0	1	1	2	3	6
	4/0	0	0	1	1	1	3	5
	250	0	0	1	1	1	1	3
	300	0	0	0	1	1	1	3
	350	0	0	0	1	1	1	3
	400	0	0	0	1	1	1	2
	500	0	0	0	0	1	1	1
	600	0	0	0	0	1	1	1
	700	0	0	0	0	1	1	1
	750	0	0	0	0	1	1	1
	1000	0	0	0	0	0	1	1
XHHW, XHHW-2	8	1	3	5	8	15	20	34
	6	1	2	4	6	11	15	25
	4	1	1	3	4	8	11	18
	2	1	1	1	3	6	8	13
	1	0	1	1	2	4	6	10
	1/0	0	1	1	1	3	5	8
	2/0	0	1	1	1	3	4	7
	3/0	0	0	1	1	2	3	6
	4/0	0	0	1	1	1	3	5
	250	0	0	1	1	1	2	4
	300	0	0	0	1	1	1	3
	350	0	0	0	1	1	1	3
	400	0	0	0	1	1	1	2
	500	0	0	0	0	1	1	1
	600	0	0	0	0	1	1	1
	700	0	0	0	0	1	1	1
	750	0	0	0	0	1	1	1
	1000	0	0	0	0	0	1	1

*Corresponds to 356.2(1).
Definition: *Compact stranding* is the result of a manufacturing process where the standard conductor is compressed to the extent that the interstices (voids between strand wires) are virtually eliminated.

Table C7 Maximum Number of Conductors or Fixture Wires in Liquidtight Flexible Metal Conduit (LFMC) *(Based on Table 1, Chapter 9)*

CONDUCTORS

Type	Conductor Size (AWG/ kcmil)	Metric Designator (Trade Size)									
		16 (1/2)	21 (3/4)	27 (1)	35 (1¼)	41 (1½)	53 (2)	63 (2½)	78 (3)	91 (3½)	103 (4)
RHH, RHW, RHW-2	14	4	7	12	21	27	44	66	102	133	173
	12	3	6	10	17	22	36	55	84	110	144
	10	3	5	8	14	18	29	44	68	89	116
	8	1	2	4	7	9	15	23	36	46	61
	6	1	1	3	6	7	12	18	28	37	48
	4	1	1	2	4	6	9	14	22	29	38
	3	1	1	1	4	5	8	13	19	25	33
	2	1	1	1	3	4	7	11	17	22	29
	1	0	1	1	1	3	5	7	11	14	19
	1/0	0	1	1	1	2	4	6	10	13	16
	2/0	0	1	1	1	1	3	5	8	11	14
	3/0	0	0	1	1	1	3	4	7	9	12
	4/0	0	0	1	1	1	2	4	6	8	10
	250	0	0	0	1	1	1	3	4	6	8
	300	0	0	0	1	1	1	2	4	5	7
	350	0	0	0	1	1	1	2	3	5	6
	400	0	0	0	1	1	1	1	3	4	6
	500	0	0	0	1	1	1	1	3	4	5
	600	0	0	0	0	1	1	1	2	3	4
	700	0	0	0	0	0	1	1	1	3	3
	750	0	0	0	0	0	1	1	1	2	3
	800	0	0	0	0	0	1	1	1	2	3
	900	0	0	0	0	0	1	1	1	2	3
	1000	0	0	0	0	0	1	1	1	1	3
	1250	0	0	0	0	0	0	1	1	1	1
	1500	0	0	0	0	0	0	1	1	1	1
	1750	0	0	0	0	0	0	1	1	1	1
	2000	0	0	0	0	0	0	1	1	1	1
TW, THHW, THW, THW-2	14	9	15	25	44	57	93	140	215	280	365
	12	7	12	19	33	43	71	108	165	215	280
	10	5	9	14	25	32	53	80	123	160	209
	8	3	5	8	14	18	29	44	68	89	116
RHH*, RHW*, RHW-2*	14	6	10	16	29	38	62	93	143	186	243
RHH*, RHW*, RHW-2*	12	5	8	13	23	30	50	75	115	149	195
	10	3	6	10	18	23	39	58	89	117	152
	8	1	4	6	11	14	23	35	53	70	91
RHH*, RHW*, RHW-2*, TW, THW, THHW, THW-2	6	1	3	5	8	11	18	27	41	53	70
	4	1	1	3	6	8	13	20	30	40	52
	3	1	1	3	5	7	11	17	26	34	44
	2	1	1	2	4	6	9	14	22	29	38
	1	1	1	1	3	4	7	10	15	20	26
	1/0	0	1	1	2	3	6	8	13	17	23
	2/0	0	1	1	2	3	5	7	11	15	19
	3/0	0	1	1	1	2	4	6	9	12	16
	4/0	0	0	1	1	1	3	5	8	10	13
	250	0	0	1	1	1	3	4	6	8	11
	300	0	0	1	1	1	2	3	5	7	9
	350	0	0	0	1	1	1	3	5	6	8
	400	0	0	0	1	1	1	3	4	6	7
	500	0	0	0	1	1	1	2	3	5	6
	600	0	0	0	1	1	1	1	3	4	5
	700	0	0	0	0	1	1	1	2	3	4
	750	0	0	0	0	1	1	1	2	3	4
	800	0	0	0	0	1	1	1	2	3	4
	900	0	0	0	0	0	1	1	1	3	3
	1000	0	0	0	0	0	1	1	1	2	3
	1250	0	0	0	0	0	1	1	1	1	2
	1500	0	0	0	0	0	0	1	1	1	2
	1750	0	0	0	0	0	0	1	1	1	1
	2000	0	0	0	0	0	0	1	1	1	1

Table C7 Continued

CONDUCTORS

Type	Conductor Size (AWG/kcmil)	16 (½)	21 (¾)	27 (1)	35 (1¼)	41 (1½)	53 (2)	63 (2½)	78 (3)	91 (3½)	103 (4)
THHN, THWN, THWN-2	14	13	22	36	63	81	133	201	308	401	523
	12	9	16	26	46	59	97	146	225	292	381
	10	6	10	16	29	37	61	92	141	184	240
	8	3	6	9	16	21	35	53	81	106	138
	6	2	4	7	12	15	25	38	59	76	100
	4	1	2	4	7	9	15	23	36	47	61
	3	1	1	3	6	8	13	20	30	40	52
	2	1	1	3	5	7	11	17	26	33	44
	1	1	1	1	4	5	8	12	19	25	32
	1/0	1	1	1	3	4	7	10	16	21	27
	2/0	0	1	1	2	3	6	8	13	17	23
	3/0	0	1	1	1	3	5	7	11	14	19
	4/0	0	1	1	1	2	4	6	9	12	15
	250	0	0	1	1	1	3	5	7	10	12
	300	0	0	1	1	1	3	4	6	8	11
	350	0	0	1	1	1	2	3	5	7	9
	400	0	0	0	1	1	1	3	5	6	8
	500	0	0	0	1	1	1	2	4	5	7
	600	0	0	0	1	1	1	1	3	4	6
	700	0	0	0	1	1	1	1	3	4	5
	750	0	0	0	0	1	1	1	3	3	5
	800	0	0	0	0	1	1	1	2	3	4
	900	0	0	0	0	1	1	1	2	3	4
	1000	0	0	0	0	0	1	1	1	3	3
FEP, FEPB, PFA, PFAH, TFE	14	12	21	35	61	79	129	195	299	389	507
	12	9	15	25	44	57	94	142	218	284	370
	10	6	11	18	32	41	68	102	156	203	266
	8	3	6	10	18	23	39	58	89	117	152
	6	2	4	7	13	17	27	41	64	83	108
	4	1	3	5	9	12	19	29	44	58	75
	3	1	2	4	7	10	16	24	37	48	63
	2	1	1	3	6	8	13	20	30	40	52
PFA, PFAH, TFE	1	1	1	2	4	5	9	14	21	28	36
PFA, PFAH, TFE, Z	1/0	1	1	1	3	4	7	11	18	23	30
	2/0	1	1	1	3	4	6	9	14	19	25
	3/0	0	1	1	2	3	5	8	12	16	20
	4/0	0	1	1	1	2	4	6	10	13	17
Z	14	20	26	42	73	95	156	235	360	469	611
	12	14	18	30	52	67	111	167	255	332	434
	10	8	11	18	32	41	68	102	156	203	266
	8	5	7	11	20	26	43	64	99	129	168
	6	4	5	8	14	18	30	45	69	90	118
	4	2	3	5	9	12	20	31	48	62	81
	3	2	2	4	7	9	15	23	35	45	59
	2	1	1	3	6	7	12	19	29	38	49
	1	1	1	2	5	6	10	15	23	30	40
XHH, XHHW, XHHW-2, ZW	14	9	15	25	44	57	93	140	215	280	365
	12	7	12	19	33	43	71	108	165	215	280
	10	5	9	14	25	32	53	80	123	160	209
	8	3	5	8	14	18	29	44	68	89	116
	6	1	3	6	10	13	22	33	50	66	86
	4	1	2	4	7	9	16	24	36	48	62
	3	1	1	3	6	8	13	20	31	40	52
	2	1	1	3	5	7	11	17	26	34	44
XHH, XHHW, XHHW-2	1	1	1	1	4	5	8	12	19	25	33
	1/0	1	1	1	3	4	7	10	16	21	28
	2/0	0	1	1	2	3	6	9	13	17	23
	3/0	0	1	1	1	3	5	7	11	14	19
	4/0	0	1	1	1	2	4	6	9	12	16
	250	0	0	1	1	1	3	5	7	10	13
	300	0	0	1	1	1	3	4	6	8	11
	350	0	0	1	1	1	2	3	5	7	10
	400	0	0	0	1	1	1	3	5	6	8
	500	0	0	0	1	1	1	2	4	5	7
	600	0	0	0	1	1	1	1	3	4	6
	700	0	0	0	1	1	1	1	3	4	5
	750	0	0	0	0	1	1	1	3	3	5
	800	0	0	0	0	1	1	1	2	3	4
	900	0	0	0	0	1	1	1	2	3	4
	1000	0	0	0	0	0	1	1	1	3	3
	1250	0	0	0	0	0	1	1	1	1	3
	1500	0	0	0	0	0	1	1	1	1	2
	1750	0	0	0	0	0	0	1	1	1	2
	2000	0	0	0	0	0	0	1	1	1	2

*Types RHH, RHW, and RHW-2 without outer covering.

Table C7 Continued

FIXTURE WIRES

Type	Conductor Size (AWG/kcmil)	16 (½)	21 (¾)	27 (1)	35 (1¼)	41 (1½)	53 (2)
FFH-2, RFH-2, RFHH-3	18	8	15	24	42	54	89
	16	7	12	20	35	46	75
SF-2, SFF-2	18	11	19	30	53	69	113
	16	9	15	25	44	57	93
	14	7	12	20	35	46	75
SF-1, SFF-1	18	19	33	53	94	122	199
RFH-1, RFHH-2, TF, TFF, XF, XFF	18	14	24	39	69	90	147
RFHH-2, TF, TFF, XF, XFF	16	11	20	32	56	72	119
XF, XFF	14	9	15	25	44	57	93
TFN, TFFN	18	23	39	63	111	144	236
	16	17	30	48	85	110	180
PF, PFF, PGF, PGFF, PAF, PTF, PTFF, PAFF	18	21	37	60	105	136	223
	16	16	29	46	81	105	173
	14	12	21	35	61	79	129
HF, HFF, ZF, ZFF, ZHF	18	28	48	77	136	176	288
	16	20	35	57	100	129	212
	14	15	26	42	73	95	156
KF-2, KFF-2	18	40	70	112	197	255	418
	16	28	49	79	139	180	295
	14	19	34	54	95	123	202
	12	13	23	37	65	85	139
	10	9	15	25	44	57	93
KF-1, KFF-1	18	48	83	134	235	304	499
	16	34	58	94	165	214	350
	14	23	39	63	111	144	236
	12	15	26	42	73	95	156
	10	10	17	27	48	62	102
XF, XFF	12	5	8	13	23	30	50
	10	3	6	10	18	23	39

Note: This table is for concentric stranded conductors only. For compact stranded conductors, Table C7(A) should be used.

Table C7(A) Maximum Number of Compact Conductors in Liquidtight Flexible Metal Conduit (LFMC) *(Based on Table 1, Chapter 9)*

COMPACT CONDUCTORS

Type	Conductor Size (AWG/kcmil)	12 (³⁄₈)	16 (½)	21 (¾)	27 (1)	35 (1¼)	41 (1½)	53 (2)	63 (2½)	78 (3)	91 (3½)	103 (4)
THW, THW-2,	8	1	2	4	7	12	15	25	38	58	76	99
	6	1	1	3	5	9	12	19	29	45	59	77
THHW	4	1	1	2	4	7	9	14	22	34	44	57
	2	1	1	1	3	5	6	11	16	25	32	42
	1	0	1	1	1	3	4	7	11	17	23	30
	1/0	0	1	1	1	3	4	6	10	15	20	26
	2/0	0	0	1	1	2	3	5	8	13	16	21
	3/0	0	0	1	1	1	3	4	7	11	14	18
	4/0	0	0	1	1	1	2	4	6	9	12	15
	250	0	0	0	1	1	1	3	4	7	9	12
	300	0	0	0	1	1	1	2	4	6	8	10
	350	0	0	0	1	1	1	2	3	5	7	9
	400	0	0	0	0	1	1	1	3	5	6	8
	500	0	0	0	0	1	1	1	3	4	5	7
	600	0	0	0	0	1	1	1	1	3	4	6
	700	0	0	0	0	1	1	1	1	3	4	5
	750	0	0	0	0	0	1	1	1	3	3	5
	1000	0	0	0	0	0	1	1	1	1	3	4
THHN, THWN,	8	—	—	—	—	—	—	—	—	—	—	—
	6	1	2	4	7	13	17	28	43	66	86	112
THWN-2	4	1	1	3	4	8	11	17	26	41	53	69
	2	1	1	1	3	6	7	12	19	29	38	50
	1	0	1	1	2	4	6	9	14	22	28	37
	1/0	0	1	1	1	4	5	8	12	19	24	32
	2/0	0	1	1	1	3	4	6	10	15	20	26
	3/0	0	0	1	1	2	3	5	8	13	17	22
	4/0	0	0	1	1	1	3	4	7	10	14	18
	250	0	0	1	1	1	1	3	5	8	11	14
	300	0	0	0	1	1	1	3	4	7	9	12
	350	0	0	0	1	1	1	2	4	6	8	11
	400	0	0	0	1	1	1	2	3	5	7	9
	500	0	0	0	0	1	1	1	3	5	6	8
	600	0	0	0	0	1	1	1	2	4	5	6
	700	0	0	0	0	1	1	1	1	3	4	6
	750	0	0	0	0	1	1	1	1	3	4	5
	1000	0	0	0	0	0	1	1	1	2	3	4
XHHW, XHHW-2	8	1	3	5	9	15	20	33	49	76	98	129
	6	1	2	4	6	11	15	24	37	56	73	95
	4	1	1	3	4	8	11	17	26	41	53	69
	2	1	1	1	3	6	7	12	19	29	38	50
	1	0	1	1	2	4	6	9	14	22	28	37
	1/0	0	1	1	1	4	5	8	12	19	24	32
	2/0	0	1	1	1	3	4	7	10	16	20	27
	3/0	0	0	1	1	2	3	5	8	13	17	22
	4/0	0	0	1	1	1	3	4	7	11	14	18
	250	0	0	1	1	1	1	3	5	8	11	15
	300	0	0	0	1	1	1	3	5	7	9	12
	350	0	0	0	1	1	1	3	4	6	8	11
	400	0	0	0	1	1	1	2	4	6	7	10
	500	0	0	0	0	1	1	1	3	5	6	8
	600	0	0	0	0	1	1	1	2	4	5	6
	700	0	0	0	0	1	1	1	1	3	4	6
	750	0	0	0	0	1	1	1	1	3	4	5
	1000	0	0	0	0	0	1	1	1	2	3	4

Definition: *Compact stranding* is the result of a manufacturing process where the standard conductor is compressed to the extent that the interstices (voids between strand wires) are virtually eliminated.

Table C8 Maximum Number of Conductors or Fixture Wires in Rigid Metal Conduit (RMC) *(Based on Table 1, Chapter 9)*

CONDUCTORS

Type	Conductor Size (AWG/kcmil)	16 (½)	21 (¾)	27 (1)	35 (1¼)	41 (1½)	53 (2)	63 (2½)	78 (3)	91 (3½)	103 (4)	129 (5)	155 (6)
RHH, RHW,	14	4	7	12	21	28	46	66	102	136	176	276	398
RHW-2	12	3	6	10	17	23	38	55	85	113	146	229	330
	10	3	5	8	14	19	31	44	68	91	118	185	267
	8	1	2	4	7	10	16	23	36	48	61	97	139
	6	1	1	3	6	8	13	18	29	38	49	77	112
	4	1	1	2	4	6	10	14	22	30	38	60	87
	3	1	1	2	4	5	9	12	19	26	34	53	76
	2	1	1	1	3	4	7	11	17	23	29	46	66
	1	0	1	1	1	3	5	7	11	15	19	30	44
	1/0	0	1	1	1	2	4	6	10	13	17	26	38
	2/0	0	1	1	1	2	4	5	8	11	14	23	33
	3/0	0	0	1	1	1	3	4	7	10	12	20	28
	4/0	0	0	1	1	1	3	4	6	8	11	17	24
	250	0	0	0	1	1	1	3	4	6	8	13	18
	300	0	0	0	1	1	1	2	4	5	7	11	16
	350	0	0	0	1	1	1	2	4	5	6	10	15
	400	0	0	0	1	1	1	1	3	4	6	9	13
	500	0	0	0	1	1	1	1	3	4	5	8	11
	600	0	0	0	0	1	1	1	2	3	4	6	9
	700	0	0	0	0	1	1	1	1	3	4	6	8
	750	0	0	0	0	1	1	1	1	3	3	5	8
	800	0	0	0	0	0	1	1	1	2	3	5	7
	900	0	0	0	0	0	1	1	1	2	3	5	7
	1000	0	0	0	0	0	1	1	1	1	3	4	6
	1250	0	0	0	0	0	0	1	1	1	1	3	5
	1500	0	0	0	0	0	0	1	1	1	1	3	4
	1750	0	0	0	0	0	0	1	1	1	1	2	4
	2000	0	0	0	0	0	0	0	1	1	1	2	3
TW, THHW,	14	9	15	25	44	59	98	140	216	288	370	581	839
THW, THW-2	12	7	12	19	33	45	75	107	165	221	284	446	644
	10	5	9	14	25	34	56	80	123	164	212	332	480
	8	3	5	8	14	19	31	44	68	91	118	185	267
RHH*, RHW*, RHW-2*	14	6	10	17	29	39	65	93	143	191	246	387	558
RHH*, RHW*, RHW-2*	12	5	8	13	23	32	52	75	115	154	198	311	448
	10	3	6	10	18	25	41	58	90	120	154	242	350
RHH*, RHW*, RHW-2*	8	1	4	6	11	15	24	35	54	72	92	145	209
RHH*, RHW*, RHW-2*, TW, THW, THHW, THW-2	6	1	3	5	8	11	18	27	41	55	71	111	160
	4	1	1	3	6	8	14	20	31	41	53	83	120
	3	1	1	3	5	7	12	17	26	35	45	71	103
	2	1	1	2	4	6	10	14	22	30	38	60	87
	1	1	1	1	3	4	7	10	15	21	27	42	61
	1/0	0	1	1	2	3	6	8	13	18	23	36	52
	2/0	0	1	1	2	3	5	7	11	15	19	31	44
	3/0	0	1	1	1	2	4	6	9	13	16	26	37
	4/0	0	0	1	1	1	3	5	8	10	14	21	31
	250	0	0	1	1	1	3	4	6	8	11	17	25
	300	0	0	1	1	1	2	3	5	7	9	15	22
	350	0	0	0	1	1	1	3	5	6	8	13	19
	400	0	0	0	1	1	1	3	4	6	7	12	17
	500	0	0	0	1	1	1	2	3	5	6	10	14
	600	0	0	0	1	1	1	1	3	4	5	8	12
	700	0	0	0	0	1	1	1	2	3	4	7	10
	750	0	0	0	0	1	1	1	2	3	4	7	10
	800	0	0	0	0	1	1	1	2	3	4	6	9
	900	0	0	0	0	1	1	1	1	3	4	6	8
	1000	0	0	0	0	0	1	1	1	2	3	5	8
	1250	0	0	0	0	0	1	1	1	1	2	4	6
	1500	0	0	0	0	0	1	1	1	1	2	3	5
	1750	0	0	0	0	0	0	1	1	1	1	3	4
	2000	0	0	0	0	0	0	1	1	1	1	3	4

Table C8 Continued

CONDUCTORS

Type	Conductor Size (AWG/kcmil)	16 (½)	21 (¾)	27 (1)	35 (1¼)	41 (1½)	53 (2)	63 (2½)	78 (3)	91 (3½)	103 (4)	129 (5)	155 (6)
THHN,	14	13	22	36	63	85	140	200	309	412	531	833	1202
THWN,	12	9	16	26	46	62	102	146	225	301	387	608	877
THWN-2	10	6	10	17	29	39	64	92	142	189	244	383	552
	8	3	6	9	16	22	37	53	82	109	140	221	318
	6	2	4	7	12	16	27	38	59	79	101	159	230
	4	1	2	4	7	10	16	23	36	48	62	98	141
	3	1	1	3	6	8	14	20	31	41	53	83	120
	2	1	1	3	5	7	11	17	26	34	44	70	100
	1	1	1	1	4	5	8	12	19	25	33	51	74
	1/0	1	1	1	3	4	7	10	16	21	27	43	63
	2/0	0	1	1	2	3	6	8	13	18	23	36	52
	3/0	0	1	1	1	3	5	7	11	15	19	30	43
	4/0	0	1	1	1	2	4	6	9	12	16	25	36
	250	0	0	1	1	1	3	5	7	10	13	20	29
	300	0	0	1	1	1	3	4	6	8	11	17	25
	350	0	0	1	1	1	2	3	5	7	10	15	22
	400	0	0	1	1	1	2	3	5	7	8	13	20
	500	0	0	0	1	1	1	2	4	5	7	11	16
	600	0	0	0	1	1	1	1	3	4	6	9	13
	700	0	0	0	1	1	1	1	3	4	5	8	11
	750	0	0	0	0	1	1	1	3	4	5	7	11
	800	0	0	0	0	1	1	1	2	3	4	7	10
	900	0	0	0	0	1	1	1	2	3	4	6	9
	1000	0	0	0	0	1	1	1	1	3	4	6	8
FEP, FEPB,	14	12	22	35	61	83	136	194	300	400	515	808	1166
PFA, PFAH,	12	9	16	26	44	60	99	142	219	292	376	590	851
TFE	10	6	11	18	32	43	71	102	157	209	269	423	610
	8	3	6	10	18	25	41	58	90	120	154	242	350
	6	2	4	7	13	17	29	41	64	85	110	172	249
	4	1	3	5	9	12	20	29	44	59	77	120	174
	3	1	2	4	7	10	17	24	37	50	64	100	145
	2	1	1	3	6	8	14	20	31	41	53	83	120
PFA, PFAH, TFE	1	1	1	2	4	6	9	14	21	28	37	57	83
PFA, PFAH, TFE, Z	1/0	1	1	1	3	5	8	11	18	24	30	48	69
	2/0	1	1	1	3	4	6	9	14	19	25	40	57
	3/0	0	1	1	2	3	5	8	12	16	21	33	47
	4/0	0	1	1	1	2	4	6	10	13	17	27	39
Z	14	15	26	42	73	100	164	234	361	482	621	974	1405
	12	10	18	30	52	71	116	166	256	342	440	691	997
	10	6	11	18	32	43	71	102	157	209	269	423	610
	8	4	7	11	20	27	45	64	99	132	170	267	386
	6	3	5	8	14	19	31	45	69	93	120	188	271
	4	1	3	5	9	13	22	31	48	64	82	129	186
	3	1	2	4	7	9	16	22	35	47	60	94	136
	2	1	1	3	6	8	13	19	29	39	50	78	113
	1	1	1	2	5	6	10	15	23	31	40	63	92
XHH, XHHW,	14	9	15	25	44	59	98	140	216	288	370	581	839
XHHW-2, ZW	12	7	12	19	33	45	75	107	165	221	284	446	644
	10	5	9	14	25	34	56	80	123	164	212	332	480
	8	3	5	8	14	19	31	44	68	91	118	185	267
	6	1	3	6	10	14	23	33	51	68	87	137	197
	4	1	2	4	7	10	16	24	37	49	63	99	143
	3	1	1	3	6	8	14	20	31	41	53	84	121
	2	1	1	3	5	7	12	17	26	35	45	70	101
XHH, XHHW,	1	1	1	1	4	5	9	12	19	26	33	52	76
XHHW-2	1/0	1	1	1	3	4	7	10	16	22	28	44	64
	2/0	0	1	1	2	3	6	9	13	18	23	37	53
	3/0	0	1	1	1	3	5	7	11	15	19	30	44
	4/0	0	1	1	1	2	4	6	9	12	16	25	36
	250	0	0	1	1	1	3	5	7	10	13	20	30
	300	0	0	1	1	1	3	4	6	9	11	18	25
	350	0	0	1	1	1	2	3	6	7	10	15	22
	400	0	0	1	1	1	2	3	5	7	9	14	20
	500	0	0	0	1	1	1	2	4	5	7	11	16
	600	0	0	0	1	1	1	1	3	4	6	9	13
	700	0	0	0	1	1	1	1	3	4	5	8	11
	750	0	0	0	0	1	1	1	3	4	5	7	11
	800	0	0	0	0	1	1	1	2	3	4	7	10
	900	0	0	0	0	1	1	1	2	3	4	6	9
	1000	0	0	0	0	1	1	1	1	3	4	6	8
	1250	0	0	0	0	0	1	1	1	2	3	4	6
	1500	0	0	0	0	0	1	1	1	1	2	4	5
	1750	0	0	0	0	0	0	1	1	1	1	3	5
	2000	0	0	0	0	0	0	1	1	1	1	3	4

Table C8 Continued

FIXTURE WIRES

Type	Conductor Size (AWG/kcmil)	16 (½)	21 (¾)	27 (1)	35 (1¼)	41 (1½)	53 (2)
FFH-2, RFH-2, RFHH-3	18	8	15	24	42	57	94
	16	7	12	20	35	48	79
SF-2, SFF-2	18	11	19	31	53	72	118
	16	9	15	25	44	59	98
	14	7	12	20	35	48	79
SF-1, SFF-1	18	19	33	54	94	127	209
RFH-1, RFHH-2, TF, TFF, XF, XFF	18	14	25	40	69	94	155
RFHH-2, TF, TFF, XF, XFF	16	11	20	32	56	76	125
XF, XFF	14	9	15	25	44	59	98
TFN, TFFN	18	23	40	64	111	150	248
	16	17	30	49	84	115	189
PF, PFF, PGF, PGFF, PAF, PTF, PTFF, PAFF	18	21	38	61	105	143	235
	16	16	29	47	81	110	181
	14	12	22	35	61	83	136
HF, HFF, ZF, ZFF, ZHF	18	28	48	79	135	184	303
	16	20	36	58	100	136	223
	14	15	26	42	73	100	164
KF-2, KFF-2	18	40	71	114	197	267	439
	16	28	50	80	138	188	310
	14	19	34	55	95	129	213
	12	13	23	38	65	89	146
	10	9	15	25	44	59	98
KF-1, KFF-1	18	48	84	136	235	318	524
	16	34	59	96	165	224	368
	14	23	40	64	111	150	248
	12	15	26	42	73	100	164
	10	10	17	28	48	65	107
XF, XFF	12	5	8	13	23	32	52
	10	3	6	10	18	25	41

Note: This table is for concentric stranded conductors only. For compact stranded conductors, Table C8(A) should be used.
*Types RHH, RHW, and RHW-2 without outer covering.

Table C8(A) Maximum Number of Compact Conductors in Rigid Metal Conduit (RMC) (*Based on Table 1, Chapter 9*)

COMPACT CONDUCTORS

Type	Conductor Size (AWG/ kcmil)	16 (1/2)	21 (3/4)	27 (1)	35 (1 1/4)	41 (1 1/2)	53 (2)	63 (2 1/2)	78 (3)	91 (3 1/2)	103 (4)	129 (5)	155 (6)
THW, THW-2, THHW	8	2	4	7	12	16	26	38	59	78	101	158	228
	6	1	3	5	9	12	20	29	45	60	78	122	176
	4	1	2	4	7	9	15	22	34	45	58	91	132
	2	1	1	3	5	7	11	16	25	33	43	67	97
	1	1	1	1	3	5	8	11	17	23	30	47	68
	1/0	1	1	1	3	4	7	10	15	20	26	41	59
	2/0	0	1	1	2	3	6	8	13	17	22	34	50
	3/0	0	1	1	1	3	5	7	11	14	19	29	42
	4/0	0	1	1	1	2	4	6	9	12	15	24	35
	250	0	0	1	1	1	3	4	7	9	12	19	28
	300	0	0	1	1	1	3	4	6	8	11	17	24
	350	0	0	1	1	1	2	3	5	7	9	15	22
	400	0	0	1	1	1	1	3	5	7	8	13	20
	500	0	0	0	1	1	1	3	4	5	7	11	17
	600	0	0	0	1	1	1	1	3	4	6	9	13
	700	0	0	0	1	1	1	1	3	4	5	8	12
	750	0	0	0	1	1	1	1	3	4	5	7	11
	1000	0	0	0	0	1	1	1	1	3	4	6	9
THHN, THWN, THWN-2	8	—	—	—	—	—	—	—	—	—	—	—	—
	6	2	5	8	13	18	30	43	66	88	114	179	258
	4	1	3	5	8	11	18	26	41	55	70	110	159
	2	1	1	3	6	8	13	19	29	39	50	79	114
	1	1	1	2	4	6	10	14	22	29	38	60	86
	1/0	1	1	1	4	5	8	12	19	25	32	51	73
	2/0	1	1	1	3	4	7	10	15	21	26	42	60
	3/0	0	1	1	2	3	6	8	13	17	22	35	51
	4/0	0	1	1	1	3	5	7	10	14	18	29	42
	250	0	1	1	1	2	4	5	8	11	14	23	33
	300	0	0	1	1	1	3	4	7	10	12	20	28
	350	0	0	1	1	1	3	4	6	8	11	17	25
	400	0	0	1	1	1	2	3	5	7	10	15	22
	500	0	0	0	1	1	1	3	5	6	8	13	19
	600	0	0	0	1	1	1	2	4	5	6	10	15
	700	0	0	0	1	1	1	1	3	4	6	9	13
	750	0	0	0	1	1	1	1	3	4	5	9	13
	1000	0	0	0	0	1	1	1	2	3	4	6	9
XHHW, XHHW-2	8	3	5	9	15	21	34	49	76	101	130	205	296
	6	2	4	6	11	15	25	36	56	75	97	152	220
	4	1	3	5	8	11	18	26	41	55	70	110	159
	2	1	1	3	6	8	13	19	29	39	50	79	114
	1	1	1	2	4	6	10	14	22	29	38	60	86
	1/0	1	1	1	4	5	8	12	19	25	32	51	73
	2/0	1	1	1	3	4	7	10	16	21	27	43	62
	3/0	0	1	1	2	3	6	8	13	17	22	35	51
	4/0	0	1	1	1	3	5	7	11	14	19	29	42
	250	0	1	1	1	2	4	5	8	11	15	23	34
	300	0	0	1	1	1	3	5	7	10	13	20	29
	350	0	0	1	1	1	3	4	6	9	11	18	25
	400	0	0	1	1	1	2	4	6	8	10	16	23
	500	0	0	0	1	1	1	3	5	6	8	13	19
	600	0	0	0	1	1	1	2	4	5	7	10	15
	700	0	0	0	1	1	1	1	3	4	6	9	13
	750	0	0	0	1	1	1	1	3	4	5	8	12
	1000	0	0	0	0	1	1	1	2	3	4	7	10

Definition: *Compact stranding* is the result of a manufacturing process where the standard conductor is compressed to the extent that the interstices (voids between strand wires) are virtually eliminated.

Table C9 Maximum Number of Conductors or Fixture Wires in Rigid PVC Conduit, Schedule 80 (*Based on Table 1, Chapter 9*)

CONDUCTORS

Type	Conductor Size (AWG/ kcmil)	16 (1/2)	21 (3/4)	27 (1)	35 (1 1/4)	41 (1 1/2)	53 (2)	63 (2 1/2)	78 (3)	91 (3 1/2)	103 (4)	129 (5)	155 (6)
RHH, RHW, RHW-2	14	3	5	9	17	23	39	56	88	118	153	243	349
	12	2	4	7	14	19	32	46	73	98	127	202	290
	10	1	3	6	11	15	26	37	59	79	103	163	234
	8	1	1	3	6	8	13	19	31	41	54	85	122
	6	1	1	2	4	6	11	16	24	33	43	68	98
	4	1	1	1	3	5	8	12	19	26	33	53	77
	3	0	1	1	3	4	7	11	17	23	29	47	67
	2	0	1	1	3	4	6	9	14	20	25	41	58
	1	0	1	1	1	2	4	6	9	13	17	27	38
	1/0	0	0	1	1	1	3	5	8	11	15	23	33
	2/0	0	0	1	1	1	3	4	7	10	13	20	29
	3/0	0	0	1	1	1	3	4	6	8	11	17	25
	4/0	0	0	0	1	1	2	3	5	7	9	15	21
	250	0	0	0	1	1	1	2	4	5	7	11	16
	300	0	0	0	1	1	1	2	3	5	6	10	14
	350	0	0	0	1	1	1	1	3	4	5	9	13
	400	0	0	0	0	1	1	1	3	4	5	8	12
	500	0	0	0	0	1	1	1	2	3	4	7	10
	600	0	0	0	0	0	1	1	1	3	3	6	8
	700	0	0	0	0	0	1	1	1	2	3	5	7
	750	0	0	0	0	0	1	1	1	2	3	5	7
	800	0	0	0	0	0	1	1	1	2	3	4	7
	1000	0	0	0	0	0	1	1	1	1	2	4	5
	1250	0	0	0	0	0	0	1	1	1	1	3	4
	1500	0	0	0	0	0	0	1	1	1	1	2	4
	1750	0	0	0	0	0	0	0	1	1	1	2	3
	2000	0	0	0	0	0	0	0	1	1	1	1	3
TW, THHW, THW, THW-2	14	6	11	20	35	49	82	118	185	250	324	514	736
	12	5	9	15	27	38	63	91	142	192	248	394	565
	10	3	6	11	20	28	47	67	106	143	185	294	421
	8	1	3	6	11	15	26	37	59	79	103	163	234
RHH*, RHW*, RHW-2*	14	4	8	13	23	32	55	79	123	166	215	341	490
	12	3	6	10	19	26	44	63	99	133	173	274	394
	10	2	5	8	15	20	34	49	77	104	135	214	307
	8	1	3	5	9	12	20	29	46	62	81	128	184
RHH*, RHW*, RHW-2*, TW, THW, THHW, THW-2	6	1	1	3	7	9	16	22	35	48	62	98	141
	4	1	1	3	5	7	12	17	26	35	46	73	105
	3	1	1	2	4	6	10	14	22	30	39	63	90
	2	1	1	1	3	5	8	12	19	26	33	53	77
	1	0	1	1	2	3	6	8	13	18	23	37	54
	1/0	0	1	1	1	3	5	7	11	15	20	32	46
	2/0	0	1	1	1	2	4	6	10	13	17	27	39
	3/0	0	0	1	1	1	3	5	8	11	14	23	33
	4/0	0	0	1	1	1	3	4	7	9	12	19	27
	250	0	0	0	1	1	2	3	5	7	9	15	22
	300	0	0	0	1	1	1	3	5	6	8	13	19
	350	0	0	0	1	1	1	2	4	6	7	12	17
	400	0	0	0	1	1	1	2	4	5	7	10	15
	500	0	0	0	1	1	1	1	3	4	5	9	13
	600	0	0	0	0	1	1	1	2	3	4	7	10
	700	0	0	0	0	1	1	1	2	3	4	6	9
	750	0	0	0	0	0	1	1	1	3	4	6	8
	800	0	0	0	0	0	1	1	1	3	3	6	8
	900	0	0	0	0	0	1	1	1	2	3	5	7
	1000	0	0	0	0	0	1	1	1	2	3	5	7
	1250	0	0	0	0	0	1	1	1	1	2	4	5
	1500	0	0	0	0	0	0	1	1	1	1	3	4
	1750	0	0	0	0	0	0	1	1	1	1	3	4
	2000	0	0	0	0	0	0	0	1	1	1	2	3

Table C9 Continued

CONDUCTORS

Type	Conductor Size (AWG/kcmil)	16 (½)	21 (¾)	27 (1)	35 (1¼)	41 (1½)	53 (2)	63 (2½)	78 (3)	91 (3½)	103 (4)	129 (5)	155 (6)
THHN, THWN, THWN-2	14	9	17	28	51	70	118	170	265	358	464	736	1055
	12	6	12	20	37	51	86	124	193	261	338	537	770
	10	4	7	13	23	32	54	78	122	164	213	338	485
	8	2	4	7	13	18	31	45	70	95	123	195	279
	6	1	3	5	9	13	22	32	51	68	89	141	202
	4	1	1	3	6	8	14	20	31	42	54	86	124
	3	1	1	3	5	7	12	17	26	35	46	73	105
	2	1	1	2	4	6	10	14	22	30	39	61	88
	1	0	1	1	3	4	7	10	16	22	29	45	65
	1/0	0	1	1	2	3	6	9	14	18	24	38	55
	2/0	0	1	1	1	3	5	7	11	15	20	32	46
	3/0	0	1	1	1	2	4	6	9	13	17	26	38
	4/0	0	0	1	1	1	3	5	8	10	14	22	31
	250	0	0	1	1	1	3	4	6	8	11	18	25
	300	0	0	0	1	1	2	3	5	7	9	15	22
	350	0	0	0	1	1	1	3	5	6	8	13	19
	400	0	0	0	1	1	1	3	4	6	7	12	17
	500	0	0	0	1	1	1	2	3	5	6	10	14
	600	0	0	0	0	1	1	1	3	4	5	8	12
	700	0	0	0	0	1	1	1	2	3	4	7	10
	750	0	0	0	0	1	1	1	2	3	4	7	9
	800	0	0	0	0	1	1	1	2	3	4	6	9
	900	0	0	0	0	0	1	1	1	3	3	6	8
	1000	0	0	0	0	0	1	1	1	2	3	5	7
FEP, FEPB, PFA, PFAH, TFE	14	8	16	27	49	68	115	164	257	347	450	714	1024
	12	6	12	20	36	50	84	120	188	253	328	521	747
	10	4	8	14	26	36	60	86	135	182	235	374	536
	8	2	5	8	15	20	34	49	77	104	135	214	307
	6	1	3	6	10	14	24	35	55	74	96	152	218
	4	1	2	4	7	10	17	24	38	52	67	106	153
	3	1	1	3	6	8	14	20	32	43	56	89	127
	2	1	1	3	5	7	12	17	26	35	46	73	105
PFA, PFAH, TFE	1	1	1	1	3	5	8	11	18	25	32	51	73
PFA, PFAH, TFE, Z	1/0	0	1	1	3	4	7	10	15	20	27	42	61
	2/0	0	1	1	2	3	5	8	12	17	22	35	50
	3/0	0	1	1	1	2	4	6	10	14	18	29	41
	4/0	0	0	1	1	1	4	5	8	11	15	24	34
Z	14	10	19	33	59	82	138	198	310	418	542	860	1233
	12	7	14	23	42	58	98	141	220	297	385	610	875
	10	4	8	14	26	36	60	86	135	182	235	374	536
	8	3	5	9	16	22	38	54	85	115	149	236	339
	6	2	4	6	11	16	26	38	60	81	104	166	238
	4	1	2	4	8	11	18	26	41	55	72	114	164
	3	1	2	3	5	8	13	19	30	40	52	83	119
	2	1	1	2	5	6	11	16	25	33	43	69	99
	1	0	1	2	4	5	9	13	20	27	35	56	80
XHH, XHHW, XHHW-2, ZW	14	6	11	20	35	49	82	118	185	250	324	514	736
	12	5	9	15	27	38	63	91	142	192	248	394	565
	10	3	6	11	20	28	47	67	106	143	185	294	421
	8	1	3	6	11	15	26	37	59	79	103	163	234
	6	1	2	4	8	11	19	28	43	59	76	121	173
	4	1	1	3	6	8	14	20	31	42	55	87	125
	3	1	1	3	5	7	12	17	26	36	47	74	106
	2	1	1	2	4	6	10	14	22	30	39	62	89
XHH, XHHW, XHHW-2	1	0	1	1	3	4	7	10	16	22	29	46	66
	1/0	0	1	1	2	3	6	9	14	19	24	39	56
	2/0	0	1	1	1	3	5	7	11	16	20	32	46
	3/0	0	1	1	1	2	4	6	9	13	17	27	38
	4/0	0	0	1	1	1	3	5	8	11	14	22	32
	250	0	0	1	1	1	3	4	6	9	11	18	26
	300	0	0	1	1	1	2	3	5	7	10	15	22
	350	0	0	0	1	1	1	3	5	6	8	14	20
	400	0	0	0	1	1	1	3	4	6	7	12	17
	500	0	0	0	1	1	1	2	3	5	6	10	14
	600	0	0	0	0	1	1	1	3	4	5	8	11
	700	0	0	0	0	1	1	1	2	3	4	7	10
	750	0	0	0	0	1	1	1	2	3	4	6	9
	800	0	0	0	0	1	1	1	2	3	4	6	9
	900	0	0	0	0	0	1	1	—	3	3	5	8
	1000	0	0	0	0	0	1	1	1	2	3	5	7
	1250	0	0	0	0	0	1	1	1	1	2	4	6
	1500	0	0	0	0	0	0	1	1	1	1	3	5
	1750	0	0	0	0	0	0	1	1	1	1	3	4
	2000	0	0	0	0	0	0	1	1	1	1	2	4

Table C9 Continued

FIXTURE WIRES

Type	Conductor Size (AWG/kcmil)	16 (½)	21 (¾)	27 (1)	35 (1¼)	41 (1½)	53 (2)
FFH-2, RFH-2, RFHH-3	18	6	11	19	34	47	79
	16	5	9	16	28	39	67
SF-2, SFF-2	18	7	14	24	43	59	100
	16	6	11	20	35	49	82
	14	5	9	16	28	39	67
SF-1, SFF-1	18	13	25	42	76	105	177
RFH-1, RFHH-2, TF, TFF, XF, XFF	18	10	18	31	56	77	130
RFHH-2, TF, TFF, XF, XFF	16	8	15	25	45	62	105
XF, XFF	14	6	11	20	35	49	82
TFN, TFFN	18	16	29	50	90	124	209
	16	12	22	38	68	95	159
PF, PFF, PGF, PGFF, PAF, PTF, PTFF, PAFF	18	15	28	47	85	118	198
	16	11	22	36	66	91	153
	14	8	16	27	49	68	115
HF, HFF, ZF, ZFF, ZHF	18	19	36	61	110	152	255
	16	14	27	45	81	112	188
	14	10	19	33	59	82	138
KF-2, KFF-2	18	28	53	88	159	220	371
	16	19	37	62	112	155	261
	14	13	25	43	77	107	179
	12	9	17	29	53	73	123
	10	6	11	20	35	49	82
KF-1, KFF-1	18	33	63	106	190	263	442
	16	23	44	74	133	185	310
	14	16	29	50	90	124	209
	12	10	19	33	59	82	138
	10	7	13	21	39	54	90
XF, XFF	12	3	6	10	19	26	44
	10	2	5	8	15	20	34

Note: This table is for concentric stranded conductors only. For compact stranded conductors, Table C9(A) should be used.

*Types RHH, RHW, and RHW-2 without outer covering.

Table C9(A) Maximum Number of Compact Conductors in Rigid PVC Conduit, Schedule 80 (*Based on Table 1, Chapter 9*)

COMPACT CONDUCTORS

Type	Conductor Size (AWG/kcmil)	Metric Designator (Trade Size)											
		16 (½)	21 (¾)	27 (1)	35 (1¼)	41 (1½)	53 (2)	63 (2½)	78 (3)	91 (3½)	103 (4)	129 (5)	155 (6)
THW, THW-2, THHW	8	1	3	5	9	13	22	32	50	68	88	140	200
	6	1	2	4	7	10	17	25	39	52	68	108	155
	4	1	1	3	5	7	13	18	29	39	51	81	116
	2	1	1	1	4	5	9	13	21	29	37	60	85
	1	0	1	1	3	4	6	9	15	20	26	42	60
	1/0	0	1	1	2	3	6	8	13	17	23	36	52
	2/0	0	1	1	1	3	5	7	11	15	19	30	44
	3/0	0	0	1	1	2	4	6	9	12	16	26	37
	4/0	0	0	1	1	1	3	5	8	10	13	22	31
	250	0	0	1	1	1	2	4	6	8	11	17	25
	300	0	0	0	1	1	2	3	5	7	9	15	21
	350	0	0	0	1	1	1	3	5	6	8	13	19
	400	0	0	0	1	1	1	3	4	6	7	12	17
	500	0	0	0	1	1	1	2	3	5	6	10	14
	600	0	0	0	0	1	1	1	3	4	5	8	12
	700	0	0	0	0	1	1	1	2	3	4	7	10
	750	0	0	0	0	1	1	1	2	3	4	7	10
	1000	0	0	0	0	0	1	1	1	2	3	5	8
THHN, THWN, THWN-2	8	—	—	—	—	—	—	—	—	—	—	—	—
	6	1	3	6	11	15	25	36	57	77	99	158	226
	4	1	1	3	6	9	15	22	35	47	61	98	140
	2	1	1	2	5	6	11	16	25	34	44	70	100
	1	1	1	1	3	5	8	12	19	25	33	53	75
	1/0	0	1	1	3	4	7	10	16	22	28	45	64
	2/0	0	1	1	2	3	6	8	13	18	23	37	53
	3/0	0	1	1	1	3	5	7	11	15	19	31	44
	4/0	0	0	1	1	2	4	6	9	12	16	25	37
	250	0	0	1	1	1	3	4	7	10	12	20	29
	300	0	0	1	1	1	3	4	6	8	11	17	25
	350	0	0	0	1	1	2	3	5	7	9	15	22
	400	0	0	0	1	1	1	3	5	6	8	13	19
	500	0	0	0	1	1	1	2	4	5	7	11	16
	600	0	0	0	1	1	1	1	3	4	6	9	13
	700	0	0	0	0	1	1	1	3	4	5	8	12
	750	0	0	0	0	1	1	1	3	4	5	8	11
	1000	0	0	0	0	0	1	1	1	3	3	5	8
XHHW, XHHW-2	8	1	4	7	12	17	29	42	65	88	114	181	260
	6	1	3	5	9	13	21	31	48	65	85	134	193
	4	1	1	3	6	9	15	22	35	47	61	98	140
	2	1	1	2	5	6	11	16	25	34	44	70	100
	1	1	1	1	3	5	8	12	19	25	33	53	75
	1/0	0	1	1	3	4	7	10	16	22	28	45	64
	2/0	0	1	1	2	3	6	8	13	18	24	38	54
	3/0	0	1	1	1	3	5	7	11	15	19	31	44
	4/0	0	0	1	1	2	4	6	9	12	16	26	37
	250	0	0	1	1	1	3	5	7	10	13	21	30
	300	0	0	1	1	1	3	4	6	8	11	17	25
	350	0	0	1	1	1	2	3	5	7	10	15	22
	400	0	0	0	1	1	1	3	5	7	9	14	20
	500	0	0	0	1	1	1	2	4	5	7	11	17
	600	0	0	0	1	1	1	1	3	4	6	9	13
	700	0	0	0	0	1	1	1	3	4	5	8	12
	750	0	0	0	0	1	1	1	2	3	5	7	11
	1000	0	0	0	0	0	1	1	1	3	3	6	8

Definition: *Compact stranding* is the result of a manufacturing process where the standard conductor is compressed to the extent that the interstices (voids between strand wires) are virtually eliminated.

Table C10 Maximum Number of Conductors or Fixture Wires in Rigid PVC Conduit, Schedule 40 and HDPE Conduit (*Based on Table 1, Chapter 9*)

CONDUCTORS

Type	Conductor Size (AWG/kcmil)	Metric Designator (Trade Size)											
		16 (½)	21 (¾)	27 (1)	35 (1¼)	41 (1½)	53 (2)	63 (2½)	78 (3)	91 (3½)	103 (4)	129 (5)	155 (6)
RHH, RHW, RHW-2	14	4	7	11	20	27	45	64	99	133	171	269	390
	12	3	5	9	16	22	37	53	82	110	142	224	323
	10	2	4	7	13	18	30	43	66	89	115	181	261
	8	1	2	4	7	9	15	22	35	46	60	94	137
	6	1	1	3	5	7	12	18	28	37	48	76	109
	4	1	1	2	4	6	10	14	22	29	37	59	85
	3	1	1	1	4	5	8	12	19	25	33	52	75
	2	1	1	1	3	4	7	10	16	22	28	45	65
	1	0	1	1	1	3	5	7	11	14	19	29	43
	1/0	0	1	1	1	2	4	6	9	13	16	26	37
	2/0	0	0	1	1	1	3	5	8	11	14	22	32
	3/0	0	0	1	1	1	3	4	7	9	12	19	28
	4/0	0	0	1	1	1	2	4	6	8	10	16	24
	250	0	0	0	1	1	1	3	4	6	8	12	18
	300	0	0	0	1	1	1	2	4	5	7	11	16
	350	0	0	0	1	1	1	2	3	5	6	10	14
	400	0	0	0	1	1	1	1	3	4	6	9	13
	500	0	0	0	1	1	1	1	3	4	5	8	11
	600	0	0	0	0	1	1	1	2	3	4	6	9
	700	0	0	0	0	0	1	1	1	3	3	6	8
	750	0	0	0	0	0	1	1	1	2	3	5	8
	800	0	0	0	0	0	1	1	1	2	3	5	7
	900	0	0	0	0	0	1	1	1	2	3	5	7
	1000	0	0	0	0	0	1	1	1	1	3	4	6
	1250	0	0	0	0	0	0	1	1	1	1	3	5
	1500	0	0	0	0	0	0	1	1	1	1	3	4
	1750	0	0	0	0	0	0	1	1	1	1	2	3
	2000	0	0	0	0	0	0	0	1	1	1	2	3
TW, THHW, THW, THW-2	14	8	14	24	42	57	94	135	209	280	361	568	822
	12	6	11	18	32	44	72	103	160	215	277	436	631
	10	4	8	13	24	32	54	77	119	160	206	325	470
	8	2	4	7	13	18	30	43	66	89	115	181	261
RHH*, RHW*, RHW-2*	14	5	9	16	28	38	63	90	139	186	240	378	546
	12	4	8	12	22	30	50	72	112	150	193	304	439
	10	3	6	10	17	24	39	56	87	117	150	237	343
	8	1	3	6	10	14	23	33	52	70	90	142	205
TW, THW, THHW, THW-2	6	1	2	4	8	11	18	26	40	53	69	109	157
	4	1	1	3	6	8	13	19	30	40	51	81	117
	3	1	1	3	5	7	11	16	25	34	44	69	100
	2	1	1	2	4	6	10	14	22	29	37	59	85
	1	0	1	1	3	4	7	10	15	20	26	41	60
	1/0	0	1	1	2	3	6	8	13	17	22	35	51
	2/0	0	1	1	1	3	5	7	11	15	19	30	43
	3/0	0	1	1	1	2	4	6	9	12	16	25	36
	4/0	0	0	1	1	1	3	5	8	10	13	21	30
	250	0	0	1	1	1	3	4	6	8	11	17	25
	300	0	0	1	1	1	2	3	5	7	9	15	21
	350	0	0	0	1	1	1	3	5	6	8	13	19
	400	0	0	0	1	1	1	3	4	6	7	12	17
	500	0	0	0	1	1	1	2	3	5	6	10	14
	600	0	0	0	0	1	1	1	3	4	5	8	11
	700	0	0	0	0	1	1	1	2	3	4	7	10
	750	0	0	0	0	1	1	1	2	3	4	6	10
	800	0	0	0	0	1	1	1	2	3	4	6	9
	900	0	0	0	0	0	1	1	1	3	3	6	8
	1000	0	0	0	0	0	1	1	1	2	3	5	7
	1250	0	0	0	0	0	1	1	1	1	2	4	6
	1500	0	0	0	0	0	1	1	1	1	1	3	5
	1750	0	0	0	0	0	0	1	1	1	1	3	4
	2000	0	0	0	0	0	0	1	1	1	1	3	4

Table C10 Continued

CONDUCTORS

Type	Conductor Size (AWG/kcmil)	Metric Designator (Trade Size)											
		16 (½)	21 (¾)	27 (1)	35 (1¼)	41 (1½)	53 (2)	63 (2½)	78 (3)	91 (3½)	103 (4)	129 (5)	155 (6)
THHN, THWN, THWN-2	14	11	21	34	60	82	135	193	299	401	517	815	1178
	12	8	15	25	43	59	99	141	218	293	377	594	859
	10	5	9	15	27	37	62	89	137	184	238	374	541
	8	3	5	9	16	21	36	51	79	106	137	216	312
	6	1	4	6	11	15	26	37	57	77	99	156	225
	4	1	2	4	7	9	16	22	35	47	61	96	138
	3	1	1	3	6	8	13	19	30	40	51	81	117
	2	1	1	3	5	7	11	16	25	33	43	68	98
	1	1	1	1	3	5	8	12	18	25	32	50	73
	1/0	1	1	1	3	4	7	10	15	21	27	42	61
	2/0	0	1	1	2	3	6	8	13	17	22	35	51
	3/0	0	1	1	1	3	5	7	11	14	18	29	42
	4/0	0	1	1	1	2	4	6	9	12	15	24	35
	250	0	0	1	1	1	3	4	7	10	12	20	28
	300	0	0	1	1	1	3	4	6	8	11	17	24
	350	0	0	1	1	1	2	3	5	7	9	15	21
	400	0	0	0	1	1	1	3	5	6	8	13	19
	500	0	0	0	1	1	1	2	4	5	7	11	16
	600	0	0	0	1	1	1	1	3	4	5	9	13
	700	0	0	0	0	1	1	1	3	4	5	8	11
	750	0	0	0	0	1	1	1	2	3	4	7	11
	800	0	0	0	0	1	1	1	2	3	4	7	10
	900	0	0	0	0	1	1	1	2	3	4	6	9
	1000	0	0	0	0	1	1	1	1	3	3	6	8
FEP, FEPB, PFA, PFAH, TFE	14	11	20	33	58	79	131	188	290	389	502	790	1142
	12	8	15	24	42	58	96	137	212	284	366	577	834
	10	6	10	17	30	41	69	98	152	204	263	414	598
	8	3	6	10	17	24	39	56	87	117	150	237	343
	6	2	4	7	12	17	28	40	62	83	107	169	244
	4	1	3	5	8	12	19	28	43	58	75	118	170
	3	1	2	4	7	10	16	23	36	48	62	98	142
	2	1	1	3	6	8	13	19	30	40	51	81	117
PFA, PFAH, TFE	1	1	1	2	4	5	9	13	20	28	36	56	81
PFA, PFAH, TFE, Z	1/0	1	1	1	3	4	8	11	17	23	30	47	68
	2/0	0	1	1	3	4	6	9	14	19	24	39	56
	3/0	0	1	1	2	3	5	7	12	16	20	32	46
	4/0	0	1	1	1	2	4	6	9	13	16	26	38
Z	14	13	24	40	70	95	158	226	350	469	605	952	1376
	12	9	17	28	49	68	112	160	248	333	429	675	976
	10	6	10	17	30	41	69	98	152	204	263	414	598
	8	3	6	11	19	26	43	62	96	129	166	261	378
	6	2	4	7	13	18	30	43	67	90	116	184	265
	4	1	3	5	9	12	21	30	46	62	80	126	183
	3	1	2	4	6	9	15	22	34	45	58	92	133
	2	1	1	3	5	7	12	18	28	38	49	77	111
	1	1	1	2	4	6	10	14	23	30	39	62	90
XHH, XHHW, XHHW-2, ZW	14	8	14	24	42	57	94	135	209	280	361	568	822
	12	6	11	18	32	44	72	103	160	215	277	436	631
	10	4	8	13	24	32	54	77	119	160	206	325	470
	8	2	4	7	13	18	30	43	66	89	115	181	261
	6	1	3	5	10	13	22	32	49	66	85	134	193
	4	1	2	4	7	9	16	23	35	48	61	97	140
	3	1	1	3	6	8	13	19	30	40	52	82	118
	2	1	1	3	5	7	11	16	25	34	44	69	99
XHH, XHHW, XHHW-2	1	1	1	1	3	5	8	12	19	25	32	51	74
	1/0	1	1	1	3	4	7	10	16	21	27	43	62
	2/0	0	1	1	2	3	6	8	13	17	23	36	52
	3/0	0	1	1	1	3	5	7	11	14	19	30	43
	4/0	0	1	1	1	2	4	6	9	12	15	24	35
	250	0	0	1	1	1	3	5	7	10	13	20	29
	300	0	0	1	1	1	3	4	6	8	11	17	25
	350	0	0	1	1	1	2	3	5	7	9	15	22
	400	0	0	0	1	1	1	3	5	6	8	13	19
	500	0	0	0	1	1	1	2	4	5	7	11	16
	600	0	0	0	1	1	1	1	3	4	5	9	13
	700	0	0	0	0	1	1	1	3	4	5	8	11
	750	0	0	0	0	1	1	1	2	3	4	7	11
	800	0	0	0	0	1	1	1	2	3	4	7	10
	900	0	0	0	0	1	1	1	2	3	4	6	9
	1000	0	0	0	0	0	1	1	1	3	3	6	8
	1250	0	0	0	0	0	1	1	1	1	3	4	6
	1500	0	0	0	0	0	1	1	1	1	2	4	5
	1750	0	0	0	0	0	0	1	1	1	1	3	5
	2000	0	0	0	0	0	0	1	1	1	1	3	4

Table C10 Continued

FIXTURE WIRES

Type	Conductor Size (AWG/kcmil)	Metric Designator (Trade Size)					
		16 (½)	21 (¾)	27 (1)	35 (1¼)	41 (1½)	53 (2)
FFH-2, RFH-2, RFHH-3	18	8	14	23	40	54	90
	16	6	12	19	33	46	76
SF-2, SFF-2	18	10	17	29	50	69	114
	16	8	14	24	42	57	94
	14	6	12	19	33	46	76
SF-1, SFF-1	18	17	31	51	89	122	202
RFH-1, RFHH-2, TF, TFF, XF, XFF	18	13	23	38	66	90	149
RFHH-2, TF, TFF, XF, XFF	16	10	18	30	53	73	120
XF, XFF	14	8	14	24	42	57	94
TFN, TFFN	18	20	37	60	105	144	239
	16	16	28	46	80	110	183
PF, PFF, PGF, PGFF, PAF, PTF, PTFF, PAFF	18	19	35	57	100	137	227
	16	15	27	44	77	106	175
	14	11	20	33	58	79	131
HF, HFF, ZF, ZFF, ZHF	18	25	45	74	129	176	292
	16	18	33	54	95	130	216
	14	13	24	40	70	95	158
KF-2, KFF-2	18	36	65	107	187	256	424
	16	26	46	75	132	180	299
	14	17	31	52	90	124	205
	12	12	22	35	62	85	141
	10	8	14	24	42	57	94
KF-1, KFF-1	18	43	78	128	223	305	506
	16	30	55	90	157	214	355
	14	20	37	60	105	144	239
	12	13	24	40	70	95	158
	10	9	16	26	45	62	103
XF, XFF	12	4	8	12	22	30	50
	10	3	6	10	17	24	39

Note: This table is for concentric stranded conductors only. For compact stranded conductors, Table C10(A) should be used.

*Types RHH, RHW, and RHW-2 without outer covering.

Table C10(A) Maximum Number of Compact Conductors in Rigid PVC Conduit, Schedule 40 and HDPE Conduit (*Based on Table 1, Chapter 9*)

COMPACT CONDUCTORS

Type	Conductor Size (AWG/ kcmil)	Metric Designator (Trade Size)											
		16 (½)	21 (¾)	27 (1)	35 (1¼)	41 (1½)	53 (2)	63 (2½)	78 (3)	91 (3½)	103 (4)	129 (5)	155 (6)
THW, THW-2, THHW	8	1	4	6	11	15	26	37	57	76	98	155	224
	6	1	3	5	9	12	20	28	44	59	76	119	173
	4	1	1	3	6	9	15	21	33	44	57	89	129
	2	1	1	2	5	6	11	15	24	32	42	66	95
	1	1	1	1	3	4	7	11	17	23	29	46	67
	1/0	0	1	1	3	4	6	9	15	20	25	40	58
	2/0	0	1	1	2	3	5	8	12	16	21	34	49
	3/0	0	1	1	1	3	5	7	10	14	18	29	42
	4/0	0	1	1	1	2	4	5	9	12	15	24	35
	250	0	0	1	1	1	3	4	7	9	12	19	27
	300	0	0	1	1	1	2	4	6	8	10	16	24
	350	0	0	1	1	1	2	3	5	7	9	15	21
	400	0	0	0	1	1	1	3	5	6	8	13	19
	500	0	0	0	1	1	1	2	4	5	7	11	16
	600	0	0	0	1	1	1	1	3	4	5	9	13
	700	0	0	0	0	1	1	1	3	4	5	8	12
	750	0	0	0	0	1	1	1	2	3	5	7	11
	1000	0	0	0	0	1	1	1	1	3	4	6	9
THHN, THWN, THWN-2	8	—	—	—	—	—	—	—	—	—	—	—	—
	6	2	4	7	13	17	29	41	64	86	111	175	253
	4	1	2	4	8	11	18	25	40	53	68	108	156
	2	1	1	3	5	8	13	18	28	38	49	77	112
	1	1	1	2	4	6	9	14	21	29	37	58	84
	1/0	1	1	1	3	5	8	12	18	24	31	49	72
	2/0	0	1	1	3	4	7	9	15	20	26	41	59
	3/0	0	1	1	2	3	5	8	12	17	22	34	50
	4/0	0	1	1	1	3	4	6	10	14	18	28	41
	250	0	0	1	1	1	3	5	8	11	14	22	32
	300	0	0	1	1	1	3	4	7	9	12	19	28
	350	0	0	1	1	1	3	4	6	8	10	17	24
	400	0	0	1	1	1	2	3	5	7	9	15	22
	500	0	0	0	1	1	1	3	4	6	8	13	18
	600	0	0	0	1	1	1	2	4	5	6	10	15
	700	0	0	0	1	1	1	1	3	4	5	9	13
	750	0	0	0	1	1	1	1	3	4	5	8	12
	1000	0	0	0	0	1	1	1	2	3	4	6	9
XHHW, XHHW-2	8	3	5	8	14	20	33	47	73	99	127	200	290
	6	1	4	6	11	15	25	35	55	73	94	149	215
	4	1	2	4	8	11	18	25	40	53	68	108	156
	2	1	1	3	5	8	13	18	28	38	49	77	112
	1	1	1	2	4	6	9	14	21	29	37	58	84
	1/0	1	1	1	3	5	8	12	18	24	31	49	72
	2/0	1	1	1	3	4	7	10	15	20	26	42	60
	3/0	0	1	1	2	3	5	8	12	17	22	34	50
	4/0	0	1	1	1	3	5	7	10	14	18	29	42
	250	0	0	1	1	1	4	5	8	11	14	23	33
	300	0	0	1	1	1	3	4	7	9	12	19	28
	350	0	0	1	1	1	3	4	6	8	11	17	25
	400	0	0	1	1	1	2	3	5	7	10	15	22
	500	0	0	0	1	1	1	3	4	6	8	13	18
	600	0	0	0	1	1	1	2	4	5	6	10	15
	700	0	0	0	1	1	1	1	3	4	5	9	13
	750	0	0	0	1	1	1	1	3	4	5	8	12
	1000	0	0	0	0	1	1	1	2	3	4	6	9

Definition: Compact stranding is the result of a manufacturing process where the standard conductor is compressed to the extent that the interstices (voids between strand wires) are virtually eliminated.

Table C11 Maximum Number of Conductors or Fixture Wires in Type A, Rigid PVC Conduit (*Based on Table 1, Chapter 9*)

CONDUCTORS

Type	Conductor Size (AWG/ kcmil)	Metric Designator (Trade Size)									
		16 (½)	21 (¾)	27 (1)	35 (1¼)	41 (1½)	53 (2)	63 (2½)	78 (3)	91 (3½)	103 (4)
RHH, RHW, RHW-2	14	5	9	15	24	31	49	74	112	146	187
	12	4	7	12	20	26	41	61	93	121	155
	10	3	6	10	16	21	33	50	75	98	125
	8	1	3	5	8	11	17	26	39	51	65
	6	1	2	4	6	9	14	21	31	41	52
	4	1	1	3	5	7	11	16	24	32	41
	3	1	1	3	4	6	9	14	21	28	36
	2	1	1	2	4	5	8	12	18	24	31
	1	0	1	1	2	3	5	8	12	16	20
	1/0	0	1	1	2	3	5	7	10	14	18
	2/0	0	1	1	1	2	4	6	9	12	15
	3/0	0	1	1	1	1	3	5	8	10	13
	4/0	0	0	1	1	1	3	4	7	9	11
	250	0	0	1	1	1	1	3	5	7	8
	300	0	0	1	1	1	1	3	4	6	7
	350	0	0	0	1	1	1	2	4	5	7
	400	0	0	0	1	1	1	2	4	5	6
	500	0	0	0	1	1	1	1	3	4	5
	600	0	0	0	0	1	1	1	2	3	4
	700	0	0	0	0	1	1	1	2	3	4
	750	0	0	0	0	1	1	1	1	3	4
	800	0	0	0	0	1	1	1	1	3	3
	900	0	0	0	0	0	1	1	1	2	3
	1000	0	0	0	0	0	1	1	1	2	3
	1250	0	0	0	0	0	1	1	1	1	2
	1500	0	0	0	0	0	0	1	1	1	1
	1750	0	0	0	0	0	0	1	1	1	1
	2000	0	0	0	0	0	0	1	1	1	1
TW, THHW, THW, THW-2	14	11	18	31	51	67	105	157	235	307	395
	12	8	14	24	39	51	80	120	181	236	303
	10	6	10	18	29	38	60	89	135	176	226
	8	3	6	10	16	21	33	50	75	98	125
RHH*, RHW*, RHW-2*	14	7	12	20	34	44	70	104	157	204	262
	12	6	10	16	27	35	56	84	126	164	211
	10	4	8	13	21	28	44	65	98	128	165
	8	2	4	8	12	16	26	39	59	77	98
RHH*, RHW*	6	1	3	6	9	13	20	30	45	59	75
TW, THW, THHW, THW-2	4	1	2	4	7	9	15	22	33	44	56
	3	1	1	4	6	8	13	19	29	37	48
	2	1	1	3	5	7	11	16	24	32	41
	1	1	1	1	3	5	7	11	17	22	29
	1/0	1	1	1	3	4	6	10	14	19	24
	2/0	0	1	1	2	3	5	8	12	16	21
	3/0	0	1	1	1	3	4	7	10	13	17
	4/0	0	1	1	1	2	4	6	9	11	14
	250	0	0	1	1	1	3	4	7	9	12
	300	0	0	1	1	1	2	4	6	8	10
	350	0	0	1	1	1	2	3	5	7	9
	400	0	0	1	1	1	1	3	5	6	8
	500	0	0	0	1	1	1	2	4	5	7
	600	0	0	0	1	1	1	1	3	4	5
	700	0	0	0	1	1	1	1	3	4	5
	750	0	0	0	1	1	1	1	3	3	4
	800	0	0	0	0	1	1	1	2	3	4
	900	0	0	0	0	1	1	1	2	3	4
	1000	0	0	0	0	1	1	1	1	3	3
	1250	0	0	0	0	0	1	1	1	1	3
	1500	0	0	0	0	0	1	1	1	1	2
	1750	0	0	0	0	0	0	1	1	1	1
	2000	0	0	0	0	0	0	1	1	1	1

Table C11 Continued

CONDUCTORS

Type	Conductor Size (AWG/kcmil)	Metric Designator (Trade Size)									
		16 (½)	21 (¾)	27 (1)	35 (1¼)	41 (1½)	53 (2)	63 (2½)	78 (3)	91 (3½)	103 (4)
THHN, THWN, THWN-2	14	16	27	44	73	96	150	225	338	441	566
	12	11	19	32	53	70	109	164	246	321	412
	10	7	12	20	33	44	69	103	155	202	260
	8	4	7	12	19	25	40	59	89	117	150
	6	3	5	8	14	18	28	43	64	84	108
	4	1	3	5	8	11	17	26	39	52	66
	3	1	2	4	7	9	15	22	33	44	56
	2	1	1	3	6	8	12	19	28	37	47
	1	1	1	2	4	6	9	14	21	27	35
	1/0	1	1	2	4	5	8	11	17	23	29
	2/0	1	1	1	3	4	6	10	14	19	24
	3/0	0	1	1	2	3	5	8	12	16	20
	4/0	0	1	1	1	3	4	6	10	13	17
	250	0	1	1	1	2	3	5	8	10	14
	300	0	0	1	1	1	3	4	7	9	12
	350	0	0	1	1	1	2	4	6	8	10
	400	0	0	1	1	1	2	3	5	7	9
	500	0	0	1	1	1	1	3	4	6	7
	600	0	0	0	1	1	1	2	3	5	6
	700	0	0	0	1	1	1	1	3	4	5
	750	0	0	0	1	1	1	1	3	4	5
	800	0	0	0	1	1	1	1	3	4	5
	900	0	0	0	0	1	1	1	2	3	4
	1000	0	0	0	0	1	1	1	2	3	4
FEP, FEPB, PFA, PFAH, TFE	14	15	26	43	70	93	146	218	327	427	549
	12	11	19	31	51	68	106	159	239	312	400
	10	8	13	22	37	48	76	114	171	224	287
	8	4	8	13	21	28	44	65	98	128	165
	6	3	5	9	15	20	31	46	70	91	117
	4	1	4	6	10	14	21	32	49	64	82
	3	1	3	5	8	11	18	27	40	53	68
	2	1	2	4	7	9	15	22	33	44	56
PFA, PFAH, TFE	1	1	1	3	5	6	10	15	23	30	39
PFA, PFAH, TFE, Z	1/0	1	1	2	4	5	8	13	19	25	32
	2/0	1	1	1	3	4	7	10	16	21	27
	3/0	1	1	1	3	3	6	9	13	17	22
	4/0	0	1	1	2	3	5	7	11	14	18
Z	14	18	31	52	85	112	175	263	395	515	661
	12	13	22	37	60	79	124	186	280	365	469
	10	8	13	22	37	48	76	114	171	224	287
	8	5	8	14	23	30	48	72	108	141	181
	6	3	6	10	16	21	34	50	76	99	127
	4	2	4	7	11	15	23	35	52	68	88
	3	1	3	5	8	11	17	25	38	50	64
	2	1	2	4	7	9	14	21	32	41	53
	1	1	1	3	5	7	11	17	26	33	43
XHH, XHHW, XHHW-2, ZW	14	11	18	31	51	67	105	157	235	307	395
	12	8	14	24	39	51	80	120	181	236	303
	10	6	10	18	29	38	60	89	135	176	226
	8	3	6	10	16	21	33	50	75	98	125
	6	2	4	7	12	15	24	37	55	72	93
	4	1	3	5	8	11	18	26	40	52	67
	3	1	2	4	7	9	15	22	34	44	57
	2	1	1	3	6	8	12	19	28	37	48
XHH, XHHW, XHHW-2	1	1	1	3	4	6	9	14	21	28	35
	1/0	1	1	2	4	5	8	12	18	23	30
	2/0	1	1	1	3	4	6	10	15	19	25
	3/0	0	1	1	2	3	5	8	12	16	20
	4/0	0	1	1	1	3	4	7	10	13	17
	250	0	1	1	1	2	3	5	8	11	14
	300	0	0	1	1	1	3	5	7	9	12
	350	0	0	1	1	1	3	4	6	8	10
	400	0	0	1	1	1	2	3	5	7	9
	500	0	0	1	1	1	1	3	4	6	8
	600	0	0	0	1	1	1	2	3	5	6
	700	0	0	0	1	1	1	1	3	4	5
	750	0	0	0	1	1	1	1	3	4	5
	800	0	0	0	1	1	1	1	3	4	5
	900	0	0	0	0	1	1	1	2	3	4
	1000	0	0	0	0	1	1	1	2	3	4
	1250	0	0	0	0	0	1	1	1	2	3
	1500	0	0	0	0	0	1	1	1	1	2
	1750	0	0	0	0	0	1	1	1	1	2
	2000	0	0	0	0	0	0	1	1	1	1

Table C11 Continued

FIXTURE WIRES

Type	Conductor Size (AWG/kcmil)	Metric Designator (Trade Size)					
		16 (½)	21 (¾)	27 (1)	35 (1¼)	41 (1½)	53 (2)
FFH-2, RFH-2, RFHH-3	18	10	18	30	48	64	100
	16	9	15	25	41	54	85
SF-2, SFF-2	18	13	22	37	61	81	127
	16	11	18	31	51	67	105
	14	9	15	25	41	54	85
SF-1, SFF-1	18	23	40	66	108	143	224
RFH-1, RFHH-2, TF, TFF, XF, XFF	18	17	29	49	80	105	165
RFHH-2, TF, TFF, XF, XFF	16	14	24	39	65	85	134
XF, XFF	14	11	18	31	51	67	105
TFN, TFFN	18	28	47	79	128	169	265
	16	21	36	60	98	129	202
PF, PFF, PGF, PGFF, PAF, PTF, PTFF, PAFF	18	26	45	74	122	160	251
	16	20	34	58	94	124	194
	14	15	26	43	70	93	146
HF, HFF, ZF, ZFF, ZHF	18	34	58	96	157	206	324
	16	25	42	71	116	152	239
	14	18	31	52	85	112	175
KF-2, KFF-2	18	49	84	140	228	300	470
	16	35	59	98	160	211	331
	14	24	40	67	110	145	228
	12	16	28	46	76	100	157
	10	11	18	31	51	67	105
KF-1, KFF-1	18	59	100	167	272	357	561
	16	41	70	117	191	251	394
	14	28	47	79	128	169	265
	12	18	31	52	85	112	175
	10	12	20	34	55	73	115
XF, XFF	12	6	10	16	27	35	56
	10	4	8	13	21	28	44

Note: This table is for concentric stranded conductors only. For compact stranded conductors, Table C11(A) should be used.
*Types RHH, RHW, and RWH-2 without outer covering.

Table C11(A) Maximum Number of Compact Conductors in Type A, Rigid PVC Conduit (*Based on Table 1, Chapter 9*)

COMPACT CONDUCTORS

Type	Conductor Size (AWG/kcmil)	Metric Designator (Trade Size)									
		16 (½)	21 (¾)	27 (1)	35 (1¼)	41 (1½)	53 (2)	63 (2½)	78 (3)	91 (3½)	103 (4)
THW, THW-2, THHW	8	3	5	8	14	18	28	42	64	84	107
	6	2	4	6	10	14	22	33	49	65	83
	4	1	3	5	8	10	16	24	37	48	62
	2	1	1	3	6	7	12	18	27	36	46
	1	1	1	2	4	5	8	13	19	25	32
	1/0	1	1	1	3	4	7	11	16	21	28
	2/0	1	1	1	3	4	6	9	14	18	23
	3/0	0	1	1	2	3	5	8	12	15	20
	4/0	0	1	1	1	3	4	6	10	13	17
	250	0	1	1	1	1	3	5	8	10	13
	300	0	0	1	1	1	3	4	7	9	11
	350	0	0	1	1	1	2	4	6	8	10
	400	0	0	1	1	1	2	3	5	7	9
	500	0	0	1	1	1	1	3	4	6	8
	600	0	0	0	1	1	1	2	3	5	6
	700	0	0	0	1	1	1	1	3	4	5
	750	0	0	0	1	1	1	1	3	4	5
	1000	0	0	0	0	1	1	1	2	3	4
THHN, THWN, THWN-2	8	—	—	—	—	—	—	—	—	—	—
	6	3	5	9	15	20	32	48	72	94	121
	4	1	3	6	9	12	20	30	45	58	75
	2	1	2	4	7	9	14	21	32	42	54
	1	1	1	3	5	7	10	16	24	31	40
	1/0	1	1	2	4	6	9	13	20	27	34
	2/0	1	1	1	3	5	7	11	17	22	28
	3/0	1	1	1	3	4	6	9	14	18	24
	4/0	0	1	1	2	3	5	8	11	15	19
	250	0	1	1	1	2	4	6	9	12	15
	300	0	1	1	1	1	3	5	8	10	13
	350	0	0	1	1	1	3	4	7	9	11
	400	0	0	1	1	1	2	4	6	8	10
	500	0	0	1	1	1	2	3	5	7	9
	600	0	0	0	1	1	1	3	4	5	7
	700	0	0	0	1	1	1	2	3	5	6
	750	0	0	0	1	1	1	2	3	4	6
	1000	0	0	0	0	1	1	1	2	3	4
XHHW, XHHW-2	8	4	6	11	18	23	37	55	83	108	139
	6	3	5	8	13	17	27	41	62	80	103
	4	1	3	6	9	12	20	30	45	58	75
	2	1	2	4	7	9	14	21	32	42	54
	1	1	1	3	5	7	10	16	24	31	40
	1/0	1	1	2	4	6	9	13	20	27	34
	2/0	1	1	1	3	5	7	11	17	22	29
	3/0	1	1	1	3	4	6	9	14	18	24
	4/0	0	1	1	2	3	5	8	12	15	20
	250	0	1	1	1	2	4	6	9	12	16
	300	0	1	1	1	1	3	5	8	10	13
	350	0	0	1	1	1	3	5	7	9	12
	400	0	0	1	1	1	3	4	6	8	11
	500	0	0	1	1	1	2	3	5	7	9
	600	0	0	0	1	1	1	3	4	5	7
	700	0	0	0	1	1	1	2	3	5	6
	750	0	0	0	1	1	1	2	3	4	6
	1000	0	0	0	0	1	1	1	2	3	4

Definition: *Compact stranding* is the result of a manufacturing process where the standard conductor is compressed to the extent that the interstices (voids between strand wires) are virtually eliminated.

Table C12 Maximum Number of Conductors in Type EB, PVC Conduit (*Based on Table 1, Chapter 9*)

CONDUCTORS

Type	Conductor Size (AWG/kcmil)	Metric Designator (Trade Size)					
		53 (2)	78 (3)	91 (3½)	103 (4)	129 (5)	155 (6)
RHH, RHW, RHW-2	14	53	119	155	197	303	430
	12	44	98	128	163	251	357
RHH, RHW, RHW-2	10	35	79	104	132	203	288
	8	18	41	54	69	106	151
	6	15	33	43	55	85	121
	4	11	26	34	43	66	94
	3	10	23	30	38	58	83
	2	9	20	26	33	50	72
	1	6	13	17	21	33	47
	1/0	5	11	15	19	29	41
	2/0	4	10	13	16	25	36
	3/0	4	8	11	14	22	31
	4/0	3	7	9	12	18	26
	250	2	5	7	9	14	20
	300	1	5	6	8	12	17
	350	1	4	5	7	11	16
	400	1	4	5	6	10	14
	500	1	3	4	5	9	12
	600	1	3	3	4	7	10
	700	1	2	3	4	6	9
	750	1	2	3	4	6	9
	800	1	2	3	4	6	8
	900	1	1	2	3	5	7
	1000	1	1	2	3	5	7
	1250	1	1	1	2	3	5
	1500	0	1	1	1	3	4
	1750	0	1	1	1	3	4
	2000	0	1	1	1	2	3
TW, THHW, THW, THW-2	14	111	250	327	415	638	907
	12	85	192	251	319	490	696
	10	63	143	187	238	365	519
	8	35	79	104	132	203	288
RHH*, RHW*, RHW-2*	14	74	166	217	276	424	603
	12	59	134	175	222	341	485
	10	46	104	136	173	266	378
RHH*, RHW*, RHW-2*	8	28	62	81	104	159	227
RHH*, RHW*, RHW-2*, TW, THHW, THW-2	6	21	48	62	79	122	173
	4	16	36	46	59	91	129
	3	13	30	40	51	78	111
	2	11	26	34	43	66	94
	1	8	18	24	30	46	66
	1/0	7	15	20	26	40	56
	2/0	6	13	17	22	34	48
	3/0	5	11	14	18	28	40
	4/0	4	9	12	15	24	34
	250	3	7	10	12	19	27
	300	3	6	8	11	17	24
	350	2	6	7	9	15	21
	400	2	5	7	8	13	19
	500	1	4	5	7	11	16
	600	1	3	4	6	9	13
	700	1	3	4	5	8	11
	750	1	3	4	5	7	11
	800	1	3	3	4	7	10
	900	1	2	3	4	6	9
	1000	1	2	3	4	6	8
	1250	1	1	2	3	4	6
	1500	1	1	1	2	4	6
	1750	1	1	1	2	3	5
	2000	0	1	1	1	3	4

Table C12 Continued

CONDUCTORS

Type	Conductor Size (AWG/kcmil)	Metric Designator (Trade Size)					
		53 (2)	78 (3)	91 (3½)	103 (4)	129 (5)	155 (6)
THHN, THWN, THWN-2	14	159	359	468	595	915	1300
	12	116	262	342	434	667	948
	10	73	165	215	274	420	597
	8	42	95	124	158	242	344
	6	30	68	89	114	175	248
	4	19	42	55	70	107	153
	3	16	36	46	59	91	129
	2	13	30	39	50	76	109
	1	10	22	29	37	57	80
	1/0	8	18	24	31	48	68
	2/0	7	15	20	26	40	56
	3/0	5	13	17	21	33	47
	4/0	4	10	14	18	27	39
	250	4	8	11	14	22	31
	300	3	7	10	12	19	27
	350	3	6	8	11	17	24
	400	2	6	7	10	15	21
	500	1	5	6	8	12	18
	600	1	4	5	6	10	14
	700	1	3	4	6	9	12
	750	1	3	4	5	8	12
	800	1	3	4	5	8	11
	900	1	3	3	4	7	10
	1000	1	2	3	4	6	9
FEP, FEPB, PFA, PFAH, TFE	14	155	348	454	578	888	1261
	12	113	254	332	422	648	920
	10	81	182	238	302	465	660
	8	46	104	136	173	266	378
	6	33	74	97	123	189	269
	4	23	52	68	86	132	188
	3	19	43	56	72	110	157
	2	16	36	46	59	91	129
PFA, PFAH, TFE	1	11	25	32	41	63	90
PFA, PFAH, TFE, Z	1/0	9	20	27	34	53	75
	2/0	7	17	22	28	43	62
	3/0	6	14	18	23	36	51
	4/0	5	11	15	19	29	42
Z	14	186	419	547	696	1069	1519
	12	132	297	388	494	759	1078
	10	81	182	238	302	465	660
	8	51	115	150	191	294	417
	6	36	81	105	134	206	293
	4	24	55	72	92	142	201
	3	18	40	53	67	104	147
	2	15	34	44	56	86	122
	1	12	27	36	45	70	99
XHH, XHHW, XHHW-2, ZW	14	111	250	327	415	638	907
	12	85	192	251	319	490	696
	10	63	143	187	238	365	519
	8	35	79	104	132	203	288
	6	26	59	77	98	150	213
	4	19	42	56	71	109	155
	3	16	36	47	60	92	131
	2	13	30	39	50	77	110
XHH, XHHW, XHHW-2	1	10	22	29	37	58	82
	1/0	8	19	25	31	48	69
	2/0	7	16	20	26	40	57
	3/0	6	13	17	22	33	47
	4/0	5	11	14	18	27	39
	250	4	9	11	15	22	32
	300	3	7	10	12	19	28
	350	3	6	9	11	17	24
	400	2	6	8	10	15	22
	500	1	5	6	8	12	18
	600	1	4	5	6	10	14
	700	1	3	4	6	9	12
	750	1	3	4	5	8	12
	800	1	3	4	5	8	11
	900	1	3	3	4	7	10
	1000	1	2	3	4	6	9
	1250	1	1	2	3	5	7
	1500	1	1	1	3	4	6
	1750	1	1	1	2	4	5
	2000	0	1	1	1	3	5

Note: This table is for concentric stranded conductors only. For compact stranded conductors, Table C12(A) should be used.
*Types RHH, RHW, and RHW-2 without outer covering.

Table C12(A) Maximum Number of Compact Conductors in Type EB, PVC Conduit (*Based on Table 1, Chapter 9*)

COMPACT CONDUCTORS

Type	Conductor Size (AWG/kcmil)	Metric Designator (Trade Size)					
		53 (2)	78 (3)	91 (3½)	103 (4)	129 (5)	155 (6)
THW, THW-2, THHW	8	30	68	89	113	174	247
	6	23	52	69	87	134	191
	4	17	39	51	65	100	143
	2	13	29	38	48	74	105
	1	9	20	26	34	52	74
	1/0	8	17	23	29	45	64
	2/0	6	15	19	24	38	54
	3/0	5	12	16	21	32	46
	4/0	4	10	14	17	27	38
	250	3	8	11	14	21	30
	300	3	7	9	12	19	26
	350	3	6	8	11	17	24
	400	2	6	7	10	15	21
	500	1	5	6	8	12	18
	600	1	4	5	6	10	14
	700	1	3	4	6	9	13
	750	1	3	4	5	8	12
	1000	1	2	3	4	7	9
THHN, THWN, THWN-2	8	—	—	—	—	—	—
	6	34	77	100	128	196	279
	4	21	47	62	79	121	172
	2	15	34	44	57	87	124
	1	11	25	33	42	65	93
	1/0	9	22	28	36	56	79
	2/0	8	18	23	30	46	65
	3/0	6	15	20	25	38	55
	4/0	5	12	16	20	32	45
	250	4	10	13	16	25	35
	300	4	8	11	14	22	31
	350	3	7	9	12	19	27
	400	3	6	8	11	17	24
	500	2	5	7	9	14	20
	600	1	4	6	7	11	16
	700	1	4	5	6	10	14
	750	1	4	5	6	9	14
	1000	1	3	3	4	7	10
XHHW, XHHW-2	8	39	88	115	146	225	320
	6	29	65	85	109	167	238
	4	21	47	62	79	121	172
	2	15	34	44	57	87	124
	1	11	25	33	42	65	93
	1/0	9	22	28	36	56	79
	2/0	8	18	24	30	47	67
	3/0	6	15	20	25	38	55
	4/0	5	12	16	21	32	46
	250	4	10	13	17	26	37
	300	4	8	11	14	22	31
	350	3	7	10	12	19	28
	400	3	7	9	11	17	25
	500	2	5	7	9	14	20
	600	1	4	6	7	11	16
	700	1	4	5	6	10	14
	750	1	3	5	6	9	13
	1000	1	3	4	5	7	10

Definition: *Compact stranding* is the result of a manufacturing process where the standard conductor is compressed to the extent that the interstices (voids between strand wires) are virtually eliminated.

This annex is not a part of the requirements of this NFPA document but is included for informational purposes only.

For the 1999 *Code*, these examples were moved from Chapter 9 following the tables to this new Annex D. In addition, Examples D11, Mobile Home, and D12, Park Trailer, were moved from Articles 550 and 552 to Annex D and are now more appropriately located in this explanatory annex.

Contents

Selection of Conductors. In the following examples, the results are generally expressed in amperes (A). To select conductor sizes, refer to the 0 through 2000 volt (V) ampacity tables of Article 310 and the rules of 310.15 that pertain to these tables.

Voltage. For uniform application of Articles 210, 215, and 220, a nominal voltage of 120, 120/240, 240, and 208Y/120 V is used in computing the ampere load on the conductor.

Fractions of an Ampere. Except where the computations result in a major fraction of an ampere (0.5 or larger), such fractions are permitted to be dropped.

Power Factor. Calculations in the following examples are based, for convenience, on the assumption that all loads have the same power factor (PF).

Ranges. For the computation of the range loads in these examples, Column C of Table 220.19 has been used. For optional methods, see Columns A and B of Table 220.19. Except where the computations result in a major fraction of a kilowatt (0.5 or larger), such fractions are permitted to be dropped.

SI Units. For metric conversions, $0.093 \ m^2 = 1 \ ft^2$ and $0.3048 \ m = 1 \ ft$.

In the following examples, loads are assumed to be properly balanced on the system. If loads are not properly balanced, additional feeder capacity may be required. The calculations are based on the standard method.

Example D1(a) One-Family Dwelling

The dwelling has a floor area of 1500 ft^2, exclusive of an unfinished cellar not adaptable for future use, unfinished attic, and open porches. Appliances are a 12-kW range and a 5.5-kW, 240-V dryer. Assume range and dryer kW ratings equivalent to kVA ratings in accordance with 220.18 and 220.19.
Computed Load *(see 220.10)*
General Lighting Load: 1500 ft^2 at 3 VA per ft^2 = 4500 VA

Minimum Number of Branch Circuits Required
[see 210.11(A)]

General Lighting Load: 4500 VA ÷ 120 V = 37.5 A

This requires three 15-A, 2-wire or two 20-A, 2-wire circuits.

Small Appliance Load: Two 2-wire, 20-A circuits [see 210.11(C)(1)]

Laundry Load: One 2-wire, 20-A circuit [see 210.11(C)(2)]

Bathroom Branch Circuit: One 2-wire, 20-A circuit (no additional load calculation is required for this circuit) [see 210.11(C)(3)]

Minimum Size Feeder Required (see 220.10)

General Lighting		4500 VA
Small Appliance		3000 VA
Laundry		1500 VA
	Total	9000 VA
3000 VA at 100%		3000 VA
9000 VA – 3000 VA = 6000 VA at 35%		2100 VA
	Net Load	5100 VA
Range (see Table 220.19)		8000 VA
Dryer Load (see Table 220.18)		5500 VA
	Net Computed Load	18,600 VA

Net Computed Load for 120/240-V, 3-wire, single-phase service or feeder

18,600 VA ÷ 240 V = 77.5 A

Sections 230.42(B) and 230.79 require service conductors and disconnecting means rated not less than 100 amperes.

Calculation for Neutral for Feeder and Service

Lighting and Small Appliance Load		5100 VA
Range: 8000 VA at 70% (see 220.22)		5600 VA
Dryer: 5500 VA at 70% (see 220.22)		3850 VA
	Total	14,550 VA

Computed Load for Neutral

14,550 VA ÷ 240 V = 60.6 A

The general lighting and general-use receptacle load is computed from the outside dimensions of the building, apartment, or other area involved. For a dwelling unit, the computed floor area is not to include open porches, garages, or, as stated in the opening paragraph, "an unfinished cellar not adaptable for future use." See 220.3(A) and 220.3(B)(10) for the requirements on how to calculate the lighting and general-use receptacle load for this occupancy.

A two-story dwelling measures 30 ft × 30 ft for the first floor and 30 ft × 20 ft for the second floor.

First-floor area: 30 ft × 30 ft =   900 ft^2

Second-floor area: 30 ft × 20 ft =   600 ft^2

Total area = 1500 ft^2

Example D1(b) One-Family Dwelling

Assume same conditions as Example No. D1(a), plus addition of one 6-A, 230-V, room air-conditioning unit and one 12-A, 115-V, room air-conditioning unit,* one 8-A, 115-V, rated waste disposer, and one 10-A, 120-V, rated dishwasher. See Article 430 for general motors and Article 440, Part VII, for air-conditioning equipment. Motors have nameplate ratings of 115 V and 230 V for use on 120-V and 240-V nominal voltage systems.

*(For feeder neutral, use larger of the two appliances for unbalance.)

From Example D1(a), feeder current is 78 A (3-wire, 240 V).

	Line A	Neutral	Line B
Amperes from Example D1(a)	78	61	78
One 230-V air conditioner	6	—	6
One 115-V air conditioner and 120-V dishwasher	12	12	10
One 115-V disposer	—	8	8
25% of largest motor (see 430.24)	3	3	2
Total amperes per line	99	84	104

Therefore, the service would be rated 110 A.

Example D2(a) Optional Calculation for One-Family Dwelling, Heating Larger Than Air Conditioning (see 220.30)

The dwelling has a floor area of 1500 ft^2, exclusive of an unfinished cellar not adaptable for future use, unfinished attic, and open porches. It has a 12-kW range, a 2.5-kW water heater, a 1.2-kW dishwasher, 9 kW of electric space heating installed in five rooms, a 5-kW clothes dryer, and a 6-A, 230-V, room air-conditioning unit. Assume range, water heater, dishwasher, space heating, and clothes dryer kW ratings equivalent to kVA.

Air Conditioner kVA Calculation

6 A × 230 V ÷ 1000 = 1.38 kVA

This 1.38 kVA [item 1 from 220.30(C)] is less than 40% of 9 kVA of separately controlled electric heat [item 6 from 220.30(C)], so the 1.38 kVA need not be included in the service calculation.

General Load

1500 ft^2 at 3 VA	4500 VA
Two 20-A appliance outlet circuits at 1500 VA each	3000 VA
Laundry circuit	1500 VA
Range (at nameplate rating)	12,000 VA
Water heater	2500 VA
Dishwasher	1200 VA
Clothes dryer	5000 VA
Total	29,700 VA

Application of Demand Factor [see 220.30(B)]

First 10 kVA of general load at 100%	10,000 VA
Remainder of general load at 40% (19.7 kVA × 0.4)	7880 VA
Total of general load	17,880 VA
9 kVA of heat at 40% (9000 VA × 0.4) =	3600 VA
Total	21,480 VA

Calculated Load for Service Size

21.48 kVA = 21,480 VA

21,480 VA ÷ 240 V = 89.5 A

Therefore, the minimum service rating would be 100 A in accordance with 230.42 and 230.79.

Feeder Neutral Load, per 220.22

1500 ft^2 at 3 VA		4500 VA
Three 20-A circuits at 1500 VA		4500 VA
	Total	9000 VA
3000 VA at 100%		3000 VA
9000 VA − 3000 VA = 6000 VA at 35%		2100 VA
	Subtotal	5100 VA
Range: 8 kVA at 70%		5600 VA
Clothes dryer: 5 kVA at 70%		3500 VA
Dishwasher		1200 VA
	Total	15,400 VA

Calculated Load for Neutral

15,400 VA ÷ 240 V= 64.2 A

The air-conditioning load is calculated at 100 percent and is calculated separately to comply with the requirements of 220.30(C)(1).

Example D2(b) Optional Calculation for One-Family Dwelling, Air Conditioning Larger Than Heating
[see 220.30(A) and 220.30(C)]

The dwelling has a floor area of 1500 ft², exclusive of an unfinished cellar not adaptable for future use, unfinished attic, and open porches. It has two 20-A small appliance circuits, one 20-A laundry circuit, two 4-kW wall-mounted ovens, one 5.1-kW counter-mounted cooking unit, a 4.5-kW water heater, a 1.2-kW dishwasher, a 5-kW combination clothes washer and dryer, six 7-A, 230-V room air-conditioning units, and a 1.5-kW permanently installed bathroom space heater. Assume wall-mounted ovens, counter-mounted cooking unit, water heater, dishwasher, and combination clothes washer and dryer kW ratings equivalent to kVA.

Air Conditioning kVA Calculation

$$\text{Total amperes} = 6 \text{ units} \times 7 \text{ A} = 42 \text{ A}$$

$$42 \text{ A} \times 240 \text{ V} \div 1000 = 10.08 \text{ kVA (assume PF} = 1.0)$$

Load Included at 100%

Air Conditioning: Included below *[see item 1 in 220.30(C)]*
Space Heater: Omit *[see item 5 in 220.30(C)]*

General Load

1500 ft² at 3 VA	4500 VA
Two 20-A small appliance circuits at 1500 VA each	3000 VA
Laundry circuit	1500 VA
Two ovens	8000 VA
One cooking unit	5100 VA
Water heater	4500 VA
Dishwasher	1200 VA
Washer/dryer	5000 VA
Total general load	32,800 VA
First 10 kVA at 100%	10,000 VA
Remainder at 40% (22.8 kVA × 0.4 × 1000)	9120 VA
Subtotal general load	19,120 VA
Air conditioning	10,080 VA
Total	29,200 VA

Calculated Load for Service

$$29,200 \text{ VA} \div 240 \text{ V} = 122 \text{ A (service rating)}$$

Feeder Neutral Load, per 220.22

Assume that the two 4-kVA wall-mounted ovens are supplied by one branch circuit, the 5.1-kVA counter-mounted cooking unit by a separate circuit.

1500 ft² at 3 VA	4500 VA
Three 20-A circuits at 1500 VA	4500 VA
Subtotal	9000 VA
3000 VA at 100%	3000 VA
9000 VA − 3000 VA = 6000 VA at 35%	2100 VA
Subtotal	5100 VA

Two 4-kVA ovens plus one 5.1-kVA cooking unit = 13.1 kVA. Table 220.19 permits 55% demand factor or 13.1 kVA × 0.55 = 7.2 kVA feeder capacity.

Subtotal from above	5100 VA
Ovens and cooking unit: 7200 VA × 70% for neutral load	5040 VA
Clothes washer/dryer: 5 kVA × 70% for neutral load	3500 VA
Dishwasher	1200 VA
Total	14,840 VA

Calculated Load for Neutral

$$14,840 \text{ VA} \div 240 \text{ V} = 61.83 \text{ A (use 62 A)}$$

Example D2(c) Optional Calculation for One-Family Dwelling with Heat Pump (Single-Phase, 240/120-Volt Service)
(see 220.30)

The dwelling has a floor area of 2000 ft², exclusive of an unfinished cellar not adaptable for future use, unfinished attic, and open porches. It has a 12-kW range, a 4.5-kW water heater, a 1.2-kW dishwasher, a 5-kW clothes dryer, and a 2½-ton (24-A) heat pump with 15 kW of backup heat.

Heat Pump kVA Calculation

$$24 \text{ A} \times 240 \text{ V} \div 1000 = 5.76 \text{ kVA}$$

This 5.76 kVA is less than 15 kVA of the backup heat; therefore, the heat pump load need not be included in the service calculation *[see 220.30(C)]*.

General Load

2000 ft² at 3 VA	6000 VA
Two 20-A appliance outlet circuits at 1500 VA each	3000 VA
Laundry circuit	1500 VA
Range (at nameplate rating)	12,000 VA
Water heater	4500 VA
Dishwasher	1200 VA
Clothes dryer	5000 VA
Subtotal general load	33,200 VA
First 10 kVA at 100%	10,000 VA
Remainder of general load at 40% (23,200 VA × 0.4)	9280 VA
Total net general load	19,280 VA

Heat Pump and Supplementary Heat*

$$240 \text{ V} \times 24 \text{ A} = 5760 \text{ VA}$$

15-kW Electric Heat:

$$5760 \text{ VA} + 15,000 \text{ VA} = 20,760 \text{ VA or} = 20.76 \text{ kVA}$$

$$20.76 \text{ kVA} \times 100\% = 20.76 \text{ kVA}$$

*If supplementary heat is not on at same time as heat pump, heat pump kVA need not be added to total.

Totals

Net general load	19,280 VA
Heat pump and supplementary heat	20,760 VA
Total	40,040 VA

Calculated Load for Service

$$40.04 \text{ kVA} \times 1000 \div 240 \text{ V} = 166.8 \text{ A}$$

Therefore, this dwelling unit would be permitted to be served by a 175-A service.

Example D3 Store Building

A store 50 ft by 60 ft, or 3000 ft², has 30 ft of show window. There are a total of 80 duplex receptacles. The service is 120/240 V, single phase 3-wire service. Actual connected lighting load is 8500 VA.

Computed Load *(see 220.10)*

Noncontinuous Loads

Receptacle Load *(see 220.13)* 80 receptacles at 180 VA	14,400 VA
10,000 VA at 100%	10,000 VA
14,400 VA − 10,000 VA = 4400 at 50%	2200 VA
Subtotal	12,200 VA

Continuous Loads

General Lighting* 3000 ft² at 3 VA per ft²	9000 VA
Show Window Lighting Load 30 ft at 200 VA per ft	6000 VA
Outside Sign Circuit *[see 220.3(B)(6)]*	1200 VA
Subtotal	16,200 VA
Subtotal from noncontinuous	12,200 VA
Total noncontinuous loads + continuous loads =	28,400 VA

*In the example, 125% of the actual connected lighting load (8500 VA × 1.25 = 10,625 VA) is less than 125% of the load from Table 220.3(A), so the minimum lighting load from Table 220.3(A) is used in the calculation. Had the actual lighting load been greater than the value computed from Table 220.3(A), 125% of the actual connected lighting load would have been used.

Minimum Number of Branch Circuits Required

General Lighting: Branch circuits need only be installed to supply the actual connected load [see 210.11(B)].

$$8500 \text{ VA} \times 1.25 = 10,625 \text{ VA}$$

$$10,625 \text{ VA} \div 240 \text{ V} = 44 \text{ A for 3-wire, } 120/240 \text{ V}$$

The lighting load would be permitted to be served by 2-wire or 3-wire, 15- or 20-A circuits with combined capacity equal to 44 A or greater for 3-wire circuits or 88 A or greater for 2-wire circuits. The feeder capacity as well as the number of branch-circuit positions available for lighting circuits in the panelboard must reflect the full calculated load of 9000 VA × 1.25 = 11,250 VA.

Show Window

$$6000 \text{ VA} \times 1.25 = 7500 \text{ VA}$$

$$7500 \text{ VA} \div 240 \text{ V} = 31 \text{ A for 3-wire, } 120/240 \text{ V}$$

The show window lighting is permitted to be served by 2-wire or 3-wire circuits with a capacity equal to 31 A or greater for 3-wire circuits or 62 A or greater for 2-wire circuits.

Receptacles required by 210.62 are assumed to be included in the receptacle load above if these receptacles do not supply the show window lighting load.

Receptacles

Receptacle Load: 14,400 VA ÷ 240 V = 60 A for 3-wire, 120/240 V

The receptacle load would be permitted to be served by 2-wire or 3-wire circuits with a capacity equal to 60 A or greater for 3-wire circuits or 120 A or greater for 2-wire circuits.

Minimum Size Feeder (or Service) Overcurrent Protection [see 215.3 or 230.90]

Subtotal noncontinuous loads	12,200 VA
Subtotal continuous load at 125% (16,200 VA × 1.25)	20,250 VA
Total	32,450 VA

$$32,450 \text{ VA} \div 240 \text{ V} = 135 \text{ A}$$

The next higher standard size is 150 A (see 240.6).

Minimum Size Feeders (or Service Conductors) Required
[see 215.2, 230.42(A)]

For 120/240-V, 3-wire system,

$$32,450 \text{ VA} \div 240 \text{ V} = 135 \text{ A}$$

Service or feeder conductor is 1/0 Cu per 215.3 and Table 310.16 (with 75°C terminations).

Example D4(a) Multifamily Dwelling

A multifamily dwelling has 40 dwelling units.

Meters are in two banks of 20 each with individual feeders to each dwelling unit.

One-half of the dwelling units are equipped with electric ranges not exceeding 12 kW each. Assume range kW rating equivalent to kVA rating in accordance with 220.19. Other half of ranges are gas ranges.

Area of each dwelling unit is 840 ft^2.

Laundry facilities on premises are available to all tenants. Add no circuit to individual dwelling unit.

Computed Load for Each Dwelling Unit (see Article 220)

General Lighting: 840 ft^2 at 3 VA per ft^2 = 2520 VA

Special Appliance: Electric range (see 220.19) = 8000 VA

Minimum Number of Branch Circuits Required for Each Dwelling Unit [see 210.11(A)]

General Lighting Load: 2520 VA ÷ 120 V = 21 A or two 15-A, 2-wire circuits; or two 20-A, 2-wire circuits

Small Appliance Load: Two 2-wire circuits of 12 AWG wire [see 210.11(C)(1)]

Range Circuit: 8000 VA ÷ 240 V = 33 A or a circuit of two 8 AWG conductors and one 10 AWG conductor as permitted by 220.22 [see 210.19(C)]

Minimum Size Feeder Required for Each Dwelling Unit (see 215.2)

Computed Load (see Article 220):

General Lighting	2520 VA
Small Appliance (two 20-ampere circuits)	3000 VA
Subtotal Computed Load (without ranges)	5520 VA

Application of Demand Factor (see Table 220.11)

First 3000 VA at 100%	3000 VA
5520 VA − 3000 VA = 2520 VA at 35%	882 VA
Net Computed Load (without ranges)	3882 VA
Range Load	8000 VA
Net Computed Load (with ranges)	11,882 VA

Size of Each Feeder (see 215.2)

For 120/240-V, 3-wire system (without ranges)

$$\text{Net computed load of } 3882 \text{ VA} \div 240 \text{ V} = 16.2 \text{ A}$$

For 120/240-V, 3-wire system (with ranges)

$$\text{Net computed load, } 11,882 \text{ VA} \div 240 \text{ V} = 49.5 \text{ A}$$

Feeder Neutral

Lighting and Small Appliance Load	3882 VA
Range Load: 8000 VA at 70% (see 220.22)	5600 VA
(only for apartments with electric range)	
Net Computed Load (neutral)	9482 VA

Calculated Load for Neutral

$$9482 \text{ VA} \div 240 \text{ V} = 39.5 \text{ A}$$

Minimum Size Feeders Required from Service Equipment to Meter Bank (For 20 Dwelling Units — 10 with Ranges)

Total Computed Load:

Lighting and Small Appliance 20 units × 5520 VA	110,400 VA
Application of Demand Factor	
First 3000 VA at 100%	3000 VA
110,400 VA − 3000 VA = 107,400 VA at 35%	37,590 VA
Net Computed Load	40,590 VA
Range Load: 10 ranges (less than 12 kVA) (see Col. C, Table 220.19)	25,000 VA
Net Computed Load (with ranges)	65,590 VA

Net computed load for 120/240-V, 3-wire system,

$$65,590 \text{ VA} \div 240 \text{ V} = 273 \text{ A}$$

Feeder Neutral

Lighting and Small Appliance Load	40,590 VA
Range Load: 25,000 VA at 70% (see 220.22)	17,500 VA
Computed Load (neutral)	58,090 VA

Calculated Load for Neutral

$$58,090 \text{ VA} \div 240 \text{ V} = 242 \text{ A}$$

Further Demand Factor (220.22)

200 A at 100%	200 A
242 A − 200 A = 42 A at 70%	29 A
Net Computed Load (neutral)	229 A

Minimum Size Main Feeders (or Service Conductors) Required (Less House Load) (For 40 Dwelling Units — 20 with Ranges)

Total Computed Load:

Lighting and Small Appliance Load 40 units × 5520 VA	220,800 VA

Application of Demand Factor *(from Table 220.11)*

First 3000 VA at 100%	3000 VA
Next 120,000 VA − 3000 VA = 117,000 VA at 35%	40,950 VA
Remainder 220,800 VA − 120,000 VA = 100,800 VA at 25%	25,200 VA
Net Computed Load	69,150 VA
Range Load: 20 ranges (less than 12 kVA) *(see Col. C, Table 220.19)*	35,000 VA
Net Computed Load	104,150 VA

For 120/240-V, 3-wire system

Net computed load of 104,150 VA ÷ 240 V = 434 A

Feeder Neutral

Lighting and Small Appliance Load	69,150 VA
Range: 35,000 VA at 70% *(see 220.22)*	24,500 VA
Computed Load (neutral)	93,650 VA

93,650 VA ÷ 240 V = 390 A

Further Demand Factor *(see 220.22)*

200 A at 100%	200 A
390 A − 200 A = 190 A at 70%	133 A
Net Computed Load (neutral)	333 A

[See Tables 310.16 through 310.21, and 310.15(B)(2) and (B)(4).]

Example D4(b) Optional Calculation for Multifamily Dwelling

A multifamily dwelling equipped with electric cooking and space heating or air conditioning has 40 dwelling units.

Meters are in two banks of 20 each plus house metering and individual feeders to each dwelling unit.

Each dwelling unit is equipped with an electric range of 8-kW name-plate rating, four 1.5-kW separately controlled 240-V electric space heaters, and a 2.5-kW, 240-V electric water heater. Assume range, space heater, and water heater kW ratings equivalent to kVA.

A common laundry facility is available to all tenants *[see 210.52(F), Exception No. 1].*

Area of each dwelling unit is 840 ft².

Computed Load for Each Dwelling Unit *(see Article 220)*

General Lighting Load:

840 ft² at 3 VA per ft²	2520 VA
Electric range	8000 VA
Electric heat: 6 kVA (or air conditioning if larger)	6000 VA
Electric water heater	2500 VA

Minimum Number of Branch Circuits Required for Each Dwelling Unit

General Lighting Load: 2520 VA ÷ 120 V = 21 A or two 15-A, 2-wire circuits, or two 20-A, 2-wire circuits

Small Appliance Load: Two 2-wire circuits of 12 AWG *[see 210.11(C)(1)]*

Range Circuit: 8000 VA × 80% ÷ 240 V = 27 A on a circuit of three 10 AWG conductors as permitted in Column B of Table 220.19

Space Heating: 6000 VA ÷ 240 V = 25 A

Number of circuits *(see 210.11)*

Minimum Size Feeder Required for Each Dwelling Unit *(see 215.2)*

Computed Load *(see Article 220)*:

General Lighting	2520 VA
Small Appliance (two 20-A circuits)	3000 VA
Subtotal Computed Load	5520 VA
(without range and space heating)	

Application of Demand Factor

First 3000 VA at 100%	3000 VA
5520 VA − 3000 VA = 2520 VA at 35%	882 VA
Net Computed Load	3882 VA
(without range and space heating)	
Range	6400 VA
Space Heating *(see 220.15)*	6000 VA
Water Heater	2500 VA
Net Computed Load	18,782 VA
(for individual dwelling unit)	

Size of Each Feeder For 120/240-V, 3-wire system,

Net computed load of 18,782 VA ÷ 240 V = 78 A

Feeder Neutral *(see 220.22)*

Lighting and Small Appliance	3882 VA
Range Load: 6400 VA at 70% *(see 220.22)*	4480 VA
Space and Water Heating (no neutral): 240 V	0 VA
Net Computed Load (neutral)	8362 VA

Calculated Load for Neutral

8362 VA ÷ 240 V = 35 A

Minimum Size Feeder Required from Service Equipment to Meter Bank (For 20 Dwelling Units)

Total Computed Load:

Lighting and Small Appliance Load	
20 units × 5520 VA	110,400 VA
Water and Space Heating Load	
20 units × 8500 VA	170,000 VA
Range Load: 20 × 8000 VA	160,000 VA
Net Computed Load (20 dwelling units)	440,400 VA
Net Computed Load Using Optional Calculation *(see Table 220.32)*	
440,400 VA × 0.38	167,352 VA

167,352 VA ÷ 240 V = 697 A

Minimum Size Main Feeder Required (Less House Load) (For 40 Dwelling Units)

Computed Load:

Lighting and Small Appliance Load	
40 units × 5520 VA	220,800 VA
Water and Space Heating Load 40 units × 8500 VA	340,000 VA
Range: 40 ranges × 8000 VA	320,000 VA
Net Computed Load (40 dwelling units)	880,800 VA

Net Computed Load Using Optional Calculation *(see Table 220.32)*

880,800 VA × 0.28 = 246,624 VA

246,624 VA ÷ 240 V = 1028 A

Feeder Neutral Load for Feeder from Service Equipment to Meter Bank (For 20 Dwelling Units)

Lighting and Small Appliance Load 20 units × 5520 VA	110,400 VA
First 3000 VA at 100%	3000 VA
110,400 VA − 3000 VA = 107,400 VA at 35%	37,590 VA
Net Computed Load	40,590 VA
20 ranges: 35,000 VA at 70% *(see Tables 220.19 and 220.22)*	24,500 VA
Total	65,090 VA

65,090 VA ÷ 240 V = 271 A

Further Demand Factor *(see 220.22)*

First 200 A at 100%		200 A
Balance: 271 A − 200 A = 71 A at 70%		50 A
	Total	250 amperes

Feeder Neutral Load of Main Feeder (Less House Load) (For 40 Dwelling Units)

Lighting and Small Appliance Load 40 units × 5520 VA	220,800 VA
First 3000 VA at 100%	3000 VA
Next 120,000 VA − 3000 VA = 117,000 VA at 35%	40,950 VA
Remainder 220,800 VA − 120,000 VA = 100,800 VA at 25%	25,200 VA
Net Computed Load	69,150 VA
40 ranges: 55,000 VA at 70% (see Tables 220.19 and 220.22)	38,500 VA
Total	107,650 VA

$$107{,}650 \text{ VA} \div 240 \text{ V} = 449 \text{ A}$$

Further Demand Factor (see 220.22)

First 200 A at 100%	200 A
Balance: 449 − 200 A = 249 A at 70%	174 A
Total	374 A

Example D5(a) Multifamily Dwelling Served at 208Y/120 Volts, Three Phase

All conditions and calculations are the same as for the multifamily dwelling [Example D4(a)] served at 120/240 V, single phase except as follows:

Service to each dwelling unit would be two phase legs and neutral.

Minimum Number of Branch Circuits Required for Each Dwelling Unit (see 210.11)

Range Circuit:

$$8000 \text{ VA} \div 208 \text{ V} = 38 \text{ A}$$

or a circuit of two 8 AWG conductors and one 10 AWG conductor as permitted by 210.19, Column B

Minimum Size Feeder Required for Each Dwelling Unit (see 215.2)

For 120/208-V, 3-wire system (without ranges),

Net computed load of 3882 VA ÷ 2 legs ÷ 120 V/leg = 16.2 A

For 120/208-V, 3-wire system (with ranges),
Net computed load (range) of 8000 VA ÷ 208 V = 38.5 A
Total load (range + lighting) = 38.5 A + 16.2 A = 54.7 A

Feeder neutral: (range) of 8000 VA × 70% = 5600 VA ÷ 208 V = 26.9 A

Total load: (range + lighting) = 26.9 A + 16.2 A = 43.1 A

Minimum Size Feeders Required from Service Equipment to Meter Bank (For 20 Dwelling Units — 10 with Ranges)

For 208Y/120-V, 3-phase, 4-wire system,

Ranges:

Maximum number between any two phase legs = 4

2 × 4 = 8.

Table 220.19 demand = 23,000 VA

Per phase demand = 23,000 VA ÷ 2 = 11,500 VA

Equivalent 3-phase load = 34,500 VA

Net Computed Load (total):

$$40{,}590 \text{ VA} + 34{,}500 \text{ VA} = 75{,}090 \text{ VA}$$

$$75{,}090 \text{ VA} \div (208 \text{ V})(1.732) = 208.4 \text{ A}$$

Feeder Neutral Size:

Net Computed Lighting and Appliance Load & Equivalent Range Load:

$$40{,}590 \text{ VA} + (34{,}500 \text{ VA} \times 70\%) = 64{,}700 \text{ VA}$$

Net Computed Neutral Load:

$$64{,}700 \text{ VA} \div (208 \text{ V})(1.732) = 179.7 \text{ A}$$

Minimum Size Main Feeder (Less House Load)
(For 40 Dwelling Units — 20 with Ranges)

For 208Y/120-V, 3-phase, 4-wire system,

Ranges:

Maximum number between any two phase legs = 7
2 × 7 = 14.
Table 220.19 demand = 29,000 VA
Per phase demand = 29,000 VA ÷ 2 = 14,500 VA
Equivalent 3-phase load = 43,500 VA

Net Computed Load (total):

$$69{,}150 \text{ VA} + 43{,}500 \text{ VA} = 112{,}650 \text{ VA}$$

$$112{,}650 \text{ VA} \div (208 \text{ V})(1.732) = 312.7 \text{ A}$$

Main Feeder Neutral Size:

$$69{,}150 \text{ VA} + (43{,}500 \text{ VA at } 70\%) = 99{,}600 \text{ VA}$$

$$99{,}600 \text{ VA} \div (208 \text{ V})(1.732) = 276.5 \text{ A}$$

Further Demand Factor (see 220.22)

200 A at 100%	200.0 A
276.5 A − 200 A = 76.5 A at 70%	53.6 A
Net Computed Load (neutral)	253.6 A

Example D5(b) Optional Calculation for Multifamily Dwelling Served at 208Y/120 Volts, Three Phase

All conditions and calculations are the same as for Optional Calculation for the Multifamily Dwelling [Example D4(b)] served at 120/240 V, single phase except as follows:

Service to each dwelling unit would be two phase legs and neutral.

Minimum Number of Branch Circuits Required for Each Dwelling Unit (see 210.11)

Range Circuit: 8000 VA at 80% ÷ 208 V = 30.7 A or a circuit of two 8 AWG conductors and one 10 AWG conductor as permitted by 210.19(B)

Space Heating: 6000 VA ÷ 208 V = 28.8 A

Two 20-ampere, 2-pole circuits required, 12 AWG conductors

Minimum Size Feeder Required for Each Dwelling Unit
120/208-V, 3-wire circuit
Net computed load of 18,782 VA ÷ 208 V = 90.3 A
Net computed load (lighting line to neutral):

$$3882 \text{ VA} \div 2 \text{ legs} \div 120 \text{ V per leg} = 16.2 \text{ amperes}$$

Line to line = 14,900 VA ÷ 208 V = 71.6 A

Total load = 16.2 A + 71.6 A = 87.8 A

Minimum Size Feeder Required for Service Equipment to Meter Bank (For 20 Dwelling Units)

Net Computed Load

$$167{,}352 \text{ VA} \div (208 \text{ V})(1.732) = 464.9 \text{ A}$$

Feeder Neutral Load:

$$65{,}080 \text{ VA} \div (208 \text{ V})(1.732) = 180.65 \text{ A}$$

Minimum Size Main Feeder Required (Less House Load) (For 40 Dwelling Units)

Net Computed Load:

$$246{,}624 \text{ VA} \div (208 \text{ V})(1.732) = 684.6 \text{ A}$$

Main Feeder Neutral Load:

$$107{,}650 \text{ VA} \div (208 \text{ V})(1.732) = 298.8 \text{ A}$$

Further Demand Factor (see 220.22)

200 A at 100%	200.0 A
298.8 A − 200 A = 98.8 A at 70%	69.2 A
Net Computed Load (neutral)	269.2 A

Example D6 Maximum Demand for Range Loads

Table 220.19, Column C applies to ranges not over 12 kW. The application of Note 1 to ranges over 12 kW (and not over 27 kW) and Note 2 to ranges over 8¾ kW (and not over 27 kW) is illustrated in the following two examples.

A. Ranges All the Same Rating *(see Table 220.19, Note 1)*
Assume 24 ranges, each rated 16 kW.
From Table 220.19, Column C, the maximum demand for 24 ranges of 12-kW rating is 39 kW. 16 kW exceeds 12 kW by 4.
5% × 4 = 20% (5% increase for each kW in excess of 12)
39 kW × 20% = 7.8 kW increase
39 + 7.8 = 46.8 kW (value to be used in selection of feeders)

B. Ranges of Unequal Rating *(see Table 220.19, Note 2)*
Assume 5 ranges, each rated 11 kW; 2 ranges, each rated 12 kW; 20 ranges, each rated 13.5 kW; 3 ranges, each rated 18 kW.

5 ranges × 12 kW	= 60 kW	(use 12 kW for range rated less than 12)
2 ranges × 12 kW	= 24 kW	
20 ranges × 13.5 kW	= 270 kW	
3 ranges × 18 kW	= 54 kW	
30 ranges Total kW	408 kW	

408 ÷ 30 ranges = 13.6 kW (average to be used for computation)
From Table 220.19, Column C, the demand for 30 ranges of 12-kW rating is 15 kW + 30 (1 kW × 30 ranges) = 45 kW. 13.6 kW exceeds 12 kW by 1.6 kW (use 2 kW).
5% × 2 = 10% (5% increase for each kW in excess of 12 kW)

45 kW × 10% = 4.5 kW increase

45 kW + 4.5 kW = 49.5 kW (value to be used in selection of feeders)

Example D8 Motor Circuit Conductors, Overload Protection, and Short-Circuit and Ground-Fault Protection *(see 240.6, 430.6, 430.22, 430.23, 430.24, 430.32, 430.52, and 430.62, Tables 430.52 and 430.150)*

Determine the minimum required conductor ampacity, the motor overload protection, the branch-circuit short-circuit and ground-fault protection, and the feeder protection, for three induction-type motors on a 480-V, 3-phase feeder, as follows:

(a) One 25-hp, 460-V, 3-phase, squirrel-cage motor, nameplate full-load current 32 A, Design B, Service Factor 1.15

(b) Two 30-hp, 460-V, 3-phase, wound-rotor motors, nameplate primary full-load current 38 A, nameplate secondary full-load current 65 A, 40°C rise.

Conductor Ampacity
The full-load current value used to determine the minimum required conductor ampacity is obtained from Table 430.150 *[see 430.6(A)]* for the squirrel-cage motor and the primary of the wound-rotor motors. To obtain the minimum required conductor ampacity, the full-load current is multiplied by 1.25 *[see 430.22 and 430.23(A)]*.
For the 25-hp motor,

34 A × 1.25 = 42.5 A

For the 30-horsepower motors,

40 A × 1.25 = 50 A

65 A × 1.25 = 81.25 A

Motor Overload Protection
Where protected by a separate overload device, the motors are required to have overload protection rated or set to trip at not more than 125% of the nameplate full-load current *[see 430.6(A) and 430.32(A)(1)]*.
For the 25-hp motor,

32 A × 1.25 = 40.0 A

For the 30-hp motors,

38 A × 1.25 = 47.5 A

Where the separate overload device is an overload relay (not a fuse or circuit breaker), and the overload device selected at 125% is not sufficient to start the motor or carry the load, the trip setting is permitted to be increased in accordance with 430.32(C).

Branch-Circuit Short-Circuit and Ground-Fault Protection
The selection of the rating of the protective device depends on the type of protective device selected, in accordance with 430.52 and Table 430.52. The following is for the 25-hp motor.

(a) Nontime-Delay Fuse: The fuse rating is 300% × 34 A = 102 A. The next larger standard fuse is 110 A *[see 240.6 and 430.52(C)(1), Exception No. 1]*. If the motor will not start with a 110-A nontime-delay fuse, the fuse rating is permitted to be increased to 125 A because this rating does not exceed 400% *[see 430.52(C)(1), Exception No. 2(a)]*.

(b) Time-Delay Fuse: The fuse rating is 175% × 34 A = 59.5 A. The next larger standard fuse is 60 A *[see 240.6 and 430.52(C)(1), Exception No. 1]*. If the motor will not start with a 60-A time-delay fuse, the fuse rating is permitted to be increased to 70 A because this rating does not exceed 225% *[see 430.52(C)(1), Exception No. 2(b)]*.

Feeder Short-Circuit and Ground-Fault Protection
The rating of the feeder protective device is based on the sum of the largest branch-circuit protective device (example is 110 A) plus the sum of the full-load currents of the other motors, or 110 A + 40 A + 40 A = 190 A. The nearest standard fuse that does not exceed this value is 175 A *[see 240.6 and 430.62(A)]*.

Example D9 Feeder Ampacity Determination for Generator Field Control *[see 215.2, 430.24, 430.24 Exception No. 1, 620.13, 620.14, 620.61, and Tables 430.22(E) and 620.14]*

Determine the conductor ampacity for a 460-V 3-phase, 60-Hz ac feeder supplying a group of six elevators. The 460-V ac drive motor nameplate rating of the largest MG set for one elevator is 40 hp and 52 A, and the remaining elevators each have a 30-hp, 40-A, ac drive motor rating for their MG sets. In addition to a motor controller, each elevator has a separate motion/operation controller rated 10 A continuous to operate microprocessors, relays, power supplies, and the elevator car door operator. The MG sets are rated continuous.

Conductor Ampacity. Conductor ampacity is determined as follows:

(a) Per 620.13(D) and 620.61(B)(1), use Table 430.22(E), for intermittent duty (elevators). For intermittent duty using a continuous rated motor, the percentage of nameplate current rating to be used is 140%.

(b) For the 30-hp ac drive motor,

140% × 40 A = 56 A.

(c) For the 40-hp ac drive motor,

140% × 52 A = 73 A.

(d) The total conductor ampacity is the sum of all the motor currents:

(1 motor × 73 A) + (5 motors × 56 A) = 353 A

(e) Per 620.14 and Table 620.14, the conductor (feeder) ampacity would be permitted to be reduced by the use of a demand factor. Constant loads are not included *(see 620.14, FPN)*. For six elevators, the demand factor is 0.79. The feeder diverse ampacity is, therefore, 0.79 × 353 A = 279 A.

(f) Per 430.24 and 215.3, the controller continuous current is 125% × 10 A = 12.5 A

(g) The total feeder ampacity is the sum of the diverse current and all the controller continuous current.

$$I_{total} = 279 \text{ A} + (6 \text{ elevators} \times 12.5 \text{ A}) = 354 \text{ A}$$

(h) This ampacity would be permitted to be used to select the wire size.

See Figure D9.

Example D10 Feeder Ampacity Determination for Adjustable Speed Drive Control [see 215.2, 430.24, 430.24(b), 620.13, 620.14, 620.61, and Tables 430.22(E), and 620.14]

Determine the conductor ampacity for a 460-V, 3-phase, 60-Hz ac feeder supplying a group of six identical elevators. The system is adjustable-speed SCR dc drive. The power transformers are external to the drive (motor controller) cabinet. Each elevator has a separate motion/operation controller connected to the load side of the main line disconnect switch rated 10 A continuous to operate microprocessors, relays, power supplies, and the elevator car door operator. Each transformer is rated 95 kVA with an efficiency of 90%.

Conductor Ampacity. Conductor ampacity is determined as follows:

(a) Calculate the nameplate rating of the transformer:

$$I = \frac{95 \text{ kVA} \times 1000}{\sqrt{3} \times 460 \text{ V} \times 0.90_{eff.}} = 133 \text{ A}$$

(b) Per 620.13(D), for six elevators, the total conductor ampacity is the sum of all the currents.

$$6 \text{ elevators} \times 133 \text{ A} = 798 \text{ A}$$

(c) Per 620.14 and Table 620.14, the conductor (feeder) ampacity would be permitted to be reduced by the use of a demand factor. Constant loads are not included *(see 620.13, FPN No. 2)*. For six elevators, the demand factor is 0.79. The feeder diverse ampacity is, therefore, $0.79 \times 798 \text{ A} = 630 \text{ A}$.

(d) Per 430.24 and 215.3, the controller continuous current is 125% × 10 A = 12.5 A.

(e) The total feeder ampacity is the sum of the diverse current and all the controller constant current.

$$I_{total} = 630 \text{ A} + (6 \text{ elevators} \times 12.5 \text{ A}) = 705 \text{ A}$$

(f) This ampacity would be permitted to be used to select the wire size.

See Figure D10.

Example D11 Mobile Home

Example D11 was relocated from 550.13(B) to Annex D for the 1999 *Code*. It is considered explanatory material and is therefore more appropriately located in an annex.

A mobile home floor is 70 ft by 10 ft and has two small appliance circuits; a 1000-VA, 240-V heater; a 200-VA, 120-V exhaust fan; a 400-VA, 120-V dishwasher; and a 7000-VA electric range.

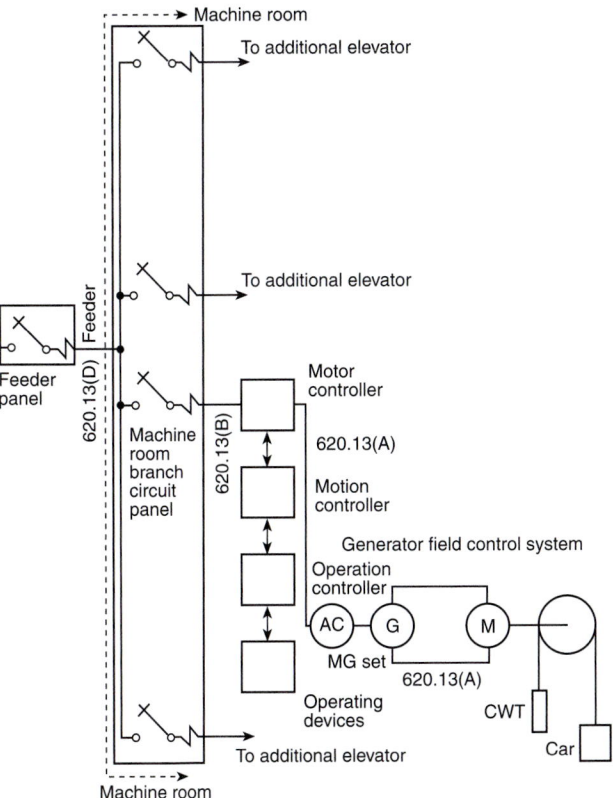

Figure D9 Generator field control.

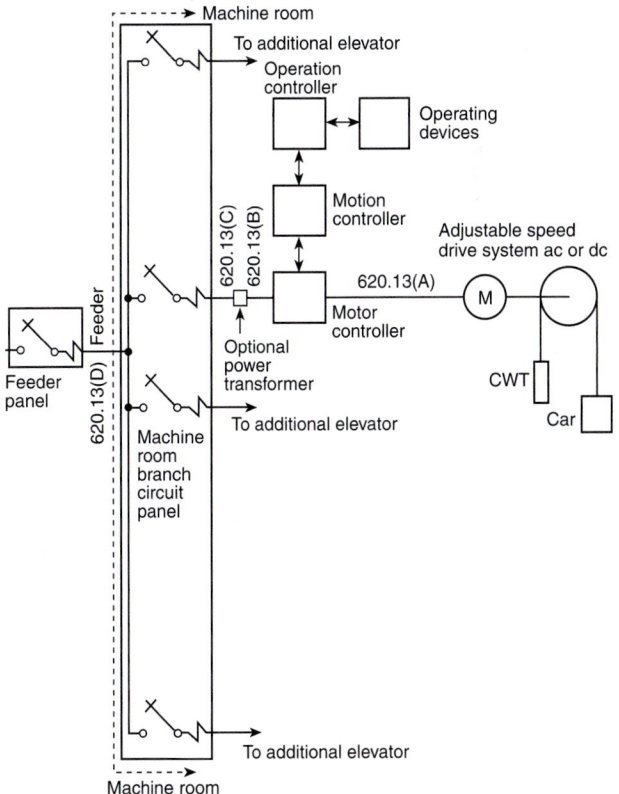

Figure D10 Adjustable speed drive control.

Lighting and Small Appliance Load

Lighting (70 ft × 10 ft × 3 VA per ft^2)		2100 VA
Small appliance (1500 VA × 2 circuits)		3000 VA
Laundry (1500 VA × 1 circuit)		1500 VA
	Subtotal	6600 VA
First 3000 VA at 100%		3000 VA
Remainder (6600 VA − 3000 VA = 3600 VA) × 35%		1260 VA
	Total	4260 VA

4260 VA ÷ 240 V = 17.75 A per leg

Amperes per Leg

	Leg A	Leg B
Lighting and appliances	17.75	17.75
Heater (1000 VA ÷ 240 V)	4.20	4.20
Fan (200 VA × 125% ÷ 120 V)	2.08	—
Dishwasher (400 VA ÷ 120 V)	—	3.30
Range (7000 VA × 0.8 ÷ 240 V)	23.30	23.30
Total amperes per leg	47.33	48.55

Based on the higher current calculated for either leg, a minimum 50-A supply cord would be required.

For SI units, 0.093 m^2 = 1 ft^2 and 0.3048 m = 1 ft.

Example D12 Park Trailer

Example D12 was relocated from 552-47(b) to Annex D for the 1999 *Code*. It is considered explanatory material and is therefore more appropriately located in an annex.

A park trailer floor is 40 ft by 10 ft and has two small appliance circuits, a 1000-VA, 240-V heater, a 200-VA, 120-V exhaust fan, a 400-VA, 120-V dishwasher, and a 7000-VA electric range.

Lighting and Small Appliance Load

Lighting (40 ft × 10 ft × 3 VA per ft^2)		1200 VA
Small appliance (1500 VA × 2 circuits)		3000 VA
Laundry (1500 VA × 1 circuit)		1500 VA
	Subtotal	5700 VA
First 3000 VA at 100%		3000 VA
Remainder (5700 VA − 3000 VA = 2700 VA) × 35%		945 VA
	Total	3945 VA

3945 VA ÷ 240 V = 16.44 A per leg

Amperes per Leg

	Leg A	Leg B
Lighting and appliances	16.44	16.44
Heater (1000 VA ÷ 240 V)	4.20	4.20
Fan (200 VA × 125% ÷ 120 V)	2.08	—
Dishwasher (400 VA ÷ 120 V)	—	3.3
Range (7000 VA × 0.8 ÷ 240 V)	23.30	23.30
Totals	46.02	47.24

Based on the higher current calculated for either leg, a minimum 50-A supply cord would be required.

For SI units, 0.093 m^2 = 1 ft^2 and 0.3048 m = 1 ft.

Types of Construction

This annex is new in the 2002 *Code*. This table is extracted from NFPA 220, *Standard on Types of Building Construction*. It explains the various construction types and has been provided for use in conjunction with 334.10.

This annex is not a part of the requirements of this NFPA document but is included for informational purposes only.

Table E.1 Fire Resistance Ratings (in hours) for Type I through Type V Construction

	Type I		Type II			Type III		Type IV	Type V	
	443	332	222	111	000	211	200	2HH	111	000
Exterior Bearing Walls —										
Supporting more than one floor, columns, or other										
bearing walls	4	3	2	1	0^1	2	2	2	1	0^1
Supporting one floor only	4	3	2	1	0^1	2	2	2	1	0^1
Supporting a roof only	4	3	1	1	0^1	2	2	2	1	0^1
Interior Bearing Walls —										
Supporting more than one floor, columns, or other										
bearing walls	4	3	2	1	0	1	0	2	1	0
Supporting one floor only	3	2	2	1	0	1	0	2	1	0
Supporting a roof only	3	2	1	1	0	1	0	1	1	0
Columns —										
Supporting more than one floor, columns, or other										
bearing walls	4	3	2	1	0	1	0	H^2	1	0
Supporting one floor only	3	2	2	1	0	1	0	H^2	1	0
Supporting roofs only	3	2	1	1	0	1	0	H^2	1	0
Beams, Girders, Trusses & Arches —										
Supporting more than one floor, columns, or other										
bearing walls	4	3	2	1	0	1	0	H^2	1	0
Supporting one floor only	3	2	2	1	0	1	0	H^2	1	0
Supporting roofs only	3	2	1	1	0	1	0	H^2	1	0
Floor Construction	3	2	2	1	0	1	0	H^2	1	0
Roof Construction	2	1½	1	1	0	1	0	H^2	1	0
Exterior Nonbearing Walls[3]	0^1	0^1	0^1	0^1	0^1	0^1	0^1	0^1	0^1	0^1

    Those members that shall be permitted to be of approved combustible material.

Source: Table 3.1 from NFPA 220, *Standard on Types of Building Construction*, 1999.

[1]See A-3-1 in NFPA 220.
[2]"H" indicates heavy timber members; see text for requirements.
[3]Exterior nonbearing walls meeting the conditions of acceptance of NFPA 285, *Standard Method of Test for the Evaluation of Flammability Characteristics of Exterior Non-Load-Bearing Wall Assemblies Containing Combustible Components Using the Intermediate-Scale, Multistory Test Apparatus*, shall be permitted to be used.

Cross-Reference Tables

This annex is not a part of the requirements of this Code but is included for informational purposes only.

Because Chapter 3 was extensively reorganized for the 2002 *Code*, Annex F is provided to make the *Code* more user friendly. This annex provides cross references between the 1999 and 2002 editions of the *Code* as well as an alphabetical cross index.

Table F.1 Chapter 3 Cross-Reference from the 2002 *NEC* to the 1999 *NEC*

2002 *NEC*	1999 *NEC*	Article Title
300	300	Wiring Methods
310	310	Conductors for General Wiring
312	373	Cabinets, Cutout Boxes, and Meter Socket Enclosures
314	370	Outlet, Device, Pull, and Junction Boxes; Conduit Bodies; Fittings; and Manholes
320	333	Armored Cable: Type AC
322	363	Flat Cable Assemblies: Type FC
324	328	Flat Conductor Cable: Type FCC
326	325	Integrated Gas Spacer Cable: Type IGS
328	326	Medium Voltage Cable: Type MV
330	334	Metal-Clad Cable: Type MC
332	330	Mineral-Insulated, Metal-Sheathed Cable: Type MI
334	336	Nonmetallic-Sheathed Cable: Types NM, NMC, and NMS
336	340	Power and Control Tray Cable: Type TC
338	338	Service-Entrance Cable: Types SE and USE
340	339	Underground Feeder and Branch-Circuit Cable: Type UF
342	345	Intermediate Metal Conduit: Type IMC
344	346	Rigid Metal Conduit: Type RMC
348	350	Flexible Metal Conduit: Type FMC
350	351 (Part A)	Liquidtight Flexible Metal Conduit: Type LFMC
352	347	Rigid Nonmetallic Conduit: Type RNC
354	343	Nonmetallic Underground Conduit with Conductors: Type NUCC
356	351 (Part B)	Liquidtight Flexible Nonmetallic Conduit: Type LFNC
358	348	Electrical Metallic Tubing: Type EMT
360	349	Flexible Metallic Tubing: Type FMT
362	331	Electrical Nonmetallic Tubing: Type ENT
366	374	Auxiliary Gutters
368	364	Busways
370	365	Cablebus
372	358	Cellular Concrete Floor Raceways
374	356	Cellular Metal Floor Raceways
376	362 (Part A)	Metal Wireways
378	362 (Part B)	Nonmetallic Wireways
380	353	Multioutlet Assembly
382	342	Nonmetallic Extensions
384	352 (Part C)	Strut-Type Channel Raceway
386	352 (Part A)	Surface Metal Raceways
388	352 (Part B)	Surface Nonmetallic Raceways
390	354	Underfloor Raceways
392	318	Cable Trays
394	324	Concealed Knob-and-Tube Wiring
396	321	Messenger Supported Wiring
398	320	Open Wiring on Insulators
404	380	Switches
408	384	Switchboards and Panelboards
527	305	Temporary Installations

Table F.2 Chapter 3 Cross-Reference from the 1999 *NEC* to the 2002 *NEC*

1999 *NEC*	2002 *NEC*	Article Title
300	300	Wiring Methods
305	527	Temporary Installations
310	310	Conductors for General Wiring
318	392	Cable Trays
320	398	Open Wiring on Insulators
321	396	Messenger Supported Wiring
324	394	Concealed Knob-and-Tube Wiring
325	326	Integrated Gas Spacer Cable: Type IGS
326	328	Medium Voltage Cable: Type MV
328	324	Flat Conductor Cable: Type FCC
330	332	Mineral-Insulated, Metal-Sheathed Cable: Type MI
331	362	Electrical Nonmetallic Tubing: Type ENT
333	320	Armored Cable: Type AC
334	330	Metal-Clad Cable: Type MC
336	334	Nonmetallic-Sheathed Cable: Types NM, NMC, and NMS
338	338	Service-Entrance Cable: Types SE and USE
339	340	Underground Feeder and Branch-Circuit Cable: Type UF
340	336	Power and Control Tray Cable: Type TC
342	382	Nonmetallic Extensions
343	354	Nonmetallic Underground Conduit with Conductors: Type NUCC
345	342	Intermediate Metal Conduit: Type IMC
346	344	Rigid Metal Conduit: Type RMC
347	352	Rigid Nonmetallic Conduit: Type RNC
348	358	Electrical Metallic Tubing: Type EMT
349	360	Flexible Metallic Tubing: Type FMT
350	348	Flexible Metal Conduit: Type FMC
351 (Part A)	350	Liquidtight Flexible Metal Conduit: Type LFMC
351 (Part B)	356	Liquidtight Flexible Nonmetallic Conduit: Type LFNC
352 (Part C)	384	Strut-Type Channel Raceway
352 (Part A)	386	Surface Metal Raceways
352 (Part B)	388	Surface Nonmetallic Raceways
353	380	Multioutlet Assembly
354	390	Underfloor Raceways
356	374	Cellular Metal Floor Raceways
358	372	Cellular Concrete Floor Raceways
362 (Part A)	376	Metal Wireways
362 (Part B)	378	Nonmetallic Wireways
363	322	Flat Cable Assemblies: Type FC
364	368	Busways
365	370	Cablebus
370	314	Outlet, Device, Pull, and Junction Boxes; Conduit Bodies; Fittings; and Manholes
373	312	Cabinets, Cutout Boxes, and Meter Socket Enclosures
374	366	Auxiliary Gutters
380	404	Switches
384	408	Switchboards and Panelboards

Table F.3 Chapter 3 Alphabetical Cross-Reference, 2002–1999 *NEC*

Article Title	2002 *NEC*	1999 *NEC*
Armored Cable: Type AC	320	333
Auxiliary Gutters	366	374
Busways	368	364
Cabinets, Cutout Boxes, and Meter Socket Enclosures	312	373
Cable Trays	392	318
Cablebus	370	365
Cellular Concrete Floor Raceways	372	358
Cellular Metal Floor Raceways	374	356
Concealed Knob-and-Tube Wiring	394	324
Conductors for General Wiring	310	310
Electrical Metallic Tubing: Type EMT	358	348
Electrical Nonmetallic Tubing: Type ENT	362	331
Flat Cable Assemblies: Type FC	322	363
Flat Conductor Cable: Type FCC	324	328
Flexible Metal Conduit: Type FMC	348	350
Flexible Metallic Tubing: Type FMT	360	349
Integrated Gas Spacer Cable: Type IGS	326	325
Intermediate Metal Conduit: Type IMC	342	345
Liquidtight Flexible Metal Conduit: Type LFMC	350	351 (Part A)
Liquidtight Flexible Nonmetallic Conduit: Type LFNC	356	351 (Part B)
Medium Voltage Cable: Type MV	328	326
Messenger Supported Wiring	396	321
Metal Wireways	376	362 (Part A)
Metal-Clad Cable: Type MC	330	334
Mineral-Insulated, Metal-Sheathed Cable: Type MI	332	330
Multioutlet Assembly	380	353
Nonmetallic Extensions	382	342
Nonmetallic Underground Conduit with Conductors: Type NUCC	354	343
Nonmetallic Wireways	378	362 (Part B)
Nonmetallic-Sheathed Cable: Types NM, NMC, and NMS	334	336
Open Wiring on Insulators	398	320
Outlet, Device, Pull, and Junction Boxes; Conduit Bodies; Fittings; and Manholes	314	370
Power and Control Tray Cable: Type TC	336	340
Rigid Metal Conduit: Type RMC	344	346
Rigid Nonmetallic Conduit: Type RNC	352	347
Service-Entrance Cable: Types SE and USE	338	338
Strut-Type Channel Raceway	384	352 (Part C)
Surface Metal Raceways	386	352 (Part A)
Surface Nonmetallic Raceways	388	352 (Part B)
Switchboards and Panelboards	408	384
Switches	404	380
Temporary Installations	527	305
Underfloor Raceways	390	354
Underground Feeder and Branch-Circuit Cable: Type UF	340	339
Wiring Methods	300	300

During the 1999 cycle, Article 250 also was extensively reorganized. Because some jurisdictions may update directly from the 1996 edition of the *Code* to the 2002 edition, the tables of cross reference between the 1996 and 1999 editions have been removed from the 2002 edition. However, those tables have been updated and placed in this commentary as Exhibits F.1 and F.2 to provide the reader with a cross index of Article 250 for the 1996, 1999, and 2002 editions of the *Code*.

Exhibit F.1 Article 250 Cross Reference, by 1996 Topic — 1996, 1999, and 2002 *NEC*

1996 Article 250 Topic	1996 *NEC*	1999 *NEC*	2002 *NEC*
Part A. General			
Scope	-1	-1	.1
Application of Other Articles	-2	-4	.3
Part B. Circuit and System Grounding			
Direct-Current Systems	-3	-162	.162
Alternating-Current Circuits and Systems to Be Grounded	-5	-20, -21	.20, .21
Portable and Vehicle-Mounted Generators	-6	-34	.34
Circuits Not to Be Grounded	-7	-22	.22
Part C. Location of System Grounding Connections			
Objectionable Current Over Grounding Conductors	-21	-6	.6
Point of Connection for Direct-Current Systems	-22	-164	.164
Grounding Service-Supplied Alternating-Current Systems	-23	-24	.24
Two or More Buildings or Structures Supplied from a Common Service	-24	-32	.32
Conductor to Be Grounded—Alternating-Current Systems	-25	-26	.26
Grounding Separately Derived Alternating-Current Systems	-26	-30	.30
High-Impedance Grounded Neutral System Connections	-27	-36	.36
Part D. Enclosure and Raceway Grounding			
Service Raceways and Enclosures	-32	-80	.80
Other Conductor Enclosures and Raceways	-33	-86	.86
Part E. Equipment Grounding			
Equipment Fastened in Place or Connected by Permanent Wiring Methods (Fixed)	-42	-110	.110
Fastened in Place or Connected by Permanent Wiring Methods (Fixed)—Specific	-43	-112	.112
Nonelectric Equipment	-44	-116	.116
Equipment Connected by Cord and Plug	-45	-114	.114
Spacing from Lightning Rods	-46	-106	.106
Part F. Methods of Grounding			
Equipment Grounding Conductor Connections	-50	-130	.130
Effective Grounding Path	-51	-2	.4
Grounding Path to Grounding Electrode at Services	-53	-24, -28	.24, .28
Common Grounding Electrode	-54	-58	.58
Underground Service Cable	-55	-84	.84
Short Sections of Raceway	-56	-132	.132
Equipment Fastened in Place or Connected by Permanent Wiring Methods (Fixed)—Grounding	-57	-119, -134	.119, .134
Equipment Considered Effectively Grounded	-58	-136	.136
Cord-and Plug-Connected Equipment	-59	-138	.138
Frames of Ranges and Clothes Dryers	-60	-140	.140
Use of Grounded Circuit Conductor for Grounding Equipment	-61	-142	.142
Multiple Circuit Connections	-62	-144	.144
Part G. Bonding			
General	-70	-90	.90
Service Equipment	-71	-92	.92
Method of Bonding Service Equipment	-72	-94	.92(B)
Metal Armor or Tape of Service Cable	-73	(removed)	(removed)
Connecting Receptacle Grounding Terminal to Box	-74	-146	.146
Bonding Other Enclosures	-75	-96	.96
Bonding for Over 250 Volts	-76	-97	.97
Bonding Loosely Jointed Metal Raceways	-77	-98	.98
Bonding in Hazardous (Classified) Locations	-78	-100	.100
Main and Equipment Bonding Jumpers	-79	-28, -102, -168	.28, .102, .168
Bonding of Piping Systems and Exposed Structural Steel	-80	-104	.104

Exhibit F.1 Continued

1996 Article 250 Topic	1996 *NEC*	1999 *NEC*	2002 *NEC*
Part H. Grounding Electrode System			
Grounding Electrode System	-81	-50	.50
Made and Other Electrodes	-83	-52	.50, .53
Resistance of Made Electrodes	-84	-56	.56
Use of Lightning Rods	-86	-60	.60
Part J. Grounding Conductors			
Material	-91	-54, -62, -64, -118	.54, .62, .64, .118
Installation	-92	-64, -120	.64, .120
Size of Direct-Current Grounding Electrode Conductor	-93	-166	.166
Size of Alternating-Current Grounding Electrode Conductor	-94	-66	.66
Size of Equipment Grounding Conductors	-95	-122	.122
Outline Lighting	-97	(600-7)	(600.7)
Equipment Grounding Conductor Continuity	-99	-124	.124
Part K. Grounding Conductor Connections			
To Grounding Electrode	-112	-68	.68
To Conductors and Equipment	-113	-8	.8
Continuity and Attachment of Equipment Grounding Conductors to Boxes	-114	-148	.148
Connection to Electrodes	-115	-70	.70
Protection of Attachment	-117	-10	.10
Clean Surfaces	-118	-12	.12
Identification of Wiring Device Terminals	-119	-126	.126
Part L. Instrument Transformers, Relays, Etc.			
Instrument Transformer Circuits	-121	-170	.170
Instrument Transformer Cases	-122	-172	.172
Cases of Instruments, Meters, and Relays Operating at Less Than 1000 Volts	-123	-174	.174
Cases of Instruments, Meters, and Relays Operating at Voltage 1 kV and Over	-124	-176	.176
Instrument Grounding Conductor	-125	-178	.178
Part M. Grounding of Systems and Circuits of 1 kV and Over (High Voltage)			
General	-150	-180	.180
Derived Neutral Systems	-151	-182	.182
Solidly Grounded Neutral Systems	-152	-184	.184
Impedance Grounded Neutral Systems	-153	-186	.186
Grounding of Systems Supplying Portable or Mobile Equipment	-154	-188	.188
Grounding of Equipment	-155	-190	.190

Exhibit F.2 Article 250 Cross-Reference, by 2002 Topic — 2002, 1999, and 1996 *NEC*

2002 Article 250 Topic	2002 *NEC*	1999 *NEC*	1996 *NEC*
Part I. General			
Scope	.1	-1	-1
Definitions	.2	N/A	N/A
Application of Other Articles	.3	-4	-2
General Requirements for Grounding and Bonding	.4	-2	-51, -1 FPN 1 & 2
Objectionable Current Over Grounding Conductors	.6	-6	-21
Connection of Grounding and Bonding Equipment	.8	-8	-113
Protection of Ground Clamps and Fittings	.10	-10	-117
Clean Surfaces	.12	-12	-118
Part II. Circuit and System Grounding			
Alternating-Current Circuits and Systems to Be Grounded	.20	-20	-5
Alternating-Current Systems of 50 Volts to 1000 Volts Not Required to be Grounded	.21	-21	-5
Circuits Not to Be Grounded	.22	-22	-7
Grounding Service-Supplied Alternating-Current Systems	.24	-24	-23, -53(a)
Conductor to Be Grounded—Alternating-Current Systems	.26	-26	-25
Main Bonding Jumper	.28	-28	-79, -53(b)
Grounding Separately Derived Alternating-Current Systems	.30	-30	-26
Two or More Buildings or Structures Supplied from a Common Service	.32	-32	-24
Portable and Vehicle-Mounted Generators	.34	-34	-6
High-Impedance Grounded Neutral Systems	.36	-36	-27
Part III. Grounding Electrode System and Grounding Electrode Conductor			
Grounding Electrode System	.50	-50	-81
Grounding Electrodes	.52	-50, -52	-81, -83
Grounding Electrode System Installation	.53	-50, -52	-81, -83
Supplementary Grounding Electrodes	.54	-54	-91(c)
Resistance of Rod, Pipe, and Plate Electrodes	.56	-56	-84
Common Grounding Electrode	.58	-58	-54
Use of Air Terminals	.60	-60	-86
Grounding Electrode Conductor Material	.62	-62	-91(a)
Grounding Electrode Conductor Installation	.64	-64	-91(a), -92
Size of Alternating-Current Grounding Electrode Conductor	.66	-66	-94
Grounding Electrode Conductor and Bonding Jumper Connection to Electrodes	.68	-68	-112
Methods of Grounding and Bonding Conductor Connection to Electrodes	.70	-70	-115
Part IV. Enclosure, Raceway, and Service Cable Grounding			
Service Raceways and Enclosures	.80	-80	-32
Underground Service Cable or Conduit	.84	-84	-55
Other Conductor Enclosures and Raceways	.86	-86	-33
Part V. Bonding			
General	.90	-90	-70
Services	.92	-92	-71
Bonding for Other Systems	.94	-92(b)	-72
Bonding Other Enclosures	.96	-96	-75
Bonding for Over 250 Volts	.97	-97	-76
Bonding Loosely Jointed Metal Raceways	.98	-98	-77
Bonding in Hazardous (Classified) Locations	.100	-100	-78
Equipment Bonding Jumpers	.102	-102	-79
Bonding of Piping Systems and Exposed Structural Steel	.104	-104	-80
Lightning Protection Systems	.106	-106	-46
Part VI. Equipment Grounding and Equipment Grounding Conductors			
Equipment Fastened in Place or Connected by Permanent Wiring Methods (Fixed)	.110	-110	-42
Fastened in Place or Connected by Permanent Wiring Methods (Fixed)—Specific	.112	-112	-43
Equipment Connected by Cord and Plug	.114	-114	-45

Exhibit F.2 Continued

2002 Article 250 Topic	2002 *NEC*	1999 *NEC*	1996 *NEC*
Nonelectric Equipment	.116	-116	-44
Types of Equipment Grounding Conductors	.118	-118	-91(b)
Identification of Equipment Grounding Conductors	.119	-119	-57(b)
Equipment Grounding Conductor Installation	.120	-120	-92(c)
Size of Equipment Grounding Conductors	.122	-122	-95
Equipment Grounding Conductor Continuity	.124	-124	-99
Identification of Wiring Device Terminals	.126	-126	-119
Part VII. Methods of Equipment Grounding			
Equipment Grounding Conductor Connections	.130	-130	-50
Short Sections of Raceway	.132	-132	-56
Equipment Fastened in Place or Connected by Permanent Wiring Methods (Fixed)—Grounding	.134	-134	-57
Equipment Considered Effectively Grounded	.136	-136	-58
Cord-and Plug-Connected Equipment	.138	-138	-59
Frames of Ranges and Clothes Dryers	.140	-140	-60
Use of Grounded Circuit Conductor for Grounding Equipment	.142	-142	-61
Multiple Circuit Connections	.144	-144	-62
Connecting Receptacle Grounding Terminal to Box	.146	-146	-74
Continuity and Attachment of Equipment Grounding Conductors to Boxes	.148	-148	-114
Part VIII. Direct-Current Systems			
General	.160	-160	(new)
Direct-Current Circuits and Systems to Be Grounded	.162	-162	-3
Point of Connection for Direct-Current Systems	.164	-164	-22
Size of Direct-Current Grounding Electrode Conductor	.166	-166	-93
Direct-Current Bonding Jumper	.168	-168	-79(d)
Ungrounded Direct-Current Separately Derived Systems	.169	-169	(new)
Part IX. Instruments, Meters, and Relays			
Instrument Transformer Circuits	.170	-170	-121
Instrument Transformer Cases	.172	-172	-122
Cases of Instruments, Meters, and Relays Operating at Less Than 1000 Volts	.174	-174	-123
Cases of Instruments, Meters, and Relays Operating at Voltage 1 kV and Over	.176	-176	-124
Instrument Grounding Conductor	.178	-178	-125
Part X. Grounding of Systems and Circuits of 1 kV and Over (High Voltage)			
General	.180	-180	-150
Derived Neutral Systems	.182	-182	-151
Solidly Grounded Neutral Systems	.184	-184	-152
Impedance Grounded Neutral Systems	.186	-186	-153
Grounding of Systems Supplying Portable or Mobile Equipment	.188	-188	-154
Grounding of Equipment	-190	-190	-155

Cable installation, 392.8

Conduits and cables supported from, 392.6(J)

Construction specifications, 392.5

Definition, 392.2

Grounding, 392.7

Installation, 392.6

Separation, 392.6(F)(2)

Uses not permitted, 392.4

Uses permitted, 392.3, 392.6(A)

Cablebus, Art. 370

Conductors, 370.4

Overcurrent protection, 370.5

Terminations, 370.8

Definition, 370.2

Fittings, 370.7

Grounding, 370.9

Marking, 370.10

Support and extension through walls and floors, 370.6

Use, 370.3

Cables

Aerial, 820.10, 830.10

Armored (Type AC). *see* Armored cable (Type AC), Art. 320

Border lights, theater, 520.44(B)

Bundled

Definition, 520.2

CATV, Art. 820

Continuity, 300.12

Definition, 800.2

Flat cable assemblies (Type FC). *see* Flat cable assemblies (Type FC)

Flat conductor (Type FCC). *see* Flat conductor cable (Type FCC)

Grouped

Definition, 520.2

Heating. *see* Heating cables

Installation in cable trays, 392.8

Installed in shallow grooves, 300.4(E)

Instrumentation tray (Type ITC). *see* Instrumentation tray cable (Type ITC)

Integrated gas spacer cable (Type IGS). *see* Integrated gas spacer cable (Type IGS)

Medium voltage cable (Type MV). *see* Medium voltage cable (Type MV)

Metal-clad cable (Type MC). *see* Metal-clad cable (Type MC)

Mineral-insulated metal-sheathed (Type MI). *see* Mineral-insulated metal-sheathed cable (Type MI)

Nonmetallic extension. *see* Nonmetallic extensions

Nonmetallic underground conduit with conductors. *see* Nonmetallic underground conduit with conductors

Nonmetallic-sheathed (Types NM, NMC, and NMS). *see* Nonmetallic-sheathed cable (Types NM, NMC, and NMS)

Optical fiber. *see* Optical fiber cables

Other types of. *see* names of systems

Point of entrance

Definition, 800.2, 820.2, 830.2

Portable. *see* Cords, flexible

Power and control tray cable (Type TC). *see* Power and control tray cable (Type TC)

Preassembled in nonmetallic conduit. *see* Nonmetallic underground conduit with conductors

Protection against physical damage, 300.4

Sealing, 501.5(D), 501.5(E), 505.16(A)(2), 505.16(B)(5), 505.16(B)(6), 505.16(B)(7), 505.16(C)(2)

Secured, 300.11, 314.17(B) and (C)

Service. *see* Service cables

Service-entrance (Types SE and USE). *see* Service-entrance cable (Types SE and USE)

Splices in boxes, 300.15

Stage, 530.18(A)

Supported from cable trays, 392.6(J)

Through studs, joists, rafters, 300.4

Underground, 230-III, 300.5, 300.50

Underground feeder and branch circuit Type UF. *see* Underground feeder and branch-circuit cable (Type UF)

Calculations, Annex D. *see also* Loads

Camping trailer

Definition, 551.2

Canopies

Boxes and fittings, 314.25

Live parts, exposed, 410.3

Luminaires (lighting) fixtures

Conductors, space for, 410.10

Cover

At boxes, 410.12

Combustible finishes, covering required between canopy and box, 410.13

Capacitors, Art. 460. *see also* Hazardous (classified) locations

Enclosing and guarding, 460.2

Induction and dielectric heating, 665.24

600 volts and under, 460-I

Conductors, 460.8

Grounding, 460.10

Marking, 460.12

Rating or setting of motor overload device, 460.9

Over 600 volts, 460-II

Grounding, 460.27

Identification, 460.26

Means for discharge, 460.28

Overcurrent protection, 460.25

Switching, 460.24

X-ray equipment, 660.36

Capacity, interrupting. *see* Interrupting capacity

Caps. *see* Attachment plugs

Carnivals, circuses, fairs, and similar events, Art. 525

Conductor overhead clearance, 525.5

Grounding and bonding, 525-V

Equipment bonding, 525.30

Equipment grounding, 525.31

Grounding conductor continuity assurance, 525.32

Power sources, 525-III

Separately derived systems, 525.10

Services, 525.11

Protection of electrical equipment, 525.6

Wiring methods, 525-III, 525.20

Concessions, 525.21

Ground-fault circuit-interrupter protection, 525.23

Portable distribution or terminal boxes, 525.22

Rides, 525.21

Tents, 525.21

Cartridge fuses, 240-VI

Disconnection, 240.40

CATV systems. *see* Community antenna television and radio distribution (CATV) systems

Ceiling fans, 680.22(B), 680.43(B)

Support of, 314.27(D), 422.18

Cell

Cellular concrete floor raceways

Definition, 372.2

Electrolytic

Definition, 668.2

Sealed, storage batteries

Definition, 480.2

Solar

Definition, 690.2

Cell line, electrolytic cells

Attachments and auxiliary equipment

Definition, 668.2

Definition, 668.2

Cellars. *see* Basements

Cellular concrete floor raceways, Art. 372

Connection to cabinets and other enclosures, 372.6

Definitions, 372.2

Discontinued outlets, 372.13

Header, 372.5

Inserts, 372.9

Junction boxes, 372.7

Markers, 372.8

Number of conductors, 372.11

Other articles, 372.3

Size of conductors, 372.10

Splices and taps, 372.12

Uses not permitted, 372.4

Cellular metal floor raceways, Art. 374

Connection to cabinets and extension from cells, 374.11

Construction, 374.12

Definitions, 374.2

Discontinued outlets, 374.7

Inserts, 374.10

Installation, 374-I

Junction boxes, 374.9

Markers, 374.8

Number of conductors, 374.5

Size of conductors, 374.4

Splices and taps, 374.6

Uses not permitted, 374.3

Chair lifts. *see* Elevators, dumbwaiters, escalators, moving walks, wheelchair lifts, and stairway chair lifts

Churches, Art. 518

Cinder fill

Intermediate or rigid metal conduits and electrical metallic tubing, in or under, 342.10(C), 344.10(C), 348.12(3)

Rigid nonmetallic conduit, 352.10

Circuit breakers, Art. 240. *see also* Hazardous (classified) locations

Accessibility and grouping, 404.8

Circuits over 600 volts, 490.21

Definition, Art. 100-I

Disconnection of grounded circuits, 404.2(B), 514.11(A)

Enclosures, 404.3

General, 110.9, 240-I

Overcurrent protection, 230.208, 240-I, 240-VII

Generators, 445.12

Motors, 430.52(A), 430.58, 430.110, 430.111

Transformers, 450.3

Panelboards, 408-III

Rating, motor branch circuits, 430.58

Rating, nonadjustable trip, 240.6(A), 240.83(C)

Service overcurrent protection, 230.90, 230.91

Services, disconnecting means, 230.70, 230.205

Switches, use as, 240.83(D), 404.11, 410.81

Wet locations, in, 404.4

Circuit directory, panelboards, 110.22

Circuit-interrupters, ground-fault. *see* Ground-fault circuit-interrupters

Circuits

Anesthetizing locations, 517.63(F)

Bonding jumper

Definition, Art. 100-I

Branch. *see* Branch circuits

Burglar alarm. *see* Remote-control, signaling, and power-limited circuits

Central station. *see* Fire alarm systems

Communication. *see* Communication circuits

Fire alarm

Definitions, 760.2

Fuel cell systems, 692-II

Grounding, Art. 250

Impedance, 110.10

Information technology equipment, 645.5

Intrinsically safe, 504.30

Definition, 504.2

Inverter input circuit

Definition, 690.2

Less than 50 volts, Art. 720

Class 1, 725-II
Grounding, 250.20(A)
More than 600 volts. *see* Over 600 volts
Services, 230-IX
Motor, 430-II
Motor control, 430-VI
Definition, 430.71
Nonincendive
Definition, Art. 100-I
Number of, in enclosures, 90.8(B)
Photovoltaic output
Definition, 690.2
Photovoltaic source
Definition, 690.2
Power-limited. *see* Remote-control, signaling, and power-limited circuits
Protectors required, 800.12, 800.30, 800.31, 800.32
Remote-control, Art. 725
Definition, Art. 100-I
Motors, controllers, 430-VI
Signal. *see* Remote-control, signaling, and power-limited circuits
Telegraph. *see* Communication circuits
Telephone. *see* Communication circuits
Underground. *see* Communication circuits
Ungrounded, 210.10, 215.7, 410.48
Circuses. *see* Carnivals, circuses, fairs, and similar events
Clamp fill, boxes, 314.16(B)(2)
Clamps, ground, 250.8, 250.10, 250.70
Class 1, 2, and 3 circuits. *see* Remote-control, signaling, and power-limited circuits
Class I, II, and III locations. *see* Hazardous (classified) locations
Clean surfaces, grounding conductor connections, 250.12
Clearances. *see also* Enclosures; Space
Antennas, 810.3, 810.18, 810.54
Community antenna systems, Art. 820
Conductors
Open, outside branch circuits and feeders, 225.18, 225.19
Service drop, 230-II
Lighting luminaires (fixtures), 410.66, 410.76(B)
Live parts
Auxiliary gutters, 366.8
Circuits over 600 volts, 110-III
Network-powered broadband communications systems, 830.11
Swimming pools, 680.8
Switchboards, 408.7, 408.8, 408.10
Climbing space, line conductors on poles, 225.14(D)
Closed-loop and programmed power distribution, Art. 780
Cables and conductors, Art. 334, 780.6

Control, 780.3
Hybrid cable, 780.6(A)
Noninterchangeability, 780.7
Power limitation, in signaling circuits, 780.5
In same cabinet, panel, or box, 780.6(B)
Clothes closets
Heating, 424.38(C)
Luminaires (lighting fixtures), 410.8
Overcurrent devices, 240.24(D)
Clothes dryers
Calculations for, 220.18
Feeder demand factors, Table 220.18
Grounding, 250.140
Mobile homes, 550.15(E)
Clothing manufacturing plants, 500.5(D), Art. 503. *see also* Hazardous (classified) locations
CO/ALR
Receptacles, 406.2(C)
Switches, 404.14(C)
Collector rings, 490.54, 675.11
Definition, 675.2
Collectors, cranes and hoists, 610.22
Color code
Branch circuits, 210.4(D), 210.5
Conductors, 310.12, 504.80(C), 647.4(C)
Grounded conductor, 200.6
Grounding conductor, 250.119, 310.12(B), 400.23
Heating cables, 424.35
Higher voltage to ground
Feeders, 215.8
Panel boards, 408.3(E)
Sensitive electronic equipment, 647.4(C)
Service-entrance conductors, 230.56
Combustible dusts, Art. 502. *see also* Hazardous (classified) locations
Commercial garages. *see* Garages, commercial
Common grounding electrode. *see* Electrodes, grounding, common
Common neutral
Feeders, 215.4
Outside wiring, 225.7(B)
Communications circuits, Art. 800
Access to electrical equipment, 800.5
Bonding, 800.40(D), 800.41(B)
Cable marking, 800.50
Cable substitution, 800.53(G)
Circuits requiring primary protectors, 800.12
Conductors
Other circuits with, 800.53(E)
Overhead, 800.10
Within buildings, 800-V
Definitions, 800.2
Essential electrical systems, 517.32(D), 517.42(D)
Fire resistance, 800.49
Grounding, 800.33, 800-IV

Hazardous (classified) locations, 800.8
Health care facilities, 517.32(D), 517.42(D), 517-VI
Hybrid power and communications cables, 800.3
Installation, 800.7, 800.52
Lightning conductors, 800.13
Listing of communications wires and cables, 800.50, 800.51
Listing of equipment, 800.4
Marking, 800.50
Mechanical execution of work, 800.6
Mobile homes, 800.41
Protection, 800-III
Devices, 800.30
Grounding, 800.40, 800.41
Requirements, 800.31, 800.32, 800.40, 800.41
Types of cable, 800.51
Underground circuits entering buildings, 800.11
Community antenna television and radio distribution (CATV) systems, Art. 820
Access to electrical equipment, 820.5
Cables
Aerial entrance, 820.10
Within buildings, 820-V
Entering buildings, 820.11
Fire resistance of, 820.49
Listing and marking, 820.50, 820.51
Outside, 820.10
Types of, 820.51
Underground, 820.11(A)
Definitions, 820.2
Energy limitations, 820.4
Grounding, 820-IV
Cables, 820.33, 820.40, 820-IV
Equipment, 820.41
Installation of cables and equipment, 820.52
Installation of systems, 820.7
Locations, 820.3
Mechanical execution of work, 820.6
Mobile homes, 820.42
Protection, 820.40(A)(6), 820-III
Compressors, refrigeration, Art. 440
Computer rooms. *see* Information technology equipment
Concealed
Definition, Art. 100-I
Knob-and-tube wiring. *see* Knob-and-tube wiring
Concentric knockouts. *see* Knockouts, bonding service equipment
Concrete
Electrodes encased in, 250.52(A)(3)
Metal raceways and equipment in, 300.6(B)
Conductive surfaces, exposed
Definition, 517.2
Conductor fill
Audio systems, 640.23(A)
Auxiliary gutters, 366.6
Boxes, 314.16(B)(1) and (5)
Cable trays, 392.9, 392.10

Cellular concrete floor raceways, 372.11
Cellular metal floor raceways, 374.5
Control circuits, 517.74(A)
Electrical metallic tubing, 358.22
Electrical nonmetallic tubing, 362.22
Elevators, 620.32, 620.33
Fixture wire, 402.7
Flexible metal conduit, 348.22, Table 348.22
Flexible metallic tubing, 358.22
General installations, 300.17
Hazardous (classified) locations, sealing fittings, 501.5(C)(6), 505.16(D)(5)
Intermediate metal conduit, 342.22
Liquidtight flexible metal conduit, 350.22
Liquidtight flexible nonmetallic conduit, 356.22
Maximum permitted, 300.17
Outlet boxes, etc., 314.16
Remote control circuits, 725.28
Rigid metal conduit, 344.22
Rigid nonmetallic conduit, 352.22
Signs, 600.31(C)
Surface raceways, 386.22, 388.22
Theaters, 520.6
Underfloor raceways, 390.5
Wireways, 376.22, 378.22
X-ray equipment, 660.8
Conductors. *see also* Cords, flexible; Fixture wires
Aluminum, conductor material, 310.14
Aluminum, properties of, Chap. 9, Table 8
Motors, 430-II
Ampacities of, 310.15, Tables 310.16 through 310.21, Tables 310.69 through 310.86, B.310.1 through B.310.10
Application, 310.13
Armored cable, Type AC. *see* Armored cable (Type AC)
Bare. *see* Bare conductors
Bending radius, 300.34
Boxes and fittings, junction, 314.16, 314.17
Branch circuits, Art. 210
Bundled, in theaters, etc.
Definition, 520.2
Buried, 310.7
Busways. *see* Busways
Cabinets and cutout boxes, 312.5 through 312.7
Cablebus. *see* Cablebus
Capacitors, 460.8
Cellular concrete floor raceways. *see* Cellular concrete floor raceways
Cellular metal floor raceways. *see* Cellular metal floor raceways
Circuit
Communications, Art. 800
Control, health care facilities, 517.74
Fire alarm, Art. 760

About the Editors

Mark W. Earley, P.E.

Mark W. Earley is Assistant Vice President and Chief Electrical Engineer at NFPA. He has served as Secretary of the *NEC* for the past 12 years and is the co-author of NFPA's reference book, *Electrical Installations in Hazardous Locations*. Prior to joining NFPA, he worked as an electrical engineer at the Factory Mutual Research Corporation. Additionally, he has served on several of NFPA's electrical committees and NEC code-making panels. Earley is a registered professional engineer and a member of IAEI, IEEE, SFPE, the UL Electrical Council, U.S. National Committee on the International Electrotechnical Commission, and the Canadian Electrical Code Committee. He serves as chair of Task Group 1—Electrical Installation Codes for the Pacific Area Standards Congress.

Joseph V. Sheehan, P.E.

Joseph V. Sheehan is Principal Electrical Engineer at NFPA and an instructor of NFPA's electrical code seminars. Prior to joining NFPA in 1993, he was employed for over 28 years in the electrical construction industry as an electrician, chief estimator and project manager in the Boston area. He also taught electrical engineering for over 15 years in the evening division of the Franklin Institute of Boston. Sheehan now serves as staff liaison to various NFPA committees that deal with industrial machinery, electrical equipment maintenance, and Electrical Workplace Safety. He is a registered professional engineer and a licensed master electrician in three states and is a member of IAEI, IEEE, and IBEW.

Jeffrey S. Sargent

Jeffrey S. Sargent is Senior Electrical Specialist for NFPA. Before joining NFPA, he worked for the New Hampshire Electricians' Licensing Board, where he served as a state electrical inspector. In addition to inspecting and approving electrical installations and providing technical assistance to municipal code enforcement officials, he also has taught electrical courses for apprentice and licensed electricians at New Hampshire Technical College. Sargent is a member of the International Association of Electrical Inspectors and a certified electrical inspector through IAEI and co-author of NFPA's *1999 NEC® Changes* and NFPA's *Electrical Inspection Manual with Checklists*.

John M. Caloggero

John M. Caloggero, Principal Electrical Specialist at NFPA, serves as staff liaison to the technical committee on lightning protection. He has contributed to seven editions of the *National Electrical Code Handbook*, has developed several *NEC* workshops, and is technical advisor to the *NEC* videos. He has served as an instructor for numerous *Code* classes for Spanish-speaking students in Latin America. Prior to joining NFPA in 1980, he was employed as a master electrician, electrical inspector, and electrical instructor. Caloggero is a member of IAEI, holds a journeyman electrician's license, and is a nationally certified electrical inspector.

Timothy M. Croushore, P.E.

Timothy M. Croushore is a product manager for Allegheny Power Service Corporation. Former Chief Electrical Engineer for the NFPA, he holds a B.S. and M.S. degree in electrical engineering from the University of Pittsburgh. He is a registered professional engineer, a certified electrical inspector, a licensed electrical contractor, an IEEE member, and an NFPA member and technical committee chairman. Croushore has authored several articles in industry trade magazines and holds a patent in distribution power line communication.

Tentative Interim Amendment

NFPA 70

National Electrical Code®

2002 Edition

Reference: 527.4(B) and 527.4(C)
TIA 02-1 (NFPA 70)

Pursuant to Section 5 of the NFPA Regulations Governing Committee Projects, the National Fire Protection Association has issued the following Tentative Interim Amendment to NFPA 70, *National Electrical Code®*, 2002 edition. The TIA was processed by the National Electrical Code® Committee, and was issued by the Standards Council on October 3, 2002, with an effective date of October 23, 2002.

A Tentative Interim Amendment is tentative because it has not been processed through the entire standards-making procedures. It is interim because it is effective only between editions of the standard. A TIA automatically becomes a proposal of the proponent for the next edition of the standard; as such, it then is subject to all of the procedures of the standards-making process.

1. Revise section 527.4(B) to read as follows:

(B) Feeders. Feeders shall be protected as provided in Article 240. They shall originate in an approved distribution center. Conductors shall be permitted within cable assemblies or within multiconductor cords or cables of a type identified in Table 400.4 for hard usage or extra-hard usage. For the purpose of this section, Type NM and Type NMC cables shall be permitted to be used in any dwelling, building, or structure without any height limitation <u>or limitation by building construction type as described in 334.10(3).</u>

2. Revise section 527.4(C) to read as follows:

(C) Branch Circuits. All branch circuits shall originate in an approved power outlet or panelboard. Conductors shall be permitted within cable assemblies or within multiconductor cord or cable of a type identified in Table 400.4 for hard usage or extra-hard usage. All conductors shall be protected as provided in Article 240. For the purposes of this section, Type NM and Type NMC cables shall be permitted to be used in any dwelling, building, or structure without any height limitation <u>or limitation by building construction type as described in 334.10 (3).</u>

Chapter 3 for 2002

♦ *Where to Find 1999 Articles in 2002 NEC*

1999	2002	ARTICLE TITLE
300	300	Wiring Methods
305	527	Temporary Installations
310	310	Conductors for General Wiring
318	392	Cable Trays
320	398	Open Wiring on Insulators
321	396	Messenger Supported Wiring
324	394	Concealed Knob-and-Tube Wiring
325	326	Integrated Gas Spacer Cable: Type IGS
326	328	Medium Voltage Cable: Type MV
328	324	Flat Conductor Cable: Type FCC
330	332	Mineral-Insulated, Metal-Sheathed Cable: Type MI
331	362	Electrical Nonmetallic Tubing: Type ENT
333	320	Armored Cable: Type AC
334	330	Metal-Clad Cable: Type MC
336	334	Nonmetallic-Sheathed Cable: Types NM, NMC, and NMS
338	338	Service-Entrance Cable: Types SE and USE
339	340	Underground Feeder and Branch-Circuit Cable: Type UF
340	336	Power and Control Tray Cable: Type TC
342	382	Nonmetallic Extensions
343	354	Nonmetallic Underground Conduit with Conductors: Type NUCC
345	342	Intermediate Metal Conduit: Type IMC
346	344	Rigid Metal Conduit: Type RMC
347	352	Rigid Nonmetallic Conduit: Type RNC
348	358	Electrical Metallic Tubing: Type EMT
349	360	Flexible Metallic Tubing: Type FMT
350	348	Flexible Metal Conduit: Type FMC
351 (Part A)	350	Liquidtight Flexible Metal Conduit: Type LFMC
351 (Part B)	356	Liquidtight Flexible Nonmetallic Conduit: Type LFNC
352 (Part A)	386	Surface Metal Raceways
352 (Part B)	388	Surface Nonmetallic Raceways
352 (Part C)	384	Strut-Type Channel Raceway
353	380	Multioutlet Assembly
354	390	Underfloor Raceways
356	374	Cellular Metal Floor Raceways
358	372	Cellular Concrete Floor Raceways
362 (Part A)	376	Metal Wireways
362 (Part B)	378	Nonmetallic Wireways
363	322	Flat Cable Assemblies: Type FC
364	368	Busways
365	370	Cablebus
370	314	Outlet, Device, Pull, and Junction Boxes; Conduit Bodies; Fittings; and Manholes
373	312	Cabinets, Cutout Boxes, and Meter Socket Enclosures
374	366	Auxiliary Gutters
380	404	Switches
384	408	Switchboards and Panelboards
410 (Part L)	406	Receptacles, Cord Connectors, and Attachment Plugs (Caps)